FIFTEENTH
INTERNATIONAL CONFERENCE ON
RAMAN SPECTROSCOPY

ORGANIZATION

International Steering Committee

R. S. Armstong (Australia)
S. S. Bukalov (Russia)
F. Cerdeira (Brazil)
P. M. Champion (USA)
J. H. Clark (UK)
James R. Durig (USA)
J. Greve (Netherlands)
H. Hamaguchi (Japan)
A.M. Heyns (South Africa)
W. Kiefer (Germany)
T. Kitagawa (Japan)
J. Koningstein (Canada)
J. Lascombe (France)
D. Long (UK)
S. Montero (Spain)
A. Otto (Germany)
W. L. Peticolas (USA)
V. V. Smirnov (Russia)
A. K. Sood (India)
G. J. Thomas (USA)
M. Tsuboi (Japan)
P. -I. Turpin (France)
N. -T. Yu (Hong Kong)
Shu-Lin Zhang (China)

Organizing Committee

Sanford A. Asher, Chairman University of Pittsburgh
PamelaM. Aker, University of Pittsburgh
Robert W. Baudoux, Sr., RWB Convention Management
Rob D. Coalson, University of Pittsburgh
John F. Jackovitz, Westinghouse
Paul W. Jagodzinski West Virginia University
Christine McCreary, Pittsburgh Energy Technology Center
Foil A. Miller, University of Pittsburgh
Linda A. Peteanu, Carnegie Mellon University
Paul B. Stein (Treasurer), Duquesne University
Sonja Straub
William Straub (Retired)
Cynthia Vanda, University of Pittsburgh
Gilbert C. Walker (Secretary), University of Pittsburgh
Robert E. Witkowski, University of Pittsburgh

Program Committee

Pamela M. Aker, University of Pittsburgh
Sanford A. Asher, (Chairman), University of Pittsburgh
Rob D. Coalson, University of Pittsburgh
Paul W. Jagodzinski, West Virginia University
Linda A. Peteanu, Carnegie Mellon University
Pal B. Stein, Duquesne University
Gilbert C. Walker, University of Pittsburgh

Scientific Advisory Committee

A. K. Arora (India)
R. Callender (USA)
M. Cardona (Germany)
P. Champion (USA)
R. K. Chang (USA)
B. Chase (USA)
R. J. H. Clark (UK)
E. da Silva (France)
J. Greve (Netherlands)
T. L. Gustafson (USA)
H. Hamaguchi (Japan)
R. E. Hester (UK)
P. Hildebrandt (Germany)
R. M. Hochstrasser (USA)
B. Hudson (USA)
P. V. Huong (France)
W. Kiefer (Germany)
J. Kincaid (USA)
T. Kitagawa (Japan)
M. Klein (USA)
N. Koroteev (Russia)
Y. Koyama (Japan)
S. Krimm (USA)
I. W. Levin (USA)
M. D. Morris (USA)
S. Mukamel (USA)
A. B. Myers (USA)
L. A. Nafie (USA)
Y. Ozaki (Japan)
W. L. Peticolas (USA)
A. Pinczuk (USA)
J. F. Rabolt (USA)
A. K. Ramdas (USA)
S. Sharma (USA)
Y. R. Shen (USA)
A. Sood (India)
T. G. Spiro (USA)
H. L. Strauss (USA)
H. Takahashi (Japan)
H. Takeuchi (Japan)
M. Tasumi (Japan)
J. Terner (USA)
G. J. Thomas, Jr. (USA)
P. Treado (USA)
M. Tsuboi (Japan)
D. Tuschel (USA)
G. Vergoten (France)
K. Yoshihara (Japan)

FIFTEENTH INTERNATIONAL CONFERENCE ON RAMAN SPECTROSCOPY

Proceedings of the Fifteenth International Conference on Raman Spectroscopy
August 11-16, 1996
Pittsburgh, PA
USA

Edited by

S. A. Asher
University of Pittsburgh
and
P. Stein
Duquesne University

JOHN WILEY & SONS

Chichester • New York • Brisbane • Toronto • Singapore

Telephone: National 01243 779777
International (+44) 1243 779777

Other Wiley Editorial Offices

John Wiley and Sons, Inc., 605 Third Avenue,
New York, NY 10158-0012, USA

Jacaranda Wiley Ltd, 33 Park Road, Milton,
Queensland 4064, Australia

John Wiley & Sons (Canada) Ltd, 22 Worcester Road,
Rexdale, Ontario M9W 1L1, Canada

John Wiley & Sons (Asia) Pte Ltd, 2 Clementi Loop #02-01,
Jin Xing Distripark, Singapore 0512

British Library Cataloguing in Publication Data

A catalogue record for this book is available from the British Library

ISBN 0 471 96747 5

Typeset by the authors
Printed and bound in Great Britain by Biddles Ltd, Guildford, Surrey

This book is printed on acid-free paper responsibly manufactured from sustainable forestation, for which at least two trees are planted for each one used for paper production

PREFACE

The Pittsburgh scientific community is proud to host the XVth International Conference on Raman Spectroscopy on the campus of the University of Pittsburgh. The conference is hosted by scientists from the entire Pittsburgh area, from the University of Pittsburgh, Duquesne University, Carnegie Mellon University and from companies in the Pittsburgh area. This is the fourth time the conference has been held in the USA. Previous conferences were held in Hong Kong (1994), Würzburg (1992), Columbia, South Carolina (1990), London (1988), Eugene, Oregon (1986), Tokyo (1984), Bordeaux (1982), Ottawa (1980), Bangalore (1978), Freiburg (1976), Brunswick, Maine (1974), Reims (1972), Oxford (1970) and Ottawa (1969).

On behalf of the International Steering Committee, the Scientific Program Committee and the Local Organizing Committee we welcome you to an exciting week of discussions of advances in Raman spectroscopy and related areas. This is an exciting time in Raman spectroscopy where advances in the fundamental understanding and instrumentation combine to create new opportunities to study molecular structure and dynamics. New applications have emerged which are having an important impact on industry and technology. The conference will include a large equipment exhibit, a workshop on new instrumental advances and short courses on Raman theory and instrumentation.

You will find that Pittsburgh is an exciting and vibrant city, where excellent science can be combined with adventure and camaraderie during walks in different ethnic neighborhoods and through its many parks, and by formal or casual dining in Pittsburgh's numerous restaurants. Pittsburgh, which was known as the "smoky city" was the center of industrialization in the United States early in this century. It is a city forged by immigrants from countries from all over the world.

We have received more than 650-camera ready manuscripts. This volume includes 579 manuscripts. Those manuscripts which could not be included due to space will be published in another volume with the postdeadline papers.

The first part of this volume contains the abridged texts of the Keynote Lecture (M. Dresselhaus) and twelve Plenary Lectures (T. Spiro, M. Klein, A. Myers, A. Ramdas, E. Da Silva, T. Cotton, N. Koroteev, A. Pinczuk, L. Ziegler, W. Kiefer, J. Valentini, and A. Heynes). The second part contains two-page summaries of invited oral papers (55) and poster papers (511). These papers are grouped according to the following sessions:

1. Raman Theory
2. Resonance Raman Theory
3. Time-Resolved Raman Scattering and Transient Species
4. Nonlinear Raman Scattering
5. Vibrational Analysis
6. Band Shapes/Band Intensities
7. Molecular Dynamics
8. Macromolecules and Polymers
9. Proteins
10. Heme Proteins
11. Lipids and Biomembranes
12. Nucleic Acids and Viruses
13. Amino Acids and Peptides
14. Biological Systems
15. Surfaces, Interfacial Phenomena
16. Surface Enhanced Scattering
17. Thin Films
18. Inorganic Materials
19. The Solid State
20. Superconductors
21. Carbon and Fullerenes
22. Semiconductors and Microstructures
23. Phase Transitions
24. Effects of Temperature and Pressure
25. Non-Crystalline Materials
26. Industrial Applications
27. Medical Applications
28. Raman Microscopy
29. Raman Imaging
30. New Techniques

The abridged texts and summaries were reproduced from camera-ready copy supplied by the authors who bear responsibility for any errrors, misprints and irregularities therein.

We express our sincere gratitude to those who provided co-sponsorship and financial support. We are especially appreciative of financial help from the Spectroscopy Society of Pittsburgh and the Society for Analytical Chemists of Pittsburgh for the generous help at the beginning of our efforts. We are indebted to the University of Pittsburgh which has generously donated its facilities. We are indebted to the Department of Chemistry of the University of Pittsburgh for financial and secretarial help. Crucial financial support was granted by the Coblentz Society and the Petroleum Research Fund. The following companies contributed to the Conference's Founding Members Fund: ChemIcon, Coherent, Dilor, EIC Laboratories, Instruments SA, Kaiser Optical Systems, Nicolet, Oriel, Princeton Instruments, Proctor and Gamble and Renishaw.

We also thank Ms. Charle-Pan Dawson and Sandy Law for their help in organizing the conference and creating this volume. We thank Cynthia Vanda, the University of Pittsburgh Director of Conference Coordination for her help and wise advice. We acknowledge and appreciate the thoughtful efforts of the Scientific Advisory Committee, the Program Committee and the Local Organizing Committee for the hard work and wisdom they showed in the selection of subjects and speakers for the conference.

CONTENTS

KEYNOTE LECTURE

PLENARY LECTURES

SECTION 1 - RAMAN THEORY

SECTION 5 — VIBRATIONAL ANALYSIS

SECTION 14 — BIOLOGICAL SYSTEMS

SECTION 15-SURFACE AND INTERFACIAL PHENOMENA

SECTION 16— SURFACE ENHANCED SCATTERING

SECTION 19-THE SOLID STATE

SECTION 20 — SUPERCONDUCTORS

SECTION 21 — CARBON AND FULLERENES

SECTION 25 — NON-CRYSTALLINE MATERIALS

SECTION 26 — INDUSTRIAL/ ANALYTICAL APPLICATIONS

SECTION 27 — MEDICAL APPLICATIONS

SECTION 29 — RAMAN IMAGING

SECTION 30 — NEW TECHNIQUES

ADDITIONAL PLENARY LECTURE

KEYNOTE LECTURE

RAMAN SPECTROSCOPY AS A TOOL FOR THE CHARACTERIZATION OF CARBONACEOUS MATERIALS

M. S. Dresselhaus

Massachusetts Institute of Technology, Cambridge MA 02139, USA

Carbon-based materials are of great scientific and practical interest because of the large variety of structures that can be prepared and the wide range of properties these materials display. The dominant forms of carbon include: two-dimensional graphite, the equilibrium semimetallic crystalline phase; the metastable three-dimensional wide bandgap semiconductor diamond lying ~20 meV/ carbon atom higher in energy; the zero-dimensional quantum dot closed-shell fullerene molecules of which C_{60} is the most common; and the one-dimensional quantum wire carbon nanotubes, which relate closely to fullerenes, graphite and carbon fibers. The talk will present an overview of the role that Raman scattering plays in the characterization of these basic forms of carbon and carbon-based materials derived from them.

The lowest energy state for carbon atom clusters in the 40-100 atom range is that of a closed shell molecule or a fullerene, where the C-C nearest neighbor distance is 1.44Å, very close to the corresponding 1.42Å distance in graphite. The C_{60} molecule is the most abundant of the fullerenes, has exceptionally high stability, and high icosahedral symmetry. Thus, despite the large number of degrees of freedom (180-6=174) of C_{60} this molecule has relatively few distinct vibrational mode frequencies (46) and a much smaller number of Raman-active modes (10). Other closed-shell fullerene molecules can be formed, all having even numbers of carbon atoms because of Euler's theorem for polyhedra. Because of the difference in symmetry and number of carbon atoms between the various fullerene molecules, Raman spectroscopy provides a highly sensitive tool to distinguish between each of the fullerene molecules, as for example C_{70} (the next most common fullerene) relative to C_{60}. In the condensed phase, Raman spectroscopy also yields valuable information on bonding between fullerene molecules, providing strong evidence for the molecular nature of the condensed phases of fullerene-derived solids. The molecular nature of crystalline C_{60} has been exploited to study the higher-order Raman spectra, as is frequently done in molecular spectroscopy, yielding mode frequencies and symmetries for the 32 silent mode frequencies of C_{60}. Study of the intermolecular modes for crystalline C_{60} as a function of temperature and pressure provides a sensitive probe for structural phase transitions. Light, however, is found to photo-transform the C_{60} molecule into a dimer or higher oligomer, which is formed through (2+2) cycloaddition intermolecular bonding in the temperature range $260 < T < 330$ K. Because of the small difference in bond length of the polymerized phase relative to the isolated molecules, Raman scattering can be used as a quantitative probe of the quantity of material that has been transformed to the polymerized phase, whether polymerization is induced by light, electron excitation, pressure, or doping with alkali metals. Raman spectroscopy has also been used successfully to study doped fullerene phases with regard to structure, electrical properties, and superconductivity. For example, Raman spectroscopy has been applied to study the magnitude of the electron-phonon coupling and to identify the modes with the largest electron-phonon coupling in M_3C_{60} phases (M = alkali metal species or alloys) which become superconducting at low temperatures. Just as Raman scattering has provided a sensitive probe of the charge transfer for doped fullerenes, this technique also provides valuable information regarding charge transfer from a metal surface to fullerenes adsorbed on the surface.

The one-dimensional carbon nanotube refers to a long carbon cylinder of nanometer diameter. This cylinder contains only a few atoms along the circumference of the cross section, is one atom thick in the radial direction and is capped by half of a fullerene at both ends of the cylinder. The atomic arrangement of carbon atoms along the cylinder is that of a rolled up single layer of crystalline graphite (called a graphene sheet). Both single-wall carbon nanotubes, as described above, and multi-wall nanotubes, consisting of the concentric stacking of these cylindrical single-wall structures, have been synthesized and characterized with regard to their structure and properties. The remarkable property which has attracted most attention is the ability of carbon nanotubes to exhibit either metallic or semiconducting electrical properties, depending on their geometry which is denoted by the tube diameter and chiral angle. The chiral angle is described by the angles that the unit vectors of the graphene sheet make with respect to the axis of the cylindrical nanotube. The one-dimensional quantum features in carbon nanotubes are associated with the periodic boundary conditions imposed upon the small number of carbon atoms in the circumferential direction of the cylinder. To observe these quantum effects it would be necessary to measure the Raman spectra for small diameter (3 nm) nanotubes of known diameter and chiral angle because of the dependence of the phonon mode frequencies on tubule geometry. Despite the large number of phonon modes in the smallest carbon nanotubes (those that are capped by C_{60} hemispheres) and the rapid increase in the number of modes with increasing tube diameter, the number of modes that are Raman active remains small and independent of tubule diameter; for example, only modes of A, E_1 and E_2 symmetries (15 in number) are Raman active for chiral nanotubes of arbitrary chiral angle and diameter. To date most Raman studies of carbon nanotubes have been carried out on poorly characterized carbon nanotube samples containing a wide distribution of tubule diameters and chiralities, and low concentrations of small diameter (<3 nm) single-wall nanotubes. Recently Raman spectra have been reported for a sample containing 40 single-wall carbon nanotubes, with average diameters of ~1.5 nm. The Raman spectra show broad features identified with contributions from nanotubes of various diameters and chiralities. Further progress on this topic depends critically on the availability of well-characterized samples of carbon nanotubes of small diameters.

Crystalline graphite is a material with unique properties, exhibiting extreme structural anisotropy, thereby giving rise to highly anisotropic physical properties. Graphite has four atoms per unit cell and crystallizes in the $P6_3/mmc$ D^4_{6h} space group with two Raman-active modes at 48 cm^{-1} E_{2g1} and 1582 cm^{-1} E_{2g2} with the low frequency E_{2g1} mode sensitively dependent on the interlayer displacements, but the higher frequency E_{2g2} mode strongly dependent on in-plane displacements. Thus the 1582 cm^{-1} mode is commonly used to characterize graphite-derived forms of carbon, many of which have important industrial applications. Raman scattering was first used by Tuinstra and Koenig in 1970 to distinguish several forms of carbon from each other, including single-crystal graphite, pyrolytic graphite, activated charcoal, and carbon blacks. In these studies, all disordered graphite samples showed two prominent Raman features, with peaks at 1580 and 1335 cm^{-1}. They assigned the upper frequency peak 1580 cm^{-1} to the in-plane E_{2g2} Raman-allowed mode vibration, and the second peak (1355cm^{-1} was identified with in-plane disorder-induced scattering from graphitic modes throughout the Brillouin zone, but emphasizing modes near the zone edge, where the phonon density of states has a strong maximum. Of particular importance for characterizing the amount of disorder in a graphite-derived material is the close correlation between the in-plane crystalline size L_a as established by x-ray diffraction and the ratio of the integrated intensity of the disorder-induced line (1355 cm^{-1}) to the first-order Raman-allowed line (1580 cm^{-1}). These two Raman features have been used to monitor the degree of disorder of graphite-derived materials and the graphitization process of a wide variety of carbons as they are heat treated at elevated temperatures (up to 3200 °C), including carbon fibers, glassy carbon and various porous carbons. For example, the properties of carbon fibers, such as their strength, modulus and thermal conductivity, which are all commercially exploited, are strongly dependent on the degree of graphitization of these carbons and this property can be conveniently monitored by Raman scattering. In this sense Raman scattering has also been used to characterize and gain physical insight into porous carbons such as activated carbons, activated carbon fibers and carbon aerogels. For the case of carbon aerogels, Raman spectroscopy was among the first techniques to show that the small particles in the "string-of-pearls" microstructure of carbon aerogels is essentially independent of the mass density of the aerogel, as this mass density varies over an order of magnitude. Disordered carbons are now being considered for many new applications, such as for lithium ion batteries. For this application, PPP (poly-paraphenylene) seems especially promising as an electrode precursor material because of its superior ability when heat treated at 700 °C for lithium uptake and hence battery charging capacity, approximately three times greater than that of single crystal graphite. Through study of the carbonization process of PPP as a function of temperature by Raman spectroscopy, new insights into the mechanism for Li uptake have been provided.

Raman scattering has been one of the main tools for studying and characterizing graphite intercalation compounds in which layers of guest species are introduced between the weakly coupled graphene layers. Through Raman spectroscopy, the integrity of the graphene sheets during the intercalation process has been verified, charge transfer between the intercalate and graphene layers has been determined leading to the use of Raman spectroscopy as a non-destructive tool for measurement of the intercalate uptake in graphite intercalation compounds. Raman spectroscopy has also been shown to provide a powerful tool for studying the intercalate layers when these layers contain molecular species that have Raman-active modes.

Because of the different crystal structures and different C-C bond lengths in two-dimensional graphite and three-dimensional diamond, their Raman spectra are very different, thus providing a non-destructive characterization tool for distinguishing between the sp^2 C-C bonding that is characteristic of graphite and the sp^3 C-C bonding associated with diamond in materials of commercial interest such as polycrystalline diamond films, and hydrogen-terminated diamond-like films. Since the Raman-active line in diamond (at 1334 cm^{-1}) has a Raman cross-section 50 times smaller than that for graphite (at 1582 cm^{-1}), Raman scattering provides a very sensitive tool for the detection of residual graphite in a polycrystalline diamond or diamond-like film.

Thus Raman spectroscopy offers an important characterization tool for a broad spectrum of carbonaceous materials, including fullerenes, carbon nanotubes, graphites and diamond. Additional insights into carbon-based materials overall have been provided by common features observed in the Raman spectra of different types of carbons, such as between fullerenes, carbon nanotubes and graphite, and also between graphite intercalation compounds and doped fullerene-based solids.

PLENARY LECTURES

HEMOGLOBIN DYNAMICS FROM TIME RESOLVED RESONANCE RAMAN SPECTROSCOPY

T. G. Spiro,[1]* V. Jayaraman,[1] K. R. Rodgers,[2] Ishita Mukerji[3]

[1]Department of Chemistry, Princeton University, Princeton, NJ 08544
[2]Department of Chemistry, North Dakota State University, Fargo, ND 58105
[3]Department of Biochemistry and Molecular Biology, Wesleyan University, Middletown, CT 06457

The advent of reliable ultraviolet lasers has made possible the study of proteins using ultraviolet resonance Raman [UVRR] spectroscopy. Vibrational modes of aromatic side chains, and of the peptide bonds, can be resonantly enhanced, and used to probe secondary and tertiary structure. When carried out in the time-resolved mode, UVRR spectroscopy can provide structural information concerning dynamical processes in proteins. Excitation at 230 nm is particularly useful, because tyrosine and tryptophan modes are selectively enhanced at this wavelength [1]. Because the frequencies and intensities are sensitive to H-bonding, hydrophobic effects, and sidechain conformation, these modes can serve as markers of conformation change in specific regions of the protein.

A fruitful application of time-resolved UVRR spectroscopy is the study of allostery, the process by which proteins change their shape upon binding small molecules or ions [2]. For several proteins, x-ray crystallography has provided atomic resolution structures of the end states of allosteric transitions, but the pathways leading from one end state to another are unknown. A prime objective of current research is to understand how protein architectures accommodate bistable structures, and what molecular motions constitute the allosteric reaction coordinates.

Hemoglobin [Hb] offers an attractive experimental system for investigating the allosteric transition, because of an extensive knowledge base upon which new techniques and hypotheses can be tested. The tertiary and quaternary structure differences between the low- and high-affinity states, R and T [2], have been analyzed in detail. The allosteric transition can be studied by kinetic methods, using the photo-dissociation of HbCO to generate unligated Hb in the R structure, from which it evolves toward the equilibrium T structure. Transient absorption spectroscopy has provided rate constants for several kinetic steps [4] subsequent to HbCO photolysis. The allosteric pathway can in principle be elucidated by determining the structures of the intermediates defined by these rate constants.

We have applied resonance Raman spectroscopy to this task, using UV excitation to monitor the aromatic side chains, and also visible excitation to enhance vibrational modes of the heme group, where the primary ligation events occur [3]. The aromatic residues are distributed through the protein, and provide useful markers of structure change throughout the allosteric transition. We have obtained time-resolved spectra from 10 to 10^5 ns following HbCO photolysis, using both visible and UV excitation, and have extracted spectra of the kinetic intermediates (Figure 1). These spectra permit construction of a working model of the allosteric transition, that involves reciprocating tertiary and quaternary motions, as illustrated in Figure 2.

UVRR Spectra and Quaternary Contacts: 230 nm - excited RR spectra of hemoglobin (Hb) were obtained with sufficient signal/noise to produce difference spectra between ligated and unligated (deoxyHb) forms of the protein, which contain well-defined and reproducible tyrosine and tryptophan bands [1]. These difference signals are attributed mainly to changes in H-bond contacts across the $\alpha_1\beta_2$ interface of the Hb tetramer between the R and T quaternary states [2]. The interfacial H-bond contacts involve the Trp β37 and Tyr α42 residues. Using the mutant Hb Rothschild, in which the Trp β37 residue is replaced by arginine, it was possible to establish that the W3 band in the static difference spectrum is entirely due to Trp β37 [1]. The two interior residues, Trp α14 and β15 make no contribution to this band, indicating that the environment of these residues is the same in the R and T states. The contributions of the different Trp residues can be distinguished because the W3 frequency differs for Trp β37 and for Trp α14 and β15 owing to different dihedral angles connecting the indole rings to the backbone.

A large bisignate signal for the Y8a and Y8b tyrosine bands in the static T-R difference spectrum is assigned to the interface residue Tyr α42, since Tyr α42 forms an H-bonded contact with the sidechain of Asp β99, and the Y8a and Y8b bands are known to be sensitive to H-bonding [1]. However, the direction of the frequency shifts is opposite to that expected if Tyr α42 is an H-bond donor to ionized carboxylate, as has generally been supposed; instead it is the expected direction for H-bond acceptance from unionized carboxylic acid. The Asp β99 side chain is exposed to solvent in the R state, but protected from solvent in the T state, consistent with the hypothesis of Asp β99 protonation in the T state. Asp β99 may therefore be an important contributor to the Bohr effect.

Time-resolved UVRR Spectra: The time course of the UVRR intensities, following HbCO photolysis with a 10 ns 419 nm laser pulse, was fit to a series of exponential decays, with time constants of 0.02, 0.50 and 17 μs at pH 7.4 [3], in good agreement with the optical transient analysis of Hofrichter et al. [4]. These time constants were used to extract the difference UVRR spectra of intermediates on the R-T reaction coordinate. The 0.50 and 17 μs intermediates, called S and T' (Figure 2), have difference spectra resembling the static T-R difference spectrum (Figure 1), but with lower amplitudes. These intermediates therefore involve formation of the quaternary contacts. This inference requires that rearrangement of the subunits occurs in 0.5 μs, although the 17 μs transient had been assumed to be the quaternary transition [4].

The 0.02 μs intermediate, called R, has an altered difference spectrum. This spectrum is attributed to a purely tertiary change, associated with the loss of ligand within the R quaternary state, on the basis of its similarity to spectra of hybrid Hb's which are in the R state but have deligated chains [5,6]. The spectrum contains large negative bands, indicating blue-shifts in the aromatic excitation profiles, as would occur upon weakening of H-bonds. The position of the W3 difference band implicates Trp α14 and β15 as the responsive residues; Trp β37 makes no contribution. The Trp signals are therefore attributed to movement of the E helix, which lines the distal side of the heme pocket, toward the heme [5]. The E helix motion weakens the H-bonds between Trp α14 and 15, which are located on the A helices, and their H-bond acceptors on the E helix, producing the negative RR difference bands.

Time-resolved visible RR spectra: 436 nm-excited RR spectra of the initial photoproduct, B, showed deoxy-heme skeletal modes at slightly depressed frequencies, indicating porphyrin core expansion [3]. Also the stretching mode of the bond between the iron and the proximal histidine ligand (Fe-His) was elevated, 228 cm^{-1}. Together with the absence of any UVRR spectral changes in the initial photoproduct [3], these features indicate that molecular strain is localized at the heme group, resulting in core expansion and a compressed Fe-His bond. The strain relaxes in the 0.02 μs transient, since the intermediate R has normal deoxy-heme skeletal mode frequencies and a lowered Fe-His frequency, 224 cm^{-1}, close to that of deoxy-myoglobin.

In addition, half the photolyzed CO molecules recombine geminately in the same 0.02 μs transient, as determined by the intensity of the CO-heme ν_4 band. It is proposed that the E helix displacement toward the heme, detected in the UVRR spectrum of R, induces geminate rebinding of CO in half the chains, and departure of the CO in the other half. Thus the intermediates S and T' have two chains bound by CO, on average. The fact that the UVRR difference intensities for T' are only half those of deoxHb is interpreted as reflecting a weakened quaternary interface in tetramers having two ligands on the same dimer unit, which is the most stable arrangement for two ligands [6]. The time-resolved UVRR spectra also reveal a recombination intermediate, U, which has the same difference bands as R (Figure 1), consistent with the expected R state structure for tetramers having three ligands.

The Fe-His frequency changes very little between R and S, but is significantly lowered (216 cm^{-1}) in T', as it is in deoxyHb. Thus, the 17 μs transient is interpreted as a tertiary motion, probably involving the F helix, which strains the Fe-His bond and locks the quaternary contacts into place.

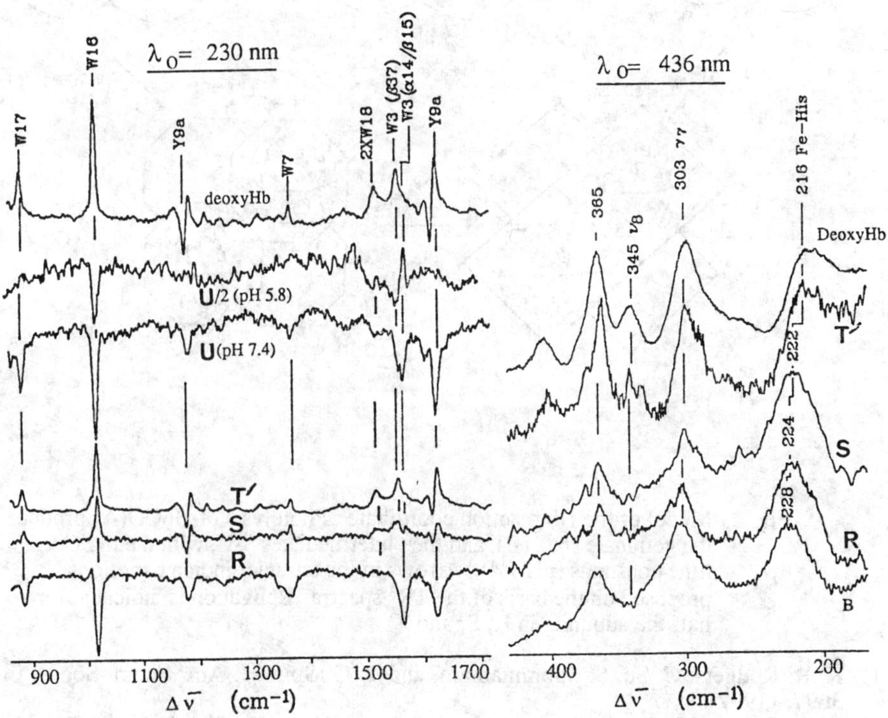

Fig. 1. Low-frequency heme RR spectra (right), and UVRR difference spectra (left) of the intermediates B-U extracted from time-resolved spectra; species populations were calculated with the experimental time constants [3]. The static spectrum (right) and difference spectrum (left) of deoxyHb are shown at the top for comparison.

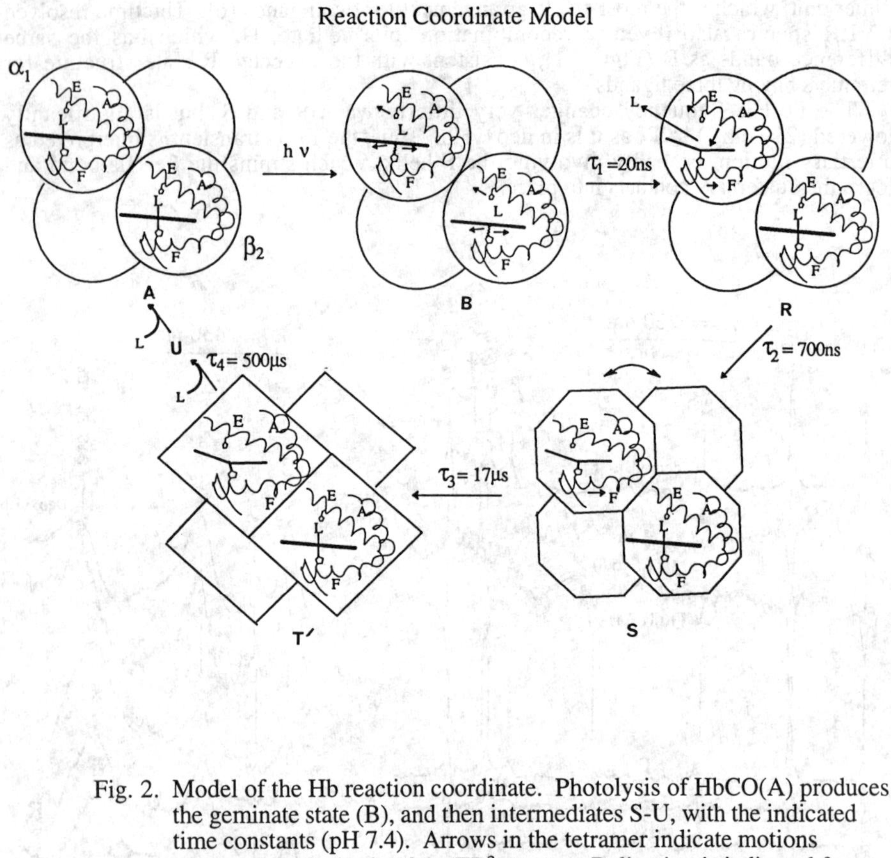

Fig. 2. Model of the Hb reaction coordinate. Photolysis of HbCO(A) produces the geminate state (B), and then intermediates S-U, with the indicated time constants (pH 7.4). Arrows in the tetramer indicate motions proposed on the basis of the TR^3 spectra. Religation is indicated for half the subunits in R, S, and T'.

1. K. R. Rodgers, C. Su, S. Subramaniam, and T. G. Spiro, J. Am. Chem. Soc., 114, 3697, (1992).
2. M. F. Perutz, *Mechanisms of Cooperativity and Allosteric Regulation in Proteins* (Cambridge University Press, Cambridge, 1990).
3. V. Jayaraman, K. R. Rodgers, I. Mukerji, and T. G. Spiro, Science, 269, 1843 (1995).
4. J. Hofrichter, J. H. Sommer, E. R. Henry, and W. A. Eaton, Proc. Natl. Acad. Sci., USA, 80 , 173 (1983).
5. K. R. Rodgers and T. G. Spiro, Science, 265, 1697 (1994).
6. V. Jayaraman and T. G. Spiro, Biochemistry, 34, 4511 (1994) .

RAMAN PROBING OF ELECTRONIC AND MAGNETIC EXCITATIONS IN CUPRATE HIGH TEMPERATURE SUPERCONDUCTORS

M. V. Klein[1] and Girsh Blumberg[1]

[1]Department of Physics and Science and Technology Center for Superconductivity, University of Illinois at Urbana-Champaign, 104 S. Goodwin Avenue, Urbana IL 61801 USA

Raman scattering was performed on high temperature superconductors (HTS) soon after their discovery, but it took several years before decent single crystals became available and reproducible results could be obtained. [1] Raman scattering has revealed much about lattice vibrations and how they are affected by the carriers and by superconductivity. Here we will concentrate on scattering from carriers (electronic Raman scattering) and from spin excitations. Before discussing experimental data, we must first mention the underlying theoretical framework. Because of the complexity of the HTS field, many aspects of the theory are still poorly understood or missing all together.

The superconducting state in HTS is characterized by BCS-like spin-singlet pairing; nevertheless, these materials are peculiar relative to conventional superconductors, and a persistent issue is the extent to which the peculiarities affect fundamental explanations of physical properties. Compared with low temperature superconductors, HTS have far fewer superconducting carriers in a coherence volume (25 or less). This allows for much larger fluctuations, e.g., of superconducting regions persisting above the transition temperature T_c. In addition, the energy gap is a much (~ 10 times) larger fraction (~ 1/25) of the width of the energy band(s) in which the carriers move, and for most interesting cuprates the density of energy states is quite non-uniform on the energy scale of the superconducting gap. Qualitative differences are very important for Raman scattering. The non-superconducting state that exists above T_c is anomalous. This abnormal "normal" state is characterized by strong inelastic scattering of carriers (electrons and holes) by other carriers. As the temperature goes to zero in an ordinary metal, carriers having wave vector k and low energy $\varepsilon=\varepsilon(k)$ with respect to the chemical potential have a lifetime τ that grows as ε^{-2} as ε tends to zero, if superconductivity does not intervene. This behavior, which is that expected for a so-called “Fermi liquid” means that the carrier energy becomes increasingly well-defined as its energy becomes smaller. For the cuprate HTS τ grows only as ε^{-1}, and the carrier energy remains somewhat ill-defined as its energy becomes smaller. This non-Fermi liquid behavior was pointed out early by Anderson [2] and forms the basis for the "marginal Fermi liquid" phenomenology [3] and the "nested Fermi liquid theory" [4]. Lifetime broadening of the carrier energy ε is comparable to ε or thermal energy k_BT, whichever is larger.

The BCS theory of superconductivity [5] allows one to calculate many properties of superconductivity, including electronic Raman spectra. [6] Although it is based on a Fermi liquid model for the normal state, BCS theory has been applied to HTS by many workers, in the hope that the greater lifetime broadening and other peculiarities make only a quantitative difference. BCS theory says that to calculate an energy spectrum of excitations in the superconducting state, one must use the superconducting quasi particle energy $E(k)= [\varepsilon(k)^2+\Delta(k)^2]^{0.5}$, instead of $\varepsilon(k)$, where $\Delta(k)$ is the superconducting gap parameter. Excitation probabilities must also be multiplied by coherence factors to account for the fact that the quasi particles are superpositions of electrons and holes.

Upon neglect of effects of lifetime broadening, one obtains from BCS theory at T=0 a differential Raman cross-section internal to the crystal that is the Thomson electron radius squared multiplied by

$$S(\gamma;\omega) = -\pi^{-1} Im\ U(\gamma^2;\omega) , \tag{1a}$$

$$U(f;\omega) = 4\sum_k f(k)\Delta(k)^2/\{E(k)[\omega^2-4E(k)^2+i\delta]\} . \tag{1b}$$

Here the sum is over all wave vectors k in the first Brillouin zone of the lattice (and, implicitly, over those different bands whose carrier energies $\varepsilon(k)$ pass through zero); ω is the Raman frequency shift in energy units; and $\gamma(k)$ is the light scattering "vertex", the amplitude for scattering of a photon with wave-vector transfer q and energy transfer ω, wherein a carrier scatters from k to k+q, in the limit as q goes to zero. $\delta = 0^+$. The coupling to photons occurs via the kinetic energy of the carriers in the band(s) and leads to the expression for $\gamma(k)$,

$$\gamma(k) = \mathbf{e}_i \cdot \mu(k) \cdot \mathbf{e}_s , \tag{2}$$

where $\mathbf{e}_{i,s}$ denote polarization vectors for incident and scattered light, and $\mu(k)$ is the symmetric, dimensionless inverse effective mass tensor at wave vector k. The difference between μ and the unit tensor is due to virtual transitions induced by light between the band under consideration and other bands. If any such transitions resonate with the frequency of the photons, the definition of μ must be appropriately generalized (and it may no longer be symmetric).

The carriers and the spins which we expect to produce most of the electronic and magnetic Raman scattering are found on the CuO_2 planes of the cuprates. We customarily use the tetragonal symmetry of these planes rather than the true orthorhombic symmetry of many cuprates HTS, and for light polarized in these planes the Raman symmetries and irreducible components of $\mu(k)$ are as follows:

$$A_{1g}:\ [\mu_{xx}(k) + \mu_{yy}(k)]/2 , \tag{3a}$$

$$B_{1g}:\ [\mu_{xx}(k) - \mu_{yy}(k)]/2 , \qquad B_{2g}:\ \mu_{xy}(k) . \tag{3b}$$

In the case of A_{1g} symmetry, one must account for screening by the Coulomb interaction. One should replace Eq. (1a) by

$$S(\gamma;\omega) = -\pi^{-1} Im\, \hat{U}(\gamma^2;\omega) , \quad (4a)$$

$$\hat{U}(\gamma^2;\omega) = U(\gamma^2;\omega) - [U(\gamma;\omega)]^2/U(1;\omega) . \quad (4b)$$

The experiments on hole-doped cuprates near optimum doping give a peak in B_{1g} symmetry at a frequency that we will label $2\Delta_0$. In B_{2g} symmetry, a peak, if present, is observed at a lower frequency and with lower intensity. The A_{1g} spectra are puzzling. They are weak, if present, in single layer cuprates such as $La_{2-x}Sr_xCuO_4$ [7] and $Tl_2Ba_2CuO_{6+x}$. [8] For $YBa_2Cu_3O_{6+x}$, which has been studied the most, the A_{1g} intensity is comparable to the B_{1g} intensity, but the peak is at a frequency that is perhaps 35% lower than $2\Delta_0$. [9] Neglecting the effects of dynamical coupling between the CuO_2 planes, one expects the screening effect [Eq. (4b)] to reduce the A_{1g} intensity very drastically. The inter-layer coupling is strong for $YBa_2Cu_3O_{6+x}$ (YBCO) at optimal doping where x is close to 1.0, but it is much weaker for $Bi_2Sr_2CaCu_2O_{8+x}$ (Bi2212); nevertheless the intensities of the spectra at optimal doping are similar to those for YBCO. [10] Theoretical treatments of interlayer coupling in the Raman context are just beginning to appear. [11,12]

The effects of finite lifetime on Raman spectra have been treated theoretically for scattering from impurities in the superconducting state, for constant gap function [13] and for the d-wave pairing case, where $\Delta(k)$ changes sign along the lines $k_x=\pm k_y$. [14] For inelastic scattering of carriers by other carriers the only treatment thus far is for the normal state using the nested Fermi-liquid model. [15]

The HTS cuprate superconductors are obtained by doping carriers into a "parent state" that is a spin 1/2 antiferromagnetic Mott-Hubbard, charge-transfer insulator. Spins localized on Cu atoms in the CuO_2 planes interact via a Heisenberg Hamiltonian on a square lattice

$$H = J\sum_{<i,j>}(\mathbf{S}_i \cdot \mathbf{S}_j - 1/4) , \quad (5)$$

with superexchange constant $J \approx 125$ meV ≈ 1000 cm^{-1} and a Néel temperature of $\approx$300K. There is much evidence for the persistence of strong electronic correlations of the parent state into the metallic, superconducting state. These effects hamper the straight-forward application of BCS theory and complicate data interpretation.

Insight into the strong correlation effects may be gained by doping-dependent Raman studies, starting with the insulators. The most dramatic feature is the two-magnon peak at about 3000 cm^{-1}, due to photon-induced excitation of two flipped spins out of the ordered Néel state. [16] Recent resonance Raman work has yielded new insights. The lowest intermediate state is the charge-transfer band wherein a hole from the singly-occupied Cu $3d_\sigma(x^2-y^2)$ band is transferred to the O $2p_{x,y}$ bands where it forms a singlet state [17] with another Cu 3d hole. Detailed calculations of resonant two-magnon Raman scattering have been performed for the Hubbard model of the insulating state using the spin-density-wave approximation to "de-couple" the repulsive U-term and obtain a band structure for the occupied and un-occupied bands. [18] With realistic parameters for the dispersion of these bands and of the magnon bands, it is found that the strongest resonance is a "triple-resonance" having three singular energy denominators. The strongest resonance occurs at energy E_g^{max} at the top of the charge-transfer band. The intensity is predicted to diverge as $1/|E_i - E_g^{max}|$, where E_i is the incident laser photon energy. Such behavior was observed in resonance Raman experiments on insulating YBCO with $E_g^{max} \approx 3.1$ eV. [19]

Oxygen doping of YBCO adds holes and produces an insulator-superconductor transition. The two-magnon peak weakens and broadens, but it is still seen in superconducting YBCO. [20] Its resonance Raman behavior is weaker, but similar to that of the insulating case. Moreover, the low-energy portion of the two-magnon band has a flat portion that is indistinguishable from the universal continuum seen in all doped cuprates. The consequences of this fact merit future exploration.

We thank our many collaborators, especially M. Kang and P. Abbamonte, and NSF for support under DMR 93-20892 and 91-20000.

1. Reviews: C. Thomsen and M. Cardona, Physical Prop. of High Temperature Superconductors I, ed. D. M. Ginsberg (World Sci., Singapore, 1989), p. 409; R. Feile, Physica C 159, 1 (1989); S. L. Cooper and M. V. Klein, Comments in Condensed Matter Physics 15, 99 (1990); C. Thomsen, Light Scattering in Solids VI, ed. M. Cardona and G. Güntherodt (Springer-Verlag, Heidelberg, 1991), p. 285.
2. P. W. Anderson, Strong Correlation and Superconductivity, ed. H. Fukuyama, S. Maekawa, and A. Malozemoff (Springer, Berlin, 1989).
3. C. M. Varma, P. B. Littlewood, S. Schmitt-Rink, E. Abrahams, and A. E. Ruckenstein, Phys. Rev. Letters 63, 1996 (1989).
4. A. Virosztek and J. Ruvalds, Phys. Rev. B 42, 4064 (1990); J. Ruvalds and A. Virosztek, Phys. Rev. B 43, 5498 (1991).
5. J. Bardeen, L. N. Cooper, and J. R. Schrieffer, Phys. Rev. 108, 1175 (1957).
6. M. V. Klein and S. B. Dierker, Phys. Rev. B 29, 4976 (1984).
7. X. K. Chen et al., Phys. Rev. Letters, 73, 3290 (1994).
8. R. Nemetschek et al., Phys. Rev. B 47, 3450 (1993).
9. For recent results see X. K. Chen et al., Phys. Rev. B 48, 10530 (1993).
10. C. Kendziora and A. Rosenberg, Phys. Rev. B 52, 9867 (1995).
11. M. Krantz and M. Cardona, J. Low Temp. Physics 99, 215 (1995).
12. T. P. Deveraux, A. Virosztek, and A. Zawadowski, preprint (1996).
13. T. P. Deveraux, Phys. Rev. B 45, 12965 (1992)
14. T. P. Deveraux, J. Phys. Chem. Solids 56, 1771 (1995).
15. A. Virosztek and J. Ruvalds, Phys. Rev. B 45 , 347 (1992).
16. K. B. Lyons et al., Phys. Rev. B 37, 2353 (1988).
17. F. C. Zhang and T. M. Rice, Phys. Rev. B 37, 3759 (1988).
18. A. V. Chubukov and D. M. Frenkel, Phys. Rev. Letters 74, 3057 (1995); Phys. Rev. B 52, 9760 (1995).
19. G. Blumberg et al., Phys. Rev. B 53 (1996), to be published.
20. G. Blumberg et al., Phys. Reb. B 49, 13295 (1994).

COMPARISON OF TIME- AND FREQUENCY-DOMAIN RESONANCE RAMAN SPECTROSCOPIES FOR TRIIODIDE IN SOLUTION

Alan E. Johnson and Anne B. Myers*

Department of Chemistry, University of Rochester, Rochester, NY 14627-0216

Raman scattering is traditionally detected in the frequency domain by using a monochromator or spectrograph to resolve the scattered radiation into its various wavelength components. Even the "Fourier transform" methods often applied to near-IR Raman spectroscopy still involve detecting the response of the material system to a continuous, quasi-monochromatic perturbation. However, the recent development of lasers reliably generating pulses shorter than many molecular vibrational periods has spurred the development of true "all-time-domain" vibrational spectroscopies such as Impulsive Stimulated Raman Scattering (ISRS),[1] in which a single light pulse excites molecular vibrations which are then probed directly in the time domain by a second pulse. The resulting time-dependent signal can be Fourier transformed to give vibrational information. These methods are most valuable for low-frequency vibrations that are difficult to separate from scattered laser light in traditional Raman experiments, but as ever-shorter laser pulses become available, the range of vibrational frequencies over which the two techniques overlap continues to expand.

Under conditions of electronic resonance, spontaneous resonance Raman (RR) spectroscopy contains specific information about the electronically resonant state *via* the ground-state RR intensities.[2] There can, however, be complications due to the electronically resonant nature of the excitation, notably interference from the fluorescence that occurs when dephasing processes allow a true population to be established in the resonant electronic state. The time-domain analog, resonant ISRS (RISRS),[3-5] similarly carries information about both ground and resonant electronic states. In general, electronically resonant excitation with a short pulse generates vibrational coherences on both surfaces, and the contributions to the pump-probe signal from the ground state (due to absorption depletion) and the excited state (due to stimulated emission and/or transient absorption) can be difficult to separate clearly. Both experiments are simplified if the upper state is directly dissociative, as excited-state oscillations will then be absent from the RISRS data and the RR scattering will exhibit little or no interfering fluorescence. But even under these idealized conditions, the Fourier transform of the RISRS signal (FT-RISRS) is not simply equivalent to a RR spectrum.[4,6,7]

In RR with a photodissociative excited state, a nearly monochromatic laser field transfers part of the initial ground-state wavefunction to the excited electronic state, and the spectrum of the spontaneously emitted light reflects the evolution of that amplitude on the dissociative surface. The cross section for resonance Raman scattering at frequency $\omega_{LS} = \omega_L - \omega_S$ can be conveniently cast in a "time-domain" form[2,8] as

$$\sigma_R \propto \omega_S^3 \omega_L \sum_{i,f} P_i \int_0^{\infty} d\omega_0\, G(\omega_0) \left| \int_0^{\infty} \langle f|i(t)\rangle \exp[i(\omega_L+\omega_0+\omega_i)t]\, dt \right|^2 \frac{\gamma_{fi}/\pi}{(\omega_{LS}-\omega_{fi})^2+\gamma_{fi}^2} \quad (1)$$

Here $G(\omega_0)$ is an inhomogeneous distribution of electronic zero-zero frequencies, γ_{fi} is the ground-state Raman linewidth, P_i is the Boltzmann population of state $|i\rangle$, and $|i(t)\rangle = \exp(-iH_e t/\hbar)|i\rangle$ is the initial ground-state wavefunction propagated by the excited-state vibrational Hamiltonian. The time-domain RISRS signal can be written in analogous fashion. Adapting eq. (17) of ref. [4] and dropping the excited-state population and coherent coupling terms, we obtain

$$S(\tau) \propto \mathrm{Re} \sum_i P_i \sum_f \int_0^\infty d\omega_0 G(\omega_0) \int_{-\infty}^{\infty} dt\, E_{pr}^*(t-\tau) \int_0^\infty dt_1\, E_{pr}(t-\tau-t_1) \int_0^\infty dt_2 \int_0^\infty dt_3$$
$$\{E_{pu}(t-t_1-t_2-t_3)\, E_{pu}^*(t-t_1-t_2) \exp[-i\omega_0(t_1+t_3)+i\omega_i(t_1+t_2+t_3)-i\omega_f t_2-\gamma_{fi}t_2]\langle i(-t_1)|f\rangle\langle f|i(t_3)\rangle$$
$$+E_{pu}^*(t-t_1-t_2-t_3)\, E_{pu}(t-t_1-t_2) \exp[-i\omega_0(t_1-t_3)-i\omega_i(t_2+t_3)+i\omega_f(t_1+t_2)-\gamma_{fi}t_2]\langle i(t_3)|f\rangle\langle f|i(t_1)\rangle\} \qquad (2)$$

where E_{pu} and E_{pr} are the fields of the pump and probe pulses. The RR and RISRS signals depend on the same material properties but in different ways. In particular, inhomogeneous broadening affects the two spectra differently because the contributions from spectroscopically distinct molecules (different initial vibrational states or zero-zero frequencies) are added in the frequency domain in RR, while in RISRS the contributions of different molecules are added in the time domain and then Fourier transformed to give a "frequency-domain" spectrum. However, even in the absence of any inhomogeneous broadening the FT-RISRS and RR spectra at the same excitation frequency can be very different.

We have measured and modeled the RR excitation profiles of triiodide, I_3^-, in several solvents[9,10] in order to gain more insight into the photodissociation dynamics that lead to the formation of coherently vibrating I_2^- products[11] and for comparison with the RISRS experiment on the parent I_3^-.[12] Fig. 1 compares the experimental RR and FT-RISRS spectra of I_3^- in ethanol at nearly the same excitation wavelength. While both spectra are dominated by the symmetric stretch fundamental (ν_1 at 112 cm^{-1}), the antisymmetric stretch (ν_3 at 145 cm^{-1}) also has considerable intensity in the RR, but is nearly undetectable in the FT-RISRS. The appearance of this nominally forbidden transition indicates that the solvent environment breaks the symmetry, making the two I-I bonds inequivalent on the Raman time scale. The intensity of this mode is highly solvent dependent, as discussed elsewhere.[10] The vibrational linewidths in the RR spectrum also appear about a factor of two broader than those obtained from both the FT-RISRS and nonresonant Raman, an observation that was initially difficult to understand since all of these techniques are expected to yield the same ground-state vibrational linewidths.

Computational modeling of the RR and RISRS data allows us to explore the source of these discrepancies. The excited-state potential surface and ground-state linewidth parameters were parameterized by fitting to the experimental absorption spectrum and absolute resonance Raman intensities at eight wavelengths. The same potential parameters, along with the experimental pulse characteristics,[12] were then employed in the RISRS calculations. We reproduce the intensity in ν_3 by employing a "skewed" LEPS potential for the excited state,[9,10] and the apparent additional line broadening on resonance is attributed to unresolved combination bands of the I_3^- modes with one or more low-frequency solvent modes coupled to the electronic transition.[7,10] No explicit inhomogeneous site broadening is needed to reproduce the absolute RR intensities, but there is significant thermal population of initial states, particularly the solvent mode (modeled as a 13 cm^{-1} frequency).

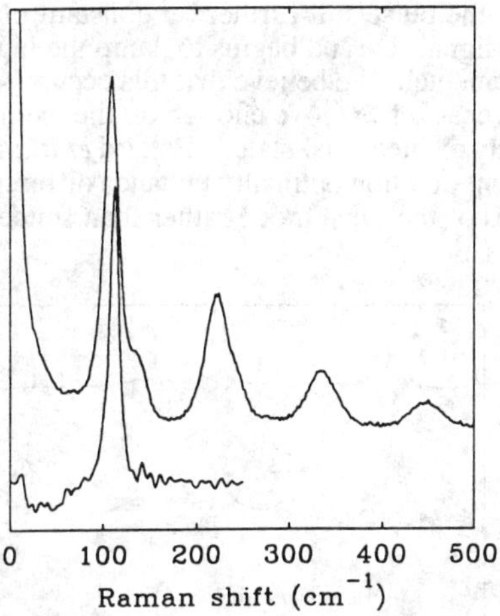

Fig. 1. Top: Experimental RR spectrum of I_3^- in ethanol obtained with excitation at 309 nm. Bottom: Experimental FT-RISRS signal with 60 fs, 308 nm pump and probe pulses (ref. [12].)

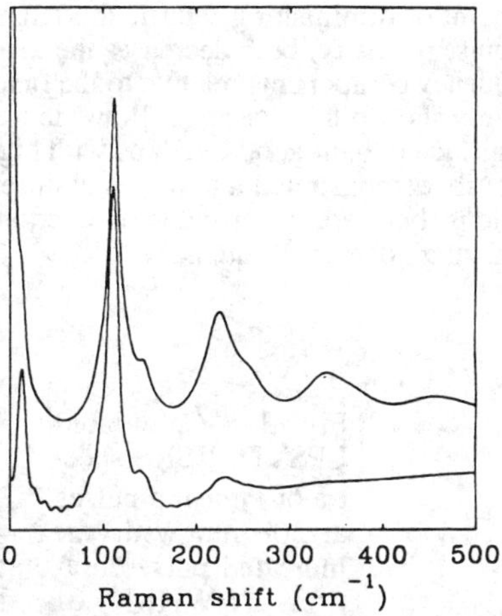

Fig. 2. RR (top) and FT-RISRS (bottom) spectra calculated from eqs. (1) and (2) for the experimental conditions of Fig. 1. Parameters used in the calculations are given in refs. [7] and [10].

Fig. 2 shows the calculated RR and FT-RISRS spectra. The RR spectra show a progression in ν_1 as well as activity in ν_3, in agreement with experiment. The calculated FT-RISRS also agrees with the corresponding experiment in clearly showing the ν_1 fundamental while barely resolving ν_3 and $2\nu_1$. Part of the reason for the apparent weakness of the $2\nu_1$ feature is that its phase differs from that of ν_1 and so it appears in the cosine transform with a partly "derivative" lineshape that tends to obscure its true amplitude. Use of Linear Prediction Singular Value Decomposition (LPSVD)[13-15] to fit the time-domain signal accounts for the different phases of different components and reveals more intensity in ν_3 and $2\nu_1$ as shown in Fig. 3. The calculated FT-RISRS spectrum also has considerable intensity in the presumed intermolecular mode at 13 cm^{-1}, while a feature near this frequency is only possibly present in the experimental spectrum. This discrepancy may reflect inaccuracies in our model for this mode or modes, whose frequency distribution, linewidth, and coupling strength are not well defined by our fit to the RR intensities.[10]

One reason to expect differences between the RR and RISRS spectra lies in the nature of the downward transition. In RR this step proceeds by spontaneous emission, so the probability of reaching a particular final vibrational state depends only on the potential surfaces and dynamics. In RISRS this step is stimulated, so vibrations having frequencies outside the bandwidth of the pulse cannot be excited. Fig. 3 shows the dependence of the calculated LPSVD-RISRS spectra on pulse duration. Shortening the pulse from 60 fs to 15 fs brings in higher harmonics of the symmetric stretch vibration and makes the intensity distribution in the $n\nu_1$ progression roughly comparable to that observed in spontaneous RR. However, there appears to be a point of diminishing return; shortening the pulse still further (at constant integrated pulse intensity) both decreases the total signal size and begins to damp the higher-frequency components relative to the fundamental. We believe that this occurs because a very short pulse does not allow the wavepacket to move enough on the excited-state surface to gain good overlaps with highly excited final states. Pollard *et al.*[4] previously demonstrated that pulses of different duration optimally enhance different vibrations, but here the effect is on overtones of the same mode rather than fundamentals having different frequencies.

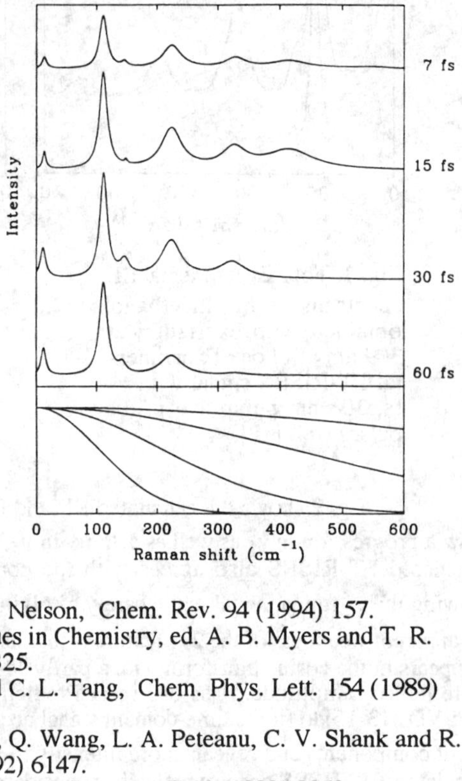

Fig. 3. Calculated LPSVD-RISRS spectra of I_3^- using pulses at 308 nm with the indicated pulse durations (FWHM of transform limited Gaussian). The parameters used in the calculations are given in ref. [7]. The bottom panel shows the power spectra of the pulses (top to bottom, 7 to 60 fs).

[1] L. Dhar, J. A. Rogers and K. A. Nelson, Chem. Rev. 94 (1994) 157.
[2] A. B. Myers, in Laser Techniques in Chemistry, ed. A. B. Myers and T. R. Rizzo (Wiley, New York, 1995) p. 325.
[3] I. A. Walmsley, F. W. Wise and C. L. Tang, Chem. Phys. Lett. 154 (1989) 315.
[4] W. T. Pollard, S. L. Dexheimer, Q. Wang, L. A. Peteanu, C. V. Shank and R. A. Mathies, J. Phys. Chem. 96 (1992) 6147.
[5] Y. Tanimura and S. Mukamel, J. Opt. Soc. Am. B 10 (1993) 2263.
[6] D. M. Jonas, S. E. Bradforth, S. A. Passino and G. R. Fleming, J. Phys. Chem. 99 (1995) 2594.
[7] A. E. Johnson and A. B. Myers, J. Chem. Phys. 104 (1996) 2497.
[8] S.-Y. Lee and E. J. Heller, J. Chem. Phys. 71 (1979) 4777.
[9] A. E. Johnson and A. B. Myers, J. Chem. Phys. 102 (1995) 3519.
[10] A. E. Johnson and A. B. Myers, J. Phys. Chem., in press.
[11] U. Banin and S. Ruhman, J. Chem. Phys. 98 (1993) 4391.
[12] U. Banin, R. Kosloff and S. Ruhman, Isr. J. Chem. 33 (1993) 141.
[13] H. Barkhuisjen, R. de Beer, W. M. M. J. Bovee and D. van Ormondt, J. Mag. Res. 61 (1985) 465.
[14] N. F. Scherer, L. D. Ziegler and G. R. Fleming, J. Chem. Phys. 96 (1992) 5544.
[15] F. W. Wise, M. J. Rosker and C. L. Tang, J. Chem. Phys. 86 (1987) 2827.

RAMAN AND BRILLOUIN SCATTERING FROM DIAMOND: LATTICE DYNAMICS, ZERO-POINT ENERGY AND ISOTOPIC COMPOSITION

A. K. Ramdas,[a]* R. Vogelgesang,[a] S. Rodriguez,[a] M. Grimsditch,[b] and T. R. Anthony[c]

[a] Dept. of Physics, Purdue University, W. Lafayette, IN 47907, U.S.A.
[b] Argonne National Laboratory, Argonne, IL 60439, U.S.A.
[c] GE Corp. Research & Development, Schenectady, NY 12309, U.S.A.

Diamond, the cubic modification of crystalline carbon, belongs to the space group O_h^7 with two atoms per primitive cell. The characteristic tetrahedral coordination underlying its structure is a feature diamond shares with Si, Ge, α-Sn, and most of the III-V and II-VI semiconductors. The Raman active, infrared inactive, triply degenerate F_{2g} zone-center optical phonon of diamond has a large frequency thanks to the large covalent bonding [1]. For the same reason, the elastic constants of diamond are some of the largest known for any material and thus lead to Brillouin components with large frequency shifts [2]. All these (and many additional) striking physical characteristics, amenable to a theoretical analysis in view of the simplicity of structure, account for its importance in condensed matter physics. The crystalline perfection and transparency make diamond eminently suited for light scattering studies.

Studies on diamond were performed in the past only with naturally occurring specimens until the synthesis of diamonds was achieved in the early 1950's by the high-temperature-high-pressure (HPHT) technique [3,4]. The low temperature growth of polycrystalline diamond from the gaseous phase by chemical vapor deposition (CVD) is another significant development [5] in this context. One can now look forward to the elimination of undesired, often unknown, impurities/imperfections, followed by the controlled introduction of known impurities. In this context it is of great interest to appreciate that the *isotopic composition* of the CVD diamond can be controlled by that of the gas employed, *e.g.*, CH_4; in turn, the isotopically controlled polycrystalline diamond thus produced can be transformed by the HPHT technique into single crystals [6]. An extraordinary scientific opportunity has thus arisen for the investigation of the properties of diamond in which its isotopic composition plays a special role. Whereas natural diamonds have 98.9% ^{12}C and 1.1% ^{13}C, reflecting the natural abundance of the two stable isotopes of carbon, one can now synthesize single crystals of $^{12}C_{1-x}{}^{13}C_x$ diamonds over the entire composition range. We have recently completed a Raman and Brillouin scattering study of a series of isotopically engineered diamonds ranging from $x = 0.003$ to 0.992 and interpreted our results in terms of a valence-force-field-model in which the effects of zero-point motion and the anharmonicity of the lattice vibration are explicitly incorporated [7,8].

We consider a valence-force-field model [9] in which the strain energy is expressed in terms of the changes in the bond lengths and bond angles. We introduce two harmonic parameters, k_1 and k_2, associated with the bond length and bond angle, respectively, and two corresponding anharmonic parameters, g_1 and g_2. It transpires that only g_1 is of significance in the present context. One can compute the zero temperature bulk modulus (B) and the frequency of the F_{2g} zone center optical phonon Ω_0 in the harmonic approximation. A variational calculation based on the zero-point motion and anharmonicity parameter enables one to deduce the *renormalized* force constant (K), bulk modulus ($\mathcal{B}$) as well as the lattice parameter (a) in terms of g_1, k_1 and $M = xM_{13} + (1-x)M_{12}$, where M_{12} and M_{13} are the masses of ^{12}C and ^{13}C, respectively. In Table I, the expressions deduced in the harmonic approximation (first row) are compared with their renormalized counterparts (second row).

Table I. Effects of zero-point motion and anharmonicity

F_{2g} Frequency	Lattice Parameter	Bulk Modulus
$\Omega_0 = \left[\frac{8(k_1+4k_2)}{3M}\right]^{1/2}$	a_∞	$B = (k_1/3a)$
$\omega_0 = \Omega_0\left(1 - \frac{32\hbar g_1^2}{81M^3\Omega_0^5}\right)$	$a = a_\infty + \frac{\hbar g_1}{\sqrt{6}k_1^{3/2}M^{1/2}}$	$\mathcal{B} = (K_1/3a)$

In Table I $K_1 = k_1\left(1 - \frac{\hbar g_1^2}{8\sqrt{2}k_1^{5/2}M^{1/2}}\right)$ and a_∞ is the lattice parameter of a diamond made up of 'infinitely' massive carbon atoms.

The experimental values of $\omega_0(x)$ from Raman spectra reported by Hass *et al.* [10], Spitzer *et al.* [11], Chrenko [12] as well as our measurements [8] are well represented by the polynomial $\omega_0(x) = 1332.82 - 34.77x - 16.98x^2$. Theory predicts

$$\omega_0(x) = \omega_0(0)(M_{12}/M)^{1/2}\left[1 + \frac{32\hbar g_1^2}{81M_{12}^3\omega_0(0)^5}\left\{1 - \left(\frac{M_{12}}{M}\right)^{1/2}\right\}\right], \quad (1)$$

keeping corrections to second power of g_1 only. Isotopic disorder produces phonon life times proportional to $x(1-x)$ accompanied by shifts in the real part of the frequency of the form $Cx(1-x)$ as shown by Hass *et al.* [10] and Spitzer *et al.* [11]. Thus Eq. (1), along with the addition of $Cx(1-x)$ on the right hand side, is in excellent agreement with the polynomial fit to $\omega_0(x)$. Since the disorder induced term vanishes at $x = 0$ and 1, one can deduce the contribution of zero-point motion over and above the $M^{-1/2}$ dependence; $\omega_0(^{13}C)$ predicted to be 1280.9 cm^{-1}, instead of 1280.6 cm^{-1} expected on the basis of

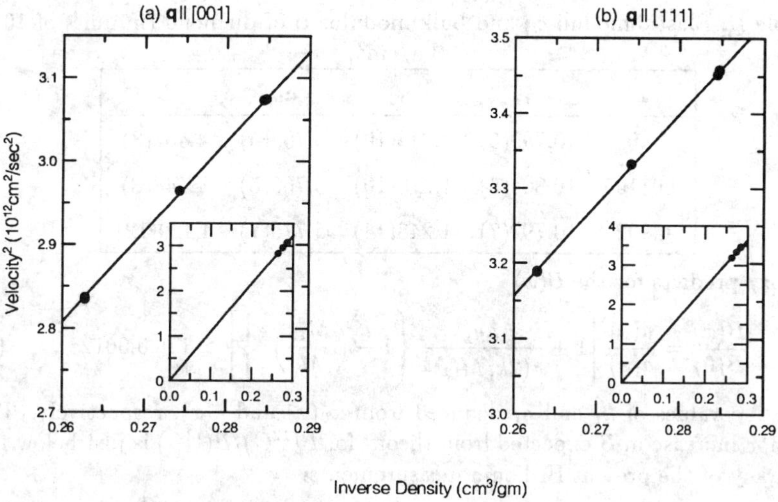

Figure 1. v_s^2 *vs.* $\rho^{-1}(x)$ for longitudinal acoustic phonons travelling along [001] and [111] in diamond specimens of varying isotopic composition. Our data points are indicated by full circles, the solid line represents a linear least squares fit passing through the origin ($\rho^{-1}(x) = 0$).

$M^{-1/2}$ dependence only, is in reasonable agreement with 1281.07 cm^{-1} deduced from the polynomial fit.

The elastic moduli c_{ij} are deduced from the Brillouin shift, $\Delta\omega = \pm 2\omega_L n(v_s/c)\sin(\theta/2)$, where ω_L is the frequency of the incident laser radiation; n the refractive index; c, the speed of light in vacuum; θ, the scattering angle; and v_s, the velocity of the sound wave responsible for the Doppler shift producing the Brillouin component. The appropriate combination of the elastic moduli (X) is given by $X = v_s^2\rho$ where ρ, the density, is a function of x. This immediately underscores the isotopic effects one might expect. The density can be determined from $a(x)$ reported by Holloway *et al.* [13]. We have investigated Brillouin scattering in the back scattering geometry using a high contrast, piezoelectrically scanned, multipassed Fabry-Perot interferometer. Backscattering by acoustic phonons with wave vectors $\boldsymbol{q}$ along [001], [110], and [111] by longitudinal acoustic phonons (LA) and by transverse acoustic (TA) phonons for $\boldsymbol{q}$ along [111] yielded c_{11}; $\frac{1}{2}(c_{11} + c_{12} + 2c_{44})$; $\frac{1}{3}(c_{11} + 2c_{12} + 4c_{44})$, and $\frac{1}{3}(c_{11} - c_{12} + c_{44})$, respectively.

Results for $\boldsymbol{q} \parallel [001]$ and [111] are displayed in Figure 1 as v_s^2 *vs.* $\rho^{-1}(x)$. The elastic moduli c_{11}, c_{12} and c_{44} deduced and the renormalized bulk modulus $\mathcal{B}$ for ^{12}C, natural, and ^{13}C diamond are displayed in Table II.

Table II. Elastic moduli c_{ij} and bulk modulus $\mathcal{B}$ of diamond (in units of 10^{12} dyn/cm^2)

x	c_{11}	c_{12}	c_{44}	$\mathcal{B}$
0.0	10.799(5)	1.248(10)	5.783(5)	4.432(8)
0.01105	10.804(5)	1.270(10)	5.766(5)	4.448(8)
0.992	10.792(7)	1.248(14)	5.776(7)	4.429(12)

Theory predicts for the $\mathcal{B}(x)$

$$\frac{\mathcal{B}(x)}{\mathcal{B}(0)} = \frac{a(0)}{a(x)}\left[1 + \frac{\hbar g_1^2}{8(2k_1^5 M_{12})^{\frac{1}{2}}}\left\{1 - \left(\frac{M_{12}}{M}\right)^{\frac{1}{2}}\right\}\right] \simeq 1 + 0.0012x\ , \qquad (2)$$

with the values of k_1 and g_1 deduced from $\omega_0(x)$ and $a(x)$, respectively. The $\sim 0.1\%$ increase in $\mathcal{B}$ expected from theory for $\mathcal{B}(^{12}C)/\mathcal{B}(^{13}C)$ is just below the accuracy of the present Brillouin measurements.

Acknowledgments: The work reported in this paper received support from the National Science Foundation (Grant No. DMR93-03186) at Purdue University and from the U.S. Department of Energy, BES Materials Sciences (Grant No. W-31-109-ENG-38) at the Argonne National Laboratory.

1. S. A. Solin and A. K. Ramdas, Phys. Rev. B **1**, 1687 (1970).
2. M. H. Grimsditch and A. K. Ramdas, Phys. Rev. B **11**, 3139 (1975).
3. F. P. Bundy, H. T. Hall, H. M. Strong, and R. H. Wentorf, Nature **176**, 51 (1955).
4. H. M. Strong and R. H. Wentorf, Am. J Phys. **59**, 1005 (1991).
5. W. G. Eversole, U.S. Patent Nos. 3030187 and 32030188, (1962); B. V. Derjaguin, D. V. Fedosev, V. M. Lukyanovich, B. V. Spitzin, V. A. Ryabov, and A. V. Lavrentyev, J. Cryst. Growth **2**, 380 (1968).
6. T. R. Anthony and W. F. Banholzer, Diamond and Related Materials **I**, 717 (1992).
7. A. K. Ramdas, S. Rodriguez, M. Grimsditch, T. R. Anthony, and W. F. Banholzer, Phys. Rev. Lett. **71**, 189 (1993).
8. R. Vogelgesang, A. K. Ramdas, S. Rodriguez, M. Grimsditch and T. R. Anthony, submitted for publication.
9. M. J. P. Musgrave, *Crystal Acoustics* (Holden-Day, San Francisco, 1970) pp. 248-257.
10. K. C. Hass, M. A. Tamor, T. R. Anthony, and W. F. Banholzer, Phys. Rev.B **45**, 7171 (1992).
11. J. Spitzer, P. Etchegoin, and M. Cardona, T. R. Anthony, and W. F. Banholzer, Solid State Commun. **88**, 509 (1993).
12. R. M. Chrenko, J. Appl. Physics **63**, 5873 (1988).
13. H. Holloway, K. C. Hass, M. A.Tamor, T. R. Anthony, and W. F. Banholzer, Phys. Rev. B **44**, 7123 (1991).

70 YEARS OF RAMAN INSTRUMENTATION : FROM MULTICHANNEL TO MULTICHANNEL

E. Da Silva

DILOR SA, 244 ter rue des Bois-Blancs, F-59000 Lille

From 70 years ago when the Raman effect was discovered up until present day, according to the recent papers presented at ICORS XV, the amount of sample material needed to record a spectrum has decreased from around ten grams to only a few picograms. In the same period, the recording time has been reduced from tens of hours to seconds in conventional measurements and even for time resolved studies down to picoseconds or femtoseconds.

It is interesting to consider this evolution from the point of view of the experimentalist since the most impressive progresses have been clearly related to the availability of new components and new technologies.

It is universally accepted that the laser light sources were at the origin of a complete renewal of Raman instrumentation but we also feel that decisive improvements have been permitted by the advent of other components such as photomultipliers, image intensifiers, solid state detector arrays, or sophisticated optical devices and evidently by the developments of computer data processing.

In this presentation, we would like to emphasize some important steps of the evolution of instrumentation by presenting the chronology of the techniques at the date at which they became available to the whole community of Raman spectroscopists.

It is important to notice that these instrumental progresses often rely on previous experimental studies pioneered sometimes 10 or 20 years before. A considerable amount of laboratory study in various countries has been the necessary basis for the evolution of commercial instruments.

Together with the present talk, an exhibition is also presented to show the main steps of the evolution of instrumentation. We would like to express our thanks to M. DELHAYE, W. KIEFER and D.A. LONG who have accepted to select and collect valuable data and to put it together in order to present this series of posters.

TABLE 1 : Principal themes of this exhibition

A) Books on Raman.

B) Light sources :

From the sun to the laser.

C) Sampling :

Large cuvette for mercury arc to microscopy sample.

D) Dispersive system for multichannel :

Photographic plate, intensifier tubes,
diode array, CCD.

E) Dispersive system for single channel :

Double triple monochromators.
Holographic gratings.
Cooled low noise PMT.

F) Non-dispersive system :

Fourier transform spectroscopy.

G) Some other important developments :

Microscopy.
Imaging.
Time resolved ...

H) Selected new fields of applications.

I) Pictures of commercial instruments from the
beginning upto a few years ago :

From the prism spectrograph to modern instruments.

TABLE 2 : Chronology of Raman instrumentation

1928	Discovery of the Raman effect.	
1930 1940	"The best gift offered by Physicists to Chemists". Raman spectrometry can be used routinely in molecular spectroscopy whereas infrared is just at the beginning (1).	Mercury lamp Prism spectrograph Photographic multichannel
1950 1960	Photoelectric recording by means of photomultipliers (2)(3).	Grating spectrometers
1965	The laser (4) : an ideal source for Raman. - single channel or intensified multichannel ?? - CW or pulsed laser ??	Lasers Photoelectric multichannel
1975	The age of the single channel (HeNe and Argon lasers). Raman microscopy (5).	Holographic gratings Microscopy
1985	The development of multichannel : microscopy and imaging.	CCD
1990	Non-dispersive techniques : Interferometry and holographic filters (6).	F.T. Raman Confocal microscopy
1995	The maturity of multichannel (7).	
2000	XVII ICORS. What's new ??	???

Commercial Raman instruments :

(1) Large prism spectrograph.
(2) high power mercury lamps.
(3) Grating spectrometers and the first integrated Raman instruments.
(4) Double and triple spectrometers.
(5) Raman microprobes.
(6) Extension of F.T.I.R. to Raman.
(7) Renewed spectrographs : - integrated compact instruments
- routine Raman microscopy
- remote industrial measurements

NEW DEVELOPMENTS IN SERS WITH APPLICATIONS TO BIOLOGICAL MOLECULES

T.M. Cotton, G. Chumanov, M. Sibbald and J. Zheng

Department of Chemistry, Iowa State University, Ames, IA 50014

Surface enhanced Raman scattering (SERS) spectroscopy is a powerful method for the study of biological molecules on metal surfaces. One aspect of our research has been concerned with the characterization of electron transfer reactions between various biological molecules and silver or gold electrodes. In the case of cytochrome c, it has been demonstrated that modification of the metal surface with organic and inorganic species serves to preserve the native structure of the protein in the adsorbed state. As a result, a reversible electrochemical response is obtained. In the presence of bifunctional modifiers, such as 4-pyridylthiol and 11-undecanoic acid, cytochrome c is strongly adsorbed on the surface. The electron transfer reaction may be followed by monitoring the change in the oxidation sensitive bands of the heme in the SERS spectrum as a function of potential. Detection of the resonance Raman spectrum of the heme prosthetic group at distances > 5Å from the metal surface is possible because of the long range electromagnetic component of the enhancement process [1].

In another example, bifunctional surface modifiers are used to orient reaction centers on the metal surface. Because of the short range enhancement of Raman scattering, SERS spectra were used to monitor different orientations of the this complex. Thiolated compounds with amino and carboxy terminated groups, as well as I^- or S^{2-} were used as modifiers of the roughened silver surfaces. Strong SERS spectra were observed which contained signature features of all chromophores present in the reaction centers, such as carotenoids, bacteriochlorophyll and bacteriopheophytin. It is necessary to emphasize that all Raman bands on modified surfaces were as sharp as those observed in the Raman spectra of reaction centers in solution, whereas all bands in the SERS spectra on bare silver electrode were noticeably broadened. Assuming that broadening of the Raman bands on the bare metal surface results from perturbation of the native structure, it can be concluded that surface modifiers stabilize membrane protein structures. In addition, the relative intensities of the pigment bands were different for different modifiers. This fact clearly implies orientational differences. For example, the intensities of the carotenoid and bacteriochlorophyll peaks relative to those of bacteriopheophytin were stronger on iodide-modified surface in comparison to a carboxylated-modified surface.

The role of the electrode modifier in the electron transfer process has also been of concern. Although bifunctional modifiers clearly are involved in orienting the protein near the surface, other types do not interact specifically with the protein. Examples of the latter include 2,2'-bipyridine and strongly adsorbed halide ions (iodide and bromide). These appear to present a more biologically hospitable surface to the protein as compared to the bare metal electrode and prevent structural changes that result in poor electron transfer kinetics. Adsorbed inorganic halide ions were initially thought to merely interact electrostatically with the protein [2], but this simple assumption has been revised based upon recent results with colloidal metals [3]. In these studies, it was discovered that iodide, bromide and sulfide-modified silver particles can reduce cytochrome c. Adsorption of these ions causes a charging of the metal particles and a shift in their Fermi potential to more negative values. At certain surface coverages, the potential is sufficient to reduce the protein. In the absence of oxidizing agents (e.g. O_2) charge is stored for an extended period of time and the particles behave as nanoelectrodes. Thus, the modifier plays an active role in the redox process.

Organic modifiers may themselves partake in redox reactions with the metal surface. The dication form of methylviologen adsorbed on a silver electrodes was found to undergo one-electron reduction to the cation radical [4]. The radical is produced photochemically with maximal charge transfer occurring under blue (413.1 nm) excitation. At liquid nitrogen temperatures the radical is stable indefinitely, whereas at room temperature the reverse electron transfer process occurs rapidly and an appreciable concentration of the cation radical is not present under steady state conditions. Because the initially adsorbed methylviologen dication is colorless at the laser excitation wavelengths used in this study, absorption of light by silver appears to be responsible for the photochemical process. Two possible mechanisms were proposed. These are plasmon assisted charge transfer from the metal to the adsorbed viologen and the formation of a resonance charge transfer complex.

The possibility of utilizing plasmon enhanced photoelectrochemistry for reduction of species in solution is currently under investigation. It has been known for some time that electron photoemission occurs upon irradiation of metals with photons of sufficient energy to match or exceed the work function of the metal. Silver is unique among metals in that it possesses a low work function (4.2 eV), a surface plasmon absorption band near 3.6 eV and is relatively oxide free under normal conditions. It has already been demonstrated that enhanced photoemission occurs in vacuum from silver surfaces with strongly adsorbed organic molecules [5]. An enhancement of the hydrogen yield in an electrochemical cell with plasmon excitation has also been reported [6].

The photocurrent response of several metal electrodes, including Pt, Au and Ag has been investigated in aqueous solutions containing 0.1 M Na_2ClO_4 as a function of electrode potential. No increase in current is observed for Pt or Au upon irradiation with 413.1 nm excitation. However, in the case of silver, an increase in current is observed which is strongly dependent upon the electrode potential. The threshold potential was determined to be 0.45 V (vs. SCE) which is in good agreement with that calculated according to the photoemission relationship. In these conditions, protons are reduced to hydrogen and the photoelectrochemical yield is quite small (ca. 0.01%).

When methylviologen is added to the above electrolyte solution, an increase in photocurrent is observed at more positive potentials (-0.2 - 0.4 V). In this potential range, adsorbed viologen undergoes reduction and transfers an electron to solution species. This behavior is only observed for roughened silver electrodes at which the excitation of surface plasmons is particularly efficient.

The possibility of utilizing methylviologen as a photocatalyst for reduction of CO_2 was explored. For this purpose, a silver electrode was roughened in electrolyte solution, removed from the electrochemical cell and immersed in a 1 x 10^{-3} M methylviologen methanol/water (50:50, v/v) solution for 1 hour. The electrode was removed, rinsed with water and placed in an electrochemical cell containing 0.5 M $NaClO_4$. The cell was purged with nitrogen and the SERS spectrum was measured with 413.1 nm excitation as a function of electrode potential. As shown in the top spectrum of Figure 1, the methylviologen cation radical was present at -0.8 V, as expected based on previous electrochemical and Raman studies. However, when CO_2 was bubbled into the cell, the spectrum of methylviologen dication was observed at all potentials (Figure 1 shows the spectrum at -0.8 V), indicating immediate oxidation of the adsorbed cation radical by CO_2. This effect is completely reversible. If the cell is purged with nitrogen, the spectrum of the cation radical is again present at -0.8 V. It should be noted that the onset of photochemical reduction of CO_2 -1.2 V at bare silver. Because the cation radical has strong absorption at this wavelength, it is likely that the additional energy required for the reduction of CO_2 at this potential is obtained from the absorbed photon. Thus, it is appears that methylviologen functions as a photocatalyst in this experiment.

In summary, recent investigations of the behavior of inorganic and organic on silver have shown that these species are not necessarily passive modifiers. The halides have been shown to function as electron donors, whereas methylviologen is an electron acceptor under photoexcitation. Excitation of the surface plasmon resonance in silver may also assist in the electron transfer processes.

References:

1. Hobara, D., Niki, K., Zhou, C., Chumanov, G. and Cotton, T.M. Colloids and Surfaces 1994, 93, 241.
2. Lu, T., Yu, X., Dong, S., Zhou, C., Ye, S. and Cotton, T.M. J. Electroanal. Chem. 1994, 369, 79.
3. Sibbald, M.S., Chumanov, G. and Cotton, T.M. J. Phys. Chem., 1996, 100, 4672.
4. Feilchenfeld, H., Chumanov, G. and Cotton, T.M. J. Phys. Chem., in press.

5. Kim,C.-W., Villagran, J.C., Even, U., Thompson, J.C. J. Chem. Phys. 1991, 94, 3974.
6. Sass, J.K., Sen, R.K., Meyer, E. and Gerischer, H. Surf. Sci. 1974, 44, 515.

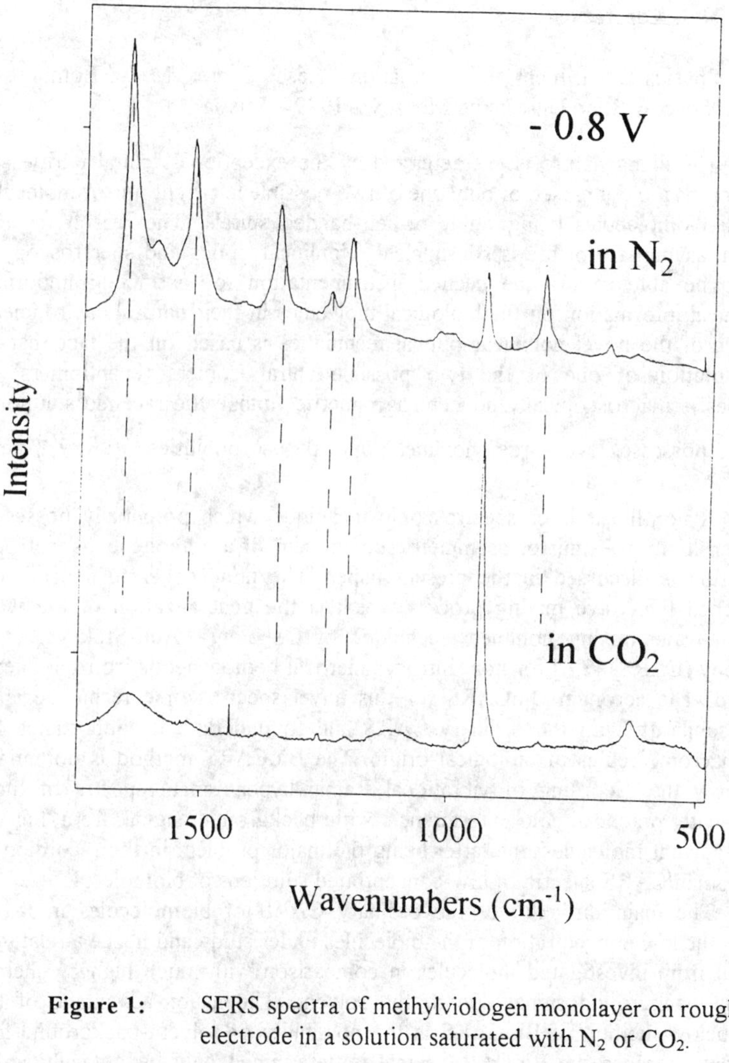

Figure 1: SERS spectra of methylviologen monolayer on roughend Ag electrode in a solution saturated with N_2 or CO_2.

BIOCARS –NOVEL HIGHER-ORDER COHERENT ANTI-STOKES RAMAN SCATTERING TECHNIQUE TO STUDY CHIRAL BIOMOLECULES IN SOLUTION

N. I. Koroteev

Physics Department and International Laser Center, M. V. Lomonosov Moscow State University, Moscow 119899, Russia

The living nature is characterized by the exceptional "chiral purity": all living organisms are composed of only one of two possible left-right mirror molecular isomers, all biomolecules being right- or left-handed spirals. The reason for this fundamental asymmetry of life is still unclear. Nonlinear optics and spectroscopy is believed to be able to provide needed instrumentation for extracting important stereochemical information on the biological molecules in their natural environment. The essence of the novel nonlinear optical techniques is based on the fact that an isotropic solution of one of the two possible chiral isomers (enantiomers) of biomolecules is macroscopically noncentrosymmetric (unlike the racemic solutions) and, hence, possesses even-order nonlinear optical susceptibilities, $\chi^{(2)}_{ijk}$, $\chi^{(4)}_{ijklm}$ etc. [1,2].

A non-linear laser spectroscopic technique which potentially possesses selective sensitivity to a mirror asymmetric component of a homogeneous isotropic liquid solution is discussed in the present paper. The nondegenerate noncollinear phase-matched five-wave mixing process which is the generalization of the well-known four-wave mixing nonlinear technique of Coherent Anti-Stokes Raman Spectroscopy (CARS) [3,4] for noncentrosymmetrical homogeneous isotropic media is described. The acronym BioCARS for this novel spectroscopic technique is to stress it's similarity with the ordinary CARS and to underline its importance for studying macromolecules of biological origin. The BioCARS method is potentially able to allow the recording of vibrational Raman/hyper-Raman spectra of these molecules with practically no interference with background signal from mirror-symmetric solvent molecules, the latter being the major problem in the recording of ordinary Raman/CARS spectra of low-concentrated solutions of biomolecules.

The main difficulty of the ordinary CARS of biomolecules in native solutions is the low concentration of the molecules under study and hence a relatively small signal from investigated molecules in comparison with much higher coherent nonresonant background signal due to the solvent. The major advantage of the described below method of BioCARS is the possibility of solvent-background-free measurements because the BioCARS signal is determined only by optically-active (chiral) molecules. However, the BioCARS signal is expected to be actually very weak. In order to enhance the signal level and still to reduce the total energy of the laser radiation affecting the biological sample in the nonlinear BioCARS scheme, it is necessary to use laser pulses with highest possible peak intensities and with lowest possible energies. The proper choice here is to use amplified pico- or femtosecond laser pulses.

BioCARS is based on measurements of the frequency dispersion of fourth order nonlinear optical susceptibility $\chi_{ijklm}^{(4)}(\omega_a;\omega_1,\omega_1,\omega_1,-\omega_2)$, a fifth-rank tensor, responsible for generation in the solution under study a new spectral "anti-Stokes" component with frequency $\omega_a=\omega_1+\omega_1+\omega_1-\omega_2$ when the solution is illuminated simultaneously with two laser beams having frequencies ω_1 and ω_2 (tunable). This five-wave mixing process can be made phase-matched (see Fig. 1) in solutions with normal dispersion (i.e. the momentum conservation condition $\vec{k}_a=3\vec{k}_1-\vec{k}_2$ can be satisfied, $\vec{k}_1$, $\vec{k}_2$ and $\vec{k}_a$ being wave vectors of light beams with frequencies ω_1, ω_2 and ω_a, respectively) in slightly noncollinear geometry – like in ordinary CARS, based on four-wave mixing process [3,4] $\omega_{CARS}=2\omega_1-\omega_2$. Phenomenologically, BioCARS can be considered as a generalization of ordinary CARS and hence a variety of experimental equipment and techniques developed to be used in CARS can be applied, with some changes, in the BioCARS scheme. By tuning the frequency of one of the laser beams, an experimenter can measure the dispersion of $|\chi^{(4)}(\omega_a)|^2$ where Raman-type $(\omega_1-\omega_2=\Omega_R)$ and/or hyper-Raman-type $(2\omega_1-\omega_2=\Omega_{HR})$ resonances will be displayed representing the vibrational eigenmodes of the chiral molecules, Ω_R, Ω_{HR} etc.

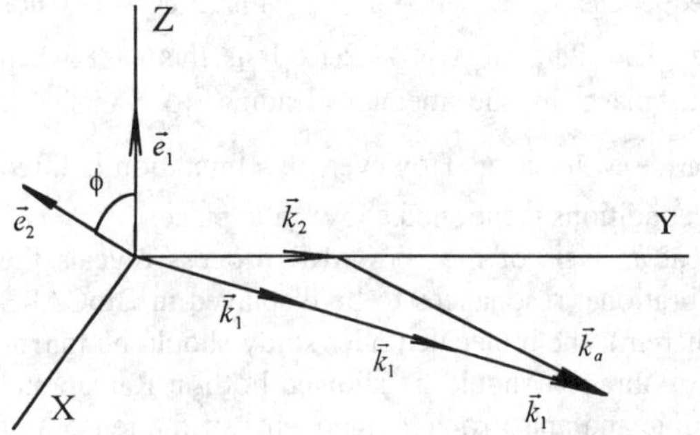

Fig. 1. Phase-matched beam arrangement for the five-wave mixing process of BioCARS: $\omega_a=3\omega_1-\omega_2$; $\vec{k}_a=3\vec{k}_1-\vec{k}_2$. All wave-vectors of interacting beams lie in the XY plane.

The phenomenological expression for the dipole fourth-order nonlinear optical polarization $\vec{P}^{(4)}(\omega_a)$ of the macroscopically acentric non-racemic solution of chiral biomolecules is as follows [6]:

$$P_i^{(4)}(\omega_a)=4\chi_{ijklm}^{(4)}(\omega_a;\omega_1,\omega_1,\omega_1,-\omega_2)E_j(\omega_1)E_k(\omega_1)E_l(\omega_1)E_m^*(\omega_2).$$

and at the output plane of the medium, at $z=L$, the intensity of the BioCARS signal $I_a=R|\chi_{31112}^{(4)}(\omega_a;\omega_1,\omega_1,\omega_1,-\omega_2)|^2$, where, at $\vec{k}_a=3\vec{k}_1-\vec{k}_2$

$$R=\frac{2^{15}\pi^5L^2\omega_a^2}{c^5n_1^3n_2n_a}I_1^3I_2|[\vec{e}_1\vec{e}_2^*]|^2|(\vec{e}_1\vec{e}_2)|^2,$$

and $\vec{e}_{1,2}=\vec{E}(\omega_{1,2})/|\vec{E}(\omega_{1,2})|$, $n_{1,2,a}$ – refraction indices at $\omega_{1,2,a}$, $[\vec{a}\vec{b}]$ and $(\vec{a}\vec{b})$ mean vectorial and scalar products, respectively, of the vectors $\vec{a}$ and $\vec{b}$.

We pointed out above, that $\chi_{31112}^{(4)}\equiv 0$ in the bulk of centrosymmetric isotropic media. However, even in non-centrosymmetrical media such as non-racemic solutions of chiral molecules there are many situations when the susceptibility $\chi_{31112}^{(4)}(\omega_a)$ is completely vanishing, namely, when the chiral medium under study is transparent at all frequencies ω_1, ω_2, ω_a and possesses neither two-photon-allowed transitions at combinations of frequencies: $2\omega_1$, $\omega_1-\omega_2$, $\omega_a+\omega_2$, $\omega_a-\omega_1$ nor three-photon-allowed transitions at $3\omega_1$, $2\omega_1-\omega_2$, $\omega_a-2\omega_1$. It is this case when Kleinman's permutation rules take place in the medium leading to vanish all components of $\chi_{ijklm}^{(4)}(\omega_a;\omega_1,\omega_1,\omega_1,-\omega_2)$ tensor. However, this limitation is lifted when any one (or more) resonance conditions mentioned above take place.

The quantum mechanical analysis of the BioCARS process reveals the following selection rules for the vibrational resonances to be displayed in BioCARS spectra (see ref. 6 and references therein): the molecule under study should be mirror asymmetrical (chiral), the molecular vibration should be allowed both in Raman and hyper-Raman scattering, only isotropic and anisotropic components of the tensors of Raman and hyper-Raman scattering being responsible for the generation of the BioCARS signal.

We have succeeded in the experimental observation of a frequency degenerate phase-matched five-wave-mixing process of the BioCARS type in an optically active water solution of d- and l-arabinose first described in ref. [5]. This process can be symbolically depicted as $\omega_a=2\omega=3\omega-\omega$ (i.e. in the BioCARS scheme $\omega_1=\omega_2=\omega$), where the first three (positive) ω's on the right hand side correspond to a laser beam with the wave vector $\vec{k}$ (along the direction of the $\vec{k}_1$ vector at fig.1) and the frequency ω, and the latter (negative) ω represents photons from the second beam of the same fundamental frequency and with the wave vector $\vec{k}'$ (along the direction of the $\vec{k}_2$ vector at fig.1) crossing the first beam at a phase matching angle $\alpha\approx 15.5°$.

Water solution of arabinose was used in our experiments as it was in those of Rentzepis et al. [7], because of its high specific rotation and availability of its d- and l- isomers. Solutions were prepared at a concentration of 1 part arabinose to 2 parts water, molarity M=2.46 and had refractive indexes 1.375, 1.394 and 1.415 at 532, 355 and 266 nm respectively. Experiments were carried out using an original picosecond spectrometer based on the actively mode-locked cw-pumped neodymium-doped yttrium aluminum garnet (Nd^{3+}:YAG) laser capable to deliver the sequence of pulse trains (each containing 30-35 pulses of 100 ps duration) at 1064 nm at a repetition rate of 1.0 kHz. The output of the laser with an average power of up to 5 W was converted into the green radiation (532 nm) in two potassium titanyl-phosphate (KTP) crystals. For second harmonic generation (SHG) we used two green beams of different intensity, one with average power 100 mW (beam $\vec{k}$) and another (beam $\vec{k}'$) of 80 mW. The beam waist diameters inside sample were 120 and 100 μm, respectively. The $\vec{k}$ beam was polarized vertically and fixed. The $\vec{k}'$ beam's polarization was varied using a Fresnel Rotator. After spatial separation, the SHG emission passed through the single monochromator was detected by a fast photomultiplier tube with microchannel plates. A time-correlated single-photon counting technique was used with the resolution of about 150 ps. We have observed SHG with wavelength $\lambda = 266$ nm and with theoretically predicted time characteristic behavior (see fig. 2). The signal demonstrates predicted polarization, intensity, spectral and time characteristics. When the equal amount of d-arabinose was added to the solution, the polarization dependence becomes flat.

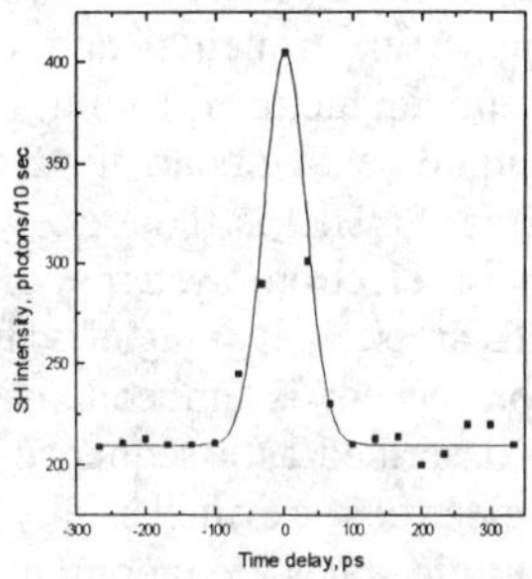

Fig. 2. Time cross-correlation function of SH signal vs. the beam $\vec{k}$ delay. The polarizations of the interacting waves are linear and perpendicular to one another. The solid line is the Gaussian fitting with pulse duration $\tau_p = 100 ps$

Taking the parameters of the beams used in our experiments we get an order-of-magnitude estimate value: $|\chi^{(4)}| \approx 10^{-23}$ esu, which is quite low — probably, due to incomplete lifting of Kleinman's type prohibition since frequency ω lies in the transparency region of the solution and the frequency 2ω does on the long-wavelength wing of arabinose's UV absorption band. Now we are in the process of the preparation for the experimental observation of the frequency-nondegenerate BioCARS process with the specially designed amplified femtosecond laser system [8].

1. N. I. Koroteev, In Frontiers in nonlinear optics. The Sergei Akhmanov memorial volume, ed. by H. Walther, N. Koroteev, M. O. Scully, Institute of Physics Publishing, Bristol, 1993, p. 228.
2. N. I. Koroteev, JETP, 79, 681 (1994).
3. S. A. Akhmanov, N. I. Koroteev, Methods of nonlinear optics in light scattering spectroscopy, Nauka Publ. House, Moscow, 1981 (in Russian).
4. G. L. Eesley, Coherent Raman spectroscopy, Pergamon Press, Oxford, 1981.
5. A. P. Shkurinov, A. V. Dubrovskii and N. I. Koroteev, Phys. Rev. Lett., 70, 1085 (1993).
6. N. I. Koroteev, Biospectroscopy, 1, 341 (1995).
7. P. M. Rentzepis, J. A. Giordmaine, and K. W. Wecht, Phys. Rev. Lett., 16, 792, (1966).
8. N.I.Koroteev. In SPIE Proceedings, 1996 (in press).

RAMAN SCATTERING BY ELECTRON QUANTUM LIQUIDS

A. Pinczuk*, B.S. Dennis, L.N. Pfeiffer, and K.W. West

Bell Laboratories, Lucent Technologies,
Murray Hill, New Jersey 07974-0636, USA

Studies of two-dimensional electron systems have uncovered novel behaviors that emerge from the combined effects of strong electron-electron interactions and the reduction in the physical dimensions. Intensive experimental and theoretical research of remarkable effects observed in semiconductor quantum structures continue to uncover intriguing electron quantum liquid behaviors arising from strong correlations[1-3]. The fractional quantum Hall effect(FQHE) is the archetype of the class of electron quantum liquids that occur in electron systems of reduced dimensions[4-5]. The transport anomalies of FQHE are seen at "magic" values of the Landau level filling factor when the 2D electron system is embedded in a large perpendicular magnetic field. The FQHE can be described as a sequence of metal-insulator transitions that appear when the free electrons are in partially occupied Landau levels. In the FQHE the electron quantum fluid states are incompressible and an energy gap separates the ground state from its excited states. The gaps of the FQHE are manifestations of the strong electron interactions that lead to condensation of the 2D system into quantum fluids[5-8].

Gaps in quantum Hall states are represented by neutral density excitations in which a particle (or quasiparticle) is promoted to an excited state leaving behind a hole (or quasihole). For excitations with wave vector $\vec{q}$ there is a displacement between the centers of the cyclotron orbits that represent the single-particle states of the particle-hole pairs given by $x_o = q\ell_o^2$, where $\ell_o = (\hbar c/eB)^{1/2}$ is the magnetic length[6]. Measurements of thermal activation of magnetotransport yield gap energies of the FQHE[9]. These gaps, as well as those determined from acceptor luminescence[10], are interpreted as energies required to create widely separated quasiparticle-quasihole pairs that have $q \rightarrow \infty$.

The study of dispersive gap excitations is the major goal of spectroscopy of the electron liquids in semiconductor quantum structures. This quest has particular significance in the fractional quantum Hall regime, in which the existence of well-defined dispersive gap excitations is among the most significant predictions of current theories of the electron quantum fluid of the FQHE. Collective gap excitations of the FQHE at $\nu = 1/3$ were observed by resonant inelastic light(Raman) scattering[11]. The success in the fractional quantum Hall regime suggests that resonant inelastic light scattering methods, proposed in 1978 for studies of 2D electron systems at semiconductor interfaces[12], will play important roles in research of free electron systems of artificial semiconductor structures.

The light scattering observations in the FQHE regime are surprising because for these *intra*-Landau-level excitations the dynamical structure factors, the functions that describe conventional inelastic scattering cross-sections, as well as dipole oscillator strengths vanish as q^4 for the small wave vectors in optical experiments[5,8]. The results are also intriguing because the structure of q=0 gap excitations of the FQHE is still a subject of conjecture[8,13,14]. Resonant inelastic light scattering experiments in the regime of the integer quantum Hall effect have also demonstrated observations of *inter*-Landau-level excitations with vanishing dynamical structure factors for q→0[15].

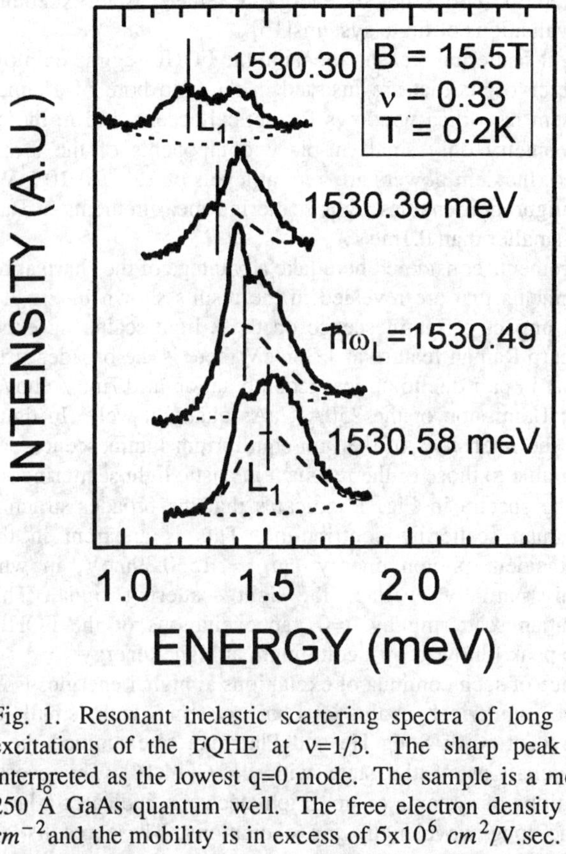

Fig. 1. Resonant inelastic scattering spectra of long wavelength gap excitations of the FQHE at ν=1/3. The sharp peak at 1.38meV is interpreted as the lowest q=0 mode. The sample is a modulation doped 250 Å GaAs quantum well. The free electron density is n=1.25x10^{11} cm^{-2} and the mobility is in excess of 5x10^6 cm^2/V.sec. The dotted line is the estimated background. The dashed line is the estimated non-equilibrium luminescence L_1. The vertical bars give the position of the peak L_1. $\hbar\omega_L$ is the incident photon energy.

In the experiments carried out in the quantum hall regime the resonant inelastic light scattering method emerges as a powerful technique that enables studies of electron excitations that at this time are not within reach of other experimental tools.

Resonant inelastic light scattering scattering by q=0 collective gap excitations of FQHE state with $\nu=1/3$ was measured in several GaAs quantum wells with free electron densities in the range $0.7 \leq n \leq 1.3\text{x}\ 10^{11} cm^{-2}$. Figure 1 shows results measured at a field of 15.5Tesla. The spectra display a sharp peak (FWHM≈0.07meV) that overlaps a broader structure(FWHM≈0.25meV). The sharp peak is observed only for magnetic fields near $\nu=1/3$ and is interpreted as the q=0 gap excitation of the FQHE state. Its energy, 1.38meV, is in excellent agreement with numerical evaluations of finite systems[16].

Inelastic light scattering measurements in the FQHE regime employ ^{3}He or dilution ^{3}He/^{4}He cryostats that are inserted in the cold bore of a superconductive magnet. The system has silica windows for optical access, and in the conventional backscattering geometry only small in-plane components of the scattering wave vector are allowed. Incident powers are kept at levels of 10^{-4} to 10^{-3}Wcm^{-2}. This experimental configuration enables light scattering measurements at Raman energy shifts that can be smaller than 0.1meV.

The measurements considered here take advantage of the sharp and strong resonant enhancements that are revealed in the results shown in Fig. 1. The figure also displays the impact of luminescence on these light scattering experiments. In addition to the sharp Raman feature at 1.38meV there is the broader structure. This structure is at least in part due to optical recombination at 0.7meV above the energy of the fundamental emission of the 250Å GaAs quantum well. In figure 1 we see that at 15.5Tesla the intensities of the non-equilibrium luminescence, labeled L_1 in Fig. 1, are comparable to those of the resonant inelastic light scattering. In fact, close examination of the spectra in Fig. 1 indicates that the broader structure may also incorporate a Raman scattering contribution. This is apparent in the spectrum measured with incident photon energy $\hbar\omega_L$ = 1530.39meV, in which the L_1 luminescence only partially overlaps the light scattering signal. This spectrum suggests that Raman scattering by q=0 gap excitations of the FQHE state may consist of a sharp peak followed by a continuum at higher energy.

The existence of such continua of excitations at higher energies is indeed predicted by the numerical evaluations of systems with small numbers of electrons[7]. In recent work Song He and Platzman re-examined the structure of collective gap excitations of the state with $\nu=1/3$[14,16]. This work, based on numerical evaluations of systems of up to ten electrons, indicates a large two-roton component in the long wavelength gap excitations. These results also uncover numerical evidence that the lowest q=0 mode consists of a weakly-bound two-roton state at energy below that of the two-roton continuum.

The resonant inelastic light scattering work reviewed here shows that the method offers unique insights into the dispersive collective excitations of electrons in the quantum Hall regimes. Clearly, these experiments are difficult and require new approaches that enhance the resonant light scattering signal in the presence of strong background due to luminescence. We anticipate that the method will find applications in studies of remarkable electron states of semiconductor quantum structures created by state-of-the-art fabrication technologies. Access to the elementary excitations of these systems could uncover intriguing new physics of electron interactions that manifest in quantum liquid states.

1. *The Quantum Hall Effect*, 2nd Ed., (R.E. Prange and S.M. Girvin, eds.), Springer, New York, 1990.
2. T. Chakraborty and P. Pietilainen, *The Fractional Quantum Hall Effect: Properties of an incompressible quantum fluid, Springer-Verlag, Berlin, 1988.*
3. *Perspectives in Quantum Hall Effects: Novel Quantum Liquids in Low Dimensional Semiconductor Structures*, (S. Das Sarma and A. Pinczuk, eds.), J. Wiley & Sons, New York, 1996, in press.
4. D.C. Tsui, H.L. Stormer, and A.C. Gossard, Phys. Rev. Lett. **48**, 1559(1982).
5. R.B. Laughlin, in Ref. 1, p. 233.
6. C. Kallin and B.I. Halperin, Phys. Rev. **30**, 5655(1984).
7. F.D.M. Haldane and E.H. Rezayi, Phys. Rev. Lett. **54**, 237(1985).
8. S.M. Girvin, A.H. MacDonald and P.M. Platzman, Phys. Rev. Lett. **54**, 581(1985).
9. H.L. Stormer and D.C. Tsui, in Ref. 3.
10. I.V. Kukushkin, N.J. Pulsford, K. von Klitzing, K. Ploog, R.N. Haug, S. Koch and V.B. Timofeev, Europhys. Lett. **18**, 63(1992).
11. A. Pinczuk, B.S. Dennis, L.N. Pfeiffer and K.W. West, Phys. Rev. Lett. **70**, 3983(1993).
12. E. Burstein, A. Pinczuk, and S. Buchner, *Physics of Semiconductors 1978*, (B.L.H. Wilson, ed.), The Institute of Physics, London, 1979, p. 1231.
13. D.-H. Lee and S.-C. Zhang, Phys. Rev. Lett. **66**, 1220(1991).
14. Song He and P.M. Platzman, Procs. of the 11th Int. Conf. on Electr. Props. of 2D Systems, Surface Science, 1996, in press.
15. A. Pinczuk, B.S. Dennis, D. Heiman, C. Kallin, L. Brey, C. Tejedor, S. Schmitt-Rink, L.N. Pfeiffer, and K.W. West, Phys. Rev. Lett. **68**, 3623(1992).
16. Song He, private communication.

MOLECULAR DYNAMICS STUDIES OF RESONANCE EMISSION: ABSORPTION LINE BROADENING AND SOLVATION DYNAMICS

L. D. Ziegler* and T. Kalbfleisch

Department of Chemistry, Boston University, Boston MA 02215

It is well established that the analysis of the spontaneous resonance emission of molecules in solution can be used as a probe of the ultrafast fluctuation dynamics of the solvent or bath surrounding the electronically resonant chromophore.[1-3] The redistribution of resonance emission into narrow Raman-like (RL) features and broad fluorescence-like(FL) features is dependent on the relative rates of excited electronic state population decay and the solvent-based pure-dephasing decay of the resonant electronic coherence. Thus, spontaneous emission studies of optical dephasing dynamics are most valuable when the lifetime of the resonant electronic state is of the order of a picosecond or less because this corresponds to the relevant time scale of solvent fluctuations and inverse electronic absorption line widths. For such short lived excited state levels the two phenomenological emission RL and FL components can be easily comparable. Here we examine the absorption and resonance emission of the B-state of methyl iodide in high pressures of methane, argon and carbon dioxide. This 6s Rydberg electronic state is short-lived owing to curve crossing to a photodissociative surface.

Previous experimental studies have employed a stochastic model of line broadening to analyze the optical band shape.[1,2,4,5] The solvation dynamics were hence characterized by an amplitude of transition energy fluctuations and a single correlation time. Here, solvent fluctuation correlation functions given by molecular dynamics (MD) simulations are used to analyze the resonance emission band shapes. These simulations provide much more detailed information concerning the coupling of the solvent dynamics to the electronic excitation and serve as a test of line broadening optical dephasing models. In addition, the molecular level description afforded by this approach helps clarify the origin of the macroscopic response time scales and their evolution from low to high densities.

In order to carry out the resonance emission MD calculations, a set of ground and excited state solute-solvent interaction potentials must be established. A set of solute-solvent interactions parameters are determined by fitting MD simulations of the density/pressure dependence of the B-state absorption spectrum to the observed experimental spectra.[6] Lennard-Jones and exp6 potentials are used to model the electronic ground and excited CH_3I states respectively interacting with the bath. Good fits to the absorption spectra in high to moderate pressures of Ar and CH_4 (55 - 140 atm, reduced densities ~ 0.08 - 0.15) are found in the so-called static limit i.e. where the decay of the dipole correlation function is dominated by the range of configuration-dependent transition energies and not their fluctuations. Thus, the absorption band shape is due to inhomogeneous broadening on the time scale of the decay of the dipole correlation function. The pressure dependence of the absorption lineshapes is attributable to the shape of the solute-solvent difference potential. Consistent with the static limit description the corresponding transition energy correlation functions also known as the solvation energy correlation function in the context of time-dependent fluorescence stokes-shift measurements, decay on a time

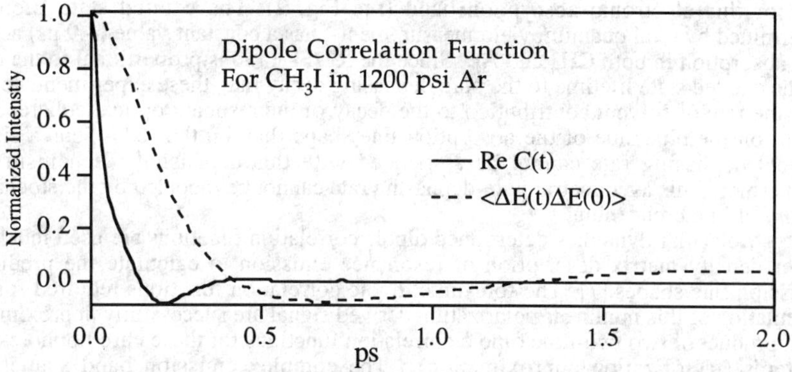

Fig. 1. The real part of the dipole correlation function (DCF) due to trajectories on the ground electronic surface, Re C(t) and the transition energy correlation function (ECF), $\langle \Delta E(t)\Delta E(0) \rangle$ are shown here for CH_3I in 1200psi Ar.

scale (~100 to 150 fs) much longer than that of the DCF or inverse absorption line width (see Fig. 1). This ECF is also a quantity of central importance in stochastic and Brownian oscillator line shape modeling.[1,2]

The spontaneous resonance emission of CH_3I in Ar and CH_4 excited in the region of the B-state origin consist of Raman-like and fluorescence-like features.[3,7] The area of the broad FL component increases relative to that of the narrow RL component as the bath pressure increases for these solvents. The labels RL and FL are used here to signify that a separation of the total complex emission band shape into two components is purely phenomenological. These features have distinguishable excitation frequency dependencies, band shapes (narrow and broad) and polarization characteristics. Thus, it is possible to separately describe the behavior of these two observable components on that basis.

The observed ratio of the RL/FL intensity in these high pressure solutions is observed to be a function of the excitation wavelength as the incident radiation is tuned through the pressure broadened electronic origin band. The relative amount of RL intensity increases as the excitation wavelength is tuned from the blue to the red

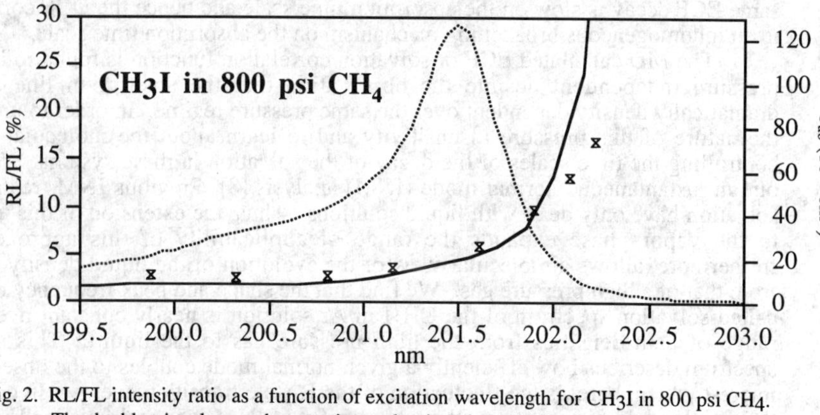

Fig. 2. RL/FL intensity ratio as a function of excitation wavelength for CH_3I in 800 psi CH_4. The double triangles are the experimental ratios, the solid line is the MD calculated RL/FL ratio. The pressure broadened absorbtion band is also shown.

side of this electronic absorption band (see Fig. 2). The excited state lifetime, determined by total quantum yield measurements, has a constant value (~.9 ps) across this absorption in both CH_4 and Ar. Since the RL/FL ratio is proportional to the ratio of the excited state lifetime to the pure dephasing decay rate, these experiments reveal that the rate of solvent contribution to the decay of the resonant optical coherence is faster on the blue side of the absorption line shape than on the red. Thus a single optical dephasing rate cannot be associated with this dephasing mechanism and furthermore, this asymmetric pure-dephasing rate cannot be modeled by the stochastic theory of line broadening.

Molecular dynamics determined dipole correlation functions are used in a third order density matrix description of resonance emission to calculate the predicted emission line shapes.[7] The four-time dipole correlation functions required for the calculation of this nonlinear polarization derived signal are successfully approximated by a product of two two-time dipole correlation functions for these chromophore-bath systems (factorization approximation). The complex emission band shapes are captured by this approach. The dispersion in the RL/FL emission redistribution is shown to be due to the multiple time scales inherent to this resonant optical coherence decay of these pressure broadened absorptions. The wavelength dependent pure-dephasing rate results from the nonlinear shape of the solute-solvent difference potential. There is simply a larger range of transition frequencies on the blue side of the absorption band than on the red side due to the shape of this difference potential and hence the pure-dephasing decay is more rapid on the blue side than the red side.

These experiments and their analyses clearly illustrate how the use of the labels homogeneous and inhomogeneous is always dependent on some reference or experimental time scale. The absorption spectra are unequivocally due to inhomogeneous broadening. The ECF decay is slower than that of the dipole correlation function (DCF) and the absorption bandshape is given by the static limit. However, the emission spectra are dominated by the FL component relative to the RL component, hence homogeneous pure dephasing must be an order of magnitude faster than the population decay. The relevant time scale for distinguishing those solvent fluctuations which are considered homogeneous or inhomogeneous for the nonlinear resonance emission experiment are the excited state lifetime (~1 ps) and the inverse Raman vibrational line width, which is also on the order of ~ 1ps. The ECF decay (~ 150 fs) is faster than this time scale and hence the DCF decay constitutes a homogeneous dephasing mechanism for the scattering experiment. However, the same ECF decay is slow on the absorption time scale and hence the DCF corresponds to an inhomogeneous broadening mechanism on the absorption time scale.

The MD calculated ECF or solvation correlation function is found to be nearly pressure independent despite the observation that the absorption line shape is dramatically density dependent over the same pressure regime. In order to understand the nature of this pressure insensitivity and to learn about the molecular motions controlling the time scales of the decay of the solvation in these systems, we carried out an instantaneous normal mode (INM) analysis.[8] Previous INM treatments of solvation have only dealt with liquid solutions. Thus, the extension of this technique to the vapor phase expands the range of applicability of this approach and, furthermore, allows a molecular view of the evolution of the liquid density response from that of a high pressure gas. We find that the shape and peak frequency of the so-called solvation spectrum of the CH_3I in Ar solution is nearly constant over a wide range of bath densities from the high pressure gas to the liquid. This solvation spectrum describes how efficiently a given normal mode couples to the observable of interest which, here, is the fluctuation of the B-state transition energy. In contrast to the weighted density of states, the total density of INMs varies dramatically over this same density range. The ECF calculated from the INM density of states agrees with

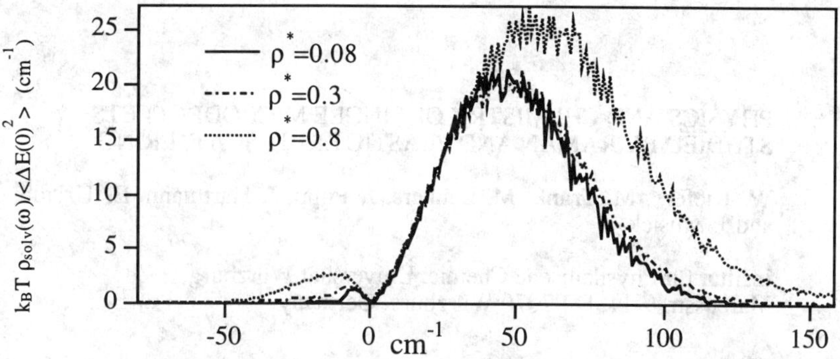

Fig. 3. Solvation spectra for CH_3I in Ar, corresponding to reduced argon densities of 0.08, 0.3, and 0.8. Each of the spectra is multiplied by the Boltzman constant and the temperature, and divided by the mean squared fluctuations of the transition energy spectrum at the corresponding density.

the MD simulated solvation correlation function at short times which is to say up to ~ 250 fs. This is well beyond the so-called inertial regime (≤ 100 fs). The ECF near pressure independence results from the cancellation of the density dependence of the solvation spectrum area and the mean squared fluctuations of the transition energy. These quantities are dependent on the total coupling strength of the energy gap fluctuations per INM and the second moment of the resonant absorption band width respectively. It is found from an analysis of the normal mode eigenvectors contributing to the solvation spectrum and an INM semi-analytical model that the largest contribution to the solvation dynamics are derived from a collection of independent pairwise interactions of the methyl iodide solute with a single Ar solvent particle at low density and with only one or two Ar particles at high (liquid) density.

Finally, the polarization of the resonance fluorescence component is given by a third order density matrix treatment.[9] It is demonstrated by the resonance emission of CH_3I in CH_4 and CO_2 that the depolarization of the FL component is controlled by the excited state lifetime and excited state (rovibronic) coherence loss decay rate. This is in contrast to the RL component, where the depolarization is controlled by the decay rate of the resonant optical coherence. Since quasi-elastic interactions are required for the appearance of the FL band, one might expect the FL component to be totally depolarized as a result of these stochastic interactions. However, no additional reorientational scrambling of the FL polarization beyond that caused by free rotation during the short excited state lifetime is found.

1. Y. Yan and S. Mukamel, J. Chem. Phys. 86, 6085 (1987).
2. S. Mukamel, Adv. Chem. Phys. 70, 165 (1988).
3. L. D. Ziegler, Acc. Chem. Res. 27, 1 (1994).
4. A. B. Myers and B. Li, J. Chem. Phys. 92, 3310 (1990).
5. E. T. J. Nibbering, K. Duppen and D. A. Wiersma, J. Chem. Phys. 93, 5477 (1990).
6. T. Kalbfleisch, R. Fan, J. Roebber, P. Moore, E. Jacobsen and L. D. Ziegler, J. Chem. Phys. 103, 7673 (1995).
7. R. Fan, T. Kalbfleisch and L. D. Ziegler, J. Chem. Phys. 104, (1996).
8. T. Kalbfleisch, L. D. Ziegler and T. Keyes, J. Chem. Phys. (in press).
9. L. D. Ziegler and R. Fan, J. Chem. Phys. (in press).

PHYSICS AND CHEMISTRY OF SINGLE MICRODROPLETS STUDIED BY RAMAN AND ELASTIC LIGHT SCATTERING

W. Kiefer*, M. Trunk, M. Lankers, J. Popp, I. Hartmann, E. Urlaub, and J. Musick

Institut für Physikalische Chemie, Universität Würzburg, Marcusstr. 9-11, D-97070 Würzburg, Germany

Spectroscopy of single microdroplets has become feasible due to the application of various optical trapping techniques. Particularly the optical levitation technique, which has been pioneered by Ashkin [1] already in 1971, has been proofed to be a powerful tool for nondistructible spectroscopic investigations. Studies on liquid droplets [2-15] as well as on solid particles [16,17] have been reported in the literature. Data derived from the elastic light scattering, e.g. the angular distribution of the scattered radiation or the radiation pressure, acting on spherical resonators like droplets give only information on the morphology of the scattering particle. Raman or fluorescence spectroscopy [18-20] additionally allows to obtain information on the chemical composition of the material under investigation. As well known, in Raman spectroscopy of bulk samples, the energy shift of the peaks in a Raman spectrum can be ascribed to a specific molecular vibration and the intensity of each peak is linearly proportional to the concentration of the specific molecule and the Raman scattering cross section of the particular vibrational transition. The above mentioned trapping techniques made it possible to obtain Raman spectra of samples whose sizes are of the order of or larger than the wavelength of the exciting light. However, if such small scattering systems have well defined geometries, e.g. spheres, cylinders, spheroids, etc., the use of Raman spectroscopy as a diagnostic probe becomes much more complicated as a result of morphology-dependent resonances (MDRs). Such resonances can actually give rise to sharp peaks in a Raman spectrum that are not present for a bulk sample of the same composition. These peaks can result from resonance-induced enhancement of the Raman scattering. The physical nature of these resonances can be described for dielectric particles by means of the well-known Lorenz-Mie theory [21]. On the other hand, these MDRs can be used together with Raman data for a comprehensive study of physical properties as well as of the time dependence of chemical reactions. Such studies have been performed in recent years on various microsystems in our laboratory. This lecture intends to give a short review and a state-of-the-art report on such combined inelastic/elastic light scattering studies. However, due to space limitation we will concentrate in this article to only a few aspects.

First we show how the generation of morphology dependent resonances can be explained in a very simple manner. Figure 1 represents theoretical calculations of the light intensity distribution applying the Mie-Debye theory for the case of a focused Gaussian laser beam (coming from below, see arrow) hitting a microdroplet of water close to its edge. As in geometrical optics, part of the light is reflected on the curved surface and another part is deflected into the sphere. The latter beam is again deflected and reflected when it hits the surface a second time. This continues until the internally reflected part of the laser light comes back to the starting region. It can be seen that part of the laser radiation is trapped by the droplet and circulates around the

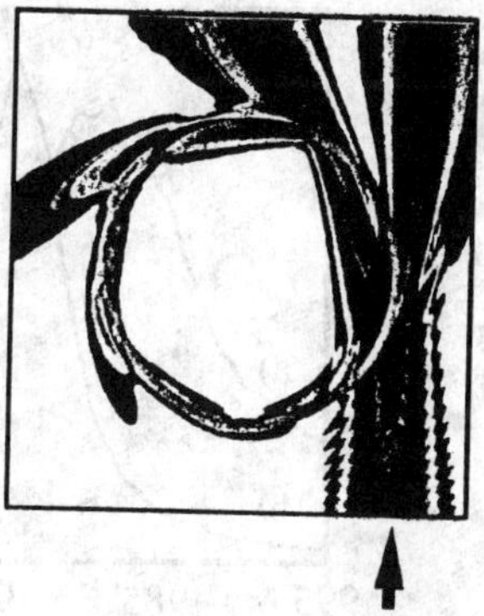

Fig. 1. A laser beam is tightly focused close to the edge of a microdroplet. For the generation of morphology dependent resonances see text.

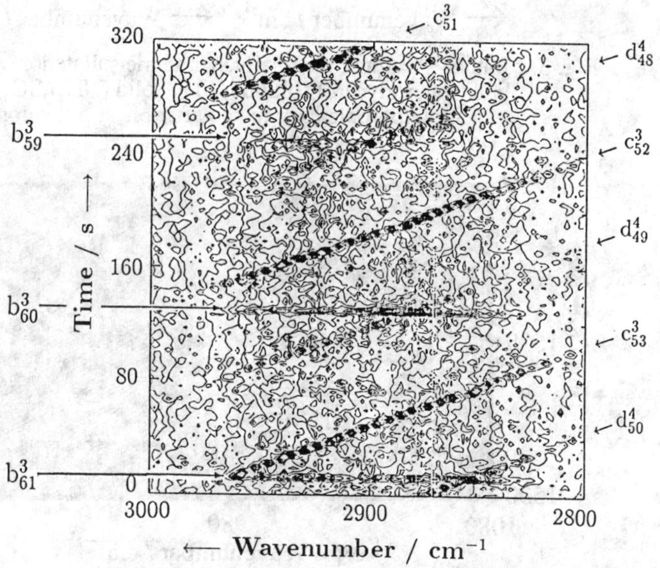

Fig. 2. Morphology dependent resonances (MDRs) in the Raman spectra of a glycerin microdroplet (a=5.9986 μm at t=0) as revealed by a contour plot.

droplet rim. If the size (radius a) and the ratio of the indices of refraction of the liquid and of the surrounding medium (m) are such that there is constructive interference between the wave which circulates around the droplet rim, the droplet acts as a high-Q optical cavity at specific wavelengths and gives rise to so-called input MDRs. A similar mechanism can happen for wavelengths which are shifted by the Raman frequency. The latter are called output MDRs. Input and output resonances can nicely be revealed when Raman spectra are taken as a function of time and displayed three-dimensionally [10] with the time as the third axis or by a two-dimensional contour plot of such three-dimensional plots. An example is given in Fig. 2 where we display a contour plot of measurements on a glycerin droplet of radius a=5.9986 μm at the start of the series (t=0). The MDR peaks, which are - based on simulations using the Mie-Debye formalism - assigned to the multipole expansion coefficients c (transverse

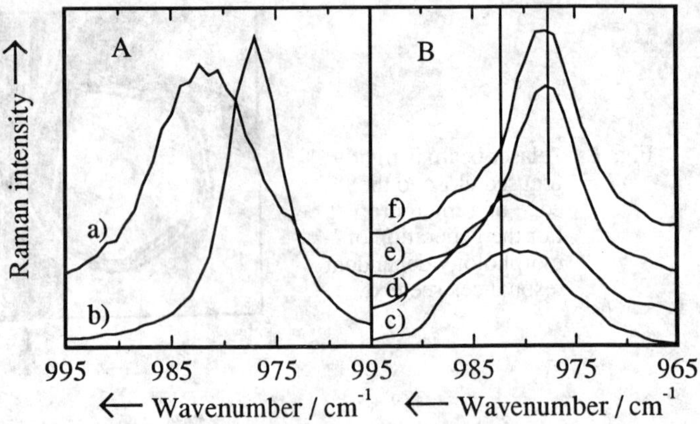

Fig. 3. Raman spectra of the totally symmetric sulfate ion vibration; a) in bulk water/glycerol solution; b) in solid $(NH_4)_2SO_4$; c)+d) and e) +f) in microdroplet before and after phase transition, respectively.

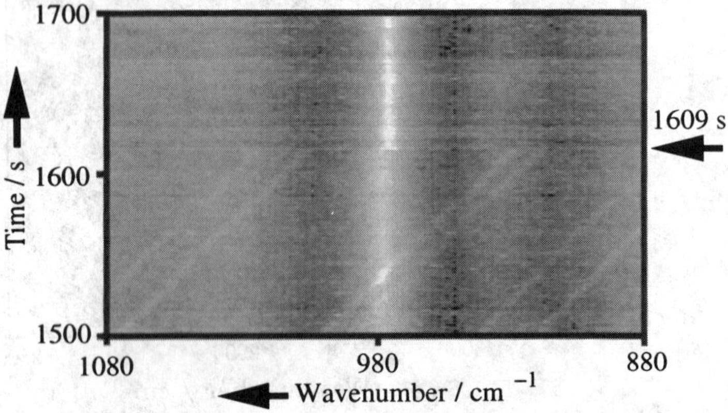

Fig. 4. Contour plot of Raman spectra of an evaporating ammonium sulfate water/glycerol microdroplet. At time t=1609 s the phase transition from the liquid to the solid state takes place.

magnetic) and *d* (transverse electric) for the electric field, can be visualized as traces. The peaks, which move from higher to lower wavenumbers with increasing time as a consequence of droplet evaporation, are the output MDRs. From these traces very accurate data for evaporation rates can be derived. Traces marked by horizontal arrows correspond to input MDRs. They can be observed only during a few seconds when the radius of the evaporating droplet corresponds to an electromagnetic eigenresonance of the microdroplet.

Next we show results of an investigation of solute nucleation in an optically levitated microdroplet. The latter was made from a water solution of ammonium sulfate with addition of glycerol to prevent too rapid evaporation. Spectra a) and b) of Fig. 3 show the totally symmetric sulfate vibration at about 981 cm^{-1} and 976 cm^{-1} observed in the water/glycerol bulk solution and in solid $(NH_4)_2SO_4$, respectively. Spectra c)+d) and e)+f) were taken from the microdroplets at times before and after crystallisation, respectively. Each spectrum was taken with one second accumulation time. In order to follow up evaporation and to determine the exact time of crystallisation we have made a contour plot similar to the one displayed in Fig. 2 exept for the time scale and the fact that white corresponds to high and black to zero intensity. Such a plot is displayed in Fig. 4 for the time period between 1500 and 1700 seconds after the start of the experiment. Up to time 1609 s we see the slightly broader sulfate band in solution. Additionally, traces from MDRs can be noticed which show that there is still evaporation taking place, since the change in wavenumbers is to lower values with increasing time. At exactly 1609 s these traces stop and there is a change of the sulfate vibrational band by about five wavenumbers as well as a decrease in band width. These observations clearly demonstrate that the phase transition from liquid to solid has taken place at this time.

Using both, data from elastic and inelastic light (Raman) scattering we have been able to follow up several chemical reactions with resolution in time of about 1 second. Examples for this are acid/base reactions in microdroplets where the formation of the solid salt could be monitored or emulsion polymerization studies. Such results will be presented at the conference.

Financial support from the Deutsche Forschungsgemeinschaft and from the Fonds der Chemischen Industrie is highly acknowledged.

1. A. Ashkin and J. M. Dziedzic, Appl. Phys. Lett. 19, 283 (1971).
2. R. Thurn and W. Kiefer, Appl. Opt. 24, 1515 (1985).
3. R. Thurn and W. Kiefer, Appl. Spectrosc. 38, 78 (1984).
4. T.R. Lettieri and R.E. Preston, Opt. Commun. 54, 349 (1985).
5. C.K. Chan, R.C. Flagan, and J.H. Seinfeld, Appl. Opt. 30, 459 (1991).
6. G. Schweiger, Opt. Lett. 15, 156 (1990).
7. G. Schweiger, J. Raman Spectrosc. 21, 165 (1990).
8. T.R. Lettieri, W.D. Jenkins, and D.A. Swyt, Appl. Opt. 20, 2799 (1981).
9. K. Schaschek, J. Popp, and W. Kiefer, J. Raman Spectrosc. 24, 69 (1993).
10. K. Schaschek, J. Popp, and W. Kiefer, Ber. Bunsenges. Phys. Chem. 97, 1007 (1993).
11. M. Lankers, J. Popp, and W. Kiefer, Appl. Spectrosc. 48, 1166 (1994).
12. M. Trunk, M. Lankers, J. Popp, and W. Kiefer, Appl. Spectrosc. 48, 1291 (1994).
13. J. Popp, M. Lankers, K. Schaschek, W. Kiefer, and J.T. Hodges, Appl. Opt. 34, 2380 (1995).
14. J. Popp, M. Trunk, I. Hartmann, M. Lankers, and W. Kiefer, J. Mol. Structure 348, 281 (1995).
15. I. Hartmann, J. Popp, M. Lankers, M. Trunk, and W. Kiefer, J. Mol. Structure 349, 203 (1995).
16. R. Thurn and W. Kiefer, J. Raman Spectrosc. 15, 411 (1984).
17. J.F. Owen, R.K. Chang, and P.W. Barber, Aerosol Sci. Techn. 1, 293 (1982).
18. S.C. Hill, R.E. Benner, C.K. Rushforth, and P.R. Conwell, Appl. Opt. 23, 1680 (1984).
19. R.E. Benner, P.W. Barber, J.F. Owen, and R.K. Chang, Phys. Rev. Lett. 44, 475 (1980).
20. H.-M. Tzeng, H.F. Wall, M.B. Long, and R.K. Chang, Opt. Lett. 9, 273 (1984).
21. for a short review see W. Kiefer, Croat. Chim. Acta 61, 474 (1988).

RESONANCE RAMAN STUDIES OF HYDROGEN BONDING IN AMIDES

James J. Valentini

Department of Chemistry
Columbia University
New York, NY 10027

The hydrogen bond represents an intermolecular, or occasionally intramolecular, interaction between an electronegative atom, B, and a hydrogen atom covalently bonded to an electronegative atom, A-H, that results in an attractive force, or "bond" between B and H, written as A-H•••B. What is implied by this picture is something between a purely electrostatic interaction and a covalent bond. The presence of such a hydrogen bond is taken to be evidenced by the modification of chemical and physical properties, such as boiling points, vibrational frequencies, dielectric strength, solubility, NMR chemical shifts, and crystal structures.[1-3] For example, in solids a relatively short, characteristic distance between A and B, or between H and B when that distance can be determined, is associated with hydrogen bond formation. In liquids, solids, and solution, hydrogen bonds are associated with reductions in the A-H vibrational band frequency and increases in the band width.

However, these features are not merely evidence for hydrogen bond formation, they effectively define it. We have worked to establish a definition of the hydrogen bond in terms of particular features in the Raman and resonance Raman spectra of molecules that can form hydrogen bonds, most extensively for the amides. We have studied a variety of compounds containing the amide group, in the gas phase, in neat liquid, in solution, both aqueous and non-aqueous, and in solids. These compounds range from the simplest amides, such as ε-caprolactam, to the polyamide polymer nylon 6,6. Hydrogen bonding in amides is possible because the amide group incorporates both a proton donating group, -N-H, and a proton accepting group, -C=O. For evidence of hydrogen bonding we naturally look to the effect on the band frequencies and widths of the N-H and C=O stretch vibrations, as we would with any vibrational spectroscopic probe. However, we also look to an effect specific to resonance Raman spectroscopy, the strong modification of the intensities of vibrational bands, particularly the intensities of the C=O stretch and C-N stretch vibrations. Modification of the intensities of these bands by interactions that we would call hydrogen bonds are much more pronounced than the changes in vibrational band frequencies and widths, and are quite distinctive.

The amides have an unstructured $\pi-\pi^*$ electronic absorption band in the far uv, with a peak at around 190 nm and a long tail extending to longer wavelengths.[4,5] There have been many reports of resonance Raman spectra of amides, peptides, and proteins in aqueous solution taken with excitation wavelengths within this absorption band.[6-9] While some of our resonance Raman spectra have been recorded at excitation wavelengths that are well into the $\pi-\pi^*$ band of the amides, the most useful spectra are those taken at wavelengths longer than 250 nm, which are perhaps better described as pre-resonant rather than resonant. We have also recorded Raman spectra of the amides at wavelengths in the visible that are well removed from the absorption band.

To take all these spectra we use harmonics of the output of a pulsed Nd:YAG laser, with Raman shifting providing further excitation wavelengths. Pulse energies of a 0.1 to 10 mJ are used, with an unfocused or weakly focused beam. Depending on the sample the Raman scattered light is collected at 90° or 180° from the incident beam direction, and dispersed by a 0.75-m double monochromator. When stray light rejection is a less significant concern, only the first stage of the monochromator is used. The dispersed light is detected with either a uv-sensitive photomultiplier or a uv-enhanced CCD array.

Our results show that hydrogen bonding results in several changes in the Raman spectra. The most important of these affect the Am I band, which is predominantly the C=O stretch, and the Am II and Am III bands, which are a mixture of C-N stretch and C-N-H in-plane bend in the trans amides, but are, respectively, predominantly C-N stretch and C-N-H bend in the cis amides. The C=O stretch frequency is reduced by hydrogen bonding, to an extent about twice as great as that due to general solvation. The width of the Am I C=O stretch band also increases upon hydrogen bonding. The frequency of the C-N stretch vibration does not appear to change. More significantly, hydrogen bonding alters the relative intensities of the C=O and C-N stretch vibrations. The ratio of these intensities, measured by the Am I:Am II ratio for the cis amides and the Am I:(Am II + Am III)/2 ratio for the trans amides, is greatly reduced, from a value significantly greater than one to a value significantly less than one, upon hydrogen bonding. To the extent that comparisons can be made, these results are consistent with previous resonance Raman measurements on N-methylacetamide in aqueous solution, in which the effects of hydrogen bonding were investigated.[10,11]

The most surprising result of our studies is the observation that these signatures of hydrogen bonding are not evident in the resonance Raman spectra of some crystalline amides for which the crystal structures would seem to indicate the presence of a hydrogen bond. For some crystalline amides the resonance Raman spectra look very much like spectra of the amide group in the low-density gas phase, where hydrogen bonding is not important, and like the spectrum of N,N-dimethylacetamide in the liquid, for which hydrogen bonding is not possible due to the absence of an amidic hydrogen. The differences between spectra of solids that show hydrogen bonding and those that do not are illustrated by the spectrum of ε-caprolactam in Fig. 1, and the spectrum of 2-azacyclanonanone in Fig. 2.

The ε-caprolactam and 2-azacyclanonanone molecules are both cyclic amides, the former having a ring size N=7 and the latter N=9. Both have crystal structures that show the distance between the O of the amidic C=O in one molecule and the N of the amidic N-H in an adjacent molecule to be relatively short and characteristic of the distances associated with amide-amide hydrogen bonding in the solid. Yet the resonance Raman spectra of these two crystalline solids are dramatically different. The ε-caprolactam spectrum shows a broad (38 cm^{-1} FWHM) Am I band and an Am I:Am II intensity ratio of 0.53(3), both indicative of hydrogen bonding. The 2-azacyclanonanone spectrum reveals an Am I band that is narrow (13 cm^{-1} FWHM), and is characterized by an Am I:(Am II + Am III)/2 ratio of 2.5(2), both indicative of the absence of hydrogen bonding. The two lactams are different in that ε-caprolactam has a cis amide configuration, while 2-azacyclanonanone has a not-quite-planar near-trans configuration,[12] but this cannot account for the differences in the resonance Raman spectra. We believe that these differences, which are evident in many other compounds as well, reflect differences in the amide-amide interaction in the solid. The absence of any signature of hydrogen bonding in crystalline solids like 2-azacyclanonanone actually suggests that the solids should not be described as hydrogen-bonded.

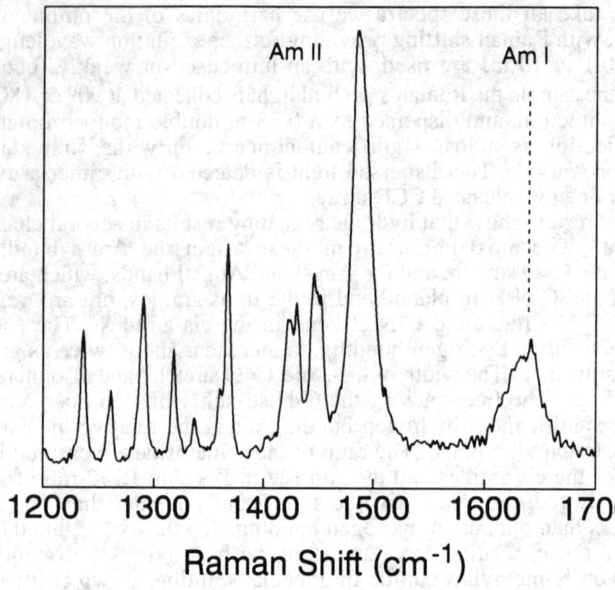

Fig.1 Pre-resonant Raman spectra of crystalline ε-caprolactam taken with 266 nm excitation.

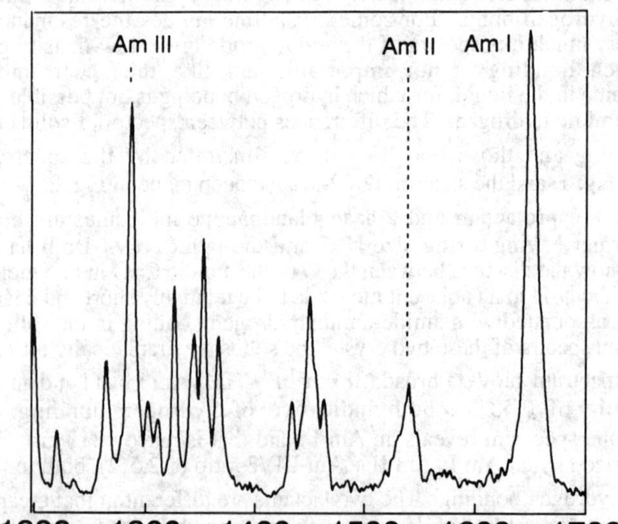

Fig.2 Pre-resonant Raman spectra of crystalline 2-azacyclononanone taken with 266 nm excitation.

It is almost always assumed that amides hydrogen bond with one another in the solid, and crystallographic evidence supports this assumption. However, while proximity of the -N-H and -C=O groups on adjacent molecules in the solid can be taken as one definition of a hydrogen bond, such proximity does not of necessity imply the kind of electron density redistribution that the term "bond" connotes. The resonance Raman spectra are sensitive to such electronic rearrangement, since it influences the vibrational frequencies, and more important, the band intensities. The resonance Raman spectra will be particularly sensitive to electron density redistribution associated with a hydrogen bond, if the intramolecular electronic perturbation caused by the hydrogen bond is different in the ground electronic state and the excited electronic state accessed by the Raman excitation frequency. Thus, although the Raman spectral features that we associate with hydrogen bonding serve only as one of many operational definitions of the hydrogen bond, this definition is a particularly important one. In particular, the absence of the signatures of hydrogen bonding in the resonance Raman spectra provide quite compelling evidence that a "bond" is not actually forming.

This work is supported by the National Science Foundation and by the Du Pont Company.

1 G.C. Pimentel and A.L. McClellan, The Hydrogen Bond, W.J. Freeman, San Francisco, 1960.
2. W.C. Hamilton and J.A. Ibers, Hydrogen Bonding in Solids, W.A. Benjamin, Reading, MA, 1968.
3. G.A. Jeffrey and W. Saenger, Hydrogen Bonding in Biological Structures, Springer-Verlag, Berlin, 1991.
4. K. Kaya and S. Nagakura, Theor. Chim. Acta 7, 124.
5. E.B. Nielsen and J.A. Schellman, J. Phys. Chem. 71, 2297 (1967).
6. I. Harada, Y. Sugawara, and H. Matsuura, J. Raman Spectrosc. 4, 91 (1975).
7. L.C. Mayne, L.D. Ziegler, and B. Hudson, J. Phys. Chem. 89, 3395 (1985).
8. J.M. Dudik, C.R. Johnson, and S.A. Asher, J. Phys. Chem. 89, 3805 (1985).
9. Y. Wang, R. Purrello, and T.G. Spiro, J. Am. Chem. Soc. 111, 8274 (1989).
10. L.C. Mayne and B. Hudson, J. Phys. Chem. 95, 2962 (1991).
11. Y. Wang, R. Purrello, S. Georgiou, and T.G. Spiro, J. Am. Chem. Soc. 113, 6368 (1991).
12. F.K. Winkler and J.D. Dunitz, Acta Crystallogr. B31, 276 (1975).

SECTION 1

RAMAN THEORY

NEW FEATURES IN COHERENT RAMAN SCATTERING DERIVED FROM NOISY LIGHT INTERFEROMETRY AND SPECTROSCOPY

Darin J. Ulness and A. C. Albrecht*

Department of Chemistry, Cornell University, Ithaca, New York, 14853

The study of ultrafast wavepacket (coherences) and population dynamics traditionally involves the skillful use of femtosecond lasers - recently tailored in both their phase and amplitude. An interesting alternative approach to such questions is offered by noisy light interferometry. Introduced just over a decade ago [1] it is based on light sources of considerable duration (usually nanoseconds) but being so "noisy' that their field autocorrelation time approaches the femtosecond domain - the transform of their spectrum. By now a variety of applications of such incoherent light, particularly in nonlinear spectroscopies, have appeared from several laboratories leading to some interesting new light/matter phenomena not seen in the real femtosecond domain.

We have recently focussed on the principal differences between these two approaches to femtosecond dynamics and extended our understanding of noisy light interferometry. Simply put, where in real femtosecond work the pulse envelope function defines the effective time domain, in noisy light experiments it is the two-point time (pair) correlator that plays the defining role. Real femtosecond light pulses necessarily contain a broad spectrum of colors (frequencies) whose relative phases are under deterministic control (not random). (This includes chirped pulses of various sorts.) When they are transform limited, the reciprocal of the bandwidth must give a good measure of the pulse duration. Different frequencies from such a pulse *can act in a coherent fashion* upon a material system. In complete contrast, noisy light having the same spectrum of colors, and having a correlation time also measured well by the reciprocal spectral width, consists of individual frequencies whose relative phases are *completely uncorrelated.* (This is exactly true if "stationarity" applies to the pair correlator.) Different frequencies from such a noisy light source pulse *cannot act coherently* on a material system. At the amplitude level noisy light produced coherences are themselves noisy. However, once the final signal is *quadrature detected each component color must appear in self quadrature ('color locking') and the noise can be completely nulled.* Incidentally because one source is phase coherent, and the other is not, one predicts that these two very different "femtosecond" light sources will differ in the nature of the wavepackets they prepare.

Formally, quadrature detected coherent spectroscopies involve the square modulus of the new field launched by the nonlinear polarization appearing in the medium. By far the dominant contribution arises from the cross terms between the polarization on different sites ('chromophores') in the medium. Stochastic averaging over the noise in the light field(s) takes place at the quadrature level and therefore is likely to involve noise correlation across chromophores pairs. Typically the time correlators that appear are higher order than two (the pair correlator). One then immediately invokes the complex Gaussian moment theorem to achieve the factoring of the higher order correlator into products of pair correlators. These products of pair correlators are readily represented by diagrams we call factorized time correlator (FTC) diagrams. This FTC diagrammatic approach was developed in the analytic treatment of interferometric noisy light four wave mixing (4WM) in a Bloch two-level system [2]. There we have learned how to use the FTC diagrams in a broadly predictive way and have applied them to a variety of noisy light interferometric versions of nonlinear spectroscopies - otherwise well known in continuous wave or real femtosecond applications.

Typically, in any experiment the noisy source is split into two beams by an interferometer to offer twin beams F and F', the latter shifted by τ over the former. These may join one or more monochromatic beams, and/or themselves multiply intervene, to produce a nonlinear polarization. Interferograms (measuring the intensity of the field launched by the polarization) exhibit signals that decay in τ (dynamics information) and may contain significant fringes which can determine (noise-free) transition frequencies. Among other things an interferogram is characterized by its contrast ratio (CR) (its value at $\tau = 0$ divided by the plateau value reached at $\tau = \infty$) and its symmetry in τ. Surprisingly, depending on the spectroscopy, the CR's may be limited to a simple numerical ratio or may in effect be 'unbounded'. Some must be symmetric, others may be asymmetric. FTC analysis can answer such questions immediately. FTC predictions for noisy light interferograms in several spectroscopies at the $\chi^{(2)}$ and $\chi^{(3)}$ levels are offered in the TABLE on the next page. Here we concentrate on the FTC diagrams for noisy light interferometry in the two coherent Raman spectroscopies - $I^{(2)}CARS$ and $I^{(2)}CSRS$.

FTC DIAGRAMS IN $I^{(2)}CRS$

These $\chi^{(3)}$ level $I^{(2)}CRS$ have in common an amplitude level stimulated Raman (SR) process (Step 1), followed by a Stokes step (in *CSRS*) or an anti-Stokes step (in *CARS*), Step 2. They are k-vector designed to involve noisy field F in one step and its twin F' - acting *with the same phase* in the second step. (F,F' permutations are always considered.) At the quadrature level the noisy fields consummate their *color-locked conjugation* through the pair correlator. Since F and F' act *in phase* on a given chromophore their conjugate partners (for conjugate action in a pair correlator) must be on a second chromophore (the bichromophore model). All pair correlators appearing in $I^{(2)}CRS$ must be *interchromophoric*. Furthermore Step 1 involves a monochromatic beam, ω_M so that with the Raman resonance (Bohr frequency ω_a) Step 1 serves to single out the unique color from broadband F (of F'): $\omega_x = \omega_M \pm \omega_a$ (+ for $I^{(2)}CARS$; - for $I^{(2)}CSRS$). Likewise Step 2 serves as a color filter provided that the signal field is monochromatically detected at ω_D. Now $\omega_y = \omega_D \mp \omega_a$ is selected (*oppositely* signed to the Step 1 filter). Stochastic averaging leads to 8 distinct products of two pair correlators. In FTC diagrammatic form these are:

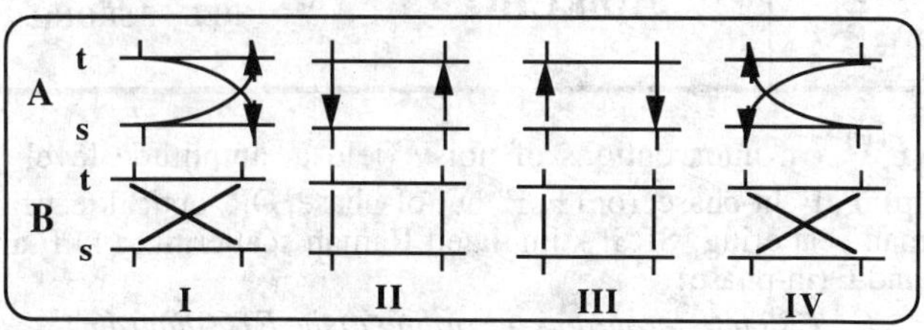

In each diagram the two ticks on the t-line correspond to the Step 1 and Step 2 noisy field interventions on one chromophore, similarly for the s-line on the second chromophore. Each connecting segment corresponds to a color locked pair correlator - the *arrow* is τ independent with F' at the head and F at the tail; the *line* is τ-independent with correlation between F and F and F' and F'. All segments are *interchromophore* as noted. When $\tau = 0$ all arrowed pair correlators reduce exactly to lines - Row A turns into Row B. At $\tau = \infty$ all arrowed diagrams disappear - Row A is gone.

TABLE*

			Noisy Light Spectroscopy	FTC predictions re: interferograms
CHI-TWO	$I^{(2)}$ ND3WM	in-phase $I^{(2)}$ ND3WM(ip)	t-dependent frequency summing, like $I^{(2)}$SHG	CONTRAST RATIO (CR): 2:1(noncolinear), 4:1(colinear) SYMMETRIC REL. WEAK
		out-of-phase $I^{(2)}$ ND3WM(op)	t-dependent frequency differencing, like optical rectification or $I^{(2)}$OR	CR: FTC UNbounded SYMMETRIC REL. STRONG (DC energy density)
CHI-THREE	$I^{(2)}$ ND4WM	in-phase $I^{(2)}$ ND4WM(ip)	$I^{(2)}$ CRS: $I^{(2)}$ CARS, $I^{(2)}$ CSRS	CR: 2:1 SYMMETRIC REL. WEAK
			$I^{(2)}$ 4WM(ip), ECHO	CR: 2:1 SYMMETRIC (?) REL. WEAK
		out-of-phase $I^{(2)}$ ND4WM(op)	$I^{(2)}$ SRS	CR:FTC UNbounded ASYMMETRIC (?) REL. STRONG
			$I^{(2)}$ 4WM(op)	CR:FTC UNbounded ASYMMETRIC REL. STRONG
	$I^{(3)}$ D4WM	$I^{(3)}$ D4WM (-)	three-beam FID/ECHO	CR: 4:1 SYMMETRIC (?) REL. STRONG
		$I^{(3)}$ D4WM (+)	two (or three)-beam FID/ECHO	CR:FTC UNbounded ASYMMETRIC REL. STRONG

*Glossary: $I^{(n)}$ - n interventions of noisy field at amplitude level; ND: nearly degenerate; (ip): F, F' in-phase; (op) F, F' out-of-phase; OR: optical rectification; CRS: coherent Raman scattering; SRS: stimulated Raman scattering; (+) F and F (out-of phase); (-) F and F (in-phase)

$I^{(2)}$CRS interferograms according to the FTC diagrams

FTC diagram predictions for $I^{(2)}CRS$ are enumerated: 1) CR must be 2:1 - since at $\tau = 0$ we have 2 times Row B; at $\tau = \infty$ one Row B only. 2) It must be

SYMMETRIC in τ – arrows in Row A are symmetrically disposed along the time axis. 3) The signal is REL. WEAK because all correlators are *interchromophore* (see [2]). 4) Diagrams II and III dominate the signal because they are free to accumulate over the dephasing time of the vibration (the Step 1 to Step 2 separation damped by memory loss). The crossed diagrams I and IV are not free to accumulate in this way (see [2, 3]). 5) The interferogram possess fringes that oscillate at the beat frequency between the Step 1 A and Step 2 filtered colors: $\Delta = \left| \omega_x - \omega_y \right|$ (called a Rabi detuning oscillation). 6) In general color-locking is favorable for II and III because the correlators are pure Step 1 (at ω_x) and pure Step 2 (at ω_y). The crossed diagrams suffer in this respect for they try to correlate *with a single color* Step 1 (at ω_x) on one chromophore with Step 2 (at ω_y) on another. This conflict is relieved provided ω_x is made experimentally equal to ω_y (by adjusting ω_M and/or ω_D). 7) However in binary materials involving two different Bohr frequencies signals from mixed chromophore pairs are principally from the crossed diagrams (relieved of conflict by different Bohr frequencies) while FTC diagrams II and III are conflicted. In binary mixtures they continue to strongly dominate the "same" chromophore pair signals.

Spectral 'poles'

The spectrum of an $I^{(2)}CRS$ is predicted to have remarkable spectral poles with "zero" material linewidth [4, 5, 6] located at $\omega_D = \omega_M \pm 2\,\omega_a$ (+,CARS; -, CSRS) exactly where the two filtered colors are identical, namely at $\Delta = 0$. In theory their width is governed by the slit function only (for truly monochromatic ω_M). As the slit is narrowed the feature becomes sharper (arbitrarily sub material linewidth), though its area diminishes even as it identifies a vibrational frequency [6] with unusual precision. It is absent at "zero" slit width. Seen in several liquids including binary systems, these 'poles' are now known to originate from the crossed FTC diagrams I and IV (not color conflicted when $\Delta = 0$). In binary systems spectral 'poles' appear from same chromophore pairs as well as from mixed (a/b) chromophore pairs - triplets are seen at $\omega_{D_a} = \omega_M \pm 2\,\omega_a$, $\omega_{D_b} = \omega_M \pm 2\,\omega_b$, and at $\omega_{D_{ab}} = \omega_M \pm \left(\omega_a + \omega_b \right)$.

Tailored noisy light sources

Recently [7] light from two independent noisy laser sources has been joined into one beam whose F and F' interferograms produce fringes having very short periods corresponding to beating between the two color components. Among other things time resolution better than the autocorrelation time of either source alone is promised.

Acknowledgement: We are grateful for support through NSF Grant CHE-9311959

1. N. Morita and T Yajima, Phys. Rev. A, 30, 2525(1984); S. Asaka et. al. Phys. Rev. A, 29, 2286(1984); R. Beach, D. Debeer, and S. R. Hartman, Phys. Rev. Lett. 53, 663(1984).
2. D. J. Ulness and A. C. Albrecht, Phys. Rev. A, 53, 1081(1996).
3. D. J. Ulness and A. C. Albrecht, Appl. Phys. B, submitted.
4. M. A. Dugan, J. S. Melinger, and A. C. Albrecht, Chem. Phys. Lett. 147, 411(1988); M. A. Dugan and A. C. Albrecht, Phys. Rev. A, 43, 3877, 3922(1991).
5. S. P. Smith, D. Tan, S. A. Schaertel, D. DeMott, and A. C. Albrecht, Laser Physics 5, 667(1995).
6. S. Smith, Ph.D. Thesis, Cornell University, 1995.
7. D. C. DeMott, D. J. Ulness, and A. C. Albrecht, Phys. Rev. A, submitted.

ULTRAFAST NON-BORN OPPENHEIMER DYNAMICS DETECTED BY RESONANCE RAMAN SPECTROSCOPY: A CASE STUDY ON PYRAZINE

Gerhard Stock, Clemens Woywod, and Wolfgang Domcke

Institut für Physikalische und Theoretische Chemie,
Technische Universität München, D-85748 Garching, Germany

The photophysics and photochemistry of excited singlet states of simple aromatic molecules have been a topic of continued interest over the past decades [1]. While the spectroscopy of low vibronic levels of the first singlet state (S_1) is well established, very little is still known about the photophysical dynamics of the S_2 state and higher excited singlet states. The spectroscopic investigation of these states is often hampered by large line broadening of the electronic spectra, rendering the appearance of the absorption and fluorescence bands typically diffuse and rather structureless.

Besides time-resolved techniques [2], resonance Raman (RR) spectroscopy has proven to be very useful to obtain specific information on the vibrational and relaxational dynamics of short-lived higher singlet states. Most theoretical investigations of RR processes assume a separable harmonic model for both ground and excited electronic states, whereby the vibrational frequencies and geometry shifts are inferred from experimental data [3]. The goal of this work is to demonstrate that high-level *ab initio* calculations and time-dependent multimode propagation techniques enable us to make *first-principles predictions* of RR spectra for vibronically coupled electronic states of an aromatic molecule.

The $S_1(n\pi^*)$ and $S_2(\pi\pi^*)$ excited states of pyrazine represent a classic example of vibronic coupling in aromatic systems [4]. Based on recent *ab initio* CASSCF/MRCI calculations [5], we have constructed a vibronic coupling model of the conically intersecting S_1 and S_2 states of pyrazine, which includes all 24 vibrational degrees of freedom of the molecule [6]. Employing a time-dependent approach that treats the intramolecular couplings in a nonperturbative manner [7], we calculate RR cross sections for this model, taking explicitly into account *seven* nonseparable vibrational modes.

As a survey of the calculated S_2 RR spectra, Fig. 1 presents the excitation functions of the most prominent Raman lines of pyrazine. For all excitation wavelengths, the fundamentals, overtones and combination levels of the totally symmetric modes ν_1 and ν_{6a} dominate the spectra, while the nontotally symmetric mode ν_{10a}, which dominates the S_1 RR spectrum, is hardly observed under resonant excitation of the S_2 state of pyrazine.

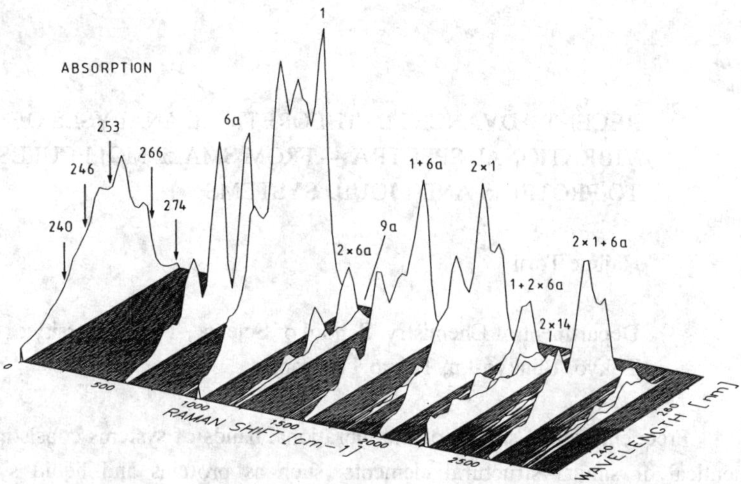

Fig. 1: Calculated absorption and RR spectra of the $S_2(\pi\pi^*)$ state of pyrazine as obtained by the *ab initio* three-state seven-mode model.

The theoretical data have been compared to experimental gas-phase RR spectra [6], which have been obtained for five different excitation wavelengths as indicated in Fig. 1. The comparison reveals that the *ab initio* predictions match the experimental results in almost every detail. Furthermore, the validity and limitations of the vibronic coupling model for the description of RR spectroscopy and the effects of the conical intersection on the Raman emission have been discussed [6,7]. For a consistent simulation of absorption, fluorescence, and RR spectra, it turned out to be essential to explicitly account for the vibronic coupling by a nonperturbative multimode wave-packet propagation [7].

In conclusion, it is encouraging that predictive calculations of RR spectra are nowadays possible for molecules of this size, even when complex non-Born-Oppenheimer dynamics associated with surface-crossings is involved. It has to be kept in mind, however, that, due to the rather short optical dephasing times in these systems ($\approx$ 10 fs for the $S_2 - S_1$ transition in pyrazine), absorption and RR spectroscopy only accounts for the very first step of the photodynamics of the molecule.

This work has been supported by the Deutsche Forschungsgemeinschaft.

1. J. B. Birks, *Photophysics of Aromatic Molecules* (Wiley, New York, 1970).
2. See, for example, papers in *Ultrafast Phenomena IX*, edited by P. F. Barbara, W. H. Knox, G. A. Mourou, and A. H. Zewail, (Springer, Heidelberg, 1994).
3. See, for example, A. B. Myers, R. A. Mathies, D. J. Tannor, and E. J. Heller, J. Chem. Phys. **77**, 3857 (1982).
4. K. K. Innes, I. G. Ross, and W. R. Moomaw, J. Mol. Spectrosc. **132**, 492 (1988).
5. C. Woywod, W. Domcke, A. L. Sobolewski and H.-J. Werner, J. Chem. Phys. **100**, 1400 (1994).
6. G. Stock, C. Woywod, W. Domcke, T. Swinney, and B. S. Hudson, J. Chem. Phys. **103**, 6851 (1995).
7. G. Stock and W. Domcke, J. Chem. Phys. **93**, 5496 (1990).

RECENT ADVANCES IN THEORETICAL ANALYSES OF VIBRATIONAL SPECTRA — FROM SMALL MOLECULES TO PROTEINS AND LIQUID SYSTEMS

Hajime Torii

Department of Chemistry, School of Science, The University of Tokyo, Bunkyo-ku, Tokyo 113, Japan

Frequencies and shapes of the vibrational bands of systems consisting of many identical or similar structural elements, such as proteins and liquid systems, are determined by the interactions between the constituting elements (off-diagonal term) as well as by the local perturbation on each constituting element (diagonal term). In this sense, analyses of the vibrational bands of proteins and liquid systems can be carried out under similar ideas.

For the purpose of clarifying the factors determining the diagonal term, studies on small subsystems are substantially useful. For example, by analyzing the structures, vibrational frequencies, and band intensities of systems consisting of *N*-methylacetamide (NMA) and water molecules, parameters characterizing the local interactions between peptide groups and water molecules in proteins and polypeptides can be obtained. By contrast, in order to clarify the factors determining the off-diagonal term, phenomena arising from delocalized nature of vibrational modes should be studied. For example, the frequencies of the isotropic and anisotropic components of a Raman band are observed to be different for some liquid systems (called "Raman noncoincidence") because the vibrational modes are delocalized due to intermolecular interactions. Differences between the IR and Raman peaks of the amide I band observed in some proteins and polypeptides are considered to arise from the same mechanism.

In the present study, vibrational spectra of species relating to proteins and liquid systems are analyzed theoretically from the above viewpoint.

(1) Vibrational analyses of the NMA–water systems

We carry out theoretical analyses of the structures, vibrational frequencies, and band intensities of the NMA–water systems by *ab initio* MO calculations [1]. Formation of hydrogen bonds with water molecules induces a low-frequency shift for the amide I band and high-frequency shifts for the amide II and III bands. There is a strong correlation between the changes in the C=O bond length and the amide I frequency. Upon formation of hydrogen bonds, the C=O bond becomes longer and the C–N bond becomes shorter. In contrast to these structural changes in the ground electronic ($1^1A'$) state, the C=O bond becomes shorter and the C–N bond becomes longer upon formation of hydrogen bonds in the excited electronic ($2^1A'$) state. These opposite structural changes in the two electronic states are rationalized by using a simple Hamiltonian based on a two-state model, and are consistent with the observed changes in the resonance Raman intensities of the amide I, II, and III bands.

(2) Raman noncoincidence effect of liquid formamide

For a vibrational mode with a strong IR intensity, vibrations of molecules (or structural elements in general) in condensed phases interact with each other by the transition dipole coupling mechanism. In the case where the liquid structure is determined by the dipole–dipole interactions and the transition dipole is parallel to the permanent dipole, it is well known that the relation among the frequencies of the isotropic and anisotropic components of the Raman band and the IR band is expressed as $\nu_{iso} < \nu_{aniso} \cong \nu_{IR}$. However, when one of the above two conditions is not satisfied, *the relation among the Raman and IR frequencies may be different even when the vibrations of molecules in the liquid system interact with each other by the transition dipole coupling mechanism.* For example, the liquid structure of methanol is determined by hydrogen bonding of the OH group. The frequency of the isotropic component of the CO stretching Raman band of liquid methanol is observed to be higher than that of the anisotropic component. We have shown that this experimental result is explained by the transition dipole coupling mechanism when the liquid structure determined by hydrogen bonding is taken into account [2].

In this study, we analyze the frequencies of the Raman and IR bands of another hydrogen-bonding liquid, formamide. Recently, it is suggested that liquid formamide mainly consists of cyclic hexamers on the basis of NMR measurements [3], contrary to the previous proposal that linear-chain structure prevails in liquid formamide. We analyze the relation between the liquid structure and the vibrational frequencies of the CO stretching band by performing *ab initio* MO calculations. It is shown that formation of cyclic hexamers in the liquid is consistent with the experimental result [4] that the frequency of the IR band of the CO stretching mode is noticeably lower than that of the anisotropic Raman band.

[1] H. Torii, T. Tatsumi, and M. Tasumi, manuscript in preparation.
[2] H. Torii and M. Tasumi, J. Chem. Phys. **99**, 8459 (1993).
[3] R. Ludwig, F. Weinhold, and T. C. Farrar, J. Chem. Phys. **102**, 5118 (1995).
[4] A. Mortensen et al., J. Phys. Chem. **98**, 5221 (1994); **99**, 4435 (1995).

Chirped-Pulse Sequences for Impulsive Vibrational Excitation

Eric M. Hiller* and Jeffrey A. Cina†

* James Franck Institute, The University of Chicago, Chicago, IL 60637
† Department of Chemistry, University of Oregon, Eugene, OR 97403

Recent calculations [1, 2] have explored the use of electronically preresonant femtosecond laser pulse sequences to induce large amplitude coherent molecular vibrations. [3,4] In order that the samples might tolerate some small amount of electronic absorption, dilute chromophores in low temperature hosts were considered. Results indicate that:

- preresonant pulses can efficiently impart vibrational momentum without undue electronic absorption or other loss processes [1]
- wave packet spreading due to intramolecular anharmonicity eventually limits vibrational excitation by fixed-center-frequency pulses by triggering near resonant electronic population transfer [1]
- frequency modulation, in the forms of variable pulse center frequency and variable chirp rate, can significantly further enhance the efficiency of impulsive stimulated Raman preparation of coherent vibrations in the electronic ground state [2]

Here we wish to illustrate the last of these three findings, by reporting on numerically exact calculations on a model system of oriented I_2 molecules at 0 K. The preresonantly driven $I_2(B \leftrightarrow X)$ transition is assumed to dominate impulsive Raman scattering by a sequence of pulses with $\tau_L = 10$ fs and $\mu E/\hbar = 734$ cm^{-1}. In keeping with the second point highlighted above, vibrational relaxation and dephasing due to interactions with the cryogenic matrix are neglected for the duration of the pulse sequence.

One pulse was applied during each half-period of outgoing vibrational motion. The pulses had one of 14 center frequencies, equally spaced between 3400 and 6000 cm^{-1} below vertical resonance, and one of 100 chirp rates, equally spaced between -70 and + 5 cm^{-1}/fs. Pulse arrival times, chirp rates, and center frequencies were chosen on a period-by-period basis in such a way as to maximize the quality function

$$Q = \delta E/E - 10^3 \Delta x \Delta p/\hbar - 10^6 P_{lost}^2 .$$

The weighting coefficients for fractional vibrational energy increase, uncertainty product, and electronic ground state population loss, respectively, were arrived at by trial and error. Pulsing was stopped when the electronic ground state depletion reached $P_{lost} = 0.05$.

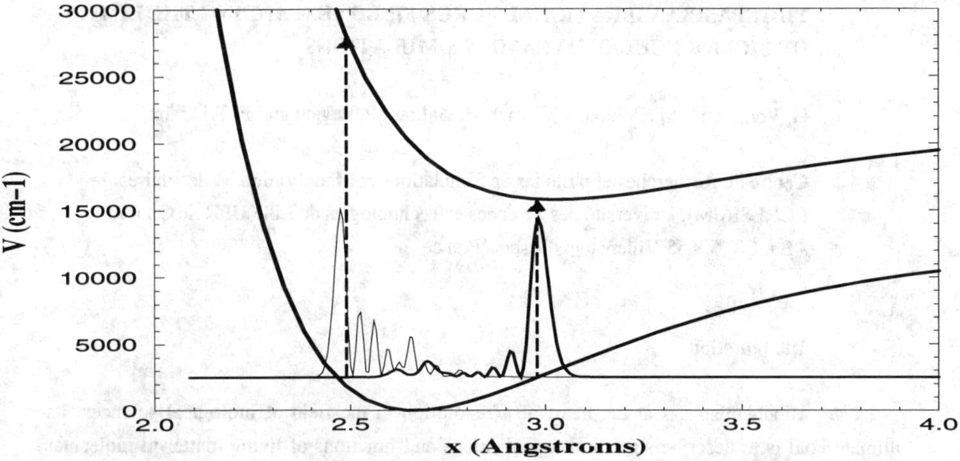

The figure shows the wave packet from the optimal pulse sequence at its inner and outer turning points on the $I_2(X)$ potential curve. The large change in effective vertical resonance frequency over a half-period of motion is evident. The initial pulses in the 20.113 ps-long sequence that generates this large-amplitude vibration were nearly unchirped and offset from vertical resonance at the X-state potential minimum by about - 3800 cm^{-1}. By the end of the sequence, the chosen pulses had chirp rates in the range - 30 to - 40 cm/fs and were offset from vertical resonance by - 6000 cm^{-1}. Pulse arrival times were clustered near the inner and outer turning points for the first half of the sequence, and near the instant of most rapid bond stretching for the second half. The final vibrational energy is 2397 cm^{-1} in excess of the 107 cm^{-1} zero point. The uncertainty product at the end of the pulse sequence is 3.9 $\hbar$. As can be seen from the figure above, nuclear displacements more than 0.2 Å larger than the equilibrium position are predicted. This displacement is roughly double that achieved when the resonance offset is fixed and the pulses are all unchirped. [1]

Research supported by the NSF and the Camille and Henry Dreyfus Foundation.

1. T. J. Smith and J. A. Cina, J. Chem. Phys. **104**, 1272 (1996).
2. E. M. Hiller and J. A. Cina, manuscript under preparation.
3. For a review of impulsive stimulated Raman scattering, see: L. Dhar, J. A. Rogers, and K. A. Nelson, Chem. Rev. **94**, 157 (1994) and references therein.
4. For recent experimental applications of ultrashort chirped pulses, see: C. J. Bardeen, Q. Wang, and C. V. Shank, Phys. Rev. Lett. **75**, 3410 (1995); B. Kohler, V. V. Yakovlev, J. Chi, J. L. Krause, M. Messina, K. R. Wilson, N. Schwentner, R. M. Whitnell, and Y. Yan, Phys. Rev. Lett. **74**, 3360 (1995).

THE SPASIBA VIBRATIONAL FORCE FIELD : EXPANDING THE LIMITS OF BIOMOLECULAR DYNAMICS SIMULATIONS

G. Vergoten*, M. Chhiba, V. Durier, H. Gallouj, P. Lagant and F. Tristram

Centre de Recherches et d'Etudes en Simulations et Modélisation Moléculaires (CRESIMM). Université des Sciences et Technologies de Lille UFR de Chimie - Bat C8- 59655 Villeneuve d'Ascq - France

Introduction

Biochemistry is at the heart of a revolution in the field of biological sciences. Its ultimate goal is to describe structural organizations and functions of living matter in molecular terms using all the laws of Chemistry and Physics. The classical view of biomolecules in their native state has been static. It is now well accepted that atoms in biomolecules undergo important motions at room temperature. This dynamical behavior is closely related to biological functions and properties.

In most cases information on three dimensional structure and dynamics of biological macromolecules may be reached using experimental techniques (X ray crystallography, NMR spectroscopy, neutron scattering and diffraction, vibrational spectroscopy). Unfortunately carbohydrates and lipids are two classes of biomolecules for which such experimental information is difficult to obtain. This is due to their intrinsic flexibility and/or their ability to be donor or acceptor of hydrogen bonds.

In these cases, in order to obtain information at the molecular level, one can take advantage of computer simulation techniques among which molecular dynamics represent one of the most promising.

The SPASIBA force field and molecular dynamics simulations

When used without experimental constraints (for example NOE distances) molecular dynamics results are highly dependent on the force field used. It has been clearly demonstrated that in order to be confidently used in molecular dynamics simulations, the force field has to be accurate over the whole potential energy surface rather than in the vicinity of local or global minima. To check a force field correct vibrational frequencies and assignments are valuable tests. The SPASIBA force field [1], a spectroscopic potential agorithm for simulating biomolecular adaptability, has been developed in our laboratory. Its greatest merit over all commonly available molecular force fields, consists in its capactity to reproduce at the same time structures, energies and vibrational frequencies with a high accuracy. It has been parameterized for a large number of model molecules with biological interest (alkanes, alkenes, carboxylic acids, alcohols, ethers, acetals, mono-and di-saccharides, aliphatic amino-acids, phospholipids). For the Π-electrons systems, aromatic amino acids and nucleic acid bases, the parameters are derived from a scaled-*ab initio* force field based on the density functional theory.

The SPASIBA force field parameterization of model compounds related to lipids of biomembranes

In order to investigate the structural and dynamical properties of phospholipids, one has first to correctly parameterize the SPASIBA force fields for some dialkylphosphate anions. Monomethylphosphate, dimethylphosphate, diethylphosphate, methylphosphate and methylpropylphosphate anions and some of their deuterated derivatives have been carely investigated.

Partial atomic charges were obtained using *ab intio* quantum mechanical calculation at the HF/6-31 G* level. Table I gives relative energies for the dimethyl phosphate anions calculated using the SPASIBA force field, the Amber and Charmm molecular mechanics force fields and an *ab initio* quantum chemistry procedure.

Table 1 : Comparison of the calculated realtive energies of dimethylphosphate (kcal/mol).

Conformation	SPASIBA	AMBER [2]	CHARMM [3]	Ab initio 6-31G*[4]
gg	0.	0.	0.	0.
gt (tg)	1.34	0.80	0.96	1.15
tt	2.73	1.70	1.99	2.78

Molecular dynamics simulation of a model of biomembrane

A hydrated bilayer model consisting of 24 dimyristoyl-2,3-D-glyycero-1-phosphatidylcholine molecules (DMPC) and 284 water molecules, in the gel phase, has been investigated using the molecular dynamics technique at constant temperature (320 K) during 400 ps. From the analysis of the trajectory, a number of results have been extracted and compared with experimental data when available (the thermodynamic order parameter: S; conformations of head groups; glycerol hinges; hydrocarbon chains: number of kinks in the aliphatic chains, ie $g^+tg^-tg^-tg^+$ arrangements, electronic density along a normal to the bilayer and distributions of N and P atoms and of water molecules). Radial distributions functions of water oxygens around N and P atoms have also been obtained. A typical example of the ability of SPASIBA to reproduce experimental data is given through the conformational behavior of the polar head group in table 2. Conformations relative to α_4 and α_5 torsion angles are well explained using a spectroscopic derived force field (see figure 1 for notation of torsion angles α_i).

Fig. 1 : Notation of torsion angles of the polar head group.

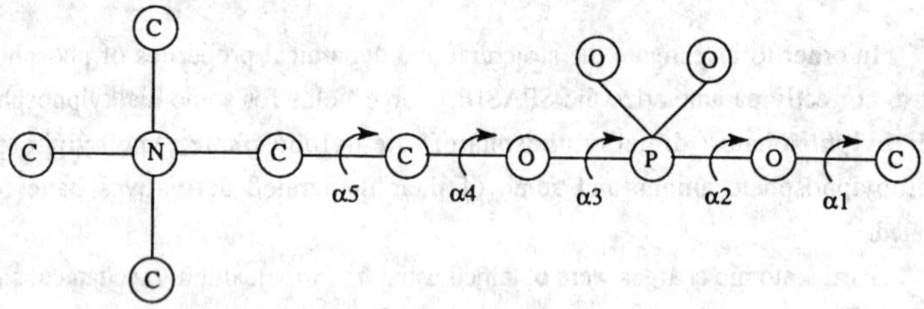

Table 2 : Headgroup conformation populations obtained by SPASIBA, AMBER and GROMOS force fields compared with the experimental data.

Torsion angle	SPASIBA	AMBER [5]	GROMOS [6]	Exptl [7-10]
α 1	t (69%)	t (66%)	t (70%)	t
α 2	g (84%)	g (81%)	g (74%)	g
α 3	g (83%)	g (78%)	g (73%)	g
α 4	t (57%)	t (76%)	t (72%)	t-g
α 5	g (93%)	t (91%)	t (57%)	g

1. P. Derreumaux, G. Vergoten, J. Chem. Phys. 102, 8586-8605 (1995).
2. S.J. Weiner, P.A. Kollman, D.T. Nguyen, D.A. Case, J. Comput. Chem. 7, 230 (1986).
3. B.R. Brooks, R.E. Buccoleri, B.D. Olagson, D. J. States, S. Swaminathan and M. Karplus, J. Comput. Chem. 4, 187 (1983).
4. C.Liang, C.S. Ewig, T. R. Stouch, A.T. Hagler, J. Am. Chem. Soc. 115, 1537 (1993).
5. J.W. Essex, M.M. Hann, W.G. Richards, Phil. Trans. R. Soc. Lond. B 344, 239 (1994).
6. E. Egberts, S.J. Marrink, H.J.C. Berendson, Euro. Biophys. J. 22 , 423 (1994).
7. J. Seelig, H.U. Gally, R. Wohlgemuth, Biochim. Biophys. Acta 467, 109-119 (1977).
8. J. Seelig, Biochim. Biophys. Acta 515, 105-140 (1978).
9. H. Akutsu, Biochemistry 20, 7359-7366 (1981).
10. J. Seelig, H.U. Gally, Biochemistry 15, 5199-5204 (1976).

PHASE RECOVERY FROM THE RAMAN EXCITATION PROFILE

Soo-Y. Lee

Department of Chemistry, National University of Singapore, Kent Ridge, S. 119260, Singapore

A Fourier transform of the Raman amplitude, $A_{n,m}(E)$, gives the correlation function, $C_{n,m}(t)$, from which the dynamics, as well as the shape of the excited state potential can be inferred [1]. The observable Raman excitation profile, REP ($\propto |A_{n,m}(E)|^2$), depends only on the modulus of the complex Raman amplitude. The phase information contained in the Raman amplitude is lost in the REP. Recovery of the phase from the REP to reconstruct the Raman amplitude is therefore an important problem. With the Raman amplitude, the transform theory [2] can also be applied to obtain the absorption spectrum, ABS. This may be useful in separating overlapping absorption bands in a situation where only one of them is responsible for the observed REP.

There are essentially two known procedures for a direct phase recovery : the dispersion method [3] and the maximum entropy method [4]. The dispersion method relies on the analyticity of $\ln A_{n,m}(z)$ in the right-half complex plane, where the Raman amplitude $A_{n,m}(E)$ has been extended in the complex plane $z = \gamma - iE$, for $\gamma > 0$, while the maximum entropy method requires the analyticity of $1/A_{n,m}(z)$ in the same region. The sum-over-states expression for the Raman amplitude [5] assures us that all the poles lie in the left-half complex plane. Therefore, for both methods to work, the Raman amplitude, $A_{n,m}(z)$, should not have any zeros in the right-half complex plane. i.e. the methods work only for minimum-phase problems. This is a severe limitation, because the Raman amplitude $A_{n,m}(z)$ can have zeros in the right-half complex plane, which makes $\ln A_{n,m}(z)$ and $1/A_{n,m}(z)$ non-analytic in the region. Also, both methods have to be used with care, because the REP is known only in a limited energy range. The minimum phase by the dispersion method can be shown [6] to be,

$$\theta_m(E) \approx -\frac{2E}{\pi}\wp\int_{-\infty}^{\infty}\frac{\ln|A_{n,m}(E')|}{E'^2 - E^2}dE' \,. \qquad (1)$$

To handle Raman amplitudes with zeros in the right-half complex plane, requires first the identification of such zeros $z_j = \gamma_j - iE_j$, $\gamma_j > 0$. By contour integration to avoid the branch cuts containing the zeros, it can be shown [6] that the non-minimum phase is given by,

$$\theta(E) \approx \theta_m(E) + 2\sum_j \Im m\left\{\ln\left[-\frac{(\gamma_j + i(E - E_j))(E + E_j)}{(\gamma_j - i(E + E_j))(E - E_j)}\right]\right\} + 2\pi\sum_j \eta(E - E_j), \quad (2)$$

where $\Im m$ denotes taking the imaginary part, and the Heaviside function

$$\eta(E - E_j) = \begin{cases} 1 & E > E_j \\ 0 & E < E_j \end{cases}, \tag{3}$$

accounts for the phase jump of 2π on crossing each of the branch cuts at E_j. The additional terms in Eq.(2) over Eq.(1) which correct for the phase in the neighbourhood of the branch cut do not depend explicitly on the *REP*. They are incorporated after the minimum phase scheme, whether given by Eq.(1) for the dispersion method or by the maximum entropy method, has been carried out. The non-minimum phase theory has been applied to the two-mode model of iodobenzene *REP* described in Ref. 4, with good recovery of the phase and correlation function for the problematic 270 cm^{-1} *REP*, as shown in Fig.1 below.

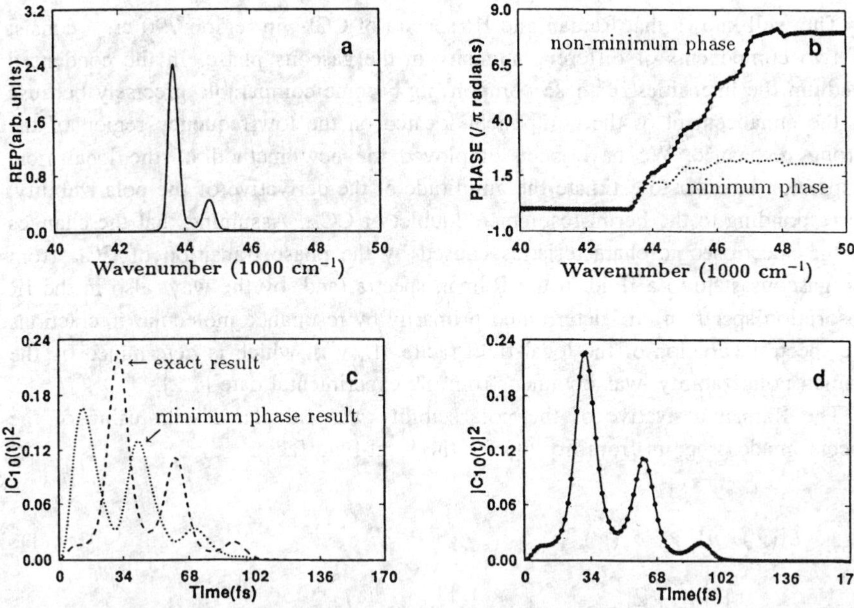

Fig. 1. (a) *REP* for the 270 cm^{-1} mode. (b) Dotted line : minimum phase $\theta_m(E)$ using Eq.(1). Solid line : non-minimum phase using Eq.(2), taking into account three zeros in the right-half complex plane. Solid circles : exact result. (c) Cross-correlation function obtained using minimum phase (dotted line) versus exact result (dashed line). (d) Cross-correlation function using non-minimum phase (solid line) versus exact result (solid circles).

1. S.-Y. Lee and E.J. Heller, J. Chem. Phys. 71, 4777 (1979).
2. J.B. Page and D.L. Tonks, J. Chem. Phys. 75, 5694 (1981).
3. T. Joo and A.C. Albrecht, J. Phys. Chem. 97, 1262 (1993).
4. F. Remacle and R.D. Levine, J. Chem. Phys. 99, 4908 (1993).
5. A.C. Albrecht, J. Chem. Phys. 34, 1476 (1961).
6. Soo-Y. Lee, Chem. Phys. Lett. 245, 620 (1995).

CALCULATION OF MOLECULAR POLARISABILITY BY LOCAL FIELD METHOD

V. S. Libov and A. Yu. Tuchomirov

S.I.Vavilov State Optical Institute, St.-Petersburg, Russia, 199034

The well-known that Raman and IR spectra of CCl_4 in region 790 см$^{-1}$ consist of two components of different intensity in the gaseous phase. In the condensed medium the intensities of these components become comparable precisely because of the enhancement of the component located on the lowfrequency region of the strong resonance. We have been employed the new method of the local field dispersion in order to estimate the amplitude of the derivative of the polarizability, corresponding to the Fermi-resonance doublet of CCl_4. Assuming that the changes in the spectroscopic characteristics caused by the phase transition of CCl_4 from the gaseous state to a fluid in the Raman spectra (and, by the way, also in the IR absorption spectrum) are determined primarily by resonance molecular interactions the spectral variation of the local-field factor ($L_\nu(\nu')$), which is determined by the Raman polarizability, was obtained from the experimental date [1,2].

The Raman derivative of the polarizability was calculated with using of the special made program from Eq. (1) for the local-field factor

$$L_\nu(\nu') = n_{\nu'}^3 \cdot \gamma_\nu \cdot \delta_\nu(\nu'), \tag{1}$$

where $\gamma_\nu = \frac{1}{n_\nu^2}\left|\frac{n_\nu^2+2}{3}\right|^2, \delta_\nu(\nu') = \frac{1}{n_\nu^2}\left|f_1/(1-\frac{f_2}{r^3}\hat{\alpha}_R(\nu'))\right|^2$, f_1, f_2 and r^3 are

the parameters of the Onsager models and

$$\hat{\alpha}_R(\nu') = \alpha_0 + \frac{2\nu_0' k\Gamma\left(\frac{\partial\alpha}{\partial q}\right)q}{\nu_0'^2 - \nu'^2 + i\Gamma\nu'}.$$

As a result, it was found that $(\partial\alpha/\partial q)q \approx 5\cdot 10^{-24}$ см^2. This significance agrees with the polarizability value obtained by independent methods [3].

1. I. I. Condilenko, G. A. Worobjva and I. F. Klassen, Optica i spectroskopia (Russ.) 27, 420 (1969).
2. V. M. Sidorenco, V.S. Libov and N. G. Bakhshiev, Optica i spectroskopia (Russ.) 37, 680, 1084 (1974).
3. J. Brandmuller und H. Moser. Einfuhrung in die Ramanspektroskopie, Darmstadt, 1962.

LOCAL FIELD METHOD IN RAMAN SPECTROSCOPY OF DISORDER CONDENSED MEDIUM

V. S.Libov

S.I.Vavilov State Optical Institute, St.-Petersburg, Russia, 199034

The new approach based on the concept of the local field dispersion is proposed and developed to interpretation of the influence of the resonance and dispersion molecular interactions on the Raman spectroscopic characteristics (shift, intensity and form) of a disorder condensed medium. The similar approach there is only in spectroscopy absorption of condensed media (see the literature in [1,2]). The general expression for the Raman band intensity with taking into account of the local field dispersion can be written in the form

$$I_\nu(\nu') \approx |\alpha_\nu(\nu')|^2 \cdot E^2_{(\nu)av} \cdot L_\nu(\nu') \qquad (1)$$

where $|\alpha_\nu(\nu')|$ is element of the polarizability tensor, $E_{(n)av}$ is the average field, n and n'are the frequencies of the excitation source and Raman line, respectively, and $L_n(n')$ is the local field factor. This factor can be expressed as

$$L_n(n') = n^3{}_{n'} \cdot \gamma_\nu \cdot \delta_n(\nu'), \qquad (2)$$

where $\gamma_\nu = \frac{1}{n_\nu^2}\left|\frac{E_{(\nu)loc}}{E_{(\nu)av}}\right|^2 = \frac{1}{n_\nu^2}\left|\frac{n_\nu^2+2}{3}\right|^2$ can be calculated with according to the Lorentz model and

$$\delta_\nu(\nu') = \frac{1}{n_{\nu'}^2}\left|\frac{\hat{E}_{loc}(\nu')}{\hat{E}_{av}(\nu')}\right|^2 = \frac{1}{n_{\nu'}^2}\left|f_1/(1-\frac{f_2}{r^3}\hat{\alpha}_R(\nu'))\right|^2$$

can be determinated more complicated way from the Onsager model. It is necessary to emphasize the magnitude $\delta_\nu(\nu')$ can be change versus from a frequency in the region of the Raman band. That is why all the parameters of the Raman band (shift, intensity and form) can be change too.

The method of the local field dispersion is used to analyse the Raman bands of liquids and solutions, for example, $CHCl_3$ (760 cм $^{-1}$) and CCl_4 (790 cм$^{-1}$) for room temperature. These results are in agreement with the experimental date, obtained for the gas-liquid phase transition, as well as with the results of the consideration IR bands absorption for the same vibrational transitions [1,2].

The general expressions (1) and (2) are valid also for resonance Raman spectrum. In this case the intensity of the Raman band for the gas - condensed medium transition is increased in addition in comparison with nonresonance Raman Spectrum. This effect can be especially significant for UV resonance Raman [3], i.e., for the strong electronic transitions.

The method of local field makes it possible not only to describe variations in the spectra of condensed media due to action of resonance forces, but also makes it possible to determine molecular parameters. Besides we obtained the preliminary data about the frequency dependence of a intensity for SERS', for which the maximum of the enhanced coefficient is shifted toward lower frequencies with respect to the maximum of the surface plasmon resonance [4].

At last the method of local field makes it possible to predict new spectroscopic effects that have not yet been realized experimentally. Thus, for example, by analyzing the Raman spectra of condensed systems consisting of two kinds of molecules by the local field method, it can be concluded that the Raman spectra are enhanced in the region of the molecular resonance corresponding to absorption. This resonance (or band) can have a different nature, i.e. it can correspond to transitions with the absorption of light not only between electronic energy levels, but also between vibrational levels. In both these cases, the optical characteristics of the medium surrounding the molecule being investigated (the Raman spectrum of which is being recorder) can experience a strong spectral variation and can thereby differ from the average value. This also leads to an increase of the intensity of the Raman spectrum.

1. V.S.Libov, Optichesky Zhurnal (Russ.), № 11, 55 (1993).
2. V.S.Libov, Zhurnal Fizicheskoi Khimii (Russ.), 54, 817 (1980).
3. S.A.Asher, Annual Review of Physical Chemistry, 39, 537 (1988).
4. Surface Enhanced Raman Scattering, by Ed. R. Chang and T.Furtak, Plenum Press, N.-Y. (1982).

THE EXTENSION OF LANDAU-PLACZEK THEORY ON MEDIA WITH CURRENT-CARRIERS

V.A. Voitenko

Technical State University 195251 St.Petersburg

We introduce the hydrodinamical approach in the Landau-Paczek theory of light scattering linewidth. Our approach enables one to determine such a rare parameter of substance as thermal diffusion coefficient.

It is always accepted that light scattering from temperature fluctuations $\langle\delta T^2\rangle = T^2/C_V$ where C_V is heat capacity is hard to observe due to very faint susceptibility -temperature dependence [1]. Nevertheless it is confirmed by parabolic temperature dependence of scattering intensity, quadratic in superionic glasses $(AgI)_x(AgPO_3)_{1-x}$ [2] and cubic in some doped semiconductors [3]. Therefore we include in the expansion of the differential extinction coefficient temperature autocorrelation function $\langle\delta T^2\rangle_{k\omega}$ and mixed correlation function $\langle\delta n\delta T\rangle_{k\omega}$ where δn is particle number fluctuation, k and ω are wave vector and frequency transfered to light during scattering process. In order to calculate all necessary functions the hydrodinamical theory is sufficient. The system of particle diffusion equations consists of diffusive flow expansion

$$\delta j - \frac{\sigma}{e} E = - D \nabla \delta n - \frac{D_T}{T} \nabla \delta T + g \qquad (1)$$

and continuity equation:

$$\frac{\partial \delta n}{\partial t} + \operatorname{div} \delta j = 0$$

Here D, D_T and σ are diffusion, thermal diffusion and conductivity coefficients, E is the electric field, g is random particle flow. Thermal conductivity equations include heat flow expansion

$$\delta q - (\mu - \Pi)\, \delta j = - \kappa \nabla \delta T + \xi \qquad (3)$$

and continuity equation

$$\frac{\partial \delta \varepsilon}{\partial t} + \mathrm{div}\ \delta q = 0 \tag{4}$$

where κ is heat conductivity, Π is Peltier coefficient, $\delta\varepsilon$ is energy density fluctuation, μ is chemical potential and ξ is random heat flow. To calculate scattering inten sity one should compose all required correlation functions from δn and δT with the help of (I)-(4) and take appropriate linear combination of them. Taking into account correlation properties of random flows, lacking of correlation between different flows, one gets the extinction coefficient for scattering of light:

$$\frac{\partial^2 h}{\partial\omega\partial\Omega} = \frac{\tau h_0}{\pi}\ \frac{P - N\,(\omega\tau)^2 + M\,(\omega\tau)^4}{[(\omega\tau)^4 - (A+1)(\omega\tau)^2 + B]^2 + (\omega\tau)^2\,[A - 2(\omega\tau)^2]^2} \tag{5}$$

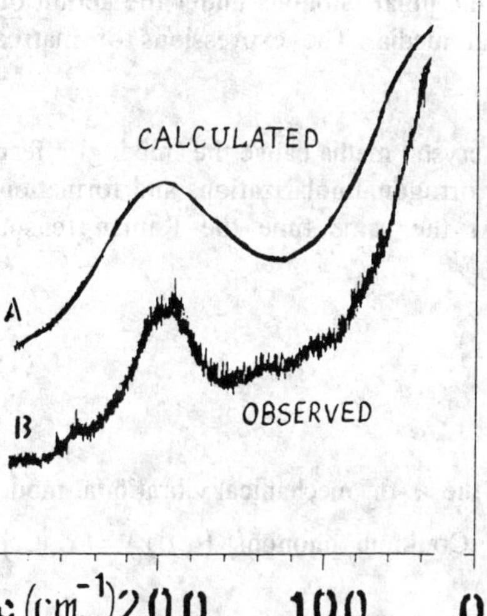

where h_0 is the integrated extinction coefficient, A-P are dimentionless constants, τ is the transport relaxation time. The result of a calculation according to (5) is shown in fig 1, spectrum A, together with the experimental spectrum B.[3] Very good agreement could be achieved for n-InP (shown in fig.1) as well as for n-GaAs (not shown here).

Fig.1. A :Calculated electronic Raman spectrum according to (5). B:Observed Raman spectrum of n-InP with n= 5,6 10^{17} cm^{-3}.

1. L.D. Landau, E.M. Lifshitz Electrodinamics of Continious media. Pergamon, Oxford, 1984.
2. P.A.M. Rodrigues, P.Benassi, and A. Fontana, J.Non Cryst. Sol. 143, 274, (1992)
3. B.H. Bairamov, I.P. Ipatova, V.A. Voitenko, Phys. Rep. 229, 223, (1993)

VARIATION PECULIARITIES OF THE RAMAN TENSOR IN THE POLAR PHONONS INTERACTION

G.N.Dmitrik[1], P.A.Korotkov[1*], and L.N.Ovander[2]

[1]Radiophysical Department, Kiev University,
prosp. Akademika Glushkova,6 ,Kiev, 252127, Ukraine
[2]Engineer-technological institute, Zhitomir, Ukraine

The phenomenological theory of the asymmetry in the intensity of Raman scattering of light in uniaxial electrooptical crystals is worked out. It is suggested that observable effect is caused by the 'mixing' of the polar phonons under the action of the macroscopic electric field of the crystal media. The expressions of matrix elements of the Raman tenzor are presented.

The macroscopic electric fields of the crystal media cause the 'mixing' effect of the normal lattice modes with the mutually ortogonal polarizations and formation of the Coulomb type vibrational states. At the same time the Raman tensor components change in the following way

$$R_{ij}^{\upsilon} = \sum_{n(\gamma)} C_{n(\gamma)}^{\upsilon} R_{ij}^{n(\gamma)} \tag{1}$$

where $C_{n(\gamma)}^{\upsilon}$ is the transition parameter of the n-th mechanical vibrational mode polarized along γ direction to the υ-th Coulomb phonons; $(i,j,\gamma)=(x,y,z)$ are the crystallographic axes [1].

In the case when normal vibrational states $n(\beta)$ and $m(\gamma)$ mixe, formula (1) takes the form

$$R_{ij}^{\upsilon} = C_{m(\gamma)}^{\upsilon} R_{ij}^{m(\gamma)} + C_{n(\beta)}^{\upsilon} R_{ij}^{n(\beta)} \tag{2}$$

The coeffcients $C_{m(\gamma)}^{\upsilon}$ and $C_{n(\beta)}^{\upsilon}$ satisfy the system of linear homogeneous equations

$$\begin{cases} \left(E_{m(\gamma)} - E_{\upsilon}\right)C^{\upsilon}_{m(\gamma)} + V_{m(\gamma),n(\beta)}C^{\upsilon}_{n(\beta)} = 0 \\ V_{n(\beta),m(\gamma)}C^{\upsilon}_{m(\gamma)} + \left(E_{n(\beta)} - E_{\upsilon}\right)C^{\upsilon}_{n(\beta)} = 0 \end{cases} \tag{3}$$

and the normalization condition

$$\left(C^{\upsilon}_{m(\gamma)}\right)^2 + \left(C^{\upsilon}_{n(\beta)}\right)^2 = 1 \tag{4}$$

Here $E_{n(\beta)}$, $E_{m(\gamma)}$ and E_{υ} are energies of "mixing" mechanical and Coulomb phonons relatively; $V_{m(\gamma),n(\beta)}$ is the matrix element of the intermode interaction operator [1, 2].

The possible values of E_{υ} are obtained from the condition of vanish of the determinant of system (3).

The simultaneous solution of (3) - (4) , on taking account of the explicit form of E_{υ}, enables one to determine $C^{\upsilon}_{m(\gamma)}$ and $C^{\upsilon}_{n(\beta)}$ and after this, accordingly (2), obtain the expressions of the Raman tensor components of the Coulomb type phonons:

$$R^{\upsilon(\pm)}_{ij} = \frac{1}{\sqrt{1+\tau^2_{mn}}}\left(\tau_{mn}R^{m(\gamma)}_{ij} \pm R^{n(\beta)}_{ij}\right) \tag{5}$$

where:

$$\tau_{mn} = \sqrt{\kappa^2_{mn} + 1} + \kappa_{mn}; \quad \kappa_{mn} = \frac{E_{m(\gamma)} - E_{n(\beta)}}{2V_{m(\gamma),n(\beta)}}$$

If energies of "mixing" lattice modes are equal , relation [5] are trasformed to the form

$$R^{n(\pm)}_{ij} = \frac{1}{\sqrt{2}}\left(R^{n(\gamma)}_{ij} \pm R^{n(\beta)}_{ij}\right) \tag{6}$$

Formulas (5) and (6) can be used to the interpretation of the experimentally registered effect of intensity asymmetry of the Raman scattering (IARS) of light [2,3]. The tensors $R^{\upsilon(+)}_{xx}$ and $R^{\upsilon(-)}_{xx}$ describe IARS of light by anisotropic phonons in crystals with the point symmetry 3 and 3m, and $R^{n(+)}_{xy}$ and $R^{n(-)}_{zy}$ discribe the IARS of light by the transverse and longitudinal phonons of E-type in crystals with the point symmetry 3, 4 and 6.

1.L.N.Ovander, Usp.Fiz.Nauk.,86,1(1965)
2.G.N.Dmitrik,P.A.Korotkov,and L.N.Ovander, Opt. i Spektr.,79, 36 (1995)
3.S.M.Kostritskiy, A.E.Semenov, E.V.Cherkasov, Opt. i Spektr.,53, 180 (1982)

NUMERICAL CALCULATION OF MORPHOLOGY DEPENDENT RESONANCES IN RAMAN SPECTRA

I. Hartmann, J. Popp, M. Trunk, M. Lankers, and W. Kiefer *

Institut für Physikalische Chemie, Universität Würzburg, Marcusstraße 9-11, D-97070 Würzburg, Germany

H. Chew et. al. [1] developed a model for Raman scattering and fluorescence from single molecules embedded within a dielectric particle whose dimensions are comparable to the wavelength of visible light. The optical levitation technique [2] enabled us to investigate a single aerosol droplet with diameters of about 5-15 microns by means of Raman spectroscopy. Besides the conventional Raman signal the so-called morphology dependent resonances (MDR) occur for a particular radius of such a particle. The appearence of MDRs is furthermore correlated to the wavelength of the incident light λ_0 (input resonances) or to the Raman wavelength λ (output resonances).

In order to simulate the experimental data we apply the Chew model, taking into account a superposition of induced scattering dipoles inside the dielectric sphere [3].

Fig. 1 presents results of recorded (panel A) compared to calculated (panel B) Raman spectra superimposed by MDRs of an evaporating tetraethylene glycol droplet. The displayed spectral range is that of the C-H stretching region for spectra taken at 10, 20, 30 and 40 seconds after the beginning of the droplet levitation. The expected movement of resonances c_n, d_n towards smaller wavenumbers becomes obvious. The time dependent behaviour of the radius and index of refraction - both are necessary for simulating the spectra - was fitted from the movement of the experimental d^3_{72}, d^4_{68}, d^3_{73} output resonance types during a time periode of 400s. Because of the small signal accumulation time (1s) applied, weak but theoretically expected output resonances such as c^4_{67}, c^3_{72}, c^4_{68} (see panel B) vanish in the noisy structure of the recorded spectra. Nevertheless, in some cases, they can be notified as weak signals. After solving several numerical problems, we are now able to calculate such spectra from spheres with diameters up to about 14 µm.

Financial support from the Deutsche Forschungsgemeinschaft and by the Fonds der Chemischen Industrie is highly acknowledged.

1. H. Chew, P. J. McNulty, and M. Kerker, Phys. Rev. A 13, 396 (1976).
2. A. Ashkin, Phys. Rev. Lett. 24, 156 (1970).
3. I. Hartmann, J. Popp, M. Lankers, M. Trunk, and W. Kiefer, J. Mol. Struct. 349, 203 (1995).

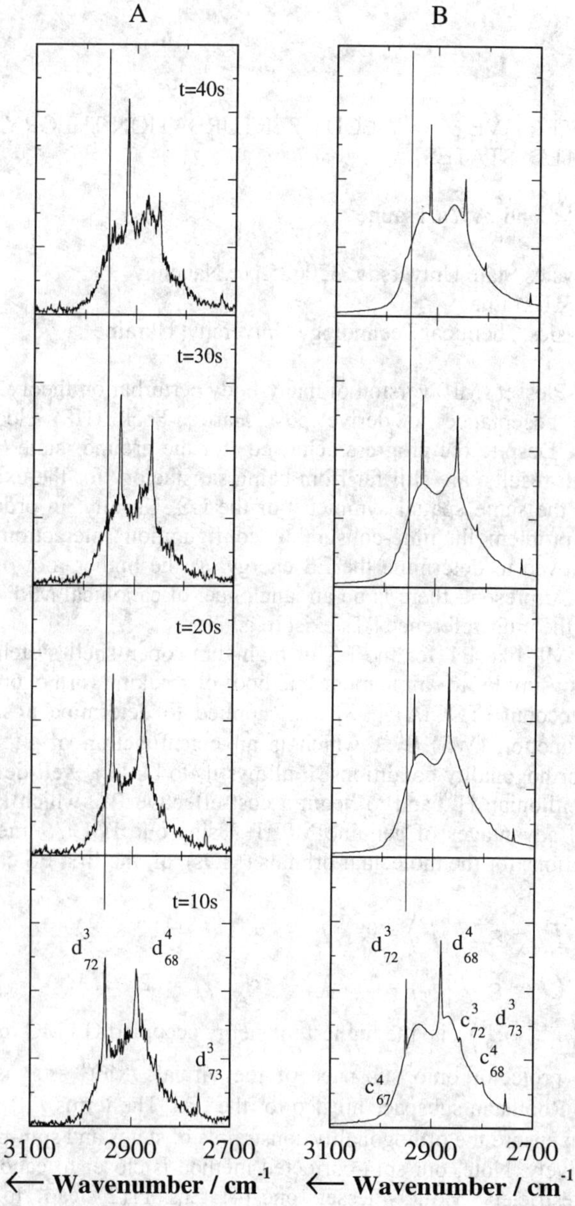

Fig. 1: Experimental (panels A) and calculated (panels B) morphology dependent resonances (MDRs) in the Raman spectra of an optically levitated tetraethylene glycol droplet. Spectra were taken every 10 s with a 1 s accumulation time.

A COST-EFFECTIVE MANY-BODY PERTURBATION THEORY FOR EXCITED STATES

V.Glushkov[1*] and A.Ya.Tsaune[2]

[1] Dept.of Physics,State University, 320625,per.Nauchny 13,Dniepropetrovsk,Ukraine
[2]Dept.of Physics,Chemical Technology University, Ukraine

The Moller-Plesset (MP) version of many-body perturbation theory (PT) has gained wide acceptance to derive post Hartree-Fock (HF) electron correlation effects. Despite the progress achieved for the ground state (GS), the computational results are still far from being satisfactory for the excited state (ES) having the same spatial symmetry of the GS. Usually, in order to avoid collapsing problem the time-consuming configuration interaction (CI) methods are employed to determine the ES energy, as the higher root of the secular equation. At present, there is no an analogue of canonical MPPT for the ESs, although the multireference PTs exist (e.g.[1]).

Here, the MP like PT for the ES of high-spin open-shell systems is proposed. A very simple to implement method of taking orthogonality constraints into account (TOCIA) [2,3] is applied to determine a single reference wave function (WF) $\Psi^{(0)}$, which is an eigenfunction of S^2 and compatible with orthogonality conditions. It allows us to build a well-defined zeroth-order Hamiltonian $H^{(0)}$ and to begin a cost-effective PT, which keeps the computational advantages of genuine MPPT. Using our TOCIA method amended HF equations for the molecular orbitals (MOs) of the first ES can be obtained

$$P(F^{\alpha}+\lambda_1 P_n-\varepsilon_i^{\alpha})P\left|i^{\alpha}\right\rangle=0,\quad \lambda_1\to\infty,\quad i^{\alpha}=1,2,\ldots,M \tag{1}$$

$$P(F^{\beta}+\lambda_2 Q^{\alpha}-\varepsilon_i^{\beta})P\left|i^{\beta}\right\rangle=0,\quad \lambda_2\to\infty,\quad i^{\beta}=1,2,\ldots,M \tag{2}$$

Here $P_n=\left|n_0^{\alpha}\right\rangle\left\langle n_0^{\alpha}\right|$, $\left|n_0^{\alpha}\right\rangle$ is the highest in energy occupied GS MO of the α set, Q^{α} is the projector onto subspace of the virtual MO^{α} s. P is the orthoprojector on Roothaan subspace ajusted to the ES. The terms $\lambda_1 P_n$ and $\lambda_2 Q^{\alpha}$ are added to ensure the orthogonality constraints of states and spin-purity condition respectively. Note, our spin-corrected method is closer to canonical HF theory than restricted Moller-Plesset one [4]. Eqs.(1),(2) lead to the uniquely defined $H^{(0)}$

$$H^{(0)}=\sum_i^{n^{\alpha}}F^{\alpha}(x_i)+\sum_i^{n^{\beta}}F^{\beta}(x_i) \tag{3}$$

with $F^{\alpha}=\sum_{i=1}^{M-1}\left|i^{\alpha}\right\rangle\varepsilon_i^{\alpha}\left\langle i^{\alpha}\right|$, $F^{\beta}=\sum_{i=1}^{M-1}\left|i^{\beta}\right\rangle\varepsilon_i^{\beta}\left\langle i^{\beta}\right|$

Singly, doubly etc. excited configurations from $\Psi^{(0)}$ are orthogonal themselves and are eigenfunctions of $H^{(0)}$. Raley-Shrodinger PT based on (3) lead to MP like PT for the ESs with the same computational efforts as for the GS. For example, the second order correction to the first ES energy is as follows

$$E^{(2)} = \sum_{i}^{n^{\beta}} \sum_{a}^{M-1} \frac{\left|\left\langle i^{\beta}\left|F^{\beta}\right|a^{\beta}\right\rangle\right|^{2}}{\varepsilon_{i}^{\beta}-\varepsilon_{a}^{\beta}} + \frac{1}{4}\sum_{i,j}^{occ}\sum_{a,b}^{virt}\frac{\left|\left(ai|bj\right)-\left(aj|bi\right)\right|^{2}}{\varepsilon_{i}+\varepsilon_{i}-\varepsilon_{a}-\varepsilon_{b}} + $$

$$+\left\langle \Psi^{(0)}\left|H\right|\Psi_{0}^{(0)}\right\rangle\left\langle \Psi_{0}^{(1)}\left|\Psi^{(0)}\right.\right\rangle \quad (4)$$

The sums run over spin-orbitals. The single replacements contribute to $E^{(2)}$ since Brillouine's theorem is not obeyed for the β set. Note, in the RMP method this theorem is not obeyed for both the α and β sets. The second term in (4) conforms to the geniune MPPT result. The last term appears because of the orthogonality constraints imposed on approximative Wfs. The calculation practice shows that this summand may be neglected compared to the second term.

For HeH the energy difference $\Delta = E(A^2\Sigma^+) - E(X^2\Sigma^+)$ across the potential energy curve has been computed to test the proposed method (see Table 1). The basis sets consisting of 10 1s Gaussians for the GS and 14 - for the ES were used. As one can see our second-order results are in good agreement with their precision [5], which have been obtained by CI method using 3190 configurations. In conclusion it should be stressed that MOs obtained by means of Eqs. (1),(2) may be applied to CI calculations, which be more reliable and accurate if based on these MOs specifically optimized for the parent configurations.

TABLE 1. The energy difference $\Delta = E(A^2\Sigma^+) - E(X^2\Sigma^+)$ as function of the internuclear distance for HeH (All quantities in atomic units)

R	1.0	1.5	2.0	2.5	3.0	4.0	10.0
Our method	0.0464	0.1484	0.2519	0.3149	0.3470	0.3697	0.3732
[5]	0.0474	0.1505	0.2509	0.3143	0.3474	0.3697	0.3752

1. K.Wolinski and P.Pulay, J.Chem.Phys. 90, 3647 (1989)
2. A.Ya.Tsaune,V.N.Glushkov and A.I.Aprasyukhin, J.Mol.Struct. THEOCHEM, 312 ,289 (1994)
3. V.N.Glushkov, Chem.Phys.Lett. 244 ,1 (1995)
4. P.J.Knowles,J.S.Andrews,R.D.Amos,N.C.Handy and J.A.Pople, Chem.Phys.Lett. 186 ,130 (1991)
5. G.Theodorakopoulos,I.D.Petsalakis,C.A.Nicoladies and R.J.Buenker, J.Phys. B20 ,2339 (1987)

SEMICLASSICAL THEORY OF SELF-CONSISTENT VIBRATION-ROTATION FIELDS FOR DESCRIBING HIGH-EXCITED MOLECULAR SPECTRA

A.S.Skalozub* and A.Ya.Tsaune

Ukrainian University of Chemical Technology,
av.Gagarin 8, 320005 Dnepropetrovsk, Ukraine

A knowledge of high-excited vibration-rotation (VR) molecular states and an understanding of the intramolecular dynamics for these states is of importance to many fields of spectroscopy, Raman spectroscopy among them.

For this purpose we have developed a semiclassical theory to calculate the high-excited states in VR spectra of polyatomic nonrigid molecules. Our method is based on introducing auxiliary time-periodical fields, which transfer interactions between rotational and vibrational molecular degrees of freedom, into molecular Hamiltonian H. This procedure leads to separating the variables in the auxiliary VR Hamiltonian $\mathscr{H}$. As a result the Schrodinger equation solution

$$H|\varphi_n\rangle = E_n|\varphi_n\rangle \tag{1}$$

reduces to a simpler problem. The Hamiltonians H and $\mathscr{H}$ are related by the following equation [1]

$$\mathrm{Tr}\, e^{-iHT} = \int Dw(t) e^{iS_0[w]}\, \mathrm{Tr}\, T_{ts} \exp\{-i\int_0^T \mathscr{H}\,[w(t)]dt\}, \tag{2}$$

where w(t)=w(t+T) are the time-periodical configurations of the auxiliary fields, $S_0[w]$ is some weight function accounting the contribution of various configurations in the path integral (2), T_{ts} is the time-order symbol in combination with symmetrizing operators at the same point in time,

$$\mathscr{H}[w(t)] = H_{vib}[w(t)] + H_{rot}[w(t)]. \tag{3}$$

The Hamiltonian $H_{vib}[w(t)]$ is a sum of usual vibrational energy and some additives, which describe a motion of the vibrational collective variables in the auxiliary fields. It doesn't depend on the angular coordinates explicitly. The Hamiltonian $H_{rot}[w(t)]$ governs a motion of the molecular angular momentum **J** in some field. It isn't a function of the vibrational coordinates.

For high-excited vibration or rotation states the path integral (2) can be evaluated in the mean-field approximation. This yields the "classical" configurations which make a main contribution to the integral value. These are indicated by underlining. As a consequence a solution of Eqn.(1) is associated with two problems:
a) the vibrational problem; it is a eigenvalue one for the time- periodical Hamiltonian $H_{vib}[w(t)]$. According to the Floquet's theorem it has time-periodical solutions

$$(i\partial_t - H_{vib}[\underline{w}(t)])\,|\varphi_{\{v\}}(t)\rangle = -\,\varepsilon_{\{v\}}[\underline{w}]|\varphi_{\{v\}}(t)\rangle\,, \tag{4}$$
$$|\varphi_{\{v\}}(t+T)\rangle = |\varphi_{\{v\}}(t)\rangle,$$

where $\varepsilon_{\{v\}}[\underline{w}]$ are quasienergies, $\{v\} = (v_1, v_2, \ldots, v_{3N-6})$ is a set of vibrational quantum numbers;
b) the rotational problem; it is formulated in terms of the spin-coherent states [2] for the time-periodical Hamiltonian $H_{rot}[w(t)]$

$$i\partial_t\ |\zeta(t)\rangle = (H_{rot}[\underline{w}(t)] - \chi'(t))|\zeta(t)\rangle\ , \qquad (5)$$
$$|\zeta(t+T)\rangle = |\zeta(t)\rangle,$$

where the function $\chi(t)$ satisfies the following equation

$$\alpha_J[\underline{w}] = \int_0^T \chi'(t)dt\ .$$

The quantity $\alpha_J[\underline{w}]$ is defined as the Floquet index of the quasienergy eigenvalue problem for the rotational Hamiltonian.

The VR energy levels follow from the quantization condition [3]

$$j\beta(C) = 2\pi(N(C) + m(C)/4) - \gamma_{\{v\}}(C), \qquad (6)$$

where $j=J+0.5$, J is the angular momentum quantum number, $\beta(C)$ is the Berry's phase [4] of the rotational problem, N(C) are integers running a series of values over the range (0,1,...J-1,J), m(C) are the number of turning points on a orbit C in the space of the field variables $\underline{w}(t)$, $\gamma_{\{v\}}(C)$ is some quantity which equals to the Berry's phase for the vibrational problem in the adiabatic approximation.

The quantization rules have been derived for various topology of the closed circuit C. This circuit divides the parameter space into four invariant regions. In two of them $m(C_{1,2})=0$ for the corresponding circuits, in other two $m(C_{3,4})=2$. Owing to orbit periodicity Eqn.(6) is the same for the first two regions and has form

$$j\beta(C_{1,2}) = 2\pi N - \gamma_{\{v\}}(C_{1,2}), \qquad (7)$$
$$N=J,J-1,\ldots,N_{min}\ .$$

Similarly, for the two another regions one has

$$j\beta(C_{3,4}) = 2\pi(N' + 1/2) - \gamma_{\{v\}}(C_{3,4}), \qquad (8)$$
$$N'=0,1,2,\ldots,N'_{max}\ .$$

The values $N_{min} = N_S+1$, $N'_{max}= N_S-1$, where N_S determines the energy of a separatrix, which divides the regions outlined. For the separatrix Eqn.(6) can be written as

$$j\beta(C_S) = 2\pi(N_S + 1/4) - \gamma_{\{v\}}(C_S), \qquad (9)$$

where C_S is the separatrix circuit.

As a result the energy levels are two-fold degenerated and their total number are J+1. The degeneration is removed if effects of tunneling and overbarrier reflection when orbiting are taken into account.

The quantization rules, Eqns.(7)-(9), the mean-field configurations of the auxiliary fields, and the quasienergy solutions to the Eqn.(4) and Eqn.(5) are related by a self-consistent procedure.

The application of this approach to a problem of bending-rotational interaction in the H_2O molecule has demonstrated its efficiency [5].

1. A.S.Skalozub and A.Ya.Tsaune, Ukrain. J. Physics, 33, 498 (1988).
2. A.M.Perelomov, Sov. Phys. Usp. 20, 703 (1977).
3. A.S.Skalozub and A.Ya.Tsaune, Proc. SPIE, 1811 ,168 (1992).
4. M.V.Berry, Proc. R. Soc. London, A392, 45 (1984).
5. A.S.Skalozub and A.Ya.Tsaune, Opt. and Spectrosc. 77, 900 (1994).

VARIANT OF BRINGING TOTAL ANGULAR MOMENTUM AND INTERNAL VARIABLES INTO MOLECULAR HAMILTONIAN

A.Tsaune[1*] and M.Dyachenko[2]

[1]Ukrainian State Chemical and Technological University, av.Yu.Gagarin 8

320005 Dnepropetrovsk, Ukraine
[2]Dnepropetrovsk State University, Dnepropetrovsk, Ukraine

Possibility of using molecular hamiltonian in the principal axes of inertia as the basis for solution of phenomenological problem of description of vibration-rotation absorption and Raman spectra was confirmed by calculations [1]. The study of internal molecular dynamics and its manifestations in spectra is not less important problem. To investigate this problem it is necessary to have an obvious expression of complete (ro-vibronic) hamiltonian in the principal axes of inertia. It turns out that various variants of such hamiltonians which differ through ways of bringing total angular momentum and internal variables are possible. As an example we refere to [2-4].

In this communication one of new variants of molecular hamiltonians is proposed. Particles forming a molecule are broken up in two subsystems. The first subsystem consists of nuclei, the second one - of both nuclei oscillating with large amplitude and electrons. Masses of the first subsystem are designated as m_τ ($\tau = 1,2,...,N_1$), masses of the second one - m_a ($a = 1,2,...,N_2$). The total number of the particles is $N = N_1 + N_2$. A mobile frame of reference is principal one for the first subsystem.

The basis of derivation of the hamiltonian is differential forms which represent infinitesimal changes of coordinates of the first ($dr''_{\tau(\alpha)}$) and the second ($dr''_{a(\alpha)}$) subsystems in laboratory frame of reference.

$$dr''_{\tau(\alpha)} = dr^c_{(\alpha)} + (\vec{\omega} \times \vec{r}_\tau)_\alpha + \sum_j N^{(\alpha)}_{\tau j} d\xi^j - \frac{1}{M}\sum_a m_a (\vec{\omega} \times \vec{r}_a + d\vec{r}_a)_\alpha ,$$

$$dr''_{a(\alpha)} = dr^c_{(\alpha)} + \sum_b \left(\delta_{ab} - \frac{m_b}{M}\right)(\vec{\omega} \times \vec{r}_b + d\vec{r}_b)_\alpha , \quad M = \sum_\tau m_\tau + \sum_a m_a .$$

Here $dr^c_{(\alpha)}$, $dr_{a(\alpha)}$, $d\xi^j$ - infinitesimal changes of grid coordinates (in the mobile frame of reference) of mass center, coordinates of particles of the second subsystem and independent variables of the first subsystem which are invariant with respect to translations and rotations, $\alpha = x, y, z$; $j = 1,2,...,3N_1-6$. ω_α - are differential forms (in tensor analysis - non-holonomic components) of angular velocity $\vec{\omega}$ of the mobile frame of reference with respect to the laboratory one.

$$N^{(\alpha)}_{\tau k} = \sum_j T_{kj} \left\{ m_\tau^{-1} B^j_{\tau(\alpha)} + 2\sum_\varsigma \left[\frac{r_{\tau(\beta)} r_{\varsigma(\beta)}}{I_{\beta\beta} - I_{\alpha\alpha}} + \frac{r_{\tau(\gamma)} r_{\varsigma(\gamma)}}{I_{\gamma\gamma} - I_{\alpha\alpha}} \right] \right\} B^j_{\varsigma(\alpha)}$$

$$B^k_{\tau(\alpha)} = \frac{\partial \xi^k}{\partial r''_{\tau(\alpha)}}, \quad G^{ij} = \sum_\tau m_\tau^{-1} \vec{B}^i_\tau \vec{B}^j_\tau, \quad \sum_j G^{ij} T_{jk} = \delta^i_k,$$

$$I_{\alpha\alpha} = \sum_\tau m_\tau (r^2_{\tau(\beta)} + r^2_{\tau(\gamma)}), \qquad \alpha \neq \beta \neq \gamma; \qquad \alpha, \beta, \gamma = x, y, z.$$

Components of the operator of total angular momentum are as follows

$$L_{(\alpha)} = -i\hbar \frac{\partial}{\partial \varpi_\alpha}, \qquad \alpha = x, y, z.$$

As was proved in [5], the use of non-holonomic components as variables does not change well-known general tensor representation of Laplacian having an effect on an absolut scalar (wavefunction ψ)

$$g^{\mu\nu} \nabla_\mu \nabla_\nu \psi = g^{-1/2} \partial_\mu g^{1/2} \partial_\nu \psi, \qquad g = det(g_{\mu\nu}).$$

Finally, the ro-vibronic hamiltonian is

$$H_{rve} = \frac{1}{2M} \sum_\alpha P^2_\alpha + H_{ve} + \frac{1}{2} \sum_\alpha \mu_\alpha L^2_\alpha - \frac{1}{2} \sum_\alpha \left(2\mu_\alpha L^{II}_\alpha + F_\alpha \right) L_\alpha,$$

where vibronic hamiltonian

$$H_{ve} = \frac{1}{2} \sum_a \sum_\alpha \frac{1}{m_\alpha} p^2_{a(\alpha)} + \frac{1}{2} \sum_j \sum_k G^{jk} p_j p_k +$$

$$+ \frac{1}{M_1} \sum_b \sum_c \sum_\alpha p_{b(\alpha)} p_{c(\alpha)} + W'.$$

Here P_α, $p_{a(\alpha)}$, p_j - are impulses: of molecular mass center with respect to the laboratory frame of reference; of particles of the second subsystem with respect to the mobile frame of reference and impulses conjugated to the vibrational coordinates of the first subsystem (j = 1,2,...,3N$_1$-6); L^{II}_α - components of angular momentum of the second subsystem with respect to the mobile frame of reference; μ_α, F_α, W' - functions of variables of both subsystems and $\mu_\alpha = \frac{I_\alpha}{\left(I_\beta - I_\gamma\right)^2}$, I_α - principal inertia momentum of the first subsystem ($\alpha \neq \beta \neq \gamma$ = x,y,z), M$_1$ - is the mass of the first subsystem.

The process of derivation and final result will be shown in poster presentation.

1. Tsaune A.Ya., Dyachenko M.P., J.Mol.Struct., 266, 241, (1992)
2. Darling B.T., Cells M.de, J.Mol.Spectrosc., 29, 66 (1969)
3. Buck B., Biedenharn L.C., Cusson R.V., Nucl.Phys., A317, 205 (1979)
4. Skalozub A.S., Tsaune A.Ya., Optika i Spektroskopia, 50, 458 (1981)
5. Tsaune A.Ya., Optika i Spektroskopia, 53, 228 (1982)

SIMULATION OF RAMAN SPECTRA FROM MD CALCULATIONS. APPLICATION TO ZEOLITES AND ADSORBED MOLECULES

D. Dumont[1], K.S. Smirnov[2] and D. Bougeard[1*]

[1]LASIR-CNRS, Université de Lille 1, Bat C5, F 59655 Villeneuve d'Ascq Cedex
[2]Institute of Physics, St Petersburg State University, St Petersburg 198904, Russia

Molecular dynamical (MD) calculations give the possibility to follow the time evolution of a system of particles. The linear response theory shows that the Raman spectrum can be derived from the Fourier transformation of the autocorrelation function of the polarizability [1]. For the simulation of the Raman spectra it is necessary to develop a model permitting the calculation of the polarizability tensor α of the MD box. A simple description of α can be obtained if the tensor is written as a sum of bond polarizabilities α_i and if each bond polarizability is developed in a Taylor series according to the bond length and the series cut after the linear term:

$$\alpha_i = \alpha_{0i} + (d\alpha_i/dr_i)_0\, r_i$$

Thus the autocorrelation function of the polarizability can be obtained from the coordinates by using one parameter $(d\alpha_i/dr_i)_0$ for the longitudinal polarizability of each bond type [2].

This formalism was applied to a series of zeolites (sodalite, faujasite, zeolite A and silicalite) and to a series of hydrocarbons adsorbed in silicalite (methane, ethane, propane, ethene and ethyne). The spectra obtained for siliceous zeolites are of good quality in comparison with the available experimental spectra. The same technique

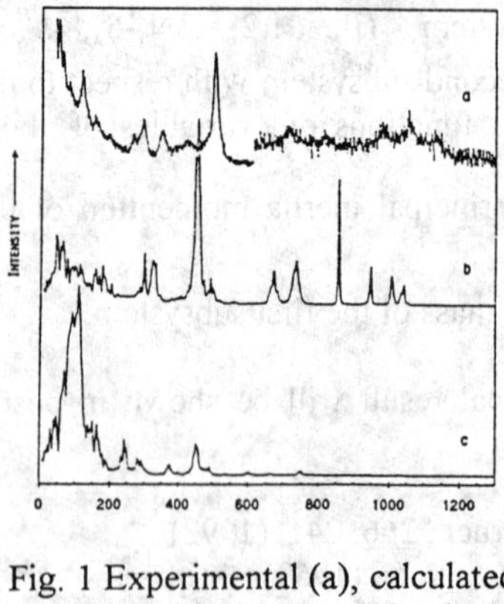

Fig. 1 Experimental (a), calculated (b) Raman spectra of Na zeolite A and calculated spectral density of window fluctuations in the structure.

was also applied to Na substituted zeolite A with an Si/Al ratio of 1, where two parameters are necessary . It turned out that the parameter for the Al-O bond has to be only slightly different from the Si-O bond parameter [3]. Further a detailed analysis of the low frequency spectra shows that the pulsation of the windows can be assigned to a Raman band observed around 100 cm^{-1} in all zeolites A [4].

The same simple model applied to five hydrocarbons in silicalite yields spectra in good agreement with the experimental spectra for the gaseous phase as well as for the adsorbed molecules. This fact confirms that the role of the cations in silicalite is not as important as in some other zeolites [5].

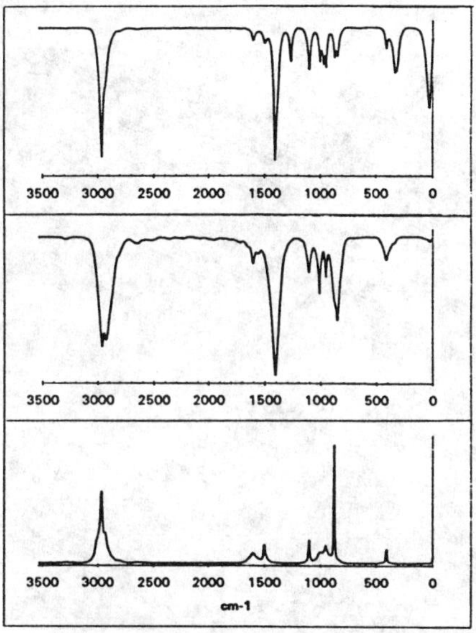

Fig. 2. Calculated density of states (a), infrared (b) and Raman (c) spectra of propane adsorbed in silicalite.

Finally, at the time scale of some picoseconds the different behaviors of the hydrocarbons in silicalite nicely reproduces the experimental values obtained by pulsed field gradient nmr spectroscopy, thus showing that the model used to simulate the Raman spectra is also able to reproduce the dynamical processes at a larger time scale [6].

Presently work is in progress in order to improve the theoretical model and to determine a corresponding set of parameters for aluminosilicates.

1. P.H. Berens, S.R. White and K.R. Wilson, J. Chem. Phys. 75 515 (1981).
2. K.S. Smirnov and D. Bougeard, J. Raman. Spectroc. 24 255 (1993).
3. K.S. Smirnov, M. Le Maire, C. Brémard and D. Bougeard, Chem. Phys. 179 445 (1994).
4. K.S. Smirnov and D. Bougeard, Zeolites 14 203 (1994).
5. D. Dumont and D. Bougeard, Spectrochim. Acta 51 1671 (1995).
6. D. Dumont and D. Bougeard, Zeolites 15 650 (1995).

SECTION 2

RESONANCE RAMAN THEORY

RESONANCE RAMAN SPECTROSCOPY OF AZODYES: EVIDENCES OF THE PARTICIPATION OF A SINGLE EXCITED ELECTRONIC STATE

Luiz F.C. de Oliveira[1], Mauro C.C. Ribeiro[2] and P.S. Santos[2*]

[1] Departamento de Química, Instituto de Ciências Exatas, Universidade Federal de Juiz de Fora, CEP 36036-330, Juiz de Fora, MG, Brazil
[2] Departamento de Química Fundamental, Instituto de Química, Universidade de São Paulo, CP 26.077, CEP 05599-970, São Paulo, SP, Brazil

The interpretation of the optical spectrum of substituted azodyes has been highly controversial [1], despite the fact that the electronic spectrum of the parent system, azobenzene is at present well understood [2].

In particular, 4-(2-pyridylazo)-resorcinol (PAR), a widely employed metallochromic indicator for Cu(II) and Zn(II), displays an optical spectrum whose interpretation is far from trivial. When coordinated to metal ions additional complications appear, as charge transfer transitions involving the metal ion must, in principle, be present [3].

In the present work, resonance Raman spectra of aqueous solutions of PAR in its monoanionic and dianionic forms were obtained, confirming the existence of a pH dependent azo-hydrazone tautomerism, as revealed by the presence of the marking bands at *ca.* 1336 and 1391 cm^{-1}, respectively. In addition, the resonance Raman spectra of the $Cu(PAR)_2$ and $Zn(PAR)_2$ chelates were obtained, revealing an azo-type coordination. The transform method and time dependent theory [4] were employed to generate the excitation profiles (Figure 1) and electronic spectra (Figure 2), strongly suggesting the participation of a single excited electronic state in the transition responsible for the strong band at *ca.* 500 nm. The more significantly enhanced modes involve mainly the N=N stretching and modes that seem to be related to the resorcinol moiety, that are the ones responsible for overtones and combination bands.

Although a substantial shift in the absorption maximum is observed for the low energy transition when one proton is removed from the neutral PAR, just a small shift is observed when its monoanionic form coordinates to the metal ion, thus suggesting that the extent of electronic delocalization is quite comparable in both systems. This seems to be confirmed by the calculation of displacements of equilibrium internal coordinates for both systems, that yelded similar values. No sign of metal-ligand modes were present in the resonance Raman spectra of the Zn(II) and Cu(II) chelates.

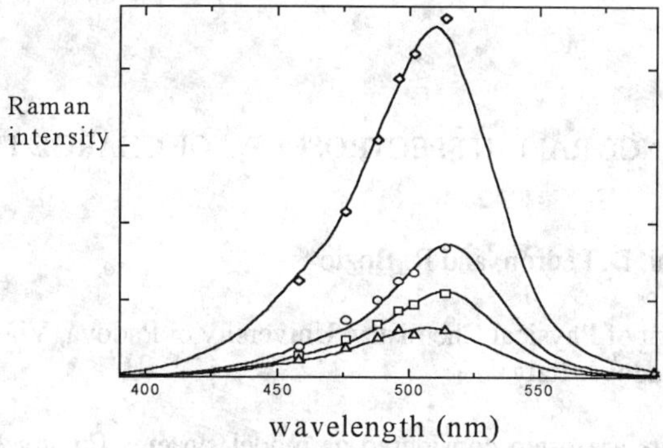

Figure 1. REP of $Cu(HL)_2$: experimental (symbols) and calculated by transform method (full line): ◊ 1339 cm^{-1}; o 1018 cm^{-1}; □ 1114 cm^{-1}; Δ 1448 cm^{-1}.

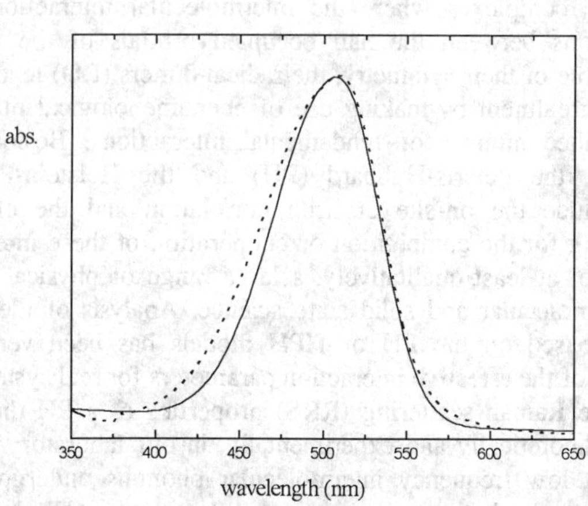

Figure 2. Electronic absorption spectra of $Cu(HL)_2$: experimental (dashed line) and calculated by the time dependent theory (full line).

Acknowledgements: Financial support from FAPESP and CNPq are highly acknowledged.

1. C. Sourisseau, P. Maraval, J. Raman Spectrosc. 25, 477 (1994).
2. S. Monti, G. Orlandi, P. Palmieri, Chem. Phys. 71, 87 (1982).
3. CRC Handbook of Organic Analytical Reagents, (K. L. Cheng, K. Ueno, T. Imamura, ed.), CRC Press, Florida, 1982, p.195.
4. M. C. C. Ribeiro, W. J. Barreto, P. S. Santos, J. Raman Spectrosc. 25, 923 (1994).

RESONANCE RAMAN SPECTROSCOPY OF CHARGE TRANSFER DIMERS

V. Mulloni, D. Pedron, and R. Bozio*

Department of Physical Chemistry, University of Padova, Via Loredan 2, I-35131 Padova, Italy

Molecular dimers are often considered as model systems for investigating the effects of the intermolecular interactions in larger aggregates, in molecular solids and in biological systems. Donor-acceptor complexes have long been investigated to clarify the basic aspects of the charge transfer (CT) interaction, which is of pivotal importance for both electron transfer processes and electrical transport in organic materials. Dimers formed by identical molecular ion radicals can be considered as a special case of CT complexes where the intermolecular interaction involves the exchange of electrons between the half occupied orbitals of the two molecular components. By virtue of their symmetry, the radical dimers (RD) lend themselves to an easy theoretical treatment by making use of phenomenological models that take into account a limited number of fundamental interactions. Beside the electron transfer interaction, the Peierls-Hubbard (PH) and the Holstein-Peierls-Hubbard (HPH) models include the on-site electron correlation and the electron-phonon coupling. Accounting for the competition or cooperation of these interactions, these models can describe, at least qualitatively, a large range of physical phenomena of great relevance in molecular and solid state science. Analysis of the spectroscopic properties of RD based on the PH or HPH models has been very succesful in providing estimates of the effective interaction parameters for real systems [1].

The resonance Raman scattering (RRS) properties of a PH dimer have been investigated, both theoretically and experimentally, in our laboratory [2,3]. The PH model predicts that low frequency intermolecular phonons undergo a remarkable intensity enhancement for laser excitation resonant with the CT absorption of the dimer. Calculation of the CT absorption profile and of the Raman excitation profiles (REP's) is carried out by solving the PH model perturbatively to second order in the electron-phonon interactions and using a time correlator technique. In this way, the needed Herzberg-Teller potential surfaces, including frequency changes and Duschinsky rotation, and the transition moments with non-Condon contributions can be calculated directly from the fundamental parameters of the model, namely the transfer integral t, the on-site correlation U, the electron-intermolecular phonon (EIP) coupling constants $g_e = (\partial t / \partial u_e)_\circ$ where u_e is the intermolecular phonon coordinate.

For comparison with the theory, experimental REP data, for resonance with the CT absorptions, have been measured for two crystalline radical ion salts containing almost isolated RD as basic structural units. $(TTF)_2(W_6O_{19})$ (TTF = tetrathiafulvalene) contains layers of $(TTF^+)_2$ dimers whereas TEA-DDQ (TEA =

tetraethylammonium; DDQ = dichloro-dicyano-p-benzoquinone) contains strongly dimerized stacks of DDQ^- anions. In both cases, the low frequency resonance Raman spectra exhibit three prominent bands which have been associated to the out-of-phase combination of translational motions of the two molecular radicals composing the dimer. This modes are, in fact, most effective in modulating the intradimer transfer integral. Quantitative analysis of the REP's data, accounting also for the finite temperature effects, allows us to extract values for the basic interaction parameters of the PH Hamiltonian which are collected in Table 1.

TABLE 1. Parameter values used for the calculation of the REP's of the low frequency modes

	$(TTF)_2(W_6O_{19})$		
	$t = 0.30$ eV	$U = 1.13$ eV	$\gamma = 750\ cm^{-1}$
mode	Ω_e[a] (cm^{-1})	ω_e (cm^{-1})	g_e (eV/Å)
T_x^-	116	116.5	−0.171
T_z^-	90	90.5	0.078[b]
T_y^-	55	55.5	−0.079
	TEA-DDQ		
	$t = 0.33$ eV	$U = 1.19$ eV	$\gamma = 800\ cm^{-1}$
mode	Ω_e[a] (cm^{-1})	ω_e (cm^{-1})	g_e (eV/Å)
1	148	151	0.37
2	174	174	0.28
3	194	195	0.34

[a] Observed Raman frequencies. [b] The sign of g_z remains indeterminate.

Very recently, the PH model has been extended to allow for the possible mixing between the CT and a low lying molecular transition and to predict the spectroscopic consequences of such a mixing. REP data for totally symmetric intramolecular modes of the $(TTF^+)_2$ dimers in $(TTF)_2(W_6O_{19})$ have been measured by tuning the exciting radiation to resonance both with the CT and with molecular excitations. A theoretical analysis of the data based on the extended model has allowed us to estimate the interaction parameter responsible for the mixing of electronic states [4].

1. R. Bozio and C. Pecile, in Spectroscopy of Avanced Materials (R.J.H. Clark and R.E. Hester, ed.) Wiley, Chichester, 1991, p.1 and references therein.
2. D. Pedron, A. Speghini, V. Mulloni, and R. Bozio, J. Chem. Phys. 103, 2795 (1995)
3. D. Pedron, S. Lavina, V. Mulloni and R. Bozio, to be published.
4. V. Mulloni, D. Pedron and R. Bozio, to be published.

RESONANCE RAMAN SCATTERING IN PHOTODISSOCIATING HALOMETHANES

F. Duschek, M. Schmitt, A. Materny, V. Engel, and W. Kiefer*

Institut für Physikalische Chemie, Universität Würzburg,
Marcusstr. 9-11, D-97070 Würzburg, Germany

The photodissociation of iodoalkanes in the A–band absorption has been extensively investigated as a prototype for direct photodissociation taking place on a purely repulsive excited state potential energy surface. The examination of the dissociation after excitation with ultraviolet light by means of resonance Raman spectroscopy gives fundamental insight into the short time dynamics of these molecules on their excited potential energy surface(s) during the reaction. Resonance Raman spectroscopic studies of methyl iodide have been carried out by several research groups [1–3]. The principle features of gas phase photodissociation of methyl iodide appear to be fairly well understood. However, not much attention has been payed to the resonance Raman scattering in vapours of iodomethan derivatives [4].

In this work we present the absorption and resonance Raman spectra of the iodomethanes CH_3I, CH_2I_2, and CH_2ICl. The CH_3I and CH_2ICl UV absorption spectra, displayed in Fig. 1 panel A, in the region from 325 to 220 nm are broad and featureless with maxima at 258 and 270 nm, respectively. In contrast to this CH_2I_2 has three absorption bands between 210 and 370 nm that can be deconvoluted into four distinct absorption bands (Fig.1 panel A). The major photochemical process after excitation within the respective A–band absorption is the breaking of the C–I–bond of each molecule.

As shown in Fig. 1 panel A, the 275.394 nm laser line of an argon ion laser is resonant with the A–band absorption transition of the three molecules. The vapour phase resonance Raman spectra of the molecules CH_3I, CH_2I_2 and CH_2ICl excited with this line are shown in Fig. 1 panel B. They are all dominated by an overtone progression of the resonance enhanced C–I stretch mode and weak combination band progressions (Fig. 1 panel B). In the case of CH_2I_2 and CH_2ICl the combination bands $9_1 3_2$, $9_2 3_1$ and $5_1 4_1$, $5_2 4_1$ are found, which indicate an electronic coupling between the chromophores C–I, C–I and C–I, C–Cl, respectively. For CH_2I_2, which has 4 absorption maxima between 230 and 370 nm, further excitation wavelengths of an argon ion laser (333.613, 351.112, 363.789 nm) have been used to record the resonance Raman spectra. All these spectra show several overtones of the symmetric and antisymmetric CI_2 stretching modes ν_3 and ν_9.

The resonance Raman intensities and the absorption spectrum of CH_2I_2 are calculated using the time–dependent theory of Raman scattering [5]. In this calculations CH_2I_2 is treated as a pseudotriatomic molecule with the symmetric and antisymmetric C–I stretching mode as vibrational coordinates. The resonance Raman intensities are calculated for different parameters of a model potential energy surface of the first excited electronic state.

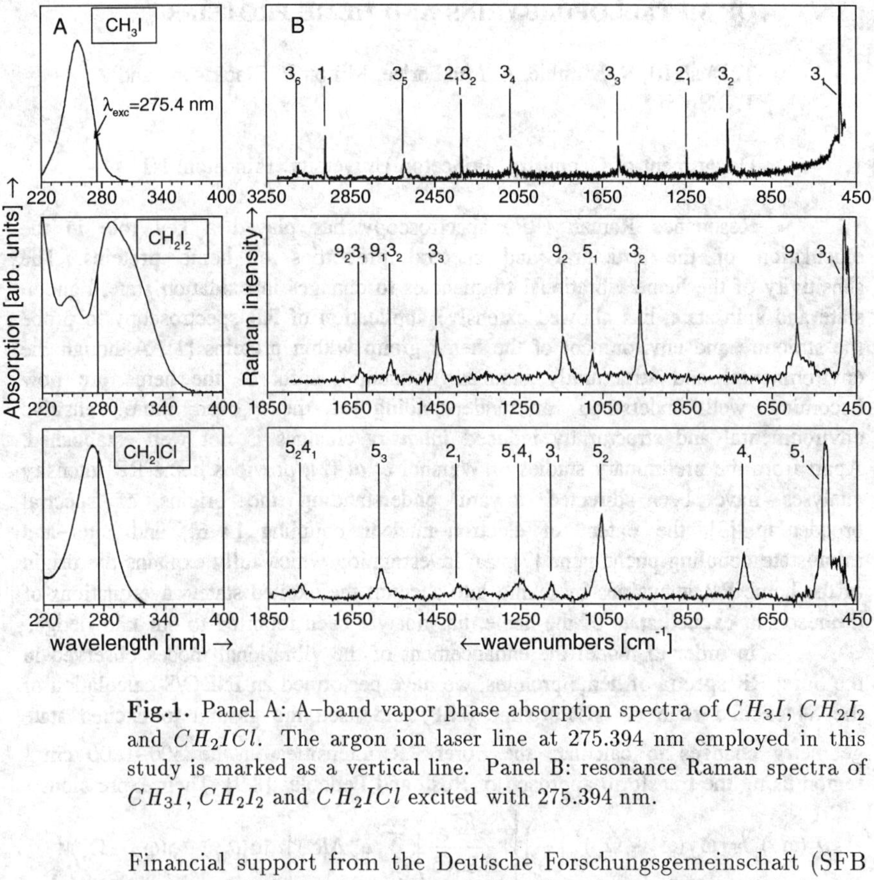

Fig.1. Panel A: A–band vapor phase absorption spectra of CH_3I, CH_2I_2 and CH_2ICl. The argon ion laser line at 275.394 nm employed in this study is marked as a vertical line. Panel B: resonance Raman spectra of CH_3I, CH_2I_2 and CH_2ICl excited with 275.394 nm.

Financial support from the Deutsche Forschungsgemeinschaft (SFB 347, projects C2 and C5) and by the Fonds der Chemischen Industrie is highly acknowledged.

1. D. Imre, J.L. Kinsey, A. Sinha and J. Krenos, J. Phys. Chem. 88, 3956 (1984).
2. K.Q. Lao, M.D. Person, P. Xayariboun and L.J. Butler, J. Chem. Phys. 92, 823 (1990).
3. P.G. Wang and L.D. Ziegler, J. Chem. Phys. 97, 3193 (1993).
4. J. Zhang, E.J. Heller, D. Huber and D.G. Imre, J. Chem. Phys. 89, 3602 (1988).
5. S.Y. Lee and E.J. Heller, J. Chem. Phys. 71, 4777 (1979).

MODELING THE SORET RESONANT RAMAN INTENSITIES OF METALLOPORHYRINS AND HEME PROTEINS

T. Rush III, R. Kumble, A. Mukherjee, Milton E. Blackwood and T.G.Spiro *

Department of Chemistry, Princeton University, Princeton, NJ

Resonance Raman (RR) spectroscopy has played a key role in the elucidation of the structural and spectral properties of heme proteins. The sensitivity of the heme vibrational frequencies to changes in oxidation state, ligation state and spin state, has allowed extensive application of RR spectroscopy to probe the structure and environment of the heme group within proteins [1]. Although the environmental and structurally induced frequency shifts of the heme are now becoming well understood, an understanding of the even more sensitive environmental and structurally induced *intensity* changes is not well established. Apart from the preliminary studies of Warshel *et al* [2], previous heme RR intensity analyses have been directed toward understanding the origins of spectral broadening [3], the extent of electron–nuclear coupling [4–6], and inter–and intra–state coupling phenomena [7]. An investigation which fully explains the origin of the heme RR intensities by taking into account the excited state wavefunctions of the resonant excited state of the heme, has not yet been reported to our knowledge.

In order to model the enhancement of the vibrational modes observed in the Soret RR spectra of heme proteins, we have performed an INDO/S calculation of the S_2 excited state of Ni Porphine (NiP) and used the ground to excited state geometry changes to calculate the Soret RR intensities in the 300–1800 cm^{-1} region using the transform approach of Rush and Peticolas [8,9]. Their expression:

$$I_k(\omega_L) \propto \omega_L(\omega_L - \Omega_k)^3 \left(\frac{1}{2}\right)\left[\left(\frac{\sqrt{\Omega_k}}{5.8065}\right)\left(\sum_{j=1}^{3N-6} L_{kj}^{-1}\Delta R_j'\right)\right]^2 |\phi(\omega_L) - \phi(\omega_L - \Omega_k)|^2$$

incorporates the usual standard assumptions of the transform theory, and the assumptions that the temperature is low enough that all of the transitions originate from the v=0 state, that Dushinsky rotations are negligible, and that non–Condon effects are also negligible. In the same spirit as this Rush and Peticolas reformulation of the transform formula for the fundamental RR intensity of a symmetric vibration, we have reformulated the expressions for the RR intensities of overtone bands and for non–totally symmetric vibrations of b_{1g} symmetry, and used these in our intensity analysis.

Initially, the INDO geometry predicted reasonable agreement to the relative intensities in this region of the spectra, but the intensity of the ν_8 overtone was predicted to be too high. This reflects an overestimation of the INDO determined excited state geometry. By comparing the experimental ν_8 fundamental to overtone intensity ratio, we were able to determine a scaling factor of 0.81 +/– 9% for the INDO values. A comparison between the experimental (406.7 nm excitation, 1mM NiP in CS_2) and the final calculated spectrum is shown in figure 1. The agreement is quite good, although the enhancement of the Jahn–Teller active modes is underestimated. This is presumably due to the fact that their enhancement also requires consideration of non–Condon mechanisms. A full analysis of the spectrum, comparisons at different levels of approximations and the excited state geometry values are not presented here, but will appear in an upcoming research article.

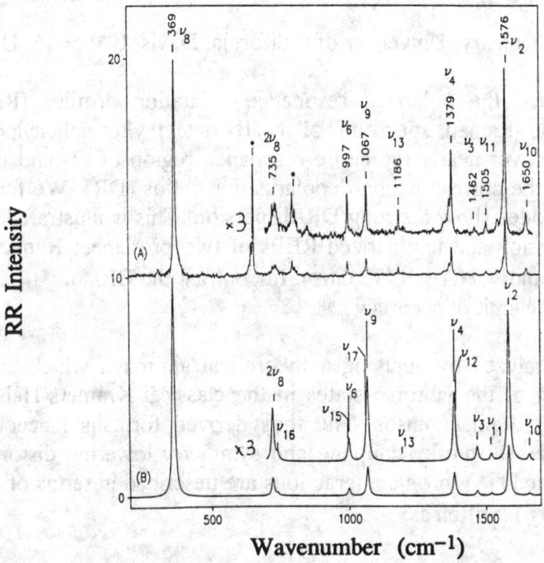

1. Spiro, T.G. in *Iron Porphyrins*, Part II; Lever, A. and Gray, H., Eds.; Addison–Wesley: Reading, MA, 1983; pp 89–160.
2. Warshel, A.; Lappicirella, A. *J. Am. Chem. Soc.* **1981**, 103, 4664.
3. Champion, P.M. in *Biological Applications of Raman Spectroscopy,* Spiro, T.G. Ed.; Wiley: New York, NY, 1988; Vol. 3, p 249.
4. Stallard, B.R.; Callis, P.R.; Champion, P.M.; Albrecht, A.C. *J. Chem. Phys.* **1984**, 80, 70.
5. Gu, Y; Champion, P.M. *Chem. Phys. Lett.* **1990**, 171, 254.
6. Morikis, D.; Li, P.; Bangcharoenpaurpong, O.; Sage, J.T.; Champion, P.M. *J. Phys. Chem.* **1991**, 95, 3391.
7. Bobinger, U.;Schweitzer–Stenner, R.; Dreybodt, W. *J. Phys. Chem.* **1991**, 95, 7625.
8. Rush III, T.; Peticolas, W.L. *J. Phys. Chem.* **1995**, 99, 14647.
9. Peticolas, W.L.; Rush III, T. *J. Comp. Chem.* **1995**, 16, 126.
10. Kozlowski, P.M.; Jarzecki, A.; Li, X.–Y.; Zgierski, M.Z.; Pulay, P. (submitted)

OUT-OF-PLANE AND IN PLANE DISTORTIONS OF NI(II)OCTAETHYLTETRAPHENYLPORPHYRIN PROBED BY RESONANCE RAMAN DISPERSION SPECTROSCOPY.

Reinhard Schweitzer-Stenner[1*], Andreas Stichternath[1], Wolfgang Dreybrodt[1], Xingzhi Song[2], John A. Shelnutt[2], Craig J. Medforth[3] and Kevin Smith[3]

[1]FB1-Institut für Experimentelle Physik, Universität Bremen, 28359 Bremen, Germany
[2]Fuel Science Department 6211, Sandia National Laboratories, Albuquerque, NM 87185, USA
[3]Department of Chemistry, University of California, Davis, CA95616, USA

We have measured the polarized resonance excitation profiles (REPs) of numerous bands in the Raman spectrum of Ni(II)-octaethyltetraphenylporphyrin (NiOETPP). The data cover nearly the entire resonance region of Q- and B-bands. From this we obtained the corresponding depolarization ratios (DR). We found that all A_{1g}- and A_{2g}-type modes show a strong DR-dispersion. This is illustrated in Fig.1 which shows DR-dispersion and unpolarized REPs of two prominent Raman bands, i.e. ν_2 (A_{1g}, 1509 cm^{-1}) and ν_{19} (A_{2g}, 1490 cm^{-1}). In contrast the DRs of B_{1g} and B_{2g} - modes are constant in the limit of accuracy.

The data were analyzed by applying a theoretical approach which invokes a non-adiabatic expansion of the vibronic states in the classical Kramers-Heisenberg-Dirac expression of the Raman tensor. The thus derived formalism accounts for multimode mixing, linear Dushinsky coupling and symmetry lowering distortions of the porphyrin macrocycle [1]. Vibronic interactions are described in terms of vibronic coupling parameters $c_{es}^{\Gamma}(r)$ written as:

$$c_{es}^{\Gamma}(r) = \langle e | \partial \hat{H}_{el} / \partial Q_r^{\Gamma} | s \rangle \langle 1_r | Q_r^{\Gamma} | 0_r \rangle \qquad (1)$$

where $|e\rangle, |s\rangle$ represent the excited electronic states Q_x, Q_y, B_x, B_y, Γ is the apparent symmetry of the Raman active normal mode Q_r^{Γ} and $\partial \hat{H}_{el} / \partial Q_r^{\Gamma}$ is the corresponding vibronic coupling operator. If we use D_{4h} as the reference point group this operator reads as:

$$\frac{\partial \hat{H}_{el}}{\partial Q_r^{\Gamma}} = \frac{\partial \hat{H}_{el}}{\partial Q_r^{\Gamma_r}} + \sum_a \frac{\partial^2 \hat{H}_{el}}{\partial Q_r^{\Gamma_r} \partial \overline{Q}_a^{\Gamma_a}} \delta \overline{Q}_a^{\Gamma_a} + \sum_a \sum_b \frac{\partial^3 \hat{H}_{el}}{\partial Q_r^{\Gamma_r} \partial \overline{Q}_a^{\Gamma_a} \partial \overline{Q}_b^{\Gamma_b}} \delta \overline{Q}_a^{\Gamma_a} \delta \overline{Q}_b^{\Gamma_b} \qquad (2)$$

Eq. (2) describes a Taylor expansion of the vibronic coupling operator with respect to normal distortions $\delta \overline{Q}_a^{\Gamma_a}, \delta \overline{Q}_b^{\Gamma_b}$ which are classified in terms of the symmetry types Γ_a, Γ_b = A_{1g}, B_{1g}, B_{2g}, A_{2g}, A_{1u}, A_{2u}, B_{1u} and B_{2u} of the D_{4h}-point group. Only the first term contributes to the Raman scattering of in-plane modes, if the macrocycle is in its ideal D_{4h}-symmetry. The second term accounts for the influence of in-plane distortions of even symmetry, while the second order term becomes relevant when two out-of-plane distortions of odd symmetries are operative. Vibronic coupling remains unaffected if only one symmetry type of out-of-plane distortions exists.

The above theory was used to fit the experimental data. The thus obtained coupling parameters suggest that B_{1u} and B_{2u}-distortions are operative, which yield ruffling and saddling of the porphyrin's macrocylce [2]. This agrees with results from INDO calculations which show that these distortions mix σ-orbitals of B_{1g}- and B_{2g}-symmetry with the classical A_{1u}- and A_{2u}-orbitals of the π–electron system. Moreover theoretical reasonings suggest that the above out-of-plane distortions can induce in-plane A_{1g}- and A_{2g}-distortions by intrastate vibronic coupling in both, the excited states and the ground state. For the ground state A_{2g}-type normal distortions read as:

$$\delta\overline{Q}_a^{A_{2g}} = -\frac{\langle g|\partial^3 \hat{H}_{el} / \partial Q_r^{A_{2g}} \partial \overline{Q}_a^{B_{1u}} \partial \overline{Q}_b^{B_{2u}}|g\rangle}{\Omega_r^2} \delta\overline{Q}_a^{B_{1u}} \delta\overline{Q}_b^{B_{2u}}$$

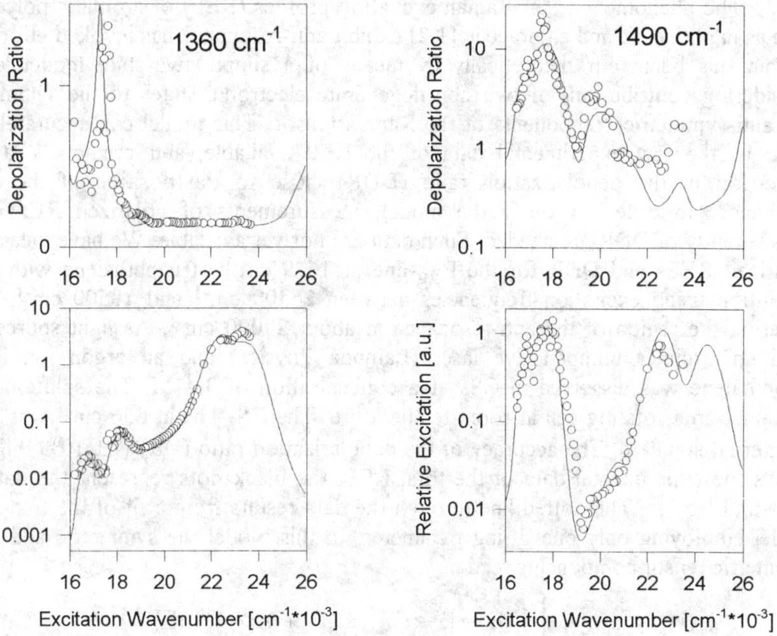

Fig. 1 DR-Dispersion and REPs of ν_4(1360 cm^{-1}) and ν_{19}(1490 cm^{-1})

1. E. Unger, U. Bobinger, W. Dreybrodt and R. Schweitzer-Stenner, J. Phys. Chem. 97, 9956(1993).
2. W. Jentzen, M.C. Simpson, J.D. Hobbs, X. Song, T. Ema, N.Y. Nelson, C.J. Medforth, K.M. Smith, M. Veyrat, M. Mazzanti, R. Ramasseul, J.-C. Marchon, T. Takeuchi, W.A. Goddard III, and J.A. Shelnutt, J. Am. Chem. Soc.117,11085(1995).

POLARIZATION DISPERSION IS NECESSARY FOR INFORMATION ON THE CONTRIBUTION OF ELECTRONIC STATES TO THE RAMAN TENSOR.

F. Ehland[1], S. Hassing[2*] and W. Dreybrodt[1]

[1]FB 1 - Institut für Experimentelle Physik, University of Bremen,P.O. Box 330 440, D-28334 Bremen, Germany
[2]Department of Electrical Engineering, The Engineering College of Odense, Niels Bohrs Alle 1, DK 5230 Odense M., Denmark

The phenomenon that Raman excitation profiles (REP) of aromatic polyenes, such as naphthalene and anthracene [1,2] exhibit anti-resonance has invoked efforts to explain this behaviour theoretically by means of a simple two-state model [3,4], considering contributions of two non degenerate electronic states to the symmetric and anti-symmetric components of the Raman tensor. This model can accurately be fitted to the few experimental data of the REP available, and predicts a strong dispersion in the depolarization ratio (DDR) close to the position of the anti-resonance. In order to verify the model, measurements of polarized REPs and consequently of DDR are needed. Such data are not yet available. We have measured polarized REPs and DDR for the B_{3g}-line at 1629 cm^{-1} of naphthalene with high resolution using excitation frequencies between 26300 cm^{-1} and 18500 cm^{-1}, thus covering the region of the anti-resonance at about 24000 cm^{-1}. As light source we used an excimer pumped dye laser (Lambda Physics) and an argon ion laser. Naphthalene was dissolved in CS_2 at a concentration of 3.4 M. The solution was contained in a rotating cell at room temperature. The CS_2-line at 656 cm^{-1} was used as internal standard. The accuracy of the depolarization ratio (DR) is ± 10%. Fig. 1a shows the experimental data of the total REP; the black dots represent the data of Ohta and Ito [1]. The dotted line through the data results from a fit of the two-state model employing only one fitting parameter. In this model the symmetric and anti-symmetric tensor components read

$$\alpha^{\pm}_{\rho\sigma} = C\cdot\Big[\left\{(E_e - E - i\Gamma)^{-1} \pm (E_e + \hbar\omega - E - i\Gamma)^{-1}\right\} + K_{e'e}\left\{(E_{e'} + \hbar\omega - E - i\Gamma)^{-1} \pm (E_{e'} - E - i\Gamma)^{-1}\right\}\Big] \quad (1)$$

where C is a constant, E_e = 34600 cm^{-1}, and $E_{e'}$ = 45400 cm^{-1} are the energies of the excited states <e|, <e'|. E is the excitation energy and ω is the vibrational frequency. $K_{e'e}$ is the fitting parameter related to vibronic coupling. Γ is the line width. This fit predicts a strong DDR as shown by the dotted line in Fig. 1b. The experimental points, however, do not show any dispersion. To resolve this

contradiction, we have extended our model by the contribution of a third state <e″| with $E_{e''} = 59800\ cm^{-1}$. This adds an extra term to eqn. (1) reading

$$C \cdot K_{e''e}\left\{(E_{e''} - E - i\Gamma)^{-1} \pm (E_{e''} + \hbar\omega - E - i\Gamma)^{-1}\right\} \qquad (2)$$

where $K_{e''e}$ provides an additional vibronic fitting parameter and Γ has to be increased with respect to the two-state model [4]. This augmented model (full line) yields similar agreement with the REPs as the two-state model, but predicts no DDR in accordance with the experimental findings.

From these data one concludes that experimental data of total REPs are not sufficient to discriminate between different theoretical models, even in simple cases. The knowledge of DDR turns out to be of utmost importance for a complete investigation.

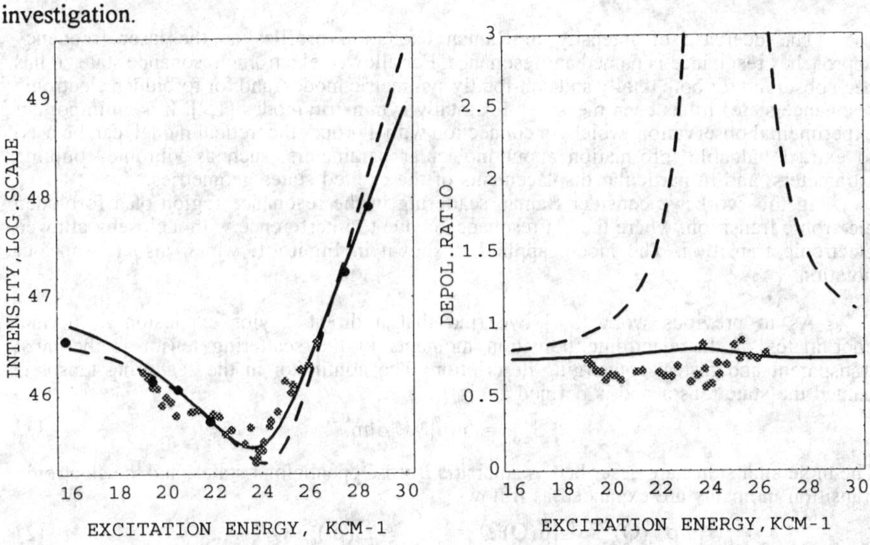

$K_{e''e} = -8.75$, $K_{e'e} = +3.3$, $\Gamma = 2400 cm^{-1}$

Fig. 1.: a) Raman excitation profiles calculated from the two-state model (----) and from the three-state model (____) in comparison to experimental data points. The full dots represent the data of [1]. b) Depolarization ratio from the two-state model (----) exhibits strong dispersion, but the prediction from the three-state model (____) is in accordance with the experimental data.

1. N. Ohta and M. Ito, J. Chem. Phys. **20**, 71-81,(1977).
2. A. Aminzadeh, V. Fawcett and D.A. Long, J. Raman Spectrosc. **9**, 219, (1980).
3. S. Hassing, P. Waage Jensen, L.T.K. Kristensen, E. N. Svendsen, Proc. of the XIV. ICORS, N.-T. Yu and X. Y. Li (eds.), 264, (1994) John Wiley and Sons
4. S. Hassing, J. Raman Spectrosc., (submitted)

ANTIRESONANCE IN RESONANCE RAMAN SCATTERING FOR FORBIDDEN ELECTRONIC TRANSITIONS

S. Hassing[1] and E. Nørby Svendsen[2]*

[1]Department of Electrical Engineering, Engineering College of Odense, Niels Bohrs Allé 1, 5230 Odense M, Denmark

[2]Department of Chemistry, Odense University, Campusvej 55, 5230 Odense M, Denmark

The decrease of intensity in Raman excitation profiles as the laser frequency approaches resonance is named antiresonance. For allowed electronic resonance states it has been observed fc- both totally and non-totally symmetric modes, and for forbidden electronic resonance states it has been measured for totally symmetric modes [1,2]. It is an important experimental observation, which, in connection with a proper theoretical model, can be used to extract valuable information about molecular parameters, such as vibronic coupling parameters, and in particular displacements of the excited states geometries.

In this work we consider Raman scattering in the resonance region of a forbidden electronic transition, where the antiresonance is due to interference with a closeby allowed electronic transition. The model applied is shown in Figure 1, which also reveals our notation.

As in previous work [3,4], we find that a direct Taylor expansion in normal coordinates of the electronic transition moments in the scattering tensor is the most transparent and fastest converging description. The nominator in the scattering tensor is named the state tensor and is defined as

$$S^{\ell}_{\rho\sigma} = \langle n|\rho|\ell\rangle\langle \ell|\sigma|m\rangle \tag{1}$$

The molecular states are described as adiabatic Born-Oppenheimer states, and the electronic transition moments are expanded as follows

$$\rho_{ge}(Q) = \langle g|\rho(Q)|e\rangle = \rho^{0}_{ge} + \sum_{i}(\rho^{i})^{0}_{ge}\, Q_i + \ldots \tag{2}$$

$$\text{where}\quad \rho^{0}_{g} = \langle g^{0}|\rho|e^{0}\rangle \quad\text{and}\quad (\rho^{i})^{0}_{ge} = \left(\frac{\partial \rho_{ge}}{\partial Q_i}\right)^{0} \tag{3}$$

Since the electronic transition is forbidden, the usual terms in the state tensor responsible for Franck-Condon and Herzberg-Teller scattering disappear and the term describing the scattering from state $|e\rangle$ is given by

$$\sum_{i,j}(\rho^{i})^{0}_{ge}(\rho^{j})^{0}_{eg}\,\langle n|Q_i|v\rangle\langle v|Q_j|m\rangle \tag{4}$$

Considering the fundamental symmetric 0-1 Raman mode, we have for symmetry reasons an asymmetric coupling mode Q_a with $(\rho^{a})^{0}_{ge} \neq 0$, giving for eq. (1) the following

$$(\rho^{a})^{0}_{ge}(\rho^{a})^{0}_{eg}(1_s v_a|Q_a|v_s v_a)\,(v_s v_a|Q_a|0,0) \tag{5}$$

Using the properties of harmonic oscillator wavefunctions, eq. (5) reduces to

$$\alpha_{\rho\rho} = |\rho^0_{ge}|^2 \sum_{v_s, v_s'} \left(\frac{\langle 1_s v_s' | v_s' 0 \rangle}{E_{ge'} + v_s' \omega_s - E - i\tau} \right) + \left(K^a_{ee'}\right)^2 \left(\frac{\langle 1 v_s | v_s 0 \rangle}{E_{ge} + \omega_a + v_s \omega_s - E - i\tau} \right) \tag{6}$$

$$\text{where} \quad K^a_{ee'} \equiv \frac{\sqrt{\frac{\hbar}{2\omega_a}} \cdot (\rho^a)_{ge'}}{\rho^0_{ge}} \tag{7}$$

is a slope parameter, similar to the ones introduced in previous work [3,4], describing the amount of vibrational coupling. The Franck-Condon factors are given by the analytical expressions

$$(1|_s v_s)(v_s|0) = M e^{-M^2} \frac{M^{2v_s}}{v_s!} \left(1 - \frac{v_s}{M^2} \right) \tag{8}$$

where M is the displacement of the electronic state |e⟩, and a similar expression for the second term in eq. (9) with a displacement of state |e'⟩ named M'.

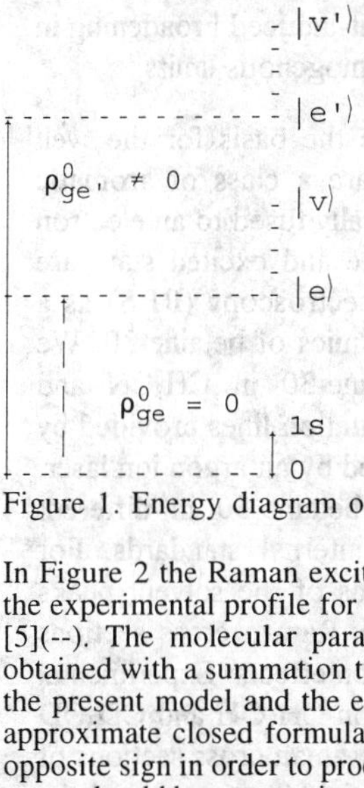

Figure 1. Energy diagram of Model.

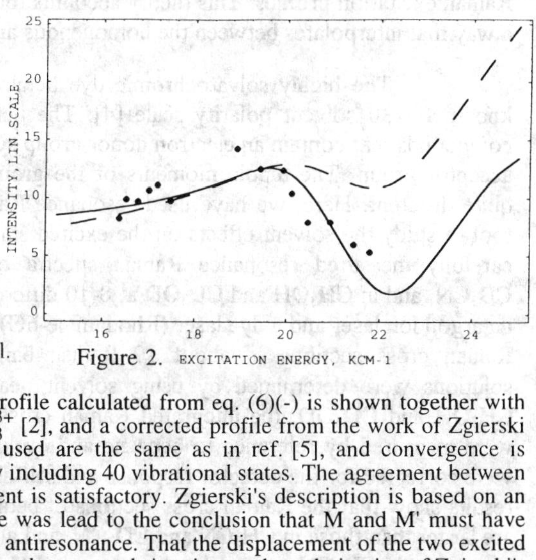

Figure 2.

In Figure 2 the Raman excitation profile calculated from eq. (6)(-) is shown together with the experimental profile for $Co(en)_3^{3+}$ [2], and a corrected profile from the work of Zgierski [5](--). The molecular parameters used are the same as in ref. [5], and convergence is obtained with a summation typically including 40 vibrational states. The agreement between the present model and the experiment is satisfactory. Zgierski's description is based on an approximate closed formula, and he was lead to the conclusion that M and M' must have opposite sign in order to produce an antiresonance. That the displacement of the two excited states should have opposite sign is a rather unusual situation, and a rederivation of Zgierski's formula shows that it contains a simple algebraic error (eq. (12) in ref. [5]). We have corrected this error in the figure, and now both models describe antiresonance with same signs of excited states displacement.

1. P. Stein, V. Miskovski, W.H. Woodruff, K.G. Werner, B.P. Garber, and T.G. Spiro, J. Chem. Phys. 64, 2159 (1976).
2. H. Okamoto, H. Hamaguchi, and M. Tasumi, Chem. Phys. Lett. 130, 185 (1986).
3. S. Hassing and E.N. Svendsen, Chem. Phys. Lett. 86, 342 (1982).
4. S. Hassing and E. N. Svendsen, J. Raman Spectrosc. 24, 675 (1993).
5. M. Z. Zgierski, J. Raman Spectrosc. 6, 53 (1977).

RESONANCE RAMAN STUDIES OF INTRAMOLECULAR CHARGE TRANSFER: BETAINE-30

Yaping Zong and Jeanne L. McHale

Department of Chemistry, University of Idaho,
Moscow, ID 83843, U. S. A.

The application of resonance Raman spectroscopy to the study of dynamics, and the utilization of Heller theory for that purpose, is largely due to the pioneering work of Mathies, Myers and their coworkers [1]. Recently, Myers et al. [2] have further refined the approach by incorporating the stochastic lineshape theory of Mukamel [3] into the direct modeling of absorption and Raman excitation profiles. This theory accounts for solvent-induced broadening in a way that interpolates between the homogenous and inhomogenous limits.

The highly solvatochromic dye betaine-30 is the basis for the well known ET-30 solvent polarity scale [4]. The betaines are a class of aromatic compounds that contain an electron donor group synthetically fused to an electron acceptor group. The dipole moments of the ground state and excited state are quite different. Here, we have used resonance Raman spectroscopy (RRS) as a tool to study the solvent effects on the excited state dynamics of betaine-30. We carefully measured resonance Raman spectra of betaine-30 in CH_3CN and CD_3CN, and in CH_3OH and CD_3OD at 8-10 different excitation lines provided by an argon ion laser and a dye laser (Rhodamine-6G) pumped by an argon ion laser. Raman cross sections of about 10 Raman bands of betaine-30 in different solutions were determined by using solvent peaks as internal standards. For CH_3OH and CD_3OD, the integrated Raman cross sections of the solvent peaks were measured by reference to benzene and cyclohexane. Raman cross sections were corrected for the detector response and self-absorption effects. Experimental results show that the Raman cross sections of betaine-30 in CH_3OH and CD_3OD are larger than those in CH_3CN and CD_3CN, and also the Raman cross sections of betaine-30 in CD_3CN are smaller than those in CH_3CN. Initial efforts to model the experimental profiles have used transform theory (see fig. 1-2). The results of using time-dependent wavepacket theory fitting will be presented. The experimental results show that the displacements along the vibrational modes between the ground and resonant electronic states of betaine-30 are larger in CH_3CN than in CD_3CN. It is necessary to include the non-Condon term in the process of fitting the experimental Raman excitation profiles. The parameters resulting from the application of time-dependent wavepacket theory will be used to clarify the solvent effects on the dynamics of the excited state of betaine-30.

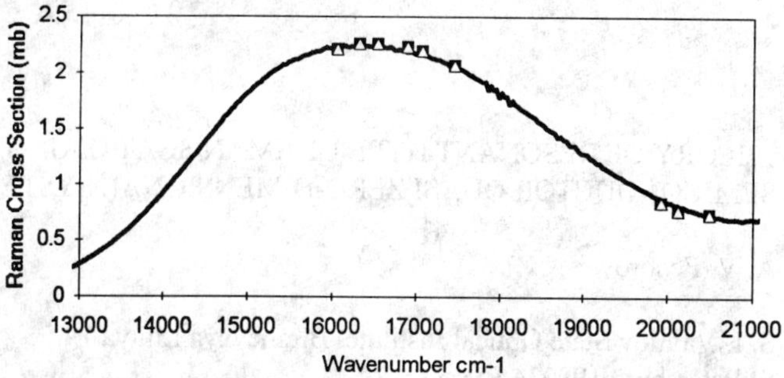

Figure 1. Experimental Raman cross section data (△) and transform theory fitting (line) of 1622 cm^{-1} mode of betaine-30 in CH_3CN.

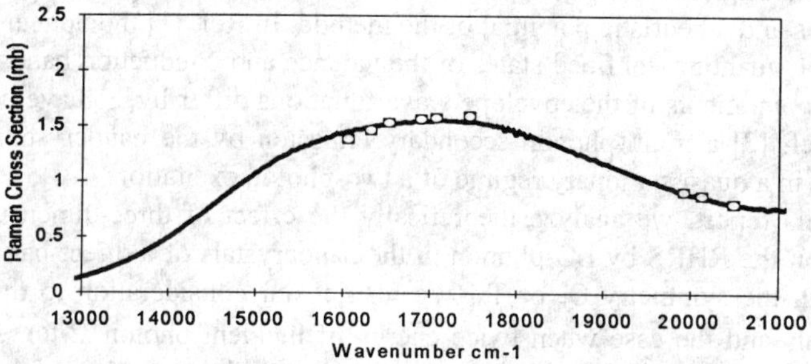

Figure 2. Experimental Raman cross section data (□) and transform theory fitting (line) of 1622 cm^{-1} mode of betaine-30 in CD_3CN.

References:

1. A. B. Myers and R. A. Mathies , "Resonance Raman Intensities: A Probe of Excited State Structure and Dynamics in Biological Applications of Raman Scattering" in Biological applications of Raman Scattering, Vol. 15, p. 1-58, (1987).
2. F. N.S. Ferris, I. R. Gould and A. B. Myers, J. Am. Chem. Soc., 114, 6208 (1992).
3. J. Sue, Y. J. Yan and S. Mukamel, J. Chem. Phys., 85, 462 (1986).
4. C. Reichardt, in Molecular Interactions , Edited by H. Ratajczak and W. J. Orville-Thomas, Vol. 3, p. 241-282, (1982).

THEORY OF RESONANT HYPER-RAMAN SCATTERING BY SEMICONDUCTOR QUASI ZERO-DIMENSIONAL SYSTEMS

A. V. Fedorov

S. I. Vavilov State Optical Institute, Birzhevaya Liniya 12,
St.Petersburg 199034 Russia

Experimental studies of semiconductor quasi zero-dimensional systems (nanocrystals embedded in a dielectric medium) by a technique of resonant hyper-Raman scattering (RHRS) [1], [2] have demonstrated both the unique spectroscopic possibilities and a heuristic potential of the method. In Ref. [1] the optical transitions between the quantum-confined states of the valence and conduction bands for which the angular moments of the envelope wave functions differ by ±1 have been identified. In Ref. [2] a multi-phonon secondary emission by the nanocrystals has been first found in a quasi stationary regime of a two-photon excitation.

In this report, we analyze theoretically the effect of three-dimensional confinement on the RHRS by LO-phonon in the nanocrystals of a direct-band semiconductor with the symmetry O_h or T_d. We restrict our consideration to the spherical nanocrystals and the case when twice energy of incident photon $2\hbar\omega$ falls into a region of the lowest interband transitions. Analytical expressions for the cross-section and excitation profile of the RHRS have been derived in the cases of a strong confinement regime (SCR) $R \ll R_{ex}$ or week confinement regime (WCR) $R \gg R_{ex}$, where R is the nanocrystal radius and R_{ex} is exciton Bohr radius of the bulk material. In the first regime an uncorrelated motion of the electrons and holes determines an energy spectrum of nanocrystal, but in the second one the Coulomb correlation leads to the rise of exciton levels. It have been shown that the RHRS process shares in either case both the common points and the essential distinctions.

Without regard to the confinement regimes, two main channels of scattering contribute to the RHRS by the nanocrystals of cubic semiconductors, e.g. CuCl or CuBr.

The first channel exists due to the fact that, in semiconductors of the O_h and T_d symmetry, one-photon transitions are allowed in the dipole approximation. For this channel the typical process of Stokes scattering may be described as follows: the first photon $\hbar\omega$ creates electron-hole pair (exciton), then the electron or hole absorbs the second photon $\hbar\omega$ and emits LO-phonon $\hbar\Omega$ by the successive "intraband" transitions, and finally the electron-hole pair (exciton) annihilates with the emission of scattered photon $\hbar\omega_S = \hbar(2\omega - \Omega)$. It is easily seen that only non-diagonal part of Fröchlih-like interaction involving LO-phonons with angular moment l_p=1 determines the first channel of the RHRS. In the SCR the dipole allowed interband one-photon transition joins the quantum-confined states of the valence and conduction bands with the same angular moment of the envelope wave functions while the one-photon intraband transition changes the angular moment by 1. It follows that the electron-hole pair will annihilate only when the angular moment turns back to its original value as a result of LO-phonon emission. In the WCR the excitons are created (annihilated) in quantum-confined states with the angular moment l=0 at the dipole allowed one-photon transition, therefore, a change of the angular moment by 1 at one-photon exciton-exciton ("intraband") transition should be compensated by emission of LO-phonon with l_p=1.

The second channel of the RHRS exists in non-centrosymmetric (e.g., T_d) semiconductors due to the fact that both one- and two-photon transitions are allowed in dipole approximation. For this channel the typical process of Stokes scattering may be represented as follows: the first and second photon $\hbar\omega$ simultaneously create electron-hole pair (exciton), then the electron or hole emits LO-phonon $\hbar\Omega$ by the "intraband" transition, and finally the electron-hole pair (exciton) annihilates with the emission of scattered photon $\hbar\omega_S = \hbar(2\omega - \Omega)$. Notice that similar process, only without LO-phonon, is responsible for a resonant second harmonic generation in non-centrosymmetric semiconductors. All features of electron-hole generation (radiative recombination) at one-photon dipole allowed transitions described in earlier section hold also at two-photon dipole allowed transitions. Consequently, both the diagonal and non-diagonal part of Fröchlih-like interaction involving LO-phonons with angular moment l_p=0 determines the second channel of the RHRS. In the SCR the RHRS looks like a resonant Raman scattering discussed in Ref.[3] and to obtain a nonvanishing value of RHRS cross-section, the spherical potential with finitely high walls should be considered. In this case, only non-diagonal electron-phonon interaction causes the RHRS. By contrast, in the WCR, the diagonal part of Fröchlih-like interaction makes an essential contribution to the RHRS. The reason is that for exciton a spatial localization of electron is distinct from that of hole. Then the contributions to the RHRS cross-section from the processes in which the electron emits LO-phonon and the hole emits LO-phonon do not cancel each other. This situation looks like the RHRS by molecules or impurity centers in crystals. It is highly probable that the diagonal part of Fröchlih-like interaction determines the multi-phonon secondary emission by the CuBr nanocrystals [2].

Finally, the derived analytical results allow ones to interpret qualitatively the main features of the RHRS by LO-phonon from the quasi zero-dimensional systems.

This work was supported by the Russian Basic Research Foundation (RBRF), Grant No. 95-02-03821a.

1. A.V. Baranov, K. Inoue, K. Toba, A. Yamanaka, V.I. Petrov, and A.V. Fedorov, Phys. Rev. B 52, (DEC 15, 1995).
2. K. Inoue, A. Yamanaka, K. Toba, A.V. Baranov, A.A. Onushchenko, and A.V. Fedorov, Phys. Rev. Lett., in press.
3. M.P. Chamberlain, C. Trallero-Giner, and M. Cardona, Phys. Rev. B 51, 1680 (1995).

THEORY AND CALCULATION OF INTENSITY DISTRIBUTION IN RESONANCE RAMAN SCATTERING SPECTRA OF POLYATOMIC MOLECULES

T.Burova

Saratov State Pedagogical Institute,
Michurin Str. 92, 410071 Saratov, Russia

The direct use of Kramers-Heisenberg expression for theoretical description and calculation of resonance Raman spectra (RRS) is associated with decision of such question as the determination of the dipole moment matrix elements and sums over vibration quantum numbers of excited electronic states (EES), i.e. the Green function matrix elements. In this paper the intensity distribution in RRS spectra is analysed using the method of direct quantum-mechanical calculation of relative intensities suggested in [1]. This method permits one to determine the relative intensities on the basis of the dipole moment matrix element representation [2], taking into account the Herzberg-Teller effect. Such technique is the extension of theory of vibronic interaction, usually applied to absorption and fluorescence spectra, to RRS as the spectra of similar physical nature. It makes possible to describe all these spectra from one point of view with the same set of parameters, such as vibrational and geometrical parameters of ground electronic state and electronic structure data. No adjusting parameters are used except for standard parametrization of electronic structure calculation method.

Herzberg-Teller effect is responsible for appearing of RRS spectrum in the case of resonance between frequency of exciting radiation and frequency of forbidden transition. This effect also explains the presence of spectral lines which correspond to excitation of single-quantum nontotally symmetric vibrations, their odd overtones and combinations and corrects the values of other line intensities.

Sums over vibrational quantum numbers of EES are calculated as functions of vibrational parameters of ground electronic state and frequency of exciting radiation without direct summation. Both frequency effect and Dushinsky effect are taken into account. It is possible due to proposed method, representing the Green function matrix element as a product of two parts. The first part is the same for all vibration lines, but the second one is dependent on the exciting vibration mode. To obtain the closed expression for total multimode Green function some admissions are needed. So the square of difference between Dushinsky matrix and the unit matrix is proposed small as well as the square of vibration frequency difference in EES and in ground electronic state related to the value of vibration frequency in ground state. Utilizing the Green function approach [3] to such situation it is possible to represent the multimode oscillatory Green function as a combination of degenerate hypergeometric functions. Vibration frequences in ground electronic state, Dushinsky parameters and the exciting radiation frequency are used in the arguments of these functions.

Such technique is used to analyse the intensity distribution in RRS spectra of pyrazine excited by lazer radiation of 337 nm wavelength (resonance with lowest singlet excited electronic state (EES)) and 266 nm wavelength (resonance with second EES). Vibrational frequences from [4] in the lowest EES are used as well as calculated frequences for the second one.

The most considerable relative line intensities of pyrazine RRS spectra calculated in accordance with proposed method together with experimental results [5] are given in Table 1. It's important, that the presence of spectral lines corresponding to excitation of nontotally symmetric vibrations (for example, 10a and 4) can be explained only by Herzberg-Teller effect. Comparison of the calculated and experimental results in Table 1 shows their good qualitative agreement. So, the systematic accounting for the vibronic interaction yields adequate description of RRS intensity distribution and the proposed method of Green function matrix elements calculation makes possible to realize this approach.

TABLE 1. Relative intensities of pyrazine RRS spectra

freq. (cm^{-1})	assignment	relative intensity			
		resonance with first EES		resonance with second EES	
		calcul.	experiment [5]	calcul.	experiment [5]
2028	2*I	0.19	-	0.43	m
1838	2*10a	0.48	m-w	0.02	-
1827	9a+6a	0.10	w	0.02	-
1788	3*6a	0.07	-	0.14	w
1610	I+6a	0.09	-	0.77	s
1578	8a	0.25	m-w	0.06	-
1515	6a+10a	0.11	w	0.08	-
1231	9a	0.72	s	0.08	-
1192	2*6a	0.16	w	0.24	w
1014	I	0.45	m	1.00	s
919	10a	1.00	v.s.	0.17	w
705	4	0.12	w	0.09	w
596	6a	0.37	m	0.89	s

1. M.V.Priyutov and T.G.Burova, Opt. Spektrosk. 64, 182 (1988).
2. M.A.Kovner and M.V.Priyutov, Dokl. Akad. Nauk SSSR 204, 634 (1972).
3. Y.J.Yan, S.Mukamel, J. Chem. Phys. 85, 5908 (1986).
4. V.I.Berezin, Opt. Spektrosk. 16, 240 (1964).
5. I.Suzuka, Y.Udagawa and M.Ito, Chem.Phys.Lett. 64, 333 (1979).

NONLINEAR ROTATION SPECTRA OF RESONANCE RAMAN SCATTERING OF INTENSE LASER LIGHT FROM DIATOMICS

A.I. Pegarkov [1*] and I.I. Tugov [2]

[1] Angewandte Laserphysik, Fakultät für Physik, Universität Bielefeld, Universitätsstraße 25, D-33615 Bielefeld, Germany
Department of Theoretical Physics, Voronezh State University, 1 University Square, Voronezh 394693, Russia
[2] General Physics Institute, Russian Academy of Sciences, 38 Vavilov Street, Moscow 117942, Russia

The excited electronic state of real diatomics is a complicated system of crossing and interacting molecular terms. An important problem of the theory is to develope the quantitative methods for calculation of the spectra including the nonadiabatic interactions of the molecular terms and rotations of the molecule. The nonadiabatic effects can be induced by following two things: 1) by motion of intramolecular particles (dynamic nonadiabatic effects), 2) by the incident laser radiation (radiative nonadiabatic effects). In this report there is presented the approach of multi-channel Green functions of nuclear motion [1,2] for the calculation.

The nonlinear rotation spectra, involved the dynamic coupling, were studied for the example of the IBr molecule

$$IBr(1\Sigma^+) + h\nu \Rightarrow IBr(B'O^+) \Rightarrow IBr(1\ \Sigma^+) + h\nu' . \qquad (1)$$

The dynamic nonadiabatic interaction couples the intermediate attractive $B(^3\Pi O^+)$ and repulsive O^+ terms. Fig. 1 displays the intensity distribution of rotation RRS overtones I_{RRS} (arb.units) for different magnitudes of the dynamic coupling V_d. Fig.2 does the intensity of various RRS transitions I_{RRS} versus strength of the dynamic coupling. Fig. 3 shows the dependence of depolarization ratio ρ on the dynamic coupling.

The nonlinear rotation RRS spectra, involved the radiative nonadiabatic coupling, were considered for the example of the Ar_2^+ molecule

$$Ar_2^+(2\Sigma_{1/2u}^+) + h\nu \Rightarrow Ar_2^+(2\Sigma_{1/2g}^+) \Rightarrow Ar_2^+(2\Sigma_{1/2u}^+) + h\nu' . \qquad (2)$$

The term $2\Sigma_{1/2g}^+$ is repulsive. Fig.4 displays the rotation RRS cross section σ_{RRS} (arb. units) in dependence on the laser strength F_r. The nonlinear behavior of the rotation RRS cross section is due to the laser-induced generation of quasi-bound states of nuclear motion in the nuclear continuum. These states appear as a result of the radiative coupling of bound nuclear states in the ground electronic term with the dissociative states in the excited repulsive term.

The work of A.I. Pegarkov was supported by the Alexander von Humboldt

Foundation and the International Center for Fundamental Physics in Moscow (INTAS Grant N 93-2492).

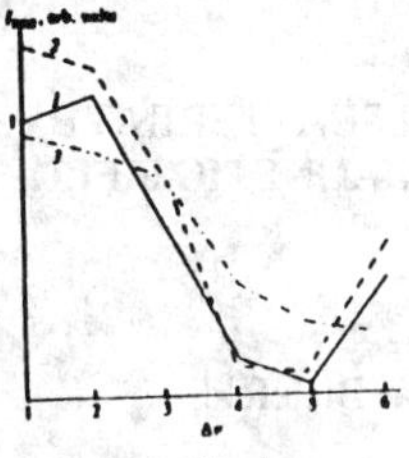

Fig. 1

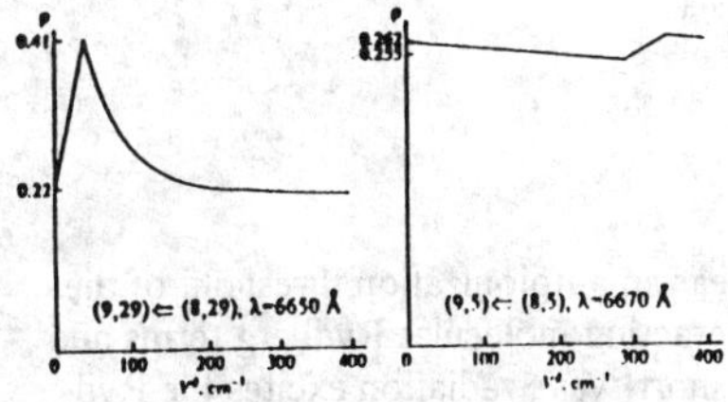

Fig. 3

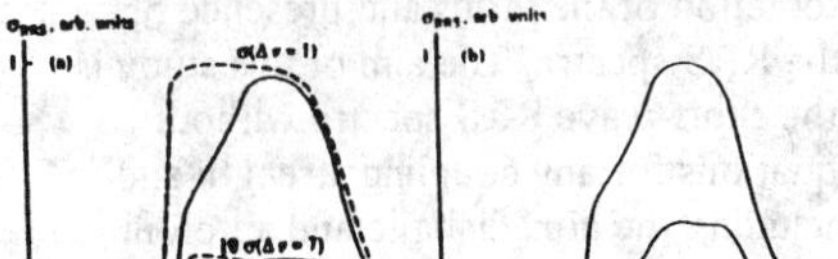

Fig. 4

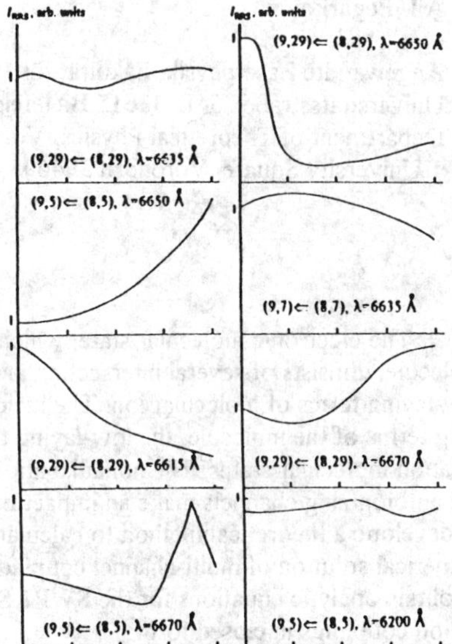

Fig. 2

Fig.1. Intensity distribution of the Q-branch overtones of the Stokes RRS from IBr molecule (1) for different values of dynamic coupling V_d. λ=6635 Å, J=5, V_d = (1) 0, (2) 100, (3) 350 cm^{-1}.

Fig.2. Intensity of Q branch of RRS from IBr molecule (1) for the $(v_2,J_2) \Leftarrow (v_1,J_1)$ RRS transitions vs. strength of dynamic coupling at different wavelengths of incident radiation λ.

Fig.3. Depolarization coefficient for RRS (1) of linearly polarized radiation.

Fig.4. Q-branch cross sections of the Stokes RRS (2) vs. the strength of incident radiation F_r for the overtones $\delta v = v_2 - v_1$ with (-) and without (- -) of molecular rotations. λ=3510 Å.

1. A.I. Pegarkov and I.I. Tugov, in Multiphoton Processes, (D.K.Evans, S.L.Chin, eds.), World Scientific, Singapore, 1994, p.89.
2. A.I. Pegarkov and I.I. Tugov, Opt.Spectrosc. 79, 44 (1995).

THEORY OF SHORT-WAVE RESONANCE RAMAN SCATTERING BY SIMPLE MOLECULES IN AN INTENSE ELECTROMAGNETIC FIELD

A.I. Pegarkov

Angewandte Laserphysik, Fakultät für Physik, Universität Bielefeld,
Universitätsstraße 25, D-33615 Bielefeld, Germany
Department of Theoretical Physics, Voronezh State University,
1 University Square, Voronezh 394693, Russia +

The electronic molecular state, which is near an autoionization threshold of the molecule, consists of several intersecting and interacting molecular Rydberg terms and low-laying terms of molecular ion. The incident short-wave radiation excites the Rydberg terms of the molecule, the low-laying terms of the molecular ion, and the autoionization molecular states. The nonadiabatic interaction of the terms and presence of the autoionizing channels make an impact on the RRS spectra. The aim of this study is to develope a theoretical method to calculate the short-wave RRS spectra without numerical solution of multi-channel coupled equations for any coupling strength and to obtain analytic equations for the SWRRS including the nonadiabatic and autoionization couplings in closed form.

In this report the resonance Raman scattering of short-wave radiation via Rydberg and autoionization states is considered within and beyond a dipole approximation. The analytic equations for electronic-rotation-vibration SWRRS tensor and cross section are obtained by the method proposed in our papers [1,2]. The contributions of molecular Rydberg, ion, and autoionization states are analyzed in detail. The theory, developed within the dipole approximation, may be applied to numerical calculations of the following RRS processes: the scattering of soft X-ray, XUV laser or synchrotron radiation via low-laying Rydberg states ($n=1\div3$) of multicharged molecular ions ($Z\geq10$), the scattering of UV, optical, and infrared radiation from high-Rydberg molecules ($n\geq10$, $Z=1$), the scattering of the XUV and longer waves from low-charged high-Rydberg molecules ($Z=2\div4$, $n\geq10$). For other radiation wavelengths and molecules one may apply the theory developed beyond the dipole approximation. Unfortunately, but an experimental observation of such RRS spectra is unknown, although the SWRRS could be a perspective method to study the spectra and structure of the Rydberg and autoionization states of the molecules.

Weak autoionization coupling. In the case of weak autoionization coupling a high-order perturbation theory is applied to calculate the SWRRS tensor and cross section. The analytic equations are obtained, which allow for the multi-term structure of the excited molecular state and autoionization coupling within the perturbation theory.

+ permanent address

Intense autoionization coupling. The perturbation theory is inapplicable if the nonadiabatic and autoionization couplings are strong or if the distance between vibration-rotation states of the interacting electronic terms is too small. Within our nonperturbative method the equations for SWRRS tensor and cross section are obtained and take into account the couplings which can be more than vibration quantum of the molecule. For a limit of weak coupling these equations take the form of the high-order perturbative results.

Exact resonance effects. Resonance fluorescence. If the incident radiation frequency is near with an electronic-vibration-rotation transition to an autoionization state, the equations for the tensor and cross section take the simpler form. If the field frequency is equal to the frequency of the transition, there is a resonance fluorescence. The cross section of the short-wave resonance fluorescence is obtained.

Discussion. The obtained analytic equations involve sums of the Clebsch-Gordan coefficients, 6-j symbols, and compound matrix elements. Besides an interference of contributions from the intermediate high-laying Rydberg terms and low-laying terms of the molecular ion, these equations allow for the nonadiabatic and autoionization couplings. The total transition integral can increase or decrease as one in the case of a second-order perturbation theory. The analytic equations for the RRS tensor and cross section beyond the perturbation theory involves the nonadiabatic and autoionization couplings through wave functions and Green functions of nuclear motion in the multi-term system without the perturbation theory restriction upon magnitudes of the couplings. In the paper [3] an analytic method to receive these functions under nonadiabatic conditions has been given. This method is applied here to calculate the analytic equations for the SWRRS spectra and intensity.

Our theoretical approach permits to obtain the analytic equations for high-order branches of the scattering and to allow for the dynamic, radiation-induced, and autoionization couplings without the numerical integration of coupled differential equations for nuclear wave functions, developed in [4]. The equations, presented in this report, may be used to evaluate the spectra of following resonance ionization-scattering process [5]:

$$AB + Nh\nu \Rightarrow AB^{+} + e^{-} + (N-1)h\nu + h\nu' .$$

The work of A.I. Pegarkov was supported by the Alexander von Humboldt Foundation and the International Center for Fundamental Physics in Moscow (INTAS Grant N 93-2492).

1. A.I. Pegarkov and I.I. Tugov, Chem.Phys.Lett., 203, 456 (1993).
2. A.I. Pegarkov and I.I. Tugov, Opt.Spectrosc., 79, 44 (1995).
3. A.I. Pegarkov, Bull.Russian Acad.Sci.Phys.Ser., 54, 1217 (1992).
4. A.D. Bandrauk, Int.Rev.Phys.Chem., 13, 123 (1994).
5. A.I. Pegarkov and I.I. Tugov, prepared for "J.Raman Spectrosc.".

INFLUENCE OF COLLECTIVE DECAY OF EXCITED ELECTRONIC STATES OF THE CHROMOPHORES UPON DISPERSION CHARACTERISTICS OF RESONANCE RAMAN SCATTERING BY CHROMOPHORE DIMERS WITH A DYNAMIC COUPLING BETWEEN CHROMOPHORES

V.A.Morozov

Institute of Organic Chemistry,
Russian Academy of Sciences,
Leninskii Prospekt 47, Moscow 117334, Russia

Theoretical treatment of Raman scattering by molecules or molecular dimers consisting of two identical chromophores - "chromophore dimers" or "bichromophores" - have been given in the literature [**1**,**2**] for the cases of two different types of interactions between the chromophores. The first type interaction described using coupling energy operator has been considered using quantum theory [**1**]; the corresponding matrix elements of this operator were supported to be real. The second type of interaction described using off-diagonal elements of a damping matrix of excited electronic states of the bichromophore (written as the product of the wavefunctions of the chromophores) has been considered using quantum electrodynamic theory [**2**]. The first type interaction (we shall call it "dynamic" to distinguis it from the second type of interaction, which we shall call "collective decay" or "common damping") causes the excitation profiles (EP) of Raman lines (RL) to separate into two components relative to the EP monomer because of doublet splitting of the degenerate electronic energy levels of the bichromophore. This interaction also leads to dispersion of the depolarization ratio of the RL in the frequency range between components of the EP, when the vectors of the transition dipoles of the chromophores are not parallel or anti-parallel. The extremes of these dispersion curves are at the frequency of the pure electronic transition, i.e. at the minima of the EP in the frequency field examined. On the other hand common damping only changes the intensity and width of the EP band of bichromophoric RL at the frequency of the pure electronic transition, and leads to the dispersion of depolarization ratio with extremes of the dispersion curves being at the same frequency, i.e. at the maxima of the EP in the frequency field examined.

Most bichromophores, however, belong to molecular compounds, in which both above type interactions act simultaneously. Here, we report some results of the theoretical study of the dispersion characteristics of RL for such bichromophores with identical chromophores.

Expressions for the EP and dispersion curves of the depolarization ratio of RL as well as for the Raman optical activity signal are obtained for freely orienting above mentioned rigid chromophore dimers. The frequency region near the frequency of the pure electronic transition of the chromophores is considered. It is shown that in such chromophores the collective decay leads to suppression and widening one of two components of EP (separeted by dynamic coupling) and to enhancement and narrowing the other. There is the same influence on intensity and width of two components of dispersion curves of Raman optical signal. The collective decay also causes shifting of the extremes of dispersion curves of the depolarization ratio of the RL towards the enhancing components of the EP and a decrease in width, when the vectors of the transition dipoles of the chromophores are not parallel or antiparallel. Calculated examples are presented.

1. A.R. Gregory, W.H. Henneker, W. Siebrand, M.Z. Zgiersky,

J. Chem. Phys. **63**, 5475 (1975).

2. V.A. Morozov, J. Raman Spectrosc. **24**, 585 (1993).

SECTION 3

TIME-RESOLVED RAMAN SCATTERING and TRANSIENT SPECIES

DYNAMIC STRUCTURE OF PHOTOEXCITED MOLECULES: THE OUTCOME OF PICO/NANOSECOND TIME-RESOLVED RAMAN STUDIES IN THE PAST FIVE YEARS

H. Hamaguchi

Department of Basic Science, The Graduate School of Arts and Sciences, The University of Tokyo, 3-8-1 komaba, Meguro, Tokyo 153 and Molecular Spectroscopy Laboratory, Kanagawa Academy of Science and Technology, KSP East 301, 3-2-1 Sakato, Kawasaki 213, Japan

In the past five years, much effort has been directed in our laboratory toward the development of new time-resolved vibrational spectroscopies in the pico- to milli-second time regimes. These include transform-limited picosecond time-resolved Raman spectroscopy [1], picosecond time-frequency two-dimensional multiplex CARS spectroscopy [2], nanosecond dispersive infrared spectroscopy [3], and millisecond multichannel Fourier-transform infrared spectroscopy [4]. With a full use of these new spectroscopies, we accumulated considerable amount of vibrational data on a variety of photoexcited active molecules in solution. Quite noteworthy among these vibrational data is the fact that the band widths (here we focus on Raman band widths) of photoexcited molecules are often much more broad than those of the ground-state molecules. As a representative example, the Raman spectra of N,N-dimethyl-4-aminobenzonitrile (DMABN) in butanol are shown for the excited triplet state (Fig. 1, upper) and for the ground state (Fig. 1, lower). On going from the ground state to the triplet, the width of the C≡N stretch band increases by a factor of two or more.

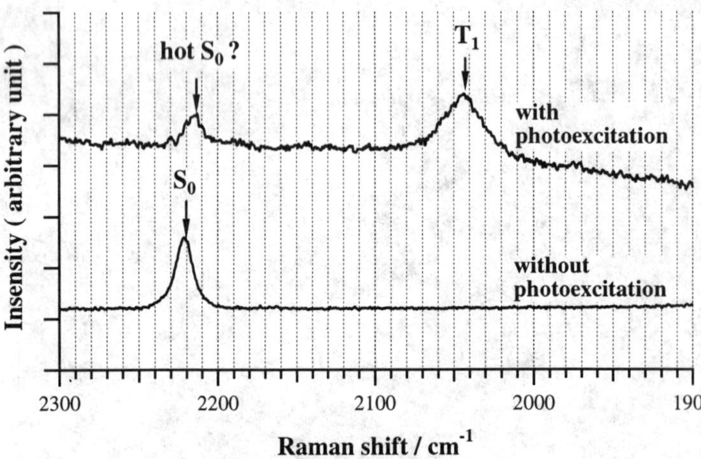

Fig. 1. Time-resolved Raman spectrum of T_1 DMABN (upper, 20 ns after photoexcitation) and Raman spectrum of DMABN in butanol.

This marked band broadening in the excited triplet state of DMABN can be attributed to a vibrational dephasing mechanism that is associated with the electronic excitation. The excited triplet state of DMABN in a polar solvent is known to have a polarized structure due to the intramolecular charge transfer from the dimethyl amino group to the benzonitrile moiety [5]. This polarized structure is expected to fluctuate according to the fluctuation of the solvent field and so is the C≡N stretch frequency. Then, the C≡N stretch frequency is modulated by the solvent fluctuation, leading to effective vibrational dephasing that causes the Raman band broadening. Note that such a mechanism of vibrational dephasing is associated only with highly polarizable molecules like DMABN in the excited triplet state and not with ordinary ground state molecules.

The Raman band broadening due to dynamic polarized structure has also been found for the excited triplet charge-transfer complex of chloranil with alkylbenzenes [6]. For this system, we previously showed that some Raman frequencies of the donor (triplet chloranil) changes gradually with the electron-donating ability of the accepter (alkylbenzenes) according to the degree of partial charge transfer from the donor to the accepter. Though not explicitly shown in our previous publication, we noticed that the Raman band widths gradually increase with increasing the degree of charge transfer. This increase of the band width indicates that, the more the complex is polarized, the more the structure is affected by the solvent fluctuation, resulting in the more effective dephasing of vibrations.

In order to quantitatively analyze the solvent-induced dynamic polarization and vibrational dephasing of electronically excited molecules described above, we constructed a theoretical model that expresses a Raman band shape in terms of two dynamic parameters, the mean frequency of the solvent fluctuation and the mean phase shift for each fluctuation [7]. The model explains very well the experimental results for the third example, S_1 trans-stilbene, for which many sets of time, solvent, and temperature dependent Raman band shape data are available. The solvent dependent shift of the peak position as well as the concomitant change of the band width are very well reproduced. The model also accounts for the solvent dependent rate of isomerization of S_1 tSB, suggesting that the dynamic polarization plays a crucial role not only in vibrational dephasing but also in the process of isomerization.

We believe that the concept of dynamic structure can be extended to many other photochemical systems to account for the structure-reactivity relationships. We note that the concept is built solely upon the accumulated time-resolved Raman data through the consideration of the observed band shapes. The Raman band shapes and the information on vibrational dephasing derived therefrom are certainly a portion of Raman resources that should be more utilized in the future time-resolved Raman spectroscopy.

1. K. Iwata and H. Hamaguchi, Rev. Sci. Instr., **64**, 2140 (1993).
2. T. Tahara and H. Hamaguchi, Rev. Sci. Instr., **65,** 3332 (1994).
3. T. Yuzawa, C. Kato, M. W. George, and H. Hamaguchi, Appl. Spectrosc., **48**, 684 (1994).
4. M. Hashimoto and H. Hamaguchi, Appl. Spectrosc., in press.
5. M. Hashimoto and H. Hamaguchi, J. Phys. Chem., **99**, 7875 (1995).
6. T. Tahara and H. Hamaguchi, J. Phys. Chem., **96**, 8252 (1992).
7. H. Hamaguchi, Mol. Phys., submitted.

PHOTOCHEMISTRY OF BIPYRIDINES AND PHENYLPYRIDINES IN ALCOHOLS

G. Buntinx*, R. Naskrecki and O. Poizat

LASIR, CNRS, USTL, 59655 Villeneuve d'Ascq, France

Aza-aromatic derivatives of biphenyl such as phenylpyridines and bipyridines are model compounds with simple and symmetric structure for studying the reactivity of the imine function involved in larger heterocyclic molecules of biological interest (e.g. porphyrins, nucleotide bases), in organometallic complexes (MLCT, mixed-valence complexes), or in polymeric conductors such as polypyrrole or polyaniline. The photochemistry of this class of compounds implicates both the lowest singlet (S_1) and triplet (T_1) states and depends strongly on the nature of the solvent. One reason for this effect is the presence of two close-lying $n\pi^*$ and $\pi\pi^*$ levels with relative energies varying with the solvent polarity and acidity. We present here an overview of results obtained by UV-visible transient absorption and time-resolved resonance Raman spectroscopies related to the photoreactivity of 4,4'-bipyridine (44BPY), 2,2'-bipyridine (22BPY) and 4-phenylpyridine (4PPY).

In all cases, transient measurements in the nanosecond range following $S_1 \rightarrow S_n$ excitation are essentially characteristic of the T_1 state and of a N-hydro radical $RH^{\cdot}$ (Figures 1-3). The T_1 Raman data are in agreement with the ab initio spectra calculated for a $\pi\pi^*$ configuration. In the case of 44BPY [1] the formation of the radical species shows a double - fast and slow - kinetics. The slow component results

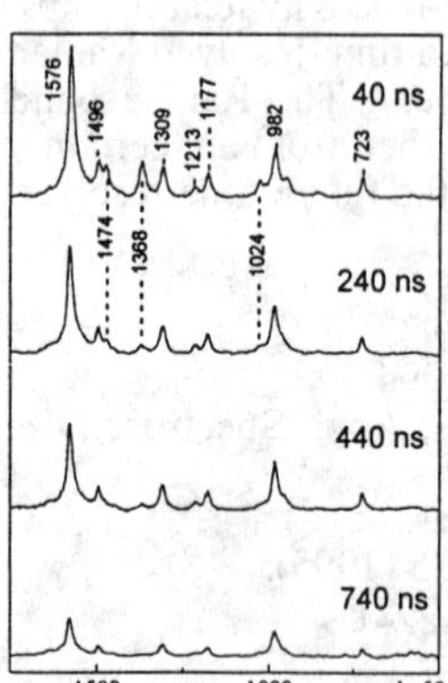

Figure 1- transient Raman spectra of 4PPY in methanol (pump 248nm, probe 370nm)

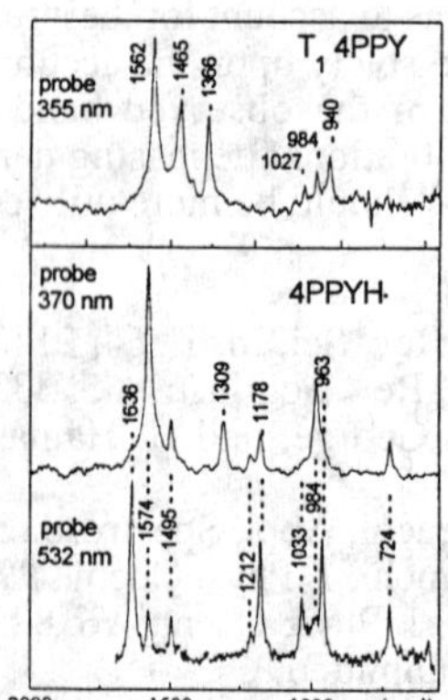

Figure 2- transient Raman spectra of T_1 4PPY and $4PPYH^{\cdot}$

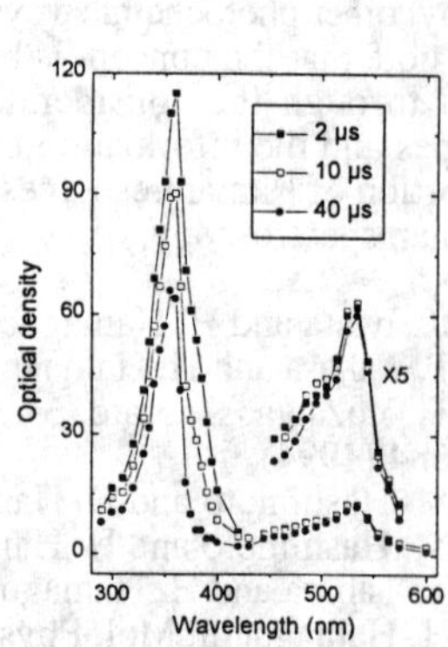

Figure 3- transient absorption spectra of 4PPY in methanol (pump 248nm)

from a reduction of the triplet state via hydrogen-atom abstraction from the alcohol. Isotopic effects on the radical Raman frequencies in deuterated alcohols indicate that this reaction arises from the homolytic break of an alcohol CH bond in position α to the hydroxyl group. The fast component is unresolved at this time resolution. The 22BPYH˙ radical yield is negligible. As shown in Figures 1 and 3 for methanol solutions, the 4PPYH˙ radical (40-μs spectrum in Fig.3) is not produced from the T_1 state but with the fast kinetics exclusively.

Transient absorption measurements using a femtosecond set-up (Ti:sapphire laser, ≈100-fs, 50 μJ pump pulses at 266 nm (third harmonic) for a repetition rate of 1 kHz, fwhm of the pump/probe intensity cross correlation at the sample ≈ 300 fs) allow to characterize the S_1 state spectrum and to follow its reactivity. In acetonitrile one observes the decay of the S_1 state and the simultaneous rise of the T_1 spectrum. The short S_1 lifetimes (Table I) compared to biphenyl (16 ns), indicating much faster ISC processes, suggests a nπ* nature for the lowest singlet state. In alcohols, comparable behavior is observed for 22BPY whereas the spectra of 44BPY (Figure 4) and 4PPY reveal a much faster S_1 decay rate (in methanol τ_{S_1} = 11.5 and 28 ps, respectively), and show the rise of the N-hydro radical from the S_1 state in parallel to the T_1 species. The T_1 yield is lowered in accordance to the reduction of the S_1 lifetime, i.e. the ISC rate constant remains comparable in acetonitrile and in alcohols. Quantitative measurements of the

Table I- S_1 properties in CH_3CN

	τ_{S_1} (ps)	λ_{max} (nm)
44BPY	41	375, 580
22BPY	64	365, 530
4PPY	100	370, 575

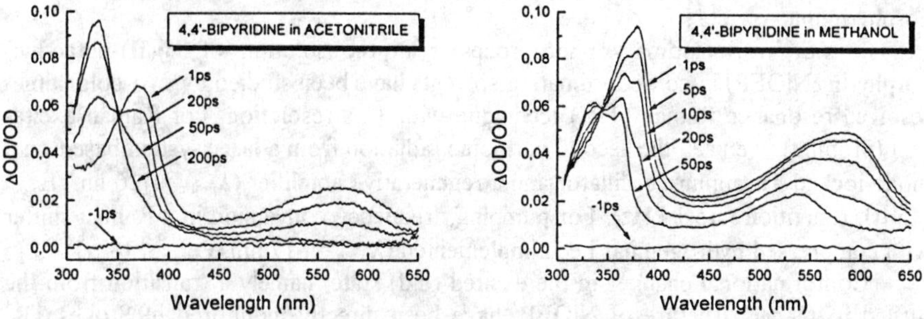

Figure 4- Transient absorption spectra of 44BPY in acetonitrile (left) and methanol (right)

S_1 lifetime and of the relative triplet and radical yields in a series of alcohols and in binary solutions such as alcohol/acetonitrile or alcohol/water leads to the following conclusions: the shortening of the S_1 lifetime in alcohols is due for a large part to quenching by the radical formation process. The rate constant estimated in the range 1×10^{10} to 5×10^{10} s^{-1} is proportional to the protic character of the alcohol. The mechanism involves probably a H-atom transfer from the hydroxyl group of an alcohol molecule H-bonded to the heterocyclic species in the ground state.

1. O. Poizat, G. Buntinx, P. Valat, V. Wintgens and M. Bridoux, *J. Phys. Chem.* 97, 5905 (1993)

TIME-RESOLVED STOKES/ANTI-STOKES RESONANCE RAMAN STUDY OF PRIMARY PHOTOPROCESSES IN THE SOLUTIONS OF Ni-PORPHYRINS

S.G. Kruglik[1,2]*, Y. Mizutani[1], V.A. Orlovich[1,2], and T. Kitagawa[1]

[1]Institute for Molecular Science, Okazaki National Research Institutes, Okazaki 444, Japan
[2]Permanent address: B.I.Stepanov Institute of Physics, Academy of Sciences of Belarus, 70 F.Skaryna Ave., Minsk 220072, Belarus

Ni-porphyrins (NiP) are considered for a long time as an exemplary system for studying both the ground-state features and the excited-state processes in heme proteins and related metalloporphyrin complexes. Photophysics of NiP has already been intensively investigated: relaxation after visible excitation exhibits biphasic kinetics, with the long-lived ($\tau \approx$ 250-300 ps) excited (d,d) state playing the key role in deactivation processes [1-3]. However, the nature of a short-lived transient species ($\tau \leq$ 10 ps) is still not clear; several controversial interpretations include either electronic relaxation within the excited states [1-3] or non-electronic relaxation due to vibrational heating and cooling [4,5] or due to conformational changes in the porphyrin macrocycle [6].

In our work, primary photoprocesses in the solution of Ni(II)-octaethyl-porphyrin (NiOEP) in non-coordinating solvents have been studied by two-color time-resolved resonance Raman (TR^3) technique with 3-ps resolution. For Raman excitation (probing), we used the second harmonic radiation from a laser system based on a mode-locked Ti:sapphire oscillator and a regenerative amplifier (λ_{probe}= 426 nm, E_{probe} $\leq$ 5 μJ, repetition rate 1 kHz). For pumping, frequency conversion in a Raman shifter with compressed hydrogen has been implemented (λ_{pump}= 517 nm, $E_{pump} \leq$ 18 μJ).

Conformational changes in the excited (d,d) state, namely a transition from the ruffled to planar structure of NiOEP, have been directly monitored in Stokes TR^3 spectra (Fig.1). Indeed, contour of the mode ν_{10}^* at first 10 ps after pumping is very broad and has two maxima at $\approx$1616 and $\approx$1625 cm^{-1}, while at longer time delays Δt it becomes narrower with the only one maximum located at 1626-1629 cm^{-1}.

Vibrational heating/cooling processes have been studied in anti-Stokes TR^3 spectra of NiOEP (Fig.2), and several conclusions can be derived from the data obtained. First, rates of anti-Stokes intensity changes for the modes ν_4^* and ν_7^* were found to be quite different, mode ν_4^* being dominant at first 5 ps after pumping, while mode ν_7^* - at longer Δt. Second, heating effect induced by the probing light (426 nm, spectrum at Δt=-10 ps) is different from that induced by the pump light (517 nm, difference spectra at $\Delta t \geq$ 0 ps). At last, we observed non-zero residual anti-Stokes TR^3 difference spectra at very long time delays of 50-200 ps.

Competition between various physical processes (electronic relaxation, vibrational heating and conformational changes) taking place at the early times following photoexcitation are discussed on the basis of the results obtained. Preliminary TR^3 studies have also been made for NiTPP in toluene.

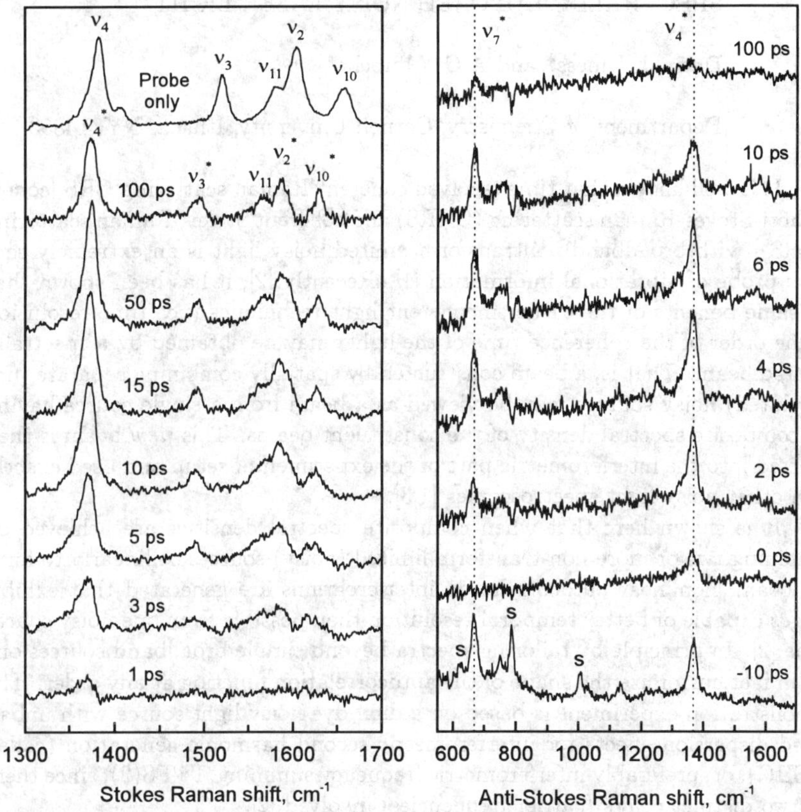

Fig. 1. (Left) Stokes TR3 difference spectra of NiOEP in toluene. Spectra were reconstructed from the experimentally measured ones recorded at different Δt.

Fig. 2. (Right) Anti-Stokes TR^3 difference spectra of NiOEP in toluene. Lower trace is an experimental spectrum recorded when probe pulse comes before pump, and heating is caused by the probing pulse itself. "S" indicates solvent bands.

1. T.Kobayashi, K.D. Straub, P.M. Rentzepis, Photochem. Photobiol. 29, 925 (1979)
2. V.S. Chirvony, B.M. Dzhagarov, A.M. Shulga, G.P. Gurinovich, Chem. Phys. Lett. 70, 79 (1980).
3. D. Kim, C. Kirmaier, D. Holten, Chem. Phys. 75, 305 (1983).
4. J. Rodriguez, D. Holten, J. Chem. Phys. 91, 3525 (1989).
5. S. Sato, T. Kitagawa, Applied Physics B59, 415 (1994)
6. S.H. Courtney, T.M. Jedju, J.M. Friedman, R.G. Alden, and M.R. Ondrias, Chem. Phys. Lett. 164, 39 (1989).

TIME RESOLVED COHERENT RAMAN SCATTERING USING SPECTRALLY TAILORED NOISY LASER LIGHT

Darin J. Ulness* and A.C. Albrecht*

Department of Chemistry, Cornell University, Ithaca, NY, 14853

It is well known that time resolved coherent Raman scattering CRS [coherent anti-Stokes-Raman scattering (CARS) and coherent Stokes-Raman scattering (CSRS)] with broadband nontransform limited noisy light is an extremely sensitive probe of vibrational information [1]. Recently [2], it has been shown that the same benefits of the earlier incoherent light techniques (i.e. time resolution on the order of the coherence time of the light) may be obtained by a spectrally tailored beam. That is, a beam constructed by spatially combining separate (uncorrelated) noisy sources may be viewed as a beam from a single source having the composite spectral density of the constituent beams. This *new* beam is then directed into the interferometric part of the experimental setup and used exactly as in other noisy light spectroscopies [1,3].

It is shown here that when composite spectral densities are achieved by combining two or more non-transform limited (noisy) sources collinearly to form one beam, nonlinear autocorrelation interferograms are generated that exhibit fringes capable of better temporal resolution then possible from one noisy source alone [2]. In principle by tailoring spectra beyond simple broadband sources one can in fact customize the shape of the autocorrelation function at any order. The demonstration experiment is based on a dual dye noisy light source with unbalanced dispersion effects for interferometric second harmonic generation (called $I^{(2)}$SHG), or, preferably interferometric frequency summing, $I^{(2)}$FS(2), since there are two distinct colors (carrier frequencies) involved [2].

A simple theory can capture the essence of the observed $I^{(2)}$FS(2) interferograms. The second-order electrical polarization is obtained for light fields containing a complex noise function; the $I^{(2)}$FS(2) interferogram (an intensity level signal) is expressed in terms of the resulting four point time correlation functions for the noisy quasi-cw light. These are written in terms of stochastically averaged noise functions, that are reduced to products of pair correlators which themselves are Fourier transforms of the composite spectral density of the tailored source. A diagrammatic technique called factorized time correlation (FTC) diagram analysis has been introduced to characterize time resolved noisy light spectroscopies directly from the resultant product of pair correlators [4]. It is believed that the FTC diagrams represent the decomposition of the noisy light spectroscopy into its most elementary components that have a 'physical' interpretation. The theory is extended to include effects of unbalanced dispersion within the two arms of the interferometer. Such differential dispersion effects prove to be significant and observable. Their control offers another tool for tailoring the structure of the interferogram.

The proposed general double noisy beam noncollinear CRS experiment called $I^{(2)}$CRS(3c) would consist of three dye lasers pumped by a single Nd:YAG laser two of which operate at zeroth order diffraction such that much of the emission spectrum of the dye is emitted from the cavity; the third operates in narrow band mode. The two separate broadband beams are made collinear to form a new beam having the composite spectral density of the two broadband dye sources. The composite beam is now used as a single source with a complicated noise spectrum. Briefly, the composite broadband beam is sent through a Michaelson interferometer. Emerging from the interferometer are twin parallel beams (F and F$'$) each having the identical composite spectral density but with F$'$ in one arm having been delayed relative to F in the other arm by a variable delay, τ. These beams along with the narrow band beam are focused into the sample in a BOX configuration [1] with the narrow band propagating along $\mathbf{k}_m$, F along $\mathbf{k}$, and F$'$ along $\mathbf{k}'$. The signal beam along $\mathbf{k}_s = \mathbf{k} + \mathbf{k}' - \mathbf{k}_m$ is sent through a monochromator and detected with a PMT.

An interesting aspect of $I^{(2)}$CRS(3c), using noisy light that has "two colors" (two spectral peaks) and a third monochromatic beam having a center frequency between the two broad spectral peaks, is that a single signal along $\mathbf{k}_s = \mathbf{k}+\mathbf{k}'-\mathbf{k}_m$ emerges, simultaneously containing both the $I^{(2)}$CARS and $I^{(2)}$CSRS frequencies along with a color that is between the former, which may be up to twice the relative intensity of either the $I^{(2)}$CARS or $I^{(2)}$CSRS frequencies. Additional dynamical information in the form of oscillations at up to twice the Bohr frequency is also contained in the $I^{(2)}$CRS(3c) signal. This is most pronounced when the 'monochromatic' detector is set near the arithmetic mean of the two component frequencies. This particular aspect of the proposed $I^{(2)}$CRS(3c) technique may be useful for up-conversion of very low frequency vibrational (or rotational) modes.

Acknowledgment: We are grateful for support from NSF grant CHE-9311959

1. M.A. Dugan, J.S. Melinger, and A.C. Albrecht, Chem. Phys. Lett. 147, 411 (1988); M.A. Dugan and A.C. Albrecht, Phys. Rev. A, 43, 3877 (1991).
2. D.C. DeMott, D.J. Ulness, and A.C. Albrecht, submitted for publication in Phys. Rev. A.
3. N. Morita and T. Yajima, Phys. Rev. A, 30 2525 (1984); S. Asaka et.al., Phys. Rev. A, 29, 2286 (1984); R. Beach, D. Debeer, and S.R. Hartmann, Phys. Rev. Lett. 53, 663 (1984).
4. D.J. Ulness and A.C. Albrecht, Phys. Rev. A, 53 1081 (1996); D.J. Ulness and A.C. Albrecht, submitted for publication in Appl. Phys. B.

COHERENT REACTION DYNAMICS OF HEME PROTEINS

Paul M. Champion*, Leyun Zhu, and Gang Zhong

Department of Physics, Northeastern University, Boston, MA 02115

Heme proteins play important roles in many biological systems. In this paper we will present recent studies on the reaction dynamics of heme proteins[1]. Two key issues will be discussed: (a) the level crossing dynamics after photodissociation of MbNO; (b) coherent reaction dynamics of HbNO.

The upper panel of Figure 1 shows the absorption spectra of MbNO and Mb. The middle panel of Figure 1 shows the pump induced transmission change at 435nm when the full (transform limited) probe bandwidth is detected. The laser pulse autocorrelation is also shown as a dashed line. Immediately after the coherent coupling signal near Δt=0, an anti-bleaching signal rapidly appears. The data used in the fitting procedure lie outside the laser pulse autocorrelation and are shown on a shifted scale, along with the best bi-exponential fit to the rapid (<ps) response. The major component of the decay is described by a time constant τ_1=80$\pm$30fs and a deoxy Mb like ground state absorption has fully developed at 250fs. The lower panel of Figure 1 shows data obtained using the spectrally dispersed detection scheme. The (75fs) laser pulse is centered at 435nm (λ_o) while the monochromator is set at 429nm (λ_m). Immediately after the coherent coupling signal is complete (150fs), the signal is seen to be modulated by 220 cm^{-1} oscillations that are a characteristic of the ground state Mb Fe-His stretching mode.

Femtosecond electronic and vibrational experiments suggest the following photodissociation dynamics: Immediately ($\sim$20fs) after photoexcitation of the (S=0) MbNO sample, the NO bond breaks and iron-histidine bond compression begins. Initially (i.e., within $\sim$20fs), the Fe-His antibonding component of the iron d_{z^2} orbital loses two electrons from NO and gains one electron from the previously occupied t_{2g} iron orbitals, leading to a S=1 iron spin state. As the Fe-His bond compresses the iron begins to move out of the heme plane and within half a period ($T_{Fe-His}/2\sim$75fs) we expect the $d_{x^2-y^2}$ orbital to lower its energy and be populated on this time scale. This process would affect the electronic absorption spectra and correspond to the creation of a S=2 deoxy Mb electronic ground state. The 80$\pm$30fs time constant observed in the electronic relaxation data is, in fact, a measure of this process. The population of the $d_{x^2-y^2}$ orbital also creates strong forces that lead to heme doming. Thus, the initial phase of heme doming is complete within $T_{Fe-His}/2+T_{dome}/4\sim$180fs.

The HbNO pump-probe data and analysis are presented in Figure 2. Trace (a) shows the raw data; (b) shows the raw data (points) and the LPSVD fits; (d) is the 230 cm^{-1} oscillatory component from the LPSVD fit; while (c) is the residual of the data after the 230 cm^{-1} component has been removed. The power spectrum from the LPSVD analysis (e) is shown in the right part of the figure together with the cw resonance Raman spectra of HbNO and Hb. We see that the reaction driven product state coherence has an iron-histidine stretching mode at 230 cm^{-1} and a doming mode at 95 cm^{-1}. Equilibrium deoxy Hb is in the T-state while the ligand bound HbNO is in the R-state. After photodissociation of the diatomic ligand, the deligated Hb is in a transient R-state, which undergoes tertiary relaxation on the ns time scale followed by a quaternary R to T transition on the μs time scale[2]. Equilibrium T-state deoxy Hb has an iron-histidine stretching mode at 214 cm^{-1}, while the transient R-state photoproduct Hb has the iron-histidine mode at 230 cm^{-1}.[2] Our results show that, as for Mb, the NO ligand in Hb is dissociated instantaneously (~20 fs) and the transient R-state deoxy Hb ground state is formed in about 100 fs. The impulsive electronic force, due to the photolytically induced changes in the d_{z^2} bond order, drives the iron-histidine mode to vibrate coherently on a product state potential energy surface with a frequency of 230 cm^{-1}, identical to that of the R-state transient[2]. The heme doming motion at ~95 cm^{-1} is also coherent, as in the case of myoglobin. This result demonstrates that no significant iron-histidine frequency change takes place in the photoproduct of HbNO between ~ 200 fs and 1 ns after photolysis.

1. L. Zhu *et al.*, *J. Raman Spectrosc.* **26**, 527 (1995).
2. D. L. Rousseau, and J. M. Friedman, in *Biological Applications of Raman Spectroscopy*, ed. by T. G. Spiro, Vol. 3, p. 134. Wiley, New York (1988).

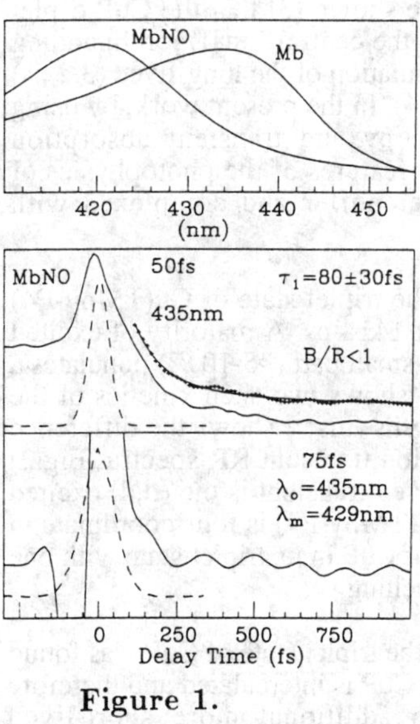

Figure 1.

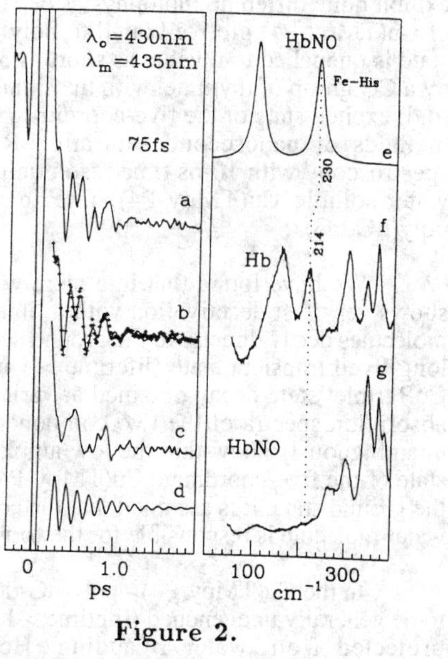

Figure 2.

PICOSECOND TRANSIENT ABSORPTION AND NANOSECOND RR STUDIES OF EXCITED ELECTRONIC STATES OF WATER-SOLUBLE CU(II) PORPHYRIN

V.A.Galievsky[1], V.S.Chirvony[1], S.G.Kruglik[2], V.V.Ermolenkov[2], V.A.Orlovich[2], C.Otto[3], P.-Y.Turpin[4*]

[1] Institute of Molecular and Atomic Physics, Academy of Sciences of Belarus, 70 F.Skaryna Ave., Minsk 220072, Belarus
[2] B.I.Stepanov Institute of Physics, Academy of Sciences of Belarus, 70 F.Skaryna Ave., Minsk 220072, Belarus
[3] University of Twente, Department of Applied Physics, P.O.217, 7500 AE Enschede, the Netherlands
[4] Université Pierre et Marie Curie, LPBC, 4 Place Jussieu, 75252 Paris Cedex 05, France

Deactivation of the Cu(II)porphyrin (CuP) lowest triplet state very strongly depends on surrounding conditions, that originates from the presence of low-lying CT state in the four-coordinate form and low-lying CT and (d,d) states in the (photoinduced) five-coordinate form [1,2]. This enables the photophysics of CuP to be used as an off-beat probe of environment.

When complexed with DNA and DNA-model compounds, water-soluble CuP exhibit quite different photophysics depending on environment. Recent investigations of Cu(TMpy-P4) groove-bound to poly(dA-dT) have shown [3] that the CuP triplet state is quenched to a value as short as 35 ps, due to the central Cu(II) ion liganding by a CO group of thymine, with the simultaneous population of the long-lived (3.2 ns) (d,d) excited state of the five-coordinate Cu(TMpy-P4). In the present work, by using methods of nanosecond saturation RR spectroscopy and transient absorption spectroscopy with 10-ps time resolution, we present features of the photophysics of water-soluble Cu(TMpy-P4), free in a mere water buffer and complexed with poly(dG-dC).

We have found that in a mere water buffer the triplet state of Cu(TMpy-P4) shows very fast deactivation with a time constant of 14±4 ps. A majority of excited molecules decay directly to the ground state but some small part (~5-10 %) populates a long-lived transient state (lifetime > 1 ns). Figure 1 shows two such kinetics of the CuP triplet state decay obtained at various wavelengths (insert shows the difference absorption spectra of the two components). Saturation transient RR spectra (Fig.2) unambiguously show that the low amplitude long-lived transient is the (d,d) excited state of the five-coordinate Cu(TMpy-P4). While Cu(TMpy-P4) is four-coordinate in the ground state, it is assumed that the coordination of CuP in its triplet state with one water molecule is responsible for the triplet state quenching.

In the Cu(TMpy-P4) - poly(dG-dC) complex the triplet state of CuP was found to be generally unquenched (lifetime > 1 ns) because CuP is intercalated and therefore protected from water liganding. However, two additional more short-lived components (~20 ps and ~600 ps) have been found. Possible origins of these short-lived features will be discussed.

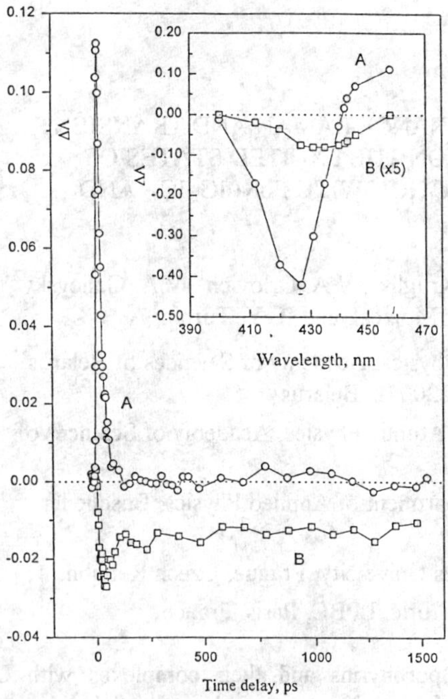

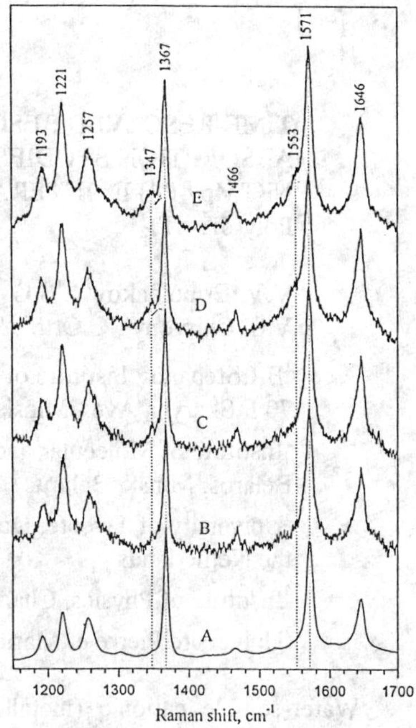

Fig. 1 (left). Kinetics of absorption changes for Cu(TMpy-P4) in a phosphate water buffer solution with excitation at 540 nm and probing at 458 nm (curve A, short-lived component) and 440 nm (curve B, mainly long-lived component). In the inset, transient difference absorption spectra are presented recorded at time delays of 5 ps (A) and 200 ps (B) following excitation.

Fig. 2 (right). Transient RR spectra of Cu(TMpy-P4) in a phosphate buffer solution recorded with excitation at 436 nm. Incident laser power density increases from spectrum B (I_{ex}=1.1x10^6 W/cm^2) to spectrum E (I_{ex}=7x10^8W/cm^2). Spectrum A was recorded with CW Kr-laser excitation at 413 nm and represents ground-state scattering of Cu(TMpy-P4). Features at 1347 and 1553 cm^{-1} originate from the (d,d) excited state population.

1. (a) V.S.Chirvony, B.M.Dzhagarov, G.P.Gurinovich, *Bull.Acad.Sci.USSR, Phys.Ser.* **1984**, *48*, 55; (b) D.Kim, D.Holten, M.Gouterman, *J.Am.Chem.Soc.* **1984,** *106*, 2793.
2. S.G.Kruglik, P.A.Apanasevich, V.S.Chirvony, V.V.Kvach, V.A.Orlovich, *J.Phys.Chem.* **1995**, *99*, 2978.
3. S.G.Kruglik, V.A.Galievski, V.S.Chirvony, P.A.Apanasevich, V.V.Ermolenkov, V.A.Orlovich, L.Chinsky, P.-Y.Turpin, *J.Phys.Chem.* **1995**, *99*, 5732.

TIME-RESOLVED RESONANCE RAMAN AND TRANSIENT ABSORPTION STUDIES ON THE EXCITED STATES OF Ni(TMpy-P4) IN INTERACTION WITH Poly(dG-dC)$_2$ AND Poly(dA-dT)$_2$

V.V. Ermolenkov[1]*, S.G. Kruglik[1], V.A. Orlovich[1], V.A. Galievsky[2], V.S. Chirvony[2], C.Otto[3], P. Mojzes[4], and P.-Y. Turpin[5]

[1]B.I.Stepanov Institute of Physics, Academy of Sciences of Belarus 70 F.Skaryna Ave., Minsk 220072, Belarus

[2]Institute of Molecular and Atomic Physics, Academy of Sciences of Belarus, Minsk, Belarus

[3]University of Twente, Department of Applied Physics, Enschede, the Netherlands

[4]Institute of Physics, Charles University, Prague, Czech Republic

[5]Université Pierre et Marie Curie, LPBC, Paris, France

Water-soluble cationic (metallo)porphyrins and their complexes with DNA attract an increased attention, in particular, because of their possible biomedical applications. Porphyrins exhibit at least three stationary modes of interaction with DNA: intercalation, outside groove bindings in the edge-on and face-on manner [1]. Recently, new transient processes of interaction between photoexcited Cu(TMpy-P4) and DNA (so-called exciplex formation) has been revealed in resonance Raman (RR) spectra under high-power irradiation [2]. Further studies [3,4] have shown the crucial role of AT sites and the flexibility of Pu-Py alternating double-helical structure for the exciplex formation. Also, participation of a CO group of thymine or uracil residues as an axial ligand attached to the central copper ion has been suggested and proven, and the electronic origin of the exciplex in terms of the excited (d,d) state of five-coordinate porphyrin species has been revealed on the basis of RR and transient absorption (TA) studies [4,5].

The purpose of the present work is to study the photoinduced behavior of another transition-metal cationic porphyrin, namely Ni(TMpy-P4), in interaction with two DNA-modeling polynucleotides, poly(dG-dC)$_2$ and poly(dA-dT)$_2$, by using jointly nanosecond RR and picosecond TA spectroscopies. For Ni(TMpy-P4)-poly(dG-dC)$_2$ complex, new transient Raman lines arise on increase of the excitation power at 1563, 1459 and 1363 cm^{-1}, pairing their ground-state counterparts ν_2 at 1585 cm^{-1}, ν_3 at 1480 cm^{-1}, and ν_4 at 1376 cm^{-1}, respectively. In TA measurements, double-exponential kinetics of excitation relaxation was found, with time constants of $\leq$10 ps and 350$\pm$20 ps; absolute absorption spectra originated from both transient species were reconstructed from the experimentally measured difference spectra. The long-lived transient species was assigned, on the basis of whole set of RR and TA data, to the intramolecular metal-centered excited (d,d) state $^3B_{1g}$ of the 4-coordinate Ni(TMpy-P4) intercalated between G-C base pairs.

More complicated picture of excitation relaxation has been revealed for Ni(TMpy-P4)-poly(dA-dT)$_2$. Transient RR spectra of this complex were found to be quite close to those of the G-C complex, with transient bands having similar power-induced behavior and values of downward-frequency shifts from their ground-state counterparts. However, some fine features, especially in the vicinity of the ν_3 band, show that RR spectra of the A-T complex contain more details than in spectra of the G-C complex. In TA measurements, at least 4 relaxation components have been separated with the time constants of ≤10 ps, ~100 ps, ~450 ps, and >1 ns. Our analysis shows that the most credible interpretation of these observations is that there are two pairs of relaxation kinetics: 1) <10 ps and 100 ps; and 2) <10 ps and 450 ps, each of them accounting for a different kind of interaction with A-T polymer. In each pair, the longest lived component likely involves the excited (d,d) state $^3B_{1g}$ of the 4-coordinate Ni-porphyrin, which in turn is responsible for the observed additional transient RR lines. The much less intense fourth TA component (>1 ns) may correspond to a small part of molecules involving an axial ligand binding in the excited state, followed by slow ligand-dissociation process. The comparative analysis of transient RR spectra of Ni(TMpy-P4) bound to A-T sequences and in phosphate buffer solution shows that no 6-coordinate $^3B_{1g}(L)_2$ transient species is photogenerated in A-T complex. Therefore, coordination of only one axial-ligand molecule, most probably from the surrounding water buffer, to the porphyrin central Ni ion is proposed to explain the whole set of experimental results.

Possible types of binding of Ni(TMpy-P4) to poly(dA-dT)$_2$ are discussed on the basis of photophysical data obtained. Concerning the major part of Ni(TMpy-P4), being 4-coordinate both in the ground and in the excited states, edge-on binding in the minor groove of poly(dA-dT)$_2$, where the porphyrin protrudes into the polynucleotide interior far enough to be screened from the surrounding water molecules, is the only type of localization consistent with all the experimental data obtained so far for this complex. Concerning the minor part of Ni(TMpy-P4) molecules participating in photoinduced ligand binding/release processes, it is plausible to suggest that Ni(TMpy-P4) is located in a face-on manner in the major groove of poly(dA-dT)$_2$ and can adopt one water molecule as an axial ligand, producing a 5-coordinate species.

1. R.J.Fiel, J. Biom. Str. & Dyn. 1989, 6, 1279.
2. P.-Y. Turpin, L. Chinsky, A. Laigle, M. Tsuboi, J.R. Kincaid, K. Nakamoto. Photochem. Photobiol. 1990, 51, 519.
3. P.-Y. Turpin, L. Chinsky, A.H.R. Al-Obaidi, S. Bell, R.E. Hester. J. Phys. Chem. 1991, 95, 5754.
4. P. Mojzes, L. Chinsky, P.-Y. Turpin. J. Phys. Chem. 1993, 97, 4841.
5. S.G. Kruglik, V.A. Galievsky, V.S. Chirvony, P.A. Apanasevich, V.V. Ermolenkov, V.A. Orlovich, L. Chinsky, P.-Y. Turpin. J. Phys. Chem. 1995, 99, 5732.

RESONANCE RAMAN AND CARS STUDIES OF Cu-PORPHYRINS: STRUCTURAL DEFORMATIONS IN THE GROUND AND EXCITED ELECTRONIC STATES

P.A. Apanasevich, V.V. Ermolenkov, S.G. Kruglik, and V.A. Orlovich*

B. I. Stepanov Institute of Physics, Academy of Sciences of Belarus
70 F. Skaryna Ave., Minsk 220072, Belarus

In the present work we carried out spectroscopic studies of Cu-porphyrins which are interesting compounds for modeling various native processes. Cu(II)-octaethylporphyrin (CuOEP) and Cu(II)-tetraphenylporphyrin (CuTPP) complexes in oxygen(O)-containing solvents tetrahydrofuran (THF) and 1,4-dioxane and in noncoordinating solvent benzene have been studied by the methods of resonance Raman (RR) and resonance coherent anti-Stokes Raman scattering (RCARS) techniques. Our attention was focused to the perturbations of porphyrin structure both in the ground and in the excited electronic states due to complexation by O-containing molecules and triplet-state excitation.

CuOEP in the ground electronic state is known to be a pure 4-coordinate species having flat structure of the porphyrin macrocycle [1]. Lowering of the temperature of a solution containing a metalloporphyrin M and a potential axial ligand L leads to the shift of equilibrium constant towards the complex ML formation because of the diminishing of the thermal energy excess. At 77 K, the absorption spectrum of CuOEP in THF/diethyl ether (1:1) mixture reveals splitting of the Soret band with maxima at 397 and 402 nm due to partial complex formation via binding of a THF molecule as an axial ligand [2]. We have carried out RR study of both complexed and noncomplexed porphyrins. RR spectrum of the complexed species recorded with excitation at 405 nm proves that porphyrin macrocycle remains flat with predomination of one of the isomeric conformations of the triclinic structure, namely, the B form. At the same time, new out-of-plane Raman bands evidenced ruffled structure of porphyrin core appear in RR spectrum of the noncomplexed species recorded with excitation at 400 nm.

At ambient temperature, CuOEP reveals ability for photoinduced binding of some O-containing molecules in the excited (d,d) state forming an excited complex (exciplex) [$(CuOEP)^*_{d,d}$-L] [2-4]. According to transient absorption kinetic study [2], the lifetime of such an exciplex lies in subnanosecond time domain. We recorded transient RR and RCARS spectra of CuOEP in THF and dioxane, and the vibrational frequency analysis on the measured spectra permits to make the following conclusions about the exciplex structure.

First, there are substantial downshifts in frequency of all the core-size marker bands in high-frequency region for the exciplex species reflecting core expansion by ca. 0.03 Å. The porphyrin excited (d,d) state involves the promotion of an electron from the highest filled d_{z^2} orbital to the half-filled $d_{x^2-y^2}$ orbital; and porphyrin core substantially expands to accommodate the added electron density in the metal $d_{x^2-y^2}$ orbital. The values of shifts of the modes ν_{10} and ν_{19} reflect also the removal of the copper ion out of the macrocycle plane towards the axial ligand.

Second, mode ν_4 known as the characteristic oxidation state and π electron density marker band also experiences large downshifts indicated the electron density donation from the axial ligand to metal orbitals.

We note the absence or negligible frequency shifts of ethyl substituent deformations and stretches. This points to the fact that perturbation of the exciplex structure does not affect remarkably the peripheral ethyl groups.

Noteworthy is also the absence in the exciplex RR spectrum of the prominent Raman bands in the low-frequency region below 500 cm^{-1}. The nature of this effect is still not clear, we can only suppose the presence of some dynamic effect leading to a significant broadening of the low-frequency modes.

Time-resolved RR study of the excited (π,π^*) triplet $^{2,4}T_1$ state of CuOEP and CuTPP in benzene [5] revealed the significant difference between RR spectra from excited triplet states and ground and excited (d,d) ones. The main features of the triplet-state spectra are: 1) low intensity and considerable broadening of the metalloporphyrin skeletal bands; 2) absence of unambiguous one-to-one correlation between excited triplet and ground state modes; and 3) absence of broadening of peripheral substituents bands and their enhancement with respect to the lines of porphyrin core. Our experimental results are quite similar to those obtained in independent study of the excited triplet states of meso-substituted copper porphyrins [6]. Probable mechanisms of the processes occurring in the excited $^{2,4}T_1$ state are analyzed in the report. It is suggested that the observed broadband and low-intense triplet-state spectra are the superposition of spectra of multiple structural conformers, having shifted skeletal and unshifted substituent Raman bands. Both Yahn-Teller effect and nonplanar distortions of porphyrin core may be proposed to define the above mentioned peculiarities in RR spectra from the excited triplet states of Cu-porphyrins.

1. L.D. Spaulding, C.C. Chang, N.-T. Yu, R.H. Felton, J. Am. Chem. Soc. 1975, 97, 2517.
2. S.G. Kruglik, P.A. Apanasevich, V.S. Chirvony, V.V. Kvach, V.A. Orlovich, J. Phys. Chem. 1995, 99, 2978.
3. P.A. Apanasevich, R. Gadonas, V.V. Kvach, V. Krasauskas, S.G. Kruglik, V.A. Orlovich, V.S. Chirvony, Bull. Acad. Sci. USSR, Phys. Ser. 1988, 52, 1089
4. P.A. Apanasevich, V.S. Chirvony, S.G. Kruglik, V.V. Kvach, V.A. Orlovich, in Laser Applications in Life Sciences, (S.A. Akhmanov, M.Yu. Poroshina, Eds.), SPIE: Bellingham, Vol. 1403, Part I, 1991, pp. 195-211
5. S.G. Kruglik, V.S. Chirvony, V.V. Ermolenkov, V.A. Orlovich, Opt. Spectrosc. 1996, 80, in press.
6. M. Asano-Someda, S. Sato, K. Aoyagi, T. Kitagawa, J. Phys. Chem. 1995, 99, 13800.

PHOTOINDUCED ELECTRON TRANSFER: TRANSIENT VIBRATIONAL SPECTROSCOPY OF SINGLET PORPHRYIN AND UBIQUINONE DONOR-ACCEPTOR RADICAL ION PAIRS.

Tione Buranda, Neal Soice, Jack Greiner, Somjai Prajakrattanakij, and Mark Ondrias*

The Department of Chemistry, The University of New Mexico, Albuquerque, NM 87131.

In the photosynthetic reaction center light induced charge separation is initiated by photoexcitation of a bacteriochlorophyll complex into its first excited singlet state, this is immediately followed by ultrafast charge migration to an intermediate acceptor (a bacteriopheophytin molecule). Rapid and efficient secondary ET to another acceptor (a ubiquinone) creates chemical potential with virtually no energy loss due to back electron transfer (BET). [1] In simple model systems however, collisional quenching of the first excited singlet state of porphyrin systems with an acceptor have left no apparent sign of redox products albeit, the inference of ET from fluorescence quenching experiments. [2] Yet quenching of the triplet excited state by the redox partner leads to primary charge separated species such as exciplexes and solvated geminate radical ion pairs in high yield. In a previous study we have used transient absorption to demostrate improved charge seperation yields between excited triplet zinc-meso-tetra (4-sulfonatophenyl) porphine (MP) and 2,3-dimethoxy-5-methyl-1,4-benzoquinone (UQ_0) through manipulation of the MP/UQ_0—$MP^{\bullet+}/UQ_0^{\bullet-}$ Franck-Condon energy gap. [3] Even though substantial yields of MP π radical cations were formed from the triplet excited state precursors, no evidence of electron transfer from the singlet manifold was observed in that study. In this work we describe our efforts to spectroscopically characterize the photoinitiated dynamics of ET from singlet excited-state free base porphyrin (P) to UQ_0. Our observations provide the strongest experimental evidence to date confirming P singlet electron transfer quenching.

Single pulse, transient resonance Raman spectra of P and UQ_o were collected using protocols described previously. [4] Raman spectra were collected in neutral aqueous solutions of P, UQ_0, and P and UQ_0 ground state spectra were taken using 0.4 mJ/pulse and excited state transients were generated from 2mJ laser pulses. A summary of the vibrational frequencies and their assignments are shown in table 1.

Oxidation of a porphyrin is commonly associated with a shift to lower frequency at "the oxidation state marker" mode ν_4. This is indeed the case where a $\Delta\nu$ of -8 cm^{-1} in this instance is typical. [5] The T_1 ν_4 mode at 1380 cm^{-1} is due to uncomplexed P. Much of the spectral interpretation of UQ_0 is based on the work of Tripathi, Brus *et al* and Parker and coworkers (table 1 B). The neutral UQ_0 (S_0) spectrum is simple with only one unresolved band at 1608 cm^{-1} corresponding to accidentally degenerate symmetric stretch of the C=C ring bonds and the symmetric combination of the two C=O stretches. Reduction of UQ_0 to form the semiquinone radical anion leads to strong electronic transitions near 420 and 440 nm compared to the single less intense band at 406 nm in the neutral species, (neutral ubiquinone $\varepsilon \approx$ 5300 M^{-1} cm^{-1} at 425 nm and semiquinone radical $\varepsilon \approx 8600$ $M^{-1}cm^{-1}$ at 440 nm). As a result, the absorption cross section of the semiquinone radical $UQ_0^{\bullet-}$ in this (436 nm) region of the Raman probe is substantially higher than the neutral species being probed in the ground state; this leads to the resonant enhancement of the modes associated with this transition, in the $UQ_0^{\bullet-}$ viz. C=C at 1618 cm^{-1} and C=O stretch at 1435 cm^{-1}. In this work, we are unable to identify some of the bands which have been observed in the literature due to inteference from the $P^{\bullet+}$ radical cation, a much better Raman scatterer than the $UQ^{\bullet-}$. However it is gratifying to note that the key UQ_0 bands (1435 cm^{-1} 1608 cm^{-1} and 1618 cm^{-1}) are easily descernible in the midst of the more prominant $P^{\bullet+}$ vibrations The 1435 cm^{-1} band which we have tentatively assigned to the C=O stretch, while similar in frequency to that of p-benzosemiquinone radical is much lower than the one obtained by Parker *et al* (1474 cm^{-1}; $UQ_0^{\bullet-}$).

Table 1 ***A.*** *Raman frequencies of S_0, T_1 and $P^{\bullet+}$ states of H_2TPPS_4 measured in this work compared to literature ground state H_2TPP frequencies.* ***B*** *Ubiquinone vibrational frequencies:*

A

	assignment[a]	typical[b] TPP	$H_2(TPPS)^c$ S_0	T_1	$P^{\bullet+}$
ν_1	$\nu(C_m\phi)$	1234	1236	1235	1230
ν_2	$\nu(C_\beta C_\beta)$	1546	1550	1539	1549
ν_3	$\nu(C_\alpha C_m)_{sym}$	1425			
ν_4	ν(pyr 1/2-ring)$_{sym}$	1360	1362	1380	1352
ν_{10}	$\nu(C_\alpha C_m)_{asym}$	1610	1595		
ν_{11}	$\nu(C_\beta C_\beta)$	1493		1495	
ν_{12}	ν(pyr 1/2-ring)$_{sym}$	1353			1320
ν_{27}	$\nu(C_m\phi)$	1270	1290		1290
ν_{28}	$\nu(C_\alpha C_m)_{sym}$	1485			1500

B

assignment[a]	UQ_0	$UQ_0^{\bullet-b}$	$UQH_0^{\bullet c}$
C=C$_{str}$(8a)	1608	1435	1506
C=O$_{str}$(7a)	1608	1608	
C=O$_{str}$(7a)		1618	1617
C=C$_{str}$(8b)		1527[1]	
C–C$_{str}$(14)		1305[1]	

A: a. Kitagawa, T.; Abe, M.; Kyogoku, Y. *J. Chem. Phys.* **1978**, *69*, 4516 **b.** Li, X.; Zgierski, M.Z. *J. Phys. Chem.* **1991**, *95*, 4268 **c.** this work. **B: a** mode assignments based on the work of (i) Tripathi, G.N.R. *J. Chem. Phys.* **1981**, *74*, 6044. (ii) Beck, S.M.; Brus, L.E. *J. Am. Chem. Soc.* **1982,** 104, 4789. (iii) Parker, A.W.; Hester, R.E.; Phillips, D.; Umapathy, S. *J. Chem. Soc. Faraday Trans.* **1992**, *88*, 2649. **b.** semiquinone frequencies measured in this work. (1) values from Parker, A.W.; Hester, R.E.; Phillips, D.; Umapathy, S. *J. Chem. Soc. Faraday Trans.* **1992**, *88*, 2649. **c.**values from Parker *et al.* see **b1**.

Vibrational spectroscopies such as resonance Raman scattering and infrared absorption play a crucial role in extracting structural information important to the understanding and ultimate control of the dynamics of electron transfer. Reaction centers succeed in suppressing BET because very little reorganization energy accompanies initial ET; and the ≈1 eV [1b] energy gap results in a poor Franck-Condon overlap to allow the system to thermally relax via BET. Most of the frequency changes undergone by the porphryin donor suggest very little molecular reorganization, due to the relatively large size of the molecule and the extended π electron system which enables the molecule to endure minor structural pertubations upon oxidation. On the other hand the ubiquinone is shown to experience large frequency shifts upon reduction, for example the C=0 stretch undergoes a 173 cm^{-1} shift. This translates into a sizable reorganizational energy which enables the conversion of 1.15 eV of electronic to vibrational energy. (Supported by the NIH (GM 33330)

References:

[1]. (a) Gouterman, M.; Holten, D. *Photochem. Photobio.* **1977**, 25, 85.
(b) M.R. Wasielewski, Chem. Rev. **1992**, 92, 435
[2]. K. Kalyasundaram, *Photochemistry of Polypyridine and Porphyrin Complexes*; Academic Press: New York. **1992**, chapters 14, and references therein.
[3]. T. Buranda, N. Soice, S. Niu, R. Larsen, M. Ondrias, *J. Phys. Chem.* submitted. **[4]**. E.W. Findsen, K. Alston, J.A. Shelnutt, and M.R.Ondrias, *J. Am. Chem. Soc.* 108, *4009*, **1989**. **[5]**. (a). Reed, R.A.; Purrello, R.; Prendergast, K.; Spiro, T.G.; *J. Phys. Chem.* **1991**, *95*, 9720-9727.(b). de Paula, J.C.; Walters, V.A.; Nutaitis; C.; Lind, J.; Hall, K.; *J. Phys. Chem.* **1992,** *92*, 10591-10594.

INFLUENCE OF *meso*-SUBSTITUENTS ON ELECTRONIC STATES OF PORPHYRIN π-CATION RADICALS

K. Jayaraj[1], J. Terner[2*], A. Gold[1*], D.A. Roberts[1], R.N. Austin[1], D. Mandon[3], R. Weiss[3*], E. Bill[4], M. Müther[4], and A.X. Trautwein[4*]

[1]Dept. of Environmental Sciences and Engineering, The University of North Carolina at Chapel Hill, Chapel Hill, NC 27599 USA

[2]Dept. of Chemistry, Virginia Commonwealth University, Richmond, VA 23284-2006, USA

[3]Institut LeBel, Université Louis Pasteur, 67070 Strasbourg, France

[4]Institut für Physik, Medizinische Universität zu Lübeck, Lübeck, Germany

Physico-chemical investigations involving the low temperature generation and stabilization of oxidized forms of model synthetic iron porphyrins have contributed towards an understanding of the active site mechanisms of heme enzymes and proteins which participate in a wide variety of metabolic process in living systems [1]. High valent iron porphyrins have been either identified or proposed as crucial intermediates in the reactions of peroxidases, catalases, cytochromes P-450, cytochrome oxidase and other heme-based enzymes found in animals and plants. The best known of the physiological high-valent hemes are the sequential intermediates of the protoheme-containing peroxidases and related catalases. These intermediates, commonly designated as compounds I and II, are two and one oxidation equivalents above the ferric resting state.

The symmetry states of the oxoferryl porphyrin π-cation radical hemes (compounds I) of peroxidases such as horseradish peroxidase and chloroperoxidase have historically been thought to be predominantly $^2A_{2u}$ [2]. Recent resonance Raman results have however favored $^2A_{1u}$ symmetry states for both of these enzymes [3,4]. In this report we discuss effects which can modulate the symmetry states in model compounds. A series of oxoferryl porphyrin π-cation radicals generated from porphyrins substituted at the *meso*-positions with highly electron-withdrawing aryl groups [5] is investigated and compared to those of oxoferryl porphyrin π-cation radical complexes with electron releasing substituents. Porphyrins with *meso*-substituents ordinarily form metalloporphyrin π-cation radicals with $^2A_{2u}$ symmetry states which exhibit strongly downshifted ν_2 porphyrin modes [6]. By contrast, resonance Raman data for some substituted *meso*-porphyrins, such as the oxoferryl tetrakis (2,6-dichlorophenyl) porphyrin π-cation radical shown in Figure 1, show a moderately upshifted ν_2 frequency [7]. The data for the porphyrins with the electron-withdrawing *meso*-substituents may be explained by admixture of A_{1u} character into the ground state. This is consistent with the hypothesis that electron-withdrawing substituents lower the energy of the a_{2u} molecular orbital, favoring a_{1u} admixture [8].

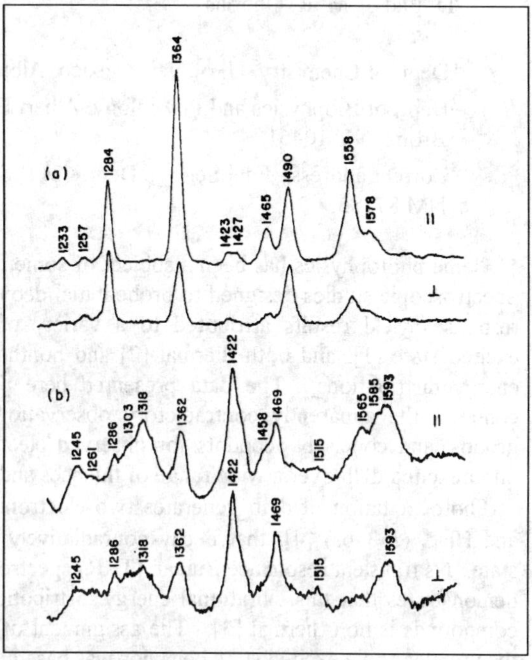

Fig. 1. Resonance Raman spectra (406.7 nm excitation at -80 °C) of (a) tetrakis (2,6-dichlorophenyl)porphinatoiron(III) triflate in methylene chloride, (b) the same compound reacted with *m*CPBA.

Acknowledgments: This work was supported in part by NIH grants GM34443 (JT) and ES03433 (AG), Centre National de la Recherche Scientifique grant UA424 (RW) and the Deutsche Forschungsgemeinschaft (AXT).

1. B. Boso, G. Lang, T.J. McMurry and J.T. Groves, J. Chem. Phys. 79, 1122 (1983).
2. D. Dolphin and R.H. Felton, Accts. Chem. Res. 7, 26 (1974).
3. V. Palaniappan and J. Terner, J. Biol. Chem. 264, 16046 (1989).
4. C.M. Hosten, A.M. Sullivan, V. Palaniappan, M.M. Fitzgerald and J. Terner, J. Biol. Chem. 269, 13966 (1994).
5. D. Mandon, R. Weiss, K. Jayaraj, A. Gold, J. Terner, E. Bill, and A.X. Trautwein, Inorg. Chem. 31, 4404 (1992).
6. R.S. Czernuscewicz, K.A. Macor, X.-Y. Li, J.R. Kincaid, and T.G. Spiro, J. Amer. Chem. Soc. 111, 3860 (1989).
7. K. Jayaraj, J. Terner, A. Gold, D.A. Roberts, R.N. Austin, D. Mandon, R. Weiss, E. Bill, M. Müther, and A.X. Trautwein, Inorg. Chem. (1996) *in press*.
8. P.J. Spellane, M. Gouterman, A. Antepas, S. Kim, and Y.C. Liu, Inorg. Chem. 19, 386 (1980).

HEME PHOTOPHYSICS: A NON-BOLTZMANN VIBRATIONAL ENERGY DISTRIBUTION IN DEOXYHEMOGLOBIN IS GENERATED AND MEASURED WITHIN 30 PICOSECOND PULSES

[1,3]M. C. Simpson, [1]C. Cheatum, [2]E. Peterson, [2]J. M. Friedman, [2]C. Shannon, [2]D. Eads, [1,*]M. R. Ondrias

[1]Dept. of Chemistry, U. of New Mexico, Albuquerque, NM 87135.

[2]Dept. of Biophysics and Physiology, Albert Einstein College of Medicine, Bronx, NY 10461.

[3]Current address: Fuel Science Dept. 6211, Sandia Nat. Labs, Albuquerque, NM 87185.

Heme photophysics has been a subject of some interest over the last two decades. Spectroscopic studies designed to probe initial deoxyhemoglobin (dHb) photoinduced responses yield results attributed to a variety of origins, including an electronic excited state [1], and both thermal [2] and nonthermal [3] ground state vibrational energy distributions. The data presented here provide a basis for a model that connects the apparently contradictory observations of three independent research groups, and concisely accounts for observed electronic and vibrational behavior of photoexcited dHb over a wide range of fluences and timescales.

Photoexcitation of dHb generates two electronic excited states, Hb_I* (τ~300 fs) and Hb_{II}* (τ~3 ps) [4], that decay nonradiatively to a vibrationally excited ground state. Ns transient resonance Raman (TRR) spectroscopy with moderate incident flux demonstrates that this vibrational energy distribution in dHb and protein-free model compounds is non-thermal [3]. The assignment of these results to vibrational rather than to electronic excited state behavior was based upon: (1) lack of saturation of the Stokes transitions, (2) all bands fit with single Lorentzian lines, (3) the Stokes ν_4 band did not shift to the lower frequency characteristic of Hb_{II}* [5], and (4) the anti-Stokes band moved to lower frequency and its relative intensity dramatically increased with laser flux in a mode specific manner. Higher pulse powers employed by Li et al. [1] produced in addition asymmetric broadening and saturation of the ν_4 Stokes line.

The present study extends these ns [1,3] TRR investigations into the ps domain. We report TRR measurements of dHb as a function of incident laser flux using 35 ps, Soret-resonant (436 nm) laser pulses. In these ps TRR studies, the heme is observed to be vibrationally excited in a non-thermal manner (Table 1). In addition, at the highest incident laser flux, evidence for an electronic excited state was found. Below

Table 1: Summary of Curvefitting Ps TRR Spectra of dHb

Flux	ν_4						ν_7					
(W/cm^2)	ω_S	Γ_S	ω_{aS}	Γ_{aS}	ratio*	T_B*	ω_S	Γ_S	ω_{aS}	Γ_{aS}	ratio*	T_B*
12×10^8	1353.8	18.4	1349.0	28.7	6.1%	730K	672.1	19.2	674.6	18.5	2.7%	280K
6×10^8	1353.6	22.0	1349.6	29.1	3.1%	590K	672.9	18.7	674.8	20.6	2.4%	270K
3×10^8	1354.1	17.2	1350.9	25.6	2.3%	540K	672.9	16.9	673.5	17.5	2.0%	260K
1.5×10^8	1355.3	19.6	----	----	----	----	672.3	15.0	675.4	15.2	1.4%	240K

Table 1: Curvefitting of TRR spectra of dHb. Bands were fit with Lorentzian lineshapes. ω and Γ are positions and FWHM values (cm^{-1}) respectively. The 'S' and 'aS' subscripts refer to Stokes and anti-Stokes scattering. *not corrected for holographic filter, resonance, or reabsorption effects

this incident flux, the anti-Stokes signal is still quite intense. Thus the intense, flux-dependent anti-Stokes scattering arises from a vibrationally excited ground state, rather than from the electronic excited state with a large anti-Stokes cross section that was invoked by Li et al. [1].

We explore the simple photophysical cycle depicted in Figure 1, from which the vibrationally excited ground state population can be described by

[1] $$[Hb(v^*)] = \frac{[Hb_I^*]_0 k_I}{k_v - k_I}\{e^{-k_v t} - e^{-k_I t}\} + \frac{[Hb_{II}^*]_0 k_{II}}{k_v - k_{II}}\{e^{-k_v t} - e^{-k_{II} t}\}.$$

Figure 1: Model Photophysical Cycle of dHb

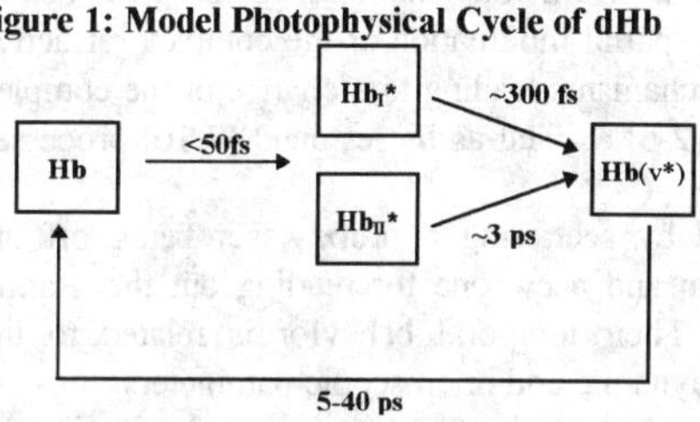

Simulation of Relative Population of Species as a Function of Time (τ_v = 10 ps)

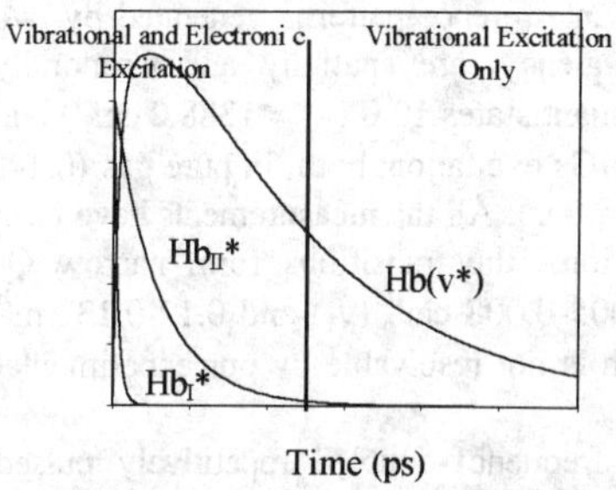

At this point, no distinction has been made between thermal and nonthermal vibrational excitation. Simulations indicate that with reasonable vibrational lifetimes (τ_v=5-40 ps), considerable vibrational excitation persists beyond the point at which Hb_{II}* has decayed completely. In addition, at all τ_v from 5-40 ps, the electronic excitation overlaps vibrational excitation. It becomes clear, then, that the Champion, Hopkins, and Ondrias groups have probed different regions of the decay process. The data of Li et al. and Lingle et al. arose from a regime where significant electronic excited state persists in the presence of a vibrationally excited ground state. Simpson et al. examined the vibrationally excited ground state in the ns TRR studies, and at the lower fluxes in the studies reported here. It now becomes clear that the observed nonthermal vibrational distribution in dHb [3] can be probed effectively by lower flux TRR methods.

References:
[1] P. Li, J. T. Sage, P. M. Champion (1992) J. Chem. Phys. 97:3214.
[2] R. Lingle, Jr., X. Xu, H. Zhu, S.-C. Yu, J. B. Hopkins (1991) J. Phys. Chem. 95:9320.
[3] M.C. Schneebeck, L. E. Vigil, M. R. Ondrias (1993) Chem. Phys. Lett. 215:251. M. C. Schneebeck et al. (1994) Proc. XIVth International Conference on Raman Spectroscopy, p 488. M.C. Simpson, et al. (1995) Proceedings of the Biophysical Society. M.C. Simpson, Ph.D. Dissertation, U. of New Mexico, (1994).
[4] J.W. Petrich, C. Poyart, J.L. Martin (1988) Biochem. 27:4049.
[5] S. Franzen, B. Bohn, C. Poyart, J.L. Martin (1995) Biochem. 34:1224.
Acknowledgements: Funded by NIH (GM33330, to MRO), and a DOE Distinguished Postdoctoral Research Fellowship (MCS), and a Minority Engineering and Mathematics Scholarship (CMC). M.C. Simpson was previously M.C. Schneebeck.

LASER-INDUCED GRATINGS IN THE GAS PHASE EXCITED VIA RAMAN-ACTIVE RESONANCES

D.N. Kozlov[1], R. Bombach[2], B.Hemmerling[2*], and W.Hubschmid[2]

[1]General Physics Institute, Russian Academy of Sciences, Vavilov-str. 38, 117942 Moscow, Russia
[2]Paul Scherrer Institute, CH-5232 Villigen PSI, Switzerland

Laser-induced gratings (LIGs) arise from the interference of two beams originating from the same laser and appear as a spatial modulation of the complex refractive index of the medium under investigation [1]. Mechanisms leading to a change of the complex refractive index are determined by nonresonant [2-6] as well as by resonant [7-10] processes of light-matter interaction.

Experimentally, LIGs are detected by scattering a probe laser beam off the grating. Using a pulsed laser for LIG excitation and a cw one for reading out the grating allows to study the time evolution of LIGs. Their temporal behavior is related to the composition of the medium, and its local thermodynamic and microscopic parameters.

Some results on LIGs in gases, resonantly excited using dipole-allowed transitions of the molecules, have been already discussed in literature [8-10]. Here, results are presented of the first observation of gratings excited via Raman-active transitions (RLIGs) of molecules in the gas phase. RLIGs appear when two interference patterns generated by two lasers, differing from each other by a Raman frequency, are spatially and temporally overlapped. Rovibrational transitions to the Fermi-resonant states 10^00 (ν_1 = 1388.2 cm^{-1}) and 02^00 (ν_2 =1285.4 cm^{-1}) of CO_2 have been used for RLIGs excitation, both, in pure gas (0.1-1 bar) and in mixtures of CO_2 (0.1 bar) with argon (0.5-4 bar). All the measurements have been carried out at room temperature. Under this conditions, the transitions form narrow Q-branches with a quasi-homogeneous linewidths of 0.005-0.008 cm^{-1} (ν_1) and 0.12-0.13 cm^{-1} (ν_2), respectively, leading to rotational structure which is not resolvable by our experimental equipment.

In the experiments the radiation of a frequency-doubled repetitively pulsed Nd:YAG laser (λ_1 = 532 nm, 40 mJ, 8 ns, 20 Hz, spectral width 0.9 cm^{-1}) and of a tunable dye laser (λ_2 = 570-575 nm, 24 mJ, spectral width 0.08 cm^{-1}) were used. In order to excited the grating two pairs of excitation beams were focused into a gas cell under a small angle (θ ~ 10). A single-line Ar^+-laser (λ_3 = 514.5 nm, 1.1 W) was employed for reading out the grating. Radiation scattered off the LIGs was detected by a photomultiplier. The time behavior of the signal was recorded using a digital oscilloscope (500 MHz, sampling rate up to 10^9 s^{-1}). Simultaneously, the intensity of coherent anti-Stokes Raman scattering (CARS) obtained from one of the pairs of excitation beams was measured.

At nonresonant LIG excitation, a damped oscillation of the grating reflectivity with a period of T ~ 60 ns was observed. This is due to the refractive index modulation of a standing acoustic wave which is electrostrictively generated by each of the lasers setting up the grating. The decay of the oscillation, which occurs in about 130-180 ns, is determined not only by viscosity and thermal conductivity but also by the divergence of the two counterpropagating components of the standing acoustic wave as well. If the frequency difference of the two lasers was adjusted to a Raman resonance, the observed time behavior of the grating reflectivity changed. It can be explained as resulting from a superposition of a standing acoustic wave (due to adiabatic compression) and a stationary thermal density gratings (due to isobaric heating). Each of these components contains contributions from electrostriction and from heat, released in the course of collisional relaxation of resonantly excited molecules. In the temporal range 0-0.5 μs after the occurrence of the laser pulses, the strong damped oscillation of the thermal LIG dominates the nonresonant electrostrictive LIG. The thermal grating exhibits a acoustic and a stationary component with equal amplitude, therefore, leading to a doubled oscillation period of the grating reflectivity, 2T ~ 120 ns [6]. The reason for the formation of such a thermal grating is supposed to be fast release of a part of CO_2 molecules excitation energy due to collisional redistribution inside the multiplet of states formed by 10^00, 02^00, and 02^20. The appearance of a broad unstructured hump in the temporal development of the grating reflectivity (t ~ 0.5-20 μs) corresponds to the formation of a stationary thermal LIG due to v-v relaxation within the ν_2 mode and subsequent slow v-T relaxation from the lowest vibrational state 01^10 of CO_2.

The decay of the spatial density modulation is governed by thermal conductivity and by diffusion of the excited molecules across the LIG fringes. The addition of the buffer at constant CO_2 partial pressure produced stronger RLIGs, because of an increased v-T relaxation rate and a decrease of both, the thermal conductivity of the mixture and the diffusion velocity of the excited CO_2 molecules. RLIG reflectivity spectra, recorded simultaneously with CARS spectra, showed no nonresonant background, had a high level of scattered light intensity, and had a signal-to-noise ratio comparable to that one of CARS.

In conclusion, the experimental realization of a new method of coherent Raman spectroscopy has been shown, and its use for local diagnostics of gaseous media has been demonstrated.

[1] H.J. Eichler, P. Günter and D.W. Pohl, Laser-Induced Dynamic Gratings, Springer, Berlin, 1986.
[2] B. Hemmerling and A.Stampanoni-Panariello, Appl.Phys. B57 (1993) 281.
[3] D.E. Govoni, J.A. Booze, A. Sinha and F.F. Crim, Chem. Phys. Lett. 216 (1993) 525.
[4] E.B. Cummings, Opt. Lett. 19 (1994) 1361).
[5] A. Stampanoni-Panariello, B. Hemmerling and W. Hubschmid, Phys. Rev. A51 (1995) 655.
[6] W. Hubschmid, B. Hemmerling and A. Stampanoni-Panariello, J. Opt. Soc. Am. B12 (1995) 1850.
[7] G. Meijer and D.W. Chandler, Chem. Phys. Lett. 192 (1992) 1.
[8] A. Dreizler, T. Dreier and J. Wolfrum, Chem. Phys. Lett. 233 (1995) 525.
[9] P.H. Paul, R.L. Farrow and P.M. Danehy, J. Opt. Soc. Am. B12 (1995) 383.
[10] P.M. Danehy, P.H. Paul and R.L. Farrow, J. Opt. Soc. Am. B12 (1995) 1564.

PICOSECOND 2D-CARS SPECTROSCOPY OF THE EXCITED STATES OF DIPHYENYLACETYLENE

Taka-aki Ishibashi[1] and Hiro-o Hamaguchi[1,2]

[1] Molecular Spectroscopy Laboratory, Kanagawa Academy of Science and Technology (KAST), KSP E301, 3-2-1 Sakado, Takatsu, Kawasaki 213, Japan
[2] Department of Basic Science, Graduate School of Arts and Sciences, The University of Tokyo, 3-8-1 Komaba, Meguro, Tokyo 153, Japan

Diphenylacetylene (DPA) is one of the fundamental aromatic molecules that has a carbon-carbon triple bond. Electronically excited states of DPA are of interest with regard to the structural change of the triple bond upon the photoexcitation. Hirata et al. observed the electronic absorption spectra of the S_1 (435 nm) and the S_2 (500 nm) states of DPA in solution. They showed that the fluorescence state of DPA is not S_1 but S_2 as a result of a lifetime analysis. Spontaneous Raman spectra of DPA were reported for the T_1 state, the cation radical, and the anion radical by Hiura and Takahashi, but no Raman spectra of the singlet excited states have been measured because of the hindrance by strong fluorescence [2]. In this paper we present the CARS spectra of the S_1 and the T_1 states of DPA in solution.

The picosecond 2D-CARS technique [3] was used to obtain time-resolved CARS spectra. CARS probing radiations were obtained from a Q-switched Nd:YLF laser (second harmonic 523 nm, ~5 ns, ~10 mW, 1 kHz) and an amplified broad band dye laser (556-608 nm, 5 ns, ~10 mW, 1 kHz). The CARS signal was analyzed with a polychromator and detected with a streak camera (time resolution: ~20 ps). The third harmonic of a mode-locked Ti:sapphire regenerative amplifier (~270 nm, < 1 ps, ~1 mW, 1 kHz) was used to photoexcite DPA (5mM in cyclohexane).

2D-CARS spectra of DPA were measured for the frequency region from 1100 cm^{-1} to 2600 cm^{-1}. Two kinds of transient CARS bands with different lifetimes were observed. The first group of bands (1127, 1556, and 1573 cm^{-1} marked by filled diamonds) appear immediately after the excitation, and decay almost completely after ~300 ps. (Fig. 1) The second kind of band (1969 cm^{-1} marked by an open diamond) increases in intensity during ~300 ps and does not decrease at least until 5 ns after the photoexcitation. We did spectral simulations based on the ordinary Lorentz-type susceptibility, assuming two transient species, and determined the peak positions and concentrations of the transient species as functions of time. The long-lived CARS band is assigned to the T_1 state, because the lifetime of this band is longer than 5 ns and the band position is identical with that of C≡C stretching band observed in the spontaneous Raman spectrum of T_1 DPA [2]. The decay curve of the short-lived species

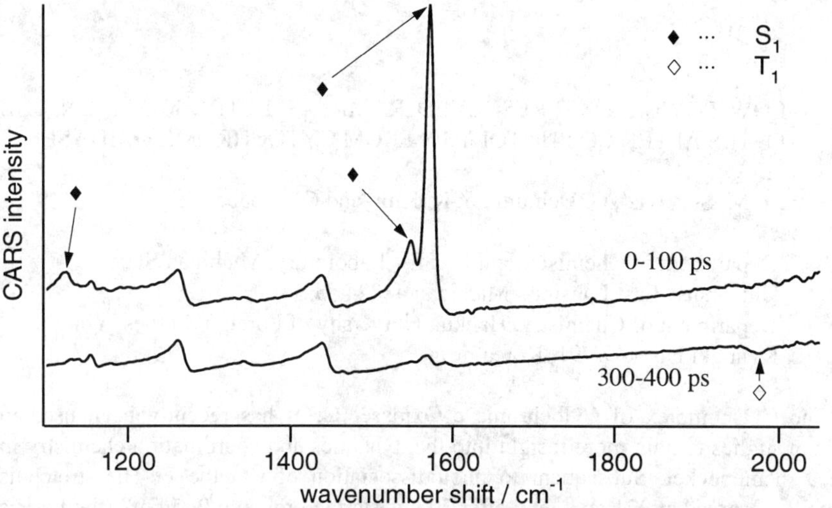

Fig. 1. Time-resolved CARS spectra of DPA

can be fitted well to a single exponential with a lifetime of ~100 ps. (Fig. 2) The short-lived CARS bands are assigned to the S_1 state, since the lifetime thus obtained is close to that of S_1 (~200 ps) rather than that of S_2 (~8 ps) obtained by time-resolved absorption spectroscopy [1]. The vibrational mode assignments of the S_1 CARS bands are not clear at present. An interesting point is that no strong band was observed in the frequency region of 1600 cm^{-1} ~ 2600 cm^{-1}, though the C≡C stretching band is usually expected to appear in this region. One possibility is that the CARS C≡C stretching band is so small in its intrinsic intensity and/or so broad that it is not observed. Another possibility is that the electronic structure of the S_1 state is markedly different from that of S_0, giving a much lower stretch frequency. The 1556 or/and 1573 cm^{-1} bands may correspond to the C≡C stretching mode. In order to clarify which is the case, we are planning to measure the CARS spectrum of ^{13}C-substituted species of DPA.

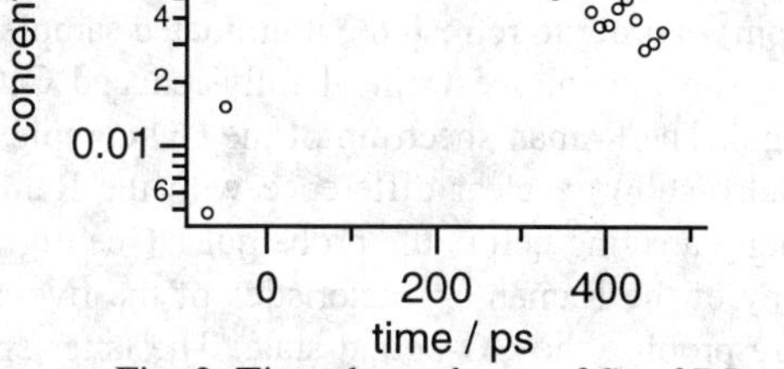

Fig. 2. Time dependence of S_1 of DPA

1. Y. Hirata et al., *J. Phys. Chem.*, 97, 9677 (1993)
2. H. Hiura and H. Takahashi, *J. Phys. Chem.*, 96, 8909 (1992)
3. T. Tahara and H. Hamaguchi, *Rev. Sci. Instru.*, 65, 3332 (1994)

LOW POWER TIME-RESOLVED RAMAN SPECTROSCOPY ON THE FE-HIS AFTER CO PHOTOLYSIS FROM CYTOCHROME OXIDASE

J.P.M. Schelvis[1], G. Deinum[1], Y-K. Kim[2] and G.T. Babcock[1]

[1]Department of Chemistry and LASER Laboratory, Michigan State University, East Lansing, Michigan 48824, U.S.A.
[2]Department of Chemistry, Hankuk University of Foreign Studies, Yongin, Kyungki Do, 449-791, Korea

The CO complex of Cytochrome c Oxidase (CcO) has recently been used in Raman studies to gain more insight into the dynamics and coordination chemistry in the a_3 hemepocket. Subsequent to photodissociation of CO the Fe-His stretch is initially observed at 221 cm^{-1}, and after 200 ns starts to relax to 215 cm^{-1} (the Fe-His stretch in the fully reduced enzyme) with a 1 μs half time. These data have been interpreted as a two step relaxation of the a_3 hemepocket to its equilibrium conformation [1]. Recently, it has been proposed that after CO photolysis a distal ligand binds to the a_3 heme Fe, breaking the Fe-His bond [2,3]. It was concluded that this new ligand must be photolabile and that the probe light used to detect this bond is sufficient to dissociate the ligand and restore the Fe-His bond, of which the time dependenc of its frequency relaxation is similar to that described above.

In this paper we present preliminary results of low-power ps resonant Raman experiments on the CO complex of CcO from bovine heart. The CcO was used in a concentration of about 150 μM. The laser system used has been described elsewhere [4]. Photolysis was initiated with 592 nm, 70 nJ pulses (23 mJ/cm^{-2}) and the Raman spectrum was taken with a 441.6 nm, 25 nJ probe pulses (8 mJ/cm^{-2}). Both dye lasers were operated at 150 kHz and the sample was contained in a spinning cell (5400 rpm) in order to refresh the illuminated sample-volume between two pump pulses.

The Raman spectrum of fully reduced CcO shows the Fe-His stretching at 212 cm^{-1}. The Raman spectrum of the CO complex of CcO obtained with only the probe pulse shows a clear difference with the Raman spectrum collected with the pump pulse arriving before the probe pulse (see Fig.1). The former spectrum does not have any of the Raman characteristics of the five-coordinate a_3 heme, indicating that we are probing the CO-bound state. The latter spectrum, however, shows a significant intensity around 217 cm^{-1} and at 364 cm^{-1}. The latter peak is due to a heme deformation mode of the 5-coordinated species and the former peak is most likely due to an Fe-His stretch. It is not clear whether this is the proximal His or a distal His. The position and the intensity of the peak at 217 cm^{-1} was obtained by normalizing both spectra to the ν3 (340 cm^{-1}), subtracting them and fitting a gaussian to the difference spectrum from 175 to 250 cm^{-1}. The position and the intensity of the peak as a function of time are shown in the inset of Fig. 1. Although the intensity of the peak rises with overlapping pump and probe pulse, the peak shows up at 217 cm^{-1} as soon as it is detectable. The 217 cm^{-1} peak persists on the ns time scale (data not shown) and relaxes back to 213 cm^{-1} at 0.8 μs.

Although the probe energy is too low to photodissociate CO to a significant extent and the pump energy is sufficient to photolyse about 30% of the CO [3], the current results are not conclusive as to whether a distal ligand binds to the a_3 heme. If the probe power in our experiments is low enough to avoid photodissociation of a possible distal ligand, this ligand is either a distal His with a Fe-His stretch at 217 cm^{-1} or not present. However, if we do photodissociate the distal ligand this might happen around t=0 ps which is the shortest time delay at which the 217 cm^{-1} peak is observed. Experiments with lower probe powers are being performed to resolve this uncertainty.

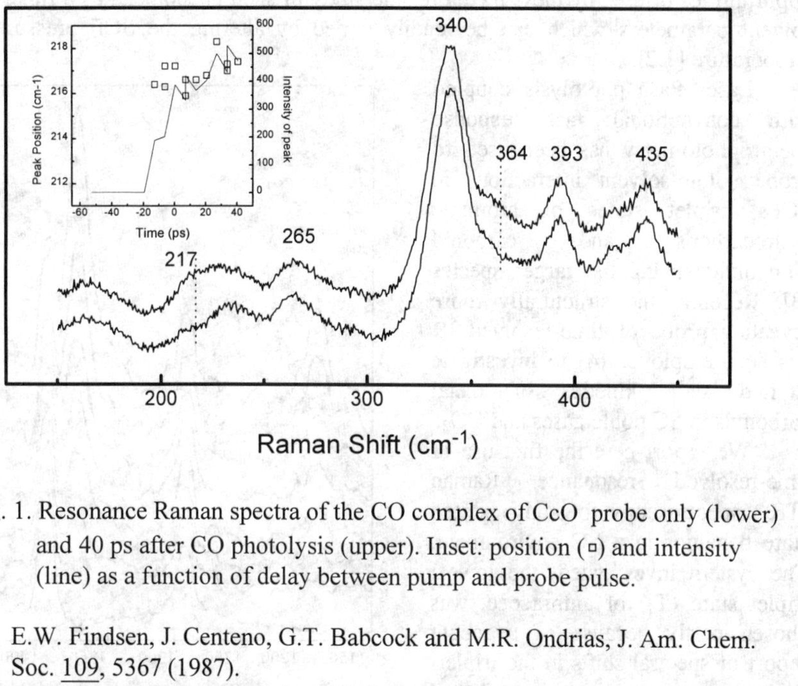

Fig. 1. Resonance Raman spectra of the CO complex of CcO probe only (lower) and 40 ps after CO photolysis (upper). Inset: position (▫) and intensity (line) as a function of delay between pump and probe pulse.

1. E.W. Findsen, J. Centeno, G.T. Babcock and M.R. Ondrias, J. Am. Chem. Soc. 109, 5367 (1987).
2. W.H. Woodruff, O. Einarsdottir, R.B. Dyer, K.B. Bagley, G. Palmer, S.J. Atherton, R.A. Goldbeck, T.D. Dawes and D.S. Kliger, Proc. Natl. Acad. Sci 88, 2588 (1991).
3. B-S. Lou, R.W. Larsen, S.I. Chan and M.R. Ondrias, J. Am. Chem. Soc. 115, 403 (1993).
4. D.H. Kreszowski, G. Deinum and G.T. Babcock, J. Am. Chem. Soc. 116, 7463 (1994).

TIME-RESOLVED RESONANCE RAMAN SCATTERING FROM THE TRIPLET STATE OF ANTHRACENE IN SUPERCRITICAL CO_2

J.N.M. Hegarty*, J.J. McGarvey , S.E.J. Bell and A.H.R. Al-Obaidi

School of Chemistry, Queen's University of Belfast, Belfast BT9 5AG, N. Ireland.

There has been a growing interest in the study of photophysical and excited state kinetic phenomena in supercritical fluids (SCFs), reflecting recognition of the opportunities offered by these media for the study of such phenomena as a function of solvent parameters which can be readily varied by altering the SCF pressure and temperature [1,2].

Laser flash photolysis coupled with conventional fast response spectrophotometry has been used to probe solute-solvent interactions in SCFs, triplet states of aromatic hydrocarbons and carbonyl compounds being the target species [3]. Recently the structurally more revealing probe of time-resolved IR has been employed [4] to investigate excited state kinetics of metal carbonyls in SC noble gases and CO_2.

We report here the first use of time-resolved resonance Raman (TR^3) spectroscopy to study excited state behavior in a SC environment. The system investigated, the lowest triplet state (T_1) of anthracene, was chosen partly because of a recent report of spectral shifts in the triplet-triplet absorption spectrum, attributed to London dispersion forces and local enhancement of solvent density effects [5].

Fig. 1 shows TR^3 spectra of the T_1 state of anthracene in SC CO_2, at 55°C recorded at a series of pump-probe delays in the range 50ns - 3μs.

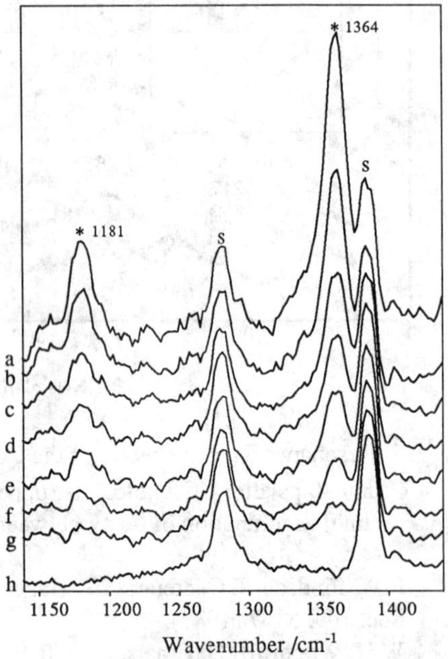

Figure 1. Time resolved resonance Raman spectra of anthracene in SC CO_2 recorded with 355nm/416nm pump/probe delays: (a) 50ns, (b) 100ns, (c) 200ns, (d) 300ns, (e) 400ns, (f) 600ns, (g)1000ns, (h) probe only. Excited state features are labeled *, CO_2 bands are marked s.

The transient features can be identified by comparison with RR spectra of the T_1 state in cyclohexane solution [6]. With increasing pump-probe delay the features decay as expected. TR^3 spectra recorded at a lower temperature (35°C) were similar to those shown in Fig. 1.

No significant frequency shifts have been noted in any of the principal T_1 anthracene bands. However, the intensity (I) of the T_1 bands relative to those of CO_2 (I_{CO_2}/I_{T_1}) increases dramatically with increasing applied pressure in the range 90-500 bar at 55°C (Fig. 2). In contrast to this study a similar experiment at 35°C over the range 78 - 170 bar showed no significant changes in I_{CO_2}/I_{T_1}. The increase in RR band intensity at 55°C could be accounted for in terms of increased resonance Raman enhancement arising from spectral shifts of the T_1 absorption band as a function of fluid pressure, [5] and/or the increased solubility [7] of anthracene in CO_2 with increasing pressure. However, measurement of the absorption spectra of the T_1 state over the same pressure range as the TR^3 measurements and at the same two temperatures, show only a small red shift (408 → 412 nm). Moreover, the pressure dependence of the spectral shifts are near identical at both temperatures. This suggests that it is solubility rather than changes in the degree of resonance enhancement which is the controlling influence on the intensity variations shown in Fig. 2.

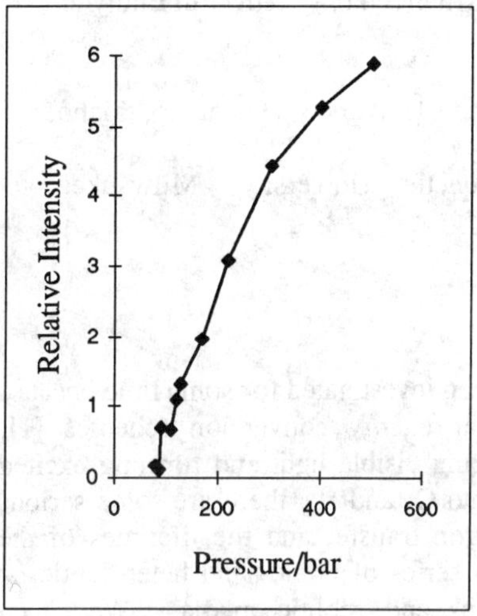

Figure 2. Relative intensity of the Raman bands of CO_2 vs. T_1 bands of anthracene measured as a function of pressure.

Acknowledgment : We thank the EPSRC for support (Grant GR/J01905)

1. M. Poliakoff, S.M. Howdle and S.G. Kazarian, Angew. Chem. Int. Ed. 34, 1275 (1995).
2. S. Akimoto and O. Kajimoto, Chem. Phys. Lett. 209, 263 (1993).
3. C.B. Roberts, J.E. Chateauneuf and J.F. Brenecke, J. Am. Chem. Soc. 114, 8455 (1992).
4. X.-Z. Sun, M.W. George, S.G. Kazarian, S.M. Nikiforov and M. Poliakoff, J. Am. Chem. Soc. submitted for publication.
5. C.B. Roberts, J.F. Brenecke and J.E. Chateauneuf, J. Chem. Soc. Chem. Commun. 868 (1993).
6. G.N.R. Tripathi and M.R. Fisher, Chem. Phys. Lett., 104, 297, (1984).
7. T.W. Zerda, B. Wiegand and J. Jonas, J. Chem. Eng. Data, 31, 274, (1986)

RESONANCE RAMAN AND TIME-RESOLVED RESONANCE RAMAN STUDIES OF 4,4' BIPYRIMIDINE COMPLEXES OF RU(II)

David J. Manuel[1], Dennis P. Strommen[1*], Milan Sykora[2], Anwarul Bhuiyan[2] and James R. Kincaid[2]

1. Department of Chemistry, Idaho State University, Pocatello, Idaho, 83209. USA
2. Department of Chemistry, Marquette University, Milwaukee, Wisconsin, 53233. USA

Tris-α-diimine complexes of Ru(II) have been investigated for some time because of their potential as photosensitizers in solar energy conversion schemes [1]. Although they are relativley efficient at capturing visible light and forming excited states with the capabilty of converting water into O_2 and H_2, there are some serious problems related to the rapidity of back-electron transfer and the lifetimes of the excited state species. Recently we prepared a series of homo- and heteroleptic α-diimine complexes of Ru(II) in both aqueous and zeolitic media . We then investigated the complexes via resonance Raman(RR) and Time-resolved resonance Raman(TR^3) [2,3]. RR and TR^3 are uniquely suited to study the electron distribution in the excited state of a molecule. Our interest has been to modify the lifetimes of the α-diimine complexes by trapping them in the supercages of zeolite Y. We were mildly succesful in this respect in the case of $Ru(bpz)_3^{2+}$ where the lifetime increased from 853 in aqueous solution to 1433 nsec in the zeolite.

We now wish to investigate the effect of the position of the N's external to the α-diimine linkage on the lifetimes within the zeolite supercage. Thus we prepared the relativley little-investigated ligand bpm(4-4'bipyrimidine) and formed the tris-homoleptic ruthenium(II) complex as well as the heteroleptic counterparts with bipyridine as the second ligand. The ligand and complexes were prepared according to literature methods [4-6].

The resonance Raman spectrum of $Ru(bpm)_3^{2+}$ is shown in Figure 1. and is typical of this type of complex. Emission is observed at 700nm. We were unable to determine the very short lifetime of the complex in aqueous solution at room temperature but obtained a value of 350nsec at 77K in methanol/ethanol(4/1) solution.

The TR^3 spectrum of the excited state of $Ru(bpm)_3^{2+}$ is shown in Figure 2. Neutral ligand bands are noticeabley absent. A spectrum taken of the complex with a 350.9 CW source showed that the intensity of the 1324cm^{-1} is comparable to that of the highest intensity band in the CW spectrum, yet it is seen as a very small peak in the TR^3 spectrum.

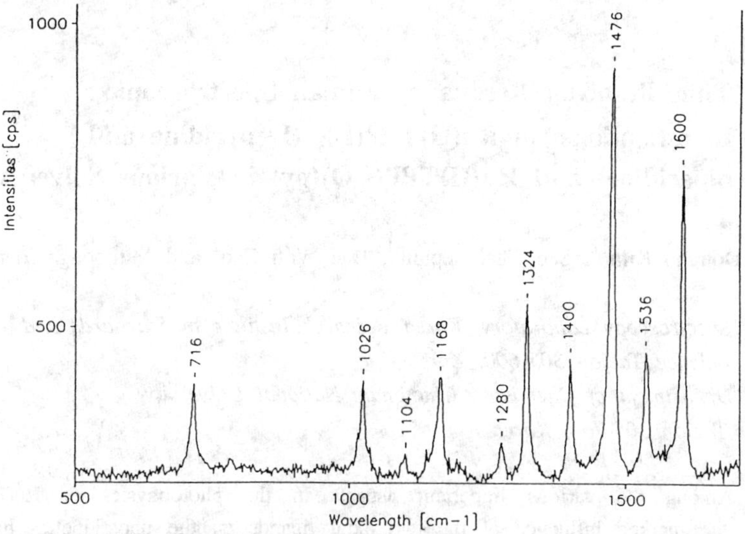

Fig. 1. Resonance Raman Spectrum of $Ru(bpm)_3^{2+}$ taken with 514.5nm excitation.

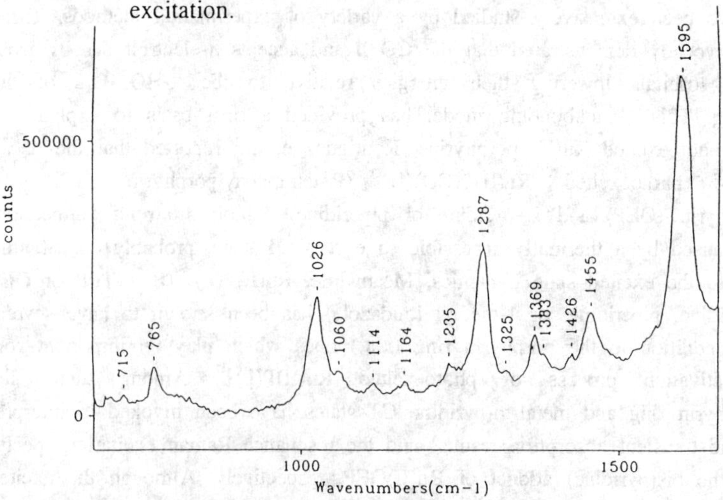

Fig. 2. Time-resolved Resonance Raman Spectrum of $Ru(bpm)_3^{2+}$ taken with 354.7nm excitation.

1. A. Juris, V. Balzani, F. Barigelletti, S. Campagna, P. Belser and A. Von Zelewsky, Coord. Chem. Rev., 1988, **84**, 85.
2. K. Maruszewski, D.P. Strommen, K. Handrich, and J.R. Kincaid, Inorg. Chem. 1991, **30**, 4579.
3. K. Maruszewski, D.P. Strommen and J.R. Kincaid, J. Amer. Chem. Soc. 1993, **115,** 8345.
4. F. Effenberger, Chem. Ber. 1965, **98**, 2260.
5. S. Ernst and W. Kaim, Inorg. Chem., 1989, **28**, 1520.

Time-Resolved Resonance Raman Spectroscopic Investigations on Ru(II)TPP$(L)_2$ (L=pyridine and piperidine) and Ru(II)TPP(CO)(py) in Various Solvents

Dongho Kim[1]*, Sae Chae Jeoung[1], Dae Won Cho[2] and Minjoong Yoon[2]

[1]*Spectroscopy Laboratory, Korea Research Institute of Standards and Science, Taejon 305-600, Korea*
[2]*Department of Chemistry, Chungnam National University Taejon 305-764, Korea*

Among the various important aspects in the photophysics of Ru(II) porphyrins, the marked influence of the two axial ligands on the photokinetics has drawn special attention. The effects of the two axial ligands on the photophysical properties have been extensively studied by a variety of experimental methods. These studies unequivocally demonstrated that the CO ligand accepts π-electron density from the metal d_π orbitals lowering their energies relative to the π-HOMO's of the porphyrin ring. This π-backbonding model has provided a firm basis to explain the dynamics of photoexcited Ru(II) porphyrins. Rodriguez et al.[1] reported that the decay dynamics of photoexcited Ru(II)P(CO)(L) (P=tetraphenylporphyrin (TPP) or octaethylporphyrin (OEP) and L=pyridine or piperidine) exhibit a strong temperature dependence caused by a thermally accessible $(d_\pi,e_g(\pi^*))$ CT state, probably contributing substantially to the excited state dynamics. Meanwhile, Ru(II)P$(L)_2$ (P = TPP or OEP and L = pyridine, piperidine, or 1-methyl imidazole) has been known to have several CT states in addition to the porphyrin ring (π,π^*) ones which play an important role in the deactivation process of photoexcited Ru(II)P$(L)_2$.[1] Among them, the metal-to-porphyrin ring and metal-to-pyridine CT states have been invoked to interpret the picosecond transient absorption results[1] and the resonance Raman excitation spectra (RRES)[2] of the bis(pyridine) adduct of Ru(II)OEP, respectively. Although the excited state dynamics of photoexcited Ru(II) porphyrins have been investigated by transient absorption and luminescence spectra,[1] the vibrational spectroscopic studies for the structural changes resulting from photoexcitation have not yet been reported.

Transient Raman spectra with an excitation of nanosecond pulses at 416 nm were first recorded for Ru(II)TPP$(L)_2$ (L=pyridine and piperidine) (Fig. 1) and Ru(II)TPP(CO)(py) in various solvents such as benzene, tetrahydrofuran, pyridine, and piperidine (Fig. 2). It was found that the electronic structure of the lowest energy charge transfer (CT) state is mainly dependent on not only the nature of axial ligand attached to the central metal but the solvent employed. As for Ru(II)TPP$(L)_2$

(L=pyridine and piperidine) in neat pyridine and piperidine solvents, the transient species were observed to be different in structure depending on the axial ligand. Several reasons for this observation were considered, including the discussion on the nature of the CT states such as (π,d), (d,d), (d_π,π^*(py)), and (d_π,π^*(ring)) states, and the configuration interaction between the two $e_g(\pi^*)$ orbitals. The present transient Raman spectroscopic studies on photoexcited Ru(II)TPP(pip)$_2$ has manifested the Jahn-Teller effects, which may result in a diamond distortion of porphyrin ring superimposed by a minor rectangular distortion. This major B_{2g} distortion accounts for the observation of a variety of unusual Raman spectral features in the photoexcited Ru(II)TPP(pip)$_2$.

Meanwhile, the introduction of a CO ligand to Ru(II) porphyrin also altered the nature of the lowest CT state in non-ligating and weakly ligating solvents. It was observed that the CO ligand of the carbonylated Ru(II) porphyrin was substantially replaced by the solvent molecules such as pyridine and piperidine in the excited state. These experimental results were interpreted in terms of the change in the charge distribution among the porphyrin ring, the central metal and the axial ligand in the metal-to-ring (d_π,$e_g(\pi^*)$) CT state. The relevance of our findings to the photodynamics observed by the previous picosecond transient absorption results of photoexcited Ru(II) porphyrins has been discussed by examining the conformational changes of photoexcited species.

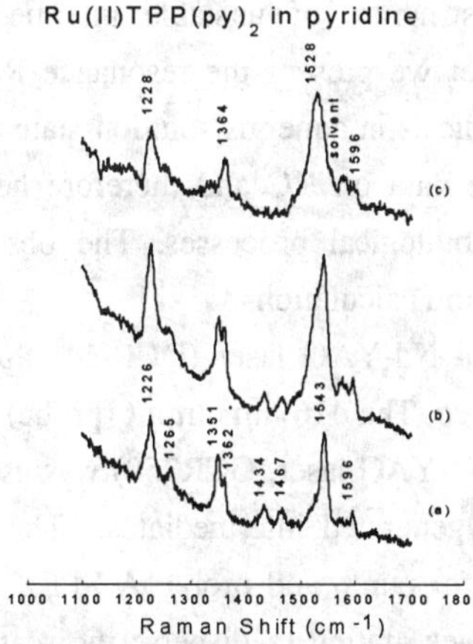

Fig. 1 Transient Raman spectra of Ru(II)TPP(py)$_2$ with 416 nm single pulse excitation; (a) the low power, (b) the high power and (c) difference Raman spectra (b)-(a) with a proper subtraction factor

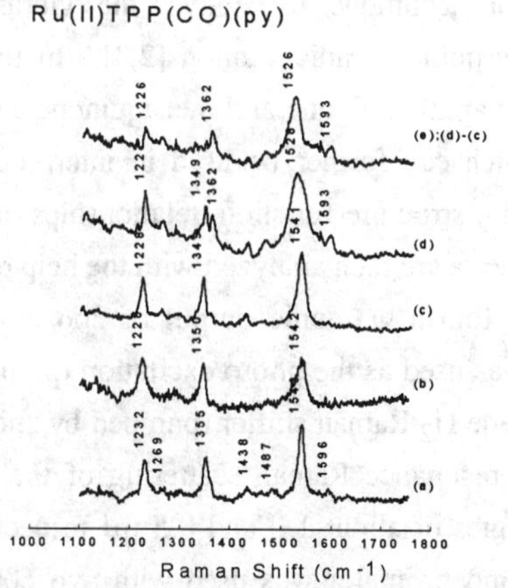

Fig. 2 Transient Raman spectra of Ru(II)TPP(CO)(py) with 416 nm single pulse excitation; (a) the low power, (b) the difference spectrum in benzene, (c) in THF, (d) in pyridine and (e) the difference spectrum (d)-(c).

1. J. Rodriguez, L. McDowell and D. Holten, Chem. Phys. Lett. 147, 235 (1988)
2. G. A. Schick and D. F. Bocian, J. Am. Chem. Soc. 106, 1682 (1984)
3. S. C. Jeoung, D. Kim, D. W. Cho and M. Yoon, J. Phys. Chem. 99, 5826 (1995)

TIME RESOLVED RESONANCE RAMAN SPECTROSCOPIC STUDIES ON THE RADICAL ANIONS OF MENAQUINONE AND NAPHTHOQUINONE

Gurusamy Balakrishnan, Pothukattil Mohandas and Siva Umapathy*

Department of Inorganic and Physical Chemistry,
Indian Institute of Science, Bangalore- 560 012, India.

Quinones and their reactive intermediates have been studied widely because of their importance to various biological redox processes and also because of their potential for applications in various industrial processes. The reactive intermediates of NQ and its derivatives participate in a number of biological electron transfer processes. For example, 2-methyl-1,4-naphthoquinone (MQ=menaquinone) radical anion is known to be an active intermediate in various biological processes, such as photosynthesis and transport of electrons to cytochrome[1].

Time resolved resonance Raman (TR3) spectroscopy has been well demonstrated to be an ideal technique to study the vibrational structure of unstable intermediates particularly quinone radical anions[2,3]. In this paper we present the resonance Raman spectra of naphthoquinone and menaquinone anion radicals in aqueous solution state using TR3S, which can further be used in interpreting the data on RC and therefore help in elucidating structure-function relationships in the biological processes. The observed Raman spectra are then analyzed with the help of ab initio calculations.

The fourth harmonic output of 266 nm from a Nd-YAG laser (DCR-11, Spectra Physics) was used as the photo excitation (pump) source. The 416 nm output (probe) from a home made H_2 Raman shifter (pumped by another Nd-YAG laser, GCR250) was used to probe the resonance Raman scattering of the photo generated intermediates. The laser pulse energies of about 1.0 and 0.5 mJ respectively, for pump and probe. A SPEX 1404 double monochromator was used with two 600 grooves grating to disperse the scattered light. A liq.N_2 cooled CCD (Princeton Instruments) with 576X378 pixels was used as the multichannel detector. The sample solutions were circulated through a capillary at a rate of about 10 ml per minute. In order to avoid possible accumulations of photoproducts, samples were replaced regularly. All the sample solutions were deoxygenated to avoid quenching of the intermediates by oxygen. A typical sample solution contained 1.0 mM of the quinone and 0.1 M of sodium nitrite in water.

Ab initio MO calculations were performed with the Gaussian-92 program on an IBM-RS6000 computer system employing the standard 6-31G split-valence basis set. Calculations for the neutral molecules and anion radicals were carried out using restricted Hartree-Fock (RHF) and unrestricted Hartree-Fock (UHF) formalisms, respectively. Ab initio studies have also been extended to investigate the effect of isotopic substitution (^{16}O to ^{18}O) on the potential energy distribution (PED) of the normal modes. The table summarises the final assignments of the experimental vibrational frequencies observed in our experiment of NQ, MQ and their anion radicals. These assignments are also consistent with the qualitative trend observed in other quinones.

Experimental Vibrational freqeuncies (cm^{-1}) and assignments of NQ, MQ and their radical anion.

NQ		MQ		Vibrational
Neutral	Anion	Neutral	Anion	assignment
1660	1441	1672	1442	sym $\nu_{C=O}$
--	--	--	1468	
--	1603	1624	1605	quin $\nu_{C=C}$
1599	1537	1599	1539	arom $\nu_{C=C}$
1445	--	1447	--	
1407	--	1385	--	arom/quin
1357	1327	1362	1339	$\nu_{C-C}/\nu_{C=C}$
1323	--	1305	--	

We have analysed the vibrational spectra of NQ and MQ in their neutral and rdaical anion forms. Ab initio and literature data of similar systems have been used to assign all the observed Raman frequencies. The asymmetric substitution present in MQ with respect to NQ shows important signatures in the radical anion spectrum. The PED values obtained from ab initio studies also reflect the effect of methyl substitution on coupling between C=O and C=C stretching modes. Further work is in progress in identifying the other intermediates, such as protonated radicals of NQ and MQ and the intermediates in vitamin K_1 (a MQ analog), and 2,3-dimethyl-1,4-Naphthoquinone.

Acknowledgments: We thank JNCASR, DST and CSIR for financial support.

References

1. (a) S. Patai, Eds. The Chemistry of the Quinoid Compounds, Wiley: New York, 1974. (b) R.A. Morton, Ed. Biochemistry of Quinons, Academic Press: London and New York, 1965.
2. (a) S. Umapathy, A.M. Cartner, A.W. Parker, R.E. Hester, J. Phys. Chem. 94, 8880, (1990) , (b) G. Balakrishnan, P. Mohandas, S. Umapathy, submitted to J. Phy. Chem.
3. G.N.R. Tripathi, In Time Resolved Spectroscopy; Clark, R.J.H.; Hester, R.E., Eds.; John wiley & Sons: New York, 1989; Vol.18, Chapter 4.

EXCITED STATE CHARGE LOCALIZATION OF METAL COMPLEXES WITH SUBSTITUTED DIPYRIDO[3,2-a:2',3'-c] PHENAZINE LIGANDS, PROBED BY TIME-RESOLVED RESONANCE RAMAN SPECTROSCOPY AND SPECTROELECTROCHEMISTRY

K. C. Gordon* and M. Waterland

Department of Chemistry, University of Otago, PO Box 56, Dunedin, New Zealand

Metal complexes containing ligands based on dipyrido[3,2-a:2',3'-c]phenazine, dppz (Figure 1), are of interest because of their properties with respect to DNA intercalation [1]. However complexes with these ligands also possess properties that may be utilized in solar energy harvesting schemes. Solar energy conversion using metal polypyridyl complexes use metal-to-ligand charge-transfer (MLCT) states for photon capture. Charge-separation is an important factor in the success of a material as a solar photosensitizer. One of the main drawbacks of the metal polypyridyl complexes is their rapid back electron transfer form the MLCT state. The dppz ligands possess poor ligand-π^* metal-dπ orbital overlap which should reduce the rate of back electron transfer. Indeed the ligand-π^* acceptor orbital tends to localize charge away form the metal center. However, the low ligand-π^* metal-dπ orbital overlap means that the MLCT transitions of these complexes have low oscillator strength. It is possible to use the ligand-centered $\pi \rightarrow \pi^*$ transition as the sensitizing chromophore for solar photon capture. The allowed transition has a high oscillator strength and with appropriate ring substituents it may be tuned to the 400 nm region of the spectrum. This accomplishes high photon absorption but to utilize the energy it is necessary to separate the charge through the population of the MLCT state. This may be the first step in a series of charge-separation stages.

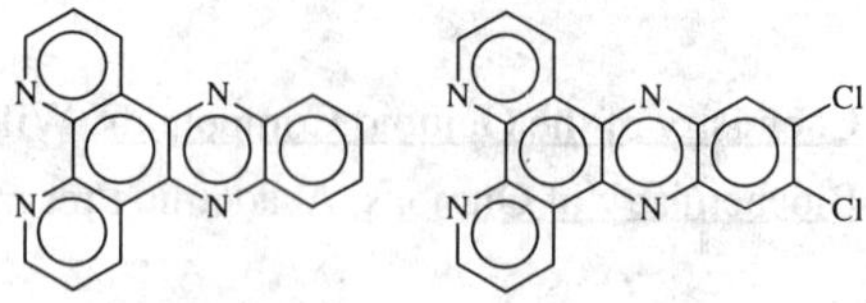

dppz dppzCl_2

Figure 1

Time-resolved resonance Raman (TR^3) spectroscopy may be used to gauge the extent of charge-separation within the dppz ligand framework [2] as a result of photoexcitation. We have used TR^3 spectroscopy to observe charge-localization in the excited states of a series of rhenium(I) complexes with substituted dppz ligands. As the extent of charge-transfer, MLCT character for the lowest excited state increases then the observed TR^3 spectrum resembles that of the radical anion of the ligand. The radical anion spectrum may be obtained using Raman spectroelectrochemical techniques.

For $Re(CO)_3Cl(dppzCl_2)$ the TR^3 spectrum and that of $dppzCl_2^{\cdot-}$ are similar indicating significant MLCT character in the THEXI state.

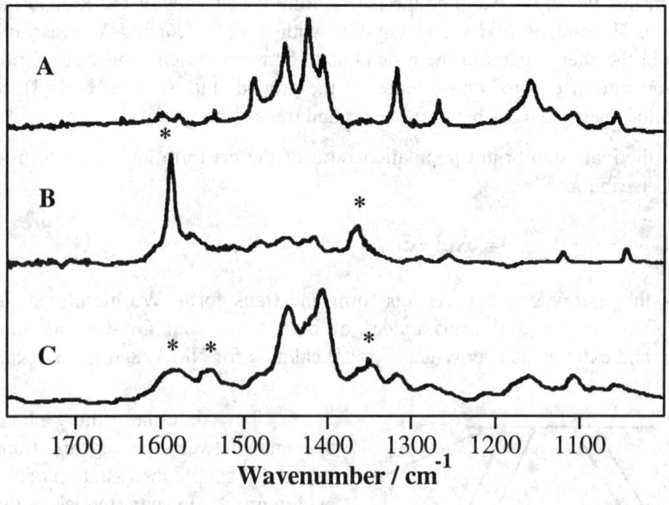

Figure 2. A. Resonance Raman spectrum of $Re(CO)_3Cl(dppzCl_2)$ in CH_2Cl_2, λ = 457.9 nm. B. Resonance Raman spectrum of electrochemically reduced $Re(CO)_3Cl(dppzCl_2)$ in CH_2Cl_2, λ = 457.9 nm. C. Time-resolved resonance Raman spectrum of electrochemically reduced $Re(CO)_3Cl(dppzCl_2)$ in CH_2Cl_2, λ = 355 nm.

Acknowledgments: Support from the New Zealand Lottery Commission and the University of Otago Research Committee for the purchase of the Raman spectrometer is gratefully acknowledged. We also thank the University of Otago for the award of a Ph.D. scholarship to M. W.

1. R. Hartshon and J. K. Barton J. Am. Chem. Soc. 116, 5919, (1994).
2. J. R. Schoonover, W. D. Bates and T. J. Meyer Inorg. Chem. 34, 6421, (1995).

Resonance Raman Studies of the Ground and Excited State Photochemical Dynamics of Amides and Peptides

Pusheng Li, Xingguo Chen, ShuLin E, and Sanford A. Asher
Department of Chemistry, University of Pittsburgh, Pittsburgh, PA 15260, U. S. A.

UV resonance Raman studies[1,2] have recently demonstrated the existence of monophotonic photoisomerization of the normally trans-amide linkage to a cis-linkage for aqueous solutions of *N*-methylacetamide (NMA), Glycylglycine (Gly-Gly), and other small peptides. In this work, we examine the temperature dependence of the Raman intensity of the cis-amide II band of NMA and Gly-Gly with a CW 206nm UV laser, in order to determine (1) the energy gaps in the ground state between the cis- and trans-forms, (2) the isomerization rate of cis- to trans-amides on the ground state surface, and (3) the ground state activation energy barrier between the cis and trans form.

The thermal equilibrium population ratio of the cis-form and trans form obeys the Boltzmann distribution:

$$\frac{N_c}{N_t} = \exp(-E_g / k_B T) \tag{1}$$

where E_g is the energy gap between cis form and trans form. We monitored the Raman intensities of cis-amides and trans-amides of both NMA and Gly-Gly as functions of temperature, and extract the energy gap E_g=2.5kcal/mile for NMA, and E_g=3.1kcal/mole for Gly-Gly.

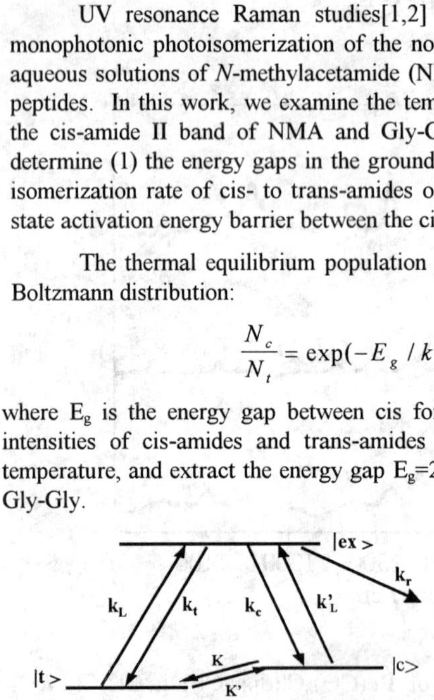

Figure 1. The three-state model

To describe the photochemical dynamics between the cis- and trans amides, we used a simple three-state model as shown in Figure 1. In our model, the amides isomerization reaction coordinate is restricted to one dimension about the CN bond axis. Figure 1 shows that the isomerization occurs by both photochemical and thermal pathways. K and K' are the thermal isomerization rates of trans and cis amides in the ground state, respectively, and k_L and k'_L are the photoexcitation rates from the trans and cis forms to the excited state, respectively. k_L and k'_L depend on the absorption cross sections and pump flux. k_L and k_c are photochemical relaxation rates back to the trans and cis forms, respectively. k_r is the photochemical rate to different chemical species[3]

In the case of pulsed laser excitation, the photoexcitation rates, k_L and k'_L, are much faster than the thermal exchange rates between ground state cis and trans amides, K and K'. Thus, the population of the cis-form only depends on the quantum yield of the photoisomerization of trans- to the cis-form. In small k_r case, we can determine the quantum yield by measuring the cis population under photo-steady state conditions with the formula:

$$R = \frac{k_c}{k_t} = \frac{\sigma_c N_c^{st}}{\sigma_t N_t^{st}} \tag{2}$$

here σ_t and σ_c are the absorption cross sections of the trans- and cis-forms, and N_c^{st} is the steady state cis concentration. In a previous paper from our group[1], 3.2% cis- NMA was observed from 220 nm excitation near photo-steady conditions by measuring the intensity of the 1496 cm^{-1} cis-Amide II mode, which suggested a quantum yield of $R \geq 0.1$

In the case of CW laser excitation, the photoexcitation rates, k_L and k'_L are comparable to the ground state thermal conversion rates, K and K', while the photochemical relaxation rates, k_t, k_c,, and k_r, are sufficiently fast that the photochemical relaxation process depends only on relative values of k_t, k_c,, and k_r .We utilized a new 206.5 nm CW source from an intra-cavity doubled Kr^+ laser to measure the ground state conversion rate between the cis and trans forms. We found these rates to be in the ms time region. We used a flowing sample technique to generate the cis form and to monitor the temperature dependence of the conversion of cis form to trans form in the ground state manifold (see Figure 2). We determined the energy barrier between the trans and cis form ground states from(see Figure 3):

$$K = K_0 \exp(-E_a / k_B T). \tag{3}$$

We measure E_a=*13.4kcal/mole* for NMA and E_a=*10.3kcal/mole* for Gly-Gly, which is consistent with theoretical calculations. We believe that our measurement is the first measurement of the activation barrier for cis-trans isomerization in a monosubstituted amide such as found in peptides.

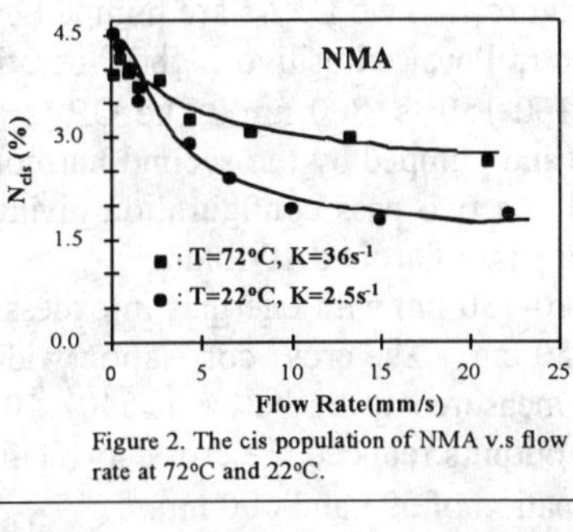

Figure 2. The cis population of NMA v.s flow rate at 72°C and 22°C.

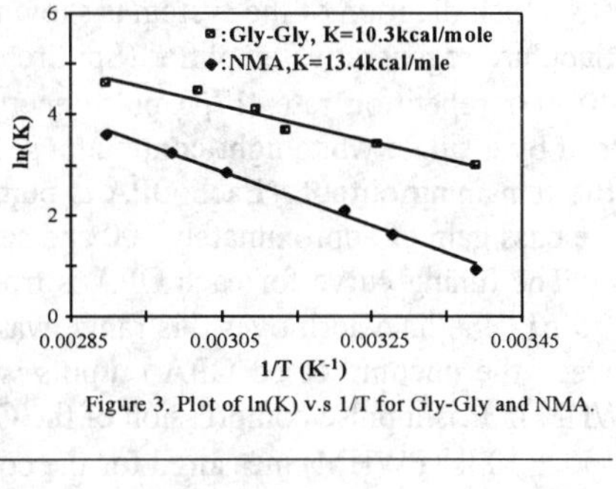

Figure 3. Plot of ln(K) v.s 1/T for Gly-Gly and NMA.

References:

1. Song, S. A. Asher, S. Krimm, and K. D. Shaw, *J. Am. Chem. Soc.* **113,** 1155 (1991).
2. Y. Wang, R. Purrello, and T. G. Spiro, *J. Am. Chem. Soc.* **111,** 8274 (1989). (b) Y. Wang, R. Purrello, S. Georgiou, and T. G. Spiro, *J. Am. Chem. Soc.* **113**, 6368 (1991).
3. R. R. Hill, J. D. Coyle, D. Birch, E. Dawe, G. E. Jeffs, D. Randall, I. Stec, and T. M. Stevenson, *J. AM. Chem. Soc.* **113**, 1805 (1991)

TWO 40 KHZ REPETITION RATE INDEPENDENTLY TUNABLE, SYNCHRONISED FEMTOSECOND PULSES GENERATED IN THE VISIBLE USING OPTICAL PARAMETRIC AMPLIFIERS

M.Towrie*, P.Matousek, A.W.Parker, P.F.Taday, W.T.Toner

Central Laser Facility, Rutherford Appleton Laboratory, Chilton, Didcot, Oxfordshire, OX11 0QX, UK. E.mail M.Towrie@RL.AC.UK, Tel. 01235 446141, Fax. 01235 445888

1. Introduction

Many time-resolved spectroscopy techniques require two broadly and independently tunable, synchronised light pulses. Recent advances in short pulse laser technology and non-linear crystal materials have lead to rapid development of optical parametric oscillators (OPOs) and travelling wave optical parametric amplifiers (OPAs) capable of meeting these requirements. Our aim was to develop a multi-kilohertz repetition rate OPA system for femtosecond/picosecond, pump and probe time-resolved absorption and resonance Raman spectroscopies. The system represents a unique concept in synchronously pumped travelling wave OPAs. The OPA design was adopted from [1].

2. Description of the System

A block diagram of the system is shown in figure 1. Two OPAs are pumped by a Ti:Sapphire regenerative amplifier (Spitfire, Spectra Physics/Positive Light) operating at 40 kHz repetition rate, 17μJ pulse energy (100-150 fs, 800 nm). The OPAs are seeded by a single white light continuum source and pumped by the second harmonic of the remaining output. Each OPA is pumped in a two pass configuration giving a single pass gain of approximately 100 and second pass of around 10 fold.

The tuning curve for each OPA is from 480-730 nm with energies in excess of 40-75 nJ. The linewidth over this range was <250 cm^{-1}. The cross correlation widths between the uncompressed OPA outputs were measured to be between 540-610 fs (FWHM). Prism pulse compression of the OPA outputs reduced the cross correlation to 150 ± 10 fs (FWHM) measured for the combination of 595 and 600 nm.

3. Preliminary TR^3 experiments

The spectra of p-quarterphenyl dye shown in figure 2 illustrate the potential of OPAs for use in ps-TR^3 spectroscopy. The dye is excited by a fixed pump wavelength of 277 nm (3rd harmonic of the regenerative amplifier) and probed by the output of OPA1 from 593 to 647 nm. A grating pulse shaper is used to spectrally filter the probe to a linewidth of ~30 cm^{-1}. The temporal resolution was estimated to be ~1 ps. All bands show strong relative intensity variation with probe wavelength. This is most evident for the band at c. 1500 cm^{-1} which has a remarkably sharp Raman resonance of about 20 nm centred at ~620 nm . This unique ability to elicit Raman excitation profiles from molecules in excited states adds a new and illuminating dimension to the ps-TR^3 technique.

FIGURE 1: A block diagram of the parametric amplifier system.

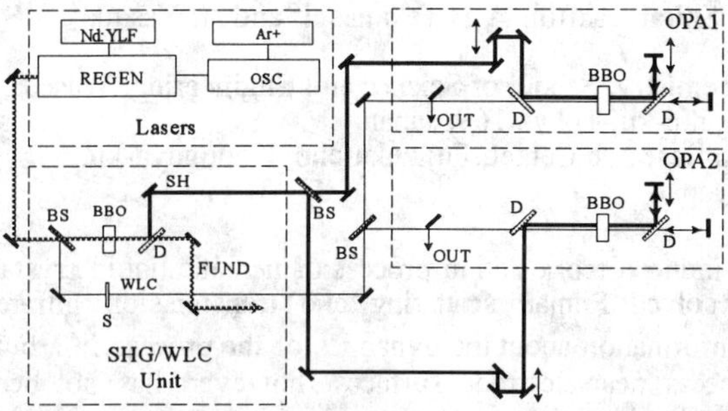

Legend: Nd:YLF - Q-switched Nd:YLF laser, Ar^+ - 10W argon ion laser, REGEN - Ti-Sapphire regenerative amplifier, OSC - femtosecond Ti:Sapphire laser, BS - beamsplitter, BBO - BBO crystal, D - dichroic mirror, S - sapphire crystal, FUND - fundamental output of the regenerative amplifier, SHG - second harmonic generator, SH - second harmonic of the fundamental, OUT - output beam, WLC - white-light continuum, OPA1, OPA2 - optical parametric amplifier units.

FIGURE 2: The TR^3 excitation profile of p-quarterphenyl dye.

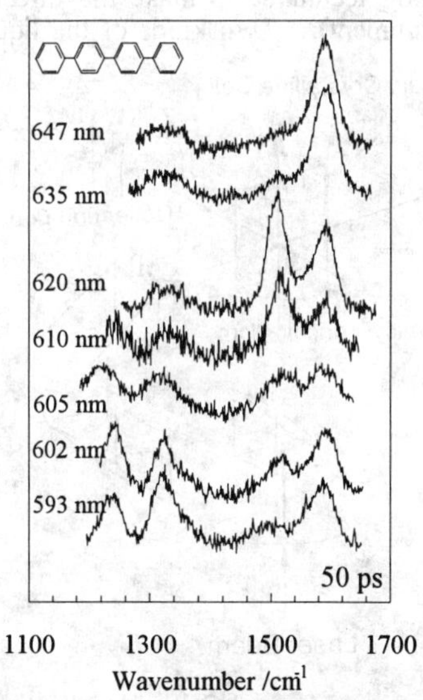

P-quarterphenyl dye pumped at 277 nm and probed after 50 ps delay over a wavelength range of 593 to 647 nm.

Reference: [1] M.K.Reed, M.K.Steiner-Shepard, D.K.Negus, IEEE Journal of Quantum Electronics 31 (1995), 1614.

TIME-RESOLVED TOTAL INTERNAL REFLECTION RAMAN SCATTERING STUDY ON ELECTRIC-FIELD INDUCED RE-ORIENTATION PROCESS OF A NEMATIC LIQUID CRYSTAL

T. Morikawa[1], E. Shirai[1], K.Itoh[1]*, H.Takanashi[2] and A. Yasuda[2]

[1]Department of Chemistry, School of Science and Engineering, Waseda University, Shinjuku-ku, Tokyo 169, Japan
[2]Sony Corporation Research Center, Fujizuka-cho, Hodogaya-ku, Yokohama 240, Japan

The electric field-induced reorientation process of nematic liquid crystals has been studied by time-resolved Raman scattering and transmission infrared spectroscopies [1,2] , giving information about the dynamics of the process in a bulk phase. The reorientation process near electrode surfaces, however, has not been studied yet, although this kind of information is of crucial importance to understand the process as a whole. In this paper we applied time-resolved total internal reflection (TIR) Raman spectroscopy to study the electric field-induced reorientation process of the liquid crystal cells of 4-*n*-hexyl-cyanobiphenyl (6CB), which exhibits the nematic phase in the temperature range of 14 - 29° C.

Fig. 1 exhibits a liquid crystal cell and the collection optics of TIR Raman scattering. The liquid crystal cell with a thickness of ca. 10 μm was fabricated at the bottom of a hemicylindrical high index glass prism (LaSF08 , 15 × 15 × 7.5 (ϕ) mm, n = 1.888) by using an oblique angle SiO sputtering technique to make the director parallel to the electrode surface (homogeneous alignment). Two kinds of the liquid crystal cells were prepared by adjusting the direction of the sputtering; one of the cells has the director parallel to the axis of the hemicylindrical prism (*p-configuration cell*) and the other has the direction perpendicular to the axis (*s-configuration cell*). By introducing a pulsed Nd-YAG laser (λ=532 nm, 15 ns, 10 Hz) from the prism to the liquid crystal cell at an angle Ψ ($>\Psi_c$), TIR Raman spectra were recorded by using a Kaiser Optical Systems HoloSpec f/1.8 spectrometer (aperture ratio; 1.8) equipped with an ICCD detector. Experimental conditions for TIR measurements are abbreviated such as PSX and PSY, where the first letter indicates the cell (p-configuration cell), the second letter the polarization of the excitation laser (s-polarization), and the last letters X and Y the electric field polarization of the collected Raman

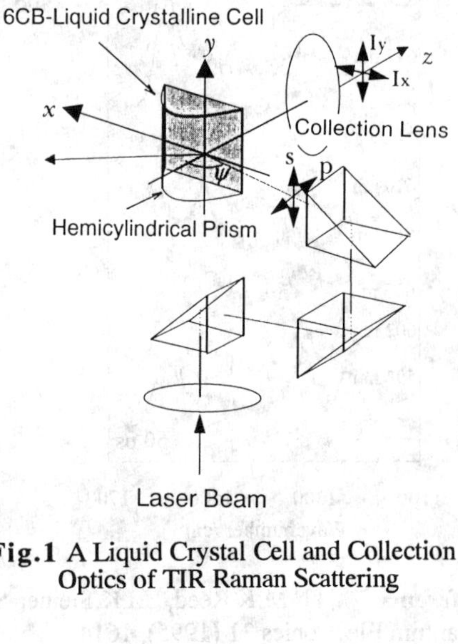

Fig.1 A Liquid Crystal Cell and Collection Optics of TIR Raman Scattering

scattering parallel to the x and y-axes, respectively. (Fig.1)

Fig. 2 illustrates the time-resolved TIR Raman spectra (PPX condition, Ψ=56°) under the application of a periodic rectangular electric field shown by the insert. The nematic liquid crystal has a uniaxial symmetry and its Raman scattering tensor can be expressed by α_1 and α_2, which are diagonal elements parallel and perpendicular to the axis, respectively, and α_3 and α_4, which are the corresponding off-diagonal elements. The relative values of these elements could be determined by measuring the polarized TIR Raman spectra of the p- and s-configuration cells. The TIR Raman scattering intensities measured under the conditions of PSX and PSY can be expressed as

$$I_{PSX}=K\left(\alpha_3\cos\theta+\alpha_4\sin\theta\right)^2\left\langle E_{sy}\right\rangle^2 \quad (1)$$

$$I_{PSY}=K\alpha_2^{\,2}\left\langle E_{sy}^{\,2}\right\rangle \quad (2)$$

, where $\langle E_{sy}^2\rangle$ is a squared amplitude of the electric field at the SiO/6CB interface and θ the angle of the director rotation in the xz-plane from the x-axis. Since $\langle E_{sy}^2\rangle$ does not depend on the angle θ for the p-configuration cell, as experimentally confirmed by the constant intensity of I_{PSY} under electric field-on conditions, eq.1 can be used to determine θ as a function of time from the time-resolved TIR Raman spectra measured under the condition of PSX. Fig.3 plots θ as a function of time after application of the rectangular electric fields (A; 8V, B; 5V) shown by the insert in Fig. 2. With the onset of the field of 8V (0 - 10 ms) the rotation angle shows a rapid rise within several milliseconds, and then gradually increases, giving a final value of ca. 60°, while on application of the field of 5V the angle stays almost constant for a few milliseconds and then increases up to ca. 40°. When the electric fields are off (t> 10 ms), the director recovers its original orientation within ca. 8 ms in both cases. These results are in marked contract to the case of a bulk phase studied by time-resolved infrared spectroscopy (not shown in this paper); e.g., the electric field-induced reorientation processes in the bulk phase proceed in a similar manner irrespective of the electric fields except that the rise time under the 8V-field (ca. 2 ms) is appreciably faster than that (ca. 4 ms) under the 5V-field.

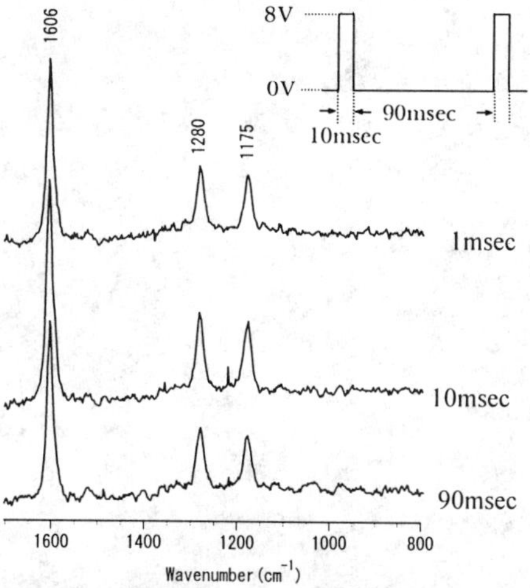

Fig.2 Time-resolved TIR Raman Spectra of the Liquid Crystal Cell (PPX, Ψ = 56°)

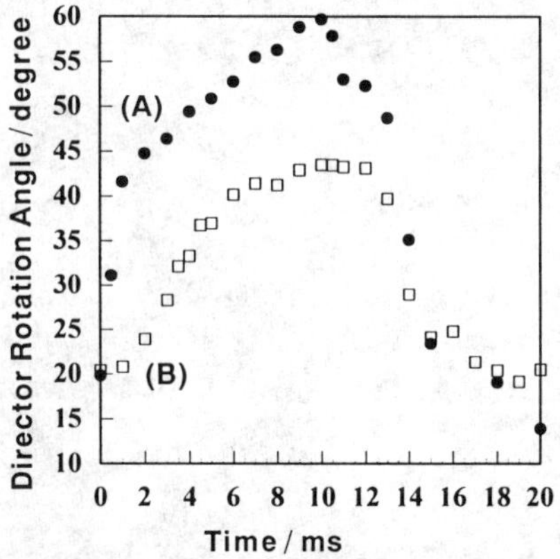

Fig.3 Time-course of the Director Rotation under 10-ms Rectangular Electric Field. (A) 8V, (B) 5V

[1] T.I.Urano and T.Hamaguchi, Chem. Phys. Lett. **195**, 287(1992).
[2] K.M.Booth and H.J.Cole, Liq. Cryst. **13**, 677 (1993).

SECTION 4

NONLINEAR RAMAN SCATTERING

SPECTROSCOPY OF PENDULAR STATES BY IONIZATION-LOSS STIMULATED RAMAN SPECTROSCOPY

Wousik Kim, Mark W. Schaeffer, and Peter M. Felker

Department of Chemistry, University of California
Los Angeles, CA 90095-1569

Recent advances in mass-selective, ionization-loss stimulated Raman spectroscopy (ILSRS) have allowed for the species-specific measurement of *inter*molecular transitions in molecular clusters across a wide spectral range with subwavenumber resolution.[1] Such measurements hold out the promise of revealing much about intermolecular potential energy surfaces and collective molecular motions.

One of the major challenges in trying to relate a spectrum of intermolecular transitions in a molecular cluster to an intermolecular potential energy surface is the assignment of observed bands to vibrations of the cluster. In ILSRS such assignments can be made, in principle, if the Raman resolution is sufficient to observe the band contours of the intermolecular vibrational resonances. With 0.03 cm^{-1} resolution we have recently acquired this capability. The band contours we have thus far observed, however, have characteristics that clearly show them to be something other than *rotational* band contours.[2] In particular, they have significant dependences on the intensities of the stimulated-Raman excitation fields. And, they have polarization dependences that are not those to be expected for the rotational band contours of anisotropic or isotropic Raman bands.

The observed contours,[2] in fact, are manifestations of transitions between "pendular" states[3] formed by interaction of one of the Raman fields with the permanent polarizability tensor of the species being studied. Pendular states can arise when gaseous species with permanent polarizability anisotropies interact with a linearly polarized optical field via the AC Stark effect. Such species in such fields have potential energies that depend on their angular position with respect to the field polarization. When the angular variation of this potential energy is large in comparison to the rotational kinetic energy of a species, it no longer rotates freely. Instead, it librates over a limited angular range about the field's polarization direction. The eigenenergies and wavefunctions corresponding to these librational (i.e., pendular) states are significantly different from the free rotational states of the same species. Hence, at the band-contour level, the spectroscopy of a sample of species in pendular states can differ markedly from the same species in free rotational states. In fact, these differences can can be such as to render the assignment of vibrational bands considerably easier from pendular band-contour information than from rotational band-contour information.

In this paper we present ILSRS spectra of intermolecular transitions in naphthalene trimer. The results show that the species is one of high symmetry – either C_{3h} or very close to C_{3h}.[4] The species thus has eight intermolecular fundamentals, four of which are doubly degenerate. Two fundamentals are of E' symmetry, two are E'', two are A', and two are A''. From pendular band-contour data we assign six of these fundamentals to observed intermolecular bands. (The two A'' fundamentals, by symmetry, are not Raman-active.) Intermolecular spectra on three other isotopomers of the trimer are also presented and assigned. In total, the results represent a significant step toward the elucidation of the intermolecular forces binding the naphthalene moieties together. The spectroscopic results are compared with results from normal coordinate analyses employing semi-empirical atom-atom intermolecular potential energy surfaces. The comparison indicates that such surfaces are successful in reproducing the qualitative nature of the intermolecular Raman spectra, yet quantitative agreement is not achieved.

1. P. M. Maxton, M. W. Schaeffer, S. M. Ohline, W. Kim, V. A. Venturo, and P. M. Felker, J. Chem. Phys. **101**, 8391 (1994).
2. W. Kim and P. M. Felker, J. Chem. Phys. **104**, 1147 (1996).
3. For example, B. Friedrich and D. Herschbach, Phys. Rev. Lett. **74**, 4623 (1995).
4. M. W. Schaeffer, W. Kim, P. M. Maxton, J. Romascan, and P. M. Felker, Chem. Phys. Lett. **242**, 632 (1995).

IMPULSIVE RAMAN SCATTERING IN $LaAlO_3$: SEEKING DOMAIN REVERSAL

R. Merlin

The Harrison M. Randall Laboratory of Physics and Center for Ultrafast Optical Science, The University of Michigan, Ann Arbor, MI 48109-1120, U. S. A.

Using ultrashort light pulses, it has become possible in recent years to generate coherent *optical* phonons through impulsive stimulated Raman scattering (ISS) [1]. This has provided a unique tool for time-domain studies of high-frequency modes and their coupling to other excitations [1]. These developments have also led to intriguing proposals bearing on the potentially large amplitude atomic motion that can be attained with ISS [2]. Here, we present a brief review of theoretical results on ISS-induced domain-reversal in soft-mode materials [2] and preliminary ISS data on $LaAlO_3$ [3].

In materials showing structural phase transitions, such as ferroelectrics, the formation of domains is associated with the instability of a normal vibrational mode (the *soft* phonon) whose frequency tends to zero as T approaches the transition temperature T_C. Below T_C, the strongly anharmonic soft modes can be visualized as vibrations about the minima of a double-well potential, each well corresponding to a different domain [4]. We have used this model to simulate the effect on the lattice of an impulsive excitation leading to a uniform shift of the velocity of each ion by an amount V (note that V is proportional to the pulse energy per unit area). In the simulations, we have considered two- and three-dimensional arrays of oscillators with nearest-neighbor coupling [4] and studied the long-time behavior of a previously thermalized system after the impulse. Shown in Fig. 1 is the final polarization of such an array as a function

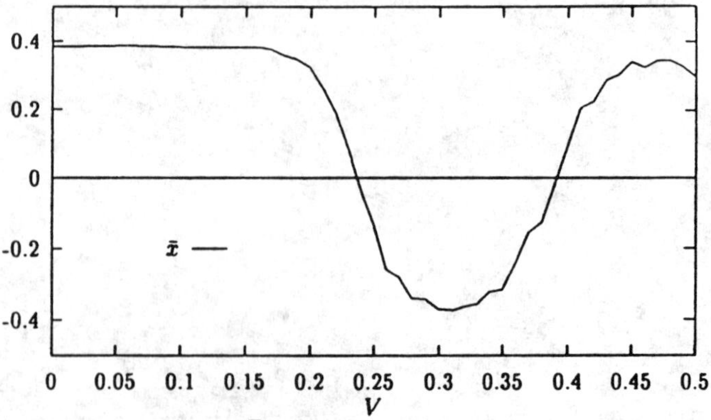

Fig. 1. Final polarization $\bar{x}$ versus the impulsively added velocity V for two-dimensional 400×400 oscillators at temperature $T/T_C \approx 0.9$.

of V. The original domain is not altered until the energy imparted by the pulse is approximately equal to the (T-dependent) barrier height. Close to this value and within the window shown in Fig. 1, we find that the entire domain undergoes a reversal following a rather sharp threshold. This behavior can be understood by mapping the problem of the center-of-mass motion of a domain fragment of size N_D to that of a single oscillator at temperature T/N_D [2].

Figure 2 shows ISS as well as conventional Raman scattering (RS) data on $LaAlO_3$ [3]. This compound undergoes a nearly second-order transition at $T_C \approx 750$ K from the cubic perovskite structure (O_h) into a rhombohedral form (D_{3d}). The transformation is characterized by the softening of a phonon at the R-point of the cubic Brillouin zone and concomitant doubling of the unit cell. Below T_C, the R-point becomes the Γ-point of the D_{3d}-phase where the soft phonon splits into Raman-allowed *gerade* modes of symmetries A_{1g} and E_g. The spectra in Fig. 2 correspond to the latter phonon. The fact that inversion symmetry is preserved makes $LaAlO_3$ a promising candidate for domain reversal studies. In non centro-symmetric substances, and at the small wavevectors probed by ISS, the polaritons created by pulses are primarily infrared photons. An important parameter that can be gained from our measurements is the ratio between the coherent phonon amplitude and the pulse energy density, which is 0.6-6 mÅ/(Jm^{-2}). This value suggests that the ISS mechanism for reversing domains should operate in a range accessible to experiments with standard lasers.

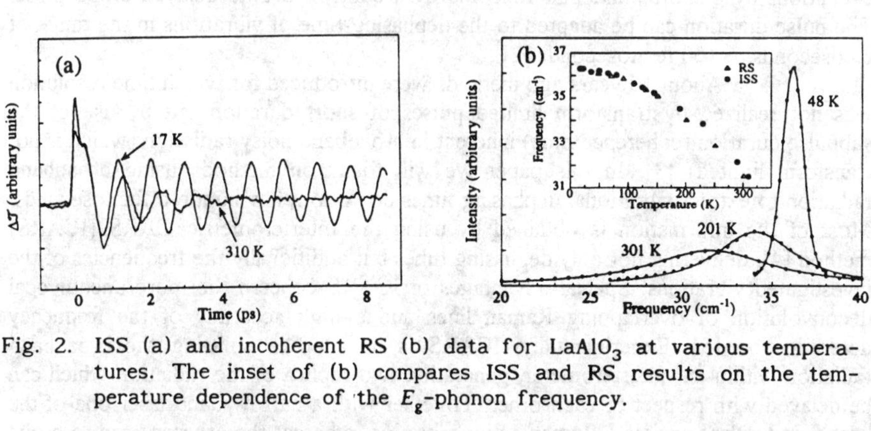

Fig. 2. ISS (a) and incoherent RS (b) data for $LaAlO_3$ at various temperatures. The inset of (b) compares ISS and RS results on the temperature dependence of the E_g-phonon frequency.

Acknowledgements - I am indebted to G. A. Garrett, Y. Liu, C. Uher, J. F. Whitaker and, in particular, to S. Fahy for contributing to this work. The support of the National Science Foundation through the Center for Ultrafast Optical Science is gratefully noted.

[1] For reviews, see: S. Ruhman, A. G. Joly and K. A. Nelson, IEEE J. Quantum Electron. **24**, 460 (1988); W. A. Kütt, W. Albrecht and H. Kurz, IEEE J. Quantum Electron. **28**, 2434 (1992).
[2] S. Fahy and R. Merlin, Phys. Rev. Lett. **73**, 1122 (1994).
[3] Y. Liu, A. Frenkel, G. A. Garrett, J. F. Whitaker, S. Fahy, C. Uher and R. Merlin, Phys. Rev. Lett. **75**, 334 (1995).
[4] See: T. Schneider and E. Stoll, Phys. Rev. B **13**, 1216 (1976).

MEASUREMENT AND INTERPRETATION OF DEPHASING TIMES OF MOLECULES IN ORGANIC SOLUTIONS

A. Lau, A. Kummrow

Max-Born-Institut für Nichtlineare Optik und Kurzzeitspektroskopie
Rudower Chaussee 6, 12489 Berlin, Germany

Line shape analysis of Raman lines has been used for a long time to extract dynamic parameters of neat liquids and solutions [1]. In comparison with stationary frequency domain Raman spectroscopy, time domain techniques provide not only more direct but also more controllable ways to reveal solvent-induced dynamics. Most time resolved Raman techniques are based on third order coherent Raman excitations by a short pulse pair followed by a controlled time delayed probe pulse. The pulse duration can be adapted to the dephasing time of vibrations in the range of picoseconds to 100 femtoseconds [2].

About 10 years ago methods were introduced for which time resolution was not realized by transform limited pulses of short duration but by use of the subpulse duration (coherence time) inherent in broadband noisy radiation (which is not transform limited) [3]. In this paper we will focus on methods using broadband radiation to extract vibrational dephasing times down to a few hundred femtoseconds. Most of the information is obtained by using the Interferometric CARS (ICARS) method [4], delivering not only dephasing times but additionally the frequencies of the investigated vibrations. Special advantages of ICARS concern the more unequivocal deconvolution of overlapping Raman lines and a high accuracy of the frequency determination [5]. Experimentally, ICARS is built up in splitting the broadband radiation within an interferometer, generating two replica of the radiation which can be delayed with respect to each other. Together with a narrow band laser one of the broadband pulses excites a Raman active vibration, whereas the other is used to probe this excitation. Using a frequency resolution of the generated ICARS signal, the damped signal is modulated by detuning oscillations. For instance in Fig. 1, an ICARS interferogram of the 1522 cm^{-1} vibration of β-carotene is depicted. From the decay of the signal a dephasing time of this vibration of 600 fs is determined in excellent agreement with corresponding measurements using a femtosecond CARS technique [6]. Dephasing times deduced from linewidth measurements of CARS give the same value. However, this can be expected due to calculations, which showed a principle equivalence of time resolved data obtained by methods based on $\chi^{(3)}$ with data in the frequency domain [7]. Methods using higher order susceptibilities can give more information, separating homogeneous and inhomogeneous contributions to the linewidth. This was already demonstrated by time resolved femtosecond methods, for example the Raman echo [8].

Are there corresponding possibilities for methods using nontransformed broadband lasers?

In a first attempt we tried to excite narrow band lines using spectral features of ICSRS [9]. Our experimental conditions for excitation of narrow band lines suggest that higher order Raman processes come into play. A closer look shows instead that the features can be explained by successive processes of third order susceptibility. As a result our technique cannot be used to resolve homogeneous and inhomogenous contributions to the width of Raman lines. Currently we are examining higher order Raman processes in analogy to corresponding femtosecond setup applying broadband laser radiation. Experiments were carried out with modified setup and the results are shown and discussed. Possibilities to reduce the background usually associated with some of the broadband techniques are also considered.

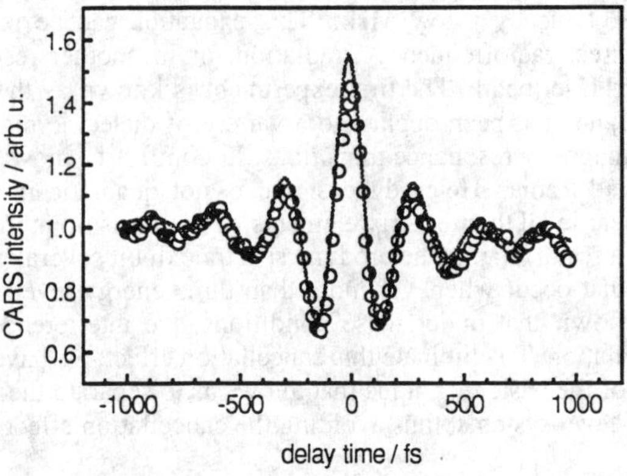

Fig. 1. ICARS interferogram of the 1522 cm^{-1} vibration of all-trans-β-carotene, 1.3 10^{-4} mol/l in cyclohexane. Circles: experimental data; solid line: fit curve.

1. A. Morresi, L. Mariani, M.R. Distefano, M.G. Giorgini, J. Raman Spectroscopy 26, 179 (1995).
2. M. Fickenscher, H.G. Purucker, A. Laubereau, Chem. Phys. Lett. 191, 182 (1992).
3. T. Hattori, A. Terasaki, T. Kobayashi, Phys. Rev. A35, 715 (1987).
4. M.A. Dugan, J.S. Melinger, A.C. Albrecht, Chem.Phys. Lett. 147, 411 (1988).
5. A. Lau, A. Kummrow, M. Pfeiffer, S. Woggon, J. Raman Spectr. 25, 607 (1994).
6. A. Kummrow, A. Lau, J. Raman Spectr. 26, 457 (1995).
7. R.F. Loring, S. Mukamel, J. Chem. Phys. 83, 2116 (1985).
8. D. Vanden Bout, L. Muller, M. Berg, Phys. Rev. Lett. 67, 3700 (1991).
9. M.A. Dugan, A.C. Albrecht, Phys. Rev. A43, 3877 and 3922 (1991).

Coherent Raman Scattering with Pulsed and CW Excitation

Dieter Suter

Fachbereich Physik, Universität Dortmund, D-44221 Dortmund, Germany

Coherent Raman scattering (CRS) was first proposed by Giodrmaine and Kaiser [1]: They showed that the generation of Raman sidebands can occur at arbitrarily low intensities if the medium with which the laser beam interacts contains a coherent excitation. They assumed that the excitation was a lattice vibration, but any type of coherence will have the same effect. Besides being linear in the incident intensity, CRS also occurs with the same probability in the anti-Stokes as in the Stokes direction.

Several realizations of CRS were demonstrated in the field of laser-assisted magnetic resonance, in particular in dilute ionic crystals. The coherent excitation of the medium is in this case a precessing magnetic moment, whose transition frequency is in the radio frequency range, typically a few MHz. This excitation can be prepared either continuously, using direct radiofrequency irradiation, or in another resonant Raman process, using a second laser beam. The first experiment is known as the Raman Heterodyne experiment [2] and has been applied to a variety of dielectric materials for the sensitive detection of magnetic resonance transitions. In contrast to most spectroscopic techniques, the observed Raman Heterodyne signal is not quadratic in the transition matrix element, but linear in all three matrix elements of the transitions involved in the scattering process. As a result, Raman heterodyne spectra exhibit several peculiar interference effects [3, 4] that occur whenever more than three energy levels contribute to the signal. It can be shown that under these conditions, the interference leads to an almost complete cancellation. To eliminate this cancellation effect, we have developed a pump-probe variation of the basic technique that allows us to separate the contributions from the various three-level systems, thus avoiding the cancellation effect.

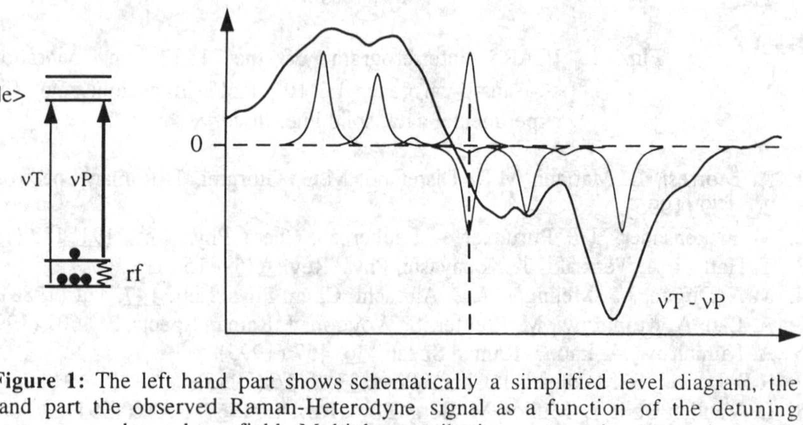

Figure 1: The left hand part shows schematically a simplified level diagram, the right hand part the observed Raman-Heterodyne signal as a function of the detuning between pump- and test laser field. Multiple contributions occur due to excited state sublevels; cancellation is observed at the position of the dashed line.

A purely optical technique for coherent Raman scattering is the Raman Beat experiment: in this case, a laser pulse excites the spin coherence, which subsequently undergoes free precession. This freely precessing magnetization can be observed through coherent Raman scattering, which yields in this case a free induction decay. This signal can be Fourier transformed to provide a spectrum of the sublevel transitions in the electronic ground state as well as in the electronically excited state. The technique is resonant with respect to both the pump- and the test laser beam: The pump laser must be resonant with respect to two optical transitions that participate in the Raman process, as shown in the left hand part of figure 2.

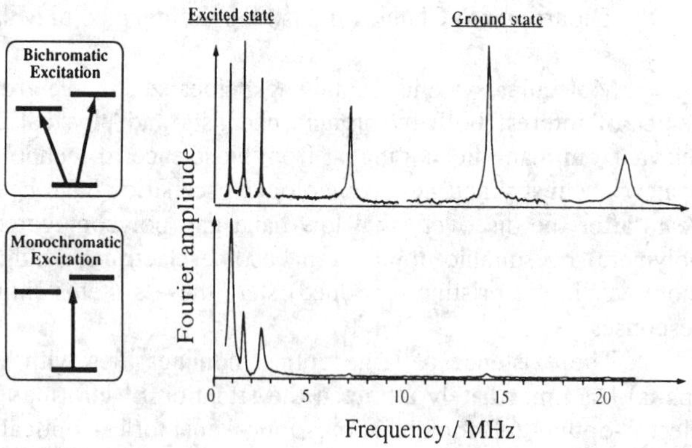

Figure 2: Sublevel spectra obtained by Fourier transformation of a Coherent Raman beat signal. The lower trace, which was obtained with a monochromatic excitation pulse, shows a decrease of the signal amplitude at higher frequencies, while the amplitudes of the upper trace, which was obtained with bichromatic excitation, are roughly equal.

For a nonvanishing sublevel splitting, this is only possible if the laser beam contains frequency components at both optical transition frequencies. In a narrow-band laser beam, this is no longer the case when the level splitting exceeds approximately 1 Mhz. As a result, the excitation efficiency decreases dramatically at higher splittings, as the lower trace on the right hand part of figure 2 demonstrates: The signal at 21 MHz, which originates from a level splitting in the electronic ground state of Pr^{3+}, is some two orders of magnitude smaller than the signals close to the origin of the frequency axis. This problem can be avoided by using a bichromatic laser beam [5], which may be derived from a monochromatic beam by amplitude modulation. If the resulting sidebands are matched to the two optical transition frequencies, the Raman process becomes highly efficient again, as shown in the upper trace on the right hand side of figure 2. Obviously, all six resonance lines have comparable intensity in this case, demonstrating that the drop-off at higher frequencies is eliminated with this technique. Besides the Pump laser beam, also the test laser beam must be resonant with one of the two optical transitions to obtain the resonant enhancement of the Raman cross section.

References:

[1] J.A. Giordmaine and W. Kaiser, Phys. Rev. 144, 676 (1966).

[2] N.C. Wong, E.S. Kintzer, J. Mlynek, R.G. DeVoe, and R.G. Brewer, Phys. Rev. B 28, 4993 (1983).

[3] M. Mitsunaga, E.S. Kintzer, and R.G. Brewer, Phys. Rev. B 31, 6947 (1985).

[4] T. Blasberg and D. Suter, Phys. Rev. B 51, 12439 (1995).

[5] T. Blasberg and D. Suter, Phys. Rev. B 51, 6309 (1995).

HYPERPOLARIZABILITIES OF MATERIALS RELEVANT IN MOLECULAR ELECTRONICS : A SPECTROSCOPIC APPROACH

Giuseppe Zerbi*, M. Del Zoppo, C. Castiglioni and P. Zuliani

Dipartimento Chimica Industriale, Politecnico di Milano, Milano, Italy

Molecular systems containing delocalized π electrons have been always the center of interest both by organic chemists and physical chemists because of their relevance in many fields ranging from bioscience to technology. A new physics, a lot of new chemistry and new technological scenarios have been opened in the past few years after the discovery that low band gap polyconjugated systems (oligomers and polymers) by suitable doping can become electrical conductors as good as metals; moreover in the pristine (undoped) state these systems show large non linear optical responses.

The existence of large optical nonlinearities with very short response times (ps-fs) has immediately attracted the attention of information technology, photonics, electro-optics, telecomunication, optical memories, optical computers and all other fields where information can be treated by optical signals. Physicists became immediately interested in the understanding of the origin of such large response while chemists have been engaged for the synthesis of new materials with improved optical properties which must also cope with the strict requirements by technology for the construction of suitable devices (long term chemical and thermal stability, processability etc).

The basic relationship which rules non linear optical responses of a macroscopic sample under the action of an external electromagnetic field (**E**) is:

$$\mathbf{P} = \chi^{(1)}\mathbf{E} + \chi^{(2)}\,\mathbf{E{:}E} + \chi^{(3)}\,\mathbf{E{:}E{:}E} + \ldots..$$

where **P** is the volume polarizability $\chi^{(1)}$, $\chi^{(2)}$ and $\chi^{(3)}$ are the bulk electrical susceptibilities. The values of $\chi^{(2)}$ and $\chi^{(3)}$ depend on the number of molecules per unit volume, on the way molecules pack in the crystal lattice and on the strength of the intermolecular interactions. One can also aim at the molecular non linear optical responses (hyperpolarizabilities)

$$\boldsymbol{\mu} = \alpha\mathbf{E} + 1/2!\,\beta\,\mathbf{E{:}E} + 1/3!\,\gamma\,\mathbf{E{:}E{:}E} + \ldots$$

where $\boldsymbol{\mu}$ is the molecular dipole moment, α, β and γ are the molecular polarizabilities of first, second and third order.

The first aim of the research in this field is to produce new molecules with the largest possible β and γ. The molecules should then be suitably packed to reach optimal bulk susceptibilities.

The measurement of bulk or molecular nonlinearities requires complex optical experiments not always easy to be carried out. In our group in Milano a great effort is being made to combine molecular engineering and vibrational spectroscopy aiming at the discovery of new polyconjugated materials suitable for technological applications.

The basic fact which underlies the physics of polyconjugated molecules is the existence of strong electron-phonon coupling which provides the way to obtain information on the electronic properties using as probes vibrational observables. It is at present well known that the Raman spectra of polyconjugated oligomers and polymers are very peculiar and show frequency and intensity dispersion with conjugation length. The origin of such peculiar spectroscopic properties is at present known [1,2] and is again related to the fact that electron phonon coupling takes place only for a few specific normal modes which describe the trajectory of atoms along a path which strongly involves electron delocalisation.

On this basis we have proven theoretically [3] and verified experimentally [4,5] that a reliable measure of β and γ is given just by the contribution of nuclear relaxations which can be expressed as function of infrared, Raman and Hyper-Raman intensities.

The "vibrational method" turns out to be easy, fast and cheap compared with the traditional laser-optical experiments (SHG, THG, EFISH, etc.). It has been applied to many classes of polyconjugated oligomers and polymers starting from the classical case of polyenes $(-CH=CH)_n$- to polyaromatic molecules (polyphenylenes, polyphenylene-vynilenes) and polyheteroaromatics molecules (polythiophenes, polypyrroles, polyfuranes etc.

We think this is a new field of vibrational spectroscopy worth to be further exploited for its relevance both for basic and applied material science. Structure property relationship can be and have already been worked helping to develop new synthetic strategies for the discover of new materials with improved optical properties. Moreover a lot of new physics and exciting spectroscopy can be learned related to the electronic polarisation and their coupling with nuclear motions.

1. M. Gussoni, C. Castiglioni, G. Zerbi in *"Advances in Material Science Spectroscopy"* (R.J.H. Clark, R.E. Hester Eds.), Wiley, New York (1991)
2. G. Zerbi, M. Gussoni, C. Castiglioni in *"Conjugated Polymers"* (J.L. Brédas, J. Silbey Eds.), Kluwer, New York (1991)
3.C. Castiglioni, M. Del Zoppo, G. Zerbi, *Phys. Rev. B* in press
4. C. Castiglioni, M. Del Zoppo, P. Zuliani, G. Zerbi, *Synth. Met.*, **74**, 171 (1995)
5. P. Zuliani, M. Del Zoppo, C. Castiglioni, G. Zerbi, S.R. Marder, W.J. Perry, *J. Chem. Phys.*, **103**, 9935 (1995)

HIGH-RESOLUTION IONIZATION-DETECTED RAMAN SPECTROSCOPY

Marshall Crew, Matthias Leuchs, and Joseph W. Nibler*

Department of Chemistry, Oregon State University,
Corvallis, Oregon 97331 USA

Ionization-detected stimulated Raman spectroscopy (IDSRS) combines the high resolution capability of stimulated Raman spectroscopy with the high detection sensitivity of resonantly enhanced multiphoton ionization (REMPI) spectroscopy. Esherick and Owyoung first demonstrated the feasibility of the technique on NO [1] and C_6H_6 [2] using a simple electrode detector. More recently, Felker and co-workers have utilized more sensitive channel-plate detectors and time-of-flight mass analysis in medium resolution Raman studies of a number of aromatic clusters formed in free jets.[3] We report here results obtained with a similar apparatus but at much higher resolution (0.0015 cm^{-1}).

The Raman-pumping step is achieved with an amplified ring dye laser output and a 50 ns 532 nm beam from a custom long-pulse Nd-YAG laser. Molecules excited to the upper state of the Raman transition are then resonantly ionized by a separate, tunable uv beam brought to the sample focus after a delay of 50 ns. Resultant ions are detected by a dual channel-plate detector using a standard time-of-flight apparatus. As an example illustrating the detection capability, Fig. 1 shows an IDSRS spectrum of a portion of the υ_{16} plus $\upsilon_2 + \upsilon_{18}$ Fermi dyad of C_6H_6. A spectrum of comparable signal to noise and resolution was recorded earlier by Esherick and Owyoung [2] but the spectrum shown here was obtained at a sample density about 600 times lower than the value of 3 x 10^{15} molecules per cm^3 which they reported.

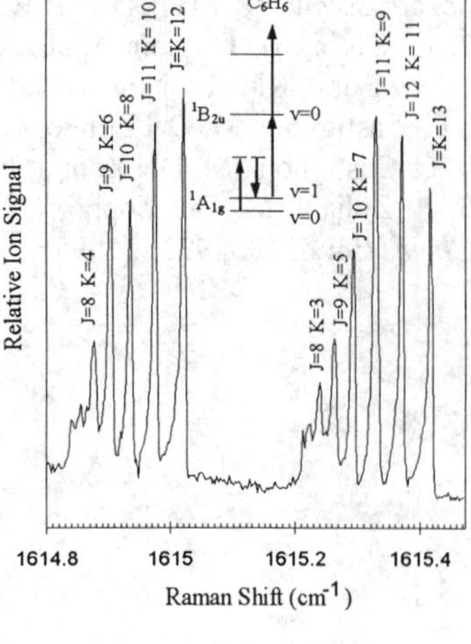

Fig. 1.
Ionization-detected Raman-gain spectrum of the v_{16} (e_{2g}) mode of benzene in a molecular beam. An energy level diagram is shown for the double-resonance experiment. The region of the Raman active ^{S}S transitions is displayed along with J and K symmetric top quantum number labels for the lower state of the transitions. The UV frequency was tuned to 36479 cm^{-1}.

The ν_{16} benzene transition has a low Raman cross section but the 1 + 1 ionization step is known to be quite efficient. An example of a case in which the Raman pumping is efficient but the ionization is more difficult in shown in Fig. 2, where 2 + 2 REMPI ionization of N_2 is employed. The bottom trace (a) is an optical Raman loss Q-branch measurement made in a jet expansion at the high density region near the nozzle. Trace (b) shows the ion signal obtained when the J = 2, v=1 level was Raman-pumped and the uv laser was scanned. The intensities of the four observed transitions are in good agreement with intensities predicted from the Hönl-London factors. The top trace shows the Raman J = 2 line that results when the uv laser is fixed at the S_2 position and the Raman-pumping lasers are tuned. Part of the linewidth (0.0015 cm^{-1}) is attributed to frequency jitter in the Nd-YAG laser and can be reduced by locking this source to an iodine line. The density at the sampling point for trace (c) was 2 x 10^{14}, about 25,000 times lower than that at the sampling point for the optical SRS trace in (a). Several applications of this high detection sensitivity and resolution capability will be presented.

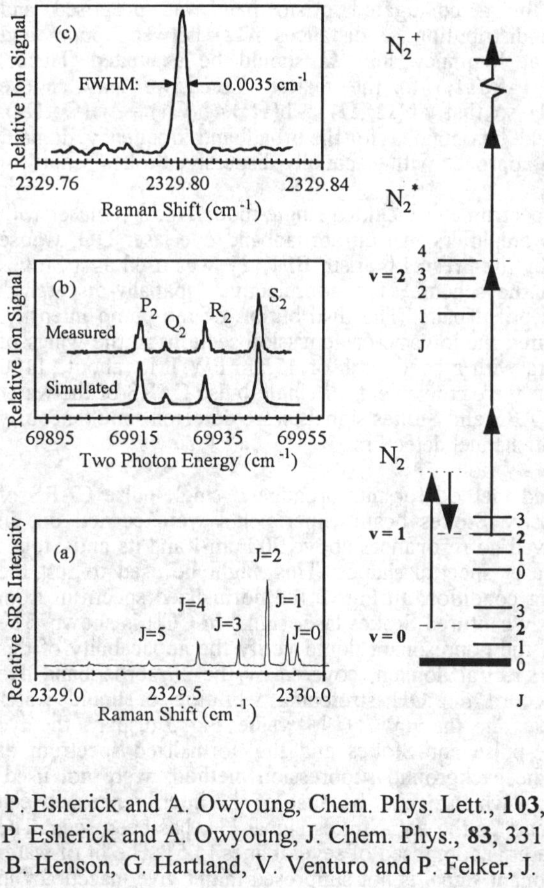

Fig. 2.
Raman-pumped 2+2 ionization spectrum of N_2 taken in a jet and compared with a simulated spectrum. The energy level diagram for the process is shown to the right of the figure.
(a) Optical SRS spectrum of the Q branch of N_2 cooled to about 27 K in a free jet.
(b) Spectrum of N_2 in a jet when the v=2, J=2 level was Raman-pumped and the ionization laser was scanned.
(c) Spectrum obtained when the uv laser was fixed at the S_2 line and the Raman-pumping lasers were scanned across the J=2 line.

1. P. Esherick and A. Owyoung, Chem. Phys. Lett., **103**, 235 (1983).
2. P. Esherick and A. Owyoung, J. Chem. Phys., **83**, 3311 (1985).
3. B. Henson, G. Hartland, V. Venturo and P. Felker, J. Chem. Phys., **91**, 2751 (1989).

SINGLE-PULSE CARS OF LIQUIDS USING AN ULTRABROADBAND SPATIALLY-DISPERSIVE STOKES BEAM

V.H. Astinov* and G.M. Georgiev

Quantum Electronics Department, Faculty of Physics, Sofia University, 5 James Bourchier Blvd., 1126 Sofia, Bulgaria

In this report we undertook to investigate liquid water by CARS. To use the single-pulse technique advantages it is necessary to have an appropriate Stokes source as well as to satisfy the phase-matching condition. In previous work [1] we proposed a scheme for phase-matching of CARS process in liquids. To satisfy simultaneously the phase-matching condition in a broad spectral domain, the Stokes frequency components should cross the pump wave at different angles, e.g. the Stokes beam should be spatially-dispersive in front of the focusing lens (Fig. 1). A scheme supplying similar distribution by a conjugated prism pair was proposed and demonstrated in [1]. First the distribution of distances $r(\Omega)$ between pump and monochromatic Stokes wave at frequincy shift Ω should be evaluated. Hence, calculating $\Delta r_S(\Omega_2,\Omega_1) = r_S(\Omega_2) - r_S(\Omega_1)$ for the matter studied, we may choose the prism glass and distance d_2 so that $\Delta b(\Omega_2,\Omega_1) = b(\Omega_2) - b(\Omega_1) = \Delta r_S(\Omega_2,\Omega_1)$. Then the phase-matching should be obtained for the broadband frequency domain $\Omega_2 - \Omega_1$ by the proposed CARS approach with a spatially-dispersive Stokes beam.

The ultrabroadband CARS spectrometer included single-mode Nd:YAG laser with an active Q switching and two amplifiers. An ultrabroadband dye laser DO, whose pump was spatially modulated by the Fresnel biprism BF1 [2], was used as a Stokes source. The laser operated on the scheme of a nonselective "spatially-dispersive" resonator with two conjugated prism pairs. The distribution of the pump intensity, appropriately chosen, compensated the lower cross-emission sections in the wings of the dye gain profile. The spectra with a bandwidth of 45 nm FWHM (about 1100 cm^{-1}), produced by the dye laser, were convenient for single-pulse CARS of the water OH stretching vibration. Both CARS and Stokes signals were detected simultaneously by two polychromators and multichannel detectors.

To demonstrate the described method for ultrabroadband single-pulse CARS of liquids using a spatially-dispersive Stokes beam experiments were carried out in carbon tetrachloride. CCl_4 showed no resonances above 900 cm^{-1} and its anti-Stokes spectra could repeat the dye laser spectral shapes. This might be used to test the fulfilment of the phase-matching condition. In Fig. 2 the normalized spectrum from the CARS and the simultaneously captured Stokes laser signal in CCl_4 is shown. The recorded anti-Stokes spectra in the nonresonant liquid verify the applicability of this CARS approach to the whole spectral domain, covered by the ultrabroadband dye laser. We used this method to record the OH stretching vibration of liquid water and to clarify the discrepancies in the $\chi^{(3)R}/\chi^{(3)NR}$ value. Fig. 3 displays the dye laser spectrum, resultant single-pulse anti-Stokes and the normalized spectrum of double-distilled liquid water. The background suppression methods were not used. The low-frequency side of a dip, which occurs because of the interference between the resonant and nonresonant part of the anti-Stokes signal [3], has been seen. The OH stretching band of liquid water is composed of several lines [4,5]. The fit of water CARS spectra in which nonresonant signal is not suppresed might give inaccuracy in determining the line parameters. For simplicity, Itzkan and Leonard assumed that in the region of interest there was only one Raman vibration [3]. Using the one-line approximation and curve fitting of the single-pulse normalized CARS spectrum with a theoretical function (the smooth line in Fig. 3), we have found the Raman mode parameters: $\chi^{(3)R}_{1111}/\chi^{(3)NR}_{1111} = 2.3$. The obtained for single shot (10 ns) value of the ratio between $\chi^{(3)}$ parts is close to that of Itzkan and Leonard determined by scanning CARS [3]. The future application of this single-pulse ultrabroadband CARS approach in the polarization technique [4,5] would allow to resolve the water OH stretching vibration, and it would clarify the ambiguity in the liquid water models.

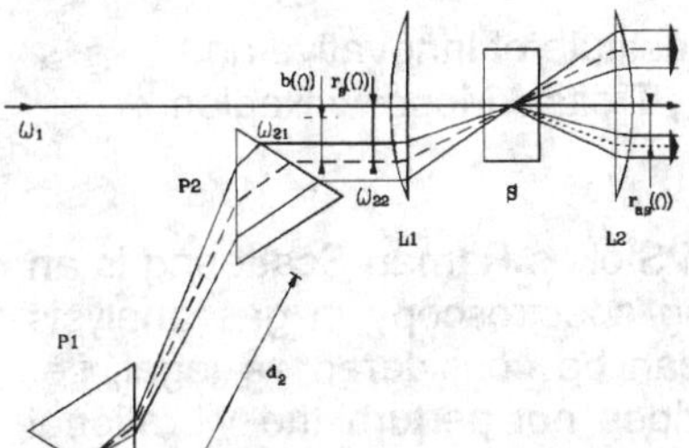

Fig. 1. Broadband CARS with a spatially-dispersive Stokes beam. P1, P2 are conjugated prisms, L1, L2 are input and output CARS lenses, and S is the sample under investigation.

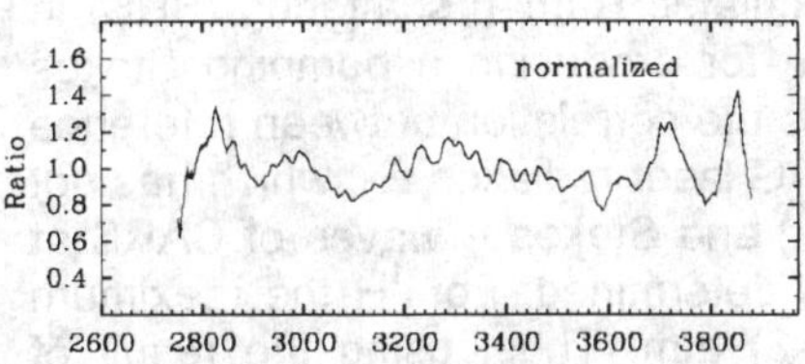

Fig. 2. Single-pulse normalized CARS spectrum in nonresonant liquid.

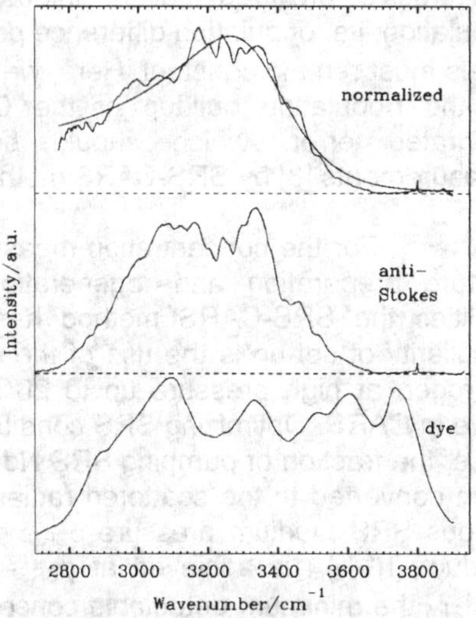

Fig. 3. Single-pulse ultrabroadband spectra of dye laser and CARS signal in liquid water.

1. V.H. Astinov, G.M. Georgiev, Appl. Phys. B, submitted for publication.
2. V.H. Astinov, Opt. Commun. 118, 297 (1995)
3. I. Itzkan, D.A. Leonard, Appl. Phys. Lett. 26, 106 (1975)
4. N.I. Koroteev, M. Endemann, R.L. Byer, Phys. Rev. Lett. 43, 398 (1979)
5. A.F. Bunkin, D.V. Vlasov, A.S. Galumian, K.O. Surskii, Opt. Spectrosk. 58, 481 (1985)

CALCULATION OF LEVELS POPULATION PERTURB EFFECT FOR CONCENTRATION MEASUREMENTS WITH SRS-CARS

A. A. Ivanov

Federal Scientific Center “Troitsk Institute of Innovative and Thermonuclear Investigation”, OIP, Troitsk, Moscow Region 142092, Russia

It is well known that Coherent anti-Stokes Raman Scattering is an alternative to conventional Raman scattering spectroscopy in gas analysis [1,2]. However, the CARS measurements can be considered as legal, i.e. nonintrusive, provided that the interaction does not perturb the vibrational population, i.e. population difference per molecule (Δ_j) between Raman active levels must remain constant. Here, we give a calculation of the pumping rate for the population buildup of the $Q_{01}(1)$ transition in protium and the integrated error, which should be expected in our concentration measurements [3] by SRS-CARS method.

For the concentration measurements of hydrogen isotopes in fuel mixture preparation and regeneration systems of thermonuclear fusion facilities the SRS-CARS method has been proposed [2,3]. An essential peculiarity of set-up is the use of the stimulated Raman scattering (SRS) in hydrogen at high pressure up to 20 atm for producing a pumping Stokes wave in CARS. Optimizing SRS conditions the correlation between reference I_1 (i.e. the fraction of pumping SRS Nd:YAG laser radiation E_0, which has not been converted to the scattered radiation) and Stokes I_2 waves of CARS at various SRS medium pressure p_{SRS} was determined. For ^1H the maximum product $(I_1^2 \cdot I_2)_m$ is achieved at $p_{SRS}=(2\div3)$ atm. Thus, using the result of $(I_1^2 \cdot I_2)_m$ the minimum detectable concentration of protium ^1H in deuterium ^2H as a buffer gas at the total mixture pressure p_t=50 Torr is 80 ppm within the limits of quadratic concentration law.

The pumping rate for the population buildup of a Raman transition according to the classical approach for monochromatic excitation [4,5] can be calculated to be:

$$r \equiv -\frac{1}{\Delta_j}\frac{\partial \Delta_j}{\partial t} = 2\left(\frac{4\pi c}{\hbar\omega_2^2}\right)^2 \left(\frac{\partial\sigma}{\partial\Omega}\right)_j g_j\, I_1\, I_2\, \frac{\gamma_j}{4\Delta\omega_j^2+\gamma_j^2} \qquad (1)$$

where $(\partial\sigma/\partial\Omega)_j$ is the spontaneous Raman scattering cross section of the mode j and g_j the weighting factor. Eqn. (1) can be used to calculate the effect of levels population perturb of the $Q_{01}(1)$ transition in ^1H as function of p_{SRS} i.e. energies of the reference and Stokes waves. For this purpose one need the values of the squared detuning $\Delta\omega_j^2=\omega_j^2-(\omega_1-\omega_2)^2$ and the transition linewidth γ_j, which can be obtained from Refs. 1, 6 and 7, as well as numerical values for the right hand side of Eqn. (1). The result of a calculation is shown in Fig. 1 together with the experimental I_1 and I_2 values. The Raman line shift for ^1H has been modeled using a second-order expansion in the pressure p_{SRS} as: $\omega_j(p)=\omega_j(0)+C \cdot p+D \cdot p^2$, where $\omega_j(0)=$ 4155.247 cm^{-1}, C and D are the temperature coefficients.

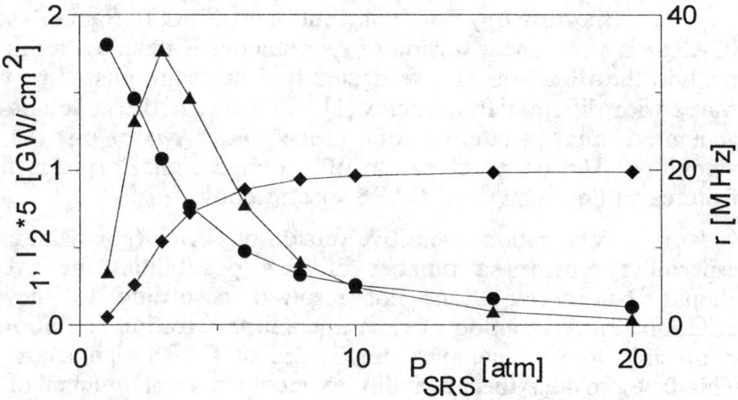

Fig. 1. Variation of power densities I_1 (●), I_2 (◆) and pumping rate r (▲) with SRS pressure. E_0=160 mJ.

Thus, taking into consideration that the Δ_j=0.67 for $Q_{01}(1)$ at T=300 ^{0}K, for ^1H on its $Q_{01}(1)$ line at p_{SRS}=3 atm (I_1=1.07 GW/cm^2 and I_2=0.11 GW/cm^2 at E_0=160 mJ) if we use 20 ns laser pulse, a 50 % reduction in Δ_j, and hence a 25 % integrated error on the concentration measurements is observed.

1. P. Regnier, and J.-P.E. Taran, in Laser Raman Gas Diagnostics, (M. Lapp, C.M. Penney, ed.), Plenum Press, N.Y.-London, 1974, p. 87.
2. A.A. Ivanov, G.A. Polyakov, and V.B. Voronin, B.R.A.S. Physics 57, 350 (1993).
3. A.A. Ivanov, G.A. Polyakov, and V.B. Voronin, Instruments and Experimental Techniques 3, 172 (1995).
4. M. Maier, W. Kaiser, and J. Giordmaine, Phys. Rev. 117, 580 (1969).
5. P. Regnier, F. Moya, and J.-P.E. Taran, AIAA Journal 12, 826 (1974).
6. W. Bischel, and M. Dyer, Phys. Rev. A 33, 3113 (1986).
7. D. Jennings, A. Weber, and J. Brault, Appl. Optics 25, 284 (1986).

THE EXPERIMENTAL POSSIBILITIES OF CARS SPECTROSCOPY IN COMPLEX RAMAN BANDS ANALYSIS

A.I.Fishman, A.I.Skvortsov

Phys. department, Kazan State University, Lenin str. 18, Kazan, 420008, Russia. E-mail: fishman@phys.ksu.ras.ru

Spectroscopy of coherent anti-Stokes Raman scattering (CARS) is a nonlinear version of spontaneous Raman scattering. In this method the vibrations of investigated molecules are phased by two laser waves with different frequencies [1]. As a result, the coherent signal is generated after scattering of "probe" laser wave on the phased vibrations. The coherent nature of scattered signal explains the main features and advantages of CARS spectroscopy.

Polarization - sensitive version of CARS (polarization CARS) especially, provides a number of new possibilities for overlapping Raman bands resolution, not resolved according to the Raleigh criterion [2]. A variation of experimental polarization conditions makes it possible to vary complex amplitudes of CARS signal components. This fact provides the possibility to record a great number of spectra, with different line shapes. At the same time all these spectra can be described with the only set of spectroscopic parameters.

An investigation of vibrational spectra of conformationally - inhomogeneous media is one of the most interesting applications of polarization CARS unique possibilities for overlapping bands resolution. In this report some examples of using polarization CARS for resolution of complex contours which components belong to different conformers are presented.

The coherent nature of CARS spectra gives the additional experimental "degree of freedom" in monitoring of spectral contour. The coherent nonresonance component resulting from the cuvette windows is superimposed on the signal from the substance under investigation. By varying the phase - mismatch we can alter the nonresonance background complex amplitude (phase - mismatching technique [3]). That is why we can reach the most contrasting interference picture of two close resonances with its minimum equal to zero.

The distinctive property of this approach is that the phase difference between complex amplitudes of resonant components is kept constant while the value of phase - mismatch is changed.

In the present report the results of application of polarization CARS in combination with phase - mismatching technique for an investigation of organic molecules in liquid are presented. The use of suggested combination allowed us to resolve the complex structure of some contours in cases with the distance between separate lines less then half - width at half - maxima and determine the values of spectroscopic parameters of separate components.

The additional possibilities in forming the spectral contours arise when the asymmetrical cuvette with different thicknesses of the windows is used.

The dependence of the interference picture on the wave mismatch value Δk significantly changes. As the Δk changes the vectors E_{3l}^{NR} and E_{3l}^{R} have not remained perpendicularly relative to each other (at frequency detuning equal to zero). The experimental CARS spectra may be significantly modified by variation of the ratio of the windows thickness. (Fig.1).

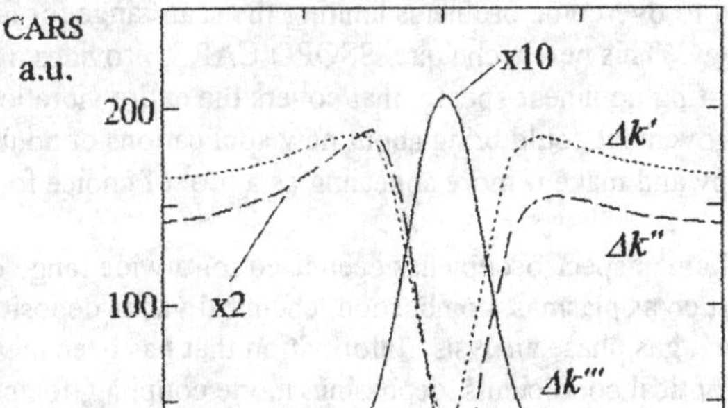

Fig.1 Polarization CARS spectra of two close resonances for different value of wave - mismatch (Δk) (asymmetrical cuvette)

1. S.A.Akhmanov, N.I.Koroteev, Methods of nonlinear optics in spectroscopy of light scattering, Nauka, Moscow, 1981.

2. N.I.Koroteev, Usp. Phys. Nauk (in Russian), 152, 493 (1987)

3. T.A.H.M.Scholten, G.W.Lucassen, F.F.M.de Mul, and J.Greve, Appl.Optics, 28, 1387 (1989).

Synchronously Scanned OPO CARS: Improved Tunability for Nonlinear Raman Spectroscopy

Peter C. Chen*

Spelman College, 350 Spelman Ln. Box 307, Atlanta, Georgia, 30314

A new method for performing Coherent Antistokes Raman Spectroscopy (CARS) using the Optical Parametric Oscillator (OPO) and synchronous scanning has been developed to overcome problems limiting the scan range for nonlinear Raman Spectroscopy. This new technique, SSOPO CARS, provides a means for conveniently generating nonlinear spectra that covers the entire vibrational region [1]. This new improvement could bring about new applications of nonlinear Raman spectroscopy and make it more appealing as a tool of choice for studying chemical species.

Nonlinear Raman spectroscopy has been used for a wide range of studies, including systems such as plasmas, combustion, chemical vapor deposition, condensed phase and gas phase analysis. Information that has been measured includes nonlinear optical coefficients, dephasing, mode coupling, temperature profiles, energy transfer, reaction mechanisms, and quantitative analysis. Most of these applications have involved producing spectra over relatively short tuning ranges (hundreds of wavenumbers). Covering broader regions of the spectrum has required frequent changing of laser dyes and the task of gluing together individual spectral pieces in order to produce a spectrum that covered the range of interest [2]. Not only has this task been inconvenient, but it has also limited the ability of the technique for real-time qualitative and quantitative analysis. Furthermore, the act of changing dyes can affect the spatial overlap, angle, and position of the laser beams, thereby changing the overlap region or signal intensity. Controlling and maintaining such overlap regions is essential for studies that involve determining spatial information.

In addition to being limited by the tunability of the dye lasers, the scan range for traditional CARS in condensed media is also plagued by phase mismatch problems. The phasematching requirements vary as the laser frequencies are changed, limiting the scan range to a few hundreds of wavenumbers. In order to overcome this long-standing problem, researchers have adjusted the angles of the input lasers [2] during the scan, a process that has been somewhat inconvenient, expensive, and limited in its success.

The basis for this new technique lies with recent commercial development of the OPO as a convenient and promising method for generating tunable coherent light. This development has benefited from improvements in injection seeding, Gaussian optics, and nonlinear crystal technology. Commercial OPO systems have narrow linewidths ($0.2cm^{-1}$) and broad tuning ranges of roughly 400-2000nm, which is almost two orders of magnitude longer than that of dye lasers.

Furthermore, the OPO can be used with synchronous scanning in order to significantly reduce the phasematching problem. The OPO has two output beams: a signal and an idler. These two beams provide a convenient means of synchronously scanning the pump beams so that phasematching is maintained throughout the scan. Synchronous scanning, which was developed for Four Wave Mixing by Nguyen and Wright [3] involves tuning two fields in the same direction at the same rate while keeping the other two field frequencies constant. For SSOPO CARS, two fields (the signal and idler) are automatically tuned at the same rate, but in *opposite* directions. For example, in CARS, with $\omega_4=\omega_1-\omega_2+\omega_3$, frequencies ω_2 and ω_4 can remain constant if ω_1 and ω_3 are synchronously scanned in opposite directions. The resulting cancellation of change in frequency can also be used to cancel changes in the corresponding k vectors. Experimental results show that problems with increasing phase mismatch during CARS scans have consequently been reduced, and that the scan range has been increased by one to two orders of magnitude.

Using SSOPO CARS, complete nonlinear vibrational spectra of molecules have been produced in a way that is fast, efficient, and simple. These new improvements help make nonlinear Raman practical as a tool for a broader range of chemical measurement experiments, since complete fingerprint-like spectra can now be produced quickly and be associated with a fixed overlap region in space. As an analytical tool, this technique will enable CARS to move from being a tool for monitoring a specific species to a tool that could be used for performing a complete qualitative and quantitative analysis.

1.Chen, P.C., "Synchronous Scanning for Coherent Antistokes Raman Spectroscopy using a Tunable Optical Parametric Oscillator," submitted to *Analytical Chemistry,* 1996.
2.Nguyen, D. C.; Wright, J. C. Appl. Spectrosc. 39, 230-242, 1985.
3.Carreira, L.A and Horovitz, M.L., in Nonlinear-Raman Spectroscopy and Its Chemical Applications, (W. Kiefer and D. A. Long, eds.), D. Riedel Publishing Co., Dordrecht, Holland, 1982, pp. 367-389.

2λ-CARS THERMOMETRY OF HYDROGEN

W.Clauß[1], D.N.Kozlov[2], R.L.Pykhov[2], V.V.Smirnov[2], O.M.Stel'makh[2], K.A.Vereschagin[2]

[1]German Aerospace Research Establishment (DLR), Institute of Chemical Propulsion and Engineering, Energy and Propulsion Section, D74239 Lampoldshausen, Germany
[2]General Physics Institute of the Academy of Sciences of Russia, Vavilov Str. 38, 117333 Moscow, Russian Federation

The aim of this work, carried out in frame of the development of H_2-based CARS temperature diagnostics of high-pressure technical combustors (P = 60-200 bar, T= 1000-3000 K) with hydrogen as the main component, was to investigate the abilities of the two-wavelength rotational CARS spectroscopy of H_2 for single-shot temperature measurements in the heated gas cell under stable laboratory conditions. At this stage the question of the optimal choice of the pair of spectral lines has been regarded, an experimental setup for two-wavelength single-shot CARS spectroscopy of hydrogen has been assembled and tested, software for experimental data processing has been developed. The results of laboratory single-shot temperature measurements in a heatable high-pressure cell with hydrogen at pressures up to 26 bar and temperatures up to 1100 K are presented. The temperature evaluation and the analysis of the accuracy of measurements and its dependence on measured temperature values have been accomplished.

In hydrogen we prefer to work with a pure rotational transitions of the S-branch because the pressure shift coefficients for pure rotational lines is an order of magnitude smaller than those for the Q-branch and in addition for the rotation lines the Dicke-narrowing effect takes place for smaller densities and for pressure higher then 5 bar lines should be broadened by collisions homogeneously and linearly with density.

The analysis of the temperature sensitivity together with consideration about intensities of S-branch lines and their detectivity in CARS spectra shows, that for the temperature region 500 -1000 K the preferable pair of the lines are S1 and S3 just as the lines S3 and S5 are more suitable for the temperature region 1500 K - 2000 K.

To take into account the pulse-to-pulse fluctuations of measured intensities due to lasers power and frequency variations a reference cell with the same resonant gas at room temperature T_0 and pressure P_0 and with the adequate pump beams interaction geometry was installed.

As a result, pure rotational S-branch transitions 1-3 (Raman shift=587cm-1) and 3-5 (Raman shift=1035 cm-1), with the corresponding Stokes laser wavelength of ~549 nm and ~563 nm, were choosen.

The laser system of the 2λ-CARS-spectrometer consists of a Nd:YAG pump laser with frequency doubling, and a dye laser module, providing Stokes radiation at two adjustable wavelengths. As the Stokes lasers we tested two different types of dye lasers: distributed feedback dye laser (DFDL) and superluminescent laser (SLDL) - corresponding to two different types of spectra structure.

To test this CARS-system in pure hydrogen, a high-pressure heatable sample cell has been designed which was supposed to provide stationary conditions at gas pressures up to 200 bar and temperatures up to 1300 K. Gas temperature inside the cell was measured by a thermocouple. The data acquisition was accomplished using the SI POSMA software. Both single-shot and accumulated spectra has been recorded. Further data processing was performed off-line with the software developed in the laboratory. Collinear, planar BOXCARS and USEDCARS pump configurations were tried in the experiments, and the measured temperatures were found to be the same within the experimental errors, which proves that the temperature field in the central part of the cell is uniform.

Temperature measurements in 2λ-CARS thermometer are based on temperature dependence of CARS signal energy, which is primarily determined by molecular population distribution function between rotational states. To exclude the dependence on gas density one should take a ratio $I_i(T)/I_j(T)$ of CARS-signals from two transitions and use the temperature dependence of this ratio for temperature measurements. When a reference cell, with the same resonant gas at known stable temperature T_{REF} and pressure P_{REF}, is used the total ratio $RAT(T) = (I_i(T)/I_j(T))/(I_i(T_{REF})/I_j(T_{REF}))$ has to be measured for evaluation of unknown temperature T in the probed volume. The random error of this ratio has been shown to be reduced even in a single-shot measurement.

The temperatures, measured with averaging over 500 laser shots (300 K < T < 1100 K, pressures of 5.5 and 11 bar), agree with the thermocouple measurements with accuracy of 3-5%. For each single-shot realisation the value of temperature T_i was determined. As a results for single shot temperature measurement Relative Standatd Deviations (RSD) 1.5% at room temperature and 8% at 1000 K were obtained.

The dependence of the Relative Standard Deviations σ_{RAT} (for CARS signals ratio) and σ_T (for temperature) on parameters of the experimental setup, pump and Stokes laser spectrum and Stokes laser configuration, as well as on sample gas temperature and density has been studied.

The observed growth of σ_T with temperature at moderate pressures is explained by simultaneous effect of two factors: decrease of sensitivity of the transition pair S(1) and S(3), and increase of the RSD for total ratio (σ_{RAT}) due to loss of correlation of CARS signals from sample and reference channels for both transitions caused by changing linewidthes with temperature and pressure variations.

On the base of the comparative analysis of data, obtained for different confugurations of dye laser module (with different spectral characteristics of Stokes lasers),we expect to reach with SLDL CARS configuration the accuracy of single-shot temperature measurements in the flame of real burner of about 5-6% at temperatures 1000-2000 K.

HIGH RESOLUTION CARS AND INFRARED SPECTRA OF ISOTOPOMERS OF AMMONIA: $^{14}NH_3$, $^{15}NH_3$, $^{14}ND_3$, and $^{15}ND_3$

S. Anders[1], J. Jonuscheit[1], D. Illig[1], D. Bermejo[2], J. Santos[2], R. Bini[3], E. Castellucci[3], P. Pracna[4], K. Sarka[5], and H.W. Schrötter[1]*

[1]Sektion Physik der LMU München, Schellingstrasse 4, D-80799 München, Germany
[2]Instituto de Estructura da la Materia, C.S.I.C., Serrano 123, E-28006 Madrid, Spain
[3]LENS, University of Florence, Largo E. Fermi 2, I-50127 Firenze, Italy
[4]J. Heyrovsky Institute, Czech Academy of Sciences, Dolejšková 3, CZ-18223 Praha, Czech Republic
[5]Faculty of Pharmacy, Comenius University, Odbojárov 10, SK-83232 Bratislava, Slovakia

In a previous investigation we have studied the CARS and FTIR spectra of the ν_1 band of ammonia $^{14}NH_3$ [1]. The CARS spectrum could only be satisfactorily simulated by introducing individual linewidths $\Delta\nu(J,k)$ depending on the quantum numbers J and k of symmetric top molecules. Later Markov et al. [2] have calculated and measured self broadening coefficients in th Q-branch of the infrared ν_1 band. In this communication we report on the CARS spectra of the ν_1 bands of $^{14}NH_3$, $^{15}NH_3$, $^{14}ND_3$, and $^{15}ND_3$ measured at pressures from 1 kPa to 80 kPa with our reconstructed cw CARS spectrometer [3]. Preliminary results have already been published [4]. To facilitate the assignment of the ND_3 spectra also difference frequency IR [5] and FTIR spectra have been recorded.

The simulation of the $^{14}NH_3$ CARS spectra in the pressure range 1 to 5 kPa using the linewidths derived from the experimental IR data [2] showed good agreement after some modifications for the k = 0 lines [4]. The CARS spectrum of $^{15}NH_3$ was assigned by combining the wavenumbers from MW [6] and IR [7] data and also 24 previously unassigned IR transitions could be identified. Pressure broadening coefficients were determined to be lower by factors between 0.6 and 0.9 for isolated lines as compared to the $^{14}NH_3$ values to achieve the best fit of the experimental and the simulated CARS spectra.

In the CARS spectra of the Q-branches of the ν_1 bands of $^{14}ND_3$ [4] and $^{15}ND_3$ (see Fig. 1) the lines are much more closely spaced and could not be so well resolved. Because no high resolution IR spectra could be found in the literature, the Q-branch region was recorded by the difference frequency IR technique in Madrid [5] at a pressure of 200 Pa in a cell with 3 m pathlenth. A section of the spectrum of $^{15}ND_3$ is shown in Fig. 2. In addition the ν_1 bands including the P- and R-branches were measured with a Bruker FTIR interferometer in Florence at a pressure of 3 kPa in a 20 cm cell. At the conference we shall report on the state of the assignment of the IR and CARS spectra.

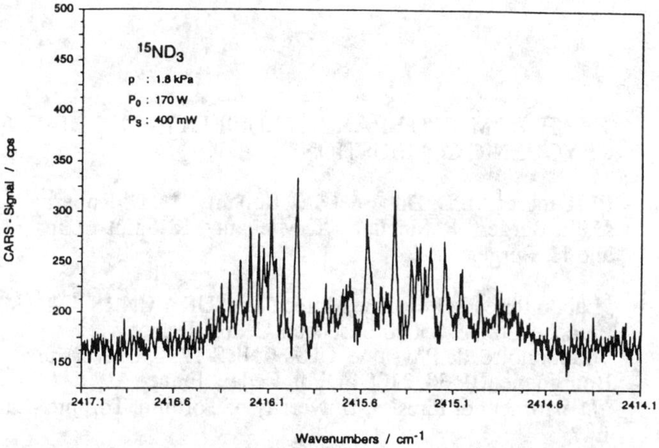

Fig. 1. CARS spectrum of the Q-branch of the ν_1 band of $^{15}ND_3$ at a pressure of 1.8 kPa. Argon laser power 170 W (intracavity), dye laser power 400 mW.

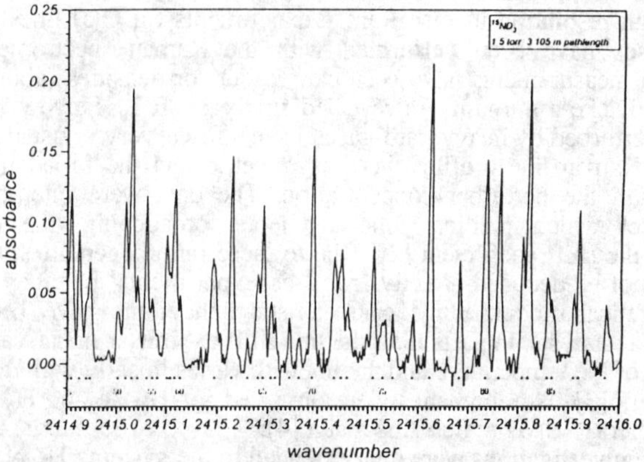

Fig. 2. IR difference frequency spectrum of part of the Q-branch of the ν_1 band of $^{15}ND_3$ at a pressure of 200 Pa in a 3 m cell.

1. R. Angstl, H. Finsterhölzl, H. Frunder, D. Illig, D. Papoušek, P. Pracna, K. N. Rao, H.W. Schrötter, and Š. Urban, J. Mol. Spectrosc. 114, 454 (1985).
2. V.N. Markov, A.S. Pine, G. Buffa, and O. Tarrini, J. Quant. Spectrosc. Radiat. Transfer 50, 167 (1993).
3. Th. S. Bican, J. Jonuscheit, U. Lehner, and H.W. Schrötter, J. Raman Spectrosc. 26, 699 (1995).
4. J. Jonuscheit, U. Lehner, D. Illig, S. Anders, K. Sarka, and H.W. Schrötter, J. Mol. Structure 349, 389 (1995).
5. J. Santos, P. Cancio, J.L. Domenech, J. Rodriguez, and D. Bermejo, Laser Chem. 12, 53 (1992).
6. H. Sasada, J. Mol. Spectrosc. 83, 15 (1980).
7. D. Papoušek and Š. Urban, private communication (1984).

THE H_2 RAMAN Q BRANCH PROFILES IN THE DIAGNOSTIC OF CRYOGENIC COMBUSTION

P. Joubert [1], M. L. Dubernet [1], J. Bonamy [1]*, D. Robert[1]
J. Ph. Berger[2], P. Sinclair[3], X. Michaut [2], R. Saint-Loup[2]
and H. Berger[2]

[1] Laboratoire de Physique Moléculaire, URA CNRS 772, Université de Franche-Comté, 25030 Besançon-Cedex, France
[2] Laboratoire de Physique, URA CNRS 1796, Université de Bourgogne, BP 38, 21004 Dijon Cedex, France
[3] Department of Physics, University of Toronto, Toronto, Canada, M5S 1A7

H_2 is of particular interest in aerospace research with regard to cryogenic combustion. Due to the simplicity of its spectrum (isolated lines) it plays a major role in the diagnostic of temperature and concentration.

High resolution inverse Raman experiments for Q(J) lines of H_2 diluted in various gases have been performed with the Raman spectrometer of Dijon University. All measurements have been carried out for densities (about 10 amagat) well above the Dicke minimum. As revealed firstly by R.L. Farrow *et al.* [1], the spectra of H_2 perturbed by heavy rare gases have shown very unusual features. The Q(J) isotropic Raman line profiles were asymmetric and the broadening increased non-linearly with the perturber concentration. The data were interpreted using a lineshape model which predicts inhomogeneous broadening due to the speed dependence of the shift coefficient [2]. Due to these inhomogeneities, the collisional linewidths cannot be deduced directly from the experiments, but they are extracted from the theoretical model. Some characteristic behaviours have been obtained through the data analysis. In particular, the line shift exhibits a linear variation versus the square root of the temperature and the linewith varies linearly with the temperature for all perturbers [3]. No anomalies were observed for H_2-He and H_2-H_2 molecular pairs, and the theory also predicts this result [4].

The investigations were then extended to the systems H_2-N_2 and H_2-H_2O for combustion applications. The experiments have been performed, as in previous studies, from room temperature(from 600K for H_2O) up to 1200K. At this higher temperature, the line profiles of H_2 diluted in N_2 were very asymmetric and again, the linewidth was non-linear versus perturber concentration. It was possible to explain these features by including speed dependent broadening in the model in addition to speed - dependent vibrational shifting[5]. In contrast, the H_2-H_2O lines present no anomalies[6].

The second step of this work consisted in the calculation of the broadening coefficients through a semiclassical approach[7]. This was done first on the simplest system H_2-He, for which a precise potential surface is available, including in particular a dependence on the intramolecular distance. We have obtained a quantitative agreement with experiment concerning the line shift and especially its linear variation versus the square root of the temperature. The consistency is not so good concerning the linewidth but the linear variation with temperature is effectively obtained. This calculation on a model system allows us to draw some conclusions concerning its extension to other H_2-X pairs. The main difficulty lies in the availability of a good potential surface, especially in the short range domain. Very precise calculations are thus only possible, to our knowledge, for H_2-He and H_2-Ar. The second problem is tied to the use of a semiclassical theory. The hydrogen molecule has a very large rotational constant (about 60 cm^{-1}), so that the inelastic energy jumps due to collisions are not negligible compared to kT. Again, this difficulty may be overcome for the same systems as cited above for which a purely quantum calculation is possible. Such a calculation is the subject of current work.

As said above, the inhomogeneous broadening is a consequence of the large variation of the vibrational dephasing shift with temperature (or with relative speed). The question thus arises concerning the existence of such collisional effects for the pure rotational transitions. In order to check this specific point, the rotational Raman spectrum of H_2 mixed with foreign gases has been investigated in a large temperature range. The results have been interpreted through quantum and/or semiclassical calculations starting from the knowledge of the intermolecular potential.

1. R.L. Farrow, L.A. Rahn, G.O. Sitz and G. J. Rosasco, Phys. Rev. Lett. 63, 746 (1989).
2. D. Robert, J.M. Thuet, J. Bonamy and S. Temkin, Phys. Rev. A 47, 771 (1993).
3. J. Ph. Berger, R. Saint-Loup, H. Berger, J. Bonamy and D. Robert, Phys. Rev. A 49, 3396 (1994).
4. J.W. Forsman, J. Bonamy, D. Robert, J. Ph. Berger, R. Saint-Loup and H. Berger, Phys. Rev. A 52, 2635 (1995).
5. P. M. Sinclair, J. Ph. Berger, X. Michaut, R. Saint-Loup, H. Berger, J. Bonamy and D. Robert, Phys. Rev. A (submitted).
6. J. Ph. Berger, Thesis, Dijon (France), July 1994.
7. D. Robert and J. Bonamy, J. Phys. (Paris) 40, 923 (1979).

OBSERVATION OF MULTIPHONON RESONANT HYPER RAMAN SCATTERING PHENOMENA IN A CdS CRYSTAL

K.Ondriash*, Yu.Polivanov, K.Prokhorov, L.Semenova

Institute of General Physics, Russian Academy of Sciences, Vavilova 38, Moscow, 117942, Russia

The scattering of light near the fundamental absorption edge of crystals has attracted considerable research interest, since such studies can provide important information on the nature and mechanisms of the electron-phonon interaction. One of the prominent and well known phenomena in resonant Raman scattering (RRS) investigations is abnormal increasing of the even-order (2LO, 4LO etc.) intensity in multiphonon RRS by longitudinal optical phonons (see, for example, [1]).

In this paper we report the first observation of the multiphonon resonant hyper Raman scattering (RHRS) of light in CdS crystal applying an optical parametric oscillator with KTP crystal as a source of the tunable pumping radiation (0.94÷1.04 μm). The Fig.1 presents some experimental spectra of RHRS we have obtained. A considerable increase of the scattering intensity by 1LO phonons has been observed as the twice excitation photon energy $2\hbar\omega_l$ approaches a region of the exciton resonances (spectra 1-5 in Fig.1.). Both theoretical and experimental dependences of the RHRS intensity from 1LO phonons on the $2\hbar\omega_l$ are plotted in the Fig.2, showing satisfactorily agreement of our calculations with experimental results. However, the most interesting features in the RHRS spectra has appeared under conditions of two-photon resonance with the 1S exciton energy level (λ_{1S}~487.5nm). In this case, the intensity of the 1LO line decreases because of the attenuation of the scattered light near the edge of fundamental absorption (spectrum 6). Simultaniously, the intensity of the two-phonon RHRS (2LO line) has drastically increased. With further increase in the twice exciting photon energy, the 1LO line has vanished because of absorption, while the intensity of the 2LO line has continued to grow up (spectrum 7).

An appearance of the 2LO line in the HR spectra, while the 3LO line has not been fixed, seems to be one of the most interesting results of this studies. Indeed, taking into account the theoretical model of resonant scattering [1], in the RHRS process in CdS an absorption of the first photon of incident radiation assumes to lead to excitation of the virtual *s*-exciton state and absorption of the second photon induces transition to the *p*-exciton state. Since the dipole intraband Frohlich exciton-lattice interaction couples only the states of different parity, we can compose the following chains of intermediate states for the multiphonon RHRS process: *s-p-s* (1LO), *s-p-s-p* or *s-p-s-s* (2LO), *s-p-s-p-s* (3LO). Therefore, the HRS by 2LO phonons should only arise whether the transition from *p*-exciton to ground state or the quadruple Frohlich interaction, coupling the states of the same parity, contribute to the scattering process. In consequence of the oscillator strength of the one-photon transition to *s*-exciton state is considerably greater than that for transition to the *p*-state, the odd parity multiphonon lines have to prevail in the RHRS spectra. So, only odd-order scattering lines (1LO and 3LO) were previously observed under near-to-resonance conditions (λ_L~1064 nm), whereas the 2LO line was not detected [2]. However, both the one-photon transition to the *p*-exciton state and the two-photon transition to the *s*-state are allowed by the selection rules. Therefore, an appearance of the forbidden 2LO line in resonant HRS spectra can be accounted for a drastic resonant enhancement of the matrix element that describes the following mechanism of RHRS: (1) two-photon resonant excitation of the 1S exciton; (2) consequent virtual transitions to the p-state and to the s-state with two phonons being created in the latter one; (3) and, finally, emission of the scattered photon (*p-s-p-s*).

This work was partially supported by the Russian Foundation of Fundamental Investigations, grant No. 95-02-05403A.

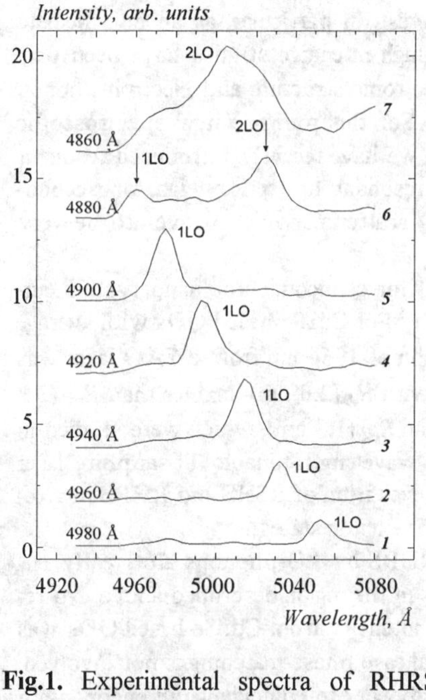

Fig.1. Experimental spectra of RHRS from 1LO and 2LO optical phonons. Near each spectrum, the wavelength of doubled incident frequency is indicated.

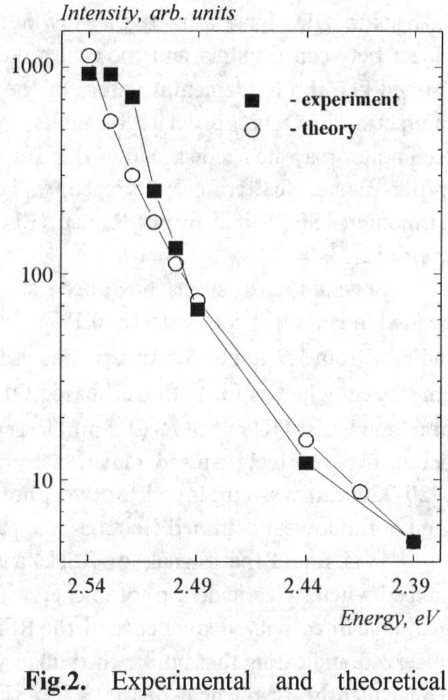

Fig.2. Experimental and theoretical dependences of the RHRS intensity from 1LO phonons on the twice excitation photon energy $2\hbar\omega_l$.

1. Klochikhin, S.A. Permogorov, and A.N. Reznitskii, Sov. Phys. JETP, 44, 1176 (1976).
2. Zubkova, K.K. Ondriash, Yu.N. Polivanov, and K.A. Prohkorov, Sov. JETP Lett., 57, 348 (1993).

RESONANT HYPER-RAMAN SCATTERING SPECTROSCOPY OF SEMICONDUCTOR QUANTUM DOT SYSTEMS

A. V. Baranov[1]* and K. Inoue[2]
[1] S. I. Vavilov State Optical Institute, St.Petersburg 199034, Russia
[2] Res. Inst. for Electronic Sci., Hokkaido Univ., Sapporo 060, Japan

Quantum dots (QDs), or semiconductor nanocrystals of sizes smaller than or comparable to the corresponding bulk exciton Bohr radius R_{ex}, are a new class of matter. Three-dimensional confinement modifies drastically electron energy spectrum and electron-phonon interaction in bulk crystal. These confinement-induced changes in QDs have attracted a growing interest, in particular, as an intermediate object between crystals and molecules. Although intensive studies have been performed [1], the fundamental nature of the electronic structure and electron-phonon interaction in QDs are yet to be understood. For this purpose new spectroscopic techniques may be needed, and in this respect we have recently introduced resonant hyper-Raman scattering spectroscopy, i.e., resonant hyper-Reyleigh, or second-harmonic (RSHS), and hyper-Raman (RHRS) scattering, which proves to be very useful [2-3].

Several QD systems have been studied for semiconductor nanocrystals embedded in glass matrices with ca. 0.1% in weight. For CdSSe-based QDs with average radii R_0 from 1.9 nm to 8.5 nm, pulsed radiation of 1064 nm from a YAG laser was employed ,whereas for both CdS-based QDs with R_0 (2.0 nm) smaller than R_{ex} (3.2 nm), and CuBr QDs with R_0 (3.2 nm) larger than R_{ex} (1.3 nm), which were studied in detail, a repetitively-pulsed radiation from a wavelength-tunable Ti-sapphire laser (920-720 nm) was employed for two-photon excitation of RSHS and RHRS in a region of the lowest confined electron-hole pair or exciton states.

It is found that signals of RSHS and RHRS by LO-phonons are greatly enhanced when twice icident photon energy falls in the region descriibed above. No remarkable directivity dependence of the RSHS intensity from CdSSe-based QDs was observed, indicating that unlike the bulk crystal case phase-matching is not involved. Consequently, resonances of the RSHS should occur in coincidence in energy with the energies of electron-hole pair states, which can be thereby probed. Note that SHG signal from glass matrix was negligibly small and thus RSHS signals from QDs at low concentration could be observed. The intensity of RHRS by LO-phonon from QDs proved to be surprisingly large as compared to the corresponding bulk crystals. It is important to note that the selection rules of optical transition accessible in both RSHS and RHRS are different from those of one-photon absorption (OPA) and resonant Raman scattering (RRS). Moreover, so-called "silent" phonon mode inactive in both RRS and IR spectra was also observed in RHRS spectrum of CdSSe-based QDs owing to the specific selection rules of RHRS.

Both two-photon excitation spectra of RSHS and RHRS from CdS QDs in the strong confinement regime where momenta of electron and hole are separately quantized, show two maxima more pronounced and shifted in energy relative to those of OPA spectrum. Analysis has shown that the maxima may be assigned to the confined electron-hole pair states with total angular momentum of odd number, unobservable in OPA, and thereby the effective masses of electron and hole in the QD can be estimated. The profile of the lowest maximum in the RHRS spectrum is similar to that of RRS, but the second peak manifests itself in the former in contrast to the RRS case where only one maximum was observed around the lowest OPA peak. This fact presumably indicates that the off-diagonal intraband Frohlich interaction governs the present RHRS process.

In the CuBr case belonging to the weak confinement regime, the RSHS and RHRS excitation spectra show one maximum in the lowest-energy OPA region due to the confined $Z_{1,2}$, *S*-exciton state. The RHRS spectrum turns out rather anomalous: up to 5th-order LO-phonon bands have appeared as incident photon energy was increased. At first glance, the RHRS spectra look like the well known spectra of RRS by LO-phonon in polar bulk crystals such as CdS, but no continuous electronic band states exist in the CuBr-based QDs in the energy region in question! The present anomalous result indicates that the strong-coupled states of Wannier exciton and LO-phonon through diagonal intraband Frohlich interaction are realized. In principle, the observed features may be interpreted in terms of either the RHRS or hot luminescence. Discussions will be presented in detail together with other experimental results.

In conclusion, we have unravelled the respective confined electronic states inherent to CdS and CuBr QDs. Furthermore, we have demonstrated that the resonant RHRS spectroscopy provides a promising and powerful tool for studying the electronic structure as well as electron (exciton)-phonon interaction in semiconductor quantum dots.

1. A.D. Yoffe, Adv. in Phys. 42, 173 (1993).
2. A.V. Baranov, Ya.S. Bobovich, and V.I. Petrov, Sov. Phys. Usp. 33, 812 (1990).
3. A.V. Baranov, K. Inoue, *et al.*, Phys. Rev. B, 53, 1721 (1996).

EXTERNAL SEEDING OF STIMULATED RAMAN SCATTERING IN A BINARY MIXTURE MICRODROPLET

Mitchell H. Fields, Jürgen Popp, and Richard K. Chang*

Department of Applied Physics and Center for Laser Diagnostics
Yale University, P. O. Box 208284, New Haven, CT 06520-8284 U. S. A.

In a multicomponent droplet, the dominant stimulated Raman scattering (SRS) signal is from the vibrational mode with the largest Raman gain. The SRS from the minority species is usually below the detection limit, even though a micron-sized droplet has low input-laser intensity threshold for SRS, mainly because of the high-Q optical feedback provided by the droplet-air interface. The droplet-cavity modes are commonly referred to as morphology dependent resonances (MDR's). SRS in a droplet or cell is amplified spontaneous Raman scattering, generated at the beginning of the pump pulse. In a cell, external "seeding" of the SRS can enhance the signal.[1] We have investigated the possibility of enhancing the SRS of the minority species in a droplet by externally seeding the droplet with broad-band radiation (at the Raman wavelength of the minority species) prior to the arrival of the pump pulse.

We have experimentally investigated external seeding of SRS in a binary mixture microdroplet (14% ethanol, 86% water). At this ratio, the SRS from water (3400 cm^{-1} band) is much stronger than the SRS from ethanol (2928 cm^{-1} mode). A pulse (532 nm, 7 ns) from a second harmonic Nd:YAG Q-switched laser is split into two beams by a beam-splitter. One green beam pumps a tunable dye laser (DCM). The red output pulse of the dye laser (the seed pulse) is recombined with the other green pulse (the pump pulse) and both are focussed onto a droplet stream. The 40 μm droplets are created by a Bergund-Liu droplet generator at a rate of 65 kHz and synchronized with the laser. The red-seed pulse (630 nm with 2 nm FWHM) arrives at the droplet ~3 ns before the pump pulse. At 630 nm, the seed pulse overlaps spectrally with the 2928 cm^{-1} Raman mode of ethanol. The illuminated droplets are imaged onto the entrance slit of a spectrograph and detected single-shot by an intensified linear photodiode array. An electronic shutter in the path of the seed pulse permits seeded and un-seeded spectra to be acquired alternately and in rapid succession. The droplet's elastic scattering spectrum of the seed pulse can be obtained by blocking the green pump pulse. Measurement of the elastic scattering spectrum is essential to distinguish SRS of the 2928 cm^{-1} mode from the elastic scattering of the seed pulse.

Figure 1 illustrates the results of a typical measurement, which consists of an average of 100 seeded and 100 un-seeded spectra. The averaged elastic scattering spectrum of the seed pulse is subtracted from the averaged seeded SRS spectrum. Two features are noteworthy. First, there is an 85X enhancement of the SRS from the 2928 cm^{-1} mode of ethanol due to external seeding of this weaker gain mode. Second, the SRS from the 3400 cm^{-1} band of water decreases as a result of seeding the weaker gain 2928 cm^{-1} mode of ethanol.

These results can be explained by a simple model for SRS in a microdroplet.[2] In a microdroplet, there is incomplete spatial overlap between the pump laser and MDR's that provide feedback for SRS. Thus, SRS can occur at several MDR's (i. e., from different Raman bands and different MDR's within the same Raman band), depending upon the relative spatial overlaps of the pump and SRS MDR's. Without seeding, SRS builds from spontaneous photons, so the vibrational modes with the highest gain will build most rapidly and deplete most of the pump before weaker gain modes can build SRS. If, however, the weaker gain modes are seeded by externally injecting photons at the Raman shifted frequency, then the weaker gain modes build from a larger initial number of photons than the stronger gain mode. Despite a smaller Raman gain, the weaker gain mode can build and partially deplete the pump before the stronger gain mode can deplete the pump. Hence, by seeding a weaker gain Raman mode, the SRS from that weaker gain mode is enhanced and the SRS from the stronger gain modes are decreased. These effects are observed in Figure 1.

In summary, we have externally seeded the SRS of the weaker gain Raman mode in a binary mixture microdroplet. By externally seeding at the wavelength of the weaker gain mode, we observed an enhancement of the SRS from the weaker gain mode and a decrease of the SRS from the stronger gain modes. This technique can be applied to reduce the concentration detection threshold of minority species in a multicomponent microdroplet.

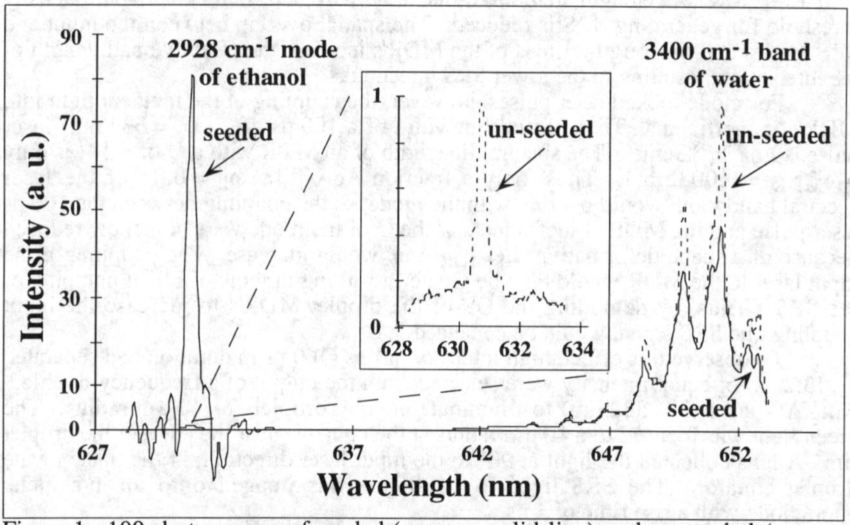

Figure 1: 100 shot average of seeded (——— solid line) and un-seeded (------ dotted line) SRS spectra (after subtraction of the elastically scattered seed-pulse spectrum) from 14% ethanol, 86% water microdroplets. The inset shows ethanol SRS from the un-seeded SRS spectrum. An enhancement of 85X is observed for the seeded ethanol SRS compared to the un-seeded ethanol SRS. SRS from the water band is diminished when the ethanol band is seeded. The sharp peaks in the SRS spectrum of the water Raman bands are associated with MDR's.

We gratefully acknowledge the partial financial support from the U.S. Air Force Office of Scientific Research (F49620-94-1-0135), the National Defense Science and Engineering Graduate Fellowship Program (for MHF), and the Deutsche Forschungsgemeinschaft (for JP).

[1] B. Zhang, W. R. Lempert, and R. B. Miles, *Opt. Lett.* **18**, 1132-1134 (1993).
[2] M. M. Mazumder, M. H. Fields, J. M. Hartings, X. Pu, A. S. Kwok, K. Schaschek, R. K. Chang, SPIE Proceedings *Nonlinear Frequency Generation and Conversion*, **2700** (1996).

LOCALIZED CAVITY Q-SPOILING FOR EFFICIENT PUMPING OF STIMULATED RAMAN SCATTERING IN MICRODROPLETS

Justin M. Hartings, Xiaoyun Pu, and Richard K. Chang*

Department of Applied Physics and Center for Laser Diagnostics
Yale University, P.O. Box 208284, New Haven, CT 06520-8284, U.S.A.

The high Q optical resonances of liquid microdroplets, known as morphology dependent resonances (MDR's), provide feedback for stimulated Raman scattering (SRS) as well as other nonlinear processes. When the incident pump wavelength, as well as the SRS wavelength are *both* resonant with droplet MDR's, the input intensity threshold for generating SRS is reduced. The spatial overlap between the input and SRS modes and the long lifetimes of the MDR's increase the effective path length of the interaction, resulting in the lower SRS threshold.

For mode-locked laser pulses, however, the coupling of the incident light into MDR's is inefficient. The spectral linewidth of a 100 ps green ($\lambda = 532$ nm) laser pulse is $\Delta\omega_L = 0.3\text{cm}^{-1}$. The spectral linewidth of an MDR with a Q of $\approx 10^7$ is only $\Delta\omega_{MDR} = 0.003\text{cm}^{-1}$. Thus, only a fraction $\Delta\omega_{MDR}/\Delta\omega_L = 0.01$ of the laser spectral bandwidth would overlap with the mode, so the coupling between the 100 ps laser pulse and the MDR is inefficient. If the Q of the mode were somehow reduced because of a shape deformation, $\Delta\omega_{MDR}/\Delta\omega_L$ would increase. The coupling of the input laser to the MDR would then be greater, resulting in more efficient pumping of the SRS. Thus, by degrading the Q's of the droplet MDR's to increase the input coupling, the SRS signal would be enhanced.

To observe this effect, high intensity pulses (100 ps in duration and separated by 13.2 ns) of equal intensity were selected from the output of a frequency-doubled, Nd:YAG laser ($\lambda = 532$ nm) to illuminate ethanol droplets of 40 μm radius. The green light was focused to a 10 μm radius at the equator, near the edge of the droplet rim. A lens collected the light at 90° to the input laser direction, in the plane of the droplet equator. The SRS light ($\lambda = 630$ nm) was imaged onto an avalanche photodiode with a rise time of 1 ns.

In the first experiment, a single input pulse was used [see Fig. 1(a)]. Oscilloscope traces of the SRS light are shown in Figure 1(b). With an input intensity of 30 GW/cm^2 (I_0), a single pulse generates SRS [see Fig. 1(b)]. The input intensity was then decreased in order to determine the input intensity threshold for generating SRS. When the input intensity is reduced to 65% I_0, no SRS is observed [see Fig. 1(b)]. The threshold for generating SRS from a single 100 ps pulse is between 65% I_0 and 75% I_0.

In the second experiment, two input pulses of equal intensity were used to illuminate the droplets [see Fig. 2(a)]. For an input intensity of 60% I_0, no SRS is observed from the first input pulse, but SRS is generated by the second input pulse [see Fig. 2(b)]. The input intensity must be reduced to 5% I_0 before the second pulse is below the SRS threshold [see Fig. 2(b)]. The input intensity threshold for SRS is a factor of ten lower for the second input pulse than that for the first input pulse.

This reduction in the SRS threshold can be explained by a localized reduction in the Q's of the droplet MDR's [1,2]. The first input pulse imparts an electrostrictive impulse in the region of droplet illumination. This electrostrictive impulse initiates shape distortions which require milliseconds to fully damp out [3,4]. In the 13.2 ns between pulses, the distortion grows into a highly localized bulge on the droplet surface [5]. For plane-wave illumination, this electrostrictive effect gives rise to shape distortions at the Descartes ring [6], which in tens of microseconds grow into the broad bulge previously observed [3,4].

When the second pulse impinges upon the droplet, the localized surface bulge locally lowers the Q's of those droplet MDR's which have some spatial overlap with the deformation. This localized Q-spoiling increases $\Delta\omega_{MDR}$, which allows greater coupling of the laser illumination to MDR's. Consequently, the increased input coupling provides more efficient pumping for the SRS. Because the distortion is localized, the deformation would not degrade the Q's to such a degree that they could not provide feedback for the SRS. Localized Q-spoiling should similarly enhance the pumping of other nonlinear processes such as lasing and stimulated Brillouin scattering.

We gratefully acknowledge the partial support of this research by U.S. Army Research Office (DAAH 04-94-G-0031) and an ASSERT Fellowship for Justin M. Hartings.

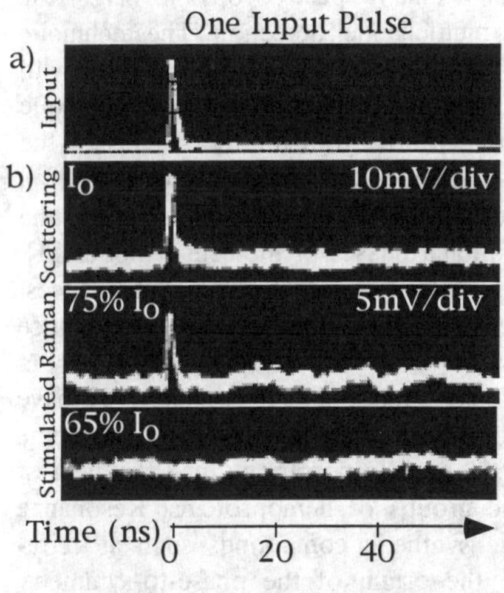

Figure 1: (a) Time trace of the 100 ps input pulse. (b) Time traces of the SRS generated by three different input intensities. The vertical scale appears in the upper right corner. The threshold for generating SRS from a single pulse is between 65% and 75% I_0.

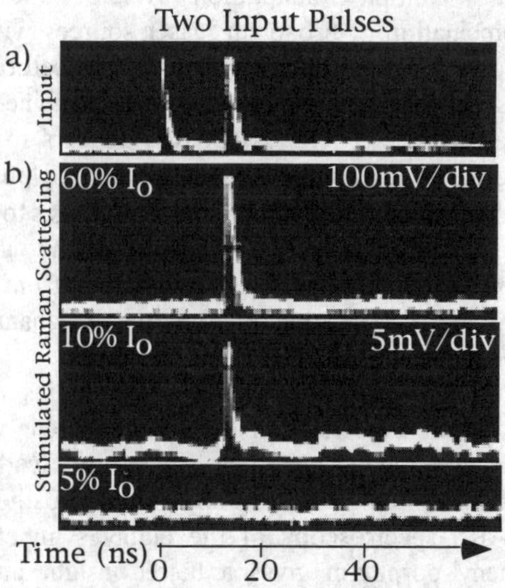

Figure 2: (a) Time trace of the two 100 ps input pulses. (b) Time traces of the SRS generated by three different input intensities. The vertical scale appears in the upper right corner. The threshold for generating SRS from the second pulse is between 5% and 10% I_0.

1. A. L. Huston, H.-B. Lin and A. J. Campillo, Opt. Lett. **15**, 1176 (1990).
2. J. L. Cheung, J. M. Hartings and R. K. Chang, Opt. Lett. **20**, 1089 (1995).
3. J.-Z. Zhang and R. K. Chang, Opt. Lett. **13**, 916 (1988).
4. H. M. Lai, P. T. Leung, P. L. Poon, and K. Young, L. Opt. Soc. Am. B **6**, 2430 (1989).
5. C. K. Ng, Ph.D. dissertation, (Chinese University of Hong Kong, Shatin, Hong Kong, 1994).
6. J.-G. Xie, T. Ruekgauer, J. Gu, R. L. Armstrong, and R. G. Pinnick, Opt. Lett. **16**, 1310 (1991).

VIBRATIONAL STRUCTURE OF Cu-TETRAPHENYL-PORPHYRIN STUDIED BY POLARISATION SENSITIVE MULTIPLEX CARS SPECTROSCOPY.

Artemy Voroshilov, Cees Otto* and Jan Greve,

Institute for Biomedical Technology (BMTI) and Applied Optics Group, Department of Applied Physics, University of Twente, P.O. Box 217, 7500 AE Enschede, the Netherlands

Multiplex adaptation of the coherent Raman spectroscopy involves the combination of broadband laser sources with multichannel detection. The technique allows for rapid data acquisition and features an improved signal-to-noise ratio with respect to scanning methods [1,2]. Combined with the polarisation sensitive technique the multiplex CARS (MCARS) appears to be a fruitful tool for the studies of the biological macromolecules in condensed phase. Besides the possibility to resolve all independent $\chi^{(3)}$ components and thus, to establish the depolarisation ratios, the method enables the determination of the vibrational phases of molecular modes [3]. The coherent phases, as originating from the manifold of electronic resonances, provide an information on the mechanisms of the mode-selective resonance enhancement.

Previous polarisation CRS studies of oxy- and deoxyhemoglobin [3] have indicated a correlation between coherent vibrational phase displayed by the heme skeleton modes and their symmetry. Synthetic metallo-porphyrins are the nearest model systems for naturally found prosthetic groups of hemoproteins. Resonance CARS spectroscopy of the highly symmetric synthetic compounds such as tetra-phenyl-porphyrin gives a better insight into the origin of the phase-to-symmetry relations. Furthermore, the primary structural information on the heme of the related native biological systems can be obtained.

The layout of our picosecond multiplex polarisation CARS spectrometer has been reported elsewhere [2]. Solution of Cu-*meso*-tetraphenylporphyrin in CH_2Cl_2 (0.9 mM) was passed through a nozzle that formed a ~300 μm jet. Using the jet allowed both to reduce the extent of phase-mismatch due to the broad Stokes spectrum $\Delta\lambda_2$ and to avoid thermal effects. The broadband Stokes laser emission enabled the coverage of typically 350-400 cm^{-1} broad Raman range. The pump wavelength λ_1 was chosen at 582.5 nm. In this fashion the resonant enhancement of the susceptibility $\chi^{(3)}$ was achieved only through the rigorous resonance of the continuum $\Delta\lambda_{CARS} = 2\lambda_1 - \Delta\lambda_2$ with the Q band of Cu-TPP absorption at 538 nm. Using an analyser four dispersion profiles of the susceptibility components χ_{1111} and χ_{1221} and their combinations were resolved in the 1300-1650 cm^{-1} range. Figure 1 shows respectively the spectra due to χ_{1111}, χ_{1221}, $(\chi_{1221}+3\chi_{1111})$ and $(3\chi_{1221}+\chi_{1111})$ as indicated by the insets *A* through *D*. Spectrum *D* was acquired with the complete suppression of the nonresonant background. On the basis of depolarisation ratios, the major bands could readily be assigned to A_{2g}, B_{1g} and A_{1g} modes of porphyrin macrocycle [4].

As can be seen from Fig. 1 the use of multiplex CARS technique resulted in a fairly high signal-to-noise ratio enabling further CARS line-shape analysis. All spectra were least-square fitted to the theoretical expression for the $\chi^{(3)}$ (Fig. 1). Eight porphyrin bands were assumed in the fit along with three bands of the solvent. Simultaneously fitting spectra [2,3] provided a set of parameters including positions, bandwidths, amplitudes, phases and the depolarisation ratios for the bands considered. As can be seen the A_{1g} modes, namely ν_2 and ν_4 (at 1565 and 1370 cm^{-1}), prevail over the non-totally symmetric B_{1g} and A_{2g} which is different from what is expected upon the resonance excitation in the Q-band range. The anti-symmetric ν_{20} mode has an untypically large bandwidth of ~25 cm^{-1} as compared to other modes (8 to 13 cm^{-1}). The extracted phases appear to correlate strongly with the symmetries of porphyrin modes. While the anti-symmetric modes ν_{19} and ν_{20} have phases of ~ -10^0, the phases of ν_2 and ν_4 (A_{1g}) are ~ 160^0 and those of the ν_{11} and ν_{12} (B_{1g}) are ~ -165^0. Such correlation conforms with the hypothesis of the symmetry dependent coupling of the vibronic transitions to the electronic ones.

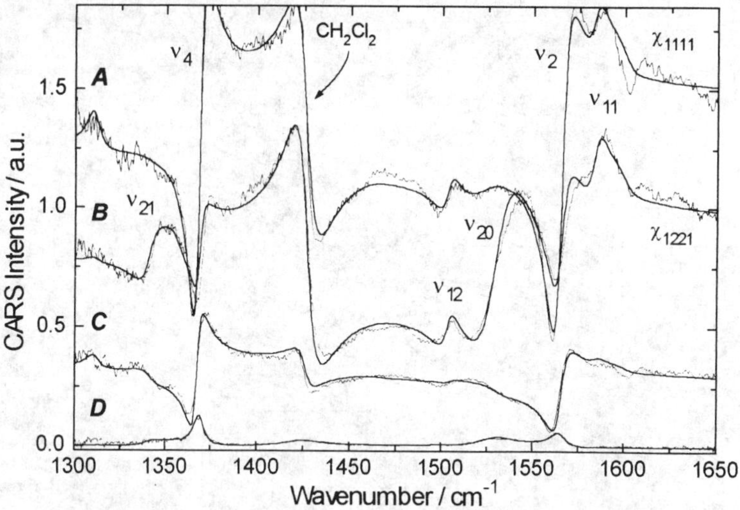

Fig.1 Polarisation MCARS spectra of Cu-*meso*-TPP in CH_2Cl_2 solution (thin lines) and the best corresponding CARS fits (bold lines). The spectra are normalised on the one of the nonresonant background

1. B. N. Toleutaev, T. Tahara, H. Hamaguchi, Appl. Phys. B59, 369, (1994).
2. A. Voroshilov, C. Otto, J. Greve, Appl. Spectrosc. 50, 78, (1996).
3. A. Voroshilov, G. W. Lucassen, C. Otto, J. Greve, J. Raman Spectrosc. 26, 443, (1995).
4. M. Atamian, R. J. Donohoe, J. S. Lindsey, D. F. Bocian, J. Phys. Chem. 93, 2236, (1989).

SECTION 5

VIBRATIONAL ANALYSIS

SPECTROSCOPICALLY DETERMINED FORCE FIELDS: A ROUTE TO VIBRATIONALLY RELIABLE POTENTIAL ENERGY FUNCTIONS

Samuel Krimm*, Kim Palmö, and Noemi G. Mirkin

Department of Physics and Biophysics Research Division
University of Michigan, Ann Arbor, MI 48109, USA

There are two main reasons for wanting to have a spectroscopically reliable molecular mechanics (MM) potential energy function, particularly for macromolecules. First, such a function, through the explicit dependence of its second derivatives (force constants) on chain conformation, can provide accurate normal mode frequencies (and hopefully intensities) for the many conformations that are possible for a macromolecule, for example the polypeptide chain of proteins. This will enhance the power of vibrational spectroscopy as a tool for studying the three-dimensional structures and interactions of macromolecules. Second, the availability of an accurate MM energy function can be expected to improve the quality of the structure and dynamics predictions that have recently made molecular mechanics and dynamics such important computational tools. Many currently used MM functions give relatively poor reproduction of spectroscopic properties, leading one to at least wonder whether their other predictions suffer from underlying weaknesses in the functional forms and parameters.

Our approach to solving this problem, in distinction to other methods, is to incorporate spectroscopic in addition to structure and energy reproducibility in the initial stages of refinement of the MM function. Furthermore, wishing to maximize the agreement with as complete as possible a spectroscopic force field, we choose an *ab initio* force field as the criterion for correspondence to the MM function. Since present *ab initio* calculations of moderate size molecules usually require scaling of force constants to experimental spectral data, we call such an MM function a spectroscopically determined force field (SDFF).

The key to the method is our development of an analytical transformation [1,2] and an associated optimization procedure [3] that make the structure and frequency predictions of the SDFF exactly the same as those of the *ab initio* calculation. In practice, we determine the SDFFs of a set of conformers of model molecules of the macromolecular system, and then optimize a common set of parameters for all structures. In this process, the set of force constant parameters is systematically reduced based on a pre-assigned limit on the frequency error. The optimized SDFF is produced by a suite of programs called SPEAR (Spectroscopic Potential Energy Analytical Refinement), which also compute other quantities such as optimized structures, conformational energies, normal mode frequencies and eigenvectors, dynamics trajectories, etc.

There are a number of advantages to the SDFF approach to refining an MM force field. By transforming a complete (i.e., *ab initio*) spectroscopic force field scaled to experimentally assigned bands, frequency agreement is assured at the start.

In contrast to a least squares fit, this not only directly incorporates highly accurate information about the minima in the potential surface, but it avoids the possibility of biasing the parameters to compensate for an incomplete formulation of the terms in the potential function. In addition, since the optimization is done in a non-redundant coordinate basis [4], the uniqueness of the force field is assured, and it is often possible to check that the MM interaction force constants are physically reasonable. Subsequent transformation to a redundant basis of choice is always possible. Finally, there is a better chance of revealing the nature of specific conformation dependences, for example in the cross-terms, since no such dependence needs to be initially imposed on the MM force constants but can be deduced from the values obtained from many different conformers.

We have implemented this procedure to produce an SDFF for the linear hydrocarbon chain [5], using scaled *ab initio* force fields for the 14 stable conformers of *n*-pentane and *n*-hexane [6]. In addition to reproducing geometries, relative energies, and barriers between conformational states, this 50-parameter SDFF gives very good frequency agreement: for example, the observed frequencies of *tt*-pentane and *ttt*-hexane below 1500 cm^{-1} are reproduced with an rms error of ~11cm^{-1}. This force field reproduces the observed chain elastic modulus of polyethylene to within ~1% [7]. We have now extended this SDFF to include branched hydrocarbon structures, with comparable results [8].

It is important that SDFFs be capable of reliably reproducing band intensities as well as frequencies, since particularly in the infrared case this determines the kind of electrostatic terms needed in the MM function, which of course influences its other predictions. Infrared intensities are determined by dipole derivatives, which at the very least require partial charges and charge fluxes for their reproduction. We now find that consistency with energies and spectroscopic properties also requires the inclusion of atomic dipoles [9]. All of these contributions are being determined for the hydrocarbons. Raman intensities can be treated in an analogous manner.

The SDFF approach holds the promise of providing a straightforward method for obtaining potential energy functions that give reliable structural, energetic, spectroscopic, and dynamic properties of macromolecular systems. We are presently using the method to develop an SDFF for the polypeptide chain.

1. K. Palmö, L.-O. Pietilä, and S. Krimm, Computers Chem. 15, 249 (1991).
2. K. Palmö, L.-O. Pietilä, and S. Krimm, J. Comp. Chem. 12, 385 (1991).
3. K. Palmö, L.-O. Pietilä, and S. Krimm, Computers Chem. 17, 67 (1993).
4. K. Palmö, L.-O. Pietilä, and S. Krimm, J. Comp. Chem. 13, 1142 (1992).
5. K. Palmö, N.G. Mirkin, L.-O. Pietilä, and S. Krimm, Macromolecules 26, 6831 (1993).
6. N.G. Mirkin and S. Krimm, J. Phys. Chem. 97, 13887 (1993).
7. K. Palmö, and S. Krimm, J. Polym. Sci.. Polym. Phys. Ed. 34, 37 (1996).
8. K. Palmö, N.G. Mirkin, and S. Krimm, this Conference.
9. W. Qian and S. Krimm, this Conference.

AB INITIO FORCE FIELDS AND VIBRATIONAL ANALYSIS OF α-PYRIDONE

Paul B. Stein*, Jianli Yang, and Iona N. L. Black

Department of Chemistry, Duquesne University, Pittsburgh, PA 15282, USA

The α-pyridone/α-hydroxypyridine system is the simplest model of nucleic acid base tautomerism, resulting in considerable interest to understand their vibrational spectra. Both tautomers are found in the vapor phase and nonpolar solvents. α-Pyridone is known to dimerize in nonprotic solvents and the vapor phase. Raman and UVrR spectra of α-pyridone and α-pyridone-ND are reported in aqueous solution, where the keto monomer is favored. Five additional isotopomers are investigated in the solid state. *Ab initio* force fields of these tautomers are determined .

Raman spectra in the 1200-1700 cm^{-1} region (λ_{ex}= 235 *nm* and 488 *nm*) are shown in Figures 1 and 2 for α-pyridone and α-pyridone-ND in aqueous solution. UVrR bands (1642 cm^{-1} and 1609 cm^{-1} in α-pyridone, and 1642 cm^{-1} and 1558 cm^{-1} in α-pyridone-ND) are enhanced by a π-π^* absorption at 224*nm* (ε=7250) .The first π-π^* transition occurs at 294*nm* (ε=5900). Bands at 1645 cm^{-1} and 1563 cm^{-1} have been reported by IR for α-pyridone in D_2O [1]. The two normal modes contain CO and CC (double bond) stretching. δ_{NH}, but not δ_{ND}, is also involved, as judged by frequency shifts with ^{15}N substitution. ν_{CO} shows sensitivity to the environment, as observed in IR studies at 1674 cm^{-1} in chloroform [2], 1676 cm^{-1} in acetonitrile [3], 1704 cm^{-1} in the vapor phase [4], and 1705 cm^{-1} in an Ar matrix [5]. Raman spectra show intense bands at 1259 cm^{-1} and 848 cm^{-1} in aqueous solution (1244 cm^{-1} and 844 cm^{-1} in the solid) with 488*nm* excitation. Shifts of 8 cm^{-1} and 3 cm^{-1} with ^{15}N substitution occur for the 1259 cm^{-1} and 1231 cm^{-1} bands. The latter band gains intensity with UV excitation. α-Pyridone-ND shows bands at 1308 cm^{-1} and 1254 cm^{-1} that shift 16 cm^{-1} and 2 cm^{-1} in the ^{15}ND isotopomer. Our normal coordinate calculations indicate that the C(6)-N and C(2)-N stretchings contribute to both modes of α-pyridone; the C(2)-N stretch is the leading term in the PED for the 1254 cm^{-1} mode, and the C(6)-N stretch, (with some ND bending), describes the 1308 cm^{-1} mode of α-pyridone-ND. The observed *frequency upshift* is the result of an alteration in the normal modes upon N-deuteration. The magnitude of the *upshift* is large but not unprecedented [6]. The NH bending mode, δ_{NH}, is assigned to the 1425 cm^{-1} band, due to its disappearance from the region upon N-deuteration and its correspondence with δ_{NH} in *cis*-lactams. The isotopomers, reported in this study, include α-pyridone-ND, α-pyridone-^{15}NH, α-pyridone-^{15}ND, 3,5-dideutero-α-pyridone, 3,5-dideutero-α-pyridone-ND, and perdeutero-α-pyridone.

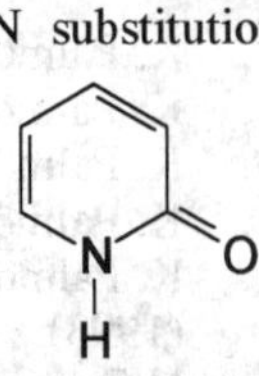

α-pyridone

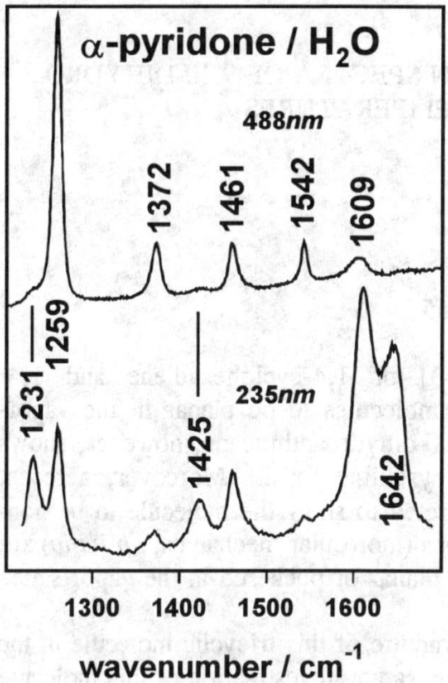

Fig.1. Raman spectra of α-pyridone in H_2O. Lower trace: 20mM α-pyridone. Upper trace: 5M α-pyridone.

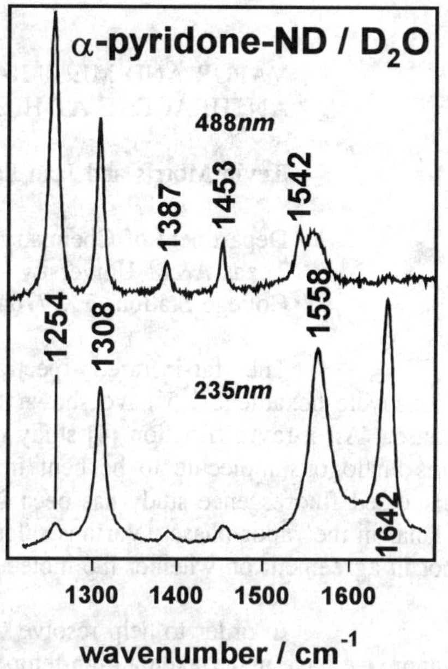

Fig.2. Raman spectra of α-pyridone-ND in D_2O. Lower trace: 20mM α-pyridone Upper trace: 0.75M α-pyridone.

The pyridone/2-hydroxypyridine system was studied theoretically by several *ab initio* methods, beginning with 6-31G*. Energies, geometries, and force fields were determined for the two tautomers as monomers and dimers. The monomer keto-enol equilibrium favored the enol tautomer, ΔH=-0.232 kcal/mole; but the dimer keto-enol equilibrium favored the keto form, ΔH=4.86 kcal/mole. The latter is attributed to dipolar interactions of the keto units. Geometry optimization gave a N-H···O distance of 2.89Å for the keto dimer (2.75Å from experiment [7]). A single scaling factor of 0.903 was applied to the frequencies. The Cartesian force field was transformed to internal coordinates so that the Wilson GF method could be used to obtain the PEDs.

Acknowledgment: We wish to thank Professor Masako Nagai for assistance in collecting the UVrR data at the Institute of Molecular Science, Okazaki, Japan.

1. K. C. Medhi *Bull. Chem. Soc. Jpn.* 64, 1944 (1991).
2. G. H. Keller, L. Bauer, C. L. Bell *Can. J. Chem.* 46 , 2475 (1968).
3. O. Bensaude, M. Chevier, J. E. Dubrois *J. Am. Chem. Soc.* 100 , 705 (1978).
4. J. Smets, G. Maes *Chem. Phys. Let.* 187 , 532 (1991).
5. M. J. Nowak, L. Lapinski, J. Fulara, A. Les, L. Adamowicz *J. Phys. Chem.* 96 , 1562 (1992).
6. A. Toyama, Y. Takino, H. Takeuchi, I. Harada *J. Am. Chem. Soc.* 115 , 11092 (1993).
7. A. Held, D. W. Pratt *J. Am. Chem.* 112, 8629 (1990).

VAPOR AND MELT RAMAN SPECTRA OF 9,10-DIHYDRO-ANTHRACENE AT HIGH TEMPERATURES

Kevin Morris and Jaan Laane

Department of Chemistry
Texas A&M University
College Station, TX 77843-3255

The far-infrared spectra [1-3] of 1,4-cyclohexadiene and 1,4-dioxacyclo-hexadiene-2,5 have shown these molecules to be planar in the vapor-phase. An x-ray diffraction [4] study of 9,10-dihydroanthracene, however, shows this analogous molecule to be bent in its crystalline form. Moreover, a recent dispersed fluorescence study has been interpreted to show the molecule to be non-planar in the vapor-phase. Various calculations (molecular mechanics, *ab initio*) are not in agreement on whether the molecule is planar or puckered in the vapor state.

In order to help resolve the structure of this tricyclic molecule in the vapor, we have undertaken a high-temperature Raman investigation of this molecule in both vapor and liquid (melt) states. Of particular interest were the low-frequency modes shown below in terms of D_{2h} (planar) symmetry.

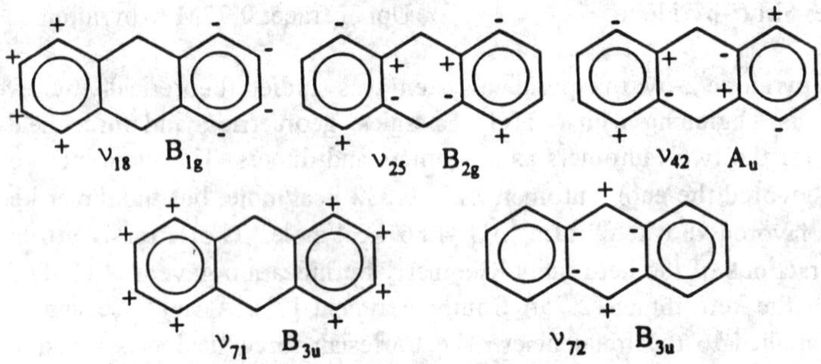

The Raman spectra of 9,10-dihydroanthracene vapor at 300° C is shown in Figure 1 and the low-frequency region is expanded in the inset. The two strongest bands at 223 and 171 cm^{-1} were assigned as the D_{2h} active ring-flapping (ν_{18}) and central ring twisting (ν_{25}) modes, respectively. The third band at 119 cm^{-1} appears to be two low in frequency to be a fundamental and, moreover, is found 52 cm^{-1} below ν_{25}, almost exactly the 51 cm^{-1} value measured for ν_{22}, the ring-puckering from its dispersed fluorescence spectra [5]. This ν_{25}-ν_{72} difference can only be Raman active if the ring-puckering potential energy function has a double minimum with nearly degenerate levels (Ag, B_{3u} pairs for D_{2h} or A_1, B_2 pairs in C_{2v} symmetry). We therefore conclude that 9,10-dihydroanthracene is bent in the vapor, in agreement with the conclusion from the fluorescence work [5].

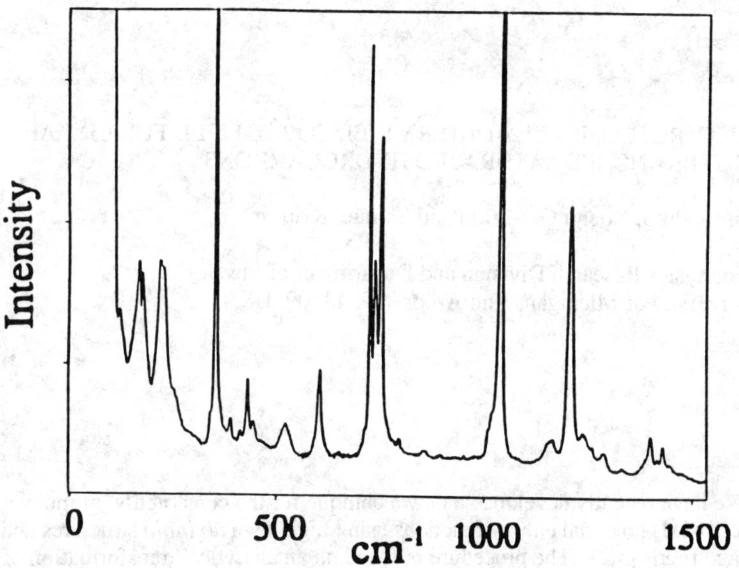

Fig. 1. Vapor-phase Raman spectrum of 9,10-dihydroanthracene at 300° C.

Table 1 summarizes the assignments in the low-frequency region for both the vapor and the melt. Values from an MM3 molecular mechanics calculations are also given. These assignments differ from previous work [6].

TABLE 1. Low-frequency vibrations of *trans*-stilbene and derivatives

Symmetry	Vibration	Description	Vapor	Melt	MM3
B_{1g}	ν_{18}	Ring-flapping (o.p.)	223	231	243
B_{2g}	ν_{25}	Central-ring twisting (i.p.)	171	182	179
A_u	ν_{42}	Central-ring twisitng (o.p.)		138 ?	116
B_{3u}	ν_{71}	Ring-flapping (i.p.)	256	267	224
B_{3u}	ν_{72}	Central-ring puckering	51[a]		
B_{2g}	ν_{25}-ν_{72}	Combination	119		

[a] Ref. [5]

1. J. Laane and R. C. Lord , J. Mol. Spectrosc. 39, 340 (1971).
2. L. A. Carriera, R. O. Carter, and J. R. Durig, J. Chem. Phys. 59, 812 (1973).
3. M. M. Strube and J. Laane, J. Mol. Spectrosc. 129, 126 (1988).
4. F. H. Herbstein, M. Kapon, and G. M. Reisner, Acta Crystallogr. B 42, 181 (1986).
5. Y.-D. Shin and H. Saigusa, M. Z. Zgierski, F. Zerbetto, and E. C. Lim, J. Chem. Phys. 94, 5 (1991).
6. M. Brigodiot and J. M. Lebas, J. Mol. Struct. 32, 297 (1976).

SPECTROSCOPICALLY DETERMINED FORCE FIELD FOR LINEAR AND BRANCHED SATURATED HYDROCARBONS

Kim Palmö, Noemi G. Mirkin and Samuel Krimm*

Biophysics Research Division and Department of Physics,
University of Michigan, Ann Arbor, MI 48109, USA

We have recently developed a new technique for the construction of molecular mechanics (MM) potential energy functions using high level *ab initio* structures and scaled force fields [1,2]. The procedure is based on an analytical transformation of spectroscopic (*ab initio*) force constants into corresponding MM parameters. We call the resulting set of energy parameters a Spectroscopically Determined Force Field (SDFF), since the (scaled) *ab initio* vibrational frequencies and modes are preserved in the transformation.

Previously, the SDFF technique was used to derive an MM force field for *n*-alkane chains [3]. This force field accurately reproduced the structures, relative energies, and vibrational frequencies of 14 conformers of *n*-pentane and *n*-hexane. In addition, crystal properties of these and longer chains were successfully computed, including the chain elastic modulus of polyethylene [3,4].

We have now extended this approach to develop an SDFF for the branched alkanes. *Ab initio* force fields were obtained for the seven stable conformers of 2-methylbutane, 3-methylpentane and 2,2-dimethylpropane. These force fields were scaled to fit the assigned vibrational frequencies. In the transformation from *ab initio* the same electrostatic and van der Waals parameters as in the linear alkanes were used. To extract the significant valence interaction force constants, all nearest neighbor and next-nearest neighbor cross terms were generated for all the molecules. Their transformed values were examined and most of them were retained in the final SDFF. After a reoptimization of some of the force constants, the rms frequency deviation with respect to the (scaled) *ab initio* results of all the alkanes, including the previously treated *n*-pentanes and *n*-hexanes, was 7.1 cm^{-1} (excluding C-H stretching vibrations).

Procedures to compute infrared and Raman intensities have also been developed. The calculation of Raman intensities is based on a model in which the derivatives of the molecular polarizability tensor are described as a sum of transferable local polarizability derivative tensors. Our procedure is somewhat similar to that of Machida et al. [5], but instead of fitting the intensity parameters to observed Raman spectra, we obtain them from *ab initio* polarizability derivatives. Figure 1 shows preliminary Raman intensity results for *tt*-pentane in the form of a comparison of intensities given by *ab initio* polarizability derivatives (circles) and MM parameters (solid line).

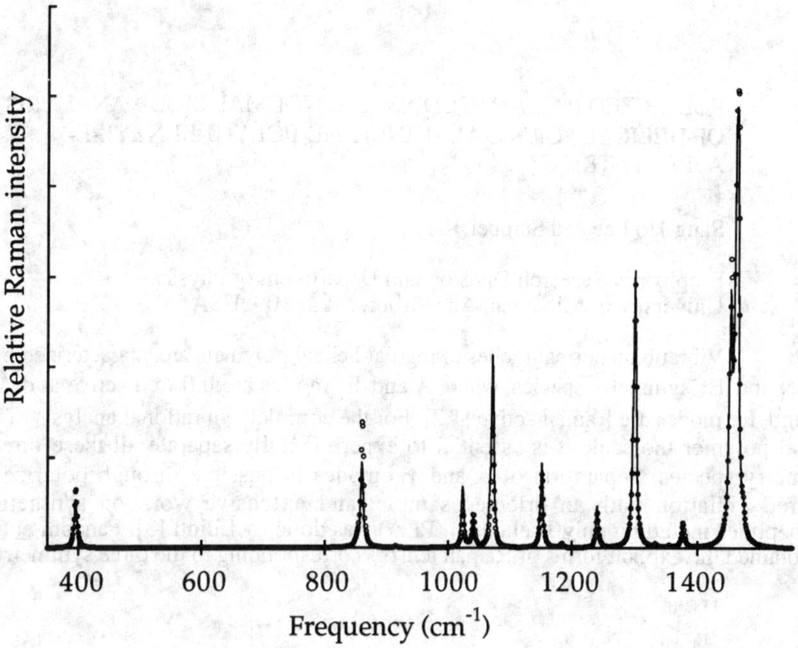

Fig. 1. Simulated Raman spectrum of *tt*-pentane at 0 K. The circles represent the intensity distribution given by the HF/6-31G polarizability derivatives. The solid line represents the intensity distribution calculated using MM parameters obtained from *tt*-pentane and *ttt*-hexane. Each band has been given a width of 5 cm^{-1}. The full (scaled) HF/6-31G force field was used to calculate the frequencies and normal modes. The intensities were calculated assuming plane polarized incident light with a wavelength of 5145 Å.

1. K. Palmö, L.-O. Pietilä and S. Krimm, J. Comp. Chem. 12, 385 (1991).
2. K. Palmö, L.-O. Pietilä and S. Krimm, Computers Chem. 17, 67 (1993).
3. K. Palmö, N.G. Mirkin, L.-O. Pietilä and S. Krimm, Macromolecules 26, 6831 (1993).
4. K. Palmö and S. Krimm, J. Polym. Sci., Polym. Phys. Ed., 34, 37 (1996).
5. K. Machida, H. Noma and Y. Miwa, Indian J. Pure & Appl. Phys. 26, 197 (1988).

POLARIZED RAMAN STUDIES AND NORMAL MODE ANALYSIS OF HELICAL POLY(L-ALANINE) AND POLY(β-BENZYL-L-ASPARTATE)

Sang-Ho Lee and Samuel Krimm*

Biophysics Research Division and Department of Physics
University of Michigan, Ann Arbor, MI 48109, USA

Vibrational normal modes of regular helical polymers are characterized by A, E_1, and E_2 symmetry species, where A and E_1 modes are infrared active and A, E_1, and E_2 modes are Raman active [1,2]. For the complete vibrational analysis of a helical polymer molecule it is essential to experimentally separate all these three symmetry species. Separation of A and E_1 modes is possible through polarized infrared radiation with an oriented sample, and extensive work on synthetic polypeptides including poly(L-alanine) (PLA) was done by Elliott [3]. Fanconi et al [4] obtained the explicit forms of Raman tensors corresponding to the three symmetry

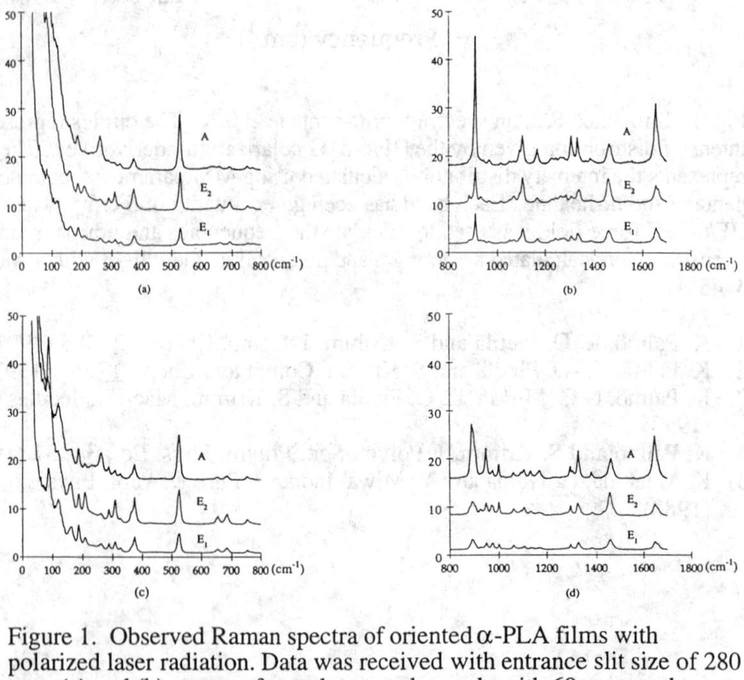

Figure 1. Observed Raman spectra of oriented α-PLA films with polarized laser radiation. Data was received with entrance slit size of 280 μm. (a) and (b): spectra for undeuterated sample with 60 scans at laser power of 0.16 W. (c) and (d): spectra for N-deuterated sample with 50 scans at laser power of 0.15 W.

species, and tried with limited success to separate each mode for α-PLA by using polarized laser radiation. We have been able to clearly separate these three symmetry species for α-PLA and helical poly(β-benzyl-L-aspartate) (PBLA) by using polarized laser Raman scattering (Figure 1). Polarized infrared spectra were also measured for uniaxially oriented α-PLA and PBLA samples. Similar Raman and infrared spectra were measured also for the N-deuterated α-PLA and PBLA samples.

In an effort to find vibrational force fields suitable for polypeptides and proteins, we had already done detailed vibrational analyses for right-handed α-PLA [5,6]. Based on our new experimental results with the availability of E_2-species Raman and polarized infrared spectra for N-deuterated α-PLA, which were not available at the time of the previous studies, rigorous normal mode analyses have been repeated. The initial force fields were taken from *ab initio* computational results for *trans*-N-methylacetamide and alanine dipeptides [7-11], these being refined to the observed frequencies of α-PLA. Transition dipole coupling for the amide I and amide II modes, which was first developed in a perturbation approximation [12,13], was treated by the introduction of computed interaction force constants between adjacent and between third-nearest-neighbor peptide groups. The observed splittings are reproduced more satisfactorily by this than by the perturbation treatment. The observed Raman frequencies at 289 cm^{-1} (E_1) and 290 cm^{-1} (E_2) and the infrared frequency at 290 cm^{-1} (A) are better assigned to the $C^{\alpha}C^{\beta}$ torsional vibrations of the side chain methyl group than, as previously [6], the medium intensity 260 cm^{-1} (Raman) band and weak 240 cm^{-1} (infrared) band. Depending on the experimental conditions of sample preparation, PBLA can exist as a left-handed α-helix (α_L-helix), a left-handed ω-helix (ω_L-helix), a right-handed α-helix (α_R-helix), and a β-sheet [14-16]. The refined force fields have been applied to reproduce the observed main chain frequencies for α_L-PBLA and ω_L-PBLA.

1. P. W. Higgs, Proc. Roy. Soc. London A133, 472-485 (1953).
2. H. Tadokoro, J. Chem. Phys. 33, 1558-1567 (1960).
3. A. Elliott, Proc. Roy. Soc. London A226, 408-421 (1954).
4. B. Fanconi, B. Tomlinson, L. A. Nafie, W. Small and W. L. Peticolas, J. Chem. Phys. 51, 3993-4005 (1969).
5. J. F. Rabolt, W. H. Moore and S. Krimm, Macromolecules 10, 1065-1074 (1977).
6. A. M. Dwivedi and S. Krimm, Biopolymers 23, 923-943 (1984).
7. N. G. Mirkin and S. Krimm, J. Mol. Struct. 242, 143-160 (1991).
8. T. C. Cheam and S. Krimm, J. Mol. Struct. (Theochem) 188, 15-43 (1989).
9. T. C. Cheam and S. Krimm, J. Mol. Struct. (Theochem) 206, 173-203 (1990).
10. T. C. Cheam, J. Mol. Struct. (Theochem) 257, 57-73 (1992).
11. T. C. Cheam and S. Krimm, J. Mol. Struct. 295, 259-271 (1993).
12. S. Krimm and Y. Abe, Proc. Nat. Acad. Sci. USA 69, 2788-2792 (1972).
13. W. H. Moore and S. Krimm, Proc. Nat. Acad. Sci. USA 72, 4933-4935 (1975).
14. E. M. Bradbury, B. G. Carpenter and R. M. Stephens, Biopolymers 6, 905-915 (1968).
15. H. Kyotani and H. Kanetsuna, J. Polymer Sci. 10, 1931-1939 (1972).
16. Y. Kondo, E. Izuka, A. Oka and T. Hayakawa, Polymer 18, 111-113 (1977).

CONFORMATIONAL STABILITY OF N_2F_4 FROM COMBINED RAMAN DATA AND *AB INITIO* CALCULATIONS

J. R. Durig* and Z. Shen

Department of Chemistry, University of Missouri-Kansas City,
5100 Rockhill Road, Kansas City, MO 64110-2499

From the initial structural study [1] of tetrafluorohydrazine, N_2F_4, it was concluded that the molecule was 97% in the *gauche* form in the gas phase at ambient temperature. However, from a vibrational study [2], it was shown that both the *gauche* and *trans* conformers were present in nearly equal amounts which lead to a re-investigation of the electron diffraction data [3]. It was concluded that the *gauche* conformer had a slightly higher abundance. Several additional vibrational studies [4,5] have appeared but the conformational stability of N_2F_4 has not yet been determined. Therefore, we have recorded additional Raman data for conformational stability determinations and carried out *ab initio* calculations utilizing the 6-311++G** basis sets at the RHF and MP2 levels [6,7].

Temperature studies of the vibrational spectrum gave an enthalpy difference of 70±10 cm^{-1} (200±29 cal/mol) with the *trans* conformer the more stable rotamer. Utilizing the predicted Raman activities from the *ab initio* calculations along with the relative intensities of five Raman lines for each conformer, Fig. 1, again gives the *trans* conformer as the more stable form (25 cm^{-1}).

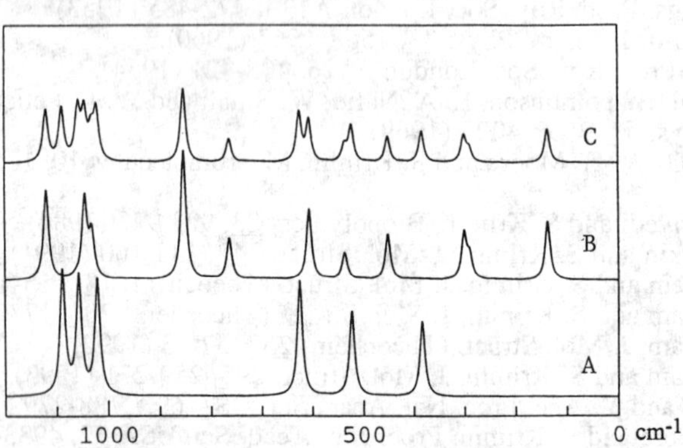

Fig. 1. Calculated Raman spectrum; A: pure *trans*; B: pure *gauche*; C: mixture of *trans* and *gauche* conformers with ΔH=70 cm^{-1} and *trans* form more stable.

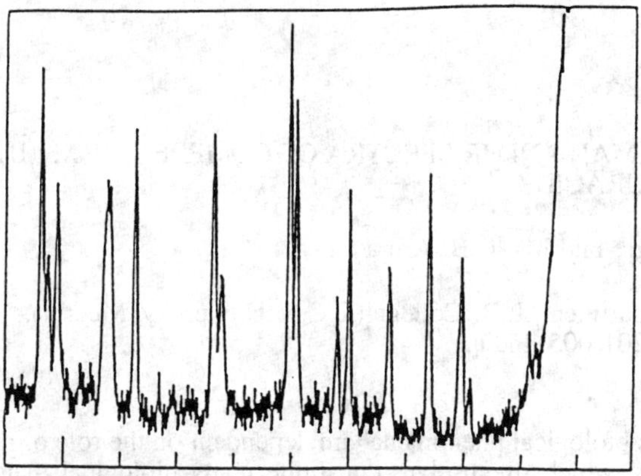

Fig. 2 Raman spectrum of glassy N_2F_4.

Therefore, the experimental spectroscopic data clearly indicate that the *trans* conformer of N_2F_4 is the more stable form of this molecule.

From the *ab initio* calculations at the MP2/6-311++G** level the *gauche* conformer was predicted to be more stable than the *trans* rotamer by 130 cm^{-1} (372 cal/mol). Smaller basis sets or at the RHF level gave even larger values of the enthalpy difference with the *gauche* conformer the more stable form. With such a small enthalpy difference *ab initio* calculations can not be expected to predict the more stable rotamer.

The predicted Raman spectrum (Fig. 1) from the *ab initio* at the RHF/6-311++G** level is in remarkably good agreement with the experimental determined spectrum, Fig. 2. The relative intensities of the bending modes which appear below 800 cm^{-1} are predicted well at this level of calculation. In fact, the Raman spectrum is predicted relatively well even at the RHF/6-31G* level of calculation. These data clearly show the value of predicting both the frequencies and the intensities of the Raman spectrum for conformational studies of small molecules. Even a very low level of *ab initio* calculation, i.e., RHF/6-31G* can be very valuable for interpreting the Raman spectra of small inorganic molecules.

1. R. K. Bohn and S. H. Bauer, Inorg. Chem. 6, 304 (1967).
2. J. R. Durig and J. W. Clark, J. Chem. Phys. 48, 3216 (1968).
3. M. J. Cardillo and S. H. Bauer, Inorg. Chem. 8, 2086 (1969).
4. J. R. Durig and R. W. MacNemee, J. Raman Spectrosc. 2, 635 (1974).
5. D. N. Shchepkin, L. A. Zhygula and L. P. Belozerskaya, J. Mol. Struct. 49, 265 (1978), and references therein.
6. M. J. Frisch, et al., Gaussian-92, Gausian, Inc. Pittsburgh, PA, 1992.
7. P. Pulay, Mol. Phys., 17, 197 (1969).

LASER RAMAN AND IR SPECTRA OF COMPLEX OF Mn(II) WITH 5-BROMOURACIL

V.K. Rastogi* and Ms. C.B. Arora

Physics Department, L.R. College (C.C.S. University, Meerut), Sahibabad-201 005, India.

Various biological phenomena are dependent on the role of metal ions in biological processes, which themselves constitute many biological materials or affect the progress of reactions [1]. Metals may be best transported into a cell in a chelated form, because the most chelates have a relatively high lipid solubility. During the last couple of years [1-3], there has been a considerable inerest in the interaction of metal ions with nucleic acid bases and their derivatives because metal-complexes function as antitumour agents against many type of tumours by inhibiting DNA and protein synthesis. 5-bromouracil(5-BU) is an interesting ligand, since coordination may occur through two nitrogen atoms of pyrimidine ring, the two carbonyl oxygens and the bromine of (C_5–Br) group. In order to understand the mechanism by which metal ions intervene in the function of nucleic acids, we in the present work, describe the preparation and characterisation of the complex of Mn(II) with 5-BU, which has the ability to bind metal or to tissues via metals.

The complex was isolated by mixing the hot , hydrated metal nitrate solution (1 m mole) prepared in ethanol (14 ml)-water(100 ml) mixture and hot alcoholic solution of 5-BU (1 m mole). The resulting mixture was heated to about 70 °C and the NaOH was added to adjust the pH to about 6-7. The brownish precipitate thus obtained was filtered, washed several times with ethyl alcohol and recrystallised from the same solvent. The residue was dried in an oven at about 70°C. The IR and Raman spectra of the complex and the ligand in solid phase have been recorded on Beckman IR-12 spectrophotometer in KBr pellet in the region 200-4000 cm^{-1} and on a Spex spectrophotometer in the region 50-3500 cm^{-1} respectively. Sample for the Raman work was sealed in a pyrex glass capillary cell, which was mounted in a sample illuminator. Raman spectrum was excited with 514.5 nm radiation of a Spectra Physics Model-165 argon ion laser, using 100 mW power at the sample.

The complex formed possesses 1:2 stoichiometry and is insoluble in common organic solvents and does not melt upto 275°C. The chemical analyses corresponds to the molecular formula $[Mn(C_4H_3N_2O_2Br)_2.4H_2O]$ and the percentage (found/calc.) as; Mn=17.28/17.27; C= 15.03/15.09; H= 3.41/3.45 and N= 8.86/8.80.

The vibrational spectra of 5-BU have been analysed and fundamental vibrations have been assigned by assuming that the molecule belongs to C_s point group. In the present study, the strong bands observed at 3205 and 3100 cm^{-1} have been assigned to N-H stretching mode. The low value of this vibration is considered to be indicative of the strong hydrogen bonding in the molecule considered. The ν(C–Br), $\nu(C_2{=}O)$ and $\nu(C_4{=}O)$ modes in the spectra of the free ligand appear at 540/550(IR/Raman) cm^{-1}, 1700 and 1655/1670 cm^{-1} (IR/Raman) respectively.

The vibrational spectra of the complex show some band shifts which reveal the coordination modes of the ligand well. The Raman spectra of the Mn-complex also show considerable intensity changes in the 2000-800 cm^{-1} region. However, the region of main interest is 3000-1600 cm^{-1} where the N–H, C–H, C=O and C=C stretching modes appear. In the spectra of the complex $\nu(C_4=O)$ band is shifted to lower frequency by 35 cm^{-1}, while $\nu(C_2=O)$ band remains unaffected, indicating that the ligand coordinates to the metal ion through $C_4=O$ group and not with $C_2=O$ group[2]. The further evidence of bonding of oxygen with metal ion is provided by the appearance of a band around 250 cm^{-1} in the spectrum of the complex. The appearance of a band around 3100/3115 cm^{-1} in association with another stretching band around 840-842 cm^{-1} confirm the presence of coordinated water molecules in the complex. The bands due to ν(N–H) and ν(C–Br) modes remain unaltered indicating the non-involvement of these groups in complexation.

Thus on the basis of these studies a octahedral structure has been suggested to this complex. Further support to this structure comes from Bohr Magneton value(5.80 BM) of the complex.

TABLE 1. Assignment of fundamental frequencies (in cm^{-1}) of 5-bromouracil

Sym. species	IR	Raman	Assignments	Sym. species	IR	Raman	Assignments
a'	3205	–	ν(N-H)	a"	940	945	γ(N–H)
a'	3100	–	ν(N-H)	a"	860	–	γ(N–H)
a'	3045	3040	ν(C-H)	a"	804	–	γ(C–H)
a'	1700	–	ν(C=0)	a"	778	785	γ-ring
a'	1655	1670	ν(C=O)	a'	750	750	ν-ring
a'	1620	1620	ν(C=C)	a'	740	–	β(C=O)
a'	1560	–	β(N-H)	a'	640	635	β-ring
a'	1438	1450	ν(C-N)	a'	610	612	β(C=O)
a'	1420	1405	β(N-H)	a'	540	550	ν(C–Br)
a'	1340	1340	ν(C-N)	a"	440	–	γ-ring
a'	1242	1240	β(C-H)	a"	418	425	γ(C=O)
a'	1224	1225	ν(C-N)	a"	–	300	γ(C=O)
a'	1058	–	ν(C-N)	a"	–	205	γ(C–N)
a'	1000	1010	β-ring				

The antitumour activity depends upon the central metal ion and the ligand. The ligand(5-BU) and its Mn-complex were also tested for their antitumour activity against Dalton's lymphoma. It was found that 5-BU and its Mn -complex have effective antitumour activity against Dalton's lymphoma.

1. V.K. Rastogi, H.P. Mital and S.N. Sharma, Indian J.Phys. B64, 312(1990).
2. V.K. Rastogi, B. Lal, C.B. Arora and S.L. Gupta, in Procd. IXV International Conf. on Raman Spectroscopy, John Wiley and Sons, 1994.
3. V.K. Rastogi, H.P. Mital and R.C. Saxena, in Spectroscopy of Biological Molecules: The State of Art, Societa Editrice Esculapio srl, Bologna, Italy, 1989, p. 381.

Raman and Infrared Spectroscopic Investigation of Some Hydrous LnTPPTS Complexes (Ln=La,Sm)

Zhen-Hua Chen, Xin-Lan Chen, Ji-Ping He, Nai-Teng Yu and Lun Zhang
(Department of Chemistry, Wuhan University, Wuhan 430072, China)
* *(Centre of Research and Analysis of Materials, Wuhan University of Technology, Wuhan 430070 China)*
* * *(Department of Chemistry, Hong Kong University of Science and Technology Clear Water Bay, Kowloon, Hong Kong)*

Larpent et al. [1] reported their studies of the highly hydrosoluble trisodium phosphinetriyltri-m-benzene-sulfonate, TPPTS=P(m-$C_6H_4SO_3Na$)$_3$, Water-soluble metallocomplexes containing the ligand tris(sodium-m-sulfonatophenyl) phosphane which were synthesized from readily accessible precursor compounds and were isolated pure by Herrmann et al. [2]but the synthesis and structural characterization of rare earth TPPTS complexes have not been reported. In the present paper, Raman and IR spectroscopic study on rare earth TPPTS complexes are described.

The hydrous tris(sodium-m-sulfonatophenyl) Phosphane (TPPTS · $4H_2O$), 3. 3′. 3″-Phosphinidynetrisbenzenesulfonate Lanthanide (Ⅲ) (LaTPPTS · $4H_2O$) and Samaride (Ⅲ) (SmTPPTS · $4H_2O$) were prepared by Xin-Lan Chen et al. The experimental and calculated Value show good agreement for the elemental analysis.

Raman spectra of the samples were analysed in the solid state using the Nicolet 910 FT-Raman spectrometer. The infrared spectra of LnTPPTS complexes were measured on a Nicolet 60 SXB FT-IR Spectrometer. The Raman and infrared spectra of LnTPPTS · $4H_2O$ complexes are shown in Fig. 1 and 2 respectively. Comparing vibrational spectroscopy of rare earth metal triflato complexes and triphenyl phosphine complexes, assignments of the LnTPPTS complexes are made on the basis of previous studies. [3]

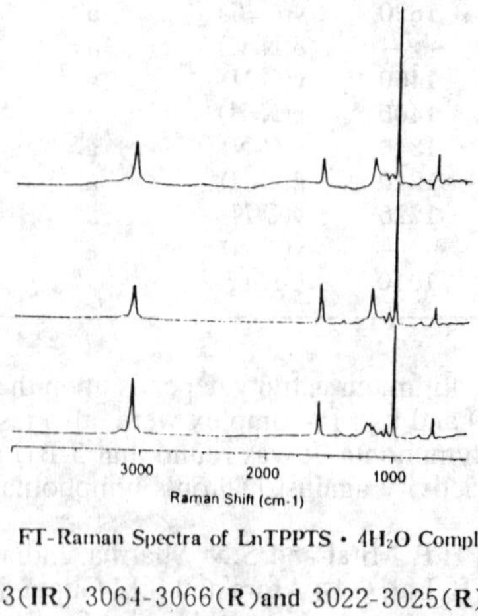

Fig. 1 FT-Raman Spectra of LnTPPTS · $4H_2O$ Complexes

For these complexes, 3063(IR) 3064-3066(R)and 3022-3025(R) cm^{-1}were observed which are attributed to C-H aromatic stretching vibrations, around 3170, 3138-3145, 2925-2926 cm^{-1}bands are assigned to combination of the C-H deformations. The aromatic C-C stretching vibrations were observed at 1589(IR), 1585-1587(R), 1553-1570 (R), 1468(IR) 1466(R), 1448 (R) cm^{-1}. The in-plane C-H deformations are observed at 1169-1170(IR),1174-1176(R); 1129-

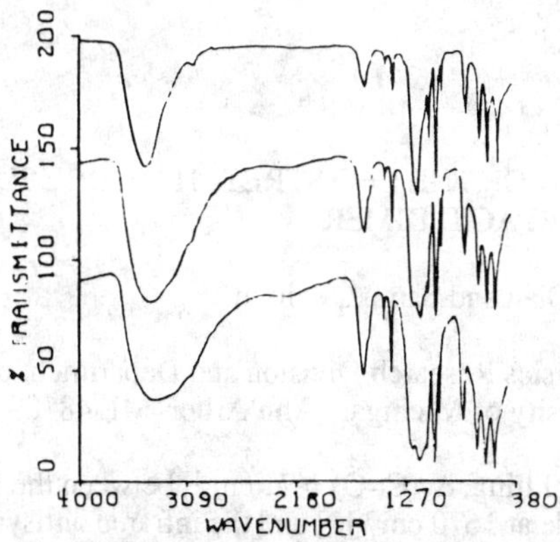

Fig. 2 Infrared Spectra of LnTPPTS • $4H_2O$ Complexes

1130(IR), 1140(R);1085(R), 1037(IR), 1043-1044(R) cm^{-1}. The in-plane C-C ring deformation is assigned to the 616-617(IR),620-625(R) cm^{-1}. The aromatic ring breathing mode vibration are seen at 993 cm^{-1} in the IR spectra and as a single very strong Raman band at 995 cm^{-1}. The C-H out of plane deformations are assigned to bands at 913(R), 801(IR), 809-810, 789(R) cm^{-1}. The out of plane C-C ring deformations are observed at 688 (IR), 688-690(sh. R). 404-406(IR) cm^{-1}. Those vibrations affected by the substituent X which are seen in the Raman and IR spectra at 1102,688, 674-675 (sh IR) 673-674, 688-690 (sh R), 535(IR),550-560(R) 460-462(sh. IR) 462-480(R),246(R) cm^{-1}. Above bands are mainly characteristic frequencies of triphenyl phosphine of ligand.

For the hydrated complexes, the ν_{OH} hydrous vibration is seen at 3287-3454 cm^{-1} (IR) broad band and another band δOH at 1640 cm^{-1}.

The SO_3 groups are bridging tridentate or bidentate. In the LnTPPTS complexes, the board IR six-bands between 1000-1240 cm^{-1} are characterization of the stretching vibrations of SO_3 group. $\nu_{s(SO_3)}$ is assigned to the strong band around 1037(IR). 1044,1085(R)cm^{-1}, while $\nu_{as(SO_3)}$ is located around 1129-1130(IR) 1169-1170(IR).1140.1174-1176(R) cm^{-1}.

Coordinaton of TPPTS to a rare earth metal ion is expected to induce marked spectral changes owing to changes in geometry and symmetry, with vibrations of the SO_3 group being particularly affected. In the Raman and Infrared spectra of LnTPPTs complexes, coordination induces a lengthening of the S-O bonds and a shortening of the C-S bond. Comparing Raman and IR spectra of TPPTS, the C-S stretching frequencies $\nu_{(C-S)}$ 799→801(IR), 799→810(R), 1220(sh)→1230, 1235 cm^{-1} should therefore increase while the position of the $\nu_{(SO_3)}$ 1040→1037 (IR), 1049→1044 (R), 1178→1172,1174(R), 1206→1169, 1170(IR), 1150→1140(R)cm^{-1} should move to a lower wavenumber in the LnTPPTS • $4H_2O$ complexes.

The TPPTS anion can be present for defferent coordinational structures; ionic Td symmetry, mono-coordinational C_{3v} structure, di-coordinational C_{2v} and mixed structure. In general, vibrational modes of metal-oxygen polyhedron $\nu_{(Ln-O)}$ band at 146,168, 253 cm^{-1} are characteristic for coordinate bonds of multicoordinate rare earth TPPTS complexes. The Ln^{3+} ion is eight coordinated by four oxygens from the water molecules, by two oxygen from P-(m-$PhSO_3$)$_2$ and by two oxygen from P-(m- PhSO(O)(O)).

Reference

1. C. Larpent, R. Dabard and H. Patin, Inorg. Chem. 26,2922(1987)
2. W. A. Herrmann, J. Kellner and H. Riepl, J. Of Org. Chem. 389,103(1990)
3. M. M. Hamidi and J. L. Pascal, Polyhedron 13,1787(1994)

ORIGIN OF THE C=O STRETCH MODE SPLITTING IN THE FORMIC ACID DIMER

Weili Qian and Samuel Krimm*

Biophysics Research Division and Department of Physics,
University of Michigan, Ann Arbor, MI 48109, USA

The large splitting, $\Delta\nu(C=O) = 71\ cm^{-1}$, between the Raman symmetric A_g C=O stretch(s) mode at 1670 cm^{-1} [1] and the infrared antisymmetric B_u C=O s mode at 1741 cm^{-1} [2] provides important insights into the intermonomer interactions responsible for this splitting in the formic acid dimer (FAD). An understanding of these interactions is essential to developing the molecular mechanics potential functions used to obtain the more accurate spectral predictions of a spectroscopically determined force field [3]. We had shown earlier [4] that charge fluxes are crucial in an electrostatic model of the interaction if the observed $\Delta\nu(C=O)$ is to be reproduced. We have extended this approach to account for the intermonomer energy, and now find that it is at least necessary to also include atomic dipoles to reproduce both the interaction energy as well as $\Delta\nu(C=O)$.

In order to base our analysis on a detailed understanding of the FAD, we have done *ab initio* computations of $\Delta\nu(C=O)$. Normal mode calculations of the harmonic frequencies were done at the MP2/6-13G** level, and anharmonic corrections [4] (cubic, quartic, and Coriolis) for the two C=O s modes were calculated by a perturbation method [5], the cubic force field in internal coordinates being transferred from a HF/6-311++G** calculation. The results are shown in Table 1, the calculated $\Delta\nu$ of 67 cm^{-1} being in reasonable agreement with the observed value of 71 cm^{-1}. (Scaling of the harmonic frequencies to give ν(anharm) closer to the observed values does not significantly change the predicted splittings.)

Table 1. Harmonic Frequencies (cm^{-1}) and Anharmonic Corrections to C=O Stretch Frequencies of FAD

	ν(harm)	Anharm. Correction			ν(anharm)	Obs
		Cubic	Quartic	Coriolis		
B_u:	1822	-34.9	-19.1	0.2	1768	1741
A_g:	1764	-41.1	-22.4	0.1	1701	1670
$\Delta\nu$:	58	6	3	0	67	71

A self-consistent model of the intermonomer interaction should be accurate to second derivatives of the energy [3] and the molecular dipole [4]. For planar molecules, such as FAD, dipole first and second derivatives uniquely define charges and charge fluxes, respectively, and atomic dipoles are determined by quadrupole derivatives, under the constraint that there are no contributions of the atomic dipoles and their derivatives to the molecular dipole and its derivatives, respectively [6]. Using HF/6-311++G** derivatives, we have computed such dipole derivative-derived charges and charge fluxes and quadrupole-derived atomic dipoles.

The electrostatic interaction energy, which is a uniquely defined quantity, was calculated from the above charges and atomic dipoles, and was compared to this energy as obtained from a B3LYP DFT/6-31G** calculation. The latter is -26.73 kcal/mole (being somewhat basis set dependent); the HF/6-311++G** charge-charge energy is -12.10 kcal/mole and the atomic dipoles contribute -12.92 kcal/mole. It is clear that atomic dipole interactions are vital in accounting for the electrostatic interaction energy.

The $\Delta\nu(C{=}O)$ predicted by this model was determined from normal mode calculations using the MP2/6-31G** intramonomer force constants and replacing the MP2/6-31G** intermonomer constants by second derivatives of nonbonded terms involving all possible charge, charge flux, and atomic dipole interactions plus a van der Waals 12-6 term (CHEM-X parameters were used). The results are shown in Table 2, and the total splitting of 65 cm^{-1} is to be compared to the *ab initio* harmonic

Table 2. Contributions to $\Delta\nu(C{=}O)$ (cm^{-1}) of FAD from Nonbonded Interactions

Interaction	$\Delta\nu_i$	$\Delta\nu_{total}$
Charge-charge	4	4
Charge-charge flux	24	28
Charge flux-charge flux	29	57
Atomic dipoles (all)	29	86
van der Waals	-21	65

splitting of 58 cm^{-1}. Considering the uncertainty in the unrefined van der Waals parameters, the agreement must be considered satisfactory. We found no significant charge fluxes through the hydrogen bonds, but the interactions between the O and H atoms account for almost half of $\Delta\nu(C{=}O)$. It is again clear that the contribution of the atomic dipoles is crucial in accounting for $\Delta\nu(C{=}O)$, and therefore that they must be included in potential energy functions designed to reproduce spectroscopic as well as energetic properties.

1. J.E. Bertie and K.H. Michaelian, J. Chem. Phys. 76, 886 (1982).
2. T. Wachs, D. Borchardt and S.H. Bauer, Spectrochimica Acta, 43A, 965 (1987).
3. K. Palmö, L.-O. Pietilä and S. Krimm, J. Comp. Chem. 12, 385 (1991).
4. J. Dybal, T.C. Cheam and S. Krimm, J. Mol. Struct. 159, 183 (1987).
5. J.F. Gaw, A. Willetts, W.H. Green and N.C. Handy, in Advances in Molecular Vibrations and Collision Dynamics, edited by Bowman, J.M., JAI Press Inc., Vol. 1B, 169 (1990).
6. U. Dinur and A.T. Hagler, J. Chem. Phys., 91, 2949 (1989).

POLARIZED RAMAN, IR, AND FORCE FIELD STUDY OF 3Cl-PYRIDINE

C.M. Pathak [1*], S.N. Singh [2], R.K. Singh [3], and B.P. Asthana [2]

[1] Department of Chemistry and Physics, Alcorn State University, Lorman, MS 39096, U.S.A.

[2] Laser and Spectroscopy Laboratory, Department of Physics, Banaras Hindu University, Varanasi - 221005, India

[3] Department of Physics, North Eastern Hill University, Shillong, 793022, India

The parent molecule, pyridine, is one of the most thoroughly investigated molecules [1-4] from different spectroscopic angles. A few years ago, Innes et al. [3] made a critical review of the electronic spectra/states of azabenzenes and discussed the assignments of their fundamental modes. Further, the vibrational frequencies and infrared intensities in a number of azabenzenes were calculated using ab initio by Wiberg [4]. It was observed in this study that even the ab initio results, without applying any scaling factor, are quite off both for vibrational frequencies and the force constants. Moreover, in a more recent Raman study of 3Cl-pyridine in a (3Cl-p) + CH_3OH mixture [5], some interesting aspects of some of the ring modes were revealed. In view of these studies [1-5], the present study on 3Cl-p was undertaken with a view to study the force field, carry out a normal coordinate analysis (NCA), and to compare the results with recent results reported on pyridine [3-4].

The polarized Raman and IR spectra of 3Cl-p were recorded in the frequency regions : 100-3600 and 600-4000 cm^{-1}, respectively. An important aspect of this study was to measure the depolarization ratio of the Raman bands after a careful polarization adjustment using the Raman lines of carbon tetrachloride. The other details of the experimental set-up have been described elsewhere [5].

The force field calculations were carried out using a software employing the well known modified least squares technique of Overend and Scherer [6]. A planar geometry for 3Cl-p was obtained using the PRECISE option of the semi-empirical MNDO method of geometry optimization. The potential energy function was expressed in terms of 14 principal and 13 interaction force constants, and 31 internal coordinates. The initial choice of most of the valence constants for the ring modes was made on the basis of an earlier study by Dongre et al. [7]. With the planar geometry, 3Cl-p belongs to the C_s point group and its genuine normal vibrations are distributed as : Γ_g = 19 a' + 8 a". The normal modes belonging to the species a' and a" were dealt with separately during the NCA. Some salient features and important results of the present study are presented in Table 1.

TABLE 1. Observed and calculated frequencies, $\nu_{obs.}$ and $\nu_{calc.}$ for some selected normal modes of 3Cl-pyridine.

$\nu_{obs.}$ (cm^{-1})	$\nu_{calc.}$ (cm^{-1})	Assignment	PED %
728(IR), 731(R,p)	729.7	Ring breathing	$K_R(29)+K_\alpha(27)+$ ...
429(R,p)	424.9	C-Cl stretch	$K_R(42)+K_\alpha*(13)+$...
287(R,dp)	278.4	Torsion	$K_\tau(65)+K_\tau'(25)+$...
200(R,dp)	200.4	C-Cl rock	$K_\beta*(93)+$...
103(R,dp)	104.9	C-Cl wag	$K_\gamma*(32)+K_\tau(24)+$...

In order to describe 8 non-planar modes, 9 internal coordinates were chosen. However, on the basis of the observed IR and Raman frequencies, only 6 fundamentals could be assigned and the remaining 2 frequencies were calculated to be 456.4 and 270.1 cm^{-1}, where the force constants corresponding to C-H and C-Cl waggings and H-C-C-H and H-C-C-Cl type torsions contribute significantly to the potential energy distribution (PED). The PED's for different modes revealed several interesting aspects, such as a heavy mixing in the ring breathing mode at 731 cm^{-1}. The substitution of one Cl atom at position 3 in pyridine is expected to reduce the frequency of the torsional modes, 16a and 16b. The depolarized Raman band of medium intensity at 287 cm^{-1} has, therefore, been assigned to a torsional mode of the ring in 3Cl-p, as against 404 and 374 cm^{-1} reported consistently in pyridine [1-3]. Further, this substitution would essentially give rise to 3 different normal modes, namely, C-Cl stretching, C-Cl in-plane bending, and C-Cl out-plane wagging. On the basis of the depolarization data, the strongly polarized band at 429 cm^{-1} was assigned to the C-Cl stretching mode. THis band also showed a composite peak nature due to C-^{35}Cl and C-^{37}Cl stretches. The two other Raman bands, one at 200 cm^{-1} having medium intensity and the other depolarized, weakest band at 103 cm^{-1}, have been assigned to C-Cl in-plane and out-plane wagging, respectively. All the observed frequencies were very nicely reproduced during NCA and also showed PED's typical of normal mode in question.

1. C.H. Kline, Jr. and T. Turkevich, J. Chem. Phys. 12, 300 (1944).
2. I. Haque and J.L. Wood, Spectrochim. Acta A23, 959(1967).
3. K.K. Innes, I.G. Ross and W.R. Moomow, J.Mol.Spectrosc. 132.492 (1988)
4. K.B. Wiberg, J. Mol. Struct. 244, 61(1990).
5. Ranjan K. Singh, B.P. Asthana, A.L. Verma and C.M. Pathak, Chem. Phys. Lett., submitted for publication.
6. J. Overend and J.R. Scherer, J. Chem. Phys., 32, 1289 (1960).
7. N.G. Dongre, B.P. Asthana, P.C. Mishra and C.M. Pathak, Spectrochim. Acta, 47A, 673 (1991).

RAMAN AND INFRARED VIBRATIONAL SPECTRA, AB INITIO CALCULATIONS AND NORMAL COORDINATE ANALYSES FOR tBUTYL SUBSTITUTED SILANES $^tBuSiX_3$ AND DISILANES $^tBuSiX_2SiX_2{}^tBu$, X = H, F, Cl, Br, I

R.Zink, K.Hassler*

Institute of Inorganic Chemistry, University of Technology, Stremayrgasse 16, A-8010 Graz, Austria.

Structural distortions and hindered dynamics of molecules containing bulky groups can be studied by a combination of theoretical and spectroscopic methods, including Raman spectroscopy. The number of sterically crowded silicon compounds studied so far is much smaller than the number of organic molecules, because Si compounds generally are much more difficult to prepare and to handle.

This work is part of a Raman spectroscopic investigation of the rotational isomerism of ^{t}butyl and $SiMe_3$ substituted disilanes $(CH_3)_3ESiX_2SiX_2E'(CH_3)_3$(E,E' = C, Si; X = H, F, Cl, Br, I). For this purpose we have recorded temperature dependent Raman spectra of $^tBuH_2SiSiH_2{}^tBu$, $^tBuF_2SiSiF_2{}^tBu$, $^tBuCl_2SiSiCl_2{}^tBu$, $^tBuBr_2SiSiBr_2{}^tBu$, $SiMe_3Cl_2SiSiCl_2SiMe_3$ and $SiMe_3Br_2SiSiBr_2SiMe_3$. All molecules can, in principle exist as mixtures of anti and gauche conformers (Fig 1).

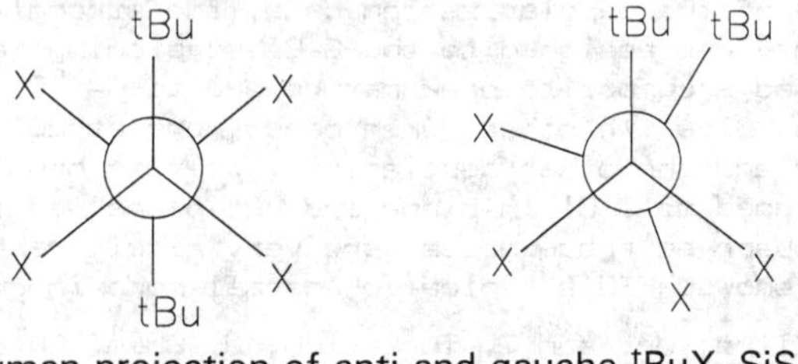

Fig.1 Newman projection of anti and gauche $^tBuX_2SiSiX_2{}^tBu$

Surprisingly, the Raman spectrum of none of these disilanes offers any evidence for the presence of more than one rotamer in the liquid as well as in the solid state, such as doubling of some bands. To illustrate this, Figures 2 gives the Raman spectrum of $^tBuH_2SiSiH_2{}^tBu$ at two different temperatures. The results are confirmed by gas phase electron diffraction experiments [1]. The single rotamer present at 340 K is anti with a torsional CSiSiC angle of 180°. Molecular mechanics

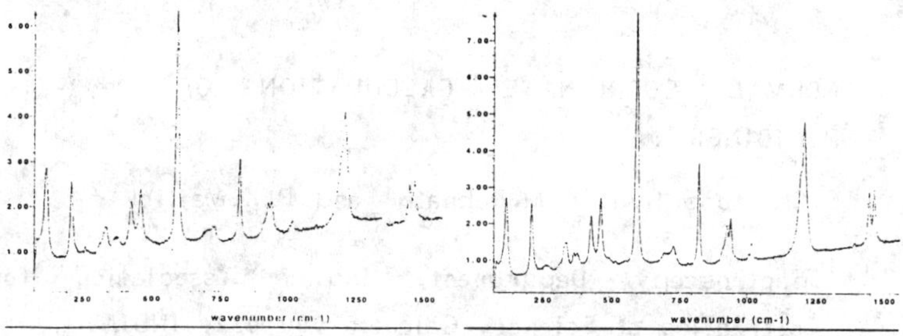

Fig.2: Raman spectra of $^tBuH_2SiSiH_2{}^tBu$ at 313 K (left) and 173 K (right)

calculations using the MM3 program package gave the result that the barrier separating gauche and anti rotamers is very small (Fig.3)

Fig.3: Potential function for the interconversion of anti and gauche $^tBuH_2SiSiH_2{}^tBu$ as calculated with molecular mechanics

Fundamental vibrations, molecular force fields and potential energy distributions have been calculated for ^{t}butylmono- and disilanes using a non relativistic core potential [2] for the heavy atoms, as this method can be applied for the whole series from fluorine to iodine. An internal force field has been developed for the ^{t}butyl group, facilitating normal coordinate analyses for silanes with many ^{t}butyl substituents

1. D.Hnyk, R.S.Fender, H.E.Robertson, D.W.H.Rankin, M.Bühl, K.Hassler and K.Schenzel, J.Mol.Struct., 346 (1995) 215.
2. W.J.Stevens, H.Basch and M.Krauss, J.Chem.Phys., 81 (1984)6026.

NORMAL COORDINATE CALCULATION OF 2-,3-,4-ACETYL PYRIDINES

S.Chattopadhyay[1], M.Debnath[1] and P.K.Mallick*[2]

[1]Spectroscopy Department, Indian Association for the Cultivation of Science, Calcutta 700 032, INDIA.

[2]Physics Department, Golapbag, Burdwan 713 104, West Bengal, INDIA.

This work reports the normal coordinate calculation (NCA) of the three isomers of acetylpyridine (AP). In order to simplify the calculation, we have considered methyl part of the acetyl group as a point mass and confined our calculation only to the in-plane modes. We have used general valence force field (GVFF) since it was pointed out [1,2] that this field is more appropriate than Urey-Bradley force field (UBFF) for large aromatic molecules as ours. The initial choice of diagonal stretching force constants was based on the existing bond length force constant relation [3] and previous works on pyridine [2] and acetophenon [4]. Initial values of all other constants were judiciously transferred from previous works [1,2,4,5] on some relevant molecules. In order to reduce the number of independent force constants, as before [1,5], we have linked certain off-diagonal elements with others through some relations incorporated in the Z-matrix. For fitting the calculated values with the observed frequencies, experimental data were taken from earlier work [6].

Structural parameters used were : ring CC, CN and CH distances 1.38, 1.34 and 1.08 Å; ring-acetyl (C_rC_a) and acetyl-methyl (C_aC_m) distances 1.5 Å. During floating procedure, values of the force constants were kept on close watch so that they do not become unrealistic. We thus constructed a force field of 25-parameters. For best fitting, error was found to be 14.6 cm^{-1} (1.5%). The final set of force field (in the usual unit) was:

Diagonal

1. f(CN)=6.756; 2. f(CC)=6.681; 3. f(CH)=5.045;
4. f(CO)=9.524; 5. $f(C_rC_a)$=4.8; 6. $f(C_aC_m)$=4.47
7. f(CCC)=0.960; 8. f(CCH/NCH)=0.491;
9. $f(CC_rC_a/NC_rC_a)$=1.652; 10. $f(C_rC_aC_m)$=1.331 and
11. $f(C_rC_a/C_mC_aO)$=0.84.

Off-diagonal

12. f(1,1)=0.846; 13. f(1,2)=0.722; 14. f(2,2)=0.899;
15. F(1,5/2,5)=0.212; 16. f(5,6)=0.612; 17. f(4,6)=0.189;
18. f(1,8/2,8)=0.124; 19. f(1,9/2,9)=0.136;
20. f(1,7/2,7)=0.199; 21. f (CC_r,CC_rC)=0.289;
22. f(5,10/6,10)=0.557; 23. f(5,9)=0.668;
24. f(5,11/6,11)=0.089 and 25. f(4,11)=0.174.

REFERENCE

1. P.K. Mallick, G.D. Danzer, D.P. Strommen and J.R. Kincaid, J. Phys. Chem. **92**, 5628 (1988).
2. G. Zerbi and S. Sandroni, Spectrochim. Acta. **24A**, 511 (1968).
3. H.B. Burgi and J.D. Dunitz, J. AM. Chem. Soc. **109**, 2924 (1987).
4. A. Gambi, S. Giorgianni, A. Passerini, R. Visinoni and S. Ghersetti, Spectrochim. Acta. **36A**, 871 (1980).
5. D.P. Strommen, P.K. Mallick, G.D. Danzer, R.S. Lumpkin and J.R. Kincaid, J. Phys. Chem. **94**, 1357 (1990).
6. K.C. Medhi, Indian J. Physics **52A**, 399 (1977).

RAMAN AND INFRARED VIBRATIONAL SPECTRA, SYNTHESES, AB INITIO CALCULATIONS AND NORMAL COORDINATE ANALYSES FOR IODODISILANES $I_nSi_2H_{6-n}$

M.Ernst[1], K.Hassler[2*], W.Köll[2]

[1]Institute of Theoretical Chemistry, Karl Franzens University, Mozartgasse 14, A-8010 Graz, Austria.
[2]Institute of Inorganic Chemistry, University of Technology, Stremayrgasse 16, A-8010 Graz, Austria.

Over the years we have developed methods for the preparation of ioddodisilanes $I_nSi_2H_{6-n}$ by the use of the reaction of arylated disilanes $Ar_nSi_2H_{6-n}$with liquid or gaseous HI (Ar = phenyl, mesityl, naphtyl)

$$Ar_nSi_2H_{6-n} + HI \rightarrow I_nSi_2H_{6-n} + ArH$$

We have prepared all members of the series except $I_2HSiSiH_3$[1,2]. The Raman spectra of IH_2SiSiH_2I,$I_2HSiSiHI_2$, IH_2SiSiI_3 and $I_2HSiSiI_3$ have been reported and assigned [3,4], and $^{29}Si^{29}Si$ one bond coupling constants for all disilanes were measured [5]. J(SiSi) increases with increasing number of iodine atoms from 83.4 Hz for IH_2SiSiH_3 to 119.2 Hz for $I_2HSiSiI_3$, indicating a strengthening of the SiSi bond with increasing n.

Fig.1 gives the Raman spectra of $I_2HSiSiH_3$ and $I_2DSiSiD_3$, that were prepared from $Ph_2HSiSiH_3$ and $Ph_2DSiSiD_3$,resp.

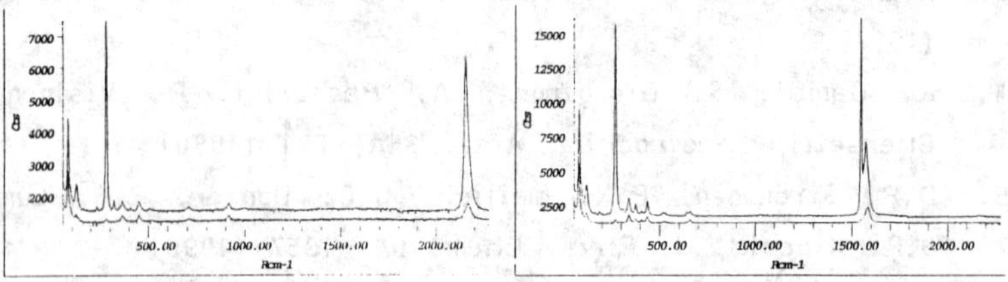

Fig.1: Raman spectra of $I_2HSiSiH_3$ (left) and $I_2DSiSiD_3$

Ab initio calculations have been performed for IH_2SiSiH_3, $I_2HSiSiH_3$, $I_3HSiSiH_3$, IH_2SiSiH_2I, $I_2HSiSiHI_2$ as well as I_3SiSiI_3. The core electrons of the heavy elements were represented by a non relativistic core potential (ECP) as a compromise between accuracy and feasibility, as the method can be used for the complete halogen series from fluorine to iodine. Unconstrained geometry optimizations were performed using standard restricted Hartree-Fock (RHF) theory, a valence double zeta basis with polarization functions on all atoms and the SBK-ECP[6].
Table 1 summarizes the observed and calculated (scaled by 0.92) fundamental vibrations of the iododisilanes mentioned above.

Table 1: Calculated and observed fundamental vibrations for selected iododisilanes $I_nSi_2H_{6-n}$

$I_2HSiSiH_3$			anti $I_2HSiSiHI_2$			I_3SiSiI_3		
species	calc.	exp.	species	calc.	exp.[4]	species	calc.	exp.[7]
A'	2158	2157	A_g	2177	2165	A_{1g}	493	510
	2151	2157		778	-		149	154
	2174	2157		467	477		56	-
	935	919		310	308	A_{2u}	264	255
	895	850		112	111		115	116
	725	712		56	-		16	-
	518	510	B_g	721	693	E_g	402	398
	443	456		385	386		98	94
	276	283		80	85		56	-
	122	128	A_u	724	679	E_u	393	388
	79	82		382	387		76	81
A''	2159	2157		84	-		36	-
	937	919		11	-			
	733	712	B_u	2180	2165			
	533	510		644	655			
	363	369		96	-			
	79	82		35	-			
	105	-						

1. K.Hassler and M.Pöschl, J.Organomet.Chem.,398 (1990) 225.
2. K.Hassler and G.Bauer, J.Organomet.Chem.,460 (1993) 149.
3. K.Hassler, Spectrochim. Acta 50A (1994) 243.
4. K.Hassler and M.Pöschl, Spectrochim Acta 47 A (1991) 439.
5. U.Katzenbeisser and K.Hassler in Organosilicon Chemistry, VCH, N.Y., 1994.
6. W.J.Stevens, H.Basch and M.Krauss, J.Chem.Phys. 81(1984) 6026.
7. F.Höfler, W.Sawodny and E.Hengge, Spectrochim.Acta 26A (1970) 819.

RAMAN AND INFRARED VIBRATIONAL SPECTRA, ROTATIONAL ISOMERISM, AB INITIO CALCULATIONS AND NORMAL COORDINATE ANALYSES FOR 1,2-DIMETHYLTETRACHLORO-DISILANE AND ITS d_6 ISOTOPOMER

M. Ernst[1], K. Hassler[2], A. Jähn[3], K. Schenzel[3*]

[1]Institute of Theoretical Chemistry, Karl Franzens University, Mozartgasse 14, A-8010 Graz, Austria
[2]Institute of Inorganic Chemistry, University of Technology, Stremayrgasse 16, A-8010 Graz, Austria
[3]Institute of Analytical Chemistry, Martin Luther University, Weinbergweg 16, D-06120 Halle, Germany.

1,2-Dimethyltetrahalodisilanes MeX_2SiSiX_2Me (X = F, Cl, Br, J) exist as mixtures of anti and gauche rotamers at room temperature (Fig.1), as was deduced from dipole moment measurements [1].

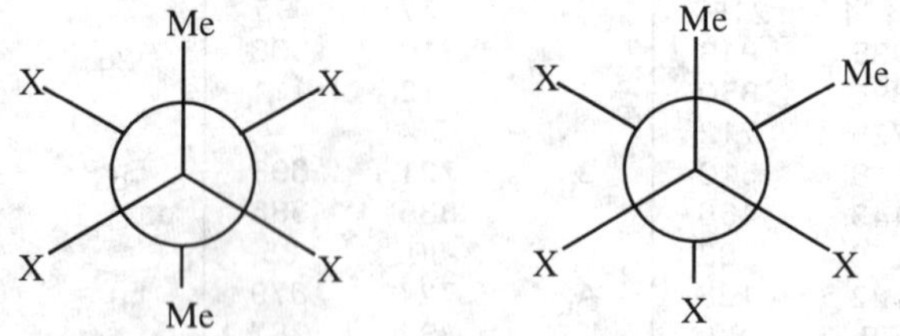

Fig.1: Newman projection of anti and gauche MeX_2SiSiX_2Me

So far, no vibrational spectroscopic study of these molecules has been reported. We have therefore prepared $CH_3Cl_2SiSiCl_2CH_3$ and $CD_3Cl_2SiSiCl_2CD_3$ and recorded their Raman spectra at various temperatures. The spectra clearly prove the presence of both rotamers at room temperature (Fig.2). In the solid state (170 K) only the anti rotamers are present. The selection rules predict five Raman active heavy atom deformations for the anti rotamers, which are clearly observed. The band most suitable for determination of ΔH values for anti - gauche interconversion is $\nu_s(SiCl_2)$ of the totally symmetric irreducible representation. If $\ln I_g / I_a$ is plotted against T^{-1} (Fig.3), ΔH values of 0.9 ± 0.2 and 1.2 ± 0.2 kJ / mol are obtained.

Ab initio calculations of vibrational spectra and molecular force fields for both rotamers have been performed. Since this work is part of a series of similar studies of halogenated disilanes MeX_2SiSiX_2Me and $Me_2XSiSiXMe_2$ with X = F, Cl, Br, I, the core electrons of all heavy atoms were represented by a core potential (ECP) to ensure comparability of the results and to keep the computational costs at a reasonable level. The ECP parameters according to [2] were chosen. The calculated and observed fundamental vibrations of anti and gauche $CH_3Cl_2SiSiCl_2CH_3$ and

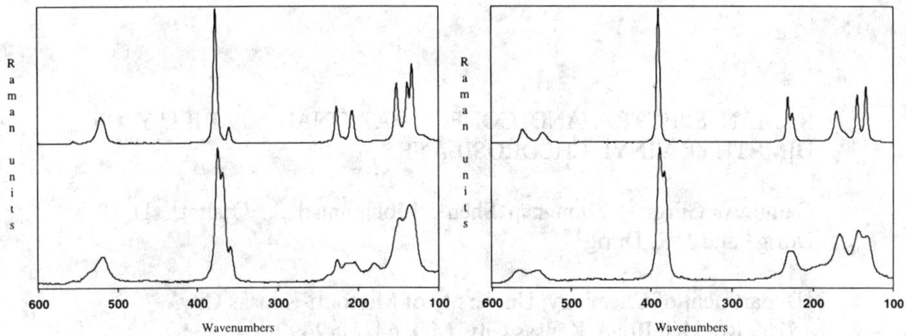

Fig.2: Solid (170 K, upper spectra) and liquid (300 K) Raman spectra of $CD_3Cl_2SiSiCl_2CD_3$ (left) and $CH_3Cl_2SiSiCl_2CH_3$ (right).

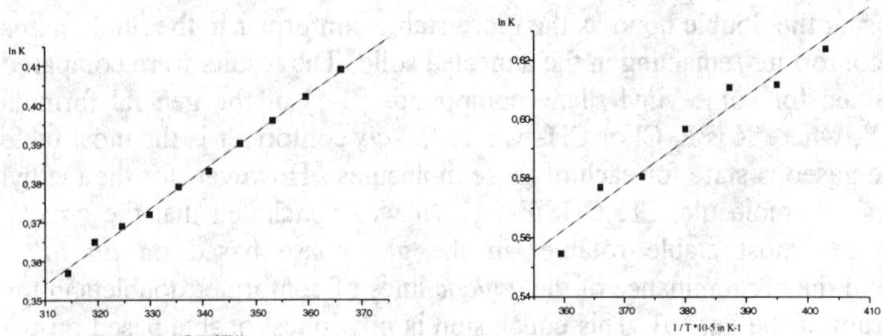

Fig.3: Plots of ln I_g / I_a of $\nu_s(SiCl_2)$ against T^{-1} for $CD_3Cl_2SiSiCl_2CD_3$ (left) and $CH_3Cl_2SiSiCl_2CH_3$ (right).

$CD_3Cl_2SiSiCl_2CD_3$ as well as harmonic force fields and potential energy distributions will be reported.

1. J.Nagy, S. Ferenczi-Gresz, E.Hengge and S. Waldhör, J. Organomet. Chem., 96 (1975) 199.
   J.Nagy, G.Zsombok and E.Hengge, J.Organomet. Chem., 182 (1979) 165.
2. W.J.Stevens, H.Basch and M. Kraus, J.Chem. Phys., 81 (1984) 6026.

RAMAN SPECTRA AND CONFORMATIONAL STABILITY OF DIMETHYL VINYL CHLOROSILANE

Gamil A. Gurgis[1], Zhongnan Shen[1], Mohammad A. Qtaitat[1], D. T. Durig[2] and J. R. Durig[1]*

[1]Department of Chemistry, University of Missouri-Kansas City, 5100 Rockhill Road, Kansas City, MO 64110-2499
[2]Department of Physics, University of the South, Sewanee, TN 37375

We [1] recently carried out a spectroscopic study of methyl vinyl dichlorosilane, $CH_2CHSiCl_2CH_3$, and found the *cis* conformer where the methyl group is eclipsing the double bond as the more stable conformer in the fluid phases and the only conformer remaining in the annealed solid. The results were compared to those obtained for some vinyl silane compounds [2-5] of the general formula CH_2CHSiH_2X, where X is F, Cl or CH_3 where the *cis* conformer is the most table rotamer in the gaseous state for each of these molecules. However, for the methyl vinyl difluorosilane molecule, $CH_2CHSiF_2CH_3$, it was concluded that the *gauche* conformer is the most stable rotamer in the gas phase based on *ab initio* calculations and the predominance of the *gauche* lines of conformer doublets in the Raman spectrum of the gas.[6] This conclusion is now questionable based on the more recent studies of the corresponding dichloride. Also, in this recent study of methyl vinyl dichloride silane, it was suggested that a conformational study of dimethyl vinyl chlorosilane, $CH_2CHSi(CH_3)_2Cl$, would be of interest since the size of the methyl group is nearly the same as that of a chlorine atom. Any differences in conformational stability of these two molecules can be attributed to the electronegativity difference between the chlorine atom and the methyl group.

Therefore, we have investigated the Raman spectra of dimethyl vinyl chlorosilane in the gaseous, liquid and solid states. We have undertaken the present study to determine if conformers are present in the fluid states and if there are conformers present, their relative stabilities. *Ab initio* calculations were carried out to obtain predictions on the conformer stability, relative structural parameters of the possible rotamers, and to aid in the vibrational assignment.

The Raman (3500 to 20 cm^{-1}) spectra of gaseous and solid dimethyl vinyl chlorosilane, $CH_2CHSi(CH_3)_2Cl$, have been recorded. Additionally, the Raman spectrum of the liquid, Fig. 1, has been recorded and qualitative depolarization values obtained. Both the *cis* and *gauche* conformers have been identified in the fluid phases but only the *gauche* conformer remains in the solid, Fig. 1. From this data the enthalpy difference has been determined to be 100±10 cm^{-1} (286±29 cal/mol) with the *gauche* conformer the more stable rotamer in the fluid phase. A complete vibrational assignment is proposed for the *cis* conformer based on relative intensities, depolarization values and group frequencies. Several of the fundamentals for the *gauche* conformer have also been identified. Complete equilibrium geometries have been determined for both rotamers by *ab initio* calculations employing the 3-21G* and 6-31G* basis sets at the levels of the restricted Hartree-Fock (RHF) and Møller-Plesset (MP) to second order.

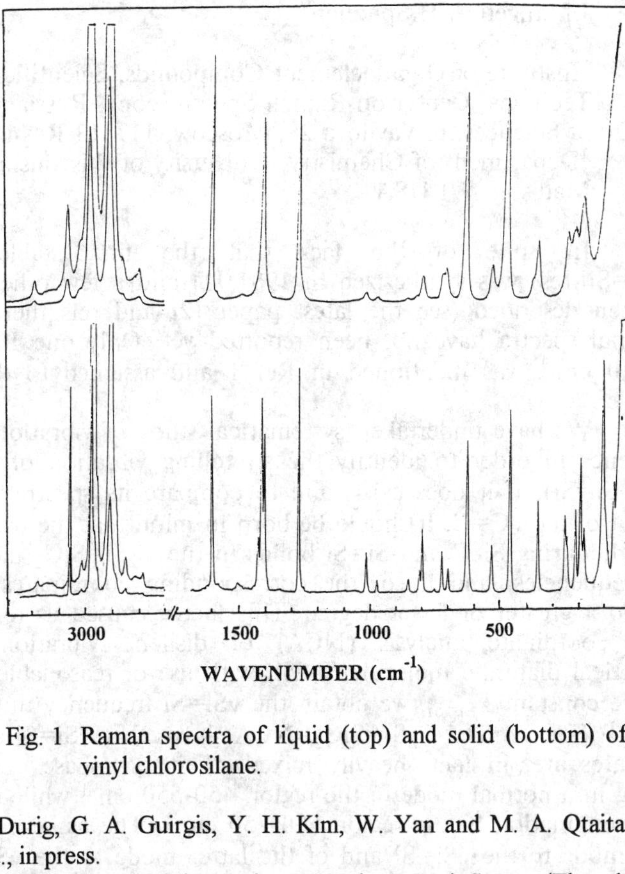

Fig. 1 Raman spectra of liquid (top) and solid (bottom) of dimethyl vinyl chlorosilane.

1. J. R. Durig, G. A. Guirgis, Y. H. Kim, W. Yan and M. A. Qtaitat, J. Mol. Struct., in press.
2. J. F. Sullivan, M. A. Qtaitat and J. R. Durig, J. Mol. Struct. (Theochem), 202, 159 (1989).
3. J. R. Durig, J. F. Sullivan, G. A. Guirgis and M. A. Qtaitat, J. Phys. Chem. 95, 1563 (1991).
4. J. R. Durig, J. F. Sullivan and M. A. Qtaitat, J. Mol. Struct. 243, 239 (1991).
5. J. R. Durig, Zhongnan Shen and M. A. Qtaitat, Chem. Phys., in press.
6. M. A. Qtaitat, G. A. Guirgis, and J. R. Durig, J. Raman Spectrosc. 26, 413 (1995).

RAMAN SPECTRA OF DISILENES - COMPOUNDS WITH THE Si=Si DOUBLE BOND. IDENTIFICATION AND NORMAL COORDINATE ANALYSIS OF THE νSi=Si STRETCHING VIBRATION.

L.A.Leites[1*], S.S.Bukalov[1], I.A.Garbuzova[1], R.West[2], J.Mangette[2]. H.Spitzner[2]

[1]Institute of Organoelement Compounds, Scientific and Technical Center on Raman Spectroscopy, Russian Academy of Sciences, ul.Vavilova 28, Moscow, 117813 Russia.
[2]Department of Chemistry, University of Wisconsin, Madison, WI USA

In spite of the fact that the first stable disilene $Mes_2Si{=}SiMes_2$ was synthesized in 1981 [1] and a few other disilenes have been described (see the latest paper [2] and refs therein), their vibrational spectra have not been reported yet. Only one Raman line near 530 cm^{-1} was mentioned in Ref 1 and assigned to the νSi=Si stretch.

We have undertaken systematical study of vibrational spectra of disilenes in order to identify the stretching vibration of the Si=Si bond (νSi=Si), if it does exist, and to compare its spectral behaviour with that of the νC=C. It should be born in mind that the oscillators of the neighbouring Si-C and Si=Si bonds in the $C_2Si{=}SiC_2$ moiety have close frequencies, and if so, the corresponding coordinates could be mixed to a greater or lesser degree. This factor caused us to carry out normal coordinate analysis (NCA) of disilene vibrations. For a hypothetical diatomic molecule Si=Si, with use of reasonable values of the force constant $F_{Si=Si}$, we obtain the νSi=Si frequency in the region 600 cm^{-1}. For a model $C_2Si{=}SiC_2$, NCA shows that νSi=Si and νSi-C coordinates are, in fact, heavily mixed, their syn-phase combination resulting in a normal mode in the region 450-550 cm^{-1} while their anti-phase combination in the region 600-750 cm^{-1}. The assignment of the former mode to the νSi=Si and of the latter mode to the νSi-C is, in essence, conventional. The results of NCA allowed us to conclude that if a molecule contains a Si=Si bond, its Raman spectrum should exhibit a band in the region 450-550 cm^{-1}, corresponding to a vibration of complex eigenvector but with significant participation of the νSi=Si stretch. All disilenes studied have centrosymmetrical structure, so this mode is IR-inactive.

Raman spectra of three disilenes:
trans-Mes(t-Bu)Si=Si(t-Bu)Mes (I), Mes=2,4,6-trimethylphenyl, and $R_2Si{=}SiR_2$ with R=Mes (II) , $CH(SiMe_3)_2$ (III) in the region of the skeleton vibrations 50-900 cm^{-1} are presented in Fig.1.

Three intense lines in the spectrum of III are assigned on the basis of the NCA results for a model $C_2Si{=}SiC\text{-}CH(SiC_3)_2$. All the stretching coordinates: those of Si=Si, =Si-C and $Si\text{-}C_3$ bonds participate in the normal vibration at 454 cm^{-1} but the contribution of the νSi=Si is predominant, so that this mode may be readily assigned as the νSi=Si stretch. The lines at 579 and 649 cm^{-1} are mostly Si-C stretching vibrations, the former being a syn-phase while the latter an anti-phase combination of the =Si-C and $Si\text{-}C_3$ stretching coordinates.

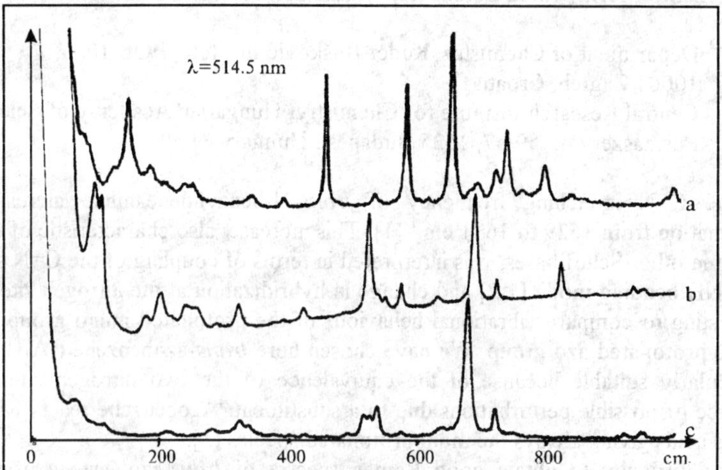

Fig. 1 Raman spectra of disilenes (solid samples) $R_2Si{=}SiR_2$
R=$CH(SiMe_3)_2$ (a), Mes (orange modification) (c) and
Mes(t-Bu)Si=Si(t-Bu)Mes (b)

In the molecules of I and II, mesityl group is present which exhibits at least two vibrations with frequencies in the region of 500 cm^{-1}. On the basis of NCA results for a model $C_2Si{=}SiC$-Mes and taking into consideration the molecular structures of I and II (the X-ray data [4,5]), we can assign an intense Raman line of I at 522 cm^{-1}, having no counterpart in the IR spectrum, to the νSi=Si, while weak lines at 547 and 575 cm^{-1} to mesityl vibrations. A triplet observed in the Raman spectrum of II in the region 550 cm^{-1} can be rationalized as an overlap of the νSi=Si and the ring vibrations. To identify the νSi=Si line among this triplet, we studied the excitation profiles of the Raman lines. When approaching resonance (orange modification of II absorbs at 460 nm) the 540 cm^{-1} component of the Raman triplet is most enhanced and thus can be assigned to the νSi=Si.

1. R.West, M.J.Fink, J.Michl; *Science* **214**, 1343. 1981.
2. M.Kira, T.Maruyama, C.Kabuto,.K.Ebata, H.Sakurai; *Angew. Chem. Int. Ed. Engl.* **33**, 1489, 1994.
3. M.J.Fink, M.J.Michalczyk, K.J.Haller, R.West, J.Michl; *Organometallics* **3**, 793, 1984.
4. B.D.Sheperd, C.F.Campana, R.West; *Heteroatom Chem.* **1**, 1, 1990.

VIBRATIONAL ANALYSIS OF THE PROTONATED AZO GROUP. RAMAN SPECTRA AND NORMAL COORDINATES OF PROTONATED *trans*-AZOBENZENE AND ITS ISOTOPOMERS

Zlatko Meić,[1] Goran Baranović,[1*] Vilko Smrečki[1] and Gábor Keresztury[2]

[1] Department of Chemistry, Ruđer Bošković Institute, POB 1016, 10001 Zagreb, Croatia
[2] Central Research Institute for Chemistry, Hungarian Academy of Sciences, Pusztaszery ut. 59/67, 1525 Budapest, Hungary

The C=N stretching frequency of *trans*-N-benzylideneaniline increases upon protonation from 1629 to 1671 cm^{-1} [1]. This increase, also characteristic of the imino group in other Schiff bases, was interpreted in terms of coupling of the C=N stretching with NH bending mode [1,2] and change in hybridization at the nitrogen site [2]. It is interesting to compare vibrational behaviour of the protonated imino group with that of the protonated azo group. We have chosen here *trans*-azobenzene (tAB), which is particularly suitable because of the equivalence of the two nitrogen sites and the absence of possible perturbations due to a substituent. Azobenzene reacts with strong Lewis acids giving always the monoprotonated species [3].

The difficulty to obtain good Raman spectra of *trans*-azobenzene using visible lasers [4] has been overcome by the use of the near-infrared Nd-YAG laser excitation, with a Nicolet Raman System 950 FT-Raman spectrometer. The strongest band in the Raman spectrum of powdered D_0-tAB at 1439 cm^{-1} (Fig. 1a) is assigned to the N=N stretching mode. Raman spectra of *trans*-azobenzene ethanolic solution protonated with different acids reveal a decrease of the N=N stretching frequency by approximately 40 cm^{-1}, with no appreciable effect from the counter ion (Table 1). It is interesting to note that Raman spectra of all protonated samples have strong bands near 1550 cm^{-1}, which are assigned to the NH bending mode. Three tAB isotopomers were used here to obtain Raman spectra of powdered, protonated and deuterated samples: D_0-tAB, ^{15}N-tAB and D_{10}-tAB (Table 2). In the Raman spectrum of D_0-tAB protonated in pure CF_3COOH, a strong broad band is found at 1547 cm^{-1} (Fig. 1b). In the Raman spectrum of D_0-tAB deuterated in the same manner, this band is missing, but a new one of a medium intensity appears at 1075 cm^{-1} (Fig. 1c), which could be ascribed to the ND bending mode.

TABLE 1. N=N stretching and NH bending frequencies (cm^{-1}) of unprotonated and protonated *trans*-azobenzene.

	unprotonated (powder)	HCl/EtOH	$HClO_4$/EtOH	H_2SO_4/EtOH	CF_3COOH/ EtOH
N=N str.	1439 vs	1399 vs	1403 vs	1398 vs	1401 vs
NH bend.	-	1548 s	1548 s	1547 s	1547 s

TABLE 2. N=N stretching and NH/ND bending frequencies (cm^{-1}) of protonated and deuterated *trans*-azobenzene isotopomers.

		unprotonated (powder)	protonated (CF_3COOH)	deuterated (CF_3COOD)
D_0-tAB	N=N str.	1439 vs	1401 vs	1446 vs
	NH/ND bend.	-	1547 s	1075 m
^{15}N-tAB	^{15}N=N str.	1428 vs	1383 vs	1435 vs
	NH/ND bend.	-	1545 s	1075 m
D_{10}-tAB	N=N str.	1466 vs	1411 vs	1460 vs
	NH/ND bend.	-	1518 s	1073 m

The most relevant Raman frequencies of unprotonated, protonated and deuterated isotopomers of tAB are given in Table 2. The normal coordinate calculations showed a substantial contribution to the N=N stretching mode from the phenyl group mode 19b [5]. Therefore the increase of the N=N stretching frequency in D_{10}-tAB is a result of a decoupling on going from undeuterated to deuterated phenyl groups. This trend is preserved also in protonated and deuterated species. In protonated D_{10}-tAB the NH bending band is shifted to the lower frequency, which also indicates a certain coupling with ring modes.

Preliminary FT infrared spectra of protonated D_0-tAB also indicate the appearance of the weak N=N stretching and NH bending bands.

^{13}C NMR spectrum of the D_0-tAB / CF_3COOH / C_6D_6 solution reveals only one signal for the C-1,1' site (144.8 ppm), which is more shielded than in D_0-tAB / C_6D_6 (153.5 ppm). This implies that fast proton transfer occurs between two nitrogens of an azobenzene molecule and surroundings.

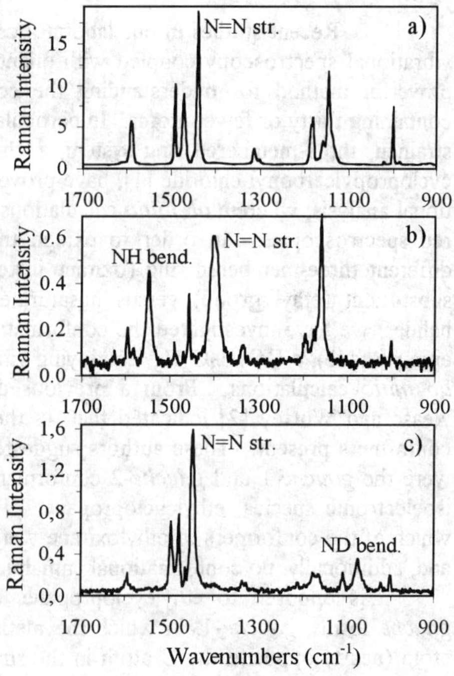

Fig. 1. FT Raman spectra of a) D_0-tAB powder, b) D_0-tAB in CF_3COOH and c) D_0-tAB in CF_3COOD

1. Z. Meić, G. Baranović and T. Šuste, J. Mol. Struct., 296, 163 (1993).
2. J. J. López-Gariga, S. Hanton, G. T. Babcock and J. F. Harrison, J. Am. Chem. Soc., 108, 7251 (1986).
3. F. Gerson, E. Heilbronner, A. van Veen and B. M. Wepster, Helv. Chem. Acta, 43, 1889 (1960).
4. B. Kellerer, H. H. Hacker and J. Brandmüller, Ind. J. Pure Appl. Phys., 9, 903 (1971).
5. G. Baranović, B. Dippel and B. Schrader, EUCMOS XXII, 1994, paper IL 18.

CONFORMATIONAL BEHAVIOR OF ETHYLOXIRANE FROM TEMPERATURE DEPENDENT RAMAN SPECTRA

C. J. Wurrey*, Shiyu Shen, T. K. Gounev and J. R. Durig

Department of Chemistry, University of Missouri-Kansas City, 5100 Rockhill Road, Kansas City, MO 64110-2499

Recent studies in our laboratories have shown that the combination of vibrational spectroscopy coupled with quantum mechanical calculations can be a powerful method for understanding the conformational behavior of molecules containing thirty or fewer atoms. In particular, even molecules containing a highly strained, three-membered ring system, such as cyclopropylcarbonyl fluoride and cyclopropylcarbonyl chloride [1], have proven amenable to this type of conformational analysis, wherein *ab initio* calculations are combined with Raman and infrared spectroscopies. In order to extend these studies to systems containing a different three-membered ring (oxirane instead of cyclopropane) and a saturated substituent (ethyl group) versus unsaturated substituents such as the carbonyl halides, we have investigated the conformational equilibrium in ethyloxirane (1,2-epoxybutane), *c*-$OC_2H_3C_2H_5$, employing vibrational spectroscopic techniques and *ab initio* calculations. From a previous conformational study of ethyloxirane Nease and Wurrey [2] indicated that, in the gas phase, there were at least two conformers present. These authors suggested that the observed rotameric forms were the *gauche*-1 and *gauche*-2 conformers by analogy with their study on the isoelectronic species, ethylcyclopropane. This previous study did not ascertain which of the conformers of ethyloxirane were present in the liquid or solid phases and, additionally, no conformational enthalpies were determined.

As opposed to ethylcyclopropane, ethyloxirane has two non-equivalent *gauche* forms, *gauche*-1, in which the methyl group is located near the oxygen atom (nearly opposite the C atom in the ring without the ethyl attachment), and *gauche*-2, in which the methyl group is nearly opposite the oxygen atom. In addition to these two *gauche* forms, a *cis* conformation (methyl group *cis* to the ring) is also possible.

Variable temperature Raman spectra (25 to -89°C) of the liquid have been recorded, Fig. 1. Spectroscopic evidence for the co-existence of the *gauche*-1 and *gauche*-2 conformers of the title compound is found in the fluid phases, along with small amounts of the *cis* rotameric form. From the temperature dependent Raman spectra of the liquid phase the enthalpy difference between the *gauche*-1 and *gauche*-2 conformers was determined to be 95±6 cm^{-1} (272±17 cal/mol.). In the annealed solid phase, only the *gauche*-1 conformer is found. *Ab*

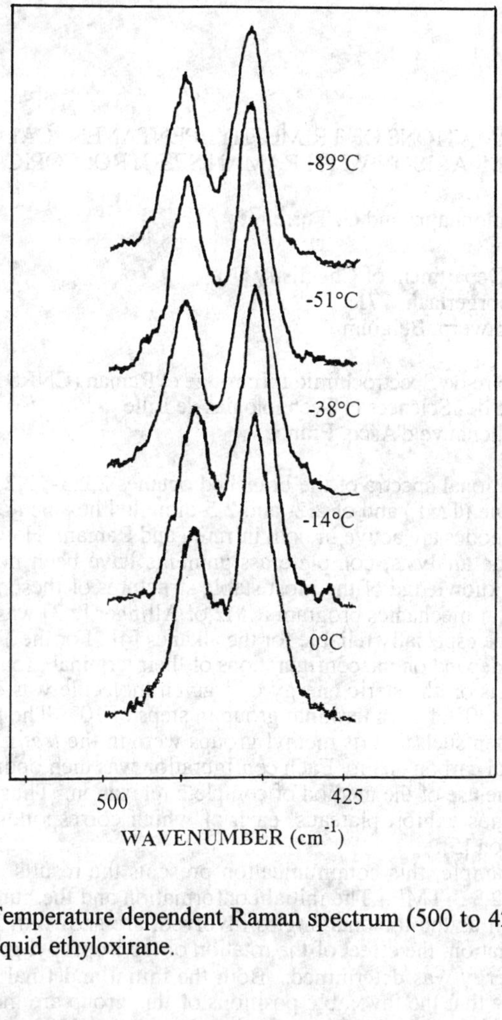

Fig. 1 Temperature dependent Raman spectrum (500 to 425 cm^{-1}) of liquid ethyloxirane.

initio calculations, have been carried with different basis sets up to MP2/6-31G* [3,4] from which conformation stabilities have been determined. These calculations support the experimental conformational conclusions. These results will be compared to the corresponding quantities for some similar molecules. The calculated Raman spectra are similar to those obtained from a mixture of the *gauche*-1, more stable, and *gauche*-2 conformers with an enthalpy difference of 200 cm^{-1}.

1. J. R. Durig, Shiyu Shen, Wayne Zhao and Lin Zhou, J. Mol. Struct., in press.
2. A. B. Nease and C. J. Wurrey, J. Raman Spectrosc. 9, 107 (1980).
3. M. J. Frisch, et al., Gaussian-92, Gausian, Inc. Pittsburgh, PA, 1992.
4. P. Pulay, Mol. Phys., 17, 197 (1969).

CONFORMATIONS OF TRIMETHYLPENTANES: CALCULATIONS AND COMPARISON WITH RAMAN SPECTROSCOPIC RESULTS

A. Aboulmouhajir[1] and G. Turrell[2*]

[1]RUCA, Department of Chemistry
Groenenborgerlaan, 171
B2020 Antwerp, Belgium

[2]Laboratoire de Spectrochimie Infrarouge et Raman (CNRS A2631L),
Université des Sciences et Technologies de Lille
59655 Villeneuve d'Ascq, France

The vibrational spectra of the branched octanes 2,2,3-; 2,2,4-; 2,3,3- and 2,3,4-trimethyl pentane (TMP) and of 2,2- and 2,5-dimethyl hexane (DMH) are very rich, as all of the 72 modes are active in both infrared and Raman. However, with the use of a normal-mode analysis, complete assignments have been made [1]. This analysis requires the knowledge of the most stable structures of these molecules. In this work the molecular mechanics program MM2 of Allinger [2,3] was employed, as it has been shown to be especially reliable for the alkanes [4]. For the TMP's the most stable conformations depend on the configurations of their terminal groups. The effect of these configurations on the steric energy of a given molecule was determined by making a rotation of 360° of each terminal group in steps of 10°. The position of the other group was chosen such that its methyl groups were in the *trans* position with respect to the principal carbon chain. Each configuration was then optimized without any constraint with the use of the method of complete relaxation. Thus, the curves of the final steric energies exhibit plateaus, each of which corresponds to a specific molecular conformation [5].

As an example, this communication presents the results of the conformational analysis of 2,3,3-TMP. The initial conformation and the numbering of the carbon atoms, as well as the torsonal angles involved, are shown in Fig. 1. In the first step in the calculation, the effect of the rotation of the isopropyl group C_1-C_2-C_6 (τ_1) on the steric energy was determined. Both the initial and final steric energy curves (Fig. 2a) show that the favorable positions of this group are the totally-*trans* conformation (p_1) and that obtained by a further rotation by 120° (p_2), the former being slightly less stable (Fig. 3). The two most favorable position of the methyl group C_5 correspond to τ_1=180°, -60°, as shown in Figs. 2b and 2c.

The results of these calculations can be compared with the temperature dependence of the Raman spectrum of this compound in the low-frequency region. Upon cooling 2,3,3-TMP, a band at 220 cm^{-1} decreases in intensity with decreasing temperature while a band at 240 cm^{-1} increases. The latter can then be assigned to the most stable conformer. In this case these bands are broad and weak. Thus, a reliable calculation of the enthalpy difference between conformers could not be made[6,7].

The normal mode analysis of the most stable conformer of 2,3,3-TMP, *viz.* cp'_1 (Fig. 3), shows that the band at 240 cm^{-1} is due principally to methyl-group torsions[5]. Complete vibrational assignments for this conformer will be presented on the poster.

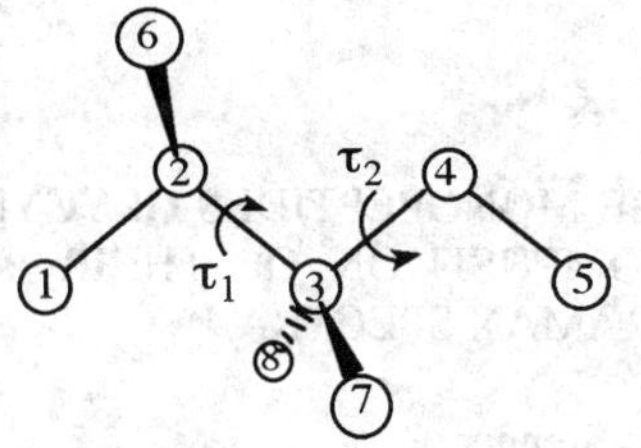

Fig. 1. Atom numbering and torsional angles in 2,3,3-TMP.

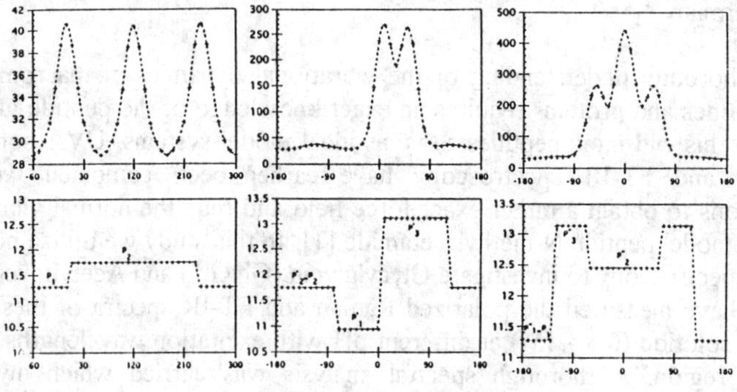

(a) (b) (c)

Fig. 2. Energy *vs.* torsional angles. (a) τ_1 ; (b) and (c) τ_2 .

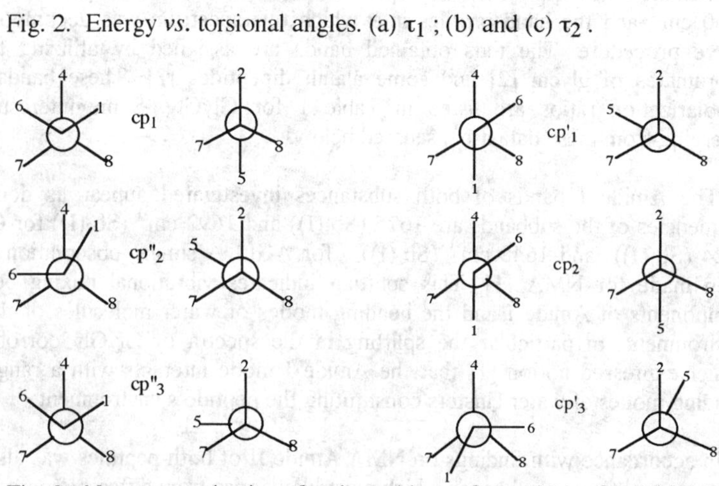

Fig. 3. Newman projections for the stable conformations of 2,3,3-TMP.

1. A. Aboulmouhajir and G. Turrell, to be published.
2. N. L. Allinger, J. Am. Chem. Soc. 99, 8127 (1977).
3. N. L. Allinger, Y. H. Yuh and J. Lii, J. Am. Chem. Soc. 111, 8551 (1989).
4. J. E. Anderson, J. Chem. Soc. Perkin Trans. 2, 299 (1991).
5. A. Aboulmouhajir, Doctoral Thesis, University of Antwerp, Belgium (1994).
6. J. R. Shearer and R. G. Snyder, J. Chem. Phys. 72, 5798 (1980).
7. A. Aboulmouhajir, A. Adil and G. Turrell, J. Mol. Structure 349, 365 (1995).

VIBRATIONAL ANALYSIS OF THE MONOPEPTIDES GLYCYLGLYCIN AND ACETYLGLYCIN BY DECOMPOSITION OF THEIR COMPLEX ISOTROPIC AND ANISOTROPIC RAMAN SPECTRA.

Guido Sieler* and Reinhard Schweitzer-Stenner

[1]FB1-Institut für Experimentelle Physik, Universität Bremen, 28359 Bremen, Germany

A thorough understanding of the vibrational dynamics of macromolecules like oligopeptides and proteins requires an exact knowledge of the peptide group's force field. To this end monopeptides serve as ideal model systems. UV-resonance Raman (UVRR) and FT-IR spectroscopy have earlier been combined with *ab-initio* calculations to obtain a rather exact force field and thus the normal vibrations of the classical model peptide N-methylacetamide [1]. In this study we utilize non-resonance Raman spectroscopy to investigate Glycylglycin (GlyGly) and Acetylglycin (AGL).

We have measured the polarized Raman and FT-IR spectra of these peptides in aqueous solution (0.5 - 1M) at different pH with excitation wavelengths in the visible and UV-region. A thorough spectral analysis was carried which involves a self consistent fitting of the same model to corresponding Raman and IR-spectra. To give an example Fig. 1 depicts the isotropic Raman scattering of GlyGly between 1200 and 1800 cm^{-1} and the band profiles into which this spectrum was decomposed by the above procedure. The thus obtained bands are assigned by utilizing the normal coordinates of glycin [2] and some alanin dipeptides [3]. These bands and their depolarization ratios are listed in Table 1 for GlyGly. Some interesting results emerging from these data are discussed below.

1. The Amide I bands of both substances investigated appear as doublett. The frequencies of the subbands are 1675 ($Sb_1(I)$) and 1692 cm^{-1} ($Sb_2(I)$) for GlyGly and 1624 ($Sb_1(I)$) and 1648 cm^{-1} ($Sb_2(I)$) for AGL. A similar observation has earlier been made for NMA [1]. This splitting indicates vibrational mixing between the components of Amide I and the bending modes of water molecules of the aqueous environment. In particular the splitting in the spectra of GlyGly corroborates the earlier expressed notion [1] that the Amide I mode interacts with a range of HOH bending modes of water clusters constituting the peptide's environment.

2. In accordance with findings on NMA Amide II of both peptides was also found to be composed of two subbands which most likely arise from different conformes.

3. The spectral region between 1370 and 1450 cm^{-1} contains a comparatively strong band at 1393 cm^{-1} (in GlyGly and AGL) the intensity of which is drastically reduced with lowering pH. It also appears in the UV-resonance Raman spectra and was earlier assigned to the overtone of Amide [4]. This seems unlikely since this band is still apparent in the corresponding spectra of DGL in D_2O (data not shown). In

accordance with Asher and coworkers [5] we assign it to the symmetric stretch of COO^-.

4. Amide III at 1280 cm^{-1} (GlyGly) and 1298 cm^{-1} (AGL) overlaps with two polarized (1235 and 1265 cm^{-1}) and a more depolarized band (1315 cm^{-1} for GlyGly, 1323 cm^{-1} for AGL). The latter most likely arises from the CH_2 w of the methin at the carboxyl group. For GlyGly it remains nearly unchanged by NH deuteration which clearly argues against vibrational mixing with NH b [3].

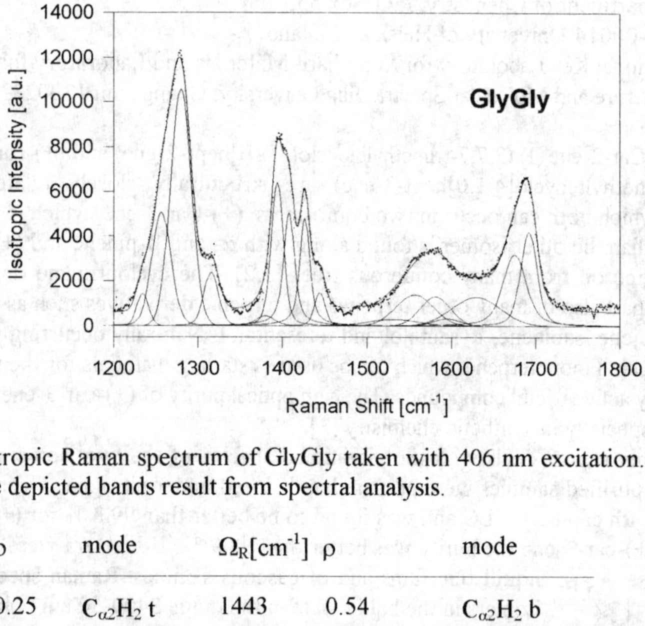

Fig. 1: Isotropic Raman spectrum of GlyGly taken with 406 nm excitation. The solid line and the depicted bands result from spectral analysis.

$\Omega_R[cm^{-1}]$	ρ	mode	$\Omega_R[cm^{-1}]$	ρ	mode
1257	0.25	$C_{\alpha2}H_2$ t	1443	0.54	$C_{\alpha2}H_2$ b
1280	0.18	Amide III	1556	0.19	Amide II_1
1315	0.47	$C_{\alpha2}H_2$ w	1581	0.16	Amide II_2
1394	0.45	COO^- ss	1594	0.56	COO^- as
1408	0.24	$C_{\alpha1}H_2$w ?	1675	0.10	Amide I_1
1426	0.30	$C_{\alpha1}H_2$ b	1691	0.15	Amide I_2

Table 1: Assignment of prominent Raman bands appearing in the spectrum of GlyGly (ρ: depolarization ratio; t: twisting, w: wagging, b: bending, ss: symmetric stretch, as: asymmetric stretch).

1. X.G. Chen, R. Schweitzer-Stenner, S.A. Asher, N.G. Mirkin and S. Krimm. J. Phys. Chem. 99, 3704(1995).
2. W. Quian and S. Krimm, J. Phys. Chem. 98,9992(1994).
3. M. Diem, O. Lee and G.M. Roberts. J. Phys. Chem. 96, 548(1992).
4. S. Krimm, S. Song and S.Asher, J. Am. Chem. Soc. 111, 4290(1989).
5. X.G. Chen and S.A. Asher, manuscript in preparation

THE VIBRATIONAL SPECTRUM OF (+)-CAR-2-ENE AND (+)-CAR-3-ENE: A COMPARISON

H. Tylli[1*] and Y. Zhao[2]

[1]Department of Chemistry, P.O.Box 55,
FIN-00014 University of Helsinki, Finland
[2]National Key Laboratory for Superhard Materials and Laboratory for Molecular Structure and Molecular Spectra, Jilin University, Changchun 130023, P.R. China

(+)-Car-2-ene (I) (3,7,7-trimethylbicyclo[4.1.0]hept-2-ene) and (+)-car-3-ene (II) (3,7,7-trimethylbicyclo[4.1.0]hept-3-ene) are structurally closely related bicyclic terpenes which both can occur in two conformers. (+)-Car-3-ene, which is much more abundant than the other isomer is found along with α- and β-pinene and (+)-car-2-ene in turpentine oil from many coniferous trees [1,2]. The cyclopropane ring forms an important part also of many other terpenes and terpenic derivatives such as thujane, α- and β-thujene, sabinene, α-santalol and teresantalol. Naturally occurring carenes are abundant chiral monoterpenes which can be used as starting materials for the preparation of optically active useful compounds. The high optical purity of (+)-car-3-ene is of great interest especially in synthetic chemistry [3].

This report deals with the vibrational spectra of the two isomers of carene. Carefully purified samples were prepared and stored under nitrogen. Their purity was checked with capillary GLC and was found to be better than 99.8 % for (+)-car-2-ene and for (+)-car-3-ene the purity was better than 98.9 %. IR spectra were obtained of liquid thin films and of gaseous samples. Raman spectra of both isomers in the liquid state in the range 3400-50 cm^{-1} obtained with a multipass cell are shown in Fig. 1. Due to the structural similarity of the two isomers more significant differences between the spectra are found only in the low-frequency region. According to MO calculations the pseudo chair conformer of (+)-car-2-ene is more stable than the pseudo boat conformer [4]. For (+)-car-3-ene the inverted boat conformer was found to be more stable than the boat conformer [5]. Because the energy difference between the conformers is small for both isomers the recorded spectra represent the spectra of conformational mixtures with both conformers present in nearly equal amounts already at room temperatures. A temperature study of the IR spectrum of liquid samples of both isomers performed in the range 22-80 °C indeed showed only band broadening and minor intensity changes at higher temperatures.

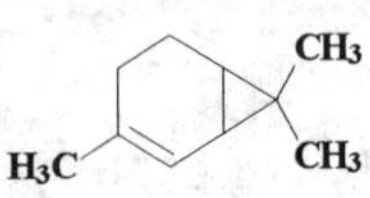

(I)

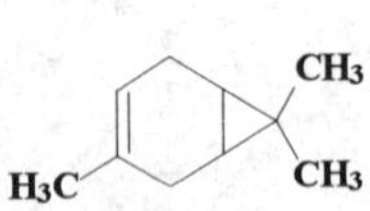

(II)

Theoretical calculations were performed in order to facilitate vibrational assignments. The calculations were carried out with the basis set 6-31G using full geometry optimization. Equilibrium structures, vibrational frequencies and IR and Raman intensities were calculated for both conformers of the two isomers. Scaling factors relating calculated vibrational transitions to the observed frequencies were also evaluated. For calculations using the 6-31G basis set an average scaling factor of 0.88 was obtained. The streching mode of the double bond was observed at 1667 cm^{-1} for (+)-car-2-ene and at 1689 cm^{-1} for (+)-car-3-ene. The experimentally observed bands are slightly assymmetric due to the existence of the two conformers for both isomers. The

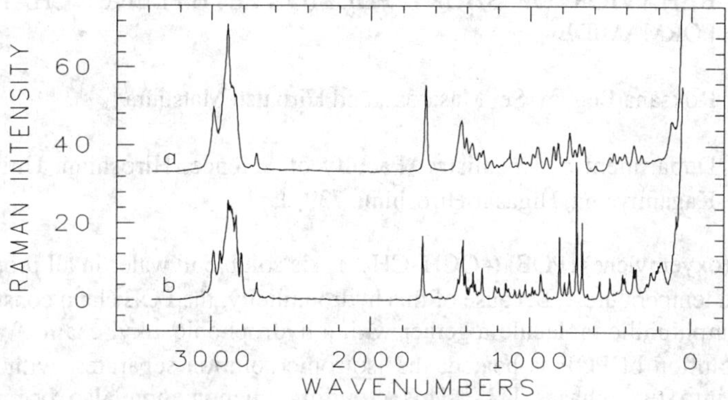

Fig. 1. Raman spectra of the two isomers of carene: curve a, (+)-car-2-ene; curve b, (+)-car-3-ene. Intensity given in units of 10^4 counts s^{-1}.

bands were resolved into two subbands in both cases using Voigt profiles. For the isomer (+)-car-2-ene components were obtained at 1664.9 cm^{-1} and 1669.3 cm^{-1}. The calculated value for the C=C stretching mode after scaling was 1667.8 cm^{-1} for the pseudo boat conformer and 1664.5 cm^{-1} for the pseudo chair conformer. The experimentally observed band for this mode in (+)-car-3-ene was resolved into components at 1686.1 cm^{-1} and 1688.1 cm^{-1}, whereas the scaled calculated values were 1683.5 cm^{-1} for the boat conformer and 1686.3 cm^{-1} for the inverted boat conformer.

The low-frequency part of the spectra contain the torsional transitions of the methyl groups and certain low-frequency modes of the carbon skeleton. The methyl torsions act as sensitive conformational probes in the carenes because the more crowded structure found in the pseudo boat form of (+)-car-2-ene and in the boat form of (+)-car-3-ene influence the torsional frequencies. For (+)-car-2-ene broad bands are seen in the experimental spectrum close to the Rayleigh line. These could be resolved into three components at 96, 122 and 171 cm^{-1}. Unscaled calculated values for the methyl torsions were 119, 124 and 171 cm^{-1} for the pseudo boat conformer and 108, 132 and 199 cm^{-1} for the pseudo chair conformer. For (+)-car-3-ene the observed low-frequency bands were resolved into three components at 101, 170 and 191 cm^{-1} compared to the unscaled calculated values 81, 169 and 189 cm^{-1} for the boat conformer. The skeletal low-frequency modes are well resolved for both isomers.

1. F. W. Semmler and H. von Schiller, Chem. Ber. **60**, 1591 (1927).
2. O. Aschan, Liebighs Ann. Chem. **461**, 1 (1928).
3. H. Sasai, T. Arai, E. Emori and M. Shibasaki, J. Org. Chem. **60**, 465 (1995).
4. H. Tylli, J. Mol. Struct. **181**, 11 (1988).
5. H. Tylli, J. Mol. Struct. **151**, 71 (1987).

RAMAN SPECTROSCOPIC STUDY OF THE CONFORMATIONAL BEHAVIOR OF SHORT POLY(OXYETHYLENE) CHAINS IN FORMAMIDE

Roksana Begum, Sei Masatoki, and Hiroatsu Matsuura*

Department of Chemistry, Faculty of Science, Hiroshima University, Kagamiyama, Higashi-Hiroshima 739, Japan

Poly(oxyethylene) (POE) $(–OCH_2CH_2–)_m$ is soluble in water in all proportions at moderate temperatures. Because of this hydrophilicity, the POE chain constitutes a nonionic amphiphilic molecule together with a hydrophobic alkyl chain. When an aqueous solution of POE is heated, the isotropic solution separates, with sudden clouding, into two phases [1]. This clouding phenomenon also occurs in a nonaqueous polar solvent of formamide. In our previous Raman and infrared spectroscopic studies [2,3], we have observed anomalous conformational behavior of short POE compounds $CH_3(OCH_2CH_2)_mOCH_3$ ($C_1E_mC_1$) with $m = 1–4$ in water. In the present work, we have studied the conformational behavior of the same compounds in formamide in order to elucidate possible factors responsible for the conformational stability of the POE chain in comparison with the conformational behavior in aqueous solution.

The Raman spectra of formamide solutions of $C_1E_mC_1$ were measured for mole fractions below 0.5 The spectral profiles in the 800–900 cm^{-1} region, where well-defined conformation-sensitive bands of the POE chain are observed [4], were analyzed by fitting with resolved components based on the Pearson function. We used the Raman bands at 848–854 and 801–816 cm^{-1} due to the gauche and trans conformations, respectively, around the OC–CO bond in order to examine the conformational behavior of the POE chain in formamide. The Raman spectra shown in Fig. 1 indicate a decreasing tendency of the trans band intensity from higher to lower concentrations. Fig. 2 shows comparative plots of the intensity ratio of the two bands, I_{gauche}/I_{trans}, in formamide and water against the mole fraction of $C_1E_mC_1$.

The plots show that, on addition of formamide, all of the $C_1E_mC_1$ compounds studied progressively prefer the gauche conformation. The plots for the aqueous solutions of $C_1E_2C_1$, $C_1E_3C_1$, and $C_1E_4C_1$ indicate a decrease of the gauche population with increasing water fraction in concentration regions below 0.02 mole fraction, where the formamide solutions show a further increase of the population of this conformation. The different tendencies of the conformational behavior in the two solvents suggest that some different factors associated possibly with solvophobic interactions are responsible for the conformational stability in those low concentration regions. The hydrophobicity of the terminal methyl groups in particular is likely to be associated with the interactions with the solvent.

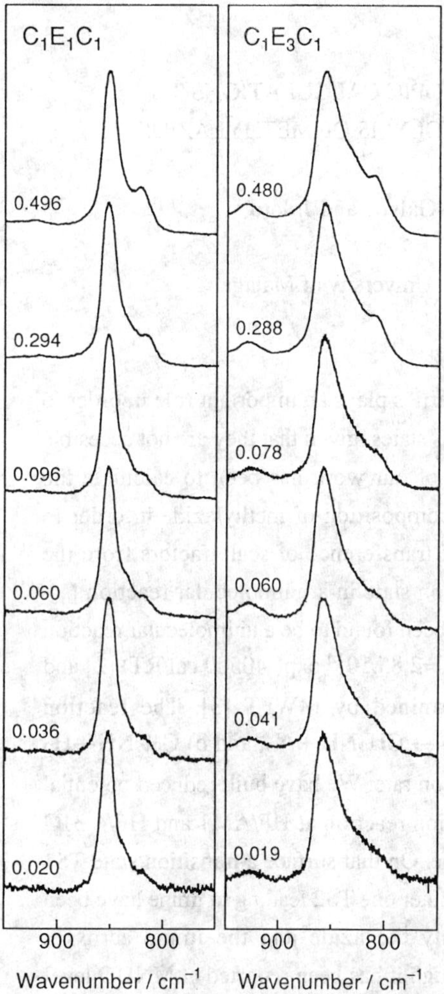

Fig. 1. Raman spectra of $C_1E_1C_1$ and $C_1E_3C_1$ in formamide for different mole fractions.

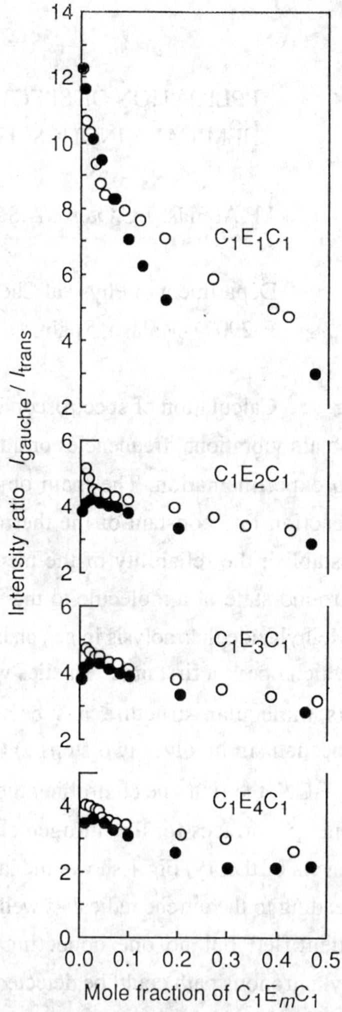

Fig. 2. Plots of the intensity ratio for $C_1E_mC_1$ (m = 1–4) in formamide (○) and water (●).

1. G. Karlström and B. Lindman, in *Organized Solutions* (S.E. Friberg and B. Lindman, eds.), Marcel Dekker, 1992, p. 49.
2. S. Masatoki, M. Takamura, H. Matsuura, K. Kamogawa, and T. Kitagawa, Chem. Lett., 991 (1995).
3. H. Matsuura and T. Sagawa, J. Mol. Liq., **65/66**, 313 (1995).
4. H. Matsuura and K. Fukuhara, J. Polym. Sci., Part B, **24**, 1383 (1986).

APPLICATION OF SPECTROSCOPIC CALCULATIONS TO CHEMICAL KINETICS: THERMOLYSIS OF METHYLAZIDE

J.F.Arenas, J.C.Otero, A.Sánchez-Gálvez and J.Soto

Department of Physical Chemistry, University of Málaga,
E-29071-Málaga, Spain.

Calculation of spectroscopic properties plays an important role in order to obtain vibrational frequencies of transitions states, given that they are not accesible to experimentation. The main objective of our work has been to calculate the reaction rate constant of the thermal decomposition of methylazide in order to establish the reliability of the method of transference of scale factors from the ground state of a molecule to the transition state in a unimolecular reaction [1]. Methylazide thermolysis in gas phase has been found to be a unimolecular reaction which obeys a first order kinetics with k_{∞} =2,85.10^{14} exp(-40500 cal/RT) [2] and its molecular structure has been determined by MWFT [3]. The reaction mechanism involves two steps a) $CH_3N_3 \rightarrow CH_2NH + N_2$, and b) $CH_2NH \rightarrow H_2 + HCN$, the first one controlling the reaction rate. We have built reduced potential energy surfaces for the nitrogen elimination reaction at HF/AM1 and HF/4-31G levels of theory, Fig.1 shows the latter one. On that surface a transition state TS1 leading to the nitrene radical as well as another one TS2 leading to imine have been identified but no one conecting directly the azide and the imine across a synchronous path could be detected although it has been reported at MNDO level by Bock [4] and found by ourselves at AM1 level. The chemically interesting points on the potential energy surface have also been calculated at HF/6-311G* level but that synchronous path was never found. Starting with the scale factors calculated by the SQMFF method [5] for the force field of methylazide, we have computed the scaled force field of TS1 and its vibrational frequencies. Table 1 collects some of the diagonal scale factors , the respective off-diagonal ones are calculated according to : $f_{ij} =(f_{ii} \cdot f_{jj})^{1/2}$. Once the geometry and the vibrational frequencies of azide and TS1 are known, the preexponential factor of the rate constant has been found to be A_{∞} = 1,35.10^{14} which compares quite well, we think, with experiment.

Table 1. Force constants (F_{ii}, in mdyn/Å) and scale factors (f_{ii}) for selected internal coordinates of methylazide.

Internal coordinate	Methylazide 4-31G F_{ii}	Methylazide 4-31G f_{ii}	Methylazide 6-311G* F_{ii}	Methylazide 6-311G* f_{ii}	TS1 4-31G F_{ii}	TS1 6-311G* F_{ii}
ν (N≡N)	17.994	0.83	17.535	0.72	23.385	21.923
ν (N=N)	8.428	0.96	9.121	0.88	-0.973	-1.076
ν (C-N)	4.913	1.00	4.301	0.79	4.656	4.279
δ (NNN)	0.642	1.15	0.586	0.85	0.329	0.181
τ (NN)	0.012	1.09	0.004	0.80	0.022	0.009
τ (CH_3)	0.020	0.91	0.018	0.72	0.106	0.009

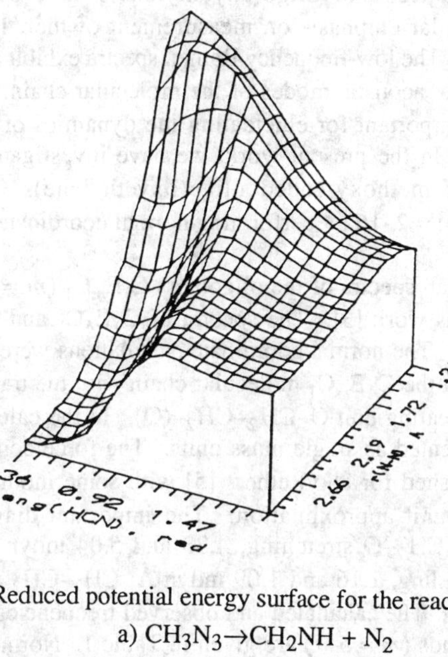

Figure 1. Reduced potential energy surface for the reaction
a) $CH_3N_3 \rightarrow CH_2NH + N_2$

1. J. F. Arenas, J. I. Marcos, A. Sánchez- Gálvez and J. Soto, J. Chem. Soc., Faraday Trans., 92, 000 (1996) in press.
2. C. L. Currie and B. deB. Darwent, Can. J. Chem., 41, 1048 (1963).
3. N. Heineking and M. C. L. Gerry, Z. Naturforsch. 44, 669 (1989)
4. H. Bock and R. Dammel, J. Am. Chem. Soc., 110, 5261 (1988)
5. G. Fogarasi and P. Pulay, J. Mol. Struct., 39, 275 (1977)

LOW-FREQUENCY RAMAN SPECTRA AND NORMAL MODES OF METHOXY-ENDED OLIGO(OXYETHYLENE)S

Hiroatsu Matsuura[1,2]*, Kyriakos Viras[2,3], Colin Booth[2], and Colin Price[2]

[1] Department of Chemistry, Faculty of Science, Hiroshima University, Kagamiyama, Higashi-Hiroshima 739, Japan
[2] Manchester Polymer Centre, Department of Chemistry, University of Manchester, Manchester M13 9PL, UK
[3] Department of Chemistry, National and Kapodistrian University of Athens, Panepistimiopolis, 157 71 Athens, Greece

Poly(oxyethylene) $(\text{–}OCH_2CH_2\text{–})_m$ is a crystalline polymer with a helical chain conformation. The Raman spectra of oligo(oxyethylene)s have been studied extensively by placing particular emphasis on measurement of their longitudinal acoustic modes (LAMs) [1–3]. The low-frequency Raman spectra exhibit a number of other bands associated with the acoustic modes of the molecular chain. Studies of these vibrational modes are important for elucidating the dynamics of the whole-molecule motion in crystals. In the present work, we have investigated the low-frequency normal modes of methoxy-ended oligo(oxyethylene)s $CH_3(OCH_2CH_2)_mOCH_3$ ($C_1E_mC_1$) with m = 2–16 by performing normal coordinate analysis of the single molecular chain.

The low-frequency Raman spectra of the oligomers $C_1E_mC_1$ (m = 4–25) have been measured in our previous work [3]. The spectra of $C_1E_6C_1$ and $C_1E_7C_1$ are shown in Fig. 1 as illustration. The normal coordinate calculations were carried out with the program MVIB [4] on the $C_1E_mC_1$ molecular chain with the trans–gauche–trans conformation for the repeating unit O–CH_2–CH_2–(O). In the calculation, the CH_2 and CH_3 groups were treated as single mass units. The force constants were transferred from those established for alkyl ethers [5] with some modifications to accord with the single-mass-unit approximation. The important diagonal force constants are: CH_2–CH_2 and CH_2–O stretching, 4.33 and 5.04 mdyn Å^{-1}; CH_2–CH_2–O and CH_2–O–CH_2 bending, 1.16 and 1.09 mdyn Å; CH_2–CH_2 and CH_2–O torsion, 0.12 and 0.10 mdyn Å. The calculated and observed frequencies below 100 cm^{-1} for the $C_1E_mC_1$ compounds (m = 5–7) are given in Table 1. Normal modes of some of the low-frequency vibrations for $C_1E_6C_1$ are depicted in Fig. 2.

The lowest two normal modes of $C_1E_6C_1$ with frequencies of 8 and 9 cm^{-1} are observed as a weak band at 11 cm^{-1}. These modes are associated with the bending of the helical chain. The calculated frequency of LAM-1, 31 cm^{-1}, for $C_1E_6C_1$ is considerably lower than the observed frequency, 50 cm^{-1}. The same trend is noted for all of the homologous oligomers. This implies that interchain forces are important in particular in this longitudinal mode [3].

TABLE 1. Calculated and observed[a] frequencies below 100 cm^{-1} for $C_1E_mC_1$ (m = 5–7)

$C_1E_5C_1$		$C_1E_6C_1$		$C_1E_7C_1$	
ν_{calc}	ν_{obs}	ν_{calc}	ν_{obs}	ν_{calc}	ν_{obs}
10		8	11	7	
13		9	11	7	
23	23	15	23	10	12
39	55	27	31	18	23
45	55	31	50	25	50
75	80	46	50	32	34
76	80	63	61	45	50
		73	75	56	61
		93	83	70	73
		97		82	83
				92	91

[a] Observed Raman frequencies are taken from Ref. [3].

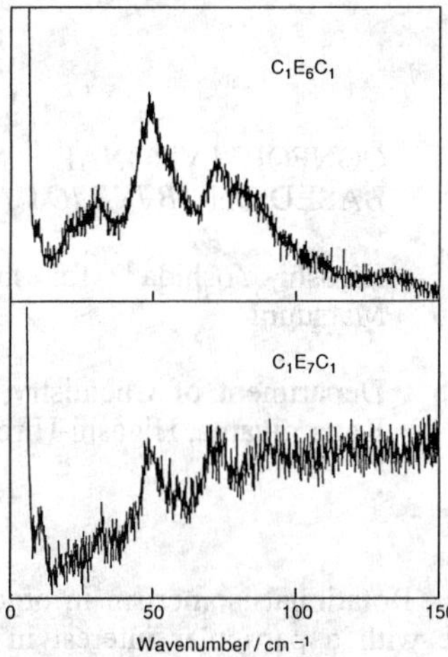

Fig. 1. Raman spectra of $C_1E_6C_1$ and $C_1E_7C_1$ at 173 K [3].

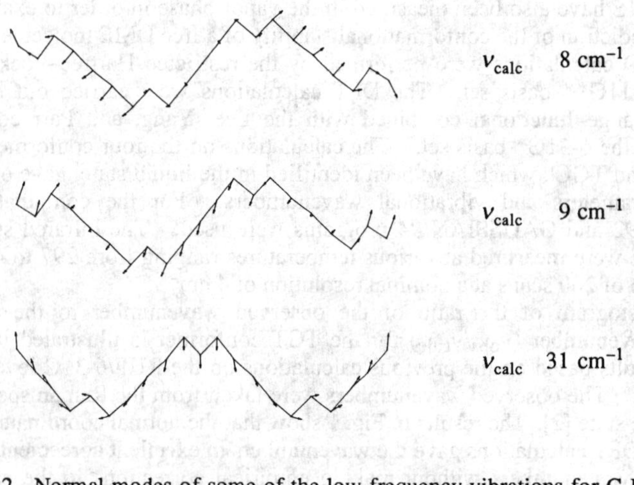

Fig. 2. Normal modes of some of the low-frequency vibrations for $C_1E_6C_1$.

1. K. Viras, H.H. Teo, A. Marshall, R.C. Domszy, T.A. King, and C. Booth, J. Polym. Sci., Polym. Phys. Ed., **21**, 919 (1983).
2. K. Viras, T.A. King, and C. Booth, J. Polym. Sci., Polym. Phys. Ed., **23**, 471 (1985).
3. C. Campbell, K. Viras, and C. Booth, J. Polym. Sci., Part B, **29**, 1613 (1991).
4. H. Matsuura, Comput. Chem., **14**, 59 (1990).
5. T. Shimanouchi, H. Matsuura, Y. Ogawa, and I. Harada, J. Phys. Chem. Ref. Data, **7**, 1323 (1978).

CONFORMATIONAL ANALYSIS OF 1,2-DIMETHOXYETHANE BASED ON *AB INITIO* CALCULATIONS

Hiroshi Yoshida*, Takeshi Tanaka, Miyoko Komoda, and Hiroatsu Matsuura

Department of Chemistry, Faculty of Science, Hiroshima University, Kagamiyama, Higashi-Hiroshima 739, Japan

Rotational isomerism in organic compounds has been investigated for a long time with a particular interest in the conformation around C–C bonds. Despite progressive development, a number of fundamental problems still remain unresolved. The conformational stability around the OC–CO bond is among these subjects. In the present work, we have studied the conformational stability of 1,2-dimethoxyethane (DME) $CH_3OCH_2CH_2OCH_3$ by normal coordinate analysis based on *ab initio* molecular orbital (MO) and density functional theory (DFT) calculations. The infrared spectra of DME have also been measured in the vapor phase in order to examine the theoretical prediction of the conformational stability of a free DME molecule.

The MO calculations were performed by the restricted Hartree–Fock method using the 6-311G** basis set. The DFT calculations were carried out by using Becke's exchange functional combined with the Lee, Yang, and Parr correlation functional on the 6-31G* basis set. The calculations on the four conformers, TTT, TGT, TTG, and TGG', which have been identified in the liquid state, gave optimized structural parameters and vibrational wavenumbers. For the computation, the GAUSSIAN 92 and GAUSSIAN 94 programs were used. The infrared spectra of gaseous DME were measured at various temperatures ranging from 297 to 400 K by the coaddition of 200 scans at a nominal resolution of 4 cm^{-1}.

The histogram of the ratio of the observed wavenumber to the unscaled calculated wavenumber (ν_{obs}/ν_{calc}) for the TGT conformer is illustrated in Fig. 1, where the results based on the previous calculations on the RHF/6-31G level [1] are also included. The observed wavenumbers were taken from the Raman spectrum in the crystalline state [2]. The results in Fig. 1 show that the normal coordinate analysis based on the DFT calculations gave the wavenumbers in excellent agreement with the experimental wavenumbers without making empirical corrections to the calculated values.

From the temperature dependence of the infrared band intensities at 987 and 946 cm^{-1} in the vapor phase (Fig. 2), the enthalpy difference between the TGG' and TTT conformers ($H_{TGG'} - H_{TTT}$) was determined to be 0.31 ± 0.04 kcal mol^{-1}. This experimental value is in good agreement with a value, 0.543 kcal mol^{-1}, calculated on a high level of MP3/6-311+G*//HF/6-311+G* [3]. The present spectroscopic study has also shown the existence of an intramolecular 1,5-CH...O interaction which is important for stabilizing the TGG' conformation, resulting in the largest population of this conformer in the vapor phase.

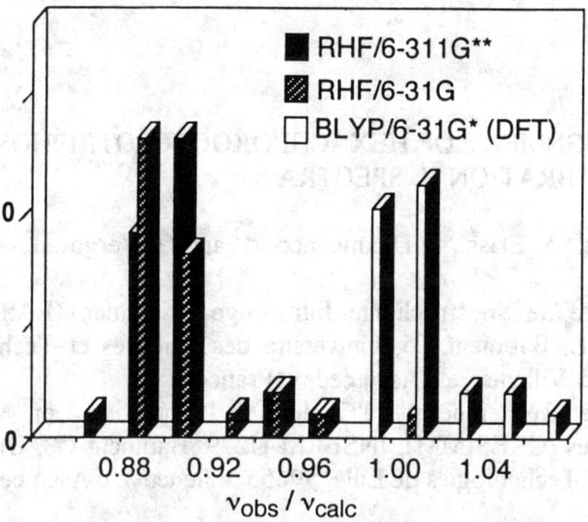

Fig. 1. Histogram of the ratio of the observed wavenumber to the unscaled calculated wavenumber (ν_{obs}/ν_{calc}) for the TGT conformer of DME.

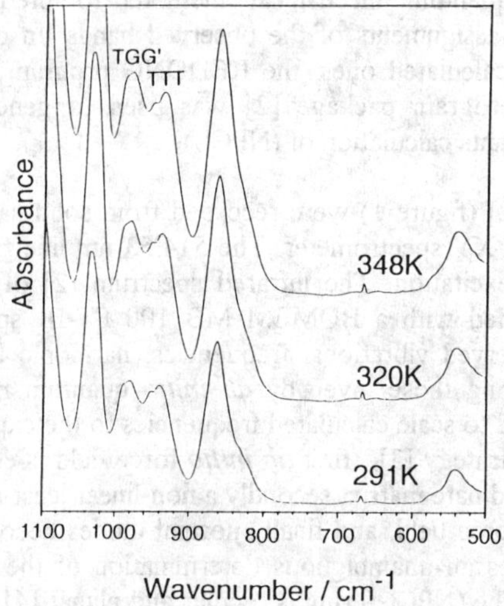

Fig. 2. Infrared spectra of DME in the vapor phase at different temperatures.

1. H. Yoshida, I. Kaneko, H. Matsuura, Y. Ogawa, and M. Tasumi, Chem. Phys. Lett., **196**, 601 (1992).
2. Y. Ogawa, M. Ohta, M. Sakakibara, H. Matsuura, I. Harada, and T. Shimanouchi, Bull. Chem. Soc. Jpn., **50**, 650 (1977).
3. S. Tsuzuki, T. Uchimaru, K. Tanabe, and T. Hirano, J. Phys. Chem., **97**, 1346 (1993).

THE ASSIGNMENT OF HEXACHLOROCYCLOTRIPHOSPHAZENE $(NPCl_2)_3$ VIBRATIONAL SPECTRA

R. Becquet[1], A. Elass[2], P. Dhamelincourt[1] and G. Vergoten[2]

[1]Laboratoire de Spectrochimie Infrarouge et Raman (LASIR), CNRS UPR 2631L, Bâtiment C5, Université des Sciences et Technologies de Lille, 59655 Villeneuve d'Ascq cedex, France
[2]Centre de Recherche et d'Etudes en Simulations et Modélisation Moléculaires (CRESIMM), INSERM U279, Bâtiment C8, Université des Sciences et Technologies de Lille, 59655 Villeneuve d'Ascq cedex, France

Since 1962, several works [1] have been done on the vibrational spectra and assignments for $(NPCl_3)_2$. To our knowledge, no attempt has not been made yet to compute an *ab initio* force field for this molecule. Therefore, we decided to explore the capacity of quantum mechanical methods to predict vibrational frequencies and to give the assignments of the observed bands. In order to fit the experimental data with the calculated ones, the REDONG program has been used [2]. The GAUSSIAN 92 program package [2] was used to generate geometry optimization and force constants calculation of $(NPCl_2)_3$.

Raman spectra [2] (figure 1) were recorded from solid samples at room temperature using a DILOR XY spectrometer. The 514.53 nm line from an ionized Argon laser was used for excitation. The infrared spectrum [2] of the sample in solution in CCl_4 was recorded with a BOMEM MB 100 FT-IR spectrometer. In order to reproduce the observed vibrational frequencies, harmonic force constants were determined by modifying those given by *ab initio* quantum mechanics. The program REDONG was used to scale calculated frequencies to the experimental ones according to the following strategy [3] : first *ab initio* force field is converted into a complete set of internal coordinate matrix, secondly a non-linear least-square fitting is used to obtain the reliable force field, and finally normal modes decomposition by a least-square method provides an unambiguous determination of the non-redundant coordinates. Assuming that the $(NPCl_2)_3$ ring is regular and planar [4] (D_{3h} symmetry for the isolated molecule), we should obtain : $4A_1'$ (Raman), $2A_2'$ (inactive), 6E' (IR, Raman), $1A_1''$ (inactive), $3A_2''$ (IR) and 4E" (Raman). Thus, an appropriate normal coordinate analysis should give an unambiguous assignment of all these modes and the corresponding PED. Most of the fundamental modes were previously reported by Painter et al. [1]. However, assignments of observed frequencies, found in the literature, are different depending on the authors and are still ambiguous concerning some of the ring modes (E', A_2'' and E"). In our studies, the standard deviation, corresponding to the difference between the observed and calculated frequencies, is 14.2 cm^{-1} [2]. To our knowledge, no quantum scaled force field for $(NPCl_2)_3$ has been published hitherto. We have defined 39 internal coordinates which can be classified into the following coordinates : stretching P-Cl and N-P coordinates, angle bending N-P-N, P-N-P, N-P-Cl and Cl-P-Cl coordinates, and finally torsional P-N coordinates. From the number of internal coordinates and assuming that theorically 3N-6 (30) frequencies are expected, 9 redundancies should be omitted. The program REDONG indicated that 3 of these redundancies correspond to PCl_2 moieties and 6 to the N_3P_3 ring. From the Potential Energy Distribution (PED) [2], the ring mode frequencies retained are : 781.3 and 665.3 cm^{-1} (A_1'), 1207.5, 879 and 518.2 cm^{-1} (E'), 612.3 cm^{-1} (A_2'') and 293.6 cm^{-1} (calculated but non observed) (E"). At last, the inactive one (A_2') was calculated at 1184.7 cm^{-1}. The validity of this PED was verified a posteriori by the good transferability of the force field calculated for the trimer $(NPCl_2)_3$, which has permitted to reproduce the geometries of the tetramer $(NPCl_2)_4$ and pentamer $(NPCl_2)_5$ [5].

Fig. 1. : Solid state Raman spectrum of $(NPCl_2)_3$ in the range 100 to 1400 cm^{-1}

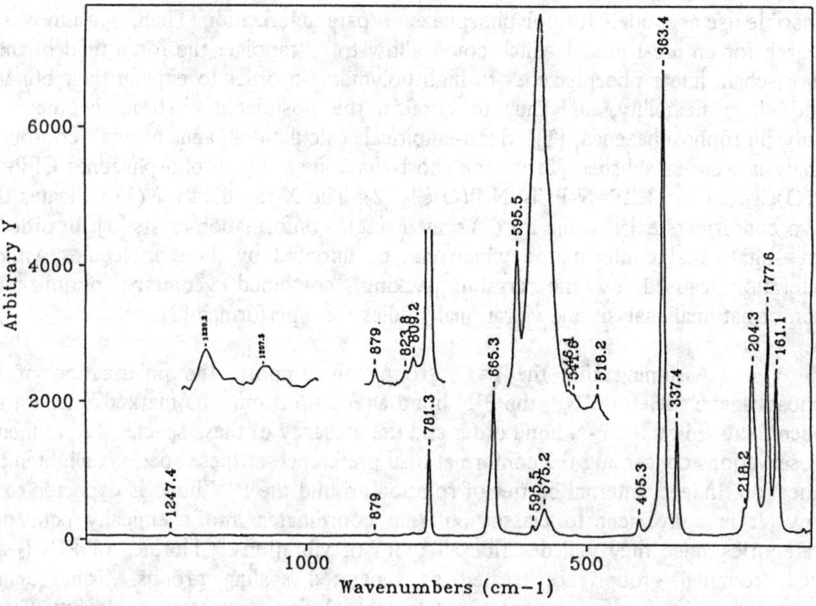

1. P. C. Painter, J. Zarian and M. M. Coleman, Appl. Spectrosc., 36 (3), 265 (1982)
2. A. Elass, P. Dhamelincourt, R. Becquet and G. Vergoten, to be published
3. A. Allouche and J. Pourcin, Spectrochim. Acta, 49A, 571 (1993)
4. J. Bullen, J. Chem. Soc., A, 1450 (1971)
5. A. Elass, P. Dhamelincourt, R. Becquet and G. Vergoten, to be published

A SEMI-EMPIRICAL SCALED FORCE FIELD FOR SIMPLE SHORT-CHAIN PHOSPHAZENE $Cl_3P{=}N{-}(PCl_2{=}N)_{n-1}{-}P(O)Cl_2$ (n=1 AND 2)

R. Becquet[1], A. Elass[2], P. Dhamelincourt[1] and G. Vergoten[2]

[1]Laboratoire de Spectrochimie Infrarouge et Raman (LASIR), CNRS UPR 2631L, Bâtiment C5, Université des Sciences et Technologies de Lille, 59655 Villeneuve d'Ascq cedex, France
[2]Centre de Recherche et d'Etudes en Simulations et Modélisation Moléculaires (CRESIMM), INSERM U 279, Bâtiment C8, Université des Sciences et Technologies de Lille, 59655 Villeneuve d'Ascq cedex, France

Current interest in the short-chain linear phosphazenes resides in their possible use as models for polyphosphazenes parameterization. Then, our aim was to search for an ideal model which could allow to extrapolate the force field of these short-chain linear phosphazenes to high polymers, in order to explain their bonding and chain flexibility, and thus to confirm the postulated cis-trans geometry of polychlorophosphazenes [1]. Semi-empirical calculations and normal coordinate analysis were established [2] for the short-chain linear chlorophosphazenes $Cl_3P{=}N{-}P(O)Cl_2$ (1) and $Cl_3P{=}N{-}PCl_2{=}N{-}P(O)Cl_2$ (2). The X-ray data for (1) indicates that two conformers exist, while for (2) only a single conformation exists [3]. In order to investigate real conformation which can be adopted by these molecules without alteration caused by the crystal packing, combined geometry optimization, conformational analysis and vibrational studies were performed [2].

Assuming that the P=O group can support the polarization of the phosphazene skeleton PNP, the P-N bond alternations must be marked. Interest has been focussed on the P-N bond order and the planarity of these species. Experimental observations concerning the conformational preference of these species enlighten this question. Indeed, internal barrier of rotation around the P-N bond is expected to be low. It is convenient to classify internal coordinates into chemically equivalent categories since they will describe all types of vibrations. Chlorines of $POCl_2$ and PCl_3 terminal groups are treated as separated scaling groups. Consequently, concerning (1), force constants given in table 1 for parameters containing Cl are slightly different from the literature [1], while mean values could be considered equivalent. Excepting bending constants of P-N-P bond angle, the remaining force constant values are in good agreement with the previous studies. The most important points are : low torsional constants around both N-P or N=P bonds (0.04 and 0.05 mdyn/Å, respectively) suggesting that (1) could exist in fast equilibrium with other conformer(s), and significant differences between the N-P (3.8 mdyn/Å) and N=P (7.76 mdyn/Å) stretching constants. This last fact is correlated to the bond length values observed in (1) for which difference between the N-P and N=P bond lengths is marked, while, no significant difference could be observed for (2) (table 1). The most important finding for this last molecule is that the bond angle P_2-N_5-P_3 is highly dependent on the intramolecular nonbonded interactions between N_4 and Cl_{11} (probably Cl_{12} too). The electrons in these atoms are distributed to minimize the Van der Waals repulsion between them. Thus, bending force constants for P_1-N_4-P_2 and P_2-N_5-P_3 angles are not equivalent, all other force constant values being unequal. If we compare bond stretching constants for these two angles, they seem well correlate with bond lengths obtained from X-ray data [3]. No neat separation into single and double bond are observed experimentally or calculated as having separated force constants. We have found that P_3-N_5-P_2-N_4 had cis conformation while P_1-N_4-P_2-N_5 had a trans one. The terminal groups are more flexible than the central cores. The cis-trans conformation maintains a planar structure for the phosphazene skeleton. Concerning the variation of the P-N-P bond angles with the nature of substituent [2], we have observed that these angles are dependant on both intramolecular and intermolecular nonbonded interactions. The P-N-P geometry is then function of the steric repulsion generated by the side groups or by the ions charges. Physical properties of these polymers are probably consigned in the capability of the P-N-P skeleton to adopt multiples forms.

Table 1 : Harmonic force constants for molecules (1) and (2)

Coordinate	Values (2)	Values (1)	Coordinate	Values (2)	Values (1)	Coordinate	Values (2)	Values (1)
Stretching			Bending			Torsion		
ν(P=O)	9.20	10.31	δ(O=P-N)	1.17	0.65	$\tau(P_1$-$N_4)$	0.16	0.04
$\nu(P_1$-$N_4)$	8.49	3.80	δ(O=P-Cl)	2.74	0.76	$\tau(P_2$=$N_4)$	0.19	0.05
$\nu(P_2$=$N_4)$	8.48	7.76	$\delta(P_1$-N_4-$P_2)$	2.3	0.33	$\tau(P_2$-$N_5)$	0.46	0.04
$\nu(P_2$-$N_5)$	8.20	3.80	$\delta(P_2$-N_5-$P_3)$	3.27	0.33	$\tau(P_3$=$N_5)$	0.35	0.05
$\nu(P_3$=$N_5)$	7.23	7.76	δ(N-P-N)	3.3	-			
$\nu(P_1$-Cl)	3.21	2.41	δ(Cl-P_1-Cl)	1.16	0.95			
$\nu(P_2$-Cl)	3.24	2.41	δ(Cl-P_2-Cl)	0.35	-			
$\nu(P_3$-Cl)	2.53	1.92	$\delta(Cl_{11}$-P_3-$Cl_{12})$	0.35	1.00			
			$\delta(Cl_{11}$-P_3-$Cl_{13})$	1.03	1.00			
			$\delta(Cl_{12}$-P_3-$Cl_{13})$	1.53	1.00			
			δ(Cl-P_1-N_4)	0.81	0.84			
			δ(Cl-P_2-N)	1.30	0.84			
			δ(Cl-P_3-N_5)	1.56	0.84			

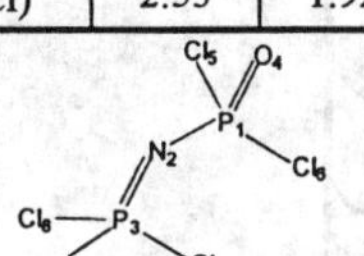

Molécule (1)

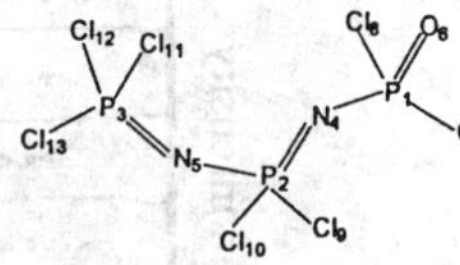

Molécule (2)

Stretching constants in mdyn/Å, bending constants in mdyn.Å/rad^2 and torsions in mdyn/Å

1. Bougeard, C. Brémard, R. De Jaeger and Y. Lemmouchi, Spectrochim. Acta, 49A, 199 (1993)
2. A. Elass, P. Dhamelincourt, R. Becquet and G. Vergoten, to be published
3. R. Hallcock, N. M. Tollefson, R. A. Arcus and R. R. Whittle, J. Am. Chem. Soc., 107, 5166 (1985)

RAMAN SPECTROSCOPY OF $AlCl_3$-SO_2Cl_2 AND $AlCl_3$-LiCl-SO_2Cl_2 SYSTEMS

M.C. Dhamelincourt*, R. De Jaeger

Laboratoire de Spectroscopie Infrarouge et Raman, UPR2631 - Bat C5
Université des Sciences et Technologies de Lille
59655 - VILLENEUVE D'ASCQ Cedex - France

Because high energy densities are permitted by the SO_2Cl_2 catholytes, a large number of investigations have been conducted on these systems. What it is essential to known is what species are present in solution and what are their structure. This is very important to better understand the reaction path of the electroreduction reactions inside the SO_2Cl_2 electrolyte. Some information have already been obtained using IR spectrometry by P. A. Mosier-Boss and al. (1). The structural aspect of the $AlCl_3$-SO_2Cl_2 and $AlCl_3$-LiCl-SO_2Cl_2 system have been examined by Raman spectroscopy.

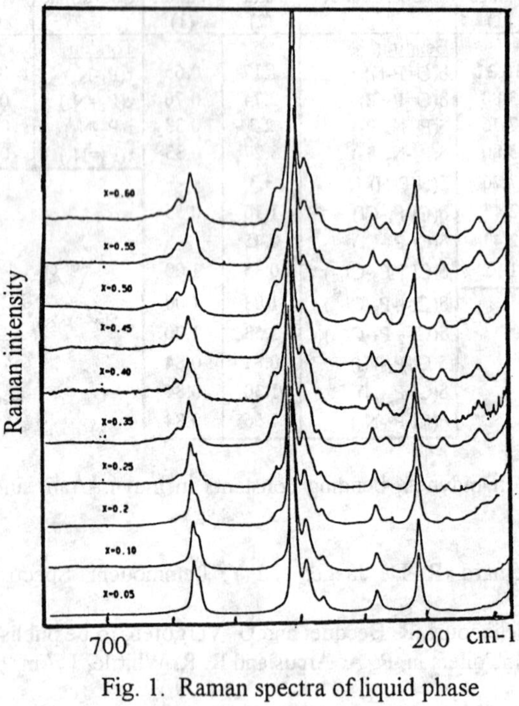

Fig. 1. Raman spectra of liquid phase

A large concentration range of $AlCl_3$- SO_2Cl_2 mixtures was studied (0.05 <X< 0.60, where X represents the molar fraction). For molar fractions up to 0.4, a solid phase appears.

In dilute solutions as well as in more concentrated phases, new bands at 1131 and 1380 cm^{-1} appear with this addition of $AlCl_3$. The shift of the S-O stretching vibration bands toward low frequencies indicates that the bonding occurs through one or two oxygen atoms of SO_2Cl_2. At the same time, modifications are observed in the O-S-O bending mode frequency range.

More over besides this modifications, new bands at 117, 172, 226, 261, 283, 435 appear (figure 1).

On the spectra of the solid phase shown on figure 2, other new bands appear.

The interpretation of these spectra will be presented.

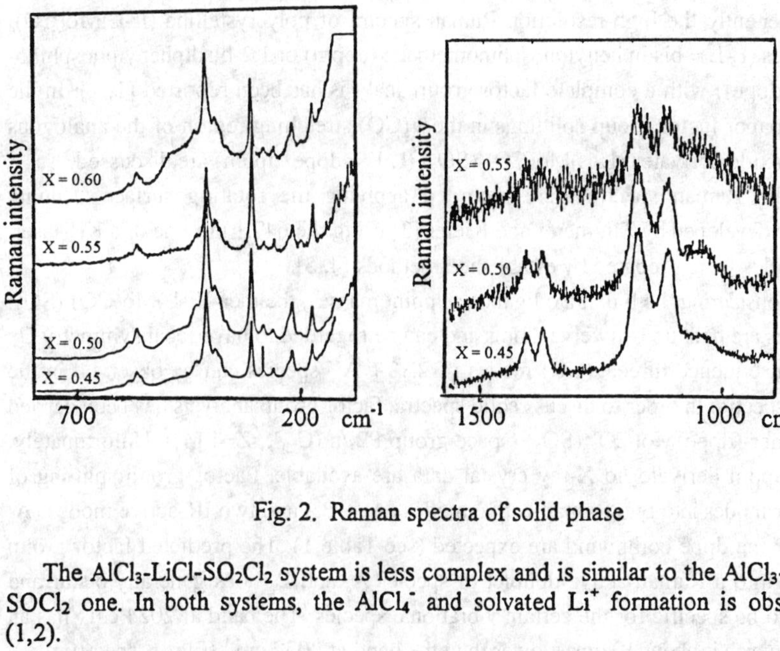

Fig. 2. Raman spectra of solid phase

The $AlCl_3$-LiCl-SO_2Cl_2 system is less complex and is similar to the $AlCl_3$-LiCl-$SOCl_2$ one. In both systems, the $AlCl_4^-$ and solvated Li^+ formation is observed (1,2).

1. P. A. Mosier-Boss and al., J. Electrochem. Soc., 136, 5,1282 (1989)
2. M. C. Dhamelincourt and al. J. Power Sources, 14, 77 (1985)

FACTOR GROUP SPLITTING IN THE ν(CO) REGION OF POLY-CRYSTALLINE TRICARBONYLSULFURDIOXIDE COMPLEXES OF MOLYBDENUM

P. Scholz[1], W. Kiefer[1]*, K. Nielsen[2], and W. A. Schenk[2].

[1]Institut für Physikalische Chemie, Universität Würzburg,
Marcusstr. 9-11, D-97070 Würzburg, Germany.
[2]Institut für Anorganische Chemie, Universität Würzburg,
Am Hubland, D-97074 Würzburg, Germany.

Recently the high-resolution Raman spectra of polycrystalline (L-L)Mo$(CO)_4$ complexes (L-L = bisdiphenylphosphinomethane (dppm) or 1,2-bisdiphenylphosphino-ethane (dppe)) with a complete factor group analysis has been reported [1, 2]. In the present paper factor group splittings in the ν(CO) stretching region of the analogous sulfur dioxide derivates (L-L)Mo$(CO)_3(SO_2)$ (L-L = dppe, dppm) are discussed.

The Raman spectra were recorded applying the rotating surface-scanning technique developed by Zimmerer and Kiefer [3] using the 647.1 nm line of a Kr^+-laser. The samples were prepared by established methods [4, 5].

Considering the bidentate ligands as point masses, the mer-(L-L)Mo$(CO)_3(SO_2)$ complexes are built up by twelve atoms and can be regarded to have local symmetry C_s. As a consequence three ν(CO) modes of 2A´ + A´´ species can be observed in the solution spectra. In order to discuss solid spectra, factor group analysis has been carried out for mer-(dppe)Mo$(CO)_3(SO_2)$ (space group $P2_1/n$ (C_{2h}^5), Z=4 [4]). Unfortunately, for the dppm derivate no X-ray crystal data are available. Factor group splitting of molecular modes into two Raman-active modes ($A_g + B_g$) and two IR-active modes ($A_u + B_u$) for the dppe compound are expected (see Table 1). The predicted factor group splitting into a Raman-active doublet of species A_g and B_g does generally occur and proved to be specific for the certain vibrational species. The band at 2024 cm^{-1} has an intensity which is about 10 times lower than the band at 2032 cm^{-1} (Fig. 1, spectrum A).

The Raman spectra of solid mer-(dppm)Mo$(CO)_3(SO_2)$ show completely different splitting patterns and band intensities. In contrast to the dppe complex, the A´´ -mode does not split whereas the $A'^{(1)}$-mode splits into four bands and has about half of the intensity of the analogous band in the dppe compound. The ν(CO) $A'^{(2)}$ band splits into two bands with equal intensities (Fig. 1, spectrum B).

From the observations it is concluded that the dppm derivative has different factor groups than those for the dppe compound.

TABLE 1. Correlation diagram for mer-(dppe)$Mo(CO)_3(SO_2)$.

point group	site group	factor group
C_s	C_1	C_{2h}^5

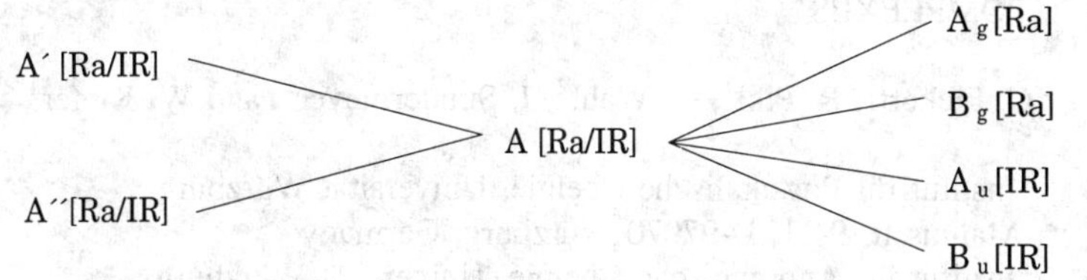

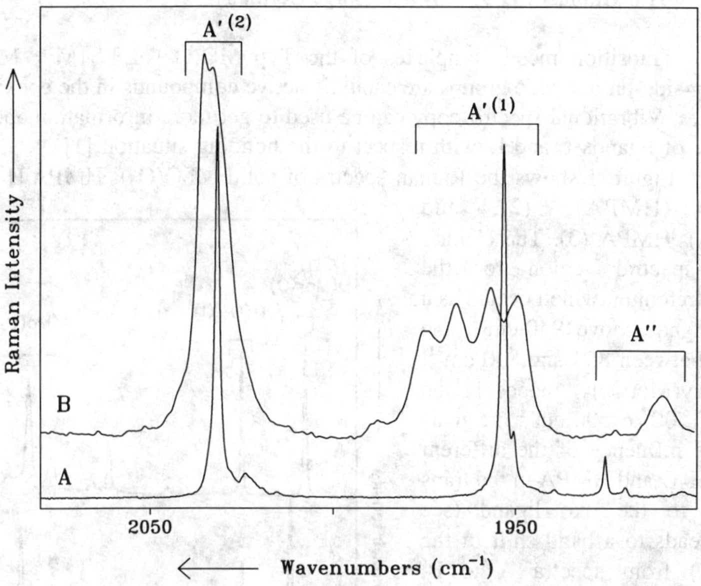

Fig. 1. Raman spectra in the CO stretching region of polycrystalline mer-(dppe)$Mo(CO)_3(SO_2)$ (A) and mer-(dppm)$Mo(CO)_3(SO_2)$ (B) at 18 K; resolution: 0.15 cm^{-1}.

Financial support from the Deutsche Forschungsgemeinschaft (SFB 347, projects B3 and C2) and the Fonds der Chemischen Industrie is highly acknowledged.

1. U. Posset and W. Kiefer, Vibrat. Spectrosc. 3, 47 (1992).
2. U. Posset and W. Kiefer, J. Mol. Struct. 349, 427 (1995).
3. N. Zimmerer and W. Kiefer, Appl. Spectrosc. 28, 279 (1974).
4. K. Nielsen, Master thesis, Universität Würzburg (1995).
5. F.-E. Baumann, Ph. D. thesis, Universität Würzburg (1985).

VIBRATIONAL ANALYSIS OF MOLYBDENUM BISPEROXO COMPLEXES

C. Fickert[1], R. Pikl[1], G. Wahl[2], J. Sundermeyer[2], and W. Kiefer[1*]

[1]Institut für Physikalische Chemie, Universität Würzburg, Marcusstr. 9-11, D-97070 Würzburg, Germany
[2]Institut für Anorganische Chemie, Universität Würzburg, Am Hubland, D-97074 Würzburg, Germany

Transition metal complexes of the Typ $MO(O_2)_2{\cdot}L{\cdot}L'$ (M = Mo, W) involving side-on peroxide groups are catalytic active compounds in the epoxidation of olefines. Vibrational spectroscopy can be used to get more information about the influence of ligands L and L' with respect to the bonding situation [1].

Figure 1 shows the Raman spectra of solid $MoO(O_2)_2{\cdot}HMPA{\cdot}H_2O$ (**1**), $MoO(O_2)_2(HMPA)_2$ (**2**), and $MoO(O_2)_2{\cdot}HMPA$ (**3**). The characteristic spectral region for the Mo=O stretching, which occurs as a strong signal above 940 cm^{-1}, the ν(O-O) between 850 and 890 cm^{-1}, and the $\nu(Mo(O_2)_2)$ modes in the range of 500 to 600 cm^{-1} are marked. The influence of the different ligands H_2O and HMPA in the trans position to the oxo ligand (see Fig. 2) leads to a band shift of the ν(Mo=O) from spectra A to B, which indicates a weaker Mo=O bond in **2**. Comparing the spectra one notices characteristic changes in the $\nu(Mo(O_2)_2)$ region for **3**, which will be discussed elsewhere [1].

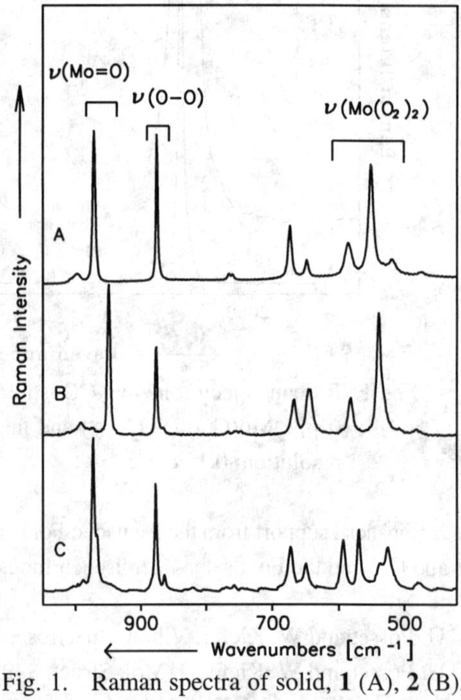

Fig. 1. Raman spectra of solid, **1** (A), **2** (B), and **3** (C); CCD detection; λ_0 = 647.1 nm.

In earlier studies Griffith *et al.* [2] discussed the metal peroxid vibrations by the concept of local symmetry as a separated $M(O_2)$ triangle under C_{2v} symmetry, the O-O stretching showing A_1 and the two metal peroxid stretchings A_1 and B_2 symmetry. However, we expect for the whole molecule interactions between vibrations of both triangles. So the vibrational spectra should show splittings into *in phase* and *out of phase* modes in the ν(O-O) and νM(O_2) regions. In order to confirm this proposal a normal coordinate analysis (NCA) has been carried out for the pentagonal bipyramidal $MoO(O_2)_2$·HMPA·H_2O (**1**) complex, from which X-ray data are available [3].

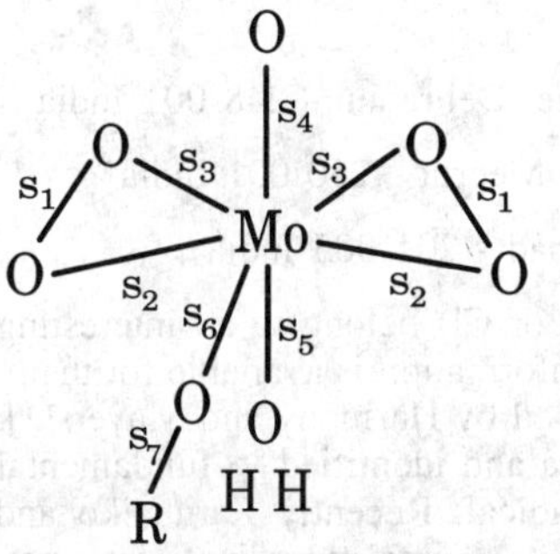

Fig. 2. Structure and stretching coordinates of **1** [R = P{N(CH_3)$_2$}$_3$].

The calculations are based on the experimental data of Raman and IR spectra (listed in Table 2). Raman spectrum A (Figure 1) shows only splitting of the Mo(O_2) modes, although the molecular symmetry is C_s and all vibrations are Raman- and IR-active. The detected IR spectrum shows a doublet in the ν(O-O) region. For the NCA some simplifications have been taken into account and only stretch coordinates have been used for the calculations (see Fig. 2). Stretch coordinates s_1, s_2, and s_3 in the Mo(O_2) triangles involved also bending character. The optimized force constants for the different metal oxo coordinates are listed in Table 1. The interactions in the Mo(O_2)$_2$ unit are relatively strong. Table 2 shows the potential energy distribution for the characteristic vibrations.

TABLE 1. Force constants of the internal coordinates and their interactions.

internal coordinate	force constant [mdyne/Å]
s_1	3.063
s_2	2.743
s_3	2.884
s_4	7.622
s_1s_2	-0.185
s_1s_3	0.224
s_2s_3	0.280

TABLE 2. Assignments and potential energy distribution (PED) of **1**.

wavenumbers [cm^{-1}]		assignments[a]	PED (%)			
observed	calculated		s_1	s_2	s_3	s_4
972	972	ν(Mo=O)				99
881	881	ν_{ip}(O-O)	81	7	12	
857	857	ν_{op}(O-O)	75	9	16	
583	583	$\nu_{s/op}$(Mo(O_2)$_2$)	27	41	32	
544	544	$\nu_{s/ip}$(Mo(O_2)$_2$)	21	44	32	
533	533	$\nu_{as/op}$(Mo(O_2)$_2$)	3	56	41	
520	520	$\nu_{as/ip}$(Mo(O_2)$_2$)	2	40	55	

[a]s: symmetric, as: antisymmetric, ip: in phase, op: out of Phase.

Financial support by the Deutsche Forschungsgemeinschaft (SFB 347, C2 and A4) and by the Fonds der Chemischen Industrie is highly acknowledged.

1. G. Wahl, J. Sundermeyer, C. Fickert, W. Kiefer, R. Stowasser, and G. Bringmann, in preparation.
2. W.P. Griffith and T.D. Wickins, J. Chem. Soc. (A), 398 (1968).
3. J.-M. Le Carpentier, R. Schlupp, and R. Weiss, Acta. Cryst. B28, 1278 (1972).

BOND PROPERTIES OF NXY (X,Y = F or Cl) RADICALS

A.K. Dublish[1]*, Y.C. Sharma[2] and V.K. Rastogi[3]

[1]Physics Deptt, S.G.R.R. (PG) College, Dehradun - 248 001 India

[2]Physics Deptt, N.A.S. (PG) College, Meerut - 250 001 India

[3]Physics Deptt, L.R. College, Sahibabad - 201 005 India

The halogen aminyl radicals NXY (X,Y = F or Cl) belong to an interesting class of free radicals, but a very little spectroscopic information is available for them. The IR spectra of NF_2 radical has earlier been recorded by Harmony and Mayers[1]. Kohlmiller and Andrews[2] have recorded IR spectra and identified ν_3 fundamental (asym. stretch) for various isotopomers of NCl_2 radical. Recently Zarubaiko and Gilbert[3] synthesised NFCl radical and performed its normal coordinate analysis. The present work aims at to study the bond properties of NXY (X,Y = F or Cl) radicals by evaluating force constants, bond order and bond charge.

The force constants for NXY radicals are calculated using Müller's and L–F approximation methods[4] and are presented in Table 1. Both methods give comparable values of force constants. The frequencies and geometry required are taken from references[1-3]. In the present calculations interbond angle for NFCl is taken as the mean of interbond angles of NF_2 and NCl_2. In case of NCl_2 , since ν_1 (sym. stretch) and ν_2 (bend) are not known, therefore ν (NCl) and half of ν (Cl_2) as reported by Gilbert and Smith[5] in their photolysis study of NCl_3 has been taken for ν_1 and ν_2 respectively. The bond order and effective bond charge (Q) have been evaluated following relations proposed by Politzer and Habibollahzadeh[6]. The results are presented in Table 2 along with bond properties of the related molecules.

TABLE 1. Valence force constants (mdyn /Å) for aminyl radicals NXY (X, Y = F or Cl)

Radical	fr	frr	$f\alpha/r^2$	$fr\alpha/r$	Ref.
NF_2	5.398	1.765	1.013	0.654	L-matrix approx.
	5.268	1.635	0.922	0.471	L–F approx.
	5.1 ± 0.5	1.5 ± 0.5	1.15 ± 0.15	0.4 ± 0.3	8
NFCl	4.259[a]	0.768[c]	0.972[d]	0.400[e]	L-matrix approx.
	3.496[b]			0.568[f]	
NCl_2	4.113	1.950	0.338	0.303	L-matrix approx.
	4.070	1.907	0.312	0.255	L–F approx.

a – fr_1 (N–F), b – fr_2 (N–Cl), c – $f_{r_1r_2}$, d – $f\alpha/r_1r_2$, e – $fr_1\alpha/r_1$, f – $fr_2\alpha/r_2$

A look in Table 1 shows that f_r (N–F) or f_r (N–Cl) decreases with increase in mass of the third atom. It is evident from the Table 2 that stretch force constants f_r(N–F) rather than bond distance r(N–F) vary consistently in radicals and molecules considered and the force constant f_r in general decreases with decrease in dissociation energy and hence force constant seems to be a correct guide for trends in (N–F) bond strength. The same type of systematic trend in force constants and anomalous behaviour of bond length are also observed by Politzer and Habibollahzadeh in (N–F) and (O–F) containing molecules. Our results also show that recently[7] proposed value (60 Kcal/mol) of dissociation energy for (N–F) bond in NFCl is in order. The bond order and bond charge also vary consistently and both changes in the same direction as dissociation energy. Mass of the atoms attached with (N–F) bond seems to affect the (N–F) bond strength. It is also observed from in Table 2 as the mass of attached atoms increases, the (N–F) bond strength decreases.

TABLE 2. Bond properties of some (N–F) containing radicals and molecules

Radical/ Molecule	Bond	Bond Dissociation Energy (Kcal/mol)	Bond Distance† r (Å)	Force constant f_r (mdyn/Å)	Bond order	Bond charge (Q)
NF	N–F	75.38 (D_0^0)	1.32	6.19	1.207	1.878
H_2NF	N–F	73.5 (De)	1.433†	4.71†	1.011	1.853
FNF	N–F	65.7(D_0^0)+	1.3994†	5.1 ± .5	1.105	1.791
HFNF	N–F	69.6 (De)*	1.400†	4.62†	1.013	1.772
ClNF	N–F	60.0 (Do)	1.37	4.26	0.983	1.672
F_2NF	N–F	58.0 (De)	1.371†	4.31†	0.988	1.659

(* Calculated at DF/GGA–DZVPP level, expected over estimated, + De > D_0^0 , † and dissociation energy – refs 6 and 9)

1. M.D. Harmony and R.J. Mayers *J. Chem. Phys.* **37** (1962) 636

2. C.K. Kohlmiller and L. Andrews *Inorg. Chem.* **21** (1982) 1519

3. Ann M. Zarubaiko and J. V. Gilbert *J. Phys. Chem.* **97** (1993) 4331

4. A.N. Pandey, D.K. Sharma, U.P. Verma, L.D. Arora, S.L. Gupta and B.P. Singh *Ind J Pure and Appl. Phys.* **14** (1976) 815 and references therein.

5. J.V. Gilbert and L.J. Smith *J. Phys. Chem.* **95** (1991) 7278

6. Peter Politzer and Dariush Habibollahzadeh *J. Chem. Phys.* **98** (1993) 7659 and references therein.

7. E. Arunan, C.P. Lin, D.W. Setser, J.V. Gilbert and R.D. Coombe. *J. Phy. Chem.* **98** (1994) 494

8. John C. Miller and L. Andrews *J. Chem. Phys.* **71** (1979) 5276

9. J. Berkowitz, J.P. Greene, K. Foropoulos Jr. and O.M. Neskovic *J. Chem. Phys.* **81** (1984) 6166.

CONSTRAINED FORCE FIELDS FOR TETRAHEDRAL IONS SiO_4^{4-}, GeO_4^{4-} AND SO_4^{2-}

A.K. Dublish[1] and Prem Singh[2] *

[1]Deptt. Physics, S.G.R.R. (PG) College, Dehradun - 248 001 India

[2]Opto-electronics Training Institute, Opto-electronics Factory, Raipur, Dehradun - 248 001 India

The force constants are a necessary part of the vibrational problem, reflecting as they do bonding characteristics of the molecules. In case of XY_4 (Td) system at least one additional constraint is required to supplement the vibrational frequencies for calculation of its symmetrized force constants. The aim of this communication is to develop two kinematic constraint methods (extended L-matrix and isotopic shift) for XY_4 system and apply them to SiO_4^{4-}, GeO_4^{4-} and SO_4^{2-}.

The force constant calculation of 2×2 F_2 symmetry block using extended L-matrix approximation[1] requires knowledge of off diagonal elements of L-matrix. The reliable force constants of XY_4(Td) type molecules are available[2-8] for twenty one molecules, which are used to obtain the ratio of off diagonal elements $K(L_{34}/L_{43})$. A graph is plotted between K and mass coupling term $T(G_{34}/\sqrt{|G|})$. A linear best fit is, then obtained giving the following equation

$$K = -0.0583\,T - 0.0741 \quad , \qquad (1)$$

It is possible to write F_2 species symmetrized force constants in terms of isotopic shift $\Delta\nu_3$. The relation[9] for $\Delta\nu_3$ for XY_4(Td) can be written as

$$\Delta\nu_3 = \frac{2\nu_3}{3m_x(\nu_3^2 - \nu_4^2)}\left[\frac{3\nu_4^2 m_y}{M} - \frac{F_{33} + 4F_{44} - 4F_{34}}{Cm_x}\right], \qquad (2)$$

where $C = 5.892 \times 10^{-7}$, m_x and m_y are the masses of X and Y atoms respectively (in amu) and $M = m_x + 4m_y$. $\Delta\nu_3$ can be obtained for molecules whose reliable force constants are available, then it is possible to express $\Delta\nu_3$ as

$$\Delta\nu_3 = -5.193\,\nu_3 m_x^{-1.873} \,, \qquad (3)$$

It is found that $\Delta\nu_3$ as observed spectroscopically be Handke[10] is very close to one predicted for SiO_4^{4-} using equation (3). A semi-empirical formula such as eqn. (3) might be useful for other XY_4(Td) system for which direct measurement of isotopic shift is difficult. In order to obtain reliable unperturbed molecular force fields for SiO_4^{4-} and GeO_4^{4-}, vibrational spectra[10] of Na_4SiO_4 and Na_4GeO_4 in water solution is employed. The symmetrized force constants are calculated using Wilson's FG matrix method[1] and using relations (1) and (3) and are summrized in Table 1. A number of workers[11,12] have reported symmetrized force constants for SO_4^{2-}. Our calculations support the results obtained by Müller *et al* [11]. It is interesting to note that Badger relation[13] for first row transition elements seems to give good results for GeO_4^{4-} ion. ($r_{exp.}$ = 1.84Å, r_{cal} = 1.81Å).

We have been able to develop two constraints for evaluating symmetrized force constants and a relation for isotopic shift for $XY_4(Td)$ molecules. The reliable and unperturbed symmetrized force constant for SO_4^{4-} and GeO_4^{4-} are reported for the first time. This study helps in chosing correct set of force constants for SO_4^{2-}.

TABLE 1. Symmetrized force constants (mdyn/Å).

Ion	F_{11}	F_{22}	F_{33}	F_{34}	F_{44}	Reference
SiO_4^{4-}	5.732	0.636	5.472	0.860	0.927	Ext. L-matrix approx.
			5.402	0.801	0.921	Isotopic shift
GeO_4^{4-}	4.749	0.342	4.063	0.238	0.421	Ext. L-matrix approx.
			4.011	0.061	0.418	Isotopic shift
SO_4^{2-}	9.07	0.64	7.73	0.92	0.99	3
	9.0729	0.6414	6.1581	0.1078	1.1393	4
	9.086	0.639	7.823	0.884	0.986	Ext. L-matrix approx.
			7.062	0.628	1.001	Isotopic shift

1. A. Müller, N. Mohan and U. Heidborn, Z Naturforsch A, **27** (1972) 129.
2. J.H. Carpenter, J.D. Muse, J.G. Smith S.W. Walters and D.H. Whiffen J. Mol Spectrosc, **118** (1986) 500.
3. K.A. Curtiss and V.A. Morine, Spectrochim Acta Part A, **41** (1985) 585.
4. N. Mohan, K.H. Schmidt and A. Müller, J. Mol Struct, **13** (1972) 155.
5. G. Cazzoli and W. Caminati, J Mol Spectrosc, **62** (1976) 1.
6. N. Mohan, S.J. Cyvin and A. Müller, Coord Chem Rev. **21** (1976) 236.
7. A. Müller and B. Krebs, J Mol Spectrosc, **26** (1968) 136.
8. I.W. Levin, J Chem Phys, **44** (1966) 2259.
9. R.S. McDowell, Spectrochim Acta Part A, **42** (1986) 1053.
10. M. Handke, J Mol Struct, **114** (1984) 187.
11. A. Müller, N. Weinotoch, N. Mohan and K. Nakamoto, Appl Spectrosc **27** (1973) 257.
12. S. Alvarez, V. Tabacik and J. Casabe, J. Mol Struct, **114** (1984) 293.
13. J. Hanuza, M. Baluka, and B. Jerowska-Trzebiatowska, Acta Phys Pol A, **42** (1972) 537.

MEAN AMPLITUDES OF VIBRATION OF SOME TETRAHEDRAL OXOANIONS

V.K. Rastogi[1], H.P. Mital[2]*, Sudhir Kumar[1]

[1]Physics Department, L.R. College, Sahibabad-201 005, India
[2]Physics Department, Meerut College, Meerut-250 001, India

Mean square amplitudes of vibration computed from spectroscopic method greatly improve the precision of structure determination by electron diffraction[1]. Recently, Lopez-Bste and Montero[2] have studied the Raman spectra of some oxoanions (SO_4^{2-}, ClO_4^-, PO_4^{-3}) in polycrystalline samples and found that fundamental wavenumbers of the internal modes are markedly influenced in different environments. Therefore, it was thought worthwhile to investigate the mean amplitudes of vibration for these oxoanions.

TABLE 1. Mean amplitude of vibration of tetrahedral oxoanions

Complex	Distance	Mean amplitude of vibration (in Å)		
		T = 0 K	T = 298.16 K	T = 500 K
Na_2SO_4	S – O	0.0378	0.0380	0.0394
	O···O	0.0530	0.0561	0.0625
K_2SO_4	S – O	0.0381	0.0383	0.0398
	O···O	0.0537	0.0566	0.0610
Rb_2SO_4	S – O	0.0383	0.0385	0.0400
	O···O	0.0539	0.0570	0.0636
Cs_2SO_4	S – O	0.0384	0.0387	0.0403
	O···O	0.0542	0.0572	0.0641
$(NH_4)_2SO_4$	S – O	0.0384	0.0386	0.0402
	O···O	0.0540	0.0569	0.0636
$SrSO_4$	S – O	0.0374	0.0376	0.0389
	O···O	0.0530	0.0557	0.0620
$BaSO_4$	S – O	0.0376	0.0378	0.0392
	O···O	0.0533	0.0561	0.0624
Ti_2SO_4	S – O	0.0386	0.0388	0.0405
	O···O	0.0547	0.0580	0.0651
$KClO_4$	Cl – O	0.0381	0.0384	0.0399
	O···O	0.0571	0.0616	0.0630
NH_4ClO_4	Cl – O	0.0383	0.0385	0.0401
	O···O	0.0538	0.0566	0.0631
Na_3PO_4	P – O	0.0396	0.0340	0.0419
	O···O	0.0518	0.0596	0.0673

The oxoanions studied in the present investigation belong to T_d symmetry point group. According to relevant symmetry and selection rules, this model gives rise to nine vibrational degrees of freedom, forming only four vibrational frequencies. In this model all the four frequencies are Raman active, whereas, the frequencies belonging to F_2 species are only IR active. The mean amplitudes of vibration for the system under investigation, have been computed at temperatures, T = 0 K, 298 K and 500 K for bonded as well as for non-bonded atoms pair following Cyvin's procedure[3,4] and the results are presented in Table 1. It is apparent from Table 1 that the mean amplitudes of vibration for bonded as well as for non-bonded distances are in increasing order with the increase in temperature as expected.

Further, one can easily see from Table 1 that the mean amplitudes increase slightly due to decreasing S–O force constants in series Na_2SO_4, K_2SO_4, $RbSO_4$ and CsO_4. Similar trend is also exhibited by the mean amplitudes in series $SrSO_4$ and $BaSO_4$. Inspite of minor variation in mean amplitudes, for isostructural sulphate ions, there is a great similarity of mean ampltiude values because of similar character of the bond.

On examining the trend in mean amplitudes of vibration in isoelectronic series : SiO_4^{4-} [Ref 5], PO_4^{3-}, SO_4^{2-} and ClO_4^-, we find that the respective values for bonded and non-bonded atoms pair (in Å) at room temperature are : (0.0437, 0.0658); (0.04, 0.0596); (0.0386, 0.0569); (0.0385, 0.056). It is evident that mean amplitudes decrease with increasing positive charge. This is because the masses are nearly the same, but the force constants increase with increasing positive charge.

$$(f_{SiO} = 4.18, f_{PO} = 6.830, f_{SO} = 7.797, f_{ClO} = 7.931).$$

Further, it is interesting to study the trend of mean amplitudes for bonded atoms pair among the following series :

SO_4^{2-} (0.0383), SO_3^{2-} (0.041), SO_3 (0.0349), SO_2 (0.03546),
ClO_4^- (0.0384), ClO_3^- (0.040), ClO_2 (0.0385), Cl_2O (0.0437)

(The mean amplitudes at room temperature are given in bracket in Å unit).

A survey of above results shows that the mean amplitudes decrease with the decrease of the oxidation state of the central atom in oxoanions and oxides. This trend is in accordance with the respective trend in bond stretching force constant and bond distances [6].

1. K. Kuchitsu and L.S. Bartell, J. Chem. Phys., 35, 1945 (1961).
2. M.A. Lopez-Bste and S. Montero, J. Raman Spectrosc., 9, 386 (1980).
3. S.J. Cyvin (Ed.), Molecular Structure and Vibrations, Elsevier Amsterdam (1972).
4. S.J. Cyvin, Spectrochim Acta, 17, 1219 (1961).
5. S.L. Gupta, U.P. Verma A.N. and Pandey, Acta Physica Polonica, A55, 153 (1979).
6. H.P. Mital, in spectral studies of some polyatomic molecules and biologically active compounds, Ph.D. Thesis, C.C.S. Univ., Meerut, India, 1995.

INTRAMOLECULAR FORCE FIELDS OF OCTAHEDRAL HEXAHALIDES OF THE ELEMENTS OF GROUPS III-A,IV-A AND V-A

A.N.Pandey, S.Bhardwaj* & Ekta Gupta

Molecular Spectroscopy Research Laboratory,
Department of Physics,
Meerut College, Meerut-250 001 (India)

Wilson GF matrix method [1] has been followed to investigate thoroughly the intramolecular force fields (GVFF, MUBFF & MOVFF) for the octahedral hexahalides of the elements of groups III-A, IV-A and V-A using latest vibrational data [3,4]. Labonville et al [5] and Pandey et al [6,9] have also applied the modified force fields for different octahedral hexahalogens species to study the trend among force constants. The secular determinant at F_{1u} species has been solved by L-F approximation method [2].

Keeping the oxidation state, stereochemistry, coordination number and central atom etc. constant, it can be demonstrated that as the mass of the legand increases, the bond stretching force constant fr (GVFF), K (MUBFF & MOVFF) and bending force constant $f\alpha$ - $f\alpha\alpha'''$ (GVFF) decreases with the corresponding increase in bond distance X-Y. A comparison of the effects of man made by keeping the halogen constant reveals that bond stretching and bending force constants (GVFF) decrease with the increase of mass of the central atom in case of hexafluoroanions. But its not true in other hexahalides. The bond stretching force constants K (MUBFF & MOVFF) show opposite trend in hexafluoroanions as compared to bond stretching force constant fr (GVFF), similar trend has also been noticed by others [5,6]. The intraction reveals that the frr (cis) > frr' (trans) in the hexahalides of the elements of groups III-A and IV-A. But this trend is not general in case of the halides of the elements of group V-A. The effect of oxidation state is exhibited by the hexahalides $SbCl_6^-$, $SbCl_6^{3-}$ and $SbBr_6^-$, $SbBr_6^{3-}$. The bond stretching and bending force constants decrease with the decrease of oxidation state of the central atom as expected. From this its inferred that relative strength of the chemical bond is in the sequence Sb(V) - Y (Y=Cl or Br) > Sb (III) - Y.*

Its informative to study the trend in bond stretching force constants in isoelectronic sequence in the following series. I. AlF_6^{3-} (1.920), SiF_6^{2-} (3.223), PF_6^-

(4.338), SF_6 (4.84) ; [7]. II. GaF_6^{3-} (1.801), GeF_6^{2-} (2.958), AsF_6^- (4.012), SeF_6 (4.97) ; [5]. III. InF_6^{3-} (1.723), SnF_6^{2-} (2.857), SbF_6^- (3.943), TeF_6(5.13) ; [5]. IV. $TlCl_6^{3-}$ (1.092), $PbCl_6^{2-}$ (1.126) ; V. $GeCl_6^{2-}$ (1.247), $AsCl_6^-$ (1.358) ; VI. $SnBr_6^{2-}$ (1.182), $SbBr_6^-$ (1.385) (values in Parenthesis are in 10^2 N/m unit). An examination of these values shows that the stretching force constant increases with the increase of the oxidation state of the central atom. Therefore, its concluded that the bond in higher oxidation state is more stable then the bond occurring in lower oxidation state.

A comparison of bond stretching force constants of the octahedral hexahalides with the corresponding values of the tetrahedral tetrahalides [8], leads us to conclude that the chemical bonds in tetrahalides are relatively stronger than in the respective bonds in hexahalides.

*The values of the force constants are available on request from authors.

1. Wilson,E.B.,Decius, J.C. & Cross, P.C., Molecular Vibrations, McGraw-Hill, New York, (1955).
2. Pandey, A.N.,Sharma, D.K.,Verma U.P.,Arora, L.D.,Singh,B.P. & Gupta, S.L.,Indian J.Pure Appl. Phys. 14, 815 (1976).
3. Nakamoto, K. "Infrared & Raman Spectra of Inorganic & Coordination compounds, John Wiley & Sons, New York (1978).
4. Geldard, J.F. & McDowell, H.K.Spectrochem. Acta 43A 439 (1978).
5. Labonaille, P.,Ferraro, J.R., Wall, M.C., S.M.C & Basile, L.,J. Coord.Chem. Rev.7, 257(1972).
6. Pandey, A.N.,Sharma, D.K. & Verma, U.P., Acta. Phys. Pol. A51, 475 (1977).
7. Pandey, A.N.,Sharma, D.K. & Kumar, V,Acta. Ciencia Indica 2, 63(1976).
8. Müller, A. & Krebs, B.,J. Mol. Spectry, 24, 180 (1967).
9. Pandey,A.N., Bhardwaj, S., Sharma, S.K. & Sharma, D.K., in Procd. XIV Int. Conf. on Raman Spectroscopy, Hong Kong, John Wiley and Sons (1994), P. 246.

SYMMETRY LOWERING OF NICKEL(II)OCTAETHYL 5,15-DI-*MESO*-NITRO-PORPHYRIN IN CS_2 DETECTED BY RESONANCE RAMAN SPECTROSCOPY

C. Lemke[1]*, J.A. Shelnutt[2], J.M.E. Quirke[3], R. Schweitzer-Stenner[1] and W. Dreybrodt[1]

[1]FB1-Institut für Experimentelle Physik, Universität Bremen, P.O. Box 330440, 28334 Bremen, Germany
[2]Fuel Science Department, Sandia National Laboratories, Albuquerque, NM 87185-0710, USA
[3]Department of Chemistry, Florida International University, Miami, Florida 33199, USA

In metalloporphyrins with ideal D_{4h}-symmetry, e.g. Ni(II)octaethylporphyrin, only vibrations with representations A_{1g}, A_{2g}, B_{1g} and B_{2g} are Raman active and the depolarization ratios (DPRs) ρ of the corresponding Raman lines are independent of the excitation wavelength: $\rho_{A_{1g}} = 1/8$, $\rho_{A_{2g}} = \infty$, $\rho_{B_{1g}} = \rho_{B_{2g}} = 3/4$ [1]. Symmetry lowering of the macrocycle by peripheral substituents causes distortions which can be classified by representations Γ_D. Their contribution to Raman scattering is accounted for by adding a term with representation $\Gamma_M = \Gamma_D \times \Gamma_R$ to the Raman tensor of the undistorted mode. This can cause dispersion of the DPR if $\Gamma_M \neq \Gamma_R$ and $\Gamma_M \in \{A_{1g}, A_{2g}, B_{1g}, B_{2g}\}$ [2].

Nickel(II)octaethyl 5,15-di-*meso*-nitroporphyrin (NiOE5,15DNP) shows D_{2d}-symmetry due to the two NO_2–groups, located at opposite meso-positions of the molecule. Accordingly this induces a B_{2g} distortion with respect to the D_{4h}-symmetry. Molecular mechanics calculations performed for this molecule [3] predict additional out-of-plane distortions of strong ruffling type (B_{1u}). These can affect DPRs only if they appear in pairs ($\Gamma_{D_1}, \Gamma_{D_2}$) with $\Gamma_{D_1} \times \Gamma_{D_2} = \Gamma_D$ [4]. Our analysis of the DPR-dispersion of structure sensitive Raman bands of NiOE5,15DNP [5] reveals, that Raman tensors $\Gamma_M = A_{2g}$ admix to modes of A_{1g} and B_{1g} representation, and correspondingly $\Gamma_M = A_{1g}$ to A_{2g} and B_{2g} vibrations. Fig. 1 represents the DPR-dispersion determined from the polarized excitation profiles of the A_{2g}-line ν_{19} and the B_{1g}-line ν_{10} and the corresponding unpolarized excitation profiles. The DPR of the ν_{19} varies between 1/8 to ∞. In the B-band region its value is close to 1/8, below 3/4, indicating admixture of A_{1g}-tensors. The DPR of the ν_{10} rises from values of 3/4 in the B-band region up to 3 in the Q-band. This indicates to an A_{2g} contribution to its Raman tensor. From the above group multiplication one finds, that only in plane distortions of type A_{2g} and B_{2g}, or out-of-plane distortions of type A_{1u}, A_{2u}, B_{1u} and B_{2u} are present. Our data of the Raman bands ν_{11}, ν_{19}, ν_2 and ν_{10} from different conformers of NiOE5,15DNP [5], yield qualitatively identical results for each conformer, though the deviations from the corresponding D_{4h}-values differ in quantitative terms. The above DPR dispersions are interpreted in terms of in plane and out-of-plane distortions, namely B_{2g} distortions due to the electron withdrawing capacity of the two NO_2-groups and a mixing of ruffling and saddling ($B_{1u} \times B_{2u} = A_{2g}$) as induced by steric interactions between the substituents.

The DPRs of all analysed Raman lines show large dispersion for all conformers. The lower frequency mode of each Raman line exhibits an excitation profile red shifted [5] with respect to the higher frequency line. Taking together our results suggest, that the porphyrin macrocycle is subject to strong out-of-plane perturbations and both conformers are nonplanar.

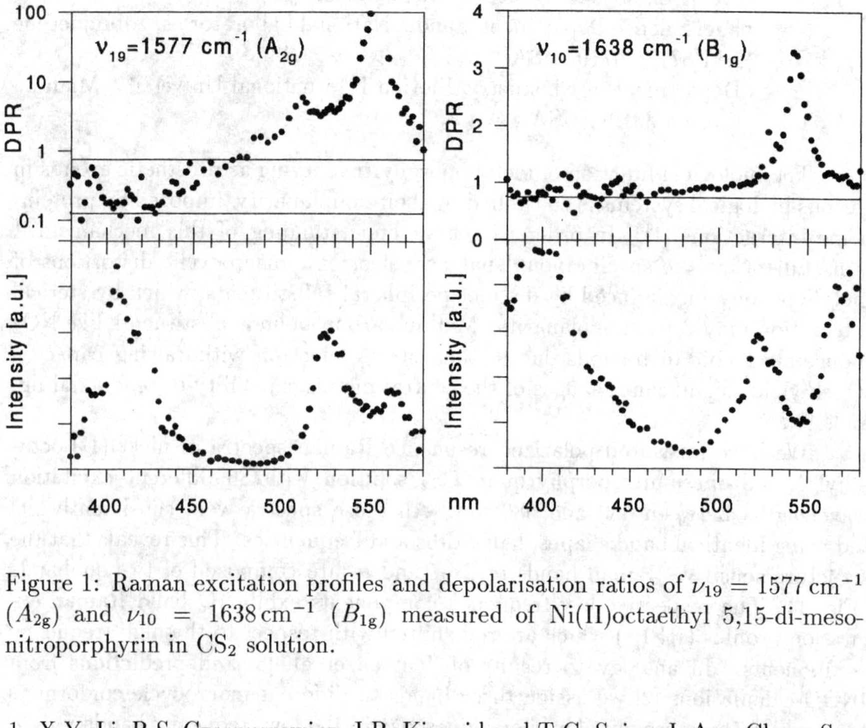

Figure 1: Raman excitation profiles and depolarisation ratios of $\nu_{19} = 1577\,\text{cm}^{-1}$ (A_{2g}) and $\nu_{10} = 1638\,\text{cm}^{-1}$ (B_{1g}) measured of Ni(II)octaethyl 5,15-di-meso-nitroporphyrin in CS_2 solution.

1. X.Y. Li, R.S. Czernuszewicz, J.R. Kincaid and T.G. Spiro, J. Am. Chem. Soc. **111**, 7012 (1989)
2. R. Schweitzer-Stenner, Q. Rev. Biophys. **22**, 381 (1989)
3. J.D. Hobbs, S.A. Majumder, L. Luo, G.A. Sickelsmith, J.M.E. Quirke, C.J. Medforth, K.M. Smith and J.A. Shelnutt, J. Am. Chem. Soc. **116**, 3261 (1994)
4. R. Schweitzer-Stenner, A. Stichternath, W. Dreybrodt, X. Song, J.A. Shelnutt, C.J. Medforth and K. Smith, this volume
5. C. Lemke, J.A. Shelnutt, J.M.E. Quirke, R. Schweitzer-Stenner and W. Dreybrodt, this volume

CONFORMATIONAL HETEROGENEITY OF NICKEL(II)OCTAETHYL 5,15-DI-*MESO*-NITRO-PORPHYRIN IN CS_2 DETECTED BY RESONANCE RAMAN SPECTROSCOPY

C. Lemke[1]*, J.A. Shelnutt[2], J.M.E. Quirke[3], R. Schweitzer-Stenner[1] and W. Dreybrodt[1]

[1]FB1-Institut für Experimentelle Physik, Universität Bremen, P.O. Box 330440, 28334 Bremen, Germany
[2]Fuel Science Department, Sandia National Laboratories, Albuquerque, NM 87185-0710, USA
[3]Department of Chemistry, Florida International University, Miami, Florida 33199, USA

The biological function of metalloporphyrins serving as prosthetic groups in various biological systems is controlled by their nonplanarity, imposed by protein-heme interactions [1]. In order to get an understanding of this mechanism it is useful to analyse specific nonplanar and electronic macrocyclic distortions of metalloporphyrins, induced by distinct peripheral substituents, which by sterical interaction may cause nonplanarity. Methin carbon bound substituents like NO_2 induce electronic distortions due to their strong electron withdrawing capacity, which yields significant changes of the central metals capability to bind axial ligands [2].

We have measured polarized resonance Raman spectra of nickel(II) octaethyl 5,15-di-*meso*-nitroporphyrin in CS_2 solution with 86 different excitation wavelengths between 380 and 580 nm. All these spectra were consistently fitted using identical band shapes, halfwidths and frequencies. This reveals that the structure sensitive Raman bands ν_{11}, ν_{19} and ν_2 are composed of two subbands (Fig. 1). The respective low frequency components exhibit Q band Raman excitation profiles (REPs) which are red shifted with respect to the high frequency components. In analogy to results of Jentzen et al. [3] and predictions from INDO-calculations [2] we assign these bands to different macrocycle conformers from which the more nonplanar one gives rise to the low frequency bands.

The above analysis of the data clearly revealed that the ν_{10} band is composed of four sublines with similar though not identical depolarization ratios [4] (i.e. $0.75 < \rho < 5$). The lower frequency modes $1631\,cm^{-1}$ and $1638\,cm^{-1}$ exhibit similar excitation profiles red shifted with respect to the higher frequency modes at $1646\,cm^{-1}$ and $1654\,cm^{-1}$ which also show similar REPs (Fig. 1). These findings indicate that multiple conformers of the macrocycle coexist in solution, in accordance with predictions from molecular mechanics calculations [2]. An alternative explanation of the larger multiplicity of ν_{10} may tentatively invoke a Fermi resonance between a fundamental and a less intense combination tone, for which possible combinations can be inferred from Ni(II)octaethylporphyrin spectra [5]. To clearify this issue, further experiments are nessessary.

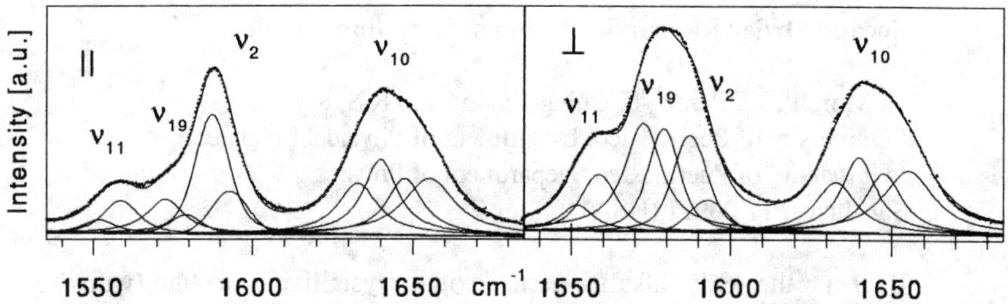

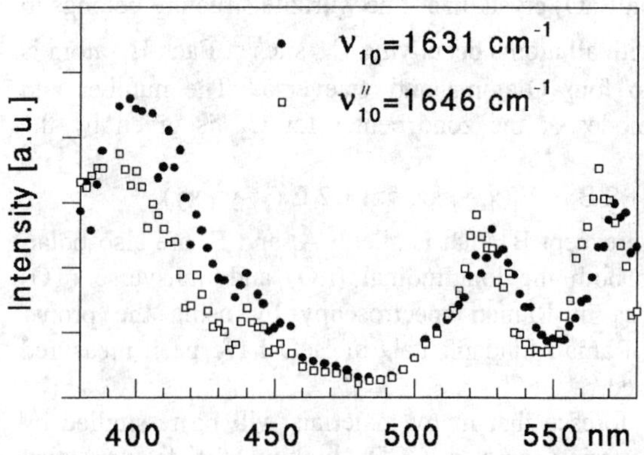

ν_{11}	1550
ν'_{11}	1557
ν_{19}	1571
ν'_{19}	1577
ν_2	1586
ν'_2	1591
ν_{10}	1631
ν'_{10}	1638
ν''_{10}	1646
ν'''_{10}	1654

Figure 1: Above: Polarized Raman spectra of Ni(II)octaethyl 5,15-di-*meso*-nitroporphyrin measured with 514.5 nm excitation wavelength. Polarization (⊥ and ||) is indicated with respect to the laser polarization. The data are represented by dots and the fit by solid lines. The plotted bands result from the spectral analysis. Their frequency positions are listed in the table below in cm^{-1}. Bottom: Raman excitation profiles of ν_{10} and ν''_{10}.

1. J.D. Hobbs and J.A. Shelnutt, J. Protein Chem. **14**, 19 (1995)
2. J.D. Hobbs, J.M.E. Quirke, L. Luo, G.A. Sickelsmith, J.M.E. Quirke, C.J. Medforth, K.M. Smith and J.A. Shelnutt, J. Am. Chem. Soc. **116**, 3261 (1994)
3. W. Jentzen, E. Unger, G. Karvounis, J.A. Shelnutt, W. Dreybrodt, and R. Schweitzer-Stenner, submitted for publication
4. C. Lemke, J.A. Shelnutt, C.J. Medforth, R. Schweitzer-Stenner and W. Dreybrodt, this volume
5. X.Y. Li, R.S. Czernuszewicz, J.R. Kincaid, P. Stein and T.G. Spiro, J. Phys. Chem. **94**, 47 (1990)

Second Order Raman Scattering of Beryllium Oxide

G. Morell[1], W. Pérez[2], E. Ching-Prado[2], and R.S. Katiyar[2]
[1] University of Puerto Rico, Department of Physical Sciences
[2] University of Puerto Rico, Department of Physics
San Juan, PR 00931 (USA)

BeO is the only alkaline-earth oxide crystallizing in the wurtzite structure, instead of the NaCl-type structure in which the other members of this family (MgO, CaO, SrO, and BaO) crystallize. The wurtzite structure belongs to space group C^4_{6v} ($P6_3mc$) with all atoms occupying C_{3v} sites[1,2]. Each Be-atom is tetrahedrally coordinated to four O-atoms and viceversa.[1] The number and symmetry of the optical modes at the zone center for C_{6v} is given by the representation:[3]

$$\Gamma = A_1(z, z^2, x^2+y^2) + 2\,B_1 + E_1(x, y, xz, yz) + 2\,E_2(x^2-y^2, xy).$$

All modes are Raman-active except B_1, that is silent. A_1 and E_1 are also polar. One can therefore observe both the longitudinal (LO) and transverse (TO) components of these modes in Raman spectroscopy by using the proper scattering geometry.[2,3] All Raman fundamentals of BeO have been measured and their polarization analyzed.[2,3,4]

It is not difficult to foresee that many materials will be re-studied by detecting their Raman scattering onto a CCD, looking for features that photomultipliers and diode-arrays might have left undiscovered. In the case of BeO, this is the first time that its second order Raman spectra are observed. Previous researchers reported the Raman spectra of BeO to be "clean" above 1100 cm^{-1}.[2,4] Infrared spectroscopy was however able to detect second order vibrational modes.[4] Here is where a serious inconsistency existed until our measurements. According to group theory, all infrared-active modes are also Raman-active. A second-order mode can be infrared-active only if the reduced representation of the product of the two (or more) modes that are combining contains at least one infrared-active species.[5] In wurtzite (C_{6v}) materials, this condition automatically guaranties that all infrared-active second orders are also Raman-active. In fact, two of us had unsuccessfully searched for the second order Raman spectra of BeO before utilizing both a photomultiplier and a diode-array detector.[3] In addition to the infrared second-order spectra, our analysis of the anharmonic behavior of the A_1(LO) mode indicated that second orders should be present.[6]

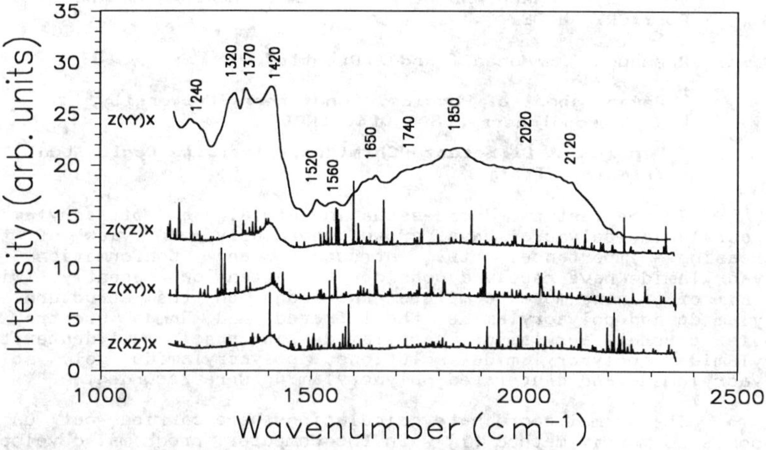

Figure 1. Polarized second-order Raman spectra of BeO. The scattering geometry is indicated on the left side of each spectrum.

The second-order Raman spectra of BeO are shown in Figure 1. Four different polarization geometries were employed, but only the yy polarized spectrum shows well defined features up to 2400 cm^{-1}. The positions of bands and shoulders are annotated on top. They generally agree with the infrared measurements of Loh[4], especially when taking into account the broad nature of second-order scattering. We did not however observe features at around 3000 cm^{-1} corresponding to four-phonon interactions. If present, these signals would be too weak to be detected.

This work was supported by NSF Grant OSR-9452893.

1. G. Vidal-Valat, J.P. Vidal, K. Kurki-Suonio, and R. Kurki-Suonio, Acta Cryst A43, 540 (1987)
2. C.A. Arguello, D.L. Rousseau, SPS Porto, Physical Review 181, 1351 (1969)
3. S Devanarayanan, G Morell, R.S. Katiyar, J Raman Spec 22, 311 (1991)
4. E. Loh, Physical Review 166, 673 (1968)
5. A. Anderson, The Raman Effect, Volume I, Marcel Dekker (1971), p.118
6. G. Morell, W. Pérez, E. Ching-Prado, and R.S. Katiyar, Physical Review B, to be published on March 1996

POLARISED RAMAN AND FTIR SPECTRA OF ACRYLAMIDE AND POLYACRYLAMIDE

R. Murugan[1], S. Mohan[1*] and A. Bigotto[2]

[1] Raman School of Physics, Pondicherry University, Pondicherry, 605 014, INDIA

[2] Department Di Scienze Chimiche, Universita Degli Studi Di Trieste, Italy.

In the last two decades the industrial use of acrylamide and acrylamide polymers have grown enormously. In view of the increasing importance, the structure and conformation of polyacrylamide have received considerable attention recently. With the aim of gaining more complete knowledge on the structure of acrylamide and polyacrylamide, the infrared and Raman spectra of single crystal acrylamide, acrylamide in nujol mull, deuterated acrylamide, polyacrylamide solutions, polyacrylamide gel, solid polyacrylamide and deuterated polyacrylamide were recorded.

The normal coordinate calculations were carried out using wilson's FG matrix method [1] with the computer programs developed by Schachtschneider[2]. Crystals of acrylamide are monoclinic, space group $P_{21}/c[C_{2h}^{5}]$, with four molecules in unit cell. The selection rule predicts that each molecular fundamental should give four components in the crystal spectrum, two being infrared active and two Raman active. A general discussion of the assignments are given on the basis of the dichroism of the infrared bands and of Raman depolarization data. The fit between the calculated and observed frequencies is good and the experimental facts are satisfactorily accounted for by the force field, indicating that the force constants obtained are reasonable. The vibrational mean amplitudes evaluated at 298.6K for the bonds C=O, C-N, C=C, C-C, C-H, NH_2 and CH_2 are 4.302, 4.611, 4.198, 4.805, 8.117, 7.581 and 7.747 X 10^{-2} A^0 respectively.

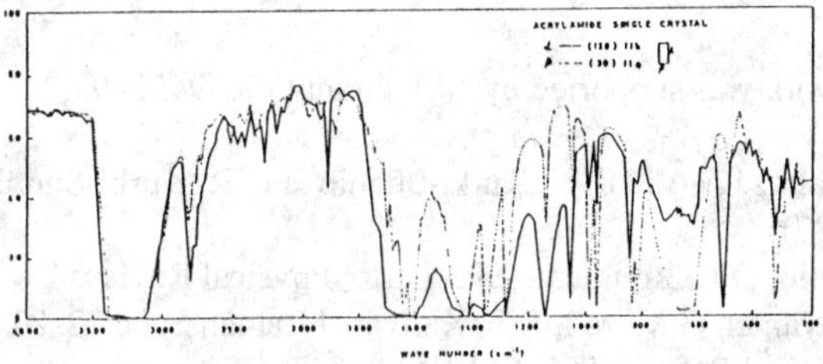

Fig. 1. Infrared spectrum of single crystal acrylamide

Based on the preferred chain confirmation of 3_1 isotactic helical structure,the 86 optically active normal vibrations of polyacrylamide are classified under the line group isomorphous to the point group C_3 and are both Raman and infrared active. The 28 A mode exhibit parallel dichroism in the infrared while the 29 doubly degenerate E modes exhibit perpendicular dichroism. No remarkable differences are observed in the Raman spectra of solid polyacrylamide and the aqueous solutions, but in gel phase, one new band appears around 1295 cm^{-1}. The methylene group vibrations are useful to monitor the extent of polymerization. The lowering of CH_2 stretching vibrations in polyacrylamide compared to monomer acrylamide is mainly due to the unsaturated linkage in monomer acrylamide. The vibrational modes of amide groups are considerably affected by the involvement of these groups in hydrogen bonding. The lowering of C-O stretching frequency in solid polyacrylamide compared to solutions indicates considerable polymer -polymer hydrogen bonding in the solid state. The assignments for the infrared and Raman spectra of polyacrylamide ,its aqueous solution and its covalently cross linked gel in the region 4000-200 cm^{-1} are discussed on the basis of a comparison with spectra of related polymer structure and acrylamide.

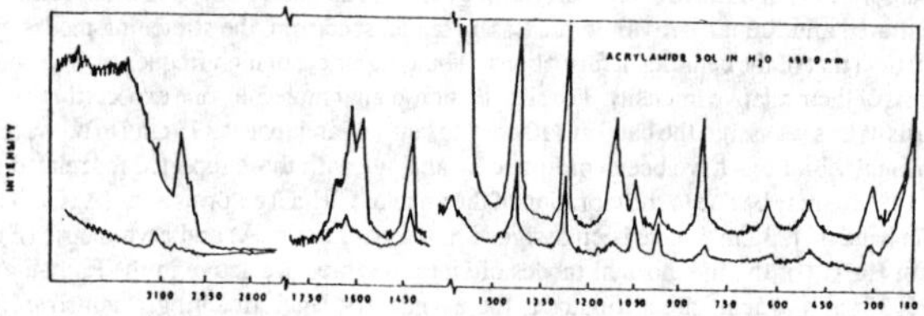

Fig. 2. Polarised infrared spectrum of acrylamide

1. E.B. Wilson, J.Chem.Phy., 7,1047(1941)
2. J.H.Schachtschneider, Vibrational analysis of Polyatomic molecules, Parts V and VI, Shell development, Emeryville, CA 1964 and 1965 ,Technical reports Nos. 231 and 57.

INFRARED AND LASER RAMAN SPECTROSCOPIC INVESTIGATIONS AND VIBRATIONAL ANALYSIS OF TRICHLOROMERCURY (II) AND MERCURY (I) CHLORIDE.

S. GUNASEKARAN [1] AND USHA DESAI [2]

1. PG and Research Department of Physics, Pachaiyappa's College Madras - 30, INDIA.
2. Department of Physics, Women's Christian College, Madras - 6, INDIA.

Mercury (I) Chloride and Trichloromercury (II) are known for their medicinal use but because of their contamination by the more soluble and poisonous $HgCl_2$, they are highly hazardous to life. Hg_2Cl_2 (colomel) and $HgCl_3$ have been known for a long time, however a complete spectral data has not been reported till now. This paper presents the Laser Raman and Infrared Spectra of Hg_2Cl_2 and $Hgcl_3$. $HgCl_2$ belongs to the point group of $D_{\infty h}$ with the seven fundamental modes of vibration represented by $\Gamma vib = 2\ \Sigma\ g + 1\pi u^{+} + 1\pi g + 1\pi u$ and $HgCl_3$ possessing a pyramidal c_{3v} symmetry with $\Gamma vib\ =\ 2A_1 + 2E$.

Spectroscopically pure sample of Hg_2Cl_2 and $HgCl_3$ were procured from SIGMA chemical company, U.S.A. The Infrared spctra were recorded on Perkin - Elmer double beam spectrometer, model - 983 in the region 4000-200 cm $^{-1}$, and the Laser Raman Spectra were recorded on DILOR-Z24 over the region 500-10 cm $^{-1}$. The spectra of these compounds are presented in figures 1 and 2 and the observed frequencies along with their assisgnment are summarised in Table 1 respectively. Four fundamental absorption bands have been observed at 375,310, 123 and 70 cm $^{-1}$. in the spectrum of $HgCl_3$. All the four frequencies are infrared and Raman active. In the Laser Raman spectrum, the stretching modes are expected to occur at frequencies higher than that of the bending deformations. The assignment of high frequency fundamentals (stretching) can be made on the basis of their relative intensity. For slightly non-planar molecule, one expects that ν_3 mode be more intense than ν_1. On the basis of this we assign the band at 310 cm $^{-1}$ to be ν_1 (A) and that at 375 cm $^{-1}$ to be ν_3 (E). Further the assignment of the fundamental vibrations have been confirmed by analogy with those reported for related molecules. [1] The bands at 70cm $^{-1}$ and 123cm $^{-1}$ may be due to the vibration of the angular ClHgCl deformations (A_1,E). The strong band at 70cm $^{-1}$ and the medium band at 123 cm $^{-1}$ have been assigned to the symmetric (A) and asymmetric (E) deformations respectively. In the case of Hg_2Cl_2 of the five normal modes of vibration three are active in the Raman and two in the infrared. The bands at 425, 354 and 275cm $^{-1}$ are attributed to the asymmetric Hgcl stretching, symmetric HgCl stretching and HgHg stretching respectively. The low frequency bands at 47 and 32 cm $^{-1}$ may be attributed to asymmetric and symmetric bending modes of the molecule [2]

A normal coordinate analysis has been carried out using Wilsons F.G. matrix method and a set of molecular constants have also been evaluated Table 2. In the case of $Hgcl_3$ the interaction force constants assume negative values. The force constant fd is in good agreement with literature. The mean amplitudes of vibration for both bonded and non - bonded distances are found in the characteristic range [4]. It is observes that $l_{cl\cdots cl}$ is > $l_{Hg\text{-}cl}$ and this in agreement with literature [5]. A similar observation can be made on the molecule Hg_2Cl_2.

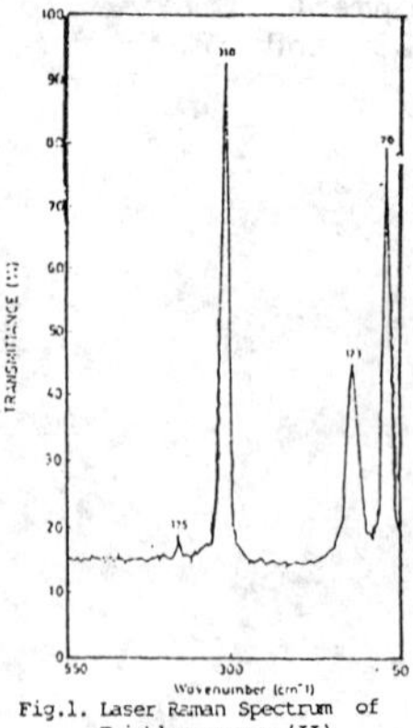

Fig.1. Laser Raman Spectrum of Trichloromercury(II)

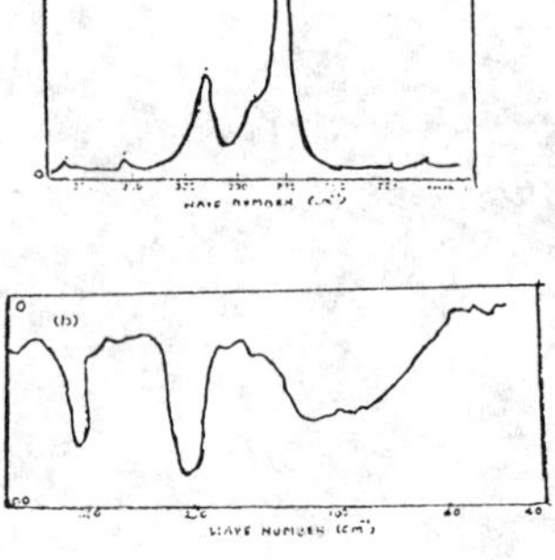

Fig.2.(a) Laser Raman and (b) Infrared Spectra of Mercury (I) Chloride.

TABLE 1

The fundamental frequencies of $HgCl_3$ and Hg_2Cl_2.

$HgCl_3$		Hg_2Cl_2
Frequency cm^{-1}	Assignment	Frequency cm^{-1} assignment
70 S	ClHgCl Sym. Bending	32 vw Sym. Bending
123 m	ClHgCl Asym.Bending	47 vw Asym. Bending
310 s	HgCl Sym. Stretching	275 w HgCl Stretching
375 w	HgCl Asym. Stretching	353 vw HgHg Stretching
		425 s HgCl Asym. Stretching.

TABLE 2 MOLECULAR CONSTANTS

	Force Constants			Mean Amplitudes of Vibration
HgCl3	fd fdd 1.5432 -0.2701	fα fdα 0.0280 -0.0040	fαα fd -0.0091 0.0062	l Hg-Cl= 4.859 l Cl... Cl= 18.795
Hg2Cl2	fd fαα 2.9079 0.0025	fD fdd 5.2583 -0.2985	fα fdD 0.0043 0.8253	l Hg-Cl = 4.4670 l Hg-Hg = 3.1584

1. S. Mohan, K.G. Ravikumar, and S. Gunasekaran. Acts Chimica Acadomide (Hung)III, 59 (1982) .
2. S. Mohan, S. Gunasekaran, and R. Venkatesan, Proc, II mtl.conf. on vibl. Spectroscopy (1985).
3. S.J. Cyvin, Molecular vibrations and mean square amplitudes, Elsevier Publishing Co, Amsterdam (1968).
4. J.C. Decius, J. Chem Phys 38, 241 (1963).
5. N.K. Sanyal, R.K. Goel and A.N. Pandey Indian . J. Phys 49, 546 (1975).

SECTION 6

BAND SHAPES / BAND INTENSITIES

RAMAN SPECTROSCOPY: A STRUCTURAL PROBE FOR SOLUTE-SOLVENT INTERACTION.

Mitambo M. Musilli and Glen R. Loppnow*

Department of Chemistry, University of Alberta, Edmonton, Alberta, Canada, T6G 2G2.

Research on covalently-linked donor-acceptor molecules has been quite intensive since the discovery of dual fluorescence in these molecules. Grabowski and co-workers [1] explained the dual fluorescence as resulting from a twisted intramolecular charge transfer (TICT) state. The development of the concept of TICT states has been reviewed in depth by Rettig [2]. Molecules with TICT states have been used extensively to characterize solvent-solute interactions, principally through the use of time-resolved fluorescence spectroscopy [3]. However, fluorescence spectroscopy doesn't allow the characterization of structurally-specific interactions between solute and solvent.

Raman spectroscopy is a structurally-specific probe of the molecular motions that are sensitive to the solvent environment. Raman spectra of ethyl 4-dimethylaminobenzoate (DMABEE) were obtained in 17 solvents with a range of solvent parameters, such as dielectric constant, refractive index, and donor and acceptor numbers [4,5]. These solvents also have a range of values on the π^* and $E_T(30)$ scales [6,7]. Typical Raman spectra of DMABEE in the C=O stretch region are shown in Fig. 1.

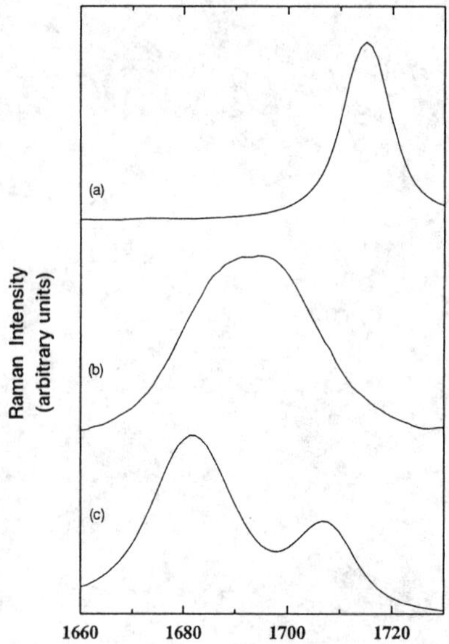

Fig. 1-The DMABEE C=O stretching vibrational mode in (a) hexane, (b) choroform, and (c) methanol.

Solvent effects on the Raman spectra were observed in three modes, localized at the ester end of the molecule. The frequencies of the C=O stretch, C-O stretch, and OCO bend were found to shift systematically with increasing solvent acceptor number (Fig. 2); the C=O stretch frequency decreases while both the C-O stretch and OCO bend frequencies increase. Additionally, the C-O and C=O stretches broaden as the acceptor number increases, with the C=O splitting into two peaks in the alcohol solvents.

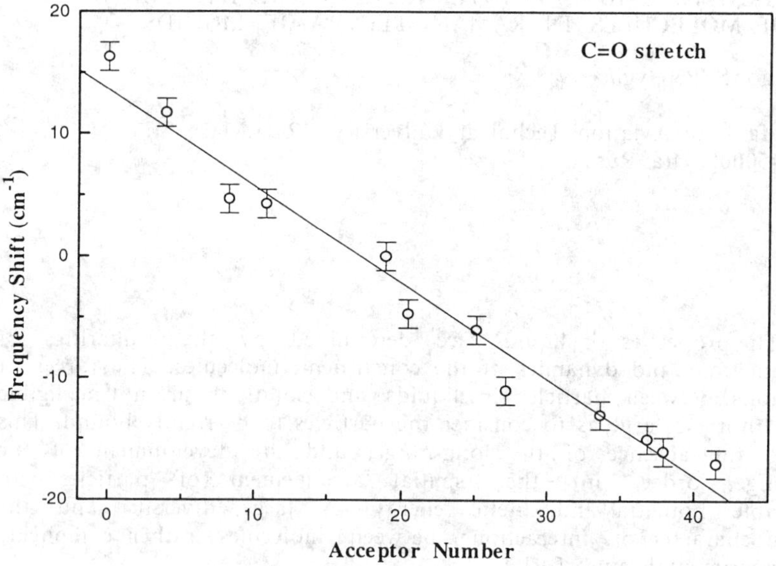

Fig. 2-Correlation of the C=O stretch frequency shift with solvent acceptor number.

A molecular model of DMABEE-solvent interaction is proposed to explain the observed behavior. The interaction between the solvent (electron acceptor) and the DMABEE molecule (electron donor) causes electron density to shift from the carbonyl double bond to solvent, as the acceptor number increases. Strengthening of the C-O bond also occurs as a result of this polarization. This shift results in the C=O bond becoming more single-bond-like and the C-O bond becoming more double-bond-like, resulting in the C=O stretching frequency to shift to lower wavenumbers and the C-O stretching frequency to shift to higher frequencies. These results indicate the strong nature of the solvent-solute interactions in this system and are discussed in terms of Fermi resonance and hydrogen bonding.

1. Z. R. Grabowski, et al., Nouv. J. Chim. 3, 443 (1987).
2. W. Rettig, Angew. Chem. Int. Ed. Engl. 25, 971 (1986).
3. C. F. Chapman, et al., J. Phys. Chem. 99, 4811 (1995).
4. V. Gutmann, The Donor-Acceptor Approach to Molecular Interactions, Plenum, New York, 1978.
5. F. L. Riddle and F. M. Fowkes, J. Am. Chem. Soc. 112, 3259 (1990).
6. J. M. Kamlet, et al., J. Am. Chem. Soc. 99, 6027 (1977).
7. K. Dimroth, et al., Liebigs Ann. Chem. 661, 1 (1963).

MANIFESTATIONS OF INTERACTIONS AND DYNAMICS OF MOLECULES IN RAMAN SPECTRA OF LIQUIDS

Igor S. Perelygin

Ufa State Aviation Technical University, 12 K.Marx Str., 450000, Ufa, Russia

The properties of liquids are determined by the peculiarities of the interactions and dynamics of the constituent molecules. The forces of interactions between particles in liquids and solutions are not negligible and yet not so large as to consider the particles to be rigidly bound. This leads to the absence of the long-range, and the development of the short-range, order in the spatial arrangement of particles with comparable potential and kinetic energy. A large diversity and the dynamic character of interactions between molecules rather complicate their experimental study [1].

Vibrational spectra of molecules are characterized by the maxima positions of bands, their intensities and shapes. Studying a change taking place in these parameters under the influence of intermolecular interactions, one can obtain information on the structure of liquids and the presence of relatively long-lived complexes and associates in them [2 - 5], and also investigate molecular dynamics. Vibrational spectra bandshapes of liquids contain extensive information on molecular processes: peculiarities of association, vibrational and orientational relaxation.

Raman spectroscopy has a certain advantage in investigation of these processes: in many cases Raman spectra are less complicated, their band intensities are in proportion to the concentration of scattering molecules, regardless of the type of the associate, while polarization measurements provide additional data on the objects involved. The method of time correlation functions, which play the role comparable to that of distribution functions in statistical physics, allows one to follow the molecular dynamics at length. Until recently, the Raman investigation of molecular dynamics in liquids was confined to the systems wherein intermolecular interactions do not produce significant disturbing effect upon intramolecular vibrations. In real liquids, the dynamics of vibrational as well as orientational motion of molecules is disturbed greatly by intermolecular interactions and environmental inhomogeneity. Such systems, however, have been studied to a very small extent only [6].

Association of polar aprotic molecules of the liquids without well-defined intermolecular hydrogen bonds shows up in the complex Raman structure of the vibrational band contours for the most polar bond in that molecule, through the atoms of which the intermolecular interaction preferentially occurs. The procedure is described for the revealing of this structure and the separation of components pertaining to each type of the associate.

The form of Raman bandshapes of monomeric or associated molecules in liquids is determined by the vibrational and orientational relaxation processes. The isotropic scattering spectrum is related to the isotropic polarizability only and thus does not depend on the orientational molecular motion conditions. The anisotropic scattering spectrum is related to both the vibrational and orientational motion. This provides the possibility to separate the effects due to the vibrational and orientational motion of molecules [7, 8].

The extraction of information on the dynamics of molecules from the spectral band contour shape involves certain difficulties. associated with the acquisition of the real spectral band contours. The most serious of them is the overlapping of spectral bands. It is therefore first necessary to separate the individual components from the complex spectrum, and determine their contours, which then provides access to the information on molecular dynamics. In the component analysis of the spectrum no firm restrictions on the component contour shapes are allowed, as this would predetermine the result. In our study, the shape of individual components was approximated by the symmetric curves of the Gaussian or Lorentzian function product shape. Such a curve does not impose firm restrictions on the band contour shapes so that one may use the analytical form of integral Fourier transformations that are necessary for calculating the molecular dynamics parameters.

The examples are given of using bandshapes of the Raman spectra of dipolar aprotic solvents for quantitative determination of the relative content of molecules as well as for studying mechanisms and parameters of their vibrational and orientational relaxation in monomeric state and in different types of associates.

1. I.S.Perelygin, in Ionic Solvation. Ellis Horwood, New York, 1994, p. 100.
2. I.S.Perelygin, I.G.Itkulov, Russ. J. Phys. Chem. 65 1064 (1991).
3. I.S.Perelygin, I.G.Itkulov, A.S. Krauze, Russ.J. Phys.Chem. 66 302 (1992).
4. I.S.Perelygin, A.S.Krauze, I.G.Itkulov, D.Makkambaev, Russ. J.Phys.Chem. 66 1578 (1992).
5. I.S.Perelygin, I.G.Itkulov, A.S.Krauze, Russ. J. Phys. Chem. 68 311 (1993).
6. Z.Kecki, J.Mol. Struct. 265 1 (1992).
7. I.S.Perelygin, A.S.Krauze, I.G.Itkulov. Sov. J. Chem. Phys. 12 573 (1994)
8. I.S.Perelygin, I.G.Itkulov, Chem. Phys. Rep. 13 326 (1994).

CONCENTRATION DEPENDENT RAMAN STUDY OF THE MIXTURE OF C_6H_5Cl + CH_3OH

Ranjan K. Singh*[1], B.P. Asthana[2] and A.L. Verma[1]

[1]Department of Physics, North Eastern Hill University, Permanent Campus, Shillong - 793022, India.

[2]Laser and Spectroscopy Laboratory, Department of Physics, Banaras Hindu University, Varanasi-221005, India.

A systematic study of the inter/intra-molecular interactions in condensed phase can be made through spontaneous Raman spectroscopy by analysing the isotropic component of the Raman spectra which has contribution purely from vibrational dephasing. A pretty good account of studies on vibrational dephasing in liquids and liquid mixtures has been presented in recent literatures [1-2].

The isotropic components of the Raman spectra of C_6H_5Cl + CH_3OH system were recorded after making a linear combination, $I_{iso} = I_{\parallel} - 4/3\ I_{\perp}$, in the frequency ranges, 680-730 cm^{-1}, 980-1060 cm^{-1} and 3050-3090 cm^{-1} on a Spex Ramalog 1403 spectrometer using 488.0 nm line of Ar^+- laser as excitation source at several concentrations with the varying mole fraction of C_6H_5Cl in the mixture, C = 0.05, 0.1, 0.2, 0.3, 0.5, 0.7, and 0.9.

A plot of the true Raman linewidths (FWHM), Γ_L, obtained after correction for the instrumental slit effect [3], versus concentration, C in mole fraction of C_6H_5Cl has been shown in Fig. 1 for the vibrational bands at ~1000 and ~1021 cm^{-1}. It is immediately obvious that the plots for both the bands as shown in Fig. 1, depict a minimum at C = 0.7. A similar effect has already been observed in our earlier study [4-5] of 3Cl-pyridine on dilution with methanol. The variation of the true Raman linewidth with concentration can be explained in terms of two competing phenomena, namely diffusion and motional narrowing. The process of diffusion causes the vibrational modes in question for the reference molecule, C_6H_5Cl to dephase at a faster rate due to increased collision between the reference and the solvent molecules. This phenomenon obviously leads to an increase in the linewidth. On the other hand, motional narrowing [6] arises from the fact that the environment around a molecule changes during the vibrational lifetime so rapidly that all the vibrators see nearly the same mean surrounding, thereby increasing the dephasing time of the ensemble, which, in effect, decreases the linewidth. At a concentration around C=0.7, it probably so happens that the effect of motional narrowing dominates. On further increasing the concentration of the solvent CH_3OH, the linewidth starts increasing due to the dominant effect of diffusion. Another minimum around C=0.2 occurs in the plot for the 1000 cm^{-1} band (see Fig.1) which may probably be caused due to yet another species, $C_6H_5Cl.....(HOCH_3)_2$ being formed at the higher concentrations of methanol.

The effect of motional narrowing is more clearly seen in the band at ~3069 cm^{-1}, which also shows an almost linear concentration dependence of the frequency shift of ~3.5 cm^{-1} towards the higher wavenumber in going from pure liquid to very high degree of dilution. The linewidth (FWHM) shows a steady decrease of ~2 cm^{-1} in going from C = 0.9 to C = 0.1. The band at ~700 cm^{-1}, however, shows only very small frequency shifts and the linewidth changes. Although the changes in linewidth and frequency for this mode are small, they may be explained using the concentration fluctuation model of Bondarev and Mardaeva [7]. A detailed study to offer some quantitative explanation is in progress.

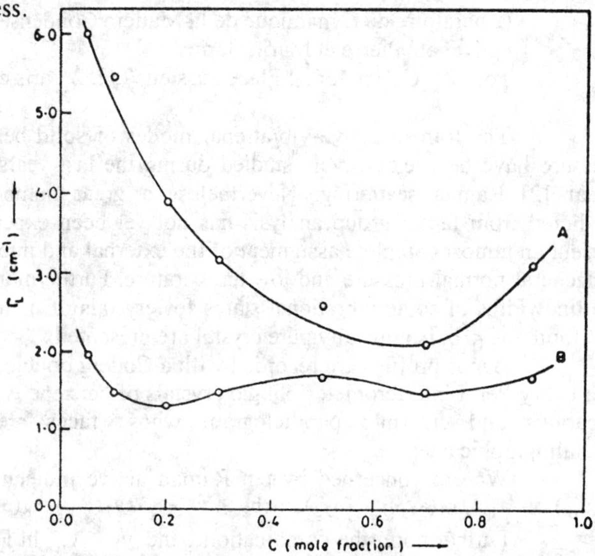

Fig.1. The true Raman linewidth, Γ_L plotted against the concentration of C_6H_5Cl in mole fraction in the (C_6H_5Cl + CH_3OH) mixture for the A : ~1021 cm^{-1} and B : ~1000 cm^{-1} bands.

1. B.P. Asthana and W. Kiefer, in Vibrational Spectra and Structure (J.R. Durig, ed.), Vol. 20, Elsevier Science Publishers B.V., Amsterdam, 1993, p. 67.
2. A. Morresi, L. Mariani, M.R. Distefano and M.G. Giorgini, J. Raman Spectrosc. 26, 179 (1995).
3. Ranjan K. Singh, S.N. Singh, B.P. Asthana and C.M. Pathak, J. Raman Spectrosc. 25, 423 (1994).
4. C.M. Pathak, B.P. Asthana, Ranjan K. Singh and A.L. Verma, Proc. XIV Int. Conf. on Raman Spectrosc. (N.-T. Yu and X.-Y. Li, eds.), John Wiley & Sons, New York, 1992, p. 335.
5. Ranjan K. Singh, B.P. Asthana, A.L. Verma and C.M. Pathak, Chem. Phys. Lett., submitted for publication.
6. D.W. Oxtoby, Adv. Chem. Phys. 40, 1 (1979).
7. A.F. Bondarev and A.I. Mardaeva, Opt. Spectrosc. 35, 267 (1973).

RAMAN LINE SHAPES OF ISOTOPICALLY PURE AND NATURAL CRYSTALS OF BENZENE

R.Ouillon, J.-P. Pinan Lucarre, and P.Ranson

Laboratoire de Dynamique de la Matière Condensée
Université Pierre et Marie Curie
Tour 22, Casier 136, 4 Place Jussieu, 75252 Paris cedex 05, France

The Raman active vibrational modes of solid benzene at normal or high pressure have been extensively studied during the last years using linear [1] or non linear [2] Raman scattering. Nevertheless a great number of vibrational states predicted from factor group analysis has not yet been experimentally analyzed. We present an almost complete assigment of the external and internal modes in crystalline benzene at normal pressure and low temperature. Furthermore a comparison between the linewidths of some vibrational states for crystals of natural isotopic composition and for a $^{12}C_6H_6$ isotopically pure crystal are presented.

Bands profiles are recorded with a Coderg double monochromator coupled to a Fabry-Perot interferometer. Single crystals of benzene were grown in a Bridgman apparatus and cut into parallelogram whose faces are perpendicular to the crystallographic axes.

We are concerned by ten Raman active molecular bands in the crystal: $\nu_1(a_{1g})$, $\nu_2(a_{1g})$, $\nu_3(a_{2g})$, $\nu_4(b_{2g})$, $\nu_5(b_{2g})$, $\nu_6(e_{2g})$, $\nu_7(e_{2g})$, $\nu_8(e_{2g})$, $\nu_9(e_{2g})$, $\nu_{10}(e_{1g})$

Furthermore, the combination band $\nu_1+\nu_6$, in Fermi resonance with ν_8, is visible in the spectra. Each non degenerate mode splits into four components $(A_g, B_{1g}, B_{2g}, B_{3g})$ and each degenerate mode into eight components $(2A_g, 2B_{1g}, 2B_{2g}, 2B_{3g})$. Besides, there are twelve Raman active librational lattice modes, three for each symmetry species. The different regions of the spectrum, corresponding to lattice vibrations and internal fundamental bands, are analyzed in detail. The underlying table is given as an exemple.

Factor group components for the $\nu_1(a_{1g})$ and $\nu_6(e_{2g})$ internal modes in natural solid C_6H_6 (T = 6 K).

	ν_1 mode	
symmetry	wavenumber (cm^{-1})	F.W.H.M. (cm^{-1})
A_g	991.20	0.124
B_{2g}	991.80	0.080
B_{3g}	992.30	0.15
B_{1g}	992.35	0.10

	ν_6 mode	
symmetry	wavenumber (cm^{-1})	F.W.H.M. (cm^{-1})
B_{2g}	605.50	0.10
B_{2g}	605.75	-----
A_g	606.75	0.125
B_{1g}	607.0	0.12
A_g	607.85	0.11
B_{3g}	608.10	0.09
B_{3g}	609.55	0.165
B_{1g}	611.75	0.04

It is of great importance to give experimental information on the vibrational dynamics in a solid. When excited in the energy region of the vibrational fundamentals, the crystal undergoes vibrational energy redistribution which may be studied in the frequency domain from the temperature dependence of the line shapes. The basic relaxation mechanisms are: a) population decay due to energy transfer processes; b) pure dephasing processes due to phono-phononscattering. The presence of mass defects (isotopic impurities) can, however, make a large quantitative contribution to the Raman linewidths through scattering and energy trapping processes [3]. We present some results concerning this problem in all its experimental aspects. We compared the linewidths for a given mode between an isotopically pure and a natural crystal (Fig.1, Fig.2, Fig.3) and of close lying factor group components in one or the other material (Fig.4).

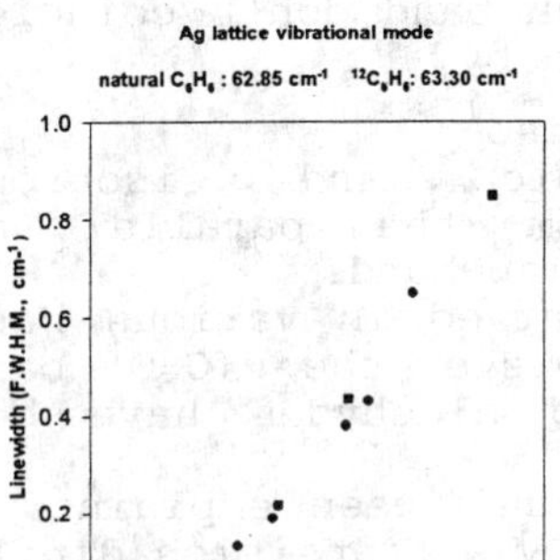

Fig.1 squares: $^{12}C_6H_6$ circles: natural C_6H_6

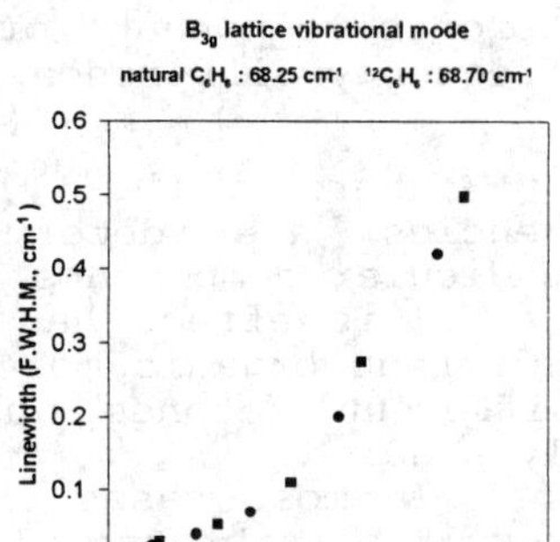

Fig.2 squares: $^{12}C_6H_6$ circles: natural C_6H_6

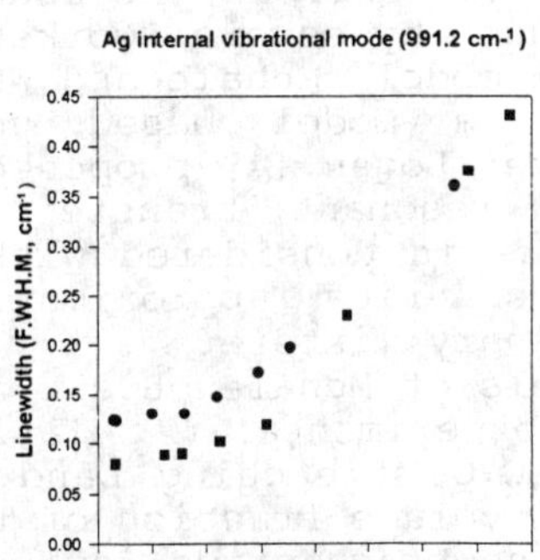

Fig.3 squares: $^{12}C_6H_6$ circles: natural C_6H_6

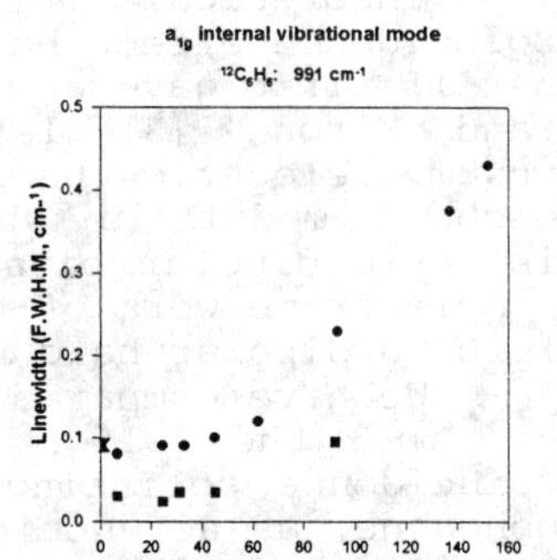

Fig.4 squares: B_{2g} 991.85 cm^{-1} circles: Ag 991.2 cm^{-1} triangles: ref.2

[1] H.Bonadeo, M.P.Marzocchi, E.Castellicci, and S.Califano J.Chem.Phys **57**, 4299 (1972)

[2] T.J.Trout, S.Velsko, R.Bozio, P.L.Decola, and R.M.Hochstrasser J.Chem.Phys **81**, 4746 (1984)

[3] A.A.Maradudin, and S.Califano Phys.Rev B **48**, 16628 (1993)

SOLVENT DEPENDENT NON-COINCIDENCE EFFECT

Debajyoti Bhattacharjee and T.N.Misra*

Department of Spectroscopy, Indian Association for the Cultivation of Science. Jadavpur, Calcutta-700032, India.

The non-coincidence effect is the phenomenon that the peak frequencies of the isotropic and anisotropic components of the Raman band don't coincide. The frequency difference is defined by;

$$\Delta\nu = (\nu_{aniso} - \nu_{iso})$$

where the anisotropic and isotropic frequencies are determined from the parallel and perpendicular components of the Raman band.

This effect has been observed in various bands of different organic molecules, however the ν(C=O) bands of different ketones, amides and aldehydes have been widely studied.

In most cases the effect has been explained as the result of vibrational coupling via transition dipoles in the presence of local quasi crystalline order arising from the interaction of permanent dipoles. Concentration studies have shown that the magnitude of the splitting decreases with increasing concentration of the solvent.

Theoretical works on the solvent dependent non-coincidence effect have been done by several workers. Mirone [1] first gave an empirical model. Mchale and Wang [2] and Mirone [3] later gave a model based upon continuum dielectric theory. However Logan [4] proposed a microscopic model in which the resonant transfer in non-isotopic dipolar binary mixture is considered within the simple frame work of the mean spherical approximation (MSA) for a dipolar hard sphere binary mixture.

We have compared the models of Mchale et.al. and Logan in the light of experimental relative non-coincidence splitting of the C=O stretching band of acetophenone, as a function of the volume fraction of the solute in chloroform and carbontetrachloride solvent systems. It is observed that Logan's theory fits better with the experimental results than that of Mchale-Mirone's model. However none of these two models include the experimentally observed threshold volume fraction (ϕ_A) term, below which the non-coincidence splitting vanishes.

Therefore considering the fact that the

non-coincidence splitting does not depend on the total volume fraction ϕ_A but on $(\phi_A - \phi_0)$ where ϕ_0 is the threshold volume fraction, we have employed an modified form of Logan's theory, following the empirical relation of Mirone [1] by including the threshold volume fraction term.

Then in a binary mixture, the relative non-coincidence splitting, for the mode α of solute (A) at a constant temperature T(K), as a function of the volume fraction ϕ_A of the solute (A) is;

$$\frac{\Delta_{\alpha,A}(\phi_A,T)}{\Delta_{\alpha,A}(1,T)} = \frac{(\phi_A - \phi_0)}{(1 - \phi_0)} \frac{[1 + (5/4)Y_0(\phi_A)]}{[1 + (5/2)Y_0(\phi_A)]^2} \frac{[1 + (5/2)\xi'_{A,0}]^2}{[1 + (5/4)\xi'_{A,0}]} \quad (1)$$

$$\text{where } Y_0(\phi_A) = \xi'_{A,0}\left[\phi_A + (\mu_B/\mu_A)^2 \frac{(1 - \phi_A)}{R}\right]$$

$$\text{and } \xi'_{A,0} = \frac{\mu_A^2\,\rho(1)}{72\varepsilon_0\,KT}$$

where μ_A and μ_B are the dipole moment of solute (A) and solvent(B) respectively. $\rho(1)$ is the number density of pure solute and ε_0 is the permittivity of the vacuum.

When this model (Equ.1) is compared with the experimental data, it is observed that in case of carbontetrachloride solvent system, the model shows excellent fit. In case of chloroform solvent system, however, although the experimental data show a more close correspondence with expression(1) than with the models of Logan [4] and Mchale and Mirone [2,3], but still the experimental data lie moderately below. This inconsistency may stem from the fact that some other local factors (viz. hydrogen bonding interaction) can also contribute to the additional lowering of the non-coincidence effect, which can not be incorporated in the model through the introduction of threshold volume fraction term in Logan's theory.

1. P.Mirone and G.Fini, J.Chem.Phys., 71(5), 2241 (1979).
2. J.L.McHale, J.Chem.Phys., 75(1), 30 (1981).
3. M.G.Giorgini, G.Fini and P.Mirone, J.Chem.Phys., 79(2), 639 (1983).
4. D.E.Logan, Chem.Phys., 131, 199 (1989).

THE LINE SHAPE OF THE LIGHT SCATTERING IN MOLECULAR CRYSTALS

Bashirov I.F., Bashirov F.I., and Gaisin N.K.
Department of Physics, Kazan State University, 420008, Kazan, Tatarstan, Russia

A molecular hindered motion (MHM) increases a line width of Reilegh and Raman scattering. The theory of the line shape in molecular crystals has been developed by application of the abstract point groups [1]. The model of fixed angular jumps has been used as basic model of the MHM. We assume that the stationary angular orientations of a molecule do not fix, but they have a continuous distribution in potential wells. Moreover, the frame of abstract groups does not allow to cover all crystallographic sites where the rotating molecules may be placed. The results of new approach in solution of such stochastic problem are given in [2]. In this report, we present an advanced description of the line shape for the internal modes of the light scattering by symmetrical top molecules in molecular crystals.

The calculated intensities I_z and I_y of polarized and depolarized lines are related to spectral densities $J_0(\Delta\omega)$ and $J_1(\Delta\omega)$ of the unitary spherical tensor components $Y_0^{(2)}(g)$ and $Y_1^{(2)}(g)$ by

$$I_z(\Delta\omega) = \frac{8\pi}{15}\left[\alpha_0^{(2)}\,E\right]^2 J_0(\Delta\omega) \qquad \text{and} \qquad I_y(\Delta\omega) = \frac{2\pi}{5}\left[\alpha_0^{(2)}\,E\right]^2 J_1(\Delta\omega) \tag{1}$$

where $\alpha_0^{(2)}$ is the principal value of the spherical tensor of molecular polarizability in the case of Reilegh scattering or the same of its increment in the case of Raman scattering, E is the amplitude of the strength of polarized incident light. We give the general form of spectral density $J_m(\Delta\omega)$ (m = 0, 1) in a monocrystal by

$$J_m(\Omega,\Delta\omega) = \frac{5}{4\pi}\sum_{\alpha,\beta} q_\alpha \chi_{\alpha E}^{-1}\left|\sum_n \psi_{\alpha\beta n}^{(2)} D_{nm}^{(2)}(\Omega)\right|^2 \frac{\omega_\alpha}{(\Delta\omega)^2+\omega_\alpha^2} \tag{2}$$

and in powder by

$$J_m(\Delta\omega) = \frac{\delta_{mm}}{2\pi}\sum_\alpha q_\alpha \frac{\omega_\alpha}{(\Delta\omega)^2+\omega_\alpha^2} \tag{3}$$

where Ω is the three-dimensional angle between the laboratory coordinate frame and that where the molecule motion symmetry group G is defined, $D_{nm}^{(2)}(\Omega)$ is Wigner function of 2nd rank, $\psi_{\alpha\beta n}^{(2)}$ is a coefficient of the decomposition of the 2nd rank basis function of the irreducible representation (IR) Γ_α upon the spherical harmonics, $\chi_{\alpha E}$ is the dimension of Γ_α, and α and β enumerate the IRs and their rows. The spectral densities (2) and (3) include two dynamical variables q_α and ω_α, respectively, the dynamic weights and the correlation frequencies, corresponding to the IRs Γ_α. The following relations may be written for q_α and ω_α

$$\sum_\alpha q_\alpha = 1 \qquad \text{and} \qquad \omega_\alpha = \omega\,(1-\chi_{\alpha E}^{-1}\sum_i p_i \chi_{\alpha i}) \tag{4}$$

where ω - the mean frequency of MHM, $\chi_{\alpha i}$ - the character of ith class, and p_i - the probability of reorientation about the axes of ith class.

There are two main peculiarities of new spectral densities (2) and (3). At first, Eq. (2) shows the anisotropy properties of the scattered lines that depend solely on the crystal installation about the direction of light polarization. That is, the spectraldensities $J_m(\Omega,\Delta\omega)$ do not depend explicitly on the orientations of the molecule as it prescribed in [1]. Secondly, our theory includes new variables q_α, the dynamic weights of IRs, which allow to describe the change of line intensities whereas the reorienting molecules occupy the crystalline positions of various site symmetry [3].

To demonstrate the effect of new dynamical variables q_α on the light scattering study in a molecular crystal, we present the graphs where the angular dependence of the theoretical intensities of polarized (left fig.) and depolarized (right fig.) lines are shown: $J_0(\varepsilon)$ and $J_1(\varepsilon)$. They are drawn for the case when a molecule exhibits a reorientation, the symmetry of that is cubic, but its site symmetry changes from the entirely tetragonal ($q_1 = 0$) up to completely trigonal one ($q_1 = 1$). The figures correspond to the crystal installation where the external field is directed perpendicularly to its axis C that is also chosen as the axis of crystal rotation. That is ε is the angle between the electric vector and the observation direction. The notation q_1 corresponds to the dynamic weight of the two-dimensional IR E of a cubic system group. In such case the dynamic weight of the three-dimensional representation F is $q_2 = 1 - q_1$.

One may see how the site symmetry affects on the anisotropy of scattering intensity for high symmetry rotating molecule. This anisotropy vanishes for $q_1 = 0.4$. It is the case of the undistorted cubic site that consistent with a behavior of 2nd rank physical quantity in cubic crystals. For a tetragonal position in the cubic unit cell - $q_1 = 0$ and the intensity of a polarized line $J_0(\varepsilon)$ is proportional to $\sin^2 2\varepsilon$. There is an experimental confirmation of such dependence by the internal mode, observed at 1008 cm^{-1} in zirconium silicate $ZrSiO_4$ [4]. Here, the tetrahedral atomic groups SiO_4 occupy the tetragonal positions. If we assume that the silicate groups exhibit random rotations and give the predominant extension of the line, the angular dependence of scattered intencities must be expressed by $\sin^2 2\varepsilon$. In the tetragonal coordinate frame the experimental normalized intensity is well described by $\cos^2 2\varepsilon$ [4] that agrees with $\sin^2 2\varepsilon$ - in the coordinate frame of the tetrahedral SiO_4-groups.

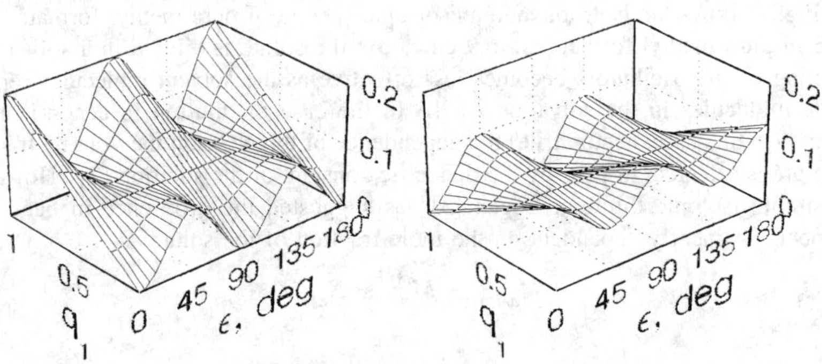

1. Ivanov E. N., Optica i Spectroscopia, **37,** 672, 1974.
2. Bashirov F. I., Magnetic Resonance and Related Phenomena. Extended Abstracts of the XXVIIth Congress AMPERE, Kazan, **1**, 282 - 283, 1994.
3. Halford R. S., J. Chem. Phys., **14**, 8, 1946.
4. Dawson P., Hargreave M., Wilkinson G. R., J. Physics C, **4**, 240, 1971.

A STUDY OF THE NON-COINCIDENCE EFFECT IN SOLUTIONS OF METHYL FORMATE

R. J. Bartholomew and D. E. Irish*

Guelph-Waterloo Centre for Graduate Work in Chemistry, Department of Chemistry, University of Waterloo, Waterloo, ON. Canada N2L 3G1

In pure liquids vibrations of neighbouring molecules may be coupled because of the interactions between them [1]. If the alignment of the molecules is not random (i.e., there is some local order), the transition dipole moments of neighbouring molecules can couple. This will lead to resonance energy transfer (RET) which most clearly manifests itself as a non-coincidence effect (NCE). The non-coincidence effect is defined as the difference between the maxima of the isotropic and anisotropic components of the spectral bands.

Methyl formate was chosen for study partially because it is closely related to another molecule of interest to us - formic acid. As such, it is useful to know something about the liquid structure of methyl formate.

Solutions of methyl formate in acetonitrile and d(formyl)-methyl formate were prepared on a mole fraction basis. Spectra were recorded with a Jarrell-Ash 25-100 spectrometer with a 1 m focal length. Excitation was with the 514.5 nm line from an argon ion laser.

Fig. 1 shows the isotropic and anisotropic spectra of pure methyl formate. The NCE in pure methyl formate is ~6.9 cm^{-1}. As the solute is diluted in a solvent the coupling of the oscillators becomes less effective as the solvent molecules replace solute molecules in the solvation shell. In the case of dilution in a non-isotopic solvent (in this case, acetonitrile) the dependence of the NCE on the volume fraction of solute is affected, in part, by the dielectric constant of the solution [2]. However, for simple isotopic dilution, Logan [3] has suggested the difference in band first moments is linearly dependent on the mole fraction of the solute:

$$M_{aniso.} - M_{iso.} = \Delta_{neat} X$$

where Δ_{neat} is the difference for the pure solute and X is the mole fraction of solute.

Fig. 2 shows the variation of $M_{aniso.} - M_{iso.}$ with X_{MF}. It is indeed linear with a correlation of r = 0.939. Δ_{neat} for this system is 4.2 cm^{-1} and within the 95% confidence interval the slope of the line equals Δ_{neat} and the intercept is zero.

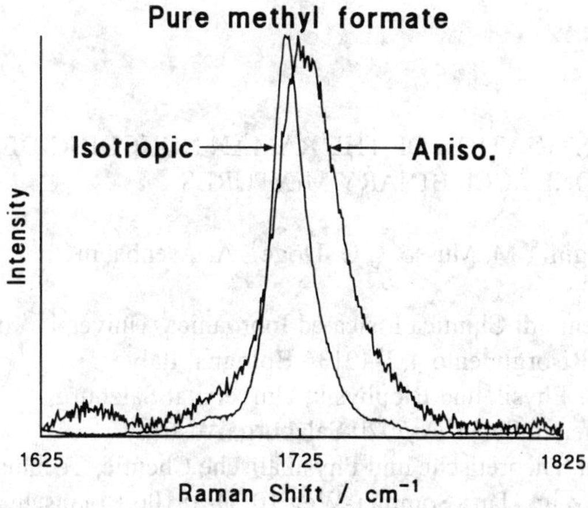

Figure 1: Isotropic and Anisotropic Peaks of Pure Methyl Formate.

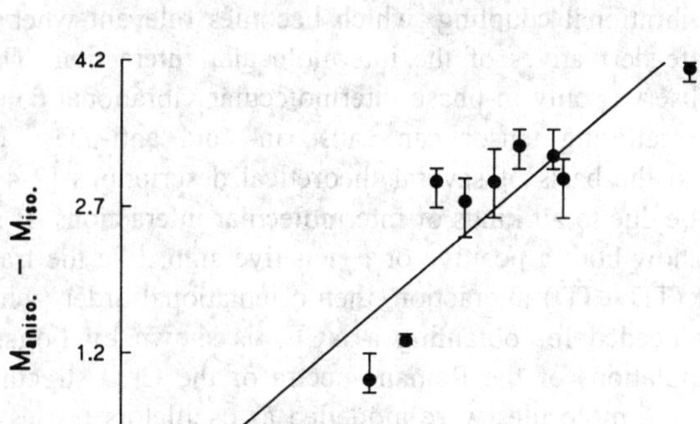

Figure 2: $M_{aniso.}$ - $M_{iso.}$ as Function of Mole Fraction of Methyl Formate

1. G. Fini, P. Mirone, *J. Chem. Soc. Faraday Trans.* **1974** 70 1776
2. P. Mirone, G. Fini, *J. Chem. Phys.* **1979** 71 2241
3. D. E. Logan, *Chem. Phys.* **1989** 131 199

AN INVESTIGATION OF THE RAMAN NONCOINCIDENCE EFFECT IN ACETONE / CCl_4 BINARY MIXTURES

M.G. Giorgini[1], M. Musso[2]*, G. Döge[3], A. Asenbaum[2]

[1]Dipartimento di Chimica Fisica ed Inorganica, Universita' di Bologna, Viale del Risorgimento 4, I-40136 Bologna, Italy
[2]Institut für Physik und Biophysik, Universität Salzburg, Hellbrunnerstraße 34, A-5020 Salzburg, Austria
[3]Institut für Theoretische und Physikalische Chemie, Technische Universität Braunschweig, Hans-Sommer-Weg 10, D-38106 Braunschweig, Germany

The noncoincidence effect (NCE) of a Raman band is the frequency difference between the positions of the anisotropic and the isotropic component of the band and was first experimentally noticed by [1]. The main reason for a NCE is intermolecular vibrational coupling, which becomes relevant when there are large normal coordinate derivatives of the intermolecular interaction. Then in isotropic scattering one observes only in-phase intermolecular vibrational couplings, whereas the anisotropic scattering tensor can cause in- and anti-phase couplings. This knowledge is also the basis of several theoretical descriptions [2-4]. The coupling mechanism can be due to all kinds of intermolecular interactions. If such a coupling interaction can show both a positive or a negative sign, like the transition dipole - transition dipole (TD - TD) interaction, then orientational order (which favours one of the signs) is needed for obtaining a NCE, as shown by Logan [4]. Recently Monte-Carlo simulations of the Raman spectra of the C=O strecth of acetone [5] were performed: the molecules were modelled as oscillators possessing a transition dipole and a coupling between these oscillators was introduced as a TD - TD intermolecular coupling. The results for neat acetone are in good agreement with experiment, but simulations for binary liquid mixtures are still not available.

Our present investigations on the NCE aim at its concentration dependence in binary mixtures, with particular attention to the low concentration region. Observations by Fini & Mirone [6] and by Giorgini et al. [7] suggested the existence of a threshold concentration for the onset of the NCE. The presently available theories predict a smooth onset of the effect, contrary to those observations. We therefore performed a series of VV-VH Raman experiments in 90° scattering geometry for the binary liquid mixture acetone + CCl_4, taking advantage of the high sensitivity of a LN2-cooled CCD camera mounted on a Jobin-Yvon U-1000 monochromator. We have simultaneously measured the shift of position of the isotropic and anisotropic components of the carbonyl stretching vibration at 1710 cm^{-1} ($\nu(^{12}C{=}O)$) and at 1680 cm^{-1} (mainly $\nu(^{13}C{=}O)$, present as impurity at 1%) of standard spectroscopy grade acetone. Figure 1 shows the currently obtained results for the concentration dependence of the NCE of acetone in CCl_4. The dependence for the 1710 cm^{-1} band, determined from the peak positions and from the first moment values (M1), is remarkably different due to bandshape changes with concentration. The NCE for the 1680 cm^{-1} band is approximately zero in the whole concentration range, since ^{13}C-acetone is isotopically highly diluted in ^{12}C-acetone. The shifts of the 1680 cm^{-1} band components are therefore solely determined by chemical shift. This fact can be exploited for the determination of the true NCE: by subtracting the $\nu(^{13}C{=}O)$ shifts from the $\nu(^{12}C{=}O)$ shifts (caused by chemical effects as well as by resonance transfer), we may extract the contribution only due to resonance transfer. Experiments are in progress with isotopic mixtures in order to improve the accuray of the results, furthermore bandshape features and the change of sign of the NCE at low concentrations will be studied in more detail.

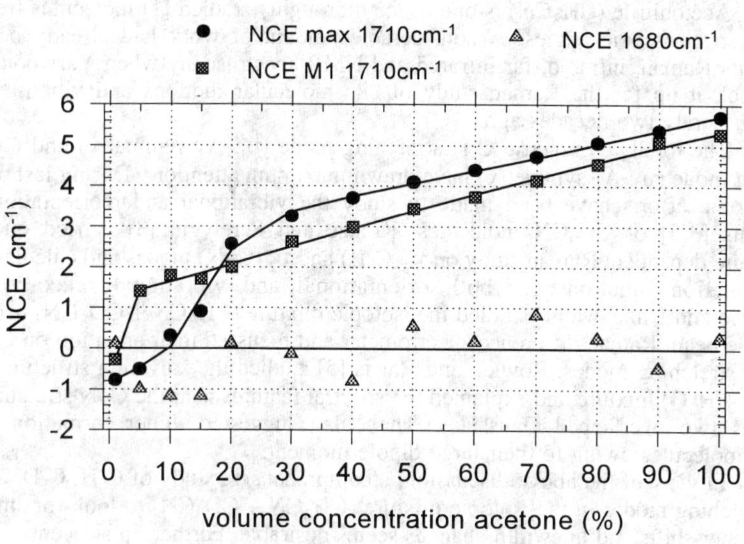

Fig. 1: Noncoincidence effect of the carbonyl stretching mode of acetone diluted in CCl_4

1. G. Fini, P. Mirone, J. Chem. Soc. Faraday Trans. II **70**, 1776 (1974)
2. G. Döge, Z. Naturforsch A **28**, 919 (1973), D. Scheibe, J. Raman Spectrosc. **13**, 103 (1982)
3. C.H. Wang, J. McHale, J. Chem. Phys. **72**, 4039 (1980)
4. D. Logan, Chem. Phys. **103**, 215 (1986); Mol. Phys. **58**, 97 (1986); Chem. Phys. **131**, 199 (1989)
5. H. Torii, M. Tasumi, J. Chem. Phys. **99**, 8459 (1993), H. Torii, J. Mol. Struct. (THEOCHEM) **311**, 199 (1994)
6. G. Fini, P. Mirone, J. Chem. Phys. **71**, 2241 (1979)
7. M.G. Giorgini, G. Fini, P. Mirone, J. Chem. Phys. **79**, 639 (1983)

ISOTOPIC DILUTION STUDY IN THE (CH_3CN + CD_3CN) MIXTURE USING THE SCANNING MULTICHANNEL RAMAN DIFFERENCE TECHNIQUE

B.P. Asthana[1*], V. Deckert[2], and W. Kiefer[2]

[1]Laser and Spectroscopy Laboratory, Department of Physics, Banaras Hindu University, Varanasi-221005, India
[2]Institut für Physikalische Chemie, Universität Würzburg, Marcusstr. 9-11, D-97070 Würzburg, Germany

Acetonitrile (CH_3CN) is one of the thoroughly studied [1] molecules from different spectroscopic angles. A considerable amount of work had already been done on its Raman, infrared, far-infrared and NMR spectroscopy when Yarwood et al. [2] took it up for the Raman study of its molecular motions and vibrational relaxation neraly two decades ago.

The totally symmetric C-H stretching mode (υ_1, A_1 symmetry) and C≡N stretching mode (υ_2, A_1 symmetry) have drawn maximum attention. During last one decade some efforts have been made to study the vibrational and reorientational relaxations [3-5] of CH_3CN molecules. Krauze and Perelygin [3] carried out a temperature dependent Raman study on υ_1 (C-H) and υ_2 (C≡N) modes and calculated the correlation functions for both orientational and vibrational relaxations. Kamogawa and Kitagawa [4] studied the isotopic mixture (CH_3CN + CD_3CN) using the double beam Raman difference spectrometer and focussed their attention on C-H and C-D stretching modes. Rowlen and Harris [5] studied the solvation structure in (CH_3CN + H_2O) mixture and explained the spectral features near the C≡N stretching region as two overlapped Gaussians. They also suggested dimer formation by CH_3CN molecules owing to their large dipole moment.

In view of the above discussion, a comprehensive study of C-H, C-D, and C≡N stretching modes in the isotopic mixture (CH_3CN + CD_3CN) to look for small wavenumber shifts and linewidth changes seems desirable. Further, in a recent work Döge et al. [6] measured the shifts in an isotopic dilution study of the C_6H_6/C_6D_6 system, basically to study the negative coincidence effect. Deckert et al. [7] have recently developed a new device for Raman difference spectroscopy (RDS) using scanning multichannel technique. The $I_{\parallel}$ and $I_{\perp}$ for C-H, C-D, and C≡N stretching modes in different mixtures with varying mole fractions (0.1, 0.2, 0.3, 0.4, 0.5, 0.6, 0.7, 0.8, and 0.9) of CH_3CN were recorded. In order to get the optimal advantage of RDS, two sets of measurements were made, taking CH_3CN as the reference in one case and CD_3CN as the reference in the other.

It is to be noted that the C≡N stretching frequency in CH_3CN is higher than that in the CD_3CN. The isotropic part of the typical spectra of CH_3CN, CD_3CN and a (CH_3CN + CD_3CN) mixture, each component having 0.5 mole fraction, are shown in Fig. 1. It is evident from these spectra that the main band has a shoulder also at a lower wavenumber. In view of this fact, the spectra A and C have to be fitted taking two Voigt profiles and the spectra B (see Fig. 1) has to be fitted taking

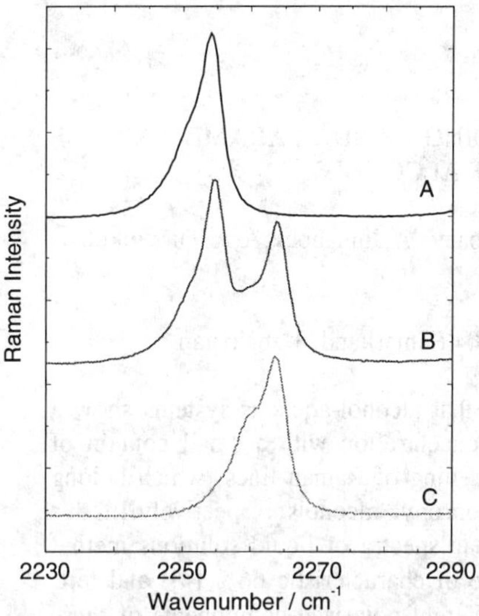

Fig. 1. Isotropic part of the Raman spectra in the C≡N stretching region; A: CH_3CN, B: (CH_3CN + CD_3CN) mixture (0.5 mole fraction, each), C: CD_3CN.

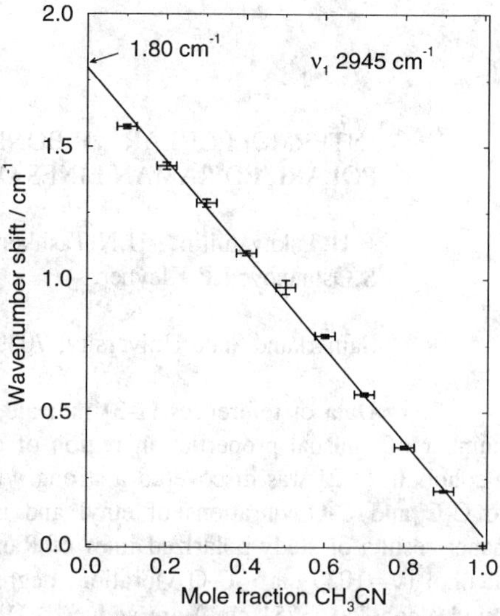

Fig. 2. Wavenumber shifts of the υ_1 (C-H) mode in the isotopic mixture (CH_3CN + CD_3CN) as a function of mole fraction of CH_3CN.

four different Voigt profiles into account. This makes the determination of the shift more complicated and they have to be fitted taking eight Voigt profiles, which is obviously quite cumbersome. The spectra in the C-H stretching region are, however, relatively less complicated. The isotropic part of the spectra were obtained as ($I_\parallel$ - 4/3 $I_\perp$) and the difference spectra could also easily be obtained applying the RDS methods [7]. The precise wavenumber shifts of υ_1 (C-H) in different mixtures with respect to pure CH_3CN were determined and they are plotted as a function of mole fraction of CH_3CN in Fig. 2 where they show a nice linear dependence. A linear dependence of wavenumber shifts indicates that static effects due to concentration fluctuations dominate and the diffusion dynamics plays a minor role. The C-H, C-D, and C≡N stretching frequencies are shifted to higher wavenumbers in the mixtures. From the preliminary analysis it appears that non-resonant interactions play a major role in the dephasing of these vibrations.

1. J. Yarwood and R. Arndt, Molecular Associations, edited by R. Foster, Academic Press, New York (1979) and references cited therein.
2. J. Yarwood, R. Arndt, and G. Döge, Chem. Phys. 25, 387 (1977).
3. A.S. Krauze and I.S. Perelygin, Zh. Prikl. Spektrosk. 45, 453 (1986).
4. K. Kamogawa and T. Kitagawa, J. Phys. Chem. 94, 3916 (1990).
5. K.L. Rowlen and J.M. Harris, Anal. Chem. 63, 964 (1991).
6. G. Döge, D. Schneider, and A. Morresi, Mol. Phys. 80, 525 (1993).
7. V. Deckert, W. Liebler, R. Eck, and W. Kiefer, Appl. Spectrosc. (submitted) and references cited therein.

INTERMOLECULAR H-BONDING AND PARAMETERS OF POLARIZED RAMAN LINES OF ALCOHOLS.

F.H.Tukhvatullin*, U.N.Tashkenbaev, A.Jumaboev,A.A.Tursunkulov, S.Osmanov, I.P.Kleyner.

Samarkand State University, 703004 Samarkand, Uzbekistan

Data of references [1-3] indicate, that alcohol-aqueous systems show a number of unusual properties in region of concentration with a small contain of alcohol. In [1,2] was discovered a strong widening of Raman lines, which belong to C-C and C-O vibrations of ethyl and isopropyl alcohols respectively.In this paper results of study polarized lines of Raman spectra of liquid solutions methyl alcohol (ν=1035 cm^{-1}, C-O vibration, degree of characteristic 66% [4]) and tert. butyl alcohol (ν=751 cm^{-1}) are adduced. Different polarized components of these lines at 293 K were obtained by spectrometer at incident light wave length 488 nm.

On fig.1 was showed concentrational dependence of line width 751 cm^{-1} for tert. butyl alcohol. We see, as in [1,2], that in aqueous solution the line width passes through maximum, and what's more this region of concentrations coincides with that for maximum of intensity Rayleigh light scattering in this solution [3]. However the maximum of lines broadening for 751 cm^{-1} was observed in the region with small contain alcohol and in solution with hexane, although it is clear, that it is impossible to be any intermolecular H-bonding between them. Any anomalies in behaviour 751 cm^{-1} line width were not observed in solution of tert. butyl alcohol with acetone, or dimethyl sulfoxide, or methanol. Displacement of line is unusually for solutions in water and in hexane - in region with small contain of alcohol the displacement of line in low frequency area sharply increases. In mixture with $CHCl_3$ the maximum of line broadening was observed at 0,6 molar fraction of $CHCl_3$. Taking into account the properties of solvents molecules, their proton acceptor and proton donor properties the above-mentioned character of behaviour of line width we may associate with availability of molecules with free and bonded atom of hydrogen of OH group, which correspond to separated near situated lines which are not solved by device. The availability of these lines in solutions of alcohols with water and cyclohexane, at near equal contents of such molecules, leads to unusual concentrational broadening of line. In solution with acetone and dimethyl sulfoxide the free from H-bonding atoms of hydrogen of OH group do not appear and accordingly of anomalies in behaviour of line width it is not observed.

It is analogous, but is more clear the picture of behaviour C-O vibrations (1035 cm^{-1}) of methyl alcohol in solutions. In pure liquid this band is asymmetrical at lower frequency side and within of band the depolarization coefficient changes not monotonously. These data show a complexity of band - a availability in band at least of two lines - lines belonging to alcohol molecules with free and H-bonded hydrogen atoms of OH group. These lines in solution with CCl_4 were recognized clear (fig. 2). In solutions with acetonitryl, in case of which H-bonded hydrogen atoms of OH group alcohol is absent, some splitting of line for components is not observed, and asymmetry observing in pure alcohol vanishes.

In our opinion, observing anomaly of behaviour line width in aqueous solutions [1,2] is caused by break of H-bonds and appearance in noticeable number of molecules with free hydrogen atoms from H-bond.

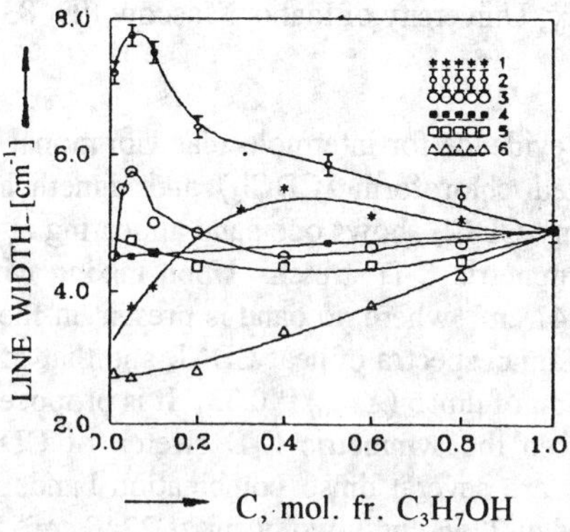

Fig. 1. The width of Raman line ν=750 cm^{-1} in solutions of tert-butanol: 1 - $CHCl_3$, 2 - H_2O, 3 - C_6H_{14}, 4 - CH_3OH, 5 - C_2H_6SO, 6 - C_3H_5O.

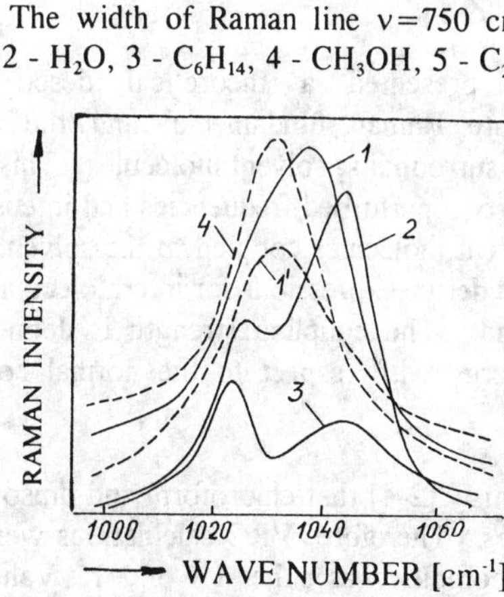

Fig. 2. C-O vibration band of methanol (1, 1' - I_{vv} (ν) and I_{VH}(ν) components), and its solutions in CCl_4 (2 - 0.1, 3 - 0.02 mol. fr.), in acetonitryl (4 - 0.1 mol. fr.).

1. Tanabe K. J.Raman Spectr. 1984, V.15, N 4, P.248.
2. Pogorelov V.E., Hodjieva E. Chem. phys. (Russia).1993, V.12, N 7, P.1032.
3. Vuks M.F. Light scattering in gases, liquids and solutions. 1977. Leningrad.
4. She Un-bo, Chulanovski V.M. Opt. and spectr. 1963. (USSR)., V.2, P.218.

INTERMOLECULAR VIBRATIONAL RESONANCE COUPLING BETWEEN d-CHLOROFORM AND DMSO

D. C. Daniel and J. L. McHale

Department of Chemistry, University of Idaho, Moscow, ID 83844-2343 USA

In this work we report evidence for intermolecular vibrational resonance coupling (VRC) between deuterated chloroform ($CDCl_3$) and dimethyl sulfoxide (dmso). The Raman spectrum of pure $CDCl_3$ shows one band appearing at 2256 cm^{-1} which corresponds to the totally symmetric C-D stretch. Upon mixing with dmso a band appears in the region 2236-2241 cm^{-1} where no band is present in the spectrum of neat dmso. Figure 1 shows the Raman spectra of neat $CDCl_3$ and that of a mixture of dmso and $CDCl_3$ with mole fraction of dmso (X_{dmso}) = 0.33. It is proposed that this new band results from VRC between the symmetric C-D stretch of $CDCl_3$ and a combination band of dmso. There are several dmso combination bands which are close in frequency to the $CDCl_3$ band at 2256 cm^{-1}; two occur at 2246 cm^{-1} and one at 2241 cm^{-1}.

McHale and Veas [1] presented a theoretical description for intermolecular VRC between a solute Raman fundamental and the collective vibrational excitation of N equivalently surrounding solvent molecules. This coupling mechanism leads to expressions for the two perturbed frequencies and intensity ratios which depend upon the number of solvent molecules coupled to the solvent and also the coupling strength. In this work, we derive expressions for intermolecular VRC of a fundamental and a combination band. The coupling strength is defined as the change in intermolecular potential energy with respect to the normal coordinates involved in the coupling.

It has been reported previously [2-4] that chloroform and dmso form 1:1 and 2:1 ($2CHCl_3$ to 1 dmso) complexes. Therefore, VRC calculations were carried out assuming the stoichiometry of the complex was either 1:1 or 2:1. Values of the coupling strength were found to be 3-8 cm^{-1} which is about 0.6-3.5% that of a typical hydrogen bond. The observed intensity ratios were compared to the calculated ratios obtained from theory assuming either a 1:1 or 2:1 stoichiometry for the complex. This was done for the case of VRC between the C-D stretch of $CDCl_3$ and the 2241 cm^{-1} and 2246 cm^{-1} modes of dmso. It was found that agreement between theory and experiment was reasonably good assuming a 1:1 complex but poor for the assumption of a 2:1 stoichiometry.

Evidence will also be presented which suggests that the coupling involves one or both of the 2246 cm^{-1} modes of dmso rather than the 2241 cm^{-1} mode. Furthermore, it appears that the VRC interaction is of a collisional nature rather than a long-lived hydrogen-bonded complex.

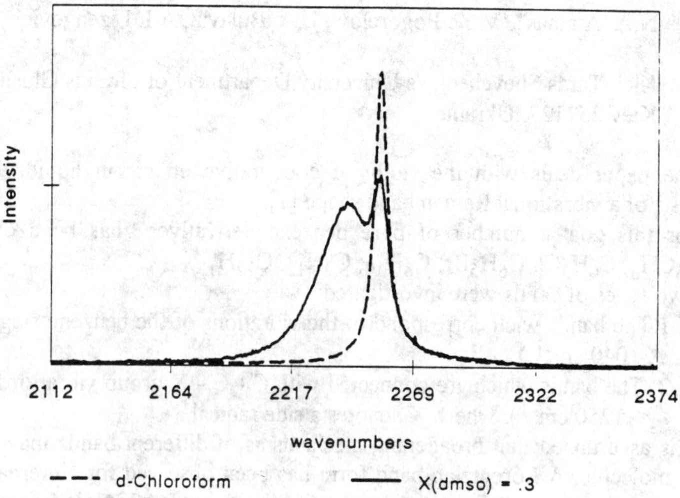

Fig. 1 Raman spectra of neat $CDCl_3$ and a mixture of dmso and $CDCl_3$ with mole fraction of dmso equal to 0.3

1. C.H. Veas and J.L. McHale, J. Am. Chem. Soc. 111, 7042-7046 (1989).
2. A.L. McClellan, S.W. Nicksic, and J.C. Guffy, J. Mol. Spectrosc. 11, 340 (1963).
3. D.V. Fenby and L.G. Helper, J. Chem. Thermodynamics 6, 185-189 (1974).
4. L. Rintoul and H.F. Shurvell, J. Raman Spectrosc. 21, 501-507 (1990).

DEVELOPMENT OF A PARTIAL ORDERING OF LIQUIDS IN RAMAN SPECTRA.

N.A .Atamas*, V.Ye.Pogorelov , L.A. Bulavin, A.I. Lizengevich

Kiev Taras Shevchenko's University,Department of Physics,Glushkov pr.6, Kiev 252127, Ukraine

The paper deals with the study of cooperative effects in liquids by means of analysis of a vibrational Raman band shape [1].

For this goal a number of pure benzene derivatives has been chosen (C_6H_6 $C_6H_5CH_3$, C_6H_5Cl, C_6H_5Br, C_8H_{10}, C_9H_{12}, $C_{12}H_{18}$).

Two types of bands were investigated:

1. The bands wich correspond to the vibrations of the benzene ring (range1020-1040 cm^{-1})
2. The bands which are connected with C - C - X group vibrations (range 1050 - 1250 cm^{-1}), where X denotes a side radical.

It is ascertained that broadening mechanisms of different bands may differ for the same molecule. A Lorentzian band form has been observed for "internal" vibrations. On the contrary, the shape of the "external" vibrational band is not Lorentzian and has noticeable Gaussian component. It concerns the vibrations of C - C links with connect benzene rings with X - radicals (X= CH_3, CH_2CH_3, $(CH_2)_5CH_3$). From the presence of a slow modulation regime for a broadening of the "external" vibrations we can make a deduction that these vibrations are more "feeling" for slow modes of a structural relaxation in a media.

The second part of the work is devoted to the investigation of the non-coincidence in positions of isotropic and anisotropic Raman bands. The observations of the positions of the parallel $I_{II}(\omega)$ and perpendicular $I_{\perp}(\omega)$ components of Raman bands have been carried out. To study these parameters we have carried out the measurement of C - H stretching bands in above - mentioned liquids and in some benzene monosubstituted. The positions of component maximums of the studied objects are presented in the Table 1. Such non-coincidence takes place in some cases where resonant vibrational exchange is significant. The value of the band shift depends on a degree of an orientational ordering in a media.

The isotropic - anisotropic band non-coincidence points out the development of the co-operative (at least three - body) processes in benzene monosubstituted. In this case particularly the procedure of a deconvolution of vibrational and orientational correlation functions is not valid.

Table 1. The positions of the parallel and perpendicular components of Raman bands for benzene monosubstituted.

substance	ω_{max} $I_{II}(\omega)$	cm^{-1} $I_{\perp}(\omega)$	substance	ω_{max} $I_{II}(\omega)$	cm^{-1} $I_{\perp}(\omega)$
C_6H_6	992.45	992.67	C_8H_{10}	1001.94	1001.97
	3062.75	3063.05		3053.05	3053.79
	1001.68	1001.48		1001.76	1000.76
$C_6H_5CH_3$	3053.47	3053.68	C_9H_{12}	3035.81	3035.60
				3951.62	3051.76
				3062.66	3062.97
	3055.24	3052.80			
C_6H_5Br	3063.90	3062.56		1001.11	1000.76
			$C_{12}H_{18}$	3062.80	3063.00
				3052.03	3052.64
C_6H_5Cl	1000.50	1001.08		3035.16	3036.18
	3066.80	3062.15			

The displacement of bands frequencies can be put down as additional component to the first spectral moment which is a characteristic of the bands maximum location (1):

$$\Delta M_{if}^{1} = \sum_{kj} \int dx dy d\Omega_x d\Omega_y \langle \alpha_{fi}(\Omega_{fi}) \alpha_{kj}^{*}(\Omega_y) \varphi_{ij}^{kf}(x,\Omega_x;y,\Omega_y)\rangle_{or} * \\ *(n_i - n_f) n_k G(x,y) \quad (1)$$

α–tensor of scattering of light, $\varphi_{ij}^{kf}(x,\Omega_x;y,\Omega_y)$–resonant intermolecular interaction of potential; ij–initial states, k,f–final ones; n_k, n_f, n_i– filling numbers of vibrational levels.

1. N.A. Atamas, V.Ye.Pogorelov, L.A.Bulavin, Ukrainian Phys. J. 40, 807 (1995)
2. N.A. Atamas, Adamenko I.I., L.A. Bulavin, V.Ye. Pogorelov. J. Phys. Chem. 70, 97 (1996).

EXCITED STATE STRUCTURES OF 2,2'-DIPYRIDYL AMINE

S.Chattopadhyay[1*], P.Jana[1], S.K.Sarkar[2], T.Ganguly[1*], A.Mitra[3] and P.K. Mallick[4*]

[1]Spectroscopy Department, Indian Association for the Cultivation of Science, Calcutta 700 032, INDIA.

[2]Vidyasagar College, Calcutta 700 006, INDIA.

[3]VECC, Salt Lake City, Calcutta 700 064, INDIA.

[4]Physics Department, Burdwan University, Burdwan 713 104, INDIA.

It is known that the totally symmetric vibrations get Raman intensity contributions from both Albrecht-A and Albrecht-B terms. A-term is proportional to the vibrational overlap integral which is dependent on the distortion of the excited electronic state with respect to the ground state. On the otherhand, the B-term arises from the vibronic coupling mainly between two excited electronic states. So critical examination of Raman excitation profile (REP) is supposed to be very helpful in drawing informations regarding structures and symmetries of molecules in the excited electronic states. In our recent investigation [1], we have seen that the structures in the ground and excited electronic states of 2,2'-Dipyridyl amine (DPA) play some important role in determining its photophysical and photochemical behaviour. In order to get further insight into its structures in various excited states, we have critically analysed the observed REP in the light of Albrecht theory using the intensity formula (for the a normal mode) given below [2]:

$$I_\alpha = \text{Constant X } (\nu_a - \nu_\alpha)^4 \left[\sum_I F_A \Delta_\alpha^I \frac{\nu_{IG}^2 + \nu_a^2}{(\nu_{IG}^2 - \nu_a^2)^2} + \sum_{I<J} \sum F_B \frac{\nu_{IG}\,\nu_{JG} + \nu_a^2}{(\nu_{IG}^2 - \nu_a^2)\,(\nu_{JG}^2 - \nu_a^2)^2} \right]^2$$

Where all the symbols have usual significance including $\Delta^{l}_{\alpha'}$ measuring distortion of the molecule in the lth electronic state along with the ath normal coordinate.

The electronic absorption spectra [1] exhibit two strong bands at 312 (^1L_a) and 256 nm (^1B_b) and one weak and broad band at 220 nm (^1B_a) corresponding to the excited electronic states S_1, S_2 and S_3 respectively. Comparing the observed profile with the theoretical one of the strong and polarised ring breathing mode (998 cm^{-1}) we see that S_3 -state is the chief contributor to the Albrecht-A term. This means that the distortion of the ring in this state is very much significant. Besides this A-term contribution, B-term is found to be equally strong for the mode corresponding to the pair of states S_2 and S_3, indicating the symmetries of the molecule in these two states to be the same. Similarly S_1 and S_2-states are found to be the chief contributors to the modes denoting mainly the CC/CN stretching vibrations at 1594 (weak) and 1610 cm^{-1} (strong) respectively. Two strong polarised Raman bands at 1260 and 1280 cm^{-1} are identified chiefly as amino nitrogen-ring stretching mode [3]. Each of these is found to get contribution only from Albrecht-B term through vibronic coupling of the S_1-state with the other two states. Since no significant A-term contribution is coming to these modes, amino nitrogen -ring distance is not expected to undergo any significant change in the excited states. Since most of the intense Raman bands are not getting appreciable diagonal contribution from the lowest state (S_1), so the molecular symmetry is not expected to change significantly in this state.

REFERENCE

1. P.Jana, A.Mitra, S.K.Sarkar, T.Ganguly and P.K. Mallick, J.Photochem. Photobiol. (Awaiting publication).
2. J.Marconi, J. Raman Spectrosc., 22, 361 (1991).
3. S.Chattopadhyay, S.K.Sarkar, S.Sarkar (nee' Chaudhury) and P.K.Mallick, Proceedings of XIII International Conference on Raman Spectroscopy, (W.Keifer et al., ed.), John Wiley & Sons, Chichester, N Y, Brisbane, 1992, p. 294.

SECTION 7

MOLECULAR DYNAMICS

MOLECULAR ORIENTATION, SPECTRA AND PHASE TRANSITIONS AT AQUEOUS INTERFACES - SUM FREQUENCY VIBRATIONAL STUDIES

D. Zhang, J. H. Gutow, and K. B. Eisenthal*

Chemistry Department, Columbia University, New York, NY 10027, USA

Studies of the orientational structure and hydrogen bonding interactions at air/aqueous-nitrile interfaces using infrared plus visible sum frequency generation spectroscopy will be presented. The unexpected structural phase transition, first observed in our laboratory for acetonitrile (CH_3CN) at the air/acetonitrile-water interface, has also been observed recently in our laboratory for propionitrile (CH_3CH_2CN) at the air/propionitrile-water interface, but not for butyronitrile ($CH_3CH_2CH_2CN$)at the air/butyronitrile-water interface. These are the first examples to our knowledge of phase transitions in interfaces composed of small mutually soluble molecules. For acetonitrile and propionitrile the phase transition is manifested by an abrupt change in the vibrational frequency of the CN chromophore and an abrupt change in its orientation as the bulk nitrile concentration, and hence nitrile surface density, is varied. FIG. 1.

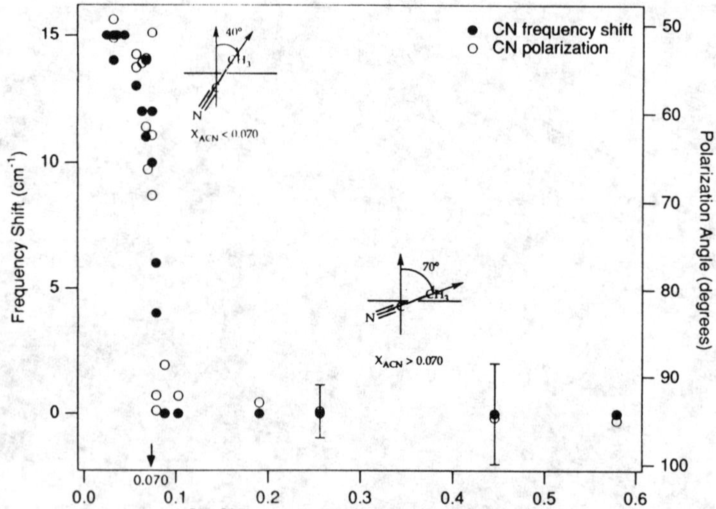

Figure 1. Compilation of CN frequency shifts of the surface acetonitrile with respect to the neat bulk acetonitrile absorption, and also the measured sum frequency polarization for the CN transition of acetonitrile in the interface vs.

mole fraction of acetonitrile in the bulk water solution. Calculation using the polarization data yield a molecular orientation with respect to the surface normal of 70° at high mole fraction and 40° at low mole fraction.

The interfacial structural phase transition is described in terms of the competing hydrogen bonding and solvation interactions of the CN head group with neighboring water molecules and the dipole-dipole interactions among the CN head groups. At low surface densities the CN chromophores are surrounded by water molecules, are hydrogen bonded, and are tilted by about 40° to the interface normal. As the nitrile surface density increases, water molecules are squeezed out of the interface, the distances separating the aligned CN dipoles decrease, and the water screening of the nitriles is diminished. At the phase transition density the reduced number of water molecules separating and thus solvating and hydrogen bonding the nitrile molecules, enables the nitrile molecules to reorient towards the flatter and energetically more favorable head to tail dipole configuration. The trigger for the structural phase transition is the weakening or rupturing of the hydrogen bonding to and solvation by neighboring water molecules as the interface nitrile density increases, together with the lowering in energy of the dipoles that can be achieved by rotating to a configuration closer to a head to tail orientation. The sum of these effects is the observed spectral red shifts in the CN vibrational frequency and the shift in molecular orientation. The different interfacial behavior for the three nitrile interfaces is attributed to the limited interfacial packing density that can be achieved for the propionitrile and especially the butyronitrile molecules.

In addition to the short chain nitrile studies we have also used SFG to obtain the vibrational spectra and orientations for both the nitrile (-CN) head group and the methyl (-CD_3) tail group of the insoluble long chain nitrile $CD_3(CH_2)_{19}CN$ at the air/water interface. These SF studies of nitriles comprise the first studies in which two chromophores are attached to the same molecule, and offer the powerful advantage of probing how different regions of the same molecule are oriented and interacting with the molecules of the interface. The results show that the orientations and spectra of the nitrile head group and the methyl tail group vary with the molecular density, but in distinctly different ways. The nitrile head group maintains the same orientation and hydrogen bonding as the surface density increases until the coexistence to liquid phase transition density is reached. At this density there is an abrupt change in orientation as well as the breaking of the hydrogen bonding and solvation of the -CN group with the neighboring water molecules. We use this finding in a model developed to explain the phase transition behavior of the short chain nitriles we have studied. With respect to the methyl tail we observed that its orientation changed continuously as density increased, and in addition underwent a further abrupt change at the phase transition density. The methyl spectral shifts are of interest in that we found the symmetric stretch did not shift with density whereas a methyl overtone, which is sensitive to local environment, underwent a shift as the density was changed. As an additional point of interest we have discovered that SFG can be used to remove the ambiguity in the ratio of the two Raman polarizability elements for a symmetric stretch vibration; an ambiguity that cannot be resolved by the linear Raman spectroscopy.

FEMTOSECOND SPECTROSCOPY OF MID-GAP STATES IN POLYPHENYLACETYLENES

Takayoshi Kobayashi*

Department of Physics, University of Tokyo
7-3-1 Hongo, Bunkyo-ku, Tokyo 113, Japan

Femtosecond pump-probe experiments have been performed on three polyphenylacetylenes poly[*o*-(trimethylsilyl)phenylacetylene] (PMSPA), poly[4-tert-butyl-2,6-dimethylphenyl)acetylene] (PMBPA), and poly[(*o*-isopropylphenyl) acetylene] (PPPA) thin films to investigate the effect of confinement of a photogenerated soliton-antisoliton pair on its ultrafast relaxation dynamics. The decay kinetics of transient photoinduced absorption due to the photoexcitation of PMSPA were found to behave as erf (t^{-n}) with $n = 0.65 \pm 0.05$ as shown in Fig. 1, implying a geminate recombination process of the soliton-antisoliton pair on a quasi-one-dimensional chain. The distance between the soliton and antisoliton could be estimated as about six repeat unit lengths, suggesting a fairly strong confinement of the pair. It is consistent with the size of a certain regular structure indicated to exist in this material by X-ray diffraction measurements. Creation of confined pairs of likely charged solitons due to interchain photoexcitation is also discussed in terms of the following energetically favored reaction between likely charged polarons on the same chain: $P^{\pm} + P^{\pm} \rightarrow S^{\pm} + S^{\pm}$.

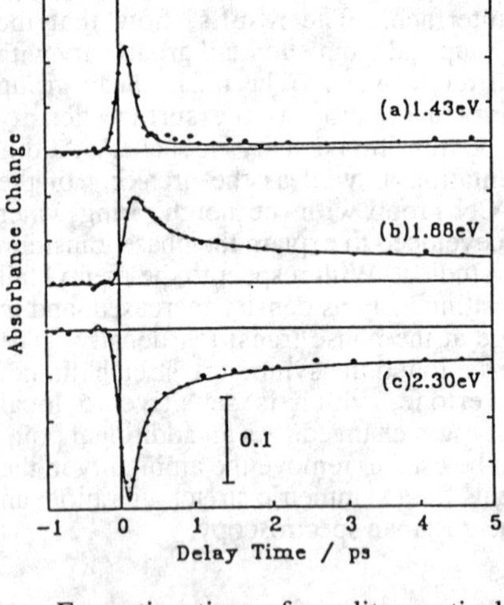

Fig. 1 Decay profiles of PMSPA probed at three different photon energies. Open circles represent the experimental data, and solid lines show the fitting curves with an exponential function for (a) and a power-law function for (b) and (c).

Formation time of a soliton-antisoliton pair has been determined as 100-50 fs experimentally by femtosecond transient spectroscopy performed on three substituted polyphenylacetylenes as shown in Fig. 2. It agrees favorably with results of the numerical simulation based on the Su-Schrieffer-Heeger model. A temporal behavior of photoinduced absorption was fitted successfully by a sum of an exponentially and a power-low decaying components (Fig. 3). This allowed one to determine separately the spectrum of the soliton-antisoliton pair and that of an exciton-polaron created by an intrachain photoexcitation.

The confinement parameter g = 0.23 ± 0.07 and on-site Coulomb repulsion U_0 = 2.8 ± 0.4 eV were determined from the results.[1] Similar experimental results were obtained for the other two polymers PMBPA and PPPA. The parameters obtained for the three polymers were compared from the viewpoint of the side chain effect.

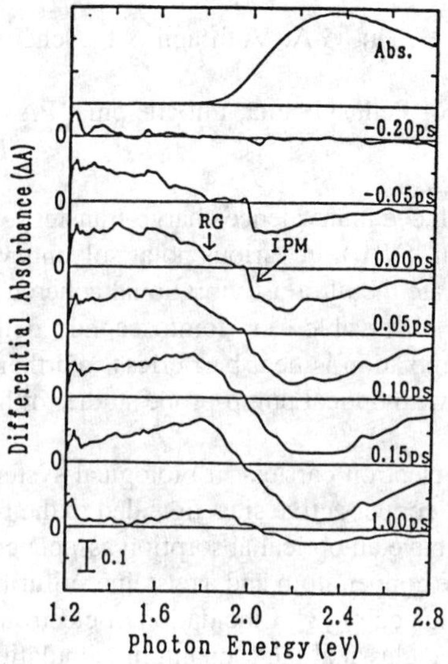

Fig. 2 Difference absorption spectra of PMSPA measured at 10 K up to 1 ps after photoexcitation at 1.98 eV. Pump duration and excitation density are 100 fs, 2x10^{16} photons/cm^2, respectively. The positive change means photoinduced absorption, while the negative one means photoinduced bleaching. The absorption spectrum (Abs.) of PMSPA and pump laser spectrum (pump) are also shown for comparison at the top. RG: peak minimum due to the Raman gain, IPM: sharp dispersion-type feature due to induced phase modulation.

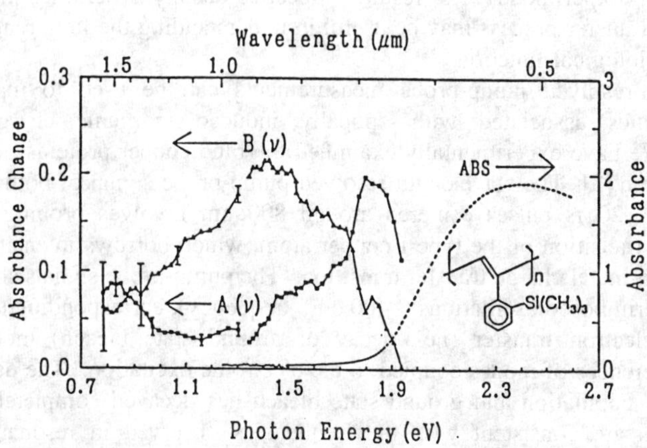

Fig. 3 The fractions of the soliton-antisoliton pair (A) and exciton-polaron (B) components plotted against probe photon energy $\hbar\upsilon$. The chemical structure of PMSPA and its stationary absorption spectrum (broken curve) are also shown.

Reference

[1] S. Takeuchi, T. Masuda, and T. Kobayashi, *Phys. Rev. B* 52(10), 7166 (1995). *IEEE J. Quantum Electron*, QE-28, 2518 (1992). *Solid State Commun.*, 87, 655 (1993).

OPTICAL AND VIBRATIONAL COHERENCE STUDIES OF LIQUIDS AND PROTEINS

D.C. Arnett, L.D. Book, L.W. Ungar, H. Hu, G.A. Voth and N.F. Scherer*

Department of Chemistry, University of Pennsylvania, Philadelphia, PA 19104-6323

The ultrafast dynamics of the optically-induced intervalence charge-transfer reaction of the mixed valence dimer[1] $(CN)_5Ru^{II}CNRu^{III}(NH_3)_5$ in various polar solvents will be presented. These results will be used to illustrate the ultrafast vibrational coherence and excitation relaxation dynamics that are observable in femtosecond nonlinear spectroscopy for a system whose electronic relaxation is near barrierless. Furthermore, these studies will be compared and contrasted with optical pump-probe studies of ligand-metal charge-transfer dynamics in proteins.

Blue copper proteins function as mobile electron carriers in biological systems by transfering electrons to and from their "type I" copper active sites liganded to the protein matrix.[2] In oxidized form, these active sites have an optical absorption associated with the ligand-to-metal charge transfer between the copper atom and a cysteine sulfur ligand, which gives the proteins their characterisitic blue color.[3] Elucidation of electronic and nuclear dynamics of these systems requires classical and quantum simulations in conjunction with experiment. The resultant spectral density describing the optically induced charge transfer process may be useful in understanding the long range electron transfer of physiological function.

Wavelength-resolved pump-probe measurements can be used to map out the ultrafast dynamics associated with optically induced inorganic charge transfer reactions.[4] We have experimentally examined the blue copper proteins plastocyanin and ceruloplasmin.[3] The wavelength resolved pump-probe signal of both proteins at 800 nm, using <20 fs pulses centered around 800 nm involves probing a Laporté forbidden $d \rightarrow d$ transition of the type I copper atom, which borrows intensity from the intense ligand-to-metal charge transfer transition. The pump-probe signals show a rapid decay with superimposed oscillations (~400 cm^{-1} frequency), corresponding to metal-to-ligand return electron transfer (i.e., decay of ground state bleach) modulated by vibrational coherences of modes coupled to the electronic excitation. The data indicate that the excited population and ground state bleach has decayed completely by 1 ps. The oscillations are consistent with the frequencies observed in resonance Raman measurements for the $Cu\text{-}S_{cysteine}$ bond.[5],[6]

On the theoretical side, we have developed a model for a generic blue copper active site and a complete plastocyanin protein that can be used in classical molecular dynamics simulations using the CHARMM molecular mechanics force field[7] modified to incorporate copper atom-protein interactions. The model was tested against resonance

Raman data[5],[6] for blue copper proteins and an x-ray crystallographic structure.[8] Simulation of optical excitation is achieved by modeling the potential difference between the ground and excited states as a change in the equilibrium distance between the copper atom and its cysteine sulfur ligand, without changing atomic partial charges.[3],[9] The Figure shows the Fourier transform of the Cu-$S_{cysteine}$ distance autocorrelation function from an equilibrium simulation of our model blue copper active site. The spectrum of the Cu-$S_{cysteine}$ frequency (at about 400 cm^{-1}) is split into three peaks and a shoulder, consistent with resonance Raman spectra.[5],[6] These classical simulations are useful in identifying the frequency, intensity and decay time of nuclear modes coupled to the optical electron transfer.

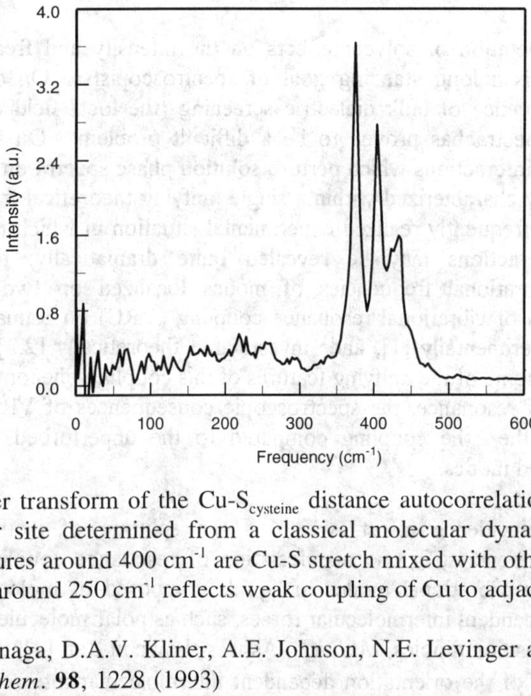

Fig. Fourier transform of the Cu-$S_{cysteine}$ distance autocorrelation function in a model blue copper site determined from a classical molecular dynamics simulation. The intense features around 400 cm^{-1} are Cu-S stretch mixed with other amino acid motions. Amplitude around 250 cm^{-1} reflects weak coupling of Cu to adjacent histidine nitrogens.

1. K. Tominaga, D.A.V. Kliner, A.E. Johnson, N.E. Levinger and P.F, Barbara, *J. Phys. Chem.* **98**, 1228 (1993)
2. R. Langen, I.-J. Chang, J.P. Germanas, J.H. Richards, J.R. Winkler and H.B. Gray, *Science* **268**, 1733 (1995)
3. E.I. Solomon, M.J. Baldwin and M.D. Lowery, *Chem. Rev.* **92**, 521 (1992)
4. D.C. Arnett, P. Vöhringer and N.F. Scherer, *J. Am. Chem. Soc.* **117**, 12262 (1995)
5. W.H. Woodruff, R.B. Dyer and J.R. Schoonover, *Biol. App. Raman Spec.* **3**, 413 (1988)
6. D. Qiu, J.A. Ybe, M.H. Hecht and T.G. Spiro, *J. Am. Chem. Soc.* **117**, 6443 (1995)
7. B.R. Brooks, R.E. Bruccoeri, B.D. Olafson, D.J. States, S. Swaminathan and M. Karplus, *J. Comp. Chem.* **4**, 187 (1983)
8. J.M. Guss and H.C. Freeman, *J. Mol. Biol.* **169**, 521 (1983)
9. S. Larsson, A. Broo and L. Sjölin, *J. Phys. Chem.* **99**, 4860 (1995)

VIBRATIONAL RESONANCE COUPLING IN RAMAN SPECTROSCOPY: A PROBE OF INTERMOLECULAR INTERACTIONS IN SOLUTION

Jeanne L. McHale

Department of Chemistry, University of Idaho, Moscow, ID 83843, USA

The interpretation of solvent effects on the intensity and frequency of vibrational transitions is a long standing goal of spectroscopists. On one hand, accounting for the influence of bulk dielectric screening (the local field effect) on infrared and Raman spectra has proven to be a difficult problem. On the other, specific intermolecular interactions which perturb solution phase spectra on a case by case basis are not easily characterized within a single unifying theoretical framework. However, there is one frequently realized experimental situation in which the effects of intermolecular interactions may be revealed quite dramatically: that of a coincidence in the vibrational frequencies of modes localized on two different molecules. The effects of vibrational resonance coupling (VRC) on Raman spectra have been verified experimentally [1], and investigated theoretically [2,3]. In this work we will consider some of the unifying features of this coupling phenomenon. As in intramolecular Fermi resonance, the spectroscopic consequences of VRC depend on the magnitude of the the coupling compared to the unperturbed frequency difference of the coupled modes.

The noncoincidence effect, which is the splitting of isotropic and anisotropic Raman frequencies, is one manifestation of resonance coupling. It is frequently observed for vibrational modes in liquids composed of molecules with strongly orientation-dependent intermolecular forces, such as polar molecules, and for modes having large transition dipoles, $d\mu / dQ$. As was shown in Ref. [4], the normal coordinate dependence of the orientation dependent intermolecular interaction leads to a difference in the peak frequencies for infrared ($\ell = 1$), isotropic Raman ($\ell = 0$), and anisotropic Raman ($\ell = 2$) spectra:

$$\omega_{\ell} = \omega_0 + \frac{N}{2m\omega_0} \langle V_{12} P_{\ell}(\cos\theta_{12}) \rangle \tag{1}$$

where $V_{12} = \left(\partial^2 V / \partial Q_1 \partial Q_2\right)_0$ is the "intermolecular force constant" connecting molecules 1 and 2, θ_{12} is the angle between their symmetry axes, ω_0 is the unperturbed vibrational frequency, and P_{ℓ} is the Legendre polynomial.

Perturbation theory has been employed to interpret the effect of VRC on the intensities and frequencies of overlapping solvent and solute vibrational modes. Two cases were considered. In the first case [2], equations were derived for the intensities and frequencies of a solute molecule solvated by N discrete solvent molecules. It was shown that VRC leads to an apparent intensity enhancement of the coupled peaks which is proportional to $N-1$. For N_A solute and N_B solvent molecules, the total intensity of the overlapping peaks was found to be:

$$I_{tot} = N_A\sigma_A + [N_B + N_A(N-1)]\sigma_B \qquad (2)$$

This theory has been successfully applied to the interpretation of VRC in the Raman spectra of cobalt porphyrin molecules [1]. The enhancement effect was demonstrated in the resonance Raman spectra of dye molecules in solutions with overlapping solvent bands [5]. Recently, we have observed VRC in the spectra of solutions of chloroform and dimethylsulfoxide [6], and the observed intensities were used to determine the stoichiometry (1:1 vs. 2:1) of the hydrogen bonded complex.

In the second case [3], long-ranged interactions are considered to be responsible for resonance coupling. It this situation it is necessary to average the solvent-solute interaction over the equilibrium structure of the liquid. It was shown that the first order (but not the second order) correction to the frequencies of the coupled modes vanishes, while the intensity perturbations are apparent at the first-order level. The perturbed intensities were shown to depend on the orientational correlation function $g_{112}(r)$.

Although the theory of VRC is straightforward, there are practical limitations on the experimental applications of this phenomenon. Nevertheless, recognition of the influence of VRC on observed spectra can avoid errors in assignments and even help to elucidate details of specific intermolecular interactions in solution.

1. a) K. Bajdor, J. R. Kincaid, and K. Nakamoto, J. Am. Chem. Soc. 106, 7741,(1981).
   b) L. M. Proniewicz, K. Nakamoto, J. R. Kincaid, J. Am. Chem. Soc. 110, 4541 (1988).
2. C. Veas and J. L. McHale, J. Am. Chem. Soc. 111, 7642 (1989).
3. J. L. McHale, Int. J. Quant. Chem. 25, 593 (1991).
4. a) C. H. Wang and J. L. McHale, J. Chem. Phys. 72, 4039 (1980).
   b) J. L. McHale J. Chem. Phys. 75, 30 (1981).
5. B. L. Rice and J. L. McHale, Proceedings of the XIIth Internat. Conf. on Raman. Spec., ed. J. R. Durig, and J. F. Sullivan (Wiley, New York, 1990) p. 574.
6. D. C. Daniel and J. L. McHale, submitted to the Journal of Physical Chemistry.

'MAGIC' FEMTOSECOND CARS IN LIQUIDS AND LIQUID MIXTURES

A. Laubereau

Physik-Department E11, Technische Universität München,
James-Franck-Strasse, D-85748 Garching, Germany

Introducing magic polarization geometries in time-resolved coherent Raman spectroscopy of liquid molecules, the three scattering contributions can be separated: the resonant isotropic, resonant anisotropic and nonresonant component.[1] This achievement is important for the observation of non-exponential vibrational dephasing on the subpicosecond time scale, described by the correlation time τ_c, since nonresonant and anisotropic scattering contribute notably in this range. More detailed information on vibrational dephasing has become directly accessible applying our polarization techniques and the high time resolution of femtosecond pulses.

Using the Abragam-Kubo theory the autocorrelation function may be written:[2]

$$\Phi_v(t) = \exp\left[-|t|/T_2 - \left(e^{-|t|/\tau_c} - 1 \right) \tau_c / T_2 \right] \qquad (1)$$

with parameters τ_c and T_2. Eq. 1 has model character; the oscillatory part of the autocorrelation function has been omitted. τ_c describes the time scale of the dephasing mechanism, while the asymptotic decay time T_2 is also governed by the strength of the phase-changing events. Φ_v can be measured by time-resolved CARS.

Our experimental system for femtosecond 3-color CARS has been discussed previously.[3] Three different input pulses are derived from a dye laser, continuum generation and dye amplifiers (frequency widths 50 - 70 cm^{-1}, durations 200-300 fs). Some results are depicted in Fig. 1. Neat bromochloromethane is studied at room temperature. The symmetric methylen stretching mode is resonantly excited. The measurements are carried out for the three magic angles while the other experimental parameters are kept constant. Fig. a shows the signal transient for analyzer orientation -30° with elimination of the resonant-isotropic scattering, while the nonresonant component clearly dominates. The small anisotropic part (dotted line) decays with a time constant of 950 ± 40 fs. A different situation is found for analyzer orientations of 60° and 49.1° (Figs. b and c) with a dominant resonant-isotropic part (dash-dotted lines). Fitting theoretical curves to the data yields a value of 320 ± 100 fs for τ_c and T_2 = 1.65 ± 0.03 ps.

Knowing T_2 precisely from the signal decay of Figs. b,c, the isotropic lineshape measured by Raman spectroscopy is readily compared with computed lineshapes deduced from Eq. 1. This analysis is highly reliable, since only one relevant parameter, τ_c, has to be determined.[4] A similar value of 350 ± 100 fs is obtained. Combining the two results we arrive at τ_c = 330 ± 70 fs. The numbers agree with the $\tau_c \simeq$ 400 fs of

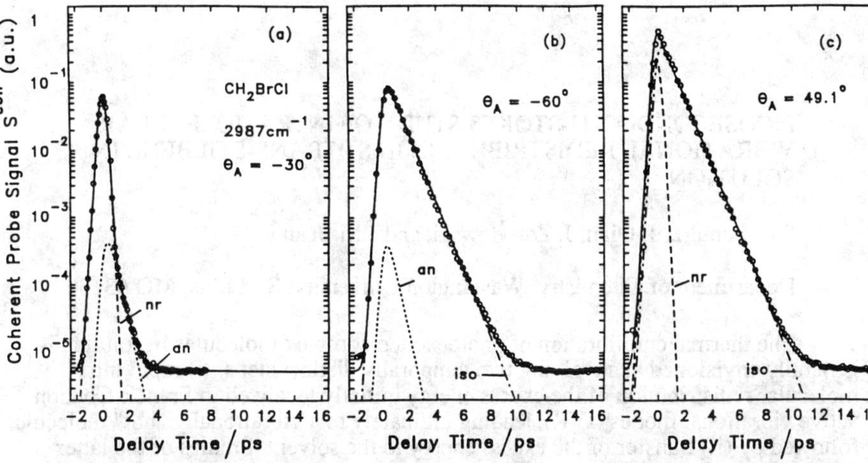

Fig. 1: Magic 3-colour CARS for the symmetric stretching vibration of CH_2BrCl at 2987 cm^{-1} with elimination respectively of the isotropic (a), nonresonant (b) and anisotropic scattering (c); experimental points, calculated curves.

Ref. 5, using a different method in ps time domain. Similar findings are obtained for other neat liquids, e.g. the symmetric mode at 2913 cm^{-1} of dimethyl sulphoxide with τ_c = 240 ± 50 fs. From the short values of τ_c, rapid modulation is indicated, representing homogeneous linebroadening. The results directly point to interaction via the (short range) repulsive part of the intermolecular potential and are consistent with estimates for a Enskog hard sphere liquid.[4] A contribution from the (long range) attractive part of the intermolecular potential is not seen.

In liquid mixtures the situation is found to be different. A considerably slower dephasing mechanism is introduced as indicated by larger values of the correlation time of a few picoseconds.[6] Experimental data will be presented for the system iodomethane:deuterochloroform. Our results are fully consistent with the theory of Knapp and Fischer[7] considering concentration fluctuations in the first solvent shell and will be compared with findings from other techniques.[8]

1. W. Li, H.-G. Purucker and A. Laubereau, Opt. Commun. 94, 300 (1992).
2. see for example, W. G. Rothschild, Dynamics of Molecular Liquids, Wiley, New York, 1983, p. 221 ff.
3. M. Fickenscher, H.-G. Purucker and A. Laubereau, Chem. Phys. Lett. 191, 182 (1992); J. Raman Spectr. 24, 453 (1993).
4. F.Lindenberger, C. Rauscher, H.-G. Purucker and A. Laubereau, J. Raman Spectrosc. 26, 835 (1995).
5. H. R. Telle and A. Laubereau, Chem. Phys. Lett. 94, 467 (1983).
6. F. Lindenberger et al, to be published
7. E.W. Knapp and S.F. Fischer, J. Chem. Phys. 76, 4730 (1982).
8. L.J. Muller, D. Vanden Bout and M. Berg, J. Chem. Phys. 99, 810 (1993).

PICOSECOND ANTI-STOKES STUDY OF INTRAMOLECULAR VIBRATIONAL REDISTRIBUTION IN S_1 TRANS-STILBENE IN SOLUTION

S.L. Schultz, J. Qian, J. Zdrodowski, and J.M. Jean*

Department of Chemistry, Washington University, St. Louis, MO 63130

The thermal equilibration of photoexcited complex molecules in solution is generally envisioned as involving two temporally distinct steps: (i) rapid intra-molecular redistribution of the excess energy initially localized in Franck-Condon active vibrational modes (IVR), leading ultimately to a vibrationally "hot" molecule, followed by (ii) transfer of the excess energy to the solvent. Studies of the latter process in the S_1 state of trans-stilbene (tS) have been carried out using picosecond Stokes scattering by our group and others [1-4]. The timescale of intermolecular vibrational energy transfer was found to be ~12 ps.

While these studies clarified the timescale of the overall thermal equilibration process, they provided little insight into the pathways for energy flow in this system. Recently our group [5] and Hester and co-workers [6] reported observation of anti-Stokes scattering from high frequency modes in S_1 tS. We report here results from picosecond pump-probe anti-Stokes experiments on the S_1 state of tS with ~1 ps resolution, which provide insight into the competition between internal and external pathways for vibrational energy transfer. Anti-Stokes spectra of tS following pumping 1600 cm^{-1} above the electronic origin are shown in Fig.1

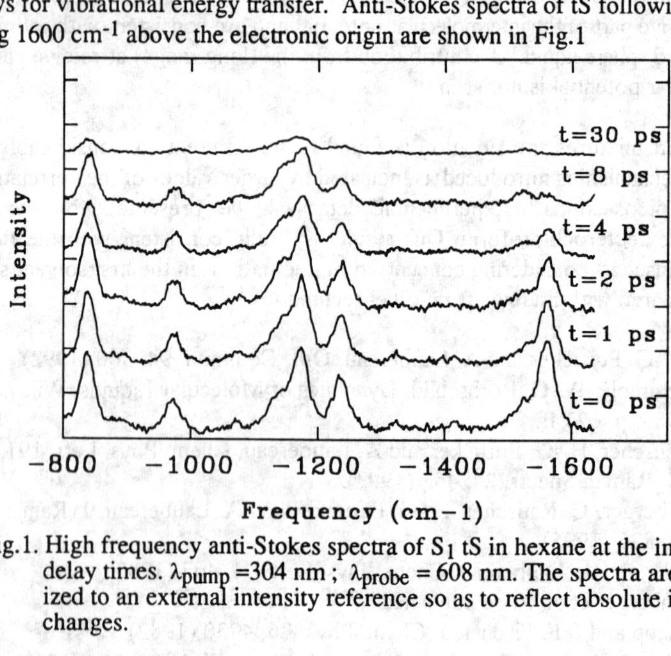

Fig.1. High frequency anti-Stokes spectra of S_1 tS in hexane at the indicated delay times. λ_{pump} =304 nm ; λ_{probe} = 608 nm. The spectra are normalized to an external intensity reference so as to reflect absolute intensity changes.

At t=0 ps, there is significant intensity in a number of high frequency bands. Much of the excess energy is initially localized in the strongly Franck-Condon active ν_7 mode (1570 cm^{-1}), which is predominately ethylenic stretch in character. Significant changes in both absolute and relative intensities in both high and low frequency regions are seen over the timescale of 30 ps.

In order to determine the extent to which these intensity changes reflect intra- and intermolecular energy transfer processes, we compare our intensity profiles with those obtained from a simple model that assumes instantaneous IVR. To apply this model, we need to calculate the transient temperature rise created by ultrafast IVR. This is done using the relation $\langle E(T)\rangle = \sum_i h\nu_i / \left\{\exp(h\nu_i / kT) - 1\right\}$ where $\langle E(T)\rangle = \langle E(295K)\rangle + E_{excess}$. Using the S_1 vibrational frequencies of tS , we find that converting the excess energy into a transient temperature rise results in an initial temperature of 385K. We then calculate the time-dependent intensities assuming that at all times the molecule is described by a Boltzmann distribution of the excess energy and that the excess energy decays exponentially with a time constant of 12 ps. To implement this idea, we fit the decay curves to the intensities at t=30 ps, which correspond to the S_1 state at ambient temperature (i.e. 295K), and extrapolate back to t=0 ps.

The experimental and predicted intensities for several prominent bands are shown in Fig.2. The discrepancy at short times between the experimental intensities and those predicted based on rapid thermalization, particularly for the 1236 cm^{-1} and 1570 cm^{-1} bands, clearly shows that the IVR process occurs on ~5 ps timescale. Thus the often-made assumption of a separation of timescales between intra- and intemolecular relaxation pathways does not hold in this sytem. In addition, we do not observe any rise times in the anti-Stokes intensities. This suggests that if IVR follows specific pathways, these pathways do not involve the resonance Raman active motions.

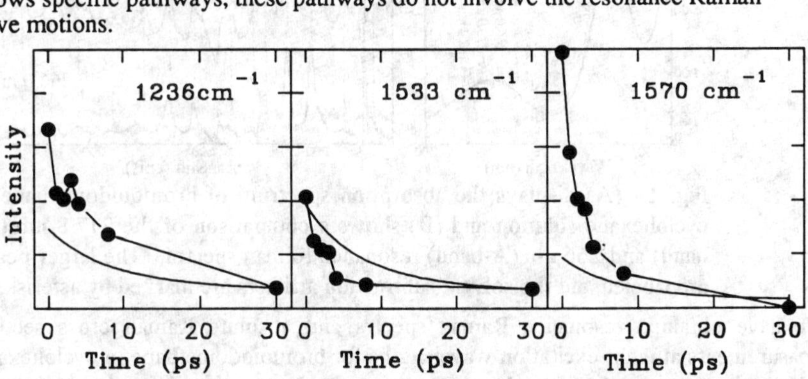

1. J.Qian, S.L.Schultz, G.Bradburn and J.M.Jean, J. Phys. Chem. **97**, 10638 (1993).
2. W.Weaver, L.Huston, K.Iwata and T.L.Gustafson, J. Phys. Chem. **96**, 895 (1992).
3. R.E.Hester, P.Matousek, J.N.Moore, A.W.Parker, W.T.Toner and M.Towrie, Chem. Phys. Lett. **208**, 471 (1993).
4. K.Iwata and H.Hamaguchi, Chem. Phys. Lett. **196**, 462 (1992).
5. J.Qian, S.L.Schultz and J.M. ean, Chem. Phys. Lett. **233**, 9 (1995).
6. P. Matousek, A. Parker, W.T.Toner, M.Towrie, D.L.A. de Faria, R.E.Hester, and J.N.Moore, Chem. Phys. Lett. **237**, 373 (1995).

RESONANCE RAMAN EXAMINATION OF BOND SELECTIVE ELECTRONIC EXCITATION IN BROMOIODOMETHANE

S.-Q. Man, W. M. Kwok, and D. L. Phillips*

Department of Chemistry, University of Hong Kong, Pokfulam Road, Hong Kong

We have investigated the nature of bond selective electronic excitation in bromoiodomethane in cyclohexane solution. Photoexcitation of gas phase bromoiodomethane has been observed to lead to bond selective photodissociation with the C-I bond predominately broken in the A-band and the C-Br bond predominately broken in the B-band [1,2]. Anisotropy measurements from molecular beam experiments [1,2] have shown that the A-band and B-band photodissociation reactions are very fast compared to molecular rotation.

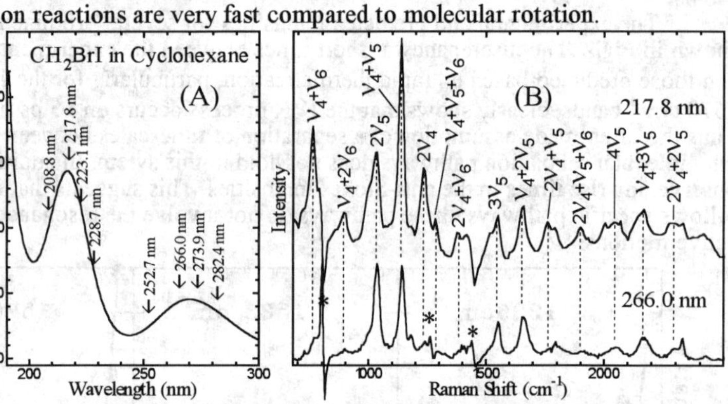

Fig. 1. (A) displays the absorption spectrum of bromoiodomethane in cyclohexane solution and (B) shows a comparison of the 217.8 nm (B-band) and 266 nm (A-band) resonance Raman spectra. The larger peaks are labeled and the solvent subtraction artifacts are marked by asterisks.

We have obtained resonance Raman spectra and absolute Raman cross section measurements at eight excitation wavelengths for bromoiodomethane in cyclohexane solution in order to elucidate the short-time photodissociation dynamics associated with this prototypical molecule at the vibrational mode-specific level.

The experimental methods have been described in previous work [3,4]. Fig. 1 (A) shows the absorption spectrum of bromoiodomethane in cyclohexane solution and Fig. 1 (B) shows a comparison of the 217.8 nm and 266 nm resonance Raman spectra. Most of the intensity in the A-band and B-band resonance Raman spectra appear in fundamentals, overtones and/or combination bands of three Franck-Condon active modes: ν_4 (the nominal C-Br stretch), ν_5 (the nominal C-I stretch), and ν_6 (the nominal I-C-Br stretch). Inspection of Fig. 1 (B) shows that the 217.8

nm B-band resonance Raman spectrum displays more intensity in the overtone of the nominal C-Br stretch mode ($2\nu_4$), in the combination bands of the nominal C-Br stretch fundamental and overtone with the fundamentals and overtones of the nominal C-I stretch and nominal Br-C-I bend vibrational modes ($\nu_4 + \nu_5$, $\nu_4 + \nu_6$, $\nu_4 + 2\nu_6$, $\nu_4 - \nu_6$, $2\nu_4 + \nu_6$, $2\nu_4 + \nu_5$, $2\nu_4 + 2\nu_5$, $\nu_4 + \nu_5 + \nu_6$ and $2\nu_4 + \nu_5 + \nu_6$) than the 266.0 nm A-band spectrum. This suggests that there is more motion along the nominal C-Br stretch and Br-C-I bend modes in the B-band than in the A-band. We have done time-dependent wavepacket calculations to simulate the resonance Raman intensities and absorption bands using harmonic oscillators with the same ground and excited state frequencies displaced by an amount Δ in dimensionless normal coordinates. The parameters for the calculations are shown below in Table I.

Table I. Parameters for Calculation of Resonance Raman Intensities

Vibrational Mode	Freq. (cm^{-1})	A-Band |Δ|	B-Band |Δ|
ν_4 (C-Br stretch)	621 cm^{-1}	2.65	2.5
ν_5 (C-I stretch)	521 cm^{-1}	4.8	3.0
ν_6 (Br-C-I bend)	151 cm^{-1}	4.5	3.6

E_0 for A-band = 27,200 cm^{-1} and E_0 for B-band = 41,100 cm^{-1}

Figure 2 shows a comparison of the experimental resonance Raman intensities with the calculated Raman intensities using the parameters from Table I. There is reasonable agreement between experiment and the simulated intensities for most of the combination bands and overtones. These results imply that the dissociation coordinate in both the A- and B-bands have a degree of multidimensional character as well as noticeable coupling of the C-Br and C-I chromophores [3,5].

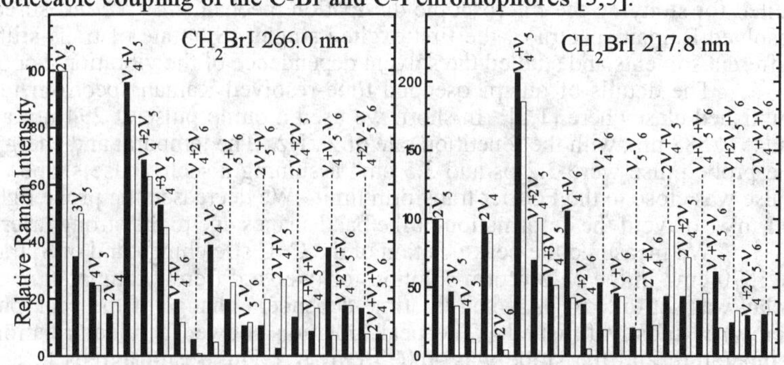

Fig. 2. Comparison of experimental (solid bars) and calculated (open bars) Raman intensities for the 266 nm A-band and 217.8 nm B-band resonance Raman spectra.

1. S. J. Lee and R. Bersohn, J. Phys. Chem. 86, 728 (1982).
2. L. J. Butler, E. J. Hintsa, S. F. Shane, and Y. T. Lee, J. Chem. Phys. 86, 2051 (1987).
3. S.-Q. Man, W. M. Kwok, and D. L. Phillips, J. Phys. Chem. 99, 15705 (1995).
4. W. M. Kwok and D. L. Phillips, J. Chem. Phys., 104, 2529 (1996).
5. S.-Q. Man, W. M. Kwok, A. E. Johnson, and D. L. Phillips, J. Chem. Phys., submitted for publication.

SOLVENT DEPENDENT EXCESS ENERGY DISSIPATION PROCESS IN SOLUTION MONITORED WITH PICOSECOND TIME-RESOLVED RAMAN SPECTROSCOPY

Koichi Iwata[1] and Hiro-o Hamaguchi[1,2]

[1]Molecular Spectroscopy Laboratory, Kanagawa Academy of Science and Technology (KAST), KSP East 301, 3-2-1 Sakato, Kawasaki 213, Japan
[2]Department of Basic Science, Graduate School of Arts and Sciences, The University of Tokyo, 3-8-1 Komaba, Meguro, Tokyo 153, Japan

Fast energy dissipation from a solute to the surrounding solvents is a process characteristic to chemical reactions in solution. The solvent molecules effectively remove the excess energy from newly formed unstable product molecules. Such a process is lacking in the gas phase reactions. Because of this difference, the same reactants sometimes give different reaction products in the gas and solution phases. Although the excess energy dissipation process is crucially important for chemical reactions in solution, our current understanding on its microscopic mechanism is quite limited. Clearly, more experimental observations are needed on this subject.

Ultrafast time-resolved vibrational spectroscopies are suitable methods to monitor the excess energy dissipation process from a photoexcited molecule, because the excess energy is first deposited in the vibrational degrees of freedom of the solute. We can monitor the energy flow from the solute to the solvent by observing the vibrational cooling process of the solute. Because a typical time scale for this vibrational cooling in solution is 10 ps, picosecond time-resolved Raman spectroscopy is best suited for studying the microscopic dissipation mechanism. We measured the time-resolved Raman spectra of the first excited singlet (S_1) state of trans-stilbene in ten different solvents and studied the solvent dependence of the vibrational cooling rate.

The details of our picosecond time-resolved Raman spectrometer has been published elesewhere [1,2]. In short, we used a pump pulse at 294 nm and a probe pulse at 588 nm, with the repetition rate of 2 kHz. The temporal and energy widths of the probe pulse were 3.2 ps and 3.5 cm^{-1} assuming a $sech^2$ pulse shape. The probe pulse was close to the Fourier transform limit. We decreased the probe light power to 0.1 mW to avoid the deformation of the band shapes due to the strong laser light field.

We measured the peak position of the C=C stretching band of S_1 trans-stilbene at 1570 cm^{-1} in the chloroform solution for twelve different temperatures. The time delay was set to t=50 ps, when the time-dependent change of the peak position (see below) is mostly finished. The peak position showed a linear relation with the temperature and the slope was -0.093 cm^{-1}/K. This result agrees to a previously reported result [4] qualitatively. It has been known that the 1570 cm^{-1} band changes its position when the time delay is changed [1-4]. By using the obtained linear relation, we can determine the temperature of the photoexcited S_1 trans-stilbene at each time delay after the photoexcitation. The result is shown in Fig. 1.

By monitoring the anti-Stokes/Stokes intensity ratio, we can also study the vibrational cooling process. The anti-Stokes/Stokes intensity ratio of the 285 cm^{-1} band is plotted against the time delay in Fig. 1, together with the peak position of the 1570 cm^{-1} band. Because the two analyses gave the same cooling curve, we conclude that we can use the time-resolved Raman spectrometer as a "picosecond thermometer".

The observed vibrational cooling rates of photoexcited S_1 trans-stilbene in ten different solvents are plotted against the thermal diffusivities of the solvents in Fig. 2.

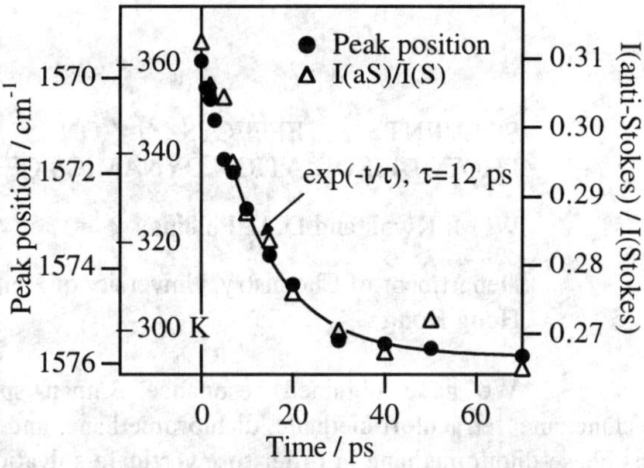

Fig. 1. Vibrational cooling of S_1 trans-stilbene in chloroform

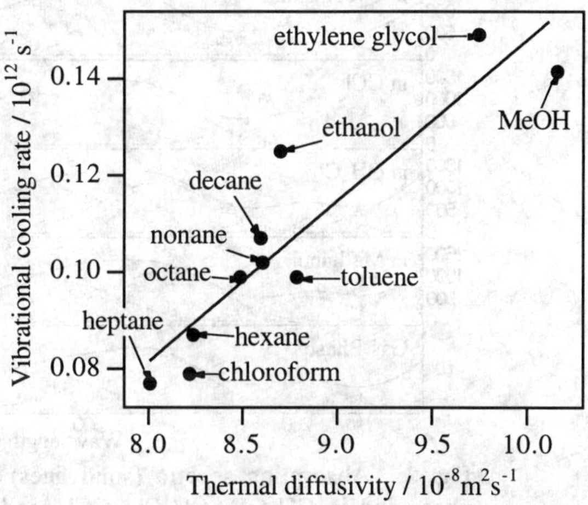

Fig. 2 Vibrational cooling rate vs thermal diffusivity

There is a strong positive correlation between the two physical quantities. However, the vibrational cooling is primarily an intramolecular process of a solute, while the thermal diffusivity is a bulk property of a solvent. These two quantities do not have any intrinsic correlation with each other. To account for the observed correlation, we assume that the energy transfer from the solute to the nearest solvent molecules are much faster than the energy flow between the solvent molecules. In this case, the rate of the whole energy dissipation process is determined only by the energy flow rate in the bulk solvent. We should also note the variety of solvents listed in Fig. 2. Properties of solvents such as polarity or dielectric constant do not seem to affect the overall energy dissipation rate in a decisive way

We made a simple numeric model to explain the observed cooling curve quantitatively [5]. With the known heat conduction property of hexane, for example, the solution of the diffusion equation of heat can explain the observed cooling curve satisfactorily. The best agreement with the observed cooling curve was obtained with an initial condition in which the heat at t=0 is distributed uniformly inside a box. The size of the box is larger than trans-stilbene by 2.5 nm for each direction. This supports our conclusion that, in the excess energy dissipation process, the energy transfer from the solute to the nearest solvents happens very fast (at least within a few picoseconds) and then the slow energy flow in the solvent follows.

1. K. Iwata and H. Hamaguchi, Chem. Phys. Lett. 196, 462 (1992).
2. K. Iwata, S. Yamaguchi, and H. Hamaguchi, Rev. Sci. Instrum. 64, 2140 (1993).
3. W. L. Weaver, L. A. Huston, K. Iwata, and T. L. Gustafson, J. Phys. Chem. 96, 8956 (1992).
4. R. E. Hester, P. Matousek, J. N. Moore, A. W. Parker, W. T. Toner, and M. Towrie, Chem. Phys. Lett. 208, 471 (1993).
5. K. Iwata and H. Hamaguchi, J. Mol. Liquids 65/66, 417 (1995).

SOLVENT EFFECTS ON THE EARLY-TIME PHOTODISSOCIATION DYNAMICS OF DIIODOMETHANE

W. M. Kwok and D. L. Phillips*

Department of Chemistry, University of Hong Kong, Pokfulam Road, Hong Kong

We have obtained resonance Raman spectra of diiodomethane in cyclohexane, tetrachloromethane, dichloromethane, and methanol solvents as well as gas phase diiodomethane in order to investigate solvation and solvent effects on the short-time photodissociation dynamics of diiodomethane.

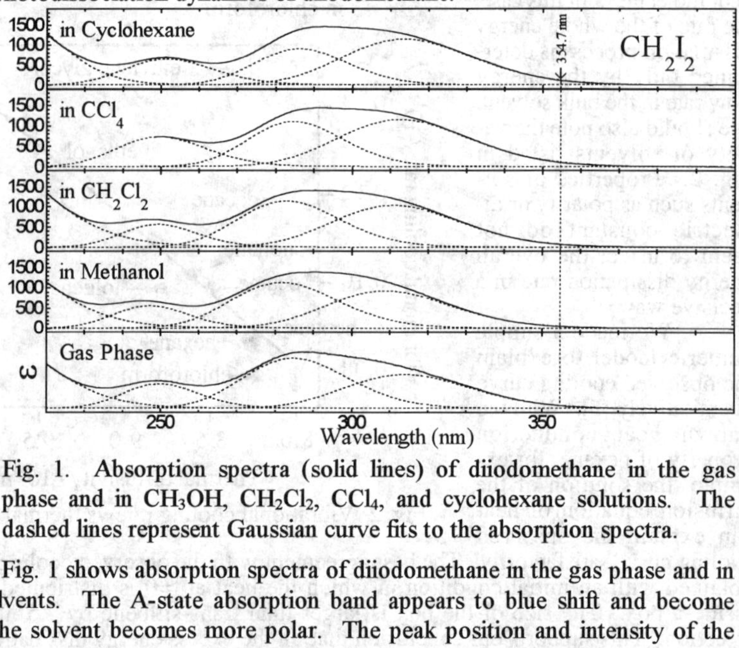

Fig. 1. Absorption spectra (solid lines) of diiodomethane in the gas phase and in CH_3OH, CH_2Cl_2, CCl_4, and cyclohexane solutions. The dashed lines represent Gaussian curve fits to the absorption spectra.

Fig. 1 shows absorption spectra of diiodomethane in the gas phase and in different solvents. The A-state absorption band appears to blue shift and become weaker as the solvent becomes more polar. The peak position and intensity of the A-state Gaussian used in fitting the diiodomethane absorption spectra are 313.6 nm and ε=1145 in cyclohexane, 312.4 nm and ε=1151 in CCl_4, 310.8 nm and ε=1030 in CH_2Cl_2, and 308.8 nm and ε=922 in CH_3OH.

Fig. 2 displays resonance Raman spectra for gas and solution phase diiodomethane and inspection of Fig. 2 reveals differences in the overall intensity patterns of the I-C-I antisymmetric stretch (ν_9), the I-C-I bend (ν_4) and their combination bands. If diiodomethane has a strict C_{2v} symmetry then the resonance Raman spectrum will not have ν_9, $\nu_9+\nu_4$, $\nu_9+2\nu_4$, and $\nu_9+3\nu_4$ peaks since these are nontotally symmetric vibrations. This is indeed the case for gas phase diiodomethane where the isolated molecules have C_{2v} symmetry and both C-I bonds lengthen at the same rate in the Franck-Condon region[1-3]. However, solvent perturbations appear to reduce the symmetry of diiodomethane in the solution phase and this symmetry breaking shows up rather dramatically as allowing the ν_9, $\nu_9+\nu_4$, $\nu_9+2\nu_4$, and $\nu_9+3\nu_4$ peaks appear with substantial intensity in the A-state resonance Raman spectra. This implies that the two C-I bonds are not equivalent in the solution phase and that the molecule chooses which C-I bond is broken in the Franck-Condon region soon after photoexcitation [3,4].

The amount of apparent symmetry breaking is very solvent dependent with the following order: cyclohexane > CCl_4 > CH_2Cl_2 > methanol. The observed solvent-induced symmetry breaking may be mostly due to mode-specific solvent-solute interactions and/or a polarization solvent-solute interaction. The time-scale of solvent fluctuations relative to the resonance Raman scattering time-scale will also likely play a role in determining the amount of symmetry breaking seen in different solvents. For instance, the solvent fluctuations in methanol may be faster than in cyclohexane so that the symmetry of the two C-I bonds is better maintained in methanol than in cyclohexane which would give rise to less symmetry breaking in methanol than in cyclohexane. We are currently modeling the resonance Raman spectra of diiodomethane in a variety of solvents in order to quantitate the solvent induced perturbation of the excited state potential energy surface in the Franck-Condon region and this will be reported in the future [5].

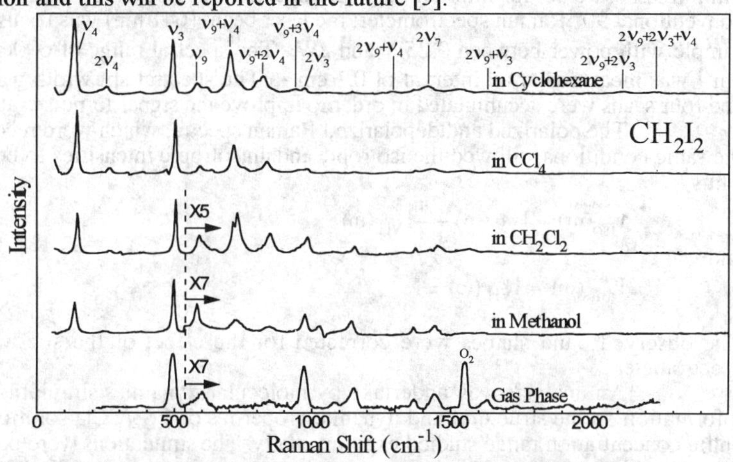

Fig. 2. Resonance Raman spectra of gas phase diiodomethane and diiodomethane in cyclohexane, CCl_4, CH_2Cl_2, and methanol solvents. The larger peak assignments are labeled.

1. J. Zhang, and D. G. Imre, J. Chem. Phys., 89, 309 (1988).
2. J. Zhang, E. J. Heller, D. Huber, D. G. Imre, and D. Tannor, J. Chem. Phys., 89, 3602 (1988).
3. W. M. Kwok and D. L. Phillips, J. Chem. Phys., 104, 2529 (1996).
4. W. M. Kwok and D. L. Phillips, Chem. Phys. Lett., 235, 260 (1995).
5. W. M. Kwok and D. L. Phillips, in preparation.

MOLECULAR DYNAMICS AND RAMAN SPECTROSCOPIC STUDY OF REORIENTATIONAL RELAXATION OF CS_2 IN DILUTE CCl_4 SOLUTIONS

A. Idrissi, I. Vercuque, F. Sokolic and G. Turrell*

Laboratoire de Spectrochimie Infrarouge et Raman (CNRS UPR A2631L), Université des Sciences et Technologies de Lille
59655 Villeneuve d'Ascq, France

The profiles of the infrared and Raman bands of a molecule in solution depend on the effective potential function which governs solute-solvent interaction. The autocorrelation function (acf) obtained by Fourier transformation of a spectral profile provides a dynamic description of the system.

In previous work [1] the vibrational and orientational relaxation of CS_2 in liquid CCl_4 solutions was investigated at room temperature over the concentration range $CS_2/CCl_4 = 0.25$ to 0.05. As the polarizability of this solute is very high, it was possible to obtain Raman spectra with relatively good signal-to-noise ratios. Thus, spectra were recorded of this system at concentrations which are sufficiently low to minimize solute-solute interaction. The spectra were obtained with the use of a conventional 90° Raman spectrometer; the laser beam (641 nm) was focused onto the sample with power between 0.25W and 2W. The spectral range of 600 cm^{-1} to 700 cm^{-1} was measured at an interval of 0.1 cm^{-1}. The spectral slit width was 0.5 cm^{-1} and four scans were accumulated in order to improve the signal-to-noise ratio.

The polarized and depolarized Raman spectra, which were recorded under the same conditions, allowed the isotropic and anisotropic intensities to be evaluated. Thus,

$$I_{iso}(\omega) = I_{VV}(\omega) - \frac{4}{3} I_{VH}(\omega) \tag{1}$$

and

$$I_{anis}(\omega) = I_{VH}(\omega) \ . \tag{2}$$

The observed band shapes were corrected for the effect of the slit width of the spectrometer.

An analysis was undertake by molecular dynamics simulation to derive information on the structural and dynamic properties of CS_2/CCl_4 solutions over the entire concentration range studied experimentally. The simulations were performed on a system of 18, 26, 38, 64 solutes molecules in interaction with 238, 230, 218, 192 molecules of CCl_4, respectively. One time step was equivalent to 5 10^{-15} s. After reaching equilibrium, the time dependent acf's for different dynamic properties of the system were obtained. A pairwise-additive potential was used for the description of the interaction, *viz.*

$$U_{LJ}(\mathbf{r}^N) = \sum_i \sum_{j>i} u(r_{ij}) \ , \tag{3}$$

the energy between molecules i and j is represented by the simple Lennard-Jones function, plus the Coulombic interaction, in the form

$$u\,(r_{ij}) = 4\sum_{\alpha}\sum_{\beta}\varepsilon_{\alpha\beta}\left[\left(\frac{\sigma_{\alpha\beta}}{r_{i\alpha j\beta}}\right)^{12} - \left(\frac{\sigma_{\alpha\beta}}{r_{i\alpha j\beta}}\right)^{6}\right] + \frac{1}{4\pi\varepsilon_0}\sum_{\alpha}\sum_{\beta}\frac{q_\alpha q_\beta}{r_{i\alpha j\beta}} \,. \qquad (4)$$

Here $r_{i\alpha j\beta}$ is the distance between site α on molecule i and site β on molecule j. The parameters $\varepsilon_{\alpha\beta}$ and $\sigma_{\alpha\beta}$ are the energy and the effective atomic diameters in the LJ potential function. These parameters are taken from ref. [2]; the q's are the effective charges on the appropriate atoms.

The autocorrelation function $C(t) = \left\langle \frac{3}{2}[\mathbf{u}_i(0)\cdot\mathbf{u}_i(t)]^2 - \frac{1}{2}\right\rangle$, where $\mathbf{u}_i$ is a unit vector along the axis of the solute molecule i, was calculated to provide a comparison with the results of the Raman spectroscopic investigation. These acf's are shown in Fig. 1. The decay becomes more rapid with increasing CS_2 concentration. The same behavior is observed experimentally. In the range of the concentration considered here, the first coordination number was determined to provide a simple description of the environment of the solute molecule. This quantity and the self-diffusion coefficient of the solute molecule is shown in Table 1.

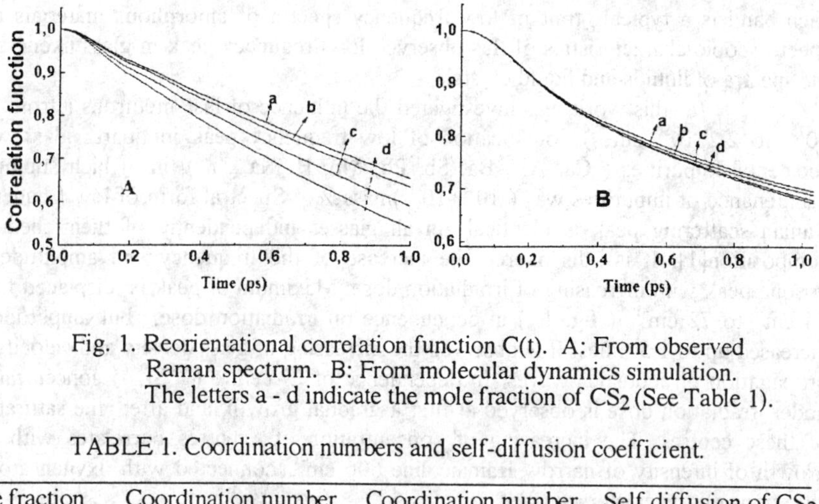

Fig. 1. Reorientational correlation function C(t). A: From observed Raman spectrum. B: From molecular dynamics simulation. The letters a - d indicate the mole fraction of CS_2 (See Table 1).

TABLE 1. Coordination numbers and self-diffusion coefficient.

Mole fraction of CS_2	Coordination number (CS_2)	Coordination number (CCl_4)	Self diffusion of CS_2 (10^{-5} cm^2/s)
a = 0.07	0.56	10.74	2.35
b = 0.10	1.23	10.35	2.59
c = 0.15	1.73	10.12	2.74
d = 0.25	2.58	9.53	2.95

The increase of the number of CCl_4 molecules around the solute molecule indicates an increase of the steric effect on the reorientational movement and explains the decay of the reorientational acf. This result is confirmed by the decrease of the self-diffusion coefficient with dilution.

The "Institut du Développement et des Ressources en Informatique Scientifique" is thanked for the allocation of CPU time on the CRAY C98.

1. A. Idrissi, I. Vercuque, F. Sokolic and G. Turrell, Poster presentation at "Fast Elementary Processes in Chemical and Biological Systems", Lille, 1995.
2. U. Mittag, Doctoral Thesis, University of Bielfeld, Germany, 1988.

SELF ORGANIZATION AND LOW FREQUENCY RAMAN SCATTERING IN QUARTZ GLASS IRRADIATED BY FAST NEUTRONS.

Erkin Nurmat, O.D.Davronov, U.N.Radjabov, A.N.Salachitdinov

Physical department, Samarkand State University, University boulevard 15, Samarkand, 703004, Uzbekistan

In all investigated glass materials [1] in low frequency region of the IR absorption and Raman scattering spectra it is observed intensive and sufficienly broad band with maximum within ~10-100 cm^{-1} and bandwidth ~ν_{max}. The availability of such band is a typical trait of low frequency spectra of amorphous materials and spectroscopic characteristics of this observed low frequency peak in glass likeness to the spectra of liquids and liquid crystals.

In this work was investigated the influence of fast neutrons (from 2,5 10^{15} to 2,2 10^{20} cm^{-2}) on location of low frequency peak in quartz glass with accidental impurities (Ca, Al , Ba, Sb, Pb, Mn, B, Na, Zn), in which summary maintenance of impurities was (10^{-3} -10^{-1})mass % . Spectral form of low frequency Raman scattering peak is identical in all glasses independently of their chemical composition[1]. It is discovered the increase of the frequency and amplitude of bosone peak with increasing of irradiation dose. Maximum of peak is displaced from 54 cm^{-1} to 72 cm^{-1} (fig. 1.) in dependence on irradiation dose, but amplitude is increased up to 1,5 times. It is observed the increasing of glass density and velocity of propagation of acoustic waves. In dependence of E'-centre ($^{28}Si^{3+}$) concentration under irradiation dose is observed at first a gradual growth, and after the saturation of these centres. The increasing of concentration of centres correlates with the growth of intensity of narrow Raman line 606 cm^{-1}, connected with oxygen atoms' vibrations on the clusters surface.

The results of this investigation with taking into account the ideas developed in [2,3] allows to develop another approach to the nature of low frequency Raman scattering in glasses.

The irradiation by fast neutron lead to the changing degree of self organization or phase correlation in glasses. It is lead to the rising of internal field of phase structure, and consequently, to the changing of wave vector of phase structure, which is displayed in the shift of frequency of bosone peak. The changing of self organization degree influence on macroscopic parameters of medium and it is displayed in the changing of glass density and velocity of propagation acoustic waves'. The increasing of internal field of phase structure equivalent to the changing of distribution energetic states because of the changing of self orga-nization degree. This process leads to the changing of density of fluctuation transitions between hierarchies of energies' states, which is displayed in the growth of E'-centres concentration and Raman line intensity.

According to our ideas [2,3] about structure-dynamical processes in non- regulated media and theirs spectral display, and also by results of this work the nature of bosone peak directly don't connect with velocity of propagation acoustic waves and hypersound data, which can be obtained from Brillouin scattering. In our opinion the nature of bosone peak is formed by more sensible processes in medium, just by phase correlation radius, which cause the main role in the forming of macroscopic parameters of medium. Frequency distribution of bosone peak displays the distribution of internal field of phase structure of nonregulated medium.

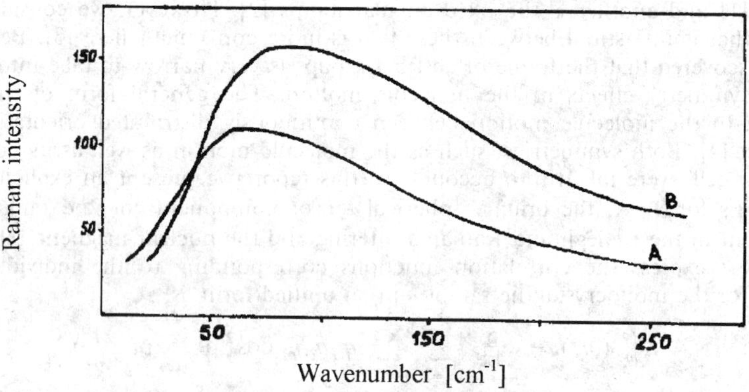

Fig. 1. Spectral form of low frequency Raman scattering in quartz glass. A - nonirradiated, B - irradiated by fast neutron (2,0 10^{20} cm^{-2})

1. T.S.Perova, Physics and chemistry of glasses. 20, 301, (1994)
2. Erkin Nurmat, Proc. XIV Int. Conf. Raman. Spectr., Hong Kong, A 371, (1994)
3. Erkin Nurmat, R.M.Ishburiev, Sh.Umidullaev. Proc.XIV Int.Conf.Raman. Spectr., Hong Kong, 1068, (1994)

ORIENTATIONAL AUTO-CORRELATION FUNCTIONS OF SECOND RANK SPHERICAL TENSORS IN CRYSTALS

Bashirov I.F., Bashirov F.I., and Gaisin N.K.
Department of Physics, Kazan State University, 420008, Kazan, Tatarstan, Russia

Time depended auto-correlation functions (ACF) are important in study of the molecule hindered motion (MHM) in condensed matter. There are two well-known models of the MHM in molecular crystals. The first one is the rotational diffusion model [1] and another is the fixed angular jumps [2]. However, we consider that a model that is interstitial between these two is more convenient here [3]. Besides, we have discovered that the frame of abstract groups is very narrow to take into account the dissymmetry effects in the molecule motion. The general form of the ACFs, relating to the molecule motion between continuously distributed orientations, was given in [4]. Both symmetries, such as the molecule motion as well as its position in the unit cell, were taken into account. In this report we present an explicit form of the ACFs for $Y_m^{(2)}$, the unitary spherical tensor components of 2nd rank that are important in the Reilegh and Raman scattering and the nuclear magnetic relaxation.

We express the correlation functions corresponding to the individual point groups for the monocrystalline samples in an unified form as

$$K_m^{(2)}(\theta, t) = \frac{5}{4\pi} \sum_{\alpha=1}^{3} \sum_{l=0}^{2} q_\alpha a_{\alpha lm} \cos^{2l}\theta \exp(-\frac{t}{\tau_\alpha}) \qquad (1)$$

where the mutual orientation between the principal axis C of a crystallographic coordinate frame and the polar axis of an axial laboratory coordinate frame is expressed by θ. The factors $a_{\alpha lm}$ relating to the different irreducible representations (IR) Γ_α are presented for some crystallographic point symmetry groups in Table. The angle of the crystal orientation about its axis C is denoted by ε.

CRYSTAL SYSTEM	ε	*l* / *m*	*0* ($\alpha = 1$)	*1* ($\alpha = 1$)	*2* ($\alpha = 1$)	*0* ($\alpha = 2$)	*1* ($\alpha = 2$)	*2* ($\alpha = 2$)
CUBIC	π/4	0	1/8	−3/4	9/8	1/4	1/2	−3/4
O, T		1	1/4	1/2	−3/4	1/6	−1/3	1/2
		2	3/16	−1/8	3/16	5/24	1/12	−1/8
HEXAGONAL	-	0	3/8	−3/4	3/8	0	3/2	−3/2
TRIGONAL		1	1/4	0	−1/4	1/4	−3/4	1
D_6, C_6, D_3, C_3		2	1/16	3/8	1/16	1/4	0	−1/4

There are two kinds of dynamical variables in Eq. 1 - q_α and τ_α, named as the dynamic weight and the correlation time symmetrized by Γ_α. A variable q_α is considered as a probability that the molecule occupies the hindered state transforming by Γ_α and obeys the normalized relation $\Sigma q_\alpha = 1$. Another dynamical variable τ_α depends on τ, the mean time between two successive molecule rearrangements, p_i, the probabilities of the MHM about the classes i, and $\chi_{\alpha i}$, the characters of group G, by

$$\tau_\alpha = \tau \left[1 - \chi_{\alpha E}^{-1} \sum_i \chi_{\alpha i} p_i(g_e) \right]^{-1}$$

where E denotes the identity class. Generally, q_α and τ_α are the experimentally defined quantities. The examples of the allowed angular dependence for the ACF amplitudes are presented in Figs. 1 and 2. Fig. 1 corresponds to the point groups of the cubic system, Fig. 2 - to that of the hexagonal and trigonal systems. It is shown how the anisotropy of the ACFs changes with the variation of the parameter q_1, one of the dynamic weight of the given group G. The subscript 1 in q_1 denotes the two-dimensional IR for groups of the cubic system and that of two two-dimensional IRs for groups of the hexagonal or trigonal systems. It should be noted that the weight of the identity representation for the groups D_6, C_6, D_3, C_3 is taken equally to 0 here.

One may see the straight lines in the curves of Fig. 1. It is the case when the site symmetry of the rotating molecule is a purely cubic. It corresponds to $q_1 = 0.4$. This fact is consistent with a behavior of any 2nd rank physical quantities in cubic crystals. The case of $q_1 < 0.4$ corresponds to the tetragonal distortion of the cubic site symmetry. Alternatively, if $q_1 > 0.4$, there is the trigonal distortion.

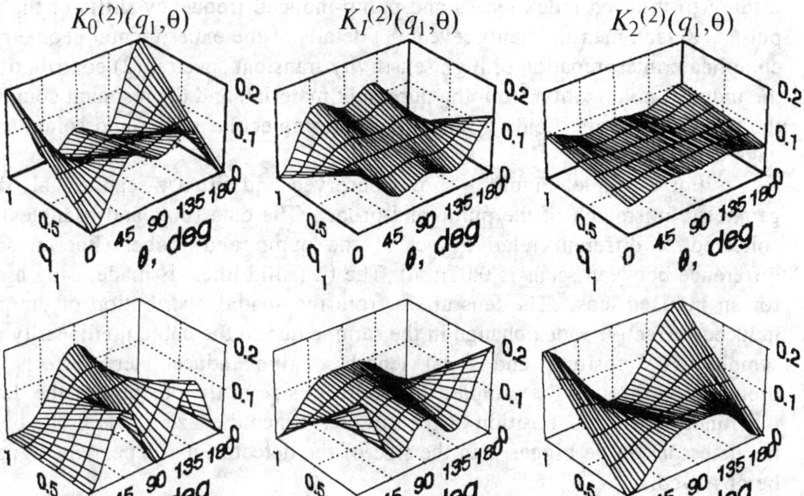

Fig. 1.

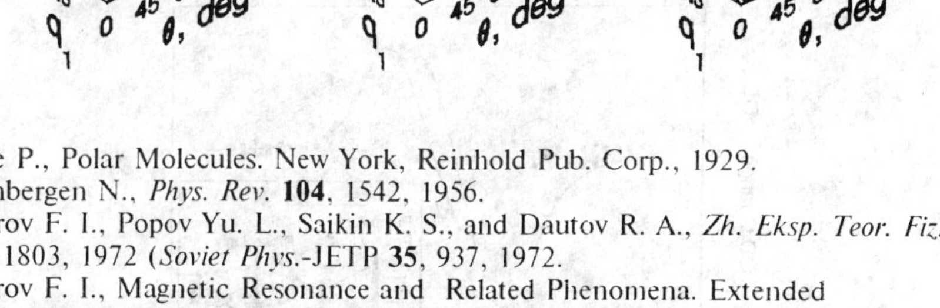

Fig. 2.

1. Debye P., Polar Molecules. New York, Reinhold Pub. Corp., 1929.
2. Bloembergen N., *Phys. Rev.* **104**, 1542, 1956.
3. Bashirov F. I., Popov Yu. L., Saikin K. S., and Dautov R. A., *Zh. Eksp. Teor. Fiz.* **62**, 1803, 1972 (*Soviet Phys.*-JETP **35**, 937, 1972.
4. Bashirov F. I., Magnetic Resonance and Related Phenomena. Extended Abstracts of the XXVIIth Congress AMPERE. Kazan, **1**, 282 - 283, 1994.

ULTRAFAST VISIBLE AND INFRARED MEASUREMENTS OF SOLVENT-SOLUTE COUPLING AND PROTON TRANSFER IN SIMPLE LIQUIDS AND PROTEINS

Boris Akhremitchev, Chengfei Wang, Brian Mohney and Gilbert C. Walker*, Chemistry Department, University of Pittsburgh, Pittsburgh, PA 15260

A solid state laser system has been constructed for the generation of pulses throughout the visible and infrared.[1] The generation of tunable 100fs IR pulses is accomplished by a continuum-seeded OPA using $AgGaS_2$, GaSe, and KTP. In the pump-probe experiments we employ HgCdTe detector arrays, permitting high sensitivity (0.6mOD/sec) over patches of the IR spectrum with varying frequency resolution.

Various material responses can contribute to the observed transient signal, most commonly a pump induced change in the material index of refraction, leading to induced index lenses and pump-induced frequency shifts of the probe pulse.[2] Z-scan measurements reveal (1) details of the experimental geometry that complicate interprepation of high sensitivity transient spectra, (2) contributions of the induced polarization from the pure bulk material, and (3) resonant coupling to the pump and probe fields to intra- and intermolecular solute and solute-solvent degrees of freedom.

For example, Figure 1 shows observed and fitted z-scan signals due to geometric mismatch of the pump and probe. The data (dots and triangles) were collected at different relative foci of the pump and probe. (Foci mismatch difference between scans is 0.07mm) The fit (solid lines) is made using a model for an induced lens. The lens arises from the spatial distribution of the pump-induced refractive index change in the sample due to the cubic nonlinearity of the sample. The position and focal length of the induced lens affects beam propogation through the sample. Probe beam size at the position of the detector is a function of the z-position of the sample. Since the size of the probe beam at the detector can be bigger than the size of the detector, a z-dependent signal can be observed.

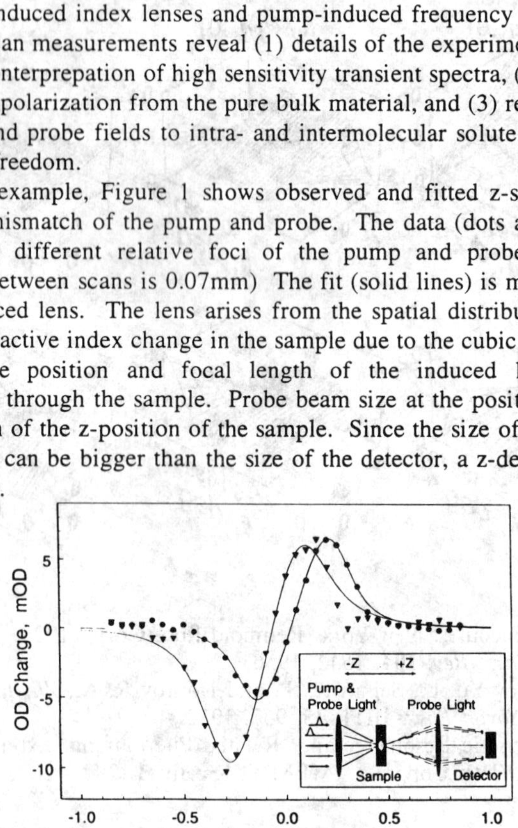

Figure 1

We have studied the excited state dynamics of acylaminoanthraquinonoidal systems, which exhibit femtosecond proton transfer processes. There are both solvent dynamical and intramolecular control of proton transfer rates, depending on the solvent polarity and dynamics. The initial proton transfer occurs in ca 100fs following optical excitation. Subsequent solvent polarization diffusion results in proton reshuffling, which depends on the characteristic time for dynamic solvation. Isotopic substitution experiments suggests that proton/deuteron tunneling contributions are small.

Figure 2 shows the excited state absorption of chloroacylaminoanthraquinone in dichloromethane. The overall spectral shape reflects an excited state electronic absorption. The NH ground state bleach may be seen at at ca 3200 cm^{-1}. Dynamics suggestive of excited state OH absorption occur at ca 3400cm^{-1}. The overall lifetime of the excited state absorption is ca 100ps, consistent with the fluorescence lifetime.

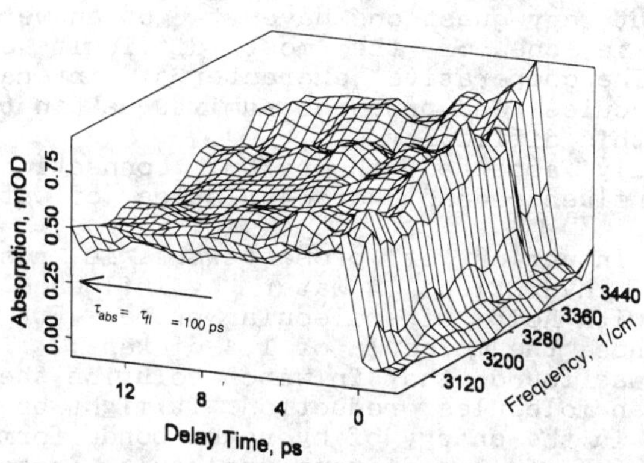

Figure 2. Transient absorption of chloroacylaminoanthraquinone in dichloromethane. Excitation at 400nm, IR probe at 70° relative polarization.

1. M. T. Asaki, C. P. Huang, G. Garvey, J. Zhou, H. C. Kapteyn, M. M. Murnane *Opt. Lett.* 1993, **18**, 977. J. Squier, F. Salin, G. Mourou *Opt. Lett.* 1991, **16**, 324.
2. E. Tokunaga, A. Terasaki, and T. Kobayashi *Phys. Rev. A,* 1993, **47**, R4581.

Acknowledgements: This research was supported by the CRDF Fund of The Unversity of Pittsburgh, the Petroleum Research Fund, and 3M Company.

RAMAN SPECTROSCOPY IN RESEARCH OF THE DISSIPATIVE STRUCTURES IN WATER INFLUENCED BY MICROWAVE RADIATION IN THE PRESENCE OF MOLECULAR PROBE

V.M. Klimovich, L.D. Bobrovnik, I.S. Gulay

Ukrainian State University of Food Technologies, Kiev,
st.Vladimirskay 68, 252601, Kiev-17, Ukraine

One of the fundamental properties of substance is creating space and temporary dissipative structures in unbalance tract. Arose in the water´s systems on molecules level the space structures were researched in sufficiently so us compare with homogeneous distribution of molecules, space temporal structures is more complex autoorganization arising in result cooperation motion of liquid molecules.

In research the properties of water was achieved some successes but any questions haven´t got answers because the water is one of the most difficult object of research. The cooperative character of interactions by water´s molecules and presence admixtures in the water-are causes this difficulty.

Now generally accepted theory of connection between external influences and characteristics of water system is absent.

Intention of present work is research the processes of hydrations in water by influence microwave radiation with help of molecular probe. For molecular probe was chose the molecule of 1,4-dioxane.

Earlier it was found, that in water solution the symmetry of 1,4-dioxan molecules reduction. It might be result of distinction in the energy of hydrogen bonds formed by two oxygen atoms of 1,4-dioxane molecule (acceptors of protons) and by water molecules (donors of protons) [1]. In the system of water activated by microwave radiation the energy of those hydrogen bonds of 1,4-dioxane equalized to each other. As a result we had an increase of symmetry of 1,4-dioxane molecule. Thus all the above mentioned findings testify to the fact that the water activated by microwave radiation makes an effect of the "structure memory", which is displayed in the process of 1,4-dioxane hydration.

For explain obtaining result necessary remember that water´s systems own non-linear properties. Microwave radiation have created powerful non-equilibrium conditions when radiation influences on this systems. In such a way have been reached threshold of stability by thermodynamic branch-bifurcation point. In this point the thermodynamic branch becomes non-stability relatively by fluctuations. Macroscopic distance molecules lose independence (the cogerent state) and appears distant correlations. The lows of fluctuations and correlations lose their universality and find strong dependence from type non-linear systems. This is take place in non-balance range as complex systems as more variety types fluctuation with threaten their stability. In compound systems where some types interact one each other the diffusion provides stability influence. It is well-known that diffusion provides stability for systems but fluctuation assists instability. Competition take place in water´s systems. Threshold´s stability dependence from result of this competition. The dissipative structures request very non-balance conditions. But the kinetic equation with diffusion has parameters which permit displacement in weak non-balance domain [2]. The system can evolve approach to equilibrium and more off one. In results this evolution the system can choice condition which can be do in itself benefit nonlinear correlations which provide stability of preceding state. We think that so as can be explain the experimental effect of structure "memory" by water system.

1. V.M. Klimovich, I.S. Gulay, L.D. Bobrovnik, Proceeding of Academy of Engineering Sci. of Ukraine, N 1, p.45, 1995.
2. I. Prigogine, I. Stengers, Order out of Chaos, Heinemann. London, 1984.

SECTION 8

MACROMOLECULES and POLYMERS

PHASE CHANGE ANALYSIS OF POLYMER BY THERMAL SCANNING RAMAN SPECTROSCOPY

N. Muraki*, G. Katagiri, H. Ishida

Material Science Laboratory, Toray Research Center, Inc., Otsu, Shiga 520, Japan

A combination of the thermal scanning stage and Raman Spectroscopy is described. This Thermal Scanning Raman Spectroscopy gives thermal and spectroscopic information of polymers simultaneously. A similar approach in combining a thermal scanning stage and FTIR analysis have been already reported in the literature [1]. In this work, the phase transitions of poly-(ethylene terephthalate) (PET) were used to illustrate the potential of Thermal Scanning Raman Spectroscopy. Undrawn and uniaxially drawn PET films were used for the experiment.

An argon ion laser operating at a wavelength 514.5 nm with a power of 20 mW on sample was used as the excitation source. The sample were heated in the FP82 (Mettler) from 30 to 260 °C at 10 °Cmin^{-1} and Raman spectra were simultaneously collected using Ramanor T-64000 (Jobin-Yvon/Atago bussan). The interval of the measurement was 15 sec. All the recorded spectra were analyzed with a commercially available software (Grams/386, Galactic Ind. Co.).

Significant changes in the spectra are seen when the material is heating through its crystallization temperature (i.e., 130°C) and melting point (i.e., 230°C). Major spectral changes are seen in the relative intensity of the 1096cm^{-1} (ethylene glycol νC-C and νC-O combination mode) and the 998cm^{-1} bands (ethylene glycol C-O stretching mode), and the width of the 1730cm^{-1} band (C=O stretching mode) [2]. These bands of ethylene glycol segment are characteristic of the *trans* conformational structure of PET [2]. It is well known that the width of the 1730cm^{-1} band have a good correlation with the density of PET [3]. Temperature dependence of the relative intensity of the 998cm^{-1} band to the 1285cm^{-1} band and the width of the 1730cm^{-1} band of undrawn PET film are shown in Fig. 1. The behavior of these two parameters are in good agreement with each other. Therefore, the conformational transition of ethylene glycol segments and the ordering of carbonyl environments may occur simultaneously during the phase transition.

We also examine the temperature dependence of the peak position of Raman bands. Figure 2 shows a temperature dependence of the 1615cm^{-1} band (benzene ring stretching mode) of undrawn PET film. There are almost linear correlation between the peak position of the 1615cm^{-1} band and the temperature. The shift of the peak position may be due to the thermal expansion of the molecular chain [4]. The peak position varies continuously even in the region of the crystallization temperature (i.e., 130°C). Therefore, the crystallization of PET may not change the length of the benzene ring segments, in other words, the PET framework is relatively rigid, even in the amorphous and melt phase. Additionally, there is a slight difference between the correlation coefficient of the peak position and temperature in amorphous phase (30-120, 220-240°C) and crystalline phase (120-220°C). The phenomenon may be explained by the difference of the thermal expansion coefficient between two phases.

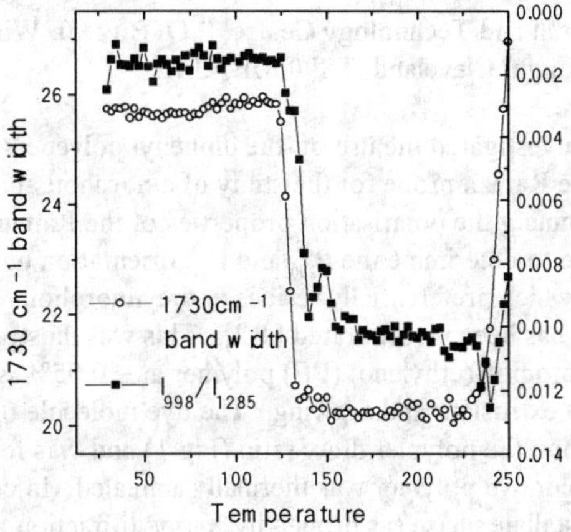

Fig. 1 Temperature dependence of the band parameters: ■, the intensity ratio of the $998cm^{-1}$ band to the $1285cm^{-1}$ band. ○, the width of the $1730cm^{-1}$ band.

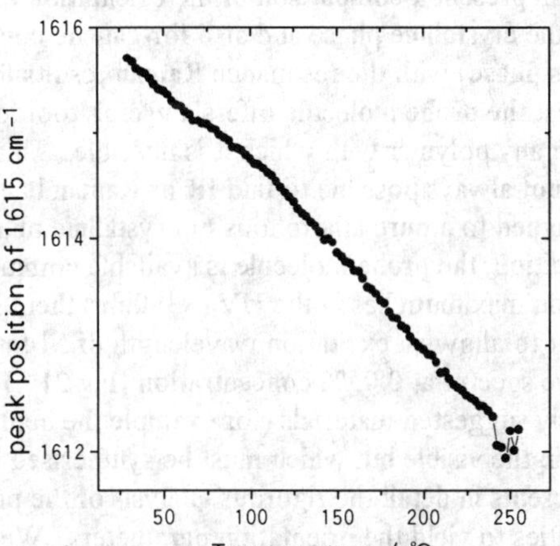

Fig. 2 Temperature dependence of the peak position of the benzene ring stretching mode ($1615cm^{-1}$ νC=C).

1. D.J. Johnson, D.A.C. Compton and P.L. Canale, Thermochim. Acta., **195**, 5 (1992)
2. J. Stokr, B. Schneider, D. Doskocilova and J. Lovy, Polymer, **23**, 714 (1982)
3. A.J. Melveger, J. Polym. Sci., **10**, 317 (1972)
4. V.I. Vettegren, L.S. Titenkov and V. Bronnikov, J. Therm. Anal., **38**, 1031 (1992)

MEASUREMENT OF AMORPHOUS AND CRYSTALLINE PHASE ORIENTATION IN POLY(ETHYLENE) USING POLARISED RESONANCE RAMAN SPECTROSCOPY AND INFRARED DICHROISM.

N Everall*, J Chalmers and P Mills

ICI Research and Technology Centre, P.O. Box 90, Wilton, Middlesbrough, Cleveland, TS90 8JE, U.K.

We have investigated the use of the diphenyl polyene (Φ-(C=C)-$_4\Phi$) as a potential Resonance Raman probe for the study of amorphous-phase orientation in polymers. By examining the polarisation properties of the Raman scattered radiation, it is possible to deterimine the P_{200} and P_{400} orientation parameters of the rigid probe molecule which preferentially resides in the amorphous phase of the polymer into which it has been incorporated [1,2]. This was illustrated by melt-mixing the dye into poly(ethylene) (PE) polymer at ~ 0.05% w/w concentration prior to extrusion and drawing. The dye molecule orientation was strongly dependent upon the polymer draw ratio (Fig 1) and was found to relax significantly when the drawn polymer was thermally annealed. In contrast, the orientation of the crystalline phase (as probed by X-ray diffraction and FTIR dichroism) was largely unaffected by thermal annealing.

This poster will present a comparison of the orientation results obtained by FTIR dichroism (for the crystalline phase and also for gauche conformers present in the amorphous phase) with the resonance Raman results for the probe molecule. It is shown that the probe molecule offers a useful tool for monitoring amorphous orientation in any polymer with which it is miscible. This is advantageous since it is not always possible to find IR or Raman bands of a polymer which can readily be assigned to a pure amorphous or crystalline phase, as we can in the case of PE. In addition, the probe molecule is available commercially, and even though its absorption maximum lies in the UV (~390nm) there is sufficient preresonant enhancement to allow an excitation wavelength of 514 nm to be used and still obtain strong dye spectra at 0.05% concentration [Fig 2]. This is an advantage over previously suggested materials (for example the hepta-ene homolog) which absorb in the visible but which must be synthesised by the user.

This work presents in detail the rigorous analysis of the polarised resonance Raman intensities to yield the orientation parameters. We show how the simplified analysis of Margulies and Stockburger [1] overestimates the orientation parameter by 10-20% compared with the rigorous analysis (which also yields the fourth-order orientation parameter P_{400}). The rigorous analysis demands a knowledge of all elements of the scattering tensor - the effect of including or neglecting the off-diagonal elements is discussed and is shown to be more important for P_{400} than P_{200} . If the off-diagonal elements are ignored, the orientation parameters are underestimated by ~ 30% and 10% respectively. A full exposition of this work will be given elsewhere [2].

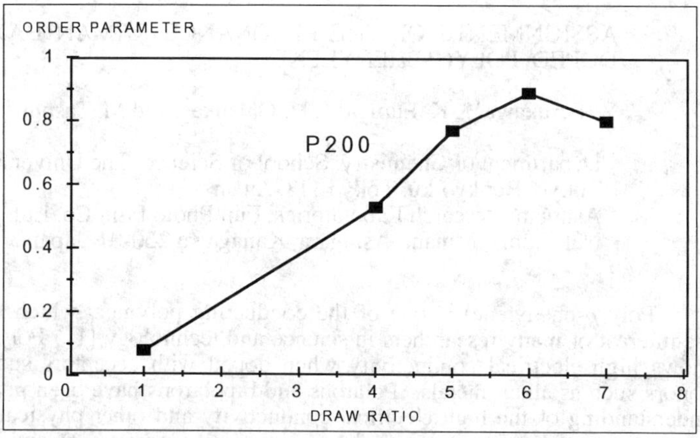

Figure 1 Plot of P_{200} versus draw ratio for polyene dopant molecule.

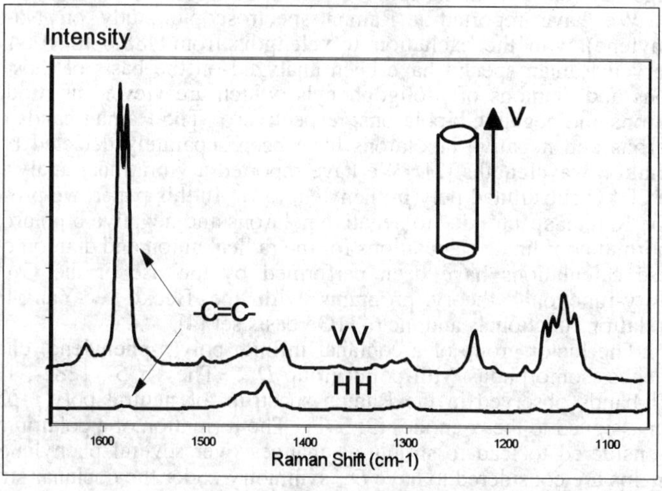

Fig. 2 Comparison of VV and HH resonance Raman spectra for a 0.05% probe molecule concentration in a 6-fold drawn PE tube (long axis held vertically). The probe molecule is strongly aligned parallel to the draw axis, as evidenced by the weakness of the HH spectrum.

1 L Margulies and M Stockburger, J Raman Spectrosc., 8, 26 (1979).
2 N Everall, J Chalmers and P Mills, submitted to Appl. Spectrosc.

ASSIGNMENTS OF THE RESONANCE RAMAN BANDS OF DOPED POLY(*P*-PHENYLENE)

Y. Furukawa[1]*, K. Furuya[1,2], H. Ohtsuka[1], and M. Tasumi[1]

[1] Department of Chemistry, School of Science, The University of Tokyo, Bunkyo-ku, Tokyo 113, Japan
[2] Ashigara Research Laboratories, Fuji Photo Film Co. Ltd., 210 Nakanuma, Minami Ashigara, Kanagawa 250-01, Japan

Poly(*p*-phenylene) is one of the conducting polymers which have attracted the interest of many researchers in science and technology [1]. Poly(*p*-phenylene) shows high electrical conductivity when doped with acceptors such as AsF_5 or donors such as alkali metals. Polarons and bipolarons have been proposed for the understanding of the high electrical conductivity and other physical properties of doped poly(*p*-phenylene). A polaron has charge $+e$ (or $-e$) and spin 1/2 with structural changes extending over several phenylene rings, whereas a bipolaron has charge $+2e$ (or $-2e$) and no spin.

We have reported a Raman spectroscopic study on Na-doped poly(*p*-phenylene) with the excitation wavelengths from 488.0 to 1064 nm [2]. The observed Raman spectra have been analyzed on the basis of those of the radical anions and dianions of *p*-oligophenyls, which are viewed as models of negative polarons and negative bipolarons, respectively. The Raman bands due to negative polarons and negative bipolarons have been separately detected by changing the excitation wavelengths [2]. We have reported a vibrational analysis based on the data of ^{13}C-substituted poly(*p*-phenylene) [3]. In this paper, we present an analysis of the Raman spectra due to negative polarons and negative bipolarons on the basis of normal coordinate calculations for the radical anion and dianion of *p*-sexiphenyl. These calculations have been performed by the use of the GAUSSIAN 92 / density-functional theory programs with the Becke-Lee-Yang-Parr exchange-correlation functionals and the 6-31G* basis set [4].

The factor group of a coplanar infinite poly(*p*-phenylene) chain at the zone center is isomorphous with point group D_{2h}. The 1595-, 1282-, 1222-, and 798-cm^{-1} bands observed in the Raman spectrum of neutral poly(*p*-phenylene) have been assigned to the a_g modes [3, 5–7]. The formation of a polaron and a bipolaron is considered to lead to structural changes over several phenylene rings. These domains are considered to have D_{2h} symmetry under the coplanar structure. The *yz* plane is taken in the phenylene-ring plane (*z* axis along the polymer chain) and *x* axis perpendicular to the phenylene-ring plane. Table 1 shows the frequencies of the observed Raman bands associated with negative polarons and negative bipolarons in Na-doped poly(*p*-phenylene) and the results of the normal coordinate calculations for the radical anion and dianion of *p*-sexiphenyl. All the observed Raman bands have been attributed to the a_g bands. Each band is labeled with the mode number of a benzene ring [8] and the phase sequence of the assigned mode (the seventh and eighth columns in Table 1, respectively). These bands can be classified into two groups: bands corresponding to the zone-center modes of an infinite chain (bands A, C, E, and H) and bands corresponding to the non-zone-center modes (bands B, D, F, and G). For example, band E is assignable to the in-phase sequence (+ + + + + +) of the 9a mode of a benzene ring, whereas band F is assignable to an out-of-phase sequence (+ + − − + +) of the 9a mode. Band B is attributable to an out-of-phase sequence (+ + + − − −) of the 19a mode. The appearance of the Raman bands belonging to the latter group can be understood by considering that the observed Raman bands arise from several phenylene rings which are changed from the neutral state by the formation of negative polarons and bipolarons. In other words, these results are spectroscopic evidence of the localized structural changes associated with the formation of polarons and bipolarons.

It has been demonstrated that the normal coordinate calculations performed by density functional theory for the radical ions and divalent ions of long oligomers are useful for analyzing the Raman spectra of doped conducting polymers.

TABLE 1. Vibrational frequencies (cm^{-1}) and assignments.

No.	obsd.[a]		calcd.[b]		sym.	assignment	
	negative polaron	negative bipolaron	RA	DA		mode (benzene ring)	phase sequence
A	1591–1590	1594–1586	1590	1591	a_g	(8a, a_g)	++++++
B		1477–1471	1482	1470	a_g	(19a, b_{1u})	+++−−−
C	1315–1314	1352–1344	1294	1315	a_g	inter-ring CC str	+++++
D		1297–1290	1278	1297	a_g	inter-ring CC str	+0−0+
E	1221–1219	1222–1219	1210	1215	a_g	(9a, a_g)	++++++
F	1179	1190–1179	1198	1202	a_g	(9a, a_g)	++−−++
G	980–974	992–952	981	972	a_g	(18a, b_{1u})	0++−−0
H	783–777		777	770	a_g	(6a, a_g)	++++++

[a] Ref. 2. [b] RA and DA mean the radical anion and dianion of *p*-sexiphenyl, respectively.

1. Conjugated Conducting Polymers, (H. Kiess, ed.), Springer-Verlag, Berlin, 1992.
2. Y. Furukawa, H. Ohtsuka, and M. Tasumi, Synth. Met. 55, 516 (1993).
3. Y. Furukawa, H. Ohtsuka, M. Tasumi, I. Wataru, T. Kanbara, and T. Yamamoto, J. Raman Spectrosc. 24, 551 (1993).
4. M. J. Frisch et al., GAUSSIAN 92/DFT, Gaussian, Inc., Pittsburgh, PA, 1993.
5. G. Zannoni and G. Zerbi, J. Chem. Phys. 82, 31 (1985).
6. J. P. Buisson, S. Krichene, S. Lefrant, Synth. Met. 21, 229 (1987).
7. L. Cuff and M. Kertesz, Macromolecules 27, 762 (1994).
8. E. B. Wilson, Jr., Phys. Rev. 45, 706 (1934).

PHOTOISOMERIZATION AND ANGULAR REORIENTATION IN DR1-DOPED PMMA FILMS : A RESONANCE RAMAN STUDY.

F.Lagugné Labarthet and C.Sourisseau *.

Laboratoire de spectoscopie moléculaire et cristalline, URA 124 CNRS, Université de Bordeaux I, 33405 Talence, France.

The azo-type dye molecules due to their reversible "trans↔cis" photoisomerization with respect to the -N=N- bond can be used as optical active centers in doped or functionalized polymer films; their use in amorphous high-Tg polymer films has received a lot of attention in the last few years [1,2]. Indeed, upon irradiation with a linearly polarized light pump, an angular dependent photoselection occurs which causes a redistribution in the orientation of the photochromic entities; the optical properties (linear dichroism) of such systems are thus modified by light-induced variations in their orders and orientations.

In the present study we have thus investigated these phenomena in DR1 dye-doped PMMA films (DR1= 4'-[(2-hydroxyethyl)ethyl)amino]-4-nitroazobenzene and PMMA=poly(methyl-methacrylate)) by dichroism measurements and polarized resonance Raman experiments.

First of all, it is shown that anisotropy and dichroism induced in low optical density thin films (with dye contents ranging from 1.3 to 5 wt %) can be created and followed in classical pump-probe c.w. laser absorption measurements; the absorbance variations are well reproduced by a simple three level model taking account of the cis population and of angular redistributions and relaxations [2,3]. Then, we have shown that the relative resonance Raman intensities of various bands of the chromophore are sensitive to the photoinduced structural anisotropy (Fig. 1). By comparing the intensities on a starting isotropic sample ("erased"), using a circularly polarized light ($I^{\frown}$), with those obtained for an anisotropic sample ("written"), using a vertically polarized light ($I^{\uparrow}$), and assuming an effective uniaxial symmetry , one gets the following Raman intensity ratios:

$$R_1 = \frac{I^{\uparrow}_{ZY}}{I^{\uparrow}_{ZZ}} = \frac{\left(7+5\langle P_2\rangle - 12\langle P_4\rangle\right)}{7\left(\dfrac{8a_1^2+4a_1+3}{(a_1-1)^2}\right) - 20\langle P_2\rangle\left(\dfrac{4a_1+3}{a_1-1}\right) + 24\langle P_4\rangle} \qquad (1)$$

$$R_2 = \frac{I^{\uparrow}_{ZY}+I^{\uparrow}_{ZZ}}{I^{\frown}_{total}} = 1 + \frac{\langle P_2\rangle}{7}\left(\frac{-75a_1^2+10a_1+65}{9a_1^2+2a_1+4}\right) + \frac{12}{7}\langle P_4\rangle\left(\frac{(a_1-1)^2}{9a_1^2+2a_1+4}\right) \qquad (2)$$

where $a_1 = \alpha_{xx}/\alpha_{zz} = \alpha_{yy}/\alpha_{zz}$ and $\langle P_2\rangle$, $\langle P_4\rangle$ are the orientational order parameters of Legendre polynomials $P_i(\cos\theta)$, with θ the angle between the molecular long axis (rod-like trans molecule) and the director [3].It is noteworthy that the a_1 molecular values can be determined from the depolarizations ratios (ρ) in solution.

Under relatively weak pumping regimes (10-15 mW/cm^2), one can thus estimate the $\langle P_2\rangle$ and $\langle P_4\rangle$ order parameters for various vibrational modes (Table I) and get a better insight into the primary intramolecular orientational mechanisms.

As expected the <P2> values are negative demonstrating that the principal long axis of some DR1 molecules are reorientating perpendicularly to the electric field direction. Nevertheless, the <P4> parameters are very weak, nearly equal to zero.

Table I: Raman results in a 3% DR1-doped PMMA film ($N=6.9 \times 10^{19}$ molecules/cm^3; $e=5\mu m$) under 15 mW/cm^2 at $\lambda=496.5$ nm.

Mode/νcm^{-1}	νsNO2 1339	νN=N 1392	ν19a 1423	ν19b+δCH2 1449	ν8a 1588
ρ solution.	0.358	0.353	0.365	0.410	0.410
<P2>	-0.120	-0.086	-0.102	-0.028	-0.066
<P4>	+0.013	+0.014	+0.025	+0.007	+0.032

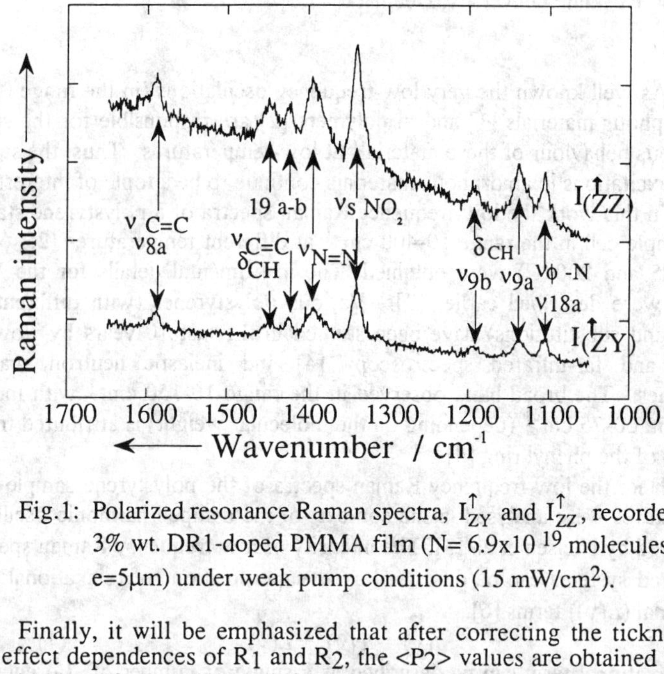

Fig.1: Polarized resonance Raman spectra, $I^{\uparrow}_{ZY}$ and $I^{\uparrow}_{ZZ}$, recorded on a 3% wt. DR1-doped PMMA film ($N= 6.9 \times 10^{19}$ molecules/cm^3; $e=5\mu m$) under weak pump conditions (15 mW/cm^2).

Finally, it will be emphasized that after correcting the tickness and absorption effect dependences of R_1 and R_2, the <P2> values are obtained in good agreement with those from visible or infrared dichroic absorption measurements[4], convincing the reliability of the resonance Raman scattering method. In addition, valuable dynamical information in such potential optoelectronic systems can be obtained from time dependent resonance Raman measurements.

1. S. Xie, A. Natansohn and P. Rochon, Chem. of Mat. 5, 403 (1993).
2. Z. Sekkat and M. Dumont, Synthetic Metals 54, 373 (1993).
3. F. Lagugné Labarthet and C. Sourisseau, J. Raman Spectrosc., in press.
4. T. Buffeteau and M. Pézolet, Applied Spectrosc., in press.

DAMPED HARMONIC OSCILLATOR MODEL ANALYSIS OF THE LOW FREQUENCY RAMAN SPECTRA OF POLYSTYRENE

T.S. Perova[1,2], O. Faurskov Nielsen[2], P.A. Perov[3], E. Praestgaard[4], U. Rasmussen[4]

1 Vavilov State Optical Institute, St.-Petersburg, Russia
2 Chemical Institute, University of Copenhagen, Denmark
3 Gertzen University, St.-Petersburg, Russia
4 Roskilde University, Denmark

As well known the very low-frequency oscillations (in the range 0-150 cm^{-1}) of amorphous materials [1] and in polymers [2] are responsible for the specific heat anomalous behaviour of these materials at low-temperatures. Thus, the study of low-energy excitations in amorphous materials continue to be a topic of interest.

In this work the low-frequency Raman spectra of a polystyrene standard UV-VIS sample cell in the range 10-400 cm^{-1} at different temperatures (25, 50, 75, 100, 125, 175 and 200°C) were obtained. The instrumental details for the VIS-Raman spectra were described earlier [3]. Various polystyrenes (with different molecular weight and substitutions) have been studied in the last 10 years by low-frequency Raman and far-infrared spectroscopy [4] and inelastic neutron scattering [5] techniques. The broad band observed in the range 10-150 cm^{-1} with the frequency maximum 60-75 cm^{-1} (depending on the molecular weight) is attributed to librational motions of the phenyl ring [4].

Here the low-frequency Raman spectra of the polystyrene sample at different temperatures were analysed in the context of the damped harmonic oscillator model introduced by Nielsen et al. [3]. The intensity of low-frequency Raman spectra $I(\nu)$ of disordered systems can be described as a superposition of relaxational ($\Psi(\nu)$) and vibrational ($L(\nu)$) terms [6]:

$$I(\nu) = \Psi(\nu) + L(\nu). \qquad (1)$$

The relaxational term can be described as a single or number of Lorentzians. For a single Lorentzian [6]

$$\Psi(\nu) = \mathrm{H}\Gamma^2/(\nu^2 + \Gamma^2), \qquad (2)$$

where Γ is the width and H is the height of the Lorentzian function centred at zero wavenumber. The $L(\nu)$ function in accordance with the damped harmonic oscillator model [3] can be written as:

$$L(\nu) = G\gamma K_o/((K_o - \nu^2)^2 + \gamma^2\nu^2) \text{ and } R(\nu) = L(\nu)\,\nu\,[1-\exp(-h\nu/kT)], \qquad (3)$$

where ν_o is the undamped frequency, G is the height at the frequency ν_o, $K_o = \nu_o^2$ and γ is the damping factor .

The fitting procedure as described in [5] is easier to carry out for the $R(\nu)$ representation of the low frequency Raman spectrum. The results of the fitting procedure for the $R(\nu)$ spectra of the polystyrene sample at 25°C is shown on Fig. 1. Despite the fact that the "experimental" $R(\nu)$ function fitts well, no good fit was obtained for the observed Raman intensities at very low frequencies (~ < 20 cm^{-1}). Possibly more Lorentzians should be taken into account for the relaxation process as was pointed out in [6], or the influence of the stray light makes this region less reliable.

The frequency maximum of the low-frequency Raman spectra (in $R(\nu)$ - representation) decreases as the temperature increases. This result is in a good agreement with the earlier observation by Zoidis et al. [4].

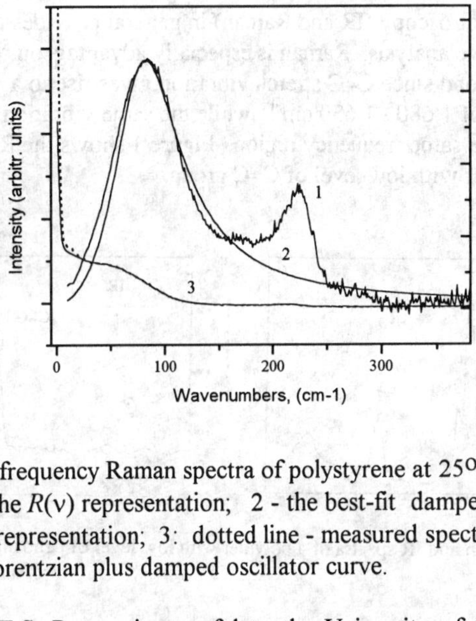

Fig. 1. Low-frequency Raman spectra of polystyrene at 25°C: 1 - measured spectrum in the $R(\nu)$ representation; 2 - the best-fit damped oscillator curve in the $R(\nu)$ representation; 3: dotted line - measured spectra $I(\nu)$, solid line - a best-fit Lorentzian plus damped oscillator curve.

Acknowledgment: T.S. Perova is greatful to the University of Copenhagen for a visiting professorship.

1. W.A. Phillips, ed., Amorphous Solids - Low Temperature Properties. Topics in Current Physics, v.24 , Springer, Berlin, 1981.
2. S.J. Spells and I.W. Shepherd, Chem. Phys. Lett., 45, 606 (1977).
3. S.E. May Colaianni and O. Faurskov Nielsen, J.Molec.Struct., 347, 267 (1995).
4. E. Zoidis, Ch. Borsdorf, F. Strehle and Th. Dorfmuller, Chem Phys., 168, 99 (1992).
5. T. Kanaya, T. Kawaguchi, K. Kaji, J.Non-Cryst.Solids, 172-174, 327 (1994).
6. O. Faurskov Nielsen, Ann. Rep.Prog. Chem.Sec. C, 90, 3 (1993).

QUANTITATIVE ANALYSIS OF PENDANT C=C GROUP IN A POLYMER BY RAMAN SPECTROSCOPY

Ken Laughlin and Wei Xie *

Research Laboratories, Rohm & Haas Co., 727 Norristown Road, Spring House, PA 19477

Vibrational spectroscopy (IR and Raman) in general provides a quick and accurate means of quantitative analysis. Raman is especially advantageous over IR in the detection of C=C bond since C=C stretch vibration gives rise to a very strong Raman band in the region of 1,680 - 1,630 cm^{-1}, while the same vibration produces a fairly weak IR band in the same frequency region. Figure 1 shows the Raman and IR spectra of a polymer with low level of C=C group.

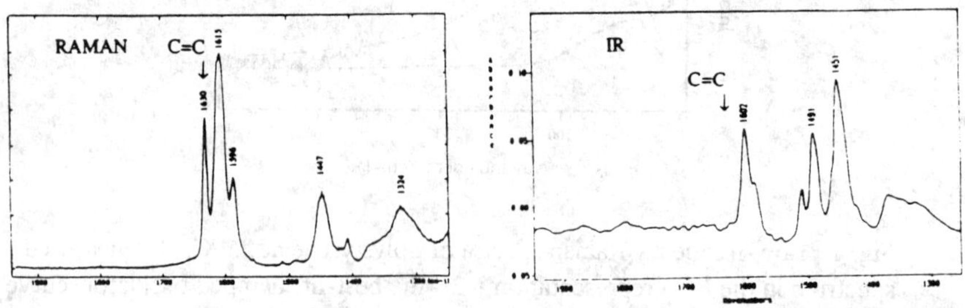

Figure 1. Raman and IR spectra of a polymer with low level of pendant C=C group.

The key in any quantitative measurement is the establishment of a calibration with standards of known composition. Problems arise when one deals with the quantitative analysis of polymers samples. For copolymer analyses it is not always easy to synthesize a series of polymer standards with different known ratios of monomers. Furthermore when a specific functional group needs to be analyzed quantitatively it is also difficult to have standards with different known amount of the targeted functional group within a monomer unit. In practice the monomer ratio or certain functional groups in a copolymer can be determined by NMR which is the primary means to study composition of a copolymer. Another alternative for standards is to use mixtures of small molecules which resemble the monomer units of the copolymer.

The example that we are presenting here is a copolymer containing divinyl benzene. We need to determine the amount of pendant C=C group left after the

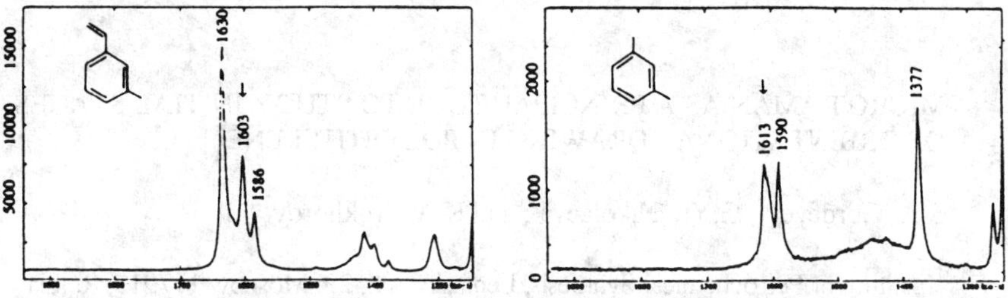

Figure 2. Raman intensity of aromatic-ring mode with and without a vinyl group.

polymerization. To establish a calibration for the quantitative analysis we used mixtures of monomers to simulate the monomer units in the polymer. An interesting observation we had during the process is that the Raman intensity of the aromatic-ring mode near 1,610 cm^{-1} is dramatically increased by about 6 times when one or more vinyl groups are attached to the ring. Figure 2 shows the Raman intensity of two aromatic rings, one with a vinyl group and the other without. Since we use the intensity of the ring mode as a reference for normalization in the calibration, the intensity change of the ring mode due to the presence of a vinyl group gives rise to a nonlinear calibration curve for the quantitative analysis of the pendant C=C group, as shown in Figure 3.

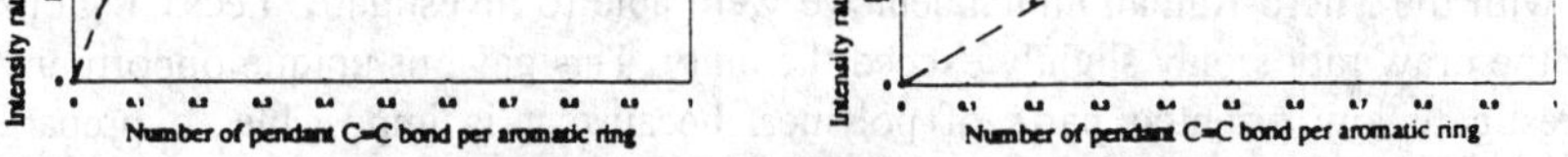

Figure 3. Calibration for pendant C=C group by Raman and IR.

Nevertheless with the established calibration by Raman spectroscopy, we were able to measure low level of pendant C=C group in the polymer to a few percent. A couple of points need to be addressed here. First the attachment of a vinyl group to an aromatic ring does not affect the IR band intensity of the ring mode. Therefore we could get a linear calibration with IR spectroscopy. Secondly although the full range calibration from Raman is a nonlinear one, the portion at the lower level of the calibration is very close to being linear, which is important since that's where Raman has the advantage over IR in detecting very low level pendant C=C groups.

MICRO RAMAN AS A PRINCIPAL TOOL TO STUDY INITIAL STAGES OF ORIENTATIONAL DRAWING IN POLYETHYLENE

S.A. Gordeyev[1], G.Yu. Nikolaeva[2], and K.A. Prokhorov[2]*

[1]Institute of Petrochemical Synthesis, Leninsky Av., 29, Moscow 117912, Russia
[2]General Physics Institute, Vavilova St. 38, Moscow 117942, Russia

Raman spectroscopy is a very informative and convenient tool for non destructive characterization of polymers. Micro-Raman technique combines advantages of Raman spectrometer and an optical microscope. In this work Micro-Raman was used to study the process of macromolecules orientation in polyethylene (PE) at the initial stages.

It is well known that deformation of PE begins through formation of a "neck". In this region the lamellar structure of nonoriented sample transforms into the fibrillar one. Subsequent drawing leads to the improvement of fibrillar structure. Studying the initial stages of transformation is very important for understanding the whole mechanism of PE deformation under the strain and for practical recommendations. Up to now only X-ray analysis was applied for studying the evolution of the structure at a "neck" region [1,2]. However, this method has a limitation in a local characterization of the structure. Raman analysis enables to investigate very small volume of a sample up to 10 mkm in diameter and to get information about macromolecule orientational order and structure's conformational and phase composition.

Polarized Raman spectra were recorded for PE samples of different origin. Among them were the tapes prepared by compression of linear and branched PE with molecular weight of 10^6 , and single crystal mats prepared by crystallization from solution of polyethylenes with molecular weights of $3 \cdot 10^5$ and 10^6 in xylene. The range of "neck" lengths was 150-600 mkm. The evolution of the spectra was studied under spatial scanning from nonoriented to oriented sample parts with a step of about 15 mkm. Thus, with the Micro-Raman attachment we were able to investigate "neck" regions where the draw ratios only slightly exceed the unity. This gave us unique opportunity to investigate low oriented parts of polymer, because it is impossible to prepare separately the series of homogeneous samples with such a low values of draw ratio.

Raman spectra of the tape prepared by compression of PE are shown in the figure 1. Polarization of the incident and scattering light was parallel to the direction of drawing. For this scattering geometry the asymmetric stretching vibration of C-C bond at 1063 cm^{-1} and twisting vibration of CH_2 group at 1295 cm^{-1} (both have $B_{2g}+B_{3g}$ symmetry) are forbidden for the case of highly oriented samples with c-texture structure. The more structure is ordered, the less intensities of these vibrations. CH_2 bending region (1400-1500 cm^{-1}) and C-H stretching bands (2800-3000 cm^{-1}) have similar behavior although their symmetries are differed. These features could be obviously seen from the figure, where spectrum **a** corresponds to nonoriented part of the sample at the beginning of the "neck" and spectrum **e** - to the end of "neck" transition region.

Analysis of the all Raman data allows to make some conclusions that could be summarized in the following way.

We consider that the principal changes in the Raman spectra occur due to continuous and monotonous transformation from lamellar to fibrillar structure along the "neck".

Intensities of the lines except ones at 1130 and at 1418 cm^{-1} are very sensitive to the macromolecules orientation. Intensities of C-H stretching bands are most strongly affected by this factor.

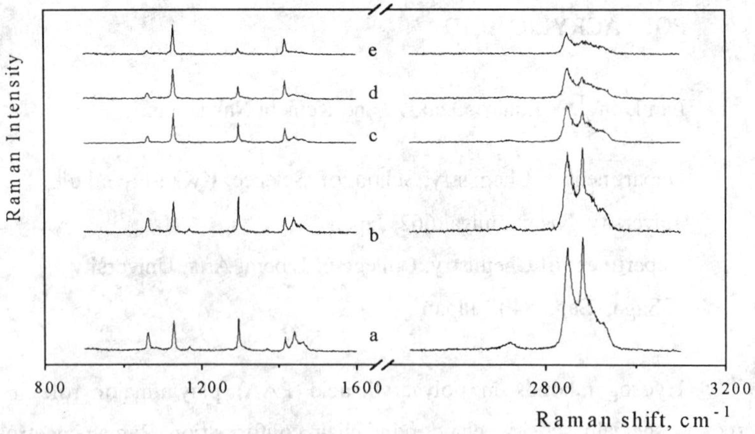

Fig. 1. Series of Raman spectra recorded at the different parts of the "neck" in the drawn polyethylene tape. **a:** the beginning of transition region; **e:** the end of transition region.

Molecules ordering in tapes prepared by compression of PE powder at the same values of draw ratio is higher than in single crystal mats. This fact can explain the high deformation ability of the single crystal mats.

The data obtained for the "neck" region have a good continuation in results obtained for independently prepared series of samples with high draw ratios [3,4]. Combined analysis of these data enables to distinguish two stages in the process of the fibrillar structure formation under the drawing. The first stage is the continuous transformation of the lamellar structure to the fibrillar one. The second stage is the evolution of the fibrillar structure which implies fibrils translation in respect to each other as well as tautening and unfolding of the tie molecules. At the first stage the main changes in the sample structure occur, so it is very important to study deformation process at the initial stages. Micro-Raman technique allows to investigate the whole transformation from isotropic to highly oriented PE structure.

1. N.A.J.M. Van Aerle and A.W.M. Braam, J. Material Sci., 23, 4429 (1988).
2. Yu.A. Zubov, S.N. Chvalun, V.I. Selihova, M.B. Konstantinopol'skaya, and N.F. Bakeev, Russian J. Phys. Chem., LXII, 2815 (1988).
3. S.A. Gordeyev, G.Yu. Nikolaeva, and K.A. Prokhorov, Polymer Science, 5 (1996).
4. S.A. Gordeyev, G.Yu. Nikolaeva, and K.A. Prokhorov, Laser Physics, 6, 121 (1996).

FT-RAMAN, INFRARED AND NEAR-INFRARED SPECTROSCOPIC EVIDENCE FOR THE EXISTENCE OF LINEARLY ASSOCIATED HYDROGEN BOND IN POLYACRYLIC ACID

Jian Dong[1], Yukihiro Ozaki[1]*, and Kenichi Nakashima[2]

[1]Department of Chemistry, School of Science, Kwansei-Gakuin University, Nishinomiya 662, Japan
[2]Department of Chemistry, College of Liberal Arts, University of Saga, Saga 840, Japan

Hydrogen bonds in polyacrylic acid (PAA) play a major role in such properties as special viscosity behavior and chain conformation. Raman spectroscopy is a useful technique in elucidating the hydrogen bond forms in carboxylic acids. In concentrated solutions of acetic, propionic, and *n*-butyric acids, the monomeric and oligomeric forms of carboxylic groups were identified by Raman spectroscopy [1,2]. Lateral hydrogen bonded acid oligomers have also been proposed for several kinds of long-chain fatty acids in monolayer films on solid surfaces [3]. There has been no report which suggests the existence of linearly associated hydrogen bonding structures for polymers containing carboxylic acid groups. Here, we report FT-Raman, FT-IR and near infrared (NIR) studies of possible forms of hydrogen bonding in PAA.

In the Raman spectrum of PAA, the broad asymmetric C=O stretching band near 1690 cm^{-1} shows 4 components, which are located at 1742, 1715, 1680 and 1660 cm^{-1}, respectively. The 1680 cm^{-1} band may be assigned to an in-phase C=O stretching vibration of the cyclic dimer form [1,2,4]; while the other sub-bands at 1742, 1715 and 1660 cm^{-1} correspond to C=O stretching modes of free COOH, terminal COOH and inner COOH in the oligomeric forms, respectively [1,2]. When the temperature is raised, the C=O stretching region in the Raman spectra illustrates that these four different structures of hydrogen bond persist even at the rubber state of PAA. The observed bands are similar to those reported from Raman studies of the association of small organic acids. The less restricted motion of the linear oligomeric COOH as compared to the cyclic dimer can account for the slightly lower frequency observed for the linear oligomers. We postulate that the hydrogen bonding structure in oligomeric form may take a face-on fashion proposed for small aliphatic acids, or a lateral fashion proposed for fatty acids adsorbed on solid surface. Both fashions involve the polarization of the electrons in the covalent CO bonds or π -bond cooperativity.

Additional evidence of various species of hydrogen bond comes from FTIR and FT-NIR spectra. In the FT-IR spectra, four resolved bands in the C=O stretching region, near 1686, 1705, 1723 and 1742 cm^{-1}, are observed over a temperature range of 40-140°C. They may be assigned to C=O stretching modes of the linear open chain hydrogen bonded groups in oligomeric forms, cyclic hydrogen bonded COOH groups in dimer form, terminal groups in the oligomeric COOH chains and free COOH groups, respectively [5,6]. In the NIR spectral region, temperature-dependent intensity changes in the first overtone of an OH stretching mode of PAA reveal that the COOH groups dissociate significantly at high temperatures. This is the first time that various possible hydrogen bond forms analogous to those in small aliphatic acids are demonstrated for polymers such as PAA. The oligomeric chains of COOH groups in PAA may justify the previous conjecture that some sort of cooperative hydrogen bonding exist in PAA or PAA blended polymer alloys [4].

1. J.B.Ng and H.F.Shurvel, J.Phys.Chem., 91, 496(1987).
2. N.Tanaka, H.Kitano and N.Ise, J.Phys.Chem., 94, 6290(1990).
3. L.Sun, L.J.Kepley and R.M.Crooks, Langmuir, 8, 2101(1992).
4. N.Tanaka, H.Kitano, and N.Ise, Macromolecules, 24, 3017(1991).
5. I.F.Wang, E.M.Pearce, and T.K.Kwei, J.Polym.Sci., Part C, 28, 317(1990).
6. M.A.Czarnecki, Y.Liu, Y.Ozaki, M.Suzuki and M.Iwahashi, Appl.Spectrosc., 47, 2162(1993).

RAMAN SPECTROSCOPY OF STRUCTURAL CHANGES IN POLYMER SOLUTIONS

S. D. Cooper and D. N. Batchelder*

Department of Physics and Astronomy, University of Leeds, Leeds, LS2 9JT, UK.

Raman spectroscopy has been used to investigate the effect of extensional flow on polymer molecules in a dilute solution. An opposed jets apparatus was used to create a uniaxial extensional flow field that aligns and extends the polymer molecules. The techniques used enabled a spatial resolution of 10 μm in the extensional flow region to be obtained **[1]**.

The polymer solution investigated was cis-polyisoprene in 1,1,1-trichloroethane. The vibrational frequency studied was the ν(C=C) stretch at 1664 cm^{-1}, which gave information on the strain within the polymer backbone. Figure 1 shows examples of the Raman spectra obtained at different strain rates.

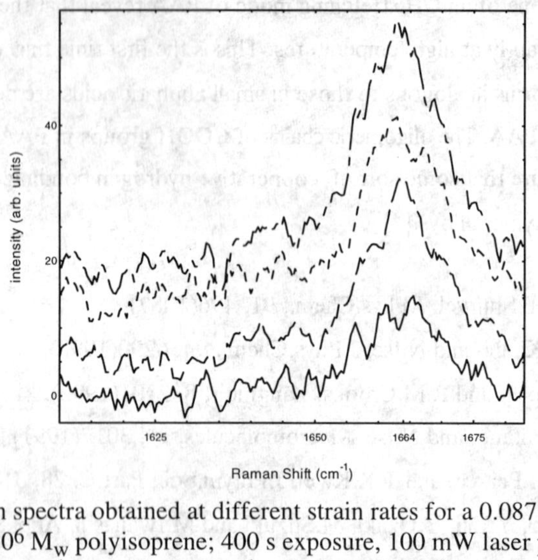

Fig. 1 Raman spectra obtained at different strain rates for a 0.087% solution of 4.4×10^6 M_w polyisoprene; 400 s exposure, 100 mW laser power; —— $\dot{\varepsilon} = 0\ s^{-1}$, – – $\dot{\varepsilon} = 3800\ s^{-1}$, - - - $\dot{\varepsilon} = 4500\ s^{-1}$, — - $\dot{\varepsilon} = 5300\ s^{-1}$.

No changes were observed in the Raman spectra until the strain rate was larger than the critical strain rate for the coil-to-stretch transition. The vibrational frequency was then observed to shift linearly to lower wavenumber with increasing strain rate until, at a high enough strain rate, the rate of decrease changed to a smaller value, as shown in figure 2. The strain rate where this occurred was related to the concentration of the solution. This is evidence that transient network effects were occurring and that the average strain in the region was reduced; a moderate level of strain, however, persisted throughout the region.

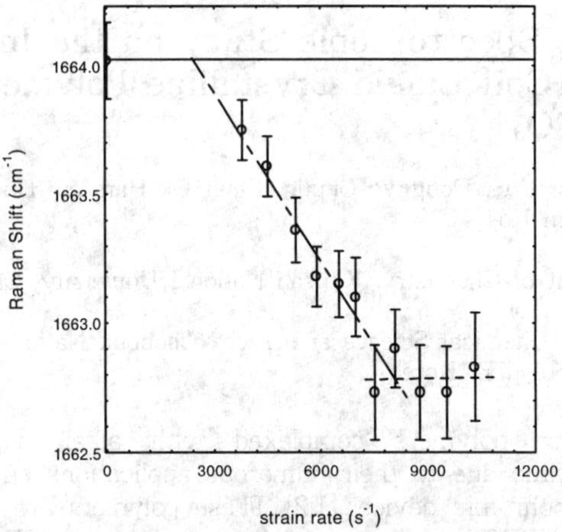

Fig. 2 The average change in frequency of the ν(C=C) vibration of a 0.087% solution of 4.4×10^6 M_w CPI with increasing strain rate, in the region of maximum strain. The solid line shows the frequency at zero strain.

Spatial variations in the Raman spectra were investigated within the extensional flow region. In dilute solutions, below the strain rate for network effects to occur, a cylindrical region of significant Raman shift was observed; at higher concentrations, more complicated structures were seen.

The shifts expected in the Raman spectra were calculated and compared to the measured values. The force on the molecule due to the flow field was calculated using Stokes' law; the stress on the bonds had a parabolic distribution, with the greatest strain on the central bond **[2]**. The shape of the expected resultant Raman band was calculated as the sum of a series of Lorentzian curves representing bonds in the molecule under different levels of strain. The calculated band showed a dramatic shift to lower frequency; the frequency shifts were more than an order of magnitude greater than those observed experimentally.

The probable reason that the shifts are so much smaller than the calculated values is that the polymer chains do not fully stretch out in the flow field, but are instead driven into a folded or kinked state **[3]**. Calculations of the effect of the folds on the expected shifts showed that if the molecule was folded into three or four parts the expected shift would be similar to the experimentally measured values.

1. S. D. Cooper and D. N. Batchelder, Proc. XIV International Conference on Raman Spectroscopy, (N.-T. Yu and X.-Y. Li , eds.) Wiley and Sons, Chichester, 1994, p 700
2. J. A. Odell and A. Keller, J. Polym. Sci. B 24, 1889 (1986)
3. J. M. Rallinson and E. J. Hinch, J. Non-Newt. Fluid Mech. 29, 37 (1988)

Raman Spectroscopic Study on the Interaction of a Rod-Coil Liquid Crystalline Polymer with $LiCF_3SO_3$

Soo-Chang Yu*, Donggyo Chung, Keun-Ok Han Yu, Ho-Seob Choi, Seuk-Beum Ko#

Department of Chemistry, Kunsan National University, Chonbuk 573-701, Korea
#School of Chemical Science and Tech., Chonbuk National University, Chonbuk 561-756, Korea

Conducting polymers complexed with alkali-metals have been studied for a long time due to their numorous applications such as solid state batteries and electrochromic devices.[1,2] These polymers are mostly based on poly(ethylene oxide)(PEO) as a spacer group because PEO shows better solvating power for salts than the others. Nowadays, the liquid crystalline(LC) polymers with two-phase mophology which have aromatic moieties as a rigid mesogenic segment and PEO as a flexible coil segment have attracted much attentions. The mesogenic unit allows variable structures while the coil unit contributes to ionic association and segmental motion of the polymers, which play major roles in the ionic transport in the amorphous regions.[2] Out of LC polymers thermotropic polyesters are one of the most well studied groups in the area of so called main-chain LC polymers.[3,4] But little work has been carried out on rod-coil diblock systems containing PEO as a coil unit. In this study we report Raman spectroscopic results of a rod-coil liquid crystalline polymer, ethyl-4′-(biphenylcarbonyloxy) biphenylcarboxylate with 12 ethylene oxides(EBBEO), complexed with $LiCF_3SO_3$ focusing on which functional groups are susceptible to metal cations. The spectral regions corresponding to both the coil and the mesogenic units are analysed in detail.

The coil unit of EBBEO comprises the repeat units of ethylene oxide, and it is well known that ether oxygens interact with salt with four oxygens bound to one metal cation. To find the binding sites of salt in EBBEO, we have observed the changes in the spectral region of 780-900cm^{-1} that belongs to the coil unit. The spectra of figures 1(a), (b), and (c) represent $[Li^+]/[EO]$= 0, 0.15, and 0.3, respectively. As can be seen in figure 1, the bands have not been changed until $[Li^+]/[EO]$ reaches 0.3. At the concentration of $[Li^+]/[EO]$=0.3 the bands are changed apparently in intensity and frequency. The four bands in figures 1(a) and (b) appeared at 801, 816, 851, and 883cm^{-1} are merged into the three apparent bands as shown in figure 1(c), implying a change in the conformational distribution. These changes account for the binding of the Li^+ to the ether oxygens in the coil unit.

To investigate the mesogenic unit we have chosen the carbonyl stretching bands in figure 2. There are two carbonyl groups in the mesogenic unit, but the carbonyl stretching bands will be different each other in frequency because both terminals centering around ester groups have different moieties. As a result the Raman bands of the carbonyl stretching mode should be different to some extent; The bands at around 1724 and 1732cm^{-1} in figure 2 are thought to be due to either of the carbonyl stretching modes. When the Li^+ binds to the mesogenic unit, the low frequency component at 1724cm^{-1} diminishes in intensity whereas the high frequency one remains unchanged. This means that the Li^+ binds to either of the carbonyl groups with lower frequency. The shoulder band at 1710cm^{-1} is thought to arise from the carbonyl stretching vibration as a result of the formation of the salt complexs.

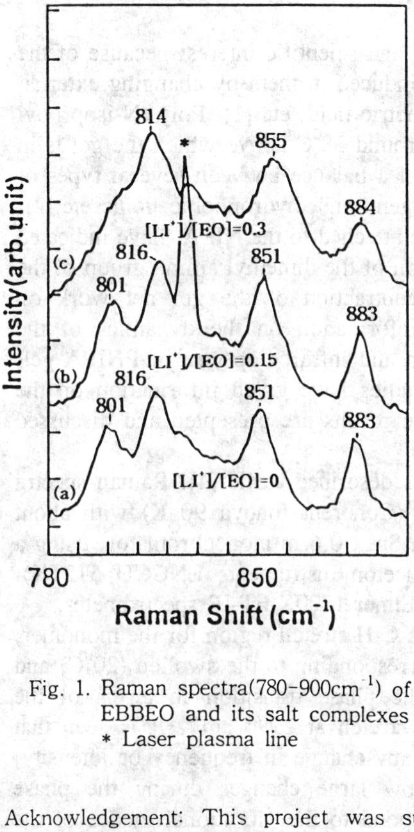

Fig. 1. Raman spectra(780-900cm^{-1}) of EBBEO and its salt complexes
* Laser plasma line

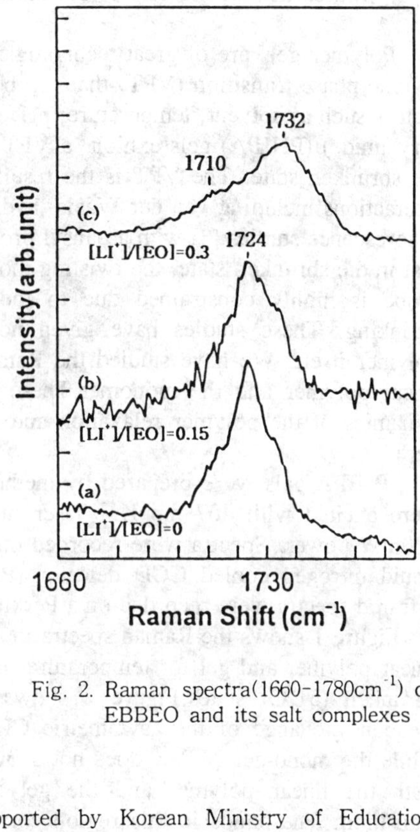

Fig. 2. Raman spectra(1660-1780cm^{-1}) of EBBEO and its salt complexes

Acknowledgement: This project was supported by Korean Ministry of Education, Project No. BSRI-95-3431

1. Polymer Electrolytes Reviews edited by MacCallum, J. R. and Vincent, C. A. Elsevier, London (1987), Vol. 1
2. Polymer Electrolytes Reviews edited by MacCallum, J. R. and Vincent, C. A. Elsevier, London (1989), Vol. 2
3. W. Huang and R. Frech, Polymer, 35, 235 (1994)
4. S. Yoon, K. Ichikawa, W. J. MacKnight, and S. L. Hsu, Macromolecules, 28, 4278(1995)

RAMAN SPECTRAL STUDIES OF POLYMER GELS

K. Tanaka[1*], V.B. Kartha[1], R.R. Dasari[1], M. Feld[1], C. Wang[2], and T. Tanaka[2]

1. G.R. Harrison Spectroscopy Laboratory
2. Department of Physics
Massachusetts Institute of Technology, Cambridge, MA 02139, U.S.A.

Polymer gels are of great technological and scientific interest because of the volume phase transition (VPT) that can be induced in them by changing external factors such as solvent, temperature, pH, electric field, etc.[1]. Poly (N-isopropyl acrylamide) (PNIPA) gels exhibit a VPT around 38^0C above which the gel is in the shrunken state. The VPT is the result of a balance between several types of interactions including van der Waals, hydrogen bond, hydrophobic, ionic etc.[2]. Fluorescence studies [3] with a dansyl probe attached to the NIPA have indicated that in the shrunken state, the twisting motion of the dimethyl amino group of the probe is highly constrained due to the contraction of the gel net work on shrinking. These studies have given no information on the dynamics of the polymer itself. We have studied the Raman and infrared spectra of PNIPA gel, linear polymer and the monomer.These studies have given information on the dynamics of the polymer relaxation and the results are presented and discussed here.

PNIPA gels were prepared by methods described earlier[4]. Raman spectra were excited with 407 nm Kr^+ laser line (Coherent Innova 90 K) with about 100mW power. Spectra were recorded on a Spex 0.6 m monochromator, using a liquid nitrogen cooled CCD detector (Princeton Instruments LNCCD 512TK). Infrared spectra were recorded on a Perkin Elmer 1700X FT-IR spectrometer.

Figure 1 shows the Raman spectra in the C-H stretch region for the monomer, linear polymer and gel for temperatures corresponding to the swollen (20^0C) and shrunken (51^0C) states.Figure 2 shows the phase transition in terms of the frequency change of the asymmetric CH_3 stretch at 2994 cm^{-1}. It is seen that while the monomer NIPA does not show any change in frequency or intensity, both the linear polymer and the gel show large changes during the phase transition. The bands in this region correspond to the vibrational modes of the isopropyl group. That the isopropyl group is undergoing a reorientation during the transition is also shown by the large intensity changes in the infrared spectra of the 1170-1150 cm^{-1} doublet, assigned to the C-C stretch of this group[5].

The Raman and infrared spectra thus clearly show that the VPT of PNIPA gel is accompanied by a change in the isopropyl group orientation and relaxation. The Amide -II band in this system comes at about 1560 cm^{-1} in all the three cases, indicating a trans conformation for the C=O and N-H. The isopropyl group can have a trans or gauche conformation with respect to the N-H. The spectral changes with temperature arise most probably from a conformational change of this group brought about by steric constraints in the polymer and gel. Such a change will not take place in the monomer because the isopropyl group in this case is free to rotate, without any constraints.

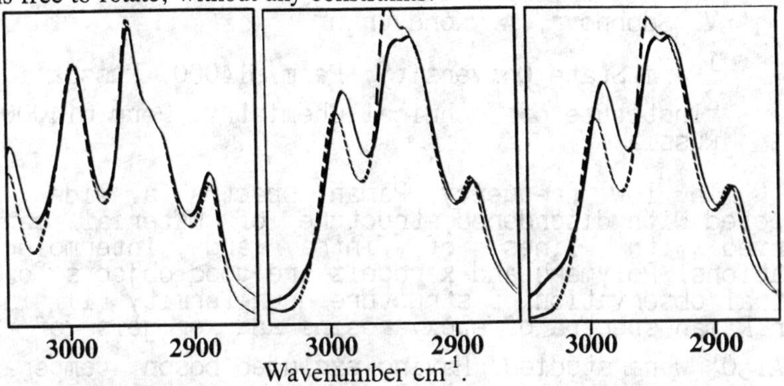

Figure 1. Raman spectra of C-H stretch region: from left to right- NIPA monomer, NIPA polymer, and NIPA gel.- - - 20^{0}C. — 51^{0}C.

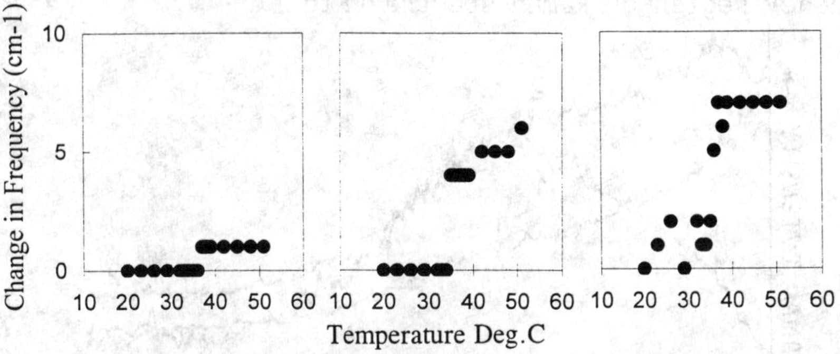

Figure 2. Change in asym. CH_3 stretch: from left to right- NIPA monomer, NIPA polymer, and NIPA gel.

1.Tanaka, T., Sci. Am, 244, 124 (1981).
2. Ilmain, F., Tanaka, T., and Kokufuta, E., Nature, 349, 400 (1991).
3. Asano, M., Horie, K., and Yamashita, T., (in press).
4. Tanaka, T., Fillmore, D.J., S.-T. Sun, I., Nishio G. Swislow, and Shah, A. Phys.Rev. Lett. 45, 1636 (1980).
5. Lin-Vien, D., Colthup, N.B., Fateley, W.G., and Grasselli, J.G., Infrared and Raman Characteristic Frequencies of Organic Molecules, Academic Press, New York (1991).

FRACTAL AND PSEUDO-RAMAN SPECTRA OF EPOXY RESINS AND XEROGELS

V. Sdobnov[1], A. Kondyurin[2*]

[1]Perm State University, Perm 614000, Russia
[2]Institute of Technical Chemistry, Perm 614000 Russia

In low frequency Raman spectra a wide peak connected with disordered structure of material can be observed with lines of intra and intermolecular vibrations. Polymers and xerogels are good objects for the spectral observation of structure ununiformity [1]. In our paper Raman spectra of epoxy resins and xerogels of SiO_2 and Al_2O_3 were studied. Having excluded boson temperature factor [2] wide peak with maximum at 80 cm^{-1} for epoxy resins and 300 cm^{-1} for xerogels were observed in low frequency region of Raman spectra (Fig.1).

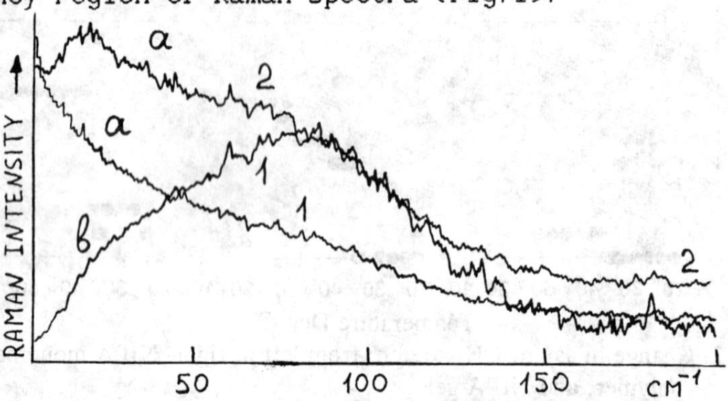

Fig.1. Raman spectra (a) and LF-peak (b): (1) ED-24 epoxy resin (MM=624), (2) ED-24 and monoethanolamin material.

Spectral distribution of peak can be explained by fractal structure of material [3]:

$$I_{vh}(\omega) \sim \omega^{\mu} [n(\omega,T) + 1] \quad (1)$$

$$\mu = d(1 + 2/D) - 2 \quad (2)$$

where D - fractal dimension and d - fracton dimension. Equations (1) and (2) can be used for calculation of fractal structure parameters of material by low frequency Raman spectra.

In Raman spectra of epoxy resins in region to 2000 cm^{-1} the same wide peak can be observed too (Fig.2). This high frequency peak is connected with electron exitation wave [4, 5]. Maximum position and intensity of peak don't correlate with exiting light wavelength and correlate with chemical structure of polymer. Parameters of peak are sensitive to temperature treatment of polymers. This spectral method can be used for study of polymer structure.

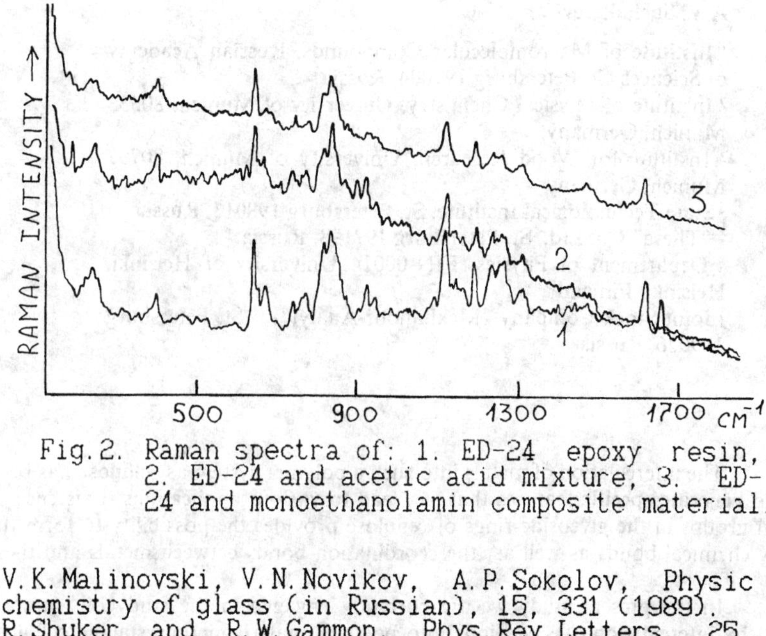

Fig.2. Raman spectra of: 1. ED-24 epoxy resin, 2. ED-24 and acetic acid mixture, 3. ED-24 and monoethanolamin composite material.

1. V.K.Malinovski, V.N.Novikov, A.P.Sokolov, Physic and chemistry of glass (in Russian), 15, 331 (1989).
2. R.Shuker and R.W.Gammon, Phys.Rev.Letters, 25, 222 (1970).
3. V.K.Malinovsky, V.N.Novikov, A.P.Sokolov, V.A.Bagryansky, Chem.Phys.Lett, 143, 111 (1988).
4. J.P.Biscar, N.Kollias, Chem.Phys.Lett, 27, 100 (1974).
5. I.F.Klassen, V.E.Pogorelov, G.I.Salivon, Ukrain's Physical J. (in Russian), 32, 1342 (1987).

NEW DATA ON RAMAN AND FTIR STUDY OF CELLULOSE WITH INTERCALATED SILVER CLUSTERS

N.E. Kotelnikova[1*], H. Knozinger[2], G. Wegener[3], M. Scheithauer [2], V.N. Demidov[4], V.P. Kotelnikov[5], T. Paakkari[6], A.V. Shchukarev[7]

[1] Institute of Macromolecular Compounds, Russian Academy of Sciences, St. Petersburg 199004, Russia
[2] Institute of Physical Chemistry, University of Munich, 80333 Munich, Germany
[3] Institute for Wood Research, University of Munich, 80797 Munich, Germany
[4] State Technological Institute, St. Petersburg 198013, Russia
[5] "Thesa" Co. Ltd., St. Petersburg 197198, Russia
[6] Department of Physics, FIN-00014 University of Helsinki, Helsinki, Finland
[7] Joint Stock Company "Mekhanobr-Analyt", St. Petersburg 199026, Russia

The intercalation of metals into such a polymer system as cellulose has been shown to generate specific features that can lead to various applications. Existence of hydroxyl groups in the glycoside rings of cellulose provides the possibility to form the ordinary chemical bonds as well as the ccordination bonds between metals and these groups.

In the present study we are currently engaged in the study of samples obtained by intercalation of silver ions into polymer matrix microcrystalline cellulose (MCC) and subsequently reduced to metal silver with zero oxidation number. The amount of silver in this sample was 5.5% in bulk (according to the elemental analysis) and about 13% on the surface (according to ESCA data).

FTIR spectra of the resulting compound obtained by Digilab FTS 40 FTIR spectrometer reveal no changes in the range of 400-1800 cm^{-1} in comparison with the spectrum of the initial MCC sample; deconvolution spectra in the range 2400-3700 cm^{-1} show only small redistribution of the absorption bands intensity of OH groups attributed to the intramolecular hydrogen bonds between O(2)H···O(6) in cellulose (3410 and 3470 cm^{-1}). Therefore, it is impossible to conclude from these data about any interaction between MCC and silver in any form.

In spite of this, great changes in the range of 1000-2000 cm^{-1} are observed in Raman spectra of MCC-Ag sample. The bands at 1360, 1450, 1520 and 1558 cm^{-1} form intense wide band with a complex structure. Its intensity strongly increases (coefficient of the increase as compared to the MCC spectrum is about 70). Character of absorbance at this range makes it possible to attribute this bands to symmetric and antisymmetric vibrations of (-COO)- groups. It means that ionised and unionised metal-carboxyl groups present in the samples. These groups appear in MCC after the reduction of a part of silver ions to silver with the oxidation number zero and subsequent oxidation of some part of MCC aldehyde groups to carboxyl groups and their complexation with silver ions.

Except this, according to WAXS data the resulting compound is a two-component system of MCC and metallic silver (oxidation number zero).

Hence, application of Raman spectroscopy allowed us to elucidate the path of the complex reaction of the preparation of cellulose-silver compounds.

RADICAL POLYMERIZATION AND COPOLYMERIZATION OBSERVED IN OPTICALLY LEVITATED UNSATURATED-POLYESTER-RESIN-MICROPARTICLES

J. Musick, J. Popp, M. Trunk, and W. Kiefer*

Institut für Physikalische Chemie, Universität Würzburg,
Marcusstr. 9-11, D-97070 Würzburg, Germany

Investigations of polymerization reactions are needed for several applications. The polymerization of microparticles consisting of synthetic materials are of great interest in building and electrical industries. Polymerization is the process of linking many monomer units into larger polymer molecules by a radical chain reaction. Copolymerization is the 3-dimensional linking-up of the polymer chains.

In situ analysis of single aerosol particles of synthetics is necessecary for understanding the fundamental polymerization reactions in single particles [1]. In this context we present measurements of polymerization and copolymerization reactions of styrene and single unsaturated-polyester-resin (UP-resin) microdroplets. UP–resin is formed by condensation reaction of maleic acid and glycol. As a radical–initiator we used dibenzoylperoxid which was cleaved with dimethylanilin.

Different spectroscopic techniques in combination [2] with optical levitation [3] were used for the investigations by recording elastic and inelastic light scattering and measurements of the laser power. Apart from the observation of the reactions by educt and product signals in the Raman spectra the polymerization and copolymerization can be followed by means of morphology dependent resonances (MDR). Also, changes in the particle size during the reaction and the refractive index can be detected by these MDRs in the inelastically and elastically scattered light as well as in the laser power necessary for stable levitation [3-5]. MDRs correlate with eigenfrequencies of the droplet and therefore depend only on the morphology (size, shape, refractive index) of the droplet. Therefore precise data can be obtained during the reaction and the course of the reaction can be investigated.

In panel A of Fig. 1 the Raman spectra in the region between 1550 and 1800 cm^{-1} is shown as a function of time. The phenyl–vibration and the doublebonding vibration of styrene are recorded at 1596 cm^{-1} and 1628 cm^{-1}, respectively. The latter decreases in a short time and describes the polymerization. The maleic acid C=C–doublebonding mode and the C–O–stretching of UP-resin are observed at 1655 cm^{-1} and 1725 cm^{-1}, respectively. In panel B the decrease of the ratio of the maleic acid doublebonding at 1655 cm^{-1} to the C–O–bonding at 1725 cm^{-1} is shown, which characterizes the course of the copolymerization.

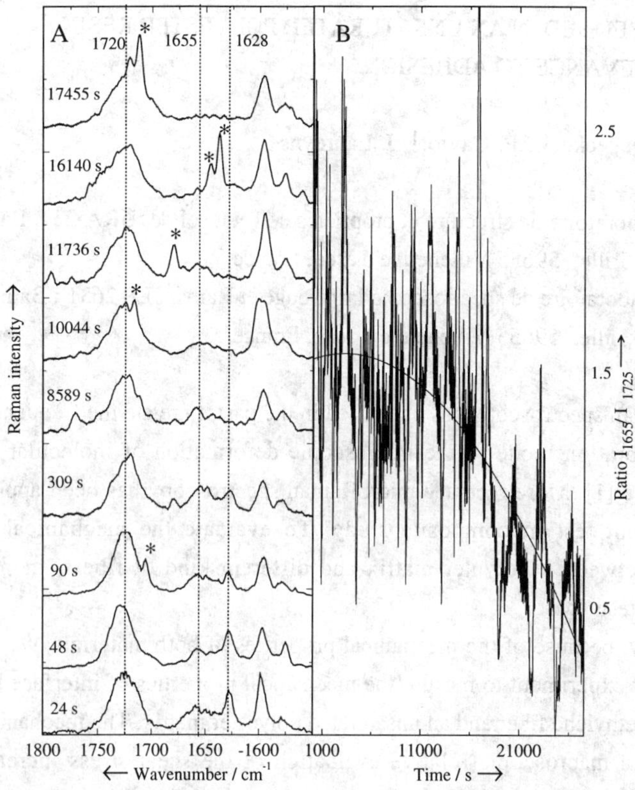

Fig.1 A: Raman spectra of the polymerization and copolymerization of styrene. Morphology dependent resonances are marked by asterisks; **B:** Intensity ratio of the maleic acid doublebonding to the C–O–bonding as a function of time.

Financial support from the Deutsche Forschungsgemeinschaft and the Fonds der Chemischen Industrie is highly acknowledged.

1. J. Musick, J. Popp, M. Trunk, and W. Kiefer, to be published.
2. J. Popp, M. Lankers, M. Trunk, I. Hartmann, E. Urlaub, and W. Kiefer, to be published.
3. A. Ashkin, Phys. Rev. Lett. 24, 156 (1970).
4. K. Schaschek, J. Popp, and W. Kiefer, J. Raman. Spectrosc. 24, 69 (1993).
5. W. Kiefer, Croat. Chim. Acta 61, 473 (1988).

RAMAN SPECTROSCOPY OF TREATED POLYETYLENE FIBERS EMBEDDED IN AN UNSATURATED POLYESTER RESIN : RELEVANCE TO ADHESION.

C. Depecker[1],J.P. Cavrot[1], J. Laureyns[2].

[1] Laboratoire de structure et propriétés de l'état solide, URA 234, Bâtiment C6 UST Lille, 59655 Villeneuve d'ascq, France.
[2] Laboratoire de spectroscopie infrarouge et Raman, LP 2631 ; Bâtiment C5 UST Lille, 59655 Villeneuve d'ascq, France.

Raman spectroscopy has been used in the past, to relate the frequency shift of polymer vibration mode to the microscopic deformation of molecular chains of polyethylene [1]. More recently microRaman spectroscopy has been applied to the monofilament test in composite study. To evaluate the mechanical interface propretiesbetween a polymer matrix and different kind of fibers, monofilament mechanical tests exist .
In our study, because of the mechanical properties of both materials, we have used pull out test experiment to mesure the mechanical properties of interface between a treated polyethylene fiber and an unsaturated polyester matrix. The mechanical results provide us a macroscopic relative evaluation of the shear stress increase of the interface in function of the fiber treatment [2]. Mechanical model as shear lag analysis [3,4] are very far from the experimental values.
To get a better understanding of this system, we have recorded Raman polyethylene spectra along the fiber axis for different points in the embedded part of the fibre in the resin matrix. The C-C symmetric stretching vibration mode is used to monitor the stress transfert along the embedded PE fiber into the polyester matrix. To eliminate the structural influence of the melt spun fiber and to take into account the compressive state of the cylindrical fiber into the polyester resin [5] a calibration curve and a corected profile (corresponding to the mesurement of the stress profil in the embedded part without any strength exerced on the external part of the fiber) is used to get access to the absolute stress versus the length between the focus point and the edge of the sample. This profile is compared with those calculated from the mechanical measurements with the stress and the shear stress relations (figures 1 and 2):

$$\sigma(x) = \frac{F}{\pi r_f^2}(\coth(\alpha l_c)\sinh(\alpha x) - \cosh(\alpha x)) \qquad (1)$$

$$\tau(x) = \frac{F}{2\pi r_f}(\coth(\alpha l_c)\cosh(\alpha x) - \sinh(\alpha x)) \qquad (2)$$

where F represent the strength exerced on the fiber, r_f is the radius of the fiber, x is the distance of the fiber in the matrix at the considered point, l_e is the critical embedded length (the maximal length for which we observe a transition between an interface rupture mode and an external fiber rupture mode), and α contains the mechanical interphase parameters developped by Grezczuk. Taking into account the Raman spectra, these calculations are in good aggreement with the mechanical properties of both materials and allow the determination of the interface parameters with more realistic considerations.

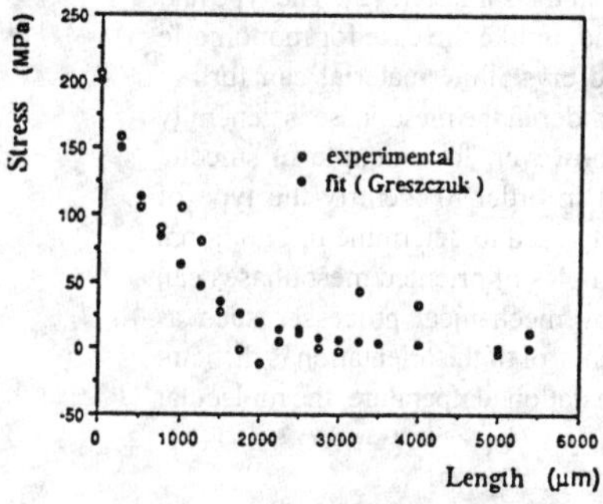

Fig 1 : Untreated fiber

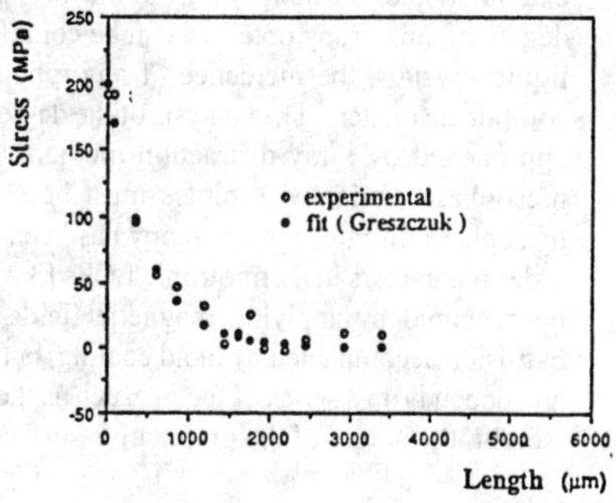

Fig 2 : Treated fiber

1- R.P. Wool, R.S. Bretzlaff ; J. Polym. Sci., Part B :Polym. Phys. **24**, 1039,(1986).

2- P. Masse, J.P. Cavrot, P. François, J.M. Lefebvre, B. Escaig ; Polym. Comp. 15,247,(1994).

3- L.B. Gřeszczuk ; Interfaces in composites ASTM STP Philadelphia PA, USA **452**,42,(1969).

4- P.S. Chua, M.R Piggott ; Comp. Sci. Tech. **22**,107,(1985). and P.S. Chua, M.R Piggott ; Comp. Sci. Tech. **22**,185,(1985).

5- P. Masse, J.P. Cavrot, C. Depecker, J. Laureyns, B. Escaig ; Macromol. Symp. **94**,249-257,(1995).

ANALYSIS OF MECHANICALLY-INDUCED ORIENTATION IN THERMOTROPIC LIQUID CRYSTAL POLYESTERS BY POLARIZED RAMAN SPECTROSCOPY

G. Ellis*, M.A.Gómez, J. del Pino and C.Marco

Department of Polymer Physics and Engineering
Instituto de Ciencia y Tecnología de Polímeros,
CSIC, Juan de la Cierva 3, 28006-Madrid, SPAIN.

Main-chain thermotropic liquid-crystal polymers (LCPs) containing a rigid aromatic mesogenic unit and a flexible aliphatic spacer are of great interest for the study of ordered and disordered states in condensed phases. The vast majority of LCPs of this type exhibit either smectic or nematic mesophases [1], normally identified by polarized thermo-optical microscopy (TOA). In some thermotropic LCPs, the ordered liquid crystal structure can be "frozen-in" allowing the examination of the morphology of the mesophase in the solid state [2]. The type and degree of anisotropy obtained can be complex and, unlike the case for monomeric liquid crystals, the incidence of amorphous and crystalline material can further complicate matters. The analysis of the degree of order in the mesophase is generally approached by x-ray diffraction methods [1], however, for example in smectic mesophases, oriented samples must be studied in order to identify the type of mesophase. Infrared spectroscopy has been widely used to determine orientational order parameters in thermotropic LCPs [3,4]. Samples of oriented mesophases can be prepared by applying magnetic fields, or by mechanical processes such as extrusion accompanied by rapid cooling, but the control of the orientation is difficult and depends on a series of factors such as the preparation temperature, the molecular weight, the nature of the preparative surface, etc.

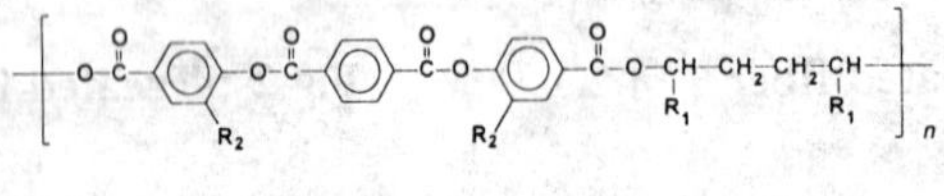

This communication describes the analysis by polarized Raman spectroscopy and x-ray diffraction of a homologous series of LCPs with the above chemical structure, where R_1 = H, CH_3 and R_2 = H, Cl, and the polymer with R_2 = H but with 7 methylene units in the aliphatic spacer, denominated P7TOB, see Table 1.

TABLE 1. Thermal transitions and inherent viscocity data [2,5]

Polymer	Tg (°C)	T_1 (°C)	T_i (°C)	η_{inh} (dl·g^{-1})	Type of mesphase
P4TOB	75	285	350	0.33	nematic
PDM4TOB	75	185	250	0.28	nematic
P4CTOB	—	274	319	0.39	nematic
PDM4CTOB	77	133	176	0.29	nematic
P7TOB	65	173	296	0.67	smectic

Oriented samples were prepared by drawing fibres from the polymers in the mesophase at a range of temperatures above T_1. The fibre forming capabilities of the samples varied significantly as a function of temperature, as did the degree of orientation induced in the materials, determined by the azimuthal thickness in the x-ray fibre diagrams. In the case of the first four polymers in the table, the Raman data supports the hypothesis of preferential alignment of the molecular axis along the fibre draw axis. Although the differences observed in the polarized spectra point towards only low levels of orientation, differences in spectral behaviour can be observed as a function of temperature. In the case of P7TOB, the polarization behaviour is anomalous. Fig. 1 shows a series of spectra recorded at different geometries for this sample. The strong depolarization of a series of bands associated with the mesogenic unit when the laser polarization is perpendicular to the fibre axis is very interesting. In particular, the variation in intensity of the quadrant stretching modes (ν_{16} Herzberg) of both the oxybenzoyl and terephthaloyl rings, previously adopted in various IR studies because they are characteristic of the molecular axis of the mesogen, seem to indicate that the molecular axes lie perpendicular to the draw direction. The ability of the mesogenic units to orient themselves perpendicular to the orientation direction has been observed by IR dichroism measurements on uniaxially oriented films of low molecular weight LCPs, and shown to be molecular weight and shear dependent [4]. Further, when P7TOB is drawn at temperatures close to T_1, the fibre forming capability of the material improves drammatically, and the observations in the Raman data become analogous to those observed for the nematic systems. These and other observations related with mesophase and crystal orientation are described in detail in the communication.

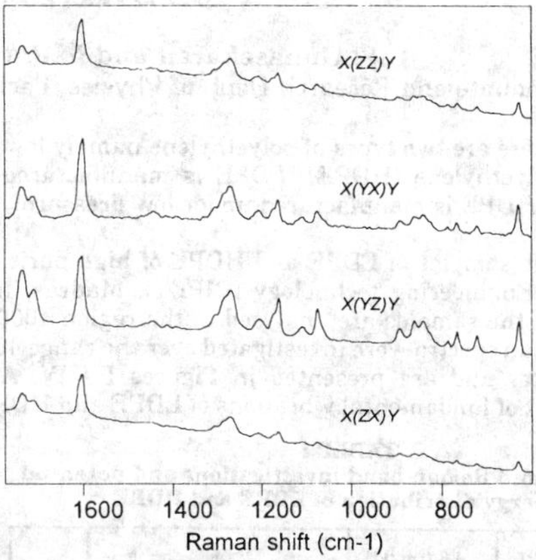

Fig.1. Polarized Raman spectra of P7TOB

1. A.Blumstein, "Polymeric Liquid Crystals", Plenum Press, NY (1985)
2. J.Lorente, Ph.D Thesis, Universidad Complutense de Madrid (1996)
3 B.Z.Volchek, N.S.Kholmuradov, A.V.Purkina, A.Yu.Bilibin, S.S.Skorokhodov, Polym. Sci. USSR, 27, 90 (1985)
4 B.Z.Volchek, D.A.Medvedeva, A.V.Purkina, S.V.Shilov, Poym. Sci. USSR, 32, 1462 (1990)
5. F.J.Del Pino, Ph.D Thesis, Universidad Complutense de Madrid (1996)

SPECTROSCOPIC INVESTIGATIONS AND VIBRATIONAL ANALYSIS OF LOW DENSITY AND HIGH DENSITY POLYETHYLENE

S. Gunasekaran and M.Rathna Kumari
Post Graduate and Research Dept. of Physics, Pachaiyappa's College, Madras, India.

There are two types of polyethylene namely low density polyethylene (LDPE) and high density polyethylene (HDPE). LDPE is manufactured under high pressure using peroxide initiators. HDPE is manufactured under low pressure.

The samples of LDPE and HDPE of high purity were obtained from Central Institute of Plastic Engineering technology (CIPET), Madras, India. The Fourier transform infrared spectra for the samples are analysed in the region 4000-400 cm^{-1} on Bruker IFS 66V and the Laser Raman spectra were investigated over the range 4000-100 cm^{-1} using DILOR Z-24 Raman spectrometer and are presented in Figures I - IV. A detailed interpretation on the band assignment of fundamental vibrations of LDPE and HDPE has been made as shown in Table I.

TABLE I
Infrared and Raman band investigations and potential energy distribution of LDPE and HDPE

FTIR (cm^{-1})		LR (cm^{-1})		Description	PED%
LDPE	HDPE	LDPE	HDPE		
-	-	484	514	$\upsilon_1 - \upsilon_8$	
504	529	-	-	$\upsilon_{12} - \upsilon_7$	
-	-	565	588	$\upsilon_1 - \upsilon_8$	
-	-	580	600	$\upsilon_1 - \upsilon_8$	
720	730	-	-	$2\upsilon_{10} - \upsilon_7$	
-	-	825	850	$(\upsilon_3 + \upsilon_8) - \upsilon_1$	
840	853	-	-	(υ_{10}) C = CH asymmetric deformation	99
-	-	930	950	(υ_8) out of plane deformation	97
925	964	-	-	(υ_7) out of plane deformation (symmetric)	98
-	-	1070	1100	$\upsilon_2 - 2\upsilon_8$	
-	-	1300	1325	(υ_6) C = CH asymmetric deformation	92
-	-	1303	1350	(υ_3) symmetric HCH deformation	90
1462	1471	-	-	(υ_{12}) HCH asymmetric deformation	95
-	-	1453	1484	(υ_1) C = C stretching	100
2018	2019	-	-	$\upsilon_9 - \upsilon_7$	
-	-	2310	2323	$\upsilon_5 - \upsilon_8$	
2330	2343	-	-	$\upsilon_{12} + \upsilon_{10}$	
-	-	2660	2679	$\upsilon_3 + \upsilon_1$	
2955	2991		-	(υ_{11}) C-H symmetric stretching	90
-	-	2950	3001	(υ_2) C-H symmetric stretching	87
-	-	2975	3050	(υ_5) C-H asymmetric stretching	91
3000	3060	-	-	(υ_8) C-H asymmetric stretching	86
3603	3604	-	-	$(\upsilon_{11} + \upsilon_{12}) - \upsilon_{10}$	

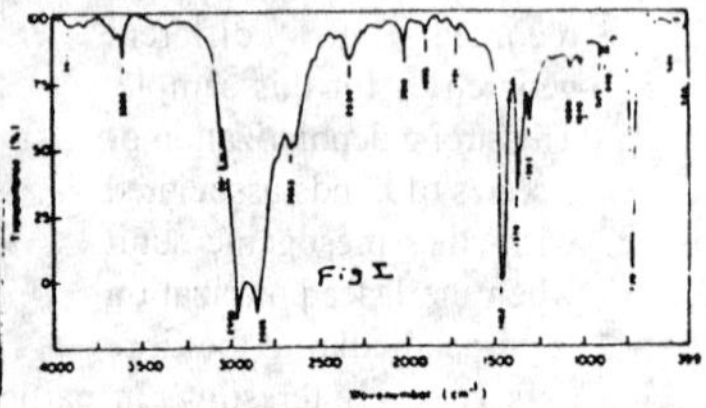

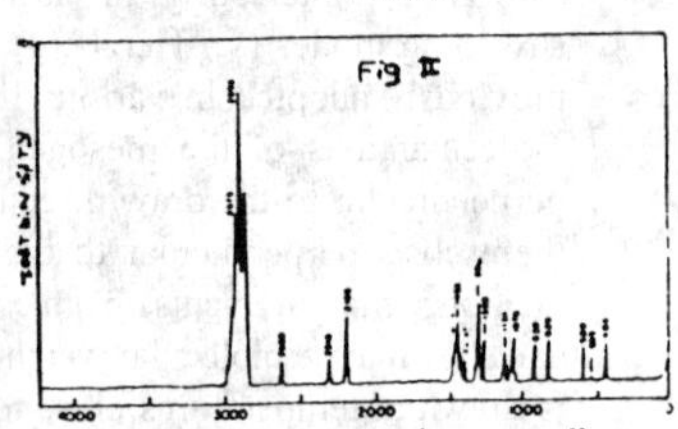

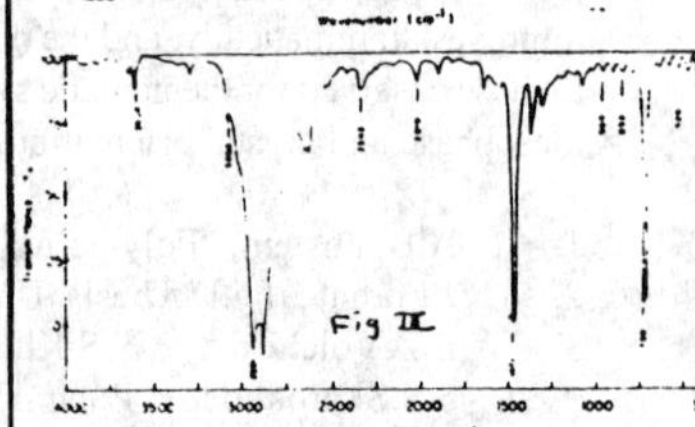

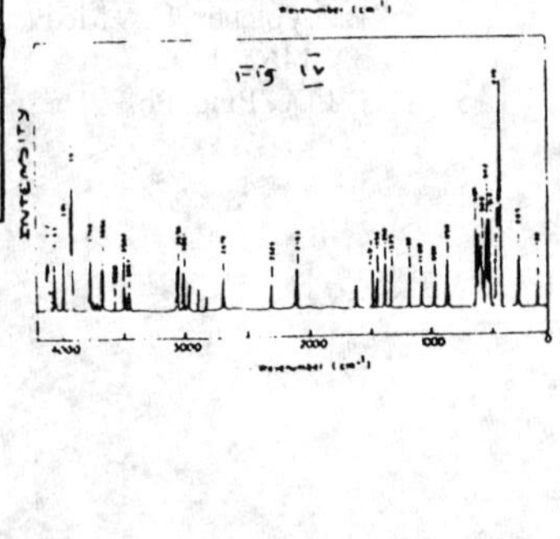

The present work is entirely devoted to make a complete vibrational analysis on polyethylene by assuming the repeat unit with D_{2h} symmetry. A normal coordinate analysis has been carried out following Wilson's FG matrix method [1] and systematic set of molecular constants has been evaluated. The potential constants, vibrational mean amplitudes, coriolis coupling constant, centrifugal distortion constants are presented in Table II, III, IV and V. The fact that the force constat F_D is higher than fd suggests that C=C bond is much stronger than C-H bond. The values of F_D, fd and f_θ agree quit well with the value reported in literature [2-5].

Comparative study on LDPE and HDPE

Differences between LDPE and HDPE can be observed most easily. The wave numbers of Raman active only chain vibrations of polyethylene have been interpreted in analogy with literature [6-8]. A good example of sensitivity of structural order is given by the 1130 cm^{-1}/1060 cm^{-1} lattice vibrations of polyethylene [7].

A systematic set of molecular constants made available on polyethylene confirms the correctness of the present assignment.

Table II
Comparison of potential constant of C_2H_4 (mdyne/Å)

Potential Constant	PW	2	3	4	5
f_D	11.70	10.51	9.93	11.43	9.40
f_d	5.20	5.14	5.91	5.38	-
f_θ	0.40	0.23	1.14	0.34	-

Table III
Bounded and non - Bounded mean amplitude of vibration (10^{-23}) of LDP4E

I	Values
1_D (C-C)	4.161
1_d (C-H)	7.734
1_p ($H-C_2$)	13.446
1_pC (H_1-H_2)	12.502
1_r (H_1 - H_3)	15.286
1_s (H_1 - H_4)	14.561

Table IV
Coriolis Coupling Constants

ξ	Values
$\xi_{1.5}$	0.3345
$\xi_{1.6}$	-0.5887
$\xi_{2.6}$	0.7808
$\xi_{3.5}$	0.9424
$\xi_{3.6}$	0.4568
$\xi_{4.10}$	-0.8820
$\xi_{4.11}$	-0.8483
$\xi_{4.12}$	0.5294
$\xi_{5.8}$	0.7717
$\xi_{6.8}$	0.0614
$\xi_{7.9}$	-0.8820

ξ	Values
$\xi_{7.10}$	-0.4711
$\xi_{4.9}$	0.4711
$\xi_{1.8}$	0.0581
$\xi_{2.8}$	-0.4847
$\xi_{3.8}$	-0.8395
$\xi_{9.4}$	-0.0671
$\xi_{9.12}$	-0.9977
$\xi_{10.11}$	0.9977
$\xi_{10.12}$	0.0671
$\xi_{7.11}$	0.5293
$\xi_{7.12}$	0.8483

TABLE V
Centrifugal Distortion Constants (HZ)

D_J	34.5
D_K	2915.5022
D_{JK}	336.5904
R_5	86.3977
R_6	0.59231
δ_J	6.9789

[1] Wilson Jr., Decius and Cross, 'Molecular Vibrations', McGraw Hill, New York, 1955.

[2] B.M.Arthur, 'Molecular Constants by G.K.F. matrix method', Ph.D. Thesis-Univ. of Madras.

[3] P.Puley and W. Meyer, J. Mol. Spectra C.40, 59 (1971).

[4] K.Ramasamy and V.Devarajan, Acta. Phys. Pol., 34, 985 (1968).

[5] J.L.Duncan, D.C. McKean, and P.D. Mallison, J. Mol. Spectra., 45, 221 (1973).

[6] Schaufele, R.F. Macromol. Rev. 4, 67 (1970)

[7] Strobel, G.R., Colloid Polym. Sci., 257, 584 (1979).

[8] Hendra,D in : Huminel,D O (ed):Polymer Spectroscopy, Weinhein:Verlag chemic 1974 p.151.

CHEMOMETRICS ANALYSIS OF FOURIER-TRANSFORM RAMAN SPECTRA OF ETHYLENE/VINYL ACETATE COPOLYMERS; DISCRIMINATION AND PREDICTION OF PERCENTAGE OF VINYL ACETATE

M. Shimoyama[1], H. Maeda[2], K. Matsukawa[3], H. Inoue[3], T. Ninomiya[1], and Y. Ozaki[2]*

[1]Forensic Science Laboratory, Hyogo Prefectural Police Headquarters, Shimoyamate-dori, Chuo-ku, Kobe 650, Japan

[2]Department of Chemistry, School of Science, Kwansei-Gakuin University, Uegahara, Nishinomiya 662, Japan

[3]Plastics Department, Osaka Municipal Technical Research Institute, Morinomiya, Joto-ku, Osaka 536, Japan

Recently, chemometrics has been introduced to the field of Raman spectroscopy for qualitative and quantitative analysis of Raman spectral data. For example, Everall et al. [1,2] carried out characterization of polymer morphology using chemometrics analysis of Raman spectra. Chemometrics aims at correlating statistically observed spectral changes with known sample properties such as concentration, density, and crystallinity.

The purpose of the present paper is to demonstrate potential of chemometrics analysis of Raman spectra in investigating the discrimination and composition analysis of copolymers. We have employed principal component analysis (PCA) and partial least squares (PLS) to produce calibration models which allow the discrimination of ethylene/vinyl acetate (EVA) copolymers with different percentage of vinyl acetate (VA) and the prediction of its percentage, respectively.

The samples used, EVA copolymers, were generous gifts from TOSOH Co. Ltd. (Tokyo) and used without further purifications. The percentage of VA in EVA changes from 6 to 42. The FT-Raman spectra of the copolymers were measured at an 8 cm^{-1} resolution with a JEOL JRS-FT 6500 Raman spectrometer equipped with an InGaAs detector. A 1064-nm cw Nd;YAG laser (Spectron SL301) was used for excitation. All the spectra were the results of the coaddition of 100 interferrograms.

The obtained FT-Raman spectra were transformed from .JDX into .UNS files. PCA and PLS were applied to the spectral data by using the Unscrambler (CAMO AS, Trondheim, Norway) software programs. All regressions were calculated for the spectral region of 2000-500 cm^{-1}. The calibration models were tested by cross-validation. All the FT-Raman data were subjected to the treatment of multiplicative scatter correlation (MSC) before proceeding with any analyses.

The Raman spectra in the 2000- 500 cm^{-1} region of EVA copolymers are characterized by bands at 1740, 1439, 1296, 1126, 1064, and 629 cm^{-1} due to C=O stretching, CH_2 bending, C-C stretching, C-C stretching, C-C stretching and O-C=O deformation modes, respectively. Fig. 1 illustrates the result of the PCA analysis which shows the variation in the percentage of VA in EVA. The twelve kinds of EVA copolymers are clearly discriminated by using three factors.

Fig. 2 shows a PLS calibration model which predicts the percentage of VA in EVA copolymers. The correlation coefficient and standard error of prediction were 0.997 and 0.911%, respectively. Loading plot for factor 1 shows that all the above-mentioned bands contribute largely to the prediction.

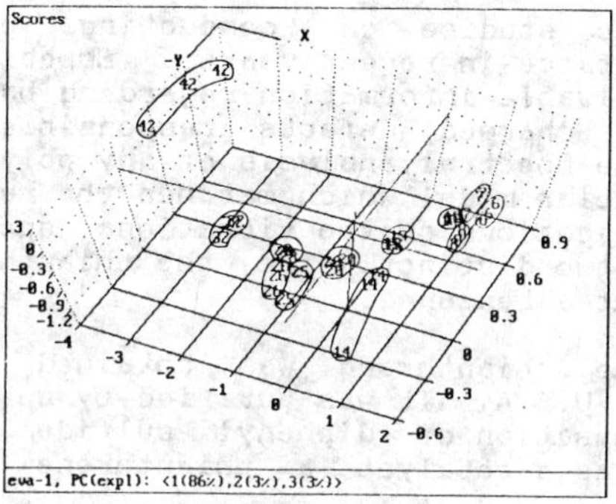

Fig. 1 Three-dimensional score plot showing the variation of the percentage of VA in EVA copolymers.

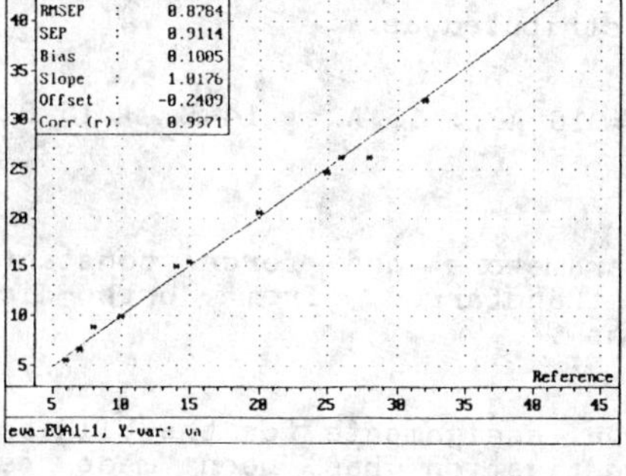

Fig. 2 Scatter plot of actual and predicted percentage of VA in EVA copolymers

1. N.Everall, P.Tayler, J.Chalmers, R.Ferwerda, and J.van der Maas, Polymer, 35, 3184 (1994).

2. N.Everall, J.Chalmers, R.Ferwerda, J. van der Maas, and P.Hendra, J. Raman Spectrosc. 25, 43 (1994).

INVESTIGATIONS ON THE FTRAMAN AND FTIR SPECTRA OF THIANTHRENE AND POLY(PHENYLENE SULFIDE) OF THIANTHRENE-TYPE POLYMER

S. Mohan* and R. Murugan
Raman School of Physics, Pondicherry University,
Pondicherry, India 605 014

Spectroscopic studies on conducting polymers received considerable importance in recent years . Especially, vibrational spectra provide valuable information regarding band structure and on the nature of charged defects responsible for doping and transport. Since the spectral analysis of any polymer is based on suitable low molecular model which matches the repeat unit. Here in this case , investigations on the vibrational spectra of two repeat units is based on two distinct units; the thianthrene and 1,2,4,5-tetrahalo substituted benzene .

Reagent grade thianthrene was obtained commercially from Aldrich Chemicals, U.S.A. It was purified by sublimation in vacuo. By bulk poly condensation of diphenyl sulfide at 225°C in the presence of $AlCl_3$ as a catalyst, a thianthrene type polymer was synthesized as described by Costa Bizzarri et al [1]. The Fourier transform infrared spectra and Fourier transform Raman spectra of thianthrene and the syntesised polymer were recorded. In order to assign all the vibrational modes unambiguously, normal coordinate analysis has been carried out using Wilson's F-G matrix method [2] with computer programs developed by Fuhrer et al [3]. The molecular model used for the normal coordinate analysis of thianthrene is of C_{2v} symmetry and consists of 22 atoms. The 60 internal modes of this system are distributed as

$$G_{vib} = 16\ A_1 + 15\ A_2 + 14\ B_1 + 15\ B_2$$

The structural parameters and force constants used for the thianthrene were transferred from ortho-dichlorobenzene and ortho-dibromobenzene .

The vibrational assignments for the FTIR spectrum of polymer in the present investigation has been made essentially on the basis of magnitudes and relative intensities of the observed peaks and the data available from the spectra of both thianthrene and 1,2,4,5-tetrachloro substituted benzene. The infrared bands observed in the region 1120-1050 cm^{-1} and 600-400 cm^{-1} in the present work are of particular interest, where stretching and bending vibrations are of phenylene groups are observed, and where drastic changes upon doping are expected.

1 D.G. Brady , J.Appl.Polym. Sci., 20, 2541(1976).
2 B.J. Tabore , E.P Magre and J.Boon , Eur.Polym.J., 7, 1127(1971).
3 P.Costa Bizzari , C.Della Casa ,M.Fiorini and W.Porzio, J.Polm.Sci.: Polym.Chem.Ed., 26, 255(1988).

INFLUENCE OF PARAFFINIC COMPONENTS ON THE FINE STRUCTURE OF ULTRA HIGH MOLECULAR WEIGHT POLYETHYLENE. A STUDY BY RAMAN SPECTROSCOPY

J.V.García-Ramos*, A. Alizadeh and J. Martínez-Salazar
Instituto de Estructura de la Materia (CSIC)
c/Serrano, 119-121, 28006-Madrid, Spain

The plasticizing effect of the paraffins is currently used for the production of high performance polyethylene (PE) fibres (1). The n-paraffins reduce considerably the density of the entanglements in the molten which, in turn, enhances dramatically the drawability of the polymeric material. The overall result is a final product with a much higher elastic modulus. Furthermore the paraffinic material employed in this process is easily recycled in comparison with other conventional solvents. Yet being this a well defined process the internal role of the paraffin upon the fine structure of the polymer is not well understood. Raman spectroscopy has been proved to be an useful technique for the characterization of the different phases in PE (2). The aim of the present work is then to study the influence of the paraffinic material on the fine structure of melt crystallized ultrahigh molecular weight PE.

Following the procedure proposed by Mutter et al, the mass fraction of the different phases that form the semicrystalline PE can be evaluated from the intensities of the observed bands. These authors have also demonstrated that the integrated intensity of the twisting band is independent of the chain conformation. Therefore it can be used as an internal standard for normalization. The mass fraction of the orthorhombic crystalline α_c can be derived from the intensity of the CH_2 bending band at 1416 cm^{-1}:

$$\alpha_c = \frac{I_{1416}/I_{Tws}}{I_{1416}^{\alpha=1}/I_{Tws}}, \quad I_{1416}^{\alpha=1}/I_{Tws} = 0.493 \qquad (1)$$

Similarly, the mass fraction of the amorphous phase can be calculated from the intensity of the CH_2 twisting vibrations according to:

$$\alpha_a = I_{1303}/I_{Tws} \qquad (2)$$

The material used in this work is a linear high-molecular weight PE (*Hostalen-Gur*, $M_w > 10^6$ $g.mol^{-1}$) supplied by Hoscht which has been melt-blended with a commercial paraffin $\mathbf{C_{28}H_{58}}$. The composition by weight of the blends were $\mathbf{PE/C_{28}H_{58}}$: 100/0, 85/15, 50/50 and 20/80. The blend materials were compression moulded into films of 0.5 mm thick under rapid crystallization conditions. To remove the paraffin from the samples, part of the films were treated with petroleum ether at room temperature during one week. This period of time was enough to remove all the paraffinic material from the mixtures. Both films before and after extraction have been examined.

The Raman spectra of the PE/$C_{28}H_{58}$ blends after paraffin extraction has been analyzed by calculating the integrated intensities of each band in the internal mode region. The different experimental bands have been separated assuming Voigt distribution functions. The relative mass fraction of the crystalline core and global amorphous phases have been calculated applying equations (1) and (2). The level of the interfacial phase, i.e. the regions where the chains adopt all-trans conformations but have lost their lateral order, has been deduced from the following relation:

$$\alpha_i = 1 - (\alpha_c + \alpha_a)$$

The results are summarized in figure 1. As the composition of the n-alkane increases, a higher crystal core fraction is obtained while the fraction of the interfacial phase remains constant. These results indicate that as the concentration of the $C_{28}H_{58}$ is increased, the crystallization capability of the high molecular weight PE increases continuously at expenses of the amorphous regions. This process occurs with a fixed interphase which is presumably determined by molecular weight and crystallization conditions.

The Raman spectra of the untreated films and the pure paraffin also has been analyzed. The mixtures before the extraction are composed of well separated crystalline paraffinic zones and semicrystalline PE regions. The Raman spectra of the n-paraffins which are the low molecular weight homologous of PE, is very similar to that of the fully crystalline polymer. For a given blend composition by subtracting the spectra of a treated sample from the untreated one in the CH_2 bending range, the amount of paraffin with orthorhombic crystalline structure has been deduced.

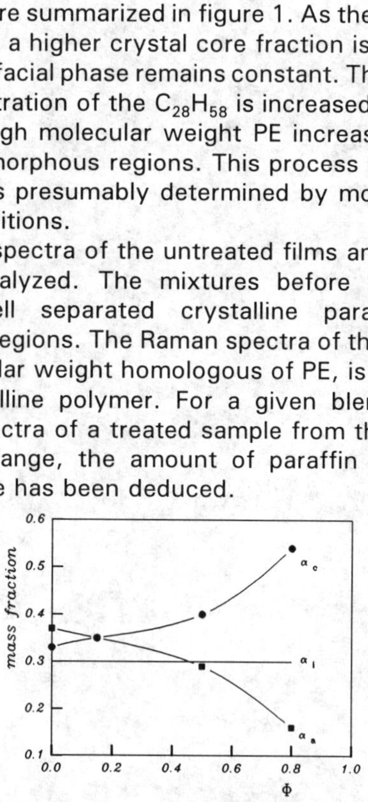

Fig. 1. Calculated mass fraction of the different phases in PE as a function of paraffin content, ϕ.

1. P.J. Lemstra, R. Kirschbaum, T. Ohta and H. Yasuda, in Developments in Oriented Polymers-2 (I.M. Ward, ed.), Elsevier App. Sc., 1987, p. 39 and ss.
2. D.I. Bower and W.F. Maddams, in The vibrational spectroscopy of polymers , Cambridge Univ. Press, 1989
3. R. Muutter, W. Stille and G. Strobl, J. of Polym. Sci., Polym. Phys. Ed., 31, 99 (1993)

SECTION 9

PROTEINS

RAMAN DIFFERENCE STUDIES OF ENZYMATIC CATALYSIS

Robert Callender*, Yong Qing Chen‡, and Hua Deng

Physics Department, City College of the City University of New York, New York, NY 10031 USA
‡Present address: Department of Chemistry, University of California, San Diego, La Jolla, CA 92093

Non-covalent interactions between an enzyme and its substrates are responsible, in most cases, for both the enormous catalytic power of enzymes as well as the fact that enzymes conduct the correct chemical reaction. Interactions that are present in the ground state are largely responsible for the specificity of the enzyme for a particular substrate. Moreover, the enzymic active site must be so designed as to lower the transition state between the interconverting species, and this too is brought about by particular interactions between the enzyme and its substrate. However, the identities of these interactions, their strengths, and the relationship of strength to the rate effects produced by the enzyme are generally not well understood, particularly the last two issues. There are several structural probes commonly used to study enzymatic catalysis. However, probes such X-ray diffraction and multi-dimensional NMR do not characterize the electronic distributions between atoms and in bonds, and these distributions are key to understanding the bond breaking/bond making steps in catalysis. It is thus evident that vibrational spectroscopy, which is sensitive to the electronic distributions between atoms (bond order, for example) as well as to geometry of the bonds and the masses of the bonded atoms, is a critical structural probe of enzymic catalysis.

Our lab has developed the technique of (non-resonance) Raman difference spectroscopy to study a wide range of problems involving enzymatic catalysis[1]. In such an experiment, a protein is 'tagged' in some way and the difference spectrum between the protein and its modified version measured. For example, an atom within a bond of interest of a bound substrate is labeled with a stable isotope, and this shifts the frequency of the modes that involve the motion of the labeled atom. Subtraction of labeled and unlabeled protein spectra yields an 'isotopically edited' difference spectrum whereby only those modes involving the labeled atom show up in the spectrum. The sensitivity of our difference spectrometer allows us to measure the spectrum of a single and specific phosphate group inside a 60 kDa protein with a signal to noise of about 20/1, at present. Recent work concerns studies of the mechanism of dihydrofolate reductase and proteins which carry out phosphoryl transfer reactions.

Dihydrofolate Reductase

Dihydrofolate reductase (DHFR) catalyzes the reduction of dihydrofolate (H_2folate) to tetrahydofolate by NADPH, and this requires that the pteridine ring of substrate be protonated at N5. A long-standing puzzle has been how, at physiological pH, the enzyme can protonate N5 in view of its solution pK_a of 2.6 and the fact that the only proton-donating group in the pteridine binding site, Asp-27, hydrogen bonds not to N5 but to the 2-amino group and N3 of the pterin ring. We have determined the pK_a of N5 of dihydrofolate in the *E. coli* DHFR/$NADP^+$/H_2folate ternary complex by Raman difference spectroscopy and found its value is 6.5[2]. In contrast, the pK_a of N5 is less than 4.0 in either the binary complex, the ternary complex with an analogue of NADPH (H_2NADPH), or the Asp27 to serine mutant DHFR (D27S) ternary complex with $NADP^+$. Thus, one needs not invoke proton donation from Asp-27 to N5 via a series of bound water molecules and/or pteridine-ring substituents. We propose instead that the N5 protonated form of H_2folate is stabilized directly at the active site in the DHFR/NADPH/H_2folate complex by specific interactions that form only in the ternary complex, involving perhaps a bound water molecule, the carboxamide moiety of the coenzyme, and/or the local electrostatic field of the enzyme molecule, to which an important contribution may be made by Asp-27. This interpretation is reinforced by our measurement of the pK_a of Asp-27, which has long been supposed to be around 6.5 from indirect measurements. With our techniques, it is possible to monitor the presence or absence of Raman bands that are characteristic of a protonated group of the carboxyl group. We find that the pK_a of Asp-27 is less than 5.0.

Phosphoryl Transfer Reactions

Perhaps one third of all proteins bind phosphate or catalyze reactions involving phosphate groups. Very little is know about how phosphate binds to these proteins, and next to nothing is known about the mechanism of phosphate hydrolysis or phosphoryl transfer in these enzymes. Under active study in the lab are phosphoglucomutase (PGM)[3], which catalyzes the interconversion of glucose 1-phosphate and glucose 6-phosphate, and two of the so-called class of G-proteins, the bacterial elongation factor protein, EF-Tu, and the Harvey *ras* p21 protein, which is involved in cell growth and mutants of which are associated with a large percentage of human tumors. The hydrolysis of GTP to GDP for bound GTP to the G-proteins is a key regulatory process of activity.

We find that the binding of phosphate to enzyme varies substantially across this set of proteins in that there is virtually no change in the spectra of bound phosphate from solution spectra in the case of PGM while a number of substantial shifts are observed in the case of the G-proteins. For phosphates and vanadates, a formulism has been worked out that relates vibrational frequencies to bond orders directly. Thus it is possible to determine how ezymic binding affects bond orders and, from this, deduce binding strengths and enzymic mechanism. From a study of a vanadate derivative of the substrate of PGM, which is believed to behave as a transition state analogue, it seems clear that the hydrolysis step in PGM invloves an essentially associative transition state. This is quite unlike that found in the solution chemistry of dianionic phosphoryl transfer reactions which are dissociative in character. The shifts in the phosphate bands for GTP bound to the p21 protein suggest that the transition state in this case is largely associative in character as well.

1. Robert Callender and Hua Deng, Ann. Rev. of Biophysics and Biomolecular Structure 23, 215 (1994).
2. Yong Q. Chen, Joseph Kraut, Raymond L. Blakley, and Robert Callender, Biochemistry 33, 7021 (1994).
3. Hua Deng, William Ray, John Burgner, and Robert Callender, Biochemistry 32, 12984 (1993).

SPECIFIC PROTEIN-PIGMENT INTERACTIONS MODIFYING THE PHYSICOCHEMICAL PROPERTIES OF THE PRIMARY DONOR IN THE BACTERIAL REACTION CENTER

A. Ivancich and T.A. Mattioli

Section de Biophysique des Protéines et des Membranes, DBCM, CEA and URA CNRS 1290, C.E. de Saclay, 91191 Gif-sur-Yvette cedex, France.

Photosynthesis involves the conversion of light energy into chemical energy via a series of electron and proton transfer reactions. In bacterial photosynthesis, the primary light reactions take place in a membrane-bound pigment-protein complex called the reaction center (RC). Transmembrane electron transfer within this complex is initiated by the photoexcitation of the primary electron donor (P) to a singlet excited state (P*). The primary donor is a dimer of excitonically interacting bacteriochlorophyll molecules (P_L and P_M) and plays a crucial role in the rapid and efficient photochemistry of the RC. The physicochemical properties (e.g. redox midpoint potential, electronic structure) of P have a profound influence on the rate and efficiency of the electron transfer reactions involved [1].

Fourier transform (FT) Raman spectroscopy has proven to be a powerful and valuable tool in assessing the protein-pigment interactions which determine the energy levels of ground state (P^o), the excited state (P*), and the oxidized radical cation state ($P^{\bullet +}$) of the primary donor [2,3]. Recently, a direct correlation was observed between the number and strengths of histidine-donated H-bonds to the π-conjugated carbonyl groups of the primary donor in mutant RCs from *Rb. sphaeroides*. For example, the rupture or formation of a single such H-bond can modify the $P^o/P^{\bullet +}$ redox midpoint potential by as much as 125 mV [4]. Specifically, the removal of the H-bond, donated by His L168, to the C_2 acetyl carbonyl of P_L lowered this redox potential by 95 mV [2,4].

The binding site and protein-pigment interactions of P in several bacterial RCs have been determined using FT Raman spectroscopy. For these RCs whose 3-dimensional structures have not been elucidated but whose protein sequence data are available, amino acid residues donating H-bonds to the carbonyl groups of P may be proposed. Table 1 lists the frequencies of the C_2 acetyl carbonyl group stretching modes of P in various RCs as observed in their FT Raman spectra. For the case of *Rb. sphaeroides*, the 3-dimensional structure of the RC is solved to atomic resolution. The 1620 cm^{-1} band in the FT Raman spectra of these RCs has been assigned to the P_L C_2 acetyl carbonyl which is H-bonded to Histidine L168 [5].

The deduced H-bonding pattern of P from *Chloroflexus aurantiacus* RCs, based on its FT Raman spectrum and sequence alignment with *Rb. sphaeroides* [6], indicates that one of the two C_2 acetyl carbonyl groups of P is engaged in a H-bond (weaker than that observed for *Rb. sphaeroides*), probably with a tyrosine residue, while the other C_2 carbonyl group is free from such interaction. His L168 in *Rb. sphaeroides* is not conserved in *Cf. aurantiacus*, thus the H-bond between P_L and this residue, seen in *Rb. sphaeroides*, is not present. The absence of this histidine-donated H-bond to P in *Cf. aurantiacus*, may be primarily responsible for the low $P^o/P^{\bullet+}$ midpoint potential, as compared to that of *Rb. sphaeroides*.

The FT Raman spectrum of P from the green sulfur bacterium *Chlorobium limicola* indicated that all of its conjugated carbonyl groups were free of H-bonding interactions [7]. The $P^o/P^{\bullet+}$ midpoint potential of this species is very much lower than that of *Rb. sphaeroides*. The fact that there are no histidine-donated H-bonds to P of *Chlorobium* appears to be consistent with its low oxidation potential but other factors are probably operational.

Table 1 also lists the percent localization of the resulting + charge on the dimer in its oxidized cationic $P^{\bullet+}$ state. The H-bonding pattern and the chemical nature of the H-bond to P appear to influence the repartition of the + charge over the dimer in its oxidized state.

TABLE 1. Observed Raman frequencies (cm^{-1}) of the conjugated carbonyl groups of P in various bacterial RCs.

species	P_L C_2=O	donor	P_M C_2=O	donor	$P^o/P^{\bullet+}$	$\%P_L^{+}$ [d)]
Rb.sphaeroides [a)]	1620	His	1653	none	500mV	80%
Cf.aurantiacus [b)]	1648	none	1632	Tyr	386mV	65%
Chlorobium [c)]	1654	none	1654	none	240mV	50%

[a)] from Reference [5], [b)] from Reference [6], [c)] from Reference [7].
[d)] % localization of the + charge of $P^{\bullet+}$ on the P_L molecule of the dimer.

1. J.P. Allen and J.C. Williams, J. Bioenerg. Biomemb. 27, 275 (1995).
2. T.A. Mattioli, J.C. Williams, J.P. Allen, and B. Robert, Biochemistry 33, 1636 (1994).
3. T.A. Mattioli, X. Lin, J.P. Allen, and J.C. Williams, Biochemistry 34, 6142 (1995).
4. X. Lin, H.A. Murchison, V. Nagarajan, W.W. Parson, J.P. Allen, and J.C. Williams, Proc. Natl. Acad. Sci. U.S.A. 91, 10265 (1994).
5. T.A. Mattioli, A. Hoffmann, B. Robert, B. Schrader, and M. Lutz, Biochemistry 30, 4648 (1991).
6. A. Ivancich, R. Feick, A. Ertlmaier, and T.A. Mattioli, Biochemistry (1996), in press.
7. U. Feiler, D. Albouy, B. Robert, and T.A. Mattioli, Biochemistry 34, 11099 (1995).

BOND LENGTHS AND COORDINATION GEOMETRIES IN COPPER-SULFUR ELECTRON TRANSFER PROTEINS

Thomas M. Loehr

Oregon Graduate Institute of Science & Technology, Portland, OR 97291

Cupredoxins are ubiquitous copper-containing electron-transfer proteins in which the metal is coordinated to the sulfur atom of a cysteine thiolate. The intense colors of these proteins arise from S→Cu(II) charge transfer transitions and these chromophores exhibit many novel spectroscopic and electrochemical properties [1]. Another unusual feature of these proteins is that their copper sites can accommodate a variety of structures and coordination geometries as the ligands are varied [2]. The classic type 1 copper proteins such as the azurin and plastocyanin are blue ($\varepsilon_{600} \approx$ 3000 $M^{-1}cm^{-1}$) and have three strongly coordinated ligands (Cys, His_2) in a trigonal planar geometry (Figure 1).

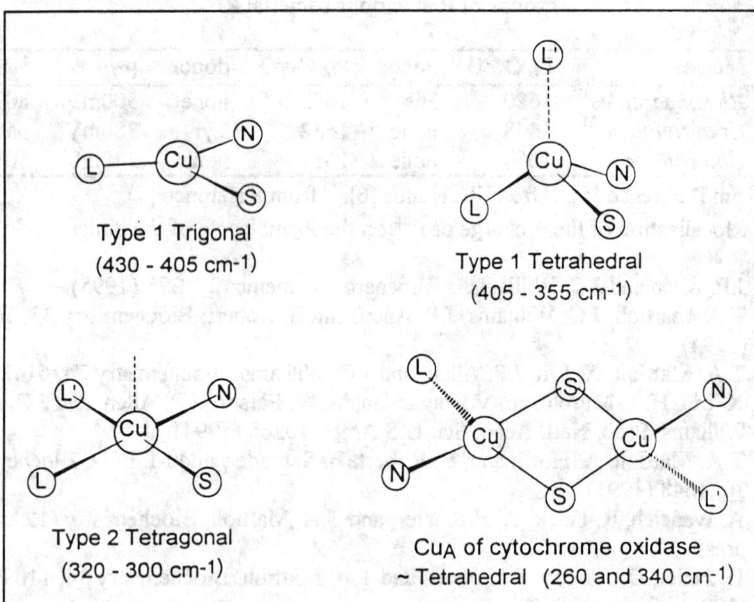

Figure 1. Coordination geometries found in copper-cysteinate proteins. The ranges of observed resonance Raman frequencies for the principal Cu-S stretching vibrations are listed below each structure [2].

Type 1 copper sites with increased rhombic distortions are green resulting from a second prominent S→Cu(II) charge transfer band at ~460 nm as in nitrite reductase [3]. Here the copper is displaced out of the trigonal plane toward a fourth ligand to adopt a distorted tetrahedral geometry. Further structural changes can be imposed on the copper by alteration of ligands using protein engineering techniques. Thus, yellow type 2 sites have been created in azurin and superoxide dismutase by site-directed mutagenesis, and, in this square-planar geometry, their S→Cu(II) CT bands are further blue shifted to ~400 nm. In the case of azurin, substitution of a histidine ligand by glycine and addition of exogenous histidine as the free amino acid generates a tetragonal site as a result of the bidentate coordination of its imidazole and amino groups [2]. The most remarkable example of modified cupredoxin structures comes from the recently identified binuclear, dithiolate-bridged Cu_A site of cytochrome *c* oxidase. The same structure has been made in engineered constructs of several type 1 proteins by insertion of an additional cysteine into the cupredoxin fold and is believed to exist in N_2O reductase [4, 5].

Excitation into these S→Cu(II) CT bands yields rich resonance Raman spectra dominated by vibrations of the Cu-cysteinate moiety. The type 1 sites have a large number of bands that arise from coupling of ν(Cu-S) and δ(Cys) modes, enhanced by the nearly coplanar arrangement of the Cu-S-C_β-C_α-N atoms [6]. Although the spectra are not yet fully assigned, isotope substitutions for Cu, total N, and total S have been extremely helpful. Most of the ~10 fundamentals exhibit S and N shifts. However, none appear to be due to Cu-N(His) vibrations as specific replacement of the histidine ligands in azurin by exogenous imidazoles (^{14}N and ^{15}N) failed to elicit any isotope shifts [2]. The most intense peak in the RR spectra has the largest S-isotope shift and is assigned as the dominant Cu-S vibration (Figure 1). In addition, this peak is the sole generator of combination bands, and its frequency may be used to estimate the Cu-S bond length. The change in ν(Cu-S) from 430 cm^{-1} in trigonal to 300 cm^{-1} in tetragonal sites corresponds to a lengthening of the Cu-S(Cys) bond by ~0.15 Å upon addition of a fourth strong ligand. The RR spectrum of Cu_A is unique in that there are two major Cu-S vibrations with large S-isotope shifts. The RR spectrum of this $Cu_2S_2N_2$ cluster has been assigned with the aid of NCA [7].

1. E. T. Adman, *Adv. Protein Chem.* **42,** 145 (1991).
2. C. R. Andrew, H. Yeom, J., S. Valentine, B. G. Karlsson, N. Bonander, G. van Pouderoyen, G. W. Canters, T. M. Loehr, & J. Sanders-Loehr, *J. Am. Chem. Soc.* **116,** 11489 (1994).
3. J. Han, T. M. Loehr, Y. Lu, J. S. Valentine, B. A. Averill, & J. Sanders-Loehr, *J. Am. Chem. Soc.* **115,** 4256 (1993).
4. C. R. Andrew, J. Han, S. de Vries, J. van der Oost, B. A. Averill, T. M. Loehr, & J. Sanders-Loehr, *J. Am. Chem. Soc.* **116,** 10805 (1994).
5. C. R. Andrew, P. Lappalainen, M. Saraste, M. T. Hay, Y. Lu, C. Dennison, G. W. Canters, J. A. Fee, C. E. Slutter, N. Nakamura, & J. Sanders-Loehr, *J. Am. Chem. Soc.* **117,** 10759 (1995).
6. J. Han, E. T. Adman, T. Beppu, R. Codd, H. C. Freeman, L. Huq, T. M. Loehr, & J. Sanders-Loehr, *Biochemistry* **30,** 10904 (1991).
7. C. R. Andrew, R. Fraczkiewicz, R. S. Czernuszewicz, P. Lappalainen, M. Saraste, & J. Sanders-Loehr, manuscript submitted.

TEMPERATURE DEPENDENCE OF RESONANCE RAMAN SPECTRA OF PHOTOSYNTHETIC REACTION CENTER PROTEINS

A. Ivancich, M. Lutz*, and T.A. Mattioli

Section de Biophysique des Protéines et des Membranes, DBCM, CEA and CNRS URA 1290, CEA/Saclay, 91191 Gif-sur-Yvette, France.

In photosynthetic reaction center proteins (RCP) from purple bacteria, a photoinduced charge separation occurs between the primary electron donor (P) - a pair of bacteriochlorophylls (BChl)- and the primary acceptor -a bacteriopheophytin (BPhe)- within 3 ps at 300 K. A species-dependent, two-fold to four-fold acceleration of this step is observed when the RCP is cooled down to 10 K [1]. Further electron transfer steps also are temperature-dependent. Understanding these properties is useful in establishing adequate models for the mechanisms of electron transfer in the RCP.

Temperature dependence of electron transfer rates may result from temperature dependence of various parameters, such as changes in donor-acceptor distances (e.g. due to protein contraction at low temperature), changes in vibrational coupling of the electron transfer, or changes in cofactor environments modifying their redox potentials.

Resonance Raman can provide detailed information about the temperature-dependent local structural changes which may occur on the electron-transferring cofactors of RCP and in their immediate surroundings. We have studied the bacteriochlorin-type cofactors in RCP from *Rhodobacter sphaeroides* in the temperature range 300 - 10 K, using both classical RR spectroscopy and FT RR spectroscopy [2]. Because 1064 nm excitation is nonactinic and induces very limited heating of the fragile RCP, the latter technique provides excellent spectra at room temperature and is particularly well fit to temperature dependence studies in this range.

Three types of resonant excitations were performed. Excitation at 363.8 nm induced Soret-resonant scattering from the six bacteriochlorin cofactors. Excitation at 546 and 537 nm selectively induced Q_x-band resonance scattering from $BPheo_L$ (the primary electron acceptor) and from $BPheo_M$ cofactors, respectively. Excitation at 1064 nm predominantly induced Q_y-band preresonance scattering from the primary donor P. Sample temperature was varied from 300 K to 10 K.

In both Soret-resonant spectra and Q_y-preresonant spectra, the methine bridge stretching band at 1609 cm^{-1} is seen to upshift by 4 to 5 cm^{-1} when the temperature is decreased from 300 K to 15 K. This change is tentatively interpreted as resulting from a change in conformation of the macrocycles of BChl cofactors. It may be non specific to the protein-bound cofactors, since a 4 cm^{-1} shift was observed for a BChl *a* solution on the same temperature range.

In Q_x-resonant spectra, the methine bridge stretching mode of $BPhe_M$ was observed to upshift by 2 cm^{-1} from 200K to 15K, while no shift was observed for the same mode of $BPhe_L$. We conclude, at variance with a previous report [3], that the macrocycle of the primary electron acceptor does not undergo any significant conformational change in the 200-15 K range.

Q_y-preresonant excitation of RCPs at 1064 nm showed that in P the stretching frequencies of the two conjugated, C_9 keto and C_2 acetyl, carbonyls of the M-branch primary donor BChl cofactor (P_M) did not significantly change between 300 and 10 K. In contrast, the C_9 keto carbonyl frequency of cofactor P_L was observed to upshift by 5 cm^{-1}, while its C_2 acetyl carbonyl frequency downshifted by 2 cm^{-1}. The latter shift indicated that the strong H-bond that the acetyl group of P_L has with His L168 may be still stronger at 10 K.

Excitation at 1064 nm of chemically oxidized RCPs selectively provided FT RR spectra of the primary donor in its radical $P^{\bullet+}$ state [2]. These spectra showed that the fraction of the unpaired charge which is localized over P_L decreased from 78% to 63% when decreasing temperature from 300 to 10K (Figure).

In conclusion, the primary electron acceptor $BPheo_L$ does not appear to undergo any significant conformational change at low temperature which could influence the electron transfer rate. Rather, the primary donor pair of BChls may undergo a significant flattening of their conjugated macrocycles, which could in part explain the change in unpaired charge localization measured at low temperature (Figure). Also, the strengthening of the H-bond between the acetyl carbonyl of P_L and His L168 could affect the oxidation potential of P.

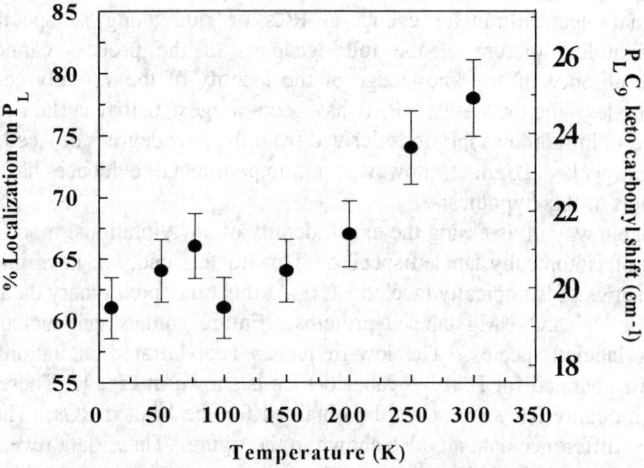

1. G.R. Fleming, J.-L. Martin, and J. Breton, Nature 333, 190 (1988).
2. T. A. Mattioli, A. Hoffmann, B. Robert, B. Schrader, and M. Lutz, Biochemistry, 30, 4648 (1991).
3. J.M. Peloquin, C.A. Violette, H.A. Frank, and D.F. Bocian, Biochemistry 29, 4892 (1990).

LOW-FREQUENCY NEAR-INFRARED-EXCITATION RESONANCE RAMAN SPECTRA OF ISOTOPICALLY LABELED BACTERIAL PHOTOSYNTHETIC REACTION CENTERS

Kazimierz Czarnecki,[1] James R. Diers,[1] Veeradej Chynwat,[2]
Joy P. Erickson,[2] Harry A. Frank,[2] and David F. Bocian[1,*]
[1]Department of Chemistry, University of California, Riverside, CA 92521
[2]Department of Chemistry, University of Connecticut, Storrs, CT 06269

The bacterial reaction center (RC) is a membrane-bound protein responsible for the primary electron-transfer events in photosynthesis [1]. Bacterial RCs contain four bacteriochlorophylls (BChls) and two bacteriopheophytins (BPhs). Two of the BChls are closely associated and form the special pair (P) which serves as the primary electron donor. Recently, the low-frequency vibrations of the cofactors in RCs have received considerable attention. Hole-burning [2] and second-derivative optical studies [3] have shown that very low-frequency modes (30-150 cm^{-1}) are strongly coupled to the P^* state from which electron transfer is initiated. The hole-burning and optical data have been interpreted in terms of only a few (one or two) strongly coupled modes. However, near-infrared-excitation resonance Raman (RR) studies have shown that there are in fact a substantial number (10-20) of strongly coupled low-frequency modes [4-7]. Recent femtosecond optical studies on the P^* state have revealed coherent oscillations in a number of low-frequency modes [8-10].

At this time, it is not certain whether strongly coupled, low-frequency modes mediate the initial electron-transfer events in RCs or ride along as spectators. Regardless, a complete picture of the full dynamics of the process cannot be developed in the absence of any knowledge of the identity of the strongly coupled low-frequency modes. In the case of P, it has been suggested that certain modes might correspond to intradimer vibrations derived from the new degrees of freedom in the supermolecule [2]. To date, however, no experimental evidence has been obtained in support of this hypothesis.

The only reliable way of assessing the exact identity of any vibrational mode is via the examination of isotopically labeled species. Toward this end, we have initiated RR studies on a series of isotopically labeled RCs. At this time, preliminary data have been acquired for ^{15}N and ^{26}Mg labeled proteins. Future studies will include the $^{15}N/^{26}Mg$ double-labeled species. The low-frequency near-infrared-excitation (865 nm) RR spectrum obtained for P in ^{15}N labeled RCs is shown in Fig. 1. These data indicate that high-quality RR spectra can be obtained for the labeled RCs. The ^{14}N minus ^{15}N isotope difference data are also shown in the figure. These data reveal that certain modes undergo ^{15}N shifts whereas others do not. The largest shifts are observed for bands in the 150-200-cm^{-1} region. At this time, we cannot offer an interpretation of the spectra; however, the apparent magnitude of the shifts observed for the bands in the 150-200-cm^{-1} region suggests that these modes are derived from motions of the BChl core. It is possible that one of these bands contains contributions from the Mg-histidine stretching vibration $\nu_{Mg\text{-}His}$. This mode has not been identified in the RR spectra of the cofactors in RCs or for model complexes. The identification

$\nu_{Mg\text{-}His}$ would be an important step in characterizing the BChl cofactors. A detailed analysis of the spectrum must await the acquisition of RR data for other isotopomers and the results of normal coordinate calculations.

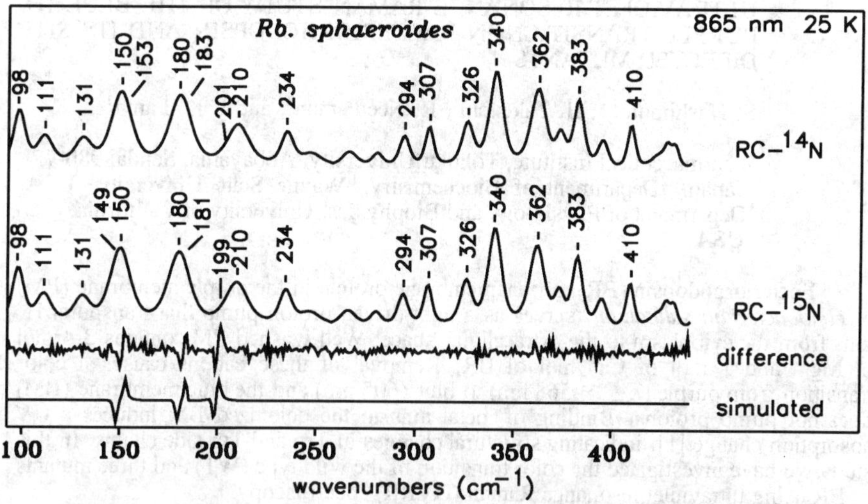

Fig. 1. Low-frequency, near-infrared-excitation RR spectra of P in ^{14}N (top trace) and ^{15}N (2nd trace) RCs along with the experimental (3rd trace) and simulated isotope difference spectra (bottom trace).

Acknowledgment. This work was supported by Grants GM39781 (D.F.B), GM36243 (D.F.B.), and GM30353 (H.A.F.) from the National Institute of General Medical Sciences; the Cooperative State Research Service, USDA, under aggreement no. 92-37306-7690 (H.A.F.); and the Human Frontier of Science Program (H.A.F.).

1. C. Kirmaier and D. Holten, Photosynth. Res. 113, 225 (1987).
2. R. Jankowiak, J. M. Hayes, and G. J. Small, Chem. Rev. 93, 1471 (1993).
3. A. V. Klevanik, A. O. Ganago, A. Y. Shkuropatov, and V. A. Shuvalov, FEBS Lett. 237, 61 (1988).
4. V. Palaniappan, M. A. Aldema, H. A. Frank, and D. F. Bocian, Biochemistry 31, 11050 (1992).
5 V. Palaniappan, C. C. Schenck, and D. F. Bocian, J. Phys. Chem. 99, 17049 (1995).
6. N. J. Cherepy, A. P. Shreve, L. J. Moore, S. Franzen, S. G. Boxer, and R. A. Mathies, J. Phys. Chem. 98, 6023 (1994).
7. N. J. Cherepy, A. Holzwarth, and R. A. Mathies, Biochemistry 34, 5288 (1995).
8. M. H. Vos, F. Rappport, J.-C. Lambry, J. Breton, and J.-L. Martin, Nature 363, 320 (1993).
9. M. H. Vos, M. R. Jones, C. N. Hunter, J. Breton, J-C. Lambry, J.-L. Martin, Biochemistry 33, 6750 (1994).
10. R. J. Stanley and S. G. Boxer, J. Phys. Chem. 99, 859 (1995).

ULTRAVIOLET RESONANCE RAMAN STUDY OF THE BLUE-TO-PURPLE TRANSITION IN BACTERIORHODOPSIN AND ITS SITE DIRECTED MUTANTS

S. Hashimoto[1]*, H. Takeuchi[1], R. Needleman[2], and J. K. Lanyi[3]

[1]Pharmaceutical Institute, Tohoku University, Aobayama, Sendai 980-77, Japan. [2]Department of Biochemistry, Wayne State University, USA. [3]Department of Physiology and Biophysics, University of California, USA.

Bacteriorhodopsin (BR), a transmembrane protein in the purple membrane (PM) of *Halobacterium salinalium*, serves as a light-driven proton pump that transports H^+ ions from the cytoplasm to the extracellular space. Well washed PM contains 3-4 mol of Mg^{2+} and 1 mol of Ca^{2+}/mol of BR. Removal of these cations causes a color transition from purple (λ_{max} = 568 nm) to blue (605 nm) and the blue membrane (BM) does not pump protons. Binding of metal ions to the deionized BM induces a UV absorption change [1], indicating structural changes in Trp and Tyr side chains. In this study we have investigated the color transition of the wild type (WT) and three mutants of BR using ultraviolet resonance Raman (UVRR) spectroscopy.

Fig. 1a shows UVRR spectrum of WT PM (in 100 mM Na_2SO_4 and 20 mM HEPES pH 7.0) excited with 244-nm line from a CW UV laser (Coherent, Innova 300 FReD). The spectrum contains several bands assignable to Trp and Tyr vibrations, which are labeled W and Y with the mode numbers proposed previously [2]. Since BR contains 8 Trp and 11 Tyr, the spectrum is a sum of the spectra of these residues. Fig. 1b shows the difference, BM - PM, the intensity being enlarged by a factor of 6. The OH stretching band of solvent water was used as a Raman intensity standard, because the addition of inorganic ions usually used as internal standards converted BM back to PM. The Trp and Tyr Raman bands show intensity decreases, which are consistent with a large decrease in absorbance at 244 nm [1].

Trp side chain. The intensity decrease of the Trp bands (Fig. 1b) is probably due to a blue shift of the B_b absorption band of Trp. The shift may reflect changes in the environment and/or the hydrogen bonding of the indole ring. Since the W17 ferquency (a measure of the strength of hydrogen bonding at the NH of indole ring [3]) shows no shift (BM, 878 cm^{-1}) upon deionization, the decrease in intensity can be ascribed to decreases in hydrophobicity of the environments of Trp side chains. Figs. 1c and d show the BM - PM difference spectra of W189F (Trp-189 being replaced with Phe) and W182F mutants, respectively. The spectra also show intensity decreases of the Trp and Tyr bands, but the spectral features of them are somewhat different from those of WT. If we assume that the decrease in intensity of the Y9a band of the Trp mutants is the same as that of WT, the decrease in intensity of the W3 band is small in these mutants as compared to WT. Thus, the W3 intensity decrease of WT is probably attributed, at least partly, to changes in environment of Trp-189 and -182. The W17 band shifts down by about 20 cm^{-1} when the indole NH proton is replaced with a deuteron. This shift can be used to monitor the Trp H/D exchange in D_2O [3]. A UVRR study of PM in D_2O has shown that the H/D exchange is very slow (half-time > 1 week) [4]. As the UVRR spectrum of BM in D_2O shows no band around 878 cm^{-1}, the NH protons of all Trp residues appear to undergo exchange with solvent water within 2 hours. The Trp H/D exchange is much faster in BM than PM. We have observed that the W17 intensity of W182F PM in D_2O is substantially weaker than that of WT PM in D_2O. This observation probably indicates that the H/D exchange of Trp-182 is slow and the side chain is buried inside the protein. According to the electron cryo-microscopy [5], Trp-182 is a part of the retinal binding pocket and surrounded by hydrophobic residues. Deionization may increase the protein fluctuation and make the solvent water accessible to Trp-182 in BM. Possibly, some salt bridges near the protein surface stabilize the native structure of PM.

Tyr side chain. The Tyr Raman bands (Y7a, Y9a) also show intensity decreases upon deionization (Fig. 1b). The Tyr intensity is largely affected by hydrogen-bonding at the phenolic oxygen [6]. Fig. 1e shows the BM - PM difference spectrum of the Y185F mutant. The spectrum is very similar to that of WT BR and probably Tyr-185 is not involved in the protein structural change associated with the color shift. The retinylidene chromophore is linked to protein through a protonated Schiff base. It has been proposed that the color transition to blue is due to the protonation of a negative charge near the Schiff base [7]. Asp-85, -212, and Arg-82 form a complex counterion in which the negative charge is not localized on a single group but delocalized over a hydrogen-bonded network among these residues. Tyr-185 is located near Asp-212 and the OH group of Tyr-185 is hydrogen-bonded to the carboxylate of Asp-212 [5]. The present result suggests that the side chain of Asp-212 remains anionic and hydrogen-bonded with Tyr-185 in BM.

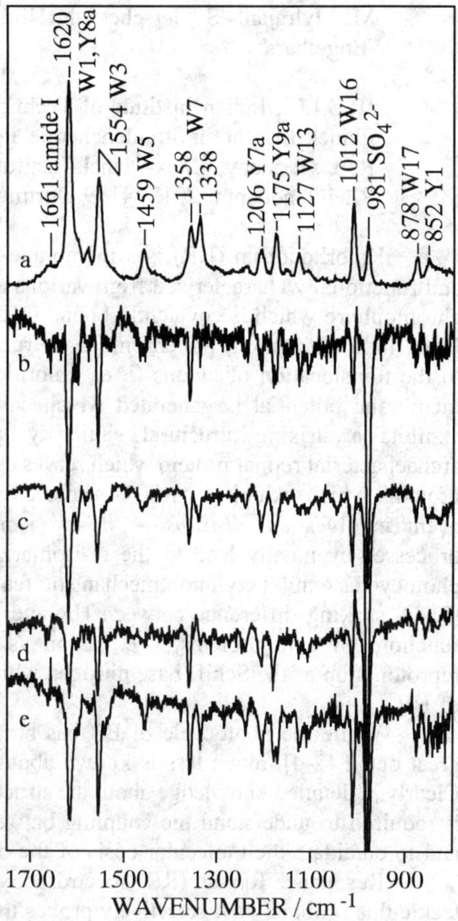

Fig. 1. 244-nm UVRR spectrum of WT PM (a) and difference spectra, BM-PM, of WT (b), W189F (c), W182F (d), and Y185F (e).

1. M. Dunach, E. Padros, M. Seigneuret, and J. L. Rigaud, J. Biol. Chem. 263, 7555 (1988).
2. I. Harada and H. Takeuchi, in Advances in Infrared and Raman Spectroscopy, (R. J. H. Clark and R. E. Hester, ed.), Vol. 13, John Wiley, New York, 1986, p. 113.
3. T. Miura, H. Takeuchi, and I. Harada, Biochemistry 27, 88 (1988).
4. S. Hashimoto, K. Miura, T. Yamagishi, H. Takeuchi, and I. Harada, Photochem. Photobiol. 56, 1097 (1992).
5. R. Henderson, J. M. Baldwin, T. A. Ceska, F. Zemlin, E. Beckmann, and K. H. Downing, J. Mol. Biol. 213, 899 (1990).
6. S. Hashimoto, T. Yabusaki, H. Takeuchi, and I. Harada, Biospectrosc., in press.
7. R. Jonas and T. G. Ebrey, Proc. Natl. Acad. Sci. USA. 88, 149 (1991).

TIME-RESOLVED RESONANCE RAMAN SPECTROSCOPIC STUDY OF THE PHOTOCYCLE OF HALORHODOPSIN

M. Mylrajan[1], S. Gerscher[2], P. Hildebrandt[2*], I. Chizov[3], J. Sydor[3], and M Engelhard[3]

[1]R.S.I.C., Indian Institute of Technology, Madras 600036, India; [2]Max-Planck-Institut für Strahlenchemie, Postfach 101365, D-45413 Mülheim, Fed. Rep. Germany; [3]Max-Planck-Institut für Molekulare Physiologie, Rheinlanddamm 201, 44139 Dortmund, Fed. Rep. Germany.

Halorhodopsin (HR) is a membrane-bound protein which has been purified and functionally characterized from various archaebacteria [1,2]. It contains a retinal chromophore which is covalently bound to the apoprotein via a Schiff base linkage. Upon light absorption, the chromophore runs through photocycle which is coupled to the translocation of anions (i. e., chloride) across the membrane. In this way, a membrane potential is generated which can be used to drive ATP synthesis. HR exhibits a striking structural similarity with bacteriorhodopsin (BR), another archaebacterial retinal protein which serves as a light-driven proton pump [2]. In both proteins which exhibit a ca 30% sequence identity [1] the primary photochemical event involves an *all-trans* → *13-cis* isomerization and the thermal relaxation processes eventually lead to the re-isomerization of the retinal chain. Thus, both photocycles exhibit common mechanistic features which trigger the ion translocation [2]. A striking difference between HR and BR which is attributed to the different functions of both proteins, i. e., anion vs proton pumping, concerns the de-and reprotonation of the Schiff base nitrogen which does not occur during the photocycle of HR.

While the photocycle of BR has been studied by a variety of techniques in great detail [2-4], much less is known about the various intermediate states of HR. Clearly, a detailed knowledge about the structural changes of the chromophore in HR is required to understand the coupling between the photocycle and the anion pump and to elucidate the molecular basis of the different functions in HR and BR.

Resonance Raman (RR) spectroscopy is a particularly powerful technique to tackle this problem as it selectively probes the vibrational bands of the chromophore [3,4]. Furthermore, the extraction of structural data from the RR spectra is significantly facilitated by previous normal mode analysis studies on model compounds [3]. As RR spectroscopy can be employed in the time-resolved domain, it is possible to measure the various intermediates of the photocycle, thereby providing simultaneously structural and kinetic information about the photocycle as successfully demonstrated in the case of BR [3,4].

In previous RR studies, HR from *Halobacterium salinarium* has been used [5-8], however, these experiments were hampered by the instability of the protein and interference by the fluorescence background. In this respect, HR from *Natronobacterium pharaonis* is significantly more appropriate for RR spectroscopic studies. This protein exhibits a strong homology with that from *H. salinarium* [9]. The kinetics of its photocycle have recently been studied by flash photolysis [10].

In this work we have studied HR from *N. pharaonis* by time-resolved RR spectroscopy. The measurements were carried out with cw laser excitation employing the rotating cell technique as described previously [4,6]. In this way high quality spectra were obtained which allowed for a detailed analysis of the effect of various anions on the chromophore structure and the kinetics of the photocycle.

In the parent state of the chloride-bound protein, $HR_{578}(Cl^-)$, the RR spectrum reveals the characteristic signature of an all-trans retinal Schiff base with far-reaching similarities to the parent state of BR, BR_{570}. Based on careful analysis of the RR spectra a structural heterogeneity of the light-adapted parent state (i. e., a mixture of ca 80% all-trans and 20% 13-cis isomers) as suggested by Varo et al. [10] can be ruled out. The main difference compared to BR_{570} refers to the region of the polyene stretching modes which in $HR_{578}(Cl^-)$ displays two bands at 1525.0 and 1536.8 cm^{-1} which are assigned to the in-phase and out-of-phase combinations of the C=C stretches of the retinal chain, respectively. In contrast, in BR_{570} no RR activity was found for the out-of-phase mode.

Exchange of chloride at the anion binding site by bromide and iodide, affects both the frequencies and the intensity distribution between the two C=C stretching modes of HR_{578} while no detectable alteration of the $C=NH^+$ stretching was observed. Thus, it is not very likely that the anion binding site is in the immediate vicinity of the Schiff base group. On the other hand, the anion is apparently sufficiently close to the retinal to alter the electron density distribution in the chromophore which might be the origin for RR-activation of the C=C out-of-phase stretching mode. A more pronounced anion-dependence is found for the C=C stretching modes in the RR spectra of the intermediates.

1. J. K. Lanyi, Physiol. Rev. 70, 319 (1990)
2. D. Oesterhelt, J. Tittor, and E. Bamberg, J. Bioenerg. Biomembr. 24, 181 (1992).
3. R. A. Mathies, S. O. Smith, and I. Palings, in Biological Applications of Raman Spectroscopy (T. G. Spiro, ed.), Vol. II, Wiley, New York, p. 59.
4. T. Althaus, W. Eisfeld, R. Lohrmann, and M. Stockburger, Isr. J. Chem., in press (1996).
5. A. Maeda, T. Ogurusu, and T. Yoshizawa, Biochemistry 24, 2517 (1985).
6. R. Diller, M. Stockburger, D. Oesterhelt, and J. Tittor, FEBS Lett. 217, 297 (1987).
7. C. Pande, Lanyi, J. K., and Callender, R. H., Biophys. J. 55, 425 (1989).
8. J. B. Ames, J. Raap, J. Lugtenburg, and R. A. Mathies, Biochemistry 31, 12546 (1992).
9. B. Scharf and M. Engelhard, Biochemistry 33, 6387 (1994).
10. G. Varo, L. S. Brown, J. Sasaki, H. Kandori, A. Maeda, R. Needleman, and J. K. Lanyi, Biochemistry 34, 14490 (1995).

QUANTITATIVE DETERMINATION OF PROTEIN SECONDARY STRUCTURE USING 206.5 NM EXCITED RESONANCE RAMAN SPECTRA

Zhenhuan Chi, X. G. Chen, Janet S.W. Holtz and Sanford A. Asher*

Department of Chemistry, University of Pittsburgh Pittsburgh, PA 15260, USA

UV resonance Raman (UVRR) spectroscopy has emerged as a powerful and selective tool to investigate protein structure and function. Excitation at 206.5 nm from a newly developed frequency-doubled CW krypton ion laser is used to excite within the $\pi \rightarrow \pi^*$ amide transitions of proteins. The 206.5 nm excitation selectively enhances the peptide vibrations (amide I, II and III bands, and 1400 cm^{-1} C_α-H bending band) [1,2]. We have examined the 206.5 nm excited resonance Raman spectra of concanavalin A (Con A), trypsin, RNase A, cytochrome *c* (cyt *c*), bovine serum albumin (BSA), hemoglobin (Hb), and myoglobin (Mb).

We analyze the amide I, II, III and 1400 cm^{-1} C_α-H bending bands in these protein spectra, using target factor analysis (TFA) [3]. We decompose the protein spectra into three secondary structural component spectra (i.e., random coil, α-helix and β-sheet structure), using the protein secondary structure compositions previously determined by X-ray crystallography and circular dichroism. Figure 1 shows the individual calculated secondary structural component spectra. The quality of the secondary structural component spectra improves as more protein spectra are used in factor analysis. The Raman cross sections, frequencies and bandwidths of the amide I, II and III bands vary with conformation. The Raman cross section of the 1400 cm^{-1} C_α-H bending band is very sensitive to α-helical content. Figure 1 shows that each secondary structural component spectrum has a unique spectral pattern.

We use a linear combination of the secondary structural component spectra to calculate each individual UVRR protein spectrum. The least-square method is used to minimize the residual between the experimental spectrum and the calculated spectrum. The fractional abundances of the three protein secondary structures are determined by the coefficients in the linear combination. Figure 2 shows three experimental protein spectra and their residuals. The residuals indicate that how closely the calculated spectrum matches its corresponding experimental spectrum. Table I compares the fractional abundances of the different protein secondary structures, as is determined by UVRR spectra and the other techniques. The 206.5 nm excited Raman spectroscopy is a new powerful tool to determine protein secondary structures in solution.

Table I Fractional Abundances of Conformation in Proteins

	α-Helix		β-Sheet		Random coil & others	
Protein	Previous	Raman	Previous	Raman	Previous	Raman
Con A	2	5	68	85	30	11
Trypsin	8	10	32	13	60	101
RNase A	23	17	40	32	37	33
Cyt *c*	39	37	0	14	61	29
BSA	68	66	18	6	14	22
Hb	67	73	0	7	33	29
Mb	79	78	0	1	21	31

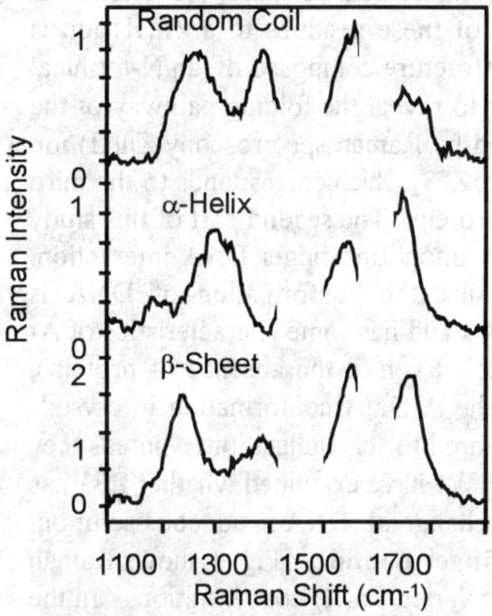

Figure 1 Three component Raman Spectra
The regions around 1450 and 1600 cm^{-1} are excluded due to overlap of C-H and ring vibrations

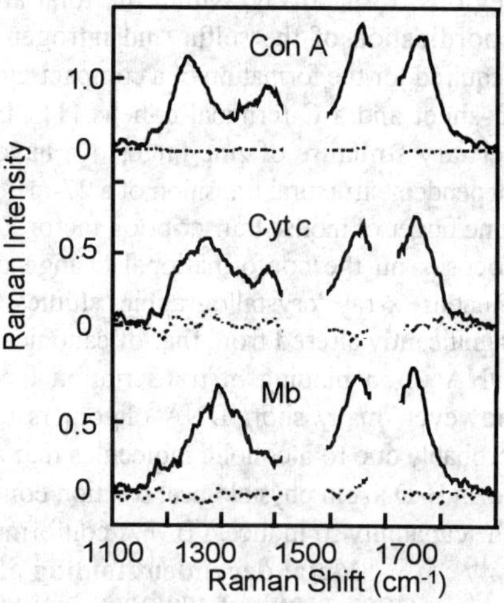

Figure 2 Protein Raman Spectra and their residuals

—— Experimental - - - - Residual

REFERENCES

[1] S. Song and S.A. Asher, *J. Am. Chem. Soc.* **111**,4295 (1989)
[2] X.G. Chen, S.A. Asher, R. Schweitzer-Stenner, N.G. Mirkin and S. Krimm *J. Am. Chem. Soc.* **117**,2884 (1995)
[3] E.R. Malinowski, Factor Analysis in Chemistry, Wiley-Interscience, New York, 1991

RAMAN SPECTROSCOPIC STUDY ON A ZINC FINGER PEPTIDE: METAL-DEPENDENT FOLDING AND INTERACTION WITH DNA

Takashi Miura*, Tamami Satoh, and Hideo Takeuchi

Pharmaceutical Institute, Tohoku University, Aobayama, Sendai 980-77, Japan

The zinc finger is a major nucleic acid-binding structural motif in eukaryotic transcription factors. One of the best characterized family of zinc finger, TFIIIA family, contains conserved two cysteines and two histidines in the N-terminal and C-terminal regions, respectively, within the total about 30 amino acid residues. A tetrahedral coordination of the sulfur and nitrogen atoms of these residues to a Zn(II) ion is required for the formation of a compact globular structure composed of an N-terminal β-sheet and a C-terminal α-helix [1]. In order to reveal the folding pathway of the tertiary structure of zinc finger, we have proved by Raman spectroscopy Zn(II) ion dependent structural transition of a 27-mer peptide, ZF3, which corresponds to the third zinc finger of mouse transcription factor Zif268 protein. The second part of this study focuses on the conformational change of DNA upon zinc finger-DNA interaction. Recent x-ray crystallographic studies show that the conformation of DNA is significantly altered from that of canonical B-DNA and has some characteristics of A-DNA upon binding of transcription factors [2]. Even in the absence of proteins, however, many short DNA oligomers exhibit the A-DNA conformation in crystals probably due to alcoholic molecules that are required for crystallization, whereas they form B-DNA in physiological solution conditions. We have examined whether ZF3 has the capability to induce a B → A conformational change of DNA in aqueous solution.

Metal dependent folding of zinc finger peptide. Fig. 1 shows Raman spectra (1750-1180 and 400-200 cm^{-1}) of ZF3 at varied Zn(II) concentrations. In the absence of Zn(II), the amide I band at 1668 cm^{-1} and the strong amide III band at 1442 cm^{-1} indicate that the secondary structure of the peptide is mainly β-sheet. As the Zn(II) concentration increases, a partial structural transition from β-sheet to α-helix takes place, which is indicated by intensity decreases at 1668 and 1242 cm^{-1} and concomitant increases at 1660 and 1266 cm^{-1}. The final form (Zn(II)/ZF3 = 1.0 or 2.0) contains both of the secondary structures. A histidine band at 1573 cm^{-1} exhibits a high frequency shift and overlaps with the 1606 cm^{-1} phenylalanine band upon coordination of histidine side chains to Zn(II). The 1573 and 1606 cm^{-1} bands of ZF3 remain unchanged between Zn(II)/ZF3 = 0 and 0.5. However, they show significant intensity decrease and increase, respectively, at higher Zn(II) concentrations. This result indicates that the Zn(II) ion first binds to the sulfur atoms of cysteine side chains at low Zn(II) concentrations and then histidines are involved in the metal coordination to form the zinc finger structure. The ligand-selective stepwise metal coordination is confirmed by the Raman spectra in the Zn-S (311 cm^{-1}) and Zn-N (343 and 242 cm^{-1}) stretching regions (Fig. 1, right); i.e. intensity increase of the 311 cm^{-1} band (Zn(II)/ZF3, 0-0.5) followed by increases of the 343 and 242 cm^{-1} bands (0.5-1.0). Either the Zn-cysteine or the Zn-histidine binding promotes the β-sheet → α-helix conversion in the peptide backbone. The pathway and mechanism of metal ion dependent folding on the basis of these results will be discussed.

Conformation of DNA bound by zinc finger peptide. The third zinc finger of Zif268 recognizes a GCG triplet in DNA. We have collected 244-nm Raman spectra of double-stranded DNA oligomers containing GCG sequences in the presence and absence of ZF3·Zn(II). Upon binding of the peptide, a drastic intensity decrease of the C=O stretching vibration of cytosine at 1630 cm^{-1} is observed, together with an intensity increase of the 1533 cm^{-1} cytosine band and a change in intensity ratio of the 1336/1320 cm^{-1} guanine doublet. These spectral changes are analogous to those observed for an ethanol-induced B → A transition of DNA. This may be the first evidence of a B/A intermediate conformation occurring in a zinc finger-DNA complex under a physiological condition.

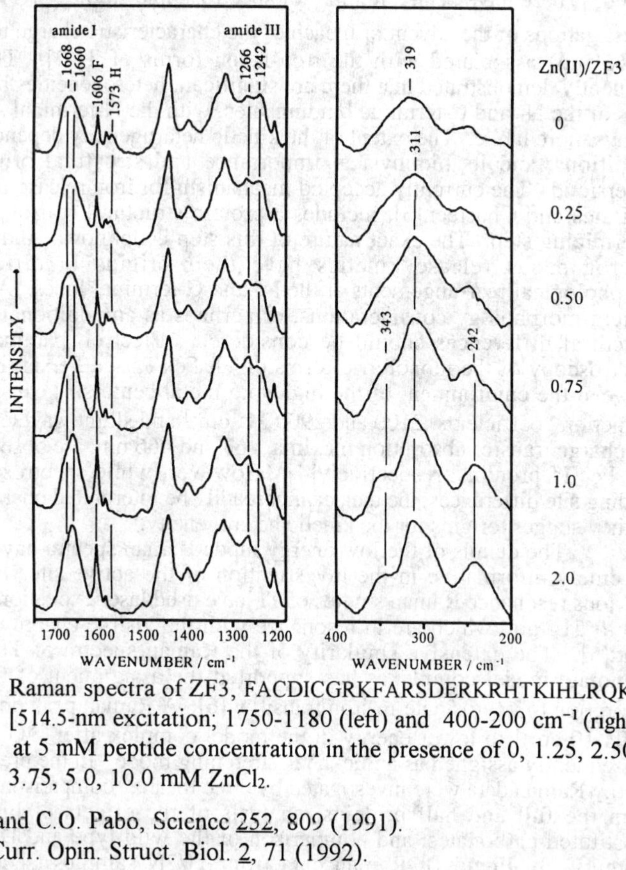

Fig. 1. Raman spectra of ZF3, FACDICGRKFARSDERKRHTKIHLRQK [514.5-nm excitation; 1750-1180 (left) and 400-200 cm^{-1} (right)], at 5 mM peptide concentration in the presence of 0, 1.25, 2.50, 3.75, 5.0, 10.0 mM $ZnCl_2$.

1. N.P. Pavletich and C.O. Pabo, Science 252, 809 (1991).
2. A.A. Travers, Curr. Opin. Struct. Biol. 2, 71 (1992).

RESONANCE RAMAN SPECTROSCOPY OF SERUM TRANSFERRIN REVEALS STRONGER METAL BINDING AT THE C-TERMINAL ACTIVE SITE

Sandra L. Mecklenburg & Robert J. Donohoe*

CST-4, Biochemistry and Biotechnology Group, MS-C345;
Los Alamos National Laboratory, Los Alamos, NM 87545

Iron translocation to proliferating cells for storage or use in biosynthesis is dependent upon transport by transferrin (Tf) and uptake of the diferric form of Tf (Fe_2Tf) into the cells via the Tf receptor protein[1]. The N- and C-terminal binding sites of Tf each include two tyrosinates, one histidine, one aspartic acid, and a carbonate that both neutralizes the site prior to metal binding and serves as a bidentate ligand. There have been a number of spectroscopic studies of Tf, including kinetic investigations of the advent or bleach of the characteristic charge transfer band (λ_{max} = 465 nm) associated with the iron-laden forms of Tf[2]. These studies have frequently demonstrated that there are significant heterogeneities between the release rates of the N- and C-terminal binding sites, with the N-terminal site usually but not always more labile. The extent of the kinetic heterogeneity depends on experimental conditions, and its biophysical importance and structural origins are not fully understood. The currently accepted mechanism for iron release from Tf, developed by Bates and Chasteen[3], includes a protein structural rearrangement as the rate determining step. The exact nature of this step is unknown, and it may be that the heterogeneous release kinetics have their origins in differences for such morphological rearrangements at the N- and C-terminal lobes. As an alternative to protein morphology considerations, or perhaps as an additional factor, active site structural differences should be considered. Electron paramagnetic resonance spectroscopy of the monoferric forms of Tf, Fe_NTf and Fe_CTf, reveals differences between the environment of the high spin metal centers[1]. In addition, disparate extinction coefficients, 2100 and 2900 $M^{-1}cm^{-1}$ and slight wavelength differences in the charge transfer absorption maxima, 463 and 466 nm, are observed for the Fe_NTf and Fe_CTf proteins, respectively[4]. However, while spectroscopy has revealed binding site differences, the data cannot readily be interpreted on a structural basis nor do they suggest origins for the kinetic heterogeneity.

The details of the low energy mode Raman spectra have been found to be extremely informative in the investigation of the active site structures in Fe_2Tf. Previous resonance Raman studies of Tf have used laser excitation wavelengths from 458 to 514 nm, which are in resonance with the visible charge transfer absorption band[5]. The extensive similarity of the Raman spectra of Tf to those of ferric phenolate model complexes has supported the assignment of the charge transfer absorption to a tyrosinate-to-iron transition[6]. A Raman peak observed at 570 cm^{-1} shifts 10 cm^{-1} to lower energy in the model complex after ^{18}O substitution and is consequently assigned as an Fe-TYR stretching mode. In the present study, the low energy Raman data were investigated by three means: comparison of the data derived from the full and half proteins, analysis of the effects of binding isotopically substituted carbonates, and comparison of the wild type and D63S mutant Tf/2N proteins. In Figure 1, Raman data from Tf/2N samples (second from top) are compared with the data from Fe_2Tf samples (top). Certain of the bands observed for the Fe_2Tf sample, such as the 698 and 336 cm^{-1} features, are absent from the data obtained for the Tf/2N sample. Subtraction of the spectrum obtained from the Tf/2N sample from the Fe_2Tf data yields a new and distinct Raman spectrum (Figure 1, Difference). This difference spectrum can be reasonably associated with the C-terminal binding site. The validity of this conclusion is supported by examination of the Raman data from a sample of Tf prepared at pH 6.8 (Figure 1, bottom). In a previous study[5], we found a strong indication from the absorption data that (at least under those particular experimental conditions), the protein at pH 6.8 is primarily monoferric Fe_CTf. As demonstrated in the poster, several of the low energy peaks can be assigned to metal-ligand modes. These are at 254 (Fe-HIS/CO_3), 420 (Fe-ASP) and 570 (Fe-TYR) cm^{-1}. Examination of the data below reveals that the metal-ligand modes in the spectrum from Fe_2Tf are comprised of higher and lower energy components from the C- and N-terminal sites, respectively, indicating stronger metal binding at the C lobe and suggesting an active-site basis for the kinetic heterogeneity.

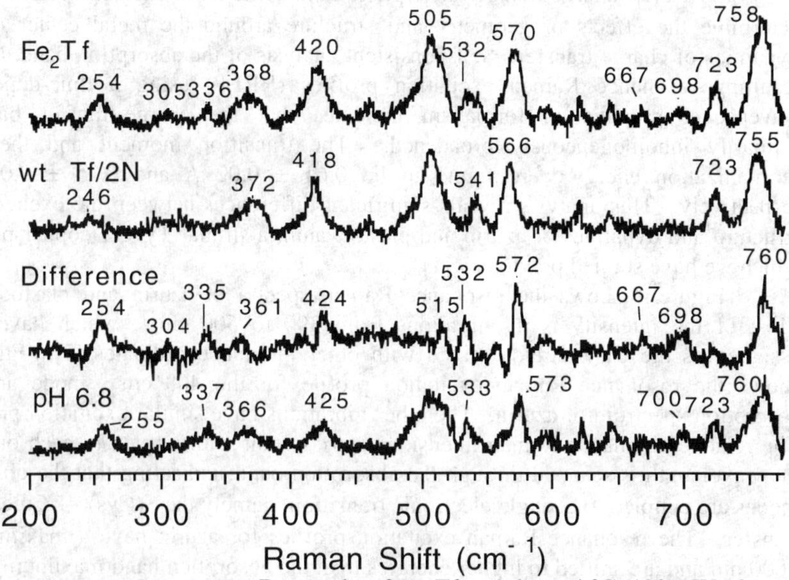

Figure 1. Low energy Raman data from Tf samples at 4 °C (100 mW of 514.5 nm excitation)

[1] Aisen, P. (1989) in *Iron Carriers and Iron Proteins* (Loehr, T. M., ed.) pp 353-372, VCH Publishers, Inc., New York.

[2] Baldwin, D. A., & de Sousa, D. M. R. (1981) *Biochem. Biophys. Res. Commun. 99,* 1101.

[3] Chasteen, N. D. (1983) *Adv. Inorg. Biochem. 5,* 201.

[4] Mecklenburg, S. L., Olah, G. A., & Donohoe, R. J. (1996) *Biochemistry,* submitted.

[5] Gaber, B. P., Miskowski, V., & Spiro, T. G. (1974) *J. Am. Chem. Soc. 96,* 6868.

[6] Pyrz, J. W., Roe, A. L., Stern, L. J., & Que, Jr., L. (1985) *J. Am. Chem. Soc. 107,* 614.

CHARGE TRANSFER DYNAMICS OF AZURIN, A BLUE COPPER PROTEIN, FROM RESONANCE RAMAN INTENSITIES

M. Adam Webb and Glen R. Loppnow*

Department of Chemistry, University of Alberta, Edmonton, Alberta, Canada T6G 2G2

The resonance Raman spectra of azurin, a 13-kDa Type 1 blue copper protein from *Pseudomonas aeruginosa*, have been measured at wavelengths throughout the S(Cys) $\rightarrow$ Cu(II) charge-transfer absorption band centered at 625 nm in an effort to determine the effects of geometry and structure around the metal center on the dynamics of charge transfer. Self-consistent analysis of the absorption band and the resulting resonance Raman excitation profiles (RREP) using a time-dependent wavepacket propagation formalism indicates that the charge-transfer band is primarily inhomogeneously broadened. The transition moment and the total reorganization energy were found to be 0.61 ± 0.02 Å and 0.25 ± 0.05 eV, respectively. This analysis reveals significant differences between the excited-state structure and dynamics of azurin and plastocyanin, a similar Type 1 copper protein, which we have studied previously [1].

Figure 1 shows the resonance Raman spectra of azurin and plastocyanin. Most of the intensity is in vibrations from 300 to 500 cm^{-1}, which have been assigned as the Cu-S stretch mixed with other internal coordinates [2]. Figure 2 shows the resonance Raman excitation profile for the 408 cm^{-1} mode and the absorption spectrum of azurin. The other vibrations have similar excitation profiles. The relative resonance Raman intensities are constant with respect to each other as the wavelength is scanned through the absorption band, indicating that the observed modes are coupled to a single electronic transition, namely the S(Cys) $\rightarrow$ Cu charge-transfer. The resonance Raman excitation profiles for azurin have a maximum at ~600 nm and are shifted to higher energies than the absorption band maximum. This is consistent with the presence of other electronic transitions under the absorption band to which the observed vibrations are not coupled [3].

A comparison of the extracted molecular parameters for plastocyanin and azurin identifies those parameters sensitive to geometry and environment. The coordination geometry of the Cu metal site is an elongated trigonal bipyramid in azurin, whereas the metal site is a distorted tetrahedron in plastocyanin [4,5]. From the analysis of the RREP and absorption spectrum, the transition moment and the reorganization energy of plastocyanin are 0.58 Å and 0.25 eV [1], respectively, and are similar to the values for azurin. Thus, the total reorganization energy seems to be independent of coordination geometry and protein composition. The charge-transfer band is primarily inhomogeneously broadened in azurin, while it is primarily homogeneous in plastocyanin, indicating that the specifics of the excited-state may be sensitive to structure. A closer comparison of the copper environment reveals two hydrogen-bonds to S(Cys) in azurin, whereas plastocyanin has only one [4,5]. The different hydrogen-bonding environment may explain the different broadening mechanisms involved in charge transfer dynamics of azurin and plastocyanin.

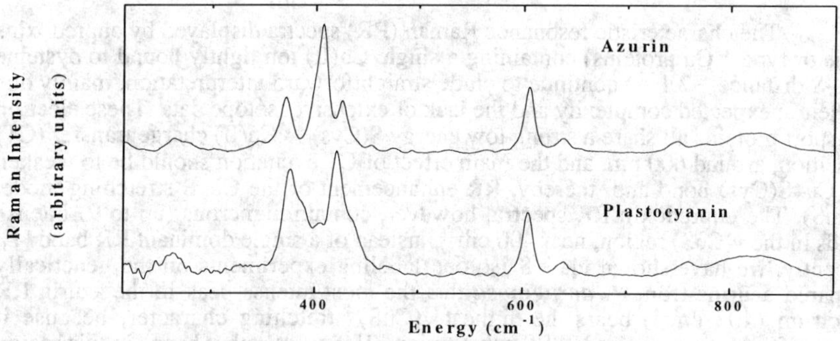

Figure 1. Resonance Raman spectra of plastocyanin and azurin. The peaks at 608 and 638 cm^{-1} are due to the cacodylate internal standard.

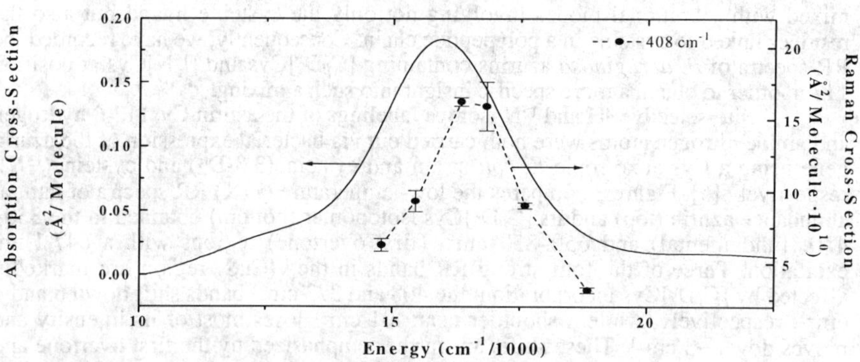

Figure 2. RREP (---) and absorption spectrum (—) of azurin.

1. E. Fraga, et al., J. Phys. Chem., in press
2. D. F. Blair, et al., J. Am . Chem. Soc. 107, 5755 (1985)
3. E. I. Solomon, et al., J. Am. Chem. Soc. 102, 169 (1980)
4. J. M. Guss and H. C. Freeman, J. Mol. Biol. 169, 521 (1983)
5. E. N. Baker, J. Mol. Biol. 203, 1071 (1988)

VIBRATIONAL MODE MIXING IN AZURIN

G. Fraczkiewicz, R. Fraczkiewicz, J. Germanas, and R. S. Czernuszewicz*

Department of Chemistry, University of Houston, Houston, Texas 77204, U.S.A.

The characteristic resonance Raman (RR) spectra displayed by cupredoxins (blue or type 1 Cu proteins) containing a single Cu(II) ion tightly bound to cysteine (Cu–S distance, ~2.1 Å) continue to elude straightforward interpretation, mainly due to their unexpected complexity and the lack of extensive isotope data. These electron transport proteins all share a strong low energy S(Cys) $\rightarrow$ Cu(II) charge transfer (CT) transition, around 600 nm, and the main effect of CT excitation should be to weaken the Cu–S(Cys) bond and, thereby, RR enhancement of the Cu–S stretching mode, ν(CuS). The cupredoxin RR spectra, however, contain numerous (up to 9) intense bands in the ν(CuS) region, near 400 cm^{-1}, instead of a single dominant RR band [1]. Recently, we have shown via ^{34}S isotope labeling experiments on the genetically prepared azurins from *P. aeruginosa* that the most intense peak in the azurin RR spectrum (408 cm^{-1}) bears the highest ν(CuS) stretching character, because it downshifts nearly 4 cm^{-1} on $^{34/32}S$ substitution [2]. Several other bands in this region also shift to lower frequency (1–2.5 cm^{-1}) for protein grown in the ^{34}S-enriched minimal medium. An even more surprising is the effect of ^{15}N isotope labeling of azurins [1,3] and plastocyanin [4] which revealed that most of the fundamental bands move 1–4 cm^{-1} to lower frequencies. It is therefore apparent that ν(CuS) is strongly mixed with vibrational modes involving not only the cysteine ligand but also the residues linked to cysteine in a polypeptide chain. Consequently, we have recorded the RR spectra of *P. aeruginosa* azurins containing [$C_\beta D_2$]Cys and [^{15}N]Cys at position 112 in order to obtain a more specific insight into such a mixing.

Site-selective ^2H and ^{15}N isotope labelings of the azurin Cys112 β-hydrogen and amide nitrogen atoms were both carried out via bacterial expression of the azurin gene using a Cys auxotropic *E. coli* strain and cysteine(3,3-D_2) and cysteine(^{15}N), respectively [5]. Figure 1 compares the low-temperature (77K) RR spectra of natural abundance azurin (top) and its [$C_\beta D_2$]Cys isotopomer (bottom) obtained in the 350–450 (fundamental) and 650–850 cm^{-1} (first overtone) regions with a 647.1-nm excitation. Three of the four strong RR bands in the ν(CuS) region are markedly affected by [$C_\beta D_2$]Cys incorporation: the 408 and 373 cm^{-1} bands shift down 6 and 4 cm^{-1}, respectively, while a shoulder near 401 cm^{-1} loses most of its intensity and moves down ~3 cm^{-1}. These shifts are further emphasized by the first overtone and combination bands as indicated in the figure. On the other hand, the specific labeling of the N atom in the Cys112 residue caused only small changes in the RR spectrum of azurin (not shown). Three shoulders at about 264, 401, and 377 cm^{-1} all shifted downward by ~1 cm^{-1}, while the ^{34}S-sensitive peaks at 408 and 428 cm^{-1} shifted less than 0.5 cm^{-1}. There were also a few small changes in the overtone region, indicating that the substantially larger ^{15}N shifts observed for azurins and plastocyanin grown in the $^{15}NH_4Cl$ media originate from mixing of ν(CuS) with deformation coordinates of several side chain residues. Our normal coordinate analysis (NCA) calculations indicate that the ν(CuS) coordinate is in fact distributed among the most prominent RR modes (408 cm^{-1}, ~40%; 401 cm^{-1}, 20%; 426 cm^{-1}, ~13%; and 373 cm^{-1}, ~5%), and that the source of ^2H shifts are heavy atom bending coordinates of Cys112 (SCC and

and bendings around C_α), while ^{15}N shifts arise from contributions of the $\delta(NC_\alpha C_\beta)$ coordinates of at least four residues, Cys112, Phe110, Phe111, and Thr113 [3].

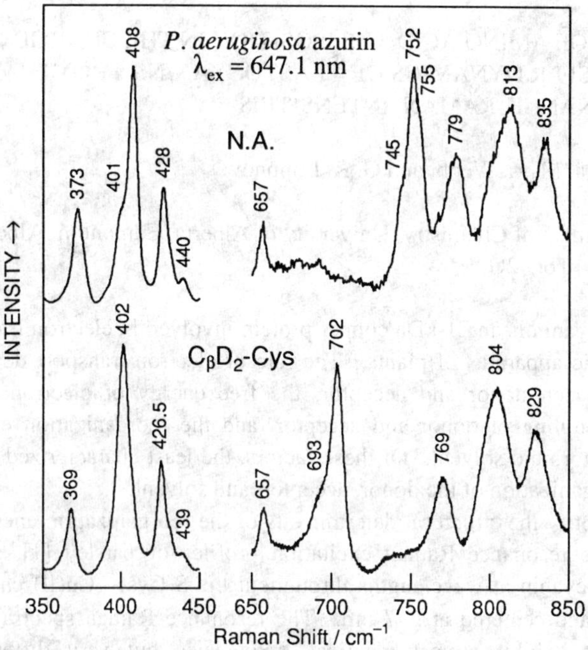

Fig. 1. Low-temperature (77K) RR spectra of *P. aeruginosa* azurin and its $[C_\beta D_2]$Cys112-labeled protein excited at 647.1 nm.

Finally, the most dramatic change in the RR spectrum caused by $[C_\beta D_2]$Cys substitution was observed in the 650-850 cm^{-1} region. A prominent ^{34}S-sensitive (–2 cm^{-1}) feature at 752 cm^{-1} in the N.A. azurin spectrum, which shifted to 702 cm^{-1} in the $[C_\beta D_2]$Cys spectrum (Fig. 1), has now been firmly assigned to cysteine C–S stretching, ν(CS). The $[C_\beta D_2]$Cys substitution also helped to assign some peaks in the region above 1000 cm^{-1} (not shown). An isolated band at 1406 cm^{-1} displayed a –289 cm^{-1} shift upon $[C_\beta D_2]$Cys incorporation, which is in excellent agreement with the calculated value for the $\rho_s(CH_2)$ scissor mode of methylene hydrogens on the cysteine 112 *β*-carbon. The azurin RR band at 1226 cm^{-1}, which occurs in other cupredoxins in the narrow range of 1225–1240 cm^{-1}, has been assigned to the cysteine amide III vibration coupled to the C_α hydrogen bending on the basis of its shift to ~1207 cm^{-1} in the $[C_\beta D_2]$Cys spectrum and NCA calculations [3]. This work was supported by National Institute of General Medical Sciences (GM48370) and Robert A. Welch Foundation (E-1184).

1. J. Sanders-Loehr, in Bioinorganic Chemistry of Copper, (K.D. Karlin and Z. Tyeklar, eds.), Chapman, New York, 1993, p. 51.
2. B.C. Dave, J.P. Germanas, and R.S. Czernuszewicz, J. Am. Chem. Soc. 115, 12175 (1993).
3. R. Fraczkiewicz, Ph. D. Dissertation, University of Houston, 1996.
4. D. Qiu, S. Dong, J.A. Ybe, M.H. Hecht, and T.G. Spiro, J. Am. Chem. Soc. 117, 6443 (1995).
5. The cysteine *E. coli* auxotropic strain was a gift from Dr. J. Markley (University of Wisconsin).

ROLE OF AMINO ACID COMPOSITION IN THE CHARGE TRANSFER DYNAMICS OF PLASTOCYANINS FROM RESONANCE RAMAN INTENSITIES

E. Fraga, M. A. Webb, and G. R. Loppnow*

Department of Chemistry, University of Alberta, Edmonton, Alberta, Canada T6G 2G2.

Plastocyanin is an 11-kDa copper protein involved in electron transport in the photosynthetic apparatus of plants. The rate of electron transport depends on the distance between donor and acceptor, the free energy of electron transfer, the electronic coupling of donor and acceptor, and the reorganization energy of the donor, acceptor, and solvent. Of these factors, the least characterized is the mode-specific reorganization of the donor, acceptor, and solvent.

To identify the structural determinants of the reorganization energy, we have measured the resonance Raman excitation profiles of parsley [1], spinach, and poplar plastocyanin at wavelengths throughout the S(Cys)→Cu(II) charge transfer absorption band centered at 597 nm. The resonance Raman spectra of the three plastocyanins exhibit almost identical frequencies but very different relative intensities (Fig. 1), indicating similar ground-state structure, but different excited-state charge transfer structure and dynamics. Self-consistent analysis of the absorption band and resulting resonance Raman excitation profiles with a time-dependent wavepacket propagation formalism [2,3] demonstrates that many of the molecular parameters for the three plastocyanins are similar, including the zero-zero energy, the transition moment, and the homogeneous and inhomogeneous linewidths. The excited-state lifetime of 20±15 fs calculated here from the observed fluorescence suggests that the charge transfer state decays rapidly via lower-lying ligand-field states. The resonance Raman excitation profiles demonstrate that the charge transfer absorption band is primarily homogeneously broadened as a result of this rapid population decay and an additional contribution from solvent-induced dephasing.

The reorganization energy contribution from resonance Raman-active modes is estimated to be 0.19 ± 0.02 eV in all three plastocyanins. If the additional reorganization energy of the protein as measured from the solvent-induced dephasing component of the homogeneous linewidth is included, the observed reorganization energy is 0.25 ± 0.03 eV in quantitative agreement with a previous upper-limit of 0.3 eV measured for the reorganization energy upon electron transport at the copper site in azurin, a similar blue copper protein. A crude comparison of the structural changes upon electron transport and charge transfer suggests that the charge transfer may be a useful model for the geometry changes upon electron transfer. The resonance Raman spectrum indicates that reorganization occurs primarily along normal modes that involve the Cu-S(Cys) stretch, but significant reorganization also occurs along specific normal modes that involve internal cysteine stretches, Cu-N(His) stretches, and protein internal motions. Although the total reorganization energy is identical in the three plastocyanins, the observation of different molecular motions as a function of protein composition demonstrates that the protein environment contributes to the reorganization energy through two mechanisms: by coupling into specific normal modes and by a solvent-induced dephasing contribution.

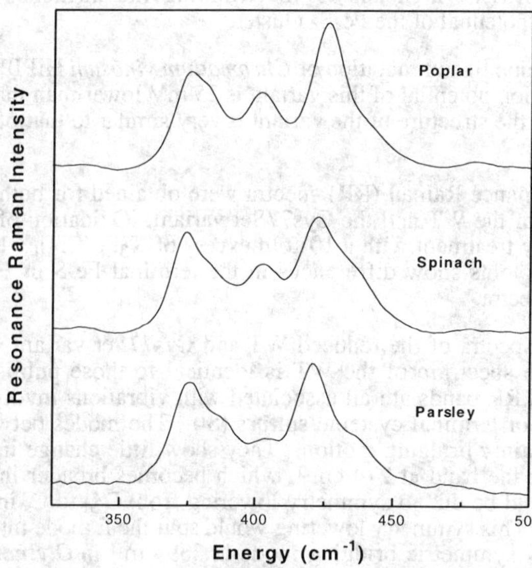

Fig. 1-The resonance Raman spectra of poplar, spinach, and parsley plastocyanin.

1. E. Fraga, M. A. Webb, and G. R. Loppnow, J. Phys. Chem., in press.
2. A. B. Myers and R. A. Mathies, in Biological Applications of Raman Spectroscopy, (T. G. Spiro, ed.), Vol. 2, Wiley, New York, 1988, p. 1.
3. F. Markel, N. S. Ferris, I. R. Gould, and A. B. Myers, J. Am. Chem. Soc. 114, 6208 (1992).

RESONANCE RAMAN STUDY OF A HIGH POTENTIAL IRON-SULFUR PROTEIN AND ITS SERINE → CYSTEINE LIGAND VARIANT

C. A. Piffat[1], L. D. Eltis[2], and R. S. Czernuszewicz[1]*

[1]Department of Chemistry, University of Houston, Houston, TX 77204-5641
[2]Department of Biochemistry, Université Laval, Québec City, Québec G1K 7P4, Canada

In the high-potential iron-sulfur proteins (HiPIP's) used for electron transfer, large differences in redox potentials occur despite nearly identical structures of the Fe_4S_4 clusters. Differences in these proteins occur in the cluster environment.[1] Changes in the cluster ligands would provide a means for determining the influence the environment has on the redox potential of the Fe_4S_4 cluster.

Recently, a serine → cysteine ligand mutation of *Chromatium vinosum* HiPIP has been prepared.[2, 3] The reduction potential of this variant is 25 mV lower than that of the wild type (WT), although the structure of the variant is very similar to that of the WT.[2]

In the present study, resonance Raman (RR) spectra were obtained for both the reduced and oxidized forms of the WT and the Cys77Ser variant. Oxidation of the proteins was accomplished by treatment with a 10-fold excess of $K_3Fe(CN)_6$. Both the reduced and oxidized proteins show differences in the terminal Fe-S or Fe-O stretching region of the RR spectra.

Figure 1 depicts the RR spectra of the reduced WT and Cys77Ser variant with excitation at 488.0 nm. The spectrum of the WT is identical to those published previously.[1, 4] The observed RR bands are all associated with vibrations involving cluster bridging sulfurs (S^b) or terminal cysteine sulfurs (S^t). The modes between 250 and 300 cm^{-1} involve mainly bridging motion. They show little change in the Cys77Ser variant, except for the band at 274 cm^{-1}, which becomes broader in the variant. This broadening could be due to symmetry lowering from D_{2d} to C_s in the cluster portion of the variant. This symmetry lowering would split the E mode into its A' and A" parts. The totally symmetric bridging mode at 339 cm^{-1} in *C.vinosum* HiPIP is upshifted 3 cm^{-1} in the variant. This higher frequency is characteristic of Fe_4S_4 clusters in which a non-cysteine ligand is coordinated to a specific iron atom.[5] The vibrational mode at 396 cm^{-1} is fairly broad and changes very little in the variant. This mode contains terminal Fe-S motion. Finally, the mode at 364 cm^{-1} loses intensity and acquires a shoulder at 373 cm^{-1} in the Cys77Ser variant. Similar spectral patterns were observed for oxidized *P.furiosus* ferredoxin when the Fe_4S_4 clusters of some of the protein molecules had been converted to Fe_3S_4 clusters. This alteration in the RR spectra of the variant could reflect a distortion of the Fe_4S_4 cluster from its slightly flattened cube form to one in which one of the Fe atoms is stretched away from the center of the cluster.

Spectra of the oxidized HiPIP's reveal that there are small differences between the WT and the Cys77Ser variant. The 3 cm^{-1} upshift of the totally symmetric bridging mode is still maintained, and the terminal vibrations show some differences in relative intensities, indicating that the Cys77Ser exchange induces similar structural and electronic changes in both oxidation states of the Fe_4S_4 cluster.

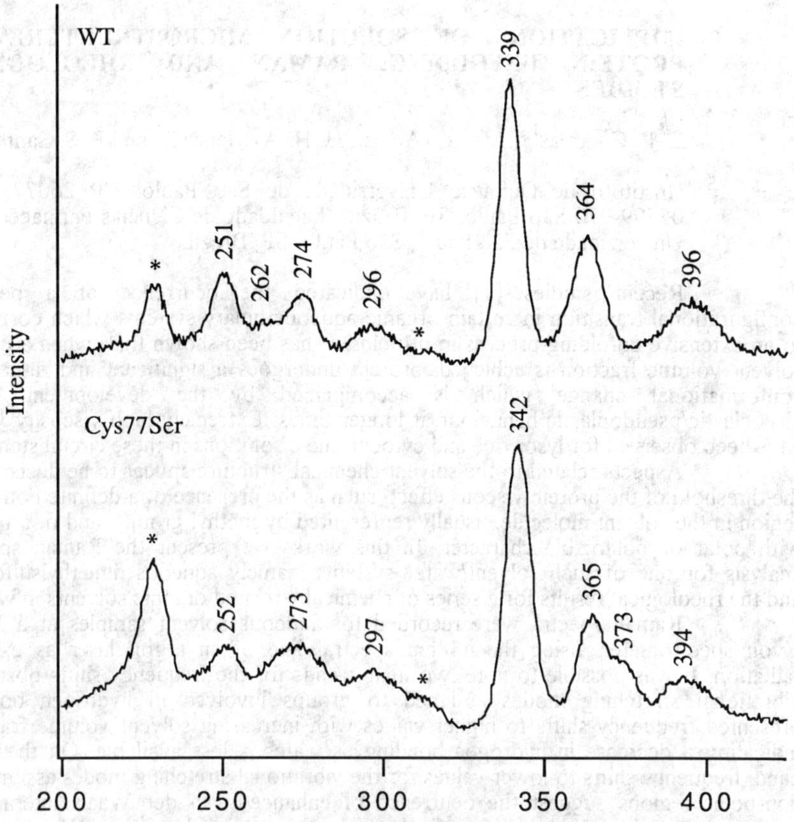

Fig. 1. Low-temperature (77 K) RR spectra of reduced *C. vinosum* HiPIP and its Cys77Ser variant. * denotes ice peaks.

Acknowledgement. This work was supported in parts by the NIGMS-NIH grant GM 48370 and the Robert A. Welch grant E-1184 (to R. S. C.).

1. G. Backes, Y. Mino, T. M. Loehr, T. E. Meyer, M. A. Cusanovich, W. V. Sweeney, E. T. Adman and J. Sanders-Loehr, J. Am. Chem. Soc. 113, 2055 - 2064 (1991).
2. E. Babini, I. Bertini, M. Borsari, F. Capozzi, A. Dikiy, L. D. Eltis and C. Luchinat, J. Am. Chem. Soc. 118, 75 - 80 (1996).
3. A. Agarwal, D. Li and J. A. Cowan, J. Am. Chem. Soc. 118, 927 - 928 (1996).
4. R. S. Czernuszewicz, K. A. Macor, M. K. Johnson, A. Gewirth and T. G. Spiro, J. Am. Chem. Soc. 109, 7178 - 7187 (1987).
5. R. C. Conover, A. T. Kowal, W. Fu, J.-B. Park, S. Aono, M. W. W. Adams and M. K. Johnson, J. Biol. Chem. 265, 8533 - 8541 (1990).

IMPLICATIONS OF SOLUTION MICROSTRUCTURE ON PROTEIN UNFOLDING: RAMAN AND RHEOLOGICAL STUDIES

E. P. G. Arêas[1]*, J. A. G. Arêas[2], H. H. A. Menezes[1] and P. S. Santos[1]

[1]Instituto de Química, Universidade de São Paulo, CP 26077, CEP 05599-970 São Paulo, SP, Brazil; [2]Faculdade de Ciências Farmacêuticas, Universidade de São Paulo, São Paulo, SP, Brazil.

Recent studies [1] have indicated the ocurrence of a peculiar configurational transition in certain organic-aqueous binary systems which correlates to an extensive unfolding process in proteins. It has been shown that, when a critical solvent volume fraction is achieved, protein undergoes a significant and rather fast conformational change which is accompanied by the development of a viscoelastic/pseudoplastic behaviour at longer times. Extremely high viscosity values have been observed for lysozyme and cytochrome c solutions in these circumstances.

Aspects related to the solvent chemical structure appear to be decisive for the threshold of the protein viscous effect, such as the presence of a definite non-polar region in the solvent molecule, usually represented by methyl groups, and of a moiety with polar or polarizible character. In this work, we present the Raman spectral analysis for one of such solvent/water systems, namely aqueous dimethylsulfoxide, and the rheological results for a series of chemically related organic solvents in water.

Raman spectra were recorded for aqueous solvent samples at a Jobin-Yvon spectrometer, using the 488nm spectral line of an argon laser as exciting radiation. It was possible to note two main trends for the frequency shifts observed: vibrational stretching modes assigned to groups involved in hydrogen bonding presented frequency shifts to higher values with increasing solvent volume fraction, indicating a decrease in hydrogen bonding as water is less available. On the other hand, frequency shifts to lower values for the vibrational stretching modes assigned to non-polar regions suggest the ocurrence of enhanced van der Waals interactions involving these molecular regions. Such a frequency dependence on solvent volume fraction can be described by two straight lines intersecting at a value that coincides with that for the viscosity critical point determined in the rheological experiments. Such observations are a further support to our model of a colapse of hydrophobic regions when a critical solvent volume fraction is achieved [1]. That would occur through a rather abrupt reversion in the configurational pattern of the organic/aqueous system at a critical solvent volume fraction, from a continuous, hydrophylic domain with non-polar microdomains dispersed in it to a continuous, non-polar milieu, with dispersed water microdomains. Such a possibility has been strongly suggested by a varied lot of physico-chemical data that we have been accumulating (^{13}C spin-lattice relaxation time, ^{13}C chemical shifts, condutimetric and refratometric data, rheological data), along with some simple geometric considerations [1]. Bending modes have not shown any appreciable dependence on solvent volume fraction, whereas a significant change in line shape has been observed for bands at low frequency, attemptively assigned to vibrations of associated molecules.

The rheological investigation of a series of chemically related solvents in water has also been undertaken in this work. Amides (formamide, dimethylformamide, hexamethylphosphortriamide), alcohols (methanol, ethanol, terc-butanol), tetrahydrofurane and acetone have been assayed, besides dimethylsulfoxide (DMSO) and tetramethylurea, previously characterized [1]. Refratometric measurements have also been performed for those systems. The results have shown the occurrence of critical points (expressed by a maximum in viscosity and an inflexion in the refractometric curves) only for solvents able to induce protein unfolding and consequent viscoelastic behavior under appropriate conditions. The DMSO/water system, spectroscopically studied here, is included in this group. Solvents without such capability showed a simple linear ratio between refraction index and solvent volume fraction and did not show the characteristic bell shaped viscosity curve, with a maximum in apparent viscosity, observed for solvents that are able to induce protein unfolding.

On the basis of both Raman and rheological data, it has been possible to start rationalizing the role played by the solvent polarized region on hydrogen bonding and the effectiveness of van der Waals interactions involving its hydrophobic regions, as a function of solvent volume fraction. These two simultaneous tendencies, when present in a same molecule, can lead to very peculiar features in solution configuration. The prevalent non-polar environment occurring after the solvent critical point would be a particularly favorable environment for protein unfolding, leading to a rather fast exposure of its inner sites, whereas allowing the protein polar groups to interact with the water microdomains.

1. E. P. G. Arêas, J. A. G. Arêas, J. Hamburger, W. L. Peticolas and P. S. Santos, *J. Coll. Interface Sci.*, in press.

HYDRATION AND STABILITY OF THE SOME GLOBULAR PROTEINS IN THE NON-POLAR MEDIUM IN THE PRESENCE OF PHOSPHATIDILHOLINE

L.D. Bobrovnik , V.M. Klimovich, I.S. Gulay

Ukrainian State University of Food Technologies, Kiev,
st.Vladimirskay 68, 252601, Kiev-17

By pressure of water steam near to pressure saturated steam in solid phase the globular proteins associate until 0,25 - 0,35gram water/gram protein [1]. Approximately the same number of water is in hydrate envelop of protein in water solution and in protein crystals [2]. Therefore the globular proteins is convenient model for study with spectral methods of the hydration, stability, dynamical structure of protein´s by it interaction with lipids, carbohydrates, pigments and other physiological-active substances.

Intention of present work is research the influence of non-polar medium and phosphatidilholine on stability of the macromolecules and hydration of cytohrom-C, tripsine and insulin by use of methods Raman and Infrared spectroscopy and isotope H/D exchange.

It was found, that non-polar medium give reversity alteration of conformation of the globule, when increase number segments with regularity structure (a-spiral, plicate layers and etc.). With it alteration is maximum-possibility formation of hydrogen bounds, ion-dipol interactions and its.

The present of phosphatidilholine give formation of associates in that electron-donor groups of phospholipide is acceptor of protons, the protons donor is peptides and other groups of protein.

The formation of associate proteins with phosphatidilholine increase number segments in the globule of protein with regularity structure and decrease mobility of their macromolecule. This can explain by effect the competition dehydration of protein globule, when some part of water earlier associated with protein, combine lipide´s phosphate group. All discover alteration of conformation protein is reversibility.

It is shown that water molecules in protein hydrate cover are in a more disordered state than in "liquid" water. However, the hydrogen bonds of protein hydrate envelope are more solid than associates of liquid water.

Thus the polarity of environment vital influence on the space structure of protein macromolecules. The non-polar medium promote increase of regulation of protein globule. The presence of water in system lead to increase number non-regular form in the polypeptide chains.

1. H. B. Bull, K. Breese, Arch. Biochim. Biophys., 128, 488 (1966).
2. J. A. Rupley, in Structure and Stability of Biological Macromolecules,(S. N. Timasheff, G. D. Fasman, ed.), MARCEL DEKKER, inc. NEW YORK, 1969, P. 255.

STRUCTURAL ANALYSIS OF MILK PROTEINS BY NEAR IR FOURIER-TRANSFORM RAMAN SPECTROSCOPY

Yih-Ming Weng and Wenlung Chen*

Department of Food Industry
National Chiayi Institute of Agriculture, Chiayi, Taiwan

Milk protein is widely utilized in the manufacture of food products, particularly for the functional and nutritional purposes. Factors that affect the functional properties of milk proteins, such as gelling, emulsifying, coagulating, and foaming ability, is ascribed to the structural changes of milk proteins during food processing. Several biochemical studies have showed that pH changes or thermal denaturations of milk proteins would lead to its conformational changes and sulfhydryl group/disulfide bond interchange resulting in affecting emulsification, gelation, and foaming properties [1-4]. However, results from the biochemical methods also presented great inconsistancy among different research groups. Therefore, in this work, The near infrared Fourier-transformed Raman spectroscopy will be used as a noninvasive probe for structural investigation of milk proteins since Raman spectroscopy is able to provide *in situ* structural information of proteins, especially in the secondary structural information, S-H/S-S conversion, and microenviornments of aromatic amino acid such as tyrosine, tryptophan, and phenylanaline [5].

Milk proteins is mainly classified as casein and whey proteins; both are important ingredients in food products. The detailed structure of milk powders, casein, and whey protein is identified by the Raman spectra showed on Figure 1 and Figure 2. The secondary structure of milk powders presented α-helix, and lactoalbumin was α-helix coupling with random coil; however, the secondary structure of casein showed mainly random coil. Preliminary results also showed that the microenvironment of disulfide bond existed in casein and in milk powder was different as evidenced by the S-S stretching at 507 cm^{-1} in lactoalbumin and 515 cm^{-1} in milk powder, respectively. Whereas, no disulfide bond existed in casein as no apparent S-S stretching around 510 cm^{-1} [Fig.2].

Based on the aforementioned results, the successive demonstration of Raman spectroscopy in milk protein enables its application extensively in understanding protein structure-function relationship in foods. Further studies on the precise relationships between changes in physicochemical properties and functional properties including foaming, gelling, emulsifying, and coagulating properties will be elucidate by Raman spectroscopy.

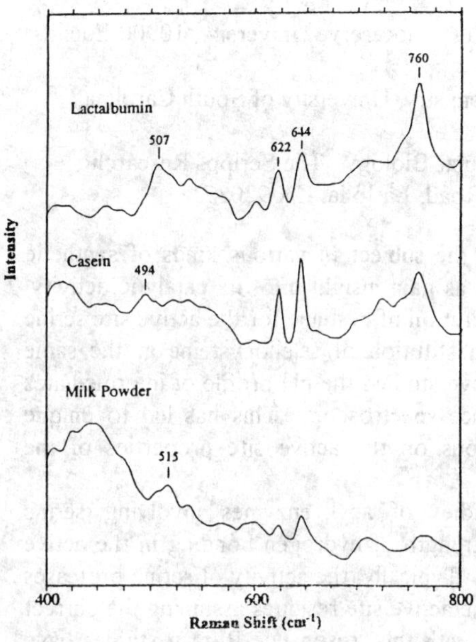

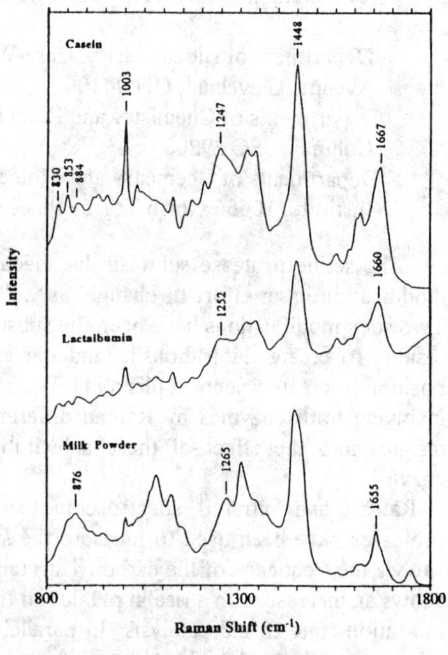

Figure 1: Raman spectra in the region of 400-800 cm^{-1} for milk proteins.

Figure 2: Raman spectra in the region of 800-1800 cm^{-1} for milk proteins.

(1) H. Zhu and S. Damodaran (1994) J. Agric. Food Chem. 42, 846-855.

(2) J. A. Hunt and D. G. Dalgleish (1994) J. Agric. Food Chem. 42, 2131-2135.

(3) F. J. Monahan, J. B. Germaan, and J. E. Kinsella (1995) J. Agric. Food Chem. 43, 46-52.

(4) K. Shimada and J. C. Cheftel (1989) J. Agric. Food Chem. 37, 161-168.

(5) W. Chen, S. Nie, J. F. Kuck and N.-T. Yu (1991) Biophys. J. 60,447-455.

RAMAN EVIDENCE FOR THE IMPORTANCE OF STERIC FACTORS IN CATALYSIS BY THE SEMI-SYNTHETIC ENZYMES THIOLSUBTILISIN AND SELENOSUBTILISIN.

D. Dinakarpandian[1], B. C. Shenoy[1], M. O'Connor[2], R B. Dunlap[2], J.D. Odom[2], E.B. Peterson[3], D. Hilvert[3], P. R. Carey[1]*

[1]Department of Biochemistry, Case Western Reserve University, 10900 Euclid Avenue, Cleveland, OH 44106
[2]Departments of Chemistry and Biochemistry, University of South Carolina, Columbia, SC 29208
[3]Departments of Chemistry and Molecular Biology, The Scripps Research Institute, 10666 North Torrey Pines Road, La Jolla, CA 92037

The serine protease subtilisin has been the subject of various kinds of synthetic modification in an effort to change, as well as gain insight into, its catalytic activity. Two such modifications have been the substitution of cysteine for the active site serine residue to create thiolsubtilisin, and the substitution of selenocysteine at the same position to create selenosubtilisin [1]. We have studied the pH profile of intermediates involving both enzymes by Raman difference spectroscopy This has led to unique insights into the effect of these substitutions on the active site properties of the enzyme.

Raman and infrared spectroscopic studies of acyl enzymes involving serine proteases have been used to measure the strength of hydrogen bonding in the active site via the frequency of the carbonyl stretch. Typically, the activity of serine proteases shows an increase with a rise in pH due to the active site residues assuming the correct ionization state to aid catalysis. In parallel with this, resonance Raman studies have shown that the carbonyl stretch shifts downwards with a rise in pH, reflecting an increase in the strength of hydrogen bonding in the oxyanion hole [2].

We have performed similar spectroscopic studies on thiolsubtilisin and selenosubtilisin. An acyl intermediate, 5-methylthienylacryloyl enzyme (5MTA) was prepared and isolated as previously described [3] A 488 nm laser and a charge coupled device detector were used to record Raman spectra Raman difference spectra were obtained by computer aided subtraction of the spectrum of the enzyme from that of the acyl enzyme intermediate. The rate of deacylation of the enzyme intermediate was measured by recording the rate of decrease in absorbance of the acyl intermediate in a spectrophotometer.

The following information was obtained (Fig 1,2) The position of the C=C stretch, at about 1600 cm^{-1} , shows only a small shift when compared to that of a corresponding model compound in carbon tetrachloride (data not shown). This suggests that the acyl group of the substrate is not subject to strong polarizing forces, as seen in the case of cysteine proteases like papain (data not shown). 5MTA is known to exhibit rotational isomerism about the C_1-C_2 single bond, and both *s-cis* and *s-trans* forms are seen in 5MTA papain. It is interesting to note, based on the peak corresponding to the C_1-C_2 stretch seen at about 1120 in selenosubtilisin or 1137 cm^{-1} in thiolsubtilisin, that this freedom to rotate is restrained in the present active sites. The *s-trans* isomer alone is observed. The most unexpected finding is the rise in frequency of the carbonyl stretch at high pH, e.g., from 1636 to 1686 cm^{-1} in the case of selenosubtilisin. The acyl-enzymes undergo a conformational change at high pH, which is marked enough to disrupt hydrogen bonding at the active site. The fact that the rate of catalysis is nevertheless enhanced at higher pH (data not shown) suggests that steric factors, such as better alignment with the catalytic apparatus at the active site, are far more important than hydrogen bonding for this particular substrate and the modified subtilisins. It is likely that the larger van der Waals radii of sulphur and selenium compared to oxygen crowd the active site and play a critical role in this 'steric control' of catalysis. Efforts are underway to generate a model that explains the conformational basis of these interesting results.

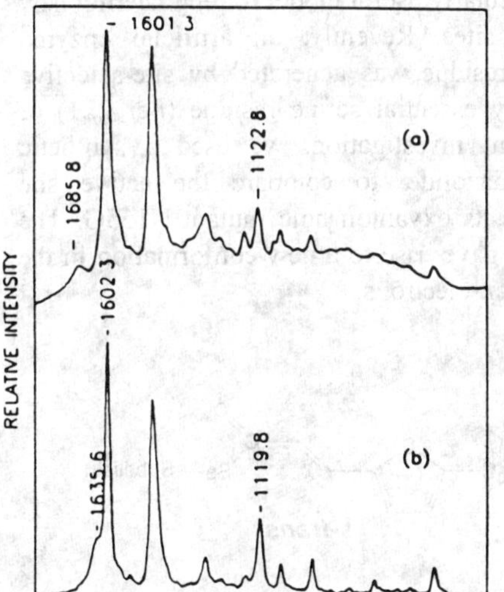

Fig. 1 Raman difference spectrum of the acyl selenosubtilisin intermediate at (a) pH 9.6 and (b) pH 6.5.

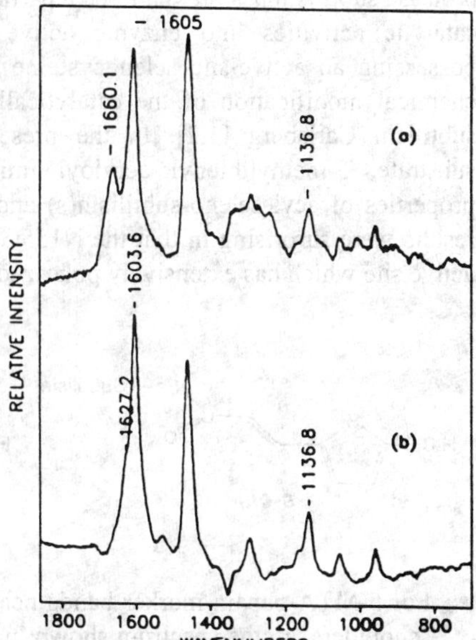

Fig.2 Raman difference spectrum of the acyl thiolsubtilisin intermediate at (a) pH 9.6 and (b) pH 6.5.

References:

1. Wu, Z. -P., & Hilvert, D. (1989) *J. Am. Chem. Soc.* **111**, 4513-4514.

2. Carey, P. R., & Tonge, P. J. (1995) *Acc. of Chem. Res.* **28**, 8-13.

3. O'Connor, M. J., Dunlap, R. B., Odom, J. D., Hilvert, D., Pusztai-Carey, M., Shenoy, B.C. and Carey, P. R. (1996) *J. Am. Chem. Soc.* **118**, 239-240.

RAMAN SPECTROSCOPY AS A UNIQUE PROBE OF NEWLY CREATED ACTIVE SITES

B. C. Shenoy[1], D. Dinakarpandian[1], M. Pusztai-Carey[1], E. B. Peterson[2], D. Hilvert[2] and P.R. Carey[1]*

[1]Department of Biochemistry, Case Western Reserve University, 10900 Euclid Ave. Cleveland, OH 44106.
[2]Department of Chemistry & Molecular Biology, The Scripps Research Institute, 10666 North Torrey Pines Rd., La Jolla, CA 92037.

There is an urgent need to develop techniques to probe the active sites of enzymes which have been genetically or chemically modified. The bacterial serine protease subtilisin has proven to be a particularly useful model for engineering new catalytic activities into enzyme active sites. Recently, an artificial enzyme possessing an active-site selenocysteine residue was generated by site-selective chemical modification of the catalytically essential serine residue (Ser 221) of subtilisin Carlsberg [1,2]. In the present investigation, we used a synthetic substrate, 5-methylthienyl acryloyl imidazolide, to compare the active site properties of acyl seleno-subtilisin(s) and its oxyanion hole mutant N155G. The results were surprising in that the N155G gave rise to a new conformation in the active site which has extensively polarized π-electrons.

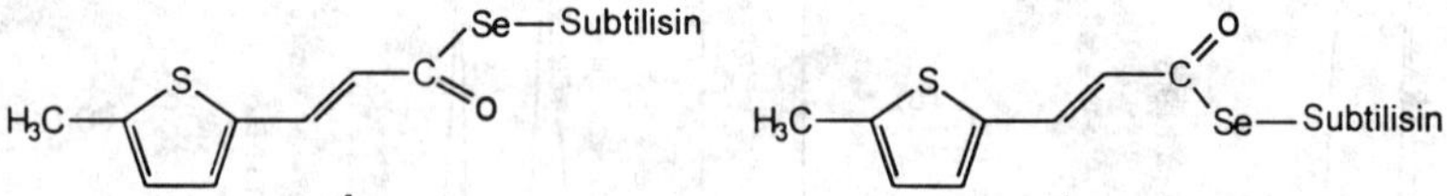

- For 5-MTA-papain marker bands near 1140 and 1030 cm^{-1} (which appear as "shoulders" in the spectrum shown in Fig. 1c) demonstrate the presence of two conformational forms for the bound acyl group. These are *s-cis* and *s-trans* about the C_1-C_2(=O) bond with the former predominating. The position of $\nu_{C=C}$ at 1563 cm^{-1} is evidence for strong electron polarization in the acyl group π electrons.

- For seleno-subtilisin (Fig. 1a) the 5-MTA group leads to a single population, *s-trans* about C_1-C_2, as evidenced by the 1109 cm^{-1} marker band. There is no polarization of the acyl group π electrons.

- For the N155G oxyanion hole mutant of seleno-subtilisin a second *s-cis* population appears (giving the 1025 cm^{-1} band, Fig. 1b) which has polarized π electrons and gives the 1569 cm^{-1} shoulder as its $\nu_{C=C}$. Thus, the mutant creates a papain like active site environment. The other population seen in "wild type" seleno-subtilisin is still present, it is *s-trans* and not polarized.

The X-ray structure of seleno-subtilisin has been determined. Molecular modeling will be able to determine the relationship between acyl group conformation in the active site and the enzyme's α-helices which are responsible for causing π-electron polarization.

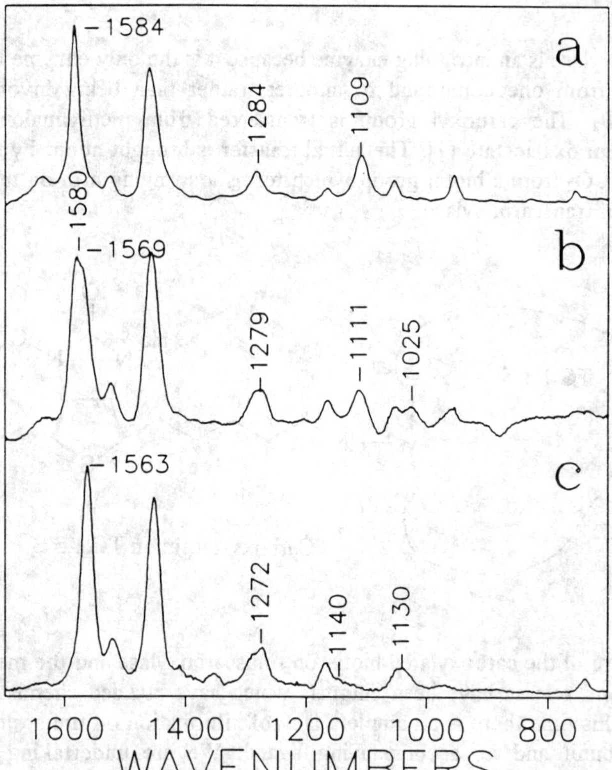

Figure 1. Comparison of the Raman difference spectra of a) 5-MTA-seleno-subtilisin b) 5-MTA-seleno-subtilisin N155G and c) 5-MTA-papain.

1. P-Z. Wu and D. Hilvert, J. Am. Chem. Soc., 111, 4513 (1989).
2. M. Philipp, and M.L. Bender, Mol. Cell. Biochem., 51, 5 (1983).

STRUCTURE AND MECHANISM OF THE CO_2 CARRYING ENZYME TRANSCARBOXYLASE

J. Clarkson, B. Shenoy, D. Samols and P. R. Carey*

Department of Biochemistry, Case Western Reserve University, 10900 Euclid Avenue, Cleveland, Ohio 44106-4935, USA

Transcarboxylase is an intriguing enzyme because it is the only enzyme known that transfers CO_2 from one compound to another, rather than being involved in the fixation of CO_2. The carboxyl group is transferred from methylmalonyl-CoA to pyruvate to form oxalacetate [1]. The actual transfer is brought about by the binding and release of CO_2 from a biotin group which forms a swinging arm on the so-called 1.3 S subunit of transcarboxylase.

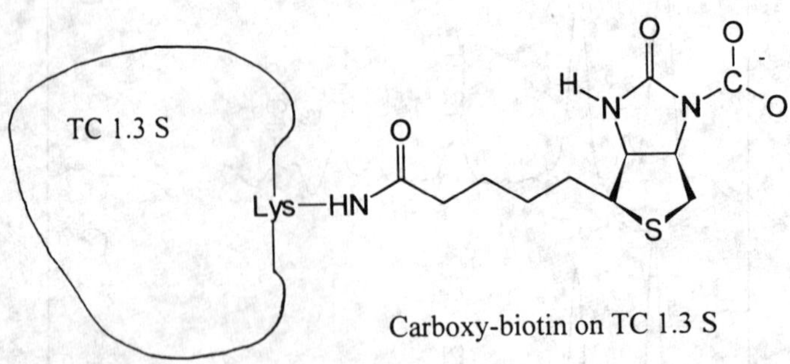

Carboxy-biotin on TC 1.3 S

The structure of the carboxylated biotin on transcarboxylase and the mechanism of CO_2 binding and release have been enigmas which have puzzled chemists for many years [2]. At this time there is a complete lack of information on the structure of the the 1.3 S subunit and its accompanying ligand. We are undertaking vibrational spectroscopic studies to define in detail the chemistry of the bound CO_2 molecule and to elucidate the mechanism of transcarboxylase by a series of structure - reactivity analyses. The approach is to undertake extensive IR and Raman studies on model biotin compounds, in the absence of protein, and to use this information to provide detailed structural information on CO_2 bound to transcarboxylase. The normal modes of the latter can be accessed by both IR and Raman difference spectroscopies.

Model compounds. Figure 1 shows the Raman spectra of unstable carboxy-2-imidazolidone. The 1332 cm^{-1} feature, due the symmetric stretching mode of the

COO^- group bound to the nitrogen of the ring decreases with time as decarboxylation occurs. Similar behaviour is seen for the 840 cm^{-1} mode (probably associated with a $-CO_2$ scissoring motion). Data such as these, for a number of biotin analogs and their isotopomers and combined with infrared spectroscopic data, allow us to develop a force field for carboxylated biotin. The time dependency of the Raman spectra provide kinetic constants for the decarboxylation process and, combined with the vibrational anlaysis sets up structure reactivity correlations for the model compounds.

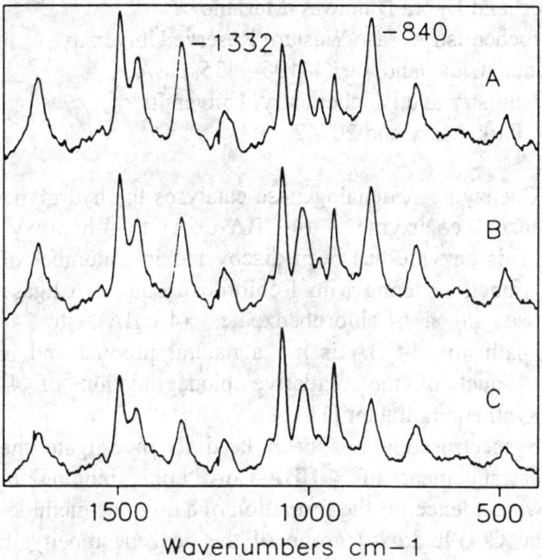

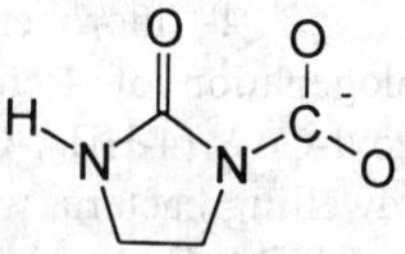

Carboxy-2-imidazolidone

Fig 1. Raman difference spectra of carboxy-2-imidazolidone A 10 mins., b 90 mins., and C 90 mins. after preparation. 488 nm excitation, 1 x 60 sec exposure. 100 mM in H_2O.

Enzyme Intermediates. We have access to a series of carboxy-biotin enzyme complexes where the protein has been modified by site selected mutagenesis [3]. This provides a series with a range of decarboxylation reactivities stretching over more than two orders of magnitude. Each member of the series is sufficiently stable to characterize by Raman and FTIR difference spectroscopies. Very high quality Raman spectra have been obtained for the biotinylated 1.3 subunit enzyme that demonstrate that it will be possible, using Raman difference spectroscopy, to observe the modes of the CO_2 molecule bound to biotin on the enzyme.

1. H. G. Wood and G. K. Kumar, Ann. N. Y. Acad. Sci. 447, 1 (1986).
2. J. R. Knowles, Ann. Rev. Biochem. 58, 195 (1989).
3. B. C. Shenoy, Y. Xie, V. L. Park, G. K. Kumar, H. Beegan, H. G. Wood and D. Samols, J. Biol. Chem. 267, 18407 (1992).

RAMAN STUDIES ON BENZOYL-Co-A COMPOUNDS BINDING TO 4-CHLOROBENZOYL-CoA DEHALOGENASE

John Clarkson[1] and Paul R. Carey*[1]
Kimberly L. Taylor[2] and Debra Dunaway-Mariano[2]
[1] Department of Biochemistry, Case Western Reserve University, 10900 Euclid Avenue, Cleveland, OH 44106-9435, USA.
[2] Department of Chemistry and Biochemistry, University of Maryland, College Park, Maryland 20742

4-Chlorobenzoyl-Coenzyme A dehalogenase catalyses the hydrolytic dehalogenation of 4-chlorobenzoyl coenzyme A (4-CBA-CoA) to 4-hydroxybenzoyl-CoA (4-HBA-CoA). This enzyme has been discovered in a number of soil dwelling bacteria where it functions along with 4-chlorobenzoate:CoA ligase and 4-HBA-CoA thioesterase in a 4-chlorobenzoate, (4-CBA), to 4-hydroxylbenzoate converting pathway. 4-CBA is not a natural product and is generated *in situ* as a byproduct of the oxidative biodegradation of 4-chlorobiphenyl, a ubiquitous, synthetic pollutant [1].

Raman difference spectroscopy has been used to investigate the nature of the electronic rearrangement of 4-HBA-CoA upon binding to dehalogenase [2] and has shown evidence for the formation of a quinone methide-like tautomer. The role of the C(4)-hydroxyl group of the benzene moeity in facilitating binding to the enzyme was examined by studying 4-methylbenzyl-CoA, a structural analog of the substrate 4-CBA-CoA for dehalogenase.

Upon binding of 4-methylbenzyl-CoA to dehalogenase the aromatic ring stretching modes (8a and 8b) observed at 1609 and 1581 cm^{-1} in buffer, are shifted to 1596 and 1570 cm^{-1} while the phenyl C-H bending modes seen at 1213 and 1182 cm^{-1} in buffer move to 1215 and 1278 cm^{-1}. The thioester C=O stretching band observed at 1651 cm^{-1} in buffer has shifted to 1610 cm^{-1} upon enzyme complexation. This data indicates that a perturbation of the benzoyl π electrons for 4-MeBA-CoA takes place, but is not as large as that observed for 4-HBA-CoA [2]. However strong electron polarization is observed localized at the thioester C=O oxygen. The 39 cm^{-1} shift in the thioester carbonyl frequency is thought to be due to strong hydrogen bonding at the active site of dehalogenase with additional effects due to the protein's α-helix dipoles being likely. The strength of this hydrogen bonding is being quantitated by FTIR studies of a simple thioester model compound of 4-MeBA-CoA, with hydrogen donors, to provide thermodynamic data. Thus, a quantitative measure of the strength of the hydrogen bonding at the thioester carbonyl of 4-MeBA-CoA in the active site will be obtained.

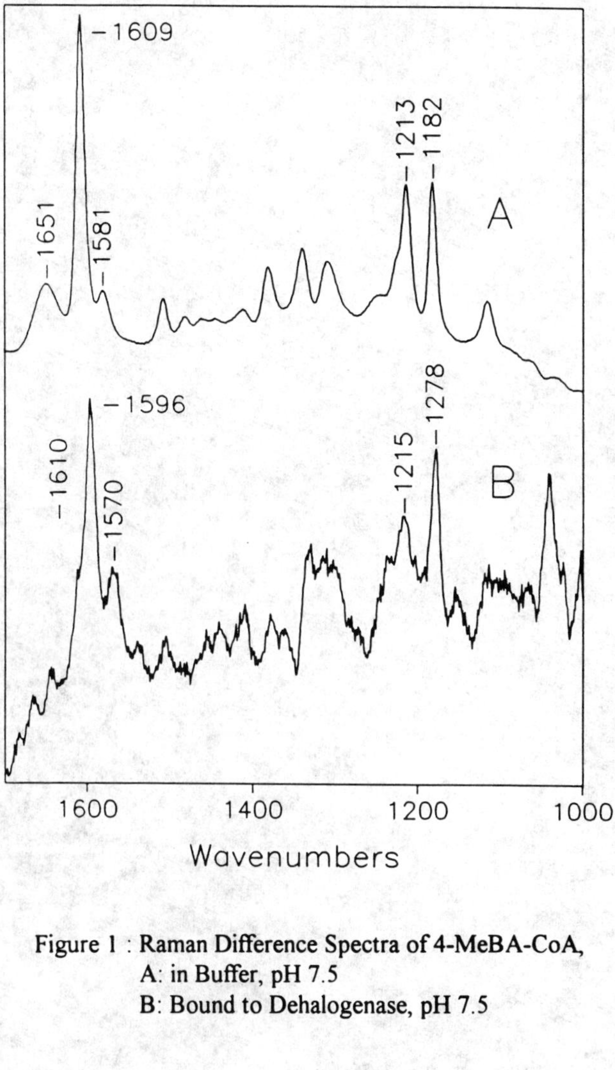

Figure 1 : Raman Difference Spectra of 4-MeBA-CoA,
A: in Buffer, pH 7.5
B: Bound to Dehalogenase, pH 7.5

1. G. Yang, P.-H. Liang and D. Dunaway-Mariano, *Biochemistry*, 33, 8527, 1994
2. K. L. Taylor, L. Rui-Qin, P.-H. Liang J. Price, D. Dunaway-Mariano, P. J. Tonge, J. Clarkson and P. R. Carey, *Biochemistry*, 34, 13881, 1995

SECTION 10

HEME PROTEINS

RESONANCE RAMAN SPECTROSCOPY OF PROTEIN DYNAMICS

T. Kitagawa,[1]* S. Nakashima,[1] Y. Mizutani,[1] M. Mukai,[1] and J. S. Olson[2]

[1]Institute for Molecular Science, Okazaki National Research Institute, Myodaiji, Okazaki, 444 Japan
[2]Department of Biochemistry and Cell Biology, Rice University, Houston, Texas 77251 U.S.A.

Proteins retain a specific three-dimensional structure as an average of time but simultaneously exhibit considerable flexibility in a short time scale. Good examples of such features are seen when a substrate is bound to an enzyme. We have been interested in elucidating rapid structural changes of proteins for performing their functions and applied time-resolved resonance Raman spectroscopy to investigate it [1]. Myoglobin (Mb), an oxygen binding protein, is one of highly characterized proteins and has served as a model molecule for studying the dynamics of proteins [2,3], since its reaction is merely binding of a substrate (O_2). Here we present the results obtained from picosecond and nanosecond pump/probe time-resolved resonance Raman (TR^3) experiments on photodissociation and recombination of carbonmonoxyMb (COMb) and its mutants.

The ps TR^3 experiments were carried out with the laser system based on a mode-locked Ti-sapphire oscillator and a regenerative amplifier operated at 1 kHz. The probe light was obtained from the second harmonic through a frequency conversion using a CH_4 Raman shifter, while the pump light was obtained through optical parametric generation by LBO crystal and optical parametric amplification by BBO crystal. The correlation width between the pump pulse (540 nm) and probe pulse (435 nm) was 1.8 ps. The ns visible TR^3 experiments were carried out with two Nd:YAG lasers ($\Delta\tau$ = 7 ns), which provide the pump- (532 nm, 4 mJ) and probe pulses (416 nm, 100 μJ), a triple polychromator equipped with an intensified photodiode array, and a slow spinning cell (25 rpm). The delay time (Δt) of the probe pulse from the pump pulse was varied between -20 ns and 1 ms, and the spectra for different Δt's were measured quasi-simultaneously [4]. In the UV TR^3 experiments, the second harmonic was generated from the XeCl excimer laser-pumped dye-laser output (470 nm) with BBO crystal. The delay times were regulated by a pulse generator and monitored with a photodiode and an oscilloscope.

In x-ray crystallographic analysis [5], it has been noted that there is no pathway for migration of CO between the buried binding site and the solvent. Therefore, rapid rearrangements of the protein structures should be accompanied by the release and recombination of CO. Upon photodissociation of CO, the photon energy absorbed by heme is partially converted to the kinetic energy of photodissociated CO but the most part is converted to vibrtional energy of heme, and then dissipates to the surrounding (protein and water) [6]. Raise of the temperature of heme by photodissociation of CO became evident from the increased anti-Stokes Raman intensity, and the transient temperatures were determined from the Stokes/anti-Stokes Raman intensity ratios of the ν_3, ν_4, and ν_7 bands. The temperature determined by the ν_3 mode was higher than those of others, while all modes exhibited higher temperatures than that expected from the equipartition law. The vibrational relaxation time determined from the time profile of anti-Stokes Raman intensity of ν_4 was ~2 ps but that of ν_3 was longer than it.

The Fe-CO stretching band ($\nu_{Fe\text{-}CO}$) of COMb sensitively reflects the CO environments in the heme pocket. Accordingly, we pursued the time-profile of this band with the ns TR³ system to monitor the structural change of the distal side of the heme pocket. In the resting state of COMb, the $\nu_{Fe\text{-}CO}$ Raman band was observed between 526 and 490 cm^{-1} depending on the residue at 68th, 64th, and 29th positions. It is generally seen that the mutants which give the $\nu_{Fe\text{-}CO}$ band around 490-495 cm^{-1} (H64Y, H64W, H64L, H64A, H64G, and H64YV68F) and at 520-526 cm^{-1} (H64WL29F, and L29F) recombined with CO much faster and slower, respectively, than the WT (wild type) with $\nu_{Fe\text{-}CO}$ at 508 cm^{-1}. Therefore, the presence of some relationship between the static structure of heme pocket and dynamics is clear.

Here we want to stress that in common to all the samples a band appears at a frequency different from that of the equilibrium state in the initial stage of recombination. For an example, WT gave it at 503 cm^{-1} for Δt = 20 ns - 100 ms, and H64Y gave it at 499 cm^{-1} for Δt = 0 ns - 100 ms. In all cases the transient band appeared at 497±2 cm^{-1} irrespective of the resting $\nu_{Fe\text{-}CO}$ frequency, and it did not exhibit the CO-isotopic frequency shift. Therefore, the spectral pattern was different from that of the equilibrium deoxy state. These observations suggest that the heme environments of all the mutants are very alike in the initial stage of rebinding and might be related with a water molecule present in the heme pocket of the equilibrium deoxy form.

In order to elucidate the transient heme pocket structure, UV TR³ experiments were carried out for H64Y and H64W to pursue the movements of the 64th residue. The WT Mb has 3 Tyr and 2 Trp residues and UVRR active bands of WT Mb did not show any detectable frequency and intensity changes during the photodissociation and recombination process. In comparison of H64Y with WT, intensity increases of the Y8a (1618 cm^{-1}) and Y9a (1179 cm^{-1}) bands were recognized in the resting state, and the same feature was observed for the W3 band (1556 cm^{-1}) in H64W. Temporal behaviors of the increased portion of the intensity will be discussed in the talk.

1. T. Kitagawa and T. Ogura, Adv. Spectrosc. (R. Clark and R. Hester, Eds) 21, Part B, 139-188 (1993).
2. H. Frauenfelder, S. G. Sligar, and P. G. Wolynes, Science, 254, 1598 (1991).
3. W. D. Tian, J. T. Sage, V. Srajer, and P. M. Champion, Phys. Rev. Lett. 68, 408 (1992)
4. S. Sato, K. Kamogawa, K. Aoyagi, and T. Kitagawa, J. Phys. Chem. 96, 10676 (1992).
5. J. Kuriyan, S. Wilz, M. Karplus, G. A. Petsko, J. Mol. Biol. 192, 133 (1986).
6. P. A. Anfinrud, C. Han, and R. M. Hochstrasser, Proc.Natl. Acad. Sci. U.S.A. 86, 8387 (1989).

STRUCTURE, FUNCTION AND DYNAMICS IN HEMOGLOBINS AND MYOGLOBINS: TIME RESOLVED AND CRYOGENIC RESONANCE RAMAN STUDIES.

J. M. Friedman and E. Peterson.

Department of Physiology and Biophysics, Albert Einstein College of Medicine, Bronx, N.Y. 10461.

Time resolved and cryogenic Raman studies in conjunction with geminate rebinding and near IR absorption measurements on several hemoglobin and myoglobin systems are used both to identify functionally relevant degrees of freedom and expose protein specific relaxation pathways. Several mutant human Hbs are used to directly expose the relationship between the iron-proximal histidine stretching mode and the barrier controlling ligand-iron formation. Proximal mutants of Mb and dimeric clam Hbs are used to expose the hydrogen bonding between the heme propionates and surrounding residues. Light induced relaxation studies at cryogenic temperatures and Raman studies of trehalose glass embedded samples are used to dissect out the hierarchy of relaxations and their temperature and viscosity dependence. The following is a partial overview of some of these results.

HbCO in trehalose. Upon photodissociation COHb exhibits a wide range of nonequilibrium relaxation phenomena starting within some fraction of a picosecond and extending out to tens of microseconds. In addition equilibrium fluctuations result in conformational averaging. All of these dynamics can have an impact on ligand rebinding. In an effort to better understand the relationship between conformational dynamics and ligand binding reactivity, we have embedded COHbA in a room temperature trehalose sugar glass [Hagen et al., *Science* **269**, 959 (1995)] in order to decouple solvent dependent and solvent independent dynamics as well as reducing the amplitude for large scale conformational fluctuations. Ligand rebinding kinetics are followed as a function of temperature (60-300 K) and compared to those observed in water-glycerol solvents (50-95% glycerol:water). Time resolved resonance Raman is used to determine the effect of the trehalose glass on the fast relaxation of the heme iron-histidine unit. The results indicate that trehalose does not impede the elastic like rapid displacement of the the iron from the heme plane. In addition, sharpening and shifting of the propionate sensitive Raman bands indicate that the trehalose is strongly interacting with the propionates and as a result imobilizing them. The same behavior is observed at cryogenic temperatures.

MbCO. Raman spectra were generated of the deoxy and the 10 ns photoproduct of wild type COMb(SW) and several mutants designed to expose the effects of systematically disrupting the hydrogen bonding network of the proximal heme pocket (in collaboration with S. Sligar and E. Chen). The following results were obtained:
a. Progrssive disruption of the hydrogen bonding to the heme-7-propionate through replacement of His FG3, Ser F7 and Leu F3 is manifested by a progressive shift of the 372 cm-1 Raman band to lower frequency. The maximum result produces a frequency of 364 cm-1, which the value observed in HbA. Both deoxy and photoproduct show this effect. This Raman band arise from a vibrational mode that is known to have a substantial propionate in plane vibrational motion. These observations along with the parallel

observations on the dimeric clam Hb s, represent the first unambiguous experimental connecton between hydrogen bond strength of the propionate and the Raman frequency.
b. As the frequency of the propionate marker band shifts to below 369 cm-1, the out of plane pyrrole tilting mode at 240 cm-1 dissappears. This correlation extends to every 5-coordinate ferrous hemeprotein we have examined. Champion claims that this mode becomes Raman active when the reflection symmetry of the porphyrin is broken. Conceivably, a strong hydrogen bond to a propionate is also associated with having the propionate locked into a bent out of the plane of the heme orientation. A weak hydrogen bond might therefore be associated with a disordered or weakly ordered propionate that is occupying on average the in plane orientation . It is likely that the 364 cm-1 value corresponds to the propionate interacting with water molecules from the hydration shell of the protein.
c. Reduction of the pH to ~ 5 for the His FG3 Phe mutant results in the partial reversal of the frequency decrease for the propionate marker band which occurs due to the loss of the hydrogen bond from the FG3 His. A possible explanation is that the heme-7-propionate is becoming partially protonated at this pH. The presence of the nonpolar phenyl ring is likely to increase the pKa for the nearby propionate.
d. Disruption of the hydrogen bonds from either Leu F3 or Ser F7 to the N of the His F8, results in an increase in the frequency of (Fe-His). Weakening of the hydrogen by itself should decrease the frequency of (Fe-His). The increase is likely due to a slight rotation of the imidazole to a less eclipsed orientation with respect to the N1-Fe-N3 axis of the heme. Such a rotation is likely to reduce the repulsive interaction between the imidazole carbons and the pyrrole nitrogens and thus result in a reduction in the iron displacement which causes the frequency to increase (Stavrov).
e. We have compared Mb(HH) and Mb(yellow fin tuna) with respect to their low frequency Raman spectra. Tuna Mb is known to have a much looser heme pocket based on NMR studies (G. Lemar, private communication). We observe that (Fe-His) is unchanged but the propionate band is at lower frequency for the tuna Mb (369 versus 372 cm-1) and the 240 cm-1 band is nearly gone. The lower frequency and reduced 240 cm-1 band are consistent with a weaker hydrogen bond to the propionates. A weakening of the hydrogen bond should result in a more mobile heme as seen from the NMR studies.
f. We have compared the photoproduct and deoxy forms of HH, SW and tuna Mb. The frequency of (Fe-His).is the same in the deoxy and 10 ns photoproduct of COMb. Small differences are observed in the propionate sensitive modes.
g. At high glycerol concentrations, the frequency of (Fe-His). for deoxy Mb decreases by ~ 2 cm-1 as also observed by Champion and coworkers. For Mb*(10 ns) the frequency for (Fe-His) is at its usual value. Thus at low glycerol or in aqueous buffer, (Fe-His) is the same for deoxy and Mb*(10 ns) whereas at high glycerol they are different due to a shift for the deoxy species. We believe that the shift in the deoxy could be due to the pulling out of water from the heme pocket by the high osmotic strength of the glycerol. Thus at least some of the glycerol dependent differences in the time the evolution of Mb* spectra could be due to the time course for water to enter the heme pocket subsequent to dissociation of CO. The lower frequency (Fe-His) could be a consequence of an empty distal pocket that causes either an actual conformational perturbation or a polarity induced effect. If these and other spectral changes are due to CO escaping and water reentering or not reentering the distal heme pocket, the relaxations should be very sensitive to the architecture and polarity of the distal pocket.

UV RESONANCE RAMAN SPECTROSCOPY STUDIES OF D_4-TYROSINE LABELED HEMOGLOBIN

Xuehua Hu and Thomas G. Spiro*

Department of Chemistry, Princeton University, Princeton, NJ 08544

Hemoglobin (Hb) is an allosteric protein showing cooperativity in oxygen binding. Even though an enormous amount of research has been carried out on its structure and function, the elementary molecular mechanism of cooperativity is still not well understood. The elucidation of a structural mechanism of allostery has been a subject of extensive research. Recent developments in Raman spectroscopy make it possible to probe the backbone and aromatic side chains via ultraviolet resonance Raman spectroscopy (UVRR) and a great deal has been learned about the nature of the transition between the states of low and high affinity, T and R, respectively. Recently we reported the T/R difference UVRR features for both Tyrosine and Tryptophan residues of deoxyHb and its carbonmonoxide adduct, recorded with 230-nm excitation [1]. These UVRR markers of R/T transition were believed to arise from conformational changes at the $\alpha 1\beta 2$ and $\alpha 2\beta 1$ interfaces, and the strong T/R difference signal for the Y8a was tentatively assigned to Tyr α-42. In this report we present the UVRR spectra of the d_4-Tyrosine labeled Hb.

Recombinant human hemoglobin (rHb) was expressed in *E. Coli* using plasmid pHE2 [2], kindly supplied by Professor Chien Ho. The rHb of natural abundance exhibits the same UV Resonance Raman features in both its deoxy and ligated carbonmonoxide forms as those of the human adult hemoglobin (HbA). The expression of the d_4-Tyrosine labeled recombinant hemoglobin (d_4Y-rHb) was made possible by using the *E. Coli* strain DL39, which is auxotrophic for Tyrosine and other amino acids. Since the H/D substitution shifts the Y9a and Y8a/Y8a' bands appreciably, the UVRR studies of this isotope labeled hemoglobin reveal some very important features. First we are now able to show that the Trp W1 band, which has very low intensity in the parent spectra, does not contribute to the UVRR difference signal in the 1615 cm^{-1} band. This supports the conclusion that the strong 1615 cm^{-1} difference signal is primarily due to the changes of Tyr environments [1].

Secondly, after the d_4-Tyrosine labeled α or β chains were separated and then recombined with complementary nonlabeled chains, the two chains were found to behave quite differently. There are a total of 12 Tyr residues in HbA, with 3 Tyr per α subunit and β subunit. While the α Y9a band upshifts about 1.6 cm^{-1} during the R/T transition, the β Y9a down shifts about 0.4 cm^{-1}. They together contribute to the about 1.2 cm^{-1} upshift in the natural abundance tetramer, as observed in HbA.

Thirdly, from the above experiments we also found that the strong T/R difference signal for the Y8a arises from the α subunit. The T/R difference signal is isotopically down shifted when the α subunit is labeled with d_4-Tyrosine but the labeling of β subunit with d_4-Tyrosine has no effect on the Y8a difference signal. This doesn't resolve the six inequivalent Tyr residues. It is, however, a strong evidence for the Tyr α-42 assignment. Of the three α-Tyr residues, Tyr α-24 is an interior residue and is not involved in hydrogen bonding. While Tyr α-140 experiences environmental changes during the R/T transition, it is actually at the same position as the other corresponding penultimate residue Tyr β-145. These two residues are expected to behave similarly. Therefore, it seems quite unlikely that the Tyr β-145 does not contribute to the strong T/R difference signal for the Y8a while Tyr α-140 does.

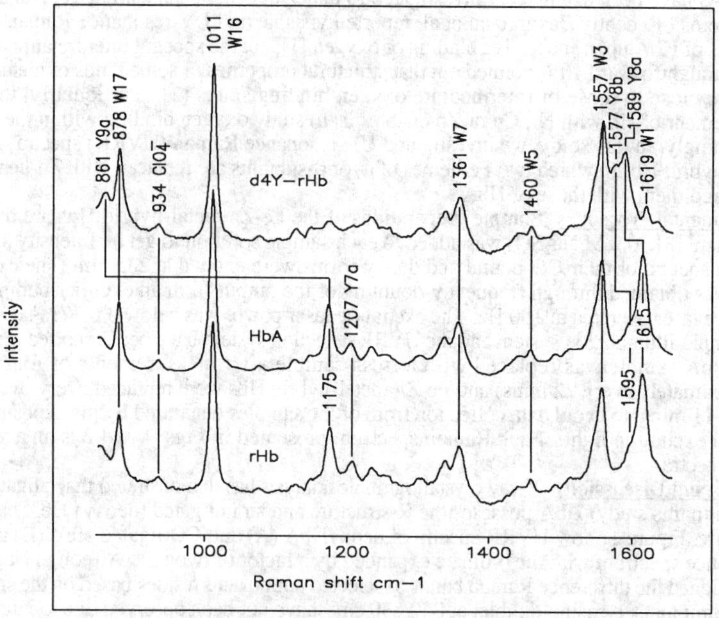

Fig. 1 230-nm excited UVRR spectra of d_4Y-rHb, HbA and rHb in their CO forms.

1. K.R. Rodger, C. Su, S. Subramaniam and T.G. Spiro, J. Am. Chem. Soc. 114, 1697(1992).
2. T-J Shen, N.T. Ho, V. Simplaneaun, M. Zou, B.N. Green, M.F. Tam and C. Ho, Proc. Natl. Acad. Sci. U.S.A. 90, 8108(1993).

ULTRAVIOLET RESONANCE RAMAN STUDY OF Fe-Zn METAL-HYBRID HEMOGLOBINS AS A T-STATE MODEL

Shoji Kaminaka[1], Antonio Tsuneshige[2], Takashi Yonetani[2], and Teizo Kitagawa[3]*

[1]Department of Virology, Kurume University School of Medicine, 67 Asahimachi, Kurume 830, Japan
[2]Department of Biochemistry and Biophysics, University of Pennsylvania School of Medicine, Philadelphia, PA 19104-6089, USA
[3]Institute for Molecular Science, Okazaki National Research Institutes, Okazaki 444, Japan

Structural changes involved in cooperative oxygen binding of human hemoglobin A (HbA) have been extensively investigated to understand a mechanism of general allostery of proteins. Recently, Jayaraman et al. reported visisble and UV resonance Raman spectral changes of HbA upon successive binding of oxygen [1], but the spectral interpretation may not be so straightforward. It is pointed out that structural properties of some kinds of metal-hybrid Hbs are close to those of intermediate oxygen binding states [2]. Particularly, the metal replacement of Hb with Ni, Cu or Zn enables us to study oxygen binding within the T-state. Accordingly, in this study, we investigated UV-resonance Raman (UVRR) spectra of Fe-Zn metal hybrid Hbs in which two Fe-hemes of a- or β-subunits are replaced with Zn-hemes, and compared them with those of HbA.

Experimental Procedues Sample preparations of the Fe-Zn metal-hybrid Hbs are described elsewhere [3]. 0.2 M Na_2SO_4 was added to each sample solution to get an intensity standard. UVRR spectra of their CO-bound and deoxy-forms were excited at 235 nm. The excitation light was obtained through frequency-doubling of the output from an excimer-pumped (308 nm) dye laser operated at 100 Hz. The excitation laser power was below 3 mJ/cm^2. Details of the sample-illumination system and the UVRR detection system have been reported previously [4]. HbA sample was replaced with a fresh sample after measurements of four spectra (approximately every 22 mins) and Fe-Zn metal-hybrid Hbs were replaced every two spectra (about 11 mins) to avoid data collection from of the samples denatured by the exposure to the Raman excitation light. Each Raman spectrum presented in Figs. 1 and 2 is an average of eight spectra.

Results and Discussion X-ray crystallographic analysis has demonstrated that a ligated (CO-ligated in this study) HbA posseses the R-structure and an unligated (deoxy) HbA has the T-structure. Figure 1 show UVRR spectra of deoxyHbA (A) and CO-ligated HbA (B) and their difference spectrum with the ordinate expanded by a factor of two (C). Although Nagai et al. [5] assigned the difference Raman bands to selected amino acid resides based on the spectra of several mutants, tyrosine-doublet at 850/830 cm^{-1} have not been observed due to interruption by pre-resonanced strong and broad Raman scattering from a quartz cell. The use of the novel spinning cell system reported previously [6], decreased its interference significantly and made it possible to observe the Tyr doublet. The difference spectrum (Fig. 1C) indicates that the 829 cm^{-1} band is slightly more intensity enhanced in the deoxy form than the 852 cm^{-1} band, although both bands are stronger in the deoxy than in the CO-ligated forms. This means that phenolic hydroxyl group of some Tyr residues have stronger hydrogen bonds in the deoxy than CO-ligated forms. Nagai et al. attributed the ligation-induced frequency shifts of Y_{8a} and Y_{9a} of HbA to the H-bond formation of α42-Tyr in the T-structure from UVRR studies with α42 mutant [7]. Considering this, the spectral change of Tyr-doublet in Figure 1(C) would also be ascribed to H-bond formation of α42-Tyr.

Figure 2 shows UVRR spectra of deoxy HbA minus CO-bound (A and B) and deoxy Fe-Zn metal-hybrid Hbs (C and D). As CO molecules bind only to the Fe subunits in the metal-hybrid HbA, the CO-bound form corresponds to the half-ligated form of normal

HbA. The bands at 979 cm^{-1} in Fig. 2(A) and (B) arise from sulfate ions as an intensity standard, meaning mismatch of its relative concentrations between the HbA (400 μM/heme) and metal-hybrid (800 μM/heme) preparations. It is apparent from Fig. 2(A) and (B) that the UVRR spectra of $COHb(\alpha Fe\beta Zn)_2$ and $COHb(\alpha Zn\beta Fe)_2$ are alike and both exhibit close resemblance to that of deoxy HbA. This means that the half-ligated Zn-Fe metal-hybrid Hbs adopt the ordinary T quaternary structure. On the other hand, UVRR spectra of deoxy $Hb(\alpha Fe\beta Zn)_2$ and deoxy $Hb(\alpha Zn\beta Fe)_2$ have different features from that of deoxy HbA. In case of deoxy $Hb(\alpha Zn\beta Fe)_2$ (C), Trp bands at 1621, 1559 and 1357 cm^{-1} decreased in intensity compared with that of deoxy HbA. The Trp band at 1559 cm^{-1} (W3) could be attributed to α14- and β15-Trps due to its frequency. This feature may suggest that replacement of α-heme Fe with Zn changes a tertiary structure of Hb but its quaternary structure remains unchanged. On the contrary, in the case of deoxy $Hb(\alpha Fe\beta Zn)_2$ (D), some Trp bands (1547, 1357, 1009, and 875 cm^{-1}) and Tyr bands (1610, 1203, 1174, and 850 cm^{-1}) increase intensities compared with those in deoxyHbA. From the frequency of W3 band (1547 cm^{-1}), the Trp bands would be ascribed to β37-residue which is known to be responsible for a significant part of the quaternary structure change. It is noted that the UVRR difference spectrum (not shown) of deoxy HbA minus ZnHb in which all of four subunits have Zn-hemes, was almost the same as the spectrum obtained by addition of Figure 2(C) and (D). These results indicate that the effects of the metal substitution from Fe to Zn on protein structures are different between the α and β subunits, although all belong to the so-called T structure.

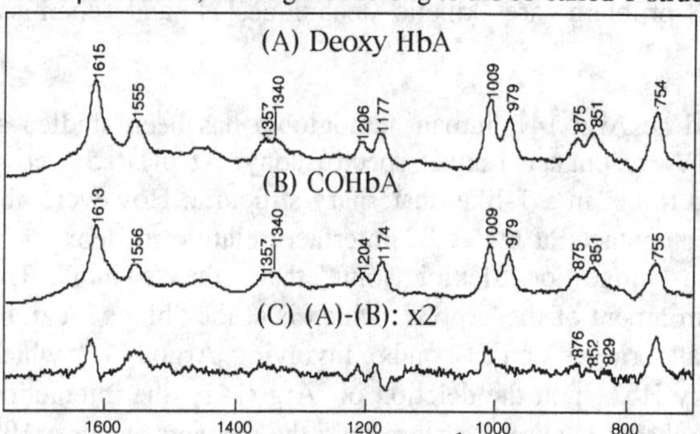

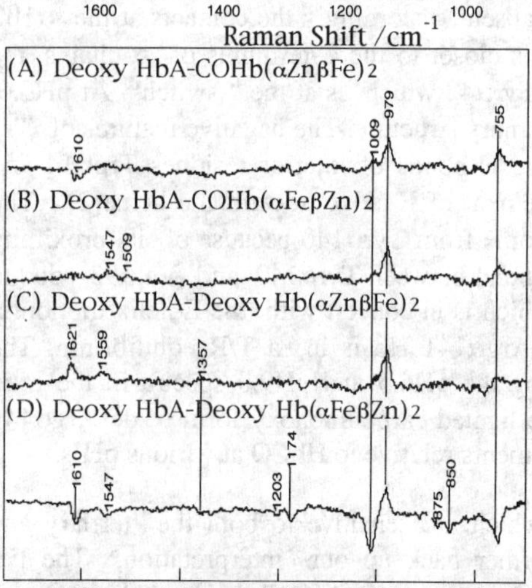

Figure 1.(Left, Upper) 235-nm excited UVRR spectra of deoxy HbA (A), CO-ligated HbA (B) and its difference (= deoxy-form - CO-form) (C).

Figure 2.(Left, Lower) 235-nm excited UVRR difference spectra of deoxy HbA minus CO-ligated and deoxy Fe-Zn metal-hybrid Hbs; (A) deoxy HbA - $COHb(\alpha Zn\beta Fe)_2$, (B) deoxyHbA - $COHb(\alpha Fe\beta Zn)_2$, (C) deoxy HbA - deoxy $Hb(\alpha Zn\beta Fe)_2$, (D) deoxy HbA - deoxy $Hb(\alpha Fe\beta Zn)_2$

References

1. V. Jayaraman, K. R. Rogers, I. Mukerji, and T.G. Spiro Scinence 269,1843 (1995)
2. I. Mukerji, and T.G. Spiro Biochemistry 33, 13132 (1994)
3. A.Tsuneshige, and T.Yonetani manuscript in preparation
4. S. Kaminaka, and T. Kitagawa Appl.Spectrosc. 46, 1804 (1992)
5. M. Nagai, S. Kaminaka, Y. Ohba, Y. Nagai, Y. Mizutani, and T. Kitagawa J. Biol. Chem. 270, 1636 (1995)
6. S. Kaminaka, and T. Kitagawa Appl. Spectrosc. 49, 685 (1995)
7. M. Nagai, K. Imai, S. Kaminaka, Y. Mizutani, and T. Kitagawa J.Mol. Struct. in press (1996)

RESONANCE RAMAN STUDIES OF DESARGα141 HEMOGLOBIN

D. Wang, V. Jayaraman, G. Heibel & T. G. Spiro*

Department of Chemistry, Princeton University, Princeton, NJ 08544

There is enduring challenge in the molecular mechanism of cooperativity in hemoglobin, as a paradigm of protein dynamics and allostery. The classical two-state model(MWC) continues to provide a productive explanation for Hb functions. In this model, hemoglobin switches between a low-affinity quaternary state T and a high-affinity quaternary state R. Because cooperativity insures that most Hb molecules are either unligated or fully ligated, intermediate states cannot readily be accessed in native Hb. Two ways around this problem are kinetic techniques[1] and chemical modifications of native Hb.

Chemically modified desArgα141 human hemoglobin has been studied at various pH conditions using UV resonance Raman spectroscopy. At pH 6.5, deoxy desArgα141 Hb has been found to be in a T-like quaternary structure. However, the UVRR spectra indicate weaker contacts at the $\alpha1\beta2$ interface relative to deoxy Hb, with larger perturbation at the "hinge" or "flexible joint" than the "switch". The marked perturbation in the environment of the Trpβ37 which is at the "hinge" can be attributed to the rupture of salt bridges and H-bonds involving Argα141 , which stabilize the T state in the deoxy Hb. Upon the deletion of Argα141, the interaction network at the C-terminus of α chains is absent; accordingly the contacts at the $\alpha1\beta2$ interface are weakened. Because Trpβ37 is closer to the C-terminus of α chains, the perturbation at the Trpβ37 is larger than Tyrα42 which is at the " switch". At pH9.0, deoxy desArgα141 Hb exists in a R quaternary structure. The negative features of Trp bands indicate tertiary perturbations. The H-bonds of interior residues Trpα14 and Trpβ15 are weakened due to the displacement of E-helix toward the heme[2]. The negative peaks of Tyr bands probably come from Tyrα140 because of its proximity to Argα141. This implies the weaker H-bond between Tyrα140 and Valα93 due to the F-helix translation across the heme, which is in concert with the E-helix motion in the R-deoxy state. At pH7.4, deoxy desArgα141 Hb is in a T/R equilibrium. The spectrum can be reproduced by co-adding the pH6.5 and pH9.0 spectra in the ratio 0.58/0.42. In contrast, the structure of the ligated carbonmonoxy form of desArgα141 Hb exhibits the same Trp and Tyr environments relative to HbCO at various pH's.

The Visible-RR spectra which are sensitive to both the tertiary and quaternary structures of hemoglobin further back up our interpretation. The Fe-N_ε(His, F8) stretching frequencies show deoxy desArgα141 Hb is in a T-like state at pH6.5, a R state at pH9.0, at pH7.4, it is intermediate, but more T-like.

The large perturbations at the α1β2 interface upon the deletion of Arg141 can account for the lower dimer-tetramer association free energy, higher oxygen affinity and lower Bohr effect for deoxy desArgα141 Hb relative to deoxyHb. This indicates the intricate network in the structure of Hb wherein an absence of one single residue causes a perturbation in the structure far from the modification site.

Several conclusions can be drawn from this study. 1. The H-bonds of Tyrα42 and Trpβ37 at the α1β2 interface are essential in stabilizing the T state in hemoglobin. The salt bridges and hydrogen bonds in the C-termini also play an important role in stabilizing the T state. 2. Unligated desArgα141 Hb can exist in T-like and R quaternary states depending on the pH. This reveals the flexibility in the Hb structure. 3. UVRR spectroscopy is proved to be useful in elucidating the quaternary and tertiary structures of hemoglobin. In UVRR difference spectra between Deoxy and CO forms, W3 features at 1550cm^{-1} come mainly from Trpβ37, while those at 1558cm^{-1} from Trpα14 and Trpβ15[3]. The Tyr features come mainly from Tyrα42 while contributions from Tyrα140 and Tyrα145 can't be ignored. 4. DesArgα141 hemoglobin can serve as a model for the study of intermediate states in the hemoglobin cooperativity.

1. V. Jayaraman, K. R. Rodgers, I. Mukerji and T. G.Spiro ***science*** 269, 1843(1995)
2. K. R. Rodgers and T. G. Spiro ***Science*** 265, 1697(1994)
3. K. R. Rodgers , C. Su, S. Subramaniam and T. G. Spiro ***J. Am. Chem. Soc.*** 114, 3697(1992)

RESONANCE RAMAN STUDIES OF AN UNUSUAL HEMOGLOBIN (HbI) FROM *Lucina pectinata.*

Y. Echevarria, J. Cerda, J. Colon, and J. Lopez-Garriga*

*Chemistry Department University of Puerto Rico at Mayagüez, Mayagüez, P.R. 00680,USA. E-mail address: J_lopez@rumac.upr.clu.edu

Three different hemoglobins (HbI, HbII and HbIII) have been isolated from *Lucina pectinata* clam which inhabits in sulfide-rich coastal sediments. Like human, sperm whale and horse myoglobins, *Lucina pectinata* has a histidine in the proximal position to the heme, however, the distal position E7, is a glutamine residue as in elephant myoglobin. HbI, the monomeric sulfide reactive specie, has in the Fe^{+3} oxidation state, an extraordinary affinity for sulfide with fast combination and very slow dissociation dynamics. The three hemoglobins present unusual dioxygen and carbon monoxide kinetics.[1,2] These properties may be related to the low distal pocket polarity and the four phenylalanyl residues which do not favor the presence of water molecules in the heme pocket.

Our FTIR spectroscopic measurements [5] for the HbICO complex showed, at room temperature, the presence of two main heme-CO conformers. For these species, the ν_{CO} frequency occurs at 1962 cm^{-1} and 1938 cm^{-1}, respectively. These results are in contrast with the ν_{CO} frequencies for the bovine hemoglobin-CO complex which present a ν_{CO} frequencies at 1970 cm^{-1} and 1952 cm^{-1}, respectively. The high intensity band of the complex is 6 cm^{-1} lower in energy than the 1944cm^{-1} found in sperm whale. Since distal histidine in heme proteins has been assigned the role of modulating the ligand reactivity towards the heme centers, it is not clear why the natural E7His→Gln substitution in HbI-CO complex showed lower ν_{CO} frequencies. One can expect in terms of removal of the histidine steric constrains the opposite effect.

Time resolved infrared spectroscopy was used to follow the 1939 cm^{-1} conformer of the HbICO moiety. The results showed that FeCO recombination time has half-life of 25 μs. The rate constant for the recombination between HbI and CO was 8.33 $\mu M^{-1}s^{-1}$. This value is higher than the rate constant reported for similar species and mutants.[4] The HbICO association rate constant is in the order of magnitude of the recombination reported for the triple mutant of 3.1$\mu M^{-1}s^{-1}$ using flash photolysis techniques. Nevertheless, a reported value of 0.78 $\mu M^{-1}s^{-1}$ was determined previously for HbI using the same technique. This suggests that the functionality of the Gln64(E7), Phe29(B10), and Phe68(E11) in HbI increases the

CO association rate relative to SM-Mb which lacks this aromatic environment in the heme pocket. Therefore, it seems that the aromatic nature of the heme pocket in HbI is the main factor contributing to the observed ligand dynamics. To explore the effect of the heme-pocket aromatic environment in the HbI chromophore structure in more detail, we carried out resonance Raman studies of HbI ligand complexes.

Our measurements show that for the aquoHbI species, there are resonance Raman bands appearing at 1622 cm^{-1}, 1560 cm^{-1}, 1485 cm^{-1}, 1426 cm^{-1}, 1372 cm^{-1}, 1128 cm^{-1}, 993 cm^{-1}, 672 cm^{-1}, 488 cm^{-1}, 408 cm^{-1}, 341 cm^{-1}, and 272 cm^{-1}. It is known that for metmyoglobin[5] the vibrational modes at 1622 cm^{-1}, 1560 cm^{-1}, and 1372 cm^{-1} correspond to the $\nu_{C=C}$ of the vinyl peripheral group, to the iron six coordinate high spin species, and to Fe^{+3} oxidation state, respectively. In our aquoHbI species, we observed these characteristics marker bands at 1622 cm^{-1}, 1560 cm^{-1}, and 1372 cm^{-1}, respectively. Similarly, for this species, the ν_9 and ν_8 vibrational modes are at the same frequencies of metmyoglobin. Our results suggest that the aromatic nature of the heme pocket in HbI, that contributes to the observed ligand dynamics, does not affect the heme oxidation state, spin state and vinyl vibrational structure of the HbI heme chromophore. In particular, Gln64(E7), Phe29(B10), and Phe68(E11) amino acid substitution in HbI, do not affect the chromophore vibrational structure. (Supported in part by MRCE-NSF and RIMI-NSF, HRD-9550705).

References.

1. E. Casale, C. Lionetti, A. Coda, A. Merli, P. Ascenzi, J.B. Wittenberg, and M. Bolognesi, *J. Mol.Biol.* 222, 447 (1991).
2. M. Rizzi, J.B. Wittenberg, A. Coda, M. Fasano, P. Ascenzi, and M. Bolognesi, *J. Mol. Biol.* 244, 86 (1994).
3. J. Lopez-Garriga, J. Colon, A. Navarro, and M. Maldonado, *J. Inor. Biochem.* 59, 447 (1995).
4. X. Zhao, K.Vyas, B.D. Nguyen, K. Rajarathnam, K., G.N. La Mar, T. Li, G.N. Phillips, R.F. Eich, J.S. Olson, J. Ling, and D. J. Bocian, *J. Biol. Chem.* 270, 20774 (1995).
5. D. Biram, and R.E. Hester, *Biochim. Biophys. Acta* 1204, 204 (1994).

EFFECT OF THE PROTEIN ELECTRIC FIELD ON THE VIBRATIONAL PROPERTIES OF THE FeCO UNIT OF CARBONYLHEME PROTEINS: THEORETICAL APPROACH

BORIS KUSHKULEY* and SOLOMON S. STAVROV

Sackler Institute of Molecular Medicine, Sackler School of Medicine, Tel Aviv University, Ramat Aviv, Tel Aviv 69978, Israel, P.O.B. 39040

Vibronic Theory of Activation (VTA) [1,2] and quantum chemical Intermediate Neglect of Differential Overlap Restricted Hartree-Fock (INDO-RHF) MO LCAO [3] calculations are applied [4] to study the activation of carbon monoxide (change of the C-O bond index and force field constant) by the heme-imidazole complex (Fe(P)Im) in dependence on the distortion of the porphyrin ring, geometry of the CO coordination, iron-carbon and iron-imidazole distances, iron displacement out of the porphyrin plane and presence of the charged groups in the heme environment. It is shown that a major contribution to the CO activation arises from the electron density transfer from the 5σ orbital of CO to the heme (σ donation) and from the heme to the 2π* orbital of CO (back-bonding).

Investigation of the effect of the different distortions of the porphyrin ring showed that the ruffling and doming of the porphyrin ring, the iron displacement out of the porphyrin plane and change of the iron-imidazole distance of reasonable magnitudes hardly change the C-O vibrational frequency (ν_{CO}) and, therefore, cannot by themselves explain wide variation of the ν_{CO} in different heme proteins (HPs) and their models.

Distortion of the carbon monoxide linear perpendicular coordination geometry or the change of the iron-carbon distance affect both σ donation and back-bonding and, consequently, can notably affect both ν_{CO} and ν_{FeC}. However, by themselves these changes in the heme geometry can explain neither wide distribution of ν_{CO}, nor inverse relationship between ν_{FeC} and ν_{CO} in HPs and model compounds.

To study the dependence of the properties of the FeCO unit on the presence of charged groups, the latter are simulated by point charges of different magnitude and location and by homogeneous electric field. The results show that the appearance of the charges in the vicinity of the heme changes both the back-bonding and σ donation and very strongly affects activation of the coordinated CO changing the C-O bond index up to $\pm$ 50 and the C-O vibrational frequency up to $\pm$ 30% of their changes upon coordination. The Fe-C and Fe-N_{Im} bonds can be also essentially affected (up to $\pm$10%) by these charges.

The axial charges (located along the C-Fe-N_{Im} bond both on the CO and Im sides of the porphyrin) and equatorial ones (located in the plane of the heme) affect mainly back-bonding and, consequently, change the Fe-C and C-O bonds strengths and corresponding vibrational frequencies in the opposite directions. The charges located in the region of the porphyrin edge on the CO side of the heme affect both σ donation and back-bonding and lead to the very complicated relationship between ν_{FeC} and ν_{CO}. The homogeneous electric field perpendicular to the porphyrin plane essentially changes back-bonding (and as the result ν_{CO}), while the field directed parallel to the porphyrin plane practically does not affect this frequency.

All the results obtained for the dependence of the Fe-C bond strength show that the protein charged groups can essentially affect also the affinity of different HPs to CO and other diatomics.

The effect of the distal histidine on the ν_{CO} in Hps was studied by location of the number of point charges (optimized for simulation of the histidine electric field [5]) in the distal region of the complex. The calculated values of ν_{CO} = -15 cm^{-1} is close to the experimentally observed [6-10].

In general, these theoretical results support the experimental data [6-10] showing that charged or polar aminoacid groups can affect Fe-C and C-O vibrational frequencies. Moreover, since the charge effect decreases slowly upon the charge displacement away from the heme, these spectroscopic properties can be notably affected even by the charges located relatively far from the heme.

REFERENCES

1. Bersuker, I. B., The Jahn-Teller Effect and Vibronic Interactions in Modern Chemistry. Plenum/New York (1984)
2. Stavrov, S.S., I.P. Decusar, and I.B. Bersuker, New. J. Chem. 17, 71 (1993).
3. Zerner, M. C., G. H. Loew, R. E. Kirchner, and U. T. Mueller- Westerhoff, J. Amer. Chem. Soc. 102, 589 (1980).
4. Kushkuley, B. and S.S. Stavrov, Biophys. J., 70, N 3 (1996).
5. Chipot, C., J.G. Angyan, B. Maigret and H. A. Scheraga, J. Phys. Chem. 97, 9788 (1993)
6. Li, T., M.L. Quillin, G.N. Phillips, and J.S. Olson, Biochemistry. 33, 1433 (1994)
7. Li, X.-Y., and T.G. Spiro, J. Amer. Chem. Soc. 110, 6024 (1988)
8 Park, K.D., K. Guo, F. Adebodun, M.L. Chiu, S.G. Sligar, and E. Oldfield, Biochemistry. 30, 2333 (1991)
9. Ray, G.B., X.-Y. Li, J.A. Ibers, J.L. Sessler, and T.G. Spiro, J. Amer. Chem. Soc. 116, 162 (1994)
10. Springer, B.A., S.G. Sligar, J.S. Olson, and G.N. Phillips, J., Chem. Rev. 94, 699 (1994)

* This work is in partial fulfillment of the requirements for the Ph. D. degree from Sackler School of Medicine at Tel Aviv University.

L-ARGININE AND TETRAHYDROBIOPTRIN BIND ON THE DISTAL SIDE OF THE HEME POCKET AND AFFECT THE LIGAND BINDING CONFORMATION IN NO SYNTHASES

Jianling Wang[1], Chaoqun Wu[2], Dipak K. Ghosh[2], Jingli Zhang[2], Dennis J. Stuehr[2], and Denis L. Rousseau[3]

[1] Pharmaceutical Division, Ciba-Geigy Corporation, Summit, NJ 07901, USA
[2] Dept of Immunology,The Cleveland Clinic, Cleveland, OH 44195,USA
[3] Dept of Physiology and Biophysics, Albert Einstein College of Medicine, Bronx, NY 10461, USA

Nitric oxide (NO) is found to play important roles as a neuronal transmitter, a vasodilator and a cytotoxic agent in physiology and is synthesized from L-arginine (L-Arg) and O_2, *in vivo*, by a family of enzymes termed NO synthases (NOSs) [1]. The catalytic function of NOSs involves two heme-activated monooxygenation reactions as reported in cytochrome P-450s and also requires the dimeric formation of two bi-domain subunits. The oxygenase domain (OX-domain) contains the binding sites for heme, tetrahydrobiopterin (BH_4) and L-Arg, and the reductase domain contains the binding sites for FAD, FMN, calmodulin and NADPH. BH_4, which is crucial for stabilizing the dimeric structure and maintaining the activity of NOSs [2-3], also affects the affinity of substrates (e.g. L-Arg). Resonance Raman spectroscopy has been used to elucidate the heme structures [4,5], to understand the mechanism for feedback inhibition of NO [6] and to clarify the functional roles that BH_4 and L-Arg play in NOSs [2]. Here, we expand our work by investigating the binding sites for BH_4 and L-arginine analogues in NOSs. Since CO forms a stable complex by coordinating to the heme iron on the distal side it is utilized as a probe for the Fe-ligand-distal environment interactions [5].

The resonance Raman data of NOS from rat brain, which contains BH_4 but not L-Arg when purified, show a high-spin heme structure in both its oxidized and reduced forms. The addition of L-Arg to the enzyme stabilizes such "active" heme structure by preserving a binding site for the catalytic ligand, O_2. The CO-adduct from the L-Arg-free enzyme exhibits broad Fe-CO ($\nu_{Fe\text{-}CO}$) and C-O stretching modes ($\nu_{C\text{-}O}$) at ~491 and ~1936 cm^{-1}, respectively, (see Fig.1) indicating multiple Fe-C-O binding conformations. The former is deconvoluted into two nearly equal contributions at 487 and 501 cm^{-1} while the latter is deconvoluted into contributions at 1949 and 1930 cm^{-1}. The addition of L-Arg (or N^{ω}-OH-L-arginine) leads to a highly homogeneous structure, termed the α-conformation, with $\nu_{Fe\text{-}CO}$ at 503 cm^{-1} and $\nu_{C\text{-}O}$ at 1930 cm^{-1}. It represents a bent Fe-C-O linkage due to the polar and steric effects exerted by substrates and also indicates the close binding of substrates directly over the heme. One of the two forms observed from the CO-adduct of the substrate free enzyme (ν_{Fe-CO}: 501 cm^{-1}; ν_{C-O}: 1930 cm^{-1}) is assigned as the α-form while the other is assigned as the β-form. In the β-form, CO coordinates to the heme iron in a nearly linear geometry suggesting an open distal environment in the absence of substrates. The correlation between the ν_{Fe-CO} and the ν_{C-O} frequencies in NOS is consistent with the existence of (a) a proximal cysteine ligand; and (b) a more polar environment in the α-form than in the β-form. Moreover, the ligation of CO to the heme iron expels both L-citrulline and L-thiocitrulline from binding on the distal pocket.

To understand why half the heme population has a bent Fe-C-O linkage (α-form) in the absence of substrate, we examined the effects of BH_4 on the heme in the OX-domain. We found that in the presence of BH_4, the OX-domain has a high-spin heme structure in both its oxidized and reduced forms as observed in the full-length dimeric enzymes. In the absence of BH_4, however, the ferric high-spin configuration is replaced by a six-coordinate low-spin heme suggesting that the O_2 binding site is blocked. We speculate that water or an internal residue may serve as the low-spin ligand. Our data also show that either BH_4 or L-Arg can displace the low-spin ligand and completely recover the "active" structure. In the absence of BH_4 and L-arginine, we observed a complete β-conformation in its CO-adduct (ν_{Fe-CO}: 484 cm^{-1}), while the addition of BH_4 causes nearly half of the heme Fe-CO complexes to convert a bent Fe-C-O geometry (α-form, ν_{Fe-CO}: 511 cm^{-1}). It demonstrates that it is BH_4 that forces CO to bind in a distorted orientation. In addition, the presence of L-arginine results in the complete conversion to a bent Fe-C-O linkage (α-form), as found in the full-length enzyme. Our data for first time provide the direct structural evidence for the binding of BH_4 on the distal side of heme pocket and suggests that BH_4 plays a functional role in the catalytic conversion of L-Arg to NO and L-citrulline.

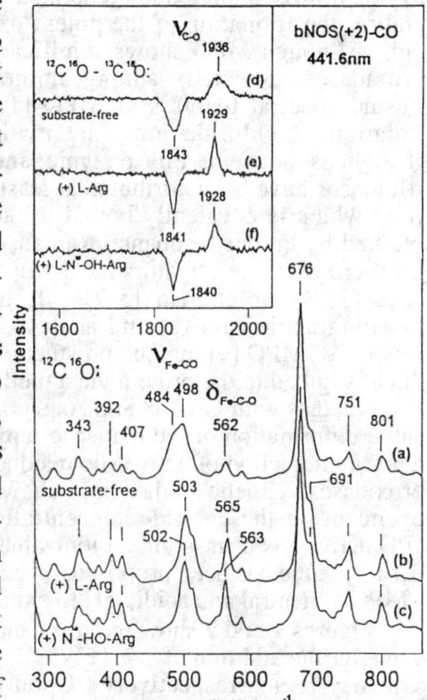

Fig.1 Resonance Raman data from CO-adducts of rat brain NOS (also containing BH_4) in the substrate-free (a,d); L-Arg (b,e) and N$^{\omega}$-OH-L-Arg bound forms (c,f).

1. P. L. Feldman, O. W. Griffith & D. J. Stuehr, Chem. Eng. News 71(51), 26-38 (1993).
2. K. J. Baek, B. A. Thiel, S. Lucas & D. J. Stuehr, J. Biol. Chem. 268, 21120-21129 (1993).
3. J. Wang, D. J. Stuehr & D. L. Rousseau, Biochemistry 34, 7080-7087 (1995).
4. J. Wang, D. J. Stuehr, M. Ikeda-Saito & D. L. Rousseau, J. Biol. Chem. 268, 22255-8 (1993).
5. J. Wang, W. S. Caughey & D. L. Rousseau, In Methods in Nitric Oxide Research, (M. Freelisch & J. Stamler, eds.), Wiley, Chichester, 1996, in press.
6. J. Wang, D.L.Rousseau, H.M.Abu-Soud & D.J.Stuehr, Proc.Natl.Acad.Sci.USA 91,10512 (1994).

RESONANCE RAMAN SPECTROSCOPY OF MYLEOPEROXIDASE AT NEUTRAL AND ACIDIC PH

K. T. Yue[1]*, K. L. Taylor[2], and J. M. Kinkade, Jr.[3]

[1]College of Life Science, Peking University, Beijing 100871, China
[2]Department of Microbiology, Emory University, Atlanta, GA 30322
[3]Department of Biochemistry, Emory University, Atlanta, GA 30322

Myeloperoxidase (MPO) is an important component of the oxygen-dependent host defense system of the human polymorphonuclear leukocyte (PMN). When released from primary granules in response to appropriate PMN stimulation, this enzyme utilizes superoxide (generated by the respiratory burst) and chloride ions to catalyze the formation of the potent oxidant and bactericidal agent, hypochlorous acid. Although MPO shows significant sequence homology to other mammalian peroxidases, especially among amino acid residues within the active site, the unusual spectral features of MPO [1, 2] and its unique ability to catalyze the oxidation of chloride ions, are evidence for significant structural/functional differences between this enzyme and other mammalian peroxidases. These differences have been attributed at least in part to the nature of the MPO prosthetic group which is covalently bound to the apoprotein. On the basis of information obtained by earlier resonance Raman, EPR and other spectroscopy studies of the native enzyme, investigators have proposed that this prosthetic group was a formyl heme [3, 4] or chlorin [5-7]. It has now been revealed by recent X-ray crystallographic data [8] and analysis of a macrocycle released during autolytic cleavage of MPO [9] that the prosthetic group of MPO is a protoheme IX derivative which is linked to the protein via a methionyl sulfonium bond with Met409 and two ester linkages with Glu408 and Asp260. This unique active site structure results in spatial deformation of the planar tetrapyrrole macrocycle [8] and may significantly alter the interaction of active site residues in MPO as compared to other mammalian peroxidases. Kinetic studies have shown that the rate of utilization of peroxidase is dependent on the chloride concentration and the pH. The physiological pH for MPO activity is near 4 and an ionizable group such as a distal histidine with a pKa near this value has been proposed. We have measured the resonance Raman spectra of MPO at neutral and acidic pH to explore the changes in the macrocycle with pH.

Figures 1 and 2 show the resonance Raman spectra of MPO at pH 7.5 (b) and soon after the addition of 1% TFA (a), with 406.7 nm excitation at the high and low frequency region respectively. Upon lowering the pH, all of the skeletal modes show an upshift in frequency of about 10 cm^{-1}. In particular, the ν_3 mode shifts from 1485 cm^{-1} to 1495 cm^{-1}. These frequencies are typical frequencies for six- and five-coordinate high spin ferric hemes respectively, indicating a loss of an axial ligand at low pH. The smaller shift of the ν_{10} mode at ca. 1614 cm^{-1} may be due to the unusual covalent bonds between the macrocycle and the protein matrix in MPO. The unusual bonding gives rise to multiple peaks in this region and can significantly alter the nature of the ν_{10} mode.

The most interesting change in the low frequency region (Figure 2) is the appearance of a weak but unambiguous peak at ca. 280 cm^{-1}. This new peak is likely due to Fe-imidazole stretching although confirmation is needed. Spiro [10] has argued that the Fe-imidazole stretching mode is not resonantly enhanced by

Soret excitation in FeIII-$(ImH)^2$ porphyrin model compound where there are two axial imidazole ligands. Since the iron in native MPO at neutral pH is six-coordinated with possibly two histidine axial ligands, we may not be able to observe this symmetric Fe-imidazole stretching mode in native MPO at neutral pH. However, at low pH, the iron is five-coordinated with the probable loss of the proximal histidine as a ligand, thus making it possible to observe the Fe-imidazole mode. In addition, two peaks at ca. 414 cm^{-1} and 332 cm^{-1} at neutral pH seem to disappear upon lowering the pH. The 414 cm^{-1} peak has been assigned to a bending mode involving the vinyl peripheral of the macrocycle, while the origin of the 332 cm^{-1} peak is not clear. The reason why these modes are not enhanced at low pH may be due to a change in the distortion of the porphyrin, but the exact nature remains unclear.

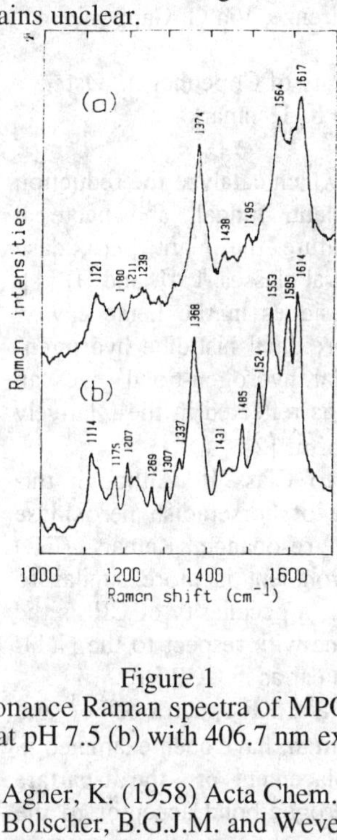

Figure 1

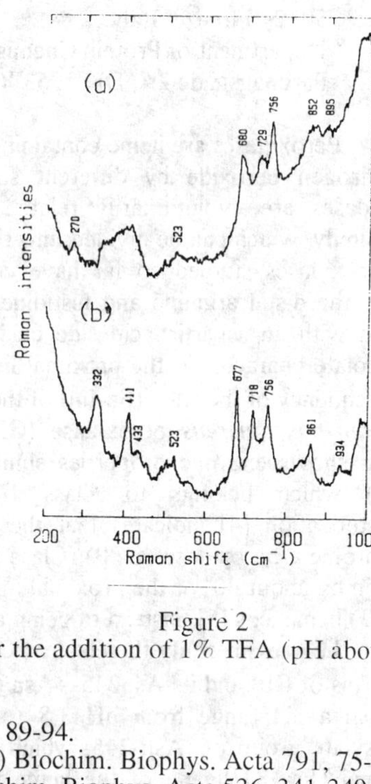

Figure 2

Resonance Raman spectra of MPO soon after the addition of 1% TFA (pH about 3) (a), at pH 7.5 (b) with 406.7 nm excitation.

1. Agner, K. (1958) Acta Chem. Scand. 12, 89-94.
2. Bolscher, B.G.J.M. and Wever, R. (1984) Biochim. Biophys. Acta 791, 75-81.
3. Harrison, J.E. and Schultz, J. (1978) Biochim. Biophys. Acta 536, 341-349.
4. Odajima, T. (1980) Biochem. J. 87, 379-391.
5. Sibbett, S.S. and Hurst, J.K. (1984) Biochemistry 23, 3007-3013.
6. Babcock, G.T., Ingle, R.T., Oertling, W.A., Davis, J.C., Averill, B.A., Hulse, C.L., Stufkens, D.J., Bolscher, B.G.J.M., and Wever, R. (1985) Biochim. Biophys. Acta 828, 58-66.
7. Andersson, L.A., Loehr, T.M. Chang, C.K., and Mauk, A.G. (1985) J. Am. Chem. Soc. 107, 182-191.
8. Taylor, K.L., Strobel, F., Yue, K.T., Ram, P., Pohl, J., Woods, A.S., and Kinkade, J.M., Jr. (1995) Arch. Biochem. Biophys. 316, 635-642.
9. Fenna, R., Zeng, J., and Davey, C. (1995) Arch. Biochem. Biophys. 316, 653-656.
10. Mitchell, M. L., Li, X.-Y., Kincaid, J. R., and Spiro, T. G. (1987) J. Phys. Chem. 91, 4690-4696.

RESONANCE RAMAN STUDIES OF COPRINUS CINEREUS PEROXIDASE AND ITS D245N MUTANT AT VARIOUS PH VALUES

G. Smulevich[1*], F. Neri[1], M. P. Marzocchi[1], and K. G. Welinder[2]

[1] Dipartimento di Chimica, Universita' di Firenze, Via G. Capponi 50121 Firenze, Italy

[2] Department of Protein Chemistry, University of Copenhagen, Øster Farimagsgade 2A, DK-1353, Copenhagen K, Denmark

Peroxidases are heme containing enzymes which catalyze the reduction of hydrogen peroxide by different substrates. Plant, fungal, and bacterial peroxidases are evolutionarily related and constitute the plant peroxidase superfamily, which can be divided into three structural classes, I, II, and III [1]. All peroxidases studied so far have conserved residues in the heme cavity, namely the distal arginine and histidine, and the proximal histidine hydrogen-bonded with an aspartic acid side chain. This polar hydrogen-bond gives an imidazolate character to the proximal ligand which is reflected in the relatively high frequency of the RR ν(Fe-Im) of the reduced forms [2].

Coprinus cinereus peroxidase (CIP) belong to Class II. Although this protein shows enzymic properties similar to those of horseradish peroxidase (HRP) which belongs to Class III [3], its resonance Raman (RR) characterization [4] indicates that the heme environment is more similar to cytochrome c peroxidase (CCP) (Class I). However, a peculiarity of CIP is the rotation by about 40° of the proximal imidazole plane with respect to the N(1)-Fe-N(3) heme axis [5], instead of being almost co-linear as in CCP [6].

In the present study the electronic absorption and RR spectra of Fe^{III} and Fe^{II} forms of CIP and its Asp245→Asn (D245N) mutant have been examined in detail in a pH range from pH 3.8 to 12.0. Replacement of the aspartate carboxylate group of Asp-245, which acts as hydrogen-bond acceptor to the proximal His-183 ligand of the heme iron, with a carboxamide group of an asparagine (Asn) residue, has a profound influence on the heme coordination of the ferric form. In fact at pH 6.0, whereas the mutant shows a 6-coordinate high spin (6-c HS) heme, CIP is 5-c HS. The sixth coordination of the iron atom of the mutant is presumably occupied by a water molecule and the change from the 5-c heme in CIP to the 6-c HS in the Asn-245 mutant is due to the loss or weakening of the His-183-Asp245 hydrogen-bond, documented in the dramatic change of the Fe-His RR band in the Fe^{II} form, as discussed below. Breaking this interaction allows the Fe atom to approach the heme plane and to bind the distal water molecule (Wat-415). The corresponding mutation (D235N) in CCP gave rise to a similar result [7].

At pH 3.8, both CIP and the mutant show a new form, coexisting with the one observed at pH 6.0, assigned to 5-c HS heme with the histidine ligand replaced by a partially protonated water molecule.

At pH 12.1 CIP converts to a 6-c low spin (LS) heme with a hydroxyl ligand coordinated in the sixth position of the heme iron and strongly hydrogen-bonded with the positively charged guanidinium group of the distal Arg-51 residue. Therefore, at alkaline pH CIP behaves as HRP-C [8], and not as CCP [9]. The pK_a of the alkaline transition of the mutant is much lower than that of the wild type. At pH 10.0 the (D245N) mutant is in its final form which markedly differs from that of the parent enzyme. The protein is 5-c HS heme with the Fe-His bond broken and the distal coordination side of the heme iron occupied by a hydroxyl group which is strongly hydrogen-bonded with distal Arg-51.

The RR spectra of the Fe^{II} form of the mutant at both neutral and alkaline pH show that the 230 and 211 cm^{-1} bands of wild type CIP, assigned to the ν(Fe-Im) stretching modes, disappear and are replaced by a band at 204 cm^{-1}. This latter band is attributed to the Fe-His stretch associated with a fairly weak or non hydrogen-bonded imidazole, as previously found for the (D235N) mutant of CCP [7].

1. K. G. Welinder, Curr. Opin. Struc. Biol., 2, 388 (1992).
2. G. Smulevich, in Advances in Spectrocopy, (R. J. H. Clark and R. E. Hester, eds), Vol. 20: Biomolecular Spectroscopy, John Wiley, London, 1993, p.163.
3. M. B. Andersen, T. Johansson, P. O. Nyman, and K. G. Welinder, in Biochemical, Molecular and Physiological Aspects of Plant Peroxidases (J. Lobarzewski, H. Greppin, C. Penel, and T. Gaspar, eds), University of Geneva, Switzerland, 1991, p. 169.
4. G. Smulevich, A. Feis, C. Focardi, J. Tams, and K. G. Welinder, Biochemistry, 33, 15425 (1994).
5. N. Kunishima, K. Fukuyama, H. Matsubara, H. Hatanaka, Y. Shibano, and T. Amachi, J. Mol. Biol., 235, 331 (1994).
6. B. C. Finzel, T. L. Poulos, and J. Kraut, J. Biol. Chem., 259, 13027 (1984).
7. G. Smulevich, J. M. Mauro, L. A. Fishel, A. M. English, J. Kraut, and T. G. Spiro, Biochemistry, 27, 5477 (1988).
8. A. Feis, M. P. Marzocchi, M. Paoli, and G. Smulevich, Biochemistry, 33, 4577 (1994).
9. G. Smulevich, M. A. Miller, J. Kraut, and T. G. Spiro, Biochemistry, 30, 9546 (1991).

RESONANCE RAMAN INVESTIGATIONS OF A PEROXIDASE ISOLATED FROM SOYBEAN SEED HULLS

M. Nissum[1], P.W. Jensen[1], and G. Smulevich[2] *

[1] Department of Chemistry, Odense University, Campusvej 55, DK-5230 Odense M, Denmark

[2] Dipartimento di Chimica, Università degli Studi di Firenze, Via G. Capponi 9, I-50121 Firenze, Italy

Peroxidases are heme containing enzymes that catalyze the oxidation of a variety of substrates by hydrogen peroxide. The superfamily of plant peroxidases is divided into three classes which show less than 20 % amino acid sequence identity [1,2]. The class I peroxidases appear to be of prokaryotic origin. Class II peroxidases comprise extracellular fungal peroxidases. Members of the class III peroxidases include secretory plant peroxidases. Members from all three classes have been investigated by resonance Raman (RR) spectroscopy and include mitocondrial cytochrome c peroxidase (CCP) (class I) [3], *Coprinus cinereus* peroxidase (CIP) (class II) [4], and horseradish peroxidase (HRP-C) (class III) [5,6]. In the present study a soybean peroxidase isolated from soybean seed hulls (SBP), which has been found to belong to class III [7], has been characterized by electronic absorption and RR spectroscopies.

RR spectra of the ferric heme at pH 7.3 were recorded with different excitation wavelengths and in polarized light. The SBP was found to exist as a mixture of four different species. On basis of the positions of the core size marker bands, the four species were assigned, respectively, to 6-coordinate high spin heme (6cHS), 5cHS, 6-coordinate low spin heme (6cLS), and to a species with core size marker band frequencies different from that of 6cHS, 5cHS, and 6cLS model heme compounds [8]. Based on results for barley peroxidase (BP) [9] this species was assigned to a quantum mixed spin heme species.

When the pH was increased a transition to complete LS was observed. At pH 11.0 SBP showed a pure 6cLS electronic absorption spectrum and the positions of the Q bands are typical of an OH ligand as the sixth ligand [10]. The high frequency RR spectrum at pH 11.0 also gave a pure 6cLS species with core size marker bands similar to HRP-C [11,12]. The ν(Fe-OH) stretching mode was identified in the low frequency spectrum around 500 cm^{-1}. The exact position of the band was difficult to determine due to band overlap. However, in D_2O a 3 cm^{-1} upshift of this composite band was observed. In HRP-C similar observations have been made and the upshift in the ν(Fe-OH) stretching mode for SBP suggests a similar H-bond interaction in the distal cavity as discussed for HRP-C [11,12].

The ferrous form of SBP at both neutral and alkaline pH gave a high frequency spectrum very similar to HRP-C and characteristic of 5cHS. In the low frequency region one band showed to be sensitive to changes in pH. The 246 cm^{-1} band observed at pH 7.3 was observed at 247 cm^{-1} at pH 6.0 and 245 cm^{-1} at pH 10.2. This band therefore was assigned to the ν(Fe-Im) stretching mode. The pH dependence of the ν(Fe-Im) stretching mode is similar to what is observed for HRP-C and is ascribed to a heme-linked ionization of an amino acid residue [13]. The ν(Fe-Im) stretching frequency for SBP is 3 cm^{-1} higher than for HRP-C at neutral pH. No structural information on SBP exists, but a plausible explanation for this higher frequency is that the N_δ atom of the imidazole fifth ligand also is H-bonded to a proximal Asp side chain in SBP, and that this H-bond is stronger in SBP, increasing the imidazolate character of the fifth ligand.

The SBP showed two ν(C=C) vinyl stretching modes. The modes were not easily identified in the RR spectra of the native enzyme at pH 7.3 due to band overlap. However, in the fluoride complex at pH 5.0 almost a complete conversion to 6cHS was observed and two vinyl stretching modes were clearly seen. This was also the case at pH 11.0 where a complete conversion to 6cLS made the identification of the two vinyl modes possible. Further evidence for splitting of the vinyl stretching modes were found in the ferrous form and in spectra obtained at low temperature. The splitting of the vinyl stretching modes has also been observed in HRP-C [5,6] and BP [9] and seems to be a general feature for class III peroxidases. In the class I peroxidase CCP and class II peroxidase CIP the two vinyl groups give rise to ν(C=C) vinyl stretching modes at the same frequency [4].

1. K.G. Welinder, Curr. Opin. Struct. Biol. 2, 388 (1992).
2. K.G. Welinder and M. Gajhede, in Plant Peroxidases: Biochemistry and Physiology, (K.G. Welinder, S.K. Rasmussen, C. Penel, and H. Greppin, eds.), University of Geneva, Switzerland, 1993, p. 35.
3. G. Smulevich, in Biomolecular Spectroscopy, Part A, (R.J.H. Clark and R.E. Hester, eds.), John Wiley & Sons, Chichester, 1993, p. 163.
4. G. Smulevich, A. Feis, C. Focardi, J. Tams, and K.G. Welinder, Biochemistry 33, 15425 (1994).
5. G. Smulevich, A.M. English, A.R. Mantini, and M.P. Marzocchi, Biochemistry 28, 772 (1991).
6. G. Smulevich, M. Paoli, J.F. Burke, S.A. Sanders, R.N.F. Thorneley, and A.T. Smith, Biochemistry 33, 7398 (1994).
7. K.G. Welinder, work in progress.
8. S. Choi, T.G. Spiro, K.C. Langry, K.M. Smith, D.L. Budd, and G.N. La Mar, J. Am. Chem. Soc. 104, 4345 (1982).
9. C.B. Rasmussen, B.D. Howes, K.G. Welinder, and G. Smulevich, to be submitted.
10. G. Smulevich, M.A. Miller, J. Kraut, and T.G. Spiro, Biochemistry 30, 9546 (1991).
11. A. Feis, M.P. Marzocchi, M. Paoli, and G. Smulevich, Biochemistry 33, 4577 (1994).
12. A.J. Sitter, J.R. Shifflett, and J. Terner, J. Biol. Chem. 263, 13032 (1988).
13. J. Teraoka and T. Kitagawa, J. Biol. Chem. 256, 3969 (1981).

RESONANCE RAMAN AND CIRCULAR DICHROISM SPECTRA OF ZINC CYTOCHROME C IN AQUEOUS SOLUTION

S. Ye, C. Shen, N. Kostic, and T. Cotton

Department of Chemistry, Iowa State University, Ames, IA 50011

Metal-substituted cytochromes c have been characterized by various techniques [1], but to date the Raman spectra of zinc cytochrome c has not been reported. The aim of this work is to examine the effect of reconstitution of the heme of cytochrome c with Zn(II) on the protein structure by resonance Raman and CD spectroscopies.

Figure 1-4 show the RR spectra of Zn(II), Fe(II) and Fe(III) Cyt c in the high and low frequency. Most bands in Fe(II) Cyt c are sharp and split, while they are broader in Zn(II) and Fe(III) Cyt c. In Figure 1, the high frequency region contains Raman lines sensitive to the core size of the porphyrin (defined as C-N, the distance from the center of the ring to the pyrrole N atoms), as determined by the properties of the central metal ion and the axial ligands [2]. The three, well-known marker lines v_3, v_{19} and v_{10} in the zinc cytochrome c spectrum are 7-15 cm^{-1} lower than those of ferrocytochrome c and 10-27 cm^{-1} lower than those of ferricytochrome c. These markers show an inverse linear dependence on the C-N distance [2-3]. From Figure 2, it can be clearly observed that the most prominent band (v_4) in the Fe(III) Cyt c and Fe(II) Cyt c is at 1375 cm^{-1} and 1364 cm^{-1}, respectively. This band is used as an oxidation-state and π-charge-density marker line in iron porphyrins and heme proteins [4]. The frequency of v_4 for Zn(II) Cyt c is between Fe(III) Cyt c and Fe(II) Cyt c.

Figure 3-4 show the RR spectra in the region below 1000 cm^{-1} for Zn(II) , Fe(II) and Fe(III) Cyt c. The frequency regions 342-421 cm^{-1} and 632-731 cm^{-1} exhibit multiple bands. Band v_{15}, corresponding to the pyrrole breathing mode, is much weaker and broader in Zn(II) and Fe(III) Cyt c than in Fe(II) Cyt c. In addition, the γ_{21} band in Zn(II) Cyt c is upshifted and much weaker than in the Fe(II) protein. It can also be seen that some of the bands shift to lower frequency in Zn(II) Cyt c. Of particular note, the band assigned to v_8, or the mixture of Fe-N(pyrrole) stretching and porphyrin-substituent bond bending[5], appears at 342 cm^{-1}, whereas in Fe(III) and Fe(II) Cyt c, this band is located at 351 and 349 cm^{-1}, respectively.

From a comparison with Fe(III) and Fe(II) Cyt c RR spectra, it can be concluded that introduction of Zn(II) into cytochrome c appears to cause very little structural perturbation of the protein conformation at the active site. The major difference in comparison with the iron form of the protein is that one of the axial ligands, most likely Met 80, does not seem to be coordinated in the Zn(II) form. Also the core size of Zn(II) Cyt c is larger than Fe(III) and Fe(II) Cyt c. Circular dichroism spectra are consistent with a similar overall protein structure and conformation as compared to the Fe(II)/Fe(III) form.

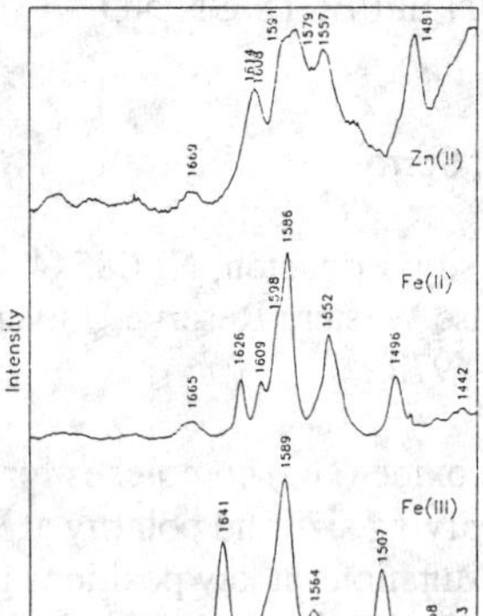

Figure 1

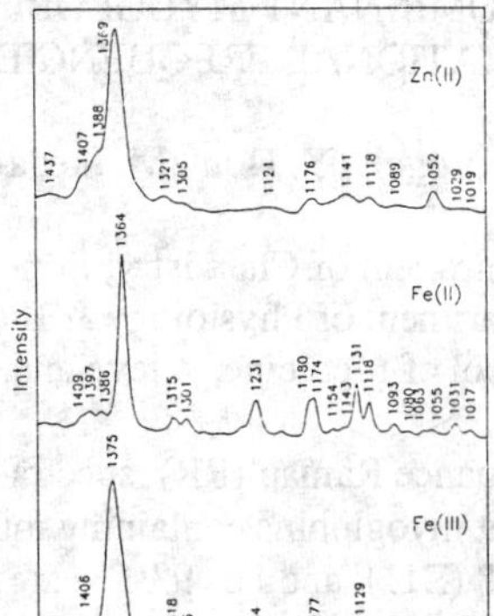

Figure 2

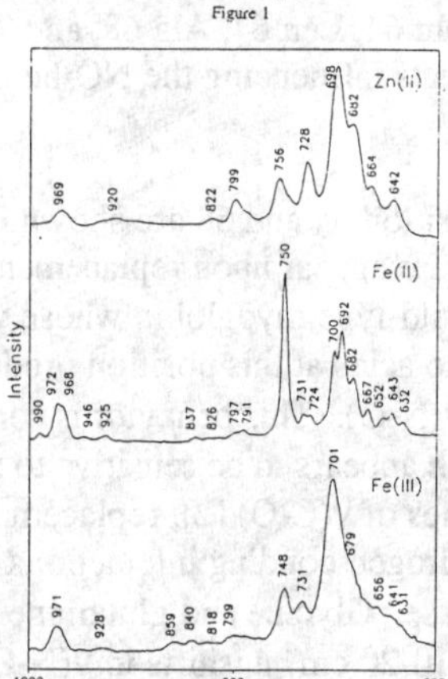

Figure 3

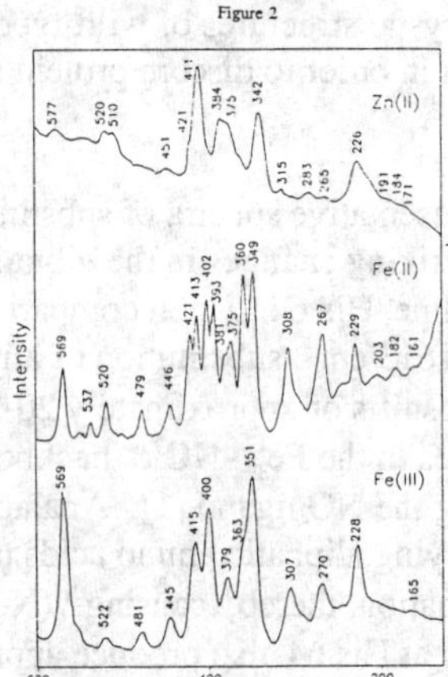

Figure 4

Figures 1-4. Raman Spectra of Zn(II), Fe(II), Fe(III) Cyt c

1. H. Anni, J. M. Vanderkooi, and L. Mayne, Biochemistry, 1995, 34, 5744
2. T. G. Spiro, Advances in Protein Chemistry, 1985, Vol 37, 111
3. K. Prendergast, T. G. Spiro, J. Am. Chem. Soc., 1992, 114, 3793
4. J. Feitlson and T. G. Spiro, Inorg. Chem. 1986, 25, 861.
5. S. Hu, Ian K. Morris; J. P. Singh; K. M. Smith; and T. G. Spiro, J. Am. Chem. Soc. 1993, 115, 12446

RESONANCE RAMAN SPECTROSCOPY OF NO BOUND TO RECOMBINANT MYOGLOBIN: PROTEIN EFFECTS ON NO VIBRATIONAL FREQUENCIES

K.M. Vogel[1], Y. Dou[2], M. Ikeda-Saito[2], T.G. Spiro[1*]

[1]Department of Chemistry, Princeton University, Princeton, NJ 08544
[2]Department of Physiology & Biophysics, Case Western Reserve University School of Medicine, Cleveland, OH 44106-4970

Resonance Raman (RR) spectra of the nitric oxide (NO) complexes for a series of recombinant myoglobins containing mutations of varying size and polarity at His 64 (E7), Val 68 (E11), and His 93(F8) are reported. Mutations at key positions in the distal pocket were selected in order to systematically change the electrostatic and steric interactions near the bound ligand and monitoring their effect on the ν(Fe-NO) and ν(N-O) vibrational frequencies. These results will be interpreted with the aid of the available crystal structures of wild-type, Gly 64, Gln 64, Leu 64, Ala 68, and Thr 68 myoglobins in order to discern protein specific effects influencing the NO-heme complex.

Representative spectra of substitutions at position 64 and 68 are shown in Figure 1. The most striking changes in the vibrational spectrum appear upon replacement of the distal histidine, His 64. Upon comparison of the wild-type myoglobin whose ν(N-O) appears at 1613 cm^{-1}, substitution of aliphatic amino acids at this position produces dramatic upshifts of approximately 20-30 cm^{-1} of ν(N-O). This behavior is consistent with changes in the $Fe_{d\pi}$-NOπ* backbonding which appears to be sensitive to the polarity surrounding the NO ligand [1]. Analagous to studies of ν(C-O) [2], replacement of His 64 with varying aliphatic amino acids prevents hydrogen bonding interactions important to back-donation, thereby causing ν(N-O) to increase. Glycine and glutamine substitution at His 64 also produce approximately 10-20 cm^{-1} upshifts in ν(N-O). In these two cases, the smaller degree of upshifts in ν(N-O) with respect to the aliphatic substituents can be explained, respectively, in terms of increased polar interactions of the bound ligand with heme pocket water molecules and multiple ligand conformations.

Figure 1 also displays the replacement of Val 68 with aliphatic and non-polar amino acids. The substitution of Val 68 with alanine and glycine produces upshifts (5-10 cm^{-1}) in the vibrational spectrum of bound NO. Such minor perturbations are consistent with those found in CO complexes where substitution of small, non-polar residues at position 68 has little effect on the IR spectrum of ν(C-O). [2] Additional studies of Thr, Phe, and Asn are underway to determine the effect of polar replacements.

Present data of ν(Fe-NO) for the series of mutants also suggests sensitivity to protein influences. The ν(Fe-NO) stretching frequency varies approximately 5-10 cm^{-1} over the span of mutations. As has been constructed for CO complexes, the extent of back-donation in each protein can be determined from a plot between ν(FeNO) and ν(N-O). Purely electrostatic effects should render an inverse correlation; deviations from this expected plot could account for kinematic effects due to the deformability of the Fe-NO unit.

The results of ν(N-O) obtained in these recombinant myoglobins indicate NO is sensitive to environmental effects near the ligand binding site in myoglobin. As NO has been implicated in an array of critical physiological responses via binding to the heme based enzyme, soluble guanylyl cyclase, a broad understanding of protein specific effects upon the NO-heme moiety is timely.

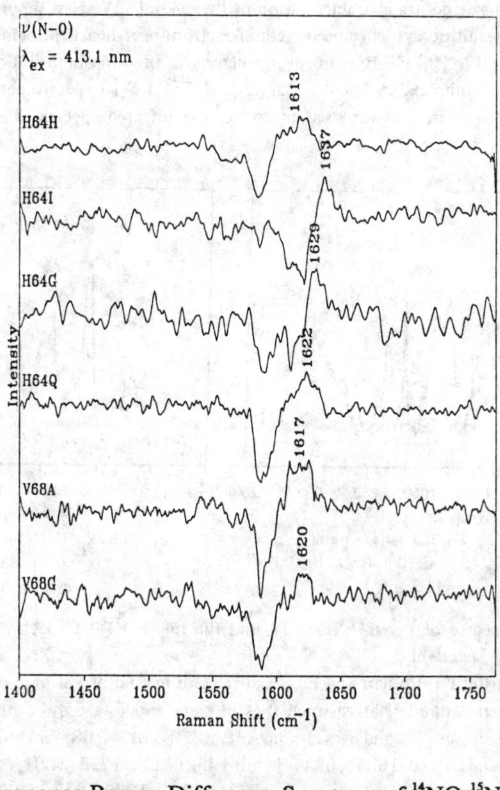

Fig.1. Resonance Raman Difference Spectrum of ^{14}NO-^{15}NO for ν(N-O) for H64H, H64I, H64G, H64Q, V68A, and V68G.

1. X.Y. Li and T.G. Spiro, J. Am. Chem. Soc. 110, 6024 (1988).
2. T. Li, M.L. Quillin, G.N. Phillips, Jr., and J.S. Olson, Biochemistry 33, 1433 (1994).

FT-Raman and FT-IR Spectra of Crystalline Cytochrome bc_1 Complex *

J-P. He[1], W-H. Yue[1], R. Xie[1], N-T. Yu[2], T. Wang[3], S-F. Weng[4], Z-H. Chen[5]

(1. Center of Research and Analysis of Materials, Wuhan University of Technology, Wuhan 430070, China. 2. Department of Chemistry, Hong Kong University of Science and Technology, Clear Water Bay, Kowloon, Hong Kong. 3. Bruker Instruments Ltd. New Chen. Bldg. Peking University, Beijing 100871, China. 4. Department of Chemistry, Peking University, Beijing 100871, China. 5. Department of Chemistry, Wuhan University, Wuhan 430072, China.)

Cytochrome bc_1 complex (Ubiquinol-cytochrome c Reductase) is an oligomeric protein W-H. Yue et al. [1] developed a simple method to obtain large crystal of cytochrome bc_1 complex. In order to study the structure characterization of the crystal, We first reported the FT-Raman and FT-IR spectra of crystalline cytochrome bc_1 complex from beef heart mitochondria.

A Bruker FRA 106 FT-Raman spectrometer was used to obtain FT- Raman Spectra of crystalline cytochrome bc_1 complex. A Nicolet Magna-IR™ 750 FT-IR spectrometer equipped with a Nicolet IR-plan infrared microscopy was used to collect the infrared spectra at room temperature. Fig. 1-4

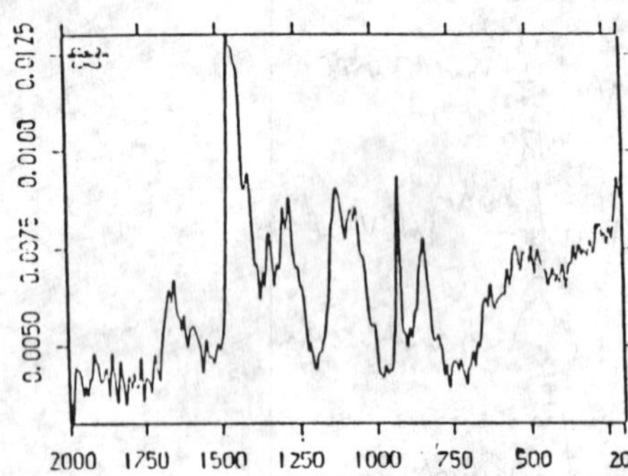

Fig. 1 FT-Raman spectra of crystal (B) of cyt. bc_1 complex

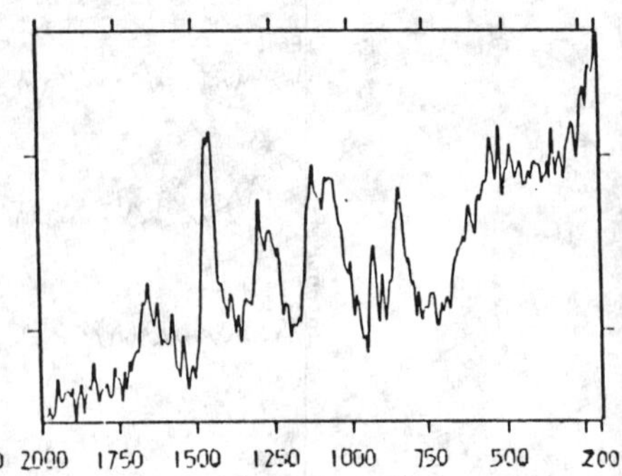

Fig. 2 FT-Raman spectra of crystal (1) of cyt. bc_1 complex

are shown the Raman spectra of crystal (B), (1) and the micro FT-IR spectra of section A, B of crystalline bc_1 complex respectively.

By comparing FT-Raman and FT-IR spectra of crystal of variable orientation and defferent section, the frequencies and relative intensities of some modes are quite different. In Raman spectra of crystal(1), 1592 cm^{-1} band was disappear, at 618 cm^{-1}, 983 cm^{-1} appear new bands. Relative intensities of bands 1670-1654-1624, 1466/1452(71), 1298/1261(>1), 925/839(<1)cm^{-1} are significantly change. These modes at 1573, 1551, 1280, 843, 547, 508, 476, 395, 358, 241, 215 cm^{-1} in Raman spectra of crystal (B) are shifted to 1577, 1542, 1261, 839, 550, 522, 486, 407, 365-341, 246, 220cm^{-1} respectively in the Raman spectra of crystal (1). In the IR spectra of section A, Bands at 1247, 1346, 1411, 1630, 2879cm^{-1} are shifted to 1242,

* This work was supported by the National Natural Science Foundation of China

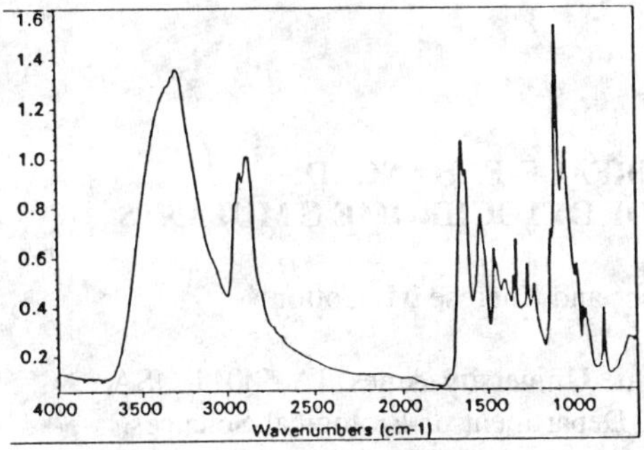

Fig. 3 Micro FT-IR spectra of crystalline(A) cyt. bc_1 complex

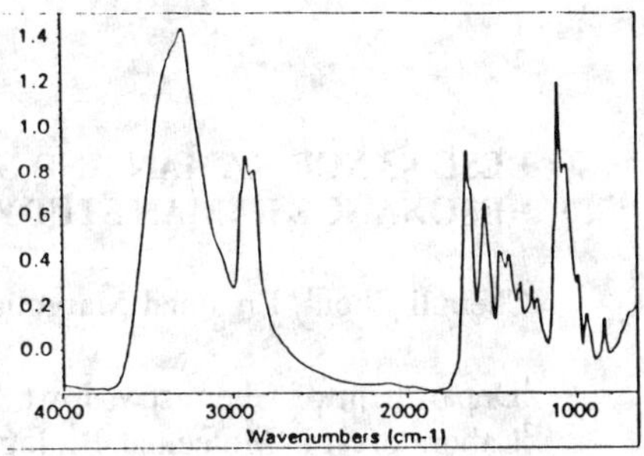

Fig. 4 Micro FT-IR spectra of crystalline (B) cyt. bc_1 complex

1343, 1408, 1627-1635, 2885cm^{-1} in the IR spectra of section B. The apparent orientation is defferent for some modes perphaps because of residues interaction and symmetry- lowering effects. FT-Raman and IR frequencies tentatively assignment of cytochrome bc_1 complex is attempted here base on assignments of previous studies[2-4]. In spectra of crystal, Bands at 1653-1654(R) 1654 (IR); 1512-1513, 1552-1554(R) 1544-1546 (IR); 1275-1282, 1253-1261(R), 1280(IR) cm^{-1} respectively are due mainly to amide I, amide II, and amide III modes of α helix structure of protein. simultaneity, those at 1663-1670, 1622-1624(R), 1627-1629, 1635(IR); 1542, 1573-1577(R); 1234-1236, 1212-1218(R) 1242-1247(IR) cm^{-1} are attributable to the amide I, amide II and amide III modes of β sheet structure respectively. The tentative assignments of other frequencies are :1452- 1468(R) 1455-1467, 1340-1360(IR) cm^{-1}- CH_2, CH_3 bending modes, 1002(R) 998-999(IR) cm^{-1}-aromatic ring mode, 925-927(R) 947-964(IR)cm^{-1}-skeletal mode of protein. 839-860, 843-865(R) shows the tyrosine doublets. 685-690(R) 697(IR)cm^{-1}-amide V of protein. 2882-2890, 2928-2931(R) 2879-2885, 2927, 2955-2960(IR) cm^{-1}-the C-H stretching 3030-3050(R) 3068-3070(IR) cm^{-1}-aromatic C-H stretching. 3160-3166,3208-3248 (R) 3290-3460 (IR) cm^{-1}- NH or OH stretching mode. These bands indicaticd that the crystal shows mixed α helix structure and β sheet formation of protein. the Raman and IR spectra shown that those bands presented at 1670-1622(R) 1654-1627(IR), 1573-1577, 1542-1551, 1512-1513(R), 1541-1544(IR), 1452-1468(R), 1455-1467(IR), 1410-1412(R) 1408-1411(IR) 1344-1367(R), 1343-1360(IR), 1126(R) 1114-1143(IR), 1064-1082(R) 1061-1078(IR) cm^{-1} are assignable to some modes of heme group. Therefore, measured results show good agreement for the FT- Raman and FT-IR spectra. they enable to investigate the structure of both heme group and apoprotein at same time. These results imply that the topolgy of crystalline cytochrome bc_1 complex has been studied by Raman and IR spectroscopy and most probably four functional units are present in an asymmetric unit. The Raman and IR spectra obtained in this study further clarify that the structure of crystalline bc_1 complex could be affected by many factors such as the distortion of the heme, variable orietation and symmetry, the geometryof protein, nature of axial ligands, amino acid residues interaction and defferences in protein environment.

Reference

1. W. H. Yue, Y. P. Zou, L. Yu and C-A. Yu, Biochemistry 1991, 30, 2303-2306
2. D. D. Hobbs, A. Kriauciunas, S. Guner et al. Biochim. Biophys. Acta, 1990, 1018, 47-54
3. W. H. Yue, J. P. He, R. Xie et al. , Science in China (series c) 1996. 39(5) in press
4. K. Zhu, Y. Zhang, J. Xu et al. , Acta Biophysica Sinica, 1995, 11, 144-148

RESONANCE RAMAN AND SURFACE-ENHANCED RESONANCE RAMAN STUDY OF CYTOCHROME C MUTANTS

Chengli Zhou[1], Emanual Margoliash[2] and Therese M. Cotton[1]*

[1]Department of Chemistry, Iowa State University Ames, IA 50011 USA
[2]Laboratory for Molecular Biology, Department of Biological Sciences, The University of Illinois at Chicago, Illinois 60680-4348 USA

Recently, site-directed mutagenesis of invariant amino acids in cytochrome c near the heme, but not directly ligated to it, have established that these substitutions can have considerable effect on the protein stability [1]. In the present study, several mutants have been characterized by RR and SERRS in order to determine the mutation effects on their spectra and in an effort to understand more about the structure/function relationships in redox proteins. Resonance Raman spectroscopy is an exquisitely sensitive technique for selectively probing the heme moiety of cytochromes. SERRS provides addition information concerning the stability and redox potential of the protein on an electrode surface [2].

Figure 1 shows the SERRS spectra of the Y67F mutant on both bare and modified silver electrodes at 0.0 mV. The spin state marker band at 1505 cm^{-1} in spectrum B indicates that the heme group of the adsorbed Y67F mutant is in the native low spin, six coordinate configuration at the 4-PyS-modified Ag electrode. In contrast, the presence of the split 1505 & 1493 cm^{-1} band at the bare Ag electrode indicates that a mixture of both low spin, six coordinate and high spin, five coordinate heme is present on the surface (spectrum A). The same behavior was also observed for other cytochrome c mutants which emphasizes that the 4-PyS modifier plays an important role in stabilizing the native configuration of cytochrome c on the electrode surface.

Figure 2 shows the resonance Raman spectra of cytochrome and the T78V mutant in the low frequency region. From the spectra it can be seen that there are considerable differences between the two, indicating that the heme environment is strongly perturbed by the mutation. The side chain of threonine 78 forms a hydrogen bond both to an interior water molecule and the anterior propionyl side chain of the heme [3,4]. When threonine is replaced by a valine in cytochrome c, both hydrogen bonds are broken, and there is a dramatic change in the properties of the protein [1].

The results from the potential dependence of the SERRS show that the so-called "water mutants" Y67F, T78V and N52I have lower redox potentials than those of other mutants which means that these mutations increase the protein stability. From the amino acid structure it can be seen that all of the three residues, tyrosine 67, asparagine 52 and threonine 78 are polar, but in the water mutants they are substituted by the non-polar residues phenylalanine, isoleucine and valine. Since three hydrogen bonds are required to hold the water molecule in the protein interior [5], the substitution of polar residues with non-polar residues in the water mutants eliminates one of these hydrogen bonds and leads to an expulsion of the water molecule from its intramolecular position. This also liberates the asparagine 52 and threonine 78 side chain from their unfavorable internal position which may account for part of the increase of the stability [6].

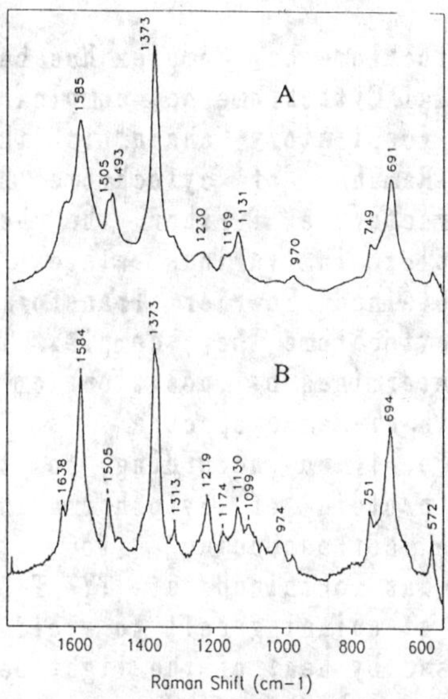

Figure 1. Surface-enhanced resonance Raman spectra of the cytochrome c Y67F mutant on (A) a bare and (B) 4-PyS-modified silver electrode at 0.0 mV vs. SCE.

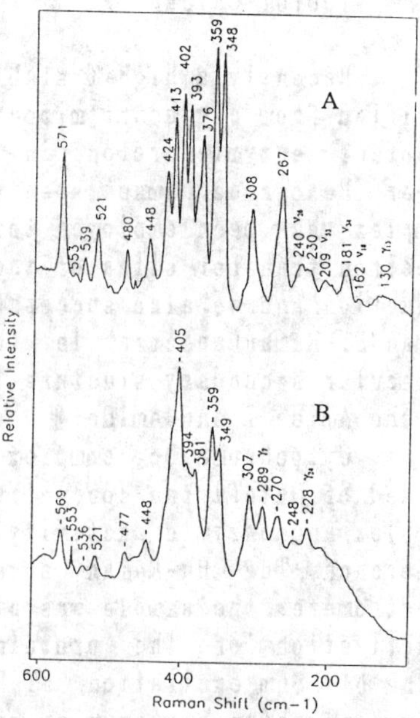

Figure 2. Resonance Raman spectra of 0.1 mM reduced cytochrome c (A) wild-type and (B) T78V mutant at 77K.

References

1. E. Margoliash, A. Schejter, T. I. Koshy, T. L. Luntz and A. E. Garber, in *Bioenergetics,* C. H. Kim and T. Ozawa, Eds., Plenum Press, New York, **1990.** 125-145.
2. T. M. Cotton, J. H. Kim and G. D. Chumanov, *J. Raman Spectrosc.* **1991,**
3. T. Takano and R. E. Dickerson, *J. Mol. Biol.* **1981,** 153, 79.
4. T. Takano and R. E. Dickerson, *J. Mol. Biol.* **1981,** 153, 95.
5. A. A. Rashin, M. Iofin and B. Honig, *Biochem.* **1986,** 25, 3619.
6. A. Schejter, T. L. Luntz, T. I. Koshy and E. Margoliash, *Biochem.* **1992,** 31, 8336.

STUDY ON THE STRUCTURAL INFORMATION OF CYTOCHROME COMPLEX AND ENERGY TRANSPORTATION WITH RAMAN SPECTROSCOPY

Zhu Keli Pan Pangxie Xu Jianxing

(Institute of Biophysics, Academia Sinica, Beijing 100101 China)

Recently a highit stable cytochrome bc_1 Complex has been purified from pig heart mitochendrial. Cytochrome bc_1 complex is impotant enzymic proton in the respiratory chain. In this paper, Resonance Raman spectra (RR-Raman) of cytochrome bc_1 complex has been employed to characterize further the heme electron transfer sites of the complex in various stages of redaction. and we also successful obtained Fowrier Transformed Raman (FT-Raman) apectra in this cytochrome bc_1 complex. The molecular secendary stucture is determines by positions (cm^{-1}) of the Amide I and Amide III bonde in FT-Raman spectre.

Cytochrome bc_1 complex was purified according to the methed of Hatefi, the concentration protein of cytochrome bc_1 complex and enzymic activity were corresponding to the referencn. The RR-Raman spectrum was obtained at TY T-800 spectrometer, the sample was placed at spinning cell to avoid the inactivation of the protein causen by heat of the light beam. using 514.5nm excitation. FT-Raman of cytochrome bc_1 complex in different redox states were measured using a Bruke IFS66 / FRA106 FT-Spectrophotometer. Near-IR excitation at 1.064μ. There are three heme groups of cytochrome bc_1 complex named as b_{566} b_{562} and c_1 in the oxidized state all the Fe of heme are in +3 valence shown as of $^{+3}b_{566}$ $^{+3}b_{562}$ $^{+3}c_1$. Ascorbate is a gentle reductant, it could only reduce cytochrome c_1 but not cytochrome b. Therefore, a state of $^{+3}b_{566}$ $^{+3}b_{562}$ $^{+2}c_1$ can be obtained by adding ascorbate to the cytochrome bc_1 complex. sodium dithionite is a strong reductant, it can reduce all the three heme groups and get the form of $^{+2}b_{566}$ $^{+2}b_{562}$ $^{+2}c_1$ state.

RR-Raman date for the cytochrome complex are germane to the determination of the roles played by the sturcture and dynamics of their heme active sites in photosynchetic electron transfer. Comparing Fig.1 with Fig.2 (a) and (b), It is clear, the different heme environment exist among the active sites the contributions from the heme c iron state and heme b iron state in the Raman spectre are different. But it has not been possible to detect the detailed secondary structure of cytochrome bc_1 complex by RR-Raman. While FT- Raman spectrum gives the information about the structure of whole protein. It was obvious that the spectra of Fig. 3 (a) (b) and (c) were almost same, the peptide CONH gromps of cytochrome bc_1 complex exhibit two characteristic group fequencies in the FT-Raman. $1660cm^{-1}$ was the signd peak of amide I band, $1266cm^{-1}$ was the signd peak of amide Ⅲ band. there are extremely weak in the normality Raman effect but quite strong in FT- Raman spectra. Clealy, indection moleculars have same conformations, it was rich of α-helical forms in the difference stages.

The purified cytochrome bc_1 complex was studied by comparing their RR-Raman spectre and FT-Raman spectra in its oxidized and reduced states, The RR-Raman spectre were extremely different, while the FT-Raman spectra do not change much with their redox states. These results indicated that the local conformation of heme group but not the conformation of whole protein molecular changed with their redox states.

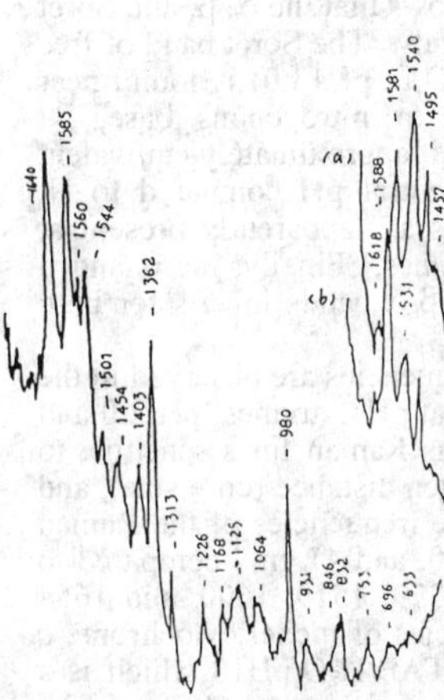

Figure 1. Raman spectra of bc_1 complex at normal oxidized state

Figure 2 Raman spectra of bc_1 complex reduced by ascorbate (a) Raman spectra of bc_1 complex reduced by dithionite (b)

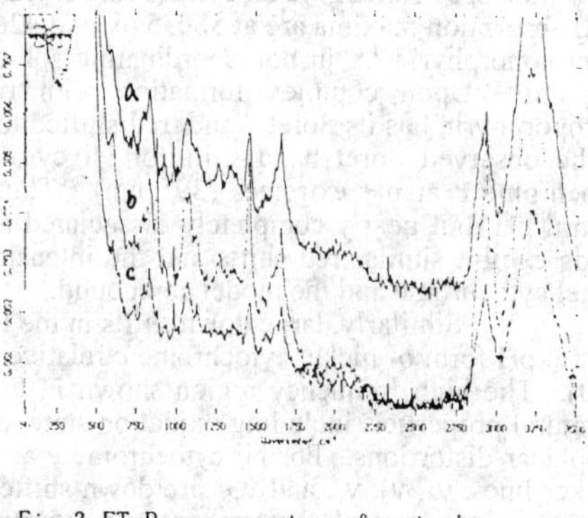

Fig.3 FT-Raman spectra of cyt. bc_1 at oxidized (a) at ascorbate-reduced state (b) at Dithionite-reduced stste (c)

1. Youssef Hatefi (1978) Methed in Enzymology vol. L Ⅲ p. 35-40 ed. bysidney Flecscher lester pacher Academic press.
2. R. J. H clark and R. E. Hester (1993) Biomoleular spectroscopy. par A.
3. Mitchell P. (1976) J. theor Biol. 62, 337-367

CONFORMATIONAL ANALYSIS OF NICKEL-RECONSTITUTED CYTOCHROME *C* BY RESONANCE RAMAN SPECTROSCOPY

J. Ma,[1,2] W. Jentzen,[1] M. Laberge,[3] J. Vanderkooi,[3] X. Song,[1,2] S. Jia,[1,2] J. D. Hobbs,[1] J. A. Shelnutt[1,2*]

[1]Fuel Science Department, Sandia National Laboratories, Albuquerque, NM 87185-0710, USA. [2]Department of Chemistry, University of New Mexico, New Mexico 87131, USA. [3]Department of Biochemistry and Biophysics, University of Pennsylvania, Philadelphia, PA 19104-6089, USA.

Catalysis, ligand binding, redox reactions, and light to chemical energy transduction are among the important functions of metalloporphyrin-containing proteins in biology. Analogous biomimetic chemistries are also of current interest. To further develop and understand these chemistries, a detailed understanding of the relationship between protein structure and metalloporphyrin reactivity for this class of protein would be useful. One method of investigating structure-reactivity relationships is by making a comparison of systematically modified enzymes. Recently, a variety of transition metals have been substituted for iron in the central core of the porphyrin ring of several heme proteins including cytochrome c, hemoglobins, and myoglobin. These metal-substituted heme proteins offer an important new method of investigating the role of the metal ion in catalytic and ligand-binding functions of heme proteins.

Nickel(II)-reconstituted cytochrome *c* and model nickel(II) porphyrins have been investigated by pH-dependent resonance Raman and absorption spectroscopy. The absorption spectrum of nickel cytochrome *c* near neutral pH shows that the α, β, and Soret band absorption maxima are at 580, 546, and 426 nm, respectively. The Soret band of free Ni protoporphyrin IX in non-coordinating solvent (CTAB/H_2O at pH13.0) is found near 400 nm. Upon complex formation with two molecules of nitrogenous base, Ni protoporphyrin has its Soret band red-shifted to 434 nm, a shift approximately equivalent to the observed Soret band shift for Ni cytochrome *c* at neutral pH compared to Ni cytochrome *c* at pH extremes (395 nm). Thus, axial ligands are apparently present at neutral pH, but nearly completely dissociated at the pH extremes. Finally, the α and β bands exhibit similar red shifts and the intensity ratio of the β/α bands inverts for both nickel cytochrome and the model compound.

Similarly, large down shifts in the Raman line frequencies are observed in the neutral pH form of nickel cytochrome *c* relative to the protein at pH extremes (pH 2.0 and 13.0). The high frequency region shown in Figure 1 contains Raman lines sensitive to structural properties, including oxidation state, center-to-nitrogen distance (core size), and nonplanar distortions. For Ni cytochrome *c* at neutral pH, the frequencies of the Raman marker lines, ν_4, ν_3, ν_2, and ν_{10}, are down-shifted by 12, 39, 20, and 41 cm^{-1} compared to Ni cytochrome *c* at pH extremes (ν_4, ν_3, ν_2, and ν_{10} are at 1379, 1519, 1600, and 1654 cm^{-1}, respectively). At pH extremes, the Raman line frequencies of nickel cytochrome *c* are very close to those observed for Ni protoporphyrin in CTAB/H_2O/pH13 which is a well-known four-coordinate model. The Raman line frequencies of the neutral form of Ni cytochrome *c* are close to those of Ni protoporphyrin in pyrrolidine which is six-coordinate. The pattern of shifts occurring upon addition of two ligands is one of large shifts to low frequency for the core size marker lines (~ -40 cm^{-1}) and a somewhat smaller shift in the oxidation-state marker line ν_4, also to low frequency (~ -12 cm^{-1}). A comparison of four-coordinate frequencies for the Raman marker lines with those for Ni cytochrome *c* at physiological pH shows that the shifts in frequency are near those expected if a σ-donating axial ligand is added [1].

On the other hand, closer inspection of the ν_4 line shows clear differences between the spectra of Ni cytochrome *c* and the model compounds. For Ni cytochrome *c* at pH 2.0, ν_4 exhibits a low-frequency shoulder near 1369 cm^{-1}, not present in the four-coordinate model. This indicates a residual concentration of axially coordinated species in the denatured proteins. This conclusion is also supported by the absorption spectra, which shows a shoulder on the low-energy side of the Soret band near 419 nm for Ni cytochrome *c* at either pH 2.0 or 13.0. For the neutral pH form of Ni cytochrome *c*, a broadened ν_4 line is observed indicating two nearly resolved ν_4-line components. The two component lines are suggested to result from a six-coordinate species (with a frequency of 1365 cm^{-1}) and a five-coordinate species (with a frequency of 1370 cm^{-1}). The latter form might also be explained by a six-coordinate species in which the axial ligands provided by the protein moiety have changed.

From the crystal structures of various cytochromes *c* and molecular modeling studies, the average pyrrole dihedral distortion angle of the heme of cytochrome *c* is found to be near 14° [2]. Thus, we expect that the macrocycle of Ni cytochrome *c* at physiological pH will be ruffled like the native Fe protein. Comparing the frequency of the ν_{10} line, the most sensitive line to nonplanar distortion of porphyrins, we find that Ni cytochrome *c* has its ν_{10} line down-shifted 4 cm^{-1}. This down-shift of the ν_{10} line is primarily a result of nonplanar distortions [3]. However, the apparent upshift of the ν_2 line for Ni cytochrome *c* at neutral pH compared to that of the six-coordination model is problematical. Further resonance Raman studies are underway to resolve the nature of the axial ligands in the two coordinated forms and the origin of the up-shift in ν_2.

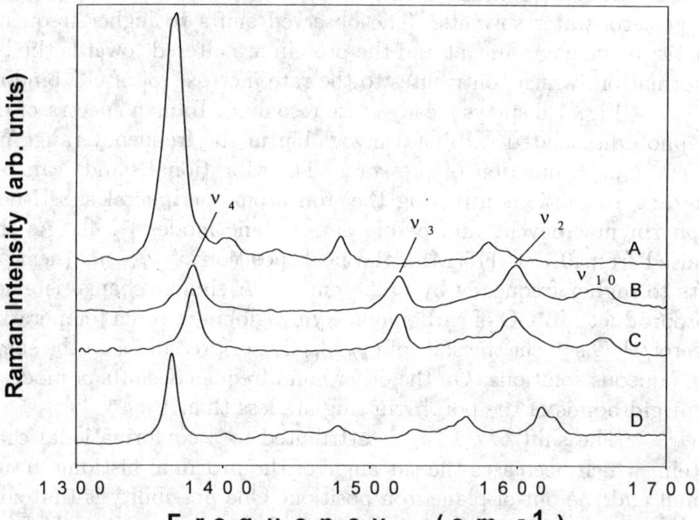

Figure 1. Resonance Raman spectra of nickel cytochrome c at (A) pH 6.8 and (B) pH 2.0, and nickel protoporphyrin IX in (C) CTAB/H2O at pH 13 and (b) pyrrolidine.

1. Shelnutt, J. A.; Alston, K.; Ho, J.-Y.; Yu, N.-T.; Yamamoto, T.; Rifkind, J. M. *Biochemistry*, **1986,** 25, 620.
2. Hobbs, J. D.; Shelnutt, J. A. *J. Protein Chem.*, **1995**, *14,* 19.
3. Jentzen, W.; Simpson, M.C.; Hobbs, J.D.; Song, X.; Ema, T.; Nelson, N.Y.; Medforth, C.J.; Smith, K.M.; Veyrat, M.; Mazzanti, M.; Ramasseul, R.; Marchon, J.-C.; Takeuchi, T.; Goddard, III, W.A.; Shelnutt, J.A. *J. Am. Chem. Soc.*, **1995**, *117*, 11085.
(This work performed at Sandia National Laboratories is supported by the U.S. Department of Energy under contract DE-AC04-94AL85000.)

HIGH PRESSURE RAMAN SPECTROSCOPY OF THE HEME-PROTEIN LINKAGE IN MYOGLOBIN

O. Galkin and A. Schulte*

Department of Physics and Center for Research and Education in Optics and Lasers, University of Central Florida, Orlando, FL 32816

The sensitivity of the iron-proximal histidine bond to variations in protein structure [1, 2] may be utilized by heme proteins to regulate the acitvity of the iron. Pressure allows one to change the molecular interactions in a controlled way and to separate thermal and volume effects on both the reaction rate and protein conformational states. Here we determine the pressure dependence of the iron-proximal histidine mode ν_{Fe-His} of horse myoglobin in H_2O and glycerol-water solvents. The observed shifts to higher frequency suggest that the heme environment and the protein are altered towards the bound state conformation, which contributes to the rate increase [3] of CO binding.

Fig. 1 displays steady state resonance Raman spectra of deoxy (Mb) and photodissociated (Mb*CO) myoglobin in the frequency range from 150 to 550 cm^{-1} as a function of pressure. The vibrational bands can be classified according to motions involving the iron atom, peripheral substituents on the porphyrin macrocycle, and pyrole out-of-plane modes [1, 4]. As the pressure is raised from 0.1 to 175 MPa the peak position of ν_{Fe-His} (near 220 cm^{-1}) shifts to higher frequency by ≈ 2.8 cm^{-1} . A similar change is evident in the photoproduct Mb*CO of carbonmonoxymyoglobin at room temperature. In 75% glycerol/H_2O solvent the shift of ν_{Fe-His} due to pressure is in the same direction as in aqueous solutions. On the other hand frequency shifts of modes associated with rigid bonds of the porphyrin ring are less than 1 cm^{-1} [5].

The shift of ν_{Fe-His} is attributed to a conformational change in the protein, which decreases the tilt angle of the proximal histidine from the heme normal and the out-of-plane iron position. One possibility is that sliding of the F-helix alters the tilt angle of the proximal histidine as well as the force constant or the length of the iron-histidine bond [5]. NIR absorption studies of band III provide additional support for iron movement. An important factor affecting the global protein conformation and the heme pocket structure is bound water. The altered water activity in a glycerol/H_2O mixture causes a 2.6 cm^{-1} downshift of ν_{Fe-His} [6] as compared to aqueous solution, though the perturbation of the Fe-His frequency is opposite to that from pressure. Molecular dynamics simulation have indeed found large pressure-induced changes in the hydration layer causing the solvation shell water to adopt a more ordered structure [7].

We thank A. Tabirian for assistance with NIR absorption measurements. Financial support by NSF (MCB-9305711) is gratefully acknowledged.

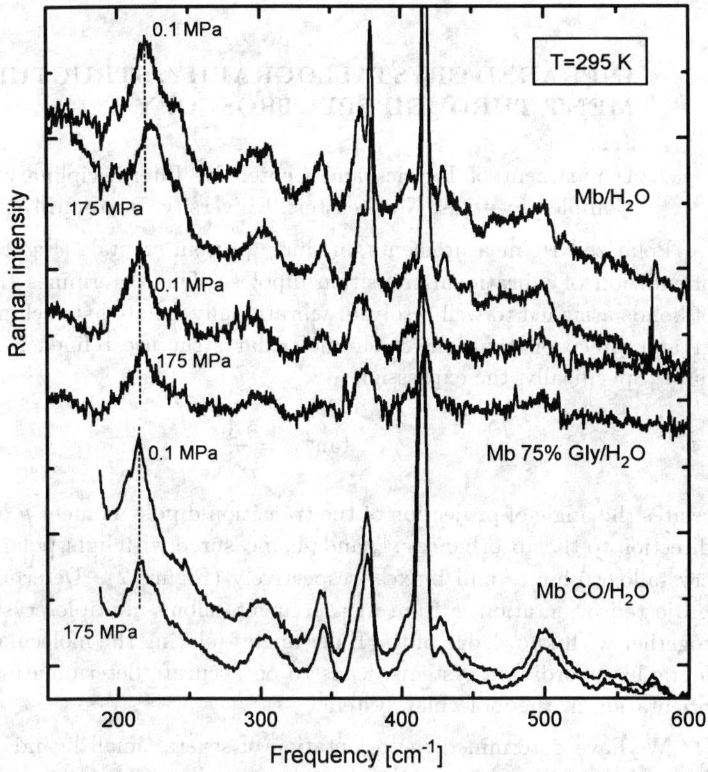

Fig. 1 Resonance Raman spectra of (horse) myoglobin at ambient and high pressure (175 MPa) obtained in potassium phosphate buffer and glycerol-water (pH 7) with 441 nm excitation. The sharp lines at 417 and 378 cm^{-1} are due to the sapphire windows of the pressure cell.

1. D. L. Rousseau and J. M. Friedman, in Biological Applications of Raman Spectroscopy, (T. G. Spiro, ed.), Vol III, Wiley, New York, 1988, p. 133.
2. T. Kitagawa, in Biological Applications of Raman Spectroscopy, (T. G. Spiro, ed.), Vol III, Wiley, New York, 1988, p. 99.
3. H. Frauenfelder et al., J. Phys. Chem. 94, 1024 (1990), and refs. therein.
4. T. G. Spiro and X. Y. Li, in Biological Applications of Raman Spectroscopy, (T. G. Spiro, ed.), Vol III, Wiley, New York, 1988, p. 133.
5. A. Schulte, S. Buchter, O. Galkin, and C. Williams, J. Am. Chem. Soc. 117, 10149 (1995).
6. J. T. Sage, K. T. Schomacker, and P. M. Champion, J. Phys. Chem. 99, 3394 (1995).
7. D. B. Kitchen, L. H. Reed, and R. M. Levy, Biochemistry 31, 10083 (1992).

INFRARED CRYSTALLOGRAPHY: STRUCTURAL REFINEMENT THROUGH SPECTROSCOPY

J.T. Sage*

Department of Physics and Center for Interdisciplinary Research on Complex Systems, Northeastern University, Boston, MA, USA

Polarized IR measurements on single protein crystals (Figure 1) yield the orientation of vibrational transition dipoles. This determines the orientation of bonds assigned to well-resolved, vibrationally isolated stretching modes with greater precision than the macromolecular structure refined from diffraction data. Specifically, the expression

$$\tan^2 \psi = \frac{\mathcal{A}_b}{\mathcal{A}_a}$$

relates the angle of projection of the transition dipole moment $\vec{\mu}$ onto the $\{001\}$ direction to the absorbances $\mathcal{A}_a$ and $\mathcal{A}_b$ measured with light polarized along the crystallographic a- and b-axes, respectively (Figure 2). Determination of the projected orientation ψ from measurements along multiple crystal directions, together with knowledge of the Euler angles relating the molecular and crystallographic coordinate systems, leads to an accurate determination of the bond orientation in the molecular system.

We have determined the orientation of several small ligands bound at the heme sites of myoglobin (Mb) crystals, including CO, CN^-, and N_3^-. As reported previously, we find that in MbCO crystals CO is displaced less than 10° from the heme normal [1], approximately 6.5° in a direction between pyrroles C and D ($\phi = 70°$). This result contradicts X-ray and neutron structures of $P2_1$ MbCO crystals [2], which show CO displaced in various directions by more than 30°. Accurate structural information allows a realistic quantitative evaluation of the contribution of steric distortion by the protein to the free energy of CO binding.

The small differences observed between crystal and solution IR spectra and the fact that the resulting orientations are consistent with polarized measurements on both $P2_1$ and $P2_12_12_1$ crystals, suggest that crystallization has minimal impact on active site structure. In fact, recent picosecond photoselection measurements [3] have confirmed that the CO lies within 7° of the heme normal in solution, but do not reveal the direction of tilt. Unlike X-ray diffraction, which leads to a time averaged electron density map, IR crystallography can independently characterize spectrally resolved bands. This allows us to observe the effect of changes in distal pocket structure and iron spin on ligand binding geometry.

1. Ivanov et al., J. Am. Chem. Soc. 116, 4139 (1994).

2. Kuriyan et al., J. Mol. Biol. 192, 133 (1986); Cheng and Schoenborn, J. Mol. Biol. 220, 381 (1991).

3. Lim et al., Science 269, 962 (1995).

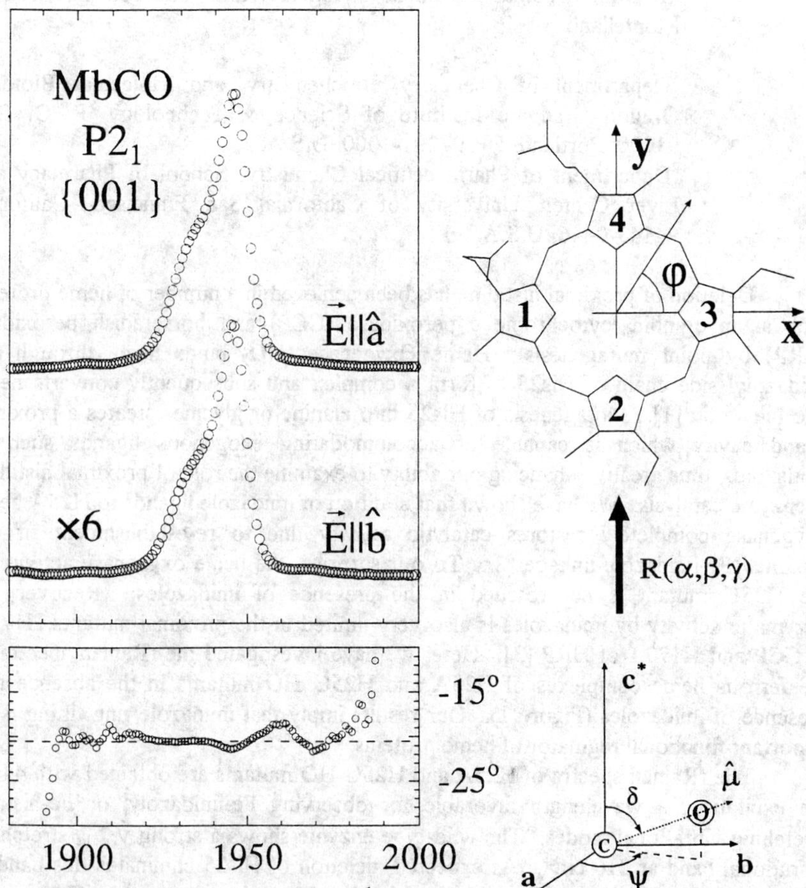

RESONANCE RAMAN STUDIES OF HEME OXYGENASE PROXIMAL HISTIDINE MUTANTS: ADDITION OF IMIDAZOLE LIGANDS.

Jie Sun[1], Thomas M. Loehr[1*], Angela Wilks[2], and Paul R. Ortiz de Montellano[2]

[1]Department of Chemistry, Biochemistry, and Molecular Biology, Oregon Graduate Institute of Science & Technology, P. O. Box 91000, Portland, OR 97291-1000, U.S.A.
[2]Department of Pharmaceutical Chemistry, School of Pharmacy and Liver Center, University of California, San Francisco, California 94143-0446, U.S.A.

Deletion of proximal histidine has been achieved in a number of heme proteins such as myoglobin, cytochrome *c* peroxidase (CCP) and horseradish peroxidase (HRP) by point mutagenesis. Heme Oxygenase (HO) binds heme through the imidazolyl side chain of His25 to form a complex and subsequently converts heme into biliverdin [1]. Mutagenesis of His25 into alanine or glycine creates a proximal ligand cavity which is capable of accommodating exogenous ligands such as imidazoles, thus greatly enhancing our ability to examine the role of proximal histidine in enzyme catalysis. We have shown that addition of imidazole ligands to H25A heme oxygenase completely restores catalytic activity due to re-establishment of the proximal Fe-imidazole linkage [2]. To our surprise, the heme oxygenase activity of the H25G mutant is not rescued in the presence of imidazoles. Recovery of enzymatic activity by imidazoles is also very limited in the proximal mutants H175G of CCP and H170A of HRP [3]. Here, we have investigated the rRaman spectra of the ferrous heme complexes of H25A and H25G HO mutants in the absence and presence of imidazoles (Figure 1). Our results imply that imidazole ring tilting is an important functional regulator of heme proteins.

The rRaman spectra of H25A and H25G HO mutants are obtained with 441.6 nm excitation, a wavelength favorable for observing Fe-imidazolyl or Fe-histidyl stretching vibrational modes. The wild-type enzyme shows a strong $\nu_{Fe\text{-}His}$ stretching vibrational band at 216 cm^{-1}. As expected, deletion of His25 eliminates this band in H25A and H25G mutants. Addition of imidazole to these mutants yields a new strong band at 225 cm^{-1} (H25A) and 220 cm^{-1} (H25G). These bands are assigned to $\nu_{Fe\text{-}Im}$ because they shift to 216 cm^{-1} (H25A) and 205 cm^{-1} (H25G) when imidazole is replaced by the heavier *N*-methylimidazole.

We interpret our results in comparison with the relaxed (R) and tense (T) states of deoxyhemoglobin (Hb). In T state Hb, the imidazole ring is tilted toward the heme plane and $\nu_{Fe\text{-}His}$ has been found at 205 cm^{-1} (α-chain) and 218 cm^{-1} (β-chain)

[4,5]. These bands move to 220 cm^{-1} (α-chain) and 222 cm^{-1} (β-chain) in R state Hb where the imidazole ring is not tilted [4,5]. In this view, the imidazole ring is more tilted in the H25G HO/imidazole complex than in the H25A HO/imidazole complex (Figure 2) as indicated by the lower $\nu_{Fe\text{-}Im}$ frequency of H25G. This 5 cm^{-1} frequency shift is not induced by weakened H-bonding to imidazolyl $N_{\delta}H$ in H25G relative to H25A, because of the larger $\nu_{Fe\text{-}Im}$ frequency difference between imidazole and *N*-methylimidazole for H25G than for H25A. The greater imidazole ring tilting in the H25G mutant may be prevented by van der Waals repulsion between the $-CH_3$ side-chain of alanine and the imidazole ring in H25A. In the wild-type enzyme, the imidazole ring tilting can be prevented by a direct tethering to C_{β} of His25.

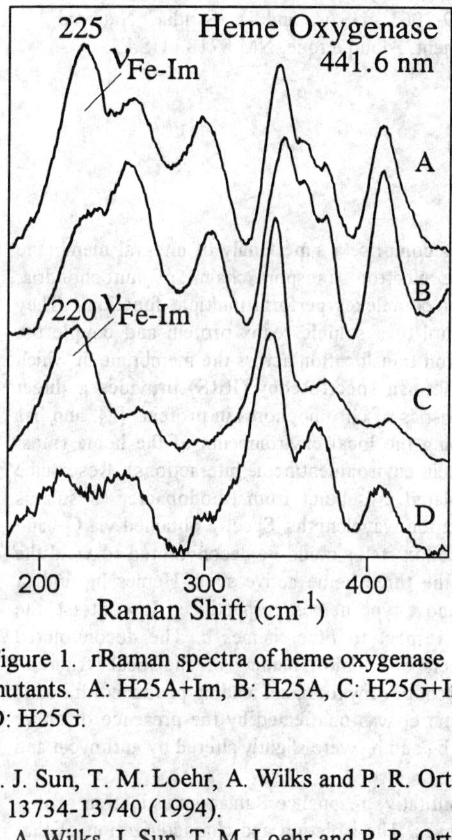

Figure 1. rRaman spectra of heme oxygenase mutants. A: H25A+Im, B: H25A, C: H25G+Im, D: H25G.

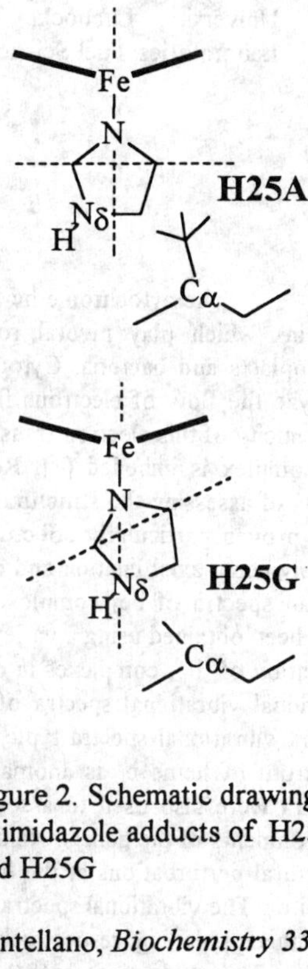

Figure 2. Schematic drawing of imidazole adducts of H25A and H25G

1. J. Sun, T. M. Loehr, A. Wilks and P. R. Ortiz de Montellano, *Biochemistry 33*, 13734-13740 (1994).
2. A. Wilks, J. Sun, T. M. Loehr and P. R. Ortiz de Montellano, *J. Am. Chem. Soc. 117*, 2915-1926 (1995).
3. D. E. McRee, G. M. Jensen, M. M. Fitzgerald, H. A. Siegel and D. B. Goodin, *Proc. Natl. Acad. Sci. U.S.A. 91*, 12847-12851 (1994).
4. K. Nagai and T. Kitagawa, *Proc. Natl. Acad. Sci. U.S.A. 77*, 2033-2037 (1980).
5. M. F. Perutz, G. Fermi, B. Luisi, B. Shaanan and R. C. Liddington, *Acc. Chem. Res. 20*, 309-321 (1987).

RESONANCE RAMAN CHARACTERIZATION OF CYTOCHROME bc_1 COMPLEXES

F.Gao[a], H.Qin[b], M.C. Simpson[c], J.A. Shelnutt[c], D.B. Knaff[b], and M.R. Ondrias[a]*

[a] Department of Chemistry, University of New Mexico, Albuquerque, NM 87131, USA, [b] Department of Chemistry and Biochemistry, Texas Tech University, Lubbock, Tx 79409-1061 USA, and [c] Sandia National Laboratories, Fuel Science Department, Albuquerque, NM 87185 USA.

The cytochrome bc_1 complexes comprise a superfamily of integral membrane proteins which play pivotal roles in the electron transport chains of mitochondria, chloroplasts and bacteria. Cytochrome bc_1 complexes perform multiple functions. They catalyze the flow of electrons from a quinol to a soluble redox protein and couple the exogenicity of this electron transfer to proton translocation across the membrane in which the complex is imbeded [1], Resonance Raman spectroscopy (RRS) provides a direct means of assessing the structural characteristics of chromophores in proteins [2], and has been proven particularly effective in probing the local environments of the heme (most notably heme axial ligation and direct protein environment/heme interactions). Resonance Raman spectra of bc_1 complexes and isolated c_1 subunit from Rhodobacter capsulatus have been obtained using a variety of excitation wavelengths. Spectra obtained via Q-band excitation of bc_1 complexes in different redox states could be deconvoluted to yield the individual vibrational spectra of each of the three heme active sites. Hemes b_H and c_1 exhibit vibrational spectra typical of b- and c-type hemes, respectively. In contrast, the spectrum of heme b_L is anomalous with respect to other hemes b. The deconvoluted spectra were also used to assess the effects of inhibitor binding on the local structural environments of the hemes. Neither antimycin nor myxothiazol binding produced dramatic structural perturbations at the hemes. Heme c_1 was unaffected by the presence of either inhibitor. The vibrational spectra of hemes b_H and b_L were slightly altered by antimycin and myxothiazol binding, respectively.

This study represents the first quantitative resonance Raman investigation of the individual heme active sites of bc_1 complexes. The isolation and characterization of the vibrational spectra of heme c_1, b_H and b_L are a necessary first step toward utilizing the structural sensitivity of RR scattering for the time-resolved investigation of the dynamics associated with the physiological functions of the complex. Several useful conclusions can be drawn from the results of these initial investigations: While B-band RR heme spectra are extremely difficult to deconvolute, the differences in Q-band absorption among heme c_1, b_H and b_L allow for the isolation of their spectra using resonance selectivity and difference techniques. The isolated RR Q-band spectra reveal that heme c_1

and b_H have "typical" c-type and b-type vibrational spectra[3,4], respectively. In contrast, heme b_L exhibits an anomalous RR spectrum, particularly for the axial ligand sensitive mode, ν_{11}. The extremely low energy (relative to other hemes b) of this mode and the low redox potential of heme b_L may be a reflection of protein-induced deprotonation (or strong hydrogen bonding) of at least one of the heme b_L histidine axial ligands. Heme c_1 appears to be structurally isolated from both the other hemes and from effects of inhibitor binding at either the Q_0 or Q_i sites of the complex. There is no evidence in the RR spectra for direct structural communication between the b-hemes. In particular, the local environment of ferrous heme b_H is unaffected by the redox state of heme b_L. Neither Q_i nor Q_0 inhibitors produce dramatic structural changes at the b-hemes, upon binding to the complex. In particular, the anomalous ligation environment of heme b_L is not dramatically changed by the binding of inhibitors. Antimycin binding does induce minor changes in the heme b_H environment, while myxithiozal binding produces similarly small perturbations of heme b_L.

Supported by the NIH (GM 33330), the USDA(93-37306-9084), the R.W Welch Foundation and the DOE Distinguished Postdoctoral Program.

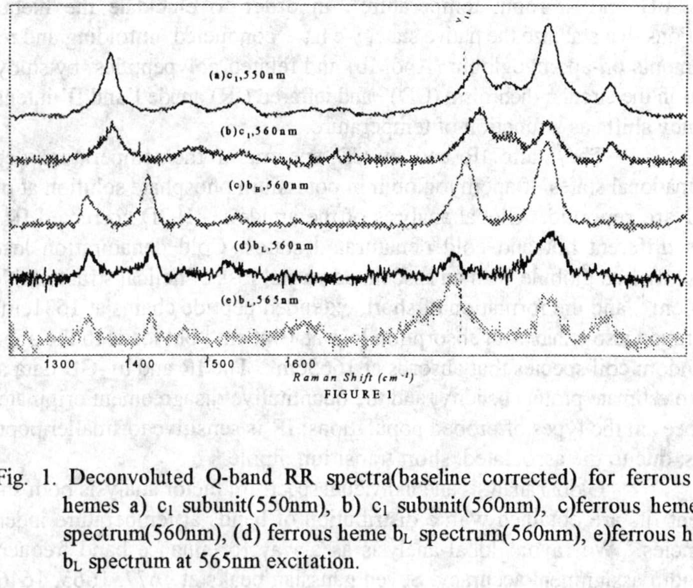

FIGURE 1

Fig. 1. Deconvoluted Q-band RR spectra(baseline corrected) for ferrous bc_1 hemes a) c_1 subunit(550nm), b) c_1 subunit(560nm), c)ferrous heme b_H spectrum(560nm), (d) ferrous heme b_L spectrum(560nm), e)ferrous heme b_L spectrum at 565nm excitation.

1. Knaff, D.B. (1993) *Photosynthesis Research 35*, 117-133.
2. Spiro, T.G. & Li, X. (1987) in *Resonance Raman Spectra of Heme and Metalloprotein.(Biological Applications of Raman Spectroscopy;v.3)*
3. Lou, B., Hobbs, J.D., Chen, Y., Yu, L., Yu, C., & Ondrias, M.R. (1993) *Biochem. et Biophys. Acta*
4. Hobbs, J.D., Kriaucuunas, A., Guner, S., Knaff, D.B., & Ondrias, M.R. (1990) *Biochem. et Biophys. Acta 1018*, 47-54.

A COMPARISON OF INFRARED AND ULTRAVIOLET CIRCULAR DICHROISM MEASURES OF APOMYOGLOBIN FOLDING STATES†

Brian K. Mohney, Victoria Uvarova, and Gilbert C. Walker*
Department of Chemistry, Chevron Science Center, University of Pittsburgh, 15260

The stability of different regions of folded globular proteins illustrates protein flexibility important in function and promises to reveal insight into protein folding paths. Protein folding is a process in which an extended polypeptide chain acquires compact packing through the formation of specific secondary and tertiary structures and hydrophobic interactions. Apomyoglobin, myoglobin with its heme group removed, exhibits a largely helical, compact, near native conformation near neutral pH and at room temperature[1]. In order to elucidate the hierarchy of interactions that stabilize the native state, we have conducted unfolding and refolding experiments on apomyoglobin (ApoMb) and related polypeptides by studying the changes in the circular dichroism (CD), and infrared (IR) amide I' and II' intensity and frequency shifts as a function of temperature.

The static IR and uv-CD spectra of the temperature dependent conformational states of apomyoglobin in potassium phosphate solution at pH*=5.0 in D_2O are reported. Global analysis of the amide I', HOD corrected IR spectra reveals different hot and cold denaturated states. Cold denaturation leads to a compact molten globule with an associated partial loss of helical structure identified at 1646cm^{-1}, and the formation of short, extended peptide chains at 1631cm^{-1}. Heat denaturation also exhibits an absorption loss at 1646cm^{-1} but leads to the appearance of a random coil species that absorbs at 1665cm^{-1}. The IR and uv-CD data are both useful to estimate protein helicity, and the quantitative disagreement originates in the differences in the types of probed populations: IR is sensitive to smaller populations subsets, due to the associated, short transition dipole.

Global analysis and individual spectrum factor analysis both show that excellent fits are obtained with a distribution of bands at temperature independent frequencies. We favor global analysis as a way to enhance band frequency and bandwidth assignment accuracy. Seven gaussian peaks at 1677, 1665, 1646, 1631, 1622, and 1570 were the minimum number necessary to provide a good fit, illustrated in the residuals which show nearly random displacements no more than 1% of the original data amplitude. The FWHM bandwidths associated with these frequencies were 10, 35, 30, 70, 26, 21, and 60cm^{-1} respectively.

1. Griko, Y.V., Privalov, P. L., Venyaminov, S. Yu. & Kutyshenko, V.P. (1988) *J. Mol. Biol. 202*, 127-138.

† This work was supported by financial support from the CRDF Fund of the University of Pittsburgh, 3M Company through an Untenured Faculty Award, and the Petroleum Research Fund.

SECTION 11

LIPIDS and BIOMEMBRANES

RAMAN STUDY OF THE EFFECT OF HIGH EXTERNAL PRESSURE ON DPPC - CHOLESTEROL MULTILAMELLAR VESICLES

Jiandong Guo, S. Medrano, and T. W. Zerda*

Physics Department, Texas Christian University, P. O. Box 298840, Fort Worth, TX 76129, USA

Incorporation of cholesterol into aqueous solutions of DPPC results in structural changes of the DPPC bilayer. In order to understand those changes we conducted a series of high pressure Raman experiments using a diamond anvil cell. The experiments were conducted within the pressure range between 1 and 30,000 bar at two temperatures, 23°C and 57°C. For comparison purposes the spectra were also recorded as a function of temperature, from 20°C to 60°C at the normal atmospheric conditions. Aqueous solutions of pure DPPC and 10, 20 and 40% DPPC-cholesterol mixtures have been studied. Raman spectral parameters, such as band frequencies, intensities, widths and shapes were analyzed. Analysis of those parameters yields information on conformer population, reorientational fluctuations, hydrogen bonding, interdigitation packing and phase transformation of the DPPC-cholesterol bilayers.

The vibrational frequency of molecular bonds is shifted when the pressure is increased. The shifts are due to changes in the conformation of the molecule or due to changes in the environment. Decreased distance between the molecules may result in increased contribution due to repulsive or attractive forces. Thus, the position of the band is a result of the balance of those two opposite effects. Repulsive forces shift the band toward higher frequencies, the so called blue shift, whereas attractive forces result in the red shift. Conformational changes are more difficult to explain, no general conclusions can be made, and frequency changes for each band need to be discussed separately.

In the region between 2800 - 3100 cm^{-1} there are several absorption bands due to symmetric and antisymmetric modes of the methylene chain and methyl groups. The positions of those bands are conformation sensitive and respond to temperature and pressure induced changes in the trans/gauche ratio in the acyl chains. In Fig. 1 we show pressure induced shifts of the 2880 cm^{-1} band for 20% cholesterol-DPPC mixture at 57°C. For a given conformation, when the pressure is increased a continuous shift of the band maximum is observed. However, during a pressure induced conformational change the slope of the frequency vs. pressure may change. As seen in Fig. 1 for pure DPPC the discontinuities in the slopes are observed at the pressure values of about 0.8, 1.7, 5.3, 9, and 20 kbar. Those pressure values correspond nicely to the previously observed pressure induced phase transitions in pure DPPC [1] and assigned to transition from the liquid crystalline phase L_α, to the ripple gel phase GI ($P_{\beta'}$), from the GI to the gel phase $L_{\beta i}$, from the $L_{\beta i}$ to the GIII, from the GIII to the Gx phase, and from the Gx to another unidentified phase G, respectively.

Similar observations were made for cholesterol - DPPC mixtures, although those discontinuities were observed always at somewhat higher pressures, compare Fig. 2. Higher pressure required to induce a phase transformation indicates that the degree of disorder of the lipid molecules increases with increased cholesterol concentration. At the highest cholesterol loading used in this experiment, all phase transformations smear out or even completely disappear. Similar conclusion can be reached from the analysis of pressure induced shifts observed for other bands. It is worth noticing that the pressure induced intensity changes correlate nicely with the observed frequency shifts.

The magnitude of temperature induced changes also depends on cholesterol concentration. For high cholesterol concentrations the phase transformations are very broad and

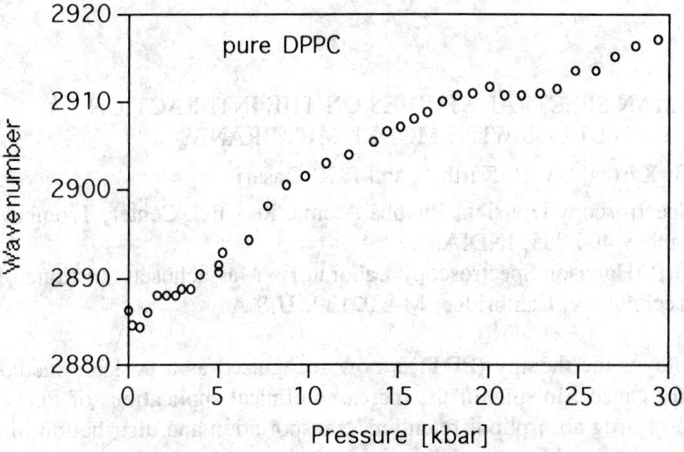

Fig. 1. Pressure dependent wavenumber of the C-H stretching mode of aqueous DPPC at 57°C.

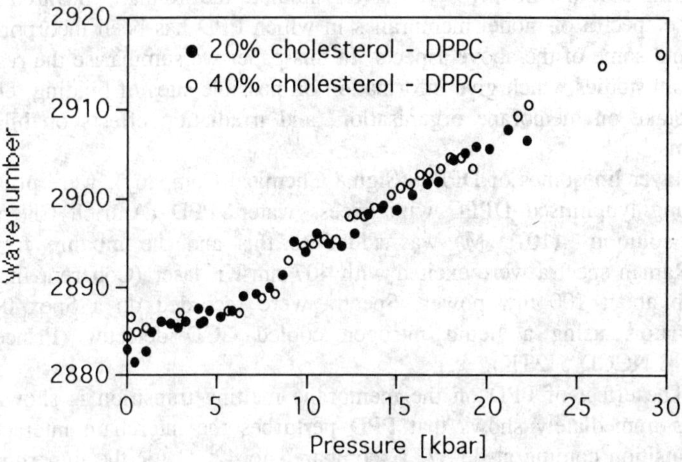

Fig. 2. The pressure dependence of the C-H stretching mode of acyl chains for DPPC-cholesterol mixtures at 57°C.

it is only possible to define the range of temperatures where the phase transition takes place.

Cholesterol increases reorientational disorder of the lipid molecules. Reorientational fluctuations of the molecules and twisting/torsion motions of the acyl chains can be hindered only by increasing the pressure. It appears that the important role of cholesterol is to make the membrane less sensitive to changes in the external environment.

1. O. Reis, R. Winter, and T. W. Zerda, Biochim. Biophys. Acta, in press (1996).

RAMAN SPECTRAL STUDIES ON THE INTERACTION OF PDT DRUGS WITH MODEL MEMBRANES.

S.B. Kartha[1*], V.B. Kartha[2], and R.R. Dasari[2]

1 Spectroscopy Division, Bhabha Atomic Research Center, Trombay, Bombay 400 085, INDIA.

2 G.R. Harrison Spectroscopy Laboratory, Massachusetts Institute of Technology, Cambridge, MA 02139, U.S.A.

Photdynamic therapy (PDT) is now recognised as a possible method of treatment for cancer. In spite of the increased clinical applications of PDT , the mechanism of drug absorption, retention, transportation and distribution in cells and tissues, relative affinity to cell components, site of primary interaction,etc. have not yet been understood clearly. Fluorescence [1], chemiluminescence [2], and various physicochemical methods [3], have been employed for this purpose. It has been suggested that Protoporphyrin (PPD) may be aggregated in membranes, may be bound to surface sites or sites inside the bilayer [1], or buried within the bilayer [3]. We have studied the Raman, infrared and fluorescence spectra of model membranes in which PPD has been incorporated to understand some of the above aspects. In this paper we summarise the results of the Raman studies which give information on possible sites of binding, effect of PPD uptake on membrane organisation, and irradiation effects on bilayer lipid systems.

Bilayer liposomes of DPPC (Sigma Chemical Company), were prpared by sonicating lyophilised DPPC with excess water. PPD (Aldrich Chemical Company) solution (10^{-5} M) was added to this and the mixture further sonicated.Raman spectra were excited with 407 nm Kr^+laser (Coherent Innova 90 K) with about 100 mw power. Spectra were recorded on a Spex 0.6m monochromator, using a liquid nitrogen cooled CCD detector (Princeton Instruments LNCCD 512TK).

The effect of PPD on the membrane melting transition is shown in Fig.1. This immediately shows that PPD perturbes the interchain interaction [4], the transition coming at 35-37^0 C compared to 42^0 C for the unperturbed membrane. Fig.1 also shows that the extent of order (disorder) is not altered much by the PPD, as seen from the initial and final intensity ratios I_{2930}/I_{2880}, indicating that the drug is not embedded in the lipid region.

Figure 2 shows the spectra for the C-C stretch (1124 cm^{-1}), CH_3 rock (890 cm -1), and C-N (717 and 874 cm^{-1}) stretch regions . The first two regions are highly perturbed even at room temperature, whereas in pure DPPC only the melting transition produces intensity changes. It has been pointed out that the intensity of the 1124 cm^{-1} band is highly sensitive to kink structures[5].The perturbatrion of the CH_3 rock shows that the PPD has entered the inter bilayer region and this has lead to the formation of the kinked structures affecting the intensity of the 1130 cm^{-1} band. As has been pointed out by Pink et al [5], this does not affect the interchain packing and we do not see any change in the C-H stretch region. Since the C-H or C-N stretch regions do not show any change at this stage, it is clear that PPD is not attached to the membrane surface and also not embedded in the inter chain lipid region. We have found no spectral changes even after 30 mts of irradiation with 100 mw laser power focussed on to an area of about 50 microns. This suggests that any damage to cell membranes may be taking place through interaction of other components of the membrane with the radiation, and not of the saturated lipids.

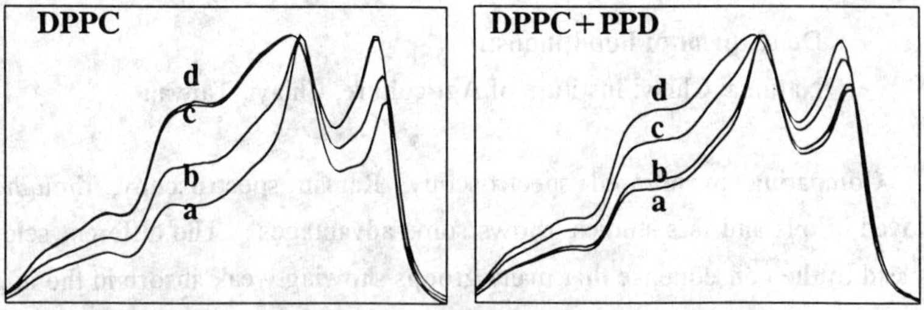

Fig. 1. Raman spectra of C-H stretch region of DPPC & DPPC+PPD.
DPPC: a-27^0; b-43^0; c-45^0; d-47^0. DPPC+PPD: a-27^0; b-33^0; c-35^0; d-37^0 C.

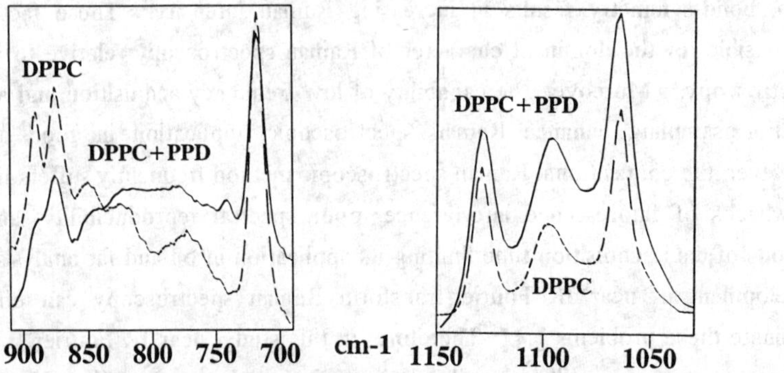

Fig. 2. Raman spectra- CH_3 rock, Head group, & C-C stretch region. 27^0 C.

1. E.B. Chernyaeva, A.G. Varanyan, N.I. Koroteev, V.F. Kamalov, and O.V. Lobanov, J. Photochem. Photobiol.,B: Biol.,10,239 (1991).
2. T. Takemura, N. Ohta, S. Nakajima, and I. Sakata, Photochem. and Photobiol., 55,137 (1992).
3. K.W. Woodburn, N.J. Vardaxis, J.S. Hill, A.H. Kaye, J.A. Reiss, and D.R. Phillips, Photochem. and Photobiol., 55 705 (1992).
4. Ira W. Levin, " Vibrational Spectroscopy of Membrane Assemblies " Chapter 1, Advances in Infrared and Raman Spectroscopy, Vol. 11 (eds. R.J.H. Clark and R.E. Hester) John Wiley & Sons, New York (1984).
5. D.A. Pink, T.J. Green, and D. Chapman, Biochemistry, 19, 349 (1980).

NEAR IR FOURIER-TRANSFORM RAMAN SPECTROSCOPIC ANALYSIS OF PEANUT OIL EXTRACTED BY SUPERCRICTICAL CARBON DIOXIDE

Kuo-Lung Ku, Robin Y. Y. Chiou and Wenlung Chen*

Department of Food Industry
National Chiayi Institute of Agriculture, Chiayi, Taiwan

Comparing to infrared spectroscopy, Raman spectroscopy, though less employed in oils and fats studies, shows some advantages. The different selection rules lead to the consequence that many groups showing weak absorb in the infrared region exhibit strong response in Raman scatters. Groups such as C = C, S — C, N = N, and S — S are observed more clearly in Raman spectra than in infrared spectra. Also, bond-symmetry results in increasing Raman intensity. These factors are responsible for the dominant character of Raman spectroscopy relative to infrared spectroscopy. Moreover, the capability of low frequency acquisition and aqueous solution sampling enhance Raman spectroscopic application in food analysis. However, the conventional Raman spectroscopic method frequently suffers from the drawbacks of fluorescence interference, poor spectral reproducibility, and long periods of data acquisition time limiting its application in oil and fat analysis. The development of near IR Fourier-transform Raman spectroscopy can efficiently eliminate these problems [1]. Therefore, in this study, near IR Fourier transform Raman spectroscopy will be employed in analyzing the peanut oil under different extraction process.

Roasted peanuts is a popular food product because of its unique texture and pleasant flavor; however, the high oil content peanuts presents a major threat for some consumers. The defatting process of roasted peanuts has received intensive attention. Supercritical fluid extraction (SFE) is a powerful technique for differential extraction and attracts much interest of food technologists [2,3]. Supercritical carbon dioxide (SC-CO_2) is one of the best choice for food extraction because of its nontoxic, inexpensive, and intrinsic physico-chemical properties. Thus, supercritical carbon dioxide is used to extract oil from roasted peanut meals and kernels. The peanut oil is then forward to compositional analysis by near IR Fourier-transform Raman spectroscopy. The Raman spectra of peanut oil extracted under different conditions is demonstrated in following figure. Results showed that the composition of peanut oils remains no change even the extraction condition is changed.

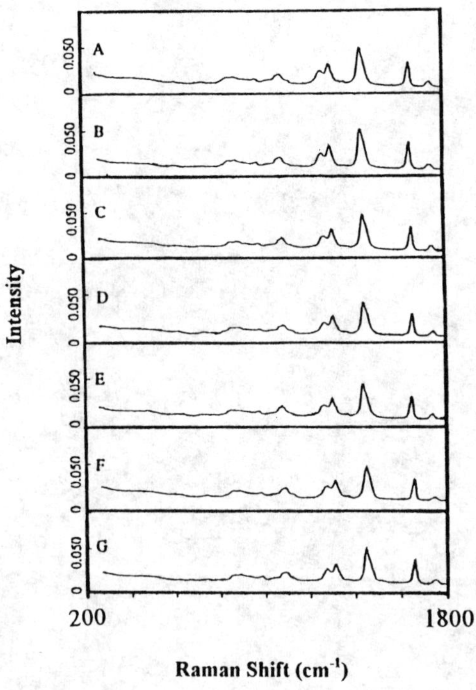

Figure: Raman spectra of peanut oils extracted from roasted, deskinned whole kernels by a hydraulic press (A) and by SC-CO_2 using SCP at 8000 psi/80 °C (B), IDP at 8000 psi/80 °C (C), IDP at 6000 psi/80 °C (D), IDP at 4000 psi/80 °C(E), IDP at 3000 psi/80 °C (F), IDP at 2000psi/80 °C (G). *SCP: Semicontinuous process, IDP: Intermittently depressurized process.

(1) W. Chen, S. Nie, J. F. Kuck and N.-T. Yu (1991) Biophys. J. 60,447-455.
(2) C. R. Santerre, J. W. Goodrum, and J. M. Kee (1994) J. Food Sci. 59, 382-386.
(3) Z. R. Yu, B. Singh, S. S. H. Rizvi, and J. A. Zollweg (1994) J. Supercrit. Fluids, 7, 51-59.

SECTION 12

NUCLEIC ACIDS and VIRUSES

RAMAN SPECTROSCOPIC STUDY OF THE MELTING OF ASSOCIATED POLYGUANYLIC ACID IN AQUEOUS SOLUTION: THERMAL AND STRUCTURAL BEHAVIOR

Simon Roy, Pascal Mercier, and Rodrigue Savoie*

Département de chimie, Université Laval, Cité Universitaire, Québec, QC, Canada G1K 7P4

Polyguanylic acid [poly(G)] is well known for its tendency to become self-associated in aqueous solution. Various methods, including X-ray diffraction on fibers [1], have shown that it forms a right-handed, four-stranded helical structure, with each base within a tetrad linked to its neighbors by N1-H...O-C6 and NH_2...N7 hydrogen bonds. This complex is extremely stable and it does not melt below 100°C, unless destabilizing species, such as Et_4N^+ ions, are added to the solution [2]. We have studied by Raman spectroscopy the thermal behavior of four-stranded poly(G) in pressurized aqueous solutions. Typical spectra below and above the melting transition are presented in Fig. 1.

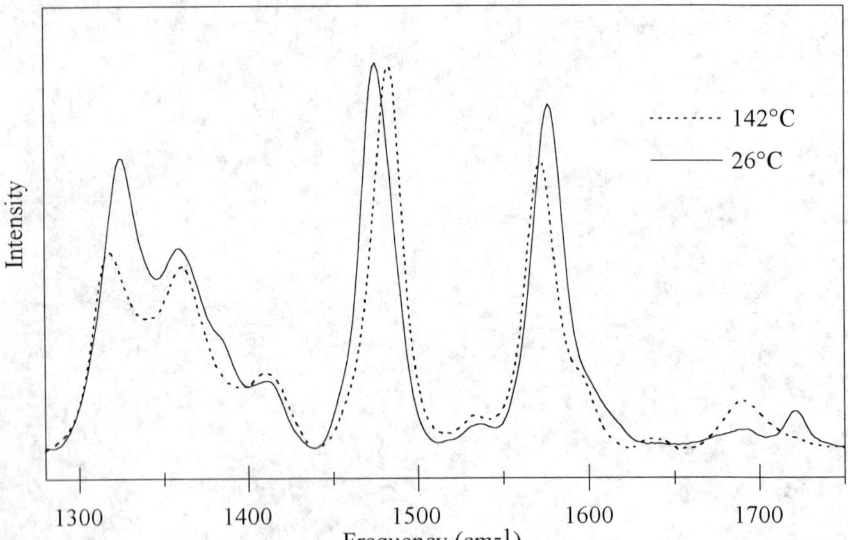

Fig. 1 Raman spectrum of an aqueous solution of the potassium salt of poly(G) (4% v/v, 0.5M NaCl, pH 7) at (A) 25°C and (B) 150°C.

Many changes in the spectrum can be used to monitor the melting transition of the four-stranded helix with increased temperature, such as the frequency of the guanine band at 1475 cm^{-1}, which shifts to higher values by about 8 cm^{-1}, and the intensity reversal of the carbonyl bands at 1691 and 1721 cm^{-1}, the latter being characteristic of the presence of the complex. Examples (obtained from the solution described in Fig. 1) are shown in Fig. 2. The melting process occurs quite cooperatively near 130°C with the type of solutions that we have used, this value being higher in KCl and lower in LiCl and alkyl ammonium salt solutions. The melting process is accompanied by some thermal degradation of the samples, as indicated by a strong fluorescence effect which does not disappear upon cooling and makes further Raman measurements impossible when excitation in the visible is used.

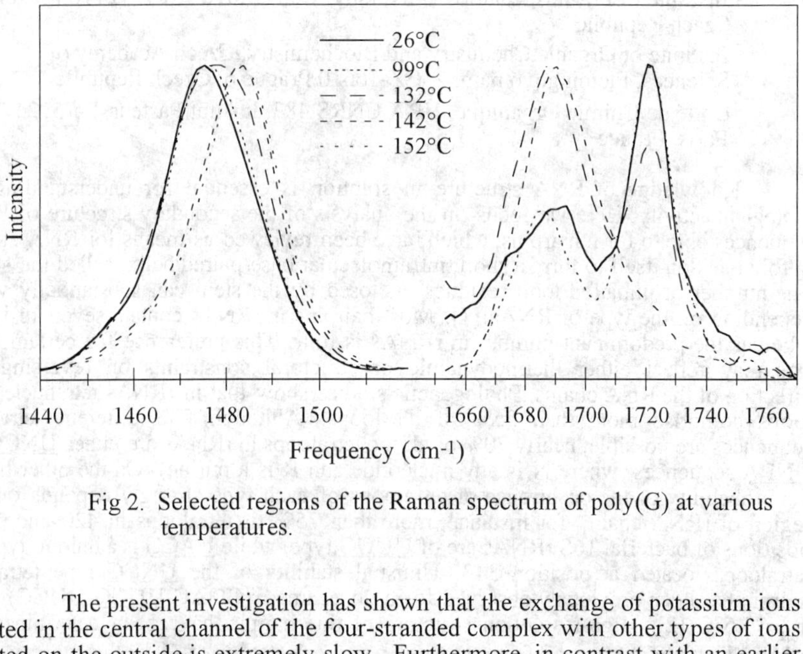

Fig 2. Selected regions of the Raman spectrum of poly(G) at various temperatures.

The present investigation has shown that the exchange of potassium ions located in the central channel of the four-stranded complex with other types of ions located on the outside is extremely slow. Furthermore, in contrast with an earlier Raman study [3] which suggested an abnormal conformation of the ribose units in self-associated poly(G), the spectral changes observed in the present study indicate a partial shift towards a normal C3'-*endo* conformation upon melting of this polynucleotide.

1. S.B. Zimmerman, G.H. Cohen, and D.R. Davies, J. Mol. Biol. 92, 181 (1975).
2. F.B. Howard, J. Frazier, and H.T. Miles, Biopolymers 16, 791 (1977).
3. C. Simard and R. Savoie, Biopolymers 34, 91 (1994).

STRUCTURAL PROPERTIES OF UNCG TETRALOOPS IN SHORT OLIGORIBONUCLEOTIDES AS REVEALED BY UV, CD, RAMAN AND INFRARED SPECTROSCOPIES

M. Abdelkafi[1], M. Ghomi[1], V. Baumruk[2], L. Bednárová[3], A. Namane[4], C. Gouyette[4], T. Huynh-Dinh[4] and P.-Y.Turpin[1]

[1] LPBC, URA CNRS 2056, Université Pierre et Marie Curie, Case Courrier 138, 4 Place Jussieu, F-75252 Paris Cedex 05, France

[2] Institute of Physics, Charles University, Ke Karlovu 5, CZ-12116 Prague 2, Czech Republic

[3] Institute of Organic Chemistry and Biochemistry, Czech Academy of Sciences, Flemingovo nám. 2, CZ-16610 Prague 6, Czech Republic

[4] Unité de Chimie Organique, URA CNRS 487, Institut Pasteur, F-75724 Paris, France

Knowledge of RNA structure in solution is essential for understanding its biological activity. Here we focus on the analysis of the secondary structure of RNA sequences able to form hairpins, which have been reviewed as means for RNA strand to fold back on itself to form a short, intramolecular base-paired helix, called the stem. The number of unpaired loop residues enclosed by the stem can substantially vary, depending on the type of RNAs [1]. Most hairpins in tRNAs contain seven residues whereas the predominant number in rRNAs is four. This preference for certain loop size may reflect either thermodynamic or structural constraints on reversing the direction of the RNA chains. Phylogenetic studies show that in rRNAs tetranucleotide loops comprise more than 55% of all loops. Although 256 different tetraloop sequences are possible, nearly 70% of all the tetraloops in rRNAs are either UNCG or GNRA sequences (where N is any nucleotide and R is a purine). On the other hand, the probability of the occurrence of tetraloops of each type strongly depends on the region of rRNA chain : for instance, more than 75% of tetraloops in 420 and 1029 positions of bacterial 16S rRNAs are of UUCG type, while UACG is a unique type of tetraloops located at position 343. Unusual stability of the UNCG type tetraloop hairpin has been previously observed for a dodecamer r(GGAC-UUCG-GUCC) with T_m=71.5°C [2]. High-resolution structure of this hairpin have been determined by NMR. The main structural features were : syn orientation of G in the loop, a G-U base-pair formed between the first and last residues of the loop, and C2'-endo ribose configuration of the remaining two residues of the loop [2].

We have studied UNCG tetraloops in octamers containg only two G-C base pairs in a stem. For instance, melting profiles obtained from UV and CD spectra confirmed high stability of the UACG loop in the octamer (T_m= 51°C) [3]. Raman and IR spectra of the octamer were studied in H_20 and D_2O buffers in the 0 - 68 °C temperature range. All UV, Raman and IR spectra were completely reversible with temperature. Vibrational spectra well below T_m correspond to the structured hairpin and those above T_m to the disordered chain (Fig. 1).

The presence of the strong Raman line at 812 cm^{-1} reveals the A form conformation of the stem which is further confirmed by the IR line at 1239 cm^{-1}. Strong Raman line at 1321 cm^{-1} is assigned to the rG residue in the tetraloop adopting a syn orientation. Raman line at 1359 cm^{-1} is also consistent with the rG residues having an N-type sugar and syn base in the loop, while the line at 1381 cm^{-1} corresponds to N-type sugar/anti rG residues in the stem. Two medium Raman lines at 671 and 636 cm^{-1} are assigned to the breathing modes of the rG residues in the stem and loop, respectively. The Raman lines at 1530 and 1252 cm^{-1} correspond to the rC residues (C3'-endo/anti) involved in the stem, while those at 1510 and 1262 cm^{-1} (shoulder) should originate from the rC residue in the loop (S-type sugar/anti base). Strong line observed in Raman spectra at 1237 cm^{-1} is assigned to the rU. However, its position corresponds well to hydrogen bonded U (probably a reverse wobble-pair with G in the loop). The two Raman lines at 725 and 715 cm^{-1} indicate possible occurrence of two conformations of the unique rA residue present in the tetraloop; the former should correspond to an rA residue with S-type sugar/anti base and the latter may be assigned to a syn base orientation. This confirms a possible rapid interconversion of the anti to syn base orientation in the rA residue.

The main structural features of the UACG tetraloop and the hairpin are in good agreement with those proposed previously for UUCG tetraloop mentioned above [2] and were further confirmed by NMR data [3].

Vibrational spectroscopy thus represents an alternative powerful tool for structural studies of hairpins in solution.

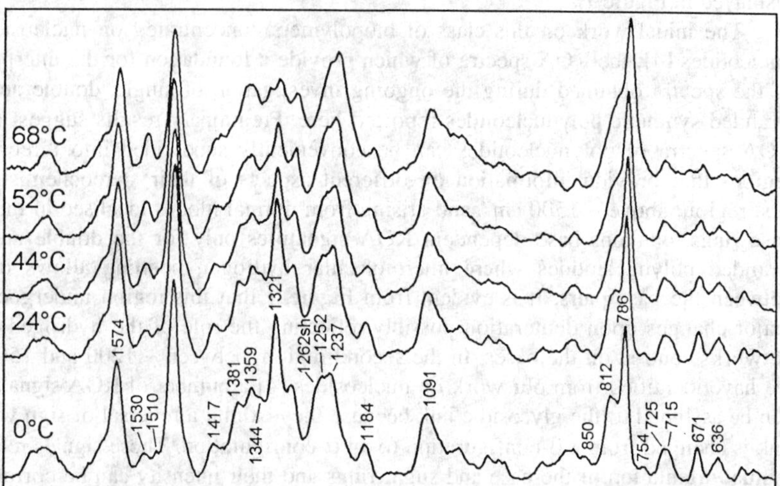

Fig. 1. Raman spectra of r(GC-UACG-GC) octamer in H_2O buffer in the 0°C - 68°C temperature range.

1. O. C. Uhlenbeck, Nature 346, 613 (1990).
2. Ch. Cheong, G. Varani, I. Tinoco Jr., Nature 346, 680 (1990).
3. M. Abdelkafi, O. Lampire, C. Hervé du Penhoat, M. Ghomi, C. Gouyette, A. Namane, T. Huynh-Dinh, V. Baumruk, L. Bednárová, P.-Y. Turpin, N. Bouchemal-Chibani, P. Goyer, submitted for publication.

VIBRATIONAL RAMAN OPTICAL ACTIVITY OF POLYNUCLEOTIDES

A. F. Bell, L. Hecht and L. D. Barron*

Chemistry Department, Glasgow University, Glasgow G12 8QQ, U.K.

Vibrational Raman optical activity (ROA) is measured in our laboratory as the intensity difference in Raman scattered light from chiral molecules in right and left circularly polarised incident light [1]. Recent instrumental advances [2] have meant that nucleosides, nucleotides and nucleic acids, which were previously regarded as difficult samples due to their weak ROA spectra [3], can now be recorded routinely. As examples the ROA spectra of poly(rG)•poly(rC), which adopts an A-type helical conformation at neutral pH, in phosphate buffer and deuterated phosphate buffer are displayed in Figure 1.

The initial work on this class of biopolymers concentrated on nucleosides and nucleotides [4], the ROA spectra of which provide a foundation for the interpretation of the spectra obtained during the ongoing investigation of single, double and triple stranded synthetic polynucleotides reported here. Preliminary results suggest that the ROA spectra of polynucleotides can be conveniently subdivided into three distinct regions that provide information on different aspects of their stereochemistry. The first region, above ~ 1500 cm^{-1} and arising from normal modes localised in the planar base rings, contains base dependent ROA signatures only for the double and triple stranded polynucleotides where intermolecular hydrogen bonding allows coupling between the base pairs. It is evident from Figure 1 that this region undergoes some major changes upon deuteration possibly reflecting the role of the hydrogen bonded networks centred on the bases. In the second region, between ~ 1200 and 1500 cm^{-1}, we have identified from our work on nucleosides [4] a number of ROA signals which can be assigned to the glycosidic link because they exhibit a reversal of sign when the link is changed from a β-configuration to an α-configuration. These signals reflect the mutual orientation of the base and sugar rings and their intensity can be correlated to changes in the degree of order of the polynucleotide. In the third region, between ~ 800 and 1200 cm^{-1}, the normal modes are mainly localised in the sugar rings and phosphate groups and generate ROA patterns that are independent of the base but which reflect sugar ring and sugar-phosphate backbone conformation. For the A-type double helices (Figure 1) the ROA sign pattern is a positive-negative-positive triplet between ~ 940 and 1100 cm^{-1} and deuteration seems to have little effect on the ROA spectra below ~ 1500 cm^{-1}. These results indicate that ROA studies may provide information on many of the key features of nucleic acid stereochemistry and that deuteration studies may provide useful additional data on the origin of the ROA signals above ~ 1500 cm^{-1}.

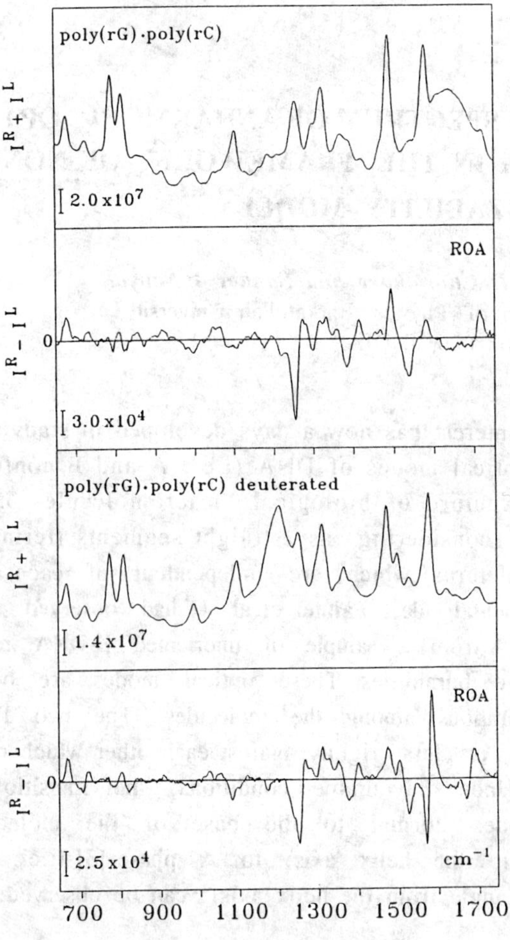

Fig.1 The Raman and ROA spectra of poly(rG)•poly(rC) in phosphate buffer pH7 (top) and in deuterated phosphate buffer pD7.4 (bottom) at 50mg/ml.

1. L. D. Barron and L. Hecht, in Circular Dichroism: Principles and Applications (K. Nakanishi, N. Berova and R. W. Woody, eds.), VCH, New York, 1994, p. 179.
2. L. Hecht and L. D. Barron, Faraday Discussion 99, 35 (1994).
3. L. Hecht, L. D. Barron, Z. Q. Wen and S. J. Ford, in Proceedings of the Thirteenth International Conference on Raman Spectroscopy, (W. Kiefer, M. Cardona, G. Schaak, F. W. Schneider and H. W. Schrötter, eds.) Wiley, Chichester,1992, p.1089.
4. A. F. Bell, L. Hecht and L. D. Barron, J. Chem. Soc. Faraday Trans., submitted for publication.

RAMAN SPECTRUM OF ANHARMONIC OPTICAL MODES OF DNA IN THE FRAMEWORK OF NONLINEAR Φ^4 POLARIZABILITY MODEL

Nabonita D. Chowdhury and Sankar P. Sanyal
Department of Physics, Barkatullah University,
Bhopal - 462 026, INDIA.

Lot of interest has now a days developed in studying the nonlinear dynamics of optical modes of DNA (i.e., A and B conformations). The complicated structure of biological macromolecule DNA is made simplified by considering as straight segments (termed as molecule) contains several turns, which are independent of each other. Looking to the experimental side, Painter et al [1] had observed a broad Raman line at 30 cm^{-1} from a sample of unoriented B-DNA molecules at two different relative humidities. These optical modes are believed to have large scale vibrations among the molecules. The two DNA modes are displaced more or less rigidly against each other which remain parallel to the helix axis. Under favourable conditions, the transition [2] from the B phase (i.e., modes normal to the base of the molecule which are nearly parallel to the helix axis) to A phase [3] (i.e., bases are tilted by an small angle from the helix axis) can be observed.

Martin [4] had presented a mathematical model which simulate the vibrations of DNA molecule in A and B phase by considering the motion in an anharmonic oscillator, arising from one-dimensional asymmetric quartic double well oscillator. Minima of this potential will represent the A and B conformations and the magnitude of asymmetry parameter will determine the lowest minima of A and B. They had obtained the depolarization ratio and the dependence of the Raman spectrum on the asymmetry of the double well [4].

In the present paper, we present an analytical study of the nonlinear excitations and Raman spectrum for the A and B modes of DNA, in the framework of nonlinear Φ^4 polarizability model [5]. This type of modelling was first proposed by Aubry [5] and by Krumhansl and Schrieffer [6] to describe the displacive type of phase transition in solids. The model we use for our treatment is an one-dimensional diatomic linear chain, in which only anions are supposed to experience an onsite asymmetric nonlinear electron-phonon coupling of the order three and four. Nearest and next nearest neighbour force constants between cation and anion are taken into consideration. We had solved the equation of motion in core-shell coordinate of anion, whose dynamics experiences an asymmetric quartic double well oscillator (observed from the energy versus displacement curve at the turning point of the double well). We had calculated numerically the frequency of the anharmonic oscillation and thus obtained two nearby close transition temperatures near the mode softening.

For calculating Raman spectrum, we had adopted the formalism developed by Mills and Maradudin [8] for anharmonic noninteracting oscillators in the high frequency region of fundamental latice absorption. We had calculated the absorption rate as a function of frequency of the incoming light frequency at three different temperatures, depending on the different magnitudes of nonlinear core-shell force constants of polarizable ion. At very low temperature, the absorption spectrum consists of a sharp localised peak at the higher frequency side, which corresponds to one of the well, whose minimum is lowest as compared to the other. As one proceeds towards the higher temperature, another localised peak of less magnitude arises from harmonic oscillation in the less depth well. In addition, the single peak at higher frequency side broadens which looses its height and a second peak appears in the low frequency side corresponding to the symmetric oscillations. At very high temperatures, only the symmetrical oscillations move towards the higher frequency side.

1. P.C.Painter, L.Mosher and C.Rhoads, *Biopolymers* **20**, 243 (1981).
2. J.M.Eyster and E.W.Prohofsky, *Phys. Rev. Lett.* **38**, 371(1977).
3. R.E.Franklin and R.G.Gosling, *Nature* **172**, 156-157 (1953)
4. A.J.Martin, *Biopolymers* **23**, 471-491 (1984).
5. N.D.Chowdhury and S.P.Sanyal, *Pramana J. Phys.* **37**, 115 (1991); **39**, 101 (1992).
6. S.Aubry, *J.Chem. Phys.* **62**, 3217 (1975); **64**, 3392 (1976).
7. J.A.Krumhansl and J.R.Schrieffer, *Phys. Rev.* **B11**, 3535 (1975).
8. D.L.Mills and A.A.Maradudin, *Phys. Rev.* **B8**, 1617 (1973).

RESONANCE RAMAN STUDIES OF DNA INTERCALATORS

C. G. Coates[1]*, L. Jacquet[2], J. J. McGarvey[1], S.E.J. Bell[1], A. H. R. Al-Obaidi[1], and J. M. Kelly[2].

[1]School of Chemistry, Queen's University of Belfast, Belfast, BT9 5AG, N.I.
[2]Chemistry Department, Trinity College, University of Dublin, Dublin 2, Ireland.

In recent years, design of small, non-radioactive molecules, capable of selectively binding to nucleic acids has been an area to which considerable attention has been devoted [1-4]. To successfully characterise the nature and photophysical implications of interactions which exist between the two, it is desirable to employ a selective spectroscopic method. Mixed ligand complexes of the type $[Ru(L)_2DPPZ]^{2+}$ (L = 2,2' bipyridine (bpy), 1,10 phenanthroline (phen) ; DPPZ = dipyrido[3,2:a-2',3':c]-phenazine) have been investigated by Barton and co-workers as luminescent DNA intercalators [1,2]. We report a thorough investigation by nanosecond pulsed laser resonance Raman spectroscopy , confirming the technique as an effective probe of these systems.

Ground state RR spectroscopy, through selection of appropriate probe wavelengths, has proven effective not only in providing information concerning the nature of the transitions involved in the ground state UV-vis electronic spectrum, but also in affording a more detailed insight into the effects of intercalation upon the transitions. These studies reveal that changes observed in the absorption spectrum upon intercalation are centered on transitions involving the intercalating DPPZ ligand.

RR spectra of the short-lived excited states of each of the above complexes in the absence of DNA were recorded using pulsed excitation at 355nm [3]. The spectra reveal features mainly attributable to the $DPPZ^{\cdot -}$ ligand. Hence, short-lived transients, undetectable by nanosecond absorption and luminescent techniques, can be effectively probed by single pulse RR scattering. DNA intercalation resulted in significant changes to these excited state spectra. Subtractions of spectra recorded of each complex in buffer from corresponding spectra recorded in presence of DNA, yielded features attributable mainly to the respective neutral ligand vibrations in the $Ru^{III}(L)_2DPPZ^{\cdot -}$ excited state (exemplified for $Ru(phen)_2DPPZ$ in Fig. 1A) . More significantly however, each subtracted spectrum shows a prominent new feature at $1526cm^{-1}$ (other new features may exist but are too weak to be confidently distinguished from the spectral noise). This band may be reasonably assigned as a mode of $DPPZ^{\cdot -}$, but one which is not present in the transient RR spectra recorded in buffer solution. It would seem that this mode arises through resonance with a π^*-π^* transition of $DPPZ^{\cdot -}$ which is distinct from that observed in the absence of DNA and attributable to the intercalative interaction of the DPPZ ligand with the binding sites.

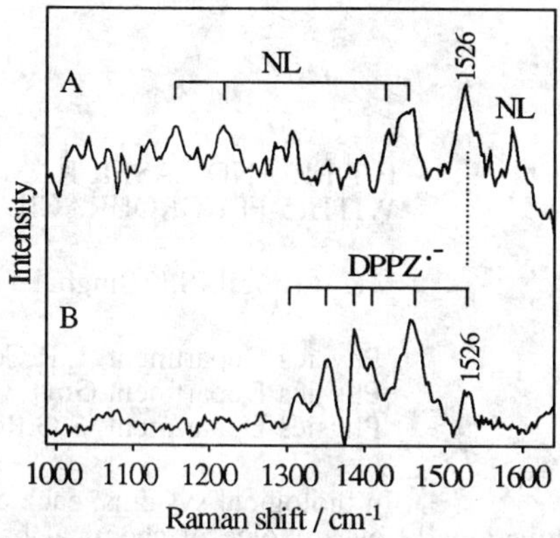

Figure1. Transient RR spectra of $Ru(phen)_2DPPZ$ recorded at λex=355nm. (A) Scaled subtraction of spectrum recorded in buffer from spectrum recorded in presence of DNA (NL denotes neutral ligand modes of the corresponding ancillary ligands) ; (B) Spectrum recorded in acetonitrile.

Transient RR spectra recorded of the complexes in acetonitrile (exemplified for $Ru(phen)_2DPPZ$ in Fig. 1B) also show the feature at 1526cm^{-1}, strengthening its assignment as a mode of $DPPZ^{\cdot-}$. It is possible that appearance of this mode is indicative of a 'non-aqueous' environment.

Investigations into the enantiomeric forms of $Ru(phen)_2DPPZ$ by Norden and co-workers [4], indicate a luminescent quantum yield for the bound Δ enantiomer 6-10 times larger than that of the bound Λ enantiomer. In addition, a slight difference in binding geometry of the two enantiomers was suggested. In line with this, we have recorded transient RR spectra of free and bound Δ and Λ enantiomers of $Ru(phen)_2DPPZ$ at an excitation wavelength of 355nm. Spectral subtraction in each case again reveals the same 1526cm^{-1} mode as in the racemic mixture. Significantly, this band is stronger in the case of the Δ enantiomer. The average of the two spectra is identical to the spectrum of the racemate. In contrast to the situation in DNA-buffer, spectra recorded at 355nm of each enantiomer in acetonitrile show identical enhancement patterns for both. Hence, it seems possible that acetonitrile provides an 'achiral' non-aqueous environment whereas DNA provides a 'chiral' non-aqueous environment.

Preliminary experiments have also been conducted with the complex $Ru(TAP)_2(DPPZ)$ (TAP = 1,4,5,8-tetraazaphenanthrene). While intercalation of the DPPZ ligand persists, electron localization is now on the TAP ligand in the MLCT excited state. Excitation of the bound complex results in a rapid electron transfer from the guanine residue of DNA to the oxidized Ru^{III} center.

Acknowledgements. We acknowledge DENI for a distinction award (C.G.C.) and EPSRC for support of the work (Grant GR/J01905).

1. E. Friedman, J.-C. Chambron, J.-P. Sauvage, N. J. Turro and J. K. Barton, J. Am. Chem. Soc., 112, 4960 (1990).
2. Y. Jenkins, A. E. Friedman, N. J. Turro and J. K. Barton, Biochemistry, 31, 10809 (1992).
3. C. G. Coates, L. Jacquet, J.J. McGarvey, S.E.J. Bell, J.M. Kelly, and A.H.R. Al-Obaidi, J. Chem. Soc. Chem. Commun., 35 (1996).
4. H. Hiort, P. Lincoln and B Norden, J. Am. Chem. Soc., 115, 3448 (1993).

F T I R AND LASER RAMAN SPECTRA OF COMPLEX OF Cu(II) WITH 5-FLUOROURACIL

V.K.Rastogi[1], S.K.Singhal[1], M Upadhyay[2] and A.K.Dublish[3]

[1]Physics Department, L.R.College, Sahibabad-201 005, India
[2]Physics Department,Govt Model Science College,Rewa,India
[3]Physics Department, S.G.R.R. (PG) College, Dehradun, India.

In biological systems, each element involved in the metabolism is tuned functionally by a choice of chemical having discriminating behaviour for different donor sites present in the organic molecules. Thus the general question of how organic molecules or molecular fragments interact with metal ions is of great importance? Vibrational spectroscopy is a powerful tool and sensitive monitor to investigate the different coordinating sites present in organic molecules and the mechanism by which metal ions interact with organic molecules because the bands observed in vibrational spectroscopy arises from all parts of the metal-ligand complex. During the last couple of years[1-3], there has been a considerable interest in the interaction of transition metal ions with nucleic acid bases and their derivatives because of their biological importance.

TABLE 1 Assignment of fundametal frequencies of 5-FU (cm^{-1})

IR	Raman	Sym sp	Assignment	IR	Raman	Sym sp	Assignments
3130	–	a′	$\nu(N_1-H)$	940	–	a″	γ(N–H)
3060	–	a′	$\nu(N_3-H)$	880	–	a′	β(C–F)
3024	–	a′	ν(C–H)	815	810	a″	γ(C–H)
1720	–	a′	ν(C=O)	752	765	a″	γ-ring
1670	–	a′	ν(C=O)	680	–	a′	β(C=O)
1558	–	a′	β(N–H)	630	635	a′	β-ring
1540	–	a′	β(N–H)	545	542	a′	β(C=O)
1508	–	a′	ν(C–H)	465	467	a″	γ(C–F)
1458	–	a′	ν(C–N)	417	410	a″	γ-ring
1450	–	a′	ν-ring	–	362	a″	γ(C=O)
1430	–	a′	β(C–H)	–	329	a″	γ-ring
1350	–	a′	ν-ring	–	170		lattice
1255	1250	a′	ν(C–F)	–	107		lattice
997	–	a′	β-ring	–	88		lattice

Like other 5-halogenated uracil derivatives , 5-fluorouracil(5-FU) is an antimetabolite that is incorporated into nucleic acids and have the ability to bind metals. In order to understand the mechanism by which the metal ions intervene in the function of nucleic acid bases, and to find the coordinating sites in 5-FU, we have undertaken the present study. In this communication , we report the isolation and characterisation of the complex of Cu(II) with 5-FU. Chemical analyses are used to determine stoichiometry and vibrational spectroscopy is used to illucidate the vibrational modes and sites of coordination.

The complex was synthesised as follows. First of all 5-FU, obtained from M/s Sigma Chem Co, U.S.A., was dissolved in ethyl alcohol. Then hot copper nitrate solution (1 m mole) prepared in ethanol (15 ml)-water (100 ml) mixture was mixed with hot alcoholic solution of 5-FU(1 m mole). The resulting mixture was heated slightly and the NaOH was added to adjust pH to about 5-7. The white precipitate thus obtained was filtered, washed several times with ethyl alcohol and recrystallised from the same solvent. The residue was dried in an oven at about 60 °C.

The FTIR and Raman spectra of the ligand and complex in the solid phase have been recorded on Nicolet-DX spectrophotometer in KBr matrix in the region 200-4000 cm^{-1} and on Jasco K-500 Raman spectrophotometer in the region 50-2000 cm^{-1} respectively. Sample for the Raman work was sealed in a pyrex glass capillary cell, which was mounted in a sample illuminator . Raman spectrum was excited with 514.5 nm radiation of a Spectra Physics Model-165 argon ion laser, using 100 mW power at the sample.

The complex formed possesses 1 : 1 stoichiometry and is insoluble in common organic solvents and does not melt upto 300 °C. The chemical analyses corresponds to the molecular formula $[Cu(C_4H_3N_2O_2F)(OH)_2.3H_2O]$ and the percentage(found/calc) as; Cu=23.00/22.42; C=17.10/17.08; H= 4.21/3.91; N=8.86/8.54.

The vibrational spectra of 5-FU have been analysed and fundamental vibrations have been assigned by assuming that the molecule belongs to C_s point group. Two N–H stretching frequencies associated with the ring is expected to lie in 2900-3500 cm^{-1} region. In the present study, the IR bands observed at 3130 and 3060 cm^{-1} have been assigned to ν(N-H) mode. The low value of this vibration is considered to be indicative of the strong hydrogen bonding in the molecule considered. The 1600-1750 cm^{-1} region of uracil and its derivatives is least crowded and hence the band identification becomes relatively easy in this region. In the present study, two bands observed at 1720 and 1670 cm^{-1} in IR spectrum of 5-FU are assigned to the C_2=O and C_4=O stretching fundamentals. The ν(C–F) mode has been identified at 1255/1250 cm^{-1} (IR/Raman). The vibrational spectra of the complex show some bands shift which confirm the coordination modes of the ligand well. In the spectra of the complex $\nu(C_4{=}O)$ band is shifted to lower frequency by 30 cm^{-1}, while $\nu(C_2{=}O)$ band remains unaffected, suggesting that the ligand coordinates to the metal ion through C_4=O group and not with C_2=O group. The further evidence of bonding of oxygen with metal is provided by the appearance of a band at 265/260 cm^{-1}. The medium broad band appearing in the region 950-940 cm^{-1} in the spectra of the complex has been assigned as OH bridging. The weak band appearing at 3320 cm^{-1} has been attributed to ν(OH) of water. The appearance of a band at 340 cm^{-1} due to $\nu(Cu{-}O)_{aquo}$ confirms the presence of coordinated water molecules in the complex. The bands due to ν(N–H) and ν(C–F) modes remain unaltered indicating the non-involvement of these groups in complexation. Thus on the basis of these studies a octahedral polymeric

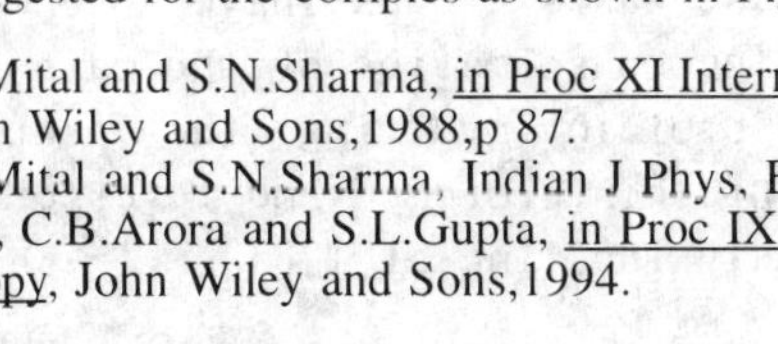

structure has been suggested for the comples as shown in Fig.

1. V.K.Rastogi, H.P.Mital and S.N.Sharma, in Proc XI International Conf on Raman Spectroscopy, John Wiley and Sons,1988,p 87.
2. V.K.Rastogi, H.P.Mital and S.N.Sharma, Indian J Phys, B64, 312(1990).
3. V.K.Rastogi,B.Lal, C.B.Arora and S.L.Gupta, in Proc IXV International Conf on Raman Spectroscopy, John Wiley and Sons,1994.

RAMAN SPECTROSCOPYIC STUDY ON HUMAN IMMUNODEFICIENCY VIRUS(HIV-1--HIV-2) SPACE STRUCTURE AND MICROCOSMIC AND PHOTOSENSITIVE DAMAGE OF HYPERICIN TO THE VIRUS

Xu Yiming and Lu Chuanzong

Institute Of Biophysics, Academia Sinica, Beijing 100101,P.R.China

Laser Raman spectra of HIV-1--HIV-2 in human sera and nagative human sera have been obtained.The prominent Raman lines in the spectra are respectively assigned to the carbohydrates of the viral glycoprotein, the main-chain and side- chain of the viral protein,the phosphate backbone,riboseand four bases of the viral double helical RNA,according to the analytical methods of Thomas et al.[1-3]. It is knowed that the photosensitive drug hypericin can prevent uninfected T-cell from being infected with the AIDS virus in cell culture[4].Animal tests show that it has low toxicity at therapeutic doses. But the mechanisms of the viral photoinactivation have not beeh understood. In this work, the microcosmic and photosensitive damage of HIV-1-- HIV- 2 caused by hypericin was studied with laser Raman spectroscopy at the molecular level.The group and mode of damage could be obtained. The results showed that the Raman characteristic frequencies and the intensities of the lines assigned to various groups of the components of the virus had already been changed which indicated that the ordered double helical RNA of virus changed, and the breakage of some interbase hydrogen bonds. Some of the line intensities of the bases groups increased which showed the occurrence of hyperchromic effect which resulted from the disruption of vertical base-base stacking interaction. Some of the line intensities of the bases groups decreased, indicating that these bases groups have been damaged in the different degrees.The conformation of the main- chain and side-chain of the viral protein,involving the protein moiety of glycoprotein,coat protein,protease and reverse transcriptase etc has changed obviously.The line at 1264 cm-1 which was assigned α -helix disappeared. The lines belonging to S-S, C-S, S-H stretching vibration, P-hydroxy phenyl ring stretching vibration of tyrosine and the series lines assigned to the indole-ring of tryptophane have also changed invarious degrees. The photosensitive damages of the carbohydrates of the viral glycoprotein may be weeker than that of viral RNA and proteins. So the characteristic and reason can be deeply investigated, and the anti-AIDS mechanism of the drug further probed. This is why short periods of photosensitization resulted in a partial loss of reverse transcriptase activity and complete inhibition of viral infectivity. These results suggest that the photodynamic effect of hypericin interferes with more than one stage inthe virus replication cycle [5].

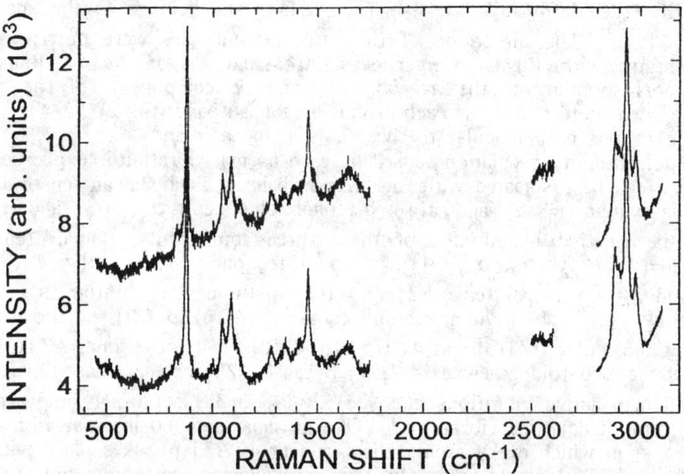

Fig. 1. Raman spectra of HIV-1--HIV-2 in human sera (lower), HIV-1--HIV-2 in human sera after the photosensitive damage caused by hypericin(upper).

1. G.J. Thomas Jr, B. Prescott, J. Mol. Biol. 102, 103-124(1976).
2. Theo.M. Theophanides(ed), Infrared and Raman Spectroscop of Biological Molecules, cooperation with NATO scientific affairs division, 1978, p.125.
3. P.R.Cary, Biochemical Applications of Raman and Resonace Raman Spectroscopies, A Subsidiary of Harcourt Brace Jovanovich, publishers, New York, 1982, p.70, p.184.
4. Jean L. Marx, Science, 244, 287(1989).
5. Susan Carpenter and George A. Kraus, Photochem. and Photobiol. 53, 169-174(1991)

POLARIZED RAMAN SPECTRUM OF A SINGLE CRYSTAL OF AZT

T. Ueda[1*], K. Ushizaswa[1], A. Kumakura[2], and M. Tsuboi[2]

[1]Dept. of Chem., Meisei University, Hodokubo, Hino-shi, Tokyo 191, Japan
[2]Dept. of Fund. Sci., Iwaki-Meisei University, Iwaki, Fukushima 970, Japan

AZT (3'-azido-3'deoxythymidine) is one of the most effective anti-HIV (Human immunodeficiency virus) agents. It is a thymidine analogue (Fig.1), and its anti-HIV activity is supposed to be based upon a competitive inhibition of reverse transcriptase. Polarized Raman scattering measurements have been made of a single crystal of AZT (purchased from YAMASA SHOYU Co. JAPAN) by the use of a Raman microscope of Jasco R-MPS-22 in connection with a Jasco NR-1100 Raman spectrophotometer with the 488.0nm exciting beam from an NEC GLG-3300 Ar^+ laser. The AZT crystal belongs to the space group $P2_1$ (monoclinic) [1]. The directions of the three crystal axes were determined by X-ray diffraction study with Rigaku Geigerflexes 2013 and CN1563A2. Raman scattering intensities, corresponding to the *aa, bb, c'c'* and *bc'* components of the crystal Raman tensor, have been determined for each prominent Raman band (Fig.2) [2-4]. Here, *c'* axis is defined as an axis perpendicular to the *a*-axis in the *ac*-plane.

The polarizability oscillation caused by each normal vibration is expressed by a Raman tensor, which can be associated with each Raman band. Each Raman tensor has its proper principal axis system (i=xyz), and three diagonal components α_{xx}, α_{yy} and α_{zz}. We deal only with the relative magnitudes of these three components, Raman tensor quotients $(r_i)=(r_1,r_2,1)=(\alpha_{xx}/\alpha_{zz}, \alpha_{yy}/\alpha_{zz}, 1)$. The direction cosine of the ℓ-th crystal axis ($\ell=a,b,c$) and i-th principal tensor axis of the p-th member in the asymmetric unit is expressed as $\Phi_{\ell i,p}$. The asymmetric unit contains AZT(p) (p=I,II) and the Bravais unit cell contains four molecules AZT(p) and AZT(p') (p'=I',II'). Of these four, AZT(p) and AZT(p') are related by a two-fold screw axis 2_1. Because AZT molecule has 32 atoms, the total number of Raman active vibrations must be 32×4×3-3=381, of which nine are translational and 12 are rotational lattice vibrations. Of the remaining 360 internal vibrations, 180 B-type vibrations, in which every atomic displacement of AZT(p) takes place with 180° phase difference from AZT(p) would give only very weak Raman scatterings and might tentatively be ignored. The Raman tensor and Raman intensity ratios $I_{\ell m,p}/I_{nn,p}$ for the A-type 90+90 vibrations are to be assigned in the following extreme cases.

Case 1: Weak intra-unit coupling

An AZT(I) vibration may take place independently of the corresponding AZT(II) vibration.

$$I_{\ell m,p}/I_{nn,p}=(\Sigma_i \Phi_{\ell i,p}\Phi_{mi,p} r_i)^2 / (\Sigma_i \Phi_{ni,p}^2 r_i)^2, \quad (\ell mn=aab,\ bbc',\ bc'c')\ \&\ (p=I,II)$$

When the observed Raman scattering is considered as a mere overlap of the Raman scatterings from AZT(I) and AZT(II), then

$$(I_{\ell m}/I_{nn})_{effective}=\Sigma_p I_{\ell m,p} / \Sigma_p I_{nn,p}$$

where Σ_i and Σ_p means the summations over x, y and z and over I and II, respectively.

Case 2: Strong intra-unit coupling

An AZT(I) vibration and the corresponding AZT(II) vibration are practically equivalent and there is a vibrational coupling between them.

$$(I_{\ell m}/I_{nn})_q=[\Sigma_i (\Sigma_p q^{p-1}\Phi_{\ell i,p}\Phi_{mi,p}) r_i]^2 / [\Sigma_i (\Sigma_p q^{p-1} \Phi_{ni,p}^2) r_i]^2, \quad (q=+1,-1)\ (p=I,II)$$

The Raman intensity of the vibrations for the in-phase coupling q=+1 is expected to be higher than that for the out-of-phase coupling q=-1.

From these experimental data and theoretical considerations, and on the basis of the known crystal structure, localized Raman tensors have been determined (Tables 1, 2).

Two bands at 2088 and 1259 cm^{-1} appear only in the $c'c'$ scattering geometry and they are considered to have only one non-zero tensor component, α_{xx}, with the principal x-axis along the N=N=N direction. They are assigned to the N=N=N anti-symmetric and symmetric vibrations, respectively. Raman bands at 1666, 1392, 1238, 771 and 495 cm^{-1} are assigned to thymine vibrations and their Raman tensors were evaluated as shown in Table 1 based on Case 2. Three bands at 870, 849 and 739 cm^{-1} are assigned to deoxyribose vibrations and their tensors were evaluated as shown in Table 2 based on Case 1.

Fig. 1

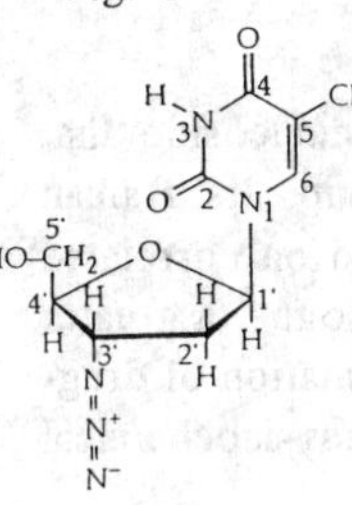

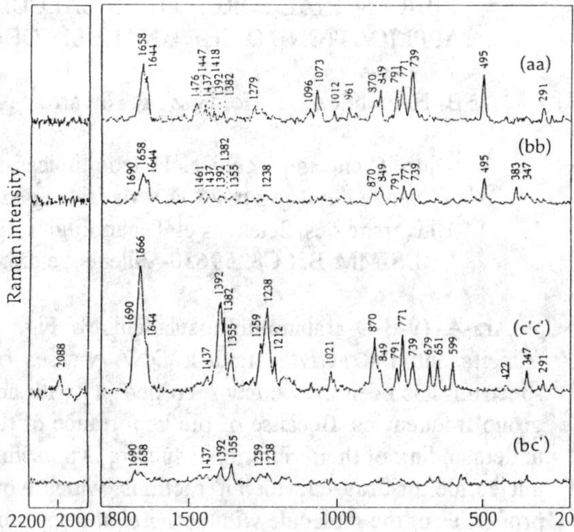

Fig. 2

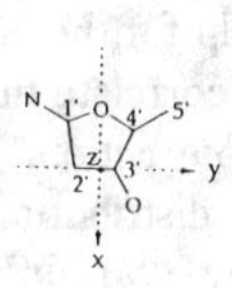

Fig.3

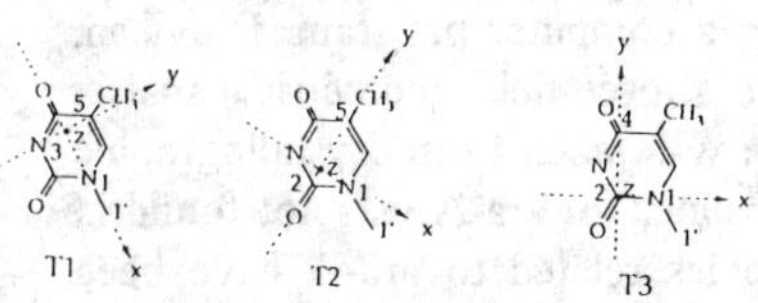

Table 2. Tensors of deoxyribose vibrations

Freq cm^{-1}	Tensor ass Axis	r_1	r_2	Raman intensity ratio I_{aa}/I_{bb}	$I_{bb}/I_{c'c'}$	$I_{bc'}/I_{c'c'}$
870	D1	5	7	1.81	0.19	0.017
				(2)	(0.2)	(0)
849	D1	5	3.2	1.80	0.77	0.007
				(2)	(0.8)	(0)
739	D1	7	5	1.88	0.59	0.009
				(2.5)	(0.6)	(0.1)

Table 1. Raman tensors assignable to thymine vibrations

Frequency cm^{-1}	mode	Tensor assumed Axis system	r_1	r_2	Raman intensity ratio calculated (observed in parentheses) I_{aa}/I_{bb}	$I_{bb}/I_{c'c'}$	$I_{bc'}/I_{c'c'}$
1690					0.08	1.94	0.62
(out[a])	C_4=O/C_5=C_6	T1	10.4	5.3	(~0.1)	(~2)	(0.6)
1666	in-phase str.				1.15	0.15	0.07
					(1.2)	(0.15)	(0.08)
1658	C_2=O str.	T2	3.7	8.6	1.78	1.24	0.09
					(1.8)	(1.2)	(~1)
1644	C_4=O/C_5=C_6	T2	1.8	2.5	1.40	0.83	0.02
	out-of-phase str.				(1.4)	(0.8)	(~0.5)
1392	C_6H in-plane				0.26	0.05	0.19
	bend. and	T1	5.8	-1.5	(0.3)	(0.1)	(0.2)
1355	ring str.				0.48	0.35	0.68
(out)					(0.5)	(0.3)	(0.7)
1238	Kekulé	T1	20.6	8.7	1.03	0.11	0.1
					(~1)	(~0.1)	(~0.1)
771	ring breathing	T3	3.8	3.4	1.86	0.34	0.01
					(1.9)	(0.3)	(~0.1)
495	in-phase ring	T2	-0.4	7.0	1.52	3800	100
	def.				(1.8)	(∞)	(-)

1. I.Dyer, J.N.Low, P.Tollin, H.R.Wilson, and R.A.Howle, Acta Cryst., C44, 767 (1988).
2. A. Kumakura, M. Tsuboi, K. Ushizawa, and T. Ueda, Biospectroscopy, submitted for publication.
3. M. Tsuboi, T. Ikeda, and T. Ueda, J. Raman Spectrosc., 22, 619 (1991).
4. S. Dijkstra, J. M. Benevides, and G. J. Thomas, J. Mol. Str., 242, 283 (1991).
5. M. Tsuboi, T. Ueda, K. Ushizawa, Y. Sasatake, A. Ono, M. Kainosho, and Y. Ishido, Bull. Chem. Soc. Jpn., 67, 1483 (1994).

VIBRATIONAL FORCE FIELD CALCULATIONS OF ARA-A. APPLICATION TO THE ANALYSIS OF RAMAN AND IR SPECTRA

B. Hernández[1], A. Hernanz[1], R.Navarro[1]*, A. Elass[2] and G.Vergoten[2]

[1]Dpto. Ciencias y Técnicas Fisicoquímicas, Fac. Ciencias, UNED, C/Senda del Rey s/n, 28040-Madrid, Spain
[2]Université des Sciences et Technologies de Lille I. CRESIMM-Bât C8. 59650-Villeneuve d'Ascq, France

Ara-A (9-β-D-arabinofuranosyladenine), Fig. 1 , is a purine nucleoside that exhibits antiviral activity against DNA viruses of the herpes group. Its Raman spectrum has been previously recorded [1] with an assignment based on correlated group frequencies. Because of the importance of this nucleoside analogue, a greater understanding of their vibrational spectra is a useful help for the elucidation of drug-nucleotide, or drug-enzyme interactions, which would correlate the physicochemical properties of the molecule with antiviral and antitumorous action.

The harmonic frequencies and potential energy distributions (PED) of the vibrational modes of ara-A have been calculated by two different methods: a classical molecular mechanics method, and the semiempirical molecular orbital method PM3. The first one was carried out using a computer program, Bioviban, developed by this group, and based on the Wilson's force field geometrical matrix (FG) method. The geometry to build the G matrix was taken from crystallographic data. As a complete set of valence force constant values for ara-A was not found in the literature, values published for smaller molecules related to ara-A have been transferred [2-6], according with the procedure suggested by Pulay [7] for large molecules. A set of selected 119 internal force constant was used in the calculation, including 29 redundancies. Semiempirical MO calculations were performed using the PM3 method. A geometry optimization was performed to obtain one minimun on the potential energy surface close to the crystalline structure. The scale factors of the PM3 harmonic force field were determined by fitting to the Raman and infrared spectra using the Redong program [8].

The FT-Raman spectrum, Fig.1 (A), has been obtained with a Bruker Raman RFS 100 spectrometer with the 1064 nm Nd^{3+}:YAG laser line as the excitation source. The nominal resolution was 8 cm^{-1}.The FT-IR spectra, Fig.1 (B), have been recorded in the mid-infrared and far infrared regions, using a Bomem-DA3 interferometer with an effective apodized resolution (Hamming apodizing function) s= 0.89 cm^{-1} and 1.77 cm^{-1}, respectively.

The experimental IR and Raman frequencies have been compared with the results of both computational methods. An assigment of the vibrational modes of ara-A is proposed on the basis of the potential energy distributions. The more intense or characteristic bands are listed in Table 1.

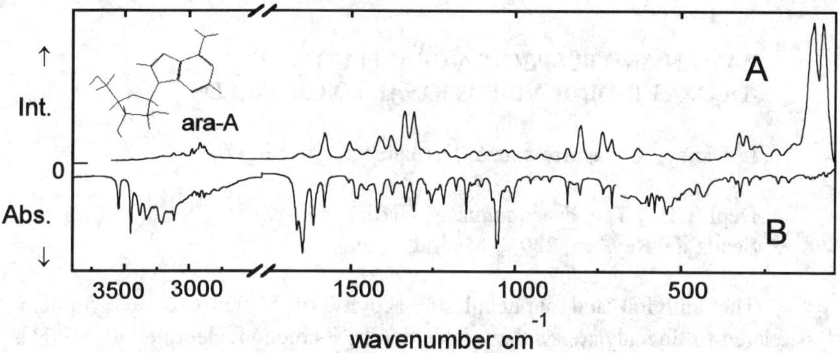

Fig. 1 A:FT-Raman spectrum (3500- 50 cm^{-1} region) of polycrystalline ara-A. B:Mid and far-IR spectra of polycrystalline Ara-A.

TABLE 1. Assignments of some computed and experimental frequencies of ara-A

FT-IR cm^{-1}	FT-Raman cm^{-1}	PM3 cm^{-1}	PED %	BioV cm^{-1}	PED %
3545 s		3537	54N6H-45N6H'(ν_aNH2)	3311	48νN6H-48νN6H'(ν_a NH2)
1644 vs	1641w	1648	42δHN6H'+20C6N6+..	1714	53δH'N6H+13δC6N6H'+..
1579 m	1578 m	1568	26δHN6H'+10δC6N6H+..	1565	50δHC5'H'+20δHC4'C5' +..
1333 m	1337 s	1332	34δC3'O3'H+18νO4'C1'+.	1326	42δC3'O3'H+10τHC4'C5'O5'+
1308 m	1308 s	1305	32νC2N3+13N7C5+..	1308	46δC5'O5'H+11τHO2'C2'H+..
1052 vs	1050 vw	1046	37νC4'C5'+8δHC4'O4'+..	1076	11νC3'O3'+11δC3'O3'H+..
706 m	706 s	707	18νC6N6+10δC6N6H+..	713	8δC6N6H'-7δC6N6H-7δN9C1'H

In general, a good agreement is observed between the experimental and calculated frequencies, as well as PED contributions, by both methods. The weak 1641 cm^{-1} Raman band (1644 cm^{-1} strong in IR) not previously assigned, is mainly due to NH_2 scissoring. Although the strong Raman band at 1337 cm^{-1} has been considered a pyrimidine ring vibration of adenine, relevant sugar contributions are calculated for the nucleoside by both methods.This vibrational coupling between the base and sugar moieties is supported by the PED calculated for many other modes.

1. T. Theophanides, S. Hanessian, M. Manfait and M. Berjot, J. Raman Spectrosc. 16(1), 32 (1985).
2. Z. Dhaouadi, M. Ghomi, J. Austin, R. Girling, R. Hester, P. Mojzes, L. Chinsky, P. Turpin, C. Coulombeau, H. Jobic, and J. Tomkinson, J. Phys. Chem., 97, 1074 (1993).
3. M. Majoube, J. Chim. Phys,. 81, 303 (1984).
4. M. Majoube, Biopolymers, 25, 1075 (1985).
5. J. Eyster and E. W. Prohofsky, Spectrochim Acta, A 30, 2041 (1974).
6. M. Ghomi and E. Taillandier, Eur. Biophys. J., 12 153 (1985).
7. P. Pulay, J. Mol. Struct., 347, 293 (1995).
8. A. Allonche and J. Pourcin, Spectrochim Acta, A 43, 591 (1993).

RAMAN AND IR SPECTRA OF 5'-FLUORO-2'-DEOXYURIDINE. VIBRATIONAL FORCE FIELDS.

L. Bailey, R. Navarro and A. Hernanz*

Dept. CC. y Téc. Fisicoquímicas, UNED
Senda del Rey, s/n, 28040-Madrid, Spain

The antiviral and antineoplastic activity of 5'-fluoro-2'-deoxyuridine, FdU, is related to thymidylate synthase inhibition by its mononucleotide FdUMP [1]. A vibrational analysis of the fluorinated nucleoside would help in the understanding of the proposed drug-target interactions [2].

Two force fields have been used for normal coordinate treatment of FdU: one transferred from simpler molecules considered as FdU fragments [3-5] and the semiempirical PM3 force field. In the first case, the molecular geometry obtained from X ray difraction [6] and the overlaid force field have been the source data for a Wilson's FG matrix calculation. In the second case, a previous PM3 geometry optimization was carried out to find a minimum of the potential energy close to the FdU structure obtained by X ray diffraction. No force constant has been refined nor scale factors have been applied in either case to compare the harmonic frequencies predicted by the different raw force fields.

The FTIR spectra, Fig 1A, in the mid and far infrared regions have been obtained in a Bomem-DA3 interferometer with effective apodized resolutions of 0.89 and 1.77 cm^{-1} respectively.The FT-Raman spectrum of FdU, Fig 1B, has been recorded with a Bruker Raman RFS 100 spectrometer, excitation line at 1064 nm of a Nd^{3+}:YAG laser and 8 cm^{-1} of nominal resolution.

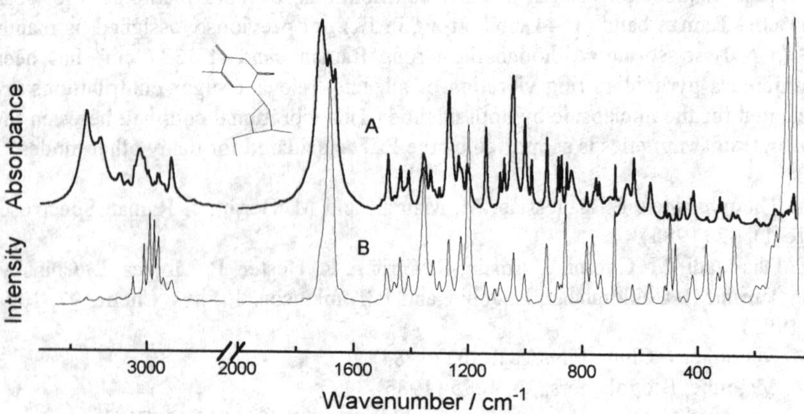

Fig. 1. Spectra of polycrystalline FdU. A: Mid and far IR, B: FT-Raman.

TABLE 1. Assignment for some relevant frequencies of FdU

FTIR /cm^{-1}	FT-Raman /cm^{-1}	PM3 /cm^{-1}	BioV /cm^{-1}	PED (BioV) %
3182	3182	3358	3102	97νN3H
1715		1946	1696	15νC4O+5νC2O
1691	1688	1849	1656	17νN1C2+16δN3C2O2-14δN1C6H- -10δC2N3H-6νC5C6-4νC2O
1271	1270	1261	1283	7δC4'C5'H+6δC5C6N1+5δC5C5F- -4νN1C1'-4νC5F
859	862	867	860	51δC2N3H+16δN1C6H- -6δC5C6H-3δC5C6N1

Some observed and calculated frequencies, with the potential energy distribution, PED, obtained are shown in Table 1. The wavenumber of the bands due to the N3H stretching vibration are strongly influenced by the hydrogen bond schemes of the amino group in solid phase [6]. According to the PEDs, the stretching vibrations of the carbonyl groups do not appear to be independent. Although a fragment method is applied to build the force field, vibrational couplings between the deoxyribose and the fluorouracyl moieties are predicted for different modes. The non scaled PM3 force field tends to overestimate the wavenumbers of bands attributed to internal modes of groups involved in hydrogen bonds.

1. A. Rossi, in Nucleoside Analogues. Chemistry, Biology and Medical Aplications (R.T. Walker, E. de Clercq and F. Eckstein, eds.), Plenum, New York, 1979, pp. 409-436.
2. J.C. Austin, A. Fitzhugh, J.E. Villafranca and T.G. Spiro, Biochemistry 34, 7678 (1995)
3. P. Lagant, A. Elass, M. Dauchez, G. Vergoten and W.L. Peticolas, Spectrochim. Acta 48A, 1323 (1992).
4. B.V. Reddy and G.R. Rao, Vib. Spectrosc. 6, 231 (1994)
5. J.M. Eyster and E.W. Prohofsky, Spectrochim. Acta 30A, 2041 (1974)
6. D.R. Harris and W.M. Macintyre, Biophys. J. 4, 203 (1964)

FTIR AND LASER RAMAN SPECTRA OF COMPLEX OF ZN(II) WITH 5-IODOURACIL

V. K. Rastogi[1], Inder Sain[1], M. Srivastava[1] and U. Awasthi[2]

[1]Physics Department, L R College, Sahibabad-201 005
[2]Department of Zoology, Govt Girls College, Rewa, India

Metal complexes play an important role in the biological activity of drugs, as the complex formation has been suggested one of the important mechanism for the drug interaction. Recent spectroscopic studies of uracil derivatives and its metal complexes have been motivated by their biological importance [1-3]. Like other 5-halogenated uracil derivatives, 5-iodouracil (5-IU) is an antimetabolite that is incorporated into nucleic acids and have the ability to bind metals or to bind to tissues via metals. In order to understand the mechanism by which the metal ions intervene in the function of nucleic acid bases, and to find the coordinating sites in 5-IU, we have undertaken the present study. Here we report the isolation and characterisation of the complex of Zn(II) with 5-IU. Chemical analyses are used to determine stoichiometry and vibrational spectroscopy is used to illucidate the vibrational modes and sites of coordination.

The compound 5-IU (mp : 274-276°C) was obtained from M/s Aldrich Chemical Co, U.S.A. and used as such without further purification. The complex was prepared by the method given in literature [4].

FTIR and Raman spectra of the ligand and complex in solid phase have been recorded on Jasco Spectrophotometer Model FTIR/7000 in KBr matrix in the region 200–4600 cm^{-1} and on a Jasco K-500 Raman spectrophotometer in the region 10–2000 cm^{-1} respectively. For recording the Raman spectrum, sample was sealed in a pyrex glass capillary sample tube of internal diameter about 2 mm, which was mounted in a sample illuminator. The Raman spectrum was excited by the 488.0 nm line of an argon ion laser of 100 mW power. The slit width at the entrance was 380 μm, the time constant was 0.5s, and the scanning speed was 5.

The complex formed possesses 1 : 1 stoichiometry and is insoluble in common organic solvents and does not melt upto 270°C. The complex is white in colour. The chemical composition corresponds to the molecular formula $[Zn(L)(OH)_2.3H_2O]$; $L = C_4H_3N_2O_2I$ and the percentage (found/calc) as; Zn = 17.00/16.62; C = 12.00/12.27; H = 3.00/2.81 and N = 7.25/7.16.

The vibrational spectra of 5-IU have been analysed and fundamental vibrations have been assigned by assuming that the molecule belongs to Cs point group. Thus the 30 fundamental modes of this molecule may be classified into 21a′ symmetry species (in-plane) and 9a″ species (out-of-plane). The wavenumbers of the observed fundamentals in IR and Raman spectra alongwith their assignments are given in Table 1. The two stretching frequencies associated with the ring is expected to lie in 2900–3500 cm^{-1} region. In the present study, the IR bands observed at 3120 and 3090 cm^{-1} have been identified as ν(N–H) mode. The low value of this vibration is considered to be indicative of the strong hydrogen bonding in the molecule considered. The 1600–1750 cm^{-1} region of uracil and its derivatives is least crowded and hence the band identification becomes relatively easy in this region. In the present study, two bands observed at 1710 and 1670/1675 cm^{-1} (IR/Raman) are assigned to C_2=O and C_4=O stretching fundamentals. The ν(C–I) mode has been observed at 780/782 cm^{-1}.

TABLE : Assignment of fundamental frequencies (in cm^{-1}) of 5-idouracil

Sym Sp	I R	Raman	Assignment	Sym Sp	I R	Raman	Assignment
a′	3120	–	ν(N–H)	a″	810	–	γ(C–H)
a′	3090	–	ν(N–H)	a′	780	782	ν(C–I)
a′	3048	–	ν(C–H)	a″	760	750	γ-ring
a′	1710	–	ν(C=O)	a″	740	732	β(C=O)
a′	1670	1675	ν(C=O)	a′	670	668	β-ring
a′	1630	1640	ν(C=C)	a′	–	640	β(C=O)
a′	1562	–	β(N–H)	a′	615	620	β(C–I)
a′	1550	–	β(N–H)	a′	540	550	β-ring
a′	1506	1490	ν(C–N)	a″	440	–	γ(C=O)
a′	1455	1472	ν(C–N)	a″	430	–	γ-ring
a′	1439	–	ν-ring	a″	420	422	γ(C=O)
a′	1420	1410	β(C–H)	a″	235	–	γ(C–N)
a′	1338	1340	ν-ring			140	Lattice
a′	1228	1215	ν-ring			122	Lattice
a″	910	–	γ(N–H)			100	Lattice
a′	880	–	β(C–N)			89	Lattice
a″	870	–	γ(N–H)			75	Lattice

The vibrational spectra of the complex show some band shifts which confirm the coordination mode of the ligand well. In the spectra of the complex ν(C_4=O) band is shifted to lower frequency by 35 cm^{-1}, while ν(C_2=O) band remains unaffected, suggesting that the ligand coordinates to the metal ion through C_4=O group and not with C_2=O group. The further evidence of bonding of oxygen with metal is provided by the appearance of a band at 270/265 cm^{-1} in the spectra of the complex. The medium band appearing at 960/955 cm^{-1} in the spectra of the complex has been assigned as OH bridging. The weak band appearing at 3325 cm^{-1} has been attributed to ν(OH) of water. The appearance of a band at 345/340 cm^{-1} due to $\nu(Zn–O)_{aquo}$ confirms the presence of coordinated water molecules in the complex. The bands due to ν(N–H) and ν(C–I) modes remain unaltered indicating the non involvement of these groups in complexation. Thus on the basis of these studies a octahedral polymeric structure has

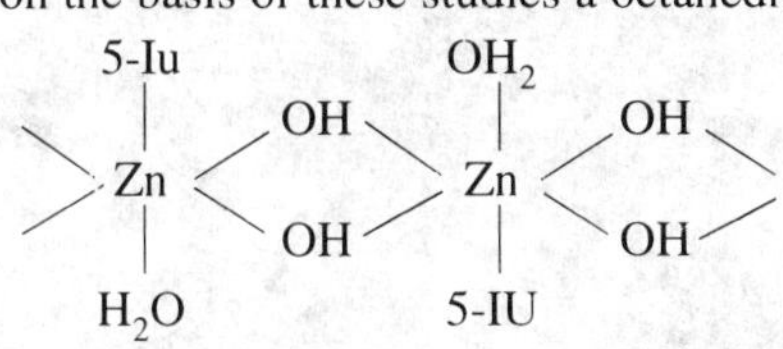

been suggested for the complex as shown in figure.

1. V.K.Rastogi, H.P.Mital and S.N.Sharma, in Proc XI International Conf on Raman Spectroscopy, John Wiley and Sons,1988,p 87.
2. V.K.Rastogi, H.P.Mital and S.N.Sharma, Indian J Phys, B64, 312(1990).
3. V.K.Rastogi,B.Lal, C.B.Arora and S.L.Gupta, in Proc IXV International Conf on Raman Spectroscopy, John Wiley and Sons,1994.
4. V.K. Rastogi, B. Lal, Y.C. Sharma & C.L. Jain, in Metal ions in Biology and Medicine, Vol 2, John Libbey Eurotext, France, 1990, p 184.

SECTION 13

AMINO ACIDS and PEPTIDES

STRUCTURE AND DYNAMICS IN HYDROGEN BONDED AMIDES

O. Faurskov Nielsen* and A. Mortensen

Department of Chemistry, University of Copenhagen, DK-2100, Copenhagen, Denmark.

The rotational and translational motions of molecules in a liquid cause that a structure present at the time scale for IR- and Raman spectroscopy (picosecond or faster) may be non-existent at the NMR-time scale (nanosecond or slower), and the NMR-spectrum will only show an average of the molecular motions. The life time of a hydrogen bond may be as short as a few picoseconds. Thus in hydrogen bonded systems the fast dynamics should be considered. NMR-spectroscopy is one of the most, or even the most, important tool to obtain the structure of biological macromolecules in solution. Our aim is to investigate whether vibrational dynamics too fast to be observed on an NMR time scale may be important in molecules of biological interest. Recently, several femtosecond laser techniques have appeared, giving a direct measure of very fast time events. Certainly, they will be very important for the study of fast dynamics in the years to come. However, it should be emphazised that the more conventional vibrational spectroscopic methods working in the frequency domain still give information about the fast dynamics.

Simple amides with the general structure RCONHR' can be used as model systems for the peptide bonds in proteins. We have performed intensive studies of formamide where R=R'=H. From low-frequency studies below 200 cm^{-1} we concluded that the intermolecular hydrogen bonded molecules in the liquid state formed a chain and not a cyclic dimer[1]. Apart from the parent compound, a number of formamide isotopomers have been investigated: $HCOND_2$, HCONDH, $DCONH_2$, $DCOND_2$, DCONDH, and $H^{13}CONH_2$[2]. Some anomalies are observed in the amide-I region for mixtures of isotopomers in the neat liquid state. Also solutions in either DMSO or water have been studied. The observed spectra could be explained in terms of two different phenomena:

1. Resonance Energy Transfer (RET).
2. Coalescence of Bands in Mixtures of Isotopomers (CBMI)

RET gives rise to a non-coincidence between isotropic and anisotropic components in the Raman spectrum. The theoretical background for RET is covered by Jeanne McHale in her contribution at this conferernce. The main issue is that there is transfer of vibrational energy between different molecules in the liquid state. The CBMI confirms that collective intermolecular effects are important in the liquid state. In a mixture of different isotopomers only one amide-I band is observed in the isotropic Raman spectrum although the individual isotopomers in the neat liquid state show different amide-I frequencies. This coupling has not been treated theoretically too well. We have tentatively used a model known as mutual entrainment of oscillators[2]. An advantage of this model is that a threshold value for the coupling parameter can explain why CBMI does not occur for all mixtures of two isotopomers.

Formamide is not the best model for the peptide bond in proteins, because both R and R' in proteins never will be hydrogens. One step further is to let R=H and R'=CH_3. The molecule is N-methylformamide (NMF). We are presently investigating this molecule, including the isotopomer $HCONDCH_3$. Unfortunately, the frequency shift for the amide-I vibration in $HCONHCH_3$ and in $HCONDCH_3$ is rather small, so it is difficult to see whether only one band is observed for the isotropic Raman component in a mixture of two isotopomers. However, the results seem to justify the presence of CBMI in liquid NMF.

An even more realistic model for the peptide bond in proteins is N-methylacetamide, NMA. For an aqueous solution of NMA a coupling between the water bending vibration and the amide-I vibration has recently been proposed[3].

Low-frequency Raman studies in the $R(\bar{\nu})$-representation[5,6] of liquid amides have shown that a deformation mode with a band maximum around 100 cm^{-1} may be important for driving conformational changes and breaking or formation of hydrogen bonds[4,5]. Although the fast dynamical fluctuations are unobservable by NMR-spectroscopy, some conclusions from NMR- and vibrational spectroscopy can be drawn regarding the structure of formamide in the liquid state. The structure reported recently from NMR-data [6] is indeed very close to the one concluded by us from low-frequency Raman and FIR data[1,4,5].

The above-mentioned fast time events point towards the importance of some collective motions in amides. Similar phenomena might easily occur in proteins where they could be of importance for transfer of information from one part of a molecule to another. Taking the coupling between water molecules and proteins through bands around 1650 cm^{-1} into account, a key understanding of the role of water in biological systems might be achieved. A coupling of this type might also influence the shape of water bands in aqueous solutions of biomolecules, making a spectral water compensation more difficult. Furthermore, dynamical couplings might explain the reason why different secondary structures of proteins show specific bands in IR and Raman spectra.

1. N. Østergaard, P.L. Christiansen and O. Faurskov Nielsen, J. Mol. Struct. Theochem. 235, 423 (1991).
2. O.Faurskov Nielsen, A. Mortensen, J. Yarwood and V. Shelley, J. Mol. Struct. in press, and references cited therein.
3. X.G. Chen, R. Schweitzer-Stenner, S. Krimm, N.G. Mirkin and S. A. Ascher, J. Am. Chem. Soc. 116, 11141 (1994).
4. O. Faurskov Nielsen, Ann. Rep. Prog. Chem. Ser. C 90 3 (1993).
5. O. Faurskov Nielsen and B.M. Colaiani, J. Mol. Struct. 347, 267 (1995).
6. R. Ludwig, F.Weinhold and T.C. Farrar, J. Chem. Phys. 102, 5118 (1995).

Resonance Raman Examination of the Electronic Excited States of Glycylglycine and Other Dipeptides: Observation of a Carboxylate→Amide Charge Transfer Transition

X. G. Chen, Pusheng Li, Janet S. W. Holtz, Zhenhuan Chi,
Vasil Pajcini, Sanford A. Asher*
Department of Chemistry, University of Pittsburgh, Pittsburgh, PA 15260
Lisa A. Kelly
Biology Department, Brookhaven National Laboratory, Upton, New York 11973

Glycylglycine (Gly-Gly) and other dipeptides have served as important model systems for theoretical, and experimental studies of the amide linkage in proteins and peptides. The study of these species has been motivated by the expectation that a fundamental understanding of the role of protein structural constraints controlling biological function requires a deep understanding of the structure and dynamics of the individual amide linkages. In this work, we use UV resonance Raman, VUV absorption spectra of aqueous glycylglycine and other dipeptides to detect and assign a new charge transfer transition in dipeptides.

We examined the VUV absorption spectra of aqueous Gly-Gly and other dipeptides, and observe a new transition band at 197nm (ε~3000$M^{-1}\cdot cm^{-1}$) in Gly-Gly (see Figure 1). However, this absorption feature is removed by the addition of a methylene spacer between the amide and the carboxylate groups (*i.e.* Gly-β-Ala and Ace-β-Ala), which indicates that this transition results from an interaction between the amide and the carboxylate groups.

Excitation at *ca.* 206.5nm results in strong resonance Raman enhancement of the amide I, II and III bands and the amide $C_\alpha H$ bending mode similar to the enhancement we observed previously with excitation within the $\pi\rightarrow\pi^*$ transition of N-methylacetamide (NMA)[1]. In addition, we observe strong resonance Raman enhancement of the ca. 1400 cm^{-1} symmetric COO^- stretching vibration, whose 206.5 nm Raman cross section is increased 20-fold compared to that of the carboxylate in sodium acetate. However, the addition of a methylene spacer between the amide and the carboxylate groups causes the resonance Raman enhancement of this symmetric COO^- stretch to disappear. We also have studied the Raman depolarization ratios of amide and carboxylate vibrations of Gly-Gly and other dipeptides, and found that the depolarization ratios of dipeptides such as Gly-Gly at pH 7 deviate from 0.33 at 206.5nm excitation (see Table 1), which indicate that at least two electronic transitions are involved in the resonance enhancement.

The data from UV resonance Raman excitation profiles, the Raman depolarization ratio dispersion, and the VUV absorption spectra of Gly-Gly and other dipeptides suggest this new transition at *ca.* 197 nm is a charge transfer band involving electron transfer from a nonbonding carboxylate orbital to a amide-like π^* orbital. This transition occurs at the penultimate carboxylate moiety of all peptides and proteins. This charge transfer transition may be important in peptide photochemistry, and serve as a conduit for electron transfer.

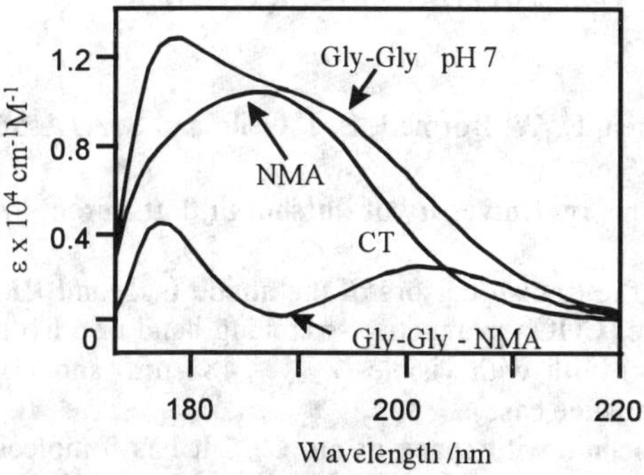

Figure 1. Absoption spectra of Gly-Gly and NMA, and their difference spectrum.

Table 1. Depolarization ratios of glycylglycine (Gly-Gly), glycyl - β -alanine (Gly-β-Ala)

Vibrations (cm^{-1})	Compounds	ρ (488 nm)	ρ (244 nm)	ρ (206.5 nm)
Amide III band				
1286	Gly-Gly pH 7	0.20	0.26	0.28
1283	Gly-β-Ala	0.25	0.33	0.33
Amide II band				
1576	Gly-Gly pH 7	0.25	0.28	0.27
1579	Gly-β-Ala	0.23	0.33	0.32
Symmetric COO^- Stretch				
1396	Gly-Gly pH 7	0.53	0.50	0.41

Reference:

1. Chen, X. G.; Asher, S. A.; Schweitzer-Stenner, R.; Mirkin, N. G.; Krimm *J. Am. Chem. Soc.*, **117**, 2884 (1995)

NORMAL AND RESONANCE RAMAN TENSOR MEASUREMENTS OF A GLYCYL-GLYCINE HYDRATED CRYSTAL

V. Pajcini, X. G. Chen, R. W. Bormett, S. J. Geib, and S. A. Asher*

Department of Chemistry, University of Pittsburgh, Pittsburgh, PA 15260.

We have measured the Raman tensors of the amide I, II, and III bands (A-I, A-II, A-III) and carboxylate (COO^-) symmetric stretching band of a hydrated crystal of Glycyl-Glycine (Gly-Gly) both with visible ($\lambda_{exc.}$ = 488 nm) and UV excitation ($\lambda_{exc.}$ = 244 nm, almost resonance case).

The crystal is monoclinic with space group $\mathbf{C_{2/c}}$. It has 8 molecules of Gly-Gly and 12 molecules of water per unit cell. It is biaxial with the optical plane perpendicular to the two-fold symmetry axis (**b** = **Y** crystallographic axis). The two optical axes, **X** and **Z**, are perpendicular in our crystal, and **XYZ** was chosen as the laboratory coordinate system. The angle between **X** and the crystallographic axis, **a** is 27^0. Vertical or horizontal plane polarized excitation was obtained using a Soleil-Babinet compensator. To avoid birefringence, the sample was excited along the optical axes. Spectra were recorded in a backscattering geometry (180^0). By using a polarization analyzer before the spectrometer we measured the polarized Raman spectra for different crystal orientations. The UV Raman spectra denoted with the Porto notation [1] are shown in Fig. 1.

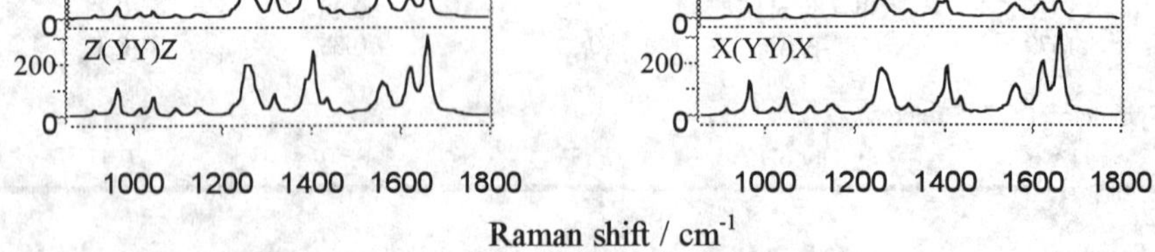

Fig. 1 The UV Raman spectra of Gly-Gly crystal.

The intensity of each Raman band in the laboratory coordinate system is given by

$$I_{\rho\sigma} = K|\alpha^{lab}{}_{\rho\sigma}|^2 \qquad \rho, \sigma = X, Y, Z. \qquad (1)$$

where K is a proportionality constant and $\alpha^{lab}{}_{\rho\sigma}$ are the elements of the Raman tensor $\alpha^{\mathbf{lab}}$ in the laboratory system. This tensor is related to the molecular Raman tensor $\alpha^{\mathbf{mol}}$ by

$$\alpha^{lab} = \mathbf{R}^T \alpha^{mol} \mathbf{R} \qquad (2)$$

where **R** is the Eulerian angles rotation matrix [1].

The molecular Raman tensor α^{mol} can be written in the diagonal frame by a suitable choice of the **xyz** coordinate system attached to the molecule; moreover the relative values of these diagonal elements can be expressed as $r_1 = \alpha_{xx}/\alpha_{zz}$, $r_2 = \alpha_{yy}/\alpha_{zz}$, and $\alpha_{zz} = 1$ [2]. The relative values of $I_{\rho\sigma}$ for each vibrational band is related to the values of r_1, r_2 and the Euler's angles (eqn. 1 and 2). A least squares fit of the measured oriented polarized crystal Raman intensities to eqn. 1 and 2 determines r_1, r_2 and the orientation of the **x,y,z** principal axes. The angles between these axes and amide plane or carbonyl bond (nearly parallel to the **b** axis) were also determined.(Table 1).

TABLE 1. Molecular Raman tensor data of Amide and Carboxylate Vibrations in Gly-Gly crystal.

	VISIBLE				UV			
	A-I	A-II	A-III	COO^-	**A-I**	**A-II**	**A-III**	**COO^-**
$r_1 = \alpha_{xx}/\alpha_{zz}$	3.65	2.92	2.81	3.03	**10.64**	**9.70**	**8.79**	**7.46**
$r_2 = \alpha_{yy}/\alpha_{zz}$	-0.057	-0.49	-0.24	-1.62	**0.41**	**-0.61**	**-0.077**	**-2.84**
ρ	0.22	0.28	0.23	0.52	**0.26**	**0.315**	**0.28**	**0.505**
γ	31^0	55^0	57^0	56^0	**41^0**	**60^0**	**60^0**	**69^0**
δ	22^0	16^0	12^0	4^0	**0^0**	**2^0**	**1^0**	**5^0**

Legend: ρ - depolarization ratio; γ - angle between principal Raman tensor axis **(x)** and crystallographic axis **(b)**; δ- angle between the plane of the **x** and **z** principal Raman tensor axes and amide plane.

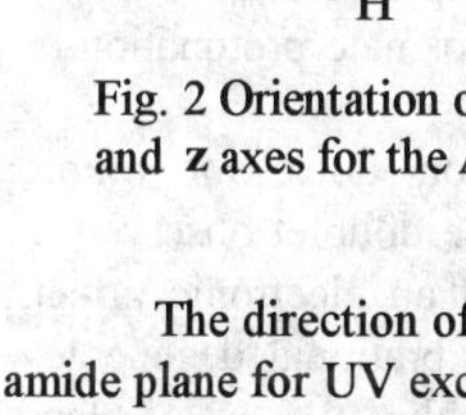

Fig. 2 Orientation of the **x** and **z** axes for the A-II vibration.

Table I shows that the major axis of the molecular Raman tensor rotates between visible and UV excitations. In the visible, numerous electronic transitions contribute; while in the UV, the resonant transition dominates. With UV excitation, the largest tensor elements (**x and z** axes) lie in the amide plane (Fig. 2). We can use the polarized crystal data to determine the molecular orientation of the electronic transition moment.

The direction of the major Raman axis of the COO^- stretching band lies in the amide plane for UV excitation (within 5^0). This enhancement was recently suggested to derive from a charge transfer transition from the carboxylate group to the amide group [3].

1. G. Turrell, Infrared and Raman Spectra of Crystals, A. P., London, 1972.
2. M. Tsuboi, T. Ikeda, and T. Ueda, J. Raman Spectrosc. 22, 619, (1991).
3. X. G. Chen, P. Li, J. S. W. Holtz, Z. Chi, V. Pajcini, S. A. Asher, and L. A. Kelly, in press.

UV RESONANCE RAMAN INVESTIGATION OF TYROSINE PHOSPHORYLATION

B.T. Cooper,[1] Z. Chi,[3] S.A. Asher,[3] * D.J. Graves,[2] T.M. Cotton[1] *

[1] Department of Chemistry
[2] Department of Biochemistry and Biophysics
Iowa State University, Ames, IA 50011 USA

[3] Department of Chemistry
University of Pittsburgh, Pittsburgh, PA 15260 USA

A wide variety of biochemical signal transduction pathways involve an enzymatic protein phosphorylation or dephosphorylation step [1]. Proteins may be phosphorylated at the hydroxyl group of serine, threonine, or tyrosine residues. In the case of tyrosine, phosphorylation may be observed directly in aqueous solutions by UV resonance Raman spectroscopy.

Phosphorylation of tyrosine mimics tyrosinate protonation with respect to the electronic environment of the substituted benzene ring. First, the UV spectrum of phosphotyrosine at all pH's is similar to that of tyrosine at physiological pH, while in tyrosinate the benzene L_a and L_b bands are shifted about 20 nm to the red. Phosphorylation also mimics protonation with respect to the Raman spectra of these species. The toluene ν_{8a} vibration (an in-plane ring stretch [2,3]) of tyrosinate shifts from 1601 cm^{-1} to 1617 cm^{-1} upon protonation; the same band occurs in phosphotyrosine at about 1618 cm^{-1} at all pH's, with resonant enhancement behavior nearly identical to that of tyrosine. Protonation of tyrosinate also shifts the toluene ν_{9a} in-plane C-H bend from 1175 cm^{-1} to 1180 cm^{-1}; this band occurs at 1178 cm^{-1} in phosphotyrosine, again independent of pH. The weak toluene ν_{7a} methyl-ring stretch near 1265 cm^{-1} and the para-benzene totally symmetric stretch near 1280 cm^{-1} also shift slightly upon tyrosinate protonation or phosphorylation.

One important way in which phosphorylation does *not* mimic tyrosinate protonation is with respect to the Fermi resonance doublet observed at 830/850 cm^{-1} for both tyrosine and tyrosinate. This is not an electronic effect, however; phosphorylation simply destroys the accidental vibrational degeneracy between the ν_1 symmetric ring-breathing fundamental and the ν_{16a} out-of-plane C-H bending overtone. A weakly enhanced band remains at 770 cm^{-1} in the UV resonance Raman spectra (see Figure 1).

There are also several bands in the normal Raman spectrum of phosphotyrosine attributable to the phosphate moiety. Two are the dianionic symmetric stretch at 983 cm^{-1} and the broader, monoanionic symmetric stretch at about 1088 cm^{-1}. Since the pK_a of the second phosphate proton of phosphotyrosine is about 6, pH-dependent spectral changes can be observed in a physiologically interesting pH range. These bands, however, are not strongly enhanced by UV excitation. Potentially more useful is a phosphate band observed at 1225 cm^{-1} for the phosphate monoanion and at 1233 cm^{-1} for the dianion, which is tentatively assigned as a phosphate-ring stretch, although this band does not acquire appreciable resonance intensity until the excitation wavelength is tuned deeper in the UV than the L_a absorption band.

As far as the practical problem of using UV resonance Raman spectroscopy to monitor tyrosine phosphorylation is concerned, two strategies present themselves. The strongest band in the UV-excited spectra is generally the toluene ν_{8a} vibration near 1618 cm^{-1}. If the pH is adjusted to 11 this band will shift to about 1601 cm^{-1} (and a second band will appear near 1556 cm^{-1}) if the tyrosine is not phosphorylated *and* if it is accessible to solvent. If the protein or peptide cannot tolerate basic conditions or if the tyrosine residues in question are not solvent-accessible, the weaker Fermi resonance doublet and the corresponding 770 cm^{-1} band of phosphotyrosine may be used. The preferred excitation wavelength (of the doubled Ar^+ lines) is 229 nm; the spectra shown below are excited with 13 mW of 229 nm light and normalized to the 0.3 M perchlorate internal standard peak at 932 cm^{-1}. (OPY: 2 mM *o*-phospho-L-tyrosine; Y: 2 mM tyrosine; Y^-: 2 mM tyrosinate.)

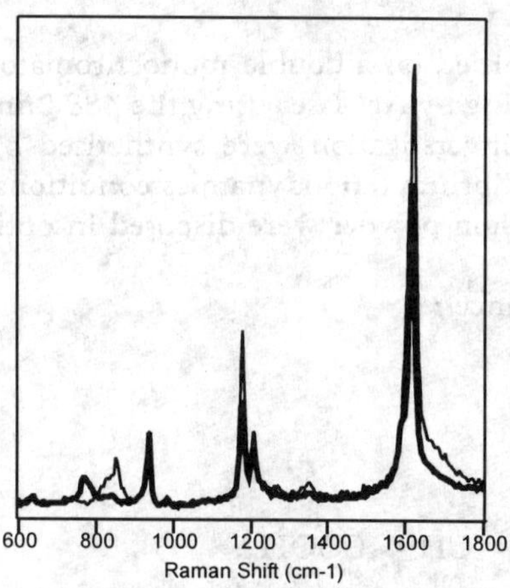

Figure 1a: UVRR spectrum of OPY (pH 5, heavy line) and Y (pH 6.5).

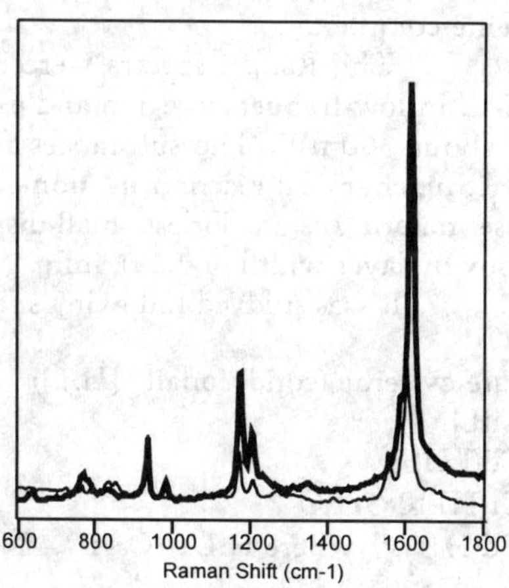

Figure 1b: UVRR spectrum of OPY (pH 8, heavy line) and Y^- (pH 11.5).

1. D.J. Graves et. al. Co- and Post-translational Modification of Proteins, New York: Oxford, 1994.
2. For assignments see: S.A. Asher, M. Ludwig, and C.R. Johnson, J. Am. Chem. Soc. 108, 3186 (1986), and [3], below.
3. R.P. Rava and T.G. Spiro, J. Phys. Chem. 89, 1856 (1985).

THE INFLUENCE OF IONS OF METALS ON LOW FREQUENCY RAMAN SPECTRA OF CYSTEINE

V.N.Moiseenko[1], I.G.Vinichenko[1], E.A.Chernushenko[1], G.D.Zegshda[1], I.P.Kolesnik[1], V.S.Gorelik[2] and V.S.Schetinkin[1]

[1]Dniepropetrovsk State University, 13, Nauchny by Street, 320625, Ukraine
[2]Physical Instityte of the Academy of Sciencies, Moscow, Russia

It is known that ions of metals exert essential influence on functioning of biological systems at molecular level [1]. On the other hand Raman scattering take a very important role in studying structural and dynamic properties of biological molecules and systems [2].

Here we report the results of recent inverstigations dealing with the analysis of singularities of lattice Raman scattering in a new solid metal-cysteine complex.

The Raman spectra were recorded on a double monochromator DFS-12 in low-frequency region and exciting by Ar^+-laser using the 488.0nm line about 300mW. The substances for inverstigation were synthetized by fast-acting chemical reaction in non-equilibrium termodynamics conditions. These materials is a colorless small-dispersion powder were disposed in optical box by layer width of 0.2-0.5mm.

It was studyed following substances:

1. Pure cysteine (conditionally $[H_2L]$)
2. [CdL]
3. $[Cd(LH)_2]$
4. $[Cr(HL)_3]•3H_2O$
5. $[Cd(HL')_2]$, where H_2L': C_6H_5 —NH—CH—COOH

                                   |
                                   CH_2—SH

Figure 1 shows some of obtained spectra at T=100K.

In spectrum of pure cysteine it is observed a number of high Q-factor lines in region of 0-150 cm^{-1}. It is possible that these lines concern to translation and libration modes of molecular cysteine and its parts in crystals lattice. In the same time $[Cd(HL')_2]$ spectra have not any lines in this region: spectra have continuum character and signal level is considerably stronger than background level. The same situation took place in other metallocomplex spectra.

Observing changes in optical vibration spectra of metallocomplexes may be interpreted on the base of conformation changes under metal ions influence model [1], taking into acount the aminoacids ambivalent properties. In case of cysteine combining it is theree coordination centers (S, N and O) provides the formation of variety types of metallocomplexes. It may lead to significant distortions of primary molecular structure and even to amorphorising of crystalline structure of pure cysteine. In this case "white" spectrum of disordering crystal appears. The normal dynamics behaviour transforms into that of chaotic system. The proof of it is the absense of band coresponding to molecular skeleton vibrations in IR transmission spectra at $\omega \approx 1600 cm^{-1}$ in $[Cd(HL')_2]$ and [CdL]. This mechanism can take place in complex of other metals, in particular, for Cr.

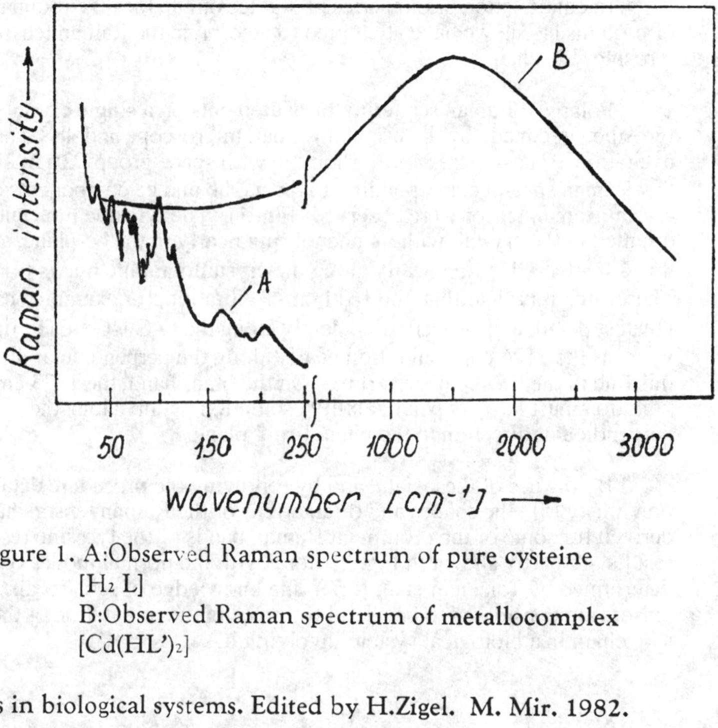

Figure 1. A:Observed Raman spectrum of pure cysteine $[H_2L]$
B:Observed Raman spectrum of metallocomplex $[Cd(HL')_2]$

1. Metal ions in biological systems. Edited by H.Zigel. M. Mir. 1982. p.168.
2. Proceeding of the XIVth International Conference on Raman Spectroscopy. Hong Kong. August 22-26. 1994.

RAMAN TENSORS OF TYROSINE SIDE CHAIN OF PROTEIN

M. Tsuboi[1], T. Ueda[2], K. Ushizawa[2], and Y. Ezaki[2]

1 Department of Fundamental Science, Iwaki-Meisei University, Iwaki, Fukushima 970, Japan
2 Department of Chemistry, Meisei University, Hodokubo, Hino-shi, Tokyo 191, Japan

In continuation of our series of studies on the localized Raman tensors of proteins [1-3], we have attempted to determine the Raman tensors of tyrosine side chain.

Polarized Raman scattering measurements on a single crystal of L-tyrosine were made by the use of a Raman microscope and 488.0 nm argon excitation. The crystal is orthorhombic with space group $P2_12_12_1$ [4], and the Raman spectra corresponding to the aa, bb and cc components of the crystal Raman tensor were observed (Fig. 1). The tyrosine molecule is oriented in the crystal, with its phenol ring nearly in the bc-plane, and with the O-C^1-C^4-CH_2 line nearly along the crystallographic b-axis [4]. Therefore, it is clear that the 1611 cm-1 vibration, for example, causes the greatest polarizability oscillation nearly along the O-C^1-C^4-CH_2 line, whereas the 1176 cm-1 vibration nearly along the perpendicular direction to this line in the six-membered ring. On the other hand, the 1433 cm-1 vibration must have its polarizability oscillation mainly along the perpendicular direction to the phenol-ring plane.

By the use of these data, and by following the procedure detailed previously [5], the shapes and orientations of the Raman tensors have been derived for some of the prominent Raman bands of the tyrosine residue. The results are partly shown in Fig. 2, along with the normal modes of vibrations determined by Takeuchi et al. [6,7]. The knowledge of such localized Raman tensors are considered to be useful for probing the orientation of the tyrosine side chain in a biological system involving it.

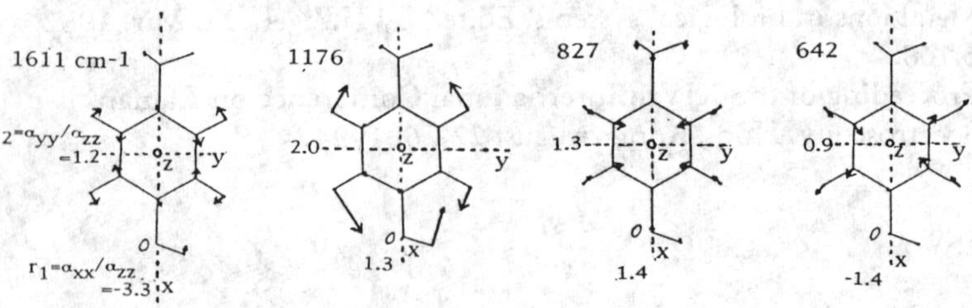

Fig. 2. Principal axes (x, y and z) and relative magnitudes of the components of some of the Raman tensors of L-tyrosine.

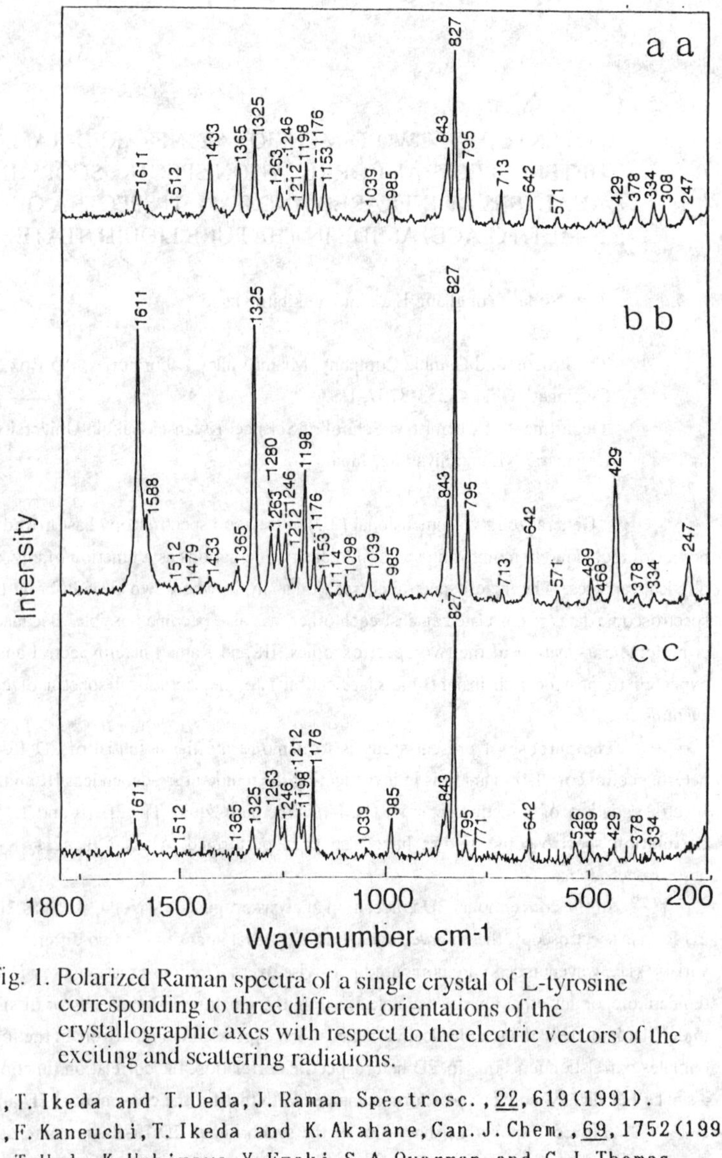

Fig. 1. Polarized Raman spectra of a single crystal of L-tyrosine corresponding to three different orientations of the crystallographic axes with respect to the electric vectors of the exciting and scattering radiations.

1. M. Tsuboi, T. Ikeda and T. Ueda, J. Raman Spectrosc., 22, 619(1991).
2. M. Tsuboi, F. Kaneuchi, T. Ikeda and K. Akahane, Can. J. Chem., 69, 1752(1991).
3. M. Tsuboi, T. Ueda, K. Ushizawa, Y. Ezaki, S. A. Overman and G. J. Thomas, J. Mol. Strtucture, in press(1996).
4. A. Mostad, H. Nissen and C. Romming, Acta Chemica Scand., 26, 3819(1972).
5. M. Tsuboi, T. Ueda and K. Ushizawa, J. Mol. Structure, 352/353, 509(1995).
6. H. Takeuchi, N. Watanabe and I. Harada, Spectrochim. Acta, 44A, 749(1988).
7. H. Takeuchi, N. Watanabe, Y. Satoh and I. Harada, J. Raman Spectrosc., 20, 233(1989).

POTENTIAL OF TWO-DIMENSIONAL INFRARED-RAMAN HETEROSPECTRAL CORRELATION SPECTROSCOPY IN ANALYSES OF INFRARED AND RAMAN SPECTRA OF N-METHYLACETAMIDE IN THE PURE LIQUID STATE

Isao Noda[1], Yongliang Liu[2], and Yukihiro Ozaki[2]*

[1]The Procter and Gamble Company, Miami Valley Laboratories, PO Box 538707, Cincinnati, Ohio 45253-8707, USA
[2]Department of Chemistry, School of Science, Kwansei-Gakuin University, Uegahara, Nishinomiya 662, Japan

Generalized two-dimensional (2D) correlation spectroscopy has proved to be a powerful technique in monitoring spectral intensity fluctuations as a function of any kinds of physical variables.[1] The heterospectral correlation analysis, where two very different types of spectroscopic data are correlated against each other, has also become possible. Because of the complementary nature of the two spectroscopies, IR and Raman heterospectral analysis is expected to provide rich insight and clarification into the vibrational spectra of complex systems.

The purpose of present study is to demonstrate the potential of 2D IR-Raman heterospectral correlation analysis in investigating the temperature-dependent IR and Raman spectral variations of N-methylacetamide (NMA) in the pure liquid. The 2D-IR and 2D-Raman correlation spectroscopic studies have also been performed to assist the 2D IR-Raman heterospectral analysis.

In the conventional 2D correlation analysis proposed by Noda,[1] such as 2D IR or 2D Raman spectroscopy, the observed variations of spectral intensities at two different spectral variable (i.e., wavenumbers) are compared pair-wise over a fixed observation interval of time, temperature, or any other physical variable affecting the spectra. The similarity or dissimilarity under the influence of such physical variable is then expressed quantitatively in the form of a complex correlation function. In 2D heterospectral correlation, the correlation function takes a similar but slightly more complex form compared with that for the conventional 2D analysis.[1]

Figure 1 shows a fishnet plot of the synchronous 2D IR-Raman heterospectral correlation spectrum in the 1750-1075 cm^{-1} region of NMA in the pure liquid. The appearance of synchronous correlation peaks observed in these spectra indicates that the thermally induced dynamic variations of spectral intensities of IR absorption and Raman scattering of NMA occur at somewhat similar temperature.

The 2D IR-Raman study reveals that the amide I Raman bands of NMA, consisting of at least three distinct bands at 1665, 1650, and 1635 cm^{-1}, are correlated with the amide I

IR bands at the same frequencies, respectively. These bands at 1665, 1650 and 1635 cm^{-1} are ascribed to dimer and medium-size and large-size oligomers of NMA, respectively. Comparison of the IR spectra in the 3500-2700 cm^{-1} region and the Raman spectra in the 1750-1050 cm^{-1} region gives the following correlation:

3275 cm^{-1} (IR, amide A)—1650 and 1635 cm^{-1} (Raman, amide I) —1305 cm^{-1} (Raman, amide III)

Combining this correlation with the correlation found for the IR amide bands in the 2D-IR, the following correlation can be proposed:

3275 cm^{-1} (IR)—1635 cm^{-1} (IR) —1570 cm^{-1} (IR) —1300 cm^{-1} (IR)

↑ ↑

1650 and 1635 cm^{-1} (Raman) —— 1305 cm^{-1} (Raman)

All the bands involved in this correlation decrease in the low temperature range. It seems, therefore, that they arise from the large or medium-size oligomers of NMA. Similar correlation has been obtained also for the dimeric NMA.

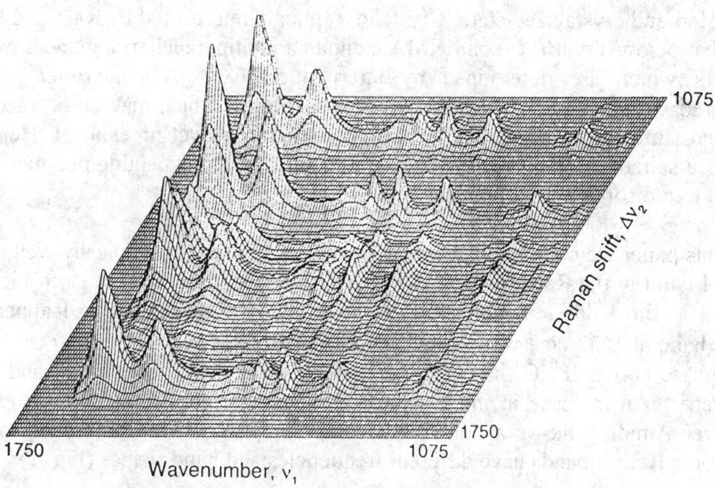

Fig. 1 Full view of the pseudo three-dimensional stacked-trace representation of the 2D IR-Raman synchronous correlation spectrum of NMA in the pure liquid in the 1750-1075 cm^{-1} region

1. Noda, I. *Appl. Spectrosc.*, **1993**, *47*, 1329.

INTERPEPTIDE COUPLING IN THE LIQUID AND SOLID PHASE OF NEAT *TRANS*-METHYLACETAMIDE PROBED BY FT-IR AND POLARIZED RAMAN SPECTROSCOPY:

R: Schweitzer-Stenner[1], G. Sieler[1], N. Mirkin[2] and S. Krimm[2]

[1]FB1-Institut für Experimentelle Physik, Universität Bremen, 28359 Bremen, Germany
[2]Biophysics Research Division, University of Michigan, Ann Arbor, MI 48109,USA

N-methylacetamide (NMA) is an ideal model system for studying the vibrational dynamics of peptides. Vibrational spectroscopy was recently combined with normal mode calculations based on an *ab initio* force field to yield the normal modes and a complete assignment of the vibrational bands [1]. The present study reports some results from non-resonance polarized Raman and FT-IR measurement on neat NMA in its liquid and crystallized state. Crystallographic data reported by Katz and Post [1] show that below T = 10^0 C solid NMA exhibits an antiparallel arrangement of NMA-polymers which are determined by interpeptide CO·······NH hydrogen. A more disordered structure is observed at higher temperature which may be a precursor of the liquid state existing above 28^0 C whose structure is yet unresolved. Hence neat NMA is a suitable model for studying the transition of a polypeptide polymer from an ordered to a disordered liquid phase.

In this paper we confine ourselves to analyzing the spectroscopically well resolved Amide I band in the Raman spectra of neat NMA. Fig. 1 shows the isotropic part of Amide I at three different temperatures. In the crystal phase Amide I appears as a narrow band at 1633 cm^{-1} with a sideband at its high energy side (≈ 1650 cm^{-1}). In the liquid phase above 28^0 C the entire band shifts up and becomes broad and clearly asymmetric. An increase in intensity and frequency occur with rising temperature. Moreover Amide I shows a non-coincidence effect (NCE) in that its isotropic and anisotropic Raman bands have different frequencies and band shapes (Fig. 1).

A thorough spectral analysis by virtue of a recently developed program MULTIFIT yields the following results:

1. Amide I of liquid NMA consists of three different gaussian subbands denoted as Sb_j(j=1,2,3). Sb_1 and Sb_3 appear at 1634 and 1673 cm^{-1} in the isotropic and anisotropic spectrum, whereas Sb_2 shows a NCE in that its frequency is at 1650 cm^{-1} in the isotropic and at 1655 cm^{-1} in the anisotropic spectrum. The subbands in the IR spectrum have frequencies which are practically identical with those of the anisotropic spectrum. Sb_2 slightly increases its intensity with rising temperature at the expense of Sb_1.

2. In the crystal phase Amide I exhibits two subbands at 1633 (Sb_1') and 1650 cm^{-1} (Sb_2'). Neither Sb_1' nor Sb_2' show any detectable NCE. A new band at 1610 cm^{-1} appears whose origin remains unclear. Sb_1' appears as a comparatively narrow lorentzian, whereas Sb_2' is still broad and gaussian.

These results can be understood as follows. In the ordered solid phase NMA has four peptides in an unit cell exhibiting D_{2h}-symmetry. The Amide I modes of peptides interact by transition dipole coupling (TDC, [3] and hydrogen bonding. This yields four delocalized normal vibrations of A_g, B_{2g}, B_{1u} and B_{3u}-symmetry. Modes with even symmetry are Raman active, whereas the odd modes may appear in the IR-spectrum. Assuming interpeptide coupling along the peptide chains to be predominant we expect $\nu(A_g) \approx \nu(B_{3u}) < \nu(B_{2g}) \approx \nu(B_{1u})$ (ν: band frequency). Consequently we assign Sb_1' to A_g and Sb_2' to B_{2g}. IR-spectra of solid NMA exhibit Amide I at 1635 cm^{-1}[4] which correspond to the above B_{3u}-mode. The IR-activity of B_{1u} is less significant.

Amide I of the liquid phase is indicative of three different polymer conformations which are in thermodynamic equilibrium. Sb_1(I) represents linear hydrogen-bonded oligomeric chains similar to the arrangements in the β-sheet like ordered phase of the crystal. Sb_2(I) now results from a more disordered structure in which NCE-splitting may arise from TDC-coupling between adjacent molecules. Similar effects have earlier been observed for other liquids [4] The third subband may represent a conformation with weak hydrogen bonding which was recently predicted by Dixon et al. [6].

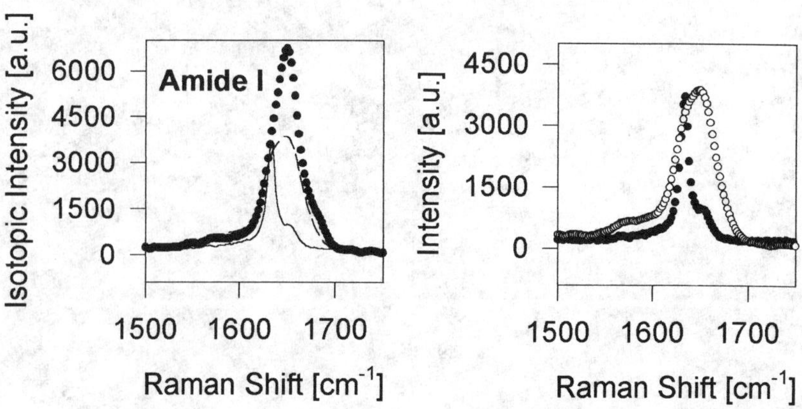

Fig. 1: *Left*: Isotropic Amide I band measured at 0^0 C (solid line), 40^0 C (dashed) and 60^0 C (dotted). *Right*: Isotropic (open circles) and anisotropic (full circles) Amide I band measured at 40^0 C.

1. X.G. Chen, R. Schweitzer-Stenner, S.A. Asher, N.G. Mirkin and S. Krimm, J. Phys. Chem. 99, 3074, 1995.
2. J.L. Katz and B. Post, Acta Crystallogr. 13, 624(1960).
3. T.C. Cheam and S. Krimm, J. Chem. Phys. 82, 1631(1985).
4. T. Miyazawa, T. Shuimanouchi and S.J. Mizushima, J.Chem. Phys. 24, 408(1956).
5. H. Torii and M. Tasumi, J. Chem. Phys.99, 8459(1993).
6. D.A. Dixon, K.D. Dobbs and J.J. Valentini, J. Phys. Chem. 98, 13435(1994).

SECTION 14

BIOLOGICAL SYSTEMS

PHOTOINDUCED REACTION CYCLE OF PHYTOCHROME STUDIED BY RESONANCE RAMAN SPECTROSCOPY

P. Hildebrandt*, J. Matysik, K. Kneip, S. E. Braslavsky, W. Gärtner, K. Smit, I. Magdo, F. Mark, and K. Schaffner

Max-Planck-Institut für Strahlenchemie, Postfach 101365, D-45413 Mülheim, Fed. Rep. Germany

Phytochrome is an ubiquitous photoreceptor in plants which controls a variety of photomorphogenic processes [1]. The protein exists in two photoconvertible forms, the physiologically active P_{fr} and the non-active P_r. The interconversion between both forms is initiated by light absorption of the chromophore, a linear tetrapyrrole which is covalently bound to the apoprotein. The primary photochemical processes which involve the isomerizations around a methine bridge is followed by a sequence of thermal reactions which also include changes in the chromophore-protein interactions. In this way, structural changes in the interaction domain of phytochrome are induced so that physiological reaction partners can distinguish between the active and the non-active state. Evidently, the elucidation of structure-function relationships in phytochrome requires a detailed knowledge about the chromophore structure and its changes during the photoinduced reaction cycle.

In this respect, resonance Raman (RR) spectroscopy is a particularly powerful technique as it selectively probes the vibrational bands of the chromophore [2-4]. Such spectra include detailed data about the configuration and conformation of the chromophore and its interactions with the immediate protein environment. However, the vibrational assignment for linear tetrapyrroles, which is a prerequisite for extracting structural information from the spectra, is far from being fully settled. Hence, we have adopted two different approaches: (i) RR spectroscopic characterization of the various states of phytochrome formed during the reaction cycle, and (ii) vibrational analysis of linear tetrapyrroles regarded as models for the phytochrome chromophore.

RR spectra of phytochrome were obtained with near-infrared (1064 nm) excitation employing the Fourier-transform technique [4,5]. Although this excitation line does not coincide with the electronic transition of the chromophore (660 - 720 nm), the preresonance enhancement of the Raman bands is still sufficient to allow for a nearly complete discrimination of the non-resonance Raman bands of the apoprotein. This experimental approach avoids the interference of the chromophore fluorescence as well as unwanted photochemical processes induced by the exciting laser beam. On the other hand, the photoinduced reaction cascade can be initiated in a controlled way by irradiating P_r and P_{fr} with 660 and 730-nm light, respectively. The intermediates which are subsequently formed can be trapped at low temperature. Using this approach, it was possible to obtain high-quality RR spectra of the parent and the intermediate states of phytochrome.

The vibrational analysis of linear tetrapyrroles focusses on biliverdin dimethyl ester (BVE) for which the crystal structure has been solved [6]. The analysis is based on quantum chemical calculations of the force field which is scaled according to Pulay's method [7]. While previous calculations were restricted to the semiempirical AM1 method [8], we have now employed ab-initio Hartree-Fock and density functional approaches which provide a more reliable force field. Furthermore, IR and RR intensities for the various modes were approximated. Based on these calculated spectral data, a plausible assignment for most of the vibrational bands in the measured IR and RR spectra of BVE was achieved. These studies were extended to a configurational and conformational isomers of BVE in order to assess the effect of double and single bond rotations on the vibrational spectra.

A comparison with the RR spectra of related linear tetrapyrroles exhibiting different ring substitution patterns suggests that some aspects of the vibrational analysis of BVE can also be extended to the phytochrome chromophore. In this way, we were able to provide a consistent albeit preliminary interpretation for the RR spectral changes of phytochrome during its photoinduced reaction cycle.

The structural changes of the chromophore during the primary photochemical process of P_r are largely restricted to the methine bridge at which photoisomerization occurs. The torsional angle around the methine bridge remote from the isomerization site remains largely the same in the primary photoproduct as in the parent state. In the subsequent thermal reactions this angle is reduced accompanied by an increased twist of the single bond at the isomerized methine bridge. This torsion is particularly pronounced in the final step of P_{fr} formation. As this molecular motion implies a significant change of the chromophore-protein interactions, it is suggested that this step of the reaction cycle is coupled to structural changes in the protein which eventually are transduced to the interaction domain of phytochrome with other cellular components. In a similar way, we interpreted the photoinduced backreactions of P_{fr} to P_r (i. e., the deactivation of phytochrome).

1. K. Schaffner, S. E. Braslavsky, and A. R. Holzwarth, in Frontiers of Supramolecular Organic Chemistry and Photochemistry (H. J. Schneider and H. Dürr, eds.), VCH Verlagsgesellschaft, Weinheim, 1991, p. 421.
2. S. P. A. Fodor, J. C. Lagarias, and R. A. Mathies, Photochem. Photobiol. 48, 129 (1988).
3. Y. Mizutani, S. Tokutomi, K. Aoyagi, K. Horitsu, and T. Kitagawa, Biochemistry 30, 10693 (1991).
4. P. Hildebrandt, A. Hoffmann, P. Lindemann, G. Heibel, S. E. Braslavsky, K. Schaffner, and B. Schrader, Biochemistry 31, 7957 (1992).
5. J. Matysik, P. Hildebrandt, W. Schlamann, S. E. Braslavsky, and K. Schaffner, Biochemistry 34, 10497 (1995).
6. W. S. Sheldrick, J. Chem. Soc., Perkin Trans. 2, 1457 (1976).
7. P. Pulay, G. Fogorasi, G. Pongor, J. E. Boggs, and A. Vargha, J. Am. Chem. Soc. 105, 7037 (1983).
8. K. Smit, J. Matysik, P. Hildebrandt, and F. Mark, J. Phys. Chem. 97, 11887 (1993).

TOWARDS *IN VIVO* CLINICAL DIAGNOSIS BY RAMAN SPECTROSCOPY

G.J. Puppels, H. van Dekken[#], E. Branderhorst, H.W. Tilanus, H.A. Bruining*

Department of General Surgery and Department of Pathology[#]
Erasmus University Rotterdam & Rotterdam University Hospital "Dijkzigt", Rotterdam, The Netherlands

One way to recognize a disease is through the changes it causes in molecular composition of body fluids, cells or tissues. Therefore, techniques that provide information about molecular compostion often play an important role in clinical diagnosis. The biomedical and clinical diagnostic application of optical spectroscopies is a hot item at the moment. In principle they are able to provide information on the physiological condition and molecular composition of tissues and organs in a non-invasive, or at least minimally invasive manner. Especially those techniques that do not depend on exogenous labels, such as diffuse reflection/transmission spectroscopy, autofluorescence spectroscopy and vibrational spectroscopies are very appealing. Raman spectroscopy appears to be a particularly promising technique due to the great information content of the spectra and due to the fact that it can be easily applied for the analysis of watery biological samples.

Which steps need to be taken to make this promise reality and turn Raman spectroscopy into a valuable clinical diagnostic tool?

First of all good applications will have to be identified that when successfully developed will meet a clinical need. Second, suitable instrumentation for *in vivo* measurements will have to be developed (read suitable Raman fiber probes). Third, databases will need to be set up. And for these databases to function it is necessary that spectra measured at different places at different times and with different equipment can be compared without hindrance from instrumental artifacts. Therefore, all measured spectra will have to be rigorously corrected for instrumental influences. A further point is that the speed with which measurements and analysis (which should preferably be real-time) can be carried out needs to be optimized as far as possible.

We have started several projects that address these points. Topics include the analysis of the molecular composition of human coronary artery wall and changes therein due to atherosclerosis and medication, detection of (pre-)malignant tissue in the oesophagus, analysis of bladder wall molecular composition and changes therein due to obstruction in the urinary tract, optimization of instumental sensitivity and the use of multivariate statistical analysis techniques for classification of tissue on the basis of their Raman spectra.

Figure 1 shows results of Raman measurements on unfixed tissue (luminal side) of a resected human oesophagus. The top spectrum is that of normal stratified squamous epithelium, the second spectrum that of Barrett's columnar epithelium and the bottom spectrum that of a small tumor in the oesophagus (adenocarcinoma).

Barrett's epithelium replaces the normal epithelium in response to chronic reflux of gastric/duodenal juices. This condition markedly enhances the risk of developing oesophageal cancer to about 1% per year [1]. Clear differences can be seen in the spectral regions 400-500 cm^{-1} and 800-1200 cm^{-1}. In fact principal components analysis of the spectra shows that features that enable correct classification of the tissues, are present in virtually every part of the spectrum. Figure 2 shows the result of a principal components analysis carried out on 75 spectra (25 normal epithelium, 25 Barrett's epithelium and 25 adenocarcinoma) using the whole measured spectral interval as input (400-1850 cm^{-1}). Using only the first two principal components already leads to a quite good separation between the different types of tissue. Before principal components analysis a fluorescent background was subtracted from the spectra and an intensity normalization was carried out, using the intensity of the 1450 cm^{-1} band as reference. These first results indicate that Raman spectroscopy in combination with endoscopy may be a useful technique for screening of patients with a Barrett's oesophagus for (pre-)malignant tissue degenarations.

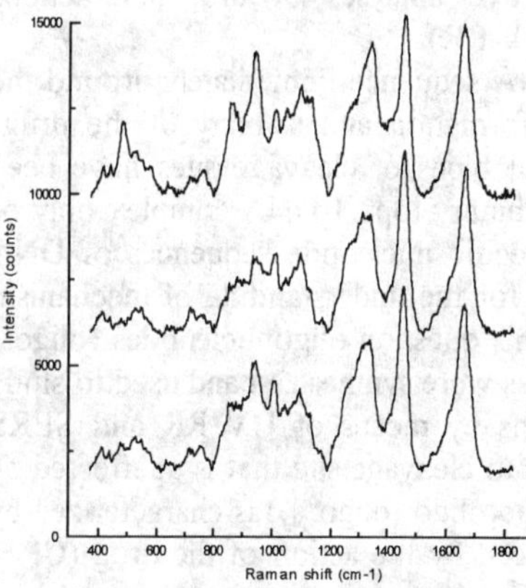

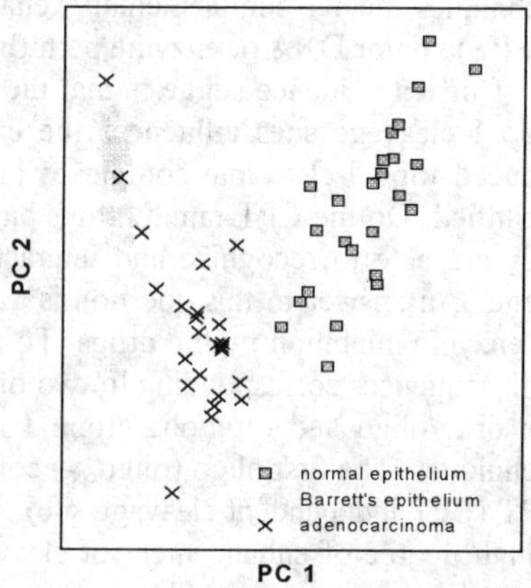

F g. 1: Raman spectra of unfixed human oesophageal tissue
NE: normal squamous stratified epithelium
BE: Barrett's columnar epithelium
AC: adenocarcinoma
experimental conditions: micro-Raman set up; x20 Nachet NIR-objective; laser 785 nm, 100 mW on sample;signal collection time 30 s; spectra shown averaged over 25 measurements

Fig. 2:75 spectra of human oesophageal tissue plotted in 2-dimensional principal component space

1. S.J. Spechler Seminars Oncol. 21, 431 (1994)

SERS AND UV RR SPECTROSCOPY STUDY OF DNA-TOPOISOMERASE I AND ITS COMPLEXES WITH DNA AND ANTITUMOR DRUGS

A.V. Feofanov[1]*, A.V. Baranov[2], F. Fleury[3], M. Manfait[3] and I.R. Nabiev[1,3]

[1]Shemyakin-Ovchinnikov Institute of Bioorganic Chemistry, Russian Academy of Sciences, ul. Miklukho-Maklaya 16/10, 117871 Moscow, Russia
[2]Vavilov State Optical Institute, St. Petersburg, Russia
[3]Université de Reims Champagne-Ardenne, Reims, France

Combined surface-enhanced Raman scattering (SERS) and UV resonance Raman (RR) spectroscopy approach has been employed to analyze molecular interactions within the binary and ternary complexes of DNA-topoisomerase I (topo I) with DNA (oligonucleotides) and anticancer drugs of camptothecin (CPT) family. Two techniques being complementary enable selective analyses of drug interactions (SERS) and/or DNA or enzyme perturbations (UV RR).

Current evidence suggests that the local DNA sequence immediately around the topo I cleavage sites influences the extent of formation and stability of the drug-induced topo I-cleavable complexes [1]. Several types of cleavage sites have been identified. Do the CPT-family drugs bind to the binary topo I-DNA complex only or they are able to recognize and interact with specific nucleotide sequences of DNA alone? An answer to this question is a keystone for the understanding of mechanism of enzyme inhibition by the drugs. To address this question oligonucleotides (oligos) with sequences corresponding to two of these sites were synthesized and used to study the drug/oligo and drug/oligo/topo I interactions by means of UV RR and SERS techniques. The first oligo ($oligo_{cpt-}$) corresponds to cleavage site that is unaffected by CPT (CPT independent cleavage site). The second oligo ($oligo_{cpt+}$) is characterized by a high degree of enhancement of cleavage induced by the action of the drug (CPT-dependent cleavage sites) [1].

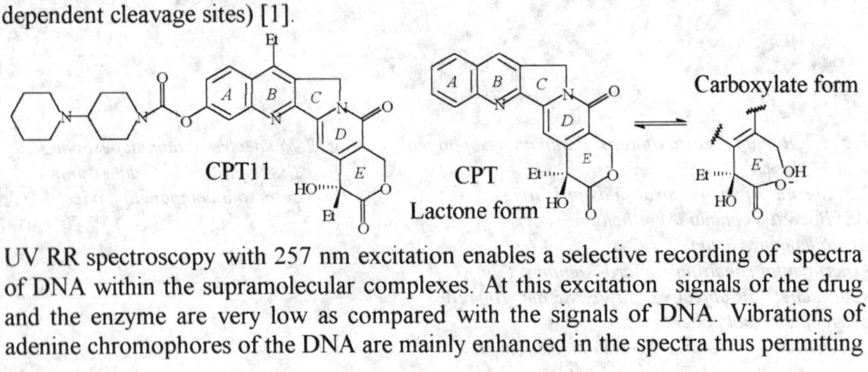

UV RR spectroscopy with 257 nm excitation enables a selective recording of spectra of DNA within the supramolecular complexes. At this excitation signals of the drug and the enzyme are very low as compared with the signals of DNA. Vibrations of adenine chromophores of the DNA are mainly enhanced in the spectra thus permitting the drug and/or enzyme-induced changes of stacking, microenvironment and interactions of adenine bases to be studied.

Two sets of SERS spectral parameters enable analysis both of the topology of supramolecular complexes and of inter- and intramolecular interactions. The first one is the SERS absolute intensity. The effect of loss of SERS intensity upon the partial penetration of the drug into the DNA or in the interior of the protein [2,3] enables us to clarify whether the substrate is bound on the surface or within the interior of supramolecular complex. The second set of parameters includes SERS frequencies and relative intensities of the bands being analyzed in the same fashion as for normal Raman spectra. The bands corresponding to vibrations of functional groups of drug involved in the interaction with the target are preferably decreased in intensity [3], thus permitting the construction of the molecular models of drug-target complexes.

UV RR and SERS study revealed interaction of CPT both with $oligo_{cpt-}$ and $oligo_{cpt+}$. UV RR spectra show drug/oligo ratio-dependent decrease of the relative intensities of the 1343 and 1489 cm^{-1} bands of adenine. Additionally, general decreasing of the oligonucleotide signal was detected when adding of CPT to $oligo_{cpt-}$. These CPT-oligo interactions are specific for the nucleotide sequence, since they were not observed in the mixture of derivative of CPT, CPT11, with $oligo_{cpt-}$. At the same time interaction of CPT11 with the $oligo_{cpt+}$ was revealed and found to be different from the CPT-$oligo_{cpt+}$ one. DNA sequence recognition by the drugs before formation of cleavable complexes could be responsible at least partially for the different cleavage pattern. SERS spectra show (i) general decrease of the intensity of the CPT signals when the drug forms the complex with both $oligo_{cpt-}$ and $oligo_{cpt+}$, and (ii) pronounced changes in the relative intensities of the drug bands upon the interaction. Detailed SERS analysis disclosed that the interaction of CPT with the oligos occurred through the quinoline moiety of the drug.

In spite of the absence of CPT11-$oligo_{cpt-}$ interaction, interactions of CPT11 within both types of ternary complexes (CPT11- $oligo_{cpt-}$ -topo I and CPT11-$oligo_{cpt+}$ -topo I) were revealed by UV RR and SERS spectroscopy. Characteristic spectral features of CPT11-$oligo_{cpt+}$ -topo I ternary complex are quite similar to those of CPT-oligo complexes. Some additional changes were observed in the spectra of CPT11- $oligo_{cpt-}$ - topo I ternary complex.

An open (carboxylate) form of the CPT11 was found to induce pronounced changes of the relative intensity of UV RR bands of oligos in the ternary complexes. Some spectral features detected in this case were observed in the spectra of the ternary complexes with a closed (lactone) form of CPT11. Confirming the UV RR data the SERS spectra indicated undoubtedly that the open form of CPT-drugs interacts with the both types of oligos mentioned above.

The preliminary models of complexes of the open and closed forms of CPT (CPT11) with oligos as well as in the ternary complexes have been proposed on the basis of the analysis of UV RR spectra of adenine derivatives and polynucleotides and of the SERS spectra of CPT-derivatives and model compounds.

1. M.Gupta, A.Fujimori, and Y.Pommier, Biochim. Biophys. Acta. **1262**, 1 (1995)
2. T.M. Cotton, in Spectroscopy of Surfaces, (R.J.H. Clark and R.E. Hester, eds.), Wiley, New York, 1988, p.91.
3. I. Nabiev, I. Chourpa, J.F. Riou, et al. Biochemistry **33**, 9013 (1994).

SPASIBA FORCE FIELD PARAMETERIZATION OF MODEL COMPOUNDS OF BIOMEMBRANES : PHOSPHOLIPIDS AND GLYCOLIPIDS

M. Chhiba, F. Tristram, V. Durier and G. Vergoten*

CRESIMM - Bat C8 - Université des Sciences et Technologies de Lille
UFR de Chimie 59655 Villeneuve d'Ascq - France -

Biomembranes play a central role in large number of life's processes. Not only do they serve as a barrier to maintain the integrity of cells, they also are the functional environment of a large number of proteins that include receptors, ion-channels and parts of the immune system. Knowledge at the molecular structure level and molecular dynamics of lipid bilayers are important for understanding many functions of biomembranes [1]. For this reason, we have extended specifically the force field SPASIBA [2] (Spectroscopic Potential Algorithm for SImulating Biomolecular conformational Adaptability) to the phospholipids and glycolipids which present the major classes of lipid molecules.

The SPASIBA force field is currently under development in our laboratory. SPASIBA derives from the Amber program [3] where a set of Urey-Bradley-Shimanouchi terms were added [4]. The force field parameters are established through different experimental results deduced from thermodynamics, crystallography and particularly vibrational spectroscopy (Infrared and Raman). SPASIBA have been parameterized for a large number of model compounds like alkanes [5] aliphatic amino acids [2], aliphatic ethers [6], alcohols [7] and acetals [8]. More recently, force field parameters of phosphate groups [9] and carbohydrates [10] were developped. With such a database, it is now possible to investigate the structural and dynamic behaviour of compounds with biological interest.

In this paper, we focus our attention on the force field development of model compounds of particular interest in biological membranes. For the phospholipids molecules, we have been mainly concerned with the parameterization of three series of compounds. The first series concerns the molecules containing phosphate and choline groups in order to have the parameters for polar head groups. The second is the aliphatic carboxylic acids linked to the glycerol moiety. Finaly, the force field SPASIBA has been extended to include linear alkenes in order to investigate the hydrocarbon region. The determination of force field parameters for sugars is rather complex. Indeed the assignment of vibrational spectra of monosaccharides is not totally precise. A set of parameters was first proposed for α-D-glucose [10] on the basis of known SPASIBA parameters. This allowed application of this model to monosaccharides units.

The SPASIBA force field development of parameters initially from simple compounds with specific function results in homogenous database for complex molecules. These force field parameters are developed in order to be transferable. They include the dynamical properties of molecules and can now be applied to glycolipids. This work extens SPASIBA to the dynamical and structural behaviour of the common part of the glycolipids (galactolipids).

1. K.V. Damodaran and K.M. Merz Jr., Review in Comput. Chem. vol. V, VCH Publishers, Inc. N.Y., 1994.
2. P. Derreumaux and G. Vergoten, J. Chem. Phys. 102, 8586 (1995).
3. S.J. Weiner, P.A. Kollman, D.T. Nguyen and D.A. Case, J. Comput. Chem. 7, 230 (1986).
4. T. Shimanouchi, Pure Appl. Chem. 7, 131 (1963).
5. P. Derreumaux, M. Dauchez and G. Vergoten, J. Mol. Struct. 295, 203 (1993).
6. F. Tristram, V. Durier and G. Vergoten, J. Mol. Struct. In Press.
7. F. Tristram, V. Durier and G. Vergoten, J. Mol. Struct.,In Press.
8. F. Tristram, V. Durier and G. Vergoten, submitted for publication to J. Mol. Struct.
9. M. Chhiba and G. Vergoten submitted for publication to J. Mol. Struct. In press.
10. V. Durier, F. Tristram and G. Vergoten, submitted for publication to J. Mol. Struct.

RESONANCE RAMAN STUDIES OF THE NONPLANAR DISTORTIONS OF NICKEL(II) 5,15-DI-SUBSTITUTED PORPHYRINS: A COMBINATION OF DISPLACEMENTS ALONG TWO OF THE LOWEST-FREQUENCY NORMAL COORDINATES

W. Jentzen,[1] X. Song,[1,2] S. Jia,[1,2] J. Ma,[1,2] L. Jaquinod,[3] D. J. Nurco,[3] C. J. Medforth,[3] K. M. Smith,[3] J. A. Shelnutt[1,2*]

[1]Fuel Science Department, Sandia National Laboratories, Albuquerque, NM 87185-0710, USA. [2]Department of Chemistry, University of New Mexico, Albuquerque, NM 87131, USA. [3]Department of Chemistry, University of California, Davis, CA 95616, USA.

Metalloporphyrins that are embedded in proteins as cofactors can be significantly nonplanar, and this nonplanarity may play a pivotal role in their biological function. A recent advancement in investigating these out-of-plane distortions allows us to identify the different distortion types of the macrocycle in crystal structures of various proteins. We have found that the displacements from the mean plane can be described quantitatively in terms of displacements along the lowest-frequency out-of-plane normal coordinates [1]. Only four normal coordinates, which in D_{4h} symmetry transform like A_{2u} (*dom*), B_{1u} (*ruf*), B_{2u} (*sad*), and E_g (*wav*), are required for an accurate description of most porphyrin structures. In other words, the total observed out-of-plane distortion is decomposed into contributions from each these four distortion types. For example, the out-of-plane distortions in the X-ray crystal structures of the heme groups in cytochromes and hemoglobins are substantially different. In ferrocytochromes *c*, a conserved, primarily *ruf* distortion is observed with minor contributions from the *wav* distortion. Further, the heme groups in the α-chains from human deoxyhemoglobin mainly consist of *ruf* and *dom* distortions, whereas the out-of-plane displacement of the hemes in the β-chains is decomposed into *ruf* and *sad* distortions.

We have been using various substituted porphyrins as models to mimic the structures of the porphyrin cofactors in these proteins. In a previous study, the ruffling in a series of nickel(II) *meso*-tetrasubstituted porphyrins served as a model for the conserved ruffling of the heme of *c*-type cytochromes [2]. In the present study, we report the resonance Raman studies of the nonplanar distortions in a series of nickel(II) 5,15-di-substituted porphyrins. These porphyrins exhibit a combination of distortion types along the lowest frequency normal coordinates of the B_{1u} (*ruf*) and A_{2u} (*dom*) distortions, modeling the ruffling and doming of the hemes in the α-chains of hemoglobin.

Nickel(II) 5,15-diphenylporphyrin (NidPP) is predicted by molecular mechanics calculations to be planar, but the broad and asymmetric line shape of the lines ν_2 and ν_8 in the resonance Raman spectra reveal the porphyrin to exist in solution as an equilibrium mixture of planar and nonplanar conformers (Figure 1a) [3].

All of the nickel(II) dialkylporphyrins, in which the alkyl group is propyl (Pr), *iso*-propyl (iPr), *tert*-butyl (tBu), are predicted to be in a roof or gabled (*gab*) conformation resulting from an αα-orientation of the substituents with respect to the mean plane of the macrocycle. This nonplanar *gab* distortion can be represented as a linear combination of distortions along the lowest-frequency macrocycle normal coordinates of B_{1u} and A_{2u} symmetry. A higher energy stable αβ-conformer is also predicted for some dialkylporphyrins, and this conformer can be represented as an equal combination of distortions along the *x*- and *y*- components of the lowest-frequency E_g normal coordinate [*wav(x)+wav(y)*]. These predictions are confirmed by the X-ray crystal structure of NidtBuP which is found to be in a *gab* αα-conformation. The resonance Raman spectra confirm this result for the other di-substituted porphyrins as well.

For NidPrP, six stable conformers resulting from all the possible orientations of the two propyl substituents with respect to the mean plane of the macrocycle and to each other are predicted. These conformers differ in energy by less than 1.4 kcal mol^{-1}. This predicted conformational heterogeneity is reflected in the line shape of a number of Raman lines, including ν_2 and ν_8, which show asymmetry and a broad line width (Figure 1b). The narrow and symmetric lines of ν_2 and ν_8 of NidiPrP (Figure 1c) agree with the molecular mechanics also, because only two conformers that have almost identical distortions are predicted by the molecular mechanics calculations. For NidtBuP, the narrow, symmetric line ν_8 suggests that only one conformer, the stable *gab* structure, exists in solution. This is expected since the *wav(x)+wav(y)* conformer is about 9 kcal mol^{-1} higher than the *gab* conformer. The origin of the doublet of ν_2 for NidtBuP is not clear at present (Figure 1d), but Raman spectra from deuterated compounds currently being synthesized and low-temperature spectra may resolve this issue.

With increasing bulkiness of the substituents in the order of phenyl, Pr, iPr and tBu, the degree of nonplanar distortion is predicted to increase from the almost planar dPP, to the slightly nonplanar dPrP, to the prominently nonplanar diPrP and dtBuP derivatives. This increasing degree of nonplanar distortion is reflected in the increasing downshifts of the structure-sensitive Raman lines of ν_2, ν_3, and ν_{10}. Figure 1e shows the frequency dependence of these Raman lines versus the degree of nonplanarity for the (*ruf+dom*) nickel(II) di-substituted porphyrins in comparison with the corresponding purely *ruf meso*-tetrasubstituted porphyrins [2]. When the degree of nonplanarity is measured solely by the out-of-plane displacement along the lowest-frequency normal coordinate of B_{1u} (*ruf*) symmetry, the slopes of the lines for the lines ν_2 and ν_3 are steeper for the dialkyl series than for the corresponding lines for the tetraalkyl series. The different slopes may result from the additional contribution of the *dom* distortion in the dialkyl porphyrin case.

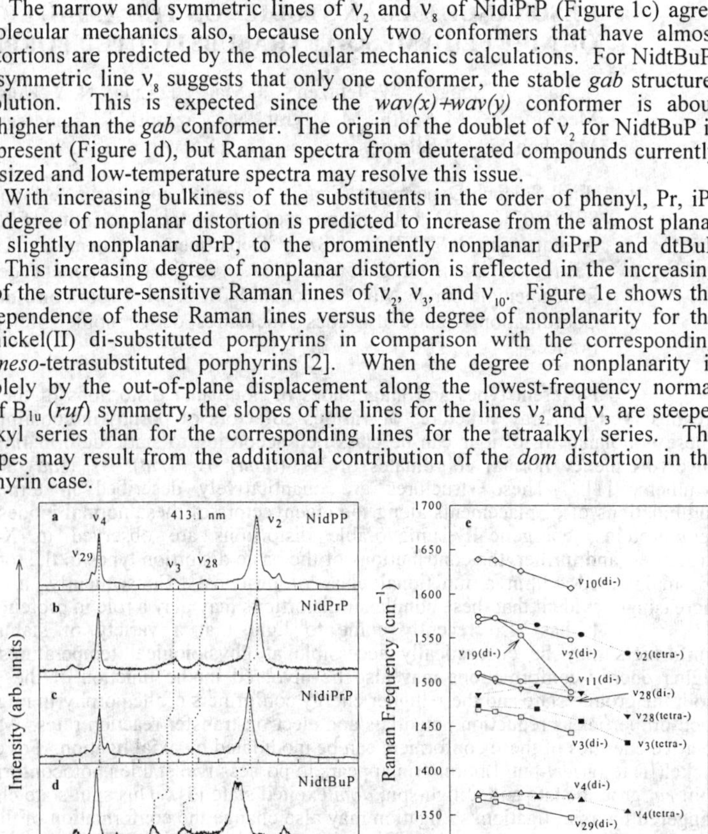

Figure 1. Resonance Raman spectra of the series of nickel(II) di-substituted porphyrins in CS_2 (a–d) and the frequency dependence of some structure-sensitive lines versus the ruf distortion for nickel(II) di- and tetra-substituted porphyrins (e).

1. Jentzen, W.; Song, X.; Shelnutt, J.A., *in preparation*.
2. Jentzen, W.; Simpson, M.C.; Hobbs, J.D.; Song, X.; Ema, T.; Nelson, N.Y.; Medforth, C.J.; Smith, K.M.; Veyrat, M.; Mazzanti, M.; Ramasseul, R.; Marchon, J.-C.; Takeuchi, T.; Goddard III, W.A.; Shelnutt, J.A. *J. Am. Chem. Soc.* **1995**, 117, 11085.
3. Jentzen, W.; Unger, E.; Turowska-Tyrk, I.; Song, X.; Nurco, D.J.; Medforth, C.J.; Smith, K.M.; Dreybrodt, W.; Scheidt, W.R.; Shelnutt, J.A., *in preparation*.
(This work performed at Sandia National Laboratories is supported by the U.S. Department of Energy under contract DE-AC04-94AL85000.)

RESONANCE RAMAN STUDIES ON THE AXIAL LIGATION OF NICKEL(II) MESO-TETRASUBSTITUTED PORPHYRINS

S. Jia,[1,2] X. Song,[1,2] W. Jentzen,[1] J. Ma,[1,2] T. Ema,[3] N. Y. Nelson,[3] C. J. Medforth,[3] K. M. Smith,[3] M. Veyrat,[4] M. Mazzanti,[4] R. Ramasseul,[4] J.-C. Marchon,[4] J. A. Shelnutt[1,2*]

[1]Fuel Science Department, Sandia National Laboratories, Albuquerque, NM 87185-0710, USA. [2]Department of Chemistry, University of New Mexico, Albuquerque, NM 87131, USA. [3]Department of Chemistry, University of California, Davis, CA 95616, USA. [4]CEA/Départment de Recherche Fondamentale sur la Matière Codensée/SESAM, Laboratoire de Chimie de Coordination, Centre d'Etudes Nucléaires de Grenoble, 38054 Grenoble, France.

Different types and magnitudes of nonplanar distortions of the macrocycle are observed in X-ray structures of various cofactors of porphyrin-containing proteins. These nonplanar distortions can be classified according to static deformations along the lowest-frequency normal coordinates of A_{2u} (*dom*), B_{1u} (*ruf*), B_{2u} (*sad*) and E_g (*wav*) symmetry [1]. These structures are quantitatively described in terms of linear combinations of displacements along the eigenvectors of these normal modes. For many hemoproteins, energetically unfavorable distortions are observed in X-ray crystal structures, and further, the contributions of the basic distortion types to the total distortion are maintained within a functional class of proteins. Consequently, it is becoming increasingly evident that these nonplanar distortions may play a role in protein function.

It has also recently come to light that a variety of stable nonplanar conformers may be energetically accessible at physiological temperatures, and these higher energy conformations may also be involved in the function of the proteins [2]. Both the ground state and these higher energy conformers of the porphyrin might influence such properties as reduction potentials and electron-transfer reaction rates. Moreover, the relative energies of these conformers can be modulated by axial ligation. For example, the nickel(II) tetra(*tert*-butyl)porphyrin appears to possess two stable photoconformers: a low-spin *ruf* ground state and a high-spin *dom* excited state [3]. This spin-state change is also triggered by axial ligation, so ligation may also change the conformation of the porphyrin. We have previously studied ruffling in a series of nickel(II) *meso*-tetrasubstituted porphyrins [1]. Here, we report the axial ligation properties of this series of porphyrins as revealed by resonance Raman spectroscopy, combined with UV-visible absorption spectroscopy and molecular modeling (MM). In combination, these techniques have proved to be one of the most useful methods for investigating the ligation state and nonplanar distortions of metalloporphyrins [1,2,4]. We use the method to address the following questions. What is the effect of ruffling on the axial ligation properties of porphyrins and what are the structural changes of the porphyrin macrocycle upon ligation?

In the series of nickel(II) *meso*-tetrasubstituted porphyrins, the degree of ruffling increases with the steric bulkiness of the substituents from phenyl (P), ethyl (Et), propyl (Pr), pentyl (Pe), and methyl (Me), to *iso*-propyl (iPr), cyclohexyl (cH), and *tert*-butyl (tBu) [1]. The resonance Raman spectra (Figure 1) and absorption spectra show that ruffling generally lowers the axial ligand affinity of the porphyrin. Specifically, the ruffled porphyrins like TiPrP, TcHP and TtBuP exhibit no axial ligation in strongly coordinating solvents such as pyrrolidine and piperidine.

Among the porphyrins which axially ligate in coordinating solvents, molecular modeling shows that TPP is planar both in low-spin and high-spin states;

although, resonance Raman spectra indicate that TPP in non-coordinating solvents like CS_2 exists in an equilibrium mixture of planar and nonplanar conformations, as evidenced by the broad and asymmetric ν_2 line. After axial ligation with pyrrolidine, ν_2 is narrow and symmetric, suggesting that NiTPP changes to one planar conformer. MM shows that TEtP, TPrP, TPeP and TMeP are nonplanar (the ruffling dihedral angles are almost the same and about 21°) in non-coordinating solvents like CS_2, as indicated by the frequencies of ν_2 which are around 1567 cm^{-1}, and ν_2 for these nonplanar alkyl derivatives is about 8 cm^{-1} lower than for planar NiTPP. Upon ligation with pyrrolidine, the degree of nonplanar distortion for these porphyrins is significantly reduced as exhibited by the behavior of ν_2; ν_2 of these porphyrins is only 1-2 cm^{-1} lower than NiTPP, indicating that they are all nearly planar in the 6-coordinate form. The decrease in the nonplanarity could be explained by the conformational change caused by axial ligation. Axial ligation changes the state of Ni(II) from low spin to high spin and expands the porphyrin core. Large metals like high spin Ni(II) favor a planar or domed macrocycle over a ruffling distortion which contracts the core. Also, doming results in a much smaller degree of nonplanarity than the ruffling distortion for the same substituted porphyrin. Thus, the decreased nonplanar distortion of the ligated porphyrins, indicated by the Raman frequencies, suggests a conformational change from ruffling to doming. This is also supported by molecular mechanics calculations which indicate that the *dom* conformers have almost the same energy as the *ruf* ones in ligated state, whereas, the *dom* conformers are much higher energies in un-ligated state.

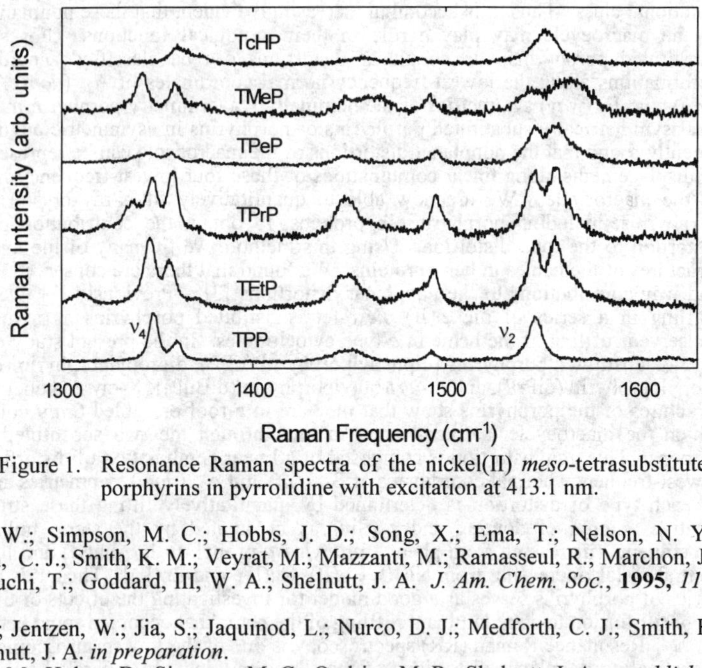

Figure 1. Resonance Raman spectra of the nickel(II) *meso*-tetrasubstituted porphyrins in pyrrolidine with excitation at 413.1 nm.

1. Jentzen, W.; Simpson, M. C.; Hobbs, J. D.; Song, X.; Ema, T.; Nelson, N. Y.; Medforth, C. J.; Smith, K. M.; Veyrat, M.; Mazzanti, M.; Ramasseul, R.; Marchon, J.-C.; Takeuchi, T.; Goddard, III, W. A.; Shelnutt, J. A. *J. Am. Chem. Soc.*, **1995,** *117*, 11085.
2. Song, X.; Jentzen, W.; Jia, S.; Jaquinod, L.; Nurco, D. J.; Medforth, C. J.; Smith, K. M.; Shelnutt, J. A. *in preparation*.
3. Drain, C. M.; Holten, D.; Simpson, M. C.; Ondrias, M. R.; Shelnutt, J. A. *unpublished results*.
4. Shelnutt, J. A.; Alston, K.; Ho, J.-Y.; Yu, N.-T.; Yamamoto, T.; Rifkind, J. M. *Biochemistry*, **1986,** 25, 620.

(This work performed at Sandia National Laboratories is supported by the U.S. Department of Energy under contract DE-AC04-94AL85000.)

RESONANCE RAMAN STUDIES OF NONPLANAR DISTORTIONS OF METAL 5,15-DI-SUBSTITUTED PORPHYRINS

X. Song,[1,2] W. Jentzen,[1] S. Jia,[1,2] J. Ma,[1,2] L. Jaquinod,[3] D. J. Nurco,[3] C. J. Medforth,[3] K. M. Smith,[3] J. A. Shelnutt[1,2*]

[1]Fuel Science Department, Sandia National Laboratories, Albuquerque, NM 87185-0710, USA. [2]Department of Chemistry, University of New Mexico, Albuquerque, NM 87131, USA. [3]Department of Chemistry, University of California, Davis, CA 95616, USA.

Substantial nonplanar distortions are present in tetrapyrrole cofactors of proteins. The X-ray crystal structures of the heme of cytochromes *c*, for example, show an energetically unfavorable and conserved distortion mainly of the ruffled type [1]. Many other heme proteins exhibit distortions that are conserved for proteins of a specific functional class. Thus, it is becoming increasingly evident that these nonplanar distortions of the macrocycle may play a role in their biological functions. For symmetrically substituted porphyrins, the nonplanar distortions can be classified according to static deformations along the lowest-frequency normal coordinates of A_{2u} (*dom*), B_{1u} (*ruf*), B_{2u} (*sad*), and E_g (*wav*) symmetries of the nominally D_{4h}-symmetry porphyrin macrocycle [2]. For asymmetrically substituted porphyrins or porphyrins in asymmetric environments, we recently found that the nonplanar distortions of the macrocycle can be represented in terms of displacements along linear combinations of these four lowest-frequency normal modes of the macrocycle. We are now able to quantitatively fit X-ray crystal structures of porphyrins, including porphyrins in proteins, to obtain the contribution of each basic distortion to the total distortion. Using this method, we fit many of the available X-ray structures of the hemes in hemoproteins. We found that there are conserved *ruf*, *sad*, *dom*, and *wav* contributions to the nonplanar distortions [3]. Previously [2], we also studied ruffling in a series of nickel(II) *meso*-tetrasubstituted porphyrins as a model for this conserved ruffling of the heme in *c*-type cytochromes. In the present study on a series of 5,15-disubstituted porphyrins [diphenylporphyrin (dPP), dipropylporphyrin (dPrP), di-*iso*-propylporphyrin (diPrP), and di-*tert*-butylporphyrin (dtBuP)], X-ray and energy-minimized structures of the porphyrins show that they are in a roof or gabled (*gab*) conformation in which the macrocycle is folded about a line through the two substituted 5, 15-*meso*-carbons. This *gab* distortion is composed of a linear combination of distortions along the lowest-frequency normal coordinates of B_{1u} (*ruf*) and A_{2u} (*dom*) symmetries and the degree of each type of distortion is determined by quantitatively fitting these structures. The relative amount of doming and ruffling is modulated by the steric bulkiness of the substituents (increasing from phenyl, propyl, *iso*-propyl, to *tert*-butyl) and the size of the central metal [increasing from Ni(II), Co(II), Cu(II), and Zn(II)]. Thus, this di-substituted series of porphyrins serves as a good model for investigating the effects of the addition of a *dom* distortion to a predominant ruffling of the heme that occurs in some hemoproteins.

Resonance Raman (RR) spectroscopy is one of the few spectroscopic methods for investigating nonplanar distortions of metalloporphyrins. In a parallel presentation, the effects of the steric bulkiness of the substituents on the relative magnitude of the *dom* and *ruf* distortions in a series of nickel derivatives of the dialkylporphyrins are studied. Here we focus on resonance Raman studies of the metal-dependence of the relative contributions of the *dom* and *ruf* distortions.

For the metal dPPs, molecular mechanics calculations show that they are all planar, but the RR spectra show that the dPPs with a small central metals such as Ni(II) and Co(II) actually exist as an equilibrium mixture of planar and nonplanar species. This is evident from the broadness and asymmetry of ν_2 and ν_8, two of the structure-sensitive

Raman lines (Figure 1a). This planar/nonplanar equilibrium results from an energetic balance between the enhanced conjugation of a planar macrocycle and the optimal M-N bond lengths in the contracted core of a nonplanar macrocycle. For dPPs with a large central metal such as Cu(II) and Zn(II), the porphyrin is planar as shown by the narrowness and symmetry of ν_2 and ν_8. Here, the optimal M-N bond distance is close to the optimal distance in a planar macrocycle, so conjugation wins out. The frequencies of the structure-sensitive lines exhibit the typical dependence on the core-size of planar porphyrins(anti-correlation).

For the metal dtBuPs, molecular mechanics calculations show that in changing the central metal from smaller to larger size, the degree of ruffling decreases significantly, and at the same time the degree of doming increases (the displacement along the *ruf* coordinate for Ni, Co, Cu, and Zn is 1.606, 1.414, 1.324, and 1.110 Å, respectively; the corresponding *dom* displacements are 0.479, 0.506, 0.520 and 0.753 Å). Thus, large central metals favor the *dom* over *ruf* distortion. This metal-dependent conformational change is reflected in frequencies of the structure-sensitive lines (Figure 1b): instead of the typical anti-correlations observed for planar porphyrins with different metals, for the dtBuPs, the frequency variations of the structure-sensitive lines (*e.g.*, ν_2) with the core-size is much smaller. One notable feature of the Raman spectra of dtBuPs is the doublet of ν_2 which disappears with increasing metal size. The doublet may be caused by Fermi-resonance or conformational heterogeneity and is under further investigation.

The RR spectra of the metal dPrPs and diPrPs agree well with predictions of molecular mechanics calculations. The dPrPs are slightly nonplanar; the degree of nonplanarity decreases with increasing metal size since a large core size disfavors ruffling over doming. For the diPrPs, the porphyrin changes gradually from considerably nonplanar for NidiPrP to almost completely planar for ZndiPrP with increasing metal size.

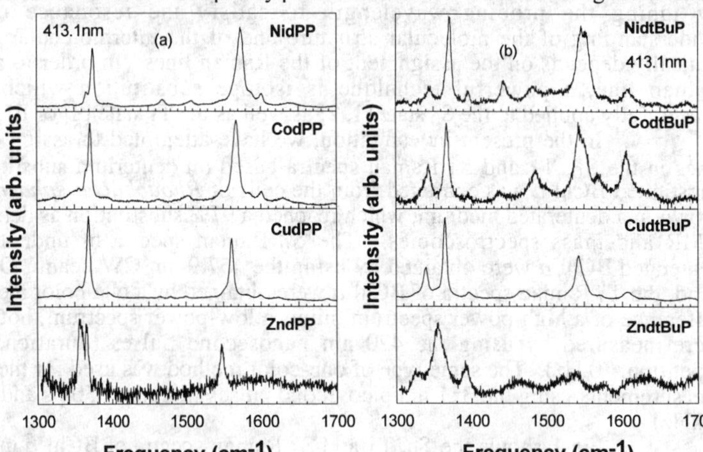

Figure 1. Resonance Raman spectra of metal derivatives of (a) dPP and (b) dtBuP in CS_2.

1. Hobbs, J. D.; Shelnutt, J. A. *J. Protein Chem.*, **1995**, *14,* 19.
2. Jentzen, W.; Simpson, M.C.; Hobbs, J.D.; Song, X.; Ema, T.; Nelson, N.Y.; Medforth, C.J.; Smith, K.M.; Veyrat, M.; Mazzanti, M.; Ramasseul, R.; Marchon, J.-C.; Takeuchi, T.; Goddard, III, W.A.; Shelnutt, J.A. *J. Am. Chem. Soc.*, **1995**, *117*, 11085.
3. Jentzen, W.; Song., X; Shelnutt, J. A. *manuscript in preparation.*

(This work performed at Sandia National Laboratories is supported by the U.S. Department of Energy under contract DE-AC04-94AL85000.)

TRANSIENT RAMAN SPECTROSCOPY OF ^{2}H-SUBSTITUTED BACTERIOCHLOROPHYLL *a*: AN EMPIRICAL ASSIGNMENT OF THE EXCITED-STATE RAMAN LINES

L. Limantara[1], M. Masci[2], I. Katheder[2], H. Scheer[2] and Y. Koyama[1]*

[1]Faculty of Science, Kwansei Gakuin University, Uegahara, Nishinomiya 662, Japan
[2]Botanisches Institut der Universität München, Munich, Germany

Bacteriochlorophyll (BChl) *a* in the excited states plays important roles in the bacterial photosynthetic systems. The S_1 state is involved in harvesting and transferring the light energy among the light-harvesting components, and also in photo-induced charge separation at the special pair of the reaction center which triggers a series of electron-transfer reactions. On the other hand, the T_1 state is involved in dissipating the excess energy through carotenoids.

Transient Raman spectroscopy is a useful means to elucidate the excited-state structures of chlorophylls because of its selective detection of a particular state by tuning the probing wavelength to satisfy the resonance conditions. Understanding of the molecular structure and of the intermolecular interaction critically depends on the assignment of the Raman lines. In order to assign each Raman line, a powerful technique is isotope substitution which has been successfully applied to the S_0 state [1, 2] as well as the T_1 state [3] of BChl *a*.

In the present investigation, we have attempted to assign the Raman lines in the S_0, T_1 and S_1 Raman spectra based on deuterium substitution. ^{2}H-substituted BChl *a* was extracted from the cells of *Rhodobacter sphaeroides* 2.4.1 grown in a deuterated medium, which reached a 97% substitution as determined by NMR and mass spectroscopies. The S_0 Raman spectra of undeuterated and deuterated BChl *a* were obtained by using the 457.9 nm CW beam. On the other hand, the T_1 Raman spectra of BChl *a* were obtained by a one-color method as the difference of a high-power spectrum minus a low-power spectrum, both of which were measured by using the 420 nm nanosecond pulses (duration, 5 ns and repetition, 10 Hz). The same type of one-color method was used for the S_1 Raman measurements using the 351 nm picosecond pulses (duration, 50 ps and repetition, 1 kHz).

Fig.1 shows the S_0, T_1 and S_1 Raman spectra of BChl *a* in the 1700 - 1000 cm^{-1} region for both undeuterated and deuterated species. The assignments for the S_0-state BChl *a* in the present investigation is in good agreement with the calculation of the normal modes of a BChl *a* model, copper methylbacteriopheophorbide *a*, calculated by using a semiempirical QCFF/PI method [4].

In general, a comparison of the S_0, T_1 and S_1 Raman lines between the undeuterated and deuterated species shows a coupling of the δC_m - H in plane (ip) bending with the double-bond stretching as well as with the single-bond stretching. Upon deuteration, the low-frequency-shift and decoupling of the δC_m - H ip bending causes a shift of the C_a - C_m double-bond stretching vibration to the lower frequencies and a shift of the C_a - N, C_a - C_b and C_b - C_b single-bond stretchings to the higher frequencies.

The empirical assignments of the T_1 Raman lines can be described as follows: The 1585 cm^{-1} line which shifts to 1574 cm^{-1} upon deuteration can be assigned to the C_a - C_m double-bond stretching. The lines at 1477 and 1407 cm^{-1} which disappeared from this region upon deuteration can be assigned to the δC_m - H ip bending. On the other hand, the 1321 cm^{-1} line which shifts slightly to the higher frequency of 1323 cm^{-1} can be assigned as the C_a - N and the C_a - C_b stretchings coupled with the δC_m - H deformations. Finally, the lines at 1185, 1100, 1060 and 1034 cm^{-1} may be assigned to the single-bond stretchings coupled with the δC_m - H deformations. All the single-bond stretchings in the 1200 - 1000 cm^{-1} region are supposed to couple strongly with one another in a complicated way, a fact which causes the changing in the spectral patterns of this region.

The bases of the set of empirical assignment for the T_1 state can be used for the S_1 state as well. The 1568 cm^{-1} line shifts to 1559 cm^{-1} upon deuteration can be assigned to the C_a - C_m stretching; the line at 1398 cm^{-1} which shifts to the higher frequency of 1428 cm^{-1} can be assigned to the C_a - C_b or C_b - C_b stretchings coupled with the δC_m - H deformations. The line at 1318 cm^{-1} is not shifted upon deuteration can be assigned to the C_a - N stretching. The rest of the Raman lines in the 1200 - 1000 cm^{-1} region can be assigned as in the case of T_1 state BChl *a*.

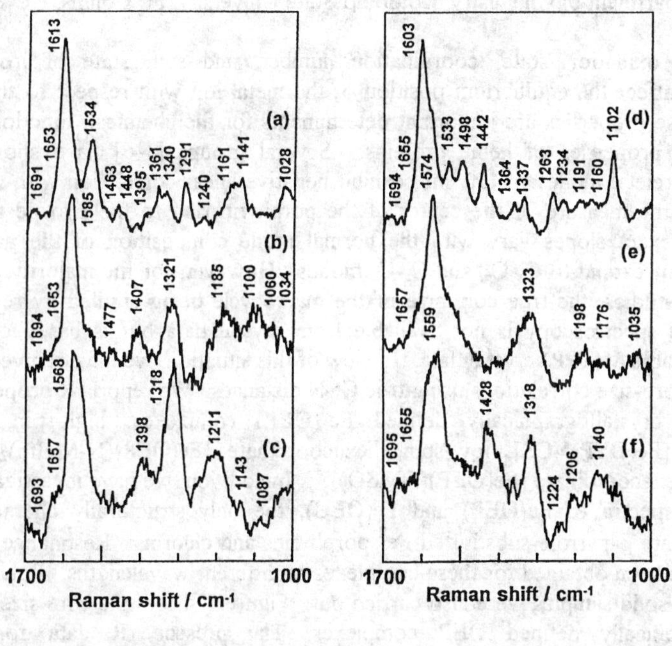

Fig. 1 The Raman spectra of undeuterated BChl *a* in the (a) S_0, (b) T_1 and (c) S_1 states; and deuterated BChl *a* in the (d) S_0, (e) T_1 and (f) S_1 states.

1. M. Lutz, in Advances in Infrared and Raman Spectroscopy, (R.J.H. Clark and R.E. Hester, eds.), Wiley, New York, 1984, p. 211.
2. T.M. Cotton and R.P. van Duyne, J. Am. Chem. Soc., 103, 6020 (1981).
3. L. Limantara, Y. Koyama, I. Katheder and H. Scheer, Chem. Phys. Lett., 227, 617 (1994).
4. R.J. Donohoe, H.A. Frank and D. Bocian, Photochem. Photobiol., 48, 531 (1988).

CORE SIZE CORRELATIONS FOR CRYSTALLOGRAPHICALLY DEFINED 4-, 5-, AND 6-COORDINATE [Fe(OEP)] COMPLEXES

M. Mylrajan[1], L. A. Andersson[2], J. Sun[3], T. M. Loehr[3], C. S. Thomas[4], E. P. Sullivan, Jr.[4], M. A. Thomson[4], K. M. Long[4], O. P. Anderson[4], and S. H. Strauss[4]

[1]RSIC, Indian Institute of Technology, Madras 600036, India
[2]Department of Biochemistry, Kansas State University, Manhattan, KS 66506
[3]Oregon Graduate Institute of Science & Technology, Portland, OR 97291
[4]Department of Chemistry, Colorado State University, Ft. Collins, CO 80523

The oxidation state, coordination number, and spin state of iron(II,III) porphyrins affect the equilibrium position of the metal ion with respect to the heme plane. These properties are important determinants for the chemical, functional, and mechanistic properties of heme proteins. Several groups [1-6] have shown that porphyrin skeletal modes >1450 cm^{-1} exhibit negative linear correlation with the core size, Ct-N, the measure of the center of the porphyrin ring to the pyrrole nitrogen atoms; the exact slopes vary with the normal mode composition of the particular frequency but extend to C_a-C_m and C_a-C_b modes. However, for the majority of these correlation studies, the true core size of the macrocycle being studied by resonance Raman (rR) spectroscopy is not available from crystal data but is based on more readily available M(TPP) crystal data. In view of this situation, we have reinvestigated frequency/core-size correlations using true Ct-N distances. We report rR properties of three new crystallographically defined Fe^{II}(OEP) complexes: high-spin, penta-coordinate [Fe(OEP)NCS]; low-spin, hexacoordinate $[Fe(OEP)(N\text{-}MeIm)_2]^+$; and high-spin, hexacoordinate $[Fe(OEP)(DMSO)_2]^+$. In addition, we have investigated the vibrational spectra of Fe(OEP) and Fe(OEC), the only structurally characterized tetracoordinate β-pyrrole-substituted Fe^{II} porphyrins and chlorins. Resonance Raman spectra have been obtained for these complexes at different wavelengths. Polarization analyses for solid samples were also carried out. Figure 1 shows the core size plot of crystallographically defined OEP complexes. The present rR data for these crystallographically characterized complexes permit the first rigorous test of earlier Ct-N *vs* frequency correlations in that the spectroscopic and structural parameters are from the same compounds in the same crystalline phase and yield a new core-size correlation for the β-pyrrole-substituted metalloporphyrins. With the present analysis as a foundation, rR spectroscopic studies may be extended to simple complexes in solution and heme proteins in determining core sizes and facilitating spectral assignments. The frequencies of the four-coordinate ferrous complex are consistently higher than those of ferric porphyrins and reflect the significantly smaller Ct-N distance in the intermediate-spin compound. Indeed, the positions of the core-size marker bands of four-coordinate Fe(OEP) are unique, being essentially identical to those of the low-spin, 6-coordinate complex. The frequencies and core-size plot parameters for OEP complexes are different from those of *meso*-tetraphenylporphyrins and reflect the strongly altered normal mode compositions induced upon a change in the substituent pattern. Although rR spectra of Fe(OEP) and Fe(OEC) show superficial similarities, the detailed spectral patterns, relative intensities, and wavelength-dependent enhancements differ. The chlorin, owing to its lower symmetry, has a larger number of rR-active bands at each excitation wavelength and exhibits a rich spectrum with red excitation that is distinct from its porphyrin counterpart.

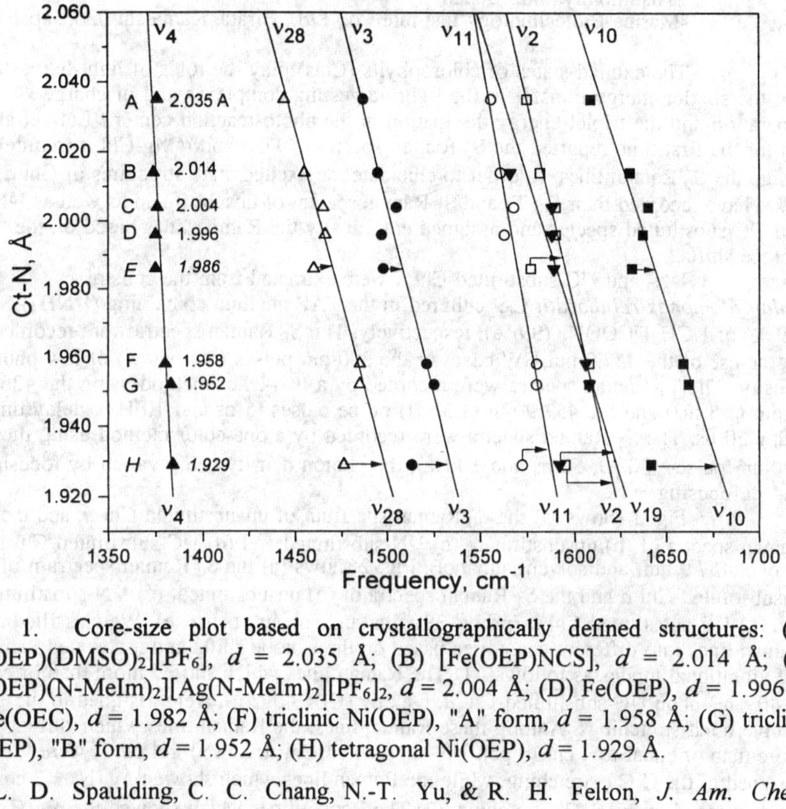

Fig. 1. Core-size plot based on crystallographically defined structures: (A) [Fe(OEP)(DMSO)$_2$][PF$_6$], d = 2.035 Å; (B) [Fe(OEP)NCS], d = 2.014 Å; (C) [Fe(OEP)(N-MeIm)$_2$][Ag(N-MeIm)$_2$][PF$_6$]$_2$, d = 2.004 Å; (D) Fe(OEP), d = 1.996 Å; (E)Fe(OEC), d = 1.982 Å; (F) triclinic Ni(OEP), "A" form, d = 1.958 Å; (G) triclinic Ni(OEP), "B" form, d = 1.952 Å; (H) tetragonal Ni(OEP), d = 1.929 Å.

1. L. D. Spaulding, C. C. Chang, N.-T. Yu, & R. H. Felton, *J. Am. Chem. Soc.*, **97**, 2517 (1975).
2. P. V. Huong, & J. -C. Pommier, *C. R. Acad. Sci., Ser. C.*, **285**, 519 (1977).
3. T. G. Spiro, J. D. Stong, & P. Stein, *J. Am. Chem. Soc.*, **101**, 2648 (1979).
4. S. Choi, T. G. Spiro, K. C. Langry, K. M. Smith, L. D. Budd, & G. N. La Mar, *J. Am. Chem. Soc.*, **104**, 4345 (1982).
5. Y. Ozaki, K. Iriyama, H. Ogoshi, T. Ochiai, & T. Kitagawa, *J. Phys. Chem.*, **90**, 6105 (1986).
6. N. Parthasarathi, C. Hansen, S. Yamaguchi, & T. G. Spiro, *J. Am. Chem. Soc.*, **109**, 3865 (1987).

RAMAN SPECTROSCOPY OF ^{15}N- AND ^{13}C-SUBSTITUTED CHLOROPHYLL *a* AND *b* IN THE S_0, T_1 AND S_1 STATES

T. Sashima[1], N. Kurano[2], S. Miyachi[2] and Y. Koyama[1]*

[1]Faculty of Science, Kwansei Gakuin University, Uegahara, Nishinomiya 662, Japan
[2]Marine Biotechnology Institute Co., Ltd., Hirata, Kamaishi 026, Japan

The excited-states of chlorophylls (Chls) play the roles of light-harvesting and the singlet-energy transfer in the light-harvesting complexes and of charge separation and the triplet-energy dissipation in the photo-reaction center. Lutz et al. [1] for the first time reported the S_0 Raman spectra of ^{15}N- and ^{26}Mg-Chl *a* in order to assign the S_0 Raman lines. In order to elucidate the excited-state structures of Chl *a* and *b*, we have recorded their S_0, T_1 and S_1 Raman spectra of unsubstituted as well as ^{15}N- and ^{13}C-substituted species and assigned empirically the Raman lines based on the isotope shifts.

^{15}N- and ^{13}C-substituted Chl *a* were extracted from the cells of *Chlamydomonas reinhardtii* C-9 cultured in the TAP medium containing $(^{15}NH_4)_2SO_4$ (99 %) or $^{13}CH_3{}^{13}COONa$ (99 %), respectively. The S_0 Raman spectra were recorded by the use of the 457.9-nm CW beam or the 390-nm pulses (see below) of low photon density. The T_1 Raman spectra were recorded by a two-color method using the 436-nm pump (1.5 mJ) and the 457.9-nm (1.5 mJ) probe pulses (5 ns and 10 Hz), delay time being 50 ns. The S_1 Raman spectra were recorded by a one-color method using the 390-nm pulses (30 μJ, 50 ps and 1 kHz), the photon density being varied by focusing and defocusing.

Fig. 1 shows (a) the S_0 Raman spectrum of unsubstituted Chl *a*, and the T_1 Raman spectra of (b) unsubstituted, (c) ^{15}N-substituted and (d) ^{13}C-substituted Chl *a* (probe, 457.9 nm; and solvent, ethanol). Fig. 2 shows (a) the S_0 Raman spectrum of unsubstituted Chl *a* and the S_1 Raman spectra of (b) unsubstituted, (c) ^{15}N-substituted and (d) ^{13}C-substituted Chl *a* (probe, 390 nm; and solvent, benzene). We classified the Raman lines into different categories based on the isotope shifts and associated them to the vibrational modes as follows: (1) The Raman lines which shifted more than or equal to 40 cm^{-1} upon ^{13}C-substituted Chl *a*, i. e., $\Delta\nu$ (C) $\geq$ 40 cm^{-1}, were assigned to the double bond stretchings. Among these Raman lines, the Raman lines which shifted more than or equal to 4 cm^{-1} upon ^{15}N-substituted Chl *a*, $\Delta\nu$ (N) $\geq$ 4 cm^{-1}, were assigned to the C_aC_m stretching, while the Raman lines which showed $\Delta\nu$ (N) $\leq$ 3 cm^{-1} were assigned to the C_bC_b stretching. (2) The Raman lines which showed the $\Delta\nu$ (C) $\leq$ 39 cm^{-1} and $\Delta\nu$ (N) $\geq$ 4 cm^{-1} were assigned to the CN stretchings. Classification of the Raman lines thus obtained are listed in the figure.

Based on this classification, the T_1 and S_1 Raman lines can be correlated with the S_0 Raman lines as follows: The T_1 Raman lines at 1611, 1564, 1540, 1598, 1499, 1239 and 1104 cm^{-1} can be correlated with the S_0 Raman lines at 1630, 1608, 1581, 1552, 1494, 1184 and 1115 cm^{-1}, respectively. The shifts upon $T_1 \leftarrow S_0$ transition are -19, -44, -41, +46, +5, +55 and -11 cm^{-1}, respectively. Thus, the average of the absolute values of the above shifts is 32 cm^{-1}. Furthermore, the S_1 Raman lines at 1598, 1542, 1495, 1338 and 1185 cm^{-1} can be correlated with the S_0 Raman lines at 1609, 1552, 1492, 1346 and 1211 cm^{-1}, respectively. The shifts upon $S_1 \leftarrow S_0$ transition are -11, -10, +3, -8 and -26 cm^{-1}, respectively. Thus, the average of the absolute values of the above shifts is 12 cm^{-1}. The above results suggest that the changes of the bond orders in the macrocycle upon the $T_1 \leftarrow S_0$ transition are larger than those upon the $S_1 \leftarrow S_0$ transition. This conclusion is consistent with ab initio CI calculation of chlorin by Petke et al. [2]; the average change in the bond order of the $C_aC_m \cdot C_bC_b \cdot CN$ upon the $T_1 \leftarrow S_0$ transition (0.042) is larger than that upon the $S_1 \leftarrow S_0$ transition (0.022).

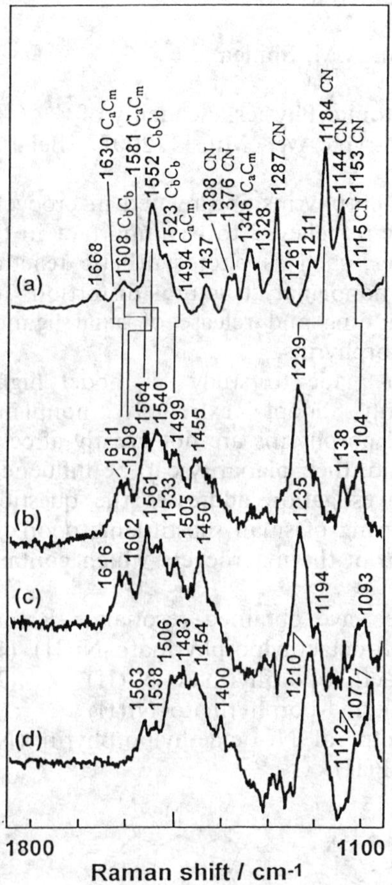

Fig. 1. The Raman spectra of (a) Chl *a* in the S_0 state, and (b) unsubstituted Chl *a* · (c) ^{15}N-Chl *a* · (d) ^{13}C-Chl *a* in the T_1 state.

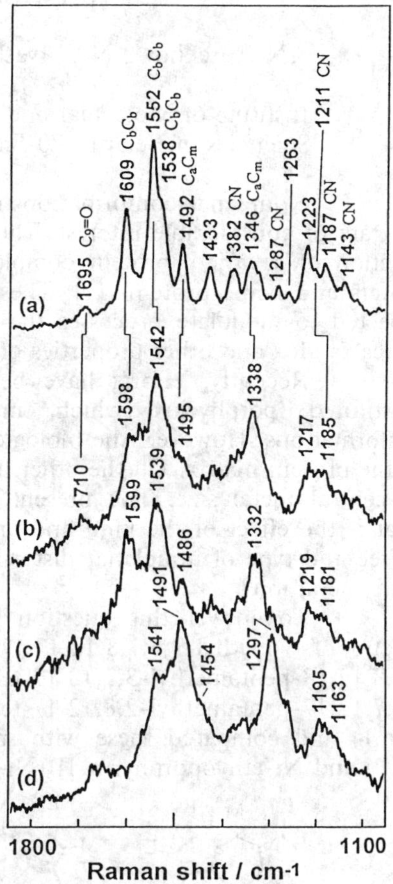

Fig. 2. The Raman spectra of (a) Chl *a* in the S_0 state, and (b) unsubstituted Chl *a* · (c) ^{15}N-Chl *a* · (d) ^{13}C-Chl *a* in the S_1 state.

1. M. Lutz, J. Kleo, R. Gilet, M. Henry, R. Plus, and J. P. Leicknam, in Proceedings of the 2nd International Conference on Stable Isotopes, (E. R. Klein and P. D. Klein, eds.), US Department of Commerce, Springfield, VA, 1975, 462.
2. J. D. Petke, G. M. Maggiora, L. L. Shipman, and R. E. Christoffersen, J. Mol. Spectrosc. 73, 311 (1978).

RESONANCE RAMAN STUDY OF THE NONPLANAR DISTORTIONS OF MESO-MONOSUBSTITUTED Ni-PORPHYRINS.

S.N. Terekhov*, N.V. Ivashin, A.M. Shulga

Institute of Molecular and Atomic Physics, Academy of Sciencies of Belarus, 70 F.Skaryna Ave., Minsk 220072, Belarus

Nonplanar conformations of porphyrins and related macrocycles now are of considerable interest. They are believed to be important in the function of a variety protein complexes, such as photosynthetic reaction centers and heme proteins [1]. These nonplanar structural distortions are expected to modulate processes the binding and release of axial ligands, optical, redox and other properties of porphyrins.

Recently, efforts have been made to study of model highly substituted porphyrins which mainly adopt extremely nonplanar conformations. However, the biological porphyrins are not greatly affected by the protein matrix. On the other hand, their planarity can be influenced by central metal size. The present investigation addresses the question, what is the effect of the kind and ordering of small substituents upon the degree and type of nonplanar distortion of the macrocycle which contains small metal ion?

To answer this question we have obtained resonance Raman spectra of 5-methyl-2,3,7,8,12,13,17,18-octaethylporphyrinato-Ni(II) (**1**), 2,5,8,12,18-pentamethyl-3,7,13,17-tetraethyl-porphyrinato-Ni(II) (**2**), 3,5,7,12,17-pentamethyl-2,8,12,17-tetraethyl-porphyrinato-Ni(II) (**3**) (Fig.1) and compared these with spectra of Ni-octaethylporphyrin (Ni-OEP) and Ni-ethyoporphyrin-II (Ni-EP(II)).

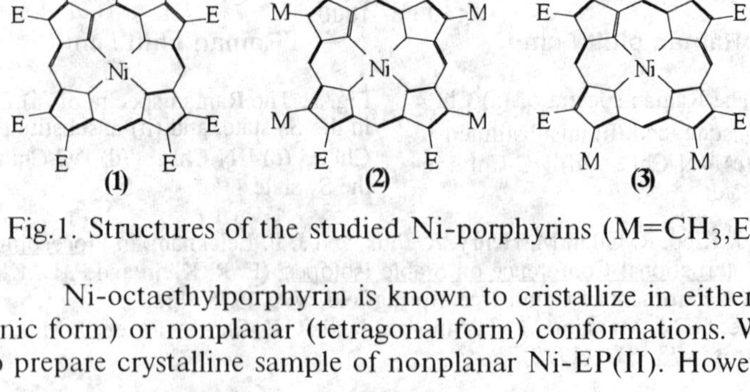

Fig.1. Structures of the studied Ni-porphyrins ($M=CH_3$,$E=C_2H_5$).

Ni-octaethylporphyrin is known to cristallize in either planar (triclinic form) or nonplanar (tetragonal form) conformations. We have tried to prepare crystalline sample of nonplanar Ni-EP(II). However, after the same as for Ni-OEP procedures of heating or quickly evaporation of a solution, Ni-EP(II) crystals have not converted to the nonplanar conformation. As shown in Table 1 frequencies of the structure-sensitive Raman lines of Ni-EP(II) are consistent with frequencies of Ni-OEP in planar form.

TABLE 1. Comparison of resonance Raman frequencies (cm^{-1}) of structure-sensitive lines for Ni-porphyrins in solids.

Porphyrin	ν_4	ν_3	ν_{11}	ν_{19}	ν_2	ν_{10}
Ni-OEP[a)] triclinic	1383	1523	1581	1609	1608	1663
Ni-EP(II)	1382	1525	1581	1608	1607	1662
Ni-OEP[a)] tetragonal	1383	1512	1572	1589	1595	1642
3	1381	1511	1570	1581	1598	1642
2	1381	1510	1564	1571	1591	1635
1	1380	1508	1561	1570	1584	1630

[a)]from ref. 3.

The introduction of only one CH_3 group into the meso-position of Ni-EP(II) significantly alters Raman spectrum from that of the unsubstituted Ni-EP(II). In particular, ν_3, ν_{11}, ν_{19}, ν_2 and ν_{10} all downshift by 15-37 cm^{-1} while the only ν_4 is not appreciable changed. It is known [2,4], the decreasing of Raman frequencies due to meso-monosubstitution in the porphyrins with large metal ions which have nearly flat structure, is about 10 cm^{-1}. Thus, the observed spectral changes for Ni-porphyrins are caused by the distortions from planarity. The similarities in the Raman spectra of Ni-OEP tetragonal form and **1,2,3** suggest that CH_3-mesosubstituted Ni-porphyrins consistent with ruffled structure. The degree of nonplanarity is modulated by the alkyl groups on the pyrrole rings and increases in series of porphyrins **3-1**. It is noteworthy, that **1,2,3** each adopt the same ruffled conformation in solids and in solution.

The molecular mechanics calculations predict the correct degree of nonplanar distortion to account for the decrease in frequencies of the structure-sensitive Raman lines that observed.

We are grateful to the Fundamental Research Foundation of the Republic of Belarus (Grant F-157) for the financial support.

1. M.O.Senge, J. Photochem. Photobiol.. B: Biol. 16, 3 (1992).
2. K.K. Anderson, J.D. Hobbs, L.Luo, K.D. Stanley, J. Martin, E. Quirke and J.A. Shelnutt, J. Am. Chem. Soc. 115, 12346 (1993).
3. R.S. Czernuszewicz, X-Y. Li and T.G. Spiro, J. Am. Chem. Soc. 111, 7024 (1989).
4. S.N. Terekhov, N.M. Ksenofontova, I.F.Gurinovich and I.V. Filatov, Zhurn. Prikl. Spektr. 52, 807 (1990) (in russian).

FT-RAMAN STUDY OF TETRAPHENYLPORPHYRINS AND PROTOPORPHYRINS

M. Mylrajan

Regional Sophisticated Instrumentation Centre
Indian Institute of Technology
Madras 600036, India

Metalloporphyrins are available with a variety of peripheral substituents, cetral metal ions and axial ligands which are used as models for heme proteins like globins, cytochromes, peroxidases etc [1-4]. Resonance Raman (RR) spectroscopy with visible excitations provide valuable information about oxidation, spin and ligation states of the central metal atom. FT-Raman spectroscopy with near infrared excitation has several advantages over conventional Raman which includes: 1) reduction in fluorescence 2. minimum photodegradation and 3. FT advantage. In this paper we report FT-Raman spectra of tetraphenylporphyrins (iron (III) tetra(pentafluorophenyl)porphyrin chloride ($Fe^{III}(F_5TPP)Cl$), $Fe^{III}(TPP)Cl$, Cu(TPP)Cl and msoTPP) and protoporphyrins ($Fe^{III}PPCl$ and protoporphyrin IX) and spectral features are compared with RR data.

FT-Raman spectra were obtained with a Raman module, Bruker FRA 106, attached to a Bruker IFS 66v spectrometer, using the 1064 nm line of a Nd-YAG laser for excitation. The spectral resolution was 4 cm^{-1}. All the spectra were obtained with an average of 1024 scans. Solid and solution spectra were obtained with back scattering geometry.

Figure 1 shows the FT-Raman spectra of TPPs. Normally freebases show strong fluorescence with visible excitation, but we have got good quality FT-Raman spectra with near-IR excitation. When the excitation is in resonance with Soret band of porphyrin RR spectrum is dominated by totally symmetric modes of porphyrin ring enhanced via A (Frank-Condon) term scattering. When the excitation is in resonance with Q band, RR spectrum is dominated by non-totally symmetric modes via vibronic scattering (B term). When the excitation in the near-IR region (1064 nm) the FT-Raman spectra are similar to that of Q-band excited RR spectra as in the case of $Fe^{III}(F_5TPP)Cl$ the FT-Raman spectrum is similar to that of 514.5 nm excited RR spectra but the relative intensities differ. FT-Raman spectrum of $Fe^{III}(F_5TPP)Cl$ show a medium intense band at 1650 cm^{-1} which can be assigned to ν_1' mode of pentafluorophenyl group [5] and this band is very weak in RR spectra. Similarly phenyl modes are seen with good intensity in FT-Raman spectra of $Fe^{III}(TPP)Cl$ and Cu(TPP) at 1596 and 1598 cm^{-1} respectively and are in good agreement with calculated values [4, 7]. These phenyl modes are very weak in RR spectra. [4, 6]. The effect of various ligands and spin states will be presented and compared with RR data. FT-Raman spectra TPPs also show bands due to CH stretching vibrations in the region 2900-3150 cm^{-1}. In addition to these bands various overtone and combinations are also seen in this region.

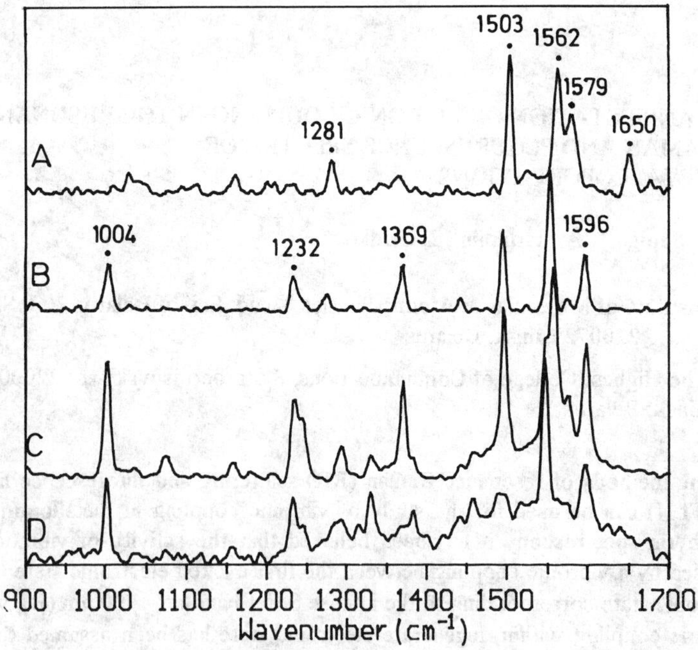

Fig. 1. FT-Raman spectra of A) $Fe^{III}(F_5TPP)Cl$ B) $Fe^{III}(TPP)Cl$ C) Cu(TPP) and D) *meso* TPP in the solid state.

FT-Raman spectra of protoporphyrins are also show similarities with Q-band excited RR spectra [8]. The main advantage of FT-Raman spectra over RR spectra of porphyrins is that one can obtain good quality FT-Raman spectra of pure solid samples without any dilution with KBr. In addition to RR bands non-resonant Raman bands are also seen in FT-Raman spectra. FT-Raman spectra of TPPs and PPs were also obtained in various solvents and also in various ligation states.

1. Biological Applications of Raman Spectroscopy (T. G. Spiro, ed.), Wiley-Interscience, New York, 1988.
2. T. Kitagawa and Y. Ozaki, Struct. Bonding, Berlin, 64, 71 (1987).
3. M. Mylrajan, L. A. Andersson, J. Sun, T. M. Loehr, C. S. Thomas, E. P. Sullivan, and S. H. Strauss, Inorg. Chem. 34, 3953 (1995).
4. X.-Y. Li, R. S. Czernuszewicz, J. R. Kincaid, Y. O. Su, and T. G. Spiro, J. Phy. Chem. 94, 31 (1990).
5. S. G. Frankiss and D. J. Harrison, Spectrochim. Acta, 31A, 1839 (1975).
6. J. M. Burke, J. R. Kinkaid, and T. G. Spiro, J. Am. Chem. Soc. 100, 6077 (1978).
7. M. Atamian, R. J. Donohoe, J. S. Lindsay, and D. F. Bocian, J. Phys. Chem. 93, 2236 (1989).
8. S. Choi, T. G. Spiro, K. C. Langry, K. M. Smith, D. L. Budd, and G. N. La Mar, J. Am. Chem. Soc., 104, 4345 (1982).

MANIFESTATION OF VIBRONIC COUPLING IN THE RESONANCE RAMAN AND FLUORESCENCE SPECTRA OF METALLOPORPHYRINS

A. Shulga[1]*, A.Starukhin[1], L.Gladkov[2]

[1]Institute of Molecular & Atomic Physics, Acad. Sci. of Belarus 70 F.Skaryna Av., 220072 Minsk, Belarus

[2]The Highest College of Communications, 8 Staroborisovsky tr., 220000 Minsk, Belarus

The methods of resonance Raman (RR) scattering and fluorescence line narrowing (FLN) can be used for the study of vibronic coupling in metalloporphyrins. For porphyrin spectroscopy it has been believed that the activity of vibrations has been caused by a vibronic coupling between the first excited electronic state (Q) and the electronic state corresponding to the intense Soret band at ~ 400 nm (B), whereas the vibronic coupling within degenerate electronic state has been assumed to be insignificant. We decided to verify this assumption.

Initially two Zn octaethylporphyrin derivatives with isocycle substitution were taken for the study. By disturbing the symmetry of the molecule, isocycle substitution results in removal of degeneracy and splitting of the long-wavelength absorption band Q (the energy gap between the $S_0 \rightarrow S_1$ and the $S_0 \rightarrow S_2$ transitions is about 300 $cm.^{-1}$). For this compounds we have observed the unique situation. The absorption spectra of these compounds are the same but excitation fluorescence polarization signs of S_0-S_2 bands are opposite. It has been found that in FLN spectra frequencies and relative lines intensities are identical practically. Nevertheless the polarization of lines with frequencies of 785, 915, 1253, 1294, 1311, 1356, 1440 and 1553 $cm.^{-1}$ is characterized by the sign, which coincide with the sign of the polarization of pure electronic $S_0 \rightarrow S_2$ transitions for every compound. It signifies that these lines borrows intensity from $S_0 \rightarrow S_2$ transition. The analysis of experimental data shows clearly, that the vibronic interaction of quasi-degenerate S_1 and S_2 levels plays the essential role in the formation of the vibrational structure in fluorescence spectra in the region of stretching vibrations.

Then, FLN spectra of symmetrical metalloporphyrins (Zn-porphine, Mg-octaethylporphyrins (Mg-OEP), Zn-OEP, In-OEP and TiO-OEP) have been studied. For compounds of interest we found that intensities of some negatively polarized lines correlate with one of Q electronic transition vibronic band in absorption spectrum. It may be concluded that these lines corresponding to nonsymmetrical vibrations borrow their intensity from the $S_0 \rightarrow S_2$ transition because this pure electronic transition namely has negative polarization also. For instance, in Mg-OEP spectrum lines with frequencies of 595, 767, 912, 1126, 1253, 1319, 1363, 1384, 1426 and 1537 cm.$^{-1}$ are very active in vibronic coupling with S_2 level. At the same time negatively polarized lines with frequencies of 934, 999, 1145, 1169 and 1401 cm.$^{-1}$ are attributed as ones borrowing their intensities from negatively polarized component of Soret band.

The vibronic coupling between the S_1 and S_2 electronic states can be a cause of a RR scattering of a nontotally symmetric vibrations. The RR study of the Cu chlorin and porphine has been proved this assumption. For compounds under investigation the energy gap between the S_1 and S_2 electronic states is substantially greater (see Fig. 1A) and RR spectra can be obtained under excitation within the range of the $S_0 \rightarrow S_1$, $S_0 \rightarrow S_2$ and Soret absorption bands [1]. As it is seen from Fig. 1B totally symmetric modes with frequencies of 366, 385, 730, 900, 1008, 1056, 1267, 1316, 1473, 1517 and 1600 cm.$^{-1}$ are most active in 632,8 nm excited RR spectra ($S_0 \rightarrow S_1$ transition). Additional polarized RR lines appear at Soret band excitation (1067, 1322, 1364 and 1553 cm.$^{-1}$).On other hand nontotally symmetric vibrations are enhanced by excitation within $S_0 \rightarrow S_2$ transition range only (λ_{exc}=514,5 nm). According to the results of normal coordinate analysis the modes of Cu chlorin with frequencies of 1325, 1355 and 1608 cm.$^{-1}$ and ones of Mg OEP 1319, 1363 and 1537 cm.$^{-1}$ have similar potential energy distributions. Thus there are a number of modes leading to vibronic coupling between the S_1 and S_2 electronic states.

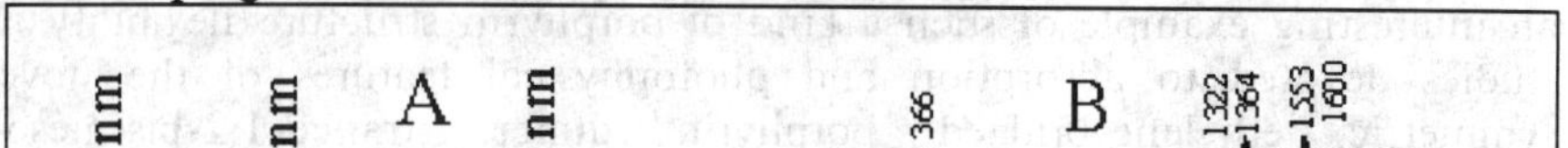

Fig. 1. A: Electronic absorption spectrum of Cu-chlorin. B: RR spectra of Cu chlorin.

1. K. N. Solovyov, L. L. Gladkov, A. S. Starukhin and S. F. Shkirman, Spectroscopy of porphyrins: Vibrational States. Nauka i Tekhnika, Minsk, 1985. (in Russian).

CONFORMATIONAL DYNAMICS OF TRANS-ETHYLENE-BRIDGED OCTHAETHYLPORPHYRIN DIMER FROM RESONANCE RAMAN AND IR SPECTROSCOPIES.

N.V. Ivashin[1]*, S.N. Terekhov[1], V.V. Ermolenkov[2], A.M. Shulga[1], V.S. Chirvony[1]

[1]Institute of Molecular and Atomic Physics, Academy of Sciencies of Belarus, 70 F. Skaryna Ave., Minsk 220072, Belarus
[2]B.I. Stepanov Institute of Physics, Academy of Sciencies of Belarus

Biological functions of porphyrin type molecules are caused in many respects by ability of such molecular systems to undergo different conformational transformations, followed by appreciable changes of their electron structure and chemical activity. Recently, one of us has revealed an interesting example of such a kind of porphyrin structure flexibility in studies devoted to absorption and photophysical features of the novel symmetric ethylene-bridged porphyrin dimer, trance-1,2-bis(meso-octaethylporphyrinyl)ethene (tbis=OEP in short) [1].

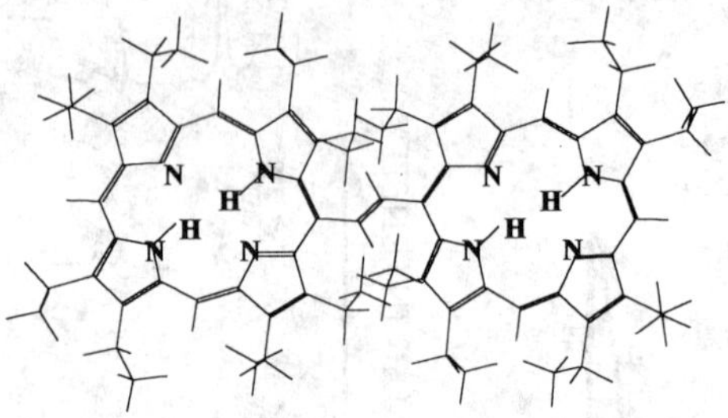

It was found that in solutions of organic solvents tbis=OEP exists as a mixture of two main forms, named P and U. Absorption spectrum of the P conformer is slightly red-shifted as compared to monomeric OEP due to excitonic interaction. However, the isomer U exhibits a quite unusual spectrum with a main band near 495 nm and a broad absorption of lower intensity in 600-900 nm region. Satisfactory explanation of this spectrum has been made possible only on the basis of the molecular model, suggesting a common π-conjugation of two macrocycles through the double-bond bridge. We report here a comparative resonance Raman, IR spectroscopic and quantum-chemical investigations of two tbis=OEP conformers and related compounds.

Resonance Raman and IR data revealed that vibrational spectra of tbis=OEP conformers as a whole are not too different from each other.

Furthermore they are close to spectra emerged for monomeric meso-monomethyloctaethylporphyrin (CH_3-OEP) and single-bond bridged OEP dimer (bis-OEP). A majority of changes in the spectra takes place in the regions innate to manifestation of νNH, δNH and $C_\alpha C_m$ vibrations.

It is well known that under sterically strained conditions porphyrin core is capable of making planar "elongated" and nonplanar "saddle", "ruffled", or "domed" structures. Because all porphyrins under this investigation have steric crowding at the meso-position, we have conducted semiempirical (PM3 method) calculations concerning geometry optimization and νNH vibrations for model porphyrin structures showing the above-mentioned distortions to understand unusual high sensitivity of νNH frequency to porphyrin structure.

The data obtained strongly suggest that a decrease of νNH frequency for CH_3-OEP (ν=3255 cm^{-1}), tbis=OEP ($\nu \leq 3290$ cm^{-1}), and bis-OEP (ν=3270 cm^{-1}), in comparison with OEP (ν=3310 cm^{-1}), originates from an elongation of the porphyrin macrocycle in direction of a meso-substitutent. Distortion of this sort leads to strenthening of the intramolecular NH······N hydrogen bond, and mainly this effect yields a high sensitivity of νNH and δNH modes to the porphyrin structure.

The optimized structures of tbis=OEP P and U conformers exhibit different value of elongation, in accord with detected frequencies of νNH vibrations (3273 cm^{-1} for U and 3290 cm^{-1} for P forms), however lengths of C-C and C-N bonds of porphyrin cores vary only slightly. Largest difference is observed for C_α-C_m bonds of methyne bridges nearest to ethylene group, connecting porphyrin macrocycles ($\Delta l \sim 0.017$ Å).

These peculiarities of tbis=OEP structure are confirmed by resonance Raman data. In the region of OEP ν_{10} mode (1615 cm^{-1}) RR spectrum of tbis=OEP P form contains not one but two lines (1614 cm^{-1} and 1603 cm^{-1}). When passing from P to U form, one of the lines (1603 cm^{-1}) experiances ~3 cm^{-1} downward frequency shift. Other structure-sensitive RR lines are unchanged under this transformation. It is interesting to note that tbis=OEP P and U conformers, as it is seen from the optimized structures, have not only different values of porphyrin elongation, but also exhibit quite different type of their small-scale nonplanar distortion ("ruffled" and "saddle"). For the present attempts to find spectroscopic reflections of these nonplanar structural features have not met with success.

We are grateful to the International Science Foundation for the financial support (Grants MWR000 and MWR300). This work has been supported also in part by the Fundamental Research Foundation of the Republic of Belarus (Grant F-157).

1. M. Chachisvilis, V.S. Chirvony, A.M. Shulga, B. Källebring, S. Larsson, V. Sundström, J. Phys. Chem., in press.

LOW TEMPERATURE RESONANCE RAMAN OF B12 MODEL COMPOUNDS: TRANS-EFFECT ON CO-C BOND STRETCH

S. Dong[1], R. Padmakumar[2], R. Banerjee[2], and T. G. Spiro[1]*

[1]Department of Chemistry, Princeton University, Princeton, NJ 08544
[2]Department of Biochemistry, University of Nebraska, Lincoln, NE 68588

Nature employs two biologically active forms of vitamin B12 derivatives to carry out two distinctive reactions in B12 dependent enzymes. Enzymes containing methylcobalamin (MeCbl) as cofactors catalyze methyl transfer reactions, with the heterolytic cleavage of the carbon-cobalt bond of the cofactor as a key step. In contrast, those using adenosylcobalamin (AdlCbl) as cofactors catalyze isomerization reactions or the de-methyl reaction from a tertiary amine, with the homolytic cleavage of the C-Co bond of AdlCbl as the initial step. The commonness of these two sets of reactions and the distinctiveness between them are closely related to the C-Co bond in the cofactors during the catalytic process of B12-dependent enzymes. As a result, information about the Co-C bond will be invaluable for the understanding of B12-dependent enzyme catalysis. Toward this end, resonance Raman spectroscopy is expected to be able to extract the structural information about the cofactors due to its high selectivity and sensitivity. However, resonance Raman study of cobalamin derivatives has been hampered by the photolability of the Co-C bond. After the initial efforts[1], no resonance Raman study has been directed to the C-Co bond study of alkylcobalamins. This undoubtedly retarded the resonance Raman studies of B12-dependent biological processes. Although FT Raman has been used to detect the Co-C bond vibration in MeCbl and model compounds[2], it can not be applied to B12-containing proteins. In this work, low temperature RR experiment has been carried out to eliminate photolysis, and Co-C bond information has been readily obtained for B12 model compounds. The effect of the coordination of the dimethylbenzimidazole(DMB), which is trans to the C-Co bond, is studied by comparing the C-Co bond stretching energy of base-on and base-off species.

Each B12 sample is dissolved in neutral solution (50mM pH7 potassium phosphate buffer) and acidic solution (0.1N HCl) respectively. For resonance Raman studies, the sample is loaded in a liquid nitrogen dewar[3] under nitrogen atmosphere in the dark and frozen with liquid nitrogen. Then the dewar is pumped to vacuum while the samples are kept frozen and in the dark. Resonance Raman spectra are collected on a Spex1877 triple monochromator and calibrated with N,N-dimethyl formamide (DMF). UV-visible spectra are taken before and after resonance Raman experiments and compared. No photolysis due to the resonance Raman experiment can be detected. Excitation at 476nm, 488nm, 520nm 530nm or 568nm is carried out for AdlCbl in neutral solution. Among these lines, 530nm is found to be the best one to give good quality spectra and is used for all the other samples. To estimate the photolysis of samples during laser exposure, the laser powers at samples are set at 5mW, 10mW, 30mW and 90mW respectively. No detectable changes can be observed in the resonance Raman spectra of all the tested samples (AdlCbl, MeCbl and neopentylcobalamin (NpCbl) in neutral and acidic solutions). Methyl-Co stretching for methylcob(III)alamin in neutral aqueous solution is identified at 508cm^{-1} by its 12cm^{-1} $^{13}CH_3$ isotope shift. For ethylcobalamin, the Co-C stretching is identified at 465cm^{-1} by its 15cm^{-1} downshift upon deuteration at β position. A prominent peak at 445cm^{-1} is observed for AdlCbl, which is tentatively assigned to Co-C stretching. NpCbl in neutral solution shows some variations in their resonance Raman profile around 400-600cm-1 from MeCbl, probably due to the weak neopentyl modes in this region. Due to the low intensity of these peaks, the assignment for the C-Co stretch can not be made. By dissolving alkylcobalamin in acidic solutions, the axial base DMB can be protionated and base-on alkylcobalamin is changed to its base-off form. For MeCbl, the Co-C stretch is upshifted by 17cm^{-1} in 0.1N HCl . This result is consistent with strengthening of Co-C bond in the base-off form as evidenced by kinetic studies[4]. It is also noteworthy that different B12 derivatives have variable relative peak intensities, which probably reflects the different corrin ring conformations when the axial alkyl ligand is changed. Direct detection of the Co-C with resonance Raman spectroscopy will facilitate the structural and functional studies of B12 dependent enzymes.

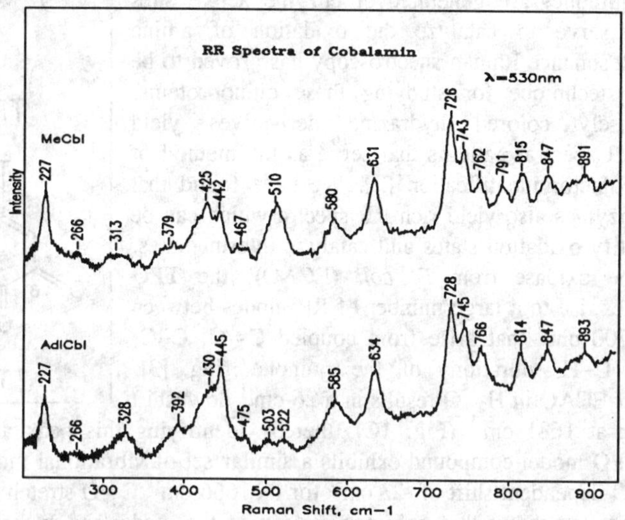

Figure 1. Resonance Raman spectra of MeCbl and AdlCbl with 530nm excitation. Laser power at sample=10mW; slit width=4cm^{-1}.

1. S. Salama and T. G. Spiro, *J. Raman Spectrosc.* **6**, 57 (1977).
2. S. Nie, P. A. Marzili, L. G. Marzili and N.-T. Yu, *J. Am. Chem. Soc.* **112**, 6084(1990).
3. R. S. Czernuszewicz and M. K. Johnson, *Appl. Spect.* **37**(3), 297 (1983).
4. B. P. Hay and R. G. Finke, *J. Am. Chem. Soc.* **109**, 8012 (1987).

QUINONE ENZYME COFACTORS DERIVED FROM TYROSINE (TPQ) AND TRYPTOPHAN (TTQ)

Pierre Moënne-Loccoz, Nobuhumi Nakamura, and Joann Sanders-Loehr

Oregon Graduate Institute of Science & Technology, Portland, OR 97291, USA

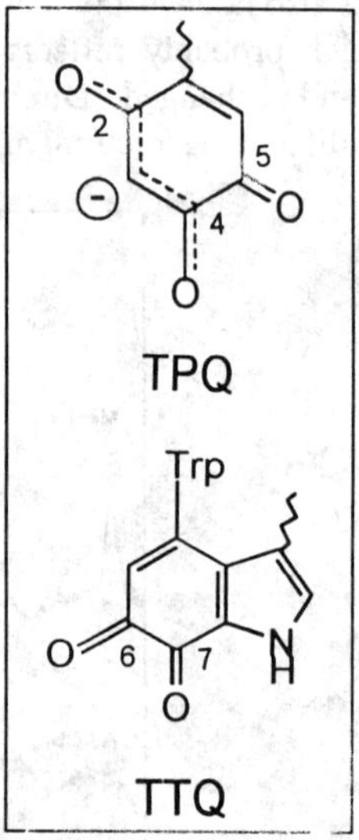

Quinones represent a newly discovered class of redox-active organic cofactors derived from the post-translational modification of protein residues [1]. TPQ (trihydroxyphenylalanine quinone) and TTQ (tryptophan-tryptophyl quinone) are generated at enzyme active sites where they serve to catalyze the oxidation of amine substrates. Resonance Raman spectroscopy has proved to be an excellent technique for studying these quinoproteins. Their intensely colored hydrazine derivatives yield distinctive RR spectral patterns that serve as the method of choice for cofactor identification [2]. We have found that the native enzymes also yield rich RR spectra which can be used to identify oxidation states and catalytic intermediates. In the amine oxidase from *E. coli* (ECAO), the **TPQ cofactor** gives rise to a large number of RR modes between 1000 and 1700 cm^{-1} that arise from coupled C=C, C–C, C=O, and C–H vibrations of the quinone ring [3]. Incubation of ECAO in $H_2{}^{18}O$ results in a 26-cm^{-1} downshift for the peak at 1681 cm^{-1} (Fig. 1C), thereby identifying this peak as a C=O stretch. A TPQ model compound exhibits a similar set of vibrational modes and a similar $H_2{}^{18}O$-dependent shift of -23 cm^{-1} for its 1666-cm^{-1} C=O stretch [3]. Since mass spectrometry shows that only one oxygen of the model has exchanged, this vibration can be assigned to the carbonyl at the **C5** position which is known to react with substrates and mechanism-based inhibitors [1]. The exchange of the **C3** proton in D_2O is indicated by similar deuterium-dependent shifts in the RR spectra of the protein (Fig. 1B) and model compounds and suggests charge delocalization from the C-O at the **C4** position. When an oxygen from $H_2{}^{18}O$ is incorporated at **C2** during cofactor biosynthesis, the primary ^{18}O-shift occurs at 1397 cm^{-1} suggesting that this C-O group also has single-bond character [4]. Thus, the resonance hybridization extends from the C4 to the C2 oxygen, thereby explaining why the oxygen at C5 is the only C=O observed in the RR spectrum.

The **TTQ cofactor** of methylamine dehydrogenase (MADH) exhibits a similarly rich set of vibrational modes with an intense peak at 1620 cm^{-1} that has been assigned to the in-phase stretch of the two C=O groups on the basis of its 25 cm^{-1} downshift in $H_2{}^{18}O$ [2,5]. In the TTQ model, 6,7-IIQ (indolyl-6,7-indole quinone), the analogous ν(C=O) occurs at 1638 cm^{-1}. Both the C=O stretching frequency and the absorption maximum of 6,7-IIQ are sensitive to the quinone environment, shifting to lower cm^{-1} and longer wavelength, respectively, as the polarity of the medium increases [5]. Inclusion of MADH in this correlation reveals that the C=O and NH groups of the TTQ cofactor are more strongly H bonded in the protein than in the model compound. Furthermore, this correlation extends to the effects of the monovalent cations, such as $(CH_3)_3NH^+$, which are known to behave as competitive inhibitors of the methylamine substrate [6]. The substantial decrease in ν(C=O) to 1588 cm^{-1} in the presence of $(CH_3)_3NH^+$ is strong evidence for an *electrostatic interaction between the cation and a carbonyl oxygen of the cofactor*. The X-ray crystal structure of MADH shows a solute molecule, probably $NH_4{}^+$, adjacent to the C6 carbonyl of TTQ. Our absorption and RR spectroscopic data [5,6] point to this being the site where the substrate amine binds and gets deprotonated, thereby facilitating its nucleophilic attack at the C6 position of the TTQ cofactor.

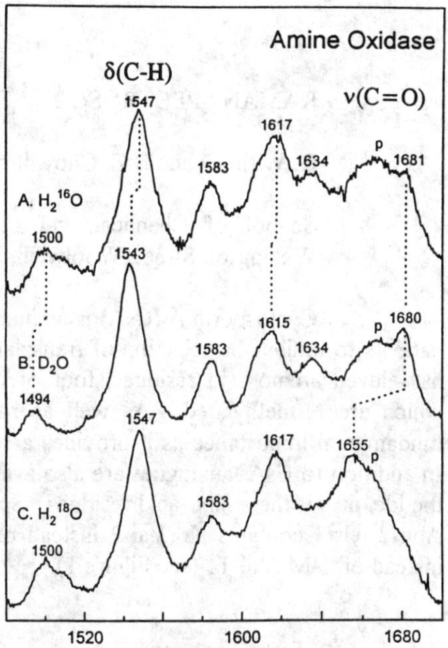

Fig. 1. Identification of C=O stretch and C–H bend in RR spectrum of ECAO.

1. J. P. Klinman & D. Mu, *Annu. Rev. Biochem.* **63**, 299-344 (1994).
2. D. M. Dooley & D. E. Brown, in *Principles and Applications of Quino-proteins*, V. L. Davidson, Ed., Marcel Dekker, New York, 1993, pp. 275-305.
3. P. Moënne-Loccoz, N. Nakamura, V. Steinebach, J. A. Duine, M. Mure, J. P. Klinman, & J. Sanders-Loehr, *Biochemistry* **34**, 7020-7026 (1995).
4. N. Nakamura, R. Matsuzaki, Y.-H. Choi, K. Tanizawa, & J. Sanders-Loehr. *J. Biol. Chem.* **271**, 4718-4724 (1996).
5. A. C. F. Gorren, P. Moënne-Loccoz, G. Backes, S. De Vries, J. Sanders-Loehr, & J. A. Duine, *Biochemistry* **34**, 12926-12931 (1995).
6. P. Moënne-Loccoz, N. Nakamura, S. Itoh, S. Fukuzumi, A. C. F. Gorren, J. A. Duine, & J. Sanders-Loehr, *Biochemistry*, **35**, in press (1996).

RAMAN SPECTROSCOPY OF CYCLOSPORIN

R. Withnall and B.Z. Chowdhry

School of Chemical and Life Sciences, University of Greenwich, Wellington Street, Woolwich, London SE18 6PF,UK

Cyclosporin A (CsA) is an immunosuppressive reagent which is given to patients to inhibit the rejection of transplanted organs. It is a cyclic peptide which has eleven amino-acid residues, four of which contain N-H groups and seven of which are N-methylated. As well as having biomedical importance, CsA is of fundamental importance as it provides a structural model of a small cyclic peptide. In addition to CsA, analogues are also available; these differ to CsA with respect to the identity of the amino-acid residues. Specifically, CsC contains Thyr-2 instead of Abu-2, CsG contains NorVal-2 instead of Abu-2, and CsH contains D-MeVal-11 instead of L-MeVal-11 (see Figure 1).

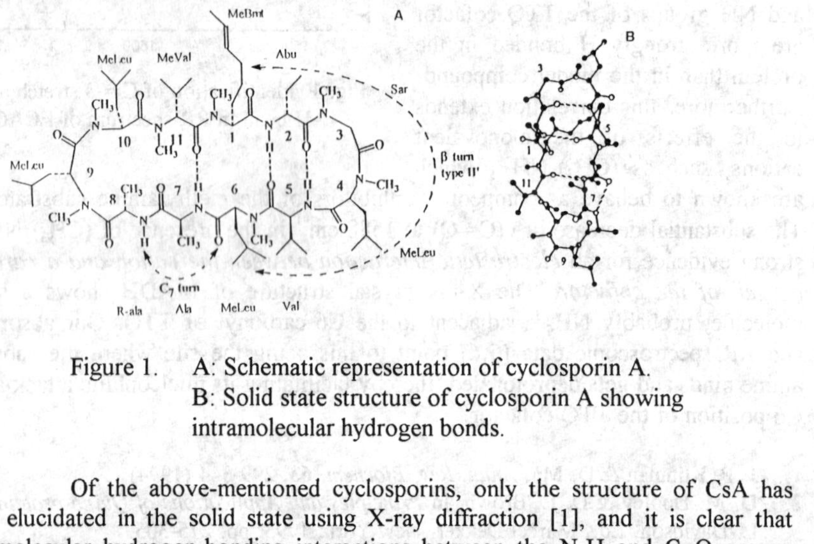

Figure 1. A: Schematic representation of cyclosporin A.
B: Solid state structure of cyclosporin A showing intramolecular hydrogen bonds.

Of the above-mentioned cyclosporins, only the structure of CsA has been elucidated in the solid state using X-ray diffraction [1], and it is clear that intramolecular hydrogen-bonding interactions between the N-H and C=O groups play a major role in determining the secondary structure. This present study has been undertaken in order to probe such interactions by Raman spectroscopy. The amide I (1610-1690 cm^{-1}) region of the spectrum is particularly informative in this respect, since the exact location of a band due to C=O stretching is dependent on the extent of hydrogen bonding. A free C=O has a stretching wavenumber in the range 1660-1690 cm^{-1}, whereas that of a C=O which is strongly hydrogen bound appears in the range 1610-1645 cm^{-1}. More weakly hydrogen-bound C=O bonds give rise to bands in the intermediate region of 1645-1660 cm^{-1}.

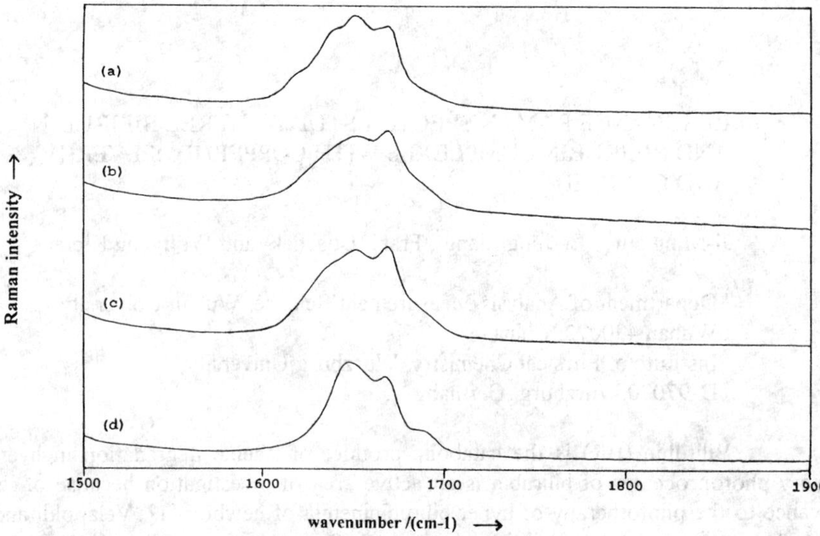

Figure 2. Raman spectra (excitation 532 nm, 4 cm^{-1} resolution) of (a) CsA, (b) CsC, (c) CsG, and (d) CsH.

The crystal structure which has been determined for CsA [1] shows that the peptide ring contains four intramolecular hydrogen bonds (shown in Figure 1). These are (i) the C_2=O ← H-N_5, (ii) the C_5=O ← H-N_2, (iii) the C_{11}=O ← H-N_7 and (iv) the C_6=O ← H-N_8 hydrogen bonds. The shoulders at ca. 1621 and 1639 cm^{-1} in the Raman spectrum of CsA (Figure 2a) are expected to be due to these H-bonded C=O groups, since they appear in the appropriate amide I region. A useful comparison can be made with a previous infrared spectroscopic study of CsA in $CDCl_3$ since it has been shown that the intramolecular H-bonding within CsA remains intact in this solution [2]. Thus the shoulder observed at ca. 1639 cm^{-1} in the Raman spectrum of CsA in the solid state (see Figure 2a) is likely to correspond to the band which was reported at 1638 cm^{-1} in the infrared spectrum of CsA in $CDCl_3$ solution. Likewise the shoulder at ca. 1621 cm^{-1} in Figure 2a most probably corresponds to the feature reported at 1625 cm^{-1} in the solution infrared spectrum.

Clearly the inversion of the MeVal-11 chiral centre which occurs on going from CsA to CsH has a significant effect on the H-bonding interactions as can be seen from a comparison of the respective spectra (see Figure 2a and 2d).

Furthermore differences are also apparent between the spectra of CsA and CsC and between those of CsA and CsG, although the spectra of CsC and CsG appear similar (Figure 2a, 2c and 2d).

1. R.B. Knott, J. Schefer and B.P. Schoenborn, Acta Crystallogr. Sect. C: Cryst. Struct. Comm. C46, 1528 (1990).
2. R.A. Shaw, H.H. Mantsch and B.Z. Chowdhry, Can. J. Chem. 71, 1334 (1993).

RESONANCE RAMAN SPECTRA STUDY OF FREE BILIRUBIN AND BILIRUBIN COMPLEXES WITH COPPER(II), SILVER(I) AND GOLD(III)

Ji-Ming Hu[1*], Er-Jung Liang[2], Frank Duschek[2] and Wolfgang Kiefer[2]

[1] Department of Analysis-Measurement Science, Wuhan University, Wuhan 430072, China
[2] Institute of Physical Chemistry, Würzburg University, D-97070 Würzburg, Germany

Bilirubin (BR) is the catabolic product of heme degradation in liver. Primary photoprocesses of bilirubin is an active area of investigation because of its relevance to the phototherapy of hyper bilirubininemia of newborn[1]. Velapoldi and Meins[2] have reported the stabilities of complexes formed between BR and metal ions. The details of structure information of the complexes of BR-metal remains largely unknown.

Raman Spectroscopy will be a useful tool to study the structure of free BR and complexes of BR with metal ions. Recently, due our own work, surface-enhanced Raman Spectroscopy of complexes of BR with iron(II), nickel(II) and cobalt(II) was the first investigations of the effects of differing metal-bonding on Raman Spectra of BR[3].

In the present paper, the absorption spectra and resonance Raman Spectra of free BR and complexes of BR with Copper(II), Silver(I) and gold(III) at different excitation wavelength are reported. The absorption peak of BR is shifted to violet region in the presence of metal ions. Some new absorption peaks are observed in the presence of Cu(II). One strong is at 340 nm, another weak is at 670 nm. The informations indicate the formation of complexes of BR with metal ions.

The resonance Raman Spectra of free BR and BR complexes with metal ions were take by using 647.1 nm, 514.5 nm, 488.0 nm, 363.8 nm, 335.8 nm, 300.3 nm and 275.4 nm for excitation. It was impossible to obtain resonance Raman Spectra of BR, BR-Ag and BR-Au with 647.1 nm excitation, because they didn't produce absorption at this wavelength. With excitation of 514.5 nm and 488.0 nm, most vibrational modes of lactam and pyrrole can be observed in BR molecules. By contrast, with UV laser excitation, there are a few bands to be observed. The new broad band appear at 1628-1631 cm^{-1} region, is also a lactam C=C, C=O, C-N and pyrrole C=C stretching mode. There are some differences appearing at band intensity and shift of complexes of BR-Ag and BR-Au in comparision with free BR. The resonance Raman Spectra of BR-Cu is entirely different from free BR, BR-Ag, BR-Au. Most of vibration modes of lactam ring disappear and some new vibration modes of pyrrole, propionic acid and big ring appear.

Since all the skeleton carbon atoms of BR bond in SP^2 orbital hybridization, the molecule is basically in a planar form, thus when adding metal ions, the complexes are also basically in a planar configuration. But the structure of complexes of BR with metal ions are dependent on their ionic charge and ionic radius. The ionic radius of silver ion (1.26) and gold ion (0.85) are bigger than copper ion (0.72). When adding silver ion or gold ion, they interact with one pyrromethenone group only. But when adding copper ion, it can interact with two pyrromethenone groups and lead to the formation of circular complex which is analogous to the structure of porphyrin big ring. The N-H band does not exist and the strong band assigned to N-H strtching mode which are at about 1268-1272 cm^{-1} and 961-964 cm^{-1} disappear. The new absorption peak appearing in UV-vis absorption spectra, most lactam vibration modes disappearing and new big ring and pyrrole vibration modes appearing in resonance Raman Spectra of complex. BR-Cu are confirmed this suggestion also.

Acknowledgement: This work was support by the Deuesch Forschungsgemeinschaft grant 446 CHV-112/15/94 (to J-M Hu and W. Kiefer) and by the National Nature Science Foundation of China grant (to J-M Hu).

1. A. F. McDonagh, L. A. Palma and D. A. Lightner, Science, 208, 145(1980)
2. R. A. Velapoldi and O. Menis, Clin. Chem., 17, 1165(1971)
3. J. Chen, J-M Hu, Z-S Xu and R-S Sheng, Spectroschim Acta, 50A, 929(1994)

FT–RAMAN AND FT–IR SPECTRA OF METAL DEOXYCHOLATE COMPLEXES

W–H Li[1], D–M Yu[1], W–D Huang[1], Q. Peng[1], W. Tian[1], Z–H Xu[1], R.D. Soloway[2] and J–G Wu[1]*

[1]Department of Chemistry, Peking University, Beijing 100871, China
[2]Department of Internal Medicine, UTMB, Galveston, TX, USA

In recent years, the study of bile acids and their complexes become a very active field because of their biological importance[1]. Chenodeoxycholic acid and ursodeoxycholic acid are chemotherapeutant of gallstone disease. Bile acids and many metal ions such as Na^+, Ca^{2+}, Mn^{2+}, Cu^{2+} and Zn^{2+} etc. are the main components in the bile of animals, and the interaction between them play an important role in physiology[2]. However, few papers were reported relating to the coordination structure of bile salts complexes. Especially, the binding between hydroxyl group and metal ions is not very clear. In this paper, FT–Raman and FT–IR spectroscopies were used to study the structure of metal deoxycholate complexes. Seven metal deoxycholate complexes ($M(DC)_2$, M^{2+}=Ca^{2+}, Sr^{2+}, Mn^{2+}, Co^{2+}, Ni^{2+}, Cu^{2+} and Zn^{2+}) were prepared in water–alcohol (1:1 in volume) medium by reacting MCl_2 aqueous solution and sodium deoxycholate solution. The FT–Raman and FT–IR spectra were performed on a Nicolet Raman 910 and a Nicolet Magna IR 750 spectrometer respectively.

In the IR spectra of complexes, the COOH stretching band of deoxycholic acid(HDC) at 1715 cm^{-1} was absent and a strong band at 1556 cm^{-1} can be observed. Similarly, the Raman band at 1711 cm^{-1} of HDC also disappeared in the spectra of $M(DC)_2$. It indicate metal deoxycholate complexes formed (see Fig. 1). In the 2800–3000 cm^{-1} region of Raman spectra of $M(DC)_2$, CH stretching band located at 2924 cm^{-1} shift to high frequency about 7–19 cm^{-1} in comparison with HDC. It suggest that after binding to the metal ions the molecular skeleton changed. IR and Raman peaks in the 900–1100 cm^{-1} region of HDC and $M(DC)_2$ were assigned to the C–O–H and C–C–H vibration modes. Both band shape and band position varied in different complexes (shown in Fig. 2), indicating the OH groups of HDC are binding to the metal ions. In other hand, IR spectra give more useful information in the range of 3000–3700 cm^{-1}. The hydroxyl O–H stretching bands in different complexes are shown in table 1. The band shifts of HDC and its complexes provided another direct evidence of the coordination between hydroxyl group and metal ions. We conclude

* The project was supported by NSF and Doctoral Foundation of National Education Commission of China

that both the carboxyl and hydroxyl groups of deoxylcholic acid binding to the metal ions in the complexes.

TABLE 1. The O–H stretching bands(IR) of HDC and $M(DC)_2$

Samples	O–H stretching bands (cm^{-1})
HDC	3551, 3434, 3334, 3212
$Ca(DC)_2$	3700*, 3616*, 3513*, 3394*, 3194*
$Sr(DC)_2$	3702*, 3615*, 3514*, 3453*, 3261*
$Mn(DC)_2$	3625, 3519*, 3437*, 3350*, 3225*
$Co(DC)_2$	3624, 3515*, 3428*, 3376*, 3218*, 3157*
$Ni(DC)_2$	3622, 3443*, 3352*, 3258*, 3181*
$Cu(DC)_2$	3600, 3444, 3361, 3338, 3318
$Zn(DC)_2$	3624, 3518*, 3429*, 3336*, 3217*

*deconvolution bands

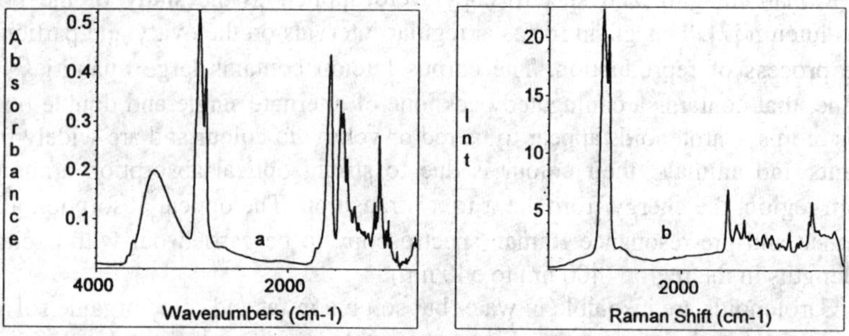

Fig. 1. The FT–IR(a) and FT–Raman(b) spectra of $Ca(DC)_2$

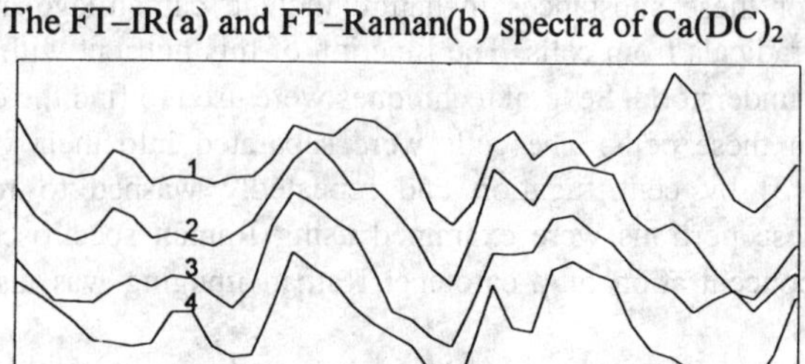

Fig. 2. The bands relating to O–H vibration modes in the Raman spectra of (1) HDC; (2) NaDC; (3) $Sr(DC)_2$; and (4) $Mn(DC)_2$.

1. M. C. Carey, in Sterols and Bile Acids, ed. by H. Danielesson and J. Sjovall, chapter 13, 345 (1985).
2. N. Rajagopalan and S. Lindenbaum, Biochimica et Biophysica Acta, 711, 66 (1982).

BIOLOGICAL APPLICATIONS OF RAMAN SPECTROSCOPY AND IMAGING: AN INVESTIGATION OF LIVE CORPUS LUTEUM CELLS

H.S. Sands[1], D.N. Batchelder[1]*, A.J. Holt[2], R.G. Rodway[2]

[1]Department of Physics and Astronomy,
[2]Department of Animal Physiology and Nutrition,
University of Leeds, Leeds LS2 9JT, UK

Raman spectroscopy makes an excellent tool for the investigation of biological materials **[1]**. Spectroscopic information can be obtained from samples containing water unlike work done in the infrared.

Raman imaging and spectroscopy were applied to the study of the bovine corpus luteum **[2]**. This gland forms at regular intervals on the ovary and participates in the process of reproduction. The corpus luteum contains large quantities of β-carotene, that contains a conjugated backbone of alternate single and double bonded carbon atoms. Carotenoids appear to be red or yellow in colour and are widely found in plants and animals; their colour is due to strong optical absorption around the 450 nm region, the energy from the π to π^* transition. The optical absorption allows resonance and pre-resonance Raman spectroscopy to be carried out with excitation wavelengths in the region (400 nm to 540 nm).

Carotenoids are insoluble in water but soluble in fat and some organic solvents. Known uses for these substances in nature include camouflage and anti-oxidants removing free radicals from cells. The function of this nutrient within this particular cell is not well understood. Several techniques were used to find the distribution of β-carotene within these cells. The cells were separated into their various organelle components **[3]** by centrifugation and repeatedly washed to remove any free carotenoid. These portions were examined using Raman spectroscopy to ascertain their relative concentration of β-carotene. Raman imaging was also performed on whole live cells.

The Renishaw Raman Imaging Microscope was used to obtain the spectra and images **[4]**. To obtain the spectra, Ar^+ (514.5 nm) laser excitation was used. At this wavelength, the resonance Raman effect occurs, allowing high signal-to-noise spectra to be taken from these low concentration samples. For the Raman imaging, HeNe (632.8 nm) laser excitation was used under pre-resonance conditions to avoid damaging the delicate cells. The Raman image was obtained by filtering out all scattered light except for that from the 1525 cm^{-1} Raman band **[5]** of β-carotene.

Figure 1 shows that, out of the fractions tested, the majority of the β-carotene was contained within the nuclei and microsome fractions. The microsome is made up of the ruptured cell wall. Figure 2 shows the white light and Raman light image of a live cell. From the Raman image, it can be seen that the β-carotene is present in discrete sites within the cell and not distributed evenly throughout it.

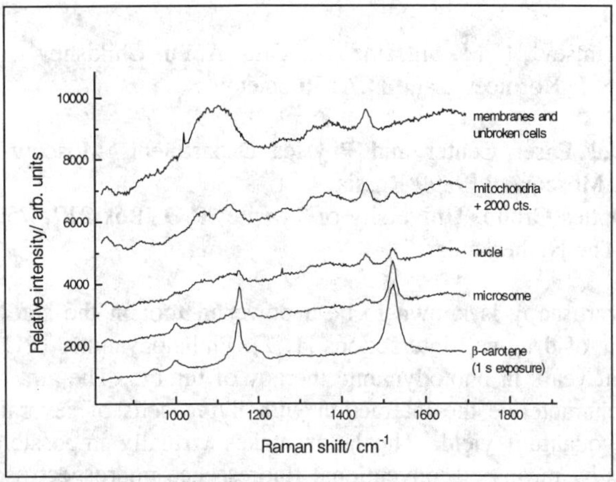

Figure 1 Raman spectra of β-carotene and separated fractions of corpus luteum cells. 514 nm excitation and 100 s exposure.

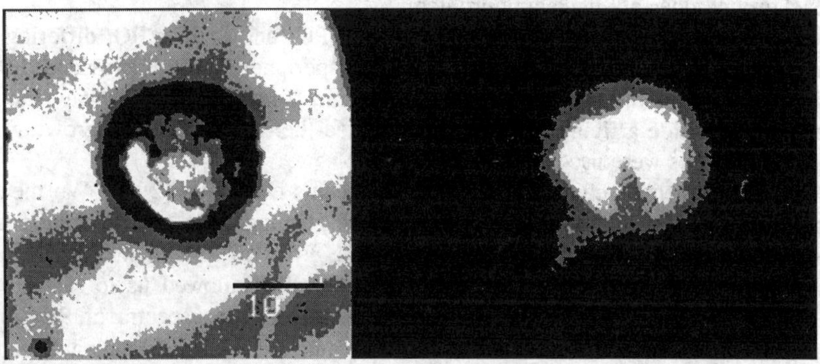

Figure 2 White light (left) and Raman light (right) images of a mature corpus luteum cell. Note the two distinct areas of β-carotene in the Raman light image. The length of the scale bar is ten microns.

1. G.J. Pupples, F.F.M. de Mul, C. Otto, J. Greve, M. Robert-Nicoud, D.J. Arndt-Jovin and T.M. Jovin, Nature, 347, 301, (1990).
2. D.N. Batchelder and R. Rodway, Opto and Laser Europe, Dec., p. 7, (1994).
3. F. Adar and M. Erecinska, Biochemistry, 17, 5484 (1978).
4. K.P.J. Williams, G.D. Pitt, B.J.E. Smith, A. Whitley, D.N. Batchelder and I.P. Hayward, Journal of Raman Spectroscopy, 25, 1, 131, (1994).
5. F. Inagaki, M. Tasumi, and T. Miyazawa, Journal of Raman Spectroscopy, 3, 335, (1975).

RAMAN SPECTROSCOPY OF PHTHALOCYANINES AND THE STUDY OF THEIR INTERACTION WITH LIVING CELLS

S.Yu. Arzhantsev[1], N.M. Sijtsema[2] ,C. Otto[2] A.Yu. Chikishev[1*], J. Greve[2], N.I. Koroteev[1], and D.A. Ruzmetov[1].

[1]International Laser Center and Physics Department, Moscow State University, Moscow, 119899 Russia
[2]Applied Optics Group, University of Twente, P.O. Box 217, 7500AE Enschede, The Netherlands

Raman microspectroscopy is known to be a powerful tool in the studies of cells and, in particular, of drug-cell interactions [1, 2]. Phthalocyanines (PCs) are widely applied in recent years in photodynamic therapy of tumors. The aim of the present study was to characterize the interaction with living cells of several PCs with low fluorescence quantum yield. The latter makes virtually impossible the study of this molecules by means of conventional fluorescence microspectroscopy. A disadvantage for the fluorescence studies becomes an advantage for the studies carried out by the methods of vibrational spectroscopy because the fluorescence background does not make any problem.

In our experiments we used three cobalt PCs and an iron PC, differing from each other by the peripheral groups. Different peripheral groups provided different hydrophoby of the molecules which, in turn, influenced the way the dyes penetrated inside a living cell and accumulated therein. Suspension growing K562 and HeLa cells were used.

Experimental setups are described in details elsewhere [2, 3]. We measured both Raman spectra of phthalocyanines inside living cells and in solutions and Raman images of the cells stained with PCs. Prior to Raman measurements the studies of absorption spectra were carried out which allowed us to select proper excitation wavelengths for the Raman studies. Absorption spectra of PCs display several strong bands: Soret band at ca. 330 nm, and two overlapping Q-bands at ca. 620 and 670 nm. Raman spectra of PCs are highly sensitive to the excitation wavelength. Though under the excitation within a single absorption band the spectra were quite similar, pronounced differences were observed in the spectra measured under pre-resonance conditions (λ_{ex} = 413 nm), off-resonance (λ_{ex} = 514 nm) and under the excitation in the Q-bands (λ_{ex} = 647, 660 nm). In the experiments with living cells and with the solutions of different pH we used the excitation in the longwavelength Q-bands.

Both absorption and Raman spectra of PCs appeared to be sensitive to pH. In Fig. 1 we present Raman spectra of cobalt phthalocyanine in solution at neutral pH (curve *1*), in solution at pH 11.1 (curve *2*) and inside a living cell (curve *3*). It is seen that the spectra are sensitive to microenvironment. The main changes take place for the spectral ranges 700–900 cm^{-1} and 1300–1450 cm^{-1}. The assignment of the corresponding vibrational bands has been given earlier (see, for example, [4]): CH out-of-plane bending mode (870 cm^{-1}) and stretching modes of the ring (higher than 1100 cm^{-1}). Note a small shift of the strongest band at ca. 1540 cm^{-1} with the change of pH along with the changes around 1200 and 1320 cm^{-1}.

The Raman images of the cells stained with PCs can roughly be divided into two major groups. The images of the cells stained with hydrophilic PCs were featured by bright "spots" inside the cell with the size of the spot about 1-2 μm. Whereas the images of the cells stained with hydrophobic PCs displayed more or less homogeneous distribution of the dye. These observations are quite in agreement with the existing hypothesis about localization of PCs in cells. Hydrophilic molecules penetrate inside the cell by means of endocytosis and the hydrophobic ones do that by diffusion. Therefore, one may expect the hydrophilic molecules to localize in lysosomes and the hydrophobic molecules may accumulate in cellular membranes and cytoplasm. Thus, the images detected may belong to the cells with the dye molecules concentrated in lysosomes or localized in membranes and cytoplasm.

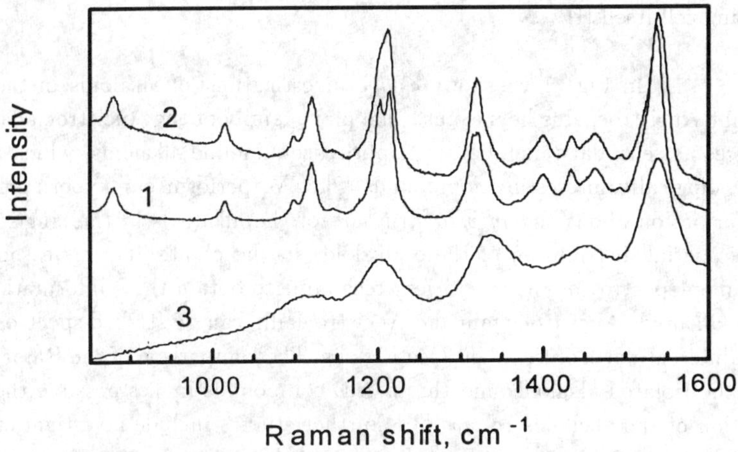

Fig. 1.Raman spectra of phthalocyanine in neutral solution (*1*), at pH 11.1 (*2*), and in a living cell (*3*)

The authors are grateful to Profs. O.L. Kaliya, E.A. Lukjanets, and Dr. M.G. Galpern who provided us with phthalocyanines. We thank Ing. I. Segers for cells preparation. The research was supported by NWO (Dutch Organization for Advancement of Scientific Research).

1. G.J. Puppels, F.F.M. de Mul, C. Otto, J. Greve, M. Robert-Nicoud, D.J. Arndt-Jovin, and T.M. Jovin, Nature, 347, 301 (1990)

2. G.J. Puppels and J. Greve, in Biomolecular Spectroscopy, (R.J.H. Clark and R.E. Hester, eds.), Vol. 20, part. A, Wiley, Chichester, 1993, p.231.

3. G.J. Puppels, M.Grond and J.Greve, Applied Spectroscopy 47 , 1256 (1993).

4. R. Simic-Glavaski, S. Zecevic, and E.B. Yeager, J. Raman Spectrosc. 14, 338 (1983).

RAMAN INVESTIGATION OF BIOLOGICAL SYSTEMS: ANALYSIS OF SINGLE CELLS AND THEIR COMPONENTS

E. Urlaub, J. Popp, M. Lankers, and W. Kiefer*

Institut für Physikalische Chemie, Universität Würzburg,
Marcusstr. 9-11, D-97070 Würzburg, Germany

Raman spectroscopy has widespread and potential applications for the analysis of isolated biological molecules and their conformation. This technique is not only non–destructive and rapid, it also yields detailed information about the biological molecules without using marker– or tracer–molecules. Thus, Raman spectroscopy even enables the investigation of biological molecules *in vivo*, in the living cell itself [1].

In Fig. 1 we show *in vivo* investigations of alkaloids in the tropical plant Ancistrocladus heyneanus. The plant family of the Ancistrocladaceae produces as secondary metabolites naphthylisoquinoline alkaloids, which are pharmacologically interesting compounds [2]. We perform our experiments to get information about the *in vivo* structure of the alkaloids in the single cells and the spatial distribution of these alkaloids in the plant. It can be shown, that Raman spectroscopy is an efficient technique to obtain these information. Thus, the alkaloids Ancistroheynin and Ancistrocladin can be studied spectroscopically in different parts of the plant (see Fig. 1). The comparison of the Raman spectra of the isolated alkaloids and the *in vivo* compounds makes possible the identification of the alkaloids *in vivo* [3]. Further studies include investigations on the production of the special alkaloids in stressed plants.

Another example for the application of Raman–spectroscopic techniques in biology is the classification of microorganisms. It can be shown, that the combination of confocal Raman spectroscopy with optical trapping techniques allows the characterization of single yeast cells in mixtures by means of their spectroscopic fingerprints. Thus, a single yeast cell is stabilized by the single beam gradient force optical trap (optical tweezers) [4] and can be investigated spectroscopically. Thereby Raman spectra of single cells with good signal–to–noise–ratios can be obtained by using Raman difference spectroscopy between cell and environmental medium. Apart from the fast identification of the microorganisms the combination of the above mentioned techniques makes it possible to investigate dynamic processes in the cell, e.g. the accumulation or the effect of harmful substances as a function of time and local position. Such studies will be discussed.

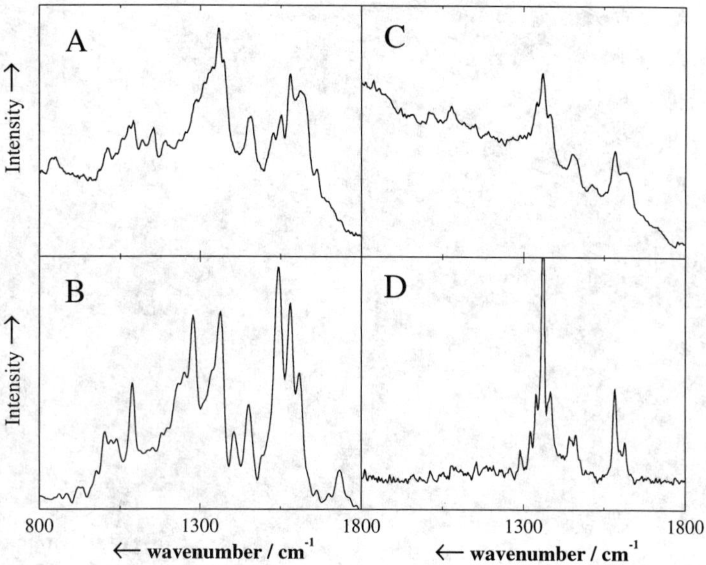

Fig.1. Characterization of alkaloids *in vivo*. **A:** FT–Raman spectra of Ancistroheynin in the shoot of the plant; **B:** FT–Raman spectra of isolated Ancistroheynin; **C:** FT–Raman spectra of Ancistrocladin in the roots of the plant; **D:** FT–Raman spectra of isolated Ancistrocladin.

Financial support from the Deutsche Forschungsgemeinschaft and by the Fonds der Chemischen Industrie is highly acknowledged. We also thank Professor G. Bringmann for supplying the tropical plant.

1. J. Greve, G. J. Puppels in Advances in Spectroscopy (Ed.: R. J. H. Clark, R. E. Hester), John Wiley and Sons, Chichester, 20 , 231 (1993).
2. G. Bringmann, F. Pokorny in The Alkaloids (Ed.: G. Cordell), Academic Press, New York, 46, 127 (1995).
3. E. Urlaub, J. Popp, W. Kiefer, G. Bringmann, D. Koppler, B. Schrader, J. Sawatzki, in preparation.
4. A. Ashkin, J. M. Dziedzic, J. M. Bjorkholm, S. Chu, Opt. Lett.,11, 288 (1986).

SECTION 15

SURFACES, INTERFACIAL PHENOMENA

RESONANCE AND NORMAL RAMAN SPECTROSCOPY OF CARBON SURFACES: RELATIONSHIPS OF SURFACE STRUCTURE AND REACTIVITY

R.L. McCreery, Y-C. Liu, M. Kagan, P. Chen, M. Fryling

Department of Chemistry, The Ohio State University, 100 West 18th Ave., Columbus, Ohio 43210

Although carbon surface chemistry is of great importance to the widespread use of carbon in energy conversion, electrochemistry, and adsorption, there are few molecular probes of carbon surface structure available. Raman spectroscopy has been used extensively for characterization of bulk carbon materials, but until recently has lacked sufficient sensitivity for the observation of monolayer or submonolayer species on carbon surfaces. Since carbon does not support electromagnetic field enhancement under any known conditions, surface analysis by Raman depends on the ability of the spectrometer to observe very low values of the product of cross section (β) and surface number density (N).

The advent of CCD detectors and efficient spectrographs has greatly increased the signal to noise ratio (SNR) for surface Raman, and decreased the minimum βN product which is detectable.[1,2] We exploited these instrumental improvements to observe monolayers adsorbed on carbon surfaces without any known surface enhancement. Monolayers of molecules with cross sections comparable to that of benzene at 514.5 nm were observable with useful SNR. Resonance Raman permitted higher SNR and lower surface coverage by increasing β. Several examples of resonance and normal Raman on carbon surfaces will be presented, including:

1. Surface quenching of fluorescence (e.g. R6G)[1]
2. Derivatization of surface oxide groups (e.g. carbonyls)[3]
3. Use of polarization ratio to determine orientation of adsorbed phthalocyanines[4]
4. Changes in cross section upon adsorption

Unenhanced surface Raman was used extensively to investigate chemisorbed species on carbon surfaces, such as nitrophenyl moities. Electrochemical derivatization of glassy carbon and pyrolytic graphite yielded monolayers of covalently bonded species, which were characterized with surface Raman.[5] After spectral characterization, the NO_2 group of chemisorbed nitrophenyl was subsequently electrochemically reduced, resulting in disappearance of the NO_2 bend (848 cm^{-1}) and stretch (1336 cm^{-1}). An example of a surface Raman spectrum is shown in Figure 1, for a monolayer azobenzene covalently bonded to glassy carbon.

Two objectives of more general interest provide the driving force for Raman spectroscopy of carbon surfaces. First, Raman spectral features correlate with the electrochemical reactivity of carbon surfaces, and provide a means to monitor surface oxidation and/or intercalation. Second, Raman provides a molecular probe of unique species at carbon surfaces which may be responsible for the highly variable behavior of carbon materials.

1. M.R. Kagan, R.L. McCreery, Reduction of Fluorescence Interference in Raman Spectroscopy via Analyte Adsorption on Graphitic Carbon, Anal. Chem. **1994**, 66, 4159.
2. M.R. Kagan, R.L. McCreery, "Quantitative Surface Raman Spectroscopy of Physisorbed Monolayers on Glassy Carbon", Langmuir, **1995**, 11, 4041.
3. M.A. Fryling, J. Zhao, R.L. McCreery, Resonance Raman Observation of Surface Carbonyl Groups on Carbon Electrodes Following Dinitrophenylhydrazine Derivatization, Anal. Chem., **1995**, 67, 967.
4. J. Zhao, R.L. McCreery, "Polarized Raman Spectroscopy of Metallophthalocyanine Monolayers on Carbon Surfaces", Langmuir, **1995**, 11, 4036.
5. Y-C. Liu, R.L. McCreery, "Reactions of Organic Monolayers on Carbon Surfaces Observed with Unenhanced Raman Spectroscopy", J. Am. Chem. Soc., **1995**, 117, 11254.

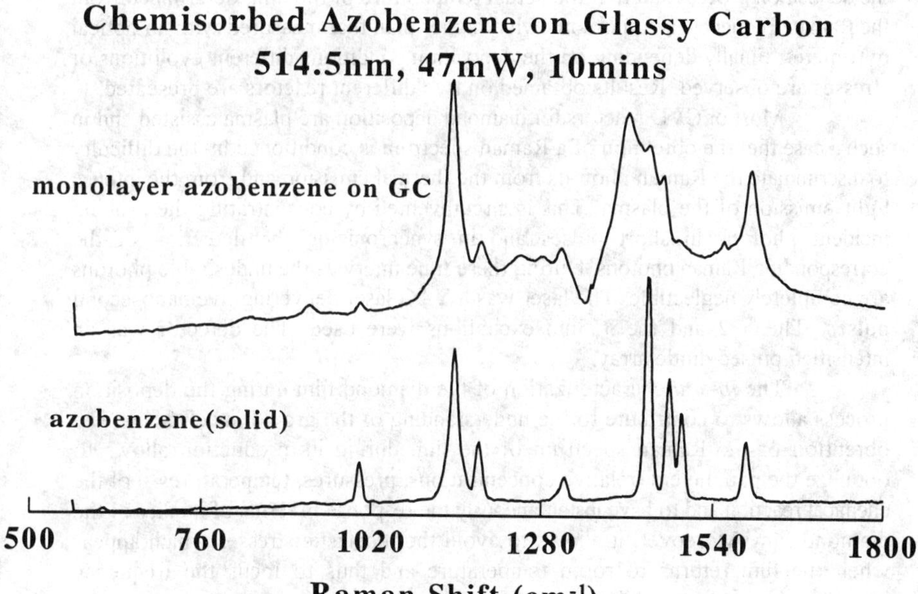

Figure 1

IN-SITU RAMAN ANALYSIS OF THE GROWTH PROCESS IN A CHEMICAL VAPOR DEPOSITION OF THIN DIAMOND FILMS

G. Lucazeau*, L. Abello, M.Mermoux

Laboratoire d'Electrochimie et de Physicochimie des materiaux et des Interfaces BP75, 38402 Saint Martin d'Häres Cedex, France

A set up designed for characterization of a diamond film during its growth in plasma assisted Chemical Vapor Deposition (CVD) reactors is described. It is shown that the chemical and crystalline quality of the film can be controlled during the deposition process and that the actual temperature of the film, determined from the Raman frequency, is found somewhat higher than that measured with an optical pyrometer. Finally depending on the deposition conditions different evolutions of stresses are observed. Results obtained on two different reactors are presented.

Most of CVD reactors for diamond deposition are plasma assisted and in such a case the the obtention of a Raman spectrum is conditioned by the difficulty to discriminate the Raman photons from the thermal emission and from the intense light emission of the plasma. This is circumvented by concentrating the probing incident photons in short pulses and in synchronizing the detection of the corresponding Raman photons. During these time intervals the undesirable photons are completely neglectible. The laser was a YAG laser delivering five nanosecond pulses. The 532 and the 355nm excitations were used. The detector was an intensified pulsed diode array.

The *in-situ* characterization of the diamond film during the deposition process allows to contribute to the understanding of the growth mechanism. The obtention of the Raman spectrum of the film during its production allows to optimize the parameters (relative concentrations, pressures, temperatures..) of the chemical reaction and to have instantaneously the response in terms of quality of the diamond film. Moreover, it allows to avoid thermoelastic stresses which appear when the film returns to room temperature and thus to focus the frequency variations of the diamond line on the internal stresses.

Figure 1 is relative to the detection of films obtained in a plasma jet reactor. In such a reactor the deposition rate is of about 1 micrometer per 10 seconds and the obtained diamond films have a poor transparency with typical penetration depths of one micrometer. In these conditions each spectrum recorded every min is thus independent from the preceding one and the amplitude of the 1310 cm^{-1} band characterizes the crystalline quality of the layer being formed. Figure 1 shows that the crystalline quality first increases then decreases when secondary nucleation takes place and the beginning of a cyclic process is observed.

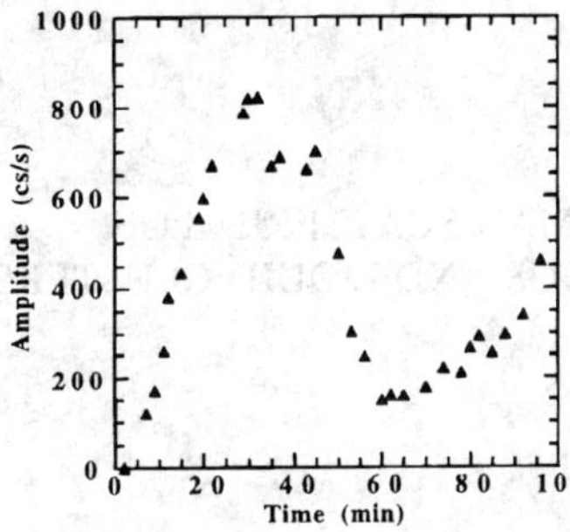

Figure 1 Amplitude variation of *in situ* Raman spectra of diamond film formed in a plasma torch assisted CVD reactor.

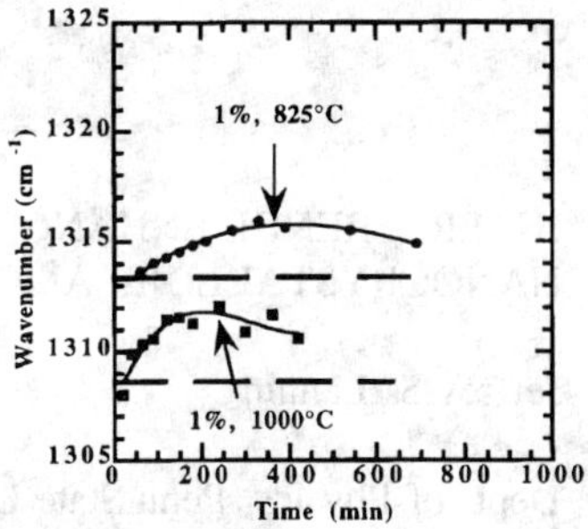

Figure 2 Evolution of the internal stresses *vs.* thickness (time of deposition) of a diamond film produced in a plasma assisted reactor.

Figure 2 represents the variation of the diamond line frequency vs time for a film obtained in a microwave plasma assisted reactor (the deposition rate is less than one micrometer per hour). From Stokes-antiStokes measurements performed on the signal of the substrate (alumina) and on the diamond band it was possible to estimate the decrease of frequency of the diamond line due to the temperature effect and thus, to obtain the frequency variation associated to the internal stresses. Figure 3 shows how the frequency of a film varies when the substrate is cooled from 800°C to room temperature depending it is silicon or alumina. On alumina thermoelastic stresses of *ca.* 2.3GPa can be estimated from the frequency variation.

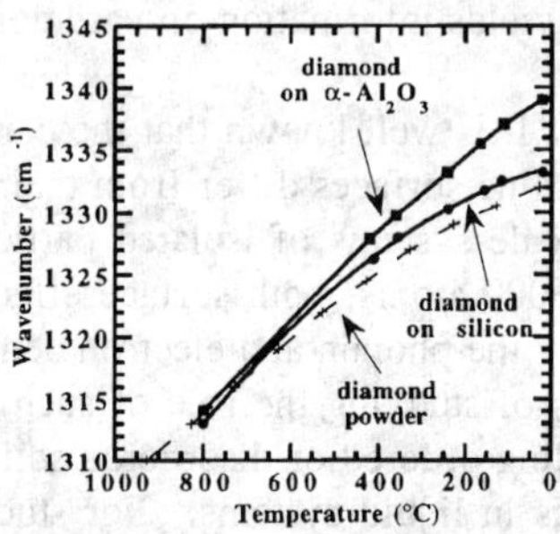

Figure 3 Difference in the evolution of the thermoelastic stresses when a diamond film deposited on silicon or alumina is cooled down to room temperature.

1. N. Rosman, L.Abello, G.Verven, J.P. Chabert, G. Lucazeau, J. Appl. Phys.**78**, 519 (1995)
2. L. Fayette, B. Marcus, M. Mermoux, N. Rosman, L. Abello, G. Lucazeau: J. Appl. Phys. **76**, 1604, (1994)
3. M.Mermoux , L.Fayette, B.Marcus, N.Rosman, L.Abello, G.Lucazeau, Physica Stat.Solidi (in press)
4. M. Mermoux , L. Fayette, B. Marcus, N. Rosman, L. Abello, G. Lucazeau, Analysis, **23**, 301-311 (1995)

INTERFERENCE ENHANCED RAMAN SCATTERING OF NANOCRYSTALLINE, AMORPHOUS AND LIQUID CLUSTERS

Jeffrey S. Lannin

Dept. of Physics, Penn State University
University Park, PA 16802, USA

The method of interference enhanced Raman scatttering (IERS) using multilayer structures provides a general means of studying ultrathin films and near surface vibrations of a variety of systems. Examples of the use of IERS for ultrahigh vacuum studies of changes in the vibrational states of disordered condensed matter systems with cluster size are presented. For such systems IERS yields information on variations in the phonon density of states.

It is well known that the electronic and vibrational properties of ordered crystalline surfaces differ from their bulk due to changes in local structure. For nanoscale systems of isolated particles or clusters, containing roughly of order 50 - 5000 atoms, both surface effects as well as quantum size effects may also modify the phonon and electron densities of states. Raman scattering provides a means of studying the role of such size effects on a diverse range of materials, including ordered or disordered solids in nanocrystalline or amorphous forms as well as in liquid systems. For studies of isolated particles or clusters, the low coverage and relatively small Raman cross sections require the use of local electric field, surface enhancement effects, such as SERS, ATR coupling and interference enhanced Raman scattering (IERS).

Of these methods IERS provides appreciable flexibility for uhv studies of a broad range of materials deposited under a diverse set of substrate conditions. We have also shown that it is possible to combine IERS and SERS using multilayer films with Ag particles. In IERS multilayer substrates with appropriate thicknesses are fabricated to yield enhanced electric fields. IERS provides a means for in situ studies of surface and near surface phenomena of molecular systems, polycrystalline solids and weaker scattering, disordered systems. The latter include amorphous, liquid and very small, less than 2nm size nanocrystalline particles. In all these systems disorder effects provide information about the density of vibrational states, as crystal momentum is not conserved.

Two types of substrate configurations have been employed for ultrathin film particle studies: either trilayer [1] or bilayer [2] structures. Both configurations utilize a lower reflective metal underlayer and a transparent optically matched layer, such as SiO_2. For the trilayer a thin top third layer such as disordered carbon has been employed. Estimates of the bilayer film thicknesses use a computer program that calculates the field distribution in the neighborhood of the surface, while for trilayers the approximate thicknesses are chosen to yield a minimum in the reflectivity for the laser wavelength of interest. Both SiO_2 and C layers are also prepared in uhv prior to deposition of the ultrathin film system of interest. The uhv system has a base pressure of $2x10^{-10}$ Torr, allowing Raman scattering measurements over extended periods of time without appreciable surface modifications. High resolution reflection EELS measurements allow monitoring of H bonding to the surface, while Auger and photoemission provide a means of studying island film growth. In Figs. 1-3 examples of IERS applied to nanocrystalline (nc), amorphous and liquid particle systems are shown. In each of these systems the spectra indicate structural changes with the distribution of particle sizes and their average.

Nanocrystalline C In Fig. 1 the high frequency Raman spectra of nanocrystalline carbon in ultrathin film form is compared to "glassy" carbon. The latter sample exhibits x-ray line widths that indicate an ~3nm in plane correlation length with graphite-like structure. Of the two distinct peaks the lower frequency one is associated with disorder induced scattering from zone boundary modes, while the higher is due to vibrations derived from the zone center allowed graphite mode. The latter is modified by wavevector mixing that shifts its centroid to higher frequency. For the ultrathin film nc-C film the two peaks essentially merge into a single peak with residual shoulders. The Raman spectrum in this case corresponds to the density of phonon states modulated by matrix elements that are assumed to be a smooth function of frequency. The position of the shoulders relative to second order scattering in glassy carbon suggests that the higher frequency density of states peak has down shifted in small nanocrystallites of C with sizes of 1-2nm. [3]

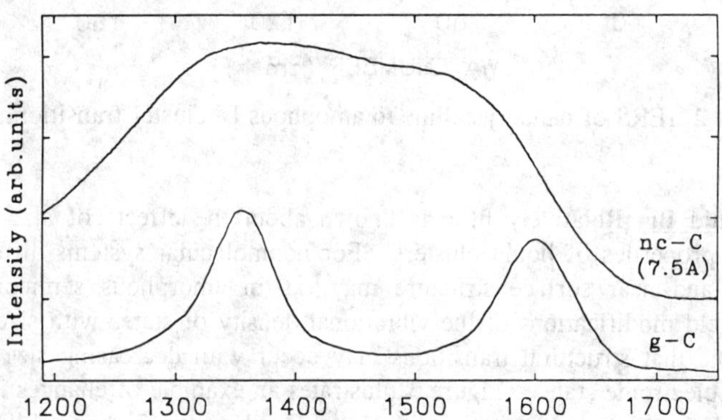

Fig. 1 Comparison of the Raman scattering of glassy (g) carbon and IERS of a 7.5A thick nanocrystalline (nc) film.

Amorphous Bi IERS of amorphous clusters has been performed on Ge [4], Bi, [5] C in the form of normal and laser polymerized [6] C_{60} and recently ultrathin random network C structures. For amorphous Ge clusters changes in the Raman spectra with the addition of atomic H to cluster surfaces demonstrated the role of surface dangling bond effects. [3] Figure 2 illustrates the changes noted as film thickness is reduced so a tranformation from nc-Bi to a new, amorphous Bi form is obtained at ~8A. In contrast, lower temperature spectra indicate a 3-fold coordination for the amorphous clusters. The increase of the amorphous Bi peak frequency with decreasing cluster size is attributed to an increase in covalent bond strength due to increased electron localization.

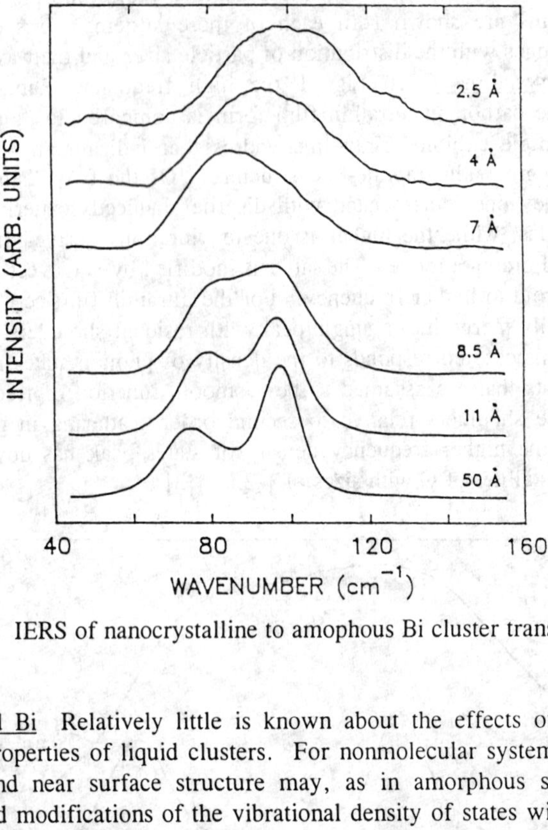

Fig. 2 IERS of nanocrystalline to amophous Bi cluster transition.

Liquid Bi Relatively little is known about the effects of size on the vibrational properties of liquid clusters. For nonmolecular systems differences in surface and near surface structure may, as in amorphous semiconductor clusters, yield modifications of the vibrational density of states with size. It is also possible that structural transitions may occur with decreasing size due to changes in electronic states. Figure 3 illustrates an example of changes in IERS of liquid Bi clusters supported on a disordered C substrate. [7] For a 140A thick film, ex situ TEM measurements indicate it is composed of hemispherical particles with a mean diameter of ~500A For these larger clusters the Raman

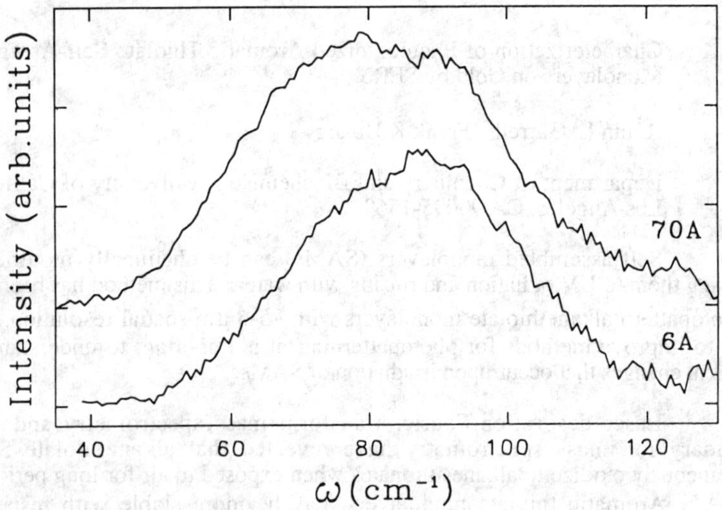

Fig. 3 Modification of liquid Bi cluster phonons with film thickness.

spectra essentially correspond to the weighted density of states of the bulk liquid metal. Neutron diffraction studies have noted that the bulk liquid coordination is ~7±1 [8]. With decreasing film thickness the IERS becomes substantially modified for thickness below ~30A. For a 6A thick film shown, corresponding to an average liquid particle size of ~45A, the peak of the IERS shifts to higher frequency. The position of the peak at ~90cm^{-1} is near to that of amorphous Bi and nc-C clusters whose coordination number is 3. Supercooling studies and an analysis of the IERS spectra, using the TEM size distributions, suggest that below a size of order 42±5A the liquid Bi clusters become 3-fold coordinated.

Supported by NSF Grant 92-24165

1. G. A. N. Connell, R. J. Nemanich and C. C. Tsai, Appl. Phys. Lett. 36, 31 (1980).
2. W. Basca and J. S. Lannin, Appl. Phys. Lett. 61, 19 (1992).
3. V. I. Merkulov, J. S. Lannin and J. M. Cowley (to be published).
4. J. Fortner and J. S. Lannin, Surf. Sci. 254, 251 (1991).
5. M. G. Mitch et. al., Phys. Rev. Lett. 67, 875 (1991).
6. W. Basca and J. S. Lannin, Phys. Rev. B49, 14750 (1994).
7. V. I. Merkulov and J. S. Lannin (to be published).
8. M. Takagi, J. Phys. Soc. Jpn. 11, 396 (1956).

Characterization of Photooxidized Aromatic Thiolate Self-Assembled Monolayers on Gold by SERS

Robin L. Garrell*, Frank K Huang

Department of Chemistry and Biochemistry, University of California, Los Angeles, CA 90095-1569

Self-assembled monolayers (SAMs) can be chemically modified by exposing them to UV radiation and rinsing with water. This method has been used to photopattern alkanethiolate monolayers with ~5-6 μm spatial resolution.[1,2] In order to improve methods for photopatterning, it is important to understand the chemical changes that occur upon irradiation of SAMs.

Laser desorption-Fourier transform mass spectrometry and static secondary ion mass spectrometry have revealed that alkanethiolate SAMs spontaneously oxidize to alkanesulfonates when exposed to air for long periods of time.[3-4] Aromatic thiolate monolayers may be more stable with respect to oxidation; using surface-enhanced Raman spectroscopy, we showed that benzenethiolate on gold does not oxidize in air.[5,6] On the other hand, exposure of *p*-chorobenzenethiolate monolayers to UV irradiation led to formation of the sulfonate,[7] suggesting that aromatic substitution may influence the tendency to oxidize, or that UV irradiation may be necessary to oxidize aromatic thiolate SAMs. Most recently, surface-enhanced Raman spectroscopy (SERS) was used to characterize hexanethiolate SAMs on silver.[8] It was proposed that upon UV irradiation, the thiolate undergoes C-S bond cleavage followed by oxidation of sulfur to SO_3^{2-} and subsequently to SO_4^{2-} on the silver surface.

We have used SERS to characterize the effects of UV irradiation on benzenethiolate, 2-phenyl-1-ethanethiolate, and *p*-chlorobenzenethiolate SAMs on roughened gold electrodes. The Figure shows SER spectra of a benzenethiolate monolayer prior to irradiation and after 1 and 2 h irradiation with a 4.5 mW/cm^2 254 nm penlight. After 1 h irradiation, a shoulder appeared at ~972 cm^{-1} that was not present in the original spectrum; after 2 h, this feature became a distinct band at 972 cm^{-1}. The benzene ring mode vibrations at 997, 1019, 1070, and 1110 cm^{-1} decreased in intensity after 1 h and even further after 2 h. The near disappearance of these modes and of the ν C-S stretching modes at 416 and 691 cm^{-1} are evidence for cleavage of the C-S bond accompanied by the loss of the phenyl groups from the monolayer. The new band at 972 cm^{-1} is assigned to the symmetric stretch of adsorbed SO_4^{2-}.[9] In contrast with the results for alkanethiolates on roughened silver[6], no SO_3^{2-} formation is observed. Analogous experiments with 2-phenyl-1-ethanethiolate and *p*-chlorobenzenethiolate SAMs on roughened gold give similar results.

In summary, UV irradiation of aromatic thiolate SAMs on gold results in C-S bond cleavage followed by oxidation of the surface-bound sulfur to SO_4^{2-} The kinetics of monolayer oxidation and the effects of ring substituents will be discussed. Results of experiments involving the displacement and removal of SO_4^{2-} from the surface will also be described.

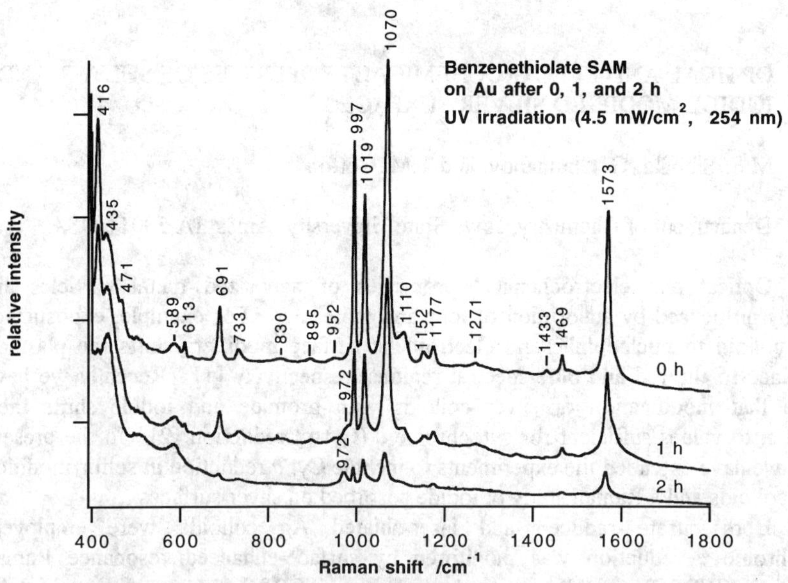

1. Huang, J.; Dahlgren, D. A.; Hemminger, J. C. *Langmuir* **1994**, *10*, 626-628.
2. Tarlov, M. J.; Donald R. F. Burgess, J.; Gillen, G. *J. Am. Chem. Soc.* **1993**, *115*, 5305-5306.
3. Tarlov, M. J.; Newman, J. G. *Langmuir* **1992**, *8*, 1398-1405.
4. Li, Y.; Huang, J.; Robert T. McIver, J.; Hemminger, J. C. *J. Am. Chem. Soc.* **1992**, *114*, 2428-2432.
5. Garrell, R. L.; Chadwick, J. E.; Severence, D. L.; McDonald, N. A.; Myles, D. C. *J. Am. Chem. Soc.* **1995**, *117*, 11563-11571.
6. Chadwick, J. E.; Myles, D. C.; Garrell, R. L. *J. Am. Chem. Soc.* **1993**, *115*, 10364-10365.
7. Huang, J.; Hemminger, J. C. *J. Am. Chem. Soc.* **1993**, *115*, 3342-3343.
8. Lewis, M.; Tarlov, M. J. *J. Am. Chem. Soc.* **1995**, *117*, 9574-9575.
9. Sasaki, Y.; Nishina, Y. *Surf. Sci.* **1991**, *242*, 549-553.

OPTICAL AND ELECTROCHEMICAL PROPERTIES OF SULFUR AND IODIDE-MODIFIED SILVER SURFACES

M.S. Sibbald, G. Chumanov, and T.M. Cotton*

Department of Chemistry, Iowa State University, Ames, IA 50011 USA

Optical and electrochemical properties of nanosized metal particles are strongly influenced by adsorption of ions and molecules. For example, exposure of silver colloid to nucleophilic and electrophilic surface modifiers shifts the plasmon resonance to the red and blue spectral regions, respectively [1]. Recently we have shown that modification of silver colloids with bromide and iodide shifts their potential to values sufficient for cytochrome *c* (Cyt *c*) reduction [2]. In the present work, we have extended the experiments to include Cyt *c* reduction in sulfur-modified silver colloids and a Raman study of iodide adsorbed on silver surfaces.

Both citrate reduced and laser-ablated Ag colloids were employed. Cytochrome *c* reduction was monitored by surface-enhanced resonance Raman scattering (SERRS) and UV-Vis spectroscopies. SERRS spectra of Cyt *c* in Ag colloid mixed with different amounts of sodium sulfide are shown in Figure 1. As the concentration of S^{2-} was increased, bands at 1374 and 1363 cm^{-1} changed in intensity. These bands are markers for the oxidation state of the heme iron and correspond to Fe^{III} and Fe^{II}, respectively. In addition, changes in SERRS spectra around 1600 cm^{-1} support Cyt *c* reduction and correlate with similar changes observed for iodide-modified Ag colloid [2]. At higher S^{2-} concentrations, the Cyt *c* is nearly all reduced.

Because of the inherent surface sensitivity of SERRS spectroscopy, it is not clear whether reduced Cyt *c* is also present in appreciable amounts in the solution. This basically concerns the reducing power of the modified particles. A related study was performed using UV-Vis spectroscopy and centrifugation. A known amount of Cyt *c* in its oxidized form was mixed with sulfur-modified colloid. After standing for 15 min, the suspension was centrifuged and the amounts of reduced and remaining oxidized Cyt *c* in the supernatant were estimated from characteristic absorption features between 500-600 nm. Reduced Cyt *c* was found in amounts which were significantly more than that required for one monolayer of protein on the Ag particles but less than that required for monolayer coverage by sulfide. Thus, it seems that several S^{2-} are needed on the Ag surface for one Cyt *c* to be reduced.

Based on these results, the following mechanism for Cyt *c* reduction is proposed. Upon adsorption of S^{2-} to the Ag surface, charge is donated to the metal particle thereby changing its potential. A certain amount of preadsorbed sulfide ions is required to achieve a potential sufficient for Cyt *c* reduction. Sulfide-modified silver particles are functioning as "microelectrodes" [1].

In order to understand the nature of the metal surface-modifier interaction, Raman studies of iodide adsorbed on Ag colloid and electrochemically roughened Ag films were performed. An intriguing new phenomenon was observed when the iodide modified Ag surface was cooled to 77 K. At room temperature, the SERS spectrum of I^- adsorbed on the silver surface (Ag_s-I) consisted of a single broad band (FWHM ca. 30 cm^{-1}) around 112 cm^{-1}. This band did not significantly change with different excitation wavelengths ranging from 406.7 to 799.3 nm. However, when the sample temperature was lowered to 77 K, the spectrum obtained with 406.7 and 413.1 nm excitations changed dramatically, Figure 2. New features include a progression with a fundamental frequency at 124 cm^{-1}, first overtone at 245 cm^{-1}, etc.

Because the Ag_s-I vibration is not expected to produce an overtone progression, iodide must be present in a different form. One possibility is that polyiodide species are formed at the silver surface. Overtone progressions of ν(I-I) have been reported in Raman spectra under resonance excitation for solutions of polyiodides such as I_2, I_2^+, I_2^- and I_3^- having fundamental frequencies at 212, 238, 114 and 113 cm^{-1}, respectively [3-6]. A second possibility is that the surface monolayer of Ag_s-I behaves optically like bulk AgI. It has been reported that wurtzite-type single crystals of AgI exhibit longitudinal optical phonons at 124 cm^{-1} with one overtone at 244 cm^{-1} when excited with 441.6 nm, near-band gap resonance [7]. In this scenario, the temperature dependence of the observed effect can be explained by simple temperature-tuning of the proper resonance in AgI to the excitation frequency. Further experiments are underway to clarify these issues.

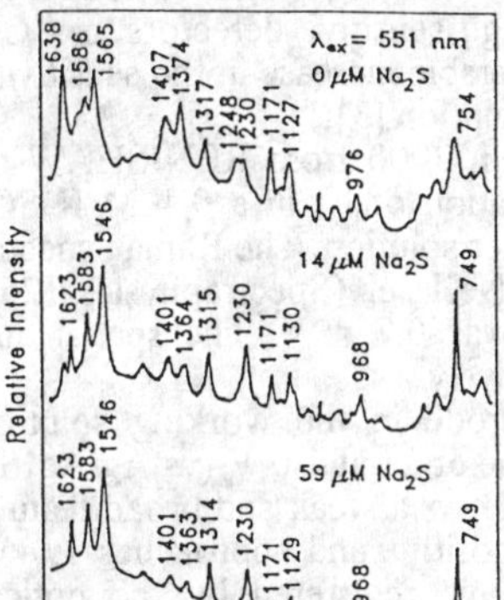

Figure 1. SERRS spectra of cytochrome *c* in aqueous Ag colloid with different concentrations of Na_2S.

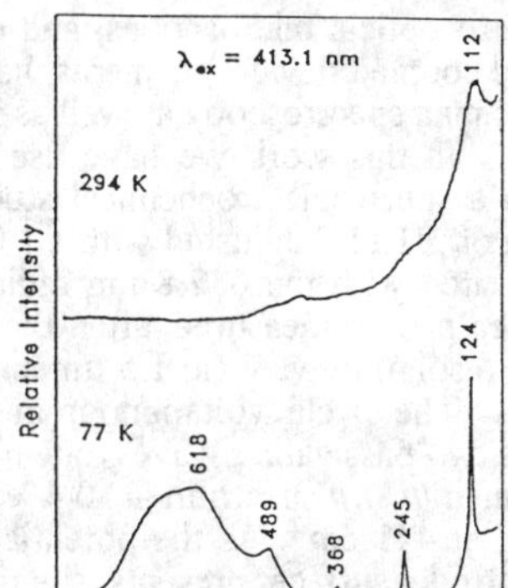

Figure 2. Raman spectra of iodide-modified Ag colloid at different temperatures.

1. A. Henglein, J. Phys. Chem. 97, 5457 (1993).
2. M.S. Sibbald, G. Chumanov, T.M. Cotton, J. Phys. Chem., in press
3. O.S. Mortensen, J. Molec. Spectrosc. 39, 48 (1971).
4. R.J. Gillespie, M.J. Morton, J. Molec. Spectrosc. 30, 178 (1969).
5. G.N.R. Tripathi et al., Chem. Phys. Lett. 113, 563 (1985)
6. W. Keifer and H.J. Bernstein, Chem. Phys. Lett. 16, 5 (1972).
7. B.Kh. Bairamov et al., Sov. Phys. Solid State 27, 333 (1985).

A *IN SITU* RAMAN SPECTROSCOPIC STUDY OF THE PASSIVATION OF NIQUEL IN ALKALINE SOLUTION

Michelle Scalabrin and Joel C. Rubim

Instituto de Química da Universidade de São Paulo,
CP. 26077, 05599-970, São Paulo, SP, Brazil.

The cross section for the Raman scattering is very low if compared to fluorescence and absorption spectroscopies. Therefore, in the absence of any kind of enhancement, e.g., resonance Raman or surface-enhanced Raman scattering, it is very difficult to observe Raman spectra of thin passive films using conventional instrumentation. For instance, in the case of passive films formed on iron electrodes, the observation of Raman spectra of the passive film was possible only for thick films [1], while for thin films it was necessary the previous deposition of a silver film to enhance the Raman signals of the passive film [2].

The development of Raman instruments with high light throughput, coupled to optical microscopes, and equipped with high sensitive detectors as CCD (charged coupled devices) cameras, has given a considerable increase in the sensitivity of the Raman spectroscopy as well as in the spatial resolution [3].

In this work we have used a Raman system 3000 from RENISHAW to perform a spectroelectrochemical study of the passivation of Ni in a 0.1 M K_2SO_4 solution of pH 11.3 adjusted with a 1.0 M NH_4OH water solution. The Raman spectra were excited with the 632.8 nm radiation from a HeNe laser (Spectraphysics 125). The laser power measured after the 80x objective was 0.7 mW. The spatial and spectral resolution were ca. 1.5 μm and 1 cm^{-1}, respectively.

The cycle voltammogram of the Ni electrode in the working solution presents two passivating waves, one near -0.6 V and the other at 1.0 V (vs Ag/AgCl). The Raman *in situ* spectrum at -0.4 V (see Figure 1) shows a weak and broad feature centered at 441 cm^{-1}. As the potential is made more positive and approaches 1.0 V, the electrode surface presents significant modifications, registered by the optical microscope, and new and relative strong Raman features appear at 484 and 558 cm^{-1} with a shoulder at 610 cm^{-1} (see Figure 1).

The Raman spectrum of the $Ni(OH)_2$ (see Figure 2) presents features at 447 and 517 cm^{-1}. Since we were not able to obtain an Ni(III) hydroxide, we performed the following experiment: a sample of $Ni(OH)_2$ was submitted to a temperature program being heated, in the presence of air, from 298 to 523 K and then cooled to 298 K. As shown in Figure 2, as the temperature increases the separation between the two main peaks of the $Ni(OH)_2$ decreases. At temperatures above 468 K only a broad band is observed close to 507 cm^{-1} and at 523 K the band becomes more defined with a main peak at 495 cm^{-1} and a shoulder at 414 cm^{-1}. If the temperature is lowered to 298 K, the spectrum remains the same.

Based on the above results we assign the Raman feature at 447 cm^{-1} on the *in situ* spectrum of the passive film to a $Ni(OH)_2$ thin layer. At potentials more positive than 0.8 V, the passive film is oxidized, i.e., a Ni(III) hydrated oxide (possibly an oxy-hydroxide) is formed given origin to the Raman features at 484, 558, and 610 cm^{-1}.

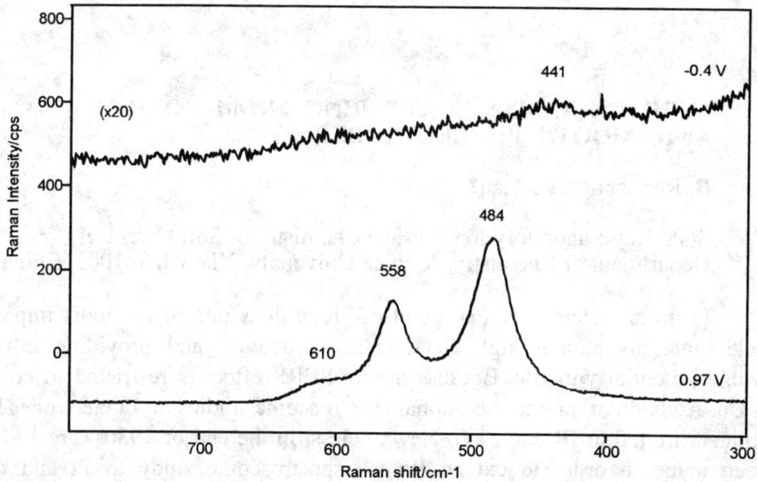

Figure 1 - *In situ* Raman spectra of a Ni electrode at the indicated potentials.

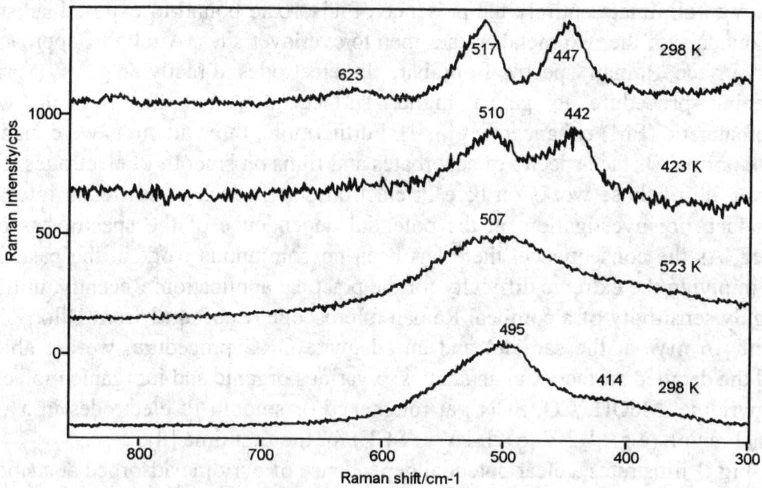

Figure 2. Raman spectra of $Ni(OH)_2$ at the indicated temperatures.

1. J. Dünnwald and A. Otto, Fresenius Z. Anal. Chem. 319 , 738 (1984).
2. J.C. Rubim and J. Dünnwald, J. Electroanal. Chem. 258, 327 (1989).
3. K.P.J. Williams, G.D. Pitt, B.J.C. Smith, A. Whitley, D.N. Batchelder and I.P. Hayward, J. Raman Spectrosc. 25, 131 (1994).

SURFACE RAMAN SCATTERING FROM ROUGHENED AND SMOOTH Pt ELECTRODES

B. Ren and Z.Q. Tian*

State Key Laboratory for Physical Chemistry of Solid Surfaces,
Department of Chemistry, Xiamen University, Xiamen 361005, China

From an electrochemical point of view, Pt is one of the most important electrode materials having high electrocatalytic activity and providing a wider polarizable potential window. Because strong SERS effect is restricted to coinage metals Cu, Ag and Au, in-situ vibrational spectroscopic studies on Pt electrodes have been almost limited to IR and SFG. Nevertheless, in the end of 1980s, great efforts had been made in order to extend Raman spectroscopic study to Pt and other transition metals. One approach involves coating SERS active electrodes with very thin Pt [1] or Ni [2] overlayer by electrochemical deposition. However, it has the inevitable limitation for practical applications mainly because the extremely difficulty to eliminate entirely the presence of adsorbate bound to exposed substrate or to boundary of the two metals rather than to overlayer sites. A reliable approach is to get surface Raman spectra from bare Pt electrodes directly, e.g., by surface roughening procedure to get a higher surface area and the possible weak electromagnetic (EM) enhancement [3, 4]. Furthermore, three attempts were made to observe normal surface spectra of adsorbates and films on smooth Pt electrodes [5-7]. However, all of those works on bare Pt electrodes were lack of sufficient intensity, and in fact, no investigation on the potential dependence of the spectra has been reported. As the consequence, there has been no continuous work in the past three years, implying the extreme difficulty for the practical application. Recently, utilizing the highly sensitivity of a confocal Raman microscope (LabRam I from Dilor, 632.8 nm with 16 mW at the sample) and an adequate ORC procedure, we are able to extend the detailed surface Raman studies to various organic and inorganic molecules (e.g., pyridine, MeOH, CO, SCN^-) at roughened or smooth Pt electrodes in a wide potential region (e.g. -1.2 V to +1.4 V *vs* SCE) for the first time [8].

Fig. 1 illustrates a clear potential dependence of pyridine adsorbed at a smooth Pt electrode, a broad band at around 1640 cm^{-1} is from bulk water. The potential profile is different remarkably from that observed from the roughened surface [8]. It seems that only the perpendicular orientation can provide the strong surface signal on the smooth surface in the negative potential range while both flat and perpendicular orientations can be observed from the roughened surface in a wide potential region.

Raman spectra with sufficiently high intensity from SCN^- adsorbed at rough platinum surfaces were obtained in a wide potential region (-1.2 V to +1.4 V). This potential range of the adsorption is much wider than that for Ag, Au and Cu electrodes. It can be divided into three regions: (i) from -1.2 V to 0 V, SCN^- is adsorbed through nitrogen and the frequency changes from 2075 to 2125 cm^{-1} with the rate of 40 cm^{-1}/V; (ii) from 0 V to +1.2 V, some SCN^- change the orientation to the sulfur end (inferred by a strong band at ca. 290 cm^{-1}) and the band width increases gradually at around 2183 cm^{-1}; (iii) at +1.4 V, SCN^- is desorbed due to the formation of the oxide film indicated by a broad band at ca. 587 cm^{-1}.

This work is supported by the National Science Foundation of China.

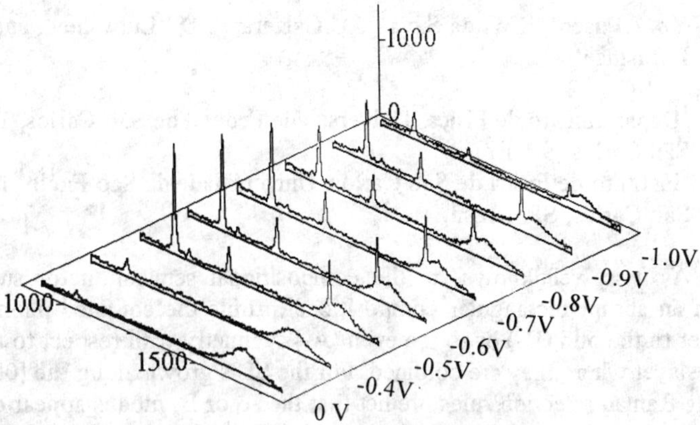

Fig. 1. Potential dependence of Raman spectra from pyridine adsorbed at a smooth Pt surface. Solution: 0.01 M pyridine and 0.1 M KCl.

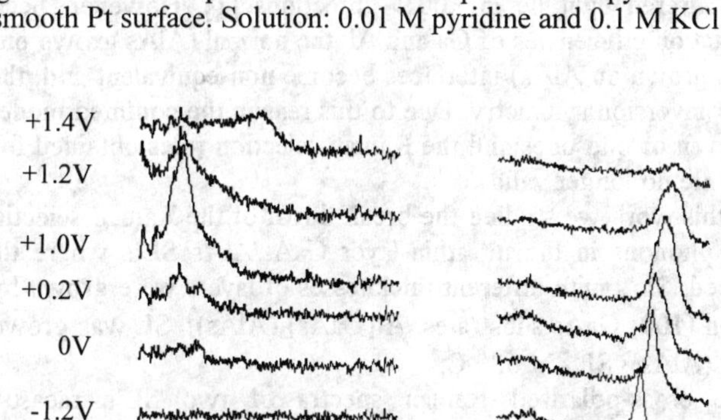

Fig. 2. Potential dependence Raman spectra from SCN^- adsorbed at roughened Pt surface. Solution: 0.1 M NaSCN.

1. L.-W.H. Leung, M.J. Weaver, J. Am. Chem. Soc., 109, 5113 (1987).
2. M. Fleischmann, Z.Q. Tian, L.J. Li, J. Electroanal. Chem., 217, 397 (1987).
3. R.P.Cooney, M.Fleischmann, P.Hendra, J.Chem.Soc.Chem.Commun.,235 (1977).
4. S.A.Bilmes, J.C.Rubim, A.Otto and A.J.Arvia, Chem. Phys. Lett, 159, 89 (1989).
5. M.Fleischmann, D.Sockalingum, Spectrochim.Acta., 46A, 285 (1990).
6. B. Pettinger, A. Friedrich, U. Tiedemann, J. Electroanal. Chem., 280, 49 (1990).
7. M.A. Bryant, S.L. Joa, J.E. Pemberton, Langmuir, 8, 753 (1992).
8. B. Ren and Z.Q. Tian, submitted to J. Electroanal. Chem.

BREAK-DOWN OF RAMAN SELECTION RULES FOR OPTIC CONFINED PHONONS CAUSED BY THE DIFFERENT ARRANGEMENTS OF THE NORMAL AND INVERTED INTERFACES IN GaAs/AlAs SUPERLATTICES

Yu.A.Pusep[1], S.W. da Silva[2], J.C.Galzerani[1], D.I.Lubyshev[2], and P.Basmaji[2]

[1]Departamento de Física, Universidade Federal de São Carlos, 13565-905 São Carlos, SP, Brasil
[2]Instituto de Física de São Carlos, Universidade de São Paulo, 13560-970 São Carlos, SP, Brasil

As it is well known, in the compositional semiconductor superlattices (SL's) with an abrupt rectangular compositional profile the confined phonon modes belong either to the odd (B_2) or to the even (A_1) symmetry with respect to the central plane of the layer where they are confined. For the SL's grown along the [001] crystal direction the Raman selection rules predict that the A_1 or B_2 modes appear only in the polarised or in the depolarised configurations respectively when the polarisations of incident and scattered light lie in <001> directions [1]. However, because of the different segregation efficiencies of Ga and Al the normal (AlAs grown on GaAs) and inverted (GaAs grown on AlAs) interfaces become non-equivalent and, therefore, the layers lose their inversion symmetry. Due to this reason the confined modes cannot be considered as even or odd ones and the Raman selection rules obtained for a SL with ideal interfaces are no longer valid.

In this work we studied the break-down of the Raman selection rules for optic confined phonons in the ultrathin-layer GaAs/AlAs SL's where this effect is more pronounced. SL's with different thicknesses of layers were grown by molecular beam epitaxy on (100) GaAs substrates. A $(GaAs)_3(AlAs)_1$ SL was grown at 580 C, while a $(GaAs)_4(AlAs)_2$ SL - at 680 C.

The cross-polarised Raman spectra of two SL's measured in the frequency range of GaAs phonons are shown in Fig.1. In order to analyse the activities of the confined modes in the Raman spectra of SL's with different normal and inverted interfaces, we used the local mode approach of the next-nearest-neighbour linear chain model where the cation intermixture in the same monolayer has been taken into account [2]. The arrows in the top panels indicate the frequencies of confined modes calculated: I - for abrupt interfaces (nominal profile), and II - for the compositional profile obtained by the best fitting. In accordance with the kinetic segregation model [3] the higher the growth temperature the stronger the deviation of the real compositional profile from the nominal one.

Our calculations give the evidence that in spite of the fact that the selection rules allow the Raman scattering for the odd modes only, in both samples we observed second LO_2 modes rather than third LO_3 ones.

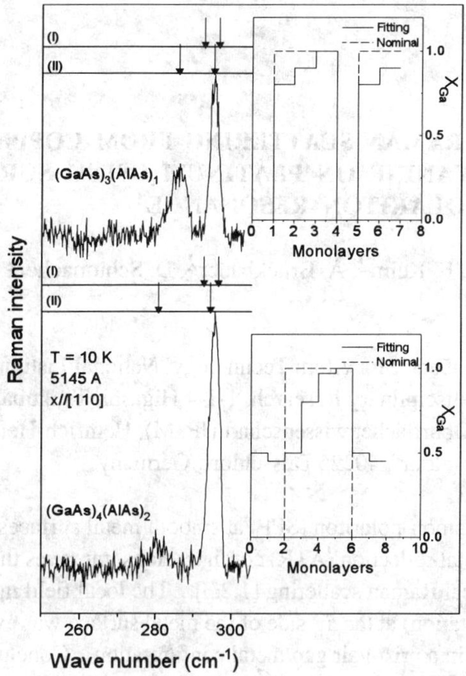

Fig.1. The depolarised Raman spectra measured in the $(GaAs)_3(AlAs)_1$ and in the $(GaAs)_4(AlAs)_2$ SL's at T=10K. The insets show the compositional profiles of corresponding SL's. Arrows are explained in the text.

Moreover, the intensities of the Raman lines corresponding to the LO_1 and LO_2 confined modes measured in the depolarized spectra with x,y || <100> were in good agreement with the calculated values of their macroscopic polarisations: in the (4,2) SL we measured $I(LO_1)/I(LO_2) \cong 3.9$, while the ratio of the corresponding polarisations gives the value 3.67; in the (3,1) SL these values were 1.9 and 1.95 respectively.

Thus our results testify the breakdown of the Raman selection rules in the GaAs/AlAs SL's due to the difference between the normal and inverted interfaces. In this case the activities of the confined modes can be estimated by the calculation of their macroscopic polarisations.

1. B.Jusserand and M.Cardona in Light Scattering in Solids V (M.Cardona and G.Güntherodt, ed.), Springer-Verlag, Berlin, Heidelberg, 1989, p.113.
2. Yu.A.Pusep, S.W. da Silva, J.C.Galzerani, A.G.Milekhin, V.V.Preobrazhenskii, B.R.Semyagin, and I.I.Marahovka, Phys.Rev.B52, 2610 (1995).
3. O.Dehaese, X.Wallart, and F.Mollot, Appl.Phys.Lett. 66, 52 (1995).

ENHANCED RAMAN SCATTERING FROM COPPER PHTHALOCYANINE ON PLATINUM USING SURFACE-PLASMON-POLARITON RESONANCE

M. Futamata[1*], E. Keim[2], A. Bruckbauer[2], D. Schumacher[2] and A. Otto[2*]

[1]Joint Research Center for Atom Technology, National Institute for Advanced Interdisciplinary Research, 1-1-4 Higashi, Tsukuba 305, Japan
[2]Lehrstuhl für Oberflächenwissenschaft (IPkM), Heinrich-Heine-Universität Düsseldorf, 40225 Düsseldorf, Germany

The surface-plasmon-polariton (SPP) at smooth metal surfaces, in- and out-coupled in attenuated total reflection (ATR) configurations improves the inherently limited sensitivity in classical Raman scattering [1,2,3]. The local field intensity $|E|^2$ due to the in-coupling (excitation) at the air side of the metal surface was evaluated for various dielectric constants in a prism/air gap/metal configuration. Conclusively the real part Re ε_{metal} is much more important than the imaginary part Im ε_{metal}, since $|E|^2$ is approximately proportional to $|\text{Re } \varepsilon_{metal}|^2$, and inversely proportional to Im ε_{metal} for $15 > |\text{Re } \varepsilon_{metal}| > 5$ or Im $\varepsilon_{metal} > 2$. Accordingly, the promising enhancement within the incident channel of more than 30 is predicted even for highly damping metals, if $|\text{Re } \varepsilon_{metal}|$ is sufficiently large. Out-coupling by the ATR configuration yields a further enhancement of about 15 [4]. Thus, theoretical evaluation predicts that even intrinsically strong damping metals give a promising enhancement of the Raman band intensity from adsorbates. Therefore, we applied the ATR method to platinum, whose dielectric constants ε_{Pt} is -8.34 + i14.2 at 514.5 nm.

The dielectric constants of the metal and of the solid CuPc, and thicknesses of the air gap and the CuPc film determine the dispersion of the SPP, and the enhancement of the electric field at the metal surface. In-coupling of the SPP was clearly observed as a sharp minimum of the reflectivity only for p-polarized light. For instance, the observed resonance angle $\alpha_{SPP}^{obs} = 34.5^\circ \pm 0.2^\circ$, optimum gap size $d_{opt} = 2.5\lambda_L/4$ and FWHM = 6.4° for t_{CuPc} = 3.5 nm are in good agreement with the theoretical values, $\alpha_{SPP}^{calc} = 34.6^\circ$, $d_{opt} = 2\lambda_L/4$ and FWHM = 8.4° within the experimental uncertainty. Raman bands from CuPc, e.g. at 1530 cm^{-1} (A_{1g}, $\nu(C_\alpha$-$N_\alpha)$+$\delta(C_\alpha$-N_β-$C_\alpha)$), display the maximum intensity at the reflectivity minimum. Consequently, the Stokes shifted light of CuPc on platinum is prominently enhanced by the excitation of the SPP using p-polarized light at $\alpha_i = \alpha_{SPP}$ and $d = d_{opt}$.

Moreover, the enhancement factor for the Raman band intensity increases with decreasing CuPc thickness due to intrinsic damping of the SPP field within the CuPc layers as shown in Table 1. Note that platinum yields a pronounced enhancement of ca. 300 at t_{CuPc} = 3.5 nm [5], almost the same as for the silver system [3], despite extremely larger internal damping than in the case of silver: Im ε_{Pt} (= 14.2) > Im ε_{Ag} (= 0.342). The observed enhancement is compatible with the theoretical values only for parallel orientation of CuPc molecular plane to the Pt surface as shown in Table 1. It suggests that one can obtain the orientation information from the anisotropy of the SPP field, and of the Raman tensor of the totally symmetric ring breathing mode of CuPc on Pt. In addition, the depolarization ratio for the 1530 cm^{-1} band diminishes with decreasing CuPc thickness: from 0.50 at 17.1 nm to 0.20 at 3.1 nm. Theoretical values for the same band are 1.00, 0.33, and 0.0 for perpendicular, random and parallel orientation of the molecular plane with respect to the surface. Therefore CuPc films increasingly well ordered with decreasing thickness, in contrast to CuPc films on evaporated silver [3] or gold films, yield the extremely large enhancement factor. Similarly, Raman bands from p-nitrothiophenol on Pt give the factor of ca. 100 under the SPP resonance. In conclusion, the ATR method improves the sensitivity level of Raman scattering from adsorbates even on highly damping transition metals.

Table 1 Enhancement factor (G) for Raman band intensity at 1530 cm^{-1} from sapphire/air gap/CuPc (t nm)/Pt sample using SPP resonance at 514.5 nm

t (nm)	Intensity (cps/mW) ATR	external	G_{exp} †	G_{theory} ‡ random	parallel	perpendicular
3.5	1891	6.56	288	47.0	169	9.3
8.6	6000	36.6	164	48.1	178	9.1
17.1	6447	73.0	88.3	36.4	133	7.2

† The peak intensities observed under optimized ATR and external geometry at the angle of incidence of β_i (= arcsin {(sin α_{SPP})/n_1} [3]. ‡ Theoretical values under the SPP resonance for various orientation of CuPc to the Pt surface.

References

1. M. Futamata, P. Borthen, J. Thomassen, D. Schumacher, and A. Otto, Appl. Spectrosc. 48 (1994) 252, and references therein.
2. M. Futamata, J. Phys. Chem. 99 (1995) 11901.
3. M. Futamata, Langmuir, 11 (1995) 3894.
4. J. Thomassen, Diplomarbeit, Universität Düsseldorf, 1988.
5. M. Futamata, E. Keim, A. Bruckbauer, D. Schumacher and A. Otto, Appl. Surf. Sci. in press.

IN SITU RAMAN STUDIES OF SURFACE-MODIFIED ELECTRODES: FeTPP IN A POLYMER MATRIX.

S.E.J. Bell, M.D. Devenny, J.G. Grimshaw, J.T. Grimshaw, S. Hara, J.J. McGarvey and J.H. Rice*.

School of Chemistry, The Queen's University of Belfast, Belfast, BT9 5AG, N. Ireland.

Many haem enzymes contain iron porphyrin complexes held within a polymeric protein matrix through coordination by an axial ligand. In order to model the structure and ultimately the function of such enzyme systems we have prepared poly(amino-acids) with imidazole functionalised pendant arms, and generated electroactive films by using these polymers as matrices to support iron porphyrins on electrode surfaces [1]. These surface modified electrodes contain iron porphyrins with one or two axial imidazole ligands and can be prepared for a wide range of porphyrins. The doping level of imidazole side arms in the poly(γ-ethyl glutamate) (shown below) and the oxidation state of the central metal ion can be controlled independently.

The films are prepared by spray coating a mixed solution of the iron porphyrin and the polymer directly onto the electrode surface, which is roughened silver in the case of surface-enhanced Raman spectroscopy. To characterise the polymer *in situ* Raman/electrochemical measurements are made in a conventional three-electrode cell. SERRS spectroscopy is used to characterise the species present near the surface while resonance Raman measurements probe the "bulk" of the films.

Spraying Fe(TPP)Cl along with a highly-doped (60% pendant arms imidazole doped) polymer gives a PEG-Im film containing six-coordinate low-spin Fe(III)TPP with the characteristic Raman bands [2] shown in Fig. 1(a). The films are electroactive and can be reduced to the Fe(II) form as shown in Fig 1(b), the changes are reversible.

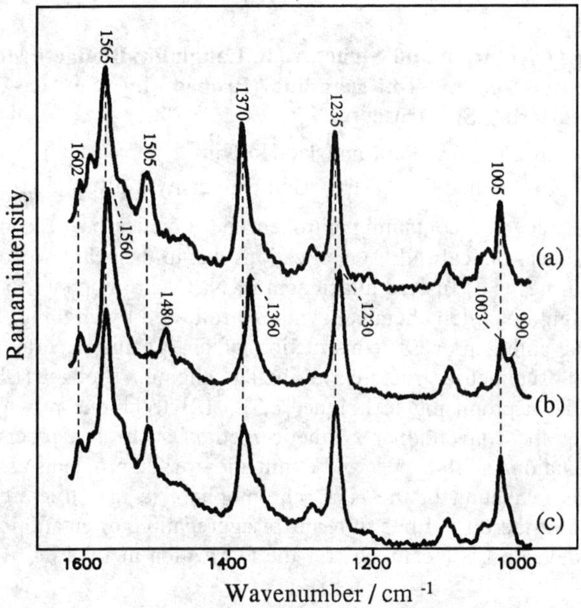

Fig.1. FeTPP.Cl in PEG-Im, resonance Raman spectra, λ_{ex}= 457.9nm. Applied potential (a) 0 mV, (b) -600 mV, (c) 0 mV, in 0.1 M NaCl/water.

The spectra shown in Fig 1 are resonance Raman spectra and therefore probe the "bulk" film. However, when surface-enhanced Raman spectra of thinner films prepared from the same solutions are recorded they show that high-spin Fe(III)TPP is present at the surface. This high-spin, presumably 5-coordinate, Fe(III) complex can again be reversibly reduced. This observation highlights the fact that surface modified electrodes are not necessarily homogenous throughout their entire thickness and rather different species can be present at different distances from the electrode surface.

The most desirable systems for applications of these electrodes as sensors or catalysts would contain exclusively 5-coordinate complexes, similar to those found near the surface, to provide binding sites for other ligands. We have been pursuing several strategies to achieve this. The most obvious route is to use a polymer with a much lower imidizole level but we have found that this leads to formation of films which contain mixtures of high- and low-spin complexes of Fe(II) and Fe(III). We now generate such films using either sterically hindered porphyrins or polymer-bound substituted imidazole ligands which prevent coordination of both axial sites.

1. S.E.J. Bell, M. D. Devenney, J.G. Grimshaw, and J. Trocha-Grimshaw, J. de Physique IV 4,157 (1992).
2. J.M. Burke, J.R. Kincaid, S. Peters, R.R. Gagne, J.P. Collman and T.G. Spiro, J. Am. Chem. Soc. 100, 6083 (1978).

Molecular Orientation and Structure in Langmuir-Blodgett Monolayers of N-docosyl-2-methyl-4-nitroaniline Probed by Surface-Enhanced Raman Scattering Spectroscopy

Jung-Hyuk Im, Sung-Ho Park and Jae-Ho Kim*

Department of Applied Chemistry, Ajou University, Suwon, 442-749, Korea

Organic materials containing nitrobenzene moieties have been extensively studied for nonlinear optical (NLO) applications due to their large second and third order susceptibilities [1]. Instead attachment of NLO chromophore to the polymer backbone through multi-step chemical synthesis followed by aligning them through the complicated poling process, construction of an organic superlattice with the required noncentosymmetric structure by utilizing Langmuir-Blodgett (LB) deposition has proven to be a promising technique [2,3]. LB technique provides a unique advantages over the conventional synthetic method such as a precise controlled molecular orientation and thickness and elimination of the problems associated with the spontaneous relaxation of the NLO chromophore in the structure. Structural characterization of these LB films in terms of aggregation, orientation and structure is essential to develop LB techniques for the fabrication method of NLO elements.

In this paper we report on surface-enhanced Raman scattering (SERS) spectroscopic studies carried out to elucidate the aggregation, orientation and structure of the n-docosyl-2-methyl-4-nitroaniline (DCMNA) LB films at the air-water interface and structural alteration upon transferring to a solid substrate. Raman and SERS spectroscopy are powerful nondestructive structural probes provide an insight to the conformation of the molecules in the LB films and to the interaction between a substrate and molecules [4]. DCMNA synthesized from 2-methyl-4-nitroaniline in out laboratory formed a well define and reproducible Langmuir monolayer at the air-water interface only under the specific condition. The optimum conditions for construction of a stable monolayer at the air-water interface and for transferring to a solid substrate were determined through systematical changes of the important parameters such as temperature, pH of the subphase and the rates of compression at the interface and transfer to the substrate. The effects of the arachidic acids on the compressibility of the DCMNA monolayer on the water surface and the structure of the DCMNA LB monolayer were also investigated. The surface pressure for collapsing DCMNA monolayer on a neutral subphase was not detectable from the subphase above 15 °C. The other hand the subphase temperature controlled at 10 °C or lower the monolayer maintained its organized structure until the pressure reached at 20 mN/m. The stability of the DCMNA monolayer at the air-water interface as well as the transfer efficiency to a solid substrate were significantly improved by forming a mixed monolayer of DCMNA and AA as shown in Fig.1. At the mole ratio of 1:1 the compressibility was increased by 25 mN/m comparing to the pure DCMNA monolayer. Fig. 1(e) shows DCMNA-AA mixed monolayer in the mole ratio of 1:10 collapsed above 50 mN/m. Figure 2 shows a typical Raman spectra of solid DCMNA (a) and SERS of DCMNA monolayers (b-d). The solid Raman spectra and SERS of DCMNA self-assembled (SA) monolayer are almost identical indicating unorganized random orientation of the molecules. The bands in the region between 1250 and 1320 cm^{-1} in fig. 2 (a) and (b) are the vibration modes associated with the $\delta(NO_2)$, $\gamma_a(NO_2)$, $\gamma_s(NO_2)$ and ν(C-C). The SERS spectra of both DCMNA and the AA mixed DCMNA LB monolayers shows strong bands from $\nu(NO_2)$, $\nu(\Phi\text{-}NO_2)$ modes indicative of the close proximity of the NO_2 group to the Ag substrate and pseudo quinoidic state of MNA moieties due to strong interaction of NO_2 group to the Ag metal. Furthermore strong bands at 1608 cm^{-1} region, a benzene ring stretching mode, suggests the vertical orientation of benzene ring on the substrate in the LB monolayer systems.

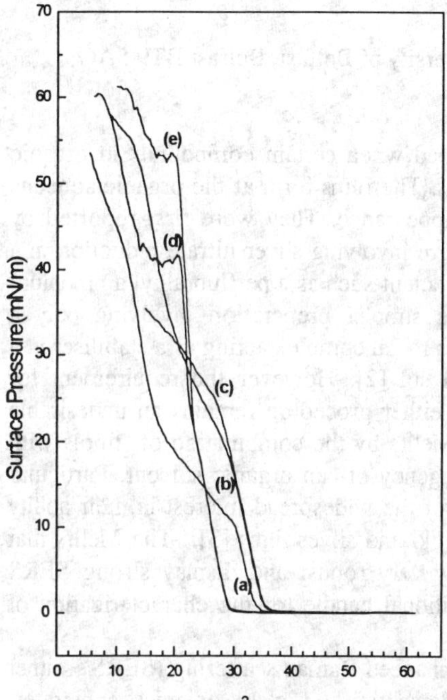

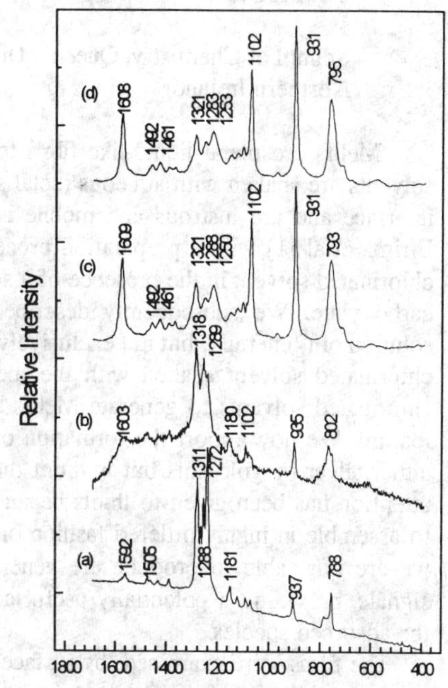

Figure 1. Surface pressure-area isotherms of pure DCMNA (a) and the mixed monolayer of DCMNA and AA in the mole ration 1:0.5 (b), 1:1 (c), 1:3 (d), and 1:10 (e). (subphase temp: 10 °C, concentration of DCMNA: 1.0 x 10^{-3} M in $CHCl_3$, compression rate: 5 cm^2/min)

Figure 2. SERS spectra of DCMNA in solid form (a), in self-assembled (SA) monolayer (b)*, in the pure LB monolayer (c) and the mixed LB monolayer (d)**. (SERS substrate: roughened Ag electrode, λ_{ex}: 514.5 nm, power: 10 mW, transfer pressure: 20 mN/m, transfer rate: 1.2 mm/min) *SA monolayer was formed by dipping SERS substrate for 5 min into the spreading DCMNA solution. **The mole ratio of DCMNA and AA (1:1)

1. H. Menzel, M. Hallensleben, A. Schmidt, W. Knoll, T. Fischer and J. Stumpe, Macromolecules, 26, 3644 (1993)
2. J. Lam and P. Stroeve, Thin Solid Films, 243, 361 (1994)
3. V. A. Howarth, N. Asai, N. Kishii and I. Fujiwara, Appl. Phys. Lett. 61(14), 1616 (1992)
4. J.-H. Kim, T. M. Cotton, R. A. Uphaus and D. Möbius, J. Phys. Chem. 93, 3713 (1989)

DIRECT FORMATION OF SILVER AND GOLD METAL LIQUID-LIKE FILMS (MELLFs) FROM THIOLS AND SOLS WITHOUT ORGANIC SOLVENT: SERS AND AFM STUDIES

A.H.R. Al-Obaidi*, S.J. Rigby, J.N.M. Hegarty, S.E.J. Bell and J.J. McGarvey

School of Chemistry, Queen's University of Belfast, Belfast BT9 5AG, Northern Ireland

Mellfs are metal liquid-like films formed when certain compounds in organic solvents are shaken with aqueous metal sols. The films form at the organic:aqueous interface and are lustrous and mobile in appearance. They were first reported by Efrima et al [1] with a preparation procedure involving silver nitrate reduction in a chlorinated solvent in the presence of a surfactant such as a perfluoralkyl ammonium carboxylate. We subsequently described a simpler preparation involving only a solution of (generally, but not exclusively) a metal complex acting as a stabiliser in a chlorinated solvent shaken with the metal sol [2]. However the requirement for chlorinated solvents to generate Mellfs by either procedure remains an undesirable feature. We now report the formation of Mellfs by the combination of thiols with either silver or gold sols but without the agency of an organic solvent. Particular attention has been given to thiols because of the widespread interest in their ability to assemble in highly ordered fashion on gold and silver films [3]. The Mellfs that we are now able to produce are generally very robust and display strong SERS signals, providing a potentially useful additional handle for the characterization of the adsorbed species.

The films were examined by surface-enhanced Raman scattering (SERS) either in the mobile interfacial state or after deposition on glass or mica substrates. Deposited films on both substrates were also examined by atomic force microscopy (AFM). SERS spectra were recorded for both gold and silver Mellfs and for comparison, resonance Raman (RR) spectra of the neat thiol. Fig. 1 shows the results for silver Mellf. To achieve plasmon resonance, the Ag Mellfs spectra (upper trace) were excited at 457.9 nm while those for the Au Mellfs were excited at 690 nm with a Ti-sapphire laser. High quality SERS spectra were obtained. Bulk spectra at the same excitation wavelength are also shown. Comparisons show several striking features of which the principal are that the intensity of the bands around 1570 and 1092 cm^{-1} double in the SERS spectra and shift down in frequency by 10 and 20 cm^{-1} respectively compared to the bulk phase. The 1570 cm^{-1} feature is presumably a phenyl ring breathing mode while the 1092 cm^{-1} band has been assigned by Pemberton et al [4] to a ring-breathing mode coupled to a C-S stretching vibration. Spectra at lower and higher frequencies have also been examined, (Fig. 1). A band at 2577 cm^{-1} due to a ν(S-H) stretch appears in the spectra of the bulk thiol. The virtual disappearance of this feature in the Mellfs spectra is as would be expected if the thiol is surface bonded via the S atom to the silver. Several strong bands appear at lower frequencies, with a mode at 367 cm^{-1} tentatively attributed to ν(Ag-S).

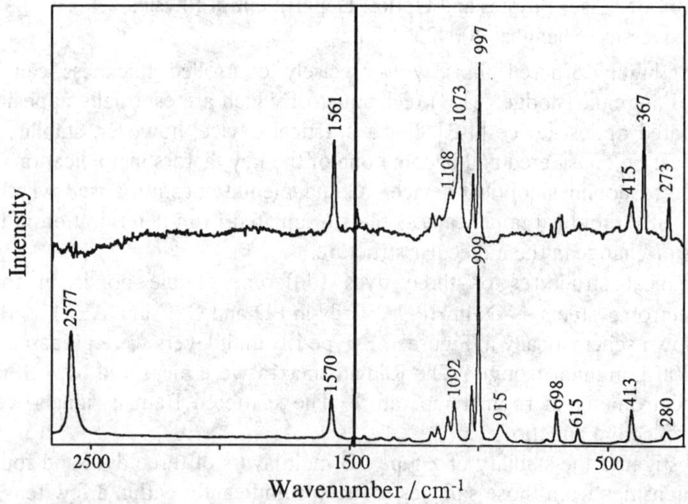

Fig. 1. Comparison of SERRS (from silver MeLLF), upper and RRS (lower) of thiophenol recorded in two regions, as indicated with λ_{ex}= 457.9 nm. laser power 25 mW; signal accumulation 3x 20 s; slit width *ca.* 5 cm^{-1}.

Evidence that SERS spectra were actually being observed on the Mellf surface was checked in the case of Au samples by recording the latter spectra also at 457.9 nm, well away from plasmon resonance. Only spectra characteristic of bulk sample were seen in this instance. SERS spectra on Mellfs surfaces prepared with and without an organic solvent were identical, showing that the absence of solvent had no influence.

AFM scans of Mellfs deposited on mica have also been recorded. They are not reproduced here but will be presented at the conference. Indicated "particle" sizes for the Ag Mellfs are < 100 nm.

The question of the orientation of the molecules at the surface requires more detailed consideration. The appearance of a strong band at 367 cm^{-1} and the shift to lower frequency of the 1092 cm^{-1} mode could be indicative [4] of an orthogonal or tilted orientation but such a conclusion remains tentative at present.

Acknowledgement We thank the EPSRC for support (Grant GR/J78969)

1. D. Yogev, S. Efrima, J. Phys. Chem. 92, 5754 (1988). For a review see S. Efrima Crit. Rev. Surf. Chem. 1, 167 (1991).
2. K.C. Gordon, J.J. McGarvey, K.P. Taylor, J. Phys. Chem. 93, 6814 (1989)
3. C.D. Bain and G.M. Whitesides, Angew. Chem. Int. Ed. 28, 506 (1989)
4. M.A. Bryant, S.L. Joa and J. E. Pemberton, Langmuir 8, 753 (1992)

STUDY OF TEMPORAL STABILITY OF THREE ORGANIC DYES WITH RAMAN SPECTROSCOPY

Shihong MA*, Xingze LU, Liying LIU, Wencheng WANG, Zhiming ZHANG

Laboratory of Laser Physics and Optics, Department of Physics, Fudan University, Shanghai 200433, China

Ultrathin, highly ordered films with precisely controlled thickness can be fabricated using Langmuit-Blodgett (LB) technique, of which are especially appealing for use in integrated optics devices[1-3]. For a practical device, however, stability of the such films must be considered and become one of the key factors in applications of the LB technique to nonlinear optical device. A fundamental question arised whether the deterioration of the optical nonlinear LB films originated from deterioration in the film frameworks or change in the molecular structure.

The chemical structures of three dyes (different double bonds in their conjugated π-electron systems, N=N in HPNA, C=N in PD and C=C in DAEP) used in this work are shown schematically in Figure 1. Z-type LB multilayers were prepared by using a KSV 5000 Langmuir trough. The Raman spectra were measured by a SPEX 1403 Raman spectrometer at room temperature. The scattered Raman signals were collected in the reflective direction.

We investigated the stability of z-type LB multilayers of three dyes and found that the SHG intensities from those samples reduced significantly within a few tens of days and then reached a saturation. The time constant for such reduction was different: 20 days for HPNA, 30 days for PD and 50 days for DAEP. On the other hand, it was found that the ratio Ipp/Isp kept almost constant in time in all three dyes samples and the small-angle X-ray diffraction patterns from those films changed little in time, suggesting invariant films structure.

The change of the Raman spectra from HPNA powder sample with the time of irradiation is shown in Figure 2. The 1610cm^{-1} peak was due to the breathing vibration of the two benzene rings in HPNA molecules, while the 1384cm^{-1} peak could be assigned to the stretching vibration of the N=N bond. It was found that the intensity of the 1610cm^{-1} line changed very little with increasing irradiation time, due to the stable structure of the benzene ring. In contrary, the intensity of the 1384cm^{-1} line decreased gradually as the time of irradiation increased, suggesting the breakdown of the N=N double bond under the UV irradiation (photobleaching). Similar behavior for C=N bond in PD and C=C bond in DAEP were also observed by Raman measurements.

Figure 3 showed the intensity reduction of the stretching modes of the double bonds 1384cm^{-1} (HPNA), 1626cm^{-1} (PD) and 1614cm^{-1} (DAEP) lines with the irradiation time. The sequence of the bleaching rate in the three dyes is different: HPNA the fastest and DAEP the slowest. Both SHG and Raman experiments deduced the same sequence of stability: N=N (HPNA) the worst and C=C (DAEP) the best, suggesting that optical bleaching effects, instead of deterioration in molecular frameworks, could possibly be responsible for the instability of optical SHG signals from LB films.

References:
[1] P. N. Prasad and D. J. Willimas, "Introduction to Nonlinear Optical Effects in Molecules & Polymers", Wiley-interscience, New York, 1991.
[2] D. Lupo, W. Prass, U. Scheunemann, A. Laschewsky, H. Ringsdorf, I. Ledoux, J. Opt. Soc. Am. B, 5(2) (1988) 300
[3] Shihong MA, Xingze LU, Jiabiao ZHENG, Wencheng WANG, Zhiming ZHANG, Thin Solid Films, 254(1-2) (1995) 263.

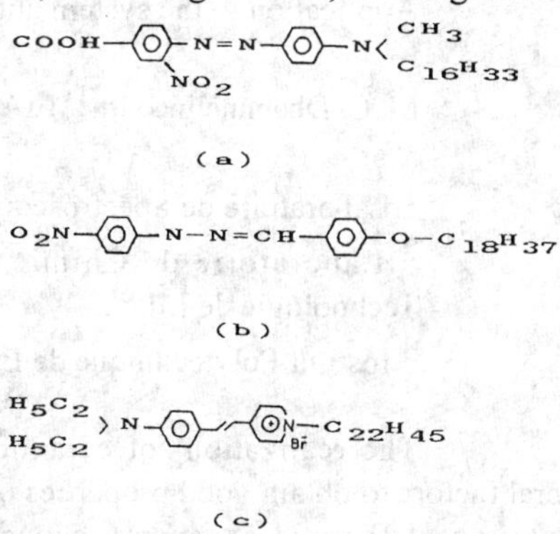

Figure 1. Molecular structures of

(a) HPNA, (b) PD, (c) DAEP.

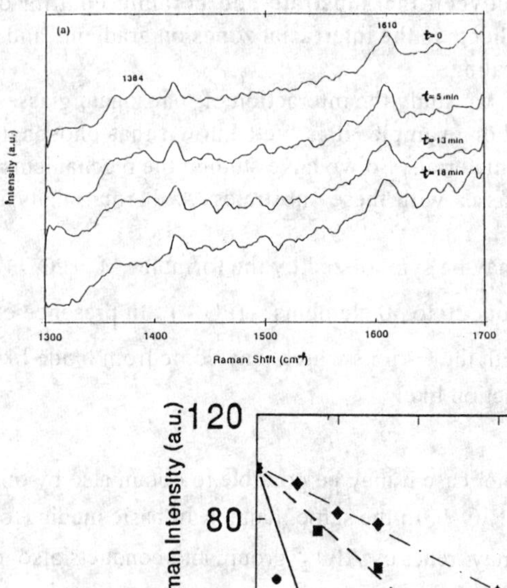

Figure 2. The Raman spectra of HPNA.

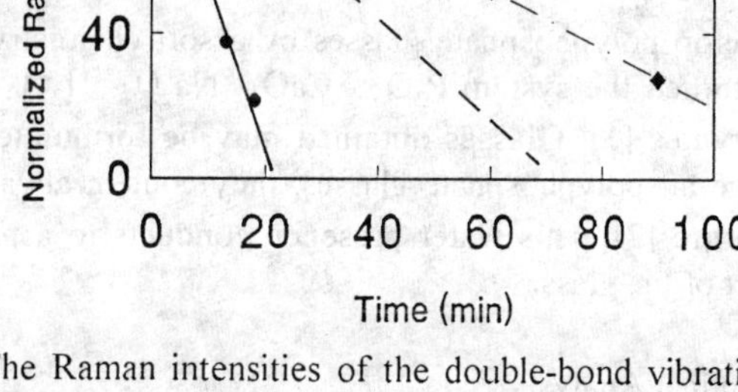

Figure 3. The Raman intensities of the double-bond vibration modes (1384cm^{-1} in HPNA, 1626cm^{-1} in PD, 1614cm^{-1} in DAEP) vs. the time of UV irradiation.

RAMAN MICROPROBE : AN IMPORTANT METHOD TO CONTROL THE STUDIES OF INTERFACE SYSTEM BY T.O.A.*

Application to the system alumina polyphosphate glasses.

M. C. Dhammelincourt [1], F. Alleene[2], O. Luczak[2], D. Ionescu [3], P. Vast[2]*

[1] Laboratoire de Spectroscopie Infrarouge et Raman UPR2631 .

[2] Laboratoire de Chimie Appliquée Université des Sciences et Technologie de Lille .

[3] Institut Polytechnique de Bucarest.

The realization of ceramic (or glass)/ceramic sealing depends on several factors to obtain good properties of adhesion. W e can give for example : the adaptation of thermal expansion between the substrate and ceramic coating or joining, the dependence of stress gradients in the interfacial zones on gradients and microstructure that form during the sealing.

In the Applied Chemistry Laboratory, we study the interaction of phosphate glasses with different ceramic substrates. For example, it is well known that phosphate glasses may be used to join alumina materials. So we have studied the mechanisms of the adhesion of polyphosphate glasses with these substrates. A thermal analysis technic was used: the T.O.A.*

Polyphosphate glasses may be symbolized by the formula $M_n^+(PO_3^-)_n$. they are formed by long chains of about 20 to 30 elements (PO_3^-) . In presence of oxides, the (PO_3^-) group may react with the O^{2-} ions which may come from oxide like Al_2O_3. In this case we may obtain reaction like:

$$(PO_3^-) + O^{2-} \text{ ---> } PO_4^{3-}$$

and obtain orthophosphate ions. In other case it may be possible to obtain also by this oxo reaction pyrophosphate groups - $P_2O_7^{4-}$. In the same manner, in basic media (for example in CaO presence), alumina may react by AlO_2^- group, and conducts also to the same species.

We develop polyphosphate glasses by a soft chemistry route, and in this paper we have studied the system P_2O_5 - CaO - Na_2O - H_2O obtained from polyphosphate coacervates [1]. Glasses obtained may be formulated in the CaO/ Na_2O ratio, but as the all polyphosphate glasses, they content always some water even at high temperature [2]. This water presence conducts to a phenomenon of crystallization of a part of the glass.

* T.O.A. : Thermo Optical Analysis

The interaction between different kinds of alumina and this polyphosphate glass has been studied by TOA* method. A little bit of alumina is put on a drop of polyphosphate glass in a microfurnace that may be heated up to 1300°C. The evolution of the biphasic system is observed with a magnifier associated with a TV camera. This TOA system permits us to observe several things.

One hand, with the camera above the system we can determine the evolution of the interface between polyphosphate glasses and different kinds of alumina. These observations are associated of course to preliminary studies by Differential Thermal Analysis which gives the thermal characteristics of glasses and, also the potential reactivity of the couple of materials.In the other hand, by an horizontal examination we can study the wettability of alumina by the polyphosphate glass. A drop of glass is deposed on a piece of alumina; the evolution of contact angle permits us to follow the wettability of alumina in relation with temperature. For the charaterization of the mechanism of wettability, we may use technics like SEM and EDX to obtain informations on diffusion phenomenon. But the information, thus obtained, doesn't exhibit the molecular aspect of the interface. This is the reason why we have also used the Raman microscopy analysis to complete information on the interfacial species.

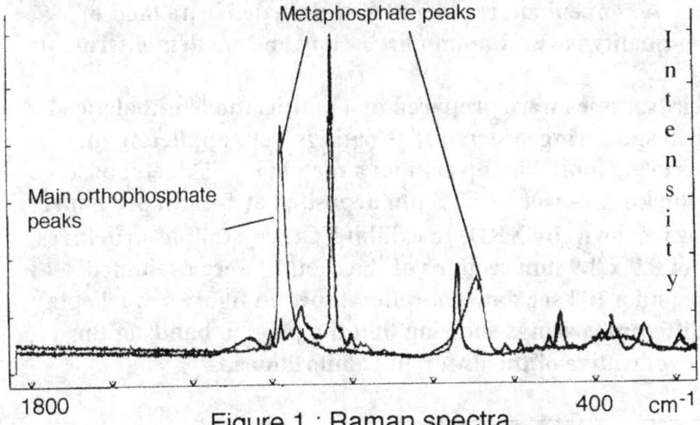

Figure 1 : Raman spectra

In fact the complexity of this study comes from the necessity to distinguish the pyrophosphate or orthophosphate species coming from alumina reaction and crystallization of glass. We have resolved this problem by using Raman maps for characteristic peaks of orthophosphates, pyrophosphates, and metaphopshates. The figure1 shows the different Raman spectra of these ions and their characteristic peaks for the symetric vibration .

We will present our results about different reactivities of alumina. Particular melted alumina doesn't react with polyphosphate glasses, presents a low wettability but may induce a crystallization of the glasses. In the other hand, on sintered alumina, we observe a good wettability , with impregnation of the alumina bulk. In this case, we observe reaction phenomenon at the interface system with or without crystallization.The role of calcium oxide will be discussed.

1. B. Deffontaines, M. Deffontaines-Fourez, P. Vast, Silicates Industriels Tome LX N°3-4 P.145-55
2. R.C. Ropp, in "Inorganics Polymerics Glass", (Elsevier ed.), 1993

RAMAN MICROSCOPIC STUDIES OF CERAMIC COATINGS BASED ON TITANIUM ALUMINIUM NITRIDE

Russell Evans[1], Iain Smith[1], W Dieter Münz[1], Ken J P Williams[2] and Jack Yarwood[1]*

[1] Materials Research Institute
Sheffield Hallam University
Pond Street
Sheffield S1 1WB, UK

[2] Renishaw Plc
OldTown
Wotton-under-Edge
Gloucestershire GL12 7DH, UK

New ceramic coatings (such as TiAlN (1), TiAlN (2) or TiAlNbN (3), etc) which may be deposited onto commercially important metal surfaces (eg drills and other cutting tool components) using modern PVD methods promise to be of crucial importance in a wide range of industrial applications. To be of maximum value they must be uniform, dense and crystalline with high quality hardness parameters. A convenient, rapid and non-destructive method of assessment of coating quality is via Raman microscopy and we demonstrate its value in this paper.

The TiN and TiAlN samples were prepared in a multicathode unbalanced magnetron closed field sputtering system [4]. Coatings were applied to high speed steel substrates using a nitride top coat on a metal (Ti, TiAl, etc) base coat. Total coating thicknesses were 1.5 - 3 μm deposited at 1-2 μm per hour. The coatings have been shown (by XRD) to exhibit FCC crystalline structures [1]. Raman spectra of 0.9 x 0.9 mm sections of the coating were examined using a x50 objective and a 100 sec data collection time. In figure 1 we display the spectra of four different coatings showing that the phonon bands of the crystalline lattice are indicative of the different compositions.

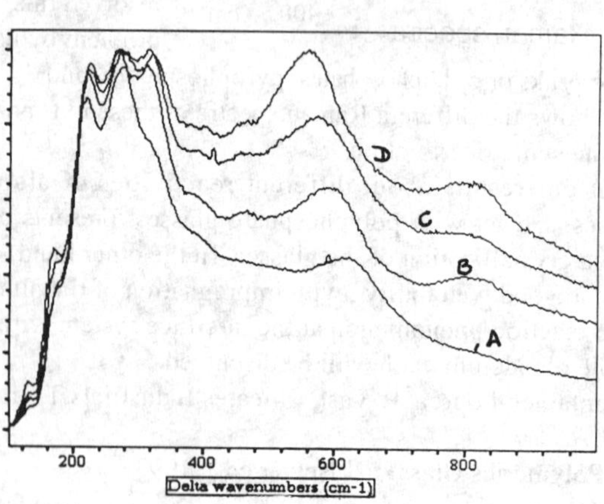

Figure 1. Raman microscopic spectra of Ti Al_xN_{4-x} coatings with increasing proportions of Al (A) x = 0 (B) x = 1 (C) x = 2 (4) x =3

For example, the band near 636 cm^{-1} from the TiAlN coating, seems to be indicative of an Al containing alloy lattice. Raman point mapping was performed by integrating over the region between 400 and 800 cm^{-1} with a 100 sec data point acquisition time (each map taking 10,000 seconds to collect). The result, shown in figure 2, demonstrates quite clearly that the coatings are uneven in the sense that material droplet formation occurs (as was already known from SEM). Such droplets are of different sizes and are distributed unevenly on a 0.5 - 15 micron scale. By examination of individual spectra on figure 1 and comparison with those of the different individual materials, it was established that the 'droplets' consisted of the metallic component of the nitride coating.

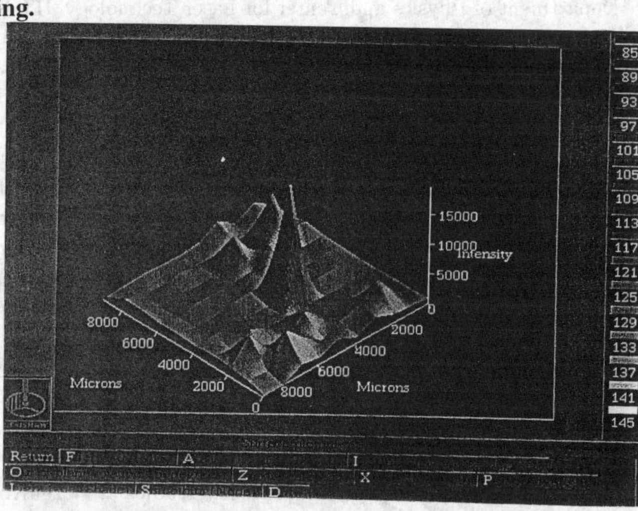

Figure 2. A Raman map of a ceramic coating of a TiAlN (0.9 x 0.9 mm surface) with spectrometer set at 636 cm^{-1}. (Shows the existence of metallic nitride 'droplets' in the coating).

Our paper will discuss these results in more detail and will include a discussion of the use of Raman spectroscopy/microscopy to study stress in these important coating materials.

1. **W D Münz, J. Vac. Sci. Technol. , 14, 2717 (1986)**
2. **L A Donohue, J Cawley, J S Brooks, Surf. Coat. Technol., 72, 128 (1995)**
3. **I Petrov, P Losbichler, V E Green, W D Münz, T Hurkmans, T Trinh. Paper presented at ICMCTFT 1994, San Diego, USA in press, Surf. Coat. Technol.**
4. **W D Münz, F J M Hauzer, D Schulze, B Buil, Surf. Coat. Technol., 49, 161 (1991)**

RAMAN AND PHOTOLUMINESCENCE SPECTROSCOPY OF SUBMICRON SURFACE OF CdTe CRYSTAL

H. D. Bist[1], P. S. Dobal[1], B. K. Rai[2], R. S. Katiyar[2], K.-T. Chen[3], and A. Burger[3]

[1]Department of Physics and Center for Laser Technology, IIT Kanpur, India
[2]University of Puerto Rico, P.O. Box 23343, Puerto Rico 00931-3343
[3]Department of Physics, Fisk University, Nashville, TN 37208, USA

Single crystals of CdTe are used for fabrication of high efficiency solar cell structures, nuclear radiation detectors, electro-optic modulators, laser windows as well as the substrate materials for HgCdTe based opto-electronic devices [1]. The band gap of CdTe changes between 1.48 to 3.35 eV and the resistivity changes from 10^4 to 10^{12} Ω-cm on controlled oxygenation [2]. We provide an example of monitoring a controlled surface oxidation on a micro-(nano-) meter scale in single crystals or wafers of CdTe using controlled laser irradiation under ambient conditions. It is established that the asymmetric TeO_3^{2-} ion is formed on the surface while Te and CdTe entities remain in the background. This is significant for basic material science and technological applications in electronic and optoelectronic devices.

Crystals of CdTe were grown using the travelling heater method (THM) [3]. Coherent Innova 99 CW Ar^+ laser was used to provide 514.5 and 457.9 nm (2 mW each) excitation for normal Raman and photoluminescence studies at $\approx$ 295K. For dynamic experiments, higher laser powers of 6.5 and 13.5 mW were used. The beam diameter on the sample was less than 2 μm using a 80X eyepiece in the Raman microprobe. Jobin-Yvon T64000 spectrometer with subtractive pre-monochromators coupled to the third spectrograph/monochromator with 1800 gr/mm grating was used for recording the spectra in backscattering geometry with the help of CCD detector (at 140 K).

Excess Te appears on the surface of single crystals of CdTe. The formation of an oxide of Te on samples standing in air for long time is expected. The changes occuring in the regions 600 to 1250 cm^{-1} on irradiations by laser powers of 6.5 and 13.5 mW were studied. The evolution of peak at 758 cm^{-1} region with the irradiation time was monitored in both the cases. To obtain the spectra under stationary conditions 2mW laser power was used to avoid possible changes occurring in the sample, while taking the spectra. The spectra obtained in the 50-800 cm^{-1} region from CdTe crystal with a 2mW Ar^+ laser falling on (a) virgin spot, and (b) a spot after irradiating with 13.5 mW for about 30 minutes were examined. The modes near 92, 103, 123 and 141 (partially) could be assigned to the presence of Te on the surface of CdTe. Te has two orders of magnitude larger scattering cross-section compared to that of CdTe. The single crystals of CdTe exhibit longitudinal optical LO phonons ($\approx$ 167 cm^{-1}) along $< 100 >$ and $< 111 >$ faces and TO phonons ($\approx$ 141 cm^{-1}) along $< 100 >$ and $< 111 >$

planes. The laser was incident over < 111 > surface, which implies that both LO and TO phonons should be present in the observed spectra. The two phonon band at 269 cm^{-1} is attributed to the anharmonic effects in the system. The anharmonic coefficient [2TO(calc.)-2TO(observed)] of 13 cm^{-1} is quite reasonable. The larger halfwidth of 269 cm^{-1} band is also consistent with its assignment as a two phonon band.

The isolated symmetric bent TeO_3^{2-} ion belongs to the pyramidal structure belonging to the C_{3v} point group. The totally symmetric Te-O stretching mode ν_1 at 758 cm^{-1} is higher than the doubly degenerate asymmetric stretching mode ν_3 at 703 cm^{-1}. Like-wise, the symmetric deformation mode ν_2 at 364 cm^{-1} is higher than the asymmetric doubly degenerate deformation mode ν_4 at 326 cm^{-1} [4].

On the basis of extra bands observed from irradiated sample the following conclusions are drawn: (a) the degenerate deformation mode ν_4, falling in the free TeO_3^{2-} at 326 cm^{-1} shows shifting of degeneracy with components at 324 and 381 cm^{-1}. Like-wise, the degenerate asymmetric stretching mode ν_3, falling at 703 cm^{-1} exhibits weaker two components at 662 and 703 cm^{-1}. There seems a remarkable coincidence of the $\nu_2(A_1)$ mode (falling exactly at 364 cm^{-1} in the CdTe-Te-TeO_3^{2-} environment) to free TeO_3^{2-} ion band at 364 cm^{-1} in solutions. The totally symmetric mode $\nu_1)A_1$ in the TeO_3^{2-} exhibits a Fermi-resonance with the upshifted position at 746.5 cm^{-1} showing a small shift (11.5 cm^{-1}) to lower frequency side as compared to the free ion position at 758 cm^{-1}.

A typical photoluminescence spectrum of CdTe crystal observed on excitation with ≈ 2mW of 457.9 nm laser light at ≈ 25 K shows a sharp peak at 1.60 eV which is unambiguously assigned to the emission maximum of near-band-gap transition of CdTe.

On the basis of a combined micro-Raman and photoluminescence studies the simultaneous existence of CdTe and Te aggregates distributed over the surface (and entire volume) of the crystal is suggested. On irradiation with an intense laser power, TeO_3^{2-} ions are created within < 2 μm diameter spot on the surface of CdTe crystal. The study is important to explore the mutual interactions of the Te aggregates, TeO_3^{2-} ions and CdTe semiconductor. One of us HDB thanks CSIR, India for the grant of an Emeritus Scientistship. This research was supported by EPSCOR-NSF and NASA grant NCCW-0088.

1. A. Partovi, J. Millerd, E. M. Garmire, M. Ziari, W. H. Steier, S. B. Trivedi and M. B. Klein, Appl. Phys. Lett. **57**, 846 (1990).
2. F. J. Espinoza Baltrán, F. Sánchez-Sinencio, O. Zelaya-Angel, J. G. Mendoza-Alejo Armenta, C. Vázquez-López, M. H. Farías, G. Soto, L. Cota-Araiza, J. L. Peña, J. A. Azamar barrios and L. Baños, Jpn. J. Appl. Phys. **30**, L1715 (1991).
3. R. Triboulet, Proc. 3rd Int. Conf. on CdTe, P.Siffert, A. Cornet (eds.), Strasbourg, p. 171, 1971.
4. H. Siebert, Z. Anorg. Allg. Chem. **275**, 225 (1955).

LIGHT INDUCED CHANGES IN RAMAN SPECTRA OF PHOTOREFRACTIVE $LiNbO_3$ WAVEGUIDES

S.M. Kostritskii

Physics Department, Kemerovo State University,
Krasnaya str. 6, Kemerovo, 650043 Russia

Proton-exchanged (PE) lithium niobate, $Li_{1-x}H_xNbO_3$, is a promising material for fabricating integrated optical waveguides and devices. We discovered previously [1] that the increase of photorefractive sensitivity (PRS) of the PE waveguides is more than of the order of three as a result of copper doping at certain conditions. Thus a new possibility in photorefractive waveguides (PRW) fabrication is provided. This improvement, a new type of PRW, requires that its properties be investigated. In recent years with development of integrated optics, it has becomes clear that the waveguide Raman technique is unique in studying the properties of thin waveguiding PE layers [2,3].

Experimental investigation indicates that doping of PE waveguides with Cu produces strong photoinduced light scattering (PLS), which is directly evident by the photorefractive effect. With He-Ne laser irradiation input into a planar (104)-cut waveguide through a wafer lateral edge, the PLS shows itself as a time-dependent spreading of the laser beam into a fan consisting of multiple rays with extraordinary polarization. Note that the intensity of the lines in Raman spectra depends crucially on the propagation direction and the polarization of laser irradiation in the crystal; these two factors determine the high effectiveness of Raman spectroscopy as an instrument for the study of PRS. Under our experimental conditions the laser irradiation that excites the PLS simultaneously excites the Raman scattering. The measurements of Raman intensity have been performed in the Stokes region on spectra frequency shifted by 152 cm^{-1} from the frequency of the excited irradiation. This frequency corresponds to the scattering on phonons of the E(TO) mode.

It is established that the Raman intensity I in Cu-doped PE waveguides depends on exposition time (Fig. 1). The characteristic times of this process and the visual effects of PLS are in agreement. The change in the Raman signal amplitude ΔI directly depends on C_{Cu} and on the input light intensity J. Thus Raman scattering can be used both the study of the kinetics of photorefraction and for the determination of PRS by comparison of the ΔI values for the waveguide in question with those for the specimen for which the value of PRS is known, and the condition of the experiment must be preset (e.g., geometry, J).

The experimental data analysis demonstrates that the Raman spectrum for any lithium niobate crystals is the overlaping of two spectra: the main structure spectrum and the spectrum of the local regions which create the polarization clusters. The Raman spectra measurements in saturated state indicate that with copper doping the strong change of the local regions spectrum is observed against the background of almost constant main spectrum (Fig. 2). It shows that at doping the copper incorporates favourably in local regions which is in accordance with the modern notations about the defect structure of the $LiNbO_3$ crystals. Moreover, the Raman intensities for additional lines are strong increasing at high level of the input laser irradiation intensity (Fig. 2). These results lead to presumption that the described effect might be caused by a change of the polarization clusters due to photoexcitation.

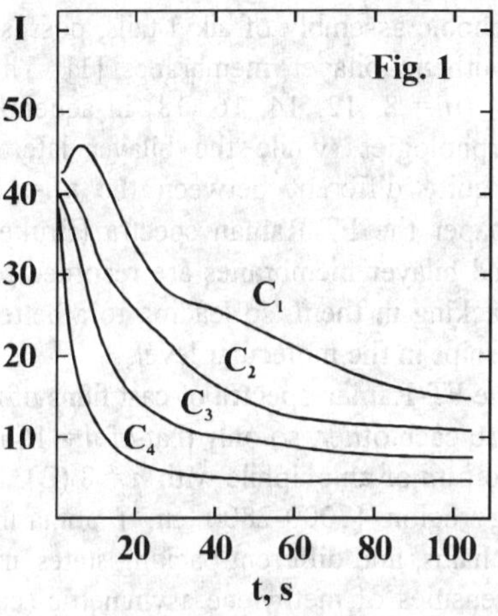

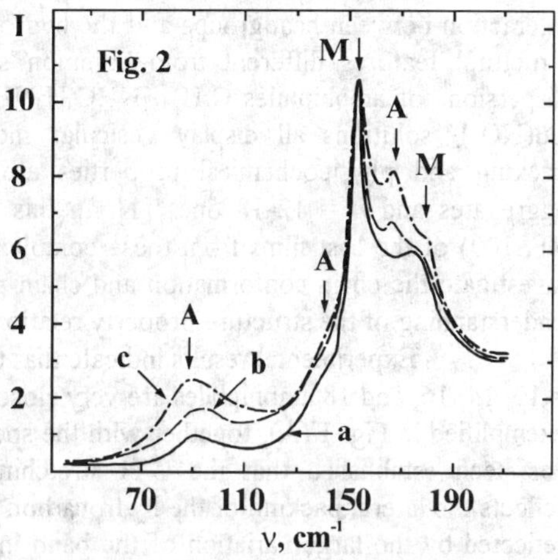

Fig. 1. Raman intensity kinetics for PE waveguides with J = 1.2 W/cm² and different Cu concentrations: 1 at.% < C_1 < C_2 < C_3 < C_4 < 4 at.%.

Fig. 2. The Raman spectra for undoped PE waveguides (a) and copper doped PE waveguides with C_{Cu} = 4 at.% measured at different levels of J: J_1 (b) and J_2 (c) at $J_2 >> J_1$. M - lines for main phonon spectrum; A - additional lines for local regions spectrum.

1. S.M. Kostritskii and O.M. Kolesnikov, J.Opt.Soc.Am. B11, 1674(1994).
2. Savatinova, S. Tonchev and M. Kuneva, Appl.Phys. A56, 81(1993).
3. Paz-Pujalt, D.D. Tuschel, G. Braunstein, T. Blanton, S. Tong Lee and L.M. Salter, J.Appl.Phys. 76, 3981(1994).

FOURIER TRANSFORM RAMAN SPECTRA OF CAST FILMS FROM COMPLEXED BILAYER MEMBRANES

Yingqiu Liang, Xianchun Lu, and Zhiqiang Zhang

Department of Chemistry, Coordination Chemistry Institute and State Key Laboratory of Coordination Chemistry, Nanjing University, Nanjing 210093, P. R. China

The complexed bilayer membranes, formed through the coordination interaction between headgroups and the hydrophobic assembly of alkyl tails, possess structural features different from common synthetic bilayer membranes [1]. The dispersions of amphiphiles $C_nH_{2n+1}NHC_2H_4NH_2$ (n = 8, 12, 14, 16, 18) in aqueous $Cu(NO_3)_2$ solutions all display vesicular morphologies, while the bilayer lateral packing and physicochemical properties are quite different between the n = 8 aggregates and n = 12-18 ones [1]. In this paper the FT-Raman spectra (Bruker RFS100) of the cast films from these complexed bilayer membranes are recorded to investigate the chain conformation and chain packing in them, so leading to a better understanding of the structure-property relationships in the molecular level.

Experimental results indicate that the FT-Raman spectra of cast films of n = 12, 14, 16, and 18 amphiphiles are very close to each other, so only that of n = 16 is exemplified in Fig. 1 (A), together with the spectrum of amphiphile with n = 8 (B). It has been established that the C-H stretching region (3000-2800 cm^{-1}) primarily reflects the lateral packing of the hydrocarbon chains, and different packing states are reflected by the large variation of the band intensities of methylene asymmetric (ca. 2880 cm^{-1}) and symmetric (ca. 2850 cm^{-1}) vibrations [2]. Peak height ratio I_{2850}/I_{2880} decreases from (B) to (A), suggesting the enhancement of chain lateral packing and increase in degree of aggregation order. Also the intramolecular chain conformation is different in these two cases, and it is known that the band intensity of the methyl symmetric stretching vibration at ca. 2932 cm^{-1} is sensitive to it [2]. The ratio I_{2932}/I_{2880} decreases from (B) to (A), indicating that the *gauche* conformers in the alkyl chain of cast films of n = 16 amphiphiles are less than those of n = 8 amphiphile, which also means the increase in degree of aggregation order. 1200-1000 cm^{-1} is the C-C skeleton stretching region, which mainly reflects the vibrational modes of different intramolecular conformation of the alkyl chains. The all-*trans* C-C stretching modes are assigned to the spectral features at ca. 1130, 1100 and 1062 cm^{-1}, while those of *gauche* conformers at ca. 1122 and 1090-1085 cm^{-1} [2]. So the increase in the ratio I_{1130}/I_{1090} from (B) to (A) as shown in Fig. 1 indicates that the population of *gauche* conformers in the alkyl chains in the n = 16 cast film is reduced, and the degree of order of the molecular chains is enhanced, compared with the condition of n = 8 cast film. The various peak height ratios calculated based on Fig. 1 are listed in

TABLE 1. It is seen from the TABLE that in the n = 16 cast film the hydrocarbon chains are arranged tightly with the degree of order approaching that of solid polyethylene. The infrared CH_2 stretching modes of the same cast film appear at ca. 2920 and 2849 cm^{-1}, suggesting the gel state of the film [3]. The ratio I_{2850}/I_{2880} of the cast film of n = 8 amphiphile is 0.97, which is much smaller than the corresponding value of its CCl_4 solution (1.5), but larger than that of the n = 16 cast film (0.7), showing that the film cast from the n = 8 complexed bilayer membranes is in a somewhat orderly state, but possesses looser chain packing and more *gauche* conformers than that of n = 16. It is consistent with the result of its FT-IR spectrum, in which the CH_2 stretching modes locate at ca. 2924 and 2854 cm^{-1} characteristic of the liquid crystalline state of the cast bilayers.

The above FT-Raman spectral data demonstrate the inter- and intra-molecular interaction in the cast films of amphiphiles with different alkyl chain length, providing structural explanation on the great disparity in the phase transition temperatures between aqueous complexed bilayers of n = 8 and those of n = 12-18 [1].

TABLE 1. Raman peak height ratios of cast films of complexed bilayer membranes of n = 8 and 16 amphiphiles, their CCl_4 solutions, and solid polyethylene (PE), respectively.

Ratio	Cast film		CCl_4 solution	Solid PE
	n = 8	n = 16		
I_{2850}/I_{2880}	0.97	0.70	1.5	0.73
I_{2932}/I_{2880}	0.97	0.32	1.0	0.18
I_{1130}/I_{1090}	0.97	2.91	0.6	11.0

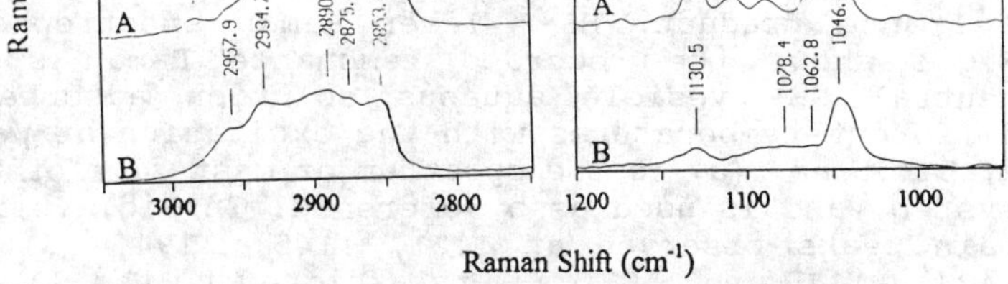

Fig. 1. FT-Raman spectra of cast films from complexed bilayer membranes of $2C_nH_{2n+1}NHC_2H_4NH_2{\cdot}Cu(NO_3)_2$. (A) n = 16; (B) n = 8.

1. X. Lu, Z. Zhang, and Y. Liang, *J. Chem. Soc., Chem. Commun.* 2731 (1994).
2. N. Yellin and I. W. Levin, *Biochem. Biophys. Acta* 489, 177 (1977).
3. R. G. Snyder, *J. Chem. Phys.* 47, 1316 (1967).

Raman Spectroscopic Evidence of Aggregation and Interaction of Methyl Orange on Surfactant Vesicle

Lixin Wu[1,2] Junhua Wei[1] and Ying Wu[1]

[1]Key Lab for Molecular Spectra & Structure, Jilin University, Changchun 130023, P.R. China
[2]Changchun Institute of Physics, Academia Sinica

As one of the important aspects of supramolecular system surfactant vesicle, a kind of simplified model system, has attracted a significant amount of scientific and technological interests due to its special structure and micro environment,especially the property in mimicking biomembrane functions[1]. It has been known that surfactant vesicles are constructed by sonicating aqueous solutions of amphiphiles, such as longchainalkyl trimethylammonium bromide incorporated a rigid segment, usually a conjugated chromophore[2]. It has also been found out that the conjugated group, stacked orderly together with molecules, performed as a sensitive probe of the variation of the vesicle structure and as an function group. We have reported that the excitation energy transfer occurred from N-benzylideneaniline, inserted in the amphiphile that formed vesicle bilayer to the dye anions adsorbed on vesicle surface[3]. The interaction between the dye and micelle carrying opposite charge has considerably been investigated by absorption and Raman spectroscopy[4]. But how the dye aggregated on vesicle surface is obscure. We herein report the Raman spectroscopic result of Methyl Orange(MO), an anionic dye, attaching and stacking on the surface of the vesicle formed by cetytrimethyl ammomnium bromide inserted a N-benzylideneaniline(NBA).

The resonance Raman spectra were carried out with a France product, HS-2 laser Raman spectrophotometer. Fig.1 shows the recorded resonance Raman spectra of neutral NBA vesicle aqueous solution adsorbed MO at different temperatures with the excitation at wavenumber of 514.5 nm. (a) is the spectrum of cast film of the mixed system, and is used as a reference. In Fig.1(b), several main peaks observed at 1121, 1145, 1194, 1316, (1396, 1424), 1445 and 1592 cm^{-1} are assigned to the SO^{3-}, C-N, S-ring, S-ring, N=N, N-ring and S-ring vibrations of MO[4], while the peak at 1412 cm^{-1} is assigned to the vibration of NBA amphiphile. Thus the spectra is basically ascribed to that of MO, but is not approaching the characteristic of it in its dilute solution, whereas is found to be rather similar to those of cetylorange in monolayer and MO mixed with cetyltrimethyl-ammonium bromide(CTAB) micelle

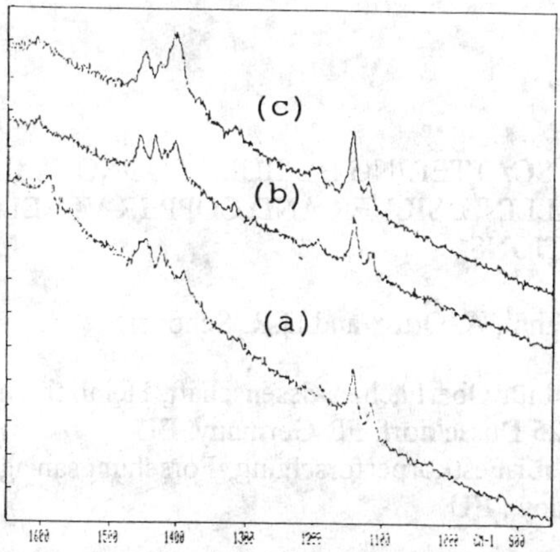

Fig.1 Resonance Raman Spectra of MO-NBA in Cast Film (a), and MO-NBA Aqueous Solution at 20°C (b) and 90°C (c) with Exitation Wavelength at 514.5 nm.

aqueous solution, indicating that MO anions aggregates on the surface of NBA vesicles.We have observed that the absorption band of MO bounded on NBA vesicle surface in visible range is at 405 nm, and shifts to around 430 nm after the temperature risen above the phase transition point of the vesicle, just like the absorption band of it in positive charged micelle solution and that of it inserted in monolayer. Much larger blue shift occurred in contrast to MO monomer at 464 nm manifest that much more ordered and tight stacking of MO emerged on vesicle surface than on micelle make the MO anions due to more smoothed and flat surface on vesicle bilayer than on micelle. However, this case is not clearly observed with Raman spectra. Fig. 1(c) recorded the spectra of MO attached to vesicle surface at 90°C. The peak at 1395 cm^{-1} is enhanced and the peak at 1420 cm^{-1} is weakened. As the Raman spectrum of NBA amphiphiles in vesicles is rather weak and basically independent of the temperature changes, spectrum (c) shows the change of the morphology of MO. By measuring the spectra of mixed system at different temperatures, it has been seen that there is a gradually but not a promptly changes of spectrum of MO occurred during the temperature increasing while the vesicle passed by the phase transition from gel phase to liquid crystal then to molecular or molecular dispersion. Even though the vesicle was destroyed up to molecular dispersion, MO is still in the aggregated state, which is confirmed from the difference between spectra (c) and (a).

References

[1] J.N.Robinson et al., Chem.Soc.Rev., 1991,20,49
[2] T.Kunitake et al.,J.Am.Chem.Soc., 1980, 102, 6642
[3] L.Wu et al.,Chem. J. of Chin. Univ.,1995,16,629
[4] K.Machida et al., J.Raman Spectro.,1980,9,198

RAMAN SCATTERING BY BULK PHONONS IN MICRO-CRYSTALLINE SILVER AND COPPER VIA ELECTRONIC SURFACE EXCITATIONS

W. Akemann[1], A. Otto* and H.R. Schober[1]

*Lehrstuhl für Oberflächenwissenschaft, Heinrich-Heine-Universität Düsseldorf, 40225 Düsseldorf, FR Germany, EU
[1] Institut für Festkörperforschung, Forschungsanlage Jülich, 52425 Jülich, FR Germany , EU

Cryocondensed silver and copper films have been prepared as described in ref.[1]. Most likely, the surface of a cryocondensed and partly annealed film consists of small facets of (111), (110) and (100) orientations,various steps, kinks and other one- and zero dimensional defects related to grain boundaries and dislocations. In figure 1 we have compiled the low frequency spectra of a silver film for various temperatures of annealing T_A for 120 seconds (solid lines). Each spectrum was recorded after re-cooling to 40 K. The quenching of the intensity by oxygen adsorption, see ref. 1, indicates a surface sensitive mechanism of exitation.

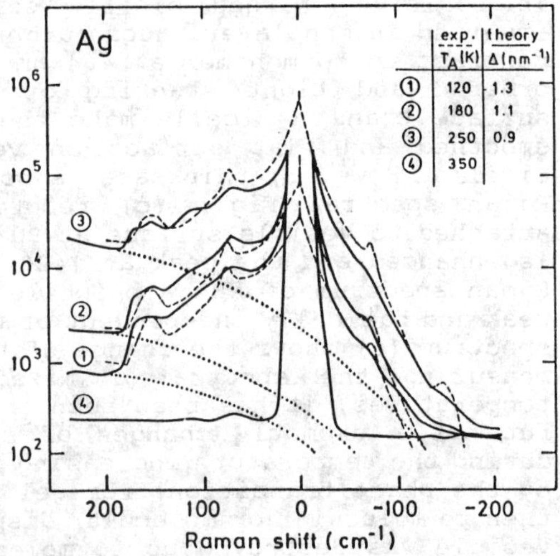

Fig. 1
Solid lines: Experimental low frequency Raman spectra of a silver film that was cryocondensed at 40 K and was subsequently annealed at the indicated temperatures T.
Dashed lines: Theoretical Raman scattering spectra as sum of the bulk phonons projected on (111) and the intra Shockley surface band transitions (dotted lines).
Parameters from bottom to top: $\Delta = 1.3, 1.1, 0.9\ nm^{-1}$.

We propose a mechanism similar to our explanation of the electronic Raman scattering above 200 cm^{-1}, as given in ref.[2], namely Raman scattering by bulk phonons via electronic surface excitations, as described in detail in ref.[3]. It involves virtual transitions between electronic surface states well known from large single crystal surfaces which still do exist on small facets. So called s/p derived crystal induced surface states at metal surfaces may penetrate several lattice constants into the bulk. An electron in these states therefore will scatter not only from surface phonons but also from bulk phonons.

One may define an electronic surface susceptibility χ modulated by lattice vibrations with normal coordinates ξ, ref. 3. Since one is off the electronic surface resonances at 514.5 nm excitation we used [4] the polarizability theory of Raman scattering of G. Placzek.

At a smooth surface, only phonons with 3 dimensional wave vector $\underline{Q}$, whose projection $\underline{q}$ onto the surface is of the order of the wavevector of the laser light may be excited. At a rough surface with a roughness spectrum $S(k_r)$ the excitation of phonons within the acoustic branches j with $\underline{q} = -\underline{k}_r$ becomes possible. The Stokes and anti Stokes phonon spectra are given, assuming the Raman tensors independent on $\underline{Q}$ and j, by

$$\begin{Bmatrix} J_{s,ph}(\omega) \\ J_{as,ph}(\omega) \end{Bmatrix} \propto \begin{Bmatrix} n(\omega)+1 \\ n(\omega) \end{Bmatrix} \frac{1}{\omega} \sum_{Q,j} S(\underline{q})\delta(\omega - \omega_{ph}(\underline{Q}, j))$$

Assuming $S(k_r)$ to be Lorentzian with Δ, the projection of the bulk phonon structure of silver on the low index facets yields the spectra $J_{s,ph}(\omega)$ in fig. 2 for Δ = 1.1 nm^{-1}.

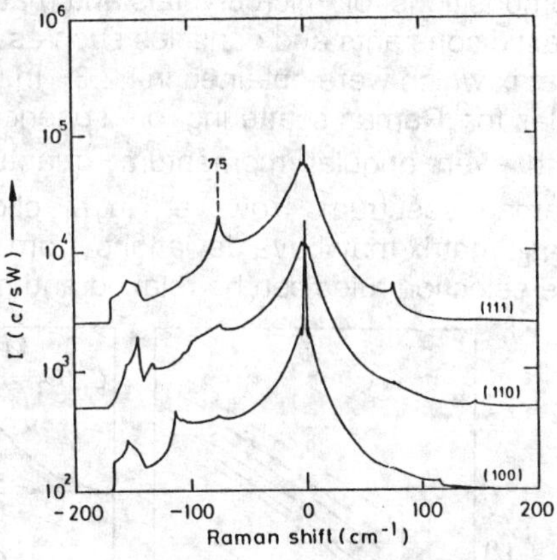

Fig. 2
Theoretical Raman scattering spectra of the bulk phonons of silver (data of Kamikatahara and Brookhouse), for Q projected onto (111), (110) and (100) faces.

To this we have added the electronic Raman scattering by intra-Shockley state excitations[3] (lines in fig 1). In fig 1 the theory for (111) facets (dotted lines) is fitted to the experimental results. The values of Δ imply that the average (111) facet size grows from 4.8 to 7.0 nm by annealing the films from 120 to 250 K. The small deviations from the experimental spectra at about 166 and 112 cm^{-1} is assigned to a weak contribution of (110) facets. We do expect a contribution of the Rayleigh phonons of Ag(111)[4] as well. They probably contribute to the experimental structure at 68 cm^{-1} in fig. 1 but are not separated from the bulk phonon signal due to the small differences in energy of bulk and Rayleigh phonons at the M - point.

---

[1] W. Akemann, A.Otto, Proceedings of ICORS XIV (1994) p. 734
[2] W. Akemann, A. Otto, Phil.Mag.B **70** (1994) 747
[3] W. Akemann, A.Otto, H.R.Schober, submitted to J. Raman Spectrosc.
[4] U. Harten, J.P. Toennies, Ch. Wöll, Faraday Discuss. Chem. Soc. **80** (1985) 137

SIZE QUANTIZATION OF ACOUSTIC PHONONS IN MICROCRYSTALS EMBEDDED IN A GLASS MATRIX

N.Ovsyuk[1]* and N.Novikov[2]
[1]Institute of Mineralogy and Petrology, Novosibirsk, 630090, Russia, e-mail: ovsyuk@uiggm.nsc.ru
[2]Institute of Automation and Electrometry, Novosibirsk, 630090, Russia, e-mail: novikov@iae.nsk.su

We consider the influence of the matrix on the spheroidal and torsional vibrational modes of microcrystals, which brought about a new interpretation of the low - frequency Raman spectra. We have studied heterophase system, consisting of Ge microcrystals embedded in GeO_2 glass matrix [1]. To gain a better understanding of the influence of the crystal - matrix boundary we have also analyzed some other chemical compositions of microcrystals and matrices with different relations between elastic constants and densities such as CdS in GeO_2 matrix and Ag in SiO_2 matrix which were obtained in [2,3]. In [4] it was shown that due to selection rules for Raman scattering on spherical particles, only spheroidal modes with even angular momentum quantum numbers l can be seen in the Raman spectrum. However, in practice the form of the microcrystals in a glass matrix may have deviations from a perfect spherical one. In this case, the selection rules can be relaxed and not only spheroidal modes with even

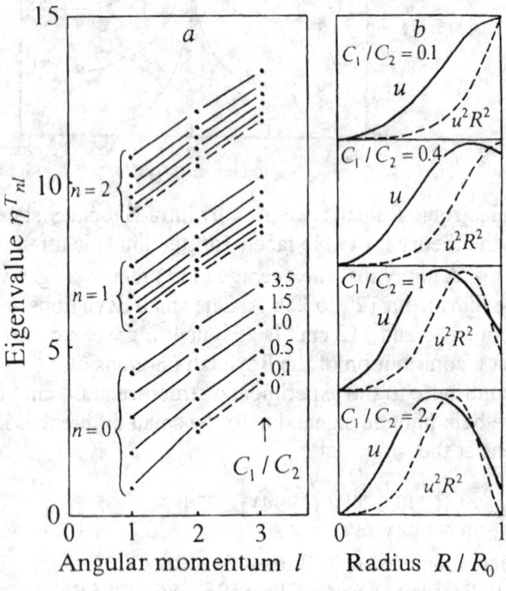

Fig.1. a.The shift of the eigenfrequencies of the torsional modes of the microcrystals for the surface mode (n = 0) and the inner modes (n = 1,2) as a function of the matrix stiffness, i.e., as a function of the ratio C_1 / C_2. The dashed lines corresponds to the case of the free microcrystal surface. b.The radial distribution of the vibrational amplitudes u and the energy density u^2R^2 of the surface vibrational mode of a microcrystal as function of the ratio C_1 / C_2.

l, but torsional vibrational modes with odd l can be seen in the Raman spectra. According to [5], eigenfrequencies of the torsional vibrational modes of the microcrystals embedded in a glass matrix can be determined from the equation $D^t_m = 0$, where D^t_m is determinant of (2x2) matrix which is given in [6]. In Fig.1 the shift of vibrational eigenfrequencies η^T_{nl} of a microcrystal is shown when the matrix stiffness increases relative to that of the microcrystal. The transverse surface mode (n = 0) with the angular momentum l = 1 is absent for microcrystals with a free surface (dashed lines). This is natural, because this mode corresponds to the rotation of the microcrystal as a whole. If a microcrystal is embedded in a matrix, then there appears a new solution for the surface mode with angular momentum l = 1. This mode arises because of the restoring force which limits the free rotation of the microcrystal. Fig.1 shows that the surface vibrational mode of the microcrystals embedded in the matrix does not disappear; its eigenfrequency increases and the maximum of the vibrational amplitude shifts inside the microcrystal with increasing of the matrix stiffness. In ref.[5] an erroneous conclusion was made that the influence of the matrix over the microcrystal leads to the opposite effect, i.e., the surface mode disappears and the inner vibrational modes softens. Hence, contrary to refs. [2,3], the frequency of the Raman peak for the surface torsional mode with the angular momentum l = 1 for the microcrystals embedded the matrix does not equal 1.83 υ_t/dc as it would be in the case of the free surface. Indeed, the existence of the matrix results in the appearance of a new solution of $D^t_m = 0$ for the surface mode with l = 1; its eigenfrequency increases, for example, from 0.2 υ_t/dc to 1.4 υ_t/dc when the matrix stiffness C_1/C_2 increases 0.1 to 3.5, (Fig.1.a). Most likely, just these vibrations with l = 1 take part in the low-frequency Raman scattering described in [2,3].

From all our experiments it is clear that the depolarized scattering is governed by the surface torsional modes with l = 1 which appear, when the crystals are embedded in a matrix. Furthermore, from the calculations follows that the influence of the matrix on the frequencies of the acoustic phonons, confined in the microcrystals, is significant even if Lame's constants λ, μ and the mass densities of the microcrystal and the matrix are quite different from each other, especially for n = 0 and 1.

1. N.Ovsyuk, V.Novikov, Phys. Rev. B**53** (1996), in press.
2. A.Tanaka, S.Onari, and T.Arai, Phys.Rev. B**47**, 1237 (1993).
3. M.Fujii, S.Hayashi, and K.Yamamoto, Phys. Rev. B**44**, 6243 (1991).
4. E.Duval, Phys. Rev. B**46**, 5795 (1992).
5. A.Tamura, K.Higeta, and T.Ichinokawa, J.Phys. C**15**, 4975 (1982).
6. N.Ovsyuk, L.Krivoputskaya, and A.Shebanin, JETP Lett. **48**, 678 (1988).

STUDY OF RELAXATION PROCESSES IN WATER-MICELLE SOLUTIONS BY RAMAN SPECTROSCOPY

R.M.Ishburiev, Sh.U.Umidullaev*, M.P.Holmuradov, A.B.Shaimanov

Samarkand State University, University Blvd. 15, 703004 Samarkand, Republic of Uzbekistan

It is known, that for many years the micelle solutions are intensely researched by various physical-chemical methods. Such interest to the water-micelle solutions is stimulated in the main with broad use of the surface active substances (SAS) in different branches of industry as well as medicine.

SAS belong to the class of substances with molecules containing both polar (hydrophilic), and nonpolar (hydrophobic) parts. Arise micelles at short region of concentrations, so-called, critic concentration of micelle formation (CCMF). Micelles in solutions of SAS can have the various shapes and structures. The shape of molecules, in dependency on concentration of SAS, in aqueous solution is changed from spherical to plate-like.

In this report we first were put into practice the molecular probe method for study of structure and basic characteristics the micelle formation solutions. The apply of this method was conditioned by first of all with lay in its foundation possibilities, that is result of particular sensibility of molecular probe to physical-chemical properties of its environment. Molecular-probe represents the original molecular sensing element of structural and dynamic information about local environment.

In the capacity, of the probe, we choosed well studied depolarized Raman scattering line ν=2235cm^{-1} of, benzonitrile (C_6H_5CN). Concentrationol dependences of vibrational and orientational relaxation times of molecules C_6H_5CN in aqueous micellar solutions of $C_8H_{17}SO_4^-Na^+$ were studied.

The apraisal of concentrational dependence of relaxational processes in aqueous SAS solutions allows to determine the position of molecular probes localization in micelle as well as to define the basic parameters of molecular systems. Concentrational dependence of vibrational and reorientational relaxation times for molecular probe in aqueous solutions of $C_8H_{17}SO_4^-Na^+$ was shown on the fig.1

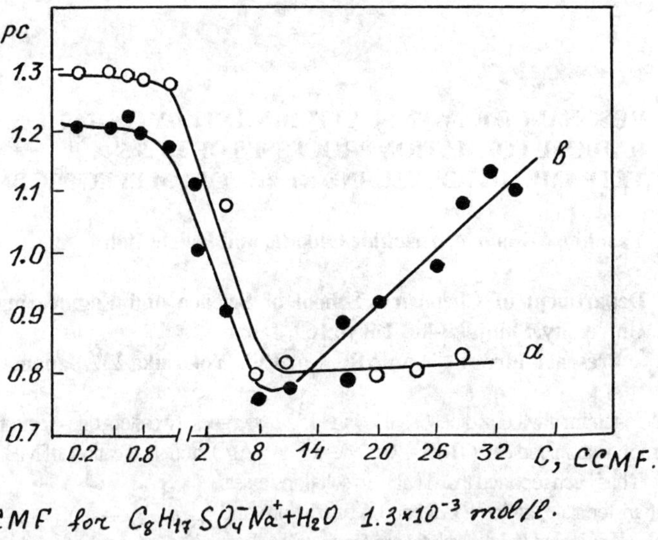

Fig.1. The concentrational dependence of vibrational (a) and reorientational (b) relaxation times of molecular probe lokalized in micellar structure at 25°C.

How the studies showed, vibrational and reorientational relaxation times depend on SAS added to water. τ_{vib} and τ_{or} remain at constant values for water, concentration 0.1-1 CCMF and is egual 1.3-1.2 pc, that it is not far from that for aquous solution, 1.25 pc. It means we have dealings with ordinary solutions at given concentrations. With increasing of concentration SAS in solution, especially beginning with CCMF1, and are noticeably decreased. This can be associated with self-organization of micellar structures. It is known, that any shotening of the time of molecule's existence in the vibrational states causes the broadening of Raman line contours. The position of these bands, their width depend on nature environment of probe's molecule. Therefore the decreasing of relaxation times for interval of concentration SAS 1-8 CCMF proves, that molecules of probe can not penetrate in micelles core and they are forced out to more peripheral parts. Further increasing of contents SAS in solution (more' 10 CCMF) leads to the increasing of τ_{or}. This may be caused only by the increasing of micelle packing density (i.e.by formation more large-scale micelle) at given concentrational intervals of SAS. Although, certainly, it is impossible of chance, that data about the increasing of meanings the orientational correlation times of molecular probe with the rise of concentration SAS can be weakened a little owing of partly force out of molecular probe to less density peripheral part of micelle.

RESONANCE RAMAN SCATTERING STUDY ON CATION RADICAL FORMATION PROCESSES OF 3,3',5,5'-TETRAMETHYLBENZIDINE AT PLATINUM ELECTRODES

Yasuhito Misono [†]*, Yasuhide Ohkata, and Koichi Itoh

Department of Chemistry, School of Science and Engineering, Waseda University, Shinjuku-ku, Tokyo 169, Japan
[†] Present address: Nissan ARC Co., Ltd., Yokosuka 237, Japan

3,3',5,5'-tetramethylbenzidine (TMB, Fig. 1.), undergoes two-step one-electron oxidation, forming a monocation radical (TMB^+, 0.60 V vs. Ag/Ag^+) and a dication (TMB^{2+}, 0.95 V), successively. It has been reported that TMB^+ in solutions exists in a charge-transfer state, giving a CT band near 650 nm and that a radical salt prepared by electrolytic oxidation of TMB at an ITO (indium tin oxide) electrode exhibits anisotropic electric conductivity [1]. Little is known, however, about electrochemical oxidation process of TMB and structures of the oxidation products, i. e., TMB^+ and TMB^{2+}. In this paper, combining resonance Raman (RR) and UV-VIS spectroscopies with a home-made thin layer flow electrochemical cell, we studied the electrochemical oxidation processes of TMB in solutions as well as at a platinum electrode surface. The points of focus in this study are as follows; (1) assignments of the RR bands of TMB^+ in free and CT states and the RR bands of TMB^{2+}, (2) effects of counter ions (BF_4^-, PF_6^-, and Br^-) and their concentrations on canonical structures of TMB^+ and on the equilibrium between the monomer and CT states, (3) electrochemical oxidation process of TMB in solutions, (4) electrodeposition process of the radical salt at a platinum electrode and the structure of the radical salt.

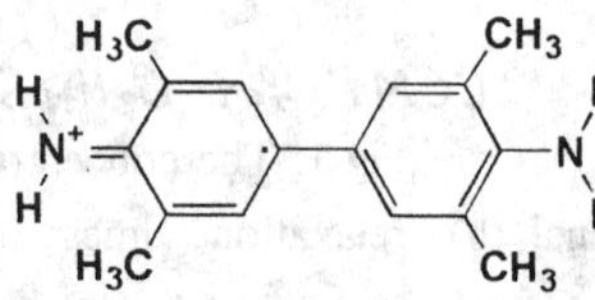

Fig. 1 3,3',5,5'-tetramethylbenzidine cation radical (TMB^+).

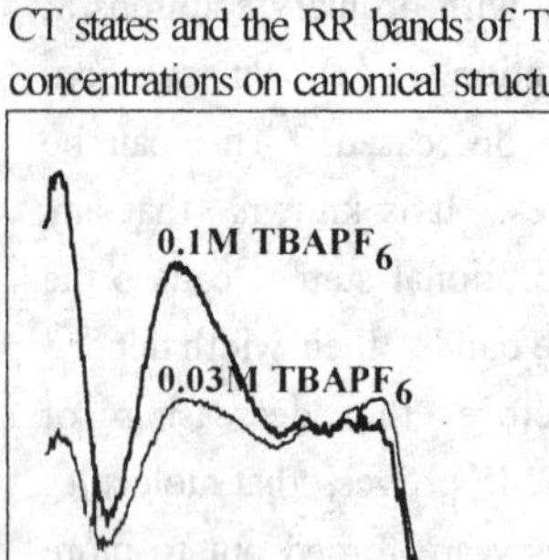

Fig. 2 UV-VIS spectra of TMB^+ in n-BuCN.

Fig. 2 exhibits the absorption spectra of TMB^+ in n-butyronitrile (n-BuCN) containing a supporting electrolyte ($TBAPF_6$), at various concentration levels (0.01-0.1 M). The spectrum measured at 0.01 M gives rise to a band near 460 nm due to the radical in a free state, while the spectrum at 0.1 M gives a band near 370 nm, which has been assigned to a CT state [1], in addition to the 460-nm band. Fig. 3, (A) and (B) , shows the RR spectra of TMB^+ in n-BuCN solution containing 0.1- and 0.01 M $TBAPF_6$, respectively. From the figure, it is clear that, on increasing the concentration of $TBAPF_6$, the intensity ratios of the 953- and 783-cm^{-1} bands to the 1335-cm^{-1} band increase. The RR spectrum of TMB^+ in an

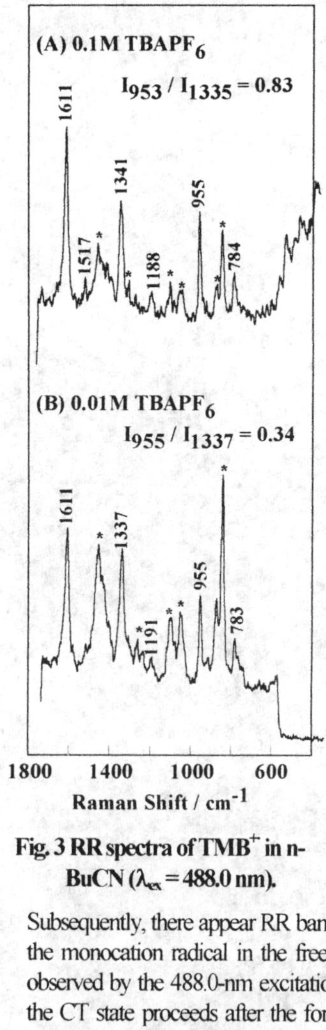

Fig. 3 RR spectra of TMB^+ in n-BuCN (λ_{ex} = 488.0 nm).

acetate buffer solution prepared by horseradish peroxidase/H_2O_2 oxidation [2], gives features almost identical with those of Fig. 3 (A). According to Josephy et al. [2], the radical exists mainly as a CT complex. Based on these results, we concluded that the prominent RR bands near 1611- and 1335 cm^{-1} in Fig. 3 are assigned to TMB^+ in both the free and CT states, while RR bands at 953- and 783 cm^{-1} are ascribable only to the CT complex. The monocation radical formation processes in n-BuCN solutions at 0.65 V were followed by UV-VIS absorption and RR spectroscopies. The results indicated that the process consists of two steps; the first step is the formation process of the radical in the free state and the second step is the formation of the CT state ion equilibrium with the monomer state.

When a TMB solution in n-BuCN containing 0.1 M $TBAPF_6$ is flowed through the electrochemical cell keeping the electrode potential at 0.95 V, there appears a film of the monocation radical salt at the platinum electrode surface. The film formation process was monitored by RR spectra excited at 488.0- and 418.5 nm. Fig. 4 illustrates the spectral changes observed by the 418.5-nm excitation. At an initial stage of the process RR bands appear at 1599-, 1459-, and 1335 cm^{-1}, which are ascribable to TMB^{2+}. Subsequently, there appear RR bands at 1610- and 1340 cm^{-1} due to the monocation radical in the free state. The RR spectral changes observed by the 488.0-nm excitation indicates that the formation of the CT state proceeds after the formation of the radical in the free state is completed. Thus, the formation of the radical salt takes place in the following manner. At first, two-electron oxidation goes on, resulting in the formation of TMB^{2+}. Then, the dication and TMB in the solution undergo coproportionation reaction to produce TMB^+. Finally, the monocation radical is partly converted to the CT complex.

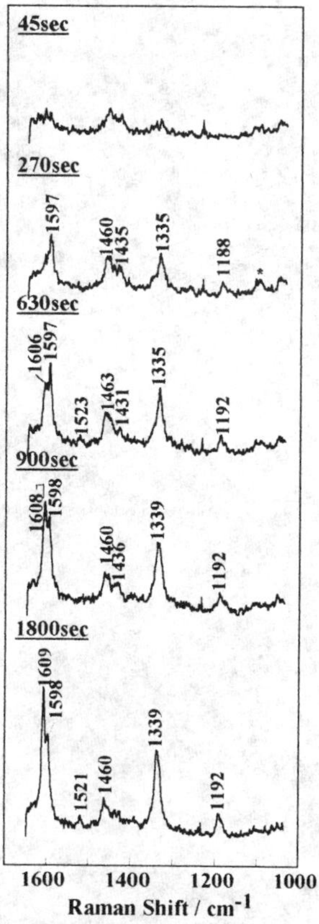

Fig. 4 RR spectra of the electrodeposition product on a Pt electrode at 0.95V (λ_{ex} = 418.5 nm).

1. H. Awano, T. Ogata, H. Murakami, H. Ohigashi, Synth. Met. 36, 263 (1990).
2. P. D. Josephy, T. Eling, R. P. Mason, J. Biol. Chem. 257, 3669 (1982).

SECTION 16

SURFACE ENHANCED SCATTERING

SERRS SPECTROSCOPY - A NEW ANALYTICAL TECHNIQUE FOR FORENSIC SCIENCE

P.C.White[1], W.E.Smith[1], C.H.Munro[2] and C.Rodger[1]
[1] Chemistry Department , University of Strathclyde, Glasgow , Scotland.
[2] Department of Chemistry , University of Pittsburg, Pittsburg,USA.

Samples submitted for forensic examinations are often minute, can be complex mixtures and contain extremely small quantities of the analyte(s) of interest which may have similar chemical structures. Since comparisons between samples eg., control and suspect, and/or positive identification of the analytes are necessary, there is a major requirement for highly discriminative and sensitive detection techniques which must be both robust and reproducible. Ideally, *in-situ* detection techniques are preferable since sample preparation and analysis times can be minimized. Raman spectroscopic techniques have previously been unable to match these performance requirements, however recent developments in SERRS have created a technique which meets the demands of forensic science.

The concept of using SERRS for the examination of dyes, especially acidic dyes, arose from previous difficulties with using other analytical techniques. Initial attempts with a citrate reduced silver colloid to obtain spectra from these anioinic dyes failed due to their poor adsorption onto the negatively charged citrate layer bound to the silver surface. However, aggregation of the colloid and intense spectra were obtained with the addition of poly(L-lysine) and a further enhancement in intensities of the siganals was achieved on the addition of ascorbic acid, due to protonation of the amino groups on the poly(L-lysine) [1,2]. Results from analyses of 22 acidic mono-azo dyes and a blind trial, performed under these conditions, demonstrated that it was possible to distinguish between and characterize all of these dyes, even geometric isomers, thereby confirming the discriminative power of SERRS spectroscopy.

Fibres, typically as single strands of only 2-5 mm in length can be transferred when objects come into contact with each other. If it can be established that any fibres recovered have a common origin this may indicate that contact has occurred. The extraction of dyes from fibres is a common approach for comparing the colourants to determine if the fibres have originated from a common source[3]. However, some dyes, notably reactive dyes, are covalently bound to cotton and wool and can only be removed by enzyme digestion but this takes 22 hours to complete [4]. Recently, an *in-situ* method for the SERRS analysis of a reactive dye covalently bound to cotton, which only takes 4 hours to complete, has been reported [5]. SERRS spectra were also obtained after enzyme digestion of the dyed cotton fibre and found to match a spectrum of reference standard of the dye. No inerferences were detected from either the cotton or the enzyme digest by-products thus establishing the selectivity of the technique. The area of the cotton fibre irradiated by the laser in this study was only 20μm in diameter, which not only confirms the sensitivity of SERRS, but also means that any variation in dyeing even along a 2 mm length of fibre can be studied. This information can be important in forensic examinations especially when fibres are coloured with a mixture of dyes.

The analyses of lipstick smears and inks on documents are typical of other problems encountered in casework which require a highly senstivite and selective detection sytsem and ideally *in-situ* methods of analysis. These have recently been achieved by SERRS with no interferences from the substrate. In the examination of one printing ink, it was established that the pH of the paper was responsible for converting a disazo printing dye from its azo to hydrazo form with a resultant change in colour from blue/green black to a red/black. When the ink was printed on alkaline paper (pH 9.9) the azo tautomer was detected at 514.5nm (Fig. 1a) , but if printed on paper with a pH of 5.5 and analysed after four weeks as shown in Fig.1b, a mixture of the azo and hydrazo tautomers can be detected. Previous SERRS studies on azo dyes [2,6] were used to confirm the presence of the hydrazone form which gives a strong band at 1305 cm^{-1} and this can be detected in Fig. 1b.

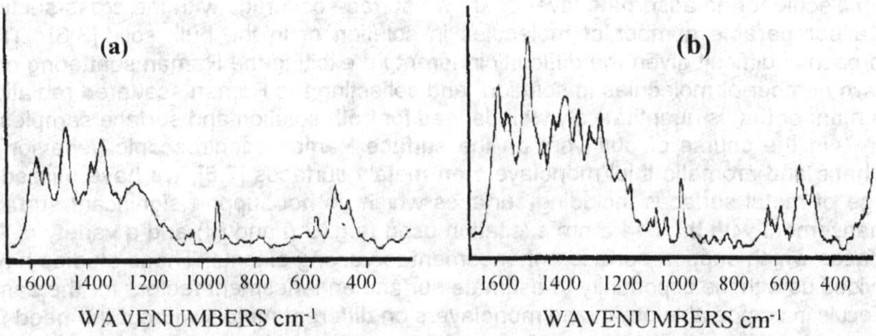

Fig. 1. SERRS spectra of a disazo dye after 4 weeks of being printed on (a) alkaline paper (pH 9.9) and (b) acidic paper (pH 5.5).

1. C.H. Munro, W.E. Smith and P.C. White, Analyst, 118 731 (1993).
2. C.H. Munro, W.E. Smith and P.C. White, Analyst 120 993 (1995).
3. P.C. White and I.R. Tebbett in The Forensic Examination of Fibres.Edited by (J. Robinson, ed.), Ellis Horwood, 1992, p.143.
4 D.R.Rendle, S.R.Crabtree, K.G.Wiggins and M.T.Salter, J.Soc.Dyers Colour. 110 338 (1994).
5. P.C.White, W.E.Smith and C.H.Munro, Analyst, In Press
6. C.H. Munro, W.E. Smith, D.R. Armstrong and P.C. White, J. Phys. Chem., 99 879 (1995).

SURFACE-ENHANCED RAMAN SCATTERING: A CONTINUUM OF POSSIBILITIES

Jeanne E. Pemberton, Chad E. Taylor, Gary G. Goodman, and Pawel Mrozek

Department of Chemistry
University of Arizona
Tucson, AZ 85721

Surface enhanced Raman scattering (SERS) has been useful for many years as a sensitive method for the study of molecules on metals surfaces. The majority of studies to date have been performed using Ag or Au surfaces, with quantitative information most prevalent for the former metal. Enhancement factors per molecule (EF) that are quoted in the literature for Ag vary depending on the nature of the surface roughness, and hence the surface preparation, and the mechanistic contributions to the overall enhancement observed for a surface, whether electromagnetic only or a combination of electromagnetic [1] and chemical [2] enhancement. In general, the enhancements that these surface types support are quite large, with EF values on the order of 10^4-10^6. EF values for surfaces which support much smaller enhancements have been both difficult to observe and difficult to quantify.

Previously reported EF values have generally been determined for a given adsorbate molecule by an attempted comparison of the Raman scattering cross-section per molecule for an adsorbate layer of known surface coverage with the cross-section for a comparable number of molecules in solution or in the bulk solid.[3-6] This approach is difficult given the difficulty inherent in exciting the Raman scattering of a known number of molecules in solution, and collecting the Raman scattered radiation in a manner that is quantitatively well-defined for both solution and surface samples.

In the course of our work on the surface Raman spectroscopic behavior of aliphatic and aromatic thiol monolayers on metals surfaces [7,8], we have utilized a range of metal surfaces including surfaces which do not support significant surface enhancement with the 514.5 nm excitation used (e.g., Au and Pt) and a variety of Ag surfaces which support surface enhancement to varying extents. These studies have provided us with the opportunity to estimate surface enhancement factors for the same molecule in comparable organized monolayers on different metals without the need for comparison with bulk Raman scattering with its associated uncertainties. The results from these EF studies for monolayers of thiophenol are the subject of this report.

The surfaces studied include polycrystalline Pt, polycrystalline Au, polycrystalline Ag, and single crystal Ag (111). Monolayers of thiophenol were formed on all surfaces by immersion into a solution of ca. 20 mM thiophenol in ethanol for 1 hour. Spectra were acquired with the samples in air using 514.5 nm excitation and a single spectrograph with a liquid nitrogen-cooled CCD detector. A holographic filter was use to reject Rayleigh scattering. Surface roughness factors were estimated electrochemically.

Spectral data from thiophenol monolayers were quantitatively analyzed in the frequency region containing the strong ring breathing vibration at ca. 1000 cm^{-1}. In analyzing the spectra, it was assumed that the surface coverage per unit area on each metal was the same. Table 1 shows a comparison of the results for different surface types. The values of EF were normalized to a value of unity for the Pt surfaces which do not support surface enhancement with 514.5 nm excitation.

Surface EF values for the different surfaces decrease as expected in the following order: roughened Ag (polycrystalline)> mechanically polished Ag (polycrystalline) > chemically polished Ag (polycrystalline) ~ chemical polished Ag (111) > mechanically polished Au (polycrystalline) > mechanically polished Pt (polycrystalline). For these surfaces, EF values range from ca. 1 (Pt) to ca. 3 x 10^4 (electrochemically roughened Ag). We believe that these numbers have a high degree of accuracy due to intensity normalization to an unenhanced surface monolayer system which is essentially chemically identical to the surface enhanced systems. The range of EF values determined for these systems clearly indicates that the phenomenon of "surface enhancement" covers a range of behaviors from very strongly to very weakly enhancing. The EF value for the electrochemically roughened Ag surface, a very strongly enhancing substrate, is smaller by approximately one to two orders of magnitude than previously reported for such surfaces. We attribute our EF values partially to the use of surfaces which largely support electromagnetic enhancement as opposed to large chemical enhancements and partially to the method used to calculate EF based on a comparison to a chemically equivalent system on a surface assumed not to support surface enhancement.

ACKNOWLEDGMENTS

The authors are grateful for support of this work by the National Science Foundation (CHE-9504345) .

REFERENCES

1. G.C. Schatz, *Acc. Chem. Res.*, 1984, 17, 370.
2. A. Otto, *J. Raman Spectrosc.*, 1991, 22, 743.
3. R.P. Van Duyne in *Chemical and Biochemical Applications of Lasers*, C.B. Moore, ed.; Academic Press: New York, 1979; Vol. 4, p. 101.
4. D./A. Weitz, S. Garoff, J.I. Gersten, A. Nitzan, *J. Chem. Phys.*, 1983, 78, 5324.
5. R.P. Van Duyne, J.C. Hulteen, D.A. Treichel, *J. Chem. Phys.*, 1993, 99, 2101.
6. W.B. Caldwell, K. Chen, B.R. Herr, C.A. Mirkin, J.C. Hulteen, R.P. Van Duyne, *Langmuir*, 1994, 10, 4109.
7. M.A. Bryant, S.L. Joa, J.E. Pemberton, *Langmuir*, 1992, 8, 753.
8. M.A. Bryant, J.E. Pemberton, *J. Am. Chem. Soc.*, 1991, 113, 8284.

Table 1.

Surface Type	Roughness Factor	Enhancement Factor
Rough Ag (polycrystalline)	4.0	3 x 10^4
Mechanically polished Ag (polycrystalline)	2.5	3 x 10^3
Chemically polished Ag (polycrystalline)	1.3	1 x 10^1
Ag (111) (chemically polished)	1.3	3 x 10^1
Mechanically polished Au (polycrystalline)	1.6	4
Mechanically polished Pt (polycrystalline)	1.6	1

SURFACE ENHANCED VIBRATIONAL SPECTROSCOPY: FROM THE VISIBLE TO THE INFRARED.

Ricardo Aroca
Materials Surface Science Group, University of Windsor,
Windsor, Ontario, N9B 3P4 Canada

Surface enhanced spectroscopy is a field with an active basic research and a broad range of analytical applications. Surface-enhanced Raman scattering [1] (SERS) is one of several well documented phenomena that is presently an established analytical technique for adsorbates on surfaces and interfaces. After the predictive power of the basic electromagnetic enhancement [2] was realized, the development of a highly efficient SERS metallic surfaces, where the radiative plasmon emission is maximized, has been a common goal [3,4]. For certain adsorbates, other effects, such as resonance Raman scattering (SERRS) or chemisorption may produce a multiplicative effect upon the SERS signal [5].

The initial SERS and SERRS studies were carried out mainly with laser lines in the visible spectral region. However, an early report on surface enhancement of infrared spectra by Hartstein [6] using gold or silver islands was an indication of the possibilities for a complementary Raman and infrared enhanced vibrational spectroscopy. The use of near-infrared lasers for excitation of the inelastic light scattering has brought about a *renaissance* in the field of *Raman Spectroscopy* and correspondingly in near-infrared-SERS. In particular, FT-Raman and FT-SERS.

Our main interest has been in the application of SERS and SERRS to the study of Langmuir-Blodgett(LB) monolayers. Since our first report of SERRS in the visible from one LB monolayer on evaporated metal island film [7], a long list of SERS applications to monomolecular films has appeared. Recently, we have extended the applications of SERS to monolayers using the 1064 nm laser excitation and gold island films to obtain FT-SERS [8]. FT-SERS of a single monolayer can also be obtained using silver island films as shown in Figure 1, for a monolayer of naphthacenedione on 6 nm silver island film. The spectra was recorded with a Bruker RFS100 equipped with and FT-Raman microscope. A simple calculation using the electromagnetic interpretation predicts that for a prolate 38nm/33nm, a number of metals could provide a hundredfold enhancement at 1064 nm. For example, Fe, Mo, Ni, Rh, Pt and Ag. Several metals are under study. Finally, closing the circle, the enhanced infrared spectra (SEIR) of LB monolayers deposited on rough metal island films of Ag and gold have been reported [9]. The nature and properties of surface enhanced infrared spectra will be discussed and compared with those of SERS. For practical applications, it is shown that a single sample can be prepared to obtain SERS, FT-SERS and SEIR spectra. The use of different metals for SERS, FT-SERS and SEIR, in keeping with the requirement of the appropriate optical constants for electromagnetic enhancement will be illustrated.

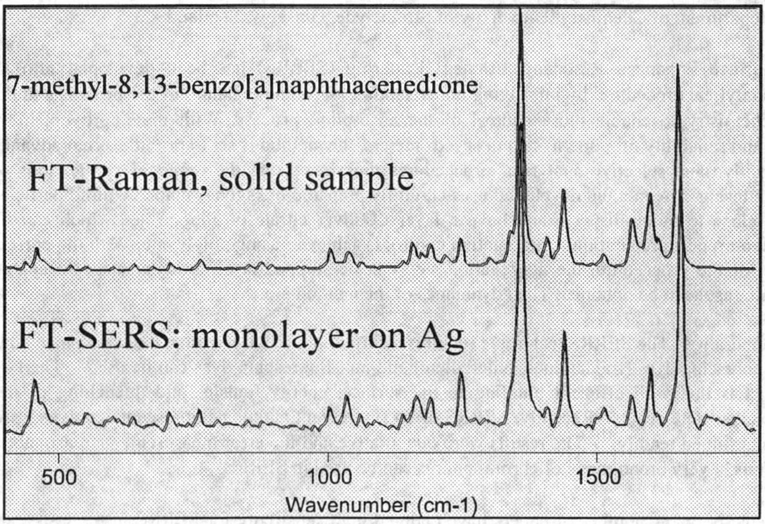

Figure 1

1. R.K. Chang, T.E. Furtak, Eds. *Surface Enhanced Raman Scattering*; Plenum: New York, 1982.
2. P.W. Barber, R.K. Chang, H. Massoudi, Phys. Rev., B, 27, 7251 (1983).
3. J.J. Laserna, W.S. Sutherland, J.D. Winefordner, Anal. Acta, 237, 439 (1990).
4. R.G. Freeman, K.C. Grabar, K.J. Allison, R.M. Bright, J.A. Davis, A.P. Guthrie, M.B. Hommer, M.A. Jackson, P.C. Smith, D.G. Walter, M.J. Natan, Science, 267, 1629 (1995).
5. A. Otto, J. Raman Spectrosc., 22, 743 (1991).
6. A. Hartstein, J.R. Kirtley, J.C. Tsang, Phys. Rev. Let., 45, 201 (1980).
7. R. Aroca, C. Jennings, G.J. Kovacs, R.O. Loutfy, P.S. Vincett. J. Phys. Chem., 89, 4051 (1985).
8. Jennings, C.A., Kovacs, G.J., Aroca, R. Langmuir, 9, 2151 (1993).
9. E. Johnson and R. Aroca, J. Phys. Chem., 99, 9325 (1995).

QUANTITATIVE AND QUALITATIVE ANALYSIS BY SURFACE ENHANCED RESONANCE RAMAN SCATTERING

W.E. Smith, P.C. White, C.H. Munro and C. Rodger
Department of Chemistry, University of Strathclyde, Glasgow, Scotland

Surface Enhanced Resonance Raman Scattering (SERRS) has been developed into an effective analytical procedure by improving understanding of both the nature of the process and the surface chemistry to enable more control of the adsorption process. With the improvement in simplicity and reliability of modern Raman spectrometers, the method is set to provide a real advance in ultrasensitive and selective analytical proceedures thus increasing the range of applications of Raman scattering The method developed uses carefully prepared hexagonal shaped colloid particles which are close to mono-disperse and have a layer of silver citrate on the surface which can be reproduced on repeat preparation. Aggregation control is achieved using polylysine. With the correct application of the proceedure, the aggregation step can be stopped at a particular stage to enable reproducible results to be obtained. The advantages of the method are,

1. Generality - The SERRS technique requires a chromophore but in contrast to resonance both fluorescing dyes (rhodamines, Nilered etc) and non-fluorescing dyes can be used . The key fact is that with efficient bonding to the surface, energy transfer and quenching of the excited state occurs. Recently, SERRS of 25 fluorescent and non-fluorescent azo dyes were determined readily.[1] The results were quantitative and the error measurements reasonable. Thus, a very broad range of chromophores can be used in SERRS.

2. Sensitivity - A number of papers have claimed high sensitivity for SERRS in the last year. In our case, we show a concentration limit of 7.8×10^{-18} indicating a much larger than expected enhancement and sensitivity comparable to or better than fluorescence. [2] This is not a detection limit. At this point, there are between 35 and 200 molecules of dye in the excitation beam. Without special facilities, the problems with contamination are serious. It is therefore essential to show the concentration dependence and reproducability of the signal to claim a final detection limit. We have shown this for Rhodamine 6G with 6 replicate analysis per concentration point.

3. Selectivity - SERRS provides sharp signals and mixtures of 4 or 5 dyes can be discriminated by eye in a blind trial.[1] Using modern methods of discrimination on more than one laser frequency, each species in much more complex mixtures could be determined quantitatively.

4. Photodecomposition - Photodecomposition is a problem with both SERS and resonance. It is much less likely with SERRS. In the first instance, very low power levels are used. For example in one experiment to obtain resonance Raman scattering with conventional macro sampling 100mW of excitation was used. SERRS with a good signal to noise ratio was obtained with 1mW. In addition, even at high powers the energy transfer between ligand and metal which quenches the fluorescence also helps to prevent photodecomposition.

5. Concentration Range -. The huge enhancement of SERRS means that concentration ranges over several orders of magnitude can be studied and self absorption does not interfere. Consequently the build up of monolayer coverage on the surface can be studied starting at less than one hundred thousandth of a monolayer. In addition, at these very low concentrations, there is little to no effect of the ligand on the aggregation process - a problem encountered with SERS.

6. Reproducability - For quantitation, the reproducability of the colloid production and the surface adsorption require to be controlled. With Lee -Meisel colloid, control of the preparation conditions leads to the production of nearly monodisperse hexagonal particles with a layer of stable silver citrate on the surface. Adsorbates are used at monolayer coverage or below and the solubility of the adsorbate in the aqueous or nonaqueous colloid medium[4] is essential for controlled adsorption.

Depending on the nature of the process, conditions can be set to favour quantitative analysis or to favour an investigation of the nature of the surface. Where excitation is at molecular resonance, the chromophore causes depolarisation of the incident light beam and mitigates against the determination of molecular orientation of the ligand on the surface. This is not yet a universal condition but certainly applies in some cases. This makes the technique less sensitive to small changes in experimental conditions and hence improves the reproducibility for quantitative analytical procedures[5] However, if off resonance excitation is used, pre-resonance still gives selectivity of the species containing the chromophore but the scattering intensities are dependant on the molecular orientation at the surface providing additional information on the orientation of the molecule and of the polarisation of the electronic structure by the surface. This has been studied in detail both for a dye (solvent yellow 14)[5] and for a protein (cytochrome C).[6] Thus, different conditions are required depending on whether the objective is ultra sensitive quantitative analysis or surface structural analysis. New calculations of normal coordinates for azo and hydrazo forms of azo dyes help to establish the effect of surface polarisation, complexing and orientation thus defining the nature of the SERRS for the study of surfaces.[3,7]

The new understanding puts SERRS in the forefront of the development of analytical techniques. Firstly, it enables the whole field of colorometric analysis to be used at attomolar concentrations or below to provide novel selective and sensitive organic and inorganic analytical procedures.[1] Secondly, it makes it possible to observe surface species from non SERS active surfaces. [9] This has been applied to metals, fibres and papers. In this particular context, the very small sample size that can be achieved by using micro Raman facilities enables applications in many fields including forensic science.

1. C.H. Munro, W.E. Smith, P.C. White, Analyst, (1995), 120, 993.
2. C.Rodger, W.E. Smith, P.C.White, G. Dent, M. Edmondson, to be published.
3. C.H. Munro. W.E. Smith, M. Garner, J. Clarkson, P.C. White, Langmuir, (1995), 11, 3712
4. J. Clarkson, C. Campbell, B.N. Rospendowski, W.E. Smith, J.Raman.Spec., (1991), 22, 771
5. C.H. Munro, W.E. Smith, D.R. Armstrong, P.C. White, J.Phys.Chem., (1995), 99, 879.
6. I.D.G. Macdonald, W.E. Smith, Langmuir (1996), accepted
7. J. Clarkson, D.R. Armstrong, W.E. Smith, J.Phys.Chem, (1995), 99, 17825
8. Wilson, H., Smith, W.E., J.Raman.Spec., (1994), 25, 899.

DETERMINATION OF CHEMICAL CONTRIBUTION IN NIR-SERS

E. J. Liang[1,2]* and W. Kiefer[2]

[1]Center of Analysis & Testing, Zhengzhou University, 450052 Zhengzhou, P. R. China
[2]Institut für Physikalische Chemie, Universität Würzburg, Marcusstr. 9-11, D-97070 Würzburg, Germany

The chemical enhancement factor in silver colloids was generally calculated by the ratio of the SERS intensities after and before addition of chloride ions [1]. We argue that chloride (also other) ions contribute not only to the chemical but also to the electromagnetic enhancement. In order to calculate the chemical enhancement factor, a correction with the absorption values at the excitation and probably also the Raman scattering wavelength is needed. The absorption of the surface plasmon excitation resonance changes with time once probe molecules or the halide ions are added to the silver colloids. In fact, the NIR-SERS intensity changes also with time. In order to discriminate the importance of the two mechanisms in NIR-SERS, it is necessary to observe the time evolution of the two processes simultaneously.

A simultaneous observation of the two processes shows that (i) the absorption value of silver colloids in the presence of pure pyridine is smaller than in the presence of Cl^- ions but larger than in the presence of Br^- ions, whereas the NIR-SERS intensity of pyridine is much stronger in the presence of halide ions than in the absence of them; (ii) though the absorption of the surface plasmon excitation resonance in the presence of Cl^- ions is larger than in the presence of Br^- ions, the NIR-SERS intensity is stronger in the presence of the latter than in the presence of the former. These results indicate that Br^- ions are able to create more surface active sites than Cl^- ions on colloidal silver surface.

The time dependence of the absorption at the excitation wavelength and corresponding NIR-SERS of pyridine in the presence of halate ions are also measured. It is found that the absorption value at the excitation wavelength caused by halate ions increases in the order $BrO_3^- < ClO_3^- < IO_3^-$. However, the total enhancement factor of the NIR-SERS does not follow this order. BrO_3^- ions lead to the largest enhancement while IO_3^- and ClO_3^- ions result in a smaller enhancement factor. These results demonstrate again that the chemical effect contribute to the NIR-SERS also in the presence of halate ions. Among the halate ions, BrO_3^- ions are the best candidates for creating surface active sites.

From the measurements we are able to determine the relationship between surface plasmon resonance and surface enhancement factor F. In Table 1 we show one example of the calculated results. A^r and $\triangle I$ in Table 1 represent the relative increment of the absorption value at the excitation wavelength and the corresponding

TABLE 1. Dependence of relative changes of SERS intensity on the relative changes of absorption A at the excitation wavelength in the presence of Cl^- ions.

	t_1 = 1.0 min. t_2 = 4.17 min.	t_1 = 4.17 min. t_2 = 7.34 min.	t_1 = 7.34 min. t_2 = 10.51 min.	t_1 =10.51 min. t_2 = 13.68 min.
$A^r = A(t_2) / A(t_1)$	2.09	1.26	1.14	1.08
$\triangle I_{SERS} = I_{SERS}(t_2)/I_{SERS}(t_1)$	2.67	1.25	1.14	1.06
$\triangle I_{SERS} / A^r$	1.28	0.99	1.00	0.98

increment of the SERS intensity from time t_1 to time t_2, respectively. From Table 1, a relationship between electromagnetic enhancement factor and the absorption value of the surface plasmon resonance at the excitation wavelength can be derived. The enhancement factors from the surface plasmon resonance and from the surface active sites can be determined seperately [2].

Based on Table 1, we get the following equation for the chemical enhancement factor:

$$F_{ch} \approx F^-_{ch} = g\,[I^-_{SERS}(t) / I_{SERS}][A / A^-(t)] \quad \text{at } t \geq 4 \text{ min.} \qquad (1)$$

where I^-, I, A^- and A being the SERS intensity and the absorption values of the surface plasmon excitation resonance at the excitation wavelength in the presence and absence of halide or halate ions, respectively. F_{ch} and F^-_{ch} are the total chemical enhancement factor and the chemical enhancement factor caused by the presence of anions, respectively and g is correction factor.

According to Eqn. (1), the chemical enhancement factor can be easily calculated from the experimental results as shown in Table 2.

TABLE 2. Chemical enhancement factors for the 1010 cm^{-1} band of pyridine caused by the presence of halide and halate ions.

ions	Cl^-	Br^-	I^-	ClO_3^-	BrO_3^-	IO_3^-
F_{Ch}	~ 69	~ 163	/	~ 27	~ 174	~ 92

1. P. Hildebrandt and M. Stockburger, J. Phys. Chem. 88 (1984) 5935.
2. E. J. Liang and W. Kiefer, to be published.

CHARGE TRANSFER PROCESSES IN SURFACE-ENHANCED RAMAN SCATTERING. FRANCK-CONDON ACTIVE VIBRATIONS OF PYRAZINE

J. F. Arenas, M. S. Woolley,* J. C. Otero, and J. I. Marcos

Department of Physical Chemistry,
Faculty of Sciences, University of Málaga, E-29071 Málaga, Spain.

SERS spectra of pyrazine on silver electrode have been recorded and analyzed, assuming a charge transfer effect and using selection rules analogous to those of resonance Raman. With the aim of predicting the effect of this mechanism on the selective enhancement of fundamentals a method has been proposed based on an analysis of the results of geometry optimizations carried out by *ab initio* calculations. For this, we have assumed the hypothesis that in a SERS-CT experiment, an electron is transferred from the metal to the adsorbate, for which the CT state is equivalent to that of the radical anion from the molecular point of view.

The strongest bands in the SERS spectra of pyrazine coincide with those assigned to the normal modes connecting the equilibrium geometries of the neutral molecule and the radical anion.

As can be seen in Figure 1, a significant enhancement of vibrations 8a, 9a, 1 and 6a can be observed with respect to the Raman spectrum of pyrazine in aqueous solution.

In all cases the strongest bands in the spectra are directly related to the largest $\Delta\mathbf{Q}$ values obtained through the transformation $\Delta\mathbf{Q} = \mathbf{L}^{-1}\Delta\mathbf{R}$. $\Delta\mathbf{Q}$ is a vector that vibrationally characterizes the chromophore and represents the displacement of the potential energy surface minimums of the two states implicated in the resonance process along each normal coordinate. $\Delta\mathbf{R}$ is a vector that contains the differences between the geometric parameters of the excited states (D_0 in SERS-CT or S_i in resonance Raman (RR)) and the ground state (S_0) and is constructed from the results of the optimized geometries.

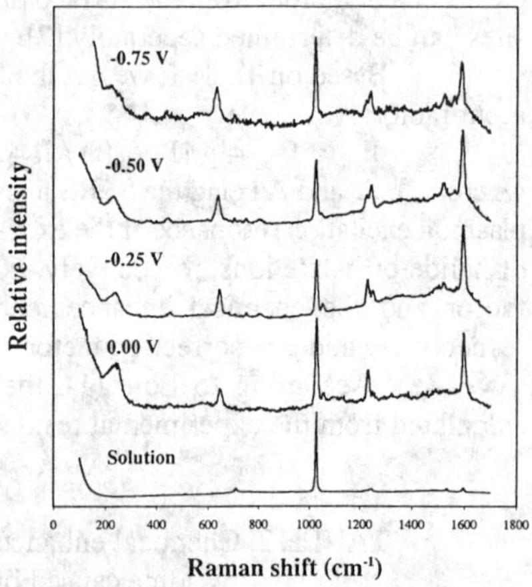

Figure 1: The SERS of pyrazine on electrode at several potentials with the Raman solution spectrum.

The L matrix of the ground electronic state needs to be known for which the force field at RHF/3-21G level has been calculated according to the SQM methodology. In the case of pyrazine, the vector $\Delta\mathbf{Q}$ will contain exclusively totally symmetric vibrations since all of the here studied equilibrium structures belong to the D_{2h} point group.

When the charge is trapped by the molecule, the nuclear structure relaxes in the direction of the potential energy surface minimum of the excited state. When the molecule comes back to its ground state, the normal modes connecting the equilibrium structures of the upper and the lower states remain excited much more probably than any other.

The origin of the selective enhancement of fundamentals that has been described is analogous to that of RR spectroscopy via Franck-Condon factors (A-term).[1] In practice, the derived selection rules of this mechanism can be summarized by the well-known empirical rule of Tsuboi [2].

If in the obtained spectra an agreement between the observed enhancement and the theoretical prediction is observed, we will be in a position to confirm the presence of the CT mechanism and to extract information related to the charge transfer process giving rise to the spectra.

As can be appreciated, such a good agreement between the ΔQ values and the selective enhancement of the A_g fundamentals has led us to conclude that a charge transfer mechanism gives rise to the SERS of pyrazine (Table 1). The prediction capability of our analysis has been checked with published RR spectra of pyrazine and pyrazine-d_4 [3] and has also successfully accounted for the observed behavior (Table 2).

These values of ΔQ can be considered as semiquantitative selection rules applicable to the SERS active modes via a mechanism analogous to the A-term in RR.

Table 1: Relative intensities of A_g fundamentals measured in the SERS of pyrazine with their respective ΔQ values. Electrode potentials measured vs. Ag/AgCl saturated electrode.

Mode	0.00 V	-0.25 V	-0.50 V	-0.75 V	ΔQ
2	19	16	12	7	0.016
8a	72	131	113	49	-0.183
9a	10	25	37	28	-0.172
1	100	100	100	100	-0.197
6a	21	48	48	36	-0.333

Table 2: ΔQ values and relative intensities for RR spectra of pyrazine.

ΔQ			Relative intensities [3]		
S_1-S_0(d_0)	S_2-S_0 (d_0)	S_2-S_0 (d_4)	S_0-S_1(d_0)	S_0-S_2(d_0)	S_0-S_2(d_4)
-0.005	-0.003	-0.012	-	-	-
0.036	-0.017	-0.026	44	-	-
-0.123	-0.030	0.128	222	-	20
-0.092	-0.244	0.236	100	100	100
0.192	-0.390	-0.376	78	100	100

1. R.J.H. Clark, T.J. Dines, Angew. Chem. Int. Ed. Engl., 25, 131 (1986).
2. D.P. Strommen, K. J. Nakamoto, Chem. Educ., 54, 474 (1977).
A.Y. Hirakawa, M.Tsuboi, Science, 188, 359 (1975).
3. I. Suzuka, Y. Udagawa, M. Ito, Chem. Phys. Lett., 64, 333 (1979).

SURFACE ENHANCED RAMAN SCATTERING (SERS) OF PYRAZINE ADSORBED ON POLYCRYSTALLINE GOLD ELECTRODES: A STUDY OF THE RELATIONSHIP BETWEEN SER INTENSITY AND BOTH SURFACE COVERAGE AND MORPHOLOGY.

Alexandre G. Brolo[1], D. E. Irish[1], G. Szymanski[2] and J. Lipkowski[2]

[1] Guelph-Waterloo Centre for Graduate Work in Chemistry, Department of Chemistry, University of Waterloo, Waterloo, ON, Canada, N2L 3G1.

[2] Guelph-Waterloo Centre for Graduate Work in Chemistry, Department of Chemistry and Biochemistry, University of Guelph, Guelph, ON, Canada, N1G 2W1.

Surface enhanced Raman scattering (SERS) has been widely used for the "in situ" study of electrochemical processes. The giant signal enhancement (10^5 - 10^6 times greater than normal Raman scattering (NRS)) makes this technique sensitive to even sub-monolayer amounts of adsorbate. However, the magnitude of the enhancement depends on several factors, such as applied potential, excitation wavelength, surface coverage, and morphology.

The dependence of the electrochemical SER intensity on the applied potential and on the energy of the incident photon has been studied [1]. The relationships are well understood and can be interpreted based on the two most accepted surface enhancement theories - the charge transfer and the electromagnetic mechanisms. However, few works have dealt with the dependence of the SER intensities on both surface coverage and morphology [2].

The study of the relationship of SER intensities with surface coverage is limited by the lack of independent electrochemical information for the surface concentration of the organic molecules. However, these electrochemical data are being produced in our Centre [3].

A previous study of the dependence of the SER intensity of the pyridine (py) ring breathing mode with surface coverage has been conducted in our laboratory[4]. A direct proportionality was found up to 2/3 of a monolayer.

The NRS of aqueous solutions of pz and its protonated forms has also been studied recently [5]. An extensive study of pz adsorbed from aqueous halide solutions on silver electrodes [6] showed that pz binds to the electrode surface via the N lone pair (end-on configuration).

In this work, the surface concentration of pz at several potentials, measured recently [7], are compared to the SER intensity of pz adsorbed on a polycrystalline Au electrode. The SER intensities of pz measured at different potentials are presented in Figure 1.

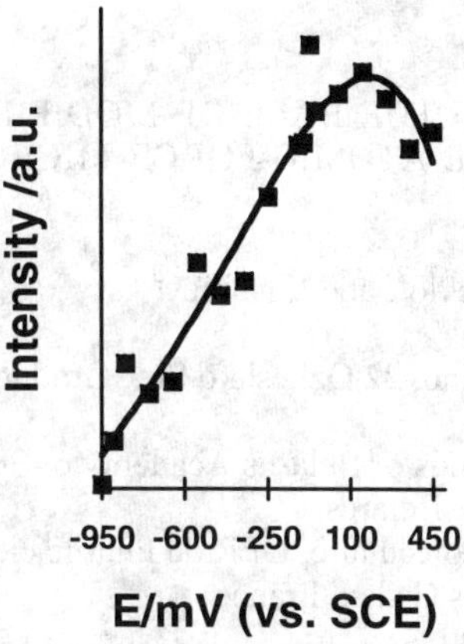

Fig. 1: Dependence of the SER intensity with the applied potential, after 20 ORC's. Au/$KClO_4$ 0.1M/pz 10^{-3} M.

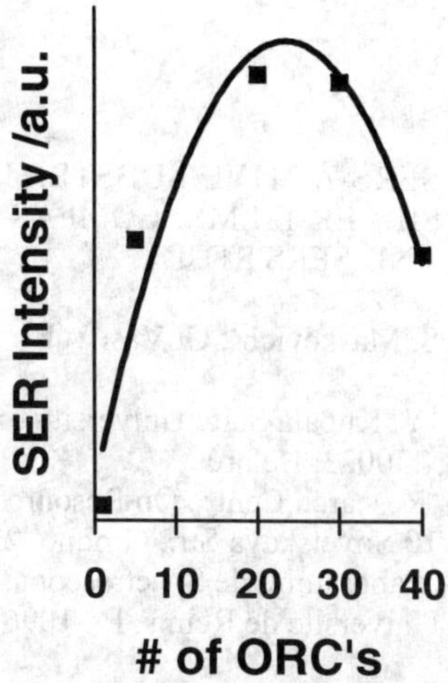

Fig. 2: Dependence of the SER intensity on the number of ORC's. Same system as Fig. 1. E = 50 mV

The polycrystalline gold electrode was activated following a sequence of oxidation reduction cycles described elsewhere [8]. This procedure yielded a stable and time independent spectrum. However, the SER intensity of the pz ring breathing mode was dependent on the number of ORC's performed. Figure 2 shows the plot of the intensity of this band at a given potential for different numbers of ORC's. These results suggest that the dependence is due to a change in the surface morphology. The surface morphology for different numbers of ORC's was monitored by probe microscopic techniques (AFM/STM). The size distribution of the surface bumps and the roughness factor were compared to the SER intensity.

1. D. A. Guzonas, D. E. Irish and G. G. Atkinson, *Langmuir*, 6, 1102 (1990).
2. D. D. Tushel, J. E. Pemberton and J. E. Cook, *Langmuir*, 2, 380 (1986).
3. J. Lipkowski and L. Stolberg, *Adsorption of Molecules at Metal Electrodes*, J. Lipkowski and P. N. Ross eds., VCH, 1992, p. 171.
4. L. Stolberg, J. Lipkowski and D. E. Irish, *J. Electroanal. Chem.,* 300, 563 (1991).
5. A. G. Brolo and D. E. Irish, *Z. Naturforsch.*, 50a, 274 (1995).
6. A. G. Brolo and D. E. Irish, *submitted to the J. Electroanal. Chem.*
7. A. Iannelli, J. Richer and J. Lipkowski, *Langmuir*, 5, 466 (1989).
8. P. Gao, D. Gosztola, L-W.H. Leung and M. J. Weaver, *J.Electroanal.Chem.*, 233, 211 (1987).

SERS-ACTIVE SUBSTRATA ON THE BASE OF MODIFIED THIN SILVER FILMS. COMPARATIVE ATOMIC-FORCE MICROSCOPY AND SERS STUDY

S. Maskevich[1], G. Vasilyuk[1*], I. Sveklo[2], and I. Nabiev[3]

[1]Y. Kupala State University of Grodno, 22 Ozheshko Str., Grodno, 230023, Belarus
[2]Research Centre On Resource Savings of Belarus Academy of Sciences, 10 Sovetskaya Str., Grodno, 230023, Belarus
[3]Laboratoire de Spectroscopie Biomoleculaire, UFR de Pharmacie, Universite de Reims, F-51096 Reims Cedex, France

Surface enhanced Raman scattering (SERS) is one of the most sensitive and informative methods for studying a structure of adsorbed molecular systems and physical chemistry of surface. Thin silver films (TFS) are widely used as SERS- active substrata. Morphology of TSF surface determines optical, Raman enhancing and adsorption properties as well as its stability. It is known that TSF post-evaporation annealing changes surface structure that is observed in change of absorption spectra (hypsochromic shift of absorption maximum from 620 to 550 nm, decreasing of optical density and narrowing of absorption band) [1-3]. On the surface topography annealed films, which were received by means of atomic- force microscope (AFM), a typical increase of silver granule dimensions and distances between them were registered [1]. We had showed earlier that high-temperature annealing (~350°C) leads to increase of relative contribution of long-range electromagnetic mechanism to the general effect of Raman scattering enhancement inducted by surface [4,5].

In the present work it is studied the possibility of modification of annealed thin silver films (TSF) by means of treatment by different solvents with the object of varying of physical and chemical properties for receiving of sers-active substrata, which are optimized for specific compounds or methods.

It was found that optical and Raman scattering enhancing properties of annealed TSF are changed at a definite sequence of applying solutions (for example, at first - acetonitrile and then - ethanol). In particular, the change of absorption spectra (narrowing of absorption band and hypsochromic shift of its maximum) after acetonitrile treatment are qualitively similar to changes occurred by annealing. It was carried out a comparative analysis of the atomic force microscopy (AFM)-images of annealed and non-annealed TSF (before and after solvents treatment) (example-in Fig.1). It was also made a comparative analysis of SERS spectra of different aromatic and heterocyclic compounds (in particular, perchlorate of 3'-metoxy-3,3-dimethyl-1-styrylphthalylium (MSPh) and 3',3'-dimethylphthalan-1'-spiro-2[2H-1-benzopyran] (PhSBP)) adsorbed on these TSF. Spectra of the present compounds adsorbed on TSF, which were modificated by such method, principally differ from spectra in case of other investigated TSF application. For instance, we received SERS spectra of cyclic (colorless) species of PhSBP (in Fig.2) (particularly testified by the presence of the band near 1640cm^{-1}) which we failed to receive by ordinary TSF application.

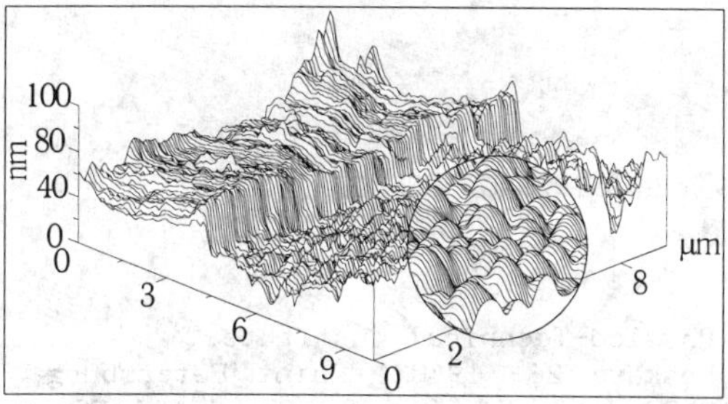

Fig.1. AFM-surface topography of the edge region of acetonitrile applied on annealed TSF. The zoom factor for the view inserted is 15.

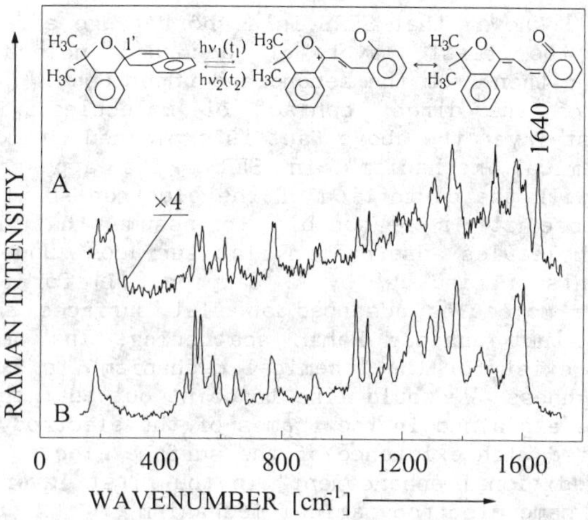

Fig.2. SERS spectra of PhSBP adsorbed on annealed TSF from acetonitrile solution at 514.5 nm excitation. A: Cyclic (colorless) species (without ethanol applying of the TSF). B: Color species (with ethanol applying of the TSF after acetonitrile solution applying).

The obtained results can be explained in the following way:
-a film solvent containing silver particles is formed after applying of acetonitrile
-posterior applying of ethanol influences on "chemical" mechanisms of Raman scattering enhancement, which are specific for different compounds.

Acknowledgments. This work was supported by INTAS project (Grant 93-1829) and by FFR of Belarus.

1. D.J. Semin, K.L. Rowlen, Anal. Chem. 66, 4324 (1994).
2. F.R. Aussenegg, A. Leitner, M.E. Lippitsch, M. Riegler, Surf. Sci. 189/190, 935 (1987).
3. R.P. Van Duyne, J.C. Hulteen, D.A. Treichel, J. Chem. Phys. 99, 2101 (1993).
4. S.A. Maskevich, I.F. Sveklo, A.I. Yanul, I.R. Nabiev, L.N. Kivach, A.I. Sviridenok, Dokl. AN B (Belarus), in press, (in Russian).
5. S. Maskevich, I. Sveklo, A. Feofanov, A. Ianoul, V. Oleinikov, O. Fedorova, S. Gromov, I. Nabiev, M. Alfimov and L. Kivach in Spectroscopy of Biological Molecules, (J. C. Merlin et al., ed.), Kluwer Academic Publishers, Dordrecht, 1995, p. 27.

WHY SERS IS A SHORT RANGE EFFECT

A.M. Polubotko

A.F. Ioffe Physico-Technical Institute,
Politechnicheskaya 26, 194021 Saint Petersburg, RUSSIA.
Fax: (812) 247-10-17, E-mail: Alex@marina.ioffe.rssi.ru

It is well known, that SERS is a short range effect. The enhancement from the first layer of adsorbed molecules is significantly larger then from the second and upper ones.

Because of the direct contact of molecules with the surface in the first layer the above fact is explained by existence of so-called "chemical mechanism" in SERS. Its appearance is usually connected with the distortion of the electron shell due to adsorption. Therefore it is reasonable to assume that it must manifests for the molecules adsorbed on flat surfaces. However the numerical experiments carried out by A. Campion, [1] for example, point out that the molecules adsorbed on flat surfaces and some similar objects do not enhance Raman scattering. The necessary factor, which must exist for the "chemical mechanism" is existence of the surface roughness. We would like to point out that the first layer effect can be explained in the frames of the electrodynamical mechanism, associated with existence of the surface electromagnetic field. Thus the "additional enhancement" in the first layer is the consequence of the same electrodynamical mechanism.

As a model of the rough surface let us consider an ideally conductive lattice of the eshelett type, such as in [2]. The electromagnetic field above the lattice can be represented in the form

$$E = g_{0,inc} \exp\left(-ik_o z \cos\vartheta_o + ik_o y \sin\vartheta_o\right) + \sum_{n=-\infty}^{+\infty} g_n \exp\left(i\alpha_n y + i\gamma_n z\right) \quad (1)$$

Here ϑ_0 is the incident angle, g_n is the amplitude of the nth space harmonic, l is the lattice period. More detailed explanation of the designations one can find in [2]. In the region of wedges the electric field is singular and can be expressed as

$$E \sim |g_{o,inc}|C_0\left(\frac{l_1}{r}\right)^{\beta} . \quad (2)$$

Here $C_o \sim 1$, $l_1 = (l, h)$ is characteristic size of the lattice (period or height), $0 < \beta < 1/2$. As it was pointed out in [2] the SERS reason is the increase of the dipole and quadrupole light-molecule interactions in the surface field strongly varying in space. Besides only the d_{ez} component of the dipole moment, which is perpendicular to the surface and Q_{eii} components of the quadrupole moments are responsible for SERS. In accordance with the results of [2] the enhancement coefficient of the cross-section, associated with the enhancement of the quadrupole interaction is estimated as

$$G = \frac{d\sigma_{surf}}{d\sigma_{vol}} \sim \frac{(Ba)^4}{16} \frac{\left|\frac{\partial E_i^{inc}}{\partial x_i}\right|^2_{surf}}{|E_{inc}|^2} \frac{\left|\frac{\partial E_i^{s_1}}{\partial x_i}\right|^2_{surf}}{|E_{s_1}|^2}, \quad (3)$$

where $\partial E_i/\partial x_i$ are derivatives of the incident and scattered fields at the surface, $B \sim 10^2$ is a numerical coefficient, a is the molecule size. Using (3) one can obtain estimation for the relation of the signals from the first and the second layers, placed at the top of the wedge. Its order is

$$\left|\left(\frac{\partial E_i}{\partial x_i}\right)_1\right|^4 : \left|\left(\frac{\partial E_i}{\partial x_i}\right)_2\right|^4 . \quad (4)$$

Here the signs 1 and 2 designate the values of the derivatives in the first and the second layers. In spite of this estimation is not correct it may give the upper limit of the difference between the enhancement coefficients. The corresponding value (4) depend on the value β and is of the order.

$$(3)^{4+4\beta} \sim (100 - 1000) . \quad (5)$$

One can see that the relation of the enhancement coefficients is very large and is sufficient for explanation of the first layer effect only by the electrodynamical mechanism. Analogous estimations in some other more realistic models corroborate this result.

REFERENCES:

1. A. Campion, J. Electron Spectroscopy and Related Phenomena 29, 397 (1983).
2. A.M. Polubotko, Phys. Lett. A 146, 81 (1990).

CHEMICAL TRANSFORMATIONS IN SURFACE-ENHANCED RAMAN SCATTERING.

J.L. Castro, J.C. Otero and J.I. Marcos*

Department of Physical Chemistry,
Faculty of Sciences, University of Málaga, E-29071 Málaga, Spain.

We report two cases where the respective SERS spectra can be explained via the interaction of two chemically different species with the silver colloid surfaces.

A) Acrylic Acid.

In previous work [1], evidence of polymerization has been found at a concentration of 6×10^{-4} M, where the ethylene plane of acrylic acid is found to lie flat with that of the surface. At the higher concentration of 2×10^{-3} M, the ethylene plane is found to be tilted with respect to the surface and no polymerization takes place.

The SERS recorded at concentrations in the order of 10^{-2} M are shown in Fig. 1 and as can seen, bands with frequencies of 750, 813, 887, 1404, 1462, 2934 and 2972 cm^{-1} are observed as well as those at 665, 902, 1276, 1633, 3007 and 3094 cm^{-1} which belong to the acrylate anion [1].

The higher frequency bands observed at 2943 and 2972 cm^{-1} are characteristic of CH stretching modes in saturated carbon molecules. This leads us to think that a saturated acid is present such as propanoic acid, the reaction product of a possible surface catalyzed hydrogenation of the ethylene group in acrylic acid.

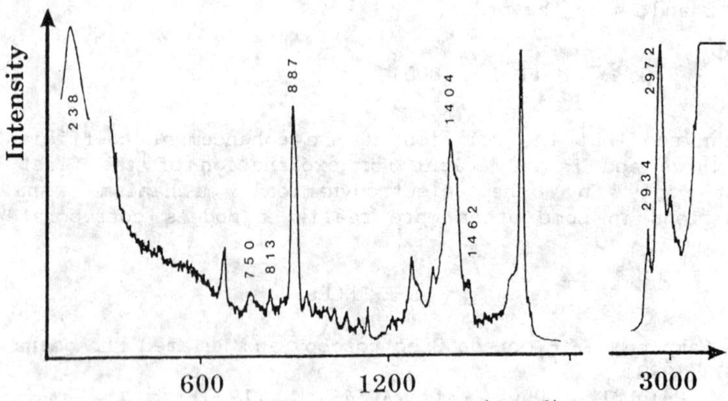

Fig. 1. SERS spectrum of acrylic acid on silver sol, 10^{-2} M.

The strong intensity of the bands recorded at 887 and 1404 cm^{-1} are due to the ν(C-CO) and the ν_s(OCO) modes respectively and imply that the propanoate ion interacts with the metal surface via the carboxylate group. The intensification of the $\nu_{as}(CH_3)$ mode at 2972 cm^{-1} corresponds with the behavior of those acids with an uneven number of carbon atoms [2].

B) Aspartic Acid.

We have recorded a SERS spectrum different to those already published for this acid [3] at concentrations in the order of 10^{-3} M. This spectrum can be seen in Fig. 2.

The difference is in the appearance of new bands at 358, 754, 1280, 1648 and 3046 cm^{-1}. The fact that the band at 3046 cm^{-1} is characteristic of an ethylenic CH stretching mode leads us to propose the hypothesis that the interactive ion $^{-}OOC\text{-}CH_2(NH_3)\text{-}COO^{-}$ can undergo a Hofmann elimination reaction on the metal surface to give the more stable $^{-}OOC\text{-}CH{=}CH\text{-}COO^{-}$ ion.

The SERS spectrum of fumaric acid studied by M. Moskovits and J.S. Suh [4] shows bands of medium intensity at 361, 753, 1274 and 1395 cm^{-1}, as well as a strong band at 1646 and another weak one at 3051 cm^{-1}, which further supports this hypothesis. In the spectrum of Fig. 2, the symmetric stretching modes of the OCO group of the aspartate and fumarate anions would be found to be overlapping in the band at 1396 cm^{-1}. The band at 1646 cm^{-1} assigned as ν(C=C) would indicate planar adsorption of the fumarate anion.

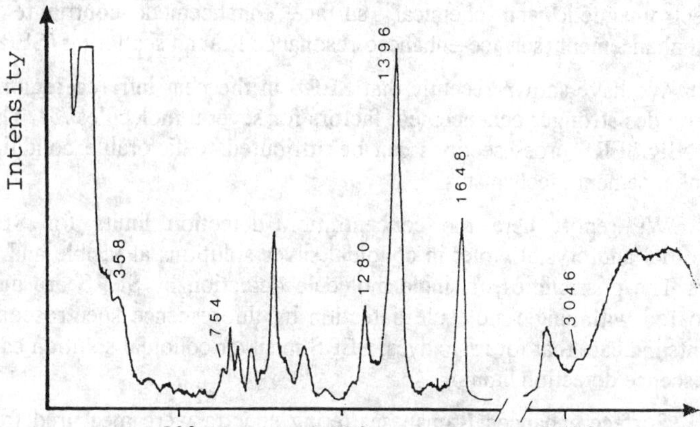

Fig. 2. SERS spectrum of aspartic acid on silver sol, 10^{-3} M.

1. J.S. Suh and K.H. Michaelian, J. Phys. Chem. 91, 958 (1987).
2. M. Moskovits and J.S. Suh , J. Am. Ghem, Soc. 107, 6826 (1985).
3. J.L. Castro, M. A. Montañez, J. C. Otero and J. I. Marcos, J. Mol. Struct. 349, 113 (1995).
4. M. Moskovits and J.S. Suh, J. Phys. Chem. 88, 1295 (1984).

APPROACH TO SINGLE MOLECULE DETECTION USING SURFACE-ENHANCED RAMAN SCATTERING (SERS)

Katrin Kneipp[1,2], Yang Wang[3], Harald Kneipp[4], Irving Itzkan[1], Ramachandra R. Dasari[1], and Michael S. Feld[1],

[1]George R. Harrison Spectroscopy Laboratory
Massachusetts Institute of Technology, Cambridge, MA 02139, USA
[2]Department of Physics, Technical University Berlin,
D-10623 Berlin, Germany
[3]PROCEPT Inc., Cambridge, MA 02139
[4]LAS- GmbH, D 14532 Stahnsdorf, Germany

The potential of surface-enhanced Raman scattering (SERS) for trace analysis has been explored actively in recent years /1/. The lowest detected concentration levels were $4x10^{-15}$ M for a cyanine dye adsorbed on a silver micro-electrode /2/, $1x10^{-12}$ M for crystal violet /3/ and $8x10^{-16}$ M for rhodamine 6G in colloidal solutions of silver /4/. All these experiments, were performed at visible excitation light. The "normal" resonance Raman enhancement of the dyes along with "electromagnetic and chemical" surface enhancement contribute to the observed enhancement (surface-enhanced resonance Raman scattering (SERRS)).

We have shown recently that SERS in the near infrared region (NIR-SERS) provides stronger enhancement factors for several molecules /5/. The large effective NIR-SERS cross sections can be attributed to favorable conditions for various enhancement mechanisms.

We report here the concentration detection limits for SERS of rhodamine 6G and crystal violet in colloidal silver solutions at visible and at NIR excitation. The possibilities of single molecule detection by SERS are discussed and compared with single-molecule detection by fluorescence spectroscopy. Our experiments indicate that for ionic dyes, SERS in silver colloidal solution can reach the fluorescence detection limits.

Surface-enhanced Raman scattering spectra were measured from dye molecules at concentrations as low as 0.8 femtomolar in colloidal silver solution activated by NaCl ions. Experiments conducted with small scattering volumes show that less than 100 molecules are sufficient to give rise to a SERS spectrum with reasonable signal to noise ratio.

Figure 1 illustrates this by SERS spectra obtained from estimated 60 rhodamine 6G molecules adsorbed on colloidal silver, using 400 mW laser excitation power at 514.5 nm (a) or 200 mW laser excitation power at 830 nm (b).

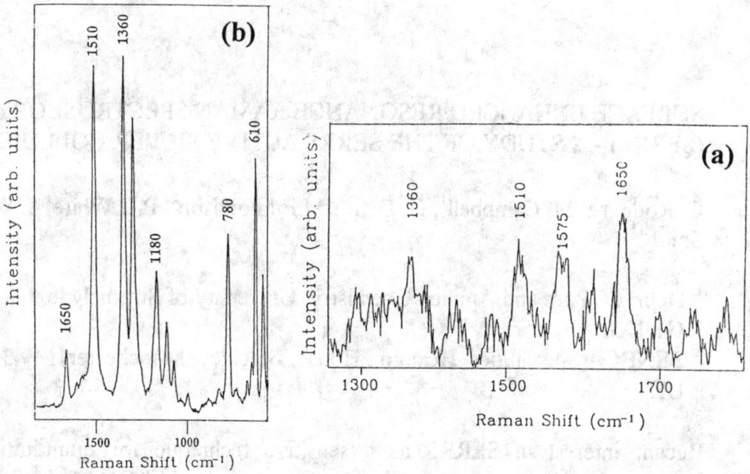

Fig. 1 : SERS spectra of estimated 60 rhodamine 6G molecules in colloidal solution of silver. Spectrum (a) was measured in 660 s collection time, spectrum (b) in 10 s.

Our experimental results demonstrate that SERS can achieve detection limits comparable to those of fluorescence spectroscopy, and at the same time provides higher structural specificity. Higher photostability of molecules due to non-resonant excitation of Raman scattering can favor SERS over fluorescence for single molecule detection

1. J. J. Laserna, Analytica Chim. Acta, 283 (1993) 7.
2. J. Macklin, in Proc. XIV Int. Conference on Raman scattering, N. T. Yu, X.-Y Li (Eds.) J. Wiley 1994, pp. 640.
3. G. T. Taylor, S. K. Sharma, and K. Mohanan, *Appl. Spectrosc.*, **44**, 635 (1990
4. K. Kneipp, Y. Wang, R. R. Dasari, and M. S. Feld, Appl. Spectrosc., 49, 780 (1995)
5. K. Kneipp, R. R. Dasari, and Y. Wang, Appl. Spectrosc., 48, 951 (1994)

SURFACE ENHANCED RESONANCE RAMAN SPECTROSCOPY (SERRS) - A STUDY OF THE SERRS ACTIVE SILVER COLLOID.

C. Rodger[1], M. Campbell[1], G. Dent[2], M.Edmondson[2], P.C.White[1] & W.E. Smith[1*]

1. Dept. of Pure and Applied Chemistry, University of Strathclyde, Glasgow, G1 1XL, UK.
2. ZENECA Specialities, Hexagon House, Blackley, Manchester, M9 3DA, UK.

Recent interest in SERRS as a sensitive technique for quantitative and qualitative analysis has prompted studies on the control of silver colloid formation and aggregation. This group has reported a comprehensive study on the formation of a modified silver Lee and Meisel colloid and on the effect of differing aggregation conditions on the SERRS spectra from adsorbed dyes[1] .

Methods of aggregation of colloidal particles include using nitric acid, poly-L-lysine or poly-L-lysine plus ascorbic acid. Scanning and Transmission Electron Microscopy (SEM and TEM)[2] have demonstrated dimensional differences in aggregates induced with these differing agents. Prior to aggregation, individual, hexagonal, colloidal particles are seen.

Nitric acid induced aggregation results in the formation of fused Ag 'clumps' which increase dramatically in size on addition of excess acid. Aggregation of the silver colloidal particles with poly-L-lysine also results in the formation of small Ag clusters. The size of these clumps increases with time. SEM studies indicate that direct physical contact between particles is prohibited by a poly-L-lysine coating which spaces neighbours at approximately 2nm apart. Colloidal aggregation with poly-L-lysine and ascorbic acid results in the formation of larger Ag clusters compared to aggregation with poly-L-lysine alone. The particles appear to be fused together indicating that the poly-L-lysine does not, in this case, separate neighbouring colloidal particles. The size and shape of the colloidal particles remains unchanged regardless of the method of aggregation employed.

Adsorbate quantification using SERRS has highlighted the requirement for a controlled aggregation procedure. A SERRS study of an azo dye yielded relative standard deviations (RSD) in peak size of 5% and 18.5% respectively for poly-L-lysine and acid induced aggregation. It was concluded that poly-L-lysine aggregation produces a more controlled and reproducible aggregation than acid.

This poly-L-lysine controlled behaviour was not observed using rhodamine 6G. This may be rationalised on the basis of surface chemistry. For acid aggregation, the colloidal surface is negative, and the positively charged rhodamine 6G is directly adsorbed onto the surface. However, when colloid is coated with poly-L-lysine the surface becomes positively charged, with the possibility of some negatively charged citrate being uncoated. Consequently, most of the rhodamine 6G will not absorb onto this positively charged surface and the amount of rhodamine 6G that does adsorb varies from sample to sample. This theory is confirmed by comparison of the fluorescent background observed from SERRS of poly-L-lysine aggregated rhodamine 6G with nitric acid induced aggregation. This is also exemplified by the high RSD (37%) calculated for the poly-L-lysine aggregated rhodamine 6G sample which compares unfavourably to the RSD of 23% calculated for the nitric acid aggregation of the rhodamaine 6G sample[3].

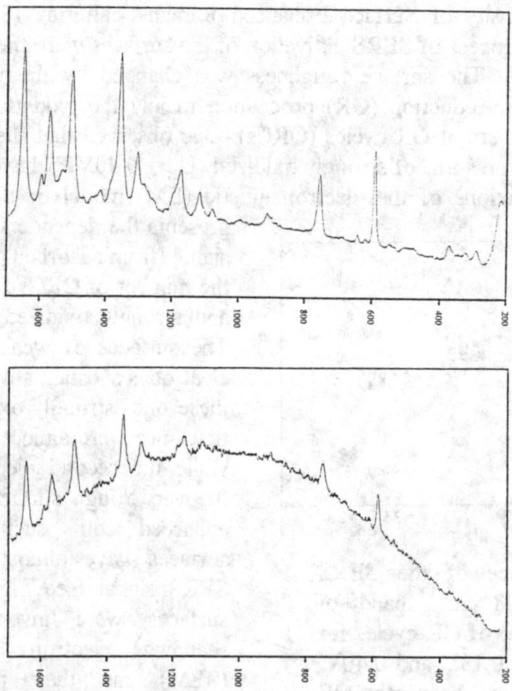

SERRS Spectra; 6.86x10^{-10}M Rhodamine 6G Sample with
(a) Nitric Acid Aggregation
(b) Poly-L-lysine and Ascorbic Acid Aggregation

---

[1] Munro C.H., Smith W.E., White P.C., *Langmuir*, 11, 3712-3720, 1995.
[2] Rodger C., Smith W.E., Dent G., Edmondson M. & White P.C. *Dalton Transactions*, 1996 (in press)
[3] Rodger C., Smith W.E. & White P.C. Unpublished work.

THE SURFACE ENHANCED RAMAN SCATTERING ON THE WEAKLY AND STRONGLY ROUGHENED SILVER ELECTRODES

Stefan Kruszewski*

Instytut Matematyki i Fizyki, Akademia Techniczno-Rolnicza,
ul. S. Kaliskiego 7, 85 - 796 Bydgoszcz, Poland

The paper presents the investigations of the influence of surface roughness variations on the intensity of Surface Enhanced Raman Scattering (SERS). The experimental and the process of SERS activation of the surfaces were the same as in the previous work [1]. The surface roughness was changed by the variation of parameters of oxidation-reduction (OR) procedure, mainly the oxidation potential limit U_a and the numbers of OR cycles (ORCs). One observed that the electrodes weakly (U_a = 0.13V) and some of strongly oxidized (U_a = 0.40V) behave seemingly contrary to the predictions of the electromagnetic (EM) model of SERS. Fig.1 presents the dependence of SERS signal (from adsorbed pyridine) on the number of ORCs from weakly and strongly oxidized electrodes. The surfaces of weakly oxidized electrodes substantially differ from these ones strongly oxidized. The first ones were almost mirror flat, while the second electrodes had the very rough surface. Pyridyne adsorbed on such different surfaces gave almost the same SERS signal (see Fig. 1). These surfaces were investigated by scanning electron microscopy (SEM) and their replicas by transmission electron microscopy (TEM). Fig. 2 presents the exemplary images of surface of weakly (a) and strongly (b) oxidized electrodes. It shows that small microobjects (grains) exist on the weakly oxidized surfaces. Their average horizontal size d is about 100nm. These microobjects are sparsely distributed on the surface. The distance between the neighbouring grains is at least 3 times larger than the grain sizes. On the other hand, as it is shown in Fig. 2b, on the

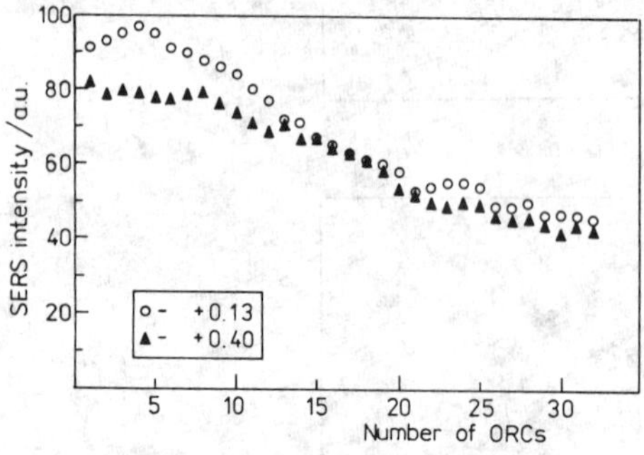

Fig.1. The dependence of the SERS intensity of the 1008 cm^{-1} band of pyridine on the number of OR cycles for electrodes oxidized to 0.13V and 0.40V. The potential sweep rate during the OR processes was 5mV/s. The SERS intensity was measured at an electrode potential of -0.60V.

Fig. 2. The SEM images of silver electrodes roughened by 1 OR cycle using the oxidation potential limit equal to 0.13V (a) and 40V (b).

strongly oxidized surfaces the large grains are formed. They are tightly distributed on the surfaces. One observe on the basis of SEM images that the increase of number of OR cycles causes the increase of the horizontal sizes of microobjects.

The shadow carbon-platinum replica of considered here surfaces were done for determination of heights of microobjects. They were investigated by transmission electron microscopy (TEM). The average height h of microobjects was determined from the length of shadows measured from TEM images. How it follows from SEM and TEM images increasing of number of ORCs cause that the horizontal sizes of grains d increase and simultaneously the ratio h/d decreases. The EM model [2] predicts that the increase of d and decrease of h/d ratio, should cause the decrease of SERS signal intensity. Experimentally, as it shows Fig.1, this decrease was observed. This model [2] also predicts that the strong SERS signal is expected from such surfaces on which the small (low h, low d) but relatively high (large h/d) microobjects exist. Why the weakly oxidized (U_a = 0.13V) and strongly oxidized (U_a = 0.40V) surfaces provide almost the same SERS signal? On the first surface the grains are substantially smaller and have the largest h/d ratio than these on the second one. The experimentally obtained results are not inconsistent with the EM model. Thought the small microobjects that exist on the electrode oxidized to U_a = 0.13V individually give the large enhancement still they are too loosely distributed on the surface to give a strong signal. One can then suppose that the SERS signal from the surface with closely distributed small microobjects will be about ten times larger than those from the investigated here weakly oxidized electrodes. The largest SERS signal is observed [3] on the surfaces on which the grains are small (10 - 100 nm) and closely packed.

1. S. Kruszewski, Surf. Interface Analysis 21,830 (1994).
2. P.W. Barber, R.K. Chang, H. Massoudi, Phys. Rev. 27, 7251 (1983).
3. J.A. Creighton, in: Specroscopy of Surfaces, (R.J.H. Clark, R.E. Hester, ed.), J. Wiley & Sons, New York 1988, p. 37.

SILVER-DOPED POLYCARBONATE FILM FOR SURFACE-ENHANCED RAMAN SPECTROSCOPY

K. Wang and Y. S. Li*

Department of Chemistry, University of Memphis, Memphis, TN 38152 USA

Silver sols have been widely used as a surface-enhanced Raman scattering (SERS)-active substrate because they can be prepared conveniently and characterized by monitoring their adsorption spectra. However, a major disadvantage in using the sols is their unstability because of the aggregation of silver particles. To avoid this disadvantage, some solid materials, such as filter paper [1-2], thin layer chromatographic plate [3-4], have been used as supports for silver particles. The technique of coating include the spraying of silver sols, high vacuum thermal evaporation, and sputtering deposition of silver particles onto the plates. These solid substrate plates are rather stable. In some cases, the aging of the substrates might affect their SERS performances and vary the Raman background. Consequently, there is a need for continuous development of SERS-active substrates.

Kurokawa and his coworkers [5, 6] have recently developed silver-doped substrates by counterdiffusing silver particles into polyvinyl alcohol or cellulose acetate films. These films could be used to activate SER spectra of pyridine, inorganic ions, and nucleic acid bases. Our recent development of SERS-active substrates have applied polycarbonate films (Poretices Corp.) for silver particle doping. The performances of the doped films as SERS-active substrates have been tested using different chemicals.

In the present work, we have used three different polycarbonate films with pore sizes of 0.01, 0.1, and 0.4 μm as well as with thicknesses of 6, 6, and 10 μm, respectively for supporting Ag particles. These films were produced by track-etching method and with an addition of polyvinyl pyrrolidone. The films are hydrophilic and have rather uniform pore size.

Ag-doped films were prepared by dipping the polycarbonate films in 0.1 M $AgNO_3$ solution and then in 0.1 M $NaBH_4$ solution. It was observed that the color of the film changed from white to yellow-brown very quickly, indicating that the Ag^+ ions were reduced to Ag in the reduction solution. Twenty minutes later, this film was taken out of the solution and washed very carefully with triple distilled water several times. After drying in air, the film was stored in a dark place until used. In the preparation, we dipped the 0.4 μm pore size in the $AgNO_3$ solution for 1 to 25 minutes, but the resulted films did not show much difference in their SERS activities. When the other two different pore-size films were used, we could not observe much Ag depositing on the films even a three hour of dipping in $AgNO_3$ solution was undertaken. This might be caused by the small pore size and the small thickness of the polycarbonate films.

When a test sample was applied onto the substrate by dropping method, an intense background appeared on the collected Raman spectrum because of the possibility of burning the film surface by laser. To eliminate or to reduce the background, a liquid solution method was used for sample applications. In this method, the film substrate was attached to a precut glass and placed in a cuvette containing about 2 ml sample solution for SER data collections.

Figure 1 shows the normal Raman and the SER spectra of p-nitrobenzoic acid (PNBA). For better comparison, all the spectra displayed in the Figure have the same intensity scale. In Figure 1-A, no normal Raman band due to PNBA can be observed except those arising from the solvent - ethanol. Displayed in Fig 1-B is the SER spectrum of PNBA adsorbed on the substrate from its 1.0×10^{-4} M solution.

To test the stability of the substrate, we have measured the SER intensity of PNBA adsorbed on the same substrate film for different lengths of time. The results indicated that the intensity remained unchange for at least two weeks. It is likely that the Ag particles in this substrate film can be stabilized by the gel network of the polymer.

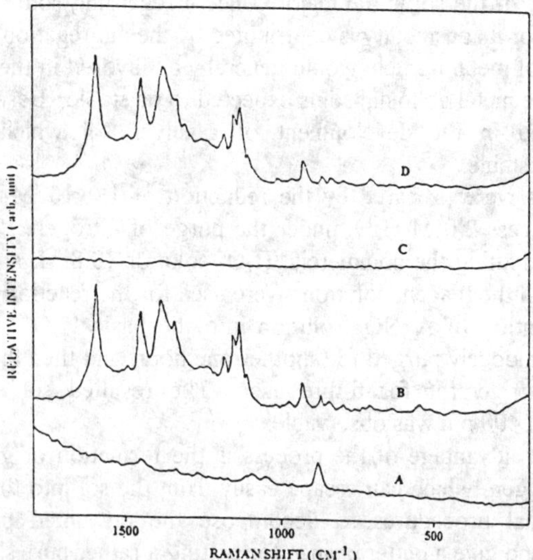

Figure 1. (A) Normal Raman spectrum of 0.1 M PNBA in ethanol solution collected in the presence of polycarbonate film. (B) SERS spectrum of PNBA absorbed on Ag-doped polycarbonate film from 1.0×10^{-4} M aqueous solution. (C) Spectrum of the used film after washing with water. (D) SERS spectrum of PNBA adsorbed on the used Ag-doped polycarbonate film from 1.0×10^{-4} M aqueous solution.

1. C. D. Trans, *Anal. Chem.* 56, 824 (1984).
2. C. D. Trans, *J. Chromatogr.* 292, 432 (1984).
3. S. A. Soper, and T. Kuwana, *Appl. Spectrosc.* 43, 1180 (1989).
4. S. A. Soper, K. L. Ratzlaff, and J. D. Winefordner, *Anal. Chem.* 62, 1438 (1990).
5. Y. Kurokawa, Y. Imai, Y. Sasaki, and T. Maeda, *Anal. BioChem.* 209, 247 (1993)
6. H. Ishikawa, Y. Imai, and Y. Kurokawa, *Vib. Spectrosc.* 8, 445 (1995).

SILVER SOLS PREPARED BY HYDROGEN PEROXIDE REDUCTION

J. Cheng and Y. S. Li*

Department of Chemistry, The University of Memphis, TN 38152 USA

Reductions of silver nitrate by sodium borohydride [1] and by citrate [2] are commonly used to produce silver sols for surface-enhanced Raman scattering (SERS). These are two prominent advantages with the use of silver sols for SERS. The first one is the simplicity of the preparation process; the second one is the convenience of monitoring the sols by visible adsorption spectroscopy. However, the unstable character of the sol is the major concern regarding some of its applications [3]. One reason for its unstability is contributed by the aggregation of metal particles. The aggregation of metal particle would generally be favored in the presence of some impurities. A pure metal colloidal sol is expected to be stable. Based on this concept, we have engaged in the development of a silver sol which is SERS-active, reproducible, and stable.

Silver sols were prepared by the reduction of $1.0\text{x}10^{-3}$M $AgNO_3$ in a basic condition with excess 2.0 M H_2O_2 under the purge of nitrogen. The H_2O_2 solution was prepared by diluting the commercial H_2O_2 solution (8.8 M, aqueous) with triple distilled water. All the reagent solutions were used for the reaction were fresh. Upon the complete addition of $AgNO_3$ solution into the basic H_2O_2 solution, the final mixture was continuously purged for another one hour, and then the solution product was stored in refrigerator for future use. The resulted sol was colorless; no absorption around 400 nm was observable.

One major advantage of the process is the formation of gaseous products - hydrogen and oxygen, which can escape easily from the sol into the air. This would eliminate additional procedures to decompose the unwanted products from the reaction system, and give a better chance for getting a rather pure silver sol. Although the sol preparation process is simple, it is important to control the reaction conditions such as temperature, time, and pH.

In the present study, we have optimized the conditions for the preparation of the Ag sols. Benzoic acid (BA) solution was adopted as a model sample for testing the SERS performances of the sols. In all the experiments, the normal Raman spectrum of ethanol solvent could be used as an internal reference for the SER intensity measurements. To optimize the reduction reaction temperature, we followed the same preparation procedures but kept the reaction temperature at 0, 21, 32, and 45°C, respectively. Relative SER intensities of BA were measured at different time using each of these prepared sols. Results indicated that the sols prepared at 0°C is more SERS-active than those prepared at higher temperatures, especially when the sols had been stored for more than three days. Thus, we fixed the reduction temperature at 0°C for the subsequent experiments.

Different rates of adding $AgNO_3$ solution into the H_2O_2 solution may affect the reduction rate, and result in different Ag particle sizes in the sols. The addition of $AgNO_3$ solution was achieved drop by drop. In the experiments, we had tried the following dropping rates: 1, 0.2, 0.1, 0.07, 0.05, and 0.03 drop/second. Shown in Figure 1 are the SER spectra of BA absorbed on the Ag sols prepared using different dropping rates. It is seen that 5 seconds of duration between the two consequent drops or 0.2 drop/see of dropping rate gives the product sol with a better SERS performance.

The pH can also significantly affect the reduction rate. In a series of experiments, we had adjusted the mole ratio between NaOH and $AgNO_3$ as a control of the solution pH. SER spectra of BA adsorbed on each of these sols prepared under different pH conditions are shown in Figure 2. From these results, it is clear that the mole rate of 0.6 provides the best pH condition for the preparation of Ag sol.

Two series of tests have been conducted for the sols prepared under the optimized conditions. In the first series experiments, we measured SER intensities of BA after adsorbing in the sols for different time. It was found that the intensity of BA increases slowly with time until day three, after then, the intensity decreases rapidly to zero on day seven. The second series experiments were performed to record the SER spectra of BA adsorbed on Ag sols which had been stored for different lengths of time since its preparation. The experimental results indicated that the sols are still SERS-active even after six months of storage.

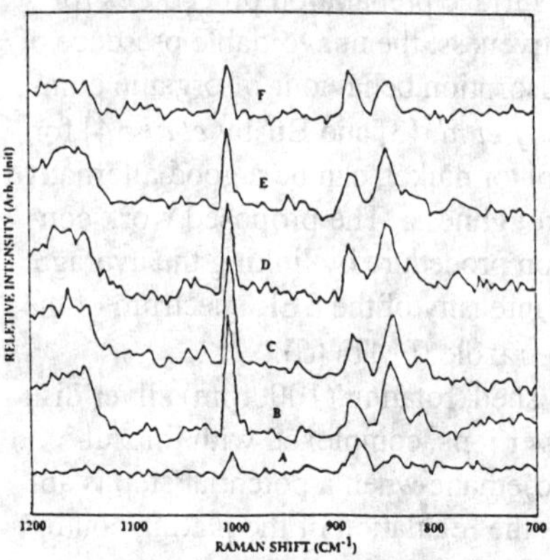

Fig. 1. SER spectrum of $2.5x10^{-3}$ M BA in Ag sols prepared with different $AgNO_3$ dropping rates, A) 1; B) .2; C) .1 D) .07; E) .05; F) .03 drop/s

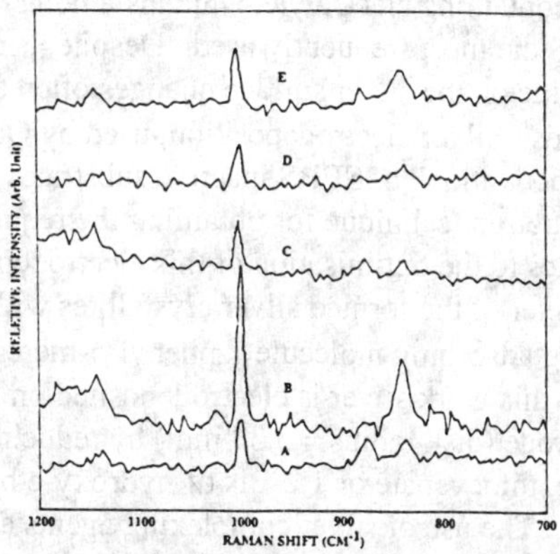

Fig. 2. SER spectrum of $2.5x10^{-3}$ M BA in Ag sols prepared using mole ratios betw. Ag NO_3 and H_2O_2: A) .4; B) .6; C) .8; D) 1.6; and E) 2.4

1. J. A. Creighton, C. G. Blatchford, and M. G. Albright, J. Chem. Soc., Faraday Trans. 75, 790 (1979).
2. P. C. Lee and D. Melsel, J. Phys. Chem. 86, 3391 (1982).
3. Y. S. Li, A. S. Lee, and Y. Wang, J. Raman Soc. 22, 191 (1919).

OPTIMISATION OF THE SURFACE ROUGHNESS OF ELECTRODES FOR SERS MEASUREMENTS BY ELECTRODEPOSITION OF SILVER

D. Gonnissen*, A. Hubin and J. Vereecken

Dept. of Metallurgy, Electrochemistry and Materials Science
Vrije Universiteit Brussel
Pleinlaan 2, B-1050 Brussels / Belgium

Surface enhanced Raman spectroscopy (SERS) has become an important tool for the *in situ* study of adsorption phenomena on silver electrodes. A strong SER spectrum can only be obtained when the electrode surface is roughened. Since the size of the surface roughness features determines the signal intensity [1,2], an accurate roughening procedure is required. Kruszewski [2] showed the influence of the surface roughness of a silver electrode on the SER spectrum of pyridine. He roughened the electrode by applying a certain number of oxidation/reduction cycles (ORC) in a chloride containing electrolyte, which is among other surface preparation procedures for SERS the most frequently used. Despite its effectiveness, the unavoidable presence of chloride on the silver surface changes often the adsorption behaviour of organic compounds. Silver electrodeposition, used by Oblonsky *et al.* [3] and Busby *et al.* [4] for the "activation" of SERS inactive substrates as iron or nickel, can be a good alternative preparation technique for obtaining the required roughness. The proposed work contributes to the optimisation of this electrodeposition procedure by linking the average diameter of the formed silver crystallites with the intensity of the SER spectrum of the strong adsorbing molecule 1-phenyl-5-mercaptotetrazole (PMT) [5].

In this work silver is electrodeposited on a polished, rotating (1000rpm) silver disk electrode (disk radius = 3.33 mm) by reducing silver ions, complexed with thiosulphate, thiocyanate or 1,2-Bis (2-hydroxy-ethylthio)ethane when a potential step is applied. The use of silver complexing agents allows the regulation of the plating conditions required to obtain a SERS active substrate. From the current response typical parameters for the electrocrystallization process (e.g. the average diameter of the formed silver crystallites) are calculated. A more detailed description of this calculation procedure will be given in future work [6]. After deposition the silver electrode is carefully rinsed with water and immersed in a 1e-5 M solution of PMT. Raman spectra of adsorbed PMT are obtained after irradiating the silver surface with an Ar^+ laser (American Laser), operating at an excitation wavelength of 514.5 nm and an output power of 60 mW, and recorded by means of DILOR microraman equipment.

The obtained crystallite diameters vary from almost 1000 down to 50 nm, depending on the composition of the electrolyte (nature and concentration of the used complexing agent) and on the step height of the applied potential. Figure 1 shows the evolution of the obtained SER spectra for PMT between 900 and 1450 cm^{-1} with the crystallite diameter. The assignment of the observed peaks has been given in earlier work [5]. The intensity of all these peaks is very low when the silver crystallites are large and rises rather weakly with decreasing diameters. Only if the preparation method results in crystallites with a diameter smaller than about 120 nm the SER spectrum of PMT becomes very intense, with a maximum intensity for diameters between 80 and 90 nm. Smaller crystallites again give rise to weaker spectra.

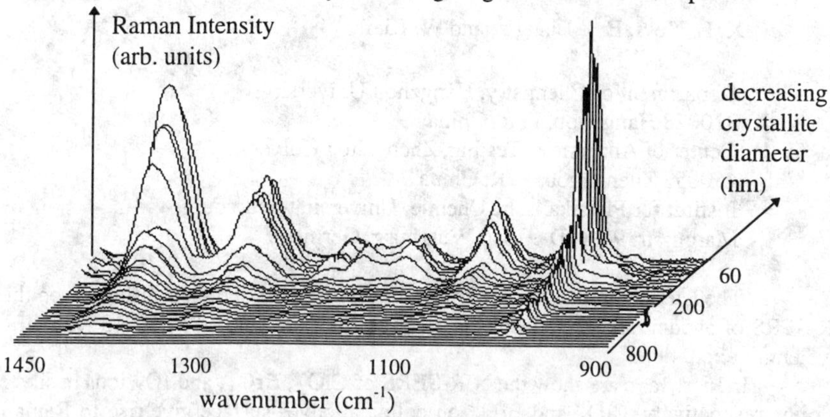

Fig. 1. SER spectra of PMT adsorbed on deposited silver crystallites in function of the average crystallite diameter

In conclusion we can state that an optimised electrode preparation technique is important for quantitative SERS measurements in future work. Next to the more known ORC procedure the electrodeposition method can be of importance when chloride must be avoided on the silver surface. Its effectiveness has already been proved in earlier work, where SERS contributed to the study of the mechanism of silver thiosulphate reduction [7] and to the study of the potential dependent influence of PMT during this process [5].

[1] H. METIU - A Survey of recent theoretical work, In : R.K. CHANG and T.E. FURTAK - Surface Enhanced Raman Scattering, Plenum Press, New York, 1982

[2] S. KRUSZEWSKI - Surf. Int. An. **21** (1994) 830

[3] L.J. OBLONSKI and T.M. DEVINE - J. Electrochem. Soc. **142**, 11 (1995) 3677

[4] C.C. BUSBY and J.A. CREIGHTON - J. Electroanal. Chem. **140** (1982) 379

[5] D. GONNISSEN, A. HUBIN and J. VEREECKEN - Conf. Proc. of ECASIA'95, 8-13 Oct. 1995, Montreux, In Press

[6] D. GONNISSEN, W. SIMONS, A. HUBIN, J. VEREECKEN and B. VAN DER LINDEN - in preparation

[7] D. GONNISSEN, S. VANDEPUTTE, A. HUBIN and J. VEREECKEN - Electrochim. Acta, in press

Research financed by the Flemish Institute for the encouragement of scientific and technological research in the industry (IWT) and AGFA-Gevaert NV.

DIFFERENT INTERACTION MANNERS OF HALATE IONS WITH COLLOIDAL SILVER SURFACE REVEALED BY SESR WITH NIR- EXCITATION

X. L. Ye[1]*, E. J. Liang[2,3], and W. Kiefer[3]

[1]Department of Chemistry, Hangzhou University,
310028 Hangzhou, P. R. China
[2]Center of Analysis & Testing, Zhengzhou University,
450052 Zhengzhou, P. R. China
[3]Institut für Physikalische Chemie, Universität Würzburg,
Marcusstr. 9-11, D-97070 Würzburg, Germany

The interaction manners of halate ions with silver and their influence on the NIR-SERS of pyridine and crystal violet in silver colloids are studied by SERS with NIR wavelength excitation.

In Figs. 1a~c are show the NIR-SERS of ClO_3^-, BrO_3^-, and IO_3^- ions in silver colloids, respectively. ClO_3^- and BrO_3^- on colloidal silver surface give rise to Raman bands at about 226 and 158 cm^{-1}, respectively. They are very similar to the cases of Cl^- and Br^- ions on colloidal silver surface [1]. A question is raised by these spectra that ClO_3^- and BrO_3^- bond to silver surface through the negatively charged O^- or the lone pair electrons in Cl and Br. One may expect the interaction of these ions with colloidal silver surface through the negatively charged O^-. However, the similarity of the spectra to those of Cl^- and Br^- ions on colloidal silver and the dependence of the band positions on the mass of the halides suggest that ClO_3^- and BrO_3^- interact with silver surface quite probably through their lone pair electrons in Cl and Br instead of through the negatively charged O^-. We assign the band at about 226 cm^{-1} to the stretching vibrations of Ag-ClO_3^- and the band at about 158 cm^{-1} to Ag-BrO_3^- stretching mode. In contrast to the spectra of ClO_3^- and BrO_3^-, the NIR-SERS of IO_3^- is much different from the spectrum of I^- ions on colloidal silver surface [1]. No Raman band in the Ag-I^- stretching vibration region could be seen. The SERS spectrum of IO_3^- is dominated by a broad and very intense Raman band at about 750 cm^{-1} and a broad band at about 330 cm^{-1}. This result indicates that the adsorption of IO_3^- on colloidal silver surface is different from those of ClO_3^- and BrO_3^-. Surface reaction of IO_3^- with silver surface occured quite possibly through their negatively charged O^-. The reaction product is expected to be $AgIO_3$. This assumption is supported by the Raman spectrum of solide KIO_3 .

In order to exclude the possibility of chemcal reactions of these ions with the reductive reagent in silver colloids, we measured the NIR-SERS of these ions in laser ablated sol and obtained similar results. This indicates that the difference of the NIR-SERS of ClO_3^- and BrO_3^- from IO_3^- does not result from the chemical reactions between these ions and the reductive reagent in the silver colloids prepared by chemical reduction methods.

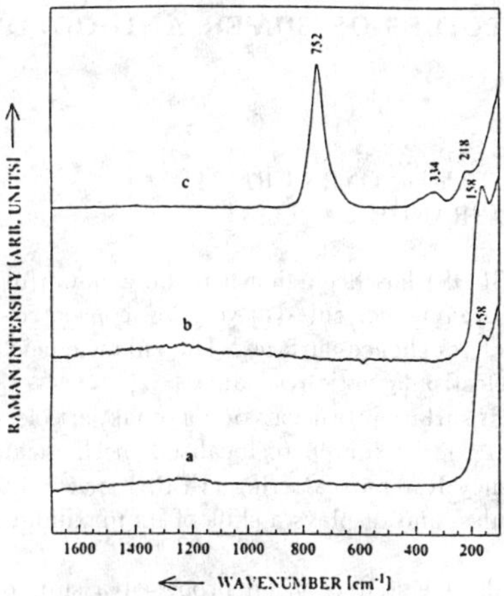

Fig. 1. NIR-SERS of (a) ClO_3^-; (b) BrO_3^- and (c) IO_3^- ions in silver colloids.

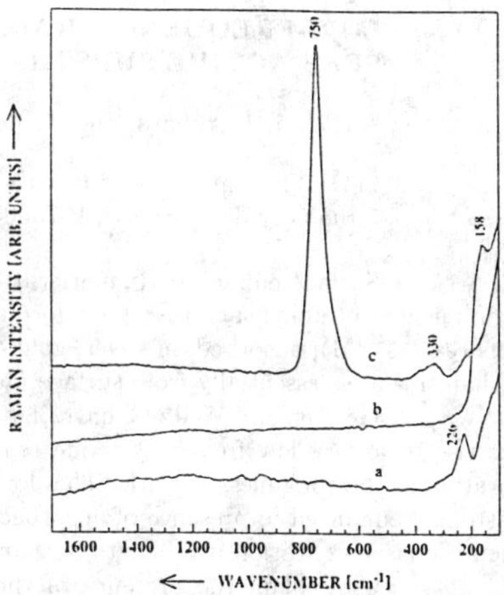

Fig. 2. NIR-SERS of (a) BrO_3^- and IO_3^-; (b) ClO_3^- and BrO_3^- and (c) ClO_3^- and IO_3^- mixtures in silver colloids.

The different adsorption manners of halate ions on colloidal silver surface may be explained by the surface potencial of the silver colloids. The surface potencial of silver colloids we used is estimated to be about -0.3 V which is very approximate to the potential -0.3551 V under which the reaction

$$Ag + IO_3^- - AgIO_3 + e$$

occurs [2]. However, this potencial seams not favourable for the formation of $AgBrO_3$ (-0.680 V) and $AgClO_3$ (no avilable data) on colloidal silver surface.

The interaction strength order between halate ions and silver is studied with the mixtures of these ions. The results are shown in Fig. 2. In both the mixture of ClO_3^- and BrO_3^- and the mixture of BrO_3^- and IO_3^-, there appears only one Raman band at about 158 cm^{-1}. As discussed above, this band is characteristic of the Ag-BrO_3^- interactions. These results indicate that in the two mixtures, BrO_3^- rather than IO_3^- and ClO_3^- interacts directly with silver surface. In the mixture of ClO_3^- and IO_3^-, besides the two Raman bands at about 752 and 334 cm^{-1}, a Raman band at about 218 cm^{-1} appears. The former two bands result from the IO_3^- ions on silver surface while the latter is due to the Ag-ClO_3^- stretching vibrations. It suggests that in the mixture of IO_3^- and ClO_3^-, both ions interact simultaneous with silver surface. The interaction strengths between the halate ions and silver surface are in the order $BrO_3^- > ClO_3^- \approx IO_3^-$.

These halate ions can enhance the Raman scattering of pyridine. The ability of these ions to enhance the ring breathing modes of pyridine decreases in the order of $BrO_3^- > IO_3^- > ClO_3^-$. Only BrO_3^- can enhance the Raman signal of crystal violet.

[1]. E. J. Liang, C. Engert and W. Kiefer, Vib. Spectrosc. 8 (1995) 435.
[2]. R. C. Weast and G. L. Tuve, eds. Handbook of Chemistry and Physics, 53rd Edition, (CRC press, 1972 - 1973) D-111.

LOW-FREQUENCY RAMAN STUDIES OF SILVER AND GOLD SERS-ACTIVE SUBSTRATES

N. Félidj, J. Aubard, and G. Lévi.

ITODYS, University of Paris 7 - Denis Diderot, CNRS URA 34,
1 rue Guy de la Brosse, F-75005 Paris, FRANCE.

Surface enhanced Raman scattering (SERS) has been shown to be a powerful technique for obtaining spectroscopic information of organic molecules, at very low concentrations (ca. 10^{-8}M), adsorbed on silver, gold or copper roughened surfaces. The enhancement mechanisms arise essentially from surface plasmon electromagnetic resonance [1,2].

Gersten and Weitz [3] have observed that surface protusions or colloidal particles give rise to a very low frequency mode (ca. 10 cm^{-1}), attributed to localized mechanical vibration of the roughness feature. This low frequency Raman scattering (LFRS) spectrum is strongly enhanced in presence of adsorbed molecules, and displays a shift of its maximum when the laser excitation wavelength is changed.

To explain the presence of these bands, Gersten *et al* [4] proposed a simple physical model where the surface roughness appears in spheroidal particles (semi major axis a and semi minor axis b). The maximum Ω_{max} of the LFRS can thus be expressed from the equation:

$$\Omega_{max} = \frac{v}{2c\,a} \quad \mathrm{cm}^{-1}$$

where v is the speed of sound in the metal and c the speed of light.

In this communication, we show how the spectral characteristics of the LFRS spectrum allow the determination of the resonant particles size distributions, by using various laser excitation wavelengths.

Different SERS-active substrates have been investigated: roughened silver electrodes and colloidal suspensions in presence of various sodium salts (NaF,...,NaI) and with molecules of different chemical nature (pyridine, benzoic acid and acridine). Figure 1 shows typical scans of both the anti-Stokes and Stokes shifts of LFRS spectrum for a silver electrode roughened in 0.1M NaI electrolyte (A), and silver colloid aggregated with 0.1 M NaI (B). If one admits there are N resonant particles of semi major axis a, the scattering intensity I can be assumed to be proportional to $I_\Omega = N \times d\sigma/d\omega$ where $d\sigma/d\omega$ is the scattering cross section for one particle. If the probability that N particles having a given value of a is supposed to be by a gaussian distribution $G(a_0, \mathcal{L})$ (where a_0 is the center of the distribution and $\mathcal{L}$ the full width at half maximum), we can simulate the experimental band shapes and determine the size distribution of resonant particles (see figure 2). The agreement between experimental and calculated profiles is satisfactory (except for the contribution of the Rayleigh line) and the a_0 and the $\mathcal{L}$ parameters, used to fit the acoustic band, give the relevant features of the resonant size distribution.

From the intensity and the frequency dependence of the band maximum on laser excitation, histogram distribution can be approached (figure 3). This histogram indicates that a stronger enhancement occurs mainly for largest particles.

A lot of microscopic techniques (AFM, STM, TEM and MEB) have been used to investigate the local topology: however, these are inadequate to reveal the microstructures deduced from the method proposed by Gersten *et al.*. Moreover, the model, based on electric dipolar plasmon resonance, appears not sufficient to explain the strong enhancement observed when Laser excitation is turned to the red [5]. Molecular mechanism via charge transfert excitations and/or magnetic scattering are then invoked [6].

Huge SERS spectra are generaly obtained on aggregating colloidal sols towards large clusters. However these large clusters are not stable enough in time. In order to overcome this difficulty, we are now investigating colloidal fims [7,8] giving rise to a stable morphology and a strong enhancement of Raman scattering.

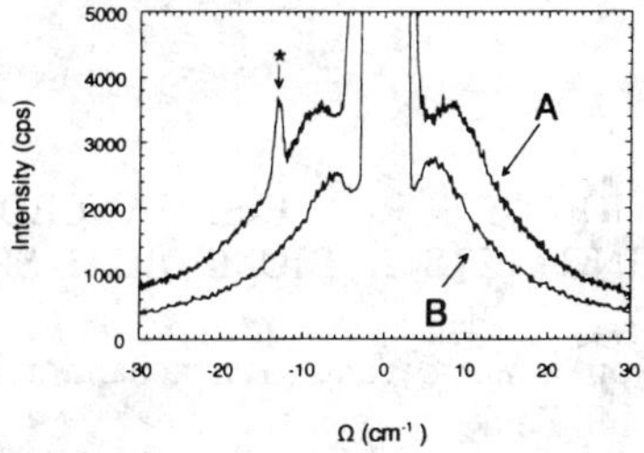

Fig.1 : LFRS spectrum of : (A) a silver electrode roughened with 0.1M NaI in deionized distilled water as electrolyte; slits: 70 μm, laser power at the sample: 24 mW, counting time: 5 s (the asterisk denotes a Ar^+ plasma line located at -13.6 cm^{-1} from the 514.5 nm Laser line. (B) a Creighton colloidal sol aggregated with 0.1 M NaI (same experimental conditions).

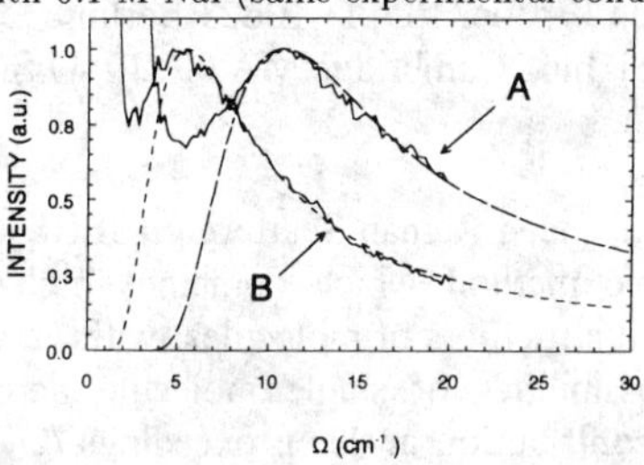

Fig.2 : comparison between experimental and calculated bands profiles for pyridine adsorbed on a roughened silver electrode at a -0.6 V potential,
A : λ_{exc}=457.9 nm,— experimental,— — calculated with a_0 = 20 Å, $\mathcal{L}$= 48 Å.
B : λ_{exc}=632.8 nm,— experimental,- - - calculated with a_0 = 32 Å, $\mathcal{L}$=100 Å.

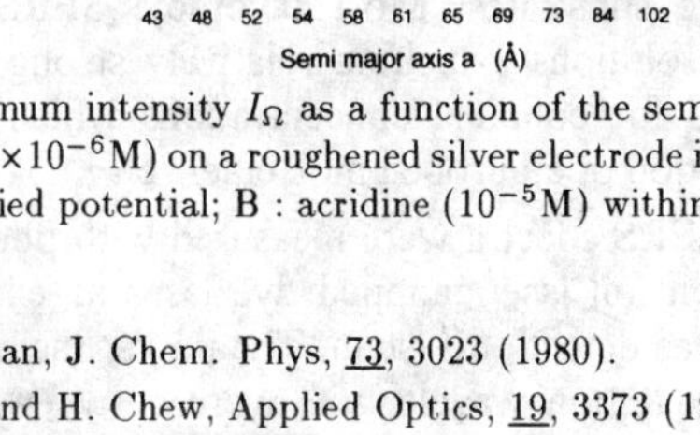

Fig. 3 : maximum intensity I_Ω as a function of the semi major axis a.
A : acridine (2×10^{-6}M) on a roughened silver electrode in 0.1 M NaCl electrolyte at -0.4 V applied potential; B : acridine (10^{-5}M) within Lee Meisel colloid with 0.1 M NaCl.

1. J.I. Gersten and A. Nitzan, J. Chem. Phys, 73, 3023 (1980).
2. M. Kerker, D.S. Wang and H. Chew, Applied Optics, 19, 3373 (1980).
3. D.A. Weitz,T.J. Gramila, A.Z. Genack and J.I. Gersten, Phys. Rev. Lett., 45, 355 (1980).
4. J.I. Gersten, D.A. Weitz, T.J. Gramila, A.Z. Genack, Phys. Rev., B 22, 4562 (1980).
5. E.J. Liang, C. Engert and W. Kiefer, J. Raman. Spectrosc., 24, 775 (1993).
6. N. Félidj, J. Aubard and G. Lévi, J. Chem. Phys., accepted for publication.
7. R.G. Freeman *et al* , Science, 267, 1629 (1995).
8. G. Chumanov, K. Sokolov, B.W. Gregory and Y.M. Cotton, J. Phys. Chem, 99, 9466 (1995).

APPLICATION OF SURFACE-ENHANCED RAMAN SCATTERING (SERS) IN BIOMEDICAL SPECTROSCOPY

Katrin Kneipp[1,2], Yang Wang[3], Ramachandra R. Dasari[1], and Michael S. Feld[1],

[1]George R. Harrison Spectroscopy Laboratory
Massachusetts Institute of Technology, Cambridge, MA 02139,U.S.A.
[2]Department of Physics, Technical University Berlin
Strasse des 17. Juni 135, D-10623 Berlin, Germany
[3]PROCEPT Inc., Cambridge MA 02139, U.S.A.

Surface-Enhanced Raman Scattering (SERS) has been established as a powerful spectroscopic method which combines high information content for characterizing chemical structures of molecules. with very low detection limits /1/. For example, the possibilities of single molecule detection by SERS will be discussed in another contribution in this proceedings /2/. In the past several years SERS has found growing applications in the field of biomedical research, as well /3/.

We have shown recently that SERS in the near infrared region (NIR-SERS) can provide strong enhancement factors, sometimes higher than that achievable with visible excitation /4/. Additionally, excitation in the near infrared avoids fluorescence background from biological materials /5/. Easy access of tunable diode lasers makes this frequency region attractive for low cost and compact Raman instrumentation.

In this work we will demonstrate applications of NIR-SERS in the field of biomedical spectroscopy. Optimized aqueous solutions of colloidal silver or gold served as SERS-active substrates. Most favorable SERS conditions could be achieved for colloidal solutions exhibiting relatively strong extinction in the NIR spectral region /4/ and for colloidal concentrations which provided just enough surface area for adsorption of all probed molecules /6/.

The NIR SERS spectra were measured with fiber optic probes using a diode laser or an argon ion laser pumped dye laser as excitation sources and a charge-coupled device as detection system. The power incident on the sample was about 100 mW, the excitation wavelengths were between 829 and 850 nm The Raman spectra were collected with accumulation times between of 25 milliseconds and 10 seconds.

We present here SERS spectra of dyes which are of interest for labeling biomolecules and spectra of molecules of biomedical interest.

For example, Fig. 1 displays SERS spectra of the neurotransmitter dopamine dissolved in colloidal silver solution at concentrations from $5x10^{-6}$ to

$5x10^{-9}$ M. Comparable spectra at the same concentration levels were recorded from the neurotransmitter norepinephrine. The low concentrations and fast acquisition from 1s to 0.025s are in the order of physiologically relevant concentrations and to the time scale of neuronal processes, respectively.

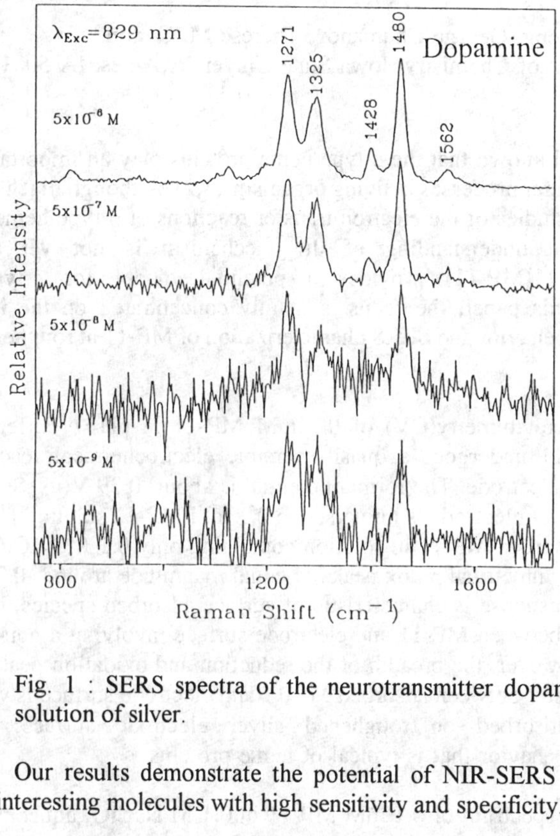

Fig. 1 : SERS spectra of the neurotransmitter dopamine in colloidal solution of silver.

Our results demonstrate the potential of NIR-SERS in probing bio-medically interesting molecules with high sensitivity and specificity.

1. J. J. Laserna, Analytica Chim. Acta, 283 (1993)
2. K. Kneipp, Y. Wang, H. Kneipp, I. Itzkan, R. R. Dasari, and M. S. Feld in this Proceedings
3. I. Nabiev, I. Chourpa, and J. Mainfait, J. Raman Spectrosc., 25, 13 (1994)
4. K. Kneipp, R.R. Dasari, M. S. Feld, Y. Wang Proceedings XIV th International conference on Raman spectroscopy, Hong Kong 1994,N.-T. Yu, X.-Y. Li (Eds.) John Wiley and Sons (Chichester, New York, Brisbane,Toronto, Singapore (1994), p. 625
5. J. J. Baraga, M. S. Feld, and R. P. Rava, Appl. Spectrosc., 46, 187 (1992)
6. K. Kneipp, Y. Wang, R. R. Dasari, and M. S. Feld, Appl. Spectrosc., 49, 780 (1995)

SERS CHARACTERIZATION AND INTERFACIAL ELECTRON TRANSFER OF MP-11 AT ROUGHENED SILVER ELECTRODES

Junwei Zheng, George Chumanov, Therese M. Cotton*
Department of Chemistry, Iowa State University,Ames, IA 50011

It is well known that the c-type heme proteins play an important role in vital electron transfer processes in living organisms(1,2). Although much effort has been devoted to studies of the electron transfer reactions of c-type heme proteins at electrodes, an understanding of the mechanism is not yet available. Microperoxidase-11(MP-11) provides a potential model for c-type heme enzymes(3). In this paper, the focus is mainly concentrated on the interfacial electron transfer behavior and SERS characterization of MP-11 at roughened silver electrodes.

Cyclic voltammetry(CV) of 0.1 mM MP-11 in pH=7 buffer solution shows that MP-11 undergoes a quasi-reversible electrochemical reaction at a roughened silver electrode. The formal potential is about -0.39 V(vs. SCE). After the roughened silver electrode is immersed in 0.1 mM MP-11 solution for 10 min, the CV of the adsorbed MP-11 in solution containing only 0.1 M $NaClO_4$ clearly shows a pair of symmetrical redox peaks of equal magnitude around -0.39 V. The electrochemical response is characteristic of surface adsorbed species, i.e., direct electron transfer between MP-11 and electrode surface involving a non-diffusing redox couple. However, the breadth of the reduction and oxidation peaks implies that electron transfer between adsorbed MP-11 and electrode surface is very slow. Thus, MP-11 adsorbed on roughened silver electrode surface processes electrochemical behavior that is typical of heme proteins

The RR spectrum of 0.1 mM MP-11 in 0.1 M $NaClO_4$ aqueous solution is characteristic of a low-spin, six-coordinate heme group, indicating that molecules of MP-11 may exist as aggregates in aqueous solution(4). The SERS of MP-11 at a roughened silver electrode reveals a mixture of five-coordinate and six-coordinate configurations, which is also similar to that of other heme proteins. However, the complete five-coordinate form of MP-11 can be obtained by switching the electrode potential between -0.2 and -0.5 V, or sweeping it continuously between these limits for several cycles. Therefore, it seems that the aggregates of MP-11 adsorbed on the electrode surface are dissociated to monomers during the redox reactions.

The time dependence of the SERS spectrum of MP-11 under open circuit conditions shows that the intensity of the bands of the oxidized form decrease, while the bands corresponding to the reduced form appear and finally become dominant. These results indicate that the oxidized form of MP-11 is reduced under laser illumination, The heme group in MP-11 is five coordinate and the reduction does not change the coordination. Therefore, it is not possible that the heme axial ligand plays a role as the electron donor, as has been suggested to explain the photoreduction of some heme proteins(5). In this case, it is proposed that electron photoemission from the silver metal may be responsible for the reduction of adsorbed MP-11. Illumination with 413 nm light excites surface plasmon resonances in the silver, which in turn can lead to electron ejection into the adsorbate.

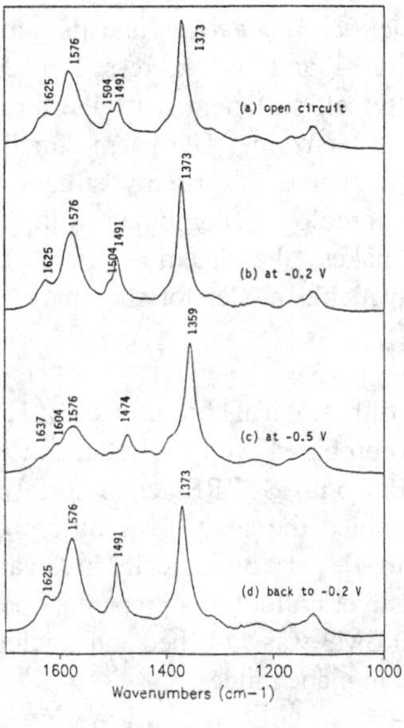

Figure 1. SERS of MP-11. (a) open circuit (b) -0.2 V (c) -0.5 V (d) back to -0.2 V

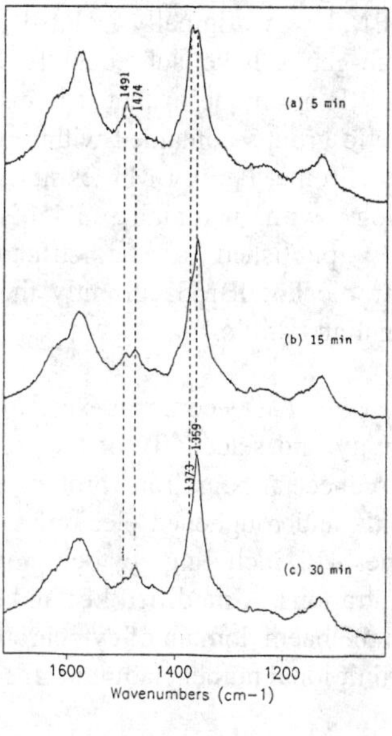

Figure 2. Time dependence of SERS of MP-11 at roughened Ag electrode under open circuit condition

REFERENCES

(1)G.R. Moore, and G.W. Pettigrew; in Cytochromes: Evolutionary, Structure and Physical chemical Aspects, Springer-Verlag, Berlin, 1990
(2)R.P.F. Gregory; in Biochemistry of Photosynthesis, Green bacteria; Summary of Photosynthetic Electron Transport, Wiley, Chichester, 3rd edn., 1989
(3)J. Aron, D.A. Baldwin, H.M. Marques, J.M. Pratt, and P.A. Adams; J. Bioinorg. Chem. 1986, 27, 227
(4)J.-S. Wang, and H.E.V. Wart; J. Phys. Chem. 1989, 93, 7925
(5)Y. Gu, P. Li, T. Sage, and P.M. Champion; J. Am. Chem. Soc. 1993, 115, 4993

SERRS OF SUBSTRATE/INHIBITOR INTERACTIONS ON THE HAEM DOMAIN OF CYTOCHROME P450 BM3

I.D.G. Macdonald[1], A.W. Munro[2], W.E. Smith[1*]

1. Department of Pure and Applied Chemistry, Thomas Graham Building, 295 Cathedral Street, Glasgow G1 1XL
2. Department of Biochemistry, Davidson Building, University of Glasgow, Glasgow G12 8QQ

The cytochrome P450 monooxygenase cycle catalyses the oxidation of thousands of compounds from metabolites to xenobiotics[1]. Cytochrome P450 BM3 was originally isolated from *Bacillus Megaterium* and its gene and sub-genes have subsequently been cloned and overexpressed in *E. Coli.*[2]. It is unique in that it is catalytically self-sufficient, with the three prosthetic groups contained within a single holoenzyme. The haem domain of cytochrome P450 BM3 exhibits a high degree of primary structural homology with mammalian P450's. This homology combined with its recently published crystal structure[3] makes the haem domain of cytochrome P450BM3 currently the best available model for the study of mammalian P450's.

SERRS has become an excellent probe of biological chromophores. The sensitivity and selectivity of the technique combined with the reduction in fluorescence, arising from protein adsorption onto SERRS-active surfaces (colloids and roughened electrodes), has enabled the study of a variety of enzymes[4], including cytochromes P450[5], at biologically relevant concentrations. A modified Lee and Meisel silver colloid was prepared, upon which the haem domain of cytochrome P450 BM3 was adsorbed and studied. The vibrational nomenclature of *Abe et al*[6] in plane modes.

SERRS of the substrate-free haem domain of cytochrome P450 BM3 has shown that a mixed spin state species (six coordinate low spin [6cls] and five coordinate high spin [5chs]) is observed for the haemoprotein adsorbed on the surface. Upon the addition of excess (~1000 x protein) substrate [palmitate] an enhanced mode appears at 1400cm^{-1}. This has been attributed to the ν_{29} haem mode and not SERS from palmitic acid[7]. The ν_{29} mode is coupled to the peripheral vinyl substituents of the ring and is therefore sensitive to slight alterations within the haem environment. Fatty acid binding within the pocket alters subtle secondary and tertiary amino acid alignments. This effect is allosterically transmitted to the immediate protein/haem environment causing an enhancement of sensitive ν_{29} mode[7].

Addition of excess metyrapone(P450 inhibitor) to substrate-free cytochrome P450 BM3 causes complete conversion to the 6cls protein species. The SERRS spectrum is presented in Figure 1. The sharp nature and frequency of the bands ν_3 ,ν_2 and ν_{10} at 1502cm^{-1}, 1583cm^{-1} and 1636cm^{-1} respectively confirm the presence of a single hexacoordinated haem surface species. The bands labelled ν_a ,ν_b and ν_c at 1682cm^{-1},1053cm^{-1} and 1032cm^{-1} are SERS contributions from metyrapone. The observation of hexacoordinated surface species infers ligation of the nitrogeneous ligand on the distal haem face.

SERRS of substrate/inhibitor/haem complex indicates that both the substrate and inhibitor are present within the haem pocket(not shown)[7]. The sensitive ν_{29} mode is enhanced confirming the binding of palmitate within the substrate pocket. Moreover, a six coordinate low spin haem species is still observed suggesting metyrapone ligation to the distal haem face. This observation of both fatty acid and metyrapone being present within the haem pocket provides evidence for the first time of a large nitrogeneous ligand and substrate both positioned within the haem pocket. Thus, SERRS provides for the view that two compounds can be present in the pocket at the same time.

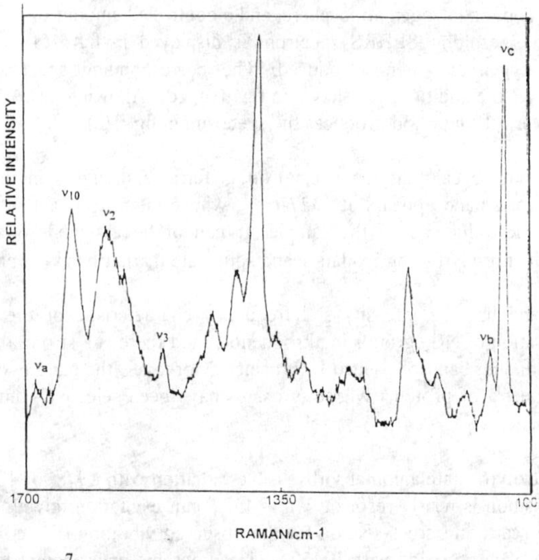

Figure 1:1.7 x 10^{-7}M haemoprotein with excess metyrapone.λ_{ex}457.9nm.

1. Guengerich, F.P. J.Biol.Chem. 266, 10019 (1991).
2. Wen, L-P., Fulco, A.J., J.Biol.Chem. 262, 6683 (1987).
3. Ravichandran, K.G., Boddupalli, S.S., Hasemann, C.A., Peterson, J.A., Deisenhofer, J.A., Science, 261, 731 (1993)
4. Macdonald, I.D.G., Smith, W.E., Langmuir, in press, (1996)
5. Rospendowski, B.N., Kelly, K., Wolf, C.R., Smith, W.E., J.Am.Chem.Soc. 113, 1217 (1991)
6. Abe, M., Kitagawa, T., Kyoguku, Y., J.Chem.Phys., 69, 4526 (1978)
7. Macdonald, I.D.G., Smith, W.E., Munro, A.W., in preparation

THE REACTION OF NITRIC OXIDE WITH CYTOCHROME P-450: *IN SITU* CHARACTERISATION BY SERRS.

L. Quaroni, W.E. Smith, A.W. Munro
Department of Chemistry, University of Strathclyde, Glasgow, G1 1XL
Department of Biochemistry, University of Glasgow, Glasgow

The details of the mechanism of nitric oxide inhibition of haemoproteins including P-450 enzymes is not well understood. The process occurs in two parts, an initial partially reversible total inhibition attributed to the formation of a haem-nitric oxide complex and a second slower non reversible process leading to a loss in activity. In the study SERRS is used to give an effective probe of the active site during the inhibition process for the soluble cytochrome P-450$_{BM3}$ from Bacillus Megaterium (haem domain).

The protein is adsorbed on a citrate-reduced silver colloid[1]. The colloid is deoxygenated under a nitrogen atmosphere, and a controlled amount of nitric oxide is introduced. The resulting SERRS spectrum is displayed in fig. 1(a). Increasing amounts of oxygen are then introduced. The corresponding spectra, recorded immediately after the addition, are shown in fig. 1(b,c,d). Allowing air to slowly leak into the cell over a 12 h period produces the spectrum in fig. 1(e).

The bands reveal that the haem iron is ferric and high-spin in all cases. However, a strong band appears at 1327 cm^{-1}, whose intensity is a function of the amount of air introduced into the sample. It cannot be assigned to any of the vibrations of the porphyrin ring oxidation and spin state marker bands ν_3 and ν_4.

The band has the intensity and frequency characteristic of the symmetric stretching vibration of NO_2 groups in nitrophenols[2] and there is a strong resemblance to the corresponding band of o-nitro-L-tyrosine. Moreover, the band is observed in the SERRS spectra of proteins whose tyrosines have been selectively nitrated with tetranitromethane.

The nitrotyrosinate anion absorbs visible radiation with a λ_{max} of 420 nm. The electronic transition is nearly resonant with 457.9 nm excitation, giving rise to the strong Raman scattering cross-section for the observed vibration. We conclude that the interaction of nitric oxide with P-450$_{BM3}$ in an aerobic environment leads to the nitration of a tyrosine. The reaction is controlled by both NO and O_2 concentrations. The sharpness of the NO_2 band suggests that the reaction is highly specific and one of tyrosine residues is affected predominately.

The electronic absorption bands of nitrotyrosines overlap with those of haems and other nitrocompounds. Therefore UV/vis absorption does not provide the fingerprinting capability required to identify nitrotyrosines. Moreover, the fluorescence emission of nitrotyrosine can hinder RR detection. Consequently SERRS is an ideal technique to identify the formation of nitrotyrosines *in situ* in complex biological systems.

SERRS is an ideal technique to identify the formation of nitrotyrosines *in situ* in complex biological systems.

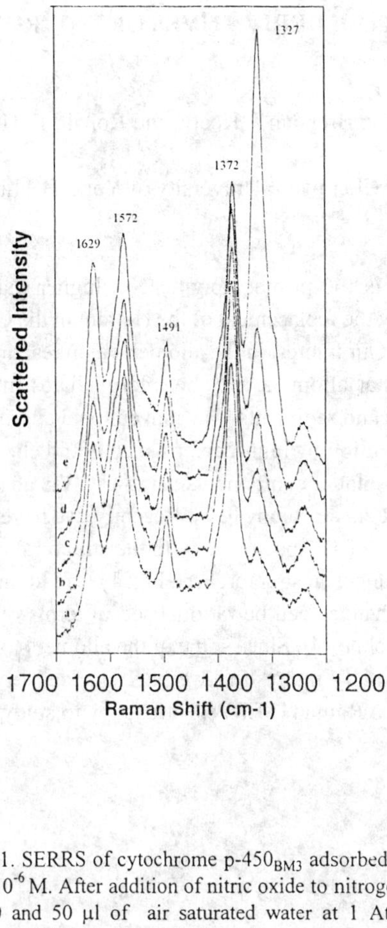

Figure 1. SERRS of cytochrome p-450_{BM3} adsorbed on silver citrate colloid. Protein conc. 10^{-6} M. After addition of nitric oxide to nitrogen atmosphere, (a). After addition of 10 and 50 μl of air saturated water at 1 Atm pressure, (b) and (c) respectively. After addition of 10 μl of air (d). After slow leakage of air for 12 h (e). Excitation at 457.9 nm, 100 mW. Slit width 5 cm^{-1}.

---

[1] Kelly, K., Rospendowski, B. N., Smith, W. E., and Wolf, C. R. (1987) *FEBS Lett.* **120**, 120-123

[2] Passingham, C., Hendra, P.J., Hodges, C., Willis, H.A. (1991) *Spectrochimica Acta* **47A**, 1235-1245

SURFACE-ENHANCED RAMAN SPECTROSCOPY OF CHLOROPHYLL DERIVATIVES OF ENVIRONMENTAL INTEREST

Paul S. Woolley, Brendan J. Keely, and Ronald E. Hester*.

Department of Chemistry, University of York, Heslington, York, YO1 5DD, UK.

Chlorophyll *a* (chl *a*) is the primary pigment in higher plants (Fig. 1). The allomers of chl *a* are formed by the replacement of the H-atom in the C-13^2 position with an oxygen containing moiety. Our interest in the allomerisation reaction and its products stems from the suggestion that allomers may be intermediates in the formation of aetioporphyrins present in oils and sediments [1]. Conventional approaches to the study of allomers often involve separation by high-performance liquid chromatography. We, and others, have studied the isolated components using UV/Vis absorption, fast atom bombardment mass, and NMR spectroscopy [2,3]. Recently, we have demonstrated the sensitivity of resonance Raman (RR) spectroscopy to the structural changes that result from allomerisation. Small changes in structure (see Fig. 1) lead to large changes in the RR spectra. These spectral changes can be rationalised in terms of structure using a normal coordinate analysis of chl *a* [4]. Since some of the allomers formed in low yield may be the most environmentally interesting we have decided to use the increased sensitivity of surface-enhanced Raman (SER) spectroscopy to study these products.

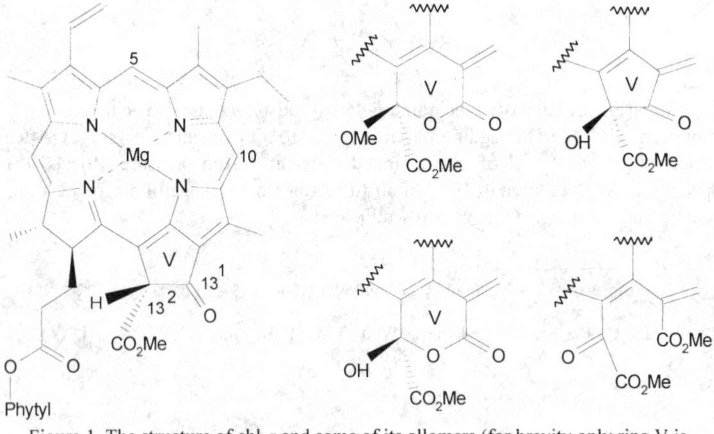

Figure 1. The structure of chl *a* and some of its allomers (for brevity only ring V is shown for the allomers).

Water insoluble compounds such as chl *a* have been studied by SER spectroscopy using silver island films [5] or silver electrodes [6]. The ease of preparation and suitability to flowing and spinning techniques of Ag hydrosols prompted us to test their suitability for obtaining SER spectra of chl *a.* Trial experiments involving the mixing of solutions of tetra-phenyl porphyrin (TPP) in 1,4-dioxane with Ag hydrosols resulted in good quality SER spectra (not shown). In previous SER studies of TPP on silver island films TPP was metallated on adsorption to the surface [7]. In this case the TPP does not undergo metallation. Bands attributed to dioxane are also seen in the SER spectrum. It seems that the dioxane is acting as a molecular spacer between the porphyrin and the Ag surface. High quality SER spectra of chl *a* have been recorded (Fig. 2) in which band positions are similar to those in the RR spectra. The molecule is protected from coordination to the surface and possible alteration from its native conformation by the molecular spacer. We are now applying this technique to allomers of chl *a*.

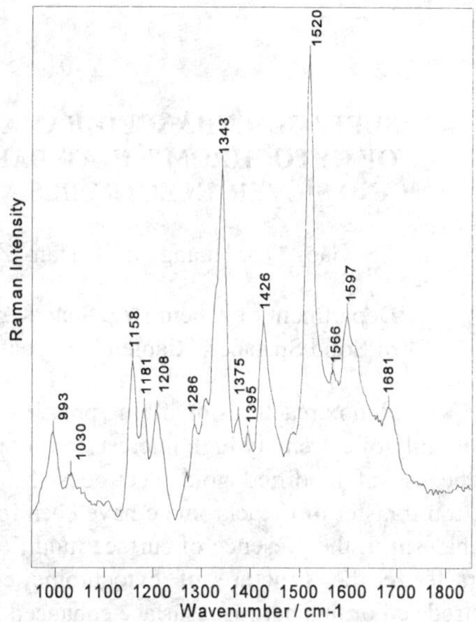

Figure 2. SER spectrum of chl *a*. Concentration 6 x 10^{-6} M, λ_{ex} 457.9 nm, laser power 40 mW, and collection time 5 minutes.

ACKNOWLEDGEMENTS

We thank B.P. Oil International (Sunbury, UK) and the E.P.S.R.C. for financial support.

REFERENCES

(1) Louda, J. W.; Baker, E. W. *Organic Marine Geochemistry* **1986**, *305*, 107-126.

(2) Kostiainen, R.; Hyvarinen, K.; Hynninen, P. H. *Rapid Communications in Mass Spectrometry* **1995**, *9*, 555-558.

(3) Hyvarinen, K.; Helaja, J.; Kuronen, P.; Kilpelainen, I.; Hynninen, P. H. *Magn. Reson. Chem.* **1995**, *33*, 646-656.

(4) Boldt, N. J.; Donohoe, R. J.; Birge, R. R.; Bocian, D. F. *J. Am. Chem. Soc.* **1987**, *109*, 2284-2298.

(5) Uphaus, R. A.; Cotton, T. M.; Mobius, D. *Thin Solid Films* **1985**, *132*, 173-184.

(6) Thomas, L. L.; Kim, J-H.; Cotton, T. M. *J. Am. Chem. Soc.* **1990**, *112*, 9378-9386.

(7) Kobayashi, Y.; Itoh, K. *J. Phys. Chem.* **1985**, *89*, 5174-5178.

SURFACE ENHANCED RAMAN SPECTROSCOPIC STUDY OF CYTOCHROME C AT BARE AND MODIFIED GOLD AND SILVER ELECTRODES

J.S. Gao, Q.J. Huang, Z.Q. Tian, Z.H. Lin

Department of Chemistry, State Key Laboratory for Physical Chemistry of Solid Surfaces, Xiamen University, Xiamen 361005, China

Redox reactions involving proteins and enzymes are a research area which is currently of extremely high interest. In protein electrochemistry , numerous studies on the use of modified gold electrodes through chemisorption to promote direct electron transfer of cytochrome c have been reported. However, the electron transfer mechanism in the presence of surface modifiers is still in question. It is essential to characterize the structure of cytochrome c and/or the promoter adsorbed at electrodes. For this purpose, surface enhanced Raman spectroscopy (SERS) would be one of the most powerful methods [1,2]. To our knowlege, no SERS of cytochrome c adsorbed on baer Au is reported. In the present study, the adsorption of cytochrome c and the influence of 2-mercaptopyrimidine (2-MP) as a new promoter at gold and silver electrodes have been investigated by using SERS and electrochemical method. The SERS spectra were acquired using a LabRam I spectrometer from Dilor with a He-Ne laser (632.8 nm, 0.4 ~ 4.0 mW at the sample).

Fig. 1 illustrates the SERS spectra of horse heart cytochrome c adsorbed directly at a bare SERS active gold electrode. The spectra are different considerably from the previous SERRS studies on Ag electrodes using 514 nm or 413 nm excitation line [2, 3]. Our result seems to be be characterize the the founction group of cytochrome c attach to the surface based on the EM enhancement mechnism and infer the structural changes and irreversible adsorption in the electrode potential region from +0.2 V to -0.4 V (vs. SCE). It is of interest that the spectra of cytochrome c adsorbed at the bare silver electrode were very changeble even with only 0.4 mW of laser power. Surface modification of the electrodes was achieved by dipping the SERS active electrodes into the solution of 0.1 M 2-MP for 15 min. Cyclic voltammograms in Fig. 2 show that 2-MP modified at a gold electrode can promote direct electron-transfer of horse heart cytochrome c with very stable activity, e.g., the curve do not changed substantially after the potential sweeping for one hour. It was impossible to observe the identical SERS spectra of cytochrome c in the 2-MP modified Au electrode systems, see Fig.3, indicating that the electron transfer of cytochrome c took place via a 2-MP adsorbed layer at the surface. The assignment and analysis of the SERS spectra and the detailed discussion on the interaction between cytochrome c, 2-MP and metal electrodes will be given.

This work is supported by the Natural Science Foundation of China.

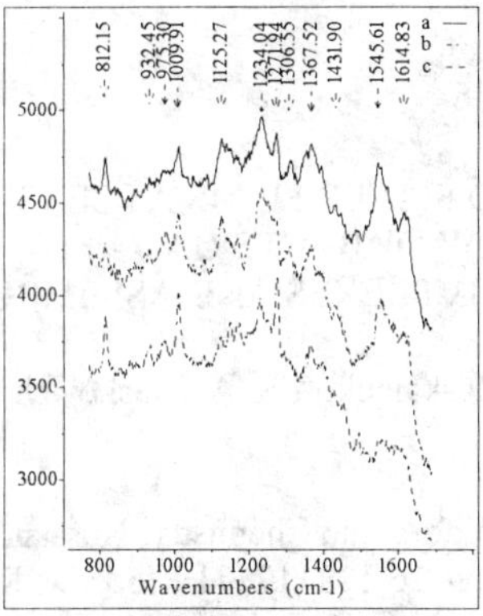

Fig.1. SERS spectra of cytochrome c in 0.02 M phosphate buffer/ 0.1 M $NaClO_4$ (pH 7.0) containing 20 μM cytochrome c at different applied electrode potentials: (a) +0.2 V, (b) -0.2 V, (c)-0.4 V vs SCE.

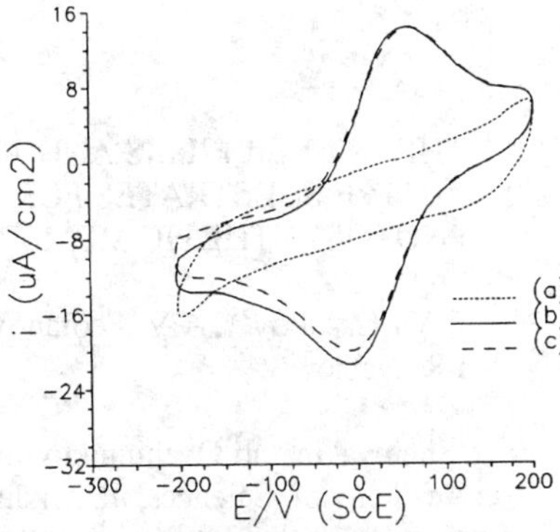

Fig.2. Cyclic voltammograms of horse heart cytochrome c (type VI) 0.38 mM using a scan rate of 50 mVs^{-1} in 0.02 M phosphate buffer / 0.1 M $NaClO_4$ (pH 7.0). (a) at bare Au, (b) at 2-MP/Au, (c) after cyclic sweep for 1 hour.

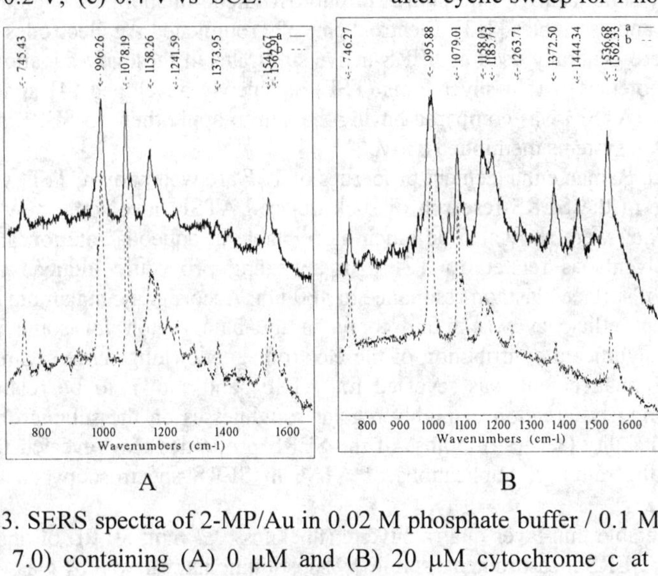

A B

Fig.3. SERS spectra of 2-MP/Au in 0.02 M phosphate buffer / 0.1 M $NaClO_4$ (pH 7.0) containing (A) 0 μM and (B) 20 μM cytochrome c at different applied electrode potentials. (a) +0.2 V; (b) -0.2 V vs. SCE.

1. D. Hobara et al., Colloids Surfaces A: Physicochem. Eng. Aspects 93, 241 (1994).
2. T.M. Cotton, S.G. Schultz, R.P. Van Duyne, J. Am.Chem.Soc. 102, 7962 (1980).
3. P. Hildebrandt, J. Molecular Structure. 242, 379 (1991).

THIN SILVER FILMS AND ANNEALED SILVER FILMS AS SERS-ACTIVE SUBSTRATES FOR THE STUDY OF BACTERIAL PHOTOSYNTHETIC MEMBRANES, COMPLEXES AND ASSEMBLES

E.Yu.Kryukov[5]*, A.V. Feofanov[1], A. A. Moskalenko[2], S. A. Maskevich[3] and I.R. Nabiev[4]

[1]Shemyakin and Ovchinnikov Institute of Bioorganic Chemistry, Russian Academy of Sciences, ul. Miklukho-Maklaya 16/10, 117871 Moscow, Russia
[2]Institue of Soil Science and Photosynthesis, RAS, Pushchino, Russia.
[3]Grodno State University, Grodno, Belarus
[4]Université de Reims Champagne-Ardenne, Reims, France
[5]Moscow Engineering-Physics Institute, Moscow, Russia

Surface-enhanced Raman scattering spectroscopy (SERS) was shown to have considerable promise as a highly selective and sensitive method for the study of functionally significant chromophores in photosynthetic membranes, pigment-protein complexes and assembles [1-3]. Electrochemically roughened Ag electrodes and silver colloids were generally used as SERS-active substrates in these investigations. In this study the potency of thin silver films (TSF) and newly developed [4] annealed thin silver films (ATSF) was compared having the aim to apply them to SERS-analysis of the biological systems mentioned above.

Although Raman enhancement properties of TSF are well known, TSF were out of general use in the SERS-research of such objects. ATSF have some new attractive features. They were shown to be much more stable in aqueous solution and several organic solvents as related to TSF. An annealing procedure induced a shift of maximum of surface plasmon resonance to 460 nm. Accordingly, maximum of Raman enhancement efficiency was found to be in the blue region for some molecules studied. A significant contribution of the electromagnetic (long-range) component to the SERS enhancement was revealed for ATSF and shown to be related to the annealing-induced creation of highly regular roughnesses on the submicron scale. A long-time stability (several months) of the SERS properties was revealed for ATSF. However, the range of applications of ATSF in SERS spectroscopy has not been defined yet.

Dismountable thin-layer quartz cuvette (thickness 0.2 mm; 30 μL of analyte) was adapted for SERS measurements simply by substituting its rear wall by glass-deposited TSF or ATSF. An absence of photo-degradation of the labile photosynthetic species used in our study was confirmed for both experimental systems. Approximately 20-times exceeding of the SERS signal over the background RR signal was easily achieved using ATSF (Fig. 1) for the membrane species at different excitation wavelengths (457.9, 488.0, 514.5, and 647.1 nm) indicating enhancement factor to be better than for an electrode. Spectra were measured from the membranes and complexes suspended in aqueous solution as well as from Langmuir-Blodgett monolayers. A relatively long-range enhancement along with blue-shifted SERS excitation maximum revealed for ATSF are good prerequisites for SERS study of chromophores buried deep into the pigment-protein complexes and membranes. The studies of acetone/methanol exctracts of the membrane are possible, since the structure of ATSF is stable in contact with these solvents. Fine conformational changes in pigments could have been traced in SERS spectra upon modifying pigment-protein complexes by adding acetone, methanol, etc.

Chromatophores as well as B890-RC and B800-850 pigment-protein complexes both of membrane and solubilized preparations for two kinds of bacteria *Chr. minutissimum* and *Rps. palustris* (sulfur and non-sulfur) were characterized by SERS in our study using ATSF and TSF. Comparative analysis of spectral features detected in the SERS and resonance Raman spectra will be presented in detail in poster.

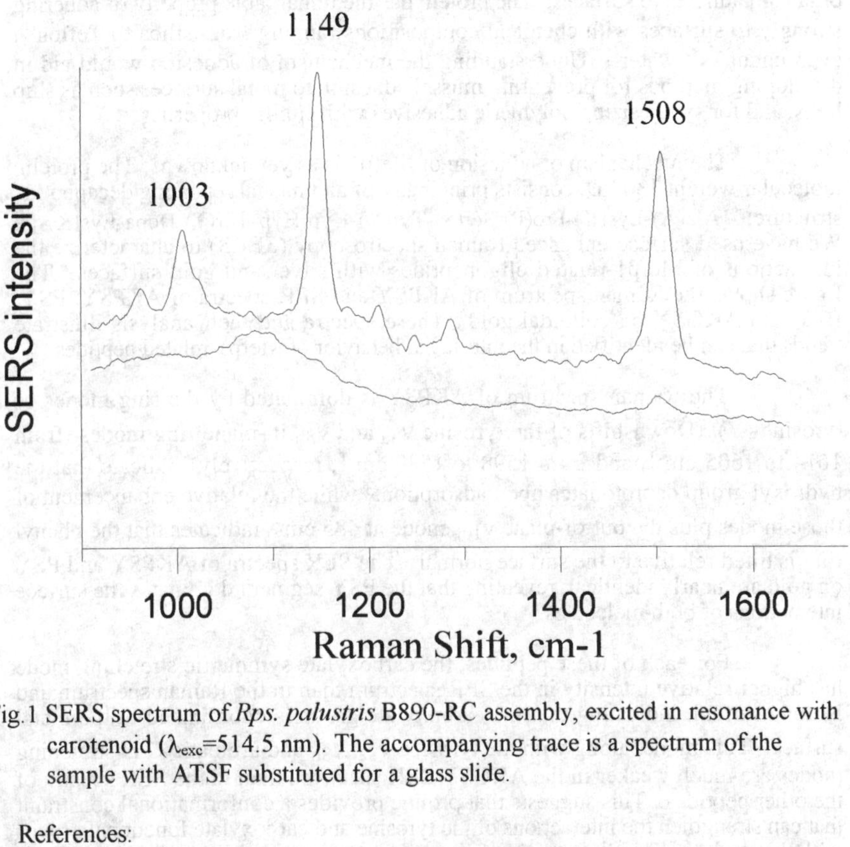

Fig. 1 SERS spectrum of *Rps. palustris* B890-RC assembly, excited in resonance with carotenoid (λ_{exc}=514.5 nm). The accompanying trace is a spectrum of the sample with ATSF substituted for a glass slide.

References:
1. R. Picorel, R.E. Holt, T.M. Cotton, M. Seibert *J. Biol.Chem.* **263**, 4374 (1988)
2. R. Picorel, T. Lu, R.E. Holt, T.M. Cotton, M. Seibert *Biochemistry* **29**, 707 (1990)
3. T.M. Cotton, J.-H.Kim, G.D. Chumanov *J. Raman Spectrosc.* **22**, 729 (1991)
4. S. Maskevich, I. Sveklo, A. Feofanov, A. Ianoul, V. Oleinikov, S. Gromov, O. Fedorova, M. Alfimov, I. Nabiev, L. Kivach *Opt.Spectrosc.* 1996, in press.

MECHANISM OF ADHESION OF MUSSELS ON METALS: SERS OF MODEL PEPTIDES OF THE PROTEIN MEFP1

Robin L. Garrell*, Namjun Cho, Siobhán S. F. Freitas

Department of Chemistry and Biochemistry, University of California, Los Angeles, CA USA 90095-1569

Mussel adhesive protein (Mefp1) from *Mytilus edulis* is secreted by mussels to form the byssus, a set of threads and adhesive plaques by which the organism adheres to surfaces. The protein has the remarkable property of adhering strongly to surfaces with chemical compositions ranging from silica to Teflon®, even under salt water.[1] Understanding the mechanism of adhesion would aid in developing methods for preventing mussel adhesion to metal surfaces such as ship hulls, and for synthesizing polymeric adhesives with similar properties.

The mechanism of adhesion of Mefp1 is as yet unknown. The protein, molecular weight 130 kD, consists principally of an unusual repeating decapeptide structure, -[Ala(A)-Lys(K)-Pro(P)-Ser(S)-Tyr(Y)-Hyp-Hyp-Thr(T)-Dopa-Lys(K)-].[2] We have used surface-enhanced Raman spectroscopy (SERS) to characterize the interactions of Mefp1-related oligopeptides with silver and gold surfaces. The figure shows the Raman spectrum of AKPSY and SER spectra of AKPSY, PSY, PTY and AKGSY on colloidal gold. These spectra and their analysis illustrate trends that can be identified in the interfacial behavior of Mefp1-related peptides.

The Raman spectrum of AKPSY is dominated by the ring modes of tyrosine (Y). Downshifts of the tyrosine ν_{8a} and ν_{8b} in-plane ring modes (from 1614 to 1605 cm^{-1}, and from 1598 to 1580 cm^{-1}, respectively) indicate that the hydroxyl group deprotonates upon adsorption,[3] while the relative enhancement of those modes plus the out-of-plane ν_{16b} mode at 485 cm^{-1} indicates that the phenyl ring is tilted relative to the surface normal. The SER spectra of AKPSY and PSY on gold are nearly identical, revealing that the PSY segment dominates the surface interactions of both molecules.

For each of these peptides, the carboxylate symmetric stretching mode has higher relative intensity in the SER spectrum than in the Raman spectrum and is downshifted, indicating that the carboxylate group is coordinated to the metal surface. This mode, along with the tyrosine ν_{8a}, ν_{8b} and aromatic C-H stretching modes are much weaker in the AKGSY SER spectrum than in the SER spectra of the other peptides. This suggests that proline provides a conformational constraint that can strengthen the interactions of the tyrosine and carboxylate functional groups with the surface. The tyrosine ring is more perpendicular to the surface in AKPSY than in AKGSY. The orientation of the ring in PTY is intermediate between the two. Trends in the interfacial behavior of other Mefp1 analogs have also been identified.

In summary, the SER spectra reveal specific interactions between Mefp1 and metals that may be responsible for the strong adhesion of the protein in submarine environments. The proline and hydroxyproline residues in the Mefp1 analogs provide conformational constraints that can promote interactions of adhesive functional groups with the surface. The significance of these results for designing adhesives for wet environments will be discussed.

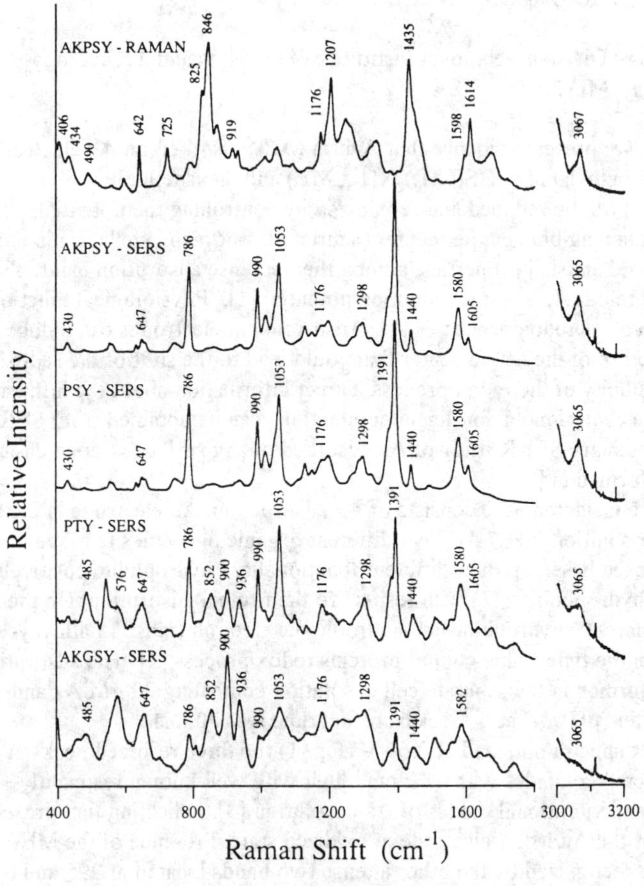

1. Crisp, D.J.; Walker, G.; Young, G.A.; Yale, A.B. *J. Coll. Interf. Sci.* (1985) *104*, 40.
2. Waite, J.H.; Tanzer, M.L. *Science* (1981) *212*, 1038.
3. Herne, T.M.; Ahern, A.M.; Garrell, R.L. *J. Am. Chem. Soc.* (1991) *113*, 846.

REDOX TRANSFORMATIONS OF ADSORBED AZURIN AS EVIDENCED BY SERRS

G. Niaura* , A. K. Gaigalas and V. L. Vilker

Biotechnology Division , National Institute of Standards and Technology, Gaithersburg, MD 20899 , USA

In this work we present evidence that Azurin (Az), adsorbed on Au electrode modified by mercaptoethylamine ($HS(CH_2)_2NH_2$; ME) and hexanethiol ($HS(CH_2)_5CH_3$; HT) can be reduced and reoxidized by controlling the potential.

Proteins containing blue copper center (azurin, plastocyanin, stellacyanin and others) exhibit several anusual properties, among them intense absorption band around 600 nm and relatively positive reduction potential [1]. Physiological function of these proteins involves redox processes [1]. Adsorption at electrodes can induce changes in the structure of the active center that could lead to the shift of the redox potential or irreversibility of the redox process. Direct information about the influence of adsorption on the central metal ion-ligand interactions can be obtained from SERR spectroscopy. Pleriminary SERR study of Az adsorbed on Ag surface at open circuit conditions was performed in [2].

The system considered here consists of Az adsorbed on Au electrode in 0.01 M phosphate buffer solution (pH 7.4). Two different organic molecules (surface modifiers) were chosen based on their different functionality (hydrophilic, positively charged MEA and hydrophobic HT) with respect to the protein adsorption. On the other hand, two different - hydrophilic and hydrophobic electron transfer pathways could be involved in the type I blue copper proteins redox process [1]. The adsorption of Az has been performed in the separate cell in solution containing 70 μM Az and 0.01 phosphate buffer (pH 7.4) at 5 ^{0}C. Adsorption time was 30 min.

In the SERR spectra obtained at 0.60 V (Fig. 1) the three resolved peaks at 371, 408, and 427 cm^{-1} coincides within ± 2 cm^{-1} limit with well-known resonantly-enhanced metal-ligand vibrational modes of Az in solution [3], indicating that protein remains adsorbed at the Au/MEA electrode in oxidized state. Presence of the MEA at the interface can be recognized by the other intense two bands located at 295 and 646 cm^{-1}; these are related mainly to the ν(Au-S) and ν(C-S) (Gauche conformers) vibrations respectively. The reduction of adsorbed Az is evidenced by the disappearence of the spectra at -0.1 V and reappearance of the spectra when potential is switched back to initial value (Fig. 1). The charge-transfer Cu(II)-S(Cys) absorption band at 625 nm is responsible for the RR effect [3]. This band completely disappears after reduction eliminating the RR effect in SERRS. The reappearance spectra at 0.60 V indicate that loss in SERRS is related with reduction process but not with the protein desorption. The approximate redox potential values were evaluated from the integrated intensity-E curves (Fig. 2) as the potential where half of adsorbed species is reduced . Near reversible intensity-E dependence is observed for modified Au electrodes (Fig. 2) . In both cases the estimated potentials 0.29 V ± 0.04 V and 0.30 V ±0.04 V for Au/HT and Au/MEA respectively is close to the potential of the native Az in solution 0.304 V [1].The near reversible intensity-E dependence, close potential value to the native form, and similarity of the RR and SERR spectra indicate that Az remains in the native form adsorbed on Au/MEA and Au/HT electrodes.

Contrary, in the case of bare Au electrode only ~ 1/5 of initially adsorbed species can be reoxidized (Fig. 2c). Two main processes can be responsible for this effect : a) desorption of reduced species; b) transformation of reduced adsorbed Az into inactive form [1]. The pleriminary attempts to observe nonresonantly surface enhanced modes from adsorbed reduced Az suggest that (b) process is more favorable.

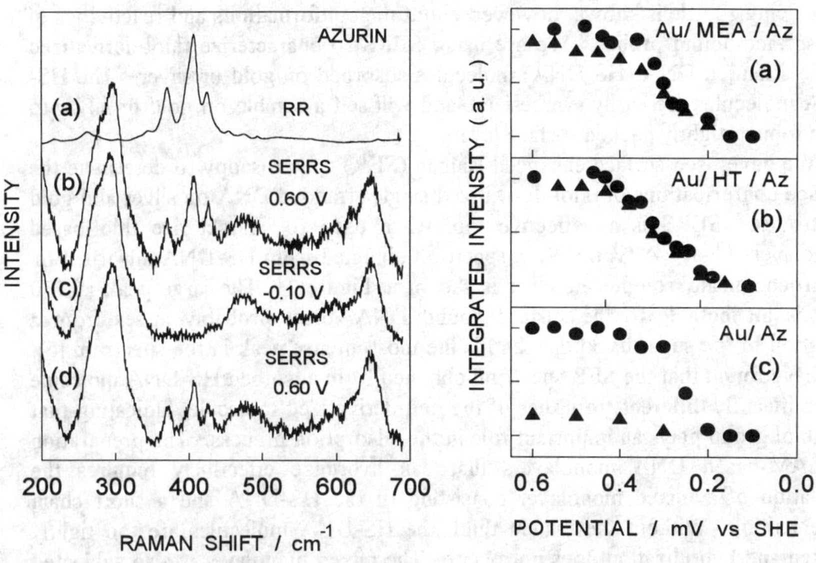

Fig. 1. (Left) Comparison of the RR (a) and SERR (b-d) spectra of adsorbed Az on Au electrode modified with MEA. Excitation wavelength - 647 nm.

Fig. 2. (Right) Potential dependence of the integrated intensity of the 408 cm^{-1} SERR band from the adsorbed Az on modified (a,b) and bare (c) Au electrodes. ● - Negative potential sweep ; ▲ - Positive potential sweep (return)

1. O. Farver and I. Pecht, in The reactivity of copper sites in the "Blue" copper proteins, (R. Lontie, ed.), Vol I, 7, CRC Press, 1984, p.183.
2. V. I. Silin and A. K. Gaigalas, J. Raman Spectrosc. 25 903 (1994).
3. T. J. Thamann, P. Frank, L.J. Willis and T. M. Loehr, Proc. Natl. Acad. Sci. USA, 79 6396 (1982).

SURFACE-ENHANCED RAMAN SPECTROSCOPY OF THIOL-DERIVATIZED DNA

T. M. Herne*, Michael J. Tarlov, and Keith McKenney

Chemical Science and Technology Laboratory, National Institute of Standards and Technology, Gaithersburg, MD 20899-0001

In our laboratory, self-assembled monolayer (SAM) technology has been exploited in the construction of novel DNA-based surfaces. DNA probes on surfaces have important potential applications in array-based DNA screening and sequencing. Little is known, however, about the conformations and reactivities of the surface-bound probes. We are using SERS to characterize thiol-derivatized single-stranded DNA (HS-DNA) molecules adsorbed on gold or silver. The HS-DNA molecules are easily synthesized, and will self assemble on gold or silver to form robust, tightly packed surface layers.

We have used surface-enhanced Raman (SER) spectroscopy to determine the surface conformations of thiol-derivatized single stranded DNA on silver and gold electrodes. SERS is an effective tool when used to characterize thiol-based monolayers [1-4]. A typical SER spectrum obtained from HS-DNA adsorbed on electrochemically roughened silver is shown in Figure 1. The large peak at 770 cm^{-1} is attributable to the surface bound DNA, most probably arises from a vibration of the sugar backbone, and is the most intense peak in the spectrum [5]. We have found that the SER spectrum obtained from adsorbed HS-DNA molecule is significantly different from that of the non-derivatized molecule, indicating that the thiol group plays an important role in the adsorption process. The preparation of SAM-based DNA monolayers that can hybridize effectively requires the formation of a mixed monolayer consisting of the HS-DNA and a short chain diluent thiol. Without the diluent thiol, the HS-DNA molecules are too tightly packed and hybridization does not occur. The mixed monolayer is also subjected to post-treatment with the diluent thiol, in order to remove the non-specifically bound DNA. For the first time, the entire process of constructing a DNA-based monolayer is monitored using SERS, and the removal/rearrangement of the HS-DNA, along with the appearance of bands attributable to the diluent thiol, can be seen.

The hybridization reaction between single strands of DNA can also be monitored using SERS. A thiol-derivatized homonucleotide is attached to the surface, and the surface bound probe is exposed to its homonucleotide complement. The appearance of vibrational bands assignable to the complement indicate that hybridization has occurred.

The results obtained from these SERS studies demonstrate that SERS is a powerful technique for characterization of surface-bound DNA. SERS is also a useful tool in characterizing the DNA mixed monolayers as they are subjected to the required post treatments.

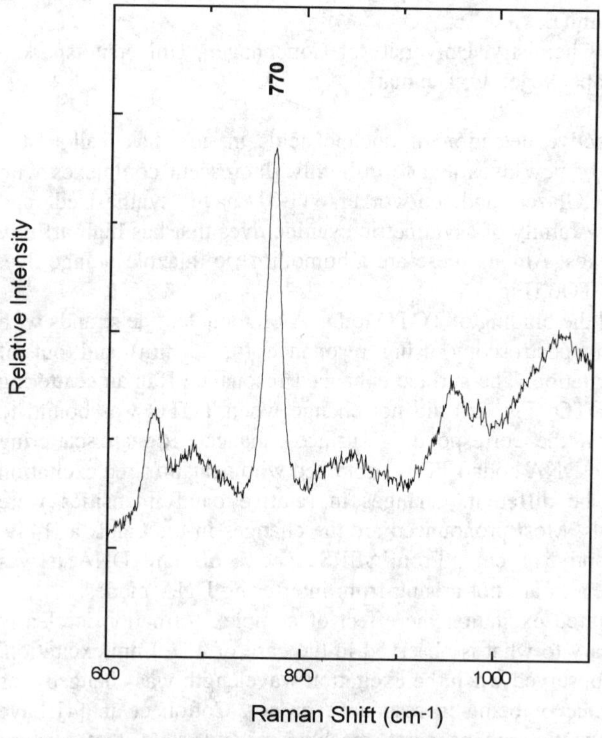

Figure 1. HS-DNA on Silver

1. M. A. Bryant and J. E. Pemberton, J. Am. Chem. Soc. 1991, 113, 8284.
2. M. A. Bryant and J. E. Pemberton, J. Am. Chem. Soc. 1991, 113, 3630.
3. J. E. Chadwick, D. C. Myles, and R. L. Garrell, J. Am. Chem. Soc. 1993, 115, 10364.
4. M. Lewis, M. Tarlov, and K. Carron J. Am. Chem. Soc. 1995, 117, 9574.
5. J. Liquiers, E. Taillandier, W. L. Peticolas, and G. A. Thomas J. Biomol. Struct. and Dynamics 1990, 8, 295.

DNA-DRUG INTERACTIONS - A COMPARISON OF SERS AND SERRS

M. Nissum[1]*, P.W. Jensen[1], and O.F. Nielsen[2]

[1] Department of Chemistry, Odense University, Campusvej 55, DK-5230 Odense M, Denmark

[2] Department of Chemistry, University of Copenhagen, Universitetsparken 5, DK-2100 Copenhagen Ø, Denmark

Interest in nonradioactive detection of nucleic acids in gels has lead to the synthesis and evaluation of new dyes that form highly fluorescent complexes with double stranded DNA. Glazer and co-workers [1-3] have synthesized and characterized a whole new family of asymmetric cyanine dyes that has high affinity for making DNA complexes. Among these are a homodimeric thiazole orange dye, TOTO, and its monomer TO6 (Fig. 1).

We have investigated the binding of TOTO to DNA oligonucleotide strands with surface-enhanced Raman spectroscopy using resonance (457.9 nm) and out of resonance (1064 nm) excitation. The surface-enhanced resonance Raman scattering (SERRS) spectrum of TOTO (Fig. 2a) did not change when TOTO was bound to DNA (Fig. 2b). However, the corresponding surface-enhanced Raman scattering (SERS) spectra of free and DNA bound TOTO obtained with near-infrared excitation (Fig. 3) were found to be different. Changes in relative band intensities were observed for several bands. Most pronounced are the changes in the bands at 1449, 1357, 1049, 637, 553, and 511 cm^{-1}. From SERS spectra of pure DNA it was concluded that the differences are not arising from interfering DNA modes.

Thus, using near-infrared excitation the effect of complex formation is clearly seen in the spectra, contrary to what is observed in the case of 457.9 nm excitation. This different behaviour observed when the excitation wavelength was changed was explained as a consequence of being in resonance or not. Cotton et al. [4] have discussed the electromagnetic enhancement mechanism and state that surface enhancement of the molecular resonance process will produce enhancement of all modes that also experience enhancement in the absence of the surface. The surface

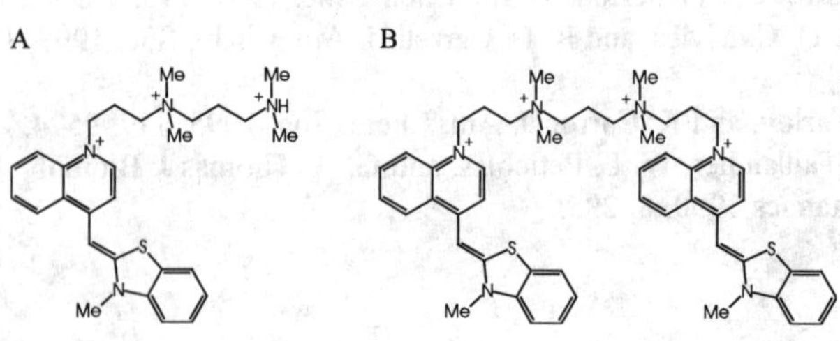

Fig. 1. Structures of a) TO6 and b) TOTO.

selection rules are modified in the resonance case so only differences in overall spectral intensities should be observed from different orientations of the adsorbate at the surface, whereas for the non-resonance process relative band intensity differences may result from different orientations. In our earlier work with TO6 and TOTO [5] we found that the surface enhancement occurred via the electromagnetic mechanism. We observed that the SERRS spectrum of TOTO is highly similar to the resonance Raman spectrum and with almost no effect of adsorption of TOTO to the surface concerning relative intensity changes. Out of resonance, using the 1064 nm laser line, large differences in relative band intensities were observed between the Raman and SERS spectrum due to specific orientation of the TOTO molecules at the colloidal surface. With near-infrared excitation the spectra therefore become much more sensitive to the specific adsorption geometry of the molecules, and the spectral differences observed in Fig. 3 were explained as due to a change in the orientation of TOTO at the surface when it is bound to DNA.

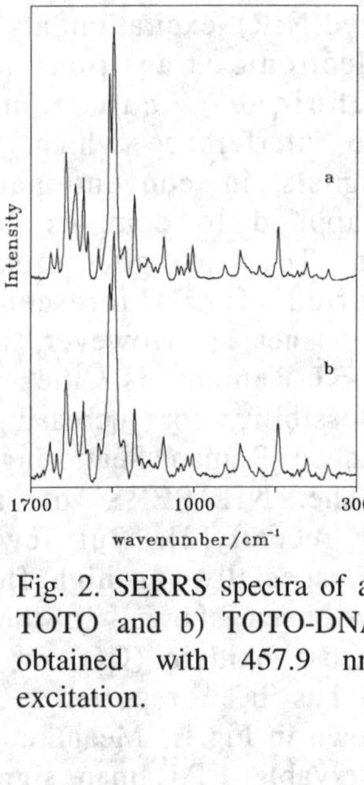

Fig. 2. SERRS spectra of a) TOTO and b) TOTO-DNA obtained with 457.9 nm excitation.

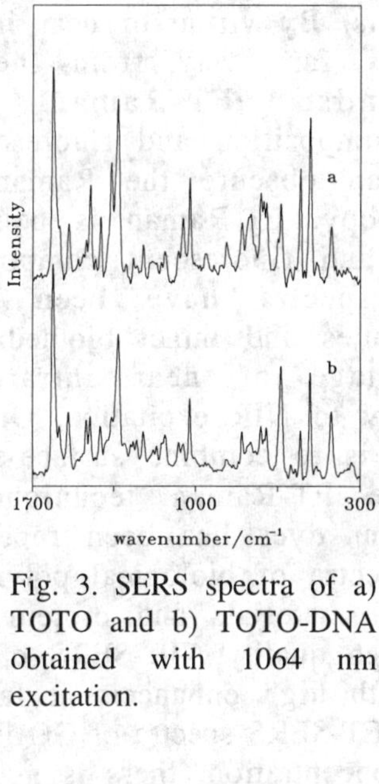

Fig. 3. SERS spectra of a) TOTO and b) TOTO-DNA obtained with 1064 nm excitation.

1. H.S. Rye, S. Yue, D.E. Wemmer, M.A. Quesada, R.P. Haugland. R.A. Mathies, and A.N. Glazer, Nucleic Acid Res. 20, 2803 (1992).
2. S.C. Benson, P. Singh, and A.N. Glazer, Nucleic Acid Res. 21, 5727 (1993).
3. S.C. Benson, R.A. Mathies, and A.N. Glazer, Nucleic Acid Res. 21, 5720 (1993).
4. T.M. Cotton, J.-H. Kim, and R.E. Holt, Adv. Biophys. Chem. 2, 115 (1992).
5. M. Nissum, O.F. Nielsen, F. Jensen, and P.W. Jensen, J. Raman Spectrosc. 26, 1009 (1995).

FOURIER TRANSFORM SURFACE ENHANCED RAMAN SCATTERING OF GLUCOSE OXIDASE ON COLLOIDAL SILVER USING Nd:YAG 1064 nm EXCITATION

F. X. Hu and S. H. Lin
Comprehansive Technology Institute, Capital Normal University
Y. Fang*
Experimental Centre,Capital Normal Univ,Beijing 100037,Chian

The development of Fourier transform (FT) Raman spectrometers in recent years has attracted the interest of researchers. By virtue of near infrared(NIR) excitation at 1064 nm which is far away from the electronic transitions of most compounds, FT-Raman technique can minimize photodecomposition and fluorescence interference which, in many cases, can obscure the Raman signals in conventional Raman spectroscopy. FT-Raman is being applied to complex biological systems and fluorescent compounds. To date, good quality FT-Raman spectra have been reported for fluorescent dyes, biomolecules and intact biomedical tissues[1]. However, a serious disadvantage of near infrared FT-Raman is the inherent insentivity to NIR excitation. One possibility for increasing Raman intensity is to combine surface-enhanced Raman Scattering (SERS) with the FT-Raman technique. The NIR-SERS of adsorbed fluorescent dyes has been reported recently[2]. But fewer NIR-SERS spectra of biological polymers, specially for high fluorescent biological materials and protein have been shown. In the present paper, high quality NIR-SERS of glucose oxidase (GO) on colloidal silver with high enhancement factor has been reported.

The FT-SERS spectra of GO is shown in Fig.1. Meantime, at the same concentration, thers is no observable FT-Raman signal could be detected in GO aqueous solution. The FT-Raman signals of GO powder could not be recorded either, even by long time scanning.

It was suggested that[3,4] the visible excitation SERS spectra of riboflavin (RF), riboflavin binding protein (RBP), old yellow enzyme (OYE) and GO were very similar, in which all the enhanced vibrational signals were from free flavin extracted from the proteins. No observable vibrational signal from proteins was detected. However, significant differences with those, either

Raman shifts or intensities, are shown in NIR-SERS of GO(Fig.1) and RF. Although many bands could be roughly attributed to the vibrations of flavin rings, the changes in intensity are markable, specially for the bands 1531 cm^{-1}, 1472 cm^{-1}, and 1422.6 cm^{-1}, which is more different from the case of the visible excitation SERS of GO. We estimated that the strongest band 1472 cm^{-1} in NIR-SERS of GO which is weak in the visible SERS of GO[4], might be attributed to both I ring of flavin and amino acid of protein in GO, and that the significant changes in Raman shift and intensities result from protein and contain the vibrational information of protein in GO.

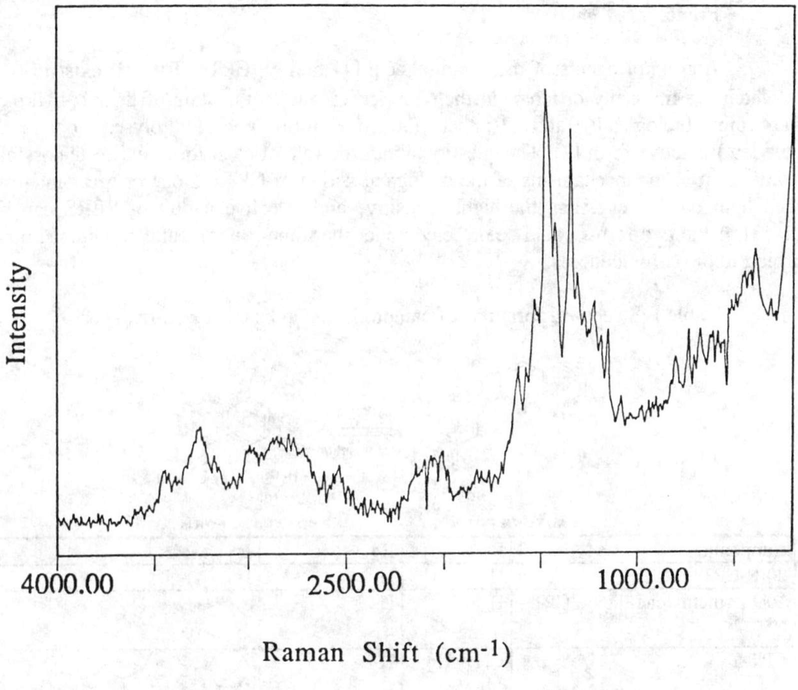

Figure1 Surface-enhanced near infrared FT-Raman scattering of GO in silver colloid.

References
[1] S.Nie, K.L.Beigbauer and J.J.Ho, Spectroscopy, 5(1990), 24.
[2] K.Kneipp, E.Roth, C.Engert and W.Kiefer, Chem.Phys.Lett., 207(1993),450.
[3] Rondall E Holt, J.Am.Chem.Soc., 109(1987), 1841.
[4] Nam-Soo Lee, J.Am.Chem.Soc., 109(1987), 1358.

SERS OF LACTONE AND CARBOXYLATE FORMS OF CAMPTOTHECINS ADSORBED ON SILVER COLLOIDS. COMPARISON WITH RAMAN AND FT-RAMAN.

I. Chourpa[1], A. Beljebbar[1], G. D. Sockalingum[1], J.-F. Riou[2], and M. Manfait[1]*.

[1]Laboratoire de Spectroscopie Biomoléculaire, UFR de Pharmacie, Université de Reims, 51 rue Cognacq Jay, 51096 Reims, France.
[2]Rhône-Poulenc Rorer, S.A., Département de Biologie, 94403 Vitry sur Seine, France.

Antitumor agents of the camptothecin [1] family (CPTs, Table I) exist in two forms which are markedly different in their properties due to the state of their α-hydroxy lactone ring: biologically active intact (lactone) form and hydrolysed, open-ring (carboxylate) inactive form [2]. The question about the role of each form and their possible interconversion in the mechanisms of the biological action of CPTs is one of great interest. To shed light on this question, the highly sensitive and selective nature of SERS can be useful [3], if the two forms could be detected under the same experimental conditions, non-perturbing to the chromophores.

Table I. Structural formulae of camptothecin and some derivatives.

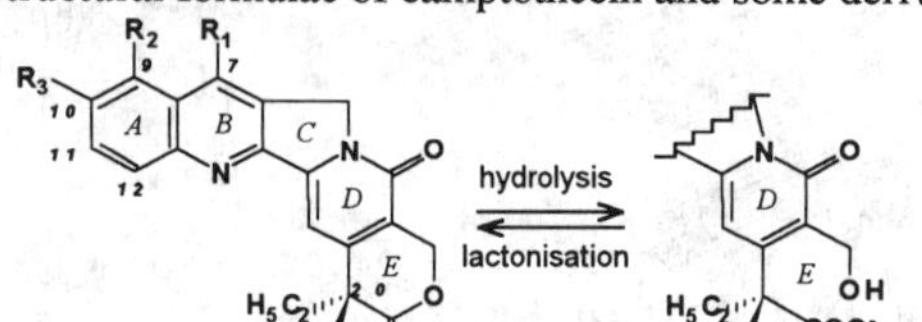

N°	compound	R_1	R_2	R_3
1	20-S-CPT	H	H	H
2	10,11-(methylene-dioxy) CPT	H	H	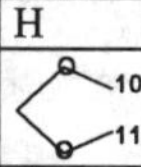
3	CPT-11	CH_2CH_3	H	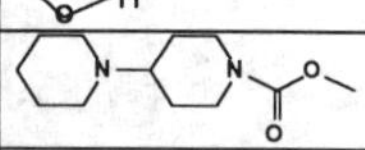
4	SN-38	CH_2CH_3	H	OH
5	topotecan	H	$CH_2NH(CH_3)$	OH

We have evaluated different types of silver colloids (reduced with sodium citrate [4] or borohydride [5], with or without activation by anions) for SERS of CPTs and concluded that for the studies of both forms of these drugs, citrate-reduced sol activated with chloride ions (CAS) was the best compromise. The interaction of the lactone and carboxylate forms of CPTs with the silver surface has been analysed by comparison of the SERS spectra with their Raman and FT-Raman counterparts. This comparison was possible because of the great similarity, both in frequencies and relative intensities, of the visible Raman spectra of CPTs with the corresponding FT-Raman spectra in the near- infrared. We suggest that in general adsorption of CPTs on CAS occurs via the nitrogen in ring B and that the adsorbate is more inclined to a flat orientation rather than to a perpendicular one. In addition, probable interactions of the substitution groups, as well as of the COO^- groups of the ring-open forms of CPTs with the CAS surface introduce some particularities in the adsorption patterns. As a result, SERS spectra are highly sensitive to substitutions and uniquely characteristic of each of the CPT derivatives. Nevertheless, pronounced hydrolysis-induced changes have been found to involve similar vibrations of the pentacyclic chromophore in the SERS spectra of different CPTs (Table II). The main features, characterising lactone-carboxylate transition are similar in the SERS and FT-Raman spectra of each CPT congener. Vibrational assignments have been proposed (Table II) for the main Raman and SERS bands of CPT, 10,11(methylenedioxy)-CPT, CPT-11, SN-38 and topotecan (Table I).

Table II. The main FT-Raman and SERS bands[(a)] affected by the lactone hydrolysis and temptative vibrational assignments for CPT (**1**) and derivatives **2-5**

FT-Raman[(b)]			SERS[(c)]					Assignments	
3	**4**	**5**	**1**	**2**	**3**	**4**	**5**	rings	modes
1616	1613	1616	1613	1602	1613	1596	1613	A-B	ring str.
–	–	–	–	–	–	–	–		
1562	1561	1561	1560	1559	1561	1559	1548	D (+C-B)	ring str.
1559	*1558*	*1558*	*1541*	*1554*	*1548*	*1557*	*1544*		
–	–	–	–	–	–	–	–		
1516	1516	1515	1516	1516	1517	1515	1515	D	ring str.
+	+	+	+	+	+	+	+		
1459	1457	1456	1469	1466	1455	1459	1471	B (+C-D)	ring str.
–	–	–	–	–	–	–	–		
1438	1440	1439	1438	1437	1436	1440	1439	B (+C-D)	ring str.
–	–	–	–	–	–	–	–		
1427	1427	1425	1426	1449	1422	1429	1418	D (+C-B)	ring str.
+	+	+	+	+	+	+	+		
			1380	1366	1373	1371	1356	B	ring str.
1401	1401	1404	+	+	+	+		+E	+ν(COO^-)
1389	1391	1400	1394	1403	1390	1392	1402	A-B	ring str.
–	–	–	–	–	–	–	–	(+C-D)	

[(a)]Wavenumbers in cm^{-1}; [(b)] and [(c)] drugs taken at 1 mM and 1 μM, respectively (pH 7.3). Underlined - wavenumbers after hydrolysis. Intensity increase or decrease upon hydrolysis is noted by "+" or "–".

1. M.E. Wall, M.C. Wani, C.E. Cook, K.H. Palmer, A.T. McPhail and G.A. Sims, J. Am. Chem. Soc. 88, 3888-3890 (1966).
2. M.E. Wall, M.C. Wani, in Natural Products and Drug Development, (Larsen, P.K., Christensen, H.K.., Eds.), Munksgaard, Copenhagen, 1984, pp. 253-266.
3. I. Chourpa, and M. Manfait, J. Raman Spectrosc. 26, 813-819 (1995).
4. P.C. Lee and D.J. Meisel, J. Phys. Chem. 86, 3391 (1982).
5. J.A. Creighton, C.G. Blachford and M.G. Albrecht, J. Chem. Soc. Faraday Trans. 2, 75, 790 (1979).

SURFACE ENHANCED RAMAN SPECTROSCOPY OF LIPIDS

M. K. Weldon, V. R. Zhelyaskov, and M. D. Morris*

Department of Chemistry, University of Michigan, Ann Arbor, MI 48109

Lipids are secreted by various glands in most mammals. Typically these are fatty acids, triglycerides and wax esters. It is important to be able to distinguish between these classes of lipids to better understand the functions and malfunctions of the glands. Because of its combination of high sensitivity and easy identification of functional groups, surface enhanced Raman spectroscopy (SERS) can be used to classify the important lipids; and to study the spatial distribution of lipid classes in glandular secretions.

We employ the silver electrode SERS microprobes previously developed in our laboratories [1,2]. These probes are used for their exquisite spatial sensitivity, which approaches or exceeds one micron, allowing study of lipid chemistry in or near glandular structures. Towards this end we have investigated archetypal skin lipids. These include oleic acid, isostearic acid, triolein, triisostearin, oleyl oleate and squalene. These compounds were used alone or in mixtures, as needed.

Spectra were obtained using a Raman microprobe described elsewhere [1, 2]. Before use a Ag wire microelectrode was activated using a standard SERS protocol [2]. It was then immersed in a lipid mixture for 30 sec - 5 min. After immersion the electrode was rinsed in chloroform to remove excess lipids from the surface, leaving only a strongly adsorbed monolayer. Spatial resolution experiments were performed in 1.5 mm i.d. glass capillaries carefully loaded with fatty acids and then triglycerides.

Figure 1 shows the SERS spectra of representative lipid classes, including mixtures. Typical spectral features are CH_2 rocking bands (700 cm^{-1} - 800 cm^{-1}), skeletal modes (1000 cm^{-1} - 1150 cm^{-1}), CH_2 twist band (1290 cm^{-1}), CH_2 bend band (1440 cm^{-1}) [3, 4].

Both the triglycerides and fatty acids adsorb on the Ag surface through their respective carbonyl and carboxyl groups, with their hydrocarbon chains directed away from the Ag surface. Evidence for this orientation is the band at ca 1380-1390 cm^{-1} observed in the spectra of both of these lipid classes. This band is characteristic for SERS of carboxylates [5]. A comparison of the SERS spectra of triglycerides and fatty acids, Figures 1A and 1B, shows a difference in the intensity of the Raman band at ca 1040 cm^{-1}. This band is strongly overlapped with another band at ca 1010-1020 cm^{-1} and it has a higher relative intensity in the fatty acid spectrum, compared to the CH_2 deformation band at 1440 cm^{-1}. The intensity of the band at 1040 cm^{-1} can be ratioed to the intensity of one of several hydrocarbon backbone bands (1010-1020 cm^{-1}, ca 1186 cm^{-1} and ca 1230 cm^{-1}) as an indicator of the presence of fatty acids and/or triglycerides. The spectral features of the fatty acids dominate in the spectrum of the total mixture, Figures 1B and 1C. This behavior indicates that the fatty acids are more strongly bound to the Ag surface, in accordance with their known strong affinities for metals.

Silver wire and silicon micro-substrate silver films have been used to spatially localize classes of lipids in synthetic mixtures mimicking glandular secretions undergoing chemical changes or bacterial transformations. The typical reaction is the hydrolysis of esters or triglycerides to the parent long-chain fatty acids. Because of the high affinity of silver for the acids and the characteristic fatty acid SERS spectra, we can localize the presence of acids to within a few microns.

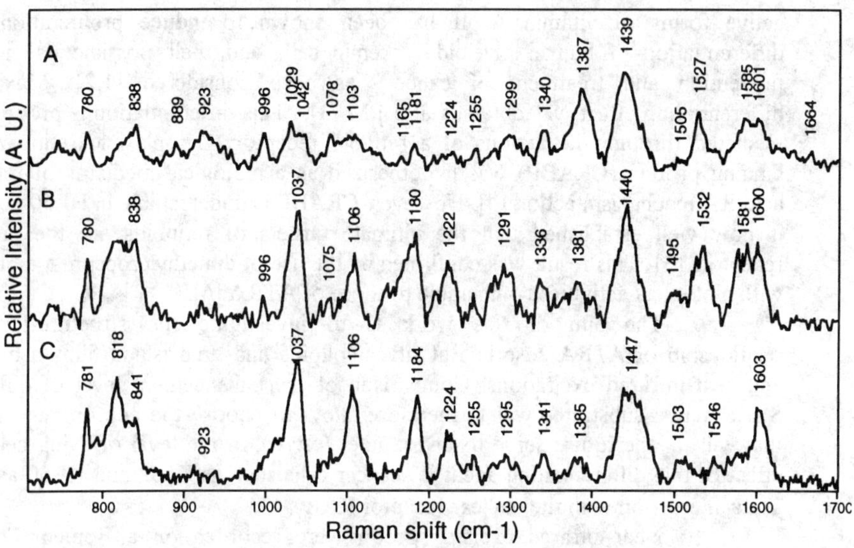

Figure 1. Silver microelectrode SERS spectra of monolayers of: (A) Triglyceride mixture and squalene; (B) Fatty acid mixture; (C) Triglycerides, fatty acids and squalene. Laser power at the sample 200μW.

1. E. A. Todd and M. D. Morris, Appl. Spectrosc. 47, 855 (1993).
2. V. R. Zhelyaskov, E. T. Milne, J. F. Hetke, and M. D. Morris, Appl. Spectrosc. 49, 1793 (1995).
3. B. P. Gaber and W. L. Peticolas, Biochim. Biophys. Acta 465, 260 (1977).
4. M. A. Bryant and J. E. Pemberton, J. Am. Chem. Soc. 113, 8284 (1991).
5. H. Bercegol and F. J. Boerio, Langmuir 10, 3684 (1994).

MICRO-SERS STUDIES OF DIFFERENTIATING AGENT ALL-TRANS RETINOIC ACID ADSORBED ON SILVER COLLOIDS AND GOLD ISLAND FILMS WITH RED AND NEAR-INFRARED EXCITATIONS.

A. Beljebbar, G.D. Sockalingum, H. Morjani, J.F. Angiboust and M.Manfait.
Laboratoire de Spectroscopie Biomoléculaire, UFR de Pharmacie, Université de Reims Champagne-Ardenne, 51 rue Cognacq Jay, 51096 Reims, France.

All-*Trans* Retinoic acid (ATRA) is one of the most potent biological active forms of vitamin A. It has been shown to induce proliferation and differentiation of human myeloid leukemic cells and their possible use in the prevention and treatment of cancer has been considered [1,2]. Maximum differentiating effect was obtained at 5 μM [3]. This differentiation is presumably mediated through the actions of a retinoid receptor. Cytoplasmic retinoic acid binding protein (CRABP) was hypothesized as a biological mediator of retinoic acid RA mechanism action [4]. However, CRABP is undetectable in HL60 cells. It is now well established that the ultimate targets of retinoids are the nuclear receptors [5]. This result was confirmed by the use of dimethylcrocetin, a molecule with biological activities comparable to those of ATRA [6].

The aim of this work is to investigate the structure-function relationship of ATRA adsorbed at silver colloids and gold island films using red and near-infrared excitations. Comparison of the enhancement level of different SERS active substrates will furthermore allow to choose the appropriate system that will in the future serve to investigate, *in vivo*, at the level of living cell, the effect of this differentiating agent in cancer cells such as K562 and HL60 as well as its interaction with the nuclear receptor RARγ.

Near-infrared excited spectra were recorded on a Bomem Fourier transform Raman spectrometer with excitation line at 1064 nm and a lateral resolution of 8 μm. A dispersive spectrometer (Labram, Dilor) equipped with a CCD detector and a He/Ne laser was used to obtain spectra in the red at 632.8 nm. The latter system gave a lateral resolution of 1 μm. Silver colloids prepared as described by Hildebrandt and Stockburger were pre-aggregated with sodium perchlorate. Gold island films were obtained by thermally evaporating the metal onto glass slides at a rate of 0.3 Å/s in a vacuum of $5x10^{-5}$ Torr. The results presented here were obtained on plates 20 Å and 40 Å thick.

Figure 1 shows the FT-SERS and SERS spectra of ATRA adsorbed on the silver colloids and gold island films. The 20 Å and 40 Å plates exhibit absorption maxima at 620 and 661 nm and these plates present good level of enhancement in the red and near infrared respectively, with drug concentration not exceeding 10^{-5} M and $5x10^{-6}$ M. The comparison between these spectra does not show any change in frequencies or relative intensities of the bands except at 1402 cm^{-1} corresponding to CH_3 deformation. Since the spectra are similar to the solid form, the adsorption on the SERS active substrate does not induce any detectable modifications on the structure of this molecule. The assignment of the other bands in the spectrum of ATRA have been described in detail before [7].

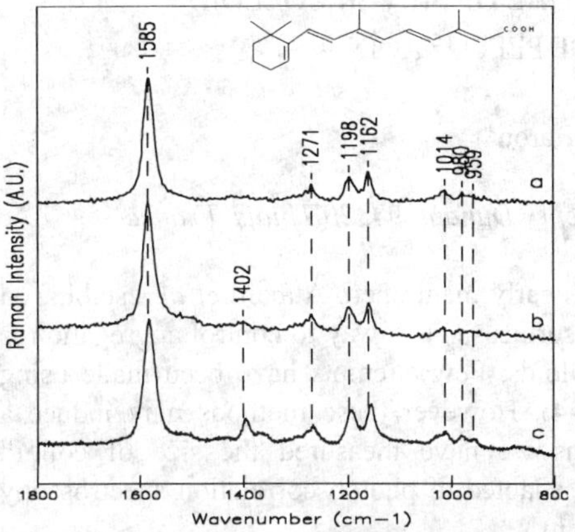

Fig.1: Micro-SERS spectra of all-*trans* retinoic acid, With (a) silver hydrosol, (b) gold island films both at 5x 10^{-6} M, 1064 nm and (c) gold island films at 10^{-5} M, at 632.8 nm.

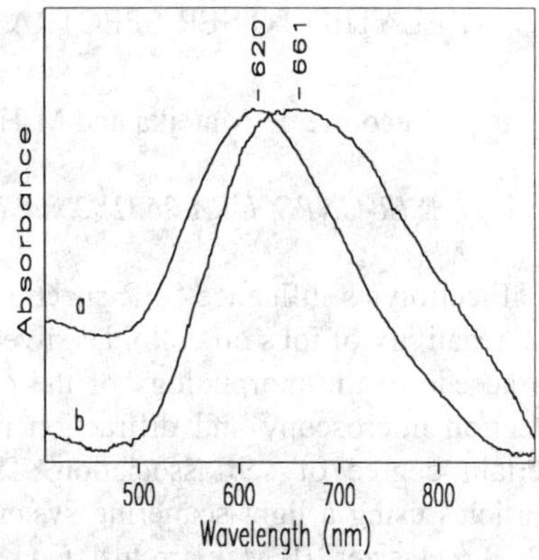

Fig.2: Absorption spectra of gold island films with evaporation rate of 0.3 Å/s (a) 20 Å thick (b) 40 Å thick

Our results indicate that the detection limit is lower with NIR radiation since no spectra were obtained below 10^{-5} M with the visible. Furthermore, no spectra could be recorded in the red with the silver colloids. All these and the absence of detectable perturbations on the structure of adsorbed ATRA suggest that gold island films associated with NIR FT-Raman spectroscopy are promising systems to study, at physiological concentrations, ATRA binding affinity to nuclear receptors (ATRA-RARγ) as well as its ternary complexes (ATRA-RARγ-DNA). These investigations will certainly give more insight into the understanding of the mechanism of intercalation of ATRA in the receptor and its mechanism of action in the K562 and HL60 cell lines, thus enabling to improve ATRA's efficiency in differentiating leukemic cells through targeting via specific carriers such as low-density lipoprotein (LDL) [8].

1. T.R Breitman, S.E. Selonick, and S.J. Collins Proc. Natl. Acad. Sci. 77, 2936-2940 (1980).
2. T.R Breitman, S.J. Collins, and B.R. Keene Blood, 57, 1000-1004 (1981).
3. P.A. Tarantilis, H. Morjani, M. Polissiou, and M. Manfait Anticancer Res. 14, 1913-1918 (1994).
4. D. Douer, and H.P. Koeffler J. Clin. Invest. 69, 277-283 (1982).
5. M. Petkovich, N.J. Brand, A. Krust, and P. Chambon Nature, 330, 444-450 (1987).
6. A. Beljebbar, G.D. Sockalingum, H. Morjani, J.F. Angiboust, M. Polissiou, and M. Manfait Spectroscopy of Biological Molecules (J.C. Merlin, S. Turrell, and J.P. Huvenne, Eds.) Kluwer Academic Publishers, 1995, pp. 465-466.
7. S.O Smith, M.S. Braiman, A.B. Myers, J.A. Pardoen, J.M.L. Courtin, C. Winkel, J. Lubtenburg, and R.A. Mathies J. Am. Chem. Soc. 109, 3108 (1987).
8. S.G. Vitols, M. Masquelier, and C.O. Peterson, J. Med. Chem., 28, 451-454 (1985).

PARTICLE SIZE, ZETA POTENTIAL OF SILVER COLLOID RELATED TO SER-SPECTRA OF PEFLOXACIN

S. Lecomte, P. Matejka and M-H. Baron*

LASIR-CNRS, URA 2631, 2 rue Henry Dunant, 94320 Thiais, France

Electrolytes influence SER-spectra as early mentioned. Aubard *et al.* established the reactivity of ions on colloidal silver surfaces (1). A way to control aggregation is to accede to the morphology of the colloids. Few attempts have been made using electron microscopy and diffraction (2-4). However these methods may induce a certain degree of self-association. Thus we have measured the size of colloid particles using a light scattering system adapted to photon correlation spectroscopy (PCS, Zetasizer III, Malvern Inst. Company).

Adsorption on Ag nanoparticles also depends on local electric fields. An access to the zeta potential of the solvated particles is possible by electrokinetics measurements. Z-potential could be regarded as a difference from a point in the uniform medium to a point in the plane of shear between the solution and the external surface of the diffuse ionic double layer (5,6). The Zetasizer III, equipped for laser velocimetry (electrophoresis), also fits to evaluate z-potentials. Pefloxacin (7,8) has been chosen to recognize the dependencies between particle-sizes, z-potentials, UV-Visible and SER-spectra.

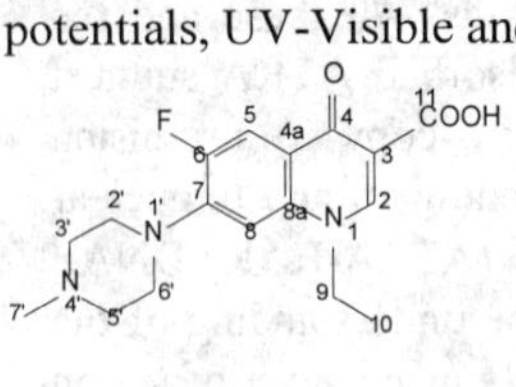

Fig. 1

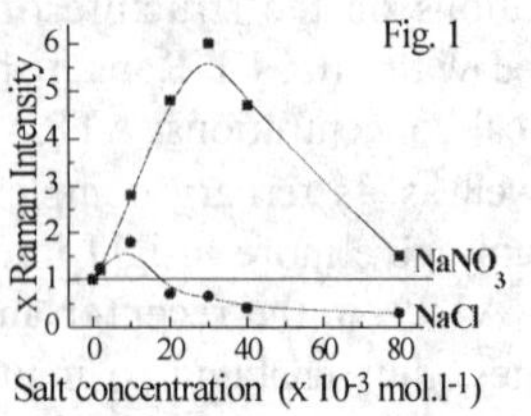

Table I

Silver colloid Creighton	Size (nm)	z-pot. (mV)
Alone	10-50	-30
+(P)	130	+9
+$NaNO_3$	900	-27
+NaCl	40-100	-10
(P)+$NaNO_3$	700	+9
P)+NaCl	40-100	-8

Fig. 2

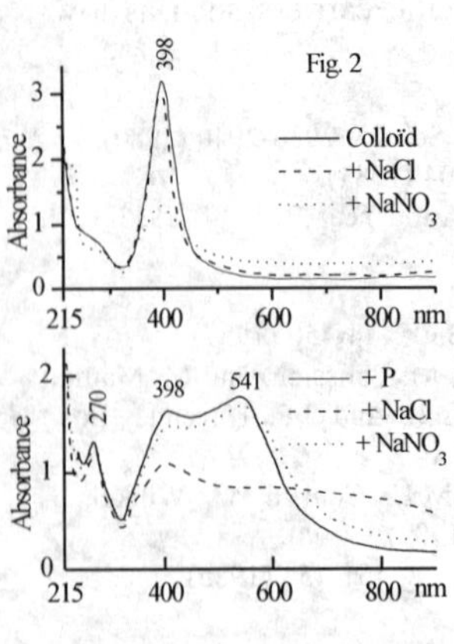

Salts effects on $\nu_s COO^-$ pefloxacin-Raman peak (1398 cm^{-1}), pefloxacin (P) 10^{-5} mol/L. (Fig. 1). Particle-sizes and z-potentials (Table I). UV-Visible spectra (Fig. 2): colloid without and with salts (up); colloid with pefloxacine, without and with salts (down) Pefloxacin SER-spectra (Fig. 3). The salt concentrations correspond to the highest Raman exaltation : $NaNO_3$ 3.10^{-2} and NaCl 10^{-2} mol/L.

$NaNO_3$ entails large self-association. The electronic transition of the plasmon surface (398 nm) deacreases. The z-potential is maintained.

Pefloxacin alone weakly inceases the particle-sizes but strongly enhances the z-potential. A new wide absorption band arises at 541 nm. The COO^- of the drug faces silver in the inner layer (8). The positive NH^+ immobilised in the outer domain, accounts for the positive potential which avoids aggregation. The 541 nm-band cannot result from a new surface plasmon related to aggregation because such spectral change was not detected with $NaNO_3$ for a much larger aggregation. This band should result from a specific charge transfer from the silver to the drug.

When **$NaNO_3$** and **pefloxacin** are both present, self-association is no more limited by the adsorbate. However, on average the positive z-potential and the charge transfer remain as for the drug alone. The colloidal pattern is a chain-like network with relatively poor contacts in between the silver particles (3,4). In absence of noticeable electronic effect of $NaNO_3$, a specific resonance with the surface plasmon or with the charge transfer cannot be hypothesis for the large Raman enhancement (×6). How self-association may influence this process is an open question.

NaCl only weakly enlarges the particle-sizes but significantly increases the z-potential. The specific Cl^--adsorption (1,8) favours the retention of Na^+ in the outer layer. The 398 nm-plasmon band is reduced and a wide background arises in the 400-900 nm range. These effects rather results from large electrostatic changes in the solvating layers, than from the weak aggregation.

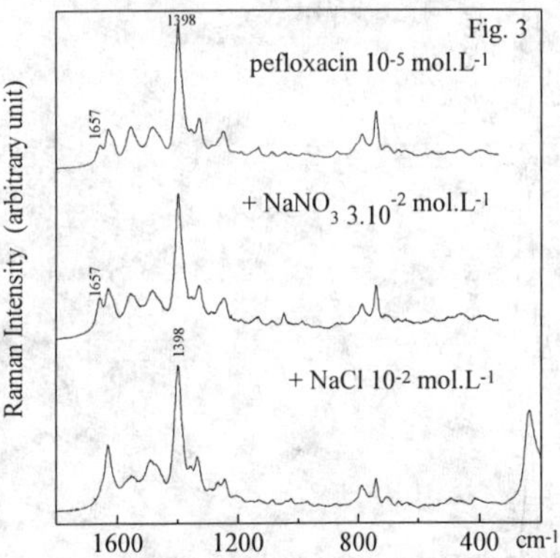

With **NaCl** and **pefloxacin** together, the silver particle-sizes, z-potential and absorption spectrum remain similar to those measured without drug. The adsorbed Cl^- cancels the charge transfer between the surface and pefloxacin. However, the pefloxacin SER-signal is ×1.8. Drug and Cl^- are co-adsorbed with large amount of Cl^- which is responsible of the negative potential. For NaCl >2 10^{-2} mol/L the adsorption sites become saturated by Cl^-.

The SER-spectra give complementary results. Without salt, or with $NaNO_3$, pefloxacin is adsorbed via the electronegative oxygens (νC4=O and $\nu_s COO^-$ at 1657 and 1398 cm^{-1} respectively) (8). With NaCl, adsorbed pefloxacin is adjacent to adsorbed Cl^-. The charge transfer cannot occurs anymore, then the 1657 cm^{-1} band cancels, meaning that the drug is no more retained via the C4=O.

1. J. Aubard, E. Bagnasco et *al.*, J. Phys. Chem. 99, 7078 (1995).
2. I.R. Nabiev, R.G. Efremov et *al.*, Sov. Phys. Ups. 31, 241 (1988).
3. D.J. Wales, AI. Kirkland et *al.*, J. Chem. Phys. 91, 603 (1989).
4. C. Rodger, W.E. Smith, et *al.*, J. Chem. Soc., in press (1996).
5. Natarajan and R.S. Schechter, AIChE Journal 33, 1110 (1987).
6. A.W. Adamson, Physical Chemistry and Surfaces, Wiley, p. 203 (1990).
7. S. Lecomte, M-H. Baron et *al.*, Antimicrob. Agents and Chemother. 38, 2810 (1994).
8 S. Lecomte, N.J. Moreau et *al.*, Biospectrosc., in press (1996).

SERS OF ISOASPARAGINE ON SILVER SOLS.

J.F. Arenas, J. L. Castro*, J. C. Otero, and J. I. Marcos

Department of Physical Chemistry,
Faculty of Sciences, University of Málaga, E-29071 Málaga, Spain.

We have studied the SERS spectrum of isoasparagine [HOOC-CH_2-CH(NH_2)-$CONH_2$] [1] correlating the results obtained with those of aspartic acid [HOOC-CH_2-CH(NH_2)-COOH] and asparagine [H_2NOC-CH_2-CH(NH_2)-COOH]. The assignment of its spectrum has allowed us to deduce which ionic species is responsible for the observed spectrum and which functional groups interact directly with the metal. Fig.1. shows the Raman spectra of the three amino acids on silver sols at approximately the same concentration and at neutral pH.

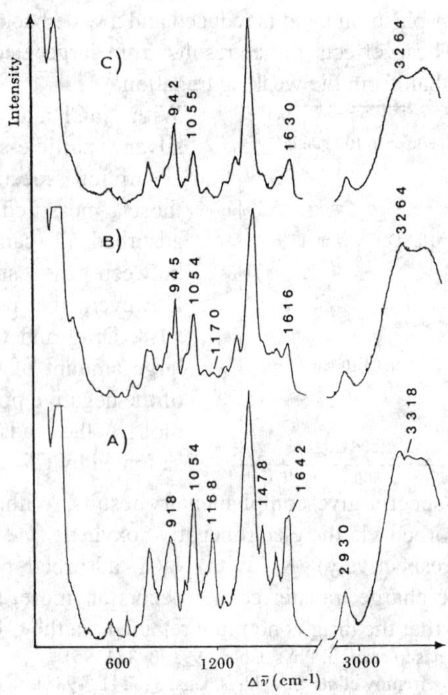

Fig. 1. SERS on silver sols at the concentration of $5x10^{-4}$ M and pH 7 of:
A) Asparagine B) Isoasparagine and C) Aspartic Acid.

As can be seen, the spectrum of isoasparagine (B) is very similar to that of aspartic acid (C) and the main difference with that of asparagine (A) is in the bands recorded in the 1100 - 1700 cm^{-1} region.

With respect to the vibrations of the carboxylate group, we have assigned the δ(OCO) to the band at 772 cm^{-1}, the ν(C-COO) to that at 945 cm^{-1} and the most intense band in all cases at 1416 cm^{-1} to the ν_s(OCO) mode, which is found to be overlapping with the δ(CH_2) vibration. We have assigned the doublet recorded at 945 and 911 cm^{-1} to carbon-carbon stretching modes in which the carboxyl group participates in the band of higher frequency, whilst the amide group is implicated in the other band of lesser intensity. In the case of aspargine, the doublet is unresolved and both vibrations are found to be included in the broad band with a maximum at 918 cm^{-1}. The carbon-nitrogen stretch of the amino group appears at the same frequency of 1054 cm^{-1} in all cases.

We have assigned the band at 1168 cm^{-1} in the spectrum of asparagine [2] to the r(NH_2) vibration, which is recorded in the spectrum of isoasparagine as a very weak band at 1170 cm^{-1}. In principle, this means that we can rule out a strong interaction between the amide group and the metal surface. This conclusion is in agreement with the absence of bands that could be assigned to the Amide I and Amide III vibrations that apear with medium intensity in the spectrum of asparagine at 1642 and 1478 cm^{-1} respectively, and that of the Amide II mode at 1616 cm^{-1} recorded with a very weak intensity. In this respect it is interesting to point out that an increase in the basicity of the system only gives rise to a relative enhancement of the bands assigned to ν(CN) and r(NH_2) of the amino group.

In the high frequency region of the spectra of the three amino acids, we have assigned the CH_2 stretching modes to the band recorded at approximately 2930 cm^{-1} and the ν(NH_2) of the amino group to that of 3264 cm^{-1}.These assignments and those previously mentioned are confirmed upon correlating the SERS spectrum of isoasaragine in acidic medium with that recorded in D_2O [1], where the bands observed correspond to vibrations belonging to the carboxylate and the ND_2 groups. The characteristic vibrations of the $COND_2$ group are not observed.

The analysis of the obtained results has lead us to the conclusion that the interaction of isoasparagine with the metal surface occurs fundamentally through the carboxylate and the amino groups of the anionic species -OOC-CH_2-CN(NH_2)-$CONH_2$, with the C=O bond a long way from or parallel to the surface.

1. J.L. Castro, Ph.D Thesis, University of Málaga, (1996).
2. J.L. Castro, M. A. Montañez, J. C. Otero and J. I. Marcos, Proc. XIV ICORS (N.T. Yu and X.Y. Li, Eds.), J. Wiley and Sons, Chichester, 1994, p. 594.

SURFACE ENHANCED RAMAN SPECTROSCOPY OF 1,5-DIMETHYLCYTOSINE ADSORBED AT A SILVER ELECTRODE

S. Sánchez-Cortés[1], J.C. Otero[2], J.I. Marcos[2] and J.V. García-Ramos[1]*

[1]Instituto de Estructura de la Materia, C.S.I.C. Serrano 121, 28006 Madrid, Spain
[2]Departamento de Química Física, Universidad de Málaga, Campus de Teatinos s/n. Málaga, Spain

1,5-dimethylcytosine(DMC) is a methylated derivative of cytosine which can be used as a model to study the electrodynamical behaviour of cytosine and 5-methylcytosine in nucleic acids. SERS spectra of this molecule on metal colloids were already studied by us [1,2].

Potential-dependence SERS measurements were carried out at two different concentrations of DMC: 10^{-2} and 10^{-3} M. The results were practically the same for these concentrations. Spectra shown in figure 1 correspond to SERS of 10^{-2} M DMC recorded at different potentials value. On changing the potential to a more negative value, a progressive enhancement of bands lying in the 1450-1700 cm^{-1} region, and this can be attributed to an increase of the double-bond character of the pyrimidine ring due to a greater transference of charge from the metal to the adsorbate at this conditions. The bands appearing at 1494, 1262, 794 and 558 cm^{-1} can be assigned to ring stretching motions and do not shift significatively their wavenumber when varying the potential. In addition, the appearance of new bands corresponding to the amino group at 1590, 1526, 1038 and 885 cm^{-1}, together with a shift of the 1580 cm^{-1} band, which have an important contribution of bending vibrations of amino group, towards 1570 cm^{-1}, indicate that when lowering the potential a reorientation of DMC is taking place in such a way that an approaching of amino group to the metal surface occurs. The greater affinitty of amino groups to a more negativelly charged surface can be accounted for by taking into account that this group is positively charged at neutral pH.

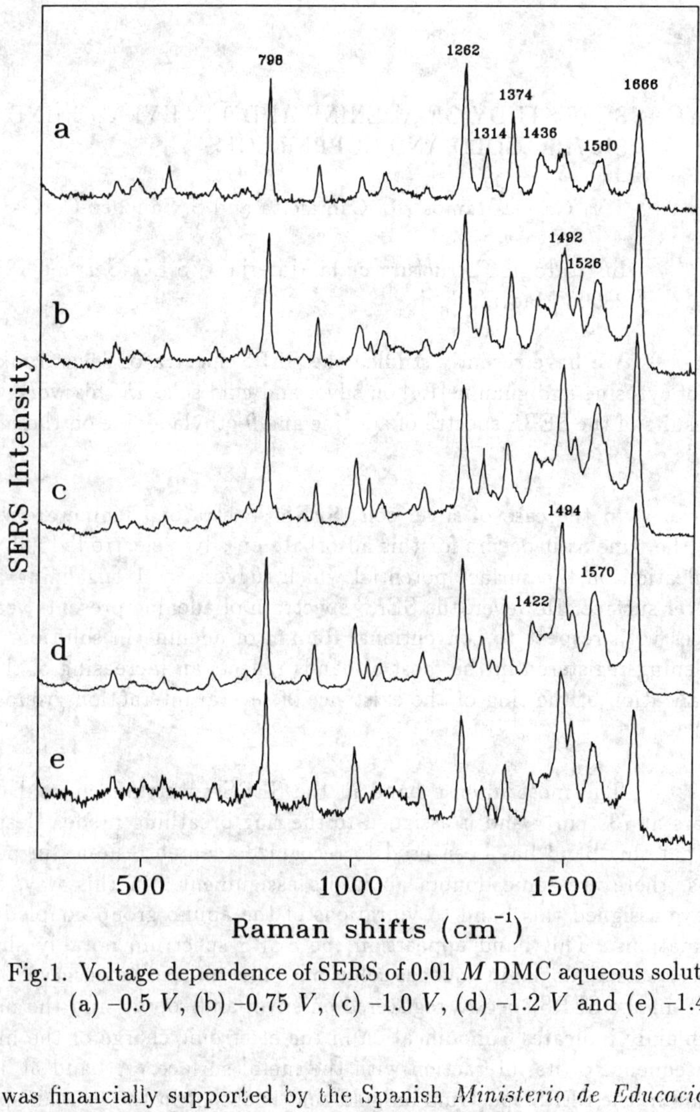

Fig.1. Voltage dependence of SERS of 0.01 *M* DMC aqueous solution: (a) −0.5 *V*, (b) −0.75 *V*, (c) −1.0 *V*, (d) −1.2 *V* and (e) −1.4 *V*.

This work was financially supported by the Spanish *Ministerio de Educación y Ciencia* (DGICYT) under project PB93-0973-02.

1. S. Sánchez-Cortés and J.V. García-Ramos, J. Raman Spectrosc. **23**, 61 (1992).
2. L.E. Camafeita, S. Sánchez-Cortés, and J.V. García-Ramos, J. Raman Spectrosc. **26**, 149 (1995).

SERS STUDY OF ADENINE AND 9-ETHYLADENINE ON SILVER, GOLD AND COPPER SOLS

J.V. García-Ramos*, L. Camafeita and S. Sánchez-Cortés

Instituto de Estructura de la Materia, C.S.I.C. Serrano 121,
28006 Madrid, Spain

We have recently studied the SERS spectra of several alkyl derivatives of cytosine and guanine[1,4] on silver and gold sols. In this work we present the results of the SERS spectra of adenine and 9-ethyladenine on these two metal colloids.

In the case of silver sols, SERS spectra of adenine are very stable along the time as it occurs for this adsorbate on silver electrode [5,6] where any modification on the surface potential which suggests a strong fixing of adenine on silver surface. However, the SERS spectrum of adenine presents very few differences with respect to conventional Raman of adenine in solution. Only the broadening registered in the most of bands reflects an increasing of the electron delocalization of the ring or the existence of several interaction groups with the metal.

The most intense band in the SERS spectrum on gold of adenine appears at 739 cm^{-1} and is assigned to the ring breathing mode. Despite of the fact that this band has been used to recognize adenine residues in polypeptide chains, there are some doubts about its assignment. In this way, Suh *et al.* [7] have assigned this band to vibrations of the amino group coupled with ring deformations. This band appears in the SERS spectrum notably shifted with respect to its analogue in aqueous solution. This can also be observed, to a lesser extent, in the SERS spectra registered on silver and copper and the pronounced shift plainly indicates a modification in the electronic charge of the molecule as a consequence of its interaction with the metal surface. A band at 1449 cm^{-1} appears in the SERS spectrum on gold that is absent in the Raman spectrum of the neutral aqueous solution. This is the frequency to which the band appearing at 1483 cm^{-1} in the Raman spectrum of the neutral aqueous solution shifts when the medium becomes alkaline. As this band has been assigned to $\nu(C{=}C)$ and $\nu(C{=}N)$ vibrations and at alkaline pH deprotonation of the N_9 atom occurs, it is possible that this deprotonation takes place in adenine as well due to its interaction with the gold surface. Thus, the new band can be assigned, taking into account the shift to lower frequencies it shows in the SERS spectrum, to stretching vibrations of the imidazole ring. Similar conclusions have been reached when studying SERS spectra on silver and copper supports.

The absence of any bands due to the NH_2 group in the 1600–1400 cm^{-1} region points out a distant position of this group with respect to the metal surface.

In the region where single–bond stretchings appear (1200–1300 cm^{-1}), the SERS spectrum on gold is less profuse in bands than the SERS on silver and copper, thus indicating a lower electronic dispersion in the ring when adenine adsorbs on the gold colloidal surface.

The metal–adsorbate stretching band appears at 214 cm^{-1} with a much lower intensity than it shows in the SERS spectra on gold of the remaining molecules hereby studied, although its intensity is even lower in the SERS spectrum of adenine recorded on silver.

In the case of 9-ethyladenine it was not possible to obtain any SERS spectrum on gold colloids. On silver sols, this adsorbate presents a SERS pattern which indicates a great participation of the imidazol ring in the interaction with the metal, whereas for copper sols the SERS spaectra obtained are very poor, possibly due to the presence of the ethyl group in the N_9 atom.

This work was financially supported by the Spanish *Ministerio de Educación y Ciencia* (DGICYT) under project PB93-0973-02.

1. S. Sánchez-Cortés and J.V. García-Ramos, J. Raman Spectrosc. **21**, 679 (1990).
2. S. Sánchez-Cortés and J.V. García-Ramos, J. Raman Spectrosc. **23**, 61 (1992).
3. L.E. Camafeita, S. Sánchez-Cortés, and J.V. García-Ramos, J. Raman Spectrosc. **26**, 149 (1995).
4. S. Sánchez-Cortés, and J.V. García-Ramos, Vibrational Spectroscopy. **4**, 185 (1993).
5. E. Koglin, J.M. Séquaris and P. Valenta, J. Mol. Struct. **79**, 185 (1982).
6. C. Otto, F.F.M. de Mul, A. Huizinga and J. Greve J. Phys. Chem. **92**, 1239 (1988).
7. J. Suh and M. Moskovits, J. Am. Chem. Soc. **108**, 4711 (1986).

ANALITYCAL SERS : DETECTION AND IDENTIFICATION OF PHOTODEGRADATION PRODUCTS IN ORGANIC PHOTOCHROMES.

J.Aubard[1], K.Karlsson[1], B.Luccioni-Houze[2], G.Lévi[1] and R.Guglielmetti[2]

[1] ITODYS, CNRS URA 34, University of Denis Diderot Paris 7, 1 rue Guy de la Brosse, 75005 Paris, France.

[2] LCMOM, CNRS URA 1320, Faculté des sciences de Luminy, 13288 Marseille Cedex 09, France.

Owing to their ability to undergo photoinduced reversible coloration in solutions or in polymer matrix films, spiropyrans and spirooxazines are very interesting organic photochromes for various applications [1] (optical filters, data imaging and storage, molecular probes...). Photochromism in these systems is initiated by UV irradiation of the colorless spiro compounds, SP, which leads to cleavage of the C - O bond to give a deeply colored metastable merocyanin type species, MC (scheme 1). However, one of the major limiting factor for the industrial availability of these compounds is their progressive degradation with the loss of their photochromism after long exposures to repeated light-dark cycles.

Scheme 1. *Photochromic Equilibrium between Closed Form and Open Form*

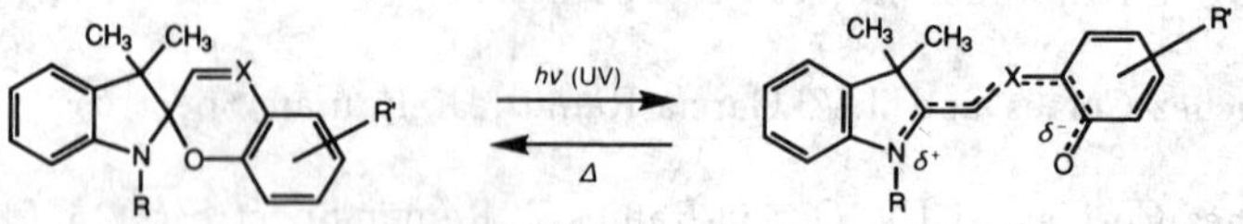

SP **MC**

Spiropyrans (BIPS) **X=CH**

Spirooxazines **X=N**

While the kinetic mechanisms and the nature of short-lived species participating to the photochromic reaction have been studied in details, to date only very limited data are available concerning the photodegradation process [2,3].

In this paper, we briefly outline a new strategy, using Surface Enhanced Raman Spectroscopy (SERS) to detect and identify the main photoproducts involved in spiropyran photoaging. The reported data were obtained on irradiating the classical nitro-methoxy-indolinospirobenzopyran (BIPS) in acetronitrile solutions.

The SERS spectrum of the non irradiated BIPS is shown in figure 1 along with the spectra recorded after 10 minutes and 4 hours of UV irradiation from a 150 W high-pressure mercury lamp. On the basis of our recent Raman assignments of indolinospiropyran derivatives [4], it appears that the SERS spectrum of the colorless BIPS (spectrum 1) is mainly dominated by the spectral features of the indoline moiety with a small involvement of the benzopyran one. This conclusion is strengthen by the absence in the SERS spectrum of any Raman line typical of the nitro-group, while very intense lines, at ca. 1320 - 1330 cm^{-1}, are observed in the NIR-FT Raman spectrum of the neat photochrome. These observations lead to the conclusion that BIPS is probably oriented with respect to the silver surface mainly on the side of its indoline part.

Comparison between SERS spectra in figure 1, shows that UV irradiation of BIPS solutions leads to important spectral changes (intensities and wavenumbers). Moreover, after 4 hours of extensive degradation it is observed that the studied compound has lossed totally its photochromic properties and the yellow color solution did not evolved anylonger. The corresponding SERS spectrum (spectrum 3) is then expected to give structural information on the main final products involved in the degradation process.

It has been shown recently that indolinospiropyran degradation in solution and in polymer matrices was mainly a photo-oxidation process which led, in the case of nitroderivatives, to a mixture of oxindols (di- and tri-) and 5-nitrovanillin. Thus, these compounds were synthetized as models and their SERS spectra recorded. Comparison of the totally degradated solution (spectrum 3) with that of the model compound, 5-nitrovanillin (spectrum 4), shows that spectra are almost identical.

This result is a clear demonstration of the usefulness of SERS for the detection and characterisation of photochromes and their main photodegradation products, even at the level of traces. The same strategy is currently underway to study the fatigue process and to identify the photoproducts in photochromic polymeric films.

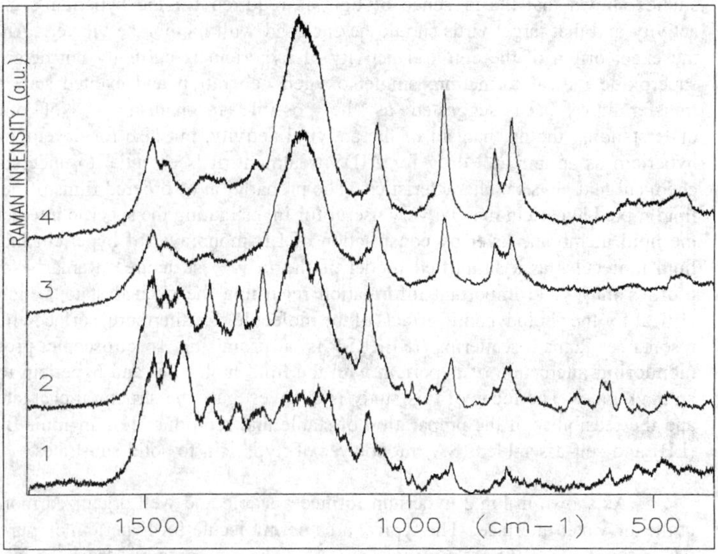

Fig.1 : SERS spectra in silver colloids of BIPS (10^{-5} M) :
(1) non irradiated solution; (2) 10 minutes UV irradiated; (3) 4 hours UV irradiated. (4) SERS spectrum of the model compound, 5–nitrovanillin (2×10^{-4}M).
Laser excitation : 514.5 nm, 100 mW.

1. R.Guglielmetti, in Photochromism: Molecules and Systems, (H.Dürr and H.Bouas-Laurent, eds), Elsevier, Amsterdam, 1990, p 855.
2. G.Baillet, G.Giusti and R.Guglielmetti, J.Photochem. Photobiol.A,70, 157 (1993).
3. V.Malatesta, R.Millini and L. Montanari, J. Am Chem. Soc.,117, 6258 (1995).
4. J.Aubard, C.M'bossa, J.P. Bertigny, R.Dubest, G.Lévi, E.Boschet and R.Guglielmetti, Mol. Cryst. Liq. Cryst., 246, 274 (1994).

Surface-Enhanced Raman Scattering from Hypericin in Langmuir-Blodgett Monolayers

Jae-Ho Kim[1*], Wan-Shik Yoon[1], Jeanne L. Wynn[2] and Therese M. Cotton[2]

[1]Department of Applied Chemistry, Ajou University, Suwon, 442-749, Korea
[2]Department of Chemistry, Iowa State University, Ames, IA 50011, USA

Hypericin is a rigid, polycyclic aromatic quinone that is known to exhibit a variety of photodynamic antiviral activity against several retroviruses including the Friends leukemia virus and the human immunodeficiency virus [1,2]. Independent studies shown that the presence of light is required for the hypericin's antiviral activity and that target virus should be enclosed with membrane bilayer. Although the exact origin of the antiviral activity of hypericin is currently not determined, superoxide radical formation, singlet oxygen generation and excited state proton transfer have been suggested as the possible mechanisms. Not only for understanding the mechanism of the antiviral activity, but also for development of hypericin as an antiviral drug for AIDS treatment, it is essential to understand its chemical and physical characteristics. The preparation of ordered films of pure and lipid mixed hypericin may be very useful for investigating the specific interaction to the lipid membrane layer by construction of LB monolayer of hypericin including lipid molecules as a simplified model of the *in vivo* structure. Raman excitation profiles may yield important information regarding the excited state, which is an critical for the photodynamic effects of the molecule. Furthermore, surface-enhanced resonance Raman scattering (SERRS) is a promising spectroscopic probe for monitoring interaction of hypericin with the lipid molecules and hypericin itself as an aggregate. The focus of this study is to investigate the structural characteristics and the feasibility of the preparation of stable and reproducible Langmuir-Blodgett (LB) and self-assembled (SA) monolayer of hypericin to solid substrates.

As shown in Fig.2 hypericin formed a stable and well organized monolayer at the air-water interface. The typical area per molecule for hypericin on pure water was estimated near 65 $Å^2$/molecule ± 3 $Å^2$. The area obtained from the isotherms suggest that hypericin orients vertically to the water surface, which is consistent with the molecular area obtained from a space filling CPK model. The pressure area isotherms obtained for hypericin were found to be very sensitive to the concentration of the spreading solution. Although monomeric hypericin film was obtained at the air-water interface by spreading pure hypericin solution in chloroform (1.0×10^{-4} M), addition of arachidic acid (AA) to the monolayer system induced further dispersion of the hypericin at the interface and lower the angle of hypericin molecule to the water surface as indicated in the area changes with the variation of AA mole ratio.

SERRS spectra were obtained for the LB films of pure hypericin and of its mixture with AA, and are displayed in Fig.1. The SERRS spectra were sharp and highly detailed. Although the spectral features for the two types of films are for the most part almost identical, subtle differences in the SERRS spectra can be seen in the 1640 cm to 1250 cm^{-1} region (the C-C stretching modes of the corner rings) due to the difference in orientation to the substrate and variation in intermolecular interaction of hypericin in monolayer structure. Since 457.9 nm excitation line is in resonance with the electronic transition along with two carbonyl groups, strong carbonyl bands of hypericin were appeared at 1608 and 1635 cm^{-1} [3]. The large increase in the relative intensities of the 1635 and 1378 cm^{-1} bands in the spectra from mixed LB films are indicative of the presence of non-hydrogen bonded carbonyl of hypericin. In summary, a stable and well ordered LB film of hypericin can be formed on the solid substrate. The SERRS studies showed that hypericin was vertically oriented to the substrate and the presence of AA in the LB monolayer induced further monomeric dispersion.

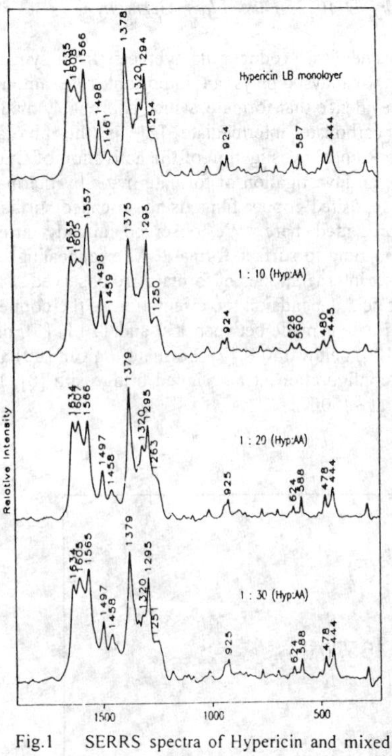

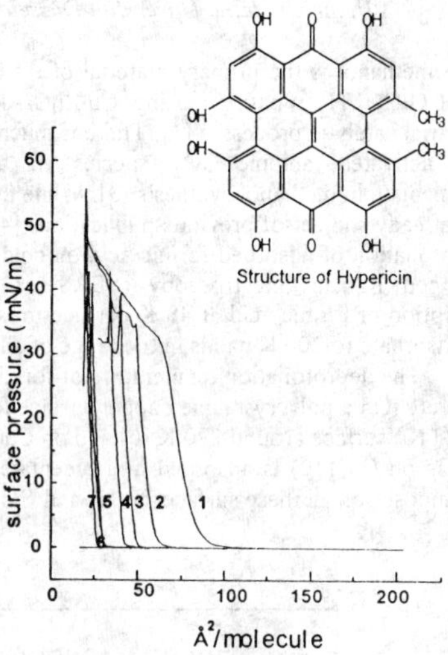

Mole ratio	1 Hyp.	2 1:1	3 1:4	4 1:10	5 1:20	6 1:30	7 AA
Area ($Å^2$/molecule)	65	44	35	29	25	22	21
Hyp.Area		67	91	109	104	82	

Fig.1 SERRS spectra of Hypericin and mixed monolayers with arachidic acid on a roughened Ag electrode. (λ_{ex}: 457.9 nm, power: 10 mW, transfer pressure: 25 mN/m, transfer rate: 1.2 mm/min)

Fig.2 π-A isotherm of pure hypericin and mixed monolayer with arachidic acid. (compression rate: 6 cm^2/min)

1. A. Giese, in Photophysiology Current Topics in Photobiology and Photochemistry, (A. Giese, ed.) Vol VI, Academic Press, New York, 1971, p.77
2. D. Meruelo, G. Lavie and D. Lavie, Proc. Natl. Acad. Sci. USA 85, 5230 (1988)
3. J. L. Wynn and T. M. Cotton, J. Phys. Chem. 99, 4317 (1995)

A Raman Study of Formic Acid and Surface Formate Adsorbed on Cold-Deposited Copper Films

M. Pohl, A. Pieck, C. Hanewinkel and A. Otto

Lehrstuhl für Oberflächenwissenschaft (IPkM), Heinrich-Heine Unversität Düsseldorf, D-40225 Düsseldorf, Germany

Since methanol is the primary material of many chemical products, its synthesis from syngas (i.e. $CO/CO_2/H_2$ mixtures) using $Cu/ZnO/Al_2O_3$-catalysts plays an important role among industrial catalytic processes [1]. There is much evidence that formate, which is probably made from activated, anionic CO_2^- species via a carbonate intermediate [2], is the pivotal intermediate in methanol synthesis [3]. While the Raman investigation of the activation of CO_2 was already subject of previous publications [4], an investigation of formate made by thermal deprotonation of adsorbed formic acid on cold-deposited copper films using so-called surface enhanced Raman Spectroscopy (SERS) is presented here. We observed already after adsorption of formic acid at 48 K peaks corresponding to surface formate. After annealing of the adsorbate to 200 K bands in the range of all six internal modes of formate are observed.

The deprotonation temperature of formic acid depends on the orientation of the copper substrate. On a polycrystalline copper surface formate is made between 150 and 170 K [5], on a Cu(110) surface around 270 K [6] and on Cu(100) below 340 K [7]. Formate on Cu(100) as well as on Cu(110) is adsorbed in a bidentate configuration at an aligned bridge site [8]. It decomposes on all these surfaces to H_2 and CO_2 at ~450 K.

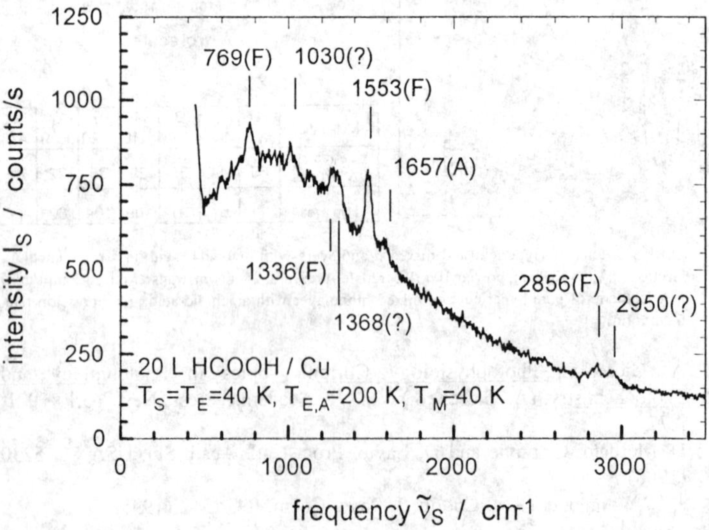

Figure 1: Raman spectrum of a copper film deposited at T_S=40 K and exposed at T_E=40 K to 20 L of HCOOH, annealed to $T_{E,A}$=200 K and recooled to T_M=40 K
(F: peak assigned to formate; A: Peak assigned to formic acid; ?: assignment to F or A unclear)

Table 1: Assignment of the observed peaks of Cu(poly) exposed to HCOOH (F: peak assigned to formate; A: peak assigned to formic acid; ?: assignment to F or A unclear)

ν(Cu-O)	δ(OCO)	π(CH)	ν(C-O)	ν_s(OCO)	δ(CH)/ δ(OH)	ν_a(OCO)	ν(C=O)	ν(CH)	ν_{comb}
290/340	769	1030	1248	1336	1368	1553	1657	2856	2950
?	F	?	A	F	?	F	A	F	?

Figure 1 shows a Raman spectrum of a cold-deposited copper film, prepared by thermal evaporation at a substrate temperature T_S=40 K, exposed at T_E=40 K to 20 L formic acid, after annealing to $T_{E,A}$=200 K and measured at T_M=40 K. The annealing temperature of 200 K has been chosen to optimize the enhancement of the substrate and hence the observed peak intensities. The assignment of the observed peaks has been carried out by comparison with literature data of surface formate, formate salts and solid formic acid. The result is summarized in table 1. This table shows that surface formate as well as adsorbed formic acid is present after annealing the adsorbate to 200 K, because modes of both species are observed. Since the bands assigned to formate are stronger and more numerous than those assigned to formic acid, it is tempting to infer that the coverage consists mainly of formate. But as long as we do not know the "SERS-cross sections" for the two species, this conclusion would be premature.

In Raman spectra recorded at T_M=46 K, without annealing of the adsorbed formic acid (not depicted here), peaks have already been observed, which can be assigned to vibrations of the formate molecule. This might be due to the special chemical reactivity of cold-deposited copper films already found for example in the activation of CO_2 to the anionic CO_2^- species [4].

It is known that the surface of these films consists of (110) and (111) facets and other unidentified parts and that the (111) facets grow in expense of the (110) facets during the annealing process [9]. Thus, still a considerable fraction of the surface consists after annealing to 200 K of facets oriented in the (110) direction. Since the annealing temperature in our experiments lies between the deprotonation temperature of the acid found on polycrystalline copper [5] and on Cu(110) surfaces [6], both, formate and formic acid species should be present.

[1] G. C. Chinchen, K. Mansfield, M. S. Spencer, *CHEMTECH*, **20**, 692 (1990)
[2] M. S. Spencer, *Surf. Sci.*, **339**, L897 (1987)
[3] G. C. Chinchen, M. S. Spencer, K. C. Waugh and D. A. Whan, *J. Chem. Soc. Faraday Trans. 1*, **83**, 2193 (1987)
[4] a) W. Akemann and A. Otto, *Surf. Sci.*, **272**, 211 (1992)
b) W. Akemann and A. Otto, *Surf. Sci.*, **287/288**, 104 (1993)
[5] R. W. Joyner and M. W. Roberts, *Proc. R.. Soc. Lond. A.*, **350**, 107 (1976)
[6] B. E. Hayden, K. Prince, D. P. Woodruff and A. M. Bradshaw, *Surf. Sci.*, **133**, 589 (1983)
[7] L. H. Dubois, T. H. Ellis, B. R. Zegarski and S. D. Kevan, *Surf. Sci.*, **172**, 385 (1986)
[8] D. P. Woodruff, C. F. McConville, A. L. D. Kilcoyne, Th. Lindner, J. Somers, M. Surman, G. Paolucci and A. M. Bradshaw, *Surf. Sci.*, **201**, 228 (1988)
[9] J. Grewe, Ü. Ertürk and A. Otto, *in preparation*

IN SITU SURFACE ENHANCED RESONANCE RAMAN SPECTROSCOPY (SERRS) OF COVALENTLY BOUND REACTIVE DYES ON COTTON FIBRES.

C.Rodger[1], W.E. Smith[1]*, P.C. White[1], C. Brennan[2], G. Dent[2].
[1]Department of Pure and Applied Chemistry, University of Strathclyde, 295 Cathedral Street, Glasgow, G1 1XL, UK.
[2]ZENECA Specialities, Hexagon House, Blackley, Manchester, M9 3DA, UK.

The identification of dyes and dye mixtures on fibres is useful in the textile industry and to forensic science. The need is for a sensitive method which will provide information on the orientation and bonding of the dye on a single fibre.

Recently surface enhanced resonance Raman spectroscopy (SERRS) has emerged as a useful probe of dyes absorbed on silver surfaces [1-3]. This study reports the development of the application of silver colloid to a fibre with a reactive dye covalently bound to the surface which was initially reported by Munro et al [4]. Chlorotriazine and Remazol dyes have been used in this study. These dyes are covalently fixed by nucleophilic addition to the fibre via a reactive group [5].

Resonance Raman, surface enhanced Raman and surface enhanced resonance Raman scattering was recorded for six dyes. *In situ* surface enhancement was also studied for each covalently bound reactive dye on cotton fibres. Raman scattering was collected using a modified Cary 81 and a Renishaw microprobe 2000 series with 514.5 nm radiation.

The cotton was pre-treated with an aqueous solution of cellosolve, (ethoxyethanol) 20% v/v, which is a strong solvent that removes any unbound dye from the cotton surface. The cotton fibres were then swollen by treatment with sodium hydroxide (3M) at 0 °C. The swelling process allows the colloid to penetrate into the fibre network and act as an antenna, reporting back the dye-cotton interacting. After this stage the sodium hydroxide was removed and the fibres washed with citrate buffer in order to increase the pH of the sample. The sample was then washed and soaked in poly-L-lysine (0.01% w/v) before a concentrated silver colloid suspension was introduced. The colloid was left in contact with the fabric for one hour before analysis with the Renishaw microprobe was attempted.

The six dyes studied were: Procion Red HE 3B, Procion Red 4BN, Procion Turquoise HA, Procion dye 5163243/1, Remazol Black B and Remazol dye 5164729. All dyes and dyed cotton samples were supplied by ZENECA Specialities, Blackley. The sodium hydroxide treatment was varied to optimise the swelling of the dye coated fibres. A sodium hydroxide treatment time of 1, 2, 4, 6, 8, 10 and 24 hours was investigated.

A sodium hydroxide treatment of between 4 to 6 hours is sufficient to observe acceptable SERRS. As the treatment time is increased the sample begins to degrade and the surface enhanced Raman signals observed deteriorate.

The solution SERRS and an *in situ* SERRS of Procion Red HE 3B is displayed below.

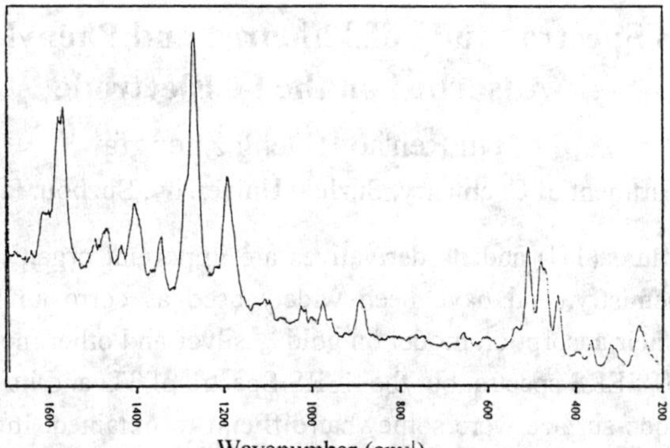

Solution SERRS of Procion Red HE 3B

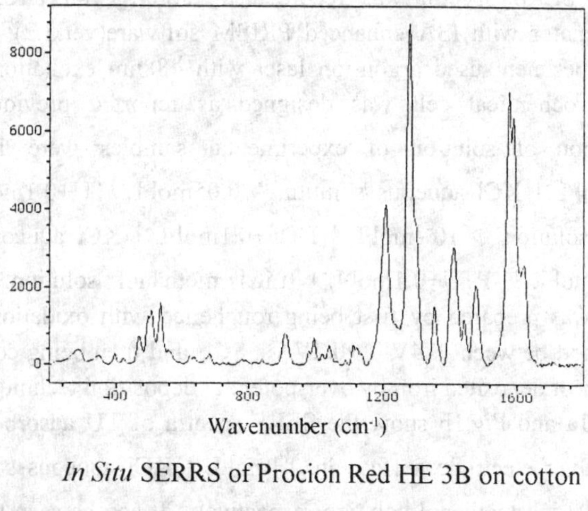

In Situ SERRS of Procion Red HE 3B on cotton

It is clear that SERRS has the sensitivity and selectivity required to obtain high quality *in situ* information from a single fibre.

1. Munro C.H., Smith W.E., White P.C. Analyst 120, 993 (1995).
2. Munro C.H., Smith W.E., Armstrong D.R., White P.C. J. Phys. Chem. 99, 879 (1995).
3. Hildebrandt P., Stockburger M. J. Phys. Chem. 88, 5935 (1984).
4. Munro C.H., Smith W.E., White P.C. unpublished work.
5. Gohl E.P.G., Vilensky L.D., in Textile Science: An explanation of fibre properties, 2nd Ed^{n}., Longman Cheshire, Melbourne, 1983.

SERS Spectra Study of Thiourea and Phenyl-thiourea Adsorbed on the Fe Electrode

Gu Ren'ao* Deng Zhengfa
Department of Chemistry, Suzhou University, Suzhou, P.R. China

Thiourea(TU) and its derivatives are important organic molecules in electrochemistry, and have been widely used as corrosion inhibitors for metals. Their adsorption model on gold 、 silver and other metals have been studied by SERS spectra, but the SERS Spectra of TU and its derivatives on Fe electrode surface were somewhat difficult to obtained. In this paper we presented the study of SERS of TU and PTU(phenyl thiourea) on iron electrode surface both in aqueous and non-aqueous electrochemical system.

The SERS spectra was recorded by Jobin-Yvon HRD-1 double monochromator with ISA enhanced PRISM software(ver.3.2). The SERS spectra experiment used argon-ion laser with 488nm excitation line. The spectroeletrochemical cell was designed as reported previously[1] The concentration of solution of experimental samples were $10^{-3}mol \cdot L^{-1}$ TU+$0.1mol \cdot L^{-1}$ KCl aqueous solution 、 $0.05mol \cdot L^{-1}$TU+$0.1mol \cdot L^{-1}$ LiCl methanol solution 、 $10^{-3}mol \cdot L^{-1}$ PTU+$0.1mol \cdot L^{-1}$ KCl aqueous solution and $0.05mol \cdot L^{-1}$ PTU+$0.1mol \cdot L^{-1}$ LiCl methanol solution. The iron electrode was prepared by first being roughened with oxidation-reduction cycle method between -0.4V ~ +0.2V vs. SCE and then being covered with a thin layer of deposited iron by over-potential deposition techniques.[2]

Fig.1a and Fig.1b show the SERS spectra of TU adsorbed on iron-covered silver electrode surface in $0.1mol \cdot L^{-1}$ KCl aqueous solution and $0.1mol \cdot L^{-1}$ LiCl methanol solution respectively. It can be seen that the TU exhibited SERS effect both in aqueous and methanol solution on iron-covered silver electrode surface, and the adsorption model of TU on electrode surface in the two solution was similar. But when comparing with the Raman spectra of solid TU, the SERS of TU on iron-covered silver electrode surface show some difference. For example, some bands were shifted($486cm^{-1}$ shifted to $575cm^{-1}$ 、 $1095cm^{-1}$ shifted to $1220cm^{-1}$ etc.) and the intensity of some bands were enhanced($388cm^{-1}$ 、 $1340cm^{-1}$ etc.). This indicated that there existed some interaction between TU and iron electrode in aqueous or methanol solution.

From comparing the SERS Spectra of TU on iron-covered silver electrode and silver electrode surface, we can know that it was by the interaction of S atom and iron that TU can adsorbed on the iron-covered silver electrode surface, and thus prevent the corrosion of iron.

It is interesting that although the Raman spectra of TU is different with that of PTU, their SERS spectra in aqueous and methanol were almost the same. This was because the molecular structure of TU and PTU was similar, and the only difference between them was that the hydrogen atom in TU is replaced by phenyl which is apart from the electrode surface and acts as steric hindrance. So their feature bands in SERS spectra was almost the same.

From the Spectra, we can also see that the intensity of SERS bands of PTU was higher than that TU, indicating that the adsorption ability of PTU was better than TU, and thus making its anti-corrosion ability better than TU. This was corresponding to our electrochemistry experiment result.[3]

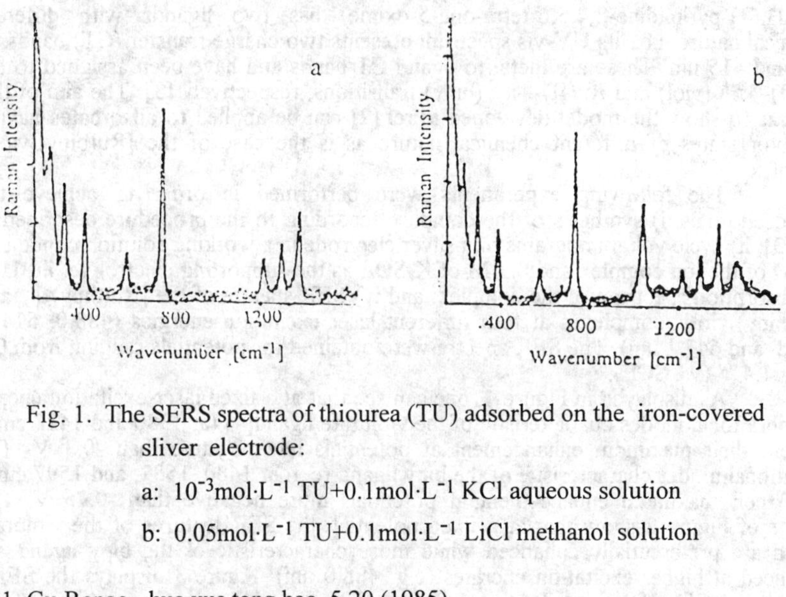

Fig. 1. The SERS spectra of thiourea (TU) adsorbed on the iron-covered sliver electrode:

a: 10^{-3}mol·L^{-1} TU+0.1mol·L^{-1} KCl aqueous solution

b: 0.05mol·L^{-1} TU+0.1mol·L^{-1} LiCl methanol solution

1. Gu Renao hua xue tong bao 5,20 (1985).
2. M.Fleischman, Z.Q.Tian,L.J.Li J.Electroanal. chem. 217,397(1987).
3.Gu Renao,Deng Zhengfa J.Suzhou University Natural science 11(4),378 (1995).

A SERS STUDY OF THE ION COMPLEX $[Ru(bipy)_2viol]^+$ ADSORBED ON A SILVER ELECTRODE

Paola Corio and Joel C. Rubim

Instituto de Química da Universidade de São Paulo,
CP. 26077, 05599-970, São Paulo, SP, Brazil

SERS (surface enhanced Raman scattering) has proved to be a powerful technique to map the density of states of molecules adsorbed on SERS-active electrodes [1,2]. Recently a model has been developed that can account for the dependence of the SERS intensity of adsorbates on the applied potential [2]. According to this model, if the laser excitation energy is maintained constant, the variation of the applied potential makes possible the excitation of photon assisted charge transfer processes involving donor states at E_F (the Fermi level of the metal electrode) and acceptor states localized on the adsorbate.

The ion complex $[Ru(bipy)_2viol]^+$ (bipy = 2,2'-bipyridine and viol = 1H,2H,5H-pyrimidine-2,4,5,6-tetra-one-5-oxime) has two ligands with different chemical nature, i.e., its UV-vis spectrum presents two charge transfer (CT) bands, at 478 and 415 nm. These are metal to ligand CT bands and have been assigned to the Ru(II)→π* (viol) and Ru(II)→π* (bipy) transitions, respectively [3]. The aim of this work is to show the model developed in ref.[2] can be applied to adsorbates having acceptor states of different chemical nature as is the case of the $[Ru(bipy)_2viol]^+$ complex.

The following experiments were performed in order to achieve the proposed goals: i) synthesis of the complex according to the procedure described in ref. [3]; ii) cycle voltammograms of a silver electrode in a working solution containing 1 mM of the ion complex and 0.1 M of K_2SO_4, as the supporting electrolyte; iii) UV-vis absorption spectrum of the complex; and iv) SER spectra of the $[Ru(bipy)_3]^{2+}$ and $[Ru(bipy)_2viol]^+$ complexes at four different laser excitation energies (488.0, 514.5, 562.8, and 647.1 nm). The SER spectra were obtained for potentials ranging from 0.0 V to -1.4 V (vs. SCE).

As displayed in Figure 1, one can see that at a fixed laser excitation energy the vibrational modes characteristic of the violurate ligand[4] (at 1380 and 1428 cm^{-1}) present their maximum enhancement at potentials less negative than -0.3 V. The vibrational modes characteristic of the bipy ligand (e.g. at 1480, 1552, and 1597 cm^{-1}) have their maximum enhancement at potentials more negative than -0.75 V. The results of Figure 2 show that at a fixed potential, the SER features of the violurate ligand are preferentially enhanced while those characteristic of the bipy ligand are enhanced at higher excitation energies (e.g. 488.0 nm). Figure 3 displays the SERS excitation profiles for two vibrational modes of the complex, one characteristic of the bipy ligand (at 1482 cm^{-1}) and the other related to the violurate ligand (at 1428 cm^{-1}). Note the presence of two potentials of maximum SERS intensity, V_{max}, close to 0.0 V and the other at -0.8 V. We have observed that both maxima shift to more negative values as the laser excitation energy decreases.

Based on these results we concluded that the dependence of the SERS intensity on the applied potential and for the existence of two different V_{max} values are due to the presence of two different metal to adsorbate photon assisted charge transfer processes. The first one involves a density of donor states at E_F and an acceptor state localized on the violurate ligand that is responsible for the less negative values of V_{max}. The second process involves the same donor states and an acceptor state localized on the bipy ligand that is responsible for the V_{max} values at more negative potentials.

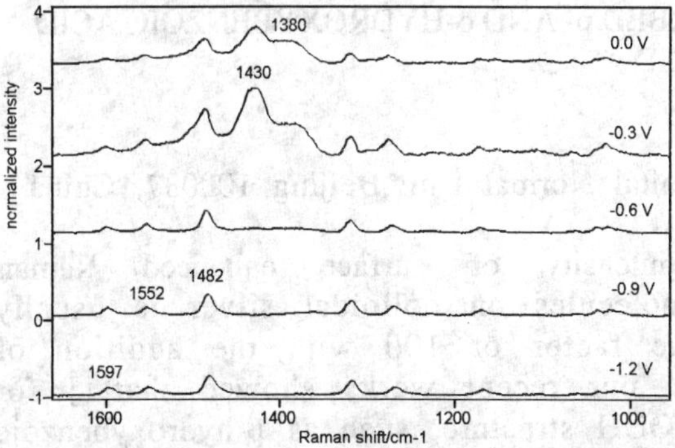

Figure1. SER spectra of the $[Ru(bipy)_2viol]^+$ complex adsorbed on a silver electrode at the indicated potentials. $\lambda_L = 647.1$ nm

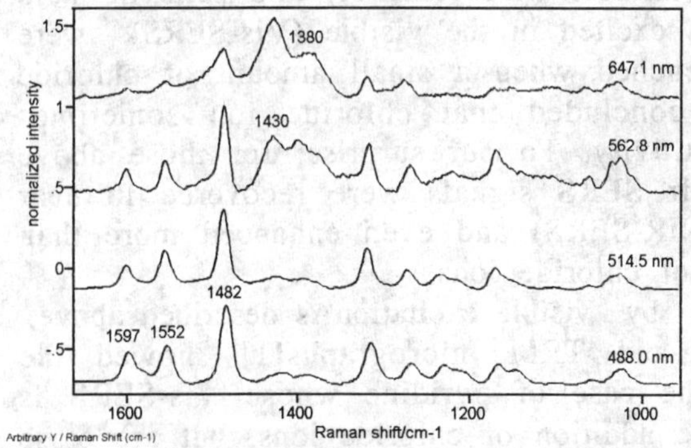

Figure 2. SER spectra at the potential of -0.1 V at the indicated excitation wavelengths.

Figure 3. SERS excitation profiles at 514.5 nm laser excitation.

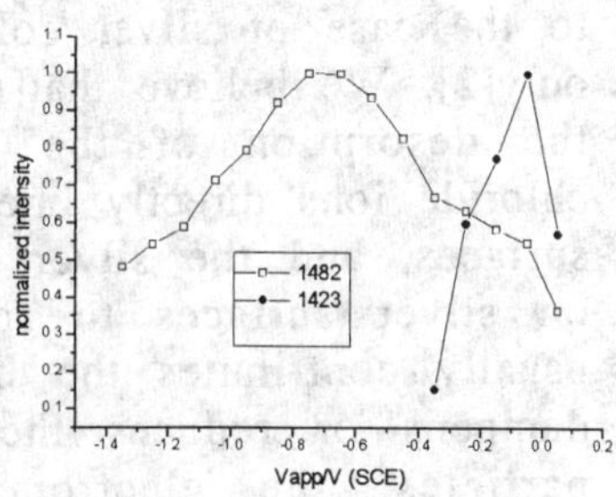

1. P. Corio and J.C. Rubim, J. Phys. Chem., 99, 13217 (1995).
2. J.C. Rubim, P. Corio, M.C.C. Ribeiro and M. Matz, J. Phys. Chem., 99, 15765 (1995).
3. V.R.L. Constantino, H.E. Toma, L.F.C. De Oliveira and P.S. Santos, J. Raman Spectrosc., 23, 629 (1992).
4. L.F.C. De Oliveira, P.S. Santos and J.C. Rubim, J. Raman Spectrosc., 22, 197 (1991).

THE NOVEL EFFECTS OF CHLORIDE ION ON VIS-SERS AND NIR-SERS OF ADSORBED p- AND o-HYDROXYBENZOIC ACID AND SALICYLIC ACID

Y. Fang* F. X. Hu
Experiment Centre,Capital Normal Univ,Beijing 100037, China

As is known, the intensity of surface enhanced Raman scattering (SERS) of molecules on colloidal silver is usually further enhanced by the factor of 100 with the addition of chloride ions. However, our recent works showed that[1] for some molecules with -COOH structure, such as p-hydroxybenzoic acid (PHBA), m-hydroxybenzoic acid (MHBA) and salicylic acid (SA), their SERS signals excited in the visible (Vis-SERS) were sharply reduced and quenched when a small amount of chloride ions were added. We concluded that chloride ion sometimes shows the SERS non-actitivity. To our surprise, for those above series of molecules, their SERS signals were recovered in near infrared (NIR) region (NIR-SERS) and even enhanced more than that before the addition of chloride ions.

For the quenching by visible excitationAs described above, The optical absorption and TEM micrographs[1] showed the much difference from the case of pyridine whose Vis-SERS is further enahnced with the addition of chloride ions, but are close to the case of silver colloid with the addition of chloride ions only[2]. We believe that[1] the addition of chloride ions leads to the desorption of the majority of adsorbed molecules. The chloride ions directly interact with silver atoms on silver particle surfaces, and the silver atoms are transported by chloride ions on silver surfaces to form a Ag-Cl clusters overlayer which usually contributes the Chemical Enhancement to SERS[2], but dampens or reduces the surface plasmon resonance of silver particles, the electromagnetic(EM) enhancement[1]. In the meantime, adsorption behavior of the molecules remaining adsorbed on surfaces would be changed because of the formation of Ag-Cl clusters overlayer. the overall effect is still a quenching in SERS intensities.

However, our recent work has shown that the SERS signals of

those molecules were recovered in the near infrared, and further enhanced or not observably changed, although the large amount of molecules were desorbed from the silver surfaces. It demonstrates that the SERS intensities of those molecules depend on the excitation wavelengths. They are SERS non-active in the visible and SERS-active in the near infrared. We ascribe them to the formation of Ag-Cl clusters overlayer which leads to the chemisorption on the SERS active sites by the structure -OH that includes a lone electron pair in -O similar to the case of the -N in pyridine. In the meantime, the chemical enhancement plays an important role in that case.

The charge transfer (CT) between the surfaces and molecules vs the Ag-Cl active sites leads to strong chemical enhancement. On the other hand, the formation of Ag-Cl clusters overlayer dampaned EM enhancement. As a result, the CT mechanism governs the NIR-SERS enhancement, and results in the strongly chemical enhancement. Hildebrandt and Stockburger's work on Rh6G showed that[3] the number of the Ag-Cl structures that could be called the SERS active sites is by an average of three active sites on each particle. We estimated that the enhancement factor of the CT in NIR-SERS is around 1000 to 10000. It desmontrates again that the chemical enhancement is more important than EM enhancement in NIR region. The difference between NIR-SERS and Vis-SERS typically reflects the difference of SERS mechanism between the visible and the near infrared.

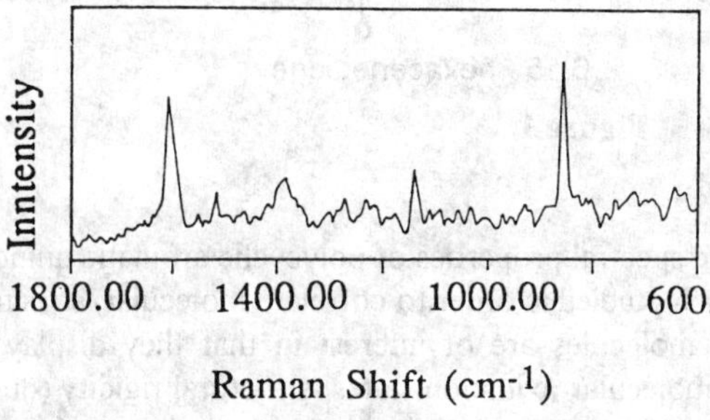

Figure 1 NIR-SERS of PHBA in Ag sol after the addition of KCl.

References:

[1] Y,Fang and S.Y.Fu et al, Chin. Phys.(USA), 12(1992)44.
[2] P.X.Zhang and Y.Fang et al, J.Raman Spectrosc., 21(1990)127.
[3] P.Hildebrandt and M.Stockburger,J.Phys.Chem.,88(1984)5935.

FT-RAMAN AND SERS ON SILVER ISLAND FILMS OF POLYCYCLIC AROMATIC QUINONES

W.J. Price[1]*, R. Aroca[1], P. Dibble[2]

[1]Department of Chemistry and Biochemistry, Materials and Surface Science Group, University of Windsor, Windsor, Canada N9B 3P4
[2]Department of Chemistry, University of Lethbridge, Alberta, Canada, T1K 3M4.

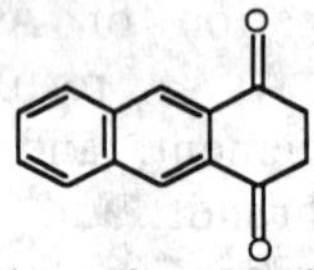

1,4 - anthracenedione

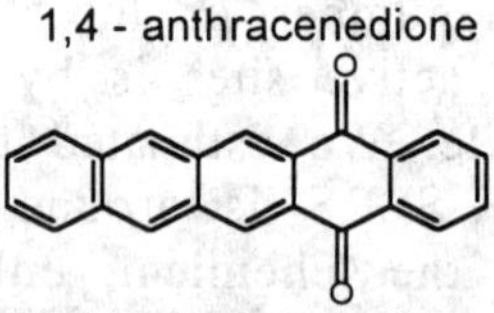

5,14 - pentacenedione

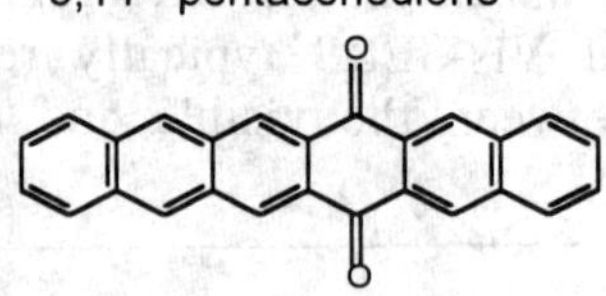

6,15 - hexacenedione

Figure 1

The electronic spectral properties of polycyclic aromatic quinones (see Figure 1) have been extensively studied in order to correlate molecular structure and spectral features [1]. These molecules are of interest in that they display diverse radical chemistry, multiple molecular oxidation states, structural rigidity (due to their planer conjugated framework) and intense colour due to their chromophore groups. These properties make the materials ideal candidates for development in organic semiconducting or electro luminescence devices containing a thin organic film of these molecules. It is our final objective to study the molecular and solid state morphologies of these films via vibrational spectroscopy. Such an endeavour would lead to a better understanding of the correlation of physical properties and solid state structure which is a necessary step in the development of new devices.

An important consideration for Raman studies is the fact that these molecules have a strong emission in the visible. Therefore, Fourier transform Raman spectroscopy (FT-Raman) is the most appropriate method for obtaining inelastic light scattering, since the 1064 nm excitation wavelength lies outside of the absorption spectrum. The infrared spectra of the 1,4-anthracenedione has been previously reported [2,3]. However, the vibrational analysis and Raman spectra of the two extended quinones (presented in Figure 1) have not been reported. The FT-Raman spectra of the three molecules are shown in Figure 2. The spectra were recorded using a Bruker RFS100 spectrometer. Another technique appropriate for studying the vibrational states of these molecules is surface-enhanced resonance Raman scattering (SERRS) on silver island films. It provides an enhanced scattered signal and a non-radiative pathway for the quenching of the phosphorescence. In the present report, using IR, Raman and SERRS, a complete analysis of the vibrational spectra of the series was carried out. An attempt is made to correlate the effect of changes in molecular structure and the observed vibrational spectra. The vibrational assignments of the observed frequencies and characteristic group frequencies in the infrared and Raman spectra are discussed. The film formation ability of these materials are also reported, and the characteristic frequencies are used to study the molecular organization in the thin solid films.

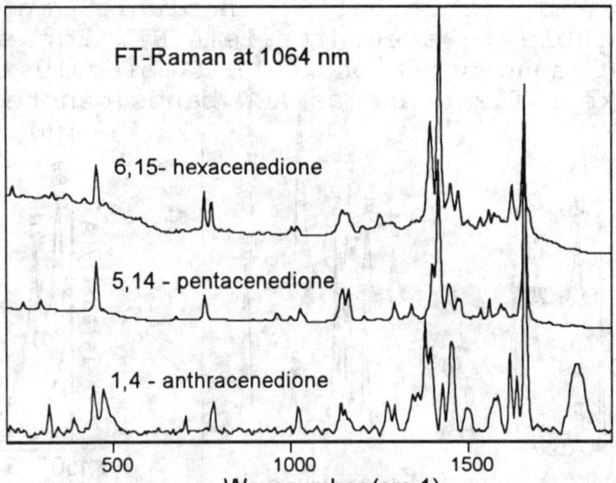

Figure 2

1. G.J. Gleicher, in The Chemistry of the Quinonoid Compounds, S. Patai, Ed. (Wiley, London, 1974), Part I.
2. L.P.Brivina, N.S.Strokach, D.N.Shigorin, M.V.Gorelik, Opt. Spectrosc. 51 376 (1981).
3. N.S.Strockach, T.V.Kainkova, D.N.Shigorin, Russian J. Phys. Chem. 60 63 (1986).

SURFACE ENHANCED RAMAN STUDY OF 2-FORMYLTHIOPHENE IN SILVER COLLIOD

Kana Mukherjee and T.N.Misra*

Department of Spectroscopy, Indian Association for the Cultivation of Science, calcutta-700032, India.

Surface enhanced Raman scattering[SERS] is a sensitive analytical tool with detection limits down to pico and femto molar concentrations. It is particularly attractive because, as a form of vibrational spectra it combines high sensitivity with high information content for establishing molecular identity and structural character. Methodology to determine the adsorbate orientation on metal surface has been developed by Moskovits and Creighton[1,2]. In this paper we report SER spectra of 2-formylthiophene[FT] adsorbed on colliodal silver particles.

2FT is practically insoluble in water but strongly soluble in acetonitrile[ACN]. For each step of reduction of concentration of 2FT molecule the addition of ACN was kept fixed and as ACN bands[denoted as * mark

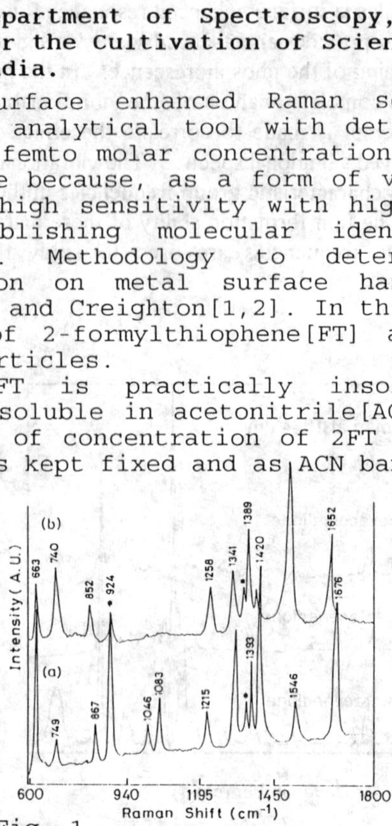

Fig. 1.

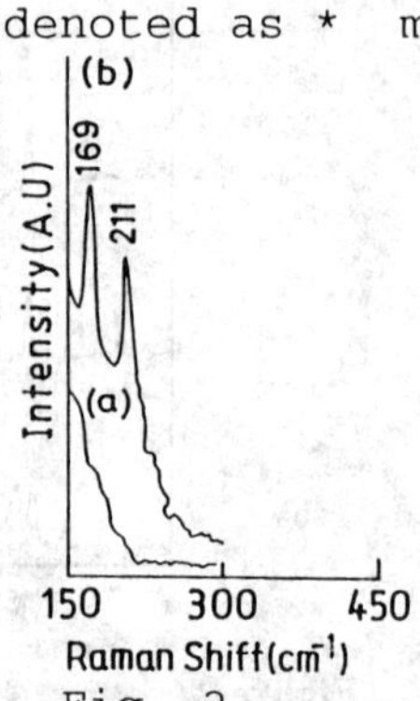

Fig. 2.

(a): NSR spectrum with concentration of 1M.
(b): SER spectrum with concentration of 10^{-18}M.

in [Fig.1] do not show[3] any surface enhancment, each band of SER spectrum is normalised with respect to ACN bands.

SER spectrum[Fig.2b] shows two strong bands at $169 cm^{-1}$ and $211 cm^{-1}$ which are not observed in the bulk and also in normal solution Raman[NSR] spectrum[Fig.2a].These bands are assigned as Ag-S and Ag-O stretching[4]. Appearance of such modes only in the SER spectrum suggests that the molecule interacts with metal surface through S and O atoms causing the molecule to adsorb through their lone pair of electrons.

Presence of ring breathing mode at 852cm^{-1} in SER spectra as well as C-H stretching modes at 3083cm^{-1} [Table 1.] indicates the vertical orientation of the thiophenyl ring on the silver surface. The C=C asymmetric stretching mode at 1546cm^{-1} is not only strongly enhanced but shows also a red shift of about 19 cm^{-1} resulting in strong conjugation of substituent group to the thiophenyl ring.

Large enhancement of C=O stretching compared to that of C-H stretching mode (table 1.) of substituent group infers the predominance of syn-rotamer form [Fig.3a] rather than anti- conformation [Fig.3b] making large projection of the C=O bond on the surface normal and for such orientation the molecule finds scope to adsorb on the metal surface through the oxygen atom of the substituent carbonyl group as being nearer to the surface. This possibillity is also substantiated by the red shift of 24cm^{-1} of carbonyl stretching mode as well as by the appearance of Ag-O stretching mode in the SER spectrum.

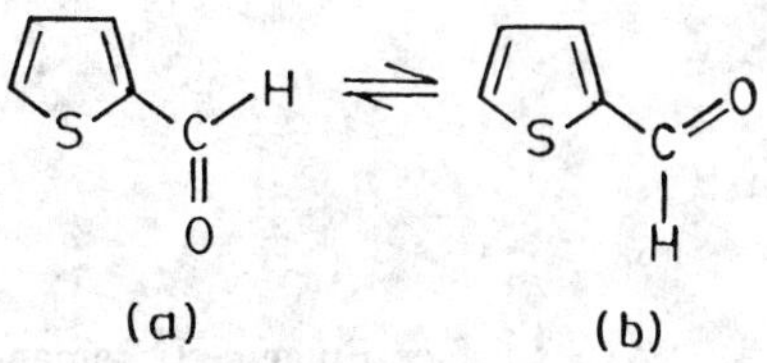

Fig.3.

TABLE 1. Raman and SER spectral data of 2FT

Raman shift in cm^{-1} solution	silver sol	Enhancement Ratio	Tentative Assignment
867	852	1.0	Ring breath
1546	1527	3.38	ν_{asym}(C=C)
1676	1652	3.02	ν(C=O)
3002	3002	1.66	ν(CH)#
3086	3083	2.15	ν(C-H)

ν: stretching, # substituent group.

REFERENCES

1. M. Moskovit and J. Suh, J. Phy. Chem., 88, 5526 (1984).
2. J. Creighton, Surf. Sci., 124, 209 (1984).
3. U. K. Sarkar, S. Chakrabarti and T. N. Misra, Chem. Phys. Lett., 190, 59 (1992).
4. S. Sanchez and J. V. Garcia-Ramos, J. Raman Spectros., 23, 61 (1992).

Surface-Enhanced Raman (SER) Spectra of Cyanopyridines Adsorbed on Aqueous Colloidal Copper

Candace M. Coyle and Paul W. Jagodzinski*

Department of Chemistry, West Virginia University
Morgantown, West Virginia USA 26505

Surface-Enhanced Raman Spectroscopy (SERS) is recognized as a fundamental technique for determining detailed information concerning bonding to surfaces. It combines the structural information of vibrational spectroscopy with the sensitivity and surface specificity that comes from the enhancement mechanism associated with SERS. The main limitation, however, is that it is only applicable to certain surfaces, specifically those of free-electron metals, although alternative methods are being developed to overcome this limitation (1). Due to this factor, colloidal metals play a distinctive role in SERS because they exhibit strong enhancement and are relatively easy to prepare by chemical reduction methods (2). SER scattering from molecules on silver and gold surfaces have been extensively studied, however, copper surfaces, specifically copper colloids, have received less attention due to their inherent colloidal instabilities. An investigation of substituted pyridines adsorbed onto aqueous copper colloids has been initiated.

Donor-acceptor substituted aromatic molecules are known to adhere to the surface of copper colloids prepared by borohydride reduction of Cu^{2+} (3). This adsorption process produces bands in the 566 nm region of the ultraviolet-visible spectrum (UV-VIS). Surface-enhanced Raman spectra and UV-VIS spectra of 2-, 3-, and 4-cyanopyridines (2-,3-,4-CP) adsorbed on copper colloids have been obtained and compared with previous studies of these molecules on silver electrodes (4). Excitation at 647.1 nm led to surface-enhancement of the spectra of these three cyanopyridines adsorbed onto copper colloids. The vibrational information contained in the Raman spectra provide the molecular specificity required to characterize the adsorbate/surface interactions, specifically, the orientation of adsorbed species on the metal surfaces. Previous studies of cyanopyridines on silver electrodes indicate a surface enhancement of the -CN stretching normal mode that is dependent upon the position of the -CN substituent on the aromatic ring. This was interpreted to indicate a dependence of the enhancement mechanism upon the orientation of the -CN group with respect to the silver surface. The spectra collected for the cyanopyridines adsorbed on the surface of a copper colloids show a complete disappearance of their -CN stretching signal (see Figure 1, for example). This would indicate a different bonding scheme than that used to explain the bonding of 4-CP and 3-CP to a silver surface in which the -CN stretching signals remained intact, but with slightly reduced intensities. However, there are similarities between the SER spectra of 2-CP adsorbed on copper and silver, in particular the disappearance of the -CN stretching signal. Studies of UV-VIS spectra will be presented to corroborate these findings. Details of the effects of concentration and time on the chemistry occurring on these copper colloids will be discussed.

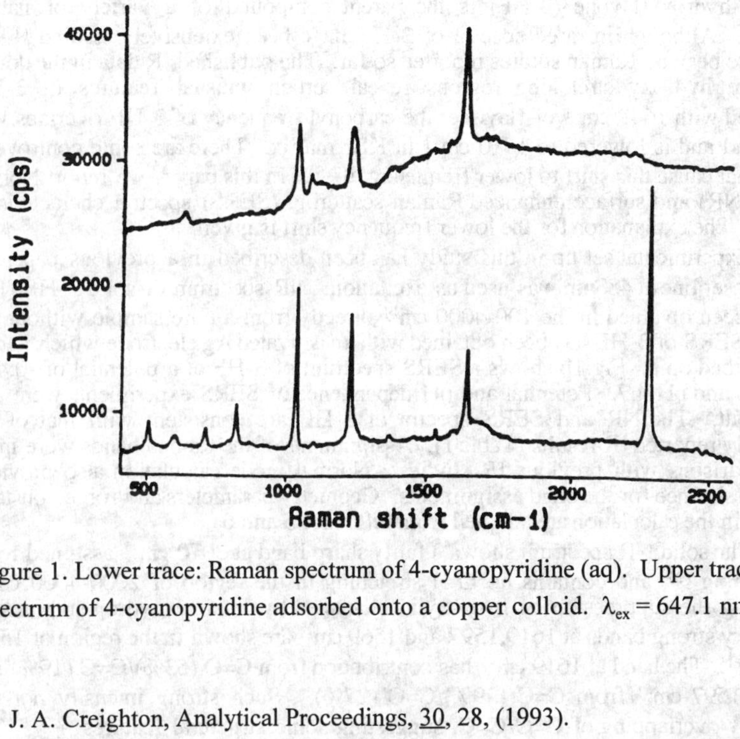

Figure 1. Lower trace: Raman spectrum of 4-cyanopyridine (aq). Upper trace: SER spectrum of 4-cyanopyridine adsorbed onto a copper colloid. λ_{ex} = 647.1 nm

1. J. A. Creighton, Analytical Proceedings, 30, 28, (1993).

2. G. Chumanov, K. Sokolov, B.W. Gregory, and T. M. Cotton, J. Phys. Chem., 99, 9466 (1995).

3. J. A. Creighton, M. S. Alvarez, D. A. Weltz, S. Garoff, and M. W. Kim, J., Phys. Chem., 87, 4793 (1983).

4. C. S. Allen, and R. P. Van Duyne, Chem. Phys. Lett., 63 (3), 455 (1979).

SURFACE ENHANCED RAMAN SCATTERING STUDIES OF 3-HYDROXYFLAVONE ON A Ag ELECTRODE

Minfa Wang, John R. Lombardi and Ronald L. Birke

Department of Chemistry, The City College of New York
Convent Avenue at 138 Street, New York, NY 10031

3-hydroxyflavone (3-HF) is the parent compound of a variety of natural products. Although infrared spectra of 3-HF have been extensively studied (1,2,3), there have been no Raman studies reported so far. The published IR data in the double bond and hydroxy stretching regions reveal certain unusual features in 3-HF. Compared with 1649 cm^{-1} of flavone, the carbonyl frequency of 3-HF overlaps with C=C band and is lowered to 1610 cm^{-1} in KBr matrix. There are some controversy about what cause this shift to lower frequency (1,2,3). In this paper, we report Normal Raman (NR) and surface enhanced Raman scattering (SERS) spectral characteristics for 3HF. The explanation for the lower frequency shift is given.

Experimental set up in this study has been described in a previous paper (4). An Ar^+ laser line at 488nm was used as excitation. NR spectrum of solid 3-HF (Fig. 1a) has been obtained in the 100-4000 cm^{-1} directly from a pure sample without any matrix. SERS of 3-HF has been obtained with an activated Ag electrode which had 3-HF adsorbed on it. Fig.1b shows a SERS spectrum of 3-HF at a potential of -0.45V (vs SCE) and pH 4.7. Potential and pH dependence of SERS experiments were also carried out. The NR and SERS spectra of 3-HF are consistent with that of the previously reported IR results (Table 1). Assignments of the Raman bands were made by comparisons with previous IR studies. Normal mode calculation also provided further assistance for the band assignments. Geometry parameters and force constants of 3-HF in the calculation are adopted from reference 5 and 6.

The solid NR spectrum shows a fairly sharp band at 3070 cm^{-1} assigned to C-H stretching (2); and contains no O-H stretching in the region of 2000-4000 cm^{-1}, suggesting the suppression of hydroxylic character by internal hydrogen bonding. Three very strong bands at 1619,1597 and 1569 cm^{-1} are shown in the region of 1500-2000 cm^{-1}. The band at 1619 cm^{-1} has contribution from C=O (53%)/C=C (19%) and band at 1597 cm^{-1} from C=C (39%)/C=O (27%). Such strong intensity not just caused by overlapping of C=O/C=C, suggesting some structural features (2,3). The SERS spectrum shows lower relative intensity for these bands; but the relative intensity of bands assigned to the phenyl ring in-plane skeletal stretching increased; whereas, the intensity of bands assigned to the benzene ring in-plane skeletal stretching and C-H stretching show little change. This suggests that main heteroatomring system is parallel to the electrode surface but the phenyl ring is at an angle to the surface. There is lower frequency shift at 1002 cm^{-1} assigned to the CCH in-plane deformation coupling with the C-O stretching in pyrone ring when applied potential is more negative,-0.75V vs SCE. The spectra obtained at different pH value (4.7 and 10.8) show no difference. This pH independence can be explained by the fact that 3-HF is a very weak acid and also a very weak base ($Ka=2.5X10^{-10}$, $Kb=1.3X10^{-17}$) (7). We conclude that the lower carbonyl frequency comes from both internal hydrogen bonding and resonance structural features.

Table 1: Calculated And Observed Frequencies of 3-HF

Vibrational assignment (main contributions)	Calc.	IR(1) (cm-1)	Solid	SERS
C–H	3070		3070m	3070m
C=O stretching coupling with C=C	1705	1627_{sh}	1619_{vs}	
C=C stretching coupling with C=O	1610	1610_s	1597_{vs}	1595_s
	1565	1564_s	1569_{vs}	
Phenyl ring in-plane stretching	1505		1495_{sh}	1498_s
Benzene ring in-plane stretching	1466	1475_s	1474_s	1457_s
Phenyl ring in-plane stretching	1426	1450_s	1447_m	1430_s
Phenyl ring in-plane stretching	1421	1418_s	1416_m	1406_s
OH in-plane deformation	1355	1351_s	1354_m	1359_m
Benzene ring in-plane stretching	1309	1305_s	1313_s	1316_s
CCH deformation coupling with C-O	989	990_s	999_s	1002_s

Figure 1

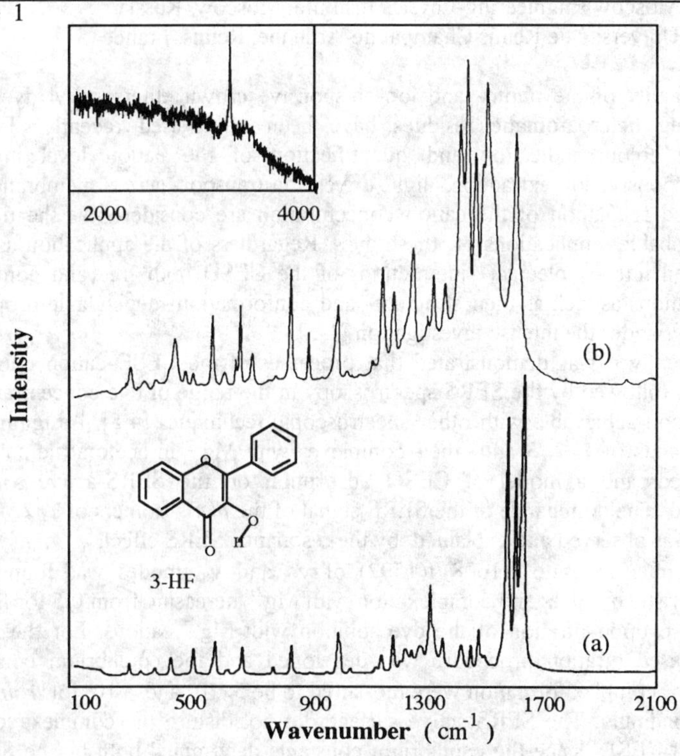

References:

[1] J. H. Looker and W. W. Hanneman; J. Org. Chem., **1962**, 27, 381
[2] C. I. Jose, P. S. Phadke & A.V. R. Rao; Spectrochem. Acta, **1974**, 30A, 1199
[3] B. L. Shaw and T. H. Simpson; J. Chem. Soc., **1955**, 655
[4] J. Xu, R. L. Birke & J. R. Lombardi; J. Am. Chem. Soc., **1987**, 109, 5645
[5] B. Dick; J. Phys. Chem., **1990**, 94, 5752
[6] Donald W. Scott; J. Molecular Spectroscopy, **1971**, 37, 77
[7] O. S. Wolfbeis, A. Knierzinger and R. Schipfer; J. Photochem., **1983**, 21, 67

SERS SPECTROSCOPY STUDY OF PHOTOCHROMIC IONOPHORES. BENZOTHIAZOLIUM CROWN-ETHER STYRYL DYES, THEIR COMPLEXES WITH Mg^{2+} AND DIMERS.

A.V. Feofanov[1], A.I. Ianoul*[3], V.A. Oleinikov[1], S.P. Gromov[2], O.A. Fedorova[2], M.V. Alfimov[2], and I.R. Nabiev[4]

[1]Shemyakin-Ovchinnikov Institute of Bioorganic Chemistry, Russian Academy of Sciences, ul. Miklukho-Maklaya 16/10, 117871 Moscow, Russia
[2]Semenov Institute of Physical Chemistry, RAS, Moscow, Russia
[3]Moscow Engineering-Physics Institute, Moscow, Russia
[4]Université de Reims Champagne-Ardenne, Reims, France

A family of the photo- and ion- responsive crown ethers styryl dyes (CESD) containing heteroaromatic residues have been synthesised recently. Type-cation sensitive colour indication and quantification of the cation level in solution, photoresponsive ion extraction, light-driven ion transport across membranes, photo-controlled regulation of the cation concentration are considered as the field of the most probable applications of these dyes. Regardless of the application aspects, the inter- and intra- molecular interactions of the CESD both free and complexed to metal cations as well as their structure- and conformation- dependable properties are currently under the intense investigation [1-5].

Recently it was demonstrated that processes of the CESD-cation complexation could be followed by the SERS spectroscopy in the range of dye concentrations 10^{-5}-10^{-8} M non-achievable with other spectroscopic techniques [4,5]. Assignment of the SERS bands of **1**, **2,** **3** and their complexes with Mg^{2+} in acetonitrile solution was performed, and a model of CESD adsorption on the SERS-active surface was proposed. The dominance of the SERS signal of the *trans*-isomers of **1, 2, 1**Mg^{2+} and **2**Mg^{2+} was observed and explained by the resonance SERS effect.

An intensity ratio I(1608)/I(1592) of ν_{8b} and ν_{8a} modes was found to be a characteristic of the ligand complexation with Mg^{2+} increasing from 0.5 for free CESD to *ca.* 1.6 upon titration of the dye solution with Mg^{2+} cations. For the first time, SERS-based titration procedure was developed and the equilibrium constants for 1L:1Mg^{2+} complex formation were measured to be 5×10^4 and $> 10^6$ for *trans*-**2** and **1**, correspondingly. The SERS-active surface did not disturb the complexation for the adsorbed CESD, since the equilibrium constants determined both by the SERS- and spectrophotometric titrations procedures were found to be in a good agreement.

Photocontroled reversible capture/release of cations was achieved by introducing sulfopropyl N-substituent in **3** and varying the proximity of anionic group to the crown ring by means of photoisomerisation [1]. The (*trans*-**3**)Mg^{2+} was found to undergo regio and stereoselective photocyclodimerization leading to the only isomer of cyclobutane derivative [2]. This unique 100% reversible photoreaction was explained by self-organisation of (*trans*-**3**)Mg^{2+} into dimers. SERS spectra of (*trans*-

3)Mg^{2+} were interpreted as indicating the formation of head-to-tail dimers **6** at the moderate Mg^{2+}/**3** molar ratios [4]. High concentration of the dimers **6** was detected by the SERS spectroscopy at concentrations 10^{-5}-10^{-7} M of the (*trans*-**3**)Mg^{2+}. On adding large excess of Mg^{2+} cations the dimers were shown to dissociate and the 1L:2Mg^{2+} complexes to be formed. The equilibrium constant for 1L:2Mg^{2+} complex formation determined for **3** by the SERS titration procedure was equal to 1.5×10^4. SERS spectroscopy is unique method for measuring this constant.

To improve the self-organisation property of the (*trans*-**3**)Mg^{2+} and to increase a release of concerted [2π+2π]-photocycloaddition reaction new CESD **4**, **5** with orto- and para- sulfobenzyl N-substituents were synthesised and studied by the SERS and absorption techniques. Assignments of bands in the SERS spectra of **4**, **5** were performed and geometry of adsorption of **4**, **5** on the SERS active surface was proposed. Spectral properties and adsorption geometry of **4**, **5** were found to be similar to those of **3**. Formation of dimers by both (*trans*-**4**)Mg^{2+} and (*trans*-**5**)Mg^{2+} was proved by the SERS spectroscopy. The equilibrium constants for 1L:2Mg^{2+} complex formation being equal to 2×10^4 and 3×10^4 were determined by SERS titration procedure for **4** and **5**, correspondingly. Relative ability of **3**, **4** and **5** to form dimers was compared by analysing the SERS titration curves measured at different (10^{-5}-10^{-7} M) concentrations of the dyes.

Among other things the data obtained demonstrate clearly both high analytical and research capabilities of SERS spectroscopy in the study of CESD.

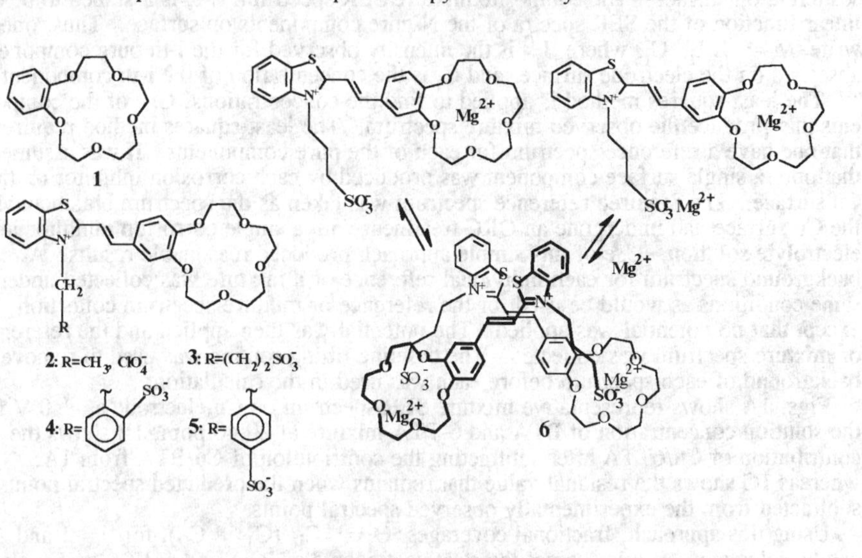

References

1. A. Barzykin, M.Fox, E. Ushakov et al. J. Am. Chem. Soc. 1992, **114**, 6381-6385.
2. M. Alfimov, S. Gromov, O. Stanislavskii et al. Russ. Chem. Bull. 1993, **42**, 1385-1389.
3. E. Ushakov, Ph.D. thesis, 1995.
4. A. Feofanov, A. Ianoul, V. Oleinikov, et al. Russ. Chem. Bull. 1995, **44**, Dec., in press.
5. A. Feofanov, A. Ianoul, V. Oleinikov, S. Gromov, et al. J Phys.Chem. 1996, Jan. in press

A QUANTITATIVE STUDY OF SURFACE COVERAGE OF BENZOTRIAZOLE AND 6-TOLYLTRIAZOLE MIXTURES ON CU ELECTRODES

B. H. Loo*, A. Ibrahim and M. T. Emerson

Department of Chemistry, University of Alabama,
Huntsville, AL 35899, USA

The least-squares method is used to generate quantitative estimates of the relative amount of each component adsorbed on the Cu electrode surface in solutions containing benzotriazole (BTA) and 6-tolyltriazole (6-TTA) from the mixture surface-enhanced Raman (SER) spectra. BTA and its derivatives are well known corrosion inhibitors for Cu and Cu alloys. The SER spectra of BTA/6-TTA on Cu electrodes were collected on a JY MOLE/U1000 Raman microprobe, using 647.1 nm excitation. Electrodeposition of the inhibitor molecules on the polished Cu electrode was carried out by an in situ oxidation-reduction cycle (ORC) using a sweep rate of 1.5 V/min. The ORC surface pretreatment was done with simultaneous laser illumination.

In the SER spectra, the intensity of the signals is proportional to the number of scatterers on surface. The composite mixture SER spectrum, M, is assumed to be a linear function of the SER spectra of the N pure components on surface. Thus, one can write $M = \Sigma I_i^o \cdot C_i$, where I_i^o is the intensity observed for the i-th pure component absorbed on the electrode surface, and C_i is the concentration of the i-th component.

The least-squares method is applied to find the concentrations, C_i, of the components that produce the observed mixture spectrum. The least-squares method requires that one have a reference spectrum for each of the pure components. It was assumed that only a single surface component was produced by each corrosion inhibitor on the Cu surface. The required reference spectrum was taken as the spectrum obtained when the Cu surface had undergone an ORC treatment with a single corrosion inhibitor in the electrolyte solution. Use of this simple approach produced reasonable results. A background spectrum for each individual reference or a mixture was collected under the same conditions as would be used for the reference or mixture spectrum collection except that no potential was applied. The potential was then applied and the reference or mixture spectrum was collected. The baseline fitting routine was used to remove the background of each spectrum before each was used in the calculation.

Figs. 1A shows representative mixture SER spectrum on Cu electrode at -1.0 V for the solution concentration of BTA and 6-TTA mixture at 20:40 ppm, 1B shows the contribution of Cu/6TTA after subtracting the contribution of Cu/BTA from 1A, whereas 1C shows the residual value that remains when the predicted spectral points are subtracted from the experimentally observed spectral points.

Using this approach, fractional coverages, $\Theta_i = C_i / (C_1 + C_2)$ for i= 1 and 2 cab be calculated. Table 1 gives the relative amounts of BTA and 6-TTA on Cu electrodes. For the 20:20 ppm solution mixture, the ratio of the fractional coverage of 6-TTA to that of BTA on Cu is 2.1:1. This result shows that 6-TTA exhibits a higher surface coverage than BTA and suggests that 6-TTA is more strongly adsorbed than BTA. When the solution concentration of 6-TTA is more than that of BTA (in the 40:20 and 100:20 ppm solution mixtures), the Cu electrode surface is predominantly covered by the 6-TTA molecules. On the other hand, when the solution concentration of BTA is more than that of 6-TTA (in the 20:40, 20:100, and 20:160 ppm solution mixtures), the BTA molecules replace the 6-TTA molecules on the Cu electrode surface, and the contribution of BTA to the mixture spectra becomes predominant. These results suggest that the absolute concentrations of the solution species affect the thermodynamics of adsorption.

In summary, we have demonstrated that least-squares analysis of the SER spectra can be used to elucidate relative surface concentration of coadsorbed species on electrode surfaces. The relative surface coverages provide important information on the thermodynamics of adsorption of adsorbates on surfaces.

TABLE 1 Fractional coverages of 6-TTA and BTA on Cu electrodes.

Solution concentration (ppm)	Fractional coverage	
6-TTA : BTA	6-TTA	BTA
20:20	0.68	0.32
40:20	0.74	0.26
100:20	0.90	0.10
20:40	0.47	0.53
20:100	0.36	0.64
20:160	0.23	0.77

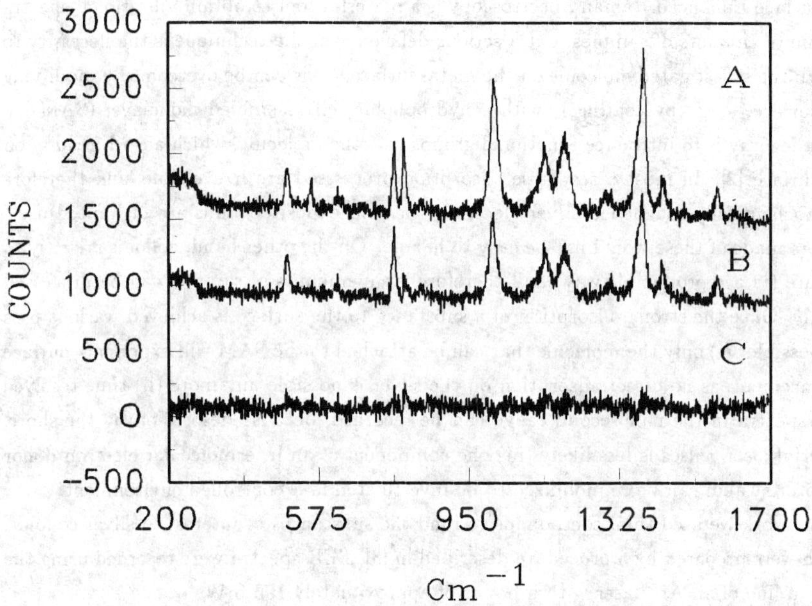

Fig. 1

SURFACE ENHANCED RAMAN SPECTROSCOPY OF COMPOUNDS ATTACHED BY THIOL GROUPS TO SILVER COLLOIDS

Georg Brehm[1*], Peter Freunscht[1] and Siegfried Schneider[1]

[1] Institut für Physikalische und Theoretische Chemie
Universität Erlangen-Nürnberg, 91058 Erlangen, Germany

Surface Enhanced Raman Spectroscopy is a powerful tool to obtain vibrational spectra of minute amounts of samples, but a serious deficiency of the technique is the necessity to adsorb the investigated molecule on the metal surface. This can be overcome by modifying the surface, e. g. by coating it with a hydrophobic self assembled monolayer (SAM)[1]. An other way is to introduce functional groups into the molecule, which adsorb readily on the surface [2]. In most cases these adsorption processsed are irreversible, and therefore electrochemically roughened silver electrodes or cold deposited films are of limited use - sensors made of these would not be easy to handle. On the other hand, colloids are easy to prepare (in a reproducible way) and therefore large amounts of samples can be processed readily. Once the strong adsorption of a substance to the surface is achieved, various uses are possible: (i) only the molecule that can be attached to the SAM will experience surface enhancement, as no direct adsorption on the silver is possible any more (ii) time resolved SER spectra in the nanosecond range can be recorded, because desorption by the short, powerful laser pulses is less likely (iii) one component of an intermolecular electron donor acceptor system can be immobilized to be investigated in a controlled environment.
In this work we used thiol compounds to bind the substances of interest to silver colloids, which were prepared by a procedure described in [3]. SER spectra were recorded using the 488 nm line of an Ar^+ laser with a power of approximately 100 mW.

Preparation of SAMs

A methanolic solution of an alkane thiol was given to the aqueous silver colloid to produce SAMs of propanethiol, hexanethiol and dekanethiol. The stability of the SAM-coated silver colloid increases with the chain length, but is also concentration dependent: If a certain concentration of thiol is exceeded the colloid begins to coagulate.

Adsorption of protein fragments

Cysteine and methionine are the two amino acids that contain a thiol group and therefore are willing to adsorb on silver. Other amino acids, like tryptophane, were coupled to them and SER spectra were taken.

The strength of the adsorption was tested by adding malachite green, which is well known to adsorb on SERS active surfaces. The sulfur containing compounds could not be removed by the dye from the surface, therefore we suggest the existence of a covalent Ag-S bond. This is supported by the disappearance of the SH stretch around 2600 cm^{-1} upon adding colloid to a dilute solution of the thiol.

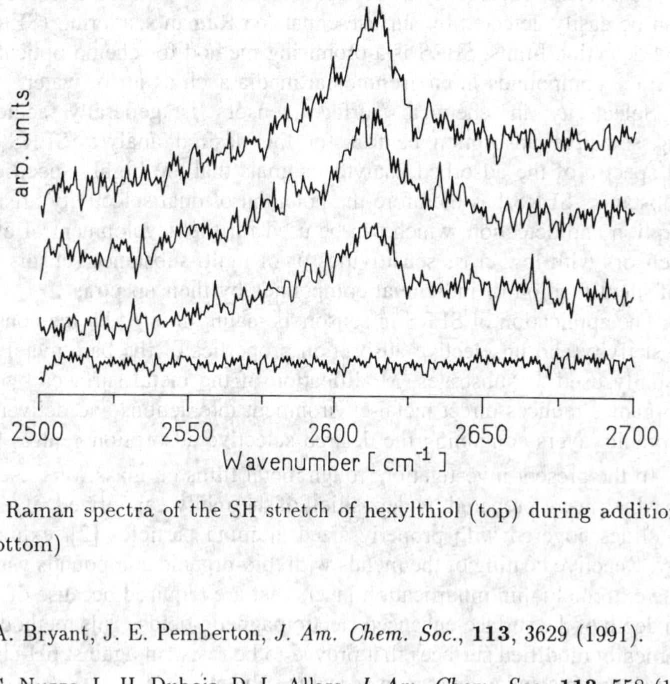

Figure 1: Raman spectra of the SH stretch of hexylthiol (top) during addition of silver colloid (bottom)

1. M. A. Bryant, J. E. Pemberton, *J. Am. Chem. Soc.*, **113**, 3629 (1991).

2. R. G. Nuzzo, L. H. Dubois, D. L. Allara, *J. Am. Chem. Soc.*, **112**, 558 (1990).

3. S. Schneider, H. Grau, P. Halbig, P. Freunscht, U. Nickel, *J. Raman Spectrosc.*, **27**, 57 (1996).

DETECTION OF AROMATICS BY SURFACE-ENHANCED RAMAN SCATTERING AT THIO-ORGANICALLY MODIFIED SUBSTRATES

W. Hill*, B. Wehling, V. Fallourd and D. Klockow

Institut für Spektrochemie und Angewandte Spektroskopie (ISAS)
PO Box 101352, 44013 Dortmund, Germany

Submonomolecular layers of organic molecules adsorbed at rough metal surfaces can be easily detected by surface-enhanced Raman scattering (SERS). Due to these low detection limits, SERS is a promising method for chemo-optical sensing of organic trace compounds in environmental media such as air or water.

Selectivity in chemical surface sensors is generally achieved by establishing selective adsorption centres for the desired analyte. SERS delivers vibrational spectra of the adsorbed analyte - signals that are highly specific for the detected substance. SERS has therefore the potential of dual selectivity arising from both, adsorption and detection, which can be used for the development of extremely selective sensors with low cross sensitivities or of multi-substance sensors with the potential of identification of individual compounds by their spectra.

The application of SERS in sensors is mainly limited by the considerable chemical reactivity and unselective adsorption properties of the bare metal surfaces that are usually used as substrates. Modification of the metal surfaces by reaction with thio-organics reduces direct metal-environment interactions and delivers stable, ultrathin organic layers containing the desired selective adsorption centres [1].

In the present investigation, rough metal films on glass were used. These substrates, which were prepared by deposition of thermally vapourized silver or gold onto glass slides covered with properly sized alumina particles [2], exhibit stable morphology. Reactive coating of the metals with thio-organic compounds was used to generate the extremely thin modification layers that are required because of the short penetration depth of the surface-enhanced electromagnetic fields. This method delivers stable organically modified surfaces that proved to be resistant against pH-changes in aqueous environments and against many organic solvents.

Thiol-derivatized cavity-like molecules are promising surface-modifiers for SERS sensor substrates because they can form inclusion complexes with various organic guest molecules. The complexation is frequently reversible and depends on shape, size and chemical structure of the adsorbed species, thus delivering the desired selectivity. Furthermore, the interaction between surface-modifier and analyte may be stoichiometric. Therefore, SERS bands of the immobilized host molecules can serve as internal standards for relative coverages with analyte molecules. Two surface-modifiers with different cavity size and chemical structure have been investigated: *p-tert*-butylcalix[4]arenetetrathiol (BCAT) and heptakis(6-thio-6-deoxy)-β-cyclodextrin (TCD).

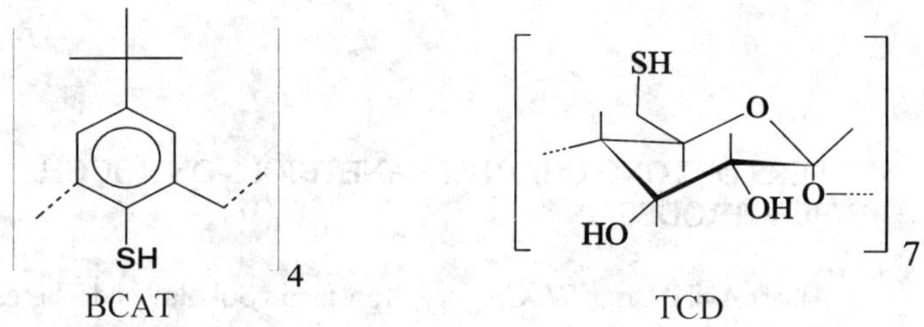

Fig. 1. Structures of thiol-derivatized substances used for modification of metal SERS substrates.

BCAT-coated substrates reversibly adsorb benzene and some of its chloro and methyl derivatives from aqueous solutions [3] and the gas phase. SERS-spectra of coated substrates with adsorbed analytes are apparently superpositions of spectra of immobilized BCAT and of the analytes. Since the band positions of the aromatics clearly depend on the substituents, SERS spectra are suited for substance identification and analysis of mixtures.

The analyte SERS peak intensities of the substances investigated depend linearly on the solution concentration and do not indicate saturation of surfaces up to the solubility limits. The dynamic range is about two orders of magnitude. This range is limited at higher concentrations by solubility and at lower concentrations by the shot noise limit of the continuous background signals from the substrates. Larger dynamic ranges may be achieved for substances with stronger affinity to the surface and with substrates exhibiting smaller background signals.

Detection limits for aromatics in water were between 5 and 100 μM. Measurements of the temperature dependence of the SERS signals showed that the adsorption enthalpy for aromatics within the small BCAT cavity was rather low, indicating van der Waals interaction.

TCD exhibits a larger cavity than BCAT, and it was therefore expected that TCD-coated substrates may complex with aromatics more efficiently than the BCAT-coated ones. Experiments on the adsorption of benzene and its chloro and alkyl derivatives from the gas phase at TCD substrates showed that some of the complexes formed are quite stable: desorption under pure nitrogen took several days in some cases. Such substrates may therefore be useful as integrating sensors and allow, for instance, the detection of lasting exposure to low concentrations.

Thanks to C. G. Gibbs and C. D. Gutsche for the synthesis of BCAT, and to the Ministerium für Wissenschaft und Forschung des Landes Nordrhein-Westfalen, the Bundesministerium für Bildung, Wissenschaft, Forschung und Technologie, and the European Commission (contract CHRXCT920039) for financial support.

1. K. Carron, L. Peitersen and M. Lewis, Environ. Sci. Technol. 26, 1950 (1992).
2. J. M. Bello, D. L. Stoke and T. Vo-Dinh, Appl. Spectrosc. 43, 1325 (1989).
3. W. Hill, B. Wehling, C. G. Gibbs, C. D. Gutsche and D. Klockow, Anal. Chem. 34, 3187 (1995).

SERS OF LONG CHAIN ALKANETHIOLS ON ROUGHENED METAL ELECTRODES

Albert Avila, Brian W. Gregory, Konstantin Sokolov and Therese M. Cotton*

Department of Chemistry and Ames Laboratory/USDOE, Iowa State Univeristy, Ames, IA 50011

Surface-enhanced Raman scattering (SERS) spectroscopy has been used in an effort to characterize self-assembled monolayers (SAM) of long chain alkanethiols on Au and Ag electrodes [1-3]. As expected, several bands are observed in the mid-frequency region arising from υ(C-S), υ(C-C) and r(CH_3) modes. However, a series of interesting new SERS bands have also appeared for these long chain alkanethiols on both Au and Ag roughened electrodes [3].

An excitation study indicates that these new bands are observed upon excitation in the red, from approximately 630 to 710 nm for octadecanethiol monolayers on roughened Au. Outside this excitation range, the SERS spectra of the alkanethiols appear as previously reported in the literature [1,2]. Figure 1 displays the spectra of an octadecanethiol monolayer on roughened Au at three different excitation wavelengths. The new bands are observed from approximately 590 to 950 cm^{-1}. These have only been observed for alkanethiols with chains consisting of 10 or more carbon groups. Many different straight chain alkanethiols, both methyl and hydroxyl-terminated versions, from many different sources all exhibit the same characteristic intense, sharp bands. In addition, isotopic labeling studies indicate that the new bands are not standard group frequency vibrations expected for this system; for example, replacement of the α-C with ^{13}C results in no shift in any of the resonantly enhanced modes.

Several experiments have been conducted to determine if impurities are the possible source of the newly observed SERS bands. Dialkyl disulfides were synthesized from the corresponding thiols as a means to purify the samples and remove non-thiol impurities; the subsequent SERS spectra of the disulfide monolayers were identical to the thiol spectra. ^1H NMR spectra of highly concentrated thiol SAM deposition solutions reveal no aromatic sulfur compound impurities. GC/MS samples of the monolayer deposition solutions indicate minor impurities, mostly of straight chain alkanes and alcohols. These latter impurities should not adsorb strongly to the metal surfaces compared to the thiols and therefore should not contribute significantly to the observed SERS spectra. Roughened bare Au and Ag electrodes (prepared in exactly the same manner as the thiol-coated electrodes except for thiol exposure) exhibit none of the new bands. Therefore, it seems unlikely that impurities are the source of the bands.

Experimental results from excitation profiles, chain length dependence studies, isotopic-labeling studies and electrochemical deposition/stripping experiments will be presented. In addition, the possible the origin of these resonantly enhanced Raman modes will be discussed.

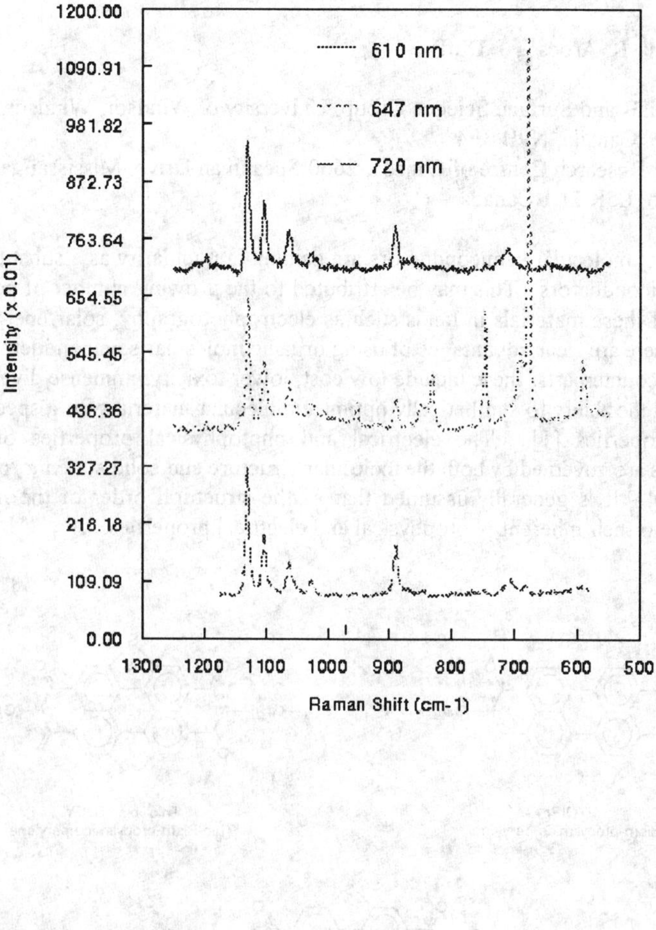

Figure 1. SERS spectra of Octadecanethiol Monolayer on Roughened Au at different excitation wavelengths

1. M.A. Bryant and J.E. Pemberton, J. Am. Chem. Soc., 113, 3629(1991)
2. M.A. Bryant and J.E. Pemberton, J. Am. Chem. Soc., 113, 8284(1991)
3. A. Avila, B.W. Gregory, K. Sokolov and T.M. Cotton, in preparation

RESONANCE AND SURFACE ENHANCED RESONANCE RAMAN SPECTRA OF THIO-PERYLENE DERIVATIVES

A. Kam[1], R. Aroca[1*], J. Duff[2]

[1.]Materials and Surface Science Group, University of Windsor, Windsor, Ontario, Canada, N9B 3P4
[2.]Xerox Research Centre of Canada, 2660 Speakman Drive, Mississauga, Ontario, L5K 2L1 Canada

Organic molecular semiconductors are gaining in popularity as a substitute for inorganic semiconductors. This may be attributed to the growing number of potential applications of these materials in fields such as electrophotography, solar energy, and electronics. There are clear advantages of using organic molecular semiconductors over their inorganic counterparts, these include low cost, lower toxicity, immense diversity of substances and the ability to synthetically optimize molecular materials for a specific set of desired properties [1]. The electrical and photophysical properties of these semiconductors are governed by both the molecular structure and solid state morphology of the material. It is generally assumed that as the structural order of the material increases so too their inherent photophysical and electrical properties.

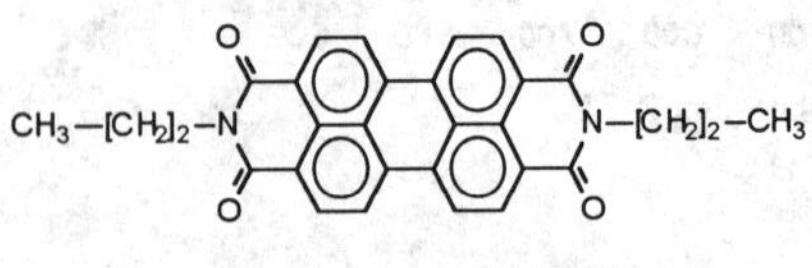

PTDIPr
Bis(n-propylimido)perylene

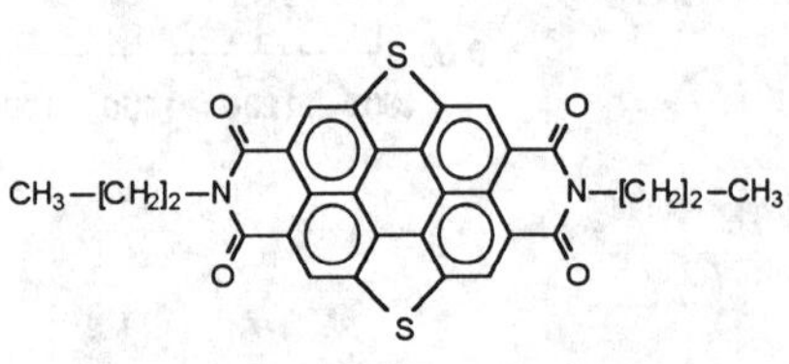

Thio- bisPTDIPr
Thio-Bis(n-propylimido)perylene

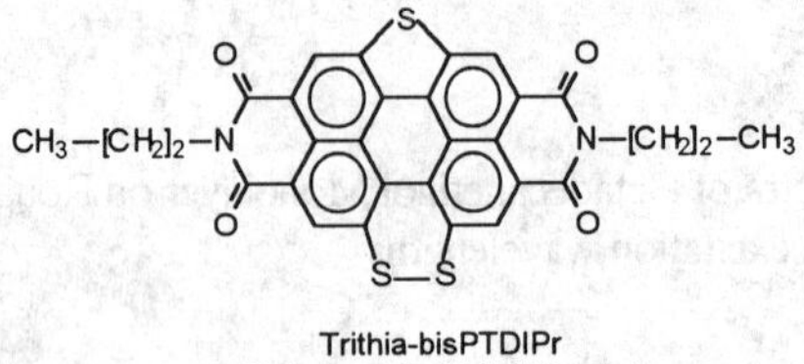

Trithia-bisPTDIPr
Bis(n-propylimido)trithiaperylene

Figure #1: Perylene derivatives: bisPTDIPr, Thio-bisPTDIPr, and Trithia-bisPTDIPr

This investigation reports three perylene derivatives:bis(n-propylimido)perylene(BisPTDIPr), thiobis(propylimido)perylene (Thio-bisPTDIPr), and trithiabis(propylimido) perylene (Trithia-bisPTDIPr). Resonance Raman scattering (RRS) and Surface-enhanced resonance Raman scatering (SERRS) were used to charcterize and monitor the interaction between the thio-perylene derivatives and silver or gold surfaces. The RRS of the two thio derivatives are shown in Figure 2. The Raman spectra of these materials are considerably different from the perylene tetracarboxylic series. Langmuir-Blodgett (LB) monolayers of the materials were also formed. Raman and infrared studies of these films will be reported in detail and effect of thio groups is discussed.

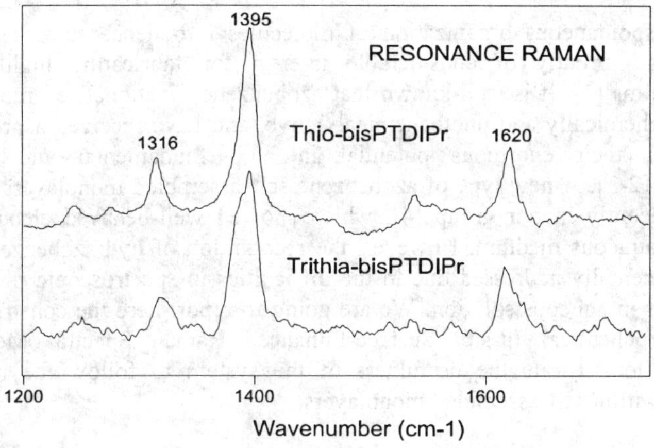

<u>Figure #2</u>: RRS of Thio-bisPTDIPr and Trithia-bisPTDIPr

Perylenes are rigid, planar, conjugated molecules. The rigidity of these molecules limits the degree of non-radiative damping effects of fluorescence due to vibrations. These properties make these molecules ideal for potential applications in organic molecular fluorescence devices. Initial investigations done indicated that the alkyl perylene derivatives show extensive fluorescence properties [2]. In our work we perform a comparative investigation into the effect of the introduction of sulphur atoms onto the initial molecular framework of peylenes. We discuss the ramifications of this structural modification on the solid state morphology of our films and their spectroscopic properties.

1. Gregg, B., J. Phys. Chem., **100**, 852, (1996)

2. Burghard, M.; Fischer, C.; Schmelzer, M.; Roth, S.; Hanack, M.; Gopel, W., Chemistry Materials, 7, 2104, (1995)

ELECTROCHEMICAL *IN-SITU* MICRO-SERS INVESTIGATIONS OF SELF-ASSEMBLED MONOLAYERS ON GOLD

T. Zhu*, H.Z. Yu, Y.Q. Wang, S.M. Cai, and Z.F. Liu
Center for Intelligent Materials Research(**CIMR**), College of Chemistry and Molecular Engineering, Peking University, Beijing 100871, China

The spontaneous organization of molecules into dense state, termed self-assembly, is currently of considerable interest for fabricating highly ordered monolayer films [1]. It is well-known that azobenzene functionalized molecules are both electrochemically and photochemical active, and have received a great deal of attention for their enormous potential in both fundamental and theoretical perspectives [2-3]. A new type of azobenzene self-assembled monolayers have been fabricated recently in our group [4], which showed well-behaved electrochemical response in aqueous medium. However, the recognition of hydrazobenzene has not been experimentally addressed due to the difficulties in spectroscopic measurement of monolayer in aqueous solution. We are going to report here the construction of a novel Electrochemical In-situ Surface-Enhanced Raman Spectral Measurement system and demonstrate the usefulness of the system to follow the reaction of azobenzene within self-assembled monolayers.

The electrochemical in-situ SERS system used in the present work consists of a Renishaw Ramascope spectrometer(System 1000), an air-cooled He-Ne laser (632.8nm), which delivered at the sample point no more than 5mW, and the electrochemical measurement system. The spectroelectrochemical cell is similar to the one we have employed elsewhere [5]. The working electrode used for SERS was a 1.0mm diameter gold disk sheathed in Teflon. We prepared azobenzene self-assembled monolayers by the chemisorption of C0AzoC2SH(shortly for C_6H_5-N=N-C_6H_4-C(O)N(H)CH_2CH_2SH) molecules onto the gold electrodes, which have been skid polished and pretreated by ORC method. Electrolyte solutions were prepared from suprapure reagent and Mill-Q water. All the potentials were reported with respect to the Ag/AgCl saturated KCl reference electrode.

Figure 1 shows the cyclic voltammogram of C0AzoC2SH SAM modified gold electrode in B-R buffer, 0.1 M $NaClO_4$, at pH 5.0. The separation between anodic peak and cathodic peak is very large, as similar as those observed in C2AzoC2SH SAM [4]. On the other hand, the film presents a stable and reproducible electrochemical behavior under the current experimental conditions.

According to the electrochemical experiments, a reversible redox process can be conducted. SERS spectra are successfully acquired by using the home-constructed electrochemical *in-situ* SERS system to follow the potential sweep between +0.25V and -0.35V, while the cyclic voltammetry is performed at a scan rate of 5mV/s (Figure 2). As the potential goes negative direction, the decrease and final diminution of the bands 1140 cm^{-1} (for benzene-N=) and 1460 cm^{-1} (-N=N-) in the in-situ spectra demonstrate that the azobenzene group was reduced into the hydrazobenzene form which has no strong bands in this region [6]. Another remarkable feature in these spectra is that by oxidation(when the potential goes positive direction), the characteristic bands mentioned above excellently recover to the original form. These observations confirmed that azobenzene molecules in the monolayer do undergo a reversible electrochemical process.

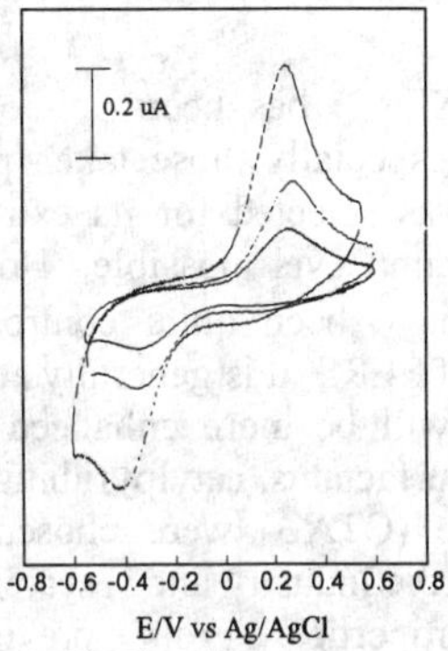

Figure 1. Cyclic voltammograms of Azo SAM obtained on gold electrodes at 100, 200, 500 mV/s.

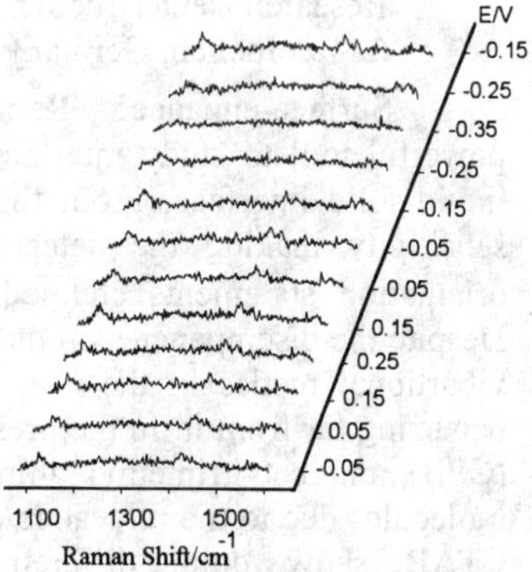

Figure 2. Sequence of SERS spectra during linear sweep voltammetric reduction of azo SAMs.

In summary, the novel electrochemical in-situ SERS system we have constructed can be used in perform real-time analysis of reactions in ultrathin films on electrodes. This made us successfully followed the redox process of the azobenzene/hydrazobenzene couple within functionalized self-assembled monolayers.

This work is jointly supported by the National Climbing Program and the NSF of China.

1. A. Ulman, An Introduction to Ultrathin Organic Films: From Langmuir-Blodgett to Self-Assembly, Academic Press, San Diego, CA, 1991.
2. E. Laviron and Y. Mungnier, J. Electroanal. Chem. 111, 337(1980).
3. Z.F. Liu, K. Hashimoto and A. Fujishima, Nature 347, 658(1990).
4. H.Z. Yu, Y.Q. Wang, J.Z. Cheng, J.W. Zhao, S.M. Cai, H. Inokuchi, A. Fujishima, and Z.F. Liu, J. Electroanal. Chem. 395, 327(1995).
5. Z.Q. Tian, Y.X. Chen, B.W. Mao, C.Z. Li, J. Wang, and Z.F. Liu, Chem. Phys. Lett. 247, 224(1995).
6. M. Jamkowshi, A. Kudelski, J. Bukowska, and K. Jackowska, J. Electroanal. Chem. 385, 177(1995).

COMBINATION OF MICRO-SERS SPECTROSCOPY AND AB INITIO CALCULATION TO DETERMINE THE ADSORPTION GEOMETRY OF CATIONIC SURFACTANTS ON SILVER ELECTRODES.

A. Tarazona*, S. Kreisig , E. Koglin, and M.J. Schwuger

Institute of Applied Physical Chemistry (IPC)
Research Center Juelich (KFA)
52425 Juelich, Germany

Surface-enhanced Raman scattering (SERS) has been pointed out as a powerful tool to study interfacial processes[1], especially those taken place at the liquid-solid interface. From the beginning it was selected for its extremely high sensitivity making the detection at pico-gramm level possible. However, the orientation statements claimed using SERS have been quite controversial [2]. Despite the discrepancies on the interpretation of SERS, it is generally accepted that vibartional modes in the near of the surface will be more enhanced than those occurring far from it. In the present work, two surfactants, cetylpyridinium bromide (CPB) and cetyltrimethyl ammonium bromide (CTAB) were chosen as model molecules due to the relevance of surfactants in the industry and research. CPB and CTAB show quite different adsorption properties. From previous SERS investigations was suggested that the CPB adsorption takes place with the pyridinium head group to the surface, leaving the hydrocarbon chain directed away from the surface. In contrast, the CTAB molecule is adsorbed with the ionic head group pointing towards the surface and the hydrophobic tail is running parallel to the surface[3]. In order to obtain more detail information for determining the adsorption geometry, experimental and computational methods were combined.

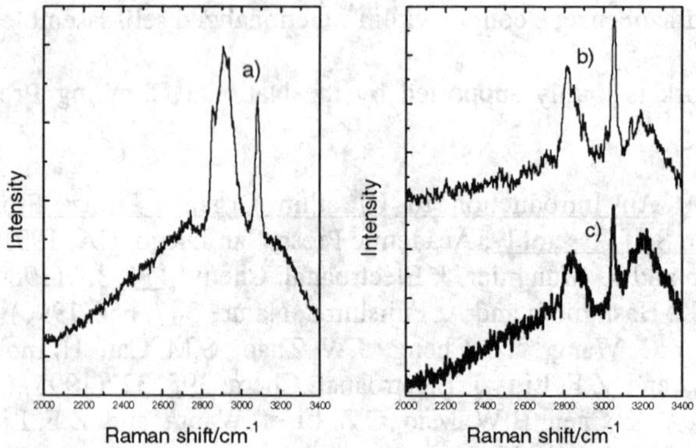

Fig. 1. Surface-enhanced Raman spectra from a) CPB, b) Pyridine and c) Mixture of pyridine and 1-Br octane (1:5). All spectra were collected at u=0.8 V (vs. sce) using a roughened Ag electrode. All solutions were 1mmol/l in 0.1 mol/l KCl.

The SERS spectra were measured from CPB and CTAB on roughened silver electrode.The corresponding spectra in the CH spectral range are shown in Fig.1a and 2c, respectively.For comparison purpose the head groups of both surfactants were modeled by pyridine in the case of CPB and by tetramethyl ammonium bromide (TMA) in the case of CTAB, Fig.1b and Fig.2a, respectively. These substances were also measured under similar conditions as for the surfactants; to this solutions 1-Br octane solution was added to this head group model substances in order to investigate the interaction of the head groups with the alkyl chains. Fig.1c shows the spectrum from the mixture of pyridine and 1-Br octane. It can be observed that no mayor changes occurred after addition of 1-Br octane. Contrary to this, when 1-Br octane was added to the TMA solution a strong band appeared at around 2858 cm^{-1}. Fig.2d shows the spectrum resulting from the summing spectra presented in Fig.2a and 2b. Here can be observed that the spectrum in Fig.2d reproduces the main spectral feature corresponding to the CTAB. From these results, it can be assured that the

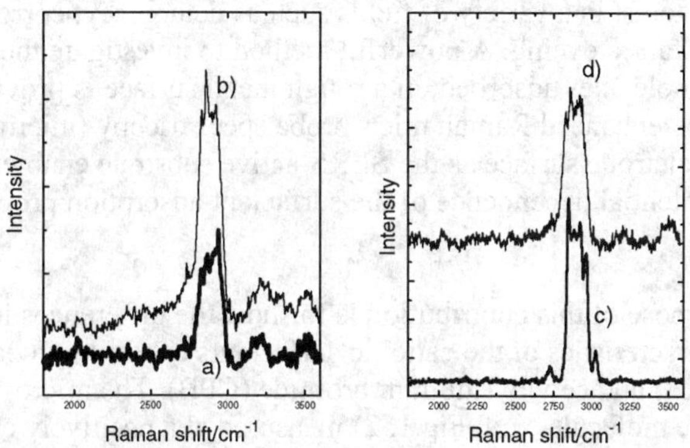

Fig. 2. Surface-enhanced Raman spectra from a) TMA, b) Mixture of TMA and 1-Br octane (1:1) c) CTAB and d) Spectra resulting from adding curves a) and b). All spectra were collected at u=-0.8 V (vs. sce) using a roughened Ag electrode, all solution were 0.1 mmol/l in 0.1 mol/l Kcl.

CTAB head group in the first layer serves as an intermediate for the alkyl chain to come close to the electrode surface. In contrast, in CPB the pyridinium head acts pushing the alkyl chain away from the electrode surface. Since a faithful assignment of the vibrational modes are necessary for the interpretation of SERS spectra, ab initio calculations were carried out using Hartree-Fock methods to compute the frequencies and intensities of the surfactants. As validation for the frequencies obtained, density functional theory methods were employed to compare the results for the head groups and short alkyl chains with the corresponding calculations from Hartree-Fock. Additionally, CPB and CTAB were marked using deuterium at the head groups and at the terminal methyl group, and the SERS investigations and calculations are compared with the undeuterated results.

[1] E. Koglin and M.J. Schwuger, Faraday Discuss., 94, 213 (1992).

[2] A. Otto, I. Mrozek, H. Grabhorn and W. Akemann, J. Phys. Condensed Matter, 4, 1143 (1992).

[3] E. Koglin, A. Tarazona, S. Kreisig, and M.J. Schwuger, Colloids Surfaces, in press.

DIFFERENT ADSORPTION PROPERTIES OF CATIONIC SURFACTANTS STUDIED BY FT-SERS MICROPROBE SPECTROSCOPY

S. Kreisig*, A. Tarazona, E. Koglin, and M.J. Schwuger
Institute of Applied Physical Chemistry (IPC)
Research Center Juelich (KFA)
52425 Juelich, Germany

Surfactants are of great importance in many different kind of both industrial and scientific applications. The mechanism of adsorption of these surface active substances at the solid/liquid interface is therefore of considerable interest in a variety of fields such as detergency, corrosion inhibition, and fabric dyeing. A powerful method to investigate the first monolayer of molecules adsorbed on a rough metal surface is provided by the NIR-FT surface-enhanced Raman microprobe spectroscopy (micro-SERS). Using a silver elctrode surface as the SERS-active substrate enables one to examine the potential dependence of the surfactant adsorption properties in detail.

The purpose of this contribution is to show the differences in the adsorption characteristics of the cationic surfactants cetyltrimethylammonium bromide (CTAB) and cetylpyridinium bromide (CPB). The micro-SERS spectra of these molecules (c.f. Fig.1, 2) measured at a positively charged silver electrode (U= -0.35 V vs. saturated calomel electrode (SCE)) reveal that inspite of the positive charge located at the nitrogene atoms of the headgroups a strong adsorption to the electrode surface is possible. This can can be explained by coadsorption of bromide ions used as electrolyte in the electrochemical cell.

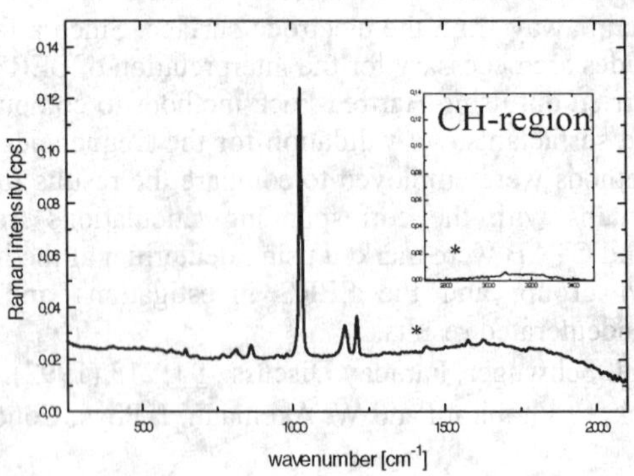

Fig. 1: NIR-FT SERS microprobe spectrum of CPB (U= -0.3 V vs. SCE)

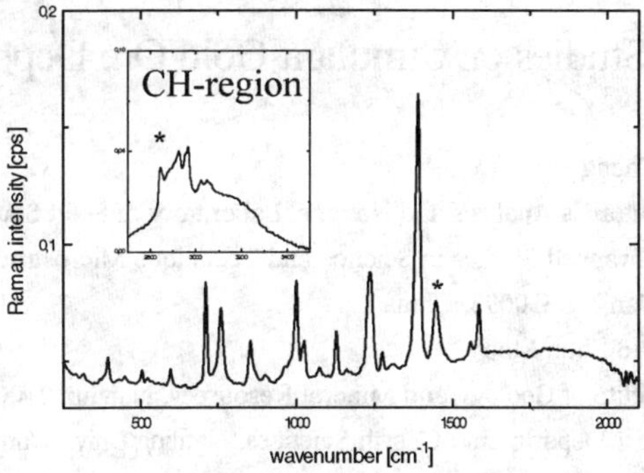

Fig. 2: NIR-FT SERS microprobe spectrum of CTAB (U= -0.3 V vs. SCE)

Investigating the whole potential range from -0.25 V up to -1.15 V one can see by the comparison of some chractaristical surfactant vibrations (1028 cm^{-1} and 1626 cm^{-1} for CPB and 712 cm^{-1} and 2843 cm^{-1} for CTAB) that CPB behaves as expected, i.e. the intensity of the SERS signals increases while changing the potential from -0.25 V up to -0.85 V. Beyond this point the signals start to decrease due to desorption of the surfactant and the Br^- counter-ions from the electrode surface. In contrast to this observations the maximum of the CTAB scattering intensity lies at about -0.45 V which means that the strongest adsorption occurs at a positively charged surface. These results are shown in Figure 3a,b.

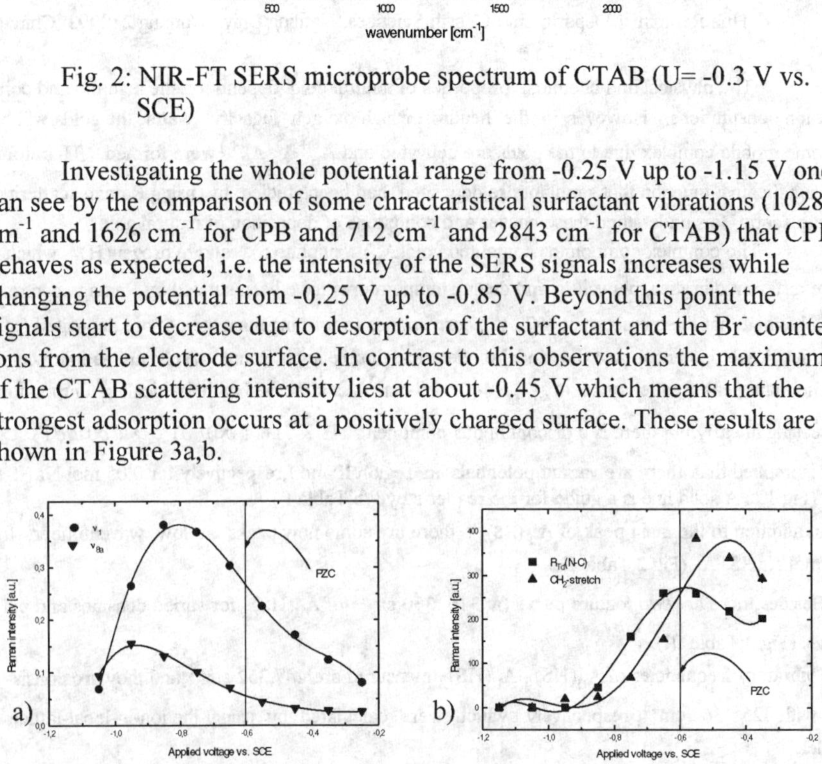

Fig. 3: Potential dependence of the signal intensity of some characteristic surfactant vibrations (a: CPB, b: CTAB)

A very important result of our investigation is indicated in the zoom windows (CH region) of the SERS spectra above. According to the short range of the SERS enhancement the occurrence or the absence of signals related to chain vibrations (marked with a *) can deliver information about the surfactants state of adsorption. Here it could be shown that in case of CPB only the headgroups attached to the surface (no CH_2 vibrations are detectable) while in case of CTAB both head- and tailparts lay flat on the surface.

Raman Studies on Simulant Gold Ore Deposition

Guang-Xu Cheng
Center of Materials Analysis and National Laboratory of Solid State Microstructures, Center for Advanced Studies in Science and Technique Microstructures, Nanjing University, Nanjing 210093, China
Cai Yuanji and Zhou Mao
Nanjing Insititute of Geology and Mineral Resources, Nanjing 210002 ,China
Hua Renmin Department of Earth Sciences, Nanjing Univ., Nanjing 210093 ,China

The physical and chemical properties of natural gold depend on the actomic and complex ion constitutions. However, in the heating or high oxygen fugacity water, the golds will be become mobile complex due to the golds are activated,and A_u^{+3}、A_u^{+1}were formed. Therefore, it is significative function that simulant ore deposited had been studied by using Raman scattering spectroscopy, for understand these causes and conditions of deposition of natural gold.

The complex compounds contented gold(A_u)、sulphur(S)、and hydrogen(H)、which were different densities, pH values and deaing temperatures have been studied by Raman spectra. It is obtained that the temperature (573 K) corresponds to forming main step of ore deposited,from Van der Waals` potential and the model of electrical double layer.Raman results are as following:

1.The main peak frequency i.e. ω_{Au-HS} (440 ± 15 cm^{-1}) is increased with the dealing temperature increasing linearly,but there is a discontinuous point near 573 K , i.e. $(\partial\omega/\partial T)_I > (\partial\omega/\partial T)_{II} > 0$, it is implied that there are variant potentials in region II and I,respectively,for 0.05 mol NaHS+ A_u (Fig.1,a: A solid line is a guide for the reader`s eye ,b;Table I).

2. In addition to the main peak of $A_u(HS)_4$, there are some new peaks at low wavenumber, for 0.1 mol NaHS+A_u (Fig.2;Table II).

3. Besides,there are two feature peaks (at 316 ,350 cm^{-1})of $A_u(HS)_2$,for varied densities and pH values (Fig.3;Table III).

The vibratory frequencies of $A_u(HS)_4$,$A_u(HS)_2$ measured are 447,352 cm^{-1},and they are consistent with 425 , 362 cm^{-1}, respectively , which were calculated by using the ion-in-local-lattice model.

Table I. Raman frequencies (cm^{-1}) of 0.05 mol NaHS, for varied temperatures.

Dealing Temp(K)	$A_u(HS)_4$ Cal.(cm^{-1})	$A_u(HS)_4$ Exp.(cm^{-1})	$A_u(HS)_2$+$A_u(HS)_4$ (cm^{-1})	*O-S-O band (cm^{-1})	HS^{-1} (cm^{-1})
300		434	802	/	/
473		447	806	/	/
573	425.4	440	810	/	/
673		452	808	1156	2332
773		464	812	1148	2372

*Ref. D.Maillard,M.Allavena and J.P.Perchard,Spectrochem Acta,31A 1523(1975).

Table.II Raman frequencies of 0.1 mol NaHS, for varied dealing temperatures.

Dealing Temperature.(K)	Low wavenumbers(cm^{-1})							$A_u(HS)_4$ (cm^{-1})		
300	109	/	148	162	178	218	244	441	491	/
473	/	128	144	/	184	219	232	451	468	483
573	/	/	148	158	184	214	230	425	/	484
673	/	122	146	/	174	212	/	431	/	482
773	/	/	/	/	/	210	/	418	476	/

Table.III Raman frequencies of NaHS, for varied densities and pH values.

Mol.& pH	$A_u(HS)_2$ (cm^{-1})		$A_u(HS)_4$ (cm^{-1})		$A_u(HS)_2$+$A_u(HS)_4$ (cm^{-1})	*$[O\text{-}S\text{-}O]^-$ band (cm^{-1})	? (cm^{-1})	HS^- (cm^{-1})
1	314	350	456	488	802	982	2224	2572
0.1	/	/	444	478	802	/	/	/
0.05	316	/	456	/	802	/	/	2792
0.01	456	/	/	472	804	/	/	/
pH = 1	/	/	448	/	804	986	/	/
pH = 6	/	/	454	490	806	/	/	/

*Ref.D.E.Millgan and M.E.Jacox,J.Chem.Phys., 55 1003(1971).

Notice: Raman shift(cm^{-1})-abscissa and Intensity(a.u)-ordinate are omitted.

Fig.1 a

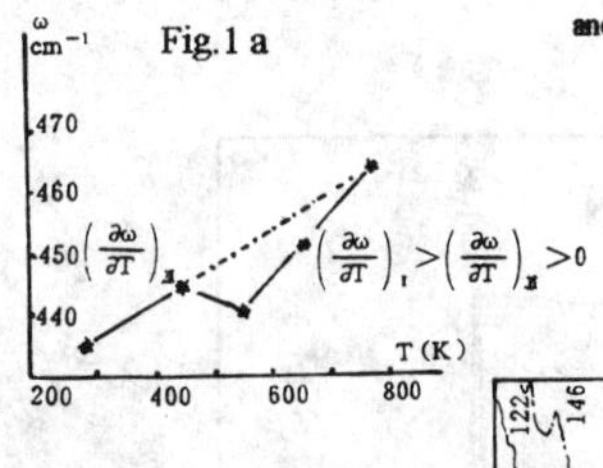

b

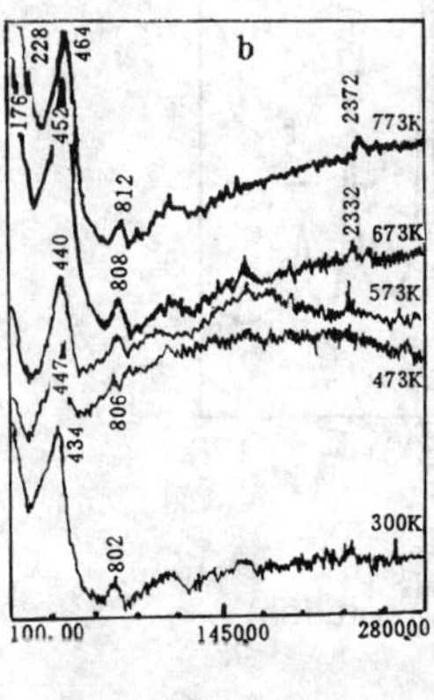

Fig.2

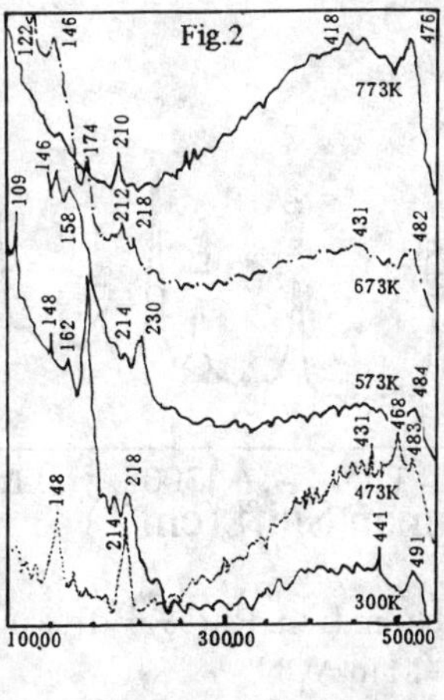

Fig.3

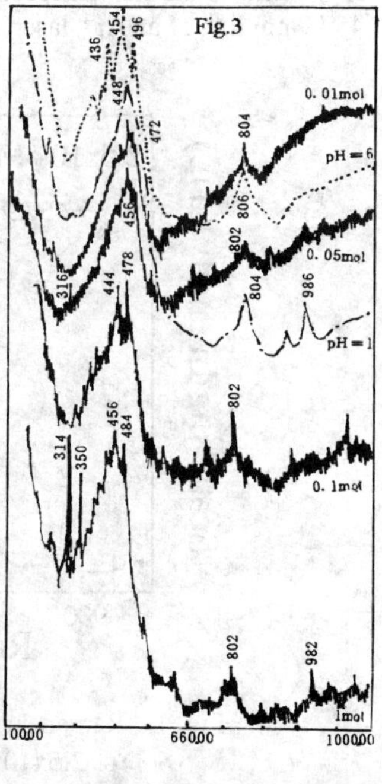

SURFACE ENHANCED RAMAN SCATTERING ON SEMICONDUCTORS.

Lucia G. Quagliano and Daniela Orani

I.M.A.I.- C.N.R., Area della ricerca di Roma, P.O.Box 10,
via Salaria km 29.300, 00016 Monterotondo Scalo, Roma, Italy

The Surface Enhanced Raman Scattering (SERS) effect is based on the enhancement of the Raman scattering cross section by several orders of magnitude of atomic and molecular species adsorbed on rough metal surfaces [1]. It is now well established that strong electric field near surfaces associated with the excitation of surface plasmon modes are responsible for the Raman enhancement on surfaces roughened by various method and on small particles including island and colloidal particles.

We present the observation of SERS from molecules adsorbed on GaAs surface. Recently, we have obtained SERS spectra of molecules adsorbed on GaAs surface onto which a silver island film has been deposited [2]. In this work we demonstrate that the SERS effect can also be obtained by using a chemical etching process.

We used tris (bipyridine) ruthenium (II) , $Ru(bpy)_3^{2+}$, as test molecule. Fig. 1 shows the Raman spectrum of the molecule free in solution obtained with the 457.9 nm line of an Ar^+ laser.

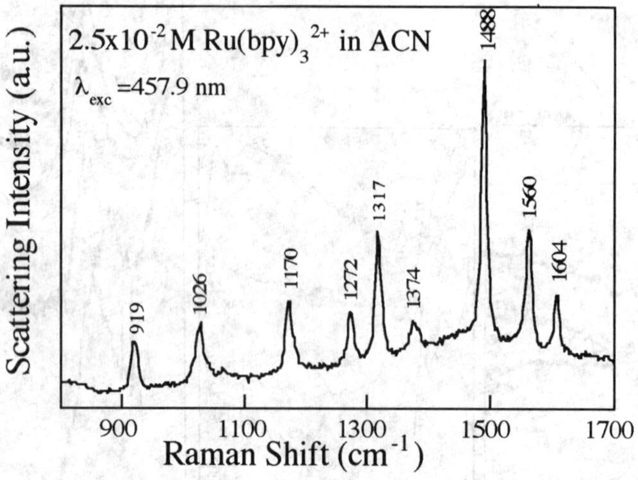

Fig. 1. Raman spectrum from $Ru(bpy)_3^{2+}$ free in a solution of $2.5x10^{-2}$M in ACN.

The GaAs samples have been subjected to an anisotropic etching containing concentrated HNO_3 and a solution of $2.5x10^{-2}$ M $Ru(bpy)_3^{2+}$ for 15 minutes. The GaAs samples with the Ag island film have been immersed in the solution of $2.5x10^{-2}$ M $Ru(bpy)_3^{2+}$ for several minutes. The Raman spectra have been performed at room temperature using the 514.5 nm line of an Ar^+ laser.

In Fig. 2 we report the SERS spectrum of $Ru(bpy)_3^{2+}$ molecule adsorbed on GaAs surface subjected to the chemical etching process. This Raman spectrum is similar to that obtained evaporating an Ag island film onto the GaAs surface. No Raman scattering could be detected from the GaAs surfaces without either the Ag film or the etching procedure. In both cases the Raman enhancement is due to the excitation of surface plasmon polaritons introduced by the surface roughness of the Ag island film as well as by the etching procedure.

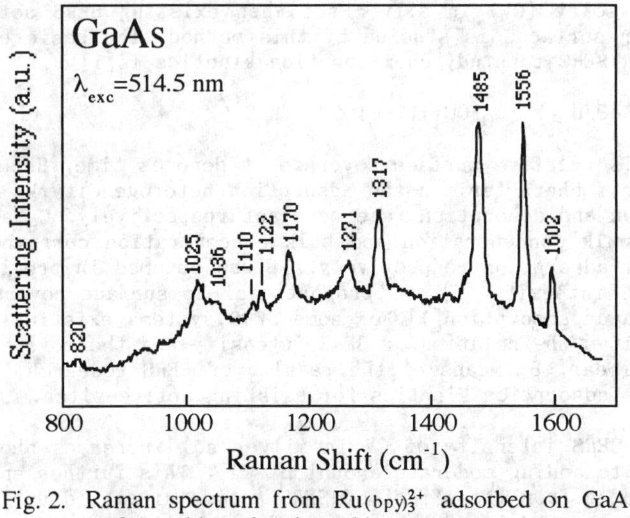

Fig. 2. Raman spectrum from $Ru(bpy)_3^{2+}$ adsorbed on GaAs surface subjected to the etching treatment.

We have demonstrated that adsorbed species on semiconductor surface can be detected by SERS spectroscopy by using a SERS-activation procedure based on the use of a chemical etching process.

The authors would like to thank P. Cafarelli for preparing the solution of $Ru(bpy)_3^{2+}$. One of us (D. O.) would like to acknowledge the CNR-Italy for its fellowship of the "Programma del Fondo Sociale Europeo".

1. for a review see R. K. Chang and T. E. Furtak (eds.), Surface Enhanced Raman Scattering , Plenum, New York 1982.
2. L. G. Quagliano, B. Jusserand and F. R. Ladan, physica status solidi (a) 152, 219 (1995).

STUDY OF CHEMISORPTION KINETICS EXISTING SERS ACTIVE-SITES

Z.F.Li[1*], S.Y.Fu[1], and P.X.Zhang[2]

[1]Department of Electronic Engineering, Daqing petroleum Institute, Heilongjing, Anda, 151400, China
[2]Institute of Physics, Chinese Academy of Sciences, Beijing, 100080, China

In our previous paper[1] an effective method was developed to study chemisorption kinetics on solid surfaces by means of SERS, (Surface Enhanced Raman Scattering). Now the adsorption kinetics of Crystal violet (CV) in silver sol when existing SERS active-sites on silver surfaces is studied by this method. The basic expression for using SERS to study chemisorption kinetics is[1]

$$d\theta/dt=k_a(C_0-\theta C_1)(1-\theta)^{1/m} - k_d\theta^{1/m} \qquad (1)$$

where θ is relative surface coverage. t denotes time. Heterogeneity parameter m characterizes the adsorption heterogeneity.k_a and k_d is adsorption and desorption rate constant respectively. C_0 and C_1 are initial bulk concentration and bulk concentration corresponding to monolayer adsorption respectively. It was assumed in previous paper that SERS intensity (I) was proportional to surface coverage θ for submonolayer adsorption[1].For some SERS systems existing so-called active-sites,the relation of SERS intensity and the coverage is not simple linear. By means of the results of SERS research, we should study the adsorption kinetics for existing active-sites.

SERS intensity of CV in silver sol increases about 10^2-fold due to adding moderate amount of Cl^-. This further enhancement mainly comes from the effect of SERS active-sites,a detail analysis will be presented in another paper. Now we discuss the influence of chloride ions to adsorption energy of CV. Adsorption energy (U) can be estimated by the dependence of CV concentration (C) and SERS intensity (I):

$$IA= \frac{NK(C-IA)}{1+K(C-IA)}, \quad K=C(H_2O)^{-1}\exp(-U/RT) \qquad (2)$$

where $C(H_2O)$ is the molarity of water. R=8.31 J/K, T=290K. A is a constant.n and N are the concentrations of adsorption molecules and adsorbed-sites on surfaces respectively, and SERS intensity comes from molecules on these adsorbed-sites.

Without existing Cl^-, the experimental data (C,I) are as follows:

(5.5×10^{-7}, 70) (9.2×10^{-7}, 95) (1.2×10^{-6},141)
(1.6×10^{-6},182) (2.1×10^{-6},225) (3.1×10^{-6},275)
(4.0×10^{-6},310)

By using the data in Eq.(2),we obtain:$U\simeq-44.6$ kJ/mol, $N\simeq2.5\times10^{-6}$M. When the concentration of chloride ions is about 10^{-3}M, the experimental data (C,I) are as follows:

(5.0×10^{-10}, 38) (8.8×10^{-10}, 57) (2.5×10^{-9}, 75)
(4.2×10^{-9}, 96) (4.7×10^{-9}, 108) (7.5×10^{-9}, 114)
(2.1×10^{-8}, 120) (3.2×10^{-8}, 130) (4.7×10^{-8}, 136)
(9.0×10^{-8}, 146) (1.2×10^{-7}, 154)

The results: $U\simeq-59.7$ kJ/mol, $N\simeq5.2\times10^{-9}$ M.

According to above results, When existing Cl^-, the adsorption energy of CV increases,but the concentration of adsorbed-sites declined by about 10^3. Therefore,SERS intensity of CV mainly originates from molecules on a few sites, these sites are so-called SERS active-sites.Because SERS intensity of CV increases 10^2-fold due to active-sites formed by means of chloride ions, the Raman section of a molecule on active-site is larger 10^5 than that on common sites. Therefore, SERS intensity of CV is proportional to the number of molecules on active-sites. Table 1 gives kinetic parameters of CV while existing chloride ions.

TABLE 1. Kinetic parameters of CV responding to various Cl^- concentration. (CV concentration: 2.5×10^{-5}M)

concentration of Cl^- (M)	m	k_a $M^{-1}s^{-1}$	k_d s^{-1}
1×10^{-4}	1.0	1.7×10^5	0
1×10^{-2}	1.0	5.1×10^5	0
5×10^{-2}	1.0	7.8×10^4	0

The parameters in Tab.1 describe the kinetic process of molecules on active-sites. m=1 shows the adsorption of CV on silver surfaces is homogeneous, all active-sites are identical. The order of adsorption rate constants on active-sites is approximately equal to that on common sites when without active-sites[2]. Since the concentration of active-sites is much smaller than that of common adsorption sites, the adsorption of molecules on active-sites would be easier than that on common sites. The desorption of molecules on active-sites is very difficult because of the high adsorption energy, it is identical with the result of $k_d=0$.

[1] Y.H.Liu,S.Y.Fu and P.X.Zhang, Acta Physic Sinica, 42(1993),157 (in Chinese)
[2] S.Y.Fu and P.X.Zhang, Acta Physic Sinica(Overseas Edition), 4(1995),721(in English)

SURFACE-ENHANCED ENERGY TRANSFER EFFECT BETWEEN MOLECULES

D. H. Pan[1]* Y. X. Huan[2]
[1]Department of Chemistry, The University of Hong Kong
Pokfulam Road, Hong Kong
[2]Department of Physics, Shaanxi Normal University
Xi' an 710062, P. R. of China

In recent years considerable research effort has been directed at undertanding a variety of optical processes that occur near a solid particle or rough surface. Much of this research is motivated by the existence of anomalous electromagnetic enhancement effect[1]. One such process of interest is the enhaned energy transfer between donor and acceptor molecules. The process of energy transfer between molecules near a solid particle was first studied in theoretically by Gerstern and Nitigan[2]. It was found that energy transfer rate between molecules may be enhanced by many orders of magnitude when the molecular transition frequencies lie in the vicinity of the resonance frequency of the particle and when the particle possesses sharp features. However, the presence of the particle also enhance the non-radiative decay of the acceptor which would tend to compete with the energy transfer. This may not lead to increased acceptor molecular radiation. For this reason, up to now no experiment work of enhanced energy fransfer between molecules on surface has been reported. Here we present the observation of sensitized fluoresence of acceptor and quenched fluoresence of donor due to enhanced energy transfer between 2,2'-cyanine and it's J-aggregate molecule adsorbed on silver colloid in the first time. The enhancement factor of energy nansfer rate between mloecules is estimated.

In previous work [3] we have shown the J-aggregates of 2,2' cyanine are characteried by an absorption band, called the J-band, at 576nm. J-aggregates are formed at 2,2' cyanine/kI solution. The SERS speectra demonstrate J-aggregate and 2,2' cyanine monomer are co-adsorbed on silver colloid. In this work 2,2' cyanine monomer and it's J-aggregate are chosen as the donor and the acceptor respectively. The experiment method was described in Ref. 3. Fig. 1 shows the fluoresence spectra of 2,2' cyanine/kI aqueous solution and 2,2' cyanine/kI adsorbed on silver colloid. The excitation wavelength is 488nm. In Fig. 1A, the fluoresence spectrum at 577. 6nm is due to 2, 2' cyaine monomer radiation, which overlaps somewhat with J-aggregate obsorption band at 576. 0nm. This characteristic feature may lead to process of non-radiative energy transfer effect between 2, 2' cyanine monomer and J-aggregate molecule. However, The J-aggregate fluoresence spectrum is not observed in Fig. 1A. When 2,2' cyanine/kI adsorbed on silver colloid, as shown in Fig. 1B, a new shonger emission band at 598. 9nm is detected and intensity of 2, 2' cyanine fluoresence is decreased. Using different exitation wavelength (488. 0, 514. 5 and 476. 5nm), we find the new emission bend is attributed to J-aggregate fluoresence from the process of energy transfer effect, which is called sensitized fluoresence of J-aggregate. It is shown that the energy transfer rate is enhanced when molecules adsorbed on silver colloid. In fact, we found the plasma resonance frequency of the silver colloid overlaps both 2,2′ cyanine and it′s J-aggregate mloecular transition frequencies [3]. It results in electromagnetic (EM) enhancement, which accelerates non-radiative energy transfer rate between molecules. Furthermore, J-aggregate has a low quantum yield ($Q=0.02$). For low quantum yield, where the intrisic non-radiative decay dominates over the surface contribution, it is important for sensitized fluoresence. Thus competing with energy transfer effect and non-rediative decay, the enhanced sensitized fluoresence of acceptor is observed in our experiment. It provids evidence for surface-enhanced energy transfer between molecules.

Let us consider the enhencement factor of energy transfer rate. According to Gerstern theory, the energy transfer rates between free molecules and adsorbed molecules are given respectively by:

$$k_0=(3c^4/2\pi r^6)\int_{-\infty}^{\infty}d\omega G(\omega)\sigma(\omega)/\omega^4 \tag{1}$$

$$k=(3c^4/2\pi r^6)\int_{-\infty}^{\infty}[d\omega G(\omega)\sigma(\omega)/\omega^4]|A(\omega)|^2, \tag{2}$$

the transfer rate enhancement factor may be defined as $\rho=k/k_0$, where $G(\omega)$, $\sigma(\omega)$ are donor emission rate and acceptor absorption cross section, respectively. $A(\omega)$ is the interaction energy implification factor because of the presence of a solid particle. Based on our experiment data, it is caculated that enhancement factor ρ is about 10^2.

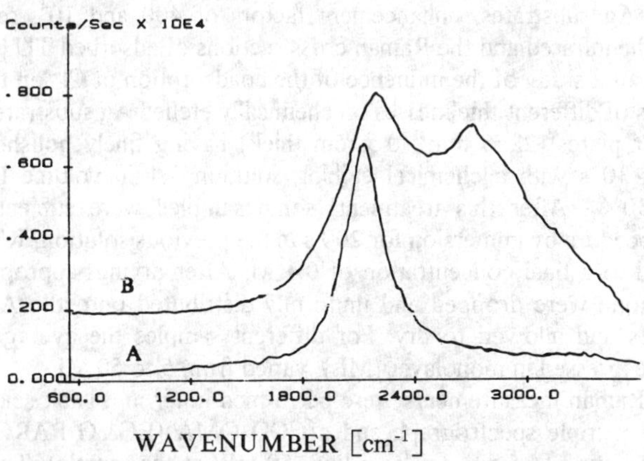

Fig. 1. Fluoresence Spectra of 2,2′ cyanine/kI in
A: aquous solution, B: silver colloid.

1. M. Moskovits, Mod. Phys. Rev., 57, 783(1985)
2. J. Gerston and A. Nitzan, Chem. Phys. Lett., 104, 31(1984)
3. D. H. Pan and P. X. Zhang, Acta Physica Sinica (Overseas Edition), 2, 925 (1993)

EFFECT OF COADSORPTION OF CHLORINE ON THE SERS SPECTRUM OF THIOUREA ADSORBED ON Ag SURFACE TREATED BY A CHEMICAL ETCHING SOLUTION

G.Mattei[1]*, Y.Mo[2], M.Pagannone[1] and V.A.Yakovlev[3]
[1] Istituto di Metodologie Avanzate Inorganiche, CNR, Area della Ricerca di Roma, C.P.10, 00016 Monterotondo Sc., Italy
[2] Institute of Physics, Chinese Academy of Sciences, P.O. Box 603,100080, Beijing, China
[3] Institute of Spectroscopy, Russian Academy of Sciences, Troitzk, Moscow reg. 142092 Russia

We have already presented and discussed [1] some results of our surface enhanced Raman scattering (SERS) study of thiourea (TU) adsorbed on silver surfaces treated with suitable chemical etching solutions according to a preparation method previously reported [2]. Recently, from a comparative investigation made by infrared reflection-adsorption spectroscopy (IRRAS) and by SERS on TU adsorbed on similar Ag substrates, enhancement factors of 400 and 10^4, respectively, were found for the infrared and the Raman cross sections of adsorbed TU [3]. In this paper, the results, of a study of the influence of the coadsorption of Cl^- on the SERS spectra of TU films of different thicknesses on chemically etched Ag substrates are presented.

Silver plates (22x6 mm^2, 0.5 mm thick) having finely polished surfaces were treated for 30 s with a chemical etching solution 1:1 in volume H_2O_2 (35%) and NH_3OH (30%). After this treatment, some samples were subjected to a second etching procedure by immersion for 200 s in the previous solution, where KCl salt had been added to a final concentration of 0.1 M. After drying, appropriate amounts of TU in acetone were dropped and uniformly distributed onto the Ag surface of the silver plates and allowed to dry. For different samples the average calculated film thickness, expressed in monolayer (ML), varied from 5 to 50 ML.

The Raman measurements were performed in air in a back scattering geometry using a Spex triple spectrograph and a CCD OMA (EG&G PARC) detector, with excitation by the 514.5 nm Ar^+ laser line (50 mW at the sample). The Raman spectra were measured in several different points of the surface and the averaged spectrum was taken as representative of the sample.

The two sets of samples (with and without KCl in the etching solution) exhibit similar spectra only for thick TU films. For thicknesses of 50 ML or less the spectra are quite different in the 650 - 1650 cm^{-1} frequency range, where three bands are present in the Raman spectra of bulk TU: at 734 cm^{-1} (S-C stretching), 1093 cm^{-1} (C-N stretching) and 1384 cm^{-1} (NH_2 rocking) [4]. For samples without KCl the line shape analysis shows that two components at frequencies, 717 cm^{-1} and 734 cm^{-1}, contribute to the S-C stretching band, with the intensity of the "bulk" component increasing with the thickness. As an example, the SERS spectrum of the 20 ML thick sample prepared without KCl is shown in Fig. 1 curve a).

The Raman spectra of the Ag substrates treated with KCl in the etching solution are characterised by a strong band at 234 cm^{-1} attributable to the SERS of the Ag-Cl vibration. After the TU film deposition a SERS spectrum similar to the one presented in Fig. 1, curve b) (case of a 20 ML thick sample), is obtained. The low frequency band shifts from 700 cm^{-1} to 710 cm^{-1} as the thickness increases. For thicker samples two other components at 717cm^{-1} and 734 cm^{-1} (those present in the other set of samples) appear with increasing intensity for higher thicknesses.

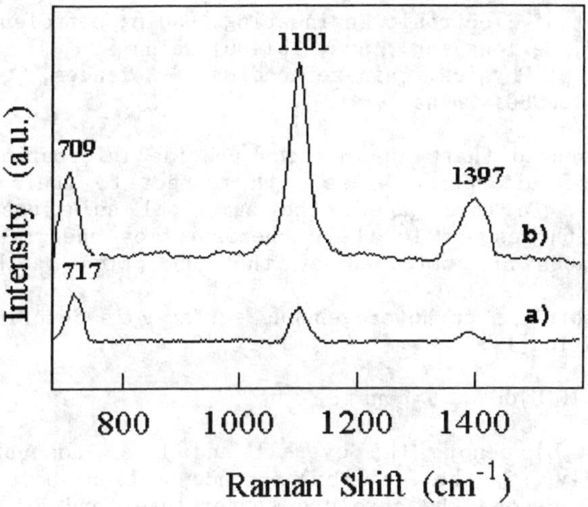

Fig. 1 SERS spectra of a 20 ML thick sample prepared without (a) and with (b) KCl salt in the etching solution.

These results could indicate that the adsorption of the TU molecules always takes place through the S atom, but to different sites on the Ag surfaces. Chlorine free sites (always present on the sample prepared without KCl) should be responsible for the 717 cm^{-1}mode being the TU molecule vertically bonded on the Ag surface. On sites present in the Ag surfaces etched with KCl in the solution, TU and Cl should interact giving rise to the 700 - 710 cm^{-1} mode and probably causing a tilt of the S-C TU axis toward the Ag surface (see the similarity of the present results with those reported for TU adsorbed on Ag electrodes [5]).

IRRAS measurements on the same sets of samples give results consistent with the SERS data.

1. G. Mattei, Y. Mo and M. Pagannone, in Proc XIV Int. Conf. on Raman Spectroscopy, eds. Y.N.-T Yu and X.-Y Li, John Wiley & Sons, 1994, p. 646.
2. G. Mattei, L. G. Quagliano and M. Pagannone, Europhys. Lett. 11, 373 (1990).
3. G. Mattei, Y. Mo, M. Pagannone and V. A. Yakovlev, Microchimica Acta, (1996, in press)
4. A. Vijay and D. N. Sathyanarayana, Spectr. Acta, 49, 1565 (1993).
5. Z. Q. Tian, Y. Z. Lian and M. Fleishman, Electrochimica Acta 35, 879 (1990).

DETERMINATION OF DISTRIBUTION FUNCTION OF CHEMISORPTION ENERGY BY SERS

S.Y.Fu[1*], S.H.Song[1], and P.X.Zhang[2]

[1]Department of Electronic Engineering, Daqing petroleum Institute, Heilongjing, Anda, 151400, China
[2]Institute of Physics, Chinese Academy of Sciences, Beijing, 100080, China

It was presented that many methods provide information about distribution of adsorption sites with respect to their adsorption energies[1]. In this paper, the numerical solutions of this distribution with respect to their chemisorption energies on liquid-solid interfaces are obtained by the adsorption isotherm measured SERS method.

Overall adsorption on heterogeneous surfaces is described by the integral equation[1]:

$$\theta_t(C)=\int(C,U)F(U)dU \qquad T=const.,$$

where $\theta_t(C)$ and $\theta(C,U)$ denote the overall and local adsorption isotherms respectively; C is the equilibrium concentration; U is the adsorption energy; T denote the absolute temperature; and F(U) is the distribution function of adsorption energy normalized to unity.

The function F(U) is defined as the derivative of the number of adsorption sites with respect to adsorption energy. The expression F(U)dU denotes the fraction of the surface with adsorption energies between U--U+dU. In this paper the local adsorption isotherm chooses Langmuir equation[1]. We assumed SERS intensity I is proportional to surface coverage θ for submonolayer adsorption, and the maximum of SERS intensity, Imax, corresponds to monolayer adsorption, so I/Imax denote the coverage. Therefore, the coverage can be obtained by measuring SERS intensity.

F(U) is expressed by m-Gaussian distributions as follows:

$$F(U)=\sum_{k=1}^{m}A_k\exp\{-[U-C_k]/B_k]^2\}$$

where A_k is amplitude; B_k is half-width; C_k is center position of Gaussian peak. When the root-mean-square relative deviation between the computed and experimental isotherms reaches a minimum, the best-fit solution is obtained.

The intensity of Crystal violet (CV) in silver sol relates to whether existing Cl^-. SERS intensity of CV may increase 10^2-fold for moderate amount of Cl^-. The adsorption energies of CV in silver sol was estimated by Langmuir equation[2]: $U_0 \simeq 44.6$ kJ/mol, without Cl^-; and U=58.6 kJ/mol corresponding to moderate amount of Cl^-. The distribution of adsorption energies of CV on silver surfaces is solved as follows.

The isotherm of CV in silver sol with existing Cl^- is given in Fig.1. First, a Gaussian peak is chosen and the initial data is: A_1=0.4, B_1=10 kJ/mol, C_1=44.6 kJ/mol, The integration region is from 20kJ/mol to 80kJ/mol. The calculation shows: Gaussian amplitude reduces, the half-width increases, the peak move to high energy. Second, adjusting fit-calculation by adding a small Gaussian peak in high energy region, the initial data is chosen again by previous calculation results: A_1=0.13, B_1=12.8, C_1=46.1; A_2=0.02, B_2=4.0, C_2=58.6; and the curve of initial data is showed in Fig.2. The calculation result is: A_1=0.041, b_1=8.550, C_1=44.59; A_2=10.6, b_2=0.366, C_2=59.18. Fig.3 gives best-fit curve.

It is found that the peak in high energy region is very narrow, similar to s-function, which is in agreement with the character of chemisorption.

Fig.3 indicates the distribution of chemisorption energies. The SERS intensity comes from molecules on the sites corresponding to these energies, because surface coverage is expressed SERS intensity.So,SERS intensity of CV in silver sol mainly originates from the molecules on sites possessing high adsorption energy. According to the results of SERS study[2], these sites are socall SERS active-sites.

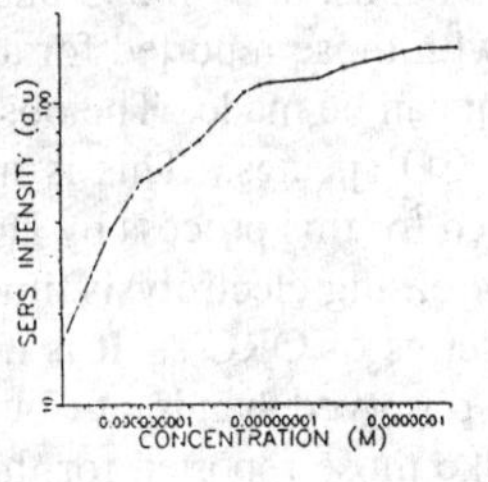

Fig.1 The isotherm of CV,

the concentration of Cl^-:10^{-3}M

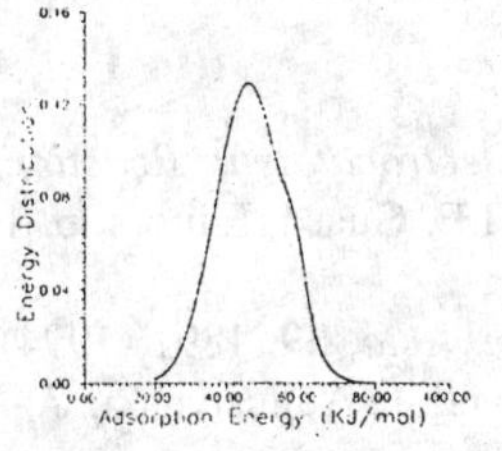

Fig.2 Initial data:
A_1=0.13,B_1=12.8,C_1=46.1
A_2=0.02,B_2=4.0, C_2=58.6

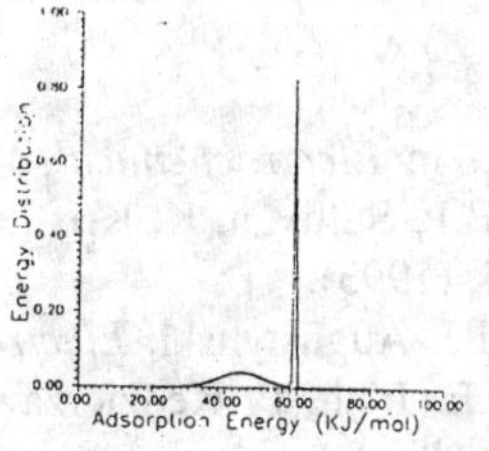

Fig.3 Result:
A_1=0.041,b_1=8.550,C_1=44.59
A_2=10.6, b_2=0.366, C_2=59.18

[1] M.Jaroniec and P.Bräuer, Surf.Sci.Rep. 6(1986)65
[2] Z.F.Li,S.Y.Fu and P.X.Zhang, in this conference.

SERS STUDY OF CO_2 REDUCTION AT THE COPPER ELECTRODE.

B.D. Smith and D.E. Irish*

Guelph-Waterloo Centre for Graduate Work in Chemistry, Department of Chemistry, University of Waterloo, Waterloo, ON, Canada, N2L 3G1

The reduction of CO_2 on a Cu electrode can provide large yields of hydrocarbon species; however the electrode is readily poisoned [1]. Jermann and Augustynski have found that the activity of the Cu electrode can be prolonged by subjecting it to a series of oxidation reduction cycles (ORC's). This removes the poisoning species from the electrode and restores the activity of the Cu for CO_2 reduction [2]. This procedure, by its nature, will roughen the electrode. This creates an ideal opportunity to employ SERS to study the reaction mechanism since the electrode is already roughened.

Spectra from a Cu electrode in CO_2 saturated 0.1 M $NaHCO_3$ have been recorded at a potential of -1.5 V (SCE) and three bands, at 270, 343 and 2078 cm^{-1}, are observed [3]. By comparing these features with those reported for adsorbed CO on Cu in the UHV environment [4] an assignment can be made. The observed bands (Fig. 1 and 2) are attributed to an intermediate CO species. This is in agreement with the intermediate which has been hypothesized for this process by Hori et al. [5]. The intensities of the CO bands decrease with increasing electrolysis time. However the intensity can be restored by an appropriate series of ORC's. It is noted that, in the case of the copper electrode, there are no observed bands attributable to the presence of adsorbed carbonaceous species, unlike those reported for the $Ag/HCO_3^-/CO_2$ system [6].

1. K.W. Frese Jr., in *Electrochemical And Electrocatalytic Reactions of Carbon Dioxide*, B.P. Sullivan, K. Krist, and H.E. Guard, Editors, p. 145, Elsevier, New York (1993).
2. B. Jermann and J. Augustynski, *Electrochim. Acta*, **39**, 1891 (1994).
3. B.D. Smith, D.E. Irish, P. Kedzierzawski and J. Augustynski, *J. Electrochem. Soc.*, (submitted).
4. W. Akemann, and A. Otto, *J. Raman Spectrosc.*, **22**, 797 (1991).
5. Y. Hori, H. Wakebe, T. Tsukamoto and O. Koga, *Electrochim. Acta*, **39**, 1833 (1994).
6. Y. Ichinohe, T. Wadayama and A. Hatta, *J. Raman Spectrosc.*, **26**, 335 (1995).

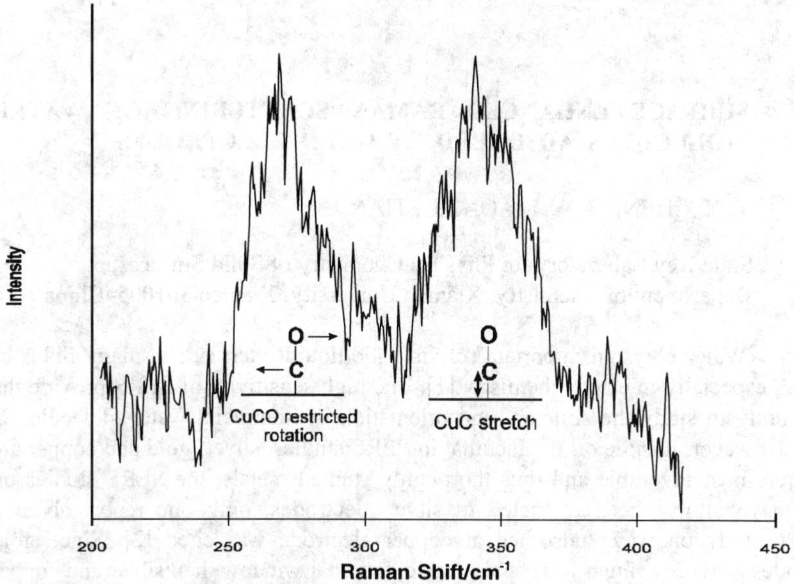

Figure 1 Raman spectrum of the Cu-C modes of adsorbed CO at a potential of -1.5 V.

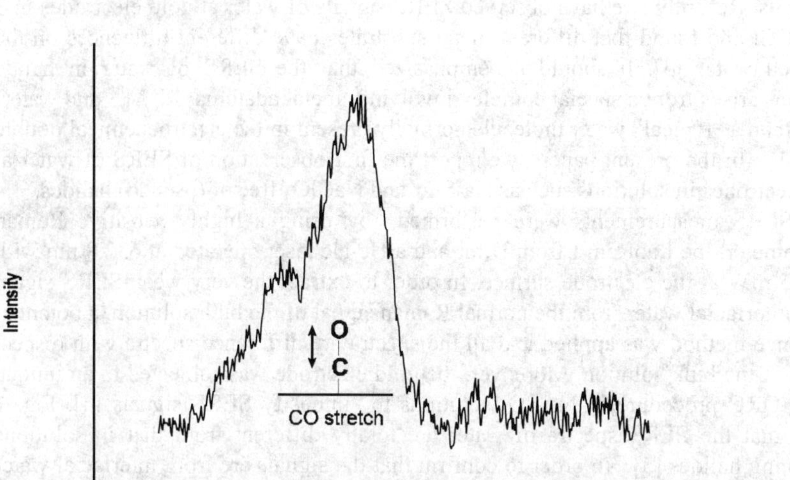

Figure 2 Raman spectrum of the CO stretching mode at a potential of -1.5 V.

SURFACE ENHANCED RAMAN SCATTERING OF WATER MOLECULES ADSORBED AT GOLD ELECTRODES

Y. X. CHEN, B. W. MAO, Z.Q. TIAN

State Key Laboratory for Physical Chemistry of Solid Surfaces,
Department of Chemistry, Xiamen University, Xiamen 361005, China

Water plays an important role in solid/liquid interfaces in many fields of science, especially in electrochemistry [1]. The high sensitivity of SERS provide the opportunity to study the structure and orientation of interfacial water molecules in detail. However, in spite of the fact that in SERS studies, silver, gold and copper are the three most favorable and thus thoroughly studied metals, the SERS studies on water have almost been restricted to silver electrodes, only one report about a primary study on water adsorbed at copper electrode was cited [2]. Since gold electrodes provide a much wider polarizable potential window than silver and copper in aqueous media, it aids the studies for a variety of electrochemical adsorption including interfacial water in the wide range of both positive and negative electrode potentials. Recently, we have observed SERS signals of water at gold electrodes in 3 mol/l KCl and found that different metal substrates have different influences on the adsorbed water [3]. It should be emphasized that the SERS of water in halide solutions arises from a special complex involving 'metal adatoms, X^-, M^{n+} and water' rather than a "typical" water molecule normally present in the electrochemical double layer [3]. In the present paper, we report the first observation of SERS of water at gold electrodes in solutions such as Na_2SO_4 and $NaClO_4$ free of (pseudo) halides.

SERS measurements were performed by using a highly sensitive Raman spectrometer, the LabRam I from Dilor and a He-Ne laser operated at 632.8 nm with about 5 mW at the electrode surface. In order to extract the very weak SERS signal of the interfacial water from the normal Raman signal of the bulk solution, a potential difference method was applied and all the spectra are difference spectra with respect to that of the bulk solution. Moreover, the gold electrode was subjected to an unique *in-situ* ORC procedure in Na_2SO_4 solutions to intensify SERS signals [4]. Fig. 1 shows that the SERS spectra of water are totally different from that in solutions containing halides [3]. In order to confirm that the signals are from interfacial water molecules rather than impurities, D_2O was used and signals corresponding to $\delta_{D\text{-}O\text{-}D}$ and $\nu_{O\text{-}D}$ were observed unambiguously as shown in Fig. 1c.

There are several interesting features observed in the present study: (i) The SERS of water can only be observed with the *in-situ* ORC treatment particularly under the laser illumination; (ii) In terms of relative peak intensity for the stretching and bending bands, the SERS signals of the interfacial water in halide free solutions are different from that in halide containing solutions; (iii) SERS of water on gold electrodes could only be observed within a very narrow potential range from -1.5 V to -1.0 V (*vs* SCE), while at more positive potentials the totally symmetric vibrational band form adsorbed SO_4^{2-} are, instead, observed due to the replacement of surface water molecules by the electrolyte anions; (iv) The intensity of the stretching mode is much weaker than the bending mode, in particular, for H_2O, the stretching mode can hardly be detected, which may be due to, at least partially, that the quantum efficiency of the CCD detector decrease dramatically in the frequency region higher than 3000 cm^{-1}. Accordingly, a model for the $_{surface}$ water with much more ordered structure and one hydrogen ending orientation is postulated. Our primary result demonstrates the possibility to detect and characterize "typical" water molecules in the electrochemical double layer of gold electrode systems. A systematic investigation including correlation of the SERS band intensity, frequency and shape with electrode potential, composition and pH of solutions as well as laser illumination is being carried out.

Acknowledgment

This work was supported by the Natural Science Foundation of China.

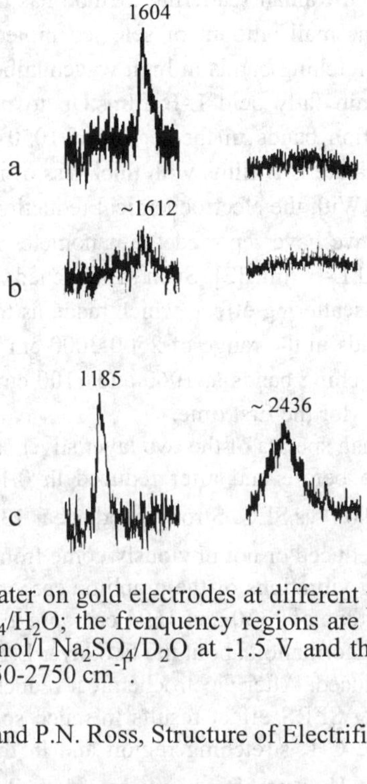

Fig. 1 SERS of water on gold electrodes at different potentials: (a) -1.5 V, (b) -1.0 V in 1 mol/l Na_2SO_4/H_2O; the frenquency regions are 1250-1750 cm^{-1} and 3000-3600 cm^{-1} . (c) in 0.1 mol/l Na_2SO_4/D_2O at -1.5 V and the frenquency regions are 1000-1500 cm^{-1} and 2250-2750 cm^{-1}.

1. J. Lipkowski and P.N. Ross, Structure of Electrified Interfaces, (VCH, New York, 1993).
2. B. Pettinger, M. R. Philpott and J. G. Gordon, Surf. Sci., 105, 469 (1981).
3 Z. Q. Tian, Y. X. Chen and B. W. Mao, Chem. Phys. Lett., 240 ,224 (1995).
4. Z. Q. Tian, Y. X. Chen, S. Z. Zou and B. W. Mao, prepared for publication.

SERS SPECTRA OF TWO-LAYER STEARIC ACID LANGMUIR-BLODGETT FILMS CONTAINING SAND-WICHED NANO-METER AG CLUSTERS IN THE FILMS

Y.H. Zhang[1], B.Z.Hua[2], Y.X. Chen[2], Z.Q. Tian[2], Y.Z. Chen[2], Q.S. Li[1]

[1]College of Chemical Engineering and Materials Science, Beijing Institute of Technology, Beijing 100081, China
[2]State Key Lab for Physical Chemistry of Solid Surfaces, Xiamen Univ.,Xiamen 361005, China

Langmuir-Blodgett (L-B) Films have found uses in a wide variety of experiments and techniques. This activity is being accompanied by an increasing understanding of the structure of these films because of improvement in instrumentation. Surface enhanced Raman scattering method has made it possible to monitor vibrational spectra of a small amount of selected molecules adsorbed on roughed Ag surface. The C-H stretching bands in high wavenumber region has been reported for one layer long chain fatty acid L-B films.Up to now, however, the skeleton (C-C) stretching vibration bands in the region of 1050-1150 cm^{-1} can be obtained only for the barium stearate L-B films with thickness of more than 7 layers for their very weak intensity[1]. With the electrochemical reduction of silver stearate L-B films on HOPG electrode, we have reported the nanometer Ag clusters in the hydrophilic layers of stearic acid L-B films[2]. Such sandwiched Ag clusters have a strong surface enhanced Raman scattering effect, which made us to observe not only the C-H stretching vibration bands in the range of 2500-3000 cm^{-1} for the two-layer L-B films, but also the C-C stretching bands at 1060 and 1100 cm^{-1} as well as other bands between 1200 to 1700 cm^{-1} for the first time.

Fig.1 shows the Raman spectra of the two layer silver stearate L-B films on glass carbon (GC) electrode before and after reduced in 0.1M H_2SO_4/ 0.1M $NaNO_3$ solution for 10 min at 0.0V vs.SEC. Strong bands near 1330 and 1604 cm^{-1} for the film-covered GC either reduced or not obviously come from the GC substrate which are due to C=C stretching vibrations of the graphitic carbon. The overtone of 1330 cm^{-1} and the combination tone of 1330 and 1604 cm^{-1} are also observed in the Fig.1. There is not any spectral information about the two layer L-B films if the film-covered GC electrode is not reduced. After electrochemical reduction of the Ag^+ to Ag in the same films, the strong SERS effect results in much spectral information concerning the L-B films in the CH_2 stretching region and in the region between 1000 and 1700 cm^{-1}. Although CH_2 stretching bands have heavily overlap with the band of the combination tone, a shoulder peak on the band of the combination tone appears at 2850 cm^{-1}, which is undoubtedly assigned to the CH_2 symmetric stretching vibration of the methylene of the stearic acid. Since the enhancement of the CH_2 and CH_3 stretching bands of the stearic acid L-B films, the band of the combination tone has a higher intensity compared with that of the overtone in contrast with the situation of the film-covered GC without reduction. On the other hand, there are two bands appeared at 1100.7 and 1128.0 cm^{-1} respectively, which are related to C-C stretching vibrations of the methylene chains. We can also observe $\delta(CH_2)$ bending vibration near 1470 cm^{-1} and CH_2 rocking vibration at 1200cm^{-1}, respectively.

Many authors have related the deformation of the phospholipid chains, of the stearic acid chain, to the appearance of a broad band at about 1090 cm^{-1} and simultaneous disappearance of the sharp lines at 1060 and 1130 cm^{-1}, respectively, assigned to the ν_s(C-C) and ν_a(C-C) vibrations of the acyl chains in the *all-trans* conformation. But in our experiment according to the SERS Raman spectra in Fig.1, a slightly different occasion appears. Only the pair to the lower wavenumber exists at 1128 cm^{-1} (due to ν_s(C-C) mode). At the same time, the line corresponding to the *gauche* conformation appears at 1100.7 cm^{-1}. Considering the physical SERS mechanism, it can be resonablely suggested that the incident electric field is enhanced in the direction of the normal line of the L-B films (usually defined Z axis). According to Harrand's analysis[3], if stearic acid molecules are in *all-trans* conformation and the molecular axis of the stearic acid is perpendicular to the films plane, the Raman scattering activity α_{zz}^2 for the ν_s(C-C) mode will be several hundred times more than that of ν_a(C-C), which is the main reason that we do not observe the line corresponding to ν_a(C-C). For the two-layer L-B films, it may be concluded that some chains have became disorder but the greater number remain *all-trans*.

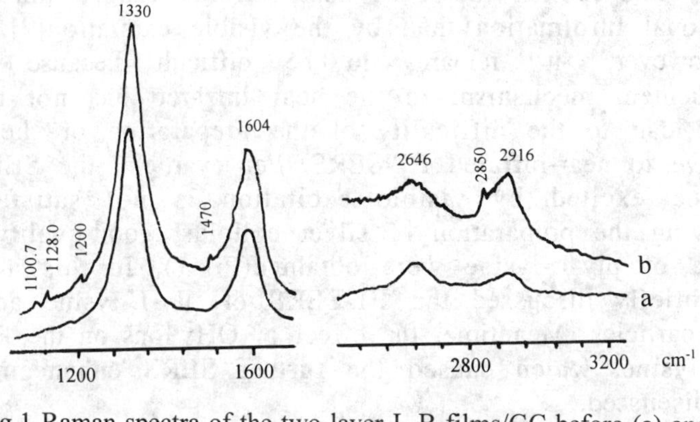

Fig.1 Raman spectra of the two-layer L-B films/GC before (a) and after (b) electrochemical reduction in in 0.1M H_2SO_4/ 0.1M $NaNO_3$ solution for 10 min at 0.0V vs.SEC.

1. M. Harrand, M. Masson, J.Chem.Phys., 87,5176(1987)
2. YH Zhang et al., Thin Solid Films, in press
3. M. Harrand, J.Chem.Phys.,79, 5639(1983)

THE STUDY OF NEAR-INFRARED FOURIER TRANSFORM SURFACE-ENCHANCED RAMAN SCTTERING OF PLY-L-LYSINE

Q. Li
Optical Engineering Dept,Beijing Inst of Techn,Beijing
H. J. Tian
National Bio-medical Centre of China, Beijing
Y. Fang*
Experimental Centre,Capital Normal Univ,Beijing 100037,China

The combination of near infrared Fourier transform Raman Spectroscopy (FT-Raman) and Surface Enhanced Raman Spectroscopy (SERS), named near-infrared FT-SERS, can provide an high sensitive and in situ analysis technique for the vibrational structures of biomolecules because of the near infrared excitation by which the lower photon energy can avoid the damage to biomolecules, can overcome those severe limitations of Raman measurement caused by chromophores in biomolecules and increase the sensitivity[1]. Near-infrared FT-SERS could obtain higher enhancement factor in near infrared range, and give more vibrational information than by the visible excitation[2].

However, it remains to be difficult because the SERS enhancement mechanism in the near infrared has not been clear, which lead to the difficulity of the preparation of the substrstes effective to near-infrared FT-SERS. For example, the SERS of ply-L-lysine excited by visible excitation is not satisfactory. By improving the preparation of silver colloids, good-quality FT-SERS spectra of ply-L-lysine were obtained(Fig.1). In this abstract, we have briefly discussed the FT-SERS of ply-L-lysine adsorbed on silver particles. Meantime, the effect of OH^- ions on the FT-SERS of ply-L-lysine, which caused the further SERS enhancement , was also discussed.

Fig.1(b) shows a satisfactory near-infrared FT-SERS spectrum of ply-L-lysine. With the addition of a small amount of OH^- ions, the FT-SERS intensities are further enhanced(Fig1.(c)). There seems no significant shift comparing with the case of that without OH^- ions added(Fig.1(b)). We believe that the surfaces of original silver colloidal particles are covered with H^+ ions before the addition of NaOH, which makes the silver colloid stable. When OH ions were added, OH^- ions could remove the H^+ ions from the particle surfaces and destroy the stability of silver colloid. In that case, the interaction between the surfaces and molecules becomes stronger than before, and more ply-L-lysine molecules could be adsorbed on the colloidal silver particles, and as a result, the SERS signals are further enhanced. By contrast with the case of Cl^- added in which no extra SERS enhancement could be recorded, it is concluded that OH^- ion can effectively enhance the near-infrared FT-SRES of ply-L-lysine.

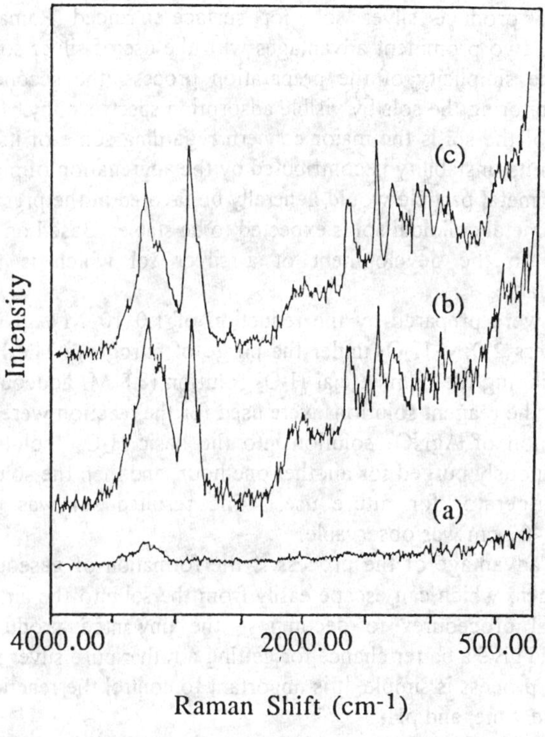

Figure1 Near-infrared FT-Raman and FT-SERS spectra of Ply-L-lysine. (a) Raman spectrum of aqueoue solution; (b) FT-SERS, and (c) after the addition of NaOH.

References:
[1] T.M.Cotton,Jae-Ho Kim and George D. Chumanov, J.Raman Spectrosc., 22(1991)729.
[2] Dider Bourgeois, Spectrochimica Acta, 46A, 295, 1990.

SILVER SOLS PREPARED BY HYDROGEN PEROXIDE REDUCTION

J. Cheng and Y. S. Li*

Department of Chemistry, The University of Memphis, TN 38152 USA

Reductions of silver nitrate by sodium borohydride [1] and by citrate [2] are commonly used to produce silver sols for surface-enhanced Raman scattering (SERS). These are two prominent advantages with the use of silver sols for SERS. The first one is the simplicity of the preparation process; the second one is the convenience of monitoring the sols by visible adsorption spectroscopy. However, the unstable character of the sol is the major concern regarding some of its applications [3]. One reason for its unstability is contributed by the aggregation of metal particles. The aggregation of metal particle would generally be favored in the presence of some impurities. A pure metal colloidal sol is expected to be stable. Based on this concept, we have engaged in the development of a silver sol which is SERS-active, reproducible, and stable.

Silver sols were prepared by the reduction of 1.0×10^{-3}M $AgNO_3$ in a basic condition with excess 2.0 M H_2O_2 under the purge of nitrogen. The H_2O_2 solution was prepared by diluting the commercial H_2O_2 solution (8.8 M, aqueous) with triple distilled water. All the reagent solutions were used for the reaction were fresh. Upon the complete addition of $AgNO_3$ solution into the basic H_2O_2 solution, the final mixture was continuously purged for another one hour, and then the solution product was stored in refrigerator for future use. The resulted sol was colorless; no absorption around 400 nm was observable.

One major advantage of the process is the formation of gaseous products - hydrogen and oxygen, which can escape easily from the sol into the air. This would eliminate additional procedures to decompose the unwanted products from the reaction system, and give a better chance for getting a rather pure silver sol. Although the sol preparation process is simple, it is important to control the reaction conditions such as temperature, time, and pH.

In the present study, we have optimized the conditions for the preparation of the Ag sols. Benzoic acid (BA) solution was adopted as a model sample for testing the SERS performances of the sols. In all the experiments, the normal Raman spectrum of ethanol solvent could be used as an internal reference for the SER intensity measurements. To optimize the reduction reaction temperature, we followed the same preparation procedures but kept the reaction temperature at 0, 21, 32, and 45°C, respectively. Relative SER intensities of BA were measured at different time using each of these prepared sols. Results indicated that the sols prepared at 0°C is more SERS-active than those prepared at higher temperatures, especially when the sols had been stored for more than three days. Thus, we fixed the reduction temperature at 0°C for the subsequent experiments.

Different rates of adding $AgNO_3$ solution into the H_2O_2 solution may affect the reduction rate, and result in different Ag particle sizes in the sols. The addition of $AgNO_3$ solution was achieved drop by drop. In the experiments, we had tried the following dropping rates: 1, 0.2, 0.1, 0.07, 0.05, and 0.03 drop/second. Shown in Figure 1 are the SER spectra of BA absorbed on the Ag sols prepared using different dropping rates. It is seen that 5 seconds of duration between the two consequent drops or 0.2 drop/see of dropping rate gives the product sol with a better SERS performance.

The pH can also significantly affect the reduction rate. In a series of experiments, we had adjusted the mole ratio between NaOH and $AgNO_3$ as a control of the solution pH. SER spectra of BA adsorbed on each of these sols prepared under different pH conditions are shown in Figure 2. From these results, it is clear that the mole rate of 0.6 provides the best pH condition for the preparation of Ag sol.

Two series of tests have been conducted for the sols prepared under the optimized conditions. In the first series experiments, we measured SER intensities of BA after adsorbing in the sols for different time. It was found that the intensity of BA increases slowly with time until day three, after then, the intensity decreases rapidly to zero on day seven. The second series experiments were performed to record the SER spectra of BA adsorbed on Ag sols which had been stored for different lengths of time since its preparation. The experimental results indicated that the sols are still SERS-active even after six months of storage.

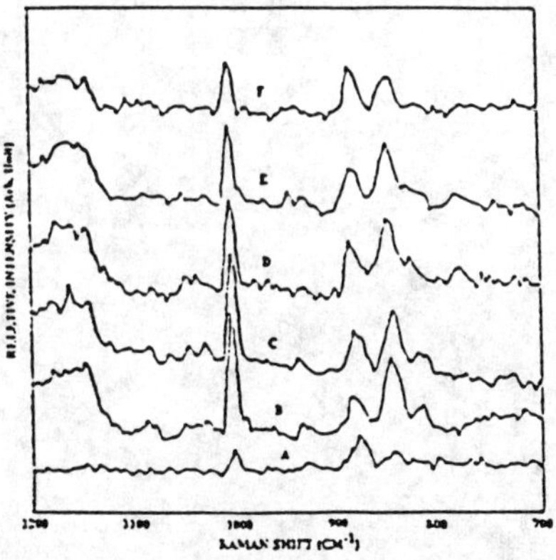

Fig. 1. SER spectrum of 2.5×10^{-3} M BA in Ag sols prepared with different $AgNO_3$ dropping rates, A) 1; B) .2; C) .1 D) .07; E) .05; F) .03 drop/s

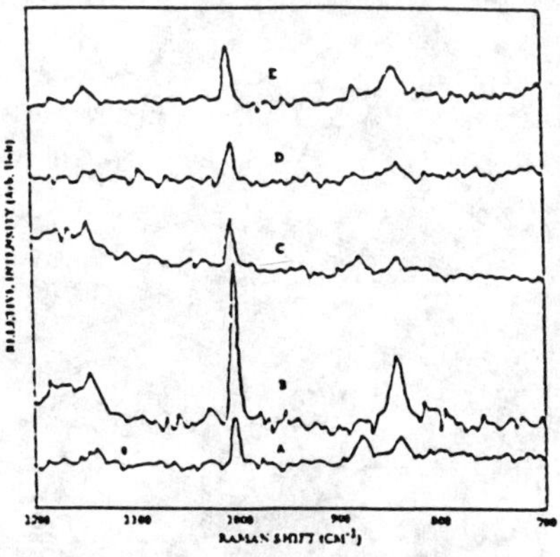

Fig. 2. SER spectrum of 2.5×10^{-3} M BA in Ag sols prepared using mole ratios betw. Ag NO_3 and H_2O_2: A) .4; B) .6; C) .8; D) 1.6; and E) 2.4

1. J. A. Creighton, C. G. Blatchford, and M. G. Albright, J. Chem. Soc., Faraday Trans. 75, 790 (1979).
2. P. C. Lee and D. Melsel, J. Phys. Chem. 86, 3391 (1982).
3. Y. S. Li, A. S. Lee, and Y. Wang, J. Raman Soc. 22, 191 (1919).

SECTION 17

THIN FILMS

RAMAN STUDIES OF CHEMICAL BONDING AND PHASE TRANSFORMATION DYNAMICS IN STRESSED OPTICAL FILMS

Gregory J. Exarhos and Nancy J. Hess

Pacific Northwest National Laboratory
PO Box 999, MS K2-44, Richland, WA 99352-1611

The availability of high intensity laser systems, intensified gated array detectors, and micro-optical imaging devices has led to a resurgence of Raman spectroscopy for rapid, non-destructive, *in situ* analysis of dielectric films. [1] Measured spectra can reveal details concerning localized atom arrangements, phase composition, crystallite orientation, and residual stress which develop in the film as a result of deposition parameters and post deposition film processing. Raman Spectroscopy is appealing particularly because measurements can be acquired in real time and the intrinsic optical properties of a thin film or multilayer coating can be used to improve sensitivity through interference and resonance enhancement.

Previous Raman studies of oxide films have shown that deposition conditions control the degree of phase homogeneity in the film and the tendency of one phase to transform to a second phase when annealed or exposed to high energy pulsed laser irradiation. [2,3] Thus, rutile or anatase nucleation centers can be incorporated into an amorphous film by altering the beam energy during ion-beam deposition of TiO_2. For example, a relatively high current density during deposition will produce films which crystallize preferentially in a dense rutile phase upon heating to 400 °C as determined from Raman spectroscopy measurements. [3] Related time-resolved measurements of sol-gel deposited amorphous titania films on silica (fig. 1) show gradual transformation of the film to an anatase phase during post deposition annealing. [4] In this instance, spheroidal crystallites evolve where the time-dependent fractional crystallinity of the film follows a sigmoidal crystallite ingrowth curve described by a modified Avrami equation. Results have been interpreted in terms of a mass diffusion limited process influenced by the solution processing conditions used to generate the film. [4]

Some insight into the degree of transient stress evolution in films undergoing transformation to a new phase also can be gleaned from measured spectra. [5] Such transformations are induced either thermally or by exposure to high energy cw or pulsed laser irradiation. [6] In these experiments, measured variations in phonon frequencies are related to film temperature and relative residual stress through correlation with vibrational data acquired as a function of pressure and temperature and derived from bulk samples in heated diamond anvil cells. Successive spectra measured during the transformation have been acquired on millisecond time scales.

Modification to film optical properties resulting from chemical interactions at interfaces or the presence of defects also can be interpreted from perturbations to the Raman spectrum. Microprobe measurements are particularly valuable here since a shallow depth of focus tends to suppress substrate contributions to the scattering thereby minimizing interference, and lateral spatial resolution on the order of several micrometers can be achieved. Regions of stress heterogeneity induced thermally or by means of an intense focused laser beam have been evaluated from phonon line shifts and mapped. [7] Recent work has shown the utility of Raman methods for *in situ* evaluation of films during deposition. Such studies are being actively pursued.

[1] GJ Exarhos, in **Handbook of Optical Properties, Vol I, *Thin Films for Optical Coatings***, Ed. RE Hummel and K H Guenther, CRC Press, Inc., Boca Raton, FL, pp. 213-241 (1995)

[2] GJ Exarhos and PM Morse, in **Laser Induced Damage in Optical Materials**, Ed. HE Bennett, AH Guenther, D Milam, BE Newnam, NIST Sp. Pub. 746, Washington, DC pp. 395-402 (1988).

[3] LS Hsu, R Rujkorakam, JR Sites, and CY She, J. Appl. Phys. **59(10)**:3475 (1986).

[4] GJ Exarhos and MJ Aloi, Thin Solid Films **193/194**:42 (1990).

[5] GJ Exarhos and NJ Hess, Thin Solid Films **220**:254 (1992).

[6] GJ Exarhos, Li-Q Wang, and T Dennis, Thin Solid Films **253**:41 (1994).

[7] PL White, GJ Exarhos, M Bowden, NM Nixon, and DJ Gardiner, J. Mater. Res. **6(1)**:126 (1991).

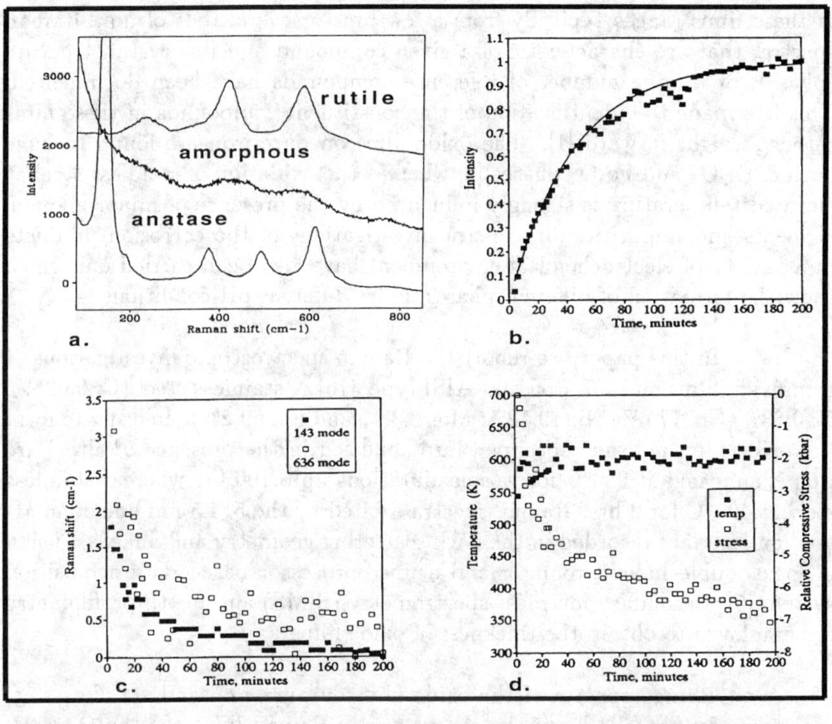

Figure 1. (a) Raman spectra of several phases of titania, TiO_2. (b)Time dependent increase in anatase mode intensity (143 cm^{-1}) during isothermal annealing of a 500 nm thick sol-gel deposited film on silica at 350 °C. (c) Time-dependent variation in anatase E_g mode frequencies during isothermal crystallization (d) Calculated film temperature and residual film stress during isothermal crystallization. [5]

RAMAN STUDIES OF THIN FILMS FORMED DURING OXIDATION AND CORROSION

Akhilesh K. Arora,[1] R.V. Subbarao,[2] T. Sakuntala[1] and R.K. Dayal[2]

[1]Materials Science Division, [2]Metallurgy Division, Indira Gandhi Centre for Atomic Research, Kalpakkam 603102 T.N. India.

Metal surfaces are extremely sensitive to attack by oxygen, halogens and other aqueous environments. As a result of oxidation and corrosion, the surface gets modified into thin films of oxides, oxyhydroxides and similar compounds depending on the environment and temperature. The thickness of these films ranges typically from a few nm to several tens of nm. Raman spectra, that are characterstic of a given compound and its crystal structure (phase), of a large number of reference compounds have been documented. This has made the identification of the constituent compounds in these films rather straight forward [1]. The oxide films on pure iron are found to have both α-Fe_2O_3 and Fe_3O_4 phases [2] whereas the oxidation of stainless steel at elevated temperature is strongly influenced by the presence of minor alloying elements and impurities [3]. In-situ investigations of the corrosion products in a variety of electrochemical environment have also been carried out which show the formation of different phases under different pH conditions [4].

In this paper, we report the Raman spectroscopic investigations of thin oxide films on pure iron and AISI type 316LN stainless steel (C: 0.025%, N: 0.68%, Cr: 17.6%, Ni: 10.12%, Mo: 2.38% and Fe: 69.2%). In order to form the oxide film, polished specimens are annealed in the presence of air. Pure iron is annealed at 250°C for various durations upto 100 hrs whereas stainless steel at 400°C for 2 hrs. Raman spectra excited by the 514.5 nm line of an argon ion laser are recorded in the back scattering geometry and analysed using a Spex double monochromator and a microprocessor based data acquisition system [5]. Secondary ion mass spectrometry (SIMS) and depth profilometry are employed to obtain the thickness of oxide films.

Raman spectra of the oxide films on pure iron and stainless steel are shown in Fig. 1. From the characteristic Raman lines of α-Fe_2O_3 (227, 245, 293, 414, 501 and 612 cm^{-1}) and Fe_3O_4 (616 and 663 cm^{-1}) we find that both these compounds are present in the oxide film on pure iron. Further it is found that the surface layer is rich in α-Fe_2O_3 whereas the inner layer in Fe_3O_4 consistent with the earlier report [2]. The oxygen deficiency of the inner layer arises due to the fact that stoichiometry of the inner layer is controlled by the diffusion of oxygen through the top oxide film. In the case of stainless steel, we find that the oxide films also have α-Fe_2O_3 and Fe_3O_4 whereas no signatures of α-Cr_2O_3 (360, 523, 718 and 1051 cm^{-1}) are detected. This is in contrast to the oxidation behavior at elevated temperature where Cr_2O_3 is reported to be the main constituent. Absence of Cr_2O_3 in these films have also been confirmed from our x-ray photoelectron spectroscopic studies [6].

Depth profiling of oxygen done using SIMS revealed the film thickness to be $\approx$100 nm in the case of pure iron. The quality of Raman spectra suggests that films that are as thin as a few tens of nm can also be studied. However, the Raman spectrum of a thin passive film ($\sim$2 nm) on stainless steel (passivated by 30% HNO_3 for 30 min at 60°C) was too weak to be detected.

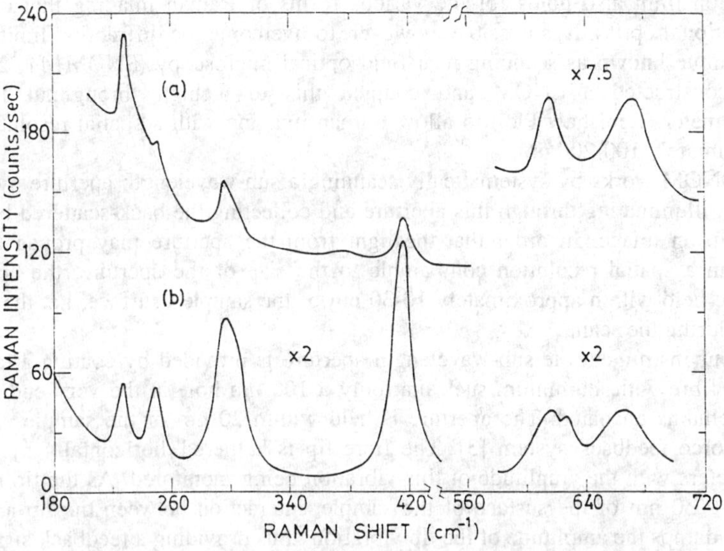

Fig. 1. Raman spectra of oxide films after annealing (a) pure iron at 250°C for 100 hrs and (b) 316LN stainless steel at 400°C for 2 hrs.

1. P.R. Graves, AERE Harwell Report AERE-R11698 (1985).
2. R.J. Thibeau, C.W. Brown and R.H. Heidersbach, Appl. Spectrosc. **32**, 532 (1978).
3. R.L. Farrow, P.L. Mattern and A.S. Nagelberg, Appl. Phys. Lett. **36**, 212 (1980).
4. C. Johnston and P.R. Graves, Appl. Spectrosc. **44**, 105 (1990).
5. A.K. Arora and T. Sakuntala, J. Phys.: Condens. Matter **4**, 8697 (1992).
6. R. V. Subbarao, T. Sakuntala, R.K. Dayal and A.K. Arora, in *Applications of Analytical Instruments* Eds. B.S. Acharya and Rajeev (Allied Publishers, New Delhi, 1993) p303.

SUB-MICRON RESOLUTION RAMAN IMAGING

S.Webster[1,*], D.A.M.Smith[2], M.N.Ayad[2], K.Kershaw[2], D.N.Batchelder[2]

[1]SOMS Centre, Department of Chemistry and [2]Department of Physics and Astronomy, University of Leeds, Leeds LS2 9JT. U.K.

The ultimate spatial resolution of optical microscopies is limited by diffraction to approximately half the wavelength of the light used to form the image. This resolution limit also holds for the various forms of Raman imaging that employ conventional optics. It is possible, however, to overcome the diffraction limit using a technique known as scanning near-field optical microscopy (SNOM) [1, 2]. We have constructed a SNOM and coupled this to a high throughput Raman spectrometer (Renishaw Plc.) to allow Raman imaging with a spatial resolution of approximately 100-200 nm.

A SNOM works by systematically scanning a sub-wavelength aperture over the sample, illuminating through this aperture and collecting the back-scattered light to build up an image. In order that the light from the aperture may propagate and maintain a spatial resolution comparable to the size of the aperture, the aperture must be held within approximately 10-30 nm of the sample's surface, i.e. the 'near-field', during the scan.

In our instrument the sub-wavelength aperture is provided by coating a tapered optical fibre with aluminium such that only a 100 nm hole at the very end of the fibre remains uncoated. The aperture is held within 20 nm of the sample using a shear force feedback system [3]. The fibre tip is 'dithered' horizontally by a few nanometers with the amplitude of this vibration being monitored. As the tip moves within ~ 30 nm of the surface of the sample, interaction between the tip and the sample damps the amplitude of the tip vibration, thus providing a feedback signal.

In our instrument the Rayleigh and Raman scattered light from the sample lying under the tip is collected with a long working distance objective at 45° to the sample and then fed into a high throughput, single stage Raman spectrometer equipped with a cooled CCD detector and holographic Rayleigh rejection filters [4].

High losses in the tapered region of the fibre mean that only a tiny fraction of the input light reaches the sample, even when the aperture is held in the near field. Further, local heating of the fibre SNOM tip restricts the amount of light it is possible to put into the fibre. These two factors, when coupled with the inherently low scattering cross section of the Raman effect, mean that Raman SNOM presents a considerable challenge.

Initial results have been obtained with the SNOM from a range of samples including conjugated polymers, CVD diamond and silicon. Figure 1 illustrates a one dimensional SNOM Raman image for a nano-crystal of a polydiacetylene. The image was recorded using 632.8 nm illumination.

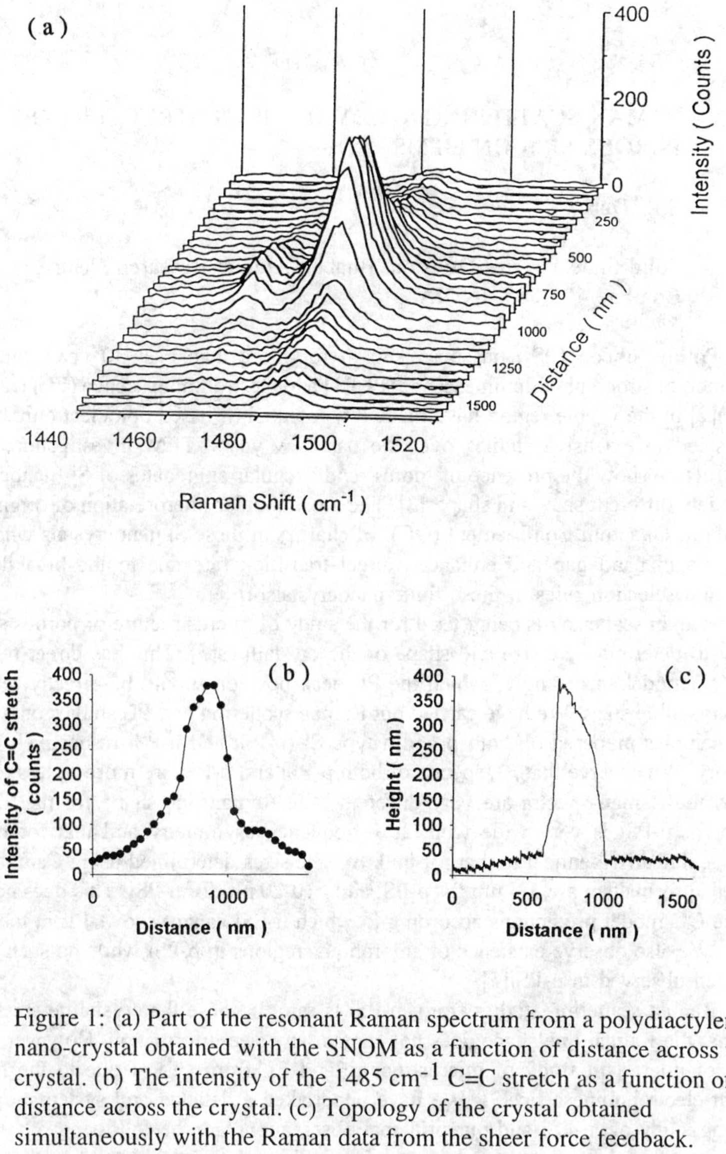

Figure 1: (a) Part of the resonant Raman spectrum from a polydiactylene nano-crystal obtained with the SNOM as a function of distance across the crystal. (b) The intensity of the 1485 cm^{-1} C=C stretch as a function of distance across the crystal. (c) Topology of the crystal obtained simultaneously with the Raman data from the sheer force feedback.

1. D.W.Pohl, W.Denk, M.Lanz. Appl. Phys Lett. 44, 651 (1984)
2. A.Lewis, M.Isaacson, A.Harootunian, A.Murry, Ultramicroscopy 13, 227 (1984)
3. E.Betzig, P.L.Finn, S.J.Weiner Appl. Phys. Lett. 60, 2484 (1992)
4. S.Webster, Ph.D thesis, University of Leeds (1994)

RAMAN SCATTERING STUDY OF MICROSTRUCTURE OF POROUS SILICON FILMS

S.K. Deb

Solid State Physics Division, Bhabha Atomic Research Centre,
Bombay 400 085, INDIA

Porous silicon (PSi) has been known for more than past 30 years but recent observation of strong photoluminescence (PL)[1] and electroluminescence (EL) (relatively weaker)[2] in the visible region have made it a potential Si based optoelectronic material and has led to extensive studies over the past few years. TEM investigations of the microstructure show the presence of 'pores' and irregular aggregates of Si 'nanocrystals' with widely different sizes and shapes[3]. The most popular interpretation of origin of PL is based on 'Quantum Confinement (QC)' of charges in these Si nanocrystals which lead to increase in band gap and enhanced direct transition rate due to the breakdown of momentum selection rules in these finite nanocrystals[4].

Raman scattering is being used for the study of microstructure of porous silicon - mainly to determine the size and shape of the crystallites[5]. This has direct relevance to the QC model according to which the PL peak position should be directly related to the Si-crystallite sizes. We have carried out Raman scattering and PL studies on a variety of PSi samples prepared on both p and n type Si (p-PSi and n-PSi respectively) in our laboratory. We observe that PL spectra of both p-PSi and n-PSi are more or less identical whereas the Raman spectra are very different. The Raman line shape for the 521 cm^{-1} phonon for p-PSi is very wide with low frequency asymmetry and that for n-PSi is narrow and nearly identical to that for bulk Si. The sizes determined using Campbell and Fauchet[6] formalism are 3-4 nm for p-PSi and $\geq$10-20 nm for n-PSi. This does not agree with the QC model predictions according to which n-PSi cannot show PL in the visible region . We also observe existence of amorphous regions in p-PSi while no such regions have been observed in n-PSi[5].

Raman scattering studies from n-PSi is rare because the crystallite sizes are so large that the Raman line shape does not contain any size information. However there is revival of interest in study of microstructure and PL from n-PSi due to the reported efficient electroluminescence[7]. We have undertaken a detailed and systematic Raman scattering study of n-PSi and our principal observations are the following:

i) At very low incident laser power the line shape is identical to that for the bulk Si, however the integrated intensity is enhanced by a factor of $\approx$12-15 compared to that for the untreated Si substrate. The intensities and line shapes for the polarized (XX) and depolarized (XY) spectra are similar with a depolarization ratio $\approx$1.

ii) With increasing laser power the peak position shifts to lower frequency and the line shape shows low frequency asymmetry with corresponding increase in the line width. However the line shape changes back to the original immediately after the power is reduced to low value.

iii) The total intensity normalized to unit laser power decreases with increasing laser power and reduces by a factor of 3 - 4 when the power is increased from 25 mW to 900 mW.

Our results can be explained by considering two layered micro-structure of n-PSi film as revealed by recent TEM studies[8,9]. The top layer consists of *nanoporous* Si with average crystallite sizes ≈ 10 - 20 nm and the bottom layer is made up of *macroporous* Si of sizes ranging from 100 nm to a few μm. The enhancement of Raman signal from n-PSi was however first reported by Andrianov et al[10] and was interpreted as due to the 'dielectric confinement'[11] of incident radiation in the *macroporous* Si particles of sizes comparable to the wavelength of light. The confinement of light also leads to a complete loss of polarization characteristics of the scattered radiation due to multiple internal reflection. To explain the changes in line shape with increasing laser power we note that, due to the presence of large number of pores and considerable oxide, the 'effective' thermal conductivity of the top layer will be much poorer than the bottom layer. At low power there is no laser heating and the line shape will be similar for both the layers and identical to bulk Si. However with increase in laser power, the laser heating induced temperature rise for the top layer will be more than that for the lower layer. So the Raman line shape for the top layer will be 'red-shifted' with respect to the lower layer giving rise to asymmetric and wider line shape. We assume lorentzian line shape for the layers and fit two lorentzians to the total intensity profile and we obtain good quantitative agreement between our observed line profiles and this two-layered model for all laser powers. Similarly the variation in normalized total intensity also can be explained within the framework of 'dielectric confinement' by considering the temperature induced changes in refractive indices of the *macroporous* Si layer. Further details will be presented in the talk and we will show that Raman scattering is indeed a very powerful nondestructive technique for probing microstructure of such systems and our results are perhaps the first unambiguous optical evidence for layered structure of n-PSi films.

The author would like to thank his collaborators Dr. A.P.Roy, Dr. S. Banerjee, Neelu Mathur, A.S. Sardesai with whom the work has been carried out and also Dr. B.M. Arora and Dr. L. Ramaniah for several illuminating discussions. Constant encouragement from Dr. K.R. Rao and Dr. B.A. Dasannacharya is also acknowledged.

1. R.L. Smith and S.D. Collins , J. Appl. Phys. 71, R1 (1992).
2. L.T. Canham, Appl. Phys. Lett. 57, 1046 (1990).
3. N. Koshida and H. Koyama., Appl. Phys. Lett. 60, 347 (1992).
4. V. Lehmann and U. Gosele, Appl. Phys. Lett. 58, 856 (1991).
5. S.K. Deb, Ind. Jr. Pure & Appl. Phys. 32, 638 (1994) and references therein.
6. I.H.Campbell and P.M. Fauchet, Solid State Commun. 58, 738 (1986).
7. F. Kozlowski, P. Steiner, W. Lang and H. Sandmaier, Sensors and Actuators, A43, 153 (1994).
8. P.C. Searson, J.M. Macauley and F. M. Ross, J. Appl. Phys. 72, 253 (1992).
9. K. Thecomata,Y.Nakamura and C.Nittono, Jpn. Jr. Appl. Phys. 33,6432(1994).
10. A.V. Andrianov, L.V. Belyakov, D.N. Goryachev, D.I. Kovalev, O.M. Sreseli, I.D. Yaroshetskii and B.Ya. Averbuch, Semiconductors 28, 1213 (1994).
11. D.V. Murphy and S.R.J. Brace, Optics Letters, 8, 494 (1983).

MICRO-RAMAN STUDY OF EPITAXY FILMS GROWN BY LASER DEPOSITION

X. B. Wang[1], Z. X. Shen[1*], C. K. Ong[1], S. Y. Xu[1], S. Z. Wang[2], S. H. Tang[1] and M. H. Kuok[1]

[1]Department of Physics, National University of Singapore, Lower Kent Ridge Road, Singapore 119260

[2]Department of Physics, Beijing University, People's Republic of China.

Raman spectroscopy is an established technique in the study of high temperature superconductors. For example, phonon anomalies have been observed in $YBa_2Cu_3O_{7-\delta}$ superconductors near Tc [1,2]; polarized Raman spectra can be used to analyze the crystal axis orientation in single crystals, epitaxy and textured films [3,4]; the oxygen content can be determined by the frequency shift of the Cu-O bridging vibration at around 500 cm^{-1} [5,6].

In the present paper, epitaxy films produced by laser deposition are studied using a Renishaw System 2000 Ramascope. The degree of axis orientation of the films are investigated using polarized Raman spectra (Fig. 1). For the c-axis oriented samples, the degree of a-b axis twinning can be determined by the relative intensities of the Raman peaks taken at different polarizations and it is found to be closely related to the preparation conditions, e.g. the substrate temperature. Systematic study of the influence of growth conditions on the film quality are being carried out.

Non-superconducting phases such as CuO, $BaCuO_2$, $Ba_2Cu_3O_6$ and $YBa_2Cu_3O_6$ are often found [7], and they are readily identified using Micro-Raman spectroscopy. CuO, $BaCuO_2$ and $Ba_2Cu_3O_6$ mainly exist as precipitates. Their distribution and size are mapped by step-scan method and correlated to the growth conditions. The distribution of the non-superconducting phase $YBa_2Cu_3O_6$ can also be mapped using the Cu-O bridging mode at about 500 cm^{-1} which shifts by 30 cm^{-1} between the non-superconducting phase and the fully oxygenated phase $YBa_2Cu_3O_7$.

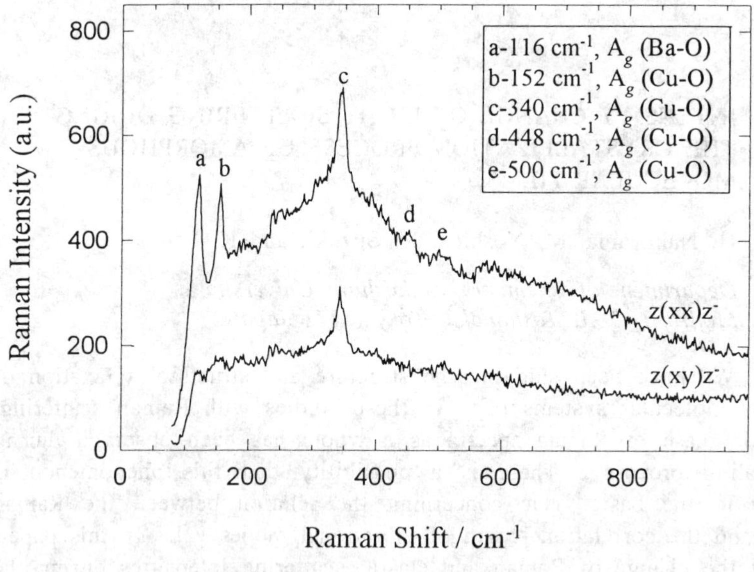

Fig. 1. Polarized Raman spectra of a laser deposited epitaxy film.

This study is supported by the National University of Singapore and the Defense Science Organization.

1. A. P. Litvinchuk, C. Thomsen and M. Cardona, Solid State Commun. 83, 343 (1992).
2. R. M. Macfalane, H Rosen and H Seki, Solid State Commun. 88, 843-846 (1993).
3. K. F. McVarty, J. Z. Liu, R. N. Shelton and H. B. Radousky, Phys. Rev. B41, 8792 (1990).
4. C. Thomsen, R. Wegerer, H. U. Habermeier and M. Cardona, Solid State Commun. 83, 199 (1992).
5. S. Fantini, L. Ulivi and M. Zoppi, Solid State Commun. 93, 519 (1995)
6. C. Thomsen, R. Liu, M. Bauer, A. Wittlin, L. Genzel, M. Cardona, E. Schonherr, W. Bauhofer and W. Konig, Solid State Commun. 65, 55 (1988).
7. N. Dieckmann, A. Bock and U. Merkt, Phsica C 244, 35 (1995)

INTENSITY CHANGE OF LIGHT SCATTERING DURING THE CRYSTALLIZATION PROCESS OF AMORPHOUS MOLECULAR FILMS.

H. Nakayama, M. Yoshida, K. Suzuki, and K. Ishii

Department of Chemistry, Gakushuin University, Mejiro 1–5–1, Toshimaku, Tokyo, 171 Japan

We have been studying the structure and structural relaxation of amorphous molecular systems [1]. In these studies with Raman scattering, intensity increase of Raman spectra as a whole has been observed during the annealing process. There is a possibility that this phenomenon is connected to the basic issue concerning the relation between the Raman intensity and the correlation length of vibrational modes [2]. In this paper, we report the change of Raman and elastic–scattering intensities during the crystallization process of amorphous molecular films kept at constant temperatures, and discuss the basic issue concerning the light scattering.

Amorphous films with the thickness of 10 μm were prepared by vacuum deposition onto gold–coated copper substrates kept at 78 K or 15 K. The 632.8 nm laser light and the spontaneous emissions from a 25 mW He–Ne laser were used as the Raman–excitation and elastic–scattering sources respectively, with the incident angle of 55°. The scattered light was collected at the right angle and analyzed with a single monochromator and a CCD. No polarization analyzer was used.

We used chlorobenzene as the sample material. Amorphous chlorobenzene crystallizes in a narrow temperature range around 105 K when the sample temperature is raised stepwise. However, structural relaxation occurs even if the sample is kept at a constant temperature slightly lower than 105 K. Figure 1 shows the evolution of the Raman spectra of amorphous chlorobenzene during the annealing at 103 K. This sample was prepared at 78 K. The times indicated in the figure represent the time lapse after the sample temperature was set at 103 K. The bands at 1001 and 1021 cm^{-1} are Raman bands arising from the breathing and in–plane–deformation modes of the benzene ring, respectively. On the other hand, the bands at 960 and 1008 cm^{-1} indicated with * marks are attributed to the elastic scattering of spontaneous emissions from the laser. The change in the integrated intensity of these bands are shown in Fig. 2. After the time lapse of 20 min at 103 K, increase in the intensities both of the Raman and elastic scattering started to take place. However, the intensity change of the elastic scattering was apparently more rapid than that of the Raman scattering. In addition, elastic–scattering intensity had a maximum value at 50 min and then showed a decrease after this point. On the other hand, the Raman–scattering intensities showed monotonous increases and were saturated finally; the behavior of the two Raman bands being slightly different.

A gradual appearance of factor–group–split bands of other vibration mode was observed along with the intensity changes mentioned above. Thus these intensity changes are related to the crystallization. The crystallization process in amorphous molecular films includes the homogeneous generation of a large number of nucleation sites and the growth of crystalline particles [3]. Therefore, to understand the above light scattering results, we must take account of the size dependence of the scattering by a single particle as well as the effect of the multiple scattering. Mie's theory accounts for the elastic scattering by a single particle. On the other hand, the Raman intensity depends on the correlation length of the vibrational mode [2]. The different changes in the Raman and elastic–scattering intensities suggest the contribution of the correlation length of vibrational mode to Raman intensity.

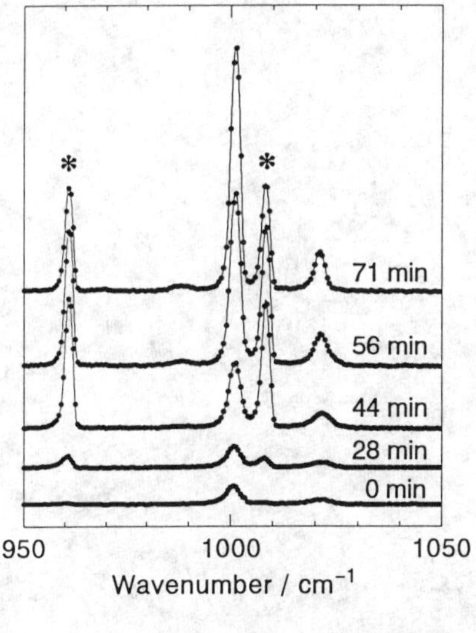

Fig. 1. Evolution of Raman spectra of amorphous chlorobenzene kept at 103 K.

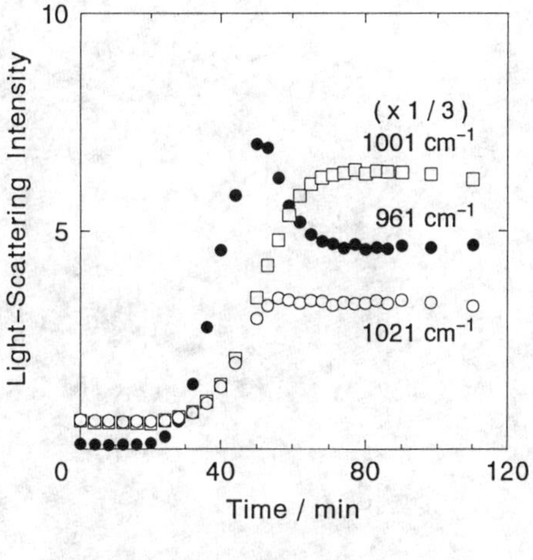

Fig. 2. Intensity change of Raman and elastic–scattering bands.

1. K. Ishii, H. Nakayama, M. Kawahara, K. Koyama, K. Ando, and J. Yokoyama, *Chem. Phys.,* **199** (1995) 245.
2. R. Shuker and W. Gammon, *Phys. Rev. lett.*, **25** (1970) 222.
3. K. Ishii and H. Nakayama, *Thin Solid Films,* in press.

SECTION 18

INORGANIC MATERIALS

RESONANCE RAMAN IN THE EXTENDED NEAR INFRARED: EXPERIMENTAL TIME DEPENDENT SCATTERING STUDIES OF MIXED-VALENCE SYSTEMS

Vladimir I. Petrov[1], Robert D. Williams, and Joseph T. Hupp*

Dept. of Chemistry, Northwestern University, Evanston, IL 60208 U.S.A.
[1]Permanent address: S. I. Vavilov State Optical Institute, St. Petersburg, Russia

Nonresonant Raman spectroscopy in the near infrared (NIR) has become popular because of its ability to eliminate fluorescence interferences. These interferences are absent because electronic transitions (usually) are not found at such low energies. Low energy electronic transitions do exist, however, for an important class of electron transfer model compounds, namely, symmetrical mixed-valence ions [1]. These ions can exist either in localized (discrete valence; class II) or delocalized (averaged valence; class III) form, where the form obtained depends on the relative strengths of localizing (vibrational and solvational) versus delocalizing (electronic) forces. In principle, Raman spectroscopy can provide unique quantitative insight into the key vibrational factors defining the electronic form. Toward that end, we have recently shown that resonance Raman spectroscopy can be implemented in the *extended* NIR (1320 nm excitation) for dilute solutions (ca. 2 to 8 *mM*) of both class II and class III chromophores [2-4], despite ν^4 effects and severe detector noise problems. Here we describe the resonance Raman spectroscopy of a delocalized mixed-valence ion (**1**) that can be reversibly converted to localized form via partial encapsulation [5] and supramolecular symmetry reduction by a polyether macrocycle such as dibenzo-36-crown-12 (**2**). To motivate the quantitative interpretation of the spectroscopy we make extensive use of a time-dependent overlap (gaussian wavepacket propagation) methodology popularized by Heller [6].

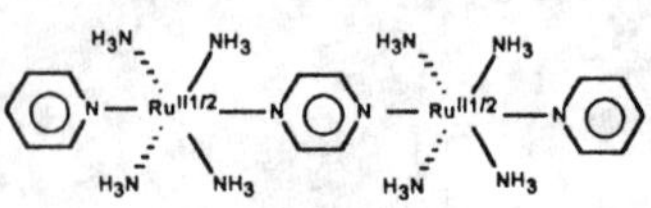

1

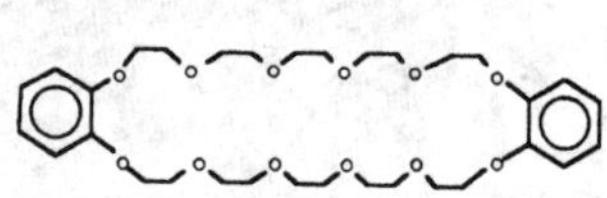

2

Figure 1 shows uncorrected resonance Raman spectra (1320 and 1337 nm excitation) for **1** (lower curve) and **1**•crown (upper curve) in nitromethane-d_3 as solvent. The former is dominated by peaks corresponding to symmetric displacements along the metal-bridging ligand (pyrazine)-metal axis. As we have noted elsewhere for a related chromophore [3,4], enhancement of pyrazine modes like ν_{6a} (674 cm^{-1}) can occur only if: a) a delocalized ground-state electronic structure exists, and b) nominal three-center bonding (Ru(dπ)/pz(π*)/Ru(dπ)) exists. (Enhancement was demonstrated by comparison to spectra obtained with λ_{exc} = 1064 nm (post resonant excitation) and by examining the fully reduced (4+) and fully oxidized (6+) forms of **1**). Consistent with this description and with the observed pattern of enhancement, excitation in the extended NIR leads to redistribution of charge from the bridge to both metal centers simultaneously.

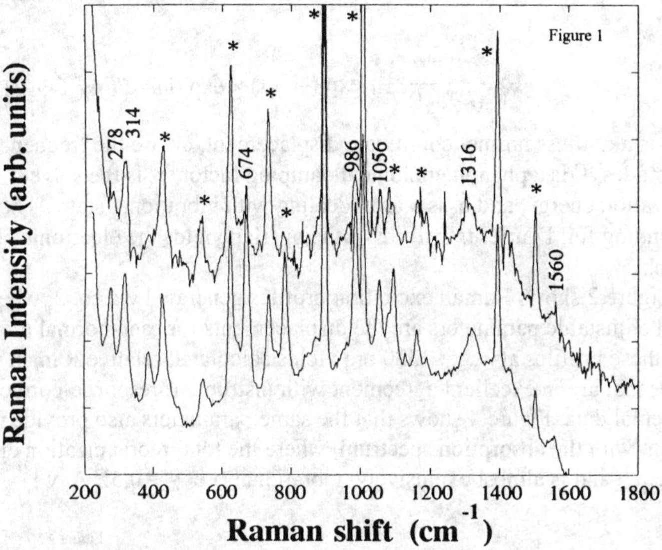

Partial crown encapsulation (via ether oxygen/ammine hydrogen interactions) largely suppresses scattering from symmetric bridging modes, but appreciably enhances modes that distort the bridging ligand in an asymmetric fashion (e.g. ν_{18a} (1056 cm^{-1})). While the latter represent nontotally symmetric modes of both free pyrazine and the delocalized mixed-valence ion, they are totally symmetric modes of the valence localized ion. Indeed, their appearance in the intervalence-enhanced Raman spectrum can be taken as *prima facie* evidence for single-metal localization (or partial localization) of the odd electron in **1**. Qualitatively, the appearance of these types of modes can be ascribed to the induction of significant metal-metal charge transfer character in the intervalence transition, as shown qualitatively in eq. 1:

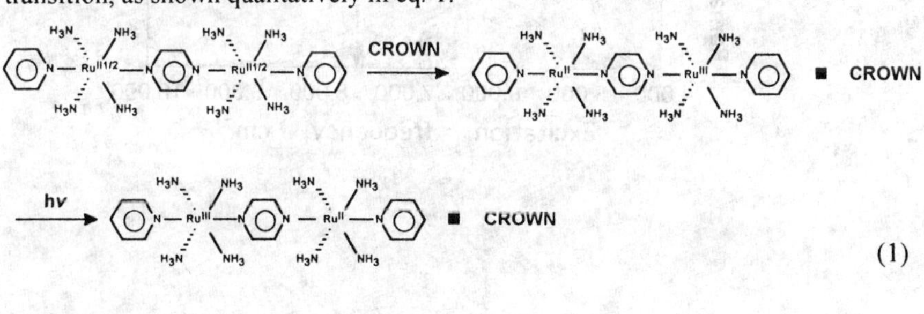

(1)

A more quantitative description of vibrational coupling to the NIR electronic transition can be obtained by using a time dependent scattering analysis. In this analysis, the scattering or polarizability tensor can be related to the half Fourier transform of the overlap of a vibrational wavepacket ($\phi(t)$) moving on the initial (i.e. upper) electronic surface and an excited vibrational wavefunction (ϕ) on the final (lower) electronic surface:

$$<\phi/\phi(t)> = \prod_k \{ \exp [-\frac{\Delta_k^2}{2} (1-\exp(-i\omega_k t)] \times (1-\exp(-i\omega_k t))^{n_k}$$

$$\times \frac{(-1)^{n_k}\Delta_k^{n_k}}{(2^{n_k} n_k!)^{1/2}} \} \exp(-i\omega_0 t) \times \exp(i\omega t - \Gamma t - i\chi_s t - \chi_s kTt^2) \qquad (2)$$

where Δ is a unitless normal coordinate displacement, ω_k are the frequencies of the normal modes, Γ is a phenomenological damping factor, χ_s is the solvent reorganization energy, and n_k is a term dealing with vibrational state. The corresponding full Fourier transform of the overlap yields the electronic absorption spectrum.

Figure 2 shows Raman excitation profiles generated via eq. 2, where the pertinent adjustable parameters are the displacements for each normal mode. A slice through these profiles at λ_{exc} = 1320 nm yields calculated relative Raman intensities (figure 3) that are in excellent agreement with instrument-response-corrected experimental data. Figure 4 shows that the same parameters also provide reasonable agreement with the absorption spectrum, where the total reorganization energy (χ) is ca. 900 cm^{-1} and is almost exclusively vibrational (i.e. $\chi = 0.5\Sigma\, \Delta_j^2 v_j$).

Figure 2

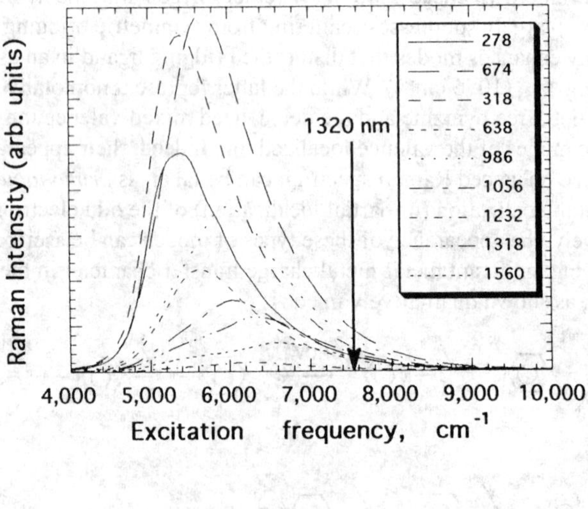

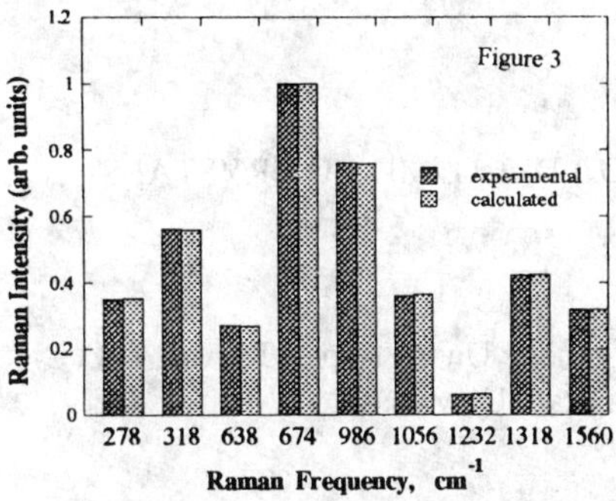

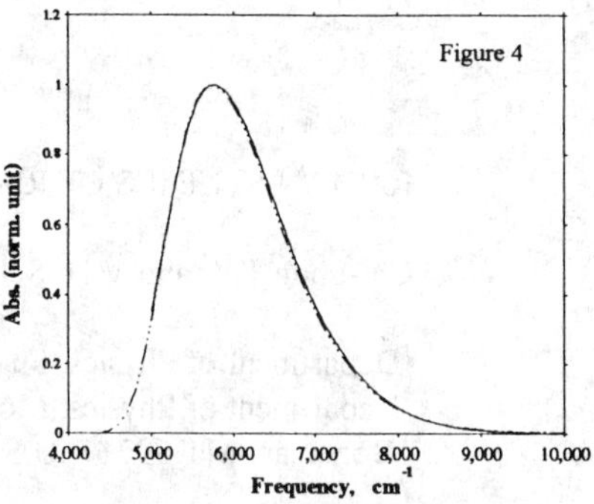

Figures 5 and 6 illustrate how the corrected scattering and absorption spectra transform following crown-induced valence localization. The time-dependent analysis shows that the total reorganization energy has: a) increased to ca. 3200 cm^{-1}, b) redistributed itself into modes that asymmetrically distort the bridge, and c) acquired a small, but significant solvent component. In summary, the Raman experiments and analysis provide the first clear experimental picture of how electronic delocalization (and subsequently localization) is induced and sustained in these unusual model compounds.

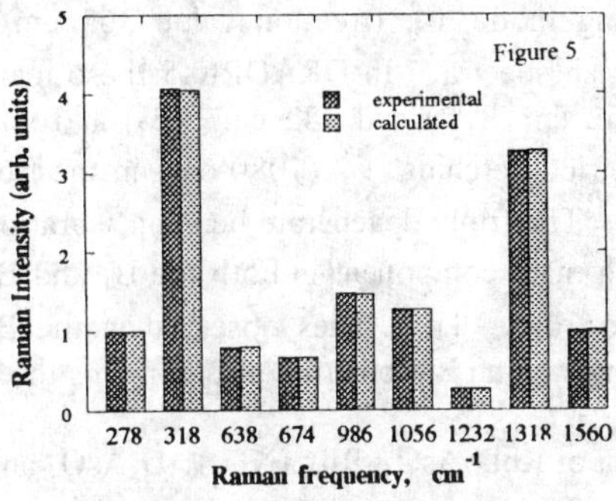

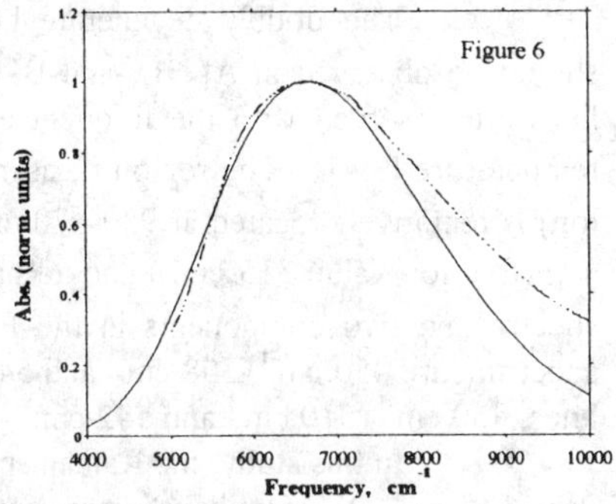

Acknowledgments: We thank the U.S. Department of Energy, Office of Energy Research, Division of Chemical Sciences (Grant No. DE-FG02-87ER13808), the NSF (graduate traineeship for R.D.W.) and the Dreyfus Foundation (Teacher-Scholar Award to J.T.H.) for support of this work.

1. R. J. Crutchley *Adv. Inorg. Chem.* **1994**, *41*, 273.
2. V. I. Petrov, R.D. Williams, H.P. Lu, J.T. Hupp *Chem. Phys. Lett.* to be submitted.
3. V. I. Petrov, J.T. Hupp, C. Mottley, L.C. Mann *J. Am. Chem. Soc.* **1994**, *116*, 2171.
ñ. H. Lu, V. I. Petrov, J. T. Hupp *Chem. Phys. Lett.* **1995**, *235*, 521.
5. Y. Dong, J. T. Hupp, D. I. Yoon *J. Am. Chem. Soc.* **1993**, *115*, 2048.
6. E. J. Heller *Acc. Chem. Res.* **1981**, *14*, 368.

RAMAN STUDIES OF $Rb_{1-x}(ND_4)_xD_2AsO_4$ MIXED CRYSTALS

Chi-Shun Tu[1*] and V.H. Schmidt[2]

[1]Department of Physics, Fu-Jen Catholic Univ., Taipei, Taiwan, R.O.C.
[2]Department of Physics, Montana State University,
Bozeman, MT 59717, U.S.A.

DRADP-x mixed crystals in the concentration range $0.3 \leq x \leq 0.7$ have tetragonal symmetry (space group $I\bar{4}2d - D_{2d}^{12}$) of the KDP type down to liquid-He temperature. The factor-group analysis of normal vibrations for such a structure where both PO_4 and ND_4 tetrahedral occupy sites S_4 local symmetry is well known [1]. According to the factor-group analysis of the ideal tetragonal KDP structure, the fully symmetric stretching vibration ν_1^P of the PO_4 ion should be observed only in the $A_1 - Y(ZZ)X$ geometry. The frequency of this mode in a free ion is 980 cm^{-1} [2]. In DRADP-0.5 this mode has a frequency of 882 cm^{-1} [3,4].

The doubly degenerate bending mode ν_2^P (free-ion value 363 cm^{-1}) should be observed in A_1, B_1, and B_2 Raman spectra. In DRADP-0.5 these main lines are observed with the frequencies 352 cm^{-1} (A_1) and 382 cm^{-1} (B_2) at room temperature [3,4]. The region of asymmetric stretching ν_3^P (1080 cm^{-1} in the free ion) is tentatively located at 930-970 cm^{-1}. The triply degenerate bending vibration ν_4^P (free-ion value 515 cm^{-1}) should have a main component in both the B_1 and B_2 spectra and two components in the E spectrum. Three lines observed in the B_1 spectrum are 450 cm^{-1}, 508 cm^{-1} and 540 cm^{-1}. The B_2 spectrum also contains three lines: 450 cm^{-1}, 510 cm^{-1} and 542 cm^{-1}.

In this study, the Raman spectra of RbD_2AsO_4, $Rb_{0.9}(ND_4)_{0.1}D_2AsO_4$ and $Rb_{0.72}(ND_4)_{0.28}D_2AsO_4$ have been investigated at room temperature. Special attention is focused on the $A_1 - \nu_1^A$ and $A_1 - \nu_2^A$ vibrations of the AsO_4 group, because they are sensitive to the ordering of the deuterons. The green light with λ=514.5 nm from an Argon ion laser was used as an excitation source. A double grating monochromator (ISA Model U1000) was used. The right-angle spectra were measured with geometry $Y(ZZ)X$. Here X, Y and Z correspond to the crystal a-, b- and c-axes, respectively. The vibrations from scattering configuration $Y(ZZ)X$ is associated with the A_1 mode. The damped harmonic oscillator model has been used to find the position and half-width of various vibrational modes, i.e.

$$S(\omega) = \frac{\chi_o \Gamma \omega \omega_o^2}{\left(\omega^2 - \omega_o^2\right)^2 + \Gamma^2 \omega^2} \cdot \frac{1}{1 - e^{-\hbar\omega/\kappa T}} \qquad (1)$$

where ω_o and Γ correspond to the mode frequency and half-width, respectively, χ_o is the susceptibility constant, k is Boltzmann's constant and T is absolute temperature.

Fig. 1 shows the Raman spectra of the three different compounds between 80 cm^{-1} and 1200 cm^{-1}. The main corresponding frequencies (labeled 1–5) are summarized in Table 1. As compared with DRADP-0.5, these frequencies of ν_1^A, ν_3^A and ν_4^A are smaller than those of the PO_4 group [3,4] and can be attributed to the difference of atomic mass (As and P). The ν_2^A frequency is larger than in DRADP-0.5, which implies that the force constant of the ν_2^A vibration mode is stronger in DRADA-x compounds.

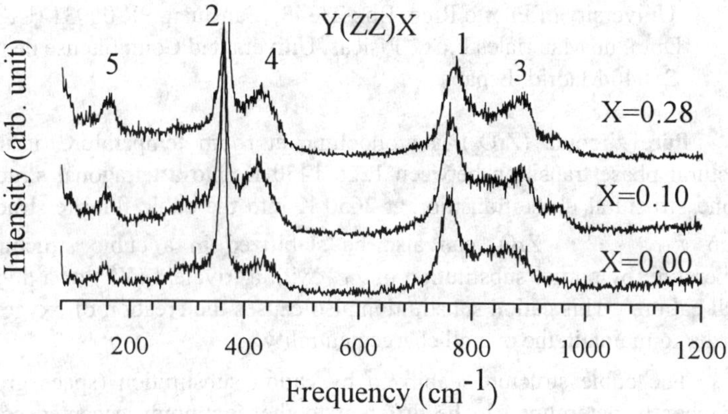

Fig. 1 Raman spectra of A_1 symmetry for DRADA-x at 300 K.

TABLE 1. Main frequencies (cm^{-1}) for internal vibrations ν_i^A of the AsO_4 group and ND_4 libration (ν_L^N).

$Rb_{1-x}(ND_4)_xD_2AsO_4$	$1\leftrightarrow\nu_1^A$	$2\leftrightarrow\nu_2^A$	$3\leftrightarrow\nu_3^A$	$4\leftrightarrow\nu_4^A$	$5\leftrightarrow\nu_L^N$
x=0.00	760	362	877	436	156
x=0.10	762	363	884	424	158
x=0.28	770	366	880	429	161

1. E. Courtens and H. Vogt, J. Chem. Phys. 82, 317 (1985).
2. G. Herzberg, *Infrared and Raman Spectra of Polyatomic Mol.*, (Princeton, 1966).
3. Y.I. Yuzyuk and I. Gregora *et al.*, J. Phys. Cond. Matter 7, 683 (1995).
4. J.L. Martinez, F. Agullo-Rueda and V.H. Schmidt, Ferroelectrics 76, 23 (1987).

RAMAN SCATTERING STUDY OF THERMOCHEMICALLY REDUCED Y-STABILIZED CUBIC ZrO_2

G. Morell[a], R. S. Katiyar[b], D. Torres[c], S. E. Paje[c], and J. Llopis[c]

[a]Department of Physical Sciences, College of General Studies
University of Puerto Rico, Box 23323, San Juan, PR 00931 (USA)
[b]Department of Physics, College of Natural Sciences
University of Puerto Rico, Box 23343, San Juan, PR 00931 (USA)
[c]Dpto. de Materiales F.CC. Físicas, Universidad Complutense de Madrid, 28040 Madrid, España

Pure zirconia (ZrO_2) is monoclinic at room temperature, it undergoes a structural phase transition between 1225-1450 K into a tetragonal structure, and a second structural phase transition at 2650 K into the cubic fluorite structure (space group $Fm\bar{3}m$).[1] ZrO_2 can also be stabilized in a cubic structure at room temperature by partial substitution of Zr^{4+} with a trivalent (Y^{3+}) or a divalent (Ca^{2+}) metal cation.[1] This cation substitution also causes the creation of oxygen vacancies in order to maintain the overall charge neutrality.

The cubic structure stabilized by cation substitution (space group $P\bar{4}3m$) was recently determined to be different to the commonly accepted cubic fluorite structure (space group $Fm\bar{3}m$).[2] Therefore, the observed bands have to be reassigned according to the character table of the new group (T_d in this case). T_{2g} becomes T_2, E_g becomes E, and A_{1g} becomes A_1. In addition, now T_2 is also infrared active, while in the O_h group T_{2g} is not.

We are considering here the room temperature polarized Raman spectra of untreated and thermochemically reduced yttrium-stabilized cubic zirconia (9.5 mol % Y_2O_3). The advantage of having the untreated and reduced cubic states frozen at room temperature is that thermal expansion and anharmonic temperature effects are equivalent in the two materials so that the spectral differences should be accounted for by intrinsic structural changes. All the experiments were done in the backscattering configuration with the laser beam pointing along the crystallographic z-axis and polarized along the x-axis . Therefore, the $Z(XY)\bar{Z}$ geometry gives the T_2 symmetry species, while the $Z(XX)\bar{Z}$ geometry gives the E + A_1. Figure 1 shows the T_2 spectra of the untreated and thermochemically reduced samples. Figure 2 shows the E + A_1 spectra of the same two samples. The bands are broad phonon density of states due to the disorder induced Raman activity of all phonons along the Brillouin zone. Although the cation sublattice shows periodicity, the oxygen sublattice is disordered since one O^{2-} vacancy is created for every two Y^{3+} entering the lattice and the oxygens are slightly displaced from their ideal positions. In addition, there is mass disorder in the cation sublattice from the Y^{3+} substitution.

The broad continuum at frequencies below 500 cm^{-1} diminishes in intensity with respect to the band 600 cm^{-1} as a function of the reducing treatment for both the XY and the XX polarization geometries. The Raman scattering in this region is predominantly due to acoustic phonons at the Brillouin zone edge, where the phonon density of states (PDOS) is maximum. The acoustic branches have linewidths proportional to q^2 due to the disorder, therefore becoming poorly defined at large wavevectors[3] and Raman active. These modes arise from the in-phase motion between the cation and oxygen sublattices.[3] On the other hand, the high-frequency region mainly contains contributions form the optical branches. Therefore, the observed reduced Raman activity in the acoustic mode region as a function of thermochemical reduction indicates that the lattice in the reduced material is less distorted. This can be understood as a result of the additional oxygen vacancies in the reduced material leaving more room for the yttrium atoms which are larger than the zirconium atoms, minimizing in this way their disruption of the cationic lattice. In addittion, oxygen vacancies can distribute themselves more uniformly, so that the oxygens staying in the lattice do not have a preferential displacement direction, therefore minimizing their static distortions.

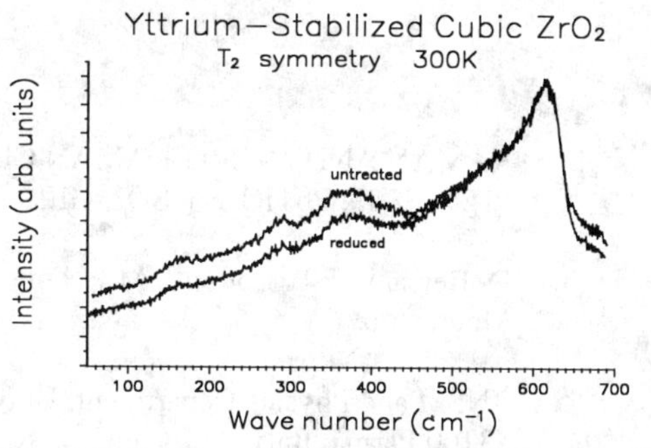

Figure 1. Spectra of the untreated and reduced samples taken in the Z(XY)Z geometry.

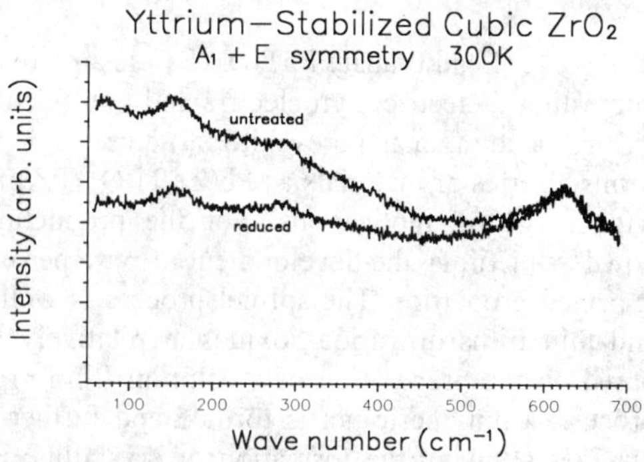

Figure 2. Spectra of the untreated and reduced samples taken in the Z(XX)Z geometry.

This work was supported by NSF Grant OSR-9452893 (USA) and DGICYT PB93-0428 (Spain).

[1] J. Cai, C. Raptis, Y.S. Raptis, and E. Anastassakis, Phys Rev B, 51, 201 (1995)
[2] K.J. McClellan, S.Q. Xiao, KPD Lagerlof, AH Heuer, Phil Mag A, 70, 185 (1994)
[3] D.W. Liu, C.H. Perry, A.A. Feinberg, R. Currat, Phys Rev B 36, 9212 (1987)

RAMAN AND MICRO-RAMAN CHARACTERIZATION OF PHASES IN $PbTiO_3$ BY SOL-GEL

D. Bersani[1], P.P. Lottici[1*], M.P. Fontana[1], M. Polli[1], G. Gnappi[2], A. Montenero[2]

[1]INFM and Physics Department, University, Viale delle Scienze, 43100 Parma, Italy
[2]Chemistry Department, University, and CORIVE, Consorzio Ricerca Innovazione Vetro, Viale delle Scienze, 43100 Parma, Italy

Lead titanate $PbTiO_3$ (PT) is a perovskite type ferroelectric material with interesting dielectric, pyroelectric and piezoelectric properties. At room temperature PT has a tetragonal perovskite structure and, when combined with other oxides, forms a series of materials as $Pb(Zr,Ti)O_3$ (PZT), $(Pb,La)(Zr,Ti)O_3$ (PLZT) having a wide range of applications. For the production of PT based materials, both as powders or films, the development of pure perovskite phase is required to obtain the expected properties. The sol-gel process is well suited to obtain powders, ceramics and thin films of various oxides at relatively low temperatures. The technique is based on the preparation of a solution from organic precursors. After a hydrolysis process, a polymeric gel is formed and further heat treatments remove the organic residues allowing the formation of crystalline phases. Sol-gel derived PT powders often crystallise directly into the perovskite structure [1] but sometimes give first cubic paraelectric pyrochlore $Pb_2Ti_2O_{7-x}$ phase [2] and subsequently transform into perovskite at higher (500-700°C) temperatures. Our previous Raman studies on sol-gel PT [3] have shown that the phase evolution is strongly influenced by the nature of the starting solution and by the hydrolysis conditions. Starting with titanium isopropoxide (TPOT), which is distilled into titanium methoxide (TMET), and lead acetate (LAT) as organic precursors and methoxyethanol as solvent, an intermediate pyrochlore phase before the final perovskite is formed. A minor $PbTi_3O_7$ phase is also present. On the other hand, a pure perovskite phase is obtained when no distillation is performed but HCl is added to the solution as catalyst [4].

To assess the role of the HCl addition, a new preparation without catalyst or distillation has been performed and investigated by XRD and Raman techniques. The Raman spectra of the powders treated at temperatures from 500°C to 700 °C, taken at 488.0 nm, are shown in Fig.1. At 500 °C the main Raman features may be attributed to lead oxide PbO [5]: the spectrum is similar to that previously reported for the pyrochlore phase [3], but the XRD patterns indicate the presence of PbO and perovskite without evidence of pyrochlore or lead trititanate phases. At higher temperatures the spectrum of the perovskite phase emerges and at 700 °C one still observes PbO and perovskite.

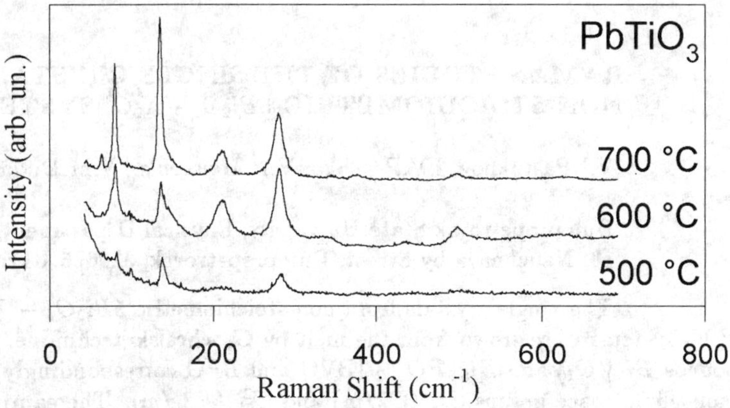

Fig.1 Raman spectra of $PbTiO_3$ powders treated at different T.

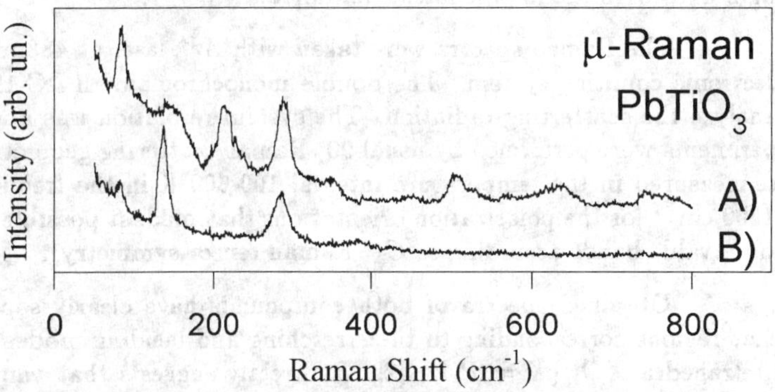

Fig.2 Micro-Raman on different grains in $PbTiO_3$ powder (700 °C).

Through the microscope, the powder treated at 700 °C appears to be made of two types of grains: A) with clear, homogeneous aspect and B) with darker and rough appearance. The B) grains often look like protuberances on bigger A) grains. Micro-Raman (magnification x100) measurements on single different A) and B) grains are reported in Fig.2. The typical spectrum of the perovskite is observed for A) grains, whereas the unambiguous PbO features are obtained in the B) case, indicating that the present preparation route of $PbTiO_3$ gives PbO segregation. The use of HCl as catalyst seems therefore to be the best way to obtain pure perovskite phase starting from lead acetate and TPOT.

1. J. B. Blum, S. R. Gurkovich, J. Mater. Science 20, 4479 (1985).
2. R. A. McCauley, J. Appl. Phys. 51, 290 (1980).
3. D. Bersani, P.P. Lottici, A. Montenero, S. Pigoni, G. Gnappi, J. Non-Cryst. Solids 192&193, 490 (1995).
4. L. Lan, A. Montenero, G. Gnappi, E. Dradi, J. Mater. Science 30, 3137 (1995).
5. N.M. Sammes, G. Tompsett, A.M. Cartner, J. Mater. Science 30, 4299 (1995).

RAMAN STUDIES OF THE SINGLE CRYSTALS FROM NON-STOICHIOMETRIC $5Bi_2O_3 - X_2O_5$ SYSTEMS

V.I.Pastukhov, I.V.Prushko, V.N.Moiseenko, V.M.Duda

Dniepropetrovsk State University, Physical Department,
13, Nauchnaya by Street, Dniepropetrovsk, 320625 Ukraine

The single crystals from non-stoichiometric $5Bi_2O_3 - X_2O_5$ ($X = P, V$) systems were grown from the melt by Czochralski technique. Both compounds Bi_5VO_{10} and $Bi_{5.8}PO_{11.2}$ (BVO and BPO correspondingly) belong to monoclinic space groups C_{2h}^3 ($C2/m$) and C_{2h}^2 ($P2_1/m$). The samples were in the form of rectangular with one edge along monoclinic z-axes and another edge along x-axes with the largest lattice parameter.

The Raman spectra were taken with Ar^+-laser at 488nm and photoelectronic counting system. The double monochromator ДФС-12 was used to analyze the scatterring radiation. The system resolution was about 2cm^{-1}. Experiments were performed by usual 90° Raman scattering geometry. Spectra were measured in the temperature interval 100-600 K in the frequency range 10-1500 cm^{-1} for the polarization orientations that made it possible to identify all observable bands according to C_{2h} Raman tensor symmetry.

Obtained spectra of both compounds have clearly separated frequency regions corresponding to the stretching and bending modes of individual tetrahedra XO_4 ($X = P, V$) [1]. It directly suggests that vanadium and phosphorus cations are coordinated by four oxygen. It is possible to expect that due to the low monoclinic symmetry of crystalline structure degeneration of nine internal vibrational modes should be removed. The sum of nine simple dumping oscillators was used for fitting the Raman spectra in bending and stretching regions for BPO and satisfactory results for every single oscillator with damping about 40-50 cm^{-1} were obtained (Fig.1.). But for BVO crystals similar fitting is impossible - another number of simple oscillators with higher damping (about 65-80 cm^{-1}) is needed (Fig.2).

Observed bands in the regions free from individual tetrahedra vibrations (400-700 cm^{-1} for BVO and 1100-1400 cm^{-1} for BPO) allow to suggest condensation of XO_4 tetrahedra into more complicated polytetrahedral complexes like X_4O_{9-12} with the bridge $X - O - X$ bonds. The possible structure of such complexes in both compounds have been discussed. The high quality mode at 21 cm^{-1} in (zy) spectrum (Fig.2) may be connected with libration of tetrahedral V_4O_{10} units as whole.

No anomalies which may be connected with any phase transitions have been observed in both compounds.

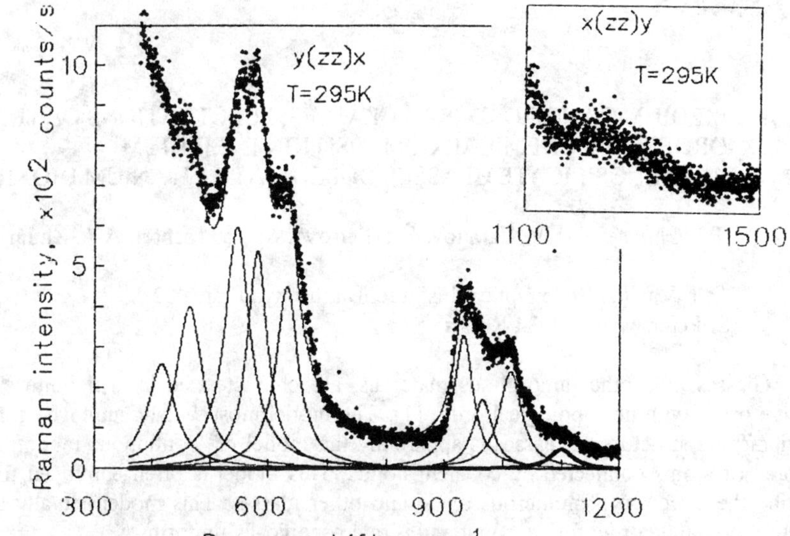

Fig.1. Raman spectra of $Bi_{5.8}PO_{11.2}$.

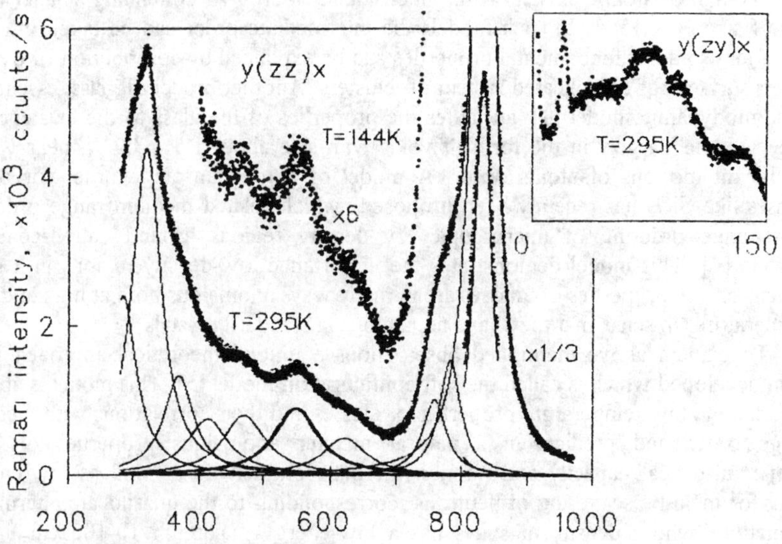

Fig.2. Raman spectra of Bi_5VO_{10}.

1.K.Nakamoto Infrared and Raman Spectra of Inorganic and Coodination Compounds, Moscow, "Mir" Publication, 1991.

MEDIUM-RANGE ORDER IN GLASSES: FLUCTUATIONS OF THE ORDER AND CHEMICAL COMPOSITION IN LITHIUM ALUMINOSILICATE GLASSES DOPED WITH TITANIUM DIOXIDE

T.I. Chuvaeva, A.V. Baranov, V.I. Petrov*, M.Ya. Tsenter, A.A. Zhilin

S. I. Vavilov State Optical Institute, Birzhevaya Liniya 12,
St.Petersburg 199034 Russia

Glasses, like other amorphous materials, are characterized by a definite short-range order with no topological order [1]. The model most widely employed for describing the structure of glasses is apparently the model of "continuous random network" of atoms connected by covalent bonds. This model is often employed to describe the structure of numerous oxide and other glasses. This model actually takes only short-range order into account and is microscopically uniform.

On the other hand, less disordered, but microscopically more nonuniform models are under discussion. Thus, in the model of Ref. 2 the glass looks more like a multi-bond system of atomic clusters with dimensions 30 - 100 Å, containing smaller clusters of size 10 - 15 Å. In such models the intermediate order can be taken into account and some of fundamental properties can be explained by contribution of overall "inner surface" of complicated system of clusters. Another model of glass, essentially a thermodynamic model [3], attributes the properties of the glass to the existence of "free volume" largely in the form of voids with size about 1.5 - 3.0 Å, near which significant motions of atoms occur. A model of a medium-range order for oxide glasses like SiO_2 has recently been proposed which related medium range order to short-range ordering of microscopic low-density regions around cation-centered clusters [4]. This model demonstrates the importance low-density regions in a glass structure. The properties of glasses are in many ways anomalous both at high and low temperatures in sense that they have no analogs in standard crystals.

To explain above mentioned observations a unified theoretical approach have been developed which is called the soft configuration model [5]. This model is able to describe the low-temperature properties of glasses and their correlations with medium range order and predicts in a natural manner anomalous properties of low-temperature heat capacity and phonon thermal conductivity as well as low-energy peak for inelastic scattering of neutrons corresponding to the quasilocal anharmonic excitations whose density of states has a low-energy maximum 10-100 cm^{-1}. This peak may be also associated with well known "Boson maximum" in IR and Raman scattering observed in non-metallic glasses.

In this work the nature of "Boson maximum" has been investigated for lithium aluminosilicate glasses doped TiO_2 (C = 4 - 10 mole %) with using low-frequency Raman spectroscopy technique. These glasses under a secondary heat treatment with temperature 600-750^0 C exhibit a phase decay with formation of very small crystalline aluminotitanate particles with sizes about 20 Å or more, which could be determined by small-angle x-ray scattering (SAXS). The Raman spectra of initial glasses without secondary heat treatment display a broad asymmetric low-frequency peak with frequency 96 cm^{-1}. The former studies show that this peak is connected with elastic vibrations of inhomogeneity regions enriched Al_2O_3 and TiO_2 with sizes of 10-15 Å [6, 7] which can appear in multicomponent glasses during heating and cooling of the melt. These regions are caused by structural and chemical fluctuations of composition. The spectrum of elastic vibrations for a spherical shape of the inhomogeneity regions and how they are affected by the surrounding matrix was studied theoretically at Ref. 8. It was shown that it is a set of spherical and torsional modes with different values of angular momentum that depends on the longitudinal and transverse velocities of sound and frequencies proportional to reciprocal sizes of particles. SAXS data display no phase decay for these glasses.

First stages of a secondary heating lead to slow transformation of low-frequency peak in Raman spectrum. We observed slow diminish of frequency and narrowing of the band. On the final stage of heating we detected only spherical and torsional vibrations of nanocrystals of pseudobrucite and anatase with frequencies 60 and 35 cm^{-1}, respectively [9]. At this stage nanocrystals with size about 19 Å are easily detected by SAXS. It is important to emphasize that the Raman spectrum of final glasses does not contain in the low-frequency region any features ("Boson maximum") besides the bands of elastic vibrations of nanocrystals. Estimates of their sizes from Raman measurements coincide with SAXS data a fairly well. Therefore this glass is somewhat unprecendent amorphous material without medium-range order.

So, our data prove that "Boson maximum" in glasses is connected with elastic vibrations of structural and chemical inhomogeneities regions with sizes of few nm and does not connected with soft configuration of atom groups.

On the other hand, low temperature measurements of heat capacity have not been performed yet for this kind of glasses. We expect that the host matrix of such glasses would lose the anomalous temperature dependence of heat capacity and has almost "normal" (crystal-like) one.

1. R. Zallen. *"Physics of amorphous solids"*, J. Willey, New York, 1983.
2. J. Phillips *"Solid State Physics"*, edited by S.H. Ehrenreich and A. Turnbull, Academic Press, New York, 1983, vol. 27.
3. M.H. Cohen, G.S Grest. *Phys. Rev. B*, 20, 1077 (1979).
4. S.R. Elliot. *Phys. Rev. Lett.* , 67, 711 (1991).
5. M.I. Klinger. *Phys. Rep.*, 94, 183 (1983).
6. Ya.S. Bobovich, V.I. Petrov, M.Ya. Tsenter. *Opt. Spectrosc.*, 66, 462 (1990).
7. V.I. Petrov, Ya.S. Bobovich, A.A. Zhilin, M.Ya. Tsenter, T.I. Chuvaeva. *Opt. Spectrosc.*, 68, 216 (1990).
8. A. Tamura, K. Higeta, T. Ichinokawa. *J.Phys. C.* 15, 4975 (1982).
9. V.I. Petrov, Ya.S. Bobovich. *Opt. Spectrosc.*, 67, 363 (1989).

RAMAN SPECTRA OF MASS-SELECTED POLYATOMIC TRANSITION-METAL MOLECULES

Huaiming Wang, Hanae Haouari, Robert Craig, Yifei Liu, John R. Lombardi, and D.M.Lindsay

Department of Chemistry and Center for Analysis of Structures and Interfaces (CASI), The City College of New York(CCNY), New York, NY 10031

Raman spectra have been reported for Sc_3, Cr_3, Ni_3, Cu_3, Ag_3 transition-metal molecules[1,2]. Well-resolved vibrational progressions have been analyzed to give accurate spectroscopic constants for these cluster molecules. Since different n-mers were generally identified by their behavior under annealing or by intensity versus concentration studies, there was often considerable controversy as to the spectral carrier, even for sizes as small as the dimer. With the recent advent of mass-selected cluster deposition techniques, it has become possible to overcome these assignment problems[3]. Recently, we have reported resonance Raman spectra of two argon matrices isolated, mass-selected polyatomic transition-metal molecules, Ta_4[4] and Zr_3[5]. Here, we report on the observation of resonance Raman spectra of Nb_3 in argon matrices.

As previously described[6], Our samples were prepared by neutralizing a mass-selected beam of the trimer ion. In the case of Nb_3, the fragmentation is estimated to be less than 0.6%. Therefore, there is little or no interference between the spectroscopic transitions of the trimer and those from atom and dimer, which provides unambiguous assignment of spectral features. The absorption spectrum shows three overlapping transitions from 425 to 650 nm. Resonance Raman spectra were observed throughout the region 458.5 ~ 614.0 nm. The average Stokes shifts found for over 60 different excitation wavelengths are listed in Table 1. Typical Raman spectra are shown in Fig.1. The progressions observed by excitation into blue band(spectrum a) were somewhat more extensive than those observed in the red region(spectrum b), but otherwise no significant difference between the spectra were obtained. All the observed Stokes lines can be explained by assuming two fundamental frequencies at 334.9(ν_1) and 227.4(ν_2)cm^{-1}, as listed in Table 1. Since the linewidth of ν_2 is almost as twice as that of ν_1 and the ratio of these two numbers is 1.47, very close to the ratio($\sqrt{2}$:1=1.41) expected for an equilateral triangular geometry(D_{3h})[7], we expect that the Nb_3 molecule is almost D_{3h} in shape with slight distortion. In this case we assign the high frequency mode as the totally symmetric stretch a'_1 and the low frequency vibration to the doubly degenerate bend of symmetry e'. A metal-metal force constant and a stretch-stretch interaction constant calculated in D_{3h} geometry were 1.95 and 0.05 mdyne/Å, respectively.

TABLE 1. Observed resonance Raman transitions and assignments for the niobium trimer in argon matrix on the Al substrate.

obsd level, cm^{-1}	$\nu_1(a_1')$	$\nu_2(e')$
227.4±2.9	0	1
334.9±2.8	1	0
561.1±2.8	1	1
669.7±3.3	2	0
895.9±4.5	2	1
1003.8±2.8	3	0
1240.6±6.2	4	0

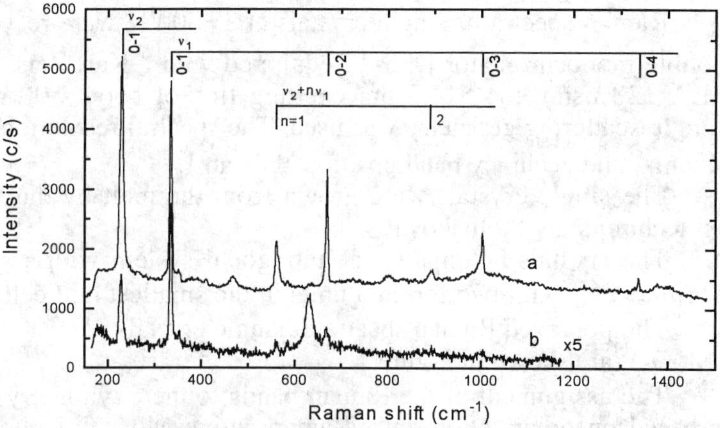

Fig.1. The resonance Raman spectra of Nb_3 in Ar matrices excited with (a)514.5nm; and (b) 588.8nm radiation at 14K on the Al substrate. The content of Nb_3 is 100nA·hours(1nA·hour = 2.25×10^{13} particles).

1. M.D. Morse, Chem.Rev. 86, 1049(1986).
2. M.M. Kappes, Chem.Rev. 88, 369(1988).
3. See, for example: W. Harbich, S. Fedrigo, F. Meyer, D.M. Lindsay, J. Lignieres, J.C. Rivoal and D. Kreisle, J.Chem.Phys. 93,8535(1990).
4. H. Wang, R. Craig, H. Haouari, J.-G, Dong, Z. Hu, A. Vivoni, J.R. Lombardi, and D.M. Lindsay, J.Chem.Phys. 103, 3289(1995).
5. H. Haouari, H. Wang, R. Craig, J.R. Lombardi, and D.M. Lindsay, J.Chem.Phys. 103, 9527(1995).
6. Z. Hu, B. Shen, Q. Zhou, S. Deosaran, J.R. Lombardi, and D.M. Lindsay, Proc. SPIE 1599, 65(1991).
7. G. Herzberg, *Infrared and Raman Spectra of Polyatomic Molecules*(Van Nostrand Princeton, 1945).

RAMAN SPECTRA OF ACOUSTOOPTIC $NaBi(MoO_4)_2$ CRYSTALS

V.N.Moiseenko, A.M.Jeryemenko, T.M.Stolpakova

Dniepropetrovsk State University, 13, Nauchnij by-street.
320625, Dniepropetrovsk, Ukraine

The properties of the single crystals of $NaBi(MoO_4)_2$ and their usefulness for acousto-optic applications have been reported previously by Kaminskii at al. [1].

The aim of this work is to study the Raman spectra of $NaBi(MoO_4)_2$ crystals in terms of spectral quality and interpretability.

Raman spectra of single crystals at T=300 K were recorded on a double monochromator DFS-12 equipped with a Karl Zeiss Ar^+ laser ILA-120 using the 514.5 nm exciting line of about 500 mW. Right angle scattering geometry was used. The spectral resolution was about 2 cm^{-1} and accuracy band position ± 1 cm^{-1}.

The single crystals were grown from the melt by the Czochralski technique by Stolpakova.

The crystals belongs to the tetragonal system with a space group $I4_1/a \equiv C^6_{4h}$ with two formula units in the smallest unit cell.

The polarized Raman spectra of single crystals of $NaBi(MoO_4)_2$ are shown in Figure 1.

The assignment of the Raman bands to their symmetry species is based on polarization data. Raman active Mo - O stretching band are localized in the 660-980 cm^{-1}: 2Ag + 2Bg +2Eg modes are expected. Raman active Mo - O bending vibrations are localized below 500 cm^{-1}: 3Ag + 3Bg + 2Eg modes expected. Raman active external modes are localized below 300 cm^{-1}: Ag + 3Bg + 4Eg translational modes and Ag + Bg + 2Eg rotational modes are expected.

The interpretation of the Raman spectra in terms of influence of defects of crystal can be facilitated by using analogy with the spectra of similar compounds [2]. There are two kind of defects in $NaBi(MoO_4)_2$ crystals.One of it is a Mo nonstoichiometry. The redundancy of rombic MoO_3 can be formed solid solution of substitution.Another kind of defects is disordered structure in disposition Na^+ and Bi^{3+} ions. It can be changed the coordination number of atoms formed the crystal.This is a reason why the Raman spectra of $NaBi(MoO_4)_2$ crystals have a continuum character and consists from a broad overlap bands.

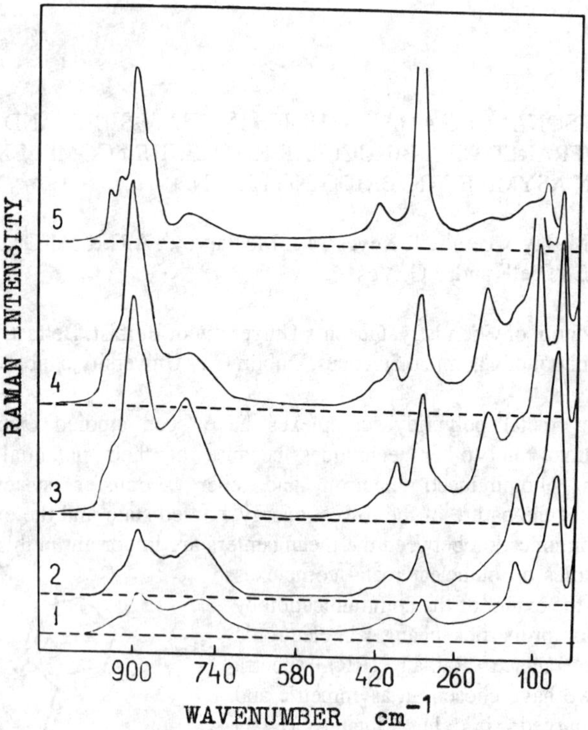

Figure 1. Raman spectra of $NaBi(MoO_4)_2$ single crystals at T=300 K: 1- Z(XZ)Y, 2 - Z(YZ)Y, 3 - Z(YX)Y, 4 - Z(XX)Y, 5 - X(ZZ)Y.

Another peculiarity of Raman spectra of $NaBi(MoO_4)_2$ is a anomaly big value of Raman scattering efficiency. It was estimated.The value of cross-section of Raman scattering for mode at ω = 323 cm^{-1} is $\sigma = 1657 \cdot 10^{-28}$ cm^2. It is in $2 \cdot 10^3$ times more than this in $CaCO_3$ crystals ($\sigma \approx 1 \cdot 10^{-28}$ cm^2). It can be a reason of high acoustooptic efficiency of this crystals.

1. A.A.Kaminskii, A.Kholov, P.V.Klevtsov, S.Kh.Khafizov, Phys/ Stat. Sol.(a).114,713(1989).
2. A.B.Kiss, S.Holly, E.Hild.Acta Chim. Acad. Sci. Hung.72, 147(1972).

RESONANCE RAMAN STUDIES IN THE VISIBLE AND NEAR-INFRARED OF A BINUCLEAR RUTHENIUM COMPLEX WITH AN ASYMMETRIC BRIDGING LIGAND

P.M. Jayaweera[1*], T. Keyes[2], J.J. McGarvey[1], A.H.R. Al-Obaidi[1], S.E.J. Bell[1] and J.G. Vos[2]

[1]School of Chemistry, Queen's University of Belfast, Belfast, N.Ireland.
[2]School of Chemical Sciences, Dublin City University, Dublin 2, Ireland.

Binuclear metal polypridyl complexes have been studied extensively by resonance Raman and other techniques because of their potential ability to participate in photoinduced electron and energy transfer systems[1]. An understanding of the nature of the lowest energy excited state and the extent of the electronic communication between the metal centers are important in this respect. A number of studies on binuclear metal complexes have assessed the extent of the communication by measuring redox properties, changes in electronic absorption, resonance Raman (RR) spectra etc[2]. Here we have chosen an asymmetric and negatively charged bridging ligand (HL1) coordinated to two ruthenium centers to produce a bi-nuclear complex $[(bpy)_2Ru\text{-}HL1\text{-}Ru(bpy)_2]^{3+}$ (**1**) where bpy = 2,2'-bipyridine, HL1 = 3 - (2-hydroxyphenyl) -5-(pyridin-2-yl) - 1,2,4-triazole in which the metal centers are not in equivalent coordination environments.

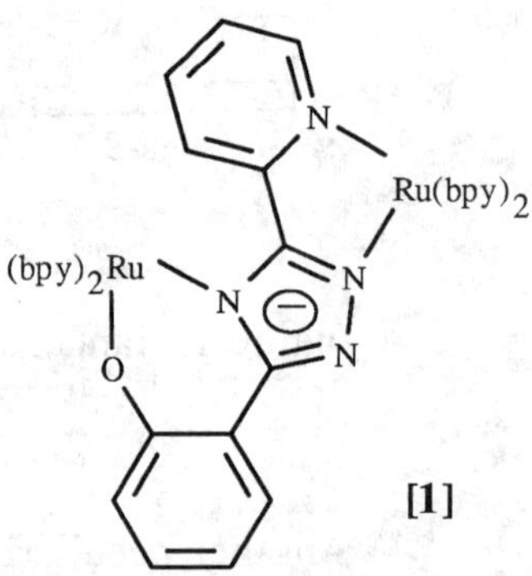

[1]

Fig.1a shows the UV-vis absorption spectrum of **1** in acetonitrile. Oxidation of one ruthenium center, chemically or electrochemically, leads to a complete loss of the MLCT bands at 537 and 650nm and the growth of a new, broad absorption band at around 814nm in the near-IR region. However, oxidation of the second ruthenium center does not cause any major additional changes in the spectrum. This suggests that there is little communication between the metal centers and that the NIR band is not due to a metal→metal charge-transfer (MMCT) transition but is possibly ligand→metal (LMCT) in character. Lack of communication between the metal centers could result from loss of planarity in the bridging ligand. In other bis-Ru systems which do exhibit strong intervalence MMCT bands the bridging ligand remains planar[3].

Resonance Raman spectra recorded using 810nm laser excitation for the parent [2,2], half oxidized[2,3] and fully oxidized[3,3] species are shown in Fig.1.

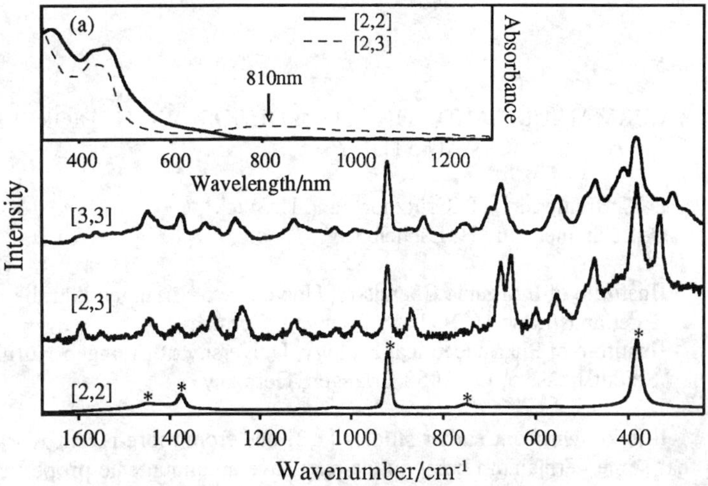

Fig.1 RR spectra of **1** in [2,2], [2,3] and [3,3] oxidation states in MeCN solutions.(Insert, UV-vis absorption spectra of [2,2] and [2,3]) * Solvent features, $\lambda_{exc.}$=810nm.

For the [2,2] state, which does not absorb in this region, only solvent features can be observed. Several new bands appear for the [2,3] and [3,3] species and the similarities in the spectra suggest that they are enhanced through the same transition. RR spectra recorded at 810 nm for fully deuterated bpy analogues of **1** show no band shifts for either the [2,3] or the [3,3] species. Thus if the NIR absorption band is LMCT in character, it appears to involve a bridging (HL1) ligand→ $Ru^{(III)}$ charge transfer transition rather than bpy→$Ru^{(III)}$. Experiments carried out with 920nm and 1064nm excitation tell a similar story. The RR spectrum of the [2,2] species recorded using 457.9nm excitation shows typical neutral bpy vibrations enhanced via the normal MLCT transition. The corresponding excited state RR spectrum recorded with 355nm pulsed laser excitation (4mJ pulse energy) shows strong modes at 1213, 1286 and 1427cm^{-1} which can be ascribed to the negatively charged bpy moiety (bpy^{-}). This confirms that the peripheral bpy ligand is negatively charged in the lowest energy MLCT excited state.

We thank the Engineering & Science Research Council for support (Grant GR/J01905) and the School of Chemistry, QUB for a studentship to PMJ.

1. V.Balzani and F.Scandola, Supramolecular Photochemistry, Ellis Horwood, Chichester, UK, 1991.
2. S.K.Doorn, J.T.Hupp, D.R.Porterfield, A.Campion and D.B.Chase. J.Am.Chem.Soc. 112, 4999, (1990), and references therein.
3. R.Hage, J.G.Haasnoot, J.Reedijk, R.Wang, J.G.Vos., Inorg. Chem. 30, 3263, (1991).

GERMANIUM AND TIN COMPOUNDS WITH BIORELEVANT LIGANDS: A RAMAN STUDY

D.K. Breitinger[1]*, T. Grützner[1], and H. Wick[1],
O. Schimmer[2] and H. Eschelbach[2]

[1]Institute of Inorganic Chemistry, University of Erlangen-Nürnberg, Egerlandstrasse 1, D-91058 Erlangen, Germany,
[2]Institute of Pharmaceutical Biology, University of Erlangen-Nürnberg, Staudtstrasse 5, D-91058 Erlangen, Germany

It is known from earlier studies [1,2] and from more recent work of our own [3] that some germanium compounds may have antimutagenic properties in the Ames test with *Salmonella typhimurium*. These findings prompted us to synthesize systematically germanium and tin compounds with biocompatible substituents in order to study their effects on the genom of this test organism. The most recent results of these tests look quite promising. Here we discuss skeletal vibrations observed in the Raman spectra of selected germanium compounds prepared in this project.

Reactions of germanium dioxide GeO_2 with the appropriate acids in aqueous media yield diaquabis(glycolato)germanium(IV) $Ge(OCH_2COO)_2(H_2O)_2$ (**1**), bis(thioglycolato)germanium(IV) $Ge(SCH_2COO)_2$ (**2**) and bis(thiolactato)germanium(IV) $Ge(SCH(CH_3)COO)_2$ (**3**). Surprisingly, on recrystallization of **2** from methanol or isopropanol tetrakis(methylthioglycolato)germanium(IV) (**4**) or tetrakis(isopropylthioglykolato)germanium(IV) (**5**) $Ge(SCH_2COOR)$ ($R = CH_3$, $CH(CH_3)_2$) are readily formed.

In **1** germanium adopts octahedral coordination [4]. In the equatorial plane two dianions derived from glycolic acid are coordinated *via* the oxygen atoms to form two chelate rings, whereas the water molecules are arranged in the axial positions. In **2** to **5** germanium obviously prefers tetrahedral environment, and the ligands coordinate to germanium in a bidentate way *via* the oxygen and sulfur donors. In the latter compounds water does not coordinate, irrespective of the same conditions kept in the syntheses of all species. Thus, the Raman spectrum of solid **1** clearly shows a band for the valence vibration $\nu(Ge\text{-}OH_2)$ at 559 cm^{-1}, properly shifted on deuteration, which is lacking with **2** and **3**. The skeletal vibrations of the octahedral GeO_6 entity of **1** as a solid are assigned as follows: ν_1 516, ν_2 336, and ν_5 297 cm^{-1}.

As to the skeletal vibrations of the tetrahedral GeO_2S_2 entities in **2** and **3** only bands of the GeS_2 fragments are observed reliably in both solid state and solution Raman spectra, since the bands of the GeO_2 fragments are weak and probably partly obscured by the strong $\nu(GeS_2)$ bands. For **2** in aqueous solution the bands of the GeS_2 group are clearly identified by means of their polarisation states, *viz* $\nu_{as}(GeS_2)$ at 449 (dp), $\nu_s(GeS_2)$ at 422 (p), and $\delta(GeS_2)$ at 237 cm^{-1} (p) (Fig. 1). Further bands are to be assigned to $\nu(CO)$ (932, p), $\rho(CH_2)$ (786, dp), $\nu(CS)$ and $\nu_s(GeO_2)$ (broad,

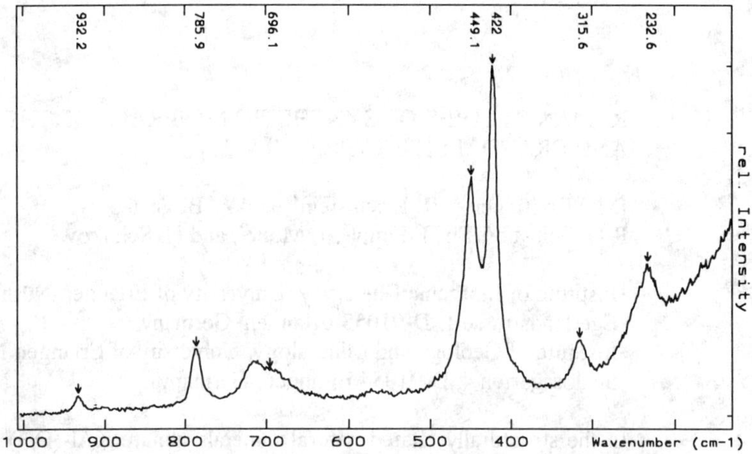

Fig. 1. Raman spectrum of $Ge(SCH_2COO)_2$ (**2**) in aqueous solution.

overlapping, ≈700, p), and δ(CCO) (316 cm^{-1}, dp). For **3** the skeletal vibrations are found at similar wavenumbers and with a similar intensity pattern as for **2**.

The Raman spectra of **4** and **5** clearly display the skeletal vibrations of the tetrahedral GeS_4 cores of the molecules. Thus, the most intense band of the Raman spectra of both compounds is lacking in the IR spectra and is therefore assigned to the valence vibration ν_1; the bands ν_3 (strong) and ν_4 (medium) appear in both Raman and IR spectra (down to 200 cm^{-1}). The frequencies of these vibrations (Table 1) are similar to those of related tetrathiometallates MS_4^{n-}, *e g* AsS_4^{3-} (*cf* [5]).

Table 1. Skeletal vibrations in the Raman spectra of the tetrakis(alkylthioglycolato)germanium compounds **4** and **5**.

	4	**5**
ν_1/cm^{-1}	357 vs	349 vs
ν_2/cm^{-1}	162 w	161 w
ν_3/cm^{-1}	406 s	393 s
ν_4/cm^{-1}	212 m	190 m

Support by the Deutsche Forschungsgemeinschaft (grant Br 422/21-1) and by the Fonds der Chemischen Industrie is gratefully acknowleded.

1. R.R. Brutkiewicz and F.Suzuki, In vivo 1,189 (1987).
2. T. Kada, H. Mochizuki, and K. Miyao, Mutation Research 125, 145 (1984).
3. M. Krüger, Doctoral Thesis, Erlangen 1991.
4. H.-Ch. Chiang, M.-H. Wang, and Ch.-H. Ueng, Acta Cryst. C49, 244 (1993).
5. H. Siebert, Z. anorg. allg. Chem. 275, 225 (1954).

RAMAN STUDIES OF SYNTHETIC MINERALS OF THE ALUNITE AND CRANDALLITE TYPE

D.K. Breitinger[1]*, R. Krieglstein[1], and A. Bogner[1], R.G. Schwab[2], Th.H. Pimpl[2], J. Mohr[2], and H. Schukow[2]

[1]Institute of Inorganic Chemistry, University of Erlangen-Nürnberg, Egerlandstrasse 1, D-91058 Erlangen, Germany,
[2]Institute of Geology and Mineralogy, University of Erlangen-Nürnberg, Schlossgarten 4, D-91054 Erlangen, Germany

In the structurally related natural minerals alunite $KAl_3(OH)_6(SO_4)_2$ and crandallite $CaHAl_3(OH)_6(PO_4)_2$ both cations and anions can be substituted to some extent with conservation of the crystal structure. In synthetic compounds of this type the substitution can be varied in a wide frame. Thus, in alunites all alkali metals, Cu(I) or Tl(I), and SeO_4 in place of SO_4 can be inserted. On simultaneous insertion of rare earth metal ions and PO_4, AsO_4 or VO_4, compounds with the same structure are obtained. Synthetic crandallites may contain earth alkali metals or lead, and oxoanions analogous to phosphate. The remarkably high chemical and thermal stability of these compounds renders them promising for fixation and deposition of toxic elements.

Synthetic minerals of this type are prepared under hydrothermal conditions (180 °C, 15 bar) from metal carbonates, hydroxides or oxides and the appropriate acids [1], also in deuterated form for spectroscopy and neutron diffraction.

X-ray diffractograms show all compounds to adopt the alunite [2] or crandallite structure [3] (space group $R\bar{3}m$, $Z = 3$, trigonal setting) with slightly variable lattice constants a and considerably scattering constants c. The structure contains layers of interconnected $AlO_2(OH)_4$ octahedra (site symmetry $2/m$) and SO_4 or PO_4 tetrahedra ($3m$), respectively, and large cavities between these layers which are occupied by K^+ or (disordered) Ca^{2+} ions. In alunite the K^+ ions ($\bar{3}m$) are regularly surrounded by six oxygens common to both Al and S, and by six OH groups, whereas in crandallite Ca^{2+} is closer to four out of six oxygens and to four OH groups. Further, half of the PO_4 particles is protonated and involved in hydrogen bonds. Hence, the $AlO_2(OH)_4$ and PO_4 groups are no longer equal, and their symmetries are distorted.

A unit cell group analysis for the alunite type helps to understand the vibrational spectra. As a result the vibrations of the unit cell are classified into the irreducible representations of D_{3d} - $\bar{3}m$ as follows (acoustic branches $A_{2u} + E_u$ omitted):

$$\Gamma_{vib} = 8\ A_{1g}(R) + 3\ A_{2g}(-) + 11\ E_g(R) + 4\ A_{1u}(-) + 10\ A_{2u}(IR) + 14\ E_u(IR).$$

In crandallite the extra proton located at one phosphate O adds 3 degrees of freedom; non-equivalence of the two PO_4 groups and perturbation of symmetry causes lifting of degeneracies and activation of originally inactive modes. Detailed analysis for alunite predicts two ν(OH) vibrations in Raman ($A_{1g} + E_g$) and IR spectra ($A_{2u} + E_u$) (the

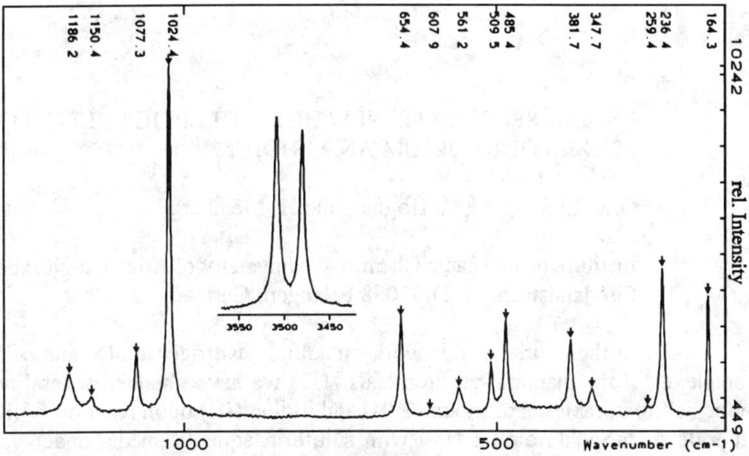

Fig. 1. Raman spectrum of alunite $KAl_3(OH)_6(SO_4)_2$; the insert shows the ν(OH) doublet (same intensity scale).

mutual exclusion rule applies), 3 $\nu(SO_4)$ and 5 $\nu(AlO_6)$ vibrations of the $AlO_2(OH)_4$ octahedra, and 3 deformation modes involving the OH groups in each of the spectra.

These predictions are found in the spectrum of alunite (Fig. 1), of deuterated isotopomer, and of isomorphic compounds. High frequencies, profiles and isotope shifts of the ν(OH/D) doublets preclude hydrogen bonds from OH/D as donors. Imbedding of the SO_4 groups in the framework rises the frequencies of the valence vibrations $\nu(SO_4)$ compared with free SO_4^{2-} (averages 1108 *vs* 1075 cm^{-1}). The $\nu(AlO_6)$ modes appear in the range 650 to 480 cm^{-1}, partly obscured by SO_4 deformations. The low-frequency domains in the Raman spectra of alunites and of crandallites show characteristic band patterns, strongly modified by the imbedded anions and slightly by the counter ions in the cavities. The extra proton in crandallites gives rise to a strong and broad IR band below the ν(OH) doublet with all features of hydrogen bonding, to spreading of the valence vibrations of the PO_4 groups between 1120 and 920 cm^{-1}, and to broadening of the low-frequency Raman bands [4].

Support by the Deutsche Forschungsgemeinschaft (grant Br 422/23-1) and by the Fonds der Chemischen Industrie is gratefully acknowledged.

1. R.G. Schwab, C. Götz, H. Herold, and N. Pinto de Oliveira, N. Jb. Miner. Mh. 1991, 97.
2. S. Menchetti and C. Sabelli, N. Jb. Miner. Mh. 1976, 406.
3. A.M. Blount, Amer. Mineral. 59, 41 (1974).
4. D.K. Breitinger, R. Krieglstein, A. Bogner, R.G. Schwab, Th.H. Pimpl, J. Mohr, H. Schukow, Vth European Conference on Solid State Chemistry, Montpellier, Abstracts, Vol. II, 1995, 596.

RAMAN SPECTRA OF NEW HETEROLEPTIC SULFITO COMPLEXES OF RHODIUM AND IRIDIUM

D.K. Breitinger*, A. Bogner, and H. Meinberg

Institute of Inorganic Chemistry, University of Erlangen-Nürnberg, Egerlandstrasse 1, D-91058 Erlangen, Germany

In the course of our work on sulfito, hydrogensulfito, and sulfur dioxide complexes of the platinum-group metals [1-3] we have obtained several new heteroleptic sulfito complexes of rhodium(III) and iridium(III) upon reaction of rhodium(III) chloride or hexachloroiridate(IV) with sulfite in aqueous media under various conditions. Thus, *e g* $Na[Rh(SO_3)_2(NH_3)_4]\cdot 3\,H_2O$ (**1**) is formed in a hydrothermal reaction, also in deuterated form (**1D**), whereas $(NH_4)_5[IrCl_2(SO_3)_3(NH_3)]\cdot 2\,H_2O$ (**2**) and $Na_3[IrCl_2(SO_3)_2(OH_2)_2]\cdot 7\,H_2O$ (**3**) are obtained under ambient conditions.

So far we have carried out X-ray structure analyses of **2** (space group $P2_12_12_1$, $Z = 4$) and **3** (*Pnma*, $Z = 4$). The complex anion of **2** in a general crystallographic site contains two chloro ligands in *trans*-position, three S-coordinated sulfito groups in *mer*-configuration perpendicular to the quasi-linear Cl–Ir–Cl fragment, and the ammine ligand *trans* to one sulfite. In **3** the complex occupies a site with mirror symmetry with two independent chloro ligands (*trans*) in the mirror plane and pairs of symmetry related sulfito and aqua ligands (*cis*), respectively. Thus, both aqua ligands experience the *trans*-influence of the S-coordinated sulfito groups (*cf* [4,5]). Work on the structure analysis of **1** with a huge unit cell (*Pbcn*, $Z = 16$) is in progress; Raman and IR spectra favour a *cis*-configuration of the anion in a centrosymmetric unit cell.

The Raman spectra of **1** and **1D** show pairs of narrow bands for the symmetric valence and deformation modes $\nu_s(SO_3)$ (995/969 cm^{-1}) and $\delta_s(SO_3)$ (651/628 cm^{-1}), respectively, indicating *cis*-position of these ligands, two strong $\nu(RhN)$ bands (495/415 cm^{-1} in **1**, 463/397 cm^{-1} in **1D**) for pairs of structurally different Rh–N bonds, and a strong, asymmetric band with a peak at 250 cm^{-1}, covering the $\nu(RhS_2)$ vibrations and the deformations of the RhN_4 fragment. Using these $\nu(RhN)$ frequencies symmetry force constants $f_r + f_{rr}'$ and $f_r + f_{rr}$ for the linear and bent $Rh(NH_3)_2$ fragments, respectively, perpendicular to each other, are estimated in a modified valence force field approximation, assuming point masses for the NH_3/ND_3 ligands. The calculations give the averaged values for **1** and **1D** $f_r + f_{rr}' = 250$ N m^{-1} and $f_r + f_{rr} =$ 153 N m^{-1}. Substantial *trans*-coupling (f_{rr}') and negligible *cis*-coupling (f_{rr}) are to be expected (*cf* [3]). The remarkably low value of $f_r + f_{rr}$ for the bent $Rh(NH_3)_2$ group is clearly due to the strong *trans*-influence of the two sulfite ligands.

The structurally different sulfite groups in **2** give rise to two $\nu_s(SO_3)$ and $\delta_s(SO_3)$ Raman bands with rather different frequencies (1002/957 and 661/635 cm^{-1}) (Fig. 1). The strong signals for the $\nu(IrN)$ (472 cm^{-1}), $\nu_s(IrCl_2)$ (308/311 cm^{-1}), and

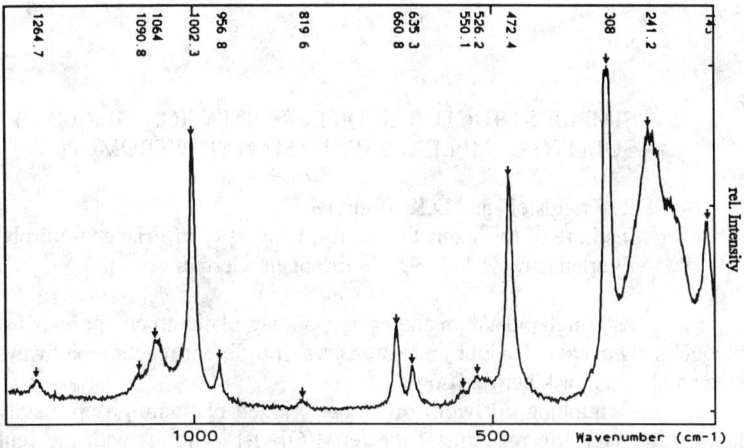

Fig. 1. Raman spectrum (section) of $(NH_4)_5[IrCl_2(SO_3)_3(NH_3)]\cdot 2\ H_2O$.

three $\nu(IrS_3)$ vibrations dominate the low frequency range of the Raman spectrum. The $\nu(IrN)$ frequency yields a force constant $f(IrN) = 205$ N m^{-1}, and $\nu_s(IrCl_2)$ gives $f_r + f_{rr}' = 199$ N m^{-1} for the quasi-linear fragment Cl–Ir–Cl. The latter value agrees with a figure derived from data for octahedral systems containing this fragment [6].

In the Raman spectrum of **3** the strongest band is unequivocally $\nu_s(SO_3)$ (1008/974 cm^{-1}), whereas $\delta_s(SO_3)$ (665 cm^{-1}) is very weak (strong in the IR spectrum). The further skeletal valence vibrations, *viz* $\nu_s(IrO_2)$ (414 cm^{-1}, weak), $\nu_s(IrCl_2)$ (319 cm^{-1}, strong), and $\nu_s(IrS_2)$ (298 cm^{-1}, strong), appear in the expected frequency regions. From $\nu_s(IrO_2)$ a force constant $f_r + f_{rr} = 164$ N m^{-1} in the bent IrO_2 fragment (angle 80.5 °) is estimated, similar to that in a RhO_3 group with a comparable bonding situation [4]. Finally, $\nu_s(IrCl_2)$ yields $f_r + f_{rr}' = 212$ N m^{-1}, the difference to the above value for **2** being explained by the different average bond lengths $\bar{r}(IrCl)$ (236.3 (**2**) *vs* 234.8 pm (**3**)).

Generous support by the Fonds der Chemischen Industrie is gratefully acknowledged.

1. D.K. Breitinger, G. Bauer, W. Haegler, and R. Krieglstein, in ICORS Hong Kong '94, XIVth International Conference on Raman Spectroscopy (N.-T. Yu and X.-Y. Li, eds.), Wiley, Chichester-New York, 1994, p. 696.
2. D.K. Breitinger, G. Bauer, M. Raidel, R. Breiter, and J. Tomkinson, J. Mol. Struct. 267, 55 (1992).
3. D.K. Breitinger and R. Breiter, J. Mol. Struct. 349, 45 (1995).
4. A. Mäurer and D.K. Breitinger, in ICORS Würzburg '92, XIIIth International Conference on Raman Spectroscopy (W. Kiefer, M. Cardona, G. Schaack, F.W. Schneider, H.W. Schrötter, eds.), Wiley, Chichester-New York, 1992, p. 330.
5. A. Mäurer, D.K. Breitinger, and R. Breiter, Z. Naturforsch. 48b, 1187 (1993).
6. K. Irmer and W. Preetz, Z. Naturforsch. 46b, 1200 (1991).

SIMPLE STRUCTURAL DIFFERENTIATION OF (μ-S,O) BRIDGED SULFITO COMPLEXES BY RAMAN SPECTROMETRY

R. Krieglstein and D.K. Breitinger*,
Institute of Inorganic Chemistry, University of Erlangen-Nürnberg,
Egerlandstrasse 1, D-91058 Erlangen, Germany

Although palladium and corresponding platinum compounds usually show analogous chemical behaviour there are cases of different reaction pathways under the same conditions for kinetic reasons.

Continuing our work on sulfite bridged platinum-group metal complexes [1] we found that on reaction of the $[enM(OH_2)_2]^{2+}$ cations with the $[enM(SO_3)_2]^{2-}$ anions (M = Pt and Pd) the dinuclear complexes with two parallel sulfite bridges in $[enPt(SO_3)_2Pten]{\cdot}3H_2O$ (**1**) and antiparallel bridges in $[enPd(SO_3)(O_3S)Pden]{\cdot}3H_2O$ (**2**), respectively, are formed, as is shown by X-ray structure analyses [2]. In contrast, the complexes $K_2[M(SO_3)_2]$ contain polymeric anions $^1_\infty[M(SO_3)_2]^{2-}$ only with parallel double bridges in both the platinum and palladium species, as suggested by vibrational spectra [3] and proved by X-ray structure analysis [4]. Thus, in the formation of the dinuclear palladium complex an isomerization step is to be assumed.

It is stated that the proof for (μ-S,O) bridges in sulfite complexes is difficult with IR spectrometry [5], however, Raman spectrometry allows to differentiate easily the products (**1**) and (**2**) from each other, and from the mononuclear educts.

Assignm. (cm^{-1})	$K_2[Pt(SO_3)_2en]{\cdot}2H_2O$	$[enPt(SO_3)_2Pten]{\cdot}3H_2O$	$K_2[Pd(SO_3)_2en]{\cdot}2H_2O$	$[enPd(SO_3)(O_3S)Pden]{\cdot}3H_2O$
$\nu_s(MN_2)$	535	574; 517	499	523
$\nu_{as}(MN_2)$	446	490; 442	428	429
$\nu(MO)$		385(ν_s); 401(ν_{as})		313
$\nu(MS_2)$	≈263(ν_s+ν_{as})	≈252(ν_s+ν_{as})	≈247(ν_s+ν_{as})	262
$\nu_s(SO_3)$	998; 962		995; 952	
$\nu_{as}(SO_3)$	1109; 1087		1109; 1077	
$\nu_s(SO_2)$		1117; 1062		1130
$\nu_{as}(SO_2)$		1154; 1178		1189
$\nu(S\text{-}O)$		923		928

The complex with an antiparallel bridge (Fig. 1 (**2**)) is centrosymmetric; thus, the mutual exclusion rule applies, in contrast to the case of (**1**). Consequently, only half the number of $\nu_s(MN_2)$ and $\nu_{as}(MN_2)$ modes is Raman active, as compared with (**1**), and there is only one $\nu(MO)$ and $\nu(MS)$ band each left in the Raman spectra.

The products differ from the educts (S-coordination only) in the existence of one (**2**) and two Raman active $\nu(MO)$ vibrations (**1**), respectively, and in the high-frequency shifts of the $\nu(SO_2)$ bands combined with the appearence of low-frequency intense ν(S-O) bands around 925 cm^{-1}. Intramolecular coupling of the $\nu_{as}(SO_2)$ and $\nu_s(SO_2)$ modes causes splitting of these bands with (**1**), which is not observed with (**2**) (mutual exclusion rule!).

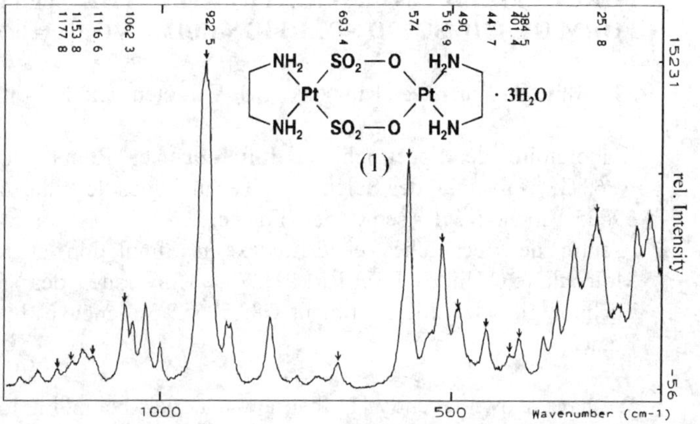

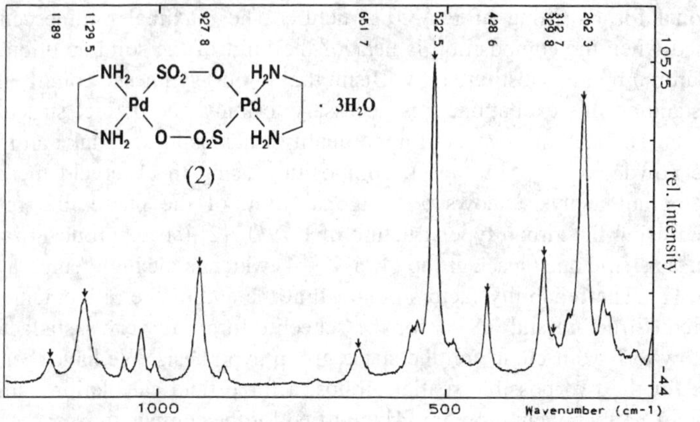

Fig. 1. Raman spectra of [enPt(SO$_3$)$_2$Pten] (**1**)
and [enPd(SO$_3$)(O$_3$S)Pden] (**2**).

Support from the Fonds der Chemischen Industrie is gratefully acknowledged.

1. D.K. Breitinger, G.Bauer, W.Haegler, and R. Kriegistein, in ICORS Hong Kong'94, XIVth International Conference on Raman Spectroscopy (N.-T. Yu and X.-Y. Li, eds.), Wiley, Chichester-New York, 1994, p. 696.
2. R. Krieglstein, D.K. Breitinger, and G. Liehr, to be published
3. G. Bauer, Doctoral Thesis, Erlangen, 1979; W. Haegler, Doctoral Thesis, Erlangen, 1979.
4. D. Messer, D.K. Breitinger, and W. Haegler, Acta Cryst. B35, 815 (1979).
5. K. Nakamoto, Infrared and Raman Spectra of Inorganic and Coordination Compounds, Wiley, Chichester-New York, 4th Ed., 1986, p. 258.

FORCE FIELD INVESTIGATION IN THE TETRAGONAL FORM OF THE SOLID SOLUTION $Bi_xEu_{1-x}VO_4$ (x = 0.72)

J.L. Blin[1], A. Lorriaux - Rubbens[1]*, G. Vergoten[2] and F. Wallart[1]

[1]Laboratoire de Spectrochimie Infrarouge et Raman, UPR CNRS A2631L, Université des Sciences et Technologies de Lille, Bâtiment C5, 59655 Villeneuve d'Ascq Cédex, France.
[2]Centre de Recherches et d'Etudes en Simulation et Modélisation Moléculaire, Unité INSERM U279, Université des Sciences et Technologies de Lille, Bâtiment C8, 59655 Villeneuve d'Ascq Cédex, France.

We have shown recently [1] that new materials belonging to the pseudo binary system $EuVO_4$ - $BiVO_4$, are good phosphors when they crystallise in the tetragonal form drifting of $EuVO_4$ structure. The maximal luminescence yield is obtained when the composition is near of the limit of the solid solution (ie 74% of the Europium are substituted by Bismuth, in our synthesis conditions) and for understanding this exaltation, it is necessary to know the perfect structure of these materials. The lack of crystals of good quality leads us to undertake an investigation on the powder of $Bi_{0.72}Eu_{0.28}VO_4$ composition using the Rietveld method. As the diffractogram analysis shows a tetragonal form of the phase, the refinement is realized using the zircon type structure of $EuVO_4$ [2] (space group $I4_1/amd$, Z = 4) and the scheelite one (space group $I4_1/a$, Z = 4) which is the high temperature form of $BiVO_4$ [3]. The reliability factors obtained at the end of the refinement : 3.84% for the zircon structure and 3.82% for the scheelite form are very sastisfactory, but do not allow to fix our choice on the space group hypothesis. We have then decided to realize for the two possible spatial groups, a force field calculation using the NCT version of Shimanouchi program [4] modified for the computation.

We have choosen a modified general force field and considered interactions as far as 5Å distance for each atom of the unit cell. In Table 1 are given the values of the different force constants used in the computation, and in Table 2, we have reported the observed and calculated frequencies for the Raman scattering. A good agreement beetween the computarised and obtained frequency values for each structure has been reached, but few anomalies have been ascertained in the force constant values of the scheelite structure. We have used principal force constant values higher than those given in the litterature [5] for $EuVO_4$; and to fit the results of the computation with the experimental ones, we must use abnormal variation of the O...O interaction force constant with the interatomic distance.
These anomalies are not established in the set of force constants used for the zircon structure hypothesis, which is similar to those of the litterature.

TABLE 1. Force constant values (in mdyn/Å) for $Bi_{0.72}Eu_{0.28}VO_4$.

Type of force constant	Force constant value in zircon form	Force constant value in scheelite form	Interatomic distance (Å)
f_r(V-O)	5.10	6.25	1.687
f_{rr}	1.20	1.56	2.591
f_α(O-V-O)	1.13	1.45	2.832 (O...O)
$f_{\alpha\alpha}$	0.30	0.22	
$f_{r\alpha}$	0.23	0.48	
f_i(M...O)*	0.07	0.50	2.405
f_i(O...O)	0.05	0.20	2.866
f_i(M...V)*	0.10	0.31	3.217
f_i(V...O)	0.05	0.30	3.568
f_i(M...M)*	0.02	0.54	3.981
f_i(V...V)	0.06	0.85	3.981
*(M = Eu or Bi)			

TABLE 2. Calculated and observed Raman frequencies (in cm^{-1}) of $Bi_{0.72}Eu_{0.28}VO_4$.

	Calculated frequency in zircon form	Calculated frequency in scheelite form	Observed frequency
$\nu_1(VO_4^{3-})$	858 (A_{1g})	860 (A_g)	860
$\nu_2(VO_4^{3-})$	370 (A_{2g})	369 (A_g)	369
	252 (B_{2g})	869 (B_g)	250
$\nu_3(VO_4^{3-})$	731 (B_{2g})	734 (B_g)	734
	772 (E_g)	766 (E_g)	767
$\nu_4(VO_4^{3-})$	250 (B_{1g})	555 (E_g)	250
	449 (E_g)	454 (B_g)	444
Lattice	688 (A_{2g})	680 (A_g)	
	199 (E_g)	251 (E_g)	
	180 (E_g)	203 (B_g)	
	120 (B_{1g})	115 (E_g)	114
modes	74 (E_g)	80 (E_g)	76
	66 (B_{1g})	60 (B_g)	59

1. A. Lorriaux - Rubbens, J. Corset, J. Ghamri and H. Baussart, Adv. Mater. Res, 1-2, 433 (1994).
2. A.T. Aldred, Acta Crystallogr., sect. B, 40, 569 (1984).
3. W.I.F. David, A.M. Glazer and A.W. Hewat, Phase Transitions, 1, 155 (1979).
4. T. Shimanouchi, Computer Programs for Normal Coordinate Treatment of Polyatomic Molecules (1975).
5. M.E. Escobar und E.J. Baran, Z. Naturforsch., 35a, 1110 (1980).

RAMAN SPECTRA OF NANOCRYSTALLINE TRANSITION METAL DICHALCOGENIDES AND THEIR INTERCALATES

A.P.Naumenko[1*], N.I.Gerasimuk[1], M.M.Bilyi[1], L.M.Kulikov[2]
A.Semjonov-Kobzar[2], , and L.G.Akselrud[3]

[1]Kyiv Taras Shevchenko University, 252601, Kyiv, Ukraine
[2]Institute for Materials Science Problems, National Academy of Science of Ukraine, 252680, Kyiv, Ukraine
[3]L'viv State University, 290602, L'viv, Ukraine

Transition Metal Dichalcogenides (TMD) and their intercalated complexes fall into the category of low-dimensional solids, which are of significant interest raised by achievements in semiconductor microelectronics. These highly anisotropic solids have layer type structures [1]. The anisotropy is evident from studies of optical properties, electron and ion transport and other physical properties. The layered materials readily form intercalated complexes with various foreign atoms or molecules that have electron-donating properties and the correct size to insert between the Van der Waals planes. Intercalation of the layered TMD by variety of transition metals provides a powerful way to the finely electron occupation of the relatively narrow "d"-bonds met in this solids.

The intercalation - deintercalation processes in TMD and their intercalated complexes in nanostate have been studied. The realization of crystal structures activated due to their intercalation and transfer into nanostate will promote obtaining of such new physical properties which don't exist in coarse grained powders or single crystals. The possibilities of nanocrystal materials on the basis of initial TMD and their intercalates (2H-$MoSe_2$, 2H- $NbSe_2$, 2H-WSe_2 , Cu_xWSe_2, Ti_xWSe_2 , etc.) produced by means of high power electrochemical and ultrasonic have been studied. It was established that powders crystal structure and composition do not undergo essential changes during there dispersion in absence of intercalation by components of final solid product solution.

The method of Raman spectroscopy was chosen as the most simple, high sensitive, express and nondestructive control methods of materials quality.

Raman spectra were obtained from several different samples which included single crystals of TMD and their intercalated compounds with various intercalates (Cr, Mn, Fe, Co, Ni and Cu) and various concentration of intercalating compounds (from 0.67 to 1.3). Powder samples only were available. Raman measurements were performed in backscattering geometry using a cylindrically focused 514.5 nm Ar-ion laser incident beam. The scattered light was dispersed with a DFS-24 model double grating monochromator.

In order to relate the first-order Raman features of nanocrystalline TMD and their intercalates the Raman spectra shown in Fig.1A and Fig.1B. This figures demonstrate the main regularities of the phonon spectra dependence on intercalating compound (Fig.1A) and concentrations of this compounds(Fig.1B).

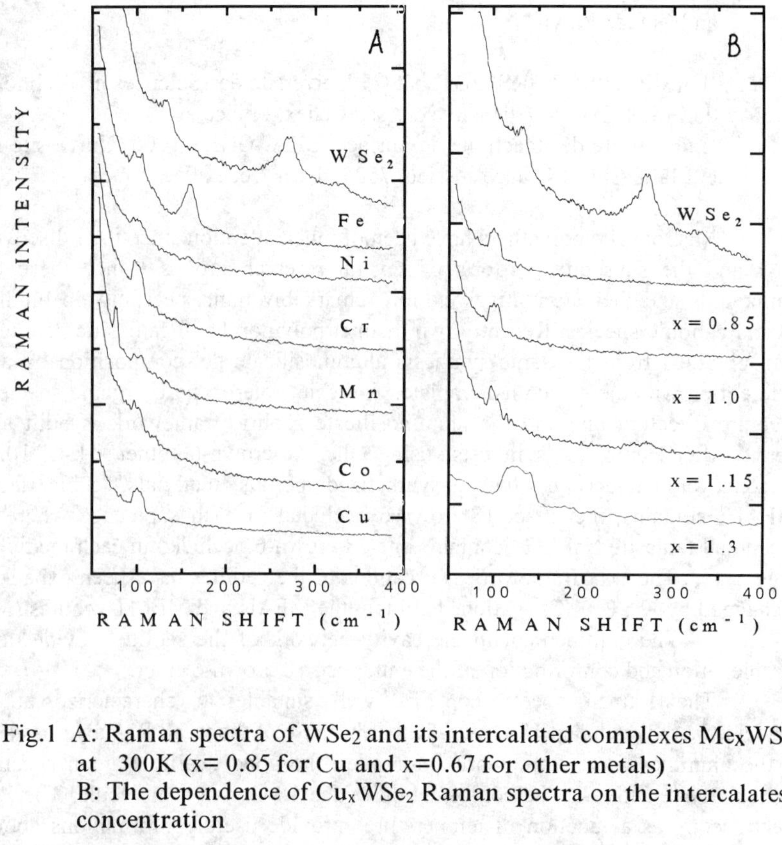

Fig.1 A: Raman spectra of WSe_2 and its intercalated complexes Me_xWSe_2 at 300K (x= 0.85 for Cu and x=0.67 for other metals)
B: The dependence of Cu_xWSe_2 Raman spectra on the intercalates concentration

The dependence of the first-order TMD Raman spectra on crystallite size also has been studied for a series of samples with different average size of particles and some empirical relations were devised.

1. J.A.Wilson and A.D.Yoffe, in Advances in Physics, vol.18, 193-335 (1969)

CONFORMATIONS OF Na^+, K^+, Rb^+ AND Cs^+ COMPLEXED WITH 18-CROWN-6 ETHER OCCLUDED IN FAUJASITIC ZEOLITES

Claude Brémard[1], Jacky Laureyns[1], Marielle Le Maire[1], Jean Marie Manoli[2] and Claude Potvin[2]

[1]LASIR, UPR-CNRS 2631, bât.C5 Université des Sciences et Technologies de Lille I, 59655 Villeneuve d'Ascq Cedex, France
[2]Laboratoire de Réactivité de Surface, URA -CNRS 1106, Université Pierre et Marie Curie, 4 Place Jussieu, 75252 Paris Cedex 05, France

Macrocyclic polyethers have been of much attention since their discovery in 1967 and the subsequent efforts of several research groups. Since these cyclic compounds and their derivatives exhibit remarkably high specificity in binding to several cationic species. Recently two distinct polymorphs of faujasitic zeolite have been obtained from the same synthesis aluminosilicate gel composition by adding different crown ethers used as template. The chief role of these organic species is a structure directing one and the aluminosilicate zeolitic framework is built around them. After template synthesis using the 18-crown-6 ether (1,4,7,10,13,16 hexaoxacycloocta-decane) the as-synthetized porous materials, cubic faujasite NaFAU[1] contains one intact 18-crown-6 occluded in each supercage whereas the hexagonal faujasite NaEMT[2] contains one 18-crown-6 occluded in each medium and large cage.[3] The extraframework Na^+ cations of the zeolites have been exhaustively exchanged by K^+, Rb^+, Cs^+ cations to obtain the MFAU and MEMT zeolites (M=K^+, Rb^+, Cs^+).[4] The confinement in the cavity network of the zeolite provide unusual complexation and conformation of the entrapped 18-crown-6 ether.

The Raman spectroscopy is well suitable to characterize at room temperature the occluded 18-crown-6 ether because of the weak Raman cross-section of the aluminosilicate framework. Moreover, the good quality Raman spectra are particularly sensitive to the conformation of the 18-Crown-6 molecule. The Raman spectroscopy as a function of temperature provide usefull informations about the dynamic of the enclosed molecules as a function of the temperature.

The most structural-indicative Raman bands are in the [900-800] cm^{-1} region. The band near 860 cm^{-1} derived from the A_{1g} mode (D_{3d} molecular symmetry) is assigned to the ring-breathing vibration. Two bands observed near 860 cm^{-1} in the spectra of MEMT samples are found to be in good agreement with the medium and large cages whereas the single feature near 860 cm^{-1} detected at room temperature in the MFAU spectra corresponds to the the single supercage type. The splitting of the E_g mode (D_{3d}) near 830 cm^{-1} indicates the low local symmetry of the occluded 18-Crown-6 in all the samples under study. The band at 550 cm^{-1} assigned to deformation vibration becomes sharper when a bulky cation reduced the void space of the cages. The prominent band at 500 cm-1 arises from the symmetric bending vibrations of the silicoaluminate framework. Upon cooling this band remains broad and provides evidence of the Si, Al positional disorder. In contrast, the broadening of the bands of the 18-Crown-6 occluded in MFAU at low temperature suggests that crown ether motions within the supercage are appreciable at room temperature. The dynamic disorder is reduced in some extent to static disorder at low temperature. The occluded 18-Crown-6 molecules in MEMT exhibit two different dynamic behaviors according to the size of the medium and large cages. In the medium cage the molecule is held more rigidily than in the large cage. The Raman results are compared to the far-IR data which are more sensitive to the dynamics of the extraframework cations. All the low temperature vibrational spectroscopic results are in good agreement with Molecular Mechanics calculations performed both on dehydrated 18-Crown-6 MFAU and MEMT systems.

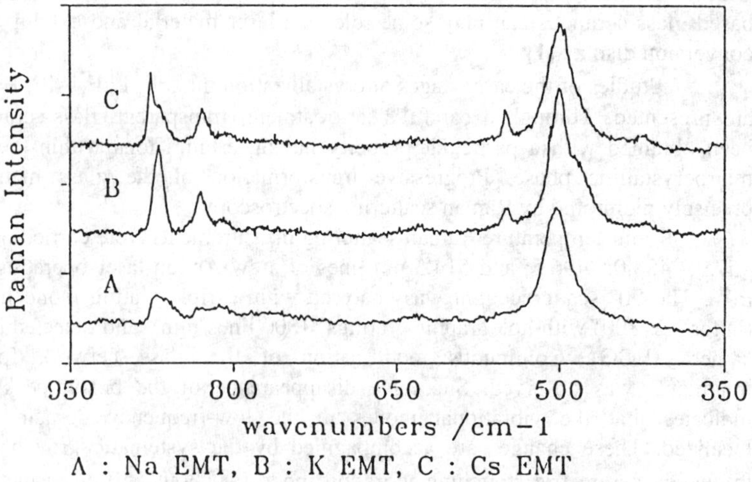

Fig. 1 Low temperature Raman spectra (77K) of MEMT zeolites (M=Na^+, K^+, Cs^+) containing occluded 18-Crown-6 ether.

1. S.L. Burkett and M. Davis, Microporous Mater., 1, 265 (1993).
2. F. Delprato, L. Delmotte, J.L. Guth and L. Huve, Zeolites, 4, 546 (1990).
3. C. Baerlocher, L.B. McCusker and R. Chiappetta, Microporous Mater., 2, 269 (1994).
4. C. Bremard, M. Le Maire, J.M. Manoli and C. Potvin, Zeolites and Microporous Materials Studies in Surface Science and Catalysis, 84, 1027 (1994).

EARLY STEPS OF CRYSTALLIZATION IN Er^{3+} DOPED FLUOROZIRCONATE GLASSES.

K.E. Lipinska-Kalita [1*], F. Auzel, and P. Santa-Cruz [2]

France Telecom, Centre National d'Etudes des Télécommunications, Paris B, Laboratoire de Bagneux, BP 107, F-92225 Bagneux, France

Fluorozirconate glasses, rare-earths doped, highly transparent from 0.3 to 0.6u are attractive for applications as active optoelectronic materials. Erbium ion is of particular interest because of its transition at 1.54u, coinciding with low-loss window in these glasses. It has been proved that rare-earths doped fluoro-oxide based glass-ceramics can play some role as a laser material and exhibit also up-conversion display [1].

Studies of the early stages of crystallization in: ZrF_4-LaF_4-AlF_4-ErF_3 glass are presented. Through a careful heat-treatment, transparent glass-ceramic has been obtained with a particular property having erbium ions within the single, microcrystalline phase. Progressive transformation of the glass matrix was precisely monitored by Raman scattering spectroscopy.

Room-temperature Raman scattering measurements were carried out using 457.9; 488.0; 496.5; and 514.5 nm lines of a cw Ar ion laser operating at 300 mW. The 90° scattered light was analysed with a triple-grating monochromator (Dilor XY 800) with holographic gratings (1800 lines/mm) and detected by CCD camera (EGG). Continuous modification of the glass network following treatments was observed. Succesive disappearance of the boson peak clearly indicates that the vibrational modes in the low-frequency domain become localized. These changes are accompagnied by the systematic growth of high frequency symmetric stretching vibration mode (at about 650 cm^{-1}) that can be directly attributed to the increasing amount of the crystalline phase. The spectra confirmed that the structure of the crystalline phase is very similar to the single phased $ErZr_3F_{15}$ reference crystal. The same conclusions were obtained from fluorescence spectroscopy (visible and infrared), transmission electron microscopy, and X-ray diffraction studies [2,5].

1
On leave from: Jagiellonian University, 30060 Krakow, Poland
Present address: University de La Rochelle, Dept. Physics, Lab. LEMMA, 170042 La Rochelle, France

2
Present address: Dpto de Quimica Fundamental, CCEN/UFPE, Cidade Universitaria, 50670-901 Receive - PE, Brasil.

We expected to observe the Raman scattering from surface vibration modes of microcrystals after observation in oxide glasses [3,4]. We failed and one of the most probable explanation is that the boundary conditions (close contact between microcrystal and glass) or elastic parameters of vitreous matrix have supressed these located at the particle surface modes.

At the end we would like to mention that our transparent glass-ceramic material has increased cross-section for 1.54 u emission comparing to untreated glass [5] and so is interesting from the point of view of future applications.

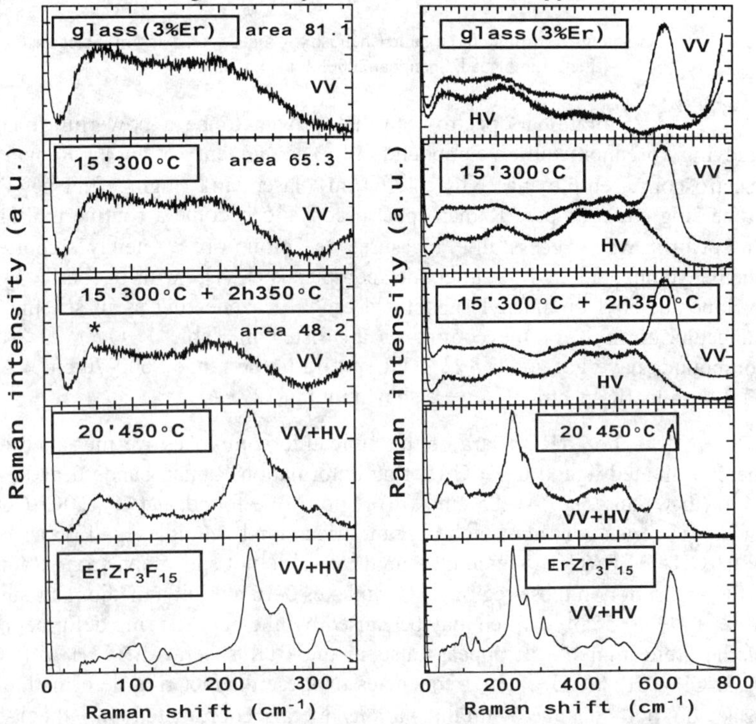

Fig. 1. Low-frequency Raman scattering spectra after different heat treatments, showing the disappearance of the boson peak (the given areas were calculated on the basis of integration between 20 and 300 cm^{-1}).

Fig. 2. Low- and high-frequency, polarized Raman scattering spectra for an untreated glass, thermaly treated samples, and for the reference crystal.

1. F. Auzel, D. Pecile, and D. Morin, J. Electrochem.Soc. 101 , 122 (1975)
2. P. Santa-Cruz, F. Auzel, A. Sadoc et al J.Non-Cryst.Solids 161, 7 (1993)
3. K.E. Lipinska-Kalita, G.Mariotto and E.Zanghellini, Phil.Mag. 71, 547 (1995)
4. K.E. Lipinska-Kalita, G. Mariotto and A. Fontana, J. Phys. C4, 469 (1994)
5. F. Auzel, K.E. Lipinska-Kalita ,and P. Santa-Cruz, Optical Materials - in press

FT-Raman Spectra of Some Heterobimetallic Complexes with Phenylcyclopentadienyl Ligand

Chong-Shi Nie*, Jianhua Guo, Changtao Qian, and Ying Tan

Shanghai Institute of Organic Chemistry, Chinese Academy of Sciences, 354 Feng-Ling Road, Shanghai 200032, P. R. China

Vibrational spectroscopy has proved to be a powerful method for studying organometallic compounds. The advent of FT-Raman (FTR) spectroscopy, employing NIR (1.064μm) laser and high efficiency in data processing has prompted Raman spectroscopy to become a routine tool [1,2]. In this work we successfully measured a group of newly reported π,π-phenylcyclopentadienyl bimetallic complexes [3,4], with a basic ligand of cyclopentadienyl or indenyl substituted on a benzene ring as in scheme 1. The molecular formula of the complexes are listed in Table 1. Over 90% of these compounds have got good S/N FTR spectra in the range 3600-70 cm^{-1}. A typical FTR with its IR spectra of **6** were shown in Fig.1.

The FTR spectra of the bimetallic complexes we measured all show the coordinated ligand C_6H_5-C_5H_4 ring deformation Raman bands near 1540 (vs) , 1490 (m-w) and 1275-1285 cm^{-1} (m) region. We found that $(C_6H_6)Cr(CO)_3$ [5,6] and a series of arene chromium tricarbonyl complexes [7] as well as $(C_6H_5CH_2OH)Cr(CO)_3$ which measured by us, do not show any strong Raman peak or IR absorption in these regions. Complexes **3-16** exhibit a medium strong Raman peak at 995 + 5 cm^{-1}, which may be caused by the benzene ring deformation mode of the arene moiety. It appears also in the FTR spectra of $(C_6H_6)Cr(CO)_3$ and $(C_6H_5CH_2OH)Cr(CO)_3$. The frequencies of these distinct ring deformation Raman peaks do not change with the various metals coordinated on the π-bonding cyclopentadienyl ring. The above mentioned characteristic ring deformation Raman peaks appear as very weak bands in IR spectra, see Fig.1. In fact, in IR spectra one could not detect these important ligand vibration modes, which are very typical for these complexes.

The FTR spectra of the complexes we studied show medium strong peaks near 480 cm^{-1} and 690-660 cm^{-1} without exception. We assigned the peaks near 480 cm^{-1} as M-CO stretching modes and the doublet near 660 and 690 cm^{-1} as M-C-O bending modes. The FTR spectra all show two medium strong peaks at 350-300 cm^{-1} and 290-270 cm^{-1}. We temporarily assigned the 350-300 cm^{-1} Raman peak of the title complexes for Cr-C_6H_5(ring) stretching mode frequency and the lower frequency one at 290-270 cm^{-1} for M-Cp stretching band [5]. The latter one may be interfered more by the mixing of the other modes.

Table 1. Molecular formula of the studied complexes

1 $(CO)_3Cr(C_6H_5)$-(C_5H_5)

2 $(CO)_3Cr(C_6H_5)$-(C_9H_7) indenyl

3 $(CO)_3Cr(C_6H_5)$-$(C_9H_6)Mn(CO)_3$

$(CO)_3Cr(C_6H_5)$-$(C_5H_4)M(CO)_n(NO)_m$ X

	M	n	m	x
4	Mn^+	3	0	
5	Co^+	2	0	
6	Cr^{2+}	2	1	
7	Mo^{2+}	2	1	

	M	n	m	x
8	W^{2+}	2	1	
9	Mo^{2+}	3	0	CH_3
10	W^{2+}	3	0	CH_3
11	Mo^{2+}	3	0	Br^-
12	W^{2+}	3	0	Br^-
13	Ti^{2+}	2	0	Cp
14	Ru^{2+}			Cl^-, 2(PPh_3)
15	Ti^{4+}			Cp, $2Cl^-$
16	Zr^{4+}			Cp, $2Cl^-$

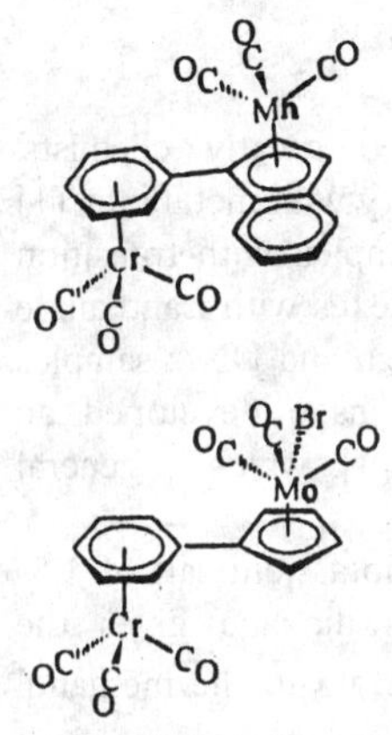

Scheme 1.

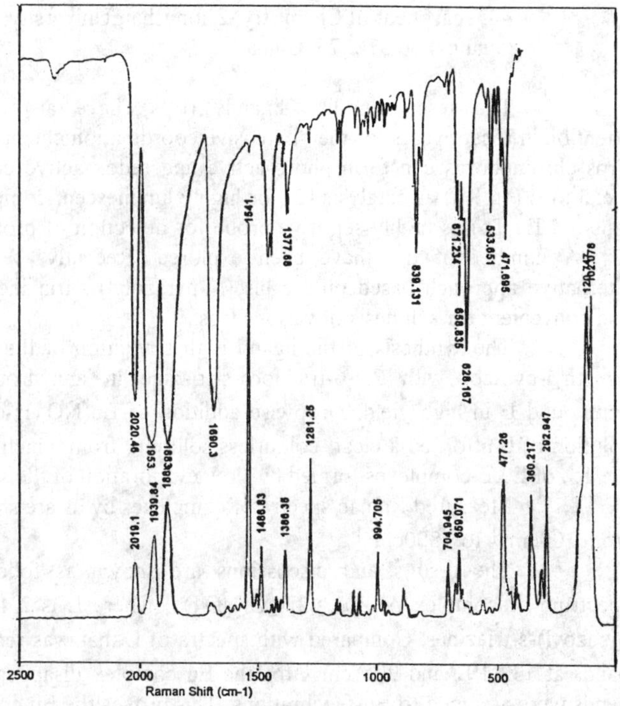

Figure 1. FT-Raman and FT-IR spectra of 6 in solid state.

[1] T. N. Day and P. J. Hendra, A. J. Rest and A. J. Rowlands, *Spectrochim. Acta*, **47A** (1991) 1251.
[2] C. -S. Nie, Z. -Y. Zhu, C. Mao and Y. Tan, *Spectrochim. Acta*, **50A** (1994) 1259.
[3] Y. K. Kang, Y. K. Chung and S.W. Lee, *Organometallic*, **14**(1995) 4905.
[4] C. Qian, J. Guo and J. Sun, submitted to *Inorg. Chem.* for publication.
[5] E. M. Bisby, G. Davidson and D. A. Duce, *J. Mol. Struct.*, **48** (1978) 93.
[6] L. J. Hyams and E. R. Lippincott, *Spectrochim. Acta*, **28A** (1972) 1741.
[7] S. M. Barnett, F. Dicaire and A. F. Ismail, *Can. J. Chem.*, **68** (1990) 1196.

RAMAN STUDY OF EUROPIUM AND L_x COMPLEXES L=2,4,6-TRI(3,5-DIMETHYL-N-PYRAZOLEYL]-S-TRIAZINES]

Zhuo-liang ZHANG[1]*, Jian CHEN[1], Wei-hong ZHANG[1], Chi YANG[2], and Yan-sheng YANG[2]

[1] Instrumentation Analysis & Research Center, Zhongshan University, Guangzhou 510275, China
[2] Department of Chemistry, Zhongshan University, Guangzhou 510275, China

The terpyridine-like ligands (trps) have attracted greatly chemists' attention in recent years for their extensive coordination chemistry with metal ions [1]. Trps chromphores can form photo-active and redox-active complex with transition metal ions (eg Ru) as catalysts [2], or highly luminescent complexes with Lanthanide ions (eg Eu,Tb) as highly sensitive probe for detection of protein and DNA samples [3]. A number of trps have been explored. Recently, We have developed an alternative approach based on 2,6-bis (N-pyrazolyl)-s-triazine [4], which is general and convenient for synthesis of various trps.

The synthesis of the ligand is that reaction of the potassium salt of 3,5-dimethyl-pyrazol with 2,4,6-trichloro-s-triazine in anhydrous dioxane gives the compound L in high yield. Dropwise addition of $Eu(NO_3)_3{\cdot}6H_2O$ to the methanol solution of L afforded a clear, colourless solution, from which transparent crystals of the 1:2 of Eu:L complexes formed by slow evaporation of the solvent.

Measured Raman spectra of complexes by us are shown in Fig.1 and Fig.2 from $100cm^{-1}$ to $1800cm^{-1}$.

The results and discussions are shown as follows. Fig.1 shows the spectrum of complex $A{=}EuL_2{\cdot}2H_2O(NO_3)_3$, where L is 2,4,6-tri (3,5-dimethyl-N-pyrazoyl)-s-triazines. Compared with spectra of L that was recorded by us, the three bands at 133,191 and 208 cm^{-1} for the Eu complex disappear for sample L. These bands were assigned to Eu-O vibrations. This proves the Eu complex existence. Eu-N vibration of the Eu complex is at 241 cm^{-1}. The X-ray structure of the Eu complex reveals it to be composed of a neutral $[EuL(NO_3)_3H_2O]$ molecule and one uncoordinated ligand L, where Eu is ten-coordinated by three nitrogen atoms of a ligand L, Six oxygen atoms of three bidentate NO_3 groups and one water molecule. Two bands at 1001 and 1223 cm^{-1} appear for Eu complex, but it were not recorded for sample L. The two bands could be assigned to nitric anion vibration.

Fig.2 shows the spectra of complex $B{=}EuL_3{\cdot}6H_2O(ClO_4)_3$.The two bands at 1001 and 1223 cm^{-1} for complex A disappear for complex B. The new three bands at 458, 624 and 927 cm^{-1} appear for compound B. The band at 927 cm^{-1} is very strong. The three bands belong to ClO_4^- vibration. In the low frequency region the band at 133 cm^{-1} for complex A shifts to 128 cm^{-1} for complex B, two bands at 191 and 208 cm^{-1} for complex A become to one band at 201 cm^{-1} for complex B. These bands were assigned to Eu-O vibration. The band at 278 cm^{-1} for complex B is 37 cm^{-1} higher frequency than for complex A could be assigned to Eu-N vibration. Additionally, Their spectra are similar for both complex A and B. Therefore their ligands are similar also.

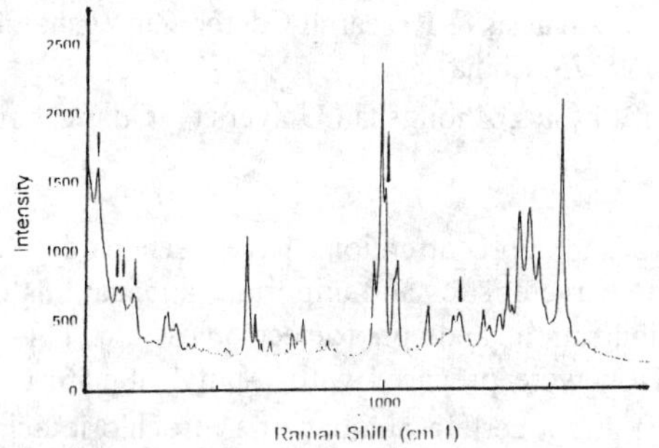

Fig.1. Raman spectrum of complex $A=EuL_2{\cdot}2H_2O(NO_3)_3$.

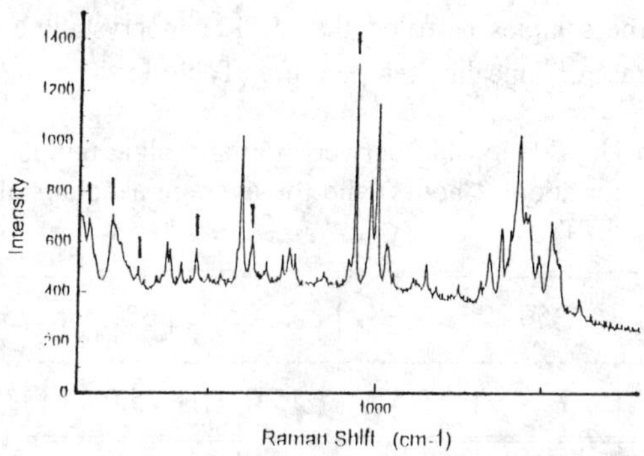

Fig.2. Raman spectrum of complex $B=EuL_3{\cdot}6H_2O(ClO_4)_3$.

1. C. Piguet, B. Bocquet, E. Muller, and A.F. Williams, Helv. Chim. Acta, 72,323(1989).
2. E.C. Constable, Adv. Inorg. Chem. Radiochem., 30,69(1986).
3. A.K. Saha, K. Kross, E.D. Kloszewski, D.A. Upson, J.L. Toner, R.A. Snow, C.D.V. Black, and V.C. Desai, J. Am. Chem. Soc., 115,11032(1993).
4. Chi YANG et al., submitted for publication.

RAMAN SPECTROSCOPIC STUDY OF NANOSTRUCTURED MATERIAL TiO_2 ANNEALED AT DIFFERENT TEMPERATURES

Zhuo-liang ZHANG[1]*, Jian CHEN[1], Wei-hong ZHANG[1], Qi-fa ZHOU[2], Shang-hua WU[2], Guang-ming LIN[2], and Jin-xiu ZHANG[2]

[1]Instrumentation Analysis & Research Center, Zhongshan University, Guangzhou 510275, China

[2]Department of Physics, Zhongshan University, Guangzhou 510275, China

Recently, more and more attentions have been paid to nanostructured materials for their unusual characteristics and important applications of new material complex, catalysis, sensibilization and photoelectronics etc [1]. In this paper, nanostructured material TiO_2 were prepared with n-butyl titanate (as raw material), ethanol (as organicsolvent) and a certain amount of hydrochloric acid by using Sol-Gel method, and the samples were annealed furthermore in air at different temperatures for several hours.

TEM of the samples revealed that TiO_2 nanocrystalline grains grew in diameter with the increase of annealing temperature (Table 1).

TABLE 1. The relationship between the annealing temperature (annealed for about 2 hours) and the average nanocrystalline grains size of TiO_2

T(℃)	350	450	550	650	750	950
ϕ(nm)	7.5	9.5	15.0	25.0	33.0	>50

Raman spectra of TiO_2 nanocrystalline grains were collected by SPEX 1403 0.85m double spectrometer equipped with 514.5 nm Ar-ion laser excitation, and they were shown in Fig.1.

It is known that there are two structural types of TiO_2 ----rutile-type and anatase-type, their space groups are D_{4h}^{14} and D_{4h}^{19}, and their Raman-active modes are $A_{1g}+B_{1g}+B_{2g}+E_g$ and $A_{1g}+2B_{1g}+3E_g$ respectively. For rutile-type TiO_2, its Raman shifts are normally at the following approximate values: 144(B_{1g}), 446(E_g), 610(A_{1g}) and 825(B_{2g}) cm^{-1}. For anatase-type TiO_2, its features are at 144(B_{1g}), 196(B_{1g}), 325(E_g), 398(E_g), 516(E_g), 638 cm^{-1}(A_{1g}).

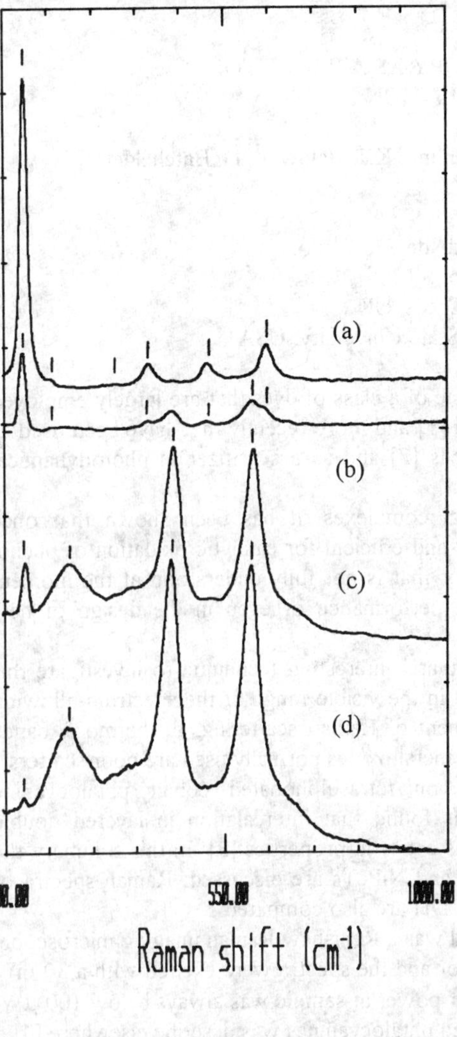

Fig.1. Raman spectra of nanostructured TiO_2 annealed at (a) 450 oC, (b) 650 oC, (c) 750 oC, and (d) 950 oC.

Fig.1(a), the sample was annealed in air at 450 ℃ for 2 hours, its Raman spectrum has six peaks identical with the features of anatase-type TiO_2, no more other peaks are found. It can be determined that after annealed at 450 ℃ , TiO_2 gel powder form anatase-type nanocrystalline grains completely. Fig.1(b), at 650 ℃, there appear seven peaks within the range of 100 ~1000 cm^{-1}, new peaks such as 232 , 450, 610 cm^{-1} are attributed to rutile-type TiO_2, and the two structural types of TiO_2 exist together.

After the sample was annealed at 750 ℃ for 2 hours, and at 950 ℃ for 1.5 hours, the spectra (seen in Fig.1(c), (d)) show four peaks same to the features of rutile-type TiO_2, and the peak at *ca.* 236 cm^{-1} is considered as a complex Raman line by S.P.S.Porto et al[2]. The peaks gain intensity except for the peak at 144 cm^{-1} (it becomes very weak), and no peaks of anatase-type are found. It can be believed that the anatase-type are transformed to rutile-type almostly, and the X-ray diffraction experiments also verified this transformation.

1. C.A.Melendres, A.Narayanasamy, V.A.Maroni, and R.W.Siegel, J. Mater. Res., 4(5), 1246(1989).
2. S.P.S.Porto, P.A.Fleury, and T.C.Damen, Phys. Rev., 154, 522(1967).

RAMAN MICROSPECTROSCOPY AS A PROBE FOR PHTALOCYANINE INTERCALATION

D.L.A. de Faria*, V.R.L. Constantino, K.J. Baldwin[1], D. Batchelder[1], T. J. Pinnavaia[2] and M. Chibwe[2]

Instituto de Química, Universidade de São Paulo,
CP 26077, 05599-970, São Paulo, SP
[1]Dept. of Physics, University of Leeds, UK
[2]Dept. of Chemistry, Michigan State University, USA

Phtalocyanine is the general name of a class of dyes that are largely employed in the pigment and petroleum industries [1] and more recently they have been used as catalyst in oxygen reduction in fuel cells [2] and as a sensitizer in photodynamical therapy .

Concerning phtalocyanine metal complexes, it has been shown that once immobilized in a matrix, both turn over and efficient for catalytic oxidation of phenol derivatives increases [3]. The reason for that is not fully understood at the moment and could help to optimize a catalyst performance or even in the design of new catalyst.

Raman spectroscopy is a particularly interesting technique to investigate this point as phtalocyanine absorbs strongly in the visible range of the spectrum allowing the observation of resonance enhancement of Raman scattering, furthermore, bands due to the support don't interfere as the metal oxides normally used are poor scatters.

The effect of immobilization on tetrasulphonated cobalt phtalocyanine (CoPcTs) has been studied and it was found that intercalation in layered double hydroxide (LDH) affects significantly its catalytic properties [3]. In this summary the Raman spectra of intercalated CoPcTs and NiPcTs are discussed; Raman spectra of CoPcTs intercalated and supported on LDH are also compared.

The Raman equipment employed was a Renishaw Raman imaging microscope, fitted with a Peltier-cooled CCD detector and the spectra were excited with a 30 mW air-cooled He-Ne laser (632.8 nm); laser power at sample was always below 100 μW. Preparation of intercalated and supported phtalocyanines was describe elsewhere [4].

In the case of CoPcTs, both intercalated and supported materials showed small shifts in band positions when compared to the pure phtalocyanine as showed in Table 1; the shifts, albeit small, are real as the spectra of all the three samples were taken before moving the grating to another spectral window.

Comparing the data for intercalated and supported samples it can be noticed that interaction with surface leads to a shift to higher wavenumber except for the band at 1034.0 cm^{-1}. On the other hand, the band at 1546.0 cm^{-1} is not affected by intercalation but this is not the case when CoPcTs is supported on the LDH surface as it shows up at 1549.5 cm^{-1}. It has been shown by X-ray analysis [4] that CoPcTs lies in a edge-on position when intercalated in LDH, with the negatively charged sulphonate groups electrostaticaly bounded to the positively charged layers of the LDH; with such an arrangement the macrocycle π electrons are only very slightly disturbed. However, when CoPcTs is not allowed in the LDH galleries, it stays on the surface in a parallel geometry where the environment effect on the π electrons are expected to be greater than in the perpendicular one. Indeed, change in some relative band intensities have been used to attribute the nature of the phtalocyanine-surface interaction [5,6] Similar behaviour is found for the system here reported and if considered together with the band shift at 1546.0 it can be used as a criteria to define the geometry of phtalocianine-LDH surface interaction.

TABLE 1. Band positions (cm^{-1}) and assignments for CoPcTs intercalated (INT) and supported (SUP) on LDH. Only bands that are sensitive to theinteraction are showed

CoPcTs	**CoPcTs INT**	**CoPcTs SUP**	**Assignment [6,7]**
688.5	690.1	691.6	Macrocycle deformation
752.5	753.0	753.0	Macrocycle deformation
1034.0	1032.3	1031.0	Macroc. def. and C-H bend.
1271.5	1273.5	1273.5	C-H bending
1460.0	1460.0	1461.5	C-N_α, C-C pyr,
1546.0	1546.2	1549.5	C=N stretching

Comparing the LDH intercalated NiPcTs with pure NiPcTs Raman spectra it could not be noticed any significant change in band intensities nor in band positions, except for the ones at ca. 690 cm^{-1}, 752 cm^{-1} and 1558 cm^{-1}. The 1558 cm^{-1} is shifted by 1 cm^{-1} what perhaps could be explained taking into account that X-rays data indicate that the molecule is tilted within the gallery.

ACKNOWLEDGMENT

D.L.A.F. and V.R.L.C. are indebted to FAPESP and CNPq (Brazil) for financial support.

[1] F.H. Moser and A.L. Thomas, The Phtalocyanines, v. II, CRC Press, Florida, 1983

[2] S. Srinivasan, J. Electrochem. Soc., **41C**, 136 (1989)

[3] M. Chibwe and T.J. Pinnavaia, J. Chem. Soc. Chem. Commun., 278 (1993)

[4] L. Ukrainczyk, M. Chibwe, T. J. Pinnavaia and S. A. Boyd, J. Phys. Chem., **98**, 2668 (1994)

[5] B.J. Palys, G.J. Puppels, D. van den Ham and D. Feil, J. Electroanal. Chem., **326**, 105 (1992)

[6] B.J. Palys, D.M.W. van den Ham, W. Briels and D. Feil, J. Raman Spectrosc., **26**, 63 (1995)

[7] R. Aroca and F. Martin, J. Raman Spectrosc., **17**, 243 (1986)

RAMAN SPECTRUM STUDY ON $LiNbO_3$:Zn SINGLE CRYSTALS

Yun-Cheng Ge* Chao-Zhong Zhao

Department of Physics, Harbin Normal University,
Harbin 150080, P.R.China

$LiNbO_3$ doped with Mg higher than the concentration threshold(4.6mol.% in the melt) was found to show a substantial reduction of the photorefractive damage(1,2). Recently, $LiNbO_3$:Zn (>6.2mol.% in the melt) has also been found to be able to withstand much higher laser intensity than pure $LiNbO_3$(3,4). It is found that the damage resistance is due to an essential increase in the photoconductivity σ_p(5). In $LiNbO_3$:Zn the increase in σ_p is most pronounced at concentrations exceeding certain threshold (6mol.% ZnO in the congruent melt), and is directly related to the reduction of the intrinsic defects Nb_{Li} because their substitution by damage-resistant-impurities.

The Raman spectra of A_1(TO) and E(TO) modes for the $LiNbO_3$:Zn cyrstals were observed, the results of A_1(TO) modes are represented in Fig.1. Comparing with the results of pure $LiNbO_3$, it is found that the number of A_1(TO) modes and E(TO) modes are essentially unchanged, also the frequency-shifts and relative intensities of them are the same as the case in pure $LiNbO_3$. However, there are some difference between them. Note the two low-frequency A_1(TO) modes in Fig.1, the border of them changes unclearly with the increasing of doping concentration, the $276cm^{-1}$-peak is difficult to observe as the concentration of Zn ions up to 6 mol.%. The same phenomenon occured for two E(TO) modes. It is likely that the Zn ions substitute for the Nb_{Li} ions first, i.e. the intrinsic impurities are substituted by the extrinsic impurities. As the doped concentration of Zn ions increasing up to 6 mol.%, they begin to substitute for the Li ions on normal lattice sites, make the concentration of impurities in the doped crystal increased. So that the lattice vibrational modes associated with Li ions show some changes as comparing to the pure $LiNbO_3$, such as the two low-frequency A_1 modes and two E modes coupled to each other. Also our results show that the A_1 modes and E modes in $LiNbO_3$ cyrstals are mainly determined by the oxygen octahedrons(NbO_6), and the inteaction between the cations and the oxygen octahedra is weak. The increase in σ_p may relate to the reduction of the intrinsic defects Nb_{Li} for their substitution by damage-resistant-impurities.

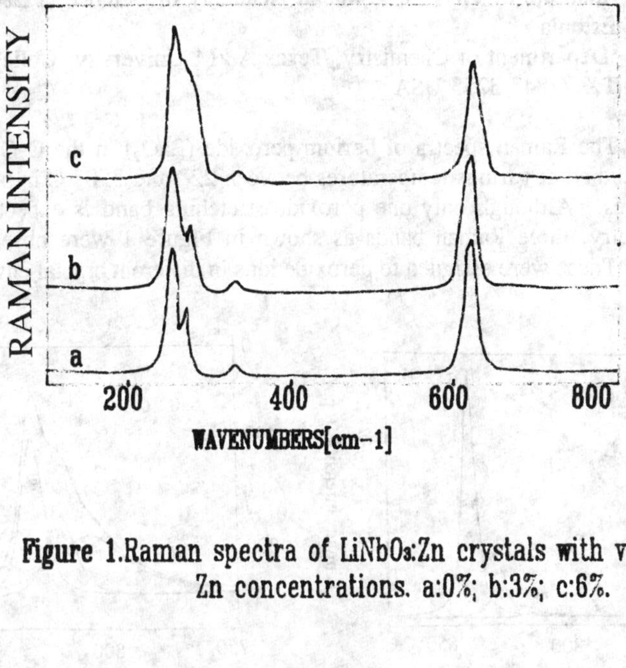

Figure 1.Raman spectra of LiNbO₃:Zn crystals with various Zn concentrations. a:0%; b:3%; c:6%.

1 G.G.Zhong, J.Jin, and Z.K.Wu, Proceedings of the 11th International Quantum Electronics Conference (Optical Society of America, Washington DC, 1980), IEEE Cat. No.80 CH1561-0, p.631.
2 D.A.Bryan, R.Gerson, and H.E.Tomaschke, Appl. Phys. Lett. 44, 847 (1984).
3 T.R.Volk, N.M.Rubinina, V.I.Pryalkin, V.V.Krasinkov, and V.V.Volkov, Ferroelectrics 109, 345 (1990).
4 T.R.Volk, V.I.Pryalkin, and N.M.Rubinina, Opt.Lett.15, 996 (1990).
5 T.R.Volk, J.Opt.Soc.Am.B.Opt.Phys.(USA), vol.11, no.9, p.1681(1994).

RAMAN SPECTRA OF BARIUM PEROXIDE AT TEMPERATURES OF 22° TO 850° C

K. Haller[1], J. H. Lunsford[2], and J. Laane[2]

[1]Institute of Physics, Estonian Academy of Science, EE2400 Tartu, Estonia
[2]Department of Chemistry, Texas A&M University, College Station, TX 77843-3255 USA

The Raman spectra of barium peroxide (BaO_2) in the O-O stretching region were studied at various temperatures between 22° and 850° C in both O_2 and He atmospheres. Although only one peroxide stretching band is expected for the crystal symmetry, three Raman bands as shown in Figure 1 were observed at all temperatures. These were assigned to peroxide ions in different crystal environments

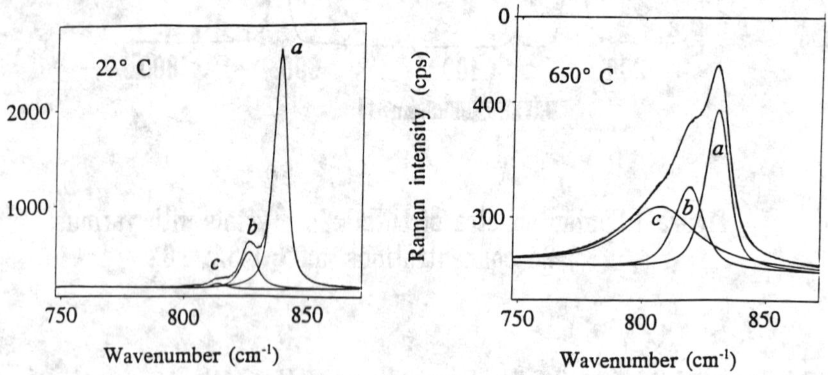

Figure 1. Raman spectra of BaO_2 at 22° and 650° C.

due to the relative locations of oxide impurities. Raman frequencies, bandwidths, and the absolute and partial intensities of each of these bands as a function of temperature were investigated in detail. The results are shown in Figure 2, where a, b, c refer to the bands labelled in Figure 1. These results can be understood [1] when peak a is assumed to be the "pure" form of BaO_2 while peaks b and c are produced through the interaction of oxide ions (present as an impurity) with Ba^{+2} ions, thus slightly altering the peroxide barium ion interactions.

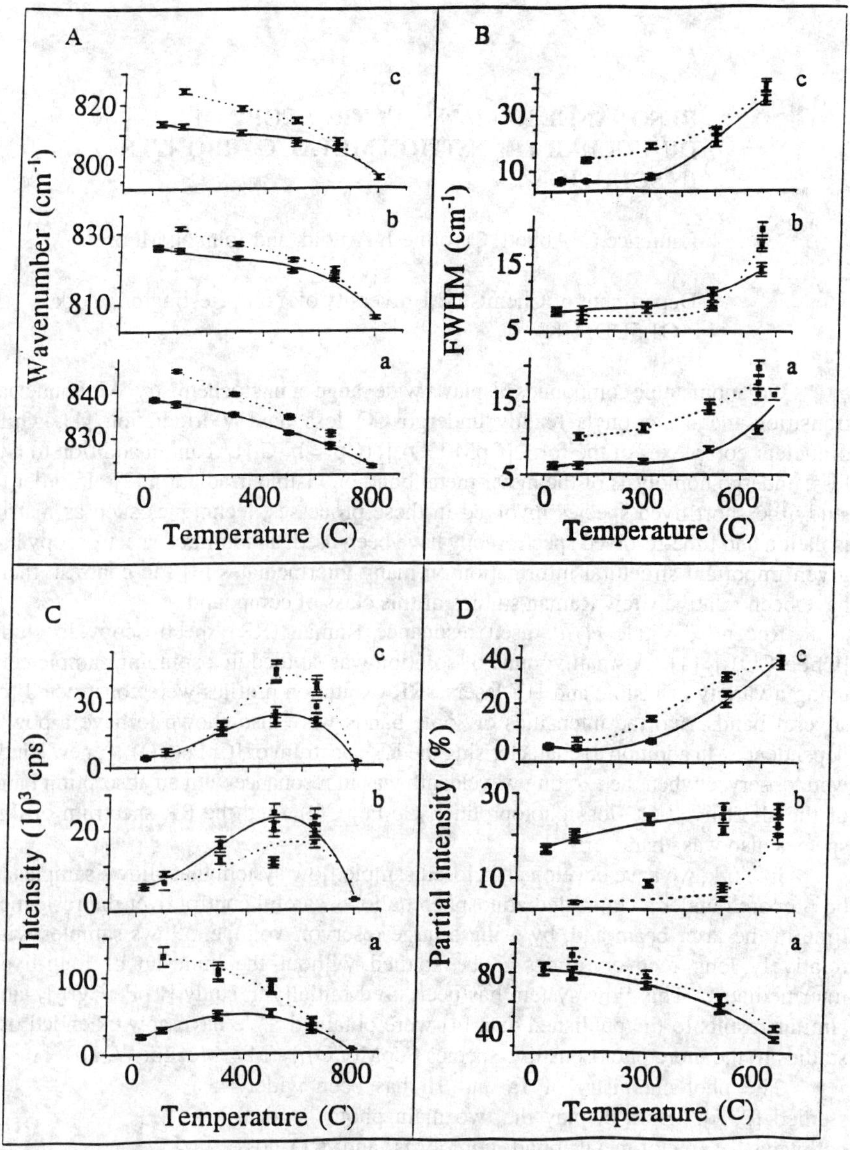

Figure 2. Temperature dependence of BaO_2 spectra.

1. K. Haller, J. Lunsford, and J. Laane, J. Phys. Chem. 100, 551 (1996).

RESONANCE RAMAN SPECTROSCOPY OF DINUCLEAR TRANSITION METAL CARBONYLS IN SOLUTION

Laurence C. Abbott, Caroline J. Arnold, and John N. Moore*

Department of Chemistry, University of York, Heslington, York
YO1 5DD, UK

Organometallic compounds display a wide range of photochemistry. Mononuclear transition metal carbonyls readily undergo CO loss on UV irradiation [1], while dinuclear complexes of the form $[CpM(CO)_n]_2$ ($Cp = \eta^5\text{-}C_5H_5$) can, in addition to CO loss, undergo homolysis of the metal-metal bond on visible irradiation [2]. In order to study the short-lived species involved in these processes, techniques such as matrix isolation and time-resolved spectroscopy have been used, and infrared spectroscopy has given important structural information on many intermediates [3]. In contrast, there have been relatively few Raman studies of this class of compound.

Recently, Vitale *et al.* used resonance Raman (RR) spectroscopy to study $[CpFe(CO)_2]_2$ [4]. A small volume of solution was studied in a spinning sample cell, using a variety of visible and UV lasers. RR excitation profiles were constructed for several bands, and the intensities of some bands were also shown to have a power dependence. In addition to characterising the RR spectrum of $[CpFe(CO)_2]_2$, new bands were observed when the exciting wavelength was in resonance with an absorption band of the short-lived CO-loss photoproduct, $Cp_2Fe_2(CO)_3$, and the RR spectrum of this species also was characterised.

In York, we have developed a closed sample flow system that allows samples to be prepared under a controlled atmosphere, allows careful control over the residence time in the laser beam and, by using a large reservoir volume, allows samples with relatively long recovery times to be studied without the build-up of long-lived intermediates. This flow system has been used initially to study $[CpFe(CO)_2]_2$, and similar results to the published data [4] were obtained. We have now extended our studies to the more photosensitive species $[CpM(CO)_3]_2$ (M = Mo, **1a**; W, **1b**).

The photochemistry of **1a** and **1b** has been widely studied [2, 5]. Both display the two main photochemical pathways of metal-metal bond homolysis and CO loss indicated above, in addition to further photoreactions of the products [5]. This diverse photochemistry can result in rapid decomposition on irradiation, and makes RR spectroscopy difficult under normal sampling conditions [6, 7]. We have applied the flow technique to study the RR spectra of **1a** and **1b** in solution, and spectra have been recorded without

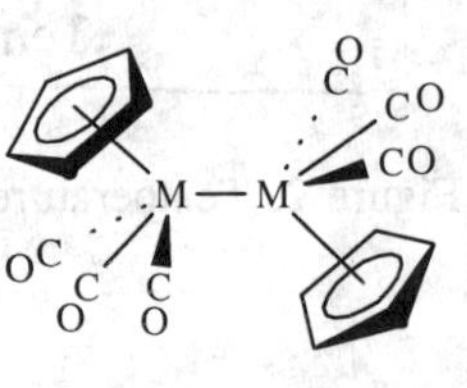

1a M = Mo
1b M = W

significant sample degradation. This is illustrated in Figure 1, which shows a RR spectrum of **1a** in *p*-xylene solution; the band positions compare well with those described in early work on solid state samples [6, 7]. RR spectra recorded at high laser powers showed a marked decrease in the sample band intensities, relative to solvent bands, allowing quantum yields of photolysis to be estimated. No new bands were observed under these conditions, attributable to the poor resonance enhancement of the photoproduct Raman spectrum at this excitation wavelength.

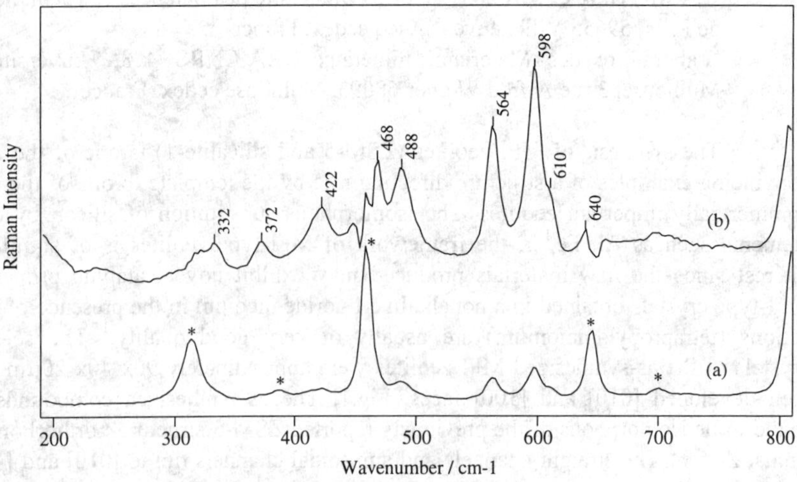

Figure 1. RR spectrum of $[CpMo(CO)_3]_2$ (8×10^{-3} mol dm^{-3}) in *p*-xylene obtained using the flow system with a linear flow rate of 400 cm s^{-1} and 514.5 nm excitation at a power of 80 mW. (a) Raw data; (b) Solvent subtracted spectrum. Asterisks denote solvent bands.

ACKNOWLEDGEMENTS

We thank the EPSRC for studentship support (LCA), for Advanced Fellowship support (JNM), and also Shell for EPSRC CASE studentship support (CJA).

REFERENCES

(1) Geoffroy, G. L.; Wrighton, M. S. *Organometallic Photochemistry*; Academic Press: New York, 1979.

(2) Meyer, T. J.; Caspar, J. V. *Chem. Rev.* **1985**, *85*, 187-218.

(3) Poliakoff, M; Weitz, E. *Adv. Organomet. Chem.*, **1986**, *25*, 277-316.

(4) Vitale, M.; Lee, K. K.; Hemann, C. F.; Hille, R.; Gustafson, T. L.; Bursten, B. E.; *J. Am. Chem. Soc.* **1995**, *117*, 2286-2296.

(5) Knorr, J. R.; Brown, T. L. *J. Am. Chem. Soc.* **1993**, *113*, 4087-4092.

(6) Parker, D. J. *J. Chem. Soc., Dalton Trans.* **1974**, 155-162.

(7) San Filippo, J. Jr.; Sniadoch, H. J. *Inorg. Chem.* **1973**, *12*, 2326-2333.

POLARIZED MICRORAMAN SPECTROSCOPY OF SMALL CRYSTALS OF SILICALITE-1, Al MFI AND Fe MFI-TYPE ZEOLITES

Claude Brémard[1], Jacky Laureyns[1], and Joël Patarin[2]

[1]LASIR, UPR-CNRS 2631,Bât.C5, Université des Sciences et Technologies de Lille, 59655 Villeneuve d'Ascq cedex, France
[2]Laboratoire des Matériaux minéraux URA-CNRS 428, Université de Mulhouse, 3 rue Alfred Werner 68093, Mulhouse cedex, France

The synthesis of MFI zeolites (ZSM-5 and silicalite-1) is one of the more convincing examples of a structure-directing role by the template to one of the most commercially important zeolites. The isomorphous substitution of silicon by other elements such as Al, Fe, in the framework of MFI-type zeolites is of significant interest since the new materials produced may exhibit novel catalytic properties. MFI-type crystals obtained in a nonalkaline fluoride medium in the presence of TPA cations (tetrapropylammonium) are usually of very good quality.[1] The selected crystals of the as-synthesized MFI zeolites were approximately 90x 40x 20μm with well developed [010] and [100] faces, Fig.1. The as-synthesized compounds are found to be isomorphous of the previously reported ZSM-5 structure : orthorhombic, Pnma, Z = 8.[2] The straight channels and sinusoidal channels run to [010] and [100], respectively.

The present work reports the polarized Raman spectroscopic study of single-crystals of four as-synthesized MFI-type samples (Si /Al > 1000, Si /Al ~ 30, 20 and Si / Fe ~ 30) using a Raman microprobe. The assignment of the Raman bands to the framework and to the occluded TPA ion templating the MFI structure was examined through the contribution of the different elements of the Raman tensors as functions of the Al or Fe content, respectively.

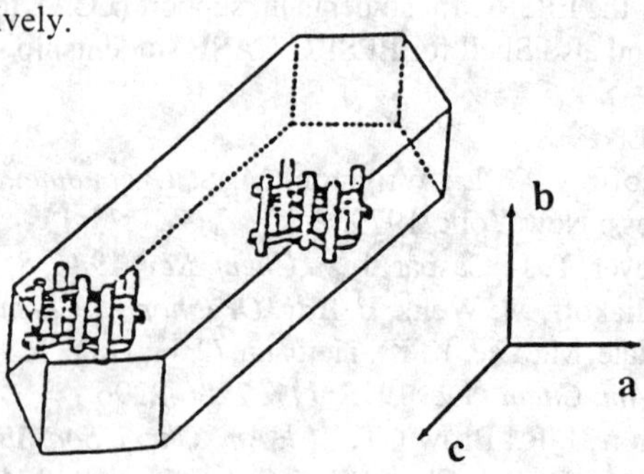

Fig. 1 Schematic representation of the MFI crystals showing the orientation of the faces and the network arrangement of intersecting straight and sinusoidal channels according to the cell parameters.

The comparison between the corresponding Raman spectra of the as-synthesized and calcined zeolites provides evidence to the assignment of the Raman bands to the zeolite framework and TPA^+ cation vibrational motions, respectively. The steric constraints in the channels induce some changes in the Raman bands attributed to the internal vibrational motions of the propyl fragments. In neither of the templating examples does the TPA cation adopt the fully extended conformation found in the bromide salt TPABr, $[N(CH_2CH_2CH_3)_4Br]$.

The weak bands derived from the intense feature at 1320 cm^{-1} observed in the spectrum of the free TPA^+ cation, depend markedly on the orientation of the crystal. The presence of five-membered rings in the framework leads to the motion of the O atom along the bisecting line of the Si-O-Si angle in the 350-400 cm^{-1} region. Some Raman bands can be resolved by the polarization study, by comparison with the broad Raman features shown by the powder spectra.[3] With increasing Al or Fe content the A_g, B_{2g} and B_{3g} modes are slightly shifted to lower or higher values. Only minimal differences are detected in the polarized spectra in the wavenumber regions of the vibrations of the template and the zeolite framework, respectively, due to the low Al and Fe content of the MFI framework. It was not possible to check the validity of recent computational and EXAFS results by polarized Raman spectroscopy.

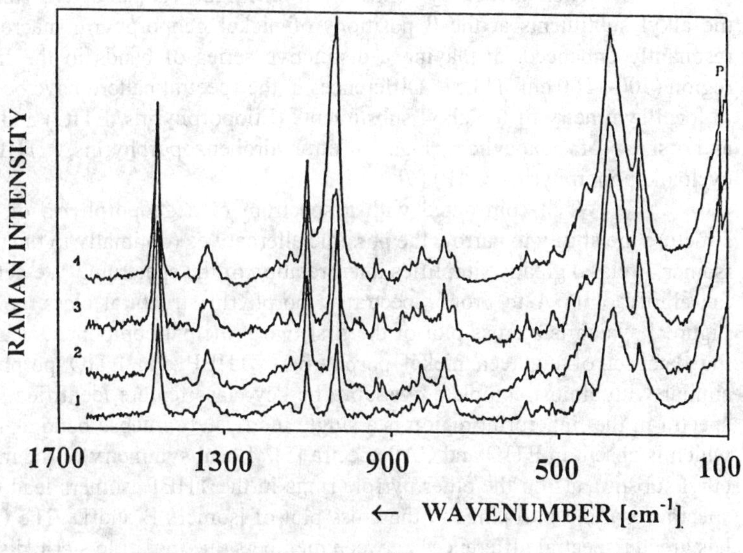

Fig.2 : **b(cc)b** polarized Raman spectra of as-synthetized MFI zeolite crystals over the range 1700-100 cm^{-1} (λ_{ex} =514.5 nm): 1- silicalite-1 (Si/ Al = 1000) ; 2)-Al MFI (Si/ Al ~ 20) ; 3- Al MFI (Si/ Al ~ 20) ; 4- Fe MFI (Si/ Fe ~30).

1. M. Soulard, S. Bilger, H. Kessler and J.L. Guth, Zeolites, 7, 463 (1987).
2. H. van Koningsveld, H. van Bekkum and J. C. Jansen, Acta Crystallogr., B43, 127 (1987).
3. P.K. Dutta and M. Puri, J. Phys. Chem., 91, 4329 (1987).
4. J. Patarin, M. H. Tuilier, J. Durr and H. Kessler, Zeolites, 12, 70 (1992).

PRINCIPAL COMPONENT ANALYSIS OF RESONANCE RAMAN SPECTRA OF NICKEL GEOPORPHYRINS

J. G. Rankin and R. C. Tucceri

Department of Chemistry, Marshall University, Huntington, West Virginia 25755-2520

Principal component analysis (PCA) has been shown to be a powerful tool for classifying vibrational spectra based on the presence or absence of certain functional groups [1]. We have applied this technique to resonance Raman spectra of synthetic nickel geoporphyrins analogous to those found in geological matrices to develop methods of classifying isolated geoporphyrins. Over 70 geoporphyrins are known from geological sources and determination of the structural types and relative abundance of these geoporphyrins can aid in the study of the formation of and search for fossil fuels [2,3].

When excited by yellow (568 nm) light, Raman active vibrations of the alkyl subtituents at the β positions of nickel geoporphyrin macrocycle are resonantly enhanced, displaying a distinctive series of bands in the fingerprint region (700-1200 cm^{-1}) [4,5]. Differences in the spectral pattern have been related to local symmetry of the alkyl substituents (Etioporphyrins, ETIO) and presence and size of exocylic rings (Tetrahydrobenzoporphyrins, THBP, and cycloalkanoporphyrins, CAP) [6].

When confronted with a spectrum of a geoporphyrin of unknown structure the ability to narrow the possible alternatives (optimally to one particular isomer or class) greatly simplifies interpretation of the spectrum. We demonstrate the ability of PCA to group spectra by porphyrin structural class as shown in Figure 1 which is a cross plot of the first two principal components (factors) for the data set of eighteen nickel porphyrins. THBP and ETIO porphyrins are similar with almost complete overlap. The key class feature identified for THBP spectra in the fingerprint region is a single moderately intense band at 1092 cm^{-1} which is absent in ETIO and CAP spectra [5]. Local symmetry of the methyl and ethyl substituents on the other pyrrole rings in the THBP isomers lead to similar spectra resulting groupings in the cross plot of isomers B and D. The CAP class has greater spectral differences between members due in part to steric distortion of the porphyrin macrocycle by the exocyclic ring and the presence of a new series of marker bands in this spectral region for exocyclic ring size [6]. In Figure 1 it can be seen that the CAP6 and CAP7 compounds (6 or 7 member exocylic ring size, respectively) are group together. The single CAP5 compound analyzed (CAP5-Et, 5 member exocyclic ring with an ethyl group at porphyrin carbon position 12) is quite removed from the group (0.61,-0.56) with a Mahalanobis distance of 5.47. The Mahalanobis distance represents the distance between group means: the greater the distance the more distinctive the groups. The discriminate analysis software used for this analysis (GRAMS/386, Galactic Industries, USA) clearly identifies CAP5-Et as belonging to a separate group. Raman spectra of additional members of this class, particularly the predominate geoporphyrin, deoxophylloerthroetioporphytin (DPEP) need to be obtained and analyzed to better define this region of factor space.

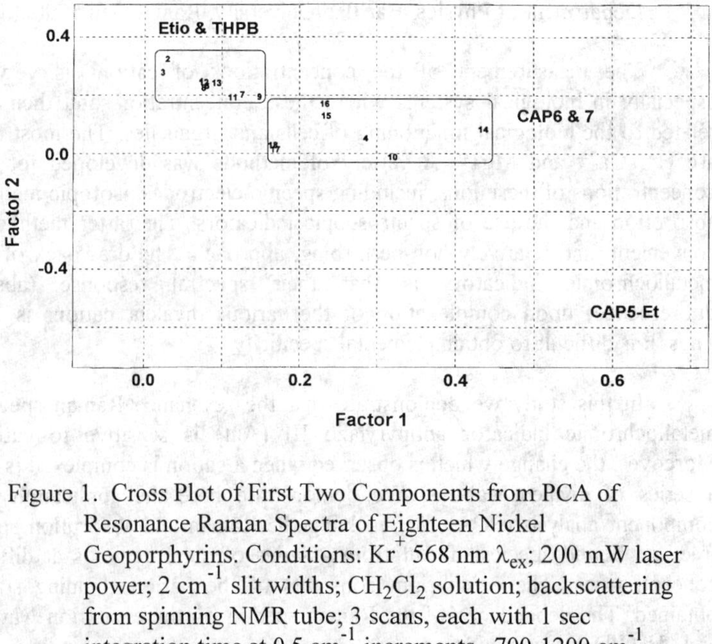

Figure 1. Cross Plot of First Two Components from PCA of Resonance Raman Spectra of Eighteen Nickel Geoporphyrins. Conditions: Kr^+ 568 nm λ_{ex}; 200 mW laser power; 2 cm^{-1} slit widths; CH_2Cl_2 solution; backscattering from spinning NMR tube; 3 scans, each with 1 sec integration time at 0.5 cm^{-1} increments. 700-1200 cm^{-1}, parallel scattering.

1. (a) Hasenoehrl, E. J.; Perkins, J. H.; Griffiths, P. R. *Anal. Chem.* **1992**, *64*, 656-663. (b) Remberg, B.; Nikiforov, A.; Buchbauer, G. *Fresenius J. Anal. Chem*. **1994**, 348, 258-263.
2. Baker, E. W.; Palmer, S. E. In *The Porphyrins*; Dolphin, D. Ed.; Academic Press: New York, 1978; vol. 1, pp. 486-552.
3. Callot, H. J.; Ocampo, R.; Albrecht, P. *Energy & Fuels* **1990**, *4*, 635-639.
4. Rankin, J. G.; Czernuszewicz, R. S. *Org. Geochem.* **1993**, 20, 521-538.
5. Rankin, J. G.; Czernuszewicz, R. S.; Lash, T. D. *Inorg. Chem.* **1995**, *34*, 3025-3037.
6. Rankin, J. G.; Czernuszewicz, R. S.; Lash, T. D. *Org. Geochem.* **1995**, 23, 419-427.

PRINCIPAL COMPONENT ANALYSIS OF ABSORPTION AND RESONANCE RAMAN SPECTROSCOPY OF ANTIPYRYLAZO, A METALLOCHROMIC INDICATOR

Benjamin Ehrenberg* and Eliyahu Pevzner

Department of Physics, Bar Ilan University, Ramat Gan 52900, Israel

The measurement of the concentrations of cations is very important, especially in biological systems where these concentration, and their changes are related to the biological functioning of cells and organelles. The most relevant ions are H^+, Ca^{2+} and Mg^{2+}. A variety of methods was developed for probing the concentrations of these ions, including specific electrodes, isotopic methods, atomic absorption and the use of spectroscopic indicators. The latter method provides a convenient, and relatively non-perturbing, approach. The drawback of most of the metallochromic indicators is that their spectral response (absorption or fluorescence) upon complexation of the various divalent cations is very similar. Thus, it is difficult to obtain elemental specificity.

In this study we demonstrate that the resonance Raman spectrum of the metallochromic indicator antipyrylazo III (Ant) is sensitive to cation binding. Moreover, the change which is observed when a cation is complexed is different for a series of divalent cations. We demonstrate here the application of principal component analysis (PCA) for resolving the Raman and absorption spectra of the free and bound species of Ant from a series of measurements at different cation concentrations. The equilibrium constants of the various binding stages is also obtained. The dependence of the Raman spectra on the excitation wavelength, pH and H_2O/D_2O composition was used to evaluate the binding modes and configurations. In a previous study we have shown that the resonance Raman of the metallochromic indicator arsenazo III can be used to monitor Ca^{2+} ions, even in the presence, and without interference, of a much higher concentration of Mg^{2+} ions[1].

The following summarizes the results which we obtained with both the absorption and resonance Raman spectroscopy of Ant upon binding of Ca^{2+}, Mg^{2+}, Sr^{2+}, Ba^{2+} and Zn^{2+} ions.

Calcium: We found that the Raman spectrum is resolvable into the free Ant and CaAnt, while the spectrum of the species $Ca(Ant)_2$ has no resonance enhancement. The resolution of the absorption spectra shows indeed its absorption band at wavelengths above 600 nm. The equilibrium constants are shown in the Table. The RR spectra are shown in Figure 1a and b.

Magnesium: Here, too, the Raman spectrum of $Mg(Ant)_2$ is not resonance enhanced and the spectra of only Ant and MgAnt are resolved. The equilibrium constants are given in the Table.

Strontium: The resolution of the Raman spectra by PCA reveals that the spectra of SrAnt and of $Sr(Ant)_2$ are almost identical.

Barium: PCA yields that the Raman spectra of Ant and $Ba(Ant)_2$ are very similar. The spectra and the equilibrium constants were resolved.

Zinc: Both resonance Raman and absorption spectra indicate that the formation of $Zn(Ant)_2$ has a very high binding constant. Even at micromolar Zn concentration it is fully bound and there is no evidence of the formation of the 1:1 complex ZnAnt.

Table: equilibrium constants for binding of metal cations to antipyrylazo

Ion	$K_1 = ([MgAnt]/([Mg]\cdot[Ant]))$	$K_1 = ([MgAnt_2]/([Mg]\cdot[Ant]^2))$
Ca	2234 M^{-1}	$8.80\cdot10^7$ M^{-2}
Mg	1820 M^{-1}	$1.19\cdot10^7$ M^{-2}
Sr	592 M^{-1}	$1.27\cdot10^7$ M^{-2}
Ba	391 M^{-1}	$0.56\cdot10^7$ M^{-2}
Zn	---	$>1\cdot10^{10}$ M^{-2}

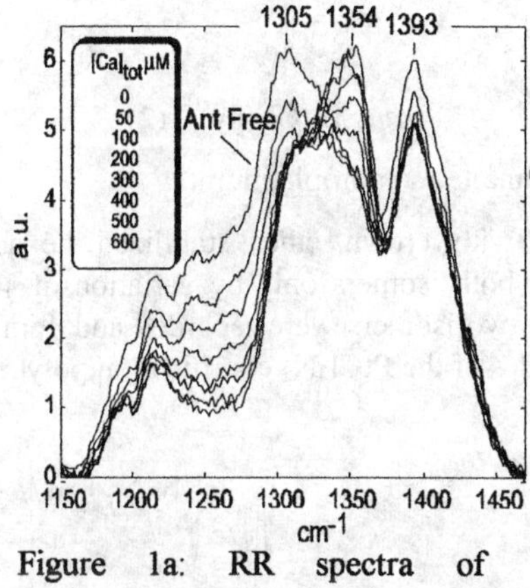

Figure 1a: RR spectra of Antipyrylazo (50 µM) with increasing concentrations of Ca^{2+}

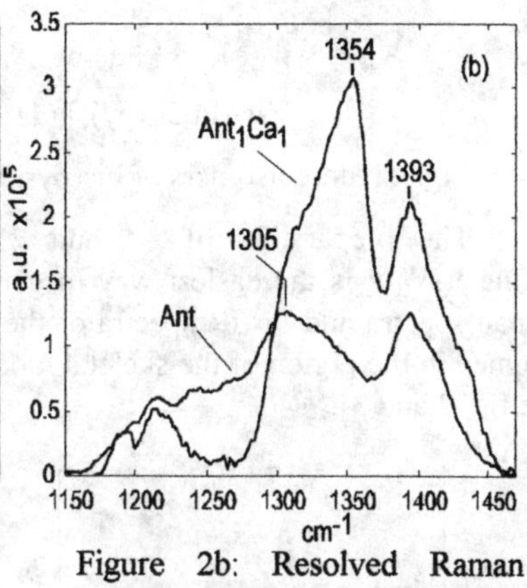

Figure 2b: Resolved Raman spectra of free Ant and the 1:1 complex with Ca^{2+}

1. B. Ehrenberg, FEBS Letters 140, 219 (1982).

Raman and IR spectroscopic study of bis(pyridine)tetrachloroplatinum(IV)

H. Junicke[1], K. Schenzel[2*]

[1]Institute of Inorganic Chemistry, Martin Luther University Halle-Wittenberg, Halle, Germany
[2]Institute of Analytical Chemistry, Martin Luther University Halle-Wittenberg, Halle, Germany

Platinum(IV) amino complexes are well known since the end of the last century[1]. In the last two decades, the discovery of platinum (IV) amino compounds which have cancerostatic properties accelerated dramatically the research activities in this field [2,3,4]. Our synthesis of new crown ether stabilized hexachloroplatinates(IV) gave the possibility to investigate the influence of ligand exchange reactions in different organic solvents. Vibrational spectroscopy was used[5] to identify the different isomers of the platinum(IV) complexes, [see fig. 1].

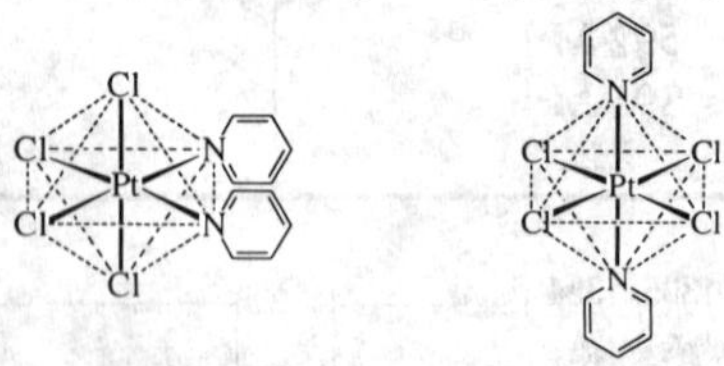

cis-[$PtCl_4(py)_2$] (**1**) *trans*-[$PtCl_4(py)_2$] (**2**)

Fig. 1: Stereoisomers of bis(pyridine)tetrachloroplatinum(IV)

The preparation of **1** and **2** with crown ether stabilized hexachloroplatinate(IV)[6] is the easiest way to get both isomers only by variation of solvent. Raman spectra and far IR spectra of the two isomers were recorded and completely assigned in the region of the skeletal modes of the $PtCl_4N_2$ coordination polyhedrons, [see fig. 2 and 3].

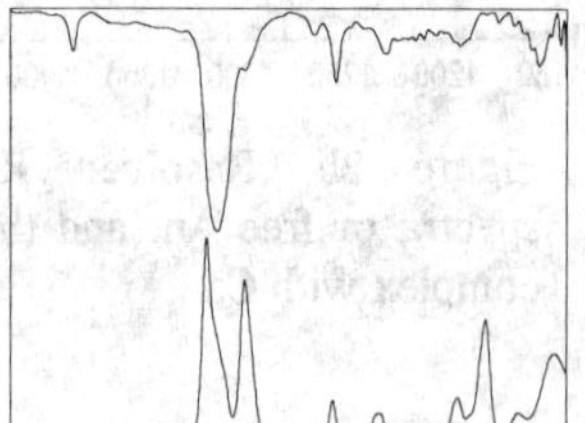

Fig. 2: IR and Raman spectra of *cis*-$PtCl_4(py)_2$

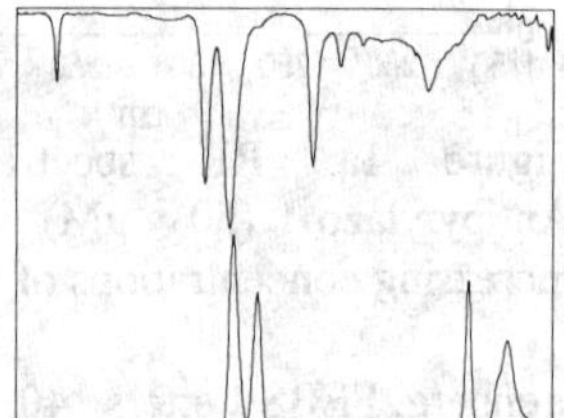

Fig. 3: IR and Raman spectra of *trans*- $PtCl_4(py)_2$

Assuming the highest possible symmetry the normal vibrations of *cis* and *trans* $PtCl_4N_2$ skeletons span the irreducible representations represented in table 1.

Table 1: Irreducible representations of nomal vibrations of $PtCl_4N_2$

isomer	*cis*-$PtCl_4(N)_2$	*trans*- $PtCl_4(N)_2$
irreducible representation	$\Gamma_{C_{2v}} = 6A_1+2A_2+4B_1+3B_2$	$\Gamma_{D_{4h}} = 2A_{1g}+B_{1g}+B_{2g}+E_g+2A_{2u}+B_{2u}+3E_u$
irreducible representation of ν(PtCl) only	$\Gamma_{C_{2v}} = 2A_1(IR/Ra)+B_1(IR/Ra)+B_2(IR/Ra)$	$\Gamma_{D_{4h}} = A_{1g}(Ra) + B_{1g}(Ra) + E_u(IR)$

The splitting of the degenerate infrared active mode ν_9 ($\nu_{as}(PtCl_4)$) of the *trans* isomer indicates a lowering of the symmetry of this coordination polyhedron. According to the correlation tables, the lowering of symmetry from point group D_{4h} to D_{2h} or C_{2h} has to be considered. The point group C_{2h} for *trans*- $PtCl_4(py)_2$ was finally confirmed by crystal structure analysis[6]. These results clearly disprove the assignment for *cis* and *trans* isomers reported by Kukushkin[2]. In solution the reaction of crown ether stabilized hexachloroplatinate(IV) leads either to *cis* or *trans* isomers depending on the solvent. Raman measurements of *cis*-$PtCl_4(py)_2$ in a solution of DMSO prove the conversion of the *cis*-isomer into the *trans*-isomer, [see fig. 4]. The process was finished after 24 h at room temperature.

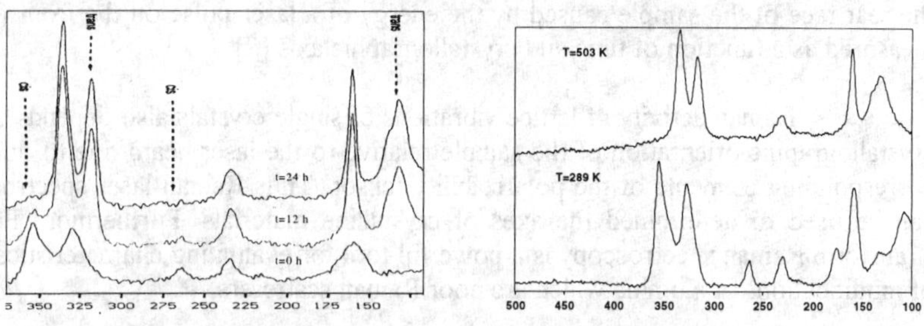

In solution of DMSO In solid state

Fig. 4 : Conversion of *cis*-$PtCl_4(py)_2$ into *trans*- $PtCl_4(py)_2$

Temperature depended Raman experiments show an immediately isomerization of the *cis*-isomer into the *trans*- $PtCl_4(py)_2$ at 503 K,[see fig. 4].

1. S. G. Hedin, *Acta Univ. Lund.*, Sect. 2, 22, 1-58, (1886).
S. M. Joergensen, S. M., *Z. Anorg. Chem.*, 25, 353-377, (1900).

2. Yu. N. Kukushkin, G. N. Sedova, A. N. Belyaev, L. V. Konovalov, *Zh. Neorg. Khim.* 25(4), 1036-43, (1980).
Yu. N. Kukushkin, V. N. Demidov, *Zh. Neorg. Khim.*, 27(11), 2839-46, (1982).

3. S. Bajusaz, T. Janaky, V. J. Csernus, L. Bokser, Fedeke, G. Srkalovic, T.W. Redding, A. V. Schally, *Proc. Nath. Acad. Sci., USA*, 86, 6313, (1989).

4. N. V. Vorob'ev-Desyatovskii, A. A. Barinov, Yu. N. Lukin, V. V. Sokolov, V. N. Demidov, A. Yu. Kuptsov, *Zh. Obshch. Khim.*, 61(3), 709-16, (1991).

5. W. Preetz; P. Erlenhöfer, *Z. Naturforsch.*, 44b, 412-418, (1989).
W. Preetz, G. Rimkus, *Z. Naturforsch.*, 37b, 579-586, (1982).

6. F. W. Heinemann, H. Junicke, K. Schenzel, D. Steinborn, in preparation.

FOURIER TRANSFORM RAMAN SPECTROSCOPY AND LASER FLASH MEASUREMENTS OF ALUMINA SINGLE CRYSTALS

H. P. Jeziorowski

Materials Research, D-85435 Erding, Germany

Single crystals of alumina are used as infrared transparent materials especially in high temperature applications where thermal diffusivity and thermal conductivity are significant properties. Single crystalline corundum (α-Al_2O_3) belongs to the trigonal crystal system D_{3d} [1]. Due to the anisotropy of the material physical properties may depend on the crystallographic directions, for example on the c-axis (threefold symmetry axis) and on the a-axis (twofold symmetry axis).

One of the most popular and widely used method for measuring the thermal diffusivity, is the laser flash technique. The relative temperature increase on the rear face of the sample caused by the energy of a laser pulse on the front face is measured as a function of time and crystallographic axis [2].

Raman activity of lattice vibrations of single crystals also depends on the crystallographic orientation of the sample relative to the laser beam due to different corresponding elements of the polarizability tensor. Thus Raman laser spectroscopy can be used to determined the axes of crystalline materials. Furthermore Fourier Transform Raman spectroscopy is a powerful tool for evaluating characteristics even of modifications of alumina which are poor Raman scatterers.

The Raman bands of polycrystalline α-Al_2O_3 observed by Bertoluzza [3] are shown in Table 1 together with the Raman bands of α-Al_2O_3 (corundum) single crystals observed at different orientations in the xyz space coordinates.

Table 1. Raman bands of polycrystalline α-Al_2O_3 and α-Al_2O_3 single crystals (a-axis and c-axis parallel to the laser beam).

	Raman bands of α-Al_2O_3 [cm^{-1}]						
Polycrystalline	750	645	578	447	428	417	377
Single crystal a_z	749					418	380
Single crystal a_x		645				418	380
Single crystal c	752					418	

The Raman spectra were recorded using a Bruker FRA 106 FT-Raman accessory mounted on an IFS 88 FT-IR optical bench and the 1064 nm line of a Nd: YAG laser for excitation in 180° scattering geometry. The Raman spectra obtained of flat samples of single crystal corundum prepared for laser flash measurements are shown in Fig. 1. The weak bands of the Raman spectrum of polycrystalline α-Al_2O_3 are not detectable in both spectra. From the Raman bands with strong and medium intensity three bands are observed in spectrum a whereas in spectrum c only two bands are observed indicating the case with the c-axis parallel to the laser beam.

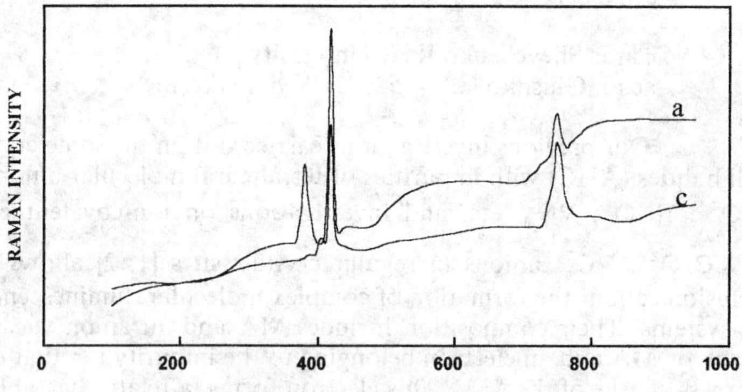

Fig. 1. Raman spectra of single crystals of α-Al_2O_3 (corundum) a: Crystal axis a parallel to the laser beam. c: Crystal axis c parallel to the laser beam (reproduced by PC graphics).

The results of measurements with the laser flash technique to determine the thermal diffusivity and thermal conductivity show that the heat transfer in single crystalline alumina samples with the c-axis normal to the surface, i.e. parallel to the laser beam is higher than in samples with other orientation. These results are in accordance with results found in the literature [4] and prove the assignment of the FT Raman spectra.

1. G. Turrell, Infrared and Raman Spectra of Crystals, Academic Press, London, 1972.
2. J.B. Henderson, W.D. Emmerich, L. Hagemann, and E. Post, High Temperatures-High Pressures 25, 693 (1993).
3. A. Bertoluzza, C. Fagnano, and A. Tinti, J. Biomedical Materials Research 23, 523 (1989).
4. Handbook of Chemistry and Physics, CRC Press, Boca Raton, 1981 Vol. 60.

INVESTIGATIONS OF THE BONDING ELECTRON - PHONON MODES IN THE LOCAL, MOLECULAR LUMINESCENCE CENTERS OF IONIC AND ION-COVALENT CRYSTALS

S. Nedelko

Taras Shevchenko Kyiv University,
acad. Glushkova. 6, 252022, Kyiv, Ukraine

Our previous investigations carried out on the some crystals such as alkali halides (AHC) with impurities of tetrahedral molecular anions (MA) like CrO_4^{2-}, ReO_4^-, VO_4^{3-} etc. and investigations on ion-covalent crystals with SO_4^{2-}, CrO_4^{2-}, VO_4^{3-} anions in regular crystals sites [1, 2] allows us to make conclusions about the formation of complex molecular luminescence centers in such systems. Their composition includes MA and an anion vacancy trapped electron or MA and an electron belonging to the impurity ion that compensates the excess charge of the MA. This electron forms a relatively stable bond with the MA. The bond is stronger in the excited state of the center . The created complex is characterized by two wells in the potential curve. We called such a complex : the ion-polaron pair. The trapped electron plays the role of localized polaron. With increasing temperature, the polaron releases itself and it can transfer the excitation energy. Complex formation manifests itself by the appearing of the long wavelength luminescence bands connected with the short wavelength bands those origin from uncomplexed luminescence centers.

The formation of such complexes is exhibited in the dynamic of high frequency internal vibrational modes of complexes and in the peculiarities of the low frequency crystal modes. These peculiarities were investigated by Raman scattering methods.

Raman spectra of pure and doped with CrO_4^{2-}, VO_4^{3-}, Ca^{2+} , Ba^{2+} ions MeHal and $Me_n AO_4$ crystals (Me = K, Cs; Hal = Cl, Br ; A = S, V, Cr) were studied at 77 K and 300 K. The samples were obtained by solid state synthesis method as well as from melting and aqueous solutions.

The strong deformation of AO_4^{n-} molecular anions and the essential changes in Raman spectra of internal modes of mentioned above complexes centers were established (Fig. 1, A). The new modes appear in the low frequency region lattice vibrations. Their frequencies lie beyond off the phonon spectrum region of "pure" crystals and they are lower than ones of internal modes of molecular groups (Fig. 1, B). Raman intensity of these modes depends on exciting light frequency (there is evidence of resonance character of the excitation) and sharply decreases at temperature decreasing. Sometimes these modes have a structure .

We have analyzed the structure and some parameters that characterize series of structure peaks were established. Obtained results allow us to interpret such low frequency modes as a bonding electron-phonon modes.

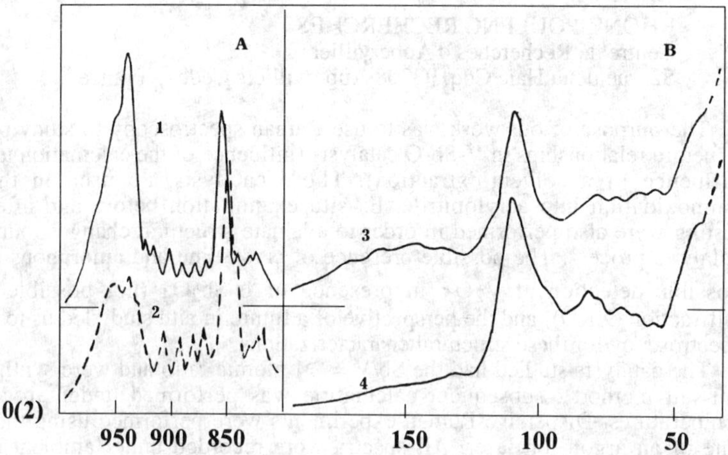

Fig. 1. Raman spectra of CsI - VO_4^{3-} crystals with La^{3+} (1, 3, 4) and Ca^{2+} (2) cations impurities . Concentration relation is: La^{3+} : VO_4^{3-} = 1.0 (1, 3), 0.01 (4); Ca^{2+}: VO_4^{3-} = 1.0 (2). λ_{ex} = 488 nm. RT.

A: Region of internal modes of VO_4^{3-} - anions.

B: Region of lattice modes of CsI crystals.

1. S.G. Nedelko, M.M. Podshivajlov, N.P. Smolyar and V. I. Sheludko, Zhurnal prikladnoj spectroskopii (Minsk, Belarus, in Russian), **55,** 838 (1991).
2. S.G. Nedelko, V. I. Sheludko, N.P. Smolyar and V.P. Koshelenko, Physics in Ukraine. Intern. Conf., June, 1993. Proceedings: Solide States Physics. Kiev, 149 - 151 (1993).

RAMAN DIFFUSION STUDY OF V-Sb-O CATALYSTS FOR PROPANE AMMOXIDATION

B. Amram*, M. Belleil, and P. Burattin

RHÔNE-POULENC RECHERCHES
Centre de Recherches d'Aubervilliers
52, rue de la Haie Coq, 93308 Aubervilliers Cedex, France

The purpose of our work was to use Raman spectroscopy to study preparation-structure relationships in V-Sb-O catalysts (influence of the calcination temperature, influence of a solvent-extraction). Those catalysts are used in the propane ammoxidation into acrylonitrile. Ex-situ examination before and after catalytic testing were also performed in order to evaluate structural changes induced by the catalytic process. The possible presence of crystalline and amorphous phases, the possible detection of V_2O_5 in presence of α-Sb_2O_4 (not possible by X-Ray Diffraction (XRD)) and the perspective of a future in-situ study led us to use Raman spectroscopy for these structural characterizations.

The catalysts studied had the Sb/V = 2.1 atomic ratio and were synthetized by a sol-salt method ; subsequent calcination was performed under air at different temperatures. Dispersive Raman experiments were performed using the 514.5 nm line of an argon ion laser. All spectra were recorded under ambient conditions. Identifications of the structural species were made according to reference spectra, literature data [1-3] and complementary XRD measurements.

Influence of the calcination temperature

V-Sb-O samples calcined at different temperatures (500, 575, 650, 750, 850°C) were examined. The principal features were the detection of V_2O_5 (998, 305, 287 cm^{-1}) at 500°C and its disappearance from 575°C, the detection of α-Sb_2O_4 (403, 197 cm^{-1}) and its transformation into β-Sb_2O_4 (632, 467, 440, 401, 209, 194 cm^{-1}) from 850°C, and the detection of two broad bands at 884 and 788 cm^{-1} increasing their intensity ratio ($R = I(884\ cm^{-1})/I(788\ cm^{-1})$) with increasing calcination temperature (see Fig.1). These two broad bands can be associated with mixed-oxide phases [2-3] and the variation of R can therefore be interpreted by structural changes occuring within these mixed-oxide phases when the calcination temperature is increased.

Influence of a solvent extraction

Extractions were performed (on the sample calcined at 500°C and containing V_2O_5) using various solvents in order to selectively modify the different structural phases present in the catalyst. Raman diffusion showed that oxalic acid (0.5 M in water) induces V_2O_5 disappearance and a modification of the mixed-oxide phase (appearance of new Raman signals at 736 and 658 cm^{-1}) ; NH_3 (0.5 M in water) and HCl (0.5 M in water) extraction induces only V_2O_5 disappearance; and no modification of the catalyst is detected after isobutanol extraction. Therefore, Raman diffusion allowed us to put into evidence different structural modifications according to the nature of the extracting solvent.

*Corresponding author.

Influence of the catalytic test

Comparisons of the Raman before and after catalytic test (sample tested after calcination at 650°C) showed a slight increase of the intensity ratio R of the 884 and 788 cm^{-1} bands indicating a structural modification of the mixed-oxide phases induced by the ammoxidation of propane.

Raman diffusion was then shown to be a very informative technique for the investigation of V-Sb-O catalysts. We were then able to follow the structural changes induced respectively by the variation of the calcination temperature, by a solvent extraction and by the catalytic test. Those above-mentioned results are now taken into account for structure-catalytic properties correlations and illustrate well that Raman spectroscopy is a powerfull tool for catalyst study and design.

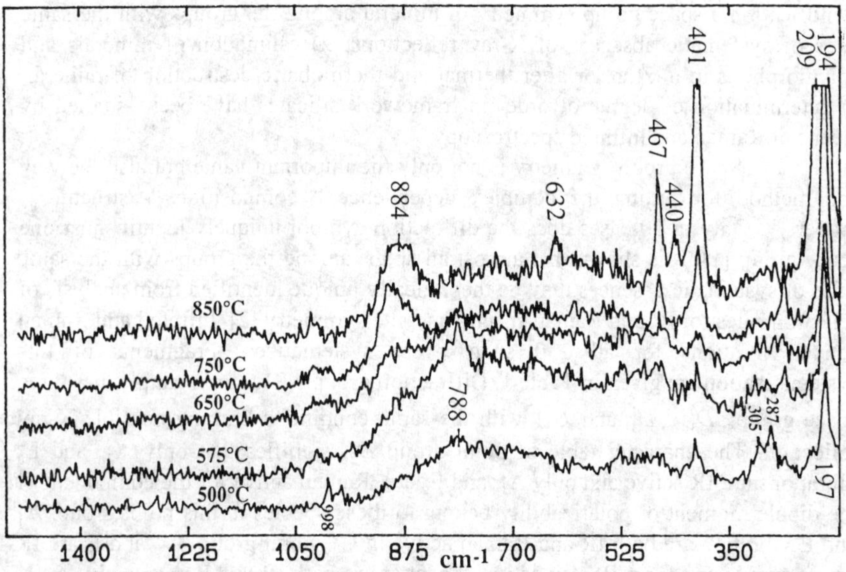

Fig.1. Raman spectra of V-Sb-O samples. Influence of the calcination temperature.

1. C.A. Cody, L. Dicarlo and R.K. Darlington , Inorg. Chem. 18(6), 1572 (1979).
2. R. Nilsson, T. Lindblad, A. Andersson, C. Song and S. Hansen, II World Cong. & IV Euro. Work. Meeting, "New Developments in Selective Oxidation", (V. Cortes Corberan and S. Vic Bellon, eds.) G.E.C., 1993, C 7-1.
3. R. Nilsson, T. Lindblad and A. Andersson, J. Catal. 148, 501 (1994).

APPLICATION OF RAMAN AND INFRARED SPECTROSCOPY TO MINERALS.

T. Moroz, D.Arkhipenko, T.Grigorieva

United Institute of Geology, Geophysics and Mineralogy, Siberian Branch RAS, 630090 Novosibirsk, Russia

Any crystallographic and mineralogical problems, namely: 1) unique identification a space group symmetry of mineral among the groups with the same rules of systematic absences of X-ray reflections; 2) estimation of minerals and polymorphous in mixture or after thermal and thermobaric destruction of mineral; 3) determination of degree of order in framework silicates have been studied by means of Raman and infrared spectroscopy.

Space group symmetry is not only the important value but also the way and method for setting the complex dependence " compositions - structure - property". Only in 50 cases does the diffraction symbol uniquely identify just one space group [1]. It is shown that almost all sp.gr. among the groups with the same rules of systematic absences may be theoretically unique identified from analysis of selection rules for IR and Raman spectra and site symmetry [2]. Infrared and Raman active vibrations for all 230 sp.gr. were systematized. Fragments of this systematization are given in Table 1. Diffraction group N81a, for example, has three space groups: D^1_{3d}, D^1_3 and C^2_{3v} with the same conditions limiting possible X-ray reflection. The character table of point group D_{3d} signifies that only A_{2u} and E_u vibrations are IR active and only A_{1g} and E_g are Raman active for the component of the dipole moment or polarizability belong to these species in this point group. A_1 and E vibration are both IR and Raman active in C_{3v} point group. A_2 and E are IR active and A_1 and E are Raman active, the latter are both IR and Raman active in D_3 point group. Thus a simple comparison between IR and Raman experimental vibrational frequencies made possible to select one space group among the groups with the same rules of systematic absences even for powder samples. The method has been applied to various minerals with of PO_4, SO_4, CO_3 anions (apatite, alunite, hydrotalcite, dawsonite, etc.).

Estimation of minerals or polymorphic modifications in mixture by X-ray diffraction is beset with problems of X-ray amorphous, of preferred orientation and disorder in minerals. Structure of minerals with volatile components after temperature treatment are often X-ray amorphous in major T-interval. In this case we can observe the transformation and destruction of mineral by means of vibrational spectroscopy. The development of the simple and convenient Raman or

IR method for quantitative determination is based on measurement of the relationship between intensities of vibrations of two phases. In specific cases IR or Raman method is the only possible.

The determination of degree of order and composition in a large collection of different genesis feldspars approached as ussually by means of X - Ray powder diffraction has been additionally carried out by vibration spectroscopy methods. The procedure of structural feature determination extended over a series of sanidine-adylar-microcline has been developed according to Raman spectra. Unpolarized Raman spectra were measured in right-angle scattering geometry using Jobin Ivon spectrometer RAMANOR U 1000 with single channel detection system. 514.5 nm line of Ar^+ ion laser was used as the excitation source. The relationship between laser produced fluorescence and factors of mineral formation can be followed. It is discussed as possible to use the halfwidth, peak position, the splitting of Raman line of feldspars as typorphic property which allows distinguishing mineral grains with different conditions of original. A comparison between data obtained by means of Raman spectroscopy and results of X-ray diffraction and IR spectroscopy estimated showed that they were in good agreement.

TABLE 1. Diffraction symbols, possible space - groups and selection rules for infrared and Raman spectra

ORTHORHOMBIC, Laue class *mmm*

	Diffraction symbols	Space - groups (number)	Active modes Infrared	Raman	*
8	*mmmP---*	14 D^1_{2h}=*Pmmm*(47)	$B_{1u}B_{2u}B_{3u}$	$A_gB_{1g}B_{2g}B_{3g}$	-
		15 C^1_{2v} =*Pmm*2(25)	$A_1B_1B_2$	$A_1A_2B_1B_2$	+/-
		D^1_2=*P*222(16)	$B_1B_2B_3$	$AB_1B_2B_3$	+/-
9	*mmmP*2_1 --	16 D^2_2=*P*$2_1$22(17)	$B_1B_2B_3$	$AB_1B_2B_3$	+/-
18	*mmmPnc-*	31 D^7_{2h} =*Pncm*(53)	$B_{1u}B_{2u}B_{3u}$	$A_gB_{1g}B_{2g}B_{3g}$	-
		32 C^6_{2v} =*Pnc2*(30)	$A_1B_1B_2$	$A_1A_2B_1B_2$	+/-
		TRIGONAL, Laue class *3m*			
81a	*3mP*-1-	121 D^1_{3d}=*P*3/1*m*(162)	$A_{2u}E_u$	$A_{1g}E_g$	-
		122 D^1_3 =*P*312(149)	A_2E	A_1E	+/-
		123 C^2_{3v} =*P*31*m*(157)	A_1E	A_1E	+

*Part vibrations are both IR and Raman active, part only IR and only Raman active (+/-); both IR and Raman active (+); only IR and only Raman active (-)

1. International Tables for Crystallography, (T. Hahn, ed.), Vol A, Kluver Academic Publishers, Dordrecht, Boston, London, 1989, p.878
2. D.Arkhipenko and T.Moroz, Kristallographi Report, 41, 1 (1996)

RAMAN SPECTRUM OF NATURAL ELLESTADITE.

D. Arkhipenko and Moroz T.

United Institute of Geology, Geophysics and Mineralogy, Siberian Branch RAS, 630090 Novosibirsk, Russia

Apatite $Ca_{10}(PO_4)_6(OH)_2$ as the main mineral constituent of hard tissues (bones, teeth) plays a very important role on earth. A series of cationic and anionic isomorphous substitutions take place in the apatite crystal lattice in nature. One of them is substitutions for P^{5+} by S^{6+} and Si^{4+}. The term ellestadite is redefined to make it a group name for all the compositions having sum (Si + S) > P. Apatite like structure of ellestadite with the following crystallochemical formula $Ca_{10}(Si_{3.14}S_{2.94}C_{0.03}P_{0.02})_{\Sigma 6.18}\,O_{24}(OH_{1.12}Cl_{0.316}F_{0.06})_{\Sigma 1.50}$ was investigated by X-ray diffractions [1]. The same samples were studied by means of Raman and infrared spectroscopy.

Unpolarized Raman spectra were measured in right-angle scattering geometry using Jobin Ivon spectrometer RAMANOR U 1000 with single channel detection system. 514.5 nm line of Ar^+ ion laser was used as the excitation source. The observed Raman spectrum superimposed on an extensive laser produse fluorescence is centered in the 2600cm^{-1} and high background noise level. The spectrum showed in Fig.1 was obtained as difference of Raman plus fluorescence spectrum and pure fluorescence spectrum, the latter is reasonable well approximated by Gaussian with experimentral parameters (FWHM=1800cm^{-1}).

Replacement of the PO_4^{3-} ions by SiO_4^{4-} and SO_4^{2-} and particulary by CO_3 causes the sp.gr. $C_{6h}^2=P6_3/m$ in apatite structure to decrease to $P2_1/m=C_{2h}^2$ which has the same rules of systematic absences with $P2_1=C_2^2$ sp.gr. This groups may be uniquely identified from the analysis of selection rules for IR and Raman spectra. The expected 126 fundamentals at k=0 split to

$\Gamma_{126}= 38A_g+ 25\,B_g+25A_u+ 38\,B_u$ (C_{2h}^2 sp.gr.) & $\Gamma_{126}= 63A + 63B$ (C_2^2 sp.gr.)

A_g, B_g vibration are only Raman active, A_u, B_u - only IR active for $P2_1/m=C_{2h}^2$ sp.gr. A,B are both Raman and IR active for $P2_1=C_2^2$ sp.gr. The low symmetry C_1 of the Si and S atoms site raises the degeneracy of the ν_2, ν_3, ν_4 modes. The correlation scheme is given in Table 1. Experimental Raman and IR spectra give evidence for $P2_1=C_2^2$ sp.gr. In the Raman spectrum the band due to ν_1 mode of SO_4^{2-} is expected to be the most intense. From the complete lifting of the bending and the stretching internal SO_4^{2-} modes from the free ion values it can be inferred that the angular distortion of SO_4^{2-} ions and the bond distortion is great.

The ratio ν_1 / ν_3 decreases for SO_4 from its free state value and decrease for SiO_4 from its free state one. This result is consistent with the crystal structure data, as the mean S-O bond (1.541-1.567Å) is increased from it free state value (1.49Å), Si-O bond is decrease (Table 2). Estimation of CO_3^{2-} ions in an ellestadite from Raman spectra data was in good agreement with microchemical analysis: C/S=0.013 (Raman), 0.0125 (micro-chemistry). A quantitative determination of the relative amounts of carbonate ions was achieved by used a calibration constant in reference to the relative intensities of the corresponding free anions [3]. Accordingly, the calibration constant determined from Raman spectrum of this sample may be used for estimation of $[SO_4]^{2-}/[CO_3]^{2-}$, $[SO_4]^{2-}/[PO_4]^{3-}$ and $[SO_4]^{2-}/[SiO_4]^{4-}$ in another natural ellestadites with (Si + S) > P.

TABLE 1.Correlation scheme for the external and internal modes of TO_4 [a)]

	Free Ion T_d	Site C_1	Factor Group C_2		
ν_1	A_1		A	$\nu_1 2\nu_2 3\nu_3 3\nu_4$	3r 3t
ν_2	E	A			
r	F_1		B	$\nu_1 2\nu_2 3\nu_3 3\nu_4$	3r 3t
t ν_3 ν_4	F_2				

a) T=S,Si,P

TABLE 2. Relation of the ratio ν_1 / ν_3 internal modes and bonds

	Bond S -O	ν_1/ν_3 SO_4^{2-}	Bond Si - O	ν_1/ν_3 SiO_4^{4-}	Bond P - O	ν_1/ν_3 PO_4^{3-}
Free ion	1.49[2]	0.89	1.61[2]	0.86	1.534 [2]	0.922
Ellestadite	1.541-1.567[1]	0.87	1.515-1.562[1]	0.9		0.924

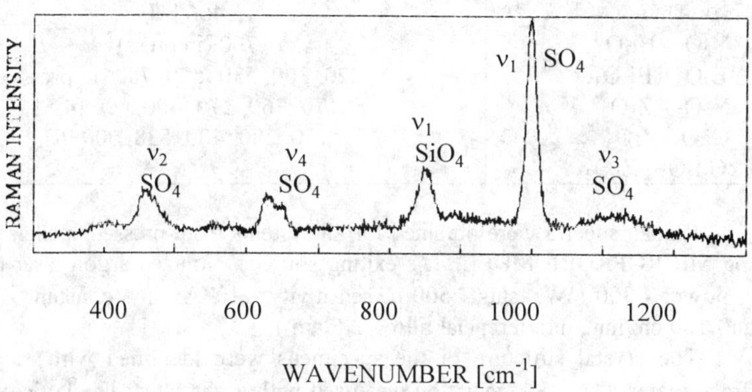

Fig.1. Raman spectrum of natural ellestadite

1. N. Organova, et al., Kristallographi Report, 39, 278 (1994)(Russia)
2. International Tables for X-Ray Crystallography. V.III Birmingham. 1968. 362p.
3. Chantry G.W. and Plane R.A., J. Chem.Phys., 32, 319 (1960)

RAMAN SCATTERING IN SOLID SOLUTION BASED ZrO_2 or Bi_2O_3 WITH YTTRIUM AND ERBIUM OXIDES

V.N.Strekalovsky

Institute High Temperature Electrochemistry of Ural Division Russia Academy Science, 620219, Ekaterinburg, Russia

The possibilities of Raman Spectroscopy for investigation oxides in systems ZrO_2 doped with oxides 2 or 3- valence cations, which are used how high temperature solid ionic conductors, were showed in survey [1]. In the last years for this aim the materials based on Bi_2O_3 began to use. The most high oxygen ion conduction is found in sintered oxides of the Bi_2O_3 - Er_2O_3 system [2].

In this report Raman scattering data in the solid solution Y_2O_3 and Er_2O_3 in Bi_2O_3 with structure type of fluorite (Ph.gr. Fm3m) and their a comparision with structure analogs in ZrO_2 - based systems are resulted.

In table 1 are given Raman - modes for individual and double oxides with above mentioned structure.

TABLE 1. Raman-modes individual and double oxides with structure type of fluotite.

Oxides	Frequency Raman-modes, 1/cm
CeO_2	465 [3]
UO_2	467 [3]
ThO_2	466 [3]
Y_2O_3 - Bi_2O_3	700 (in press)
E_2O_3 - Bi_2O_3	220, 300, 530, 620, 780 (in press)
Y_2O_3 - ZrO_2	170, 260, 270, 400, 600 [4,5]
GaO - ZrO_2	150, 280, 400, 538, 600 [7]
Gd_2O_3 - ZrO_2	

Raman spectra were obtained for all systems from pressed pellets or powder using MICROPROBE MOLE. The exiting source - ionized argon laser (l = 514,5 nm, power - 300 mW, slits - 500 μ, sensitivity - 10^{-9}A, time constant - 0,5 c, rate scan - 100 cm/min, interferencial filtre - 2,5 nm)

The crystal structure of the specimens were identified with a DRON -3 diffractometer. Cu - K_α radiation was used with a Ni filter. The lattice parameters were calculated from all diffraction angles automatic (external standard - quartz).

The comparison of Raman - spectra (Tab.1) are allowed to note :

1. In accordance with factor-group analysis Raman spectra of the individual oxides with phase group Fm3m must have 1 peak T_{2g} only. Really oxides cerium, thorium and uranium have one sharp intensive line in region 465-467 cm^{-1} .
2. Raman-spectra of the solid solution with the same structure are distinguished considerable : it have a few broad lines, high background.

3. Raman-spectra of solid solution ZrO_2 or Bi_2O_3 -based are discerned considerable.

The Raman scattering of cubic zirconia due to the breakdown of the wave vector k = 0 selection rule was firstly observed by Keramidas and White [6]. The Raman scattering of CaO-stabilized zirconia showed the broad Raman bands with some structures which were ascribed due to the breakdown of wave vector selection rule by translational disorder.

This experimental fact are obtained in another oxides systems [7-10].

The lattice disorder in solid solution will relax the selection rules relating with simmetries. The breakdown of selections rules give disorder-induced light scattering, while the degree of of the breakdown of selection rules will depend on each material [10]. The broad bands are explained by present O^{2-} - defects and their interaction with nearest environment.

The difference of spectra in systems with Bi_2O_3 or ZrO_2 are connected with nature isomorphism of the solid solutions - isoor heterovalence, accordingly.

The further investigations will useful for understanding of real structure oxides with phase group Fm3m.

1. V.N.Strekalovsky, E.G.Vovkotrub, Ogneupori.3, 27,(1985)
2. M.J.Verkerk, K.Keizer, A.J.Burggraaf, J.Applied Electrochemistry , 10, 81(1980)
3. V.G.Keramidas, W.B.White, J. Chem. Phys., 59, 1561 (1973)
4. A.Feinberg, C.H., J.Phys.Chem. Solids, 42, 513, (1981)
5. V.N.Strekalovsky, Yu.N.Makurin, E.G.Vovkotrub, Izv.AN SSSR, Neorgan. materiali, 19, 925 (1983)
6. V.G.Keramidas, W.B.White, J Amer.Ceram.Soc., 57, 22, (1974)
7. D.Mishel, H.Perez y Jorba, R. Collongues, Mater. Res.Bull. 9, 1457,(1974)
8. V.N.Strekalovsky, Yu.M.Polezaev, S.F.Pal'guev, in "Oxides with the admix-ture conductivity", Moscow,: Science, 20, (1987).
9. E.G.Vovkotrub, V.N.Strekalovsky , Z. pricladnoi spectrosc. 49, 50, (1988). 10. M. Ishigame, S.Shin, T.Suemoto, J.Solid State Ionics, 47, 1, (1991).

RAMAN SPECTRA COMPLEX COMPOUNDS Me_2MCl_6

(Me - alkali metal, M-element of 4 group)

A.B.Salyulev, E.G.Vovkotrub

Institute High Temperature Electrochemistry of Ural Division Russia Academy Science, 620219, Ekaterinburg, Russia

Raman spectra for the little number of the complex compounds Me_2MCl_6 (Me-Cs, Rb, K, Na, Li , M - Ti, Zr, Hf, Si, Ge, Sn, Pb) are known. It is connected with the complexity their synthesis and analysis. The method ways of synthesis (the number of complexes had been produced first) and the registration of spectra are developed by us. It is allowed to extend the information about Raman spectra of above- mentioned compounds, to generalize and systematize the results. The some investigations have been published by us earlier [1-4].

The complex compounds were obtained by alloying the initial chlorides in sealing quartz ampules (Me_2ZrCl_6 and Me_2HfCl_6 [1]), saturating melts MeCl by vapours MCl_4 at pressure up to 8-16 MPa (Cs_2SiCl_6 [2], Me_2GeCl_6 , Me_2SnCl_6 and Me_2 $TiCl_6$ [3]), passing the fusion MeCl with MCl_2 in liquid chlorine at 300 K (Me_2MCl_6, M - Ti, Sn, Pb) or 900-1100 K and P_{Cl_2} 3-11 MPa (Me_2 $PbCl_6$ [3]).

Raman spectra of salt fusion were registrated directly in the sealed reaction ampules at presence of abundant MCl_4 and liquid Cl_2 , and also after their evacuation. We used Raman Mikroprobe MOLE (France) with Ar^+ - laser power to 600 mW , line for excitation 514,5nm. The investigations are realized in macrochamber and under microscope of spectrometer.

In the first Raman - spectra of the compounds: Na_2MCl_6 and Li_2MCl_6 (M - Ti, Sn, Pb), Rb_2GeCl_6, K_6GeCl_6, Cs_2SiCl_6 ,Tl_2MCl_6 and Ag_2MCl_6 (M - Ti, Sn) are received. Furthermore, the all known Raman spectra in this class complex compounds were checked.

In the table (for example) frequencies of fundamental vibration groupings [$TiCl_6^{2-}$] and [$SnCl_6^{2-}$] in their compounds with chlorides alkali metal are given.

Compound	Frequencies, cm^{-1}	Compound	Frequencies, sm^{-1}
Cs_2TiCl_6	324 278 189	Cs_2SnCl_6	309 235 168
Rb_2TiCl_6	330 280 189	Rb_2SnCl6	318 240 170
K_2TiCl_6	332 v.w. 189	K_2SnCl_6	323 246 172
Na_2TiCl_6	331 v.w. 192	Na_2SnCl_6	316 246 178
Li_2TiCl_6	333 v.w. 188	Li_2SnCl_6	322 256 170

1. A.B.Salyulev, E.G.Vovkotrub, V.N.Strekalovsry, J.Inorg. Chem. (USSR) 35, 902 (1990)
2. A.B.Salyulev, E.G.Vovkotrub, V.N.Strekalovsky, Melts (USSR), 6, 63 (1992)
3. A.B.Salyulev, E.G.Vovkotrub, V.N.Strekalovsky, Melts (USSR), 3,103, (1989)
4. A.B.Salyulev, E.G.Vovkotrub, V.N.Strekalovsky, J.Inorg.Chem. (USSR), 37, 243 (1992)

INFLUENCE OF THE NEAREST NEIGHBORHOOD ON THE VIBRATIONAL MODES OF TETRAHEDRAL OXYANIONS IN CRYSTALS AND AQUEOUS SOLUTIONS

M. Bilij, O. Chukova
Taras Shevchenko Kyiv University,
6, av. acad. Glushkova, 252022, Kyiv, Ukraine

Increasing practical employment together with simple scientific interest to the crystals with tetrahedral molecular anions in their lattice at last time stimulate a wide investigation of these objects.

We have studied influence of ions and defects in the nearest neighborhood on vibrational modes of CrO_4^{2-} chromate anion in aqueous solutions and pure and doped crystals of Me_2CrO_4 salts (Me = Na, K, Cs).

Formerly Raman spectra of such a type anions were studied in pure matrices and doped alkali halide crystals [1, 2]. The spectra of pure alkali metal chromate crystals agree with the data on local symmetry of chromate anion in the crystals. Implantation of M^{2+} dopant in the Me_2CrO_4 crystals leads the formation of covalent bonds of CrO_4^{2-} anion with M^{2+} cation. This effect is displaying in corresponding characteristics of inner vibrational modes of distorted chromate anion and in appearance of new lowfrequency modes. Such doped crystals contain both the CrO_4^{2-} anions of host matrix and complexes of CrO_4^{2-}*M^{2+} type with distorted chromate anion. Second are destroyed under powerful laser irradiation.

The dependence of inner modes characteristics of chromate anion on the Me_2CrO_4 salt concentration in aqueous solution were studied. Comparison of Raman spectra of solutions with ones for corresponding crystals was carried out. Four peaks founded in Raman spectra of aqueous solutions are corresponded to the ν_1 - ν_4 modes of CrO_4^{2-} anion. Any splitting of Raman modes were no founded. Therefore, the existence of CrO_4^{2-} anion of T_d symmetry in the solutions was established. At the same time the very broad band in the region of ν_3 vibrational mode allows us to suppose that complexes of Me^+ - CrO_4^{2-} type with decreased symmetry of chromate anion are present in solution too. Changes in geometric configuration of CrO_4^{2-} anion and role of dynamic interaction of this anion with water molecules and Me^+ cations were discuussed on the basis of concentration dependences of Raman mode parameters. Observed dependence of ν_1 mode peak position on the K_2CrO_4 and Cs_2CrO_4 salts concentration is the same to one for nitrite salts [3]. But, for Na_2CrO_4 ν_1 mode frequency decreases and tends out from the ν_1 mode value of CrO_4^{2-} anion in T_d symmetry with concentration decreasing. The fact that Na_2CrO_4 forms three types of crystallhydrates at room temperature [4] may causes formation of corresponded complexes of chromate anion with hydrate shell. Relative amount of these complexes in aqueous solution increases with decreasing Na_2CrO_4 concentration and leads observed concentration shifting in Raman spectra. The character of polarization of Raman peaks also indicates some degree of order like to the crystals.

So, our investigations of Raman effects evidence the great influence of nearest neighborhood on the symmetry of chromate anion even in solutions. These results are in a good agreement with previous conclusions [5] about essential role of nonadiabatic effects in centers of secondary light scattering in pure and doped ion-covalent crystals.

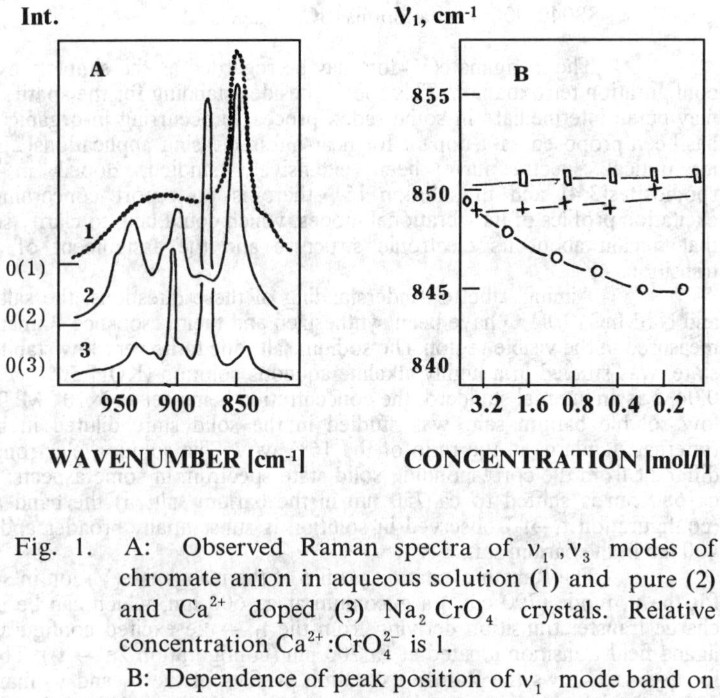

Fig. 1. A: Observed Raman spectra of ν_1, ν_3 modes of chromate anion in aqueous solution (1) and pure (2) and Ca^{2+} doped (3) Na_2CrO_4 crystals. Relative concentration $Ca^{2+}:CrO_4^{2-}$ is 1.

B: Dependence of peak position of ν_1 mode band on concentration of Na_2CrO_4 (o), K_2CrO_4 (+) and Cs_2CrO_4 ([]) salts in aqueous solution.

1. W. Kiefer and H.J. Bernstein, Mol. Phys. 23, 835 (1972).
2. R. L. Carter and C. E. Bricer, Spectrochimica Acta, 27A, 569 (1972).
3. D.E. Irish and M.H. Brooker, Trans. Farad. Soc. 67, 1916 (1971).
4. M. Geslin, M. Garntier, G. Pannetier. Bull. Soc. Chim. France, 3, 939 (1968).
5. M.U. Bilij, V.G. Kravets, I.Ya. Kushnirenko, S.G. Nedelko and M.M. Pod shivajlov, Dopovidi Akad. Nauk Ukraini, N4, 49 (1992).

RESONANCE RAMAN SPECTROSCOPY OF THE MANGANATE(V) ION IN THE SOLID STATE AND AQUEOUS SOLUTION

Oswaldo Sala[1*], Norberto S. Gonçalves[2] and Lucia K. Noda[1]

[1]Instituto de Química da Universidade de São Paulo,
CP. 26077, 05599-970, São Paulo, SP, Brazil
[2] Depto de Química, Universidade Federal de Santa Catarina,
88040-900, Florianópolis, SC, Brazil

The manganate(V) ion may be regarded as the simplest example of a d^2 configuration tetroxoanion. This specie, besides standing for the apatite blue color[1], may be an intermediate in some redox processes occurring in organic chemistry and has been proposed as a dopant for near-infrared lasing applications[2]. Even though its optical spectra have been extensively studied, doped in apatites and spodiosites[3,4] and in solution [5], there is no report concerning the Raman excitation profiles of its vibrational modes, which could help to clarify some questions that remain about its electronic structure and the assignment of the electronic transitions.

Aiming a better understanding of these questions, the salts $Ba_3(MnO_4)_2$ and $Na_3MnO_4.10H_2O$ have been synthesized and their resonance Raman spectra were measured in the visible region. The sodium salt, due to its very low stability in the solid state, was studied in a highly alkaline aqueous solution (KOH 50% w/w) and KNO_3 0.04M as an internal standard, the concentration being set at $3x10^{-3}$ M. The stable, but low soluble barium salt, was studied in the solid state diluted in barium sulfate (internal standard) in the ratio of the 16% w/w. The solution electronic spectrum is different from the corresponding solid state spectrum in some aspects: i) the band at ca 680 nm is shifted to ca 750 nm in the barium salt; ii) the band at ca 316 nm (configuration $t_1 \rightarrow t_2$) observed in solution is substantially broader and shifted to ca 400 nm in the barium salt.

The Raman excitation profile of the manganate(V) ion in solution shows for the v_1 mode (790 cm^{-1}) a maximum at ca 600 nm, which can be indicative of a charge transfer transition deriving from the $t_1 \rightarrow 2e$ excited configuration, near the ligand field transition located at ca 700 nm (configuration $2e \rightarrow t_2$). The barium salt, for its turn, shows the Raman excitation profiles of the v_1 and v_2 maximizing at ca 530nm, which was interpreted as resulting from the blue shift of the aforementioned charge transfer transition in the barium salt caused by a descending in symmetry in the solid state. This charge transfer band is probably overlapped by the apparent intense band observed around 400 nm. Overtones and combination bands have not been observed in both cases, indicating that the displacement of the excited state potential minimum with respect to that of the ground state is probably very small. This is consistent with the fact that in the $t_1 \rightarrow 2e$ electronic transition, an electron is transferred from an essentially non-bonding orbital localized on the ligand atoms to a metal non-bonding orbital.

These results are interesting in the sense that they provide an alternative assignment for the optical spectra of the manganate(V) ion, since some authors[6,7] do not consider the existence of charge transfer transitions below 30000 cm^{-1}.

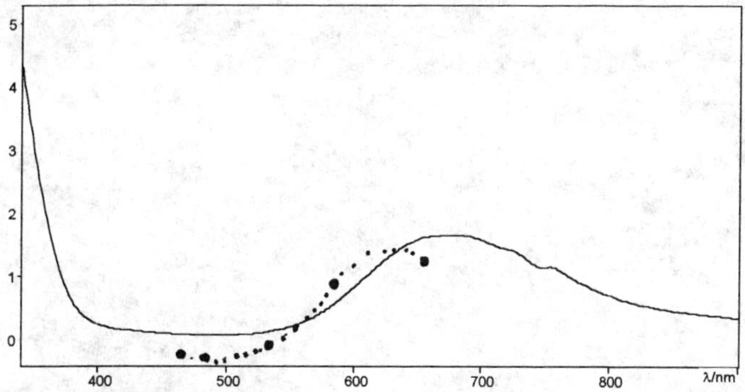

Fig.1: Absorption spectrum (solid line) of the manganate(V) solution and corresponding Raman excitation profile (dots) for the ν_1 mode (780 cm^{-1}).

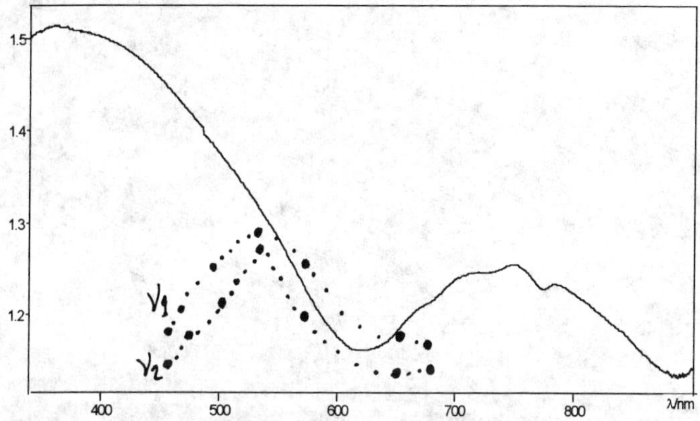

Fig.2: Absorption spectrum (solid line) of the barium manganate(V) in a KBr pellet and corresponding Raman excitation profile (dots) for the ν_1 (789 cm^{-1}) and ν_2 (305 cm^{-1}).

References

[1] P.D.Johnson, J.S.Prener and J.D.Kingsley, Science, 141, 1179 (1963).
[2] U.Oetliker, M.Herren, H.U.Güdel, U.Kesper, C.Albrecht and D.Reinen, J.Chem.Phys., 100, 8656 (1994).
[3] J.D.Kingsley, J.S.Prener and B.Segall, Phys.Rev, 137(1A), 189 (1965).
[4] J.B.Milstein, J.Ackerman, S.L.Holt and B.R.McGarvey, Inorg.Chem., 11(6), 1178 (1972).
[5] G.de Michelis, L.Oleari, L.di Sipio and E.Tondello, Coord.Chem.Rev., 2, 53 (1967).
[6] R.Borromei, L.Oleari and P.Day, J.Chem.Soc., Faraday Trans.2, 77, 1563 (1981).
[7] H.Lachwa and D.Reinen, Inorg.Chem., 28, 1044 (1989).

SECTION 19

THE SOLID STATE

RAMAN SCATTERING IN BISMUTH DOPED YTTRIUM-IRON GARNET (BI^{3+}: YIG)

G. G. Siu[1]* and Yulong Liu[2]

[1]Department of Physics and Materials Science, City University of Hong Kong, Tat Chee Avenue, Kowloon, Hong Kong
[2] Institute of Physics, Chinese Academy of Sciences, P. O. Box 603, Beijing 100080, P. R. China

Synthetic yttrium-iron garnet $Y_3Fe_5O_{12}$ crystal (YIG) is ferrimagnetic with important applications in microwave techniques. Its performances are enhanced by proper impurity doping. Bi^{3+}: YIG have unique magnetic and magneto-optic properties [1,2] and are characterized by a variety of techniques [3-6]. Here we study the effect of Bi doping by Raman scattering.

The unit cell of YIG contains four molecules and is made up with connected oxygen tetrahedrons and octahedrons with Fe^{3+} at centers, denoted as the d site and the a sites respectively. In their interstices are distorted oxygen cubics (dodecahedrons) with Y^{3+} at centers, denoted as the c sites [7]. Doped B^{3+} ions replace Y^{3+} of similar size at c site. Bi^{3+}: YIG can be written as $\{Y_{3-x}Bi_x\}^c[Fe_2]^a(Fe_3)^dO_{12}$.

Bi^{3+}:YIG sigle crystals were grown by slowly cooling the melt of constituent oxides [8]. Compositions of the samples were estimated from lattice parameters determined by X-ray diffraction and four samples with x = 0.00, 0.14, 0.36 and 0.54 were prepared. They all exhibit natural rhomboid facets, (110) planes.

An INNOVA 100 Ar ion laser was used as excitation source. The spectrometer was SPEX RAMALOG 1403 system, used at room temperature. Any fluorescence from unidentified trace impurities was excluded by matching spectra with different excitation wavelengths. Raman scattering was detected at a pseudo-backscattering geometry with incident angle less than 30° with the normal [110] to minimize stray light and to make the refracted beam almost along [110] inside crystal.

The crystal structure of Bi^{3+}: YIG belongs to space group O_h, isomorphic to that of YIG. Following irreducible representation of the factor group there are 25 Raman active phonon modes: $3A_{1g}$, $8E_g$ and $14T_{2g}$ at wave vector $k \equiv 0$, which can be identified umabiguously by specific polarization configuration. The plane polarization along the direction [001] is denoted as V and that parallel to the incident plane as H. Polarization Raman spectra of the four samples were measured at V-V and V-H configurations respectively, which denote incident and scattered polarizaion consecutively. The V-V configuration allows scattering from modes A_{1g} and E_g and the V-H allows T_{2g} to be detected. Figs. 1 and 2 show the polarized Raman spectra of the four samples with excitation wavelength 514.5 nm, where the subscripts of modes are neglected. No fluorescence line exists.

The spectra at x = 0 is just those of YIG [8]. In Bi^{3+}: YIG, YIG modes are affected by increaing x of Bi^{3+}: line's position redshifts with moderate width broadening, intensity decreases and relative intensities change drastically. These changes

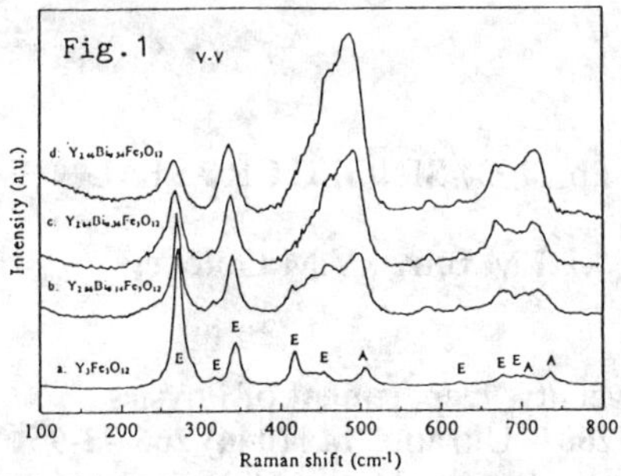

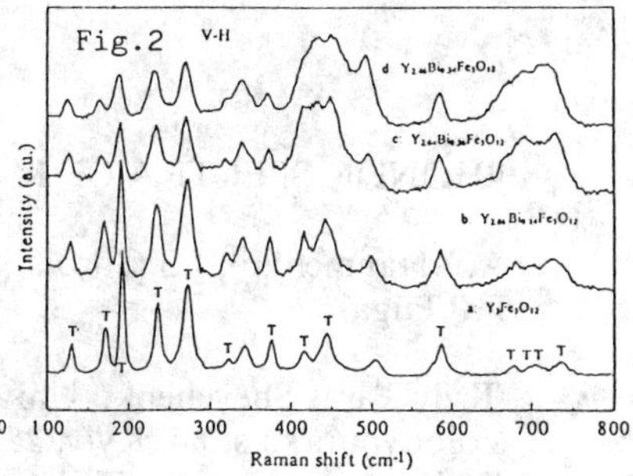

can be attributed to the disruption of the long-range translational symmetry of YIG crystal by reorientation or tilting of the oxygen dodecahedron to accommodate the larger Bi^{3+} substitution for Y^{3+}. Besides, new type of bands in the ranges of [400, 500] (band I) and [690, 720] (band II), all in cm^{-1}, appear with oppostie tendency of intensity variation with x and with anomalous broadening (Table 1), which are likely related to magnetic or electronic excitations.

Table 1. Frquencies (cm^{-1}) and assignments of Raman active phonons of Bi^{3+}: YIG

Assign.	YIG (ref. 8)	YIG x=0	x=0.14	Bi^{3+}:YIG x=0.36	x=0.54
T_{2g}	131	130	128	126	122
T_{2g}	174	174	170	166	162
T_{2g}	194	192	190	188	186
T_{2g}	238	236	234	232	230
$E_g + T_{2g}$	274	274	270	268	266
E_g	319	320	320	318	/
T_{2g}	324	322	320	318	316
E_g	346	344	340	338	336
T_{2g}	378	376	374	372	368
E_g	416	416	416	414	band I
T_{2g}	419	418	418	band I	band I
T_{2g}	445	444	442	band I	band I
E_g	456	454	458	band I	band I
A_{1g}	504	504	498	492	486
T_{2g}	592	588	586	584	582
E_g	624	622	620	620	620
$E_g + T_{2g}$	685	678	676	672	668
$E_g + T_{2g}$	692	692	690	band II	band II
A_{1g}	704	704	702	band II	band II
T_{2g}	711	710	/	band II	band II
A_{1g} or E_g	736	736	732	730	728

1. P Hansen, W Tolksdorf, K Witter and J M Robertson, IEEE Trans. Magn. MAG-20 1009 (1984)
2. P Hansen, K Witter and W Tolksdorf, Phys. Rev. B27 6608 (1983)
3. S Geller and A A Colville, Proc. AIP Conf. 24 Magn. and Magn. Mat. 372 (1974)
4. Q F Ling, X Z Xu and C X Liu, Acta Phys. Sinica 31 1680 (1982)
5. Yulong Liu, P X Zhang, Y J Mo and A Tu, Acta Phys. Sinica 36 651 (1987)
6. P X Zhang, D J Lockwood and H J Labbé, Phys. Rev. B48 6099 (1993)
7. J H van Vleck, J. Phys. Chem. Solids 27 1047 (1966)
8. X Z Xu, W Y Jia and C X Liu, Acta Phys. Sinica 30 1554 (1981)

PHONON SPECTRA OF $K_2Pb_2Si_2O_7$ SINGLE CRYSTAL

A.P.Naumenko[1], I.S.Gorban[1]*, M.M.Biliy[1], V.M.Golovei[2], P.P.Puga[2]

[1]Kyiv Taras Shevchenko University, Department of Physics, Volodymirs'ka st.64, Kyiv, 252601, Ukraine; fax (044) 266-23-95
[2]Uzhgorod Institute of Electronic Physics National Academy of Science of Ukraine, Uzhorod, 294016, Ukraine

The vibration spectra of "pyro"-compounds (pyro-silicates, -phosphates,-vanadates, etc.) were investigated by many authors [1]. But the most complete results were obtained by P.Tarte *et al.* [2] who investigated both i.r. and Raman spectra. The experimental results were explained with using of quasi-molecular model and group-theoretical analysis. In this model anion Si_2O_7 was considered as quasimolecule. In general the authors ascribed vibrational frequencies correctly but they couldn't solve the problem in full because of absence of the single crystal samples. Furthermore they carried out the experiments at room temperature and some important details escaped their notice

In a previous study we reported the first-order Raman spectrum of potassium pyrosilicate single crystal [3]. The focus of that preliminary report was to show the differences between Raman spectra of powder and single crystal of this compound. The main purpose of present work is to make up for the pointed deficiencies especially in the experiments on Raman scattering which should allow to understand more deeply the phonon spectra of investigated crystals and as far as possible to establish the spectral features upside the quasi- molecular approximation.

We obtained polarized Raman spectra from $K_2Pb_2Si_2O_7$ single crystal with size 20*20*4 mm. The single crystal of potassium pyrosilicate has been grown from melt with the direct Czochralski method. The crystal growth was carried out in air in a platinum crucible along the direction [1000] or [0001]. Raman measurements were performed in classical 90°-geometry using 488.0 and 514.5nm Ar-ion laser incident beam. The scattered light was dispersed with a DFS-24 model double grating monochromator. Raman spectra were detected under temperature of liquid helium.

The observed Raman spectrum (Fig.1) allows to take notice of two main peculiarities. The first is that the spectrum under low temperature is richer then one determined earlier [2] and we detected some new lines. Two of new low frequency lines (13171 and 99.5 cm-1) can be corresponded to the anion rotations since they at the convenient frequency region and have "pure" polarization. The nature of new high-frequency line (687.9 cm-1) can be explained with taking into account mixing of the valence banding modes with vibrations of potassium ion. The second peculiarity is in absence of "pure" polarization of the majority of lines, in their completely character indicating of the strong mixing of vibrations in different polarization under convenient polarizations.

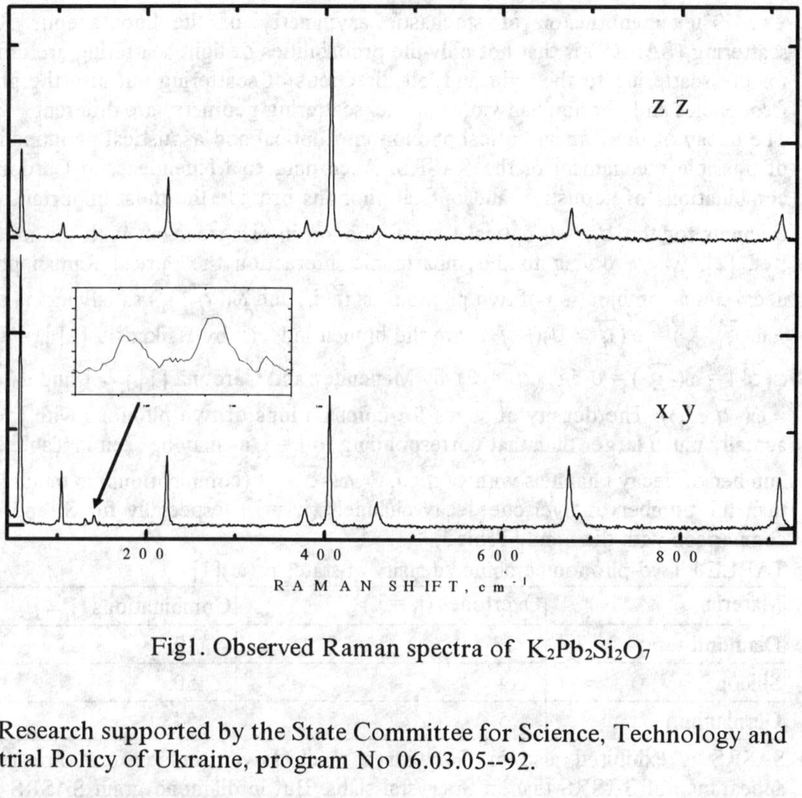

Fig1. Observed Raman spectra of $K_2Pb_2Si_2O_7$

Research supported by the State Committee for Science, Technology and Industrial Policy of Ukraine, program No 06.03.05--92.

1. Lasarev A.N.,"Vibrational spectra and structure of silicates", Leningrad, Nauka, 1968
2. P.Tarte, M.J.Potier and A.M.Proces, Spectrochimics Acta, 1973, vol. 29A, p.1017-1027
3 .A.P.Naumenko, M.M.Biliy, I.S.Gorban, in Proc. of the XIV-th Intern. Conf. on Raman Spectr., Hong Kong, 1994 (Wiley, Chichester,1992), vol.A, p.253-254

ASYMMETRICAL DECAY OF OPTICAL PHONONS IN SI, GE AND DIAMOND CRYSTAL MICROSLABS

A. Je. Semjonow
GOS, Rudower Chaussee 5, 12489 Berlin, Germany

The phenomenon of stochastic asymmetry of the spontaneous Raman scattering (SASRS) is that not only the probabilities of light scattering are different for the scattering to the right and left directions of scattering but also the phonon frequencies and phonon halfwidths in backscattering geometry are different.

The decay of the Raman optical phonon into optical and asoustical phonons is one of possible mechanism of the SASRS. According to Menendez and Cardona [1] combinations of acoustical and optical phonons provide the most important decay channels for the $\vec{q} = 0$ optical Raman phonon in disagreement with the model of Ref. [2], where owing to the anharmonic interaction the optical Raman phonon decays in a combination of two phonons $\omega(\vec{q}\ j_1)$ and $\omega(-\vec{q}, j_2)$ satisfying $\omega(\vec{q}, j_1) + \omega(\vec{q}, j_2) = \omega(\vec{q} = 0)(j_1, j_2)$ are the branch indecis; by Balkanski [2] $j_1 = j_2$ and $\omega(\vec{q}) = \omega(-\vec{q}) = 0{,}5\ \omega_0(\vec{q} = 0)$, by Menendez and Cardona [1] $j_1 \neq j_2$ and $\omega(\vec{q}, j_2) \neq \omega(-\vec{q}, j_2)$). The density of states for combinations of two phonons with $j_1 \neq j_2$ is actually much larger than that corresponding to $j_1 = j_2$ as may be seen in Table I. The number of decay channels with $\omega(\vec{q}, j_1) \neq \omega(-\vec{q}, j_2)$ (combinations) is much higher than the number of overtone decay channels ($j_1 = j_2$), especially for Si and Ge in comparison with diamond (Table I).

TABLE I Two-phonon combined density of states $\rho^2(\omega)$[1]

Material	Overtones ($j_1 = j_2$)	Combinations ($j_1 \neq j_2$)
Diamond	13	19
Silicon	4	50
Germanium	5	54

SASRS is exhibited also in Ge with a much lower temperature threshold of appearance of SASRS than in Si crystal slabs. But in diamond grain SASRS is not observed. An examination of the phonon dispersion relations suggests that this contribution arises from pairs (LA-LO) of phonons belonging to the Q branches (along L-W line) and from nearby regions of the Brillouin zone [1]. The probabilities of the scattering for (LA-LO) pairs in directions W- and L-points are not equal. The approximate relation for (LA-LO) pairs has form

$$\omega(\vec{q}, j_1) = 0{,}35\ \omega(\vec{q} = 0),\ \omega(-\vec{q}, j_2) = 0{,}65\ \omega(\vec{q} = 0). \qquad (1)$$

The concrete mechanism of the interaction between each intact optical phonons $\omega(\vec{q} = 0)$ and decay phonons $\omega(\vec{q}, j_1)$, $\omega(-\vec{q}, j_2)$ is unknown.

As the illustrative model for an understanding of SASRS, can be represented by 4-phonons interaction between decay phonons and intact "side" phonons. The scheme of these process is next. As the temperature in the centre of the laser beam of the crystal surface is maximal, then at first phonons from centre of laser beam with wave vectors $\vec{k}_0$ (normal to surface) are decayed into two asymmetrical optic-acoustic pairs

$$\vec{k}_o = \vec{q}(j_1) + \vec{q}(j_2), \qquad (2)$$

and then $\vec{q}(j_1)$ and $\vec{q}(j_2)$-phonons interact with "side" phonons $\vec{k}_s$ ($\vec{k}_s$ phonons have different directions for left and right scattering). As result of this 4-phonons interaction we have $\vec{q}_L$ not equivalent $\vec{q}_R$ in direction and absolute value of wave vectors

$$\vec{q}_R = \vec{k}_S^{(R)} + \vec{q}_1 + (-\vec{q}_2),\ \vec{q}_L = \vec{k}_S^{(L)} + (-\vec{q}_1) + \vec{q}_2. \quad (3)$$

In 4-phonon processes interactions between phonons with $k_{ph} = 0$ and phonons with $k_{ph} \neq 0$ are permitted [3].

Will be denoted ASRS as stochastical asymmetry? At first, we have a series of independent collisions, the probability for N collisions has Poisson distribution, furthermore, there are several symmetrically equivalent combinations of vibration modes associated with a given $\vec{q}_j$ in the reduced zone which contribute simultaneously to the scattered intensity at a given frequency. (The polarization effects in second-order Raman spectra are much more complicated than those in the first-order-spectra. In first-order-spectra the wave vectors of the incident and scattered radiation completely determine the wave vector of the vibration modes of a given branch which can take part in the scattering process. In second-order-spectra the combination of vibration modes that take part in the scattering processes are largely determined by density of states factors and are relatively independent of the wave vectors of the incident and scattered radiation.) The state of polarization of the total scattered radiation at a given frequency will involve a superposition of I_{yy} and I_{xy} components of the radiation scattered by each of the symmetry equivalent combination of modes whose orientations relative $\vec{k}_i$ (and $\vec{k}_S$) may be different [4]. From the above, it might be assumed that ASRS has stochastical character.

Note that the frequency position of the asymmetrical optical phonon (AOP) changes drastically with the decreasing of temperature: for T < 280 K $\omega_{aop} > \omega_{op}$, while for T >280 K $\omega_{aop} < \omega_{op}$, where ω_{op} is the frequency of the optical phonon, and in low temperature region (70 K ≤ T ≤ 280 K) is disappeared asymmetry of Raman scattering for ω_{aop} [5]. We interpret experimental results with the aid of inhomogeneous phonons (solutions of the system for the linear chain of connected anharmonic oscillators according to the work [6] and using linewidths and frequencies shifts of AOP as calculated parameters for anharmonicity and nonlinearity consequently. These results are in resonably good agreement with the experimental results, including the low temperature behaviour of $\omega_{aop}(T)$.

1. J. Menendez and M. Cardona, Phys. Rev. B 29 (1984) 2051
2. M. Balkanski, R. F. Wallis, and E. Haro, Phys. Rev. B 28 (1983) 1928
3. J. M. Zieman "Electron and phonons", Oxford at the Clarendon Press, 1960
4. E. Burstein, Lattice dynamics, Pergamon Press, Oxford (1964) p. 315
5. A. J. Semjonow and A. P. Sokolow, Proc. 14th ICORS, p. 778 (1994)
6. M. Y. Valakh et al, Phys. stat. sol. (b) 113 (1982) 635

TEMPERATURE GRADIENTS IN SI CRYSTAL MICROSLABS

A. Je. Semjonow
GOS, Rudower Chaussee 5, 12489 Berlin, Germany

Effect of temperature gradients on the first-order Raman spectrum of Si was studied by Raptis et al and Semjonow [1]. Under conditions of large temperature inhomogeneities generated by a critically focused cw laser beam they found that the Stokes (S) and anti-Stokes (A) bands are extensively deformed due to large asymmetric broadenings which develop on the side of the low frequencies. The asymmetrical broadenings are thus attributed to temperature gradients ($\vec{\nabla} T$). We observed asymmetric broadening in our measurements [2].

Asymmetry of Raman lines induced by the temperature gradient ∇T has a few special features which can be used as criterions for the existence of $\vec{\nabla} T$: a) The shape of the tail low-frequency is different for S and A band. When the spectra are normalized to their corresponding peak intensities, the tails of the A bands are more intense than those of the corresponding S bands. Furthermore, the apparent A peaks are shifted more than the S peaks (Fig. 4 in [2]). The S-A frequency shift can achieve up to $\Delta\omega = \omega_s - \omega_A = 6\ \text{cm}^{-1}$ in our measurements. b) With increasing $\vec{\nabla} T$ the integrated intensity decreases or increases for S bands, but not so for the A band. c) Dependencies of the integrated intensities for S and A bands on laser intensity have nonlinear character. d) The low-frequency sides represent scattering from regions of higher temperatures.

In Si such effects have been observed and manifest themselves through the decreasing of the Stokes / anti-Stokes ratio as the laser power is increased [1]. We applied the results of the work [3] for discription of experimental results (Fig.1) by influence of $\vec{\nabla} T$ in the directions parallel (Y) and perpendicular (Z) to the surface.

As was shown by M. Rasolt [3] that with the availability of enough large $\vec{\nabla} T$ the phonon occupation $n_o = 1/(e^{h\omega/kT} - 1)$will have transformation to the next form

$$\lim_{q\to 0} n(\vec{q}) = \lim_{q\to 0}(n_o(\vec{q}) - \gamma(\vec{q}) \bullet \vec{\upsilon}_q \bullet \tfrac{\vec{\nabla}_T}{T}), \qquad (1)$$

$$\text{where } \gamma(\vec{q}) = \omega_{\vec{q}} \bullet \tau(\vec{q}) \bullet \frac{\delta n_o(\vec{q})}{\delta\omega_{\vec{q}}}, \qquad (2)$$

$\upsilon_{\vec{q}}$ is the velocity of the optical phonons with wave vector $\vec{q} \neq 0$, $\tau(\vec{q})$ is lifetime of the optical phonon, $\omega_{\vec{q}}$ is phonon frequency.

Using the eq. (1, 2) and taking into consideration that the wave vectors (or $\vec{\upsilon}_o$) of S and A phonons have opposite direction we get for S and A intensities

$$I_S \approx (n_o + 1) - \gamma\vec{\upsilon}_z \bullet \frac{\vec{\nabla}_z T}{T} + \gamma\vec{\upsilon}_y \bullet \frac{\vec{\nabla}_y T}{T}, \qquad (3)$$

$$I_A \approx n_o + \gamma\vec{\upsilon}_z \bullet \frac{\vec{\nabla}_z T}{T} - \gamma\vec{\upsilon}_y \bullet \frac{\nabla_y T}{T}. \qquad (4)$$

$\vec{\nabla} T$ introduces asymmetry in the Raman cross section if $\gamma(\vec{q})$, $\vec{\upsilon_{\vec{q}}}$, and $\vec{\nabla} T$ are nonzero. In ideal crystal the effect of $\vec{\nabla} T$ on the optical phonons occupation cannot be, as $\vec{\upsilon}(\vec{q}=0)=0$. With the help of eq. (3, 4) can understand behaviour of the temperatures ratio T_R/T_L measured with $I_{2.0rd}/I_{1.0rd}$ and with I_S/I_A (Fig. 1). Second order of Raman scattering is not sensitive to $\vec{\nabla}$ T as in scattering participate two phonons with $\vec{\upsilon_1}=-\vec{\upsilon_2}$, therefore $T_L=T_R$. For small laser intensity we have weak second order and large experimental errors (Fig. 1).

For first order Raman scattering we have different I_{st}, I_{Ast} for left and right scattering as $\vec{\nabla_y} T \neq \vec{\nabla_z} T$. With scheme of scattering and expressions 3, 4 and taking into account parallel and antiparallel orientation of Stokes and anti-Stokes phonons wave vectors relatively

$\vec{\nabla} T$ can be obtained $I_{St}^{(L)} \sim n_o + 1 + \gamma \frac{\vec{\nabla_z T} \bullet \vec{\upsilon_z}}{T}$, $I_{Ast}^{(L)} \sim n_0 - \gamma \frac{\vec{\nabla_z T} \bullet \vec{\upsilon_z}}{T}$ and

$I_{St}^{(R)} \sim n_0 + 1 + \gamma \frac{\vec{\nabla_y T} \bullet \vec{\upsilon_y}}{T}$, $I_{Ast}^{(R)} \sim n_0 - \gamma \frac{\vec{\nabla_y T} \bullet \vec{\upsilon_y}}{T}$, and as result $T_L \neq T_R$.

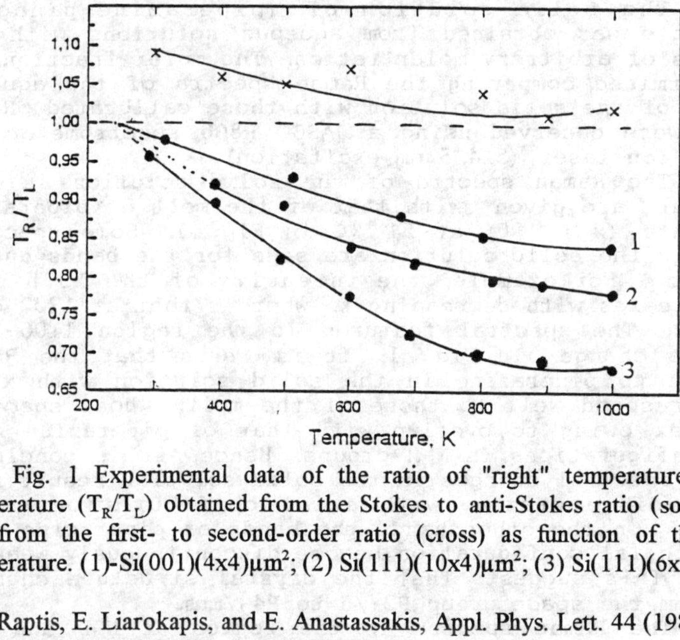

Fig. 1 Experimental data of the ratio of "right" temperature to "left" temperature (T_R/T_L) obtained from the Stokes to anti-Stokes ratio (solid circles) and from the first- to second-order ratio (cross) as function of the surface temperature. (1)-Si(001)(4x4)μm^2; (2) Si(111)(10x4)μm^2; (3) Si(111)(6x5)μm^2.

1. J. Raptis, E. Liarokapis, and E. Anastassakis, Appl. Phys. Lett. 44 (1984) 125; A. Je. Semjonow, SPIE vol. 2648 (1995) 470
2. A. J. Semjonow, L. Holz, A. Lau, and W. Richter, Phys. Stat. Solidi 171 (1992) 505
3. M. Rasolt, Phys. Rev. B 27 (1983) 6066

VIBRATIONAL SPECTRA OF PIPERAZINE$_X$-PINACOL$_{1-X}$ HEXAHYDRATE AND POLARONS

Isao Kanesaka, Satomi Yonezawa and Mizuho Sakanaka

Faculty of Science, Toyama University,
Gofuku, Toyama 930, JAPAN

The crystal structures of piperazine hexahydrate [1] and pinacol hexahydrate [2] are almost isomorphous with the space group $P2_1/n$ and $P4_2/mnm$, respectively. This suggests the formation of the solid solution, whose vibrational study is reported. We are also interested in polarons in the solid solution, which have been found in N-X stretches (X = H or D) of piperazine hexahydrate [3].

The solid solution of (piperazine-pinacol) hexahydrate was obtained from aqueous solutions with two compounds of arbitrary molar ratios. The molar fraction, x, was determined comparing the Raman spectra of the aqueous solution of the solid solution with those calibrated. Raman spectra were observed using a JASCO R800 spectrometer and an argon ion laser (514.5 nm excitation).

The Raman spectra of the solid solution at room temperature are given with that of the melt of piperazine hexahydrate (x = 1.0) at 51 °C in Fig. 1. Some spectral changes in the solid solution are seen for the bands due to piperazine. For example, the intensity of the 1325 cm^{-1} band increases with decreasing x, whereas that at 1302 cm^{-1} decreases. The spectral features in the region 1100-1000 cm^{-1} also change. In Fig. 1, it is seen that the Raman bands due to piperazine in the solid solution with x = 0.71 correspond well to those of the melt, whose spectrum is complex owing to overlap with that of piperazine with axial configurations in N-H groups. Hence, it is concluded that the spectral changes in the solid solution result from increase of piperazine with axial configurations as x decreases. On the other hand, the bands of piperazine with the equatorial configuration become discontinuously weak at x $\approx$ 0.5. This suggests that the crystal structure changes there from the space group $P2_1/n$ to $P4_2/mnm$.

The Raman spectra in the region of N-D and O-D stretches of the D system with x = 0.80 are given for various temperatures in Fig. 2. It is seen that the Raman intensity of the band B increases with lowering temperature. The similar feature has been found in the D system with x = 1.0 and the band B is assigned to polarons [3]. Hence, the band B in Fig. 2 is considered to be polarons. The relative intensity, $I_R(T) = I_B/(I_A+I_B)$, is plotted relatively to $I_R(100)$ against T^2 in Fig. 3, where the same plot reported [3] is also given. The fact that $I_R(T)$ in x = 0.80 is larger than that in x = 1.0 indicates that polarons are formed more in the solid solution than the parent compound, which is reasonable, because solitary states, that is polarons, are formed inevitably by mixing pinacol for piperazine. Hence, the assignment of the band B is confirmed.

1. D. Schwarzenbach, J. Chem. Phys. 48, 4134 (1968).
2. H. S. Kim. and G. A. Jeffrey, J. Chem. Phys. 53, 3610 (1970).
3. I. Kanesaka, K. Kuwano and T. Ishioka, J. Raman Spectrosc. 24, 889 (1993); 26, 283 (1995).

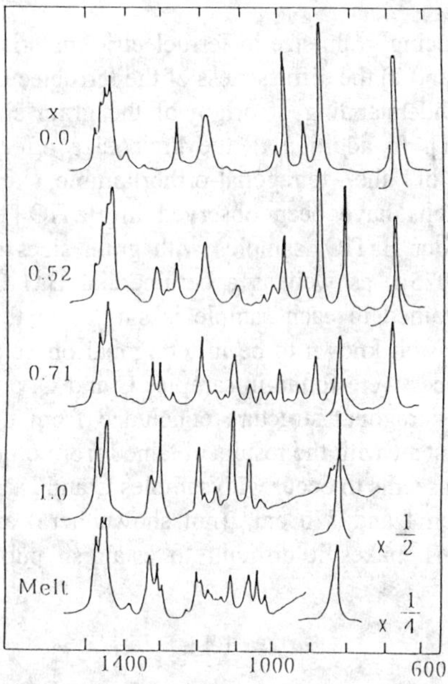

Fig. 1. Raman spectra of the solid solution and the melt at room temperature.

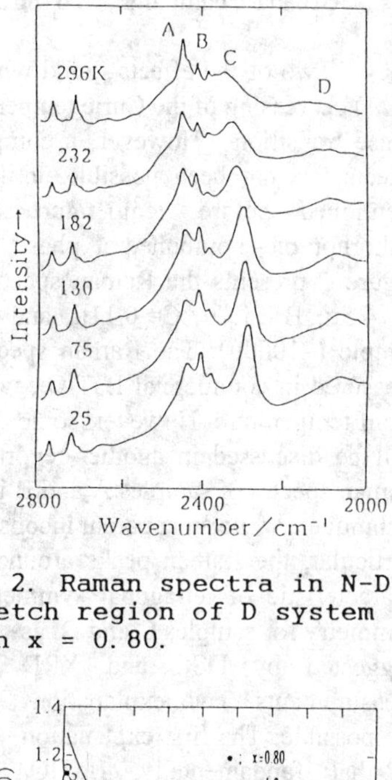

Fig. 2. Raman spectra in N-D stretch region of D system with x = 0.80.

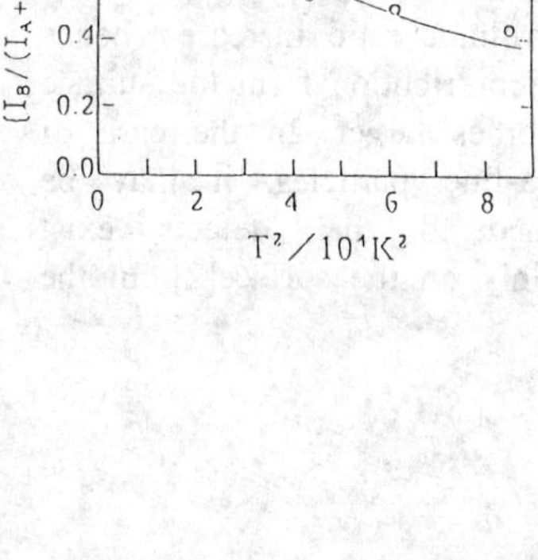

Fig. 3. $I_R(T)$ vs. T^2 plot.

GRAIN SIZE STUDY OF $BaTiO_3$ GROWN BY SOL-GEL TECHNIQUE

E. Ching-Prado[1], W. Pérez[1], S. Sen[1], R. S. Katiyar[1*], M. H. Frey[2], and D. A. Payne[2]

[1]Department of Physics, University of Puerto Rico, Río Piedras, P.R. 00931.
[2]Department of Materials Science,and Beckman Institute,University of Illinois, Urbana-Champaign, Il. 61801.

Two of the effects found when reducing grain size in ferroelectric materials are 1) decreasing of the Curie temperature, and 2) the diffusseness of the ferroelectric phase transition. However, a complete understanding of origin of the grain size effects has not been possible until now[1]. In addition to the grain size effects mentioned before, temperature shifts of the tetragonal-orthorhombic, and orthorhombic-rhombohedral phase transitions have been observed in $BaTiO_3$[1]. Figure 1 presents the Raman spectra of four $BaTiO_3$ samples with grain sizes of A= 4.5μ, B= 1.7μ, C= 0.11μ, and D= 0.025μ, as well as a commercial $BaTiO_3$ sample (~1000μ). The Raman spectra obtained in each sample is similar to that measured in commercial $BaTiO_3$, which is well known to be in tetragonal phase at room temperature. However, some differences were found in samples C and D, and will be discussed in another report. The tetragonal structure concluded from the Raman spectra of samples A and B is consistent with the results obtained from other techniques, while an apparent inconsistency seems to occur with samples C and D. In particular, the Raman peaks around 309 cm^{-1} and 720 cm^{-1} (not shown here) are characteristic of tetragonal symmetry. This makes it difficult to establish cubic symmetry for samples C and D, as suggested by DSC and XRD measurements. Two explanations are possible: The first explanation is that fundamentally, $BaTiO_3$ possesses defects, where the amount of defects depend on the method of preparation. As the particle size decreases, the surface to volume ratio increases, hence the contribution from the surface becomes larger. In the case of ultra-fine particles, it may be considered that defects exist mainly on the surface[2]. In the

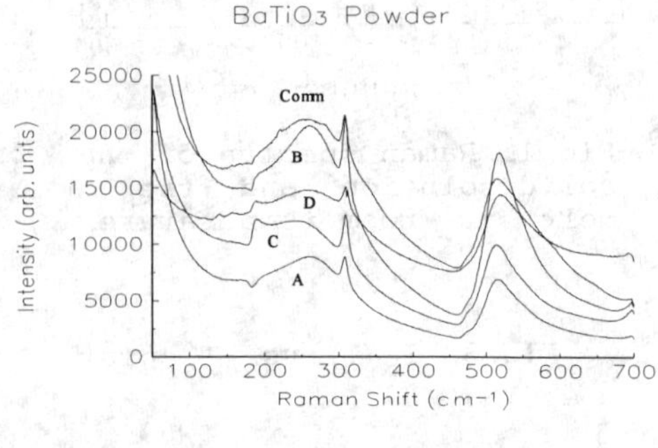

Figure 1 :Raman spectra of $BaTiO_3$ with different grain sizes at RT.

case of samples C and D the lattice shrinks considerably due to these defects, and a pseudocubic phase transformation takes place[2]. These defects appear to be similar to the effects of neutron irradiation upon tetragonal barium titanate, where XRD indicated an apparent cubic symmetry with the introduction of a large number of defects[3]. In this case, the large number of defects may cause a breakdown in translational symmetry and the wave vector selection rules are relaxed. Hence, the first order phonons need not be restricted to the zone center. Thus, contributions from phonons at other symmetry points in the Brillouin zone may be observed. As a result, the Raman spectrum is a weighted average of the density of phonon states and, due to large number of defects, the halfwidth of the phonon bands are expected to be practically temperature-independent.

The second explanation is the space charge layer at the grain boundary. X-ray and electron diffraction experiments with very small $BaTiO_3$ particles have been found to exhibit discrepancies in symmetry between the surface layer and the bulk[4]. The difference has been found to be smaller below the Curie temperature of the bulk. Thus, the structure of the surface layer is closer to the tetragonal structure than to the cubic structure. Moreover, the tetragonal surface strain does not vanish if the crystal is heated above the Curie temperature of the bulk[4]. Thus, the short-range order Raman spectra of samples C and D contain a contribution of the tetragonal surface layer, while XRD measurements for the average long-range order show the cubic structure of the bulk. In this case, the first-order Raman spectrum is, therefore, due to the space charge layer at the surface of the grain. Figure 2 shows the Raman spectra of samples C and D compared to that of commercial $BaTiO_3$ at 93K. The rhombohedral phase can be obtained from sample C, while no similar conclusions can be obtained from sample D. The results suggest that the high defect concentration is responsible for grain effects.

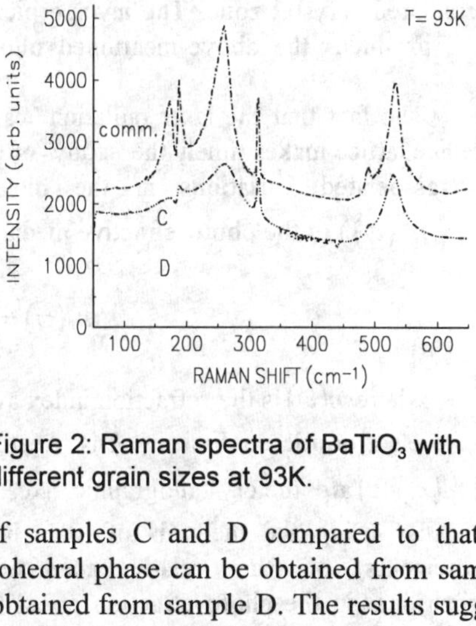

Figure 2: Raman spectra of $BaTiO_3$ with different grain sizes at 93K.

1- A. J. bell, Proceedings of the 9th ISAF- catalog CH3416-5, IEEE 1995.
2- T. Yamamoto, K. Urabe, and H. Banno, Jpn. Appl. Phys., 32, 4272 (1993).
3- M. C. Wittels, and F. A. Sherrill, J. Appl. Phys., 28, 606 (1957).
4- W. Känzig, Phys. Rev., 98, 549 (1955).

OBSERVATION OF THE PHOTODEFORMATION EFFECT IN THE RAMAN SPECTRA OF $LiNbO_3$:Fe

V.N.Malnev, G.N.Dmitrik [*], P.A.Korotkov

Radiophysics Department , Kiev University,
Prosp.Acad.Glushkova 6,Kiev, 252127,Ukraine

The action of the laser radiation on the photorefractive crystal $LiNbO_3$:Fe is accompanied by local changes in the ion lattice structure. The nature of this photoinduced phenomena can be described in the following way. A light beam initiates the recharge of the Fe^{2+} , Fe^{3+} impurity centers that results in the formation of the quasistationary electric field of a strength up to $E=10^5$ V/cm in the irradiated crystal zone. The inverse piezoelectric effect which induced by this field E produces the above-mentioned photodeformation process.

The fact that the laser radiation can produce the local deformation of a crystalline lattice makes timely the study of the peculiarities of the vibration states and associated variations in the dielectric constant and refraction index ($\varepsilon(\varpi) = n^2(\varpi)$) of the photorefractive media

$$\Delta\varepsilon_i(\varpi) = -n^4(\varpi) r_{ij} E_j$$

where $n(\varpi)$ is the refraction index averaged with respect to the crystal bulk, r_{ij} is the electrooptic coefficient, E is vector of the photoinduced electric field, $(i,j) \equiv (x,y,z)$ are the crystallographic axes.

By using the methods of the Raman spectroscopy we can carry out the measurements of the relative magnitudes of the lattice (ion) and electron contributions in the photorefraction effect (the photoinduced measurement of the refraction index).

In our experiments the photodeformation of the $LiNbO_3$:Fe crystalline lattice ($c_{Fe} = (0.01 \div 0.05)$ mass per cents) was accompanied by a relatively small $\left(3 \div 5\ cm^{-1}\right)$ in the vibration modes frequency. The band intensity in the polarized Raman spectra depended on the power density of the exciting (laser) radiation and concentration of alloying impurities.

It is known that the dielectric constants of the uniaxial crystals depend on frequencies of the transverse $\left(\varpi_{ni}^{T}\right)$ and longitudinal $\left(\varpi_{ni}^{L}\right)$ optical phonons, and , neglecting a damping, can be written in the form [1,2]

$$\frac{\varepsilon_i(\varpi)}{\varepsilon_i^{\infty}} = \prod_{n=1}^{N} \frac{\left(\varpi_{ni}^{L}\right)^2 - \varpi^2}{\left(\varpi_{ni}^{T}\right)^2 - \varpi^2}$$

where ϖ is the pumping frequency, N is number of the dipole-active modes.

The calculated values of the photoinduced variation in $\varepsilon_i(\varpi)$ must be as much as $\Delta\varepsilon_i(\varpi) = 10^{-5}$ that considerably differ from the experimental values ($\Delta\varepsilon = 10^{-2}$). Therefore, the lattice contribution in the process of the optical damage of samples does not exceed 0.1%.

Accordingly the above reported one can assume that the photorefraction effect in the $LiNbO_3$:Fe crystals is associated with the photoinduced perturbation of the electron subsystem, i.e., the variation of the electron polarization of the medium in the irradiation zone.

Determing the relative intensity of the Raman bands of optical phonons, it is possible to evaluate the magnitudes of ion-electron contribution in the tensor of quadratic nonlinear polarization and linear electrooptic effect [3,4] and analyze the dependence of these quantities on the irradiation intensity, and alloying degree of the $LiNbO_3$:Fe crystals.

1.T.Kurosava, J.Phys.Soc.Jap.,16. 1288 (1961)
2.R.Claus, Phys.Stat.Sol.,100b, 9 (1980)
3.I.P.Kaminov,W.D.Johnston, Phys.Rev.,160, 519 (1967)
4.W.D.Johnston, Phys.Rev.B, 4, 3494 (1970)

SPATIAL DISPERSION EFFECTS IN RAMAN SPECTRUM OF PARATELLURITE TeO_2 UNIAXIAL GYROTROPIC CRYSTALS

A.V.Slobodyanyuk*, S.G.Garasevich, P.I.Korenyuk

Department of Physics, Kiev University, Prospect Glushkova 6, 252127 Kiev, Ukraine

Recently we have reported E-modes splitting completely caused by spatial dispersion in Raman spectra of ZnP_2 and CdP_2 uniaxial gyrotropic crystals [1]. Main spatial dispersion effects as Raman Circular Dichroism (RCD) and splitting increasing with magnitude q of phonon wave vector **q** may be described with simple model of isolated polar mode in gyrotropic crystal based on effective Hamiltonian taking into account anisotropy of short-range forces, long-range Coulomb interaction and linear with q contribution of spatial dispersion [2].

Well-known for their unique acoustooptical properties paratellurite TeO_2 crystals belong to the same space groups $P4_12_12$ or $P4_32_12$ as zinc and cadmium diphosphides. The RCD was observed for components of the lowest frequency E-doublet in Raman spectrum of TeO_2 near 124 cm^{-1} [3] but unlike ZnP_2 and CdP_2 unexpected *decrease* of splitting with q instead its *increase* was found before in experiments with excitation by light of different wavelengths. It was in drastic contradiction with basic ideas of the first order spatial dispersion effects confirmed experimentally for all gyrotropic crystals studied before. Here we report preliminary results of new study of this violation of general trend.

Raman spectra of paratellurite samples immersed in LN were obtained in unchanged right-angle geometry with propagation of incident light from He-Ne or Ar^+ lasers operating at wavelengths 633, 514, 496, 488, 476, 458 nm along optical axis ($z(e^Iz)x$, where e^I is for polarization of incident light).

Fig.1a shows presence of the third line L3 overlapping high frequency component L2 of 124 cm^{-1} E-doublet in spectrum obtained with 633 nm linearly polarized incident radiation. Dashed lines display result of decomposition of original spectrum in sum of three Lorenz peaks. Fig.1b shows the same part of spectrum obtained with left and right circularly polarized laser radiation and RCD of main components L1 and L2 of doublet. If frequency of laser light increases line L3 shifts closer to line L2 and at λ=488 nm they become practically completely overlapped .

Plots of splitting of main components L1 and L2 and of splitting L3-L2 and L3-L2 against wave vector magnitude q are presented in Fig.1c. Values of q were carefully calculated for every laser wavelength taking in consideration dispersion of refraction indexes for both normal waves of incident and scattered light. Evidently that separation of L3 line almost completely eliminates anomalous decrease of L1-L2 splitting. Consequently there is no sufficient reasons to revise picture of spatial dispersion effects in Raman spectra of uniaxial gyrotropic crystals although nature of additional line L3 is not clear now. Line L3 was observed in spectra of all TeO_2 samples of different origin studied up to now with the same intensity relative to main components of doublet L1 and L2. This indicates its intrinsic character.

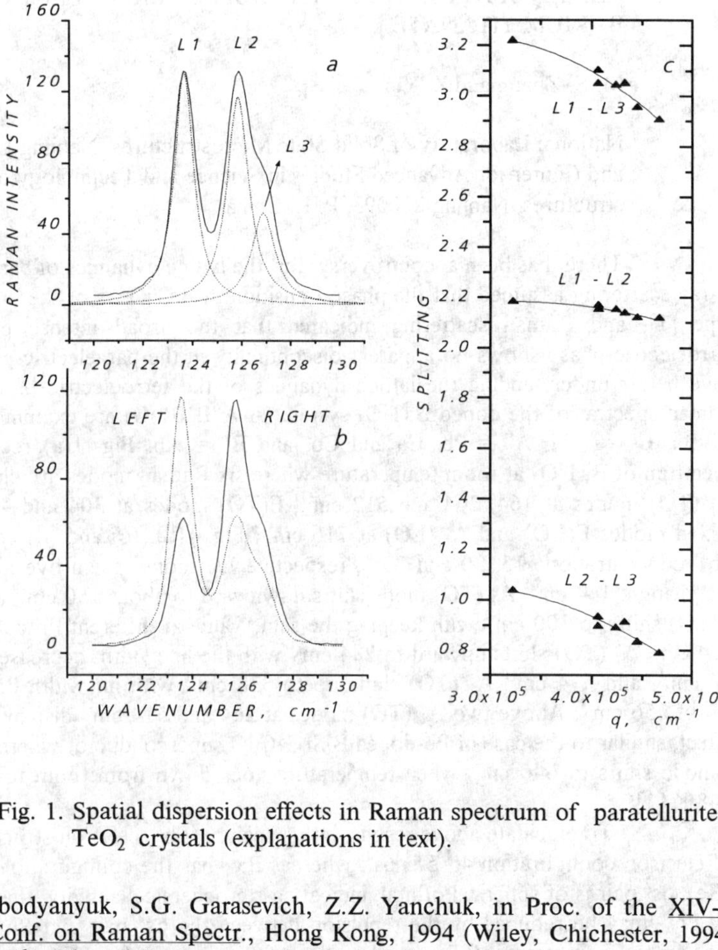

Fig. 1. Spatial dispersion effects in Raman spectrum of paratellurite TeO_2 crystals (explanations in text).

1. A.V. Slobodyanyuk, S.G. Garasevich, Z.Z. Yanchuk, in Proc. of the XIV-th Intern. Conf. on Raman Spectr., Hong Kong, 1994 (Wiley, Chichester, 1994), vol. A, p.261; Physics Letters A 197, 238 (1995).
2. A.V. Slobodyanyuk, Sov. Phys.-Solid State 28, 1845 (1986); Opt. & Spectr. (USA) 61, 626 (1986): transl. of Opt. Spektrosk. (USSR) 61, (1986).
3. L.P. Avakyants, N.N. Vorivonchik, V.P. Grishchuk, L.P. Naulik, A.V.Slobodyanyuk, Opt. Spectrosc. (USA) 66, 341, (1989): transl. of Opt. Spektrosk. (USSR) 66, 583, (1989).
4. V.P. Grischuk, L.R. Naulik, A.V. Slobodyanyuk, Ukrainskii Fizicheskii Zhurnal 34, 1670 (1989) (in Russian).

RAMAN SCATTERING OF THE DOPED FERROELECTRIC BARIUM TITANATE

M. S. Zhang and Z. Yin

National Laboratory of Solid State Microstructures, Nanjing University, and Center for Advanced Studies in Science and Technology of Microstructures, Nanjing 210093, P. R. China

There has been a controversy for the lattice dynamics of barium titanate. X-ray scattering assumed that the phase transition in $BaTiO_3$ is of the order-disorder type [1], and Raman scattering indicated that the broad intense peaks in the ferroelectric phase shows no apparent discontinuity in the paraelectric phase [2]. To have better understanding the lattice dynamics of the ferroelectric barium titanate, Raman spectra of the doped $BaTiO_3$ system (A'A"B'B")O_3 are examined where A' = Ba, B' = Ti, A" = Pb, Ca and Cu, and B" = Nb. Fig. 1(a) records Raman spectrum of $BaTiO_3$ at room temperature where six Raman modes are clearly shown: A_1 (TO) modes at 164, 264 and 512 cm^{-1}; E(TO) modes at 304 and 476 cm^{-1} and mixed modes E(LO) and A_1 (LO) at 716 cm^{-1}, Fig. 1(b), (c) and (d) correspond to Pb^{2+} concentrations 15, 20 and 25%, respectively. Among them five modes exhibit big change: 164 cm^{-1} A_1 (TO) mode shifts downward to about 150 cm^{-1} and 304 cm^{-1} E(TO) shifts to 300 cm^{-1} with keeping the same value at different Pb concentrations; 264 cm^{-1} A_1 (TO) shifts upward to 284 cm^{-1} with the linewidth decrease from 92 to 50 cm^{-1}, and 514 cm^{-1} A_1 (TO) shifts up to 538 cm^{-1} with linewidth decrease from 67.4 to 56 cm^{-1}. Above two A_1 (TO) modes at 264 and 514 cm^{-1} display "hardening" effect, similar to the case of Pb-doped BiSrCaCuO superconductor where the 628 cm^{-1} mode shifts to 640 cm^{-1} when temperature goes down from room temperature to 148 K [3].

To elucidate above results, let us analyse Pb^{2+} electron structures. Pb^{2+} is of electron configuration $4f^{14}5d^{10}6s^2$, whereas Ba^{2+} has the configuration of inert gas (Xe). $6s^2$ pair is of spherical orbital, indicating sterochemical activity, the deformation by Pb^{2+} may be reduced by the repulsion between the $6s^2$ pair and Pb^{2+}-O^{2-} bonds, leading to an increase of T_c and decrease of the volume V_c. The increase of Pb^{2+} concentrations indicate that the tetragonal ferroelectric-cubic paraelectric transitioon temperature is increased and the tetragonal-orthorhombic and orthorhombic-rhombohedral temperatures are decreased [4]. Since the displacements from the centrosymmetric position of Pb^{2+} are from non-bonded $6s^2$ electron, the off-center effects from Pb^{2+} doping become stronger. Partial substitution of Ca^{2+} does not cause detectable change of the Raman spectra since Ca^{2+} is of electron configuration of inert gas (Ar), causing less off-center effects which is consistent with the results that the tetragonal ferroelectric-cubic paraelectric transition temperature shows no Ca concentration dependence [4]. Dielectric measurements of the Nb-doped $BaTiO_3$

show that the tetragonal-cubic transition disappear when the Nb concentration varies from atomic weight of 0.2 to 1% [5]. It is relative to that the heterogeneity coefficient increase with the doping concentration as does the diffuse phase transition.

Six Raman lines at 164, 266, 306, 474, 518 and 718cm^{-1} are clearly measured in the Cu-doped $BaTiO_3$. The features are consistent with those from Didomenico et al. [2]. It is worth to note that two modes at 266 and 518 cm^{-1} undergo big change. Their intensities are almost threefold decreased comparing with those in pure sample, and correspondingly the linewidths are 110 and 65 cm^{-1} in the doped, wider than 90 and 50 cm^{-1} in the pure sample, respectively. The broadening of these two lines by Cu doping is involved in local disorder by Ti ion sitting off the center of the unit cell. The doping enhances the off-center effect of the Ti ions, which provides a longer relaxation time for Ti ion tunelling. When the relaxation time is longer than time for the Raman process, a broad-band Raman scattering would happen which is consistent with EPR results obtained by Muller [6] where the spin relaxation rates showed that the Ti ions are located at off-center sites in cubic phase. Ti ions move from one off-center site to another. This indicates that the possibility that the broadening lines near 260 and 520 cm^{-1} were previously considered owing to the determining presence of impurities or second-order scattering would be ruled out.

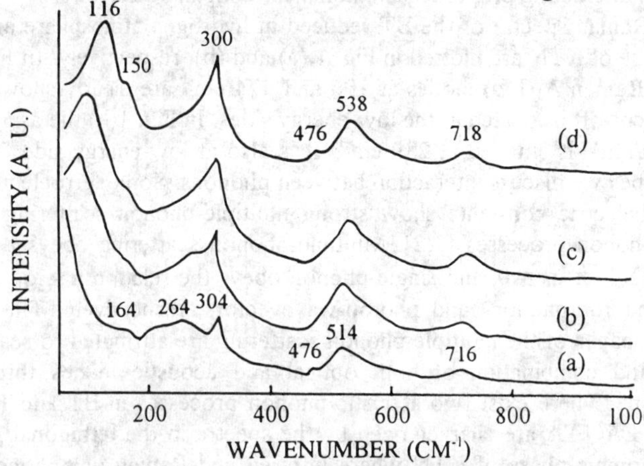

Fig. 1 (a) Raman spectrum of barium titanate; (b)-(d) Raman spectra of Pb-doped barium titanate at the various concentrations 15, 20, and 25%, respectively.

1. M. Lambert and R. Comes, Solid State Commun. 7, 305 (1969).
2. M. Jr. Didomendico, S. H. Wemple, S. P. S. Porto, and R. P. Bauman, Phys. Rev. B**174**, 522 (1968).
3. M. S. Zhang, Q. Chen, G. J. Hu, X. Jin, Z. Yin, and Y. F. Hsia, Ferroelectrics **112,** 195 (1990).
4. G. Godefroy, B. Jannot, and F. M. Michel-Calendini, Phase Trans. **33**, 3 (1991).
5. G. Godefroy, P. Chatry, and L. Benguigui, Ferroelectrics **34**, 169 (1981).
6. K. A. Muller, Solid State Commun. **61**, 21 (1989).

INTERACTION BETWEEN ONE- AND TWO-PHONON IN BARIUM TITANATE AND POTASSIUM TANTALATE NIOBATE

Z. Yin and M. S. Zhang

National Laboratory of Solid State Microstructures, Center for Materials Analysis, Nanjing University, and Center for Advanced Studies in Science and Technology of Microstructures, Nanjing 210093, P. R. China

Interaction between one-phonon and two-phonon is characteristic of broadening and distorting of the absorption, emission and scattering line shapes, weakening of emission lines to lower energy states from single-phonon state etc. By studying Raman spectroscopic lineshapes, we examine and report the resonant interference in ferroelectric barium titanate(BT) and potassium tantalate niobate (KTN): their interacting properties and nonlinear coupling behaviors.

Raman spectra of the BT reduced in hydrogen atmosphere and in orthorhombic phase of KTN are plotted in Fig. 1 (a) and (b), respectively. In Fig. 1(a) two asymmetric Raman A_1(LO) modes at 196 and 474 cm^{-1} are clearly shown where the asymmetric cut-off is located at the low-energy sides. In Fig. 1(b) the asymmetric cut-off of Raman line is situated at 239 cm^{-1}. It is also at low-energy side. To elucidate their lineshapes we discuss interaction between phonons. Some ferroelectric materials with high dielectric constants show strong multiple-phonon scattering, and rather sharp one-phonon processes [1,2]. Multiple-phonon scattering obeys a wavevector relationship $\Sigma\, \mathbf{k}_i \cong \mathbf{q}_o \cong 0$, and single-phonon obeys the relation $\mathbf{k} \cong \mathbf{q}_o \cong 0$ where $\mathbf{k}$ and $\mathbf{q}$ stand for phonon- and photon-wavevectors, respectively. The asymmetric broadening shapes of the multiple-phonon scattering are attributed to scattering from overtones and combination of both optical and acoustic modes throughout the Brillouin zone. There exist two-acoustic-phonon processes in BT and KTN, that is 2LA, 2TA, LA +TA are characterized by the spectra in the tetragonal phase of BT and orthorhombic phase of KTN where infrared and Raman one-phonon processes are allowed [3]. A Fano-type interference with an asymmetric lineshape can be caused by an anharmonic coupling in the energy expansion. As shown in Fig. 1, the anharmonic terms in the potential function lead to the discrete state interacting with the continuum state, and the transition rate between these two states is proportional to the anharmonicity of the potential function measured by the linewidth of the discrete state. The continuum process are shown at two broad bands at about 240 and 530cm^{-1} in BT and at about 240 and 280 cm^{-1} in KTN. These board bands do not disappear in paraelectric cubic phase in BT or in tetragonal phase in KTN which were interpreted as two-phonon scattering [3]. However the polariton scatering did not support the interpretation [4]. The quasi-mode-spectra measurements in both tetragonal and cubic phases for $BaTiO_3$ with different impurities indicated that the broad peaks at about 240 and 530 cm^{-1} are not owing to the determining presence of impurities, probably to some kinds of intrinsic disorder which existed in both phases [5]. The continuum consisted of two broad bands in both BT and KTN come most likely from rather large number of scattering centers which is quite similar to the case of intervalley scattering in the doped indirect gap semiconductors [6]. This argument is well confirmed in the measurement of BT thin films prepared by sol-gel technique where no apparent continuum state was found due to small number of scattering centers.

The interaction between one- and two-phonon may causes a dip on the low- or high-energy sides of the Raman lines. Such interaction does distort the lineshape and provide a radiationless decay channel for the phonon. The sign of the asymmetric coupling coefficients in BT and KTN is correlated to the interaction parameter. Positive or negative sign of the coupling coefficient is closely related to which one of low- or high-energy side is cut-off. In BT and KTN the Raman lines at 196, 474 and 239cm^{-1} are all cut-off low-energy sides, showing positive coefficients. This is same as the case of BiSrCaCuO superconductor where 290 cm^{-1} is also sharply located at low-energy side, and oppsite to the case of YBaCuO where the 116 and 340 cm^{-1} lines are cut-off on the high-energy sides [8].

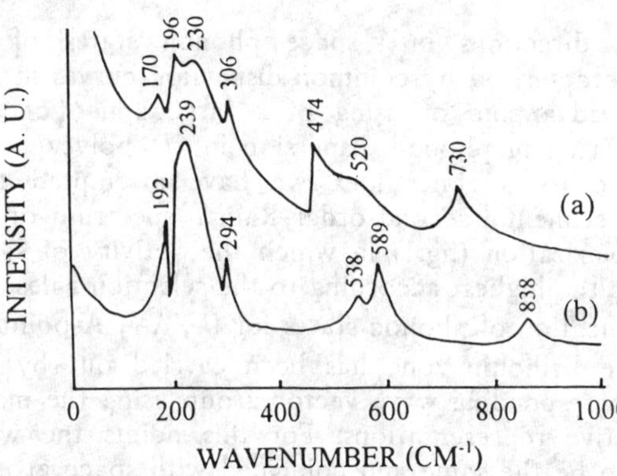

Fig. 1 Raman spectra of (a) the reduced BT, (b) in the orthorhombic phase of KTN

1. C. H. Perry and D. B. Hall, Phys. Rev. Lett. **15**, 700 (1965).
2. G. Shrane, B. C. Frazer, V. J. Minkiewiczz, and J. A. Leake, Phys. Rev. Lett. **19**, 234 (1967).
3. M. DiDomenico, Jr. S. H. Wemple, S. P. S. Porto, and R. P. Bauman, Phys. Rev. **174**, 522 (1968).
4. H. J. Benson and D. L. Mills, Solid State Commun. **8**, 1387 (1970).
5. M. P. Fontana and M. Lambert, Solid State Commun. **10**, 1 (1972).
6. P. J. Colwell and M. V. Klein, Phys. Rev. B**6**, 498 (1972).
7. S. L. Cooper, M. V. Klein, B. G. Pazol, J. P. Rice, and D. M. Ginsberg, Phys. Rev. B**37**, 5920 (1988).

LARGE ZONES CONCEPTION FOR DETERMINATION OF PHONON STATES DISPERSION IN ZnO CRYSTALS

N.I.Gerasimuk*, V.A.Gubanov, M.M.Biliy

Physics Department, Kyiv Taras Shevchenko University,
Prosp. Academic Glushkov 6, 252127, Kyiv, Ukraine

The first-order Raman spectra of ZnO single crystals have been studied and interpreted before [1-3]. But the interpretation of second-order Raman spectrum has been proved to be very complicated problem as the knowledge of phonon frequencies of Brillouin zone edge was required. In [1] the frequencies of the most prominent multiphonon processes were listed. In [3] it has been done an attempt of approximative interpretation of several second-order bands. We have also tried to find the solution of this problem.

For some directions in k-space phonon states of crystalline structures can be characterized by common dispersion curves in large zones - Jones zones. The advantage of using of a large zone conception was illustrated by considering of phonon dispersion in 2H polytype of SiC [4]. On the basis of the Jones zones idea we have attempted to give an interpretation of experimental second-order Raman spectrum of ZnO single crystals in x(zz)y polarization (fig.1) in which the activity of second-order Raman spectrum was the highest according to the selection rules.

The classification of phonon states for Γ-, Δ- , A-points along the Γ-A direction in the Brillouin zone has been carried out by irreducible representations of corresponding wave vector group using the mathematical apparatus of projective representations. For this points the wave vector groups were found to be the same and coincided with space group. As well their factor-systems coincided with standard factor-system of 6mm-group of K_0-class, all elements of which were equal 1.

Taking into the consideration the invariance of phonon states to time inversion, in other words after application of Herring criterion, all irreducible projective representations characterizing states with integral spin united in pairs at A-point in contrast to Γ- and Δ- points: $A_1 + A_4$, $A_2 + A_3$, $A_5 + A_6$. This fact gave us the possibility to pass on to new classification of phonon states in ZnO crystals in Jones zone which is twice extended in Γ-A direction in comparison with Brillouin zone and in which the quantity of phonon branches decreases twice. This extended zone corresponds to "relative" primitive cell which stretches along the optical axis for only one double closely packed layer consisted of Zn and O atoms and corresponds only to half of the primitive cell parameter in this direction.

Fig.2 shows the dispersion curves of phonons for ZnO crystal which were built according to experimental frequencies of the first-order

Raman scattering lines. The bands of the second-order spectrum are caused by overtones and combined frequencies of phonons of Γ- and Δ-points and overtones of A-point. In particular we consider that the band of 327 cm^{-1} is overtone of A-point and the bands of 209, 541, 789 cm^{-1} are overtones of Γ-point. The band of 1154 cm^{-1} is overtone of the longitudinal phonon Γ_1.

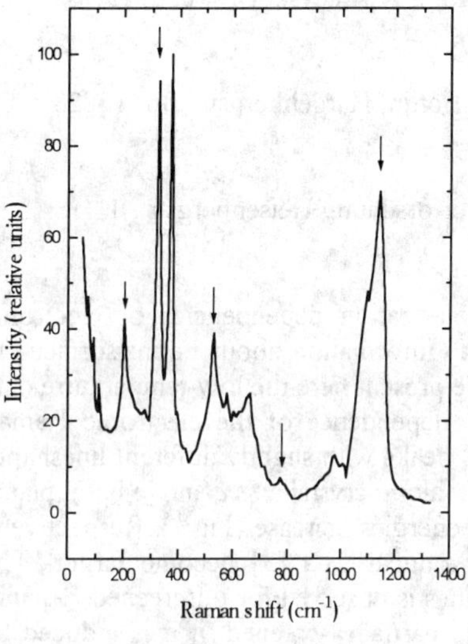

Fig.1. Raman spectrum of ZnO crystal in x(zz)y polarization, T=300K.

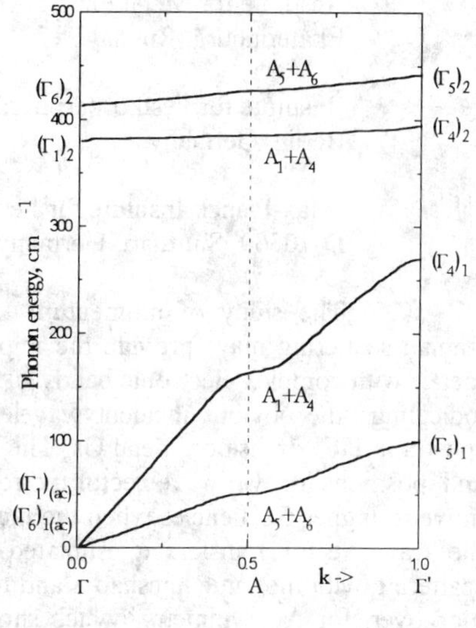

Fig.2. Dispersion curves of phonon states along the Γ-A direction in large zone for ZnO crystal.

1. T.C. Damen, S.P.S. Porto, B. Tell, Phys. Rev. 142, 570 (1966).
2. C.A. Arguello, D.L. Rousseau, S.P.S. Porto, Phys. Rev. 181, 1351 (1969).
3. E.A. Vinogradov, N.N. Mel'nik, A.E. Tsurkan, L.V. Kicherman, Zhurnal Pricladnoy Spectroscopii 26, 1059 (1977) (in Russian).
4. Lyle Patrick, D.R. Hamilton, W.J. Choyke, Phys. Rev. 143, 526 (1966).

SPATIAL DISPERSION OF THE DIELECTRIC RESPONSE IN TRANSITION METALS: MANIFESTATION IN RAMAN SPECTRA.

Yu. S. Ponosov[1*], G. A. Bolotin[1], C. Thomsen[2], and M. Cardona[3]

[1]Institute for Metal Physics UD RAS, S.Kovalevskoi 18, 620219, Ekaterinburg, Russia

[2]Institute für Festkörperphysik TU Berlin, Hardenbergstr., 36, 10623, Berlin, Germany

[3]Max-Planck Institute für Festkörperforschung Heisenbergstr., 1, D-70569, Stuttgart, Germany

The study of momentum and polarization dependencies of electronic Raman scattering may provide the important information about Fermi surface in metals with complex electronic bands [1,2]. We present here the low-temperature data indicating the obvious incident wavelength dependence of the electronic Raman spectra in HCP transition metal Os. The broad peaks with slightly different lineshapes and positions for two wave vector q directions: along crystal axis *c* and in basal plane, move to higher frequencies when the incident energies increase (Fig.1). Respectively, the wave vector transfers q estimated from optical data [3] become larger. The scattering with the same lineshapes and intensities is observed for unscreened E_{2g} and, moreover, for A_{1g} symmetry, which should be partially screened, if it is induced by charge fluctuations. Superimposed on the electronic background E_{2g} symmetry 170 cm^{-1} phonon line demonstrates clear q-dependent renormalization [4] and for 6471 Å exciting wavelength its energy coincides with continuum peak position.

The observed anisotropy of the electronic scattering in respect to q-direction and to polarization geometry shows only small changes for used exciting energies (1.5-2.4 eV) in spite of the intensity growth starts above 1.7 eV. So we tried to model the found effects in terms of intraband mechanism assuming the collisionless case and taking Raman vertex to be related to the curvature of the energy band dispersion as usually one supposes. The tight-binding expression for E(k):

$$E(k) = E_1 \cos(ak_\perp) + E_2 \cos(ck_\parallel) + E_3 \cos(ak_\perp)\cos(ck_\parallel)$$

was used to fit the form and dimensions of two main Fermi surface sheets of Os; the ratio of electron velocities for investigated directions of q was taken concerning to experimental findings. To calculate the lineshape of the effective electronic scattering cross section we completed also the integration over field distribution of the incident and scattered waves in opaque material [5]. The results of such modelling in Fig.1 show both large difference of the peak energies for two q-directions and linear q-behaviour contrary to the experimental data. This discrepancy indicates the importance of the resonant contributions to Raman vertex and, possibly, to electronic response. Their substantial k-dependence over Fermi surface might be the reason of the unscreened A_{1g} scattering also and should be explained to understand the close correlation between electronic continuum and found in Os and other HCP transition metals anomalies of the optical phonon self-energies.

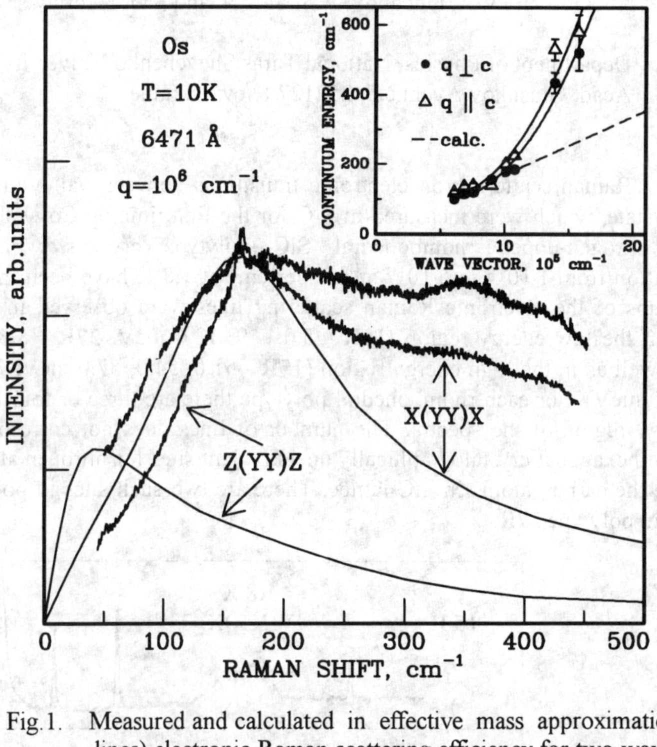

Fig.1. Measured and calculated in effective mass approximation (solid lines) electronic Raman scattering efficiency for two wave vector directions q ∥ c - Z(YY)Z and q ⊥ c - X(YY)X. Insert shows measured and calculated dependencies of the continuum energies on q value.

This work was supported in part by ISF and RF Government Grants No. NMF000 and NMF300.

1. A. A. Abrikosov and V. M. Genkin, Sov. Phys. JETP 38, 417, (1974).
2. I. P. Ipatova, M. I. Kaganov, V. A. Subashiev., Sov. Phys. JETP 57, 1066, (1983).
3. V.V. Nemoshkalenko et al, Sov. Phys. JETP, 20, 201, (1986).
4. Yu. S. Ponosov, C. Thomsen, M. Cardona, Physica C, 235-240, 1153, (1994).
5. A. Dervisch and R. Loudon, J. Phys. C: Sol. State Phys., 9, L569, (1976).

RAMAN SCATTERING ON ELECTRONIC TRANSITIONS IN RHOMBOHEDRAL SiC POLYTYPES

I.S. Gorban', V.A. Gubanov, A.S. Skirda, and S.G. Kudriavtsev

Department of Physics, National Taras Shevchenko University,
Acad. Glushkov Avenue 6, 252127 Kiev, Ukraine

Raman scattering on electronic transitions between valley-orbit splitted donor **1s**-state, which were measured in SiC for the first time by Colwell and Klein [1], in nitrogen-doped rhombohedral SiC polytype crystals with impurity concentration from $1{\cdot}10^{18}$ to $5{\cdot}10^{19}$ atoms per cm^3 at 1.8 K have been investigated. Two groups of the electronic Raman scattering lines were observed in all spectra (Fig. 1) in the low energy region (**15R** - 5.0, 7.9, 12.0 meV, **27R** - 8.5, 9.9, 22.6 meV) as well as in the high energy region (**15R** - 51.0, 54.0, 57.0 meV, **27R** - 65.0, 82.6, 84.0 meV). For each rhombohedral polytype there are three or four lines in the low energy region of the spectra. The number of lines does not correspond to the number of hexagonal crystalographically unequivalent sites for nitrogen atoms which substitute the carbon atoms in SiC lattice. There are two such sites in polytype **15R** and four in polytype **27R**.

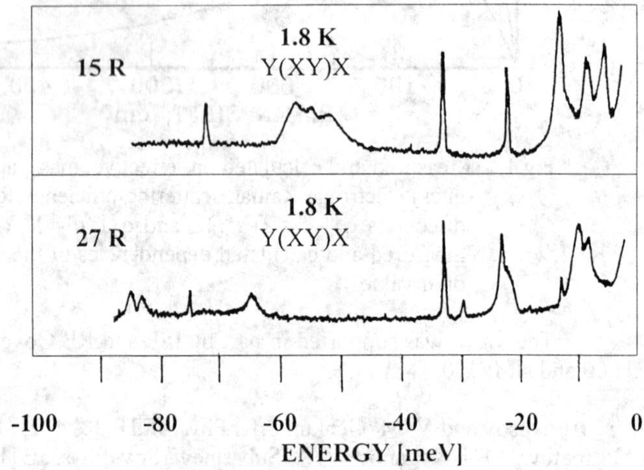

Fig. 1. Raman scattering spectra in crystals of 15R- and 27R-SiC:N.

The character of splitting for every line of electronic Raman scattering in uniaxially stressed at 1.8 K crystals in dependence on value and direction of applied pressure is described in [2]. The splitting of electronic Raman scattering lines and appearance of their components in various polarizations prove their electronic origin.

The change intensities of electronic Raman lines and their spectral positions in undeformed **15R**-SiC crystals in different geometry is demonstrated on Fig. 3. The number of electronic Raman lines in different geometry changes from 3 to 4.

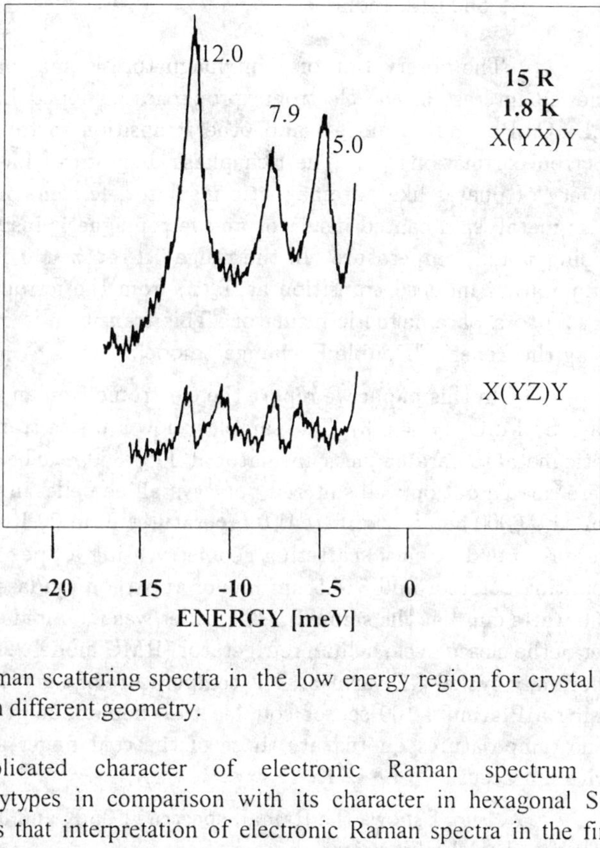

Fig. 3. Raman scattering spectra in the low energy region for crystal of undeformed 15R-SiC in different geometry.

The complicated character of electronic Raman spectrum in rhombohedral SiC polytypes in comparison with its character in hexagonal SiC polytypes [1] witnesses that interpretation of electronic Raman spectra in the first case is not so simple as in the second one. It can be connected to different character of disposition of equivalent valleys for rhombohedral SiC structure, and obviously requires further study.

1. P.J. Colwell and M.N. Klein, Phys. Rev. B 6, 498 (1972).
2. I.S. Gorban', V.A. Gubanov, V.D. Kulakovskii, A.S. Skirda, and B.N. Shepel', Physica Tverdogo Tela, 30, 1605 (1988).

ELECTRONIC RAMAN SCATTERING FROM CORRELATED METALLIC $La_{0.7}Sr_{0.3}MnO_3$

Rajeev Gupta' and A.K. Sood

Department of Physics, Indian Institute of Science, Bangalore 560 012, India

The observation of giant magnetoresistance has resulted in intense renewed interest in the electronic properties of doped $La_{1-x}A_xMnO_{3-\delta}$ (A = Sr,Ba,Ca,Pb and vacancies) and other transition metal oxides having strong electron correlations [1]. The rich phase diagram of $La_{1-x}Sr_xMnO_3$ shows a variety of phases like paramagnetic insulator, ferromagnetic metal, paramagnetic metal, spin canted insulator and ferromagnetic insulator as a function of doping x and temperature. In the range $0.18 < x < 0.30$ the system shows a temperature induced transition at $T_c(x)$ from the ferromagnetic metal (at $T < T_c$) to a paramagnetic insulator. This transition is qualitatively explained using the Zener's "Double Exchange" model.

In this paper we report the electronic Raman scattering from doped $La_{1-x}Sr_xMnO_3$ ($x = 0.3$). The samples show a phase transition from ferromagnetic metal to paramagnetic insulator at $T_c = 360$ K. The Raman measurements were carried out on well sintered polycrystalline pellet in the spectral range 200 cm^{-1} to 6000 cm^{-1}, at different temperatures from 20 K to 300 K. The spectra were recorded in back scattering geometry using a Spex Ramalog with photon counting detector and 514.5 nm line of an Argon ion laser (power density of $\sim$ 150 watts cm^{-2} at the sample). The pellet was mounted on the copper cold finger of the closed cycle helium refrigerator (RMC model 22C CRYODYNE) using thermally cycled GE varnish. The temperature of the cold finger was measured using a Platinum 100 sensor coupled to a home made temperature controller. The temperatures quoted are those of the cold finger and were measured to within a accuracy of $\sim$ 2 K.

Fig. 1 shows the Raman spectra at 35 K and 300 K. The broad band was fitted to the function

$$I(\omega) = A(n(\omega) + 1)Im[-1/\epsilon(\omega)] \tag{1}$$

where $\epsilon(\omega)$ is taken in the Drude model,

$$\epsilon(\omega) = \epsilon_\infty[1 - \omega_p^2/(\omega + i\Gamma\omega)^2]$$

The fitted curve is shown by the solid line in Fig. 1 with values of the plasmon frequency $\omega_p = 1817$ cm^{-1} and the relaxation rate $\Gamma = 4007$ cm^{-1}.

These values of Γ and ω_p can be used to obtain the resistivity $\rho(T)=4\pi\Gamma(T)/\omega_p^2\epsilon_\infty$. The value of resistivity thus obtained is 14.7 mohm-cm which is very close to the measured value [2]. Using $\omega_p^2 = 4\pi ne^2/(\epsilon_\infty m^* m_e)$ where n is the number density of the carriers of effective mass $m^* m_e$ and $\epsilon_\infty = 4.9$, $n/m^* = 1.81 \times 10^{20}$ cm^{-3}. Raman spectra measured at different temperatures have been fitted to equation (1) to obtain the temperature dependence of Γ and ω_p.

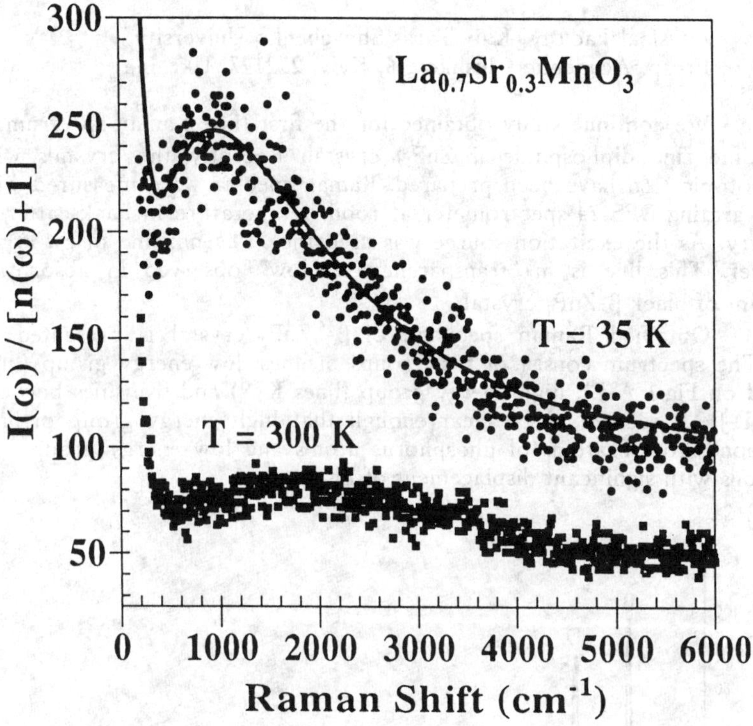

Fig. 1. Raman spectra of $La_{0.7}Sr_{0.3}MnO_3$ at 35 K and 300 K. The solid line shows the fit to the experimental data using equation (1).

We thank Prof. C.N.R. Rao for the samples and R. Mahendiran and Prof. A.K. Raychaudhuri for discussions. AKS thanks Department of Science and Technology and RG thanks CSIR for the financial assistance.

1. S.Jin, T.H. Tiefel, M. McCormack, R.A. Fastnacht, R. Ramesh, and L.H. Chen, Science **264**, 413(1994).

2. R. Mahendiran, S.K. Tiwary, A.K.Raychaudhuri, T.V. Ramakrishnan, R. Mahesh, N. Rangavittal and C.N.R. Rao, Phys. Rev. B (1996).

ISOTOPIC SHIFTS IN RAMAN SPECTRUM OF MONOCLINIC ZINC DIPHOSPHIDE CRYSTALS

I.M.Dmitruk*, M.M.Bilyi, I.S.Gorban, and Z.Z.Yanchuk

Physical Faculty, Kyiv Taras Shevchenko University,
Prosp. Academic Glushkov, 6, Kyiv, 252127, Ukraine

We continue study obtained for the first time Raman spectrum of monoclinic zinc diphosphide (β-ZnP_2) crystal. Recently the crystals with monoisotopic ^{64}Zn have been prepared. Raman spectra were measured with double-grating DFS-24 spectrometer at room temperature in backscattering geometry. As the excitation source was used the 514.5 nm line of cw argon ion laser. This line is in "transparency window" observed in absorption spectrum of black β-ZnP_2 crystal.

Obtained Raman spectrum of β-$^{64}ZnP_2$ crystal is presented on Fig.1. The spectrum consist of two groups of lines: low energy group (lines marked on Fig.1 A-E), high energy group (lines K-N) and five lines between them (F-J). It seemed to be clear enough that high energy group of lines corresponds to vibrations of phosphorus atoms and low energy group - to vibrations with significant displacement of Zn atoms.

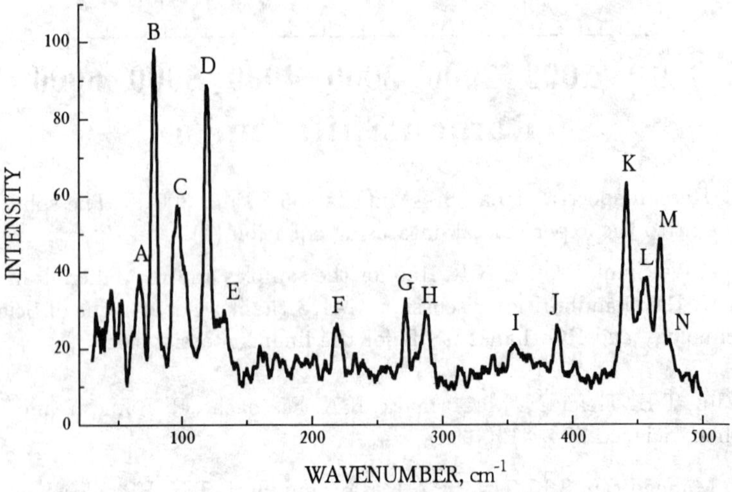

Fig.1. Raman spectrum of β-$^{64}ZnP_2$ single crystal. Excitation 514.5 nm. T=300 K.

The nature of intermediate frequency lines need additional investigation. We have measured the changes in phonon energies when natural isotope mixture zinc with mean mass number 65.46 was substituted by monoisotopic ^{64}Zn. These shifts are presented in Table 1 for the narrow lines of Raman spectrum for which ones can be determined with acceptable accuracy.

TABLE 1. Isotopic shifts in Raman spectrum of β-$^{64}ZnP_2$ crystal.

line	ω, cm^{-1}	$\Delta\omega$, cm^{-1}	$\frac{\Delta\omega}{\omega} \cdot 10^3$
B	78	0.2	2.6
D	119	0.6	5.0
F	218	0.9	4.1
G	272	0.5	1.8
H	287	0.5	1.7
J	388	0.6	1.5
K	441	0.4	0.9
L	456	0	0
M	467	0.7	1.5

The results of these measurements proved the interpretation of high frequency group of lines. Low energy lines have greater isotopic shifts, except line B for which absolute frequency shift is close to experiment accuracy. The origin of such low frequency line with little isotopic shift is not clear enough yet. We suppose that this line may correspond to spring-like vibrations of phosphorus quasi-spirals or to rotation of 4 nearest phosphorus atoms around Zn atom. Isotopic shift of line D correlates well with expected for Zn-P_2 or (ZnP_2)-(ZnP_2) relative motion and shift of line F with expected one for Zn-P oscillation.

RAMAN ACTIVE LATTICE VIBRATIONS AND THE TRAP ENERGY SPECTRA OF CRYSTALS

A. Gumenjuk, S. Kutovoy*

Kiev University, Physics Faculty,
Volodymirska, 64, 252601 Kiev, Ukraine

At investigation carefully by the thermoluminescence methods of the trap energy spectra in complex oxide and halide crystals the oscillator regularity of the thermal activation trap energy was found: $E_n = h\omega_i (n+1/2)$ with the energy of the vibration quantum $h\omega_i$ typical for lattice vibrations. In different crystals from one to five such series were observed (experimental data in detail for certain of crystals see [1]). By comparison of the TL-quants with spectroscopic data it turns out that for each $h\omega_{TL}$ the certain Raman line is corresponding. Discrepancy is no more than 3%. Most of $h\omega_{TL}$ correspond to the high-frequency lines of the first-order Raman spectra, others correspond to the second order Raman lines or to the weak Raman lines caused by impurities (local modes) or by symmetry violation (IR-modes). The values of the vibration TL-quants and corresponding positions of the Raman spectra lines are given in Table 1.

The oscillator regularity is observed in ion crystals and points to the strong interaction of a charge and the lattice, therefore one way of looking at the trap model is in terms of the polaron conception of charge carriers. The main features of it are below.

The polaron state function is usually written as $\Psi \sim \phi(\mathbf{r},\mathbf{R})*\chi(\mathbf{R})$, where $\phi(\mathbf{r},\mathbf{R})$ is an electron (hole) wave function in a tight-binding approach, and $\chi(\mathbf{R})$ is is the state function of the phonon system. The polaron energy states are considerably determined by the lattice states; in facts, interaction of a charge with the lattice forms equidistant polaron states in the parabolic potential pit at the distorted lattice. Since the bandwidth is sharply decreased to account for the polaron effect, the polaron movement in these crystals, as it follows from the estimation, even at comparatively small temperatures is realised by incorrelated jumping (activated by phonons) from one site of localisation to another. A self-trapped polaron or a polaron stabilized by the field of the defect can be activated by interaction with a definite type of phonons and jump as long as it will be captured by the excited state of the luminescence center; after that the recombination with giving of light may occur. The thermal activation energy is determined by the vibration level from which the tunneling of the charge in the excited state of the ionized recombination center can be possible. The level energy from what the jumping occurs depend on the distance between the trap and the recombination center, and increases with distance.

All investigated crystals have rather complicated lattices where it is possible to separate out comparatively isolated ion groups (here tetrahedra or octahedra), so that the interaction inside them is stronger than between the groups. Phonons which are the activators of jumping correspond, as a rule, to the high-frequency internal vibrations of such ion groups-quasimolecules. This demonstrates, in part, that the polaron movement is a result of electron (hole) localization just at the quasimolecule and a following jump to the adjacent one. The crystal defects as a whole are mostly determined by the defects of the quasimolecule.

There are certain troubles with the model, but despite that, it gives opportunity to be well on the way of the problem of chemical identification of the traps in ion crystals. Now there can be little doubt that the oscillator dependence of the trap activation energy is rather common regularity typical for ion crystals. Notice that although thermoluminescence is one of the basic methods of investigation of the defects energy structure, the analysis of Raman spectra (and IR-absorption) which makes it possible to determine unambiguosly the activation phonons values and symmetry, the order of process and others characteristics of the phonon system - and far more exactly than one from the TL-experiments - is very important.

TABLE 1. Values of the TL- and Raman quants for some investigated crystals (cm^{-1}).

Crystal	$h\omega_{TL}$	$h\omega_{Ram}$	Crystal	$h\omega_{TL}$	$h\omega_{Ram}$
$Ba_2NaNb_5O_{15}$	249	252	$CsCdCl_3$	113 250	115 248
$YAlO_3$	1048	-	$CdWO_4$	494 515	495 521
$Y_3Al_5O_{12}$	782 901 968 1274 1347	786 912 982 1309 1382	$ZnWO_4$	550 581 628	545 578 631

1. I. S. Gorban, A. F. Gumenjuk, S. Yu. Kutovoy, SPIE Proc., v.2113, p.173 (1994).

RAMAN SPECTRUM OF A TUNGSTEN BRONZE FAMILY MATERIAL KNSBN:Cu.

A. Savvinov[1*], I. Siny[1], R.S. Katiyar[1] and H.-C. Chen[2].

[1]Department of Physics, University of Puerto Rico, P.O.Box 23343
San Juan, Puerto Rico 00931-3343, U.S.A.
[2]Institute of Crystal Materials, Shandong University, Jinan 250100, PRC

The present work is devoted to the Raman scattering measurements in a single crystal from the tungsten bronze family of ferroelectrics, namely $(K_{0.5}Na_{0.5})_{0.2}(Sr_{0.75}Ba_{0.25})_{0.9}Nb_2O_6$ doped with CuO (KNSBN:Cu). The stability of the crystal structure does not appear to be disturbed if the doping concentration is controlled to be under 0.1 wt.% [1], in our sample it is 0.04 wt.%. The crystal was grown by Czochralski method. There were several reasons for growing such a complicated material. Some ferroelectrics from the tungsten bronze family like $Sr_xBa_{1-x}Nb_2O_6$ (SBN) have large electro-optic coefficients. However, the Curie temperatures of the most promising materials around $x \approx 0.75$ are rather low to prevent undesirable effects of depolarization in course of operating with a laser beam. A new ferroelectric crystal KNSBN has the advantage of a high Curie temperature, large electro-optic coefficients and good mechanical properties. However, the reported self-pumped phase conjugation reflectivity is relatively low [2] and the response time is much longer than that of $BaTiO_3$ and SBN [3]. The Cu^{2+} doping enhances the photorefractive properties of KNSBN [1]. The Raman scattering in the present work was used to find some differences in the phonon spectra of this new material, KNSBN:Cu, in comparison with the data obtained in other compounds of the tungsten bronze type.

The tungsten bronze structure consists of a framework of sharing corner oxygen octahedrons forming tetragonal (A1), pentagonal (A2) and trigonal (C) interstitials. A general formula is $(A1)_2(A2)_4(C)_4Nb_{10}O_{30}$. The unite cell of $Ba_2NaNb_5O_{15}$ (BNN) contains a large number of atoms so that 138 vibrational modes are expected in an ordered compound. The Raman spectra of KNSBN as well as those of analogous SBN and BNN show many lines which are rather broadened due to disordering and overlapping one another. Only the main features of the spectra will be pointed out here. Two sets of strong Raman lines of the $A_1(z)$ symmetry were observed around 260 and 640 cm^{-1}. These modes have been identified in BNN [4] and SBN [5] as internal modes of the NbO_3 octahedrons. Comparing those frequencies with our data on KNSBN, we can conclude that the binding energy of the oxygen framework is closer to that in SBN and lower than in the ordered BNN. The shape of two most intense line in the Raman spectra of KNSBN is complicated due to overlapping and, probably, disordering. An approximated width is about two times more than in BNN and is also close to that in SBN. In spite of a formal similarity between BNN and KNSBN (all and almost all the A1 and A2 sites are occupied, respectively), the vibrational characteristics of an oxygen framework in KNSBN are closer to those in SBN, where the A1 sites are only partially occupied. Also one can conclude that the vibrations of an oxygen framework in the tungsten bronze structure are considerably affected by disordering in the arrangement of cations in the A1, A2 and C interstitial channels, the latter is occupupied by the Cu ions.

More interesting is a comparison of the low-frequency parts of the Raman spectra with external modes due to the lattice vibrations involving cation motions relative to the oxygen octahedron framework. Three spectra of KNSBN obtained in the most important geometries are shown in Fig.1. There is a large difference between (ZZ) and (XX) spectra. Only the $A_1(z)$ photons contribute to the (ZZ) spectra. An additional contribution from the B_1 phonons is expected in the (XX) and (YY) spectra if the point group of crystal is 4mm. The low-frequency E line is seen in the (XZ) spectrum. The SBN crystal exhibits a similar spectrum although the low-frequency B_1 mode has not been found [5]. The BNN crystal belongs to the mm2 point group at room temperature, so there are only the A_1 modes in the diagonal spectra and the initial E mode from the 4mm high-temperature phase appears to be split in the off-diagonal spectra like (YZ) or (XZ) in the mm2 phase. So, one can conclude from the external mode analysis, that the KNSBN crystal has the tetragonal structure characteristic of the tungsten bronze type ferroelectrics in spite of a formal similarity to BNN. The BNN crystal has a rather unusual sequence of seven phases including incommensurate structures [6]. Even slight deviations from stoichiometry may produce some discommensurations. The stoichiometric formula of KNSBN implies the arrangement of the K, Na, Sr and Ba cations without any regular order. However, it seems that the structure of KNSBN does not predispose to discommensurations as in the latter case. The present study shows that the structure of KNSBN remains to be stable in a wide temperature interval below the Curie point.

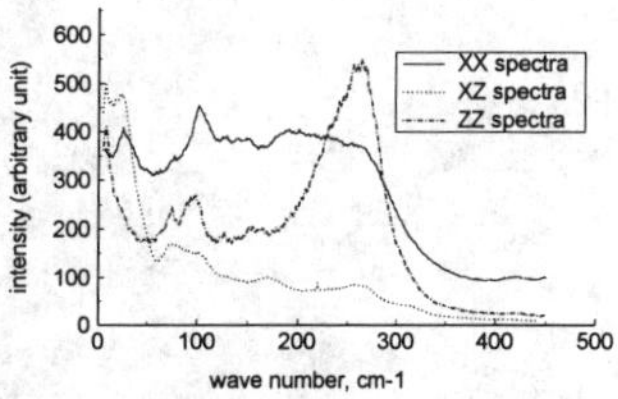

Fig.1. Raman spectra of the KNSBN crystal

Acknowledgments. This work was supported by NSF Grant OSR-9452893 and DOE Grant DE-FG02-94-ER75764.

1. Xuefeng Yue et al., Appl. Phys. B **53**, 319 (1991)
2. J. Rodriguez et al., Appl. Opt. **26**, 1732 (1987)
3. M.J. Miller et al., Opt. Lett. **12**, 340 (1987)
4. A. Boudou and J. Sapriel, Phys. Rev. B **21**, 61 (1980)
5. R. E. Wilde, J. Ram. Spectr. **22**, 321 (1991)
6. A.Shawabkeh and J.F. Scott, Phys. Rev. B **43**, 10999 (1991)

SECTION 20

SUPERCONDUCTORS

SYMMETRY-DEPENDENT POWER LAWS OF THE ELECTRONIC RAMAN RESPONSE IN CUPRATE SUPERCONDUCTORS

R. Nemetschek, M. Opel, B. Stadlober, and R. Hackl

Walther-Meissner-Institut, 85748 Garching, Germany

Inelastic scattering of light from carrier excitations in normal- and superconducting copper oxides have been studied in some detail since the discovery of this material class with particular emphasis placed on the temperature range below T_c in order to determine the **k** dependence of the superconducting order parameter $\Delta(\mathbf{k})$ [1]. Recently a study on overdoped $Bi_2Sr_2CaCu_2O_{8+\delta}$ (Bi2212) with $\delta \approx 0.24$ was proposed [2]. As a general trend the spectra for overdoped samples look more isotropic than those for materials with maximal T_c.

The Raman spectra of cuprate superconductors consist in general of a broad continuum and superimposed phonon lines. Below T_c the continuum is redistributed exhibiting gap-like structures with a reduced intensity at low energies and a pile up at an energy $\hbar\omega_p^\mu$ due to the breaking of Cooper pairs which may depend on the symmetry μ selected by the light polarizations. However, in contrast to the BCS prediction the intensity increases continuously from zero energy in the cuprates. This fundamental difference is still present in heavily overdoped Bi2212 with $\delta \approx 0.25$ and a T_c of 55 K although the peak frequencies have merged together (Fig. 1). The peak frequencies seem to suggest that the material is isotropic. A well defined onset of the scattering at an energy given approximately by $2\Delta_o$ which is then to be expected is however missing. The question arises how these apparent inconsistencies are to be interpreted.

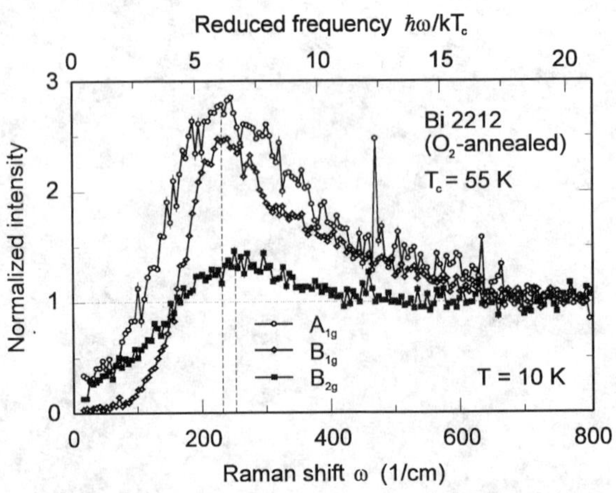

Fig.1: Electronic Raman spectra for overdoped Bi2212 (T_c=55K).

It has been shown recently that the Raman data of optimally doped materials can be explained quantitatively if a superconducting energy gap with $d_{x^2-y^2}$ symmetry, $\Delta(\mathbf{k}) = \Delta_0(\mathbf{k}_x^2 - \mathbf{k}_y^2)$, is assumed [3]. The theory even in its simplest approximation predicts the symmetry dependence of the peak frequencies and, in addition, power laws for the low energy part of the spectra at $T = 0$ and for the intensity at $\omega = 0$ at finite temperatures. The basic idea is to separate the **k** dependence of the Raman vertex γ from its variation with the incoming and outgoing light frequency, ω^I and ω^S,

$$\gamma\left(\mathbf{k};\omega^I,\omega^S\right)=\sum_{L,\mu}\gamma_L^\mu\left(\omega^I,\omega^S\right)\Phi_L^\mu(\mathbf{k})$$

where $\Phi_L{}^\mu(\mathbf{k})$ is a complete set of functions which transform according to the point group symmetry of the crystal and are, in the most general case, orthonormal over the entire Brillouin zone. $\gamma_L{}^\mu(\omega^I,\omega^S)$ determines the intensities in the different symmetries μ and depends in a complicated way on the band structure. Due to the limited frequency range the power-law behavior is completely given by the interplay between the **k** dependences of the gap and that of the functions $\Phi_L{}^\mu(\mathbf{k})$, whereas the position of the peaks can vary strongly with the band structure [4]. The low-frequency part of the spectra (see Fig. 1) is therefore believed to contain the most relevant information. It is shown in a log-log plot in Fig. 2. The observed frequency power laws are in close agreement with the predictions for slightly disordered d-wave superconductors, in particular the crossover from a linear to a cubic frequency dependence of the intensity at the B_{1g} symmetry [5]. The theoretical prediction for the temperature power laws are displayed in Fig. 3 along with the data from an optimally doped sample. Similar behavior is observed in the overdoped material. It is concluded, that overdoped Bi2212 fits also in the d-wave scenario.

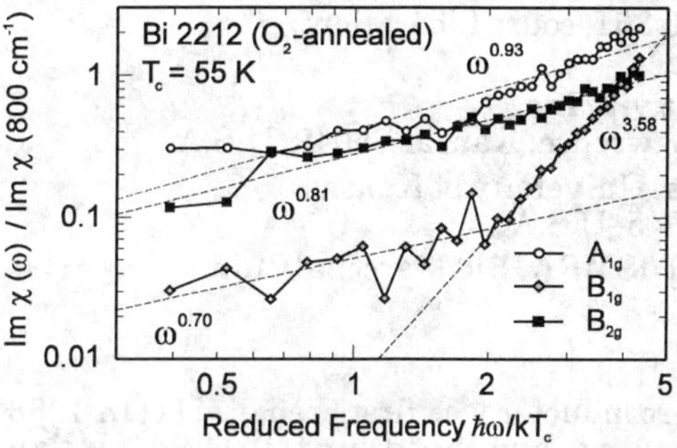

Fig.2: Low energy part of the electronic Raman spectra for overdoped Bi2212 at 10K.

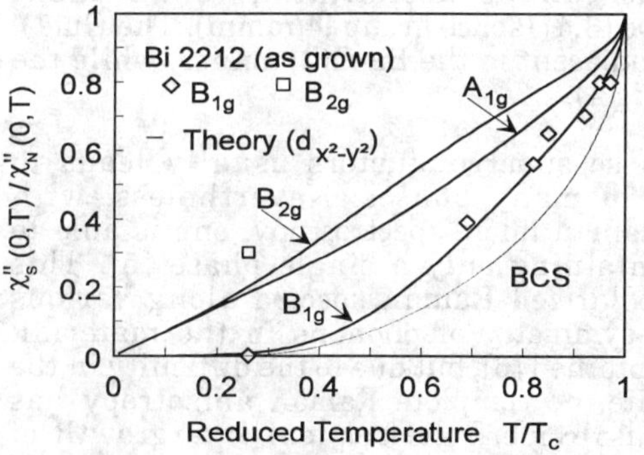

Fig.3: Theoretical curves (lines) and experimental data (symbols) for the normalized response.

Support by the Bayerische Forschungsstiftung via the Forschungsverbund Hochtemperatur-Supraleiter (FORSUPRA) is gratefully acknowledged.

1. D. Einzel and R. Hackl, J. Raman Spectroscopy **27**, #3/4 (1996).
2. C. Kendziora and A. Rosenberg, Phys. Rev. B **52**, R9867 (1995).
3. T.P. Devereaux et al., Phys. Rev. Lett. **72**, 396 (1994).
4. T.P. Devereaux, Proc. of the SPIE meeting in San Jose, Jan. 27-Feb. 2, 1996.
5. T.P. Devereaux, Phys. Rev. Lett. **74**, 4313 (1995).

PHONONS AND SUPERCONDUCTIVITY IN $T\ell_{2223}$ AND $(Hg/T\ell)_{2223}$

Pham V. Huong[a]*, Hui Wang[a], Ying Xin[b], Kai Wai Wong[c], C.Martin[d], and C.Michel[d]

[a] Laboratoire de Spectroscopie Moléculaire et Cristalline, Université Bordeaux I, 351, cours Libération, 33045 Talence, France

[b] Midwest Superconductivity Inc., 1315 Wakarusa Dr., Lawrence, Kansas 66049, U.S.A.

[c] Department of Physics, University of Kansas, Lawrence, Kansas 66045, U.S.A.

[d] Laboratoire CRISMAT/ISMRA, Bld Maréchal Juin, 14050 Caen, France

$T\ell_2Ca_2Ba_2Cu_3O_{10}$ superconductor was first prepared [1] in 1988 and has a Tc = 124 K. This record hold until 1994, when $(Hg/T\ell)_2Ca_2Ba_2Cu_3O_{10}$ with Tc = 135 K is discovered [2].

All these compounds can be assimilated to an ideal body-centered-tetragonal structure [3,4] (space group I4/mmm). The Cu(1), O(1) planes are the Cu-O planes adjacent to the Ba-O(2) planes, while the other Cu-O plane is Cu(2), O(4) (Fig.1).

The preparation of these superconductors usually leads to crystals containing a mixture of many phases. Nevertheless, with micro-physical techniques like Raman micro-spectroscopy, one is able to illuminate small crystallites containing only a single phase [5]. This technique also offers to record polarized Raman spectra along various crystal axes and to identify the symmetry of phonons in the material. Partial Raman data are already obtained [6], but due to the difficulty in the orientation of the micro-crystallite, no complete Raman anisotropy has been obtained. The difficulties also come from frequent intergrowth of different phases in the crystal.

As the insertion of mercury in thallium based superconductor, $T\ell_{2223}$, increases sharply its critical temperature, it could be interesting to record and compare the phonons of the two materials and to deduce possible correlations between phonons and superconductivity.

The single crystals of $T\ell_{2223}$ and $(Hg/T\ell)_{2223}$ were prepared with methods described earlier [2]. The polarized Raman spectra were recorded at room temperature on a micro-spectrometer Dilor-OMARS equipped with a CCD detector and an ion argon laser source Spectra-Physics emitting at 5145 nm. The spatial resolution was better than 0.5 μm^2.

The Raman spectrum of $(Hg/T\ell)_{2223}$ is very similar to that of $T\ell_{2223}$ and presents strong anisotropy between the *c* axis and the two others (Fig.1). All phonons have nearly the same frequencies except for those involving the Cu(1) motion in opposition of phase with Ba which is at 160 cm^{-1} in $(Hg/T\ell)_{2223}$ instead of 153 cm^{-1} in $T\ell_{2223}$, and the $T\ell$ motion in phase with Ba which appears at 96 cm^{-1} in the former and at 107 cm^{-1} in undoped compound.

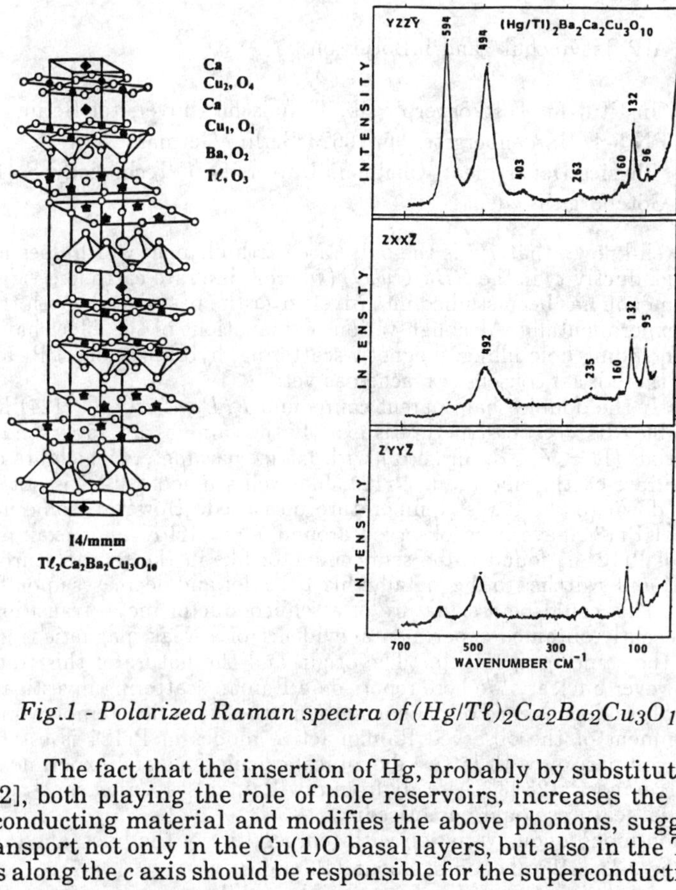

Fig.1 - Polarized Raman spectra of $(Hg/T\ell)_2Ca_2Ba_2Cu_3O_{10}$

The fact that the insertion of Hg, probably by substitution on $T\ell$ sites [2], both playing the role of hole reservoirs, increases the Tc of the superconducting material and modifies the above phonons, suggests that the transport not only in the Cu(1)O basal layers, but also in the $T\ell$OCu(1) groups along the *c* axis should be responsible for the superconductivity.

[1] Sheng L.Z., Hermann A.M., Nature, **332**, 55 (1988).
[2] Goutenoire F., Maignan A., Van Tendeloo G., Martin C., Michel C., Hervieu M., Raveau B., Solid State Commun., **90**, 47 (1994).
[3] Shimakawa Y., Kubo Y., Hanako T., Nakabayashi Y., Igarashi H., Physica C, **156**, 97 (1988).
[4] Kulkarni A.D., de Wette F.W., Prade J., Schroder U., Kress W., Phys.Rev. B, **41**, 6409 (1990).
[5] Huong P.V., Verma A.L., Phys.Rev. B, **48**, 9869 (1993).
[6] McCarty K.F., Morosin B., Ginley D.C., Boehme D.R., Physica C, **157**, 135 (1989).

PHONON ANOMALIES AROUND A SEMICONDUCTOR-METAL LIKE TRANSITION IN $PrBa_2Cu_4O_8$

A.P. Litvinchuk[a] and L.Börjesson[b*]

[a]Institut für Festkörperphysik, Technische Universität Berlin, Sekr. PN 5-4, Hardenbergstr. 36, 10623 Berlin, Germany
[b]Physics Department, Chalmers University of Technology, S-412 96 Göteborg, Sweden

It is well-known that Pr is the only rare-earth element which does not exhibit superconductivity in the $RBa_2Cu_3O_7$ (where R is a rare-earth ion) structure. This phenomenon has been studied intensively over the past decade both theoretically and experimentally. Although various explanations of the effect have been proposed, including hole-filling, magnetic scattering, hybridization of Pr and Cu bands etc., there is no consensus reached as yet.

Recently the double-chain parent compound $PrBa_2Cu_4O_8$ (Pr124) has become available. Its crystal structure is exactly the same as for superconducting R124 materials (R = Y, Nd, Sm, etc.) with lattice parameters which are dependent on the rare-earth ionic radii. Pr124, like well studied Pr123, is not superconducting down to the lowest temperature measured. However, some unusual properties of Pr124 have been observed around $T^* \approx 160K$. For example, the resistivity of Pr124 is found to be semiconductor-like in the temperature range above 160 K and switches to the metallic-like behavior and decreases upon further cooling [1]. This could be a signature of a semiconductor-metal transition. We have also recently obtained experimental evidence of a weak magnetic transition [2] around the same temperature. The origin and the nature of this transition remains however unclear. We here report on a Raman scattering investigation of the temperature dependence of phonons around the transition temperature T^*.

Assignment of the observed Raman-active modes in Pr124 is established based on a comparison with lattice dynamical calculations and experimental results for Y124 and Pr123 [3]. Already from raw Raman scattering spectra in Fig. 1 two unusual features could be noticed:

i) a non-monotonous variation with temperature of the lineshape (and the width) of the z-axis Ba-phonon at about 110 cm^{-1};

ii) upon cooling there is only negligibly small narrowing of the B_{1g}-like plane-oxygen mode at about 300 cm^{-1}.

In Fig.2 we display as an axample the temperature dependence of the phonon parameters of the Ba-mode as obtained by fits of the lineshape to an asymmetric Fano profile. By dotted lines the expected behavior of the phonon frequency and the linewidth is shown in case usual anharmonicity-related processes predominate. A rather abrupt change of the slope in the temperature dependence of the frequency, linewidth and also line asymmetry q is observed around T^*. *All other* Raman-active phonons also exhibit anomalous behaviour around T^*. These facts clearly indicate the existence of an additional channel for phonon decay beside usual anharmonicity.

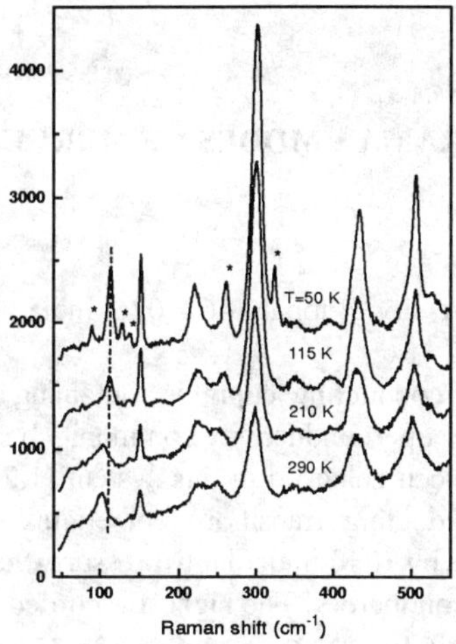

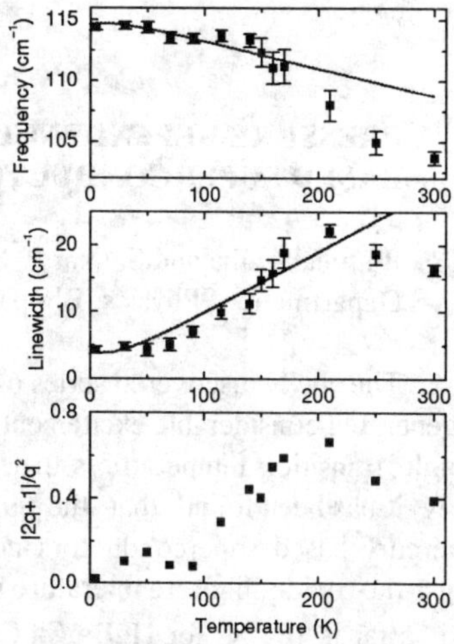

Fig. 1. Raman scattering spectra of Pr124 at various temperatures. Asterisks mark the position of the phonon peaks due to $BaPrO_3$ impurities, which appear in the low temperature spectra only.

Fig. 2. (a) Temperature dependence of the frequency, linewidth and lineshape asymmetry parameter q of the Ba-mode. Dashed curves represent fits to anharmonic phonon decay processes.

It is important to note that similar anomalies of phonon parameters for Ba and B_{1g}-like modes, as well as for some of the infrared active phonon modes have been observed in *superconducting* Y124 at almost the same temperature [4,5]. They were interpreted in terms of a spin gap opening and phonon coupling to spin excitations. A similarity in the temperature behavior of phonons for Pr124 and Y124 may suggest that the effects in the two compounds are of similar origin. It could be that the driving force of these unusual phenomena is a structural instability of the R124 lattice around T^*. In fact, weak structural changes around T^* has been observed for superconducting Ca-doped Y124 [6].

References

1. N. Seiji et al., Physica C**227**, 377 (1994).
2. A.P. Litvichuk, L. Börjesson and P. Svedlindh, to appear.
3. M. Käll et al., Phys. Rev. **B53**, 3570 (1996).
4. A.P. Litvinchuk et al., Solid State Commun. **83**, 343 (1992); Z. Phys. B**92**, 9 (1993).
5. M. Käll et al., Physica C**225**, 317 (1994).
6. H. Schwer et al., Physica C**235**, 801 (1994).

PRESSURE DEPENDENCE OF RAMAN MODES OF MERCURY BASED SUPERCONDUCTORS

Prafulla K. Jha and Sankar P. Sanyal
Department of Physics, Bhopal University, Bhopal - 462 026, India

The newly discovered series of superconducting cuprates containing mercury have generated considerable excitement in the superconductivity community because of the high transition temperatures that have been achieved in this system [1,2]. More recently, it has been found that the superconducting transition temperature (T_C) of the mercury based superconductors increases by 30 K under high pressure, the largest amongst the oxide high temperature superconductors. The highest recorded T_C reported so far is 164 K for $HgBa_2Ca_2Cu_3O_{8+\delta}$ (Hg -1223) at 30 GPa [3]. Hence, it is important to investigate the mechanism of the pressure induced increase in T_C to understanding the nature of high T_C superconductivity. Very recently, Yang et al [4] have studied the effect of pressure on Raman modes of $HgBa_2CuO_4$ (Hg-1201) and $HgBa_2Ca_2Cu_3O_8$ (Hg-Hg1223) compounds up to 10 GPa. Their results show that the Raman frequency of O_A - A_{1g} mode increases from 592 cm^{-1} at 1 atm to 626 cm^{-1} at 7.4 GPa for Hg- 1201 and 585 cm^{-1} at 1 atm to 649 cm^{-1} at 9.0 GPa for Hg-1223 compounds.

In the present paper,we report for the first time the calculated results for the effect of pressure on frequency of Raman O_A-A_{1g} mode in Hg-1201 and Hg-1223 by using a simple lattice dynamical model theory. The study of lattice vibrations of the high T_C superconductors is of importance not only for the understanding of the physical nature and interatomic forces in these compounds, but also for the role played by the phonons in the superconducting phenomena.

We consider an unscreened rigid ion model, in which the ions are considered as rigid during the vibrations[5]. This model has been used successfully earlier for several high temperature oxide superconductos[6] and gives a very qualitative account of many of the observed structural and dynamical properties. The crystal energy is given as

$$V_{kk'} = \frac{e^2 Z(k)Z(k')}{4\pi\varepsilon_0 r(kk')} + a \exp \frac{-b\, r(kk')}{R(k)+R(k')} \qquad (1)$$

Here $e^2/4\pi e_0$ = 144 eV/nm, a = 1822 eV, b= 12.364 and $r_{kk'}$ = is the distance between the two atoms of species k and k'. Z(k) and Z(k') are the effective charges situated at the ionic positions. The R(k) and R(k') are reffered as effective radii and treated as adjustable parameters along with the effective charges. while the relative values of effective charges and radii are determined from the requirement of producing a stable crystal structure, they have been scaled to ensure the maximum of the phonon frequencies to be positive and approximately of right magnitudes. Chrage neutrality condition for the unit cell was imposed as required in the rigid ion model.

The calculated results for the effect of pressure on Raman frequency of O_A-A_{1g} mode of Hg-1201 and Hg-1223 are plotted in Figs. 1and 2 respectively and compared with experimental Raman messurments [4]. The Figs. 1 and 2 reveal that the frequency of Raman mode increases as pressure is increased for both the compounds. This suggests that the same mechanism is involved in the enhancement of T_c in Hg-1201 and Hg-1223. A simple explanation for this can be thought as when pressure is applied, lattice shrinks and hence the change in Hg- O_A bond strength which results in the enhancement of Raman frequency of O_A-A_{1g} mode of these componds. A remarkable feature observed in the case of Hg-1223 is that this compound shows a sudden increase of the rate above 4.5 Gpa (in experiemnt 5.0 Gpa). In fact, at this pressure the Hg-O_A orbital energy crosses Fermi energy E_F [7] which leads to a large charge transfer from CuO_2 planes to Hg-O_A layers as the pressure is increased and hence enhancement in the Raman frequency of Hg-1223.

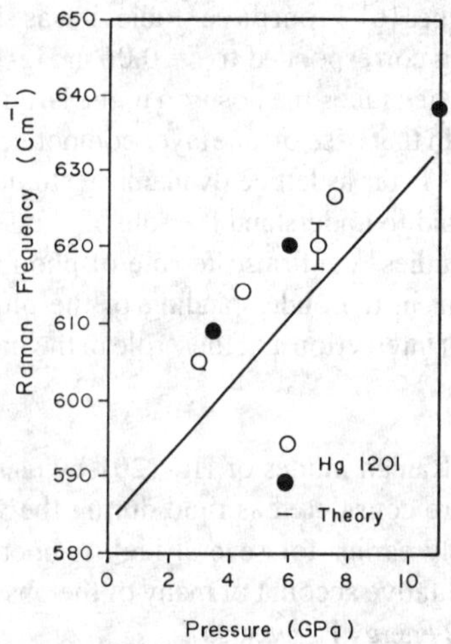

Fig. 1: Raman frequency of O_A- A_{1g} mode of of O_A-A_{1g} mode Hg-1201 as function of pressure

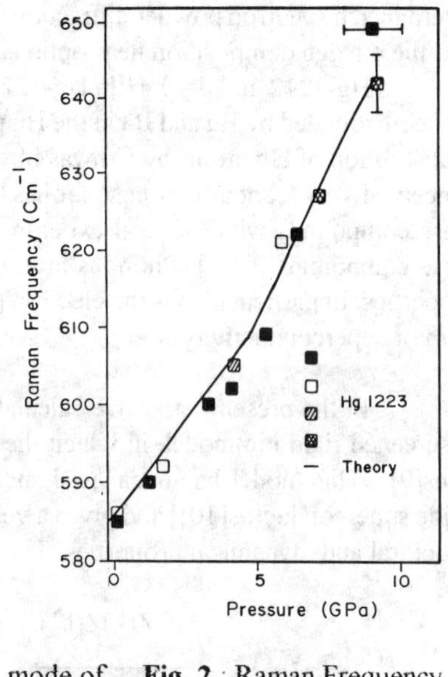

Fig. 2 : Raman Frequency Hg-1223 as funcion of pressure

1. S. N. Putlin, E.V. Angtipov, O. Chamissen and M. Merezic, Nature **362,** 226 (1993).
2. A. schilling, M. Cantoni, J. D. Guo and H. R. Ott, Nature **363**, 56 (1993).
3. C. W. Chu, L.Gao, F. Chen, Z. J. Huang, R. L. Meng and Y. Y. Xue, Nature **365**, 323 (1993).
4. In Sang Yang et al. (to be published)
5. S. L. Chaplot, Phys. Rev. **B37**, 7435 (1988). Phys. Rev. **B42**, 2149(1990)
6. P. K. Jha and S. P. Sanyal, Physica **C** (In Press).
7. D. L. Novikov, O. N. Mryasov and A. J. Freeman, Physica **C222**, 38(1994).

RAMAN MODES OF $HgBa_2CuO_4$ HIGH TEMPERATURE SUPERCONDUCTORS

Prafulla K. Jha and Sankar P. Sanyal

Department of Physics, C. V. Raman Building, University of Bhopal, Bhopal - 462 026, INDIA.

Since the discovery of Superconductivity in mercury based oxide high temperature superconducting compounds [1,2], considerable efforts have been made to understand the mechanism of superconductivity in these compounds [3-5]. The great interest in the mercury based oxide superconductors is due to not only their remarkably high critical transition temperature but also their structural similarity with thallium and bismuth - based compounds. At present there is interest in understanding their structural and physical properties. The crystal structure for the whole series has been recently determined by neutron powder diffraction techniques [6]. From these studies it was shown that the oxygen composition near optimal doping corresponded to $\delta = 0.06$ in Hg-1201, $\delta = 0.2$ in Hg-1212 and d= 0.41 in Hg-1223. Oxygen takes the position in the large open space surrounded by Hg and Ba in the Hg plane. In that case of one layer compound only , substitution of Hg atoms by Cu was observed. As far as lattice dyanamical studies are concerned very scant attention so far has been paid to understand the role of phonons in these compounds while several experimental studies confirms the role of phonons in these compounds [7-8]. Phonons are important in the understanding of the physical properties. In particular so is the electron phonon interaction and their role in the mechanism of superconductivity.

In the present paper we calculate the Raman modes of Hg-1201 by using an unscreened rigid ion model, in which the ions are considered as rigid during the vibrations[9].. This model has been used successfully earlier for several high temperature oxide superconductos[10]] and gives a very qualitative account of many of the observed structural and dynamical properties. The crystal energy is given as

$$V_{kk'} = \frac{e^2 Z(k)Z(k')}{4\pi\varepsilon_0 r(kk')} + a \exp \frac{-b\, r(kk')}{R(k)+R(k')} \tag{1}$$

Here $e^2/4\pi e_0 = 144$ eV/nm, a = 1822 eV, b= 12.364 and $r_{kk'}$ = is the distance between the two atoms of species k and k'. Z(k) and Z(k') are the effective charges situated at the ionic positions. The R(k) and R(k') are referred as effective radii and treated as adjustable parameters along with the effective charges. while the relative values of effective charges and radii are determined from the requirement of producing a stable crystal structure, they have been scaled to ensure the maximum of the phonon frequencies to be positive and approximately of right magnitudes. Charge neutrality condition for the unit cell was imposed as required in the rigid ion model. Model parameters are listed in table I.

TABLE I. Interatomic potential parameters for Hg- 1201 compound.

Atom	Z	R(nm)
Hg	0.77	1.05
Ba	1.63	2.04
Cu	1.87	1.39
O	-1.46	1.88

Our calculated results for the Raman active modes for $HgBa_2CuO_4$ are given in table II and compared with the available Raman data from recent Raman Scattering experiment. There is generally a fair agreement between the calculated results with the observed modes. We have obtained four Raman active modes at frequencies 105, 130, 535, and 583 cm^{-1} which are assigned to Ba and O(2) atoms by using a factor group analysis.

TABLE II. Raman active phonon frequencies of Hg-1201 compound.

Raman Mode	Present work cm^{-1}	Experimental Values cm^{-1} [7]
Ba A_{1g}	105	108
Ba E_g	130	158
O(2) E_g	535	570
O(2) A_{1g}	583	592

1. S. N. Putlin, E.V. Angtipov, O. Chamissen and M. Merezic, Nature **362,** 226 (1993).
2. A. schilling, M. Cantoni, J. D. Guo and H. R. Ott, Nature **363**, 56 (1993).
3. C. W. Chu et al Physica **C213**, 261(1993).
4. J. L. Wagner, P. G. Radaelli, D. G. Hinks, GJ. D. Jorgensen, J. F. Mitchell, B. Dabrowski, G. S. Knapp and M. A. benu, Physica **C210,** 447 (1993).
5. C. W. chu, L. Gao, F. Chen, Z. J. Huang, R. L. Meng and Y. Y. Xue , Nature**365**, 323(1993)
6. M. Marezio, E. V. Antipov, J. J. Capponi, C. Chaillout, S. Loureiro, S. N. Putlin, A. Santoro, and J. L. Tholene, Physica **B197** , 570 (1994).
7. N. H. Hur, H. G. Lee, J.H. park, H.S. shin and I. S. Yang, Physica **C218**, 365 (1994).
8. I. S. Yang, H. G. Lee, H. S. Shin, J.H. Park, S. I.Lee and S. Lee, Physica **C222**, 386 (1994).
9. S. L. Chaplot, Phys. Rev. **B37**, 7435 (1988). Phys. Rev. **B42**, 2149(1990)
10. P. K. Jha and S. P. Sanyal, Physica **C** (In Press).

RAMAN STUDY OF THE ROLE OF OXYGEN ATOMS IN THE MERCURY-BASED SUPERCONDUCTORS

In-Sang Yang*

Department of Physics, Ewha Womans University, Seoul, 120-750, Korea

The role of the oxygen at the center of the four mercury atoms (O_δ) and the apical oxygen between the mercury and the plane copper (O_A) in the mercury-based superconductors are investigated using Raman spectroscopy. These oxygens are believed to play an important role in high T_c superconductivity. Our results indicate that the O_δ atoms affect only the local environment of O_A, not changing the Hg-O_A bondstrength.

Mercury-based superconductors $HgBa_2Ca_{n-1}Cu_nO_{2n+2+\delta}$ (n=1,2,3), hereafter Hg-1201, Hg-1212, and Hg-1223, respectively, were synthesized by sintering stoichiometric mixture of appropriate precursors as previously described.[1] Bulk samples of $Hg_{1-x}Tl_x$-1223 (x=0.0, 0.25, 0.5, 0.75, 1.0) were obtained from the similar procedure.[2] The quality and purity of the samples were checked by XRD and SQUID magnetometry. Micro-Raman measurements show typical spectra for the mercury-based superconductors, confirming the high quality of the samples. Raman measurements were done at room temperature using a Jobin Yvon U1000 double monochromator in a back-scattering geometry. The spectra were excited with the 514.5 nm line of an argon-ion laser focused on freshly-exposed surfaces of the polycrystalline samples through a metallurgical microscope.

Raman spectra of all the mercury-based superconductors show two strong peaks at $\sim$ 570 cm^{-1} and $\sim$ 585 cm^{-1}.[3,4] The 585 cm^{-1} mode is unambiguously from the A_{1g}-type vibration of the apical oxygen O_A. The origin of the 570 cm^{-1} mode, however, had been debated.[5,6] We investigated the origin of the 570 cm^{-1} mode by measuring the Raman spectra of the Tl-doped Hg-1223 system, $Hg_{1-x}Tl_x$-1223 (Fig. 1). Three distinctive Raman peaks are observed at $\sim$ 520, 570, and 585 cm^{-1}, and determined to have the A_{1g} symmetry from the polarization-dependence measurements. It is noted that the relative intensity of the 570 cm^{-1} mode to that of the 585 cm^{-1} mode increases as the Tl-doping level increases, while the Raman frequencies of these modes show only slight changes. We found that the sum of the intensities of the 570 and 585 cm^{-1} modes is proportional to the Hg content, i.e. $1-x$ in $Hg_{1-x}Tl_x$-1223. This result strongly supports that both 570 and 585 cm^{-1} modes are due to the A_{1g}-type vibration of the O_A in the Hg-O_A-Cu bond of the samples.

It is believed that Tl-doping controls the amount of the O_δ in the Hg/Tl-O_δ plane. Therefore, the number of the neighboring O_δ atoms around a particular O_A atom should increase as Tl doped. It is proposed that the 585 cm^{-1} mode is from the O_A without immediate O_δ neighbors, while the 570 cm^{-1} mode is from the O_A with the O_δ neighbors. To test our proposal, we measured Raman spectra of the samples annealed in vacuum at relatively low temperatures. The relative intensity of the 570 cm^{-1} mode to that of the 585 cm^{-1} mode decreased, as expected. After annealing in vacuum, it is expected that the amount of the O_δ decrease and, consequently, the number of the O_A with the immediate O_δ neighbors decrease. From these Raman measurements of the Tl-doped Hg-1223 and annealing treatments, we suggest that these peaks are due to two A_{1g}-type Raman frequencies for the O_A atoms in the Hg-1223 system.[7] These two-peak behavior of the O_A atoms is common in the Raman spectra of the Hg-12(n-1)n systems. Only the relative intensity of the 570 cm^{-1} mode to that of the 585 cm^{-1} mode is different for different Hg-12(n-1)n systems.

Under pressure, the T_cs of the mercury-based superconductors are known to increase significantly ($\sim$ 30 K). The effect of pressure on the Raman frequencies of the apical oxygen O_A is observed (Fig. 2). The increase rate of the force constants $\Delta(\omega^2/\omega_0^2)/\Delta P$ is found to be strongly correlated with $\Delta T_c/\Delta P$. For Hg-1223, especially, there is an abrupt change in $\Delta(\omega^2/\omega_0^2)/\Delta P$ at $\approx$5 GPa, at which the T_c versus P curve changes the slope.[8] The Raman study on the Tl-doped Hg-1223 supports that the content of the O_δ affects only the relative intensity of the 570 cm^{-1} mode to that of the 585 cm^{-1} mode, without changing the frequencies of the two modes. These results indicate that the Raman frequency of the O_A mode is independent of the O_δ content. Therefore, the increase of the O_A A_{1g} mode under high pressure seems to be irrelevant with the possible change in the O_δ content under pressure.

These results inform us about the role of O_δ and O_A in the mercury-based superconductors. The occupancy of the O_δ affects the local environments of O_A without changing the strength of the Hg-O_A bond. The main role of the O_δ seems to be to supply charges directly into the CuO_2 planes. High pressure Raman measurements show that there is a close connection between the increase of T_c and the increase of the bondstrength of the Hg-O_A bond in the mercury-based superconductors. Under high pressure, the Hg-O_A bond strengthens, which might lead to supplying holes into the CuO_2 planes toward optimal values via charge transfer between the Hg-O_A bond and the CuO_2 planes.

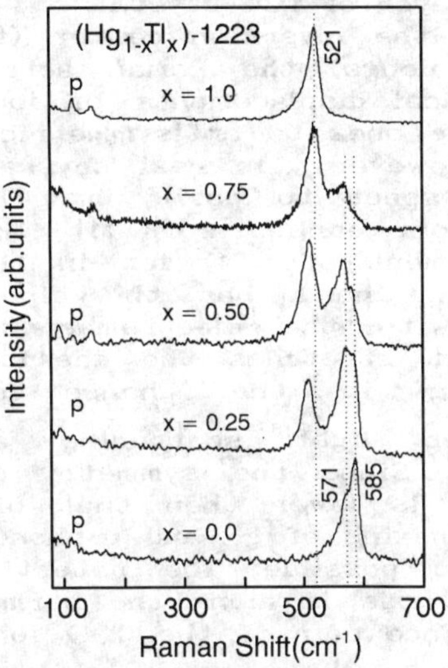

Fig. 1. Raman spectra of $Hg_{1-x}Tl_x$-1223.

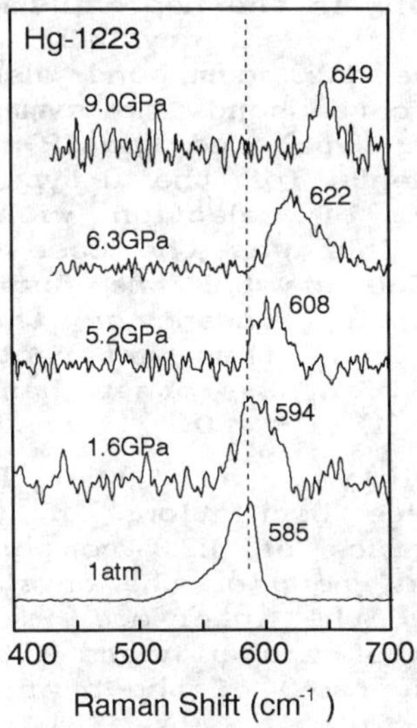

Fig. 2. Raman spectra of Hg-1223 at high pressures.

1. S. Lee, M.O. Mun, M.K. Bae, and S.I. Lee, J. Mater. Chem. 4, 991 (1994).
2. N.H. Hur, N.H. Kim, K.W. Lee, K.H. Yoo, Y.K. Park, and J.C.Park, Physica C 234, 19 (1994).
3. Y.T. Ren, H. Chang, Q. Xiong, Y.Q. Wang, Y.Y. Sun, R.L. Meng, Y.Y. Xue and C.W. Chu, Physica C 217, 273 (1993).
4. In-Sang Yang, H.G. Lee, H.S. Shin, J.H. Park, S. Lee, S.I. Lee, Physica C 222, 386 (1994).
5. Y.T. Ren, H. Chang, Q. Xiong, Y.Y. Xue, and C.W. Chu, Physica C 226, 209 (1994).
6. M.C. Krantz, C. Thomsen, Hj. Mattausch, and M. Cardona, Phys. Rev. B 50, 1165 (1994).
7. In-Sang Yang, H.G. Lee, Nam H. Hur, and Jaejun Yu, Phys. Rev. B 52, 15 078 (1995).
8. In-Sang Yang, H.S. Shin, H.G. Lee, S.J. Jeon, H.S. Ahn, Jaejun Yu, S. Lee, S.I. Lee, and Nam H. Hur, Phys. Rev. B 51, 644 (1995).

THE RAMAN SPECTROSCOPY OF 1:2:3-CRYSTAL NEAR-SURFACE LAYER

V.V.Rumyantsev* and E.Ya.Shtaerman

Donetsk Physico-Technical Institute, Ukrainian Academy of Sciences, 340114 Donetsk, Ukraine.

Because of the small coherence length ξ of the YBaCuO-type high-temperature superconductors the experimental data are often determined by layer of width $d \sim \xi$. The Raman light scattering is a good means for diagnostics of the state of such near-surface region. The paper theoretically studied both spontaneous and stimulated light scattering at surface polaritons propagating in the near-surface layer of 1:2:3-crystal.

In ideal crystals with the inversion center (to which the 1:2:3 compound also belongs) the Raman active phonons corresponds to symmetrical displacements of ions (of the g-type) and the IR-active ones to antisymmetrical displacements (of the u-type). However , in real crystals the rules of selection with respect to parity may be violated. This was the case of our earlier work [1] - in the YBaCuO crystal the displacement of O(4) ion in the bridge having vacancy,on the one hand, has the dipole moment (and, thus interacts with the electromagnetic field) and on the other hand, it stretches the chemical bond Cu(1) - O(4) resulting in the change of polarizability $\hat{\chi}$ and the Raman light scattering at dipoleactice oscillations of O(4). Since the symmetry of (110) surface of 1:2:3 compounds is lower then that off the space group of the crystal, mixing of g- and u-phonon modes at the interface is also possible. The potential energy U of z-oscilations (the z-axis is along the normal to the surface) of the near-surface pair of the O(4) ions is:

$$U=m(\omega_0^2+\Delta\omega^2)/2(q_u+\alpha q_g)^2/2+m(\omega_g^2+\Delta\omega^2)/2(q_g-\alpha q_u)^2/2 \qquad (1)$$

where m is the mass of the O(4) ion. In case of even small correction $\Delta\omega$ to the freguency of surface O(4) (because of the distortion of lattice symmetry), symmetrical $\vec{q}_g$ and antisymmetrical $\vec{q}_u$ combinations of O(4) displacements with corresponding frequences ω_q and ω_u have no sense, $\alpha=\Delta\omega^2/2 \quad (\omega_g^2-\omega_u^2)$. We have studied the spontaneous Raman light scattering at surface polaritons (frequency Ω) due to like mixing of g- and u-modes in YBaCuO single crystal. The intensity of scatted light flow generated at Stocks and anti-Stocks frequencies:

$\omega_s=\omega_i\pm\Omega$ (ω_i is the frequency of incident light) is found by the formula:

$$I_s=I_i 4\pi^2 q_g{}^2(0)(\delta\chi_{zz}/\delta q_g)^2(\omega_s d/c)^2 tg^2\theta_s \qquad (2)$$

where q(0) is the addition of g-oscillations to displacement of ions under the action of polariton, d is the thickness of the near-surface layer, θ_s is the angle between the direction of scattering and the z-axis. Account of the anharmonicity summand $\mu q_g q_u^2$ added to formula (1) has made it possible to consider other consequences of mixing u- and g-electromagnetic modes in the near-surface layer of YBaCuO crystal: second-harmonic generation and stimulated Raman light scattering.

When considering the stimulated Raman light scattering due to parametric bound of nuclei and electrons we used the model of bound oscillator which is described by the potential:

$$U(x;Q)=m\omega_0^2x^2/2+M\Omega_0^2Q^2/2-\mu x^2Q-eEx \qquad (3)$$

here x_0 and ω are coordinate and eigen frequency of the electron, Q and Ω_0 are the same for nuclei ($q_{g,u}$ is leaner combination of normal displacements of electrons $x_{g,u}$ and nuclei $Q_{g,u}$), the binding parameter $\mu\approx(\delta x/\delta Q)m^2\omega_0^4/2e^2$. If the incident field is the biharmonic one $\omega_1-\omega_2\approx\Omega$ (frequency of polariton) then O(4) displacements coherent with the incident light will cause, due to the cubic anharmonism $\mu q_g q_u^2$ in energy operator, additional coherent scattering at $2\omega_2-\omega_1$ Stocks and anti-Stocks $2\omega_1-\omega_2$ frequencies. The approach under consideration is a rough one (it studies only z-oscillations of the apex oxygen O(4) since it does not take into account the mixing, in full-symmetrical high-frequency modes, of oscillationes of the oxygen ions belonging to CuO_2 plane and the bridge. But it allows us to elucidate the peculiarities of the Raman light scattering because of O(4) symmetry violation (vacancies, surface, ets.). E.Sodtke and H.Munder [2] also note a peculiar role of the apex O(4) in disordering of the oxygen subsystem. The above optical methods of investigation are good as the theoretical ground for non-destructive diagnostics of the oxygen subsystem state in the near-surface layer of a HTSC-crystal.

1. V.Rumyantsev, V.Shunyakov, Superconductivity 5, 832 (1992)
2. E.Sodtke,H.Munder, Prepr.Res.Center Julich Gmbh (1992)

SECTION 21

CARBON and FULLERENES

A RAMAN STUDY OF C_{60} PHASE TRANSITIONS

H. Kuzmany, B. Burger, and J. Winter

Inst. f. Festkoerperphysik, Universität Wien,
Strudlhofg. 4, A-1090 Wien, Austria

Raman scattering was from the begin on a key experimental technique to study fullerenes and fullerene derived materials. This is largely due to the resonance character for the excitation spectrum [1] but also due to the sensitivity of the vibrational modes to structural transitions and changes in the electronic system. (For a review on this subject see Ref.[2].) Both, resonance enhancement and response to changes in the structural and electronic system are particularly characteristic for the total symmetric pinch mode at 1469 cm^{-1}. However, recently also other modes like the A_g breathing mode at 490 cm^{-1} and the fivefold degenerated $H_g(1)$ mode at 270 cm^{-1} were found to provide valuable information on the state of the various fulleren materials.

We report recent results of Raman studies in which the various phase transitions were analysed. The most frequently investigated transition is the order-disorder transition in pristine C_{60} where the nearly spherical molecules start to undergo a diffuse rotational motion. The investigation of this transition is not trivial since simultaneously with the diffuse rotation a 2+2 cycloaddition starts by a photoreaction. The Raman spectra of the photoreacted phase depend on the reaction temperatur. This is shown as an example for the A_g breathing mode in Fig. 1a. The Raman response for this mode is retained if the reaction temperature is close to the limit where the photoreaction can take place. A theoretical analysis indicates a dimeric phase in this case. On the other hand, if the crystal is photoreacted at room temperature the Raman response for this mode is nearly lost. In this case a clustered phase is expected from a comparison with calculations. Raman spectra from pressure polymerized C_{60} phases are different to those after photopolymerization [3]. However, for low pressure and short time polymerized material [4] the spectra become similar to those which represent the dimeric phase. Interestingly, after long time low pressure transformation the Raman spectra become equivalent to those of moderately doped C_{60} material. This means both polymers have the same backbone structure.

Phase transitions in the moderately doped C_{60} compounds where the stoichiometry between alkali metal and C_{60} is 1:1 are of particular interest. Such phase transitions have been observed for A = K, Rb, Cs but not for A = Na, Li. Figure 1b shows as an example the Raman response of the pinch mode for the phase transition from the high temperature rocksalt state to the low temperature orthorhombic state. The phase transition is reversible with a hysteresis of more than 10 K. In the low temperature phase the material is metallic and the individual C_{60} molecules are linked by a 2+2 cycloaddition process [5]. The situation is different for KC_{60} where the polymeric phase is only obtained for moderately fast cooling. For very slow cooling the high temperature phase is not stable but

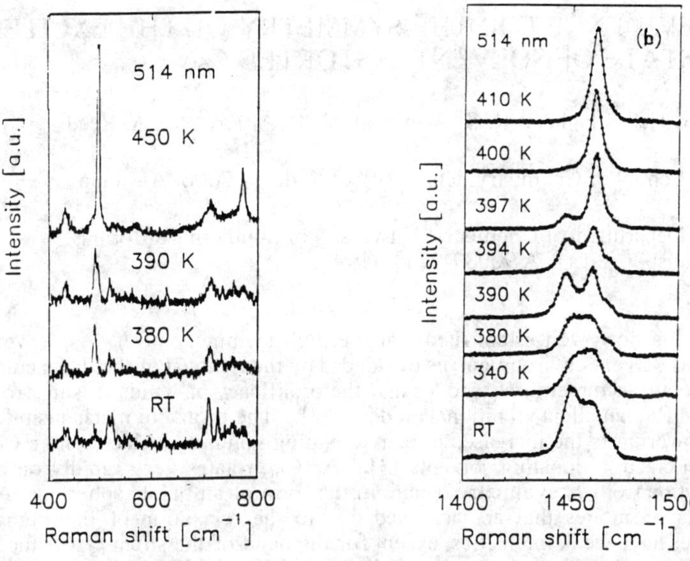

Figure 1: Raman response for the A_g breathing mode after phototransformation at different temperatures (a) and fcc to orthorhombic transition for RbC_{60} (b).

rather separates into the metallic phase K_3C_{60} and undoped C_{60}. Very recent Raman experiments showed that even in the case of the Rb compound the polymeric phase can not be obtained directly by doping below the phase transition temperature. It rather needs to go first to the high temperature rocksalt phase and then cool the sample to room temperature.

The examples given demonstrate again the key role of Raman spectroscopy for the discovery and for the analysis of new phases in the fullerenic materials.

1. M. Matus, H. Kuzmany, W. Krätschmer, Solid State Communic., 80, 839(1991)
2. H. Kuzmany et al., Advanced Materials 6, 731(1994)
3. Y. Iwasa et al., Science 264 1570(1994)
4. A. Soldatov et al., Int. Winterschool on Electronic Properties of Novel Materials, Kirchberg 1996
5. S. Pekker et al., Solid State Communic. 90, 349(1994)

EVIDENCE FOR THE SYMMETRY OF THE EXCITED STATE OF SOLVENT-DISTORTED C_{60}

S. H. Gallagher[1], R. S. Armstrong[1*], P. A. Lay[1], C. A. Reed[2]

[1]School of Chemistry, University of Sydney, 2006, Australia

[2]Department of Chemistry, University of Southern California, Los Angeles, CA 90089-0744, USA

It is now well-established that the high symmetry of I_h C_{60} is very sensitive to its solvent environment, as evidenced by the removal of the degeneracy of the non-totally symmetric H_g modes and the prolificacy of bands arising from IR-active and Raman-silent vibrational modes [1,2]. The probable mechanism for symmetry lowering is the increased effective rotating volume of electrophilic C_{60} that arises in electron-donating solvents [3]. As C_{60} rotates very rapidly on an axis, a reduction from I_h symmetry occurs in the form of an oblate spheroid. All of the bands of modes that are activated due to the relaxation of the Raman selection rules have weak intensities, except for the band of the silent $G_g(4)$ mode at 1140 cm^{-1} (Fig. 1). Upon resonance with the HOMO-LUMO transition of C_{60}, the $G_g(4)$ mode band intensity is comparable to that of the most intense bands that arise from the Raman-active fundamental A_g and H_g modes of C_{60}. This is indeed remarkable and provides a unique opportunity to investigate the nature of the symmetry of the excited state of C_{60} in a solvent environment.

The band at 1140 cm^{-1} was assigned based on the comparison of the inelastic electron tunnelling spectroscopy of C_{60} and theoretical data [4]. Further evidence of the degeneracy of the $G_g(4)$ mode is that it is split with a small band at 1136 cm^{-1} appearing in the deconvolution. This suggests that not only does the distortion of symmetry of C_{60} allow for the Raman-activation of this mode, but removes its degeneracy as well.

From the work done on the solvent effects on the electronic spectrum of C_{60} [5], there is evidence from the electronic distribution in the excited state that the symmetry of C_{60}* may closely approximate the suggested symmetry of C_{60}^{1-}. Initially, Koga and Morokuma predicted that, for the monoanion, D_{5d}, D_{2h}, and D_{3d} were the most favourable symmetries [6]. Recent work has shown that D_{5d} and D_{2h}, are the most plausible symmetries on the EPR spectroscopy and NIR electronic spectroscopy timescales [7]. However, as will be discussed, the intensity enhancement of the $G_g(4)$ vibration in the RR spectra of C_{60} in this work is suggestive of a D_{5d} symmetry for C_{60}*.

Table 1 gives the correlations of the vibrational G_g and electronic T_{1u} symmetries under the I_h point group with D_{5d}, D_{2h}, and D_{3d} point groups. The predicted on-resonance depolarisation ratio for each non-totally symmetric mode is shown in brackets. The depolarisation ratio for the band arising from the $G_g(4)$ vibrational mode in I_h symmetry was measured to be $\rho = 0.6 \pm 0.1$. Immediately, it becomes clear that, for a non-totally symmetric mode, this experimental value for ρ can only correspond to an E_{2g} vibration in the D_{5d} symmetry point group. Therefore, the symmetry correlations in Table 1 indicate that D_{5d} is the best approximation to the symmetry of C_{60}*. Hence, the T_{1u} LUMO, and the excited states of other allowed electronic transitions, split into A_{2u} + E_{1u} pairs. From group theory, an E_{2g} vibration most likely couples the E_{1u} components of the excited states of the weak HOMO-LUMO ($1^1\mathbf{T}_{1u}$-$1^1\mathbf{A}_g$) transition with the strongly-allowed $3^1\mathbf{T}_{1u}$-$1^1\mathbf{A}_g$ electronic transition in D_{5d} symmetry.

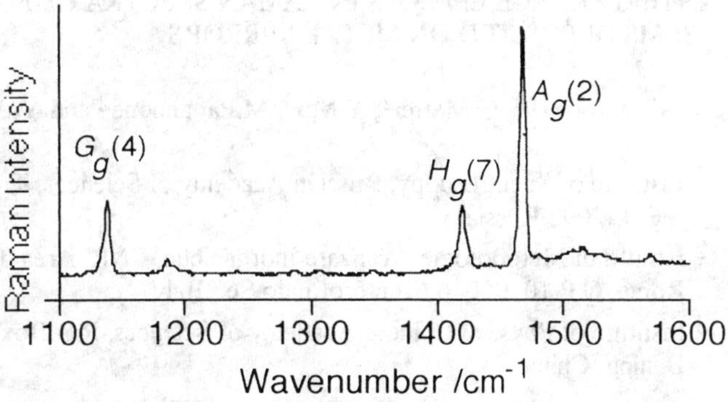

Fig. 1. The resonance Raman spectrum of C_{60} in CS_2 at 413.1-nm excitation in the range 1100-1600 cm^{-1}.

TABLE 1.Correlations of the T_{1u} electronic and G_g vibrational symmetries of I_h point group with D_{5d} , D_{2h} , and D_{3d} point groups. The bracketed numbers correspond to the on-resonance depolarisation ratio predicted for that symmetry.

I_h Point group	T_{1u} Electronic	G_g Vibrational
D_{5d}	$A_{2u} + E_{1u}$	E_{1g} (>2) + E_{2g} (0.75)
D_{2h}	$B_{1u} + B_{2u} + B_{3u}$	B_{1g} (>2) + B_{2g} (>2) + B_{3g} (>2)
D_{3d}	$A_{2u} + E_u$	A_{2g} (∞) + E_g (>2)

1. S.H. Gallagher, R.S. Armstrong, P.A. Lay and C.A. Reed, J. Am. Chem. Soc. 116, 12091 (1994).
2. S.H. Gallagher, R.S. Armstrong, P.A. Lay and C.A. Reed, Chem. Phys. Lett. 248, 353 (1996).
3. I.V. Rubstov, D.V. Khudiakov, V.A. Nadtochenko, A.S. Lobach and A.P. Moravskii, Chem. Phys. Lett. 229, 517 (1994).
4. J. Brousseau, K. Tian, S. Gauvin, R.M. Leblanc and P. Delhaes, Chem. Phys. Lett. 202, 521 (1993).
5. S.H. Gallagher, R.S. Armstrong, P.A. Lay and C.A. Reed, J. Phys. Chem. 99, 5817 (1995).
6. N. Koga and K. Morokuma, Chem. Phys. Lett. 196, 191 (1992).
7. R. Bolskar, S.H. Gallagher, R.S. Armstrong, P.A. Lay and C.A. Reed, Chem. Phys. Lett. 247, 57 (1995); and references therein.

INTERFERENCE EFFECTS IN RAMAN SPECTRA OF FULLERENE FILMS DEPOSITED ON METAL MIRRORS

V.A.Yakovlev[1]*, G.Mattei[2], Y.Mo[3], M.Pagannone[2] and Yu.Ye.Petrov[1]

[1] Institute of Spectroscopy, Russian Academy of Sciences, Troitzk, Moscow reg. 142092 Russia

[2] Istituto di Metodologie Avanzate Inorganiche, CNR, Area della Ricerca di Roma, C.P.10, 00016 Monterotondo Sc., Italy

[3] Institute of Physics, Chinese Academy of Sciences, P.O.Box 603,100080, Beijing, China

Thin film deposited on a metal substrate is one of the most popular subjects of Raman studies. But if the film thickness is comparable with the wavelength of light, optical interference strongly affecting absorption and reflection spectra of such films appears also in Raman spectra. We have shown [1] that a small change of a film thickness can cause a strong change of intensities of Raman bands of the film. This effect should be taken into account in quantitative analysis of Raman spectra. On the other hand it gives an opportunity of non-destructive thickness profiling of thin films. Now we discuss the influence of the interference on the Raman spectra of fullerene films on metal mirrors

Fullerene C_{60} thin films were deposited on 200-300 nm thick metal (gold or nickel) coated glass substrates by thermal evaporation in vacuum. A mask was placed a few cm ahead the sample to obtain rapid but smooth decrease of the film thickness near the edges. Infrared reflectivity spectra (recorded in the region 5000-400 cm^{-1}) of these samples were used to obtain IR refractive index and thickness of these films. The thicknesses of films on both gold and nickel appear to be equal to 650 nm in the central part (approx. 0.5 cm^2). All four IR active bands were present and dominant in the spectra, but weak bands belonging to C_{70} and organic impurities were also seen.

The Raman measurements were carried out using a SPEX triple spectrometer, equipped with an OMA detector, with excitation from 10 mW Ar^+-Kr^+ laser operating on the 514.5 nm, 488 nm and 568.2 nm lines. The backscattering geometry was used. The spatial resolution was 0.1 mm. The Raman spectra were taken at different points along the sample in the spectral region 160-1800 cm^{-1} .

The intensities of all the observed bands do not decrease smoothly from the centre to the edges of the film, as could be expected because of the decrease of the film thickness, but oscillate. Fig. 1 presents the Raman spectrum as a function of a probing point position x along the fullerene film on gold. Film centre is placed approximately at x=8 mm. These oscillations correspond to the fringes produced by the interference of the excitation laser light as well as of scattered light within the film. For the film on nickel the behaviour is similar. The interference of laser light was used by Connell at al.[2] to enhance Raman spectra of very thin overcoatings.

The condition of maximum laser intensity inside the film producing maximum Raman scattering coincides with the condition of reflectivity minimum and depends on the wavelength, film thickness, its refractive index and metal optical constants. The same interference effects are experienced by the scattered light inside the film. If the frequency shift between the laser and the scattered light is small, both waves are in phase, and maximum intensity contrast is expected for the modulation in the Raman spectra. Otherwise, the interference will be damped, and the contrast reduced especially for thicker films. Such qualitative consideration was confirmed by direct calculations of integrated fields of laser and scattered light inside the film using literature data on the optical constants of fullerene and gold.

We report in Fig. 2 the intensities of the 1465 cm^{-1} (2 scans) and 495 cm^{-1} Raman bands, as functions of the point position for 514.5 nm excitation. The oscillations are clearly evident also for other excitation laser lines, but minima and maxima are observed at different *x* values depending on the excitation wavelength. Using thicknesses corresponding to maxima of calculated field and maxima positions in Fig. 2 we obtained the film thickness profile from the Raman measurements. The thickness in the point closest to the film centre is in a good agreement with IR data.

Thus we have done non-destructive Raman measurements of fullerene film thickness profile. In this particular case of single component film the changes of relative intensities of different Raman bands are not so important but for multicomponent films like in [1] this problem should be also taken into account. Note that these interference effects are most significant for weakly absorbing films with high refractive index deposited on the metal exhibiting not too high reflectivity.

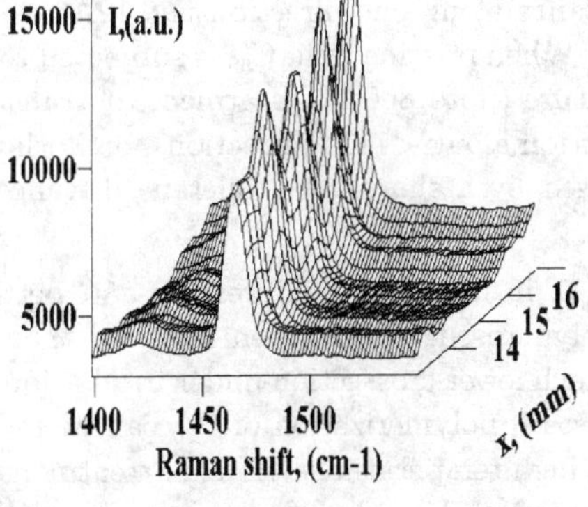

Fig. 1. Raman spectra of C_{60} film at different point positions x for 514.5 nm laser line.

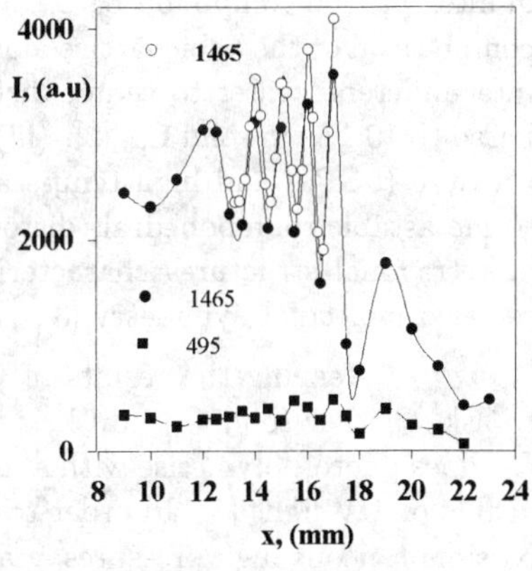

Fig. 2. Dependence of Raman intensity of different bands on the point position x.

1. V.A.Yakovlev, G.Mattei, A.Iembo, F.Fuso, E.Arimondo, M.Allegrini, F.Leccabue and B. E.Watts. Journal of Applied Physics, 78, 6321 (1995).
2. G.A.Connell, R.J.Nemanich and C.C.Tsai, Appl. Phys. Lett. 36, 31 (1980).

HIGH PRESSURE HIGH TEMPERATURE INDUCED POLYMERIZATION OF C_{60} AND C_{70}

A.K. Sood[1*], P.Ch. Sahu[2], V.S. Sastry[2], G.V.N. Rao[2], V. Sridharan[2], M. Premila[2], A. Bharathi[2], Y. Hariharan[2], T.S. Radhakrishnan[2], D.V.S. Muthu[1] and C.S. Sundar[2]

[1] Department of Physics, Indian Institute of Science, Bangalore 560 012, India

[2] Materials Science Division, Indira Gandhi Centre for Atomic Research, Kalpakkam 603 202, India

During the last three years, it has been shown that C_{60} molecules in a crystal can be brought closer along one direction to form fullerene "polymers". This can be achieved by three routes: (i) photo-induced polymerization [1] –laser or UV light irradiation of C_{60} films resulted in toluene -insoluble product which had oligomers as large as $(C_{60})_{20}$ indicated by laser desorption mass spectrometry. The proposed mechanism of photopolymerization is associated with the (2+2) cycloaddition reaction by the conversion of double bonds on adjacent molecules to a four membered ring. (ii) The room temperature phase of intercalation compound RbC_{60} and KC_{60} has an orthorhombic distortion (i.e. compression of the cubic lattice along one of the <110> directions) in which the inter-fullerene center to center distance along one direction is 9.12A^o as compared to 10.1A° in solid C_{60} [2]. (iii) When pristine solid C_{60} is subjected to high pressure (5GPa) and high temperature (upto 800°C) treatment, it transforms to metastable rhombohedral (distortion along <111> direction), orthorhombic or tetragonal structures characterised by a short inter-fullerene distance and reduced molecular symmetry [3].

Keeping the results of C_{60} in mind, it is interesting and pertinent to ask whether solid C_{70} can be polymerised. It has been reported that solid C_{70} does photo-polymerise with a much lower crossection under irradiation with visible or UV light [4]. In order to see if polymerization of C_{70} can be induced by simultaneous high pressures and high temperatures (HPHT) treatments, we have carried out the present Raman, infrared, x-ray diffraction and photoluminescence studies on solid C_{70} subjected to pressures upto 7.5 GPa at various temperatures from ambient upto 800 C [5].

Pellets of 3mm diameter of C_{60} and C_{70} powders were subjected to high pressures (upto a maximum of 7.5 GPa) over a period of 10 minutes and the temperature was increased upto maximum of 800C [5]. After keeping the sample under these conditions for about 6 hours, the temperature was lowered then pressure was released slowly. Raman measurements were carried out under ambient conditions using DILOR-XY spectrometer and 5145A° line (about 10 Watts/cm^2).

A comparison of the spectra of the HPHT treated and the starting C_{60} samples shows that all the Raman modes soften. In particular, $A_g(2)$ Raman mode at 1469 cm^{-1} shifts to 1460cm^{-1} and develop a shoulder. However, the large Raman shift of 39cm^{-1} of the $A_g(2)$ mode reported by Iwasa et al [3] for the case of C_{60} sample treated at 5GPa, 800C is not observed in the present experiments. We also note that the relative intensities of the modes also change significantly in polymerised samples. The red shift and splitting of the $A_g(2)$ mode, similar to that in photopolymerised C_{60}, can be qualitatively understood in terms of stretching of the carbon bonds. X-ray diffraction (XRD) measurements of HPHT treated samples show that the various treatements of (5GPa, 400C); (5GPa, 750C), (7.5GPa,500C) and (5GPa, 750C) were indexable to an orthorhombic structure.

The XRD pattern of the starting C_{70} powder could be index to hcp structure with a=10.60A° and c=17.26A°. The HPHT treatments resulted in broad diffraction lines and the XRD patterns were closest to the rhombohedral structure (space group R3m). Surprisingly, the Raman spectra of the HPHT samples are same as the starting C_{70} powder. Infrared and photoluminescence spectra of the HPHT samples are also unaltered [5]. These observations lead us to suggest that even though the structure changes from the starting hcp to a rhomohedral structure, C_{70} is not polymerised under the HPHT treatments. The smaller crossection for polymerization as compared to C_{60} can be attributed to a reduction in the number of reactive double bonds from 30 in C_{60} to only 5 in each polar cap of a C_{70} molecule [4].

1. A.M. Rao, P. Zhou, K. Wang, G.T. Hager, J.M. Holden, Y. Wang, W.T. Lee, X. Bi, P.C. Eklund, D.S. Cornett, M.A. Duncan and I.J. Amster, Science **259**, 955(1993).

2. S. Pekker, L. Forro, L. Mihaly and A. Janossy, Sol.State Communc. **90**, 349(1994).

3. Y.Iwasa, T. Arima, R.M. Fleming, T. Siegrist, O. Zhou, R.C. Haddon, L.J. Rothberg, K.B. Lyons, H.L. Carter, A.F. Hebard, R. Tycko, G. Dabbagh, J.J. Krajewski, G.A. Thomas and T. Yagi, Science **264**, 1570 (1994).

4. A.M. Rao, M. Menon, K. Wang, P.C. Eklund, K.R. Subbaswamy, D.S. Cornett, M.A. Duncan and I.J. Lester, Chem. Phys. Lett. **224**, 106(1995).

5. C.S. Sundar, P.Ch. Sahu, V.S. Sastry, G.V.N. Rao, V. Sridharan, M. Premila, A. Bharathi, Y. Hariharan, T.S. Radhakrishnan, D.V.S. Muthu and A.K. Sood (To be published).

THE VARIATION OF FREQUENCY AND WIDTH OF C60'S RAMAN SPECTRAL LINES AT LOW TEMPERATURE

Chang-Lian Gao[1], Jian-Mo Wang[1], Xin-Min Xiao[1*], Hong-Lei Luo[3], Zhen-Dong Liu[2], Xiang-Yang Tao[1], Jian- Zhi Xing[1]
1.Jiangxi Normal University, Nanchang 330027, CHINA
2.Dept. Physics, Zhejiang University, Hangzhou 310027, CHINA
3.Sparty Co. Ltd., Hangzhou 310023, CHINA

The Raman Spectra of the solid C60 at 284-10 K were measured by a SPEX1877B triplemate and an OMA III optical multichannel analyzer with a 1421B Reticon detector.The samples, whose purity was better than 99.9%,were connected with the cooling head of a LTS-21 cryogennic system (RMC)with a 805 temperature controller (Lake Shore). These samples were excited by the 514.5 nm line of an Ar laser (Coherent) at actual ca. 3 mW.

The temperature was adjustable continuously and the accuracy of control and measurement was 0.1 K.

We fixed the grating without restarting the triplemate during the whole course of measuring the group data. The experiment data were fitted by repeatedly measuring, so the repetition of spectra was 0.1cm^{-1} and the resolution was actually ca. 0.2 cm^{-1}.

TAB.1 Raman shifts, line-width (FWHM), intensity of C60's all spectral lines at 284 K and 10 K

Spec.line,cm^{-1}		270	429	498	716	785	1128	1426	1470	1580	1692
Raman	284K	270.1	428.9	497.9	716.1	785.1	1128.1	1425.8	1470.2	1580.2	1692.0
sh.cm^{-1}	10K	270.3	430.1	498.0	715.8	784.6	1127.0	1427.0	1471.0	1578.0	1692.6
FWHM	284K	profile	8.9	7.2	9.5	9.1	14.7	22.6	7.0	29.2	profile
cm^{-1}	10K	unclear	9.3	7.7	8.0	7.4	14.5	22.0	7.5	25.7	unclear
Inten,	284K	7	10	22	12	12	9	12	132	11	4
arb.unit	10K	8	11	19	12	12	12	16	121	14	7

Tab.1 shows the Raman shifts, line-width (FWHM), and the intensity of all spectral lines at 284 K and 10 K. The 496, 1470 cm^{-1} were Ag mode, and the others were Hg mode. Fig.1 illustrates the variation of frequency and width of some spectral lines.

We observed that the line-width, variation of frequency and line-width were small at Ag mode, but large at Hg mode.

A remarkable tendency of line-width variation curve was opposite to the frequency curve, i.e. FWHM decreased when the spectral line presented red-shift (Stokes line, Raman shift increased) and FWHM increased when the spectral line presented blue-shift.

For all curves, mutation appear at ca. 250 K, which is the first order phase transition point[1].

Above 250K, the orientations of C60 molecules in fcc lattice are disorderly with rotational thermal motion.As temperature decreases, the energy of rotational thermal motion decreases too,so the vibrational energy band broadens diffusely and decreases height slightly due to the perturbation from vibration-rotation coupling. That is why spectral line presents blue-shift and the line-width broadens between 284 to 250 K.

Below 250 K, first order phase transition has taken place and the orientations of C60 molecules in sc lattice are orderly. The rotation was frozen and thermal motion was translational motion only, so the broadening of the energy band due to the perturbation will reduce gradually, and the energy band will shrink to somewhat higher level. Then the spectral line presents red-shift and line-width decreases.

Below 50 K, the spectral line presents blue- shift once again and broadens obviously. We think that mode-splitting causes this phenomina.

We had thought 150 K may be a new phase transition point[2]. In this work,mutation or peak was not obviously present at 150 K. However,as fig.2 indicates,for temperature's rising and falling, the line-width curves of 1470 cm^{-1} fail to meet each other at 150 K.We think perhaps this is a Curie-point of a certain second order phase transition.

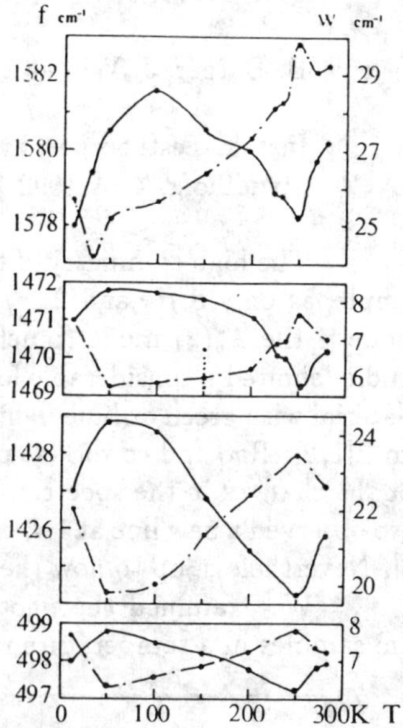

Fig.1 The variation of frequency and width
—— frequency, cm⁻¹, left scale
—·—· width, cm⁻¹, right scale

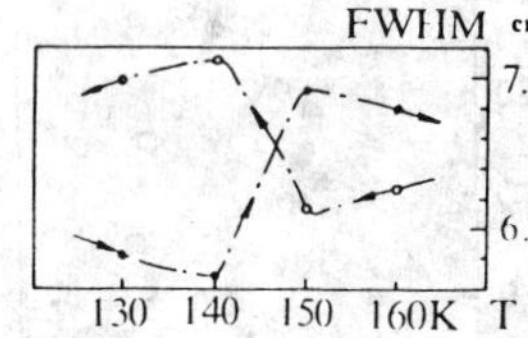

Fig.2 FWHM – T curve of 1470 cm⁻¹
⇄ direction of temperature variation

Financial support was received from the foundation of natural science of Jiangxi province.

1. P.H.M. van Loosdrecht, et al, Phys. Rev. Lett. 68 1176 (1992)
2. Xin-Min Xiao, et al, Proceedings of the 14th Inter.Conf.on Raman Spec. (Aug. 1994,Hong Kong) p308, John Wiley & Sons Ltd. England

RAMAN EVIDENCE FOR DIFFERENT PHASES IN PHOTOTRANSFORMED C_{60}

B. Burger, J. Winter and H. Kuzmany

Inst. f. Festkoerperphysik, Universität Wien,
Strudlhofg. 4, A-1090 Wien, Austria

The high symmetry of the C_{60} molecule makes the Raman spectrum very simple, as only 8 H_g and 2 A_g modes are active. The most intensively studied mode is the $A_g(2)$ mode (pinch mode) at 1469 cm^{-1}. Early Raman scattering studies showed a considerable loss of intensity for this line upon irradiation of the material with green or blue light. Simultaneously an other mode develops at 1458 cm^{-1}[1, 2]. Rao and coworkers proposed a 2+2 cycloaddition as an interpretation for the changes in the spectrum. This explanation is now widely accepted. They also observed a new line at 118 cm^{-1}, which was assigned to an interball stretching [3]. Nevertheless, up to now the structure of the the resulting product is unclear.

We examined the phototransformation of C_{60} single crystals and thin film samples in a temperature range from 200 K to 450 K.

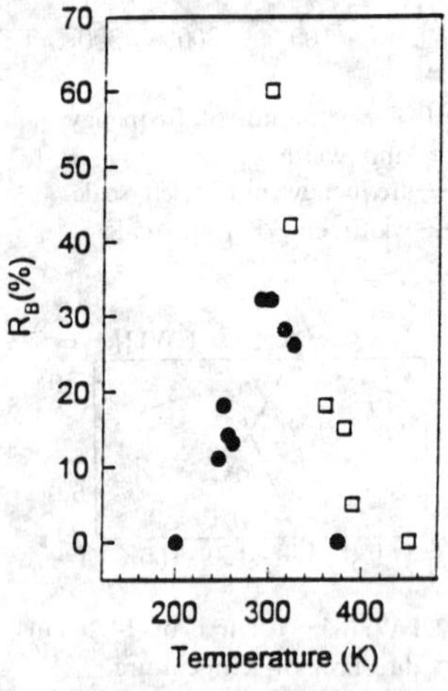

Figure 1: Bleaching rate of the $A_g(2)$ mode at various temperatures. The rate is defined as the loss of intensity within the first 30 seconds of measurement. Open squares: single crystal, full circles: thin films, power density 60 W/cm^2.

Within the temperature range studied strong differences in the dynamics of the reaction were observed. The cycloaddition process takes place between 260 and 400 K. Below 260 K the reaction is inhibited by the low temperature structure of C_{60}, which has no parallel lying double bonds. The high temperature limit is due to the higher kinetic energy of the molecules. In Fig. 1 the bleaching rate is given versus temperature.

Analyzing the spectra obtained during these dynamical studies we found striking differences. In Fig. 2 the spectra in the region of the $H_g(1)$ and of the $A_g(2)$ mode are shown for various transformation temperatures. The strongest differences are observed by a comparison of the spectra of material transformed at 300 K and 380 K. In the region of the $H_g(1)$ mode only one slightly asymmetric line is observed at 271 cm^{-1} in the spectrum of the sample transformed at 380 K, while in the 300 K transformed material this mode splits into three components at 252, 269 and 293 cm^{-1}. In the region below the $H_g(1)$ we observe a mode at 117 cm^{-1} in the spectrum of the 300 K transformed material, which is in good agreement to the work of Rao et al. In the sample irradiated at 380 K a strong line at 96 cm^{-1} is observed. Similar characteristic differences are observed for the pinch mode.

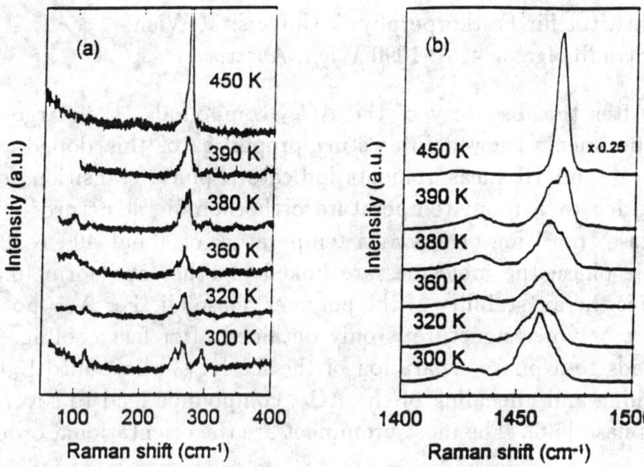

Figure 2: Spectra of C_{60} phototransformed at the temperatures indicated (a) the region of the $H_g(1)$ mode and (b) in the region of the $A_g(2)$ mode. The uppermost spectrum for untransformed C_{60} is shown for comparison

To assign the measured spectra to different structures we compared them to ab initio calculations from Porezag et al. [4]. The best fit is obtained between the sample transformed at 380 K and the calculation for a C_{60} – dimer. In the case of the sample irradiated at 300 K, none of the calculated spectra gives a reasonable good match. Accordingly we conclude that 3 d clusters are formed at this temperatures [5].

References

[1] S.J. Duclos et al., Sol State Comm., 80, 481 (1991).
[2] M. Matus et al., Phys. Rev. Lett., 68, 2822 (1992).
[3] A.M. Rao et al., Science, 259, 955 (1992).
[4] D. Porezag et al., Phys. Rev. B, 52, 14963 (1995).
[5] B. Burger et al., Z. Phys. B, to be published.

RAMAN SCATTERING OF QUENCHED AND ANNEALED RbC_{60} DOPED SINGLE CRYSTALS

J. Winter and H. Kuzmany

Institut für Festkörperphysik,Universität Wien,
Strudlhofgasse 4, A–1090 Wien, Austria

After the discovery of the AC_{60} compounds [1] a large number of different experiments showed interesting properties of this doped phase. For example X-ray and IR measurements indicate a phase transition from a high temperature fcc to a room temperature orthorhombic structure for A = Rb, Cs. The phase transition occurs at a temperature of about 400 K [2,3]. In the orthorhombic phase the molecules are linked together and form linear chains. Remarkable is the air stability of the polymer phase [4]. For A = potassium the orthorhombic polymer structure is only obtained after fast cooling. Annealing at 410 K leads to a phase separation of the fcc KC_{60} phase into K_3C_{60} and α-C_{60}. Quenching and annealing of the AC_{60} compounds lead to several different metastable phases[5,6]. The most prominent are the orientational ordered simple

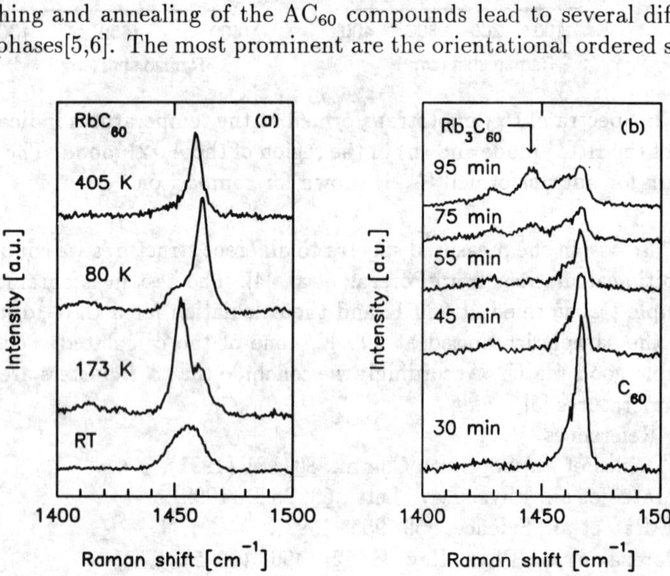

Fig. 1. (a): Raman spectra of the $A_g(2)$ mode at 405 K, after quenching at 80 K, at 173 K and at room temperature.
(b): Raman response of the $A_g(2)$ mode during the doping process at 380 K.

cubic phase after quenching and the dimer phase below 220 K after annealing. In this dimer phase two C_{60} molecules are linked only by a single bond in contrast to the linking by a four membered ring in the case of photopolymerized C_{60}.

We present Raman measurements of quenched and annealed RbC_{60} single crystals. In addition Raman results of the doping process for C_{60} performed below the fcc-orthorhombic phase transition are shown.

The C_{60} single crystals were grown by the sublimation technique [7] and intercalated with Rb by a combination of a doping and equilibration process [8]. The RbC_{60} fcc phase was quenched from 405 K to 80 K within in 7 minutes. Fig. 1a shows the Raman spectra of the frequency range of the $A_g(2)$ pinch mode taken at different temperatures. At 80 K the Raman response of the monomer phase could be retained. After annealing to 173 K a remarkable shift of 8 cm^{-1} of the pinch mode is detected. The mode at 1453 cm^{-1} is interpreted as the pinch mode of the dimer phase. After annealing to room temperature the well known Raman response of the orthorhombic phase is observed. The broad features at the low and high frequency shoulder of the pinch mode at 80 K and 173 K are interpreted as a contamination by the polymer phase due to a not high enough quenching rate (46 K/min).

As shown by many different experiments slow cooling of the high temperature fcc RbC_{60} phase leads always to a stable orthorhombic phase at room temperature. In a doping experiment we checked the possibility to obtain the orthorhombic RbC_{60} phase directly from the undoped C_{60}. Fig. 1b shows the Raman response of C_{60} during continuous doping process at 380 K just below the fcc-orthorhombic transition temperature. The spectra show a strong decrease of the $A_g(2)$ mode of C_{60} with exposure to Rb vapour and an increase of a new mode at 1448 cm^{-1}. The new mode is identified as the pinch mode of Rb_3C_{60}. The broad features around 1460 cm^{-1} result from phototransformed C_{60} material.

Summarizing the presented Raman measurements confirm the metastable low temperature phases obtained during annealing after quenching the sample. Also the results show clearly that doping below the fcc-orthorhombic phase transition temperature leads to the A_3C_{60} phase. This means the polymeric phase can only be obtained from the high temperature rock salt structure.

1. J. Winter et al., Solid State Commun. 84, 935 (1992).
2. P. Stephens et al., Nature 370, 636 (1994).
3. T. Pichler et al. Phys. Rev. B 49, 15879 (1994).
4. R. Winkler et al. Appl. Phys. Lett. 66, 1211 (1995).
5. G. Oszlanyi et al. Phys. Rev. B 51, 12228 (1995).
6. M. Kosaka et al. Phys. Rev. B 51, 12018 (1995).
7. M. Haluska et al. Appl. Phys. A 56, 161 (1993).
8. J. Winter et al. Phys. Rev. B 52, 7115 (1995).

206.5 NM BANDGAP EXCITED RAMAN SPECTRA OF CVD AND NATURAL DIAMOND EXAMINES ENHANCEMENT OF SECOND AND THIRD ORDER PHONON BANDS

Richard W. Bormett, Calum H. Munro, Robert E. Witkowski, Sanford A. Asher*

Department of Chemistry, University of Pittsburgh, Pittsburgh PA 15260

We have measured the 206.5 nm excited Raman and UV photoluminescence spectra of natural and chemical vapor deposition (CVD) diamond. We have shown previously that 228.9 nm and 244 nm excited Raman spectra of CVD diamond does not contain luminescence bands that interfere or obscure the first and second order phonon bands of diamond[1]. In fact, our UV excitation near the diamond bandgap shows that resonance enhancement allows the observation and characterization of the third order phonon bands of CVD diamond. By increasing the Raman excitation frequency to 206.5 nm which is deep within the diamond bandgap we observe for the first time a higher relative enhancement of the ca. 2200 cm^{-1} second order phonon band compared to other second order phonon bands.

The 206.5 nm excitation also produces luminescence bands in the 230 to 255 nm region of the spectrum that are not observed with 228.9 nm excitation. These bands, which were initially reported as third order phonon bands, occur at wavelengths of 247 and 251 nm and are actually photoluminescence (PL) bands associated with the diamond N9 absorption system[2]. The N9 absorption and photoluminescence occurs in natural diamond when nitrogen is the major impurity.

The Raman spectra of natural and CVD diamond excited close to the bandgap at 228.9 nm are shown in Figure 1. The diamond spectrum can be separated into three regions based on the number of phonons involved in the scattering process. The first order diamond phonon band (1332 cm^{-1}) results from the scattering of a single optical phonon at the maximum in the diamond lattice-dispersion curve[3]. Two phonon scattering in diamond produces a complex band with a maximum at 2467 cm^{-1} and a sharp cutoff at 2667 cm^{-1}. The second order phonon bands of diamond are resonantly enhanced with UV excitation and their relative intensity with respect to the first order phonon band increases ten-fold compared to visible excitation[4]. Three phonon scattering in diamond produces a band with maxima at ca. 3690 cm^{-1} in the CVD diamond. The bands observed in the third order phonon region of natural diamond are not Raman bands, but rather luminescence bands. Based on our 228.9 nm excited Raman spectra of different CVD diamond, the intensity ratio of the diamond third order phonon band to the second order phonon band appears to be sensitive to either the diamond crystallite size or crystallite orientation.

The Raman spectrum of CVD diamond excited with 206.5 nm is shown in Figure 2. The 206.5 nm excited Raman spectrum is similar to that excited at 228.9 nm (Figure 1), but shows a stronger second order phonon band at 2200 cm^{-1} with respect to the other second order phonon bands. Like the 228.9 nm excited diamond Raman spectra, the 206.5 nm excited CVD diamond Raman spectrum has a vanishingly small luminescence intensity in the wavelength region associated with the diamond's first, second and third order phonon bands.

1. R.W. Bormett, S.A. Asher, R.E. Witkowski, W.D. Partlow, R. Lizewski, and F. Pettit *J. Appl. Phys.* **77**, 5916 (1995).
2. D.R. Wight, P.J. Dean *Phys. Rev.* **154**, 689 (1967).
3. S.A. Solin and A.K. Ramdas *Phys. Rev. B* **1**, 1687 (1970).
4. J.M. Calleja, J. Kuhl, M. Cardona *Phys. Rev. B* **17**, 876 (1978).

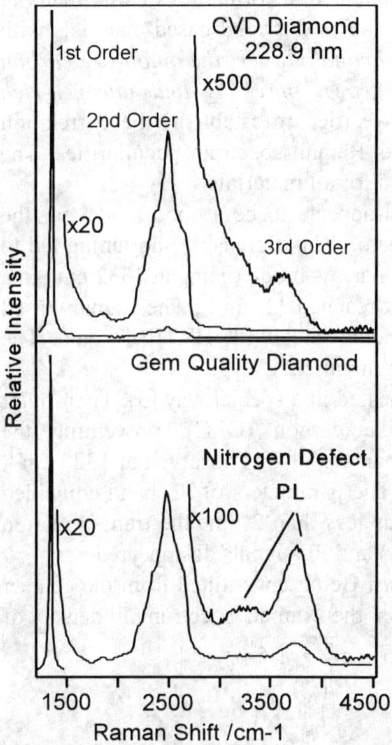

Figure 1 228.9 nm excited Raman spectrum of A) CVD diamond Film B) Large gem quality diamond. Laser spot size was ca. 50 μm with an incident power of ca. 7-10 mW. Spectral resolution was ca. 25 cm^{-1}.

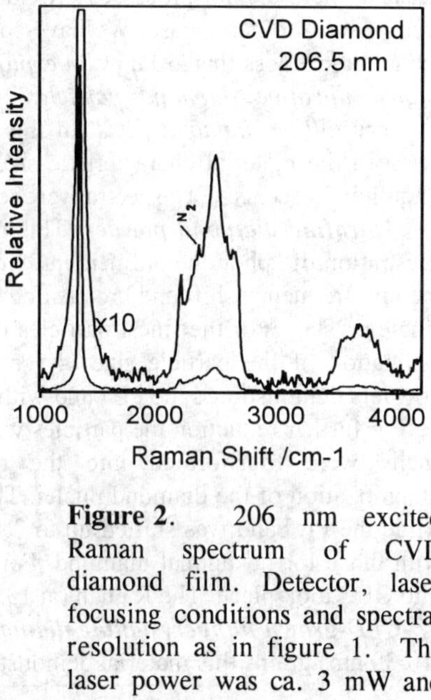

Figure 2. 206 nm excited Raman spectrum of CVD diamond film. Detector, laser focusing conditions and spectral resolution as in figure 1. The laser power was ca. 3 mW and the total integration time was 9 min.

RAMAN STUDY of DIFFERENT MODIFICATIONS of NANOCRYSTALLINE DIAMOND

E.D. Obraztsova[1]*, S.M. Pimenov [1], V.G. Pereverzev[1], V.L. Kuznetsov[2]

[1] General Physics Institute, RAS, 38 Vavilov street, 117942, Moscow, Russia
[2] Boreskov Inst. of Catalysis, Lavrentieva 5, Novosibirsk 6300090, Russia

The dimensional effects in the Raman spectra of diamond are difficult to observe because there are no processes provided a gradual transformation of this material from macro- to nanophase. We have collected a few diamond-based materials with the grain size less than 30 nm: *an explosive diamond powder; the onion-like carbon coated ultrafine diamond particles; CVD-grown pure* [1] *and nitrogen-rich nanocrystalline diamond films.* In this work we tried to establish the correlation between the material characteristic size and its Raman spectrum peculiarities. The Photoluminescence (PL) spectra were registered for all materials.

Ultrafine diamond powder. This material appears to be a good model for the illustration of "phonon confinement" mechanism. The dimension shortening led to the low-frequency shift and broadening of one narrow mode (being at 1332 cm^{-1} in a single crystal). For the finest particles the shift reached 11 cm^{-1}. The Raman-based estimation of the particle size (4.5 nm) coincide well with HRTEM data. The powders demonstrated the PL band with maximum at 2.2 eV [2].

For further reduction the particles were annealed in a special way [3]. Their outer shells were transformed into the onion-like carbon (OLC), preventing the graphitization of the diamond nuclei. The Raman signal was observed at 1320 cm^{-1}, while the PL band was shifted up to 2.4 eV. The parameters of PL band coincided with those for hexagonal diamond. For nucleus less than 2 nm the transformation into OLC took place. The Raman (at 1320 cm^{-1}) and PL signals disappeared.

CVD-grown nanocrystalline diamond films. Being constituted from the coarser (10-30 nm) grains this material demonstrated in the Raman spectrum all density of

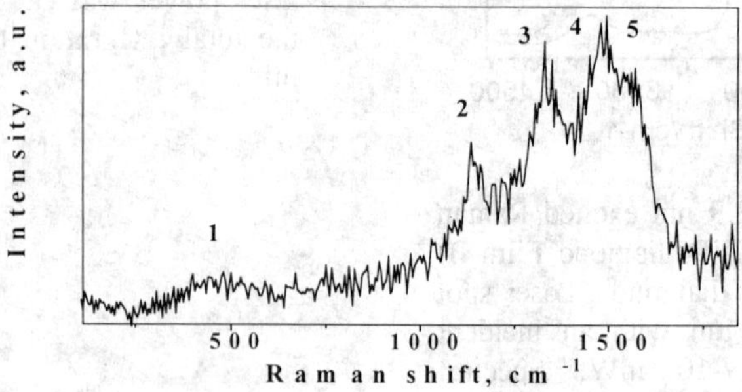

Fig.1. Typical Raman spectrum of nanocrystalline diamond film CVD-grown from methane-rich gaseous mixture CH_4/H_2

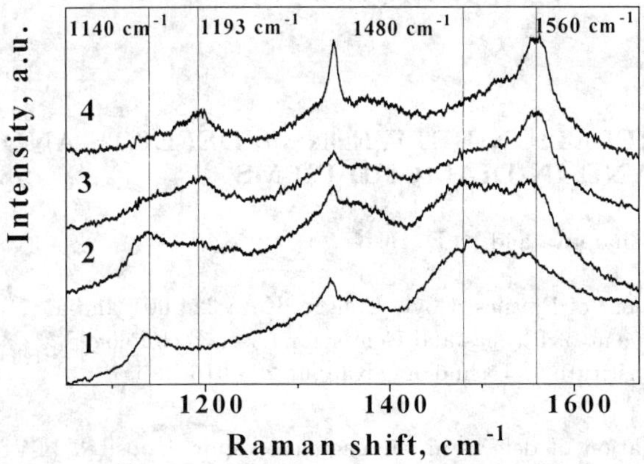

Fig.2. Raman spectra of the nanocrystalline diamond films grown from CH_4/N_2 gaseous mixture with the different N_2 contents

one-phonon diamond states (Fig.1). In addition to TO-like band at 1140 cm^{-1} (2) the acoustical modes at 500 cm^{-1} appeared (1). The similar spectrum was observed for *a*-diamond [4]. In our case it was completed by the bands of sp^2-bonded carbon (3,4,5). The ultrafine (1.5-2 nm size) deformed tetrahedral carbon particles, stabilized by the mechanical stresses in the space between the coarser grains, are supposed to be the source of such complicate spectrum. The mechanical stresses presence were confirmed by the Raman and X-ray data. The films demonstrated a strong PL with the maximum at 2.2 eV.

Nitrogen-rich nanocrystalline diamond films. Using the methane-rich gaseous mixture N_2/CH_4 instead of H_2/CH_4 we deposited the nanocrystalline diamond films with the enhanced nitrogen content. Their Raman spectra (Fig.2) displayed a specific band at 1193 cm^{-1} ascribing to C-N vibrations [5]. The intensities of the bands at 1193 cm^{-1} and 1560 cm^{-1} correlated. The competition between modes at 1140 cm^{-1} and 1193 cm^{-1} took place. For the pure diamond film (1) only first one was observed. As the incorporated nitrogen content increased (2,3) the contributions of two modes became comparable, and at last (4) only mode at 1193 cm^{-1} dominated in the spectrum. This fact is supposed to confirm the similar origin of both modes. The PL spectrum reflected the increase of the nitrogen content.

The described research was supported in part by Grants N4Q000 and N4Q300 from the International Science Foundation and Russian Government.

1. V.I. Konov, A.A. Smolin, V.G. Ralchenko, S.M. Pimenov, E.D. Obraztsova, E.N. Loubnin et al.., Diamond and Related Materials 4(1995) 1073.
2. E.D. Obraztsova, K.G. Korotushenko, S.M. Pimenov, V.G. Ralchenko, A.A. Smolin, V.I. Konov, E.N. Loubnin , Nanostructured Materials 6 (1995) 827.
3. V.L.Kuznetsov, A.L. Chuvilin, Yu.V. Butenko, I.Yu. Mal'kov, V.M. Titov, Chem. Phys. Lett., 222 (1994) 343.
4. V. Paillard, P. Melinon, V. Dupuis, A. Perez et al.,Phys. Rev. B 49 (1994)11433.
5. L.Bergman, M.McClure,J.T. Glass,R.J. Nemanich, J. Appl. Phys.76(1994) 3020.

CORRELATION BETWEEN sp^2 DEFECTS AND PL BAND IN DIAMOND FILMS

Sanjay Bhargava and H. D. Bist*

Department of Physics, Govt. College, Kota 324 001, India
*Department of Physics and Center for Laser Technology,
Indian Institute of Technology, Kanpur 208 016 India.

Investigation of defects in the chemical vapor deposited (CVD) diamond films is important for the effective use of such films in potential electronic applications. In this paper we show the use of micro-Raman and photoluminescence (PL) spectroscopy for the characterization of defects and study of their behavior on successive heat treatment in air.

The samples were deposited using the CVD technique [1]. The room temperature Raman and PL spectra were excited with 514.5 nm line fron an Ar^+ laser. The backscattered radiation passing through Spex triplemate was detected by a CCD/PMT. The spectra were analyzed using the Jandel scientific Peakfit programme [2]. The samples were subjected to heat in air at a required temperature for 30 minutes and were allowed to come to room temperature by furnace cooling.

We have found a correlation between the integrated intensity of non diamond Raman bands ($I^R_{nondiamond}$) and broad PL abnd at 2.08 eV ($I_{broadPL}$) in diamond films. These areas have been normalized by the area of the Raman band for cubic diamond ($I^R_{diamond}$). Fig 1(a) shows a plot between the ratios $(I^R_{nondiamond})/(I^R_{diamond})$ and $(I_{broadPL})/(I^R_{diamond})$ for the as grown diamond films. The linear relationship between these indicates that the defects which cause this PL band are the amorphous sp^2 type bonds, which also give rise to nondiamond features in the Raman spectrum. In good quality diamonds this PL is absent, and the Raman spectra also show no nondiamond features.

To further support the idea successive heat treatment of two films was carried out to study the behavior of these defects. The 2.08 eV PL band and nondiamond Raman band vanishes at nearly the same temperature. The ratio $(I_{broadPL}/I^R_{diamond})$ follows a linear decrease with a rise in heat treatment temperature. A similar decrease of this ratio indicates a common type of defect as origin of nondiamond Raman and PL bands. Fig. 1(b) shows a plot between ratio $(I_{broadPL}/I^R_{diamond})$ and $(I^R_{nondiamond}/I^R_{diamond})$ obtained at different treatment temperatures. The linear relationship once again indicates that both originate from the same type of amorphous sp^2 bonded carbon structures. These arguments are consistent with Bergman et.al. who find that 2.08 eV PL band and nondiamond Raman intensities follow a similar time evolution [3]. The amorphous carbon hydrogenated films which have same sp^2 bond configuration as in diamond films exhibit a broad PL band similar to that observed in our PL spectra. These observations also suggest that amorphous sp^2 bonding configuration is the origin of 2.08 eV PL band. Thus in conclusion we have shown that amorphous sp^2 bonding is responsible for the origin of 2.08 eV PL band.

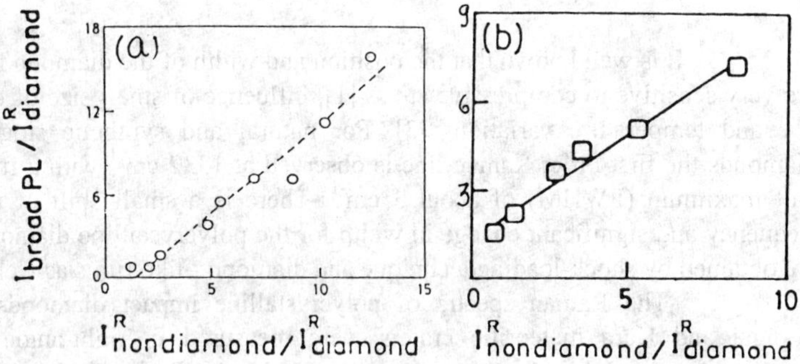

Figure 1. Variation of integrated intensity ratiosus $I^R_{nondiamond}/I^R_{diamond}$ and $I_{broadPL}/I^R_{diamond}$ for (a) as-grown diamond films, and (b) a diamond film at different heat treatment temperatures.

1. Sanjay Bhargava, H. D. Bist, S. sahli, M. Aslam and H. B. Tripathi, Appl. Phys. Lett. **67**, 1706 (1995).
2. Sanjay Bhargava, H. D. Bist, A. V. Narlikar, S. B. Samanta, J. Narayan and H. B. Tripathi, J. Appl Phys. **79**, 0000 (1996).
3. L. Bergman, B. R. Stoner, K. F. Turner, G. T. Glass and R. J. Nemanich, J. Appl. Phys. **73** 3951 (1993).

RAMAN LINESHAPE OF THE NATURAL IMPACT DIAMONDS

E. N. Fedorova[1], T. N. Moroz[2], V. P. Afanasiev[1] and G.M. Rylov[1]

[1]Institute of Mineralogy and Petrography, 630090 Novosibirsk, Russia
[2]Institute of Geology, 630090 Novosibirsk, Russia

It is well known that the position and width of the diamond Raman line are very sensitive to compressive stress [1], influence of small size of the crystals [2], and temperature variations [3]. For natural and synthetic single crystals diamonds the first-order Raman line is observed at 1332 cm^{-1} with a full width at half maximum (FWHM) of about 3 cm^{-1}. There is a small shift of the Raman frequency and significant change in width for the polycrystalline diamonds, which are obtained by shock-loading technique and diamond - like films as well.

The Raman spectra of polycrystalline impact diamonds from the Popigai and Kara meteoritic craters were measured in right-angle scattering geometry using Jobin Ivon spectrometer RAMANOR U 1000 with a single channel detection system (Fig.1). Exciting was perform with 514.5 nm line of Ar ion laser using cylindrical focalized lens for the heat decrease.

The observed Raman spectra of impact diamonds are superimposed by intensive luminescence background. The pure luminescence spectrum is reasonable well approximated by the Gaussian [4]. The spectrum at Fig.2 reflects a difference between the Raman plus luminescence spectrum of polycrystalline impact diamond and background of Gaussian form with a peak at 900 cm^{-1}, FWHM = 2000 cm^{-1}. We revealed anomalies in Raman lines of impact diamonds compared to those observed for the natural monocrystalline diamond (Fig.1). The Raman spectrum of the impact diamonds shows an intensive assymetric broading with a superposition of several lines.We used the bandfitting algorithms for separation of the components. Three Lorentzian curves were required to fit the spectrum. Position and width of lines were optimized by nonlinear least squares techniques for all spectra. The results of the fit are presented in Fig.2. We observed a splitting of the diamond Raman line in three lines of 1323, 1329.5, 1336 cm^{-1} for Kara impact diamond and 1326, 1331 and 1336 cm^{-1} for Popigai impact diamond.
Knight and White reported the small Raman band in 1315-1326 cm^{-1} range for diamond shock prepareted powder [5]. The appearance of additional peak of weak intensity may be caused by two reasons. Firstly, when the diamond crystals have small size ($d < 200$ Å) the position of the Raman line peak shifts to lower frequencies and the intensity decreases significantly [2]. In this case there is a mode in 1150 -1180 cm^{-1} range characterizing the frequency of the Brilluin zone

boundary. We observe a small shoulder in 1150 cm^{-1} range for the Popigai samples. On the other hand, the low-frequency band possibly can be exploited to irregular distribution of stacking faults on {111} planes of diamond grains [6]. These stacking faults in the cubic diamond have a structure of hexagonal diamond (lonsdaleite).

In principle, the numerous randomly distributed boundaries of twinning at the surface of diamond shock - produced grains may be considered as thin layers with hexagonal symmetry. The existence of such randomly oriented thin regions of the hexagonal diamond may influence the first-order Raman peak. Some authors reported that the Raman line of diamonds with lonsdaleite has a significant broading up to FWHM = 30 cm^{-1}. The high-frequency splitting (peak at 1336 cm^{-1}) is likely to be related with internal stress [5]. The Raman peak shifts towards the high wavenumber with increase of one-axial pressure and shift values $\delta\nu_{[100]}$=0.073 cm^{-1}/kbar and $\delta\nu_{[111]}$= 0.22 cm^{-1}/kbar were determined by Grimsditch et al. [1]. Thus, it is possible to estimate internal stress from the shift of Raman line. It seems the majority of stresses are related to the grain boundaries in which the density of defects increases. The possiblity is discussed as to the use of changes of the Raman lines shape for the investigation of the natural diamonds formed at different conditions.

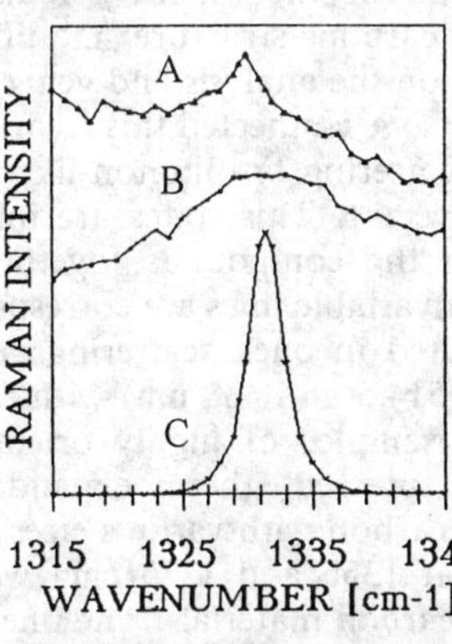

Fig. 1. Raman spectra of natural diamonds: A,B- impact samples from Kara and Popigai crater, respectively; C - single crystal

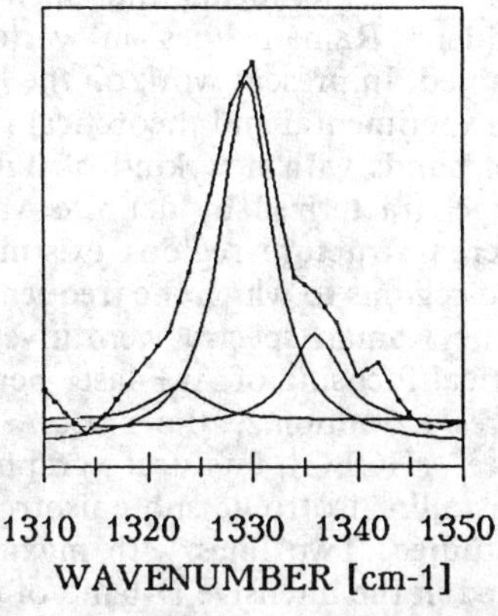

Fig. 2. Resolved spectra of Kara diamond: experimental line and three components of the fit with maxima at 1323,1329.5,1336 cm^{-1}

1. M.H.Grimsditch, E. Anastassakis and M. Cardona, Phys. Rev. B18, 901 (1978)
2. L.P. Avakyants, V.S. Gorelik et al, Kratkie soobshenya po physike 4, 28 (1994)
3. Y. Sato, New Diamond 1, 30 (1988)
4. C.K.Mann and T.J. Vickers, Appl.Spectr. 41, 427 (1987)
5. D.S. Knigh and W.B. White, J. Mater.Res. 4, 385 (1989)
6. S.A.Stuart, S. Prawer and P.S. Welser, Diamond and relat. mater., 2, 753 (1993)

PHONON-LIKE SCATTERING ON INTERMEDIATE CARBON STRUCTURES

M.E.Kornienko*, A.P.Naumenko

Department of Physics, Kyiv Taras Shevchenko University, Volodymirs'ka St.64, 252017, Kyiv, Ukraine

How it was noted by some authors [1-4] the position of the corresponding to fundamental modes spectral bands (~1350, 1380, 1550 and 1600 cm^{-1}) and some of the overtones depends on the exciting beam frequency and the technology of the sample preparing. And at the same time the frequencies of other lines (1582, 1332 cm^{-1}) stay practically unchanged. There is one of the most interesting peculiarity of the Raman scattering in carbon materials (disordered graphite, diamond-like films, glassy carbon, etc.). The experimental investigations which we have carried out earlier [5,6] revealed complicated structure of vibration bands of graphite and diamond- like materials. Moreover due to this experiments the phenomena of the intensity redistribution of "forbidden" Raman lines in various carbon structure modifications was determined. In present work on the base on the analysis and general conclusion of the experimental and theoretical results we connected this frequency variable Raman bands with new kind of light scattering on phonon-like states, which doesn't characterized by definite wave vector. This states are inherent in the disordered structure regions existing in the compounds together with more ordered regions to which the frequency invariable lines are correspondent.

The Raman spectra were investigated in back scattering geometry with cylindrical focusing of Ar+-laser beam (514.5 and 488 nm) with taking all due measures to minimize the noises. The samples of highly oriented pyrolytic graphite (HOPG), natural graphite, pressed thermoexpanded graphite, polycristalline isotropic and anisotropic carbon with various sizes of cristallites were studied. Two lines with maxima at 1350 and 1375 cm^{-1} were distinctly observed in the intensive D-band of the carbon materials. The line 1350 cm^{-1} is more intensive in the anisotropic and some isotropic samples of graphite , but the band at 1375 cm^{-1} prevailing the disordered carbon and in the case of the detection Raman spectra from the face perpendicular to the basal planes (which is demonstrated on fig.1a) . In spite of existence the bands with 1350 cm^{-1} in carbon and diamond with isolated Nitrogen-centers we connect this band with Raman scattering on graphite-like structure characterizing with maximum of vibratinal states density (VSD) in this region. And the band 1375 cm^{-1} is caused by vibration modes of the carbon structure which have different types of chemical coupling including sp^2 and sp^3 hybridization, or in the other words, this structure is intermediate one between the graphite and diamond-like coordination. Maybe this band correlates with infrared band 1368 cm^{-1} in diamond with high contents of the plate formations. The asymmetric band 1540cm^{-1} with low frequency shadow 1510 cm^{-1} either corresponds to intermediate structure. The existence of intensive bands 1375 and 1550 cm^{-1} in diamond-like films [3-5] testify to essential contents of environ tetrahedral regularity in intermediate structures. The observed doublet structure seems cannot be explained by the conception of double coupling resonance [2] since the intensities of components change in essential limits and the band-shapes depend on frequency of exciting radiation (fig.1a,b) and each band component can be registered separately. To separate the shape of the overlapping spectral bands we used the method of spectra difference.The dependence of intensity of the intermediate structure bands on frequency of exciting radiation can be caused for resonant character of Raman scattering and photoexcited carrier coupling with lattice.

In the majority of publications devoted to Raman scattering in disordered materials all the vibration states considered as localized and the VSD is under discussion. But for numerous intermediate states of disorder between states with long-range order and full chaos the vibration modes still keep their pronounced wave characteristics and they delocalized in some space region. The essential vibrational zone structure disturbances characterized to phonon-like excitations of this intermediate structures lead to possibility of the observing spectral band position variation. In the real materials with practically unchanged substance density the space inhomogeneity in the types of chemical coupling between atoms can exist and the sharp lines with fixing frequencies may be occurred equally with frequency variable bands. On Fig.1. we show the vibrational bands shapes in the environs of "forbidden" and Raman-allowed modes for different structure modifications of graphite: highly oriented pyrolytic graphite (1), anisotropic (2) and some different isotropic (3) policrystalline samples, thermoexpanded (4) and natural (5) graphite. The spectra obtained from face perpendicular to basal plane are marked with *t*.

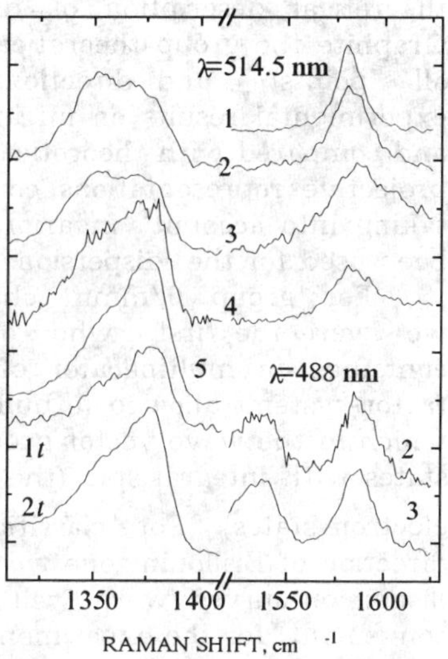

1. Vidano R.P. Fishbach D.B., Willis L.J., Loehr T.M., Sol.St.Commun., vol.39, 341 (1981)
2. Baranov A.V., Bechterev A.N., Bobovich Ya.S., Petrov V.P., Opt.& Spectr., vol.62, 1036, (1987)
3. M.Yoshikawa, G.Kalagiri, H.Ishida, A.Ishitani, T.Akamatsu, J.Appl.Phys., vol.64, 6464 (1988)
4. N.E.Kornienko, A.P.Naumenko, A.G.Gontar, S.A.Ivanov, M.A.Voronkin, in Using Prospects of Diamonds in Electronics and Electronic Techniques, Moscow, 1991
5. N.E.Kornienko, A.P.Naumenko, in press

DISPERSION OF PHONON AND ELECTRON EXCITATIONS IN GRAPHITE SINGLE CRYSTALS

V.A.Gubanov*, A.P.Naumenko

Departmrnt of Physics, Kyiv Taras Shevchenko University, Prosp.Akademik Glushkov, 6, Kyiv-22, Ukraine

Because of existing in literature [1-3] some differences in theoretical description of dispersion of phonon and electron states in Graphite the group-theoretical analyze of elementary excitations for all points and directions in Brillouin zone has been done. The experimental results on infrared and Raman spectroscopy discussed and compared with theoretical predictions. New original method of the projective representations construction of wave vector groups with taking into account invariance of energy states to time inversion has been used for the dispersion curves calculation in graphite.

For group 6/mmm characterizing crystal class of graphite we were the first, who obtained standard factor systems, which contained in multiplicator of projective classes K_1, K_3 and K_6. This factor-systems allowed to build irreducible projective representations by which in the wave vector groups it could be realized the classification of states with integral spin (the phonon states) and half-integral spin (the electron states). For classifications of elementary excitations in Γ-A direction of Brillouin zone we used the double Jones zone in which the dispersion curves were well described theoretically and especially were convenient for the experimental results interpretation.

The same calculations fulfilled for other directions of Brillouin zone (M-U-L, K-P-H, etc.) allowed to interpret exactly the second order Raman spectrum. The developed method also gives possibility to investigate elementary excitations dispersion in graphite intercalated compounds and even such complicated structures as fullerens and fullerites.

The dispersion curves on phonon states in graphite for direction Γ-A of Brillouin zone and Jones zone are given on fig.1a and fig.1b respectively. This curves were built with according to the experimental data of the first-order Raman spectra and were used for interpretation the second-order Raman spectrum bands. Thus, the band near 3181.6 cm^{-1} observed earlier seems to be overtone caused by scattering on A-point phonons with symmetry $A^{(6)}$. The second-order spectrum bands 2710, 2450 and 3250 cm^{-1} are caused by contribution of Γ-M direction and especially of point M. In present work the dispersion curves shape for the electron states in graphite is also under discussion. It has been established the expediency of the Jones zone conception application. The qualitative comparisons of the electron states dispersion dependencies in graphite crystals and hexagonal boron nitride has been done.

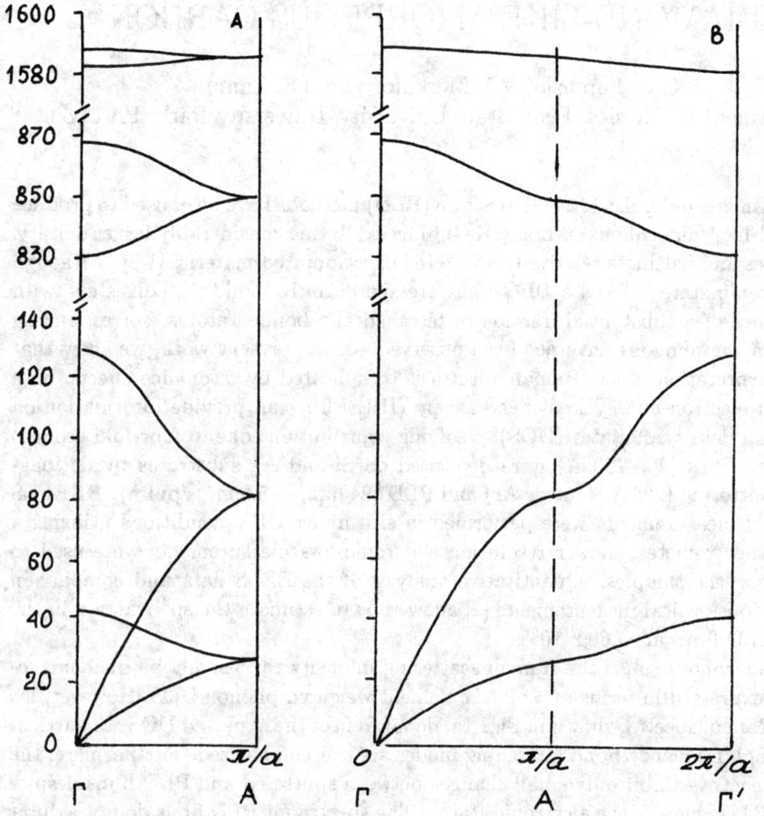

Fig.1. The dispersion curves on phonon states in Graphite for direction Γ–A in Brillouin zone (a) and Jones zone (b)

1 J.A.Young, and J.U.Koppel, J.Chem.Phys.42, 357 (1965)
2 R.Nicklow, N.Wakabayashi, and H.G.Smith, Phys.Rev. B5, 4951 (1972)
3 R.J.Nemanich and S.A.Solin, Phys.Rev. B20, 392 (1979)

RAMAN SCATTERING AND HREELS OF DIAMOND-LIKE AMORPHOUS CARBON

G.P. Lopinski, V.I. Merkulov and J.S. Lannin
Department of Physics, Penn State University, University Park, PA 16802

Vacuum arc and pulsed laser deposition (PLD) methods have been used to produce "diamond-like" amorphous carbon (DLC) films exhibiting considerably larger density, resistivities and hardness relative to sputtered or evaporated material [1,2]. Although the electronic states of these DLC films are significantly modified, consistent with the presence of a substantial fraction of tetrahedrally bonded atoms, corresponding changes in the phonons have not been observed. In the present work, we show that while interpretation of the Raman spectra is complicated by resonance effects, high resolution electron energy loss spectroscopy (HREELS) can provide information on the phonon density of states (DOS), resolving contributions due to four-fold atoms.

Carbon films (400-700 Å) were deposited on Si and Ag substrates by dc magnetron sputtering (225W, 5mtorr Ar) and PLD (248nm, $\sim$ 5Jcm^{-2}/pulse). HREELS vibrational measurements were performed in situ under UHV conditions utilizing a LK2000 spectrometer. Lower resolution electronic loss measurements were used to characterize the samples. Quantitative analysis of the EELS data and comparison with previous optical measurements [2] allowed us to estimate the sp^3 fraction in the current PLD films at $\sim$ 60$\pm$ 10%.

For amorphous solids the Raman scattering intensity can usually be phenomenologically expressed in terms of a matrix element weighted phonon DOS. However, for a-C the Raman spectra, shown in Fig. 1a, do not reflect the expected DOS. In particular, the high frequency bond-stretching modes are overemphasized. Furthermore, the Raman spectra exhibit only small changes between sputtered and PLD films despite substantial changes in the electronic states. The spectra for PLD films do not exhibit contributions of four-fold bonded atoms expected to have a peak in the density of states at $\sim$1300cm^{-1} [3]. In fact, the high frequency peak exhibits a slight shift to higher frequency. The scattering below 1000cm^{-1} also resembles the three-fold DOS. This is attributed to resonant enhancement of the three-fold contributions due to the much smaller local gap associated with $\pi - \pi^*$ transitions.

For impact scattering from an amorphous solid, the HREELS response should correspond to a matrix element weighted DOS. In order to extract the vibrational contributions from the measured spectra, the quasi-elastic peak was removed by reflecting the peak about zero loss energy, subtracting the anti-Stokes from the Stokes scattering. A background due to low-lying electronic transitions was then subtracted and the data multiplied by frequency to yield the weighted DOS shown in Fig. 1b.

The HREELS determined DOS for sputtered a-C is in good agreement with theoretical calculations for a random network model of a-C with 92% sp^2 bonding [4]. In

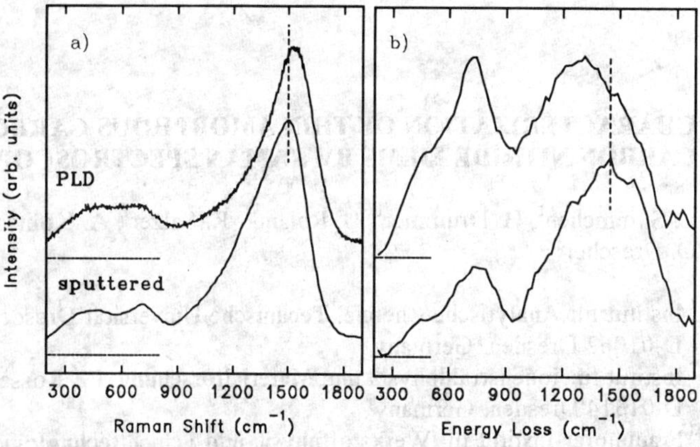

Figure 1: a)Raman spectra (5145nm excitation) for a-C films prepared by sputtering and PLD methods. b) Weighted phonon DOS extracted from HREELS data (E_i=10eV) as described in the text.

contrast to the Raman spectra, substantial differences are noted for the DLC films. In particular, the high frequency peak is observed to broaden, its centroid shifting downwards by $\sim$ 150 cm^{-1}. This shift is attributed to contributions of tetrahedrally bonded atoms as the broadened DOS of crystalline diamond exhibits a peak at $\sim$1300cm^{-1}[3]. A downward shift of the high frequency peak with increasing sp^3 content is also consistent with theoretical phonon DOS obtained in random network models[4] and molecular dynamics simulations[5].

The short-range nature of the impact scattering mechanism implies that the weighted DOS measured in HREELS is characteristic of the first few layers of material. Agreement with theoretical calculations of the bulk DOS, noted above, suggests that the surface phonons are not dramatically different from that of the bulk for both predominantly three-fold and DLC films. Although the structure of crystalline diamond surfaces are significantly modified due to surface reconstruction, disorder and strain effects in a-C make such a rearrangement unlikely in the DLC films.

This work was supported by NSF Grant DMR 92241665.

[1] D.R. McKenzie, D. Muller and B.H. Pailthorpe, Phys. Rev. Lett. **67**, 773 (1992).
[2] D.L. Pappas et al., J. Appl. Phys. **71**, 5675 (1992).
[3] W.S. Bacsa, J.S. Lannin, D.L. Pappas, and J.J. Cuomo, Phys. Rev. **B47**, 10931 (1993).
[4] D. Beeman, J. Silverman, R. Lynds, and M.R. Anderson, Phys. Rev. **B30**, 870 (1984).
[5] C.Z. Wang and K.M. Ho, Phys. Rev. Lett. **71**,1184 (1993).

CHARACTERIZATION OF THIN AMORPHOUS CARBON AND CARBON NITRIDE FILMS BY RAMAN SPECTROSCOPY

L. Sümmchen[1], H. Drummer[1], U. Roland[1], R. Salzer[1], A. Kolitsch[2] and D. Drescher[3]

[1] Institut für Analytische Chemie, Technische Universität Dresden, D-01062 Dresden, Germany
[2] Institut für Ionenstrahlphysik und Materialforschung, FZ Rossendorf, D-01314 Dresden, Germany
[3] Fraunhofer-Institut für Werkstoffphysik und Schichttechnologie, Helmholtzstr. 20, D-01069 Dresden, Germany

Increasing interest is devoted to the investigation of thin amorphous carbon and carbon nitride films due to their importance in materials science and technology. Amorphous carbon films show diamond-like properties such as high hardness, chemical resistance, special optical properties and a very high electrical resistivity [1]. The interest in the synthesis of carbon nitride films is focused on the β-C_3N_4 structure theoretically predicted by Liu and Cohen [2,3]. The β-C_3N_4 phase is expected to have a hardness comparable to or higher than diamond.

The thin amorphous carbon films studied in the present contribution were prepared by the laser-arc and the IBAD (ion beam assisted deposition) method [4,5], the carbon nitride films were synthesized using the IBAD technique.

Raman spectroscopy is a suitable tool for the investigation of carbon and carbon nitride films since it allows fast and non-invasive measurements. It provides information about structure, phase purity and crystallinity. The thin film samples were measured using a microscope connected to the dispersive Raman spectrometer Dilor XY 800. The excitation wavelength was 514 nm (Ar^+ ion laser). A laser power of 3.7 to 10 mW (on the sample) was applied. Structural modifications of the measured sample did not occur under these conditions.

The spectra of amorphous carbon films as well as carbon nitride films are characterized by a broad band in the spectral region between 1000 and 1800 cm^{-1}. Carbon nitride films show an additional weak band at about 2225 cm^{-1}. This band corresponds to the carbon-nitrogen triple bond and indicates that nitrogen is chemically bound to carbon. Nevertheless, Fig. 1 shows that both spectra are very similar in respect to the dominance of the broad peak between 1000 and 1800 cm^{-1} related to the carbon framework. This peak can be quantitatively described by the ratio of the intensities of the so-called „G“ (graphite) and „D“ (disorder) peaks defined as I_D/I_G. The concerning simulation has been carried out using the program GRAMS (Galactic Industries).

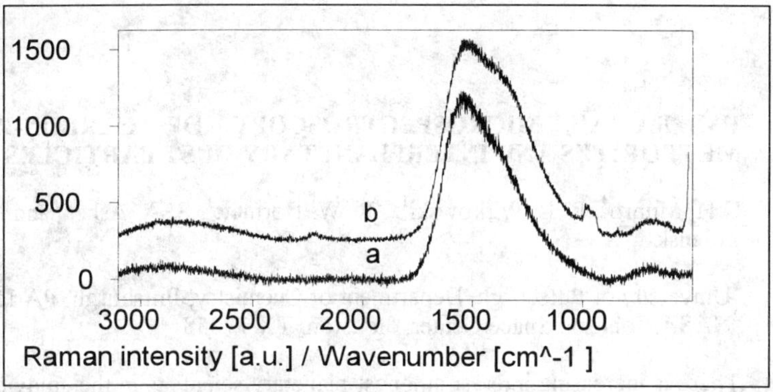

Fig. 1 Raman spectra of ***(a)*** an amorphous carbon and ***(b)*** a carbon nitride film

According to a model calculation by Beeman et al. [6] changes of the bond angle and length as well as an increasing sp^3 content lead to a shift of the Raman peaks to lower wavenumbers. Especially, the shift of the „G" peak indicates a change of the bond angle and the shift of the „D" peak corresponds to an increase of the sp^3 content or to increasing disorder. Changes of the I_D/I_G ratio can be interpreted by changes of the number and/or the size of graphitic clusters [7]. Such changes are closely connected to changes of mechanical properties like micro-hardness [8].

Our Raman spectroscopic investigation of carbon and carbon nitride films was carried out in order to correlate preparation parameters with information available from the Raman spectra. Carbon and carbon nitride films prepared at different temperatures between 30^0C and 200^0C were investigated. The temperature was kept constant during the deposition process. It was shown that the ratio I_D/I_G increased with increasing preparation temperature due to an increasing number and/or size of the graphitic clusters. A shift of the „G" peak to higher wavenumbers and a decrease of its band width indicate the formation of graphite structures. The „D" peak was also shifted to higher wavenumber but its width increased in contrast to the „G" peak. This behaviour can be interpreted by a decrease of the content of sp^3 coordinated carbon atoms. The formation of graphitic clusters at higher preparation temperatures is due to the thermal modification of the film. Obviously, the structure of the film is disordered when the preparation is carried out at lower temperatures, e.g., room temperature. Graphite formation due to healing is favoured by a sufficiently high substrate temperature which can be observed for carbon films as well as for carbon nitride films.

1. N.N. Cho, K.M. Krishnan, D.K. Veirs, M.D. Rubin, C.B. Hopper, B. Bhushan and D.B. Bogy, J. Mater. Res. 5, 2543 (1990).
2. A.Y. Liu and M.L. Cohen, Science 245, 841 (1989).
3. A.Y. Liu and M.L. Cohen, Phys. Rev. B 41, 10727 (1990).
4. H.-J. Scheibe, P. Siemroth, B. Schöneich and A. Mucha, Diamond and Relat. Mater. 1, 98 (1992).
5. A. Kolitsch, E. Hentschel and E. Richter, Diamond and Relat. Mater. 2, 1175 (1993).
6. D. Beeman, J. Silverman, R. Lynds and M.R. Anderson, Phys. Rev. B 30, 870 (1984).
7. R.O. Dillon, J.A. Woollam and V. Katkanant, Phys. Rev. B 29, 3482 (1984).
8. R. Salzer, U. Roland, H. Drummer, L. Sümmchen, A. Kolitsch and D. Drescher, Z. Phys. Chemie 191, 1 (1995)

UV RAMAN MICROSPECTROSCOPY OF CARBON-RICH METEORITES AND INTERPLANETARY DUST PARTICLES

C.H. Munro[1], R. E. Witkowski[1]*, R. W. Bormett[1], S. A. Asher[1] and M. E. Zolensky[2]

[1]University of Pittsburgh, Department of Chemistry, Pittsburgh, PA 15260
[2]NASA, Johnson Space Center, Houston, TX 77058

There is increasing interest amongst planetary scientists in the study of the carbon and carbonaceous species present in carbon-rich meteorites and interplanetary dust particles (IDP). Recent work by Zinner et al.[1] showed that much can be learned about carbon species in meteorites such as Murchison using ion microprobe analysis techniques and visible Raman spectroscopy systems. However, extensive mechanical and chemical separation of the phases of interest must be carried out prior to analysis. In addition, the long accumulation times (>8 hours) and relatively high excitation powers (up to 5 mW) required to produce adequate signal-to-noise (S/N) visible Raman spectra may result in physical and chemical changes to the sample that compromise the measurement.

UV Raman instrumentation has evolved rapidly over the last few years and we have demonstrated the superiority of new continuous wave (CW) UV laser sources[2,3]. These intracavity frequency doubled laser sources have several discrete output wavelengths between 206.5 nm and 257.3 nm. In addition, the ability to focus the CW laser spot to a size that can be efficiently imaged into the narrow entrance slits of a high resolution Raman spectrometer permits much higher S/N ratios. Most importantly, we have demonstrated that the CW laser can be used to examine thermally sensitive samples including strongly absorbing solids and the beam can be focused to small diameters without saturation and non-linear phenomena and with comparatively small photodegradation of the sample. Most recently we demonstrated the high sensitivity, selectivity, spatial resolution and ease of operation of UV Raman microspectroscopy using a new highly efficient UV Raman microspectrometer with excitation at 244 nm[4]. We demonstrated the ability to selectively excite high S/N Raman spectra from spatially resolved areas of solid, absorbing samples and biological samples using low CW excitation powers and short accumulation times to minimize sample damage.

To demonstrate the application of UV Raman microspectrometry to study meteorite and IDP samples, we examined two meteorite fragments from Ivuna and Murchison, and two IDP samples identified as L2006, cluster 4 and cluster 14. The meteorite fragments were approximately 2 x 2 mm in size, while the IDP samples comprised grains of the order of a few tens of microns in size. The IDP samples were pressed into gold foil for handling, otherwise UV Raman microspectroscopic measurements were made "as-received" without mechanical or chemical separation. Sample results of this study are presented in Figures 1 and 2. Figure 1 shows the intense band at 1588 cm^{-1} observed from one of the IDP particles (L2006, cluster 4). This may be assigned to glassy carbon. Glassy carbon is a semi-crystalline phase described by Robertson [5] as entangled ribbons of graphitic polymeric molecules. A measure of the disorder present in the sample is further suggested by the broad band observed at approximately 1355 cm^{-1}[6]. Figure 2 shows the UV Raman spectrum obtained from one area of the meteoritic fragment Murchison. The band can be fitted to known carbon bands at 1355 cm^{-1}, 1553 cm^{-1}, 1582 cm^{-1} and 1622 cm^{-1}. We note a close agreement between our data and those of Zinner [1] but further suggest the possible contribution by amorphous carbon at ca. 1553 cm^{-1} to the overall carbon band envelope.

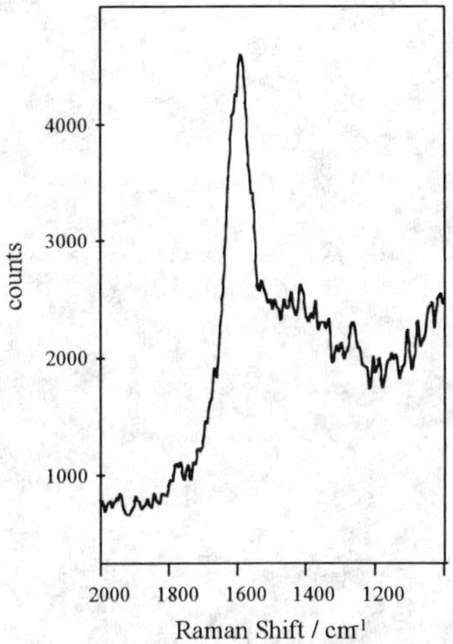

Fig. 1. UV Raman spectrum of Interplanetary dust particle L2006, cluster 4 (244 nm excitation: Spot diameter 10 μm: average power 1.5 mW: accumulation time of 100 s.)

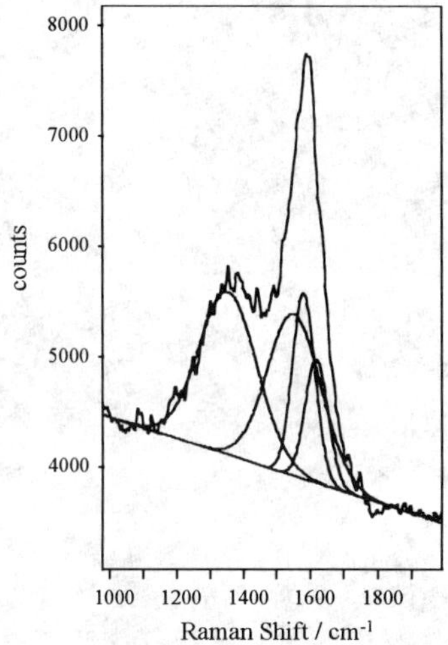

Fig. 2. UV Raman spectrum from spatially resolved area of meteorite fragment, Murchison (244 nm excitation: Spot diameter 5 μm: average power 0.5 mW: accumulation time 180 s.)

1. E. Zinner, S. Amari, B. Wopenka and R. S. Lewis, *Meteoritics*, 30, 209 (1995).
2. S. A. Asher, R. W. Bormett, X. G. Chen, D. H. Lemmon, N. Cho., P. Peterson, M. Arrigoni, L. Spinelli and J. Cannon, *Appl. Spectrosc.*, **47,** 628-633 (1993).
3. J. S. W. Holtz, R. W. Bormett, Z. Chi, N. Cho, V. Pajcini, S. A. Asher, M. Arrogoni, P. Owen and L. Spinelli, *Appl. Spectrosc.*, submitted.
4. V. Pajcini; C. H. Munro; R. W. Bormett; R. E. Witkowski and S. A. Asher, *Appl. Spectrosc.*, submitted.
5. J. Robertson, in *Advances in Physics*, Vol 35, 4 (1986) p. 317.
6. Y. Wang, D. C. Alsmeyer and R. L. McCreery, *Chemistry of Materials*, 2, 557 (1990).

SECTION 22

SEMICONDUCTORS and MICROSTRUCTURES

RAMAN SCATTERING CHARACTERIZATION OF WIDE-BAND GAP SEMICONDUCTORS

S. Nakashima and H. Harima

Department of Applied Physics, Osaka University
2-1Yamadaoka, Suita, Osaka 565 Japan

Recently, wide band gap semiconductors have attracted much attention from view point of basic physics and application to optoelectronic devices. However, their electrical properties have not been well understood as for Si and GaAs. Raman scattering measurement has been proved to be a powerful tool for studying the electric transport properties of the polar semiconductors. This technique is very convenient, because it does not need electrodes and non-destructive analysis is possible. In this paper, we will present results of the Raman scattering analysis of electrical properties in wide band gap semiconductors such as SiC, GaN and II-VI semiconductors.

Free carriers in semiconductors form plasmons which are coupled with LO phonons through their macroscopic polarization fields. Since the line shape of the plasmon-LO phonon coupled modes depends on the carrier concentration and carrier damping, the fitting of the experimental line shape to the theoretical one enables us to determine these quantities. The plasmons in SiC are usually overdamped ($\omega_p\tau < 1$) and the frequency shift of the upper branch relative to that of the pure LO phonon is small [1], while those in n-type GaN are underdamped and the frequency shift by the coupling is large [2]. In Fig. 1 we compare the spectra of n-SiC and n-GaN crystals. The lower branch of the coupled mode is observed in GaN like GaAs. The free carrier(electron) concentration and mobility determined from Raman scattering analysis of α,β-SiC[1,3] and GaP[4] are in a close agreement with results of Hall measurements.

Highly doped SiC shows Raman scattering components in low frequency region which arises from intraband and intervalley transitions of free electrons, and less heavily doped crystals show bound electron transitions which are ascribed to electronic transitions between valley orbit splitting states of the donor 1s level[5]. These electronic states form continuos electronic transition bands which interfere with folded transverse acoustic phonon modes lying in a low frequency region.

Figure 2 shows Fano-interference between single particle excitation and sharp phonon bands in nitrogen doped SiC polytypes. The distortion of the phonon bands varies systematically with the carrier concentration. Accordingly, the free carrier concentration is estimated from the interference patterns. We have attempted to determine the carrier concentration from the interference effect for 4H and 6H SiC.

The behavior of the plasmon-LO phonon coupled mode in p-type SiC is quite different from that of n-type SiC. The peak frequency of the coupled mode seems to be almost independent of the hole concentration, although its peak intensity decreases with increasing carrier concentration. This may be caused by heavy damping of the carriers. Heavily doped p-type SiC also shows the Fano interference effect for TO modes as well as folded transverse acoustic modes in low frequency region. This fact suggests that the broad background in p-SiC arises from inter valence band transition.

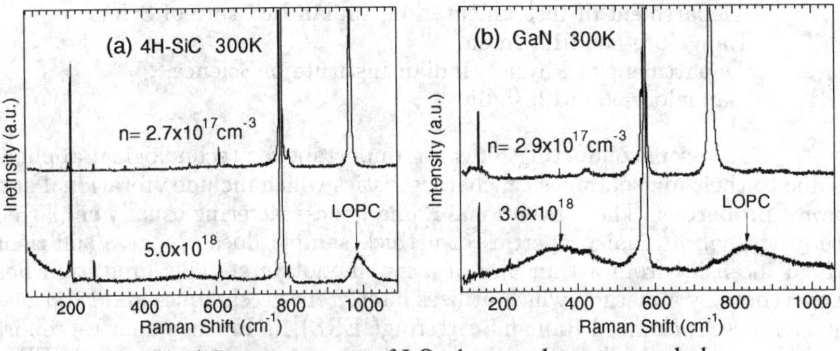

Fig.1 Raman spectra of LO-phonon plasmon coupled mode in n-type 4H-SiC (a), and n-GaN (b) (arrows).

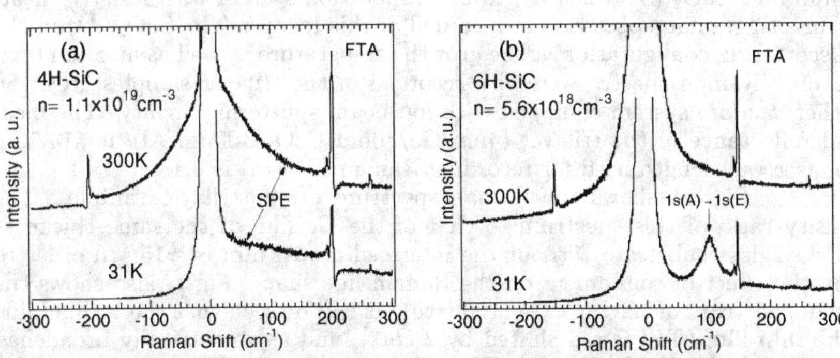

Fig.2 Raman profiles of folded transverse acoustic mode interfered with single particle excitation (a), and valley-orbit transition band of donors (b).

1. Harima, S. Nakashima and T. Uemura, J. Appl. Phys. 78, 1996(1995)
2. Harima *et al.* to be published
3. H. Yugami, S. Nakashima, A. Mitsuishi, A. Uemoto, M. Shigeta, K. Furukawa, A. Suzuki and S. Nakajima, J. Appl. Phys. 61, 354(1987)
4. G. Irmer, V. V. Toporov, B. H. Bairamov and J. Monecke, Phys. Status Solidi B119, 595(1983)
5. M. V. Klein, B. N. Gangly and P. J. Colwell, Phys. Rev. B6, 2380(1972)

.

INTERFERENCE ENHANCED RAMAN SCATTERING OF VERY THIN Ge FILMS

S.Kanakaraju[1*], A.K.Sood[2] and S.Mohan[1]

[1]Department of Instrumentation, Indian Institute of Science, Bangalore-560 012, India.
[2]Department of Physics, Indian Institute of Science, Bangalore-560 012, India.

Thin semiconducting films are important for technological applications due to their unusual physical characteristics which include vibrational and electronic properties. The conventional Raman backscattering usually employed to study the vibrational properties of a thick sample does not give sufficient scattered intensity from a thin film of a few monolayers. This limitation has been overcome by a method which utilizes interference techniques and is termed as Interference Enchanced Raman Scattering(IERS)[1]. In this paper we report Raman study of confined optical phonon in in-situ grown Ge films using IERS.

Films of Ge were deposited on a glass substrate with layers of aluminium and CeO_2 (T = 235°C.) in an evaporation system with resistive heating. In-situ Raman spectra were recorded of this trilayer (Ge/CeO_2/Al) in the backscattering configuration at the growth temperature as well as at room temperature. Raman spectra were also recorded of the trilayer samples grown in another vacuum system equipped with ion beam sputtering facility. The measured reflectance of the trilayer (4nm Ge/30nm CeO_2/100nm Al) at 514.5nm (the laser wavelength used for recording Raman spectra) is 0.1.

Fig. 1 shows the Raman spectrum of the trilayer sample. The intensity ratio of this spectrum to that of the Ge film of the same thickness on CeO_2/glass substrate without the intermediate Al film is ≈15. In order to study the effect of confinment on the Raman line shape, Fig.1. also shows the Raman spectrum of the bulk Ge (100 face). It can be seen that the Raman line in the thin film (TF) is red shifted by 2 cm^{-1}, and asymmetrically broadened as compared with the bulk Ge: ω(TF)= 296.3 cm^{-1}, ω(bulk) = 298.3 cm^{-1}, full width at half maximum FWHM(TF)= 10 cm^{-1}, FWHM(bulk) = 8 cm^{-1}. The asymmetric, red shifted Raman line arising from the phonon confinment in a thin film of thickness L is given by [2]

$$I_c(\omega) = D \int_0^{q_{max}} \frac{|C(0,q)|^2 \, dq}{[\omega - \omega(q)]^2 + (\Gamma/2)^2} \qquad (1)$$

where C(0,q) is the Fourier coefficient of the the phonon confinment function, D is a constant, Γ is FWHM of the bulk Ge which also includes instrumental broadening, ω(q) is the phonon dispersion function of the corresponding bulk material. Taking the confinment function to be Gaussion[2],

$$|C(0,q)|^2 = exp\left(\frac{-q^2L^2}{16\pi^2}\right)\left|1 - erf\left(\frac{iqL}{\sqrt{32}\pi}\right)\right|^2 \quad (2)$$

The measured phonon dispersion[3] for the LO mode in bulk Ge has been fitted to

$$\omega^2(q) = A + [A^2 - B(1 - \cos(\pi q^{0.6}))]^{1/2} \quad (3)$$

with $A = 4.45 \times 10^4\,cm^{-2}$ and $B = 9.5 \times 10^8 cm^{-4}$

We find that $I_c(\omega)$ alone does not fit the data. However $I(\omega) = I_c(\omega) + I_L(\omega)$ can fit the data very well as shown in Fig.2. Here $I_L(\omega)$ is a Lorentian function centerd at $275.5cm^{-1}$ and FWHM $= 34.1cm^{-1}$ This additional component $I_L(\omega)$ is similar to that observed in Raman spectrum of Ge nanoparticles and has been attributed to the presence of Ge with other than diamond lattice and/or amorphous Ge.

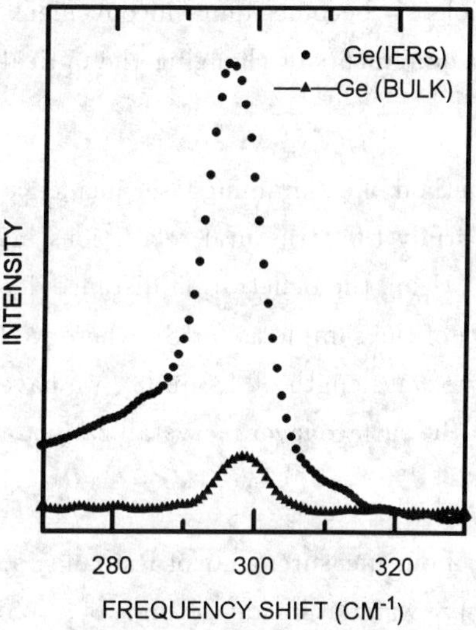

Fig.1.Raman spectrum of Trilayer and bulk Ge

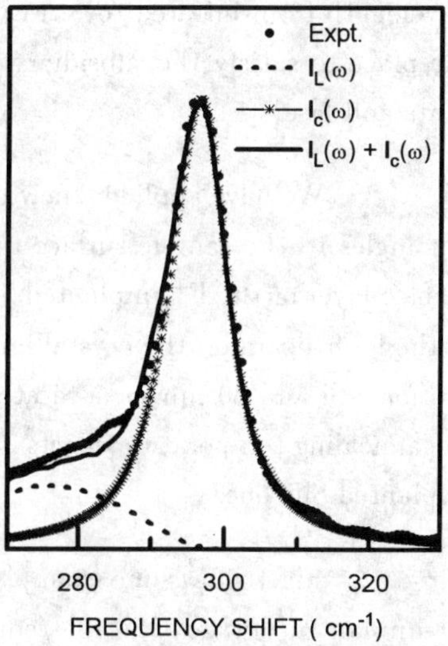

Fig.2. Experimental and fitted Raman Spectra

1. G .A. Connell, R.J. Nemanich and C.C. Tsai, Appl. Phys. Lett. 36, 31 (1980).

2. P.M. Fauchet and I.H. Campbell, Critical Reviews in Solid state and Materials Science, 14, S79(1988).

3. G.Nilsson and G.Nelson Phys. Rev. B3, 364(1971).

4. Y. Sasaki and C. Horie Phys. Rev. B47,3811(1993).

RAMAN SCATTERING FROM F^+-IMPLANTED Si

Yong-Jik Park, Yujun Mo[a)], Jong-Jean Kim*
Physics Department, KAIST
373-1 Kusung-dong, Yusung-ku, Taejon 305-701, Korea

Although Raman scattering has recently been used for depth profile studies of the ion-implanted surface layers of Si by changing the exciting laser wavelength from infrared to ultra violet, it becomes quite inconvenient to do at one place to satisfy the subsidiary requirements of changing the optical components and lasers[1,2].

We have applied a new method of a skimming laser incidence at varying angles to the sample surface to study the structural relaxations in the thin surface layer of the F^+-implanted Si. Using the penetration distance dependence of the laser beam on the crystallinity of the sample as for Si, where we have 485 nm for c-Si and 30 nm for a-Si at the wavelength of 488nm[3], we have studied the annealing temperature effects on the surface layer recrystallization of the F^+-implanted Silicons.

Fig.1 shows one example of our measurements of Raman spetra of the F^+-implanted(dosage of $5\times10^{14}/cm^2$) Si wafer at incidence angle of $\theta=55°$ for various annealing temperatures from no annealing to 700 °C. We have also analyzed the amorphous Raman band of the F^+-implanted Si by a Lorentzian band fit to obtain the peak frequency, bandwidth and bond angle variance[4], all depending on the annealing temperatures. The annealing behaviours will be discussed in terms of competition between thermal effects of annealing and bond-breaking F^+ ions.

1. V.V. Artamonov et al., Sov. Phys. Semicon. 25, 1025 (1992).
2. P. X. Zhang et al., J. Raman Spectroscopy 25, 515 (1994).
3. L. Ley, in HYDROGENATED AMORPHOUS SILICON(II), edited by joannopoulos and Lucovsky, Springer(New York,1984), p.61.

4. R. Tsu et al., Sol. St. Commun. 54, 447 (1985).

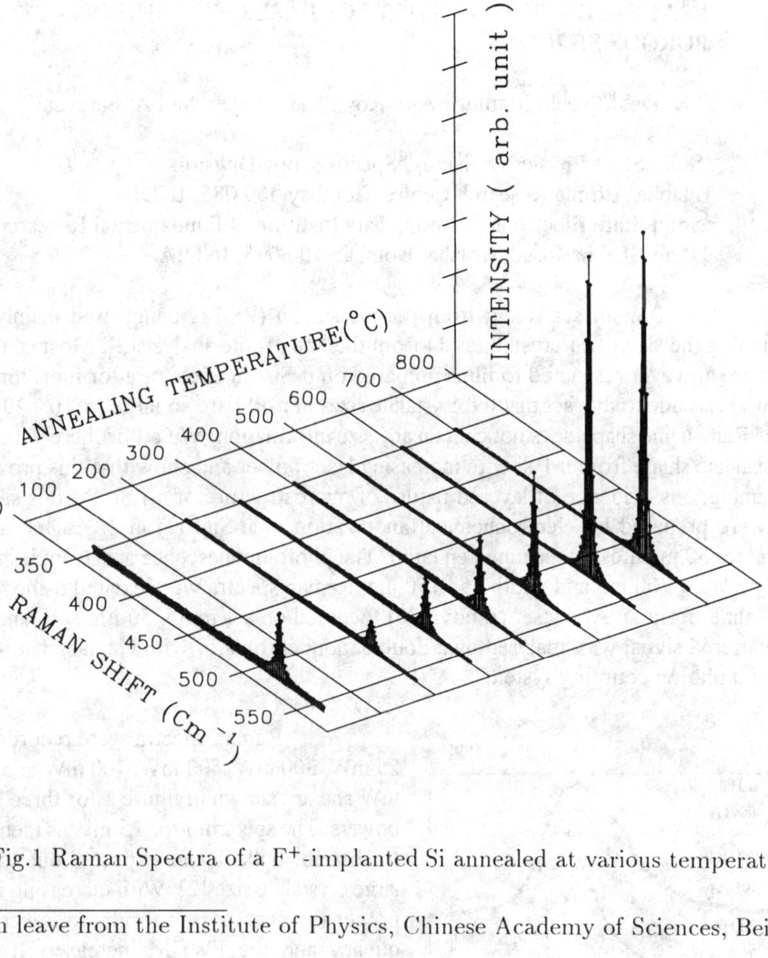

Fig.1 Raman Spectra of a F^+-implanted Si annealed at various temperatures.

---

a) on leave from the Institute of Physics, Chinese Academy of Sciences, Beijing, China

RAMAN SCATTERING STUDY OF LASER HEATING IN n- TYPE POROUS SILICON

S.K. Deb[1], Neelu Mathur[1], A.P. Roy[2], S. Banerjee[3] and A. Sardesai[3]

[1]Solid State Physics Division, [2]Spectroscopy Division
Bhabha Atomic Research Centre, Bombay 400 085, INDIA
[3]Solid State Electronics Group, Tata Institute of Fundamental Research, Homi Bhabha Road, Colaba, Bombay 400 005, INDIA

Raman scattering from porous silicon (PSi) is being used mainly for determining the size of Si crystallites[1] from the line profile analysis[2]. Most of these studies are however restricted to films prepared on p-Si while it is rare for films formed on n-Si This is due to the fact that the crystallite sizes in n-PSi are so large (≥ 10 - 20 nm) that the Raman line shape does not contain any size information. We report here variation in Raman line shape from n-PSi with increasing laser power and show that this provides an unambiguous evidence of layered nature of microstructure of n-PSi. Porous silicon films were prepared by electrochemical anodization of n-Si(100) in 48 % HF at 18 mA/cm^2 for 50 minutes under tungsten lamp. The photoluminescence spectrum is broad with a peak at ≈700 nm and width of ≈150 nm. Raman spectra were excited using 4880 A laser line from an Ar$^+$ laser (Innova-70) focussed on a 1 mm x 50 μm spot and the backscattered signal was analysed by a double monochromator with Hamamatsu R649 PMT and photon counting system.

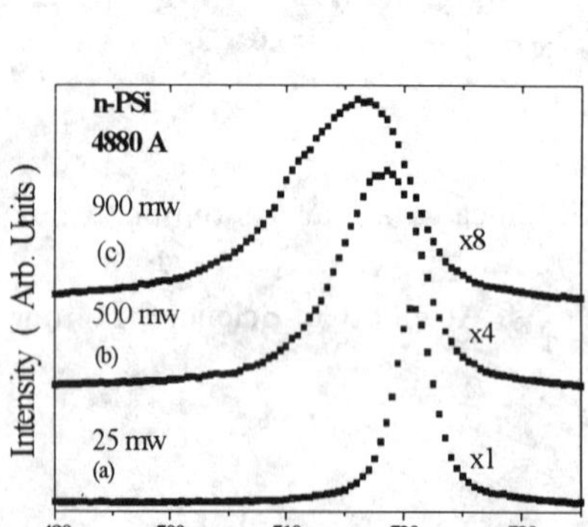

Figure 1 Raman line shape for laser powers shown.

Raman spectra were recorded at 25 mW, 300 mW, 500 mW, 700 mW and 900 mW and are shown in figure 1 for three laser powers. The spectrum for 25 mW is identical to that for bulk Si - as expected due to the large crystallite sizes[2]. With increasing laser power the peak position shifts to lower frequency and the FWHM increases. It also acquires a low frequency asymmetry which increases with laser power. These changes are due to laser heating, but the observed peak shifts donot agree with those expected for temperatures[3] obtained from Stokes/anti-Stokes ratio. The line shape has been found to change back to original when the laser power is reduced to its initial value.

The observed changes in the line shape can be explained by considering two layered structure of the PSi film as revealed by recent TEM investigations[4,5]. The top layer consists of a complicated network of pores and Si nanocrystallites of average sizes ≈ 10 - 20 nm similar to 'sponge' like morphology in p-PSi. The bottom layer is made up of 'rod' like Si crystallites of dimension 100 nm to a few microns. Because of presence of large number of pores and considerable amount of oxide the 'effective' thermal conductivity of the top layer will be 'poorer' than that of the bottom layer. At low laser power the laser heating is negligible and crystallite sizes for both the layers being ≥ 10 nm the line shape will be almost identical to the bulk Si. As the laser power increases there is considerable laser heating and the temperature of the upper 'sponge' like layer will be larger than the bottom layer due to its poorer thermal contact with the substrate. The Raman line position shifts to lower frequency and the FWHM increases with increase in temperature[3] due to anharmonicity and so the contribution from the top layer will be at a lower frequency and of larger width than that for the bottom layer. This leads to an asymmetric line shape with increase in FWHM which is seen in figure 1. The contribution from the two layers can be taken to be lorentzians and we fit the total line shape to experimental data at different powers and such a fit for 900 mW is shown in figure 2.

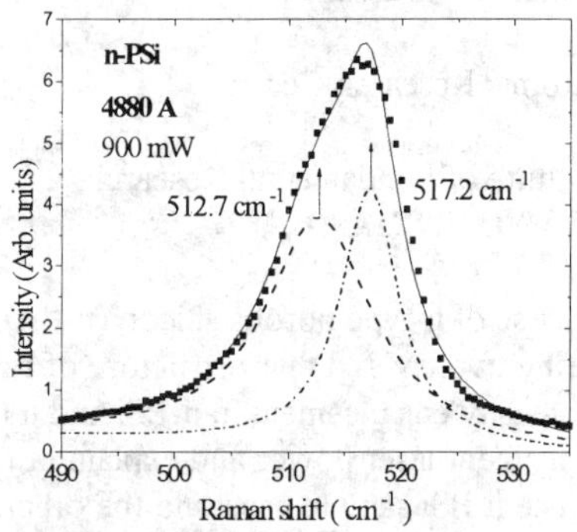

Figure 2 Fit to line shape at 900 mW with two lorentzians.

The position and width of the components from the bottom layer are 517.2 cm^{-1} and 6.2 cm^{-1} respectively while the corresponding values for the top layer are 512.7 cm^{-1} and 11.5 cm^{-1}. We obtain fairly good agreement with the observed data for all laser powers and observe that the peak position decreases and FWHM increases with increasing laser power for both the components. On reducing the laser power to the original low value the temperature of both the layers will decrease resulting in bulk Si like line shape. Our results differ significantly from similar results of laser heating on p-PSi[6] where the line shape remains lorentzian, which is due to the fact that the microstructure for p-PSi consists of one 'sponge' like homogeneous layer. This is a further justification for our model for the observed line shape changes in n-PSi. This is perhaps the first unambiguous optical evidence for layered structure of n-PSi.

1. S.K. Deb, Ind. Jr. Pure & Appl. Phys. 32, 638 (1994) and references therein.
2. I.H. Campbell and P.M. Fauchet, Solid State Commun. 58, 738 (1986).
3. M. Balkanski , R.F. Wallis and E. Haro, Phys. Rev, B28, 1928 (1983).
4. P.C. Searson, J.M. Macauley and F. M. Ross, J. Appl. Phys. 72, 253 (1992).
5. K. Takemoto, Y. Nakamura and C. Nittono, Jpn. Jr. Appl. Phys. 33, 6432 (1994).
6. E.D. Obraztsova, L.P. Avakyants and G.B. Demidovich, J. Elec. Spec & Rel. Ph. 64/65, 857 (1993).

RAMAN INTENSITY ENHANCEMENT IN n-TYPE POROUS SILICON :VARIATION WITH TEMPERATURE AND LASER POWER.

Neelu Mathur[1], S.K. Deb[1], S. Banerjee[2] and A. Sardesai[2]

[1]Solid State Physics Division, Bhabha Atomic Research Centre, Bombay 400 085, INDIA
[3]Solid State Electronics Group, Tata Institute of Fundamental Research, Homi Bhabha Road, Colaba, Bombay 400 005, INDIA

We have reported changes in Raman line shape of n-type porous silicon (n-Psi) with laser power and interpreted the results considering two layered microstructure of n-PSi[1]. Here we report our observation of Raman intensity enhancement in n-PSi and its variation with sample temperature as well as with incident laser power and explain our results within the 'dielectric confinement' model. The n-PSi sample used and the other

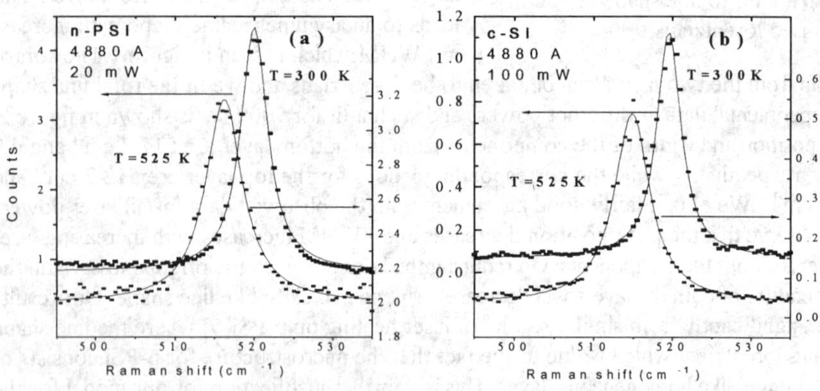

Figure 1 Raman spectra at 300 K and 525 K for n-PSi(a) and c-Si(b).

experimental arrangements are same as in ref. [1] and the sample temperature can be varied in a home made high temperature cell. The laser power used for n-PSi is 20 mW to avoid laser heating while 100 mW has been used for c-Si. The Raman spectra at 300K and 500K for n-PSi and c-Si substrate are shown in figure 1(a) and (b) respectively .The observed enhancement factors at 300K and 500K are ≈ 18 and 12 respectively indicating that the enhancement in intensity reduces by a factor of ≈ 1.6 when the sample temperature is changed from 300K to 500K. We further observe that at 300K the normalized integrated intensity under the Raman line profile (counts/mW/sec) reduces by a factor of about 3 when the incident laser power is increased from 25 mW to 900 mW as shown in figure 2.

These observations can be explained from the 'dielectric confinement' of electromagnetic fields in semiconducting microparticles[2]. Here a microparticle can support a series of 'normal modes' of wavelength ≈ particle size and for incident radiation matching with these modes the Raman polarizability tensor is enhanced considerably. According to the two layer model of n-PSi microstructure the size of the 'rod' like bottom layer Si particles ranges from a fraction to a few microns and the incident light at 0.488μm can excite these modes with enhancement of Raman intensity. On the other hand, the average sizes of Si crystallites in the top layer being ≈ 10 - 20 nm these modes cannot be excited by 0.488 μm radiation and this layer does not show enhanced Raman intensity.Our observed values agree quite well with those reported earlier in n-PSi and Si microparticles[3,4]. To explain the observed variation with temperature we note that the enhancement factor is a complicated function of real and imaginary parts of refractive indices (*n* and *k* respectively)[3] and we must consider the variation in these quantities with temperature. The reflectivity and absorption coefficient of Si for 4880Å changes from 0.374 to 0.384 and 1.2×10^{4} cm^{-1} to 2.2×10^{4} cm^{-1} respectively when the temperature changes from 300K to 500K[5]. The corresponding variation in *n* and *k* can be used to calculate the enhancement factors[3] and it reduces by a factor of ≈ 1.8 which agrees quite well with our observed value. To explain the variation with laser power we note that the enhancement factors estimated in this case is with respect to the c-Si signal at 300K and 25 mW. Thus at higher power when there is considerable laser heating we have to take into account the variation in absorption coefficient and phonon population factor at high temperature on the observed intensity. The rise in temperature of the bottom layer for laser power of 900 mW can be estimated from the peak position and FWHM corresponding to this layer and it is ≈ 425 K. The absorption coefficient and phonon population changes correspond to a reduction in intensity by a factor of about 1.6 and hence in effect the enhancement factor reduces by ≈ 3 relative to that at 25 mW. This agrees quite well with our experimental observation Thus we have shown that the origin of the intensity enhancement as well as its variation with temperature and laser power can be explained within the two layered model of n-PSi microstructure and 'dielectric confinement' of electromagnetic field in a semiconducting particles of size of the order of the wavelength of light.

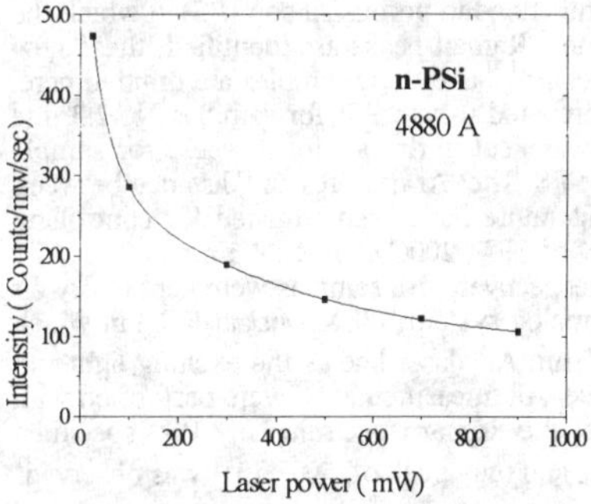

Figure 2 Variation of normalized intensity with incident laser power.

[1] S.K. Deb, Neelu Mathur, A.P. Roy, S. Banerjee and A. Sardesai , (this conference)
[2] Y. Wang, J. Phys. Chem. 95, 1119 (1991).
[3] D.V. Murphy and S.R.J. Brueck, Optics letters 8, 494 (1983).
[4] A.V. Andrianov , L.V. Belyakov, D.N. Goryachev, D.I. Kovalev, O.M. Sreseli, I.D. Yaroshetskii and B. Ya. Averbuch ,. Semiconductors 28, 1213 (1994).
[5] Landolt-Bornstein Tables, Physics of Group IV elements and III-V compounds (O. Madelung)Springer, Berlin, 1982, Vol 17a.

Raman Spectroscopy of Silicon Cluster Embedded In Porous Silicon

CHENG Guangxu(Guang-Xu Cheng)*+, LIU Fengqi+, LIAO Liangsheng+, Wei Zhang,+, and WANG Guanghou+

Center of Materials Analysis*, National Laboratory of Solid State Microstructures+, and Center for Advanced Studies in Science and Technology of Microstructures, Nanjing University, Nanjing 210093, Jiangsu, CHINA

CHU Haoyu
The Engineering Institute of the Engineer Corp, Nanjing 210002, Jiangsu, CHINA

For cluster-embedded materials or nano-crystals,the Raman study is also one of the important methods in obtaining information on vibrational states,in this letter,we shall report observation of Raman scattering from silicon cluster-embedded in porous silicon (PS),in which the second-order Raman scattering are represented and new Raman peaks are identified; these new peaks are thought as the cluster-induced TA modes. Details[1,2] about the samples are omitted here. For sample No.1, the clusters were deposited on as-prepared n-type PS; for samples No.2,3, the clusters were deposited after as-prepared p-type PS was kept in dry air for 3 weeks;for sample No.4,the clusters were deposited on as-prepared n-type PS. The Ar gas pressure,distance between molybdic boat and substrate,and the evaporation temperature have been adjusted for controlling the cluster size. The cluster size was determined by way of TEM-200CX(JEM): 4.5 nm, 4.9 nm, 5.1 nm,and 6.0 nm for samples No.1,No.2,No.3,and No.4,respectively. All samples were kept in dry air for 8 days before taking the Raman measurement (Ramalog system SPEX-1403 at R.T.) in which is backscattering configuration with 488.0 nm or 514.5 nm Ar^+ laser line as the exciting light,and the laser power is no more than 30mW at sample surface. All measurements were performed with identical experimental condition. In order to compare, the Raman spectrum of PS specimen without Si clusters embedded had been measured too, and only peak of 500 c m^{-1} was observed.

Shown in Fig.1 a,b,c,and d are the Raman spectra for samples of cluster-embedded No.1, No.2, No.3, and No.4 , respectively. For sample No.1, in addition to 500 cm^{-1} , two new peaks are located at 218 and 302 cm^{-1} , for sample No.2, besides the common peak of 502 cm^{-1} ,the 218 cm^{-1} is also shown,but the intensity reduced,and weak peaks at 280 and 292 cm^{-1} overlap on a broad background can be identified.For sample No.3,the 218 cm^{-1} disappear ,while 284, 294 cm^{-1} overlap on a broad background remain,and the main peak shifts to 496 cm^{-1} .For sample No.4 only 276 and 496 cm^{-1} were exhibited. It is also found that the peaks of sample with p-type substrate are very small and the background are more broader. As above Raman features of cluster-embedded PS originate from quantum comfinement effects and the difference in the surface boundary condition [3]. The 218 cm^{-1} was shown in sample No.1,No.2 can be identified as the TA(W) mode [4,5] ,samples No.3 and No.4 have no any trace of this feature. Since the size of introduced Si cluster in samples No.3 and No.4 are larger than samples No.1 and No.2, we conclude that the quantum size effect plays a key role in the scattering process, The most marked feature is that each sample shows a broad band between 250 and 310 cm^{-1} ,the broad band is very similar to the amorphous -like Raman signals. It is implied that the Raman signals of amorphous are come from the surface-layers of cluster in which atomic arrangement is more or less in disorder. The 302 cm^{-1} in sample No.1 can be identified as 2TA (X) made which are second-order Raman scattering. A peak at 301 cm^{-1} in microcrystalline Si had been observed and assigned to 2TA mode of bulk Si crystal [6].The 280 and 292 cm^{-1} for No.2,or the 284 and 294 cm^{-1} for No.3 may result from the splitting of 2TA (X),the discretness of wave vector can lead to fine structures in Raman spectrum. The two small peaks overlapped onto a broad background indicating that the competition between the wave vector may exist in the scattering process, the 276 cm $^{-1}$ can be assigned to the second -order feature of TA+TA(X) (see Table I).

Table I. Features of modes for Si cluster embedded in porous silicon.

Sample	Observed Peaks (cm^{-1})	Possible modes and assignment	Grain size(nm) Results from TEM-200CX	Grain size(nm) Results from the Formula(*)[7]
No. 1	218 302	TA(X) 2TA(X)	4.5	2.1
No. 2	218 280 292	TA(W) 2TA(W) splitting 2TA(X) splitting	4.9	1.9
No. 3	284 294 292	2TA(W) splitting 2TA(X) splitting	5.1	1.9
No. 4	276	TA+TA(X)	6.0	2.2

In this table, TA(X)=150 cm^{-1} ,TA(W)=210 cm^{-1} , TA=130 cm^{-1} , corresponding to vibration frequencies of X, W points in Brillouin zone, and the mean values, respectively.

* The formula is $d=2\pi(B/\Delta\omega_{TO})^{1/2}$ (nm), where B is 20.0 cm^{-1} $(nm)^2$, and the $\Delta\omega_{TO}$ is some broadening of TO-like mode with its single crystal counterpart on Raman spectra, this formula is only available to spherical model for these grains.

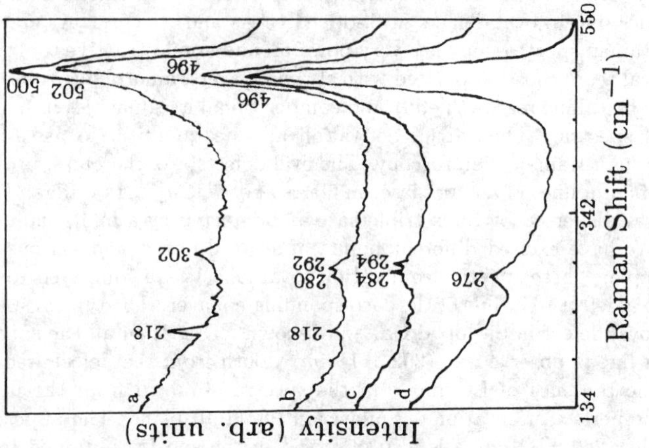

Fig.1 The Raman spectra of silicon cluster embedded in porous silicon at R.T.

1. Zhifeng Sui, Patrick P.Leong, and Irving P.Herman; Gregg S.Higashi and Henryk Temkin Appl.Phys.Lett. **60** 2086 (1992) .
2. F.Q.Liu , M.Han,J.J.Zhao, X.S.Chen, Q.Wang, and G.H.Wang, Appl.Phys.Lett. **66** 523 (1995) .
3. M.Fujii, S.Hayashi, and K.Yamamoto, Appl.Phys .Lett. **57** 2692 (1990).
4. F.A.Johnson, Proc.Phys.Soc.**73** 265 (1959) .
5. F.A.Johnson and R.Loudon, Proc.Roy.Soc., **A281** 274 (1964) .
6. Z.Iqbal and S.Veprek, J.Phys.C: Solid State Phys.**15** 377 (1982).
7. M.Cardona and G.Guntherodt Light Scattering in Solids. Vol .II Springer-Verlag Berlin Heidelberg New York. 1982(p.80).

MICRO-RAMAN STUDY OF INHOMOGENEITIES IN MBE $Al_XGa_{1-X}As$

P. S. Dobal*[1] and H. D. Bist

Department of Physics and Center for Laser Technology,
Indian Institute of Technology, Kanpur 208 016 India.

Molecular beam epitaxy is a well known effective epitaxial growth technique for fabricating high speed electronic and opto-electronic devices having ultra-thin $GaAs/Al_xGa_{1-x}As$ layers. The surface morphology of these layers exhibit different types of inhomogeneities, namely micro segregation, whiskers and oval defects [1]. Among these, oval defects are the most undesirable for the realization and performance of low dimensional devices based on $GaAs/Al_xGa_{1-x}As$ heterostructure. The origin, effect on devices and methods of eliminating oval defects in MBE $GaAs - Al_xGa_{1-x}As$ have been extensively reviewed by Chand *et. al.*[2]. Here, we summarise our Raman spectroscopic study of the morphological, compositional and carrier concentration inhomogeneities due to oval defects in Si-doped GaAs and $Al_xGa_{1-x}As$ epitaxial layers.

Raman spectroscopy is a very powerful and nondestructive tool to investigate the structural (*e.g.*, those associated with strained layers, amorphous zones or damaged layers, and crystalline regrowth etc.) and compositional irregularities in semiconductors *via* phonon spectra. The carrier activation and its quantitative assessment is also possible through Raman spectroscopy. The oval defects are, therefore, probed utilizing 488 and 514.5 nm lines of an Ar^+ laser in GaAs and $Al_xGa_{1-x}As$ wafers. The scattered radiation was dispersed by Spex triplemate and detected by a multichannel CCD.

We have excited different points of an oval defect and various oval defects in each sample. These spectra from various oval defects are compared to those corresponding to a defect free zone of the corresponding epilayer. The typical spectra excited around an oval defect in Si-doped GaAs are shown in Fig.1. In all the spectra, a sharp LO phonon line is observed at $(292 \pm 1)\ cm^{-1}$ both from the defect free and faulted regions. The presence of LO mode in the spectra obtained from the normal region indicates the perfect orientation of epilayer in the (100) plane. The appearance of TO modes at the $(269 \pm 1)\ cm^{-1}$ is also observed in the spectra obtained from the oval defects; which has been attributed to the deviation from the normal crystallographic orientation. As we approach towards the center of an oval defect the TO mode gains intensity, whereas the relative intensity of LO mode diminishes. Thus the surface of an oval defect can not be considered as homogeneous. A large number of oval defects were studied in these samples and the systematic variation of crystalline orientation was found common to all the defects. By varying the incident photon energy (using 488, 514.5 and 607 nm excitations), we notice considerable changes in LO and TO mode intensities at the defects. These results clearly show that the oval defects also display inhomogeneities along the depth. Similar results of changed crystalline orientation were observed in $Al_xGa_{1-x}As$ epitaxial layers.

[1]Permanent address: Department of Physics, V S S D College, Kanpur, India

The strain induced effects in GaAs can shift the LO and TO mode frequencies. Additionally, amorphization leads to the broadening and shifting of these modes towards low frequency side [3]. We neither observe any such broad structure nor any major downshift of LO modes in the spectra obtained from GaAs oval defects. However, the tensile and compressive stress has been inferred, respectively, in the oval defects of direct ($0 < x < 0.4$) and indirect ($0.4 < x < 1$) gap $Al_xGa_{1-x}As$ alloys.

In order to investigate the carrier concentration inhomogeneity the centers of various oval defects in GaAs were excited by $\approx 1KW/cm^{-2}$ fluence of 514.5 nm. On the basis of peak positions of LO-phonon plasmon coupled modes, the carrier concentration of oval defects and defect free regions was studied. Though the carrier density at oval defects does not vary systematically, it was found relatively lower as compared to the defect free regions. A detailed study of the carrier concentration has been done using the surface depletion layer induced changes in the LO mode intensity [4]. The oval defects were also investigated under scanning electron microscope. The overall size of the defects decreases, while the defect density increases with increase in Si-doping in GaAs. As x increases, the dimensions of the oval defects generally diminishes while the overall defect density increases for $Al_xGa_{1-x}As$ alloys. More atomic percent of As was found along the periphery and in the balled up portion of the defects by energy dispersive analysis of x-rays in GaAs. The stoichiometry at the defects was found unchanged in $Al_xGa_{1-x}As$ alloys. Thus, the properties of devices based on $GaAs - Al_xGa_{1-x}As$ heterostructures can not remain unaffected by these kinds of inhomogeneities. Authors are thankful to Dr. S. K. Mehta and Dr. R. K. Jain for providing samples.

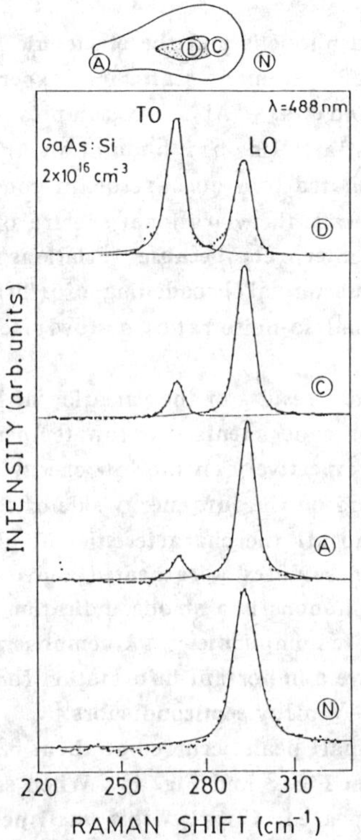

Figure 1. Raman spectra from the various spots (shown in the top) in Si doped GaAs sample. A, C, and, D are the defect spots and N is from defect free region.

1. Y. H. Wang *et. al.*, Jpn. J. Appl. Phys. **24**, 628 (1985).
2. N. Chand and S. N. G. Chu, J. Crystal Growth **104**, 485 (1990).
3. M. H. Brodsky in Light Scattering in Solids (M. Cardona ed.) Vol **1**, p.205 (1983).
4. P. S. Dobal, H. D. Bist, S. K. Mehta and R. K. Jain J. Appl. Phys. **77** 3934 (1995)

STUDY OF BISTABILITY OF DONORS IN $Ga_{1-x}Al_xAs$:Te BY RAMAN SCATTERING AT LOW TEMPERATURES

Shi-Yang Lian

Department of Physics, Xiamen University,
Xiamen 361005, Fujian, CHINA

The first observation of the bound phonons and the electronic Raman scattering in $Ga_{1-x}Al_xAs$: Te is reported. Raman scattering experiments performed on Te-doped $Ga_{0.34}Al_{0.66}As$ and $Ga_{0.39}Al_{0.61}As$ samples at low temperutures changed from 20K to 300K. A laser line 514.5 nm of an Ar^+ laser was used for these experiments, it approximated to a quasi-resonant conditionn resulting in a large spectral signal. In general, the vibrational spectra of alloy semiconductors are extremely difficult to interpret, because variations in the local environment of the defect lead to substantial broadening of vibrational spectra. In order to achieve a sufficient signal-to-noise ratio, a slowly scanning method was used in these experiments.

Fig. 1 to Fig. 4 represent the essential results of the variation of Raman spectra with temperature obtained from our experiments. At low temperatures (such as 82K), novel structures appeared respectively on the low energy side of the LO_1 line for GaAs-like mode (Fig. 1) and on the low energy side of the LO_2 line for AlAs-like mode (Fig. 2) which had all the characteristics of phonons bound to neutral donor impurities[1]. When samples were heated above 100K, they disappeared. The existence of bound phonons is a strong indication of the effective-mass-like character of the associated impurities[1]. It emphasizes the polar character of the host lattice, which give a important information for probe into the physical origin of DX centers in III-V alloy semiconductors[2].

In addition to bound phonons, a small peak occurred at about 320cm^{-1} at low temperatures (20K, 40K and 82K in Fig. 3 and Fig. 4). When samples were heated over 100K, the small peak disspeared. Clearly, this small peak had all the characteristics of the electronic Raman scattering due to shallow donors in indirect band semiconductors. Temporarily, regardless of DX level being in III-V alloy semiconductors, shallow donor states may be introduced into $Ga_{1-x}Al_xAs$ by Te substituting for As atoms. The As site donors have S-like symmetry. This threefold degeneracy of the 1S donor ground state in $Ga_{1-x}Al_xAs$ ($x>0.45$) due to multiple valleys is lifted by valley-orbit interactions, producting a lowest singlet 1S(A_1) and a doublet 1S (E). The transitions between these states gave rise to the electronic Raman scattering.

In reality, it is well known that DX centers exist as a general call of defects inherent to $Ga_{1-x}Al_xAs$ ($x>0.22$)[3]. However, it is observed in our experiments that the existence, at low temperatures, of bound phonons and the electronic Raman scattering are associated with electrons trapped to an effective-

mass-like level. Obviously, it shows that the effective-mass-like state is of photon-induced at low temperatures.

In conclussions, these investigations of Raman scattering experiments performed on Te-doped $Ga_{1-x}Al_xAs$ samples as a function of temperature show that the optical induced metastable state of donors at low temperatwres is the usual effective-mass-like state giving rise to bound phonons and the electronic Raman scattering. This metastable state disappears at higher temperatures to the benefit of a stable DX-like state. These results indicate a shallow-deep bistable character of donors in $Ga_{1-x}Al_xAs$ $(x>0.22)$

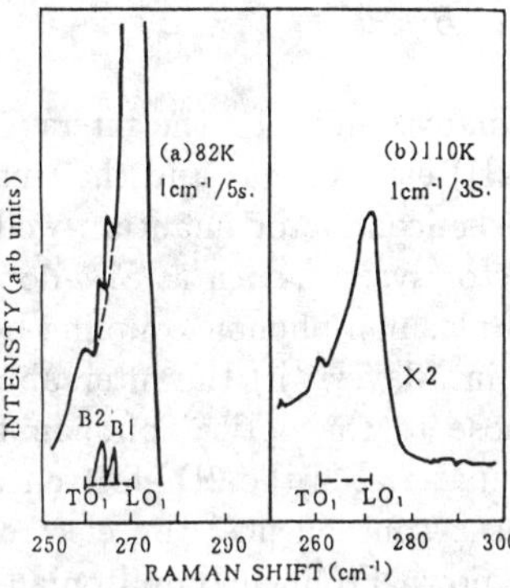

Fig. 1. Temperature dependence of Raman spectra for GaAs-like phonons in $Ga_{0.34}Al_{0.66}As:Te$.

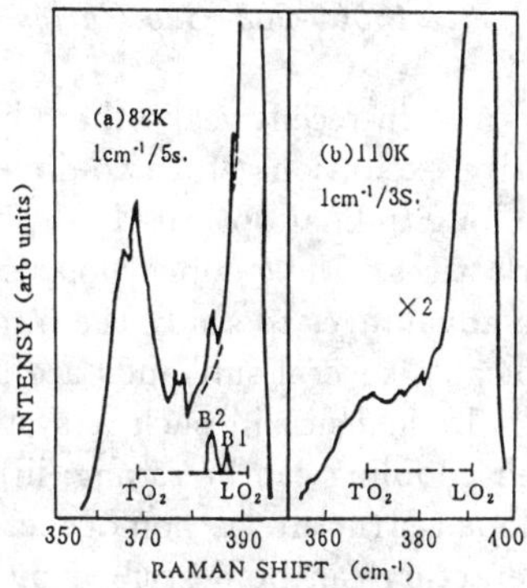

Fig. 2. Temperature dependence of Raman spectra for AlAs-like phonons in $Ga_{0.34}Al_{0.66}As:Te$

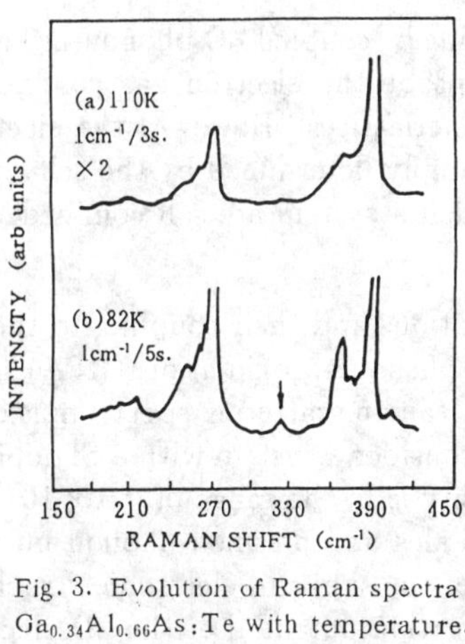

Fig. 3. Evolution of Raman spectra for $Ga_{0.34}Al_{0.66}As:Te$ with temperature.

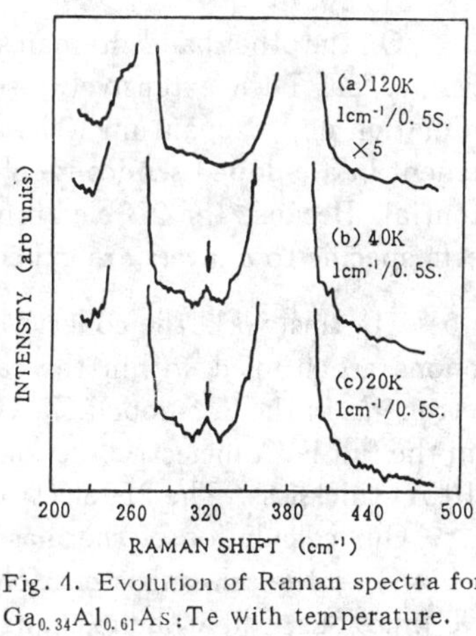

Fig. 4. Evolution of Raman spectra for $Ga_{0.34}Al_{0.61}As:Te$ with temperature.

1. **P. Seguy, M. Zigone, and G. Martinez, Phys. Rev. Lett.** 68,518(1992).
2. **S. Y. Lian and J. Xu, (to be published).**
3. **P. M. Mooney, J. Appl. Phys.** 67, **R1**(1990).

RAMAN SCATTERING DUE TO COUPLED INTERSUBBAND PLASMON-PHONON MODES IN Si DELTA-DOPED GaAs

G.-Q. Hai, G. E. Marques, and N. Studart

Departamento de Física, Universidade Federal de São Carlos, 13565-905, São Carlos, São Paulo, Brazil

In recent years, there have been many studies on the intersubband collective excitations of a two-dimensional (2D) electron gas and their coupling to the longitudinal optical (LO) phonons in semiconductor quantum wells and superlattices. The δ-doped polar semiconductor system, such as Si δ-doped GaAs, has advantages to study the intersubband plasmon-phonon coupling because: i) typically, several subbands are populated in δ-layers; ii) the intersubband plasmon frequencies in such a system are close to the optical phonon frequency, their coupling can be strong; iii) the host material of the 2D electron gas in the δ-doped structure is homogeneous and, as a consequence, the electron-phonon interaction can be described by the well known Fröhlich Hamiltonian, which is different in other 2D systems, such as heterojuctions and quantum wells, due to the presence of interfaces; and iv) due to the spatial symmetry of the system, intra-intersubband mode coupling becomes unimportant.

On the other hand, Raman scattering by coupled LO phonon-intersubband plasmons has been extensively used to probe the electron gas confinement in modulation-doped quantum wells and superlattices. However, the electric confinement in a δ-doped semiconductor is simply determined by the self-consistent potential. Because the 2D electrons in such a system are self-confined, peculiar effects specific to δ-layers are expected.

In this work, the collective excitations and their coupling to the optical phonons are studied within the random-phase approximation (RPA) for a 2D electron gas in the Si δ-doped GaAs. The Raman scattering spectrum is obtained from the full RPA dielectric fuction. We consider a system with a Si doping layer of 10 Å thickness. The 2D electron density is in the range of 0.5×10^{12} to 10^{13} cm^{-2}. Our calculation of the plasmon modes and plasmon-phonon interactions are based on the self-consistent solution of the subband electronic structure and wave functions, where the exchange-correlation potential of the 2D electron gas and the nonparabolicity of the conduction band are included.

Fig. 1 shows the electron density dependence of the frequencies (at $q = 0$) of the intersubband plasmon modes (thin curves) and the splitted intersubband plasmon-phonon modes in Si δ-doped GaAs. The different intersubband modes are indicated by the solid (2,3), dashed (2,4), and the dash-dotted (1,2) curves.

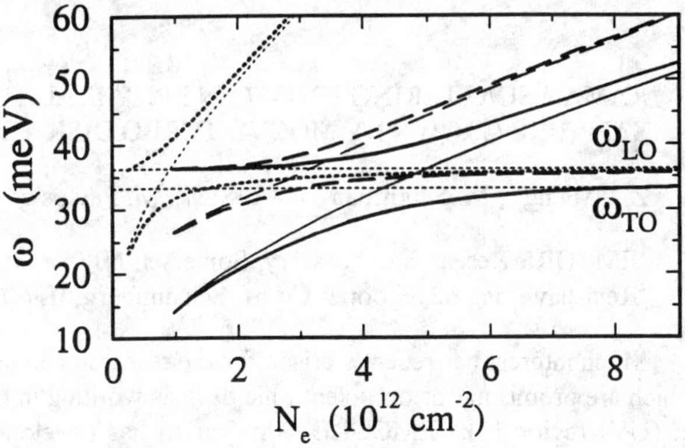

Fig. 2 shows the Raman scattering spectrum due to the coupled intersubband (1,2) plasmon-phonon modes in the Si δ-doped GaAs system of $N_e = 2 \times 10^{12}$ cm^{-2}.

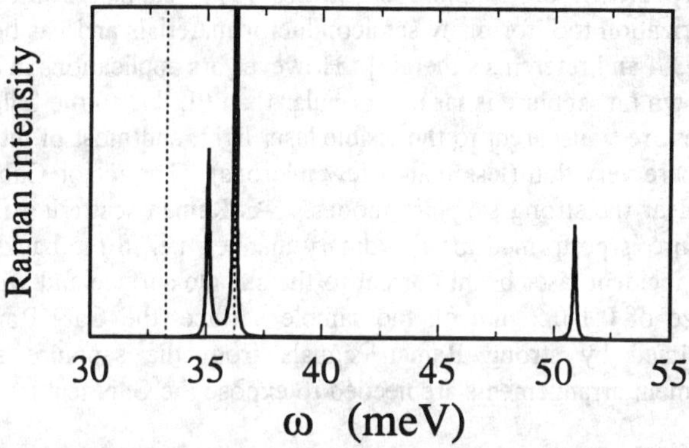

Our calculation shows that the high electron density leads to a large depolarization shift of the intersubband plasmon frequencies between the two adjacent subbands. Furthermore, both intrasubband plasmon and intersubband plasmon modes are strongly coupled to the optical phonon modes. The frequencies of the coupled intrasubband plasmon-phonon modes do not enter in the reststrahlen region of GaAs. However the frequencies of the coupled intersubband plasmon-phonon modes split around the LO-phonon frequency. In a wide range of the 2D electron density (donor concentration), the frequencies of the phonon-like branch of the intersubband plasmon-phonon modes due to 1 and 2 subbands and 2 and 4 subbands are in the reststrahlen region of GaAs.

1. A. Mlayah, R. Carles, E. Bedel, and A. Muñoz-Yagüe, Appl. Phys. Lett. **62**, 2848 (1993); J. Appl. Phys. **74**, 1072 (1993).
2. M. P. Chamberlain, M. Cardona, and B. K. Ridley, Phys. Rev. B. bf 48, 14356 (1993).

RAMAN SCATTERING FROM GaN EPITAXIAL THIN FILMS ON SAPPHIRE GROWN BY MOCVD TURBO DISK TECHNOLOGY

Z. C. Feng[1*], M. Schurman[1], R. A. Stall[1], M. Pavlosky[2] and A. Whitley[2]

[1]EMCORE Research Laboratory, Somerset, NJ 08873, USA
[2]Renishaw Inc., 623 Cooper Court, Schaumburg, IL 60173, USA

Much interest has recently arisen for the research and application of III-V nitrides which are promising for optoelectronic devices working in the blue, violet and ultraviolet (UV) region [1]. EMCORE Corporation has developed and applied an advanced Turbo Disk technology, with a vertical growth configuration and a high speed rotating disk, for the metalorganic chemical vapor deposition (MOCVD) of the large size growth of new nitrides materials, including GaN, InGaN and AlGaN epitaxially grown on sapphire substrates [2]. Raman scattering is a powerful characterization tool for many semiconductor materials and has been applied to III-V nitrides [3,4 and references therein]. However, its applications in epitaxial nitride thin films grown on sapphire is far less popular than PL due to the following reasons. III-V nitrides are transparent to the visible laser light, and most of nitride films grown on sapphire are very thin (less than a few microns). The major GaN Raman modes are located near the strong sapphire modes. As Raman scattering from GaN thin films on sapphire is performed in an ordinary manner, i.e. in the backscattering geometry with the incident laser beam normal to the sample surface and with the focused laser beam size of 0.1-0.2 mm on the sample surface, the GaN Raman modes will be overwhelmed by strong Raman signals from the sapphire substrate. Special measurement arrangements are needed to expose the GaN features [3,4].

In this study, we show two techniques to make Raman technique a convenient tool for the characterization of GaN thin films grown on sapphire. Two set of Raman facilities were used. Figure 1 show Raman spectra from a 2.2 μm thick GaN thin film, grown on sapphire by the MOCVD Turbo Disk technology, described in [2], measured by a SPEX 1877 triple spectrometer - optical multichannel analyzer (OMA) system, excited by an Ar^+ laser 488 nm line and with an ordinary camera lens focusing the laser beam size to 0.1-0.2 mm on the sample surface. We made the laser beam incident direction perpendicular to the optical axis of the spectrometer and had the sample mounted on a stage able to rotate to vary the incident angle, θ. As seen, when $\theta=40^\circ$ or less, we observed only sapphire Raman features (Fig. 1a). As $\theta=60^\circ$, the GaN E_2 mode appeared at 568 cm^{-1} (Fig. 1b). At $\theta=85^\circ$, it is stronger than the sapphire mode at 575 cm^{-1} and the GaN A_1(LO) mode is observed at 735 cm^{-1}, comparable in intensity with the sapphire mode at 750 cm^{-1}. They became dominant as $\theta=89^\circ$ was employed.

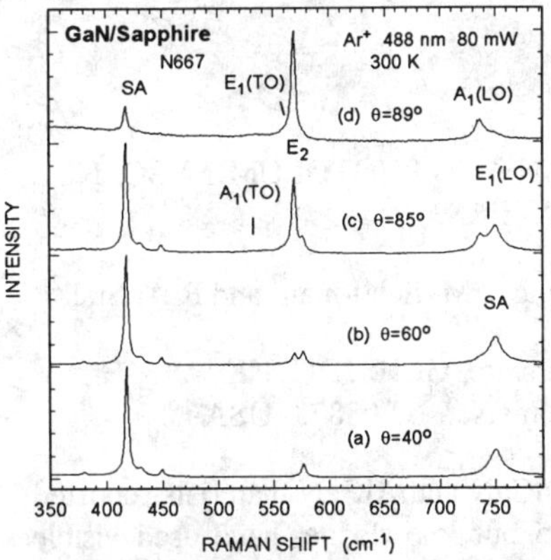

Figure 1 Raman spectra of a MOCVD-grown GaN/sapphire, N667, with different incident angles θ.

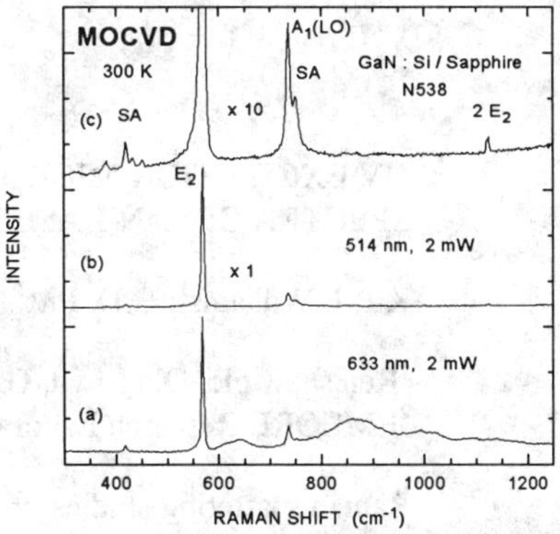

Figure 2 Micro-Raman spectra of a Si:GaN/sapphire, N538, under excitation of 633 nm in (a), and 514 nm in (b) with (c) magnified by 10 with respect to (b).

Figure 2 shows Raman spectra of a Si-dope GaN 1.8 μm thick film grown on sapphire by MOCVD, using a Renishaw Raman microscope under excitations of 633 and 514 nm. The sharp GaN E_2 Raman mode with a full width at half maximum of 5 cm^{-1} is dominant in both cases. The GaN E_1(LO) mode is stronger than the sapphire mode at 750 cm^{-1} also. The second order feature, $2E_2$, appears at 1123 cm^{-1}. These are characteristic of good crystalline quality of the GaN film. The micro-Raman system focused the laser beam to a spot size of 1-2 μm, corresponding to the studied film thickness. Therefore, the laser light entering into the substrate after leaving the film was de-focused, causing much weaker Raman scattering from substrate only. This demonstrates that the microscope focusing, in contrast to the case using an ordinary optical lens, leads to the Raman features from the GaN thin film much stronger than that from the sapphire substrate. In addition, Fig. 2a shows some broad bands beyond 580 cm^{-1}, which are not seen under the 514 nm excitation in Figs. 2b and 2c. These are PL recombinations and can be detected at the same energy positions under 514 nm excitation (not shown here) as seen under the 633 nm excitation. The origins of these deep center related PL bands are currently being studied. The authors are grateful to Prof. S. Perkowitz for his kindness and help in the use of Raman facility at Emory Univrsity.

1. S. Strite and H. Morkoc, J. Vac. Sci. Technol. B10, 4 (1992).
2. C. Yuan, T. Salagai, A. Gurary, P. Zawadzki, C. S. Chern, W. Kroll, R. A. Stall, Y. Li, M. Schurman, C.-Y. Hwang, W. E. Mayo, Y. Lu and S. J. Pearton, J. Electrochem. Soc. 142, L163 (1995).
3. T. Kozawa, T. Kachi, H. Kano, Y. Taga, M. Hashimoto, N. Koide and K. Manabe, J. Appl. Phys. 75, 1098 (1994).
4. H.-R. Kuo, M.-S. Feng, J.-D. Guo and M.-C. Lee, Jpn. J. Appl. Phys., part 1, 34, No. 10 (1995).

UV EXCITATION MICRO-RAMAN AND PHOTOLUMINESCENCE SPECTRA OF GaN-BASED EPILAYERS

K.P.J. Williams[1], G.D. Pitt[1], Z.C. Feng[2*], M. Schurman[2] and R.A. Stall[2]

[1]Renishaw plc, OldTown, Gloucestershire, GL12 7DH, UK
[2]EMCORE Research Laboratory, Somerset, NJ 08873, USA

Raman scattering studies on GaN, InGaN and AlGaN materials reported in the literature [1-4 and references therein], to our knowledge, have used visible light excitation. We present here for the first time the report of ultra violet (UV) laser excited Raman scattering study from these GaN-based epitaxial materials. The samples are grown on sapphire substrates by the low pressure metalorganic chemical vapor deposition (MOCVD) with the advanced turbo disk technology [5] and analyzed by a novel UV-Raman microscope designed by Renishaw. Figure 1 shows room temperature (RT) Raman spectra from four MOCVD-grown nitride epitaxial thin films of 1-2 μm thick on sapphire substrates, in (a) un-doped GaN film, (b) Mg-doped p-type GaN, (c) InGaN/GaN with 6% In, and (d) AlGaN/GaN with 3% Al. In contrast to the Raman studies using visible laser excitation [1-4] where GaN or GaN-like Raman E_2 mode at 565-570 cm^{-1} is dominant and much stronger than $A_1(LO)$ at 735-740 cm^{-1} and other features, we have the $A_1(LO)$ mode stronger than the E_2 mode. In additions, we also have observed a much strong second order Raman feature near 1470 cm^{-1}, assigned to $2A_1(LO)$, for all the nitride films. This is attributed to resonance Raman scattering corresponding to the proximity of the 325 nm laser with the energy band gap of these materials.

Figure 2 exhibits the corresponding photoluminescence (PL) spectra, ranging between 340-500 nm, of these GaN-based films, recorded using 325 nm excitation. As seen, the RT PL spectrum of an un-doped GaN film shows a dominant band near 364 nm, which is due to the band-to-band recombination of GaN. This energy band can also be seen from the p-type Mg:GaN and InGaN/GaN in Figs. 2b and 2c, respectively. It appears as a shoulder for the AlGaN/GaN in Fig. 2d, which shows the major band to the short wavelength side. The major band of the InGaN (6% In) film is shifted to the long wavelength side in comparison with that of GaN, while the Mg:GaN sample has its intense broad band, which is due to Mg-complex recombinations, located at even longer wavelength with modulated structures due to the multiple reflectance inside the film. We see, the wavelength positions of these major PL bands affect the resonance Raman scattering behavior in Fig. 1. The AlGaN film possesses the major PL band closest to the laser line at 325 nm and has its second order Raman features immersed inside the strong PL band tail. The un-doped GaN and InGaN (6% In) have their main PL bands close to the 325 nm secondly and thirdly among these four samples. The Mg:GaN film has its resonance Raman features seen best because its major PL band being far from 325 nm than that of other three nitride samples. In Fig. 1b, we also observe a band located near 1300 cm^{-1}, which may be due to the E_2+A_1(LO) combination.

The 325 nm excitation from a He-Cd laser provided 2 mW of power focused on the sample surface is above but close to the energy band gap of these materials. Therefore, the resonance enhance or outgoing resonance Raman scattering occurs for all the cases in Figs. 1 a-d, and makes the modes with A_1 symmetry greatly enhanced. The reasons why other GaN Raman phonon modes are not enhanced and the A_1(LO) mode enhancement behaviors under UV excitation are under further studies. These may provide further understanding on the GaN-based materials for blue-UV optoelectronic devices.

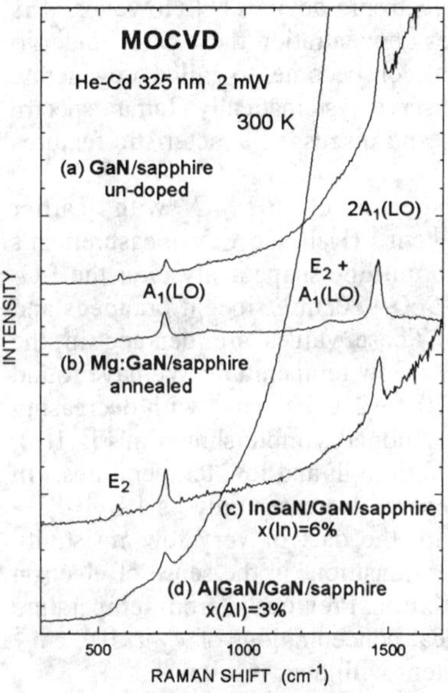

Figure 1 UV 325 nm excited Raman spectra of five nitride films on sapphire.

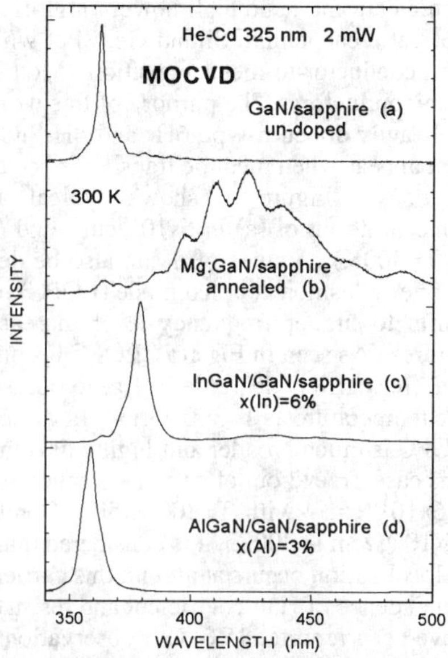

Figure 2 RT PL of five nitrides on sapphire grown by MOCVD Turbo Disk technology.

1. T. Kozawa, T. Kachi, H. Kano, Y. Taga, M. Hashimoto, N. Koide and K. Manabe, J. Appl. Phys. 75, 1098 (1994).
2. S. J. Sheih, K. T. Tsen, D. K. Ferry, A. Botchkarev, B. Sverdlov, A. Salvador and H. Morkoc, Appl. Phys. Lett. 67, 1757 (1995).
3. H.-R. Kuo, M.-S. Feng, J.-D. Guo and M.-C. Lee, Jpn. J. Appl. Phys., part 1, 10, (1995).
4. Z. C. Feng, M. Schurman, R. A. Stall, S. Perkowitz, M. Pavlosky, and A. Whitley, in this conference.
5. C. Yuan, T. Salagai, A. Gurary, A. G. Thompson, W. Kroll, R. A. Stall, C.-Y. Hwang, M. Schurman, Y. Li, W. E. Mayo, Y. Lu, S. Krishnankutty, I. K. Shmagin, R. M. Kolbas and S. J. Pearton, J. Vac. Sci. Technol. B13, 2075 (1995).

ELECTRONIC RAMAN SCATTERING FROM HEAVILY DOPED SIC

H.Harima*, S.Nakashima and T.Hamamoto

Department of Applied Physics, Osaka University,
Suita, Osaka 565, Japan

Recently, 6H- and 4H-SiC crystals with low resistivity have gathered much attention as electrically conductive substrates for short wavelength optoelectronic-, and high-power / high-frequency electronic devices.[1] However, it is not yet clear, unlike Si and Ge,[2] at which carrier concentration these SiCs undergo semiconductor-to-metal transition (Mott transition), or become so-called degenerate semiconductors. The purpose of this work is to observe systematically Raman spectra of heavily doped n-type 6H- and 4H-SiC crystals, and discuss characteristic features that appear when metallic transitions occur.

Figure 1 shows typical Raman spectra of 6H-SiC with carrier concentrations of (a) n=6x10^{18}cm^{-3} and (b) 2x10^{19}cm^{-3} (Hall and C-V measurements at T=300K). Values of n can also be deduced from a line shape analysis of the LO-phonon-plasmon coupled mode (LOPC) peaked at 960~980cm^{-1}, since it broadens and shifts to higher frequency as n increases.[3,4] These values are denoted in the figures. As seen in Fig.1(a), LOPC becomes sharp at low temperature: We have found that free carriers decrease or freeze out as n=5.6x10^{18}→2.1x10^{18}cm^{-3} with decreasing the temperature as T=300→35K. In a more heavily doped sample shown in Fig.1(b), LOPC is much broader and highly asymmetric at both high and low temperatures. In this case, freeze-out of carriers is much less efficient (n decreases only as 1.8x10^{19} → 1.6x10^{19} cm^{-3} with T=300→35K). Combining with the data of very low resistivity (5x10^{-3}Ωcm at 300K), it is considered that "metallic transition" in the sense of electron delocalization occurs at around this carrier concentration. Previously, from temperature dependence of Hall coefficient and resistivity, critical concentrations of n_c~2x10^{19}cm^{-3} have been reported.[5,6] Our observation is consistent with their results.

The sharp peaks observed at 35K in Fig.1(a) (shown by arrows) are 1s(A)→1s(E) valley-orbit transitions of electrons bound to nitrogen donors substituting three inequivalent carbon sites.[3] In Fig.1(b) at 35K, the two valley-orbit peaks at ~500cm^{-1} which merge by broadening are still clearly observed. This means that, although metallic character has been suggested from the temperature dependence of LOPC line shape, electrons still have some localized nature. According to Colwell and Klein,[3] electrons retain localized character up to n>6x10^{19}cm^{-3}. The valley-orbit transition at ~100cm^{-1} is smeared out in a strong Rayleigh tail. It is considered that the donor wavefunctions for this impurity site heavily overlap with each other, thus localized character of electrons is reduced, and contribution of single-particle excitation of free carriers [7] increases in the low frequency spectrum.

For comparison, we show in Fig.2 low frequency spectra of 4H-SiC at 35K. Free carrier concentrations at 300K are shown in the figure. At n=1x10^{18}cm^{-3} a sharp peak appears at 57cm^{-1}. It can be assigned to 1s(A)→1s(E) transition, because it has E_2 symmetry as in 6H-SiC, and the transition energy agrees with a proposed level scheme deduced from IR absorption.[8] At n=6x10^{18}cm^{-3}, however, this transition appears only as a weak shoulder superimposed on a continuum band which is due to single-particle excitations. This is very different from 6H-SiC with the same doping level as shown in Fig.1(a), where bound-electronic character is strongly retained at low temperatures. In 4H-SiC at $n=1\times10^{19}cm^{-3}$, the low frequency spectrum is clearly dominated by the free carrier excitation, and typical Fano-interference profile with sharp phonon peaks is observed at ~200cm^{-1}. The fact that electrons in 4H-SiC lose localized character at less donor concentration than in 6H-SiC can be roughly understood by the difference of donor binding energy E_d , or equivalently by the spread of donor wavefunctions a_H, using the well-known Mott criterion for the critical concentration; n_c~$(0.25/a_H)^3$. Although it is rude to apply effective mass approximation to donor 1s level of SiC, since it is severely affected by the chemical shift and valley-orbit splitting, we have used it here by assuming that the shallowest donor level among inequivalent donor sites determines n_c. Putting E_d = 80 (50) meV for 6H- (4H-)SiC,[8,9] we get a_H=9 (14) Å and $n_c=2\times10^{19}$ (5×10^{18}) cm^{-3}, which agree well with our observation.

We thank Advanced Technology Materials Inc. for supplying part of the heavily doped samples together with the data of Hall coefficient and resistivity.

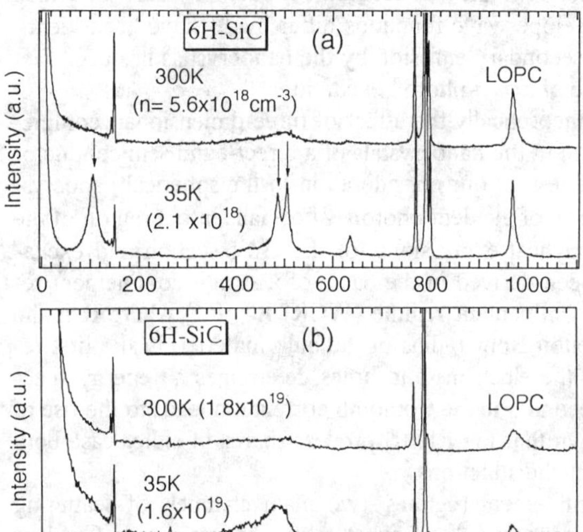

Fig.1 Raman spectra of 6H-SiC with different carrier concentrations. Arrows show valley-orbit transitions.

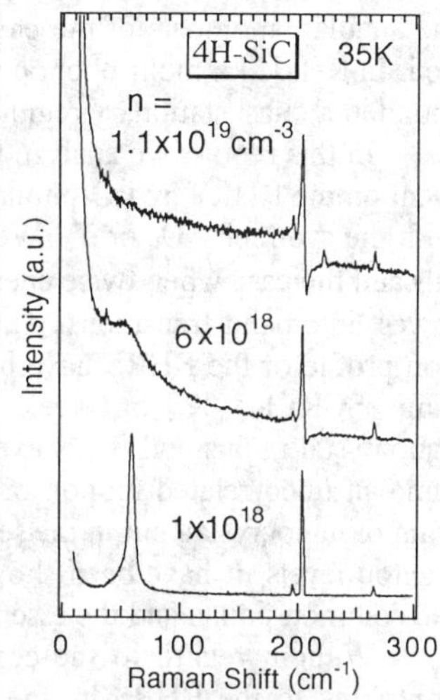

Fig.1 Raman spectra of 4H-SiC with different carrier concentrations. The concentration at 300K is shown.

1. H.Morkoç, S.Strite, G.B.Gao, M.E.Lin, B.Sverdlov and M.Burns, J.Appl.Phys. 76, 1363 (1994).
2. M.N.Alexander and D.F.Holcomb, Rev.Mod.Phys. 40, 815 (1968).
3. P.J.Colwell and M.V.Klein, Phys.Rev. 6, 498 (1972).
4. H.Harima, S.Nakashima and T.Uemura, J.Appl.Phys. 78, 1996 (1995).
5. J.A.Lely and F.A.Kröger, in *Proceedings International Colloquium Partenkirchen*, ed. by M.Schön and H.Welker (Interscience Publishers,Inc. NY, 1958), p.525.
6. G.N.Violina, Yeh Liang-hsiu and G.F.Kholuyanov, Sov.Phys.Solid State 5, 2500 (1964).
7. M.V.Klein, in *Light Scattreing in Solids*, ed. by M.Cardona (Splinger, Berlin, 1975), p.147.
8. W.Gotz,A.Schöner,G.Pensl,W.Suttrop,W.J.Choyke,R.Stein and Leibenzeder, J.Appl.Phys. 73, 3332 (1993).
9. W.Suttrop,G.Pensl, W.J.Choyke,R.Stein, and S.Leibenzeder, J.Appl.Phys. 72, 3708 (1992).

THEORY OF RESONANT HYPER-RAMAN SCATTERING BY SEMICONDUCTOR QUASI ZERO-DIMENSIONAL SYSTEMS

A. V. Fedorov

S. I. Vavilov State Optical Institute, Birzhevaya Liniya 12,
St.Petersburg 199034 Russia

Experimental studies of semiconductor quasi zero-dimensional systems (nanocrystals embedded in a dielectric medium) by a technique of resonant hyper-Raman scattering (RHRS) [1], [2] have demonstrated both the unique spectroscopic possibilities and a heuristic potential of the method. In Ref. [1] the optical transitions between the quantum-confined states of the valence and conduction bands for which the angular moments of the envelope wave functions differ by ±1 have been identified. In Ref. [2] a multi-phonon secondary emission by the nanocrystals has been first found in a quasi stationary regime of a two-photon excitation.

In this report, we analyze theoretically the effect of three-dimensional confinement on the RHRS by LO-phonon in the nanocrystals of a direct-band semiconductor with the symmetry O_h or T_d. We restrict our consideration to the spherical nanocrystals and the case when twice energy of incident photon $2\hbar\omega$ falls into a region of the lowest interband transitions. Analytical expressions for the cross-section and excitation profile of the RHRS have been derived in the cases of a strong confinement regime (SCR) $R \ll R_{ex}$ or week confinement regime (WCR) $R \gg R_{ex}$, where R is the nanocrystal radius and R_{ex} is exciton Bohr radius of the bulk material. In the first regime an uncorrelated motion of the electrons and holes determines an energy spectrum of nanocrystal, but in the second one the Coulomb correlation leads to the rise of exciton levels. It have been shown that the RHRS process shares in either case both the common points and the essential distinctions.

Without regard to the confinement regimes, two main channels of scattering contribute to the RHRS by the nanocrystals of cubic semiconductors, e.g. CuCl or CuBr.

The first channel exists due to the fact that, in semiconductors of the O_h and T_d symmetry, one-photon transitions are allowed in the dipole approximation. For this channel the typical process of Stokes scattering may be described as follows: the first photon $\hbar\omega$ creates electron-hole pair (exciton), then the electron or hole absorbs the second photon $\hbar\omega$ and emits LO-phonon $\hbar\Omega$ by the successive "intraband" transitions, and finally the electron-hole pair (exciton) annihilates with the emission of scattered photon $\hbar\omega_s = \hbar(2\omega - \Omega)$. It is easily seen that only non-diagonal part of Fröchlih-like interaction involving LO-phonons with angular moment l_p=1 determines the first channel of the RHRS. In the SCR the dipole allowed interband one-photon transition joins the quantum-confined states of the valence and conduction bands with the same angular moment of the envelope wave functions while the one-photon intraband transition changes the angular moment by 1. It follows that the electron-hole pair will annihilate only when the angular moment turns back to its original value as a result of LO-phonon emission. In the WCR the excitons are created (annihilated) in quantum-confined states with the angular moment $l=0$ at the dipole allowed one-photon transition, therefore, a change of the angular moment by 1 at one-photon exciton-exciton ("intraband") transition should be compensated by emission of LO-phonon with $l_p=1$.

The second channel of the RHRS exists in non-centrosymmetric (e.g., T_d) semiconductors due to the fact that both one- and two-photon transitions are allowed in dipole approximation. For this channel the typical process of Stokes scattering may be represented as follows: the first and second photon $\hbar\omega$ simultaneously create electron-hole pair (exciton), then the electron or hole emits LO-phonon $\hbar\Omega$ by the "intraband" transition, and finally the electron-hole pair (exciton) annihilates with the emission of scattered photon $\hbar\omega_s = \hbar(2\omega - \Omega)$. Notice that similar process, only without LO-phonon, is responsible for a resonant second harmonic generation in non-centrosymmetric semiconductors. All features of electron-hole generation (radiative recombination) at one-photon dipole allowed transitions described in earlier section hold also at two-photon dipole allowed transitions. Consequently, both the diagonal and non-diagonal part of Fröchlih-like interaction involving LO-phonons with angular moment $l_p=0$ determines the second channel of the RHRS. In the SCR the RHRS looks like a resonant Raman scattering discussed in Ref.[3] and to obtain a nonvanishing value of RHRS cross-section, the spherical potential with finitely high walls should be considered. In this case, only non-diagonal electron-phonon interaction causes the RHRS. By contrast, in the WCR, the diagonal part of Fröchlih-like interaction makes an essential contribution to the RHRS. The reason is that for exciton a spatial localization of electron is distinct from that of hole. Then the contributions to the RHRS cross-section from the processes in which the electron emits LO-phonon and the hole emits LO-phonon do not cancel each other. This situation looks like the RHRS by molecules or impurity centers in crystals. It is highly probable that the diagonal part of Fröchlih-like interaction determines the multi-phonon secondary emission by the CuBr nanocrystals [2].

Finally, the derived analytical results allow ones to interpret qualitatively the main features of the RHRS by LO-phonon from the quasi zero-dimensional systems.

This work was supported by the Russian Basic Research Foundation (RBRF), Grant No. 95-02-03821a.

1. A.V. Baranov, K. Inoue, K. Toba, A. Yamanaka, V.I. Petrov, and A.V. Fedorov, Phys. Rev. B 53, 1721 (1996).
2. K. Inoue, A. Yamanaka, K. Toba, A.V. Baranov, A.A. Onushchenko, and A.V. Fedorov, Phys. Rev. Lett., in press.
3. M.P. Chamberlain, C. Trallero-Giner, and M. Cardona, Phys. Rev. B 51, 1680 (1995).

RESONANT HYPER-RAMAN SCATTERING SPECTROSCOPY OF SEMICONDUCTOR QUANTUM DOT SYSTEMS

A. V. Baranov[1]* and K. Inoue[2]

[1] S. I. Vavilov State Optical Institute, St.Petersburg 199034, Russia
[2] Res. Inst. for Electronic Sci., Hokkaido Univ., Sapporo 060, Japan

Quantum dots (QDs), or semiconductor nanocrystals of sizes smaller than or comparable to the corresponding bulk exciton Bohr radius R_{ex}, are a new class of matter. Three-dimensional confinement modifies drastically electron energy spectrum and electron-phonon interaction in bulk crystal. These confinement-induced changes in QDs have attracted a growing interest, in particular, as an intermediate object between crystals and molecules. Although intensive studies have been performed [1], the fundamental nature of the electronic structure and electron-phonon interaction in QDs are yet to be understood. For this purpose new spectroscopic techniques may be needed, and in this respect we have recently introduced resonant hyper-Raman scattering spectroscopy, i.e., resonant hyper-Reyleigh, or second-harmonic (RSHS), and hyper-Raman (RHRS) scattering, which proves to be very useful [2-3].

Several QD systems have been studied for semiconductor nanocrystals embedded in glass matrices with ca. 0.1% in weight. For CdSSe-based QDs with average radii R_0 from 1.9 nm to 8.5 nm, pulsed radiation of 1064 nm from a YAG laser was employed ,whereas for both CdS-based QDs with R_0 (2.0 nm) smaller than R_{ex} (3.2 nm), and CuBr QDs with R_0 (3.2 nm) larger than R_{ex} (1.3 nm), which were studied in detail, a repetitively-pulsed radiation from a wavelength-tunable Ti-sapphire laser (920-720 nm) was employed for two-photon excitation of RSHS and RHRS in a region of the lowest confined electron-hole pair or exciton states.

It is found that signals of RSHS and RHRS by LO-phonons are greatly enhanced when twice icident photon energy falls in the region descriibed above. No remarkable directivity dependence of the RSHS intensity from CdSSe-based QDs was observed, indicating that unlike the bulk crystal case phase-matching is not involved. Consequently, resonances of the RSHS should occur in coincidence in energy with the energies of electron-hole pair states, which can be thereby probed. Note that SHG signal from glass matrix was negligibly small and thus RSHS signals from QDs at low concentration could be observed. The intensity of RHRS by LO-phonon from QDs proved to be surprisingly large as compared to the corresponding bulk crystals. It is important to note that the selection rules of optical transition accessible in both RSHS and RHRS are different from those of one-photon absorption (OPA) and resonant Raman scattering (RRS). Moreover, so-called "silent" phonon mode inactive in both

RRS and IR spectra was also observed in RHRS spectrum of CdSSe-based QDs owing to the specific selection rules of RHRS.

Both two-photon excitation spectra of RSHS and RHRS from CdS QDs in the strong confinement regime where momenta of electron and hole are separately quantized, show two maxima more pronounced and shifted in energy relative to those of OPA spectrum. Analysis has shown that the maxima may be assigned to the confined electron-hole pair states with total angular momentum of odd number, unobservable in OPA, and thereby the effective masses of electron and hole in the QD can be estimated. The profile of the lowest maximum in the RHRS spectrum is similar to that of RRS, but the second peak manifests itself in the former in contrast to the RRS case where only one maximum was observed around the lowest OPA peak. This fact presumably indicates that the off-diagonal intraband Frohlich interaction governs the present RHRS process.

In the CuBr case belonging to the weak confinement regime, the RSHS and RHRS excitation spectra show one maximum in the lowest-energy OPA region due to the confined $Z_{1,2}$, S-exciton state. The RHRS spectrum turns out rather anomalous: up to 5th-order LO-phonon bands have appeared as incident photon energy was increased. At first glance, the RHRS spectra look like the well known spectra of RRS by LO-phonon in polar bulk crystals such as CdS, but no continuous electronic band states exist in the CuBr-based QDs in the energy region in question! The present anomalous result indicates that the strong-coupled states of Wannier exciton and LO-phonon through diagonal intraband Frohlich interaction are realized. In principle, the observed features may be interpreted in terms of either the RHRS or hot luminescence. Discussions will be presented in detail together with other experimental results.

In conclusion, we have unravelled the respective confined electronic states inherent to CdS and CuBr QDs. Furthermore, we have demonstrated that the resonant RHRS spectroscopy provides a promising and powerful tool for studying the electronic structure as well as electron (exciton)-phonon interaction in semiconductor quantum dots.

References

1. A.D. Yoffe, Adv. in Phys. 42, 173 (1993).
2. A.V. Baranov, Ya.S. Bobovich, and V.I. Petrov, Sov. Phys. Usp. 33, 812 (1990).
3. A.V. Baranov, K. Inoue, et al., Phys. Rev. B, 53, 1721 (1996).

LO PHONON ASYMMETRY IN SEMICONDUCTOR DOPED GLASSES

Alka Ingale

Non Linear Optics Group, Laser Programme,
Centre for Advanced Technology, Indore-452 013, India

Semiconductor doped glasses (SDG) are borosillicate glasses, doped with nanocrystals of CdS_xSe_{1-x} of about 100A size. In this paper, we present line shape analysis of the asymmetry of the LO phonon, in the light of disorder effect due to alloying in addition to confinement effect due to small size crystallites. The first order Raman spectra of OG515 (x≈.92, d≈56A), OG530 (x≈0.87,d≈56A), OG590 (x≈0.5, d≈60A), OG715 (x≈0, d≈110A) Schott filters [1] and $CdS_{.62}Se_{.38}$ single crystal were recorded in back scattering geometry.

Asymmetry of LO/TO modes in alloys has been extensively investigated and is mainly attributed to Fano effect produced by an anharmonic decay of LO phonons into a continuum of other phonons [2]. In this case Raman intensity profile is given by ,

$I(\omega)\ \alpha\ |Q+\eta|^2/(1+\eta^2)$, where Q is the asymmetry parameter and η represents

the reduced frequency $(\omega-\omega(q))/\Gamma$, where Γ is broadening. In these nanocrystals, phonons are confined to the dimensions of ≈100 A, which leads to relaxation of q = 0 selection rule for Raman scattering. Raman intensity profile using Gaussian confinement function is given by [3],

$$I(\omega) \propto \int \frac{|C(o,q)|^2}{(\omega-\omega(q))^2+(\Gamma/2)^2} d^3q ,$$

Where $\omega(q)$ gives phonon dispersion and Γ is FWHM of corresponding bulk material.

We find that neither of these shapes fit the experimental data of SDG's very well. In fact, best fit to experimental data {fig. 1} is obtained by fitting two Lorentzians to each LO mode. The background was kept same in all three fitting procedures. We observed this additional phonon (asymmetry), also in single crystal data of $CdS_{0.62}Se_{0.38}$. Therefore, we believe that this additional phonon corresponds to disorder activated zone edge LO (ZE-LO) phonon. Earlier this phonon has been attributed to surface phonon l=2(n=0) by Malaya et al [5]. Whereas, l=1(n=0) surface phonon, a Frolich mode, has uniform polarisation over volume and is expected to appear with larger crosssection than l=2 mode, which is not observed [6]. In addition, the width of surface phonon obtained in the spectrum (Malaya &

our) is quite large ≈ 20-30 cm^{-1}, whereas, the damping of surface phonon is expected to be same as that of bulk LO/TO phonons [6]. On the other hand, in mixed crystals, we expect co-existance of peaks 1) at LO and TO phonon frequencies of the mixed crystal, obeying the usual Raman selection rules, and 2) broad, disorder activated spectral feature often associated with zone edge phonon [7]. In principle, disorder activated Raman scattering from phonon modes belonging to essentially all wave vectors is possible, however, phonon density of states is large at the zone edge. It is important to note here, that lower frequency spectrum of these glasses mimics phonon density of states below CdSe like LO phonon pretty well. We have observed additional mode at 268 cm^{-1} in OG715, which we believe to be a local mode due to small S concentration [8]. This explains, the appearence of ZE-LO phonon in the same. We have calculated ZE-LO phonon frequencies as a function of concentration, the way LO frequency for mixed crystal is calculated, taking ZE-LO for (x=1) = 293 cm^{-1} and for (x=0) = 203.5 cm^{-1} [4,5,8]. Fig. 2 shows, that this curve gives reasonable fit to ZE-LO phonon frequencies obtained from the fit, taking into account the shift of LO frequencies of SDG's from calculated values.

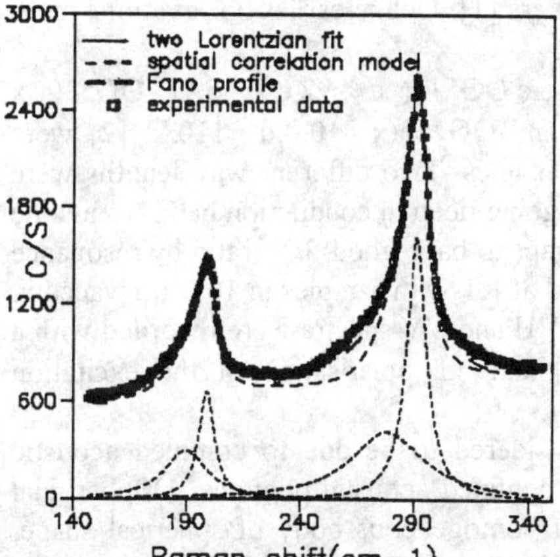

fig.1 LO modes of OG590 schott filter, excited by 5145A(100mW,line focus) of Ar ion laser and fit to experimental data by different model. (---) shows two lorentzians obtained from the fit.

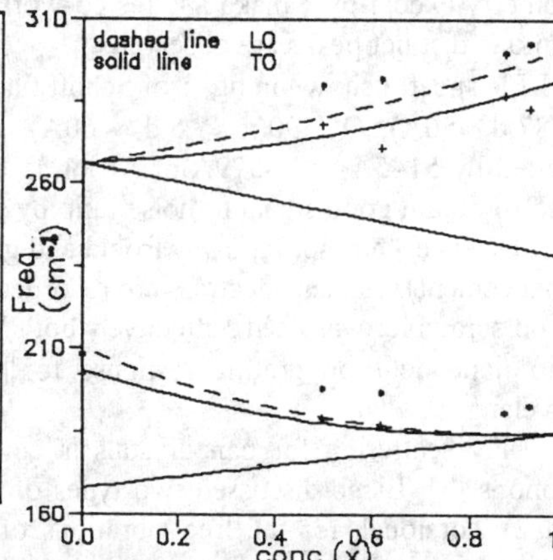

fig.2. LO_1,TO_1 and LO_2,TO_2 modes as function of concentration(x). Solid line between LO and TO shows, ZE–LO phonon freq. variation with x. (•) are LO and (+) are ZE–LO obtained from fit.

1. Gian Piero Banfi and Uittorio Degiorgio ; JAP 74 (11), 6925 (1993).
2. D.J. Olego, P.M. Raceah and J.P. Faurie; PRB 33 (6), 3819(1986).
3. J.H. Campbell and P.M. Fanchet,; SSC, 58 (10), 739 (1986).
4. M. Zigone, R. Beserman and H.D. Fair; Light Scattering in Solids,ed. by Balkaski M., R.C.C. Leite, SPS Porto p-597, Flammarion, Paris (1976).
5. A. Malayah, A.M. Brugman, R. Carles J.B. Renucci, M.Ya. Valakh and A.V. Pogorelov; SSC 90 (9), 567 (1994).
6. R. Ruppin; J. Phy. C:SSP. 8,19691(1975)
7. T.C. McGlinn, T.N. Krabach and M.V. Klein, G. Bajor, J.S. Greeve,B. Kramer, S.A. Barnett, A. Lastras and S. Gorbatkin; PRB 33 (12),8396(1986).
8. R. Beserman; SSC 23, 323 (1977).

LOW FREQUENCY RAMAN SCATTERING FROM CdS_xSe_{1-x} NANOCRYSTALS IN GLASS MATRIX

Alka Ingale

Nonlinear optics group,Laser programme,
Centre For Advanced Technology , Indore 452 013, India

Semiconductor doped glasses (SDG) contain CdS_xSe_{1-x} nanocrystals of about 100 A size, embedded in borosilicate glasses. In this paper we report low frequency modes observed in SDG's in backscattering geometry. Most of the earlier observations of low frqeuency Raman (LFR) scattering are in 90^0 scattering geometry except by Tanaka and his co-workers [1]. Unlike earlier observations more than one distinct peaks are observed.

LFR spectra shown in fig. 1 of Schott filters OG515($x \approx .92$,$d \approx 56$A), OG530($x \approx .87$,$d \approx 56$A), OG590($x \approx .5$,$d \approx 60$A), and OG715($x \approx 0.0$,$d \approx 110$A) [2] were excited by 5145A and 4579A of Ar ion laser lines. Two different wavelengths were used to obtain good signal to noise ratio by going deep in conduction band (to avoide Fluorescence), but not far away from band gap(to have good S/N ratio by resonance enhancement). Raman spectra were recorded at RT with samples in 10^{-3} torr vaccum. As no scrambler was used , effectively both VH and VV spectra were recorded with a ratio dependent on grating response to $\|$ and $\perp$ polarisation of the excitation wavelength .

LFR scattering for nanocrystals is considered to be due to confined acoustic phonons [3]. Lamb discussed two types of confined acoustic phonons, 1)Spheroidal and 2) Torsional, for the free vibrations of homogeneous body of spherical shape, under stress free boundary conditions [4]. Eigenfrequencies of torsional modes depends only on transverse acoustic velocity υ_t in the material , while the spheroidal mode eigenfrequencies involve the ratio of longitudinal to transverse acoustic velosities (υ_l/υ_t) in the material. As both for CdS (3570m/s/1540m/s) and CdSe(4277m/s/1862m/s), $\upsilon_l/\upsilon_t \approx 2.3$, eigenvalues (in cm^{-1}) for ω_{ln}^s and ω_{ln}^T for entire x range of $CdS_xSe_{(1-x)}$ can be given as [1],

$\omega_{00}^s = .78\ (\upsilon_t/dc)$ $\qquad \omega_{01}^s = 1.9\ (\upsilon_t/dc)$ $\qquad \omega_{02}^s = 2.04\ (\upsilon_t/dc)$

$\omega_{20}^s = .85\ (\upsilon_t/dc)$ $\qquad \omega_{2,1}^s = 1.60\ (\upsilon_t/dc)$ $\qquad \omega_{22}^s = 2.35\ (\upsilon_t/dc)$

$\omega_{10}^T = 1.83\ (\upsilon_t/dc)$ $\qquad \omega_{20}^T = .80\ (\upsilon_t/dc)$ $\qquad \omega_{30}^T = 1.23\ (\upsilon_t/dc)$

Where, c is the velocity of light and d is nanocrystal diameter. We have calculated these eigenfrequencies as a function of nanocrystal diameter for x = .92, .87, .5 and

0.0, taking $\upsilon_t = x\upsilon_{t(CdS)} + (1-x)\upsilon_{t(CdSe)}$. These were plotted as function of diameter for each x and matched to the experimentally observed frequencies in corresponding SDG's, as shown in fig. 2 for OG530. From fig. 2 the lowest freqency mode in OG530 at 8.5 cm^{-1} is attributed to ω_{20}^s. Corresponding modes observed in OG515 and OG590 (8-12cm^{-1}) also lie close to ω_{20}^s. For OG590, it is not shown in fig 1. as that region is not included here. This mode is not observed in OG715, probably because, it is much closer to Rayleigh line. The next LFR mode of SDG's including OG715 (12-20 cm^{-1}), lies between ω_{21}^s and ω_{10}^T. Since, The spheroidal modes for even ι and torsional modes for odd ι are Raman active, both are allowed by selection rule [1]. This mode as seen in fig. 1 is indeed broad and supports the view that both of these modes may be present. However, for clear identification line shape fitting and polrisation dependent data is required. The next higher mode in OG530 can be identified as ω_{22}^s. However in other smples, it does not appear as clear peak, but if taken to be at the position of change of slope, as shown in fig. 1 by arrows, it can be attributed to ω_{22}^s in all SDG's. The broad LFR spectrum of OG515 unlike OG530(in both d ≈ 56A) may be due to large size distribution of the sample. The structure seen above 30 cm^{-1} in fig. 1, we believe to be disorder activated phonon density of states, as they appear also in single crystal data of $CdS_{.62}Se_{.32}$. Further studies of confined acoustic phonon, using polarisation dependent LFR in back and 90^0 scattering geometry is in progress.

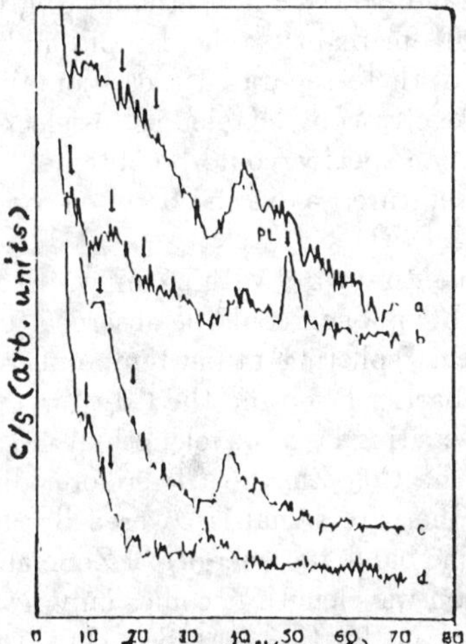

fig.1 LFR spectrum of a) OG515, b) OG530, c)OG590, and d) OG715 filters in backscattering geometry using 5145A and 4579A (100mW,line focus) of Ar ion laser at RT in vaccum.

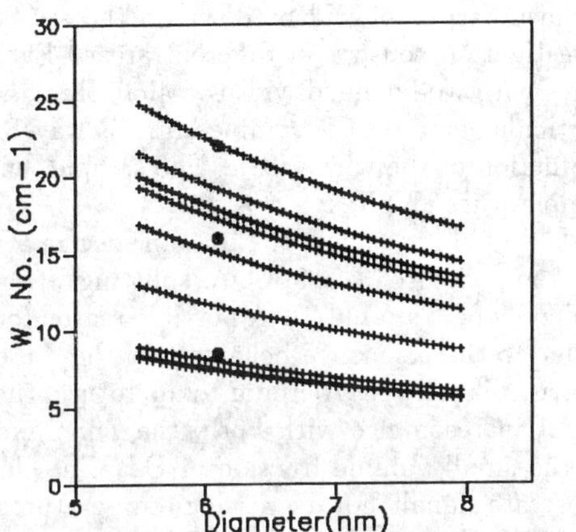

fig. 2 Eigenfrequencies of spheroidal and torsional modes of confined acoustic phonons calculated for x = .87 (OG530), as function of nanocrystal diameter. (●) shows frequencies obtained from OG530 LFR data.

1. Akinori Tanaka, Seinosuke Onari and Toshihiro Arai; PRB 47, 1237 (1993).
2. Gian Piero and Vittorio Degiorgio; JAP 74, 6925 (1993).
3. E. Duval, A. Boukenter and B. Champagnon; PRL 56, 2052 (1986).
4. H. Lamb; Proc. Lond math. Soc.13, 189 (1882).

MICRO RAMAN INVESTIGATIONS ON MBE GROWN ZNSE/ZNCDSE QUANTUM WIRES

G. Lermann, T. Bischof, A. Materny, and W. Kiefer*
Institut für Physikalische Chemie der Universität Würzburg,
Marcusstraße 9-11, D–97070 Würzburg, Germany

G. Landwehr, G. Bacher, and A. Forchel
Physikalisches Institut der Universität Würzburg,
Am Hubland, D–97074 Würzburg, Germany

Recently, quantum wires of different semiconductors have attracted considerable interest. This is due to the restriction of the vibrational and electronic (vibronic) system to only one dimension. Therefore new physical phenomena are expected which strongly depend on the width of the wires. Up to now, only few information on light scattering from II–VI–semiconductor quantum wires can be found in literature, mainly reporting on the dependence of the luminescence on the wire width [1,2].

$ZnSe/Cd_{0.2}Zn_{0.8}Se$ quantum wells grown by molecular beam epitaxy were structured by electron beam lithography and wet chemical etching to give quantum wires of widths down to the sub–100 nm region. The sample under investigation consists of different areas. Each of these separated areas contains wires with well defined widths. Micro Raman spectroscopy enabled us to select a particular area on the sample. In this way Raman spectra could be obtained as a function of the wire width. The sample temperature was varied between room temperature and 9 K.

Fig. 1 shows the Raman spectra of quantum wires with 25 nm width at low temperature (T = 9 K). A 'splitting' of the LO phonon could be observed (labelled by peaks 1 and 2). A possible reason for this 'splitting' at low temperatures is due to the resonance behaviour of the ZnSe–barrier layer and the $Cd_{0.2}Zn_{0.8}Se$ intermediate layer. At room temperature, the exciting laser wavelength at 457.9 nm is in resonance with both, the ZnSe and the $Cd_{0.2}Zn_{0.8}Se$. Therefore, due to the small volume fraction of $Cd_{0.2}Zn_{0.8}Se$, the wire signal is covered by the substrate signal. For decreasing temperatures the band gap energies of ZnSe and $Cd_{0.2}Zn_{0.8}Se$ increase. As a result, the excitation wavelength becomes only near or even off resonance for ZnSe but remains resonant for $Cd_{0.2}Zn_{0.8}Se$. Thus, the signal of the ZnSe–like Raman peak of the $Cd_{0.2}Zn_{0.8}Se$ layer should increase relatively to that of the ZnSe layer for decreasing temperature. Since the ZnSe–like mode of the intermediate layer is additionally shifted by 7 cm^{-1} with respect to the ZnSe mode of the barrier layer [3] a seperation of these modes at low temperatures is possible.

Low temperature Raman spectra were also recorded for other wire widths [see Fig. 2]. However, the splitting could be resolved only for wires with widths smaller than 70 nm. For larger widths, the LO signal becomes asymmetric.

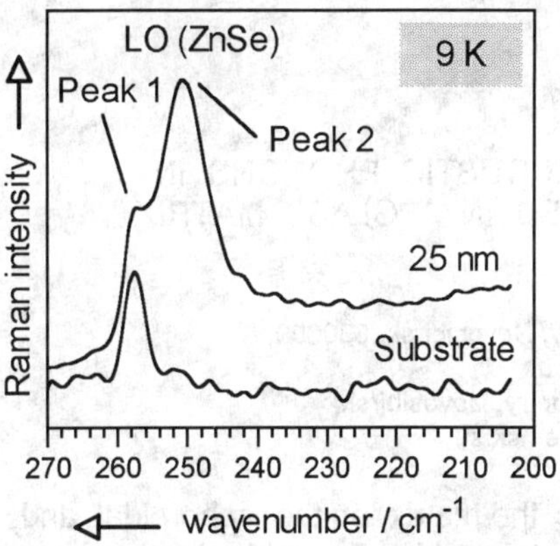

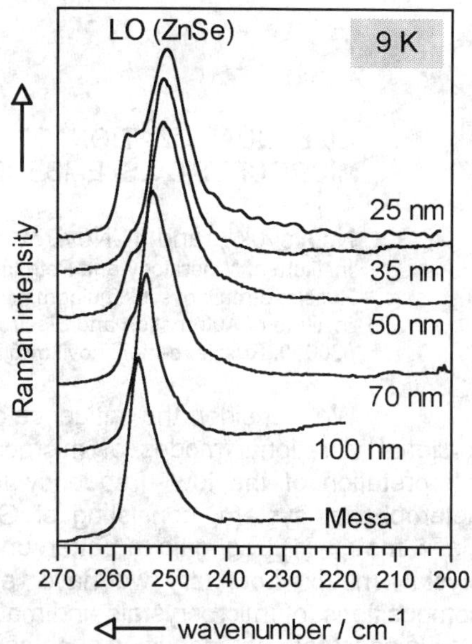

Fig. 1: Raman spectrum of 25 nm wires of CdZnSe / ZnSe and the substrate at T = 9 K.

Fig. 2: Raman spectra of CdZnSe / ZnSe wires for different wire widths at T = 9 K.

Fig. 2 indicates that the wavenumber of the Raman signal assigned to the wire is a function of the wire width. An explanation for the 'splitting' of the LO mode as a function of the wire width is possible, when the effects of strain are taken into consideration. Considering the edge zone of the intermediate layer ($Cd_{0.2}Zn_{0.8}Se$) a relaxation of the strain due to the smaller lattice constant of ZnSe may occur. In this regard, one may assume that the influence of the strain relaxed edges of the intermediate layer is dependent on the wire width.

Nakashima *et al.* [4] reported that the ZnSe LO is shifted by about 1 cm^{-1} with respect to the bulk material when a thin ZnSe 'film' is grown on GaAs. Thus a change of 0.6 % of the lattice constant shifts the LO phonon by about 1 cm^{-1}. According to this, we suppose that the wire width dependent Raman shift observed in Fig. 2 is also due to a change of the lattice constant caused by strain in the intermediate layer.

We gratefully acknowledge financial support from the Deutsche Forschungsgemeinschaft (Sonderforschungsbereich 410, Teilprojekte A1, C1, C3). Four of us (G. L., T. B., A. M., W. K.) also acknowledge support from the Fonds der Chemischen Industrie.

1. C. M. Sotomayor Torres, A. Ross, Y. S. Tang, A. Ribayrol, S. Thoms, A. S. Bunting, H. P. Zhou, K. Tsutsui, H. McLelland, H. P. Wagner, B. Lunn, and D. E. Ashenford, European II–VI–Semiconductor Workshop (1994).
2. C. M. Sotomayor Torres, A. P. Smart, M. Watt, M. A. Foad, K. Tsutsui, and C. D. W. Wilkinson, J. Electron. Mater. 23, 289 (1994).
3. R. G. Alonso, E.-K. Suh, A. K. Ramdas, N. Samarth, H. Luo, and J. K. Furdyna, Phys. Rev. B 40(6), 3720 (1989).
4. S. Nakashima, A. Fujii, K. Mizoguchi, A. Mitsuishi, and K. Yoneda, Jap. J. Appl. Phys. 27(7), 1327 (1988).

SIZE QUANTIZATION OF ACOUSTIC PHONONS IN MICROCRYSTALS EMBEDDED IN A GLASS MATRIX

N.Ovsyuk[1,*] and N.Novikov[2]
[1]Institute of Mineralogy and Petrology, Novosibirsk, 630090, Russia, e-mail: ovsyuk@uiggm.nsc.ru
[2]Institute of Automation and Electrometry, Novosibirsk, 630090, Russia, e-mail: novikov@iae.nsk.su

We consider the influence of the matrix on the spheroidal and torsional vibrational modes of microcrystals, which brought about a new interpretation of the low-frequency Raman spectra. We have studied heterophase system, consisting of Ge microcrystals embedded in GeO_2 glass matrix [1]. To gain a better understanding of the influence of the crystal-matrix boundary we have also analyzed some other chemical compositions of microcrystals and matrices with different relations between elastic constants and densities such as CdS in GeO_2 matrix and Ag in SiO_2 matrix which were obtained in [2,3]. In [4] it was shown that due to selection rules for Raman scattering on spherical particles, only spheroidal modes with even angular momentum quantum numbers l can be seen in the Raman spectrum. However, in practice the form of the microcrystals in a glass matrix may have deviations from a perfect spherical one. In this case, the selection rules can be relaxed and not only spheroidal modes with even

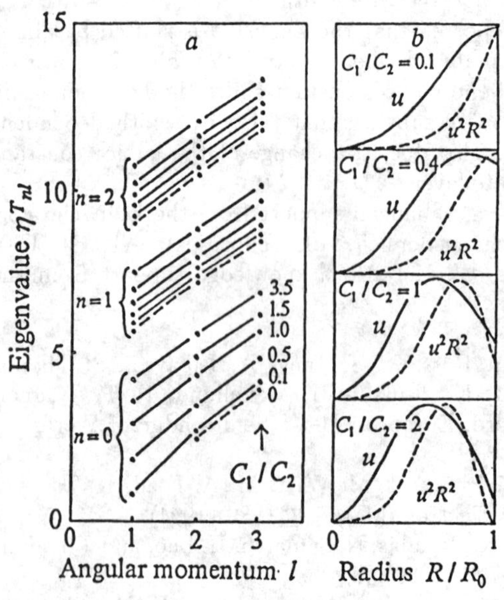

Fig.1. a.The shift of the eigenfrequencies of the torsional modes of the microcrystals for the surface mode (n = 0) and the inner modes (n = 1,2) as a function of the matrix stiffness, i.e., as a function of the ratio C_1/C_2. The dashed lines corresponds to the case of the free microcrystal surface. b.The radial distribution of the vibrational amplitudes u and the energy density u^2R^2 of the surface vibrational mode of a microcrystal as function of the ratio C_1/C_2.

l, but torsional vibrational modes with odd l can be seen in the Raman spectra. According to [5], eigenfrequencies of the torsional vibrational modes of the microcrystals embedded in a glass matrix can be determined from the equation $D^t_m = 0$, where D^t_m is determinant of (2x2) matrix which is given in [6]. In Fig.1 the shift of vibrational eigenfrequencies η^T_{nl} of a microcrystal is shown when the matrix stiffness increases relative to that of the microcrystal. The transverse surface mode ($n = 0$) with the angular momentum $l = 1$ is absent for microcrystals with a free surface (dashed lines). This is natural, because this mode corresponds to the rotation of the microcrystal as a whole. If a microcrystal is embedded in a matrix, then there appears a new solution for the surface mode with angular momentum $l = 1$. This mode arises because of the restoring force which limits the free rotation of the microcrystal. Fig.1 shows that the surface vibrational mode of the microcrystals embedded in the matrix does not disappear; its eigenfrequency increases and the maximum of the vibrational amplitude shifts inside the microcrystal with increasing of the matrix stiffness. In ref.[5] an erroneous conclusion was made that the influence of the matrix over the microcrystal leads to the opposite effect, i.e., the surface mode disappears and the inner vibrational modes softens. Hence, contrary to refs. [2,3], the frequency of the Raman peak for the surface torsional mode with the angular momentum $l = 1$ for the microcrystals embedded the matrix does not equal $1.83\, \upsilon_t/dc$ as it would be in the case of the free surface. Indeed, the existence of the matrix results in the appearance of a new solution of $D^t_m = 0$ for the surface mode with $l = 1$; its eigenfrequency increases, for example, from $0.2\, \upsilon_t/dc$ to $1.4\, \upsilon_t/dc$ when the matrix stiffness C_1 / C_2 increases 0.1 to 3.5, (Fig.1.a). Most likely, just these vibrations with $l = 1$ take part in the low-frequency Raman scattering described in [2,3].

From all our experiments it is clear that the depolarized scattering is governed by the surface torsional modes with $l = 1$ which appear, when the crystals are embedded in a matrix. Furthermore, from the calculations follows that the influence of the matrix on the frequencies of the acoustic phonons, confined in the microcrystals, is significant even if Lame's constants λ, μ and the mass densities of the microcrystal and the matrix are quite different from each other, especially for $n = 0$ and 1.

1. N.Ovsyuk, V.Novikov, Phys. Rev. B**53** (1996), in press.
2. A.Tanaka, S.Onari, and T.Arai, Phys.Rev. B**47**, 1237 (1993).
3. M.Fujii, S.Hayashi, and K.Yamamoto, Phys. Rev. B**44**, 6243 (1991).
4. E.Duval, Phys. Rev. B**46**, 5795 (1992).
5. A.Tamura, K.Higeta, and T.Ichinokawa, J.Phys. C**15**, 4975 (1982).
6. N.Ovsyuk, L.Krivoputskaya, and A.Shebanin, JETP Lett. **48**, 678 (1988).

TEMPERATURE EFFECTS IN NONLINEAR RAMAN SPECTRA OF II-VI-SEMICONDUCTOR NANOCRYSTALLITES

T. Bischof, M. Ivanda, G. Lermann, A. Materny and W. Kiefer*

Institut für Physikalische Chemie der Universität Würzburg, Marcusstraße 9-11, D–97074 Würzburg, Germany

Semiconductor crystallites with spatial dimensions less than a few tens nanometers show certain optical properties which are quite different from those of the bulk semiconductor. In such systems, so–called quantum confinement occurs, if the crystal size is comparable to or smaller than the exciton Bohr radius [1,2]. This is reflected in a size–dependent change of the electronic structure with respect to the bulk semiconductor. Up to now, a general picture of the optical properties, especially the relaxation dynamics of II-VI semiconductor crystallites does not exist. Raman spectroscopy turned out to be an important tool for the investigation of the vibrational and electronic properties of semiconductor quantum dots.

The samples discussed here are commercially available long pass filter glasses (from SCHOTT Glass Inc.). The glasses consist of CdS_xSe_{1-x} nanocrystallites with diameters of a few nanometers embedded in a borosilicate glass matrix. The volume fraction of the nanocrystallites in the matrix is about 1%. Low frequency Raman spectroscopy was applied on a variety of filter glasses with band gap energies ranging from about 1.75 eV to 2.5 eV in order to determine the size of the semiconductor crystallites. The average size of the nanocrystallites was determined using a method established by Duval *et al.* [3]. Assuming homogeneous, elastic, spherical-shaped particles, the vibrational eigenfrequency of the lowest-energy spherical mode is shown to be inversely proportional to the particle diameter. For crystallites of a few nanometers in diameter, the Raman wavenumbers of these modes are typically 10–30 cm^{-1} as shown in Fig. 1 for different filter glasses.

The large optical nonlinearities observed in low-dimensional semiconductor crystallites have attracted considerable attention due to their applicability for optoelectronic devices. Nonlinear optical properties of low-dimensional semiconductors have been extensively studied by degenerate four-wave mixing [4,5]. We report on the temperature dependence of nonlinear Raman spectra recorded by means of CARS spectroscopy. In Fig. 2 CARS spectra of the OG550 filter glass are shown for different temperatures ranging from room temperature to 10 K. The Raman resonances of the CdSe– and the CdS–like phonon modes at 200 cm^{-1} and 300 cm^{-1}, respectively, are observed to be coherently superimposed by a strong despersive non–Raman–resonant background contribution. The background obviously increases as the wavelength detuning between the pump and the Stokes laser decreases.

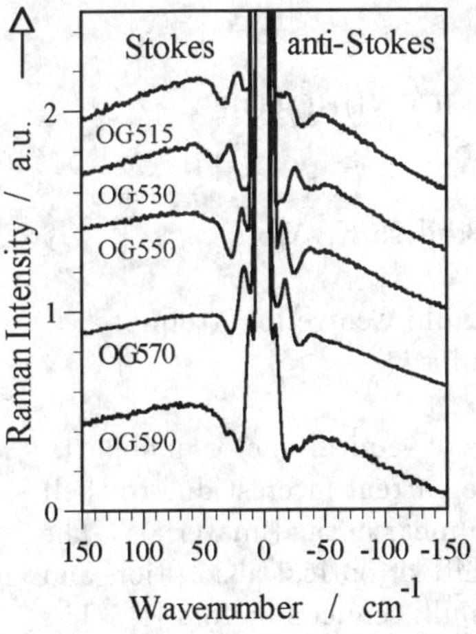

Fig. 1:
Low frequency Raman spectra of differently coloured glasses.

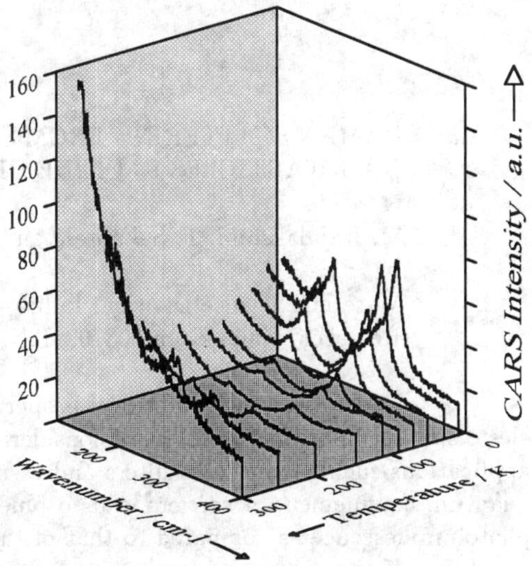

Fig. 2: Resonance CARS spectra of a semiconductor doped glass for different temperatures. The pump laser wavelength was 520 nm.

The same behaviour has been observed in bulk semiconductors, like InSb, GaAs, HgCdTe or PbSe [6]. This behaviour may be attributed to interband-excited electron dynamics. From studies of the background dispersion the homogeneous linewidth of the resonantly excited state can be deduced, which is correlated to the relaxation time of the excited carriers. In this way fast relaxation times on the picosecond time scale may be infered. The CARS spectra in Fig. 1 show a considerable decrease of the frequency dispersion from 300 K to about 110 K, followed by a moderate change down to about 10 K, indicating a similar behaviour of the linewidth of the resonantly excited state. A simulation of the non–resonant background allows a separation of the Raman–resonant part of the CARS susceptibility for line shape analysis.

We gratefully acknowledge financial support from the Deutsche Forschungsgemeinschaft (Sonderforschungsbereich 410, Teilprojekt C3) and from the Fonds der Chemischen Industrie. One of us (M. I.) acknowledges financial support from the Alexander von Humbolt Foundation.

1. Al. L. Efros and A. L. Efros, Sov. Phys. Semicond. 16, 772 (1981).
2. L. E. Brus, J. Chem. Phys. 80, 4403 (1984).
3. E. Duval, A. Boukenter, and B. Champagnon, Phys. Rev. Lett. 56, 2052 (1986).
4. Y. Wang and W. Mahler, Optics Comm. 61, 233 (1986).
5. S. H. Park, M. P. Casey, and J. Falk, J. Appl. Phys. 73, 8041 (1993).
6. S. Y. Yuen and P. A. Wolff, Appl. Phys. Lett. 40, 457 (1982).

RAMAN SCATTERING FROM SEMICONDUCTOR NANOPARTICLES IN OXIDE GLASS

M. Rajalakshmi, T. Sakuntala and Akhilesh K. Arora*

Materials Science Division, Indira Gandhi Centre for Atomic Research, Kalpakkam 603102 T.N. India

The optical and vibrational properties of semiconductor nanoparticles dispersed in oxide glasses are of considerable current interest due to their applications such as long pass filters and as nonlinear optical materials. The quantum confinement of carriers lead to blue shift of optical absorption and photoluminescence as compared to that of the bulk semiconductors [1]. The small size of the particle also causes wavevector selection rule to be relaxed in the Raman scattering process from phonons leading to modification of the line shapes [2]. Surface acoustic modes of the particle are also observed in the low frequency Raman scattering [3]. Annealing of these semiconductor doped glasses at suitable temperatures has been used to control the particle size and hence its optical and vibrational properties. In this work we report the results of annealing treatment on Raman scattering from low frequency particle modes and optical phonons and their overtones in commercial glass filter GG475 obtained from Schott Glaswerke.

Samples were annealed for 5 hrs at different temperatures between 500 and 700 C. After annealing, Raman scattering measurements are made at room temperature in back scattering geometry using Spex double monochromater and a PMT in the photon counting mode. Several lines of argon and krypton lasers are used to excite the Raman and photoluminescence spectra. Figure 1 shows the low frequency Raman spectra of the untreated sample and after annealing at 600 C. The change in the frequency of the spheroidal mode from 20 to 15 cm^{-1} suggests a growth of particle size from 58 to 76 Å when appropriate sound velocities are used. Particle size was also estimated from the x-ray diffraction using the Scherrer formula.

Figure 2 shows the Raman spectra of the filter in the region of optical phonons and their overtones. The frequency of the observed 1-LO Raman peak at 315 cm^{-1} being significantly higher than that corresponding to LO phonon in pure CdS suggests the presence of Zn also in the particle. From the one mode behaviour of phonon frequencies in $Zn_xCd_{1-x}S$ and the composition of Se (5% of the total chalcogenide anion estimated from mass spectroscopy), the approximate composition of the particle turns out to be $Zn_xCd_{1-x}S_ySe_{1-y}$

(x=0.22, y=0.95). A small peak seen at 196 cm^{-1} in the CdSe-like LO phonon region is consistent with the fraction of the Se. Annealing causes the LO peak to shift from 315 to 320 cm^{-1} and appearance of a shoulder at 310 cm^{-1}. This may be due to an increase in the value of x. ZnO is usually present in borosilicate glass to reduce the melting point. A similar behaviour of increase of phonon frequency due to increase in x has been recently reported in $Zn_xCd_{1-x}S$ nanoparticles [4]. The present samples also exhibit, subsequent to annealing, a blue shift of photoluminescence (PL) which was excited using 4067 Å line of krypton ion laser. In general, a red shift of optical absorption and PL is expected due to particle size growth, whereas a blue shift would occur due to increase in x. It appears in these samples that increase in x more than compensates the red shift, thus leading to a net blue shift. On the other hand annealing above 650 C causes a red shift of PL. Annealing also results in a significant decrease in the intensities of 1-LO and 2-LO phonons. This is again because of shifting of the electronic transition in the annealed samples away from the "in" resonance condition satisfied for the untreated samples.

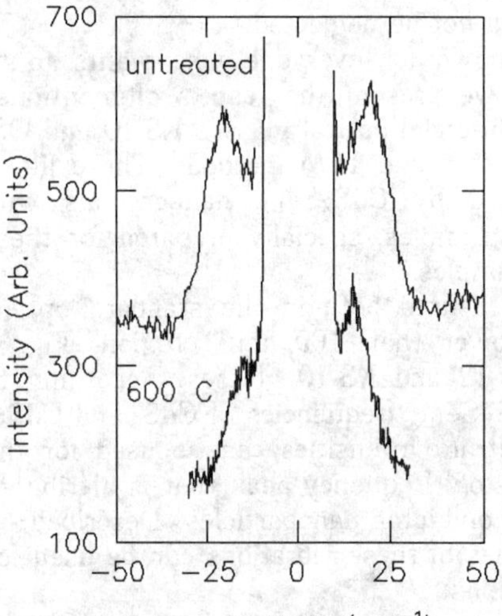

Fig. 1. Low frequency Raman spectra of untreated sample using 4762 Å line and after annealing at 600 C using 5145 Å line.

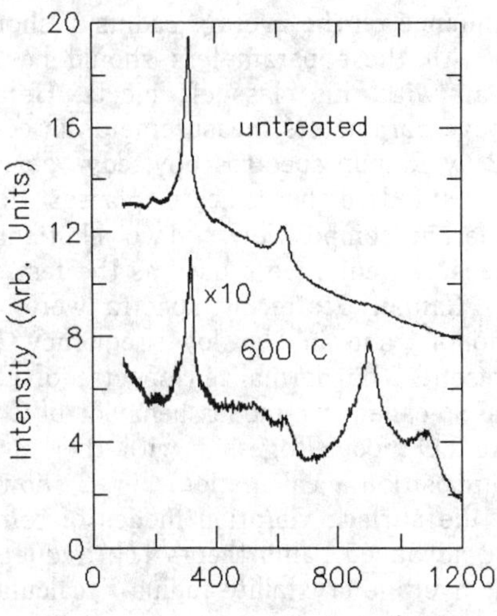

Fig. 2. 1-LO and 2-LO spectra of untreated glass sample and after annealing at 600 C using 4579 Å line.

1. L.E. Brus, J. Chem. Phys. **80**, 4403 (1984)
2. G. Mei, J. Phys.: Condens. Matter **4**, 7521 (1992)
3. A. Tanaka, S. Onari and T. Arai, Phys. Rev. B **45**, 6587 (1992)
4. H. Yukselici, P.D. Persans and T.M. Hayes, Phys. Rev. B **52**, 11763 (1995)

LOW FREQUENCY AND PHONON RAMAN SCATTERING FOR PARAMETERS DEFINITION OF CDSSE NANOPARTICLES EMBEDDED INTO GLASS MATRIX

N.R. Kulish, V.P. Kunets, A.V. Pogorelov*, M.Ya. Valakh

Institute of Semiconductor Physics Ukrainian Academy of Sciences, Kiev, Ukraine

Semiconductor crystallites embedded into glasses are known for years because of their optical properties: variation of the composition x of CdS_xSe_{1-x} small crystals in glasses leads to the changes in the absorption spectrum what is used for commercial filters production.

In the last 10 years these objects have attracted a great attention of scientists because of their intense nonlinear properties. The greatest influence on these nonlinearities can be made by changing the average radius $\bar{r}$ of semiconductor nanoparticles. Of course, the host matrix pressure factor **P** influence on the average radius $\bar{r}$ should not be neglected.

All these parameters should be known for everyone who wants to be dealt with this class of objects. Here we present the results of mentioned above parameters measurements in commercial optical glasses KS-10 and OS-12 by Raman spectroscopy, edge absorption and TEM methods. These filters are standard borosilicate glasses doped by CdS_xSe_{1-x} nanocrystals with different composition x. Two additional samples, specially prepared for these measurements, were used as the test samples.

Raman Scattering spectra were measured both in the standard optical phonon, and in the low frequency (lower than 100 cm^{-1}) region. Fig. 1 presents the normalized spectra of OS-12 and KS-10 glasses. According to the so-called two-modes behavior of CdS_xSe_{1-x}, frequencies of CdS- and CdSe-like LO-modes together with their arbitrary intensities can be used for the composition x calculation. Fig. 2 shows low frequency peak that is attributed to the surface vibration modes of semiconductor nanoparticles. Described by the advanced Lamb theory [1], frequencies of these vibrations can be used for the average crystallite radius $\bar{r}$ calculations:

$\omega_o^s = \frac{SF v_l}{2c\bar{r}}$ — for breathing modes characterized by the angular momentum $l=0$, and

$\omega_2^s = \frac{SF v_t}{2c\bar{r}}$ — for ellipsoidal modes with $l=2$. Here ω_o^s and ω_2^s are the breathing and ellipsoidal mode energies respectively, v_l and v_t are longitudinal and transverse sound velocities with $v_l / v_t = 2.3$, SF — shape factors for $l=0$ and $l=2$ modes (0.91 and 0.85 respectively), and c — is the velocity of the light in vacuum.

The numeric values of $\bar{r}$ and x obtained from Raman measurements were compared with those taken from edge absorption and TEM studies. It is shown that:

i) nanocrystal size distribution is described by the Lifshits-Slezov equation. It makes possible to define the average nanocrystalline radius $\bar{r}$ what is the one variable parameter of this equation;

ii) edge absorption spectra reveal size quantization peaks. Using the Efros theory [2], information about $\bar{r}$, band gap, and composition x, corrected on the matrix pressure to the nanocrystals, can be obtained from the energy position of these peaks.

Numeric values of all parameters, taken by presented above different methods, were compared with each other taking into account the real possibilities and limitations of each technique. It is shown that electron microscopy investigations can provide $\bar{r}$ data with good accuracy for the samples where characteristic nanocrystal size is $\bar{r} \geq 5$nm. Edge absorption method can be used for average radius $\bar{r}$ estimation only if $\bar{r} \leq a_b$, where a_b— exciton Bohr radius, but for x definition — for radii $\bar{r}$ both greater and lower a_b. Raman spectroscopy data, in principle, isn't affectable by the nanoparticle average radius value, but the possibility to perform a provable measurements is depending on technical level of the setup having been used. It also can be concluded that Raman data themselves can be used for all these parameters definition.

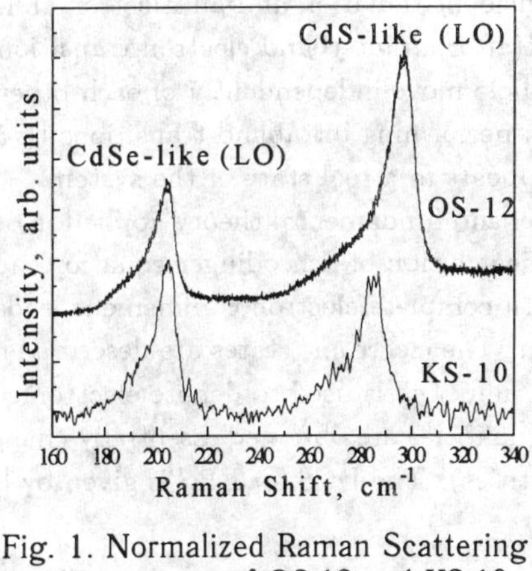

Fig. 1. Normalized Raman Scattering spectra of OS-12 and KS-10 glasses.

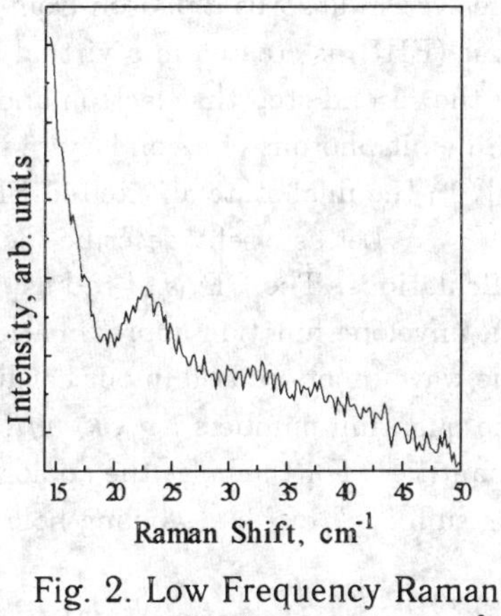

Fig. 2. Low Frequency Raman scattering spectrum of the OS-12 glass.

1. E. Duval, A. Boukenter, and B. Champagnon, Phys. Rev. Lett, 56, 2052 (1986).

2. Al. L. Efros and A. L. Efros, Fiz. Tekh. Poluprovodn. 16, 1209 (1982); [Sov. Phys. Semicond. 16, 772 (1982)].

ELECTRON RAMAN SCATTERING IN A SPHERICAL QUANTUM DOT

R.Riera, J.L. Marin and J.M Bergues

[1]Dept. of Physics-CIFUS, Sonora University, A.P. 1626, 83000-Hermosillo Sonora MEXICO

The Differential-Cross-Section (DCS) for an Electron Raman Scattering (ERS) process in semiconductors Quantum-Dots (QD) of spherical shape is calculated for T=0 K and neglecting phonon assisted transitions. Electron states are considered assuming complete confinement within the QD. We also assume a single parabolic conduction and valence bands. Two kinds of spectra are discussed: emission spectra (DCS as a function of emitted photon energy) and excitation spectra (DCS as a function of incident photon energy). In both cases we analyse DCS for different scattering configurations. We study selection rules for the processes. Singularities in the spectra are found and interpreted. The ERS here studied can be used to provide direct information about the electron band structure of the system.

ERS is qualitatively explained as a two-step process: in the first step the system absorbs a photon from the incident radiation and an Electron-Hole-Pair (EHP) is created in a virtual state (after an interband electron transition); in the second step the electron and the hole move independently of each other and emit photons of secondary radiation performing intraband transitions [1, 2]-[3] . In the final state an excited EHP appears in a real state of the system.

Let us briefly describe the model and fundamental theory applied in our calculations. The QD is of radius r_0. The solution of Schrödinger equation in the envelope function approximation with complete electron confinement leads the wave function used in our calculation. The electronic states are described by the quantum numbers : n_e, k_e, m_e. The suffix "e" is used to denote electronic quantities. For the holes the analogous quantities are obtained essentially changing suffix "e" by "h" (labeling hole quantities). The DCS for ERS is given by [1]

$$\frac{d^2\sigma}{d\omega_s d\Omega} = \frac{V^2\omega_s^2 n(\omega_s)}{8\pi^3 c^4 n(\omega_l)} W(\omega_s\, , \vec{e}_s)$$

where V is the normalization volume, $n(\omega)$ is the refraction index as a function of the radiation frequency, ω_s and $\vec{e}_s$ is the unit polarization vector for the emitted secondary radiation. c is the light velocity and ω_l is the frequency of the incident radiation and $W(\omega_s\,,\,\vec{e}_s)$ is the transition rate calculated by means of the Fermi's Golden Rule. Using the theory depicted above we can obtain, after cumbersome calculations, explicit expressions for the DCS of the ERS process. We just write the final results that is expressed in the figures. We have computed emission and excitation spectra of the ERS process for a given polarization $\vec{e}_{sz}$ of the emitted radiation. The physical parameters entering in our formulas were taken for the GaAs case; i.e., E_g=1.43 eV ; μ_e=0.0665 μ_0 ; μ_h=0.45 μ_0.

In figure (1a) we show the emission spectra for the QD. We have chosen $r_0 = 4$ nm. It can be observed that for $\hbar\omega_l = 2.3$ eV the differential cross section display only two singularities at $\omega^e\,(0,1,1,1)$, and $\omega^h\,(1,1,0,1)$, while for $\hbar\omega_l = 2.9$ eV four singularities are observe at $\omega^e(0,1,1,1, \omega^h\,(1,1,0,1)\,, \omega^e(1,1,2,1)$, and $\omega^h\,(2,1,1,1)$. We should note that the ω_s values for which we find singularities do not depend on the incident radiation frequency ω_l and only depend on the energy differences between the valence and conduction subbands for the fixed value r_0. For higher energies $\hbar\omega_l$ of the incident radiation photon we shall observe a large number of singular peaks in the emission spectra. For fixed value of ω_l we can also observe a certain step-like behavior at given values of ω_s. The position of the steps in the spectra depend on the value chosen for ω_l.

In figure (1b) we show the excitation spectra for a QD. In this case, $\hbar\omega_s$ is fixed and $\hbar\omega_l$ is a variable quantity. The other parameters coincide with those of figure (1a). We observe the same behavior in the spectra for different values of r_0 with the same $\hbar\omega_s$ as those different values of $\hbar\omega_s$. The resonant transitions are indicated only in one curve, because they are same in all curves.

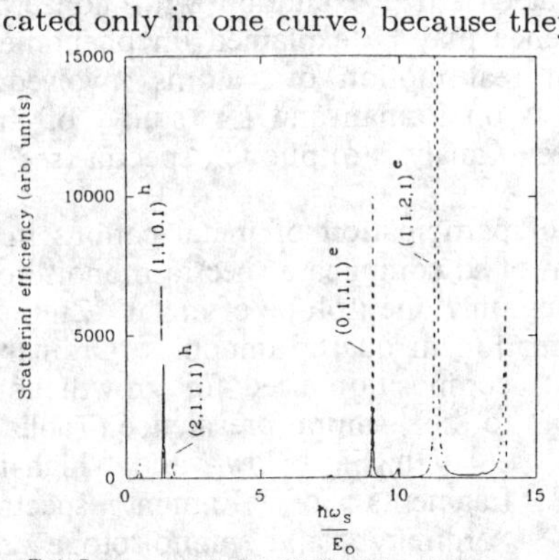

Fig.1a Emission spectra for GaAs QD. The solid curve corresponds to $\hbar\omega_l$ =2.3 eV, the dashed curve to $\hbar\omega_l$ =2.9 eV, in both case r_0 = 4 nm.

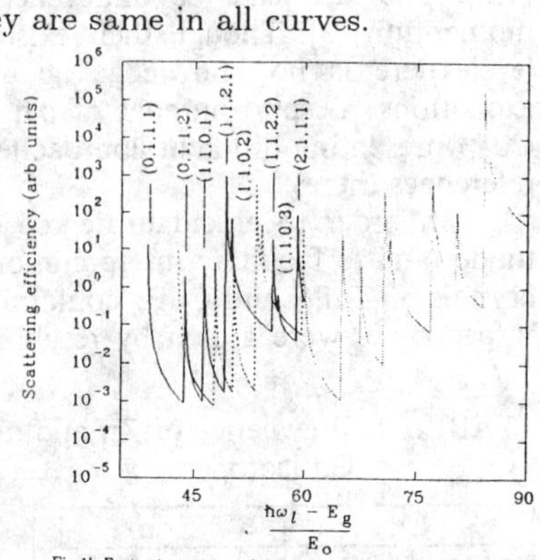

Fig.1b Excitation spectra. Solid curve: r_0 = 3 nm and $\hbar\omega_s$ =2 eV. Medium dashed: 3 nm and 2.3 eV. Short dashed: 4 nm and 2 eV.

1. R.Riera, F.Comas, C.Trallero and S.Pavlov, Phys. Stat. Sol.b **148** 533 (1988)
2. J.M. Bergues, R.Riera, C.Trallero-Giner and F.Comas, J. Phys. Cond. Matter. 7, 7273, (1995)
3. J.M.Bergues, A.Villares and R.Riera, submitted to Phys. Stat. Sol. b (1996).

Raman scattering from tetragonal $^{64}ZnP_2$ and $^{114}CdP_2$ monoisotope crystals

S.G.Garasevich, P.I.Korenyuk, A.V.Slobodyanyuk, Z.Z.Yanchuk

Department of Physics, Kiev University, Prospect Glushkova 6, 252127 Kiev, Ukraine

Pair of isostructural ion-covalent wide band gap semiconductors ZnP_2 and CdP_2 transparent in red region gives an unique chance for Raman study of fine peculiarities in lattice dynamics due to such features as strong anisotropy of mechanical properties (easy sliding fracture in XY-plane), very strong gyrotropy (specific rotatory power at λ=633 nm is more one order of magnitude greater than in quartz), very narrows lines even at room temperature, and opportunity to compare spectra of both crystals with very close physical properties.

Both ZnP_2 and CdP_2 crystals may exist in "right" and "left" enantiomorphic structures corresponding to space groups $P4_12_12$ and $P4_32_12$, respectively, with 24 atoms per unite cell. In Γ-point 69 optical phonons can be classified according irreducible representations of 422 point group as follows: $9A_1$,, $8A_2$,, $9B_1$,, $9B_2$, and 17E. Modes of A_2 symmetry are only IR active. So 61 Raman lines may be expected, taking into account possible splitting of double degenerated E-modes due to long range Coulomb interaction or spatial dispersion [1,2]. The last leads to undegenerated lattice excitation with nonzero angular momentum [3]. Though their existence may be explained on phenomenological level there is no knowledge about real motion of atoms involved in such excitations despite number of papers on Raman and IR study of ZnP_2 and CdP_2 presenting different approaches to analysis of phonon spectra (see [1-4] and references there).

In order to elucidate degree of participation of metal cations in different modes and to facilitate interpretation of so complicate spectra monoisotope single crystals of ZnP_2 and CdP_2 containing only one isotope of metal ^{64}Zn or ^{114}Cd, respectively, were grown by resublimation in quartz ampoules. Ordinary isotope composition used for growth corresponds to their natural prevalence (Table 1).

TABLE 1. Prevalence of Zn and Cd isotopes.

Cd		Zn	
a.u.	%	a.u.	%
106	1.4	**64**	**50.9**
108	1.0	66	27.3
110	12.8	67	3.9
111	13.0	68	17.4
112	24.2	70	0.5
113	12.3		
114	**28.0**		
116	7.3		

In Fig.1 low- and high-frequency fragments of Raman spectra from "ordinary" and monoisotope ZnP_2 are shown. In the Table 2 positions ν_O of few prominent lines in spectra of "ordinary" ZnP_2 and CdP_2 crystals are given as well as their isotope shifts ν_O-ν_M, bandwidths $\Delta\nu$, and symmetry according [4]. Shifts of lines in ZnP_2 and CdP_2 have opposite signs in agreement with fact that ^{64}Zn is the lightest isotope and ^{114}Cd is one of the most heavy among Cd isotopes. Magnitudes of shifts reflect involving of metal atoms in corresponding mode. High frequency lines are supposed due to motion of P atoms because they manifest no shifts and have a very close position in both crystals. The most low frequency lines correspond to motion of heavy parts of unit cell. Relative change of their masses due to isotope enrichment is small and shifts also are small. The largest shifts of lines near 100 cm^{-1} indicate that metal ions contribute significantly in corresponding modes.

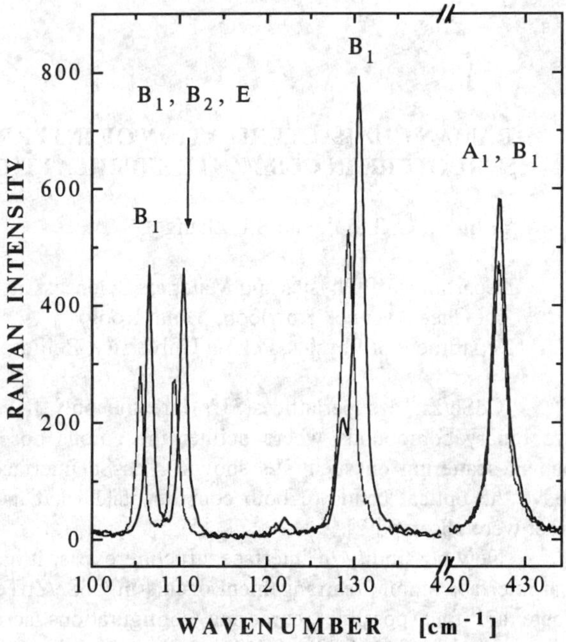

Fig.1. Raman spectra from "ordinary" (dashed line) and monoisotope (solid line) $^{64}ZnP_2$.

TABLE 2. Isotopic shifts of some lines in Raman spectra of ZnP_2 and CdP_2

ν_O, cm^{-1}	ν_O-ν_M, cm^{-1}	$\Delta\nu$, cm^{-1}	Line symmetry	ν_O, cm^{-1}	ν_O-ν_M, cm^{-1}	$\Delta\nu$, cm^{-1}	Line symmetry
	ZnP_2				**CdP_2**		
57.4	-0.3	0.6	A_1	72.2	0.3	0.6	A_1
84.3	-0.5	0.7	E	73.9	0.3	1.3	B_2
105.5	-1.0	0.9	B_1	90.0	0.5	1.0	A_1
109.4	-1.1	1.0	B_1,B_2,E	449.3	0	1.7	B_2
129.2	-1.3	1.5	B_1				
426.7	0	2.2	A_1,B_1				
459.7	0	2.0	E				

1. A.V. Slobodyanyuk, Sov. Phys.-Solid State 28, 1845 (1986) and Opt. & Spectr. (USA) 61, 626 (1986).
2. A.V. Slobodyanyuk, S.G.Garasevich, Z.Z.Yanchuk, Physics Letters A 197, 238 (1995).
3. A.V. Slobodyanyuk, in Proc. of the XIII-th Intern. Conf. on Raman Spectr., Wurzburg, Aug.-Sept. 1992 (Wiley, Chichester, 1992). p.790.
4. I.S.Gorban', V.A.Gubanov, G.I.Salivon, Z.Z.Yanchuk, Ukrainskii Fizicheskii Zhurnal 30, 202 (1985) (in Russian).

RAMAN CHARACTERIZATION OF INTERFACE ATOMIC STRUCTURE IN CDSE/ZNTE SUPERLATTICES

Y. Jin[1,2], G.G.Siu[1], and S.L.Zhang[2]

[1]Department of Physics and Materials Science, City University of Hong Kong
Tat Chee Avenue, Kowloon, Hong Kong
[2]Department of Physics, Peking University, Beijing, P.R.China

CdSe/ZnTe superlattices(SL's) are the only II-VI system formed between two binary compounds where neither the anions nor the cations are common. Raman scattering on such SLs shown a Zn-Se interface vibrational mode (IFM) above the optical bands of both constituents[1] and its multiphonons up to fifth order were observed[2].

New possibility of interface structure exists. It has recently been shown [3] that interface atomic rearrangement occurs in CdSe/ZnTe samples grown by MBE. There are four possible reordering configurations across the interfaces [3]. A schematic plot of a $(CdSe)_4/(ZnTe)_8$ SL with all possible interfaces is shown in Fig. 1, following the convention of linear chain. It is readily seen that Zn-Se and Cd-Te single wrong bonds (WB's) alternatively appeared at interfaces in the structure with sharp interfaces (S). On the other hand, possible exchange of the interface atomic planes in all reordering structures (R_i, i=1-4) introduces not only single WB's but also double WB's such as Zn-Se-Zn or Se-Zn-Se (Zn-Se double WB), Cd-Te-Cd or Te-Cd-Te (Cd-Te double WB) to SL.

This interface reordering affects vibrational properties of SL. Longitudinal phonon spectra of $(CdSe)_4/(ZnTe)_8$ SL along [001] growth direction are calculated with linear-chain model. As in previous study [1], Zn-Se single WB in S structure produces a localized IFM at 222cm^{-1} (LIF), lying above the optical bands of both constituents. In all reordering R_i structures, beside the IFM at 222cm^{-1}, an additional IFM at 235cm^{-1} (LIF_2) are generated by Zn-Se double WB's at interfaces, which is the characteristic of interface atomic reordering.

LIF has been observed in previous Raman measurements on ALE-grown CdSe/ZnTe SLs[1], and the Raman investigation on WB-originating IFM have been widely employed in characterization of interface structure in semiconductor SL's[4]. For ascertaining the Raman identifiability of LIF_2, we have calculated the relative Raman scattering intensity at excitation line of 501.7nm with bond polarizability model. This model has already been applied to II-VI heterostructure (ZnSe/ZnTe SL) [5]. The simulated spectra near the ZnTe LO confined mode (LO_1) for both S and R_i (i=1-4) structures are displayed in Fig.2. Noted that in all reordered R_i structures, LIF_2 has similar intensity to LIF, which make it possible to probe the interface reordering by Raman scattering.

On the bottom of Fig.2, the Raman spectrum of the ALE-grown $(CdSe)_4/(ZnTe)_8$ is also shown for comparison. In this experimental spectrum, the LIF_2 at $235cm^{-1}$ is completely absent, which indicates that the numbers of Zn-Se double WB arising from atomic rearrangement in this ALE grown SL should be much smaller than Zn-Se single WB. The decisive factor for low possibility of interface reordering might be the relatively lower substrate-buffer temperature, 220 °C, at ALE growth.

Fig.1

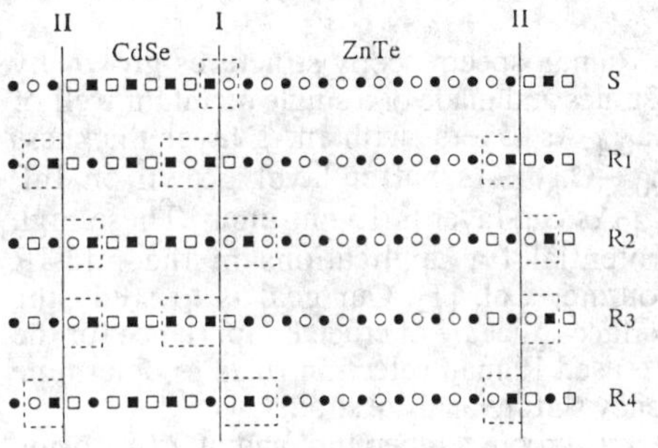

Fig.2

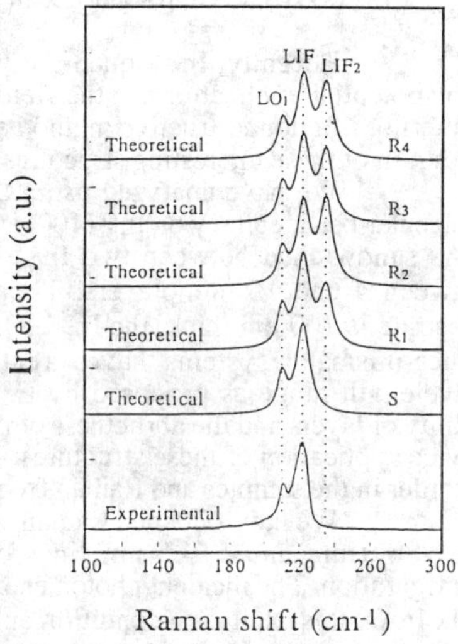

Fig. 1 Schematic plot of $(CdSe)_4/(ZnTe)_8$ SL: S for sharp interface, R_i (i=1-4) are four configuration of atomic rearrangement. Zn-Se WB's are boxed with dashed line.

Fig. 2 Raman spectra of $(CdSe)_4/(ZnTe)_8$ SL in $z(x',x')\bar{z}$ configuration: five spectra from top downward are simulated ones for s and R_i (i=1-4) configuration; the bottom spectrum is the experimental one of the ALE grown SL.

1. Y.Jin, Y.T.Hou, Shu-Lin Zhang, J. Li, S.X.Yuan and G.G.Qin, Phys. Rev. B**45**, 12141(1992)
2. S.L.Zhang, C.L.Yang, Y.T.Hou, Y.Jin, Z.L.Peng, J.Li, and S.X.Yuan, Phys. Rev. B **52**, 1477(1995)
3. K.M.Kemner, B.A.Bunker, A.J.Kropf, H.Luo, N. Samarth, J.K.Furdyna, M. R. Weidmann and K.E.Newman, Phys. Rev. B**50**, 14327(1994).
4. S.G.Lyapin, P.C.Klipstein, N.J. Mason and P.J.Walker, Phys. Rev. Lett. **74**, 3285 (1995)
5. L.A.Farrow, J.M.Worlock, F. Turco-sandroff, R.E.Nahory, R. Beserman and D.M. Hwang, Phys. Rev. B **45** 1231(1992)

RAMAN STUDY OF HIGHLY STRAINED QUANTUM WELLS

L. G. Quagliano, D. Orani and A. Ricci

I.M.A.I.- C.N.R., P.O.Box10, via Salaria km 29.300
00016 Monterotondo Scalo, Roma, Italy

Recently, InAs quantum wells have received considerable interest for their potential applications in the field of infrared optoelectronics devices. Raman scattering is a nondestructive method sensitive to study the quality, the disorder and the strain of these interesting structures.

We have analyzed using the Raman spectroscopy structures grown by molecular beam epitaxy on InP (100) substrates and made of a single quantum well of InAs sandwiched between two $In_{0.53}Ga_{0.47}As$ layers, with InAs layer thickness between 4 and 15 monolayers. The $In_{0.53}Ga_{0.47}As$ buffer layer grown on InP substrate is 100 nm thick, the $In_{0.53}Ga_{0.47}As$ cap layer is 15 nm thick. These high lattice-mismatch systems have great potential for applications in the mid-IR wavelength range as proposed by E. Tournié et al. [1]. Our goal is to study the quality of layers and the abruptness of interfaces, that are of crucial importance for the device application of these structures. We used Raman selection rules to determine disorder in the samples and Raman frequency shift to analyze strain.

We have obtained Raman spectra at room temperature and at 77 K, using the 488.0 nm and 514.5 nm lines of an Argon ion laser in the backscattering configuration. The incident photon energy is close to the resonance with the E_1 gap of bulk InAs, this excitation condition enhances the Raman scattering cross section and facilitates a more detailed study of vibrational modes. In these experiments the penetration depth of the laser light is approximately 20 nm into the $In_{0.53}Ga_{0.47}As$ and 10 nm into the InAs, so we expected to obtain significant signal only from the quantum well and from the alloy cap layer.

In the Raman spectra of Fig. 1 we can observe the GaAs-like LO phonon from $In_{0.53}Ga_{0.47}As$ alloy and the InAs LO phonon from the quantum well. The dominant and sharp features of these vibrational modes characterize the good quality and homogeneity of these structures. In addition we observe distinct peaks increasing with the well thickness. The peak in the figure associated with the InAs TO phonon should be forbidden in our scattering configuration and it can be seen only through crystal imperfections increasing with the well thickness. We can also observe another peak close to the InAs LO phonon attributed to disorder at the interface between the InAs and the cap layer, with intensity increasing with the well thickness.

The $In_{0.53}Ga_{0.47}As$ buffer layer is lattice-matched with the InP substrate, while due to the different lattice constants the InAs layer grown on the $In_{0.53}Ga_{0.47}As$ buffer layer is subject to a 3.2 % compressive biaxial strain. This in-plane strain shifts the vibrational modes to higher frequencies with respect to the bulk values and we observe an experimental energy shift of the InAs LO phonons. This experimental shift decreases with the increase of the well thickness indicating a smaller strain. The $In_{0.53}Ga_{0.47}As$ cap layer grown on the InAs layer is subject to a tensile biaxial strain, so we observe an energy shift of the GaAs-like LO phonons to lower frequencies with respect to those observed in the unstrained alloy. In this case these modes remain constant with the increase of the well thickness.

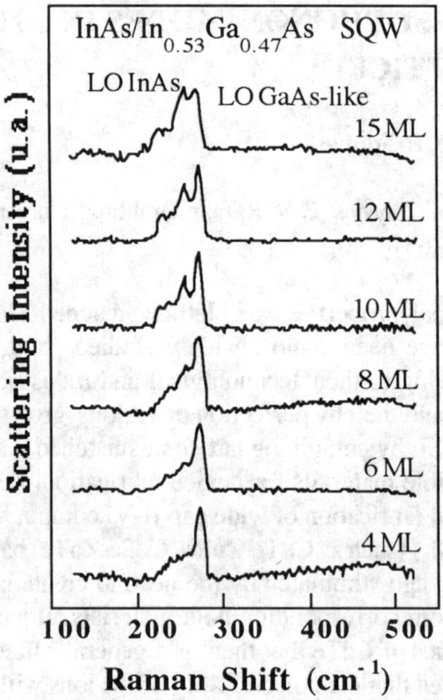

Fig. 1. Raman spectra of $InAs/In_{0.53}Ga_{0.47}As$ SQW obtained at 80 K with λ_{exc} =488 nm.

In conclusion, we have used Raman spectroscopy to characterize and study a series of $InAs/In_{0.53}Ga_{0.47}As$ single quantum wells grown by MBE on InP substrates. The observed Raman spectra consist of many peaks. The intense and narrow lines corresponding to the GaAs-like LO mode of the $In_{0.53}Ga_{0.47}As$ cap layer and to the LO phonon mode of the InAs are dominant. In addition to these modes we have observed two different features indicative of a slight deterioration of the structural perfection of the sample with the increase of the well thickness.

This study demonstrates that Raman scattering is an useful technique to characterize the local structures of interfaces and the quality of these systems that are of considerable interest for device applications in the long wavelength optical communications.

The authors would like to thank the M.B.E. of ICMAT-CNR . One of us (D. O.) would like to acknowledge the CNR-Italy for its fellowship of the "Programma del Fondo Sociale Europeo".

1. E. Tournié and K. H. Ploog, Journal of Crystal Growth **135**, 97 (1994).

TRANSVERSE PHONON MODES IN CdTe-ZnTe SUPERLATTICE

*S. Rath and S.P. Sanyal**

Department of Physics, C.V. Raman Building, University of Bhopal
Bhopal-462026 (India).

The vibrational properties of lattice matched GaAs-AlAs semiconductor superlattice (SSL) have been most widely studied both theoretically [1] and experimentally [2] owing to their technological and industrial importance. However, with the development and thereby perfection of various growth techniques, it has been possible to fabricate SSL by combining lattice mismatched materials thereby widening the choice of compatible materials for device fabrication[3].

The successful fabrication of wide gap II-VI compound semiconductor strained layer superlattice (SLSL) (such as CdTe-ZnSe, CdSe-ZnTe) based on zinc chalcogenides was achieved not long ago stimulated by the need to create metastable thin films with properties different from corresponding bulk materials. It has been observed from the lattice dynamical studies of CdTe that there is a general softening of the optical modes [4]. Hence it is expected that the polarizability of the ions will play an important role to bring out the different features of phonon dispersion curves (PDC). In view of these facts and also with the availability of experimental results [5] we report in the present paper, the effect of strain on the transverse phonon modes of CdTe-ZnTe SLSL using a simple one-dimensional chain model taking into account the effect upto the next nearest neighbour (NNN) and also by considering anion to be polarizable. Such a model with the effect of nearest neighbour (NN) interaction for both strained and unstrained cases has been successfully used in our earlier work in periodic[6] as well as quasi-periodic [7] systems for the analysis of longitudinal modes. The dispersion relation has been calculated by following the transfer matrix method explained in previous papers [6,7].

In the present paper, we have studied the effect of strain on the highest TO phonon frequency of CdTe and ZnTe for $(CdTe)_4(ZnTe)_4$ SLSL. It is observed that besides the usual features of superlattice vibrations such as folding and confinement effects, there occurs the downward and upward shifting of the highest TO modes of ZnTe and CdTe respectively. This is due to the accomodation of lattice mismatch by compressive strain in CdTe and tensile strain in ZnTe so as to attain a common lattice parameter.

We have plotted in Fig. 1, the highest TO phonon frequencies for $(CdTe)_n(ZnTe)_n$ SSL for the thickness n varying from 1 to 5 and compared them with the measured data. This figure also reveals that the calculated results are in good agreement with the experimental measured data. It can also be noted from this figure that the TO phonon modes of CdTe are very sensitive to the variation of thickness in comparison to ZnTe.

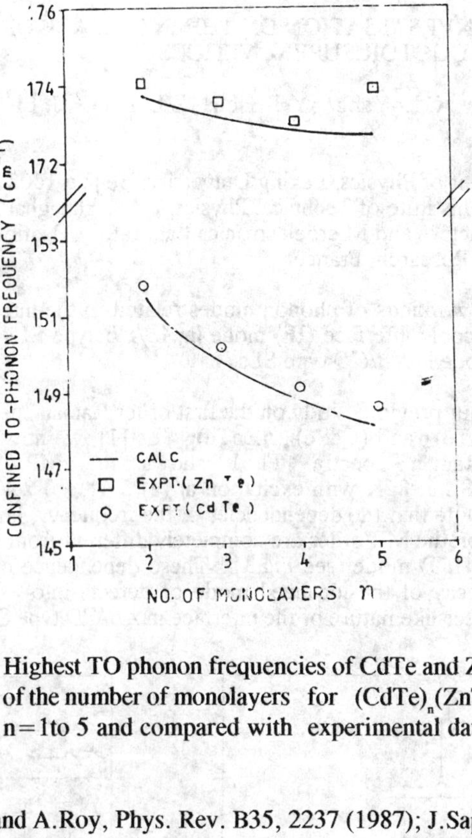

Fig.1 Highest TO phonon frequencies of CdTe and ZnTe as a function of the number of monolayers for $(CdTe)_n(ZnTe)_n$ SLSL where n=1to 5 and compared with experimental data.

1. A.Kobayashi and A.Roy, Phys. Rev. B35, 2237 (1987); J.Sapriel, Surf. Sci. Rep. 126, 197 (1983).
2. M.Cardona in Spectroscopy of Semiconductor Microstructures (Plenum, New York, 1990).
3. R.F.Davis, Z.Sitar, B.E.Williams, H.S.Kong, H.J.Kim, J.W.Palmour, J.A.Edmond, J.Ryn, J.T.Glass and C.H.Carter, Mater. Sci. and Engg. B1, 77 (1988).
4. J.M.Rowe, R.M.Nicklow, D.L.Price and K.Zanio, Phys. Rev. B10, 671 (1974).
5. T.Fromherz, F.Hauzenberger, W.Fascinger, M.Helm, P.Juza, H.Sitter and G.Bauer, Phys. Rev. B47, 1998 (1993).
6. S.Rath and S.P.Sanyal, Pramana- J.Phys. 41, 21 (1993).
7. S.Rath, P.K.Jha and S.P.Sanyal, Pramana - J.Phys. 45, 25 (1995).

RAMAN INVESTIGATION ON THE INTERFACE OF SHORT-PERIOD SEMICONDUCTOR SUPERLATTICES

S.L. Zhang[1], C.L. Yang[1], Y.T. Hou[1], Z.L. Peng[2], J. Li[2], S.X. Yuan[2], and R. Planel[3]

[1]Department of Physics, Peking University, Beijing 100871, China
[2]Shanghai Institute of Technical Physics, CAS, Shanghai 200083, China
[3]Microstructure and Microelectronics Laboratory, National Center of Scientific Research, France

There are two kinds of phonon modes related to the interface of superlattices (SLs), the macroscopic interface (IF) mode in *AB/CB* type SLs and the microscopic interface (MIF) mode in *AB/CD* type SLs.

Based on our previous study on the first order Raman spectrum of MIF modes with $(CdSe)_4/(ZnTe)_4$ and $(CdSe)_4/(ZnTe)_8$ SLs [1], we now report their multiple phonon (MP) Raman spectra. Fig.1 shows the MP Raman spectra of $(CdSe)_4/(ZnTe)_8$ SL at 82K with excitation at (a) 2.41, (b) 2.47, (c) 2.50, (d) 2.54, and (e) 2.71 eV. Note that the dependencies of the frequency, linewidth, and intensity of the MIF mode on the MP order are completely different from those of the confined longitudinal optic (LO) mode (see Fig.3). These dependence can be understood by the anharmonic decay of the localized mode of defects into extended band modes, suggesting the defect-like nature of the interface in *AB/CD* type SLs [2].

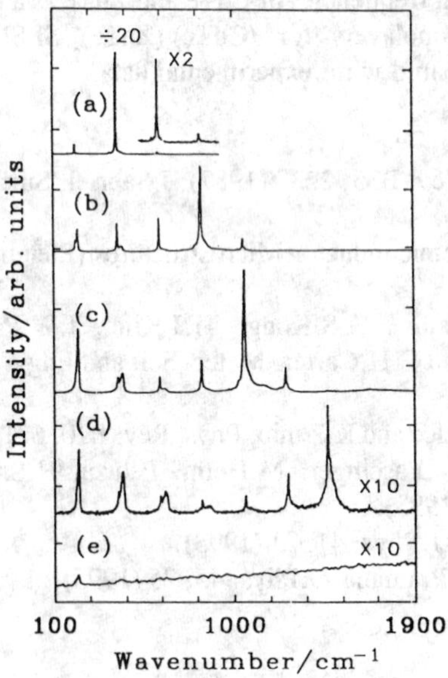

Fig.1

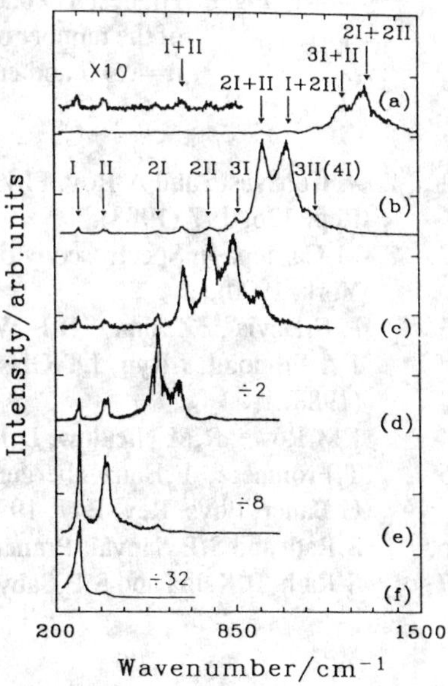

Fig. 2

In $(GaAs)_2/(AlAs)_4$ and $(GaAs)_4/(AlAs)_2$ *AB/BC* type SLs, the single phonon and MP Raman scattering for the GaAs-like (labeled by I) and the AlAs-like (labeled by II) IF modes are measured. In the single phonon spectrum, there is no sharp peak superimposed on the broad GaAs-like IF mode, confirming the new point--view on the property of the IF mode [3]. The MP Raman spectra of $(GaAs)_4/(AlAs)_2$SL at 82K with excitation at (a) 2.07, (b) 2.04, (c) 2.01, (d) 1.99, (e) 1.96 , and (f) 1.94 eV are shown in Fig. 2. No MP LO Raman peak can be seen in Fig 2. This is attributed to the penetration of electronic wavefunction into the barrier. Furthermore, the frequency and intensity behavior of the MP spectra of the IF modes are similar to those of the LO modes while the dependence of the linewidths on the MP order is reminiscence of that of the MIF mode. These dependencies are shown in Fig.3. These behaviors indicate that the IF modes contain characteristics of both the confined LO mode and the localized defect mode.

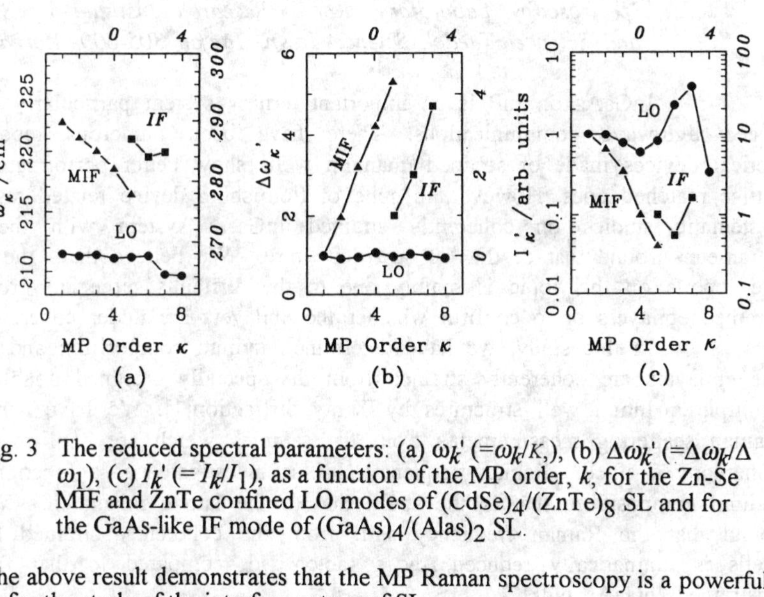

Fig. 3 The reduced spectral parameters: (a) ω_k' ($=\omega_k/\kappa$,), (b) $\Delta\omega_k'$ ($=\Delta\omega_k/\Delta\omega_1$), (c) I_k' ($=I_k/I_1$), as a function of the MP order, *k*, for the Zn-Se MIF and ZnTe confined LO modes of $(CdSe)_4/(ZnTe)_8$ SL and for the GaAs-like IF mode of $(GaAs)_4/(Alas)_2$ SL.

The above result demonstrates that the MP Raman spectroscopy is a powerful technique for the study of the interface nature of SLs.

This work is supported by the State Committee of Science and Technology and the NSF of China.

1. Y. Jin, Y. T. Hou, S. L. Zhang, J. Li, S. X. Yuan, and G. G. Qin, Phys. Rev. B45, 12141 (1992).
2. S. L. Zhang, C.L. Yang, Y.T. Hou, Y.Jin, Z.L. Peng, J. Li, and S.X. Yuan, Phys. Rev. B52, 1477(1995).
3. A.J. Sheilds, M. Cordona, and K. Eberl, Phys. Rev. Lett. 72, 412(1994).

RAMAN FREQUENCY SHIFTS FROM COHERENTLY STRAINED $In_xGa_{1-x}As/InP$ QUANTUM WELLS

E. -j. Shin[1,3], D. Lee[1], J. E. Zucker[2], C. H. Joyner[2], and D. Kim[3]

[1]*Department of Physics, Chungnam National University, Taejon 305-764, Korea*

[2]*AT&T Bell Laboratories, Holmdel, New Jersey 07733, U.S.A.*

[3]*Spectroscopy Laboratory, Korea Research Institute of Standards and Science, Taedok Science Town, Taejon 305-600, Korea*

InGaAs on InP is an important ternary system particularly for high speed lightwave communications. There have been numerous reports that optical devices made of strained quantum wells show better performances than lattice matched ones. However, in spite of flourishing device related researches, systematic studies on coherently strained InGaAs system with the lattice parameter around that of the InP are quite rare. We believe that the lack of the reports on this topic is mainly due to the difficulty either to grow good strained epilayers or to confirm whether the epilayers are under coherent strain.

In this study, we have determined indium composition and whether the epilayers are coherently strained from the specially designed double period multiple quantum well structures by X-ray diffraction[1]. We have carried out Raman scattering measurements from the samples which are confirmed to be coherently strained. Raman frequency shift is investigated as a function of the indium composition, that is the strain due to the lattice mismatch. We have found that the Raman frequency shift from the coherently strained quantum wells is dramatically reduced, by a factor 1/6, compared to that from the unstrained InGaAs bulk[2].

Strained InGaAs/InP quantum well samples are grown on [100] InP substrate by metalorganic vapor phase epitaxy (MOVPE). Special care is applied to make sure that each structure is under coherent strain. Sample parameters determined by the X-ray diffraction with the simulation[1] are written in Table 1. The sample with the indium mole fraction(x) = 0.53 has the same lattice parameter with InP substrate and has no lattice mismatch strain. The sample with x = 0.39 has the extreme tensile strain and the one with x = 0.71 has the other extreme compressive strain. As x deviates from 0.53, the amount of the strain increases and the critical thickness decreases.

Raman frequencies of GaAs-like LO phonon, presented as solid circles in Figure 1, are measured from the coherently strained InGaAs/InP quantum wells with x ranging from 0.39 to 0.71. We have found from the coherently strained structures a drastically reduced shift of the Raman frequency, only 1/6 of the shift reported from the unstraied inGaAs bulk[2]. The Raman frequency difference due to strain at the same composition is calculated by the following equation[3]:

$$\Delta\omega^{LO} = \left[\frac{\{p^{LO}/(\omega_0^{LO})^2\}\, S_{12}}{S_{11} + S_{12}} + \{q^{LO}/(\omega_0^{LO})^2\}\right] \varepsilon\, \omega_0^{LO}$$

where ω_0^{LO} is the Raman frequency from unstrained epilayers and ε is the in-plane strain($\varepsilon = [a_{InP} - a_{InGaAs}]/a_{InGaAs}$). S_{11} and S_{12} are elastic compliance constants, and p^{LO} and q^{LO} are phenomenological parameters measuring changes in spring constants due to an applied stress. We adopted the values for the above parameters from the collection in Ref. 3. However, the measured value is still smaller than the number calculated with the known parameters.

For the InGaAs epilayer system, as x changes, the strain can change as well. In this case, we should know the Raman frequency precisely at a given strain and composition, to reliably determine them. This information can be critical to figure out whether the corresponding epilayer is under coherent strain or not.

Table 1. Indium composition(x) and thickness of the quantum well(L_w) and the barrier(L_b, L_{2b}) from the coherently strained InGaAs/InP double period quantum well structures determined by X-ray diffraction[1].*(down left)*

sample	x	L_w(nm)	L_b(nm)	L_{2b}(nm)
2321	0.71	1.52	17.35	34.70
2303	0.65	4.46	9.85	19.70
2299	0.62	4.90	8.50	17.00
2295	0.60	4.44	9.83	19.66
2175	0.53	5.50	8.80	-
2306	0.48	4.77	6.37	12.74
2304	0.46	1.67	7.15	14.30
2323	0.39	2.24	16.98	33.96

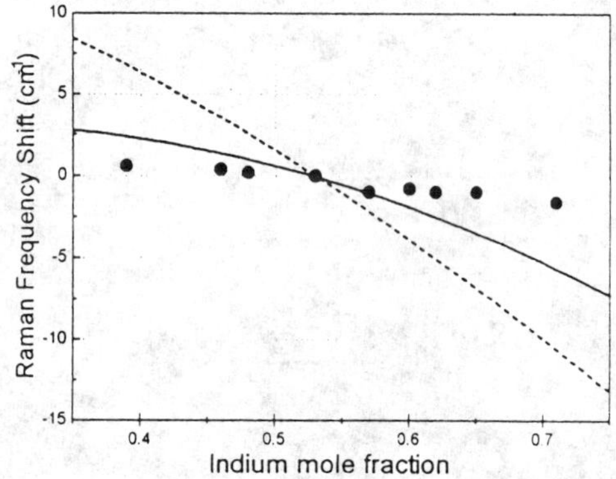

Figure 1. GaAs-like LO phonon frequency shifts($\Delta\omega^{LO}$) as a function of indium composition. Solid circles are the experimental data from the coherently strained InGaAs/InP quantum wells and the dotted line is a fit for unstrained InGaAs [2]. Solid line is a calculated curve for strained InGaAs layers with the given parameter values[3].*(upper right)*

[1]. J.E.Zucker, C.H.Joyner, and A.G.Dentai, Photo. Tech. Lett. 4, 432 (1992)
[2]. S.Emura, S.Gonda, Y.Matsui, and H.Hayashi, Phys. Rev. B32, 3280 (1988)
[3]. P.Bhattacharya, "Properties of Lattice-Matched and Strained Indium Gallium Arsenide", INSPEC, IEE, London (1993) p. 192

SECTION 23

PHASE TRANSITIONS

NEW EVIDENCES OF ORDER-DISORDER BEHAVIOR IN THE PHASE TRANSITION OF $BaTiO_3$

Y. J. Jiang*, L. Z. Zeng

Department of Applied Physics, Beijing Polytechnic University, Beijing, 100022, People's Republic of China

The phase transitions in the ferroelectric perovskites have long been considered prototypical examples of "soft-mode" displace transitions. However, in recent years, "order-disorder" behavior in $BaTiO_3$ has also been confirmed by EPR, XRD, NRD, XAFS, TEM, Raman scattering, hyper-Raman scattering, Infrared reflectivity spectra and femtosecond spectrocopy measurements [1]. The Raman spectra of $BaTiO_3$ have been studied for thirty years [2-6]. Numerous independent investigations [2,4] have demonstrated that two broad asymmetric bands at 275 and 514 cm^{-1} in the A_1(TO) Raman spectrum persist well into the cubic paraelectric phase where first-order Raman scattering is forbidden by symmetry. One of the explanations of this fact is that these two bands are second-order Raman spectra [2]. On the other hand, the polariton and oblique phonon dispersion characteristics of $BaTiO_3$ show that the A_1 phonon spectrum has the first-order character. In this paper, we report some new results on Raman spectra of $BaTiO_3$ which support the order-disorder phase transition model [7,8].

The $BaTiO_3$ single crystal used in this work was grown at the Institute of Physics, Chinese Academy of Sciences, by the top-seeded growth method using a TiO_2-$BaTiO_3$ solution. The crystal was prepared in single-domain form and then was cut and polished into a rectangular block (5.18mm*5.19mm*5.27mm) with edges along a, b and c directions. The sample was mounted on a sample holder (high-thermal-conductivity copper) heated by an electric furnace capable of operating between 290 K and 690 K, Raman spectra were collected using a JY U1000 Raman spectrometer. The sources for excitation were Argon laser radiation with 200 mw at 514.5 and 488.0 nm. The effective spectral slit width is about 2 cm^{-1}.

Polarized Raman spectra of a $BaTiO_3$ single crystal have been taken over the frequency range of 100 — 1600 cm^{-1}. Besides all the A_1 and E fundamental vibration modes reported previously, three new week bands at 986, 1204 and 1480 cm^{-1} in the X(ZZ)Y, Z(YY)X, Z(XY)Z and X(ZY)X geometries are observed. According to the temperature dependence of Raman scattering intensity and band-width of these bands, the weak bands at 986, 1204 and 1480 cm^{-1} are attributed to the second-order Raman spectra and the strong bands at 275 and 514 cm^{-1} in the X(ZZ)Y spectrum are confirmed to be of first-order. It is not easy to measure accurately the temperature dependence of the Raman scattering intensity in $BaTiO_3$ because the absorption of light in the sample varies with temperature. However, one can obtain the temperature dependence of the relative intensity ratio of two bands in one spectrum without any difficulty. Here, we chose the strong band at 514 cm^{-1} and the weak band at 1480 cm^{-1} to compare [4].

With the configurations of X(ZZ)Y, X(ZZ)X+ΔZ, Y(ZX)Y+ΔX and Y(XZ)Y+ΔZ, the frequencies of all the A_1(TO), A_1(LO), E(TO) and E(LO) modes in $BaTiO_3$ are determined accurately. Using four types of optical geometry, Raman spectra of oblique phonons in tetragonal $BaTiO_3$ single crystal have been obtained for different values of the angle which the phonon wavevector makes with the c-axis. On the basis of Merten's equation, the directional dispersion curve were fitted and plotted. The experimental points agree with calculated curves perfectly, which means it is correct to consider the bands at 275 and 514 cm^{-1} coming from the first-order scattering process [5].

Temperature dependence of the Raman spectra of both stokes (s) and antistokes (as) process in a tetragonal $BaTiO_3$ single crystal have been studied [6]. According to the macroscopic polarization and fluctuation dissipation theory of Raman scattering cross-section, the (Is - Ias) of the first-order scattering does not change with temperature, while the (Is - Ias) of the second-order scattering do changes. The experimental results up to 403K show (Is - Ias) is temperature independent within experimental error.

The appearance of first-order Raman bands above Tc can be understood on the basis of the disorder structure model [7,8]. In the cubic phase of $BaTiO_3$, Ti ions is located at one of the eight equivalent positions shifted from the center of the along the body diagonals of the cell, Ba is also displaced along each of the 12 equivalent <110> directions and O ions are displace with the {001} planes on which they are sited [8]. These distortions result in a net displacement of the cations with respect to the oxygen octahedra and give rise to the observed spontaneous polarization. This model is also supported by the recent report of Laabidi and co-workers [9], where evidence for underdamped soft phonons and central peak relaxors related to the order-disorder has been found in $BaTiO_3$.

This research was supported by the National Natural Science Foundation of China and the Project of Science and Technology New Stars of Beijing Municipality.

1. M. E. Lines, A. M. Glass, Principles and Applications of Ferroelectrics and Related Materials. Clarendon Press, Oxford(1977).
2. M. Didomenico, S. H. Wemple, S. P. S. Porto and R. P. Bauman, Phys. Rev. 174, 522(1968).
3. A. Pinczuk, E. Burnstein and S. Ushioda, Solid State Commun.7, 139(1969).
4. Y. J. Jiang, L. Z. Zeng, R. P. Wang, Y. Zhu, Y. L. Liu, J. Raman Spec.27, 31(1996).
5. R. P. Wang, Y. J. Jiang, L. Z. Zeng, Y. L. Liu, Y. Zhu, J. Raman Spec.26, 295(1995).
6. L. M. Li, Y. J. Jiang, L. Z. Zeng, J. Raman Spec. 27, to be published.
7. R. Comes, M. Lambert, A. Guinier, Solid State Commun. 6, 715(1968).
8. K. Itoh, L. Z. Zeng, E. Nakamura, N. M. Shima, Ferroelectrics 63, 29(1985)
9. K. Laabidi, M. D. Fontana, M. Maglione, B. Jannot and K. A. Muller, Europhys. Lett. 26, 309(1994).

RAMAN, UV AND FLUORESCENCE STUDY OF THERMOCHROMIC PHASE TRANSITIONS IN POLY(DIALKYLSTANNANE)S

S.S.Bukalov[1], L.A.Leites[1], T.D.Tilley[2], V.Lu[2].
[1]Institute of Organoelement Compounds, Russian Academy of Sciences, 28 Vavilova str., Moscow 117813, Russia
[2]Department of Chemistry, University of California at Berkeley Berkeley, CA 94720-1460, USA

Raman spectra of recently synthesized polymers [1], poly(di-n-alkylstannanes) $[Alk_2Sn]_n$ Alk=n-Bu (I), n-Hex (II), n-Oct (III), are reported for the first time and compared to those of analogous polygermanes and polysilanes. (see Fig.1 where the Raman spectrum of $[n\text{-}Hex_2Sn]_n$ is given along with those for $[n\text{-}Hex_2Ge]_n$ [2] and $[n\text{-}Hex_2Si]_n$ [2,3]). Band assignment is proposed based on data for corresponding alkylstannane compounds.

The most intense lines in the spectra of the polystannanes are in the low-frequency range 100-150 cm^{-1} and correspond to νSn-Sn stretching vibrations. The lines in the region 550-650 cm^{-1} correspond to the νSn-C stretching vibrations, and are less intense than analogous νGe-C and νSi-C lines in the spectra of polydialkyl-germanes and -silanes.

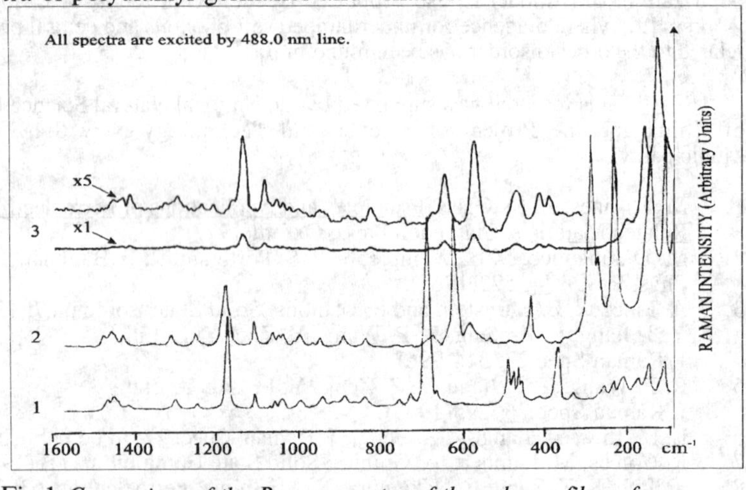

Fig.1 *Comparison of the Raman spectra of the polymer films of $[n\text{-}Hex_2M]_n$ M=Si (1), Ge (2), Sn (3).*

Poly(dialkylstannane)s absorb in the UV-visible region at 380-405 nm [1], the absorption band being red-shifted compared to those of corresponding polysilanes and polygermanes (310-370 nm) and also temperature-dependent. Thermochromic phase transitions have been observed for II and III by UV-vis and DSC techniques [1]. The manifestation of these transitions in the Raman spectra differs substantially from that in the spectra of analogous polysilanes [2,3] but is close to what is observed for polygermanes [2].

Fig.2 presents Raman spectra of III before and after the phase transition which takes place between 30 and 40°C, according to DSC data. The phase transition clearly manifests itself in the region of νSn-Sn, where a broad band at 104 cm^{-1} is replaced by a narrow line at 147 cm^{-1} on cooling from 50°C to room temperature, and also in the νSn-C region where a weak band at 657 cm^{-1} grows in intensity. The changes in the 900-1500 cm^{-1} region where the vibrations of the alkyl tails are situated are less prominent. Surprisingly, no band narrowing is observed in this region. Analogous results were obtained for the phase transition of II, which takes place at about the same temperature.

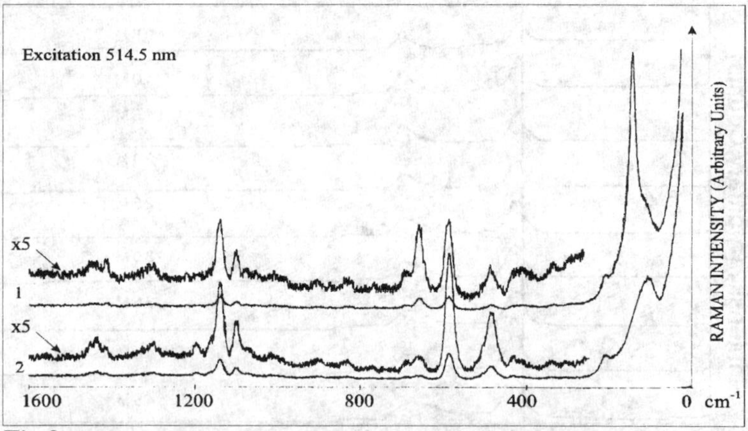

Fig.2 *Raman spectra of [n-Oct$_2$Sn]$_n$ before (1) and after (2) the thermochromic phase transition at about 40°C.*

The variable-temperature Raman studies of II and III allow us to conclude that the thermochromic phase transitions of polystannanes are of epy order-disorder type, and primarily include the conformational changes in the main polymer chain consisting of Sn atoms. Unlike their Si analogs, the polystannanes exhibit no distinct ordering of the side alkyl chains through the phase transition.

The phase transitions in II and III can be detected also by the change in fluorescence emission in the region 450-600 nm.

1. T.Imori, V.Lu, H.Cai, T.D.Tilley; *J.Am.Chem.Soc.* 1995, **117**, 9931
2. V.M.Hallmark, C.G.Zimba, R.Sooriyakumaran, R.D.Miller, J.F.Rabolt; *Macromolecules* 1990, **23**, 2346.
3. L.A.Leites, S.S.Bukalov, R.West, R.Menescal; to be published.

A RAMAN SPECTROSCOPIC AND IMAGING STUDY OF AQUEOUS KDP AND NaDP FREEZING PHENOMENA IN A BUFFER SYSTEM

Kenneth A. Christensen and Michael D. Morris*

Department of Chemistry, University of Michigan, Ann Arbor, MI 48109-1055, USA.

Raman spectroscopy has been used to study the dihydrogen phosphate structure in solutions ranging in concentration from dilute to supersaturated [1,2,3,4]. However, the Raman spectral bands broaden with increasing concentration as shown in Figure 1.

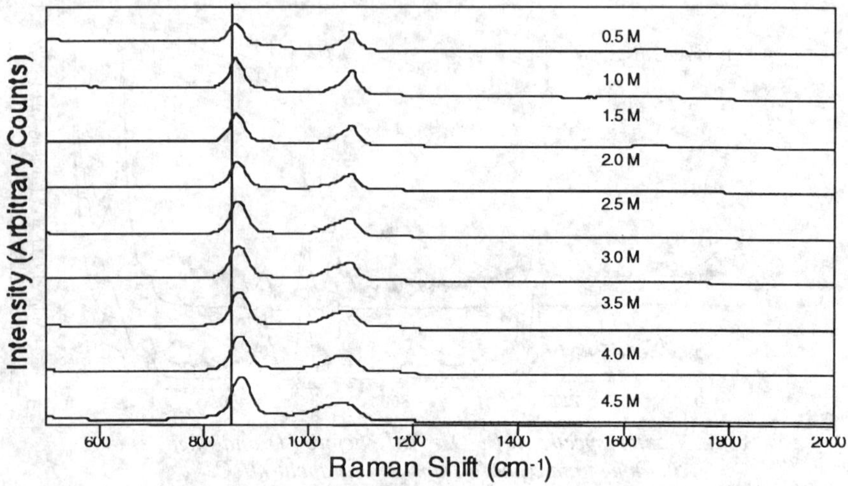

Figure 1. Raman spectra of NaH_2PO_4 solutions obtained at 20°C, 532.4 nm excitation, 5 second integration, 10 frames averaged.

The broadening of the $P{=}O_2$ and $P\text{-}(OH)_2$ symmetric stretches of the phosphate moiety have been attributed to the formation of several solution species whose Raman bands are unresolved even at high spectral resolution. Curve-fitting [2,4] and Fourier self-deconvolution [3] have been used to determine the number of species present in solution; however, these techniques require some knowledge of the system *a priori*. We will show the feasibility of using Principal Components Analysis (PCA) to deconvolve the overlapped bands from data acquired by simultaneously diluting a 6.0 M solution of NaH_2PO_4 and recording a Raman spectrum every 10 seconds.

These dihydrogen phosphate salts are also important as components of phosphate buffers, which are standard models to study freeze-drying of pharmaceuticals. The differing solubilities of the two phosphate salts, accompanied by freeze concentration, lead to large pH shifts during the initial freezing of buffer solutions. These concentration changes occur as pure water freezes and the remaining solution is freeze concentrated. The hydrogen phosphate precipitates while the more soluble dihydrogen phosphate is further freeze concentrated. We have used a Raman microprobe to monitor the concentrations of the two buffer components during freezing. Figure 2 shows how the measured peak center of mass for the $P{=}O_2$ symmetric stretch shifts during freeze concentration. The peak center of mass shift responds linearly with the concentration of the $H_2PO_4^-$. Correspondingly, the peak center of mass change implies an increase in $H_2PO_4^-$ concentration from 250 mM to about 6 M and a pH shift of nearly three pH units! Our completely non-invasive measurements are consistent with results obtained using conventional pH electrodes [5].

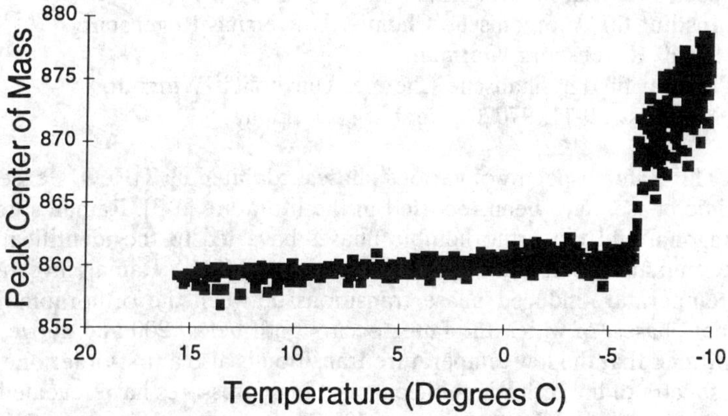

Figure 2. Plot of $P{=}O_2$ stretch of $H_2PO_4^-$ peak center of mass vs. temperature during freezing of 250 mM phosphate buffer solution.

Previous techniques have not been able to measure the spatial component associated with the freezing process. We will demonstrate that Raman microscopy can be used to simultaneously monitor the buffer component concentration and its spatial distribution in a multiphase/multicomponent sample. Raman images were acquired using either a 11-12 cm^{-1} dual-cell Fabry-Perot liquid-crystal filter (DFPF) [6] or a flexure-stage-based line-imaging apparatus which are both described previously [7]. Progress in using PCA of the image data sets to deconvolve the overlapped spectral bands will also be reported.

1. C.M. Preston and W.A. Adams, J. Phys. Chem. 83, 814 (1979).
2. M.K. Cerreta and K.A. Berglund, J. Cryst. Growth 84, 577 (1987).
3. F. Rull, A.D. Valle, F. Sobron, and S. Veintemillas, J. Raman Spectrosc. 20, 625 (1989).
4. G. Liu and G. Wu, Spectrochim. Acta 44A, 1007 (1988).
5. G. Gomez,Crystallization-related pH Changes During Freezing of Sodium Phosphate Buffer Solutions, Ph.D. Thesis, University of Michigan, Ann Arbor, MI (1995).
6. K.A. Christensen, N.L. Bradley, M.D. Morris, and R.V. Morrison, Appl. Spectrosc. 49, 1120 (1995).
7. N.L. Bradley, S.W. Dyer, J. Slater, and M.D. Morris, submitted to J. Raman Spectrosc. (1996)

RAMAN STUDY OF THE LOW TEMPERATURE PHASE TRANSITIONS IN MONOCLINIC $TlReO_4$

D. de Waal[1]*, K.- J. Range[2] and W. Kiefer[3]

[1]Department of Chemistry, University of Pretoria, 0002 Pretoria, South Africa
[2]Institut für Anorganische Chemie, Universität Regensburg, 93040 Regensburg, Germany
[3]Institut für Physikalische Chemie, Universität Würzburg, Marcusstr. 9-11, 97070 Würzburg, Germany

The Raman spectra of various phases exhibited by $TlReO_4$, except the monoclinic phase, have been reported in the literature [1-3]. Raman studies of the tetragonal and the orthorhombic phases have led to the identification of both external and internal modes of the compound [1]. A Raman investigation of the temperature-induced phase transitions between the orthorhombic and tetragonal phases, of which the latter occurs both below 200 and above 400K, has confirmed that the low temperature transition is also a first-order one [2, 5]. Raman spectra of the orthorhombic phase at high pressures have revealed three more pressure-related phase transitions [3]. The existence of a monoclinic form of $TlReO_4$ has been reported recently [4]. Pressure applied to the monoclinic phase in the form of the grinding of the sample with pestle and mortar leads to the formation of the orthorhombic phase [6]. The low temperature behaviour of the Raman active modes of the monoclinic modification of $TlReO_4$ have been followed in the present work between room temperature and 80K. Monoclinic $TlReO_4$ was prepared according to the method described by Rögner & Range [4], and it has been determined to crystallize in space group $P112_1/a$ (no.14) at room temperature [4]. A single crystal X-ray investigation has also shown that monoclinic $TlReO_4$ undergoes reversible phase transitions to result in the tetragonal structure at 170 and 419K [5]. The monoclinic sample has been cooled down to 80K and then heated back to room temperature. Spectra were recorded every 10K in the regions where phase changes could be observed. The spectra in Fig.1 represent the phase transitions as observed upon heating the sample back to room temperature. The spectrum at 210K could be identified to represent the orthorhombic phase by comparison with the literature [2]. For the orthorhombic phase five bands (339, 334, 329, 324 and 319 cm^{-1}) are observed in the tetrahedral $[ReO_4]^-$ deformational region instead of four (338, 330, 325 and 320 cm^{-1} in monoclinic) and a distinct third band at 895 cm^{-1} in the asymmetrical stretching region between 850 and 900 cm^{-1}. The spectrum at 180K corresponds to the one reported for the tetragonal phase in the literature [2]. Upon cooling monoclinic $TlReO_4$ has been observed to undergo a transition to the orthorhombic phase at 180K and to the tetragonal phase at 170K. It remains in the latter phase down to 80K. Upon heating back to room temperature the sample retained the tetragonal phase up to 200K. At 210K the orthorhombic phase could be identified again and above 220K the Raman spectrum characteristic for the monoclinic modification was observed.

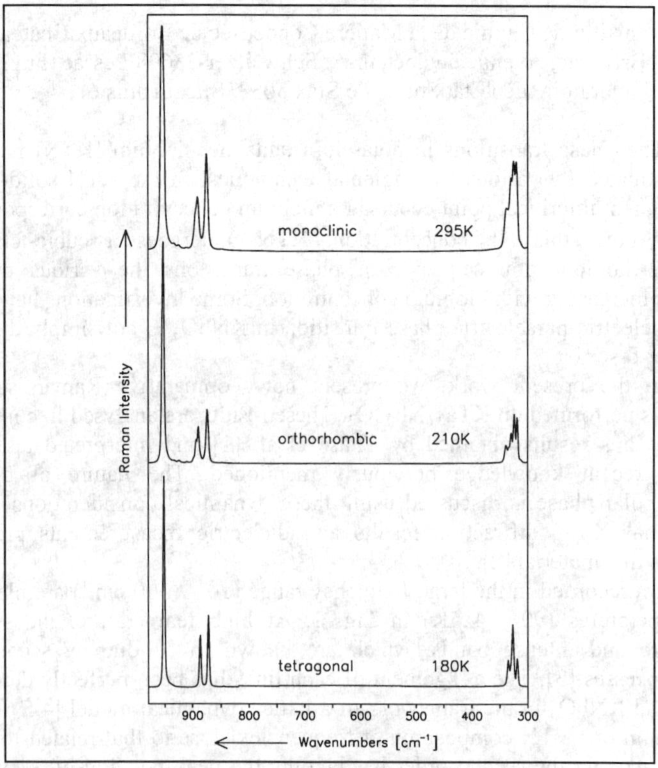

Fig. 1. Raman spectra of the phases observed after cooling monoclinic $TlReO_4$ down to 80K and heating back to the temperatures indicated

D. de Waal thanks the Alexander von Humboldt Foundation, Germany, for a research fellowship.

1. K. Ulbricht and H. Kriegsmann, Z. Anorg. Allg. Chem. 358, 193 (1968).
2. A. Jayaraman, G. A. Karoukis, R. M. Flemming and L. G. van Uitert, Phys. Rev. B37, 664 (1988).
3. A. Jayaraman, G. A. Karouklis and L. G. van Uitert, Phys. Rev. B36, 8547 (1987).
4. P. Rögner and K.-J. Range, Z. Anorg. Allg. Chem. 619, 1017 (1993).
5. P. Rögner, Dissertation, University of Regensburg, 1993.
6. M. Götz, Dissertation, University of Regensburg, 1996.

RAMAN SPECTROSCOPY MEASUREMENTS ON $KTa_{0.6}Nb_{0.4}O_3$ [KTN].

H. Khemakhem[1*,2], T. Mhiri[2], Z. Fakhfakh[2], J. Ravez[1] and A. Daoud[2]

[1 *]Institut de Chimie de la Matière Condensée de Bordeaux,Château Brivazac, Avenue du docteur A. Schweitzer-33608 Pessac (France).
[2] Laboratoire de l'Etat Solide de Sfax 3038, Sfax (Tunisie).

The phase transitions in potassium tantalate niobium [KTN] have been widely investigated by various experimental techniques. In the KTN solid-solution $KTa_{1-x}Nb_xO_3$, a multicritical point occurs at a niobium concentration corresponding to $x \approx 0.05$. Above this Nb concentration, KTN undergoes a cubic-tetragonal-orthorhombic-rhombohedral sequence of phase transitions the various transition temperatures increasing with niobium concentration. Some investigations have shown that the ferroelectric-paraelectric phase transition in $KNbO_3$ is not simply displacive but also order-disorder.

In the present work, we present new comparative Raman scattering measurements performed on $KTa_{0.6}Nb_{0.4}O_3$. These results are analysed in comparison with the previous results obtained by Chase et al. [1] and interpreted taking into account the recent knowledge previously mentioned. The nature of the low-temperature polar phase is discussed using these dynamical considerations together with additional X-ray diffraction results and dielectric measurements previously performed on this material [2].
Raman spectra recorded in the large frequency range (10 - 1300 cm^{-1}) are plotted for various temperatures (90 - 425K) in Fig. 1. At high temperature, the spectrum displays large and intense bands, which are known to be due by second-order scattering processes [3]. The assignment of each lines has been perfectly determined for $KTaO_3$ and $KNbO_3$ in the framework of a lattice dynamical model [4,5,6]. These bands appear in our KTN composition at frequencies close to that related to $KTaO_3$ and $KNbO_3$. Among the second-order bands, only the peak position of S_1, which is attributed to the two transverse acoustic phonon scattering, displays a weak temperature dependence. However all the other bands present a remarkable decrease in intensity when the temperature is lowered, especially below $T_c = 365K$, which corresponds to the ferroelectric-paraelectric phase transition according to our dielectric, pyroelectric and piezoelectric measurements on this composition [2,7,8]. The spectrum at high temperature shows the presence of intense second-order scattering making it difficult to observe first-order Raman lines which could be induced by any disorder or structural change. However, these second-order scattering bands were determined by Kugel et al. [9]. The Raman spectra reported in Fig. 1 as a function of temperature can be analysed as follows. At temperatures less than 360K, (90 to 345K), in addition to the remaining second-order bands S_2 and S_4, the spectrum exhibits first-order lines. However, these first-order lines can be seen more clearly when the temperature is lowered. According to the notation used for $KNbO_3$ [10], the lines denoted F_1, F_2, F_3 and F_4 correspond to phonons issued from the soft modes $F_{1u}(TO1)$, $F_{1u}(2)$, F_{2u} and $F_{1u}(TO3)$ respectively of the cubic phase.

At high temperatures (405 and 425 K), the spectra do not exhibit any first-order lines induced by disorder and show only well assigned second-order bands, which follow the normal light polarization effect. The peak (F_4) presents a nearly constant value in the low-temperature phase from 90 to 345K and then exhibits a steep fall at about T_c. Above T_c, it has a significant value which decreases with increasing temperature, so that it vanishes at around 425K. The peaks F_1, F_2 and F_3 decrease in intensity and become more largest when the temperature increases. At high temperatures (T > 400K) the structure is cubic and the niobium and/or tantalate atoms are on sites without any motion. The Raman spectrum exhibits only second-order phonon bands.

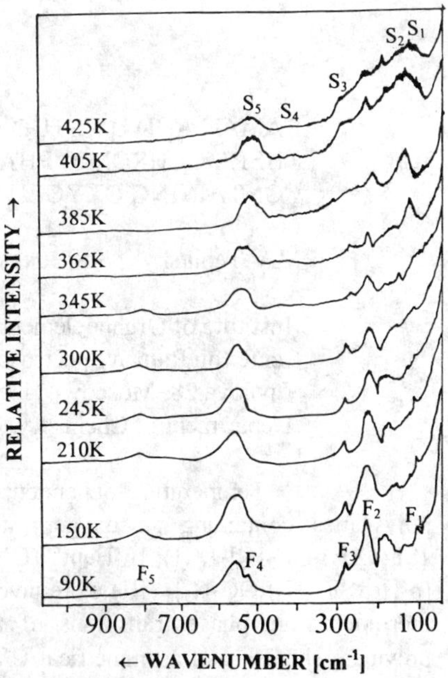

Fig.1. Temperature dependence of the Raman spectrum of $KTa_{0.6}Nb_{0.4}O_3$ ceramic. F denotes the first-order lines and S the second-order peaks.

When the temperature decreases and approaches the phase transition, either the niobium and/or tantalate ions hop between several off-center positions, as revealed by the appearance of quasi-elastic scattering. The dipolar clusters which form around relaxing niobium and/or tantalate locally break the selection rules and thus lead to activation above T_c of the first-order phonon lines which are normally forbidden in the cubic phase. These lines concern the soft-phonon mode and the three hard modes associated with the KTN lattice. These phenomena characterise an intermediate state which on average retains cubic symmetry and is locally a precursor of the rhombohedral structure of the low-temperature phase as announced by Fontana [6].

1. G.E. Kugel, H. Mesli, M. D. Fontana and D. Rytz, Phys. Rev. B37, 5619 (1988).
2. H. Khemakhem, J. Ravez and A. Daoud, Ferroelectrics in press.
3. R. Migoni, H. Bilz and D. Bäuerle, Phys. Rev. Lett., 37,115 (1979).
4. G. Nilsen and J. G. Skinner, J. Chem. Phys., 47, 1413 (1967).
5. D. Fontana, G. Métrat, J. L. Servoin and F. Gervais, J. Phys. C: Solid State Phys. 16,483 (1984).
6. M. D. Fontana, E. Bouziane and G. E. Kujel, J. Phys.: Condens. Matter, 2, 8681 (1990).
7. H. Khemakhem, R. von der Mühl, J. Ravez and A. Daoud, Ferroelectrics submitted for publication.
8. H. Khemakhem, J. Ravez, P. Dubernet and A. Daoud, Phase Trans.in press.
9. G.E. Kugel, H. Mesli, M. D. Fontana and D. Rytz, Phys. Rev. B37, 5616 (1988).
10. M. D. Fontana, A. Ridah, G. E. Kujel and C. Carabatos-Nedelec, J. Phys. C: Solid State Phys. 21 5853 (1988).

RAMAN AND UV STUDY OF THE NATURE OF COMPLEX THERMOCHROMIC PHASE TRANSITIONS IN POLYSILANES CONTAINING OXYGEN ATOM IN THE SIDE ALKYL CHAIN

M.V.Teplitsky[1*], S.S.Bukalov[1], L.A.Leites[1], C.-H.Yuan[2], R.West[2]

[1]Institute of Organoelement Compounds, Scientific and Technical Center on Raman Spectroscopy, Russian Academy of Sciences, ul. Vavilova 28, Moscow, 117813 Russia.
[2]Department of Chemistry, University of Wisconsin, Madison, USA.

Temperature dependence of the UV and Raman spectra of three polysilanes containing an oxygen atom in one or both alkyl side chains, $\{Si[(CH_2)_5OC_2H_5]_2\}_n$.(I); $[n\text{-}HeptSi(CH_2)_5OC_2H_5]_n$ (II) and $[n\text{-}HexSi(CH_2)_5OC_2H_5]_n$ (III) were investigated in order to elucidate the nature of thermochromic phase transitions of the polymers. At room temperature the polymers show broad asymmetrical UV bands with λ_{max} near 320 nm, typical of hexagonal columnar mesophases (*hcm*), and Raman spectra consisting of broad featureless bands pointing to a disordered structure of both main and side chains.

On lowering the temperature from 10° to -60°C, dramatic changes in the both types of spectra are observed. In the UV spectra the band at ca.320 nm is gradually replaced by a red shifted band of a more ordered phase. This new band has a complex profile with two or three peaks whose intensity ratio is not a single-valued function of temperature but depends on the sample prehistory and the rate of cooling (Fig. 1). These peaks lie in the region 350-375 nm which is known to be characteristic for the all-*trans* conformation of the silicon chain. Thus, it can be concluded that during the phase transitions studied the *hcm* gradually converts to a modification with the all-*trans* backbone conformation. However, the nature of several peaks observed is not understood yet. Phase transition for all polymers studied reaches its completion at about -60°C.

Results obtained in the Raman spectra agree well with those in the UV. The changes occurring on cooling are much the same as for $[n\text{-}Hex_2Si]_n$ [1]: a sharp resonantly-enhanced band of the νSi-C stretch at ca.690 cm^{-1}, known to be diagnostic of the all-*trans* backbone conformation, emerges and grows in intensity, and other narrow lines typical of the all-*trans* form appear (Fig.2).

To elucidate the changes in the side chain conformations, the temperature dependence of the Raman spectrum of a model monomeric compound $Cl_2Si(CH_2)_5OC_2H_5$ was investigated in detail.

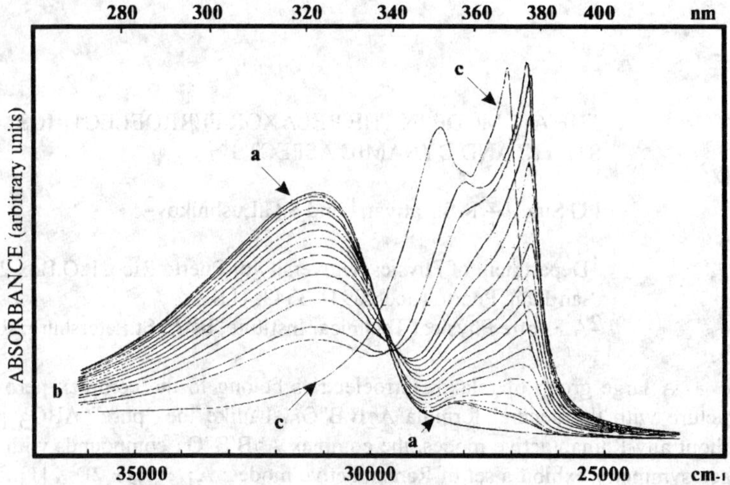

Fig. 1 Temperature evolution on cooling of the UV spectrum of the polymer II
a. room temperature just after the film preparation;
b. the interval of cooling from room temperature to -20°C
c. -248°C

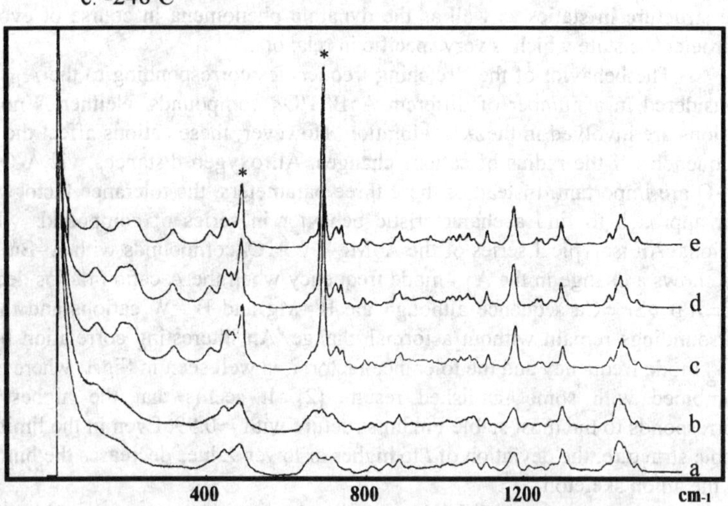

Fig. 2 Temperature evolution on cooling of the Raman spectrum of II.
a. room temperature;
b. -12°C; c. -50°C; d. -70°C; e. -160°C;
*. the lines of the all-*trans* form.

1. H.Kuzmany, J.F.Rabolt, B.L.Farmer, R.D.Miller, J. Chem. Phys., 85, 7413 (1986); V.M.Hallmark, R.Sooriyakumaran, R.D.Miller, J.F.Rabolt, J. Chem. Phys., 90, 2486 (1989).

THE A_{1g} MODE IN THE RELAXOR FERROELECTRICS: STATIC AND DYNAMIC ASPECTS

I.G.Siny,[1,2] R.S.Katiyar,[1] and S.G.Lushnikov[2]

[1]Department of Physics, University of Puerto Rico, P.O.Box 23343
San Juan, Puerto Rico 00931-3343, U.S.A.
[2]A.F.Ioffe Physical Technical Institute, RAS, St.Petersburg, Russia

A large group of relaxor ferroelectrics belong to the complex perovskite-type structure with the general formula $A_2B'B''O_6$. Unlike the "pure" ABO_3 perovskites without any Raman active modes, the complex $A_2B'B''O_3$ compounds with the Fm3m space symmetry exhibit a set of Raman active modes: A_{1g}+ E_g+ $2F_{2g}$ [1]. The group theoretical analysis show that the A_{1g} and E_g modes represent the stretching vibrations of an oxygen octahedron around the B cations. Both the oxygen atoms and the A cations are involved in vibrations determined by a pair of the coupled F_{2g} modes. The A_{1g} mode is a simple motion of the oxygen atoms like the breath-type mode of a free oxygen octahedron. However, this mode reflects clearly the effect of subtle changes in the structure in statics as well as the dynamic phenomena in course of evolution to a ferroelectric state which is very specific in relaxors.

The behavior of the stretching frequencies corresponding to the A_{1g} mode was considered in a number of different $A_2B'B''O_6$ compounds. Neither A nor B′ or B″ cations are involved in the A_{1g} vibration. However, these cations affect the stret-ching frequencies if the radius of cations changes. All oxygen distances, viz A-O, B′-O and B″-O are important. Instead of these three parameters, the tolerance factor was used in our approach to find a characteristic behavior in series of compounds with varying cations. An isotypical series of the $A_2Mg^{2+}W^{6+}O_6$compounds with A=Ba, Pb, Sr and Ca shows a change in the A_{1g} mode frequency when the A cation radius decreases in a Ba⇒Pb⇒Sr⇒Ca sequence although the B′=Mg and B″=W cations and their oxygen surroundings remain without a formal change. An interesting correlation between the A_{1g} mode frequency and the tolerance factor, t, is well seen in Fig.1, where our data are combined with some published results [2]. It seems that the highest frequency corresponds to the most stable Fm3m structure with $t \approx 0.99$. Even in the limit of a stable cubic structure, the deviation of t to higher or lower values decreases the binding energy of the anion skeleton.

The dynamic modification of the A_{1g} mode spectrum was studied on an example of Pb_2ScTaO_6 (PST) with ferroelectric state occurring slightly below room temperature [3]. At high temperatures far above the transition region the Raman spectrum of PST is well consistent with the Fm3m symmetry and the A_{1g} mode has the shape of a singlet line (Fig.2). An evident structure of the initially singlet line appears when the temperature is lowered down to the vicinity of the ferroelectric phase transition (Fig.2). No evidence for a change in the crystal structure has been known in this temperature region. The structure around the A_{1g} mode is more

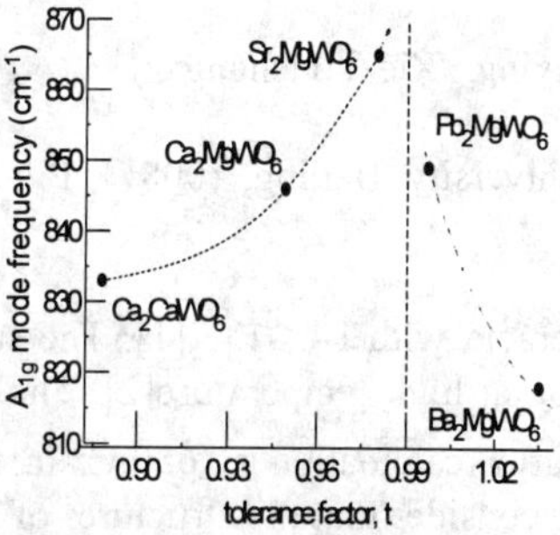

Fig.1. The A_{1g}-mode frequency vs. tolerance factor in a series of $A_2B'B''O_6$ compounds.

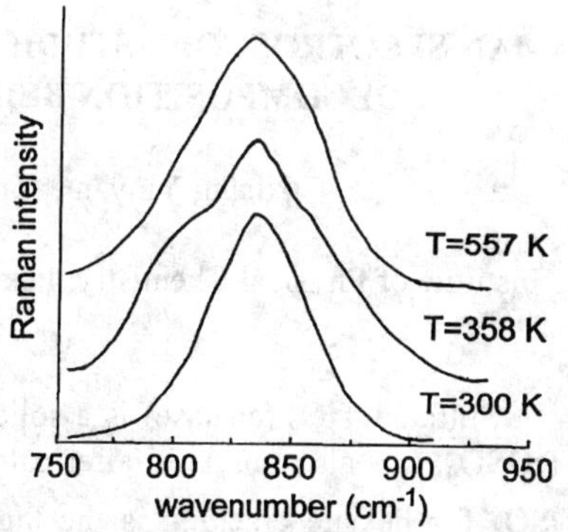

Fig.2. The complex structure of the A_{1g} line in PST above T_c (T=358 K) and two "singlet" spectra at lower and higher temperatures.

pronounced in the samples with higher disordering on the B sites. Fig.2 shows the behavior of the A_{1g} mode in a sample with the higher degree of ordering between all studied materials. Here the A_{1g} line takes a singlet shape again in the ferroelectric phase. One can suppose that the complex structure of the A_{1g} mode is connected with a break down in the wave-vector selection rules, so some symmetry points elsewhere in the Brillouin zone contribute to the Raman scattering in the vicinity of the initial singlet line.

The light scattering therefore gives evidence of a special state in PST which precedes the transition to a ferroelectric phase from above. This conclusion correlates with some other suggestions on PST like an incommensurate phase [4] or an additional nonpolar phase [3]. The latter model imply that in some temperature range there is a competition between two phases, namely between the ferroelectric phase and a postulated nonpolar phase. The authors assumed [3] that it was possible to have intensive heterophase fluctuations between these two states. The present work gives the experimental evidence in support of such a model. We suppose that the loss of translational symmetry in PST occurs in a dynamic process initiated by heterophase fluctuations.

Acknowledgments. This work was supported in parts by NSF Grant OSR-9452893, DOE Grant DE-FG02-94ER75764, ISF Grant #NP5F300 and RFBR Grant #93-02-14156.

1. G.A.Smolensky, I.G.Siny et al., Ferroelectrics **12**, 135
2. G.Blasse and A.F.Corsmit, J. Solid State Chem. **6**, 513 (1973).
3. F.Chu, N.Setter, and A.K.Tagantsev, J.Appl.Phys. **74**, 5129 (1993).
4. C.A.Randall, S.A.Markgraf, A.S.Bhalla, K.Baba-Kishi, Phys.Rev.B**40**, 413 (1989)

RAMAN SPECTROSCOPY STUDIES ON THE DISPERSION AND THERMO-DECOMPOSITION BEHAVIOUR OF $TiOSO_4$ ON SiO_2

Huang Yinyan*, Zhao Biying, Xie Youchang

Institute of Physical Chemistry, Peking University, Beijing, 100871, P.R.China

Sulfated TiO_2 (anatase) is a solid superacid with $H_o \leq -13.7$[1]. The superacidity of $TiOSO_4$ was also observed after calcination at high temperature[2]. The activity of $TiOSO_4$ for hexane cracking is the highest after calcination at 650°C; in this case a mixture of TiO_2(a) and $TiOSO_4$ is formed. We consider that the structures of the surface superacid sites on sulfated TiO_2 and calcined $Ti(SO_4)_2$ may be of the same in nature; they are depicted as SO_4^{2-} adsorbed on the surface of TiO_2 (a). When the calcination temperature is not high enough (650°C), transformation of $TiOSO_4$ into TiO_2(a) will be incomplete. This will limit the superacidity of $TiOSO_4$ prepared by thermo-decomposition. On the other hand, while $TiOSO_4$ decomposes completely at elevated temperature, SO_4^{2-} will be released from the surface of TiO_2(a). In this paper, it is shown by Raman spectroscopy that this limitation can be avoided by supporting $TiOSO_4$ onto SiO_2 of large surface area.

(1) $TiOSO_4$ can be dispersed on SiO_2 by impregnation and calcination at 500°C, as confirmed by the appearance of its characteristic Raman peaks and their line widths (see Fig. 1). The relative amounts of dispersive and crystalline $TiOSO_4$ can be obtained by deconvoluting the band at 784cm^{-1} and measure their respective areas, then calculating the area ratios to that of the internal standard KNO_3 at 1050cm^{-1}. From Fig. 2 the conclusion can be drawn that the utmost dispersion capacity of $TiOSO_4$ on SiO_2 with a specific surface area of 469 m^2/g is about 1.2 g/g SiO_2, i.e., 0.25 g/100 m^2 SiO_2.

(2) The decomposition temperature of $TiOSO_4$ is, obviously, lowered by dispersion on SiO_2. Fig. 3 shows that after calcination at 650°C there is no residual $TiOSO_4$ even in the sample of 1.5 g$TiOSO_4$/gSiO_2. It decomposes completely and is converted into TiO_2(a) with the characteristic Raman peaks at 144, 395, 518, 639 cm^{-1}. The apparent broadening of the XRD peaks and Raman peaks of TiO_2(a) demonstrates the small crystallite size and large accessible surface area. The decrease in decomposition temperature is advantageous for the retention of SO_4^{2-}; the large surface area of TiO_2(a) is favourable to increasing the SO_4^{2-} amount which can be adsorbed on it. As a result, the number of surface superacid sites will increase.

[1] K. Arata, *Adv. Catal.*, **37**, 165 (1991).
[2] K. Arata, M. Hino and N. Yamagata, *Bull. Chem. Soc. Japan*, **63**, 244 (1990).

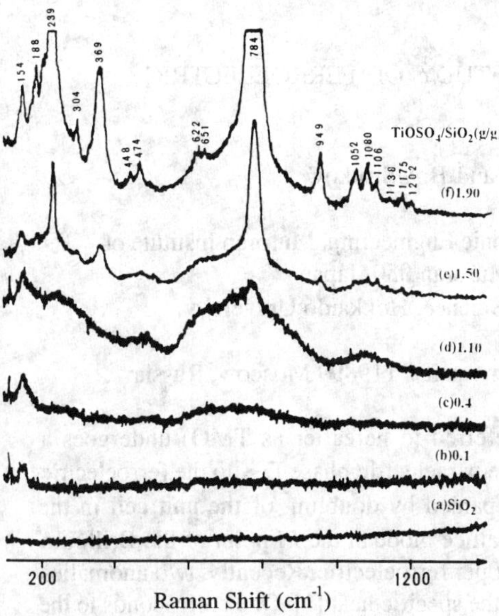

ig.1 Raman spectra of SiO_2 supported $TiOSO_4$ that was calcinated at 500°C

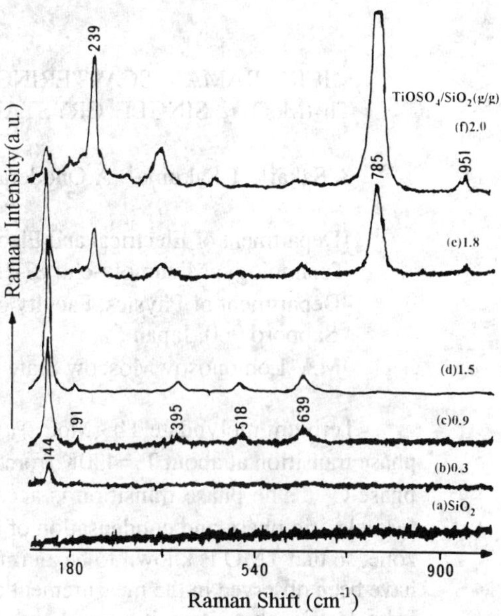

Fig.3 Raman spectra of SiO_2 supported $TiOSO_4$ that was calcinated at 650°C

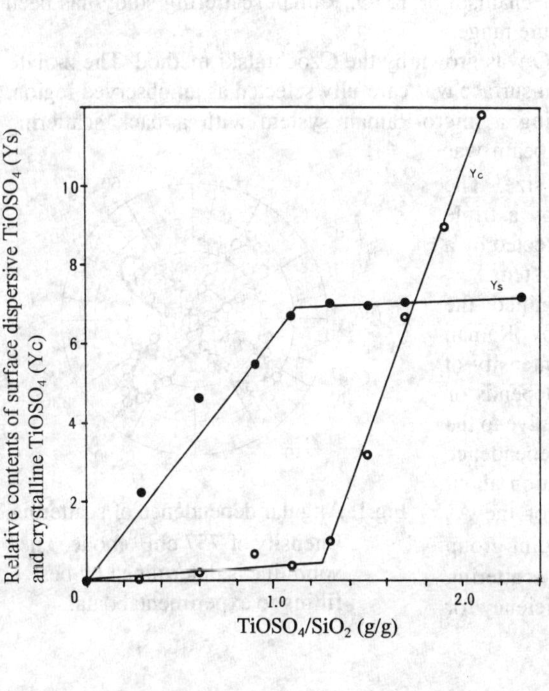

Fig 2 Quantitative Raman results

MICRO-RAMAN SCATTERING STUDY OF FERROELECTRIC $Tb_2(MoO_4)_3$ SINGLE CRYSTAL

A.Sakai[1], T.Sakuma[1], A.Onodera[2], and B.A.Strukov[3]

[1]Department of Electrical and Electronic Engineering, Muroran Institute of Technology, Mizumoto-cho 27-1, Muroran 050, Japan
[2]Department of Physics, Faculty of Science, Hokkaido University, Sapporo 060, Japan
[3]M.V.Lomonosov Moscow State University, 119899 Moscow, Russia

Terbium molybdate $Tb_2(MoO_4)_3$ (referred to hereafter as TMO) undergoes a phase transition at about T_c=430K from the paraelectric phase D_{2d} to the ferroelectric phase C_{2v}. The phase transition is accompanied by doubling of the unit cell in the ferroelectric phase and condensation of a lattice mode at the *M* point of the Brillouin zone, so that TMO is known to be an improper ferroelectric. Recently, two anomalies have been observed in the measurement of the specific heat [1]. One corresponds to the ferroelectric phase transition and the other appears near 270K. Because the later anomaly has not been reported by any other experiments, it is necessary to elucidate whether the anomaly exits or not. In order to confirm above point and to study the ferroelectric phase transition mechanism of TMO, Raman scattering study has been carried out in the wide temperature range.

The single crystal of TMO was grown by the Czochralski method. The z-plate samples were used and the clean surface was carefully selected as an observed region. Spectra were obtained by using a micro-Raman system with a back scattering configuration [2,3]. The laser beam was focused to 10 μm in spot size. The scattered light was analyzed by a triple grating monochromator and detected by a CCD (charge coupled device) system.

Firstly, we have determined the crystal axes of the sample by Raman scattering measurement. The intensity of the Raman scattering spectra depends on the directions of observation relative to the crystal axes [4]. The angular dependence of the scattering gives information about the symmetry of the lattice. For the A_2 phonon symmetry in the C_{2v} point group under the condition of the back scattering along the z-axis, the Raman efficiency for

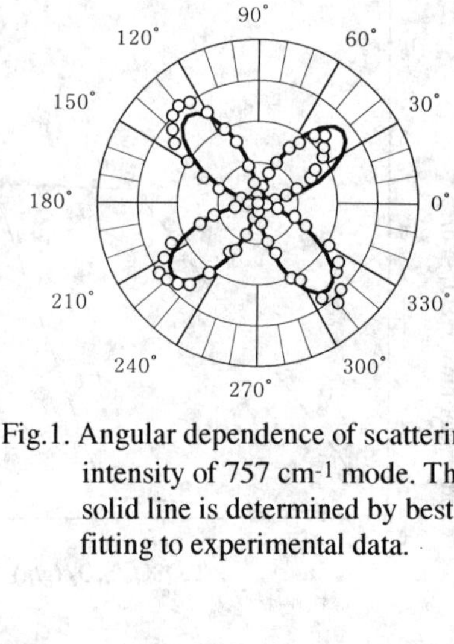

Fig.1. Angular dependence of scattering intensity of 757 cm^{-1} mode. The solid line is determined by best fitting to experimental data.

the polarized scattered-light is given as $S=A\sin^2 2\theta$. Here, A is an arbitrary factor and θ is defined as an angle between the x-axis and the direction of the polarization of the incident light. Fig.1 shows the angular dependence of the scattering intensity of 757 cm^{-1} mode. The solid line in Fig.1 is fitted to observed data by a least-squares method.

Fig.2 shows the temperature dependence of the peak position of Raman spectra in the $z(xx)\bar{z}$ geometry. As the temperature decreases, some of new lines are appeared at about T_c because of the symmetry change from paraelectric to ferroelectric. Though the peak positions of the appeared modes moves slightly, the peak intensities show remarkable temperature dependence. Thus, these modes are considered to become Raman active owing to the doubling of the unit cell.

In the case of the high frequency region, new broad modes have been appeared as shown in Fig.3 in the $z(xx)\bar{z}$ geometry from 1000 to 1300 cm^{-1}. The intensity of the mode has a maximum at about 460K and disappears at about 230K. The anomalous temperature region is agreement with that of the specific heat measurement. Therefore, it is concluded that these modes are strongly related to the specific heat anomaly. For the next step, the knowledge of the micro origin of the modes is necessary to understand the anomalous behavior.

This work is partially supported by the New Energy and Industrial Technology Development Organization in japan.

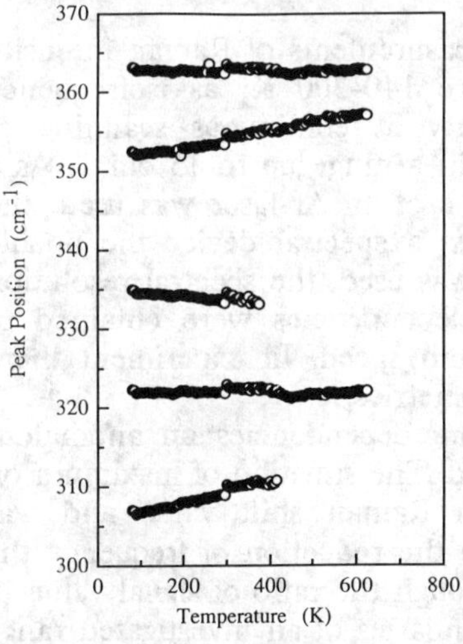

Fig.2. Temperature dependence of peak frequency of Raman spectra of TMO in the $z(xx)\bar{z}$ geometry.

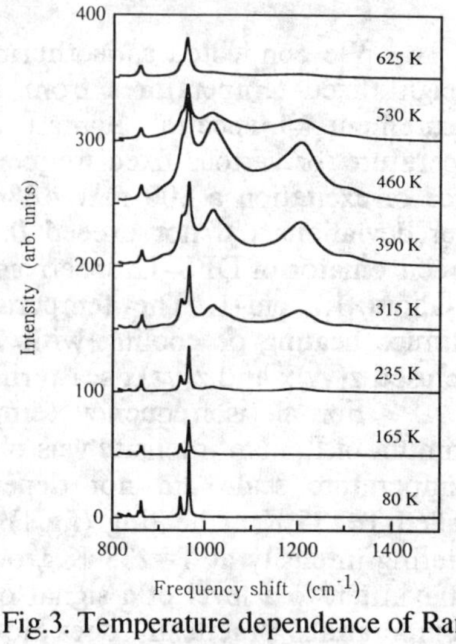

Fig.3. Temperature dependence of Raman scattering spectra of TMO in the $z(xx)\bar{z}$ geometry.

1. B.A. Strukov, S.A.Taraskin, I.V.Shnaidshtein, A.Onodera, H.Haga, and B.S.Red'kin, Sov. Phys. JETP 81, 202 (1995).
2. A.Sakai, H.Kawamura, and Y.Hidaka, Ferroelectrics 170, 87 (1995).
3. A.Sakai and A.Araya, Physica B (1996), in press.
4. R.Loudon, Adv.Phys. 13, 423 (1964).

ABOUT AN OPPORTUNITY OF PHASE TRANSITION IN $Li_2B_4O_7$ PYROELECTRIC CRYSTALS

V.N.Moiseenko[1*], A.V.Vdovin[1], Ya.V.Burak[2]

[1]Dniepropetrovsk State University, Department of Physics
13 Nauchny by Street, Dniepropetrovsk, 320625 Ukraine
[2]Institute of Physical Optics,
23 Dragomanov Street, Lvov, 290005 Ukraine

This work is devoted to studying the temperature dependencies of low-frequency Raman spectral intensity in $Li_2B_4O_7$ crystals near the point T=235 K. In a vicinity of this point a number of anomalies of macroscopic physical properties were observed [1-3]. However any anomaly in Raman spectra was not earlier found out, though an opportunity of phase transition of order - disorder type was not excluded [4].

We conducted as isothermal measurements of Raman intensity for eight fixed temperatures from a range 140-300 K, as isofrequency measurement of spectral Raman intensity at continuous scanning of temperature for various fixed frequencies in a range up to 15 cm-1. As a source of excitation a 300 mW 488nm line of an Ar-laser was used, the power deviations did not exceed 0.1%. As a spectral device the double monochromator of DFS-12 spectrometer was used, the spectral resolution was about 1.2 cm-1. The temperature dependencies were obtained in continuos heating or cooling with 2 K/min speed. In experiment there were used z(yy)x and z(yz)x scattering geometries.

For all isofrequency temperature dependencies an anomalous maximum of Raman intensity was observed. The situation of maximum on a temperature scale did not depend on Raman shift value and was equalled T≈235 K at heating (fig.1). With the reduction of frequency the scattering intensity at T=235 K grows, though the ratio of signal value in its maximum to a level of a signal on boundaries of an investigated range of temperatures practically does not vary at change of shifted frequency and scattering geometry. Width of a contour of quasi-elastic scattering at various temperatures remains practically constant, and its height displays an anomaly with relative size of increase of intensity within the limits of an error of measurements not distinguished from the value of an anomaly on shifted frequencies. Therefore the behaviour of Raman intensity on shifted frequencies is probably called by temperature change of a wing of an elastic scattering.

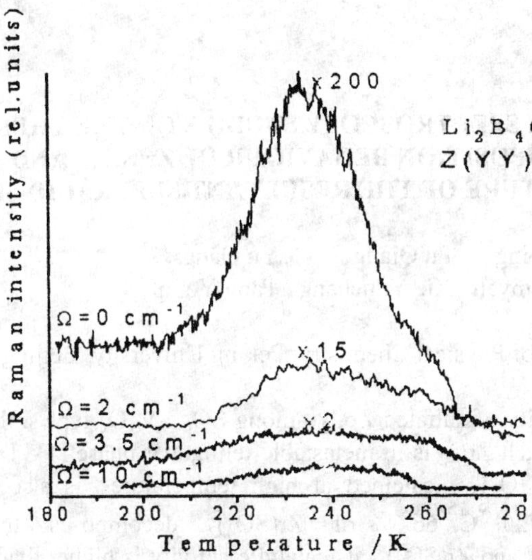

Fig.1. The temperature dependencies of Raman spectral intensity at different fixed shifts in $Li_2B_4O_7$ crystals

In order to explain this anomaly the model of phase transition with conservation of crystal class was used. The analysis of crystal structure of $Li_2B_4O_7$ gives the possibility to connect such phase transition with displacement of Li+ ions along [001] direction. It can lead to diffuse scattering of light at T=235K.

The intensity of a peak grows with lowering of initial temperature of heating. Besides at cooling the form of isofrequency temperature dependence was not repeated and the maximum in an interval 300-150 K was not observed. Such behaviour can testify to availability of metastable conditions in lithium sublattice.

1. A.E.Aliev, Ya.V.Burak, I.T.Lyseyko, Izvestiya AN SSSR, ser.fizich., (1990), p.1991
2. N.P.Tehanovich, A.U.Sheleg, Ya.V.Burak, Fizika tverdogo tela 32, 2513 (1990)
3. Ya.V.Burak, V.M Gaba, I.T.Lyseyko et.al., Ukrainsky fizicheskiy zhurnal, 36, 1638 (1991)
4. S.Furusawa, S.Tange, Y.Ishibashi, K.Miwa, J.of Phys. Society of Japn. 59, 1825 (1990).

RAMAN SPECTROSCOPY STUDIES ON THE THERMO-DECOMPOSITION BEHAVIOUR OF $Zr(SO_4)_2$ AND THE STRUCTURE OF THE RESULTANT SUPERACID SITES

Zhao Biying* Fu Qiang Yu Gang
Huang Yinyan Xie Youchang Tang Youqi

Institute of Physical Chemistry, Peking University, Beijing, P.R.China

The acidity of sulfated ZrO_2 among SO_4^{2-}-M_xO_y type solid superacids is the strongest, in which ZrO_2 is in metastable tetragonal phase [1]. The followings were reported[2]: $Zr(SO_4)_2$ calcined at high temperature is also a superacid; after calcination at 725 ^{o}C, 66.6% of $Zr(SO_4)_2$ decomposes to be a mixture of monoclinic ZrO_2 and $Zr(SO_4)_2$; its catalytic activity is higher than those of $Zr(SO_4)_2$ calcined at lower or higher temperatures. However, the structure of its superacid sites and thermo-decomposition behaviour have not been studied in detail so far.

The structure of $Zr(SO_4)_2$ is very complex, in which SO_4^{2-} combines with Zr^{4+} in various ways and their structure environments and symmetries are very different. Fig.1 (a) is the Raman spectrum of the $Zr(SO_4)_2$ calcined at 600 ^{o}C. Obviously, it has not decomposed and the 17 Raman peaks of SO_4^{2-} can be observed between 900 and 1400 cm^{-1}. After calcination at elevated temperature, the number of peaks due to SO_4^{2-} decreases considerably while $Zr(SO_4)_2$ decomposes to be ZrO_2. The peaks at 998, 1092, 1148, 1202 and 1222 cm^{-1} are observed (see the Fig.1(c)) and their wavenumbers are corresponding to 980-990, 1040, 1130-1150 and 1210-1230 cm^{-1} of SO_4^{2-} in IR spectra of sulfated ZrO2,TiO_2, and SnO_2. Those peaks of IR spectra were assigned to the bridging bidentate sulfate coordinated to metal elements[3].

ZrO_2 obtained by calcinating $Zr(SO_4)_2$ at 750 ^{o}C is ZrO_2(m) mainly. However, the characteristic bands of ZrO_2 (t) at 268, 314, 472 and 642 cm^{-1} can also be detected in the Raman spectrum of $Zr(SO_4)_2$ calcined at 750 ^{o}C (see Fig 1(b)).The existence of ZrO_2 (t) is also proved by its XRD pattern in Fig. 2.

In addition, we have prepared and studied the $Zr(SO_4)_2$ /SiO_2 samples. The preliminary results show that the $Zr(SO_4)_2$/SiO_2 has the following characteristics as compared with calcined $Zr(SO_4)_2$: 1) the decomposition temperature decreases obviously; 2) all of ZrO_2 obtained by decomposing $Zr(SO_4)_2$ /SiO_2 is ZrO_2(t) only if the $Zr(SO_4)_2$ content is equal to or lower than 1.2 g /g SiO_2 (see Fig. 3); 3) its catalytic activity per gram of $Zr(SO_4)_2$ for the conversion of n-C_6H_{14} is doubled.

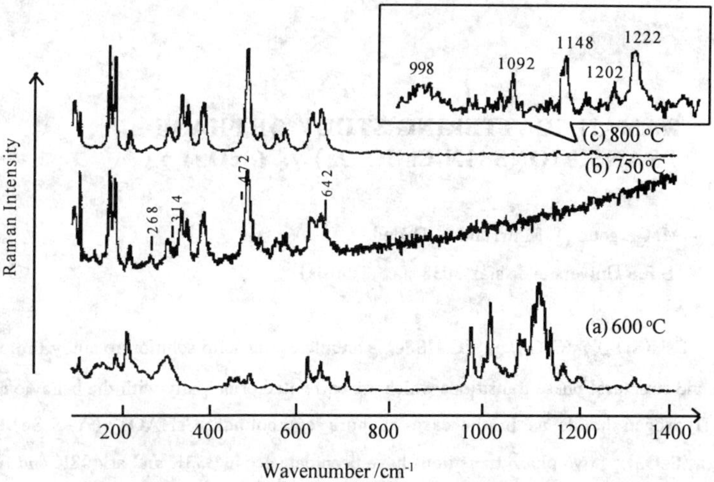

Fig. 1 Raman spectra of $Zr(SO_4)_2$ calcined at different temperatures.

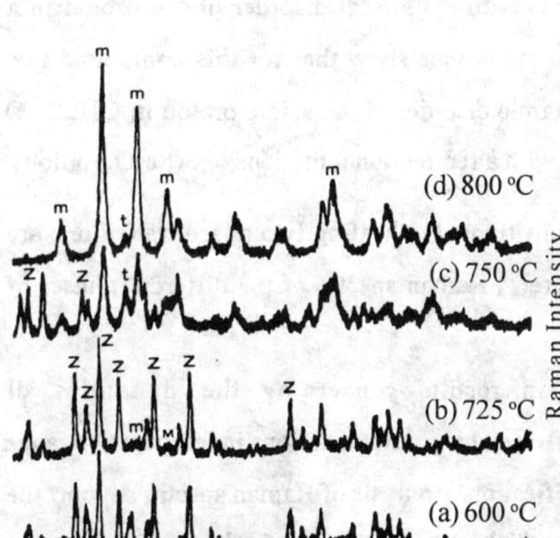

Fig. 2 XRD patterns of $Zr(SO_4)_2$ calcined at different temperatures.
Z: $Zr(SO_4)_2$, m: ZrO_2(m); t: ZrO_2(t)

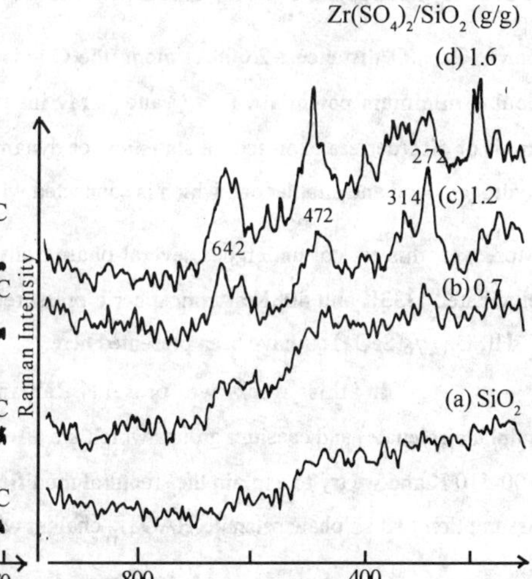

Fig. 3 FT-Raman spectra of $Zr(SO_4)_2/SiO_2$ calcined at 750 °C.

The conclusions are the followings: 1) ZrO_2(t) is necessary for forming the superacid sites; 2) the structures of their superacid sites are the same, i.e., SO_4^{2-} adsorbed on ZrO_2(t) in bridging bidentate state; 3) The acidity of calcined $Zr(SO_4)_2$ can be improved by supporting it on SiO_2.

1 K.Arata, *Adv. Catal.*, **37**,165(1991).
2 K. Arata, M. Hino and N. Yamagata, *Bull. Chem. Soc. Japan*, **63**, 244 (1990).
3 T.Yamaguchi, K.Tanabe, *J. Phys. Chem.*, **90**, 3148(1986).

RAMAN SCATTERING STUDY OF PHASE TRANSITIONS IN $CsH(SO_4)_{.76}(SeO_4)_{.24}$

M. Gargouri ,T. Mhiri and A. Daoud

L.E.S Universite de sfax 3038 sfax (Tunisia).

$CsH(SO_4)_{.76}(SeO_4)_{.24}$, (CsHSSe) is an interesting solid solution to study Protonic conduction and structural phase transitions which are correlated principally with the behaviour of OH......O Hydrogen bonds as in the case of pure compounds $CsHAO_4$ (A=S,Se).For $CsH(SO_4)_{.76}(SeO_4)_{.24}$ two phase transitions have been detected at 333K and at 408K and two phases have been described. At room temperature the monoclinic structure , space group $P2_1/a$ [1], is built of mixed layers $S(Se)O_4^{2-}$ tetraheders and Cs^+ ions which are joined by a strong hydrogen bond (O......O distance =2.608Å) along the C axis with a statiscal disorder of the proton in a double minimum potential . D S C and X-ray investigations show that for this compound two kinds of disorder are expected : a statistical or dynamic disorder of the acidic proton in OH.......O hydrogen bond and another one which is connected with a reorientional motions of $S(Se)O_4^{2-}$ anions. Moreover, this crystal undergoes several phase transitions. On heating two phase transitions are elucidated :333K and 408K at atmospheric pressure[1].Raman spectra of the different phases of $CsH(SO_4)_{.76}(SeO_4)_{.24}$ have been presented here.

In this work we present Raman results concerning the dynamics of sulphate(selenate) and caesium groups which are affected by phase transitions in temperature range 100-510 K and we try to explain the stuctural modification. Analysis of Raman spectra support the assumption that sulphate selenate$(HAO_4)_n^-$ chains, which are connected by hydrogen bonds ,were partially brooken during the I------> II transition. This transition shows some structural modifications. Figs.1,2 Show some modifications of sulfate selenate and caesium bands. Plots of the ν (A-O) and ν(A-OH) frequencies versus temperature, Fig.2, evidence the pretransitionnal effects. At the 408K phase transition we observe a continious decrease of ν(A-OH) frequency which traduce a weakening of the hydrogen bonding and thus superprotonic phase is occur. Intense Rayleigh wings is observed A broadening of S-OH and Se-OH stretching band is occur above 875 and 715 Cm^{-1} respectively [3].The superionic phase transition at 408K is indicated, at the plot of the full width at half height Fig.1b, by the increase of broadening of S-OH and Se-OH stretching toward the melt. The behaviour of intensity, FWHH and Frequency of the thermosensitive bands

Figs.1,2 indicates that the degree of disordering increases as the crystal is heating. It is found to be the highest in the superionic state . This result is consistent with earlier work[2,4].

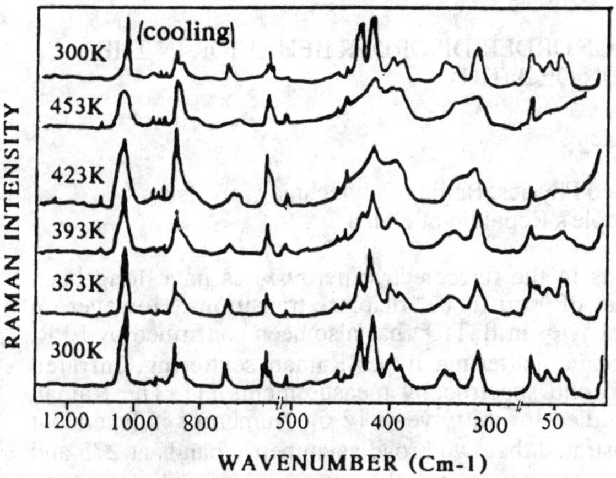

Fig.1a : Raman spectra in the range 10-1300 Cm-1 of CsHSSe at various temperatures.

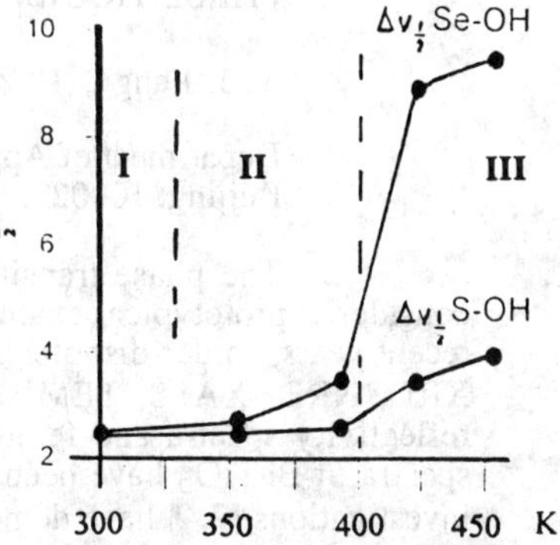

Fig.1b : Temperature dependences of the FWHH of νS-OH and νSe-OH for CsHSSe.

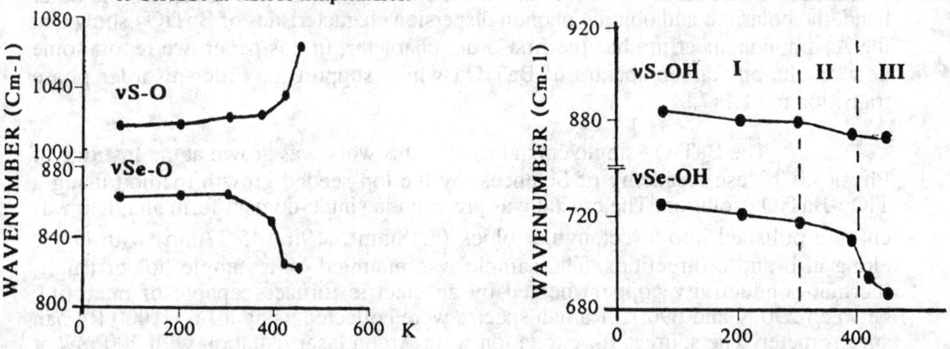

Fig.2 :Plots of the νS-O , νSe-O , νS-OH and νSe-OH Raman frequencies as a fuction of temperature for CsHSSe.

Acknowledgements: The authors would like to thank Drs Ph. Colomban and F. Romain from L.A.S.I.R CNRS 2 rue Henri Dunant 94320 Thiais-France

REFERENCES

1. M. Gargouri, T. Mhiri, A. Daoud and T. Jouini; to be published.
2. T. Mhiri and Ph. Colomban in Proc. XII th ICORS, J.r. Durig and J.f. Sullivan Eds.Wiley and Sons (1990)484.
3. M. Gargouri, T. Mhiri, A. Daoud and Ph. Colomban, to be published.
4. T. Mhiri, A. Daoud, M. Bouachir and F. Romain , J.Alloys Comp.188 (1992) 215-220.

NEW EVIDENCES OF ORDER-DISORDER BEHAVIOR IN THE PHASE TRANSITION OF $BaTiO_3$

Y. J. Jiang*, L. Z. Zeng

Department of Applied Physics, Beijing Polytechnic University, Beijing, 100022, People's Republic of China

The phase transitions in the ferroelectric perovskites have long been considered prototypical examples of "soft-mode" displace transitions. However, in recent years, "order-disorder" behavior in $BaTiO_3$ has also been confirmed by EPR, XRD, NRD, XAFS, TEM, Raman scattering, hyper-Raman scattering, Infrared reflectivity spectra and femtosecond spectrocopy measurements [1]. The Raman spectra of $BaTiO_3$ have been studied for thirty years [2-6]. Numerous independent investigations [2,4] have demonstrated that two broad asymmetric bands at 275 and 514 cm^{-1} in the A_1(TO) Raman spectrum persist well into the cubic paraelectric phase where first-order Raman scattering is forbidden by symmetry. One of the explanations of this fact is that these two bands are second-order Raman spectra [2]. On the other hand, the polariton and oblique phonon dispersion characteristics of $BaTiO_3$ show that the A_1 phonon spectrum has the first-order character. In this paper, we report some new results on Raman spectra of $BaTiO_3$ which support the order-disorder phase transition model [7,8].

The $BaTiO_3$ single crystal used in this work was grown at the Institute of Physics, Chinese Academy of Sciences, by the top-seeded growth method using a TiO_2-$BaTiO_3$ solution. The crystal was prepared in single-domain form and then was cut and polished into a rectangular block (5.18mm*5.19mm*5.27mm) with edges along a, b and c directions. The sample was mounted on a sample holder (high-thermal-conductivity copper) heated by an electric furnace capable of operating between 290 K and 690 K, Raman spectra were collected using a JY U1000 Raman spectrometer. The sources for excitation were Argon laser radiation with 200 mw at 514.5 and 488.0 nm. The effective spectral slit width is about 2 cm^{-1}.

Polarized Raman spectra of a $BaTiO_3$ single crystal have been taken over the frequency range of 100 — 1600 cm^{-1}. Besides all the A_1 and E fundamental vibration modes reported previously, three new week bands at 986, 1204 and 1480 cm^{-1} in the X(ZZ)Y, Z(YY)X, Z(XY)Z and X(ZY)X geometries are observed. According to the temperature dependence of Raman scattering intensity and band-width of these bands, the weak bands at 986, 1204 and 1480 cm^{-1} are attributed to the second-order Raman spectra and the strong bands at 275 and 514 cm^{-1} in the X(ZZ)Y spectrum are confirmed to be of first-order. It is not easy to measure accurately the temperature dependence of the Raman scattering intensity in $BaTiO_3$ because the absorption of light in the sample varies with temperature. However, one can obtain the temperature dependence of the relative intensity ratio of two bands in one spectrum without any difficulty. Here, we chose the strong band at 514 cm^{-1} and the weak band at 1480 cm^{-1} to compare [4].

With the configurations of X(ZZ)Y, X(ZZ)X+ΔZ, Y(ZX)Y+ΔX and Y(XZ)Y+ΔZ, the frequencies of all the A_1(TO), A_1(LO), E(TO) and E(LO) modes in $BaTiO_3$ are determined accurately. Using four types of optical geometry, Raman spectra of oblique phonons in tetragonal $BaTiO_3$ single crystal have been obtained for different values of the angle which the phonon wavevector makes with the c-axis. On the basis of Merten's equation, the directional dispersion curve were fitted and plotted. The experimental points agree with calculated curves perfectly, which means it is correct to consider the bands at 275 and 514 cm^{-1} coming from the first-order scattering process [5].

Temperature dependence of the Raman spectra of both stokes (s) and antistokes (as) process in a tetragonal $BaTiO_3$ single crystal have been studied [6]. According to the macroscopic polarization and fluctuation dissipation theory of Raman scattering cross-section, the (Is - Ias) of the first-order scattering does not change with temperature, while the (Is - Ias) of the second-order scattering do changes. The experimental results up to 403K show (Is - Ias) is temperature independent within experimental error.

The appearance of first-order Raman bands above Tc can be understood on the basis of the disorder structure model [7,8]. In the cubic phase of $BaTiO_3$, Ti ions is located at one of the eight equivalent positions shifted from the center of the along the body diagonals of the cell, Ba is also displaced along each of the 12 equivalent <110> directions and O ions are displace with the {001} planes on which they are sited [8]. These distortions result in a net displacement of the cations with respect to the oxygen octahedra and give rise to the observed spontaneous polarization. This model is also supported by the recent report of Laabidi and co-workers [9], where evidence for underdamped soft phonons and central peak relaxors related to the order-disorder has been found in $BaTiO_3$.

This research was supported by the National Natural Science Foundation of China and the Project of Science and Technology New Stars of Beijing Municipality.

1. M. E. Lines, A. M. Glass, Principles and Applications of Ferroelectrics and Related Materials. Clarendon Press, Oxford(1977).
2. M. Didomenico, S. H. Wemple, S. P. S. Porto and R. P. Bauman, Phys. Rev. 174, 522(1968).
3. A. Pinczuk, E. Burnstein and S. Ushioda, Solid State Commun.7, 139(1969).
4. Y. J. Jiang, L. Z. Zeng, R. P. Wang, Y. Zhu, Y. L. Liu, J. Raman Spec.27, 31(1996).
5. R. P. Wang, Y. J. Jiang, L. Z. Zeng, Y. L. Liu, Y. Zhu, J. Raman Spec.26, 295(1995).
6. L. M. Li, Y. J. Jiang, L. Z. Zeng, J. Raman Spec. 27, to be published.
7. R. Comes, M. Lambert, A. Guinier, Solid State Commun. 6, 715(1968).
8. K. Itoh, L. Z. Zeng, E. Nakamura, N. M. Shima, Ferroelectrics 63, 29(1985)
9. K. Laabidi, M. D. Fontana, M. Maglione, B. Jannot and K. A. Muller, Europhys. Lett. 26, 309(1994).

RAMAN SPECTRA OF $K_{1-x}(NH_4)_xH_2PO_4$ PELLETS

Jae-Hyeon Ko, and Jong-Jean Kim*

Physics Department, KAIST

373-1 Kusung-dong, Yusung-ku, Taejon 305-701, Korea

$K_{1-x}(NH_4)_xH_2PO_4$(KADP-x) is one of the dipole glass systems where a spin glass analogy is well studied in more varieties of experiments. We have studied the Raman spectra of KADP-x pellets of various mixing concentrations from x=0(pure KDP) to x=1(pure ADP). With respect to the reference Raman spectra of KDP(x=0) and ADP(x=1) pellets observed at T=290 K(tetragonal) and T=19 K(orthorhombic), we have examined the Raman spectral changes observed in KADP-x(x=0.52) during the glass freezing. We have observed a global symmetry conservation of one to one correspondence between the Raman bands during the temperature change from 290 K to 19 K. However, a close examination of the temperature dependence of the Raman bands can give significant informations on the freezing dynamics.

We have analyzed the Raman spectra of KADP-x(x=0.52) in the region of 250 cm^{-1} to 450 cm^{-1} including NH_4-librational band and two internal modes of PO_4 as depicted in Fig.1(a,b).

From the curve fitting analysis of the observed spectra in terms of three Lorentzian bands we could obtain the details of temperature dependence for peak frequencies and bandwidths of $\nu_2(A_1)$ and $\nu_2(B_2)$ bands of PO_4. The temperature dependence for the three Raman bands can be correlated with the dielectric anomalies associated with the phase transitions and glass freezing of the respective systems.

* This work was supported in part by Korea Science and Engineering Foundation (RCDAMP-1995).

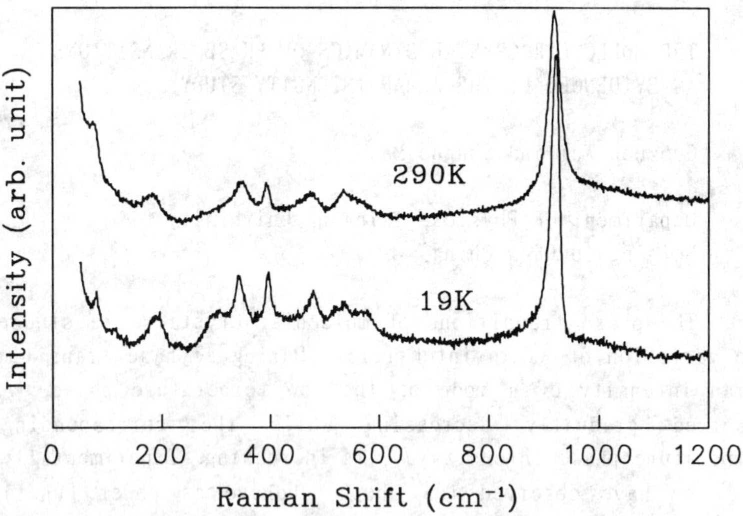

Fig.1(a) Raman Spectra of KADP-x(x=0.52) pellet at T=290K and 19K.

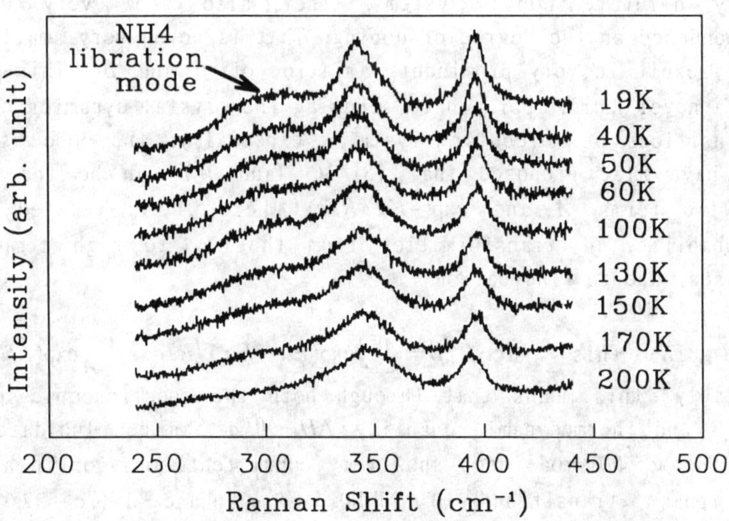

Fig.1(b) Temperature Dependence of NH_4-Librational and PO_4-Vibrational Raman bands in KADP-x(x=0.52).

THE MOLECULAR CRYSTAL DYNAMICS OF PHASE TRANSITIONS AS EVIDENCED BY THE RAMAN INTENSITY STUDY

Guozhen Wu* and Shuguo Ma

Department of Physics, Tsinghua University
Beijing 100084, China

The phase transitions of molecular crystals are studied from the variation of Raman intensities. During a phase transition, the Raman intensity of a mode of the low temperature phase, I_R , is observed gradually decreasing while the corresponding high temperature phase intensity I_R' is increasing. Experimentally [1, 2, 3] , we have observed that both behave as power functions of $|T-Tc|$, i.e. we have $I_R \sim (Tc-T)^{\beta}$, $I_R' \sim (T-Tc')^{\beta'}$ where Tc and Tc' are the temperatures at which I_R and I_R' disappear, respectively and β, β' are the critical exponents. Notably, we observed that β and β' are mode-dependent. In particular, for very dilutely doped systems, they also show very delicate dependence on the degree of doping. That is, even very small doping may result in very prominent variation of β and β'. This ensures us a novel Raman approach to studying the crystal dynamics of phase transitions of molecular crystals, especially, the doped systems. We have also proposed that [3] β (and β') can be the product of two terms A and $\exp(-Ea/RTc)$ where A is related to the probability of transformation from the low to high temperature phases, due to fluctuation.

In this way, $\beta'/\beta = A'/A \exp(Ea-Ea')/RTc \approx A'/A \exp(\Delta H)/RTc$. Notably, this means that through both the spectroscopic quantity β'/β and thermodynamic quantity ΔH, A'/A can be elucidated. A'/A of the ν_1 NO_3 mode does show very consistent behavior in a series of phase transitions of NH_4NO_3 from phase IV($Tc \sim 32.6^oC$) to I($Tc \sim 127.7^oC$). Recently, [4] the deuteriated $NH_{4-x}D_xNO_3$ system was studied with results showing that β'/β behaves rather differently for the ν_1 and ν_4 modes of NO_3 as X varies. The mode dependence of β'/β suggests that it could be a very important parameter (so is A'/A) characterizing the critical behavior of the crystal dynamics. The power law dependence of $I_R(\nu_1)/I_R(\nu_4)$ and $I_R'(\nu_1)/I_R'(\nu_4)$ on temperature has also been observed showing that the size effect of microdomains which fluctuate randomly by the thermal agitation, is significant and in a fractal way.

[1] G. Wu and J. Yu, Spectrochimica Acta 44A 1457(1988)
[2] Y. Zhou and G.Wu, Spectrochimica Acta 49A 581(1993)
[3] X. Ding and G.Wu, Spectrochimica Acta 51A 709(1995)
[4] S. Ma and G.Wu, to be published.

NEW HIGH-PRESSURE LOW-TEMPERATURE PHASE OF METHANE.

A. F. Goncharov*, M. Somayazulu, V. V. Struzhkin,
R. J. Hemley, H. K. Mao

Geophysical Laboratory and Center for High Pressure Research,
Carnegie Institution of Washington
5251 Broad Branch Road NW, Washington DC 20015-1305.

The phase diagram of methane is extremely rich at moderate pressures (≤10 GPa) and low temperatures. Six solid phases are known to exist in this P,T region [1,2]. Phase I is rotationally disordered (plastic), while phases II, III, IV are expected to be partially ordered with substantial difference in rotational potential fields between molecules in different crystallographic positions [1,3]. The transition from the phase V to fully ordered phase VI was reported at 0.9 GPa and 4.2 K [2]. Here we report low-temperature Raman measurements of methane to 11 GPa.

Measurements were done using a helium continuous-flow cryostat and diamond anvil cell. Ruby was used for pressure determination [4]. Spectra were measured using a Dilor X-Y spectrometer with CCD multichannel detector. Representative Raman spectra in different frequency ranges are shown in Fig. 1. Our spectra at 0.99 GPa are very similar to those reported in Ref. 1 for phase V. At higher pressure, in contrast with the results reported earlier [1], we do not observe substantial changes of spectra up to 2.2 GPa. However, at 3.0 GPa spectra change abruptly in all the measured frequency ranges. The ν_1 and ν_2 doublets, merge into singlet bands, while concomitantly, the ν_3 doublet evolves into a quartet. Low-frequency lattice translational and librational modes also change markedly. Most interestingly, the lowest frequency librational mode, corresponding to almost freely rotating molecules, disappears and other excitations (marked by asterisks in Fig. 1) appear. Surprisingly, we did not find any sign of the transition reported in Ref. [2] at 1.0 GPa, and our spectra in high pressure phase are different (especially in ν_3 region). Raman spectra above 3 GPa show continuous broadening of all bands and further splitting of the ν_3 multiplet. We also observed growing asymmetry of ν_1 and ν_2 fundamentals, and, finally, the second component of the ν_2 vibration becomes visible at 9.4 GPa.

The V-VI' transformation described above is anticipated transition to the molecular ordered state. The difference between our results and those reported in Ref. [2] is, probably due to problems of pressure determination in [2] or, since this transition might involve some degree of hysteresis, the transition pressure and temperature could be dependent on the P,T paths.

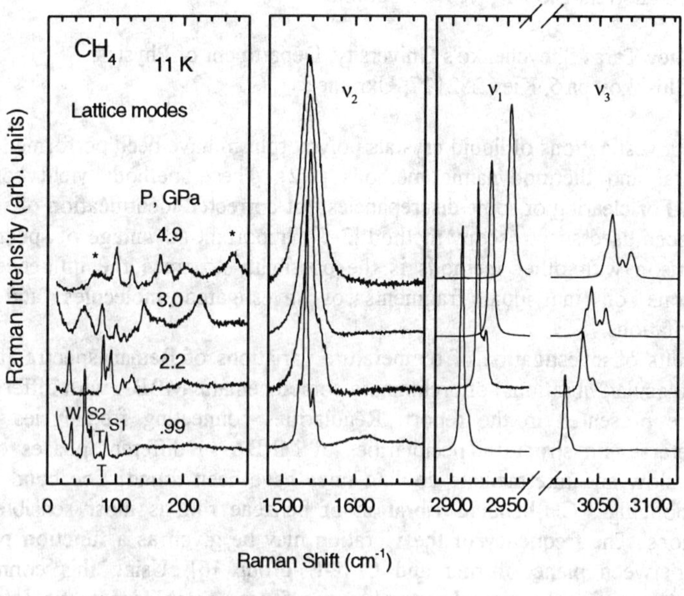

Fig. 1. Raman spectra of CH_4 at different pressures. W, S2, S1 are librational modes, corresponding to different types of molecules, T-translational modes [1]. ν_1, ν_2, ν_3 are first order Raman-active fundamentals.

1. M.-M. Thiéry, D. Fabre, K. Kobashi, J. Chem. Phys. 76, 4817 (1982).
2. M.-M. Thiéry, D. Fabre, K. Kobashi, J. Chem. Phys. 83, 6165 (1985).
3. W. Press, J. Chem. Phys., 56, 2597 (1972).
4. W. L. Vos, J. A. Schouten, J. Appl. Phys., 1, 6744 (1991)

RAMAN STUDY OF MOLECULAR STRUCTURE OF DIFFERENT SOLID PHASES OF LIQUID CRYSTALS.

I.B. Estrela-Llopis*, V.Ye. Pogorelov

Kiev Taras Shevchenko's University, Department of Physics,
Glushkov pr.6, Kiev 252127, Ukraine

The investigations of liquid crystals polymorphism have been performed by various structural and thermodynamic methods [1,2]. These methods yield contradictory results. For clearing of some discrepancies, for corrected identification of solid phases it has been used spectroscopy method [3-5]. Important advantage of spectroscopy in comparison with other methods is the possibility to trace the influence of phase transitions on individual fragments of investigated molecules and on their conformations.

Results of investigation of temperature variations of Raman spectra (the field of intramolecular vibrations) of preliminary cooled nematic MBBA with different cooling rate are presented in the report. Regularities connecting frequencies and shape parameters with structural peculiarities of MBBA in different phases obtained at shock, slow and intermediate cooling rates have been found. The band 1164 cm^{-1} corresponding CCH-bending vibration of benzene ring is most sensible to phase transitions. The frequency of the vibration may be given as a function of the twist angle between plane of ring and C=N-C group [6]. Using this connection the conformational angles have been calculated for supercooled nematic (s/c N), glass (G), nematic (N) and four crystalline phases obtained at different thermal pre-history. The calculated twist-angles of MBBA in various states are presented in the Table 1. For example, at shock cooling the low-temperature crystalline phase (M) have conformers with twist angle 50.2° and high-temperature S-crystal consist of mixture of two conformers with angles 82.7° and 43.3°. At slow cooling low-temperature and high-temperature phases (S_1 and S_2) consist of mixture of another two conformers having twist angles of 59° and 82.7°. These phases are distinguished by quantitative relationship between the two conformers. Besides it has been found that initial nematic phase (N_0) comprise three fractions of conformers. And nematic phase obtained after cooling and heating cycles (N) have twist-angles values and quantitative relationship between fractions depending on thermal pre-history of sample. Only for case of shock cooling the initial N_0-phase and the final N-phase are similar as on conformational composition as and on packing.

Determined connection of spectra with structure it has been allowed to make diagnostics of phase transitions and identification of phase states for liquid crystals having benzeliden-aniline nucleus.

TABLE 1. The values of twist angles of MBBA in various states.

States	ν, (cm^{-1})	twist-angles, (°)
N, s/c N	1160.0	44.4
	1164.0	50.2
	1168.0	59.0
G	1165.5	53.0
M	1164.0	50.2
S	1159.0	43.3
	1172.0	82.7
S_1, S_2	1168.0	59.0
	1172.0	82.7

According to temperature variations of the widths, intensities and band shifts it has been obtained information about energetic barriers between different conformers, order of non arrangement of various phases. The dynamics and kinetics of phase transitions such as supercooled nematic - glass, glass - M-crystal and M-crystal - S-crystal have been investigated.

The investigation of dynamics of phase transitions have allowed to study their mechanism. The phase transitions are accompanied by conformational conversions determining other packing of molecules. This fact is corroborated by the appearance and the increase of integrated intensities of some bands corresponding conformers of future phase at the expense of other bands corresponding conformers of previous phase. Besides the integrated intensity for all bands in process of certain phase transition is constant.

The study of spectral developments of different phases have given the possibility to determine quantitative phase composition of complicated states obtained at different pre-history.

1. L. Rosta, N. Kroo, V.K. Dolganov, Mol. Cryst. Liq. Cryst. 144, 297 (1987).
2. P.C. Jain, S.R.S. Kafle, Mol. Cryst. Liq. Cryst. 129, 199 (1985).
3. V.Ye. Pogorelov, Solid State Physics. 33, 1906 (1991).
4. V.Ye. Pogorelov, I.B. Estrela-Llopis, J. Mol. Struct. 348, 269 (1995).
5. V.Ye. Pogorelov, I.B. Estrela-Llopis, Mol. Cryst. Liq. Cryst. 265, 237 (1995).
6. M. Mizuno, T. Shinoda, Mol. Cryst. Liq. Cryst. 69, 103 (1981).

RAMAN STUDY OF THE PHASE TRANSITIONS IN LIQUID CRYSTALS.

V. Ye. Pogorelov*, I. B. Estrela -Llopis, I. P. Pinkevich, V. P. Bukalo.

Kiev Taras Shevchenko's University, Department of Physics,
Glushkov pr.6, Kiev 252127, Ukraine.

The vibrational spectra investigation of mesomorphous materials have been allowed to obtain important information about peculiarities of phase transitions in such objects [1-3].

The aim of work was to expose the development of structure peculiarities of different states MBBA in Raman spectra, to perform the theoretical analysis of Raman bands forming in various phases and to estimate structural relaxation times of samples from thermodynamic non-equilibrium states. The experimental investigations of temperature transformations of Raman bands frequencies and widths corresponding intramolecular vibrations have been performed. These vibrations are sensible to phase transitions of MBBA. Such vibrarions are conformational sensitive vibrations in field 1160-1172 cm^{-1} and also some ones of benzene ring in field 1570-1630 cm^{-1}.

It is known at fast cooling at T<203 K MBBA generate amorphous phase (A) presenting a mixture supercooled nematic (s/c N) and glasses (G) [4]. It has been found the bands widths for A-phase and nematic (N) phase are coincided in spite of the difference of heterogeneous broadening nature for the phases. It has been made the analysis of forming mechanisms of Raman bands with using of theory of Raman effect in condensed phase [5]. This theory results in the relationship discribing bands shape (1).

$$I(w) = \frac{w_o^4}{2\pi c^3} \sum_{\alpha\beta\gamma\delta} n_\alpha n_\beta E_\gamma^- E_\delta^+ i_{\beta\delta\alpha\gamma}(w) ,$$

$$i_{\beta\delta\alpha\gamma}(w) = \frac{1}{2\pi} \int_{-\infty}^{\infty} dt\, e^{-i(w - w_o)t} \left\langle P_{\beta\delta}(t) P_{\alpha\gamma}^*(0) \right\rangle \qquad (1)$$

where $E^{\pm}$ - amplitudes at exponents describing incident radiation with negative and positive frequencies, w_o- a frequency of incident light, n- a vector of polarization of scattering light, $P_{\beta\delta}(t)$- a electronic polarizability of molecules in Hejzenberg's idea. The averaging was made on states of molecule. From the analysis it follows that the proximity of Raman bands widths in A and N phases is stipulated one of parameters of corresponding averaging statistic distributions.

It has been detemined the temperature transformations of frequencies and widths correlate with peculiarities of structural conversions of MBBA at preliminary shock cooling of sample in heating cycle . The significance width of Raman bands in A-phase (12 cm^{-1}) is stipulated structurally-heterogeneous broadening. The fixed transition in heating cycle result in removal of structure heterogeneous. The widths are about 6 cm^{-1}. This transition is one to metastable crystalline state (M).

The structural evolution of MBBA under shock, slow, and intermediate cooling rates have been investigated in the report. The phase diagrams of MBBA at various pre-history are presented in Table 1, where S_1 , S_2 are low-temperature and high-temperature crystalline phases obtained at slow cooling, S is high-temperature phase obtained at shock cooling in heating cycle, N_o and N are initial and final nematic phases. In dependence on thermal pre-history of sample it may be obtained a few coexisting fractions (supercooled nematic, glass and crystalline phases). Proportional correlation of these fractions is determined temperatural pre-history. In dependence on their correlation it is possible realization of one or another phase diagram.

Model of structural transformations, possibility of prognostic of structural evolution and identification of phase transitions for wide class of liquid crystals in dependence on their thermal pre-history are discussed in the report.

TABLE 1.The phase diagram of MBBA at different cooling rate Vcool

Shock cooling	Vcool=25 K/min	Vcool=13 K/min	Slow cooling,	Vcool=2K/min
			cooling cycle	heating cycle
A	A	S_2	I_o	S_1
↓ 203 K	↓ 233 K	↓ 243 K	↓ 317 K	↓ 203K
M	M	M	N_o	S_2
↓ 253 K	↓ 263 K	↓ 253 K	↓ 263 K	↓ 293K
S	S	S	S_2	N
↓ 293 K	↓ 293 K	↓ 293 K	↓ 203 K	↓ 317K
N_o	N	N	S_1	I

1. V.Ye. Pogorelov, Solid State Physics. 33, 1906 (1991).
2. V.Ye. Pogorelov, I.B. Estrela-Llopis, J. Mol. Struct. 348, 269 (1995).
3. V.Ye. Pogorelov, I.B. Estrela-Llopis, Mol. Cryst. Liq. Cryst. 265, 237 (1995).
4. P.S. Jain, S.R.S. Kafle, Mol. Cryst. Liq. Cryst. 129, 199 (1985).
5. A.G. Tretjakov, Optics and Spectroscopy, 48, 1121 (1980).

RAMAN STUDIES OF STRUCTURAL RELAXATION IN AMORPHOUS FILM OF ADAMANTANE

L. Bistričić[1], G. Baranović[2*], and K. Furić[2]

[1] Faculty of Electrical Engineering and Computing, University of Zagreb, Croatia
[2] Rugjer Bošković Institute, P.O.B. 1016, 10001 Zagreb, Croatia

We show that small cage-like molecules like adamantane form amorphous films easily. Adamantane films were prepared by vacuum evaporation and deposition onto cold metal substrate at temperature 10 K. The structure and structural relaxation of amorphous films of adamantane have been studied by Raman spectroscopy. The characteristic spectral changes which reflect the structural change and transition from amorphous to the ordered crystalline state are observed during the annealing process.

The change of integrated Raman intensity and band width of the 760 cm^{-1} band with temperature is shown in Fig.1. and 2. This band is assigned to the totally symmetric CC stretching mode of the adamantane cage [1]. The effect of Bose factor $n(\omega,T) = \left[\exp(\hbar\omega / kT) - 1\right]^{-1}$ was accounted for in these spectra. The 760 cm^{-1} band observed in amorphous film at 10 K was weak and significantly broader than that of crystalline films at the same temperature. When the temperature was raised from 10 K to 180 K, the increase in the intensity and decrease in the band width were observed for all the measured bands. The intensity and the band width after the crystallisation remained temperature independent. The weak and broad bands are observed in amorphous systems in the low frequency region because the translational symmetry is destroyed, the usual selection rule $\vec{q} = 0$ breaks down and all vibrational modes become Raman active. According to ref. [2], reduced intensity of light scattering in disordered material is given by

$$I_R(\omega) \propto I(\omega)\cdot\omega\cdot\left[n(\omega) + 1\right]^{-1} \propto g(\omega)C(\omega)$$

where $C(\omega)$ describes the coupling between light and vibrational modes, $g(\omega)$ is the density of states. The reduced Raman intensity $I_R(\omega)$ for amorphous adamantane films is shown in Fig.3. At low frequencies the reduced intensity $I_R(\omega)$ exhibits an ω^2 dependence (Fig.4). Such a frequency dependence of $I_R(\omega)$ has been observed in other amorphous systems as well. Assuming that the Debye-like ω^2 behaviour of the density of states is also characteristic of amorphous materials, $C(\omega)$ parameter is frequency-independent in the low frequency region in amorphous adamantane films. However, accepting the propotionality $C(\omega) \propto \omega$ [3], the density of states $g(\omega)$ becomes also linearly dependent on frequency. The band shape analysis in the whole region from 180 to 10 cm^{-1} and at various temperatures from 10 to 120 K strongly indicates that the observed characteristics could be accounted for by the damped oscillatory - mechanism [4].

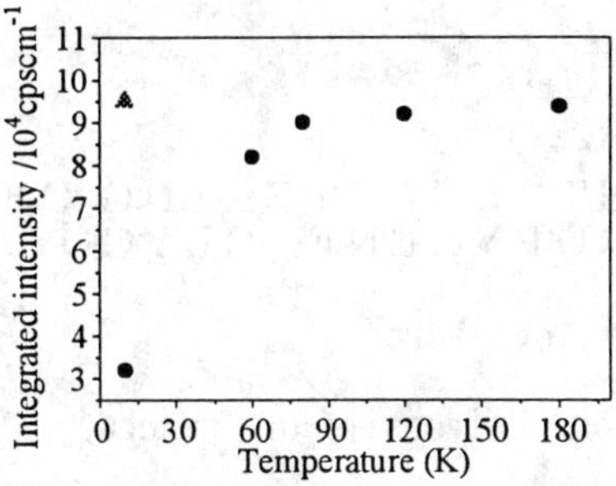

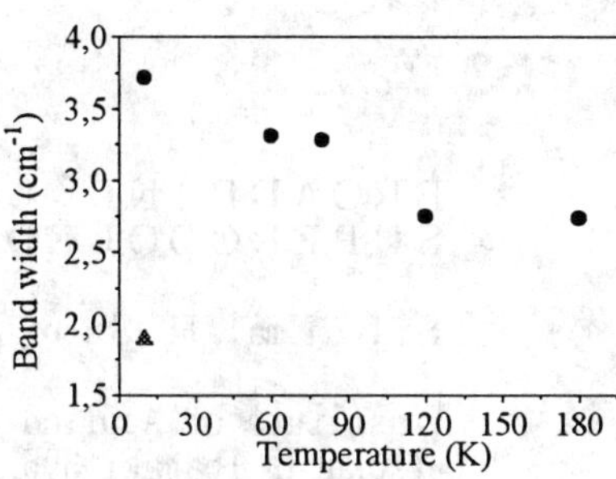

Fig. 1. Change of the integrated intensity of the 760 cm^{-1} Raman band of adamantane film deposited at 10 K. Circles: data observed during the heating of the film from 10 K to 180 K. Triangle: data at 10 K after the crystallization of the film at 180 K.

Fig. 2. Temperature evolution of the width of the 760 cm^{-1} band of the adamantane film deposited at 10 K. The data marks are the same as in Fig. 1

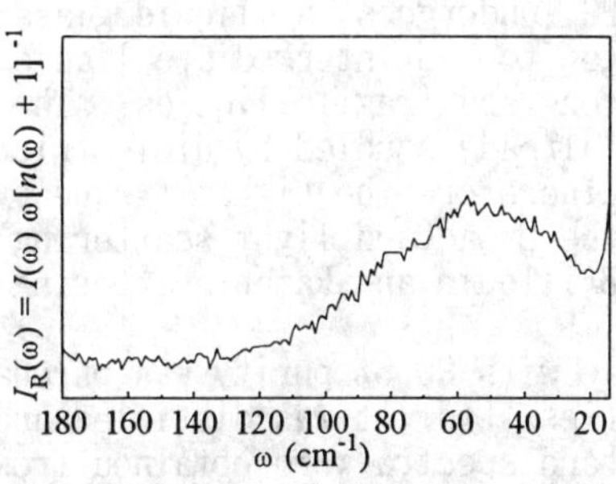

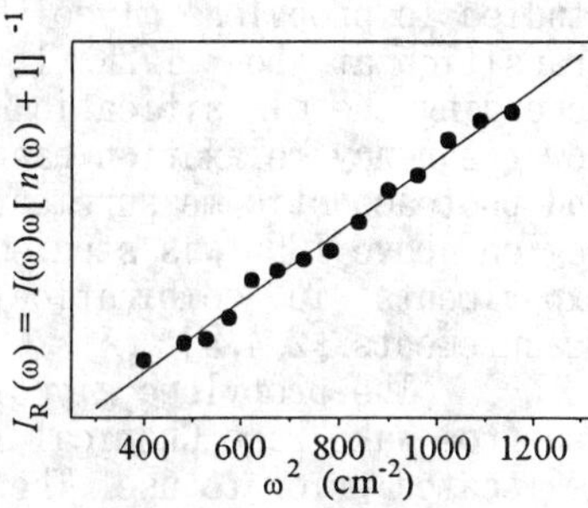

Fig. 3. The reduced Raman spectra $I_R(\omega)$ of the adamantane film deposited at 10 K.

Fig. 4. The reduced Raman intensity $I_R(\omega)$ in the low frequency region of the adamantane film deposited at 10 K

1. L. Bistričić, G. Baranović and K. Mlinarić-Majerski, Spectrochim. Acta., 51, 1643 (1995)
2. R. Shuker and R. W. Gammon, Phys. Rev. Letters 25, 222 (1970)
3. T. Achibat, A. Boukenter, and E. Duval, J. Chem. Phys. 99, 2046 (1993).
4. C. H. Wang and R. B. Wright, J. Chem. Phys. 55, 1617 (1971)

BROADBAND LIGHT SCATTERING OF SUPERCOOLED PROPYLENE GLYCOL

S. Kojima[1], H. Sato[2] and A. Yosihara[2]

[1]Institute of Applied Physics, University of Tsukuba, Tsukuba, Ibaraki 305, Japan
[2]Research Institute of Science Measurements, Tohoku University, Katahira, Aoba-ku, Sendai, Miyagi 980, Japan

Various kinds of supercooled liquids transform into glassy states, while the mechanism of this liquid-glass transition is still unknown. Thus the dynamical properties of glass-forming liquids are today a subject of increasing interest in connection with the poorly understood liquid-glass transition phenomena.

In the present study, the dynamical properties were studied in propylene glycol(PG). PG undergoes a liquid-glass transition at about 172K. It belongs to an intermediate liquid according the classification of strong and fragile liquids. The low frequency relaxation has been already studied by dielectric and photoacoustic measurements.[1] Therefore the higher frequency region above 1GHz was studied by the broadband light scattering experiments in combination of Brillouin and Raman scattering measurements.[2,4,5]

The propylene glycol sample with 99.5% purity was purchased from Wako Pure Chemical Industries, Ltd. It was distilled and desiccated prior to use. The broadband spectra were obtained from 1GHz to 10THz, where the region below 500GHz was measured utilizing the Sandercook-type Fabry-Perot interferometer.[2,5] The region above 100GHz was obtained using a triple-grating spectrometer of the additive dispersion (Jobin Yvon T64000). The output signals were detected by the photon counting system in both cases.

The dynamical susceptibility was determined from the depolarized scattering spectra in a large frequency range between 1GHz and 10THz by the combination of Brillouin and Raman scattering measurements. The obtained spectra were analyzed by the mode coupling theory at first, however it is found that the agreement is not sufficient as similar as the case of glycerol.[6] Next the phenomenological approach was attempted. Figure 1 shows the combined broadband spectra and calculated results. The spectrum at 363K reveals the two peaks. With the decrease of temperature the lower α peak shifts to lower, and the minimum position ω_{min} also becomes lower gradually, while the boson peak does not change remarkably . The lower one is attributed to the α relaxation.

According to the previous results of the dielectric and thermal dispersion, the α relaxation is well reproduced by the Davidson-Cole formula, and its relaxation frequency obeys the Vogel-Fulcher law.[1] The higher peak corresponds to the boson peak, of which lineshape is well fitted by the disorder-induced model with an exponential correlation function.[4] However, in fact, the another contribution is required to reproduce the middle region between two peaks. Thus the another relaxation mode of the Debye type was introduced. Consequently the observed spectra were well reproduced by these three contributions.

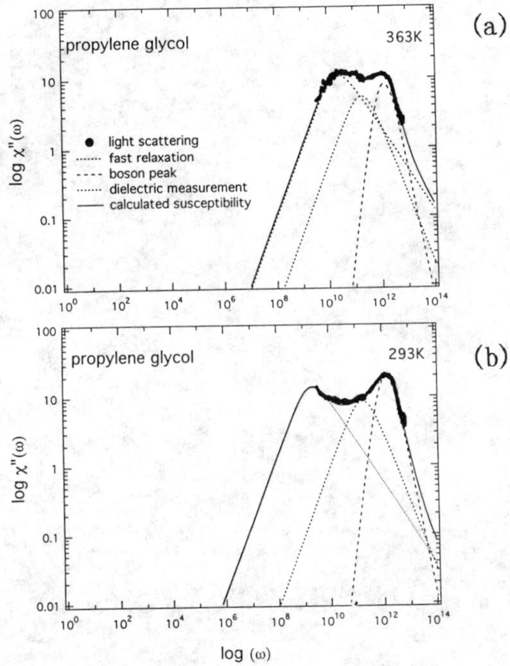

Fig. 1 Broadband dynamical susceptibility spectra, (a)363K, (b)293K.

1. S.Kojima: Ferroelectrics : 137, 315 (1993).
2. A.Yoshihara: Jpn. J. Appl. Phys. 33, 3100 (1994).
3. S.Kojima: Jpn. J. Appl. Phys. 32, 2223 (1993).
4. S.Kojima: Phys. Rev. B47, 2924 (1993).
5. K.Takanashi, A.Yoshihara and S.Kojima : Jpn. J. Appl. Phys. 33, 2867 (1994).
6. S.Kojima, K.Takanashi and A.Yoshihara: Jpn. J. Appl. Phys. 34, 2552 (1995).

SECTION 24

EFFECTS of TEMPERATURE and PRESSURE

HIGH-PRESSURE RAMAN SPECTROSCOPY: NEW WINDOWS ON MATTER UNDER EXTREME CONDITIONS

Russell J. Hemley* and Ho-kwang Mao

Geophysical Laboratory and Center for High-Pressure Research, Carnegie Institution of Washington, 5251 Broad Branch Rd. N.W. Washington, D.C. 20015, U. S. A.

High-pressure Raman scattering is a crucial technique for characterizing matter at ultrahigh pressures. As a result of breakthroughs in diamond anvil cell techniques, pressures above 300 GPa (3 megabars) have been reached [1]. These pressures correspond to compression of from a factor of two (*e.g.*, for incompressible metals) to well over an order of magnitude (*e.g.*, solidified gases). Under these conditions, the physical and chemical properties of materials can be radically altered. The high spatial resolution provided by micro-Raman is well matched to the small samples sizes (< femtoliter volumes) required for reaching these very high pressures in diamond-anvil cells. Vibrational Raman has been important for identifying and understanding the mechanism of high-pressure phase transitions, including both discontinuous and continuous crystallographic transitions, electronic and magnetic transitions, and transitions involving localization of excitations. The technique has also been useful for identifying new phases, revealing structural order-disorder, and determining thermodynamic properties of materials under extreme pressures. As such, the technique has been useful for addressing problems spanning fundamental physics, chemistry and materials science, and Earth and planetary sciences. Here, we review examples of recent key results.

Dense Hydrogen. Raman scattering continues to play a major role in the study of highly condensed hydrogen, an archetypal problem in condensed matter physics and high-pressure research. Experiments demonstrate that the molecular bond is stable to at least ~250 GPa, but several intriguing changes in bonding and physical properties are observed in this pressure range [1]. The changes can be understood in terms of a series of symmetry-breaking transitions with increasing pressure. Raman measurements reveal a phase diagram of unexpected complexity, with evidence for both classical and quantum mechanical orientational ordering in different high-pressure phases [2]. At the highest pressures, there is a significant drop in Raman vibron frequency and a large increase in infrared vibron absorption, although samples remain transparent in visible and mid-infrared portions of the spectrum [3].

Vibrational Localization Transitions. High-pressure Raman scattering has also revealed unusual transitions associated with the localization and delocalization of vibrational excitations. This was done to explore a novel transition in which double vibrational excitations transforms from being localized on a single molecule to being delocalized throughout the crystal to give a bound-unbound transition. Raman scattering measurements of the Raman vibron overtones in deuterium showed the first example of such a transition [4]. The results of parameter-free theoretical calculations were in remarkable agreement with the experimentally observed transition. Another study showed that Raman spectra of the vibrons in ortho-para hydrogen samples exhibit localization phenomena and parallel the physics of a random binary alloy [5]. Here, the disorder is associated with vibrational excitations on the J=0 and J=1 molecules. The spectral signature of the J=0 and J=1 vibron peaks agree with the results of supercell calculations using a tight-binding model Hamiltonian.

New Molecular Compounds. The newly accessible pressure domain provides the prospect of greatly enlarging the number of known materials. An example is the discovery of high-pressure stoichiometric compounds, or ordered alloys, in simple molecular mixtures (e.g., N_2, H_2, H_2O CH_4, rare gases; see Ref. [6]). Raman scattering has played an important role in the characterization of these materials. A particularly exciting result is the recent observation of four new compounds in the CH_4-H_2 system [6]. In addition, novel high-pressure clathrates have been observed in H_2O bearing systems, including a compound in which the H_2O and H_2 molecules form two interlocking diamond networks stable to at least 30 GPa [7]. Such molecular compounds may exist in icy bodies of the outer solar system.

Amorphous Materials and Amorphization. Other examples of both an unusual high-pressure transition and new materials are those connected with structural disorder under pressure. Pressure-induced amorphization transformations have been observed in a growing number of materials [8]. Such transformations may access unusual metastable states with varying degrees of disorder and possibly unique properties. Because Raman scattering provides distinct signatures for both crystalline and amorphous phase, it has played an important role in identifying and understanding the nature of these transitions. The techniques has also been important in understanding the mechanism of irreversible compaction of amorphous solids (*e.g.*, silica glass), which can be compressed into densified states that can be retrieved on quenching to ambient conditions [9]. This suggests the possibility of synthesizing amorphous solids under pressure with a range of optical properties, densities, and chemical properties.

High-Pressure Minerals. High-pressure experiments offer the only means to examine directly the chemical and physical properties of materials that comprise deep interior of our planet. Vibrational Raman spectroscopy has been used extensively to characterize the high pressure minerals important in the Earth's mantle [9]. The results of Raman studies provide structural information on six-coordinated silicon in high-density phases of the mantle. Such measurements provide detailed information on phase transitions and on the uptake of volatile components in minerals under pressure. Recent Raman scattering experiments demonstrate that stishovite, the rutile-structured phase of SiO_2 stable in the Earth's mantle, transforms to an orthorhombic form near 50 GPa, which is the pressure at 1200 km depth in the lower mantle [10]. This transition is driven by an elastic instability that is coupled with a soft Raman mode. Such high-pressure experiments are providing a unique window on the nature of planetary interiors.

This work was supported by the NSF, NASA, and DOE.

1. H. K. Mao and R. J. Hemley, Rev. Mod. Phys. **66**, 671 (1994).
2. A. F. Goncharov, *et al.*, Phys. Rev. Lett. **75**, 2514 (1995).
3. R. J. Hemley, *et al.*, Phys. Rev. Lett. **76**, 1667 (1996).
4. J. H. Eggert, H. K. Mao and R. J. Hemley, Phys. Rev. Lett. **70**, 2301 (1993).
5. J. L. Feldman, *et al.*, Phys. Rev. Lett. **74**, 1379 (1995).
6. M. Somayazulu, L. W. Finger, R. J. Hemley and H. K. Mao, Science, in press.
7. W. L. Vos, *et al.*, Phys. Rev. Lett. **71**, 3150 (1993).
8. K. J. Kingma, *et al.*, Science **259**, 666 (1993).
9. R. J. Hemley, C. T. Prewitt and K. J. Kingma, in *Reviews of Mineralogy, Vol. 29*, edited by P. J. Heaney et al. (MSA, Washington, D.C., 1994) pp. 41-81.
10. K. J. Kingma, *et al.*, Nature **374**, 243 (1995).

HIGH PRESSURE RAMAN STUDIES ON SOME RARE-EARTH MOLYBDATES AND COPPER GERMANATE

A. Jayaraman, S.K. Sharma, and S.Y. Wang
Hawaii Institute of Geophysics and Planetology
University of Hawaii at Manoa, Honolulu, HI 96822, USA

We have studied the high pressure behavior of a number of tri- and di-molybdates involving rare earth cations viz. $R_2(MoO_4)_3$, $RA(MoO_4)_2$ where A is an alkali metal ion, as well as simple molybdates with divalent cations, viz., $SrMoO_4$ using Raman spectroscopy as the probe. We have found pressure-induced phase transitions in all of them. A few representative cases are discussed here.

For the rare earth trimolybdate the behavior of $Tb_2(MoO_4)_3$ (TMO) crystallizing in the so called orthorhombic β'-phase (Pba2) is typical. The Raman data [1] reveal that the system undergoes an irreversible pressure-induced amorphization near 6GPa. The behavior is illustrated in Fig 1. The results of the present study as well as previous high pressure - high temperature quenching experiments strongly support that the pressure - induced amorphization in TMO is a consequence of the kinetically - impeded β' to α-phase transition in TMO. The thermodynamics of the system is as such that it has a built in instability with respect to pressure. The latter favors the formation of the α-phase. However, the transition is kinetically impeded. In frustration the system disorders. When thermal activation is provided, ordering is promoted and the system goes to the stable α-phase. Thus we see an important driving force at play for pressure-induced amorphization, namely the presence of a sluggish phase transition involving a large volume change up in front of pressure.

We have found a reversible pressure-induced amorphization near 27GPa both by IR and Raman study [2], in the double molybdate system $NaLa(MoO_4)_2$ crystallizing in the orthorhombic pseudo-scheelite structure. The ready reversibility suggests a different mechanism for amorphization, such as uncorrelated tilting of the (MoO_4) tetrahedra. In the case of a simple molybdate such as $SrMoO_4$ [3] $CdMoO_4$ [4] no amorphization is observed up to 35GPa. From the different behavior of tri-, di- and simple molybdates, we believe that structural complexity and large volume change in front play an important role in pressure-induced amorphization. In $CdMoO_4$, we have observed for the first time the sequence of pressure-induced transitions, scheelite →wolframite→sixfold coordinated $[MoO_6]$ structures, using Raman scattering as the probe.

Copper germanate $CuGeO_3$ is an interesting system, for it undergoes the so called Spin Peierls transition when cooled below 14K under ambient pressure. It is a layer structured material, cleaving like mica, and is structurally related to the pyroxene family minerals such as enstatite ($MgSiO_3$). By recording the Raman spectra of $CuGeO_3$ as a function of pressure we have found a rather remarkable phase transition near 7GPa when pressurized in methanol-ethanol mixture as the pressure medium [5]. This is shown in Fig. 2 by the distinct changes in the Raman spectrum. Remarkably, the single crystal visibly undergoes a 20% length change in the direction parallel to the b-axis of the crystal, retaining the single crystal nature. Since the same type of transition was not observed in powdered sample or in argon as pressure medium, we have suggested pressure-induced intercalation as a possible explanation for the observed features.

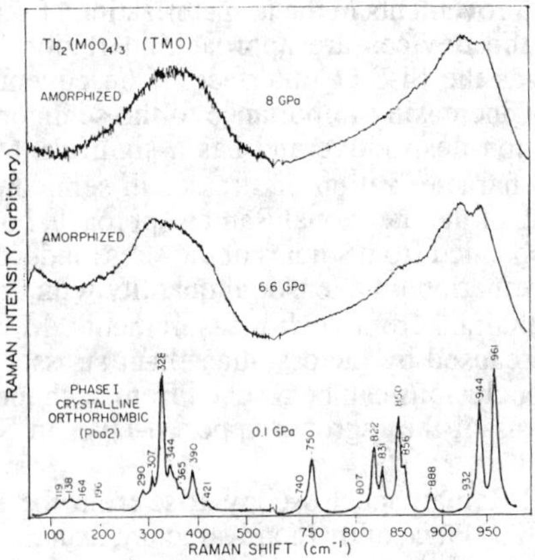

Fig. 1 Amorphization of TMO

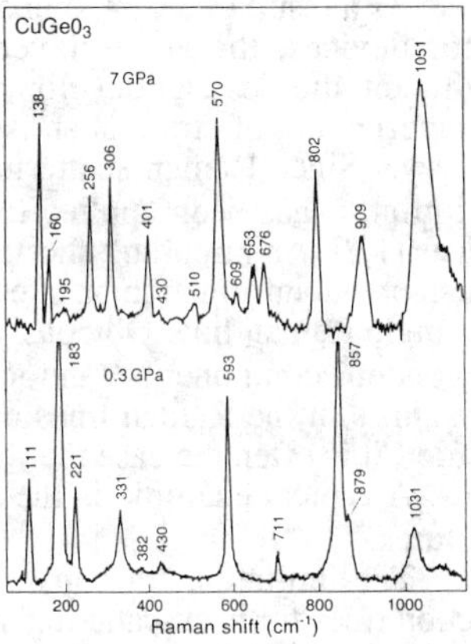

Fig. 2. Transition in $CuGeO_3$

We have observed a rather spectacular pressure-induced phase transition in $KTb(MoO_4)_2$ KTMO near 3GPa in which the Raman spectrum undergoes a drastic change, signifying a structural phase transition. Remarkably, the color of the sample changes to bright yellow in transmission, indicating a change in the electronic structure in the material. This is connected with a change in the electronic state of Tb caused by the change in the ligand field at the structural transition.

In summary, Raman scattering has proven to be a powerful tool to probe pressure-induced phase transitions and has yielded valuable information.

1. A. Jayaraman, S.K. Sharma, Z. Wang, S.Y. Wang, L.C. Ming and M.H. Manghnani, J. Phys. Chem. Solids, 54, 327 (1993).
2. A. Jayaraman, S.Y. Wang and S.K. Sharma, Solid State Commun. 93, 885 (1995).
3. A. Jayaraman, S.Y. Wang, S.R. Shieh, S.K. Sharma and L.C. Ming, J. Raman Spectrosc. 26, 451 (1995).
4. A. Jayaraman, S.Y. Wang and S.K. Sharma, Phys. Rev. B 52, 9886 (1995).
5. A. Jayaraman, S.Y. Wang, L.C. Ming and S-W. Cheong, Phys. Rev. Lett., 75, 2356 (1995).
6. A. Jayaraman, S.Y. Wang, S.K. Sharma, S.W. Cheong. Curr.Sci. (In Press, 1996).

RAMAN MEASUREMENTS OF STRESS IN SILICON-BASED STRUCTURES

José Menéndez

Department of Physics and Astronomy, Arizona State Univeristy, Tempe, AZ 85287-1504

As a result of recent dramatic improvements in the miniaturization of semiconductor devices, the active layers of the devices are approaching regions where significant stresses can develop, such as the Si-SiO_2 interface. Consequently, the characterization of stress is an issue of increasing importance to the semiconductor industry. Since Raman scattering is non-destructive and has a significant spatial resolution, it has been applied to the characterization of stresses in semiconductor devices.[1,2] An important shortcoming of the traditional Raman approach, however, is the fact that information on stress is obtained from studies of the stress-induced shift of a single Raman line. Obviously, information on a tensorial quantity with up to six independent components cannot be obtained from such measurements. Moreover, since shifts in the Raman lines can be caused by factors other thant stress, such as disorder, it is often the case that few conclusions can be reached from such measurements. A typical example is the analysis of the regrown upper Si-layer in SIMOX structures.

The limitations of conventional Raman spectroscopy arise from the Raman selection rules for backscattering at the (001) face of diamond-structure materials. This is the prefered orientation of most single-crystal semiconductor wafers used by the electronics industry. Backscattering at these surfaces allows for the observation of only one Raman peak corresponding to a longitudinal optical vibration propagating in the (001) direction. An arbitrary stress tensor in a diamond-structure semiconductor will split the three-fold degenerate Raman-active phonons at the center of the crystal's Brillouin zone. The frequency of these three phonons and their Raman intensities relative to the Raman intensity in bulk Si provide the six independent quantities needed for a full experimental determination of the stress tensor. In principle, information on these additional phonon lines can be obtained by performing experiments where the incident light is not at normal angles to the surface. However, this has not been attempted in the past because the high index of refraction makes the incident angle very small *inside* the sample (Snell's law), even for light incident on the sample at a glancing angle. In spite of this limitation, we have recently demonstrated that off-axis Raman spectroscopy can indeed be used to determine the components of the stress tensor.[3]

We constructed a device to produce a known stress on an 10 cm Si wafer by displacing the center of the wafer a certain distance D relative to the edge of the sample, which is kept at a fixed position. We use our Raman method to study the stress generated at the center of the wafer. We fit an arbitrary stress tensor to the data and obtain excellent agreement with the known stress tensor, demonstrating that our method is capable of determining stress tensors. Figure 1 shows representative Raman spectra and Figure 2 shows the fit results compared with the known stress tensor. We believe this is the first time that an externally induced LO-TO splitting has been observed in Si using backscattering Raman spectroscopy in the visible.

Previous studies of stress based on conventional Raman spectroscopy relied on theoretical modeling of the stress tensor in order to overcome the limitations of the technique. While significant advances have been made in this area,[2] our method opens up the possibility of a fully experimental determination of the stress tensor which does not rely on any theoretical model.

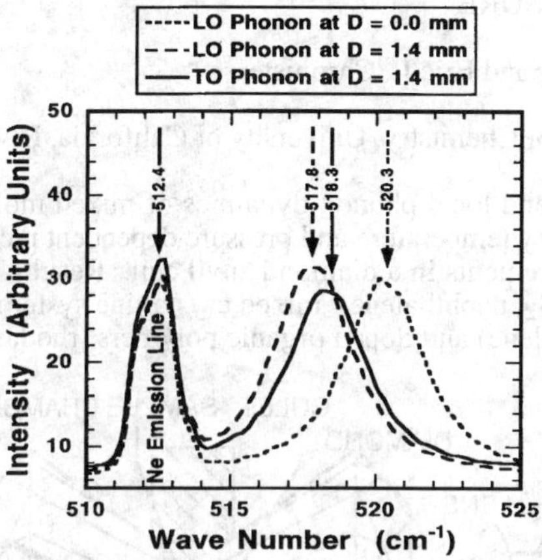

Figure 1 The Raman spectrum of a Si wafer under stress. Notice the LO-TO splitting, induced solely by stress.

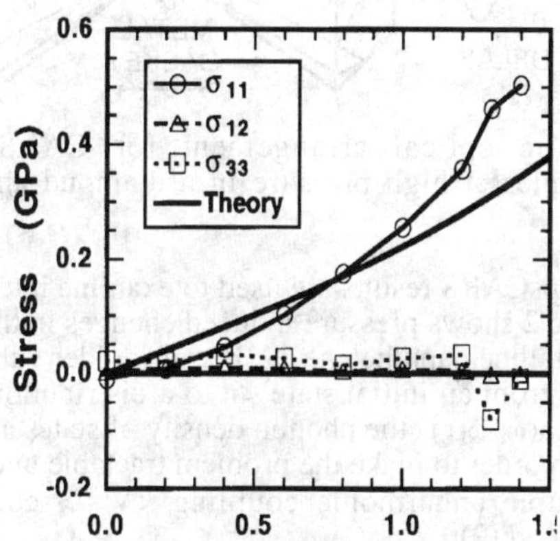

Figure 2 Comparison between expected and measured stress components for a Si-wafer under stress.

1. G. Kolb, T. Salbert and G. Abstreiter, J. Appl. Phys. **69**, 3387 (1991).
2. I.D. Wolf, J. Vanhellemont, A. Romano-Rodríguez, H. Norström and H.E. Maes, J. Appl. Phys. **71**, 898 (1992).
3. G.H. Loechelt, N.G. Cave, and J. Menéndez, Appl. Phys. Lett. **66**, 3639 (1995).

VIBRATIONAL DYNAMICS IN MOLECULAR SOLIDS UNDER HIGH PRESSURE

Bruce J. Baer and Eric L. Chronister*

Department of Chemistry, University of California, Riverside, CA 92521

The vibrational and local phonon dynamics of mixed molecular crystals and glasses are investigated by temperature and pressure dependent picosecond CARS and photon echo and measurements in a diamond anvil cell. Results have been obtained for neat crystals (N_2, CS_2, naphthalene), mixed crystalline systems (pentacene doped p-terphenyl and naphthalene) and doped organic polymers (rhodamine and pentacene doped PMMA).

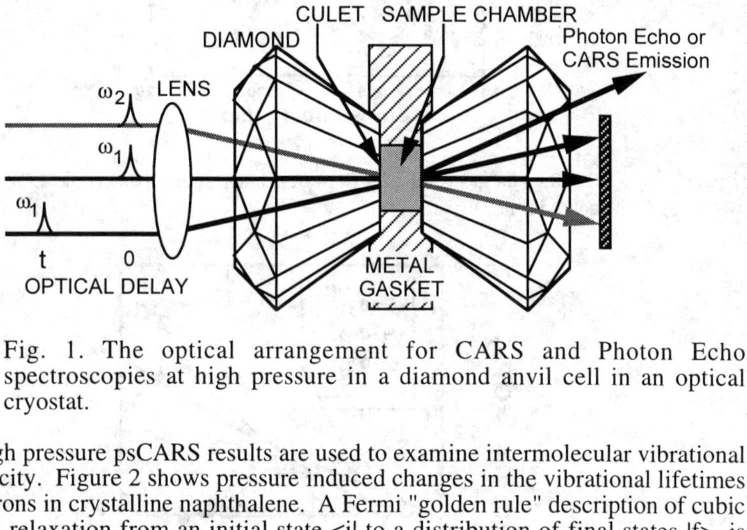

Fig. 1. The optical arrangement for CARS and Photon Echo spectroscopies at high pressure in a diamond anvil cell in an optical cryostat.

High pressure psCARS results are used to examine intermolecular vibrational anharmonicity. Figure 2 shows pressure induced changes in the vibrational lifetimes of two vibrons in crystalline naphthalene. A Fermi "golden rule" description of cubic vibrational relaxation from an initial state <i| to a distribution of final states |f>, is given in eq (1), where $\rho(\omega_i-\omega_f)$ is the phonon density of states at the frequency needed to conserve energy. In order to make the problem tractable one can examine the case where an average cubic anharmonic coupling $\langle V^{(3)}\rangle$ characterizes the decay pathways, as shown in eq (2).

$$\frac{1}{T_1} = \frac{2\pi}{\hbar}\sum_f \left\langle i\middle|V^{(3)}\middle|f\right\rangle^2 \rho(\omega_i - \omega_f) \tag{1}$$

$$\frac{1}{T_1} \cong \frac{2\pi}{\hbar}^2 \left\langle V^{(3)}\right\rangle^2 \sum_f \rho(\omega_i - \omega_f) = \frac{2\pi}{\hbar}\left\langle V^{(3)}\right\rangle^2 \cdot \rho_{2P}(\omega_i) \tag{2}$$

Figure 2b shows that pressure induced vibrational relaxation can in some cases be correlated with phonon frequency shifts (i.e. changes in the density of vibrational states), however, the pressure induced increase in the vibrational relaxation rate of the 511cm^{-1} mode is attributed to pressure induced anharmonicity, as shown in figure 2a.

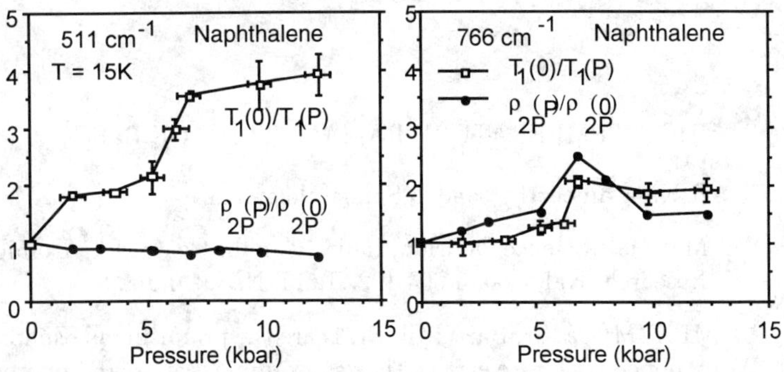

Fig. 2. The pressure induced changes in vibrational lifetimes $T_1(0)/T_1(P)$ in naphthalene can be correlated with either pressure induced anharmonicity $V^{(3)}$ or changes in the vibrational density of states $\rho(P)/\rho(0)$.

The anharmonicity of local phonon dynamics in mixed molecular crystals has been examined using temperature and pressure dependent photon echo measurements. The frequency of the local phonons are determined by Arrhenius plots of thermally activated optical dephasing, as shown in figure 3a. The pressure dependent pseudolocal phonon frequency provides a probe of the anharmonic vibrational motion of the pentacene molecule in its p-terphenyl "cage".

Similar high pressure photon echo measurements on doped glasses have been performed to provide a test of tunneling two level system (TLS) models. The relatively unchanged optical dephasing rate of rhodamine 101 in PMMA for >20% compression at 30 kbar will provide a stringent test of TLS models of optical dephasing.

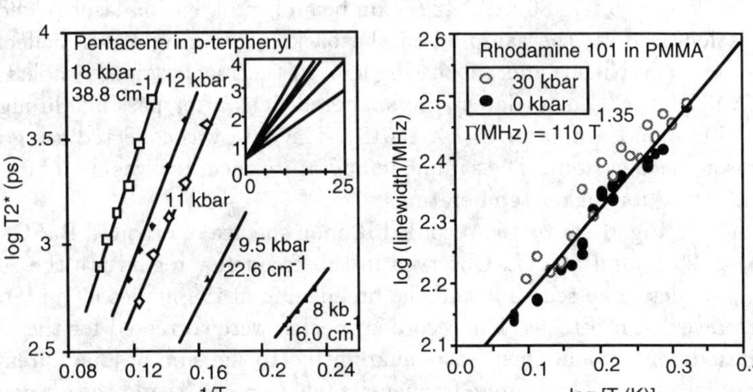

Fig. 3. a) Pressure induced increases in the pseudolocal phonon frequency of pentacene in p-terphenyl. b) Pressure independent optical dephasing of rhodamine doped PMMA.

We acknowledge support from National Science Foundation grant CHE-9400542.

HIGH TEMPERATURE RAMAN STUDY OF BaFCl

R.Kesavamoorthy* and B.Sundarakkannan

Materials Science Division, Indira Gandhi Centre for Atomic Research, Kalpakkam 603 102, Tamil Nadu, India.

MFX (M: Ca, Sr, Ba; X: Cl, Br, I) are the promising phosphor host crystals. When doped with rare earths, they show luminescence and photostimulated luminescence. These alkaline earth dihalides crystallise in tetragonal structure with P4/nmm symmetry at RTP [1]. The group theoretical analysis indicates six Raman active modes, namely, $2A_{1g}$, B_{1g} and $3E_g$ [1]. High temperature Raman investigation of these crystals has not been carried out so far.

This report deals with the Raman study of BaFCl single crystals from 25 to 700^oC. Optically transparent and dendrites-free BaFCl single crystals (5X5X0.5 mm^3) were grown by flux method using the LOBA GR grade powders of $BaCl_2.2H_2O$ and KF. Its room temperature XRD powder pattern, obtained in SIEMENS D-500 powder diffractometer, matches well with the ICDD-240096 pattern of BaFCl. The Laue pattern of BaFCl, recorded with Huber make Laue camera, reveals single crystal nature.

Unpolarised Raman spectra of the oriented BaFCl single crystal with ac-face perpendicular to the incident beam were recorded at different temperatures from 25 to 700 ^oC. 80 mW, 482.5 nm beam from a Kr-ion laser (Coherent, USA) was focussed on the sample and the backscattered light was collected using a camera lens(Nikon) and a focussing lens. A thermoelectrically cooled PMT(ITT-FW130) was used to detect the scattered light after passing through a double grating monochromator(SPEX 14018). The data was collected using a microprocessor based system. The sample temperature accuracy was $\pm 1^oC$ for T $\leq 300^oC$ and $\pm 2^oC$ for higher temperatures.

Fig. 1 shows the recorded Raman spectra of oriented BaFCl single crystal at 25^oC and 500^oC. Observed in this frequency region are the A_{1g}, B_{1g} and $2E_g$ modes. The red shift and the broadening of the modes at high temperature are revealed in Fig. 1. The recorded spectra were corrected for the phonon population density and then least square fitted to the sum of four Larentzian and a baseline curve. The Raman frequencies (ω_{oi} ; i= 1 to 4) and the observed FWHM, Γ_{ei}, were obtained. The true FWHM, Γ_i, of the Raman modes were deduced from Γ_{ei} and the instrument resolution function, $\Gamma_s = 4.2cm^{-1}$ [2]. Fig. 2 shows the contineous increase of Γ_i with temperature. The data is fitted to the sum of cubic and quartic anharmonic terms. The result indicates that the ratio of quartic to cubic anharmonicity contributions to Γ_i increases with the mode frequency. The mode frequencies, ω_{oi}, decrease contineously with the increase of temperature as shown in Fig. 3. The data fits well with the straight line. This decrease carries quasiharmonic contribution and the cubic and quartic anharmonicities contributions. The temperature behaviour of Γ_i and ω_{oi} was contineous and there was no new mode or disappearence of mode indicating the absence of phase transition in the range of 25^oC to 700^oC.

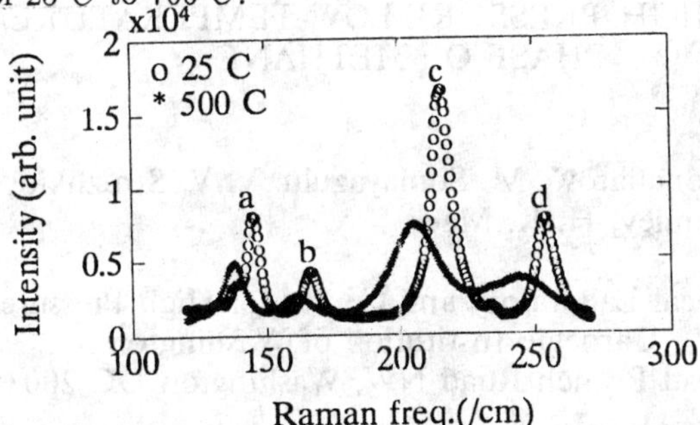

Fig. 1. Raman spectra of BaFCl at 25^oC (o) and 500^oC (*). Raman modes are marked in the spectra at 25^oC. a : E_g ($145.3cm^{-1}$); b : A_{1g} ($166.8cm^{-1}$); c : B_{1g} ($216cm^{-1}$); d : E_g ($254.3cm^{-1}$).

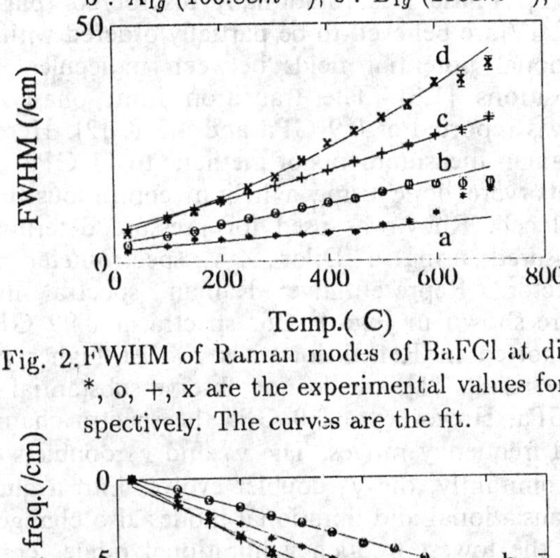

Fig. 2. FWHM of Raman modes of BaFCl at different temperatures. *, o, +, x are the experimental values for a, b, c, d modes respectively. The curves are the fit.

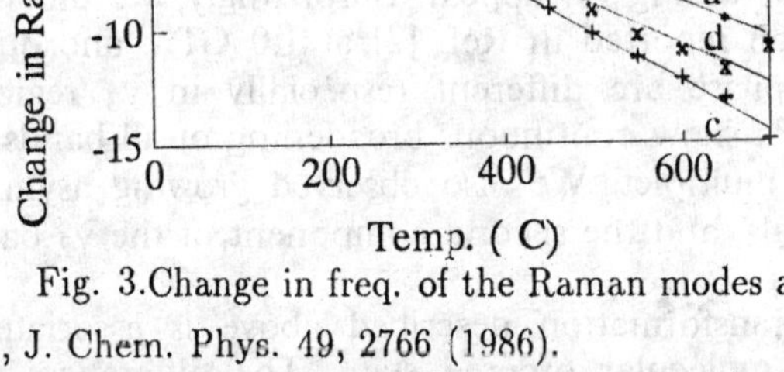

Fig. 3. Change in freq. of the Raman modes as a function of temp.

1. J.F.Scott, J. Chem. Phys. 49, 2766 (1986).
2. A.K.Arora, and V.Umadevi, App. Spectros. 36, 424 (1982).

NEW HIGH-PRESSURE LOW-TEMPERATURE PHASE OF METHANE

A. F. Goncharov*, M. Somayazulu, V. V. Struzhkin, R. J. Hemley, H. K. Mao

Geophysical Laboratory and Center for High Pressure Research, Carnegie Institution of Washington
5251 Broad Branch Road NW, Washington DC 20015-1305.

The phase diagram of methane is very rich at moderate pressures (≤10 GPa) and low temperatures. Six solid phases are known to exist in this P,T region [1,2]. Phase I is rotationally disordered (plastic) phase, while phases II, III, IV are believed to be partially ordered with substantial difference in rotational potential fields between molecules in different crystallographic positions [1,3]. The transition from phase V to fully ordered phase VI was reported at 0.9 GPa and 4.2 K [2]. Here we report low-temperature Raman measurements of methane to 11 GPa.

Measurements were done using a helium continuous-flow cryostat and diamond anvil cell. Ruby was used for pressure determination [4]. Spectra were measured using a Dilor X-Y spectrometer with CCD multichannel detector. Representative Raman spectra in different frequency ranges are shown in Fig. 1. Our spectra at 0.99 GPa are very similar to those reported in Ref. 1 for phase V. At higher pressure, in contrast with earlier results [1], we do not observe substantial changes in spectra up to 2.2 GPa. However, at 3.0 GPa the spectra change abruptly in all the measured frequency ranges. The ν_1 and ν_2 doublets merge into singlet bands; concomitantly, the ν_3 doublet evolves into a quartet. Low-frequency lattice translational and librational modes also change markedly. Most interestingly, the lowest frequency librational mode, corresponding to almost freely rotating molecules, disappears and other excitations (marked by asterisks in Fig. 1) appear. Surprisingly, we did not find any sign of the transition reported in Ref. [2] at 1.0 GPa, and our spectra in the high-pressure phase are different (especially in ν_3 region). Raman spectra above 3 GPa show continuous broadening of all bands and further splitting of the ν_3 multiplet. We also observed growing asymmetry of ν_1 and ν_2 fundamentals, and the second component of the ν_2 band becomes visible at 9.4 GPa.

The V-VI transformation described above is associated with the transition to the molecular ordered state. The difference between our results and those reported in Ref. [2] is probably due to problems of pressure determination in the latter or, since this transition might involve some degree of hysteresis, the observed transition pressures and temperatures could be dependent on the P,T paths.

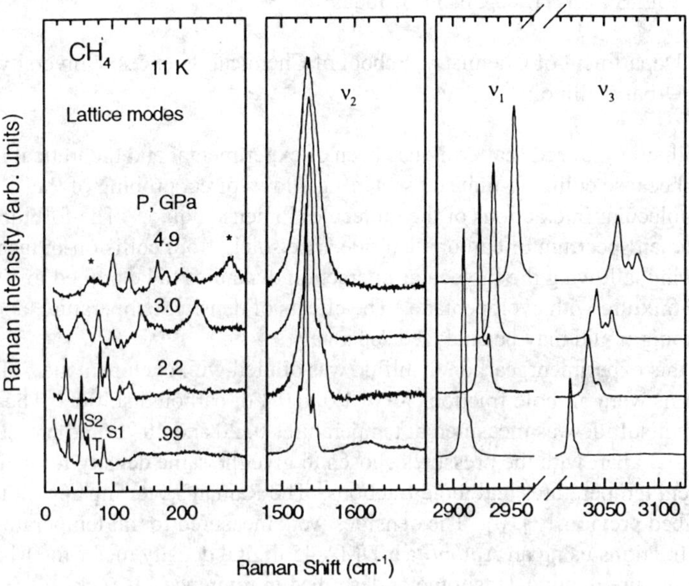

Fig. 1. Raman spectra of CH_4 at different pressures. W, S2, S1 are librational modes, corresponding to different types of molecules, T-translational modes [1]. ν_1, ν_2, ν_3 are first order Raman-active fundamentals.

1. M.-M. Thiéry, D. Fabre, K. Kobashi, J. Chem. Phys. 76, 4817 (1982).
2. M.-M. Thiéry, D. Fabre, K. Kobashi, J. Chem. Phys. 83, 6165 (1985).
3. W. Press, J. Chem. Phys., 56, 2597 (1972).
4. W. L. Vos, J. A. Schouten, J. Appl. Phys., 1, 6744 (1991)

A HIGH PRESSURE RAMAN STUDY OF COLLISION-INDUCED SCATTERING IN BINARY LIQUID MIXTURES

V.L. Slager, H.-C. Chang, J. Jonas*

Department of Chemistry, School of Chemical Sciences, University of Illinois, Urbana, Illinois

Collision-induced scattering has been of experimental and theoretical interest [1-9], because collision-induced scattering allows direct probing of the intermolecular interactions of the molecules in dense liquids. The forbidden peaks of the Raman spectrum of carbon disulfide arise solely from collision-induced scattering, allowing the molecular interaction dynamics to be probed in a binary liquid mixture with cyclopentane. The effects of density, temperature, and amount of perturbing liquid may be analyzed separately.

In this experiment, carbon disulfide was diluted with cyclopentane. These solutions were of mole fraction (X) of 1.0 to 0.3 of carbon disulfide. The ν_2 peak of carbon disulfide was measured at temperatures 0, 20 and 40 °C and pressures from 1 bar to 5 kbar, with the pressures chosen to give the same density for solutions at different temperatures and mole fractions. The Raman scattering apparatus has been described previously [10]. The densities were measured for all temperatures and mole fractions using an Anton/Paar DMA 45 digital density meter for 1 bar, and a high pressure relative densitometer described in reference 11.

The ν_2 peak of carbon disulfide was analyzed from 17 to 57 cm^{-1} from the center on the high frequency side of the peak. The data was analyzed using a method first described by Madden and Cox [2], and used in other experiments [4,5]. The line shape of the peak can be described by an exponential for collision-induced forbidden Raman peaks, such as the ν_2 peak of carbon disulfide. The line is fit for the parameter Δ using:

$$I(\omega) = A \cdot \exp(-\omega / \Delta) \quad (1)$$

The Δ values determined in this experiment are shown in figure 1.

While there has been no theoretical work done on the line shapes of collision-induced scattering Raman peaks for binary liquid mixtures, the multipole theory has been used to describe the intensities of these peaks [3,6] and the line shape of bulk carbon disulfide collision-induced scattering peaks [1,2]. These studies support the dipole induced quadrupole mechanism as the source of the forbidden bands of carbon disulfide.

In this experiment, the Δ values were found to be insensitive to temperature and the amount of cyclopentane. This agrees with a previous experiment [7], which also found little variation in the local environmental symmetry when varying temperature or the concentration of a perturbing molecule. The Δ values that were comparable agreed with previous experiments using bulk carbon disulfide [2,4]. The average of the Δ for all temperatures and mole fractions was taken for each density. When this average is plotted (figure 2), Δ is found to be a linear function of density, except for the two highest density values for pure carbon disulfide.

Translational effects dominate the mechanism for the collision-induced scattering [8,9]. The effect of the cyclopentane is reflected only in the density dependence of the collision induced scattering.

This material is based upon work supported under a National Science Foundation Graduate Research Fellowship, and in part by the NSF under grant CHE 90-176 49.

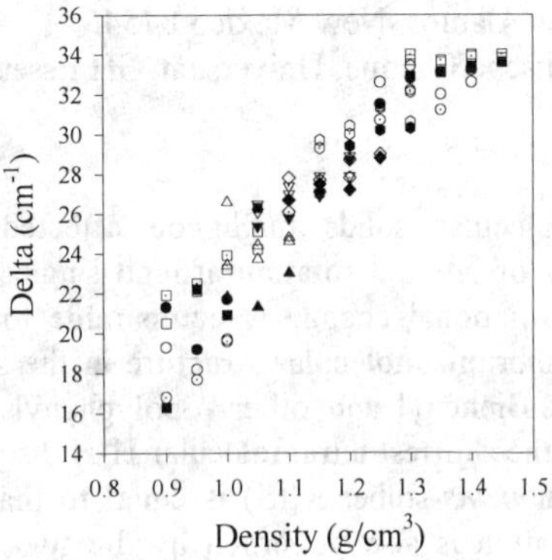

Fig. 1. Experimental line shape parameter Δ for X=0.3 to 1.0 for temperatures 0°C (filled shapes), 20°C (dotted shapes) and 40°C (outlined shapes)

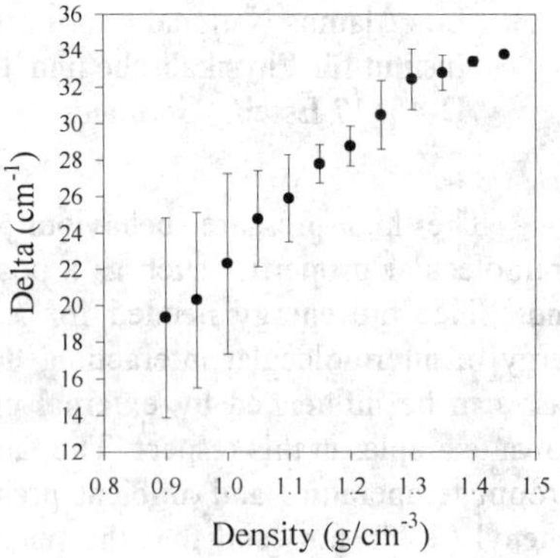

Fig. 2: Average of Δ for all mole fractions and temperatures at each density. Error bars are the standard deviation of the average.

1. T.I. Cox, and P.A. Madden, Molec. Phys. 39, 1487 (1980)
2. P.A. Madden, and T.I. Cox, Molec. Phys. 43, 287 (1981)
3. T.I. Cox, and P.A. Madden, Molec. Phys. 43, 307 (1981)
4. B. Hegemann, and J. Jonas, J. Chem. Phys. 82, 2845 (1985)
5. F. Strehle, T. Dorfmüller, and J. Samios, Molec. Phys. 72, 993 (1991)
6. E. Dayan, E. Dervil, J. Loisel, J.P. Pinan-Lucarre, and G. Tarjus, Chem. Phys. 119, 107 (1988)
7. E. Dayan, E. Dervil, J. Loisel, J.P. Pinan-Lucarre, and G. Tarjus, Molec. Phys. 76, 1457 (1992)
8. P.A. Madden, and D.J. Tildesley, Molec. Phys. 49, 193 (1983)
9. P.A. Madden, and D.J. Tildesley, Molec. Phys. 55, 969 (1985)
10. S.L. Wallen, L. Nikiel, J. Yi, and J. Jonas, J. Phys. Chem. 99, 15421 (1995)
11. J.A. Akai, Ph.D. Thesis, University of Illinois, Urbana, Illinois, 1977

NEAR-INFRARED FOURIER TRANSFORM RAMAN MICROSPECTROSCOPY AT HIGH EXTERNAL PRESSURE. CRYSTALLINE *TRANS*-STILBENE AND N- SALICYLIDENEANILINE

G. Baranović [1]*, D. S. Moore [2] and B. Schrader [3]

[1] R. Bošković Institute, P.O.Box 1016, 10001 Zagreb, Croatia
[2] Los Alamos National Laboratory, Los Alamos, New Mexico 87545
[3] Institut für Physikalische und Theoretische Chemie, Universität-GH Essen, D-45117 Essen, Germany

The high-pressure behaviour of molecular solids might be affected by intramolecular properties such as a possibility of internal rotation around single C-C bond. Since the energy needed for a conformational change is comparable to the energy of intermolecular interaction, the equilibrium molecular structure in the solid phase can be influenced by external pressure.Biphenyl and other p-polyphenyls are known examples in this respect. The fact that the shortest intramolecular H-H distance at room temperature and ambient pressure in *trans*-stilbene (tS) is equal to that of biphenyl (0.206 nm) and that the phenyl rotation is also governed by the two-well potential has prompted us to undertake high-pressure vibrational study of tS. A change in molecular conformation is also expected for various derivatives of N-benzilideneaniline which are, unlike tS and biphenyl, twisted in the crystal phase with the dihedral angle of the aniline ring equal to 40° . Semiempirical AM1 frequency calculations performed for tS predict a 1.5 % increase in the C_e=C_e stretching frequency due only to the change of the twist angle from 0° to 90°. Another geometrical factor influencing significantly the double bond stretching which might be altered by external pressure is the Ph-C=C/N bond angle.

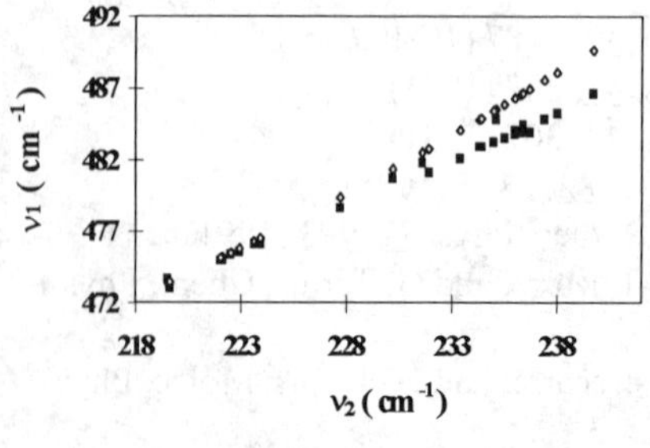

Fig. 1. By eliminating pressure from $\nu_1=\nu_1(P)$ and $\nu_2=\nu_2(P)$ [1] the curve is obtained (◊) that should be closely followed by experimental points (■).

A gasketed diamond anvil cell was used to generate high pressure. The FT Raman spectra (resolution 4 cm^{-1}) were recorded on a Bruker IFS 66 spectrometer equipped with a FRA 106 FT Raman module. The Raman microscope was coupled to the module by two fiber-optic cables. The main obstacle in using NIR FT Raman microspectroscopy at high external pressures is the lack of an internal pressure standard. One is thus forced to use a secondary standard or to perform IR microspectroscopy prior to Raman measurements. Orthorhombic sulphur is an efficient NIR Raman scatterer. The pressure dependence of its strongest Raman bands at 473 and 219 cm^{-1} from 0 to 5 GPa. [1]. was used to measure the pressure. It is evident from Fig. 1. that a satisfactory accuracy can be achieved only up to 3 GPa . Therefore the arithmentic mean of the two pressure values was used in the entire region from 0 to 5 GPa. The linear behaviour of the four anthracene bands thus reproduced supports the conclusion that the tS and N-salicylideneaniline (N-S) bands do behave linearly.

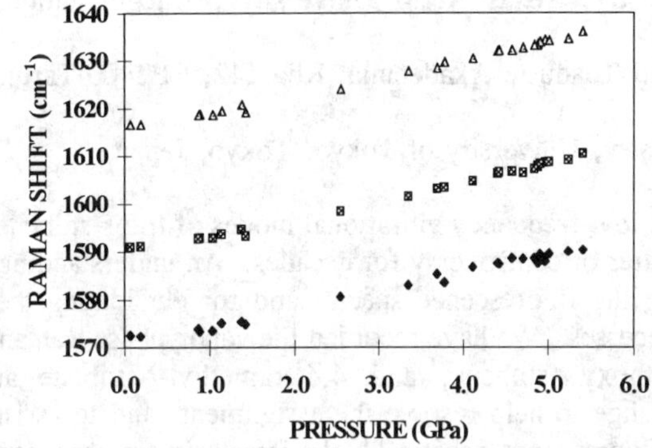

Fig. 2. The pressure dependence of internal modes in crystalline N-salicylideneaniline.

The most intense tS bands occur at 1639 cm^{-1} (C_e=C_e stretching), 1594 cm^{-1} (ring 8a), 1193 cm^{-1} (Ph-C_e stretching) and 998 cm^{-1} (ring 12) and those of N-S at 1617 cm^{-1} (C=N stretching), 1591 cm^{-1} (ring 8a), 1572 cm^{-1} (dominated by the conjugated HOCCCN region), 1457 cm^{-1} (Ph-N stretching), 1189 cm^{-1} (C-O stretching and O-H bending) and 1001 cm^{-1} (ring 12) [2]. All the bands were monitored in the entire pressure range and their frequencies change approximately 1-2 % from the ambient pressure values. No extra pressure sensitivity could be observed. Vibrational scaling law (correlation between $1/\nu_i * d\nu_i / dP$ and ν_i) [3] applies to the systems considered as well. The Raman solution band of tS at 960 cm^{-1} not seen in the crystal spectrum at ambient pressure and considered as an evidence for the molecular nonplanarity in solution is not induced by pressure changes. One may conclude that tS molecules remain planar up to 5 GPa. Contrary to expectations N-S molecules do not become planar with increasing pressure (Fig.2). Otherwise the observed C=N stretching frequency would show reduced or no pressure sensitivity, as is the case for the benzophenone C=O stretching [4] (the frequency of this mode decreases by 2.3 % when the phenyl dihedral angle is changed from 50° to 30° [5]).

1. B.Eckert, H.J.Jodl, H.O.Albert and P.Foggi, in Frontiers of High-Pressure Research, (H.D.Hochheimer and R.D.Etters, eds.), Plenum Press, New York, 1991, p. 143.
2. W. Turbeville and P.K.Dutta, J. Phys. Chem. 94, 4060 (1990).
3. R. Zallen, Phys. Rev. B, 9, 4485 (1974).
4. D.Zhang, G.Lan, S.Hu, H.Wang and J.Zheng, J. Phys. Chem Solids, 56, 27 (1995).
5. V.Volovšek, G.Baranović and L.Colombo, Spectrochim. Acta Part A, 49A, 2071 (1993).

HIGH-TEMPERATURE VAPOR-PHASE RAMAN SPECTRA OF *trans*-STILBENE AND ITS 4-METHOXY-*t*-STILBENE AND 4,4'-*t*-DIMETHYLSTILBENE DERIVATIVES

A. Del Rosario[1], K. Haller[2], W.-Y. Chiang[1], A. Sakamoto[3], M. Tasumi[3], J. Laane[1]

[1]Department of Chemistry, Texas A&M University, College Station, Texas 77843-3255
[2]Fuusika Instituut, Eesti Teaduste Akadeemia, Riia 142, EE2400 Tartu, Estonia
[3]Department of Chemistry, University of Tokyo, Tokyo, Japan

The assignments of the low-frequency vibrational modes of trans-stilbene and its derivatives have been a matter of controversy for decades. An understanding of these is essential for analyzing the fluorescence spectra and for elucidating the *trans* ↔ *cis* photoisomerization processes. We have recorded the vapor-phase Raman spectra of *trans*-stilbene, 4-methoxy-*t*-stilbene, and 4,4'-dimethyl-*t*-stilbene at temperatures in the 330-360° C range to help resolve the assignments and to assist us with the interpretation of our fluorescence spectra [1,2]. Spectra were recorded in a pyrex glass cell of the type shown in Figure 1.

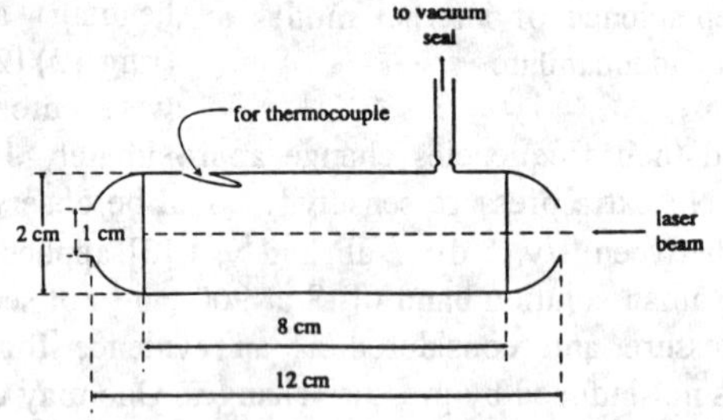

Fig. 1. Raman cell for vapors at high temperatures.

Figure 2 shows the eight low-frequency modes of *trans*-stilbene which are also present in its derivatives along with additional torsional and bending modes. Figure 3A shows the low-frequency Raman spectra (parallel and perpendicular polarization) of trans-stilbene recorded at 330° C and Figure 3B presents the full spectrum of 4,4'-dimethyl-*t*-stilbene. Table I summarizes the assignment of the low-frequency modes. In addition to these, the dimethyl compound also shows two methyl wagging vibrations at 201 cm^{-1} (out-of-plane) and 292 cm^{-1} (in-plane).

Dispersed fluorescence spectra of the dimethyl compound will be collected to complete the assignments for this species.

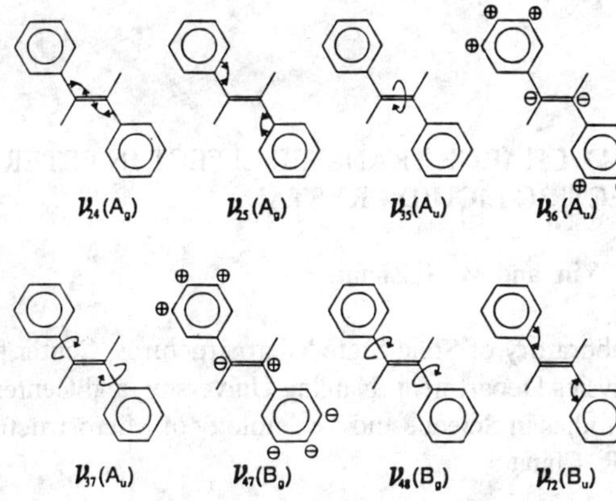

Fig. 2. Low-frequency modes of *trans*-stilbene.

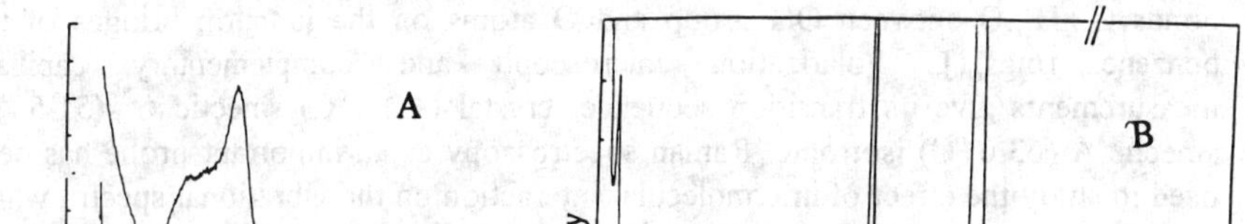

Fig. 3. Raman spectra of *t*-stilbene (A) and 4,4'-dimethyl-*t*-stilbene (B).

TABLE I. Vibrational assignments (cm^{-1}) for *trans*-stilbene, 4-methoxystilbene, and 4,4'-dimethyl-*t*-stilbene.

Description	*trans*-stilbene	4-methoxy-stilbene	4,4'-dimethyl-stilbene
υ_{24} C_e-Ph bend	273	234	247
υ_{25} phenyl wag (i.p.)	152	144	126
υ_{35} C=C torsion	101	86	-
υ_{36} phenyl flap	58	59	-
υ_{37} phenyl torsion	9	8	-
υ_{47} phenyl flap	211	179	-
υ_{48} phenyl torsion	118	123	-
υ_{72} phenyl wag (o.p.)	76	68	-

1. W.-Y. Chiang and J. Laane, J. Chem. Phys. 100, 8755 (1994)
2. W.-Y. Chiang and J. Laane, J. Phys. Chem., 99, 11823 (1995)

ELECTRICAL CHARGE TRANSFER EFFECT IN ESTER FERROELECTRIC LIQUID CRYSTAL

P. Zhang, Z. Yin, and M. S. Zhang

National Laboratory of Solid State Microstructures, Center for Materials Analysis, Physics Department, Nanjing University, and Center for Advanced Studies in Science and Technology of Microstructures, Nanjing 210093, P. R. China

Ester ferroelectric liquidcrystal (EFLC)D-4-(2-methyl-butoxy)phenyl 4-decyloxybenzoate (MBOPDOB) is of a hightend stability due to the strong hydrogen bonds O...H...O between OH group and O atoms on the junction bridges of the benzene rings[1]. Polarization microscope and complementary capillary measurements give its transition sequence: crystal (45.3 °C) smectic C* (51.5 °C) smectic A (63.6 °C) isotropic. Raman spectroscopy as an important probe has been used to study the effect of intermolecular interaction on the vibrational spectra which is strongly concerned with phase transitions in various phases in liquid crystals [2, 3]. In the low spectral region three Raman features at 15, 49 and 93 cm^{-1} display special natures: with increasing temperature, 15 and 49 cm^{-1} Raman lines vanish at smectic C* without any detectable frequency shift, at the same time they indicate thermal hysteresis during cooling and heating, and 93 cm^{-1} continuously goes through the smectic A and isotropic phase, which shows the nature of first order order-disorder type of the solid-smectic C* transition and evidence of continuous transition of Sm C*- Sm A[4]. Here we report results on temperature dependence of Raman spectra in higher spectral region where energy transfer exists between two pairs of Raman modes indeed for which we discuss microscopic dynamics process.

In the spectral region 800-1800 cm^{-1} there are four Raman lines recorded: 1166, 1178, 1710 and 1738 cm^{-1}. We have examined temperature dependence of these four lines from room temperature to isotropic phase temperature which are shown in Fig. 1(a) and (b). All the intensities are normalized to that of the 1606 cm^{-1} mode, a vibrational mode of the benzene ring skeleton, since its intensity does not show temperature dependence in the whole temperature region. In the solid phase both the 1178 and 1710 cm^{-1} Raman are very strong, and the 1166 and 1738 cm^{-1} are very weak. With increasing temperature the intensities of 1178 and 1710 cm^{-1} become weaker, those of 1166 and 1738 cm^{-1} become stronger. In the Sm C* phase the intensities of the former two lines (1178 and 1710 cm^{-1}) almost equal those of the latter two lines (1166 and 1738 cm^{-1}), respectively. The intensities of 1166 and 1738 cm^{-1} keep almost unchanged in smectic A phase, and then are dramatically increased when entering isotropic phase.

Infrared measurements show that the ester C-O-C symmetric vibration is at 1178 cm^{-1} and C=O vibration is at 1710 cm^{-1} at room temperature. We interpret above temperature behaviors of the two pair Raman modes as follows. By the action of the group, the carbonyl oxygen can drawn electrons from the other oxygen, which stiffens the C=O bond and weakens the C-O bond, thus resulting in a higher C=O frequency of 1738 cm^{-1} and a lower C-O-C frequency of 1166 cm^{-1}, respectively [5]. Enhancement of both modes at 1166 and 1738 cm^{-1} in the isotropic phase is ascribed to that the whole liquid crystal optically displays transparent system, the scattering cross section is subsequently increased. Their temperature behaviors show that the energy transfer process between these two pair modes is completed in ferroelectric smectic C^* phase. As known, in ferroelectric smectic C^* the transverse moment formed by the C=O bond results in dipolar ordering. In fact electrons could be transfered between the carbonyl oxygen and neighboring oxygen to alter the polarization due to change of the dipolar moment. This is consistent with the experimental results that both the stiffening and weakening are of remarkable temperature effects and of a short-range interaction character.

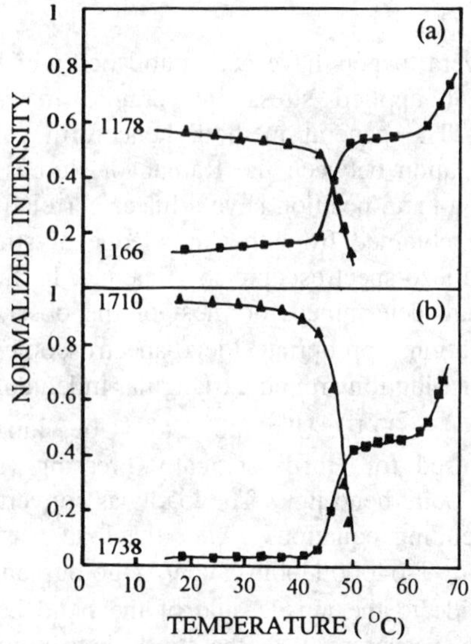

Fig. 1 Temperature behaviors of the normalized intensities of Raman modes (a) at 1178 and 1166 cm^{-1} (b) at 1710 and 1738 cm^{-1}.

1. B. I. Ostrovskii, A. Z. Rabinovich, A. S. Sonin, E. L. Sorkin, B. A. Strukov, and S. A. Taraskin, Ferroelectrics **24**, 309 (1980).
2. N. M. Amer, Y. R. Shen, and H. Rosen, Phys. Rev. Lett. **24**, 718 (1970).
3. W. J. Borer, S. S. Mitra, and C. W. Brown, Phys. Rev. Lett. **27**, 379 (1971).
4. Z. Yin, P. Zhang, Q. Chen, and M. S. Zhang, Ferroelectrics **168**, 139 (1995).
5. N. B. Colthup, L. H. Daly, and S. E. Wiberley, Introduction to Infrared and Raman Spectroscopy, Academic Press, New York, 1975, p.288.

MEASUREMENT OF THE RESIDUAL STRESS DISTRIBUTIONS IN β-Si_3N_4 USING RAMAN SPECTROSCOPY

N. Muraki[1]*, G. Katagiri[1], V. Sergo[2], G. Pezzotti[3], T. Nishida[3]

[1] Toray Research Center, Inc., Sonoyama 3-3-7, Otsu-shi, Shiga 520, Japan

[2] Materials Engineering and Applied Chemistry Department, University of Trieste, Via Valerio 2, Trieste 34127, Italy

[3] Department of Materials, Kyoto Institute of Technology, Matsugasaki, Kyoto 606, Japan

In the recent years, several papers have been published which deal with the determination of residual or applied stress in ceramic materials using spectroscopic techniques [1-3]. The general method, henceforth called Piezo-spectroscopy (PS), utilizes the relation between the Raman or fluorescence band position and the stress. Usually, a band position have a linear correlation with the stress. The residual stresses are obtained from the knowledge of stress-induced shift of the band position and the Piezo-spectroscopic coefficient. In this work, the Piezo-spectroscopic coefficients are determined for most of the observed Raman bands of β-Si_3N_4. As an application, appropriate Piezo-spectroscopic coefficient was used to determine the stress distribution around a triangular indentation.

High-purity Si_3N_4 powder (E-10, Ube Ind. Ltd., Ube, Japan) without addition of sintering aids was used for Hot-Isostatically Pressing (HIP). The sample bars were mounted on a 4-point bend jig. The load was converted to stress with the 4-point elastic beam bending equation. After the load was applied, a spectrum was recorded by Ramanor T-64000 (Jobin-Yvon/Atago bussan) every 200 μm step from the compressive side to the tensile side of the bend bar. Raman microprobe was employed for the measurement of the stress distribution around a triangular indentation. The indentation weight was 5N. The diameter of the laser spot was about 2 μm.

All the Piezo-spectroscopic coefficients of β-Si_3N_4 are summarized in Table 1. The largest absolute Piezo-spectroscopic coefficient is 2.22cm^{-1}/GPa for the band at 862cm^{-1} and this band has a relatively small standard deviation, therefore, the 862cm^{-1} band is the most advantageous mode to be used in Piezo-Spectroscopic analysis.

Profiles of the stress distribution around the indentation along the A and B lines are shown in Fig. 1. Compressive stress can be seen around the indented area and the maximum absolute compressive stress is nearly 1 GPa at about 10 μm from the indentation center. The behaviors of the compressive stress are quite similar between A and B directions, except for the region at distance 30 to 40 μm from the center. In the region from 30 to 40 μm along the B-direction, a tensile stress of about 200 MPa is detected. This is the stress which generally induces a small crack whose length is used for measuring toughness in engineering ceramics [4].

Table 1 Experimental results of stress dependence of the Raman bands of sintered β- Si_3N_4. For each band, the peak center, the width, the Piezo-spectroscopic coefficient, Π, and the standard deviation of coefficient are shown.

Center / cm^{-1}	183.23	206.08	226.75	448.92	617.62
Width / cm^{-1}	4.20	3.70	3.80	4.10	6.57
Π cm^{-1}/GPa	0.56	0.26	0.27	-0.51	-1.15
St. Dev.	0.013	0.029	0.021	0.059	0.089
	729.65	862.54	926.51	937.32	1044.91
	4.79	7.12	7.07	6.33	5.64
	-0.99	-2.22	-2.15	-2.08	-1.89
	0.067	0.051	0.076	0.088	0.054

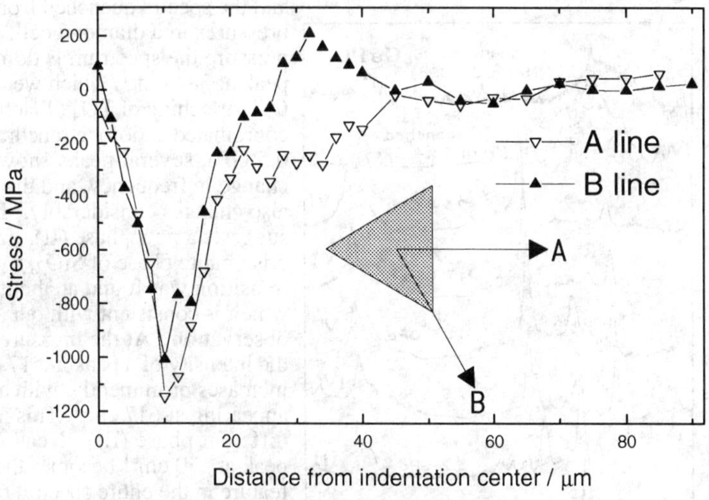

Fig. 1 Residual stress distribution around a triangular indentation: Raman microprobe measurements were performed along the A-line and B-line.

1. S.E. Molis and D.R. Clarke, J. Am. Ceram. Soc., **73**, 3189 (1990).
2. Q. Ma and D.R. Clarke, **41**, 1817 (1993).
3. Q. Ma, W. Pompe, J.D. French, and D.R. Clarke, Acta metall. mater. **42**, 1673 (1994).
4. G.R. Anstis, P. Chantikul, B.R. Lawin, and D.B. Marshall, J. Am. Ceram. Soc., **64**, 533 (1981).

High Pressure Raman Spectral Study of Low-Cristobalite Phase of $GaPO_4$: Phase Transitions and Amorphization

Zifu Wang, S. K. Sharma, A. Jayaraman, and L. C. Ming
Hawaii Institute of Geophysics and Planetology
University of Hawaii, Honolulu, HI 96822

High pressure studies of low-cristobalite phase of gallium phosphate, designated c-$GaPO_4$, have been reported recently[1, 2]. Molecular dynamics calculations show a crystal-crystal transformation at about 4 GPa [2]. A Raman study combined with x-ray investigation observed another phase transformation occurring around 15.9 GPa [1]. It is not clear yet whether there are any phase transitions between 4 and 16 GPa. In this paper, we report on a high pressure Raman study of c-$GaPO_4$ up to 30 GPa. We have observed three additional high pressure phases, and a disordered phase after quenching the sample from >23 GPa. The sample quenched from 30 GPa was found to be completely amorphous.

Figure 1 shows the Raman spectra of $GaPO_4$ recorded at high pressures and the spectra quenched from different pressures in a diamond cell. At ambient pressure the spectrum is dominated by a peak at 384 cm^{-1}, which we assign to T-O-T stretching of TO_4 (T=tetrahedrally coordinated cation) tetrahedra [3]. At 0.5 GPa, several peaks show a dramatic change in frequency, and the topology also changes considerably. This suggests a new phase (II). In cristobalite phase of SiO_2, a phase transition was found at about 1 GPa, which is consistent with our observation. At the pressure of 3.2 GPa the intensity of a peak at 517 cm^{-1} increases prominently, with a new peak appearing at 617 cm^{-1}. This indicates a different phase (III). Near 7.9 GPa, the peak at 570 cm^{-1} becomes the strongest feature in the entire spectral region. We suggest that the 570 cm^{-1} peak is associated with a new phase (IV) and with the vibrations of six-coordinated MO_6 units[4]. The peak at 428 cm^{-1}, originally appearing at 384 cm^{-1} at 1 atm., becomes weak, indicating that most of the Ga ions become six-fold coordinated. It appears that the transition between 3.2 GPa and 7.9 GPa is gradual, in which the MO_6 units grow with increasing pressure. As pressure is

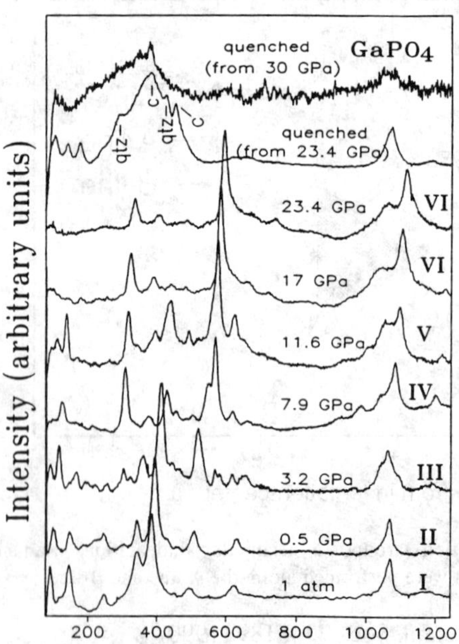

Fig. 1. Transitions in $GaPO_4$

increased to about 11.6 GPa, the two peaks in the high frequency region abruptly shift to 1093 and 1217 cm^{-1} from 1079 and 1198 cm^{-1} at 9.7 GPa. In addition, the relative intensities of peaks in the 300 and 600 cm^{-1} region also change. For this we suggest the occurrence of another phase (V) transformation around 10 GPa. Near 17 GPa, two lattice modes below 140 cm^{-1} disappear. At the same time, the peak at 436 cm^{-1} becomes very weak. A phase transition from phase V to VI in this pressure region is consistent with the observation of Robeson et al. [1], in which they found a phase transition at about 16 GPa. Up to 23.4 GPa, the sample remains in phase VI.

On releasing the pressure to 1 atmosphere, the spectrum (Fig. 1) shows a very different feature from that in the spectrum before compression. A very broad feature dominates in the range between 200 and 500 cm^{-1}, indicating strong disordering in the structure of the sample quenched from high pressure. Meanwhile, some sharp peaks in the spectrum demonstrate the coexistence of certain crystalline units in the structure. The strongest peak at 378 cm^{-1} indicates the presence of cristobalite phase of $GaPO_4$. This peak is at 384 cm^{-1} before compression. It is very interesting to note the appearance of the peaks at 292 and 423 cm^{-1}. These peaks were not present in the spectrum before compression. By comparison with the Raman spectrum of α-$GaPO_4$ [5], which is isomorphic to α-quartz, we believe that these may represent the vibrational modes of the α-$GaPO_4$ crystal phase. Therefore, in the quenched sample, two crystalline structures, i.e., c-$GaPO_4$ and α-$GaPO_4$ coexist with amorphous phase. In another experiment, we compressed the sample from 1 atm. directly to 30 GPa and quenched it back to 1 atm. (Fig. 1). The Raman spectrum of the sample quenched from 30 GPa shows a completely amorphous spectral features, which are very similar to those found in the spectrum of SiO_2 glass.

We propose a mechanism for such an unusal pressure-induced amorphization. As we know, cristobalite phase is metastable at ambient pressure and temperature, in contrast to the quartz phase. When the pressure is released to 1.5 GPa, the sample still remains in the high pressure phase (VI). On complete release of pressure, the sample tends to revert to its original cristobalite phase as revealed by the presence of the peaks at 378 and 451 cm^{-1}. At the same time, some of the sample transformed to α-$GaPO_4$ phase, as indicated by the presence of 292 and 423 cm^{-1} peaks, because the quartz type structure is more stable at ambient pressure. However, part of the sample is disordered as a result of competition between the two phases discussed above. When quenched from much higher pressure (e.g., 30 GPa), the disorder feature dominates the spectrum. This could be due to a structural change to a phase whose reversion behavior is quite different from that of phase VI.

1. J. L. Robeson, R. R. Winters, and W. S. Hammack, Phys. Rev. Lett., 73, 1644 (1994).
2. V. V. Murashov, L. S. Dubrovinsky, J. S. Tse, and Y. LePage, J. Phys., 7, 8279 (1995).
3. J. Etchepare, M. Merian, and P. Kaplan, J. Chem. Phys. 68, 1531 (1978).
4. R. J. Hemley, H. K. Mao, and E. C. T. Chao, Phys. Chem. Minerals, 13, 285 (1986).
5. M. Nakamura, H. Orihara, . Ishibashi, P. Kim, and S. Hirano, J. Phys. Soc. Japan, 59, 1831 (1990)

RAMAN SPECTROSCOPY OF PRESSURE AMORPHIZED MATERIALS

Akhilesh K. Arora

Materials Science Division, Indira Gandhi Centre for Atomic Research, Kalpakkam 603102 T.N. India.

Subsequent to the first report of pressure induced amorphizaion (PIA) in crystalline ice in 1984 [1], there has been considerable interest in understanding the mechanism and the resulting structure. By now a variety of systems with widely different types of bonding such as covalent, ionic, van der Walls and hydrogen bonding are found to exhibit this phenomenon. Though the transition is usually identified from the disappearence of sharp x-ray diffraction peaks, there are little efforts to analyse the structure of the disordered state. In this context, Raman spectroscopy has been extensively used to study the evolution of the disorder in the system. As most of the systems which exhibit PIA, have strongly bound molecules or molecular ions, growth of disorder also results in the broadening of their internal vibrational modes [2]. This is expected across amorphization because the bond lengths and bond angles now have a distribution instead of being unique. An increase in the width of Raman lines associated with the internal modes by a factor of about five is often observed across amorphization. As the amorphous state lacks periodicity, disappearence of the lattice and external vibrational modes in the Raman spectra has also been used to identify this transition.

In many compounds some of the molecular ions are often found to be misoriented in an energetically less favourable distinct orientation, e.g., 30% of the sulphate tetrahedra in $KAl(SO_4)_2.12H_2O$ point their S-O bonds towards K rather than towards Al. This orientational disorder results in the splitting of nondegenerate internal modes and appearence of more number of lines than predicted by the group theory. Recent studies on a number of binary sulphates [3-5] reveal that the high pressure crystalline phases of $LiKSO_4$ also have orientational disorder which grows with increasing pressure and eventually leads to PIA. Figure 1 shows the Raman spectra of $LiKSO_4$ in crystalline and amorphous phases. The spectra are analysed to estimate the S-O bond lengths at different pressures from the knowledge of the ν_1 internal mode frequency of the sulphate ions and the corresponding S-O bond lengths in a number of sulphates. In the three distinct orientations in $LiKSO_4$ the S-O bond lengths are thus found to be 1.446, 1.455 and 1.466 $\AA$ and in the pressure amorphized state the width of their distribution turns out to be 0.015 $\AA$.

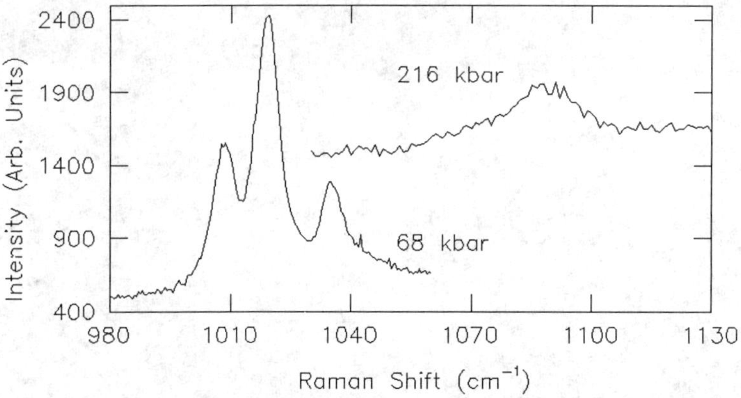

Fig. 1. Raman spectra of $LiKSO_4$ in high pressure crystalline and amorphous states.

Although both the pressure amorphized and the glassy states (obtained by quenching of melts) lack long range order, Raman spectroscopic measurements have revealed *subtle* differences between their structures in terms of short and medium range order. Early results on ice showed that pressure amorphized ice has an asymmetric distribution of O-H–O bond lengths which is broader than that of the glassy ice [6]. On the other hand fused silica, which undergoes an irreversible densification around 100 kbar, showed a decrease in the inter-tetrahedral Si-O-Si bond angles at high pressure [7]. The similarity of the Raman spectra of the pressure amorphized quartz and that of the densified silica suggests that its structure is closer to densified silica rather than to fused silica. It is found that in β-$BaZr_2F_{10}$ the pentagonal bipyramids are still intact in the amorphous state [8] and the distortions of these units are less severe than that in glassy $BaZr_2F_{10}$.

It is a pleasure to acknowledge Ms. T. Sakuntala for a fruitful collaboration.

1. O. Mishima, L.D. Calvert and E. Whalley, Nature **310**, 393 (1984).
2. A.K. Arora and T. Sakuntala, in *Recent Trends in High Pressure Research* Ed. A.K. Singh (Oxford & IBH, New Delhi, 1992) p.262.
3. A.K. Arora and T. Sakuntala, J. Phys.: Condens. Matter **4**, 8697 (1992).
4. T. Sakuntala and A.K. Arora, in *Advances in High Pressure Science & Technology* Ed. A.K. Singh (Tata McGraw-Hill, New Delhi, 1995) p.103.
5. A.K. Arora and T. Sakuntala, in *XIV Int. Conf. Raman Spectroscopy* Eds. N.-T. Yu and X.-Y. Li (John Wiley & Sons, Chichester, 1994) p.277.
6. D.D. Klug, O. Mishima and E. Whalley, J. Chem. Phys. **86**, 5323 (1987).
7. R.J. Hemley, H.K. Mao, P.M. Bell and B.O. Mysen, Phys. Rev. Lett. **57**, 747 (1986).
8. G.C. Serghiou, R.R. Winters and W.S. Hammack, J. Phys.: Condens. Matter **4**, 7617, (1992).

SECTION 25

NON-CRYSTALLINE MATERIALS

LOW-FREQUENCY RAMAN SPECTRA AND STRUCTURAL CORRELATIONS IN BINARY OXIDE GLASSES

C. Duverger[1], F. Romain[1], S. Khatir[1], M. Bouazzaoui[2] and S. Turrell[1]*

[1]Laboratoire de Spectrochimie Infrarouge et Raman (CNRS UPR A2631L), Bât. C-5 and
[2]Laboratoire de Spectroscopie Hertzienne (CNRS URA 249), Bât. P-5

Centre d'Etudes et de Recherches Lasers et Applications, Université de Lille I, 59655, Villeneuve d'Ascq Cedex (France)

Glasses in the tellurite system assume a network structure based on a three-dimensional arrangement of TeO_4 trigonal bipyramids linked in infinite chains by shared vertices. Glass formation leads to a breaking of some of the Te-O bonds thus creating TeO_3 groups. The addition of a second oxide, MO (M= Mn, Pb, Mg, Zn, for example) serves to increase the TeO_3/TeO_4 ratio.

Information on these short-order structural changes can be obtained by observing the evolution of the two components of the bands at around 700 cm^{-1}. A typical Raman spectrum of a sample in the $(1-x)TeO_2$-xMgO system ($15< x <35$) is shown in Fig 1(a). The characteristic frequencies of the Te-O vibration can be found at around 680 cm^{-1} for the TeO_3 groups and around 760 cm^{-1} for the TeO_4 groups. With an increase in the proportion of TeO_3/TeO_4, one observes a corresponding change in the relative intensities of their respective vibrational bands (1). Accompanying changes occur around 450 cm^{-1} where a band related to bending motions of the TeO_4 groups decreases in intensity with increasing concentration of the metal oxide, MO.

The accompanying medium-range spatial changes which must occur give rise to features in the low-wavenumber region of the spectrum. The broad shoulder stretching from about 100 to about 150 cm^{-1} can be correlated to external translatory, rotatory and torsional modes of the polyhedra.

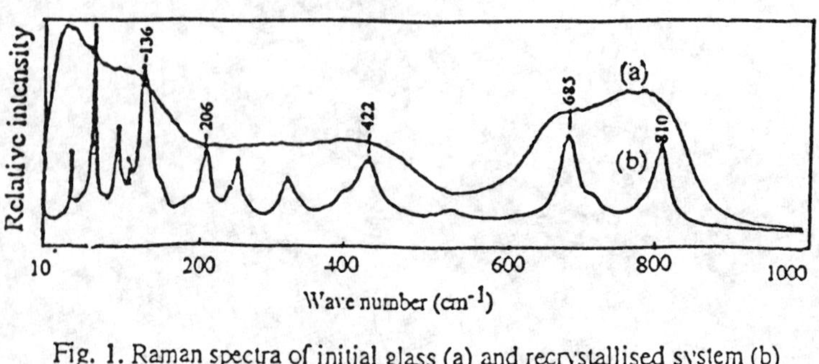

Fig. 1. Raman spectra of initial glass (a) and recrystallised system (b) obtained for the 70 TeO_2-30MgO.

The final, very intense, and lowest band observed in the spectra of glasses (observed in Fig. 1 at around 40 cm^{-1}), the Boson peak, can be related to a structural correlation, a kind of localized accumulation of disorder. This structural disorder results in a change in the previous selection rules, thus allowing the activity of acoustical modes in the Raman spectrum. The band maximum of the Boson peak is important because, according to certain models, it can be related to the distance of spatial correlation (2,3). But one might wonder whether all the features under the broad envelope of this band are indeed 'acoustic' in origin. In order to answer this question, Raman studies were performed on the glasses after they had been recrystallized. The principle is that thermal energy should enable the polyhedra to rearrange themselves into a certain order. When the system is recooled, the resulting structure should block in an arrangement which is relatively ordered but very close to that of the parent glass.

In the example of Fig. 1b, the basic spectral contour corresponds to the band envelope of the initial glass. Only two of the observed bands (114 and 68 cm^{-1}) correspond to ones characteristic of TeO_2, the others being all characteristic of the glass system. The region between 10 and 200 cm^{-1} is of particular interest. In general, for the TeO_2-MgO system, decomposition of the broad band yielded 4 components (See Figure 2). Of these, three bands corresponded to peaks observed in the spectrum of the recrystallized systems, while one was a new feature (See Table I). Though not the lowest-wavenumber component, it was always the strongest feature and dominated the band mass.Accordingly it would seem that this component, absent in the spectra of the crystalline systems, can be assigned to the Boson peak. The other components, which contribute so greatly to the broadness of the Boson peak can be actually related to bands due to external modes of the polyhedron lattice-work.

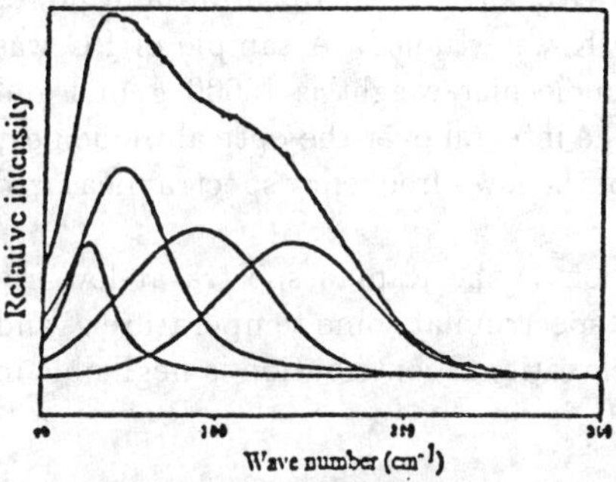

Fig. 2. decomposition of Raman spectrum of 70 TeO_2-30 MgO glass.

Decomposition for initial glass	Recrystallised system
35,2	36,2
52,1	-
-	68,1
90,5	94,4
-	114,0
-	132,7
140,7	136,5

TABLE 1. Comparaison between bands of the recrystallised system with those obtained by decomposition of glass spectrum (cm^{-1}).

The Centre d'Etudes et de Recherches Lasers et Applications (CERLA) is supported by the Ministère chargé de la Recherche, the Région Nord/Pas de Calais and the Fonds Européen de Développement Economique des Régions.

References:

1. S. Khatir, F. Romain, P. Conflant, M. Drache, B. Tanguy, J. Portier and S.Turrell, J. Mol. Structure 298 (1992) 163.
2. E. Duval, A. Boukenter and T. Achibat,J. Phys. Condens. Matter 2 (1990) 10227.
3. A. Sokolov, A. Kisluik, M. Soltwisch and D. Quitmann, Phys. Rev. Lett. 69 (10) (1992) 1540.

QUASIELASTIC LIGHT SCATTERING IN GLASSES: FREE–VOLUME APPROACH

V.N. Novikov[1]*, N.V. Surovtsev[1,2] and E. Duval[2]

[1]IA&E, Russian Academy of Sciences,
Novosibirsk, 630090, Russia
[2]Université Claude Bernard - Lyon 1,
LPCML, Bât. 205, 69622 Villeurbanne Cedex, France

Quasielastic light scattering (QLS) in glasses reveals a fast relaxation process with a characteristic time of the order of picoseconds [1, 2, 3]. The nature of QLS is still not clear. Here we present experimental data on QLS in amorphous polycarbonate (PC) and a model which connects the QLS intensity with the free-volume concentration. The latter can be determined by the positron annihilation lifetime (PAL) spectroscopy [4]. A good agreement is obtained between QLS and PAL data.

Raman spectra were measured by five-grating monochromator Z40 in right angle geometry using krypton laser light with wavelength 752.5 nm and power 330mW. A temperature range 15.6 - 450 K was studied. A sample of PC was purchased from Bayer, number average molecular weight is 15600 g/mol and glass transition temperature T_g=418K. The integral over the optical mode near 890 cm^{-1} was used for the normalization of the low - frequency spectra measured in the range 2 - 180 cm^{-1}.

We defined the quasielastic contribution to the Raman spectra at low frequencies I_{QLS} as a difference between the spectrum at some temperature T and one at very low temperature where the relaxational contribution is negligible in comparison with the vibrational one:

$$I_{QLS}(\omega, T) = I_r(\omega, T) - I_r(\omega, T = 15.6K)$$

where $I_r(\omega, T) = I(\omega, T)/(n(\omega) + 1)$ is the reduced Raman spectrum. In the spectral range of the fast relaxation, $I_{QLS}(\omega, T)$ looks like a peak at ω = 8-10 cm^{-1}, Fig.1. The width and the position of the peak are practically temperature independent while its amplitude increases with temperature as it is shown in Fig.2 by the opened circles. For the description of the QLS spectra we proposed a model which is a further development of the model [1]. In this model QLS arises due to dumping of vibrations by a fast relaxational process with a single relaxation time τ and a relaxation strength $\delta(T)$. There is no direct interaction of the light with the relaxators. At $T < T_g$, in a good approximation, $I_{QLS} \propto$

$\delta^2(T)$. We estimated the parameter δ in a free volume model; the result is $\delta^2(T) \propto V_f/V$, where V_f/V is the concentration of the free volume. The latter can be found out as a product of the ortopositronium lifetime τ_3 and intensity of the annihilation radiation I_3, $V_f/V \propto \tau_3 \cdot I_3$. Comparison of I_{QLS} with the PAL data for the free volume in PC [4] (Fig. 2) shows a good agreement with the predictions of the model.

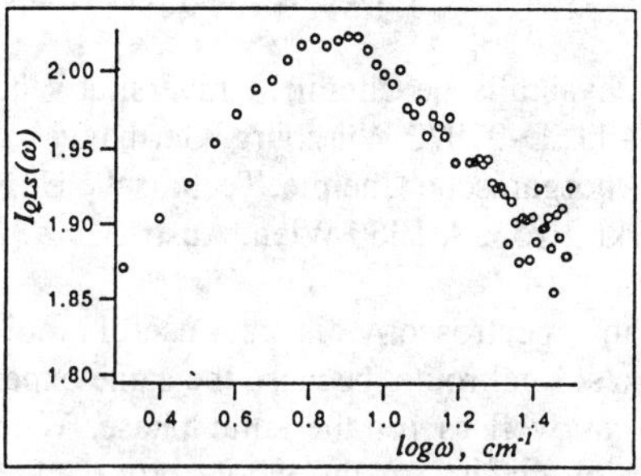

Fig. 1. Relaxational part of the Raman spectrum $I_{QLS}(\omega, T = 316K)$ in PC.

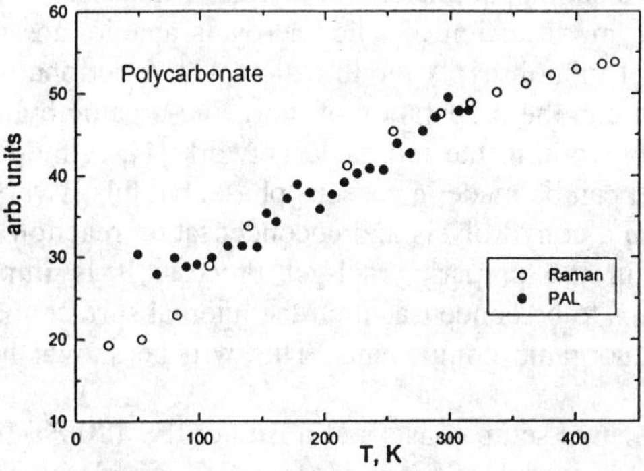

Fig. 2. Comparison of Raman and PAL data. o - intensity of QLS at 8 cm^{-1}, •- concentration of the free volume [4].

[1] V.Z. Gochiyaev, V.K. Malinovsky, V.N. Novikov, A.P. Sokolov, Phil. Mag. B63, 777 (1991).

[2] E. Rössler, A.P. Sokolov, A. Kisliuk, and D. Quitmann, Phys. Rev. B49, 14967 (1994).

[3] N.V. Surovtsev, E. Duval, T Achibat, A. Mermet, V.N. Novikov, J. Phys.: Sol. St. 7, 20223 (1995).

[4] J.Krištiak, J. Bartoš *et al*, Phys. Rev. B49, 6601 (1994).

ANALYSIS OF THE SOL-GEL PROCESS OF ORGANICALLY MODIFIED AEROGELS BY FT-IR RAMAN SPECTROSCOPY

B. Riegel[1], W. Kiefer[1]*, N. Hüsing[2], and U. Schubert[2]

[1] Institut für Physikalische Chemie, Universität Würzburg, Marcusstr. 9-11, D-97070 Würzburg, Germany
[2] Institut für Anorganische Chemie, Technische Universität Wien, Getreidemarkt 9/153, A 1060 Wien, Austria

FT-IR Raman spectroscopy is an useful tool to investigate the preparation of gels via the sol-gel route, because the same experimental setup can be used for the fluid phase as well as for the solid phase. With the non-destructive Raman technique an *in situ* analysis of the whole process is possible. The sol-gel process of silica gels and aerogels and other inorganic systems is well known. Our purpose is to understand the chemical reactions in mixed inorganic/organic systems in order to optimize the reaction parameters for aerogel synthesis.

An organic modification of silica aerogels ameliorates the mechanical properties of the aerogel monoliths. A modification with functional organic groups opens new applications, e.g. the preparation of inorganic-organic hybrid aerogels by organic cross connections within the inorganic network [1]. A modification of the readily prepared aerogel can be made in gaseous phase, but this is very difficult. The better way is to perform a cohydrolysis and cocondensation reaction of organic and inorganic compounds in the preparing sol-gel process. It is important for the functional organic groups to be condensated on the internal surface that these groups react slower than the inorganic components. This will be shown here by Raman investigations.

The experimental setup consists of a Bruker IFS 120 FT-IR spectrometer equipped with a Raman modul and a tempered cell of glass. The organically modified aerogels were prepared by NH_4OH-catalyzed hydrolysis and condensation (sol-gel process) of a tetramethoxysilane (TMOS) and R-alkyl-trimethoxysilane (R= CN, NH_2, ...) mixture, followed by supercritical drying [2]. In some cases tetraethoxysilane (TEOS) and R-alkyl-triethoxysilanes were used. The sol-gel process was started in the sample cell and Raman spectra were recorded in increasing time intervalls during the reaction (several days).

Fig. 1 shows Raman spectra for the system phenyl-trimethoxysilane and TMOS (molar ratio 20:80). Isopropanol was added as standard. The 640 cm^{-1} TMOS peak of the symmetric $Si(OCH_3)_4$-tetrahedral vibration as well as the shifted peak of the phenyl-trimethoxysilane at 656 cm^{-1} are decreasing during the reaction time . The methanol peak (CO-stretch) at 1031 cm^{-1} is increasing. Through the intensity changes of these peaks the velocity of hydrolysis reaction can be observed. It can therefore be seen in Fig. 1 that TMOS reacts faster than the phenyl-trimethoxysilane, as the 640 cm^{-1} TMOS peak has disappeared after 7 hours, whereas the shifted peak at 656 cm^{-1} of phenyl-trimethoxysilane remains much longer. The 619 cm^{-1} band of phenyl-trimethoxysilane is not affected at all. It also becomes obvious, that the 999 cm^{-1} peak of phenyl-ring breathing is not decreasing. So the ring is not destroyed during reaction and the 619 cm^{-1} peak can be attributed to a Si-(phenyl-ring) stretching vibration. By means of Raman spectroscopy it has been shown, that the hydrolysis of the organic compound is slower than that of the inorganic compound.

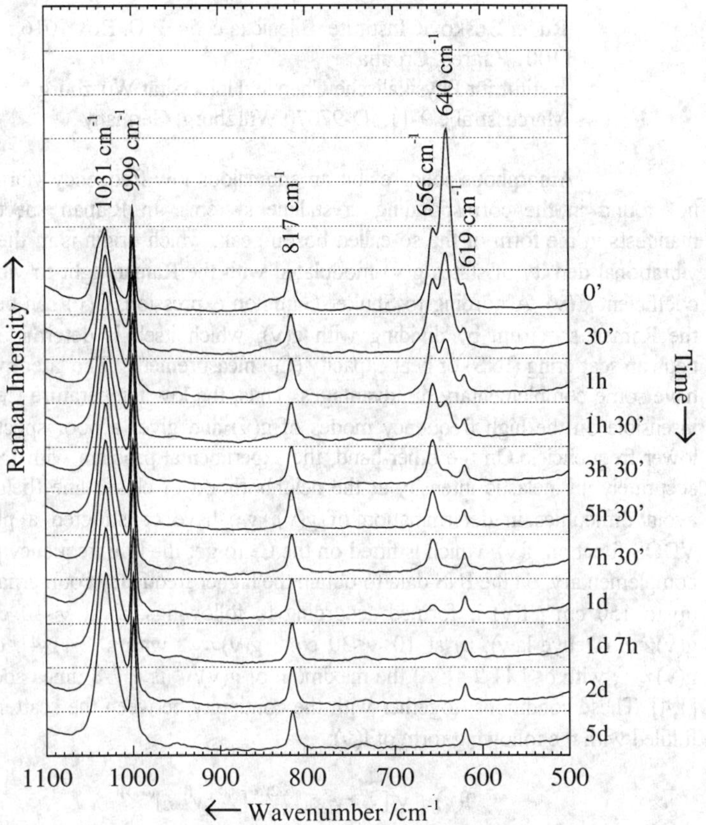

Fig. 1: Raman spectra of the sol-gel reaction of a TMOS:Phenyltrimethoxysilane 80:20 mixture at 50°C.

Financial support from the Deutsche Forschungsgemeinschaft is highly acknowledged.

1. U. Schubert, F. Schwertfeger, N. Hüsing, E. Seyfried, Mat. Res. Soc. Symp. Proc., 346 (1994) 151.
2. N. Hüsing, F. Schwertfeger, W. Tappert, U. Schubert, J. Non-Cryst. Solids 186 (1995) 37-43.

RAMAN LIGHT TO VIBRATIONS COUPLING COEFFICIENT OF VITREOUS SILICA

M. Ivanda[1] and W. Kiefer[2]

[1]Ruđer Bošković Institute, Bijenička c. 54, P.O. Box 1016, 11001 Zagreb, Croatia
[2]Institut für Physikalische Chemie, Universität Würzburg, Marcusstraße 9-11, D-97070 Würzburg, Germany

Amorphous solids exibit an anomalous low frequency vibrational behaviour not found in the corresponding crystalline systems. In Raman spectra the anomaly manifests in the form of the so-called boson peak, which origin is in the "excess" of the vibrational density of states g(ν) modulated with the Raman light-to-vibrations coupling coefficient C(ν). According to Shuker-Gammon expression, C(ν) can be evaluated from the Raman spectrum by dividing with g(ν), which itself is determined from inelastic neutron scattering (INS) or heat capacity (C_P) measurements. The later methods, however have some complementary disadvantages. Thus, the low temperature C_P measurement is insensible on the high frequency modes of g(ν) and gives a poor spectral resolution at lower frequencies. On the other hand, the experimental problem with INS is to measure accurately an inelastic intensity at the neighborhood of elastic line (below 50 cm^{-1}). To avoid difficulties in determination of g(ν), we have constructed a phenomenological VDOS function, f(ν), which is fitted on the C_P to get the low frequency part of g(ν) and, complementary, on the INS data to obtain the higher frequency counterpart. For the range up to 150 cm^{-1}, f(ν) is formed according to followings: a) at $\nu<10$ cm^{-1} we suppose $g(\nu)\propto\nu^2$ (Debye law); b) at $10<\nu<30$ cm^{-1}, $g(\nu)\propto\nu^x$, with $x>2$ [1-4]; c) at $\nu>60$ cm^{-1}, $g(\nu)\propto\nu^\delta$, with $\delta<1$ [1,2,4]; d) the maximum of $g(\nu)/\nu^2$ is ≈ 7-8 times above Debye value [1,4]. These conditions, together with the continuity between the scattering regimes, are fulfiled with the following form of f(ν):

$$f(\nu) \sim \nu^2[\nu^\alpha+\nu^\alpha_{co1}]^{(x-2)/\alpha}\,[\nu^\beta+\nu^\beta_{co2}]^{(\delta-x)/\beta} \qquad (1)$$

where ν_{co1} and ν_{co2} are the crossover frequencies from a) to b) and b) to c) scattering regimes respectively, α and β are the exponents which describe the behaviour of f(ν) in the crossover ranges. The best fit is obtained with: $\alpha=4$, $\beta=3$, $\nu_{co1}=6.5$ cm^{-1}, $\nu_{co2}=30$ cm^{-1}, $x=3.88$, $\delta=0.1$. The resulted f(ν) presented on Fig.1 shows a good agreement with the INS data of Buchenau et al. [1], and the phenomenological VDOS of Ahmad [4]. The disagrement with the INS data of Malinovsky et al. [2] at lower frequencies could be due to overestimated contributions of elastic neutron scattering and/or multiphonon scattering which lead to the lower values in g(ν) due to subtraction procedure. Fig. 2 shows C(ν) evaluated from the HV polarised and temperature reduced Raman spectrum I^R by relation

$C(\nu)=\nu I^R/[n(\nu)+1]$. The observed deviation below 20 cm^{-1} is due to quasielastic light scattering [3]. The approximately linear behaviour at frequencies from 20 to 100 cm^{-1} is in agreement with recent observation on other amorphous solids [3]. Inset of Fig. 2 shows I^R around the maximum of the boson peak together with $C(\nu)$ and $g(\nu)$ curves normalized with the same weight factor $\nu^{-0.5}$. This shows that the maximum of the boson peak is due to crossover frequency of $C(\nu)$ at 100 cm^{-1}. The crossover frequency of $g(\nu)$ at 30 cm^{-1} (maximum in $g(\nu)/\nu^2$ function) only alter the slope of $I^R(\nu)$ around 50 cm^{-1}. This completely changes the explanation of the maximum of the boson peak, which is before interpreted as a peak in $g(\nu)/\nu^2$ function [2].

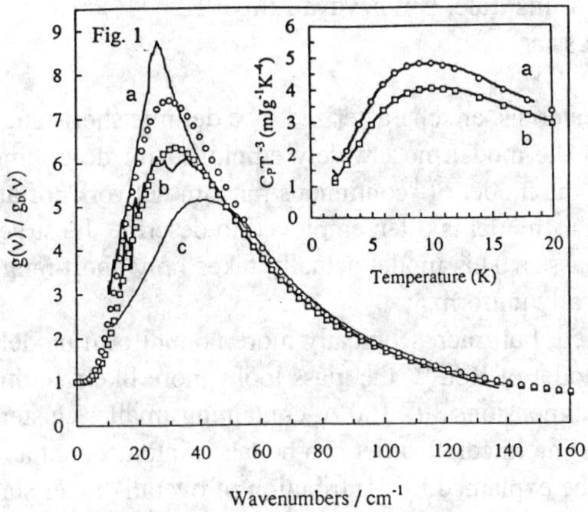

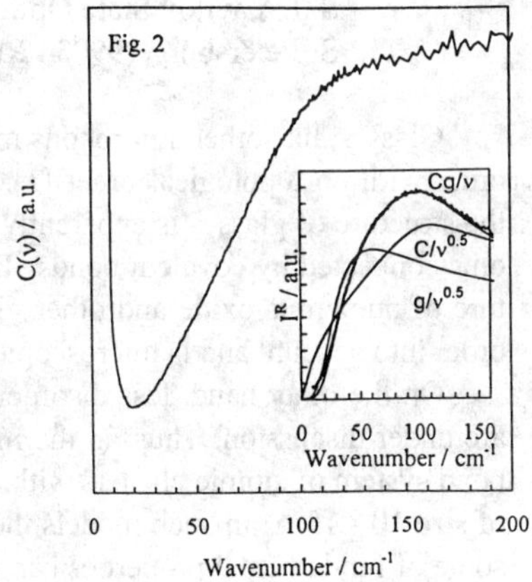

Fig. 1. VDOS of v-SiO_2 normalised on Debye value: a - phenomenological data for Spectrosil-B [4], b - INS data for Heralux [1], c - INS data for vitrous silica [2]. Our fitted function $f(\nu)/\nu^2$: O - for Spectrosil-B, □ - for Heralux. Inset - the corresponding simbols show the results of $f(\nu)$-fit on heat capacities from Ref. [5] and [1]: curves a and b, respectively.

Fig. 2. $C(\nu)$ of v-SiO_2. Inset - $I^R(\nu)$ with $C(\nu)$ and $g(\nu)$ curves normalised with $\nu^{-0.5}$.

This work is supported by Croatian Ministry of Science and Technology under a contract No. 1-03-066. M.I. is grateful to the Alexander von Humbolt Foundation for a research fellowship.

1. U. Buchenau, M. Prager, N. Nücker, A. J. Dianoux, N. Ahmad and W. A. Philips, Phys. Rev. B 34, 5665 (1986).
2. V. K. Malinovsky, V. N. Novikov, P. P. Parshin, A. P. Sokolov and M. G.Zemljanov, Europhys. Lett. 11, 43 (1990).
3. A. P. Sokolov, A. Kisliuk, D. Quitmann, E. Duval, Phys. Rev. B 48, 7692 (1993).
4. N. Ahmad, Phys. Rev. B 48, 13512 (1993).
5. Malinovsky and A. P. Sokolov, Solid State Commun. 57, 757 (1986).

MEDIUM-RANGE ORDER IN GLASSES: FLUCTUATIONS OF THE ORDER AND CHEMICAL COMPOSITION IN LITHIUM ALUMINOSILICATE GLASSES DOPED WITH TITANIUM DIOXIDE

T.I. Chuvaeva, A.V. Baranov, V.I. Petrov*, M.Ya. Tsenter, A.A. Zhilin

S. I. Vavilov State Optical Institute, Birzhevaya Liniya 12,
St.Petersburg 199034 Russia

Glasses, like other amorphous materials, are characterized by a definite short-range order with no topological order [1]. The model most widely employed for describing the structure of glasses is apparently the model of "continuous random network" of atoms connected by covalent bonds. This model is often employed to describe the structure of numerous oxide and other glasses. This model actually takes only short-range order into account and is microscopically uniform.

On the other hand, less disordered, but microscopically more nonuniform models are under discussion. Thus, in the model of Ref. 2 the glass looks more like a multibond system of atomic clusters with dimensions 30 - 100 Å, containing smaller clusters of size 10 - 15 Å. In such models the intermediate order can be taken into account and some of fundamental properties can be explained by contribution of overall "inner surface" of complicated system of clusters. Another model of glass, essentially a thermodynamic model [3], attributes the properties of the glass to the existence of "free volume" largely in the form of voids with size about 1.5 - 3.0 Å, near which significant motions of atoms occur. A model of a medium-range order for oxide glasses like SiO_2 has recently been proposed which related medium range order to short-range ordering of microscopic low-density regions around cation-centered clusters [4]. This model demonstrates the importance low-density regions in a glass structure. The properties of glasses are in many ways anomalous both at high and low temperatures in sense that they have no analogs in standard crystals.

To explain above mentioned observations a unified theoretical approach have been developed which is called the soft configuration model [5]. This model is able to describe the low-temperature properties of glasses and their correlations with medium range order and predicts in a natural manner anomalous properties of low-temperature heat capacity and phonon thermal conductivity as well as low-energy peak for inelastic scattering of neutrons corresponding to the quasilocal anharmonic excitations whose density of states has a low-energy maximum 10-100 cm^{-1}. This peak may be also associated with well known “Boson maximum” in IR and Raman scattering observed in non-metallic glasses.

In this work the nature of “Boson maximum” has been investigated for lithium aluminosilicate glasses doped TiO_2 (C = 4 - 10 mole %) with using low-frequency

Raman spectroscopy technique. These glasses under a secondary heat treatment with temperature 600-750^0 C exhibit a phase decay with formation of very small crystalline aluminotitanate particles with sizes about 20 Å or more, which could be determined by small-angle x-ray scattering (SAXS). The Raman spectra of initial glasses without secondary heat treatment display a broad asymmetric low-frequency peak with frequency 96 cm^{-1}. The former studies show that this peak is connected with elastic vibrations of inhomogeneity regions enriched Al_2O_3 and TiO_2 with sizes of 10-15 Å [6, 7] which can appear in multicomponent glasses during heating and cooling of the melt. These regions are caused by structural and chemical fluctuations of composition. The spectrum of elastic vibrations for a spherical shape of the inhomogeneity regions and how they are affected by the surrounding matrix was studied theoretically at Ref. 8. It was shown that it is a set of spherical and torsional modes with different values of angular momentum that depends on the longitudinal and transverse velocities of sound and frequencies proportional to reciprocal sizes of particles. SAXS data display no phase decay for these glasses.

First stages of a secondary heating lead to slow transformation of low-frequency peak in Raman spectrum. We observed slow diminish of frequency and narrowing of the band. On the final stage of heating we detected only spherical and torsional vibrations of nanocrystals of pseudobrucite and anatase with frequencies 60 and 35 cm^{-1}, respectively [9]. At this stage nanocrystals with size about 19 Å are easily detected by SAXS. It is important to emphasize that the Raman spectrum of final glasses does not contain in the low-frequency region any features ("Boson maximum") besides the bands of elastic vibrations of nanocrystals. Estimates of their sizes from Raman measurements coincide with SAXS data a fairly well. Therefore this glass is somewhat unprecendent amorphous material without medium-range order.

So, our data prove that "Boson maximum" in glasses is connected with elastic vibrations of structural and chemical inhomogeneities regions with sizes of few nm and does not connected with soft configuration of atom groups.

On the other hand, low temperature measurements of heat capacity have not been performed yet for this kind of glasses. We expect that the host matrix of such glasses would lose the anomalous temperature dependence of heat capacity and has almost "normal" (crystal-like) one.

1. R. Zallen, Physics of amorphous solids, J. Willey, New York, 1983.
2. J. Phillips, Solid State Physics , edited by S.H. Ehrenreich and A. Turnbull, Academic Press, New York, 1983, vol. 27.
3. M.H. Cohen, G.S Grest, Phys. Rev. B, 20, 1077 (1979).
4. S.R. Elliot, Phys. Rev. Lett., 67, 711 (1991).
5. M.I. Klinger, Phys. Rep., 94, 183 (1983).
6. Ya.S. Bobovich, V.I. Petrov, M.Ya. Tsenter, Opt. Spectrosc., 66, 462 (1990).
7. V.I. Petrov, Ya.S. Bobovich, A.A. Zhilin, M.Ya. Tsenter, T.I. Chuvaeva, Opt. Spectrosc., 68, 216 (1990).
8. A. Tamura, K. Higeta, T. Ichinokawa, J.Phys. C, 15, 4975 (1982).
9. V.I. Petrov, Ya.S. Bobovich, Opt. Spectrosc., 67, 363 (1989).

SECTION 26

INDUSTRIAL APPLICATIONS

RAMAN SPECTROSCOPY FROM THE LABORATORY TO THE PROCESS: ORIENTATION STUDIES OF POLYMERIC FIBERS UNDER DRAW

FIXED PATTERN NOISE IN CCD DETECTION

Bruce Chase

Central Research Department
DuPont Experimental Station
Wilmington, DE 19880-0328

The development of the scientific grade Charge Coupled Device (CCD) Detector has clearly revolutionized Raman spectroscopy. The high sensitivity and multichannel nature of a CCD based spectrometer when used with holographic filters and single monochromators has dramatically extended the range of accessible samples. When lasers operating in the red or near infrared are used, the problem of fluorescence is often minimized, and due to the high quantum efficiency of the CCD, good results can still be obtained. However, there is still a loss in scattering cross-section to be dealt with. In some cases, it is advantageous to operate with a visible laser and attempt to deal with the fluorescence background. By recording a large number of consecutive scans at short integration times and averaging them, one can hope to obtain a low noise spectrum, which can then be baseline corrected to remove the effects of the fluorescence. Of course, the shot noise associated with the large background can not be removed, but the averaging process should reduce this contribution in proportion to the square root of the time. In our attempts to obtain on-line Raman spectra of polymeric fibers, a diode laser operating at 532 nanometers was used. When exciting in the visible, the fluorescence background is often considerable for many polymeric fibers.

Raman spectra were obtained with a Kaiser Holoprobe® spectrometer equipped with an EEV 1511 CCD chip installed in a Photometrics camera. The CCD was operated at -40 C. The fibers were illuminated using a Holoprobe® remote fiber optic probe and a Coherent DPSS-532 laser. Under these conditions, polyethylene terephthalate fibers exhibited considerable fluorescence background when examined in a fiber spinning operation. This fluorescence was observed to quench out via a photobleaching mechanism if the fibers were made stationary. A typical Raman spectrum is shown in figure 1. The insert shows the spectra obtained in two successive experiments on an expanded scale. It is quite clear that the "noise" is completely reproducible, and as such, cannot be averaged out. It represents a coherent noise source and limits the available S/N. This fixed pattern noise arises from the variation in quantum efficiency from pixel to pixel in the CCD. The obvious approach to elimination of this noise source is to correct the spectrum with a flat field spectrum. This is ordinarily done using a white light source recorded at high S/N. Unfortunately in this experimental setup we found that simply illuminating the sample with a white light resulted in a different fixed pattern noise, which was itself reproducible. When recording spectra through a fiber optic probe and binning a column of pixels in the detection process, the nature of the fixed pattern noise is a function of both variation in QE and intensity distribution across the binned column. Any variation in illumination geometry can change the effective illumination across the CCD. The answer to this problem lies in creating a white light source which has exactly the same illumination geometry as the sample in the Raman experiment.

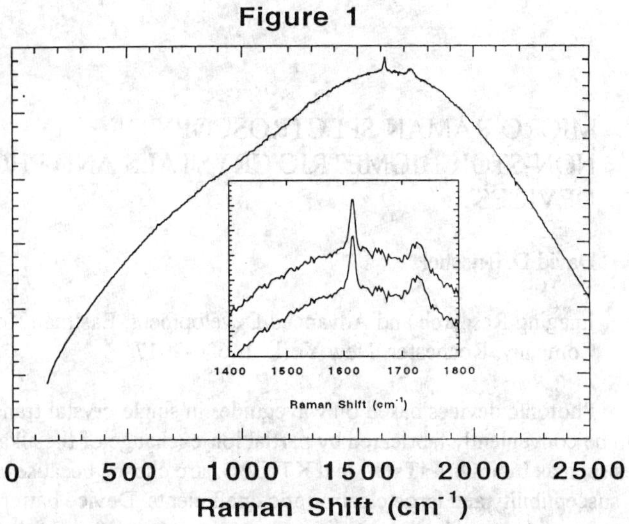

We have found that a suitably fluorescent sample inserted at the sampling point can produce such a white light spectrum. When using excitation at 532 nanometers, a very strong fluorescent signal which does not photobleach can be obtained from the yellow paper used to package Kodak lens paper. This fluorescent signal is extremely strong, can be obtained with less than 1 milliwatt of power, and shows no photobleaching. By recording a spectrum of this fluorescent card, and then fitting the continuum spectrum to a smooth polynomial function, one can obtain the relative pixel response curve. This response curve can then be used to correct succeeding spectra. Results are shown in figure 2 where the upper trace is before pixel response correction. There is a clear reduction in the amplitude of the fixed pattern noise. The correction obtained in this fashion allows signal averaging to proceed to a higher level, but there is still incomplete removal of all the fixed pattern noise.

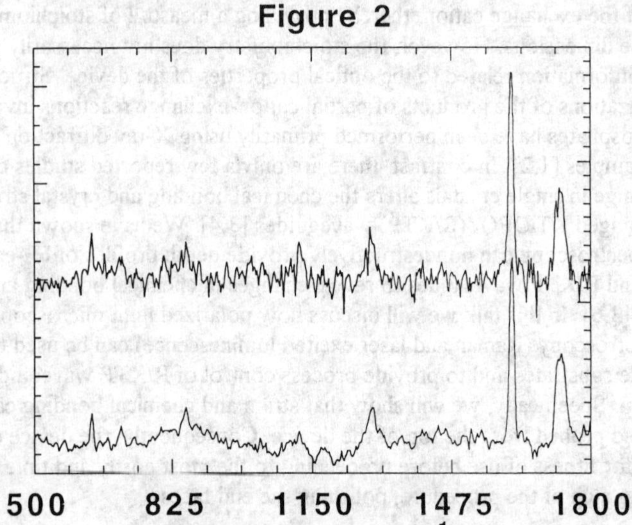

MICRO-RAMAN SPECTROSCOPY OF NON-STOICHIOMETRIC CRYSTALS AND PHOTONIC DEVICES

David D. Tuschel

Imaging Research and Advanced Development, Eastman Kodak Company, Rochester, New York 14650-2017

Photonic devices based on waveguides in single-crystal transition metal oxides can be conveniently fabricated by partial ion-exchange of the alkali cation. Substrates such as $LiNbO_3$, $LiTaO_3$, and $KTiOPO_4$ are chosen because of their high nonlinear susceptibility and large electrooptic coefficients. Device patterns are created by photolithographically depositing a mask on the surface of the crystal. Channel waveguides are then fabricated by carrying out a partial ion-exchange of H^+ for Li^+ in $LiNbO_3$ and $LiTaO_3$ or Rb^+ for K^+ in $KTiOPO_4$. A primary effect of the ion-exchange reaction is to increase the refractive index sufficient to achieve laser beam confinement; i.e., form a waveguide. However, in forming a waveguide one does not want to lose the high optical nonlinearity of the substrate, the very reason for having chosen the material. Therefore, there is a need to understand the effect of ion-exchange on the chemical bonding and crystal structure of photonic device substrates in order to design fabrication conditions that produce waveguides with optimal performance.

An analytical technique such as dynamic SIMS can generate depth profiles of the exchange cation, thereby providing a measure of stoichiometry and waveguide dimensions. However, the stoichiometry does not necessarily provide chemical information related to the optical properties of the device. Structural characterizations of the products of partial cation-exchange reactions involving titanyl phosphates have been performed primarily using X-ray diffraction on powder samples [1,2]. In contrast, there are only a few reported studies on how ion-exchange in single crystals alters the chemical bonding and crystal structure of Rb^+-exchanged $KTiOPO_4$ (R/KTP) waveguides [3,4]. We have shown that micro-Raman spectroscopy can nondestructively provide depth profiles of H^+-exchanged $LiNbO_3$ and R/KTP waveguides to reveal changes in chemical bonding and crystal structure [3,6]. In this talk we will discuss how polarized light microscopy and microspectroscopy (Raman and laser excited luminescence) can be used to evaluate waveguide substrates and to provide process control of R/KTP waveguide fabrication. Specifically, we will show that strain and chemical bonding can be imaged and probed from the top of the device. Consequently, the device can be analyzed for fitness of use before proceeding to the most costly and time consuming step of the procedure, polishing the end faces.

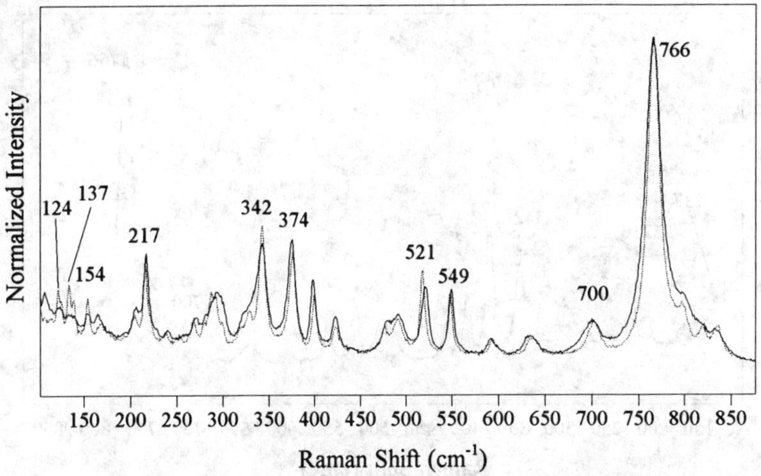

Fig. 1. Micro-Raman spectra of the center of an R/KTP test pattern (solid) and an unexchanged KTP region (dotted) from device U941123.14. The value of S_{342}/S_{375} for the R/KTP center is 0.963 whereas that for the KTP is 1.28. Band positions are for the R/KTP spectrum.

Segmented R/KTP waveguides positioned parallel to the Y-axis were prepared by first depositing an Al mask on the Z-face of a KTP crystal. Channel waveguides were then fabricated by carrying out a partial ion-exchange of K^+ in a melt containing $RbNO_3$ (92% molar) and $Ba(NO_3)_2$ (8%). Micro-Raman spectra were collected by focusing a 488.0 nm laser beam with a 0.95 NA microscope objective at various positions on the Z-face of the device. The experimental arrangement was $Z(X, XY)\overline{Z}$. A pinhole aperture was positioned in front of the spectrometer entrance slit to enhance axial spatial resolution. The aperture is essential for excluding the signal from the substrate KTP when probing the shallow R/KTP waveguides from the top.The high spatial (lateral and axial) resolution of micro-Raman spectroscopy makes it particularly well suited for the characterization of photonic devices. Furthermore, in the case of ion-exchange devices, reference spectra are obtained from the device being probed; i.e., those areas not subjected to ion-exchange. Therefore, absent physical defects such as surface roughness, differences in micro-Raman spectra obtained from different locations on the device can be interpreted in terms of chemical and crystal structure.

KTP is orthorhombic and belongs to the crystal class C_{2V}. The structure consists of TiO_6 octahedra linked at two corners to form helical chains which are separated by PO_4 tetrahedra. The Raman spectrum obtained from any face of a KTP crystal consists primarily of bands caused by the vibrational motions of the TiO_6 octahedra and PO_4 tetrahedra. Raman bands below 200 cm^{-1} are

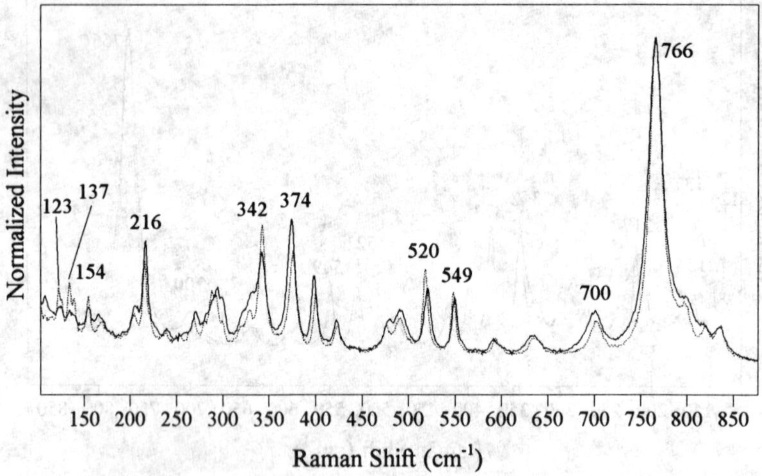

Fig. 2. Micro-Raman spectra of the corner of an R/KTP test pattern (solid) and an unexchanged KTP region (dotted) from device U941123.14. The value of S_{342}/S_{375} for the R/KTP corner is 0.715 whereas that for the KTP is 1.28. Band positions are for the R/KTP spectrum.

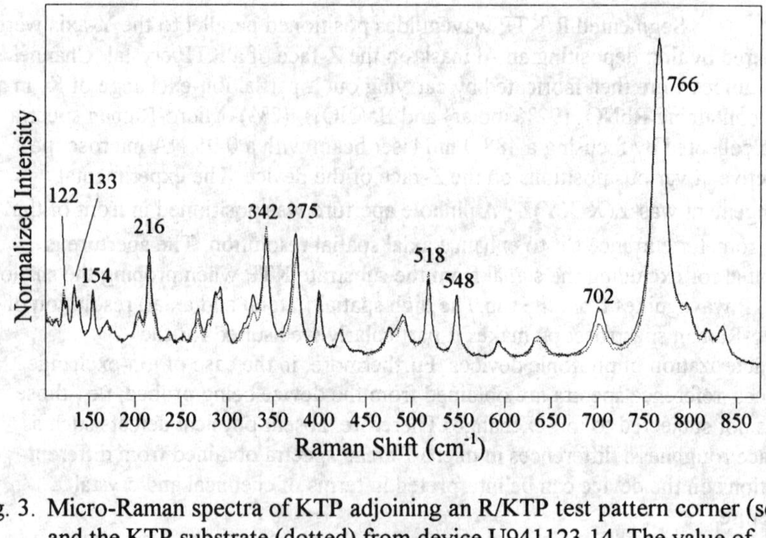

Fig. 3. Micro-Raman spectra of KTP adjoining an R/KTP test pattern corner (solid) and the KTP substrate (dotted) from device U941123.14. The value of S_{342}/S_{375} for the KTP adjoining the R/KTP is 0.961 whereas that for the KTP substrate is 1.28. Band positions are for the KTP spectrum adjoining the R/KTP.

assigned to alkali cation translational modes, and therefore manifest the most significant changes subsequent to Rb^+-exchange. Bands in the 200-400 cm^{-1} and 600-850 cm^{-1} region are attributed to TiO_6 octahedral torsional and stretching modes, respectively. Bands in the 900-1200 cm^{-1} region are associated with the PO_4 tetrahedra. The spectra of R/KTP and KTP shown in Figs. 1 through 3 illustrate how micro-Raman spectroscopy can be used to probe the structural (chemical and crystal) changes induced by Rb^+-exchange.

Micro-Raman spectra of the center of an R/KTP test pattern and an unexchanged KTP region from a channel waveguide device are shown in Fig. 1. Rb^+-exchange causes an attenuation of KTP bands below 200 cm^{-1} due to alkali cation translational modes and the emergence of a new band at 107 cm^{-1}. Also, the bands arising from TiO_6 octahedral motions, those between 200 and 850 cm^{-1}, are affected. Some bands shift and broaden and there are changes in relative band strength. In particular, the relative scattering strength of the 342 and 375 cm^{-1} bands (S_{342}/S_{375}) are reversed. The value of S_{342}/S_{375} for the R/KTP center is 0.963 whereas that for the KTP is 1.28. Clearly, changes in the refractive index of R/KTP arise from alkali cation substitution and changes in the TiO_6 sublattice. When viewing the Z-face of the device under crossed polarization with an optical microscope, many of the waveguide end segments and corners of test patterns appear bright, presumably due to strain birefringence. The Micro-Raman spectrum of the test pattern corner in Fig. 2 confirms the structural difference revealed by polarized light microscopy. The spectral differences between the R/KTP corner and KTP substrate are even more pronounced and the R/KTP value of S_{342}/S_{375} at 0.715 is significantly lower than that of the test pattern center. Perhaps most interesting of all is the spectrum in Fig. 3 of the KTP substrate adjoining an R/KTP test pattern corner. Note that there is no significant difference between the bands below 200 cm^{-1}, thereby indicating the absence of Rb^+-exchange. However, there are differences in position and relative strength for bands associated with the TiO_6 octahedra. Specifically, the adjoining KTP value of S_{342}/S_{375} is 0.961, comparable to that detected in the center of the R/KTP test pattern.

Effects similar to those described here have been observed in R/KTP waveguides. In our continuing work on photonic devices, we are trying to determine if strain in the waveguide end segment, as observed by polarized light microscopy and micro-Raman spectroscopy, can be correlated to the waveguide coupling efficiency.

1. M.E. Hagerman and K.R. Poeppelmeier, Chem. Mater., 7, 602 (1995).
2. M.L.F. Phillips, W.T.A. Harrison, G.D. Stucky, E.M. McCarron III, J.C. Calabrese, and T.E. Gier, Chem. Mater., 4, 222 (1992).
3. D.D. Tuschel, G.R. Paz-Pujalt, and W.P. Risk, Appl. Phys. Lett., 66, 1035 (1995).
4. I. Savatinova, S. Tonchev, T. Popov, E. Liarokapis, and C.C. Ziling, J. Phys. D: Appl. Phys., 27, 1384 (1994).
5. D.D. Tuschel and G.R. Paz-Pujalt, Ferroelectrics, 151, 85 (1994).

Theoretical and Experimental Evaluation of Fiber Optic Raman Probes and Diode Lasers for Industrial Applications

S. M. Angel,* Thomas F. Cooney, and H. Trey Skinner

Department of Chemistry & Biochemistry, The University of South Carolina Columbia, SC 29208, USA

THEORETICAL MODEL OF FIBER-OPTIC RAMAN PROBES

In order to obtain semiquantitative chemical information from fiber-optic Raman probes, a framework for discussion of analytical chemical issues—such as sensitivity, signal-to-noise, scattering volume, and spatial resolution—is required. The theoretical analysis of Raman probe performance for non-absorbing, non-scattering samples is based upon straightforward geometric considerations and was originally described by Schwab and McCreery [1] and Plaza et al [2]. Recently, we have re-examined this analysis to specifically include scattering volume and experimentally-determined laser irradiance, and have extended it to some more-recently-described probe designs [3].

In all probe designs considered here, one optical fiber is used to deliver the laser light to the sample and one or more additional optical fibers are used to collect scattered light from the sample. At the sample end, the fibers can be flat or beveled, or else lenses can be used to physically isolate the probe tip from the sample. In some cases, optical filters are incorporated at the probe end to minimize optical fiber silica Raman background which can interfere with analyte Raman scattering. We have developed models describing the sensitivity and sampling volume of various probes, and have compared predicted probe responses to the experimental measurements [3]. The models incorporate fiber numerical aperture, overlap geometry of illumination and excitation light cones, fiber radii, refractive indices of fiber and immersion medium, separation between excitation and collection fibers, number of fibers, and fiber bevel angle and/or lens focal length.

There is good agreement among experimental Raman sensitivity measurements and the modeled Raman sensitivity curves (signal per unit distance from the probe tip). The model accurately predicts the relative positions of sensitivity maxima of the various probes and the actual positions of the maxima to within approximately 15%. Also, the shapes of the measured profiles are very similar to, although slightly broader than, those predicted. The most interesting aspects of the fiber-optic Raman probe measurements on solid samples are the differences between the intensity profile shape (total Raman signal versus sample distance) for sample Raman signal and fiber Raman background. For non-lensed probes, the fiber-Raman background assumes an intensity profile (versus probe tip distance) resembling model-derived sensitivity profiles or white-light backscatter measurements whereas the sample Raman intensity assumes a profile resembling the integrated sensitivity profile. The fact that fiber-Raman background closely follows the sensitivity profiles shows that the primary locus of scattering is the sample surface. Yet, the fact that the sample Raman intensity profiles closely follow the integrated form of the sensitivity profiles shows that the bulk of the sample Raman signal originates from a volume below the surface.

A description of the model will be presented along with comparisons between predicted and experimentally measured probe sensitivities.

EXTERNAL CAVITY DIODE LASERS FOR RAMAN SPECTROSCOPY

Because of slight temperature and injection current fluctuations and perturbations, the output wavelength of a non-stabilized diode laser can discontinuously "hop" between modes when a different mode corresponds to the center of the gain profile. As the temperature of a single mode laser is slowly varied, the mode will first be observed to shift continuously. At some point, however, there will be a "mode hop" to an adjacent longitudinal mode caused by the different rates at which the bandgap energy distribution (gain curve) and modes shift with temperature. Thus, although the diode laser can be tuned within a range of approximately 20-30 nm, the tuning is not continuous and is punctuated by a series of gaps of approximately 0.3 nm width. The tuning gaps are not a serious concern for Raman applications although the related phenomenon of unwanted mode-hopping is. A temperature- and current-stabilized diode laser can be forced to lase in a single mode, but slight perturbations, such as back reflections from external optics, can still lead to mode hopping, which can lead to split Raman bands and calibration problems.

Laser diodes can be mounted in an external cavity configuration to yield improved wavelength stability. Because these techniques employ separate external components such as intracavity optics and an output coupler, some of the compactness of the unmodified laser diode is sacrificed. Commercially available tunable, external-cavity diode lasers incorporate the diode laser into an external optical cavity which includes a diffraction grating for wavelength selection. However, important Raman spectroscopy issues such as laser operational lifetime and frequency, power, and alignment stability characteristics have yet to be explored for these devices.

In order to determine the suitability of external cavity diode lasers for Raman spectroscopy high power diode lasers were tested for short- and long-term frequency and power stability [4]. The laser power and wavelength were measured at 1 to 2 second intervals for up to 14 hours. Raman spectra were measured with exposure times up to 12 hours. In long-term stability tests the wavelength of the tunable laser, operating at a power of ~430 mW, did not drift by more than 0.34 Å. The measured power was also relatively stable during these experiments. The results of the laser stability tests will be presented along with Raman spectra acquired under different operating conditions.

ACKNOWLEDGMENTS

This work has been supported, in part, by Westinghouse Hanford Company (Dr. F. Reich, Program Manager), by Lawrence Livermore National Laboratory (Dr. K. Kyle, Program Manager), and by UNISYS Corp.

1. S. D. Schwab and R. L. McCreery, Anal. Chem., **56**, 2199 (1984).
2. P. Plaza, Nguyen Quy Dao, M. Jouan, H. Fevrier, and H. Saisse, Appl. Opt., **25**, 3448 (1986).
3. Thomas F. Cooney, H. Trey Skinner and S.M. Angel, "Comparative Study of Some Fiber-Optic Remote Raman Probe Designs: I. Model for Liquids and Transparent Solids," and "II. Tests of Single-Fiber, Lensed, and Flat- and Bevel-Tip Multi-Fiber Probes," Appl. Spectrosc. (in press).
4. Thomas F. Cooney, H. Trey Skinner and S.M. Angel, Appl. Spectrosc., **49**, 1846-1851 (1995).

RAMAN SPECTROSCOPY: EXPLORING THE WORLD OUTSIDE THE LABORATORY

M. J. Pelletier*

Kaiser Optical Systems, Inc., 371 Parkland Plaza, Ann Arbor, MI 48103

Until recently, Raman spectroscopy was confined primarily to the laboratories of Raman specialists. It took a lot of effort from highly trained specialists to operate and maintain the equipment. Raman instruments also had special requirements, such as high power electrical service, cooling water, liquid nitrogen, and/or a clean, vibration-free environment. In general, Raman spectroscopy was too difficult and expensive for most technically-feasible applications.

Raman instrumentation has changed. It is now easy for a non-specialist to set up and operate a high performance Raman system in an office, hotel room, or plant floor. Raman spectroscopy is rapidly growing as people with no Raman expertise are discovering that Raman spectroscopy can be a cost-effective solution to their problems. Improving instrumentation, and a rapidly growing population of Raman users with little Raman expertise, are shifting the needs of modern applied Raman spectroscopy away from instrument development and toward application development.

In this paper I will review the new components and instrumentation that are making Raman spectroscopy viable in the world outside the laboratory. I will then briefly describe a number of successful industrial Raman applications to illustrate aspects of the new Raman instrumentation that are critical outside the laboratory. Finally, I will describe two Raman applications in detail in order to illustrate how the essential elements for success outside the laboratory are very different from those for laboratory research.

One successful Raman application that will be covered in more depth is quantitative analysis of silicon wafer cleaning solutions [1]. The ubiquitous SC-1 silicon wafer cleaning solution has been used in the semiconductor industry for more than 25 years. It is an unstable, bubbling mixture of H_2O_2 and NH_4OH at elevated temperature. As feature sizes of advanced semiconductors continue to shrink, product yeilds become more sensitive to the composition of the SC-1 solution. Until recently, there was no real-time analytical method for controlling the reagent concentrations in SC-1 solutions. Raman spectroscopy solves the measurement problem. Simple univariate analysis produces straight line calibration curves with R values of 0.9993 and 0.995 for NH_4OH and H_2O_2 respectively. There was a small dependence of the H_2O_2 peak area on NH_4OH concentration, indicating that multivariate analysis should yield even better results. For the standard SC-1 solution (1 : 1 : 5 volume ratio of NH_4OH : H_2O_2 : water) the 95% confidence level based on replicate measurements was 0.17% for NH_4OH and 0.05% for H_2O_2. Detection limits for NH_4OH and H_2O_2 were 0.02% and 0.01% respectively. Raman spectroscopy can also be used to analyze other wet baths used by the semiconductor industry.

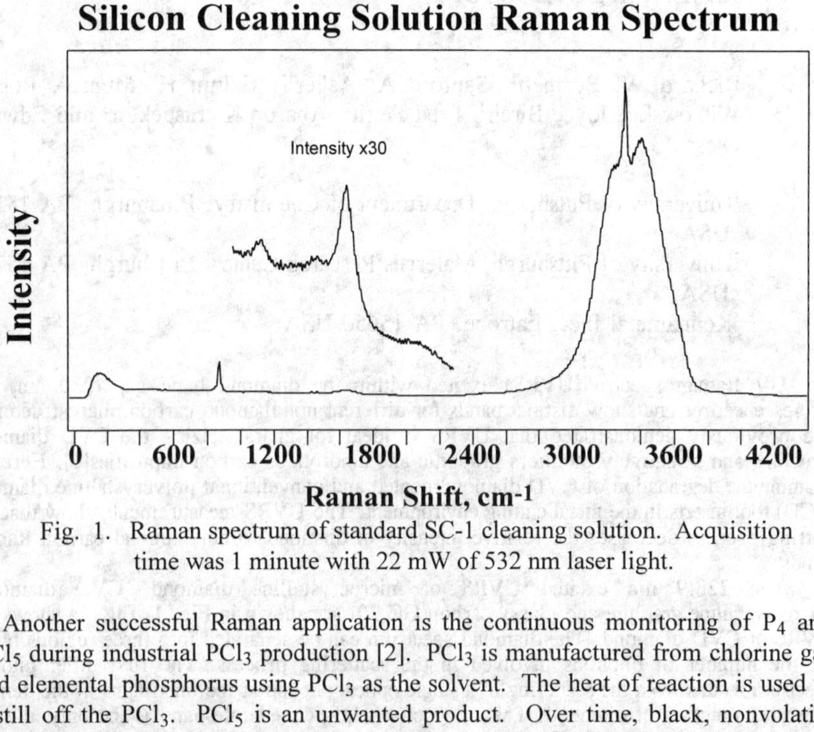

Fig. 1. Raman spectrum of standard SC-1 cleaning solution. Acquisition time was 1 minute with 22 mW of 532 nm laser light.

Another successful Raman application is the continuous monitoring of P_4 and PCl_5 during industrial PCl_3 production [2]. PCl_3 is manufactured from chlorine gas and elemental phosphorus using PCl_3 as the solvent. The heat of reaction is used to distill off the PCl_3. PCl_5 is an unwanted product. Over time, black, nonvolatile impurities from the P_4 accumulate. As a result, the reactor needs to be shut down for cleaning periodically. A knowledge of the P_4 and PCl_5 concentrations is critical to the safe and efficient operation of the PCl_3 reactor. Continuous, on-line 785 nm Raman spectroscopy has proven to be the best way to monitor these concentrations.

1. M. J. Pelletier, K. L. Davis, and R. A. Carpio, Applications of Raman Spectroscopy for SC-1 Processing Bath Concentration Control, 187th Society Meeting of the Electrochemical Society, Reno, Nevada, May 21-26, 1995.

2. G. J. Gervasio and M. J. Pelletier, On-Line Raman Anallysis of PCl_3 Reactor material, Tenth International Forum Process Analytical Chemistry, Orlando, Fl., January 21-24, 1996.

EXAMINATION OF DIAMOND FILM DETERIORATION USING UV RAMAN SPECTROSCOPY

Richard W. Bormett[1], Sanford A. Asher[1*], Calum H. Munro[1], Robert Witkowski[1], Joyce Birch[2], Fred Pettit[2], Aharon K. Inspektor[3] and Edward Oles[3]

[1]University of Pittsburgh, Department of Chemistry, Pittsburgh, PA 15260 USA
[2]University of Pittsburgh, Materials Research Center, Pittsburgh, PA 15261 USA
[3]Kennametal Inc., Latrobe, PA 15650 USA

UV Raman spectra (UVRS) excited within the diamond band gap (230 nm) are fluorescence free and show distinct bands for different nondiamond carbon microstructures. We previously demonstrated that UVRS is ideal for characterizing the CVD diamond structure and sensitively monitors graphitic and amorphous carbon impurities[1]. Here we examine the degradation of CVD diamond coated and conventional polycrystalline diamond (PCD) tool inserts in the metal cutting environment. The UVRS measurements show that the cutting process decreases the relative intensity of diamond to nondiamond carbon Raman bands.

The 228.9 nm excited UVRS of microcrystalline diamond, CVD diamond, microcrystalline graphite and glassy carbon(GC 20) are shown in Fig. 1. Fig. 1a shows the UVRS of CVD diamond. The diamond spectrum can be separated into three regions based on the number of phonons involved in the scattering process. The first order phonon spectrum results from the scattering of a single optical phonon at the maximum in the density of states function of diamond. This 1332 cm^{-1} triply degenerate band is the only allowed Raman fundamental band. Two phonon scattering in diamond which produces a complex band with a maximum at 2467 cm^{-1} and a sharp cutoff at 2667 cm^{-1} consists primarily of overtone and combination bands of phonons at high symmetry points in the Brillouin zone. Careful measurement of the second order phonon spectrum allows accurate determination of phonon energies and symmetries of high symmetry points and critical points in the Brillouin zone. The UVRS second order phonon spectrum of diamond is ten-fold stronger than for visible excitation.

CVD diamond often contains appreciable amounts of nondiamond carbon which exhibit Raman bands characteristic of graphite or highly disordered sp^2 and sp^3 carbon. The UVRS of glassy carbon and microcrystalline graphite excited with 228.9 nm excitation is shown in Figs. 1b and 1d, respectively. The graphite 1582 cm^{-1} G band results from an in-plane graphite vibration of E_{2g} symmetry. The graphite Raman spectrum is strongly affected by graphitic ordering. Absent from the 228.9 nm excited Raman spectra of graphite materials is the D band at ca. 1355 cm^{-1} and D band overtones (2450 cm^{-1}, 2710 cm^{-1}, and 2945 cm^{-1}) [2]. Since the D bands from graphitic carbon do not appear in the UVRS the diamond first and second order bands are easily measured without interference even for poor quality CVD diamond films with relatively high nondiamond concentrations. The UVRS excited second order spectra of HOPG, and microcrystalline graphite show the 3175 cm^{-1} G band overtone and the 3250 cm^{-1} overtone of a Raman forbidden band (ca. 1620 cm^{-1}). The UV second order spectrum of glassy carbon shows only a broad single band for the 3175 cm^{-1} and 3250 cm^{-1} overtones. The absence of the graphite D band overtones with UV excitation allows the diamond second order phonon band (2467 cm^{-1}) and the carbon-hydrogen (C-H) stretching bands of nondiamond carbon (ca. 2930 cm^{-1}) to be monitored in CVD diamond films.

Figure 2 shows an SEM micrograph of a CVD diamond coated tungsten carbide-cobalt tool after 4 minutes of machining. Part of the CVD film is completely worn to the substrate (Region 1 of Fig. 2). Figure 3 shows the 228.9 nm excited UVRS of the CVD coated tool through the tool wear area and Figure 4 shows the ratio of the diamond to nondiamond peak areas. The relative amount of nondiamond to diamond carbon measured in the CVD tool increases after cutting in the abraided areas (Figure 2, Region 2). A large decrease occurs in the diamond band intensity in the exposed area of the CVD diamond tool.

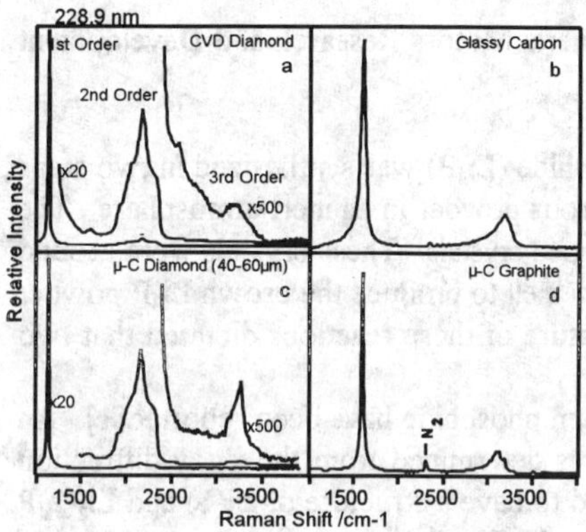

Figure 1

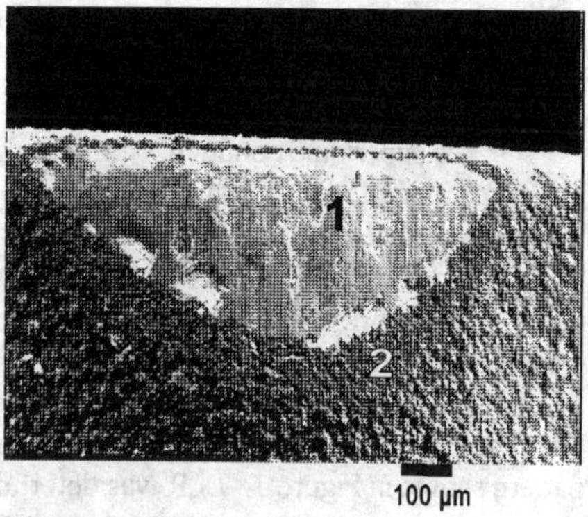

Figure 2

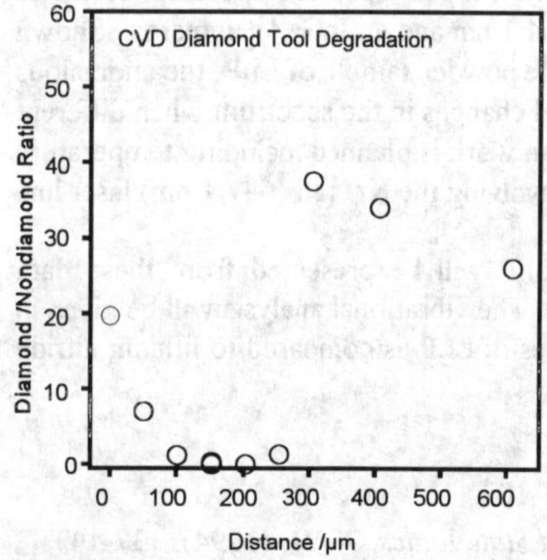

Figure 3

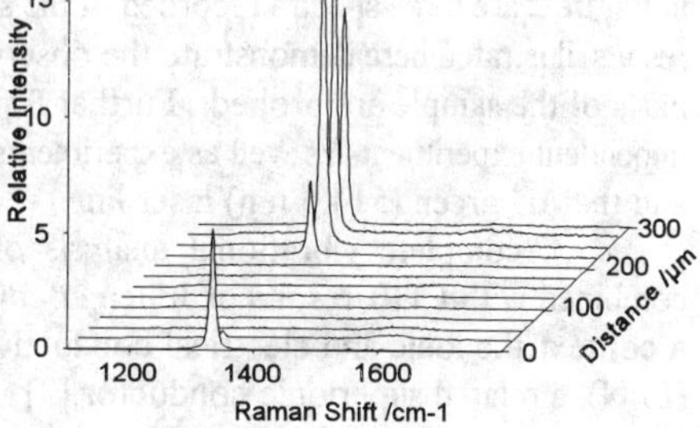

Figure 4

1. R.W. Bormett, S.A. Asher, R.E. Witkowski, W.D. Partlow, R. Lizewski, and F. Pettit *J. Appl. Phys. 77*, 5916 (1995).
2. Y. Wang, D.C. Alsmeyer and R.R. McCreery *Materials* **2**, 557 (1990).

THE RAMAN AND FT-RAMAN SPECTRA OF THE SUPERIONIC CONDUCTOR Li_3P

Bob Berno[1], Ricardo Aroca[1*], and Gholamabbas Nazri.[2]

[1] Department of Chemistry and Biochemistry, University of Windsor, Windsor, ON, N9B 3P4, CANADA.

[2] Physical Chemistry Department, General Motors Research and Development Labs, Warren, MI 48090, USA.

The superionic conductor lithium phosphide (Li_3P) was synthesized in two steps by the reaction of molten lithium and phosphorous powder in an inert atmosphere. The first step produced LiP, giving black, needle-shaped crystals. These crystals were ground into a fine powder and added to a second lithium melt to produce the brown Li_3P powder used in this work. The highly exothermic nature of these reactions dictated that two steps were preferable to one.

The structure and conductivity of lithium phosphide have been reported[1]. An hexagonal structure for lithium phosphide was determined from the x-ray diffraction pattern shown in Figure 1. Li_3P was determined to have a structure of Li(1) and Li_2(2)P alternating layers as shown in the inset to Figure 1. The lattice constants were found to be: a = 4.273 Å and c = 7.594 Å.

FT-Raman spectra were recorded on a Bomem Ramspec 150 spectrophotometer equipped with an Nd:YAG laser emitting at 1064.1 nm and an InGaAs detector. Shown in Figure 2 are three spectra recorded of the same powder sample of Li_3P; the anomalous results illustrated here demonstrate the observed changes in the spectrum when different parts of the sample are probed. Further Raman work is planned including temperature dependent experiments as well as experiments involving the Kr^+ red (647.1 nm) laser line and the Ar^+ green (514.5 nm) laser line.

A complete vibrational analysis of Li_3P will be presented from these data combined with FTIR results of Julien *et. al*[2]. The vibrational analysis will be given in a context the ionic and electrical conductivities of Li_3P as compared to lithium nitride (Li_3N), a related superionic conductor.[3]

1. G. Nazri, *Solid State Ionics*, **34**, (1989), 97-102.

2. G.A. Nazri, C. Julien, and H.S. Mavi, *Solid State Ionics*, **70/71**, (1994), 137-143.

3. H.R. Charndrasekhar, G. Bhattacharya, R. Migoni, and H. Bilz, *Phys. Rev. B*, **17**, (1978), 884-893.

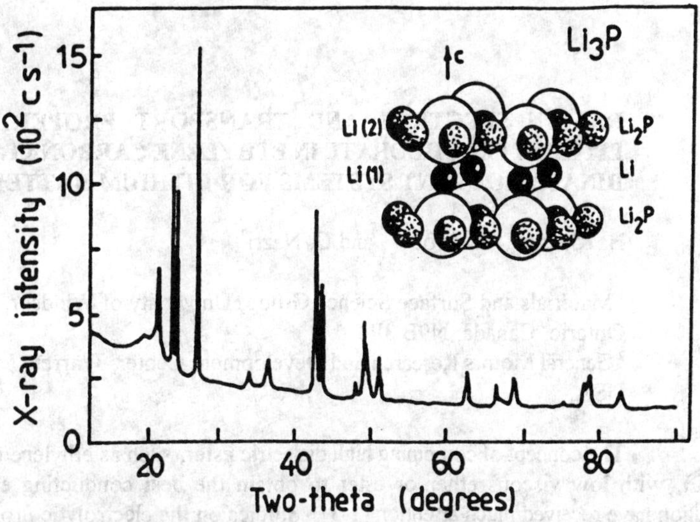

Figure 1: The x-ray diffraction pattern of Li_3P powder. Inset is an illustration of the Li(1) and $Li(2)_2P$ alternating layered structure.

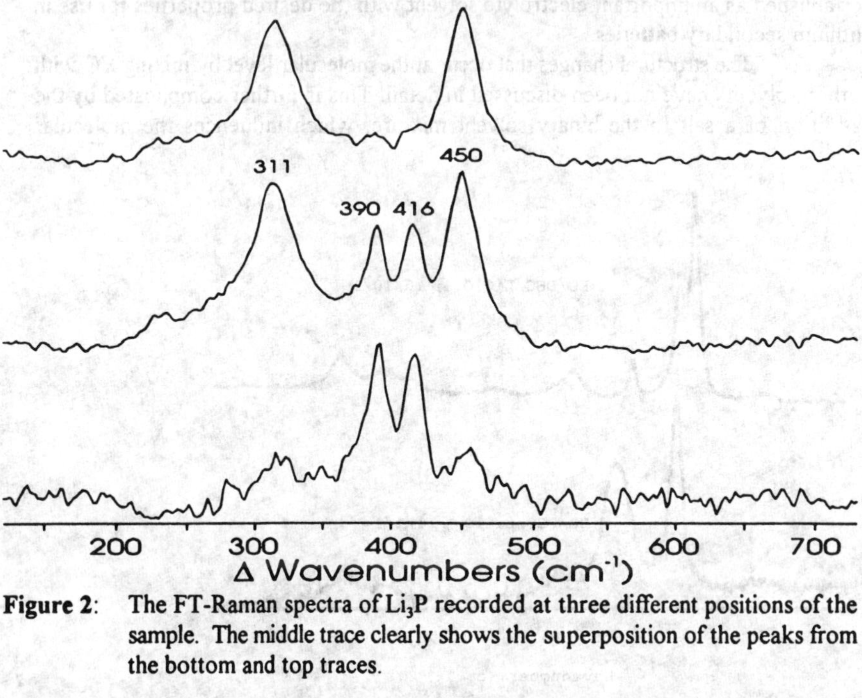

Figure 2: The FT-Raman spectra of Li_3P recorded at three different positions of the sample. The middle trace clearly shows the superposition of the peaks from the bottom and top traces.

RAMAN SPECTRAL AND TRANSPORT PROPERTIES OF LITHIUM PERCHLORATE IN ETHYLENE CARBONATE BASED BINARY SOLVENT SYSTEMS FOR LITHIUM BATTERIES

B. Klassen[1], R. Aroca[1*], and G. Nazri[2]

[1.]Materials and Surface Science Group, University of Windsor, Windsor, Ontario, Canada, N9B 3P4
[2]General Motors Research and Development Center, Warren, Mi. 48090, USA.

The concept of combining high dielectric ester, such as ethylene carbonate (EC), with low viscous ether or ester to obtain the best conducting electrolyte solution have received much attention [1-3]. Studies on the electrolytic properties of ethylene carbonate based solutions and the charge-discharge cycling behaviour of lithium in the these solutions have been performed [4-6] these studies EC has been established as an important electrolyte solvent with the desired properties for use in lithium secondary batteries.

The structural changes that occur at the molecular level by mixing EC with other solvents have not been discussed in detail. This is further complicated by the addition of a salt to the binary solvent mixture, which influences the molecular

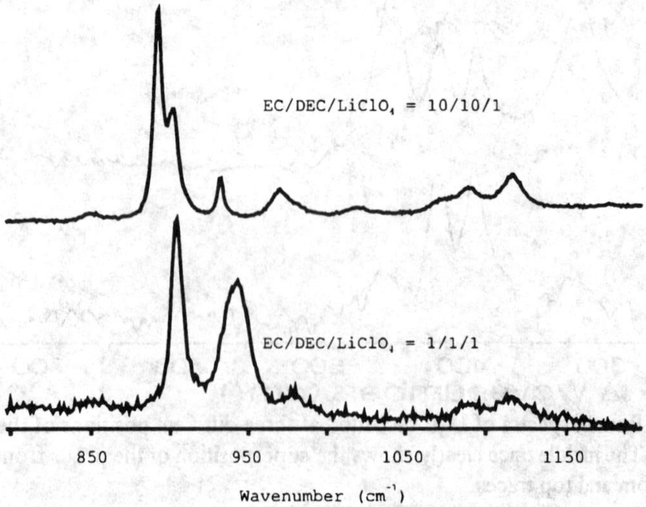

Figure 1: Raman spectra of EC/DEC/$LiClO_4$ electrolyte solutions with varying concentrations of lithium perchlorate over the range 800-1200 cm^{-1}

arrangements. Structural changes are reflected in the spectroscopy (Figure 1), conductivity, and viscosity behaviour of EC based binary solvent electrolytes. With these considerations in mind, the spectroscopic, conductivity, and viscosity studies of EC based binary electrolytes were performed. Co-solvents included: propylene carbonate (PC), diethyl carbonate (DEC), and dimethyl carbonate (DMC).

The intense band located at 891 cm^{-1} belongs to the symmetric ring breathing mode of EC, and is shifted by 10 cm^{-1} to 904 cm^{-1} with the addition of lithium perchlorate. The assignment of the bands correspond to the free solvent and the coordinated solvent molecules. The symmetric stretch $v_1(a_1)$ of the perchlorate anion at 933 cm^{-1} along with the formation of bands that include the solvent-shared ion-pairs (938 cm^{-1}) and contact ion-pairs (944 cm^{-1}) in the most concentrated electrolytes.

Viscosity for the EC/PC/$LiClO_4$ electrolyte increases with higher lithium perchlorate concentration. For lithium perchlorate concentrations from 0.33 M (EC/PC/$LiClO_4$ = 20/20/1) to 3.3 M (EC/PC/$LiClO_4$ = 2/2/1), viscosity changes from 3.82 cP to 147 cP. The electrolyte having EC/PC/$LiClO_4$ = 1/1/1 molar ratio (6.6 M) was a solid at room temperature thus viscosity data could not be collected. The viscosity data for the EC/DEC/$LiClO_4$ electrolyte behaved in the same manner as the EC/PC/$LiClO_4$ electrolyte, where the viscosity increased with increased lithium perchlorate concentration. For lithium perchlorate concentrations from 0.26 M (EC/DEC/$LiClO_4$ = 20/20/1) to 5.3 M (EC/DEC/$LiClO_4$ = 1/1/1), viscosity changes from 1.84 cP to 29.0 cP.

Conductivity increases with increasing lithium perchlorate concentration in the EC/PC/$LiClO_4$ electrolyte until a maximum of 6.1 $k\Omega^{-1}cm^{-1}$ is reached at 0.82 M $LiClO_4$ (EC/PC/$LiClO_4$ = 8/8/1). After which the conductivity decreases rapidly between 0.82 M and 3.3 M to a minimum of 1.4 $k\Omega^{-1}cm^{-1}$. In a similar fashion, the conductivity of EC/DEC/$LiClO_4$ increases with increasing lithium perchlorate concentration until a maximum of 5.1 $k\Omega^{-1}cm^{-1}$ is obtained at a lithium perchlorate concentration of 0.88 M (EC/DEC/$LiClO_4$ = 6/6/1). Once again the conductivity drops off rapidly between the lithium perchlorate concentrations of 0.88 M and 5.32 M reaching a minimum conductivity of 0.57 $k\Omega^{-1}cm^{-1}$.

1.Morita, M.; Tachihara, F.; Matsuda, Y. *Electrochim. Acta,* **33**, 239, (1988).

2.Tobishima, S.; Arakawa, M.; Yamaki, J. *Electrochim. Acta,* **33**, 317, (1988).

3.Tobishima, S.; Okada, T. *J. Appl. Electrochim.*, **15**, 467, (1985).

4.Tobishima, S.; Yamaji, A.; Okada, T. *Electrochim. Acta,* **29**, 1471, (1984).

5.Tobishima, S.; Okada, T. *Electrochim. Acta,* **30**, 1715, (1985).

6.Arakawa, M.; Tobishima, S.; Hirai, T.; Yamaji, J. *J. Power Sources*, **20**, 293, (1987).

DISSOLUTION OF ORGANIC MOLECULES IN OIL FILMS

C. F. Windisch Jr.*[1], G. J. Exarhos[1], S. K. Sharma[2], and N. J. Hess[1]

[1]Pacific Northwest National Laboratory+, Box 999, Richland, WA 99352
[2]University of Hawaii, 2525 Correa Rd., Honolulu, HI 96822

The dissolution of volatile organic molecules in a nonvolatile oil layer has relevance to several industrial processes, including scrubbing of gas mixtures in exhaust streams, the degradation of lubricating oil films in environments containing organic vapors, and the "hide-out" of unburned hydrocarbon fuel in lubricating oil films on the interior cylinder wall surfaces of internal combustion engines. The analysis of this dissolution process usually assumes equilibrium solubility is achievable [1]. While this assumption is valid in many instances, it clearly is not when liquid phases of both the volatile and nonvolatile components are present at all times during the dissolution. For two liquids that are infinitely miscible in each other, no equilibrium will be achieved over a practical time frame as long as the two liquids are always present and one or both of the liquids can freely transport through the vapor phase and dissolve in the other. For this reason, it is important in some cases to understand the *kinetics* of the vapor phase transport and dissolution of a volatile liquid into a nonvolatile oil layer.

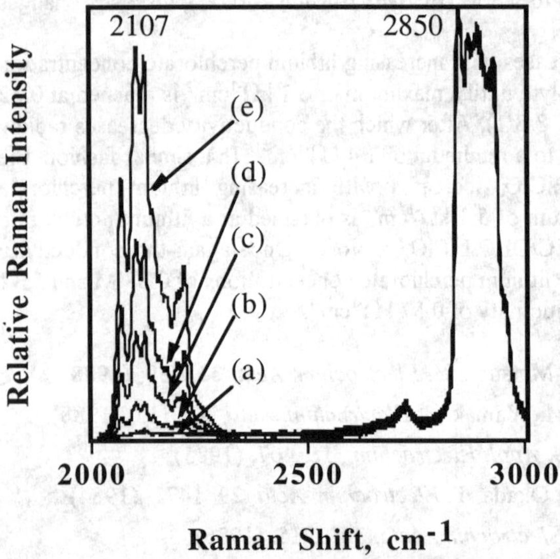

Fig. 1. Normalized Raman spectra of oil film at various times after injection of deuterated n-octane into test cell: (a) 0 s, (b) 1011 s, (c) 2910 s, (d) 5335 s, (e) 14,625 s.

This paper reports on the results of using laser Raman microprobe spectroscopy to study the kinetics of dissolution of n-octane in a 10 μm-thick film of a poly-α-olefinic hydrocarbon that is a base stock for a commercial synthetic motor oil [2]. Completely deuterated n-octane was used to distinguish vibrational bands from the volatile liquid and the oil film. A specially designed optical test cell was used to determine dissolution rates as a function of temperature and pressure.

Fig. 1 shows the growth of bands from C-D stretching modes (n-octane)

+Operated by Battelle for the U. S. Department of Energy under Contract DE-AC06-76RLO 1830.

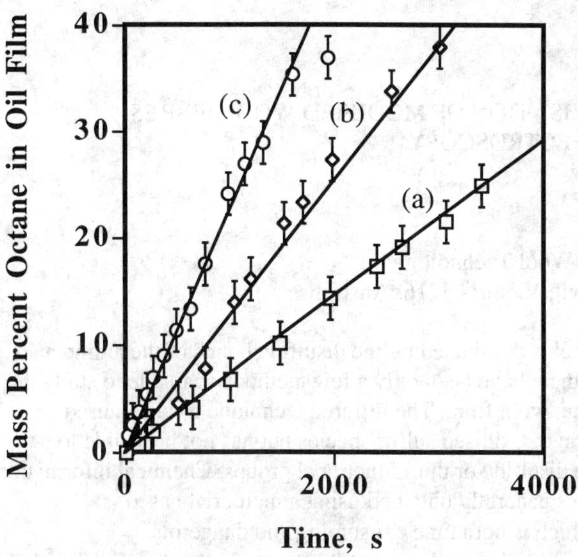

Fig. 2. Variation of oil film composition as a function of time at 1.0 atm and (a) 20°C, (b) 40°C, (c) 60°C.

relative to C-H stretching modes (oil) following the injection of excess deuterated n-octane into the cell containing the oil film (without allowing direct liquid-liquid contact). The ratio of intensities at 2107 cm^{-1} (C-D) and 2850 cm^{-1} (C-H) were determined at each time, converted to oil film composition using a previously determined calibration curve, and plotted versus time (Fig. 2). In all cases, the concentration of n-octane in the oil film increased linearly at least up to 40 % (by weight) and showed no tendency to reach an "equilibrium solubility" at less than 100 %. Predicting dissolution rates for infinitely miscible liquids, when the components are always present somewhere in the system in the liquid form, clearly requires knowledge of not only equilibrium partial pressures and vapor phase solubilities but also transport rates involving the vapor/oil film interface. This work demonstrates that the latter can be measured non-intrusively and in real time using Raman spectroscopy.

Dissolution rates increased with increasing temperature and decreasing total pressure (using N_2 gas). Arrhenius plots of the dissolution rates were linear between 20°C and 60°C. The activation energy decreased with increasing total pressure: from 25.2 kJ•mol^{-1} at 1.0 atm to 16.0 kJ•mol^{-1} at 12.6 atm. At 20°C and 1.0 atm total pressure, the dissolution rate was approximately 5 x 10^{-8} g of n-octane per cm^2 of a 10-µm thick oil film per second. In contrast to the dissolution behavior when both liquids were present, equilibrium solubility of n-octane in the oil film was observed at temperatures above the boiling point of n-octane. The solubility followed an inverse temperature dependence, which was consistent with the literature [3]. Dissolution kinetics for several simple aromatic compounds, including benzene and toluene, in oil films were also determined using this method. The aromatics were found to dissolve faster than n-octane but followed the same trends with respect to temperature and pressure. Dissolution studies of mixtures of volatile liquids are currently underway.

1. J. H. Perry (Ed.), *Chemical Engineers' Handbook*, Fourth Edition, McGraw-Hill Book Company, New York, 1963, pp. 14-17 to 14-19.

2. Red Line Synthetic Oil, Martinez, CA 94553.

3. T. Sugiyama, T. Takeuchi and Y. Suzuki, *J. Chromatogr.*, **105**, 265 (1975).

THE CHARACTERISATION OF MODIFIED WOOL FIBRES BY FT-RAMAN SPECTROSCOPY

J. S. Church

CSIRO, Division of Wool Technology
P.O. Box 21, Belmont, Victoria 3216, Australia

The chemistry of wool is dominated by the disulfide bonds of the amino acid cystine.[1] Until recently, there have been only a few methods available to study the sulfur chemistry of the intact wool fibre. The infrared technique has been used extensively for the detection of oxidised sulfur species but has not been able to provide information on the disulfide or thiol functional groups. Chemical information on these latter two species is generally obtained using a mercurial based electrochemical method which is both time consuming and dangerous.

The Raman technique provides a direct method to study the disulfide bonding as well as the amino acid residues with aromatic side chains within proteins but, until recently has seen only limited use in wool science as data acquisition suffered from fluorescence (a problem common to many biological substrates) and thermal damage problems. With the development of FT-Raman instrumentation these problems have been minimised and high-quality spectra can routinely be obtained from keratin samples.[2] In this work the technique of FT-Raman spectroscopy has been used to study the chemical modifications that occur to the wool fibre during suflitolysis (bisulfite) and reductive (dithiothreitol) treatments.

Bisulfite, a reagent commonly used in wool processing, cleaves the disulfide bond of a cystine residue to produce a thiol and a S-sulfonate group according to the reaction:

$$\text{Cy-S-S-Cy} + \text{NaHSO}_3 \rightarrow \text{Cy-S-SO}_3\text{Na} + \text{Cy-SH}.$$

The 2700 to 300 cm^{-1} region of the Raman spectra obtained from wool fabric treated with 10% on the weight of the wool (oww) sodium bisulfite at 25°C for 12 hours (B) and an untreated control sample (A) are shown in Figure 1. Several significant differences are observed when these spectra are compared, the most obvious being in the region between 550 and 350 cm^{-1} and involves the S-S stretching modes. The S-S stretching mode of cystine, observed at 512 cm^{-1}, has decreased in intensity signifying the cleavage of S-S bonds in the treated samples. A new peak appearing at 402 cm^{-1} can be assigned to the S-S stretch mode of cysteine S-sulfonate salts.[3] Two more subtle features, the weak S-H stretching mode, 2567 cm^{-1}, and the symmetric SO_3^- stretching mode associated with the formation of S-sulfonate salts, 1025 cm^{-1}, are also apparent in the Raman spectra obtained from the treated fibre sample. The spectroscopic results are consistent with those expected.

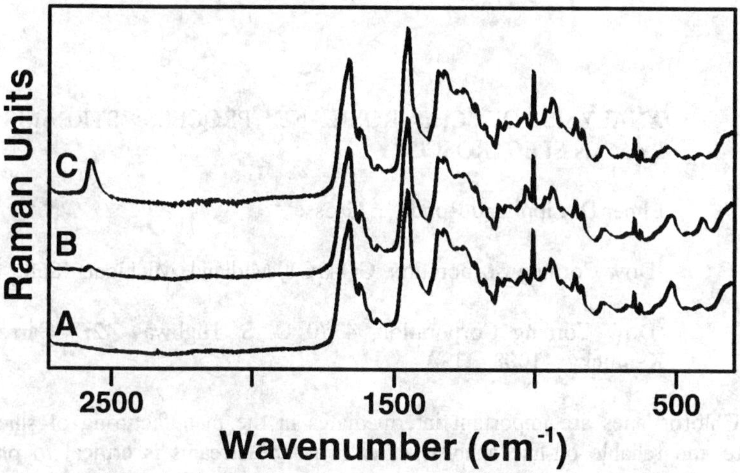

Fig. 1. The Raman spectra obtained from A: Untreated wool, B: Wool treated with 10% oww sodium bisulfite and C: Wool treated with 9.6 x10^{-3} M dithiothreitol.

The reaction of dithiothreitol (DTT) with wool results in the cleavage of the disulfide bonds of the cystine residues and the formation of -SH groups. The Raman spectrum obtained from wool treated with 9.6 x10^{-3} M DTT at 30°C for 8 hours is shown as Figure 1C. Under these conditions one would expect reduction of 85% of all disulfide bonds. The S-H stretching line is observed as a moderate feature at 2567 cm^{-1}. This peak is undetected in the spectrum obtained from the untreated wool sample where the cysteine (free SH) concentration levels are generally in the range of 20 - 40 μmol.g^{-1}.[1] The S-S stretching line which is moderately intense in the spectrum obtained from the untreated wool is greatly reduced in the spectrum obtained from the DDT treated wool sample. It should be noted that the thiol groups of reduced wool are readily oxidised back to disulfides upon exposure to air.

This work demonstrates the usefulness of the FT-Raman technique for monitoring changes in both the disulfide and thiol functional groups within the wool fibre as caused by reactions that typically occur during commercial processing.

The author recognises the support of the Australian wool growers and the Australian Government who fund research and development through the IWS.

1. J. A. Maclaren and B. Milligan, Wool Science: The Chemical Reactivity of the Wool Fibre, Science Press, Sydney, 1981.
2. E. A. Carter, J. S. Church, R. J. Denning and P. M. Fredericks, Spectrochim. Acta, 50A(11), 1927 (1994).
3. J. S. Church and K. R. Millington, Biospectroscopy, submitted for publication.

ANALYSIS OF CHLOROSILANE PROCESS STREAMS BY RAMAN SPECTROSCOPY

Elmer D. Lipp[1] and Ronda L. Grosse*[2]

[1]Dow Corning Corporation, CO41C1, Midland, Michigan 48686-0994, USA
[2]Dow Corning Corporation, 4770 U. S. Highway 42E, Carrollton, Kentucky 41008, USA

Chlorosilanes are important intermediates in the manufacturing of silicones. Accurate and reliable on-line analysis of chlorosilane streams is critical to process operations. Raman spectroscopy has recently been applied to the analysis of chlorosilane process streams. A Raman spectrometer comprised of a frequency doubled Nd:YAG laser, a holographic imaging spectrograph, and a CCD detector was assembled for this purpose. Remote sampling is accomplished with the use of fiber optic probes. A portable NEMA enclosure was constructed to house the Raman spectrometer and permit data collection in process areas.

An evaluation of Raman spectroscopy in the plant environment demonstrated that it is highly suited to on-line analysis of methylchlorosilane streams within distillation processes. Significant advantages over other analytical techniques are obtained since Raman spectroscopy is non-invasive and allows rapid analyses (seconds to minutes acquisition times). Optical fibers are utilized for transmission of laser light and collection of Raman scattering. A liquid cell was built for direct sampling of the chlorosilane monomers. A slip stream from the process is directed into the cell and the probe is immersed in the solution. Due to the corrosive nature of chlorosilanes, acid resistent windows are required at the probe tip to prevent contact with the silica fibers. A total of fourteen different distillation streams have been analyzed. On-line data are generally collected every five minutes, with actual integration times of 20 - 30 seconds. Excellent signal/noise is observed on these time scales, and fluoresence interference has not been found to be a problem using the visible laser source. Concentrations of numerous chlorosilanes can be measured in a single analysis, as the monomers show distinct Raman spectra. CH_3- and H-containing chlorosilanes have unique spectral features that can be readily differentiated. Figures 1 and 2 show examples of typical chlorosilane spectra. Various Si-H and Si-Cl vibrational bands are monitored. The real-time measurements show high accuracy and precision, and can reveal large variations in stream compositions over short sampling intervals.

Results of both laboratory experiments and on-line studies will be presented. Raman spectroscopy has great potential to be used for distillation process control. It

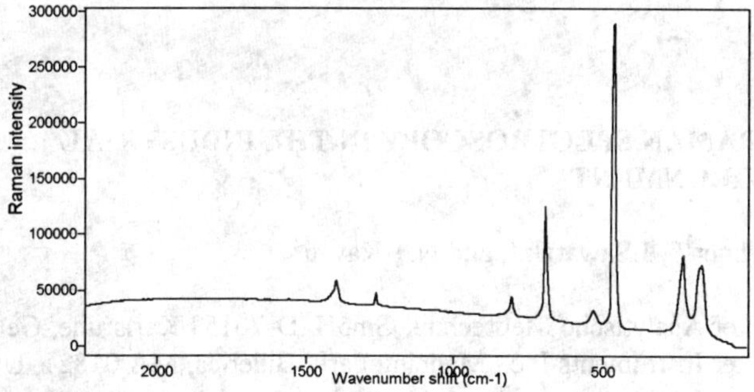

Figure 1. Raman spectrum of Me_2SiCl_2 obtained using 532 nm excitation; 10 second acquisition time.

has been demonstrated to be a powerful tool for monitoring the composition of chlorosilane process streams, and can provide rapid identification and correction of dynamic or unstable conditions in distillation columns for improved process control. Multiplexing of probes will allow simultaneous monitoring of several streams. Numerous applications also exist in the area of Quality Assurance. With the use of external probes, exposure to corrosive chemicals is reduced and analyses can be completed quickly and safely. Furthermore, this Raman system has the capability to do mobile testing within manufacturing processes.

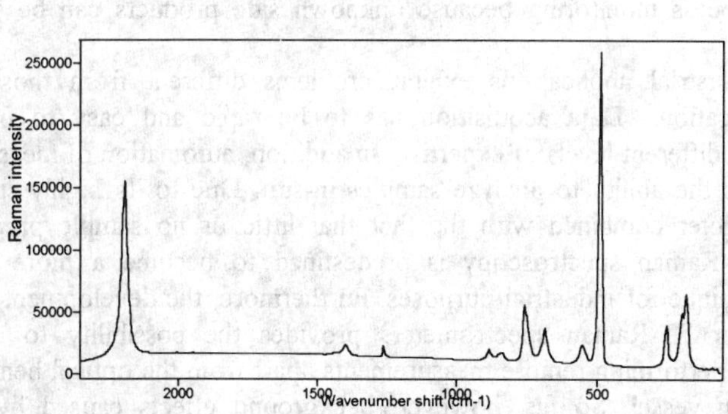

Figure 2. Raman spectrum of $MeHSiCl_2$ obtained using 532 nm excitation; 10 second acquisition time.

FT-RAMAN SPECTROSCOPY IN THE INDUSTRIAL ENVIRONMENT

C.Lehner[1]*, J.Sawatzki[1], and N.T.Kawai[2]

[1]Bruker Analytische Meßtechnik GmbH, D-76153 Karlsruhe, Germany
[2]Bruker Instruments Inc., Manning Park, Billerica, MA 01821, USA

Since several years Fourier Transform Raman spectrometers have been established as alternatives to dispersive systems. By employing highly stable near-infrared lasers in combination with the advantages of Fourier instrumentation most samples can be analyzed quickly without the ubiquitous fluorescence when excitation in the visible is performed.

NIR-FT Raman spectroscopy has been successfully applied to practical applications such as process investigation of a chemical reaction, quality control of food and *in vivo* medical diagnosis [1-3].

Recent advances in instrumental sensitivity have provided further impact in the applicability of FT-Raman spectroscopy. Especially the availability of high sensitive Raman detectors allows short measurement times just in a few seconds. Besides this the so called multiplex advantage of Fourier Transform spectroscopy provides spectroscopic information about the entire spectral range. This can be very helpful for process monitoring because unknown side products can be recognized immediately.

Industrial applications exhibit problems different from those of pure research applications. Data acquisition has to be rapid and easy to perform by operators with different levels of expertise. In addition, automation of measurement is desirable, as is the ability to analyze samples in-situ. Due to its highly specific and selective character combined with the fact that little or no sample preparation is necessary, FT-Raman spectroscopy is predestined to become a more and more attractive technique for industrial purposes. Furthermore, the development of flexible waveguides for FT-Raman spectrometers provides the possibility to do on-line monitoring by performing remote measurements apart from the optical bench directly in the reaction vessel. So far, however, background effects caused by the fiber material limited the length of the waveguide and therefore the feasabililty of the method. For this reason, recently a new probehead was designed, which eliminates this spectral contribution of the fiber material.

Fig.1 shows the result of monitoring an ester hydrolysis with a probe head which avoids any interferences by the quartz material of the fiber. A fiber as long as 15 metres was utilized to follow the reaction occuring in a glass flask.

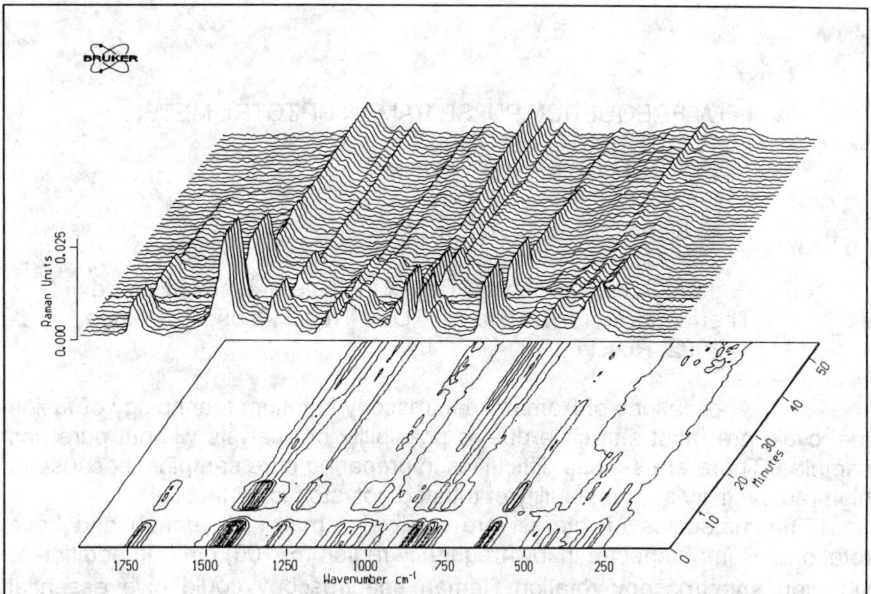

Fig.1: FT-Raman fiber optic measurement of an ester hydrolysis. Excitation: Nd:YAG, 1064 nm, ca. 400 mW; Resolution: 8cm^{-1}; Number of scans per spectrum: 20

1. M.Claybourn, T. Massey, J. Highcock and D. Gogna, J. Raman Spectrosc. 25, 123 (1994)
2. S. Keller , T. Löchte, B. Dippel and B. Schrader, Fresenius Anal. Chem. 346, 863 (1993)
3. B. Schrader, S. Keller, T. Löchte, S. Fendel, D. S. Moore, A. Simon and J.Sawatzki, J. Mol. Struct. 348, 293 (1995)

LOW-FREQUENCY PULSE RAMAN SPECTROMETER FOR TRITIUM TECHNOLOGY. SPECTRUM INTERPRETATION

A. A. Ivanov *, and G. A. Polyakov

Federal Scientific Center "Troitsk Institute of Innovative and Thermonuclear Investigation", OIP, Troitsk, Moscow Region 142092, Russia

Applications of Raman spectroscopy in tritium technology of fusion fuel cycle are most attractive due to possibility of analysis without pure gas samples. There are serious difficulties in preparing pure samples, because of high radioactivity and unstability of tritiated polyatomic molecules.

The molecules of interest are relatively small (2-5 atoms) and have rotational Raman spectra in low-frequency region (5÷100) cm^{-1}. In addition to vibration spectroscopy rotation Raman spectroscopy could give essential information on structure of such a molecule. But to obtain more reliable information one need to have high spectral resolution of the order of 0.1 cm^{-1}, and sensitivity high enough to analyze gas with pressure of the order of 0.1 bar. To develop a real time and *in-situ* process gas analyzer, application study of pulse laser Raman spectroscopy was performed by measurement of low-frequency spectra.

Details of the Raman spectrometer have been described previously [1]. As a Raman excitation source, a frequency doubled pulsed Q-switched Nd:YAG laser have been used. A $\lambda/2$ plate could be used to rotate the laser polarization to minimize the intensity of the recorded Rayleigh scattered light. The Raman spectrum was detected by self designed photon counting system with cooled photomultiplier tube which is mounted at the output of double grating monochromator. Photon counts have been processed by an on-line personal computer and all the spectra have been corrected for dark background counts and laser power.

As an example, the measured low-frequency Raman spectrum of the air at the STP is shown in Fig. 1A. The j=6 $\rightarrow$ j=8 and j=9 $\rightarrow$ j=11 transitions of the $^{14}N_2$ and $^{16}O_2$, respectively, is observed in the vicinity of 60 cm^{-1}. For the simulation of the band shape we have used Gaussian function with the same width for each of these transitions (I and II). The experimental data (◊) was numerically fitted with sum of two Gaussian functions (Fig. 1B, curve 2). Target parameters of fitting was amplitudes and FWHM of the functions. Procedure of fitting was based on minimization of sum of squares of deviation between experimental data and calculated curve.

Essential improvement was obtained with application of Fourier self-deconvolution [2]. The deconvolution filter is a simple exponential filter of the form: $\exp(2\pi\gamma X)$, where γ is the constant to change the filter shape and X is the data array whose X range is normalized between 0 and 1. This function is multiplied by the Fourier transformed trace, and the data is then reverse Fourier transformed to give the result. The deconvolution line shape with 0.755 cm^{-1} (FWHM) is shown in Fig.1B curve 1. The results of the numerical fitting and the Fourier self-deconvolution are given in Table 1. Thus, good agreement could be achieved for spectrum interpretation.

TABLE 1. Raman shift and intensities ration for the $^{14}N_2$ j=6 → j=8 (I) and $^{16}O_2$ j=9 → j=11 (II) transitions.

Line	Raman Shift, [cm^{-1}]		I/II Intensities ratio	
	num. fitt.	deconv.	num. fitt.	deconv.
I	58.354	58.311	1.645	1.583
II	60.080	60.117		

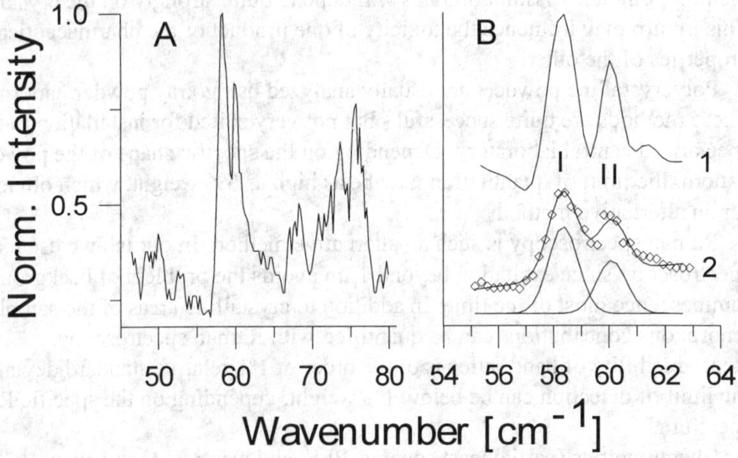

Fig. 1. A: Observed Raman spectrum of the air at the STP. B: 1 - Deconvolved spectrum; 2 - Numerically fitted experimental data (◊) with sum of Gaussian functions.

1. A.A. Ivanov, and V.B. Voronin, Instruments and Experimental Techniques 2, (1996), in press.
2. D. Massart, B. Vandeginste, S. Deming, Y. Michotte, L. Kaufman, in Data Handing in Science and Technology, (B.Vandeginste, L.Kaufman, ed), Vol. 2, Elsevier, 1988, p. 215.

QUANTITATION OF CRYSTAL POLYMORPHISM USING RAMAN SPECTROSCOPY AND CHEMOMETRICS

A. Vit and A. Zilian*

Ciba Research Services, Central Analytical Department
CIBA-GEIGY AG, CH- 4002 Basel, Switzerland

Many industrial chemical products have an active principle which is based on molecules. If the molecules are sold in crystalline form, we must consider that for one type of molecule different crystal forms might exist. Be it that a varying number of solvent molecules are incorporated in the crystal lattice or that several crystal modifications corresponding to different lattice symmetries are formed. This can be relevant from an application point of view since important parameters such as melting point and dissolution rates will depend quite strongly on the crystal type. This in turn may influence the toxicity of one product or the pharmaceutical properties of the other.

Polycrystalline powders are usually analyzed using xray powder pattern methods. These methods are quite successful - but not very suited for installation outside a specialized central laboratory. Depending on the specific shape of the powder pattern, the limit of quantitation may be as high as 5% weight, which obviously calls for an alternative method.

Raman spectroscopy is such an alternative method. In our lab we use FT-Raman spectroscopy since excitation beyond 1μm avoids the problem of background luminescence most of the time. In addition to crystalline areas of the sample, amorphous contributions can be quantified with Raman spectroscopy. Reproducibility of band ratios is of the order of 1% relative standard deviation and the limit of detection can be below 1% weight, depending on the specific Raman spectrum.

Chemometrics (partial least squares, PLS, and classical least squares, CLS [1]) is used to analyze the data. In both models a linear combination of spectral intensities is correlated in a linear model with concentration. This is expected to be valid for our physical mixture of powders with negligible interaction between grains. In a CLS calibration pure component spectra are determined which can be directly compared to the measured pure component spectra. On the other hand, a PLS calibration generates abstract factors which show optimal correlation with concentration, but which are not always easy to interpret spectrally.

We used a synthetic sample set: a binary mixture of 2 modifications of a pharmaceutically active substance. The concentrations ranged from pure α to pure β modifications including many samples in the range 80-98% β. The spectra were measured with a Bruker RFS-100 spectrometer using 300 mW laser power, 4 cm^{-1}

resolution, an acquisition time of 120 secs and a liquid N_2 -cooled Ge detector. The spectra are corrected for system response using a tungsten lamp.

The system was calibrated as follows. Once the powder mixtures were homogenized enough, a linear relationship between Raman intensity and nominal concentration was found. Both PLS (using 1 factor) and CLS (using 2 components) methods were used with comparable results. Typical RMSEE values were of the order of 0.5%, which correlates nicely with the reproducibility of band ratios. We prefer CLS, since the standards obtained in the calibration step can be directly compared to the spectra of the pure components. Fig. 1 (left) shows the spectral region with the highest information content. Fig.1 (right) shows the calibration, the line being the ideal y = x curve. A quantitative analysis of an unknown mixture yields the concentrations of the two components and a residual giving information about the quality of the fit. If the residual is high, the residual spectrum should be inspected. Any structure in the residual spectrum gives us insight about an additional component present in the sample. The analysis of a non homogenized sample yields concentration values different from the nominal concentration, but all spectral intensity can be explained using the two standards for the pure components.

Quantitation of polymorphism is an ideal application of Raman spectroscopy in an industrial laboratory: In spite of the high concentrations of the constituents the intensity - concentration relationship is strictly additive, since the different grains do not interact. The usefulness of the method will depend on the size of the spectral differences between the crystal modifications. Other important parameters are spectrometer related: The reproducibility of band ratios and the ability to detect a small signal on zero background. These spectrometer parameters currently limit the precision of the analysis.

1. H. Martens and T. Naes, *Multivariate Calibration* (Wiley, Chichester, U.K.,1989)

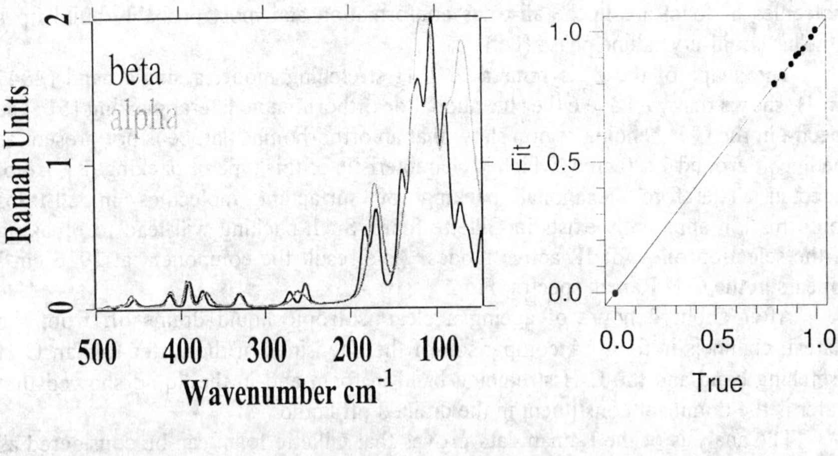

Fig. 1. left: Raman spectra of the pure α and β components. right: CLS calibration

RAMAN STUDIES OF THREE-DIMENSIONAL FOAM

N. Goutev*, Zh.S. Nickolov

Quantum Electronics Department, Faculty of Physics,
Sofia University, 1164 Sofia , Bulgaria

Foam is a random dispersion of gas bubbles separated by thin liquid films. The parameters of foam (dispersity, film thickness, rate of drainage, stability, etc.) are determined mainly by the physicochemical properties of the liquid films [1]. The latter depend on the molecular structure of the films. Raman spectroscopy can supply information about molecular conformations and lateral molecular packing in complex surfactant solutions and biological systems [2,3]. Therefore it should be suitable for studies of foam molecular structure.

We have studied the Raman spectra of shaving foam (Gillette) in the region of the C-H stretching vibrations (2800-3000) cm^{-1}, in the region of the skeletal C-C stretching modes (1050-1150) cm^{-1} and in the (1400-1500) cm^{-1} methelene deformation region. In these regions the main contributions come from the vibrational modes of the long hydrocarbon "tails" of the surfactant molecules.

The Raman spectra in the C-H stretching region change weakly with ageing, Fig.1, and show preservation of the molecular structure of the foam liquid films. The average value of the intensity ratio $R=I_{2882}/I_{2847}$ is around 1.65, Fig.2, and is characteristic of vibrationally decoupled hydrocarbon chains in the all-trans conformation. Similar values of R are observed in ordered multilayer or crystal-like structures [4,5]. The detection of well defined bands at 1460 cm^{-1} in the methelene deformation region and at 1063 and 1131 cm^{-1} in the skeletal C-C stretching region is also characteristic of surfactant molecules in all-trans conformation. The 1090 cm^{-1} component, indicating presence of gauche conformers is not observed. The appearance of the Raman spectra in the investigated regions suggests that surfactant molecules in foam are in the all-trans conformation and most probably build up a lamellar liquid-crystalline phase (gel).

The shape of the d^+ symmetrical C-H stretching mode, a single band (2847 cm^{-1}), shows that we have either hexagonal or orthorhombic lateral packing [5]. The spectra in the C-H bending region show that an orthorhombic lattice is not present as the line at around 1420 cm^{-1}, which is characteristic of this type of packing [3], is not detected. Therefore hexagonal packing of surfactant molecules in all-trans conformation apparently exists in Gillette foam. Such packing will lead to breaking of the selection rules for IR active modes. As a result the component at 2916 cm^{-1} appears in the C-H Raman spectra, Fig.2.

After about 2 hours of ageing a clear isotropic liquid drains off from the Plateau channels in foam. A comparison of the intensities of the water Raman O-H stretching band and the C-H stretching band in foam and in the liquid showed that water is the dominant constituent in the drained off liquid.

The analysis of the Raman data proves that Gillette foam can be considered as comprising two phases - isotropic liquid and a liquid crystalline phase (gel). Our

Raman studies of ageing demonstrate that foam has the ability to retain the liquid crystalline phase while the isotropic liquid is gradually released along Plateau's channels. This makes us believe that the gel phase is strongly connected to the surface of the bubbles. The origin of the gel phase could be explained by the gradual orientation of the small lamellae, initially dispersed in the liquid films, parallel to the bubbles interface and their subsequent organization in large bilayer structures. The process of lamellae growth can be followed by changes in R. When we have a dispersion of small lamellae, part of the Raman signal is due to the molecules at their curved ends. These molecules do not take part in long-range interactions characteristic of the main crystal-like part of the lamellae. As a result their C-H stretching band will have smaller ratio R than that of the C-H band, characterizing the vibrationally coupled molecules from the main body. When the lamellae group together the number of the molecules at their curved ends will decrease and R will increase. A similar effect has been observed in dispersions of biological membranes [6]. The increase of R from 1.55 for fresh foam to 1.75 for foam aged 72 hours, Fig.2, confirms this model about the molecular structure of the foam films.

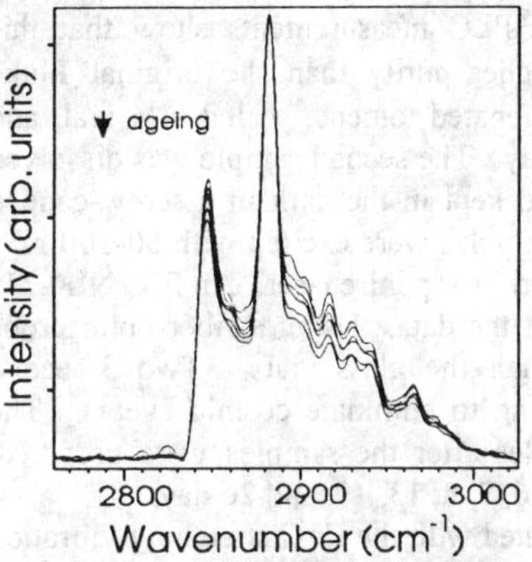

Fig.1. Deconvoluted Raman spectra in the C-H stretching region at different stages of ageing a sample of foam in a cell up to 72 hours.

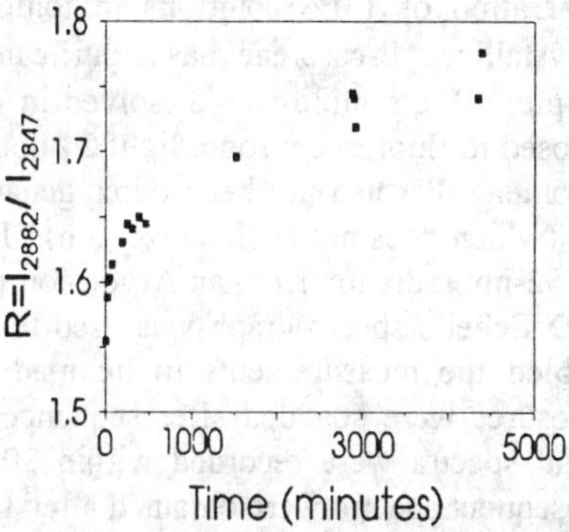

Fig.2. Dependence of the intensity ratio R of the C-H stretching band on the time elapsed from formation of foam.

1. P.M. Kruglyakov in Thin Liquid Films, Fundamentals and Applications, (I.B. Ivanov, ed.), Dekker, New York, 1988, Vol. 11, Chapter 11.
2. M. Picquart, J. Phys. Chem., 90, 243 (1986).
3. I.W. Levin Advances in Infrared and Raman Spectroscopy, (R.J.H. Clark and R.E. Hester ed.), John Wiley & Sons, London, 1984, Vol. 11, Chapter 1.
4. M. Harrand and M. Masson, J. Chem. Phys., 87, 5176 (1987).
5. R.G. Snyder, S.L. Hsu, and S. Krimm, Spectrochim. Acta, A34, 395 (1978).
6. M. Lafleur, J.L. Dasseux, M. Pigeon, J.Dufourcq, and M. Pezolet, Biochemistry, 26, 1173 (1987).

NEAR-INFRARED RAMAN STUDIES OF β-CAROTENE DEGRADATION

R. C. Reeder and C. Marcott*

The Procter & Gamble Company, Miami Valley Laboratories, Cincinnati, OH, 45253-8707, USA

β-Carotene is notoriously sensitive to degradation when exposed to light, oxygen, or especially to both. In this paper we explore the use of near-infrared Raman spectroscopy to monitor β-carotene degradation. Exciting at 752 nm, outside the main absorption bands of β-carotene between 380 and 520 nm, should minimize decomposition during measurements of Raman spectra. The β-carotene (Fluka, >97%) used was recrystallized three times from deoxygenated toluene prior to preparation of 1.0% solutions in toluene. HPLC measurements show that this recrystallized β-carotene has significantly higher purity than the original Fluka sample. One sample was dissolved in deoxygenated toluene, sealed in a vial, and exposed to fluorescent room light 24 hours a day. The second sample was dissolved in toluene that had not been deoxygenated and kept in the dark in a screw-capped vial (which does not exclude oxygen). Both samples were excited with 80-100 mW of 752-nm radiation from an Argon-ion-pumped Ti-sapphire laser. An EIC NIR-775 CCD Echelle spectrograph was used to collect the data. Using a fiber-optic probe enabled the measurements to be made through the glass vials. Two 3-second exposures were coadded after sequence filtering to eliminate cosmic events. The initial spectra were recorded within 30 minutes after the samples were prepared. Subsequent spectra were obtained after 1, 2, 5, 6, 7, 8, 13, 19, and 26 days.

Figure 1 shows the Raman signal associated with the C=C stretching vibration of β-carotene for the sample continuously exposed to room light, but not oxygen. All spectra were scaled to equalize the intensity of the toluene bands. Toluene has no absorption features in the region displayed. The arrows indicate that the intensity of the 1522-cm^{-1} band decreases as a function of time, while a shoulder at 1540 cm^{-1} grows in. Similarly, Figure 2 shows the Raman signal in the same spectral region of the β-carotene sample exposed to oxygen in the dark. The intensity of the 1522-cm^{-1} band in this sample decreases significantly faster than it does for the oxygen-free sample stored in the dark (Fig. 1). However, there is no evidence for the formation of the 1540-cm^{-1} shoulder in the sample exposed to oxygen. We believe the 1540-cm^{-1} band may be due to a photoisomerization product (*e.g.*, trans→cis). The decomposition products formed in the sample exposed to oxygen in the dark appear to be oxidation products containing carbonyl compounds, easily seen by IR absorption spectroscopy, but apparently not strong Raman scatterers.

We gratefully acknowledge significant contributions to this work from R. G. Laughlin and G. M. Bunke.

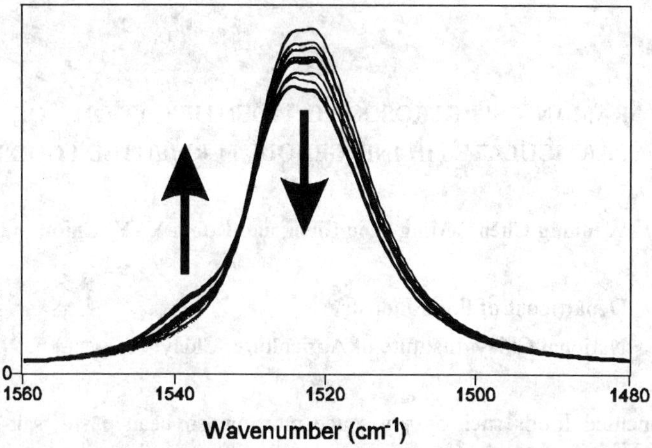

Fig. 1. Raman intensity of a 1.0% solution of β-carotene in toluene continuously exposed to room light, but not oxygen. The last spectrum shown was recorded 26 days after the first. All spectra were scaled to equalize the intensity of the toluene bands (not shown). The arrows indicate bands increasing/decreasing as a function of time.

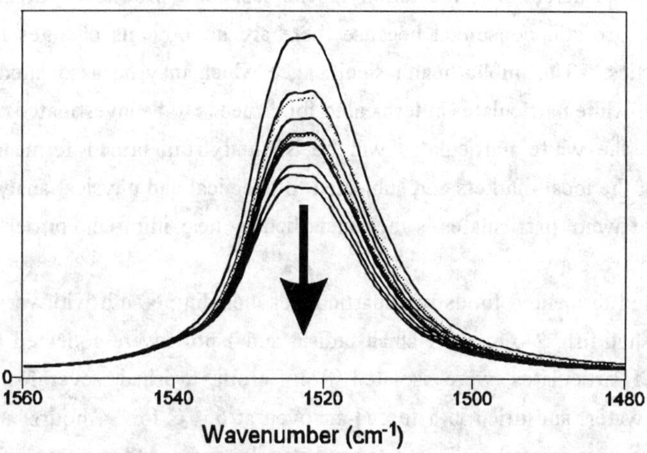

Fig. 2. Raman intensity of a 1.0% solution of β-carotene in toluene exposed to oxygen, but kept in the dark. The last spectrum shown was recorded 26 days after the first. All spectra were scaled to equalize the intensity of the toluene bands. The arrow indicates the major band is decreasing as a function of time.

RAMAN SPECTROSCOPIC IDENTIFICATION OF WHITE PARTICULATES IN INDIGENOUS FERMENTED FOODS

Wenlung Chen*, Ming-Jane Chen, and Robin Y.-Y. Chiou

Department of Food Industry
National Chiayi Institute of Agriculture, Chiayi, Taiwan

Fermented foods such as soy sauce moromi, soybean paste, salt-preserved shell fish, finfish, and pork are traditional and popular oriental foods. The principal advantage of food fermentation is to enhance flavor and to prevent spoilage [1-3]. Salt is usually added to fermented foods to extend its shelf-life and to protect it from microbiological spoilage. However, certain white particulates, usually attached to the surface of solid particles, is occasionally observed in brined fermented foods. Most consumers and some producers believe that the appearance of white particulates is a defect of fermentation. Whereas, some products containing white particulates are still consumed because there are no obvious changes in sensory characteristics. The public health significance which may be associated with the presence of white particulates in fermented foods needs to be investigated. Thus, in this study, the white particulates will be isolated from brined fermented foods available in the local markets and subjected to chemical and physical analysis. The structure of white particulates is then identified by near infrared Fourier-transform Raman spectroscopy.

Brined fermented foods with particulates including 6 sufu with various brand names, 3 shell fish, 2 oysters, 1 small finfish, and 1 pork were collected from local markets. Particulates were isolated from broth, washed several times with deionized water, and dried in a forced-air oven at 65 °C for 24 hours. The dried particulates were ground with a mortar into powder and used for amino acid analysis. The amino acid analysis showed that the major constituent of white particulates was tyrosine. Based on the Raman spectroscopic study, it also proved that tyrosine was the mainly composition of white particulates as demonstrated by the identical Raman spectra in Figure 1 and Figure 2.

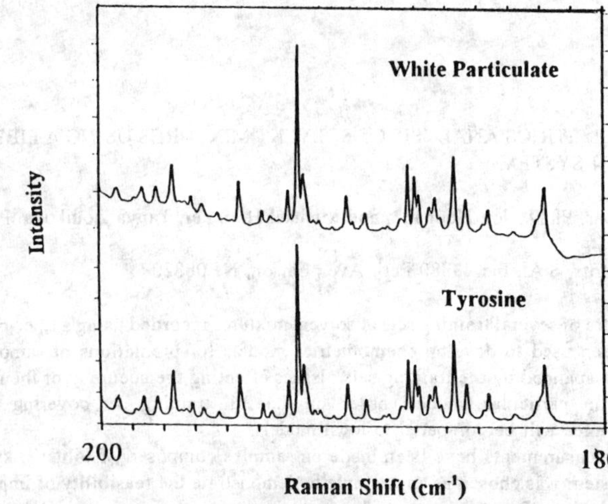

Figure 1: Raman spectra of white particulate and tyrosine between 200-1800 cm^{-1}.

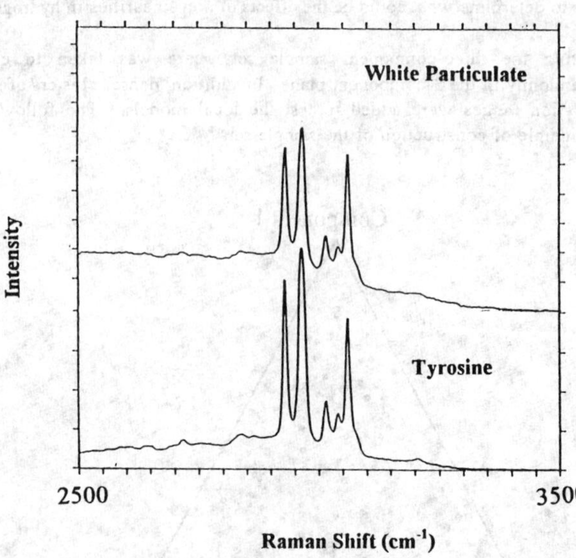

Figure 2: Raman spectra of white particulate and tyrosine between 2500-3500 cm^{-1}.

(1) L. R. Beuchat (1987) In L.R. Beuchat (ed.) Food and beverage mycology, 2nd ed. pp. 269-306, Van Nostrand Reihold, New York.

(2) C. W. Hesseltine and H. L. Wang (1967) Biotechnol. Bioeng. 9, 275-288.

(3) D. Fukuushima (1989) In K. H. Steindraus (ed.) Industrialization of indigenous fermented foods, pp. 1-88, Marcel Dekker, Inc.,, New York and Basel.

CHEMOMETRICS ANALYSIS OF SOLVENT MIXTURES USING A FIBER OPTIC BASED RAMAN SYSTEM

Fran Adar, Ph.D., Jon Noonan*, and Michael Hommer, Tanya Zouikova, Ph.D.

Instruments, S.A., Inc., 3880 Park Ave., Edison, NJ 08820

Raman spectra of several training sets of solvent mixtures recorded using a fiber-optic based system have been used to develop chemometrics models for predictions of samples whose compositions are spanned by the training sets. Issues effecting the accuracy of the model will be addressed. In particular, models based on the full training set covering the entire concentration ranges will be compared to local models.

One set of measurements have been made on samples composed of toluene, xylene, and decane. This system was chosen as an example to demonstrate the feasibility of implementing multivariate analysis to Raman detectors for analyzing solvent mixtures and solvent extracts during manufacturing and/or re-cycling. Another chemical system including alcohols have been modeled in order to determine what could be the effects of non-linearities in hydrogen-bonded systems.

In constructing the three-component sample sets, care was taken to cluster the concentrations randomly in the 3-component plane. In addition, denser clusters of samples in limited concentration ranges were added to test the local models. The following figure illustrates the principle of construction of the sample set.

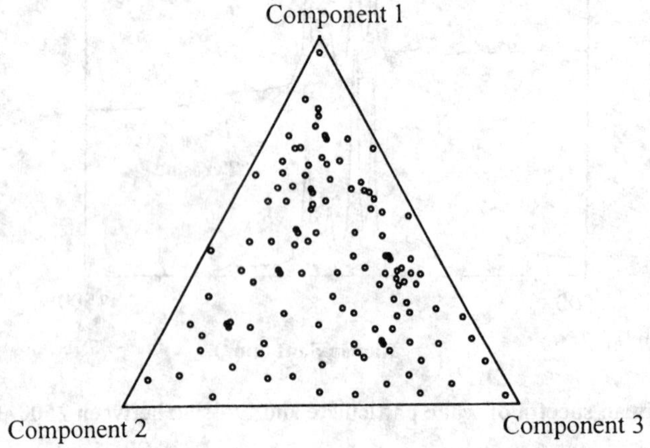

The samples were interrogated with a Dilor Superhead in standard scintillation vials. The spectra were analyzed with GRAMS PLS-Plus Chemometrics software. Multiple models were developed from a given data set in order to optimize the accuracy of the predictions. Spectral pretreatments were selected from numerous possibilities including taking derivatives, stripping baselines, performing mean centering (MC), applying multiplicative scattering correction (MSC), and applying variance scaling (VC). PCR (Principal Component Regression), PLS-1 and PLS-2 (Partial Least Squares regressing one on component, or all components at a time) were used to develop the models. Cross-validation was used as the criterion for the quality of the model; visual inspection of the results as well statistical tests aided in the selection of the best model. It was empirically found that taking the derivative can be useful in eliminating the deleterious effects of random baselines on the quality of the model. It was also observed that MC and MSC improved the model whereas VC usually did not. In general, the larger the number of samples in the training set, the better was the model. The quality of the model was also affected by the design of the sample set. Samples with concentrations clustered in limited ranges in the vicinity of the validation samples produced better results than samples with widely scattered concentration values.

A FOURIER TRANSFORM RAMAN SPECTROSCOPIC STUDY OF THE IODINE β-CYCLODEXTRIN COMPLEXES

Huijun Tian*[1], Bingyu Liu[2], Zhanjun Song[2], Yongquan Lu[2]; Hongbei Chang[3], Tingsen Shi[3]

[1]Department of Chemistry, University of Pittsburgh, Pittsburgh, PA15260, USA
[2]National Center of Biomedical Analysis; Beijing 100850, P. R. China
[3]Beijing Four Ring Pharmaceutical Factory, Beijing 100850, P. R. China

Pure iodine is volatile, hydrophobic, irritant and cannot be used directly as an antiseptic in the pharmaceutical industry. Many iodophors have been made in order to increase its solubility in water, improve its stability, mask its smell and suppress its irritation[1]. The iodine--β-cyclodextrin(β-CD) is one of these important iodophors used in clinics and widely used as the antiseptic in China. We found the stability of the iodine--β-CD complex depends strongly on the nature of the medium used for the complexation, and that the potassium ion affects the stability too. The iodine contents are varied with the medium in the complexation, and the state of iodine in the iodine--β-CD complexes is uncertain.

In this paper, we reported the FT-Raman spectroscopic study results of five iodine--β-cyclodextrin(CD) complexes obtained in different media of complexation. Elemental analysis results of five iodine--β-CD complexes are shown in table 1. In the range of 250-60 cm^{-1}, the Raman spectra of complexes A, B, C and D(formed in ethanol or mixed ethanol-water media) are very similar to each other, having three peaks 110, 141, 166 cm^{-1}, however, the intensities of these peaks in each spectrum are not the same (illustrated in table 2); and the spectrum of the complex E (formed in the water solution) is different, having four peaks 110, 141, 166, 180 cm^{-1} after curve fitted. The relative intensity of 166 cm^{-1} to 110 cm^{-1} is roughly proportional to the iodine contents in the complexes. We obtained the FT-Raman spectra of solid I_2 and KI_3, and the Raman bands are in good agreement with the results obtained by resonance Raman spectrometer. In general, bands near 105 cm^{-1} have also been observed for the pentaiodides as in references [2, 3].

A comparison of our results with the previous studies of polyiodides in the literature[2-5] indicates the strong band at 110 cm^{-1} is related with the triiodide I_3^-. When the KI was not added in the reactant solution, we still got the complex with a weaker Raman band at 141 cm^{-1}. The band at 141 cm^{-1} may be assigned to the assymmetrical I_3^-. We concluded that the iodine--β-CD complex E prepared in water contains the components of I_2, I_3^- and I_5^-, those components correspond to the Raman bands at 180, 110, 166 cm^{-1} respectively; while the complexes A, B, C, D contain the components of I_3^- and I_5^-, corresponding to the Raman bands at 110, 166 cm^{-1}. As the elemental analysis results are different from the theoretical values of $(\beta\text{-CD})_2\text{-KI}_3$(C 37.48 %, H 5.20 %, I 9.43 %) or $(\beta\text{-CD})_2\text{-KI}_5$ (C 34 %, H 4.75 %, I 17.2 %), we think the included form of iodine in these five complexes has more than one single state. The iodine content in the complex can be roughly determined by the FT-Raman relative intensity.

Table 1 The elemental analysis results of iodine--β-CD complexes

Complexes	E:W(v:v)*	Color	C(%)	H(%)	Iodine(%)
A	100 : 0	yellow	33.3	4.75	12.31
B	75 : 25	yellow	32.2	5.10	8.48
C	50 :50	brown	32.1	4.65	10.58
D	30 : 70	brown	30.4	4.65	14.45
E	0 : 100	black	31.3	4.56	18.68

* E: ethanol, W: water, v: volume

Table 2 Raman band intensities of iodine--β-CD complexes

Complexes	110 cm^{-1}	141 cm^{-1}	166 cm^{-1}	180 cm^{-1}	$I_{166}/(I_{166}+I_{110})$
A	6.65	5.01	3.94		0.372
B	10.3	2.33	2.08		0.168
C	6.87	4.23	6.61		0.490
D	3.35	4.74	10.31		0.755
E	29.4	11.3	55.4	39.4	0.638

1. L. Song and Q. Meng, *J. Inorg. Chem.* **10(4)**, 370(1994)
2. W. Yao; S. Feng; X. Wang and et al, *Journal of Academy of Military Medical Sciences*, **15(4)**,240 (1991)
3. M. Noltemeyer and W. Saenger, *Nature* **259**, 629(197)
4. 4. I. Harada; Y. Furukawa and M. Tasumi, *J. chem. Phys.* **73**, 4746(1980)
5. E. M. Nour; L. H. Chen and J. Laane, *J. Phys. Chem.* **90**, 2841(1986)

RAMAN SPECTROSCOPY STUDIES ON SHAPE-SELECTIVE CATALYSTS OF Sb_2O_3/HZSM-5

Kaixuan Wang*, Gang Yu, Biying Zhao, Youchang Xie

Institute of Physical Chemistry, Peking University, Beijing, 100871, China

Zeolite HZSM-5 due to its special structure is very important to improve selectivity of catalysis. Modification of HZSM-5 with Sb was used in synthesis of para-oxyane. The reasons were thought as narrowing of HZSM-5 channel or poisoning of strong acidic sites, respectively. It has been reported [2] that zeolites HZSM-5 with different content of Sb_2O_3 calcinated at 550°C could cause different catalystic activity and selectivity of reaction of toluene with methanol. It was found that when the content of Sb_2O_3 on zeolite HZSM-5 was increased to a certain level, the activity of this reaction was lowest, but its selectivity was highest. This "conversion point" could be corresponded to the threshold of Sb_2O_3 monolayer dispersion on zeolite HZSM-5[1].

In this paper, we provide the method to determine the threshold by Raman scattering spectra. Raman Spectroscopy can be used to measure the crystalline Sb_2O_3 remained on HZSM-5, and then its relative content can be obtained by calibrating with peaks of internal standard KNO_3 or other standard such as plasma lines.

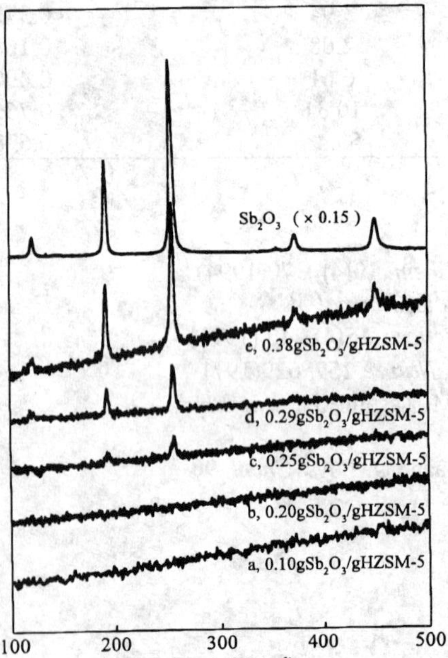

Fig.1 Laser Raman spectra of samples prepared by heating Sb_2O_3 and HZSM-5 mixtures at 500°C for 2 hours.

The samples of Sb_2O_3/HZSM-5 were prepared by mixing Sb_2O_3 with HZSM-5(SiO_2/Al_2O_3=50), then grinding them with ball mill for 20 minutes and finally calcinating them at 500°C in atmosphere of N_2 for 2 hours.

Pure Sb_2O_3 crystal powders cause very strong Raman scattering peaks in 85, 121, 192, 256, 375, 453 cm^{-1} (Fig 1). But there are no obvious signals of Raman scattering spectrum of pure HZSM-5 in this wavelength range. We have found that after the mixtures of Sb_2O_3 and HZSM-5 were calcinated in flow N_2 gas at 500°C, in their Raman spectra there are no Raman peaks of other Sb oxides except for Sb_2O_3. It is shown in Fig. 1 that when the content of Sb_2O_3 on HZSM-5 is lower than 0.25 gSb_2O_3/gHZSM-5, no obvious Raman peaks are observed, but when larger than 0.25 gSb_2O_3/gHZSM-5 the Raman scattering intensities of residual crystalline Sb_2O_3 are enlarged with increasing of the Sb_2O_3 to HZSM-5 weight ratio..

In addition, we have used two method to quantify the crystalline Sb_2O_3 which remained on zeolite HZSM-5 . One is using KNO_3 which has a strong Raman scattering at 1050 cm^{-1} as internal standard . The relative intensities of Sb_2O_3 Raman peaks at 192

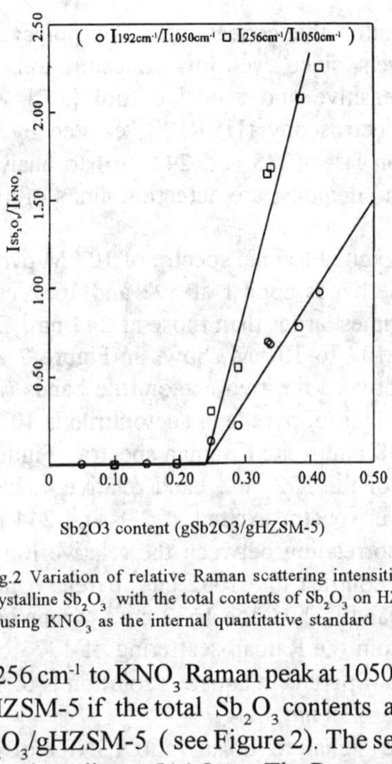

Fig.2 Variation of relative Raman scattering intensities of crystalline Sb_2O_3 with the total contents of Sb_2O_3 on HZSM-5 using KNO_3 as the internal quantitative standard

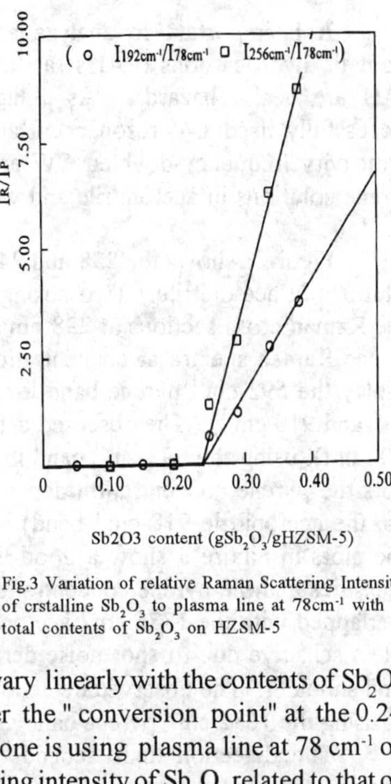

Fig.3 Variation of relative Raman Scattering Intensities of crstalline Sb_2O_3 to plasma line at 78cm^{-1} with hte total contents of Sb_2O_3 on HZSM-5

and 256 cm^{-1} to KNO_3 Raman peak at 1050 cm^{-1} vary linearly with the contents of Sb_2O_3 on HZSM-5 if the total Sb_2O_3 contents are over the " conversion point" at the 0.24 gSb_2O_3/gHZSM-5 (see Figure 2). The second one is using plasma line at 78 cm^{-1} of exiting laser line at 514.5 nm. The Raman scattering intensity of Sb_2O_3 related to that of plasma line at 78 cm^{-1} is shown in Fig. 3. The same threshold (0.24 gSb_2O_3/gHZSM-5) can be obtained by these two quantitative methods. The latter seems more simple and accurate since the homogeniousity of mixing internal standard with samples can be ignored. In order to confirm the results of quantitative Raman analysis, XRD experiments have been done and give the same threshold (0.24 gSb_2O_3/gHZSM-5) as Raman measurement, and if the total Sb_2O_3 contents are over 0.24 gSb_2O_3/gHZSM-5 the intensities of XRD peaks of crystalline Sb_2O_3 are observed clearly and increase linearly with the total Sb_2O_3 contents.

1. Xie, Y. C., Tang, Y.Q., Advances in Catalysis, 37, 1-43 (1990)
2. Li, G.Y., Zhao, J.C., Petroleum Chem. Engin. (Chinese), 16(4), 276(1987)

UV RESONANCE RAMAN DETECTION OF PYRENE AT TRACE LEVELS

Zhenhuan Chi and Sanford A. Asher*

Department of Chemistry, University of Pittsburgh, Pittsburgh, PA 15260

It is important to analyze and monitor the concentration of polycyclic aromatic hydrocarbons (PAHs) at trace levels, since even low concentrations of PAH are health hazards. As a highly sensitive and selective tool [1,2], we successfully used UV resonance Raman spectroscopy (UVRRS), excited by an intracavity frequency-doubled CW argon ion laser (238 and 244 nm) to analyze pyrene solutions in acetonitrile and water and demonstrate detection limits of 10^{-9} M.

Figure 1 shows the 238 and 244 nm excited Raman spectra of 10^{-5} M pyrene solutions in acetonitrile. Two strong pyrene bands appear at 592 and 1632 cm^{-1}. The Raman cross sections at 238 nm are 8 times larger than those at 244 nm. The pyrene Raman spectra at concentrations of 10^{-7} to 10^{-9} M shown in Figure 2 also display the 592 cm^{-1} pyrene band located between the two acetonitrile bands (376 cm^{-1} and 918 cm^{-1}). The observed detection limit of pyrene in acetonitrile is 10^{-9} M (200 ppt), using the 592 cm^{-1} band in the 238 nm excited Raman spectra. Figure 3 plots the pyrene concentration dependence of the 592 cm^{-1} band relative intensity (vs. the acetonitrile 918 cm^{-1} band) in Raman spectra excited at 238 and 244 nm. The plots in Figure 3 show a good linear correlation between the relative Raman intensities and the pyrene concentration. In contrast, the 1632 cm^{-1} pyrene band is overlapped with the 1376 cm^{-1} acetonitrile band. Thus, the 1632 cm^{-1} pyrene band is less sensitive due to shot noise derived from the Raman scattering of 1376 cm^{-1} band shoulder. The observed detection limit of pyrene in aqueous solution is 6×10^{-9} M, using the 1632 cm^{-1} pyrene band excited at 238 nm.

The detection limit depends on the signal to noise ratio (SNR) of the spectra. Compared to excitation with a YAG pulsed laser[1], a CW Ar^+ laser is able to provide 10^7 fold lower peak power at the same average power and dramatically decreases the saturation and other nonlinear effects in the Raman excitation process.

The dominant noise shown in the Raman spectra (Figure 2) is shot-noise derived from Raman scattering of the solvent. Therefore, in principle we can improve SNR by increasing the accumulation time. We also examined the simulated spectral signal to determine the possible ultimate detection limit for pyrene in our measurement. We find that a very small frequency shift of the maximum signal intensity (peak-shift) occurs for a measurement time of greater than 500 seconds. The peak-shift is attributed to thermal fluctuations in the spectrometer environment. The noise associated with the peak-shift decreases the sensitivity of the Raman measurements and determines the ultimate detection limit.

Our results demonstrate that the CW Ar^+ laser used for UVRRS is a superb excitation source to analyze PAHs at trace levels.

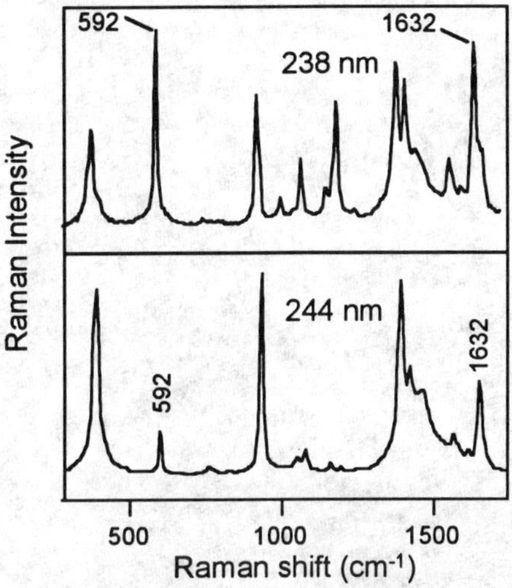

Figure 1 Pyrene Raman spectra in acetonitrile. The shade bands are acetonitrile vibrations.

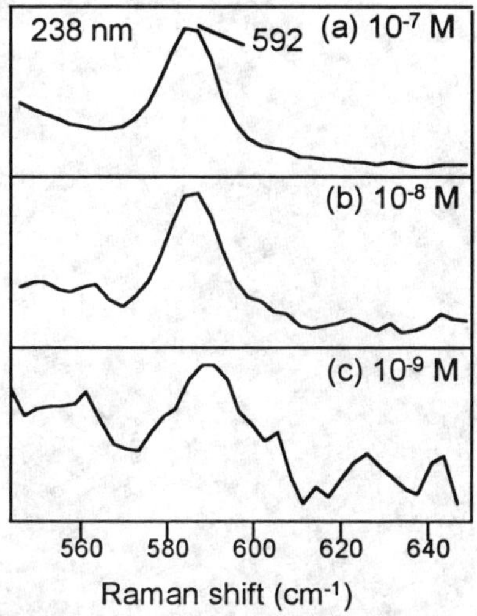

Figure 2 Pyrene Raman spectra In (b) and (c) a CH_3CN spectrum is subtracted.

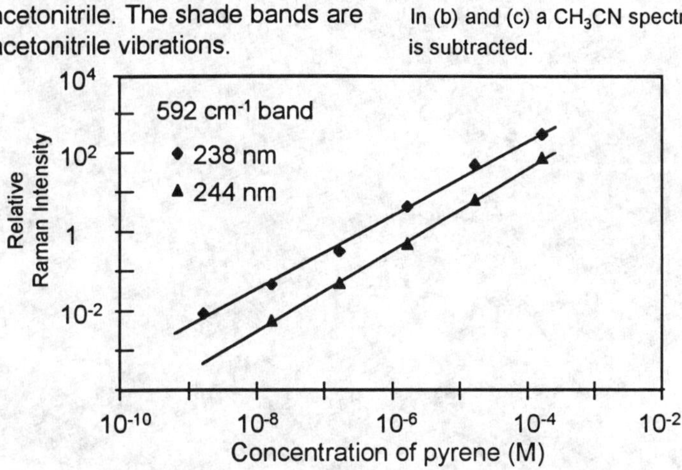

Figure 3 Concentration dependence of Raman intensity

REFERENCES

[1] S.A. Asher, *Anal. Chem.* 1984, **56**, 720.

[2] S.A. Asher, *Anal. Chem.*,(a) 1993, **65**, 59A. (b) *Anal. Chem.* 1993, **65**, 201A.

SECTION 27

MEDICAL APPLICATIONS

THE STUDY OF PARKINSON'S DISEASE BY MICRO-RAMAN SPECTROSCOPY

L. Qin[1], Z. X. Shen[1*], T. Lee[2], F. Watt[1], S. H. Tang[1] and M. H. Kuok[1]

[1]Department of Physics, National University of Singapore, Lower Kent Ridge Road, Singapore 119260.

[2]Department of Surgery, National University of Singapore, Lower Kent Ridge Road, Singapore 119260

Parkinson's disease is a common disease of the elderly which involves disruption of the motor system with symptoms of tremor and rigidity. The disease is associated with a deficiency of the neurotransmitter dopamine, following the loss of dopamine-producing neurons in the substantia nigra region of the brain. Despite intense worldwide research into this disease, the mechanism for dopamine cell death is still not known. Very recent work by the Nuclear Microscopy group in collaboration with the NUH has shown that high trace iron levels are associated with dopaminergic cell death [1]. This work reinforces the theory that in its unregulated form, iron promotes redox reactions and increases the probability of free radical cellular damage [2]. In order to confirm this theory however, we need to know whether or not the excess iron found in the lesioned substantia nigra is free or is bound. Vibrational spectroscopy is ideally suited for this study and Raman spectroscopy has demonstrated its usefulness in the study of biological samples [4,5]

In this study, 6-OHDA was injected into the medical aspect of the substantia negra (SN) of the male adult Spraque Dawley rats. The rats were tested for the adequency of the lesioning two weeks after the injection. Brain samples were prepared by cryosectioning, and contain both lesioned SN section and control SN section.

Preliminary results using the Renishaw System 2000 Ramascope are encouraging. As shown in Fig.1, these results show a difference between brain tissue exhibiting Parkinsonian symptoms (lesioned SN) and normal tissue (control SN). Two excitation sources from a He-Ne laser (λ = 632.8 nm) and a silicon diode laser (λ = 784 nm) are used in the experiments to reduce fluorescence in different regions and to assess the repeatibility of the results. This early result shows definite potential, particularly when the instrument offers micron resolutions thereby allowing study at the cellular level.

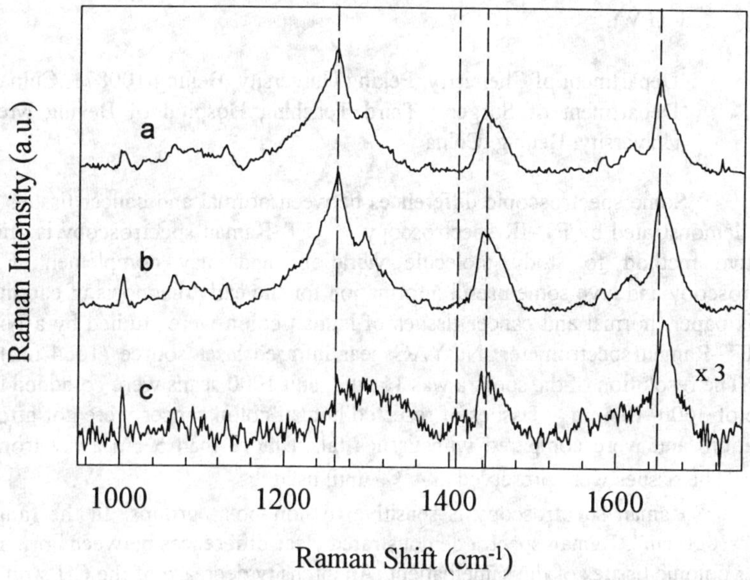

Fig. 1 Raman spectra of a rat brain sample recorded with a HeNe laser. The power used is 3 mW, and integration time 10 seconds × 180. (a) spectrum of control SN; (b) spectrum of lesioned SN; and (c) difference spectrum of b-a. Spectra a and b are background subtracted but unsmoothed.

1. F. Watt T. Lee, P. Thong and S. M. Tang, Nuclear Instr. and Meths. B 104, 361(1995).
2. C.W. Olanow, Ann. Neurol. 32 S2 (1992)
3. D. T. Dexter et al, J. Neurochem. 52, 381 (1988).
4. A. Mizuno, T. Hayashi, K. Tashibu, S. Maraishi, K. Kawauchi and Y. Ozaki, Neurosci. Let. 141, 47 (1992).
5. C. J. Frank, R. L. McCreery and D. C. B. Redd, Anal. Chem. 67, 777 (1995).

FT–RAMAN SPECTROSCOPIC CHARACTERIZATION OF THE NORMAL AND CANCER TISSUES OF HUMAN COLON

W–H Li[1], Q. Peng[1], D–M Yu[1], D–W Zhao[2], S–F Weng[1], X–S Zhou[2] and J–G Wu[1]*

[1]Department of Chemistry, Peking University, Beijing 100871, China
[2]Department of Surgery, Third Teaching Hospital of Beijing Medical University, Beijing, China

Some spectroscopic differences between normal and cancer tissues have been demonstrated by FT–IR spectroscopy [1,2]. FT–Raman spectroscopy is another effective method to study molecule vibration, and may complement FT–IR spectroscopy and give some useful information for the early diagnosis of carcinoma. In this paper, normal and cancer tissues of human colon were studied by a Nicolet 910 FT–Raman spectrometer. Nd:YAG near infrared laser source (1064 nm) was used. The resolution of the spectra was 16 cm^{-1}, and 1000 scans were co-added in the range of 3600–100 cm^{-1}. Tissues of resected human colon cancer were got after the operation, and were compared with tissues taken more than 5 cm away from the tumor. The tissues were preserved at 4 ℃ until used.

Raman spectroscopy is sensitive to non–polar groups. In the range of 2800 –3000 cm^{-1}, Raman spectra demonstrated clear differences between normal and cancer colonic tissues of the same patient. An intensity decrease of the CH stretching band was found at 2934 cm^{-1} in malignant tumour. In addition, the relative intensity of 2934cm^{-1} and 3270 cm^{-1} (I_{2934}/I_{3270}) is 1.2 in the spectrum of cancer tissue and is 2.6 in that of normal (shown in Fig. 1). There is a weak band located at 1744 cm^{-1} which is assigned to the COOH stretching vibration of fatty acid in the Raman spectra of normal tissue. But it disappeared in the cancer tissues. The result is in correspondence with our previous IR study of normal and cancer tissues[3], and it suggest that there may be fewer phospholipids or fatty acids present in cancer tissues than normals. The amide I band which relate to the C=O stretching vibration of protein is also present at 1650–1660 cm^{-1} in Raman spectra. But the bandwidth of amide I in cancer tissues is wider than normals, indicating the fragments of secondary structure of protein in cancer tissues are more than normals, and some denatured protein may exist in malignant tissues. The differences of bandwidth of three patients are given in table 1. Peaks in the 1000–1300 cm^{-1} region in cancer tissues are diffuser than normals, suggesting the structure of phospholipid also changed in malignance.

* The project was supported by NSF and Doctoral Foundation of National Education Commission of China

We conclude that FT–Raman spectroscopic detection of the differences between normal and cancer tissues provide useful results. Raman and IR spectroscopies can complement each other and develop the method of early screening for cancer.

TABLE 1. Bandwidth($w_{1/2}$) of amide I of normal and cancer tissues in FT–Raman spectra.

	patient 1		patient 2		patient 3	
	peak(cm^{-1})	$w_{1/2}$ (cm^{-1})	peak(cm^{-1})	$w_{1/2}$ (cm^{-1})	peak(cm^{-1})	$w_{1/2}$ (cm^{-1})
normal	1652.8	30.0	1654.8	60.0	1653.5	37.5
cancer	1660.1	67.5	1654.8	75.0	1657.8	75.0

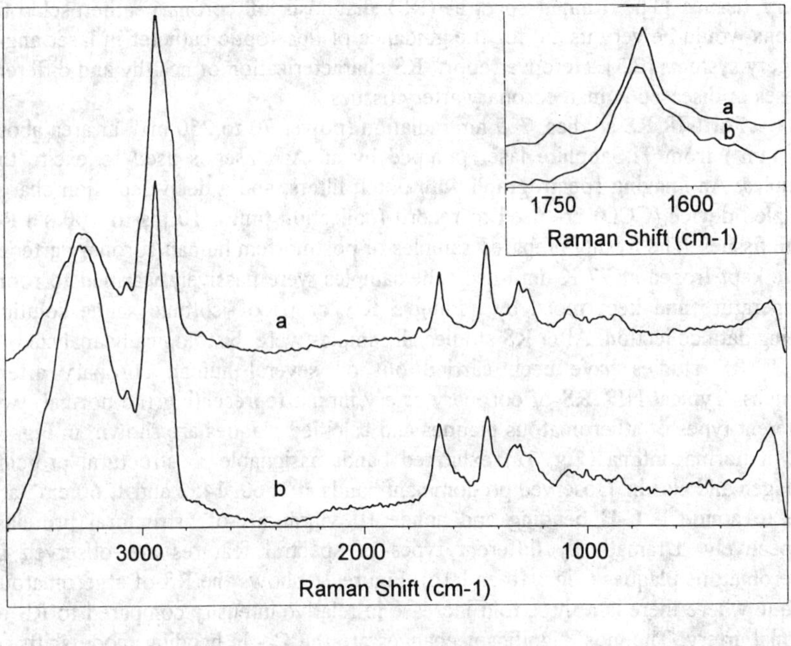

Fig. 1. FT–Raman spectra of normal(a) and cancer(b) colonic tissues

1. Wong P. T. T., Papavassilion E. O., Ring B., Applied Spectroscopy, 45, 1563 (1991).
2. Q. Peng, R. D. Soloway, J–G Wu and X–S Zhou et. al., Annual Meeting of American Gastroenterological Association (AGA) and American Association for the Study of Liver Diseases, May 19-22 (1996).
3. D–W Zhao, X–S Zhou, R. D. Soloway, Q. Peng and J–G Wu et. al., ibid.

DIAGNOSIS OF HUMAN CORONARY ARTERY WITH NEAR INFRARED RAMAN SPECTROSCOPY

S. Sathaiah[1]*, L. Silveira Jr.[1], C. A. G. Pasqualucci[2], R. A. Zângaro[1], M. C. Chavantes[2], and Marcos T. T. Pacheco[1]

[1]Center for Bio-Molecular Spectroscopy (NEBIOM),
Universidade do Vale do Paraíba, São José dos Campos, SP, Brazil.
[2]Instituto do Coração, São Paulo, SP, Brazil.

Near infrared (NIR) Raman spectroscopy provides a powerful non destructive method for repetitive analysis of biochemical composition of human artery tissues [1]. Raman spectral (RS) diagnosis of coronary atherosclerotic lesions would be very useful for the guidance of fiber-optic catheter in laser angio surgery systems [2,3]. Here we report, RS characterization of healthy and different phases of diseased human coronary artery tissues.

For NIR RS studies, 785 nm radiation (power 70 to 250 mW at area about 0.5 mm^2) from Ti:Sapphire laser pumped by an Ar^+ laser is used to excite the samples. An imaging spectrograph with notch filters; and a deep-depletion charge coupled device (CCD) are used to record (collection times: 10 ms to 100 s) RS from tissues. The freshly prepared samples of postmortem human coronary arteries were kept frozen at 77 K until use. The samples were passively warmed to room temperature and kept moist by adding a few drops of isotonic saline solution during data collection. After RS studies all samples were histologically analyzed.

RS studies have been carried out on several human coronary artery samples. Typical NIR RS of coronary artery intima representing the normal, two different types of atheromatous plaques and calcified plaques are shown in Fig. 1. RS of normal intima (Fig. 1a) exhibited bands assignable to structural proteins collagen and elastin. Observed predominant bands at 1660, 1452 and 1260 cm^{-1} are due to amide I, C-H bending and amide III vibrations of structural proteins, respectively. Dramatically different types of spectral features are observed in atheromatous plaques (Figs. 1b and 1c). Figure 1b shows the RS of atheromatous plaque where there is a three fold increase in relative intensity compared to RS of normal artery. The most significant changes are the C - H bending mode shifts to lower frequency and appears as an intense band at 1441 cm^{-1} which can be attributed to the accumulated lipids in the necrotic core. Further, by comparing the RS with that of several pure lipids, the appearance of a relatively sharper band at 1670 cm^{-1} with a bandwidth around 40 cm^{-1} in contrast to the broader (bandwidth 60 cm^{-1}) amide I band at 1661 cm^{-1} gives a clear evidence that the accumulated lipids are mainly due to cholesterol. Figure 1c shows a typical RS of another type of atheromatous plaque in which an anomalous (eight fold) increase in the relative intensity compared to RS of normal artery has been observed. The major bands at 1442, 1665, 1301, 1265, 700 cm^{-1} and several other features between 600 to 1200 cm^{-1} are assignable to cholesterol and cholesterol esters by comparing the RS of several pure lipids. The unique bands at 1518 and 1155 cm^{-1} are pre-resonance enhanced RS bands [4] of small traces of carotenoids present in the plaque.

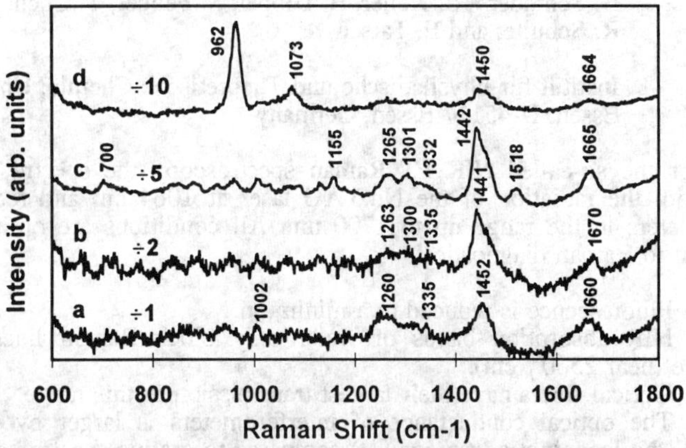

Fig. 1. Raman spectra of (a) normal ; (b) & (c) two different types of atheromatous plaques; and (d) calcified plaques of human coronary arteries. Excitation 785 nm; Power 70 mW and collected in 100 s.

Figure 1d shows a typical RS of calcified plaques. The intense bands at 962 and 1073 cm^{-1} are due to phosphate and carbonate symmetric stretching vibrations of accumulated calcium hydroxyapatite and carbonate apatites. The spectral features at 1450 and 1664 cm^{-1} are presumably due to the presence of normal tissue components that are embedded in the calcified plaque.

In conclusion, RS can clearly discriminate between healthy and diseased artery wall and thereby provides the histochemical composition of human coronary arteries with collection times as small as 10 ms which make it's clinical use feasible.

1. R. Manoharan, Y. Wang and M. S. Feld, Spectrochimica Acta A, in press, (1996); and references therein.
2. J. F. Brennan, T. J. Römer, A. M. Tercyak, Y. Wang, M. Fitzmaurice, R. S. Lees, J. R. Kramer, R. R. Dasari and M. S. Feld, Proc. SPIE, USA, 2388, 105 (1995).
3. J. R. Kramer, J. F. Brennan, T. J. Römer, Y. Wang, R. R. Dasari and M. Feld, Proc. SPIE, USA, 2395, 376 (1996).
4. J. J. Baraga, M. S. Feld and R. P. Rava, Proc. Natl. Acad. Sci. USA, 89, 3473 (1992).

NIR FT RAMAN SPECTROSCOPY IN MEDICAL DIAGNOSTICS

B. Schrader*, S. Keller, B. Dippel, S. Fendel, T. Löchte, R. Schulte, and E. Tatsch

Institut für Physikalische und Theoretische Chemie, Universität Essen, D-45117 Essen, Germany

For the so-called NIR FT Raman spectroscopy the spectra are usually excited with the radiation of the Nd:YAG laser at 1064 nm and recorded with interferometers in the range up to 1700 nm. All conditions are optimal for the application to Raman diagnostics:

- Fluorescence is reduced to a minimum.
- NIR absorption bands of water only block Raman lines of minor importance (near 2500 / cm).
- Optical fibers have their largest transmission in this range.
- The optical conductance of interferometers is larger by a factor of about 300 (the Jacquinot advantage [1]) compared to grating spectrometers.
- Ge and InGaAs detectors have their largest sensitivity in this range.

The quantum yield of the Raman effect in the NIR is, due to the ν^4 - factor, somewhat smaller than the already very small one in the visible range. For medical diagnosis it is necessary to record the Raman spectra in a limited time *in-vivo* without denaturation. These conditions force to use spectrometers and sample facilities of the largest possible figure of merit.

This can be accomplished by using multiple reflection cells and aspheric optics at the normal sample facility, and transformers of the cross-section in order to connect tissues to fiber bundles. Endoscopes are coupled to the interferometer for the recording of Raman spectra *in-vivo* and an inverse microscope for the observation of the Raman spectra of cell cultures.

We have collected Raman spectra of many different tissues, microorganisms and blood cells [2 - 5] and arrived at the following conclusions:

- We could not detect differences in the Raman spectra of living and - freshly prepared - dead tissues.
- There are reproducible differences between healthy and pathological tissues, especially cancers. However, they are small and distributed over the whole spectrum. We do not think that it is possible to tell malignant and benign tumors from benign tissues just by evaluating frequencies and intensities of distinct Raman lines [6].
- Methods of multivariate analyses may be used to develop the necessary criteria for routine Raman diagnostics [7, 8].
- It is necessary to set up a spectral data bank - hopefully in international cooperation - in which Raman spectra of any kind of tissues and cells are collected under well defined conditions in order to have a solid base for the development of programs for routine diagnostics.

Figure 1 shows the Raman spectrum of **a**, a tumor of the brain, **c**, the spectrum of the gray brain matter, and **b**, its difference spectrum. It shows typical changes in the protein structure and composition, a decrease of a cholesterol band and an increase in bands of ribonucleic acid in the tumor.

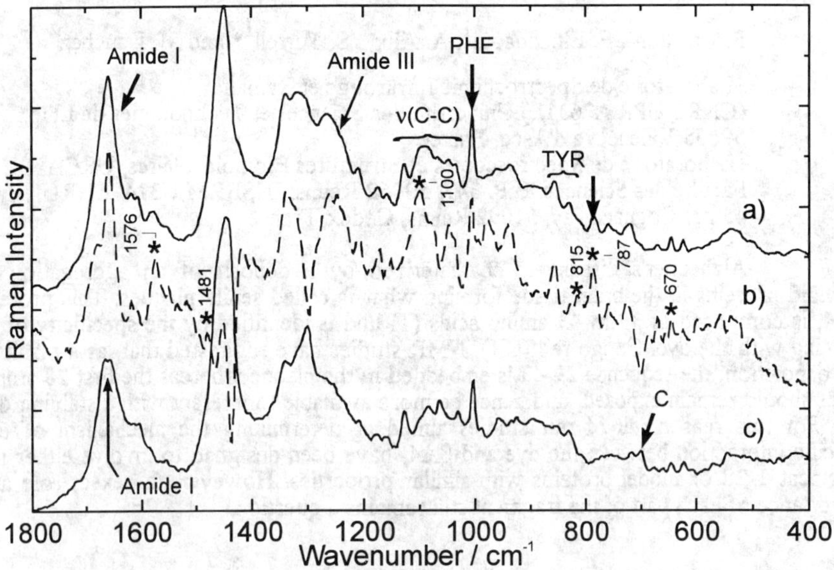

Fig. 1. NIR FT Raman spectra of **a**, a brain tumor, **c**, the gray brain matter and **b**, the difference spectrum, **a** - **c**. Positive bands describe material which is in the tumor in higher concentration, * mark bands of ribonucleic acid, **C** a band of cholesterol.

We like to thank Prof. H. de Groot, Essen, for cooperation, the Deutsche Forschungsgemeinschaft for financial and Bruker Analytische Meßtechnik, Karlsruhe, for technical help.

1. B. Schrader (ed.), Infrared and Raman Spectroscopy, Methods and Applications, VCH Verlagsgesellschaft Weinheim, **1995**.
2. S. Keller, B. Schrader, A. Hoffmann, W. Schrader, K. Metz, A. Rehlaender, J. Pahnke, M. Ruwe and W. Budach, Application of Near-Infrared-Fourier Transform Raman Spectroscopy in Medical Research, J. Raman Spectrosc. 25, 663 **(1994)**.
3. B. Schrader, S. Keller, T. Löchte, S. Fendel, D.S. Moore, A. Simon, J. Sawatzki, NIR FT Raman Spectroscopy in Medical Diagnosis, J. Mol. Struct. 348, 293 **(1995)**.
4. D. Naumann, S. Keller, D. Helm, Ch. Schultz, and B. Schrader, FT-IR Spectroscopy and FT-Raman Spectroscopy are powerful analytical tools for the non-invasive characterization of intact microbial cells, J. Mol. Struct. 347, 399 **(1995)** .
5. B. Schrader, B. Dippel, and R. Schulte, unpublished results.
6. R.R. Alfano, C.H. Liu, and W.S.Glassman, United States Patent 5 262 410, Nov. 16, **1993**.
7. M. Riedl, and B. Schrader, unpublished results.
8. G.J. Puppels, personal communication.

PREPARATION FOR THE DOCKING OF CONGO RED TO ALZHEIMER'S AMYLOID

F. Cavillon[1], E. Elhaddaoui[1], A. Alix[2], S. Turrell[1]* and M. Dauchez[2]

[1]Laboratoire de Spectrochimie Infrarouge et Raman
(CNRS UPR A2631L),Université des Sciences et Technologies de Lille,
59655 Villeneuve d'Ascq, France
[2]Laboratoire de Spectroscopies et Structures Biomoléculaires, URCA,
Faculté des Sciences, B.P. 347, 50062 Reims; INSERM U314, CHRU,
45 rue Cognacq Jay, 51092 Reims, Cedex, France

Alzheimer's disease is characterized by a deposit of an aggregation of amyloid proteins in the brain tissue forming what is called senile plaques. This protein, βA4, is composed of 39 to 43 amino acids (1) and is identified by the specificity of its staining with the dye, Congo red (CR). NMR studies have suggested that, as a result of the deposition, the sequence 29-43 is embedded in the plaque whereas the first 28 amino acids should remain exposed, and hence be more available to interact with a staining dye (2). For this reason, all of our studies aimed at determining the mechanism of this specific interaction between the dye and βA4, have been designed to involve either the fragment 1-28 or model proteins with similar properties. However, the exact role and importance of each part of the fragment still remains a question.

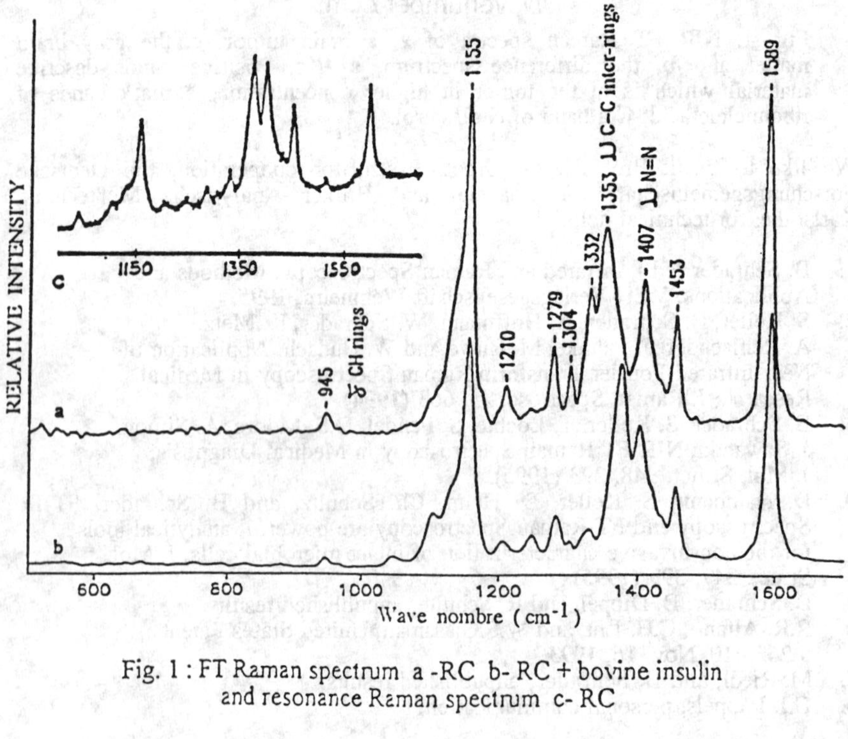

Fig. 1 : FT Raman spectrum a -RC b- RC + bovine insulin
and resonance Raman spectrum c- RC

In the present work, dye-protein interactions have been continued and centered on FT Raman studies of the CR-insulin system. In Figure 1 we have grouped the FT Raman spectra of CR alone (a), CR in interaction with insulin (b) as well as the resonance Raman spectrum of the dye alone (c). Comparison of the spectra of (a) and (c) shows that while the bands at 1155 cm^{-1} (C-N=) and 1451 cm^{-1} (C-C ring) do not change much between the resonance and non-resonance conditions, significant variations are observed in the 1360 cm^{-1} region. In effect, while the component at 1375 cm^{-1} (assigned to ν N=N) is enhanced under resonance conditions, when these conditions are removed the shoulder at 1353 cm^{-1} becomes apparent as a band. The resulting interpretation can be that the associated vibrational modes must involve vibrations of a less conjugated nature, a conclusion which is compatible with the assignment of this component to the inter-ring C-C vibration.

In Figure 1b it can be noted that the FT Raman spectrum of RC after its complexation with the protein, is identical to that of (c). The molecule is in its azoic form but all resonance effects have been removed This observation then makes it obvious that the dye-protein binding involves not only interaction between the sulfonyle groups of RC and positively charged ionic sites in the protein (3), but also van-der-Waals-type attractions due to the extensive aromatic system.

Since a model of the mechanism of the dye-protein interaction must involve these two types of interactions, the complexing site must provide a distance between positive entities which is comparable to the distance separating the two SO_3^- groups of CR, and a β intersheet region which can accommodate the flat aromatic part of the molecule.

In order locate such sites and to elucidate the tertiary structure of βA4, various models were constructed using the BIOSYM program. An α-helical form was constructed using the NMR data (2) obtained for β1-28 in solution in trifluoroethanol (which promotes intramolecular hydrogen bonding, thus stabilizing the formation of an α-helical structure). In addition, an α-helical structure (which could represent the soluble form of the protein) and an all beta strand conformation (which could represent the deposited form) were obtained using chronological sequencing.

	RMN	α	β
ε = 1	Arg 5 - Lys 16	His 14 - Lys 28	
ε = 80	Arg 5 - Lys 16		
ε = 4r	Arg 5 - Lys 16	Arg 5 - Lys 16	Arg 5 - His 13 His 6 - His 14

TABLE 1 : Distances comparable to the SO_3^- separation in RC as a function of the starting structure and the dielectric constant employed.

Observation of the three final structures obtained by energy minimization reveals that there are only a few sequences containing positively-charged sites separated by the appropriate distance. The possibilites are assembled in Table I and one can note immediately the repetition of the Arg 5- Lys16 distance.This result is coherent with the NMR results which indicated that Lysine 16 should always be solvent exposed, and confirms the importance assigned to its choice of being a possible anchoring point in βA4.

References

1. G.G. Glenner, C.W. Wong, Biochem. Biophys. Res. Commun., **120**, 885, 1984
2. M.G. Zagorski , C.J. Barrow, Biochemistry, **31**, 5621-5631, 1992
3. E. Pigorsh, A. Elhaddaoui and S. Turrell, Spectrochimica Acta, **12**, 2145-2152, 1994

RAMAN SPECTROSCOPY AS A TOOL FOR INVESTIGATION OF HUMAN AND ANIMAL SKIN

M. Gniadecka[1], H.C. Wulf[1], D.H. Christensen[2], O.F. Nielsen[2], A.D.T. Jacobsen[3], and P.W. Jensen[3]*

[1]Department of Dermatology, Danish National University Hospital, 2100 Copenhagen, Denmark
[2]Department of Chemistry, University of Copenhagen, 2100 Copenhagen, Denmark
[3]Department of Chemistry, Odense University, 5230 Odense, Denmark

In recent years, Raman spectroscopy has been used more and more for the investigation and characterization of biological tissue [1]. The technique has been used as a diagnostic tool in biomedicine since samples obtained from breast cancer [2], brain tumours [3,4] and gynaecological tract malignancies [5] have Raman spectra which differ from those of normal tissue.

Skin is the most accessible organ of the human and the animal body and therefore an obvious candidate for investigation by Raman spectroscopy. The first Raman study of human skin was made by Barry et al. [6]. Since most tissues show a strong autofluorescense when laser wavelengths in the visible region are used for excitation, it is necessary to excite with a 1064 nm Nd:YAG laser and to use the Fourier Transform Raman technique (Figure 1). The spectra were recorded using 180° scattering configuration. No background correction was performed.

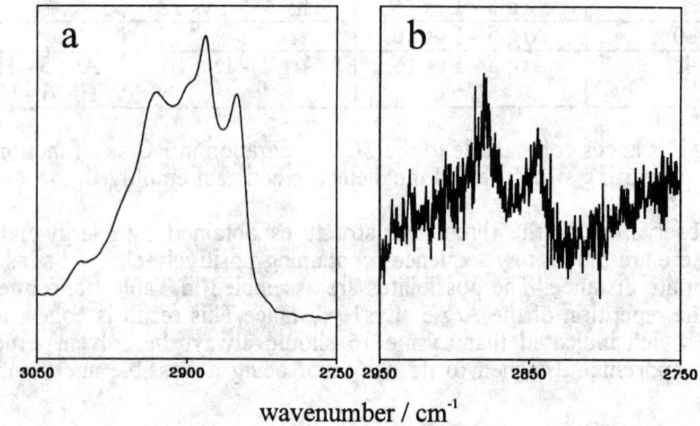

Figure 1. The C-H stretching region of spectra of mouse skin recorded with a) 1064 nm and b) 514.5 nm excitation.

Human skin consists of an outer horny layer stratum corneum, epidermis and dermis. A crucial parameter in studies of skin samples is the penetration depth. Experiments performed on skin samples of different thickness placed on a polystyrene background show that the laser beam penetrates the tissue to a depth of 2.0 - 2.5 mm.

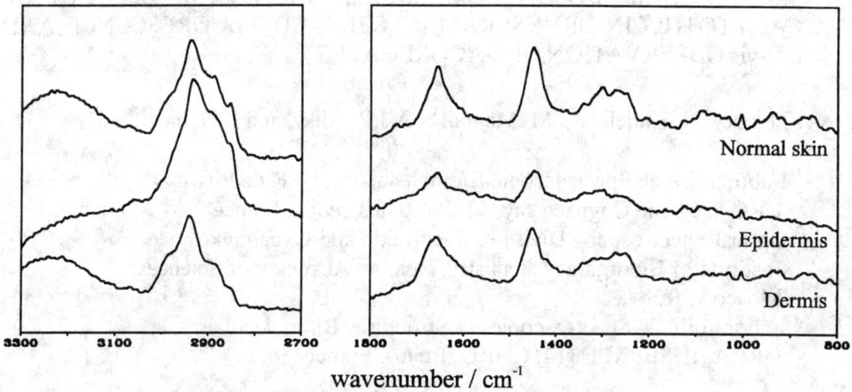

Figure 2. Spectra of human skin, epidermis and dermis.

Figure 2 shows the comparison of the Raman spectra of human dermis, epidermis and whole skin. Although most of the main bands of the spectra are quite similar, showing an amide I band with a maximum at 1660 cm^{-1}, CH_3 and CH_2 deformation at 1450 cm^{-1}, the most intense of the amide III bands at 1270 cm^{-1} and a phenylalanine ring breathing at 1004 cm^{-1}, there are also some differences. The largest is the intensity of the O-H stretch from water at 3250 cm^{-1}, which is much stronger for dermis than for epidermis. However, to perform a more quantitative study of the water content, more precise experimental conditions must be taken into account in order to prevent water evaporation. Also in the fingerprint region the spectra are different.

In progress is work on the use of other excitation frequencies and CCD detection as well as the study of the influence of different skin diseases on the Raman spectra.

1. G.J. Puppels, in Spectroscopy of Biological Molecules (J.C. Merlin, S. Turrell, and J.P. Huvenne, ed.) Kluwer, Dordrecht 1995, 457.
2. C.J. Frank, R.L. McCreery, Anal. Chem. 67, 777 (1995).
3. S. Keller, B. Schrader, A. Hoffmann, W. Schrader, K. Metz, A. Rehlaender, J. Pahncke, M. Ruwe, and W. Budach, J. Raman Spectrosc. 25, 663 (1994).
4. A. Mizuno, H, Kitajima, K. Kawanchi, S. Muraishi, and Y. Ozaki, J. Raman Spectrosc. 25, 25 (1994).
5. C.H. Liu, B.B. Das, W.L. Sha-Glassman, G.C. Tang, K.M. Yoo, H.R. Zhu, D.L. Akins, S.S. Lubicz, J. Cleary, and R. Prudente, J. Photochem. Photobiol. B16, 187 (1992).
6. B.W. Barry, H.G.M. Edwards, and A.C. Williams, J. Raman Spectrosc. 23, 641 (1992).

SELF - AGGREGATION INCREASES ANTITUMOR EFFECTS OF CAMPTOTHECIN DRUGS: RAMAN, SERS, CD, FLUORESCENCE AND UV-vis OBSERVATION OF J-AGGREGATES

F. Fleury[1], I. Kudelina[2], M. Manfait[1], A.J.P. Alix[3], and I. Nabiev[1]*

[1] Laboratoire de Spectroscopie Biomoléculaire, UFR de Pharmacie, URCA, 51 rue Cognacq Jay, 51096 Reims cedex, France
[2] Optical Spectroscopy Division, Shemyakin and Ovchinnikov Institute of Bioorganic Chemistry, Russian Academy of Sciences, Moscow, Russia
[3] Laboratoire de Spectroscopies et Structures Biomoléculaires, URCA, INSERM U314, CHRU, Reims, France

The recent identification of topoisomerase (topo) I as a principle intracellular target for the antitumor drugs has renewed interest in the development of topo I poisons, as such drugs may be valuable additions to combination chemotherapeutic regiments. Antitumor plant alkaloid, camptothecin (CPT), as well as derivatives from the CPT-family of drugs control proliferation of cancer cells by inducing topo I cleavable complex formation between the drug, topo I and DNA. Some of the CPT-drugs (20(*S*)CPT, 10,11-methylenedioxy-CPT, SN38) form, in aqueous buffer solutions, the small and very stable aggregates. The aggregates appear when solutions of CPTs in DMSO are diluted to 10^{-4}-10^{-6} M with aqueous buffer: at the conditions typical for the clinical trials [1]. The aggregates penetrate inside the cell with much higher efficiency as compared with the monomeric form of the drugs. An effectiveness of cellular uptake of aggregated and non-aggregated species of the CPT-family was found to correlate well with the inhibition of growing of the cancer cells. These effects were found to be the most important for 10,11-CPT derivative applied for trials (10 times higher intracellular penetration of the aggregated form as compared with the monomeric one) [1]. Hence, the molecular basis of the formation of the aggregates should be analysed if the mode of the drug biological action need to be refined.

The following spectral features are known to be characteristic for self-aggregation:

(i) Changes in the absorption spectrum reflecting formation of aggregated structures: both hypsochromic and bathochromic bands can occur for aggregated species. These are, respectively, the H-aggregate and J-aggregate absorption.

(ii) Fluorescence intensity diminishes upon aggregation of molecules.

(iii) Abnormally strong CD signal may appear reflecting resonance excitonic energy transfer within the aggregated species.

(iv) Appearance of the characteristic Raman (and SERS) bands upon aggregation [3].

In the case of CPT-drugs, “red” absorption bands were formed when the stock solutions of the 20(S)CPT, 10,11-CPT or SN38 were used for preparation of the buffer solutions. This finding, together with the diminishing of absorption and fluorescence intensity as well as with the appearance of the abnormally high signals in the CD spectra and characteristic bands in the Raman, SERS and fluorescence spectra of the drugs undoubtedly indicates the J-type of self-aggregation.

Due to the $1/N^{1/2}$ dependence (N is the number of monomeric units) of the line width of aggregates, it is clear that only larger aggregates give rise to narrow lines, whereas short oligomers have rather broad spectral features. In our case, we can not find considerable narrowing of the J-band, hence, the aggregated species don’t include significant amount of monomeric units. There are certainly not the aggregates including hundreds of the molecules like in the typical cases of the cyanine dyes exhibiting a tendency to form the aggregates of the J-type.

The abnormally strong signals in the CD spectra can obviously be attributed to the stacking interactions of the quinoline rings of the CPT-molecules in the polar solutions. The second possibility - self-association through formation of the hydrogen bonds is less probable, because this type of interaction must be easily revealed in non-polar solutions of the drug, but the CD spectra of CPT in DMSO are characteristic for the monomeric form of the molecule.

Raman and SERS spectra of the both monomeric and aggregated drugs have been recorded and the spectral assignments have been proposed. SERS spectra being recorded from the extremely low (10^{-8} M) concentrations of the drugs enabled to detect as low as 10% of the aggregates in solution. The data suggest that the formation of aggregates involves stacking interaction between quinoline rings of the CPT chromophores with the inverse position of the nitrogen atoms. It is known that quinoline groups of cyanine have a tendency to be overlapped as followed from the crystallographic data. The mode of dimer formation for quinolone also suppose transverse relationship between nitrogens in overlapping quinoline groups.

The role of the hydroxyl at position 20 in the self-aggregation of the drugs seems to be very important: the aggregation was totally abolished when the 20(*R*)CPT CPT-stereoisomer has been used. Hence, the process of aggregation is found to be stereospecific and one of the factors for aggregation of CPT is the relative orientation of the ethyl and hydroxyl groups in position 20. Orientation of the ethyl group may play a role of a steric hindrance for formation of the aggregates from the 20(*R*)CPT, whereas the hydroxyl group may participate in formation of the extramolecular bond and to stabilise 20(*S*)CPT self-aggregates.

Self-aggregation of 20(*S*)CPT and some other CPT-drugs influence the cellular uptake and viability of cancer cells. Hence, the self-aggregation should be regarded as an additional (and in this case favourable) factor influencing the biological activity of medications. Raman, SERS, CD and UV-vis identification of the molecular structure of the aggregates enable to propose an optimal conditions for the drugs applications in the clinical trials.

1. F. Fleury, I. Kudelina, F. Charton, M. Manfait, A.J.P. Alix and I. Nabiev, Biochemistry, submitted for publication.
2. A.S. Davydov, Theory of Molecular Excitons. McGraw-Hill: New York, 1962.
3. E.W. Knapp, Chem. Phys. 85, 73 (1984).

SERS-ACTIVE SURFACE AS A PROBE OF BIOLOGICAL ACTIVITY OF SIALYLATED MOLECULES

V.A.Oleinikov*[1], A.V.Feofanov[1], E.Yu.Kryukov[2], A.I.Ianoul[2], A.B.Tuzikov[1], S.D.Shiyan[1], N.V. Bovin[1], and I.R. Nabiev[3]

[1]Shemyakin and Ovchinnikov Institute of Bioorganic Chemistry, Russian Academy of Sciences, ul. Miklukho-Maklaya 16/10, 117871 Moscow, Russia
[2]Moscow Engineering-Physics Institute, Moscow, Russia
[3]Université de Reims Champagne-Ardenne, Reims, France

N-Acetylneuraminic (sialic) acid is the most typical terminal monosaccharide of cell glycoproteins and glycolipids. Being situated on the periphery of cell surface glycoconjugates of sialic acids are the primary target in the cell-cell (both endogenous and alien ones) and cell-virus interactions, and play a key role in biorecognition processes. A new approach to anti-influenza chemotherapy is based on the search of natural sialosides and development of synthetic sialylated biopolymers inhibiting virus attachment to host cells. A search, development and production of new potent antiviral agents require a method for quick testing and evaluation of their ability to bind to viruses and inhibit viral infections. In this report we present the data pointing out that SERS spectroscopy can be considered as such a method and discuss the physical basis of the correlations found between the intensity of SERS spectra of different sialylated molecules and their influenzavirus inhibiting properties.

Sialylated molecules were found to give intense characteristic SERS spectra originating from the sialic acid (SA) residue adsorbed on the silver surface and to be detected by this method at concentrations of SA residues as low as 10^{-5} M [1,2]. Good correlation between SERS intensity of signals of sialylated molecules and their antiviral or lymphocytes prolyferation activity (v) was established in many cases including: (i) different glycosides of SA and their derivatives, (ii) synthetic polymers with different content of SA residues, (iii) set of synthetic polymers with different carriers, (iv) polymeric conjugates of native sialooligosaccharide chains digested from glycoproteins, (v) glycoforms of α_1-acid glycoprotein (AGP) and recombinant erytropoietin and seems to be a general property of sialylated molecules. Binding affinity of a specific sialoside to the receptor-binding sites of different influenza virus strains varies sometimes significantly due to the fine variations of the binding site structure [2]. These variations of course cannot be derived from the SERS measurements, but the absence of characteristic signals in the SERS spectra of sialylated molecules reliably predicts reliably an absence or loss of antiviral activity for the molecules under study.

Similarity of the SA residue conformation being necessary both for its adsorption on the SERS-active surface and for the binding with hemagglutinin receptor of influenza viruses was found to explain this correlation. In particular, the importance of the carboxylic, N-acetyl, C8-, and C7-hydroxylic groups of SA residue in the adsorption on the silver surface as well as in ligand binding was confirmed in the SERS study.

Multipoint cooperative character of interaction of virus particles with cell surface receptors is well known. To inhibit effectively this interaction a competitive multipoint cooperative interaction of synthetic multivalent sialoside inhibitor with virus should be provided. SERS spectroscopy enables estimation of the ability of the neighboring SA residues to adopt the conformation necessary for the binding of the synthetic polymer with a receptor and therefore to provide the multipoint interaction. In this sense, the primary adsorption of the synthetic polymer on the surface of virus particle seems to be a necessary condition for the following specific interaction with receptor binding site, and SERS spectroscopy can be used as an "express" probe of specific adsorption properties of newly developed sialoside polymers.

In contrast to biochemical binding assay the SERS-probe enables to study the changes of SA residue conformation as a function of introduced substituents, monosaccharide sequences, conformation of asialic portions of sialosides, etc. For example, a shift of the *cis-trans* equilibrium toward *cis*-conformers of the acetamide moiety of SA residues was observed for a number of the sialoside inhibitors adsorbed on the SERS-active surface as compared to the aqueous solution. At the same time the most active fraction of α_1-acid glycoprotein is adsorbed on the SERS-active surface with the acetamide moiety of SA residues in *trans*-conformation.

AGP glycoform	Number of sialylated carbohydrate chains (%) Antennae			Relative content of glycoforms in AGP (%)
	Tetra-	Tri-	Di-	
AGP-A	50	50	-	40
AGP-B	40	40	20	40
AGP-D	20	20	60	1÷2

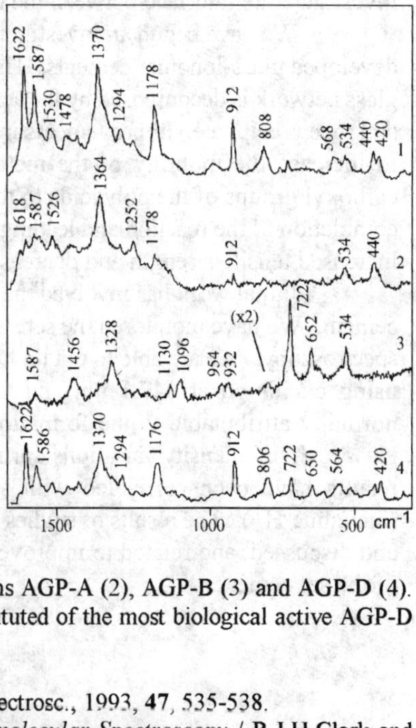

Figure. SERS spectra of AGP (1) and its glycoforms AGP-A (2), AGP-B (3) and AGP-D (4). The AGP spectrum is predominantly constituted of the most biological active AGP-D glycoform.

References

1. Sokolov K.V., Byramova N.E., et al. Applied Spectrosc., 1993, **47**, 535-538.
2. Nabiev I.R., Sokolov K.V., Manfait M. in *Biomolecular Spectroscopy* / R.J.H.Clark and R.E.Hester eds., vol. 20, Part A, Wiley, London, 1993, Chapter 7, p.325-333.

Raman Investigation of the Setting of Glass-Ionomer Dental Cements

Zheng Ouyang*, Sharon Sneckenberger*, Elizabeth Kao**,
and Paul W. Jagodzinski*

* Department of Chemistry,
West Virginia University
And
** School of Dentistry
Health Sciences Center
West Virginia University
Morgantown, West Virginia USA 26505

A number of chemical and structural techniques have been applied to the reaction that occurs when polyacrylic acid interacts with silicates in the formation of glass-ionomer dental cements. Infrared techniques have been utilized for over twenty years (1) with mirco Raman techniques having been applied more recently to investigate the interface between the dentin and the cement (2).

We have begun an investigation of the setting process chemistry of newly developed glass-ionomer cements. These types of adhesives typically harden as the glass network is decomposed by the acid, and calcium and aluminum salts are formed and then precipitate. Improvements in glass-ionomer formulations address the need to increase the mobility of the metal ions and increase the accessibility of the carboxyl groups of the polyacids to those metal ions, thus improving the degree of completion of the reaction and leading to more desirable physical properties such as increased tensile strength and decreased brittleness (3).

Initial work has involved the commercially available Fuji II glass ionomer cement. We have monitored the setting reaction as a function of time using Raman spectroscopy. We are able to obtain high S/N spectra as a function of reaction time using excitation at 647.1 nm and CCD detection. Figure 1 shows that bands normally attributable to the formation of silica gel (due to decomposition of the glass) exhibit intensity variations during the first 60 minutes of the reaction. These results have been correlated with similar trends in soluble ion concentrations (including H+). The results of studies of newly developed cements will be presented and discussed, and related to improved reaction conditions and physical properties of the set cements.

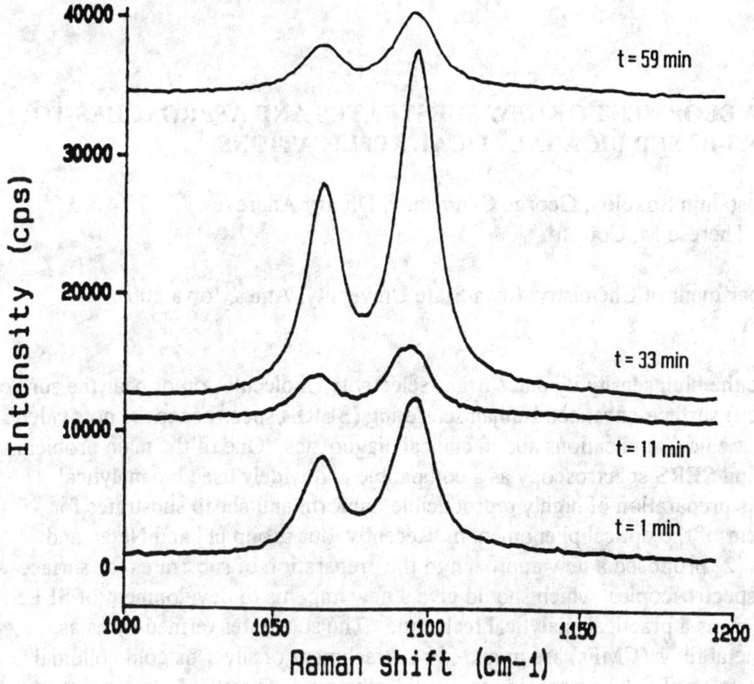

Figure 1. Raman spectra of Fuji II glass ionomer dental cement as a function of reaction time. The time dependent signals are typically due to silica gel formed during the reaction. λ_{ex} = 647.1 nm

1. S. Crisp, M. A. Pringuer, D. Wardleworth, and A. D. Wilson, J. Dent. Res., 53, 1414 (1974).

2. B. Van Meerbeek, H. Mohrbacher, J. P. Celis, J. R. Roos, M. Braem, P. Lambrechts, and G. Vanherle, J. Dent. Res., 72, 1423 (1993).

3. D. Xie, M.S. Thesis, College of Dentistry, Ohio State University, 1993.

DEVELOPMENT OF NEW SUBSTRATES AND APPROACHES TO SERS-BASED BIOANALYTICAL APPLICATIONS.

Konstantin Sokolov, George Chumanov, Dmitry Andreyev, and Therese M. Cotton*

Department of Chemistry, Iowa State University, Ames, Iowa 50011 USA

Despite the high sensitivity and surface selectivity (molecules on or near the surface are detected) surface-enhanced Raman scattering (SERS) spectroscopy is not widely used in biomedical applications and in clinical diagnostics. One of the main problems in developing SERS spectroscopy as a compatible and widely used bioanalytical technique is preparation of highly reproducible, uniform and cheap substrates for enhancement of the optical phenomenon. Recently, our group [1] and Natan and coworkers [2] proposed a new approach to the preparation of substrates for surface-enhanced spectroscopies, which should give a new impetus to development of SERS spectroscopy as a practical analytical technique. The substrates termed by us as colloidal metal films (CMFs) are prepared by attachment of silver or gold colloidal particles from a colloidal suspension to a solid substrate. The attachment occurs due to the strong specific interaction between the metal particles and different functional groups immobilized on the solid phase. Functional groups with high affinity to metals such as thiol, amino, pyridyl, etc. can be used in the substrate's preparation.

The CMFs as characterized by UV-Vis spectroscopy, electron microscopy, and atomic force microscopy revealed high reproducibility and uniformity of the their structure. As a result, highly reproducible SERS spectra were obtained from CMFs. Self-assembled monolayers of alkyl thiols, Langmuir-Blodget monolayers of fluorescein labeled phospholipids, as well as physisorbed nucleotides were used to monitor the enhancement of Raman scattering on CMFs. The observed SERS signals from the films were comparable or higher than from roughened electrodes or metal island films. CMFs are stable in different organic solvents and in water solution even in the presence of high salt concentrations. A variety of solid phases can be used for the preparation of the proposed films. We have prepared stable, reproducible and uniform CMFs based on glass slides, optical fibers, silica and polystyrene beads, and ELISA plate wells.

Our current efforts are focused on utilizing CMFs in bioanalytical assays employing SERS spectroscopy as a detection method. The assays are based on specific interaction between biomolecules immobilized on the surface (e.g. antibodies, antigens, or DNA probes) and solution species. An immobilization of "probe" molecules on the CMF's surface is accomplished by physisorption or through formation of a strong bond between metal particles and thiol groups previously attached to the "probe" molecules. The assays can be performed in "sandwich" or "competitive" format using a specific label with a high Raman scattering cross-section. Another type of analysis is to monitor changes occurring in a SERS spectrum due to specific interaction between binding partner molecules. An example of this approach is shown in Figure 1 in which an oligonucleotide $p(dT)_8$ physisorbed on silver metal particles was heated up to ca. 80 °C and then slowly cooled down to room temperature in the absence and presence of the complementary $p(dA)_8$ sequence. A decrease in the SERS signal intensity of $p(dT)_8$ indicates formation of double stranded DNA fragments in the presence of complementary $p(dA)_8$ sequence [3, 4].

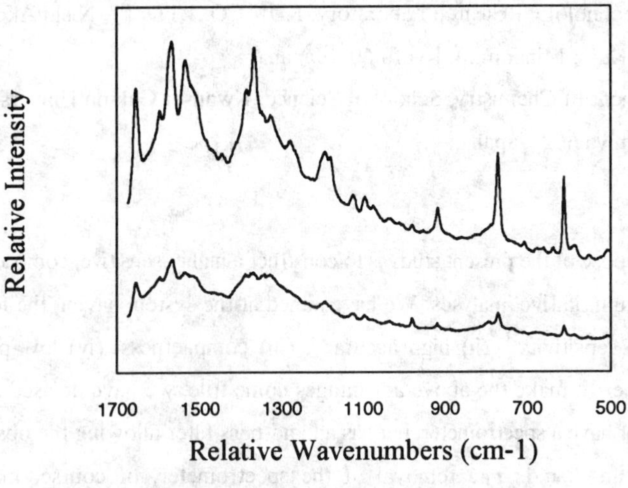

Figure 1. SERS spectra of a physisorbed on silver metal particles oligonucleotide $p(dT)_8$ annealed in the absence (top) and presence (bottom) of a complementary $p(dA)_8$ sequence.

The described approaches may be used in development of one-step DNA probe techniques. One-step means that detection and all biochemical reactions take place in one chamber without any intermediate washing steps. That is possible because of surface selectivity of SERS spectroscopy mentioned above. Another advantage of using SERS spectroscopy as a detection method is the possibility of monitoring defects such as mispairs in formation of double-stranded DNA. The later is feasible due to the fact that only a single-stranded DNA and small regions of destabilization in double-stranded DNA give a strong SERS signal [3, 4].

1. G. Chumanov, K. Sokolov, B.W. Gregory, T.M. Cotton, J. Phys. Chem. 99, 9466 (1995).
2. K.C. Grabar, R.G. Freeman, M.B. Hommer, M.J. Natan, Anal Chem. 67, 735 (1995).
3. E. Koglin and J.-M. Sequaris, Top. Curr. Chem. 134, 1 (1986).
4. I.R. Nabiev, K. Sokolov, and O. Voloshin, J. Raman Spectrosc. 21, 333 (1990).

A COMPACT RAMAN SYSTEM WITHOUT A SPECTROMETER FOR QUANTITATIVE ANALYSIS OF BIOLOGICAL SAMPLES

X. DOU[1]*, Y.YAMAGUCHI[1], S. DOI[1], H. YAMAMOTO[1], H.UENOYAMA[1], and Y.OZAKI[2]

[1]Basic Technology Research Laboratory, KDK CO.,LTD. 57, Nishi Aketa-cho, Higashi Kujo, Minami-ku, Kyoto 601, Japan

[2]Department of Chemistry, School of Science, Kwansei-Gakuin University, Nishinomiya 662, Japan

The purpose of the present study is to construct a highly-sensitive, compact Raman system suitable for quantitative analyses. We have aimed at the system having the following advantages. (i) high sensitivity, (ii) high accuracy, (iii) compactness, (iv) low price, (v) practicability. In order to make the above advantages come true, we have devised a Raman system which does not have a spectrometer, but has a band-pass filter allowing the observation of a particular Raman band. The removal of the spectrometer, of course, makes the measurements of Raman spectra impossible, but the estimation of the intensity of a Raman band becomes much more efficient. It should be also emphasized that the system without a spectrometer is highly sensitive, compact, and inexpensive. We have used this system for determining the concentrations of glucose in urine containing glucose artificially.

Figure 1 shows the proposed system. We used the semiconductor laser as a light source. The excitation wavelength was 980 nm. Biological samples, particularly, urine emit strong fluorescence, so that the excitation wavelength in the near-infrared region is essentially important. The use of the semiconductor laser diode makes the Raman system inexpensive and compact. However, the intensity of radiation of semiconductor lasers is, in general, unstable compared with that of gas lasers such as Ar laser. We compensated this fluctuation of the laser power by monitoring the intensity of the laser light and normalizing the intensity of Raman scattering light by it.

A band-pass filter for the 980 nm excitation was placed output of the laser. Full width at half maximum (FWHM) of laser beam through the band-pass filter was about 1-nm. The filtered laser beam was split into two by a beamsplitter, and one was used as a reference for the power of the excitation laser light and another was employed as an excitation beam.

The excitation beam was focused on the center of a quartz flow system (See inset in Fig.1). This cell was kept in a unique inner spherical integration type cell-holder whose inside surface was overlaid with gold to collect Raman scattering from a solid angle of 360 degree.

Raman scattering light was collected at 90 degree and condensed efficiently with a F=1.2 camera objective (Nikon Co., Ltd.) lens. The collected Raman scattering was directly introduced to a band-pass filter. The band-pass filter allows the scattering light in the particular very narrow wavelength region to pass through. We selected the wavelength regions corresponding to particular Raman bands which were useful for quantitative analyses of glucose and acetone in urine. The band-pass filters used show a spectroscopic character of Gaussian function (BARR ASSOCIATES, INC.). Their FWHM were 1 or 5 nm. The filtered Raman scattering light reached the near-infrared sensitive photodiode.

A Raman band at 1130.5 cm^{-1} due to glucose has been used to determine the concentration of glucose in urine. The correlation coefficient between the concentration and the Raman intensity have been found to be 0.991.

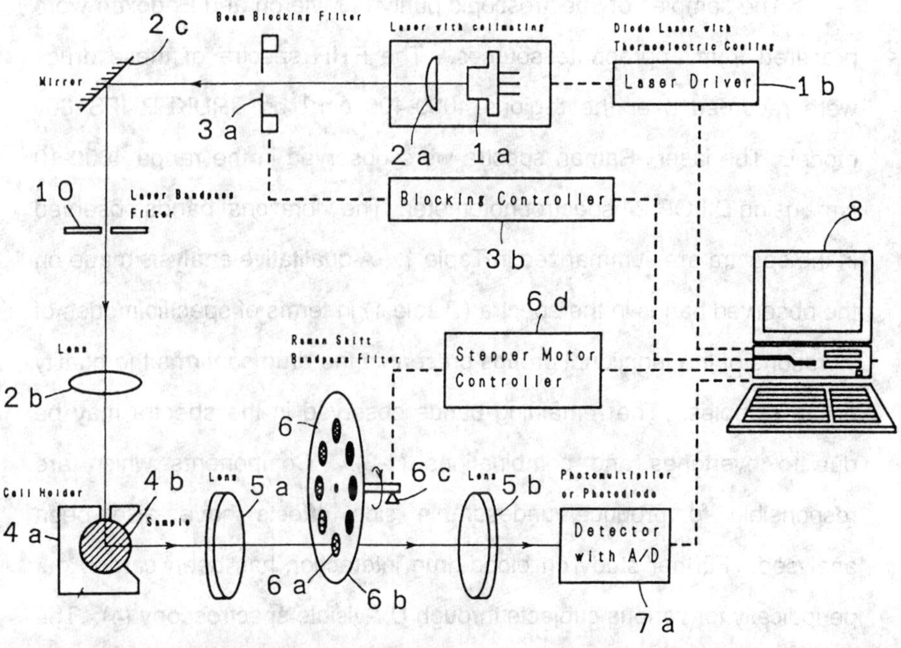

Fig. 1 A proposed Raman system dedicated for quantitative analysis.

FTIR, LR, AND UV-VISIBLE SPECTROSCOPIC METHODS IN THE ANALYSIS OF MYLERON AND ENDOXAN

S.Gunasekaran, S.Shreemathi, R. Nagendran
Post Graduate and Research Department of Physics, Pachaiyappa's College, Madras, INDIA.
and
V.K.Rastogi, LR College, Ghaziabad (India).

FTIR and LR spectroscopic methods have been employed on the qualitative analysis of some selected samples of pharmaceutical active compounds prescribed for cancer treatment. Components producing undesirable effect in these drugs have also been analysed. Further the effectiveness of these drugs on human blood is made through UV-Visible spectroscopic method.

The samples of spectroscopic purity of Myleron and Endoxan were procured from appropriate sources. The FTIR spectra of these drugs were recorded over the regions 4000-400 cm^{-1} on BRUKER IFS 66v model. The laser Raman spectra were observed in the range 4000-10 cm^{-1} using DILOR 24 spectrophotometer. The vibrational bands observed in the spectra are summarized in Table 1. A qualitative analysis made on the observed bands in the spectra (Table 1) in terms of specific modes of vibrations of the functional groups present in the drug confirms the quality of the samples. The remaining bands observed in the spectra may be due to overtones and combinations (1-3). Components which are responsible to produce undeserable side effects have also been analysed. Further study on blood-drug interaction has been carried out periodically for various subjects through UV-visible spectroscopy (4). The results are briefly discussed.

TABLE - 1 :- FTIR AND LR SPECTRA AND VIBRATIONAL BAND, INVESTIGATIONS ON MYLERON AND ENDOXAN.

MYLERON		ENDOXAN		VIBRATIONAL
FREQUENCIES (cm^{-1})		FREQUENCIES cm^{-1}		ASSIGNMENTS
FTIR	LR	FTIR	LR	
555-630 (w)	--	540-630 (w)	--	C-C-C in plane deformation.
670-920 (w)	666-813 (w)	680-940 (w)	697-948 (w)	C-H out of plane deformation.
990 (s)	987 (w)	990 (s)	969 (ms)	C-C stretchag.
1075(vs)	--	1020 (vs)	--	C-H in plane on stretch.
1100-1205 (ms)	1061 (w) (w)	1060-1210 (ms) (w)	1062-1196 (s)	C-H in plane deformation.
1255 (vw)	--	1230 (w)	--	C-H in plane deformation.
1300 (vw)	--	1330 (w)	--	C-H in plane bending.
1350 (w)	--	--	--	C-H Symmetric bending.
1370 (w)	--	1370-1480 (w)	1383 (ms)	C-H Symmetric bending.
1390 - 1470 (vw)	--	--	--	C-H Symmetric stretching.
1535-1590 (vw)	--	1580 (vw)	--	C=0 Stretching.
2360 (nw)	2596 (ms)	2380 (mw)	2373 (ms)	Co_2 stretching.

References:

1. Conley R.T., Infrared spectroscopy, 2nd edition, Boslon, Allvin and Bacon.
2. Colthup, N.P, Daly and Wiberly, H., S.E., introduction to Infrared and Raman spectroscopy, 2nd edition, New York and London, Academic Press (1975).
3. Vibrational spectral studies on pseudophedrine Hydrochloride by S. Xavier Jesu Raja and S. Gunasekaran (Asia Journal of Physics, Vol.1, No.4, 1992, 194-199).
4. UV-visible spectroscopic study of Blood in Health and Disease by S. Gunasekaran and Marshell (Asian Journal of Chemistry, Vol.5, No.1, 1993, 99-105).

THE LOW FREQUENCY RAMAN SPECTRA OF SOLID PHARMACEUTICAL PREPARATIONS WITH AROMATIC GROUPS

V.N.Moiseenko[1], T.V.Shvets[1], V.S.Gorelik[2], Bilal Abu-sal[1]

[1] Dniepropetrovsk State University, Dniepropetrovsk, UKRAINE
[2] Physical Institute of the Academy of Sciencies, Moscow, RUSSIA

The solid pharmaceutical preparations with aromatic groups find an application as an antimicrobical remedy (sulginum, streptocidum, phthalazolum), anaesthetic (novocainum), vaso-dilating (themisalum) etc.

In the present work the analysis of the low frequency Raman spectra in such preparations has been carried out in order to work out new methods of testing their quality as well as to clear the microscopic mechanism of their biological effect.

To excite the Raman spectra Ar^+-laser (λ=488 nm) was used. The registration of the low frequency spectra has been performed at forward scattering geometry by using double monochromator, photon counting system and computing analysis of the spectrogram observed. The thin layer with thickness of about 0,5 mm of the fine-graind powder placed between two plane-parallel quarts glass plates into optical cuvette served as a sample for studying in our measurements. The spectra were obtained at T = 300 and 100 K.

The objects for investigation were the following: 1) - sulfanilamidum - sulginum; 2) - anaesthetic preparation - novocainum; 3) - broadening blood-vessels remedy - themisalum. Moreover, analysis was performed for the low frequency spectra of the aromatic solid proteunogenic amino-acid - tyrosinum.

The obtained spectra of objects investigated as well as the structural chemical formulae are presented in Fig.1. As can be seen from this figure, for all the preparations a number of distinct maxima are revealed over a range of 0 - 150 cm^{-1} . For some structures (novocainum, sulginum, tyrosinum) the sharp intensive peaks appear in the immediate vicinity of the excitation line (about 30 cm^{-1}). In the case of sulginum the intensity of all the peaks are approximately equal. Finally , the spectrum of themisalum becomes essentially more diffuse and does not contain peaks in the lowest frequency region.

The bands in the region of 50 - 150 cm^{-1} are typical for all the structures; they can be attributied to the libration modes of benzol ring. The high-quality modes in the region of 70 cm^{-1}, typical for novocainum and sulginum can be naturally explained by the translation motion of aromatic ring relatively " tailing " groups which are comparatively weakly bound to aromatic ring. Thus, obvious correlation between the vibration frequencies is revealed for structures studied in the region of 50 - 150 cm^{-1} as well as for the lowest frequency modes in case of novocainum, sulginum (marked by stars in Fig. 1) .

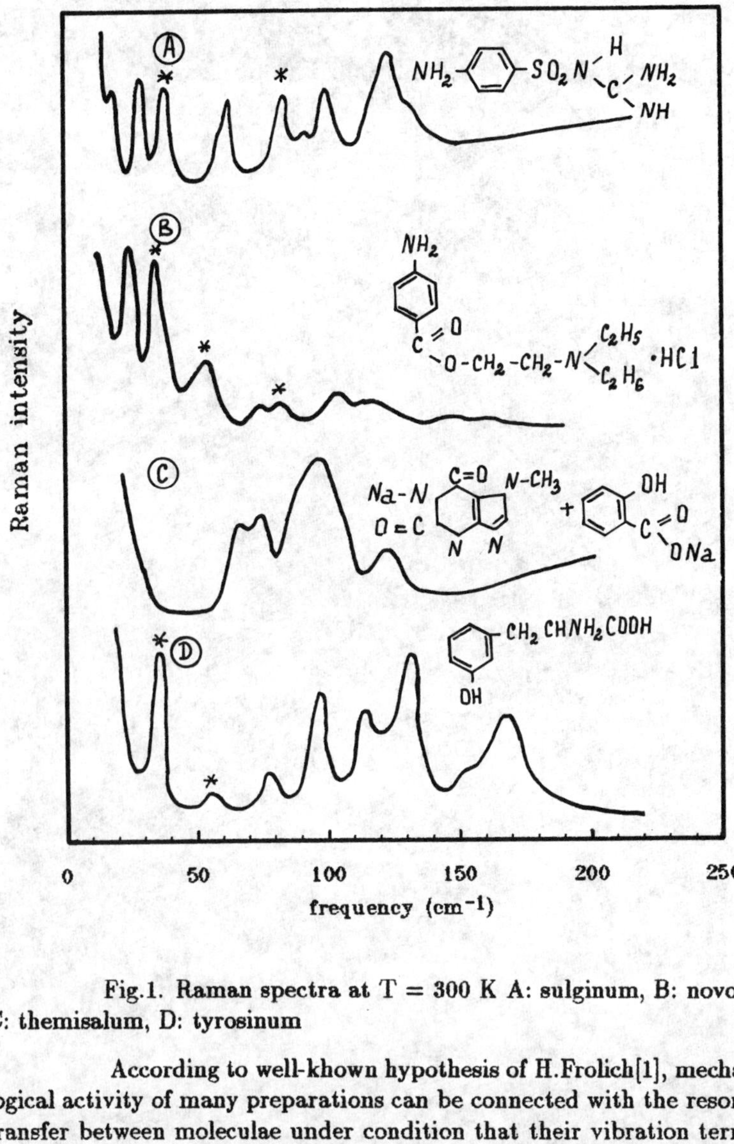

Fig.1. Raman spectra at T = 300 K A: sulginum, B: novocainum, C: themisalum, D: tyrosinum

According to well-khown hypothesis of H.Frolich[1], mechanism of biological activity of many preparations can be connected with the resonance energy transfer between moleculae under condition that their vibration terms are close. Therefore, the resonance coincidence of the lattice polar mode frequencies in this compound with the frequencies of some polar modes of the preparations in question suggests the possibility of their effective influence on processes in the protein structures and nervous tissue. The presence of sharp peaks in the lattice Raman spectra open the possibilities of controling the conformation state of corresponding pharmaceutical preparations and analysis of their composition .

[1] H.Frolich Phys.Lett., 39A, N 2, p. 153-154 (1972)

SECTION 28

RAMAN MICROSCOPY

APPLICATIONS OF RAMAN SPECTROSCOPY IN MATERIALS SCIENCE

George Turrell* and Paul Dhamelincourt

Laboratoire de Spectrochimie Infrarouge et Raman (CNRS A2631L),
Université des Sciences et Technologies de Lille
59655 Villeneuve d'Ascq, France

The first part of this communication presents a summary of the most recent developments in micro-Raman instrumentation – point analysis and imaging techniques – including the use of confocal optics and line illumination, as currently available in certain commercial instruments. It is followed by a brief review of some applications of Raman microscopy in the field of materials science. It is shown that this technique provides an extremely valuable tool for quantitative, as well as qualitative, microanalysis on a wide variety of samples.

The major development in micro-Raman spectroscopic instrumentation during the last few years is based on the extension of the principles of confocal microscopy. Applied to Raman microprobing the confocal technique provides an efficient way to obtain interference-free Raman spectra, as well as two- or three-dimensional, selective images of small specimens embedded within strong scatterers or fluorescent transparent media. Unlike the conventional microscope, where the entire field of view is uniformly illuminated and observed, the confocal configuration benefits from the use of spatial filtering by optically conjugated, pinhole diaphragms. Such a system isolates the light originating from a small region of the sample coincident with the diffraction-limited, focused laser spot, and efficiently eliminates the contributions from out-of-focus zones [1]. This technique leads to a number of significant improvements in Raman instrument performance, in particular a dramatic improvement in the depth discrimination, which allows an optical sectioning of transparent specimens to be made [2].

A new concept of Raman imaging that benefits from the advantages of confocal microscopy has been developed in this laboratory [3,4]. It is the Confocal Raman Line Scanning (CORALIS) optical configuration, an improved design which employs the line-scanning technique. This optical system allows line illumination of both the sample and the detector and at the same time the use of an adjustable, confocal pinhole diaphragm for spatial filtering of the scattered signal for any position of the laser beam on the sample. With the use of a CCD detector, a set of wavenumber-dependent intensity profiles is obtained, each of which depicts the spatial distribution of a given molecular species along the scanned laser line. Two-dimensional molecular maps of the sample are acquired line-by-line by moving the sample step-by-step and storing all of the data for each line. The construction of three-dimensional images requires displacement of the sample along the optical axis. Finally, dedicated software is used to automate the successive phases of data collection [3,4].

Raman microspectrometry and imaging have been extensively used for the identification of contaminants and defects, as well as for the characterization of materials themselves. Contamination and the formation of defects create problems which are constantly encountered in the commercial production of materials. The chemical identification of these species is of particular interest, because it can lead to properly selected materials and processes.

In semi-conductor production, for example, because circuit-board geometries and components approach micrometer dimensions, contaminants of this size often create reliability problems. Inorganic contaminants, as well as those of organic origin (*e.g.*, from etching or packing materials), can be readily detected *in situ*. In addition to the identification of foreign contaminants on devices, micro-Raman spectrometry can also provide characterization of the materials used in the fabrication of these devices. For example, with the development of very-large-scale-integration (VLSI) technology, considerable emphasis has been placed on devising methods for fabricating high-quality silicon films on a variety of insulating substrates (vapor-phase epitaxy, thermal annealing *etc.*). A good understanding of silicon crystal growth is essential in the development of materials of acceptable quality. The Raman technique is often used as a rapid, nondestructive method for the control of processed silicon, based on the observation of the position of the characteristic band at approximately 520 cm^{-1}.

Further examples of this application are the control of the stress distribution or of the temperature of devices, since stress can exceed the limit of plastic deformation and reduce the production yield. By reference to the shift of the 520 cm^{-1} band, compressive or tensile stress can be evaluated; this parameter can be determined as a function of position. In the same way, the local increase of lattice temperature in a n-MOSFET transistor has been evaluated when the device is operated in the saturation regime. Again, this measurement is based on the shift of the 520 cm^{-1} band.

Another example of these applications is the characterization of industrial polymers in terms of defect identification and characterization of molecular orientation and crystallinity. Films, fibers, or copolymers and polymer blends, often exhibit defects that reflect the non-uniform distribution of the components of the polymers. Defects inside films and fibers are usually caused by inclusions which can be directly identified through the host polymer. Substances present as part of the polymer recipe or produced during polymerization (internal contaminants), as well as substances acquired during the material handling processes (external contaminants), have thus been identified· Polarized Raman spectra of fibers provide data which are indicative of the orientation and crystallinity of the material spun from the melt, as a function of the take-up speeds. Optical birefringence measurements allow the overall orientation of fibers to be determined, but they cannot differentiate between amorphous and crystalline phases. It is important in the understanding of the mechanical performance of fibers to distinguish the molecular orientation and the crystallization of the polymer.

A last example is the characterization of ceramics such as silicon nitride (phase transformation during the hot pressing process), zirconia alloys (phase-transformation toughening) and diamond-like coatings.

1. T. Wilson, Confocal Microscopy, Academic Press, London, 1990.
2. Confocal Laser Raman Data Sheet, DILOR , Lille, France, 1992.
3. P. Dhamelincourt, J. Barbillat, M. Delhaye, Spectrosc. Eur. 5, 16 (1993).
4. J. Barbillat, P. Dhamelincourt, M. Delhaye, E. Da Silva, J. Raman Spectrosc. 25, 3 (1994).

SUB-MICRON RESOLUTION RAMAN IMAGING

S.Webster[1,*], D.A.M.Smith[2], M.N.Ayad[2], K.Kershaw[2], D.N.Batchelder[2]

[1]SOMS Centre, Department of Chemistry and [2]Department of Physics and Astronomy, University of Leeds, Leeds LS2 9JT. U.K.

The ultimate spatial resolution of optical microscopies is limited by diffraction to approximately half the wavelength of the light used to form the image. This resolution limit also holds for the various forms of Raman imaging that employ conventional optics. It is possible, however, to overcome the diffraction limit using a technique known as scanning near-field optical microscopy (SNOM) [1, 2]. We have constructed a SNOM and coupled this to a high throughput Raman spectrometer (Renishaw Plc.) to allow Raman imaging with a spatial resolution of approximately 100-200 nm.

A SNOM works by systematically scanning a sub-wavelength aperture over the sample, illuminating through this aperture and collecting the back-scattered light to build up an image. In order that the light from the aperture may propagate and maintain a spatial resolution comparable to the size of the aperture, the aperture must be held within approximately 10-30 nm of the sample's surface, i.e. the 'near-field', during the scan.

In our instrument the sub-wavelength aperture is provided by coating a tapered optical fibre with aluminium such that only a 100 nm hole at the very end of the fibre remains uncoated. The aperture is held within 20 nm of the sample using a shear force feedback system [3]. The fibre tip is 'dithered' horizontally by a few nanometers with the amplitude of this vibration being monitored. As the tip moves within ~ 30 nm of the surface of the sample, interaction between the tip and the sample damps the amplitude of the tip vibration, thus providing a feedback signal.

In our instrument the Rayleigh and Raman scattered light from the sample lying under the tip is collected with a long working distance objective at 45° to the sample and then fed into a high throughput, single stage Raman spectrometer equipped with a cooled CCD detector and holographic Rayleigh rejection filters [4].

High losses in the tapered region of the fibre mean that only a tiny fraction of the input light reaches the sample, even when the aperture is held in the near field. Further, local heating of the fibre SNOM tip restricts the amount of light it is possible to put into the fibre. These two factors, when coupled with the inherently low scattering cross section of the Raman effect, mean that Raman SNOM presents a considerable challenge.

Initial results have been obtained with the SNOM from a range of samples including conjugated polymers, CVD diamond and silicon. Figure 1 illustrates a one dimensional SNOM Raman image for a nano-crystal of a polydiacetylene. The image was recorded using 632.8 nm illumination.

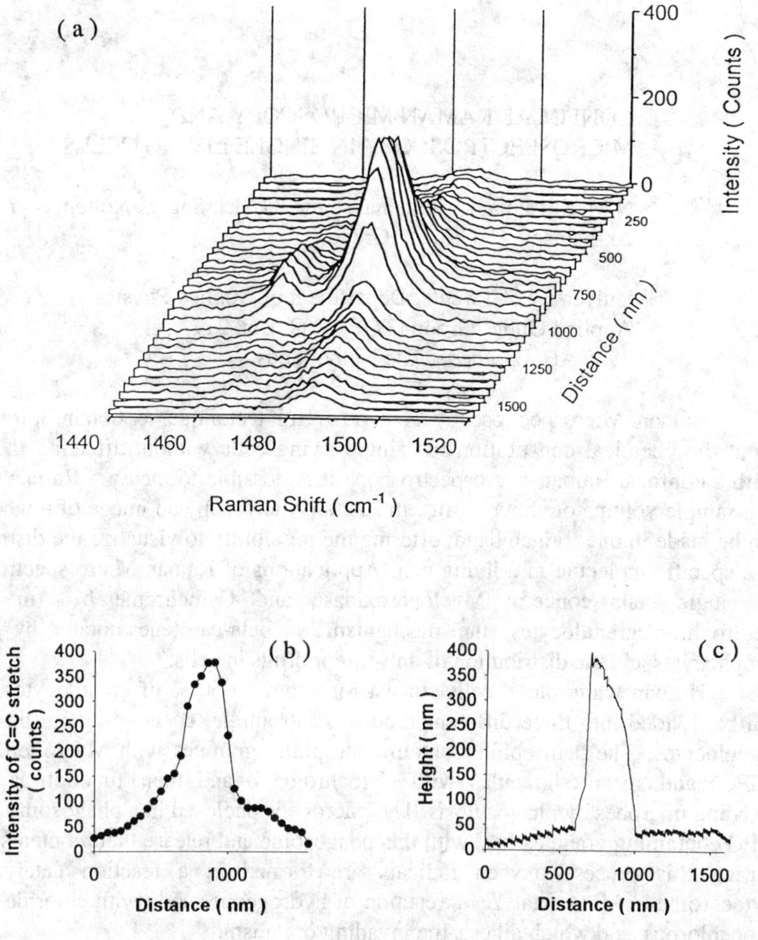

Figure 1: (a) Part of the resonant Raman spectrum from a polydiactylene nano-crystal obtained with the SNOM as a function of distance across the crystal. (b) The intensity of the 1485 cm^{-1} C=C stretch as a function of distance across the crystal. (c) Topology of the crystal obtained simultaneously with the Raman data from the sheer force feedback.

1. D.W.Pohl, W.Denk, M.Lanz. Appl. Phys Lett. 44, 651 (1984)
2. A.Lewis, M.Isaacson, A.Harootunian, A.Murry, Ultramicroscopy 13, 227 (1984)
3. E.Betzig, P.L.Finn, S.J.Weiner Appl. Phys. Lett. 60, 2484 (1992)
4. S.Webster, Ph.D thesis, University of Leeds (1994)

CONFOCAL RAMAN MICROSCOPY AND MICROSPECTROSCOPY IN SINGLE LIVING CELLS

N.M. Sijtsema, R. Ramanauskaite, G.M.J. Segers-Nolten, S. Yu. Arzhantsev, C. Otto*, J. Greve

University of Twente, Department of Applied Physics,
Applied Optics Group (TOP), P.O. Box 217,
7500 AE Enschede, The Netherlands

Raman Microspectroscopy is a powerful technique to obtain information about the chemical constitution of single living cells without affecting them [1]. With a confocal Raman microspectroscope it is possible to measure Raman spectra of a sample volume of about 1 μm^3. In Raman Microscopy an image of a whole cell can be made in one Raman band, offering the possibility to visualize the distribution of a specific molecule in a living cell. Applications of Raman Micro(spectro)scopy on single cells concern: Myeloperoxidase and Cytochrome b_{558} in human Neutrophilic granulocytes, the mechanism of beta-carotene uptake by human lymphocytes and the distribution of anti-tumor drugs in cells.

Human white blood cells consist for about 50-60 % of granulocytes, which can be divided into three different groups: neutrophilic, eosinopilic and basophilic granulocytes. The neutrophils contain azurophilic granules with MyeloPerOxidase (MPO) and specific granules with Cytochrome b and their function is to kill invading microbes by phagocytosis. The microbe is enclosed in a phagosome and the MPO containing granules fuse with this phagosome and release their contents into it. During this process oxygen radicals are formed in a reaction catalyzed by Cytochrome b. MPO catalyzes a reaction of hydrogenperoxide with chloride ions to hypochlorous acid which digests the invading organism.

We have measured Raman spectra of a group of granules in living human neutrophils and of isolated specific and azurophilic granules with Soret excitation. The Raman spectra of living neutrophils mainly comprise bands that can be attributed to MPO and Cytochrome b (figure 1). Furthermore, neutrophils were activated by adding opsonized polystyrene beads or zymosan and measured afterwards. In some cases the Raman spectra had changed after activation.

The chemopreventive action against cancer of beta-carotene is well known from epidemiological, experimental and clinical studies based on serum levels of beta-carotene [2]. It has been discovered recently that also beta-carotene is present in lymphocytes especially in $CD4^+$ cells [3] in which it is concentrated mainly in the so-called Gall bodies. Therefore, we are interested in the metabolic action of beta-carotene in lymphocytes. Carotenoid concentrations in isolated lymphocytes of cancer patients and healthy individuals have been compared. Lymphocytes were incubated with liposomes containing beta-carotene for different periods and their carotene content was measured to get more insight in the mechanism of carotene uptake by the cells. Additionally, Raman images of these lymphocytes were made to visualize the intra-cellulair distribution during uptake. Finally, a model for the beta-carotene uptake by human lymphocytes in vitro has been designed which is being applied to a study to the details of beta-carotene transport into human lymphocytes of healthy individuals as well as cancer patients.

Phthalocyanines and porphyrins are used as photosensitizers in Photodynamic therapy of cancer. The method is based on selective accumulation of photosensitizers in tumor tissue. After light irradiation a cytotoxic compound is formed which kills the tumor cells. To understand the killing mechanism of these photosensitizers it is important to know how they are distributed in a cell both before and after irradiation. Many photosensitizers fluoresce and therefore they can be imaged with a fluorescence microscope. Some phthalocyanins however show hardly any fluorescence and they can only be imaged by Raman microscopy. We have imaged cells that were incubated in phthalocyanin for 12 hours. The images show a granular distribution of the phtalocyanins.

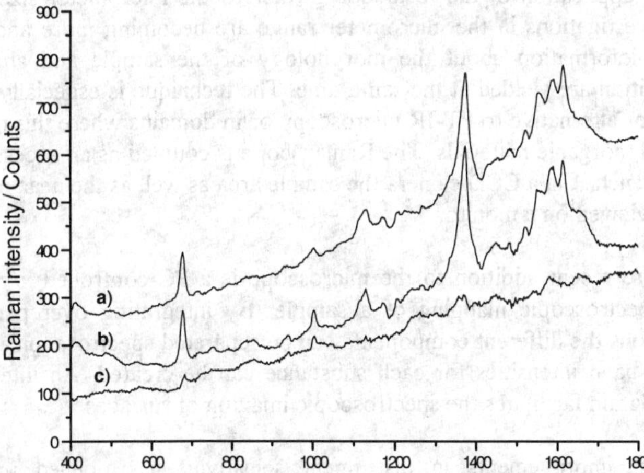

Figure 1 a) Raman spectrum of neutrophils (λ = 413 nm; 0.5 mW; 10 s); b) Fit of neutrophil spectrum with spectra of specific and azurophilic granules in a ratio of 7:4; c) Difference of a) and b)

[1] GJ Puppels, FFM de Mul, C Otto and J Greve, Studying Single Living Cells and Chromosomes by Confocal Raman Microspectrosocpy, *Nature* **347** (1990) 301-303.
[2] JA Olson Carotenoids, Vitamin A and Cancer. *J. Nutr.* **116** (1986) 1127-1130.
[3] GJ Puppels, HSP Garritsen, JA Kummer and J. Greve, *Cytometry* **14** (1993) :251-256.

CURRENT TRENDS AND FUTURE ASPECTS IN FT-RAMAN MICROSPECTROSCOPY

N.T.Kawai[2], C.Lehner[1], and J.Sawatzki[1]

[1]Bruker Analytische Messtechnik GmbH, D-76153 Karlsruhe, Germany
[2]Bruker Instruments Inc., Manning Park, Billerica, MA 01821, USA

Since one decade Fourier Transform Raman spectroscopy has been established in analytical and research laboratories [1]. By using near-infrared lasers for the excitation, high-quality spectra can be obtained from a large variety of samples while avoiding unwanted fluorescence backgrounds. In addition the method stands out for its simplicity to use and the minimal amount of sample required.

The development of a microscope accessory for FT-Raman spectrometers has extended the application of the technique further to the micrometer size scale. Spectroscopic investigations in the micrometer range are becoming more and more important, since information about the morphology of the sample and about its chemical composition are yielded at the same time. The technique is especially useful as a complement or alternative to FT-IR microscopy or in domains where this method fails, eg. for small inorganic materials. The Ramanscope is coupled as an accessory to the main optical bench. Via a CCD camera the sample area as well as the near-infrared laser spot can be viewed on a monitor.

A more recent addition to the microscope is a PC-controlled x/y-stage, which enables spectroscopic mapping of a sample. By integrating over particular wavenumber regions the different components can be separated spectroscopically and contour maps of band intensities for each substance can be created. An interactive video assisted program facilitates the spectroscopic imaging of surfaces.

Recent improvements in instrument sensitivity has opened up new application areas which were not thought to be accessible with FT-Raman microspectroscopy. As an example FT-Raman microspectroscopy has proofed to be a valuable technique for the nondestructive identification of intramineral fluids e. g. hydrocarbons and gases. Figure 1 shows the FT-Raman spectrum of a hydrocarbon fluid inclusion in a natural calcite matrix. The diameter of the inclusion was 15 μm. Even smaller inclusions down to about 5 μm containing gas under high pressure could be investigated. After subtraction of the host matrix a spectrum search in a commercial FT-Raman library was performed. Unexpectedly a similar spectrum (fig. 1, lower spectrum) was found although this library did not contain special spectra of fluid inclusions.

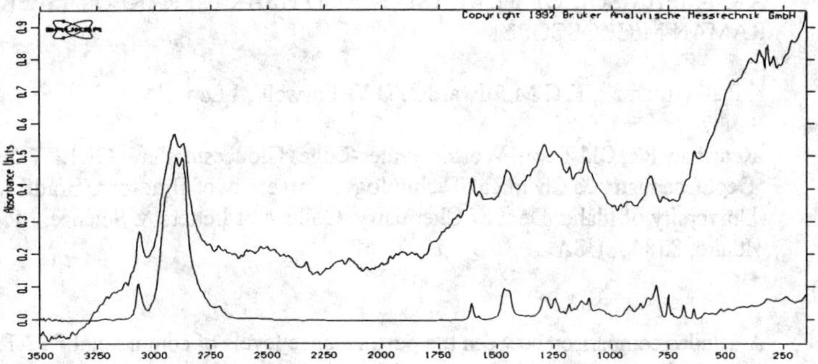

Fig. 1 Upper spectrum: FT-Raman microscope spectrum of a hydrocarbon fluid inclusion in a natural mineral, size: 15 µm
Lower spectrum: Result spectrum of a FT-Raman library search showing octylphenol decaethylene glycol ether

This contribution will focus on recent scientific and technical applications with emphasis on potentials and limitations. Specific tasks such as the minimum lateral and depth resolution as well as the use of appropriate optics e.g. for polarized micro-Raman measurements are discussed.

1. T. Hirschfeld and B. Chase, Appl. Spectrosc. 40, 155 (1986)

A COMPARISON OF FT - RAMAN AND NEAR INFRARED DISPERSIVE RAMAN MICROSCOPY

K P J Williams[1] , H G M Edwards[2], D W Farwell[2], I Lewis[3]

[1]Renishaw Plc, Old Town, Wotton-under-Edge, Gloucestershire, GL12 7DH, UK
[2]Dept Chemistry & Chemical Technology, University of Bradford, Bradford, UK
[3]University of Idaho, Dept of Chemistry, College of Letters & Science, Moscow, Idaho, 83843, USA

A detailed comparison between the performance levels of commercial FT - Raman and near infrared dispersive Raman microscope systems has been undertaken. The sample set used in this work has been chosen to be of particular interest to those working in the petrochemical industry. The application of Raman microscopy in this field has been significant. It has benefited greatly from the development of both the FT-Raman and near infrared dispersive Raman technology[1,2]

The systems compared have been used without further modification or optimisation above that normally available. Identical samples have been studied on both instruments and have included polymers, catalysts and fibres. The added benefit of optimum fluorescence rejection using the Nd : YAG laser source at 1.064 μm was seen from some samples. However, the extra sensitivity afforded by using a diode laser at 782 nm, with a dispersive spectrograph and charge couple device detector, provided data from other samples with far superior signal to noise ratio. This was achieved using both a lower laser power and a fraction of the data acquisition time.

It is well known from a theoretical point of view that the v^4 factor provides a benefit of ca 3 times when using the diode laser. The use of a CCD detector should provide in the order of 50 to 100 times more sensitivity in comparison to the liquid nitrogen cooled germanium detector commonly used in FT - Raman systems.

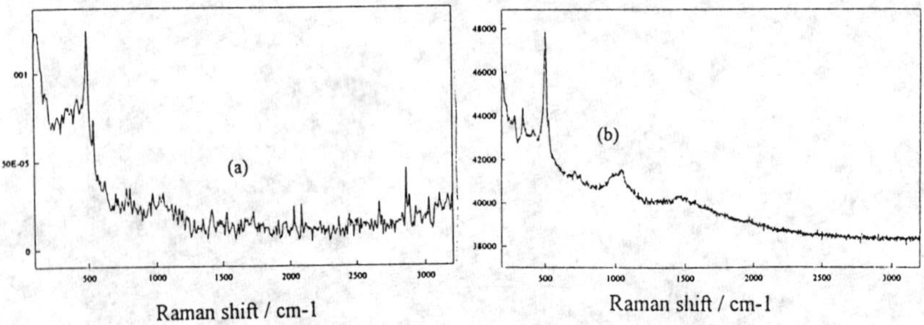

FIGURE 1 (a) FT-Raman spectrum of Zeolite A (b) 780nm dispersive Raman spectra of Zeolite A.

Figure 1 illustrates data obtained from a commonly used catalyst, Zeolite A. Both the FT - Raman method (Figure 1 (a)) and the dispersive system (Fig 1 (b)) provide clearly discernable Raman spectra with the presence of a distinct band at ca 450 cm^{-1} . However, a measurement of the $^{S}/_{N}$ ratios of this peak provides 5 : 1 and 50 : 1, respectively.

The data acquisition conditions of 2 hours, 50 mw laser power and 60 seconds, 2.5 mw laser power, respectively, leads to a relative superior sensitivity of over 2000, in favour of the dispersive based microscope system.

Figure 2, illustrates the Raman data obtained from a polyurethane foam sample using FT - Raman microscopy (Fig 2 (a)) and near infrared dispersive Raman microscopy (Fig 2 (b)). Both data sets represent raw data using (Fig 2 (a)) and 10 seconds, 2.5 mw (Fig 2 (b)). These data illustrate the almost complete fluorescence suppression achieved with the lower energy laser used in FT - Raman. However, with a background subtraction (Fig 2 (c)) the data achieved with the dispersive method is of a better $^{S}/_{N}$ ratio when laser powers and acquisition times are taken into account.

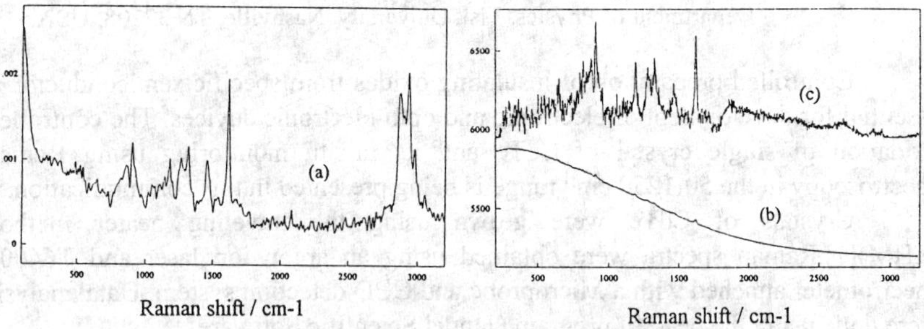

FIGURE 2 (a) FT-Raman spectrum and (b) Dispersive Raman spectrum at 780 nm, with a background subtraction (c) from a polyurethane foam.

Further issues involved in this comparative study, such as detector response, analysis area and volume will also be considered and illustrated with experimental data. In addition other sample types such as, single fibres and highly coloured materials will be used to show the strengths and weaknesses of the different approaches to fluorescence suppression in Raman microscopy.

References

1. A J Sommer and J E Katon, Appl Spectrosc, 45, 529 (1991)

2. S M Mason, N Conroy, N M Dixon and K P J Williams Spectrochimica Acta, 49A, 633 (1993)

Acknowledgements

The authors wish to thank Renishaw plc for permission to publish.

A MICRO-RAMAN STUDY OF LASER IRRADIATED CONTROLLED MICRO-OXIDATION OF CDTE SURFACE.*

Brajesh K. Rai[1], H. D. Bist[1], R. S. Katiyar[1]*, K. -T. Chen[2] and A. Burger[2]

[1]Department of Physics, University of Puerto Rico, P.O. Box 23343, Rio Piedras, Puerto Rico 00931-3343, USA.
[2]Department of Physics, Fisk University, Nashville, TN 37208, USA.

Controlled preparation of insulating oxides from specific semiconductor is essential for development of electronic and opto-electronic devices. The controlled oxidation of single crystal of CdTe and its in-situ monitoring using Raman spectroscopy in the 50-1250 cm^{-1} range is being presented in this communication.

Crystals of CdTe were grown using the traveling heater method (THM)[1].Raman spectra were obtained using an argon ion laser and T64000 spectrometer attached with a Microprobe and CCD detection system. Data analysis were done using the peak fit program (Jandel Scientific Software).

Oxidation process on the surface of CdTe crystals has been monitored in-situ and controlled with a reasonable degree of accuracy. The main product of oxidation has been identified as TeO_3^{2-}, unambiguously for the first time to the best of our knowledge. Strong new Raman bands evolve in the 600-850 cm^{-1} regions after irradiating the CdTe sample with higher powers of 514.5nm laser line.

It has been inferred that even the as-grown crystals have random distribution of the Te precipitate over whole volume. Our efforts to restore the pure CdTe surface by using 5% (by volume) solution of bromine in methanol for few minutes and 1N KOH in methanol for 20 minutes did not give Te-free spectrum from the CdTe crystal even at low laser power. Owing to the presence of Te in the bulk of CdTe crystals, Te modes near 92, 103, 123, and 141 cm^{-1}(partially) have been observed. Laser beam falling over <111> surface of the sample gives rise to TO and LO modes of CdTe at 141 and 167 cm^{-1} respectively. In addition, the 2TO(CdTe) mode at 269 cm^{-1} has been attributed to the anharmonic affects in the system. The bands at 324 and 381 cm^{-1} have been assigned to the splitted components of degenerate deformation mode $\nu_4(E)$ of TeO_3^{2-}. Likewise, the degenerate asymmetric stretching mode $\nu_3(E)$ of TeO_3^{2-} exhibits weaker two components at 662 and 703 cm^{-1}. The band at 352 cm^{-1} has been assigned to $\nu_2(A_1)$ mode of TeO_3^{2-}, while the totally symmetric mode $\nu_1(A_1)$ in the TeO_3^{2-} exhibits a fermi resonance with the overtone of ν_4' falling at 381 cm^{-1}(see Fig. 2)

Whenever two vibrational levels belonging to different vibrations for combinations of vibrations) having nearly the same energy, and symmetry species, fall near each other, only the perturbed levels are observed in the spectrum[2]. The unperturbed separations δ between the two positions could be calculated by using the simple relation between the observed intensities I_1 and I_2 of the two bands and the observed (perturbed) separation Δ of the bands by simple expression $I_1/I_2 = (\Delta + \delta)/(\Delta - \delta)$. Using the above expression, the unperturbed separation δ, for the Fermi-pair 732-777 cm^{-1} (with $I_1/I_2 = 2.009$ and $\Delta = 45$ cm^{-1}), is calculated as 15.9 cm^{-1}. The ν_1 mode in the TeO_3^{2-} system would have been in position ii(a) at 746.5 cm^{-1}. However due to the Fermi resonance it is displaced further to position ii(b) at 732 cm^{-1}. The mean position of Fermi-pair 732-777 cm^{-1} is 754.5 cm^{-1} and it is very close to the position of this band in dilute solutions at 758 cm^{-1}. With the higher excitation power, the two peaks obtained in the 720-800 cm^{-1} region show half width broadening and a shift to lower frequency side.

On the basis of micro-Raman study the simultaneous existence of CdTe and Te aggregates distributed over the surface (and entire volume) of the crystal is suggested. The study is important to explore the mutual interactions of the Te aggregates, TeO_3^{2-} ion and CdTe semiconductor.

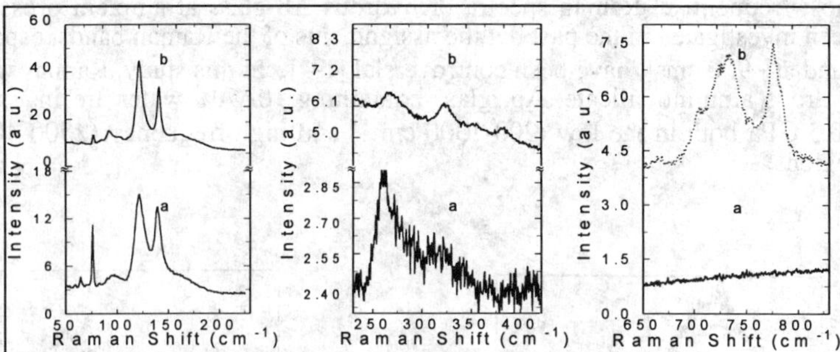

Fig. 1. Relevant portions of the micro-Raman spectra excited by 514.5 nm (2 mW) laser from a sample of CdTe having a spot diameter of < 2 μm: (a) virgin spot and (b) spot irradiated with 13.5 mW power for about 15 minutes.

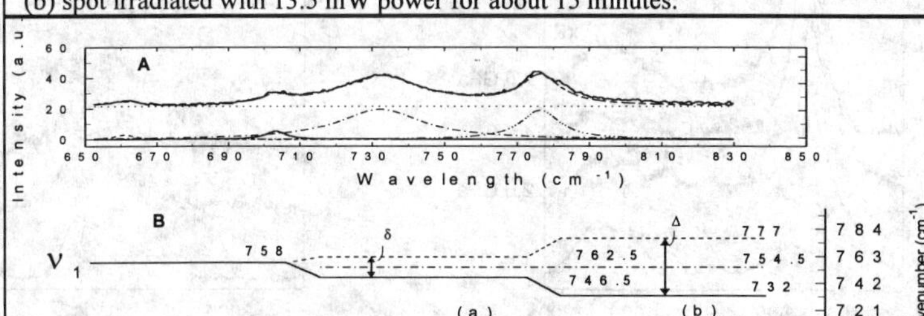

Fig. 2. The main bands 732 and 777 cm^{-1} in the Raman spectrum of TeO_3^{2-} on CdTe: [A] observed spectrum and Lorentzian peak fit analysis in the 650-830 cm^{-1} range. [B] ν_1 band (————) in (i) dilute solution and (ii) unperturbed calculated positions (a), and perturbed (due to Fermi Resonance) positions (b) of ν_1 band (————) and $2\nu_4$ band (·········), on the surface of CdTe-Te-TeO_3^{2-}. The 754.5 cm^{-1} (–·–·–·–) represents the mean of the two band positions.

1. R. Triboulet, Proc. 3rd Int. Conf. On CdTe, P. Siffert, A. Cornet (eds.), Strasbourg, p. 171, 1971.

2. G. Herzberg, Molecular Spectra & Molecular Structure II IR and Raman Spectra of Polyatomic Molecules, Van-Nostrand, New York (1945), pp 215.

*Supported by NSF grant OSR-9452893.

Laser Confocal Micro-Raman Study of Hydrous Alumino-silicate Glass at high pressure.

Shiv K. Sharma[1], Zifu Wang[1] and Sieger van der Laan[1,2]
[1] University of Hawaii, Honolulu, HI 96822, USA
[2] University of Utrecht, 3508 TA Utrecht, Netherlands

The role of water on physical properties of oxide glasses and melts is of great interest in materials science as well as in earth science. Recently, it has been found that OH containing SiO_2 glasses when electrically polarized give rise to strong second harmonic generation (SHG). In the studies of water dissolution in alumino-silicate melts, mineral albite ($NaAlSi_3O_8$, Ab) composition has been used most frequently. Raman spectra of hydrous Ab-glass at ambient pressures have been investigated in the past but the assignments of the Raman bands, especially the band at ~900 cm^{-1}, have been controversial [1-4]. In this study, Raman spectra of a hydrous alumino-silicate, Ab, glass containing 10 wt% water are investigated to 10.5 GPa both in the low (200-1300 cm^{-1}) and high frequency (2500-4000 cm^{-1}) regions.

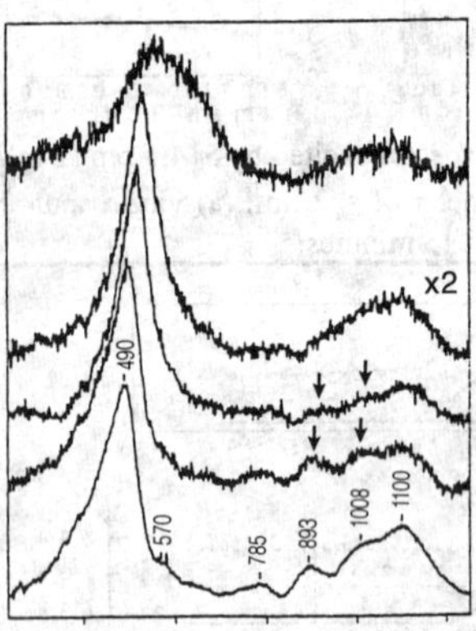

Fig. 1. Raman spectra of albite glass at pressures (200-1300 cm^{-1})

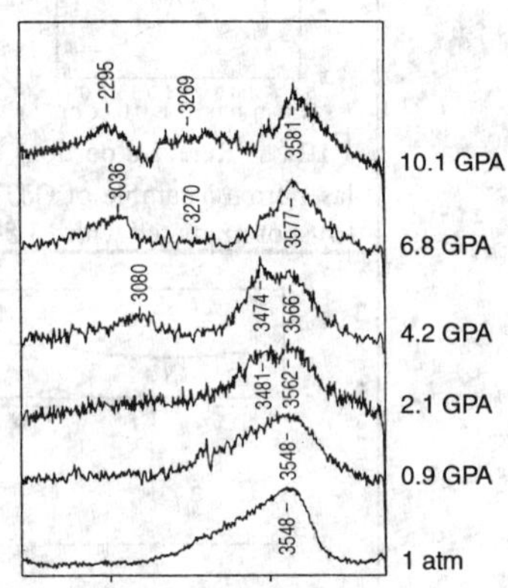

Fig. 2. Raman spectra of albite glass at pressures (2500-4000cm^{-1})

Figure 1 shows Raman spectra of hydrous Ab glass at various pressures in the 200-1300 cm^{-1} region. In agreement with previous work [2-4] at 1 atm. the spectrum of hydrous albite glass shows a broad band near 900 cm^{-1} and a doublet consisting of broad bands at ~1000 and 1100 cm^{-1}. The 900 cm^{-1} band shifts to the higher frequency with increasing pressure and its intensity drops significantly, and it finally disappears at 7.5 GPa. The two components of the T-O-T antisymmetric stretching mode shift to higher frequency with pressure and the intensity of the 1000 cm^{-1} shoulder decreases relative to the 1100 cm^{-1} band. At 7.5 GPa, the doublet is replaced by a broad symmetric band at ~1095 cm^{-1}. At higher pressure the broad band moves to lower frequency while its intensity drops significantly. It is interesting to note that in the Raman spectra of hydrated Ab glass, the 900 cm^{-1} band disappears at the same pressure as the 1000 cm^{-1} band. This indicates that these bands are inter-related. We propose that the 900, 1000 and 1100 cm^{-1} bands of the glass result from the distortion of TO_4 tetrahedra leading to lifting of degeneracy of the triply degenerate υ_{as}(T-O-T) band. The proposed assignment is also consistent with the fact that the 900 cm^{-1} does not show isotope shift in deuterated glasses[1,3].

In the OH stretching region of the spectra of hydrous albite glass (Fig. 2), several changes are observed with increasing pressure. The asymmetric band observed at about 3548 cm^{-1} in the 1 atm. spectrum splits into two bands with peaks at ~3481 and 3562 cm^{-1} at 2.1 GPa. With increasing pressure, the 3481 cm^{-1} band moves to lower frequency and the 3562 cm^{-1} band moves to the higher frequency. At pressure ≥ 4.2 GPa, additional weak features appear at lower frequencies, indicating the presence of a number of hydrous species in various environments. The decrease in frequency of Raman band indicates the formation of strong hydrogen bond in the structure of the glass. The increase in the frequency of other Raman band may result from the hydrogen-hydrogen repulson, or the different orientation of these hydroous species in the glass. In addition to the O-H stretching band, a weak band is also observed at ~1630 cm^{-1} in the 1 atm. Ab glass spectrum. This is the H-O-H bending vibration indicating the presence of molecular H_2O in the glass [2,3]. On the basis of NMR data, Kohn el al. [5] proposed that, in hydrous Ab glass, Na^+ ions are replaced by H^+ ions, and OH^- ions are associated with Na^+ ions and that most of the network remains intact with molecular water present in the open network. Our high-pressure Raman data basically support Kohn's model, except the proton must be present as hydrated H_3O^+ ions, charge balancing the Al^{3+} ions in the network. The results of present investigation show that high-pressure Raman spectroscopy is a powerful tool for investigating hydrous glass structure.

[1] B. O. Mysen, D. Virgo, W. T. Harrision and C. M. Scarfe, Amer. Mineral. 65, 900 (1980)

[2] P. F. McMillan, S. Jackobsson, J. R. Holloway and L. Silver, Geochim. Cosmochim. Acta 47, 1937 (1983).

[3] P. F. McMillan, B. T. Poe, T. R. Stanton and R. L. Remmele, Phys. Chem. Min. 19, 454 (1993).

[4] D. Sykes and J. D. Kubicki, Geochim. Cosmochim. Acta 57, 1039 (1993).

[5] S. C. Kohn, R. Dupree, and M. G. Mortuza, Geochim. Cosmochim. Acta 53, 2925 (1989).

EXAMINATION OF GLASSY MATERIALS BY RAMAN MICROSPECTROSCOPY AND IMAGING

N. L. Bradley, M. D. Morris*

Department of Chemistry, University of Michigan, Ann Arbor, Michigan 48109

Depth-resolved Raman microspectroscopy and imaging are used to examine selected high alkali silica glasses which have been treated to increase their impact resistance and flexural strength. The glasses are candidate restorative dental materials. The strengthening methods include autoclaving the glasses with steam in different combinations of time, elevated temperature, pressure, and percent humidity or refluxing the glasses in weak acetic acid solution for controlled times. The acetic acid reflux is also the standard test for chemical durability in the oral environment.

While these approaches are known to increase the strength of the glasses [1,2], the mechanisms are not clearly understood. One proposed mechanism for strengthening by autoclaving includes the penetration of water several microns into the glass network where it severs the bond between a bridging oxygen and one of its anchoring silicon atoms. As the glass network is broken, the sample occupies a larger volume and becomes more plastic. With the plastic surface, crack propagation is more difficult since deformities flow and the material effectively is stronger. There has been less investigation of the mechanism of strengthening of dental glasses by acetic acid refluxing.

A confocal Raman microprobe fitted with a 60X/1.4NA oil-immersion objective or a 50X/1.0NA water-immersion objective is used to investigate chemical changes through the first 10-25 micron depth of the glass as a function of autoclaving or refluxing time. Filter-based Raman and fluorescence imaging are used to confirm the chemical homogeneity of the glasses on the micron scale, thus insuring that single point microspectroscopy results are representative of the glass composition.

An example of the chemical effects of acid refluxing is shown in the Raman spectrum obtained after 5 hours refluxing (Figure 1.) The band at 2350 cm^{-1} is characteristic of strongly hydrogen-bonded water. The 3350 cm^{-1} band is characteristic of water weakly hydrogen bonded to silicic acid. [3] There is no evidence for substantial amounts of free water, because any 3580 cm^{-1} intensity is below our limits of detection.[4] Figure 2 shows the ratio of 2350 cm^{-1} intensity to the 3350 cm^{-1} intensity as a function of penetration depth, after five hours refluxing.

A preliminary interpretation of the acid refluxing experiments is that they result in leaching of soluble alkali components of the glass, leaving in place a strong silica network with some pendant silicic acid groups, capable of forming weak hydrogen bonds. While the structural changes are confined to the first few microns of the glass structure, the strength increases are great enough to allow the material to survive in the oral environment. The mechanism is entirely different from the plasticization promotion which is thought to occur upon autoclaving.

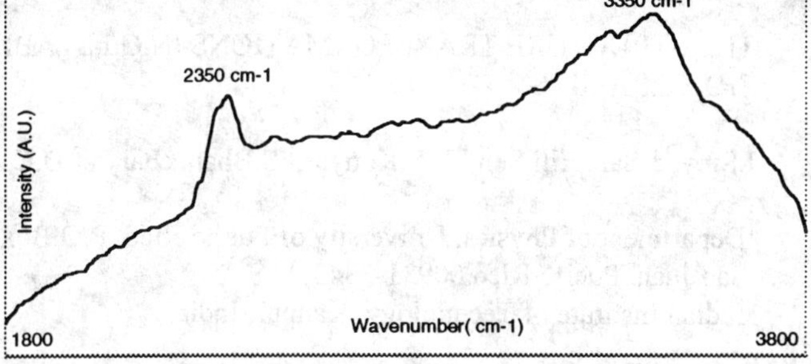

Fig. 1. Raman spectrum of a glass sample treated by refluxing in weak acetic acid for 5 hours.

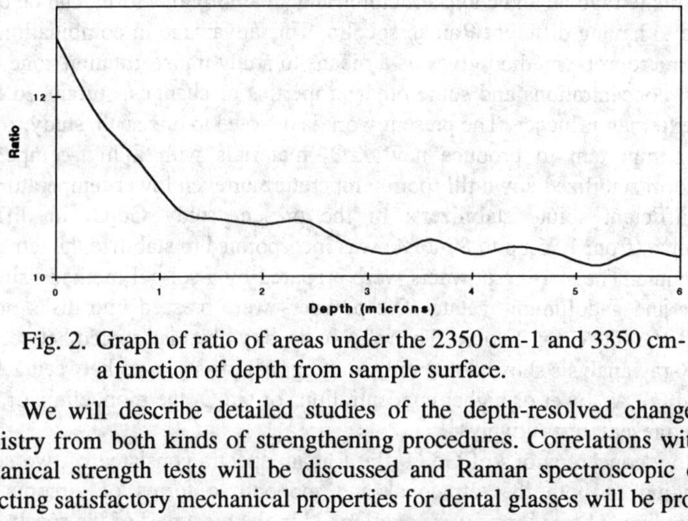

Fig. 2. Graph of ratio of areas under the 2350 cm-1 and 3350 cm-1 bands as a function of depth from sample surface.

We will describe detailed studies of the depth-resolved changes in glass chemistry from both kinds of strengthening procedures. Correlations with standard mechanical strength tests will be discussed and Raman spectroscopic criteria for predicting satisfactory mechanical properties for dental glasses will be proposed.

1. R. F. Bartholomew, J. Non-Cryst. Solids 38&39, 637 (1980).
2. W. J. O'Brien, W. Nakfoor, P. I. Mullen, and K. M. Boenke, Presentation at IADR (International Association for Dental Research) Meeting, San Francisco, March 1996.
3. T. Uchino, T. Sakka, and M. Iwasaki, J. Am. Ceram. Soc. 74, 306 (1991).
4. P. F. McMillan and R. L. Remmele, Jr., Am. Mineralog. 71, 772 (1986).

THE STRUCTURE TRANSFORMATIONS IN Gd-STABILIZED ZrO_2 CERAMICS#

I.Siny,[1]* Sampriti Sen,[1] R.S.Katiyar,[1] S.Bhattacharya,[2] D.C.Agrawal[2]

[1]Department of Physics, University of Puerto Rico, P.O.Box 23343
San Juan, Puerto Rico 00931-3343, U.S.A.
[2]Indian Institute of Technology, Kanpur, India

Raman scattering has been found very suitable for studying the crystallization and phase transformations in zirconia (ZrO_2)[1,2]. Four well-defined polymorphs, the monoclinic,tetragonal and cubic structures and an amorphous form, can be distinguished due to having different Raman spectra. This advantage in combination with the Raman microprobe method gives us a means to study transformation zone sizes, composition concentrations and some other properties or changes, produced by an internal or external influence. The present work is devoted to one such study.

It is important to produce new ZrO_2 materials with a high-temperature tetragonal form stabilized down till room temperature or even lower temperatures by adding insufficient solute stabilizers. In the present study Gd_2O_3 in different concentrations (from 1.75 up to 8 mol%) was incorporated to stabilize the tetragonal form of zirconia. The starting powders were prepared by a sol-gel method using Zr-n-propoxide and gadolinium acetate. The powders were pressed into disks or bars and sintered at 1400°C for 2 hours. The relative density obtained varied between 90-98%. The X-ray analysis showed that only a tetragonal phase exists between 2.5 to 4 mol% of Gd_2O_3. At lower or higher concentrations of Gd_2O_3 the monoclinic or cubic phases were present, respectively.

At low concentrations of Gd_2O_3 the Raman spectra consisted of two sets of lines belonging to both the tetragonal and monoclinic forms of zirconia. The characteristic lines of two forms were overlapped in the most part of the spectra. The initial Raman spectra were deconvoluted using a software program to follow their evolution in a series of samples with variable concentration of the Gd_2O_3 stabilizer. The change in both the peak intensity and the line width for "tetragonal" ($\Delta\nu \sim 146$ cm^{-1}) and "monoclinic" ($\Delta\nu \sim 177$ cm^{-1}) lines is shown in Fig.1. The hand ground samples were used in the study.

As one can see, the peak intensity exhibits some jumps in both cases which, probably, are connected with a quality of samples to provide a sufficient penetration for the laser light. The line width seems to be rather independent of the sample quality. Both dependencies confirming each other show a continuous suppression of the monoclinic phase by increasing concentration of Gd_2O_3. The line width of a "monoclinic" line increases continuously (in comparison with an initial value for ZrO_2 powder) even at low concentrations of the Gd_2O_3 stabilizer in the range from 1.75 to 3 mol% of Gd_2O_3 while the line width of a "tetragonal" line remains constant.

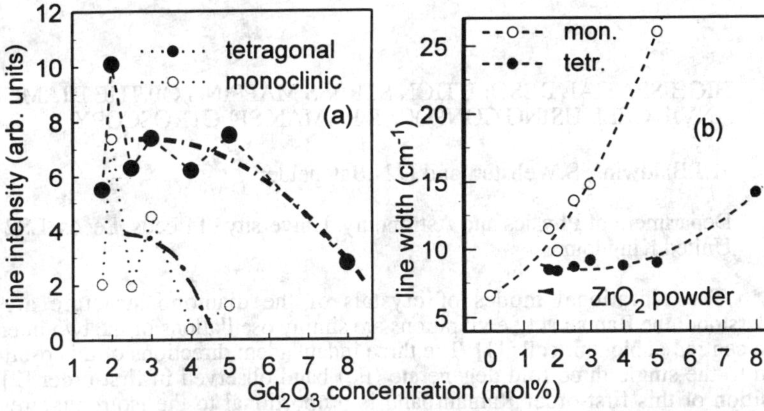

Fig.1. The intensity (a) and the width (b) of two lines, characteristic of the tetragonal and monoclinic phases in zirconia, vs. the stabilizer concentration.

Large anharmonicity of phonons, characteristic of the monoclinic phase of zirconia, is connected with a small size of the monoclinic grains impregnated in the tetragonal rather stable matrix. One can suppose that the behavior of the "monoclinic" line width reflects a continuous decrease in sizes of the monoclinic impregnations. Even at the 5 mol% concentration of Gd_2O_3 the Raman scattering evidences traces of a monoclinic phase although a size of the monoclinic regions seems to be small enough to be detected by X-ray diffraction which gives an average pattern. Probably, this is the reason of some discrepancies between the results of phase analysis obtained by X-ray and Raman measurements. Also, a predominant contribution of the surface structure is obvious in the case of Raman scattering. The hand grinding of samples may provide an additional amount of the monoclinic grains just on the surface or in the undersurface layer of the samples. When the Gd_2O_3 concentration increases, both the monoclinic and tetragonal forms disappear one after the other. The difference in the line width behavior shows that the tetragonal form is stable in some interval of concentrations of the stabilizer. According to the Raman scattering measurements, there is a very promising gap on the concentration scale where the tetragonal form is still stable and the monoclinic form is already suppressed.

The second part of the present work dealt with the measurements of the crystallographically transformed zones produced by fracture in zirconia samples with the different Gd_2O_3 concentrations. The Vickers indented samples were used in this study. Raman microprobe technique enable us to study step by step the topography of the monoclinically transformed regions adjacent to the indentation-induced cracks on the sample surface. The extent of the monoclinic region was found to be a function of the concentration of the Gd_2O_3 stabilizer.

#Supported by NSF Grant OSR-9452893

1. V.G.Keramidas and W.B.White, J. Am. Ceram. Soc. 57, 22 (1974).
2. D.R.Clarke and F.Adar, J. Am. Ceram. Soc. 65, 284 (1982).

HIGH SPATIAL RESOLUTION STRESS MAPPING OF THE DIAMOND ANVIL CELL USING CONFOCAL RAMAN SPECTROSCOPY

K.J.Baldwin*, S.Webster, and D.N.Batchelder

Department of Physics and Astronomy, University of Leeds, Leeds. LS2 9JT, United Kingdom.

The vibrational modes of crystals of the diamond structure are well understood: the Raman active vibrations are simply oscillations of the two interlaced face centred cubic unit cells [1]. The three independent directions of this oscillation lead to the single three fold degenerate (F_{2g}) band observed in first-order [2]. The position of this first-order Raman band is proportional to the isotropic stress [3]. When there is a non-uniform stress distribution within the crystal, the symmetry of the crystal structure will be lowered, and the degeneracy of the Raman band will be lifted.

Gasketed diamond anvil cells (DAC) are used extensively as a method for studying the behaviour of micro-samples at high pressures [4]. The pressure in the sample volume is usually measured using a pressure calibration such as the fluorescence from a piece of ruby [5]. However it has been suggested that the first-order Raman line could be used to measure the pressure in the sample volume [6,7,8], although care is needed to identify the feature within the band envelope that corresponds to the portion of diamond at the diamond-sample interface.

Studies of the stress distribution within the diamond anvils, especially at the base, are important since the most probable mechanism for failure of the diamonds is a concentration of tensile stress in this region. Theoretical [9,10] and experimental [6,7] studies have demonstrated the tensile stress at the base, and a local concentration of compressive stress in the vicinity of the culet. The best spatial resolution in these studies was approximately 10 μm diameter by 50 μm deep [6].

The results presented here were obtained using a Renishaw Raman Imaging Microscope, with Ar^+ laser (514.5 nm) excitation, operating in a confocal microprobe geometry [11]. A ×50 ULWD microscope objective was used for both sample illumination and collection of the 180° back-scattered light. The spatial resolution of this system was estimated to be 10 μm diameter by 20 μm deep within the diamonds of the DAC. This superior resolution, together with careful consideration of the polarisations of both the illumination and collection optics, enabled a more detailed analysis of the stress distribution within the diamond anvils to be performed.

Figure 1(A) shows the Raman spectra in the vicinity of the first-order band as the confocal volume was scanned through the axis of the diamond anvil. The shift to larger wavenumber shifts of the Raman band is clearly evident. This is attributed to an increase in the compressive stress along the axis: in an idealised back-scatter geometry, and with a stress distribution symmetrical about the axis, the only observable Raman signal would come from the 'axial' vibration. A plot of the variation of band position with depth through the anvil shows this in more detail (figure 1(B)). Both the localised compressive stress near the culet, and the tensile stress at the base are clearly evident. There is an isostatic point approximately 1/4 of the way through the anvil. Also, a decrease in the stress at the culet face can be observed. In the case of the DAC loaded to 40 kbar, the decrease in stress is approximately 3 kbar (1 cm^{-1}). The variation in measured band width (including

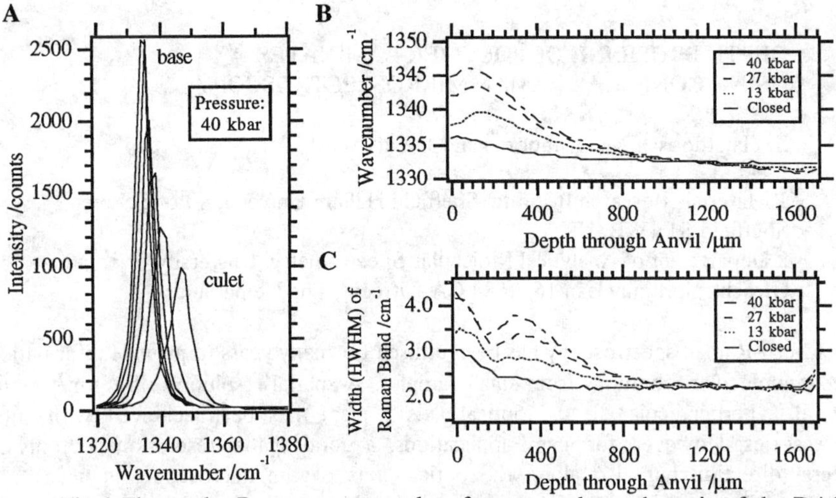

Fig.1 First order Raman spectra taken for a scan down the axis of the DAC (A). Variation of band position (B) and width (C) through the DAC.

instrumental broadening) is plotted in figure 1(C). The structure in this plot is due to the depth of focus of the instrument effectively integrating over a range of stress levels, and hence providing a measure of the gradient of the band position plot. There is also a general increase in the band width near the culet, which can be attributed to splitting of the Raman band in the large stress fields in the region.

We have also performed scans across the face of the culet. These show a general decrease in the compressive stress away from the axis of the diamond, and also a slight decrease of stress in the vicinity of the sample volume. There is also increased splitting of the Raman bands away from the axis, which can be attributed to an increased anisotropy of the local crystal unit cell, and hence a relaxation of the Raman selection rules. At the edge of the culet, there are two well resolved bands, whose absolute intensity is dependent upon the state of polarisation of the excitation and collection optics, but their relative intensities are almost independent of polarisation. The consequences of these observations are discussed in terms of the effects of morphing the diamond crystal lattice on the expected first-order Raman vibrations.

1. H.M.J.Smith, Phil. Trans. A **241** p105 (1948)
2. S.A.Solin and A.K.Ramdas, Phys.Rev. B **1** (4) p1687 (1970)
3. B.J.Parsons, Proc.R.Soc.Lond. A. **352** p397 (1977)
4. A.Jayaraman, Rev.Mod.Phys. **55** (1) p65 (1983)
5. R.A.Forman, G.J.Piermarini, J.D.Barnett and S.Block, Science **176** p284 (1972)
6. S.K.Sharma, H.K.Mao, P.M.Bell and J.A.Xu, J.Raman Spectr. **16** (5) p350 (1985).
7. M.Hanfland and K.Syassen, J.Appl.Phys. **57** (8) p2752 (1985).
8. R.D.Markwell and I.S.Butler, Can.J.Chem. **73** p1019 (1995).
9. D.M.Adams, A.G.Christy and A.J.Norman, Meas.Sci.Technol. **4** p422 (1993).
10. D.M.Adams and A.C.Shaw, J.Phys.D:Appl.Phys. **15** p1609 (1982).
11. S.Webster, Ph.D. Thesis, University of Leeds (1994).

DEPTH PROFILING OF POLYMER LAMINATES USING CONFOCAL RAMAN MICROSPECTROSCOPY

S. Hajatdoost[1], M. Olsthoorn[2] and J. Yarwood[1]*

1 Materials Research Institute, Sheffield Hallam University, Pond Street, Sheffield S1 1WB, UK.
2 Department of Analytical Molecular Spectrometry, University of Utrecht, Sorbonnelaan 16, 3584 CA, Utrecht, The Netherlands.

Micro-Raman spectroscopy has been utilised for many years to provide information about molecular vibrations from small samples. A spatial resolution of 1 μm^2 in the lateral (perpendicular to the optical axis) plane has been achieved with such accessories. However, for many applications, a good spatial resolution not only in lateral direction but also along the optical axis of the microscope, is necessary. Confocal Raman microspectroscopy already has been utilised by other researchers[1] using a pinhole at the back focal plane of the microscope. One of the disadvantages of this way of confocal arrangement is that optical alignment is difficult and time consuming.

We have used another approach[2] for confocal arrangement using a Renishaw Ramascope spectrometer with a Peltier cooled CCD detector (384 x 576 pixels). The spectral dimension is described by the long axis of the CCD (576 pixels) and the image height is described by the shorter axis (384 pixels). This system also comprises two Raman holographic edge filters which prevent the intense backscatter laser radiation. The 12 µm slit used as the primary aperture in the system provides an additional spatial filter for the confocal experiment. As a result of this arrangement, the depth resolution is determined as 2.0 µm using the full-width at half maximum (FWHM) criterion[3].

In this paper we illustrate how confocal Raman microspectroscopy can be used to focus on the interface layer of polymer laminates. The system we have chosen are PMMA/PVOH, PAN/PVOH and PAN/PAA laminates. It is expected that interdiffusion or interpenetration will occur at the interface between the two constituent polymers due to their functional groups.

Molecular depth profiling study of a PMMA/PVOH laminate on quartz shows hydrogen bonding between the ester and alcohol groups [4] near the interfacial region. The carbonyl υ(C=O), band of PMMA shows significant broadening in the interfacial region. Various PMMA/PVOH laminates with different PMMA molecular weights have been studied and it is demonstrated that the PMMA layers with lower molecular weight show a greater degree of interpenetration for a given annealing time. This is shown in figure 1 and 2. Similarly the nitrile υ(C≡N), band of PAN shows significant broadening in the interfacial region in PAN/PVOH laminate which is due to hydrogen bonding in this laminate. However the chemical interaction has not been observed in PAN/PAA laminate. One explanation for this is strong intermolecular bonding between carbonyl group of PAA with its own hydroxyl group. As a result of this the PAN nitrile, υ(C≡N), shows no band broadening (fig 3) due to hydrogen bonding.

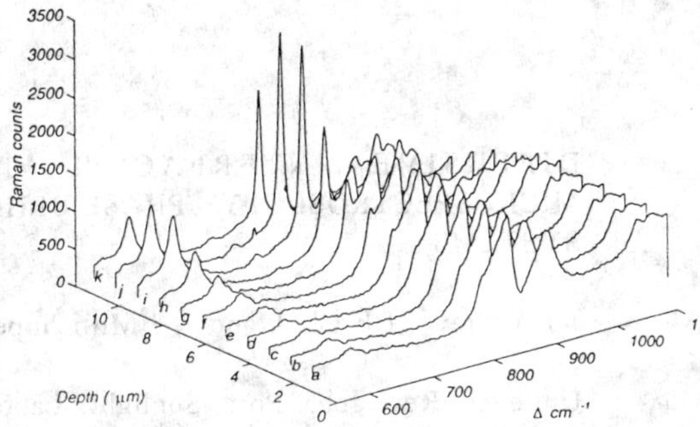

Fig. 1. 3-D representation of the depth profile of PMMA(III)-PVOH-quartz laminate obtained by focusing the microscope stepwise (1 μm) through the laminate.

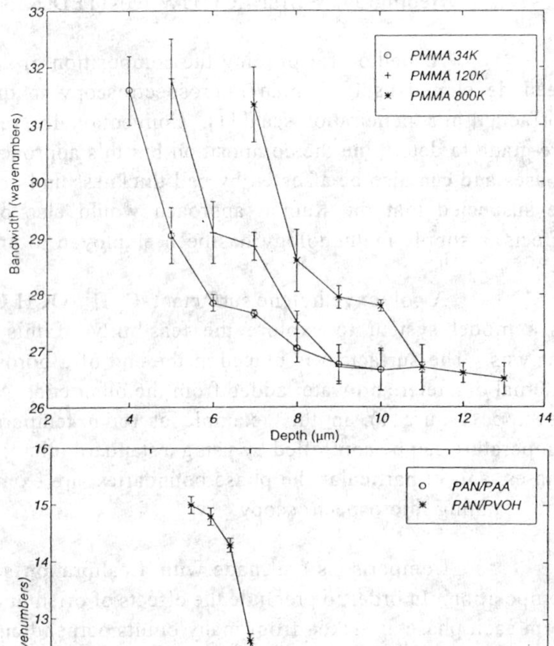

Fig. 2. Plot of υ(C=O) bandwidth changes as a function of depth through the PMMA-PVOH laminate.

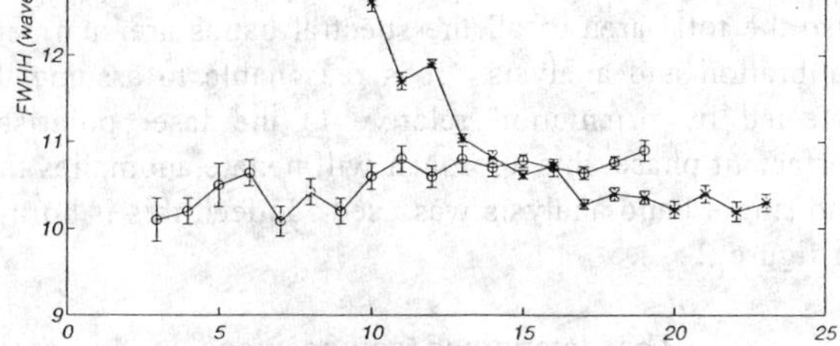

Fig. 3. Plot of band width of υ(C≡N) band changes as a function of depth through the PAN/PVOH or PAN/PAA laminate.

1. R. Tabaksblat, R. J. Meier, and B. J. Kip, Appl Spec. 46, 60 (1992).
2. K. P. J. Williams, G. D. Pitt, D. N. Batchelder, and B. J. Kip. Appl spec. 48, 232 (1994).
3. S. Hajatdoost and J. Yarwood, In press (Appl Spec, May 1996).
4. J. Yarwood, in *Spectroscopy and Structure of Molecular Complexes*, J. Yarwood, Ed. (Plenum, New York, 1973), p. 180.

DIFFUSIVE INTERFACIAL TRANSPORT RAMAN MICROSPECTROSCOPY - PHASE BEHAVIOUR IN SURFACTANT SYSTEMS

J J Andrew*, I E Clark and A J Millichope

Unilever Research, Port Sunlight Laboratory, Quarry Road East, Bebington, Wirral L63 3JW, UNITED KINGDOM.

A method for probing the composition of surfactant phase diagrams has been developed using Raman microspectroscopy to quantify the concentration of surfactant in a 'penetration scan' [1]. Conventionally, refractive index measurements are made to determine the composition but this approach is not valid for birefingent phases and can also be affected by cell thickness and non-flatness. Although it may be suspected that the Raman approach would also be influenced by orientation effects, a simple methodology has been employed to preclude this.

A polyoxyethylene surfactant, $C_{12}H_{25}(OCH_2CH_2)_8OH$, ($C_{12}EO_8$), was used as a model system to explore the feasibility of this approach for more routine analysis. The surfactant is placed at the end of a borosilicate microslide of 0.4mm internal diameter and water added from the other end. After several hours the phases develop (Figure 1), in this example at room temperature. More generally, the temperature can be controlled by using a standard microscope temperature stage. The phases, and in particular the phase boundaries, are examined by confocal point and line scanning microspectroscopy.

Comparisons are made with a calibration series to determine the phase composition. In order to preclude the effects of orientation on the calibration, spectra from each phase are taken from many points perpendicular to the diffusion flow and also the total area of all the spectral bands are summed to provide the composition calibration and analysis. It is reasonable to assume that, even if a single band is affected by orientation relative to the laser polarisation within an area of the surfactant phase, this approach will negate anomalies that could occur if single point and single band analysis was used. Indeed this is born out by the calibration shown in figure 2.

The determined compositions in this feasibility study matched those expected from viscometry and refractive index methods [1]. In the future, accurate temperature control and internal calibration between each run will be included to improve the accuracy of determination of the composition. The sensitivity of the instrumentation has also been improved by an order of magnitude since these preliminary studies which has improved the accuracy of the measurement.

Table 1: Phase Composition data as determined by DIT-RM

Phase	%Surfactant
L_1	35
H_1	54
V_1	72
L_α	77
L_1*	77

It is clear that this approach has the potential for rapid screening of surfactant systems. The spectra also contain further information, such as the nature of packing of the alkyl chains [see for example references 2 and 3]. We expect this approach to be a useful addition to the conventional methods of studying phase diagrams.

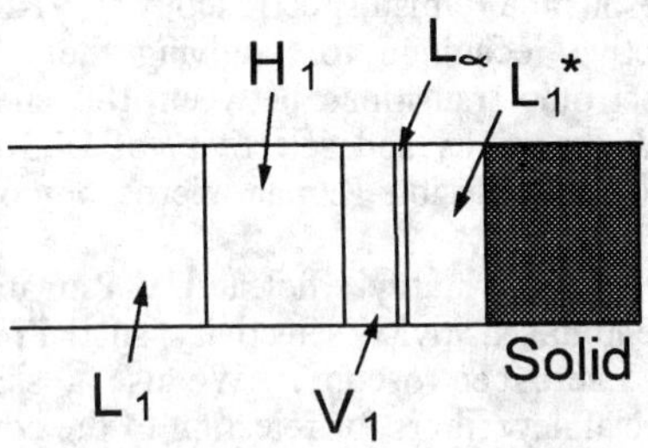

Figure 1: Phases of $C_{12}EO_8$ in water at room temperature

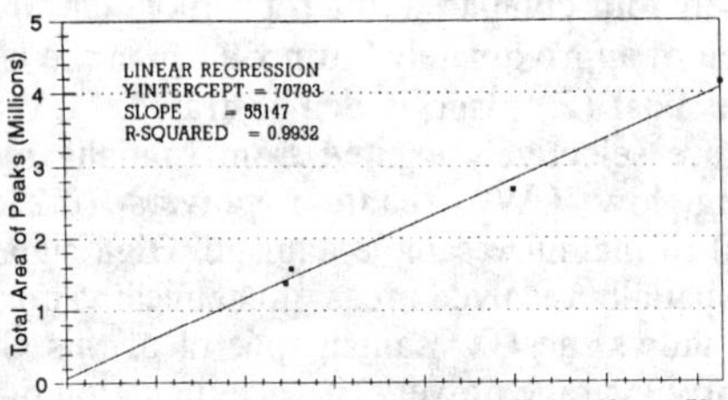

Figure 2: Calibration Graph for $C_{12}EO_8$

1. D. Mitchell, G.J.T. Tiddy, L. Waring, T. Bostock and M.P. McDonald, J.Chem.Soc.Faraday Trans., 89 4079 (1993).
2. I.W. Levin and E.N. Lewis, Anal.Chem. 62 1101A (1990).
3. J.B. Rosenholm, P. Stenius and I. Danielsson, J.Colloid and Interface Sci. 57 551 (1976).

UV RAMAN MICROSPECTROSCOPY: SPECTRAL AND SPATIAL SELECTIVITY WITH SENSITIVITY AND SIMPLICITY

C. H. Munro, V. Pajcini, R. W. Bormett, R. E. Witkowski and S. A. Asher*

University of Pittsburgh, Department of Chemistry, Pittsburgh, PA 15260

During the last 20 years, visible Raman microspectroscopy and Raman imaging have been utilized successfully in chemical analysis, geology, industrial quality control [1-4], and in biological studies including the examination of single cells[5]. In parallel, UV resonance Raman spectroscopy (UVRRS) has emerged as a highly sensitive and selective technique for studying the vibrational spectra of molecules which have electronic transitions between 180 and 300 nm[6-8]. The ability to combine the high sensitivity and selectivity of UVRRS with the ease of operation and spatial resolution of visible Raman microscopy or microspectroscopy is highly desirable.

We have developed a new highly efficient UV Raman microspectrometer, and have demonstrated the high sensitivity, selectivity, spatial resolution and ease of operation of UV Raman microspectroscopy. We use a single monochromator dispersion system with special new filters for rejection of the continuous wave (CW) 244 nm Rayleigh scattering. The throughput efficiency of this system is improved by a factor of approximately four compared to a triple monochromator.

Spatial resolution of approximately 3 μm x 9 μm in the lateral (X-Y) plane, and 10 μm or less in the axial (Z) plane is demonstrated. UV resonance Raman spectra of nucleic acids are selectively excited from spatially resolved areas of a single Paramecium using low CW excitation powers (0.2 mW) and short accumulation times (20 s) to minimize sample damage. High signal-to-noise Raman spectra are excited from spatially resolved areas of chemical vapor deposited (CVD) diamond films. Figure 1 shows the UV Raman spectra of this CVD diamond film excited at 244 nm (~1.5 mW average power: 10 s accumulation time). The diamond UV Raman spectra were recorded with the laser spot centered on the (100) face of this single diamond crystallite (Fig. 1a), or at the grain boundaries between diamond crystallites (Fig. 1b). The absolute intensity of the diamond first order phonon band at 1332 cm^{-1} was approximately the same at the (100) face and at the grain boundaries. However, the UV Raman spectrum taken from the grain boundaries showed a broad band at ca. 1600 cm^{-1}, assignable to non-diamond carbon impurities. This band was not present in the spectrum from the (100) crystallite face. These results demonstrate the ability to determine the spatial distribution of non-diamond impurities in CVD diamond films. Figure 4b illustrates that using UV Raman microspectrometry we can resolve underlying components of the non-diamond carbon band. In this instance, a sharp low energy feature, fitted to the ca. 1553 cm^{-1} amorphous carbon band, is resolved from the broad ca. 1603 cm^{-1} non-diamond carbon band (fit correlation, $R^2 = 0.97$). In other instances, the sharp ca. 1580 cm^{-1} graphite band dominates the non-diamond carbon band. The limited spatial area probed enables us to speciate the different non-diamond carbon species which make up the normally broad non-diamond carbon band.

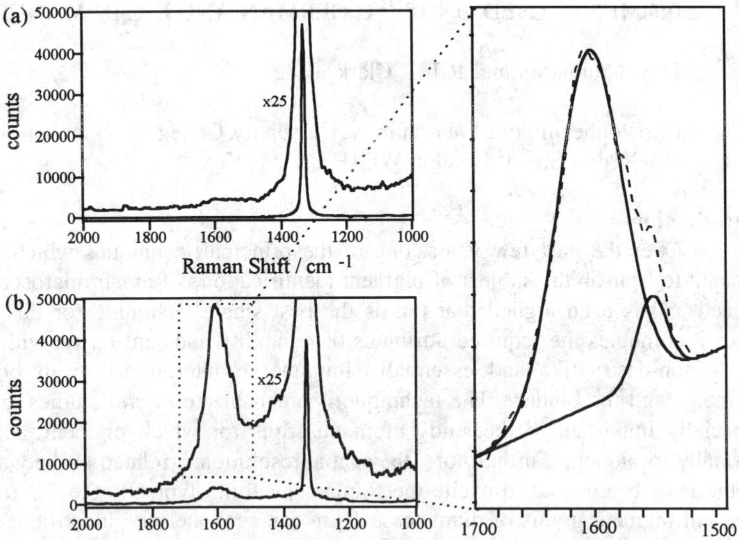

Fig. 1. Raman spectra from CVD diamond film (a) at the (100) face of a single crystallite, and (b) at the interstices between diamond crystallites (244 nm: ~1.5 mW average power: 10 seconds total accumulation time: 100 μm entrance slit).

1. M. Delhaye and P. Dhamelincourt, *J. Raman Spectrosc.*, **3**, 33 (1975).
2. P. J. Treado and M. D. Morris, *Appl. Spectrosc. Rev.*, **29,** No 1, 1 (1994).
3. J. Barbillat, P. Dhamelincourt, M. Delhaye and E. Da Silva, *J. Raman Spectrosc.*, **25**, 3 (1994).
4. K. P. J. Williams, G. D. Pitt, D. N. Batchelder and B. J. Kip, *Appl. Spectrosc.*, **48**, No.2, 232 (1994).
5. J. Puppels, F. F. M. De Mul, C. Otto, J. Greve, M. Robert-Nicoud, D. J. Arndt-Jovin, T. Jovin, *Nature*, **347**, 301 (1990).
6. S. A. Asher, *Anal Chem.* **65**, 59A (1993).
7. S. A. Asher, *Anal Chem.* **65**, 201A (1993).
8. S. A. Asher, C. R. Johnson and J. Murtaugh, *Rev. Sci. Instrum.*, **54**, 1657 (1983).

THE IDENTIFICATION BY RAMAN MICROSCOPY OF THE PIGMENTS USED ON 15TH C GERMAN AND LATIN MANUSCRIPTS

D.A. Ciomartan and R.J.H. Clark*

Christopher Ingold Laboratories, University College London,
20 Gordon Street, London WC1H 0AJ, U.K.

Introduction

Over the past few years, one of the principal techniques which has been brought to bear on the subject of pigment identification is Raman microscopy [1,2]. Indeed, it has been argued that this is the best single technique for this purpose, since it combines the required attributes of reliability and sensitivity, with those of being non-destructive and essentially immune to interference from both other pigments and the binder. The technique is amenable to *in situ* studies and this is especially important in the study of manuscripts for which pigment sampling is normally forbidden. Furthermore, the spatial resolution is related to the wavelength of the laser beam used to excite the Raman spectrum, typically 0.5-0.7 μm. Thus adjacent pigment grains of diameter ≥ 1 μm can separately be identified, a facility of obvious value for establishing the composition of pigment mixes.

There are many reasons for wishing to identify the pigments used on artefacts of all sorts. They are largely to do with restoration, conservation, dating and authentication. Clearly, the correct identification of a pigment is critically important in restorative work, since restoration with incorrect pigments could lead, at one level, to chemical reactions between incompatible pigments or, at another, simply to the incorrect hue. Conservation studies can require the identification of pigment degradation products and eventually the establishment of procedures for, or environments in which, given pigments may be preserved. Authentication and dating are closely linked to the values placed on paintings and manuscripts. Minerals and natural dyes cannot be dated by Raman microscopy, but synthetic ones clearly can. The date of the first manufacture is the marker. These matters are of concern not only to museums and galleries but also, obviously, to auction houses.

The medieval illustrator preferred in general to use inorganic pigments - usually minerals - to illuminate manuscripts and paintings, since such pigments tend to be more stable than organic ones with respect to photodegradation. Inorganic pigments are usually coloured by ligand-field transitions, charge-transfer transitions, intervalence charge-transfer transitions, and/or metallic reflectance, *e.g.* from silver or gold. The depth of colour is affected by particle size, since this relates directly to the relative importance of specular and diffuse reflectance. It is obviously also affected by the absorption coefficients and widths of bands in the visible region of the spectrum. Clearly, the dating of artefacts in this way is negative; the *absence* of a synthetic pigment of known date of manufacture does not prove that the manuscript is medieval, but the *presence* of one does define the earliest possible date for the artefact.

Case Studies

Many manuscripts have now been studied by Raman microscopy at University College London, and the pigments identified thereon. These include a 13th century North Italian antiphonal (displaying ultramarine blue and malachite) [1,3], a 13th century North Italian choir book (displaying the technique of layering of lapis lazuli over ultramarine blue) [1,3], a 16th century German choir book (displaying white lead, carbon, azurite, vermilion, red lead, massicot, and lead tin yellow type I) [1,3], a Paris bible *c.* 1275 (displaying white-lead, vermilion, red lead, lapis lazuli, azurite, orpiment, realgar and malachite) [4], the Skard copy of the Icelandic Book of Law *c.* 1360 (displaying vermilion, orpiment, realgar, red ochre, azurite, bone white, and verdigris or basic verdigris possibly mixed with green earth) [5], and many Korans from different countries. Similar studies have been carried out by a number of other groups, notably in Paris [2].

A recent study of the yellow pigments removed from the paintings at the National Gallery, London, has led to the identification of lead tin yellow type I on one painting ("Death of Acteon", Titian) and lead tin yellow type II ("Allegory of Love", Paolo Veronese) on the other. Type I is well known to be Pb_2SnO_4, but type II (which has a pyrochlore structure) had not previously been well defined. Powder diffractometry in conjunction with Rietveld analysis has now resolved the structural problem; type II is $PbSn_{0.76}Si_{0.24}0_3$, 24% of the tin sites having been randomly substituted by silicon in the lattice [6].

Recent extensions to this work include studies of the following manuscripts, which will be discussed in detail.

(1) Three 15th C German manuscripts. The pigments vermilion HgS, iron oxide Fe_2O_3, azurite $2CuCO_3.Cu(OH)_2$, malachite $CuCO_3.Cu(OH)_2$, lampblack (essentially carbon), white lead $2PbCO_3.Pb(OH)_2$, lead tin yellow type I Pb_2SnO_4, and lazurite $Na_8[Al_6Si_6O_{24}]S_n$ (but only in admixture with Pb_2SnO_4) have been identified.

(2) Three 15th C Latin manuscripts. The pigments vermilion, kermes, azurite, malachite, ivory black (a pigment based on amorphous carbon), white lead and lead tin yellow type I have been identified, but not a second black pigment. Gold illuminations were used. Some of the vermilion has blackened, due to a transformation from the red α- to the black α'-form of HgS.

1. R.J.H. Clark, Chem. Soc. Rev., 1995, 24, 187-196.
2. J. Corset, P. Dhamelincourt, and J. Barbillat, Chem. Br., 1989, 612-616.
3. S.P. Best, R.J.H. Clark and R. Withnall, Endeavour, 1993, 16, 66-73.
4. S.P. Best, R.J.H. Clark, M.A.M. Daniels, and R. Withnall, Chem. Br., 1993, 118-122.
5. S.P. Best, R.J.H. Clark, M.A.M. Daniels, C.A. Porter, and R. Withnall, Studies in Conservation, 1995, 40, 31-40.
6. R.J.H. Clark, L. Cridland, B.M. Kariuki, K.D.M. Harris, and R. Withnall, J. Chem. Soc. (Dalton), 1995, 2577-2582.

INVESTIGATION OF CRYPTOENDOLITHIC LICHENS FROM THE ANTARCTIC DESERTS USING FT-RAMAN MICROSCOPY.

N.C.Russell [1], H.G.M. Edwards [1]* and D.D.Wynn-Williams [2].

1. Chemistry and Chemical Technology, University of Bradford, Bradford, West Yorkshire. BD7 1DP. UK
2. British Antarctic Survey, High Cross, Madingley Road, Cambridge, CB3 0ET. UK

* Author to whom correspondence should be addressed.

Lichens have been investigated in the context of weathering of their substrata and the formation of soil. Recently, Raman microscopy has been used to investigate the biodeterioration of different substrata such as cement. limestone and brickwork by a variety of lichens e.g. *Xanthoria elegans, Acarospora oxytoma* and *Lecanora muralis*[1,2]

This work has now been extended to include studies on the biodeterioration of Antarctic cryptoendolithic lichen-dominated communities. Cryptoendolithic lichens are found inside the topmost layers of the rock surface and consist of fungal and algal growth in the interstitial spaces of the substratum. Vestal and Johnson [3] have examined these lichen communities and broadly analysed four zones in the rock infra-structure. These zones are often only a few millimetres in thickness and the method of analysis involved the rather crude removal of material by dental pick. Using Raman microscopy, it has now been possible to carry out detailed, non-destructive *in situ* analysis of the zones and hence overcome the problem of zonal material contamination.

A cross-section through Beacon sandstone, inhabited by cryptoendolithics, from East Beacon, Antarctica, was analysed to observe the chemical composition of the zones. Vestal and Johnson [3] grouped the rock layers into four groups:- a crustal zone; the top surface of the rock, a lichen zone; the lichen-dominated, an accumulation zone; the iron accumulation layer and the inner zone; below the accumulation layer. The light orange crustal zone spectra exhibited little spectroscopic evidence for a biological presence and the main vibrational bands were of α-quartz with bands at 465cm^{-1}and 264cm^{-1}. The lichen zone, composed of a black upper layer and a white layer showed the presence of calcium oxalate monohydrate and dihydrate, with representative bands at 1462cm^{-1} and 1488cm^{-1} for the former and 1472cm^{-1} for the latter. The accumulation zone was red in colour and showed the downward movement of leached iron oxide. The inner zone, a pale orange, produced similar spectra to that obtained for the crust zone, showing only the presence of α-quartz.

Other cryptoendolithic samples show a green zone (an algal colony) between a the lichen zone and the accumulation zone characterised by a band at 1525cm^{-1} indicative of carotenoid pigments and a broad feature from 1350cm^{-1} to 1300cm^{-1} indicative of the presence of chlorophyll.

Further studies are now being undertaken to see if substrata, location and micro-climate influence the growth and colonisation of cryptoendolithic communities.

References.

1.H.G.M.Edwards, D.W.Farwell, M.R.D.Seaward, Spectrochim. Acta, Part A, **47**, 1521 (1991).

2. H.G.M.Edwards, N.C.Russell, M.R.D.Seaward, D.Slarke, Spectrochim. Acta, Part A, **51**, 2091 (1995).

3. C.G. Johnson, J.R.Vestal, J.Microbial Ecology, **25**, 305 (1993).

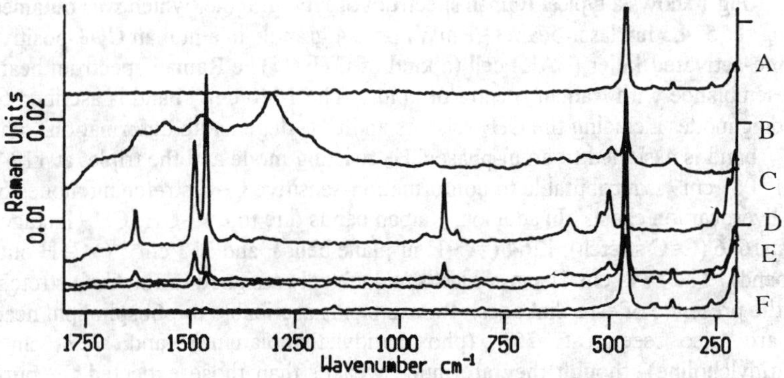

Raman spectra of a cross-section of a crytoendolithic community showing the differences between the zones. A. Crustal Zone, B, C & D lichen Zone, E Accum layer, F Inner Zone.

RAMAN MICROSPECTROSCOPIC ANALYSIS OF THE STRUCTURE OF LYTIC GRANULES IN CYTOTOXICT T LYMPHOCYTES

Hideo Takeuchi*, Yoshiaki Takai, and Takashi Masuko

Pharmaceutical Institute, Tohoku University, Aobayama, Sendai 980-77, Japan

Exocytosis of lytic granules is one of the major mechanisms by which cytotoxic T lymphocytes (CTLs) kill tumor and virus-infected cells. The granule-mediated "lethal hit" is delivered in the following way [1]: Upon immunological recognition of an appropriate target cell, lytic granules indside the CTL move toward the plasma membrane, fuse with the membrane, and then secrete their contents into the intercellular space between the CTL and terget cell. The proteins contained in the granule are perforin, which forms pores in the target cell membrane, and a variety of proteases known as granzymes, which may penetrate into the target cell through the pore. Although immunological and biochemical aspects of the lethal hit have been studied extensively, detailed mechanisms of the granule-mediated cell lysis have not been fully revealed yet. One of the enigmas remaining to be resolved is why cytotoxic granules of a CTL are not cytotoxic to the CTL itself. The structure of granules may be related to the protection of CTL from its own lytic granules.

Electron microscopy has revealed the ultrastructure of lytic granules in CTLs. A typical cell of a human CTL clone contains 15-25 granules of about 0.7 μm in diameter and each granule consists of an electron dense core, many small internal vesicles, and membrane cortices [2]. The dense core is a major pool of secretory proteins and shows a hexagonal structure attributable to highly organized lipid molecules [3]. In this study, we have investigated the structure of lytic granules in human CTL by Raman microspectroscopy.

Fig.1 shows a typical Raman spectrum of lytic granule, which was obtained by focusing the 514.5 nm laser beam (~1 mW) onto a granule in a human CD3-positive lymphokine-activated killer (LAK) cell (a kind of CTL). The Raman spectrum bears strong resemblance with Raman spectra of lipids. The 1440 cm^{-1} band is ascribed to C−H bending modes including the CH_2 scissors and CH_3 degenerate deformation. The 1302 cm^{-1} band is assigned to an in-phase CH_2 twisting mode and the triplet at 1124, 1079, and 1063 cm^{-1} is attributable to conformation-sensitive C−C stretching modes of saturated hydrocarbon chains. In addition, Raman bands due to *cis* −CH=CH− linkages are seen at 1656 (C=C stretch), 1264 (=C−H in-plane bend), and 973 cm^{-1} (=C−H out-of-plane bend). The 1744 cm^{-1} band, which is undoubtedly assigned to the C=O stretch, indicates the presence of ester linkages. Raman bands ascribable to phospholipid head groups are also seen at 751 (phosphatidylethanolamine) and 717 cm^{-1} (phosphatidylcholine), though they are much weaker than those expected for pure phospholipids. A sharp band at 1003 cm^{-1} is due to Phe residues of proteins and a weak band overlapped by the 1656 cm^{-1} C=C stretching band may be ascribed to amide I modes. The predominance of lipid Raman bands over protein Raman bands was also observed for the granules examined in the same and other LAK cells. The Raman spectrum gives evidence that the main component of the granule is lipid and a less amount of protein is contained in the lytic granule of LAK cell. Another interesting finding is that the granulous lipid is not rich in phosphatidylcholine and phosphatidylethanolamine, which are usually major components of plasma membranes.

The Raman intensities of the 1656 and 1264 cm^{-1} bands (arising from C=C bonds) relative to those of the 1656 (C=O), 1440 and 1302 cm^{-1} (C–H) bands differed from granule to granule. Generally, granules aggregated in a region within a LAK cell showed higher contributions of the C=C Raman scattering than the granules isolated in the same LAK cell. This observation indicates that the degree of unsaturation of the lipid acyl chain depend on the location of the granule within the cell. By using the relationship between the Raman intensity and the number of C=C bonds derived from Raman spectra of a series of triacylglycerols, aggregated granules are found to have ~2 C=C bonds per acyl chain, while isolated granules have 1.2 C=C bonds, on the average, as has the plasma membrane. The highly unsaturated acyl chains of the granules aggregated, possibly near the Golgi complex that supplies lytic proteins, may play a role in protecting the CTL itself from granulous lytic proteins.

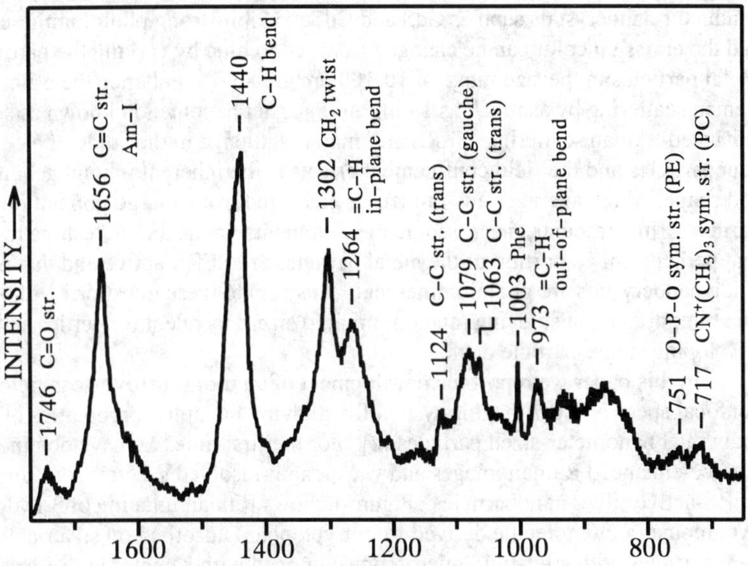

Fig. 1 Raman spectrum of a lytic granule in a LAK cell recorded on a Raman microspectrometer equipped with a x100 objective and a back-thinned CCD detector.

1. G. Berke, Cell 81, 9 (1995).
2. P.J. Peters, J. Borst, V. Oorschot, M. Fukuda, O. Krähenbühl, J. Tschopp, J.W. Slot, and H.J. Geuze, J. Exp. Med. 173, 1099 (1991).
3. J.P. Caulfield, A. Hein, R.E. Schmidt, and J. Ritz, Am. J. Pathol. 127, 305 (1991).

NEAR-FIELD AND CONFOCAL RAMAN SPECTROSCOPY OF SINGLE NANOPARTICLES

Steven R. Emory and Shuming Nie*

Department of Chemistry, Indiana University, Bloomington, IN 47405

The development of advanced methods for preparation and characterization of nanometer-scaled materials and devices is of considerable interest in many fields including microelectronics, optics, catalysis, and ceramics [1]. Nanometer-sized particles can display optical and electronic properties that are strongly dependent on the particle size and shape. These novel properties are often not available in either isolated molecules or bulk materials. For example, quantum-confined semiconductor nanocrystals such as CdS and CdSe exhibit strong photoluminescence, and the emission color can be changed from red to blue by varying the particle size. Metal particles in the size range of 10-100 nm are able to enhance the efficiency of Raman scattering by as much as 1 million times, a phenomenon known as surface-enhanced Raman scattering (SERS). A major challenge in the study of nanoparticles and the field confinement effects is the inherent inhomogeneity in size and shape, which averages any properties measured from a large population of particles. In particular, macroscopic measurements are unable to test the hypothesis that perhaps only a portion of the metal particles are SERS-active and that not all CdSe nanocrystals are photoluminescent. Answers to these questions will be of great importance in studying (and optimizing) size-dependent properties and in developing single-particle devices.

In this paper we report the development of an ultrasensitive near-field and confocal spectroscopy/imaging system for studying the optical properties of individual nanometer-sized particles [2]. For the first time, we have obtained surface-enhanced Raman images and wavelength-resolved spectra from single, SERS-active silver nanoparticles. Figure 1 shows Raman imaging of 3-hydroxy-kynurenine (a biomolecule derived from tryptophan) adsorbed on silver colloidal nanoparticles with a far-field microscope equipped with Koehler laser illumination and a cooled CCD detector. The sizes of these imaged particles are ca. 50-100 nm in diameter as revealed by scanning electron microscopy, but they appear as diffraction-limited spots of 250-350 nm. This dimension corresponds to the resolution of our far-field optical microscope.

By positioning a near-field fiber probe or by focusing a laser beam on one of these Ag particles, we have also obtained surface-enhanced Raman spectra of the adsorbed molecules with a laser excitation as low as 1 nanowatt (10^{-9} W). Figure 2 compares the single-particle Raman spectra of 3-hydroxykynurenine acquired in the near-field and the confocal far-field. The number of adsorbate molecules on such a nanoparticle is estimated to be 100 to 10,000. As will be presented in detail, these results demonstrate that surface- and resonance-enhanced Raman spectroscopy is a powerful tool for obtaining molecular information from a single nanoparticle and ultimately from a single adsorbate molecule. In contrast to laser-induced fluorescence (LIF) that has been widely employed in ultrasensitive chemical analysis and single molecule detection [3], Raman spectroscopy is capable of providing molecular/structural information and does not suffer from the problem of photobleaching.

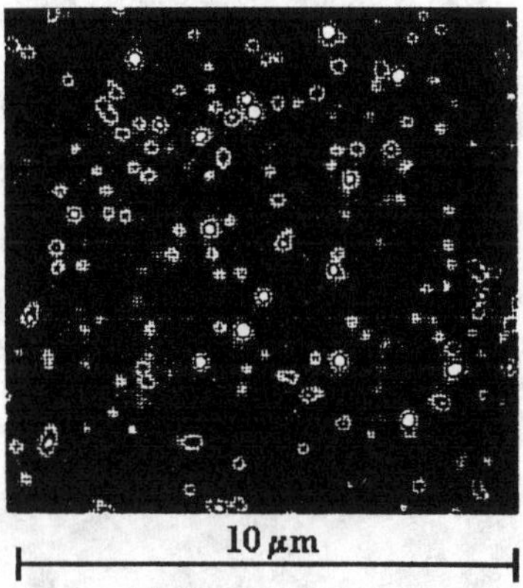

Figure 1. Raman imaging of individual silver colloidal nanoparticles. Laser wavelength = 514.5 nm; excitation power = 1.5 mW; data acquisition time = 1 second; bandpass filter = 560DF30 (passing signals from 1088 cm^{-1} to 2045 cm^{-1}).

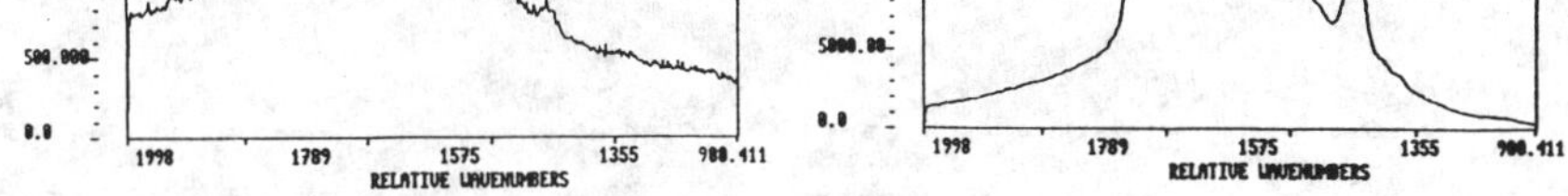

Figure 2. Near-field and confocal Raman spectra of 3-hydroxykynurenine obtained from single silver nanoparticles. Laserwavelength = 514.5 nm.

1. W. A. Reed, "Quantum dots," *Scientific American*, 118-123 (Jan., 1993).
2. S. R. Emory and S. Nie, "Near-field and confocal Raman spectrscopy of individual nanoparticles," manuscript in preparation.
3. S. Nie, D. T. Chu, and R. N. Zare, "Probing single molecules by confocal fluorescence microscopy," *Science*, 266, 1018 (1994).

SECTION 29

RAMAN IMAGING

RAMAN MICROSCOPES AND MICROPROBES: IMAGING AND SPECTROSCOPY OF COMPLEX MATERIALS

E. N. Lewis* , L. H. Kidder and I. W. Levin

Laboratory of Chemical Physics,
National Institute of Diabetes and Digestive and Kidney Diseases
National Institutes of Health, Bethesda, MD 20892-0511

Although traditional Raman spectroscopy provides information at the molecular level for determining the composition of multicomponent mixtures, it does not generate specific information pertaining to the *spatial distribution* of these components. The conventional Raman microprobe, while suitable for Raman microspectroscopy, also provides limited imaging capability [1]. The functionality and properties of complex systems are often predicted by knowledge of the microscopic spatial distribution of components. Therefore, the ability to map these distributions is crucial in activities ranging from the design and fabrication of advanced materials to the elucidation of biomolecular functionality.

Coupling Raman spectroscopy with microscopic imaging represents a powerful approach in attaining these goals, as it combines the analytical capabilities of spectroscopy with the ability to visualize the spatial distribution of the chemical constituents within a system [2,3]. For example, the technique can detect microscopic chemical heterogeneities in samples which, when using conventional optical microscopies and bulk sample spectral analysis, appear spatially and spectrally homogeneous. As with traditional Raman spectroscopy, these images can be recorded without the use of potentially invasive contrast enhancing agents, providing a highly specific, non-destructive means to analyze chemical distributions within heterogeneous samples. We describe a technique for performing high-fidelity Raman spectroscopic imaging, integrating a commercial optical microscope with laser epi-illumination, holographic filters, an acousto-optic tunable filter (AOTF) and a liquid nitrogen cooled charge-coupled device detector (CCD) [4]. The AOTF contains an optically transparent birefringent crystal, which can be electronically controlled to pass a selectable and narrow band of optical wavelengths [5].

Figure 1 is a scale drawing of the Raman imaging microscope's major aspects. It is built around an infinity-corrected metallurgical microscope (Olympus BH-2). The optical components are mounted to a superstructure attached to the microscope vertical port and fastened to the underlying optical bench. The frame provides stable and reproducible mounting for the optical components and was designed to shield, from room light, the components located after the AOTF in the optical path.

In the laser scanning mode, light from a krypton ion laser (Coherent Laser Group, Innova 300) is filtered by a 647.1 nm holographic laser bandpass filter (Kaiser Optical Systems, HLBF-647) and directed to a computer controlled x-y galvanometer scanner (two Cambridge Instruments, 6850 optical scanners). Lenses L1 and L2 (f = 300 mm) constitute an afocal unity magnification relay which images the "center" of the scanner into the objective rear entrance pupil, providing scanning epi-illumination. The laser beam is approximately 2 μm in diameter at the microscope objective, resulting in a larger than diffraction limited spot size .

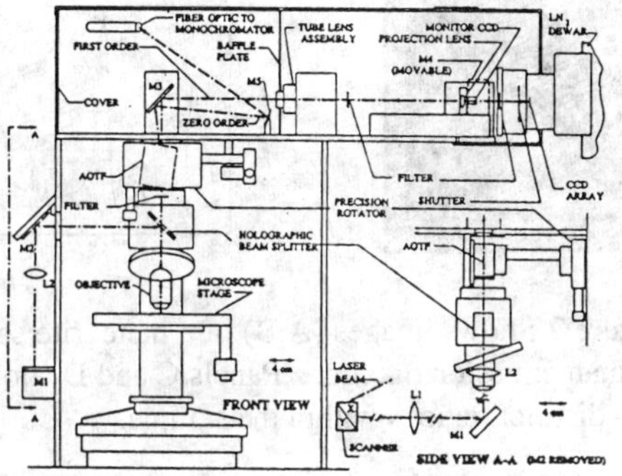

Figure 1 Schematic diagram of the Raman imaging microscope and microprobe. Laser illumination is provided with a 647.1 nm krypton ion laser and xy scanning galvonometer. The AOTF is used to continuously tune the wavelength of the Raman emission image, which is collected by the CCD array. Elements include mirrors (M) and Lenses (L).

Additionally we have provided a laser fiber optic wide field illumination system to be used with more fragile specimens that might be damaged by the higher power densities of the scanning configuration. The wide-field fiber illumination is also more convenient for sample alignment and system focusing. The system also has the potential to perform sub-micron resolution Raman microspectroscopy. The laser scanning galvanometer can be stopped at a predetermined position, corresponding to a small region of the sample. A spherical mirror, M5, placed below and in front of the tube lens focuses the zero order light from the AOTF into a fiber optic bundle. This light is coupled into a single stage monochromator, (Chromex, 500IS), equipped with a holographic filter (Kaiser Optical Systems, HNF-647) and CCD (Spex, Spectrum One). This allows a more rapid determination of the Raman spectra of the selected micro-region at a higher spectral resolution than that provided by the AOTF.

As an example of the technique, Figure 2 shows images takes from a thin section of meteorite, which highlight the compositional heterogeneities of the sample. The corresonding spectra of these distinct regions are shown, emphasizing the overall versatility and utility of the vibrational spectroscopic imaging technique.

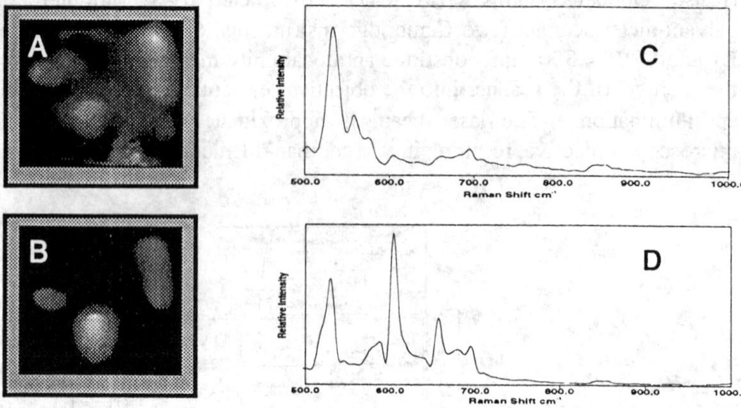

Figure 2 Shows images (A,B) of a meteorite sample at two distinct Raman emission maxima. Panels C and D show spectra corresponding to two distinct regions within the sample.

1. M. Delhaye and P. Dhamelincourt, *J. Raman Spectrosc.* **3**, 33 (1973).
2. P.J. Treado, I.W. Levin, and E. Neil Lewis, *Appl. Spectrosc.* **46**, 1211 (1992).
3. M.D. Schaeberle, V.F. Kalasinsky, J.L. Luke, E.N. Lewis, I.W. Levin, and P.J. Treado, *Anal. Chem.*, 1996 (in press).
4. S.R. Goldstein, L.H. Kidder, T.M. Herne, I.W. Levin, E.N. Lewis, *J. of Microscopy*, 1996 (in press).
5. I.C. Chang, *Optical Engineering*. **20**, 824 (1981).

HADAMARD TRANSFORM RAMAN IMAGERY

T. Mangold[1], A.N. Mortensen[2], M.K. Bellamy[1], R.M. Hammaker[1], and W.G. Fateley[1]

Departments of Chemistry[1] and Electrical and Computer Engineering[2], Kansas State University, Manhattan, KS 66506, USA

Hadamard transform spectrometry (HTS) has been of interest to investigators at Kansas State University for more than a decade and our efforts are continuing [1,2]. HTS is a unique combination of dispersive and multiplexing spectrometries where the choice of transparent or opaque for the elements of a multi-slit array (i.e. a mask), located on a focal plane provides an encoding of spectral information that may be decoded into a conventional spectrum by Hadamard mathematics [3]. As in a dispersive spectrometer, the radiation from a source is collected and separated into its individual spectral resolution elements by a grating, prism , or other spectral separator. This spectrally separated radiation is collected and focused by a mirror or lens, and spatially presented on a focal plane. Unlike a dispersive spectrometer, which possesses a single exit slit, the Hadamard transform spectrometer employs a multi-slit array. This arrangement allows for the simultaneous measurement of a multitude of spectral resolution elements and produces a multiplexing spectrometer. If the conventional one dimensional (1D) Hadamard encoding mask were to be folded in some manner, the result would be the generation of a two-dimensional (2D) Hadamard encoding mask. Similar to the 1D masks that were used for spectral multiplexing, 2D masks could be used for spatial multiplexing.

Two choices for the Hadamard encoding mask are the movable or mechanical mask and the stationary or electro-optic mask. The advantage and disadvantage of the moving mask are no optical transmission problems and potential moving parts problems [4], respectively. The advantage and disadvantage of the stationary mask are no moving parts problems and potential optical transmission problems [5], respectively . Our contribution has been the introduction of the stationary 1D and 2D Hadamard encoding masks based on liquid crystal technology . A major focus of our efforts up through 1990 has been the application of Hadamard techniques to visible [6] and near-infrared [7] Raman spectrometry. It is convenient to classify our past, present and future efforts by the dimensionality. There are three spatial dimensions (x_i, y_i, and z_i, coordinates) and one spectral dimension (frequency, v_j or wavelength, λ_j) for a total of four. Thus the work in the application of Hadamard techniques to visible [6] and near-infrared [7] Raman spectrometry is 1D spectrometry. A form of 2D spectrometry would be our experiments [2] in imaging a thin-layer chromatography (TLC) plate by surface-enhanced Raman scattering (SERS). The addition of the spectral dimension to imaging by SERS at a single frequency would provide a form of 3D spectrometry. A summary of all our work in 1D and 2D Raman spectrometry is becoming available [1].

For more than a decade we have utilized the stationary 1D and 2D Hadamard encoding masks based on liquid crystal technology in our application of Hadamard transform techniques in spectrometry, depth profiling and imaging, conveniently classified as 1D, 2D, 3D, or 4D spectrometry by the total number of spatial and spectral dimensions involved [2,8,9]. Consequently, investigations were restricted to the visible and near-infrared spectral regions. Recently, we decided to see if advances in the technology of motional devices and mask fabrication since the 1970's would overcome the problems that impeded the initial development of HT techniques with moving (mechanical) Hadamard encoding masks [3]. The use of a moving 2D Hadamard encoding mask with all mask elements either completely open or completely closed makes imaging and spectrometry possible in any spectral region for which an appropriate spectral separator and detector are available. Thus, we chose to use a Fourier transform infrared (FT-IR) spectrometer to combine the frequency precision, spectral multiplex, and throughput advantages of an FT-IR spectrometer with the spatial multiplex advantage of the moving 2D Hadamard encoding mask to obtain images in the mid-infrared (IR) spectral region and mid-IR spectra of individual pixels of the image with usable signal-to-noise ratios (SNR) in a reasonable data acquisition time. The moving 2D Hadamard encoding mask, which is based on the cyclic simplex S_{255} matrix [3] as a 15 x 17 array and was etched out of a metal substrate, was mounted vertically and was moved from one encodement pattern to the next one using a translation stage.

We are currently adapting this moving 2D Hadamard encoding mask to do a 2D Raman imaging study(single-frequency) in the visible spectral region using a Spex 14018 double monochromator as a preliminary to a 3D Raman imaging study (multi-frequency) in the near -infrared spectral region using a Midac MD2400 series Fourier transform near-infrared spectrometer. A progress report through the first half of 1996 will be given.

1. R.M. Hammaker, A.P. Bohlke, R.D. Freeman, J.M. Jarvis, K.J. Latas, D. Lin Vien, A.N. Mortensen, E.A. Orr, J.D. Tate, D.C. Tilotta, J.S. White, S.L. Wright, J.V. Paukstelis, and W.G. Fateley, "Hadamard Transform Raman Spectrometry," Chapter 5 in Modern Techniques in Raman Spectroscopy, Edited by J.J. Laserna, Wiley and Sons, Ltd., Chichester, pp. 143-226 (1996).
2. Reference 62 in Reference 1.
3. Reference 21 in Reference 1.
4. Reference 26 in Reference 1.
5. References 27, 28, 38-41, 44, 45, 55, 57, 68 in Reference 1.
6. References 30-34 in Reference 1.
7. References 35, 38, 47-50, 52, 54, 56, 57 in Reference 1.
8. Reference 66 in Reference 1.
9. R.M. Hammaker, A.N. Mortensen, E.A. Orr, M.K. Bellamy, J.V. Paukstelis, and W.G. Fateley, J. Molec. Struct., 348, 135-138 (1995).

LCTF RAMAN CHEMICAL IMAGING OF THERMOPLASTIC OLEFIN (TPO) ARCHITECTURE

Hannah R. Morris and Patrick J. Treado*

Department of Chemistry, University of Pittsburgh,
Pittsburgh, PA 15260

Thermoplastic olefins (TPO) represent an important class of impact resistant polymeric materials that are widely utilized in the automotive industry as bumpers and fascia.[1] To enhance the cosmetic appearance of TPO is challenging because base coats do not adhere to the TPO due to its low surface energy, poor wettability and resistance to solvent penetration.[2] Chlorinated polyolefin (CPO) is typically applied to the surface of the substrate to promote adhesion of the paint. Understanding the molecular architecture of TPO provides insight into the chemical and mechanical properties of the material, including paint adhesion.

We are employing Raman chemical imaging using liquid crystal tunable filters (LCTFs) to characterize the three dimensional architecture of TPO. The Raman LCTF provides diffraction limited spatial resolution and 7 cm^{-1} spectral resolution.[3] This surpasses the capabilities of previous approaches used for Raman imaging. With LCTFs it is possible to generate image contrast based on subtle spectral features which is often necessary when characterizing complex multicomponent systems such as coated polymers.

Adhesion promoter coated TPO is comprised of several components, namely CPO, polypropylene (PP) and ethylene-propylene rubber. Raman image analysis begins with collection of the spectra of the individual components. Image contrast can be derived based on the unique spectral features of the components. Raman microspectra of the pure components which comprise coated TPO, are shown in Figure 1.

We have initiated studies to characterize TPO crystallinity architecture. TPO surface architecture is influenced by thermal processing that occurs during fabrication. As a result of processing, PP crystallinity at the surface of TPO differs from the degree of PP crystallinity within the bulk. Raman microspectra of PP model systems at low and high crystallinity are shown in Figure 2. Two subtle spectral features are being investigated as a means to generate chemical image contrast. Specifically, the Raman band at 1175 cm^{-1} becomes more pronounced as the degree of PP crystallinity increases. Similarly, the ratio of the 850 to 814 cm^{-1} bands decreases as the degree of PP crystallinity increases. We are investigating the source of these subtle spectral changes, as well as using these signatures to generate image contrast attributable to crystallinity content.

References

1. R. J. Clemens et. al., Progress in Organic Coatings, **24**, 43, (1994).
2. R. A. Ryntz, A. C. Ramamurthy, Technical Report CTIS0008584.
3. H. R. Morris et. al., Applied Spectroscopy (1995) submitted.
4. This research was supported by Grant No. 47-P-W78514 from Ford Motor Company.

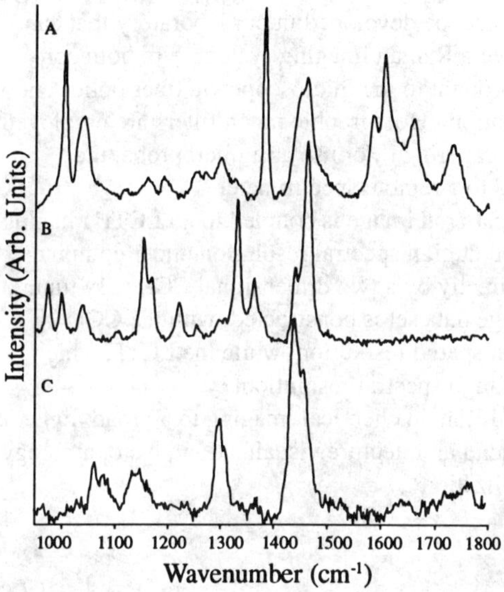

Figure 1. Raman microspectra of pure components comprising adhesion promoter coated TPO. A) CPO, B) Polypropylene, and C) Ethylene-propylene rubber. Ar^+ ion laser excitation (514.5 nm), 20X (N.A. 0.46) objective.

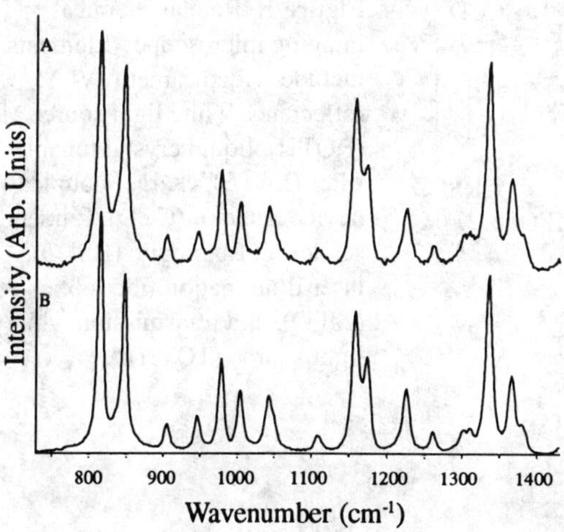

Figure 2. Raman microspectra of varying crystallinities of polypropylene: A) low crystallinity, B) high crystallinities.

LCTF RAMAN CHEMICAL IMAGING OF SEMICONDUCTORS

Michael D. Schaeberle and Patrick J. Treado*

Department of Chemistry, University of Pittsburgh, Pittsburgh, PA 1526

The implementation of liquid crystal tunable filters (LCTF) has dramatically enhanced the utility of Raman chemical imaging for materials analysis. Figure 1 shows a Raman chemical imaging microscope developed in our laboratory that can be operated as an efficient microprobe or a Raman imaging system. In both configurations, remote laser excitation is brought to the microscope via fiber optics and directed to the sample for 180° illumination. Holographic notch filters are employed to discriminate Rayleigh from Raman scattering. For use as a microprobe, the Raman output is fiber optically coupled to a remote spectrometer.

For use in imaging, the Raman scattered image is coupled to an LCTF imaging spectrometer where the image is filtered at high spectral resolution under computer control. A Raman image is recorded directly by a two dimensional CCD. By tuning the LCTF, a high fidelity chemical image data set is constructed, with the CCD recording the image information at high spatial resolution, while the LCTF, which functions as a bandpass filter, provides high spectral resolution.

In our laboratory we have applied Raman chemical imaging to a broad range of materials studies, including polymer blend architecture visualization[1], histopathology analysis[2], and corrosion products visualization[3].

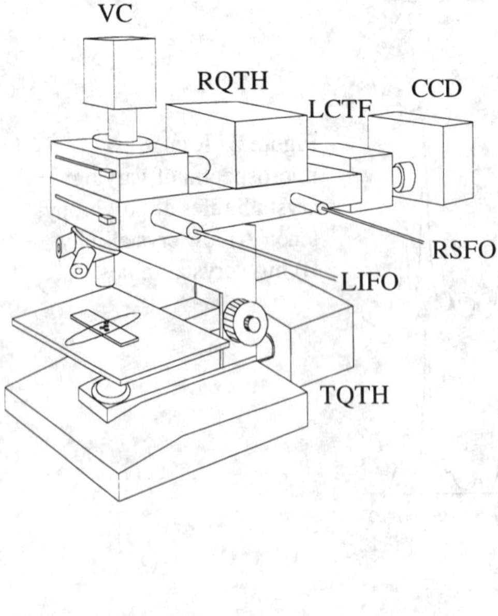

Figure 1. Raman chemical imaging microscope. Elements include video camera (VC), reflectance white light source (RQTH), liquid crystal tunable filter (LCTF), charge- coupled device detector (CCD), Raman scattering fiber optic (RSFO), laser illumination fiber optic (LIFO), and transmission white light source (TQTH).

We are currently applying our Raman chemical imaging microscope to the study of silicon semiconductors. A goal of this research is to visualize stress in silicon hetero-structures. Results of a preliminary study are shown in Figure 2. A model system comprised of an SEM test target has been analyzed. The test target consists of an aluminum pattern deposited on a silicon wafer. Fig. 2A shows a brightfield image of a 50 μm outer diameter circle of aluminum deposited on a silicon wafer. Fig. 2B is a Raman image taken at 520 cm^{-1}, the first phonon band of silicon. The high degree of visual clarity observed in Fig. 2B is matched by the quality of the silicon Raman spectrum collected through the LCTF as shown in Fig. 2C. A similarly high quality spectrum collected through a dispersive spectrometer is shown for comparison.

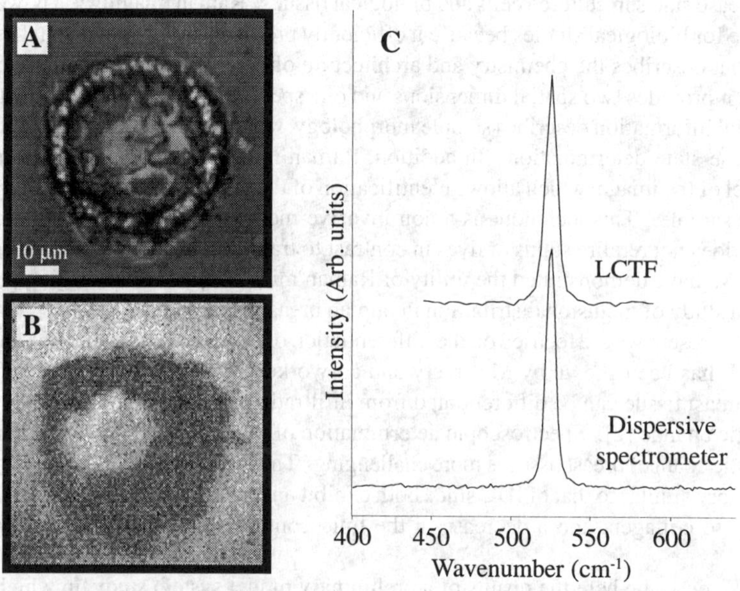

Figure 2. Image of a 50 μm outer diameter circle of aluminum deposited on a silicon substrate using a 20X (NA=0.46) objective and a Nd:YAG laser (532 nm) for excitation. (A) White light reflectance image recorded with LCTF and CCD. (B) Silicon Raman image at 520 cm^{-1}. (C) Spectra acquired using an LCTF Raman imaging spectrometer and a conventional dispersive spectrometer.

1. M. D. Schaeberle, C. G. Karakatsanis, C. C. Lau, and P. J. Treado *Anal. Chem.* 67, 4316-4321 (1995).
2. M. D. Schaeberle, V. F. Kalasinsky, J. L. Luke, E. N. Lewis, I. W. Levin, and P. J. Treado *Anal. Chem.*, submitted for publication (1995).
3. H. R. Morris and P. J. Treado *Appl. Spec.* submitted for publication (1995).

RAMAN CHEMICAL IMAGING OF DISEASE STATES

Nicole J. Kline and Patrick J. Treado*

Department of Chemistry, University of Pittsburgh,
Pittsburgh, PA 15260

Raman chemical imaging utilizing tunable filter technology is being employed to study disease states in cultured cells and biological tissues. Raman imaging is a powerful technique for biological studies because it efficiently provides multi-dimensional information that describes the chemistry and architecture of heterogeneous materials. Specifically, it provides two spatial dimensions and one spectral dimension of information. The spatial information describes sample morphology, which is one traditional means of cell disease state determination. In addition, Raman imaging provides a spectrum at each pixel of the image, which allows identification of the molecular chemical composition of a sample. This technique is a non-invasive method because Raman chemical imaging does not require stains or dyes in contrast to traditional histopathological techniques. We have demonstrated the utility of Raman microscopy for bioimaging previously in a study of inclusion distribution in human breast tissue [1].

At present we are focused on the differentiation of normal versus cancerous breast tissues. It has been shown by McCreery and co-workers that the Raman spectrum of normal breast tissue can be differentiated from infiltrating ductal carcinoma (IDC) and fibrocystic change [2]. Spectroscopic determination of malignant (IDC) versus benign (fibrocystic change) breast tissue is more challenging. The Raman spectrum of fibrocystic tissue is very similar to that of IDC since both exhibit an increase in protein content, due primarily to collagen, and a decrease in the lipid content when compared to normal tissue.

We describe here the results of a preliminary model system study, in which we have employed Raman microscopy to differentiated breast tissue components based on their spectra and based on component morphology. The Raman spectra of oleic acid methyl ester and type 1 collagen are shown in Figure 1A and B, respectively. Oleic acid methyl ester has been proposed as the major lipid component of breast tissue [3] while collagen is the primary structural protein. The lipid spectrum is readily differentiated from the collagen spectrum. These spectral differences allow for Raman imaging of the lipid distribution in a mixture of type I collagen and oleic acid methyl ester as shown in Figure 2. A Raman chemical imaging system utilizing a LCTF, as previously described [4], was used to obtain chemical images of the lipid distribution. The brightfield image shown in Figure 2A reveals the collagen fiber which appears opaque, while the lipid is translucent. In the Raman image shown in Figure 2B, the lipid distribution is observed in the region outside of the collagen fiber. Obtaining a complementary Raman image of the collagen will be more difficult due to the absence of unique, isolated Raman bands.

However Raman image contrast can be generated based on the collagen peak widths which are significantly broader then in lipid.

In real systems, differentiation of benign versus IDC tissues will be investigated by evaluating the sample morphology in conjunction with Raman spectroscopy. Raman chemical imaging's capacity to provide rapid, noninvasive differentiation of these two tissues types will be evaluated. With unambiguous identification of these tissues, *in vivo* use of Raman spectroscopy as a diagnostic tool becomes feasible.

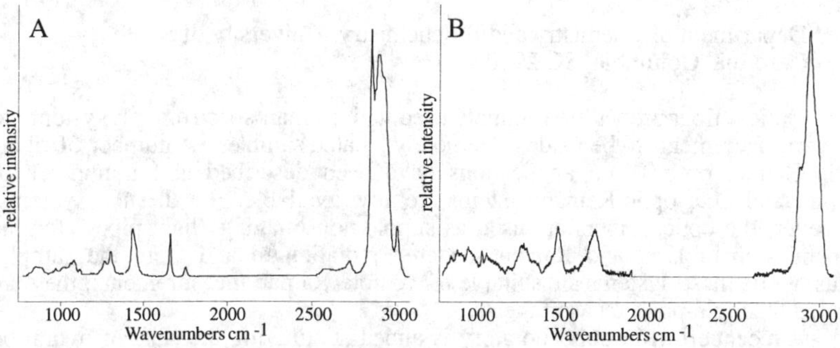

Figure 1 A) Raman spectrum of oleic acid methyl ester collected through the storage ampule; 60 seconds, 5 mW 532 nm laser power. B) Raman spectrum of type 1 collagen; 60 seconds, 150 mW at the sample.

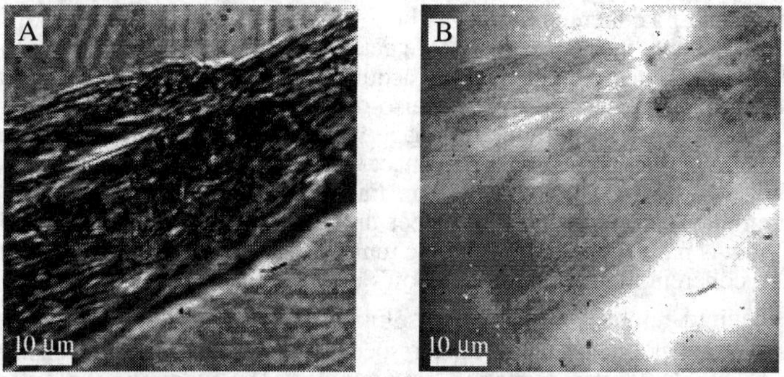

Figure 2. A) Brightfield image of an oleic acid methyl ester and type I collagen mixture. B) Raman image of the lipid content collected at 2853 cm $^{-1}$, the symmetric CH_2-stretching mode [5].

1. Schaeberle, M.D.; Karakatsanis, C.G.; Lau, C. C.; Treado, P.J. *Anal. Chem.* 67, 4316 (1995)
2. McCreery, R.L.; Frank C.J. *Anal. Chem.* 67, 777 (1995).
3. Frank, C.J.; Redd, D.C.; Gansler, T.S.; McCreery, R.L. *Anal. Chem.* 66, 319 (1994).
4. Morris, H. R.; Clifford, C.H.; and Treado, P.J. *Appl. Spectrosc.* 48, 857 (1994).
5. Verma, S. P.; Wallach, D. F. *Biochim. Biophys. Acta.* 486, 217 (1977).

Raman Microscopy Without the Microscope: Raman Imaging Using a Coherent Optical Fiber

S. Michael Angel,* H. Trey Skinner, and Thomas F. Cooney

Department of Chemistry and Biochemistry, University of South Carolina, Columbia, SC 29208

Optical fiber probes are routinely used with Raman spectrograph systems to allow measurements to be made on remotely located samples. A number of fiber-optic Raman probes and applications have been described and a number of commercial fiber-optic Raman systems are now available. In all such systems, however, the optical fibers are used as simple non-imaging "light pipes" for the transmission of laser and Raman scattered radiation to and from the sample. Thus, while these systems are suitable for remote Raman measurements, they are limited to single-point measurements.

A recent effort in our laboratory is aimed at studying the state of hydration of Nafion membranes as a function of various experimental parameters *in situ*. For this purpose it is desirable to make Raman measurements at high spatial resolution—not possible using current techniques. Recently, Walt and co-workers described a novel multi-analyte fiber-optic chemical sensor [1-4]. The sensor was constructed by attaching up to six fluorescent indicator dyes to the sampling end of a coherent microfiber bundle and transferring an image of the dye spots through the bundle to a CCD detector. Although not a microprobe, this system demonstrates the potential for acquiring fluorescence images remotely at ~4 micron-size spatial resolution. It also demonstrates the concept of proximity focusing—acquiring a high-quality image without external optics by placing the end of the imaging bundle in direct contact with the sample.

Recently, we developed a Raman microimaging probe suitable for in-situ measurements. The coherent microfiber probe is used with an imaging Raman system that is based on an acosuto-optic tunable filter (AOTF) and a CCD detector [5]. The coherent microfiber probe allows Raman, luminescence, or NIR images to be acquired remotely at ~4 μm resolution. We have tested the system by obtaining Raman images and Raman spectra of model samples consisting of pellets of highly Raman scattering compounds. The utility of the system for measuring real samples is demonstrated by imaging phase boundaries in a sample of graphitized diamond. The use of a coherent microfiber probe has not previously been demonstrated for acquiring Raman images remotely.

The figure below shows data for a sample consisting of a highly scattering organic molecule, 1,4-bis(2-methylstyryl)-benzene (BMSB), and TiO_2. This sample was prepared by pressing a pellet with a layer of BMSB and a layer of TiO_2 separated by a thin aluminum foil spacer to create a distinct boundary between the two compounds. The brightfield image (inset) taken through the AOTF at 832 nm with the coherent optical fiber clearly shows the boundary between the BMSB (lower darker region) and TiO_2 (upper bright region). The Raman spectra shown were acquired by binning pixels in each region. For the spectra, the AOTF was scanned in 1 nm increments with 20 sec exposures at each wavelength. Spectrum (a) was acquired by binning pixels corresponding to the TiO_2 (bright upper) region of the image and spectrum (b) was acquired by binning pixels corresponding to the BMSB (dark lower) region of the image. Although the background signal is high, the Raman spectra clearly show TiO_2 and BMSB bands with little overlap between the two spectra indicating low crosstalk in the coherent imaging fiber. The Raman spectrum in (c) is shown for comparison. This spectrum was acquired for a similar sample using a conventional dispersive Raman spectrograph with about 8 cm^{-1} resolution.

We have investigated several optical configurations for coupling light between the samples and the coherent microfiber probe. The most exciting uses no coupling optics (e.g., proximity focusing) and the tip of the probe is simply placed on the sample to be measured. In this case index matching gel on the probe tip improves image clarity. Good quality images are also achieved using a single graded-index rod lens at the fiber tip, but at the sacrifice of spatial resolution.

This work has been supported by the U.S. Department of Energy's South Carolina EPSCoR Program through Cooperative Agreement Number DE-FCO2-91-ER75666.

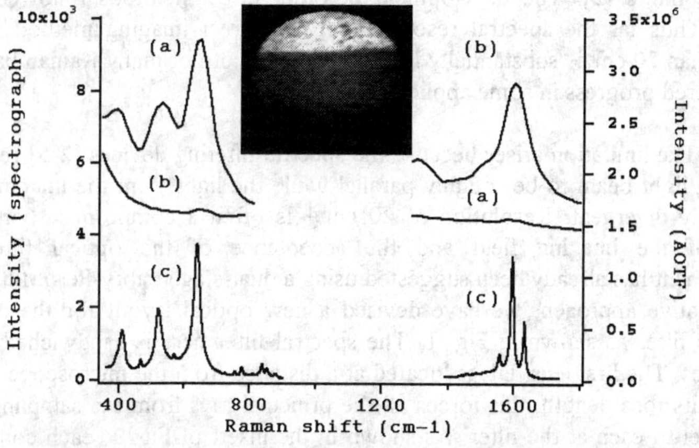

Figure 1. Inset: brightfield image of a sample containing TiO_2 (top bright region) and BMSB (lower dark region). Raman spectra acquired through the AOTF by binning pixels only in the TiO_2 region (a) and only in the BMSB region (b). The spectra are offset for clarity. (c) Raman spectra of TiO_2 and BMSB using a dispersive 0.25-m f/4 Raman spectrometer.

1. D.R. Walt, K.S. Bronk, P. Pantano and K.L. Michael, SPIE Proc. **2388**, 554 (1995).
2. K.S. Bronk and D.R. Walt, Anal. Chem. **66**, 3519 (1994).
3. B.G. Healey and D.R. Walt, SPIE-Proc. **2388**, 519 (1995).
4. B.G. Healey and D.R. Walt, SPIE-Proc. **2388**, 568 (1995).
5. H. Trey Skinner, Thomas F. Cooney, S.K. Sharma and S.M. Angel, "Remote Raman Microimaging Using an AOTF and a Spatially Coherent Microfiber Optical Probe," submitted to Appl. Spectrosc., November 27, 1995.

DIRECT RAMAN BAND IMAGING WITH IMPROVED SPECTRAL RESOLUTION AND UNIFORM INTENSITY

C. Cheng[1], D. N. Batchelder[1] and G. D. Pitt[2]

[1]Department of Physics & Astronomy, University of Leeds, Leeds LS2 9JT, UK
[2]Renishaw plc., Old Town, Wotton-under-Edge, Gloucestershire GL12 7DH, UK

Raman band imaging through a light microscope and a spectrometer was first demonstrated by M. Delhaye and P. Dhamelincourt in 1971 [1]. In 1990, we demonstrated that direct imaging was possible using rotation-tuneable multilayer dielectric filters [2]. The development of other direct methods followed quickly [3,4,5]. Thus far the spectral resolution of the direct imaging method has been greater than 20 cm^{-1}, substantially larger than the width of many Raman bands; this has hindered progress in some applications.

The limitation arises because the spectral filtering devices [2-5] require the incoming light beam to be roughly parallel while the light from the imaging area is necessarily divergent. Resolution of 20 cm^{-1} is often a compromise between the extent of the imaging field and the acceptance of the optical filters. One improvement has already been suggested using a dual-stage Fabry-Perot filter [6]. In an alternative approach, we have devised a new optical layout for the multilayer dielectric filter as shown in Fig. 1. The spectral filter here is sandwiched between two lenses. The first lens, L1, is located at a distance from the microscope objective equal to its focal length; this forces all the principal rays from the sampling area to be parallel to each at the filter. As shown in the insert of Fig. 1, each cone of rays about the principal ray has an identical divergence angle, $\Delta\beta = D / 2F$, where D is

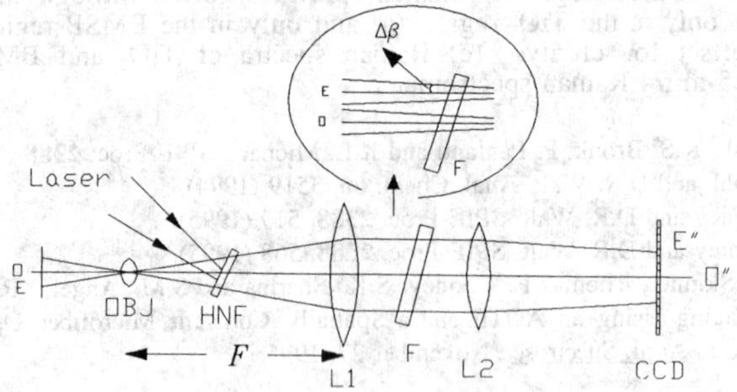

Fig. 1. Optical layout of the filter-based Raman band imaging system.

the diameter of the exit pupil of the objective and F the focal length of L1. The second lens, L2, is there to reset the system focal length, which will be shorter than F. By moderately extending F, the beam divergence can be reduced with the sampling area remaining constant. We have constructed a test system with $F = 1$ metre, giving an imaging system with 8 cm^{-1} resolution. This approach has the advantage that illumination is uniform over the image; reduction of the band-width simply reduces the intensity.

Data have been taken from Si wafer and other samples to verify the design. Fig. 2 shows two spectra obtained from the wafer sample by angle-tuning the filter shown in Fig. 1. The laser beam was defocused to illuminate an area 100 μm in diameter. Spectrum (a) was collected from the centre spot of approximately 4 μm in diameter, thus producing a nearly parallel beam; (b) was from the whole area, thus producing a diverging beam. The two spectra give the same band width. Fig. 3 shows the Raman images resulting from tuning the filter about the 520 cm^{-1} band.

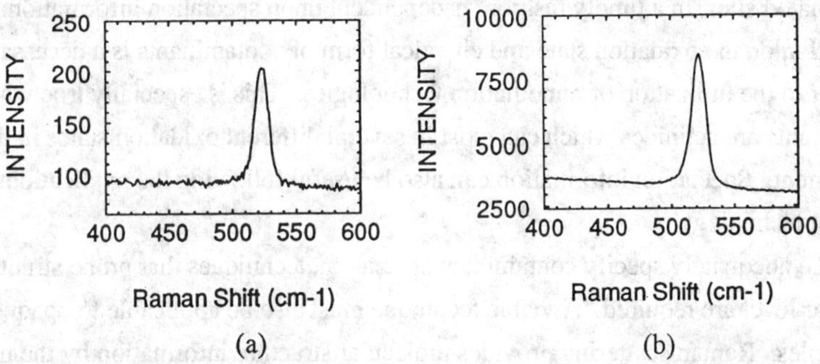

Fig. 2. Raman spectra of the Si sample: a) centre spot; b) extended area.

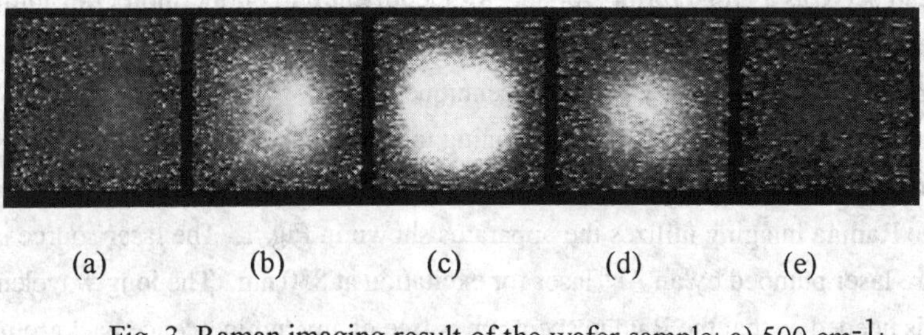

Fig. 3. Raman imaging result of the wafer sample: a) 500 cm^{-1}; b) 510 cm^{-1}; c) 520 cm^{-1}; d) 530 cm^{-1}; e) 540 cm^{-1}.

1. M. Delhaye and P. Dhamelincourt, J. Raman Spectrosc. 3, 33 (1975).
2. D.N. Batchelder, C. Cheng, W. Müller, and B.J.E. Smith, Makromol. Chem. Macromol. Symp. 46, 171, (1991).
3. P. J. Treado, I. W. Levin, and E. N. Lewis, Appl, Spectrosc. 46, 1211, (1992).
4. H.R. Morris, C. C. Hoyt, and P. J. Treado, Appl. Spectrosc. 48, 857 (1994).
5. D.M. Pallister, A. Govil, M.D. Morris, and W.S. Colburn, Appl. Spectrosc. 48, 1015, (1994).

RAMAN IMAGING OF URANIUM COMPLEXES IN SOIL

Jon R. Schoonover*, C. Drew Tait, and David E. Morris

Chemical Science and Technology Division, Los Alamos National Laboratory, Los Alamos, New Mexico 87545

A significant problem facing the United States Department of Energy is contamination of soils at various production or testing facilities. Successful remediation of contaminated sites, in a timely fashion, is dependent upon speciation information. Information on the oxidation state and chemical form of contaminants is a necessary precursor to the utilization of remediation technologies. This is especially true when the contaminants are actinides which can exist in several different oxidation states in the environment. Speciation information can also be useful following the application of a treatment technology.

To adequately specify contaminant speciation, techniques that probe structure on the molecular level are required. A viable technique must also be applicable to opaque, solid soil samples. Raman scattering provides molecular structure information by the number, energy, and intensity of the observed Raman bands. The distinctive nature of the Raman spectrum can serve as a "fingerprint" for species identification. Furthermore, the ability to spatially resolve the components makes the application of Raman imaging in the characterization soils a potentially valuable technique.

We have applied confocal Raman imaging to study speciation and spatial distribution of uranium complexes in a variety of contaminated soil samples. Our initial approach to Raman imaging utilizes the apparatus shown in Fig. 1. The laser source is a Ti: Sapphire laser pumped by an Ar^+ laser for excitation at 830 nm. The long wavelength excitation is needed so that the Raman spectrum is free of any luminescence background signals. The base microscope is a Ziess Axiovert inverted microscope adapted with a Cell Robotics programmable x, y, z stage. Detection is accomplished with a Photometrics CCD camera attached to a HoloSpec f/1.8 imaging spectrograph equipped with a SuperNotch-Plus for filtering of the 830 nm laser light. A Raman spectrum is measured at a particular point and the stage moved to another point for the next spectrum. In this manner, complete spectra are obtained at each point and Raman images built by selecting a Raman band of interest.

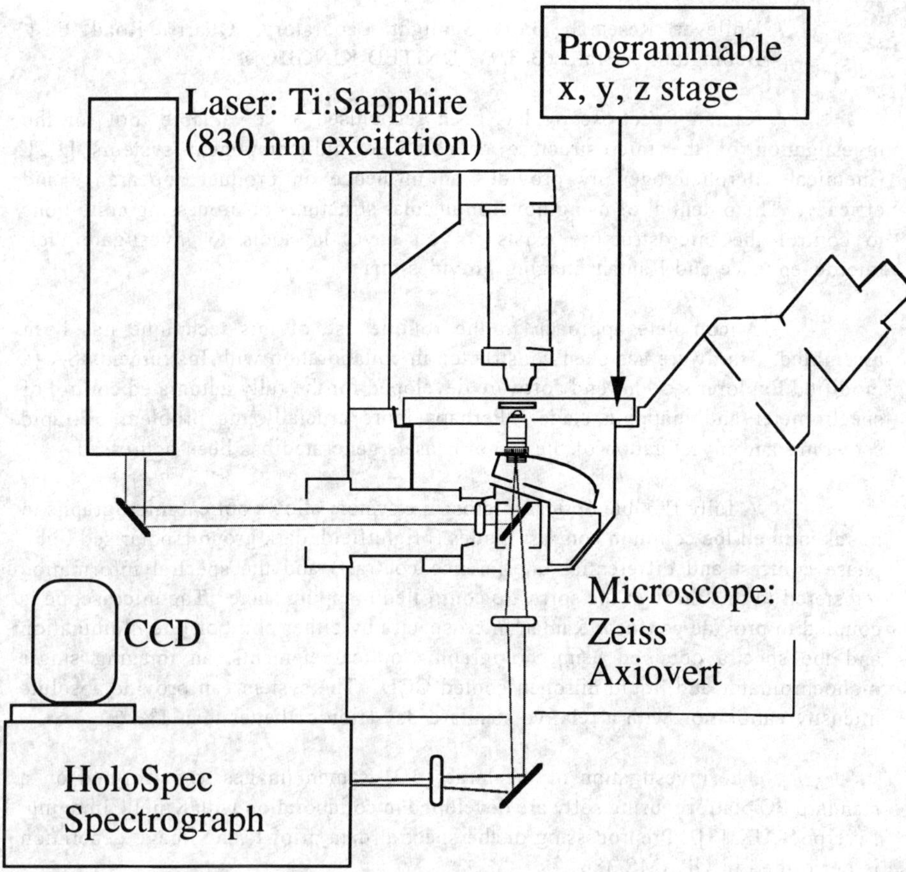

Fig. 1. Instrumentation for Raman Imaging.

INVESTIGATIONS OF MULTIPHASE SYSTEMS USING RAMAN MICROSPECTROSCOPY - ROUTINE RAMAN SPECTRAL IMAGING

J J Andrew*, I E Clark and A J Millichope

Unilever Research, Port Sunlight Laboratory, Quarry Road East, Bebington, Wirral L63 3JW, UNITED KINGDOM.

Raman Spectroscopy has been recognised as a valuable tool for the investigation of the microstructure of complex multicomponent systems [1,2]. Chemical microheterogeneity provides an influence on product appearance and efficacy. The potential of using novel molecular structures or processing conditions to control the microstructure exists if you have the tools to investigate their interdependence and Raman imaging provides this.

A complete approach to the routine use of this technique has been assembled. Hardware has been constructed in collaboration with Instruments SA (a modified Explorer system) and software developed for the fully automated control of spectrometer and mapping stage. Perhaps more crucially the problem of rapid screening and investigation of the large datasets generated has been addressed.

A fully flexible optical microscope system allows optical micrographs to be taken in all the common contrast modes (bright field, dark ground, polarised light, phase contrast and differential interference contrast) and the spectral information registered to this through the software controlled mapping stage. The microscope is coupled to provide confocal Raman microspectra by either point or line illumination and the spectra obtained using holographic optical elements, an imaging single monochromator and liquid nitrogen cooled CCD. The system can provide absolute intensity calibration with a relative standard deviation of better than 1%.

The investigation of the large [4D]-Raman images is performed on a standard PC platform using software developed in collaboration with Kinetic Imaging, Liverpool UK [3]. Preprocessing of the spectral data prior to the image generation is performed in GRAMS/386.

Several examples will be shown. Here we present the quality of information that is obtained from a seven component oil in water emulsion system. In practice, the effects of processing conditions on the chemical microstructure has been elucidated and intelligent formulation and processing changes can now be proposed to improve product quality.

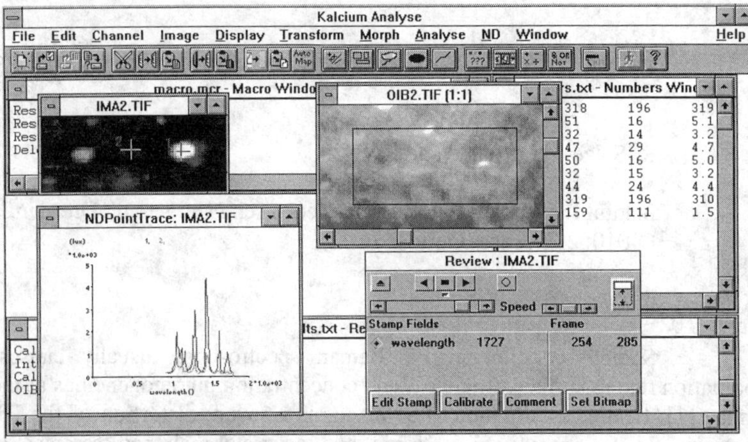

Figure 1: Spectral Image analysis.
Top right Brightfield optical image and area of spectral map 65x35 μm
Top left 4D Raman image dataset - 1μm^2 pixels
Btm right Interactive image review tool
Btm left Point spectra from dataset

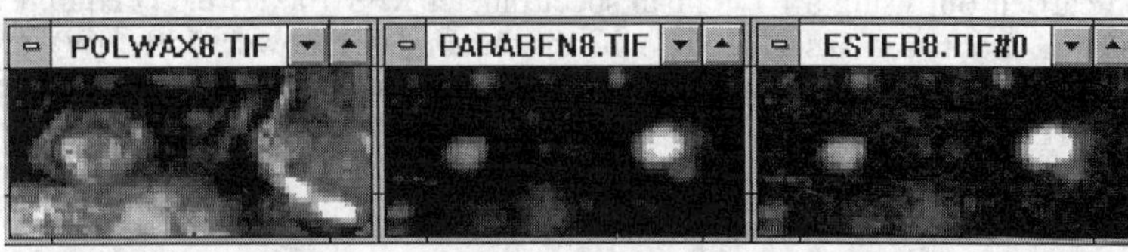

Figure 2: Images of three components from the oil in water emulsion shown above.
The distribution of polawax, parabens derivatives and a preservative ester are shown.

1. J.D. Louden, H.K. Patel and R.C. Rowe, Inter.J.Pharmaceutics 25,179 (1985).
2. J.J. Andrew, Analusis 22, M15 (1994).
3. The work of Mark Browne and his team is gratefully acknowledged.

FT RAMAN MEASUREMENTS OF ROTATING SAMPLES

R. Salzer, U. Roland and R. Born

Institut für Analytische Chemie, Technische Universität Dresden,
D-01062 Dresden, Germany

Sample rotation in FT Raman spectroscopy usually causes double modulation due to surface roughness and concentration inhomogeneities at the sample surface [1]. Double modulation often results in a severe distortion of the FT Raman spectrum. So far only very isotropic and accurately aligned samples could be measured using sample rotation in FT Raman spectroscopy [2]. That is why FT Raman spectroscopy is not regarded to be applicable for routine measurements of thermally sensitive samples where sample rotation would have to be carried out.

The step scan technique allows to overcome the problem of double modulation in FT Raman measurements of rotating samples. Our investigations have been carried out using a FT Raman spectrometer RFS 100 (Bruker) equipped with a Nd-YAG excitation laser (1064 nm, maximum laser power 750 mW) and a Ge detector. The samples were situated on a special sample holder coupled to a light barrier (Fig. 1).

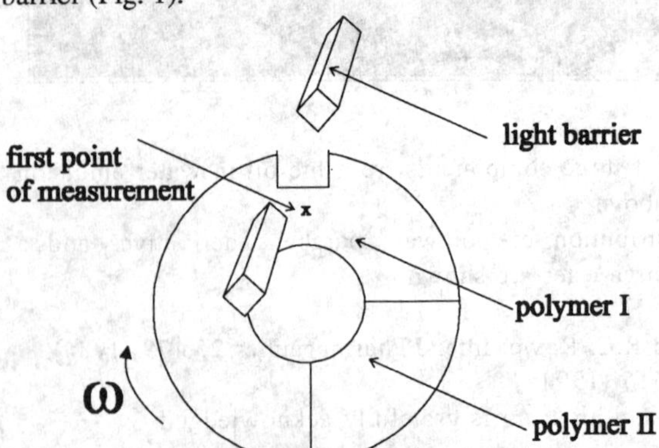

Fig. 1 Sample rotation equipment allowing FT Raman measurements without double modulation

Using this equipment is was possible to start the measurement at a distinct mirror position at constant rotation angle. Typically, a rotation frequency f of 30 Hz and a time resolution of 500 μs (time between two measuring points at one mirror position) were applied. These parameters led to an angular resolution of 5°. Significantly higher values up to 0.05^{o} can be achieved.

The obtained set of step scan FT Raman spectra represents the (angle-dependent) mapping of the sample surface, e.g., the concentration profile of the Raman scattering substance. An example is shown in Fig. 2, where two different polymers can easily be identified.

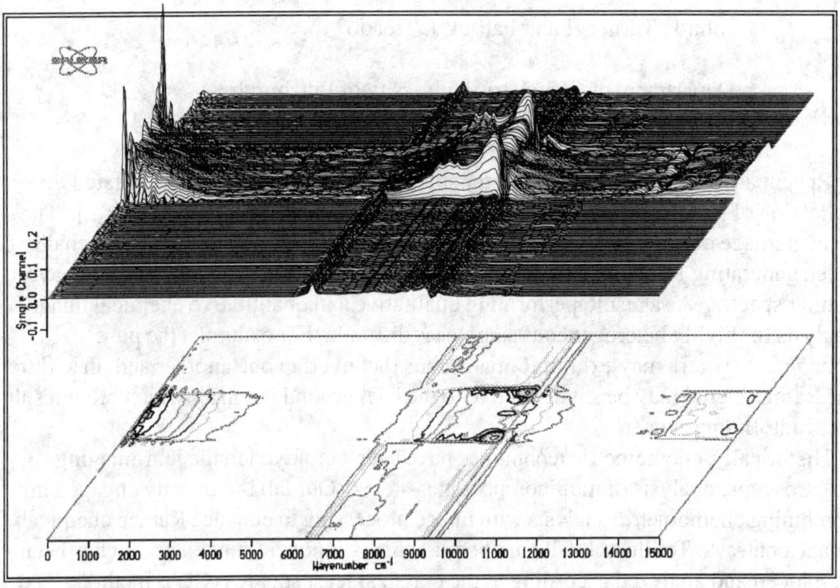

Fig. 2 Mapping of the two component polymer sample of fig. 1:
Spectrum (top) and contour plot (bottom)

The main advantages of the new method can be summarized as follows :

(i) FT Raman measurements of absorbing, sensitive sample can be carried out successfully.

(ii) A surface imaging of the rotating sample is achieved. The angular resolution depends on the rotation frequency of the sample and on the scan time resolution.

(iii) The occurrence of additional spinning bands as observed for the sample rotation in the rapid scan mode [1] is avoided.

The routine evaluation of thermally sensitive, inhomogeneous samples by FT (NIR) Raman spectroscopy can be accomplished by combining the spinning sample arrangement with the step scan technique. The new method can be used to characterize a wide range of sensitive sample such as biological systems.

The authors thank Dr. J. Sawatzki and Dr. A. Simon (Bruker) for helpful discussions. Financial support by the Sächsisches Ministerium für Wissenschaft und Kunst, Deutsche Forschungsgemeinschaft and Fonds der Chemischen Industrie is gratefully acknowledged.

[1] U. Roland and R. Salzer; "Interpretation of Spinning Bands in the FT Raman Spectra of Rotating Samples", Microchimica Acta *in press*

[2] S.H.R. Brienne, R.D. Markwell, S.M. Barnett, I.S. Butler and J.A. Finch; Appl. Spectroscopy **47** (1993) 1131

THE APPLICATION OF CHEMOMETRICS TO RAMAN CHEMICAL IMAGING

John F. Turner II and Patrick J. Treado*

Department of Chemistry, University of Pittsburgh,
Pittsburgh, PA 15260

Recent advances in electronically tunable filter technologies have facilitated the collection of high resolution Raman images of multicomponent materials [1,2]. The Raman image data are rich with spatial/spectral information which presents a challenge when generating Raman image contrast specific to individual components that have similar spectra. As a result, performing qualitative and quantitative chemical image analysis for highly heterogeneous samples is difficult. For example, the pure component spectra may exhibit Raman bands that overlap one another and, in addition, the Raman signal may be small relative to the background intensity which often results from autofluorescence.

Historically, chemometric techniques have been employed in the non-imaging spectroscopic analysis of multicomponent systems. Our lab is currently engaged in combining chemometric analysis with image processing to enhance Raman chemical image contrast. To illustrate the method, a three component model system has been developed and analyzed according to the classical least squares (CLS) method. The Raman intensity at each wavelength is related to the concentration of a training set by invoking a K-matrix calibration [3,4]. The calibration is then used to estimate the relative concentration of the pure components for each image pixel. Figure 1 shows a brightfield image of a model system comprised of three components. The pure component spectra are shown in Figure 2.

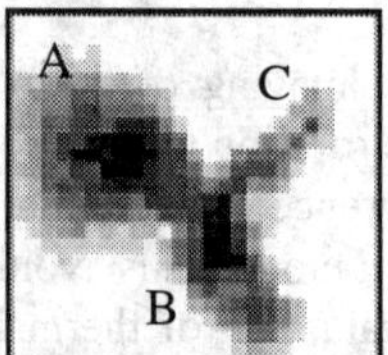

Figure 1. Brightfield image of the model three component Raman system.

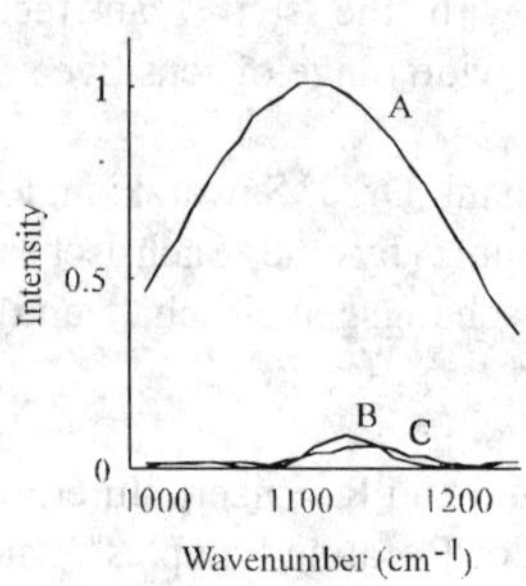

Figure 2. Pure component spectra showing significant overlapping of the Raman bands.

Pure component images can be produced by replacing each pixel intensity with the estimated relative concentration of the pure component and are shown in Figure 3. Less image contrast is produced between pure components B and C due to the high degree of overlap in their Raman spectra.

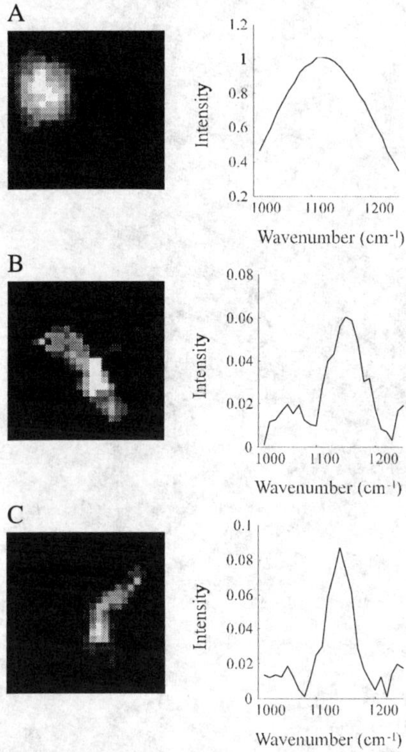

Figure 3. Pure component images estimated from K-matrix calibration are shown with representative pure component spectra.

The application of multivariate chemometric techniques to Raman chemical imaging is not limited to K-matrix analysis. Other multiple linear regression (MLR) techniques, as well as principle component analysis (PCA) and principle component regressions (PCR), can be employed to enhance image contrast [5].

1. M. D. Schaeberle, J. F. Turner II and P. J. Treado, SPIE, 2173 (1994).
2. H. R. Morris, C. C. Hoyt, P. Miller and P. J. Treado, Appl. Spec., (1996) (submitted).
3. R. C. Graham, Data analysis for the Chemical Sciences, (VCH Publishers, New York, 1993).
4. M. A. Sharaf, D. L. Illman and B. R. Kowalski, Chemometrics, (John Wiley & Sons, New York, 1986).
5. C. A. Drumm and M. D. Morris, Appl. Spec., 49, 1331 (1995).

SECTION 30

NEW TECHNIQUES

THE RAMAN SPECTRUM OF H_2 AS A PROBE OF DYNAMICS IN WATER AND ZEOLITE

Herbert L. Strauss and Zhan Chen

Department of Chemistry, University of California,
Berkeley, CA 94720-1460

The hydrogen molecule provides a unique probe of condensed phase structure. It provides strong scattering in both Raman and incoherent neutron spectroscopies and so can be examined at relatively low concentrations. In the Raman, the rotations can be seen with "tank pressure" (about 2000 lbs/sq. in.) of H_2 over water.[1] Frozen solutions form hexagonal ice and contain enough H_2 to be seen in both Raman and neutron spectroscopy.[2,3] Some zeolites such as sodalite have similar structures and will also absorb H_2.

Remarkably, the rotations of H_2 remain quantized in water and the rotational transitions show up at the same positions they have in the gas phase. The perturbations due to the water produce a substantial linewidth for these transitions. The linewidths provide interesting tests of intermolecular potentials in the condensed phase and of the perturbation dynamics. Modelling the dynamics is particularly challenging because the water molecules move on the same time scale as do the H_2 molecules. This obviates the time dependent theories that hold in the fast and slow limits. A suitable theory has been worked out [4,5] and agrees with experiment for H_2. New experiments have now been done for D_2. The D_2 rotational spacing is, of course, smaller than that for H_2. This suggests that the linewidths for the D_2 lines would be larger than those for the H_2 lines. However, we have now taken spectra of D_2 in both water and ice and find linewidths similar to those for H_2.

Since the D_2 transitions are difficult to observe, we have developed a sophisticated signal enhancement program similar to those used in NMR spectroscopy.[6] We have tested this program by comparing the results of the enhanced spectra with those obtained using long runs (many hundreds of hours!). The results are satisfactory and illustrate the possibilities for signal enhancement.

The H_2 and D_2 molecules show vibrational shifts. These are to lower frequency for H_2 or D_2 in water, as expected. However, they are in the opposite direction in ice. Calculations of the shift are compared to experiment and define the position of H_2 in the ice structure.

Neutron scattering is, of necessity, for the system H_2 in D_2O. The rotation of H_2 in ice provides spectra that can be compared to the Raman results. The neutron spectra unavoidably contain contributions from the translational modes of the H_2 (recoil). For the ice systems, analysis of the recoil contribution suggests an H_2 undergoing hindered rotation-translation. On the other hand, both the Raman results and the neutron diffusion results suggest that the H_2 motion is relatively free. Indeed, the diffusion of the H_2 is remarkably free -- the H_2 moves through ice at a rate comparable to that in pure H_2 liquid at 20 K.

The contrast with the motion of H_2 in sodalite is considerable. The rotational transitions in sodalites are narrow and show no recoil shifts -- suggesting molecules which are rotationally free, but translationally fixed. The diffusion of the H_2 in sodalite occurs only at temperatures above 300 K.

1. D. G. Taylor and H. L. Strauss, J. Chem. Phys. 90, 768 (1989).
2. D. G. Taylor and H. L. Strauss, J. Chem. Phys. 96, 3367 (1992).
3. H. L. Strauss, Z. Chen, and C. K. Loong, J. Chem. Phys. 101, 7177 (1994).
4. J. E. Hunter, D. G. Taylor and H. L. Strauss, J. Chem. Phys. 97, 50 (1992).
5. L. Xiao and D. F. Coker, J. Chem. Phys. 102, 1107 (1995).
6. Z. Chen, Y. Y. Lin, and H. L. Strauss, to be published.

RAMAN SPECTROSCOPY OF OPTICAL LATTICES

A. Hemmerich[1*] and T. Hänsch[1,2]

[1]Sektion Physik der Universität München,
Schellingstr. 4/III, 80799 München, Germany

[2]Max-Planck-Institut für Quantenoptik,
Hans-Kopfermannstr. 1, 85748 Garching, Germany

Optical lattices are ultracold samples of atoms localized by dipole forces inside optical standing waves [1]. The trapping potential arises from the spatially varying AC-Stark-shifts of the atomic ground state Zeeman levels and provides potential minima at maximum light intensity. Thus the antinodes of the light field act as microscopic light traps which are arranged on a regular three-dimensional lattice with a lattice constant of typically half an optical wavelength. Probe transmission spectra of such atomic matter bound by light involve many unexpected spectral features arising from multi-wave mixing processes as e.g. phase conjugation or hyper-Raman transitions [1-3].

The atomic vibrations in the microscopic traps of "optical lattices" can be investigated via stimulated Raman spectroscopy [1]. A weak probe laser beam is directed through the trapped atomic sample and its transmission is recorded. The interference between this probe beam and the light field which binds the atoms in the lattice shakes the microtrap potentials at the frequency detuning between the two fields. This yields a coherent excitation of atomic vibrations leading to an exchange of photons between the two light fields. This is observed in the probe transmission signal.

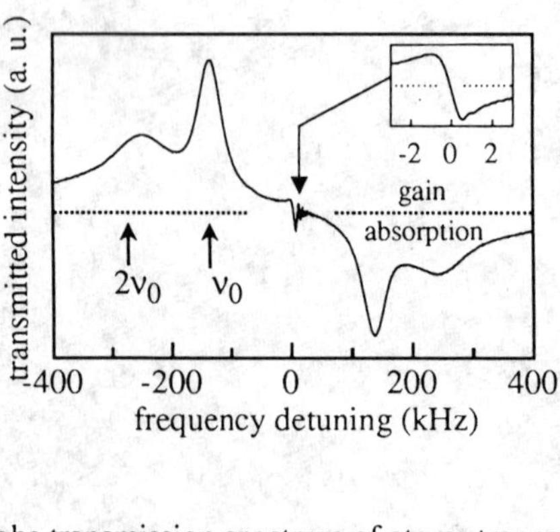

Fig. 1 Probe transmission spectrum of atoms trapped in an optical lattice at low incident probe intensity.

In fig. 1 we show the transmitted probe signal recorded versus the probe frequency detuning. We clearly recognize two pairs of Raman resonances at ±136 kHz and ± 250 kHz corresponding to transitions involving one and two vibrational quanta. An additional narrow resonance occurs in the center of the spectrum where the frequency detuning equals zero. This resonance is shown with much increased resolution in the upper right corner of fig.1. Its origin is a four-wave mixing process including the generation of a long range atomic density hologram [2]. At higher probe intensities additional resonances (fig. 2) are observed at subharmonic frequencies which can be interpreted in terms of Hyper-Raman transitions [3].

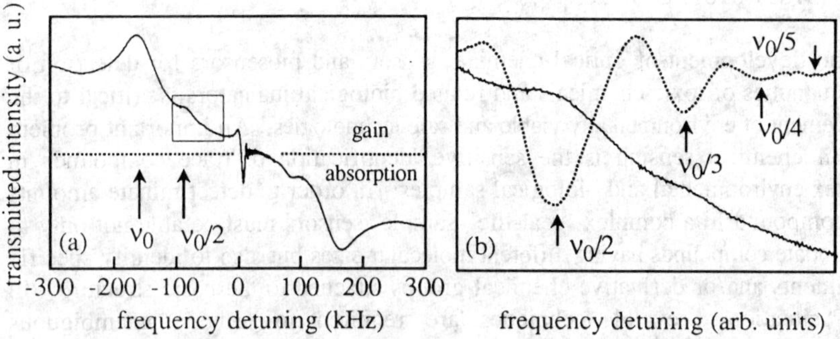

Fig. 2 Resonances arising at higher probe intensities at subharmonic frequencies ν_0/i, i = 2,3,4,5 due to hyper-Raman transitions. The solid graph in (b) shows an extension of the portion of the spectrum in (a) enclosed by the rectangle. The dotted line represents the second derivative.

1. A. Hemmerich and T. Hänsch, Phys. Rev. Lett. 70, 410 (1993). A. Hemmerich et al., Europhys. Lett. 22, 89 (1993). See also P. Verkerk et al., Phys. Rev. Lett. 68, 3864 (1992). G. Grynberg et al., Phys. Rev. Lett. 70, 2249 (1993).
2. A. Hemmerich et al., submitted to Europhys. Lett. 27, 427 (1994).
3. A. Hemmerich et al., Phys. Rev. Lett. 72, 625 (1994).

CHEMICAL SENSORS AND BIOSENSORS USING SURFACE-ENHANCED RAMAN SCATTERING (SERS)

Tuan Vo-Dinh

Advanced Monitoring Development Group, Health Sciences Research Division, Oak Ridge National Laboratory, Oak Ridge, Tennessee 37831-6101, U.S.A.

The development of optical chemical sensors and biosensors for detection of trace quantities of toxic chemicals and related biological indicators is critical to the achievement of environmentally viable and safe technologies. An important problem area in chemical sensing is the sensitive identification of trace compounds in complex environmental and biological samples. In order to detect minute amounts of a compound in a complex "real-life" sample, sensors must be able not only to differentiate compounds having different molecular sizes but also to identify specific substituents and/or derivative chemical groups attached to the basic structure. In many situations, several techniques are required to provide unambiguous identification and accurate quantification.

This lecture presents some recent advances in the development of surface-enhanced Raman scattering (SERS) substrates for use in chemical sensors and biosensors. The development of various SERS-active materials for optical probes is discussed. The SERS probes consist of substrates covered with silver-coated microstructured particles. The use of SERS probes using bioreceptors such as antibodies and enzymes for bioremediation, is presented. The results of SERS-active substrates for *in-situ* analysis in liquid and vapor samples are also presented to illustrate the usefulness of SERS sensors in environmental and biological analysis. The Raman technique is well known for its high selectivity. With the advances of fiberoptic technology, the SERS technique, which can amplify the Raman signal by several orders of magnitude, can provide a useful and versatile sensing technique with the added merit of improved sensitivity due to the surface-enhanced effect. The development of SERS-active probes will open new horizons for remote *in situ* chemical sensors and biosensors for environmental and biomedical applications.

This research is sponsored by the Office of Health and Environmental Research, U.S. Department of Energy under contract DE-AC05-96OR22464 with Lockheed Martin Energy Research Corporation, and the Biosrnsor LDRD Project. The sponsorship of BMDO managed by the Air Force Office of Scientific Research (AFOSR-MIPA-96-011) is acknowledged.

RAMAN OPTICAL ACTIVITY: AN INCISIVE PROBE OF CHIRALITY AND BIOMOLECULAR STRUCTURE AND DYNAMICS

L. D. Barron*, L. Hecht, A. F. Bell and G. Wilson

Chemistry Department, The University, Glasgow G12 8QQ, U.K.

Determination of the absolute configuration and conformation of small chiral molecules, and of the solution structure and dynamics of biopolymers, remain central problems in structural chemistry and in biochemistry. The potential value of vibrational spectroscopy in these areas has been greatly enhanced by adding the new dimension of optical activity [1]. The Raman optical activity (ROA) approach, which measures a small difference in the intensity of Raman scattered light in right and left circularly polarized incident light (or a small circular component in the scattered light) [2-4] has been transformed by recent technical advances [5] into a powerful technique for tackling these problems.

We currently have two ROA instruments, both based on a single-grating spectrograph with a backthinned CCD detector and a holographic notch filter to block the Rayleigh line [5]. ROA is measured by synchronizing the spectral acquisition with an electro-optic modulator used to switch the polarization of the incident argon ion laser beam between right and left circular. One instrument employs 90° scattering and provides ROA spectra of small chiral molecules from which conformational details and absolute configuration can be deduced, together with enantiomeric excess to within 0.1% [6]. The other employs backscattering and provides aqueous solution ROA spectra of biopolymers such as peptides, proteins, carbohydrates, glycoproteins and nucleic acids which give new information on their structure and dynamics [7].

The ROA spectra of polypeptides and proteins can be divided into four main regions: the backbone C_{α}—C and C_{α}—N stretch region ~ 870-1150 cm^{-1} which is dominated by α-helix and β-sheet bands; the extended amide III region ~ 1230-1350 cm^{-1} which contains bands characteristic of loops and turns together with helix and sheet all originating in highly local coupled backbone N—H and C_{α}—H deformations which are particularly sensitive to the Ramachandran Φ angle; the region ~ 1400-1620 cm^{-1} which contains bands originating in aliphatic and aromatic sidegroups; and the amide I region ~ 1630-1700 cm^{-1} which contains bands from both secondary and loop structure. As an example, Fig. 1 shows the Raman and ROA spectra of native α-lactalbumin. Also shown are the spectra of reduced lysozyme which illustrate the signature of disordered mobile heteropolypeptide structure, namely a broad extended amide III ROA couplet originating in a superposition of signals from the huge number of distinct conformers present in the unfolded state, and which is proving invaluable in studies of protein dynamics [8,9]. As well as the individual secondary structure and loop ROA bands, we are beginning to discern in protein ROA spectra local band patterns which might be associated with structural motifs, and overall patterns characteristic of the superfold type; so we are hopeful that ROA might ultimately provide an approximate idea of the complete solution structure of the peptide backbone.

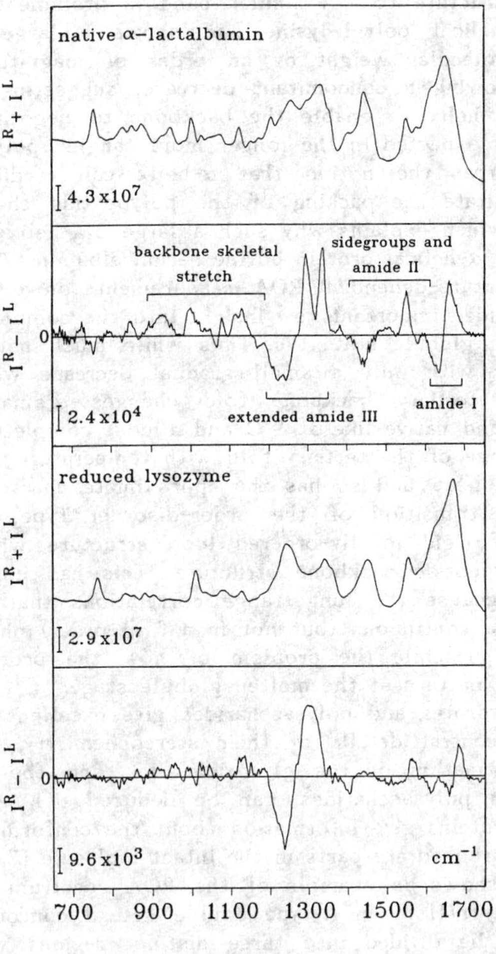

Fig. 1. Backscattered Raman ($I^R + I^L$) and ROA ($I^R - I^L$) spectra of native α-lactalbumin and reduced lysozyme.

Comparison of ROA spectra in H_2O and D_2O is proving very useful in our peptide and protein studies [7] because the rates of exchange of hydrogen atoms on the amide NH groups of the peptide backbone depend on the relative accessibility of different types of structure to the solvent. The rate of exchange for secondary structure is many orders of magnitude slower than for surface loop structure, which among other things enables ROA bands from these two types of structure to be distinguished.

The strong sharp positive ROA band at ~ 1338 cm^{-1} in α-lactalbumin (Fig. 1) is particularly interesting since it originates in 3_{10}-helix loop structure [8]. A similar band is prominent in the ROA spectrum of α-helical poly-L-lysine and shows a large increase on increasing the molecular weight by an order of magnitude, with the α-helix bands showing a concomitant decrease, suggesting that α-helix converts into 3_{10}-helix to enable the backbone to negotiate the many more sharp bends expected in the longer more tangled polypeptide chain [10]. This reinforces the notion that α-helix can readily change to 3_{10}-helix to facilitate the packing of the polypeptide chains in native folded proteins, which explains why such a large 3_{10}-helix ROA band is seen in the highly α-helical protein bovine serum albumin [7].

Temperature-dependent ROA measurements on α-lactalbumin at low pH have provided important new insight into the complexity of order within the molten globule state [9]. Thus while much native secondary structure persists with only a small gradual decrease with increasing temperature, the tertiary backbone fold changes dramatically, being almost complete and native-like at 2°C and almost completely disordered at 35°C. The change of the tertiary fold with temperature is cooperative but has no latent heat, and so has the approximate characteristics of a continuous phase transition of the order-disorder type involving the interconversion of rigid locally-ordered loop structure with disordered mobile heteropolypeptide backbone structure. This has implications for protein folding because the longe-range correlations that exist in the critical region of a continuous (but not in a first-order) phase transition could resolve in principle the problem of how the protein finds its native-like folding pattern at the molten globule stage.

Carbohydrates and polysaccharides give excellent ROA spectra from which the central details of their stereochemistry, including the nature and conformation of the glycosidic link and the existence of extended order in polysaccharides, can be deduced [7]. And the ROA spectra of glycoproteins give information about the conformation of both the protein and carbohydrate parts in the intact molecule [7].

Fig. 2 shows an example of the ROA spectrum of a nucleic acid, Poly(rA)·Poly(rU) in an A-type double helical conformation. Such ROA spectra can be divided into three distinct regions which provide information on some key features of nucleic acid stereochemistry. The first, above ~ 1600 cm^{-1} and arising from localized base ring vibrations, contains base-dependent ROA signatures only for the double and triple stranded polynucleotides in which intermolecular hydrogen bonding allows coupling between the pairs. The second, ~ 1200-1600 cm^{-1}, contains ROA signals which depend on the stereochemistry of the C—N glycosidic link and which can be correlated with the degree of order in the polynucleotide. The third, ~ 800–1200 cm^{-1}, arises from vibrations localized on the sugar rings and phosphate groups and contains sign patterns reflecting sugar ring and phosphate backbone conformation.

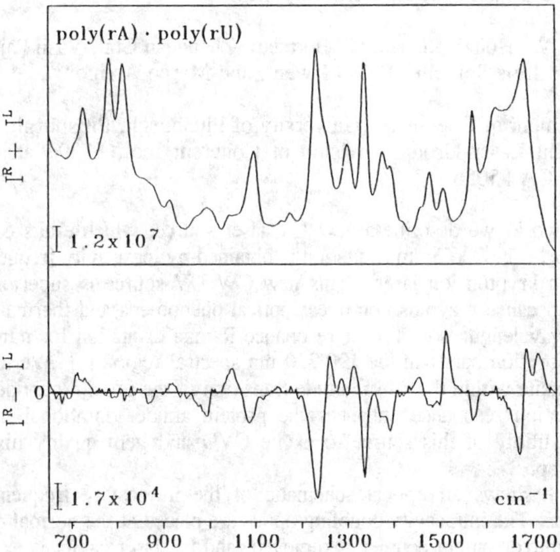

Fig. 2. Backscattered Raman and ROA spectra of A-type double helical Poly(rA)·Poly(rU).

1. Vibrational Optical Activity: From Fundamentals to Biological Applications, Faraday Discussion **99** (1994).
2. L. D. Barron, Molecular Light Scattering and Optical Activity, Cambridge University Press, Cambridge, 1982.
3. L. D. Barron and L. Hecht, in Circular Dichroism Principles and Applications, (K. Nakanishi, N. Berova and R. W. Woody, eds.), VCH Publishers, New York, 1994, p. 179.
4. L. A. Nafie and D. Che, Adv. Chem. Phys. **85**(3), 105 (1994).
5. L. Hecht and L. D. Barron, Faraday Discuss. **99**, 35 (1994).
6. L. Hecht, A. L. Phillips and L. D. Barron, J. Raman Spectrosc. 26, 727 (1995).
7. L. D. Barron, L. Hecht, A. F. Bell and G. Wilson, Appl. Spectrosc., in press.
8. G. Wilson, S. J. Ford, A. Cooper, L. Hecht, Z. Q. Wen and L. D. Barron, J. Mol. Biol. 254, 747 (1995).
9. G. Wilson, L. Hecht and L. D. Barron, J. Mol. Biol., submitted for publication.
10. G. Wilson, L. Hecht and L. D. Barron, J. Chem. Soc. Faraday Trans., in press.

A NOVEL CW 206.5 NM LASER SOURCE FOR UV RAMAN DETERMINATION OF PROTEIN SECONDARY STRUCTURE AND CVD DIAMOND MATERIAL PROPERTIES

Janet S. W. Holtz[1], Richard W. Bormett[1], Zhenhuan Chi[1], Vasil Pajcini[1], Sanford A. Asher*[1], Luis Spinelli[2], Philip Owen[2], and Marco Arrigoni[2]

[1]Department of Chemistry, University of Pittsburgh, Pittsburgh, PA 15260
[2]Coherent Laser Group, Division of Coherent Inc., 5100 Patrick Drive, Santa Clara, CA 95056

In this work, we describe a new UV laser source, which emits ca. 4 mW of CW light at 206.5 nm. The 206.5 nm excitation is obtained by intracavity frequency doubling the 413 nm line of a krypton ion laser. This new CW UV source is superior to the previous pulsed sources, because it avoids non-linear optical phenomena and thermal sample damage. The 206.5 nm wavelength is useful for resonance Raman excitation for numerous molecular species with absorption bands in the 190-220 nm spectral region.[1] We use this 206.5 nm wavelength to excite within the $\pi \rightarrow \pi^*$ amide transition of the protein peptide backbone. The 206.5 nm excitation resonance enhances the protein amide vibrational modes. We also demonstrate the utility of this source to excite CVD and gem quality diamond within its electronic bandgap.

Figure 1 shows an optical schematic of the intracavity frequency doubled CW krypton ion laser. The intracavity doubling optics are placed at the normal output port of the laser; the front mirror output coupler is removed, and the laser cavity is extended to include the frequency doubling optics. The curved folding mirror M2 directs the fundamental beam towards the Brewster-angle wavelength selection prism, P1. P1 both selects lasing for the fundamental Kr^{++} line at 413 nm and separates the second harmonic generated (SHG) beam at 206.5 nm from the fundamental intracavity beam. The fundamental beam is focused by M2 into a β-barium borate (BBO) doubling crystal. Since this is a standing-wave cavity, two counterpropagating second harmonic beams are generated. The 206.5 nm beam directed toward M3, which is a total reflector at 413 nm, is partly absorbed and partly reflected. The other 206.5 nm beam, traveling in the opposite direction, is refracted by P1 into the collimating cylindrical lens which focuses the beam outside the doubler. P1 has a very low reflectivity for the p-polarized fundamental intracavity beam. However, the s-polarized SHG beam is partly reflected (ca. 20%) at both P1 surfaces. The reflection off the first surface of P1 is directed to a photocell through a second prism, P2, which eliminates any residual 413 nm light. The photocell detector is the sensor for a control system, which regulates the output power of the SHG beam by adjusting the plasma tube current to stabilize the laser SHG output power.

We typically utilize a 206.5 nm beam output power of 3 mW, and focus the beam to spot size of ca. 100 μm at the sample. Operating powers of 3-4 mW give a BBO crystal life span in excess of 400 hours.

Protein Secondary Structure Studies

Figure 2 shows the CW 206.5 nm excited UVRR spectra of con A, cyt *c*, and Mb. These proteins show quite different secondary structures; especially, differing in their α-helical content. The con A is only ca. 2% α-helix, cyt *c* is ca. 39% α-helix, and Mb is ca. 79% α-helix. The 932 cm^{-1} bands in the Raman spectra are due to perchlorate, added as an internal standard, while the other bands mainly derive from amide vibrations of the peptide backbone. Figure 2 shows the resonance Raman spectra of these proteins dissolved in H_2O

and D_2O, left and right column, respectively. Deuterium exchange at the amide N-H bond dramatically changes the normal mode composition of the amide vibrations.[2]

UV Raman Diamond Structural Studies

The Raman spectrum of CVD diamond excited at 206.5 nm, deep within the electronic bandgap, is shown in Figure 3. The 206.5 nm excited Raman spectrum is similar to the spectrum excited at 228.9 nm, but the second order phonon band intensity at ca. 2200 cm^{-1} increases relative to that of the first order phonon band at ca. 1332 cm^{-1}. No broad band luminescence appears between 206.5 nm and 225 nm with the 206.5 nm excitation.

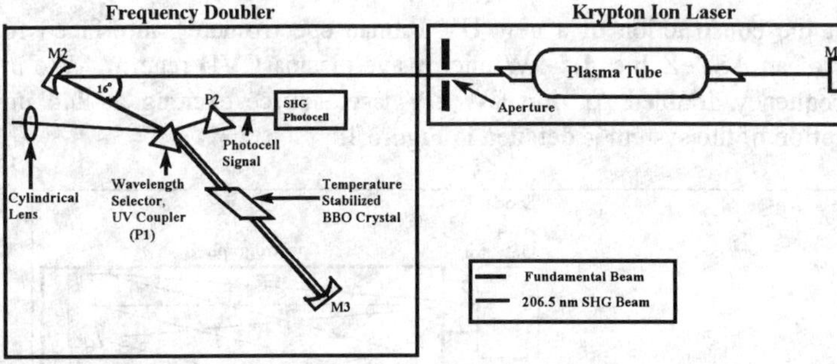

Fig. 1. Schematic diagram of the Coherent Innova 400 intracavity frequency-doubled krypton ion laser.

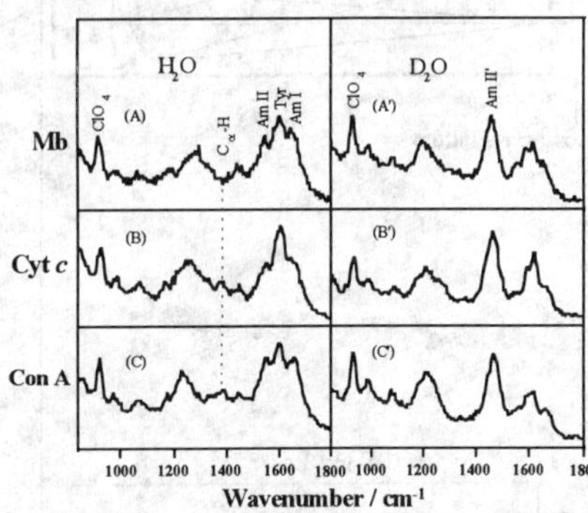

Fig. 2. The 206.5 nm excited Raman spectra of proteins with different secondary structures, dissolved in H_2O (left) and D_2O (right).

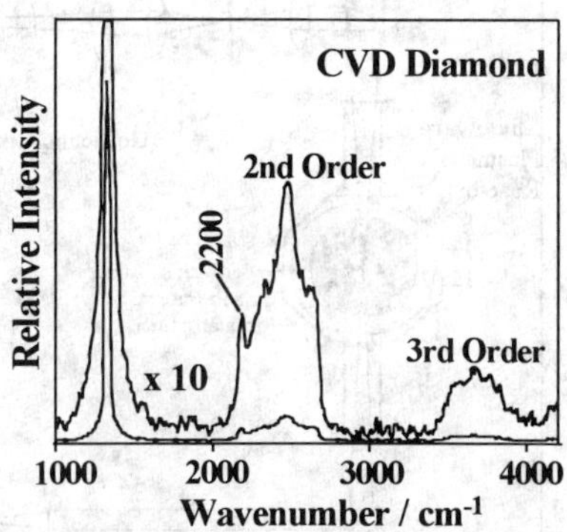

Fig. 3. The 206.5 nm excited Raman spectrum of CVD Diamond.

1. M. B. Robin, in Higher Excited States of Polyatomic Molecules (Academic Press, Orlando, 1979), Vol. I & II.
2. X. G. Chen, S. A. Asher, R. Schweitzer-Stenner, N. G. Mirkin, and S. Krimm, J. Am. Chem. Soc. 117, 2884 (1995).

CONSTRUCTION OF A CW UV RAMAN SPECTROMETER FOR *IN-SITU* CVD DIAMOND FILM GROWTH CHARACTERIZATION

James C. Worthington, Richard W. Bormett, Calum H. Munro, Robert E. Witkowski, Sanford A. Asher*

University of Pittsburgh, Department of Chemistry
Pittsburgh, Pennsylvania 15260

We report the construction of a new UV Raman spectrometer, interfaced to the growth chamber of an ASTeX Inc. 1.5 kW microwave plasma CVD reactor. We utilize an intracavity frequency doubled Ar^+ ion CW UV laser source exciting at 244 nm. A schematic description of the system is detailed in Figure 1.

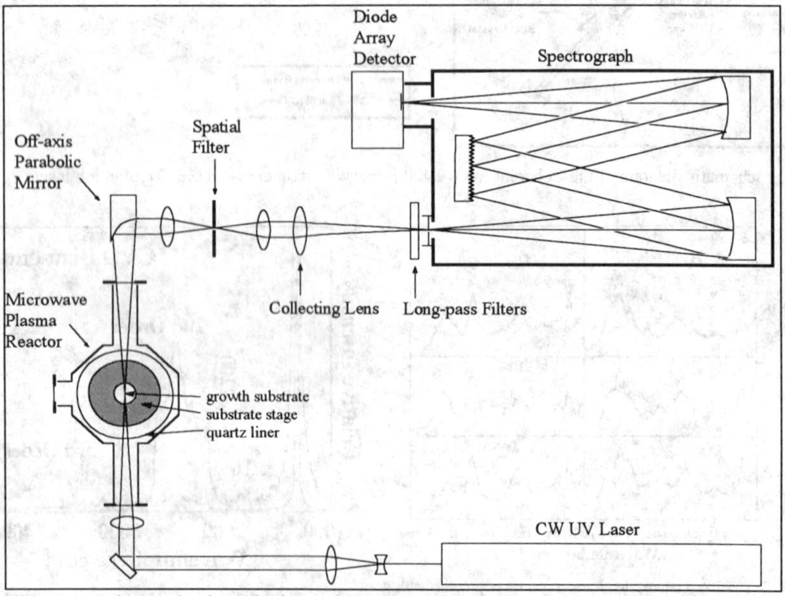

Fig. 1. Schematic description of the UV Raman Spectrometer interfaced to the MPCVD reactor; key optical elements are detailed.

Our excitation source is based on a previously described [1] frequency doubled Coherent Inc. Ar^+ ion laser. CVD diamond films and coatings are monitored *in-situ* by exciting their Raman spectra, at 244 nm or 257 nm, or at 228.9 nm near the diamond bandgap. Long pass dielectric stack filters are used for rejection of the Rayleigh scattered radiation. The Raman scattered radiation is imaged onto the entrance slit of a SPEX 1702 single spectrograph, equipped with an EG&G PARC blue intensified photodiode array and multichannel analyzer. Substrate temperatures are monitored and controlled by two methods: two-color (wavelength) optical pyrometry and direct reading thermocouples.

CVD diamond deposition is a non-equilibrium process whereby a carbon-containing feed gas, such as methane, is introduced through a gas manifold system and diluted with H_2, then ionized in a high energy environment; nucleation and subsequent carbon species layer growth occur on a heated, temperature-controlled substrate, driven by a decrease in the chemical potential.

In-situ UV Raman diagnostic measurements will allow us to monitor the diamond and non-diamond species concentration as a function of growth conditions. High S/N ratios result from resonance enhancement and because UV Raman spectra are immune from both fluorescence interference and broad band blackbody emission that occurs from the heated substrate and plasma [2].

In-situ observation of the growth surface layer has been hampered previously by the small concentrations of surface layer species, which have been difficult to measure directly, or have been near limits of detection; hence surface dynamics are not well understood. In particular, there is a need for more information on the structure and composition of species present at the growth interface. We will detect and characterize the surface species, including those which may be important to nucleation and growth.

Interest in *in-situ* monitoring of diamond film growth is evident from the recent studies by other workers who have used visible wavelength pulsed laser sources for measurement of diamond growth [3-5]. Unfortunately, pulsed Raman excitation is fraught with artifacts, since the high energy, short duration laser pulses induce photochemical changes in the non-diamond impurities and gas phase precursors, and are liable to induce saturation effects. The use of CW UV excitation is superior to pulsed sources, with higher S/N ratios, and minimal sample degradation and nonlinear saturation effects.

We will use the information obtained to characterize in real time CVD diamond films grown under different conditions, observing, for example, variation of the diamond crystallite size, stress and strain, and effects of film temperature. This information will allow us to control and optimize growth conditions for the reproducible production of high quality diamond films with various desirable physical properties.

1. S.A. Asher, R.W. Bormett, X.G. Chen, D.H. Lemmon, N. Cho, P. Peterson, M. Arrigoni, L. Spinelli, and J. Cannon, *Appl. Spectrosc.*, **47**, 628 (1993).
2. R.W. Bormett, S.A. Asher, R.E. Witkowski, W.D. Partlow, R. Lizewski, and F. Pettit, *J. Appl. Phys.*, **77**, 11 (1995).
3. L. Bernardez, K.F. McCarty, N. Yang, *J. Appl. Phys.*, **72** (5), 2001 (1992).
4. N. Rosman, L. Abello, J.P. Chabert, G. Verven, G. Lucazeau, *J. Appl. Phys.*, **78** (1), 519 (1995).
5. L. Fayette, B. Marcus, M. Mermoux, N. Rosman, L. Abello, G. Lucazeau, *J. Appl. Phys.*, **76** (3), 1604 (1994).

OPTICAL FIBER RAMAN SPECTROSCOPY COMBINED WITH SCANNING TUNNELING MICROSCOPY FOR SIMULTANEOUS MEASUREMENTS

B. Ren, W.H. Li, B.W. Mao, J.S. Gao, Z.Q. Tian*

State Key Laboratory for Physical Chemistry of Solid Surfaces, Xiamen University, Xiamen 361005, China

Being a powerful tool for studying surfaces, especially providing topographic information at the atomic level, the scanning probe microscopy (SPM) such as STM and AFM has been making unique contribution to material and surface sciences. On the other hand, Raman spectroscopy has been used extensively for structural studies on electrode materials, adsorbates, surface complexes, organic film and biologic systems. These two techniques are obviously complimentary to each other, particularly for probing surfaces or solid/liquid and solid/gas interfaces. Indeed, SPM has been combined with Raman spectroscopy for electrochemical studies [1,2]. In these studies, however, the SPM and the spectroscopic measurements were taken separately, i.e., the sample investigated had to be transferred from the sample chamber of the spectrometer to the SPM instrument. Therefore, the combined results from the two techniques remain some questions or uncertainties as it is very difficult to ensure that the interface being investigated remains unchanged after being stopped the experiment for moving the sample from one instrument to the other. Therefore, it is highly desirable to enable the two measurements to be carried out simultaneously [3].

A combined base of a scanning tunneling microscopy (STM) and a Raman spectroscopy has been set up utilizing the optical fiber technique, which enables the simultaneous Raman and STM measurements, Fig. 1. The most serious problem for the simultaneous measurement is that the SPM measurement is vitally affected by the mechanical, thermal and electric interferences. The STM head is specially designed so that the limited spaces surrounding the STM tip can be fully used for introducing the fibers. Five collection fibers and one excitation fiber are fixed in a fiber fixer of a hemisphere cap through which the fibers can be positioned closely surrounding the STM tip above the sample without any mechanical touch. Moreover, to suppress the thermal interference, the laser illumination on the tip and the electrode surface during the simultaneous measurement should be minimized. With an optical multichannel detector the time for recording a Raman Spectrum, i.e., the time of illumination ,can be reduced to 0.1 - 5 s. The graded refractive index (GRIN) lens is fixed at the end of each fiber to ensure efficient light collection. The scattering light is collected and sent to the next room through five 12 m long fibers which are then coupled into the Raman spectrometer. Fig. 2. shows the STM images of a smooth HOPG surface before and after the electrodeposition of rough silver film as well as a simultaneously measured SERS spectrum of thiourea as the organic additive for the electroplating. The improvement of the resolution of the present STM instrument would probably make it possible to observe the silver clusters with atomic resolution. Because both the STM tip and the incident laser can be utilized to induce electro-deposition and polymerization at nano- or micro-scale, the base can be applied further as a multiple tool for electrochemical nanofabrications and, in consequence, in-situ measurements for both the topography and structure of the pattern just fabricated. The set up details of the combined base, the correct correlation of the combined results and future developments of the base will be presented.

Acknowledgments

This work was supported by the Natural Science Foundation of China.

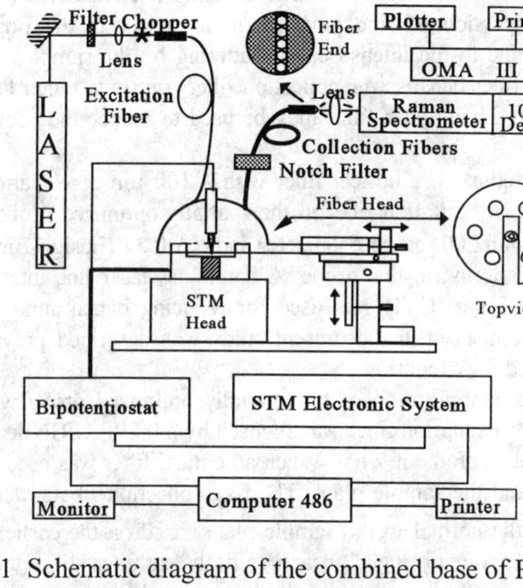

Figure 1 Schematic diagram of the combined base of Raman and STM.

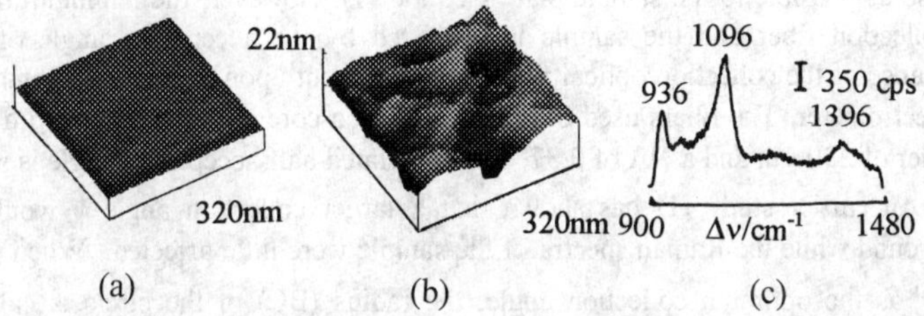

Figure 2 STM images of the HOPG surface (a) before and after (b) Ag deposition, and (c) a SERS spectrum of thiourea adsorbed on the electrode.

1. C. Stuhlmann, J. Villegas, M. J. Weaver, Chem. Phys. Lett., 219 319 (1994).
2. J.S. Chen, T.M. Devine, D.F. Ogletree, M.Salmeron, Surf. Sci.,258 346 (1991).
3. Z.Q. Tian, W.H. Li, B. Ren, B.W. Mao, J. Electroanal. Chem., (1996) in press.

A SPATIALLY OPTIMIZED PROBE WITH MULTIPLE COLLECTION FIBERS

J. Ma and Y. S. Li*

Department of Chemistry, University of Memphis, Memphis, TN 38152 USA

Our earlier study of Raman fiber probes has shown that the background is spatial dependent while the sample's Raman scattering is rather homogeneously distributed [1]. Based on this finding, we have set up a spatially optimized probe with multiple collection fibers to increase the collection efficiency. The drawback of using a Raman fiber probe to detect a powder sample or a dispersion of fine particles in a matrix is the possibility of strong background interference due to the intense back scattering by the samples. Most background arise from the excitation fiber because the collection fiber transmits much less laser light than the excitation fiber. Thus, a band-pass filter may be used to reduce the background scattering from the excitation fiber.

A silica-cladding-silica-core optical fiber with a 100 μm core diameter and NA 0.22 (General Fiber) was used as excitation fiber in the spatially optimized probe; silica-cladding-silica-core optical fibers with 200 μm core diameter and NA 0.37 (Ensign-Bickford Optic) were used to collect Raman scattering in this probe. A band-pass laser line interference filter with center wavelength at 514.5 nm (CVI) was used for reducing background interference. The Raman sprctrometric system used in the present study was described previously[2]; a green laser at 514.5 nm was used for excitation.

Fig. 1 shows the configuration of the spatially optimized probe with five collection fibers. Laser light from the excitation fiber was focused by a 0.29P GRIN lens onto the sample plate with a imagine distance of about 7 nm. When an optical filter was used, the filter was put between the GRIN lens and the sample plate. The five collection fibers were arranged in the position 60° away from the normal to the sample plate based on the earlier study. The other ends of the collection fibers were arranged linearly to fit the entrance slit of the monochromator

To collect Raman scattering signals using bare optical fibers, the fibers should be put as close as possible to the sample plate (see Fig. 1). However, the minimum distance between the collection fiber and the sample is restricted by the acceptance angle of the fiber. The acceptance of the collection optical fiber was dependent upon the NA of the monochromator of the spectrometer. The fibers used in this set up has a core diameter of 200 μm with a cladding diameter of 230 μm and a NA of 0.37. The calculated half acceptance angle is slightly less than 22°. An earlier study [1] has shown that a larger collection angle ϕ would give a lower background while the Raman spectra of the sample were little affected. When ϕ was chosen to be 60° as the optimum collection angle, the radius (BC) of the circle is calculated to have enough room for holding about 5 collection fibers. Thus, five collection optical fibers were chosen in the set up of the optimum probe.

Fig. 2 shows the SER spectra of p-nitrophenol (PNP) collected by the probe using the same laser power at 514.5 nm for excitation. Spectrum A was collected without using the band-pass filter; Spectrum B was recorded with the use of a band-pass filter. The lower signal intensity shown in Fig. 2-B is due to the optical power loss caused by the insertion of the optical filter. It appears that the signal to background ratio does not improve much with the use of the filter. Similar conclusion can be drawn from the SERS spectra of p-nitrobenzoic acid.

One disadvantage of using optical filters in the fiber Raman probe is the optical power loss. In the present study, using the band-pass filter in the spatially optimized probe does not improve the quality of the spectra very much. This may indicate that the spatially optimized method is rather efficient in reducing the fiber Raman background in this probe configuration. The main reason for the appearance of the background at the use of the filter would be the non-ideal use of the band-pass filter. For the interference band-pass filter, an ideal filtering has to have a collimated light beam. In the present experiments, the filter was positioned after the imaging GRIN lens and therefore, the light passing through the filter is a focused light. This arrangement allows small amount of the light outside the passing band going through the filter. In the configuration of spatially optimized probe, better filtering can be achieved using collimated light but this will bring additional power loss and complexity of the probe head.

The spatially optimized probe is suitable for detecting Raman and SERS spectra of film samples. Using optical fibers in Raman sampling, one may construct an ultra sensitive probe because more light can be collected by arranging fibers in all the spatial directions. It should be pointed out that the current probe does not has the expected high sensitivity because of the difficulty in the optical alignment. The five collection fibers with core diameters of 200 μm have to be positioned very carefully to have an overlapped collection area which is located exactly over the sample excitation area. To achieve this alignment requirement, very fine mechanical tuning components should be very helpful in the adjustment of the probe.

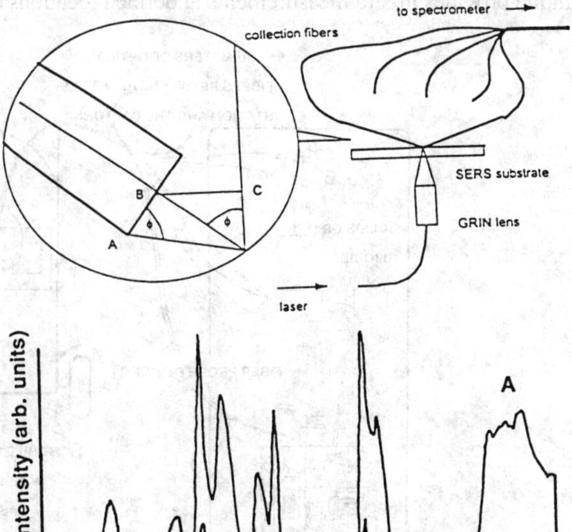

Fig.1. Spatially optimized fiber Raman probe

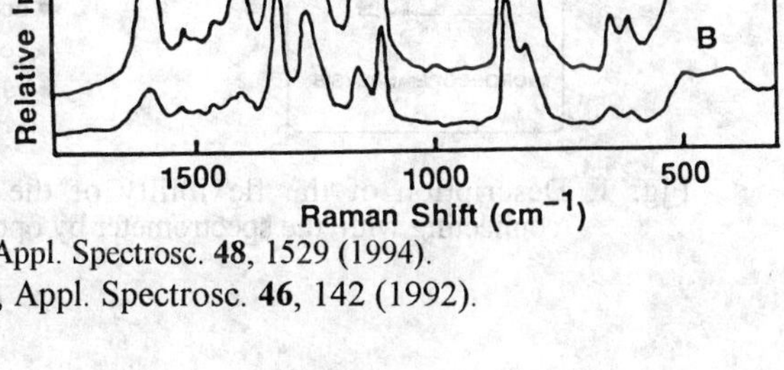

Fig.2. SER spectra of PNP (1.0×10^{-3} M) collected by the optimized probe. A: without filter; B: with filter.

1. J. Y. Ma and Y. S. Li, Appl. Spectrosc. **48**, 1529 (1994).
2. Y. S. Li and Y. Wang, Appl. Spectrosc. **46**, 142 (1992).

A MOBILE STATION USING A NEW RAMAN FIBER-PROBE TO INVESTIGATE IN-SITU AND SIMULTANEOULSLY SEVERAL INORGANIC COMPOUNDS IN THE POLLUTED GROUNDS

M. Fourmentin[1], A. Lorriaux[2], F. Adar[3] and F. Wallart[2]*

[1] Centre de Recherche sur l'Environnement Industriel de Dunkerque Université du Littoral, BP 5526 - 59379 Dunkerque Cedex 1, France
[2] LASIR - UPR CNRS A2631L, USTL, Bât. C5 - 59655 Villeneuve d'Ascq Cedex, France
[3] JY/SPEX/DILOR Raman Spectroscopy and Ellipsometry Systems, ISA Inc., 3880 Park Ave., Edison, New Jersey 08820, USA

The Northern section of France wich includes the "Nord and Pas de Calais" regions has nearly 100 km^2 of polluted sites and grounds due to old industries. The Raman fiber-probe analyzer is a promising instrument for detection of the major pollutants at these sites contaminated by chemical manufacturing. The technique provides in-situ measurements at defined locations on these sites.

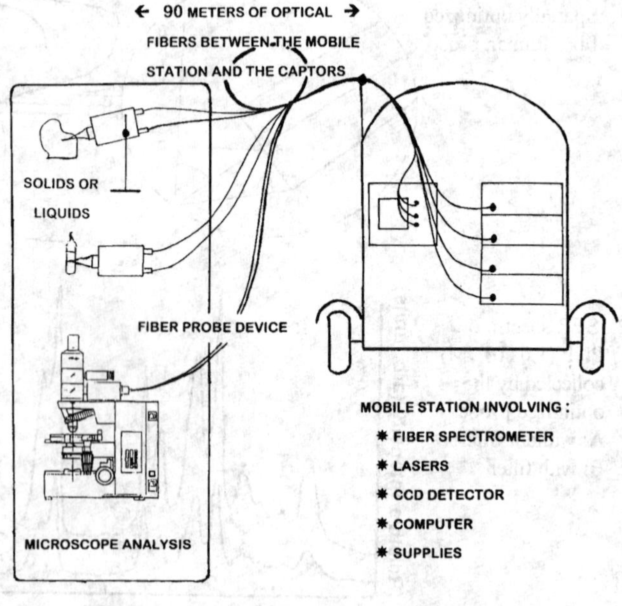

Fig. 1. Description of the flexibility of the sample stage and its connecting with the spectrometer by optical fibers

The mobile station (truck) is also equipped with a mass spectrometer providing the identification of those organic compounds which would be difficult to detect because of high fluorescence levels. The combination of these two techniques makes it possible to have a global view of the ground pollution.

The vibrational Raman spectra of reference samples are used to aid in the identification of the principal components, and to aid in estimating their concentrations. The first step is to make measurements in real time at known locations and depths. It is subsequently necessary to improve these results using more conventional spectrometers.

The first results obtained with this new equipment will be presented ; measurement points are up to 90 meters from the mobil station. Analyses can be made directly on solids, or liquids, or solutions in a vessel, or under a microscope as shown in Figure 1.

The modular nature of the contruction in the mobile station allows one the choice of laser excitation wavelength by selecting the appropriate fiber-probe head (632.8 nm of a HeNe, 514.5 nm of Ar, or 780 nm of a diode laser). The detector unit includes a stigmatic spectrometer with two interchangeable optical gratings (1800 or 600 grooves/mm) which allows the user to record both Raman and fluorescence spectra. The apparatus is assembled in a rack and is equipped with a CCD detected cooled to -65C. Four optical fibers are coupled to the entrance of the spectrograph. As a result of the stigmatic properties of the spectrograph and the two dimensional character of the CCD, this enables the simultaneous acquisition of information from four different samples. Computer control provides immediate spectral treatment and sample decomposition.

1. M.M. Carrabba, J.W. Haas III, J.M. Bello, K.M. Speccer and R.W. Forney, EIC Labs Inc, Advances in Fieldable Fiber Optic Raman Spectroscopy, FACSS XXI 1994, p. 233.

A DISPERSIVE RAMAN MICROPROBE WORKING IN THE NEAR INFRARED PART OF THE SPECTRUM

Fran Adar, Ph.D. * and Jacques Barbillat, Ph.D.+

Instruments, S.A., 3880 Park Ave., Edison, NJ 08820, USA
+LASIR CNRS, USTL Flandres Artois Bat. C5, 59655 Villeneuve d"Ascq, France

The developments of the Raman microprobe and the remote Superhead fiber probe has proven to greatly enhance the usefulness of Raman spectroscopy to materials analysis and manufacturing quality control. One of the major disadvantages of the implementation of Raman spectroscopy, however, is sample fluorescence. The development of the Fourier Transform Raman systems took advantage of fluorescence avoidance by exciting the Raman spectrum with the near IR YAG laser. However, the Raman microprobe coupled to a YAG laser and interferometer can not provide the same level of spatial resolution as a microprobe on a dispersive system.

Since the implementation of FTRaman systems, there has been interest in demonstrating the performance of a dispersive Raman microprobe with a multichannel detector in order to take advantage of the full potential for spatial resolution of the microprobe, as well as very low frequency spectra when used with a triple spectrograph. The first system built by Dilor using a YAG laser demonstrated spatial resolution better than 2μm. Using a standard CCD, the anti-Stokes Raman, and the Stokes Raman at shifts less than about 500 cm^{-1}, were detected. However, the CCD multichannel detector limited the system's performance because of its spectral range relative to that of the Raman spectrum excited by the YAG laser. The advantage of the system was to combine the avoidance of fluorescence afforded by the YAG laser with the spatial resolution of the microprobe and the wavelength multiplexing of the low noise CCD. However, it was severely limited in application because of the low sensitivity to anti-Stokes lines at frequency shifts above 1000 cm^{-1} and the total insensitivity to Stokes lines about 500 cm^{-1}.

Subsequent developments in the field of multichannel near IR detectors provided the ability to engineer a second dispersive system with a Ge multichannel detector (diode array type - 256 50 μm pixels). Because the germanium band-gap (0.67eV) is significantly lower in energy than silicon (1.11eV) on which the CCD is built, this detector can, in principle, detect Raman bands much farther from the laser line than a silicon detector. In reality the sensitivity is highest in the fingerprint region because of the decreasing quantum efficiency to detect photons close to the bandgap which is indirect. Consequently, the fingerprint region is most detectable. In this system, the 2 μm spatial resolution is maintained, as is the low frequency performance, if the system is built with a triple spectrograph. And the throughput advantage is operative because the tight spot at the microscope is imaged to a value less than the typical 100 μm entrance slit.

Raman spectra can also be acquired with the Dilor Superhead fiber probe using near IR optical fibers. In this case the throughput advantage is also operative with 100μm fibers.

Further evolution of near IR dispersive Raman instrumentation will depend on better multichannel detectors. Clearly InGaAs provides better physical properties: its bandgap can be selected by engineering the composition, and because the bandgap is direct, sensitivity will be better at the end of the range. Implementation of such devices depends only on the availability of reliable commercial devices. Results recorded with the most current InGaAs devices will be included.

The following figure is an example of a spectrum of rhodamine recorded on the 0.5m XY. Using a 60mw YAG laser, a 600 groove/mm grating, an entrance slit of 300 μm, and a 256 element Ge array, the spectrum was acquired in 3 regions, 20 seconds per region.

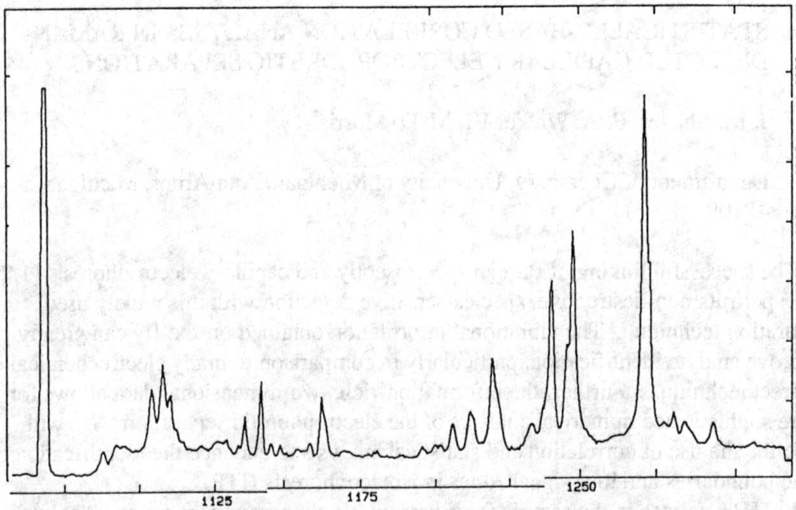

STATISTICALLY-BASED CORRELATION ANALYSES IN RAMAN-DETECTED CAPILLARY ELECTROPHORETIC SEPARATIONS

J. M. Shaver, P. A. Walker III, M. D. Morris*

Department of Chemistry, University of Michigan, Ann Arbor, Michigan, 48109

The successful linking of Raman spectroscopy and capillary electrophoresis [1,2] (CE) permits non-destructive, species-sensitive detection with this widely used separation technique. The vibrational information obtained on-the-fly can greatly improve analyte identification, particularly in comparison to many electrochemical or indirect techniques. Further, the information-rich, two-dimensional data allows for more sophisticated numerical analysis of the electrophoretic separation. We will describe the use of correlation and statistical analysis to enhance the identification of zone boundaries and low-signal zones in isotacophoresis (ITP).

In ITP, analytes in the sample form narrow, single-component zones. Analytes present in low concentration in the original sample may form temporally short zones which are more difficult to detect. The detection of these zones can be further complicated if the analyte is a poor Raman scatterer. Figure 1 shows several of the Raman spectra collected during the elution of a low concentration of a pesticide under ITP conditions.

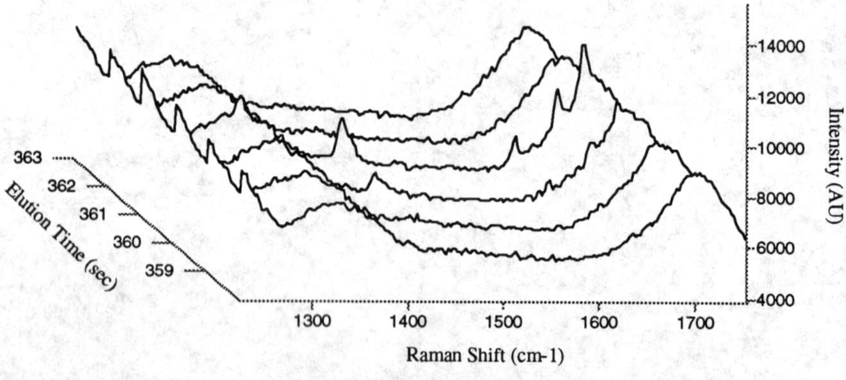

Figure 1. Raman Spectra collected on-line during the ITP analysis of a low-concentration pesticide. The peaks at 1437, 1616, 1661, and 1689 cm^{-1} are representative of the analyte.

Correlation techniques can be useful for analyzing such weak, transient signals. For example, a simple paired-covariance has been applied to CE/mass spectroscopy data.[3] Because of the nature of the line shapes, background signals and noise profiles encountered in Raman spectroscopy, there is a need for a more statistically-sound approach for the analysis of Raman spectra.

The use of correlation followed by statistical analysis will be demonstrated for CE/Raman data. This approach has been previously applied to liquid chromatography with UV absorbance detection by Castledine and co-workers.[4] In our application of this approach, the Raman signal in any individual time frame can be statistically compared to that of the background or of past frames. An example analysis of the data shown in Figure 1 is presented in Figure 2. The extension of this technique to the interpretation of multi-dimensional Raman data will be described as well as its specific use in the evaluation of ITP zone analysis.

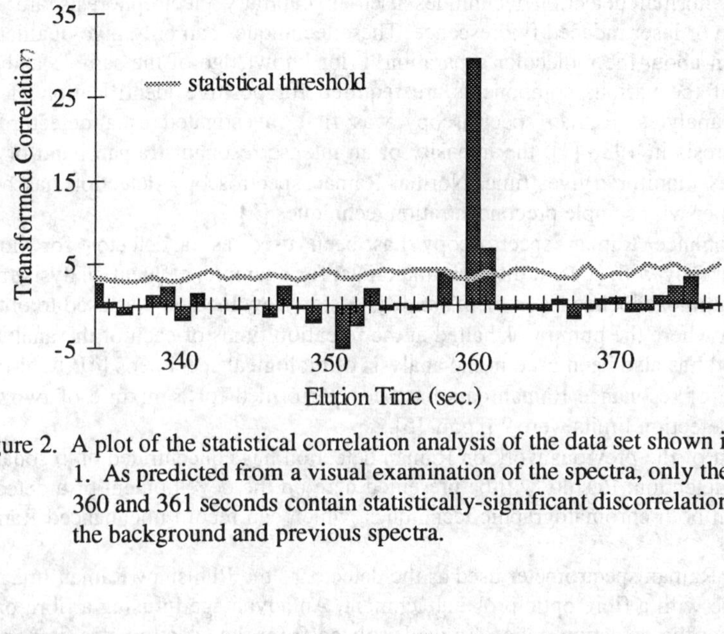

Figure 2. A plot of the statistical correlation analysis of the data set shown in Figure 1. As predicted from a visual examination of the spectra, only the data at 360 and 361 seconds contain statistically-significant discorrelation with the background and previous spectra.

1. Kowalchyk, W. K., Walker III, P. A., Morris, M. D., *Appl. Spectrosc.*, **1995**, 49, 1183-1188.
2. Walker III, P. A., Kowalchyk, W. K., Morris, M. D., *Anal. Chem.*, **1995**, *67*, 4255-4260.
3. Muddiman, D. C., Rockwood, A. L., Gao, W., Severs, J. C., Udseth, H. R., Smith, R. D., and Proctor, A., *Anal. Chem.*, **1995**, *67*, 4371-4375.
4. Castledine, J. B., Fell, A. F., Modin, R., and Sellberg, B., *J. Chromatog.*, **1992**, *592*, 27-36.

A UNIVERSAL RAMAN DETECTOR FOR CAPILLARY ELECTROPHORESIS AND MICROCHROMATOGRAPHY

S. D. Cooper [1], D. N. Batchelder [1*], M. M. Robson [2] and K. D. Bartle [2]

[1] Department of Physics and Astronomy,
[2] School of Chemistry,
University of Leeds, Leeds, LS2 9JT, UK

The normal detection techniques used in capillary electrophoresis are UV absorbance or laser induced fluorescence. These techniques can only give qualitative information about the molecular separation. Prior knowledge of the sample and the mobility of the various components are required for positive identification of the separated analytes. Raman spectroscopy was first investigated as a detector for electrophoresis in 1988 **[1]**; the intensity of an intense resonant Raman band of the sample was monitored over time. Normal Raman spectroscopy detection has been used together with sample preconcentration techniques **[2]**.

Resonance Raman spectroscopy has been used as a detector for liquid chromatography with several different materials; for example, effluent analysis from HPLC of aminoazobenzenes in methanol **[3]**. A stop flow method was used to obtain the spectra where the pump was halted at the retention times of each of the analytes; this method has also been used in the analysis of biological specimens **[4]**. Real time acquisition of resonance Raman data has been performed for a mixture of two azo dyes; the detection limits were 2 p.p.m. **[5]**.

Most of the previous work on Raman detection has concentrated on resonance enhanced detection. Results will be presented here on the development of a detector usable with both chromatographic techniques, which can record unenhanced Raman spectra.

The Raman spectrometer used as the detector is the Renishaw Raman Imaging Microscope with a fibre optic probe attachment. An advantage in using a fibre optic probe is that the detection optics are well isolated from the spectrometer (important for safety reasons because of the high voltages used in electrophoresis experiments) and can be easily attached to different chromatography apparatus.

Light of wavelength 633 nm from a HeNe laser is coupled into a optical fibre and focused onto the sample by a microscope objective lens. The scattered light from the sample is collected by the same lens, then is directed back to the spectrometer via another optical fibre which is mounted on the microscope stage of the Raman Imaging Microscope. A ×10 microscope objective lens was used to image the fibre into the spectrometer, where the Raman spectrum is analysed.

Figure 1 shows Raman spectra obtained from injecting a test sample of methanol into a HPLC column. The column used was silica capillary tubing with an internal diameter of 250 μm. The tubing was used without column packing and the mobile phase used was water. The sample volume injected was approximately 60 nL, and the pump flow rate was 1 μL/min. Raman spectra were recorded in the C-H vibrational region; the exposure time for each spectrum was 5 s. Figure 2 shows the change in the Raman scattering intensity over time. The intensity of the Raman bands of methanol increased to a maximum, then decreased again as the sample flowed past the detector.

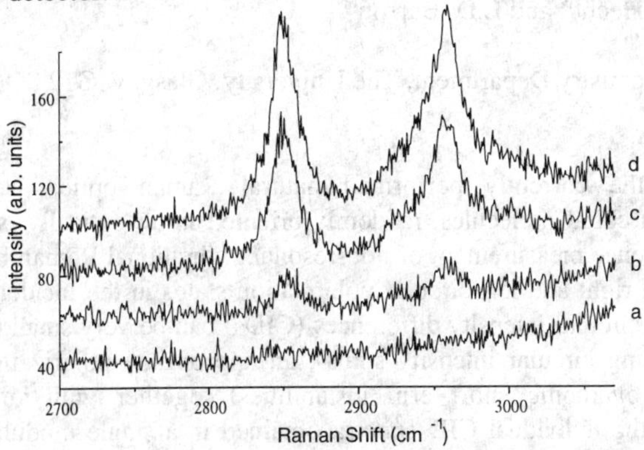

Fig. 1 Raman spectra of the HPLC capillary at time t = (a) 1100 s, (b) 1115 s, (c) 1136 s, (d) 1157 s. Methanol was injected at t = 0.

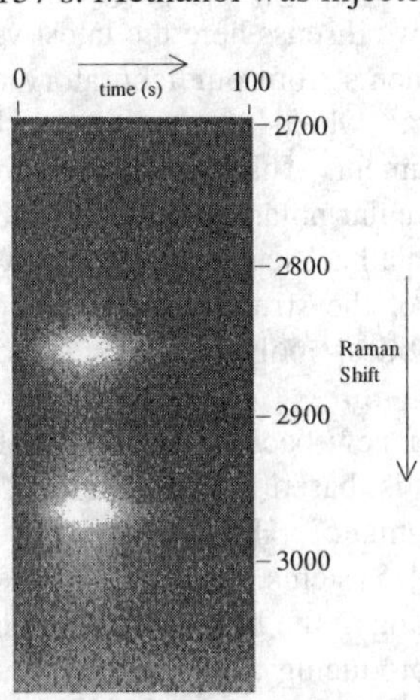

Fig. 2 A 2-dimensional map of Raman scattering intensity against time.

1. C.-Y. Chen and M. D. Morris, Appl. Spectrosc. 42, 515 (1988)
2. W. K. Kowalchyk, P. A. Walker and M. D. Morris, Appl. Spectrosc. 49, 1183 (1995)
3. S. Saito, N. Terame and S. Tanaka, Nippon Kagaku Kaishi 9, 1363 (1980)
4. K. Iriyama, Y. Ozaki, K. Hibi and T. Ikeda, J. Chromatogr. 254, 285 (1983)
5. C. K. Chong, C. K. Mann and T. J. Vickers, Appl. Spectrosc. 46, 249 (1992)

RECENT DEVELOPMENTS IN RAMAN OPTICAL ACTIVITY INSTRUMENTATION

L. Hecht* and L.D. Barron

Chemistry Department, The University, Glasgow G12 8QQ, UK

Most of the currently performed natural Raman optical activity (ROA) experiments on chiral molecules, randomly oriented in an optically isotropic phase, involve a difference measurement of non-resonant vibrational Raman band intensities associated with right and left circular polarization states in the incident radiation [1-3]. Since these circular intensity differences (CIDs) can be very small (10^{-6} to 10^{-3} of the corresponding circular intensity sums), an appropriate polarization modulation technique (to overcome short-term instabilities) together with an accumulation procedure for the individual CID spectra obtained in a single modulation cycle (to build up an acceptable signal-to-noise ratio) is required for their unequivocal detection.

As the result of our rich and long pioneering experience with different types of ROA instrumentation [4-11] we discuss here the latest version of the new generation of multichannel ROA instruments from our laboratory and present a representative selection of ROA spectra which demonstrate the excellent performance characteristics of our instrument. This novel instrument has been designed to accommodate the incident circular polarization (ICP) modulation approach for ROA measurements implemented in a backscattering geometry and has been optimized for detailed biochemical studies of the structure and dynamics of biopolymers such as polysaccharides, polynucleotides, polypeptides, proteins and glycoproteins in aqueous solution.

The optical layout of our new backscattering ICP ROA instrument is depicted schematically in Fig. 1. It is based on the actual spectrograph stage of the revolutionary imaging (stigmatic) Holospec F/1.8i on-axis single grating spectrometer (Kaiser Optical Systems, Inc.) the key element of which is a new holographic transmission grating in the beam of parallel light between the collimating and refocusing camera lens producing a 90^0 fold of the optical path. However, the original F/1.8 collimating lens has been replaced by a corresponding F/1.4 camera lens so that a F/1.4 acceptance cone is maintained throughout the entire spectrometer section of our new backscattering ICP ROA instrument (which probably represents the fastest Raman spectrograph on this planet). This spectrometer module has been further equipped with a holographic supernotch plus filter (Kaiser Optical Systems, Inc.) featuring an absorbance (optical density) in excess of 8 for a sufficient Rayleigh line rejection and with a cooled backthinned charge coupled device (CCD) detector (Wright Instruments AT1) with a quantum efficiency in excess of 80% for a virtually shot noise limited detection of Raman scattered photons. A calcite Lyot depolarizer (Leysop) and a subsequent F/1.0 collimating lens also need to be placed in the optical path of the backscattered Raman light since the 45^0 mirror would otherwise introduce large ROA polarization artefacts which are caused by nearly orthogonal linear contaminants in the two CP modulation half-cycles. The necessary ICP modulation in our instrument is achieved with a thermally stabilized potassium dideuteriumphosphate longitudinal electro-optic modulator (EOM, Leysop) eliminating any birefringence drifts caused by external temperature fluctuations.

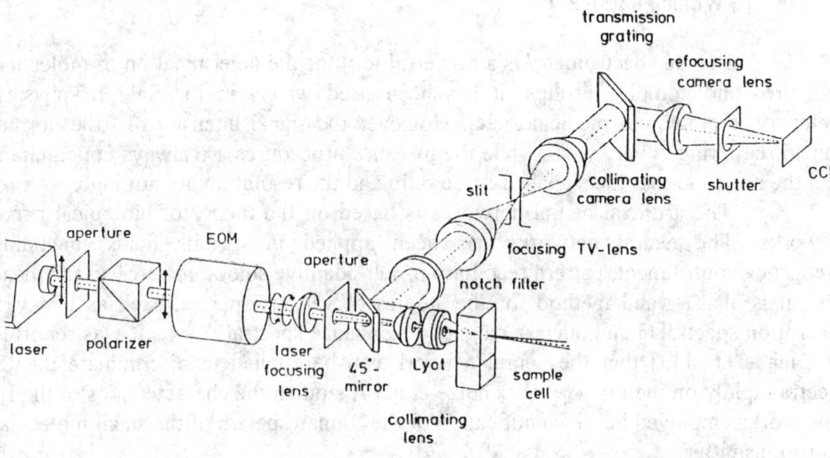

Fig. 1. Optical layout of the current backscattering ICP Glasgow University ROA spectrometer GUROAS3

1. L. Hecht and L.D. Barron, in Modern Techniques in Raman Spectroscopy, (J.J. Laserna, ed.), Wiley, Chichester, 1996, p. 265, in press.
2. L.D. Barron, L. Hecht, and A.F. Bell, in Circular Dichroism: Conformational Analysis of Biomolecules, (G.D. Fasman, ed.), Plenum, New York, 1996, p. 653, in press.
3. L.A. Nafie and D. Che, Adv. Chem. Phys. 85(3), 105 (1994).
4. L.D. Barron and A.D. Buckingham, Ann. Rev. Phys. Chem. 26, 381 (1975).
5. L. Hecht, Doctoral Thesis, Essen University (1987).
6. L. Hecht, D. Che, and L.A. Nafie, Appl. Spectrosc. 45, 18 (1991).
7. L. Hecht, L.D. Barron, A.R. Gargaro, Z.Q. Wen, and W. Hug, J. Raman Spectrosc. 23, 401 (1992).
8. A.F. Bell, L. Hecht, and L.D. Barron, J. Raman Spectrosc. 24, 633 (1993).
9. L. Hecht and L.D. Barron, J. Raman Spectrosc. 25, 443 (1994).
10. L. Hecht and L.D. Barron, Faraday Discuss. 99, 35 (1994).
11. L. Hecht and L.D. Barron, J. Mol. Struct. 347, 449 (1995).

DATA PROCESSING OF RAMAN SPECTRUM WITH ARTIFICIAL NEURAL NETWORKS

Jiming Hu*, Hailin Liu, Zhisan Xu, and Rongsheng Sheng

Department of Analysis-Measurement Science, Wuhan University, Wuhan, 430072, China

Raman spectrometry is a powerful tool for the determination of molecular structures and complex groups, it has been used widely in the field of physics, chemistry, biology and medicine, etc.. However the signal intensity of the normal Raman scattering is very weak while the measure procedures are always concomitant with the noise, so that the signal-to-noise ratio and the resolution are not high.

The artificial neural networks is based on the theory of biological nerve networks. The neural networks has been applied in various fields, including intellgency controlment, pattern recognition, self-adaptive smoothing etc.. In the mean time, it is also a valid method for the studies of various spectra, such as UV-VIS absorption spectra[1] and nuclear magnetic resonance spectra[2] etc.. It was reported by Tanabe et al.[3] that the neural network can be applied to discriminate the IR spectra rapidly on the PC. Georg Schulze et al.[4] studied the characteristics of the B-P networks employed in the identification of the Raman spectra of the small molecular neurotransmitter.

Considering the disadvantages of the normal Raman spectra as well as applications of artificial neural networks in the information processing and the pattern recognition, we employed the multi-layer B-P network of the self-adaptive smoothing to process the data of the Raman spectra.

The network consists of the input layer, output layer, and several hidden layers as well. The state of each layer will affect the state of the next layer. Once the output wanted can not met in the output layer, the system will turn into backpropagation. The error signals will be backpropagated through the same connection and then minimized by modifying the weights of the neurons in every layer.

The input signals and the goal output are both pressed on the networks. Adjusting the weight continuously through the self-adaptive algorithm so that input signal approaches the training signal in some optimum criterion. Since the error measuring function used is the mean error square, $E = E(\varepsilon_k^2)$, the minimum criterion used mostly is the gradient criterion. The mostly used self-adaptive algorithm in the data processing is Least Mass Square (LMS).

The output of the networks can be calculated to be

$$y_k = f(x_k^T w_k)$$

Where the $f(\ )$ is the nonlinear transfer function, x_k is the input signal vector, w_k is the weight vector.

The error function is defined as subtract between the goal output $\hat{y}_k$ and the output y_k, namely

$$\varepsilon_k = \hat{y}_k - y_k$$

Then the MSE will be

$$E = E(\varepsilon_k^2) = E\{[\hat{y}_k - f(x_k^T w_k)]^2\} = E[\hat{y}_k^2 - 2\hat{y}_k^2 f(x_k^T w_k) + f^2(x_k^T w_k)]$$

It is shown in the equation above that MSE is a quadratic function of weight vector, and it has a shape like a non-negative convex super-parabolic camber, so that the MSE could approach an unique least value by adjusting the weight in gradient criterion. The gradient criterion is shown as

$$w_{k+1} = w_k + \mu(-\nabla_k)$$

where w_k is the current value of the weight vector, w_{k+1} is the next value, μ is the learning time coefficient, ∇_k is the gradient on the MSE camber when $w = w_k$. Defining the gradient as

$$\nabla_k = \frac{\partial E(\varepsilon_k^2)}{\partial w_k}$$

Then the instantaneous gradient will be shown as below

$$\hat{\nabla}_k = \frac{\partial \varepsilon_k^2}{\partial w_k} = -2\varepsilon_k f'(x_k^T w_k)\frac{\partial(x_k^T w_k)}{\partial w_k} = -2\varepsilon_k x_k f'(x_k^T w_k)$$

Substituting the instantaneous gradient for the real gradient, then the gradient criterion will be adjusted as below.

$$w_{k+1} = w_k + \mu(-\nabla_k) = w_k + 2\mu\varepsilon_k x_k f'(x_k^T w_k)$$

During the study the spectrum data were pretreated and then supplied to the input layer of the neural networks in array for learning and training. By comparison with the original spectrum, it is found that the system operated optimumly if the value of μ is 0.5, either the signal-to-noise ratio or the discrimination were improved, so that the purpose of self-adaptive was realized.

From the discussion, the B-P neural network used in this study can improve the S/N and the resolution of Raman spectrum. It is just a try of the application of the artificial neural networks in Raman spectra, but it illustrated that the artificial neural networks could be used in Raman Spectra. A further research is been doing now.

Acknowledgment: Financial support for this work was provided by the National Natural Science Foundation of China and Science Foundation of Hubei Province.

1. C. W. Lin, J. C. LaManna, and Y. Takefuji, Biol. Cybern, 67, 303(1992)
2. L. S. Anker and P. C. Jurs, Anal. Chem., 64, 1157(1992)
3. K. Tanabe, T. Tamura, and H. Uesaka, Appl. Spectrosc., 46, 807(1992)
4. H.. Georg Schulze, Michael W. Blades, Alan V. Bree, Boris B. Gorzalka, L. Shane Greek, and Robin F. B. Turner, Appl. Spectrosc., 48, 50(1994)

ON-LINE NORMAL RAMAN SPECTROSCOPIC DETECTION OF ISOTACHOPHORETICALLY SEPARATED QUANTERNARY CATIONIC HERBICIDES

P. A. Walker III, J. M. Shaver, M. D. Morris*

Department of Chemistry, University of Michigan, Ann Arbor, MI 48109-1055, USA.

Normal Raman spectroscopy is used to monitor capillary isotachophoretic (CITP) separations of the quaternary cationic herbicides paraquat and diquat. Preconcentration by field amplified injections and isotachophoresis easily brings ion concentrations into the useful range of normal Raman spectroscopy, ca. 10^{-2} *M*. The Raman spectra of each analyte zone are taken with an unmodified Raman microprobe using 400-500 mW of green laser excitation and 1 sec. CCD integration.

In previous communications we have shown that field amplified injection into an electrophoresis capillary can be used to preconcentrate ions at sub-ppm concentrations to levels applicable to normal Raman spectroscopy[1]. We have also demonstrated the on-line normal Raman spectroscopy of adenosine phosphates isotachophoretically separated in biological buffers[2].

Capillary isotachophoresis is ideal for Raman spectroscopy. The micron spatial resolution of the Raman microprobe allows detection of the contiguous zones. Fluorescence problems common to normal Raman spectroscopy, which are largely due to matrix impurities, are minimized. Because the analytes are spatially segregated, fluorescence interference is insignificant even with 532 nm laser excitation.

The steady state concentration of the ions in CITP analyte zones are described by the following equation:

$$C_a = C_l\left(\frac{\mu_a}{\mu_a + |\mu_r|}\right)\left(\frac{\mu_l + |\mu_r|}{\mu_l}\right)\left(\frac{Z_l}{Z_a}\right)$$

where μ_l, μ_a, μ_r are the mobilities of the leading electrolyte, sample analyte and counter ion, respectively; C_l and C_a are the concentrations of the leading electrolyte and sample analyte; Z_l and Z_a are the leading and analyte ion charge. Analyte concentration increases can be 10^3x-10^4x, with a corresponding decrease in zone length. The zones do not undergo band broadening from longitudinal diffusion.

Electric field changes at the zone boundaries counteract the effect of diffusion so they remain sharp.

The interface between the capillary and the Raman microprobe is simple. Minor modifications are made to the microscope stage to hold a capillary in position and to provide high voltage insulation.

Paraquat and diquat are contact herbicides that damage plant tissue causing cell damage and reduced photosynthesis. In contact with soil, they bind tightly and are rendered inactive. Analysis is done of aqueous samples that contain the active form of the herbicides.

Figure 1 shows the Raman detected capillary isotachopherogram. The starting concentrations were 1 x 10^{-5} *M*. The Raman spectra of the water and sulfate counter ion have been subtracted. A fluorescent impurity found in the commercial diquat follows the diquat zone and does not interfere with the measurement. The herbicides are detectable down to at least 7 x 10^{-7} *M* (180 ppb paraquat, 241 ppb diquat).

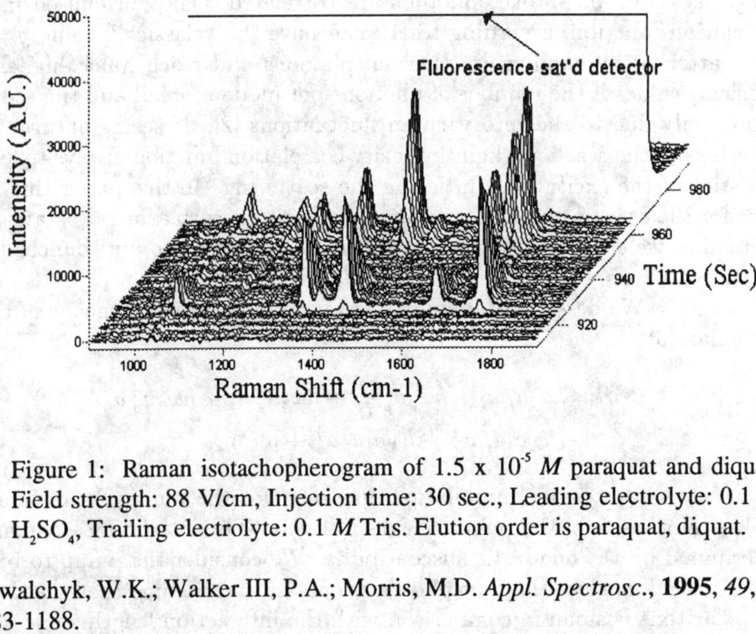

Figure 1: Raman isotachopherogram of 1.5 x 10^{-5} *M* paraquat and diquat. Field strength: 88 V/cm, Injection time: 30 sec., Leading electrolyte: 0.1 *M* H_2SO_4, Trailing electrolyte: 0.1 *M* Tris. Elution order is paraquat, diquat.

1. Kowalchyk, W.K.; Walker III, P.A.; Morris, M.D. *Appl. Spectrosc.*, **1995**, *49*, 1183-1188.
2. Walker III, P.A.; Kowalchyk, W.K.; Morris, M.D. *Anal. Chem.*, **1995**, *67*, 4255-4260.

PHOTON CORRELATION IN RAMAN SCATTERING BY POLARITONS

A. V. Belinsky and M. V. Chekhova*

Dept. Quantum Radiophysics, Moscow State University,
119899 Moscow, Russia

Traditionally, in the Raman spectroscopy of polaritons one measures the frequency-angular distribution of the scattered light intensity. The question is: does measurement of the fourth-order field moment (intensity correlation function) for the scattered light provide in this case any new information? In [1] it was shown that in the case of spontaneous Raman scattering the Stokes and anti-Stokes photons are correlated. The normalised intensity correlation function, according to [1], can have the "classical" value $g = 2$ for the scattering by equilibrium thermal phonons and reach much higher (non-classical) values if the number of phonons per mode is small and the scattering occurs only due to the zero vacuum fluctuations [2]. It seems natural to find out whether the scattered light intensity correlation function always reflects the statistics of the excitations initiating the scattering. In this paper this will be done for the case of Raman scattering by polaritons with arbitrary statistics. In particular, we consider the cases of coherent, thermal, and anti-bunched states of the polaritons.

We describe Raman scattering by polaritons by means of the Hamiltonian [3]

$$\begin{aligned} H =& \hbar\omega_l a_l^+ a_l + \hbar\omega_p a_p^+ a_p + \hbar\omega_s a_s^+ a_s + \hbar\omega_a a_a^+ a_a \\ & -(\hbar\chi a_l a_p^+ a_s^+ + \hbar\chi a_l a_p a_a^+ + \text{h.c.}), \end{aligned} \tag{1}$$

a_l, a_s, a_a and a_p being annihilation operators for the pump, Stokes and anti-Stokes photons, and the polariton, respectively. The interaction constant χ is determined by the quadratic susceptibility. We consider the pump to be given and classical, neglect the four-photon processes and assume that the damping for polaritons is small in comparison with the interaction length.

Introducing slowly varying amplitudes for the Stokes, anti-Stokes, and polariton waves (which of course implies the phase matching condition), we solve the Heisenberg equations expanding the solution in powers of the "interaction time" t.

The initial state of the system is chosen as

$$|\psi\rangle = |0\rangle_s |0\rangle_a | \quad \rangle_p , \tag{2}$$

which corresponds to the vacuum state for the scattered modes and an arbitrary state of polaritons. To find the intensity and the second intensity moment, we should average the corresponding operators over the state (2). Taking the solution of the Heisenberg equations to the accuracy of t^2, we calculate the normalized cross-correlation function of intensity for the Stokes and anti-Stokes components:

$$K_{as} \equiv \frac{\langle N_a N_s \rangle}{\langle N_a \rangle \langle N_s \rangle}, \tag{3}$$

where $N_\alpha \equiv a_\alpha^+ a_\alpha$.

We obtain:

$$K_{as} = \frac{4\langle :N_{p0}^2: \rangle + 8\langle N_{p0} \rangle + 1}{4\langle N_{p0} \rangle (\langle N_{p0} \rangle + 1)}, \tag{4}$$

where $N_{p0} \equiv a_{p0}^+ a_{p0}$ and the sign ":" means normal ordering.

For an arbitrary state of the polariton system, we introduce $K_{pp} \equiv \frac{\langle :N_{p0}^2: \rangle}{\langle N_{p0} \rangle^2}$, and

$$K_{as} = \frac{4K_{pp}\langle N_{p0} \rangle^2 + 8\langle N_{p0} \rangle + 1}{4\langle N_{p0} \rangle (\langle N_{p0} \rangle + 1)} \tag{5}$$

The dependence of K_{as} on N_{p0} is shown in Fig. 1 for the special cases of thermal and coherent polaritons. If $N_{p0} \ll 1$ then the scattered light has superclassical correlation, due to the existence of two subsequent Stokes and anti-Stokes processes. For $N_{p0} \gg 1$, the normalized correlation functions for the scattered light and for polaritons coincide.

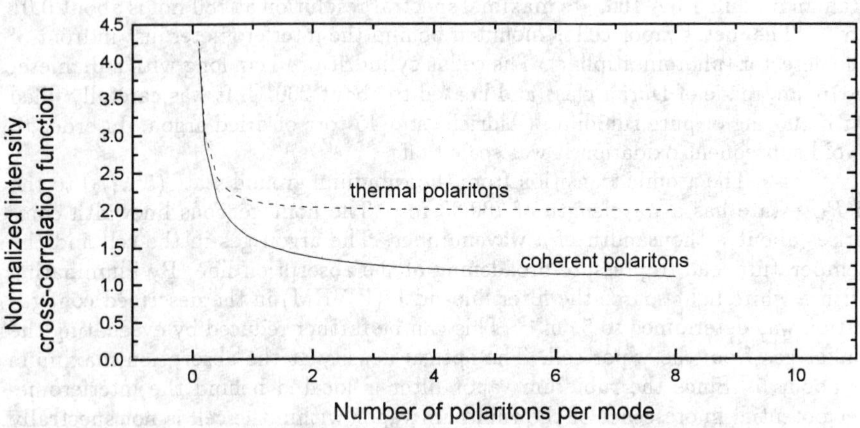

1. D.N.Klyshko, *Sov.J.Quantum Electronics*, **7**, 1977, pp.755.
2. D.N.Klyshko, Photons and Nonlinear Optics, Gordon and Breach Publ. Co., 1988.
3. J.Perina, Quantum statistics of linear and nonlinear optical phenomena, Moscow: Mir, 1987, (Russian transl.)

HIGH RESOLUTION/LOW WAVENUMBER FOURIER TRANSFORM RAMAN SPECTROSCOPY WITH A RUBIDIUM VAPOR FILTER AND A TI:SAPPHIRE LASER

A. Nohe[1], G. Lermann[1], H. Schwoerer[1], W. Kiefer[1]*,
J. Sawatzki[2], and G. Surawicz[2]
[1]Institut für Physikalische Chemie, Universität Würzburg,
Marcusstr. 9-11, D-97070 Würzburg, Germany
[2]Bruker Analytische Metechnik GmbH, 76189 Karlsruhe, Germany

The very strong and spectrally narrow absorption bands of metal vapors open the possibility to effectively suppress the Rayleigh line in Fourier transform Raman spectroscopy [1–4]. The wavelength of a narrow band and tunable excitation laser is set to the maximum of an atomic transition of the metal which features a large quantum yield. By adding the right amount of a non reactive gas as a buffer to the metal vapor cell and by changing the temperature of the vapor the bandwidth of the atomic absorption is adjusted to the laser line width. This filtering method allows to collect low wavenumber Raman spectra.

In our setup the excitation source is a linear titanium:sapphire laser (Coherent MIRA) which is additionally equipped with two intracavity étalons and a three–plate Lyot filter to reduce the linewidth to approximately 0.05 cm^{-1}. The Fourier transform spectrometer is a BRUKER IFS120 HR with an integrated Raman module FRA 106. Its maximal spectral resolution at 780 nm is about 0.01 cm^{-1}. The metal vapor cell is mounted behind the interferometer just in front of the detector (photomultiplier). The cell is cylindrical, 15 cm long with a diameter of 10 cm, made of Duran glass and heated to about 200°C. It was carefully filled with 100 mg of pure rubidium (Aldrich) and 45 torr of dried argon. In order to avoid subsequent oxidation it was sealed off.

The atomic transition from the rubidium ground state ($5^2S_{1/2}$) to the $5^2P_{3/2}$–state has a wavelength of 780.03 nm. The homogeneous linewidth comprises about a thousandth of a wavenumber. The argon gas in the cell and the temperature lead to pressure broadening of the absorption line. By illuminating with a white light source the filter linewidth (FWHM) in the described configuration was determined to 5 cm^{-1}. This can be further reduced by evacuating the buffer gas from the vapor cell. The optical density at the absorption maximum is about 5. Since the rubidium vapor filter is located behind the interferometer potential fluorescence of the rubidium vapor within the cell is not spectrally resolved but contributes to the general noise level. A cut–off filter suppresses emission below 720 nm.

Figure 1 shows as an example the Raman spectrum of crystalline sulfur. The sulfur sample was illuminated with only 5 mW CW laser power. The interferometer resolution was set to 0.5 cm^{-1} in this case.

The Rayleigh line appears strongly suppressed with a bandwidth of about 1 cm^{-1}. The lowest wavenumber Raman signal of crystalline sulfur, the phonon mode at 27 cm^{-1}, is clearly separated from the Rayleigh line on both Stokes and anti–Stokes side. The spectrum in the close vicinity of the Rayleigh line and the bandwidth of the rubidium filter let expect, that lines lying even closer can be resolved. This was verified in preliminary measurements at several scattering media. Systematic investigations, including those obtained with a potassium vapor filter, will be reported during the conference.

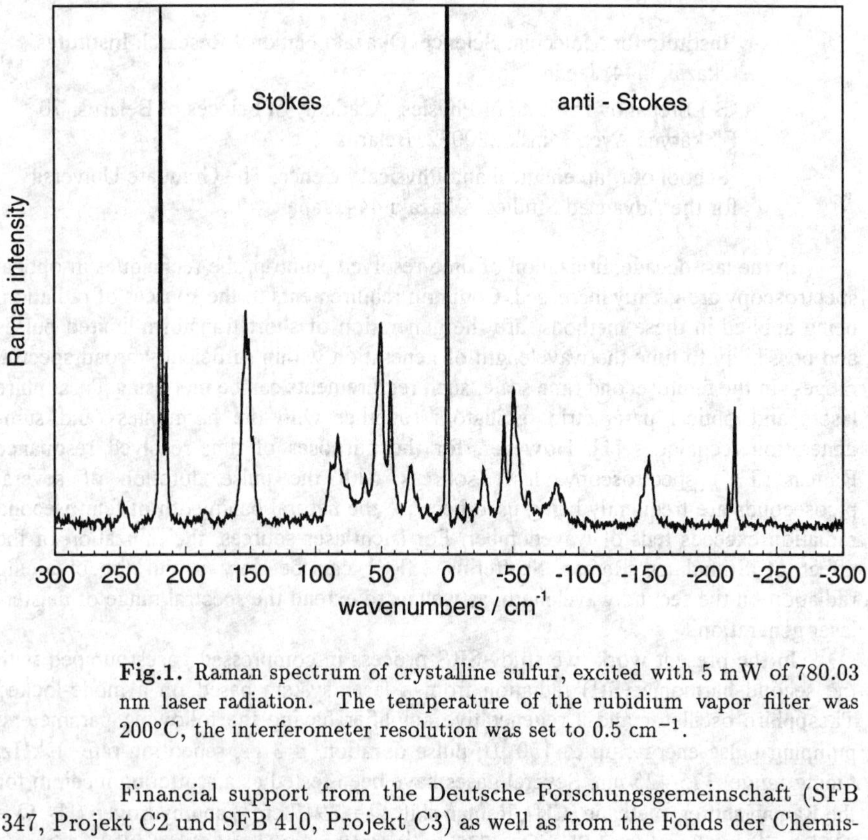

Fig.1. Raman spectrum of crystalline sulfur, excited with 5 mW of 780.03 nm laser radiation. The temperature of the rubidium vapor filter was 200°C, the interferometer resolution was set to 0.5 cm^{-1}.

Financial support from the Deutsche Forschungsgemeinschaft (SFB 347, Projekt C2 und SFB 410, Projekt C3) as well as from the Fonds der Chemischen Industrie is highly acknowledged.

1. F. Rasetti, Nuovo Cimento 7, 261 (1930).
2. M. J. Pelletier, Applied Spectroscopy 46, 395 (1992).
3. M. J. Pelletier, Applied Spectroscopy 47, 69 (1993).
4. P. J. Horoyski, M. L. W. Thewalt, Applied Spectroscopy 48, 843 (1994).

SRS-BASED SOURCE OF 3 ps, 1 kHz PULSES IN NEAR-UV-VISIBLE SPECTRAL RANGE FOR TR^3 APPLICATIONS

V.A. Orlovich[1,2]*, Y. Uesugi[3], Y. Mizutani[1], S.G. Kruglik[1,2], A.G. Shvedko[2], and T. Kitagawa[1]

[1]Institute for Molecular Science, Okazaki National Research Institutes, Okazaki 444, Japan

[2]B.I.Stepanov Institute of Physics, Academy of Scieces of Belarus, 70 F.Skaryna Ave., Minsk 220072, Belarus

[3]School of Mathematical and Physical Science, The Graduate University for the Advanced Studies, Okazaki 444, Japan

In the last decade, utilization of time-resolved pump-probe techniques in optical spectroscopy are greatly increased. Common requirements to the sources of radiation, being applied in these methods, are the generation of short transform-limited pulses and possibility to tune the wavelength of generation within sufficiently broad spectral ranges. In the femtosecond time scale, such requirements can be met using Ti:sapphire lasers and optical parametric oscillators together with the harmonics- and sum-generation techniques [1]. However, for the purposes of time-resolved resonance Raman (TR^3) spectroscopy, laser sources with the pulse duration of several picoseconds are frequently being used because the natural bandwidth of femtosecond radiation exceeds tens of wavenumber. For such laser sources, the utilization of the effect of Stimulated Raman Scattering (SRS) can be very useful for obtaining radiation on the second wavelength, as well as to extend the spectral range of master-laser generation.

In the present work, we study SRS process in compressed gases pumped with the second harmonic (SH) radiation from a laser system based on a mode-locked Ti:sapphire oscillator and a regenerative amplifier having the following parameters: pumping pulse energy, up to 150 μJ; pulse duration, ≤ 3 ps; repetition rate, 1 kHz; tuning range, 375-425 nm. Several gases have been tested as a scattering medium for the Raman shifter: methane (CH_4, Raman shift $\Omega = 2917\ cm^{-1}$) and hydrogen (H_2, $\Omega = 4155\ cm^{-1}$) at 50 atm, and SF_6 ($\Omega = 775\ cm^{-1}$) at 18 atm. Threshold of SRS excitation was ≈ 10 μJ in CH_4 and varied, depending on focusing conditions and pumping wavelength, in the range 40-70 μJ in H_2; while in SF_6 stable SRS-generation was not obtained. At high pumping energy of ≈ 120-150 μJ, several higher-order Stokes and anti-Stokes scattering components have been efficiently excited, up to 6 components in CH_4 and up to 3 in H_2.

Fig.1 gives wavelength ranges of laser radiation available using SH of master laser and several most stable and efficient SRS-components. Fig.2 shows typical energy dependencies for the 1st Stokes components of SRS in CH_4 and H_2; quite high conversion efficiencies of up to 20-25 % can be easily achieved. In Fig.3, the optimization curve for energy of the 2nd Stokes component in CH_4 on gas pressure is presented. Optimization has also been made on the Raman cell length and geometry of focusing. Statistical properties of energy of the 1st and 2nd Stokes pulses in CH_4 and 1st Stokes pulses in H_2 have been investigated using specially constructed multichannel amplitude analyzer based on an original software. Generated 1st Stokes pulses were characterized in terms of temporal and spectral widths.

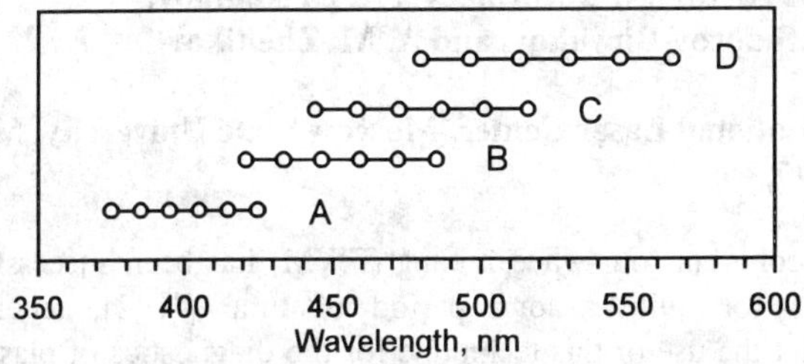

Fig. 1. Wavelength chart for the second harmonic radiation of master laser (A) and corresponding 1st Stokes components in compressed CH_4 (B), H_2 (C), and 2nd Stokes component in CH_4 (D).

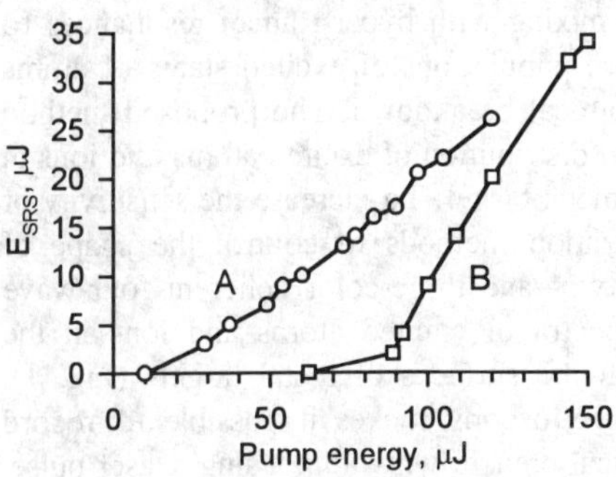

Fig. 2

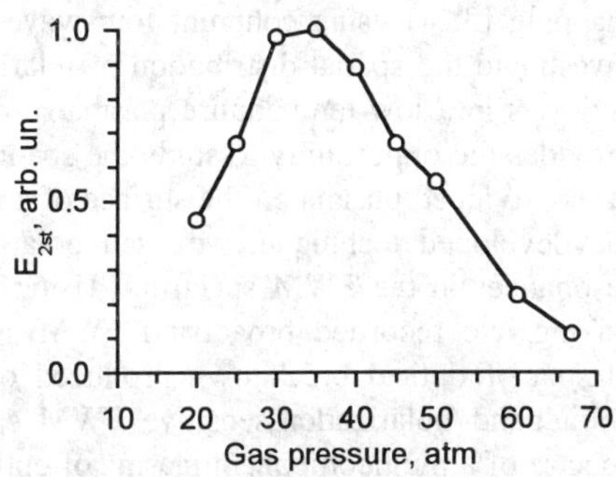

Fig. 3

Fig. 2. Energy dependence for the1st Stokes component in CH_4 (A) and H_2 (B) on the energy of pump pulses. Experimental conditions: Raman cell length, $l = 1$ m; pressure, $p = 50$ atm; focusing lens, $f = 83$ cm.

Fig. 3. Dependence of energy of the 2nd Stokes component in CH_4 on gas pressure. Experimental conditions: $l = 50$ cm; $f = 50$ cm, $E_{pump} = 120$ μJ.

On the basis of the results obtained, quite simple and reliable SRS-based source of picosecond radiation tunable in the spectral range 375-565 nm has been constructed. The possibility of its application in TR^3 spectroscopy has been successfully tested in study of primary photoprocesses in the solutions of Ni-porphyrins.

1. R. Laenen, K. Wolfrum, A. Seilmeier, A. Laubereau, JOSA 10, 2152 (1993).

TWO-DIMENSIONAL MAPPING OF SPATIAL DISTRIBUTION OF EXCITED ATOMS AND IONS IN LASER-PRODUCED PLASMAS USING THREE-COLOR FOUR-WAVE MIXING

A. B. Fedotov, N. I. Koroteev*, A. N. Naumov, D. A. Sidorov-Biryukov, and A. M. Zheltikov

International Laser Center, Moscow State University, Moscow, 119899, Russia

Although coherent four-wave mixing (FWM) has been successfully applied to different objects for quite a long period of time [1, 2], recent experiments demonstrated that the use of this technique for the diagnostics of plasma and excited gases requires certain modifications in the methodology and specific investigations in the theory of coherent FWM. Specifically, recent experiments have revealed a high efficiency of nonlinear optical interactions of a strong laser radiation with excited gases and plasma.

In this study, we apply the experimental technique of two-dimensional mapping [3, 4] using coherent four-wave mixing with hyper-Raman resonances to investigate the spatial distribution of relative populations of excited states of atoms and ions in a low-temperature plasma of optical breakdown. The proposed method provides the opportunity to study the spatial distribution of excited atoms and ions in laser-produced plasma on the surface of a metal target. To increase the sensitivity of the developed technique, we used polarization methods to control the shape of resonances in the FWM spectrum. Using two- and three-color coherent four-wave mixing, we recorded broadband FWM spectra of excited atoms and ions in the plasma of optical breakdown produced on the surfaces of metal targets (Fig. 1). Broadband polarization-sensitive FWM spectroscopy makes it possible to record spectra of a multicomponent plasma of optical breakdown within a single laser pulse, which provides an opportunity to eliminate the influence of inevitable fluctuations of plasma parameters from pulses to pulse on the accuracy of spectroscopic measurements and allows one to extract information concerning simultaneous spatial distribution of different plasma components using a single pulse of a laser spectrometer. A polarization technique in FWM spectroscopy of plasma permits an active control of the shape of the FWM spectrum, which makes it possible to suppress or enhance certain components in FWM spectra (see Fig. 1). Because of these properties, the proposed experimental technique provides an opportunity to investigate spatial distributions of the intensities of interfering spectral lines independently of each other.

Each FWM spectrum contains information on simultaneous distributions of several plasma components. Figure 2 shows the 2D mapping of the relative intensity of the FWM signal from (a) Pb atoms (the wavelength of the corresponding spectral line is 0.3683 μm) and (b) N atoms (at the wavelength of 0.3650 μm) in laser-produced plasma on a lead target. The radial distance in the figure is the horizontal displacement of the interaction region from the plasma center and the vertical distance is the distance to the target surface.

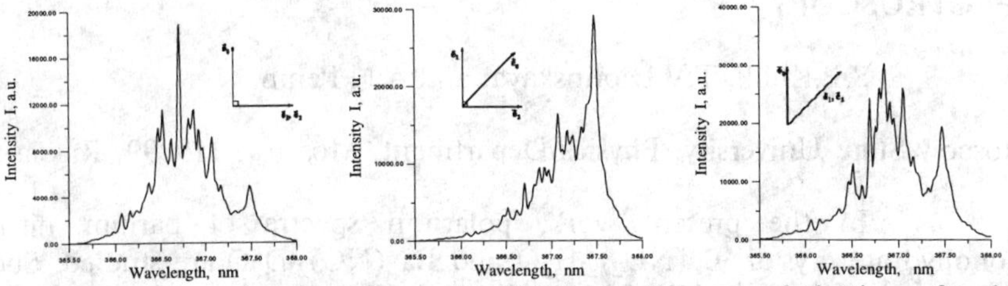

Fig. 1. Typical spectra of broadband FWM for different polarizations of probe waves

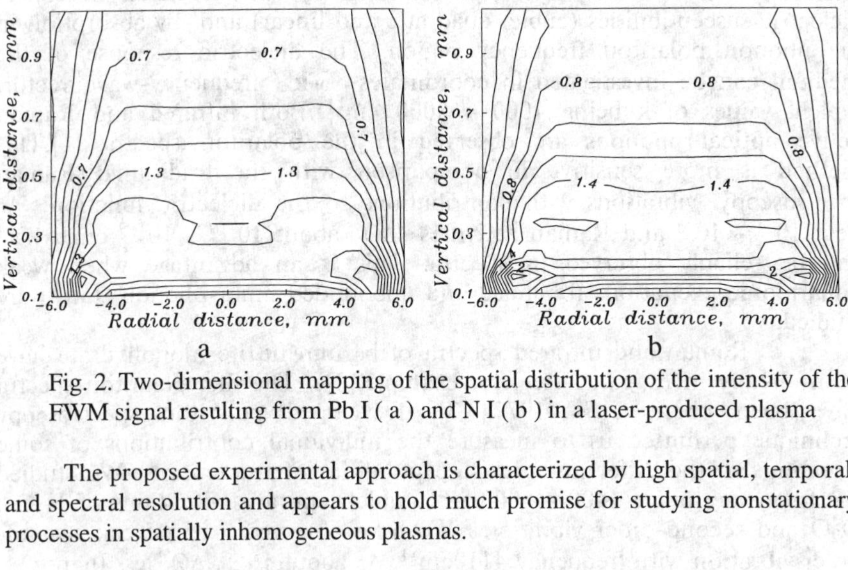

Fig. 2. Two-dimensional mapping of the spatial distribution of the intensity of the FWM signal resulting from Pb I (a) and N I (b) in a laser-produced plasma

The proposed experimental approach is characterized by high spatial, temporal, and spectral resolution and appears to hold much promise for studying nonstationary processes in spatially inhomogeneous plasmas.

1. S.A. Akhmanov and N.I. Koroteev, Methods of Nonlinear Optics in Spectroscopy of Light Scattering, Nauka, Moscow, 1981.
2. S.A.J. Druet and J.-P.E. Taran, Progress Quantum Electron., 7, 1 (1981).
3. M. Alden, S. Wallin, and W. Wendt, Appl. Phys. B, 33, 205 (1984).
4. K. Nyholm, R. Fritzon, and M. Alden, Opt. Lett., 18, 1672 (1993).

STUDY OF THE VIBRATIONAL SPECTRA OF BARIUM NITRITE MONOHYDRATE CRYSTAL BY MEANS OF POLARITON SPECTROSCOPY

S.P.Kulik, T.V.Laptinskaya, and A.N.Penin

Moscow State University, Physics Department, Moscow, 119899, Russia

In the present work, polariton spectra of barium nitrite monohydrate crystal, $Ba(NO_2)_2 \cdot H_2O$ and Ba $(NO)_2 \cdot D_2O$ are studied. Such spectra are the result of near-forward Raman scattering of light (spontaneous parametric scattering by polaritons).The frequency-angular distribution of scattered light is determined by the dispersion of crystal dielectric susceptibilities (cubic, quadratic and linear) and by absorption in the phonon polariton frequency region. The dielectric response of the medium can be investigated in coordinates ω-**k** (frequency-wave vector, typical values of k being 1000 - 8000 cm^{-1}).Both infrared and Raman active optical phonons are observed in the polariton spectra [1].This method is more sensitive in comparison with the traditional Raman spectroscopy: vibrations with contributions to the dielectric function $\Delta\varepsilon$ near 10^{-4} - 10^{-6} and Raman activities $\Delta\Theta$ about 10^{-10} - 10^{-15} $cm \cdot dyne^{-1}$ can be reliably observed in spectra. This is an advantage when weak anharmonic excitations (bounded and unbounded multi-phonon states) are studied.

Raman and infrared spectra of barium nitrite monohydrate have been already investigated (see, for example,[2]). The first polariton spectra have been obtained in 1977 [3]. Developed polariton spectroscopy technique permitted us to measure the individual contributions of some vibrations to the different-order susceptibilities of the crystals. We studied the frequency range 1400 - 5000cm^{-1}, which contains vibrations of H_2O , D_2O and second-order vibrations. We have observed a new weak second-order vibration with frequency 4110cm^{-1}, $\Delta\varepsilon$ about 10^{-5} , $\Delta\Theta$ less than 10^{-15}. The very weak IR-active vibration at frequency 3956 cm^{-1} [2] is reliably observed in our spectra ($\Delta\varepsilon$ is near $5 \cdot 10^{-5}$, the frequency width is 50 cm^{-1}).The interval 2450 - 2580 cm^{-1} in the spectrum of $Ba(NO_2)_2 \cdot H_2O$ looks like containing a single wide vibration, but there are three wide vibrations with overlapping contours, because the same frequency interval in the spectrum of the deuterated crystal includes two strong modes of D_2O (2593 and 2480 cm^{-1}) and weak overtone of D_2O (2385 cm^{-1}).

The dielectric functions calculated from the polariton spertra are shown in Fig.1.

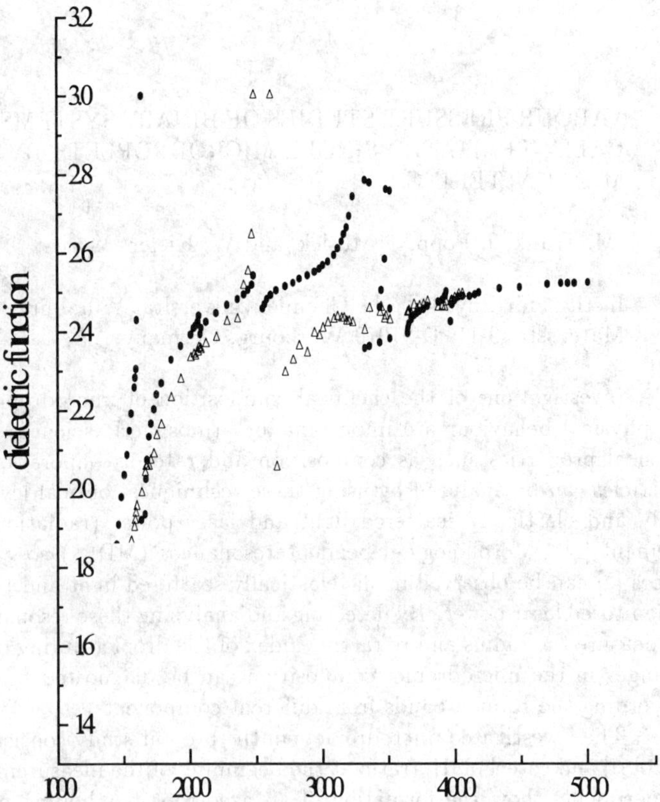

Fig.1. Dielectric function of $Ba(NO_2)_2 \cdot H_2O$ (•) and $Ba(NO_2)_2 \cdot D_2O$ (Δ) in 1400 - 5000 cm^{-1}.

1. D.N.Klyshko, Photons and Nonlinear Optics. Gordon and Breach, Amsterdam (1989).
2. O.P.Lamba and H.D.Bist., J.Phys.Chem.Solids, p.445 (1983).
3. A.A.Abdullaev, A.V.Vasilyeva, G.F.Dobrzhanskiy, Yu.N.Polivanov, Soviet Quantum Electronics, v.4, n.1, p.108 (1977).

VAPOUR PRESSURE STUDIES OF BINARY SYSTEMS IN OPTICALLY LEVITATED SINGLE MICRODROPLETS BY RAMAN-MIE SCATTERING

M. Trunk, J. Popp, J. Musick, and W. Kiefer*

Institut für Physikalische Chemie, Universität Würzburg,
Marcusstr. 9-11, D-97070 Würzburg, Germany

Investigations of the chemical composition of microdroplets as well as their physical behaviour are important for atmospheric sciences. Chemical and physical properties such as composition and rate of evaporation of single microparticles can be analyzed by using these techniques: optical levitation, inelastically and elastically scattered light and laser power (radiation pressure) measurement [1]. Morphology dependent resonances (MDRs) so-called input resonances [2] can be observed in the elastically scattered light and the data set of the monitored laser power. By detecting and analyzing these resonances we are able to measure the radius and refractive index of the droplet during evaporation [3]. Changes in the microdroplet composition can be diagnosted by simultaneously recording the Raman bands from different components.

The investigated microdroplets in the present study consist of diethylene glycol [I] and catechol [II] (ratio at the beginning of the measurement 50:50). Bulk experiments show the formation of an azeotrope. A boiling point maximum can be determined for a ratio of 46[I]:54[II], b.p.=259.5°C [4]. Here the evaporation behaviour of a single microdroplet (environment-/microdroplet temperature about 25°-30°C) is investigated. In panel A of Fig.1 the integrated peak intensity of the CH-stretching vibration of diethylene glycol [I] is displayed as a function of time. The diminution of the signal can be explained by evaporation of the droplets with time. The same behaviour can be observed for the CH-stretching vibration of catechol [II] (panel B). This is confirmed with the data of the monitored laser power (panel D), which also decreases with time, because of the decrease of the weight of the microdroplet. In panel C of Fig.1 the ratio of the Raman peak integral [II] to [I] is displayed. A decrease is observed, because catechol [II] evaporates faster than diethylene glycol [I] in the droplet. In panel E the elastically scattered light, superimposed by MDRs as function of time is shown. By analyzing these MDRs using the laser power (see insert in panel D) as well as the elastically scattered light data, the radius and temperature (refractive index) of the microdroplet can be described as function of time. In combination with Raman spectroscopy (see panel C) it is indicated that this mixture does not act as an azeotrope for temperatures much lower than the boiling point.

We have developed computer programs to simulate the changes of the

MDRs due to the radius and refractive index of multi component systems. Results of these simulations will be reported and discussed.

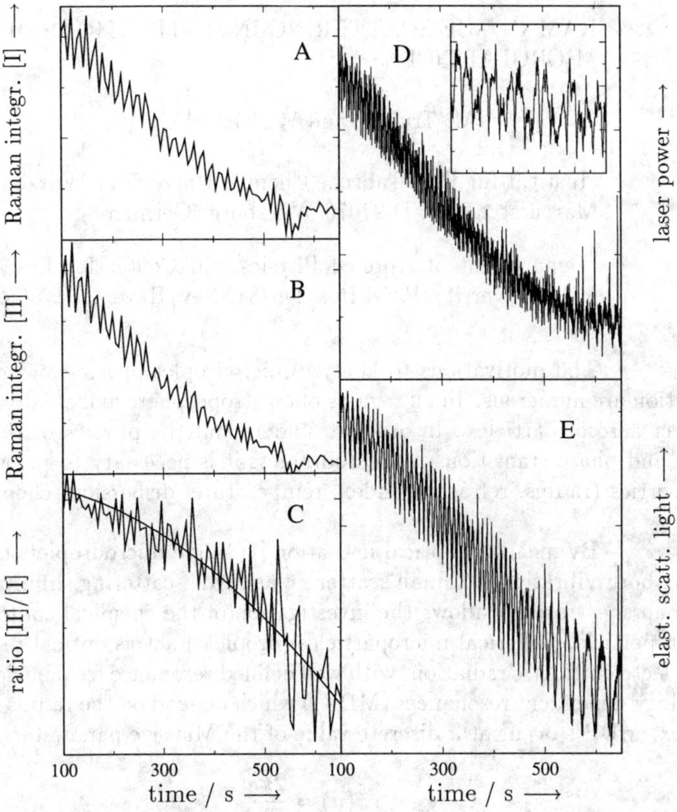

Fig.1. A: Integrated Raman band intensity of the CH–stretching vibration of diethylene glycol [I] (2840–2976 cm^{-1}). B: Same but for the CH–stretching vibration of catechol [II] (3045–3091 cm^{-1}). C: Ratio [II] to [I]. D: Laser power. E: Elastically scattered light intensity. All as a function of time (100–650 s).

Financial support from the Deutsche Forschungsgemeinschaft and by the Fonds der Chemischen Industrie is highly acknowledged.

1. M. Trunk, J. Popp, I. Hartmann, M. Lankers, E. Urlaub, and W. Kiefer, Fresenius J. Anal. Chem. 354 (1996), in press.
2. W. Kiefer, Croat. Chim. Acta 61, 473 (1988).
3. K. Schaschek, J. Popp, and W. Kiefer, J. Raman Spectrosc. 24, 69 (1993).
4. CRC Handbook of Chemistry and Physics, (R.C. Weast, ed.), CRC Press, Boca Raton, Florida, 1989.

RAMAN/MIE SCATTERING INVESTIGATIONS OF SINGLE MICROPARTICLES

J. Popp[1,2], M. Trunk[1], and W. Kiefer[1*]

[1]Institut für Physikalische Chemie, Universität Würzburg, Marcusstr. 9-11, D-97070 Würzburg, Germany

[2]Department of Applied Physics and Center for Laser Diagnostics, Yale University, P. O. Box 208284, New Haven, CT 06520–8284, USA

The motivations to analyze microdroplet physics and chemical composition are numerous. In all regions of our troposphere and stratosphere one can detect aerosol particles. In order to understand the physics of the evaporation rate and phase transitions in microdroplets it is necessary to investigate several properties (radius, refractive index, temperature, dispersion, chemical composition).

By means of optical levitation [1] single microdroplets can be studied using both vibrational Raman scattering and Mie scattering simultaneously [2,3]. Raman spectroscopy allows the investigation of the chemical composition of the microparticle. Spherical microparticles (droplets) act as optical microresonators for electromagnetic radiation, with well defined resonance frequencies, called morphology dependent resonances (MDRs) which depend on the radius and refractive index. MDR's occur at a discrete value of the Mie size parameter

$$x(n) = \frac{2\pi a}{\lambda}$$

where a is the radius, λ the wavelength and n the refractive index as function of the wavelength: $n(\lambda) = m(\lambda) + i\alpha$. The application of the Lorenz–Mie theory to the MDRs in either the Raman spectra or in the elastically scattered light permits the evalution of the particle size and of the refractive index of the microsphere as a function of time.

The purpose of this work is to present a technique which allows the general investigation of single microparticles. Dependent on the droplet material properties, Raman spectroscopy, fluorescence, or chemiluminescence can be used for analysis of the chemical composition of the microparticles. Results obtained from homogenoues droplets or droplet mixtures will be discussed.

The droplet parameter, like radius and refractive index, as well as the dependence of the refractive index on light wavelength and temperature can be determined by the evaluation of output as well as input MDRs. If only input MDRs should be used for these investigations, it is necessary to investigate the Mie–scattering of two or three different laser wavelengths. An efficient algorithm which compares the spectral locations of the MDR's in the elastic and Raman (or fluorescence, chemiluminescence, etc) spectra with the precalculated resonance position of MDR's was used. The MDR position, stored in the RAM of the computer for comparison with the observed spectra, were calculated considering both the change in refractive index as well as the dependence of the MDR's on wavelength dispersion.

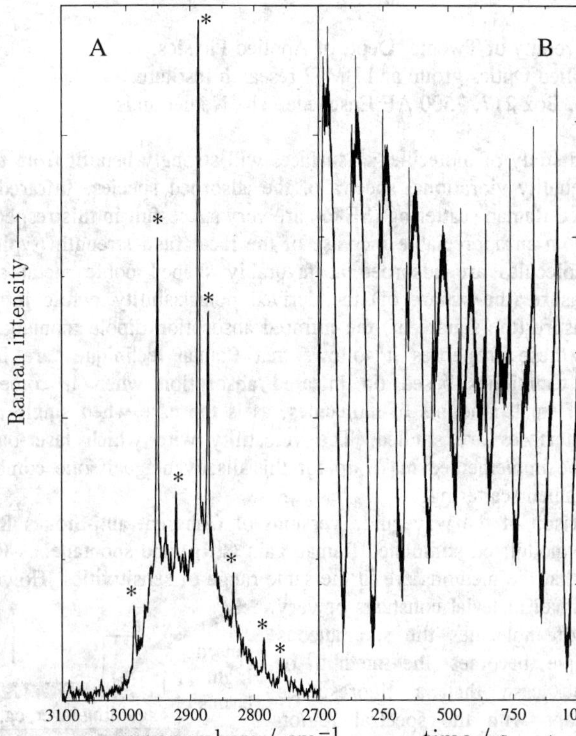

Fig.1. A: Raman spectrum of the CH–stretching vibration (2900 cm^{-1}) of triethylene glycol superimposed by MDRs (labelled by an asterisk; integration time 2 s, spectrum taken 350 s after the start of the measurement). B: Elastically scattered light intensity of the same levitated microdroplet as function of time.

Financial support from the Deutsche Forschungsgemeinschaft and the Fonds der Chemischen Industrie is highly acknowledged.

1. A. Ashkin, Phys. Rev. Lett. 24, 156 (1970).
2. W. Kiefer, Croat. Chim. Acta 61, 473 (1988).
3. J. Popp, PHD–Thesis, University of Würzburg, (1994).

BIOLOGICAL MONOLAYERS WITH WAVEGUIDE RAMAN TECHNIQUES

C. Otto*, T.W. Tukker, J.S. Kanger and J. Greve

University of Twente, Dept. of Applied Physics,
Applied Optics group and BMTI research institute,
P.O. Box 217, 7500 AE Enschede, The Netherlands

A detailled study of molecules at surfaces will strongly benefit from the possibility to acquire good quality vibrational spectra of the adsorbed species. Infrared absorption and surface enhanced Raman scattering (SERS) are very succesfull in this respect. SERS obtains it sensitivity from an appreciable increase of the local field strength (typically a factor of 100), when molecules are adsorbed at favorably shaped noble metal surfaces. Raman techniques measure (the square of) the derived polarizability, while infrared absorption techniques measure (the square of) the infrared absorption dipole moment. Comparing the magnitudes of these properties it follows that Raman techniques are in principle less sensitive than techniques based on infrared absorption when it comes down to the measurement of small amounts of molecules, as is the case when single molecular layers adsorbed at interfaces are studied. The versatility with which laser-based techniques, however, can be implemented make up for this disadvantage if one combines waveguide systems with Raman scattering.

A comparison of the waveguide versions of Coherent anti-Stokes Raman scattering (CARS), phase modulated stimulated Raman gain (stRg) and spontaneous Raman scattering (spRs) reveals that the methods are in the same range of sensitivities. However in the limit of either very small material constants or very low numbers of molecules the spontaneous Raman technique becomes the method of choice. In the case that a fluorescence emission overlaps with the spectral region where the Raman information is present, the spontaneous Raman scattering must be performed using short laser (few picoseconds) pulses and electronics in the detection system with the same temporal resolution. In this situation CARS or stRg using picosecond pulses will intrinsically lead to a suppression of the fluorescent background. The advantages of Phase modulated stRg above CARS are that the absence of a phase-matching condition enables collinear propagation of the pulses, which ensures a large interaction length and only the imaginary part of the derived polarizability contributes to the measured signal and not both the real and imaginary parts, as is the case in CARS.

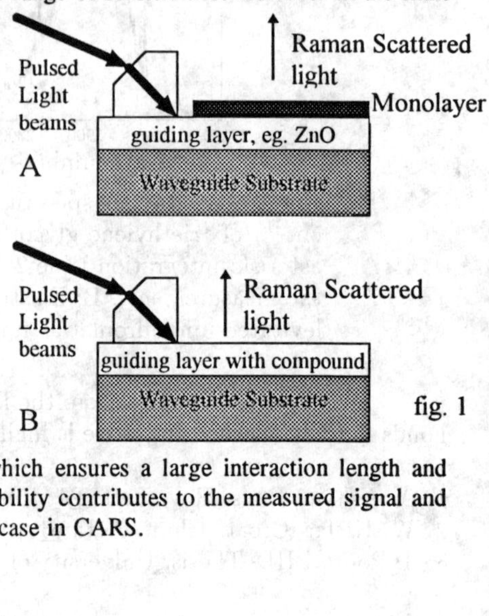

fig. 1

"Nature" provides "men" with examples of how molecular systems perform various tasks within an organism. Over the last 40 years detailled insight has been obtained in how some natural devices work and the molecular systems that can now be investigated succesfully with the various spectroscopic methods range from isolated and complexed biomacromolecules to whole organisms like bacteria and cells. It becomes feasible to address the question: Can we redesign molecular systems so that their performance is copied or improved in an environment that serves the purposes of new technologies. It must then be understood that these new devices do not suffer from the constraint of self-reproducibility that is inherent in evolution. Combinations of soft condensed matter and solid state materials will not be exceptional.

Figure 1 shows two possibilities of how Raman techniques are combined with waveguide technology. In "A" the waveguide package consists of Si and SiO_2, together mentioned as the substrate, next a thin layer of 130 nm. of ZnO is deposited. On top of this layer a monolayer of molecules is applied by dipping, coating or Langmuir-Blodgett techniques. The upper layer in the package may be air, but also water is possible. The distribution of the electrical amplitude of the exciting field depends on the refractive indices and thicknesses of all layers involved. In "B" the waveguide itself contains the molecules of interest. The waveguide material may be build from any organic polymer that has sufficient transmission for the exciting - and scattered light.

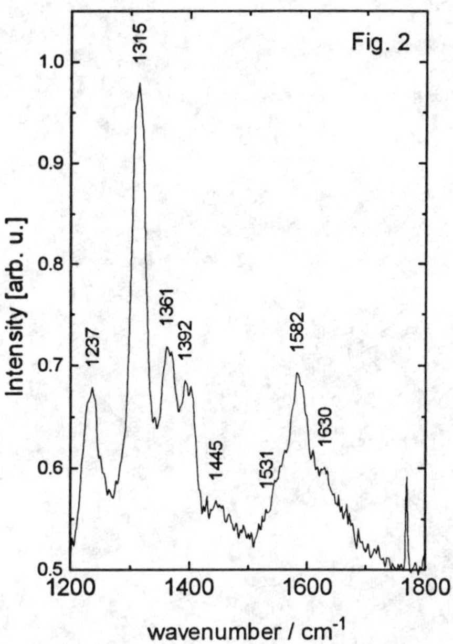

Using waveguide techniques it proved possible to measure monolayers of non-resonantly excited proteins [1]. An adequate prove of monolayer sensitivity was the detection of a monolayer of DiMyristoylPhosphatidylEthanolamine (DMPE), that was deposited on a ZnO waveguide by Langmuir-Blodgett techniques. This in spite of the fact that in the chosen scattering geometry [1] the strongest component of the polarizability tensor in such a system (α_{zz}) cannot be observed. In figure 2 the results are shown of the electron transferring protein cytochrome c. The excitation wavelength is 514.5 nm. near the Q-bands of the Fe-porphyrine system that serves as the active site in the protein. The Raman scattering is resonantly enhanced. From the spectrum the redox state of the protein can be obtained. The strong band at 1315 cm^{-1} is indicative of the reduced Fe^{2+}-state.

[1] J.S. Kanger, C. Otto, M. Slotboom and J. Greve, "Waveguide Raman Spectroscopy of Thin Polymer Layers and Monolayers of Biomolecules using High Refractive Index Waveguides", J. Phys. Chem. accepted for publication (1996).

ADDITIONAL PLENARY LECTURES

THE RAMAN SPECTRA OF SOME CERAMIC PIGMENTS

Anton M. Heyns

Institute of Applied Materials, University of Pretoria,
0002 Pretoria, South Africa

Although inorganic pigments have been subjected to numerous studies over many years, all properties of these coloured substances are still not fully understood. Toxic substances such as lead- and cadmium containing yellow pigments are banned in many countries as are zircon and zirconium based pigments containing high levels of the radio-active elements uranium and thorium. New pigments had to be developed to replace these and new methods had to be found to get rid of radio-active substances. Raman spectroscopy, in conjunction with other techniques such as electron microscopy, powder diffraction X-ray, X-ray spectroscopies, thermal analysis, electron spin resonance, solid-state nuclear magnetic resonance and electronic spectroscopy are very powerful tools which can be used in these investigations. Raman spectroscopy is a well-established and effective technique in the elucidation of structures of complex transition-metal oxides present as either bulk phases [1] or surface-supported phases [2]. However, it is a unique technique in the sense that Resonance Raman Spectroscopy enables one to enhance the intensities of certain vibrations selectively. This makes the detection of species at low levels of concentration possible, making it an ideal tool to study coloured dopants in host lattices of inorganic materials.

The following problems are often encountered with inorganic pigments:

- The structures of dopants and the positions that they occupy in the host lattice are unknown.
- The oxidation states of the transition metal oxides which act as dopants have sometimes not been unequivocally determined.
- Data on the solubility of the dopant in the host lattice are not readily available.
- The physical and chemical properties of the dopants are not completely known in the host lattice, and therefore the origin of the colour in pigments and their colour stability/instability cannot be determined.

Resonance Raman Spectroscopy can play a very important rôle in these studies and the information obtained can be used to design and synthesize new non-toxic materials with superior qualities. In this review a few selected inorganic pigments are discussed, that were studied in our laboratories with Raman Spectroscopy and other techniques. Our main aims were to obtain information on the structure and oxidation states of the dopants and to determine the origin of the colour of the pigments.

Zircon- and Zirconia-based pigments

Low concentrations of vanadium can be incorporated into white zircon, $ZrSiO_4$, to produce a bright blue solid that is commonly used in the ceramics industry. Even with this very common pigment there is still a great deal uncertainty about the sites that the vanadium ions occupy in the zircon lattice [3,4,5]. An analysis of the Resonance Raman spectra of $ZrSiO_4$:V has led to the conclusion [5] that vanadium is most probably present in the (V) oxidation state and that these ions occupy the 2b sites of Si^{4+} of D_{2d}-symmetry in the lattice of the host material, space group $I4/_{amd}$-D_{2h}^{19} [6]. Various other dopants such as Fe, Cr and Pr have a profound effect on the Raman modes of zircon [5] showing that they probably also occupy Si^{4+} sites. However, it is unlikely that Fe and Cr are present in the (IV) oxidation state in these pigments and their Resonance Raman spectra indicated that a combination of different oxidation states most probably occur in $ZrSiO_4$:Fe and $ZrSiO_4$:Cr. The origin of the colour in these pigments probably lies in mixed-valency behaviour of the Fe and Cr. In the case of $ZrSiO_4$:Pr, the Pr is in the (IV) oxidation state but it is very difficult to see how the Pr^{4+} with a radius of 1.10 Å can substitute Si^{4+} with a radius of 0.40 Å in the crystal lattice [5].

Pyrochlore-based pigments

Two pyrochlores were investigated, viz yttrium tin pyrochlore ($Y_2Sn_2O_7$) and yttrium titanium pyrochlore ($Y_2Ti_2O_7$). Yellow pigments with the composition $Ca_xY_{2-x}V_xM_{2-x}O_7$ (M = Ti, Sn) have been synthesized and are important since they do not contain the toxic Cd and Pb elements.

The vibrational spectra and the force fields have been reported by Vandenborre *et al* for various pyrochlore compounds including the yttrium/tin and yttrium/titanium pyrochlores [7,8]. Facer *et al* [9] have determined the structure of the yttrium/tin pyrochlore and found that the tin atoms are in an approximately octahedral coordination state. The yttrium ion is eightfold co-ordinated. A group theoretical analysis of the vibrational modes of the pyrochlore structure reveals that the following modes should occur, viz $\Gamma_{opt} = A_{1g}(R) + E_g(R) + 2F_{1g}(-) + 4F_{2g}(R) + 3A_{2u}(-) + E_u(-) + 7F_{1u}(IR) + 4F_{2u}(-)$. In accordance with most pyrochlore compounds, the Raman spectra of yttrium/tin and yttrium/titanium pyrochlores show only three intense bands, as is shown in Fig. 1. Seven infrared bands have been observed. The Raman spectrum of $Y_2Sn_2O_7$ doped with Ca and V is shown in Fig. 2 and it is immediately evident that the bands at 806 and 755 cm^{-1} must be due to the dopants, in this case they can be assigned to V-O stretching modes.

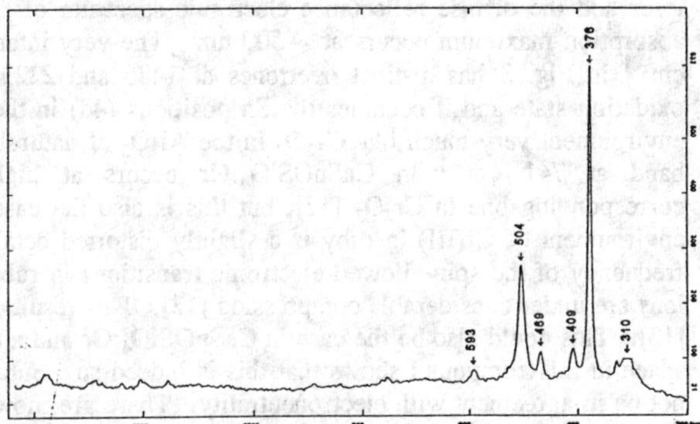

Fig. 1 The Raman spectrum of undoped $Y_2Sn_2O_7$.

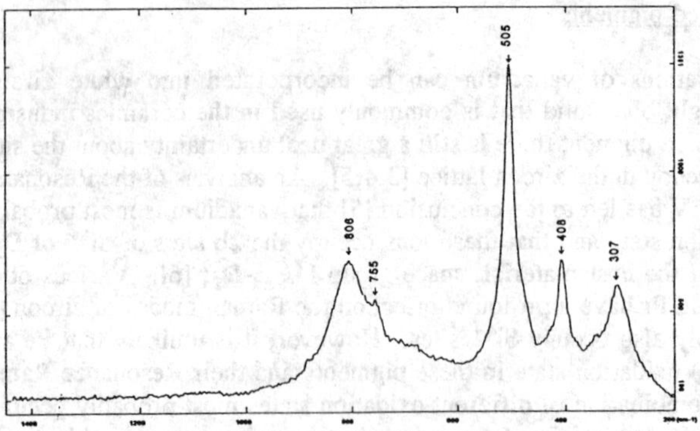

Fig. 2 The Raman spectrum of Y_2SnO_7 doped with Ca and V

The bond lengths calculated for the V-O bands in $Y_2Sn_2O_7$ are 1.71 and 1.74 using the relationship proposed by Hardcastle *et al* [10] $v = 21349 \exp(-1.917\,R)$ where v = position of the band in cm^{-1} and R = bond length in Å. The corresponding ones are 1.70 and 1.75 in $Y_2Ti_2O_7$. A consideration of these results reveals that vanadium is in the +5 oxidation state.

Malayite-based pigments

Malayite (calcium tin silicate) is a naturally occurring mineral with the general formula $CaSnOSiO_4$. It has a titanium analogue called titanite with the formula $CaTiOSiO_4$. Malayite crystallizes in the monoclinic space group 15 ($C2/_c$) [11] and consists of corner sharing SnO_6 octahedra cross linked by silicate tetrahedra to form a $(SnOSiO_4)^{2-}$ framework that accommodates Ca in irregular 7-co-ordination polyhedra [11]. Two malayite-based pigments have been investigated in the present study, viz the chromium doped maroon/pink and the cobalt doped dark blue ones. A group-theoretical analysis of the malayite structure was done and the rather limited number of Raman bands could be assigned.

The Raman spectra of undoped and chromium doped malayite are shown in Fig. 3. It is clear that resonance effects occur in the spectrum of the pigment as could be expected since these spectra have been recorded with the 514.5 nm line of an Ar^+-laser and the diffuse reflectance electronic spectrum of $CaSnOSiO_4$:Cr shows that an absorption maximum occurs at ~500 nm. The very intense Cr-O Raman band at 741 cm^{-1} in Fig. 3 has distinct overtones at 1486 and 2223 cm^{-1}. If Cr is in the +3 oxidation state and it occupies the Sn positions (4d) in the lattice, it has an octahedral environment very much like Cr_2O_3 in the Al_2O_3 of naturally occurring ruby. The Cr-O band at 741 cm^{-1} in $CaSnOSiO_4$:Cr occurs at higher wavenumbers than the corresponding one in Cr_2O_3 [12], but this is also the case in ruby crystals [13]. The environment of Cr(III) in ruby is a slightly distorted octahedron of oxide ions and the frequency of the spin-allowed electronic transitions in ruby indicate that the chromium ions are under considerable compression [13]. This results in the maroon colour of ruby [13]. This could also be the case in $CaSnOSiO_4$:Cr and a consideration of the available space in this compound shows that this is indeed so. Substituting Sn^{4+} with Cr^{3+} will not be in agreement with electroneutrality. There are, however, clear indications in the Raman spectra of $CaSnOSiO_4$:Cr that Cr-O bands occur at 977 and 938 cm^{-1} respectively (Fig. 3) which can be ascribed to Cr(VI)-ions, which are tetrahedrally co-ordinated and possibly located in the Si^{4+}-positions. This could then substitute the silicon in malayite and the problem of electroneutrality will be solved by having one Cr(VI) for every two Cr(III) present.

In the case of $CaSnOSiO_4$:Co the Raman and electronic spectra indicate that Co is in the (II) oxidation state, substituting the Si ions and the tetrahedrally co-ordinated Co(II) therefore explains the deep blue colour of the pigment.

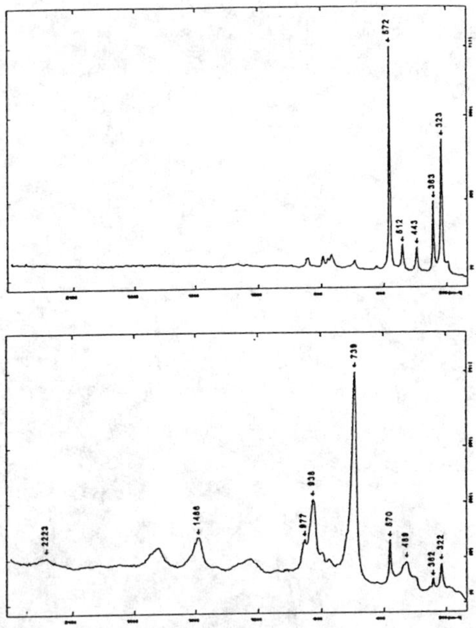

Fig. 3 The Raman spectra of undoped (top) and chromium doped malayite.

Conclusion

Resonance Raman Spectroscopy plays a unique rôle in the study of inorganic pigments and can give information on the symmetry of the dopants that is not possible with other techniques. It is an essential tool in any material science laboratory where inorganic pigments are investigated.

1. L. Dixit, D.L. Gerrard and H.J. Bowley, Appl. Spectrosc. Rev. 22, 189 (1986).
2. F.D. Hardcastle, I.E. Wachs, J.A. Horsley and G.H. Via, J. Mol. Catal. 46, 15 (1988).
3. A. Siggel and M. Jansen, Z. f. anorg. u. allgem. Chemie 583, 67 (1990).
4. P. Chandley, R.J.H. Clark, R.J. Angel and G.J. Price, J. Chem. Soc. Dalton Trans. 1579 (1992).
5. D. de Waal, A.M. Heyns, G. Pretorius and R.J.H. Clark, J. Raman Spectroscopy, 1996 (submitted).
6. T. Demiray, D.K. Nath and F.A. Hummel, J. Amer. Chem. Soc. 63, 1 (1970).
7. M.T. Vandenborre and E. Husson, J. Solid State Chem. 50, 362 (1983).
8. M.T. Vandenborre, E. Husson, J.P. Chatry and D. Michel, J. Raman Spectroscopy 14, 2 (1983).
9. G.R. Facer, C.J. Howard and B.J. Kennedy, Powder Diffraction Data 8, 245 (1993).
10. F.D. Hardcastle and I.E. Wachs, J. Phys. Chem. 95, 5031 (1991).
11. J.B. Higgins and P.H. Ribbe, American Mineralogist 62, 801 (1977).
12. D.A. Brown, D. Cunningham and W.K. Glass, Spectrochimica Acta, Vol. 1., 24A, 965 (1968).
13. A.M. Heyns and P.M. Harden, to be published.

AUTHOR INDEX

L

S

T

X

Y

Z